D1426588

Organic Electronic Spectral Data

Volume VII 1964-1965

Organic Electronic Spectral Data, Inc.

BOARD OF DIRECTORS

Officers, 1964–1965

Joseph C. Dacons, President
U. S. Naval Ordnance Laboratory
Silver Spring, Maryland

Rip G. Rice, Secretary
W. R. Grace & Co.
Clarksville, Maryland

J. P. Phillips, Treasurer and Managing Editor
University of Louisville
Louisville, Kentucky

Editorial Board

J. C. Craig, University of California, San Francisco (1963-1966)
H. Feuer, Purdue University (1962-1965)
L. D. Freedman, North Carolina State University (1963-1966)
P. R. Jones, University of New Hampshire (1962-1965)
P. M. Laughton, Carleton University, Ontario, Canada (1964-1967)
B. S. Thyagarajan, University of Idaho (1963-1966)

Organic Electronic Spectral Data

Volume VII 1964-1965

JOHN P. PHILLIPS, JOSEPH C. DACONS
& RIP G. RICE

EDITORS

CONTRIBUTORS

O. Achmatowicz
J. C. Craig
H. Feuer
W. W. Fike
L. D. Freedman
V. Gaind
K. Genest
C. Hanna
M. K. Hrenoff
P. M. Laughton
C. M. Martini
R. A. McIvor
F. C. Nachod
J. P. Phillips
E. Sawicki
B. S. Thyagarajan
H. E. Ungnade (deceased)
O. H. Wheeler

WILEY-INTERSCIENCE
A Division of John Wiley & Sons, Inc.
New York London Sydney Toronto

Copyright © 1971, by John Wiley & Sons, Inc.

All rights reserved. Published simultaneously in Canada.

No part of this book may be reproduced by any means, nor transmitted, nor translated into a machine language without the written permission of the publisher.

Library of Congress Catalog Card Number: 60-16428

ISBN 0-471-68798-7

Printed in the United States of America.

10 9 8 7 6 5 4 3 2 1

INTRODUCTION TO THE SERIES

In 1956 a cooperative effort to abstract and publish in formula order all the ultraviolet-visible spectra of organic compounds presented in the journal literature was organized through the enterprise and leadership of M.J. Kamlet and H.E. Ungnade. Organic Electronic Spectral Data was incorporated in 1957 to create a formal structure for the venture, and coverage of the literature from 1946 onward was then carried on by chemists with special interests in spectrophotometry through a page by page search of the major chemical journals. After the first two volumes (covering the literature from 1946 through 1955) were produced, a regular schedule of one volume for each subsequent period of two years was instituted. Seven volumes have now been published.

Altogether, more than fifty chemists have searched a group of journals totalling more than a hundred titles during the course of this sustained project. Additions and subtractions from both the list of contributors and of journals have occurred from time to time, and it is estimated that the effort to cover all the literature containing spectra may not be more than 95% successful. However, the total collection is by far the largest ever assembled, amounting to well over 150,000 spectra through the volumes so far published.

Volume VIII, edited by Henry Feuer and B.S. Thyagarajan, is in preparation and additional volumes are contemplated.

PREFACE

Processing of the data provided by the contributors to Volume VII was performed at the University of Louisville. The editors are grateful to Becky Sprowles, Nancy Phillips and Margaret Bell for assistance with this work. The final manuscript was typed by Mary Ann Abeles and others.

John P. Phillips
Joseph C. Dacons
Rip G. Rice

ORGANIZATION AND USE OF THE DATA

The data in this volume were abstracted from the journals listed in the reference section at the end. Although a few exceptions were made, the data generally had to satisfy the following requirements: the compound had to be pure enough for satisfactory elemental analysis and for a definite empirical formula; solvent and phase had to be given; and sufficient data to calculate molar absorptivities had to be available. Later it was decided to include spectra even if solvent was not mentioned.

All entries in the compilation are organized according to the molecular formula index system used by Chemical Abstracts. Most of the compound names have been made to conform with the Chemical Abstracts system of nomenclature.

Solvent or phase appears in the second column of the data lists, often abbreviated according to standard practice; there is a key to abbreviations on the next page.

The numerical data in the third column present wavelength values in millimicrons (or nanometers) for all maxima, shoulders and inflections, with the logarithms of the correspondin molar absorptivities in parentheses. Shoulders and inflections are marked with a letter s. In spectra with considerable fine structure in the bands a main maximum is listed and labelled with the letter f. Numerical values reported by authors are given to the nearest 0.1 millimicron for wavelength (if possible) and 0.01 unit for the logarithm of the molar absorptivity. Spectra that change with time are labelled 'changing' and temperatures are sometimes indicated.

The reference column contains the code number of the journal, the page number of the paper, and in the last two digits the year. A letter is added for journals with more than one volume in a year. The complete list of all articles with authors is found in the References at the end of the volume.

Several journals that were abstracted for previous volumes in this series have been omitted and a few new ones have been added. In a very few instances a code number for an omitted journal has been assigned to a new one in this volume.

ABBREVIATIONS

s	shoulder or inflection
f	fine structure
n.s.g.	no solvent given in original reference
C_5H_5N	pyridine
C_6H_6	benzene
C_6H_{12}	cyclohexane
DMF	dimethylformamide
DMSO	dimethylsulfoxide
THF	tetrahydrofuran

Other solvent abbreviations generally follow the practice of Chemical Abstracts.

Underlined data were estimated from graphs.

JOURNALS ABSTRACTED

Journal	No.
Acta Chem. Scand.	1
Analyst	96
Angew. Chem.	89
Ann. Chem. Liebigs	5
Ann. Chim. (Paris)	6
Ann. chim. (Rome)	7
Appl. Spectroscopy	9
Arch. Biochem. Biophys.	10
Arch. Pharm.	83
Arkiv Kemi	11
Australian J. Chem.	12
Ber. Bunsengesellschaft Phys. Chem.	61
Biochem. Z.	15
Boll. sci. fac. chim. ind. Bologna	17
Bull. soc. chim. Belges	20
Bull. soc. chim. France	22
Can. J. Chem.	23
Chem. Ber.	24
Chem. and Ind. (London)	25
Chemical Communications	77
Chem. Pharm. Bulletin(Japan)	94
Chimia	27
Coll. Czech. Chem. Comm.	73
Compt. rend.	28
Discussions Faraday Soc.	29
Doklady Akad. Nauk S.S.S.R.	30
Experientia	31
Gazz. chim. ital.	32
Helv. Chim. Acta	33
Indian J. Chem.	2
Izvest. Akad. Nauk S.S.S.R.	70
J. Am. Chem. Soc.	35
J. Biol. Chem.	37
J. Chem. Phys.	38
J. Chem. Soc.	39
J. Heterocyclic Chem.	4
J. Indian Chem. Soc.	42
J. Med. Chem.	87
J. Mol. Spect.	82
J. Optical Soc. Am.	43
J. Org. Chem.	44
J. Organometallic Chem.	101
J. Pharm. Soc. Japan	95
J. Phys. Chem.	46
J. Structural Chem.	67
Lloydia	100
Monatsh. Chem.	49
Nature	50
Proc. Chem. Soc.	77
Rec. trav. chim.	54
Roczniki Chem.	56
Science	57
Spectrochim. Acta	59
Steroids	13
Tetrahedron	78
Tetrahedron Letters	88
Trans. Faraday Soc.	60
Zhur. Obshchei Khim.	65

Journal	No.	Journal	No.
Acta Chem. Scand.	1	J. Indian Chem. Soc.	42
Indian J. Chem.	2	J. Optical Soc. Am.	43
J. Heterocyclic Chem.	4	J. Org. Chem.	44
Ann. Chem. Liebigs	5	J. Phys. Chem.	46
Ann. Chim.(Paris)	6	Monatsh. Chem.	49
Ann. chim. (Rome)	7	Nature	50
Appl. Spectroscopy	9	Rec. trav. chim.	54
Arch. Biochem. Biophys.	10	Roczniki Chem.	56
Arkiv Kemi	11	Science	57
Australian J. Chem.	12	Spectrochim. Acta	59
Steroids	13	Trans. Faraday Soc.	60
Biochem. Z.	15	Ber. Bunsengesellschaft Phys. Chem.	61
Boll. sci. fac. chim. ind. Bologna	17	Zhur. Obshchei Khim.	65
Bull. soc. chim. Belges	20	J. Structural Chem.	67
Bull. soc. chim. France	22	Izvest. Akad. Nauk S.S.S.R.	70
Can. J. Chem.	23	Coll. Czech. Chem. Comm.	73
Chem. Ber.	24	Proc. Chem. Soc.(1964)	77
Chem. and Ind. (London)	25	Chemical Communications	77
Chimia	27	Tetrahedron	78
Compt. rend.	28	J. Mol. Spect.	82
Discussions Faraday Soc.	29	Arch. Pharm.	83
Dokladt Akad. Nauk S.S.S.R.	30	J. Med. Chem.	87
Experientia	31	Tetrahedron Letters	88
Gazz. chim. ital.	32	Angew. Chem.	89
Helv. Chim. Acta	33	Chem. Pharm. Bulletin(Japan)	94
J. Am. Chem. Soc.	35	J. Pharm. Soc. Japan	95
J. Biol. Chem.	37	The Analyst	96
J. Chem. Phys.	38	Lloydia	100
J. Chem. Soc.	39	J. Organometallic Chem.	101

Organic Electronic Spectral Data

Volume VII 1964-1965

Compound	Solvent	$\lambda_{max}(\log \epsilon)$	Ref.
CBr_4			
Carbon tetrabromide	hexane	229(3.67)	46-1992-65
$CClFN_2$			
3H-Diazirine, 3-chloro-3-fluoro-	gas	351.3f(2.27)	4-0371-65
$CClF_2N_3$			
3H-Diazirine, 3-chloro-3-(difluoroamino)-	gas	331.3f(1.42)	4-0371-65
CCl_4O_2S			
Methanesulfonyl chloride, trichloro-	EtOH	204(3.49)	78-1613-64
CF_3N_3			
3H-Diazirine, 3-(difluoroamino)-3-fluoro-	gas	325.7f(2.36)	4-0371-65
$CHBrN_2O_4$			
Methane, bromodinitro-, anion	H_2O	385(4.16)	44-3587-64
	at 10°	385(4.17)	44-3587-64
$CHClN_2O_4$			
Methane, chlorodinitro-, anion	H_2O	386(4.19)	44-3587-64
	at 9°	386(4.21)	44-3587-64
$CHFN_2O_4$			
Methane, fluorodinitro-	H_2O	280(1.9)	70-2063-65
anion	base	250(3.6),385(4.28)	70-2063-65
CHN_3O_6			
Methane, trinitro-, anion	H_2O	350(4.17)	44-3587-64
	at 9.6°	350(4.18)	44-3587-64
	n.s.g.	350(4.14)	70-1066-65
CH_2HgI_2			
Iodomethylmercuric iodide	n.s.g.	227(4.34),244(4.44)	101-0097-65A
$CH_2Hg_2I_2$			
Methylenebis(mercuric iodide)	n.s.g.	237(3.72),262(3.64)	101-0097-65A
$CH_2N_2O_4$			
Methane, dinitro-	pH 5.5	360(4.32)	70-2220-65
CH_2O_2			
Formic acid	gas	159(3.40)	82-0001-64B
CH_3I			
Methane, iodo-	gas	181f(--),197f(--)	60-0597-65
CH_3NO_2			
Methane, nitro-	hexane	278(1.23)	46-3225-64
	H_2O	269(1.18)	
	MeOH	272.5(1.20)	
	EtOH	272.5(1.23)	
	PrOH	273(1.24)	
	iso-PrOH	273(1.24)	
	BuOH	273.5(1.25)	
	tert-BuOH	273.5(1.26)	
	$C_3F_7CH_2OH$	268(1.19)	

Compound	Solvent	λ_{max}(log ϵ)	Ref.
Methane, nitro- (cont.)	MeCN	272(1.18)	46-3225-64
	dioxan	272(1.36)	
	Et_2O	275(1.23)	
	($MeOCH_2$-CH_2)$_2$O	276(1.30)	
	($ClCH_2CH_2$)$_2$O	276(1.29)	
anion	H_2O	232(3.95)	44-2674-65
CH_4N_2S			
Urea, thio-	EtOH	240(4.91)	65-0372-65
	n.s.g.	241(4.08)	65-0886-65
salt of dipentyl phosphate	EtOH	220(4.84)	65-0372-65
salt of dihexyl phosphate	EtOH	221(4.08)	65-0372-65
salt of dioctyl phosphate	EtOH	222(4.55)	65-0372-65
salt of dinonyl phosphate	EtOH	223(4.51)	65-0372-65
salt of didecyl phosphate	EtOH	224(4.30)	65-0372-65
CKN_3O_6			
Methane, trinitro-, potassium deriv.	H_2O	350(4.16)	46-0053-65
CN_4			
Cyanogen azide	C_6H_{12}	220(3.33),275(2.01)	35-4506-64

Compound	Solvent	λ max (log ε)	Ref.
C_2ClF_3O Acetyl chloride, trifluoro-	gas	253.0(1.23)	59-1437-64
C_2FN_3 3H-Diazirine, 3-cyano-3-fluoro-	gas	337.9f(2.18)	4-0371-65
$C_2F_4N_2O_4$ Ethane, tetrafluoro-1,2-dinitro-	gas	282.5(1.93)	46-1758-65
	C_6H_{12}	286(2.07)	46-1758-65
	hexane	286(2.04)	46-1758-65
	MeOH	283(2.14)	46-1758-65
	Et_2O	250s(2.49)	46-1758-65
	dioxan	254(2.81)	46-1758-65
	CCl_4	286(2.07)	46-1758-65
	Et_3N	297(2.59)	46-1758-65
	Pr_2NH	290(2.65)	46-1758-65
	$C_6H_{13}NH_2$	275(2.69)	46-1758-65
$C_2F_6HgSe_2$ Mercury, bis[(trifluoromethyl)-selenyl]-	MeOH	219(4.30),274(3.04)	39-0940-65
C_2HCl_3O Chloral	heptane	222(1.2),287(1.5)	24-3322-64
	MeOH	none	24-3322-64
$C_2HCl_3O_2$ Acetic acid, trichloro-, lead salt	C_6H_{12}	223(3.87)	88-2847-65
$C_2H_2N_2OS$ 1,2,5-Thiadiazol-3-ol	H_2O	276(3.72),300s(--)	35-2861-64
$C_2H_2N_2S$ Formamide, 1-cyanothio-	EtOH	301(3.91),422(1.49)	1-1059-64
	CH_2Cl_2	438.0(1.51)	1-1059-64
	heptane-0.5% CH_2Cl_2	294.0(3.87)	1-1059-64
$C_2H_2N_6$ Formamidine, 1-azido-N'-cyano-	EtOH	219(4.02),242(3.90)	44-0650-64
C_2H_3BrO Acetyl bromide	C_6H_{12}	213(2.52),253(1.98)	22-0204-64
	H_2SO_4	206(2.13),235(1.95)	22-0204-64
$C_2H_3Br_3Ge$ Germane, tribromo-vinyl-	n.s.g.	180s(4.3),198(4.2)	70-2203-64
C_2H_3ClO Acetyl chloride	C_6H_{12}	198(1.38),242(1.61)	22-0204-64
	H_2SO_4	213(1.59)	22-0204-64
$C_2H_3Cl_3Si$ Silane, trichloro-vinyl-	n.s.g.	177(4.1)	70-2203-64
$C_2H_3FN_2O$ 3H-Diazirine, 3-fluoro-3-methoxy-	gas	357.3f(1.92)	4-0371-65
C_2H_3FO Acetyl fluoride	C_6H_{12}	212(1.53)	22-0204-64

Compound	Solvent	λ_{max}(log ε)	Ref.
$C_2H_3F_3Si$			
Silane, trifluoro-vinyl-	n.s.g.	170(4.1)	70-2203-64
C_2H_3I			
Ethylene, iodo-	gas	179f(--),199f(--)	60-0597-65
$C_2H_3KOS_2$			
Potassium methyl xanthate	H_2O	226(3.90),302(4.28), 382(1.71)	1-1113-65
	EtOH	228(3.97),304(4.31), 383(1.71)	1-1113-65
$C_2H_3N_3O_2$			
1,2,4-Triazolidine-3,5-dione	EtOH	211(3.54)	22-0500-64
	base	218.5(3.58)	22-0500-64
$C_2H_4ClNO_2$			
Ethane, 1-chloronitro-, anion	H_2O	233(3.98)	44-2674-65
$C_2H_4NNaO_2$			
Ethane, nitro-, sodium salt	H_2O	228(3.96),340(2.35), 375(2.50)	44-1687-65
$C_2H_4N_2$			
3H-Diazirine, 3-methyl-	gas	341f(2.67)	39-1700-65
$C_2H_4N_2OS$			
Oxamide, thio-	EtOH	229(3.66),301(3.82), 411(1.46)	1-1059-64
$C_2H_4N_2O_2$			
Ethylnitrosolic acid	n.s.g.	284(3.87),720(1.08)	28-0207-64A
$C_2H_4N_2O_4$			
Ethane, dinitro-, anion	n.s.g.	380(4.23)	70-1283-65
Ethane, 1,1-dinitro-	n.s.g.	382(4.25)	44-3587-64
	at 9.6°	382(4.24)	44-3587-64
$C_2H_4N_2S_2$			
Oxamide, dithio-	EtOH	312(4.05),483(1.34)	1-1059-64
$C_2H_4N_4$			
Dicyandiamide	pH -2.9	none	65-1930-65
(or cyanoguanidine)	pH 15.25	220(4.0)	65-1939-65
$C_2H_4N_4O_2$			
Formaldoximeazocarbonamide	n.s.g.	265(3.87),425(1.78)	28-0207-64A
anion	n.s.g.	340(4.22)	28-0207-64A
Methylazaurolic acid	n.s.g.	295(4.30),400(2.23)	28-0207-64A
anion 1	n.s.g.	345(4.45)	28-0207-64A
anion 2	n.s.g.	360(4.51)	28-0207-64A
$C_2H_4O_2$			
Acetic acid	gas	160(3.62)	82-0001-64B
	C_6H_{12}	211(1.58)	22-0204-64
Formic acid, methyl ester	isooctane	215.2f(1.85)	35-2384-64
	H_2O	207(1.80)	35-2384-64
	MeOH	211.4(1.77)	35-2384-64
	EtOH	213(1.89)	35-2384-64
	MeCN	211.7(1.85)	35-2384-64

Compound	Solvent	$\lambda_{max}(\log \epsilon)$	Ref.
Formic acid, methyl ester (cont.)	$C_3H_4F_4O$	207.7(1.80)	35-2384-64
$C_2H_5NO_2$			
Ethane, nitro-	hexane	278(1.29)	46-3225-64
	MeOH	275(1.28)	46-3225-64
anion	H_2O	229(3.95)	44-2674-65
$C_2H_5N_3O_2$			
Aminoacetamide, α-isonitroso-	EtOH	254(3.63)	33-0033-64
C_2H_5NaOS			
Dimethyl sulfoxide, sodium deriv.	DMSO	272(4.60),330s(4.20)	77-0108-64
$C_2H_6N_2$			
Formaldehyde, methylhydrazone	MeOH	233(3.69)	35-2395-64
$C_2H_6N_2O$			
Dimethylamine, N-nitroso-	C_6H_{12}	232(3.77)	35-4373-64
$C_2H_6N_2O_2$			
N-Methyl-N'-methoxydiimide, N-oxide	C_6H_{12}	238(3.99)	44-0311-64
Methanol, (methylazoxy)-	H_2O	215(3.94),275s(1.67)	10-0373-65B
$C_2H_6N_2S$			
Pseudourea, 2-methyl-2-thio-	n.s.g.	300(2.25)	65-0886-65
$C_2H_6S_2$			
Methyl disulfide	EtOH	256(2.64)	44-4008-65
$C_2H_2O_2$			
Cyanogen, N,N'-dioxide	hexane at 0°	256(3.9),288(4.0), 312(3.9)	5-0191-65G
C_2N_4			
Diimide, dicyano-	MeCN	240(3.89),445(1.26)	35-1819-65
sodium deriv.	MeCN	222(3.72),372(3.66), 525(2.08)	35-1819-65

Compound	Solvent	λ_{max}(log ϵ)	Ref.
$C_3Cl_2OS_2$			
3H-1,2-Dithiol-3-one, 4,5-dichloro-	MeOH	250(3.63),318(3.76)	5-0109-64I
$C_3Cl_4S_2$			
3,4,5-Trichloro-1,2-dithiol-1-ium chloride	MeOH	237(6.63),242(6.66), 248(6.81),254(6.90), 260(6.79),268(6.26) [sic]	5-0109-64I
C_3F_6S			
Propane, 1,2-epithio-1,1,2,3,3,3-hexafluoro-	n.s.g.	234(2.15)	44-4188-65
$C_3HCl_2N_3O_3$			
Cyanuric acid, dichloro-	pH 1.3	none	59-0397-64
	pH 12.0	190(4.5)	59-0397-64
	pH 8.4	215(4.0)	59-0397-64
C_3HCl_5O			
Propionaldehyde, pentachloro-	heptane	222(1.4),294(1.4)	24-3322-64
	MeOH	222(1.8)	24-3322-64
C_3HF_6N			
Ethylidenimine, 2,2,2-trifluoro-1-(trifluoromethyl)-	n.s.g.	248(1.97)	44-1398-65
C_3HN_3OS			
1,2,5-Thiadiazole-3-carbonitrile, 4-hydroxy-	H_2O	213(3.94),334(3.93)	35-2861-64
	EtOH	297(3.88),342s(--)	35-2861-64
acid salt with C_3KN_3OS	H_2O	213(3.97),335(3.94)	35-2861-64
C_3H_2BrNS			
Isothiazole, 4-bromo-	EtOH	211(3.39),256(3.87)	39-0446-64
Isothiazole, 5-bromo-	EtOH	241(3.85)	39-0446-64
$C_3H_2ClN_3O_2$			
Imidazole, 5-chloro-4-nitro-	17M H_2SO_4	267(3.86)	44-0862-64
	dil. H_2SO_4	304(3.82)	44-0862-64
	pH 13	356(4.01)	44-0862-64
C_3H_2INS			
Isothiazole, 4-iodo-	EtOH	264(4.02)	39-0446-64
$C_3H_2N_2$			
Malononitrile	EtOH-base	226(4.32)	23-1225-65
$C_3H_2N_2O_2S$			
Isothiazole, 4-nitro-	EtOH	215(4.06),272(4.03)	39-0446-64
$C_3H_2N_2O_3$			
Isoxazole, 4-nitro-	isooctane	216(3.77),239(3.75)	32-0915-64
	H_2O	217(3.69),250(3.70)	32-0915-64
	acid	218(3.70),251(3.70)	28-6379-65A
	base	220s(4.00),329(4.15)	28-6379-65A
$C_3H_2N_4$			
Malononitrile, hydrazono-	MeCN	233(3.17),282(4.13)	44-4198-65

Compound	Solvent	λ_{max}(log ϵ)	Ref.
$C_3H_2N_4O_4$			
Imidazole, 2,4(5)-dinitro-	dil. H_2SO_4	304(4.05)	44-0862-64
	pH 13	354(4.09)	44-0862-64
$C_3H_2OS_2$			
1,2-Dithia-4-cyclopenten-3-one	C_6H_{12}	310(3.6)	73-3016-65
1,3-Dithia-4-cyclopenten-2-one	C_6H_{12}	257(3.5)	73-3016-65
	EtOH	240(3.37),267(3.41)	24-1298-64
$C_3H_2S_3$			
1,2-Dithia-4-cyclopentene-3-thione	C_6H_{12}	415(3.8)	73-3016-65
1,3-Dithia-4-cyclopentene-2-thione	C_6H_{12}	363(4.2)	73-3016-65
	EtOH	231(3.89),367(4.19)	24-1298-64
$C_3H_3ClO_4S_2$			
1,2-Dithiolium perchlorate	EtOH-$HClO_4$	288(3.6)	73-3016-65
1,3-Dithiolium perchlorate	EtOH-$HClO_4$	254(3.6)	73-3016-65
$C_3H_3Cl_3O$			
2-Propanone,1,1,1-trichloro-	EtOH	286(1.45)	44-2784-65
C_3H_3N			
Acrylonitrile	cyclopenta-diene	none	70-1290-65
	$CHCl_3$	none	70-1290-65
C_3H_3NO			
Oxazole	n.s.g.	205(3.59)	24-1414-64
C_3H_3NOS			
4-Isothiazolin-3-one	H_2O	264(3.80),275(3.81)	44-2660-65
	EtOH	257(3.81),280s(3.13)	35-5307-64
	EtOH	257(3.81),280s(3.13)	88-1471-64
	EtOH-base	279(3.83)	35-5307-64
	Et_2O	256(3.82)	44-2660-65
$C_3H_3NOS_2$			
3H-1,2-Dithiol-3-one, 4-amino-	EtOH	251(3.46),342(3.81)	12-0447-64
$C_3H_3NO_2S$			
2,4-Thiazolidinedione	n.s.g.	220(3.52),295(1.76)	65-0886-65
$C_3H_3NO_4$			
Malonaldehyde, nitro-, as sodium enolate	n.s.g.	227(3.89),268(3.93), 326(3.98)	28-6379-65A
C_3H_3NS			
Thiazole	1.82N HCl	237(3.59)	59-1881-65
	pH 5	235(3.51)	59-1881-65
	H_2O	206(3.34),232(3.53)	59-1881-65
	1.92N NaOH	234(3.54)	59-1881-65
$C_3H_3NS_3$			
3H-1,2-Dithiole-3-thione, 4-amino-	EtOH	230(4.00),306(3.61), 434(4.05)	12-0447-64
$C_3H_3N_3O$			
6-Azauracil	Me_3PO_4	182(3.90),195(3.80), 229(3.38),261(3.72)	35-0011-65

Compound	Solvent	$\lambda_{max}(\log \epsilon)$	Ref.
$C_3H_3N_3O_2$			
Imidazole, 2-nitro-	8M H_2SO_4	298(3.91)	44-0862-64
	dil. H_2SO_4	325(3.95)	44-0862-64
	pH 13	372(4.13)	44-0862-64
Imidazole, 4-nitro-	8M H_2SO_4	264(3.90)	44-0862-64
	pH 7.38	298(3.86)	44-0862-64
	pH 13	350(4.01)	44-0862-64
Pyrazole, 4-nitro-	H_2O	274(3.92)	32-0915-64
	71% H_2SO_4	238(3.89)	39-1051-65
	pH 13	320.0(4.07)	39-1051-65
	EtOH	275.0(3.91)	39-1051-65
$C_3H_3N_3O_3$			
Cyanuric acid	pH 4.4	none	59-0397-64
	pH 9.7	213(4.0)	59-0397-64
	pH 12.2	192(4.4),222(3.9)	59-0397-64
	pH 14.0	222(3.9)	59-0397-64
$C_3H_3N_3O_4$			
Acetonitrile, (methyl-aci-nitro)-nitro-	HOAc	315(3.99)	70-1491-65
	MeCN	315(4.05)	70-1491-65
$C_3H_4BrN_3$			
Pyrazole, 5-amino-4-bromo-	EtOH	236.5(3.63)	95-1113-64
$C_3H_4F_2O$			
2-Propanone, 1,3-difluoro-	n.s.g.	285.0(1.46)	46-1786-64
$C_3H_4N_2$			
Imidazole	n.s.g.	206(3.70)	24-1414-64
$C_3H_4N_2O$			
Acetone, diazo-	MeOH	246(3.96),270(3.97)	88-2427-64
	at 0°	247(3.95),271(3.97)	88-2427-64
	at -78°	248(3.90),272(4.02)	88-2427-64
$C_3H_4N_2OS$			
Pseudothiohydantoin	n.s.g.	249.5(3.80)	65-0886-65
$C_3H_4N_2O_4$			
2-Isoxazoline, 3-nitro-, 2-oxide	H_2O	320(3.92)	70-1491-65
$C_3H_4N_2O_5$			
2-Isoxazolin-4-ol, 3-nitro-, 2-oxide	H_2O	315(3.92)	70-1491-65
$C_3H_4N_4$			
as-Triazine, 3-amino-	EtOH	228(4.09),324(3.41)	94-1329-64
$C_3H_4N_4O$			
6-Azacytosine	N H_2SO_4	202(3.87),237(3.61), 280(3.79)	73-1394-64
	H_2O	258(3.78)	73-1394-64
Guanidine, N-formyl-N'-cyano-	H_2O	238(4.24)	73-2783-65
s-Triazin-2-ol, 4-amino-	pH 1	236s(3.68),247(3.74)	20-0585-64
	H_2O	228s(3.74)	20-0585-64
	pH 11	253(3.63)	20-0585-64
	MeOH	230s(3.77)	20-0585-64

Compound	Solvent	$\lambda_{max}(\log \epsilon)$	Ref.
$C_3H_4N_4O_2$			
4-Furazancarboxamide, 3-amino-	pH 1	287(3.32)	4-0253-65
	pH 11	284(3.32)	4-0253-65
$C_3H_4N_6$			
Formamidine, 1-azido-N'-cyano-N-methyl-	EtOH	224(4.06),247(3.89)	44-0650-64
Tetrazoline, 5-cyanimino-1-methyl-	EtOH	221(3.98)	44-0650-64
C_3H_4O			
Acrolein	cyclopentadiene	320(1.3)	70-1290-65
	$CHCl_3$	330(1.4)	70-1290-65
Ketene, methyl-	n.s.g.	355(2.8)	37-1793-64
$C_3H_4O_2$			
Acrylic acid	H_2O	193.9(4.02)	23-0724-64
	49.2% H_2SO_4	196.6(3.99)	
	67.7%	199.2(3.95)	
	73.2%	201.0(3.94)	
	77.8%	203.8(3.91)	
	80.4%	207.3(3.90)	
	82.2%	211.0(3.92)	
	86.1%	214.2(3.95)	
	90.3%	215.6(4.02)	
	96.2	216.3(4.04)	
	100% H_2SO_4	217.5(4.02)	
	cyclopentadiene	none	70-1290-65
	$CHCl_3$	none	70-1290-65
Propiolactone	isooctane	207.2(1.73)	35-2384-64
	H_2O	201.2(1.72)	
	MeOH	204.5(1.69)	
	80% MeOH	202.9(1.72)	
	EtOH	204.5(1.72)	
	40% EtOH	201.6(1.7)	
	MeCN	205.6(1.68)	
	$C_3H_4F_4O$	201.4(1.7)	
$C_3H_4S_3$			
1,3-Dithiolane-2-thione	EtOH	318(4.03)	94-0253-65
	$CHCl_3$	318(4.17)	49-0631-65
$C_3H_5Br_3Ge$			
Germane, allyltribromo-	n.s.g.	189(4.1),225s(3.9)	70-2203-64
$C_3H_5Cl_3Ge$			
Germane, allyltrichloro-	n.s.g.	183s(4.2),208(3.8)	70-2203-64
$C_3H_5Cl_3Si$			
Silane, allyltrichloro-	n.s.g.	180s(4.0),186(4.1)	70-2203-64
$C_3H_5F_3N_2S$			
Acetic acid, trifluorothio-, S-methyl ester, hydrazone	isooctane	241(3.68)	44-3739-65
$C_3H_5F_3Si$			
Silane, allyltrifluoro-	n.s.g.	180(3.6)	70-2203-64

Compound	Solvent	$\lambda_{max}(\log \epsilon)$	Ref.
C_3H_5NO			
Isocyanic acid, ethyl ester	n.s.g.	207(1.81)	39-6421-65
$C_3H_5NOS_2$			
2-Thiazolidinethione, 4-hydroxy-	MeOH	244(3.82),277(4.19)	44-2146-64
C_3H_5NS			
Isothiocyanic acid, ethyl ester, iodine complex	heptane	288(4.51)	1-1897-65
	CH_2Cl_2	296(4.46)	1-1897-65
	CCl_4	292(4.45)	1-1897-65
$C_3H_5NS_2$			
1,3-Dithiolane, 2-imino- hydrochloride	H_2O	239(3.95),267(3.55)	44-0738-64
	MeOH	242(4.04),268(3.81)	44-0738-64
$C_3H_5N_3O$			
Acetonitrile, N-methylamino-α-iso-nitroso-	EtOH	255(3.92)	33-0033-64
Isoxazole, 3,5-diamino-, hydro-chloride	MeOH	244(4.40)	44-2862-65
	MeOH-base	228(4.11)	44-2862-65
$C_3H_5N_3O_6$			
Methane, (ethyl-aci-nitro)dinitro-	CH_2Cl_2	315(3.78)	70-1491-65
$C_3H_5N_5O$			
s-Triazole-2-carboxamide, 3-amino-	EtOH	236(3.85)	35-1980-65
	0.1N HCl	222(3.93)	35-1980-65
$C_3H_5N_5O_2$			
4-Furazancarboxamidoxime, 3-amino-	pH 1	280(3.72)	4-0253-65
	pH 11	289(3.78)	4-0253-65
$C_3H_5N_5S$			
Ammeline, thio-	H_2O	242s(3.75),279(4.33)	39-6296-65
1,2,5-Triadiazole-3-carboxamidine, 4-amino-, hydrochloride	pH 1	216(3.99),324(3.75)	44-2488-65
	pH 7	216(3.99),324(3.77)	44-2488-65
	pH 13	319(3.87)	44-2488-65
$C_3H_6N_2O$			
Ether, 2-diazoethyl methyl	Et_2O	427(<u>4.1</u>)	5-0042-64I
$C_3H_6N_2OS$			
Oxamide, N°-methylmonothio-	EtOH	240(3.74),303(3.79), 407(1.53)	1-1059-64
Oxamide, N^S-methylmonothio-	EtOH	223s(3.68),298(3.86), 394(1.40)	1-1059-64
	CH_2Cl_2	300(3.84),399(1.40)	1-1059-64
$C_3H_6N_2O_4$			
Methane, (ethyl-aci-nitro)nitro-	CH_2Cl_2	310(3.85)	70-1491-65
$C_3H_6N_2S_2$			
Oxamide, methyldithio-	heptane-0.1 % CH_2Cl_2	310.0(4.01)	1-1059-64
	heptane-1% CH_2Cl_2	486(1.28)	1-1059-64
	EtOH	307(4.04),464(1.28)	1-1059-64

Compound	Solvent	λ_{max}(log ε)	Ref.
$C_3H_6N_4S$			
Hydrazine, N-methyl-2-thiadiazolyl-	EtOH	265(3.75)	1-0871-64
$C_3H_6N_4S_2$			
Hydrazine, 5-(methylthio)-2-thia-diazolyl-	EtOH	282(3.91)	1-0871-64
$C_3H_6N_6$			
v-Triazole-4-carboxamidine, 5-amino-	pH 1	225s(--),271(3.88)	44-2488-65
	pH 7	230s(--),274(3.88)	44-2488-65
	pH 13	262(3.89)	44-2488-65
C_3H_6O			
Acetone	C_6H_{12}	280(1.16)	59-0529-65
	hexane	279(1.16)	
	H_2O	264(1.27)	
	MeOH	270(1.23)	
	EtOH	272(1.22)	
	PrOH	272(1.20)	
	BuOH	272(1.19)	
	pentanol	273(1.19)	
	hexanol	274(1.19)	
	heptanol	274(1.18)	
	Et_2O	277(1.16)	
	dioxan	277(1.16)	
	HCOOH	260(1.29)	
	HOAc	268(1.24)	
	EtCOOH	270(1.23)	
	PrCOOH	271(1.22)	
	BuCOOH	272(1.20)	
	HCOOMe	282(1.18)	
	$ClCH_2COOEt$	276(1.19)	
	CF_3COOH	252(1.37)	
	MeCN	274(1.19)	
	EtCN	274(1.18)	
	PrCN	275(1.16)	
	$HCONH_2$	271(1.23)	
	HCONHMe	274(1.20)	
	DMF	275(1.20)	
	$MeCONMe_2$	275(1.20)	
	CH_2Cl_2	276(1.20)	
	$CHCl_3$	284(1.26)	
	CCl_4	279(1.30)	
	Me_2SO	274(1.22)	
	$HOCH_2CH_2Cl$	270(1.20)	
$C_3H_6O_2$			
Acetic acid, methyl ester	isooctane	209.7(1.76)	35-2384-64
	H_2O	202.6(1.79)	35-2384-64
	MeOH	206.1(1.74)	35-2384-64
	80% MeOH	205(1.75)	35-2384-64
	EtOH	207.2(1.76)	35-2384-64
	40% EtOH	202.3(1.77)	35-2384-64
	$C_3H_4F_4O$	203.6(1.7)	35-2384-64
	MeCN	208.6(1.65)	35-2384-64
Formic acid, ethyl ester	isooctane	212(1.91),216(1.91), 220(1.89)	35-2384-64
	H_2O	207.7(1.72)	35-2384-64
	MeOH	212.6(1.81)	35-2384-64
	EtOH	213.2(1.87)	35-2384-64

Compound	Solvent	λ_{max}(log ε)	Ref.
Formic acid, ethyl ester (cont.)	$C_3H_4F_4O$	208.4(1.81)	35-2384-64
	MeCN	213.4(1.68)	35-2384-64
$C_3H_6S_2$			
1,3-Dithiolane	C_6H_{12}	248(2.43)	78-0437-64
	EtOH	246(2.46)	78-0437-64
$C_3H_7NO_2$			
Propane, 1-nitro-	hexane	279(1.34)	46-3225-64
	MeOH	275(1.30)	46-3225-64
anion	H_2O	231.5(4.00)	44-2674-65
Propane, 2-nitro-, anion	H_2O	223(4.04)	44-2674-65
sodium salt	H_2O	222(4.02)	44-1687-65
$C_3H_7N_3S$			
Acetaldehyde, thiosemicarbazone	H_2O	260(4.3)	25-0546-64
$C_3H_7N_5O_3$			
1,3-Propanediamine, 1,2,3-tri-oximino-	pH 1	276(3.60)	4-0253-65
	pH 11	267(4.08)	4-0253-65
$C_3H_8N_2$			
Formaldehyde, dimethylhydrazone	MeOH	236(3.71)	35-2395-64
$C_3H_8N_2O$			
Ethylamine, N-methyl-N-nitroso-	C_6H_{12}	233(3.76)	35-4373-64
$C_3H_8N_4$			
Pyruvaldehyde, dihydrazone	pH 1	292(4.00)	44-0560-65
	pH 11	266(4.16)	44-0560-65
C_3KN_3OS			
1,2,4-Thiadiazole-3-carbonitrile, 4-hydroxy-, potassium salt	H_2O	213(3.93),334(3.92)	35-2861-64
C_3N_4			
Malononitrile, diazo-	MeCN	241(4.09),313(2.03), 370s(1.70)	35-0652-65

Compound	Solvent	$\lambda_{max}(\log \epsilon)$	Ref.
$C_4Br_2N_2S$			
4-Isothiazolecarbonitrile, 3,5-dibromo-	CH_2Cl_2	253(3.80),267(3.84), 274s(3.78)	44-0660-64
$C_4Br_2O_3$			
Maleic anhydride, dibromo-	dioxan	211(4.09),275(3.90)	24-0764-65
$C_4ClF_2KO_2$			
1,3-Cyclobutanedione, 4-chloro-2,2-difluoro-, K deriv.	H_2O	243(4.16)	44-1629-65
$C_4Cl_2F_4$			
1,3-Butadiene, 1,4-dichloro-tetrafluoro-	gas	218.9(3.07)	46-1768-64
	hexane	230.0(4.07)	46-1768-64
1,3-Butadiene, 2,3-dichloro-tetrafluoro-	gas	198(3.86),216(3.50)	46-1768-64
	hexane	220.0(3.79)	46-1768-64
$C_4Cl_2N_2S$			
4-Isothiazolecarbonitrile, 3,5-dichloro-	isooctane	240(3.67),263(3.85), 270s(3.80)	44-0660-64
$C_4Cl_2O_3$			
Maleic anhydride, dichloro-	dioxan	211(3.96),266(3.85)	24-0764-65
$C_4Cl_3F_3$			
1,3-Butadiene, 1,2,3-trichloro-1,4,4-trifluoro-	gas	201(2.87),227(2.40)	46-1768-64
	hexane	201(4.05),228(3.71)	46-1768-64
$C_4Cl_4F_2$			
1,3-Butadiene, 1,2,3,4-tetrachloro-1,4-difluoro-	gas	207(4.21),233(3.66)	46-1768-64
	hexane	207(4.22),240(3.82)	46-1768-64
C_4Cl_4Se			
Selenophene, tetrachloro-	C_6H_{12}	266(3.97)	89-0260-65
C_4Cl_4Te			
Tellurophene, tetrachloro-	C_6H_{12}	295(4.05)	89-0260-65
C_4Cl_6			
Butadiene, hexachloro-	gas	217(3.34),256(3.11)	46-1768-64
	hexane	222.0(4.27),247(3.8)	46-1768-64
$C_4F_3KO_2$			
1,3-Cyclobutanedione, 2,2,4-trifluoro-, K deriv.	H_2O	243(4.31)	44-1629-65
C_4F_6			
Butadiene, hexafluoro-	gas	203.0(3.56)	46-1768-64
	hexane	202.0(3.68)	46-1768-64
C_4HBrN_2S			
3-Isothiazolecarbonitrile, 4-bromo-	EtOH	273(3.86)	39-3114-64
$C_4HBr_2NO_2$			
Maleimide, 2,3-dibromo-	dioxan	240(4.22),300(3.18)	24-0764-65
$C_4HCl_2NO_2$			
Maleimide, 2,3-dichloro-	dioxan	229(4.22),234(4.23)	24-0764-65

Compound	Solvent	$\lambda_{max}(\log \epsilon)$	Ref.
$C_4HCl_2NO_2S$			
4-Isothiazolecarboxylic acid, 3,5-dichloro-	EtOH	259(3.82)	44-0660-64
$C_4HF_2KN_2O$			
2-Pyrimidinol, 4,6-difluoro-, potassium salt	pH 7	217(4.18),263(3.72)	65-1433-65
	pH 13	217(4.15),263(3.67)	65-1433-65
$C_4HF_3N_2$			
Pyrimidine, 2,4,6-trifluoro-	MeOH	233(3.86)	65-1433-65
$C_4H_2BrClN_2O$			
3(2H)-Pyridazinone, 5-bromo-4-chloro-	EtOH	216(4.24),299(3.58)	22-2124-64
	NaOH	222(4.25),308(3.65)	22-2124-64
C_4H_2BrNOS			
3-Isothiazolecarboxaldehyde, 4-bromo-	EtOH	259(3.90)	39-3114-64
5-Isothiazolecarboxaldehyde, 4-bromo-	EtOH	217(3.55),260(3.81)	39-0446-64
$C_4H_2BrNO_2S$			
3-Isothiazolecarboxylic acid, 4-bromo-	EtOH	267(3.82)	39-3114-64
$C_4H_2Br_4O$			
Cyclobutanone, 2,2,4,4-tetrabromo-	C_6H_{12}	243(3.55),348(1.60)	22-1976-64
	dioxan	242(3.63)	22-1976-64
	MeCN	238(3.36),340(1.60)	22-1976-64
	CCl_4	347(1.61)	22-1976-64
$C_4H_2ClFN_2O$			
4-Pyrimidinol, 2-chloro-5-fluoro-	pH 1	229(3.72),259(3.70)	87-0253-65
C_4H_2ClNOS			
5-Isothiazolecarboxaldehyde, 4-chloro-	EtOH	214(3.65),256(3.88)	39-0446-64
$C_4H_2ClN_3O_3$			
4-Pyrimidinol, 6-chloro-5-nitro-	pH 1	285(3.51)	44-0829-65
$C_4H_2ClN_3S$			
4-Isothiazolecarbonitrile, 5-amino-3-chloro-	CH_2Cl_2	259(3.90)	44-0660-64
$C_4H_2ClN_5$			
v-Triazolo[4,5-c]pyridazine, 6-chloro-	EtOH	209(4.31),280(3.62), 316s(3.17)	4-0247-64
v-Triazolo[4,5-c]pyridazine, 7-chloro-	EtOH	210(4.18),275(3.95), 292s(3.84),305s(3.80)	4-0247-64
$C_4H_2Cl_2N_2O$			
3(2H)-Pyridazinone, 4,5-dichloro-	EtOH	215(4.29),299(3.59)	22-2124-64
	NaOH	220(4.34),306(3.66)	22-2124-64
$C_4H_2Cl_2N_2OS$			
4-Isothiazolecarboxamide, 3,5-dichloro-	EtOH	256(3.82)	44-0660-64
	CH_2Cl_2	259(3.83)	44-0660-64

Compound	Solvent	λ_{max}(log ϵ)	Ref.
$C_4H_2F_3N_3O_2$			
6-Azauracil, 5-(trifluoromethyl)-	pH 1	262.5(3.76)	4-0495-65
	pH 13	292(3.84)	4-0495-65
C_4H_2INOS			
5-Isothiazolecarboxaldehyde, 4-iodo-	EtOH	204(3.53),218(3.40), 265(3.74)	39-0446-64
$C_4H_2N_2S$			
3-Isothiazolecarbonitrile	EtOH	221(3.67),256(3.92)	39-3114-64
$C_4H_2N_2OS$			
[1,2,3]Thiadiazolo[5,4-d]pyrimidin-5(4H)-one	EtOH	236(3.57),287(3.92)	44-2121-64
[1,2,3]Thiadiazolo[5,4-d]pyrimidin-7(6H)-one	pH 1	233(3.84),286(3.72)	44-2121-64
	pH 13	237(3.87),258s(--), 310(3.80)	44-2121-64
$C_4H_3BrN_2OS$			
3-Isothiazolecarboxamide, 4-bromo-	EtOH	271(3.74)	39-3114-64
C_4H_3BrSe			
Selenophene, 2-bromo-	n.s.g.	246(3.96)	65-0841-64
Selenophene, 3-bromo-	n.s.g.	258(3.88)	65-0841-64
$C_4H_3ClN_2O$			
4-Pyrimidinol, 6-chloro-	pH 1	226(3.75),272(3.61)	87-0005-64
	pH 10	232(4.01),267(3.55)	87-0005-64
$C_4H_3ClN_4S$			
4-Isothiazolecarbonitrile, 3-chloro-5-hydrazino-	EtOH	268(3.99)	44-0660-64
$C_4H_3ClN_6$			
Tetrazolo[1,5-b]pyridazine, 7-amino-8-chloro-	EtOH	214(4.37),264(4.18), 332(4.01)	4-0042-64
Tetrazolo[1,5-b]pyridazine, 8-amino-7-chloro-	EtOH	205(4.19),273(3.96), 299(4.04)	4-0042-64
$C_4H_3Cl_2N_3$			
Pyrimidine, 2-amino-4,6-dichloro-	pH 1	233(4.18),298(3.76)	87-0792-64
	pH 10	233(4.17),299(3.75)	87-0792-64
$C_4H_3Cl_2N_3O$			
s-Triazine, 2,4-dichloro-6-methoxy-	MeOH	240(3.33)	20-0585-64
$C_4H_3FN_2O$			
2-Pyrimidinol, 5-fluoro-	pH 1	213(4.00),319(3.65)	73-3895-65
	H_2O	214(4.01),316(3.66)	73-3895-65
	pH 13	219(3.99),309(3.67)	73-3895-65
4-Pyrimidinol, 1-fluoro-	pH 1	227(3.79),251(3.70)	73-3895-65
	H_2O	226(3.79),256(3.71)	73-3895-65
	pH 13	227(3.94),266(3.69)	73-3895-65
$C_4H_3FN_2O_2$			
Uracil, 6-fluoro-	pH 1	247(3.96)	65-1433-65
	pH 7	266(4.13)	65-1433-65
	pH 13	266(4.00)	65-1433-65
	MeOH	247(3.90)	65-1433-65

Compound	Solvent	λ max(log ε)	Ref.
potassium salt	N HCl	248(3.93)	87-0207-64
	pH 6.5	267(4.12)	87-0207-64
	N NaOH	258(3.90)	87-0207-64
$C_4H_3F_2N_3O_3$			
Cyclobutanetrione, difluoro-, trioxime	EtOH	216(3.99),293(3.98)	44-3698-65
$C_4H_3F_7N_2S$			
Butyric acid, heptafluorothio-, hydrazide	isooctane	284(3.92)	44-3739-65
C_4H_3N			
Butadienenitrile	n.s.g.	207(4.0)	77-0321-65
C_4H_3NOS			
3-Isothiazolecarboxaldehyde	EtOH	205(3.59),245(3.89)	39-3114-64
5-Isothiazolecarboxaldehyde	EtOH	214(3.66),246(3.83)	39-0446-64
$C_4H_3NO_2S$			
3-Isothiazolecarboxylic acid	EtoH	233(3.53),256(3.91)	39-3114-64
5-Isothiazolecarboxylic acid	EtOH	229(3.95),263(3.76)	39-0446-64
4,5-Thiazolidinedione, 2-methylene-	dioxan	305(3.8)	24-2954-65
$C_4H_3N_3OS$			
1,2,5-Thiadaizole-3-carbonitrile, 4-methoxy-	H_2O	296(4.06)	35-2861-64
$C_4H_3N_3O_4$			
Oxonic acid	pH 8.5	253(3.87)	37-2491-65
Violuric acid, copper complex	pH 5.04	415(3.0)	28-0972-65B
$C_4H_3N_5$			
Tetrazolo[1,5-c]pyrimidine	pH 1	266(4.29)	44-0829-65
	pH 7	253(4.30)	44-0829-65
	pH 13	261(4.28)	44-0829-65
v-Triazolo[4,5-c]pyridazine	EtOH	204(4.04),270(3.90), 300s(3.42)	4-0247-64
$C_4H_3N_5O$			
5-Azahypoxanthine	pH 1	247(3.72)	35-1980-65
	pH 13	244s(3.82),257(3.85)	35-1980-65
v-Triazolo[4,5-c]pyridazin-7-ol	EtOH	208(4.20),215(4.18), 261s(3.86),278(3.89), 327s(3.91),364(4.02)	4-0247-64
$C_4H_3N_5OS$			
v-Triazolo[4,5-d]pyrimidin-5-ol, 7-mercapto-	pH 1.0	272(4.00)	44-0408-65
	pH 12.5	246(4.05),313(3.98)	44-0408-65
$C_4H_3N_5O_2$			
Furazano[3,4-d]pyrimidine, 7-hydroximino-6,7-dihydro-	pH 1	286(3.40)	4-0253-65
	pH 11	287(3.34)	4-0253-65
5H-v-Triazolo[4,5-d]pyrimidine-5,7(6H)-dione, 1,4-dihydro-	pH 2.0	263(3.81)	24-1060-65
	pH 7.0	265(3.93)	24-1060-65
	pH 13	235s(3.73),285(3.81)	24-1060-65

Compound	Solvent	λ_{max}(log ϵ)	Ref.
$C_4H_3N_5O_3$			
v-Triazolo[4,5-d]pyrimidine-5,7-diol, 4-oxide	pH 1.2	269(3.72)	44-0408-65
	pH 12.5	302(3.74)	44-0408-65
$C_4H_3N_5S$			
[1,2,3]Thiadiazolo[5,4-d]pyrimidine, 7-amino-	EtOH	244(3.79),263s(3.51), 272s(3.46),313(3.69)	44-2121-64
C_4H_4BrNOS			
3-Isothiazolemethanol, 4-bromo-	EtOH	216(3.13),258(3.88)	39-3114-64
$C_4H_4BrN_3O$			
Pyrazole-1-carboxamide, 4-bromo-	EtOH	245(3.90)	28-0606-65A
$C_4H_4Br_2O$			
Cyclobutanone, 2,2-dibromo-	C_6H_{12}	316(2.24),326(2.22)	22-1976-64
	dioxan	315(2.36)	22-1976-64
	MeCN	314(2.26)	22-1976-64
	CCl_4	316(2.30),325(2.29)	22-1976-64
Cyclobutanone, 2,4-dibromo-	C_6H_{12}	322(2.34)	22-1976-64
	dioxan	317(2.38)	22-1976-64
	MeCN	316(2.37)	22-1976-64
	CCl_4	320(2.38)	22-1976-64
C_4H_4ClNS			
Isothiazole, 4-chloro-3-methyl-	EtOH	213(3.40),255(3.92)	39-0446-64
$C_4H_4ClN_3$			
Pyrimidine, 4-amino-6-chloro-	pH 1	239(3.92),251s(3.82)	87-0005-64
	pH 10	237(4.10),272(3.52)	87-0005-64
$C_4H_4ClN_3O$			
2-Pyrimidinol, 4-amino-6-chloro-	pH 1	279(4.12)	87-0005-64
	pH 10	281.5(3.95)	87-0005-64
4-Pyrimidinol, 5-amino-6-chloro-	pH 1	265s(3.87),288(3.93)	44-0829-65
	pH 7	265s(3.88),288(3.93)	44-0829-65
	pH 13	258(3.84),279(3.89)	44-0829-65
$C_4H_4ClN_3OS$			
4-Isothiazolecarboxamide, 5-amino-3-chloro-	EtOH	222(4.39),267(3.98)	44-0660-64
$C_4H_4ClN_3O_2$			
Imidazole, 4-chloro-1-methyl-5-nitro-	14M H_2SO_4	272(3.82)	44-0862-64
	dil. H_2SO_4	312(3.93)	44-0862-64
Imidazole, 5-chloro-1-methyl-4-nitro-	14M H_2SO_4	272(3.88)	44-0862-64
	dil. H_2SO_4	308(3.85)	44-0862-64
$C_4H_4Cl_2$			
1,3-Butadiene, 1,4-dichloro-, cis-cis	n.s.g.	231s(4.15),238(4.34), 246(4.42),255(4.26)	24-0602-64
cis-trans	n.s.g.	243.2(4.46)	24-0602-64
trans-trans		234s(4.45),240(4.51), 248s(4.37)	24-0602-64
$C_4H_4Cl_5IrK_2N_2$			
Potassium pentachloro(pyrazine)-iridate(III)	H_2O	268(3.41),300s(3.47), 325(3.52),374s(3.27)	28-3420-65B

Compound	Solvent	$\lambda_{max}(\log \epsilon)$	Ref.
C_4H_4INS			
Isothiazole, 4-iodo-3-methyl-	EtOH	203(3.77),264(3.99)	39-0446-64
$C_4H_4N_2$			
Pyrazine	gas	324f(--),373f(--)	29-0192-63
Pyridazine	C_6H_{12}	335(2.6)	46-1999-64
	H_2O	300(2.6)	46-1999-64
	dioxan	330(2.5)	46-1999-64
Pyrimidine	$C_6H_{11}Me$	190(3.78),210s(3.00), 242(3.30),310(1.54)	35-0011-65
$C_4H_4N_2O$			
4-Pyrimidinol	pH 13.0	227(4.05),263(3.56)	12-0559-65
$C_4H_4N_2OS$			
3-Isothiazolecarboxamide	EtOH	216(3.83),256(3.94)	39-3114-64
4-Pyrimidinol, 6-mercapto-	pH 1	244(3.96),305(4.22)	87-0005-64
	pH 10	228s(4.14),238(4.19), 294(4.20)	87-0005-64
Thiocyanic acid, 2-carbamoylvinyl	H_2O	253(4.04)	44-2660-65
ester, cis	Et_2O	253(4.06)	44-2660-65
trans	EtOH	242(4.11)	44-2660-65
$C_4H_4N_2O_2$			
3-Pyrazolecarboxylic acid	n.s.g.	214(3.96)	39-3037-65
4,6-Pyrimidinediol	pH -2.8	242(3.92)	12-0567-64
	7.5N HCl	243(3.83)	65-3182-64
	pH 2.0	253(3.98)	12-0567-64
	neutral	253(3.98)	65-3182-64
	pH 7.8	252(3.88)	12-0567-64
	base	251(3.91)	65-3182-64
	EtOH	235(3.47),254(3.51), 256(3.52)	12-0567-64
Uracil	pH 7.0	181(4.19),202(3.94), 259(3.91)	35-0011-65
	pH 14.0	276(3.81)	35-0011-65
	Me_3PO_4	181(4.07),203(3.91), 258(3.89)	35-0011-65
$C_4H_4N_2O_2S_2$			
3H-1,2-Dithiol-3-one, 4-(methyl-nitrosamino)-	EtOH	248(3.70),323(3.85)	12-0061-65
$C_4H_4N_2O_3$			
Isoxazole, 3-methyl-4-nitro-	isooctane	225(3.78),243s(3.72)	32-0915-64
	H_2O	226(3.72),250(3.73)	32-0915-64
Isoxazole, 5-methyl-4-nitro-	isooctane	253(3.75)	32-0915-64
	H_2O	263(3.74)	32-0915-64
$C_4H_4N_2O_3S$			
1,2,5-Thiadiazole-3-carboxylic acid, 4-methoxy-	EtOH	288(3.93)	35-2861-64
$C_4H_4N_2S$			
2-Pyrimidinethiol	pH 13.0	231(3.69),270(4.24)	12-0559-65
4-Pyrimidinethiol	pH 13.0	293(4.04)	12-0559-65

Compound	Solvent	$\lambda_{max}(\log \epsilon)$	Ref.
$C_4H_4N_4$			
Carbonyl cyanide, N-methylhydrazone	MeCN	228(2.95),300(4.20)	44-4198-65
$C_4H_4N_4O_3S$			
1,2,5-Thiadiazole-3-carboxylic acid, 4-ureido-	pH 1	231(4.08),320(3.86)	44-2141-64
	pH 7 and 13	224(4.14),297(3.91)	44-2141-64
$C_4H_4N_4O_4$			
Imidazole, 1-methyl-2,4-dinitro-	5M H_2SO_4	305(4.06)	44-0862-64
Uracil, 6-amino-5-nitroso-, 1-oxide	pH 1.2	318(3.96)	44-0408-65
	pH 12.5	314(4.19)	44-0408-65
$C_4H_4N_6$			
5-Azaadenine	pH 1	261(3.89)	35-1980-65
	pH 13	249(4.06),274s(3.78)	35-1980-65
Tetrazolo[1,5-b]pyridazine, 7-amino-	EtOH	218(4.29),262(4.17), 334(3.87)	4-0042-64
Tetrazolo[1,5-b]pyridazine, 8-amino-	EtOH	209(4.05),294(4.29)	4-0042-64
v-Triazolo[4,5-c]pyridazine, 7-amino-	EtOH	210(4.25),273(4.08), 325(4.21),331(4.21), 339s(4.13)	4-0247-64
$C_4H_4N_6O$			
8-Azaisoguanine	pH 1	276(3.92)	44-2674-64
	pH 11	248(3.97),277(4.01)	44-2674-64
$C_4H_4N_6S$			
[1,2,3]Thiadiazolo[5,4-d]pyrimidine, 7-hydrazino-	EtOH	247(3.89),275s(3.58), 328(3.84)	44-2121-64
C_4H_4S			
Thiophene, S-oxide	MeOH	216(3.08),264(2.70)	73-1158-65
$C_4H_4O_2$			
1,3-Cyclobutanedione	EtOH	237(4.07)	12-0765-64
Δ^2-Furenid-4-one	Et_2O	251(4.03)	33-1322-65
$C_4H_4O_2S$			
Thiophene, S,S-dioxide	MeOH	220(3.30),289(2.94)	73-1158-65
$C_4H_4O_4$			
Oxalacetic acid	EtOH	260(3.7)	23-2806-64
$C_4H_4O_6$			
Maleic acid, dihydroxy-	n.s.g.	292(3.90)	28-1512-64A
$C_4H_5BN_2O_4$			
5-Uracilboronic acid	pH 1	262(4.03)	35-1869-64
	pH 11	284(3.75)	35-1869-64
$C_4H_5BrN_4$			
Pyrimidine, 2,4-diamino-5-bromo-	pH 1.0	282(3.64)	39-0755-65
	pH 9.4	230(4.06),295(3.83)	39-0755-65
C_4H_5BrO			
Cyclobutanone, 2-bromo-	C_6H_{12}	308(2.15)	22-1976-64
	dioxan	305(2.19)	22-1976-64
	MeCN	303(2.16)	22-1976-64
	CCl_4	310(2.15)	22-1976-64

Compound	Solvent	λ_{max}(log ε)	Ref.
$C_4H_5BrO_4$			
Succinic acid, bromo-	H_2O	220s(2.77)	39-3928-65
$C_4H_5Br_2NO$			
2-Pyrrolidinone, 3,3-dibromo-	MeOH	215(3.64)	24-1970-64
$C_4H_5ClN_4$			
Pyridazine, 3,4-diamino-5-chloro-	EtOH	216(4.31),264(4.05), 295(4.05)	4-0042-64
Pyridazine, 3,4-diamino-6-chloro-	EtOH	210(4.28),226(4.26), 262(4.11),303(4.14)	4-0247-64
Pyrimidine, 2,4-diamino-6-chloro-	pH 2	227(3.98),299(3.87)	87-0792-64
	pH 10	229s(3.97),282(3.89)	87-0792-64
$C_4H_5ClN_4O$			
s-Triazine, 2-amino-4-chloro-6-methoxy-	MeOH	218(4.14),248(3.42)	20-0585-64
C_4H_5ClO			
3-Buten-2-one, 4-chloro-	EtOH	229(4.20)	44-0385-64
$C_4H_5ClO_5S_3$			
4-Hydroxy-2-(methylthio)-1,3-dithiol-1-ium perchlorate	70% $HClO_4$	221(3.29),305(4.20)	44-1708-64
C_4H_5F			
1,3-Butadiene, 1-fluoro-, cis	n.s.g.	215(3.96),220(3.98), 226s(3.83)	24-0598-64
trans	n.s.g.	215(4.00)	24-0598-64
$C_4H_5FN_4$			
Pyrimidine, 2,4-diamino-5-fluoro-	pH 7	289(3.77)	87-0253-65
C_4H_5NO			
Acetoacetonitrile	EtOH	228(3.37)	33-1424-64
Acrylonitrile, 3-methoxy-, cis	n.s.g.	224(4.13)	44-1800-64
C_4H_5NOS			
3-Isothiazolemethanol	EtOH	213(3.47),248(3.95)	39-3114-64
4-Isothiazolin-3-one, 2-methyl-	EtOH	275(3.86)	44-2660-65
	Et_2O	281(3.82)	44-2660-65
	n.s.g.	272(3.86)	88-1477-64
$C_4H_5NOS_2$			
3H-1,2-Dithiol-3-one, 4-(methylamino)-	EtOH	348.5(3.79)	12-0061-65
$C_4H_5NO_2S$			
2,4-Thiazolidinedione, 5-methyl-	n.s.g.	293(1.83)	65-0886-65
$C_4H_5NO_2SSe$			
2-Selenophenesulfonamide	n.s.g.	247(4.07),265(3.96)	65-2201-64
C_4H_5NS			
Acrylonitrile, 3-(methylthio)-, cis	n.s.g.	274(4.08)	44-1800-64
Isothiazole, 4-methyl-	EtOH	208(3.55),251(3.82)	39-0446-64
Isothiazole, 5-methyl-	EtOH	213(3.64),243(3.95)	39-0446-64
Isothiocyanic acid, allyl ester, iodine complex	CCl_4	291(4.34)	1-1897-65

Compound	Solvent	λ_{max}(log ϵ)	Ref.
$C_4H_5NS_2$			
2H-1,3-Thiazine-2-thione, 3,6-di-hydro-	H_2O	300(4.08),325s(--)	39-4008-64
	EtOH	315(4.04)	39-4008-64
$C_4H_5N_3O$			
Cytosine	pH 1	275(4.03)	87-0005-64
	pH 10	267.5(3.81)	87-0005-64
	Me_3PO_4	185(4.09),204(4.08), 237s(3.54),277(3.88)	35-0011-65
sulfate	N H_2SO_4	275(4.05)	39-1515-65
photoisomer	EtOH	305(4.08)	28-0196-64A
Isocytosine	pH 6	203(4.9),264s(4.2), 284(4.3)	28-4387-64B
	pH 7.0	280(3.64)	12-0559-65
	EtOH	205(4.6),220(4.6), 286(4.6)	28-4387-64B
	10% EtOH in Et_2O	221(4.5),286(4.6)	28-4387-64B
Pyrazole-1-carboxamide	EtOH	233(3.98)	28-0606-65A
4-Pyrimidinol, 6-amino-	pH -2.0	217s(4.02),257(4.07)	12-0559-65
	pH 7	213(4.42),258(3.88)	12-0559-65
	pH 12.5	213(4.54),254(3.60)	12-0559-65
$C_4H_5N_3OS$			
Cytosine, 6-mercapto-	pH 1	244(3.90),318(4.58)	87-0005-64
	pH 10	222(4.22),305(4.41)	87-0005-64
6-Pyrimidinol, 4-amino-2-mercapto-	pH 1	246(3.89),277(4.28)	87-0005-64
	pH 10	243(4.24),259s(3.95), 289(4.12)	87-0005-64
Thiazole-4-carboxamide, 5-amino-	EtOH	279(3.96)	94-1319-65
$C_4H_5N_3O_2$			
Cytosine, 3-oxide	pH 7	221(4.28),270(3.66)	44-2766-65
Cytosine, N^6-hydroxy-	pH 1	216(3.94),276(4.07)	39-0208-65
	pH 7	233(4.03),273(3.72)	39-0208-65
Imidazole, 1-methyl-2-nitro-	pH 4.63	325(3.93)	44-0862-64
	5M H_2SO_4	300(3.89)	44-0862-64
Imidazole, 1-methyl-4-nitro-	pH 4.63	300(3.90)	44-0862-64
	8M H_2SO_4	266(3.87)	44-0862-64
Imidazole, 1-methyl-5-nitro-	pH 7.38	305(3.81)	44-0862-64
	2M H_2SO_4	266(3.70)	44-0862-64
4,6-Pyrimidinediol, 2-amino-	pH 1	257(4.05)	87-0792-64
	pH 10	261(4.12)	87-0792-64
2,4(1H,3H)-s-Triazinedione, 1-methyl-	EtOH	245(3.20)	73-2060-64
Uracil, 6-amino-	pH 1	264(4.24)	87-0792-64
	pH 10	263(4.23)	87-0792-64
$C_4H_5N_3O_2S$			
4-Pyrimidinesulfinic acid, 6-amino-	pH 1	249(4.12)	87-0005-64
	pH 10	235(4.00),280(3.57)	87-0005-64
1,2,3-Thiadiazole-4-carboxylic acid, 5-amino-, methyl ester	EtOH	267(3.93),285s(3.82)	44-2121-64
$C_4H_5N_3O_3$			
Uracil, 6-amino-, 1-oxide	pH 1.1	266(4.26)	44-0408-65
	pH 12.9	276(3.95)	44-0408-65
Uracil, 6-(hydroxyamino)-	pH 2	264(3.80)	87-0884-65

Compound	Solvent	λ_{max}(log ε)	Ref.
$C_4H_5N_3S$			
2-Pyrimidinethiol, 4-amino-	pH 0.2	224(4.15),277(4.30), 315s(3.74)	12-0559-65
	pH 7.0	242(4.26),269(4.26)	12-0559-65
4-Pyrimidinethiol, 2-amino-	pH -0.2	259(3.72),325(4.13)	12-0559-65
	pH 5.4	236(3.67),258(3.65), 342(4.11)	12-0559-65
	pH 10.4	262(3.77),312(4.16)	12-0559-65
4-Pyrimidinethiol, 6-amino-	pH 1	240(4.23),304(4.33)	87-0005-64
	pH 2.5	238(4.21),304(4.29)	12-0559-65
	pH 10	226(4.27),243(4.24), 291(4.16)	87-0005-64
$C_4H_5N_3S_2$			
3,4-Pyridazinedithiol, 5-amino-	EtOH	212(4.30),255(4.26), 278(4.30),314(4.08)	4-0042-64
	5N NaOH	227(4.32),262(4.34), 319(4.12)	4-0042-64
$C_4H_5N_5O$			
Tetrazolo[1,5-c]pyrimidin-5-ol, 5,6-dihydro-	pH 1	265(4.31)	44-0829-65
	pH 7	262(4.29)	44-0829-65
	pH 13	261(4.29)	44-0829-65
$C_4H_5N_5O_3$			
4-Pyrimidinol, 6-hydrazino-5-nitro-	pH 1	228(4.16),280s(3.58), 327(3.80)	44-0829-65
	pH 7	229(4.10),295s(3.51), 352(3.81)	44-0829-65
$C_4H_5N_7$			
3H-v-Triazolo[4,5-d]pyrimidine, 3,7-diamino-	pH 1	262(4.03)	44-0829-65
	pH 7	278(4.02)	44-0829-65
	pH 13	278(4.00)	44-0829-65
C_4H_6			
Butadiene	C_6H_{12}	218.5(4.36)	44-3527-64
C_4H_6BrNO			
2-Pyrrolidinone, 3-bromo-	MeOH	217(3.66)	24-1970-64
$C_4H_6BrN_3$			
Pyrazole, 5-amino-4-bromo-3-methyl-	EtOH	232(3.66)	95-1113-64
s-Triazine, 2,4-diamino-6-(bromomethyl)-	MeOH	270(3.51)	4-0340-65
$C_4H_6ClN_5$			
Pyridazine, 4-amino-5-chloro-3-hydrazino-	EtOH	219(4.29),266(4.06), 291(4.00)	4-0042-64
Pyrimidine, 2,4,5-triamino-6-chloro-	pH 1	228(4.10),306(3.71)	87-0792-64
	pH 10	236s(--),302(3.75)	87-0792-64
C_4H_6ClOP			
2-Phospholene, 1-chloro-, 1-oxide	C_6H_{12}	200(3.56)	78-1593-64
3-Phospholene, 1-chloro-, 1-oxide	C_6H_{12}	192(3.47),237(2.31)	78-1593-64

Compound	Solvent	$\lambda_{max}(\log \epsilon)$	Ref.
$C_4H_6N_2O$			
Furazan, 3,4-dimethyl-	EtOH	212(3.5)	94-1445-65
5-Pyrazolinone, 1-methyl-	N HCl	224(3.84)	5-0134-65F
	H_2O	241(3.88)	5-0134-65F
5-Pyrazolinone, 3-methyl-	C_6H_{12}	220(3.57)	78-1693-65
	H_2O	238(3.88)	78-1693-65
$C_4H_6N_2OS$			
Hydantoin, 3-methyl-2-thio-	EtOH	233(3.97),263(4.20)	65-0049-65
Pseudothiohydantoin, 5-methyl-	n.s.g.	245(3.96),298s(2.71)	65-0886-65
$C_4H_6N_2O_3$			
2-Pyrrolidinone, N-nitro-	hexane	233(3.78)	4-0308-65
3-Pyrrolidinone, N-nitro-	hexane	234(3.70)	4-0308-65
$C_4H_6N_2O_4$			
Oxazoline, 4-methyl-3-nitro-, 1-oxide	H_2O	320(3.92)	70-1491-65
Oxazoline, 5-methyl-3-nitro-, 1-oxide	H_2O	320(3.93)	70-1491-65
$C_4H_6N_2S$			
Cyanothioformamide, N,N-dimethyl-	heptane	309(3.97),433(1.32)	1-1059-64
	EtOH	312(3.97),417(1.36)	1-1059-64
$C_4H_6N_4$			
Pyrimidine, 2,4-diamino-	pH 1.0	266(3.73)	39-0755-65
	pH 10.4	228(3.97),282(3.84)	39-0755-65
$C_4H_6N_4O$			
6-Azacytosine, 6-methyl-	N H_2SO_4	210(3.90),241s(3.53), 291(3.85)	73-1394-64
	H_2O	263s(--),283(3.82)	73-1394-64
4-Pyrimidinol, 2,6-diamino-	pH 1	265(4.30)	87-0792-64
	pH 10	267(4.14)	87-0792-64
s-Triazine, 2-amino-4-methoxy-	MeOH	217(4.11),249(3.55)	20-0585-64
as-Triazin-3(2H)-one, 4,5-dihydro-5-imino-4-methyl-	N H_2SO_4	202(3.79),240(3.64), 277(3.74)	73-1394-64
	H_2O	211(3.93),270(3.74)	73-1394-64
s-Triazin-2(1H)-one, 4-amino-1-methyl-	pH 5.0	247(3.74)	73-2060-64
1,2,4-Triazole, 3-acetamido-	pH 7	220(3.9),226s(3.8)	70-1475-64
	pH 12	225(3.6)	70-1475-64
$C_4H_6N_4OS$			
s-Triazine, 2-amino-4-mercapto-6-methoxy-	MeOH	243(3.82),278(3.81)	20-0585-64
$C_4H_6N_4O_2S$			
4-Pyrimidinesulfinic acid, 2,6-diamino-	pH 1	272.5(3.82)	87-0792-64
	pH 10	229(4.06),292(3.85)	87-0792-64
$C_4H_6N_4S$			
4-Pyrimidinethiol, 2,6-diamino-	pH 1	244(3.99),323(4.49)	87-0792-64
	pH 10	226(4.24),304(4.29)	87-0792-64
$C_4H_6N_6$			
N-Cyano-N,N'-dimethylguanyl azide	EtOH	231(3.98),253(3.97)	44-0650-64
Tetrazoline, 1,4-dimethylcyanimino-	EtOH	226(4.15)	44-0650-64

Compound	Solvent	$\lambda_{max}(\log \epsilon)$	Ref.
$C_4H_6N_6O$			
as-Triazine-6-carboxamide, 3,5-diamino-	H_2O	213(4.48),242(4.09), 317(3.87)	35-1976-65
$C_4H_6N_6O_2$			
Pyrimidine, 4-amino-6-hydrazino-5-nitro-	pH 1	237(4.28),335(3.83)	44-0829-65
	pH 7	338(3.92)	44-0829-65
$C_4H_6N_6O_2S$			
1,2,5-Thiadiazole-3-carboxylic acid, 4-ureido-, hydrazide	pH 1	235(4.17),306(3.90)	44-2141-64
	pH 13	315(3.88)	44-2141-64
C_4H_6O			
Cyclobutane, 1,2-epoxy-	isooctane	285(1.47),300s(1.38), 311s(1.18),322s(0.83)	22-2755-65
Cyclobutanone	C_6H_{12}	272(1.18),279(1.26), 288(1.23)	22-1976-64
	MeOH	277(1.32)	22-1976-64
	dioxan	277(1.30),284(1.29)	22-1976-64
	MeCN	275(1.27),282(1.26)	22-1976-64
$C_4H_6O_2$			
Acrylic acid, methyl ester	cyclopentadiene	none 270-350 nm	70-1290-65
	$CHCl_3$	none 270-350 nm	70-1290-65
2,3-Butanedione	EtOH	435(1.30)	23-0113-64
Butyrolactone	isooctane	214(1.4)	35-2384-64
	H_2O	203s(1.62)	35-2384-64
	MeOH	207.8(1.60)	35-2384-64
	80% MeOH	207(1.59)	35-2384-64
	EtOH	209.8(1.59)	35-2384-64
	40% EtOH	205.1(1.61)	35-2384-64
	MeCN	212.3(1.54)	35-2384-64
	$C_3H_4F_4O$	204.6(1.45)	35-2384-64
Crotonic acid	H_2O	208.3(4.09)	23-0724-64
	38.0% H_2SO_4	211.0(4.12)	23-0724-64
	49.2%	212.1(4.10)	23-0724-64
	58.9%	214.1(4.09)	23-0724-64
	63.8%	215.9(4.07)	23-0724-64
	67.7%	218.4(4.06)	23-0724-64
	70.4%	222.1(4.08)	23-0724-64
	73.2%	224.2(4.07)	23-0724-64
	77.8%	229.2(4.12)	23-0724-64
	82.2%	232.3(4.15)	23-0724-64
	90.3%	234.5(4.19)	23-0724-64
	96.2%	235.2(4.21)	23-0724-64
	100% H_2SO_4	235.5(4.21)	23-0724-64
$C_4H_6O_2S$			
Thiophene, 2,3-dihydro-, 1,1-dioxide	H_2O	194.5	73-1158-65
$C_4H_6O_2S_4$			
Dimethyl dixanthogen	heptane	240(4.31),278(3.90), 361(1.94)	1-1113-65
	EtOH	239(4.34),279(3.91), 368(1.94)	1-1113-65
	MeCN	239(4.31),282(3.92), 358(1.92)	1-1113-65

Compound	Solvent	λ_{max}(log ϵ)	Ref.
$C_4H_6O_5$			
Malic acid	H_2O	203s(2.18)	39-3928-65
C_4H_6S			
Thiophene, 2,3-dihydro-	H_2O	202.5(3.27)	73-1158-65
$C_4H_6S_3$			
1,3-Dithiane-2-thione	EtOH	292(3.84),338(3.93)	94-0253-65
	$CHCl_3$	292(3.95),340(4.02)	49-0631-65
1,3-Dithiolane-2-thione, 5-methyl-	$CHCl_3$	320(4.10)	49-0631-65
$C_4H_7F_3N_2S$			
Acetic acid, trifluorothio-, 2,2-dimethylhydrazide	MeCN	269.5(3.96)	44-3739-65
$C_4H_7F_3N_4$			
Acetone, 1,1,1-trifluoro-, guanylhydrazone	pH 1	226(4.20)	87-0806-64
	pH 11	248(4.20)	87-0806-64
$C_4H_7NOS_2$			
Propionaldehyde, β-(thiocarbamoylthio)-	H_2O	242(3.85),287(4.07)	39-4004-64
	EtOH	240(3.84),292(4.05)	39-4004-64
2-Thiazolidinethione, 4-hydroxy-4-methyl-	MeOH	245(3.85),276(4.22)	44-2146-64
$C_4H_7NO_2$			
Cyclobutane, nitro-, anion	H_2O	229(4.10)	44-2674-65
Methane, cyclopropylnitro-, anion	H_2O	242.5(4.11)	44-2674-65
$C_4H_7NO_2S$			
Malonamic acid, 3-thio-, methyl ester	EtOH	270.0(4.05)	39-1262-65
Oxamic acid, 2-thio-, ethyl ester	heptane	301.0(3.86)	1-1059-64
	heptane-10% CH_2Cl_2	454.0(1.42)	1-1059-64
	EtOH	290(3.86),390s(1.45)	1-1059-64
$C_4H_7NS_2$			
Imidocarbonic acid, dithio-, cyclic trimethylene ester	MeOH-HCl	258(4.20)	44-0738-64
$C_4H_7N_3O_4$			
Pyruvic acid, hydroxy-, semicarbazone	pH 2	250(4.0)	28-2827-65A
$C_4H_7N_5$			
s-Triazine, 2,4-diamino-6-methyl-	MeOH	257(3.52)	4-0340-65
$C_4H_7N_5O$			
4-Pyrimidinol, 5-amino-6-hydrazino-	pH 1	264(3.74)	44-0829-65
	pH 7	273(3.78)	44-0829-65
1H-1,2,4-Triazole-1-carboxamide, 5-amino-N-methyl-	EtOH	235(3.90)	35-1980-65
$C_4H_7N_5S$			
4-Pyrimidinethiol, 2,5,6-triamino-	pH 1	225s(4.29),310(4.44)	87-0792-64
	pH 10	224(4.18),325(4.14)	87-0792-64
4(3H)-Pyrimidinethione, 2,5,6-triamino-	pH 1	310(4.35)	44-3370-64

Compound	Solvent	$\lambda_{max}(\log \varepsilon)$	Ref.
s-Triazine-2(1H)-thione, tetrahydro-4,6-diimino-1-methyl-	EtOH	282(4.28)	39-6296-65
$C_4H_8CuN_2O_4$			
Glycine Cu complex	H_2O	238(3.76)	59-1227-64
$C_4H_8N_2$			
Propene, 2-(methylazo)-	EtOH	232(3.78),386(1.74)	35-4576-65
$C_4H_8N_2O$			
2-Pyrrolidinone, 3-amino-	pH 9.2	none	46-0978-65
	dioxan	231s(2.3)	46-0978-65
	MeCN	227s(2.3)	46-0978-65
$C_4H_8N_2OS$			
Oxamide, N,N'-dimethylthio-	heptane	240(3.81),298(3.84)	1-1059-64
	heptane-10% CH_2Cl_2	400(1.45)	1-1059-64
	EtOH	238(3.79),298(3.80), 393(1.52)	1-1059-64
Oxamide, N',N'-dimethyl-1-thio-	heptane	290(3.98)	1-1059-64
	EtOH	238(3.72),279(4.07), 360s(1.72)	1-1059-64
			1-1059-64
	CH_2Cl_2	370s(1.74)	1-1059-64
Oxamide, N^2,N^2-dimethyl-1-thio-	heptane	278(3.86)	1-1059-64
	EtOH	230s(3.73),275(3.97), 358(1.63)	1-1059-64
	CH_2Cl_2	375(1.62)	1-1059-64
$C_4H_8N_2O_2$			
Glyoxime, dimethyl-	hexane	224.0(4.11)	22-1895-65
	pH 1	225.5(4.11)	22-1895-65
	H_2O	224.5(4.15)	22-1895-65
	EtOH	226.0(4.23)	22-1895-65
	n.s.g.	220(4.1)	1-0653-65
Pyrrolidine, N-nitro-	hexane	235(3.85)	4-0308-65
$C_4H_8N_2O_3$			
Methanol, (methylazoxy)-, acetate	H_2O	215(3.93),270s(1.65)	10-0373-65B
N-Methyl-N-nitrosourethane	H_2O	238.0(3.85)	28-5470-64A
$C_4H_8N_2S$			
4H-1,3-Thiazine, 2-amino-5,6-dihydro-	EtOH-HBr	211(4.06)	1-0766-65
	EtOH-HCl	214(4.00)	1-0766-65
$C_4H_8N_2S_2$			
Oxamide, N,N-dimethyldithio-	heptane	269(3.95),301(3.78)	1-1059-64
	heptane-0.5 % CH_2Cl_2	370s(2.65)	1-1059-64
	EtOH	270(4.22),350s(2.64)	1-1059-64
Oxamide, N,N'-dimethyldithio-	heptane	304(4.09)	1-1059-64
	heptane-10% CH_2Cl_2	462(1.23)	1-1059-64
	EtOH	303(4.06),440s(1.30)	1-1059-64
$C_4H_8N_4O_3$			
Cytosine, 5,6-dihydro-N-hydroxy-6-(hydroxyamino)-	pH 7	220(4.03)	39-0208-65

Compound	Solvent	λ_{max}(log ϵ)	Ref.
$C_4H_8N_4S_2$			
1,3,4-Thiadiazole, 2-(1-methyl-hydrazino)-5-(methylthio)-	EtOH	287(4.03)	1-0871-64
Δ^2-1,3,4-Thiadiazolin-5-one, 4-methyl-2-(methylthio)-, hydrazone	EtOH	289(3.74)	1-0871-64
$C_4H_8N_6$			
Pyrimidine, 4,5-diamino-6-hydra-zino-, dihydrochloride	pH 7	280(3.90)	44-0829-65
$C_4H_8O_2$			
Ethyl acetate	gas	164(3.56)	82-0001-64B
	H_2O	204.5(1.72)	35-2384-64
	MeOH	207.9(1.75)	
	80% MeOH	205.9(1.78)	
	EtOH	209.7(1.81)	
	40% EtOH	204.9(1.75)	
	MeCN	210.2(1.76)	
	$C_3H_4F_3O$	204.5(1.74)	
Formic acid, isopropyl ester	MeOH	213.4(1.81)	35-2384-64
	EtOH	213.6(1.83)	
	MeCN	213.7(1.81)	
	$C_3H_4F_4O$	208.3(1.78)	
Formic acid, propyl ester	isooctane	212(1.89),216f(1.9), 220(1.87)	35-2384-64
	MeOH	212.4(1.85)	
	EtOH	212.4(1.88)	
	MeCN	213.4(1.85)	
	$C_3H_4F_4O$	204.7(1.73)	
Isobutyric acid, lead salt	C_6H_{12}	232(4.18)	88-2847-65
Propionic acid, methyl ester	isooctane	211(1.79)	35-2384-64
	H_2O	204.4(1.70)	
	MeOH	207.7(1.79)	
	EtOH	208.7(1.78)	
	MeCN	209.5(1.72)	
	$C_3H_4F_4O$	205.7(1.72)	
$C_4H_8O_2S_2$			
2,3-Butanediol, cyclic thiosulfite	isooctane	255(3.40)	35-3891-64
isomer	isooctane	257(3.39)	35-3891-64
$C_4H_8S_2$			
m-Dithiane	C_6H_{12}	251(2.69)	78-0437-64
	EtOH	232s(2.56),250(2.67)	78-0437-64
$C_4H_9NO_2$			
Butane, 1-nitro-	hexane	279(1.36)	46-3225-64
	MeOH	276(1.37)	46-3225-64
anion	H_2O	232(4.00)	44-2674-65
Propane, 2-methyl-1-nitro-, anion	H_2O	233.5(3.98)	44-2674-65
$C_4H_9N_3O_2$			
Isobiuret, 1,4-dimethyl-	EtOH	221(4.18)	73-2060-64
$C_4H_9N_3S$			
Acetaldehyde, 2'-methylthiosemi-carbazone	pH 1	211(3.94)	44-2234-65
	pH 13	211(3.92),239(3.86), 265(4.30)	44-2234-65
	MeOH	211(3.90),239(3.82), 271(4.40)	44-2234-65

Compound	Solvent	λ_{max}(log ε)	Ref.
$C_4H_{10}N_2$			
Acetaldehyde, dimethylhydrazone	MeOH	238(3.71)	35-2395-64
Acetaldehyde, ethylhydrazone	MeOH	232(3.68)	35-2395-64
$C_4H_{10}N_4O_2$			
Propionaldehyde, α-amino-β-hydroxy-, semicarbazone	H_2O	228(3.91)	28-3059-64A
$C_4H_{11}FSn$			
Tin, (fluoromethyl)trimethyl-	hexane	211.5(1.38)	30-0969-65
$C_4H_{12}Ge$			
Germane, tetramethyl-	n.s.g.	none 175-250 nm	70-2203-64
$C_4H_{12}N_4$			
Tetrazene, tetramethyl-	MeOH	277(3.92)	35-2395-64
$C_4H_{12}Si$			
Silane, tetramethyl-	n.s.g.	none 170-250 nm	70-2203-64
$C_4H_{12}Sn$			
Tin, tetramethyl-	hexane	215.2(0.68)	30-0969-65
	n.s.g.	184(4.3)	70-2203-64
$C_4H_{13}N_2O_2PS_2$			
Ethanol, 2-amino-, O,O-diester with phosphorodithioic acid	H_2O	220s(3.54)	28-1551-65B
$C_4K_2N_2S_2$			
Malononitrile, di(potassiomercapto)-methylene-	H_2O	310s(3.98),345(4.27)	44-0660-64
	EtOH	272(3.84),313s(3.96), 341(4.33)	44-0660-64
$C_4N_2Na_2S_2$			
Malononitrile, di(sodiomercapto)-methylene-	H_2O	271(3.45),307(4.20), 341(4.32)	44-0660-64
	EtOH	271(3.63),313(4.15), 341(4.36)	44-0660-64

Compound	Solvent	$\lambda_{max}(\log \epsilon)$	Ref.
$C_5ClF_4KO_2$			
1,3-Cyclopentaendione, 2-chloro-4,4,5,5-tetrafluoro-, K deriv.	EtOH	273(4.46)	44-1629-65
C_5ClF_4N			
Pyridine, 3-chloro-2,4,5,6-tetrafluoro-	hexane	261(3.51)	39-0594-65
	EtOH	260(3.51)	39-0594-65
$C_5Cl_2F_3N$			
Pyridine, 3,5-dichloro-2,4,6-trifluoro-	hexane	265(3.53)	39-0594-65
	EtOH	265(3.52)	39-0594-65
$C_5Cl_4O_2$			
Cyclopentene-1,2-dione, tetrachloro-	hexane	226(4.09),283(3.89), 330s(1.88),463(1.45), 475(1.28)	44-3657-65
C_5Cl_5N			
Pyridine, pentachloro-	hexane	261(2.57)	39-0594-65
	EtOH	260(2.53)	39-0594-65
C_5Cl_6			
Cyclopentadiene, hexachloro-	iso-PrOH	323(3.18)	25-0709-64
$C_5F_5KO_2$			
1,3-Cyclopentanedione, 2,4,4,5,5-pentafluoro-, K deriv.	EtOH	272(4.51)	44-1629-65
$C_5HCl_2F_3N_2$			
Pyrimidine, 2,4-dichloro-5-(trifluoromethyl)-	MeOH	213(3.97),266(3.39)	44-0835-65
$C_5HCl_2F_4NO$			
2-Cyclopenten-1-one, 2,3-dichloro-4,4,5,5-tetrafluoro-, oxime	EtOH	247(4.33)	44-3698-65
$C_5HCl_3N_4$			
Purine, 2,6,8-trichloro-	pH -5.05	210(4.45),227s(3.76), 245s(3.54),279(4.11), 287(4.09)	39-3017-65
	pH 1.0	213(4.43),248(3.65), 280(4.11),288s(4.00)	39-3017-65
C_5H_2BrNO			
2-Furonitrile, 5-bromo-	EtOH	213s(3.43),253(4.19)	39-6057-65
$C_5H_2ClN_3$			
4-Pyrimidinecarbonitrile, 2-chloro-	EtOH	269(3.62)	4-0130-64
5-Pyrimidinecarbonitrile, 2-chloro-	EtOH	228(4.10),260(3.39)	44-1740-64
$C_5H_2Cl_2N_2O_3$			
Pyridine, 2,6-dichloro-4-nitro-, 1-oxide	pH 1	245(3.91),313(3.94)	4-0196-65
	pH 11	245(3.89),313(3.92)	4-0196-65
$C_5H_2Cl_2N_4$			
Purine, 2,6-dichloro-	pH -2.28	235(3.63),242s(3.49), 273(3.98)	39-3017-65
	pH 4.0	211(4.35),249s(3.55), 275(3.95)	39-3017-65
	pH 9.5	219(4.33),280(3.93)	39-3017-65

Compound	Solvent	λ_{max}(log ε)	Ref.
$C_5H_2F_3N_3$			
Malononitrile, (1-amino-2,2,2-trifluoroethylidene)-	EtOH	283(4.18)	44-0707-64
$C_5H_2N_2S_2$			
Dicyanomethylene-1,3-dithietane	EtOH	253(3.71),303(4.19)	44-0497-64
1,3-Dithiole-4,5-dicarbonitrile	MeOH	267(3.78),383(3.93)	24-1298-64
$C_5H_2O_4S_3$			
1,3-Dithiole-4,5-dicarboxylic acid, 2-thiono-	EtOH	362(4.14)	24-1298-64
$C_5H_2O_5S_2$			
1,3-Dithiole-4,5-dicarboxylic acid, 2-oxo-	EtOH	223(3.83),301(3.69)	24-1298-64
$C_5H_3BrF_2$			
1-Penten-3-yne, 1-bromo-1,2-difluoro-	MeOH	224(4.15),234(4.02)	44-2009-64
$C_5H_3Br_2NO_2$			
Maleimide, 2,3-dibromo-N-methyl-	dioxan	210(3.73),246(4.06), 311(2.73)	24-0764-65
$C_5H_3ClN_2OS$			
4-Isothiazolecarbonitrile, 3-chloro-5-methoxy-	EtOH	247(3.80)	44-0660-64
$C_5H_3ClN_2O_2$			
Pyridine, 2-chloro-3-nitro-	pH -4.9	239(3.66),278(3.76)	39-2150-64
	pH 7.0	209(4.12),240s(3.56), 281(3.45)	39-2150-64
$C_5H_3ClN_2O_3$			
Pyridine, 2-chloro-4-nitro-, 1-oxide	pH 1	237(4.04),312(4.09)	4-0196-65
	pH 11	237(4.06),312(4.12)	4-0196-65
$C_5H_3ClN_2S_2$			
4-Isothiazolecarbonitrile, 3-chloro-5-(methylthio)-	EtOH	222(4.10),288(4.02)	44-0660-64
$C_5H_3ClN_4$			
7H-Imidazo[4,5-c]pyridazine, 7-chloro-	EtOH	212(4.30),260(4.06), 272(4.08),305s(3.54)	4-0042-64
Purine, 2-chloro-	pH -1.38	221s(3.52),271(3.91)	39-3017-65
	pH 5.0	207s(4.31),273(3.87)	39-3017-65
	pH 10.5	213(4.33),278(3.89)	39-3017-65
Purine, 6-chloro-	pH -1.85	233(3.65),257s(3.91), 262(3.97),269s(3.85)	39-3017-65
	pH 4.0	245s(3.64),265(3.95)	39-3017-65
	pH 10.0	210(4.33),274(3.92)	39-3017-65
	Me_3PO_4	193(4.38),205(4.28), 245s(3.59),265(3.91)	35-0011-65
Purine, 8-chloro-	pH -0.26	269(3.97)	39-3017-65
	pH 4.0	216s(4.15),269(3.98)	39-3017-65
	pH 9.0	275(4.02)	39-3017-65

Compound	Solvent	$\lambda_{max}(\log \varepsilon)$	Ref.
$C_5H_3ClN_4S$			
Imidazo[4,5-c]pyridazine-2-thiol, 7-chloro-	EtOH	205(4.18),238(4.23), 250(4.27),333(4.34)	4-0042-64
$C_5H_3Cl_2F$			
1-Penten-3-yne, 1,1-dichloro-2-fluoro-	MeOH	228(4.10),233(4.14), 245(4.08)	44-2009-64
$C_5H_3Cl_2NO$			
Pyridine, 2,6-dichloro-, 1-oxide	pH 1	260(4.14)	4-0196-65
	pH 11	260(4.12)	4-0196-65
$C_5H_3Cl_2NO_2$			
Maleimide, 2,3-dichloro-N-methyl-	dioxan	238(4.27),304s(2.56)	24-0764-65
$C_5H_3Cl_2NO_2S$			
4-Isothiazolecarboxylic acid, 3,5-dichloro-, methyl ester	EtOH	258(3.79)	44-0660-64
$C_5H_3Cl_2N_3O$			
Pyrazine, 2-amino-5,6-dichloro-3-formyl-	pH 7.0	223(3.95),273(4.00), 388(3.93)	39-1666-64
$C_5H_3Cl_2N_3O_2$			
Pyridine, 4-amino-2,6-dichloro-3-nitro-	pH 1	243(4.08),281(3.32), 351(3.22)	4-0196-65
	pH 11	243(4.08),281(3.32), 351(3.22)	4-0196-65
Pyridine, 2,6-dichloro-4-nitramino-	pH 1	270(4.08)	4-0196-65
	pH 11	229(4.03),302(4.05)	4-0196-65
$C_5H_3FN_2O_4$			
Orotic acid, 5-fluoro-	pH 1	282(3.88)	65-4184-64
	pH 7	280(3.83)	65-4184-64
	pH 13	285(3.84)	65-4184-64
$C_5H_3F_3N_2O_2$			
Uracil, 5-(trifluoromethyl)-	pH 1	257(3.90)	87-0001-64
	pH 7	257(3.85)	87-0001-64
	pH 8.1	278(3.85)	87-0001-64
$C_5H_3F_4N_3$			
Pyridine, 2,3,5,6-tetrafluoro-4-hydrazino-	hexane	238(4.01)	39-0575-65
	EtOH	248(4.13)	39-0575-65
$C_5H_3F_4N_3O_3$			
1,2,3-Cyclopentanetrione, tetrafluoro-, trioxime	EtOH	228(4.25),282(3.89)	44-3698-65
C_5H_3I			
1,2-Pentadien-4-yne, 1-iodo-	EtOH	213(3.81),258(3.70)	39-4348-65
$C_5H_3NO_3Se$			
2-Selenophenecarboxaldehyde, 5-nitro-	n.s.g.	310(3.96)	65-2201-64
$C_5H_3NO_4Se$			
2-Selenophenecarboxylic acid, 4-nitro-	n.s.g.	240(4.29),295(3.88)	65-2201-64
2-Selenophenecarboxylic acid, 5-nitro-	n.s.g.	310(4.06)	65-2201-64

Compound	Solvent	$\lambda_{max}(\log \epsilon)$	Ref.
$C_5H_3N_3O$			
5-Pyrimidinecarbonitrile, 1,2-dihydro-2-oxo-	EtOH	257(4.40),300s(2.97)	44-1740-64
$C_5H_3N_3OS$			
Thiazolo[5,4-d]pyrimidin-7-ol	pH 1	251(3.79),257(3.78), 276(3.70)	94-1319-65
$C_5H_3N_3O_2S$			
Thiazolo[5,4-d]pyrimidine-5,7-diol	H_2O	256.5(4.49)	94-1319-65
$C_5H_3N_5O_3$			
6-Azalumazine, 7-hydroxy-	H_2O	306(3.75)	24-0005-64
C_5H_4			
1,3-Pentadiyne	hexane	205(2.20),213(2.29), 225(2.41),230(2.36), 236(2.44),249(2.24)	22-3518-65
$C_5H_4BNO_2S$			
4,5-Boroxarothieno[2,3-c]pyridin-4-ol	EtOH	209(3.91),278(4.01)	1-1271-65
$C_5H_4BrClN_2$			
Pyrimidine, 2-bromo-4-chloro-6-methyl-	n.s.g.	217(3.85),260(3.68)	44-0943-64
C_5H_4BrClO			
2-Cyclopenten-1-one, 4-bromo-2-chloro-	EtOH	234(3.97)	44-3503-64
C_5H_4BrNOS			
Isothiazole, 3-acetyl-4-bromo-	EtOH	211(3.73),279(4.10)	39-3114-64
$C_5H_4BrNO_2S$			
4-Isothiazolecarboxylic acid, 5-bromo-3-methyl-	EtOH	208(3.90),258(3.97)	39-0446-64
$C_5H_4ClF_2NO$			
2-Cyclobuten-1-one, 2-chloro-4,4-difluoro-3-(methylamino)-	EtOH	255(4.39)	44-3698-65
	pH 1	259(4.40)	44-3698-65
C_5H_4ClNO			
Pyridine, 4-chloro-, 1-oxide	EtOH	274(4.29)	39-2096-65
2-Pyridinol, 5-chloro-	pH -2.0	223(3.85),294(3.78)	39-5542-65
	pH 5.0	233(4.02),307(3.70)	39-5542-65
	pH 14.0	238(4.12),307(3.64)	39-5542-65
Pyrrole-2-carboxaldehyde, 5-chloro-	EtOH	245(3.60),294(4.32)	39-0459-65
	+base	265s(3.33),320(4.43)	39-0459-65
C_5H_4ClNOS			
5-Isothiazolecarboxaldehyde, 4-chloro-3-methyl-	EtOH	217(3.82),257(3.60)	39-0446-64
$C_5H_4ClNO_2S$			
5-Isothiazolecarboxylic acid, 4-chloro-3-methyl-	EtOH	233(3.80),279(3.79)	39-0446-64

Compound	Solvent	$\lambda_{max}(\log \epsilon)$	Ref.
$C_5H_4ClNO_3S$			
4-Isothiazolecarboxylic acid, 3-chloro-5-methoxy-	EtOH	243(3.80)	44-0660-64
$C_5H_4ClN_3O$			
Pyrazine, 2-amino-5-chloro-3-formyl-	pH 7.0	217s(3.83),270(3.96), 388(3.80)	39-1666-64
Pyrazine, 2-amino-6-chloro-3-formyl-	pH -3.6	253(3.89),267s(3.81), 374(3.77)	39-1666-64
	pH 7.0	211(3.98),232(3.90), 268(3.90),372(3.97)	39-1666-64
4-Pyrimidinecarboxamide, 2-chloro-	pH 1	267(3.63)	4-0130-64
	pH 11	268(3.63)	4-0130-64
$C_5H_4ClN_3O_2$			
Pyridine, 4-amino-2-chloro-3-nitro-	pH 1	239(4.21),262s(3.84), 341(3.37)	4-0196-65
	pH 11	239(4.22),280(3.50), 350(3.42)	4-0196-65
Pyridine, 4-amino-2-chloro-5-nitro-	pH 1	236(4.42),264s(3.84), 341(3.57)	4-0196-65
	pH 11	236(4.38),271(3.77), 354(3.64)	4-0196-65
Pyridine, 2-chloro-4-nitramino-	pH 1	276(4.13),331(3.49)	4-0196-65
	pH 11	299(4.08)	4-0196-65
$C_5H_4ClN_3S$			
4-Isothiazolecarbonitrile, 3-chloro-5-(methylamino)-	EtOH	220(4.32),270(4.03)	44-0660-64
$C_5H_4ClN_5$			
Triazolo[4,3-b]pyridazine, 7-amino-8-chloro-	EtOH	208(4.26),225s(4.24), 248(3.84),335(3.61)	4-0042-64
$C_5H_4Cl_2N_2$			
Pyridine, 2-amino-3,5-dichloro-	pH 0.5	211(4.28),215s(4.24), 242(3.94),248s(3.85), 326(3.81)	39-5542-65
	pH 7.0	240(4.04),312(3.65)	39-5542-65
Pyridine, 4-amino-2,6-dichloro-	pH 1	252(4.00),265(3.98)	4-0196-65
	pH 11	247(4.09)	4-0196-65
Pyridine, 5-amino-2,4-dichloro-	pH -1.38	228(4.37),264(3.83), 336(3.59)	39-2150-64
	pH 7.0	209(4.41),245(4.01), 305(3.50)	39-2150-64
$C_5H_4Cl_2N_2O$			
3(2H)-Pyridazinone, 4,5-dichloro-2-methyl-	EtOH	220(3.94),302(3.64)	22-2124-64
	NaOH	218(4.30),295(3.70)	22-2124-64
$C_5H_4Cl_2N_2OS$			
4-Isothiazolecarboxamide, 3,5-dichloro-N-methyl-	EtOH	256(3.86)	44-0660-64
$C_5H_4Cl_2O_2$			
2-Cyclopenten-1-one, 2,4-dichloro-3-hydroxy-	n.s.g.	272(4.07)	88-2411-64

Compound	Solvent	$\lambda_{max}(\log \epsilon)$	Ref.
C_5H_4FNO			
Pyridine, 3-fluoro-, 1-oxide	EtOH	266(4.10)	39-2096-65
$C_5H_4F_3NO$			
2-Cyclobuten-1-one, 2,4,4-trifluoro-3-(methylamino)-	EtOH	253(4.51)	44-3698-65
	pH 1	256(4.49)	44-3698-65
$C_5H_4F_7NS$			
Butyrimidic acid, heptafluorothio-, methyl ester	C_6H_{12}	244(3.76)	44-3739-65
C_5H_4INOS			
5-Isothiazolecarboxaldehyde, 4-iodo-3-methyl-	EtOH	266(3.84)	39-0446-64
$C_5H_4INO_2S$			
5-Isothiazolecarboxylic acid, 4-iodo-3-methyl-	EtOH	204(3.76),248(3.59), 288(3.75)	39-0446-64
$C_5H_4N_2O$			
4-Pyrimidinecarboxaldehyde	EtOH	245(3.54),273s(3.58)	44-2398-65
$C_5H_4N_2OS$			
2H-1,3-Thiazine-5-carbonitrile, 3,6-dihydro-2-oxo-	EtOH	241(3.75),285(3.74)	44-1740-64
$C_5H_4N_2O_2S$			
4-Pyrimidinecarboxylic acid, 2-mercapto-	pH 1	284(4.23)	4-0130-64
	pH 11	230(3.76),272(4.19)	4-0130-64
4-Pyrimidinecarboxylic acid, 6-mercapto-	pH 1	285(4.04),363(3.85)	4-0130-64
	pH 11	280(4.00),312s(3.85)	4-0130-64
$C_5H_4N_2O_2S_3$			
1,3-Dithiole-4,5-dicarboxamide, 2-thiono-	EtOH	359(4.23)	24-1298-64
$C_5H_4N_2O_3$			
Pyridine, 3-nitro-, 1-oxide	EtOH	279(3.98)	39-2319-64
Pyridine, 4-nitro-, 1-oxide	EtOH	330(4.14)	39-2319-64
2,3,6(1H)-Pyridinetrione, 3-oxime	EtOH	269(4.20),305s(3.79)	39-1423-64
2-Pyridinol, 3-nitro-	17.5M H_2SO_4	240(3.54),303(3.87)	39-2150-64
	pH 5.0	257(3.40),362(3.85)	39-2150-64
	pH 12.0	217(4.23),260(3.53), 392(3.82)	39-2150-64
4-Pyridinol, 3-nitro-	pH -2.8	218(4.34),279(3.33), 335s(2.30)	39-2150-64
	pH 4.0	225(4.22),249(3.90),	39-2150-64
		257s(3.81),326(3.54)	39-2150-64
	pH 10.0	228(4.25),265(3.50), 361(3.61)	39-2150-64
4-Pyrimidinecarboxylic acid, 2-hydroxy-	pH 1	325(3.78)	4-0130-64
	pH 11	224(3.90),306(3.66)	4-0130-64
$C_5H_4N_2O_3S_2$			
1,3-Dithiole-4,5-dicarboxamide, 2-oxo-	EtOH	220(3.87),254(3.63), 294(3.48)	24-1298-64

Compound	Solvent	$\lambda_{max}(\log \epsilon)$	Ref.
$C_5H_4N_2O_4$			
Orotic acid	pH 1	280(3.88)	65-1299-65
	pH 7	278(3.89)	65-1299-65
	pH 13	285(3.75)	65-1299-65
	H_2O	278(3.87)	70-1887-65
5-Uracilcarboxylic acid	pH 1	217(4.10),273(4.06)	65-1304-65
	pH 7	215(4.04),270(3.93)	65-1304-65
	pH 13	235(3.93),290(4.07)	65-1304-65
$C_5H_4N_4$			
Purine	Me_3PO_4	188(4.32),200(4.26), 240s(3.48),265(3.84), 290(2.78)	35-0011-65
1H-Pyrazolo[3,4-c]pyridazine	EtOH	206(4.46),253s(3.31), 262s(3.37),269s(3.40), 292(3.57)	44-2542-64
Pyrimidine-5-carbonitrile, 2-amino-	EtOH	257(4.46),296(3.54)	44-1740-64
$C_5H_4N_4OS$			
6-Purinethiol, 3-N-oxide	pH 1	250(3.65),268(3.67), 335(4.08)	87-0190-65
	pH 6.9	256(3.80),343(4.33)	87-0190-65
	pH 12.6	215s(4.15),248(4.07), 334(4.34)	87-0190-65
[1,2,3]Thiadiazolo[5,4-d]pyrimidine, 7-methoxy-	EtOH	234s(3.73),261(3.83), 280(3.70)	44-2121-64
Thiazolo[5,4-d]pyrimidine, 7-amino-, 6-N-oxide	pH 1.2	224(4.39),256(3.84), 264(3.84),285s(3.53)	87-0190-65
	pH 6	239(4.39),265s(3.84), 315(3.42)	87-0190-65
$C_5H_4N_4O_2$			
Carbazic acid, 3-(dicyanomethylene)-, methyl ester	MeCN	280(4.26)	44-4198-65
Xanthine	pH 7.5	271(3.94)	37-3407-64
$C_5H_4N_4O_3$			
Xanthine, 3-N-oxide	pH 1.2	230(4.21),284(4.01)	44-0408-65
	pH 12.5	216(4.34),295(4.08)	44-0408-65
$C_5H_4N_6O$			
Pyrimido[4,5-e]-as-triazin-8(7H)-one, 3-amino-	EtOH	219(4.15),246(4.34), 336(3.78)	35-1976-65
$C_5H_4N_6O_2$			
Pyrimido[4,5-e]-as-triazine-6,8(5H,7H)-dione, 3-amino-	H_2O	215(4.51),244(4.13), 329(3.92)	35-1976-65
	H_2O	216(4.50),244(4.16), 330(4.01)	24-3095-65
$C_5H_4O_2$			
4-Cyclopentene-1,3-dione	hexane	218(4.09),320s(1.23), 385(1.34)	65-0813-64
	EtOH	218.5(4.09)	65-0813-64
Furfural	EtOH	226(3.60),272(4.18)	95-0981-65
$C_5H_4O_2Se$			
2-Selenophenecarboxylic acid	n.s.g.	257(3.95),282(3.76)	65-0841-64
3-Selenophenecarboxylic acid	n.s.g.	260(3.94)	65-0841-64

Compound	Solvent	$\lambda_{max}(\log \epsilon)$	Ref.
$C_5H_4O_4SSe$			
2-Selenophenesulfonic acid, 5-formyl-	n.s.g.	250(3.95)	65-2201-64
$C_5H_4O_5SSe$			
2-Selenophenecarboxylic acid, 5-sulfo-	n.s.g.	272(4.09)	65-2201-64
$C_5H_5BF_4N_2O_2$			
N-Nitropyridinium tetrafluoroborate	n.s.g.	202(3.84),252s(3.63), 256(3.68),263s(3.54)	88-2117-64
$C_5H_5BN_2OS$			
4,5-Borazarothieno[2,3-c]pyridine, 4-hydroxy-	EtOH	213(4.09),221s(3.96), 262s(3.73),274s(3.97), 285(4.14),293(4.05)	1-1271-65
7,6-Borazarothieno[3,4-c]pyridine, 7-hydroxy-	EtOH	222(4.37),295(3.86)	1-1271-65
C_5H_5BrO			
2-Cyclopenten-1-one, 4-bromo-	n.s.g.	215(4.01)	44-3503-64
$C_5H_5ClN_2$			
Pyridine, 4-amino-2-chloro-	pH 1	264(3.93)	4-0196-65
	pH 11	244(3.84)	4-0196-65
$C_5H_5ClN_2O$			
Pyrazine, 2-chloro-3-methyl-, 4-oxide	N HCl	223(4.15),265(4.03)	44-1645-64
Pyridazine, 5-chloro-3-methyl-, 1-oxide	EtOH	268(4.04),318(3.60)	94-0228-64
4-Pyrimidinol, 6-chloro-5-methyl-	pH -3.6	241(3.96)	39-3204-64
	pH 5.0	232(3.74),272(3.72)	39-3204-64
	pH 11.1	235(3.93),267(3.69)	39-3204-64
$C_5H_5ClN_2OS$			
4-Pyrimidinol, 6-chloro-2-(methylthio)-	pH 1	244(3.85),293(3.98)	87-0005-64
	pH 10	221(4.26),249(3.94), 279(3.83)	87-0005-64
$C_5H_5ClN_2O_2S$			
4-Isothiazolecarboxamide, 3-chloro-5-methoxy-	EtOH	244(3.80)	44-0660-64
Pyridazine, 3-chloro-6-(methyl-sulfonyl)-	EtOH	255(3.14)	4-0001-64
Pyrimidine, 4-chloro-6-(methyl-sulfonyl)-	EtOH	255(3.62)	4-0001-64
$C_5H_5ClN_2S$			
Pyridazine, 3-chloro-6-(methylthio)-	pH 1, 11	257(4.10)	4-0001-64
Pyrimidine, 4-chloro-6-(methylthio)-	EtOH	269(3.99)	4-0001-64
C_5H_5ClO			
2,4-Pentadienal, 2-chloro-	EtOH	271(4.25),337(2.0)	78-2091-64
$C_5H_5ClO_2$			
2,4-Pentadienal, 5-hydroxy-2-chloro-	EtOH-NaOH	368(4.5)	88-0821-65
$C_5H_5ClO_4S$			
Thiopyrylium perchlorate	MeCN-1% $HClO_4$	245(3.76),284(3.54)	32-0203-64
	1% $HClO_4$	284(3.6)	73-3016-65

Compound	Solvent	λ max(log ε)	Ref.
$C_5H_5ClO_4Se$			
Seleninium perchlorate	MeCN-1% $HClO_4$	267(3.86),300(3.50)	32-0203-64
$C_5H_5ClO_5$			
Pyrylium perchlorate	MeCN-1% $HClO_4$	219(3.21),270(3.97)	32-0203-64
$C_5H_5Cl_2N_3$			
Pyridine, 3,4-diamino-2,6-dichloro-	pH 1	295(3.67)	4-0196-65
	pH 11	256(3.75),292(3.65)	4-0196-65
$C_5H_5Cl_3O_3$			
Acetoacetic acid, 4,4,4-trichloro-, methyl ester	EtOH	287(1.68)	44-2784-65
$C_5H_5Cl_5IrK_2N$			
Potassium pentachloro(pyridine)-iridate(III)	H_2O	278(3.65),320(3.32), 543(1.04)	28-3420-65B
C_5H_5DO			
3,4-Pentadien-2-one-5-d	n.s.g.	203(3.88),211s(3.82), 318(1.41)	28-1530-64B
$C_5H_5D_2NO$			
N-Methyl-4-pyridone-2,6-d_2	MeOH	262(4.21)	35-3365-65
$C_5H_5FN_2O$			
2-Pyrimidinol, 5-fluoro-6-methyl-	pH 1	208(3.95),321(3.71)	73-3895-65
	H_2O	218(3.88),313(3.62)	73-3895-65
	pH 13	221(3.82),305(3.64)	73-3895-65
$C_5H_5FN_2O_2$			
Uracil, 5-fluoro-1-methyl-	pH 1.4	273.5(3.84)	87-0253-65
	pH 7	274(3.84)	87-0253-65
	pH 12.1	271.5(3.70)	87-0253-65
$C_5H_5F_7N_2S$			
Butyric acid, heptafluorothio-, S-methyl ester, hydrazone	isooctane	242(3.70)	44-3739-65
C_5H_5N			
Pyridine	hexane	251(3.30)	65-0298-65
	octane	300(0.18)	65-0298-65
	EtOH	255f(2.5)	61-0448-65
	sulfolane	199(3.94),247s(3.30), 252(3.36),256(3.34), 263s(3.23)	88-2117-64
Cupy$_2$	pyridine-HNO_3	670(1.52)	1-0843-64
Cupy$_3$	pyridine-HNO_3	630(1.77)	1-0843-64
Cupy$_4$	pyridine-HNO_3	590(1.89)	1-0843-64
NiI_2 complex	$CHCl_3$	585s(2.51),631(2.42), 931(1.95),1015(2.00)	33-1265-64

Compound	Solvent	$\lambda_{max}(\log \epsilon)$	Ref.
C_5H_5NO			
Pyridine, 1-oxide	heptane	282(4.2),345f(3.0)	61-0448-65
	H_2O	256(4.0)	61-0448-65
	N HCl	259(3.5)	61-0448-65
	EtOH	265(4.09)	39-2319-64
	EtOH	263(4.10)	39-2096-65
complex with titanyl perchlorate	MeCN	268(3.5)	23-2685-65
uranyl complex	MeCN	432(1.86),484(1.18), 505(0.84)	23-0856-64
Pyrrole-2-carboxaldehyde	MeOH	237(3.88),287(4.23)	39-2579-64
C_5H_5NOS			
Isothiazole, 3-acetyl-	EtOH	205(3.55),238(3.78), 262(3.93)	39-3114-64
5-Isothiazolecarboxaldehyde, 4-methyl-	EtOH	217(3.46),255(3.84)	39-0446-64
$C_5H_5NOS_3$			
3H-1,2-Dithiole-3-thione, 4-acetamido-	EtOH	236(4.09),268(3.58), 294(3.43),413(4.03)	12-0447-64
$C_5H_5NO_2$			
Crotonic acid, 2-cyano-	EtOH	217.5(3.99)	28-2859-64B
Furfural, oxime, anti	EtOH	269(4.21)	22-2724-65
syn	EtOH	265(4.26)	22-2724-65
2,6-Pyridinediol	pH 1	225(3.68),300(3.83)	37-2285-64
	pH 7	234(3.93),322(4.13)	37-2285-64
2-Pyrrolecarboxylic acid	EtOH	222(3.65),258(4.10)	23-1279-64
$C_5H_5NO_2S$			
5-Isothiazolecarboxylic acid, 4-methyl-	EtOH	231(3.75),271(3.69)	39-0446-64
2H-1,3-Oxazine-2,4(3H)-dione, 6-methyl-2-thio-	EtOH	268(3.98)	70-1076-65
Thiophene, 2-methyl-5-nitro-	EtOH	330(4.0)	70-2055-64
$C_5H_5NO_2S_2$			
3H-1,2-Dithiol-3-one, 4-acetamido-	EtOH	230(4.10),327(3.98)	12-0447-64
$C_5H_5NO_3$			
Acetoacetic acid, α-cyano-	dioxan	255(4.01)	33-1424-64
2H-1,3-Oxazine-2,4(3H)-dione, 6-methyl-	EtOH	230(3.91)	70-1076-65
$C_5H_5N_3O_2$			
2-Pyrazinecarboxaldehyde, 3-amino-5-hydroxy-	pH 4.2	220s(3.66),297(4.06), 341(4.24)	39-1175-65
	pH 8.8	215(4.08),254(3.87), 289(3.96),344(4.27)	39-1175-65
5-Pyrazinecarboxylic acid, 2-amino-	H_2O	262(4.15),319(3.86)	33-0873-64
Pyridine, 2-amino-3-nitro-	pH 0.4	264(3.66),360(3.85)	39-2150-64
	pH 5.0	218(4.30),238s(3.80), 260s(3.57),389(3.82)	39-2150-64
Pyridine, 2-amino-5-nitro-	pH 0.5	212(4.03),307(4.16)	39-5542-65
	pH 7.0	220(3.92),350(4.12)	39-5542-65
Pyridine, 4-amino-3-nitro-	pH 2.0	227(4.27),259(3.98), 265s(3.92),331(3.57)	39-2150-64
	pH 8.0	231(4.27).269(3.56), 362(3.64)	39-2150-64

Compound	Solvent	λ_{max}(log ε)	Ref.
4-Pyrimidinecarboxaldehyde, 2-hydroxy-, oxime	pH 1	220(3.99),251(3.93), 330(3.76)	4-0130-64
	pH 11	239(3.87),274(4.04), 324(3.93)	4-0130-64
4-Pyrimidinecarboxylic acid, 2-amino-	pH 1	224(4.15),322(3.61)	44-0115-65
	pH 1	320(3.62)	4-0130-64
	pH 11	306(3.55)	4-0130-64
4-Pyrimidinecarboxylic acid, 6-amino-	pH 1	224(3.97),280(3.58)	4-0130-64
	pH 11	230(4.06),284(3.55)	4-0130-64
$C_5H_5N_3O_3$			
Orotamide	EtOH	296(3.61)	65-0159-64
Pyridazine, 3-methyl-4-nitro-, 1-oxide	EtOH	300(3.9),345(4.1)	94-0228-64
Pyridazine, 3-methyl-5-nitro-, 1-oxide	EtOH	270(4.0),375(3.2)	94-0228-64
$C_5H_5N_3O_4$			
Uracil-6-hydroxamic acid	pH 2	274(3.87)	87-0884-65
$C_5H_5N_3O_4S$			
4,6-Pyrimidinediol, 2-(methylthio)-5-nitro-	pH -0.2	217(4.31),246(3.92), 328(4.01)	39-3770-65
	pH 7.0	235s(4.00),267s(3.75), 337(3.96)	39-3770-65
$C_5H_5N_5$			
Adenine	gas	249(4.08)	46-3615-65
	Me_3PO_4	185(4.20),208(4.27), 260(4.10)	35-0011-64
Imidazo[4,5-c]pyridazine, 7-amino-	EtOH	213(4.30),260(4.07), 300(4.20)	4-0042-64
s-Triazolo[4,3-b]pyridazine, 7-amino-	EtOH	214(4.29),234(4.29), 252s(4.13),343(3.83)	4-0042-64
$C_5H_5N_5O$			
5-Azahypoxanthine, 1-methyl-	pH 1	254(3.81)	35-1980-65
	EtOH	257(3.54)	35-1980-65
1,2,4-Triazole, 5-cyanoacetamido-	EtOH	220(4.12)	24-1373-64
1,2,4-Triazolo[2,3-a]pyrimidine, 4-amino-6-oxo-	pH 7	266(4.15)	70-1475-64
	pH 12	222(4.54),254(3.94)	70-1475-64
1,2,4-Triazolo[4,3-a]pyrimidine, 4(6)-amino-6(4)-oxo-	pH 7	222(4.55),262(4.01)	70-1475-64
	pH 12	222(4.55),262(4.01)	70-1475-64
s-Triazolo[1,5-a]pyrimidin-5-one, 7-amino-	EtOH	211(4.33),270(4.10)	24-1373-64
s-Triazolo[1,5-a]pyrimidin-7-one, 5-amino-	EtOH	220(4.42),266(4.20), 300(3.30)	24-1373-64
$C_5H_5N_5O_2$			
v-Triazolo[4,5-d]pyrimidine-5,7-dione, tetrahydro-1-methyl-	pH 5.0	274(3.78)	24-1060-65
	pH 10	232(3.73),302(3.76)	24-1060-65
v-Triazolo[4,5-d]pyrimidine-5,7-dione, tetrahydro-2-methyl-	pH 4.0	272(3.94)	24-1060-65
	pH 10.0	233(3.73),299(3.81)	24-1060-65
3H-v-Triazolo[4,5-d]pyrimidine-5,7-dione, tetrahydro-3-methyl-	pH 2.0	232(3.75),255(3.96)	24-1060-65
	pH 8.0	247(3.92),278(3.88)	24-1060-65
1H-v-Triazolo[4,5-d]pyrimidine-5,7-dione, tetrahydro-4-methyl-	pH 2.0	270(3.73)	24-1060-65
	pH 7.0	268(3.85)	24-1060-65
	pH 14.0	245(3.94),271(4.03)	24-1060-65

Compound	Solvent	λ max(log ε)	Ref.
1H-v-Triazolo[4,5-d]pyrimidine-5,7-dione, tetrahydro-6-methyl-	pH 2.0	230s(3.62),263(3.80)	24-1060-65
	pH 7.0	230s(3.57),265(3.91)	24-1060-65
	pH 13.0	235s(3.69),285(3.78)	24-1060-65
$C_5H_5N_5S$			
Purine-6-thiol, 2-amino-	pH 1.0	258(3.94),347(4.33)	10-0379-64C
[1,2,3]Thiadiazolo[5,4-d]pyrimidine, 7-(methylamino)-	EtOH	247(3.91),267s(3.57),274s(3.54),325(3.77)	44-2121-64
v-Triazolo[4,5-c]pyridazine, 7-(methylthio)-	EtOH	226(4.19),281(3.80),328(4.16)	4-0247-64
C_5H_6			
Cyclopentadiene	n.s.g.	400(3.48)	46-2475-65
$C_5H_6BF_4N$			
Pyridinium tetrafluoroborate	sulfolane	202(3.53),252s(--),257(3.76),262s(--)	88-2117-64
$C_5H_6BNO_3S$			
2-Thiopheneboronic acid, 5-formyl-, oxime	EtOH	208(3.13),287(3.35)	1-1271-65
C_5H_6BrNS			
Isothiazole, 4-bromo-3,5-dimethyl-	EtOH	220(3.57),258(3.94)	39-0446-64
$C_5H_6BrN_3O$			
Pyrazole-1-carboxamide, 4-bromo-3-methyl-	EtOH	248(4.03)	28-0606-65A
$C_5H_6Br_2O$			
Cyclobutanone, 2,2-dibromo-3-methyl-	C_6H_{12}	320(2.25)	22-1976-64
	MeOH	318(2.20)	22-1976-64
Cyclobutanone, 2,4-dibromo-2-methyl-	C_6H_{12}	324(2.22)	22-1976-64
	MeOH	318(2.15)	22-1976-64
Cyclobutanone, 2,4-dibromo-3-methyl-	C_6H_{12}	321(2.29)	22-1976-64
$C_5H_6ClN_3$			
Pyridine, 2,3-diamino-6-chloro-	pH -3.0	233(3.92),317(3.94)	39-2150-64
	pH 1.0	232(3.82),258(3.46),312(3.71),325s(3.69)	
	pH 7.0	246(3.85),311(3.84)	39-2150-64
Pyridine, 3,4-diamino-2-chloro-	pH 1	296(3.92)	4-0196-65
	pH 11	244(3.47),284(3.53)	4-0196-65
Pyridine, 4,5-diamino-2-chloro-	pH -2.28	218(4.17),224s(4.08),265(4.20)	39-2150-64
	pH 1	290(3.84)	4-0196-65
	pH 2.5	233(4.40),291(3.88)	39-2150-64
	pH 7.0	218(4.46),249(3.81),290(3.54)	39-2150-64
	pH 11	241s(3.94),295(3.59)	4-0196-65
$C_5H_6ClN_3S$			
Pyrimidine, 4-amino-6-chloro-2-(methylthio)-	pH 1	245(4.23),280(3.76)	87-0005-64
	pH 10	230(4.29),253(4.01),287(3.78)	87-0005-64
$C_5H_6Cl_2N_2O$			
Pyrrole-2-carboxamide, 4,5-dichloro-1-methyl-	EtOH	274(3.83)	39-2641-64

Compound	Solvent	$\lambda_{max}(\log \epsilon)$	Ref.
$C_5H_6Cl_2O_2Te$			
2H-Tellurapyran-3,5(4H,6H)-dione, 1,1-dichloro-1,1-dihydro-	$CHCl_3$	243(4.80),308(4.15)	39-0688-64
$C_5H_6FN_3O$			
Cytosine, 5-fluoro-1-methyl-	pH 1.3	293.5(3.99)	87-0253-65
	pH 7	282(3.84)	87-0253-65
	pH 12.2	282(3.84)	87-0253-65
$C_5H_6N_2$			
Cyclopentadienone, hydrazone	hexane	276(3.77),307(4.16)	5-0039-64H
Pyridine, 2-amino-	n.s.g.	235(4.10),294(3.60)	25-0542-64
Pyridine, 4-amino-	n.s.g.	244(4.15),265(3.50)	25-0542-64
$C_5H_6N_2O$			
Furfural, hydrazone	$CHCl_3$	287(4.07)	44-0417-65
Pyrrole-2-carboxamide	H_2O	263(2.81)	23-2328-65
	H_2O	263(2.92)	23-1957-64
	HCl	289(4.28)	23-2328-65
	56% H_2SO_4	288(4.32)	23-1957-64
3-Pyrrolidinecarbonitrile, 2-oxo-	MeOH	203(3.40)	24-1970-64
$C_5H_6N_2OS$			
4-Pyrimidinol, 6-mercapto-5-methyl-	pH -3.6	227(3.90),266(3.84)	39-3204-64
	pH 2.0	218(3.88),235(3.87), 308(3.99)	39-3204-64
	pH 9.0	233(4.11),297(3.98)	39-3204-64
	pH 14.0	237(4.14),283(3.90)	39-3204-64
Thiazole, 5-acetamido-	EtOH	269(4.04)	94-1319-65
Thiocyanic acid, 2-(methylcarbamoyl)-vinyl ester	H_2O	254(4.10)	44-2660-65
	Et_2O	254(4.10)	44-2660-65
$C_5H_6N_2O_2$			
3-Pyrazinol, 2-methyl-, 1-oxide	H_2O	219(4.23),266(3.79), 315(3.72)	44-2623-64
Pyrazole-3-carboxylic acid, 5-methyl-	EtOH	218(3.96)	87-0350-65
Pyridazine, 4-methoxy-, 2-oxide	EtOH	227(4.3),265(4.0)	95-0344-65
4(1H)-Pyridazinone, 1-methyl-, 2-oxide	EtOH	239(4.50),273(3.92), 303(3.87)	95-0344-65
4,6-Pyrimidinediol, 2-methyl-	pH -3.0	242(3.98)	12-0567-64
	pH 2.5	253(4.11)	12-0567-64
	pH 9.4	252(3.95)	12-0567-64
	EtOH	248s(3.64),253(3.72), 256(3.72)	12-0567-64
4,6-Pyrimidinediol, 5-methyl-	EtOH	245(3.70),262(3.73)	12-0567-64
4-Pyrimidinol, 6-methoxy-	pH -2.4	241(3.90)	12-0567-64
	pH 2.0	237(3.54),254(3.56)	12-0567-64
	pH 10.8	230(3.77),266s(3.30)	12-0567-64
	EtOH	229(3.53),267(3.46)	12-0567-64
6(1H)-Pyrimidone, 4-hydroxy-1-methyl-	7.5N HCl	243(3.79)	65-3182-64
	neutral	253(3.87)	65-3182-64
	base	251(3.73)	65-3182-64
Uracil, 3-methyl-	pH 7.2	255(3.9)	49-1698-64
	pH 9.5	260(3.8)	49-1698-64
	pH 10.1	285(3.8)	49-1698-64
	pH 12.0	290(4.0)	49-1698-64

Compound	Solvent	$\lambda_{max}(\log \epsilon)$	Ref.
$C_5H_6N_2O_2S$			
4,6-Pyrimidinediol, 2-(methylthio)-	pH 2.6	243(3.83),277(3.96)	39-3770-65
	pH 7.4	207(4.35),221s(4.14), 260(3.92),270s(3.88)	39-3770-65
Uracil, 5-(methylthio)-	pH 1	212(3.92),233s(3.77)	39-3987-65
	pH 7.2	212(3.92),233s(3.77)	39-3987-65
	pH 13	235(3.85),295(3.82)	39-3987-65
$C_5H_6N_2O_3$			
Isobarbituric acid, 1-methyl-	pH 1	206(3.94),217s(3.92), 285(3.91)	73-2980-64
	pH 13	218(4.18),239s(3.88), 304(3.80)	73-2980-64
Isobarbituric acid, 3-methyl-	pH 1	206(4.04),278(3.83)	73-2980-64
	pH 13	243(3.83),322(3.76)	73-2980-64
Isoxazole, 3,5-dimethyl-4-nitro-	isooctane	259(3.76)	32-0915-64
	H_2O	267(3.75)	32-0915-64
5-Pyrimidinecarboxylic acid, 1,2,3,4-tetrahydro-2-oxo-	EtOH	288(4.02)	94-0681-65
4,6-Pyrimidinediol, 2-methoxy-	50% H_2SO_4	251(3.92)	12-1309-64
	pH 9.1	255(4.07),315(1.78)	12-1309-64
	EtOH	225(3.41),257(3.91), 317(1.97)	12-1309-64
$C_5H_6N_2O_3S$			
4-Pyrimidinol, 6-(methylsulfonyl)-	pH 1	247(3.60)	4-0001-64
	pH 11	230s(3.58),263(3.00)	4-0001-64
1,2,5-Thiadiazole-4-carboxylic acid, 3-methoxy-, methyl ester	EtOH	291(3.97)	35-2861-64
Uracil, 5-(methylsulfinyl)-	pH 1	214(4.02),267(3.88)	39-3987-65
	pH 7.2	218(3.96),280(3.87)	39-3987-65
	pH 13	237(3.98),284(3.97)	39-3987-65
$C_5H_6N_2O_4$			
Acrylic acid, 3-glyoxylamido-, oxime	EtOH	226(4.1),284(4.19)	44-2330-65
Malonic acid, diazo-, dimethyl ester	C_6H_{12}	225s(3.86),250(3.89), 352(1.34)	44-4366-65
Muscazone	pH 2-7	212(3.94)	88-2075-65
	pH 6	212(3.94),300(1.81)	33-0927-65
	pH 12	220(3.88)	88-2075-65
	pH 12	220(3.87)	33-0927-65
$C_5H_6N_2O_4S$			
Uracil, 5-(methylsulfonyl)-	pH 1	208(4.06),262(3.98)	39-3987-65
	pH 7.2	223(4.07),281(4.07)	39-3987-65
	pH 13	237(4.08),276(3.98)	39-3987-65
$C_5H_6N_4$			
Mesoxalonitrile, dimethylhydrazone	MeCN	232(3.13),300(4.24)	44-4198-65
$C_5H_6N_4O$			
Hypoxanthine, 9-methyl-	pH 6.1	200(4.33),249(4.05), 265s(--)	35-0011-65
	pH 13.8	255(4.07)	35-0011-65
	Me_3PO_4	185(4.23),198(4.31), 247(3.97),278(3.58)	35-0011-65
Pyridine, 2,6-diamino-3-nitroso-	EtOH	267(3.89),322(4.02), 377(4.25)	39-1423-64

Compound	Solvent	$\lambda_{max}(\log \epsilon)$	Ref.
$C_5H_6N_4O_2$			
Acrylamide, β-amino-N-carbamoyl-α-cyano-	H_2O	278(4.05)	39-1642-65
Imidazole-5-carboxamide, 4-formamido-	pH 1	257(3.93)	4-0253-65
Pyridine, 2,4-diamino-5-nitro-	pH 1	242(4.23),268(4.51), 313(3.82)	4-0196-65
	pH 11	233(4.14),263(3.94), 349(4.20)	4-0196-65
$C_5H_6N_4O_3$			
Acrylamide, N-carbamoyl-α-cyano-β-hydroxyamino-	H_2O	241(3.99),267(4.18)	39-1642-65
$C_5H_6N_4O_4$			
Formamide, N-(4-amino-2,6-dihydroxy-5-pyrimidinyl)-, 3-oxide	pH 1	266(4.23)	44-0408-65
	pH 12.5	276(3.93)	44-0408-65
$C_5H_6N_6$			
s-Triazole, 3-[4(5)-aminoimidazol-5(4)-yl]-	pH 1	244(4.00),260(4.00)	44-3601-65
	pH 7	265(4.09)	44-3601-65
C_5H_6O			
Ether, 1-buten-3-ynyl methyl	n.s.g.	236(4.19)	70-1318-64
C_5H_6OS			
4-Thiolen-2-one, 5-methyl-	EtOH	265(3.36)	11-0211-64
$C_5H_6O_2$			
But-2-enolide, 3-methyl-	H_2O	211.5(4.11)	39-0594-64
2-Cyclobuten-1-one, 3-hydroxy-2-methyl-	0.5N HCl	244(3.98)	12-0765-64
	2% NaOH	255(4.19)	12-0765-64
	EtOH	247(4.02)	12-0765-64
1,2-Cyclopentanedione	EtOH	253(3.95)	44-2513-65
	pH 1	244(4.00)	94-1300-65
	neutral	249(3.99)	94-1300-65
	pH 10	269(4.15)	93-1300-65
Δ^2-Furenid-4-one, 2-methyl-	EtOH	258(4.10)	33-1322-65
2,4-Pentadienal, 5-hydroxy-	EtOH-NaOH	362.5(4.7)	88-0821-65
Valerolactone, α-oxo-γ-hydroxy-	base	262(3.68)	37-1284-64
$C_5H_6O_2S$			
2(5H)-Thiophenone, 3-hydroxy-4-methyl-	EtOH	246(3.99)	39-0766-64
	EtOH-NaOH	246(3.80),304(3.79)	39-0766-64
4H-Thiopyran, 1,1-dioxide	H_2O	265(3.58)	88-2941-65
$C_5H_6O_2Te$			
2H-Telluropyran-3,5(4H,6H)-dione	$CHCl_3$	243(4.97),306(4.71)	39-0688-64
$C_5H_6O_3$			
Crotonic acid, β-formyl-, cis	H_2O	211(4.13)	39-0594-64
2(5H)-Furanone, 3-hydroxy-4-methyl-	EtOH	232(4.08)	39-0766-64
	EtOH-NaOH	268(3.90)	39-0766-64
Valerolactone, 4-hydroxy-2-oxo-	aq. base	228(3.72)	37-1284-64
$C_5H_6O_3S$			
2(5H)-Furanone, 3-hydroxy-4-mercaptomethyl-	EtOH	242(3.81)	39-0766-64
	EtOH-NaOH	278(3.54),330(3.15)	39-0766-64

Compound	Solvent	$\lambda_{max}(\log \epsilon)$	Ref.
$C_5H_6O_5$			
2-Cyclopenten-1-one, 2,3,4,5-tetra-hydroxy-	n.s.g.	290(4.15)	28-1512-64A
$C_5H_6S_3$			
3H-1,2-Dithiole-3-thione, 4-ethyl-	EtOH	224(4.08),241s(3.90), 271(3.80),315(3.81), 410(4.00)	24-0654-64
$C_5H_7BrN_4$			
Pyrimidine, 2-amino-5-bromo-1,4-dihydro-4-imino-1-methyl-	pH 7.0	210(4.44),286(3.76)	39-0755-65
Pyrimidine, 2-amino-5-bromo-4-(methylamino)-	pH 1.0	214(4.47),243s(4.07), 280(3.76)	39-0755-65
	pH 9.4	238(4.03),296(3.87)	39-0755-65
Pyrimidine, 4-amino-5-bromo-2-(methylamino)-	pH 1.0	216(4.39),225(4.39), 289(3.53)	39-0755-65
	pH 9.4	231(4.19),303(3.78)	39-0755-65
C_5H_7BrO			
Cyclobutanone, 2-bromo-2-methyl-	C_6H_{12}	320(2.15)	22-1976-64
	MeOH	313.5(2.08)	22-1976-64
Cyclobutanone, 2-bromo-3-methyl-, cis	C_6H_{12}	308(2.09)	22-1976-64
trans	C_6H_{12}	309(2.13)	22-1976-64
	MeOH	302(1.98)	22-1976-64
$C_5H_7ClN_4$			
Pyrimidine, 2-amino-4-chloro-1,6-dihydro-6-imino-1-methyl-	pH 7.0	234(3.88),296(4.07)	39-0755-65
	pH 12.0	243(3.98),297(3.82)	39-0755-65
Pyrimidine, 2-amino-4-chloro-3,6-dihydro-6-imino-3-methyl-	pH 7.0	237s(3.98),276(3.85)	39-0755-65
	pH 14.0	232(4.29),303(3.63)	39-0755-65
Pyrimidine, 2-amino-4-chloro-6-(methylamino)-	pH 1.0	212(4.26),238s(4.06), 275(3.82),302(3.73)	39-0755-65
	pH 6.4	241s(3.97),286(3.95)	39-0755-65
C_5H_7ClO			
Crotonaldehyde, 3-chloro-2-methyl-	EtOH	248(4.14)	44-1126-65
Cyclopentanone, 2-chloro-	C_6H_{12}	306(1.56),314(1.56)	28-2374-65B
	hexane	306(1.50),314(1.50)	28-2374-65B
	EtOH	302(1.54)	28-2374-65B
	DMSO	302(1.54)	28-2374-65B
1-Penten-3-one, 1-chloro-	EtOH	229(4.12)	44-0385-64
$C_5H_7ClO_2$			
Crotonic acid, 4-chloro-, methyl ester	EtOH	206(4.16)	44-3327-64
$C_5H_7Cl_3Ge$			
Germane, trichloro-2-cyclopenten-1-yl-	n.s.g.	226(3.9)	70-2203-64
$C_5H_7F_5N_2S$			
Propionic acid, pentafluorothio-, 2,2-dimethylhydrazide	isooctane	276(3.74)	44-3739-65
$C_5H_7F_6PS$			
1-Methylthiophenium hexafluoro-phosphate	H_2O	225(3.49),269(2.75)	35-5360-64

Compound	Solvent	λ_{max}(log ε)	Ref.
$C_5H_7KN_2O_6$			
2-Pentanone, 3-hydroxy-5,5-dinitro-, 5-potassium salt	dil. KOH	378(4.22)	44-2256-64
$C_5H_7KN_2O_7$			
Butyric acid, 2-hydroxy-4,4-dinitro-, 4-potassium deriv., methyl ester	dil. KOH	378(4.23)	44-2256-64
C_5H_7N			
Pyridine, 1,4-dihydro-	n.s.g.	278(3.40)	35-3283-65
C_5H_7NO			
Acrylonitrile, 3-ethoxy-, trans	n.s.g.	222(4.15)	44-1800-64
Oxazole, 3,5-dimethyl-	98% H_2SO_4	225(3.92)	70-0936-64
C_5H_7NOS			
4-Isothiazolin-3-one, 2-ethyl-	EtOH	277(3.80)	44-2660-65
	Et_2O	280(3.82)	44-2660-65
2(5H)-Thiophenone, 3-amino-4-methyl-	EtOH-HCl	230(3.70),250(3.66), 279s(3.42)	39-0766-64
	EtOH	250(3.79),288(3.59)	39-0766-64
$C_5H_7NO_2$			
2(5H)-Furanone, 3-amino-4-methyl-	H_2O	248.0(3.93)	39-0783-64
	EtOH	249(3.9)	44-3560-64
	EtOH	255(3.88)	39-0766-64
	EtOH-HCl	208(4.02),248(3.15)	39-0766-64
	HCl	207(3.85),249(3.67)	44-3560-64
Succinimide, N-methyl-	EtOH	204(4.11),242(2.04)	24-1548-64
	THF	222(2.37),242(2.06)	24-1548-64
$C_5H_7NO_2S$			
2,4-Thiazolidinedione, 5-ethyl-	n.s.g.	290s(2.00)	65-0886-65
$C_5H_7NO_4$			
Acrylic acid, 3-nitro-, ethyl ester	EtOH	222(3.98)	70-2093-64
C_5H_7NS			
Isothiazole, 3,5-dimethyl-	EtOH	217(3.76),246(3.92)	39-0446-64
$C_5H_7NS_2$			
2H-1,3-Thiazine-2-thione, 3,6-dihydro-3-methyl-	H_2O	298(4.04),320s(--)	39-4008-64
	EtOH	308(4.01)	39-4008-64
2H-1,3-Thiazine-2-thione, 3,6-dihydro-4-methyl-	H_2O	298(4.04),320s(--), 360s(--)	39-4008-64
	EtOH	310(3.97)	39-4008-64
2H-1,3-Thiazine-2-thione, 3,6-dihydro-5-methyl-	H_2O	295(4.01),318(3.92), 360s(--)	39-4008-64
	EtOH	322(3.93)	39-4008-64
2H-1,3-Thiazine-2-thione, 3,6-dihydro-6-methyl-	H_2O	302(4.07),320s(--)	39-4008-64
	EtOH	320(4.09)	
$C_5H_7N_3$			
Pyridine, 2,3-diamino-	pH -2.3	227(3.94),305(3.80)	39-2150-64
	pH 3.5	249(3.66),318(3.91)	39-2150-64
	pH 10.0	238(3.78),301(3.77)	39-2150-64
Pyridine, 3,4-diamino-	pH -1.85	208(4.13),262(4.19)	39-2150-64
	pH 5.0	225(4.28),286(3.93)	39-2150-64
	pH 12.0	245(3.80),283(3.57)	39-2150-64

Compound	Solvent	λmax(log ε)	Ref.
Pyrimidine, 1,2-dihydro-2-imino-1-methyl-	pH 13+	237(4.22),347(3.45)	12-0763-65
Pyrimidine, 1,4-dihydro-4-imino-1-methyl-	pH 14.8	260(4.23)	39-0755-65
$C_5H_7N_3O$			
Cytosine, 1-methyl-	n.s.g.	274(3.91)	94-1240-64
	acid	283(4.09)	94-1240-64
Cytosine, N^4-methyl-	N HCl	277(4.07)	44-1762-64
	H_2O	267(3.93)	44-1762-64
	pH 13	285(3.97)	44-1762-64
Pyrazole-1-carboxamide, 3-methyl-	EtOH	237.5(4.10)	28-0606-65A
Pyrazole-1-carboxamide, 5-methyl-	EtOH	232.5(3.84)	28-0606-65A
Pyrimidine, 4-amino-6-methoxy-	pH 7.0	235(3.88)	12-0559-65
4-Pyrimidinol, 6-(methylamino)-	pH 4.8	219(4.40),261(4.06)	12-0559-65
4(1H)-Pyrimidinone, 2-amino-1-methyl-	pH 13.0	260(3.74)	12-0559-65
4(3H)-Pyrimidinone, 2-amino-3-methyl-	pH 9.8,13.3	225(3.86),284(3.96)	12-0559-65
4(3H)-Pyrimidinone, 6-amino-3-methyl-	pH 4.8	216(4.54),257(3.80)	12-0559-65
$C_5H_7N_3OS$			
Cytosine, 6-(methylthio)-	pH 1	235s(3.82),295(4.34)	87-0005-64
	pH 10	287(4.10)	87-0005-64
4-Pyrimidinol, 6-amino-2-(methylthio)-	pH 1	236(4.25),270(3.98)	87-0005-64
	pH 10	264.5(3.92)	87-0005-64
4-Thiazolecarboxamide, 5-amino-2-methyl-	EtOH	281(3.94)	94-1319-65
$C_5H_7N_3O_2$			
Cytosine, N^6-hydroxy-3-methyl-	pH 1	221(3.90),284(4.09)	39-0208-65
	pH 7	234(4.04),279(3.84)	39-0208-65
Cytosine, N^6-hydroxy-5-methyl-	pH 1	217(4.02),283(4.08)	39-0208-65
	pH 7	232(3.95),272(3.82)	39-0208-65
Cytosine, N^6-methoxy-	pH 1	216(3.86),279(4.00)	39-0208-65
	pH 7	237(3.96),273(3.78)	39-0208-65
Isoxazole, 5-acetamido-3-amino-	MeOH	239(4.23)	44-2862-65
Pyrimidine-5-carboxamide, 1,2,3,4-tetrahydro-2-oxo-	EtOH	240(4.38),276(4.38)	44-1740-64
s-Triazine-2,4(1H,3H)-dione, 1,3-dimethyl-	H_2O	244(3.70)	88-2587-64
s-Triazin-2(1H)-one, 4-methoxy-1-methyl-	MeCN	254(3.34)	73-2060-64
$C_5H_7N_3O_3$			
Acrylamide, 3-glyoxylamido-, oxime	EtOH	223(4.19),285(4.18)	44-2330-65
Cytosine, N^6-hydroxy-5-(hydroxymethyl)-	pH 1	217(2.97),279(4.01)	39-0208-65
	pH 7	230(3.97),273(3.75)	39-0208-65
Imidazole, 1-(β-hydroxyethyl)-2-nitro-	8M H_2SO_4	300(3.90)	44-0862-64
	dil. H_2SO_4	326(3.93)	44-0862-64
$C_5H_7N_3S$			
Pyrimidine, 2-amino-4-(methylthio)-	pH 0.2	275s(4.01),299(4.06)	12-0559-65
	pH 7.0	234s(3.90),300(4.02)	12-0559-65
Pyrimidine, 4-amino-6-(methylthio)-	pH 7.0	235(4.39),274(3.82)	12-0559-65
4(3H)-Pyrimidinethione, 2-amino-3-methyl-	pH -0.2	262(3.67),323(4.16)	12-0559-65
	pH 7.0	258(3.68),337(4.21)	12-0559-65

Compound	Solvent	λ_{max}(log ϵ)	Ref.
$C_5H_7N_5O_2$			
Pyrimidine, 4-amino-1,6-dihydro-6-imino-1-methyl-5-nitro-	pH 4.0	234(4.36),293(3.55), 329(3.80)	39-5542-65
1H-1,2,4-Triazole-1-carboxamide, 5-formamido-N-methyl-	pH 1	222(4.06)	35-1980-65
	pH 13	239(3.83)	35-1980-65
C_5H_8			
Isoprene	C_6H_{12}	223(4.42)	44-3527-64
	EtOH	222.5(4.36)	44-3527-64
1,3-Pentadiene	C_6H_{12}	224(4.42)	44-3527-64
	EtOH	223(4.4)	44-3527-64
C_5H_8BrNO			
2-Piperidone, 3-bromo-	MeOH	211(3.84)	24-1970-64
C_5H_8ClNO			
Crotonaldehyde, 3-chloro-2-methyl-, oxime	EtOH	245.5(4.29)	44-1126-65
$C_5H_8ClNO_5S_2$			
2-(Dimethylamino)-4-hydroxy-1,3-dithiol-1-ium perchlorate	70% $HClO_4$	242(4.19)	44-1708-64
$C_5H_8N_2$			
Acrylonitrile, 3-(dimethylamino)-, cis	n.s.g.	258(4.30)	44-1800-64
Pyrazole, 3,5-dimethyl-	EtOH	213(3.67)	87-0350-65
$C_5H_8N_2O$			
5-Pyrazolinone, 1,3-dimethyl-	C_6H_{12}	251(3.61)	78-0299-64
	20N H_2SO_4	222.5(3.90)	78-0299-64
	pH 5	241(3.90)	78-0299-64
	H_2O	241(3.90)	78-1693-65
	pH 13	233(3.81)	78-0299-64
5-Pyrazolinone, 1,4-dimethyl-	N HCl	233(3.85)	5-0134-65F
	H_2O	248(3.94)	5-0134-65F
5-Pyrazolinone, 2,3-dimethyl-	C_6H_{12}	227(3.74)	78-1693-65
	H_2O	245(3.83)	78-1693-65
5-Pyrazolinone, 3,4-dimethyl-	C_6H_{12}	227(3.49)	78-1693-65
	H_2O	246(3.97)	78-1693-65
Pyrazol-3-ol, 1,5-dimethyl-	C_6H_{12}	227.5(3.74)	78-0315-64
	20N H_2SO_4	226.5(3.95)	78-0315-64
	pH 5.5	245.5(3.83)	78-0315-64
	pH 13	239(3.74)	78-0315-64
$C_5H_8N_2OS$			
Hydantoin, 1,3-dimethyl-2-thio-	97% HOAc	238(--),265(4.12)	65-0554-65
Hydantoin, 3,5-dimethyl-2-thio-	EtOH	235(4.00),265(4.23)	65-0049-65
Pseudothiohydantoin, 5-ethyl-	n.s.g.	249(3.92),300s(2.00)	65-0886-65
$C_5H_8N_2O_2$			
1,3,4-Oxadiazole, 2-ethoxy-5-methyl-	EtOH	217(--)	5-0145-65F
2-Oxazolin-4-one, 2-(dimethylamino)-	MeOH	227(4.43)	44-0370-64
$C_5H_8N_2O_2S$			
2H-1,2,6-Thiadiazine, 3,5-dimethyl-, 1,1-dioxide	EtOH	250(2.9),318(3.8)	25-0182-65

Compound	Solvent	$\lambda_{max}(\log \epsilon)$	Ref.
$C_5H_8N_2O_3S$			
Pseudothiohydantoin-5-acetic acid	n.s.g.	249(3.95)	65-0886-65
$C_5H_8N_2O_5$			
Premuscimol	pH 4	210(3.79)	33-0927-65
	pH 6-7	208(3.80),240(3.38)	33-0927-65
	pH 12	212(3.80)	33-0927-65
$C_5H_8N_4$			
Pyridine, 2,4,5-triamino-	pH 1	267(3.83)	4-0196-65
	pH 11	296(3.54)	4-0196-65
Pyrimidine, 2-amino-1,4-dihydro-	pH 9.0	235s(4.02),270(3.83)	39-0755-65
4-imino-1-methyl-	pH 14.8	297(3.44)	39-0755-65
Pyrimidine, 2-amino-4-(methyl-	pH 1.0	235(4.07),265(3.85)	39-0755-65
amino)-	pH 10.4	234(3.99),286(3.90)	39-0755-65
as-Triazine, 3-amino-5,6-dimethyl-	EtOH	229(3.82),324(3.21)	94-1329-64
$C_5H_8N_4O$			
6-Azacytosine, 3-(dimethylamino)-	N H_2SO_4	208(3.82),246s(3.58), 287(3.95)	73-1394-64
	H_2O	218s(4.08),289(3.88)	73-1394-64
2(1H)-Pyrimidinone, 4-(1-methyl-	pH 1	283(4.12)	5-0134-65F
hydrazino)-	H_2O	275(3.99)	5-0134-65F
	pH 13	288(4.00)	5-0134-65F
as-Triazin-3(2H)-one, 4,5-dihydro- 5-imino-2,4-dimethyl-	N H_2SO_4	208(3.84),240(3.58), 288(3.85)	73-1394-64
	H_2O	217(3.98),283(3.85)	73-1394-64
$C_5H_8N_4OS$			
1,2,5-Thiadiazole-3-carboxamide,	pH 1,7	219(4.08),345(3.77)	44-2141-64
N-methyl-4-(methylamino)-	pH 13	344(3.77)	44-2141-64
s-Triazine, 2-amino-4-methoxy- 6-(methylthio)-	MeOH	219(4.27),244(4.18)	20-0585-64
$C_5H_8N_4S$			
Acetone, 1,3,4-thiadiazol-2-yl-	EtOH	226(3.70),279(4.13)	1-0871-64
hydrazone	anion	246(3.89),325(4.09)	1-0871-64
C_5H_8O			
Cyclobutanone, 2-methyl-	C_6H_{12}	282(1.32),290(1.34), 291(1.20)	22-1976-64
	MeOH	284(1.45)	22-1976-64
Cyclobutanone, 3-methyl-	C_6H_{12}	279(1.18),283(1.27), 291(1.20)	22-1976-64
	MeOH	279.5(1.36)	22-1976-64
Cyclopentanone	hexane	268(0.95),279(1.11), 288(1.20),300(1.26), 311(0.95),323(0.85)	28-2374-65B
	hexane	281(1.18),290(1.26), 300(1.28),312(1.19), 324(0.88)	59-0529-65
	C_6H_{12}	269s(0.95),279s(1.11), 289(1.20),300(1.26), 311(0.95),323(0.78)	28-2374-65B
	EtOH	287(1.27)	28-2374-65B
	EtOH	275(1.18),288(1.26), 299(1.22),308(1.02), 318(0.54)	59-0529-65
	DMSO	274s(1.18),286s(1.29), 294(1.31),305s(1.22), 315s(0.93)	28-2374-65B

Compound	Solvent	$\lambda_{max}(\log \varepsilon)$	Ref.
Cyclopropanecarboxaldehyde, 1-methyl-	isooctane	285(1.38),289s(1.36),299s(1.23),311s(0.91),324s(0.31)	22-2755-65
1-Penten-3-one	EtOH	215(3.74),320(1.48)	22-0704-65
3-Penten-2-one, cis	EtOH	224(4.00),320(1.59)	22-2198-65
trans	EtOH	221(4.08),312(1.58)	22-2198-65
$C_5H_8O_2$			
Acrylic acid, β,β-dimethyl-	H_2O	220.3(4.08)	23-0724-64
	100% H_2SO_4	243.6(4.12)	
	98.3%	251.8(4.25)	
	96.2%	251.6(4.25)	
	90.3%	250.8(4.22)	
	82.2%	249.2(4.18)	
	77.8%	247.1(4.18)	
	73.2%	244.4(4.14)	
	70.4%	242.0(4.12)	
	67.7%	239.3(4.10)	
	65.8%	236.7(4.09)	
	63.8%	235.0(4.08)	
	58.9%	230.1(4.07)	
	49.2%	225.5(4.06)	
	38.0%	223.0(4.07)	
$C_5H_8O_4$			
Succinic acid, methyl-	H_2O	203(2.04)	39-3928-65
Valeric acid, 4-hydroxy-2-oxo-	n.s.g.	end absorption only	37-1284-64
$C_5H_8O_4S$			
Succinic acid, (methylthio)-	H_2O	202s(3.27),238(2.64)	39-3928-65
$C_5H_8O_5$			
Succinic acid, methoxy-	H_2O	206s(2.23)	39-3928-65
C_5H_8S			
Cyclopentanethione	C_6H_{12}	230(--)(changing)	59-0299-64
2H-Thiopyran, 3,4-dihydro-	EtOH	229(3.73),249(3.45)	44-2211-64
2H-Thiopyran, 3,6-dihydro-	EtOH	none above 220 mμ	44-2211-64
$C_5H_9BrN_2S$			
2,3,6,7-Tetrahydro-5H-imidazo-[2,1-b]thiazolium bromide	n.s.g.	213(4.05)	39-3456-65
$C_5H_9F_3N_8$			
Trifluoromethylglyoxal bis(guanyl-hydrazone)	pH 1	304(4.24)	87-0806-64
	pH 11	348(4.44)	87-0806-64
$C_5H_9KOS_2$			
Xanthic acid, butyl ester, potassium salt	H_2O	229(3.90),301(1.82),381(1.82)	1-1113-65
	EtOH	229(4.00),304(4.30),386(1.80)	1-1113-65
$C_5H_9KS_3$			
Carbonic acid, trithio-, butyl ester, potassium salt	H_2O	235(3.60),310(3.82),340(3.94),425(1.80)	1-1113-65
	EtOH	235(3.60),308(3.85),340(3.92),430(1.80)	1-1113-65
	MeCN	236(3.60),302(3.89),340(3.95),446(1.60)	1-1113-65

Compound	Solvent	$\lambda_{max}(\log \epsilon)$	Ref.
C_5H_9N			
1-Pyrroline, 2-methyl-	hexane	227(2.33)	39-2313-65
	EtOH	216(2.30)	39-2313-65
$C_5H_9NOS_2$			
2-Butanone, 4-(thiocarbamoylthio)-	H_2O	241(3.87),284(4.09)	39-4004-64
	EtOH	240(3.81),288(4.00)	39-4004-64
Butyraldehyde, β-(thiocarbamoyl-thio)-	H_2O	242(3.82),287(4.10)	39-4004-64
	EtOH	240(3.86),292(4.09)	39-4004-64
Propionaldehyde, α-methyl-β-(thio-carbamoylthio)-	H_2O	242(3.93),287(4.10)	39-4004-64
	EtOH	240(3.88),292(4.10)	39-4004-64
Propionaldehyde, β-(N-methylthio-carbamoylthio)-	H_2O	245(3.98),282(4.10)	39-4004-64
	EtOH	244(3.98),285(4.10)	39-4004-64
2-Thiazolidinethione, 4-hydroxy-3,4-dimethyl-	MeOH	252s(3.96),272(4.19)	44-2146-64
$C_5H_9NO_2$			
Cyclobutane, (nitromethyl)-, anion	H_2O	239(4.06)	44-2674-65
Cyclopentane, nitro-, anion	H_2O	226(4.12)	44-2674-65
	50% dioxan	228(4.13)	44-2674-65
Cyclopropane, 1-(nitroethyl)-, anion	H_2O	235(4.11)	44-2674-65
$C_5H_9NO_2S$			
Cysteine, S-vinyl-	50% EtOH	223(3.72)	24-0781-65
Oxamic acid, N-methyl-2-thio-, ethyl ester	heptane	303.0(3.91)	1-1059-64
	heptane-2% CH_2Cl_2	439(1.26)	1-1059-64
	EtOH	296(3.83),382(1.43)	1-1059-64
$C_5H_9N_3O$			
α-Isonitroso-N-isopropylamino-acetonitrile	EtOH	255(3.92)	33-0033-64
$C_5H_9N_3O_3S$			
Acetimidic acid, 2-diazo-N-methyl-sulfonyl)-, ethyl ester	n.s.g.	271(4.24)	24-0623-65
$C_5H_9N_3O_4$			
Acrylic acid, N-(oximinoacetyl)-3-amino-, ammonium salt	EtOH	233(4.17),285(4.12)	44-2330-65
$C_5H_9N_3S$			
5-Mercapto-2,3,4-trimethyl-s-triazolium hydroxide inner salt	n.s.g.	242(4.15)	4-0105-65
$C_5H_9N_5$			
s-Triazine, 2,4-diamino-6-ethyl-	EtOH	256(3.56)	44-0707-65
$C_5H_9N_5S$			
1,2,5-Thiazole-2-carboxamidine, 4-amino-N-ethyl-	pH 1	321(3.78)	44-2135-64
	pH 7	321(3.79)	44-2135-64
	pH 13	318(3.91)	44-2135-64
	EtOH	265(3.40),325(3.96)	44-2135-64
$C_5H_9O_2P$			
2-Phospholene, 1-methoxy-, 1-oxide	MeOH	200(3.64),231(2.01)	78-1593-64
3-Phospholene, 1-methoxy-, 1-oxide	MeOH	200(2.65),230(2.73)	78-1593-64

Compound	Solvent	λ max(log ε)	Ref.
$C_5H_{10}N_2$			
2-Butene, 2-(methyazo)-	EtOH	248(3.92)	35-4576-65
3H-Diazirine, 3-tert-butyl-	gas	346f(2.46)	39-3101-65
Isobutylene, 1-(methylazo)-	EtOH	252(3.87)	35-4576-65
$C_5H_{10}N_2O$			
2-Imidazoline, 1-(2-hydroxyethyl)-	EtOH	234(3.82)	4-0188-64
$C_5H_{10}N_2OS$			
Oxamide, N^o,N^o,N^s-trimethylthio-	heptane	276(3.84)	1-1059-64
	heptane-50% CH_2Cl_2	369.0(1.64)	1-1059-64
	EtOH	233(3.76),271(4.01), 352(1.63)	1-1059-64
Oxamide, N^o,N^s,N^s-trimethylthio-	heptane	290(3.95)	1-1059-64
	heptane-10% CH_2Cl_2	391.0(3.83)	1-1059-64
	EtOH	232(3.77),279(4.11), 362(1.72)	1-1059-64
$C_5H_{10}N_2O_2$			
Hydroxylamine, N-cyclopentyl-N-nitroso-	C_6H_{12}	230(3.80)	7-0485-65
$C_5H_{10}N_2O_3$			
Carbamic acid, ethylnitroso-, ethyl ester	H_2O	240.5(3.8)	28-5470-64A
$C_5H_{10}N_2S$			
2-Thiazoline, 2-(2-aminoethyl)-	EtOH	232(3.39),244(3.33)	94-0180-65
	N HCl	211(3.37),266(3.59)	
$C_5H_{10}N_4O_3$			
Cytosine, 4,5-dihydro-N^6-hydroxy-4-(hydroxylamino)-3-methyl-	pH 7	223(4.08)	39-0208-65
$C_5H_{10}N_4O_4$			
Cytosine, 4,5-dihydro-N^6-hydroxy-4-(hydroxylamino)-5-hydroxymethyl-	pH 7	220(3.96),270s(2.99)	39-0208-65
$C_5H_{10}OS_2$			
Xanthic acid, butyl ester, potassium salt	H_2O	229(3.90),301(1.82), 381(1.82)	1-1113-65
	EtOH	229(4.00),304(4.30), 386(1.80)	1-1113-65
$C_5H_{10}O_2$			
Acetic acid, isopropyl ester	H_2O	204.1(1.79)	35-2384-64
	MeOH	208.8(1.77)	35-2384-64
	EtOH	209.1(1.78)	35-2384-64
	MeCN	210.1(1.76)	35-2384-64
	$C_3H_4F_4O$	204.8(1.77)	35-2384-64
Acetic acid, propyl ester	H_2O	203.3(1.71)	35-2384-64
	MeOH	209.0(1.74)	35-2384-64
	EtOH	208.1(1.76)	35-2384-64
	MeCN	210.3(1.7)	35-2384-64
	$C_3H_4F_4O$	204.4(1.75)	35-2384-64

Compound	Solvent	λ_{max}(log ϵ)	Ref.
Formic acid, tert-butyl ester	H_2O	211s(1.74)	35-2384-64
	MeOH	210s(1.86)	35-2384-64
	EtOH	213s(1.87)	35-2384-64
	MeCN	210s(1.84)	35-2384-64
Isobutyric acid, methyl ester	isooctane	212.8(1.89)	35-2384-64
	H_2O	206.3(1.90)	35-2384-64
	MeOH	210.6(1.89)	35-2384-64
	EtOH	210.5(1.90)	35-2384-64
	MeCN	213.1(1.86)	35-2384-64
	$C_3H_4F_4O$	207.1(1.94)	35-2384-64
Pivalic acid, lead salt	C_6H_{12}	236(4.33)	88-2847-65
Propionic acid, ethyl ester	isooctane	205(1.82),208(1.83)	35-2384-64
	H_2O	204.2(1.79),207(1.78)	35-2384-64
	MeOH	205(1.86),207(1.87)	35-2384-64
	EtOH	205(1.85),208(1.85)	35-2384-64
	MeCN	205(1.9),208(1.9)	35-2384-64
	$C_3H_4F_4O$	203(1.78),206(1.77)	35-2384-64
$C_5H_{10}S_2$			
1,3-Dithiane, 2-methyl-	EtOH	232s(2.58),252(2.65)	78-0437-64
1,2-Dithiepane	EtOH	259(2.69)	44-4008-65
1,3-Dithiolane, 2-ethyl-	EtOH	246(2.50)	78-0437-64
$C_5H_{10}S_3$			
Carbonic acid, trithio-, butyl ester, potassium salt	H_2O	235(3.60),310(3.82), 340(3.94),425(1.80)	1-1113-65
	EtOH	235(3.60),308(3.85), 340(3.92),430(1.80)	1-1113-65
	MeCN	236(3.60),302(3.89), 340(3.95),446(1.60)	1-1113-65
$C_5H_{11}NO_2$			
Butane, 2-methyl-3-nitro-, anion	H_2O	226.5(4.05)	44-2674-65
Pentane, 2-nitro-, anion	H_2O	228(4.05)	44-2674-65
$C_5H_{11}NS$			
Butyramide, α-methyl-thiono-	MeOH	267(4.07),328(1.68)	35-0051-65
$C_5H_{11}N_3S$			
Acetone, 2'-methylthiosemicarbazone	pH 1	237(3.92)	44-2234-65
	pH 13	220(3.97),239(4.06)	44-2234-65
	MeOH	243(4.10)	44-2234-65
	$CHCl_3$	254(4.13),275(3.97)	44-2234-65
$C_5H_{12}Ge$			
Germane, trimethylvinyl-	n.s.g.	180(4.3)	70-2203-64
$C_5H_{12}OS$			
Sulfoxide, n-butyl methyl	EtOH	211s(3.04)	35-1958-65
	C_6H_{12}	206(3.55),225s(3.03)	35-1958-65
	isooctane +CF_3COOH	207s(2.98)	35-1958-65
	MeCN	200(3.58),217s(3.08)	35-1958-65
$C_5H_{12}S_2$			
Methane, bis(ethylthio)-	EtOH	237(2.74)	78-0437-64
$C_5H_{12}Si$			
Silane, trimethylvinyl-	n.s.g.	176(4.3)	70-2203-64

Compound	Solvent	$\lambda_{max}(\log \epsilon)$	Ref.
$C_5H_{12}Sn$			
Tin, trimethylvinyl-	n.s.g.	185(4.4)	70-2203-64
$C_5H_{13}N_3S$			
Semicarbazide, 4-sec-butyl-S-thio-, nickel complex	n.s.g.	440(1.51)	1-1239-65
$C_5N_2OS_2$			
1,3-Dithiol-2-one, 4,5-dicyano-	C_6H_{12}	235(3.98),302(3.86), 314(3.90)	24-1298-64
$C_5N_2S_3$			
1,3-Dithiole-2-thione, 4,5-dicyano-	EtOH	232(4.08),354(4.16)	24-1298-64

Compound	Solvent	λ_{max}(log ε)	Ref.
$C_6ClF_6KO_2$			
1,3-Cyclohexanedione, 2-chloro-4,4,5,5,6,6-hexafluoro-, K deriv.	EtOH	329(4.39)	44-1629-65
$C_6Cl_4O_2$			
Chloranil	50% EtOH	290(4.17)	88-3043-64
	CCl_4	370(2.4)	60-2189-64
hydroxide adduct	50% EtOH	285(3.83)	88-3043-64
indole complex	CCl_4	405(3.1),495(3.2)	60-2189-64
C_6F_5I			
Benzene, pentafluoroiodo-	EtOH	226(4.02),252(2.94)	30-1135-64
C_6F_7N			
2-Picoline, heptafluoro-	hexane	256(3.37)	39-2720-65
	EtOH	256(3.35)	39-2720-65
3-Picoline, heptafluoro-	hexane	255(3.43)	39-2720-65
4-Picoline, heptafluoro-	hexane	280(3.54)	39-2720-65
	EtOH	278(3.60)	39-2720-65
C_6F_8O			
2-Cyclohexen-1-one, octafluoro-	EtOH	300.0(3.43)	39-7358-65
$C_6HCl_3O_3$			
p-Benzoquinone, trichlorohydroxy-	H_2O	294(4.07)	88-3043-64
hydroxide adduct	H_2O	250(3.75),365(3.70)	88-3043-64
C_6HCl_5O			
Phenol, pentachloro-	EtOH	305.0(3.5)	39-1692-65
$C_6HF_7O_2$			
2-Cyclohexen-1-one, 2,4,4,5,5,6,6-heptafluoro-3-hydroxy- (hydrate)	EtOH	296.0(4.47)	39-5748-64
$C_6H_2Cl_2N_4$			
Pteridine, 6,7-dichloro-	pH 1.1	264(3.70),319(4.10)	39-1666-64
anhydrous	pH 7.0	225(4.09),302s(3.80), 315(4.02),328(4.01)	39-1666-64
hydrated	pH 7.0	243(3.69),281(3.82), 337(4.02)	39-1666-64
$C_6H_2Cl_2O_2$			
p-Benzoquinone, 2,5-dichloro-	H_2O	272(4.25)	88-3043-64
hydroxide adduct	H_2O	250(4.00)	88-3043-64
$C_6H_2Cl_3N_3$			
Imidazo[1,2-c]pyrimidine, 2,5,7-trichloro-	EtOH	230(4.27),280(3.86), 304s(3.70)	4-0034-64
$C_6H_2F_3N_9O$			
Tetrazolo[1,5-c]pyrimidine, 7-azido-8-(2,2,2-trifluoroacetamido)-	pH 1	233(4.43),287(4.21)	44-0829-65
	pH 7	222(4.11),296(4.18)	44-0829-65
$C_6H_2F_5N_3$			
Malononitrile, (1-amino-2,2,3,3,3-pentafluoropropylidene)-	EtOH	287(4.14)	44-0707-64
$C_6H_2F_7NO$			
2-Cyclohexen-1-one, 3-amino-2,4,4,5,5,6,6-heptafluoro-	EtOH	301.0(4.45)	39-5748-64

Compound	Solvent	λ_{max}(log ε)	Ref.
$C_6H_2N_4O_3$			
Furazanobenzofuroxan	n.s.g.	254(4.21),306(3.97)	39-5958-65
$C_6H_2N_4O_4$			
Furoxanobenzofuroxan	n.s.g.	282(4.29),308(4.14), 322(4.09),346(3.72)	39-5958-65
$C_6H_2N_4O_4S$			
2,1,3-Benzothiadiazole, 4,5-dinitro-	EtOH	212(4.15),298(4.39), 352(3.88),410(3.62)	7-0476-64
$C_6H_2N_4O_6$			
Benzofuroxan, 4,6-dinitro-	aq. HCl	259(4.00),278(4.03), 416(3.88)	44-2407-65
	H_2O	263(4.00),302(3.84), 463(4.33)	44-2407-65
	aq. NaOH	263(4.05),302(3.90), 463(4.40)	44-2407-65
	$CHCl_3$	258(4.09),272(4.11), 420(3.88)	44-2407-65
$C_6H_2N_6O_{10}$			
Aniline, 2,3,4,5,6-pentanitro-	dioxan	316(3.88),400(3.68)	77-0232-64
$C_6H_3Br_2D_2NO$			
4(1H)-Pyridone-2,6-d_2, 3,5-dibromo-1-methyl-	MeOH	280(4.08)	35-3365-65
$C_6H_3Br_2NO_3$			
Phenol, 2,6-dibromo-4-nitro-	hexane	287(3.85)	70-1666-64
$C_6H_3ClN_4$			
Pteridine, 6-chloro-	C_6H_{12}	216(4.21),244s(3.56), 285s(3.46),291s(3.61), 298s(3.77),303(3.84), 310(3.94),316(3.88), 324(3.90)	39-4920-64
	$CHCl_3$	304(3.85),311(3.96), 317(3.87),325(3.93)	39-4920-64
Pteridine, 7-chloro-	$CHCl_3$	297(3.91),304(4.00), 310(3.96),318(3.98)	39-4920-64
$C_6H_3ClO_2$			
p-Benzoquinone, chloro-	H_2O	257(4.22)	88-3043-64
hydroxide adduct	H_2O	267(4.00)	88-3043-64
$C_6H_3Cl_2NOS_2$			
3H-1,2,3-Benzodithiazole, 4,6-dichloro-, 2-oxide	EtOH	222(4.43),232s(4.39), 298(3.35),307s(3.33)	44-2763-65
$C_6H_3Cl_2NO_2$			
Pyrrole-2,5-dicarboxaldehyde, 3,4-dichloro-	EtOH	218(3.70),234s(3.96), 241(4.09),246(4.04), 307(4.20)	39-0459-65
	EtOH-base	240s(3.96),262(4.16), 330(4.34)	39-0459-65
$C_6H_3Cl_2NS_2$			
6-Chloro-1,3,2-benzothiazathiolium chloride	CF_3COOH	366(4.1),417(3.6)	44-2763-65

Compound	Solvent	λ_{max}(log ε)	Ref.
$C_6H_3Cl_2N_3$			
Imidazo[4,5-c]pyridine, 4,6-dichloro-	pH 1	240(3.62),277(3.71)	4-0196-65
	pH 11	255(3.53),286(3.69)	4-0196-65
$C_6H_3Cl_2N_3O_2$			
Pyrazine, 5,6-dichloro-2-formamido-3-formyl-	pH 7.0	209(4.02),245(4.01), 277(3.93),325(3.87)	39-1666-64
$C_6H_3F_2N_3$			
Benzotriazole, 4,6-difluoro-	pH 1	259(3.76)	87-0737-65
	pH 7	266(3.84)	87-0737-65
	pH 13	267(3.87)	87-0737-65
	EtOH	263s(--),266(3.73)	87-0737-65
$C_6H_3F_5$			
2-Hexen-4-yne, 1,1,1,2,3-pentafluoro-	MeOH	221(4.17),228s(--)	44-2009-64
$C_6H_3F_7N_2$			
1-Cyclohexen-1-ylamine, 2,4,4,5,-5,6,7-heptafluoro-3-imino-	EtOH	291.0(4.44)	39-5748-64
$C_6H_3N_5O_8$			
Aniline, 2,3,4,6-tetranitro-	dioxan	320(4.02),402(3.78)	77-0232-64
C_6H_4BrClO			
Phenol, 4-bromo-2-chloro-	HCl	218(3.92)	44-2678-65
	NaOH	244(4.05)	44-2678-65
$C_6H_4BrCl_3N_2$			
Pyrimidine, 5-(bromomethyl)-4,6-dichloro-2-(chloromethyl)-	MeOH	261(3.72)	87-0808-64
$C_6H_4BrFO_2S$			
Benzenesulfonyl fluoride, p-bromo-	$C_2H_4Cl_2$	240(4.12),270(2.95)	30-0815-65
$C_6H_4BrNO_2$			
Benzene, 1-bromo-3-nitro-	C_6H_{12}	254.5(3.77)	44-2088-64
	MeOH	258.5(3.73)	44-2088-64
	iso-PrOH	259(3.74)	44-2088-64
	dioxan	259.5(3.81)	44-2088-64
Benzene, 1-bromo-4-nitro-	C_6H_{12}	270.5(4.15)	44-2088-64
	MeOH	275.7(3.98)	44-2088-64
	iso-PrOH	274.5(4.12)	44-2088-64
	dioxan	276.5(4.22)	44-2088-64
$C_6H_4BrN_3O_4$			
Aniline, 2-bromo-4,6-dinitro-	H_2O	350(4.08)	23-1681-64
	sulfolane	395(4.37),527(4.10)	23-1681-64
$C_6H_4Br_3N$			
Pyridine, 4-(tribromomethyl)-	EtOH-HBr	268(3.64)	23-0698-64
	EtOH	269(3.37)	23-0698-64
$C_6H_4Br_3NO_2$			
Pyrrole-2-carboxylic acid, 3,4,5-tribromo-, methyl ester	EtOH	279(4.18)	39-0459-65
	EtOH-base	295(4.27)	39-0459-65

Compound	Solvent	$\lambda_{max}(\log \epsilon)$	Ref.
$C_6H_4ClFO_2S$			
Benzenesulfonyl fluoride, p-chloro-	$C_2H_4Cl_2$	232(4.18),270(2.95)	30-0815-65
C_6H_4ClCsO			
Cesium, (p-chlorophenoxy)-	DMF	337.5(3.61)	35-1857-65
	$(MeOCH_2)_2$	327(3.38)	35-1857-65
$C_6H_4ClF_4NO$			
2-Cyclopenten-1-one, 2-chloro-4,4,5,5-tetrafluoro-3-(methylamino)-	EtOH	289(4.48)	44-3698-65
	pH 1	295(4.48)	44-3698-65
C_6H_4ClKO			
Potassium,(p-chlorophenoxy)-	$(MeOCH_2)_2$	325.5(3.42)	35-1857-65
C_6H_4ClLiO			
Lithium, (p-chlorophenoxy)-	DMF	324(3.49)	35-1857-65
	$(MeOCH_2)_2$	307(3.43)	35-1857-65
$C_6H_4ClNOS_2$			
3H-1,2,3-Benzodithiazole, 6-chloro-, 2-oxide	EtOH	214(4.46),294(3.37)	44-2763-65
$C_6H_4ClNO_3$			
Phenol, 2-chloro-4-nitro-	HCl	315(3.53)	44-2678-65
	NaOH	400(3.83)	44-2678-65
Phenol, 5-chloro-2-nitro-	HCl	342(3.65)	35-4942-64
	NaOH	410(3.78)	35-4942-64
$C_6H_4ClN_3$			
1H-Imidazo[4,5-c]pyridine, 4-chloro-	pH 1	268(3.79)	4-0196-65
	pH 7	252(3.66),267(3.78), 274(3.70)	94-0866-64
	pH 11	275(3.74)	4-0196-65
1H-Imidazo[4,5-c]pyridine, 6-chloro-	pH 1	234(3.65),273(3.58), 280(3.51)	4-0196-65
	pH 11	250(3.53),275(3.58)	4-0196-65
$C_6H_4ClN_3O_2$			
Pyrazine, 5-chloro-2-formamido-3-formyl-	pH 7.0	245(4.00),272(3.89), 316(3.73)	39-1666-64
Pyrazine, 6-chloro-2-formamido-	pH 7.0	211(4.12),227(4.01), 273(4.01),318(3.92)	39-1666-64
$C_6H_4ClN_5$			
Pteridine, 4-amino-7-chloro-	pH 0.4	237(4.04),272s(3.41), 338(3.98),348s(3.93)	39-1175-65
	pH 5.0	252(4.21),280s(3.36), 348(3.79)	39-1175-65
Pteridine, 7-amino-4-chloro-	pH -1.5	228(4.14),238s(4.05), 340(3.91)	39-1175-65
	pH 3.0	238(4.27),262s(3.77), 339(3.96)	39-1175-65
C_6H_4ClNaO			
Sodium p-chlorophenoxide	DMF	336(3.58)	35-1857-65
	$(MeOCH_2)_2$	320(3.51)	35-1857-65

Compound	Solvent	$\lambda_{max}(\log \epsilon)$	Ref.
$C_6H_4Cl_2N_2O_2$			
Aniline, 2,5-dichloro-4-nitro-	pyridine	368(4.07)	23-1681-64
anion	pyridine	458(4.57)	23-1681-64
Aniline, 2,6-dichloro-4-nitro-	pyridine	368(4.14)	23-1681-64
anion	pyridine	467(4.59)	23-1681-64
$C_6H_4Cl_2O$			
Phenol, 2,4-dichloro-	HCl	218(3.86)	44-2678-65
	NaOH	245(4.00)	44-2678-65
$C_6H_4Cl_2O_4$			
1-Cyclopentene-1-carboxylic acid, 3,5-dichloro-2-hydroxy-4-oxo-	n.s.g.	276(3.94)	88-2411-64
$C_6H_4Cl_3NO_2$			
Pyrrole-2-carboxylic acid, 3,4,5-trichloro-, methyl ester	EtOH	245s(3.80),271(4.13)	39-0459-65
	EtOH-base	292(4.14)	39-0459-65
$C_6H_4Cl_3N_3O$			
Pyrimidine, 4-amino-5-chloroacetyl-2,6-dichloro-	EtOH	224(4.24),249(3.82), 318(3.76)	4-0034-64
$C_6H_4CsNO_3$			
Cesium, (p-nitrophenoxy)-	DMF	434(4.52)	35-1857-65
	$(MeOCH_2)_2$	418(4.45)	35-1857-65
C_6H_4FI			
Benzene, 1-fluoro-2-iodo-	gas	202f(--)	60-0597-65
Benzene, 1-fluoro-3-iodo-	gas	200f(--)	60-0597-65
Benzene, 1-fluoro-4-iodo-	gas	206f(--)	60-0597-65
$C_6H_4FNO_3$			
Phenol, 5-fluoro-2-nitro-	HCl	341(3.58)	35-4942-64
	NaOH	405(3.73)	35-4942-64
$C_6H_4F_2N_2O_2$			
Aniline, 2,5-difluoro-4-nitro-	EtOH	355(4.17)	87-0737-65
$C_6H_4F_2N_4$			
Benzotriazole, 4-amino-5,7-difluoro-	pH 1	258.5(3.53)	87-0737-65
	pH 7	272(3.58),293(3.70)	87-0737-65
	pH 13	272(3.61),293(3.73)	87-0737-65
$C_6H_4F_5NO$			
2-Cyclopenten-1-one, 2,4,4,5,5-pentafluoro-3-(methylamino)-	EtOH	285(4.54)	44-3698-65
	pH 1	291(4.54)	44-3698-65
$C_6H_4KNO_3$			
Potassium, (p-nitrophenoxy)-	$(MeOCH_2)_2$	415(4.43)	35-1857-65
$C_6H_4K_2O_4$			
1,1,2,2-Ethanetetracarboxaldehyde, potassium salt	n.s.g.	270(4.60)	44-3046-64
$C_6H_4LiNO_3$			
Lithium, (p-nitrophenoxy)-	DMF	435(4.53)	35-1857-65
	$(MeOCH_2)_2$	397(4.38)	35-1857-65

Compound	Solvent	λ_{max}(log ε)	Ref.
$C_6H_4NNaO_3$			
Sodium, (p-nitrophenoxy)-	DMF	435(4.54)	35-1857-65
	$MeCONMe_2$	435(4.55)	35-1857-65
	$(MeOCH_2)_2$	410(4.41)	35-1857-65
$C_6H_4N_2$			
2,4-Hexadienedinitrile, cis-cis	EtOH	260(4.46),271(4.33)	88-1433-65
trans-trans	EtOH	259(4.58),270(4.49)	88-1433-65
$C_6H_4N_2O$			
Benzofurazan	6N HCl	265(3.50),269(3.62), 276(3.71),281(3.65), 288(3.67),304s(3.53)	17-0033-64
	EtOH	268(3.53),274(3.63), 280(3.58),286(3.65), 300(3.51)	17-0033-64
2,5-Cyclohexadien-1-one, 4-diazo-	aq. EtOH	252(3.60),347(4.55)	30-1101-64
	+ acid	313(4.50)	30-1101-64
$C_6H_4N_2O_5$			
Phenol, 2,4-dinitro-, piperidine complex (other amine complexes also given)	C_6H_5Cl	360(4.23),410s(4.02)	39-6984-65
Phenol, 2,5-dinitro-	HCl	363(3.58)	35-4942-64
	NaOH	442(3.64)	35-4942-64
$C_6H_4N_2S$			
Acrylonitrile, 3,3'-thiodi-, cis	n.s.g.	292(4.32)	44-1800-64
1,2,3-Benzothiadiazole	12N HCl	233(4.02),299(3.75), 331(3.41)	39-6061-64
	H_2SO_4	233(4.03),298(3.78), 334(3.43)	39-6061-64
	pH 7.0	214(4.20),266(3.72), 313(3.40)	39-6061-64
2,1,3-Benzothiadiazole	N HCl	222(4.13),305(4.12), 312(4.11),330s(3.47)	17-0033-64
	EtOH	222(4.16),304(4.14), 310(4.14),330s(3.39)	17-0033-64
$C_6H_4N_2Se$			
2,1,3-Benzoselenadiazole	N HCl	231(3.66),333(4.18), 360s(3.41)	17-0033-64
	EtOH	232(3.66),331(4.22), 360s(3.32)	17-0033-64
$C_6H_4N_4OS$			
6-Pteridinol, 7-mercapto-	pH -1.5	227s(4.12),270(3.62), 352s(4.32),363(4.39)	39-0027-65
	pH 3.5	257(3.52),340s(4.09), 352(4.28),369(4.31), 388(4.05)	39-0027-65
	pH 7.8	232s(4.08),266(3.61), 375(4.35),390s(4.28)	39-0027-65
	pH 12.0	236(4.28),270(3.74), 362s(4.17),377(4.34), 393(4.29)	39-0027-65

Compound	Solvent	$\lambda_{max}(\log \epsilon)$	Ref.
7-Pteridinol, 6-mercapto-	pH -1.5	235(4.15),260s(3.78), 327(3.98),355s(4.17), 367(4.21)	39-0027-65
	pH 3.2	246(3.70),292(3.66), 354(4.23),371(4.27)	39-0027-65
	pH 7.5	258(3.89),308(3.70), 367(4.22)	39-0027-65
	pH 12.0	218(4.41),238(4.26), 261s(3.88),311(3.80), 363(4.24)	39-0027-65
$C_6H_4N_4O_2$			
6,7-Pteridinediol	pH -1.5	222(4.22),281s(3.88), 306(4.16)	39-0027-65
$C_6H_4N_4O_2S$			
2,1,3-Benzothiadiazole, 5-amino-4-nitro-	EtOH	232(4.24),257s(--), 315(4.06),389(4.04), 406(4.03)	7-0476-64
$C_6H_4N_4O_2Se$			
2,1,3-Benzoselenadiazole, 4-amino-5-nitro-	EtOH	225(3.98),255s(3.5), 324(4.47),360(3.87), 430(3.48)	7-0476-64
2,1,3-Benzoselenadiazole, 5-amino-4-nitro-	EtOH	224(4.23),259(3.62), 338(4.16),409(4.08)	7-0476-64
$C_6H_4N_4O_3$			
Lumazine, 5-oxide	pH 2.0	234(4.26),244(4.27), 270(3.73),345(3.84)	89-1136-65
	pH 8.0	245(4.10),266(4.23), 285s(3.95),387(3.86)	89-1136-65
$C_6H_4N_4O_6$			
Aniline, 2,4,6-trinitro-	MeOH	318(4.08),407(3.90)	35-4018-64
	dioxan	318(4.08),408(3.89)	77-0232-64
	sulfolane	328(4.04),417(3.78)	23-1681-64
anion	H_2O	412(4.37)	23-1681-64
$C_6H_4N_4S$			
Pteridine-6-thiol	pH 1.0	230(3.91),242s(3.82), 332(4.28)	39-0027-65
	pH 4.4	242(3.87),331(4.24)	39-0027-65
	pH 11.0	237(4.03),299(4.17), 411(3.83)	39-0027-65
$C_6H_4N_4S_2$			
6,7-Pteridinedithiol	pH -1.5	277(4.19),368(4.10), 382s(4.07),410s(4.00)	39-0027-65
	pH 3.0	256(4.26),290s(3.81), 378s(4.07),394(4.14), 414(4.14),433s(3.99)	39-0027-65
	pH 6.6	260(4.27),298(3.91), 380s(4.05),400s(4.13), 412(4.15),440s(3.98)	39-0027-65
	pH 12.0	247(4.22),271(4.32), 302s(3.90),396s(4.18), 409(4.22),427s(4.10)	39-0027-65

Compound	Solvent	λ_{max}(log ϵ)	Ref.
$C_6H_4N_6$			
s-Triazolo[3,4-i]purine	pH 1	258(3.97),264(3.98)	44-3601-65
	pH 7	257(3.89),263(3.92), 284(3.68)	44-3601-65
	pH 13	263(3.88),272(3.90), 301(3.76)	44-3601-65
s-Triazolo[5,1-i]purine	pH 1	261s(3.83),273(3.85)	44-3601-65
	pH 7	262(3.79),277(3.88)	44-3601-65
	pH 13	290(3.95)	44-3601-65
$C_6H_4O_2$			
p-Benzoquinone	95% THF	243(4.32),435(1.26)	44-2602-65
	n.s.g.	244(4.37),290(2.53), 438(1.36)	28-5487-65B
$C_6H_4O_3$			
2,5-Furandicarboxaldehyde	EtOH	223(3.66),280(4.27)	95-0981-65
Quinone, hydroxy-	$CHCl_3$	256(4.14),369(3.07)	44-4107-65
$C_6H_4O_4$			
p-Benzoquinone, 2,5-dihydroxy-	EtOH	285(4.26)	44-3573-65
α-Pyrone-5-carboxylic acid	pH 1.4	245(3.90),288(3.61)	88-3431-64
	pH 5.0	239(3.81),291(3.68)	88-3431-64
	pH 12.4	323(3.96)	88-3431-64
α-Pyrone-6-carboxylic acid	pH 1.4	229(3.34),301(3.89)	88-3431-64
	pH 5.0	227(3.51),303(3.88)	88-3431-64
	pH 12.4	350(4.37+)	88-3431-64
γ-Pyrone-2-carboxylic acid	pH 5.0	260(4.0)	88-3431-64
	EtOH-HCl	260(3.99)	39-2251-65
	EtOH-NaOH	260(3.74),350(4.14)	39-2251-65
$C_6H_4O_5S$			
Benzoquinonesulfonic acid	H_2O	253(4.20)	88-3043-64
hydroxide adduct	H_2O	253(3.86)	88-3043-64
$C_6H_4S_2$			
Thieno[3,2-b]thiophene	n.s.g.	213(3.7),257(4.1), 268(4.1),277(4.0)	70-0510-65
$C_6H_5BrCl_2N_2$			
Pyrimidine, 5-(bromomethyl)-4,6-dichloro-2-methyl-	MeOH	259(3.67)	87-0808-64
C_6H_5BrHg			
Mercury, bromophenyl-	90% dioxan	282s(2.6)	70-0240-65
sodium iodide complex	90% dioxan	303(3.72)	70-0240-65
$C_6H_5BrNO_5P$			
Phosphonic acid, (5-bromo-2-nitrophenyl)-	EtOH	264.5(3.72)	87-0891-65
$C_6H_5BrO_2$			
2-Penten-4-ynoic acid, 5-bromo-, methyl ester, cis	Et_2O	258.5(4.18)	24-1736-65
trans	Et_2O	257(4.25)	24-1736-65
$C_6H_5Br_2NO$			
4(1H)-Pyridone, 3,5-dibromo-1-methyl-	MeOH	280(4.04)	35-3365-65

Compound	Solvent	λ_{max}(log ε)	Ref.
$C_6H_5Br_2NO_2$			
Pyrrole-2-carboxylic acid, 4,5-dibromo-, methyl ester	EtOH	233(3.72),279(4.15)	39-0459-65
	EtOH-base	233(3.45),295(4.27)	39-0459-65
$C_6H_5ClF_4N_2O$			
Hydroxylamine, N-[2-chloro-4,4,5,5-tetrafluoro-3-(methylimino)-1-cyclopenten-1-yl]-	EtOH	286(4.48)	44-3698-65
$C_6H_5ClN_2OS$			
4-Isothiazolecarbonitrile, 3-chloro-5-ethoxy-	EtOH	248(3.83)	44-0660-64
$C_6H_5ClN_2O_2$			
Aniline, 4-chloro-2-nitro-	MeOH	417(3.54)	23-1681-64
	sulfolane	516(3.84)	23-1681-64
4-Pyrimidinecarboxylic acid, 6-chloro-, methyl ester	EtOH	265(3.53)	4-0130-64
$C_6H_5ClN_2O_3$			
Acetic acid, [(6-chloro-3-pyridazinyl)oxy]-	1% EtOH	278(3.27)	20-0532-64
1(6H)-Pyridazineacetic acid, 3-chloro-6-oxo-	1% EtOH	300(3.47)	20-0532-64
$C_6H_5ClN_2S_2$			
4-Isothiazolecarbonitrile, 3-chloro-5-(ethylthio)-	EtOH	223(4.06),288(3.98)	44-0660-64
$C_6H_5ClN_4$			
9H-Purine, 2-chloro-9-methyl-	pH -1.6	271(3.94),278s(3.72)	39-3017-65
	pH 2.9	272(3.89)	39-3017-65
9H-Purine, 6-chloro-9-methyl-	pH -2.0	238s(3.66),244s(3.72), 263(4.01),270s(3.88)	39-3017-65
	pH 2.4	266(4.05)	39-3017-65
9H-Purine, 8-chloro-9-methyl-	pH -0.2	270(3.90)	39-3017-65
	pH 4.2	267(4.03)	39-3017-65
$C_6H_5ClN_4O$			
Purine, 2-chloro-6-methoxy-	pH 1	259(4.01)	44-1110-65
	pH 7	259(4.00)	44-1110-65
	pH 13	267(4.00)	44-1110-65
$C_6H_5ClN_4S$			
Imidazo[4,5-c]pyridazine, 7-chloro-2-(methylthio)-	EtOH	207(4.23),227(4.27), 297(4.29)	4-0042-64
$C_6H_5ClN_6O$			
Pyrimidine, 5-acetamido-4-azido-6-chloro-	pH 1	244(3.88),272(3.92), 279s(3.91)	44-0829-65
	pH 7	244(3.88),273(3.91), 280s(3.90)	44-0829-65
3H-v-Triazolo[4,5-d]pyrimidine, 3-acetamido-7-chloro-	pH 7	262(3.83)	44-0829-65
C_6H_5ClO			
Phenol, 2-chloro-	HCl	213(3.86)	44-2678-65
	NaOH	236(3.94)	44-2678-65
Phenol, 4-chloro-	DMF	284(3.30)	35-1857-65
	$(MeOCH_2)_2$	283(3.30)	35-1857-65

Compound	Solvent	λ_{max}(log ε)	Ref.
$C_6H_5ClO_2$			
Ketone, 5-chloro-2-furyl methyl	EtOH	278(4.17)	25-1425-65
6(7H)-Oxepinone, 3-chloro-	Et_2O	315(3.96),370(3.58)	25-0184-65
2-Pyrone, 4-chloro-6-methyl-	MeOH	300(3.70)	54-0039-64
$C_6H_5Cl_2NO_2$			
Pyrrole-2-carboxylic acid, 3,4-di-	EtOH	250s(3.98),264(4.09)	39-0459-65
chloro-, methyl ester	EtOH-base	286(4.23)	39-0459-65
Pyrrole-2-carboxylic acid, 3,5-di-	EtOH	266(4.20)	39-0459-65
chloro-, methyl ester	EtOH-base	287(4.27)	39-0459-65
Pyrrole-2-carboxylic acid, 4,5-di-	EtOH	233(3.68),274(4.17)	39-0459-65
chloro-, methyl ester	EtOH-base	240(3.45),294(4.32)	39-0459-65
$C_6H_5Cl_3Ge$			
Germane, trichlorophenyl-	heptane	248(2.22),253(2.40), 257(2.59),263(2.72), 270(2.63)	28-3931-65A
$C_6H_5Cl_3N_2$			
Pyrimidine, 4,6-dichloro-2-(chloro- methyl)-5-methyl-	MeOH	261(3.71)	87-0808-64
Pyrimidine, 2,4,6-trichloro-5-ethyl-	MeOH	268(3.77)	87-0808-64
Pyrimidine, 4,5,6-trichloro-2-ethyl-	MeOH	270(3.73)	87-0808-64
$C_6H_5Cl_3Sn$			
Tin, trichlorophenyl-	heptane	247(2.77),257(2.69), 263(2.72),270(2.60)	28-3931-65A
C_6H_5CsO			
Cesium phenoxide	DMF	326(3.69)	35-1857-65
	$(MeOCH_2)_2$	317.5(3.54)	35-1857-65
$C_6H_5FO_2S$			
Benzenesulfonyl fluoride	$C_2H_4Cl_2$	220(3.93),267(3.11)	30-0815-65
$C_6H_5F_5N_2O$			
Hydroxylamine, N-[2,4,4,5,5-penta- fluoro-3-(methylimino)-1-cyclopen- ten-1-yl]-	EtOH	285(4.52)	44-3698-65
$C_6H_5F_7O$			
3-Hexanone, 4,4,5,5,6,6,6-hepta- fluoro-	n.s.g.	313(1.32)	46-1001-65
C_6H_5I			
Benzene, iodo-	gas	204f(--)	60-0597-65
$C_6H_5I_2NO$			
Pyrrole, 2-acetyl-4,5-diiodo-	EtOH	259(3.74),310(4.12)	39-0459-65
	EtOH-base	225(3.91),333(4.32)	39-0459-65
C_6H_5KO			
Potassium phenoxide	$(MeOCH_2)_2$	314.5(3.61)	35-1857-65
C_6H_5LiO			
Lithium phenoxide	DMF	309(3.49)	35-1857-65
	$(MeOCH_2)_2$	296.5(3.20)	35-1857-65

Compound	Solvent	$\lambda_{max}(\log \epsilon)$	Ref.
$C_6H_5NOS_2$			
3H-1,2,3-Benzodithiazole, 2-oxide	EtOH	208(4.5),276(3.25)	44-2763-65
$C_6H_5NO_2$			
Benzene, nitro-	gas	164(4.44),193(4.24), 240(3.93),288(2.70)	82-0174-64B
	C_6H_{12}	253(3.94)	44-2088-64
	MeOH	260.5(3.89)	44-2088-64
	EtOH	260(4.18)	65-0781-65
	iso-PrOH	258(3.82)	44-2088-64
	dioxan	260(3.95)	44-2088-64
2,4-Pentadienoic acid, 5-cyano-, 2-cis-5-trans	EtOH	259(4.37)	44-0203-65
2-trans-5-trans	EtOH	260(4.48)	44-0203-65
$C_6H_5NO_2S_2$			
p-Dithiin-2,3-dicarboxylic acid imide, 5,6-dihydro-	EtOH	250(4.02),405(3.52)	56-1713-65
$C_6H_5NO_3$			
Phenol, 2-nitro-	HCl	350(3.52)	35-4942-64
	hexane	350(3.82)	65-0428-65
	H_2O	274(3.59),418(3.49)	65-0428-65
	pH 13	228(4.15),348(3.60), 448(3.48)	65-0428-65
	NaOH	418(3.67)	35-4942-64
	MeOH	274(4.07),351(3.80)	44-2088-64
	EtOH	274(3.79),348(3.51)	65-0428-65
	dioxan	276(4.25),351(3.96)	44-2088-64
	cyclohexanol	273(4.30),348(3.76)	44-2088-64
	acetone	348(3.55)	65-0428-65
	+base	432(3.76)	65-0428-65
	cyclohexanone	275(3.75),346(3.45)	65-0428-65
	+base	324(3.65)	65-0428-65
	aniline	352(3.52),440(3.11)	65-0428-65
	$HOCH_2CH_2$-NH_2	278(3.64),428(3.77)	65-0428-65
Phenol, 3-nitro-	C_6H_{12}	260(4.03),315(3.40)	44-2088-64
	hexane	260(3.54),316(3.27)	65-0428-65
	H_2O	230(3.89),274(3.75)	65-0428-65
	pH 13	226(4.20),290(3.66), 390(3.20)	65-0428-65
	MeOH	270(3.77),332(3.46)	44-2088-64
	EtOH	270(3.79),330(3.34)	65-0428-65
	iso-PrOH	270(3.77),334(3.40)	44-2088-64
	dioxan	270(3.77),330(3.32)	44-2088-64
	acetone	330(3.50)	65-0428-65
	+NaOH	446(3.16)	65-0428-65
	cyclohexanone	330(3.34)	65-0428-65
	$HOCH_2$-CH_2NH_2	260(--),428(4.27)	65-0428-65
Phenol, 4-nitro-	C_6H_{12}	287(4.09)	44-2088-64
	hexane	288(3.80)	65-0428-65
	H_2O	226(4.01),314(4.13)	65-0428-65
	MeOH	313(3.92)	44-2088-64
	EtOH	314(4.06)	65-0428-65
	iso-PrOH	313(3.92)	44-2088-64
	0.1N NaOH	224(3.92),294(3.11), 378(4.27)	65-0428-65
	dioxan	306(3.96)	44-2088-64

Compound	Solvent	λ_{max}(log ϵ)	Ref.
Phenol, 4-nitro- (cont.)	acetone	308(4.20)	65-0428-65
	+NaOH	416(4.51)	65-0428-65
	cyclohexanone	324(3.85)	65-0428-65
	+NaOH	324(3.75),394(3.85)	65-0428-65
	aniline	358(3.76)	65-0428-65
	DMF	319(4.04)	35-1857-65
	$(MeOCH_2)_2$	308.5(4.04)	35-1857-65
	$C_2H_4Cl_2$	300(4.0)	38-0895-64
$C_6H_5NO_4$			
Resorcinol, 4-nitro-	C_6H_{12}	232(3.95),298(4.11), 337(4.03)	44-1821-64
$C_6H_5NO_4S$			
4,5-Isothiazoledicarboxylic acid, 3-methyl-	EtOH	252(3.73),278(3.55)	39-0446-64
$C_6H_5N_3$			
Benzotriazole	hexane	248(3.8),256(3.8), 280f(3.7),294(3.3)	61-0716-65
Nicotinonitrile, 6-amino-	pH 1.2	256(4.27),303(3.72)	39-5542-65
	pH 6.0	264(4.32),294s(3.76)	39-5542-65
1,2,3a-Triazaindene	pH 1.0	266s(3.74),275(3.80), 286s(3.57)	39-2778-65
	pH 7.0	256s(3.49),259s(3.53), 266(3.60),278(3.62), 287s(3.57)	39-2778-65
1,2,7a-Triazaindene	pH -2.0	223s(3.72),230s(3.47), 271(3.78),288s(3.67), 299(3.36)	39-2778-65
	pH 4.0	217s(4.05),223(3.72), 278(3.81)	39-2778-65
1,3a,4-Triazaindene	pH 2.0	211(4.20),244s(3.40), 251(3.42),261s(3.73), 291(3.66)	39-2778-65
1,3a,5-Triazaindene	C_6H_{12}	251(3.52),256(3.55), 260(3.63),264s(3.54), 270(3.64),300(3.31), 309(3.31),322(3.20), 339(2.83)	39-2778-65
	pH 1.7	261(3.93)	39-2778-65
	pH 7.0	215(4.29),260(3.73), 270(3.73),294(3.39), 298(3.39)	39-2778-65
1,3a,7-Triazaindene	C_6H_{12}	224s(4.26),228(4.30), 232(4.28),277(3.30), 287(3.29),326(3.41), 333(3.42),340s(3.38), 349(3.30),356s(3.08), 369(2.91)	39-2778-65
	pH 2.0	214(4.29),276(3.66), 290s(3.61)	39-2778-65
	pH 7.0	211(4.15),223(4.30), 275(3.45),283(3.47), 317(3.46)	39-2778-65

Compound	Solvent	λ_{max}(log ε)	Ref.
1,4,6-Triazaindene	pH 3.2	225(4.3),270(3.7), 330(3.5)	94-1030-64
	pH 10.3	220s(4.4),270(3.7), 300s(3.4)	94-1030-64
	MeOH	220s(4.3),270(3.7), 330s(3.4)	94-1030-64
$C_6H_5N_3O$			
Benzofurazan, 4-amino-	N HCl	264(--),269(3.47), 274(3.57),390(3.42)	17-0033-64
	3N HCl	269(3.50),274(3.59), 280(3.55),286(3.60), 305(3.54)	17-0033-64
	6N HCl	270(3.56),275(3.65), 280(3.61),287(3.65), 305(3.60)	17-0033-64
	0.02N NaOH	224(4.33),274(2.98), 390(3.42)	17-0033-64
	EtOH	227(4.39),273(3.68), 404(3.54)	17-0033-64
Benzofurazan, 5-amino-	N HCl	275(3.67),286(3.67), 300s(3.5),368(2.64)	17-0033-64
	3N HCl	275(3.70),287(3.81), 300s(3.56),358(2.19)	17-0033-64
	6N HCl	275(3.68),287(3.68), 300(3.54)	17-0033-64
	0.02N NaOH	222(4.25),277(3.58), 368(3.57)	17-0033-64
	EtOH	228(4.29),278(3.42), 377(3.55)	17-0033-64
$\Delta^{2,\alpha}$-Oxazolidinemalononitrile	EtOH	250.8(4.38)	95-0387-65
1H-Pyrazolo[4,3-c]pyridin-3-ol	10N H_2SO_4	220(4.2),283(3.55)	12-0379-65
	H_2O	226(4.29),289(3.51)	12-0379-65
	N NaOH	227(4.25),283(3.43)	12-0379-65
5-Pyrimidinecarbonitrile, 1,2-dihydro-1-methyl-2-oxo-	EtOH	262(4.04),312(2.75)	44-1740-64
1,4,6-Triazainden-7(6H)-one	pH 2	240(4.3),260(4.0)	94-1030-64
	pH 12	225(4.5),270(3.8), 285s(3.7)	94-1030-64
	MeOH	225(4.5),270(3.7)	94-1030-64
$C_6H_5N_3OS$			
5H-Pyrimido[4,5-b][1,4]thiazin-6(7H)-one	pH 1	217(4.06),242(4.05), 302(3.70),333s(--)	44-1247-65
	pH 7	213(4.07),242(4.13), 300(3.80)	44-1247-65
	pH 13	254(4.14),297(3.92)	44-2147-65
Thiazolo[5,4-d]pyrimidin-7-ol, 2-methyl-	pH 1	251(3.88),258(3.89), 279(3.77)	94-1319-65
$C_6H_5N_3O_2$			
Isoxazolo[5,4-d]pyrimidin-4(5H)-one, 3-methyl-	EtOH	235(3.81),242s(3.78), 267(3.73)	44-2117-64
1H-Pyrazolo[4,3-c]pyridin-3-ol, 5-oxide	10N H_2SO_4	232(4.26),292(3.6)	12-0379-65
	H_2O	255(4.33),295(3.6)	12-0379-65
	N NaOH	265(4.53),320(3.35)	12-0379-65

Compound	Solvent	$\lambda_{max}(\log \epsilon)$	Ref.
$C_6H_5N_3O_2S$			
Thiazolo[5,4-d]pyrimidine-5,7-diol, 2-methyl-	H_2O	212(4.28),257(3.91)	94-1319-65
$C_6H_5N_3O_3$			
Imidazo[1,2-c]pyrimidine-2,5,7-(1H,3H,6H)-trione	pH 1	267(4.35)	4-0034-64
	pH 11	284(4.37)	4-0034-64
Pyrrolo[2,3-d]pyrimidine-2,4,5-trione	pH 1	258(4.02)	4-0034-64
	pH 11	278(4.01)	4-0034-64
$C_6H_5N_3O_4$			
Aniline, 2,4-dinitro-	MeOH	336(4.17)	23-1681-64
	MeOH	336(4.16),390s(3.81)	35-4018-64
	CCl_4	316(4.15)	23-2674-64
anion	pyridine	388(4.32),535(4.18)	23-1681-64
Aniline, 2,6-dinitro-	MeOH	411(3.96)	35-4018-64
Pyrazine-5-carboxylic acid, 2-amino-3-formyl-6-hydroxy-	pH 1.2	230(3.85),291(4.12), 363(4.38)	39-3356-64
	pH 5.7	223(4.19),290(4.07), 358(4.26)	39-3357-64
	pH 10.1	220(4.13),256(3.81), 295(4.00),350(4.27)	39-3357-64
$C_6H_5N_3S$			
2,1,3-Benzothiadiazole, 4-amino-	pH 2	224(4.15),240s(4.01), 305(4.05),311(4.02), 398(3.09)	17-0033-64
	pH 1	222(4.11),245s(3.17), 304(4.07),309(4.06), 325s(3.50),395(2.27)	17-0033-64
	N HCl	221(4.16),304(4.13), 309(4.13),330s(3.30)	17-0033-64
	0.02N NaOH	238(4.28),300(3.89), 306(3.00),313(3.98), 398(3.35)	17-0033-64
	EtOH	248(4.29),300(3.77), 305(3.77),313(3.83), 424(3.34)	17-0033-64
2,1,3-Benzothiadiazole, 5-amino-	pH 2	221(4.21),305(4.18), 311(4.17),380(2.75)	17-0033-64
	N HCl	222(4.18),306(4.17), 311(4.16),330s(3.30)	17-0033-64
	0.02N NaOH	233(4.24),306(3.83), 481(3.74)	17-0033-64
	EtOH	236(4.29),304(3.81), 400(3.76)	17-0033-64
Imidazo[4,5-c]pyridine-4-thione	pH 1	336(4.17)	4-0196-65
	pH 11	229(4.12),246(4.07), 320(4.22)	4-0196-65
1H-Imidazo[4,5-c]pyridine-4(5H)-thione	pH 1	222(4.10),285s(--), 338(4.15)	87-0708-65
	pH 7	224(4.10),283(3.54), 325(4.18)	87-0708-65
	pH 13	225(4.18),246(4.06), 317(4.16),325s(--)	87-0708-65
3H-Imidazo[4,5-b]pyridine-7(4H)-thione	pH 1	244(3.90),285(3.99), 339(4.25)	87-0708-65
	pH 7	285(4.04),336(4.09)	87-0708-65
	pH 13	228(4.16),302(4.23)	87-0708-65
4-Pyrimidinecarbonitrile, 2-(methylthio)-	pH 1, 11	258(3.73)	4-0130-64

Compound	Solvent	λ max(log ε)	Ref.
$C_6H_5N_3Se$			
2,1,3-Benzoselenadiazole, 4-amino-	pH 2	233(3.85),331(4.17), 435(2.81)	17-0033-64
	pH 1	233(3.68),331(4.19), 360s(3.37),425(2.16)	17-0033-64
	N HCl	231(3.65),331(4.21), 360s(3.40),425(2.16)	17-0033-64
	0.02N NaOH	234(4.10),324(4.08), 331(4.14),338(4.11), 435(3.26)	17-0033-64
2,1,3-Benzoselenadiazole, 5-amino-	pH 2	234(4.02),333(4.08), 415(3.50)	17-0033-64
	pH 1	235(3.84),333(4.09), 447(3.42)	17-0033-64
	N HCl	235(3.88),334(4.22), 457(3.43)	17-0033-64
	0.02N NaOH	233(4.19),328(3.97), 411(3.77)	17-0033-64
	EtOH	237(4.23),325(3.92), 430(3.77)	17-0033-64
$C_6H_5N_4NaO_4S$			
Pteridine-6-sulfonic acid, 5,6-dihydro-7-hydroxy-, sodium salt	pH 1.0	226(4.45),284(3.78), 348(3.78)	39-6930-65
	pH 5.5	216(4.47),270(3.68), 325(3.82)	39-6930-65
$C_6H_5N_5$			
Pteridine, 7-amino-	pH 0.0	219(4.28),230s(4.15), 300s(3.97),328(4.20)	39-1175-65
	pH 5.1	228(4.26),262(3.80), 334(4.03)	39-1175-65
$C_6H_5N_5O$			
4-Pteridinol, 7-amino-	pH -1.0	220s(4.23),227(4.25), 239s(4.16),283s(3.85), 290(3.86),340(4.00)	39-1175-65
	pH 5.0	230(4.30),235s(4.30), 280(3.74),288(3.72), 334(3.95)	39-1175-65
	pH 11.0	228s(4.33),235(4.35), 250s(4.08),337(4.01)	39-1175-65
7-Pteridinol, 6-amino-	pH -2.4	212(4.41),230s(4.02), 281s(4.00),301s(4.18), 310(4.20),325s(3.86)	39-0027-65
$C_6H_5N_5O_2$			
Isoxanthopterin	pH 1	286(3.95),340(4.08)	37-0560-64
	pH 13	255(4.01),340(4.12)	37-0560-64
$C_6H_5N_5O_3$			
Guanine-8-carboxylic acid	pH 7.5	268(4.1),295(4.1)	37-3407-64
$C_6H_5N_5S$			
4(3H)-Pteridinethione, 2-amino-	pH 1	207s(4.39),249(3.87), 283(3.78),372(3.99)	44-3370-64

Compound	Solvent	$\lambda_{max}(\log \epsilon)$	Ref.
$C_6H_5N_9O$			
Tetrazolo[1,5-c]pyrimidine, 8-acet-	pH 1	232(4.33),288(4.18)	44-0829-65
amido-7-azido-	pH 7	231(4.32),287(4.18)	44-0829-65
C_6H_5NaO			
Sodium phenoxide	DMF	323(3.62)	35-1857-65
	$(MeOCH_2)_2$	302.5(3.57)	35-1857-65
C_6H_6			
Benzene	MeOH	264.1(1.66)(Ham band)	59-0993-64
	EtOH	264.1(1.55)(Ham band)	59-0993-64
	CH_2Cl_2	264.5(1.82)(Ham band)	59-0993-64
	$CHCl_3$	264.7(1.88)(Ham band)	59-0993-64
	CCl_4	265.6(2.02)(Ham band)	59-0993-64
	n.s.g.	256(2.35)	49-0285-65
silver complex	H_2O	233(3.2),255f(2.5)	39-6185-64
Fulvene	C_6H_{12}	241(4.15),244(4.15), 362(2.40)	33-0955-65
	C_6H_{12}	223s(3.80),228s(3.94), 231s(3.95),234s(4.00), 236s(4.04),239s(4.08), 241(4.10),244(4.10), 245s(4.08),248s(4.04), 252s(3.94),254s(3.81), 258s(3.60),362(2.33)	33-1211-64
	EtOH	242(4.14),360(2.41)	33-0955-65
	EtOH	242(4.06),360(2.35)	33-1211-64
2,4-Hexadiyne	n.s.g.	203(2.54),213(2.45), 224(2.57),236(2.52), 250(2.25)	22-1525-65
$C_6H_6BrNO_2$			
Pyrrole-2-carboxylic acid, 4-bromo-,	EtOH	233(4.04),269(4.24)	39-0459-65
methyl ester	EtOH-base	280(4.19)	39-0459-65
Pyrrole-2-carboxylic acid, 5-bromo-,	EtOH	228(3.65),271(4.30)	39-0459-65
methyl ester	EtOH-base	291(4.33)	39-0459-65
$C_6H_6Br_2$			
1,3-Cyclohexadiene, 2,3-dibromo-	EtOH	260(3.3),271(3.3), 278(3.3)	65-0453-65
$C_6H_6Br_4$			
Bicyclopropyl, 2,2,2',2'-tetrabromo-	hexane	204(3.73)	44-2951-64
$C_6H_6ClFN_2O$			
Pyrimidine, 2-chloro-4-ethoxy-5- fluoro-	H_2O	261.5(3.72)	87-0253-65
C_6H_6ClNO			
Phenol, 2-chloro-4-amino-	HCl	233(3.83)	44-2678-65
	NaOH	242(3.94)	44-2678-65
cation	HCl	216(3.83)	44-2678-65
2-Picoline, 4-chloro-, 1-oxide	EtOH	270(4.19)	39-2096-65
3-Picoline, 4-chloro-, 1-oxide	EtOH	274(4.26)	39-2096-65
2(1H)-Pyridone, 5-chloro-1-methyl-	pH 7.0	235(3.97),312(3.67)	39-5542-65

Compound	Solvent	$\lambda_{max}(\log \epsilon)$	Ref.
$C_6H_6ClNO_2$			
Pyrrole-2-carboxylic acid, 4-chloro-,	EtOH	233(3.80),269(4.11)	39-0459-65
methyl ester	EtOH-base	232(3.60),288(4.10)	39-0459-65
Pyrrole-2-carboxylic acid, 5-chloro-,	EtOH	236s(3.33),269(4.21)	39-0459-65
methyl ester	EtOH-base	291(4.30)	39-0459-65
$C_6H_6ClNO_3S$			
4-Isothiazolecarboxylic acid, 3-chloro-5-methoxy-, methyl ester	EtOH	243(3.84)	44-0660-64
$C_6H_6ClN_3O_3$			
2,6-Pyrimidinedione, 4-amino-5-chloroacetyl-	pH 1	260(4.16)	4-0034-64
	pH 11	228(4.10),279(4.30)	4-0034-64
$C_6H_6ClN_3S$			
4-Isothiazolecarbonitrile, 3-chloro-5-(dimethylamino)-	EtOH	221(4.31),277(4.06)	44-0660-64
$C_6H_6Cl_2N_2$			
Pyridine, 3,5-dichloro-1,2-dihydro-2-imino-1-methyl-	pH 7.0	211(4.33),216s(4.28), 242(3.91),246s(3.88), 324(3.81)	39-5542-65
	pH 12.4	214(4.24),249s(3.92), 257(4.03),263s(3.98), 350(3.63),365s(3.58)	39-5542-65
Pyridine, 3,5-dichloro-2-(methylamino)-	pH 0.8	213(4.30),248(4.07), 253s(4.01),311(3.81), 345s(3.63)	39-5542-65
	pH 7.0	250(4.15),322(3.64)	39-5542-65
Pyrimidine, 4,6-dichloro-2,5-dimethyl-	MeOH	263(3.78)	87-0808-64
Pyrimidine, 4,6-dichloro-2-ethyl-	MeOH	257(3.68)	87-0808-64
$C_6H_6Cl_3NS_2$			
2H-1,3-Thiazine-2-thione, 3,6-dihydro-4-methyl-6-(trichloromethyl)-	H_2O	291s(3.76),320(4.03)	39-4008-64
	EtOH	290s(--),327(4.04), 360s(--)	39-4008-64
$C_6H_6FNO_2S$			
Benzenesulfonyl fluoride, p-amino-	$C_2H_4Cl_2$	270(4.31)	30-0815-65
$C_6H_6F_2N_2$			
p-Phenylenediamine, 2,5-difluoro-	pH 1	231(3.95),284(3.43)	87-0737-65
	pH 7	231(4.06),299(3.51)	87-0737-65
	pH 13	231(4.11),297(3.49)	87-0737-65
	EtOH-HCl	238(4.09),295(3.53)	87-0737-65
C_6H_6IN			
Aniline, 2-iodo-	gas	210f(--)	60-0597-65
C_6H_6INO			
2-Picoline, 5-iodo-, 1-oxide	EtOH	269(4.07)	39-2096-65
$C_6H_6N_2$			
3H-Pyrazolo[1,2-a]pyrazol-4-ium hydroxide, inner salt	EtOH	257(3.7),465(3.5)	35-5256-65

Compound	Solvent	λ_{max}(log ε)	Ref.
$C_6H_6N_2O_2$			
Aniline, 2-nitro-	MeOH	404(3.72)	35-4018-64
	MeOH	275(3.43),398(3.44)	44-2088-64
	EtOH	231(4.22),276(3.69), 403(3.72)	65-0781-65
	cyclohexan-ol	270(3.44),379(3.54)	44-2088-64
	dioxan	276(3.27),400(3.28)	44-2088-64
	pyridine	410(3.65)	23-1681-64
C^{14} labelled	EtOH	230(4.22),277(3.68), 399(3.71)	87-0399-64
anion	pyridine	515(3.92)	23-1681-64
Aniline, 3-nitro-	C_6H_{12}	229(4.20),348(3.38)	44-2088-64
	MeOH	235(4.08),376(3.06)	44-2088-64
	iso-PrOH	235(4.07),378(3.08)	44-2088-64
	dioxan	235(4.06),370(3.02)	44-2088-64
Aniline, 4-nitro-	C_6H_6	226(3.90),321(4.17)	23-2674-64
	C_6H_{12}	324(4.34)	44-2088-64
	MeOH	370(4.12)	44-2088-64
	MeOH	371(4.20)	35-4018-64
	iso-PrOH	379(4.29)	44-2088-64
	dioxan	354(4.12)	44-2088-64
	Et_2O	348(4.21)	23-2674-64
	$C_2H_4Cl_2$	227(3.82),350(4.17)	30-1017-64
	pyridine	378(4.23)	23-1681-64
anion	sulfolane	467(4.51)	23-1681-64
4-Pyrimidinecarboxylic acid, methyl ester	EtOH	257(3.79),300s(2.58)	44-2398-65
$C_6H_6N_2O_3$			
4-Pyrimidinecarboxylic acid, 6-hy-droxy-, methyl ester	pH 1	285(3.61)	4-0130-64
	pH 11	294(3.54)	4-0130-64
$C_6H_6N_2O_4S_3$			
4-Isothiazolecarbonitrile, 3,5-bis-(methylsulfonyl)-	EtOH	270(3.79)	44-0665-64
$C_6H_6N_2S$			
4-Isothiazolecarbonitrile, 3,5-dimethyl-	EtOH	257(3.59)	39-0446-64
3-Thiophenecarbonitrile, 2-amino-4-methyl-	MeOH	219(4.62),287(3.93)	24-3571-65
$C_6H_6N_2S_2$			
Maleonitrile, bis(methylthio)-	C_6H_{12}	268(3.58),334(4.17)	24-1298-64
$C_6H_6N_2S_3$			
4-Isothiazolecarbonitrile, 3,5-bis-(methylthio)-	EtOH	215(4.05),230(4.07), 284(4.12)	44-0660-64
$C_6H_6N_4$			
Benzotriazole, 1-amino-	n.s.g.	211(3.84),263(3.76), 280s(3.65)	77-0192-65
Benzotriazole, 2-amino-	n.s.g.	216(4.12),285(4.08)	77-0192-65
Imidazo[4,5-b]pyridine, 7-amino-	pH 1	263(4.08),280(4.26), 287s(--)	87-0708-65
	pH 7	263(4.08),277(4.12), 286s(--)	87-0708-65
	pH 13	275(4.14)	87-0708-65

Compound	Solvent	λ_{max}(log ϵ)	Ref.
Imidazo[4,5-c]pyridine, 4-amino-	pH 1	260(3.93),275(3.95)	87-0708-65
	pH 7	258(3.95),267(3.95)	87-0708-65
	pH 13	274(3.88)	87-0708-65
Imidazo[4,5-c]pyridine, 6-amino-	pH 1	267(3.78),295(3.56)	4-0196-65
	pH 11	262(3.69)	4-0196-65
Pyrrolo[2,3-d]pyrimidine, 4-amino-	pH 1	224(4.24),274(4.01)	35-1995-65
	pH 13	272(4.02)	35-1995-65
5H-Pyrrolo[3,2-d]pyrimidine, 4-amino-	pH 1	234(4.18),273(4.20)	44-1528-65
	pH 7	228(4.34),274(4.04)	44-1528-65
	pH 13	229(4.37),274(3.96)	44-1528-65
s-Triazolo[4,3-a]pyrazine, 3-methyl-	C_6H_{12}	214(4.11),231s(3.08), 257(3.14),270s(3.05), 298(3.10),308(3.09), 322(2.98),338(2.64)	44-2542-64
	EtOH	212(4.38),255s(3.26), 262(3.27),270(3.26), 300(3.48)	44-2542-64
s-Triazolo[1,5-a]pyrimidine, 6-methyl-	EtOH	282(3.59)	94-0204-64
$C_6H_6N_4O$			
Hypoxanthine, 9-methyl-	pH 1	251(3.96)	44-3597-65
	pH 13	254(4.01)	44-3597-65
	MeOH	248(3.96)	44-3597-65
Isoxazolo[5,4-d]pyrimidine, 4-amino-3-methyl-	EtOH	249(3.89),273(3.76)	44-2117-64
7-Pteridinol, 5,6-dihydro-	pH 1.0	223(4.47),284(3.74), 352(3.71)	39-6930-65
	pH 6.0	211(4.47),271(3.58), 319(3.70)	39-6930-65
Purine, 6-methoxy-	Me_3PO_4	190(4.25),200(4.22), 251(3.86)	35-0011-65
6H-Pyrrolo[3,2-d]pyrimidin-6-one, 4-amino-5,7-dihydro-	pH 1	240(3.91),286(4.01)	44-1528-65
	pH 7	252(4.19),307(3.95)	44-1528-65
	pH 13	237(4.34),297(4.20)	44-1528-65
s-Triazole, 3-amino-5-(2-furyl)-	pH 2	265(4.22)	1-1191-65
	pH 12	271(4.14)	1-1191-65
$C_6H_6N_4OS$			
Purine, 6-(methylthio)-, 3-oxide	pH 1.5	250s(3.83),318(4.31)	87-0190-65
	pH 6.9	245(4.18),312(4.24)	87-0190-65
	pH 12.6	245(4.21),312(4.23)	87-0190-65
$C_6H_6N_4O_2$			
1,2,4-Oxadiazole, 3-methyl-5-(3-methyl-1,2,4-oxadiazol-5-yl)-	Et_2O	242(4.10)	33-0942-64
s-Triazolo[1,5-a]pyrimidine-5,7-diol, 6-methyl-	EtOH	271.5(4.11)	94-0204-64
Xanthine, 7-methyl-	H_2O	268(4.00)	4-0275-64
	pH 12	288(3.92)	4-0275-64
Xanthine, 9-methyl-	H_2O	237(3.92),264(4.01)	4-0275-64
	pH 12	244(3.98),278(3.97)	4-0275-64
$C_6H_6N_4O_3$			
Xanthine, 1-hydroxy-7-methyl-	H_2O	268(4.01)	4-0275-64
	pH 12	283(3.97)	4-0275-64

Compound	Solvent	$\lambda_{max}(\log \epsilon)$	Ref.
$C_6H_6N_4O_3S$			
1,2,4-Triazolo[2,3-a]pyrimidine-4-sulfonic acid, 6-methyl-, K salt	H_2O	274(3.83)	65-0504-64
$C_6H_6N_4O_5$			
Oxamic acid, (6-amino-1,2,3,4-tetrahydro-2,4-dioxo-5-pyrimidinyl)-	pH 7.4	263(4.20)	37-4272-64
$C_6H_6N_4S$			
2,1,3-Benzothiadiazole, 4,5-diamino-	EtOH	245s(4.18),263(4.27), 325(3.85),453(3.35)	7-0476-64
Purine, 6-mercaptomethyl-	N HCl	266(3.86)	87-0667-65
	pH 7.65	265(3.91)	87-0667-65
	N NaOH	273(3.96)	87-0667-65
1,2,4-Triazolo[2,3-a]pyrimidine-4-thione, 6-methyl-	pH 12	230(4.01),324(4.32)	65-0504-64
	MeOH	232(3.87),334(4.32)	65-0504-64
$C_6H_6N_6$			
Pteridine, 4,7-diamino-	pH 2.0	233(4.12),253(4.11), 258(4.11),284(3.54), 335s(4.00),342(4.03), 355s(3.89)	39-1175-65
	pH 7.0	241(4.38),254s(4.15), 263s(4.00),339(4.05), 350s(3.97)	39-1175-65
Pteridine, 6,7-diamino-	pH -2.4	219(4.41),244(4.00), 324s(4.17),337(4.35), 353(4.28)	39-0027-65
	pH 1.5	228(4.38),258(4.10), 342(4.25),352(4.25)	39-0027-65
	pH 7.0	221(4.45),242(4.08), 268s(3.59),278s(3.56), 327s(4.18),337(4.27), 352(4.13)	39-0027-65
$C_6H_6N_6O$			
Formamide, N-[4(5)-s-triazol-3-yl-imidazol-5(4)-yl]-	pH 1	248(4.18)	44-3601-65
	pH 7	251(4.08)	44-3601-65
	pH 13	251(4.06)	44-3601-65
Pteridine, 2-amino-4-(hydroxy-amino)-	pH 1	217(4.36),340(3.97)	44-3370-64
4-Pteridinol, 2,6-diamino-	pH 1	271(4.33),377(3.84)	39-4769-64
	pH 13	260(4.20),392(3.71)	39-4769-64
Pyrimido[4,5-e]-as-triazin-8(7H)-one, 3-amino-6-methyl-	H_2O	207(4.23),242(4.33), 343(3.83)	35-1976-65
$C_6H_6N_8S_2$			
Disulfide, bis(6-methyl-1,2,4-triazolo[2,3-a]pyrimidin-4-yl)	MeOH	332(4.42)	65-0504-64
C_6H_6O			
4,5-Hexadien-2-yn-1-ol	EtOH	221(3.56)	39-4659-65
Oxepin	isooctane	271(3.16),305s(2.95)	88-0609-65
	MeOH	269(3.3)	88-0609-65
	15% MeOH	269(3.5)	88-0609-65
Phenol	DMF	275(3.32)	35-1857-65
	$(MeOCH_2)_2$	274(3.32)	35-1857-65

Compound	Solvent	$\lambda_{max}(\log \epsilon)$	Ref.
C_6H_6OSe			
Selenophene, 2-acetyl-	n.s.g.	270(4.02),300(3.70)	65-0841-64
Selenophene, 3-acetyl-	n.s.g.	267(4.03)	65-0841-64
$C_6H_6O_2$			
2-Cyclohexene-1,4-dione	isooctane	219(4.19),365(1.76)	35-4971-65
	EtOH	233(4.18),352(1.81)	35-5971-65
4-Hexen-2-yn-1-al, 6-hydroxy-	Et_2O	263(4.06),274(4.01)	24-0369-65
5-Hexen-2-ynoic acid	MeC_6H_{11}	215(3.76)	5-0026-64I
Hydroquinone	95% THF	298(3.54)	44-2602-65
2-Pyrone, 5-methyl-	H_2O	297(3.73)	54-0039-64
2-Pyrone, 6-methyl-	MeOH	298(3.84)	54-0039-64
$C_6H_6O_2S$			
Thiophene, 2-acetyl-3-hydroxy-	Et_2O	270(3.74),301(3.52)	24-2109-64
$C_6H_6O_3$			
Acrolein, 3,3'-oxydi-	Et_2O	229(4.12),268(4.53), 279(4.45)	24-1959-64
$C_6H_6O_4$			
2-Furoic acid, 3-methoxy-	MeOH	258(4.16)	44-0776-64
Muconic acid, cis-trans	EtOH	259.5(4.63)	33-0157-65
$C_6H_6O_5$			
Muconic acid, α-hydroxy-	pH 1	304(4.25)	37-0740-65
	pH 13	350(4.37)	37-0740-65
$C_6H_6O_8S_2$			
Tiron, 1:1 ferric complex	aq. KNO_3	673(3.28)	23-1917-64
1:2 ferric complex	aq. KNO_3	561(3.66)	23-1917-64
1:3 ferric complex	aq. KNO_3	480(3.79)	23-1917-64
C_6H_6S			
Benzenethiol (varies with concn.)	C_6H_{12}	236(3.86),272(2.9), 280(2.88),288(2.70)	2-0143-65
$C_6H_6S_2$			
1H,3H-Thieno[3,4-c]thiophene	EtOH	232(3.79)	44-1919-64
$C_6H_6S_3$			
Cyclopenta[d]-1,2-dithiole-3(4H)-thione, 5,6-dihydro-	EtOH	221(3.96),241(3.74), 283(3.74),330(3.73), 410(4.01)	24-0654-64
$C_6H_6S_4$			
3H,4H-1,2-Dithiolo[4,3-c]thiopyran-3-thione, 6,7-dihydro-	EtOH	228(4.09),244s(3.73), 274(3.71),313(3.67), 408(4.01)	24-0654-64
C_6H_7Br			
1,3-Cyclohexadiene, 1-bromo-	C_6H_{12}	269(3.9)	44-3205-65
1,3-Cyclohexadiene, 2-bromo-	C_6H_{12}	268(3.88)	44-3205-65
$C_6H_7BrN_2$			
Pyrimidine, 2-bromo-4,6-dimethyl-	n.s.g.	217(3.83),256(3.62)	44-0943-64

Compound	Solvent	$\lambda_{max}(\log \epsilon)$	Ref.
$C_6H_7BrO_2$			
Cyclobutenone, 2-bromo-3-hydroxy-2,4-dimethyl-	Et_2O	228(4.10)	35-0673-64
$C_6H_7BrO_3$			
α-Tetronic acid, β-methyl-, methyl ether, N-bromosuccinimide reaction product	C_6H_{12}	239(3.79)	44-3560-64
$C_6H_7Br_3N_2$			
2,6-Dibromo-2,3-dihydro-1H-pyrazolo[1,2-a]pyrazol-4-ium bromide	n.s.g.	244(3.5)	35-4393-65
C_6H_7Cl			
1,3,5-Hexatriene, 1-chloro-	n.s.g.	256(4.51),265(4.64), 276(4.54)	44-2218-65
1,3,5-Hexatriene, 3-chloro-	n.s.g.	244s(4.23),253(4.47), 263(4.64),274(4.55)	44-2218-65
$C_6H_7ClF_2N_2$			
1-Cyclobuten-1-ylamine, 2-chloro-4,4-difluoro-N-methyl-3-(methylimino)-	EtOH	265(4.43)	44-3698-65
	0.1N HCl	277(4.49)	44-3698-65
	0.1N NaOH	267(4.46)	44-3698-65
$C_6H_7ClN_2$			
Pyrazine, 2-chloro-3,6-dimethyl-	MeOH	214(4.90),281(3.89), 295s(3.74)	44-2623-64
Pyridine, 5-chloro-1,2-dihydro-2-imino-1-methyl-	pH 7.0	240(4.06),315(3.70)	39-5542-65
	pH 14.5	258(4.16),264s(4.10), 344(3.51)	39-5542-65
Pyridine, 5-chloro-2-(methylamino)-	pH 3.0	245(4.16),320(3.71)	39-5542-65
	pH 7.5	248(4.17),315(3.50)	39-5542-65
$C_6H_7ClN_2O$			
Pyrazine, 3-chloro-2,5-dimethyl-, 1-oxide	H_2O	221(4.21),267(4.04), 308(3.74),312(3.72)	44-2623-64
$C_6H_7ClN_2O_2S$			
Benzenesulfonamide, 2-amino-5-chloro-	HCOOH	244(3.9),308(3.5)	20-0344-65
$C_6H_7ClN_4O$			
Acetamide, N-(4-amino-6-chloro-5-pyrimidinyl)-	pH 1	238(3.93),277(3.67)	44-0829-65
	pH 7	237(4.00),278(3.62)	44-0829-65
	pH 13	256(3.83),284(3.82)	44-0829-65
C_6H_7ClO			
1-Cyclopentene-1-carboxaldehyde, 2-chloro-	EtOH	257.5(4.12)	44-1126-65
2H-Pyran, 3-chloro-5,6-dihydro-2-methylene-	EtOH	231(3.86),251(3.73)	88-0821-65
C_6H_7DO			
4,5-Hexadien-3-one-6-d	n.s.g.	201(4.18),211(3.80), 320(1.41)	28-1530-64B
$C_6H_7FN_2O_2$			
Pyrimidine, 4-fluoro-2,6-dimethoxy-	MeOH	245(3.83)	65-1433-65

Compound	Solvent	$\lambda_{max}(\log \epsilon)$	Ref.
$C_6H_7F_3N_2$			
1-Cyclobuten-1-ylamine, 2,4,4-tri-fluoro-N-methyl-3-(methylimino)-	EtOH	263(4.43)	44-3698-65
	pH 1	275(4.51)	44-3698-65
	pH 13	265(4.44)	44-3698-65
$C_6H_7F_7N_2S$			
Butyric acid, heptafluorothio-, 2,2-dimethylhydrazide	isooctane	276(3.88)	44-3739-65
C_6H_7N			
Aniline	EtOH	235(4.14),285(3.35)	65-0781-65
	$HCONH_2$	290(3.30)	20-0741-64
hydrochloride	n.s.g.	250(2.18)	20-0741-64
iodine complex	$CHCl_3$	348(--),355s(--)	60-0064-64
2-Picoline, NiI_2 complex	CCl_4	320(--),405(--), 635(1.65)	33-1265-64
3-Picoline, NiI_2 complex	$CHCl_3$	589s(2.52),630(2.42), 930(1.98),1010(2.02)	33-1265-64
4-Picoline, NiI_2 complex	$CHCl_3$	590s(2.51),631(2.42), 921(1.95),1000(2.00)	33-1265-64
Pyrrole, 2-vinyl-	n.s.g.	274(4.16)	12-0875-65
C_6H_7NO			
2-Azabicyclo[3.2.0]hept-6-en-3-one	n.s.g.	none above 220 mμ	5-0001-65B
2H-Azepin-2-one, 1,3-dihydro-	MeOH	225(3.69)	5-0001-65B
Cyclopentanecarbonitrile, 2-oxo-	EtOH	235(3.38)	23-2512-65
	EtOH-base	262(4.23)	23-2512-65
Phenol, 3-amino-	MeOH	234(3.79),285(3.34)	33-1988-65
2-Picoline, 1-oxide	EtOH	262(4.08)	39-2319-64
4-Picoline, 1-oxide	EtOH	266(4.21)	39-2319-64
Pyrrole, 2-acetyl-	MeOH	237(3.83),286(4.22)	39-2579-64
C_6H_7NOS			
Butyric acid, 2-cyano-4-mercapto-2-methyl-, γ-(thiolactone)	MeOH	234(3.66)	24-1970-64
Valeric acid, 2-cyano-5-mercapto-, δ-(thiolactone)	MeOH	238(3.79)	24-1970-64
$C_6H_7NO_2$			
Acetoacetonitrile, 2-acetyl-	heptane	203(4.00),281(3.99)	59-0931-65
	pH 1	204(3.97),276(4.01)	59-0931-65
	pH 7	223(4.02),281(4.24)	59-0931-65
	95% H_2SO_4	203(4.10),288(4.28)	59-0931-65
Crotonic acid, 2-cyano-3-methyl-	EtOH	227.5(4.04)	28-2859-64B
Ketone, 2-furyl methyl, oxime, anti	EtOH	265(4.16)	22-2724-65
syn	EtOH	262(4.22)	22-2724-65
2-Pentenoic acid, 2-cyano-	EtOH	219(3.97)	28-2859-64B
Pyridine, 4-methoxy-, 1-oxide	EtOH	269(4.22)	39-2319-64
4(1H)-Pyridone, 3-hydroxy-2-methyl-	MeOH-acid	242(3.49),274(3.93)	44-0776-64
	MeOH	278(4.13)	44-0776-64
	MeOH-KOH	247(3.75),275(3.73), 306(3.61)	44-0776-64
Pyrrole-2-carboxylic acid, methyl ester	EtOH	234(3.72),266(4.21)	23-1279-64
	EtOH-base	238s(3.63),263(4.14)	39-0459-65
Pyrrole-2-carboxylic acid, 4-methyl-	EtOH	236(3.75),270(4.08)	44-2727-64
	EtOH-NaOH	260(4.05)	44-2727-64
$C_6H_7NO_2S$			
4-Isothiazolecarboxylic acid, 3,5-dimethyl-	EtOH	257(3.98)	39-0446-64

Compound	Solvent	λ_{max}(log ϵ)	Ref.
$C_6H_7NO_2S_2$			
Maleimide, bis(methylthio)-	EtOH	250(3.84),399(3.60)	24-1298-64
$C_6H_7NO_2S_3$			
4-Isothiazolecarboxylic acid, 3,5-bis(methylthio)-	EtOH	214(4.34),237(4.45), 282(4.47)	44-0665-64
$C_6H_7NO_3$			
4-Oxazolecarboxylic acid, ethyl ester	n.s.g.	214(3.88)	24-1414-64
$C_6H_7NO_4S$			
2(1H)-Pyridone, 1-[(methylsulfonyl)oxy]-	EtOH	226(3.64),298(3.72)	35-5186-65
$C_6H_7N_3$			
1,3a,4-Triazaindene, 2,3-dihydro-	pH 7.0	209(4.18),215s(4.05), 242(3.85),326(3.24)	39-2778-65
	pH 12.4	218(4.09),258(3.90), 264s(3.87),273s(3.59), 381(3.06)	39-2778-65
1,3a,5-Triazaindene, 2,3-dihydro-	pH 5.0	238(4.04),301(3.45)	39-2778-65
	pH 11.0	251(4.11),258s(4.04), 329(3.16)	39-2778-65
1,3a,6-Triazaindene, 2,3-dihydro-	pH 7.0	240(4.09),354(3.62)	39-2778-65
	pH 11.5	258(4.09),264s(4.05), 394(3.94)	39-2778-65
1,3a,7-Triazaindene, 2,3-dihydro-	pH 7.0	233(4.13),333(3.42)	39-2778-65
	pH 12.3	249(4.14),390(3.24)	39-2778-65
1,4,6-Triazaindene, dihydro-, hydrochloride	MeOH	210(4.1),310(3.1)	94-1030-64
$C_6H_7N_3O$			
Furo[2,3-d]pyrimidine, 2-amino-5,6-dihydro-	EtOH	280(3.57)	65-2599-64
hydrochloride	EtOH	282(3.59)	65-2599-64
Isonicotinic acid, hydrazide	pH 1	268(3.50)	56-1807-64
	H_2O	260(3.71)	56-1807-64
	EtOH	263(3.72)	56-1807-64
	dioxan	261(3.73)	56-1807-64
5-Pyrimidinecarbonitrile, 1,2,3,4-tetrahydro-1-methyl-2-oxo-	EtOH	213(3.92),286(3.97)	44-1740-64
5-Pyrimidinecarbonitrile, 1,2,3,4-tetrahydro-3-methyl-2-oxo-	EtOH	217(3.96),279(3.90)	44-1740-64
$C_6H_7N_3O_2$			
Pyrazine-2-carboxaldehyde, 5-hydroxy-3-(methylamino)-	pH 4.2	232(3.72),301(4.03), 352(4.25)	39-1175-65
	pH 8.8	218(4.07),258(4.21), 267(4.03),289(4.03)	39-1175-65
Pyridine, 1,2-dihydro-2-imino-1-methyl-3-nitro-	pH 4.0	230s(3.66),263(3.53), 357(3.81)	39-5542-65
	pH 5.0	215(3.96),304(4.12)	39-5542-65
Pyridine, 2-(methylamino)-3-nitro-	pH 0.2	212(4.29),269(3.78), 376(3.81)	39-5542-65
	pH 5.0	223(4.35),266(3.65), 423(4.35)	39-5542-65
Pyridine, 2-(methylamino)-5-nitro-	pH 0.5	214(4.04),313(4.16)	39-5542-65
	pH 7.0	223(3.97),305s(3.60), 272(4.21)	39-5542-65

Compound	Solvent	$\lambda_{max}(\log \epsilon)$	Ref.
Pyridine, 4-(methylamino)-3-nitro-	pH 2.0	231(4.21),267(4.08), 350(3.59)	39-2150-64
	pH 8.0	237(4.28),272s(3.55), 386(3.71)	39-2150-64
4-Pyrimidinecarboxaldehyde, 2-hy-	pH 1	260(3.93),342(4.01)	4-0130-64
droxy-6-methyl-, oxime	pH 11	244s(3.96),275(4.14)	4-0130-64
$C_6H_7N_3O_2S$			
4-Thiazolecarboxamide, 5-acet-amido-	EtOH	282.5(4.12)	94-1319-65
$C_6H_7N_3O_3$			
2-Imidazoline-$\Delta^{5,\alpha}$-acetic acid, 2-	pH 1	236(3.86),304(4.14)	4-0162-65
amino-α-methyl-4-oxo-	pH 7	269(4.10),312(3.98)	4-0162-65
	pH 13	263(4.00),320(3.97)	4-0162-65
Pyridazine, 3,6-dimethyl-4-nitro-, 1-oxide	EtOH	225(3.9),300(3.8), 345(4.1)	94-0228-64
Pyridazine, 3,6-dimethyl-5-nitro-, 1-oxide	EtOH	270(4.0),375(3.4)	94-0228-64
4-Pyrimidinecarboxylic acid, 2-	pH 1	283(3.72)	4-0162-65
amino-6-hydroxy-5-methyl-	pH 7	291(3.75)	4-0162-65
	pH 13	282(3.72)	4-0162-65
$C_6H_7N_3O_4$			
6-Azauracilpropionic acid	MeOH	211(3.80),254(3.71)	39-0868-64
$C_6H_7N_5$			
Adenine, 9-methyl-	gas	252(3.89)	46-3615-65
Purine, 2-(methylamino)-	pH 1.8	223(4.55),244s(3.81), 327(3.57)	39-3770-65
	pH 7.0	219(4.42),240(3.88), 319(3.74)	39-3770-65
	pH 12.5	226(4.37),272(3.48), 316(3.72)	39-3770-65
Pyrazolo[3,4-d]pyrimidine, 4-amino-	pH 1	259(3.97)	4-0215-64
1-methyl-	pH 7	261(3.95),277(3.96)	4-0215-64
	pH 13	262(3.98),275(3.98)	4-0215-64
Pyrazolo[3,4-d]pyrimidine, 4-amino-	pH 1	268(3.99)	4-0215-64
2-methyl-	pH 7	270(3.94),287(4.04)	4-0215-64
	pH 13	287(4.04)	4-0215-64
1,2,4-Triazolo[2,3-a]pyrimidine, 4-amino-6-methyl-	MeOH	286(4.16)	65-0204-64
$C_6H_7N_5O$			
Adenine, 1-methoxy-	pH 1	258(4.06)	94-1017-65
	pH 13	269(4.15)	94-1017-65
Glycinonitrile, N-(4-amino-6-hy-	pH 7.0	214(4.36),267(3.89)	39-1175-65
droxy-5-pyrimidinyl)-	pH 12.0	266(3.81)	39-1175-65
6-Purinone, 2-(methylamino)-	pH 1	250(3.80),256(4.28)	35-3752-65
	pH 11	244(3.98),278(3.86)	35-3752-65
4(1H)-Pyrimidinone, 2-azido-1,6-	pH 1	247(3.6),284(4.19)	44-2395-65
dimethyl-	pH 7	247(3.58),284(4.18)	44-2395-65
	EtOH	246(3.65),281(4.15)	44-2395-65
Tetrazolo[1,5-a]pyrimidin-5(4H)-	pH 1	244(3.83),261s(3.67)	44-2395-65
one, 4,7-dimethyl-	pH 7	244(3.83),261s(3.67)	44-2395-65
	pH 13	245(3.77),261s(3.64)	44-2395-65
	EtOH	244(3.8),261s(3.61)	44-2395-65

Compound	Solvent	$\lambda_{max}(\log \epsilon)$	Ref.
$C_6H_7N_5O_2$			
s-Triazine-2,4-dicarboxaldehyde,	MeOH	251(4.50)	44-0678-64
6-methyl-, dioxime	EtOH	252.5(4.58)	44-0678-64
1H-v-Triazolo[4,5-d]pyrimidine-	pH 6.0	279(3.80)	24-1060-65
5,7(4H,6H)-dione, 1,4-dimethyl-	pH 11.0	250(3.85),282(3.86)	24-1060-65
1H-v-Triazolo[4,5-d]pyrimidine-	pH 5.0	230(3.59),274(3.77)	24-1060-65
5,7(4H,6H)-dione, 1,6-dimethyl-	pH 10.0	235s(3.69),303(3.74)	24-1060-65
1H-v-Triazolo[4,5-d]pyrimidine-	pH 2.0	230s(3.60),271(3.80)	24-1060-65
5,7(4H,6H)-dione, 4,6-dimethyl-	pH 7.0	230s(3.62),269(3.93)	24-1060-65
2H-v-Triazolo[4,5-d]pyrimidine-	pH 7.0	277(3.91)	24-1060-65
5,7(4H,6H)-dione, 2,4-dimethyl-	pH 12.0	251(3.93),282(4.00)	24-1060-65
2H-v-Triazolo[4,5-d]pyrimidine-	pH 5.0	234(3.55),274(3.93)	24-1060-65
5,7(4H,6H)-dione, 2,6-dimethyl-	pH 10.0	236(3.67),300(3.82)	24-1060-65
3H-v-Triazolo[4,5-d]pyrimidine-	pH 5.0	239(3.85),259(3.94)	24-1060-65
5,7(4H,6H)-dione, 3,4-dimethyl-	pH 11.0	264(4.02)	24-1060-65
3H-v-Triazolo[4,5-d]pyrimidine-	pH 3.0	230s(3.52),257(3.97)	24-1060-65
5,7(4H,6H)-dione, 3,6-dimethyl-	pH 8.0	247(3.75),279(3.89)	24-1060-65
3H-v-Triazolo[4,5-d]pyrimidine-	pH 2.0	232(3.76),255(3.97)	24-1060-65
5,7(4H,6H)-dione, 3-ethyl-	pH 8.0	247(3.94),278(3.90)	24-1060-65
Xanthopterin, 7,8-dihydro-	pH 1	278(4.14),315(4.01)	39-4769-64
	pH 13	223(4.24),278(4.08), 310s(3.95)	39-4769-64
$C_6H_7N_5O_3$			
Guanidine, [(5-nitrofurfurylidene)-amino]-, as HCl salt	PG	247(4.04),356(4.23)	95-0001-64
3H-v-Triazolo[4,5-d]pyrimidine-5,7-	pH 2.0	233(3.77),256(3.97)	24-1060-65
(4H,6H)-dione, 3-(2-hydroxyethyl)-	pH 8.0	247(3.94),278(3.90)	24-1060-65
$C_6H_7N_5S$			
[1,2,5]Thiadiazolo[3,4-d]pyrimidine, 7-(ethylamino)-	pH 1	220(4.12),263(3.64), 339(4.21)	44-2135-64
	pH 7	224(4.13),272(3.69), 355(4.04)	44-2135-64
	EtOH	224(4.14),274(3.71), 283(3.64),360(4.02)	44-2135-64
$C_6H_7N_7$			
Pteridine, 2-amino-4-hydrazino-	pH 1	236(4.08),285(3.70), 318s(3.89),334(3.99), 347s(3.93)	44-3370-64
$C_6H_7N_7O$			
Pterin, 8-hydrazino-	pH 2	262(4.2),355(3.8)	33-2195-64
	pH 7	282(4.2),397(3.7)	33-2195-64
	pH 12	387(3.8)	33-2195-64
Tetrazolo[1,5-c]pyrimidine, 8-acet-	pH 1	237(4.22),292(4.05),	44-0829-65
amido-7-amino-	pH 7	216(4.27),233(4.19), 272(3.84),282s(3.82)	44-0829-65
C_6H_8			
1,3-Cyclohexadiene, $PdCl_2$ olefin complex from	EtOH	211(4.21),231s(4.02), 321(3.17)	24-2037-64
Cyclopentadiene, 1-methyl-	MeOH	249(3.62)	78-2313-65
Cyclopentadiene, 2-methyl-	MeOH	242(3.57)	78-2313-65
Cyclopropane, 1-ethynyl-1-methyl-	n.s.g.	205(2.51)	22-1525-65
1-Penten-3-yne, 2-methyl-	n.s.g.	221(4.03),228(4.03)	22-1525-65

Compound	Solvent	λ_{max}(log ϵ)	Ref.
C_6H_8BrN			
1-Methylpyridinium bromide	CH_2Cl_2	260(3.8),320(3.2)	46-3791-65
C_6H_8BrNS			
Isothiazole, 4-bromo-5-ethyl-3-methyl-	EtOH	244(3.58),258(3.88)	39-0446-64
$C_6H_8BrN_3$			
Pyrimidine, 2-amino-5-bromo-4,6-dimethyl-	pH 1.0	235(4.19),315(3.73)	39-5542-65
	pH 6.0	235(4.17),301(3.60)	39-5542-65
Pyrimidine, 5-bromo-2-(ethylamino)-	pH -0.5	243(4.37),340(3.47)	39-5542-65
	pH 7.0	248(4.36),327(3.37)	39-5542-65
Pyrimidine, 5-bromo-1-ethyl-1,2-dihydro-2-imino-	pH 7.0	237(4.25),326(3.55)	39-5542-65
$C_6H_8BrN_3O$			
Pyrazole-1-carboxamide, 4-bromo-3,5-dimethyl-	EtOH	246.5(3.94)	28-0606-65A
$C_6H_8Br_2N_2$			
2-Bromo-2,3-dihydro-1H-pyrazolo-[1,2-a]pyrazol-4-ium bromide	n.s.g.	224(3.65)	35-4393-65
$C_6H_8Br_2O$			
Cyclobutanone, 2,2-dibromo-3,3-dimethyl-	C_6H_{12}	317(2.26)	22-1976-64
	dioxan	317(2.29)	22-1976-64
	CCl_4	317(2.23)	22-1976-64
	MeCN	315(2.24)	22-1976-64
Cyclobutanone, 4,4-dibromo-2,2-dimethyl-	C_6H_{12}	334(2.24)	22-1976-64
	MeOH	330.5(2.20)	22-1976-64
	dioxan	331(2.24)	22-1976-64
	CCl_4	330(2.25)	22-1976-64
C_6H_8ClNO			
1-Cyclopentene-1-carboxaldehyde, 2-chloro-, oxime	EtOH	252(4.19)	44-1126-65
$C_6H_8ClNO_5$			
1-Methoxypyridinium perchlorate	H_2O	258(3.61)	78-2205-65
$C_6H_8ClN_3$			
Pyridazine, 3-chloro-6-(dimethyl-amino)-	pH 1	245(3.78)	4-0001-64
	pH 11	257(3.95)	4-0001-64
$C_6H_8ClN_3O$			
4-Pyrimidinol, 2-amino-5-(2-chloro-ethyl)-	EtOH	262(3.88)	65-2599-64
$C_6H_8ClN_3O_2$			
3(2H)-Pyridazinone, 4-chloro-5-(β-hydroxyethylamino)-	EtOH	231(4.38),305(3.80)	22-2124-64
	NaOH	224(4.55),290(3.82)	22-2124-64
$C_6H_8Cl_2N_2$			
2,3-Diazabicyclo[2.2.2]oct-2-ene, 1,4-dichloro-	EtOH	333(1.36),350(1.88), 366(2.11)	5-0236-65G
$C_6H_8Cl_2N_2O$			
2,3-Diazabicyclo[2.2.2]oct-2-ene, 1,4-dichloro-, 2-oxide	EtOH	230(3.85),280(2.5)	5-0236-65G

Compound	Solvent	λ_{max}(log ε)	Ref.
$C_6H_8Cl_3NOS_2$			
2-Butanone, 4-(trichloromethyl)-4-(thiocarbamoylthio)-	H_2O	236(3.79),286(4.14)	39-4004-64
	EtOH	233(3.96),290(4.09)	39-4004-64
$C_6H_8F_4O_2$			
Propionic acid, 2,2,3,3-tetrafluoropropyl ester	isooctane	210.1(1.81)	35-2384-64
	MeCN	208.2(1.76)	35-2384-64
C_6H_8IN			
1-Methylpyridinium iodide	H_2O	258(3.68)	25-0782-65
	H_2O	220(4.0),236(4.2), 260(3.8)	46-3791-65
	EtOH	220(4.2),270(3.8), 325(2.0)	46-3791-65
	CH_2Cl_2	240(4.1),315(3.2), 380(3.1)	46-3791-65
$C_6H_8N_2$			
1-Cyclopentene-1-carbonitrile, 2-amino-	EtOH	265(4.12)	23-2512-65
Pyridine, 1,2-dihydro-2-imino-1-methyl-	pH 7.0	230(3.86),299(3.73)	39-5542-65
$C_6H_8N_2O$			
Acetoacetonitrile, 2-(1-aminoethylidene)-	H_2O	218(4.13),290(4.17)	59-0931-65
Cyclohexano[c]furazan	EtOH	217(3.52)	94-1445-65
3-Piperidinecarbonitrile, 2-oxo-	MeOH	203(3.69)	24-1970-64
Pyrazine, 2-ethoxy-	MeOH	212(3.93),279(3.61), 295s(--)	44-2623-64
2-Pyrazinol, 3,5-dimethyl-	MeOH	228(4.06),327(3.92)	44-2623-64
2-Pyrazinol, 3,6-dimethyl-	H_2O	223(3.97),321(3.95)	44-2623-64
	EtOH	226(3.78),320(3.75)	39-1507-64
3-Pyrrolidinecarbonitrile, 3-methyl-2-oxo-	MeOH	207(3.24)	24-1970-64
$C_6H_8N_2OS$			
5H-Imidazo[2,1-b][1,3]thiazin-5-one, 2,3,6,7-tetrahydro-, hydrochloride	H_2O	238(4.12)	44-1720-64
	MeOH	229(4.06)	44-1720-64
4-Isothiazolecarboxamide, 3,5-dimethyl-	EtOH	254(3.79)	39-0446-64
4-Pyrimidinol, 5-methyl-6-(methylthio)-	pH -2.8	237(3.88),284(3.98)	39-3204-64
	pH 5.0	237(4.16),290(3.97)	39-3204-64
	pH 11.0	223(4.29),277(3.90)	39-3204-64
4-Pyrimidinethiol, 6-methoxy-5-methyl-	pH -3.6	217(4.29),281(4.13)	39-3204-64
	pH 5.0	224(4.20),277s(3.89), 308(4.10)	39-3204-64
	pH 11.1	227(4.14),286(4.20)	39-3204-64
Thiazole, 5-acetamido-2-methyl-	EtOH	264(3.99)	94-1319-65
Thiocyanic acid, 2-(dimethylcarbamoyl)vinyl ester	EtOH	263(4.09)	44-2660-65
	Et_2O	263(4.11)	44-2660-65
Thiocyanic acid, 2-(ethylcarbamoyl)-vinyl ester	H_2O	255(4.14)	44-2660-65
	Et_2O	255(4.15)	44-2660-65
Threonine, methylthiohydantoin deriv.	EtOH	320(4.38)	65-0988-65
$C_6H_8N_2OS_3$			
4-Isothiazolecarboxamide, 3,5-bis-(methylthio)-	EtOH	212(4.02),235(4.05), 283(4.06)	44-0665-64

Compound	Solvent	λ_{max}(log ϵ)	Ref.
$C_6H_8N_2O_2$			
4(5)-Imidazolecarboxylic acid, ethyl ester	n.s.g.	236(4.00)	24-1414-64
Pyrazine, 3-ethoxy-, 1-oxide	H_2O	218(4.31),261(4.05), 305(3.68)	44-2623-64
3-Pyrazinol, 2,5-dimethyl-, 1-oxide	H_2O	225(4.18),272(3.79), 328(3.74)	44-2623-64
Pyridazine, 5-methoxy-3-methyl-, 1-oxide	EtOH	225(4.3),275(4.0)	94-0228-64
Pyrimidine, 4,6-dimethoxy-	pH -2.0	247(3.85)	12-0567-64
	pH 4.2	242(3.50)	12-0567-64
	EtOH	239(3.48)	12-0567-64
4(1H)-Pyrimidinone, 6-methoxy-1-methyl-	pH -4.7	248(3.44),280s(3.13)	12-0567-64
	pH 1.0	217(4.11),253(3.34), 287(3.09)	12-0567-64
	EtOH	217(4.11),252(3.23), 295(3.00)	12-0567-64
4(3H)-Pyrimidinone, 6-methoxy-3-methyl-	EtOH	230s(3.56),269(3.60)	12-0567-64
Uracil, 1,3-dimethyl-	Me_3PO_4	184(4.15),206(3.81), 264(3.85)	35-0011-65
	$C_6H_{11}Me$	182(4.29),203(3.89), 262(3.90)	35-0011-65
$C_6H_8N_2O_2S$			
Benzenesulfonamide, 2-amino-	H_2O	239(3.81),303(3.49)	20-0136-65
	1% HOAc	241(3.79),299(3.45)	20-0136-65
4-Pyrimidinol, 5-(2-hydroxyethyl)-2-mercapto-	pH 1	275.5(4.20)	44-2670-64
	pH 13	259.5(4.10)	44-2670-64
4-Pyrimidinol, 2-methoxy-5-(methylthio)-	pH 1	206(3.88),252(3.77), 293(3.70)	39-3987-65
	pH 7.2	252(3.79),288(3.75)	39-3987-65
	pH 13	245(3.73),286(3.74)	39-3987-65
Uracil, 5-(ethylthio)-	pH 1	212(4.01),230s(3.82)	39-3987-65
	pH 7.2	212(4.01),230s(3.82)	39-3987-65
	pH 13	235(3.95),294(3.93)	39-3987-65
$C_6H_8N_2O_2S_2$			
3(2H)-Pyridazinethione, 2-(hydroxymethyl)-6-[(hydroxymethyl)thio]-	THF	310(4.30),390(3.36)	49-0631-65
$C_6H_8N_2O_3$			
γ-Butyrolactone, α-(ureidomethylene)-	EtOH	270(4.43)	65-2599-64
2-Furanamine, N,N-dimethyl-5-nitro-	MeOH	226(3.96),435(4.27)	65-0710-64
Pyrazine, 2-ethoxy-, 1,4-dioxide	H_2O	212(4.05),234(4.30), 256(3.74),296(4.31), 338(4.08)	44-2623-64
6(1H)-Pyridazinone, 1,3-dimethoxy-	EtOH	310(3.4)	95-0344-65
Uracil, 3-(2-hydroxyethyl)-	pH 7.25	255(3.9)	49-1698-64
	pH 9.4	260(3.8)	49-1698-64
	pH 9.93	270(3.7)	49-1698-64
	pH 12.0	290(4.0)	49-1698-64
Uracil, 5-(2-hydroxyethyl)-	pH 1	264(3.87)	44-2670-64
	pH 13	229(3.62),289(3.77)	44-2670-64
Uracil, 5-methoxy-1-methyl-	pH 1	207(4.02),217s(3.98), 282(3.92)	73-2980-64
	pH 13	278(3.79)	73-2980-64
Uracil, 5-methoxy-3-methyl-	pH 1	206(3.07),275(3.80)	73-2980-64
	pH 13	296(3.98)	73-2980-64

Compound	Solvent	$\lambda_{max}(\log \epsilon)$	Ref.
$C_6H_8N_2O_3S$			
5-Hydantoinacetic acid, 3-methyl-2-thio-	97% HOAc	235(3.96),265(4.25)	65-0554-65
Uracil, 5-(ethylsulfinyl)-	pH 1	214(3.99),268(3.84)	39-3987-65
	pH 7.2	219(3.92),281(3.84)	39-3987-65
	pH 13	235(3.94),285(3.94)	39-3987-65
$C_6H_8N_2O_4$			
Hydantoin-5-carboxylic acid, ethyl ester, sodium deriv.	H_2O	292(3.74)	44-2003-64
Uracil, 6-(β-hydroxyethoxy)-	pH 7	260(4.00)	65-1433-65
$C_6H_8N_2O_4S$			
Uracil, 5-(ethylsulfonyl)-	pH 1	207(4.01),262(4.06)	39-3987-65
	pH 7.2	223(3.96),282(4.01)	39-3987-65
	pH 13	231(4.00),275(3.98)	39-3987-65
$C_6H_8N_4$			
5H-Pyrrolo[3,4-d]pyrimidine, 4-amino-6,7-dihydro-	pH 1	251(3.97)	44-0194-65
	EtOH	234(3.95),269(3.66)	44-0194-65
$C_6H_8N_4O$			
Pyrimidine, 4-amino-5-formamido-methyl-	EtOH	236(3.99),274(3.63)	94-0393-64
s-Triazine-2-carboxaldehyde, 4,6-dimethyl-, oxime	MeOH	247(4.18)	44-0678-64
$C_6H_8N_4O_2$			
Acrylamide, N-carbamoyl-α-cyano-β-(methylamino)-	pH 1	291(4.29)	39-1642-65
	H_2O	292(4.22)	39-1642-65
	pH 13	284(4.06)	39-1642-65
1,2,3-Benzenetriamine, 4-nitro-	EtOH	218(4.47),267(3.90), 393(4.29)	7-0476-64
1,2,3-Benzenetriamine, 5-nitro-	EtOH	235s(4.23),265s(3.88), 419(3.99)	7-0476-64
1,2,5-Benzenetriamine, 4-nitro-	EtOH	211(4.29),240(4.06), 268s(3.68),381(4.02), 448(3.87)	7-0476-64
p-Benzoquinone, tetraamino-	methyl cellosolve	225(3.97),458(4.17)	78-1889-64
4-Imidazolecarboxamide, 5-formamido-1-methyl-	pH 1	217(4.05)	44-3597-65
	pH 13	265(3.94)	44-3597-65
	MeOH	245s(3.9)	44-3597-65
Pyrazine-2-carboxamide, 5-hydroxy-3-(methylaminc)-	pH -3.1	266(4.10),370(4.10)	39-3770-65
	pH 5.3	227(3.54),277(4.21), 362(4.19)	39-1175-65
	pH 9.7	211(4.28),277(4.19), 355(4.12)	39-3770-65
$C_6H_8N_4O_3$			
s-Triazine-2-carboxaldehyde, 4,6-dimethoxy-, oxime	MeOH	245(4.14)	44-0678-64
$C_6H_8N_6$			
1,2,4-Triazolo[2,3-a]pyrimidine, 4-hydrazino-6-methyl-	MeOH	260s(--),290(4.15)	65-0504-64

Compound	Solvent	$\lambda_{max}(\log \epsilon)$	Ref.
$C_6H_8N_6O$			
4-Pteridinol, 2,6-diamino-7,8-dihydro-	pH 1	274(4.13),319(3.78)	39-4769-64
	pH 13	222(4.17),280(4.03), 310s(3.83)	39-4769-64
C_6H_8O			
2-Cyclohexen-1-one	n.s.g.	224(4.06)	65-0631-65
1-Cyclopentanone, 2-methylene-	MeOH	220s(3.8),230(3.9)	83-0326-65
4-Hexen-2-yn-1-ol	Et_2O	226(4.06)	24-2118-64
C_6H_8OS			
3-Thiolen-2-one, 5-ethyl-	EtOH	265(3.45)	11-0211-64
4-Thiolen-2-one, 5-ethyl-	EtOH	268(3.14)	11-0211-64
$C_6H_8O_2$			
2-Butenolide, 3-ethyl-	H_2O	212.5(3.91)	39-3075-65
2-Cyclobuten-1-one, 3-ethoxy-	EtOH	233(4.10)	12-0765-64
Cyclobutenone, 3-hydroxy-2,4-dimethyl-	EtOH	252(4.20)	35-0673-64
1,2-Cyclohexanedione	n.s.g.	266(3.39)	23-0712-64
1,2-Cyclopentanedione, 4-methyl-	EtOH	257(3.75)	39-2187-64
1,3-Cyclopentanedione, 2-methyl-	pH 1	253(4.24)	94-1300-65
	neutral	251(4.22)	94-1300-65
	pH 10	271(4.41)	94-1300-65
	EtOH	253(4.19)	70-1648-64
	EtOH-NaOH	270(4.45)	70-1648-64
2-Cyclopenten-1-one, 2-hydroxy-3-methyl-	EtOH	258(2.85)	44-1050-65
2-Hexenal, 4-oxo-, trans	MeOH	215(3.87)	39-2955-65
$C_6H_8O_3$			
Cyclobutenone, 2,3-dihydroxy-2,4-dimethyl-	H_2O	248(4.23)	35-0673-64
	Et_2O	238(4.15)	35-0673-64
2(5H)-Furanone, 3-methoxy-4-methyl-	EtOH	223-234(3.95)	39-0766-64
$C_6H_8O_3S$			
2(5H)-Furanone, 3-methoxy-4-(mercaptomethyl)-	EtOH	231(3.80)	39-0766-64
$C_6H_8O_4$			
Meldrumic acid	MeOH	212(1.90)	88-0721-65
	MeOH-NaOH	258(4.15)	88-0721-65
$C_6H_8O_6$			
Ascorbic acid	H_2O	265(4.00)	39-0766-64
	EtOH	245(3.90)	39-0766-64
C_6H_8S			
3-Penten-1-yne, 4-(methylthio)-, cis	Et_2O	264(4.18)	24-1736-65
trans	Et_2O	273.5(4.10)	24-1736-65
$C_6H_9BrN_4$			
Pyrimidine, 2-amino-5-bromo-4-(dimethylamino)-	pH 1.0	220(4.36),257s(3.95), 294(3.88)	39-0755-65
	pH 9.4	220(4.30),307(3.91)	39-0755-65
Pyrimidine, 4-amino-5-bromo-2-(dimethylamino)-	pH 1.0	235(4.45),290(3.38)	39-0755-65
	pH 9.4	240(4.28),309(3.74)	39-0755-65
Pyrimidine, 5-bromo-1,2-dihydro-2-imino-1-methyl-4-(methylamino)-	pH 7.0	215(4.41),246s(4.02), 285(3.83)	39-0755-65

Compound	Solvent	λ_{max}(log ε)	Ref.
Pyrimidine, 5-bromo-1,4-dihydro-4-imino-1-methyl-2-(methylamino)-	pH 7.0	214(4.35),229(4.28),293(3.66)	39-0755-65
C_6H_9BrO			
Cyclobutanone, 2-bromo-3,3-dimethyl-	C_6H_{12}	305(2.18)	22-1957-64
	dioxan	301(2.20)	22-1976-64
	CCl_4	301.5(2.21)	22-1976-64
	MeCN	296.5(2.16)	22-1976-64
Cyclobutanone, 2-bromo-4,4-dimethyl-	C_6H_{12}	315.5(2.29)	22-1957-64
	MeOH	310.5(2.26)	22-1976-64
	dioxan	314(2.4)	22-1976-64
	CCl_4	315.4(2.25)	22-1976-64
	MeCN	309.5(2.25)	22-1976-64
$C_6H_9ClN_2O$			
Butyramide, 4-chloro-2-cyano-2-methyl-	MeOH	207(2.47)	24-1970-64
$C_6H_9ClN_4$			
Pyrimidine, 2-amino-4-chloro-6-(dimethylamino)-	pH 1.0	234(4.07),280(3.96),322s(3.38)	39-0755-65
	pH 7.7	239(4.06),292(3.97)	39-0755-65
Pyrimidine, 4-amino-6-chloro-2-(dimethylamino)-	pH 1.0	233(4.24),310(3.95)	39-0755-65
	pH 7.7	241(4.19),294(3.76)	39-0755-65
Pyrimidine, 4-chloro-2,6-bis-(methylamino)-	pH 1.0	221(4.26),280(3.72),308(3.75)	39-0755-65
	pH 6.4	216(4.44),242s(4.05),291(3.90)	39-0755-65
Pyrimidine, 4-chloro-2,3-dihydro-2-imino-3-methyl-6-(methylamino)-	pH 7.0	214(4.37),243s(4.08),278(3.94)	39-0755-65
	pH 15.0	239s(4.29),302s(3.58)	39-0755-65
Pyrimidine, 4-chloro-3,6-dihydro-6-imino-3-methyl-6-(methylamino)-	pH 7.0	216(4.33),282(3.81)	39-0755-65
	pH 15.0	239(4.32),310(3.49)	39-0755-65
$C_6H_9ClN_4O$			
Pyrimidine, 2-amino-4-chloro-6-[(β-hydroxyethyl)amino]-	pH 1	231s(4.08),273(3.89),298s(3.58)	87-0182-65
	pH 11	237(4.02),285(3.97)	87-0182-65
C_6H_9ClO			
Cyclohexanone, 2-chloro-	hexane	304(1.56)	28-2374-65B
	C_6H_{12}	304(1.57)	28-2374-65B
	EtOH	293(1.43)	28-2374-65B
	EtOH	295(1.32)	59-0529-65
	HOAc	292(1.40)	59-0529-65
	MeCN	292(1.32)	59-0529-65
	dioxan	300(1.40)	59-0529-65
	Et_2O	302(1.46)	59-0529-65
	CCl_4	304(1.59)	59-0529-65
	$HCONH_2$	282(1.32)	59-0529-65
	DMF	290(1.32)	59-0529-65
	Me_2SO	287(1.40)	28-2374-65B
	Me_2SO	288(1.32)	59-0529-65
	CF_3COOH	279(1.52)	59-0529-65
1-Hexen-3-one, 1-chloro-	EtOH	230(4.09)	44-0385-64
2-Pentenal, 3-chloro-2-methyl-	EtOH	247.5(4.13)	44-1126-65
1-Penten-3-one, 1-chloro-4-methyl-	EtOH	232(4.05)	44-0385-64

Compound	Solvent	λ max(log ε)	Ref.
$C_6H_9Cl_2O_2P$			
3-Phospholene, 3-chloro-1-(2-chloroethoxy)-, 1-oxide	MeOH	200(3.72)	78-1593-64
C_6H_9FO			
2-Hexenal, 2-fluoro-	EtOH	228(3.96),316(1.26)	22-2258-64
C_6H_9N			
1H-Azepine, 2,3-dihydro-	C_6H_{12}	296(3.81)	5-0001-65B
Pyridine, 1,2-dihydro-1-methyl-	EtOH	335(3.32)	44-1647-64
C_6H_9NO			
2-Cyclohexen-1-one, oxime, anti	n.s.g.	230(4.32)	65-0631-65
syn	n.s.g.	232(3.79)	65-0631-65
2(1H)-Pyridone, 3,4-dihydro-6-methy-	MeOH	248(3.65)	5-0001-65B
2(1H)-Pyridone, 5,6-dihydro-6-methyl-	MeOH	241(3.20)	5-0001-65B
2-Pyrrolidinone, 3-ethylidene-, cis	H_2O	227(4.04)	39-4591-64
trans	H_2O	222(4.11)	39-4591-64
2-Pyrrolidinone, 3-vinyl-	H_2O	196(3.69)	39-4591-64
$C_6H_9NO_2$			
Cyclohexene, 1-nitro-, anion	H_2O	278(4.21)	44-2674-65
Cyclopentene, 1-nitromethyl-, anion	H_2O	284(4.23)	44-2674-65
Glutarimide, N-methyl-	THF	215(3.91),233(2.72), 241(2.47)	24-1548-64
3-Isoxazolol, 4-ethyl-5-methyl-	C_6H_{12}	214(3.82)	78-2835-64
	0.01N H_2SO_4	224(3.80)	78-2835-64
4-Isoxazolin-3-one, 2,4,5-tri-methyl-	C_6H_{12}	241(3.64)	78-2835-64
	5% EtOH	230(3.88)	78-2835-64
2-Pentenal, 2-methyl-4-oxo-, oxime	isooctane	262(4.31)	54-1233-65
	MeOH-acid	270(4.28)	54-1233-65
	MeOH	270(4.28)	54-1233-65
	MeOH-base	320(4.28)	54-1233-65
$C_6H_9NO_2S$			
2,4-Thiazolidinedione, 5-isopropyl-	n.s.g.	240s(3.50),340(2.21)	65-0886-65
2,4-Thiazolidinedione, 5-propyl-	n.s.g.	290s(2.00)	65-0886-65
$C_6H_9NO_3$			
Butyrolactam, N-(hydroxyacetyl)-	dioxan	220(3.86)	78-3537-65
$C_6H_9NS_2$			
2H-1,3-Thiazine-2-thione, 3,6-di-hydro-3,5-dimethyl-	H_2O	293(3.98),312s(3.88)	39-4008-64
	EtOH	270s(3.34),312(3.70)	39-4008-64
2H-1,3-Thiazine-2-thione, 3,6-di-hydro-3,6-dimethyl-	H_2O	298(4.01),320s(--)	39-4008-64
	EtOH	310(4.03)	39-4008-64
2H-1,3-Thiazine-2-thione, 3,6-di-hydro-4,6-dimethyl-	H_2O	297(4.01),317s(--)	39-4008-64
	EtOH	318(4.04)	39-4008-64
2H-1,3-Thiazine-2-thione, 3,6-di-hydro-5,6-dimethyl-	H_2O	294(3.96),317(3.99), 360s(--)	39-4008-64
	EtOH	323(4.06)	39-4008-64
$C_6H_9N_3$			
Pyrazine, 2-methyl-3-(methyl-amino)-	H_2O	239(4.06),321(3.75)	44-0415-64
hydrochloride	H_2O	239(4.06),321(3.75)	44-0415-64
Pyrazine, 2-methyl-6-(methyl-amino)-	H_2O	242(4.38),330(3.76)	44-0415-64

Compound	Solvent	$\lambda_{max}(\log \epsilon)$	Ref.
Pyridine, 3-amino-4-(methylamino)-	pH -2.28	210(4.00),216s(3.85), 271(4.23)	39-2150-64
	pH 5.0	226(4.19),291(4.08)	39-2150-64
	pH 12.0	214(4.35),258(3.90), 280s(3.72)	39-2150-64
Pyridine, 4-amino-3-(methylamino)-	pH -2.3	208(4.12),262(4.19)	39-2150-64
	pH 5.0	228(4.35),294(3.85)	39-2150-64
	pH 12.0	216(4.36),252(3.74), 284(3.58)	39-2150-64
$C_6H_9N_3O$			
Ethanol, 2-(pyrazinylamino)-	pH 1.0	238(4.07),332(3.70)	39-2778-65
	pH 7.0	243(4.17),286s(3.09), 330(3.69)	39-2778-65
Pyrazole-1-carboxamide, 3,5-dimethyl-	EtOH	235(4.03)	28-0606-65A
Pyrimidine, 4-methoxy-6-(methylamino)-	pH 7.0	243(4.05)	12-0559-65
4-Pyrimidinol, 6-(dimethylamino)-	pH -1.3	265(4.19)	12-0559-65
	pH 7	224(4.37),267(4.11)	12-0559-65
4(1H)-Pyrimidinone, 1-methyl-6-(methylamino)-	pH -2.0	220(4.34),262(4.01)	12-0559-65
	pH 7	223(4.44),260(3.93)	12-0559-65
$C_6H_9N_3O_2$			
Cytosine, 5-(hydroxyethyl)-	pH 1	284(3.97)	44-2670-64
	pH 13	287(3.86)	44-2670-64
5-Pyrimidineethanol, 2-amino-4-hydroxy-	EtOH	262(3.85)	65-2599-64
	pH 1	261.5(3.86)	44-2670-64
	pH 13	233(3.87),278(3.84)	44-2670-64
$C_6H_9N_3O_2S$			
5-Hydantoinacetamide, 3-methyl-2-thio-	97% HOAc	235(3.89),265(4.25)	65-0554-65
4-Pyrimidinol, 6-amino-5-(2-hydroxyethyl)-2-mercapto-	pH 1	241(3.89),282(4.30)	44-2670-64
	pH 13	248(4.10),280(4.03)	44-2670-64
$C_6H_9N_3O_3$			
Imidazole, 1-(2-hydroxyethyl)-2-methyl-5-nitro-	2M H_2SO_4	277(3.81)	44-0862-64
	pH 6.24	319(3.97)	44-0862-64
1,2,4-Oxadiazolin-5-one, 3-carbamoyl-4-isopropyl-	EtOH	214(3.74)	33-0033-64
2,4-Pyrimidinediol, 6-amino-5-(2-hydroxyethyl)-	pH 1	271(4.28)	44-2670-64
	pH 13	274(4.20)	44-2670-64
$C_6H_9N_3O_4$			
Enteromycin carboxamide	MeOH	230(4.09),270s(4.14), 300(4.18)	78-0267-65
$C_6H_9N_3S$			
2-Pyrimidinethiol, 4-(dimethylamino)-	pH 0.2	240(4.18),278(4.32), 314(3.89)	12-0559-65
	pH 7.0	222(3.86),267(4.48)	12-0559-65
	pH 13.3	248(4.42),306(3.83)	12-0559-65
4-Pyrimidinethiol, 2-(dimethylamino)-	pH 0.2	222(4.24),266(3.76), 330(4.13)	12-0559-65
	pH 5.4	225(4.23),315(3.93), 351(3.94)	12-0559-65
	pH 13.3	234(4.27),276s(3.90), 322(3.94)	12-0559-65

Compound	Solvent	$\lambda_{max}(\log \epsilon)$	Ref.
4-Pyrimidinethiol, 6-(dimethylamino)-	pH -3.0	236(4.09),280(4.12), 340s(3.38)	12-0559-65
	pH 7.0	258(4.28),315(4.32)	12-0559-65
	pH 12.0	259(4.38),300(4.02)	12-0559-65
4(3H)-Pyrimidinethione, 3-methyl-6-(methylamino)-	pH -3.0	234(4.16),271(4.06), 328(3.94)	12-0559-65
	pH 7.0	257(4.31),312(4.32)	12-0559-65
$C_6H_9N_5$			
5H-Pyrrolo[3,4-d]pyrimidine, 2,4-diamino-6,7-dihydro-	pH 1	271(3.77),300s(3.18)	44-0194-65
	EtOH	284(3.90)	44-0194-65
$C_6H_9N_5O$			
Formamide, N-[4-amino-2-(methylamino)-5-pyrimidinyl]-	pH 3.0	216(4.39),276s(4.53)	39-3770-65
	pH 9.0	232(4.13),295(3.78)	39-3770-65
Pyrazine-2-carboxamide, 5-amino-3-(methylamino)-	pH 0.2	275(4.19),379(4.24)	39-1175-65
	pH 4.8	213(4.26),275(4.20), 363(4.11)	39-1175-65
$C_6H_9N_5O_2$			
Pyrimidine, 2-amino-4-(dimethylamino)-5-nitro-	pH 1.0	251(4.31),330(3.44)	39-5542-65
	pH 7.0	232(4.06),273(4.06), 375(3.89)	39-5542-65
$C_6H_9N_5O_2S$			
1,2,5-Thiadiazole-3-carboxamide, N-ethyl-4-ureido-	pH 1,7	232(4.16),304(3.91)	44-2141-64
1,2,5-Thiadiazole-3-carboxamide, 4-(3-ethylureido)-	pH 1,7	234(4.22),307(3.87)	44-2141-64
C_6H_{10}			
1,3-Butadiene, 2,3-dimethyl-	C_6H_{12}	228(4.34)	44-3527-64
	EtOH	226.5(4.31)	44-3527-64
Cyclobutene, 1,2-dimethyl-	isooctane	194(3.83),200(3.53), 210(2.96),220(2.68), (end absorptions)	78-1001-65
Cyclobutene, 1,3-dimethyl-	isooctane	190(3.83) (end absorption)	78-1001-65
Cyclohexene, $PdCl_2$ olefin complex from	EtOH	212(4.29),234s(4.07), 321(3.23)	24-2037-64
Cyclopropane, 1,1-dimethyl-2-methylene-	hexane	208(2.92)	28-1695-65B
	EtOH	213(2.65)	28-1695-65B
Cyclopropene, 1,3,3-trimethyl-	hexane	206(3.40)	28-1695-65B
	EtOH	215(3.09)	28-1695-65B
1,4-Hexadiene, cis	C_6H_{12}	none	5-0028-65H
trans	C_6H_{12}	none	5-0028-65H
2,4-Hexadiene	C_6H_{12}	226.5(4.38)	44-3527-64
	EtOH	226(4.38)	44-3527-64
1,3-Pentadiene, 4-methyl-	C_6H_{12}	234(4.36)	44-3527-64
	EtOH	232.5(4.35)	44-3527-64
$C_6H_{10}BrOP$			
3-Phospholene, 1-bromo-3,4-dimethyl-, 1-oxide	n.s.g.	211(3.34),264(2.82)	88-2855-64
$C_6H_{10}ClNO$			
2-Pentenal, 3-chloro-2-methyl-, oxime	EtOH	245(4.29)	44-1126-65

Compound	Solvent	$\lambda_{max}(\log \epsilon)$	Ref.
$C_6H_{10}ClNO_4S_2$			
2-(Dimethylamino)-4-methyl-1,3-dithiol-1-ium perchlorate	70% $HClO_4$	245(2.90),303(3.49)	44-1703-64
$C_6H_{10}N_2O$			
Imidazol-1-ol, 2,4,5-trimethyl-	n.s.g.	224.0(3.94)	25-1837-64
Pyrazole, 5-ethoxy-3-methyl-	C_6H_6	217(3.68)	78-1693-65
	H_2O	none	78-1693-65
3-Pyrazolinone, 1,2,5-trimethyl-	C_6H_{12}	257.5(3.97)	78-0315-64
	pH 5	246.5(3.98)	78-0315-64
	20N H_2SO_4	227.5(3.95)	78-0315-64
5-Pyrazolinone, 3-propyl-	C_6H_{12}	250(3.00)	78-1693-65
	H_2O	245.5(3.95)	78-1693-65
5-Pyrazolinone, 1,2,3-trimethyl-	C_6H_{12}	258(3.97)	78-0299-64
	H_2O	247(3.98)	78-1693-65
	29N H_2SO_4	227.5(3.94)	78-0299-64
5-Pyrazolinone, 3,4,4-trimethyl-	C_6H_{12}	241(3.71)	78-1693-65
	H_2O	238(3.68)	78-1693-65
$C_6H_{10}N_2OS$			
Formamide, N-[2-(2-thiazolin-2-yl)-ethyl]-	N HCl	265(3.69)	94-0180-65
	EtOH	232(3.36),245(3.32)	94-0180-65
Pseudohydantoin, 5-isopropyl-thio-	n.s.g.	249(3.94),300s(2.40)	65-0886-65
Pseudohydantoin, 5-propyl-thi-	n.s.g.	247(3.94),300s(2.35)	65-0886-65
$C_6H_{10}N_2O_2S$			
1,2,6-Thiadiazine, 2,4,5-trimethyl-, 1,1-dioxide	EtOH	248(2.9),323(4.2)	25-0182-65
1,2,6-Thiadiazine, 3,4,5-trimethyl-, 1,1-dioxide	EtOH	251(2.9),329(3.8)	25-0182-65
$C_6H_{10}N_2O_4S_3$			
Thiazoline, 3-ethylsulfonyl-2-(methylsulfonylimino)-	MeOH	226(3.44),280(4.03)	39-1219-65
Thiazoline, 2-(ethylsufonylimino)-3-methylsulfonyl-	MeOH	226(3.40),279(4.03)	39-1219-65
$C_6H_{10}N_4$			
Pyrimidine, 2-amino-4-(dimethylamino)-	pH 1.0	246(4.03),274(3.96)	39-0755-65
	pH 10.4	238(4.04),291(3.95)	39-0755-65
Pyrimidine, 4-amino-2-(dimethylamino)-	pH 1.0	225(4.45),279s(3.43)	39-0755-65
	pH 10.4	236(4.20),294(3.72)	39-0755-65
Pyrimidine, 1,2-dihydro-2-imino-1-methyl-4-(methylamino)-	pH 7.0	209(3.40),239(4.13), 270(3.96)	39-0755-65
	pH 15.0	234(4.35),297s(3.57)	39-0755-65
Pyrimidine, 1,4-dihydro-4-imino-1-methyl-2-(methylamino)-	pH 10.0	213(4.37),225s(4.26), 277(3.77)	39-0755-65
	pH 15.6	236(4.36),310(3.25)	39-0755-65
$C_6H_{10}N_4O$			
6-Azacytosine, 3-(dimethylamino)-6-methyl-	N H_2SO_4	215(3.87),251s(3.50), 302(4.03)	73-1394-64
	H_2O	222(4.10),299(3.97)	73-1394-64
4-Imidazolecarboxamide, 1-methyl-5-(methylamino)-	pH 1	252(3.84),268(3.81)	44-3597-65
	pH 13	267.5(3.94)	44-3597-65
	MeOH	269(3.96)	44-3597-65

Compound	Solvent	λ_{max}(log ε)	Ref.
$C_6H_{10}N_4O_3$			
Uracil, 5-amino-6-[(2-hydroxy-ethyl)amino]-	pH 1	267(4.40)	37-3493-64
$C_6H_{10}N_4S$			
2-Thiadiazolinone, 3-methyl-, azine with acetone	EtOH	257(3.93),305(3.88)	1-0871-64
$C_6H_{10}N_4S_2$			
Acetone, [5-(methylthio)thiadiazolyl]-hydrazone	EtOH	298(4.20)	1-0871-64
anion	EtOH	216(3.74),342(4.12)	1-0871-64
$C_6H_{10}O$			
Cyclobutane, 1,2-epoxy-1,2-di-methyl-	isooctane	267s(1.49),273(1.58), 280(1.57),289s(1.28)	22-2755-65
Cyclobutanone, 2,2-dimethyl-	C_6H_{12}	294(1.38)	22-1957-64
	C_6H_{12}	294(1.38)	22-1976-64
	MeOH	290(1.41)	22-1976-64
	dioxan	291(1.38)	22-1976-64
	CCl_4	293(1.38)	22-1976-64
	MeCN	291(1.39)	22-1976-64
Cyclobutanone, 2,3-dimethyl-	C_6H_{12}	293.2(1.45)	22-2755-65
	C_6H_{12}	293.2(1.45)	88-0979-65
Cyclobutanone, 3,3-dimethyl-	C_6H_{12}	277(1.21),284(1.26), 292(1.23)	22-1957-64
	dioxan	281(1.23),289(1.22)	22-1976-64
	CCl_4	282(1.37),290(1.37)	22-1976-64
	MeCN	280(1.27),286(1.27)	22-1976-64
Cyclohexanone	C_6H_{12}	290(1.19)	28-2374-65B
	hexane	291(1.18)	28-2374-65B
	EtOH	282(1.23)	28-2374-65B
	EtOH	283(1.22)	59-0529-65
	HOAc	280(1.24)	59-0529-65
	CCl_4	292(1.30)	59-0529-65
	CF_3COOH	266(1.42)	59-0529-65
	MeCN	286(1.18)	59-0529-65
	Me_2SO	286(1.22)	59-0529-65
Cyclopropanecarboxaldehyde, 1,2-dimethyl-, cis	isooctane	274s(1.30),281(1.32), 290s(1.29),298s(1.17)	22-2755-65
	isooctane	280.9(1.32)	88-0979-65
trans	isooctane	283s(0.79),288(0.80), 295s(0.77),304s(0.57), 316s(-0.14)	22-2755-65
	isooctane	288.3(0.80)	88-0979-65
2-Hexenal	hexane	216(4.31)	39-2372-65
1-Hexen-3-one	EtOH	216(3.75),322(1.47)	22-0704-65
3-Hexen-2-one	EtOH	222(3.97)	70-1070-65
4-Hexen-2-one, trans	MeOH	280(1.90)	88-3653-64
2-Pentenal, 2-methyl-	heptane	222(4.24),327(1.49)	33-0567-64
	n.s.g.	229(4.11),315(1.49)	39-5815-64
1-Penten-3-one, 4-methyl-	EtOH	217(3.70),326(1.54)	22-0704-65
	EtOH	212(4.09)	39-2340-65
3-Penten-2-one, 4-methyl-	H_2O	243.5(4.08)	28-5470-64A
	EtOH	233(4.05),328(1.58)	22-2198-65
	$ClCH_2CH_2Cl$ +acid	280(3.18)	39-1761-65
4-Penten-2-one, 4-methyl-	EtOH	286(1.90)	22-2198-65
1-Pentyn-3-ol, 3-methyl-	EtOH	226(3.68),245(3.72)	39-4348-65

Compound	Solvent	$\lambda_{max}(\log \epsilon)$	Ref.
$C_6H_{10}O_2$			
2-Hexenoic acid	H_2O	213(4.0)	39-2372-65
$C_6H_{10}O_3$			
1,2-Cyclohexanediol, 1,2-oxide	n.s.g.	none	20-0076-64
Verrucarinic acid, lactone	EtOH	none	33-0157-65
$C_6H_{10}S$			
1-Cyclohexene-1-thiol	C_6H_{12}	219(3.75)	59-0299-64
$C_6H_{10}S_3$			
2,6,7-Trithiabicyclo[2.2.2]octane, 4-methyl-	C_6H_{12}	252(3.13)	78-0437-64
	EtOH	251(3.09)	78-0437-64
$(C_6H_{10}S_3)_n$			
Polypentamethylenetrithiocarbonate	$CHCl_3$	310(4.15)	49-0631-65
$C_6H_{11}BN_2$			
Hydrazine, phenyl-, compound with BF_3	EtOH	233(4.04),282(3.45)	44-3401-64
$C_6H_{11}BrN_2S$			
1,2,3,5,6,7-Hexahydroimidazo-[2,1-b][1,3]thiazinium bromide	n.s.g.	225(4.05)	39-3456-65
$C_6H_{11}KOS_2$			
Xanthic acid, isopentyl ester, K salt	H_2O	229(3.96),303(4.30),	1-1113-65
	EtOH	229(3.98),304(4.25) 387(1.75)	1-1113-65
$C_6H_{11}N$			
1-Pyrroline, 5,5-dimethyl-	hexane	231(2.16)	39-2313-65
	EtOH	225(2.17)	39-2313-65
$C_6H_{11}NO$			
Cyclohexanone, oxime	isooctane	192(3.84),310(1.18)	88-2124-64
2-Pyrrolidinone, 3-ethyl-	H_2O	192(3.83)	39-4591-64
1-Pyrroline, 3,3-dimethyl-, 1-oxide	hexane	248(3.98)	23-2717-65
	EtOH	235(3.93)	23-2717-65
$C_6H_{11}NOS$			
2-Butanone, 4-thioacetamido-	EtOH	264.0(4.13)	39-0788-64
Thioacetamide, N-isobutyryl-	heptane	238(3.90),285(4.15), 436(1.62)	24-1556-65
	MeOH	240(3.82),290(4.06), 433(1.51)	24-1556-65
$C_6H_{11}NOS_2$			
2-Butanone, 3-methyl-4-(thio-carbamoylthio)-	H_2O	241(3.92),284(4.17)	39-4004-64
	EtOH	240(3.87),290(4.12)	39-4004-64
Butyraldehyde, α-methyl-β-(thio-carbamoylthio)-	H_2O	243(3.72),286(4.10)	39-4004-64
	EtOH	241(3.85),291(4.09)	39-4004-64
Butyraldehyde, β-(N-methylthio-carbamoylthio)-	H_2O	246(3.96),282(4.12)	39-4004-64
	EtOH	244(3.95),285(4.09)	39-4004-64
Carbamic acid, dimethyldithio-, ester with mercapto-2-propanone	MeOH	246(3.96),274(4.00)	44-2146-64
	EtOH	245(3.97),275(4.05)	44-1703-64
Propionaldehyde, α-methyl-β-(N-methylthiocarbamoylthio)-	H_2O	246(3.91),282(4.13)	39-4004-64
	EtOH	245(3.95),286(4.09)	39-4004-64
2-Thiazolidinethione, 4-hydroxy-4,5,5-trimethyl-	MeOH	248(3.81),274(4.24)	44-2146-64

Compound	Solvent	λ_{max}(log ε)	Ref.
$C_6H_{11}NO_2$			
Cyclohexane, nitro-, anion	H_2O	231(4.08)	44-2674-65
	50% dioxan	233(4.08)	44-2674-65
Cyclopentylnitromethane, anion	H_2O	235(4.04)	44-2674-65
$C_6H_{11}NO_2S$			
Lactamide, N-methylthio-, acetate	MeOH	263(4.06),323(1.62)	35-0051-65
Oxamic acid, N,N-dimethyl-2-thio-, ethyl ester	heptane	239(3.93),282(4.11), 378(1.74)	1-1059-64
	EtOH	238(3.83),280(4.06), 355s(1.79)	1-1059-64
$C_6H_{11}N_3O$			
Glycinonitrile, N-butyl-N-isonitroso-	EtOH	256(3.93)	33-0033-64
2-Pentenal, semicarbazone, cis	EtOH	267(4.46)	39-2988-65
$C_6H_{11}N_3O_5$			
Cyclopentanone, tetrahydroxy-, semicarbazone	n.s.g.	226(4.00)	28-1512-64A
$C_6H_{11}N_5OS$			
Ethanol, 2-[(2,5-diamino-6-mercapto-4-pyrimidinyl)amino]-	pH 1	233(4.27),310(4.38)	87-0182-65
	pH 11	317(4.12)	87-0182-65
$C_6H_{11}N_5O_8$			
Dipropylamine, 2,2,2',2'-tetranitro-	pH 1	270(2.18)	44-0354-65
	EtOH	250s(2.94),378(2.45)	44-0354-65
	MeCN	248s(2.92)	44-0354-65
$C_6H_{11}O_2P$			
2-Phospholene, 1-ethoxy-, 1-oxide	MeOH	199(3.66)	78-1593-64
	MeOH	200(3.57)	88-3039-64
3-Phospholene, 1-ethoxy-, 1-oxide	MeOH	200(2.48),230(2.35)	78-1593-64
	MeOH	200(2.63)	88-3039-64
C_6H_{12}			
1-Butene, 3,3-dimethyl-	n.s.g.	177(4.1)	70-2203-64
$C_6H_{12}CuN_2O_4$			
DL-Alanine, copper complex	H_2O	238(3.79)	59-1227-64
$C_6H_{12}N_2$			
2-Butene, 2-(methylazo)-3-methyl-	EtOH	262(3.86)	35-4576-65
2-Pyrazoline, 3,5,5-trimethyl-	H_2O	221(3.60)	88-1175-65
$C_6H_{12}N_2O$			
2 -Imidazoline-1-ethanol, 2-methyl-	EtOH	231(3.79)	4-0188-64
$C_6H_{12}N_2OS$			
Oxamide, tetramethylthio-	heptane	238s(3.72),281(4.01)	1-1059-64
	EtOH	235(3.83),280(4.05), 355(1.69)	1-1059-64
	CH_2Cl_2	359(1.72)	1-1059-64
$C_6H_{12}N_2O_2S$			
Carbamic acid, dimethyl-, ethyl ester, compound with isothiocyanic acid	MeCN	238(2.12)	44-2086-65

Compound	Solvent	λ_{max}(log ε)	Ref.
$C_6H_{12}N_2S_2$			
Oxamide, tetramethyldithio-	heptane	262s(4.21),275(4.27)	1-1059-64
	heptane-0.5% CH_2Cl_2	365.0(2.57)	1-1059-64
	EtOH	272(4.29),345s(2.70)	1-1059-64
$C_6H_{12}N_2S_4$			
Tetramethylthiuram disulfide	heptane	223(4.17),250s(--), 278s(3.93)	1-1113-65
	EtOH	222(4.15),250s(--), 275s(3.96)	1-1113-65
$C_6H_{12}N_4O_3$			
Cytosine, 4,5-dihydro-N^6-methoxy-4-methoxyamino-	pH 7	225(4.08)	39-0208-65
$C_6H_{12}N_6O$			
Glyoxal, methyl-, 1-hydrazone 2-(4-acetylguanylhydrazone)	pH 1	290(4.17)	44-0560-65
	pH 11	298(4.02)	44-0560-65
$C_6H_{12}O$			
2-Butanone, 3,3-dimethyl-	C_6H_{12}	288(1.32)	59-0529-65
	hexane	288(1.32)	
	H_2O	278(1.48)	
	MeOH	282(1.40)	
	EtOH	283(1.39)	
	PrOH	284(1.38)	
	BuOH	284(1.37)	
	pentanol	284(1.37)	
	hexanol	284(1.37)	
	heptanol	284(1.36)	
	Et_2O	286(1.32)	
	dioxan	286(1.32)	
	HCOOH	275(1.52)	
	HCOOMe	285(1.34)	
	HOAc	280(1.42)	
	EtCOOH	281(1.39)	
	PrCOOH	282(1.38)	
	BuCOOH	282(1.37)	
	CF_3COOH	270(1.63)	
	$Cl_2CHCOOH$	274(1.60)	
	MeCN	284(1.34)	
	EtCN	285(1.33)	
	PrCN	285(1.33)	
	$HCONH_2$	282(1.35)	
	HCONHMe	284(1.34)	
	$MeCONMe_2$	285(1.34)	
	DMF	285(1.34)	
	Me_2SO	285(1.35)	
	CH_2Cl_2	285(1.35)	
	$CHCl_3$	286(1.37)	
	CCl_4	288(1.38)	
$C_6H_{12}O_2$			
Acetic acid, butyl ester	isooctane	202(1.71),214(1.7)	35-2384-64
	MeOH	208.2(1.75)	35-2384-64
	EtOH	209(1.76)	35-2384-64
	MeCN	210(1.85)	35-2384-64
	$C_3H_4F_4O$	204(1.8)	35-2384-64

Compound	Solvent	λ_{max}(log ϵ)	Ref.
Acetic acid, isobutyl ester	isooctane	208(1.7),214(1.75)	35-2384-64
	MeOH	207.7(1.76)	35-2384-64
	EtOH	208.7(1.78)	35-2384-64
	MeCN	210(1.78)	35-2384-64
	$C_3H_4F_4O$	203.9(1.77)	35-2384-64
Acetic acid, tert-butyl ester	MeOH	213.3(1.74)	35-2384-64
	80% MeOH	212(1.78)	35-2384-64
	EtOH	213.6(1.78)	35-2384-64
	40% EtOH	211(1.78)	35-2384-64
	MeCN	214.9(1.77)	35-2384-64
	$C_3H_4F_4O$	207.1(1.81)	35-2384-64
Formic acid, neopentyl ester	isooctane	212(1.88),216(1.89)	35-2384-64
	MeOH	211.7(1.85)	35-2384-64
	EtOH	216(1.86)	35-2384-64
	MeCN	212.2(1.84)	35-2384-64
Isobutyric acid, ethyl ester	isooctane	212.3(1.93)	35-2384-64
	MeOH	210.3(1.95)	35-2384-64
	EtOH	210.5(1.95)	35-2384-64
	MeCN	211.2(1.93)	35-2384-64
	$C_3H_4F_4O$	207.5(1.99)	35-2384-64
Pivalic acid, methyl ester	isooctane	213.7(1.97)	35-2384-64
	MeOH	211.5(2.00)	35-2384-64
	80% MeOH	209.9(2.01)	35-2384-64
	EtOH	211.6(2.00)	35-2384-64
	40% EtOH	209.4(2.01)	35-2384-64
	MeCN	213.8(1.95)	35-2384-64
	$C_3H_4F_4O$	208.2(2.04)	35-2384-64
2-Pentanone, 4-hydroxy-4-methyl-	EtOH	282(1.41)	59-0529-65
	HOAc	278(1.42)	59-0529-65
	CF_3COOH	none	59-0529-65
	MeCN	283(1.39)	59-0529-65
	CCl_4	284(1.42)	59-0529-65
$C_6H_{12}S_2$			
m-Dithiane, 2,2-dimethyl-	EtOH	252(2.82)	78-0437-64
m-Dithiane, 2-ethyl-	EtOH	231s(2.71),250(2.79)	78-0437-64
o-Dithiane, 3,6-dimethyl-	EtOH	286(2.40)	78-1067-65
$C_6H_{13}NO_2$			
Butane, 2,2-dimethyl-1-nitro-, anion	H_2O	235.5(3.99)	44-2674-65
Butane, 3,3-dimethyl-1-nitro-, anion	H_2O	231.5(3.99)	44-2674-65
Butane, 3,3-dimethyl-2-nitro-, anion	H_2O	229(3.93)	44-2674-65
$C_6H_{13}NS$			
Butyramide, N,2-dimethylthio-	MeOH	261(4.07),324(1.61)	35-0051-65
$C_6H_{13}N_3O_4S$			
L-Arabinose, thiosemicarbazone	H_2O	235(4.10)	25-0546-64
D-Xylose, thiosemicarbazone	H_2O	236(4.12)	25-0546-64
$C_6H_{14}Ge$			
Germane, allyltrimethyl-	n.s.g.	173(4.0),194(4.0)	70-2203-64
$C_6H_{14}N_2$			
Acetone, isopropylhydrazone	MeOH	228(3.70)	35-2395-64
$C_6H_{14}N_2O$			
Diisopropylamine, N-nitroso-	C_6H_{12}	235(3.80)	35-4373-64

Compound	Solvent	$\lambda_{max}(\log \epsilon)$	Ref.
$C_6H_{14}Si$ Silane, allyltrimethyl-	n.s.g.	189(4.0)	70-2203-64
$C_6H_{14}Sn$ Tin, allyltrimethyl-	n.s.g.	181(4.4),207s(4.0)	70-2203-64
$C_6H_{15}N_3S$ Semicarbazide, 4-sec-butyl-2-methyl-3-thio-, nickel complex	n.s.g.	431(2.03)	1-1239-65
$C_6H_{18}BrNSi_2$ Disilazane, N-bromohexamethyl-	C_6H_{12}	230(2.48),344(1.95)	101-0430-65B
$C_6H_{18}ClNSi_2$ Disilazane, N-chlorohexamethyl-	C_6H_{12}	295.0(1.79)	101-0430-65B
$C_6H_{18}INSi_2$ Disilazane, N-iodohexamethyl-	C_6H_{12}	344(2.28),408(2.13), 513(1.78)	101-0430-65B
$C_6H_{18}Si_2$ Disilane, hexamethyl-	C_6H_{12} n.s.g.	198(3.93) none	101-0369-64B 25-1492-64

Compound	Solvent	λ_{max}(log ε)	Ref.
$C_7ClF_3N_4S$			
Acrylimidoyl chloride 2,3,3-tri-cyano-N-[(trifluoromethyl)thio]-	MeCN	223(3.87),258(3.87), 355(3.94)	44-1194-64
C_7F_9I			
2-Norbornene, nonafluoro-1-iodo-	EtOH	256(2.72)	78-2997-65
$C_7HCoF_4K_3N_5$			
Potassium pentacyano(1,1,2,2-tetra-fluoroethyl)cobaltate(III)	H_2O	311(2.40)	39-6629-65
C_7HF_5O			
Benzaldehyde, pentafluoro-	EtOH	236(4.38),242(4.35), 286(3.4)	30-1135-64
$C_7HF_5O_2$			
Benzoic acid, pentafluoro-	EtOH	218(4.58),261(2.81)	39-1135-64
$C_7H_2F_7N_3$			
Malononitrile, (1-amino-2,2,3,3,-4,4,4-heptafluorobutylidene)-	EtOH	289(4.14)	44-0707-64
$C_7H_2F_8$			
Cyclohexene, octafluoro-3-methylene-	hexane	219.0(4.15)	39-3035-64
$C_7H_2F_9NO$			
2-Cyclohexen-1-one, 3-amino-hexafluoro-2-(trifluoromethyl)-	EtOH	280.0(4.49)	39-5748-64
$C_7H_3ClN_2O_6$			
Tropone, 2-chloro-3-hydroxy-5,7-dinitro-	C_6H_{12}	367(4.19)	88-1041-65
$C_7H_3Cl_2N_3O_5$			
Benzamide, 2,5-dichloro-3,5-dinitro-	44.5% H_2SO_4	216s(4.06)	23-1957-64
	86.1% H_2SO_4	235(4.26)	23-1957-64
$C_7H_3F_3N_2$			
Benzimidazole, 4,5,7-trifluoro-	pH 1	244(3.52)	87-0737-65
	pH 7	240(3.71)	87-0737-65
	pH 13	255(3.68)	87-0737-65
$C_7H_3F_3N_2O_2$			
Nicotinonitrile, 1,2-dihydro-6-hy-droxy-2-oxo-4-(trifluoromethyl)-	EtOH	254(4.1),342(4.21)	44-3377-65
$C_7H_3F_9N_2$			
1-Cyclohexen-1-ylamine, 4,4,5,5,6,6-hexafluoro-3-imino-2-(trifluoro-methyl)-	EtOH	275.0(3.74)	39-5748-64
C_7H_3IOS			
2-Thiophenepropiolaldehyde, 5-iodo-	n.s.g.	264(3.88),324(4.31), 331(4.32)	88-0297-65

Compound	Solvent	λ_{max}(log ϵ)	Ref.
C_7H_4BrNS			
Benzothiazole, 2-bromo-	gas	211(4.27),216(4.26), 250(3.86),256(3.86), 293(3.12)	17-0079-65
	C_6H_{12}	221(4.37),233s(4.07), 256(3.96),265(3.78), 296(3.18)	17-0079-65
	EtOH	257(3.96),263s(3.92), 296(3.22)	17-0079-65
$C_7H_4ClF_3O$			
Anisole, p-chloro-α,α,α-trifluoro-	hexane	269(2.61)	65-2749-64
	EtOH	273(2.4)	65-2749-64
	Et_2O	270(2.55)	65-2749-64
	dioxan	274(2.6)	65-2749-64
	H_2SO_4	216(3.72),268(2.34)	65-2749-64
$C_7H_4ClF_3OS$			
Sulfoxide, p-chlorophenyl trifluoromethyl	hexane	225(4.04),251(3.76), 278(2.96)	65-2749-64
	EtOH	226(4.00),248(3.84), 278(2.96)	65-2749-64
	Et_2O	226(4.11),250(3.86)	65-2749-64
	dioxan	225(4.02),249(3.80), 278(2.98)	65-2749-64
	H_2SO_4	255(3.94)	65-2749-64
$C_7H_4ClF_3O_2S$			
Sulfone, p-chlorophenyl trifluoromethyl	hexane	234(4.26),258(2.93), 278(2.93)	65-2749-64
	EtOH	236(4.23),270(3.07), 278(2.99)	65-2749-64
	H_2SO_4	241(4.18),270(3.25), 278(3.12),310s(2.58)	65-2749-64
$C_7H_4ClF_3S$			
Sulfide, p-chlorophenyl trifluoromethyl	hexane	223(4.10),246(3.66), 268(2.88)	65-2749-64
	EtOH	223(4.03),244(3.70), 271(2.81)	65-2749-64
	Et_2O	223(4.00),244(3.60), 265(2.82)	65-2749-64
	dioxan	223(4.08),244(3.72), 270(2.70)	65-2749-64
	H_2SO_4	222(3.80),240(3.53), 263(2.83)	65-2749-64
$C_7H_4ClHgN_3O$			
Benzoyl azide, p-(chloromercuri)-	THF	261(4.4)	70-0111-64
$C_7H_4ClNO_3$			
Tropone, 2-chloro-7-nitro-	EtOH	241(4.08),325(3.43)	88-1041-65
C_7H_4ClNS			
Benzothiazole, 2-chloro-	gas	212(4.26),216(4.26), 248(3.88),258(3.72), 293(3.10)	17-0079-65
	C_6H_{12}	221(4.30),234s(4.03), 254(3.94),263s(3.82), 296(3.18)	17-0079-65

Compound	Solvent	$\lambda_{max}(\log \epsilon)$	Ref.
Benzothiazole, 2-chloro- (cont.)	EtOH	256(3.92),260s(3.89), 295(3.13)	17-0079-65
$C_7H_4ClN_3O_2$			
Benzimidazole , 5-chloro-4-nitro-	EtOH	303-312(3.74)	44-0476-64
	pH 2	277(3.87)	44-0476-64
	pH 12	234(3.83),374(3.88)	44-0476-64
Imidazo[1,2-a]pyridine, 2-chloro-3-nitro-	pH 1	256(4.07),266(3.97), 303(3.60),366(4.11)	44-4085-65
	pH 11	256(4.12),266(4.04), 303(3.7),366(4.13)	44-4085-65
1,2,3,4-Oxatriazolone, 3-(p-chlorophenyl)-	EtOH	218(4.00),275(4.13)	39-0906-64
$C_7H_4Cl_2HgO$			
Benzoyl chloride, p-(chloromercuri)-	THF	253(4.3)	70-0111-64
$C_7H_4Cl_2N_2$			
Benzimidazole, 5,6-dichloro-	pH 2	245(3.50),281(3.84), 289(3.74)	44-0476-64
	pH 12	290(3.55)	44-0476-64
	EtOH	252(3.61),284(3.80), 293(3.70)	44-0476-64
Imidazo[1,2-a]pyridine, 2,3-dichloro-	pH 1	282(3.85)	44-4085-65
	pH 11	272s(3.69),281(3.74), 302(3.65)	44-4085-65
$C_7H_4Cl_2S$			
Benzoyl chloride, p-chloro-thio-	C_6H_{12}	233(4.02),239(4.01), 254s(3.83),324(4.15), 532(1.83)	24-0829-65
	MeCN	232(3.89),255(3.71), 327(4.21),522(1.85)	24-0829-65
$C_7H_4Cl_3NO$			
Benzamide, 2,3,6-trichloro-	HCl	258(3.63)	23-2328-65
	34% H_2SO_4	233s(3.00)	23-1957-64
	82% H_2SO_4	258(3.87)	23-1957-64
$C_7H_4Cl_4O_2$			
1,4-Dioxaspiro[4.4]nona-6,8-diene, 6,7,8,9-tetrachloro-	iso-PrOH	313(3.38)	25-0709-64
	iso-PrOH	313(3.38)	39-2305-65
C_7H_4FNS			
Benzothiazole, 2-fluoro-	gas	210(4.25),222s(3.96), 236(3.62),283(2.60)	17-0079-65
	C_6H_{12}	215(4.37),240(3.73), 286(2.79)	17-0079-65
	EtOH	239(3.72),286(2.81)	17-0079-65
$C_7H_4F_2N_2$			
Benzimidazole, 4,6-difluoro-	pH 1	240(3.58),264(3.62), 271(3.54)	87-0737-65
	pH 7	238(3.72),263(3.51), 268(3.49),272(3.50)	87-0737-65
	pH 13	257(3.71),269(3.74)	87-0737-65

Compound	Solvent	$\lambda_{max}(\log \epsilon)$	Ref.
$C_7H_4F_3NO_2S$			
Sulfide, p-nitrophenyl trifluoromethyl	hexane	242(3.91),272(3.87), 317(2.72)	65-2749-64
	EtOH	246(3.88),274(3.91), 332(2.68)	65-2749-64
	Et_2O	250(3.90),270(3.95), 315(2.98)	65-2749-64
	dioxan	250(3.82),270(3.84), 330(2.66)	65-2749-64
	H_2SO_4	212(3.57),298(3.82), 390(2.62)	65-2749-64
$C_7H_4F_3NO_3$			
Anisole, α,α,α-trifluoro-p-nitro-	hexane	254(4.03),310(2.78)	65-2749-64
	EtOH	258(3.90),319(2.65)	65-2749-64
	Et_2O	255(3.93),306(2.66)	65-2749-64
	dioxan	254(3.85),316(2.85)	65-2749-64
	H_2SO_4	216(3.16),290(3.78), 375(2.25)	65-2749-64
$C_7H_4F_3NO_3S$			
Sulfoxide, p-nitrophenyl trifluoromethyl	hexane	240(3.97),276(3.90), 318(2.68)	65-2749-64
	EtOH	242(3.89),274(3.90), 322(2.60)	65-2749-64
	Et_2O	241(3.96),275(3.92)	65-2749-64
	dioxan	243(3.86),278(3.88), 324(2.58)	65-2749-64
	H_2SO_4	264(3.70),330(2.58)	65-2749-64
$C_7H_4F_3NO_4S$			
Sulfone, p-nitrophenyl trifluoromethyl	hexane	242(4.16),284(3.32), 298(3.16),316(2.72)	65-2749-64
	EtOH	244(4.11),288(3.36), 324(2.58)	65-2749-64
	Et_2O	244(4.15),288(3.35), 315(2.90)	65-2749-64
	dioxan	244(4.15),289(3.48), 328(2.59),333s(2.55)	65-2749-64
	H_2SO_4	252(3.81),290(3.16), 322(2.63)	65-2749-64
C_7H_4INS			
Benzothiazole, 2-iodo-	C_6H_{12}	224(4.40),260(4.06), 299(3.41)	17-0079-65
	EtOH	262(4.06),296(3.46)	17-0079-65
$C_7H_4N_4$			
3,5-Pyridinedicarbonitrile, 2-amino-	EtOH	269.4(4.45)	44-2903-64
$C_7H_4N_4O_5$			
as-Triazine-3,5(2H,4H)-dione, 6-(5-nitro-2-furyl)-	H_2O	365(4.26)	87-0819-64
$C_7H_4N_4O_7$			
Benzamide, 2,4,6-trinitro-	52% H_2SO_4	224(4.36)	23-1957-64
	88% H_2SO_4	224(4.45)	23-1957-64

Compound	Solvent	λ max(log ε)	Ref.
$C_7H_4OS_2$			
3H-1,2-Benzodithiol-3-one	C_6H_{12}	350(3.5)	73-3016-65
$C_7H_4O_6$			
4H-Pyran-2,6-dicarboxylic acid, 4-oxo-	pH 1.4	223(4.06),273(3.94)	88-3431-64
	pH 5.0	223(4.10),274(3.99)	88-3431-64
	pH 12.4	385(4.38)	88-3431-64
	MeOH	270.5(3.93)	95-1227-64
$C_7H_4S_3$			
Benzo-1,2-dithia-4-cyclopentene-3-thione	C_6H_{12}	429(3.9)	73-3016-65
Benzo-1,3-dithia-4-cyclopentene-2-thione	C_6H_{12}	367(4.0+)	73-3016-65
$C_7H_5BrN_2$			
Benzimidazole, 4(7)-bromo-	n.s.g.	253(3.87),271(3.66), 280(3.56)	39-0915-64
Benzimidazole, 5(6)-bromo-	n.s.g.	235(3.99),280(3.78), 286(3.79)	39-0915-64
Imidazo[1,2-a]pyridine, 2-bromo-	pH 1	281(3.87)	44-4085-65
	pH 11	270s(3.67),274(3.68), 280(3.72),300(3.61)	44-4085-65
$C_7H_5Br_2NO_3$			
2,5-Cyclohexadien-1-one, 2,6-dibromo-4-methyl-4-nitro-	hexane	280(4.00)	70-1666-64
$C_7H_5ClHgO_2$			
Benzoic acid, p-(chloromercuri)-	91% THF	232(4.3)	70-0111-64
$C_7H_5ClN_2$			
Benzimidazole, 4(7)-chloro-	n.s.g.	248(3.63),253(3.86), 272(3.63),280(3.55)	39-0915-64
Benzimidazole, 5(6)-chloro-	n.s.g.	246(3.68),279(3.69), 284(3.63)	39-0915-64
Benzimidazole, 5-chloro-	EtOH	246(3.76),281(3.92), 286(3.83)	44-0476-64
	pH 2	240(3.59),275(3.93), 284(3.90)	44-0476-64
	pH 12	286(3.69)	
Imidazo[1,2-a]pyridine, 2-chloro-	pH 1	269s(3.96),275(3.99), 281s(3.92)	4-0053-65
	pH 11	272s(3.72),278(3.76), 292s(3.70)	4-0053-65
Imidazo[1,2-a]pyridine, 5-chloro-	pH 1	273s(3.85),283(3.94), 292s(3.84)	4-0053-65
	pH 11	273(3.73),281(3.78), 296s(3.76)	4-0053-65
Imidazo[1,2-a]pyridine, 7-chloro-	pH 1	280(3.94)	4-0053-65
	pH 11	274s(3.63),281(3.68), 303(3.60)	4-0053-65
$C_7H_5ClN_2OS$			
Xanthic acid, ethyl-, ester with 3-chloro-5-mercapto-4-isothiazolecarbonitrile	EtOH	269(3.93),277(3.92), 304(4.05)	44-0660-64

Compound	Solvent	λ max (log ε)	Ref.
$C_7H_5ClN_2O_2S$			
2H-1,2,4-Benzothiadiazine, 6-chloro-, 1,1-dioxide	HCOOH	269(3.9)	20-0344-65
$C_7H_5ClO_2$			
Benzaldehyde, 2-chloro-4-hydroxy-	aq. acid	207(4.15),229(4.11), 281(4.15)	44-2693-64
Salicylaldehyde, 3-chloro-	aq. acid	214(4.31),257(4.05), 333(3.48)	44-2693-64
Salicylaldehyde, 4-chloro-	aq. acid	214(4.26),264(4.22), 322(3.64)	44-2693-64
Salicylaldehyde, 5-chloro-	aq. acid	222(4.38),255(3.96), 337(3.51)	44-2693-64
Salicylaldehyde, 6-chloro-	aq. acid	218(4.14),266(3.99), 337(3.50)	44-2693-64
$C_7H_5ClO_3$			
Benzoic acid, 3-chloro-4-hydroxy-	HCl	255(4.09)	44-2678-65
anion	NaOH	276(4.17)	44-2678-65
4-Cyclopentene-1,3-dione, 2-acetyl-4-chloro-	C_6H_{12}	245(4.33),265s(4.19), 320(2.72)	1-0441-64
$C_7H_5ClO_4S_2$			
1,2-Benzothiolium perchlorate	HOAc-1% $HClO_4$	381(3.6)	73-3016-65
C_7H_5ClS			
Benzoyl chloride, thio-	C_6H_{12}	224s(3.90),228s(3.9), 243s(3.70),260s(3.6), 272s(3.5),313(4.14), 530(1.82)	24-0829-65
	MeCN	225(3.74),248s(3.5), 317(4.08),518(1.78)	24-0829-65
$C_7H_5Cl_2NO_3$			
Pyrrole-2-carboxylic acid, 3,4-dichloro-5-formyl-, methyl ester	EtOH	226(4.24),231(4.22), 293(4.36)	39-0459-65
	EtOH-base	246(4.23),323(4.44)	39-0459-65
$C_7H_5Cl_2N_3S$			
Imidazo[1,2-c]pyrimidine, 2,7-dichloro-5-(methylthio)-	EtOH	242(4.32),283(3.87), 312(3.78)	4-0034-64
$C_7H_5Cl_3O$			
m-Cresol, 2,4,6-trichloro-	C_6H_{12}	286(3.44),295(3.48)	2-0417-64
$C_7H_5FN_2$			
Benzimidazole, 4(7)-fluoro-	n.s.g.	242(3.71),263(3.37), 273(3.30)	39-0915-64
Benzimidazole, 5-fluoro-	pH 1	271(3.82),278(3.80)	87-0737-65
	pH 7	242(3.73),275(3.73)	87-0737-65
	pH 13	280(3.81)	87-0737-65
Benzimidazole, 5(6)-fluoro-	n.s.g.	242(3.72),275(3.73), 280(3.67)	39-0915-64
$C_7H_5F_2N_3$			
Benzimidazole, 4-amino-5,7-difluoro-	pH 1	267(3.51),285(3.51)	87-0737-65
	pH 7	251(3.76)	87-0737-65
	pH 13	261(3.81)	87-0737-65

Compound	Solvent	$\lambda_{max}(\log \epsilon)$	Ref.
$C_7H_5F_3O$			
Anisole, α,α,α-trifluoro-	hexane	259f(2.2)	65-2749-64
	EtOH	245-267(2.32-2.08)	65-2749-64
	Et_2O	245-268(1.98-2.28)	65-2749-64
	dioxan	250-268(2.47-2.25)	65-2749-64
	H_2SO_4	224(2.98),266(2.4)	65-2749-64
$C_7H_5F_3OS$			
Sulfoxide, phenyl trifluoromethyl	hexane	248(3.48),270(3.05), 286s(2.42)	65-2749-64
	EtOH	246(3.74),273(3.20), 286s(2.36)	65-2749-64
	Et_2O	248(3.60),272(3.02), 289s(2.42)	65-2759-64
	dioxan	246(3.55),273(3.05), 288s(2.25)	65-2749-64
	H_2SO_4	248(3.76),277(3.22), 300s(2.47)	65-2749-64
$C_7H_5F_3O_2S$			
Sulfone, phenyl trifluoromethyl	hexane	220(4.09),267(3.28), 290s(2.55)	65-2749-64
	EtOH	222(4.00),269(3.14)	65-2749-64
	Et_2O	222(4.06),269(3.22), 281s(2.60)	65-2749-64
	dioxan	222(4.02),269(3.20)	65-2749-64
	H_2SO_4	226(4.06),274(3.31), 303s(2.54)	65-2749-64
$C_7H_5F_3S$			
Sulfide, phenyl trifluoromethyl	hexane	240(3.35),268f(2.88)	65-2749-64
	EtOH	242(3.54),268(2.82)	65-2749-64
	Et_2O	250(3.27),268(2.68)	65-2749-64
	dioxan	240(3.38),268(2.82)	65-2749-64
	H_2SO_4	220(4.06),260(3.80), 303(2.82),323(2.30)	65-2749-64
$C_7H_5IN_2$			
Benzimidazole, 4(7)-iodo-	n.s.g.	255(4.00),274(3.82), 282(3.65)	39-0915-64
Benzimidazole, 5(6)-iodo-	n.s.g.	249(3.69),281(3.70), 288(3.65)	39-0915-64
C_7H_5NO			
2,1-Benzisoxazole (or anthranil)	pH -4.0	220(4.28),256(3.82), 276s(3.14),312(3.32)	39-5360-65
	pH 3.0	215s(3.58),247(3.07), 252s(3.12),257(3.22), 262(3.29),267(3.33), 273(3.32),279(3.44), 306(3.63)	39-5360-65
Cyanic acid, phenyl ester	C_6H_{12}	216(3.21),256(2.58), 262(2.75),268(2.67)	24-2689-64
Isocyanic acid, phenyl ester	C_6H_{12}	227(4.02),264(2.69), 271(2.75),278(2.68)	24-2689-64
$C_7H_5NO_2$			
Benzoxazolinone	EtOH	273(3.80),279s(3.72)	1-1051-65

Compound	Solvent	λ_{max}(log ϵ)	Ref.
$C_7H_5NO_3$			
Benzaldehyde, p-nitro-	C_6H_{12}	260.5(4.10)	44-2088-64
	MeOH	265(4.01)	44-2088-64
	iso-PrOH	263.5(4.13)	44-2088-64
	dioxan	264(4.08)	44-2088-64
$C_7H_5NO_4$			
Benzaldehyde, 3-hydroxy-4-nitro-	HCl	359(3.54)	35-4942-64
	NaOH	436(3.68)	35-4942-64
Benzene, 1,2-(methylenedioxy)-4-nitro-	EtOH	340(--)	95-0857-65
Benzoic acid, m-nitro-	C_6H_{12}	250(3.78)	44-2088-64
	MeOH	257(3.80)	44-2088-64
	iso-PrOH	256(3.86)	44-2088-64
	dioxan	255(3.92)	44-2088-64
Benzoic acid, p-nitro-	C_6H_{12}	253(3.69)	44-2088-64
	N HCl	262(4.14)	73-0940-65
	borate	272(4.04)	73-0940-65
	N NaOH	272(4.04)	73-0940-65
	MeOH	263.5(3.83)	44-2088-64
	MeOH	259(4.14)	73-0940-65
	iso-PrOH	259.5(4.15)	44-2088-64
	dioxan	257(3.70)	44-2088-64
Salicylaldehyde, 3-nitro-	EtOH	278(3.82),311(3.32), 350(3.57)	39-2816-64
Salicylaldehyde, 5-nitro-	EtOH	230(3.92),258(3.32), 310(4.00)	39-2816-64
$C_7H_5NO_4S$			
Benzoic acid, 2-nitro-5-mercapto-	pH 7	412(4.13)	28-3212-65A
$C_7H_5NO_5$			
Benzoic acid, 3-hydroxy-4-nitro-	HCl	356(3.53)	35-4942-64
	NaOH	426(4.38)	35-4942-64
C_7H_5NS			
Benzothiazole	gas	212(4.17),224s(3.92), 244(3.65),252(3.53), 291(3.33)	17-0079-65
	C_6H_{12}	227(4.25),228s(4.11), 258(3.62),286(3.79), 295(3.23)	17-0079-65
	H_2O	196(4.37),216(4.20), 250(3.70),256s(3.63), 276s(3.18),283(3.25), 292(3.16)	59-1881-65
	EtOH	251(3.77),258s(3.68), 295(3.16)	17-0079-65
$C_7H_5N_3$			
3,5-Pyridinedicarbonitrile, 1,2-dihydro-	EtOH	213(4.27),255(3.85), 382(3.66)	73-1654-64
3,5-Pyridinedicarbonitrile, 1,4-dihydro-	EtOH	206(4.18),352(3.81)	73-1654-64

Compound	Solvent	$\lambda_{max}(\log \epsilon)$	Ref.
1,3a,4-Triazaindene	C_6H_{12}	223s(4.26),226(4.28), 230(4.38),250s(3.15), 268s(3.08),278(3.02), 322s(3.46),326s(3.48), 329s(3.50),333(3.54), 337(3.55),346(3.49), 350(3.48),355s(3.35), 366(3.32)	39-2778-65
	pH 7.0	221(4.20),254s(3.24), 260s(3.20),271s(3.10), 325(3.59)	39-2778-65
$C_7H_5N_3O$			
4,7-Diazaindole-3-carboxaldehyde	EtOH	257(4.03),292(4.09)	44-3454-65
$C_7H_5N_3O_2$			
Benzimidazole, 5-nitro-	pH 2	228(3.95),280(3.95)	44-0476-64
	pH 12	249(3.93),363(3.94)	44-0476-64
	EtOH	236(3.99),306(3.89)	44-0476-64
Imidazol[1,2-a]pyridine, 3-nitro-	pH 1	241s(4.00),253(4.05), 257s(4.03),328(3.80), 364(3.77)	44-4085-65
	pH 11	248s(4.01),253(4.03), 261s(3.99),294(3.68), 366(4.18)	44-4085-65
Toluene, α-diazo-p-nitro-	EtOH	370(4.29)	44-1242-65
$C_7H_5N_3O_6$			
Anthranilic acid, 3,5-dinitro-	EtOH	237(4.09),336(4.11), 392(3.87)	44-0947-64
$C_7H_5N_5$			
7H-Pyrrolo[2,3-d]pyrimidine-5-carbonitrile, 4-amino-	pH 1	239(4.14),274(4.06)	35-1995-65
	pH 13	245(4.16),289(4.00)	35-1995-65
	EtOH	226(4.03),277(4.14), 287(3.99)	35-0951-64
	EtOH	226(4.03),277(4.14), 287(3.98)	35-1995-65
$C_7H_5N_5O_3$			
6-Pteridinecarboxylic acid, 2-amino-4-hydroxy-	pH 1	235(3.92),260(3.92), 310(3.87),330(3.86)	37-0560-64
	pH 13	261(4.21),360(3.86)	37-0560-64
$C_7H_5N_5S$			
Thiocyanic acid, 6-purinylmethyl ester	N HCl	265(3.90)	87-0667-65
	pH 7.65	272(3.90)	87-0667-65
	N NaOH	275(3.90)	87-0667-65
C_7H_6			
Benzocyclopropene	C_6H_{12}	270f(3.3)	88-3625-65
$C_7H_6ClF_4NO$			
2-Cyclopenten-1-one, 2-chloro-3-(ethylamino)-4,4,5,5-tetrafluoro-	EtOH	289(4.47)	44-3698-65

Compound	Solvent	$\lambda_{max}(\log \epsilon)$	Ref.
$C_7H_6ClNO_2$			
Benzohydroxamic acid, p-chloro-	MeOH	233(4.09)	73-0940-65
	50% MeOH-N HCl	239(4.15)	73-0940-65
	50% MeOH-borate	225(4.12),271(4.03)	73-0940-65
$C_7H_6ClNO_4S$			
Thiazolo[3,2-a]pyridinium perchlorate	n.s.g.	208(4.00),224(4.08),228s(4.02),295(4.11),306(4.24)	25-1801-64
$C_7H_6ClN_3S$			
7H-Pyrrolo[2,3-d]pyrimidine, 4-chloro-2-(methylthio)-	EtOH	222(4.00),249(4.46),272s(3.82),310(3.85)	4-0034-64
$C_7H_6ClN_3S_3$			
Carbamic acid, dimethyldithio-, ester with 3-chloro-5-mercapto-4-isothiazolecarbonitrile	EtOH	272(4.27),295(4.02)	44-0660-64
$C_7H_6ClN_5O$			
1,2,4-Triazolo[2,3-a]pyrimidine, 4-acetamido-6-chloro-	H_2O	293(4.04)	70-1475-64
$C_7H_6Cl_2N_4$			
Purine, 6-chloro-9-(β-chloroethyl)-	EtOH	263(4.00)	87-0182-65
$C_7H_6Cl_2O$			
Phenol, 4,6-dichloro-2-methyl-	C_6H_{12}	283(3.36),290(3.36)	2-0417-64
Phenol, 4,6-dichloro-3-methyl-	C_6H_{12}	284(3.50),293(3.49)	2-0417-64
$C_7H_6Cl_2O_2$			
Cyclopropanepropiolic acid, 2,2-dichloro-1-methyl-	n.s.g.	220(3.73)	22-1518-65
$C_7H_6Cl_3NO_2$			
Pyrrole-2-carboxylic acid, 3,4,5-trichloro-1-methyl-, methyl ester	EtOH	245s(3.83),272(4.10)	39-0459-65
$C_7H_6Cl_4O_2$			
2,4-Cyclopentadienone, 2,3,4,5-tetrachloro-, dimethyl acetal	iso-PrOH	308(3.37)	25-0709-64
	iso-PrOH	308(3.37)	39-2305-65
$C_7H_6F_2$			
1,3-Heptadien-5-yne, 3,4-difluoro-	MeOH	260(4.39)	44-2009-64
$C_7H_6F_3NO$			
p-Anisidine, α,α,α-trifluoro-	hexane	235(4.06),292(3.37)	65-2749-64
	EtOH	240(3.97),294(3.22)	65-2749-64
	Et_2O	241(3.98),298(3.23)	65-2749-64
	dioxan	242(4.05),297(3.30)	65-2749-64
	H_2SO_4	212(3.30),261(2.41)	65-2749-64
$C_7H_6F_3NOS$			
Sulfoxide, p-aminophenyl trifluoromethyl	hexane	243(3.68),266(4.18)	65-2749-64
	EtOH	280(4.21),290(4.18),295(4.11)	65-2749-64
	Et_2O	247(3.55),275(4.30)	65-2749-64
	dioxan	246(3.48),276(4.20)	65-2749-64
	H_2SO_4	236(3.80),267(3.19),272(3.16),297s(2.43)	65-2749-64

Compound	Solvent	λ_{max}(log ε)	Ref.
$C_7H_6F_3NO_2$			
Pyridine, 2,3,5-trifluoro-4,6-dimethoxy-	hexane	210s(4.0),271(3.59)	39-0575-65
	EtOH	209s(4.0),270(3.58)	39-0575-65
$C_7H_6F_3NO_2S$			
Benzenesulfonamide, p-[(trifluoromethyl)thio]-	hexane	266(4.37)	65-2749-64
	EtOH	286(4.23)	65-2749-64
	dioxan	280(4.38)	65-2749-64
	Et_2O	217(4.06),284(4.34)	65-2749-64
	H_2SO_4	218(4.04),250(2.61), 274(3.18)	65-2749-64
$C_7H_6F_3NS$			
Aniline, p-[(trifluoromethyl)thio]-	hexane	232(3.62),258(4.30), 290(3.32)	65-2749-64
	EtOH	262(4.23),290(3.55)	65-2749-64
	Et_2O	264(4.32),290(3.50)	65-2749-64
	dioxan	262(4.11),290(3.30)	65-2749-64
	H_2SO_4	246(3.48),276(2.98)	65-2749-64
$C_7H_6F_5NO$			
2-Cyclopenten-1-one, 3-(ethylamino)-2,4,4,5,5-pentafluoro-	EtOH	287(4.54)	44-3698-65
$C_7H_6N_2$			
Benzimidazole	$C_6H_{11}Me$	201(4.63),243(3.79), 275(3.67)	35-0011-65
	MeOH-HBr	244(3.72),248(3.69), 267(3.68),273(3.72), 278(3.58)	65-0632-64
	EtOH	243(3.68),272(3.82), 279(3.81)	44-0476-64
	pH 2	266(3.90),274(3.88)	44-0476-64
	pH 12	241(3.40),272(3.66), 278(3.67)	44-0476-64
	n.s.g.	242(3.72),270(3.67), 277(3.65)	39-0915-64
3a,6-Diazaindene	pH 2.0	217(4.42),222(4.36), 242(4.15),249s(4.02), 296(3.61),302(3.62), 358(3.54)	39-2778-65
	pH 10	225(4.39),235s(4.30), 274s(3.33),283(3.49), 293(3.48),339(3.43)	39-2778-65
Imidazo[1,2-a]pyridine	pH 1	250s(3.56),274(3.80)	4-0053-65
	pH 11	268s(3.67),277(3.70), 292(3.64)	4-0053-65
	EtOH	221(4.45),226(4.37), 278(3.59),298(3.53)	44-2403-65
hydrobromide	EtOH	218(4.32),226s(4.13), 278(3.75)	44-2403-65
Indazole	EtOH	284(3.73)	32-0814-65
Pyrrolo[1,2-c]pyrimidine	EtOH	229(4.44),272(3.74), 283(3.77),345(3.03)	44-2398-65
$C_7H_6N_2O$			
2,1-Benzisoxazole, 3-amino-	EtOH	367(3.81)	24-1562-65
	dioxan	364(3.79)	24-1562-65
	3:2 EtOH: dioxan	367(3.83)	24-1562-65

Compound	Solvent	λ_{max}(log ϵ)	Ref.
Benzonitrile, 2-hydroxylamino-	3:2 EtOH:-dioxan	249(3.85),320(3.53)	24-1562-65
Imidazo[1,2-a]pyridin-5(1H)-one	pH 1	237(3.87),296(4.08), 320s(3.23)	4-0053-65
	pH 11	280s(3.81),313(4.09)	4-0053-65
Nicotinonitrile, 1,6-dihydro-1-methyl-6-oxo-	EtOH	206(4.27),254(4.32), 306(3.89)	25-1524-64
$C_7H_6N_2OS$			
2,1,3-Benzothiadiazole, 4-methoxy-	EtOH	234(4.28),295(3.95), 301(3.95),307(4.02), 358(3.40)	7-0462-64
2,1,3-Benzothiadiazole, 5-methoxy-	EtOH	306(3.93),337(3.85)	7-0462-64
	50% EtOH-6N HCl	310(3.95),343(3.79)	7-0462-64
Benzothiazole, 2-amino-, 3-oxide	EtOH	285(4.00)	25-0368-64
5H-Thiazolo[3,2-a]pyrimidin-5-one, 7-methyl-	EtOH	260(3.95),315(4.16)	95-1201-64
$C_7H_6N_2O_2S$			
2H-1,2,4-Benzothiadiazine, 1,1-dioxide	H_2O	261(3.87)	20-0136-65
	1% HOAc	261(3.87)	20-0136-65
2H-1,3-Thiazine-5-carbonitrile, 3-acetyl-3,6-dihydro-2-oxo-	EtOH	233(4.09)	44-1740-64
$C_7H_6N_2O_3$			
Benzamide, 3-nitro-	H_2O	215(4.06)	23-2328-65
	HCl	230(4.24)	23-2328-65
	12% H_2SO_4	215(3.97)	23-1957-64
	72% H_2SO_4	230(4.33)	23-1957-64
Benzamide, 4-nitro-	MeOH	261(4.05)	73-0940-65
$C_7H_6N_2O_4$			
Benzohydroxamic acid, p-nitro-	MeOH	263(4.02)	73-0940-65
	50% MeOH-HCl	267(4.06)	73-0940-65
	50% MeOH-borate	233(4.02),337(3.86)	73-0940-65
	50% MeOH-NaOH	230(3.85),261(3.86), 341(3.70)	73-0940-65
$C_7H_6N_2O_4S_2$			
3H-1,2,3-Benzodithiazole, 6-methoxy-4-nitro-, 2-oxide	EtOH	212(4.41),307(3.57), 398(3.58)	44-2763-65
$C_7H_6N_2O_5$			
Anisole, 2,4-dinitro-, monoanion	MeOH-KOMe	340(4.12),500(4.41)	28-5783-65A
dianion	MeOH-KOMe	305(4.26)	28-5783-65A
Anisole, 2,6-dinitro-, monoanion	MeOH-KOMe	300(3.90),350(3.48), 595(4.4)	28-5783-65A
dianion	MeOH-KOMe	305(4.26)	28-5783-65A
$C_7H_6N_2S$			
2,1,3-Benzothiadiazole, 4-methyl-	EtOH	225(4.21),304(4.07), 311(4.10)	7-0462-64
2,1,3-Benzothiadiazole, 5-methyl-	EtOH	225(4.15),309(4.09)	7-0462-64
Benzothiazole, 4-amino-	H_2O	238(4.0),260s(3.7), 315(3.4)	39-2258-65

Compound	Solvent	λ_{max}(log ε)	Ref.
$C_7H_6N_4O_2$			
Benzotriazole, 1-methyl-4-nitro-	EtOH	215(3.88),304(3.98)	78-0211-64
Benzotriazole, 2-methyl-4-nitro-	EtOH	308(3.98)	78-0211-64
Benzotriazole, 2-methyl-5-nitro-	EtOH	248(4.30),289(4.04)	78-0211-64
Lumazine, 8-methyl-	0.1N H_2SO_4	256(4.19),392(3.95)	37-3493-64
	0.1N NaOH	227(4.27),280(4.09), 305(3.89),400(3.08)	37-3493-64
Pteridin-4(7)-one, 1,4(or 7)-dihydro-7(or 4)-hydroxy-1-methyl-	pH 1.3	230(4.28),239s(4.25), 295(3.88),325(3.97)	39-3770-65
	pH 5.7	226(4.35),244(4.30), 250s(4.25),280(3.56), 324s(4.05),333(4.11), 346(3.98)	39-3770-65
5H-Pyrrolo[3,2-d]pyrimidine-7-carboxylic acid, 4-amino- (also shoulders, not listed)	pH 1	233(4.43),270(4.16)	44-1528-65
	pH 7	232(4.36),272(4.11)	44-1528-65
	pH 13	244(4.24)	44-1528-65
7H-Pyrrolo[2,3-d]pyrimidine-5-carboxylic acid, 4-amino-	pH 1	228(4.01),240s(3.96), 274(4.05)	35-0951-64 and 35-1995-65
	pH 13	226(4.01),259(3.98), 280(4.04)	35-0951-64
$C_7H_6N_4O_3$			
Lumazine, 1-methyl-, 5-oxide	pH 6.0	239(4.30),281(3.77), 344(3.71)	89-1136-65
	pH 11.0	245(4.33),283(3.89), 348(3.77)	89-1136-65
$C_7H_6N_6O_2$			
6-Pteridinecarboxylic acid, 2,4-diamino-	pH -1.5	248s(4.03),336(3.96)	39-1530-65
	pH 7.7	260(4.26),382(4.13)	39-1530-65
7-Pteridinecarboxylic acid, 2,4-diamino-	pH -1.5	248(3.94),295s(3.64), 350(3.98)	39-1530-65
	pH 8.0	260(4.25),380(3.90)	39-1530-65
C_7H_6O			
Bicyclo[3.2.0]hepta-3,6-dien-2-one	C_6H_{12}	208(3.83),349(1.99)	35-1623-65
2,4-Heptadien-6-ynal	Et_2O	283(4.61),295(4.59)	24-0809-64
Tetracyclo[2.2.1.0^{2,6}.0^{3,5}]heptan-7-one	C_6H_{12}	297(1.60)	35-1270-64
	H_2O	293(1.92)	35-1270-64
	EtOH	296(1.82)	35-1270-64
$C_7H_6O_2$			
Benzaldehyde, 4-hydroxy-	aq. acid	221(4.06),283(4.20)	44-2693-64
	acid	284(4.18)	44-0603-65
	base	330(4.44)	44-0603-65
Benzoic acid	pet ether	230(4.14),274(3.04), 283(2.95)	54-0949-64
Salicylaldehyde	aq. acid	210(4.24),255(4.05), 324(3.48)	44-2693-64
	10% MeOH	255(4.0),322(3.5)	20-0518-65
Toluquinone	EtOH	246(4.14),314(2.77), 429(1.28)	44-4107-65
	$CHCl_3$	249(4.33),315(2.80), 436(1.38)	44-4107-65
p-Toluquinone	EtOH	246(4.24),309(2.79), 400s(1.48)	24-1926-64

Compound	Solvent	λ max (log ε)	Ref.
$C_7H_6O_3$			
2-Buten-4-olide, 4-acetonylidene-	C_6H_{12}	290(4.06)	1-0441-64
4-Cyclopentene-1,3-dione, 4-acetyl-	C_6H_{12}	223(4.15),259(4.14), 317(2.91)	1-0441-64
1,4-Toluquinone, 6-hydroxy-	$CHCl_3$	268(4.21),380(2.87)	24-2774-65
$C_7H_6O_4$			
Benzoic acid, 2,3-dihydroxy-	EtOH	245(3.93),316(3.59)	78-1725-64
7-Oxabicyclo[4.1.0]hept-3-ene-2,5-	C_6H_{12}	307(3.82)	39-6587-65
dione, 3-hydroxy-4-methyl-	pH 3	315.5(3.82)	39-6587-65
	pH 8.5	254(3.80),311(3.67), 386(3.77)	39-6587-65
Patulin	pH 1	276(4.17)	10-0156-64A
	pH 13	291(4.13)	10-0156-64A
α-Pyrone, 3-carboxy-6-methyl-	EtOH	317(3.93)	78-2701-64
α-Pyrone, 6-carboxy-3-methyl-	pH 1.4	239(3.48),301(4.06)	88-3431-64
	pH 5.0	234(3.53),302(4.01)	88-3431-64
	pH 12.4	330(4.13+)	88-3431-64
α-Pyrone, 6-carboxy-4-methyl-	pH 1.4	230(3.36),300(3.90)	88-3431-64
	pH 5.0	228(3.34),300(3.89)	88-3431-64
	pH 12.4	258(3.92)	88-3431-64
α-Pyrone, 6-carboxy-5-methyl-	pH 1.4	232(3.43),305(3.87)	88-3431-64
	pH 5.0	228(3.57),308(3.87)	88-3431-64
	pH 12.4	345(4.17+)	88-3431-64
α-Pyrone-2-carboxylic acid,	EtOH-HCl	259(3.90)	39-2251-65
methyl ester	EtOH-NaOH	260(3.81),352(4.22)	39-2251-65
$C_7H_6O_5S_2$			
1,3-Dithiole-4,5-dicarboxylic acid, 2-oxo-, dimethyl ester	EtOH	256(3.61),294(3.67)	24-1298-64
$C_7H_6S_2$			
Thieno[3,2-b]thiophene, 3-methyl-	n.s.g.	209(3.4),259(4.1), 270(4.1),280(4.0)	70-0510-65
$C_7H_7^+$			
Tropylium ion	H_2SO_4	274(3.7)	73-3016-65
$C_7H_7BCl_4$			
Cycloheptatrienylium tetrafluoro-borate	96% H_2SO_4	274(3.64)	35-5511-64
$C_7H_7BF_4$			
Cycloheptatrienylium tetrafluoro-borate, 1:1 complex with carbazole	MeCN	430(3.55)	77-0064-65
$C_7H_7BI_4$			
Cycloheptatrienylium tetraiodoborate	96% H_2SO_4	274(3.64)	35-0539-65
	H_2O	275(3.64)	35-0539-65
$C_7H_7BO_2$			
Boronophthalide	EtOH	263(2.85),268(3.03), 275(3.05)	44-3229-64
$C_7H_7BrN_2S$			
Methyl-1,2,3-benzothiadiazolium bromide	H_2O	233(4.11),298(3.81), 331(3.48)	39-6061-64
	$CHCl_3$	247(3.86),301(3.87), 340(3.73)	39-6061-64

Compound	Solvent	$\lambda_{max}(\log \epsilon)$	Ref.
$C_7H_7BrN_4$			
Pyrazolo[1,5-a]pyrimidine, 7-amino-3-bromo-5-methyl-	EtOH	226(4.38),281s(3.78), 291(3.91),307(3.82)	95-1113-64
C_7H_7BrO			
Anisole, m-bromo-	EtOH	275(--),282(--)	95-0858-65
Anisole, p-bromo-	EtOH	281(--),288(--)	95-0858-65
$C_7H_7BrO_3$			
4H-Pyran-4-one, 2-(bromomethyl)-3-methoxy-	C_6H_{12}	220(3.76),268(3.92)	78-0093-65
$C_7H_7Br_2N$			
Pyridine, 4-(1,1-dibromoethyl)-	EtOH-HBr	264(3.43)	23-0698-64
$C_7H_7Br_2NO_2$			
Pyrrole-2-carboxylic acid, 4,5-dibromo-1-methyl-, methyl ester	EtOH	240(3.81),279(4.11)	39-0459-65
$C_7H_7ClF_4N_2$			
1-Cyclopenten-1-ylamine, 2-chloro-4,4,5,5-tetrafluoro-N-methyl-3-(methylimino)-	pH 1	318(4.57)	44-3698-65
	pH 13	293(4.40)	44-3698-65
	EtOH	290(4.45)	44-3698-65
$C_7H_7ClNO_2$			
Norcamphor, 7-chlorodiazo-	EtOH	249(4.04),298(3.54)	44-3469-64
$C_7H_7ClN_2$			
Benzaldehyde, p-chloro-, hydrazone	$CHCl_3$	279(4.29)	44-0417-65
$C_7H_7ClN_2O$			
Imidazo[1,2-a]pyridine, 3-chloro-, hydrate	pH 1	280(3.87)	44-4085-65
	pH 11	269s(3.66),273(3.67), 280(3.72),302(3.58)	44-4085-65
$C_7H_7ClN_2S$			
Methyl-1,2,3-benzothiadiazolium chloride	12N HCl	234(4.10),301(3.76), 333(3.49)	39-6061-64
	H_2O	233(4.13),297(3.76), 330(3.49)	39-6061-64
C_7H_7ClO			
Norcamphor, dehydro-3-chloro-, endo	EtOH	312(2.35)	35-4074-64
exo	isooctane	317(2.23)	35-4074-64
	EtOH	312(2.40)	35-4074-64
Phenol, 4-chloro-2-methyl-	n.s.g.	281(3.21),289(3.15)	2-0417-64
Phenol, 4-chloro-3-methyl-	C_6H_{12}	281(3.33),289(3.28)	2-0417-64
Phenol, 6-chloro-2-methyl-	C_6H_{12}	273(3.26),278(3.30)	2-0417-64
Phenol, 6-chloro-3-methyl-	C_6H_{12}	278(3.31),284(3.35)	2-0417-64
$C_7H_7Cl_2NO$			
3-Pyrrolin-2-one, 5-(dichloromethylene)-3,4-dimethyl-	EtOH	286.5(4.29)	39-5999-64
$C_7H_7Cl_2NO_2$			
Pyrrole-2-carboxylic acid, 3,5-dichloro-1-methyl-, methyl ester	EtOH	266(4.18),272s(4.15)	39-0459-65
Pyrrole-2-carboxylic acid, 4,5-dichloro-1-methyl-, methyl ester	EtOH	237(3.76),274(4.10)	39-0459-65

Compound	Solvent	$\lambda_{max}(\log \epsilon)$	Ref.
$C_7H_7Cl_2N_3OS$			
Pyrimidine, 4-chloro-6-(2-chloroacetamido)-2-(methylthio)-	EtOH	242(4.39),293(3.68)	4-0034-64
$C_7H_7Cl_3N_2$			
Pyrimidine, 2,4,6-trichloro-5-isopropyl-	MeOH	268(3.75)	87-0808-64
Pyrimidine, 4,5,6-trichloro-2-isopropyl-	MeOH	268(3.73)	87-0808-64
Pyrimidine, 2,4,6-trichloro-5-propyl-	MeOH	268(3.77)	87-0808-64
Pyrimidine, 4,5,6-trichloro-2-propyl-	MeOH	268(3.74)	87-0808-64
C_7H_7CsO			
Cesium p-cresoxide	DMF	337(3.62)	35-1857-65
	$(MeOCH_2)_2$	328(3.49)	35-1857-65
$C_7H_7CsO_2$			
Cesium, (p-methoxyphenoxy)-	DMF	348.5(3.64)	35-1857-65
	$(MeOCH_2)_2$	338(3.51)	35-1857-65
$C_7H_7FN_2O_4$			
Hydantoin, 5-(ethoxycarbonylfluoromethylene)-	pH 1	235(3.74),302(4.08)	65-4184-64
	pH 7	295(3.99)	65-4184-64
	pH 13	264(4.07)	65-4184-64
$C_7H_7FO_2S$			
p-Toluenesulfonyl fluoride	$C_2H_4Cl_2$	230(3.99),270(3.00)	30-0815-65
$C_7H_7FO_3S$			
Benzenesulfonyl fluoride, p-methoxy-	$C_2H_4Cl_2$	245(4.20)	30-0815-65
$C_7H_7F_3N_2O_2$			
Crotonic acid, 3-amino-2-cyano-4,4,4-trifluoro-, ethyl ester	EtOH	284(4.26)	44-0707-64
Pyrimidine, 2,4-dimethoxy-5-(trifluoromethyl)-	MeOH	216(3.99),257(3.73)	44-0835-65
$C_7H_7F_5N_2$			
1-Cyclopenten-1-ylamine, 2,4,4,5,5-pentafluoro-N-methyl-3-(methylimino)-	EtOH	287(4.34)	44-3698-65
	pH 1	316(4.58)	44-3698-65
	pH 13	292(4.52)	44-3698-65
C_7H_7I			
Toluene, 2-iodo-	gas	209f(--)	60-0597-65
Toluene, 3-iodo-	gas	207f(--)	60-0597-65
Toluene, 4-iodo-	gas	211f(--)	60-0597-65
$C_7H_7IN_2S$			
Methyl-1,2,3-benzothiadiazolium iodide	H_2O	230(4.39),297(3.79), 330(3.52)	39-6061-64
	$CHCl_3$	250(4.03),294(3.93), 332(3.77),495(3.44)	39-6061-64
Methyl-2,1,3-benzothiadiazolium iodide	EtOH	214(4.16),316s(4.03), 324(4.05),357s(3.42)	39-6061-64
	$CHCl_3$	246(3.99),311(4.05), 369(3.73),535(3.94)	39-6061-64

Compound	Solvent	$\lambda_{max}(\log \epsilon)$	Ref.
C_7H_7IO			
Anisole, p-iodo-	gas	200f(--)	60-0597-65
$C_7H_7IO_3$			
Furan, 2-acetyl-3-hydroxy-4-iodo-5-methyl-	EtOH	290(4.18)	39-2543-65
C_7H_7KO			
Potassium p-cresoxide	$(MeOCH_2)_2$	323(3.57)	35-1857-65
$C_7H_7KO_2$			
Potassium, (p-methoxyphenoxy)-	$(MeOCH_2)_2$	334(3.59)	35-1857-65
C_7H_7LiO			
Lithium, p-cresoxide	DMF	315(3.40)	35-1857-65
	$(MeOCH_2)_2$	304(3.04)	35-1857-65
$C_7H_7LiO_2$			
Lithium, (p-methoxyphenoxy)-	DMF	323(3.42)	35-1857-65
C_7H_7N			
Cyclopentadieneacetonitrile	EtOH	244(3.55)	44-3430-64
C_7H_7NO			
Benzamide	C_6H_{12}	233(3.9),268f(2.7)	56-0789-64
	H_2O	225(4.0),268(2.8)	56-0789-64
Formanilide	$HCONH_2$	282s(3.0)	20-0741-64
Pyridine, 2-acetyl-	0.9N HCl	270(3.9),280s(3.8)	32-0533-65
	0.1N NaOH	270(3.7),280s(3.6)	32-0533-65
Pyridine, 3-acetyl-	0.9N HCl	260s(3.6),265(3.7), 270s(3.6)	32-0533-65
	pH 7.5	270(3.5),280s(3.5)	32-0533-65
Pyridine, 4-acetyl-	0.9N HCl	280(3.6)	32-0533-65
	0.01N KOH	285(3.4)	32-0533-65
C_7H_7NOS			
Benzyl thionitrite	hexane	340(3.01),560(1.42)	77-0248-65
$C_7H_7NO_2$			
Aniline, 3,4-(methylenedioxy)-	EtOH	305(--)	95-0857-65
Anthranilic acid	H_2O	244(3.7),327(3.3)	24-1127-64
anion	H_2O	240(3.85),310(3.45)	24-1127-64
Azepine-N-carboxylic acid, potassium salt	MeOH	339(2.73)	88-1733-64
Toluene, 2-nitro-	C_6H_{12}	250(3.84)	44-2088-64
	MeOH	255(3.69)	44-2088-64
	dioxan	254(3.73)	44-2088-64
Toluene, 3-nitro-	C_6H_{12}	257(3.85)	44-2088-64
	MeOH	264(3.89)	44-2088-64
	iso-PrOH	263.5(3.90)	44-2088-64
	dioxan	264.5(3.91)	44-2088-64
Toluene, 4-nitro-	C_6H_{12}	266(4.02)	44-2088-64
	MeOH	274.5(4.00)	44-2088-64
	iso-PrOH	273.5(4.01)	44-2088-64
	dioxan	274.5(3.99)	44-2088-64

Compound	Solvent	$\lambda_{max}(\log \varepsilon)$	Ref.
$C_7H_7NO_2S$			
Sulfide, methyl p-nitrophenyl	hexane	225(3.88),265(3.03), 325(4.26),378s(2.50)	65-2749-64
	EtOH	226(3.80),268(2.88), 341(4.11),400s(2.88)	65-2749-64
	Et_2O	224(3.86),333(4.19)	65-2749-64
	dioxan	222(3.79),336(4.12)	65-2749-64
	H_2SO_4	235(3.70),305(3.53), 385(3.78),520(4.10)	65-2749-64
$C_7H_7NO_2S_3$			
1,2-Dithiole-3-thione, 4-aceto-acetamido-	EtOH	212(4.16),237(4.08), 277(3.89),416(4.13)	12-0447-64
$C_7H_7NO_3$			
Anisole, 2-nitro-	C_6H_{12}	250(3.67),306(3.53)	44-2088-64
	MeOH	257(3.73),321(3.50)	44-2088-64
	dioxan	255.5(3.75),320(3.55)	44-2088-64
	EtOH	317(--)	95-0858-65
Anisole, 3-nitro-	C_6H_{12}	263(3.85),320(3.32)	44-2088-64
	MeOH	267(3.48),325(3.06)	44-2088-64
	EtOH	325(--)	95-0858-65
	iso-PrOH	267(3.45),321(3.05)	44-2088-64
	dioxan	269(3.67),331(3.30)	44-2088-64
Anisole, 4-nitro-	C_6H_{12}	294(4.09)	44-2088-64
	MeOH	305.5(3.98)	44-2088-64
	iso-PrOH	305(4.12)	44-2088-64
	dioxan	305(4.20)	44-2088-64
Anthranilic acid, 3-hydroxy-	pH 7.3	314(3.51)	37-0740-65
Nicotinic acid, 1,6-dihydro-1-methyl-6-oxo-	EtOH	206(4.14),255(4.20), 300(3.72)	25-1524-64
Phenol, 5-methyl-2-nitro-	HCl	348(3.62)	35-4942-64
	NaOH	420(3.73)	35-4942-64
2(1H)-Pyridone, 1-acetate	EtOH	228(3.83),302(3.74)	35-5186-65
Pyrrole-2-carboxylic acid, 5-formyl-, methyl ester	EtOH	220(4.02),224(4.00), 296(4.38)	39-0459-65
	EtOH-base	245(4.02),326(4.43)	39-0459-65
$C_7H_7NO_3S_2$			
Acetoacetamide, N-(3-oxo-3H-1,2-dithiol-4-yl)-	EtOH	234(3.61),327(3.52)	12-0447-64
$C_7H_7NO_4$			
Phenol, 3-methoxy-4-nitro-	EtOH	285s(3.68),328(3.84)	44-1821-64
	EtOH-NaOH	263(3.61),394(4.33)	44-1821-64
Phenol, 5-methoxy-2-nitro-	C_6H_{12}	307(3.93),340(3.95)	44-1821-64
	HCl	345(4.04)	35-4942-64
	0.25N NaOH	317(3.80),403(3.88)	44-1821-64
	NaOH	408(3.89)	35-4942-64
Pyrrole-2,4-dicarboxylic acid, methyl ester	EtOH	264(4.14)	23-0409-64
Pyrrole-2,5-dicarboxylic acid, methyl ester	EtOH	274(4.39)	23-0409-65
$C_7H_7NO_4S$			
2-Thiazoline-2-acetic acid, 4,5-dioxo-, ethyl ester	dioxan	210(4.1),243(4.1), 315(4.1)	83-0124-65
2-Thiophenecarboxylic acid, 5-ethyl-4-nitro-	EtOH	245(4.3),270s(4.0)	70-2055-64

Compound	Solvent	λ_{max}(log ε)	Ref.
$C_7H_7NO_5$			
4,5-Oxazoledicarboxylic acid, dimethyl ester	n.s.g.	241.5(3.95)	24-1414-64
$C_7H_7N_3$			
Imidazo[1,2-a]pyridine, 5-amino-	pH 1	264(3.81),269s(3.80), 312(4.12)	4-0053-65
	pH 11	300(4.04)	4-0053-65
Imidazo[1,2-a]pyrimidine, 2-methyl-	EtOH	232(4.21),290(3.49), 323(3.56)	94-0813-64
Nicotinonitrile, 1,6-dihydro-6-imino-1-methyl-	pH 7.0	256(4.27),302(3.70)	39-5542-65
	pH 13.0	267s(4.31),273(4.41), 282(4.34),325(3.40)	39-5542-65
1,3,3a-Triazaindene, 2-methyl-	pH 0.0	237(3.67),246s(3.62), 264(3.49),269(3.52), 275(3.48),291(3.10)	39-2778-65
	pH 7.0	238s(3.45),260(3.56), 273s(3.53),283s(3.34)	39-2778-65
1,4,6-Triazaindene, 2-methyl-	pH 3	230s(4.2),270(3.7), 325(3.6)	94-1030-64
	pH 11	220s(4.3),280(3.8), 310s(3.5)	94-1030-64
	MeOH	220s(4.2),280(3.7), 310s(3.5)	94-1030-64
$C_7H_7N_3O$			
1H-Pyrazolo[4,3-c]pyridin-3-ol, 1-methyl-	N NaOH	230(4.22),288(3.41)	12-0379-65
1,4,6-Triazainden-7(6H)-one, 2-methyl-	pH 2	240(4.4),275s(3.9)	94-1030-64
	pH 12	230(4.5),270(4.0)	94-1030-64
	MeOH	235(4.5),270(4.0)	94-1030-64
Triazolo[4,3-a]pyridin-3-one, 1-methyl-	EtOH	235(4.15),281(3.54), 341(3.55)	7-0935-65
$C_7H_7N_3OS$			
2,1,3-Benzothiadiazole, 4-amino-5-methoxy-	EtOH	254(4.30),314(3.87), 320(3.87),437(3.29)	7-0462-64
	50% EtOH-6N HCl	223(4.20),315(4.05), 345(3.81)	7-0462-64
2,1,3-Benzothiadiazole, 5-amino-4-methoxy-	EtOH	240(4.26),311(3.83), 412(3.63)	7-0462-64
	50% EtOH-6N HCl	233(4.28),298(3.96), 304(3.98),310(4.04) 344(3.38)	7-0462-64
2,1,3-Benzothiadiazole, 6-amino-4-methoxy-	EtOH	245(4.39),320(3.78), 412(3.67)	7-0462-64
	50% EtOH-6N HCl	236(4.33),295(3.98), 301(3.99),307(4.04)	7-0462-64
2,1,3-Benzothiadiazole, 7-amino-4-methoxy-	EtOH	257(4.28),303(3.54), 453(3.26)	7-0462-64
	50% EtOH-6N HCl	236(4.26),295(3.95), 300(3.96),307(4.02)	7-0462-64
4H-Pyrrolo[2,3-d]pyrimidin-4-one, 3,7-dihydro-2-(methylthio)-	pH 1	273(4.14)	4-0034-64
	pH 11	229(4.30),275(4.15)	4-0034-64

Compound	Solvent	$\lambda_{max}(\log \epsilon)$	Ref.
$C_7H_7N_3O_2$			
Benzaldehyde, p-nitro-, hydrazone	$CHCl_3$	332(4.10)	44-0417-65
Imidazo[1,2-c]pyrimidine-2,5(1H,3H)-dione, 3-methyl-	pH 1	237(3.67),299(4.16)	44-1762-64
	pH 4.2	303(4.27)	44-1762-64
	pH 9.73	318(4.30)	44-1762-64
1H-Pyrazolo[4,3-c]pyridin-3-ol, 1-methyl-, 5-oxide	N NaOH	265(4.3),321(3.55)	12-0379-65
2,6-Pyridinedicarboxaldehyde, dioxime	pH <1	227(4.35),301(3.86)	39-1149-65
	pH 4-8	227(4.38),296(3.90)	39-1149-65
	pH over 12	208(3.86),275(4.35), 310s(4.13)	39-1149-65
ferrous complex	pH 5.8	453(3.98),580(3.83)	39-1149-65
	pH 10	467(4.03),560(3.95)	39-1149-65
Pyrimido[1,2-c]pyrimidine, 1,2,3,4-tetrahydro-2,6-dioxo-	pH 1	240(3.82),304(4.18)	44-1762-64
	pH 6.17	312(4.23)	44-1762-64
	pH 9.17	227(3.99),325(4.37)	44-1762-64
$C_7H_7N_3O_2S$			
Imidazo[1,2-c]pyrimidine-2,7(1H,3H)-dione, 5-(methylthio)-	pH 1	234(4.24),276(4.08)	4-0034-64
	pH 11	232(4),289(4)	4-0034-64
5H-Pyrimido[4,5-b][1,4]thiazin-6(7H)-one, 4-methoxy-	EtOH	236(4.18),246s(4.13), 283s(3.70),295(3.74)	44-2121-64
Thiazolo[5,4-d]pyrimidine-5,7(4H,6H)-dione, 4,6-dimethyl-	H_2O	283(4.67)	94-1319-65
$C_7H_7N_3O_3$			
Aniline, N-methyl-4-nitro-2-nitroso-	n.s.g.	287(4.25),358(4.11), 449(3.76),685(1.82)	39-2829-64
Benzamidoxime, p-nitro-	MeOH	232(3.92),257(3.93), 328(3.71)	73-0940-65
	50% MeOH-N NaOH	265(4.01),440(3.97)	73-0940-65
$\Delta^{2,\alpha}$-Imidazolidineacetic acid, α-cyano-4-oxo-, methyl ester	EtOH	274.5(4.43)	95-0387-65
Pyrazinepyruvic acid, oxime	pH 1	268(3.88)	4-0001-65
	pH 11	255(4.01)	4-0001-65
3-Pyridazinepyruvic acid, oxime	pH 1	243(3.62)	4-0001-65
	pH 11	245(4.02)	4-0001-65
4-Pyridazinepyruvic acid, oxime	pH 11	245(4.03)	4-0001-65
2-Pyrimidinepyruvic acid, oxime	pH 1	248(3.79)	4-0001-65
	pH 11	245(4.01)	4-0001-65
4-Pyrimidinepyruvic acid, oxime	pH 1	242(3.79)	4-0001-65
	pH 11	245(4.04)	4-0001-65
$C_7H_7N_3O_4$			
Aniline, N-methyl-2,4-dinitro-	MeOH	348(4.21),415s(3.80)	35-4018-64
2-Furanacrylic acid, 5-nitro-, hydrazide	propylene glycol	240(4.20),272(4.02), 356(4.28)	95-0212-64
$C_7H_7N_3O_5$			
Glycine, 5-uracoyl-	pH 1	220(4.06),271(4.03)	87-0001-64
	pH 13	240(4.01),291(4.16)	87-0001-64
$C_7H_7N_3O_6$			
1(2H)-Pyrimidinepropionic acid, 3,4-dihydro-5-nitro-2,4-dioxo-	pH 1	240(3.86),307(4.00)	87-0187-65
	pH 7	242(3.86),313(3.99)	87-0187-65
	pH 13	241s(3.82),324(3.99)	87-0187-65

Compound	Solvent	$\lambda_{max}(\log \epsilon)$	Ref.
$C_7H_7N_3S$			
2,1,3-Benzothiadiazole, 4-amino-5-methyl-	EtOH	251(4.32),304(3.85), 309(3.83),317(3.93), 425(3.28)	7-0462-64
	50% EtOH-6N HCl	224(4.22),311(4.20), 316(4.20)	7-0462-64
2,1,3-Benzothiadiazole, 5-amino-4-methyl-	EtOH	237(4.28),308(3.86), 403(3.71)	7-0462-64
	50% EtOH-6N HCl	225(4.24),306(4.12), 311(4.13)	7-0462-64
2,1,3-Benzothiadiazole, 6-amino-4-methyl-	EtOH	238(4.31),311(3.81), 400(3.75)	7-0462-64
	50% EtOH-6N HCl	226(4.21),306(4.07), 311(4.09)	7-0462-64
2,1,3-Benzothiadiazole, 7-amino-4-methyl-	EtOH	253(4.26),303(3.73), 315(3.74),433(3.33)	7-0462-64
	50% EtOH-6N HCl	225(4.22),303(4.08), 309(4.11)	7-0462-64
$C_7H_7N_5$			
Pteridine, 1,7-dihydro-7-imino-1-methyl-	pH 6.0	218(4.24),222s(4.21), 233s(4.01),260s(3.60), 349(4.12)	39-1175-65
Pteridine, 3,4-dihydro-4-imino-3-methyl-	pH 7.0	236(4.03),311s(3.78), 320(3.79),330s(3.62)	39-3770-65
Pteridine, 3,7-dihydro-7-imino-3-methyl-	pH 4.6	215(4.22),233s(4.01), 296(3.93),335(4.17)	39-1175-65
Pteridine, 7-(methylamino)-	pH 0.2	215(4.21),235s(3.91), 310s(4.03),337(4.14)	39-1175-65
	pH 4.8	223(4.19),232s(4.19), 270(3.92),278s(3.88), 343(4.05)	39-1175-65
$C_7H_7N_5O$			
7-Pteridinol, 2-(methylamino)-	pH -0.2	228(4.16),264(3.86), 286s(3.97),336(3.89)	39-3770-65
	pH 5.0	215(4.33),238(4.05), 294(3.71),355(4.21)	39-3770-65
	pH 10.3	230(4.58),278(3.83), 351(4.18),359s(4.15)	39-770-65
4(1H)-Pteridinone, 7-amino-1-methyl-	pH 0.1	237(4.36),243s(4.33), 288s(3.92),293(3.92), 340(4.00)	39-1175-65
	pH 4.6	231(4.36),248(4.21), 254s(4.18),279(3.57), 340(4.12),355s(3.95)	39-1175-65
4(8H)-Pteridinone, 7-amino-8-methyl-	pH 4.0	219(4.33),227s(4.22), 250(3.81),258s(3.73), 283(3.65),292(3.68), 340(4.00)	39-1175-65
	pH 8.5	220(4.32),226s(4.22), 257(3.90),261s(3.88), 286(3.54),296(3.56), 353(3.97)	39-1175-65
7(1H)-Pteridinone, 4-amino-1-methyl-	pH -0.5	220(4.17),242(4.19), 251s(4.04),289(3.73), 316s(3.98),326(4.08), 338(4.05)	39-3770-65
	pH 7.0	231(4.27),252(4.30), 257s(4.29),329s(4.08), 338(4.13),351(4.00)	39-3770-65

Compound	Solvent	$\lambda_{max}(\log \epsilon)$	Ref.
Purine-6-acetamide	pH 1	264(3.88)	44-1528-65
	pH 7	264(3.94)	44-1528-65
	pH 13	273(3.92)	44-1528-65
$C_7H_7N_5O_2$			
6-Pteridinemethanol, 2-amino-4-hydroxy-	pH 1	247(4.01),322(3.89)	44-3610-64
	pH 1	245(4.0),320(3.9)	37-2259-64
	pH 7	235(4.1),245(4.0), 275(4.1),350(3.8)	37-2259-64
	pH 13	253(4.36),362(3.85)	44-3610-64
	pH 13	255(4.4),365(3.9)	37-2259-64
1,2,4-Triazolo[2,3-a]pyrimidine, 4-acetamido-6-oxo-	pH 7	222(4.39),264(3.90)	70-1475-64
	pH 12	222(4.41),276(3.92)	70-1475-64
$C_7H_7N_5O_2S$			
6-Azalumazine, 7-(ethylthio)-	H_2O	232(4.24),266(4.20), 346(3.99)	24-0005-64
$C_7H_7N_5S$			
Pteridine, 2-amino-4-(methylthio)-	EtOH	211(4.19),233(4.18), 269(4.13),309(3.53), 383(3.88)	44-3370-64
$C_7H_7N_5S_2$			
Carbamic acid, dithio-, purin-6-yl-methyl ester	N HCl	240(3.82),267(4.01)	87-0667-65
	pH 7.65	245s(3.88),271(4.10)	87-0667-65
C_7H_7NaO			
Sodium p-cresoxide	DMF	332(3.56)	35-1857-65
	$(MeOCH_2)_2$	311(3.53)	35-1857-65
$C_7H_7NaO_2$			
Sodium, (p-methoxyphenoxy)-	DMF	342(3.61)	35-1857-65
	$(MeOCH_2)_2$	323.5(3.58)	35-1857-65
C_7H_8			
Cyclopentene, 1-ethynyl-	hexane	225(4.05)	22-3518-65
Fulvene, 6-methyl-	EtOH	255(4.24),357(2.50)	89-0862-64
3,5-Heptadien-1-yne	n.s.g.	226(3.89),247s(3.75)	5-0062-65B
2,4-Heptadiyne	n.s.g.	208(2.77),213(2.70), 224(2.77),231(2.72), 236(2.76),251(2.51)	22-1525-65
$C_7H_8BrNO_2$			
Pyrrole-2-carboxylic acid, 5-bromo-1-methyl-, methyl ester	EtOH	233(4.02),271(4.50)	39-0459-65
$C_7H_8BrNO_2S$			
3-Isothiazolecarboxylic acid, 4-bromo-, isopropyl ester	EtOH	205(3.72),270(4.00)	39-3114-64
C_7H_8BrNS			
2,3-Dihydrothiazolo[3,2-a]pyridinium bromide	pH 1	253(3.92),323(3.76)	4-0097-65
	pH 7	253(3.92),322(3.76)	4-0097-65
	pH 13	252(3.90),323(3.76)	4-0097-65
	MeOH	253(3.92),323(3.76)	4-0097-65
C_7H_8ClN			
2,5-Lutidine, 3-chloro-	EtOH	275(3.61)	32-0083-65

Compound	Solvent	λ_{max}(log ϵ)	Ref.
C_7H_8ClNO			
2,6-Lutidine, 3-chloro-, 1-oxide	EtOH	265(4.1)	94-0963-65
2,6-Lutidine, 4-chloro-, 1-oxide	EtOH	268(4.18)	39-2096-65
	EtOH	270(4.3)	94-0963-65
$C_7H_8ClNO_2$			
Pyrrole-2-carboxylic acid, 5-chloro-1-methyl-, methyl ester	EtOH	236s(3.76),269(4.24)	39-0459-65
$C_7H_8ClNO_3$			
3-Isoxazolecarboxylic acid, 5-chloro-4-methyl-, ethyl ester	C_6H_{12}	204(3.39),248(3.49)	17-0203-65
	MeOH	206(3.34),250(3.52)	17-0203-65
4-Isoxazolecarboxylic acid, 5-chloro-3-methyl-, ethyl ester	C_6H_{12}	222(3.89)	17-0203-65
	MeOH	224(3.84)	17-0203-65
$C_7H_8ClNO_4S_2$			
3,5-Dimethylthiazolo[2,3-b]thiazolium perchlorate	pH 11	221(3.84),292(4.06), 298(4.04)	88-1723-65
$C_7H_8ClN_3O_2$			
Ketone, 5-chloro-2-furyl methyl, semicarbazone	MeOH	294(4.41)	25-1425-65
$C_7H_8ClN_3O_5S_2$			
Formamide, N-[(2-amino-4-chloro-5-sulfamoylphenyl)sulfonyl]-	pH 13	227(4.60),265(4.14), 315(3.52)	95-0971-64
$C_7H_8Cl_2N_2$			
Pyrimidine, 4,6-dichloro-2-isopropyl-	MeOH	254(3.66)	87-0808-64
Pyrimidine, 4,6-dichloro-2-propyl-	MeOH	257(3.68)	87-0808-64
$C_7H_8Cl_2O$			
1-Cyclobutanone, 2-chloro-4-(chloromethylene)-3,3-dimethyl-	hexane	251(4.04),350s(--), 368(1.62),382s(--)	28-0827-64B
$C_7H_8Cl_3NS_2$			
2H-1,3-Thiazine-2-thione, 3,6-dihydro-3,4-dimethyl-6-(trichloromethyl)-	H_2O	267s(3.61),315(3.90)	39-4008-64
	EtOH	268(3.64),318(3.96)	39-4008-64
2H-1,3-Thiazine-2-thione, 3,6-dihydro-4,5-dimethyl-6-(trichloromethyl)-	H_2O	290s(3.74),320(4.03)	39-4008-64
	EtOH	290s(--),326(4.07), 360s(--)	39-4008-64
C_7H_8FNO			
2,6-Lutidine, 3-fluoro-, 1-oxide	EtOH	260(4.08)	39-2096-65
$C_7H_8INO_2$			
3-Carboxy-1-methylpyridinium iodide	pH 0.73	264(3.65),271(3.56)	37-2137-64
	pH 10.5	265(3.64)	37-1237-64
$C_7H_8N_2$			
Benzaldehyde, hydrazone	$CHCl_3$	275(4.29)	44-0417-65
1,3a-Diazaindene, 2,3-dihydro-	pH 7.0	239(4.01),324(3.58)	39-2778-65
	pH 14.5	261(4.08),368(3.37)	39-2778-65
$C_7H_8N_2O$			
Aniline, N-methyl-N-nitroso-	C_6H_{12}	274(3.85)	35-4373-64
Benzamide, 2-amino-	EtOH	249(3.81),331(3.56)	24-1562-65
	3:2 EtOH: dioxan	251(3.86),332(3.62)	24-1562-65

Compound	Solvent	λ_{max}(log ε)	Ref.
2-Propen-1-ol, 3-(4-pyrimidyl)-	EtOH	242(4.04),272(4.05)	44-2398-65
$C_7H_8N_2O_2$			
Aniline, N-methyl-2-nitro-	EtOH	429(3.7)	35-4018-64
Pyridine, 2-acetyl-, 1-oxide, oxime, anti	EtOH	266(4.04)	95-0451-65
syn	EtOH	237(4.17),272(4.11)	95-0451-65
2(1H)-Pyridone-3-carboxamide, 6-methyl-	EtOH	237(4.12),329(4.30)	44-3593-65
$C_7H_8N_2O_3$			
Pyridine, 4-ethoxy-3-nitro-	pH 0.0	221(4.28),239s(3.95), 277(3.35),332s(2.45)	39-2150-64
	pH 7.0	213(4.23),252(3.67), 294(3.41)	39-2150-64
5-Pyrimidinecarboxylic acid, 1,2-dihydro-2-oxo-, ethyl ester	EtOH-HBr	257(4.19)	94-0804-64
$C_7H_8N_2O_3S$			
Benzanilide, o-sulfonamido-	1% HCOOH	276(3.4)	20-0136-65
$C_7H_8N_2O_4$			
Imidazole-4,5-dicarboxylic acid, dimethyl ester	n.s.g.	258(3.98)	24-1414-64
1(2H)-Pyrimidinepropionic acid, 3,4-dihydro-2,4-dioxo-	pH 1	265(3.96)	87-0187-65
	pH 7	267(4.00)	87-0187-65
	pH 13	265(3.84)	87-0187-65
1(2H)-Pyrimidinepropionic acid, 3,6-dihydro-2,6-dioxo-	pH 4-7	259(3.85)	44-1762-64
	pH 14	284(4.02)	44-1762-64
$C_7H_8N_4$			
Benzotriazole, 4-amino-2-methyl-	EtOH	225(4.45),280(3.35), 325(4.45)	78-0211-64
4-Pyrimidinecarbamonitrile, 2,6-dimethyl-	MeOH	208(3.79),293(4.23)	39-3357-65
Pyrrolo[2,3-d]pyrimidine, 4-amino-	pH 1	231(4.32),284(3.93)	35-1995-65
	pH 13	278(3.90)	35-1995-65
	EtOH	279(3.94)	35-1995-65
s-Triazolo[4,3-a]pyrazine, 5,6-dimethyl-	C_6H_{12}	213(4.50),218s(4.42), 231s(3.36),261s(3.45), 278s(3.51),292s(3.59), 299(3.61),308s(3.57), 321s(3.28)	44-2542-64
s-Triazolo[4,3-a]pyrazine, 3-ethyl-	C_6H_{12}	212(4.47),219s(4.41), 260(3.41),268s(3.35), 290s(3.39),299(3.42), 308(3.42),322(3.30), 339(2.94)	44-2542-64
s-Triazolo[1,5-a]pyrimidine, 5,6-dimethyl-	EtOH	278(3.64)	94-0204-64
$C_7H_8N_4O$			
Isoxazolo[5,4-d]pyrimidine, 4-amino-3-ethyl-	EtOH	249(4.03),272(3.89)	44-2116-64
Isoxazolo[5,4-d]pyrimidine, 3-methyl-4-(methylamino)-	EtOH	252(3.99),283(3.89)	44-2116-64
1H-1,2,4-Triazole, 3-amino-5-(2-furyl)-1-methyl-	pH 2	284(4.15)	1-1191-65
	EtOH	252s(3.98),278(4.05)	1-1191-65
2H-1,2,4-Triazole, 3-amino-5-(2-furyl)-2-methyl-	pH 2	270(4.22)	1-1191-65
	EtOH	264(4.22)	1-1191-65

Compound	Solvent	$\lambda_{max}(\log \epsilon)$	Ref.
4H-1,2,4-Triazole, 3-amino-5-(2-furyl)-4-methyl-	pH 2	263(4.18)	1-1191-65
	EtOH	274(4.11)	1-1191-65
s-Triazole, 3-(2-furyl)-5-(methylamino)-	pH 2	268(4.18)	1-1191-65
	pH 12	274(4.06)	1-1191-65
$C_7H_8N_4OS$			
9H-Purine-9-ethanol, 6-mercapto-	pH 1	224(3.98),320(4.34)	87-0182-65
	pH 11	232(4.18),308(4.36)	87-0182-65
Thiocyanic acid, 1-carbamoyl-3,5-dimethylpyrazol-4-yl ester	EtOH	238(3.92)	94-0023-64
$C_7H_8N_4O_2$			
1H-Pyrazolo[4,3-d]pyrimidine-5,7(4H,6H)-dione, 4,6-dimethyl-	MeOH	285(3.69)	44-0199-65
$C_7H_8N_4O_3$			
Pyrazinecarboxamide, 5-amino-6-formyl-3,4-dihydro-N-methyl-3-oxo-	pH 3.5	229(3.98),295(4.11), 360(4.39)	39-3357-64
	pH 8.5	229(4.36),295(4.12), 359(4.36)	39-3357-64
$C_7H_8N_4O_3S$			
Xanthine, 1-hydroxy-7-methyl-8-(methylthio)-	EtOH	288(4.03)	4-0275-64
	90% EtOH-NaOH	303(3.85)	4-0275-64
$C_7H_8N_4S$			
Purine, 6-[(methylthio)methyl]-	N HCl	267(3.78),305s(2.76)	87-0667-65
	pH 7.65	268(3.89)	87-0667-65
	N NaOH	278(3.92)	87-0667-65
$C_7H_8N_6$			
Pteridine, 7(4)-amino-1,4-dihydro-4(7)-imino-1-methyl-	pH 9.8	237(4.27),256(4.22), 261(4.22),345(4.15), 360s(4.00)	39-1175-65
Pteridine, 7-amino-4-(methylamino)-	pH 3.0	234(4.17),258(4.23), 262s(4.21),289(3.78), 296s(3.71),340s(4.11), 351(4.16),365s(4.02)	39-1175-65
	pH 7.5	240(4.34),259s(4.14), 267s(4.04),346(4.05), 360s(3.94)	39-1175-65
$C_7H_8N_6O$			
6-Pteridinemethanol, 2,4-diamino-	pH 1	243(4.20),283(3.66), 337(3.99)	44-3610-64
	pH 13	227(4.05),257(4.32), 368(3.86)	44-3610-64
7-Pteridinmethanol, 2,4-diamino-	pH 3	243(4.02),287(3.67), 334(4.02)	39-1530-65
	pH 9	256(4.34),368(3.92)	39-1530-65
Pterin, 8-N-methylamino-	pH 2	277(4.3),385(3.7)	33-2195-64
	pH 7	283(4.3),400(3.7)	33-2195-64
	pH 12	263(4.3),402(3.8)	33-2195-64
Purine-6-acetic acid, hydrazide	pH 1	263(3.89)	44-1528-65
	pH 7	264(3.95)	44-1528-65
	pH 13	273(3.96)	44-1528-65

Compound	Solvent	$\lambda_{max}(\log \epsilon)$	Ref.
$C_7H_8N_6S$			
Pseudothiourea, 2-(6-purinylmethyl)-	N HCl	267(3.79)	87-0667-65
	pH 7.65	268(3.83)	87-0667-65
C_7H_8O			
Anisole	C_6H_{12}	220(3.91),224s(3.81), 265(3.15),271(3.32), 278(3.32)	23-2603-65
	EtOH	270(--),277(--)	95-0858-65
Benzyl alcohol	EtOH	222(3.18),243s(2.75), 248s(2.86),254(2.95), 259(3.00)	44-2251-65
p-Cresol	DMF	282(3.34)	35-1857-65
	$(MeOCH_2)_2$	281(3.32)	35-1857-65
2,6-Cycloheptadien-1-one	isooctane	226(4.04),235s(3.92), 244s(3.66),258(3.37), 357(1.28),366(1.32), 384(1.11)	44-2109-65
	EtOH	235(4.03),266s(3.4), 339(1.52)	44-2109-65
Cyclopentadiene, acetyl-	hexane	230(3.82),278(2.21)	5-0039-64H
4,5-Heptadien-2-yn-1-ol	EtOH	220(4.07)	39-4659-65
2-Norbornen-7-one	EtOH	273(1.63)	44-0160-64
	isooctane	274(1.49)	44-0160-64
	$CHCl_3$	273(1.61)	44-0160-64
2-Penten-4-yn-1-al, 2,3-dimethyl-	EtOH	271(4.21)	70-0546-65
Tricyclo[1.1.1.$0^{2,4}$]pentan-5-one, 2,4-dimethyl-	n.s.g.	225s(2.11),252s(1.63)	88-0961-64
C_7H_8OS			
2-Propen-1-ol, 3-(2-thienyl)-	Et_2O	276(4.13)	24-2118-64
$C_7H_8O_2$			
Benzyl alcohol, 2-hydroxy-	acid	273(3.3)	44-0603-65
	base	294(3.58)	44-0603-65
Benzyl alcohol, 3-hydroxy-	pH 1	272(3.23)	10-0156-64A
	pH 13	238(3.92),297(3.53)	10-0156-64A
Benzyl alcohol, 4-hydroxy-	acid	222(3.9)	44-0603-65
	base	244(4.09)	44-0603-65
2,4-Hexadienoic acid, 4-hydroxy-5-methyl-, gamma lactone	EtOH	289(4.27)	44-2785-64
Methane, bis(2-propynyloxy)-	MeOH	none	70-1349-64
Phenol, 2-methoxy-	EtOH	275(--),281(--)	95-0858-65
Phenol, 3-methoxy-	EtOH	275(--),280(--)	95-0858-65
Phenol, 4-methoxy-	EtOH	292(--)	95-0858-65
	DMF	295.5(3.51)	35-1857-65
	$(MeOCH_2)_2$	293.5(3.49)	35-1857-65
2H-Pyran-2-one, 4,6-dimethyl-	EtOH	295(3.88)	54-0039-64
4H-Pyran-4-one, 2,6-dimethyl-	EtOH	240(4.1)	94-0018-64
Toluene, 2,4-dihydroxy-	EtOH	220s(3.84),281(3.44), 286s(3.38)	24-1926-64
	NaOH	235s(3.91),297(3.62)	24-1926-64
Toluene, 3,4-dihydroxy-	EtOH	274(3.23),281(3.23)	1-1677-65
$C_7H_8O_2S$			
Sulfone, 2,4-hexadiynyl methyl	Et_2O	219(2.81),231(3.04), 243(3.12),256(2.93)	24-3015-65
3-Thiophenecarboxylic acid, 4-methyl-, methyl ester	Et_2O	208(4.07),238(3.80)	24-2109-64

Compound	Solvent	λ_{max}(log ε)	Ref.
$C_7H_8O_3$			
1,3-Cyclopentanedione, 2-acetyl-	MeOH-HCl	225(4.39),261(4.23)	20-0628-64
	MeOH-NaOH	248(4.57),265s(4.46)	20-0628-64
1,2,4-Cyclopentanetrione, 3-ethyl-	pH 1	280(3.95)	94-1300-65
	neutral	280(4.10)	94-1300-65
	pH 10	225(4.00),327(4.00)	94-1300-65
1-Cyclopentene-1-carboxylic acid, 2-methyl-3-oxo-	EtOH	243(4.1)	33-2234-64
Furan, 2-acetyl-3-hydroxy-5-methyl-	EtOH	288(4.29)	39-2543-65
Δ^2-Furenid-4-one, 3-acetyl-2-methyl-	pH 1	232(4.03),266(4.06)	33-1322-65
	N H_2SO_4	232(4.00),266(4.04)	33-1322-65
	H_2O	232(4.02),265(4.06)	33-1322-65
	EtOH	228(4.00),266(4.04)	33-1322-65
	Et_2O	224(3.99),264(4.04)	33-1322-65
Puntiol, desoxy-	$CHCl_3$	280(3.77)	78-0093-65
2H-Pyran-2-one, 4-hydroxy-3,6-dimethyl-	EtOH	288(3.92)	35-1264-64
4H-Pyran-4-one, 3-methoxy-2-methyl-	MeOH	259(3.99)	44-0776-64
$C_7H_8O_3S$			
3-Thiophenecarboxylic acid, 4-hydroxy-, ethyl ester	Et_2O	203(4.02),243(4.00), 297(3.44)	24-2109-64
$C_7H_8O_4$			
2-Heptenoic acid, 4,6-dioxo-	C_6H_{12}	212(--),240(--), 312(--)	1-0441-64
	N HCl	213(3.72),250(3.73), 313(4.00)	1-0441-64
	EtOH	212(3.81),240(3.71), 312(3.49)	1-0441-64
Opuntiol	$CHCl_3$	280(3.85)	78-0093-65
$C_7H_8O_4S$			
α-Tetronic acid, beta-acetylmercapto-methyl-	EtOH	240(4.16)	44-3560-64
$C_7H_8O_5$			
3-Furancarboxylic acid, 2,5-dihydro-4-hydroxy-5-oxo-, ethyl ester	EtOH	255(4.07),302(3.19)	39-0766-64
	EtOH-NaOH	303(4.19)	39-0766-64
Tetronic acid, 5-carboxymethyl-, methyl ester	EtOH-acid	223(4.16)	7-0170-64
C_7H_8S			
Sulfide, ethyl 1,3-pentadiynyl	n.s.g.	242(3.15),253(--), 268(--),284(--)	28-2847-65A
Sulfide, methyl phenyl	hexane	254(3.99),280(3.00)	65-2749-64
	EtOH	253(3.96),280(3.02)	65-2749-64
	Et_2O	253(3.27),280(3.05)	65-2749-64
	dioxan	254(4.05),280(3.00)	65-2749-64
	H_2SO_4	221(3.76),265f(3.36), 313(2.57),328s(2.13)	65-2749-64
$C_7H_8S_3$			
2H-1,2-Benzodithiole-3-thione, 4,5,6,7-tetrahydro-	EtOH	225(3.99),240s(3.76), 279(3.83),314(3.72), 408(4.02)	24-0654-64
$C_7H_9BrF_2O_2$			
Crotonic acid, 4-fluoro-3-(bromo-fluoromethyl)-, ethyl ester	MeOH	212(4.16)	5-0021-65A

Compound	Solvent	$\lambda_{max}(\log \epsilon)$	Ref.
C_7H_9ClIN			
Aniline, N-methyl-, ICl complex	$CHCl_3$	354(--),370s(--)	60-0062-64
2-Toluidine, ICl complex	$CHCl_3$	358(--)	60-0062-64
3-Toluidine, ICl complex	$CHCl_3$	356(--),362(--)	60-0062-64
4-Toluidine, ICl complex	$CHCl_3$	350(--),370s(--)	60-0062-64
$C_7H_9ClN_2O$			
4-Pyrimidinol, 6-chloro-5-isopropyl-	pH -3.6	242(3.95)	39-3204-64
	pH 5.0	234(3.72),274(3.75)	39-3204-64
	pH 11.1	237(3.93),270(3.72)	39-3204-64
$C_7H_9ClN_2O_5$			
3-Carbamoyl-1-methylpyridinium perchlorate	MeOH	265(3.64)	37-1237-64
$C_7H_9ClN_4O_2$			
4-Pyrimidinecarboxylic acid, 5,6-diamino-2-chloro-, ethyl ester	pH 5.2	232(3.89),260(3.77), 344(3.90)	39-3221-64
C_7H_9ClO			
1-Cyclohexene-1-carboxaldehyde, 2-chloro-	EtOH	255(4.26)	44-1126-65
2-Cyclopenten-1-one, 2-chloro-4,4-dimethyl-	MeOH	234(3.98),310(1.69)	20-0081-64
2-Norbornane, 3-chloro-, endo	EtOH	300(1.68)	35-4074-64
exo	isooctane	305(1.60)	35-4074-64
	EtOH	299(1.56)	35-4074-64
$C_7H_9ClO_2$			
2-Cyclopenten-1-one, 3-chloro-2-hydroxy-4,4-dimethyl-	EtOH	268(4.0)	5-0100-64I
$C_7H_9ClO_4S$			
Maleic acid, chloro(methylthio)-, dimethyl ester	MeOH	290(4.14)	24-1581-65
$C_7H_9Cl_2N$			
2H-Pyrrole, 2-(dichloromethyl)-2,5-dimethyl-	EtOH	233(3.38)	32-0083-65
C_7H_9DO			
1,2-Heptadien-4-one-1-d	n.s.g.	203(3.86),212s(3.71), 322(1.41)	28-1530-64B
$C_7H_9FN_2O_2$			
2(1H)-Pyrimidinone, 4-ethoxy-5-fluoro-1-methyl-	H_2O	282(3.90)	87-0253-65
$C_7H_9F_2N_3O_3$			
Cyclobutanetrione, difluoro-, tris(O-methyl oxime)	EtOH	226(4.06),313(4.23)	44-3698-65
$C_7H_9F_3O_2$			
Crotonic acid, 2,4-difluoro-3-(fluoromethyl)-, ethyl ester	MeOH	212(4.14)	5-0001-64D

Compound	Solvent	$\lambda_{max}(\log \epsilon)$	Ref.
$C_7H_9I_2N$			
Aniline, N-methyl-, iodine complex	$CHCl_3$	350s(--)	60-0062-64
2-Toluidine, iodine complex	$CHCl_3$	350(--)	60-0062-64
3-Toluidine, iodine complex	$CHCl_3$	355(--)	60-0062-64
4-Toluidine, iodine complex	$CHCl_3$	355(--)	60-0062-64
C_7H_9N			
Fulvene, 6-amino-6-methyl-	hexane	316(4.30)	5-0039-64H
3,4-Lutidine, NiI_2 complex	$CHCl_3$	590s(2.57),631(2.49), 925(2.02),1010(2.06)	33-1265-64
Pyrrole, 1-methyl-2-vinyl-	n.s.g.	214(3.77),282(4.08)	12-0875-65
p-Toluidine	EtOH	235.5(4.04)	78-0861-64
C_7H_9NO			
o-Anisidine	EtOH	286(--)	95-0858-65
3-Anisidine	EtOH	286(--)	95-0858-65
3H-Azepine, 2-methoxy-	MeOH	257(3.72)	5-0001-65B
2H-Azepin-2-one, 1,3-dihydro-1-methyl-	MeOH	260(3.62)	5-0001-65B
Benzyl alcohol	EtOH	241(4.08)	78-0861-64
Cyclopentadiene, acetyl-, oxime	MeOH	240(3.98)	5-0039-64H
2,4-Hexadienoic acid, 4-amino-5-methyl-, gamma-lactam	EtOH	300(4.12)	44-2785-64
4-Hexenenitrile, 4-formyl-	MeOH	226(4.09),325(1.61)	22-0550-64
2,6-Lutidine, 1-oxide	EtOH	259(4.00)	39-2319-64
4-Pyridinemethanol, alpha-methyl-	EtOH	251(3.30),256(3.35), 262(3.24)	44-2898-64
C_7H_9NOS			
Sulfoximine, S-methyl-S-phenyl-	N HCl	221(4.00),261s(3.11), 266(3.22),274(3.16)	25-1261-64
	EtOH	216(3.87),258(2.85), 264(2.94),271(2.84)	25-1261-64
$C_7H_9NO_2$			
1-Azabicyclooctane-2,8-dione	THF	232(3.15),243(2.56), 253(2.23),267(1.97), 276(1.90),287(1.64)	24-1548-64
	H_2O	222(3.94),267(1.92)	24-1548-64
Crotonic acid, 2-cyano-, ethyl ester	EtOH	218(3.97)	28-2859-64B
2-Hexenoic acid, 2-cyano-	EtOH	222(4.00)	28-2859-64B
Ketone, ethyl 2-furyl, oxime, anti	EtOH	266(4.14)	22-2724-65
syn	EtOH	263(4.19)	22-2724-65
Ketone, 3-methoxy-2-pyrrolyl methyl	MeOH	290(4.33)	44-0776-64
Ketone, (5-methyl-2-furyl) methyl, oxime, anti	EtOH	273(4.19)	22-2724-65
syn	EtOH	272(4.27)	22-2724-65
2,4-Pentanedione, 3-cyano-, O-methyl ether	heptane	232.0(3.90)	59-0931-65
	H_2O	235.0(3.89)	59-0931-65
2-Pentenoic acid, 2-cyano-3-methyl-	EtOH	229(4.05)	28-2859-64B
2-Pentenoic acid, 2-cyano-4-methyl-	EtOH	218.5(4.05)	28-2859-64B
3-Pyridinol, 4-methoxy-2-methyl-	MeOH	270(3.70),310(3.10)	44-0776-64
	MeOH-H_2SO_4	242(3.52),277(3.99)	44-0776-64
	MeOH-KOH	252(3.90),290(3.85)	44-0776-64
2(1H)-Pyridone, 1-ethoxy-	H_2O	225(3.79),295(3.77)	44-1650-64
4(1H)-Pyridone, 3-methoxy-2-methyl-	MeOH	266(4.12)	44-0776-64
	MeOH-H_2SO_4	241(3.75),259(3.73)	44-0776-64
	MeOH-KOH	245(4.00)	44-0776-64
Pyrrole-2-carboxylic acid, 1-methyl-, methyl ester	EtOH	240s(3.84),265(4.18)	39-0459-65

Compound	Solvent	λ_{max}(log ε)	Ref.
Pyrrole-3-carboxylic acid, 2-methyl-, methyl ester	EtOH	225(3.90),255(3.84)	39-2411-65
$C_7H_9NO_2S$			
3-Isothiazolecarboxylic acid, isopropyl ester	EtOH	225(3.69),256(3.94)	39-3114-64
2(5H)-Thiophenone, 3-acetamido-4-methyl-	EtOH	240(3.89)	39-0766-64
$C_7H_9NO_2S_2$			
2H-1,3-Thiazine-5-carboxylic acid, 3,6-di-hydro-2-thioxo-, ethyl ester	EtOH	340(4.27)	44-2290-65
$C_7H_9NO_2S_3$			
4-Isothiazolecarboxylic acid, 3,5-bis-(methylthio)-, methyl ester	EtOH	215(4.05),238(4.18), 284(4.17)	44-0665-64
$C_7H_9NO_3$			
Crotonic acid, 2-cyano-3-methoxy-, methyl ester	dioxan	257(4.22)	33-1424-64
cis isomer	EtOH	284(4.17)	44-0695-65
trans isomer	EtOH	284(4.07)	44-0695-65
3-Isoxazolecarboxylic acid, 5-methyl-,	C_6H_{12}	226.9(3.87)	32-1478-65
ethyl ester	EtOH	226.9(3.90)	32-1478-65
Glutarimide, 2,2-dimethyl-4-oxo-	EtOH	264(3.55)	12-0154-64
	EtOH-KOH	237s(2.75),316(3.18)	12-0154-64
3-Pyridinemethanol, 4,5-dihydroxy-	pH 1	244(3.59),271(3.78)	4-0144-65
6-methyl-	pH 13	223(4.27),303(3.96)	4-0144-65
	EtOH	281(4.12)	4-0144-65
4-Pyridinemethanol, 3,5-dihydroxy-	pH 1	250(3.30),299(3.92)	4-0144-65
2-methyl-	pH 13	315(3.83)	4-0144-65
	EtOH	258(3.36),297(3.86)	4-0144-65
3-Pyrroline-1-carboxylic acid, 2-oxo-, ethyl ester	hexane	200(4.01),225(3.74)	88-2185-64
$C_7H_9NO_3S$			
2H-1,3-Thiazine-5-carboxylic acid, 3,6-di-hydro-2-oxo-, ethyl ester	EtOH	240(3.77),292(3.96)	44-2290-65
2H-1,4-Thiazine-5-carboxylic acid, 3,4-	0.03N NaOH	229(4.12),328(3.79)	12-1071-65
di-hydro-3-oxo-, ethyl ester	n.s.g.	225(3.89),239(3.82), 320(3.83)	12-1071-65
3-Thiophenecarboxylic acid, 4-amino-2,5-dihydro-5-oxo-, ethyl ester	EtOH	271(3.68),320(3.93)	39-0766-64
$C_7H_9NO_4$			
2,4-Pentadienoic acid, 5-nitro-, ethyl ester	EtOH	281(4.20)	70-2093-64
C_7H_9NS			
Aniline, p-(methylthio)-	hexane	238(3.88),264(4.03), 300(3.18),320(2.80)	65-2749-64
	EtOH	241(3.90),263(4.18), 300(3.26),321(2.90)	65-2749-64
	Et_2O	240(3.83),265(4.13), 303(3.23),320(2.90)	65-2749-64
	dioxan	240(3.81),265(4.16), 303(3.23)	65-2749-64
	H_2SO_4	261f(3.18)	65-2749-64
2(1H)-Pyridinethione, 5,6-dimethyl-	EtOH	230s(3.51),283(4.14), 371(3.92)	94-0087-64

Compound	Solvent	$\lambda_{max}(\log \epsilon)$	Ref.
$C_7H_9N_3O$			
Pyrimidine, 2-acetamido-4-methyl-	pH 1	231(4.23),275(3.70)	44-0115-65
	H_2O	233(4.18)	44-0115-65
5-Pyrimidinecarbonitrile, 1,2,3,4-tetra-hydro-1,3-dimethyl-2-oxo-	EtOH	219(3.93),290(3.92)	44-1740-64
$C_7H_9N_3O_2$			
Cytosine, N^4-acetyl-N^4-methyl-	N HCl	213(4.02),245(3.67), 309(4.11)	44-1762-64
	H_2O	212(4.16),256(3.97), 296(3.81)	44-1762-64
Pyrazinealanine	pH 1	264(3.89)	4-0001-65
	pH 11	267(3.89)	4-0001-65
3-Pyridazinealanine	pH 1	244(3.30)	4-0001-65
	pH 11	251(3.30)	4-0001-65
4-Pyridazinealanine	pH 1	239(3.48)	4-0001-65
	pH 11	246(3.43)	4-0001-65
2-Pyrimidinealanine	pH 1	247(3.27)	4-0001-65
	pH 11	248(3.27)	4-0001-65
4-Pyrimidinealanine	pH 1	244(3.85)	4-0001-65
	pH 11	245(3.85)	4-0001-65
$C_7H_9N_3O_2S$			
Pyridazine, 3-aziridinyl-6-(methyl-sulfonyl)-	pH 1	250(4.09),306(3.20)	4-0001-64
	pH 1	249(4.08)	4-0001-64
Pyrimidine, 4-aziridinyl-6-(methyl-sulfonyl)-	pH 1	246(4.13),295(3.60)	4-0001-64
	pH 11	241(3.93),277(3.68)	4-0001-64
	EtOH	242(3.98),284(3.59)	4-0001-64
4-Thiazolecarboxamide, 5-acetamido-N-methyl-	EtOH	280(4.29)	94-1319-65
$C_7H_9N_3O_3$			
Cytosine, 3-(2-carboxyethyl)-	pH 4.72	276(3.94)	44-1762-64
	pH 10.58	297(4.03)	44-1762-64
	3N NaOH	233(3.85),294(3.93)	44-1762-64
Isoxazole, 3,5-diacetamido-	MeOH	227(4.06),246(4.06)	44-2862-65
$C_7H_9N_3O_4$			
1(2H)-Pyrimidinealanine, 3,4-dihydro-2,4-dioxo-	pH 1	262(3.957)	65-0411-64
	pH 12	266(3.813)	65-0411-64
2-Pyrimidinealanine, 4,6-dihydroxy-	pH 1	257(3.73)	4-0049-65
	pH 11	252(3.91)	4-0049-65
4-Pyrimidinealanine, 2,6-dihydroxy-	pH 1	260(3.85)	4-0049-65
	pH 11	281(3.85)	4-0049-65
5-Pyrimidinealanine, 2,4-dihydroxy-	pH 1	262(3.85)	4-0049-65
	pH 11	288(3.76)	4-0049-65
as-Triazine-6-acetic acid, 3,4,5,6-tetra-hydro-3,5-dioxo-, ethyl ester	MeOH	224(3.99),261(3.72)	39-0868-64
as-Triazine-6-butyric acid, 3,4,5,6-tetrahydro-3,5-dioxo-	MeOH	211(3.81),254(3.76)	39-0868-64
$C_7H_9N_5$			
Adenine, N,9-dimethyl-	pH 1.9	209(4.25),264(4.23), 271s(4.17)	39-3770-65
	pH 6.5	209(4.27),267(4.20)	39-3770-65

Compound	Solvent	λmax(log ε)	Ref.
$C_7H_9N_5O$			
Adenine, 1-methoxy-9-methyl-	pH 1	260(4.08)	94-1017-65
	pH 13	257(4.09),264s(4.05)	94-1017-65
Guanine, N,N-dimethyl-	pH 1	256(4.28),288(3.81)	35-3752-65
	pH 11	245s(4.11),282(3.88)	35-3752-65
Guanine, 9-ethyl-	pH 6.6	188(4.42),205s(4.30), 252(4.13),275s(3.98)	35-0011-65
	pH 9.6	190(4.32),208(4.37), 252(4.09),270(4.03)	35-0011-65
	pH 11.1	210(4.47),252s(4.00), 269(4.06)	35-0011-65
	Me_3PO_4	190(4.44),203s(4.30), 256(4.19),275s(3.97)	35-0011-65
6(3H)-Pyrimidone, 4-amino-5-cyanomethyl-amino-3-methyl-	pH 0.6	270(3.91)	39-1175-65
	pH 5.0	213(4.39),268(3.97)	39-1175-65
1,2,4-Triazole, 5-(cyanoacetamido)-3-ethyl-	EtOH	224(3.79)	24-1373-64
s-Triazolo[1,5-a]pyrimid-5(4H)-one, 7-amino-2-ethyl-	EtOH	212(4.36),270(4.10)	24-1373-64
$C_7H_9N_5O_2$			
1H-v-Triazolo[4,5-d]pyrimidine-5,7-(4H,6H)dione, 1,4,6-trimethyl-	pH 7.0	230s(3.56),280(3.74)	24-1060-65
2H-v-Triazolo[4,5-d]pyrimidine-5,7-(4H,6H)dione, 2,4,6-trimethyl-	pH 7.0	230(3.51),276(3.87)	24-1060-65
3H-v-Triazolo[4,5-d]pyrimidine-5,7 (4H,6H)dione, 3,4,6-trimethyl-	pH 7.0	242s(3.76),259(3.93)	24-1060-65
$C_7H_9N_5O_4S$			
Theophylline, 8-sulfamoyl-	MeOH	282(4.09)	54-1215-64
$C_7H_9N_5O_8P_2$			
6-Pteridinemethanol, 2-amino-4-hydroxy-, 6-(trihydrogen pyrophosphate)	pH 1	250(4.1),320(3.9)	37-2259-64
	pH 7	235(4.1),245(4.0), 275(4.1),350(3.8)	37-2259-64
	pH 13	255(4.4),365(3.9)	37-2259-64
$C_7H_9O_3P$			
Phosphonic acid, methyl-, di-2-propynyl ester	EtOH	237(2.8),242(2.8), 250(3.0),256(3.0), 264(2.8)	70-1349-64
Phosphonic acid, o-tolyl-	EtOH	215(3.87),262(2.65), 268(2.76),275(2.67)	44-2382-64
C_7H_{10}			
1,3-Cycloheptadiene, $PdCl_2$ olefin complex from	EtOH	211(4.27),257(4.29), 290s(3.77),365s(3.35)	24-2037-64
Cyclopentadiene, 1,3-dimethyl-	MeOH	248.5(3.44)	78-2313-65
	EtOH	252(3.44)	35-4533-65
1,3,5-Heptatriene, 5,6-cis-3,4-trans	C_6H_{12}	263(4.67),273(4.57)	23-2781-64
3,4-trans-5,6-trans	C_6H_{12}	251(4.61),261(4.73), 272(4.67)	23-2781-64
2-Hepten-4-yne	EtOH	225(4.06)	28-0594-64B
1,5-Hexadiene, 3-methylene-	heptane	210s(4.18),223(4.31), 230s(4.25)	35-4506-65
1,3,5-Hexatriene, 1-methyl-	n.s.g.	250(4.52),260(4.65), 270(4.55)	44-2218-65
1,3,5-Hexatriene, 3-methyl-	n.s.g.	251(4.44),261(4.57), 271(4.45)	44-2218-65

Compound	Solvent	$\lambda_{max}(\log \epsilon)$	Ref.
$C_7H_{10}BF_4NO$			
4-Methoxy-1-methylpyridinium tetrafluoroborate	MeOH	245(4.14)	35-3365-65
$C_7H_{10}BrFO_2$			
Crotonic acid, 3-(bromofluoromethyl)-, ethyl ester	MeOH	217(4.12)	5-0021-65A
Crotonic acid, 3-(bromomethyl)-2-fluoro-, ethyl ester	MeOH	228(4.11)	5-0020-64I
$C_7H_{10}BrN_3$			
Pyrimidine, 5-bromo-1,2-dihydro-2-imino-1,4,6-trimethyl-	pH 5.0	236(4.15),316(3.72)	39-5542-65
Pyrimidine, 5-bromo-4,6-dimethyl-2-(methylamino)-	pH -0.5	241(4.29),327(3.67)	39-5542-65
	pH 7.0	245(4.25),314(3.53)	39-5542-64
$C_7H_{10}Br_2$			
Norcarane, 7,7-dibromo-	hexane	203.5(3.48)	44-2951-64
$C_7H_{10}Br_2O$			
Cyclobutanone, 2,2-dibromo-3,4,4-trimethyl-	C_6H_{12}	331(2.19),338(2.22)	22-1968-64
	MeOH	335.5(2.10)	22-1968-64
	MeOH	335.5(2.10)	22-1976-64
$C_7H_{10}ClFO_2$			
Crotonic acid, 4-chloro-3-(fluoromethyl)-, ethyl ester	MeOH	210(4.07)	5-0021-65A
$C_7H_{10}ClNO$			
1-Cyclohexene-1-carboxaldehyde, 2-chloro-, oxime	EtOH	247.5(4.15)	44-1126-65
$C_7H_{10}ClNS$			
1-[(Methylthio)methyl]pyridinium chloride	MeOH	258(3.58)	35-5661-65
$C_7H_{10}Cl_3NOS_2$			
2-Butanone, 3-methyl-4-(trichloromethyl)-4-(thiocarbamoylthio)-	H_2O	237(3.80),286(4.14)	39-4004-64
	EtOH	235(3.97),290(4.09)	39-4004-64
2-Pentanone, 5,5,5-trichloro-4-(N-methylthiocarbamoylthio)-	H_2O	246(3.91),275(3.93)	39-4004-64
	EtOH	245(3.93),284(3.88)	39-4004-64
$C_7H_{10}N_2$			
Cyclopentadiene, acetyl-, hydrazone	MeOH	258(3.98)	5-0039-64H
Cyclopentadienone, dimethylhydrazone	hexane	268(3.31),327(4.41)	5-0039-64H
Toluene, α,4-diamino-	EtOH	241(4.02)	78-0861-64
$C_7H_{10}N_2O$			
ε-Caprolactam, α-cyano-	MeOH	203(3.63)	24-1970-64
Cyclohexa[c]isoxazole, 3-amino-4,5,6,7-tetrahydro-	EtOH	243(3.87)	88-2151-64
2,6-Lutidine, 3-amino-, 1-oxide	EtOH	325(3.5)	94-0963-65
2,6-Lutidine, 4-amino-, 1-oxide	EtOH	274(4.4)	94-0963-65
2-Piperidone, 3-cyano-1-methyl-	MeOH	209(3.68)	24-1970-64
2-Piperidone, 3-cyano-3-methyl-	MeOH	211(3.67)	24-1970-64
Pyrazine, 2-ethoxy-3-methyl-	MeOH	214(3.94),277s(--), 293(3.75)	44-2623-64
Pyrazine, 2-methoxy-3,6-dimethyl-	MeOH	214(3.93),295(3.87)	44-2623-64
Pyridine, 2-(dimethylamino)-, N^1-oxide	pH 1	251(4.0),328(3.7)	54-0249-64
Pyridine, 2-(dimethylamino)-, N^2-oxide	pH 1	254f(3.3)	54-0249-64

Compound	Solvent	$\lambda_{max}(\log \epsilon)$	Ref.
2-Pyrrolidone, 3-cyano-1,3-dimethyl-	MeOH	209(3.63)	24-1970-64
$C_7H_{10}N_2OS$			
Hydantoin, 3-methyl-1,5-trimethylene-2-thio-	97% HOAc	243(--),270(4.16)	65-0554-65
5H-Imidazo[2,1-b][1,3]thiazin-5-one, 2,3,6,7-tetrahydro-6-methyl-	MeOH	231(4.04)	44-1720-64
hydrochloride	H_2O	238(4.11)	44-1720-64
4-Isothiazolecarboxamide, 5-ethyl-3-methyl-	EtOH	254(3.80)	39-0446-64
4-Pyrimidinol, 5-isopropyl-6-mercapto-	pH -2.7	229(3.93),268(3.88)	39-3204-64
	pH 2.0	236(3.96),307(3.97)	39-3204-64
	pH 9.0	233(4.09),297(3.97)	39-3204-64
	EtOH	231(4.07),291(3.78)	39-3204-64
Sulfoximine, S-(p-aminophenyl)-S-methyl-	N HCl	282(4.25)	25-1261-64
	EtOH	269(4.30)	25-1261-64
$C_7H_{10}N_2OS_3$			
4-Isothiazolecarboxamide, N-methyl-3,5-bis(methylthio)-	EtOH	233(4.06),283(4.06)	44-0665-64
$C_7H_{10}N_2O_2$			
1,2-Propanediol, 3-(4-pyrimidinyl)-	EtOH	245(3.54),270(2.65)	44-2398-65
Pyrazine, 2-ethoxy-3-methyl-, 4-oxide	MeOH	217(4.22),264(3.97), 303(3.54),306(--)	44-2623-64
2-Pyrazinemethanol, 3-ethoxy-	H_2O	215(4.11),294(3.90)	44-2623-64
4-Pyridinemethanol, 5-amino-3-hydroxy-2-methyl-	pH 1	227(4.20),273(3.72), 316(3.79)	4-0144-65
	pH 13	212(4.27),316(3.85)	4-0144-65
	EtOH	218(4.24),243(3.83), 306(3.71)	4-0144-65
4,6-Pyrimidinediol, 5-isopropyl-	pH -2.7	250(3.98)	39-3204-64
	pH 3.5	261(4.11)	39-3204-64
	pH 8.6	259(4.02)	39-3204-64
	EtOH	244(3.66),264(3.66)	12-0567-64
4,6-Pyrimidinediol, 2-propyl-	pH -3.0	242(3.97)	12-0567-64
	pH 2.5	253(4.11)	12-0567-64
	pH 9.4	252(3.94)	12-0567-64
	EtOH	238s(3.53),254s(3.68), 256(3.69)	12-0567-64
4-Pyrimidinol, 5-(ethoxymethyl)-	EtOH	224(3.8),270(3.62)	94-0393-64
$C_7H_{10}N_2O_2S$			
Hydrazine, [o-(methylsulfonyl)phenyl]-	EtOH	247(3.95),318(3.61)	95-0158-65
Hydrazine, [p-(methylsulfonyl)phenyl]-	EtOH	275(4.18)	95-0158-65
5-Pyrimidineethanol, 4-hydroxy-2-(methylthio)-	pH 1	250(3.97),273(3.98)	44-2670-64
	pH 13	248(3.97),280(3.90)	44-2670-64
4-Thiazolecarboxamide, 5-ethoxy-2-methyl-	EtOH	284(3.80)	39-1190-64
Uracil, 5-(propylthio)-	pH 1	212(3.98),229s(3.80)	39-3987-65
	pH 7.2	212(3.98),229s(3.80)	39-3987-65
	pH 13	235(3.93),294(3.88)	39-3987-65
$C_7H_{10}N_2O_3$			
γ-Butyrolactone, α-[(3-methylureido)-methylene]-	EtOH	272(4.45)	65-2599-64
Isoxazole, 3-tert-butyl-4-nitro-	isooctane	224(3.73),244s(3.67), 326s(2.01)	32-0915-64
	MeOH	220(3.74),247s(3.67), 325(2.17)	32-0915-64

Compound	Solvent	λ_{max}(log ε)	Ref.
Isoxazole, 5-tert-butyl-4-nitro-	isooctane	213(3.80),261(3.77), 322s(2.20)	32-0915-64
	H_2O	269(3.72),337s(2.36)	32-0915-64
Pyridazine, 3,6-dimethoxy-4-methyl-, 1-oxide	EtOH	252(3.95),334(3.93)	95-0344-65
Pyridazine, 3,6-dimethoxy-5-methyl-, 1-oxide	EtOH	254(3.81),322(3.83)	95-0344-65
5-Pyrimidinecarboxylic acid, 1,2,3,4-tetrahydro-2-oxo-, ethyl ester	EtOH	214(3.92),288(3.98)	94-0804-64
2,4-Pyrrolidinedione, 3-acetamido-1-methyl-	H_2O	266(3.85)	44-2085-64
3-Pyrroline-3-carboxylic acid, 4-amino-5-oxo-, ethyl ester	EtOH	233(3.61),292(4.16)	39-0766-64
	EtOH-NaOH	292(4.15)	39-0766-64
Uracil, 5-(2-hydroxyethyl)-1(or 3)-methyl-	EtOH	264(3.92)	65-2599-64
Uracil, 5-(methoxymethyl)-6-methyl-	EtOH	265(3.929)	65-2171-64
$C_7H_{10}N_2O_3S$			
5-Hydantoinpropionic acid, 3-methyl-2-thio-	97% HOAc	235(3.95),266(4.25)	65-0554-65
4-Pyrimidinecarboxaldehyde, 1,2-dihydro-6-hydroxy-2-thioxo-, dimethyl acetal	pH 7.2	217(4.2),266(4.1), 310s(3.9)	87-0337-64
	pH 12.0	216(4.1),235(3.8), 262(3.9),315(4.1)	87-0337-64
$C_7H_{10}N_2O_3S_2$			
Acetic acid, [[(1-methyl-5-oxo-2-thioxo-4-imidazolidinyl)methyl]thio]-	EtOH	235(3.91),265(4.19)	65-0988-65
$C_7H_{10}N_2O_4$			
Malonic acid, diazo-, diethyl ester	C_6H_{12}	252(3.87),352(1.36)	44-4366-65
1,3,4-Oxadiazole-5-carboxylic acid, 2-ethoxy-, ethyl ester	EtOH	237(3.99)	5-0145-65F
$C_7H_{10}N_2O_5$			
Fumaric acid, ureido-, ethyl ester	H_2O	260(2.73)	65-0159-64
	pH 13	284(3.69)	65-0159-64
$C_7H_{10}N_2O_6$			
Δ^2-1,3,4-Oxadiazoline-4,5-dicarboxylic acid, 2-methoxy-, dimethyl ester	CH_2Cl_2	229(3.45)	5-0138-65J
$C_7H_{10}N_2S$			
Diethylsulfonium dicyanomethylide	EtOH	273(3.76)	44-2384-65
$C_7H_{10}N_4$			
5H-Cyclopentapyrimidine, 2,4-diamino-6,7-dihydro-	pH 1	278(3.91)	44-1837-65
	pH 10	232(3.90),285(3.98)	44-1837-65
5H-Pyrrolo[3,4-d]pyrimidine, 4-amino-6,7-dihydro-2-methyl-	pH 1	248(3.90)	44-0194-65
	EtOH	234(3.86),269(3.63)	44-0194-65
5H-Pyrrolo[3,4-d]pyrimidine, 4-amino-6,7-dihydro-7-methyl-	pH 1	250(3.93)	44-0194-65
	EtOH	235(3.96),269(3.61)	44-0194-65
$C_7H_{10}N_4O$			
as-Triazine, 3-acetamido-5,6-dimethyl-	EtOH	235(4.30),293(3.71)	94-1329-64
$C_7H_{10}N_4O_2$			
Lathyrine, hydrochloride	pH 1	222(4.18),298(3.64)	44-0115-65
	pH 13	226(4.11),292(3.64)	44-0115-65

Compound	Solvent	$\lambda_{max}(\log \epsilon)$	Ref.
as-Triazine, 2,3-dihydro-2-hydroxy-3-imino-5,6-dimethyl-, acetate	EtOH	276(4.07),364(3.88)	94-1329-64
$C_7H_{10}N_4O_2S$			
4-Pyrimidinecarboxylic acid, 5,6-diamino-2-mercapto-, ethyl ester	pH -0.5	234(3.83),297(4.41),402(3.63)	39-3221-64
	pH 5.9	239(3.89),285(4.41),375(3.61)	39-3221-64
Pseudourea, 2-[(2,4-dihydroxy-6-methyl-5-pyrimidinyl)methyl]-2-thio-	pH 1	264(3.93)	65-2774-64
$C_7H_{10}N_4O_3$			
4-Pyrimidinecarboxylic acid, 5,6-diamino-2-hydroxy-, ethyl ester	pH -4.9	307(3.92)	39-3221-64
	pH 0.0	234(4.01),365(3.91)	39-3221-64
	pH 6.3	235(4.08),348(3.89)	39-3221-64
$C_7H_{10}N_4O_5$			
Pyruvic acid, (4,6-dihydroxy-2-pyrimidinyl)-, oxime, ammonium salt	pH 1	254(3.95)	4-0049-65
	pH 11	250(4.15)	4-0049-65
$C_7H_{10}N_4S$			
Pyrimidine, 2,4-dimethyl-6-thioureido-	MeOH	207(3.99),238(3.97),291(4.43)	39-3357-65
$C_7H_{10}N_6O_5$			
Croconic acid, dihydro-, disemicarbazone	n.s.g.	250(4.11),327(4.04)	28-1512-64A
$C_7H_{10}O$			
3,5-Cycloheptadien-1-ol	EtOH	244(3.83)	35-0321-65
2-Cyclohepten-1-one	EtOH	228(4.01)	78-0031-65
Cyclopentadiene, methoxymethyl-	C_6H_{12}	247(3.60)	33-0955-65
Cyclopentadieneethanol	EtOH	246(3.10)	44-3430-64
2-Cyclopenten-1-one, 2,3-dimethyl-	EtOH	235(4.03)	70-0164-65
2,4-Hexadienal , 2-methyl-	EtOH	259(4.27),269(4.37)	22-0161-64
7-Norbornanone	isooctane	293(1.26)	44-0160-64
	EtOH	287(1.51)	44-0160-64
	$CHCl_3$	292(1.34)	44-0160-64
2-Penten-4-yn-1-ol, 2,3-dimethyl-	EtOH	229(4.13)	70-0546-65
$C_7H_{10}OS$			
4-Thiolen-2-one, 5-isopropyl-	EtOH	268(3.37)	11-0211-64
3-Thiolen-2-one, 5-propyl-	EtOH	265(3.46)	11-0211-64
4-Thiolen-2-one, 5-propyl-	EtOH	268(3.18)	11-0211-64
$C_7H_{10}O_2$			
2-Cyclobuten-1-one, 3-ethoxy-2-methyl-	EtOH	244(4.06)	12-0765-64
1,3-Cycloheptanedione	EtOH	265(3.54)	78-0031-65
	EtOH-NaOH	288(4.38)	78-0031-65
1,2-Cyclohexanedione, 3-methyl-	EtOH	272(3.1)	94-0951-65
1,3-Cyclopentanedione, 2-ethyl-	N HCl	252(4.25)	1-0441-64
	pH 1	253(4.20)	94-1300-65
	neutral	250(4.19)	94-1300-65
	pH 10	271(4.40)	94-1300-65
	N NaOH	268(4.44)	1-0441-64
	EtOH	250(4.24)	39-4472-64
	EtOH-NaOH	270(4.42)	39-4472-64
2-Cyclopenten-1-one, 2-ethoxy-	EtOH	252(3.91)	44-2513-65
2-Cyclopenten-1-one, 3-methoxy-2-methyl-	EtOH	252(4.29)	70-1648-64
	neutral	250(4.20)	94-1300-65
Furan, 2-ethoxy-4-methyl-	EtOH	221(3.7)	35-5264-65

Compound	Solvent	λ max(log ε)	Ref.
3(2H)-Furanone, 2,2,5-trimethyl-	EtOH	261(4.04)	70-0729-65
2-Heptynoic acid	EtOH	207(4.84)	44-1973-65
3-Hexyn-5-one, 2-hydroxy-5-methyl-	n.s.g.	220(3.80)	70-1318-64
Resorcinol, dihydro-2-methyl-	EtOH	262(4.23)	70-1648-64
	EtOH-NaOH	292(4.47)	70-1648-64
$C_7H_{10}O_2S$			
Valeric acid, 2-acetyl-5-mercapto-, δ-(thiolactone)	n.s.g.	300(3.86)	24-1963-64
$C_7H_{10}O_3$			
3-Butenoic acid, 3-methyl-2-oxo-, ethyl ester	EtOH	224(3.83)	39-0766-64
2-Cyclopenten-1-one, 2,3-dihydroxy-4,4-dimethyl-	EtOH	268(4.1)	5-0100-64I
3(2H)-Furanone, 4-methoxy-2,5-dimethyl-	EtOH	278.8(3.88)	25-1629-65
2-Pentenoic acid, 5-formyl-,methyl ester	heptane	204(4.16)	44-1061-65
Valeric acid, 2-(2-hydroxyethyl)-3-oxo-, γ-lactone	n.s.g.	243(4.00)	22-0651-64
$C_7H_{10}O_3S$			
Acrylic acid, 3-(acetonylthio)-, methyl ester	Et_2O	269(4.18)	24-2109-64
$C_7H_{10}O_5S$			
Acrylic acid, 3-(acetonylsulfonyl)-, methyl ester	Et_2O	206(3.99)	24-2109-64
$C_7H_{10}S$			
Dimethylsulfonium cyclopentadienylide	dioxan	283.7(4.01)	88-1757-65
$C_7H_{11}Br$			
1-Butene, 4-bromo-1-cyclopropyl-	EtOH	206.5(3.96)	22-3218-64
1,3-Hexadiene, 6-bromo-3-methyl-	EtOH	231(4.33)	22-2533-64
$C_7H_{11}BrN_4$			
Pyrimidine, 5-bromo-4-(dimethylamino)-1,2-dihydro-2-imino-1-methyl-	pH 7.0	220(4.34),255(3.94), 299(3.92)	39-0755-65
Pyrimidine, 5-bromo-2-(dimethylamino)-4-(methylamino)-	pH 1.0	229(4.50),293s(3.51)	39-0755-65
	pH 9.4	220(4.31),242(4.26), 307(3.81)	39-0755-65
Pyrimidine, 5-bromo-4-(dimethylamino)-	pH 1.0	226(4.39),304s(3.79)	39-0755-65
	pH 9.4	229(4.35),313(3.86)	
$C_7H_{11}BrO$			
Cyclobutanone, 4-bromo-2,2,3-trimethyl-,	C_6H_{12}	313.5(2.20)	22-1976-64
cis	MeOH	310(2.13)	22-1976-64
trans	C_6H_{12}	315.5(2.21)	22-1968-64
	MeOH	310.5(2.18)	22-1976-64
$C_7H_{11}ClN_2O$			
Butyramide, 4-chloro-2-cyano-N,2-dimethyl-	MeOH	209(3.19)	24-1970-64
Valeramide, 5-chloro-2-cyano-N-methyl-	MeOH	209(3.19)	24-1970-64
Valeramide, 5-chloro-2-cyano-2-methyl-	MeOH	207(2.47)	24-1970-64

Compound	Solvent	λ_{max}(log ϵ)	Ref.
$C_7H_{11}ClN_4$			
Pyrimidine, 6-chloro-4-(dimethylamino)-	pH 7.0	215(4.35),284(4.04)	39-0755-65
1,2-dihydro-2-imino-1-methyl-	pH 15.0	274(3.98)	39-0755-65
Pyrimidine, 4-chloro-2-(dimethylamino)-	pH 1.0	231(4.38),315(3.64)	39-0755-65
6-(methylamino)-	pH 7.7	222(4.38),248s(4.16), 296(3.86)	39-0755-65
Pyrimidine, 4-chloro-6-(dimethylamino)- 2-(methylamino)-	pH 1.0	222(4.32),280(3.96), 326s(3.34)	39-0755-65
	pH 7.2	222(4.40),296(3.94)	39-0755-65
$C_7H_{11}ClN_4O$			
Pyrimidine, 5-amino-4-chloro-6-(2-hy-	pH 1	305(4.12)	87-0502-65
droxypropylamino)-	pH 7	263(3.97),290(3.97)	87-0502-65
	pH 13	263(4.00),290(4.00)	87-0502-65
$C_7H_{11}ClN_4O_2$			
Pyrimidine, 5-amino-4-chloro-6-(2,3-di-	pH 1	302(4.07)	87-0502-65
hydroxypropylamino)-	pH 7, 13	260(3.95),290(3.95)	87-0502-65
$C_7H_{11}ClO$			
Cycloheptanone, 2-chloro-	hexane	294s(1.52),301(1.54), 312s(1.48),322s(1.26)	28-2374-65B
	C_6H_{12}	295s(1.52),301(1.54), 311s(1.48)	28-2374-65B
	EtOH	292(1.50)	28-2374-65B
	Me_2SO	292.5(1.46)	28-2374-65B
2-Heptenal, 2-chloro-	C_6H_{12}	235(4.09)	28-1015-65B
1-Hexen-3-one, 1-chloro-5-methyl-	EtOH	231.5(4.06)	44-0385-64
1-Penten-3-one, 1-chloro-4,4-dimethyl-	EtOH	232.5(4.09)	44-0385-64
$C_7H_{11}Cl_2N_3$			
s-Triazine, 4,6-dichloro-2,2-diethyl- 1,2-dihydro-	EtOH	290(3.27)	44-0702-65
$C_7H_{11}FO_2$			
Butenoic acid, 2-fluoro-3-methyl-, ethyl ester (isomeric mixture)	MeOH	222(3.74)	5-0001-64D
Crotonic acid, 2-fluoro-3-methyl-, ethyl ester	MeOH	222(4.12)	5-0001-64D
$C_7H_{11}I$			
Spiro[3.3]heptane, 2-iodo-	C_6H_{12}	258(2.91)	39-2281-65
$C_7H_{11}N$			
1H-Azepine, 2,3-dihydro-1-methyl-	C_6H_{12}	304(3.62)	5-0001-65B
Pyridine, 1,2-dihydro-1,4-dimethyl-	EtOH	330(3.34)	44-1647-64
$C_7H_{11}NO$			
2H-Azepin-2-one, 1,5,6,7-tetrahydro- 4-methyl-	EtOH	217.5(4.19)	94-0951-65
2-Cyclohexen-1-one, 2-amino-3-methyl-	EtOH	235(3.35),290(3.63)	94-0951-65
2-Cyclopenten-1-one, 2,3-dimethyl-, oxime	EtOH	239(4.18)	70-0164-64
Pyridine, 3-acetyl-1,4,5,6-tetrahydro-	EtOH	301(4.32)	44-2895-64
Pyrrolidinone, N-allyl-	EtOH	none	70-0371-65
Pyrrolidinone, N-propenyl-	EtOH	235(4.2)	70-0371-65
3-Pyrrolin-2-one, 3,4,5-trimethyl-	EtOH	211(4.17)	39-5999-64

Compound	Solvent	λ_{max}(log ε)	Ref.
$C_7H_{11}NO_2$			
Hexanimide, N-methyl-	THF	220.5(3.95)	24-1548-64
Methane, dicyclopropylnitro-, anion	H_2O	244(4.08)	44-2674-65
$C_7H_{11}NO_2S$			
2,4-Thiazolidinedione, 5-butyl-	n.s.g.	240(3.57)	65-0886-65
3-Thiophenecarboxylic acid, 2-amino-4,5-dihydro-, ethyl ester	MeOH	225(3.98),300(4.55)	24-1970-64
$C_7H_{11}NO_3$			
Butyrolactam, N-(β-hydroxypropionyl)-	dioxan	219(3.95)	78-3537-65
2-Piperidone, 3-carboxy-1-methyl-	MeOH	205(3.86)	24-1970-64
α-Tetronic acid, β-dimethylaminomethyl-, zwitterion	H_2O	267(3.95)	44-3560-64
$C_7H_{11}NO_3S_2$			
4-Thiazolidineacetic acid, 4-hydroxy-2-thioxo-, ethyl ester	MeOH	244(3.84),277(4.14)	44-2146-64
5-Thiazolidinecarboxylic acid, 4-hydroxy-4-methyl-2-thioxo-, ethyl ester	MeOH	244(3.74),277(4.14)	44-2146-64
$C_7H_{11}NO_5$			
Acrylic acid, 3-ethoxy-2-nitro-, ethyl ester	EtOH	380(4.52)	78-1051-64
$C_7H_{11}NS$			
Pyrrolidine-2-thione, N-allyl-	n.s.g.	270(4.0)	30-0734-64
	n.s.g.	269(3.98)	70-0702-65
	n.s.g.	270(4.0)	70-2064-64
Pyrrolidine-2-thione, 3-allyl-	n.s.g.	269(4.1)	30-0734-64
Pyrrolidine-2-thione, N-propenyl-	n.s.g.	302(4.3)	70-2064-64
1-Pyrroline, 2-(allylthio)-	n.s.g.	219(3.9)	30-0734-64
	n.s.g.	216(3.91)	70-0702-65
$C_7H_{11}NS_2$			
2H-1,3-Thiazine-2-thione, 3,6-dihydro-3,5,6-trimethyl-	H_2O	294(3.94),312s(3.93)	39-4008-64
	EtOH	270(3.45),315(3.78)	39-4008-64
2H-1,3-Thiazine-2-thione, 3,6-dihydro-4,5,6-trimethyl-	H_2O	290(3.92),318(3.93),360s(--)	39-4008-64
	EtOH	320(4.02),380s(--)	39-4008-64
2H-1,3-Thiazine-2-thione, 3,6-dihydro-4,6,6-trimethyl-	H_2O	298(4.01),311s(--)	39-4008-64
	EtOH	318(4.09)	39-4008-64
$C_7H_{11}N_3$			
Methylamine, N-(1-cyanoethyl)-N-(2-cyanoethyl)-	pH 1	259(4.10)	39-4546-65
	pH 13	263(4.09)	39-4546-65
Pyrazine, 2-(dimethylamino)-6-methyl-, hydrochloride	H_2O	247(4.04),339(3.68)	44-0415-64
Pyrimidine, 2-(isopropylamino)-	pH 1.0	230(4.27),317(3.50)	39-5542-65
	pH 7.0	236(4.25),308(3.39)	39-5542-65
$C_7H_{11}N_3O$			
Pyrazole-1-carboxamide, 3,4,5-trimethyl-	EtOH	244(3.96)	28-0606-65A
Pyrimidine, 4-(dimethylamino)-6-methoxy-	pH 0.2	262(4.16)	12-0559-65
	pH 7.0	257(4.14)	12-0559-65
2(3H)-Pyrimidinone, 4-(dimethylamino)-3-methyl-	pH 9.6	233(3.99),277(3.86)	12-0199-65
4(3H)-Pyrimidinone, 6-(dimethylamino)-3-methyl-	pH 7.0	228(4.45),270(4.11)	12-0559-65

Compound	Solvent	$\lambda_{max}(\log \epsilon)$	Ref.
$C_7H_{11}N_3OS$			
Cysteine, S-(methylthiocarbamoyl)-, methylthiohydantoin derivative	EtOH	288(4.38),320(4.35)	65-0988-65
$C_7H_{11}N_3O_2$			
Cytosine, N-(2-hydroxyethyl)-1-methyl-	neutral	275(4.00)	94-1240-64
	acid	287(4.14)	94-1240-64
Cytosine, 3-(2-hydroxyethyl)-1-methyl-	neutral	276(3.9)	94-1240-64
	acid	283(4.04)	94-1240-64
Sydnone imine, N-acetyl-3-isopropyl-	pH 1.0	280(4.10)	65-2064-64
	pH 7.00	236(3.80),310(4.10)	65-2064-64
s-Triazine, 2-ethyl-4,6-dimethoxy-	EtOH	233(3.46)	44-0702-65
$C_7H_{11}N_3O_2S$			
4-Imidazolidinepropionamide, 1-methyl-5-oxo-2-thioxo-	97% HOAc	235(3.92),265(4.24)	65-0554-65
Pyridazine, 3-(dimethylamino)-6-(methylsulfonyl)-	pH 1	257(4.22)	4-0001-64
	pH 11	275(4.28)	4-0001-64
Pyrimidine, 4-(dimethylamino)-6-(methylsulfonyl)-	pH 1	254(4.12),308(3.60)	4-0001-64
	pH 11	254(4.22),312(3.53)	4-0001-64
Uracil, 6-amino-5-(2-hydroxypropyl)-2-thio-	pH 1	242(3.89),282(4.30)	44-2670-64
	pH 13	241(4.19),264(4.05), 294(4.05)	44-2670-64
$C_7H_{11}N_3O_3$			
γ-Butyrolactone, α-acetyl-, semicarbazone	n.s.g.	222(4.07),274(3.76)	39-0141-64
Propionic acid, 2-oximino-3-(2-pyrimidinyl)-, reduction product	pH 11	251(3.96)	4-0001-65
Sydnone imine, N-carboxy-3-isopropyl-, methyl ester	pH 1.0	276(4.06)	65-2064-64
	pH 7.00	226(4.02),308(4.12)	65-2064-64
Uracil, 6-amino-5-(2-hydroxypropyl)-	pH 1	272(4.31)	44-2670-64
	pH 13	273(4.24)	44-2670-64
$C_7H_{11}N_3O_5S$			
D-Glucuronolactone, thiosemicarbazone	H_2O	231(3.89),267(4.35)	25-0546-64
$C_7H_{11}N_3S$			
Pyrimidine, 2-(dimethylamino)-4-(methylthio)-	pH 0.2	234(4.25),282(4.16)	12-0559-65
	pH 9.8	249(4.42),317(3.72)	12-0559-65
Pyrimidine, 4-(dimethylamino)-6-(methylthio)-	pH 0.0	247(4.25),289(4.19)	12-0559-65
	pH 7.0	247(4.24),289(4.17)	12-0559-65
6(1H)-Pyrimidinethione, 2-(dimethylamino)-1-methyl-	pH -0.2	237(4.08),277(3.90), 330(4.09)	12-0559-65
	pH 7.0	235(4.09),271(3.87), 351(4.15)	12-0559-65
6(1H)-Pyrimidinethione, 4-(dimethylamino)-1-methyl-	pH -3.0	236(4.15),281(4.16), 337s(3.53)	12-0559-65
	pH 7.0	259(4.31),317(4.31)	12-0559-65
$C_7H_{11}N_5$			
5H-Pyrrolo[3,4-d]pyrimidine, 2,4-diamino-6,7-dihydro-7-methyl-	pH 1	272(3.75),300s(3.31)	44-0194-65
	EtOH	283(3.88)	44-0194-65
$C_7H_{11}N_5O$			
Formamide, N-[4,6-bis(methylamino)-5-pyrimidinyl]-	pH 2,6	228(4.43),272(4.14)	39-3770-65
	pH 7.5	228(4.60),264(3.78)	39-3770-65

Compound	Solvent	λmax(log ε)	Ref.
$C_7H_{11}N_5OS$			
Morpholine, 4-(4-amino-1,2,5-thiadiazole-3-carboximidoyl)-	pH 1	317(3.78)	44-2488-65
	pH 7	316(3.78)	44-2488-65
	pH 13	306(3.86)	44-2488-65
$C_7H_{11}N_5O_5$			
Pyrazole-4,5-dione, 3-(1,2,3-trihydroxy-propyl)-, semicarbazone	n.s.g.	298(4.22)	28-5873-64A
	pH 9.30	380(3.00)	28-5873-64A
sodium salt	n.s.g.	380(4.03)	28-5873-64A
$C_7H_{11}N_5S$			
Pyrrolidine, 1-(4-amino-1,2,5-thiadia-zole-3-carboximidoyl)-	pH 1	315(3.78)	44-2488-65
	pH 7	315(3.78)	44-2488-65
	pH 13	305(3.87)	44-2488-65
C_7H_{12}			
Cycloheptene, $PdCl_2$ olefin complex from	EtOH	231s(4.13),325(3.30)	24-2037-64
1-Cyclohexene, 1-methyl-, $PdCl_2$ olefin complex from	EtOH	210(4.24),231s(4.08), 323(3.21)	24-2037-64
Cyclopentene, 1,2-dimethyl-	96% H_2SO_4	288(3.63+)	23-2768-64
Cyclopropane, 1-ethyl-1-methyl-2-methylene-	hexane	204s(--),208(2.93)	28-1695-65B
	EtOH	213(2.70)	28-1695-65B
Cyclopropene, 3-ethyl-1,3-dimethyl-	hexane	210(3.23)	28-1695-65B
	EtOH	217(3.00)	28-1695-65B
1,3-Pentadiene, 2,3-dimethyl-	C_6H_{12}	232.5(4.36)	44-3527-64
	EtOH	231.5(4.28)	44-3527-64
1,3-Pentadiene, 2,4-dimethyl-	C_6H_{12}	231(3.90)	35-5244-64
	C_6H_{12}	230.5(3.99)	44-3527-64
	EtOH	231(4.00)	44-3527-64
	Et_2O	229(4.00)	44-3527-64
1,3-Pentadiene, 3,4-dimethyl-	hexane	239(4.44)	44-3527-64
$C_7H_{12}ClNO_4S_2$			
2-(Dimethylamino)-4,5-dimethyl-1,3-di-thiol-1-ium perchlorate (hydrate)	70% $HClO_4$	246(3.5),312(3.89)	44-1703-64
$C_7H_{12}N_2$			
Cyclohexene-1-azomethane	EtOH	249(3.98)	35-4576-65
$C_7H_{12}N_2O$			
Pyrazole, 3-ethoxy-1,5-dimethyl-	C_6H_{12}	225.5(3.74)	78-0315-64
	20N H_2SO_4	229.5(4.01)	78-0315-64
	pH 7	221.5(3.80)	78-0315-64
Pyrazole, 5-ethoxy-1,3-dimethyl-	C_6H_{12}	219(3.68)	78-1693-65
	20N H_2SO_4	224(3.89)	78-0299-64
	pH 5	none	78-0299-64
	H_2O	219(3.77)	78-1693-65
Pyrazole, 5-ethoxy-2,3-dimethyl-	C_6H_{12}	225(3.74)	78-1693-65
	H_2O	221(3.80)	78-1693-65
5-Pyrazolinone, 1,3,4,4-tetramethyl-	C_6H_{12}	250(3.63)	78-0299-64
	29N H_2SO_4	260.5(3.59)	78-0299-64
	pH 5	248(3.56)	78-0299-64
	H_2O	248(3.57)	78-1693-65
5-Pyrazolinone, 3,4-tetramethylene-	C_6H_{12}	229(3.52),250(3.32)	78-1693-65
	H_2O	246(3.97)	78-1693-65
2(1H)-Pyrimidinone, 3,4-dihydro-4,4,6-trimethyl-	MeOH	236(3.51)	77-0439-65

Compound	Solvent	$\lambda_{max}(\log \epsilon)$	Ref.
$C_7H_{12}N_2OS$			
Hydantoin, 5-isopropyl-3-methyl-2-thio-	EtOH	235(4.00),266(4.23)	65-0049-65
Pseudothiohydantoin, 5-butyl-	n.s.g.	248(3.90),300s(1.93)	65-0886-65
2-Thiazoline, 2-(2-acetamidoethyl)-	EtOH	233(3.49),248(3.45)	94-0180-65
$C_7H_{12}N_2OS_2$			
Hydantoin, 3-methyl-5-[2-(methylthio)-ethyl]-2-thio-	EtOH	235(3.98),267(4.26)	65-0049-65
$C_7H_{12}N_2O_2$			
1,3-Cyclohexanedione, 2-methyl-, dioxime	EtOH	280(2.66)	70-1648-64
2-Piperidone, 3-carbamoyl-1-methyl-	MeOH	206(3.87)	24-1970-64
1-Pyrazolin-3-ol, 3,5-dimethyl-, acetate	EtOH	330(2.27)	44-1379-64
$C_7H_{12}N_2O_3S$			
Malonamic acid, 2-acetamido-3-thio-, ethyl ester	EtOH	271(4.00)	39-1190-64
$C_7H_{12}N_2S_2$			
3-Pyrroline-3-thiocarboxamide, 4-mercapto-1,5-dimethyl-	pH 13	248(3.52),291(3.48), 368(4.33)	39-4546-65
	EtOH	214(4.06),296(3.26), 373(4.37)	39-4546-65
$C_7H_{12}N_4$			
Pyrimidine, 2-(dimethylamino)-1,4-dihydro-4-imino-1-methyl-	pH 7.0	211(4.34),250s(4.02), 279(3.94)	39-0755-65
	pH 15.0	268(3.92)	39-0755-65
Pyrimidine, 2-(dimethylamino)-4-(methylamino)-	pH 1.0	226(4.50),280s(3.61)	39-0755-65
	pH 10.4	219(4.36),245s(4.15), 296(3.82)	39-0755-65
$C_7H_{12}N_4S_2$			
2-Thiadiazolinone, 3-methyl-5-(methylthio)-, azine with acetone	EtOH	262(3.98),316(4.07)	1-0871-64
$C_7H_{12}O$			
Cyclobutanone, 2,2,3-trimethyl-	C_6H_{12}	287(1.38),295(1.43)	22-1968-64
	MeOH	255(1.28),257(1.32), 264(1.38),291(1.49)	22-1968-64
Cycloheptanone	hexane	289(1.18),300s(1.15), 310s(1.00)	28-2374-65B
	C_6H_{12}	290(1.18),300s(1.15), 310s(0.98),322s(0.48)	28-2374-65B
	EtOH	283.5(1.32)	28-2374-65B
	Me_2SO	288(1.26)	28-2374-65B
Cyclohexanone, 2-methyl-	EtOH	285(1.24)	59-0529-65
	MeCN	287(1.20)	59-0529-65
	HOAc	282(1.27)	59-0529-65
	CF_3COOH	268(1.48)	59-0529-65
	CCl_4	291(1.32)	59-0529-65
3,5-Hexadien-1-ol, 4-methyl-,	EtOH	230.3(4.41)	22-2533-64
cis	EtOH	233(4.23)	22-2533-64
trans	EtOH	229(4.37)	22-2533-64
3,5-Hexadien-2-ol, 2-methyl-	n.s.g.	230(3.90),232(3.87)	70-0174-64
2-Pentenal, 2,4-dimethyl-	EtOH	226(3.93)	22-0161-64
1-Penten-3-one, 4,4-dimethyl-	EtOH	217(3.73),327(1.56)	22-0704-65

Compound	Solvent	λ_{max}(log ε)	Ref.
$C_7H_{12}O_2$			
Cyclohexanecarboxylic acid, lead salt	C_6H_{12}	243(4.38)	88-2847-65
$C_7H_{12}O_3$			
2-Furanacetic acid, tetrahydro-, methyl ester	EtOH	259.5(2.06)	44-1061-65
Glutaconaldehyde, 5-(dimethyl acetal)	EtOH	220(4.05)	70-2097-64
$C_7H_{12}O_3S_2$			
Xanthic acid, butyl-, ester with mercaptoacetic acid	H_2O	221(3.92),281(4.20), 343(1.80)	1-1113-65
	EtOH	223(3.90),280(4.21), 354(1.75)	1-1113-65
$C_7H_{12}S$			
Sulfide, 1-cyclohexen-1-yl methyl	C_6H_{12}	224(3.81),249s(3.4)	59-0299-64
isomer	C_6H_{12}	224(3.89),248s(3.2)	59-0299-64
$(C_7H_{12}S_3)_n$			
Polyhexamethylenetrithiocarbonate	$CHCl_3$	310(4.11)	49-0631-65
$C_7H_{13}N$			
1-Pyrroline, 2,4,4-trimethyl-	hexane	230(2.29)	39-2313-65
	EtOH	221(2.32)	39-2313-65
$C_7H_{13}NO$			
1-Pyrroline, 2,3,3-trimethyl-, 1-oxide	MeOH	229(3.96)	23-2717-65
$C_7H_{13}NOS$			
Acrylamide, 3-(tert-butylthio)-, cis	EtOH	215(3.32),282(4.11)	35-5307-64
	EtOH	215(3.31),282(4.11)	88-1471-64
Valeramide, 4,4-dimethyl-2-oxothio-	EtOH	269.0(4.06)	39-1262-65
$C_7H_{13}NOS_2$			
Butyraldehyde, α-methyl-β-(N-methylthiocarbamoylthio)-	H_2O	247(3.93),281(4.12)	39-4004-64
	EtOH	245(3.96),286(4.11)	39-4004-64
Carbamic acid, dimethyldithio-, ester with 3-mercapto-2-butanone	EtOH	248(3.90),277(3.93)	44-1703-64
2-Pentanone, 3-methyl-4-(thiocarbamoylthio)-	H_2O	242(3.90),283(4.17)	39-4004-64
	EtOH	240(3.87),289(4.13)	39-4004-64
2-Pentanone, 4-methyl-4-(thiocarbamoylthio)-	H_2O	242(3.80),285(4.24)	39-4004-64
	EtOH	241(3.88),290(4.15)	39-4004-64
2-Thiazolidinethione, 4-hydroxy-3,4,5,5-tetramethyl-	MeOH	255s(3.97),274(4.22)	44-2146-64
$C_7H_{13}NO_2$			
Cycloheptane, nitro-, anion	H_2O	229(4.08)	44-2674-65
	50% dioxan	230(4.11)	44-2674-65
Methane, cyclohexylnitro-, anion	H_2O	234.5(4.04)	44-2674-65
$C_7H_{13}NO_3$			
Valerolactam, N-(2-methoxyacetyl)-	dioxan	220(3.98)	78-3537-65
$C_7H_{13}NO_4S_2$			
Nereistoxin hydrogen oxalate	H_2O	320(2.18)	94-0253-65
$C_7H_{13}N_3O$			
3-Penten-2-one, 4-methyl-, semicarbazone	EtOH	267(4.19)	39-6072-65
2-Pyrazoline-1-carboxamide, 3,5,5-trimethyl-	EtOH	240(4.05)	39-6072-64

Compound	Solvent	λ_{max}(log ε)	Ref.
$C_7H_{13}N_5$			
Pyrimidine, 4,5,6-tris(methylamino)-	pH 3.5	228(4.12),230(4.12) 280(4.14)	39-3770-65
	pH 8.5	227(4.36),275(3.98)	39-3770-65
$C_7H_{13}N_5S$			
1,2,5-Thiadiazole-3-carboxamidine, 4-amino-N-butyl-	pH 1	215(3.98),320(3.79)	44-2135-64
	pH 7	215(3.97),320(3.79)	44-2135-64
	pH 13	317(3.93)	44-2135-64
	EtOH	270(3.46),325(3.98)	44-2135-64
hydrochloride	EtOH	322(3.81)	44-2135-64
s-Triazine-2-thiol, 4-amino-6-(butylamino)-	EtOH	286(4.30)	39-6296-65
s-Triazine-2(1H)-thione, 1-butyltetrahydro-4,6-diimino-	EtOH	240s(3.79),286(4.35)	39-6296-65
$C_7H_{13}O_2P$			
2-Phospholene, 1-ethoxy-3-methyl-, 1-oxide	MeOH	197(3.95)	78-1593-64
3-Phospholene, 1-ethoxy-3-methyl-, 1-oxide	MeOH	200(3.40),222(1.94)	78-1593-64
3-Phospholene, 1-isopropoxy-, 1-oxide	MeOH	200(2.65)	78-1593-64
3-Phospholene, 1-methoxy-3,4-dimethyl-, 1-oxide	MeOH	200(3.81),232(2.22)	78-1593-64
C_7H_{14}			
1-Pentene, 4,4-dimethyl-	n.s.g.	179(4.0)	70-2203-64
$C_7H_{14}ClNO$			
Trimethyl(3-oxo-1-butenyl)ammonium chloride	EtOH	207(4.86)	44-0385-64
$C_7H_{14}NNaS_2$			
Carbamic acid, diisopropyldithio-, sodium salt	H_2O	257(4.12),283(4.21), 350(1.84)	1-1113-65
	EtOH	258(4.10),292(4.21), 362(1.80)	1-1113-65
$C_7H_{14}N_2O$			
2-Imidazoline, 1-(2-hydroxyethyl)-2-ethyl-	EtOH	231(3.75)	4-0188-64
2-Piperidone, 3-(aminomethyl)-1-methyl-	MeOH	205(2.74)	24-1970-64
$C_7H_{14}N_2O_2$			
2-Pentanone, 4-methyl-4-(N-methyl-N-nitrosamino)-	H_2O	228.5(3.98)	28-5470-64A
$C_7H_{14}N_4$			
3-Penten-2-one, 4-methyl-, amidinohydrazone	EtOH	235(4.00)	39-6072-64
$C_7H_{14}O_2$			
Acetic acid, neopentyl ester	MeOH	206.3(1.79)	35-2384-64
	EtOH	206.7(1.79)	35-2384-64
	MeCN	207.4(1.77)	35-2384-64
Heptanoic acid, lead salt	C_6H_{12}	230(4.40)	88-2847-65
3-Heptanone, 4-hydroxy-	EtOH	280(1.64)	59-0529-65
	MeCN	276(1.63)	59-0529-65
	HOAc	276(1.66)	59-0529-65
	CF_3COOH	270(1.81)	59-0529-65
	CCl_4	274(1.65)	59-0529-65

Compound	Solvent	λ max(log ε)	Ref.
$C_7H_{14}O_3S$			
Butyric acid, 2-hydroxy-4-mercapto-3-methyl-, ethyl ester	EtOH	238(2.57)	39-0766-64
$C_7H_{14}S$			
4-Heptanethione	C_6H_{12}	215(3.71),230(3.80), 503(0.95)	59-0299-64
	C_6H_{12}	450s(--),503(0.95)	89-0157-64
	EtOH	213(3.70),231(3.79), 492(0.90)	59-0299-64
Sulfide, butyl propenyl	EtOH	225(3.80),244s(3.54)	44-2211-64
$C_7H_{14}S_2$			
m-Dithiane, 2-isopropyl-	C_6H_{12}	250(2.84)	78-0437-64
	EtOH	249(2.84)	78-0437-64
1,3-Dithiepane, 2,2-dimethyl-	EtOH	250(2.63)	78-0437-64
1,3-Dithiepane, 2-ethyl-	EtOH	230s(2.74),245s(2.63)	78-0437-64
$C_7H_{15}ClN_2O$			
Trimethyl(3-oxo-1-butenyl)ammonium chloride, oxime	EtOH	232.5(4.10)	44-1129-65
$C_7H_{15}NOS$			
2H-1,5-Thiazocine, hexahydro-5-methyl-, 1-oxide	C_6H_{12}	211(3.70),225(3.58)	88-4259-65
$C_7H_{15}NO_2$			
Heptane, 4-nitro-, anion	H_2O	231(4.05)	44-2674-65
$C_7H_{15}NS$			
Butyramide, N,N,2-trimethylthio-	MeOH	272(4.11),337(1.67)	35-0051-65
2H-1,5-Thiazocine, hexahydro-5-methyl-	C_6H_{12}	212(3.58)	88-4259-65
$C_7H_{15}NS_2$			
Carbamic acid, diisopropyldithio-, sodium salt	H_2O	257(4.12),283(4.21), 350(1.84)	1-1113-65
	EtOH	258(4.10),292(4.21), 362(1.80)	1-1113-65
$C_7H_{15}N_3O_4$			
Propylamine, N,N-diethyl-2,2-dinitro-	hexane	280(2.57)	44-0354-65
	pH 1	275(1.65)	44-0354-65
	H_2O	380(4.12)	44-0354-65
	pH 13	380(4.25)	44-0354-65
	MeOH	377(4.16)	44-0354-65
	EtOH	377.5(4.21)	44-0354-65
	CCl_4	280(2.56)	44-0354-65
$C_7H_{15}N_3O_4S$			
L-Rhamnose, thiosemicarbazone (hydrate)	H_2O	237(4.11)	25-0546-64
$C_7H_{15}N_3O_5S$			
D-Galactose, thiosemicarbazone	H_2O	236(4.10)	25-0546-64
D-Glucose, thiosemicarbazone	H_2O	236(4.13)	25-0546-64
D-Mannose, thiosemicarbazone	H_2O	235(4.11)	25-0546-64
$C_7H_{16}S_2$			
Acetone, diethyl mercaptole	EtOH	239(2.85)	78-0437-64
Propionaldehyde, diethyl mercaptal	EtOH	231(2.92)	78-0437-64

Compound	Solvent	$\lambda_{max}(\log \epsilon)$	Ref.
$C_7H_{16}S_3$			
Orthoformic acid, trithio-, triethyl ester	C_6H_{12}	237(3.01)	78-0437-64
	EtOH	235(2.98)	78-0437-64
$C_7H_{16}Si$			
Silane, 3-buten-1-yltrimethyl-	n.s.g.	175(4.3)	70-2203-64
$C_7H_{18}N_8O_7S$			
D-Threo-pentulose, bis(amidinohydrazone), monosulfate	EtOH	289(4.57)	44-3074-64

Compound	Solvent	$\lambda_{max}(\log \epsilon)$	Ref.
$C_8ClF_9O_2S$			
Benzenesulfonyl chloride, 2,4,5-tri-fluoro-3,6-bis(trifluoromethyl)-	n.s.g.	225(4.24),258s(3.88), 287(3.94)	39-2975-64
C_8Cl_8			
Perchlorobutenyne dimer	heptane	245(4.21),286(3.56)	88-1107-64
C_8F_9KS			
Potassium, [(nonafluoro-2,5-xylyl)thio]-	EtOH	287.0(3.94)	39-2975-64
$C_8HCl_4F_3N_2$			
Benzimidazole, 4,5,6,7-tetrachloro-2-(trifluoromethyl)-	n.s.g.	223(4.92),273(3.94), 290(3.85),301(3.81)	89-0814-65
C_8HF_9			
Benzene, 1,2,5-trifluoro-3,4-bis(tri-fluoromethyl)-	EtOH	201(4.38),237(3.30), 274(3.54)	39-2975-64
C_8HF_9S			
Thiophenol, 2,3,6-trifluoro-4,5-bis-(trifluoromethyl)-	EtOH	234(4.37),300(4.13)	39-2975-64
Thiophenol, 2,4,5-trifluoro-3,6-bis-(trifluoromethyl)-	EtOH	235(3.97),279(3.76), 375(3.58)	39-2975-64
$C_8H_2Cl_2N_2S_2$			
Benzo[1,2-d:5,4-d']bisthiazole, 2,6-di-chloro-	EtOH	250(4.64),258(4.62), 307(3.38)	7-0080-64
$C_8H_2Cl_4O_4$			
1,4-Dioxaspiro[4,5]deca-7,9-diene-2,6-dione, 7,8,9,10-tetrachloro-	CCl_4	368(3.63)	44-1657-65
$C_8H_2F_5NO_2$			
Styrene, 2,3,4,5,6-pentafluoro-β-nitro-	EtOH	264(2.59)	30-1135-64
$C_8H_2F_8S_2$			
p-Benzenedithiol, 2,5-difluoro-3,6-bis-(trifluoromethyl)-	hexane	217(4.64),236(4.57), 261(4.71),333(4.08)	39-2975-64
Benzenedithiol, difluoro-o-bis(tri-fluoromethyl)-	n.s.g.	234(4.78),323(5.22)	39-2975-64
$C_8H_3F_5$			
Styrene, 2,3,4,5,6-pentafluoro-	EtOH	240(4.12),274(2.76)	30-1135-64
$C_8H_3F_5O$			
Acetophenone, 2',3',4',5',6'-penta-fluoro-	EtOH	227(4.31),274(3.46)	30-1135-64
$C_8H_3F_5O_2$			
Acetic acid, (pentafluorophenyl)-	EtOH	263(2.6)	70-1798-65
$C_8H_3N_3O_6$			
Indole-2,3-dione, 5,7-dinitro-	EtOH	239(4.07),257s(4.02), 323(4.06),393(3.42)	44-0947-64
$C_8H_4BrMnO_3$			
Manganese, (bromocyclopentadienyl)tri-carbonyl-	EtOH	332(3.05)	101-0269-65A

Compound	Solvent	$\lambda_{max}(\log \epsilon)$	Ref.
$C_8H_4BrNO_4$			
Bicyclo[4.2.0]octa-1,3,5-trien-7-one, 2-bromo-3-hydroxy-4-nitro-	EtOH	258(4.36)	39-5343-64
$C_8H_4Br_2O$			
Benzocyclobutenone, 2,2-dibromo-	EtOH	213(4.78),247(4.08), 292(3.63),298(3.63), 330s(2.34),344s(2.08)	44-2947-64
$C_8H_4Br_2S$			
Cyclopenta[c]thiopyran, 5,7-dibromo-	hexane	232(4.15),253(4.04), 261(4.07),283(4.41), 293(4.53),330(3.48), 339(3.47),357(3.40), 498(3.20),582s(2.67), 597s(2.48),608s(2.19)	35-0708-64
$C_8H_4ClMnO_3$			
Manganese, tricarbonyl(chlorocyclopentadienyl)-	EtOH	332(3.01)	101-0269-65A
C_8H_4ClNOS			
Benzoic acid, p-chlorothio-, anhydride with isocyanic acid	n.s.g.	527(2.11)	24-2954-65
$C_8H_4ClNO_2S$			
Cyclopenta[c]thiopyran, 5(7)-chloro-7(5)-nitro-	Et_2O	244(4.07),256(4.11), 264(4.12),332(4.23), 360(3.70),375(3.72), 400(3.57),495(3.62)	35-0708-64
$C_8H_4Cl_2S$			
Cyclopenta[c]thiopyran, 5,7-dichloro-	hexane	234(4.18),252(4.15), 260(4.15),281(4.43), 291(4.56),329(3.57), 333(3.57),339(3.56), 356(3.41),497(3.19), 586s(2.67),601s(2.46), 617s(2.19)	35-0708-64
$C_8H_4FMnO_3$			
Manganese, tricarbonyl(fluorocyclopentadienyl)-	EtOH	330(3.09)	101-0269-65A
$C_8H_4F_3N_5$			
1,3-Pentadiene-1,1,3-tricarbonitrile, 2,4-diamino-5,5,5-trifluoro-	EtOH	240(4.60),252(4.41), 301(3.79),330(3.79)	44-0707-64
$C_8H_4F_4O_2$			
Acetic acid, (2,3,5,6-tetrafluorophenyl)-	EtOH	226(2.38),266(3.14)	70-1798-65
$C_8H_4F_6$			
Bi-1-cyclobuten-1-yl, 2,2',3,3,3',3'-hexafluoro-	n.s.g.	237.5(3.92)	44-1445-64
$C_8H_4F_6N_2O_2$			
Aniline, p-nitro-N,N-bis(trifluoromethyl)-	isooctane	247(2.97),340s(2.08)	35-4341-65

Compound	Solvent	$\lambda_{max}(\log \epsilon)$	Ref.
$C_8H_4F_6O_2$			
Benzene, 1,4-bis(trifluoromethoxy)-	hexane	264-270(2.76-2.68)	65-2749-64
	EtOH	262-267(2.55-2.47)	65-2749-64
	Et_2O	263-268(2.63-2.55)	65-2749-64
	dioxan	265(2.47)	65-2749-64
	H_2SO_4	265(2.29),285s(1.75)	65-2749-64
$C_8H_4F_6O_4S_2$			
Benzene, 1,4-bis[(trifluoromethyl)-sulfonyl]-	hexane	222(4.11),270(3.26), 286(3.33)	65-2749-64
	EtOH	224(4.18),282(3.53), 290(3.45)	65-2749-64
	Et_2O	223(4.17),275(3.40), 288(3.45)	65-2749-64
	dioxan	225(4.14),283(3.45)	65-2749-64
	H_2SO_4	225(4.27),283(3.57), 290(3.48),310s(2.90)	65-2749-64
$C_8H_4F_6S_2$			
Benzene, 1,4-bis[(trifluoromethyl)thio]-	hexane	216(4.00),256(3.77), 278(3.12)	65-2749-64
	EtOH	255(3.85),277(3.17)	65-2749-64
	Et_2O	214(4.18),254(3.95), 279(3.27)	65-2749-64
	dioxan	254(3.90),277(3.26)	65-2749-64
	H_2SO_4	216(3.82),254(3.63), 280(3.07)	65-2749-64
$C_8H_4F_8$			
Cyclohexane, octafluoro-1,2-dimethylene-	EtOH	208.0(3.79)	39-3035-64
$C_8H_4IMnO_3$			
Manganese, tricarbonyl(iodocyclo-pentadienyl)-	EtOH	333(3.04)	101-0269-65A
$C_8H_4I_2S_2$			
2,2'-Bithiophene, 5,5'-diiodo-	EtOH	246(3.83),324(4.30)	39-5134-65
$C_8H_4N_2O_2S_2$			
Benzo[1,2-d:5,4-d']bisthiazole-2,6-diol	EtOH	230(4.67),264s(3.68), 314(3.95)	7-0080-64
$C_8H_4N_2S_2$			
Thiazolo[4,5-f]benzothiazole	EtOH	246(4.56),252(4.55), 284(3.85),312(3.13)	7-0080-64
Thiazolo[5,4-g]benzothiazole	EtOH	228(4.28),249(4.34), 288(3.11),294(3.04), 300(3.16),312(3.02)	7-0080-64
$C_8H_4N_2S_4$			
Benzo[1,2-d:5,4-d']bisthiazole-2,6-di-thiol	EtOH	226(3.97),264(4.22), 296(4.32),352s(4.45), 365(4.61)	7-0080-64
$C_8H_4N_4O_6$			
Indole, 3,5,7-trinitro-	EtOH	216(4.30),286(4.11), 413(--)	44-0947-64

Compound	Solvent	λ_{max}(log ε)	Ref.
$C_8H_5BrN_2O$			
4-Cinnolinol, 5-bromo-	MeOH	216(4.33),234(3.96), 265(3.89),270(3.86), 292(3.61),304(3.69), 346(4.09)	4-0063-65
4-Cinnolinol, 6-bromo-	MeOH	215(4.40),243(4.31), 249(4.31),290(3.74), 301(3.81),348(4.10), 363(4.10)	4-0063-65
4-Cinnolinol, 7-bromo-	MeOH	217(4.36),243(4.11), 257(4.19),265(4.16), 289(3.41),339(4.09), 352(4.10)	4-0063-65
4-Cinnolinol, 8-bromo-	MeOH	213(4.38),234(3.84), 261(3.91),267(3.84), 286(3.53),294(3.49), 344(4.17)	4-0063-65
$C_8H_5BrN_2S$			
4-Cinnolinethiol, 6-bromo-	MeOH	231(4.56),273(3.95), 284(3.85),299(3.38), 340(3.41),361(3.53), 429(4.28)	4-0063-65
4-Cinnolinethiol, 8-bromo-	MeOH	227(4.46),263(3.66), 271(3.65),281(3.61), 320(3.36),426(4.27)	4-0063-65
C_8H_5BrO			
Benzocyclobutenone, 2-bromo-	EtOH	210(4.47),245(3.92), 288(3.55),295(3.55), 320s(2.41),345s(2.10)	44-2947-64
$C_8H_5BrO_2$			
Bicyclo[4.2.0]octa-1,3,5-trien-7-one, 2-bromo-5-hydroxy-	EtOH	235(4.25),275(4.00), 297(4.00),335(3.28)	39-5343-64
$C_8H_5BrS_2$			
2,2'-Bithiophene, 5-bromo-	EtOH	249(3.73),311(4.13)	39-5134-65
$C_8H_5Br_3N_2$			
Imidazo[1,2-a]pyridine, 3-bromo-2-(dibromomethyl)-	pH 1	235s(3.87),283(3.96)	44-4085-65
	pH 11	230(4.46),239(4.47), 280(3.8),317(3.82)	44-4085-65
$C_8H_5ClN_2O$			
4-Cinnolinol, 5-chloro-	MeOH	214(4.36),232(3.90), 259(3.90),266(3.84), 289(3.62),301(3.72), 342(4.09)	4-0063-65
4-Cinnolinol, 6-chloro-	MeOH	214(4.41),240(4.23), 245(4.23),288(3.68), 299(3.77),345(4.09), 360(4.09)	4-0063-65
4-Cinnolinol, 7-chloro-	MeOH	214(4.47),240(4.14), 253(4.20),262(4.13), 285(3.38),295(3.34), 336(4.12),350(4.13)	4-0063-65
4-Cinnolinol, 8-chloro-	MeOH	212(4.35),229(3.99), 258(3.89),266(3.84), 285(3.55),294(3.52), 342(4.14)	4-0063-65

Compound	Solvent	$\lambda_{max}(\log \epsilon)$	Ref.
1,3,4-Oxadiazole, 2-chloro-5-phenyl-	EtOH	253(4.28)	23-1607-65
$C_8H_5ClN_2O_2$			
6,7-Benzimidazoledione, 2-chloro-1-methyl-	$CHCl_3$	302(3.6),460(3.0)	65-0949-64
$C_8H_5ClN_2S$			
4-Cinnolinethiol, 5-chloro-	MeOH	231(4.48),272(3.53), 281(3.53),328(3.51), 421(4.28),430(4.26)	4-0063-65
4-Cinnolinethiol, 6-chloro-	MeOH	229(4.63),271(3.94), 281(3.85),322(3.37), 430(4.26)	4-0063-65
4-Cinnolinethiol, 7-chloro-	MeOH	227(4.60),267(3.84), 278(3.77),296(3.50), 427(4.23)	4-0063-65
4-Cinnolinethiol, 8-chloro-	MeOH	227(4.50),262(3.64), 269(3.63),279(3.60), 315(3.32),425(4.30)	4-0063-65
C_8H_5ClOS			
Benzo[b]thiophen-3(2H)-one, 7-chloro-	EtOH	239(4.30),300(3.59), 310(3.70)	39-4939-65
$C_8H_5ClO_2$			
3(2H)-Benzofuranone, 4-chloro-	EtOH	252(3.92),259(3.92), 326(3.72)	39-4939-65
3(2H)-Benzofuranone, 5-chloro-	EtOH	234(3.81),243(3.81), 248(3.81),340(3.64)	39-4939-65
3(2H)-Benzofuranone, 6-chloro-	EtOH	252(4.10),256(4.12), 323(3.82)	39-4939-65
3(2H)-Benzofuranone, 7-chloro-	EtOH	250(3.90),256(3.93), 327(3.73)	39-4939-65
$C_8H_5ClO_3$			
Benzoyl chloride 2,3-(methylenedioxy)-	C_6H_{12}	229(4.27),243(3.98), 248(4.00),256(4.12), 337(3.61)	78-1725-64
$C_8H_5ClS_2$			
2,2'-Bithiophene, 5-chloro-	EtOH	247(3.77),311(4.14)	39-5134-65
$C_8H_5Cl_5O_4S$			
2-Norbornene-1-methanesulfonic acid, 2,3,4,5,5-pentachloro-6-oxo-	Et_2O	235(3.85)	44-1110-64
$C_8H_5D_6NO$			
4(1H)-Pyridone, 1-methyl-, 2,6-di-(methyl-d_3)	MeOH	261(4.27)	35-3365-65
$C_8H_5FN_2O$			
4-Cinnolinol, 6-fluoro-	MeOH	210(4.17),234(4.11), 241(4.12),283(3.54), 294(3.60),342(4.15), 357(4.14)	4-0063-65
4-Cinnolinol, 7-fluoro-	MeOH	208(4.34),225(4.02), 248(4.06),256(3.96), 280(3.42),288(3.39), 331(4.14),345(4.11)	4-0063-65

Compound	Solvent	λ_{max}(log ε)	Ref.
4-Cinnolinol, 8-fluoro-	MeOH	209(4.29),228(4.05), 255(3.93),284(3.63), 294(3.68),336(4.12)	4-0063-65
$C_8H_5FN_2S$			
4-Cinnolinethiol, 6-fluoro-	MeOH	221(4.55),267(3.83), 313(3.33),424(4.25)	4-0063-65
4-Cinnolinethiol, 7-fluoro-	MeOH	221(4.53),270(3.87), 306(3.41),420(4.28)	4-0063-65
4-Cinnolinethiol, 8-fluoro-	MeOH	220(4.50),266(3.59), 276(3.56),316(3.40), 417(4.31)	4-0063-65
$C_8H_5F_3$			
Benzocyclobutene, 1,1,5-trifluoro-	isooctane	209(3.74),260(3.14), 263(3.19),265(3.21), 268(3.36),273(3.36)	78-1625-64
Styrene, p,β,β-trifluoro-	isooctane	207(3.86),237(4.15), 274(2.83),278(2.83), 281(2.86),284(2.79), 291(2.70)	78-1625-64
$C_8H_5F_3N_2$			
Benzimidazole, 2-(trifluoromethyl)-	EtOH	249(3.71),266(3.76), 274(3.79),281(3.72)	44-0476-64
$C_8H_5F_3N_2O$			
Nicotinonitrile, 1,2-dihydro-6-methyl-2-oxo-4-(trifluoromethyl)-	EtOH	237(3.75),355(4.04)	44-3377-65
$C_8H_5F_3O_2Se$			
1,3-Butanedione, 4,4,4-trifluoro-1-selenophene-2-yl-	n.s.g.	277(4.04),320(3.80),	67-0379-65
copper salt	n.s.g.	302(4.02),350(4.20)	67-0379-65
$C_8H_5F_5$			
p-Xylene, α,α,α,α',α'-pentafluoro-	isooctane	209(3.79),213s(3.67), 257s(2.76),263(2.92), 269(2.90)	78-1625-64
$C_8H_5F_5O$			
Ethanol, α-(2,3,4,5,6-pentafluorophenyl)-	EtOH	259(3.72)	30-1135-64
$C_8H_5F_6N$			
Aniline, N,N-bis(trifluoromethyl)-	isooctane	249(2.32),253(2.48), 258(2.56),265(2.44)	35-4341-65
$C_8H_5IS_2$			
2,2'-Bithiophene, 5-iodo-	EtOH	248(3.73),314(4.20)	39-5134-65
$C_8H_5MnO_3$			
Manganese, tricarbonyl(cyclopentadienyl)-	EtOH	327(3.06)	101-0269-65A
C_8H_5NOS			
Benzoic acid, thio-, anhydride with isocyanic acid	n.s.g.	518(1.87)	24-2954-65

Compound	Solvent	λ_{max}(log ε)	Ref.
$C_8H_5NO_2$			
Phthalimide	HCl	238(4.06),297(3.44)	44-3151-64
	NaOH	304(3.46)	44-3151-64
	EtOH	238(4.20),290(3.51)	44-3151-64
$C_8H_5NO_2S$			
Cyclopenta[c]thiopyran, 5(or 7)-nitro-	Et_2O	223(4.13),249(4.19), 259(4.17),321(4.23), 350(3.81),364(3.81), 390(3.53),478(3.72)	35-0708-64
$C_8H_5NO_3$			
2H-1,3-Benzoxazine-2,4(3H)-dione	MeOH	205(4.53),236s(3.90), 293(3.28)	4-0037-65
sodium deriv.	MeOH	206(4.48),236(3.89), 293(3.18)	4-0037-65
2H-1,4-Benzoxazine-2,3(4H)-dione	EtOH	303(3.86)	1-1051-65
Isocyanic acid, 2,3-(methylenedioxy)-phenyl ester	C_6H_{12}	210(4.59),232s(4.0), 287(3.39)	78-1725-64
Phthalimide, 4-hydroxy-	EtOH	234(4.50),264s(3.44), 280s(3.31),324(3.46)	7-0128-64
Phthalimide, 5-hydroxy-	EtOH	218(4.56),241s(3.82), 260(3.11),340(3.65)	7-0128-64
$C_8H_5N_3$			
Pyridinium dicyanomethylide	MeCN	214(4.21),238(3.59), 243(3.61),393(4.33)	35-3651-65
$C_8H_5N_3O_3$			
Benzoyl azide, 2,3-(methylenedioxy)-	EtOH	228(4.27),261(4.21), 340(3.53)	78-1725-64
$C_8H_5N_3O_4$			
1,3,4-Oxadiazole, [2-(5-nitro-2-furyl)-vinyl]-	propylene glycol	250(4.24),362(4.36)	95-0225-64
$C_8H_5N_3O_5$			
1,3,4-Oxadiazol-2-ol, 5-(5-nitro-2-furyl)vinyl-	propylene glycol	285(4.00),374(4.22)	95-0225-64
$C_8H_5N_3S_2$			
Benzo[1,2-d:5,4-d']bisthiazole, 2-amino-	EtOH	228(4.27),257(4.33), 300s(3.10),325(3.33), 334s(3.26)	7-0080-64
C_8H_6			
Benzocyclobutene	EtOH	260(3.09),266(3.28), 272(3.27)	5-0055-64B
5,6-Octadiene-1,3-diyne	EtOH	210(4.78),238(4.07), 249(4.12),263(4.16), 279(4.05)	39-4659-65
$C_8H_6BrNO_3$			
Acetophenone, 2-bromo-2-nitro-	base	365(3.88)	28-3865-64A
$C_8H_6BrN_3O_2$			
1,2,4-Triazolidine-3,5-dione, 1-(p-bromophenyl)-	NaOH	222(4.08),286(4.17)	22-0500-64
	EtOH	212(3.97),258(4.24)	22-0500-64

Compound	Solvent	$\lambda_{max}(\log \epsilon)$	Ref.
$C_8H_6Br_2O_2$			
2,4,6-Cyclooctatrien-1-one, 2,6-dibromo-7-hydroxy-	n.s.g.	234(4.04),266(4.15), 470(3.72)	25-1917-64
C_8H_6ClN			
Acetonitrile, (m-chlorophenyl)-	EtOH-Me_2SO-base	348(4.26)	23-1225-65
$C_8H_6ClNO_2$			
Styrene, α-chloro-β-nitro-, cis	C_6H_{12}	267(4.07)	94-0118-65
trans	C_6H_{12}	277(3.68)	94-0118-65
$C_8H_6ClNO_3$			
Acetophenone, 2-chloro-2-nitro-	base	360(3.92)	28-3865-64A
C_8H_6ClNS			
Benzothiazole, 5-chloro-2-methyl-	50% EtOH	222(4.46),254(3.78), 290(3.22),300(3.20)	39-0954-65
$C_8H_6ClN_3$			
1,2,3-Triazole, 2-(m-chlorophenyl)-	EtOH	265(4.24)	39-2306-64
1,2,3-Triazole, 2-(p-chlorophenyl)-	EtOH	266(4.26)	39-2306-64
$C_8H_6Cl_2O_2$			
1,3-Cyclohexadiene-1,4-dicarbonyl chloride	Et_2O	329(4.22)	23-2852-64
$C_8H_6Cl_4$			
o-Xylene, 3,4,5,6-tetrachloro-	C_6H_6	280(2.4),290s(2.31)	78-2217-64
$C_8H_6Cl_4O_2$			
1,3-Dioxolan-2-spirocyclopenta-2',4'-diene, 2',3',4',5'-tetrachloro-4-methyl-	iso-PrOH	313(3.12)	39-2305-65
$C_8H_6F_2$			
4-Octene-2,6-diyne, 4,5-difluoro-	MeOH	249(4.51),260(4.48)	44-2009-64
$C_8H_6F_2N_2O_2$			
Formamide, N,N'-bis(2,5-difluoro-p-phenylene)bis-	pH 7	253(4.19),288(3.97)	87-0737-65
	pH 13	258(4.09),293(3.93)	87-0737-65
	EtOH	257(4.28),291(4.09)	87-0737-65
$C_8H_6F_2N_4O$			
Benzotriazole, 4-acetamido-5,7-difluoro-	pH 1	265(3.83),270s(3.82)	87-0737-65
$C_8H_6F_4$			
o-Xylene, 3,4,5,6-tetrafluoro-	hexane	262.0(2.67)	39-3035-64
p-Xylene, α,α,α',α'-tetrafluoro-	isooctane	210(3.81),216s(3.68), 258s(2.64),264(2.79), 270(2.74)	78-1625-64
$C_8H_6Li_2$			
Lithium, (dihydropentalenylene)di-	$(MeOCH_2)_2$	295(3.8)	35-0249-64
	THF	296(3.7)	35-0249-64
$C_8H_6MnNO_3$			
Manganese, (aminocyclopentadienyl)tricarbonyl-	EtOH	324(3.14)	101-0269-65A

Compound	Solvent	λ_{max}(log ε)	Ref.
$C_8H_6N_2$			
1,3-Cyclohexadiene-1,4-dicarbonitrile	EtOH	295(4.06),311(4.03)	23-2852-64
2,6-Naphthyridine	MeOH	247(3.34),255(3.42), 264(3.33),318(3.34), 330(3.30)	88-1117-65
Quinoxaline	EtOH	315(3.80)	39-2319-64
$C_8H_6N_2O$			
4-Cinnolinol	MeOH	211(4.32),231(4.03), 238(4.06),253(3.94), 283(3.48),294(3.50), 338(4.15),352(4.09)	4-0063-65
	EtOH	228(4.01),236(4.04), 251(3.92),260s(3.80), 282(3.45),292(3.48), 336(4.10),352(4.07)	39-5391-65
1,2,4-Oxadiazole, 3-phenyl-	Et_2O	238(4.14)	33-0942-64
1,2,4-Oxadiazole, 5-phenyl-	Et_2O	250(4.20)	33-0942-64
Quinoxaline, 1-oxide	EtOH	325(3.97)	39-2319-64
$C_8H_6N_2OS$			
4H-1,3-Benzoxazine-4-thione, 2,3-di-hydro-2-imino-	pH 5	277(3.93),334(4.12)	78-3019-65
	pH 11	277(3.79),321(3.97)	78-3019-65
2-Quinazolinol, 4-mercapto-	pH 1	240(4.24),305(3.78), 320(3.80),358(4.11)	44-2674-64
	pH 11	242(4.32),304(3.81), 315(3.82),350(3.98)	44-2674-64
$C_8H_6N_2O_2$			
6,7-Benzimidazoledione, 1-methyl-	$CHCl_3$	307(3.6),450(3.0)	65-0949-64
Indole, 5-nitro-	EtOH	254s(4.20),264(4.24), 323(3.90)	35-3796-64
	H_2SO_4	211s(4.00),278(3.99)	35-3796-64
Indole, 6-nitro-	EtOH	248(4.03),326(3.92), 358(3.88)	35-3796-64
	H_2SO_4	254(3.08)	35-3796-64
1(2H)-Phthalazinone, 4-hydroxy-	2% EtOH	260(3.22),296(3.34)	20-0091-65
Phthalimide, 3-amino-	dil. HCl	255s(3.87),291s(3.36), 383(3.71)	44-3151-64
	dil. NaOH	304(3.46)	44-3151-64
	EtOH	235(4.17),256(3.94), 385(3.78)	44-3151-64
Phthalimide, 4-amino-	dil. HCl	254(4.23),298(3.58), 370(3.51)	44-3151-64
	dil. NaOH	267(4.03)	44-3151-64
	EtOH	242(4.50),252(4.51), 308(3.73),373(3.70)	44-3151-64
4H-Pyrido[4,3-d][1,3]oxazin-4-one, 2-methyl-	MeOH	215(4.46),222(4.45), 245(3.83),288(3.38)	87-0722-65
Quinoxaline, 1,4-dioxide	EtOH	385(4.09)	39-2319-64
$C_8H_6N_2O_2S_3$			
Acetic acid, thio-, S,S-diester with 3,5-dimercapto-4-isothiazolecarbo-nitrile	CH_2Cl_2	277(4.08)	44-0665-64
$C_8H_6N_2O_3$			
1,4-Phthalazinedione, 2,3-dihydro-5-hydroxy-	pH 13	315(3.82),332(3.85)	46-0752-64

Compound	Solvent	λ max(log ε)	Ref.
$C_8H_6N_2O_4S_3$			
Acetic acid, [(4-cyano-3,5-isothiazole-diyl)dithio]di-	EtOH	227(4.12),285(4.08)	44-0665-64
$C_8H_6N_2S$			
4-Cinnolinethiol	MeOH	224(4.58),264(3.82), 275(3.75),308(3.39), 422(4.30)	4-0063-65
1,2,5-Thiadiazole, 3-phenyl-	isooctane	228(3.99),280f(4.13)	89-0262-65
$C_8H_6N_3NaO_8$			
1,4-Dioxaspiro[4.5]decadiene, 6,8,10-trinitro-, sodium derivative	acetone	414(4.48),490(4.35)	54-0516-65
$C_8H_6N_4O$			
Pyrazolo[1,5-a]pyrimidine-6-carbonitrile, 4,7-dihydro-5-methyl-7-oxo-	EtOH	215(4.52),270(4.15), 302(4.10)	95-0442-65
$C_8H_6N_4O_3S$			
1,2,4-Thiadiazole, 5-amino-3-[2-(5-nitro-2-furyl)vinyl]-	propylene glycol	246(4.16),290(3.99), 382(4.20)	95-0948-65
$C_8H_6N_4O_4$			
1,3,4-Oxadiazole, 2-amino-5-(5-nitro-2-furyl)vinyl-	propylene glycol	300(4.02),390(4.23)	95-0212-64
1,2,4-Triazolidine-3,5-dione, 1-(p-nitrophenyl)-	NaOH	221(3.80),260(3.86), 415(4.02)	22-0500-64
	EtOH	222(3.90),332(3.94)	22-0500-64
$C_8H_6N_4S_2$			
Benzo[1,2-d:5,4-d']bisthiazole, 2,6-diamino-	EtOH	253(4.69),273s(4.15), 315s(3.83),325(3.95)	7-0080-64
$C_8H_6N_6$			
Tetrazoline, 5-cyanimino-1-phenyl-	EtOH	273(4.00)	44-0650-64
C_8H_6O			
Benzocyclobutenone	EtOH	208(4.19),242(4.17), 285(3.60),293(3.60)	44-2947-64
Benzofuran	EtOH	245(4.07),274(3.36), 281(3.48)	56-0589-65
	EtOH	244(4.06),274(3.38), 281(3.43)	56-0757-65
Furan, 2-(1-buten-3-ynyl)-, trans	n.s.g.	294(4.32),308s(4.27)	5-0062-65B
C_8H_6OS			
Benzothiophene, 4-hydroxy-	aq. base	324.0(3.84)	22-1464-65
Benzo[b]thiophen-3(2H)-one	EtOH	231(4.40),236(4.46), 258(3.74),263(3.64), 305(3.18)	39-4939-65
Phthalide, 1-thio-	n.s.g.	219(--),223(--), 244(--),286(--), 305(4.22)	20-0491-64
C_8H_6OSe			
Phthalide, 1-seleno-	n.s.g.	247f(--),338(4.15)	20-0491-64

Compound	Solvent	λ_{max}(log ε)	Ref.
$C_8H_6O_2$			
Benzocyclobutene-4,5-quinone	EtOH	270(3.63),384(3.12), 528(1.49)	78-2281-65
Benzocyclobutenone, 4-hydroxy-	EtOH	226(3.92),267(3.90), 301(3.98),337(3.94)	39-1390-65
4-Benzofuranol	aq. base	295.5(3.81)	22-1464-65
$C_8H_6O_2S_3$			
Thiophene, 2,2'-sulfonyldi-	EtOH	204(3.67),244(3.89), 276(3.83)	78-2413-65
Thiophene, 2,3'-sulfonyldi-	EtOH	206(3.76),245(4.04), 264(3.98)	78-2413-65
Thiophene, 3,3'-sulfonyldi-	EtOH	208(4.15),245(4.21)	78-2413-65
$C_8H_6O_3$			
Cyclohexadiene-1,2-dicarboxylic anhydride, cis	MeCN	258(3.59),293(2.98), 302(2.99)	35-1925-65
Piperonal	EtOH	313(--)	95-0857-65
$C_8H_6O_3S$			
3-Thiopheneacrylic acid, β,2-dihydroxy-5-methyl-, δ-lactone	n.s.g.	233(4.04),241(4.08), 300(4.43)	28-0464-65B
$C_8H_6O_4$			
Benzoic acid, 2,3-(methylenedioxy)-	EtOH	233s(3.92),312(3.56)	78-1725-64
Terephthalic acid	EtOH	240(4.30)	23-2852-64
$C_8H_6O_5$			
o-Phthalaldehyde, 2,3,4-trihydroxy-	EtOH	263(4.42),336(3.96)	23-1595-64
C_8H_6S			
Benzo[b]thiophene	MeOH	227(4.42),257(3.79), 289(3.25),297(3.50)	56-0681-65
	MeOH	227(4.45),258(3.80), 289(3.24),297(3.50)	56-0757-65
	MeOH	227(4.43),258(3.82), 289(3.23),297(3.51)	56-0939-65
	EtOH	227(4.43),258(3.78), 289(3.26),296(3.50)	56-0589-65
Isothialene	hexane	542(2.5)	73-3016-65
Thialene	C_6H_{12}	528(--)	73-3016-65
$C_8H_6S_2$			
Phthalide, 1,2-dithio-	n.s.g.	203(--),224(--), 290(4.14),343(--)	20-0491-64
$C_8H_6S_3$			
Thiophene, 2,2'-thiodi-	hexane	200(3.90),238(4.06), 264(3.88)	78-2413-65
Thiophene, 2,3'-thiodi-	hexane	204(4.08),235(4.02), 268s(3.93)	78-2413-65
Thiophene, 3,3'-thiodi-	hexane	212(4.13),272(3.77)	78-2413-65
$C_8H_6Se_2$			
Phthalide, 1,2-diseleno-	n.s.g.	205(--),237(--), 308(4.19),365(--)	20-0491-65
$C_8H_7AgO_4$			
Dehydroacetic acid, silver salt	EtOH	229(4.21),297(3.92)	44-1255-65

Compound	Solvent	$\lambda_{max}(\log \epsilon)$	Ref.
C_8H_7Br			
Benzocyclobutene, 4-bromo-	EtOH	267(3.21),273(3.34), 281(3.31)	78-0245-65
$C_8H_7BrN_2$			
Benzimidazole, 4(6)-bromo-2-methyl-	EtOH	243(3.70),248(3.75), 253(3.60),275(3.75), 280(3.75),287(3.71)	49-0614-65
Benzimidazole, 4(7)-bromo-2-methyl-	EtOH	250s(3.86),257(3.84), 279s(3.68),287(3.71)	49-0614-65
Benzimidazole, 5-bromo-2-methyl-	EtOH	250s(3.71),254(3.71), 260(3.68),287(3.77), 295(3.74)	49-0614-65
Imidazo[1,2-a]pyridine, 3-bromo-2-methyl-, hydrochloride	pH 1	223(4.37),283(3.91)	44-4085-65
	pH 11	231(4.37),274(3.66), 283(3.72),306(3.66)	44-4085-65
Imidazo[1,2-a]pyridine, 5-bromo-3-methyl-	pH 1	294(3.92)	44-4085-65
	pH 11	278s(3.71),287(3.79), 300s(3.69)	44-4085-65
$C_8H_7BrO_2$			
Acetic acid, (p-bromophenyl)-	H_2O	265(2.422)	65-1474-64
Acetophenone, 3'-bromo-2'-hydroxy-	C_6H_{12}	228(3.99),250(4.02), 335(2.75)	49-0450-65
Acetophenone, 5'-bromo-2'-hydroxy-	C_6H_{12}	220(3.92),250(3.76) 350(3.48)	49-0450-65
Acetophenone, 5'-bromo-4'-hydroxy-	C_6H_{12}	229(4.23),262(3.75), 357(3.54)	49-0450-65
$C_8H_7BrO_4$			
2,4-Hexadienedioic acid, 3-(bromomethyl)-2-hydroxy-, δ-lactone, methyl ester	50% EtOH	305(3.82)	88-3431-64
$C_8H_7Br_3O$			
Bicyclo[5.1.0]oct-4-en-3-one, 4,8,8-tribromo-	EtOH	258(3.74)	39-5343-64
C_8H_7Cl			
Benzocyclobutene, 4-chloro-	EtOH	267(3.17),274(3.30), 281(3.30)	78-0245-65
$C_8H_7ClN_2$			
Benzimidazole, 5-chloro-2-methyl-	pH 2	236(3.56),275(3.89), 283(3.86)	44-0476-64
	pH 12	243(3.42),280(3.69), 285(3.70)	44-0476-64
	EtOH	246(3.74),281(3.86), 287(3.76)	44-0476-64
1H-Pyrrolo[2,3-b]pyridine, 6-chloro-4-methyl-	EtOH	244(4.1),292(3.98)	65-1454-64
$C_8H_7ClO_2$			
Acetic acid, (p-chlorophenyl)-	H_2O	265(2.342)	65-1474-64
Acetophenone, 3'-chloro-2'-hydroxy-	EtOH	227(3.71),256(3.77), 340(2.50)	49-1214-65
Acetophenone, 5'-chloro-2'-hydroxy-	EtOH	229(3.81),250(3.70), 349(3.48)	49-1214-65
Acetophenone, 5'-chloro-4'-hydroxy-	EtOH	228(4.07),276(4.07)	49-1214-65

Compound	Solvent	λ_{max}(log ε)	Ref.
$C_8H_7ClO_3$			
Benzoic acid, 3-chloro-4-hydroxy-, methyl ester	HCl	253(4.26)	44-2678-65
	NaOH	283(4.21)	44-2678-65
$C_8H_7ClO_4S_2$			
5-Methyl-1,3-benzodithiolium perchlorate	20% $HClO_4$	319(3.8),352s(3.6)	73-3016-65
C_8H_7ClS			
m-Toluoyl chloride, thio-	C_6H_{12}	232(3.90),243s(3.74), 251s(3.69),263s(3.55), 315(4.09),525(1.88)	24-0829-65
	MeCN	230(3.75),251s(3.43), 319(4.08),515(1.93)	24-0829-65
o-Toluoyl chloride, thio-	C_6H_{12}	230(3.97),310(3.89), 520(2.19)	24-0829-65
	MeCN	228(3.90),312(3.79), 509(2.11)	24-0829-65
p-Toluoyl chloride, thio-	C_6H_{12}	231(3.85),236(3.86), 250(3.69),328(4.17), 527(1.85)	24-0829-65
	MeCN	231s(3.78),236(3.79), 259(3.66),332(4.17), 516(1.92)	24-0829-65
$C_8H_7Cl_2NO_4$			
Pyrrole-2,5-dicarboxylic acid, 3,4-dichloro-, dimethyl ester	EtOH	217(4.22),220(4.23), 274(4.24)	39-0459-65
	EtOH-base	237(4.20),299(4.35)	39-0459-65
C_8H_7FO			
Acetophenone, 4'-fluoro-	EtOH	243(4.07)	44-2165-65
$C_8H_7F_3O_2$			
Benzene, 1-methoxy-4-(trifluoromethoxy)-	hexane	220(3.94),278(3.27)	65-2749-64
	EtOH	221(3.96),277(3.29)	65-2749-64
	Et_2O	220(3.97),277(3.28)	65-2749-64
	dioxan	221(3.90),277(3.23)	65-2749-64
	H_2SO_4	223(3.88),291(3.59)	65-2749-64
$C_8H_7F_3O_2S$			
Sulfone, p-tolyl trifluoromethyl	hexane	229(4.13),271(3.04), 293s(2.64)	65-2749-64
	EtOH	231(4.11),272(3.09), 300s(2.30)	65-2749-64
	Et_2O	230(4.14),268(3.04)	65-2749-64
	dioxan	232(4.08),273(3.08), 290s(2.55),294s(2.60), 300s(2.60)	65-2749-64
	H_2SO_4	236(4.14),268(3.39), 276(3.32),315(3.16)	65-2749-64
$C_8H_7F_3S$			
Sulfide, p-tolyl trifluoromethyl	hexane	218(3.89),240(3.46), 272(2.80)	65-2749-64
	EtOH	220(3.90),236(3.55), 271(2.77)	65-2749-64
	Et_2O	220(3.87),240(3.46), 272(2.73)	65-2749-64

Compound	Solvent	$\lambda_{max}(\log \epsilon)$	Ref.
Sulfide, p-tolyl trifluoromethyl (cont.)	dioxan	220(3.90),236(3.55), 267(2.72)	65-2749-64
	H_2SO_4	240(3.16),293(2.59)	65-2749-64
$C_8H_7IN_2$			
1H-Pyrrolo[2,3-b]pyridine, 6-iodo-4-methyl-	EtOH	228(4.42),290(3.88)	65-1454-64
C_8H_7N			
Indole	MeOH	219(4.48),266(3.85), 288(3.59)	56-0757-65
	EtOH	216(4.54),266s(3.76), 270(3.77),276(3.76), 278(3.76),287(3.68)	35-3796-64
	EtOH	219(4.47),266(3.85), 288(3.58)	56-0589-65
	EtOH	220(4.5),263(3.7), 288(3.6)	56-1709-64
	H_2SO_4	233(3.59),238(3.58), 280(3.68)	35-3796-64
chloranil complex	CCl_4	405(3.1),495(3.2)	60-2189-64
o-Tolunitrile	EtOH	228(3.96),276(3.06), 284(3.08)	88-2659-64
C_8H_7NO			
1,2-Benzisoxazole, 3-methyl-	EtOH	237(3.95),271(3.41), 277(3.68)	4-0385-65
2,1-Benzisoxazole, 3-methyl-	pH -4.0	218(4.24),252(3.80), 274s(2.77),312(3.19)	39-5360-65
	pH 3.0	215s(3.73),252s(3.29), 256(3.33),260(3.32), 266(3.31),271s(3.24), 278(3.26),318(3.62)	39-5360-65
Phthalimidine	EtOH	228(3.80),264s(3.46), 271(3.51),278(3.50)	44-2251-65
$C_8H_7NO_2$			
Benzocyclobutene, 4-nitro-	heptane	267(3.92)	78-2185-64
Indole-5,6-diol	EtOH	209(4.27),275(3.60), 302(3.79)	4-0387-65
$C_8H_7NO_2S$			
Benzene, 1-(epithioethyl)-4-nitro-	hexane	266(4.08)	35-4628-64
$C_8H_7NO_3$			
Acetophenone, 2-nitro-	acid	250(4.13)	28-3865-64A
	base	350(4.35)	28-3865-64A
Benzene, 1-(epoxyethyl)-4-nitro-	hexane	265(4.08)	35-4628-64
Benzocyclobutene, 4-hydroxy-5-nitro-	EtOH	237s(3.48),282(3.62), 346(3.29)	78-2281-65
Ether, p-nitrophenyl vinyl	nonane	292(4.13)	67-0375-65
$C_8H_7NO_3S$			
1,2,3-Benzoxathiazine, 4-methyl-, 2,2-dioxide	EtOH	264(3.99),308(3.23)	44-3960-65

Compound	Solvent	λ_{max}(log ε)	Ref.
$C_8H_7NO_4$			
Acetic acid, (m-nitrophenyl)-	H_2O	270(3.697)	65-1474-64
Acetic acid, (p-nitrophenyl)-	H_2O	280(3.773)	65-1474-64
Acetophenone, 2'-hydroxy-3'-nitro-	EtOH	238(4.23),289(3.13), 335(3.79)	39-2816-64
Acetophenone, 2'-hydroxy-5'-nitro-	EtOH	235(4.24),264(3.53), 302(4.03)	39-2816-64
o-Anisaldehyde, 3-nitro-	EtOH	236(4.02),304(3.26)	39-2816-64
o-Anisaldehyde, 5-nitro-	EtOH	239(4.10),304(4.04)	39-2816-64
Glutaconic acid, 4-(1-cyanoethylidene)-	EtOH	209(3.93),256(3.96)	44-0203-65
$C_8H_7NO_5$			
Benzoic acid, 3-hydroxy-4-nitro-, methyl ester	HCl	356(3.58)	35-4942-64
	NaOH	426(3.73)	35-4942-64
C_8H_7NS			
Benzo[b]thiophen-2-amine	EtOH	281(3.95)	44-4074-65
$C_8H_7NS_2$			
Imidocarbonic acid, dithio-, cyclic 4-methyl-o-phenylene ester, hydrochloride	MeOH	233(4.03),240(3.78), 274s(3.63),303s(3.89)	44-0738-64
$C_8H_7N_3$			
3,5-Pyridinedicarbonitrile, 1,2-dihydro-2-methyl-	EtOH	214(4.32),253(3.96), 378(3.70)	73-0143-64
3,5-Pyridinedicarbonitrile, 1,2-dihydro-4-methyl-	EtOH	219(4.37),256(3.95), 378(3.68)	73-1654-64
3,5-Pyridinedicarbonitrile, 1,4-dihydro-2-methyl-	EtOH	211(4.33),349(3.78)	73-3711-65
3,5-Pyridinedicarbonitrile, 1,4-dihydro-4-methyl-	EtOH	208(4.28),346(3.85)	73-0143-64
1,4,6-Triazanaphthalene, 3-methyl-	EtOH	231(4.39),309(3.61), 317(3.59)	39-1558-65
$C_8H_7N_3O$			
Glycinonitrile, N-nitroso-N-phenyl-	EtOH	273(4.08)	33-0033-64
1,3,4-Oxadiazole, 2-amino-5-phenyl-	EtOH	277(4.2)	28-4579-64A
1,3,4-Oxadiazole, 2-anilino-	n.s.g.	252(4.36)	28-4538-65A
5-Pyrimidinecarbonitrile, 1-allyl-1,2-dihydro-2-oxo-	EtOH	264(4.11),315(2.53)	94-1418-64
5H-Pyrrolo[2,3-b]pyrazine-7-carboxaldehyde, 5-methyl-	EtOH	259(4.12),292(4.05)	44-3454-65
2-Quinazolinol, 4-amino-	pH 1	328(3.65)	44-2674-64
	pH 11	320(3.64)	44-2674-64
4-Quinazolinol, 2-amino-	pH 1	305(3.56)	44-2674-64
	pH 11	265(3.89),270(3.87), 314(3.46)	44-2674-64
6-Quinazolinol, 2-amino-, hydrochloride	H_2O	233(4.34)	78-2059-65
	pH 13	245(4.47)	78-2059-65
1,2,4-Triazolin-5-one, 4-phenyl-	H_2O	228(3.9)	5-0123-65B
$C_8H_7N_3O_2$			
1,4-Phthalazinedione, 5-amino-2,3-dihydro-	pH 13	315(3.86),348(3.90)	46-0752-64
5,6-Quinazolinediol, 2-amino-	pH 1	228(4.14),265(4.28), 310(3.85)	78-2059-65
	pH 13	228(4.13),309(4.09)	78-2059-65
1,2,4-Triazolidine-3,5-dione, 1-phenyl-	NaOH	218(4.06),278(4.08)	22-0500-64
	EtOH	211(3.94),251(4.14)	22-0500-64

Compound	Solvent	$\lambda_{max}(\log \epsilon)$	Ref.
$C_8H_7N_3O_3$			
Benzimidazole, 4-methoxy-5-nitro-	pH 2	243(3.91),300(3.81)	44-0476-64
	pH 12	245(3.91),370(3.89)	44-0476-64
	EtOH	234(3.97),300(3.81)	44-0476-64
Benzimidazole, 5-methoxy-4-nitro-	pH 2	280(3.90),351(3.54),	44-0476-64
	pH 12	390(3.6)	44-0476-64
	EtOH	308(3.67),369(3.83)	44-0476-64
$C_8H_7N_3O_4$			
Indole, 5,7-dinitro-	EtOH	218(4.03),263(4.00), 364(4.15),404s(3.90)	44-0947-64
$C_8H_7N_3O_5$			
Benzamide, 3,5-dinitro-4-methyl-	HCl	235(4.20)	23-2328-65
	24% H_2SO_4	217(4.21)	23-1957-64
	76% H_2SO_4	235(4.36)	23-1957-64
	H_2O	217(4.09)	23-2328-65
$C_8H_7N_3S$			
3-Methyl-1,2,3-benzothiadiazolium cyanide	$CHCl_3$	248(3.88),267(3.89), 385(3.85)	39-6061-64
1,3,4-Thiadiazole, 2-anilino-	EtOH	242(3.85),283(4.17)	28-0766-65B
$C_8H_7N_3S_2$			
3-Methyl-1,2,3-benzothiadiazolium thiocyanate	H_2O	230(4.00),297(3.85), 328(3.67)	39-6061-64
	$CHCl_3$	245(3.85),299(3.84), 331(3.72),437(3.14)	39-6061-64
$C_8H_7N_5$			
Pyrazolo[1,5-a]pyrimidine-6-carbonitrile, 7-amino-5-methyl-	EtOH	226(4.49),300(4.14)	95-0442-65
$C_8H_7N_5S$			
7H-Pyrrolo[2,3-d]pyrimidine-5-carbonitrile, 4-amino-6-(methylthio)-	EtOH	231(4.22),301(4.25)	35-0951-64
	EtOH	231(4.22),301(4.25)	35-1995-65
$C_8H_7NaO_4$			
Dehydroacetic acid, sodium salt	EtOH	230(4.22),296(3.92)	44-1255-65
C_8H_8			
Benzocyclobutene	C_6H_{12}	260(3.1),267(3.3), 273(3.3)	88-3625-65
Bicyclo[4.2.0]octa-2,4,7-triene	n.s.g.	273(3.49),277(3.48)	89-0432-64
Fulvene, 6-vinyl-	C_6H_{12}	277(4.30),285(4.48), 295(4.52),305(4.30), 397(2.28)	33-1022-64
	EtOH	286(4.46),293(4.48), 395(2.28)	33-1022-64
1-Heptene-3,5-diyne, 2-methyl-	hexane	205(4.48),211(4.67), 225(3.30),237(3.70), 248(4.02),263(4.20), 278(4.08)	78-1357-65
Pentalene, dihydro-	pentane	268(3.64)	35-0249-64
Styrene	hexane	244(4.60),275(3.08)	65-2088-65
	EtOH	244(4.55),275(3.08)	65-2088-65
	EtOH	251(4.10),288(2.94), 298(2.88)	56-1437-65
	EtOH-HCl	245(4.33),275(3.08)	65-2088-65
	EtOH-NaOH	245(4.60),275(3.08)	65-2088-65

Compound	Solvent	λ_{max}(log ε)	Ref.
Styrene (cont.)	CH_2Cl_2-$HClO_4$	309(3.22),427(3.59)	39-4765-65
silver complex	H_2O	270(3.9)	39-6185-64
$C_8H_8BrNO_3$			
Anisole, 2-(bromomethyl)-4-nitro-	hexane	288(4.01)	37-0722-65
	MeOH	304(3.98)	37-0722-65
	$CHCl_3$	307(4.00)	37-0722-65
	CCl_4	296(4.02)	37-0722-65
	dioxan	303(4.00)	37-0722-65
	95% dioxan	305(4.01)	37-0722-65
	50% dioxan	315(3.96)	37-0722-65
	50% dioxan-NaOH	315(3.96)	37-0722-65
	Me_2SO	317(3.97)	37-0722-65
$C_8H_8BrN_5O_3$			
2-Furanacrolein, α-bromo-5-nitro-, amidinohydrazone, hydrochloride	propylene glycol	295(4.09),385(4.13)	95-0001-64
$C_8H_8Br_2$			
Benzene, 1-bromo-4-(2-bromoethyl)-	EtOH	222(4.13)	44-2109-64
$C_8H_8Br_2O$			
2,5-Cyclohexadien-1-one, 4,4-dibromo-2,6-dimethyl-	n.s.g.	248(4.30)	70-0336-65
2,6-Xylenol, 3,4-dibromo-	n.s.g.	288(3.26)	70-0336-65
$C_8H_8Br_2O_4$			
Cyclohexanecarboxylic acid, 3,5-dibromo-2-methyl-4,6-dioxo-	EtOH	265(3.94),313(3.77)	44-3566-65
$C_8H_8ClNO_2$			
Benzohydroxamic acid, p-chloro-N-methyl-	MeOH	231(3.88)	73-0940-65
	50% MeOH-N HCl	232(4.04)	73-0940-65
	50% MeOH-borate	226(4.11)	73-0940-65
Acetanilide, 3'-chloro-4'-hydroxy-	HCl	245(4.04)	44-2678-65
	NaOH	253(4.07)	44-2678-65
$C_8H_8ClN_3$			
Pyrazolo[1,5-a]pyrimidine, 7-chloro-2,3-dimethyl-	EtOH	238(4.69),283s(3.17), 294(3.24),302s(3.10), 343(3.21)	94-1207-65
$C_8H_8ClN_3O_4S_2$			
1,2,4-Benzothiadiazine-7-sulfonamide, 6-chloro-3-methyl-, 1,1-dioxide	EtOH	226(4.49),276(4.08)	95-0095-65
$C_8H_8ClN_3O_5S_2$			
1,2,4-Benzothiadiazine-7-sulfonamide, 6-chloro-3-(hydroxymethyl)-, 1,1-dioxide	EtOH	225(4.54),276(4.06)	95-0095-65
$C_8H_8ClN_5O_3$			
Guanidine, [[2-chloro-3-(5-nitro-2-furyl)allylidene]amino]-, hydrochloride	propylene glycol	296(4.07),384(4.13)	95-0001-64

Compound	Solvent	$\lambda_{max}(\log \epsilon)$	Ref.
$C_8H_8Cl_2N_2$			
2,6-Naphthyridine, 5,7-dichloro-1,2,3,4-tetrahydro-	MeOH	218(3.95),273(3.61)	88-1117-65
$C_8H_8Cl_2N_2O_2$			
p-Benzoquinone, 2,5-dichloro-3,6-bis-(methylamino)-	methyl cellosolve	226(4.28),355(4.43), 526(2.44)	78-1889-64
$C_8H_8Cl_2O$			
3,4-Xylenol, 2,6-dichloro-	C_6H_{12}	282(3.38),290(3.43)	2-0417-64
3,5-Xylenol, 2,4-dichloro-	C_6H_{12}	282(3.39),286(3.39), 290(3.42)	2-0417-64
$C_8H_8Cl_2O_2$			
2-Cyclopenten-1-one, 3-(2,2-dichloro-1-methylvinyl)-2-hydroxy-	EtOH	275(3.94)	77-0539-65
	base	336(4.56)	77-0539-65
2-Cyclopenten-1-one, 4-(2,2-dichloro-1-methylvinyl)-5-hydroxy-	EtOH	211(4.19)	77-0539-65
$C_8H_8F_2N_2O_2$			
p-Benzoquinone, 2,5-difluoro-3,6-bis-(methylamino)-	methyl cellosolve	224(4.46),353(4.42), 544(2.34)	78-1889-64
C_8H_8INS			
N-Methylbenzothiazolium iodide	H_2O	276(3.76)	22-2868-64
	0.5N NaOH	272(3.81)	22-2868-64
	N NaOH	270(3.92)	22-2868-64
	1.5N NaOH	269(4.00)	22-2868-64
	2 N NaOH	268(4.08)	22-2868-64
	EtOH	278(3.73)	22-2868-64
	NaOEt	306(3.63)	22-2868-64
$C_8H_8N_2$			
Acetonitrile, (p-aminophenyl)-	EtOH	242.4(4.02)	78-0861-64
7-Azaindole, 4-methyl-	EtOH	220(4.22),288(4.06)	65-1454-64
Benzimidazole, 2-methyl-	EtOH	274(3.83),281(3.88)	94-0773-64
	EtOH	245(4.10),247(3.77), 270(3.68),280(4.18)	49-0614-65
Benzimidazole, 5-methyl-	pH 2	272(3.91),280(3.89)	44-0476-64
	pH 12	243(3.53),277(3.79), 283(3.79)	44-0476-64
	EtOH	244(3.75),278(3.91), 284(3.89)	44-0476-64
Imidazo[1,2-a]pyridine, 2-methyl-	pH 1	275(3.97)	4-0053-65
	pH 11	272(3.65),280(3.72), 294(3.65)	4-0053-65
hydrochloride	EtOH	222(4.27),232(3.97), 281(3.82)	44-2403-65
Imidazo[1,2-a]pyridine, 3-methyl-	pH 1	232(3.90),287(3.69)	4-0053-65
	pH 11	228s(4.25),280(3.62), 310(3.71)	4-0053-65
$C_8H_8N_2O$			
Benzimidazole, 5-methoxy-	pH 2	284(3.93)	44-0476-64
	pH 12	288(3.71)	44-0476-64
	EtOH	245(3.57),288(3.79)	44-0476-64
2,5-Cyclohexadien-1-one, 4-diazo-2,6-dimethyl-	aq. EtOH	255(3.79),362(4.61)	30-1101-64
	+ acid	323(4.30)	30-1101-64
Imidazo[1,2-a]pyridine, 5-methoxy-	pH 1	237(3.93),292(4.08)	4-0053-65
	pH 11	290(4.04),300s(3.92)	4-0053-65

Compound	Solvent	$\lambda_{max}(\log \epsilon)$	Ref.
1H-Indazole, 3-methoxy-	EtOH	299(3.67)	7-0583-65
Indazolinone, 1-methyl-	EtOH	221(4.44),260(3.15), 314(3.67)	33-1986-64
Indazolinone, 2-methyl-	EtOH	218(4.30),235(4.04), 310(3.70)	33-1986-64
	EtOH-NaOH	214(4.26),235(4.40), 351(3.69)	33-1986-64
$C_8H_8N_2O_2$			
Benzene, 1-aziridinyl-4-nitro-	n.s.g.	340(4.11)	5-0109-65I
Benzocyclobutene, 4-amino-5-nitro-	EtOH	228(4.20),261(3.86), 289(3.74),417(3.81)	78-2281-65
Methylamine, p-nitrobenzylidene-	$C_2H_4Cl_2$	285(4.20)	30-1017-64
$C_8H_8N_2O_2S$			
1H-2,1,3-Benzothiadiazine, 4-methyl-, 2,2-dioxide	EtOH	223(4.4),264(3.8), 347(4.40)	44-3960-65
$C_8H_8N_2O_3$			
Acetanilide, 2'-nitro-	C_6H_{12}	275(3.90),354(3.80)	44-2088-64
	MeOH	340(3.73)	44-2088-64
	dioxan	276(3.89),355(3.78)	44-2088-64
Acetanilide, 3'-nitro-	C_6H_{12}	236(3.94),316(2.83)	44-2088-64
	MeOH	242(4.06),328(2.74)	44-2088-64
	iso-PrOH	242(4.04),330(2.70)	44-2088-64
	dioxan	242(3.92),328(2.75)	44-2088-64
Acetanilide, 4'-nitro-	C_6H_{12}	300(3.74)	44-2088-64
	MeOH	314(4.11)	44-2088-64
	iso-PrOH	314(4.21)	44-2088-64
	dioxan	315(4.15)	44-2088-64
Acetophenone, α-nitro-, oxime	acid	250(4.02),300(4.00)	28-3865-64A
	base	260(4.06),305(4.12)	28-3865-64A
Indole-2,3,7(4H)-trione, 5,6-dihydro-, 7-oxime	EtOH	249(3.97),294(3.81), 426(3.50)	39-0991-64
Salicylaldehyde, 5-nitro-, anil with methylamine	MeOH	257(4.29),348(4.10)	24-1631-64
	MeOH-NaOH	398(4.19)	24-1631-64
$C_8H_8N_2O_4$			
Acetanilide, 4'-hydroxy-3'-nitro-	HCl	353(4.05)	35-4942-64
	NaOH	413(3.84)	35-4942-64
Anthranilic acid, N-methyl-5-nitro-	EtOH	213(4.26),243s(3.92), 320(4.09)	44-3457-65
Benzohydroxamic acid, N-methyl-p-nitro-	MeOH	269(4.00)	73-0940-65
	50% MeOH-N HCl	271(4.00)	73-0940-65
	50% MeOH-borate	269(4.00)	73-0940-65
	50% MeOH-N NaOH	268(3.99)	73-0940-65
Benzohydroxamic acid, p-nitro-, methyl ester	MeOH	263(4.04)	73-0940-65
	50% MeOH-N HCl	267(4.07)	73-0940-65
	50% MeOH-borate	236s(3.88),263(3.89), 341(3.69)	73-0940-65
	50% MeOH-N NaOH	263(3.86),345(3.68)	73-0940-65

Compound	Solvent	$\lambda_{max}(\log \epsilon)$	Ref.
$C_8H_8N_2S$			
Nicotinonitrile, 1,2-dihydro-5,6-dimethyl-2-thioxo-	EtOH	240s(3.74),311(4.30), 408(3.64)	94-0087-64
$C_8H_8N_2S_2$			
[2,2'-Bithiophene]-3,5'-diamine, tin salt of dihydrochloride	20% HCl	305(3.29)	65-0674-64
	EtOH	306(3.75),346(3.72), 371(3.68)	65-0674-64
[2,2'-Bithiophene]-5,5'-diamine, tin salt of dihydrochloride	20% HCl	244(3.78),304(4.08), 407(3.53)	65-0674-64
	EtOH	352(4.18),470(4.01), 528(3.96),598(3.50), 626(3.78)	65-0674-64
$C_8H_8N_2Se$			
2,1,3-Benzoselenadiazole, 5,6-dimethyl-	THF	225s(3.9),240s(3.6), 340f(4.1)	46-1773-65
$C_8H_8N_4$			
Quinazoline, 2,4-diamino-	pH 1	226(4.59),230s(4.57), 247s(4.11),315(3.69), 322s(3.60)	44-1837-65
	pH 10	231(4.65),266(3.94), 273s(3.89),332(3.65)	44-1837-65
$C_8H_8N_4O$			
Imidazo[4,5-b]pyridine, 7-acetamido-	pH 1	286(4.34)	87-0708-65
	pH 7	273(4.24)	87-0708-65
	pH 13	282(4.21)	87-0708-65
Pyrimido[4,5-b][1,4]diazepin-6-one, 5,9-dihydro-8-methyl-	pH 1	284(3.96)	4-0110-65
	pH 11	240(4.00),310(3.57)	4-0110-65
	EtOH	229(4.15),254(3.98), 306(3.48)	4-0110-65
$C_8H_8N_4OS$			
Acetic acid, thio-, S-purin-6-ylmethyl ester	N HCl	266(3.87)	87-0667-65
	pH 7.65	231(3.77),267(3.98)	87-0667-65
$C_8H_8N_4O_2$			
1,2,4-Oxadiazole, 3-methyl-5-[2-(3-methyl-1,2,4-oxadiazol-5-yl)vinyl]-	Et_2O	280(4.36)	33-0942-64
1,2,4-Oxadiazole, 5-methyl-3-[2-(5-methyl-1,2,4-oxadiazol-3-yl)vinyl]-	Et_2O	240(4.40)	33-0942-64
1,2,4-Triazolidine-3,5-dione, 1-(p-aminophenyl)-	NaOH	216(4.09),272(4.09)	22-0500-64
	EtOH	210(4.08),261(4.21)	22-0500-64
$C_8H_8N_4O_2S$			
Carbonic acid, thio-, O-ethyl S-purin-6-yl ester	pH 1	321(4.24)	87-0010-64
	pH 7	322(4.28)	87-0010-64
Carbonic acid, thio-, O-ethyl S-purin-8-yl ester	pH 1	318(4.23)	87-0010-64
	pH 7	318(3.98)	87-0010-64
Carbonic acid, thio-, O-methyl S-9-methyl-9H-purin-6-yl ester	pH 1	283(4.02)	87-0010-64
	pH 7	280(4.13)	87-0010-64
9H-Purine-9-carboxylic acid, 6-(methylthio)-, methyl ester	pH 1	287(4.35)	87-0010-64
	pH 7	287(4.33)	87-0010-64
9H-Purine-9-carboxylic acid, 8-(methylthio)-, methyl ester	pH 1	313(4.41)	87-0010-64
	pH 7	299(4.18)	87-0010-64
Thiazole, 5-acetamido-4-(5-methyl-1,2,4-oxadiazol-3-yl)-	pH 1.2	281(4.13)	87-0190-65
	pH 7.6	283(4.13)	87-0190-65
	pH 11.0	225(3.99),310(4.04)	87-0190-65

Compound	Solvent	λ_{max}(log ϵ)	Ref.
$C_8H_8N_4O_3$			
Lumazine, 6,7-dimethyl-, 5-oxide	pH 3.0	239(4.37),276s(3.67), 342(3.86)	89-1136-65
	pH 9.0	240(4.13),267(4.14), 285s(4.11),380(3.89)	89-1136-65
Lumazine, 8-(2-hydroxyethyl)-	0.1N H_2SO_4	256(4.19),397(3.97)	37-3493-64
	pH 13	231(4.32),281(4.13), 305(3.95)	37-3493-64
Pteridine-2,4-diol, 6,7-dimethyl-, 1-oxide	pH 1.2	337(3.99)	44-0408-65
	pH 12.5	272(4.39),394(3.93)	44-0408-65
$C_8H_8N_4O_4S$			
Benzenesulfonamide, p-(3,5-dioxo-1,2,4-triazolidin-1-yl)-	NaOH	227(3.99),302(4.19)	22-0500-64
	EtOH	210(4.00),269(4.15)	22-0500-64
$C_8H_8N_4S$			
Pyrrole-3,4-dicarbonitrile, 2-amino-5-(ethylthio)-	EtOH	224(4.16),256(3.79), 300(4.02)	35-1995-65
Thiadiazole, 2-hydrazino-5-phenyl-	EtOH	306(4.16)	1-0871-64
$C_8H_8N_6O_2$			
2(1H)-Pyrimidinone, 4,4'-hydrazodi-	pH 1	327(4.33)	5-0134-65F
$C_8H_8N_6S_2$			
Pyrimidine, 4,4'-dithiobis[6-amino-	pH 1	233(4.49),281(4.15)	87-0005-64
C_8H_8O			
Acetophenone	gas	190(4.60),230(4.19), 276(3.34),320(2.18)	35-0011-65
	C_6H_{12}	235(3.72),255(2.56), 279(2.66),286(2.61), 315(1.94)	73-3462-65
	EtOH	241(4.09)	44-2165-65
	EtOH	200(3.97),241(4.05), 279(2.94),284(2.98)	73-3462-65
	40% iso-PrOH	245(4.1),280s(3.0)	61-0296-64
Benzocyclobutene, 4-hydroxy-	EtOH	283.5(3.71)	78-2281-65
Bicyclo[3.2.1]octa-3,6-dien-2-one	n.s.g.	230(3.73)	25-0424-65
Ethane, 1,2-epoxy-1-phenyl-	EtOH	254(2.24),260(2.28)	12-0168-65
Ether, phenyl vinyl	nonane	270(3.09)	67-0375-65
Tetracyclo[3.3.0.0^{2,8}.0^{4,6}]octan-3-one	EtOH	205(3.73),281(1.69)	35-4301-65
	EtOH	205(3.73),281(1.70)	35-1876-64
C_8H_8OS			
3-Buten-2-one, 4-(2-thienyl)-	EtOH	323(3.80)	65-3645-64
	66% H_2SO_4	410(4.27)	65-3645-64
$C_8H_8O_2$			
Acetic acid, phenyl-	H_2O	260(2.138)	65-1474-64
	EtOH	260(2.1)	70-1798-65
Acetophenone, 2'-hydroxy-	EtOH	253(4.04),324(3.52)	28-5614-64A
Acetophenone, 3'-hydroxy-	0.3N $HClO_4$	310(3.4)	60-1053-64
	pH 8.80	310(3.3)	60-1053-64
	pH 9.10	315(3.2),345s(3.0)	60-1053-64
	pH 9.27	315(3.2),350s(3.1)	60-1053-64
	pH 9.54	325s(3.2),345(3.2)	60-1053-64
	pH 9.94	350(3.3)	60-1053-64
	0.12N NaOH	350(3.4)	60-1053-64

Compound	Solvent	$\lambda_{max}(\log \epsilon)$	Ref.
Acetophenone, 4'-hydroxy-	0.2N $HClO_4$	275(3.8)	60-1053-64
	pH 7.52	280(3.7),320(3.4)	60-1053-64
	pH 7.88	285(3.6),325(3.7)	60-1053-64
	pH 8.25	285s(3.6),330(3.9)	60-1053-64
	pH 8.53	330(3.9)	60-1053-64
	0.12N NaOH	330(4.1)	60-1053-64
	pH 13	325(4.37)	44-2165-65
	EtOH	275(4.16)	28-5614-64A
	EtOH	199(4.15),219(4.03), 276(4.22)	33-1775-64
	EtOH	277(4.17)	44-2165-65
Anisaldehyde	acid	284(4.21)	44-0603-65
p-Benzoquinone, ethyl-	95% THF	248(4.26),315(2.90), 435(1.40)	44-2602-65
Fulvene, 6-acetoxy-	hexane	266(4.36),273(4.29), 361(2.54)	5-0039-64H
Furfuryl alcohol, α-2-propynyl-	n.s.g.	none	5-0062-65B
4,6-Heptadiynoic acid, methyl ester	EtOH	225(3.54),238(3.56), 252(3.60),280(3.91)	70-1237-65
Toluene, 3,4-(methylenedioxy)-	EtOH	285(--)	95-0857-65
p-Toluic acid	EtOH	233(4.2),273s(2.8)	70-1798-65
$C_8H_8O_2S$			
1,3-Butanedione, 1-(2-thienyl)-	n.s.g.	277(4.24),325(4.48)	67-0379-65
copper salt	n.s.g.	284(4.32),345(4.50)	67-0379-65
$C_8H_8O_3$			
p-Benzoquinone, 3-hydroxy-2,5-dimethyl-	MeOH	265(4.19),405(3.04)	24-2774-65
	MeOH-KOH	223(4.33),270(3.96), 525(3.45)	24-2774-65
	$CHCl_3$	266(4.24),402(2.94)	24-2774-65
p-Benzoquinone, 2-methoxy-6-methyl-	EtOH	265(3.94),363(2.82)	94-0236-64
Benzyl alcohol, 3,4-(methylenedioxy)-	EtOH	286(--)	95-0857-65
1,2-Butanedione, 1-(2-furyl)-	EtOH	284(3.94)	78-2951-64
4-Cyclopentene-1,3-dione, 2-acetyl-4-methyl-	C_6H_{12}	231(4.25),261(4.18), 320(2.83)	1-0441-64
4,6-Octadiynoic acid, 8-hydroxy-	EtOH	234(3.67),242(3.63), 256(3.41),282(2.98), 301(2.91)	70-1237-65
1,2-Propanedione, 1-(5-methyl-2-furyl)-	MeOH	296(3.98)	78-2951-64
$C_8H_8O_4$			
Acetic acid, (3,5-dihydroxyphenyl)-	EtOH	276(3.31),282(3.31)	1-1677-65
Benzoic acid, 2,4-dihydroxy-6-methyl-	EtOH	260(3.97),298(3.81)	44-3566-65
p-Benzoquinone, 2,3-dihydroxy-4,5-dimethyl-	EtOH	277(4.07),445(2.61)	1-2303-64
p-Benzoquinone, 2,6-dihydroxy-3,5-dimethyl-	EtOH	297(4.26),426(2.26)	78-2319-64
	EtOH	297(4.26),426(2.26)	94-0511-65
Comanic acid, ethyl ester	EtOH-HCl	260(3.88)	39-2251-65
	EtOH-NaOH	260(3.91),352(4.21)	39-2251-65
1,3-Cyclohexadiene-1,4-dicarboxylic acid	EtOH	309(4.11)	23-2852-64
1,2,4-Cyclopentanetrione, 3-acetyl-5-methyl-	EtOH-HCl	253(4.34),273(4.21)	39-2251-65
	EtOH-NaOH	275(4.41),303(4.42)	39-2251-65
3-Cyclopentene-1,1-dicarboxylic acid, 2-methylene-	EtOH	231(4.18)	70-1460-65
Dehydroacetic acid	EtOH	226(3.99),311(4.05)	44-1255-65
Phthalic acid, 1,4-dihydro-	EtOH	284(2.32)	35-1925-65
Phthalic acid, 2,3-dihydro-	EtOH	288(3.89)	35-1925-65
2-Pyrone, 3-carbomethoxy-6-methyl-	EtOH	317(4.00)	78-2701-64

Compound	Solvent	λ_{max}(log ϵ)	Ref.
Terreic acid, methyl ester	C_6H_{12}	306(3.73)	39-6587-65
	aq. MeOH	304(3.77)	39-6587-65
$C_8H_8O_5$			
Benzaldehyde, 2,3,4-trihydroxy-6-(hydroxymethyl)-	EtOH	241(4.03),304(4.18)	23-1595-64
p-Benzoquinone, 2,3-epoxy-6-hydroxy-5-methoxy-2-methyl-	EtOH	330(3.8)	94-0935-65
$C_8H_8O_5S$			
2,5-Thiophenedicarboxylic acid, 3-hydroxy-, 2-ethyl ester	NaOH	277(3.98),355(3.83)	35-0107-64
	EtOH	273(4.11),310(3.83)	35-0107-64
C_8H_8S			
Benzene, (epithioethyl)-	hexane	226(3.93)	35-4628-64
$C_8H_8S_2$			
4,9-Dithiabicyclo[5.3.0]deca-1(10),2,7-triene	EtOH	290(4.20)	39-1154-64
1,6-Dithiacyclodeca-3,8-diyne	EtOH	none	39-1154-64
1,3-Dithiolane, 2-(cyclopentadienylidene)-	$CHCl_3$	278(3.62),343s(4.16),355(4.21),367s(4.07)	24-2825-65
$C_8H_8S_4$			
1,4,5,8-Tetrathiafulvalene, 2,7-dimethyl-	THF	298(3.88),453(2.26)	89-0453-65
$C_8H_9BN_2$			
4,3-Borazaroisoquinoline, 3-methyl-	EtOH	215(4.5),270(3.8),272s(3.8),274(3.8),278(3.7),295(3.8),300(4.0)	35-0433-64
C_8H_9BO			
Boronophthalide, B-methyl-	EtOH	264(2.85),271(3.01),278(3.04)	44-3229-64
$C_8H_9BrN_4$			
Pyrazolo[1,5-a]pyrimidine, 7-amino-3-bromo-2,5-dimethyl-	EtOH-HCl	228(4.51),273s(3.61),291(3.83),312(3.80)	95-1113-64
C_8H_9BrO			
Anisole, m-(bromomethyl)-	hexane	212(3.89),283(3.36)	78-0861-64
Anisole, o-(bromomethyl)-	hexane	233(3.82),286(3.51)	78-0861-64
Anisole, p-(bromomethyl)-	hexane	246(4.08),286(3.26)	78-0861-64
Phenethyl alcohol, p-bromo-	EtOH	221(4.00)	44-2109-64
$C_8H_9Br_2N_5$			
Adenine, N-(2,3-dibromopropyl)-	EtOH-HBr	277(4.22)	23-1599-64
	EtOH	280(4.10)	23-1599-64
C_8H_9Cl			
2,4-Heptadiyne, 6-chloro-6-methyl-	EtOH	223(2.79),234(2.99),247(3.09),261(2.97)	78-1357-65
$C_8H_9ClN_2$			
1H-Pyrrolo[2,3-b]pyridine, 6-chloro-2,3-dihydro-4-methyl-	EtOH	256(3.86),312(3.82)	65-1454-64

Compound	Solvent	λ_{max}(log ϵ)	Ref.
$C_8H_9ClN_2O_3$			
Acetic acid, [(6-chloro-3-pyridazin-yl)oxy]-, ethyl ester	1% EtOH	275(3.25)	20-0532-64
1(6H)-Pyridazineacetic acid, 3-chloro-6-oxo-, ethyl ester	1% EtOH	275s(3.39),300(3.38)	20-0532-64
$C_8H_9ClN_2O_4$			
1-Methylimidazo[1,2-a]pyridinium perchlorate	EtOH	214(4.28),218s(4.14), 284(3.85)	4-0331-65
$C_8H_9ClN_4$			
5-Pyrimidinepropionitrile, 2-amino-4-chloro-6-methyl-	pH 1	314(3.91)	4-0263-64
	pH 8.4	235(4.29),300(3.76)	4-0263-64
	pH 13	300(3.86)	4-0263-64
$C_8H_9ClN_4O$			
9H-Purine-9-ethanol, 6-chloro-	pH 1	265(3.99)	87-0502-65
α-methyl-	pH 7, 13	265(3.99)	87-0502-65
Purine-7-propanol, 6-chloro-	pH 1	268(3.93)	87-0033-65
	pH 7	269(3.94)	87-0033-65
	pH 13	269(3.94)	87-0033-65
$C_8H_9ClN_4O_2$			
1,2-Propanediol, 3-(6-chloro-9H-purin-9-yl)-	pH 1	260(3.99)	87-0502-65
	pH 7	260(3.99)	87-0502-65
	pH 13	262(4.06)	87-0502-65
$C_8H_9ClN_4O_4$			
2-Furaldehyde, 5-nitro-, 2-(2-chloro-ethyl)semicarbazone	5% EtOH	378(4.21)	44-2582-64
$C_8H_9ClN_4S$			
9H-Purine, 9-(2-chloroethyl)-6-(methylthio)-	EtOH	283(4.25)	87-0182-65
C_8H_9ClO			
Anisole, m-(chloromethyl)-	hexane	222(3.87),278(3.32)	78-0861-64
Anisole, o-(chloromethyl)-	hexane	222(3.80),277(3.40)	78-0861-64
Anisole, p-(chloromethyl)-	hexane	234(4.11),276(3.24), 280(3.21)	78-0861-64
Phenol, 4-chloro-2,3-dimethyl-	C_6H_{12}	280(3.37),284(3.35), 289(3.36)	2-0417-64
Phenol, 4-chloro-2,5-dimethyl-	C_6H_{12}	282(3.65),289(3.61)	2-0417-64
Phenol, 4-chloro-2,6-dimethyl-	C_6H_{12}	280(3.23),288(3.22)	2-0417-64
Phenol, 4-chloro-3,5-dimethyl-	C_6H_{12}	279(3.32),287(3.32)	2-0417-64
Phenol, 6-chloro-3,4-dimethyl-	C_6H_{12}	282(3.51),288(3.51)	2-0417-64
$C_8H_9ClO_2$			
Resorcinol, 2-chloro-4,5-dimethyl-	MeOH	220(4.02),284(3.19)	20-0081-64
$C_8H_9ClO_3$			
1,2,4-Cyclohexanetrione, 3-chloro-6,6-dimethyl-	MeOH-HCl	290(4.11)	20-0081-64
	MeOH-NaOH	238(3.88),340(4.04)	20-0081-64
$C_8H_9Cl_3O_3$			
Cyclopentadienone, 2,3,5-trichloro-4-methoxy-, dimethyl acetal	iso-PrOH	307(3.336)	39-4744-65

Compound	Solvent	$\lambda_{max}(\log \epsilon)$	Ref.
$C_8H_9D_2NO$			
4(1H)-Pyridone-2,6-d$_2$, 1,3,5-trimethyl-	MeOH	275(4.25)	35-3365-65
$C_8H_9F_2NO_3$			
Pyridine, 3,5-difluoro-2,4,6-tri-	hexane	230s(3.3),279(3.76)	39-0575-65
methoxy-	EtOH	229s(3.4),281(3.74)	39-0575-65
$C_8H_9F_4N_3O_3$			
1,2,3-Cyclopentanetrione, tetrafluoro-, tris(O-methyloxime)	EtOH	237(4.23),299(3.98)	44-3698-65
C_8H_9IO			
Anisole, m-(iodomethyl)-	hexane	217(4.05),286(3.33)	78-0861-64
Anisole, o-(iodomethyl)-	hexane	243(3.81),293(3.71)	78-0861-64
Anisole, p-(iodomethyl)-	hexane	260.5(4.01)	78-0861-64
C_8H_9N			
Aniline, p-vinyl-	n.s.g.	274(4.21)	44-1926-65
3H-3a-Azaindene, 1,2-dihydro-	pH 7.0	258s(3.61),264(3.70), 271(3.57)	39-2778-65
Methylamine, N-benzylidene-	$C_2H_4Cl_2$	246(4.29)	30-1017-64
1,6-Iminocyclodecapentaene	C_6H_{12}	243(4.3),275(4.5), 310s(3.9),398f(2.7)	88-3613-65
C_8H_9NO			
Acetophenone, p-amino-	EtOH	317(4.32)	44-2165-65
3-Buten-2-one, 4-pyrrol-2-yl-, trans	n.s.g.	208(3.76),269(3.23), 353(4.37)	12-0875-65
Hydroxylamine, O-(1-benzocyclobutenyl)-	EtOH	254s(2.79),261(3.05), 267(3.24),273(3.23)	87-0732-65
Phenol, o-(N-methylformimidoyl)-	EtOH	315(3.65)	59-1625-65
	acid	350(3.73)	59-1625-65
	base	355(3.87)	59-1625-65
Pyridine, 3-methoxy-2-vinyl-	EtOH	237(3.99),308(3.85)	44-1834-64
o-Toluamide	EtOH	269(2.81)	88-2659-64
Tropone, 2-amino-7-methyl-	isooctane	250(4.4),340(4.0), 395(4.0)	49-0402-64
$C_8H_9NO_2$			
Acetic acid, (m-aminophenyl)-	H_2O	286(2.657)	65-1474-64
Acetic acid, (p-aminophenyl)-	H_2O	284(2.634)	65-1474-64
Acetophenone, 3'-amino-2'-hydroxy-	EtOH	236(4.18),275(3.88)	39-2816-64
hydrochloride	EtOH	246(3.96),319(3.53)	39-2816-64
Acetophenone, 5'-amino-2'-hydroxy-	EtOH	245(3.59),324(3.62)	39-2816-64
hydrochloride	EtOH	233(4.33)	39-2816-64
Anthranilic acid, methyl ester	H_2O	244(3.6),327(3.3)	24-1127-64
Anthranilic acid, N-methyl-	10% MeOH	249(3.37),349(3.05)	24-1127-64
anion	10% MeOH	252(3.84),326(3.50)	24-1127-64
Azepine-2,5-dione, 4,6-dimethyl-	n.s.g.	228(3.26),288(2.53)	88-1071-65
Benzamide, 4-methoxy-	H_2O	253(3.21)	23-1957-64
	H_2O	253(3.23)	23-2328-65
	HCl	283(4.17)	23-2328-65
	58.7% H_2SO_4	283(4.22)	23-1957-64
2,4-Hexadienoic acid, 5-cyano-3-methyl-, cis-cis	EtOH	269(4.25)	44-0203-65
Indole-2,3-dione, 4,5,6,7-tetrahydro-	EtOH	278(4.28)	39-0991-64
Oxaziridine, 3-(p-methoxyphenyl)-	EtOH	230(4.1),275(3.23), 285(3.2)	44-3427-65
4-Pyridineacetic acid, methyl ester	N H_2SO_4	216(3.67),254(3.70), 260s(3.61)	12-0455-64

Compound	Solvent	$\lambda_{max}(\log \epsilon)$	Ref.
4-Pyridineacetic acid, methyl ester (cont.)	pH 10.0	252s(3.34),256(3.41), 263s(3.32)	12-0455-64
2(1H)-Pyridone, 1-(allyloxy)-	H_2O	226(3.74),296(3.73)	44-1650-64
p-Toluhydroxamic acid	MeOH	233(4.08)	73-0940-65
	50% MeOH-HCl	239(4.13)	73-0940-65
	50% MeOH-borate	221(4.10),265(4.08)	73-0940-65
Tropone, 2-amino-7-methoxy-	EtOH	250(4.44),262(4.24), 272(4.12),338(4.15), 398(3.99)	94-0457-65
Tropone, 3-amino-2-methoxy-	EtOH	264(4.44),274(4.39), 315(3.88),370(3.32)	94-0457-65
2,6-Xylenol, 4-nitroso-, sodium salt	H_2O	393(--)	96-0730-64
$C_8H_9NO_2S$			
Nicotinic acid, 1,2-dihydro-5,6-dimethyl-2-thioxo-	EtOH	240s(3.59),305(4.28), 390(3.66)	94-0087-64
$C_8H_9NO_3$			
Ketone, 2-furyl methyl, O-acetyloxime, anti	EtOH	273(4.17)	22-2724-65
syn	EtOH	268(4.21)	22-2724-65
Nicotinic acid, 1,2-dihydro-5,6-dimethyl-2-oxo-	EtOH	238(3.86),341(3.96)	94-0087-64
Pyridine, 3-acetyl-2,4-dihydroxy-6-methyl-	EtOH	230(3.85),265(4.22), 325(3.92)	70-1855-65
3-Pyrrolin-2-one, 3-acetyl-4-hydroxy-1-methyl-5-methylene-	EtOH	228(3.92),267(4.37)	35-5654-64
2,5-Xylenol, 4-nitro-	MeOH	312(3.88)	35-3877-64
2,6-Xylenol, 4-nitro-, sodium salt	H_2O	432(--)	96-0730-64
$C_8H_9NO_4$			
Pyrrole-1,3-dicarboxylic acid, dimethyl ester	EtOH	208(4.37),233(4.08)	23-1279-64
Pyrrole-2,5-dicarboxylic acid, dimethyl ester	EtOH	273(4.30),282(4.25)	39-0459-65
	EtOH-base	255(4.03),303(4.34)	39-0459-65
$C_8H_9NO_4S$			
Nicotinic acid, 1,2,5,6-tetrahydro-4-(methylthio)-2,6-dioxo-, methyl ester	EtOH	256(4.35),360(4.17)	95-0387-65
$C_8H_9NO_4S_2$			
Malonamic acid, N-(3-oxo-3H-1,2-dithiol-4-yl)-, ethyl ester	EtOH	234(3.73),327(3.64)	12-0447-64
$C_8H_9NO_6$			
Pyrrole-2,5-dicarboxylic acid, 3,4-dihydroxy-, dimethyl ester	EtOH	285(4.40)	44-0859-65
$C_8H_9N_3$			
Glutacononitrile, 4-(1-aminoethylidene)-3-methyl-	EtOH	219(4.05),254(3.85), 328(4.24)	23-0332-65
Glutacononitrile, 4-[(dimethylamino)-methylene]-	n.s.g.	320(4.61)	44-1800-64
Imidazo[1,2-a]pyridine, 5-amino-2-methyl-	pH 1	264(3.62),269s(3.61), 312(4.05)	4-0053-65
	pH 11	301(3.96)	4-0053-65

Compound	Solvent	λ max(log ε)	Ref.
Nicotinonitrile, 6-amino-2,4-dimethyl-	EtOH	210(4.05),271(4.24), 295(3.87)	23-0332-65
Pyrazolo[1,5-a]pyrimidine, 2,3-dimethyl-	EtOH	238(4.40),284(2.93), 290(2.94),304s(2.75), 344(2.79)	94-1207-65
1,3a,7-Triazaindene, 4,6-dimethyl-	pH 3.0	215(4.44),274(3.83), 290s(3.65)	39-2778-65
	pH 8.0	224(4.43),230s(4.27), 272(3.57),281(3.58), 306(3.53)	39-2778-65
1,4,6-Triazaindene, 2,5-dimethyl-	pH 3.1	230s(4.2),280(3.7), 330(3.7)	94-1030-64
	pH 10.3	280(3.8)	94-1030-64
	MeOH	280(3.8)	94-1030-64
$C_8H_9N_3O$			
Formamide, 1-(o-tolylazo)-	n.s.g.	293(4.00),440(2.17)	49-1314-65
Formamide, 1-(p-tolylazo)-	n.s.g.	226(4.00),302(4.14), 430(2.18)	49-1314-65
5-Pyrimidinecarbonitrile, 1-allyl-1,2,3,4-tetrahydro-2-oxo-	EtOH	213(3.92),285(4.03)	94-1418-64
5-Pyrimidinecarbonitrile, 3-allyl-1,2,3,4-tetrahydro-2-oxo-	EtOH	218(3.94),280(4.16)	94-1418-64
5H-Pyrrolo[2,3-b]pyrazine-7-methanol	EtOH	227(4.24),307(3.93)	44-3454-65
5H-Pyrrolo[3,2-d]pyrimidine, 6-ethoxy-	pH 1	240(4.03),279(3.72), 311(4.00)	44-1528-65
	pH 7	225(4.18),294(3.97)	44-1528-65
	pH 13	232(4.34),290(3.78), 305s(3.70)	44-1528-65
$C_8H_9N_3OS$			
5-Pyrimidinepropionitrile, 4-hydroxy-2-mercapto-6-methyl-	pH 1	280(4.32)	4-0263-64
	pH 8,4	280(4.35)	4-0263-64
	pH 13	262(4.20),310s(3.89)	4-0263-64
$C_8H_9N_3O_2$			
Imidazo[1,2-c]pyrimidine-2,5(1H,3H)-dione, 3,3-dimethyl-	N HCl	236(3.72),300(4.18)	44-1762-64
	pH 4.27	213(3.93),303(4.28)	44-1762-64
	pH 13	318(4.31)	44-1762-64
2H-Pyrrolo[2,3-d]pyrimidine-2,4(3H)-dione, 1,7-dihydro-1,3-dimethyl-	pH 1	243(3.78),275(3.82)	4-0034-64
	pH 11	230(4.09),277(3.83)	4-0034-64
5H-Pyrrolo[3,2-d]pyrimidin-4(3H)-one, 6-ethoxy-	pH 1	228(4.19),286(3.86), 321(3.72)	44-1528-65
	pH 7	250(4.28),297(3.81)	44-1528-65
	pH 13	247(4.35),296(3.82)	44-1528-65
$C_8H_9N_3O_2S$			
Thiazolo[5,4-d]pyrimidine-5,7(4H,6H)-dione, 2,4,5-trimethyl-	H_2O	282(4.32)	94-1319-65
$C_8H_9N_3O_3$			
Salicylaldehyde, 5-nitro-, methylhydrazone	MeOH	287(4.43)	24-1631-64
	MeOH-NaOH	345(4.11),416(4.10)	24-1631-64
$C_8H_9N_3O_4$			
Aniline, N,N-dimethyl-2,4-dinitro-	MeOH	368(4.23)	35-4018-64
Aniline, N-ethyl-2,4-dinitro-	MeOH	348(4.21),415s(3.81)	35-4018-64
2-Furanacrylic acid, α-methyl-5-nitro-, hydrazide	propylene glycol	240(4.16),272(3.94), 362(4.11)	95-0212-64

Compound	Solvent	λ_{max}(log ε)	Ref.
5-Pyrazinecarboxylic acid, 2-amino-3-formyl-6-hydroxy-, ethyl ester	pH 3.5	230(3.89),293(4.14), 361(4.40)	39-3357-64
	pH 8.5	229(4.30),293(4.14), 360(4.34)	39-3357-64
$C_8H_9N_3O_7S$			
Taurine, N-(2,4-dinitrophenyl)-	H_2O	359(4.23)	96-0788-64
	acetone	359(4.27)	96-0788-64
$C_8H_9N_5$			
Guanidine, (2-benzimidazolyl)-	MeOH-KOH	293(4.36),303(4.33)	4-0288-64
Pyrido[3,4-b]pyrazine, 5,7-diamino-3-methyl-	pH 1	244(4.25),314(4.23)	44-0734-64
	pH 7	265(4.35),312(3.86)	44-0734-64
	pH 13	266(4.37),312(3.86)	44-0734-64
$C_8H_9N_5O$			
7(1H)-Pteridinone, 4-amino-1,6-dimethyl-	pH -0.20	219(4.09),242(4.21), 248s(4.15),290(3.86), 317s(4.06),325(4.10), 335s(4.06)	39-3770-65
	pH 5.0	231(4.28),248(4.30), 251s(4.29),326s(4.15), 336(4.21),350(4.08)	39-3770-65
$C_8H_9N_5O_2$			
Triazene, 1,3-bis(5-methyl-3-isoxazolyl)-	EtOH	291(4.25)	78-0461-64
$C_8H_9N_5O_2S$			
6-Azalumazine, 7-(ethylthio)-3-methyl-	H_2O	234(4.26),261(4.20), 344(3.99)	24-0005-64
$C_8H_9N_5O_3$			
L-Alanine, N-(1,6-dihydro-6-oxo-purin-2-yl)-	pH 1	249(4.20),277(3.86)	35-3752-65
	pH 11	244(3.98),277(3.86)	35-3752-65
Guanidine, [[3-(5-nitro-2-furyl)allylidene]amino]-, hydrochloride	propylene glycol	290(4.15),388(4.18)	95-0001-64
$C_8H_9N_5S$			
[1,2,5]Thiadiazolo[3,4-d]pyrimidine, 7-(1-pyrrolidinyl)-	pH 1	227(4.13),268(3.63), 347(4.23)	44-2135-64
	pH 7	234(4.23),275(3.60), 368(4.10)	44-2135-64
	pH 13	234(4.23),275(3.61), 368(4.10)	44-2135-64
$C_8H_9O_4P$			
Phosphonic acid, (p-acetylphenyl)-	EtOH	249(4.20),284(3.20)	87-0891-65
C_8H_{10}			
1-Cyclohexene, 1-ethynyl-	hexane	222(4.15)	22-3518-65
	EtOH	223.5(4.02)	22-2541-64
Cyclopentene, 1-propynyl-	n.s.g.	226(4.20)	22-1525-65
2,3-Heptadien-5-yne, 2-methyl-	hexane	220(4.24)	78-1357-65
2,4-Octadiyne	n.s.g.	203(2.87),214(2.95), 225(2.98),237(2.86), 252(2.53)	22-1525-65
1,3,5,7-Octatetraene, cis-trans	isooctane	256(4.00),266(4.31), 277(4.59),289(4.76), 302(4.71)	24-1427-65

Compound	Solvent	λ_{max}(log ε)	Ref.
Tricyclo[3.2.1.0^{2,4}]oct-6-ene, endo	n.s.g.	192(4.00)	88-3301-65
exo	n.s.g.	192(3.95)	88-3301-65
$C_8H_{10}BrN_3O_5$			
as-Triazine-3,5(2H,4H)-dione, 6-bromo-2-(2-deoxy-β-D-ribofuranosyl)-	pH 1	279(3.81)	4-0495-65
	pH 13	267(3.76)	4-0495-65
$C_8H_{10}Br_2O$			
Bicyclo[3.2.1]octan-3-one, 2,4-dibromo-, cis	C_6H_{12}	237(2.77),342(2.29)	22-0844-64
	MeCN	237(2.78),341(2.28)	22-0844-64
trans	$CHCl_3$	232(2.68),304(2.10)	22-0844-64
	dioxan	235(2.77),306(2.05)	22-0844-64
$C_8H_{10}Br_4$			
Bicyclopropyl, 2,2,2',2'-tetrabromo-1,1'-dimethyl-	hexane	202(3.70)	44-2951-64
Bicyclopropyl, 2,2,2',2'-tetrabromo-3,3'-dimethyl-	hexane	208.5(3.73)	44-2951-64
	MeOH	208.5(3.73)	44-2951-64
Ethane, 1,2-bis(2,2-dibromocyclopropyl)-	hexane	199.5(3.65)	44-2951-64
$C_8H_{10}Br_5N_5$			
6-Amino-3-(2,3-dibromopropyl)purinium bromide, compound with bromine	EtOH	278(4.28)	23-1599-64
$C_8H_{10}ClN_3O_2$			
3(2H)-Pyridazinone, 4-chloro-5-morpholino-	EtOH	239(4.19),320(3.83)	22-2124-64
	NaOH	229(4.30),304(3.77)	22-2124-64
$C_8H_{10}ClN_3O_4S_2$			
2H-1,2,4-Benzothiadiazine-7-sulfonamide, 6-chloro-3,4-dihydro-3-methyl-, 1,1-dioxide	EtOH	226(4.58),270(4.34)	95-0095-65
$C_8H_{10}ClN_3O_5S_2$			
2H-1,2,4-Benzothiadiazine-7-sulfonamide, 6-chloro-3,4-dihydro-3-(hydroxymethyl)-, 1,1-dioxide	EtOH	225(4.59),270(4.30)	95-0095-65
$C_8H_{10}ClN_5S$			
9H-Purine, 2-amino-9-(2-chloroethyl)-6-(methylthio)-	EtOH	243(4.17),307(4.06)	87-0182-65
$C_8H_{10}Cl_2O_2$			
1,3-Cyclohexanedione, 2,4-dichloro-5,5-dimethyl-	MeOH-HCl	273(4.08)	20-0081-64
	MeOH-NaOH	301(4.24)	20-0081-64
$C_8H_{10}Cl_2O_4$			
1,2-Cyclobutanedicarboxylic acid, 1,2-dichloro-, dimethyl ester, cis	C_6H_{12}	190(3.32),222(2.40)	24-0764-65
$C_8H_{10}Cl_3NS_2$			
2H-1,3-Thiazine-2-thione, 3,6-dihydro-3,4,5-trimethyl-6-(trichloromethyl)-	H_2O	270s(3.62),314(3.89)	39-4008-64
	EtOH	270(3.63),316(3.99)	39-4008-64
$C_8H_{10}INO$			
3-Acetyl-1-methylpyridinium iodide	n.s.g.	267(3.59),274(3.49)	37-1237-64
$C_8H_{10}NO_6P$			
Pyridoxal, phosphate	pH 1	278(3.9),324(3.8)	70-0680-64
	pH 7	358(3.7)	70-0680-64

Compound	Solvent	λ_{max}(log ε)	Ref.
$C_8H_{10}N_2$			
7-Azaindoline, 4-methyl-	EtOH	248(3.80),306(3.68)	65-1454-64
Ethane, phenylazo-	hexane	261(3.98),405(2.10)	39-2788-65
Indoline, 5-amino-, dihydrochloride	MeOH	246(3.93),300(3.32)	44-2589-65
Indoline, 7-amino-	MeOH	246s(3.80),295(3.39)	44-2589-65
dihydrochloride	MeOH	241(4.05),286(3.33)	44-2589-65
1-Pyrroline, 2-pyrrol-1-yl-	EtOH	245(4.15)	23-1073-64
1-Pyrroline, 2-pyrrol-2-yl-	EtOH	279(4.21)	23-1073-64
	EtOH	241s(3.61),280(4.20)	39-0888-64
	EtOH-HCl	268s(3.48),318(4.44)	39-0888-64
$C_8H_{10}N_2O$			
Acetophenone, α-amino-, oxime, anti	EtOH-HCl	231(3.97)	24-0567-65
syn	EtOH-HCl	245.5(4.08)	24-0567-65
Aniline, N,N-dimethyl-p-nitroso-	pH 13	275(3.73),428(4.44), 685(1.79)	20-0843-64
Aniline, N-ethyl-N-nitroso-	C_6H_{12}	275(3.83)	35-4373-64
1,3-Cyclopentadiene-1-carboxaldehyde, 5-oxo-, 5-(dimethylhydrazone)	MeOH	252(4.18),374(4.35)	35-4373-64
Indol-2(4H)-one, 3-amino-5,6-dihydro-	n.s.g.	284(4.20)	39-0991-64
2(1H)-Quinoxalinone, 5,6,7,8-tetra- hydro-	EtOH	229(3.81),337(3.84)	44-0579-64
	10% HCl	228(4.06),368(3.99)	44-0579-64
	10% NaOH	233(3.95),331(3.90)	44-0579-64
$C_8H_{10}N_2O_2$			
Aniline, N,N-dimethyl-p-nitro-	C_6H_{12}	356(4.36)	44-2088-64
	MeOH	390(4.28)	35-4018-64
	MeOH	390.5(4.25)	44-2088-64
	iso-PrOH	390(4.25)	44-2088-64
	dioxan	382(4.32)	44-2088-64
	n.s.g.	408(4.30)	44-2088-64
p-Benzoquinone, 2,5-bis(methylamino)-	EtOH	337(4.47),490(2.42)	39-5569-64
1,3-Cyclohexadiene-1-carbonitrile, 2-amino-5-hydroxy-6-(hydroxymethyl)-	n.s.g.	330(3.76)	44-3252-64
1,3-Cyclohexadiene-1,4-dicarboxamide	EtOH	305(4.07)	23-2852-64
1,3-Cyclohexanedione, 2-diazo-5,5- dimethyl-	n.s.g.	227(4.24),256(3.95)	5-0101-64F
Ethane, 1-hydroperoxy-1-phenylazo-	hexane	268(3.97),409(2.10)	39-2788-65
Nicotinamide, 1,2-dihydro-5,6-dimethyl- 2-oxo-	EtOH	239(3.94),340(4.04)	94-0087-64
4-Pyrimidinepropionic acid, methyl ester	EtOH	245(3.54),270(2.51)	44-2398-65
$C_8H_{10}N_2O_3$			
$\Delta^{2,\alpha}$-Oxazolidineacetic acid, α-cyano-, ethyl ester	EtOH	255.5(4.38)	95-0387-65
5-Pyridoxamide	EtOH	290(3.76)	44-0574-64
	pH 1	293(3.96)	44-0574-64
	pH 13	250s(3.71),314(3.82)	44-0574-64
5-Pyrimidinecarboxylic acid, 1,2-di- hydro-1-methyl-2-oxo-, ethyl ester	EtOH	246(4.25),274s(3.67), 306(3.52)	94-1418-64
1-Pyrrolidinecarboxylic acid, 2-cyano- 3-oxo-, ethyl ester	MeOH	270(2.18)	35-5293-64
	MeOH-KOH	285(4.00)	35-5293-64
3-Pyrrolin-2-one, 3-acetamido-4-hy- droxy-1-methyl-5-methylene-	EtOH	268(4.31),342(3.24)	35-5654-64
$C_8H_{10}N_2O_3S$			
5-Pyrimidinepropionic acid, 4-hydroxy- 2-mercapto-6-methyl-	pH 1	280(4.35)	4-0263-64
	pH 13	262(4.27),310s(4.01)	4-0263-64

Compound	Solvent	λ_{max}(log ε)	Ref.
Uracil, 5-(2-hydroxyethyl)-4-thio-, 5-acetate	pH 1	243(3.58),333.5(4.20)	44-2670-64
$C_8H_{10}N_2O_4$			
Glutaconic acid, 3-amino-2-cyano-, dimethyl ester	EtOH	278.5(4.22)	95-0387-65
Morpholine, 4-(5-nitro-2-furyl)-	MeOH	226(4.08),432(4.30)	65-0710-64
Uracil, 5-(2-hydroxyethyl)-, 5-acetate	pH 1	263.5(3.89)	44-2670-64
	pH 13	289(3.77)	44-2670-64
$C_8H_{10}N_2O_4S$			
Uracil, 5-(carbethoxymethylthio)-	pH 1	205(3.82),277(3.91)	39-3987-65
	pH 7.2	205(3.90),283(3.93)	39-3987-65
	pH 13	220(4.13),275(3.85)	39-3987-65
$C_8H_{10}N_2O_5$			
1(2H)-Pyrimidinepropionic acid, 5-carboxy-3,4-dihydro-2-oxo-	EtOH	293(4.07)	94-0681-65
$C_8H_{10}N_2S$			
2-Pyrrolidine-3-carbonitrile, 1,5,5-trimethyl-4-thioxo-	pH 1	202(3.83),266(3.87), 382(4.34)	39-4546-65
	pH 13	268(3.85),382(4.35)	39-4546-65
$C_8H_{10}N_4$			
Malononitrile, (piperidinoimino)-	MeCN	238(3.28),305(4.33)	44-4198-65
Pyrazolo[1,5-a]pyrimidine, 7-amino-2,3-dimethyl-	EtOH	222(4.58),285(3.85), 310(3.83)	94-0142-65
Pyrazolo[1,5-a]pyrimidine, 7-amino-3,6-dimethyl-	EtOH	227(4.62),277s(3.75), 285(3.80)3,19(3.85)	94-1042-65
3H-Pyrrolo[2,3-d]pyrimidine, 4,7-dihydro-4-imino-3,5-dimethyl-	EtOH	231(4.11),273(3.91)	35-1995-65
	pH 1	231(4.31),286(3.90)	35-1995-65
	pH 13	271(3.99)	35-1995-65
Pyrrolo[2,3-d]pyrimidine, 5-methyl-4-(methylamino)-	EtOH	282(4.01)	35-1995-65
	pH 1	234(4.23),282(4.02)	35-1995-65
	pH 13	281(4.01)	35-1995-65
s-Triazolo[4,3-a]pyrazine, 3,5,6-trimethyl-	C_6H_{12}	217(4.49),223s(4.39), 263s(3.35),269s(3.38), 280(3.39),302s(3.53), 309(3.56),319s(3.50), 333s(3.23)	44-2542-64
s-Triazolo[1,5-a]pyrimidine, 5,6,7-trimethyl-	EtOH	278(3.72)	94-0204-64
$C_8H_{10}N_4O$			
Guanidine, (salicylidenamino)-	pH 1	276(4.26),316(3.96)	73-2607-64
	pH 3.91	276(4.29),316(3.99)	73-2607-64
	pH 7.35	276(4.25),316(4.01)	73-2607-64
	pH 9.53	285(4.10),323(4.07)	73-2607-64
	pH 13	222(4.22),276(4.00), 354(4.08)	73-2607-64
Pyrido[2,3-d]pyrimidin-7(6H)-one, 2-amino-5,8-dihydro-4-methyl-	pH 1, 8.4	305(3.83)	4-0263-64
	pH 13	310(4.07)	4-0263-64
5H-Pyrrolo[3,4-d]pyrimidine, 6-acetyl-4-amino-6,7-dihydro-	pH 1	256(4.15)	44-0194-65
	EtOH	237(4.09),268(3.68)	44-0194-65
1,2,4-Triazole, 3-(dimethylamino)-5-(2-furyl)-	pH 2	270(4.20)	1-1191-65
	pH 12	277(4.06)	1-1191-65
1,2,4-Triazole, 3-(ethylamino)-5-(2-furyl)-	pH 2	268(4.22)	1-1191-65
	pH 12	274(4.09)	1-1191-65

Compound	Solvent	$\lambda_{max}(\log \epsilon)$	Ref.
2H-1,2,4-Triazole, 5-(2-furyl)-2-methyl-3-(methylamino)-	pH 2	272(4.23)	1-1191-65
	EtOH	268(4.11)	1-1191-65
1,2,4-Triazole, 5-(2-furyl)-4-methyl-3-(methylamino)-	pH 2	265(4.19)	1-1191-65
	EtOH	280(4.17)	1-1191-65
$C_8H_{10}N_4OS$			
Histidine, methylthiohydantoin deriv.	EtOH	235(3.98),265(4.25)	65-0988-65
9H-Purine-9-ethanol, 6-(methylthio)-	pH 1	293(4.25)	87-0182-65
	pH 11	289(4.30)	87-0182-65
9H-Purine-9-propanol, 6-mercapto-	pH 1	322(4.34)	87-0502-65
	pH 7	320(4.37)	87-0502-65
	pH 13	310(4.33)	87-0502-65
$C_8H_{10}N_4O_2$			
9H-Purine-9-ethanol, 6-hydroxy-α-methyl-	pH 1	249(4.13)	87-0502-65
	pH 7	249(4.13)	87-0502-65
	pH 13	253(4.14)	87-0502-65
Purine-7-propanol, 6-hydroxy-	pH 1	254(3.97)	87-0710-65
	pH 7	257(3.97)	87-0710-65
	pH 13	263(4.00)	87-0710-65
1H-Pyrazolo[4,3-d]pyrimidine-5,7(4H,6H)-dione, 1,4,6-trimethyl-	MeOH	291(3.75)	44-0199-65
2H-Pyrazolo[4,3-d]pyrimidine-5,7(4H,6H)-dione, 2,4,6-trimethyl-	MeOH	288(3.74)	44-0199-65
1,2,4-Triazole, 3-amino-5-(2-furyl)-4-(2-hydroxyethyl)-	pH 2	264(4.13)	1-1191-65
	EtOH	273(4.10)	1-1191-65
$C_8H_{10}N_4O_2S$			
1,2-Propanediol, 3-(6-mercapto-9H-purin-9-yl)-	pH 1	323(4.29)	87-0502-65
	pH 7	321(4.33)	87-0502-65
	pH 13	310(4.31)	87-0502-65
$C_8H_{10}N_4O_4$			
4-Pyrimidineacetic acid, 6-amino-5-nitro-, ethyl ester	pH 1	217(4.20),252(3.88), 330(3.49)	44-1528-65
	pH 7	217(4.27),250(3.68), 285(3.32),345(3.60)	44-1528-65
	pH 13	295(4.16),332(3.75)	44-1528-65
$C_8H_{10}N_4S$			
Purine, 6-ethylthiomethyl-	N HCl	266(3.83),305s(3.09)	87-0667-65
	pH 7.65	267(3.92)	87-0667-65
	N NaOH	277(3.94)	87-0667-65
5-Pyrimidinepropionitrile, 2-amino-4-mercapto-6-methyl-	pH 1	265s(3.84),347(4.27)	4-0263-64
	pH 13	265s(3.96),317(4.22)	4-0263-64
$C_8H_{10}N_6$			
Pteridine, 7-amino-1,4-dihydro-1-methyl-4-(methylamino)-	base	238(4.23),259(4.25), 354(4.19),368s(4.08)	39-1175-65
1,2,4-Triazole, 3-anilino-5-hydrazino-	EtOH	257(4.30)	39-3912-65
$C_8H_{10}N_6O$			
Pterin, 8-(dimethylamino)-	pH 12	260(4.2),275s(4.2), 413(3.7)	33-2195-64
$C_8H_{10}N_8S_2$			
Pyrimidine, 4,4'-dithiobis[2,6-diamino-	pH 1	214(4.51),230(4.52), 301(4.26)	87-0792-64
	MeOH	287(4.14)	87-0792-64

Compound	Solvent	λ_{max}(log ε)	Ref.
$C_8H_{10}O$			
Anisole, 2-methyl-	C_6H_{12}	212(3.85),216(3.87), 221(3.86),265s(3.18), 273(3.32),278(3.32)	23-2603-65
	hexane	219(3.78),272(3.30), 278(3.28)	78-0861-64
	EtOH	219.5(3.76)	78-0861-64
	EtOH	271(--),277(--)	95-0858-65
Anisole, 3-methyl-	hexane	220(3.75),273(3.18), 279(3.18)	78-0861-64
	EtOH	220(3.74)	78-0861-64
	EtOH	273(--),279(--)	95-0858-65
Anisole, 4-methyl-	hexane	224(3.87),279(3.36), 283(3.43)	78-0861-64
	EtOH	224(3.83)	78-0861-64
	EtOH	277(--),284(--)	95-0858-65
Benzyl alcohol, α-methyl-	CH_2Cl_2-$HClO_4$	310(--),429(3.58)	39-4765-65
Benzyl alcohol, o-methyl-	EtOH	262(2.39),272(2.25)	46-0457-65
Bicyclo[3.2.1]oct-2-en-8-one	EtOH	293(1.62)	78-0215-64
Bicyclo[3.3.0]oct-3-en-2-one, cis	EtOH	221(3.99)	35-0321-65
2,5-Cyclohexadienone, 4,4-dimethyl-	isooctane	225(4.17),229s(4.14), 237s(3.96),246s(3.59), 319(0.9),332(1.0), 345(1.14),360(1.08), 370(0.8),385(0.6)	44-2109-65
	EtOH	235(4.16),323(1.38)	44-2109-65
2,7-Cyclooctadienone	isooctane	236(4.03),339(1.62), 353(1.79),366(1.90), 383(1.84),398(1.53)	44-2109-65
	EtOH	244(4.03),347(1.98)	44-2109-65
Ether, benzyl methyl	C_6H_{12}	237(2.24),243(2.26), 249(2.30),259(2.31), 265(2.19),280(1.10), 289(0.85)	39-5957-64
4,5-Heptadien-2-yn-1-ol, 6-methyl-	EtOH	221(4.21)	39-4659-65
1,3-Hexadien-5-yne, 1-methoxy-3-methyl-	EtOH	274(4.42)	54-1113-65
2,6-Octadien-4-yn-1-ol	Et_2O	262(4.24)	24-2118-64
4,5-Octadien-2-yn-1-ol	EtOH	220(4.11)	39-4659-65
Oxepin, 2,7-dimethyl-	C_6H_{12}	297(3.25)	89-0535-64
Phenetole	nonane	272(3.30)	67-0375-65
p-Xylenol	EtOH	216(3.83),275(3.29), 283(3.26)	94-0064-65
$C_8H_{10}OS$			
Ketone, 2,5-dimethyl-3-thienyl methyl	EtOH	250(3.9),255(3.9), 277(3.4)	70-2055-64
Sulfoxide, methyl p-tolyl	isooctane	210s(4.02),217s(3.97), 249(3.68),269s(3.23), 278s(2.81)	35-1958-65
	EtOH	219s(3.91),236(3.80), 264s(3.01),275s(2.64)	35-1958-65
	isooctane-CF_3COOH	218s(3.90),235(3.75), 265s(3.06),275s(2.63)	35-1958-65
	EtOH-CF_3COOH	219s(3.94),237(3.82), 266(3.07),286(2.73)	35-1958-65
$C_8H_{10}O_2$			
Acetaldehyde, di-2-propynyl acetal	MeOH	none	70-1349-64

Compound	Solvent	λ max(log ε)	Ref.
Benzene, 1,2-dimethoxy-	EtOH	281(--)	95-0858-65
	MeCN	227(3.89),276(3.40), 282s(3.33)	35-2591-64
Benzene, 1,3-dimethoxy-	EtOH	274(--),280(--)	95-0858-65
Benzene, 1,4-dimethoxy-	EtOH	290(--)	95-0858-65
	95% THF	226(4.00),290(3.51)	44-2602-65
Benzyl alcohol, m-methoxy-	hexane	219(3.81),273(3.20), 280(3.15)	78-0861-64
Benzyl alcohol, o-methoxy-	hexane	221(3.76),271(3.24), 277(3.24)	78-0861-64
	EtOH	271(--),276(--)	95-0858-65
Benzyl alcohol, p-methoxy-	hexane	227(3.88),276(3.23), 282(3.15)	78-0861-64
	EtOH	276(--),282(--)	95-0858-65
Cyclopentadienemethanol, acetate	C_6H_{12}	245.5(3.65)	33-0955-65
5-Hexen-2-ynoic acid, ethyl ester	MeOH	212(3.68)	5-0026-64I
4-Hexen-1-yn-3-one, 1-ethoxy-	EtOH	228(4.07),249(3.96)	35-0471-64
	EtOH	228(4.14),250(4.05)	39-4939-65
Hydroquinone, ethyl-	95% THF	298(3.59)	44-2602-65
2-Norbornene-2-carboxylic acid	EtOH	229(3.85)	44-3234-64
3,5-Octadiene-2,7-dione	n.s.g.	276(4.54)	5-0014-65D
5-Octen-7-yn-4-one, 2-hydroxy-	n.s.g.	238(4.06)	70-2215-65
2,4-Pentadienoic acid, allyl ester	heptane	244(4.51)	44-1061-65
Toluhydroquinone, 2-methyl ether	EtOH	217(3.78),225(3.80), 290(3.51)	24-1926-64
	NaOH	237(3.95),303(3.51)	24-1926-64
Toluhydroquinone, 5-methyl ether	EtOH	216s(3.81),226(3.81), 290(3.52)	24-1926-64
	NaOH	237(3.96),306(3.57)	24-1926-64
$C_8H_{10}O_2S$			
2-Furaldehyde, 5-[(ethylthio)methyl]-	EtOH	285(4.25)	70-1281-65
$C_8H_{10}O_3$			
Benzene, 3,4,5-trihydroxy-1,2-dimethyl-	EtOH	265(3.07)	39-4672-65
Benzene, 3,4,6-trihydroxy-1,2-dimethyl-	EtOH	207(4.03),273(3.99)	39-4672-65
1-Cyclohexene-1-carboxylic acid, 2-methyl-5-oxo-	EtOH	222(3.88)	44-3524-64
1,3-Cyclopentandione, 2-acetyl-4-methyl-	C_6H_{12}	221(4.10),265(3.89)	1-1368-64
	MeOH-HCl	226(4.29),262(4.15)	20-0628-64
	MeOH-NaOH	247(4.51),264s(4.39)	20-0628-64
2-Cyclopenten-1-one, 3-acetyl-2-hydroxy-5-methyl-	EtOH	270(3.90),305(3.71)	44-3520-64
	EtOH-base	357(4.10)	44-3520-64
2-Cyclopenten-1-one, 3-acetoxy-2-methyl-	EtOH	234(4.10)	70-1648-64
4-Hexen-2-ynoic acid, 5-ethoxy-	EtOH	278(4.18)	28-0594-64B
2-Octenoic acid, 7-hydroxy-4-oxo-, lactone	EtOH	220(4.05)	39-3239-64
$C_8H_{10}O_3S$			
3-Thiophenecarboxylic acid, 4-hydroxy-2-methyl-, ethyl ester	Et_2O	211(4.06),248(4.04), 304(3.37)	24-2109-64
$C_8H_{10}O_4$			
1-Cyclobutene-1,2-dicarboxylic acid, dimethyl ester	C_6H_{12}	228(4.10)	24-0764-65
	hexane	228(4.08)	24-2953-64
1-Cyclobutene-1,2-dicarboxylic acid, 3,3-dimethyl-	MeOH	236(4.1)	24-3672-65
3(2H)-Furanone, 4-acetoxy-2,5-dimethyl-	EtOH	268.8(4.01)	25-1629-65
Muconic acid, methyl ester, trans-trans	MeOH	262.5(4.54)	35-2095-64

Compound	Solvent	λ_{max}(log ϵ)	Ref.
$C_8H_{10}O_5$			
Acrylic acid, 3,3'-oxydi-, dimethyl ester	Et_2O	248(4.52)	24-1952-64
$C_8H_{10}O_6$			
2-Cyclohexene-1,4-dione, 3,5,6-tri-hydroxy-2-methoxy-5-methyl-	EtOH	305(4.1)	94-0935-65
$C_8H_{10}Pd$			
Palladium, allylcyclopentadienyl-	heptane	260(4.47),320(3.86), 470(2.33)	88-2881-64
	MeOH	247s(3.88),305(3.54), 400s(2.92)	88-2881-64
$C_8H_{10}S$			
Benzenethiol, o-ethyl-	MeOH	251(4.07),280(3.72)	56-0681-65
$C_8H_{10}S_2$			
Benzene, p-bis(methylthio)-	hexane	275(4.25),300(3.30)	65-2749-64
	EtOH	274(4.36),300(3.35)	65-2749-64
	Et_2O	275(4.37),300(3.40)	65-2749-64
	dioxan	275(4.30),300(3.40)	65-2749-64
	H_2SO_4	222(3.84),280(3.47), 370(4.26),720(2.60)	65-2749-64
4,9-Dithiabicyclo[5.3.0]deca-1(10),7-diene	EtOH	244(3.72),250s(3.60)	39-1154-64
$C_8H_{10}S_3$			
Cyclohepta[c]-1,2-dithiole-3(4H)-thione, 5,6,7,8-tetrahydro-	EtOH	229(4.10),243s(3.92), 280(3.93),310(3.83), 412(4.04)	24-0654-64
$C_8H_{10}Si$			
2-Silaindan	n.s.g.	264(2.9),270(3.0), 277(3.1)	30-0064-64
$C_8H_{11}BF_4O_3$			
2,4-Dimethoxy-6-methylpyrylium tetra-fluoroborate	CH_2Cl_2	272(4.02)	78-0831-64
$C_8H_{11}BrO$			
Bicyclo[3.2.1]octan-3-one, 2-bromo-	C_6H_{12}	215(2.61),315(2.02)	22-0844-64
	MeCN	215(2.58),313(1.98)	22-0844-64
2-Cyclohexen-1-one, 3-bromo-5,5-di-methyl-	EtOH	246(4.13)	44-1129-65
$C_8H_{11}ClIN$			
Aniline, N,N-dimethyl-, ICl complex	$CHCl_3$	370s(--)	60-0062-64
$C_8H_{11}ClN_2O$			
4-Pyrimidinol, 2-butyl-6-chloro-	pH -3.6	233(3.92)	39-3204-64
	pH 4.0	227(3.75),274(3.75)	39-3204-64
	pH 10.8	230(3.96),267(3.62)	39-3204-64
4-Pyrimidinol, 5-butyl-6-chloro-	pH -3.6	243(3.96)	39-3204-64
	pH 5.0	234(3.73),274(3.76)	39-3204-64
	pH 11.1	237(3.93),269(3.73)	39-3204-64

Compound	Solvent	λ max(log ε)	Ref.
$C_8H_{11}ClN_4$			
Pyridazine, 3-[2-(1-aziridinyl)ethyl-amino]-6-chloro-	pH 1	240(3.90),308(3.30)	4-0001-64
	pH 11	250(4.11),320(3.25)	4-0001-64
$C_8H_{11}ClN_4O_2$			
Pyrimidine, 4-chloro-6-(diethylamino)-5-nitro-	pH 1	254(4.17)	44-3597-65
	pH 13	251(4.07)	44-3597-65
	MeOH	253(4.16)	44-3597-65
$C_8H_{11}ClO$			
1-Cycloheptene-1-carboxaldehyde, 2-chloro-	EtOH	260(4.06)	44-1126-65
2-Cyclohexen-1-one, 3-chloro-5,5-di-methyl-	EtOH	238(4.13)	44-0385-64
	EtOH	238(4.13)	44-1129-65
$C_8H_{11}FN_2O_2$			
Pyrimidine, 2,4-diethoxy-5-fluoro-	EtOH	269(3.72)	87-0140-65
$C_8H_{11}I_2N$			
Aniline, N,N-dimethyl-, iodine complex	$CHCl_3$	355s(--)	60-0062-64
$C_8H_{11}KO_4$			
Malonic acid, ethyl-, cyclic isopropyli-dene ester, potassium derivative	$(MeOCH_2)_2$	265.5(4.25)	35-1857-65
$C_8H_{11}LiO_4$			
Malonic acid, ethyl-, cyclic isopropyli-dene ester, lithium derivative	$(MeOCH_2)_2$	265(4.25)	35-1857-65
	DMF	267.5(4.28)	35-1857-65
$C_8H_{11}N$			
Aniline, N,N-dimethyl-, tetracyano-ethylene complex	$CHCl_3$	675(3.51)	88-0189-64
4,6-Heptadienenitrile, 5-methyl-	EtOH	229(4.40)	22-2533-64
cis	EtOH	230.8(4.26)	22-2533-64
$C_8H_{11}NO$			
Benzylamine, m-methoxy-	hexane	273(3.28),280(3.26)	78-0861-64
Benzylamine, o-methoxy-	hexane	220(3.78),272(3.30), 278(3.26)	78-0861-64
	EtOH	220(3.79)	78-0861-64
Benzylamine, p-methoxy-	hexane	226(3.97),276(3.20), 283(3.11)	78-0861-64
	EtOH	226(3.86)	78-0861-64
Cyclopentanecarbonitrile, 3,3-di-methyl-2-oxo-	EtOH	235(4.21)	23-2512-65
	EtOH-base	264(4.02)	23-2512-65
2-Norbornene-2-carboxamide	EtOH	222(3.75)	44-3234-64
Phenethylamine, 2-hydroxy-	pH 1	216(3.72),273(3.30), 277s(3.27)	87-0368-65
	pH 13	239(3.91),291(3.56)	87-0368-65
Phenethylamine, 3-hydroxy-	pH 1	217(3.72),273(3.23), 278s(3.18)	87-0368-65
	pH 13	239(3.91),291(3.46)	87-0368-65
Phenethylamine, 4-hydroxy-	pH 1	222(3.84),275(3.20), 280s(3.15)	87-0368-65
	pH 13	239(4.02),294(3.42)	87-0368-65
4(1H)-Pyridone, 1,2,6-triemthyl-	MeOH	261(4.23)	35-3365-65
2H-1-Pyridin-2-one, 1,3,4,5,6,7-hexa-hydro-	EtOH	246(3.70)	44-1435-64
Pyrrole, 2-acetyl-3,4-dimethyl-	EtOH	296(4.29)	39-1518-65

Compound	Solvent	λ_{max}(log ε)	Ref.
$C_8H_{11}NOS$			
Morpholine, 4-(3-thienyl)-	C_6H_{12}	272(3.62)	54-1160-64
Thioanisole, 4-amino-3-methoxy-	EtOH	240(3.85),261(3.86), 304(3.48)	32-1137-64
Thioanisole, 6-amino-3-methoxy-	EtOH	240(3.94),242(3.93), 315(3.53)	32-1137-64
$C_8H_{11}NO_2$			
Benzyl alcohol, α-(aminomethyl)-m-hydroxy-	pH 1	217(3.70),274(3.26), 278s(3.23)	87-0368-65
	pH 13	239(3.91),292(3.48)	87-0368-65
Benzyl alcohol, α-(aminomethyl)-o-hydroxy-	pH 1	216(3.71),274(3.32), 278s(3.29)	87-0368-65
	pH 13	239(3.90),293(3.57)	87-0368-65
Benzyl alcohol, α-(aminomethyl)-p-hydroxy-	pH 1	224(3.86),274(3.15), 279s(3.08)	87-0368-65
	pH 13	242(4.09),292(3.38)	87-0368-65
	EtOH-H_2SO_4	275.5(3.20)	54-0521-65
Crotonic acid, 2-cyano-3-methyl-, ethyl ester	EtOH	229.5(4.06)	28-2859-64B
2-Heptenoic acid, 2-cyano-	EtOH	222(4.05)	28-2859-64B
2-Hexenoic acid, 2-cyano-3-methyl-	EtOH	231(4.06)	28-2859-64B
2-Hexenoic acid, 2-cyano-5-methyl-	EtOH	221.5(4.05)	28-2859-64B
Isoxazole, 3-hydroxy-4,5-pentamethylene-	C_6H_{12}	213(3.82)	78-2835-64
	0.01 H_2SO_4	226(3.80)	78-2835-64
2-Pentenoic acid, 2-cyano-3,4-dimethyl-	EtOH	229(4.05)	28-2859-64B
2-Pentenoic acid, 2-cyano-3-ethyl-	EtOH	231(4.06)	28-2859-64B
Phenethylamine, 2,4-dihydroxy-	pH 1	222(3.84),278(3.39), 283s(3.35)	87-0368-65
anion	n.s.g.	237(3.83),292(3.57)	87-0368-65
dianion	n.s.g.	224(4.09),299(3.66)	
Pyrrole-2-carboxylic acid, 3,4-dimethyl-, methyl ester	EtOH	285(4.40)	44-2727-64
Pyrrole-2-carboxylic acid, 4-isopropyl-	EtOH	237(3.77),273(4.15)	23-1279-64
Pyrrole-2-carboxylic acid, 5-isopropyl-	EtOH	231(3.52),277(4.02)	23-1279-64
Pyrrole-3-carboxylic acid, 2-methyl-, ethyl ester	EtOH	225(3.86),257(3.80)	39-2411-65
$C_8H_{11}NO_2S$			
3-Thiophenecarboxylic acid, 2-amino-4,5-dimethyl-, ethyl ester	MeOH	227(4.47),309(3.75)	24-3571-65
$C_8H_{11}NO_2S_2$			
6H-1,3-Thiazine-5-carboxylic acid, 2-(methylthio)-, ethyl ester	EtOH	240(4.11),330(4.06)	44-2290-65
$C_8H_{11}NO_3$			
1,2,3-Cyclohexanetrione, 5,5-dimethyl-, 2-oxime	EtOH	285(3.83)	25-1183-65
Ethanolamine, 2-(3,4-dihydroxyphenyl)-	pH 1	208(3.84),224(3.77), 280(3.44),285s(3.39)	87-0368-65
	pH 13	243(3.85),295(3.65)	87-0368-65
Ketoxime, methyl (5-methyl-2-furyl), acetate, anti	EtOH	285(4.27)	22-2724-65
syn	EtOH	283(4.31)	22-2724-65
2-Pyrroline-3-carboxylic acid, 2,4-dimethyl-5-oxo-, methyl ester	EtOH	219(3.58),280(4.02)	39-5999-64

Compound	Solvent	λ max(log ε)	Ref.
$C_8H_{11}NO_3S$			
2H-1,3-Thiazine-5-carboxylic acid, 3,6-dihydro-3-methyl-2-oxo-, ethyl ester	EtOH	241(3.74),298(3.86)	44-2290-65
$C_8H_{11}NS$			
Pyrrolidine, 1-(3-thienyl)-	C_6H_{12}	291(3.56)	54-1160-64
$C_8H_{11}N_2O_3P$			
Phosphonic acid, (phenylazo)-, dimethyl ester	dioxan	300(4.13),492(1.97)	24-2844-65
$C_8H_{11}N_3O_2S$			
Thiazole-4-carboxamide, 5-acetamido-N,N-dimethyl-	EtOH	274(4.04)	94-1319-65
Thiazole-4-carboxamide, 5-acetamido-N,2-dimethyl-	EtOH	215(4.18),287(4.16)	94-1319-65
$C_8H_{11}N_3O_2S_2$			
Sulfanilamide, N^1-[(methylthio)aminomethylene]-	EtOH	274.0(4.29)	95-0391-65
$C_8H_{11}N_3O_3$			
DL-Alanine, N-(1,2-dihydro-1-methyl-2-oxo-4-pyrimidinyl)-	pH 1	217(3.93),288(4.13)	44-1762-64
	pH 9.72	233(3.85),275(4.00)	44-1762-64
2-Imidazoline-$\Delta^{5,\alpha}$-acetic acid, 2-amino-α-methyl-4-oxo-, ethyl ester	pH 1	237(3.88),306(4.17)	4-0162-65
	pH 7	277(4.15),322(4.03)	4-0162-65
	pH 13	279(3.98),341(3.87)	4-0162-65
5-Pyrimidinealanine, 4-hydroxy-2-methyl-	pH 1	228(3.85),260(3.69)	65-2240-65
	pH 12	232(3.85),271(3.74)	65-2240-65
5-Pyrimidinepropionic acid, 2-amino-4-hydroxy-6-methyl-	pH 1	227(3.93),265(3.88)	87-0024-64
	pH 8.4	275(3.70)	87-0024-64
	pH 13	279(3.83)	87-0024-64
$C_8H_{11}N_3O_4$			
5-Pyrimidinealanine, 2,4-dihydroxy-6-methyl-	EtOH	265(3.533)	65-2171-64
as-Triazine-6-propionic acid, 3,4,5,6-tetrahydro-3,5-dioxo-, ethyl ester	MeOH	210(3.80),252(3.74)	39-0868-64
as-Triazine-6-valeric acid, tetrahydro-3,5-dioxo-	MeOH	211(3.82),254(3.77)	39-0868-64
$C_8H_{11}N_3O_6$			
5-Azauracil, 1-β-D-ribopyranosyl-	EtOH	240(3.24)	73-0260-64
$C_8H_{11}N_5$			
Adenine, N,N,1-trimethyl-	pH 1	221(4.13),293(4.09)	35-5320-64
	pH 7	215(4.15),298(4.10)	35-5320-64
	pH 12	232(4.13),301(4.13)	35-5320-64
Adenine, N,N,3-trimethyl-	pH 1	290(4.31)	35-5320-64
	pH 7	222(4.05),292(4.22)	35-5320-64
	pH 12	293(4.21)	35-5320-64
Adenine, N,N,7-trimethyl-	pH 1	223(3.95),293(4.26)	35-5320-64
	pH 7	222(4.11),291(4.16)	35-5320-64
	pH 12	291(4.16)	35-5320-64
Adenine, N,N,9-trimethyl-	pH 1	269(4.20)	35-5320-64
	pH 7	214(4.19),276(4.20)	35-5320-64
	pH 12	276(4.21)	35-5320-64
Adenine, N,7,9-trimethyl-	pH 7.0	270s(4.10),280(4.15)	39-3770-65
s-Triazolo[1,5-c]pyrimidine, 5,7-dimethyl-2-(methylamino)-	MeOH	232(4.60),260s(3.58), 294s(3.26)	39-3357-65

Compound	Solvent	λ max(log ε)	Ref.
$C_8H_{11}N_5O$			
9H-Purine-9-ethanol, 6-amino-α-methyl-	pH 1	260(3.87)	87-0502-65
	pH 7, 13	261(3.91)	87-0502-65
Purine-7-propanol, 6-amino-	pH 1	272(4.06)	87-0710-65
	pH 7	270(3.97)	87-0710-65
	pH 13	270(3.96)	87-0710-65
3H-Purine-3-propanol, 6-amino-	pH 1	274(4.27)	87-0710-65
	pH 7	273(4.18)	87-0710-65
	pH 13	273(4.14)	87-0710-65
Purine-9-propanol, 6-amino-	pH 1	254(3.97)	87-0033-65
	pH 7	257(3.97)	87-0033-65
	pH 13	263(4.00)	87-0033-65
5H-Pyrrolo[3,4-d]pyrimidine, 6-acetyl-	pH 1	272(3.70),305s(2.76)	44-0194-65
2,4-diamino-6,7-dihydro-	EtOH	282(3.61)	44-0194-65
$C_8H_{11}N_5OS$			
9H-Purine-9-ethanol, 2-amino-6-	pH 1	247(4.03),319(4.08)	87-0182-65
(methylthio)-	pH 11	244(4.11),308(4.09)	87-0182-65
$C_8H_{11}N_5O_2$			
1,2-Propanediol, 3-(6-amino-9H-	pH 1	260(4.13)	87-0502-65
purin-9-yl)-	pH 7	261(4.14)	87-0502-65
	pH 13	261(4.15)	87-0502-65
4,6-Pteridinediol, 2-amino-7,8-di-	pH 1	280(4.09),305(4.00)	39-4769-64
hydro-7,7-dimethyl-	pH 13	222(4.26),280(4.12)	39-4769-64
$C_8H_{11}N_5O_3$			
4-Pyrimidineacetimidic acid, 6-amino-5-nitro-, ethyl ester (shoulders, not listed)	pH 1	217(4.20),253(3.73), 296(3.76),324(3.71)	44-1528-65
	pH 7	216(4.20),302(4.19), 340(3.59)	44-1528-65
	pH 13	230(4.09),294(4.19)	44-1528-65
$C_8H_{11}N_5O_4S$			
Caffeine, 8-sulfamoyl-	MeOH	286(4.02)	54-1215-64
$C_8H_{11}N_5S$			
Adenine, N,9-dimethyl-2-(methylthio)-	pH 0.8	211(4.21),253(4.11), 274(4.15),282s(4.14)	39-3770-65
[1,2,5]Thiadiazolo[3,4-d]pyrimidine, 7-(butylamino)-	pH 1	222(4.11),264(3.63), 342(4.22)	44-2135-64
	pH 7	223(4.14),272(3.69), 357(4.06)	44-2135-64
	pH 13	272(3.67),357(4.03)	44-2135-64
	EtOH	224(4.14),274(3.71), 283(3.63),362(4.04)	44-2135-64
$C_8H_{11}NaO_4$			
Malonic acid, ethyl-, cyclic isopropyl-	DMF	267.5(4.27)	35-1857-65
idene ester	$(MeOCH_2)_2$	265.5(4.28)	35-1857-65
C_8H_{12}			
Bicyclo[4.2.0]oct-7-ene	heptane	228(3.75)	24-2339-65
1-Buten-3-yne, 2-tert-butyl-	n.s.g.	218(3.86)	22-1525-65
Cyclobutene, 1,2,3-trimethyl-4-methylene-	EtOH	230(4.16)	35-1600-64
1,3-Cyclohexadiene, 1,6-dimethyl-	n.s.g.	263(3.81)	88-0385-65
1,3-Cyclohexadiene, 5,6-dimethyl-, cis	n.s.g.	261(3.61)	88-0385-65
	n.s.g.	261(3.69)	88-0391-65
trans	n.s.g.	263(3.60)	88-0385-65

Compound	Solvent	λ_{max}(log ε)	Ref.
1,3-Cyclooctadiene, $PdCl_2$ olefin complex from	EtOH	214(4.43),253(4.33), 284s(3.96),357s(3.31)	24-2037-64
Cyclopentadiene, 1,2,3-trimethyl-	hexane	253(3.48)	70-0864-64
	EtOH	247(3.62)	70-1653-64
Cyclopentene, 2,4-dimethyl-3-methylene-	hexane	236.5(4.15)	70-0864-64
1,3,5-Heptatriene, 2-methyl-, trans	C_6H_{12}	254(4.50),264(4.65), 274(4.55)	23-2781-64
1,3,5-Heptatriene, 4-methyl-	C_6H_{12}	259(4.53),269(4.66), 280(4.52)	23-2781-64
1,4,6-Heptatriene, 2-methyl-	C_6H_{12}	226.5(4.43)	23-2781-64
2,4,6-Heptatriene, 3-methyl-	n.s.g.	256(5.32),267(5.43), 275(5.36)	88-3173-64
2,3,4-Hexatriene, 2,5-dimethyl-	hexane	230(3.85),263(4.23)	88-2175-65
	hexane	230(3.85),263(4.23)	78-1357-65
1,3,6-Octatriene, trans	C_6H_{12}	228(4.57)	44-1661-65
2,4,6-Octatriene, all-cis	n.s.g.	258(4.51),268(4.54), 277(4.40)	88-0385-65
cis-cis-trans	n.s.g.	257(4.51),265(4.59), 276(4.47)	88-0385-65
trans-cis-trans	n.s.g.	255(4.52),267(4.61), 277(4.48)	88-0385-65
	n.s.g.	256(4.51),265(4.61), 276(4.49)	88-0391-65
$C_8H_{12}BrNO$			
2-Cyclohexen-1-one, 3-bromo-5,5-dimethyl-, oxime, anti	EtOH	242.5(4.18)	44-1129-65
syn	EtOH	244(4.02)	44-1129-65
$C_8H_{12}Br_2$			
Bicyclo[5.1.0]octane, 8,8-dibromo-	heptane	206(3.54)	44-2951-64
1-Cyclobutene, 3,4-dibromo-1,2,3,4-tetramethyl-	MeOH	210(4.13),232s(3.47)	24-1811-64
$C_8H_{12}Br_2O$			
Cyclohexanone, 2,2-dibromo-3,3-dimethyl-	C_6H_{12}	302(2.02)	24-1858-65
Cyclohexanone, 2,2-dibromo-5,5-dimethyl-	C_6H_{12}	304(2.03)	24-1858-65
Cyclohexanone, 2,6-dibromo-3,3-dimethyl-	$CHCl_3$	304(2.16)	24-1858-65
$C_8H_{12}ClNO$			
1-Cycloheptene-1-carboxaldehyde, 2-chloro-, oxime	EtOH	255(4.24)	44-1126-65
2-Cyclohexen-1-one, 3-chloro-5,5-dimethyl-, oxime, anti	EtOH	242(4.06)	44-1129-65
syn	EtOH	241(4.11)	44-1129-65
$C_8H_{12}ClNO_2$			
Butyric acid, 4-chloro-2-cyano-2-methyl-, ethyl ester	MeOH	211(2.29)	24-1970-64
2-Pyrrolidinone, 1-(4-chlorobutyryl)-	EtOH	216(4.08)	35-2003-65
$C_8H_{12}ClN_3$			
Pyrazine, 2-chloro-5-(dimethylamino)-3,6-dimethyl-, hydrochloride	EtOH	260(4.05),332(3.76)	44-1645-64
$C_8H_{12}ClN_3O$			
Pyridazine, 3-chloro-6-[2-(dimethylamino)ethoxy]-	EtOH	280(3.30)	44-1751-64

Compound	Solvent	λ_{max}(log ϵ)	Ref.
$C_8H_{12}Cl_3NOS$			
2-Pentanone, 5,5,5-trichloro-3-methyl-4-(N-methylthiocarbamoylthio)-	H_2O	248(3.88),273(3.90)	39-4004-64
	EtOH	246(3.92),284(3.85)	39-4004-64
$C_8H_{12}NNaO_3$			
Pyruvaldehyde, cyano-, 1-(diethyl acetal), sodium derivative	n.s.g.	259(3.70)	28-1418-64B
$C_8H_{12}N_2$			
1-Cyclopentene-1-carbonitrile, 2-amino-3,3-dimethyl-	EtOH	263(4.10)	23-2512-65
3,4-Diazatricyclo[3.3.2.0^{4,6}]dec-3-ene	isooctane	345(2.42),349(2.52)	44-2425-65
	DMF	351(2.47)	44-2425-65
2-Picoline, 4-amino-5-ethyl-	n.s.g.	248(3.97),267(3.92)	25-0542-64
2-Picoline, 6-amino-5-ethyl-	n.s.g.	235(3.94),297(3.81)	25-0542-64
$C_8H_{12}N_2O$			
2-Piperidone, 3-cyano-1,3-dimethyl-	MeOH	210(3.75)	24-1970-64
Pyrazine, 3-ethoxy-2,5-dimethyl-	MeOH	216(3.97),297(3.89)	44-2623-64
$C_8H_{12}N_2OS$			
4-Isothiazolecarboxamide, 3-methyl-5-propyl-	EtOH	254(3.92)	39-0446-64
4-Pyrimidinol, 2-butyl-6-mercapto-	C_6H_{12}	227(4.24),282(3.75)	39-3204-64
	pH -3.6	230(4.12),262(3.95)	39-3204-64
	pH 2.0	242(4.04),306(4.23)	39-3204-64
	pH 8.0	222(4.21),235(4.21), 294(4.24)	39-3204-64
	EtOH	218(4.11),251s(3.76), 304(4.18)	39-3204-64
4-Pyrimidinol, 5-butyl-6-mercapto-	C_6H_{12}	230(4.26),282(3.75)	39-3204-64
	pH -3.6	227(4.01),270(3.99)	39-3204-64
	pH 2.0	217(4.05),242(4.00), 310(4.22)	39-3204-64
	pH 9.0	235(4.22),300(4.21)	39-3204-64
	EtOH	235(4.12),303(4.17)	39-3204-64
$C_8H_{12}N_2O_2$			
Pyrazine, 3-ethoxy-2,5-dimethyl-, 1-oxide	MeOH	220(4.34),262(4.01), 309(3.76)	44-2623-64
4,6-Pyrimidinediol, 5-butyl-	pH -2.7	256(4.00)	39-3204-64
	pH 3.5	261(4.13)	39-3204-64
	pH 8.6	259(4.03)	39-3204-64
	EtOH	245(3.69),264(3.70)	12-0567-64
$C_8H_{12}N_2O_2S$			
5-Pyrimidinecarboxylic acid, 1,2,3,4-tetrahydro-1-methyl-2-thioxo-, ethyl ester	EtOH	315(4.14)	44-2290-65
5-Pyrimidinecarboxylic acid, 1,2,3,6-tetrahydro-1-methyl-2-thioxo-, ethyl ester	EtOH	312(4.12)	44-2290-65
6H-1,3-Thiazine-5-carboxylic acid, 2-(methylamino)-, ethyl ester	EtOH	240(3.66),266(3.77), 330(4.14)	44-2290-65
$C_8H_{12}N_2O_3$			
2-Furanamine, N,N-diethyl-5-nitro-	MeOH	227(4.28),430(4.45)	65-0710-64
Primicarpin	EtOH	253(3.56)	88-1897-64
5-Pyrimidinecarboxylic acid, 1,2,3,4-tetrahydro-1-methyl-2-oxo-, ethyl ester	EtOH	214(3.91),296(4.01)	94-0804-64

Compound	Solvent	$\lambda_{max}(\log \epsilon)$	Ref.
5-Pyrimidinecarboxylic acid, 1,2,3,6-tetrahydro-1-methyl-2-oxo-, ethyl ester	EtOH	219(3.93),289(3.94)	94-0804-64
Pyrrolidine-2,4-dione, 3-acetamido-1,5-dimethyl-	H_2O	268(3.76)	44-2085-64
$C_8H_{12}N_2O_4$			
4-Pyrimidinecarboxaldehyde, 1,2,5,6-tetrahydro-1-methyl-2,6-dioxo-, 4-dimethyl acetal	pH 7.2	262(3.8)	87-0337-64
	pH 12.0	218(3.8),291(4.0)	87-0337-64
$C_8H_{12}N_2S_2$			
4-Pyrroline-3-thione, 1,2,2-trimethyl-4-thiocarbamoyl-	pH 1	268(3.87),326(4.06)	39-4546-65
	pH 13	260(3.71),327(4.10), 380(4.33)	39-4546-65
$C_8H_{12}N_3O_9P$			
6-Azauridine-5-phosphate	pH 1	263(3.76)	70-1236-64
$C_8H_{12}N_4$			
Formamidine, N,N-dimethyl-N'-(3-methyl-2-pyrazinyl)-	EtOH	218(3.52),266(4.27), 333(4.11)	44-3454-65
5H-Pyrrolo[3,4-d]pyrimidine, 4-amino-6,7-dihydro-2,7-dimethyl-	pH 1	255(3.87)	44-0194-65
	EtOH	230(3.86),267(3.61)	44-0194-65
Quinazoline, 2,4-diamino-5,6,7,8-tetrahydro-	pH 1	275(3.89)	44-1837-65
	pH 10	226s(4.01),284(3.88)	44-1837-65
$C_8H_{12}N_4O_2$			
Glutarimide, N-(3-azidopropyl)-	H_2O	208(4.16)	35-2003-65
4-Pyrimidineacetic acid, 5,6-diamino-, ethyl ester	pH 1	294(3.99)	44-1528-65
	pH 7	249(3.78),290(3.85)	44-1528-65
	pH 13	245(3.81),290(3.84)	44-1528-65
5-Pyrimidinealanine, 4-amino-2-methyl-	pH 1	247(3.96)	65-2240-65
	pH 12	235(3.88),276(3.73)	65-2240-65
2-Pyrrolidinone, 1-(4-azidobutyryl)-	EtOH	215(4.06)	35-2003-65
$C_8H_{12}N_4O_4$			
5-Azacytosine, 1-(2-deoxy-D-ribofuranosyl)-	H_2O	242(3.87)	73-2576-64
$C_8H_{12}N_4O_4S$			
4,7-Diazatryptamine sulfate	H_2O	222(4.2),308(3.95)	44-3454-65
$C_8H_{12}N_4O_5$			
5-Azacytidine	pH 5.0	246(3.78)	73-2060-64
5-Azacytosine, 1-β-D-ribopyranosyl-	pH 5.0	243(3.80)	73-2060-64
$C_8H_{12}N_4S_2$			
2H-Pyrimido[1,2-a]-s-triazine-2,4(3H)-dithione, 1,6,7,8-tetrahydro-1,3-dimethyl-	n.s.g.	271(4.3),290(4.38)	32-1342-64
$C_8H_{12}O$			
Bicyclo[3.2.1]octan-3-one	C_6H_{12}	280(1.28)	22-0844-64
Cycloheptanone, 2-methylene-	MeOH	220s(3.9),230(4.1)	83-0326-65
2-Cyclohexen-1-one, 2,3-dimethyl-	EtOH	247(4.07)	42-0242-64
2-Cyclohexen-1-one, 2,4-dimethyl-	heptane	229(4.11),323(1.60)	33-0725-64
2-Cyclohexen-1-one, 3,4-dimethyl-	EtOH	236(4.2)	42-0242-64
2-Cyclohexen-1-one, 3,6-dimethyl-	EtOH	234(4.1)	42-0242-64
1-Cyclopentene, 1-acetyl-2-methyl-	EtOH	251(3.95),309(1.86)	22-0722-65

Compound	Solvent	$\lambda_{max}(\log \epsilon)$	Ref.
2-Cyclopenten 1-acetyl-2-methyl-	EtOH	211(3.36),276(1.46)	22-0722-65
Furan, 2,3-dihydro-3-methyl-2-propenyl-	EtOH	225(3.18)	88-2983-65
2,4-Hexadienal, 2-ethyl-	heptane	270(4.08)	33-0567-64
2,7-Octadien-4-one	EtOH	221(3.90)	22-0315-65
$C_8H_{12}OS$			
3-Thiolen-2-one, 5-butyl-	EtOH	265(3.41)	11-0211-64
3-Thiolen-2-one, 5-tert-butyl-	EtOH	268(3.36)	11-0211-64
4-Thiolen-2-one, 5-butyl-	EtOH	268(3.31)	11-0211-64
4-Thiolen-2-one, 5-tert-butyl-	EtOH	270(3.33)	11-0211-64
4H-Thiopyran, 3-acetyl-5,6-dihydro-2-methyl-	n.s.g.	208(3.48),300(4.14)	24-1963-64
$C_8H_{12}O_2$			
Butyrolactone, α-isobutylidene-	EtOH	221(3.97)	65-3135-64
Cyclobutanone, 3-acetyl-2,2-dimethyl-	C_6H_{12}	289(1.78)	28-0600-64A
1,2-Cyclohexanedione, 3,5-dimethyl-	MeOH	271(4.30)	24-2111-65
1,2-Cyclohexanedione, 3,6-dimethyl-	MeOH	270(3.96)	24-2111-65
2-Cyclohexen-1-one, 3-methoxy-2-methyl-	EtOH	266(4.24)	70-1648-64
1,5-Cyclooctanedione	H_2O	287(1.40)	35-2003-65
1,3-Cyclopentanedione, 2-isopropyl-	pH 1	253(4.20)	94-1300-65
	neutral	251(4.2)	94-1300-65
	pH 10	271(4.40)	94-1300-65
	EtOH	250(4.28),270(4.40)	39-4472-64
1,3-Cyclopentanedione, 2-propyl-	EtOH	251(4.23),270(4.39)	39-4472-64
4,6-Heptadienoic acid, 5-methyl-	EtOH	229.3(4.31)	22-2533-64
cis	EtOH	232.5(4.18)	22-2533-64
4-Hexen-1-yn-3-ol, 1-ethoxy-	EtOH	none	39-4939-65
5-Hexen-3-yn-2-ol, 6-methoxy-2-methyl-	n.s.g.	236(4.21)	70-1318-64
	n.s.g.	239(4.21)	70-1318-64
5-Hexyn-2-one, 4-ethoxy-	EtOH	248(3.48)	70-1070-65
$C_8H_{12}O_3$			
2,3,4-Heptanetrione, 6-methyl-	n.s.g.	220(2.65),464(1.44)	35-5342-64
Octanoic acid, 7-hydroxy-4-oxo-, lactone	EtOH	275(1.79)	39-3239-64
2-Octenoic acid, 4,7-dihydroxy-, 7-lactone	EtOH	207(3.87)	39-3239-64
2(6H)-Pyrone, 4-methoxy-5,5-dimethyl-	n.s.g.	237(4.04)	22-0651-64
$C_8H_{12}O_3S$			
Thiopyran-3-carboxylic acid, tetrahydro-2-oxo-, ethyl ester	n.s.g.	238(3.65)	24-1936-64
$C_8H_{12}O_4$			
Fumaric acid, diethyl ester	MeOH	210(4.28)	5-0066-64F
$C_8H_{12}O_4S$			
Acrylic acid, 3-[(carboxymethyl)thio]-, 1-ethyl 3-methyl ester	Et_2O	268(4.24)	24-2109-64
Dimethylsulfonium 2,2-dimethyl-4,6-dioxo-m-dioxan-5-ylide	MeOH	247(4.04)	88-0721-65
$C_8H_{12}O_5S$			
(1,2-Dicarboxy-2-hydroxyvinyl)dimethylsulfonium hydroxide, inner salt, dimethyl ester	MeOH	260(4.14)	24-1581-65

Compound	Solvent	$\lambda_{max}(\log \epsilon)$	Ref.
$C_8H_{12}O_5S_2$			
Acetic acid, thio-, S-(tetrahydro-4-hydroxy-3-thienyl) ester, acetate, 1,1-dioxide	EtOH	229(3.54)	39-1298-65
$C_8H_{12}O_6S$			
Acrylic acid, 3-[(carboxymethyl)sulfonyl]-, 1-ethyl 3-methyl ester	Et_2O	210(3.89)	24-2109-64
$C_8H_{12}S$			
Thiophene, 2-tert-butyl-	EtOH	234.0(3.94)	22-2635-65
$C_8H_{12}S_2$			
1,6-Dithiecin, 2,5,7,10-tetrahydro-, cis-cis	EtOH	none	39-1154-64
Thiophene, 2-(butylthio)-	hexane	204(3.59),238(3.77), 272(3.52)	78-2413-65
Thiophene, 3-(butylthio)-	hexane	212(3.92),268(3.45)	78-2413-65
$C_8H_{13}BrN_4$			
Pyrimidine, 5-bromo-4-(dimethylamino)-1,2-dihydro-1-methyl-2-(methylimino)-	pH 7.0	225(4.34),261s(3.98), 303(3.89)	39-0755-65
$C_8H_{13}BrO$			
Cyclohexanone, 2-bromo-3,3-dimethyl-	C_6H_{12}	310(1.95)	24-1858-65
	MeOH	304(1.70)	24-1858-65
	EtOH	309(1.72)	24-1858-65
	$CHCl_3$	306(1.87)	24-1858-65
	CCl_4	308(1.93)	24-1858-65
	EtOAc	308(1.77)	24-1858-65
	MeCN	306(1.67)	24-1858-65
Cyclohexanone, 2-bromo-5,5-dimethyl-	C_6H_{12}	310(1.97)	24-1858-65
	MeOH	308(1.84)	24-1858-65
	EtOH	310(1.83)	24-1858-65
	EtOAc	310(1.78)	24-1858-65
	MeCN	311(1.76)	24-1858-65
$C_8H_{13}ClN_2O$			
Valeramide, 5-chloro-2-cyano-N,2-dimethyl-	MeOH	210(3.03)	24-1970-64
$C_8H_{13}ClN_4$			
Pyrimidine, 4-chloro-2,6-bis(dimethylamino)-	pH 1.0	232(4.42),266(4.05)	39-0755-65
	pH 7.2	229(4.42),300(3.89)	39-0755-65
Pyrimidine, 4-chloro-6-(dimethylamino)-2,3-dihydro-3-methyl-2-(methylimino)-	pH 7.0	221(4.43),294(3.97)	39-0755-65
$C_8H_{13}ClO$			
Cyclooctanone, 2-chloro-	hexane	302.5(1.54)	28-2374-65B
	C_6H_{12}	303(1.56)	28-2374-65B
	EtOH	295(1.49)	28-2374-65B
	Me_2SO	293(1.46)	28-2374-65B
2-Octen-1-al, 2-chloro-	C_6H_{12}	236(4.09)	28-1015-65B
$C_8H_{13}ClO_2$			
2-Pentenoic acid, 3-chloro-4-methyl-, ethyl ester	n.s.g.	225(4.69)	39-0865-64

Compound	Solvent	$\lambda_{max}(\log \epsilon)$	Ref.
$C_8H_{13}ClS$			
Sulfide, butyl 1-chloro-1,3-butadienyl	EtOH	234(4.16),280(4.06)	44-2214-64
Sulfide, tert-butyl 1-chloro-1,3-butadienyl	EtOH	246(4.27),285(3.65)	44-2214-64
$C_8H_{13}Cl_2N_7S$			
Adenin-9-ylethylisothiuronium hydrochloride	pH 1	256(4.13)	65-2774-64
$C_8H_{13}FO_2$			
2-Hexenoic acid, 2-fluoro-, ethyl ester	C_6H_{12}	217.5(--)	22-2258-64
2-Pentenoic acid, 2-fluoro-4-methyl-, ethyl ester	C_6H_{12}	214.5(3.98)	22-2258-64
cis	MeOH	214(3.87)	5-0001-64D
trans	MeOH	215(3.84)	5-0001-64D
$C_8H_{13}N$			
Pyridine, 1,2-dihydro-1,2,3-trimethyl-	EtOH	330(3.62)	44-1647-64
Pyridine, 1,2-dihydro-1,3,4-trimethyl-	EtOH	322(3.40)	44-1647-64
$C_8H_{13}NO$			
2H-Azepin-2-one, 1,3,4,7-tetrahydro-4,6-dimethyl-	MeOH	239(3.88)	5-0001-65B
2H-Azepin-2-one, 1,5,6,7-tetrahydro-4,6-dimethyl-	MeOH	217(3.18)	5-0001-65B
1-Cyclobutene-1-carboxamide, 2-ethyl-3-methyl-	EtOH	223(4.11)	39-2165-64
Cyclohexanone, 2-[(methylamino)-methylene]-	EtOH	327(4.25)	70-1054-64
3-Pyrrolin-2-one, 4-ethyl-3,5-dimethyl-	EtOH	211.5(4.14)	39-5999-64
$C_8H_{13}NOS$			
Thiophene, tetrahydro-2-oxo-, morpholine enamine	C_6H_{12}	223(3.74+)	54-1160-64
$C_8H_{13}NO_2$			
Nicotinic acid, 1,4,5,6-tetrahydro-1-methyl-, methyl ester	EtOH	240(3.28),295(4.37)	44-2534-65
$C_8H_{13}NO_2S$			
2,4-Thiazolidinedione, 5-pentyl-	n.s.g.	300s(2.82)	65-0886-65
3-Thiophenecarboxylic acid, tetrahydro-2-imino-3-methyl-, ethyl ester	MeOH	233(3.75)	24-1970-64
4H-Thiopyran-3-carboxylic acid, 2-amino-5,6-dihydro-, ethyl ester	MeOH	227(4.06),292(4.31)	24-1970-64
$C_8H_{13}NO_2S_2$			
Pipecolinic acid, N-dithiocarbomethoxy-	dioxan	280(4.07),346(1.87)	78-0407-65
$C_8H_{13}NO_3$			
2H-Azepin-2-one, 1-glycoloylhexahydro-	dioxan	214(3.23)	78-3537-65
Glutarimide, N-(3-hydroxypropyl)-	H_2O	211(4.16)	35-2003-65
	H_2O	211(4.16)	35-3397-64
5H-Oxazolo[3,2-a]pyridin-3(2H)-one, tetrahydro-8a-methoxy-	dioxan	221(2.20)	78-3537-65
$C_8H_{13}NO_3S_2$			
2H-Thiazine-4-carboxylic acid, tetrahydro-5-methyl-2-thioxo-, ethyl ester	EtOH	240(3.92),290(4.16)	39-0766-64

Compound	Solvent	λ_{max}(log ε)	Ref.
$C_8H_{13}NO_4S_3$			
Carbonic acid, trithio-, cyclic [(dimethylamino)methyl]ethylene ester, oxalate	H_2O	318(4.15)	94-0253-65
$C_8H_{13}NS$			
2-Piperidinethione, N-allyl-	n.s.g.	276(4.07)	70-0702-65
Thiophene, tetrahydro-3-oxo-, pyrrolidine enamine	C_6H_{12}	229(3.82)	54-1160-64
$C_8H_{13}NS_2$			
2H-1,3-Thiazine-2-thione, 3,6-dihydro-3,4,6,6-tetramethyl-	H_2O	297(4.02)	39-4008-64
	EtOH	273s(3.59),310(3.97)	39-4008-64
$C_8H_{13}N_3$			
Pyrazolo[1,5-a]pyrimidine, 4,5,6,7-tetrahydro-2,3-dimethyl-	EtOH	245(3.82)	94-1207-65
3-Pyrroline-3-carbonitrile, 4-amino-1,5,5-trimethyl-	pH 1	259(4.09)	39-4546-65
	pH 13	263(4.06)	39-4546-65
$C_8H_{13}N_3O$			
Pyrazinemethanol, 3-(dimethylamino)-5-methyl-, hydrochloride	H_2O	239(3.74),260(3.01), 318(3.91)	44-1645-64
Pyrazinol, 5-(dimethylamino)-3,6-dimethyl-, hydrochloride	H_2O	230(3.94),323(3.76)	44-1645-64
s-Triazine, 2,4-diethyl-6-methoxy-	EtOH	236(3.34)	44-0702-65
$C_8H_{13}N_3O_2$			
Pyrazole-4-carboxylic acid, 5-amino-3-ethyl-, ethyl ester	EtOH	257(3.80)	95-0442-65
Sydnone imine, N-acetyl-3-butyl-	pH 1.0	280(4.12)	65-2064-64
	pH 7.00	236(3.82),312(4.16)	65-2064-64
$C_8H_{13}N_3O_3$			
Sydnone imine, 3-butyl-N-carboxy-, methyl ester	pH 1.0	278(4.08)	65-2064-64
	pH 7.00	226(4.02),310(4.10)	65-2064-64
γ-Valerolactone, α-acetyl-, semicarbazone	n.s.g.	222(4.07),274(3.73)	39-0141-64
$C_8H_{13}N_5OS$			
Pyrimidine, 5-formamido-4,6-bis-(methylamino)-2-(methylthio)-	pH 1.9	225(4.29),245(4.43), 281(4.11)	39-3770-65
	pH 7.0	228(4.62),275(4.00)	39-3770-65
$C_8H_{13}N_5O_2$			
Pyrimidine, 4-amino-6-(butylamino)-5-nitro-	pH 1	241(4.38),294(3.56), 342(3.85)	44-2135-64
	pH 7	342(3.98)	44-2135-64
	pH 13	344(3.98)	44-2135-64
	EtOH	340(3.99)	44-2135-64
$C_8H_{13}N_5O_2S$			
1,2,5-Thiadiazole-3-carboxamide, N-butyl-4-ureido-	pH 1, 7	232(4.17),304(3.92)	44-2141-64
$C_8H_{13}O_5PS$			
Crotonic acid, 3-hydroxy-2-(2-hydroxyethyl)-, γ-lactone, O-ester with O,O-dimethyl phosphorothioate	MeOH	229(3.83)	5-0010-65E

Compound	Solvent	λ_{max}(log ϵ)	Ref.
$C_8H_{13}O_6P$			
Acetoacetic acid, 2-(2-hydroxyethyl)-2-phosphono-, intramolecular P-ester, P-ethyl ester	MeOH	218(2.11)	5-0010-65E
Crotonic acid, 3-hydroxy-2-(2-hydroxyethyl)-, γ-lactone, dimethyl phosphate	MeOH	220(3.90)	5-0010-65E
C_8H_{14}			
1-Cycloheptene, 1-methyl-, $PdCl_2$ olefin complex from	EtOH	230s(4.18),328(3.28)	24-2037-64
Cyclopentene, 1,3,4-trimethyl-	96% H_2SO_4	279.5(4.09)	23-2768-64
Cyclopentene, 1,4,4-trimethyl-	96% H_2SO_4	279(3.98+)	23-2768-64
2,5-Heptadiene, 2-methyl-	96% H_2SO_4	375(4.78)	23-2768-64
2,4-Hexadiene, 2,3-dimethyl-, trans	EtOH	240(4.31)	39-2465-64
2,4-Hexadiene, 2,5-dimethyl-	C_6H_{12}	244(4.38)	44-3527-64
	EtOH	242.5(4.42)	44-3527-64
	Et_2O	242.5(4.38)	44-3527-64
$C_8H_{14}Br_2N_8S$			
Adenin-8-ylaminoethylisothiuronium hydrobromide	pH 1	292(4.15)	65-2774-64
$C_8H_{14}ClNO_4S_2$			
2-(Dimethylamino)-4-ethyl-5-methyl-1,3-dithiol-1-ium perchlorate	70% $HClO_4$	245(3.03),312(3.83)	44-1703-64
$C_8H_{14}ClN_3S$			
Trimethyl[2-(methylthio)-4-pyrimidinyl]-ammonium chloride	pH 1	257(4.20)	4-0130-64
	pH 11	257(4.18)	4-0130-64
$C_8H_{14}ClN_5O_2S$			
2,6-Dihydroxypyrimidin-5-ylmethyl-aminoethylisothiuronium hydrochloride	pH 1	258(4.14)	65-2774-64
$C_8H_{14}ClO_2P$			
3-Phospholene, 1-(2-chloroethoxy)-3,4-dimethyl-, 1-oxide	MeOH	200(3.74)	78-1593-64
3-Phospholene, 1-(2-chloro-1-methylpropoxy)-, 1-oxide	MeOH	200(2.82)	78-1593-64
$C_8H_{14}Cl_2S$			
Cyclopropane, 2-(butylthio)-1,1-dichloro-3-methyl-	EtOH	226(2.42)	44-2211-64
Cyclopropane, 2-(tert-butylthio)-1,1-dichloro-3-methyl-	EtOH	225(2.64)	44-2211-64
Sulfide, butyl 1,3-dichloro-1-butenyl	EtOH	208(3.79),225s(3.48), 254(3.38)	44-2214-64
$C_8H_{14}INO_2$			
Cyclohexanecarbamic acid, 2-iodo-, methyl ester, trans	MeOH	260(2.88)	44-3640-64
	MeOH	260(2.88)	78-1037-64
$C_8H_{14}INS_2$			
1-(1,3-Dithiolan-2-ylidene)piperidinium iodide	EtOH	301(3.74)	24-1298-64

Compound	Solvent	$\lambda_{max}(\log \epsilon)$	Ref.
$C_8H_{14}NO_3P$			
Phosphonic acid, pyrrol-2-yl-, diethyl ester	EtOH	211(3.82),297(1.65)	44-0091-65
$C_8H_{14}N_2$			
2,3-Diazabicyclo[3.3.0]oct-3-ene, 1,4-dimethyl-	EtOH	236(3.57)	22-0722-65
$C_8H_{14}N_2OS$			
Hydantoin, 5-sec-butyl-3-methyl-2-thio-	EtOH	235(3.99),265(4.25)	65-0049-65
Hydantoin, 5-isobutyl-3-methyl-2-thio-	EtOH	235(3.97),265(4.23)	65-0049-65
Hydantoin, 5-isopropyl-1,3-dimethyl-2-thio-	97% HOAc	239(--),267(4.16)	65-0554-65
Pseudothiohydantoin, 5-pentyl-	n.s.g.	248(3.94),300s(2.47)	65-0886-65
$C_8H_{14}N_2O_2$			
Glutarimide, N-(3-aminopropyl)-	H_2O	208(4.16)	35-3397-64
hydrochloride	H_2O	208(4.16)	35-2003-65
Imidazole, 2-ethyl-2,4,5-trimethyl-, 1,3-dioxide	n.s.g.	347(3.98)	77-0524-65
1-Pyrazolin-3-ol, 3,5,5-trimethyl-, acetate	EtOH	330(2.29)	44-1379-64
2-Pyrrolidinone, 1-(4-aminobutyryl)-, hydrochloride	H_2O	217(4.07)	35-2003-65
$C_8H_{14}N_2S_2$			
3-Pyrroline, 3-mercapto-1,2,2-trimethyl-4-thiocarbamoyl-	EtOH	210(4.10),299(3.39), 375(4.37)	39-4546-65
	pH 13	256s(3.49),296(3.49), 369(4.33)	39-4546-65
$C_8H_{14}N_4$			
Guanidine, 2-(2-cyanovinyl)-1,1,3,3-tetramethyl-, cis	MeOH	299(4.37)	44-1278-65
trans	MeOH	298(4.42)	44-1278-65
hydrochloride	MeOH	258(4.43)	44-1278-65
Pyrimidine, 4-amino-5-(aminomethyl)-2-propyl-, dihydrochloride	EtOH	235(3.95),262(3.73)	94-0393-64
Pyrimidine, 2,4-bis(dimethylamino)-	pH 1.0	228(4.47),262s(4.06)	39-0755-65
	pH 10.4	226(4.47),300(3.89)	39-0755-65
Pyrimidine, 2-(dimethylamino)-1,6-dihydro-1-methyl-6-(methylimino)-	pH 7.0	232(4.38),280s(3.70)	39-0755-65
	pH 13.0	239(4.31),290s(3.74)	39-0755-65
Pyrimidine, 4-(dimethylamino)-1,2-dihydro-1-methyl-2-(methylimino)-	pH 7.0	217(4.35),255s(3.99), 281(3.99)	39-0755-65
	pH 15.6	267(4.02)	39-0755-65
$C_8H_{14}N_6$			
Piperidine, 1-(5-amino-v-triazole-4-carboximidoyl)-	pH 1	227(4.10),272(3.83)	44-2488-65
	pH 7	277(3.81)	44-2488-65
	pH 13	234(4.03)	44-2488-65
$C_8H_{14}O$			
Cyclobutanone, 2-tert-butyl-	C_6H_{12}	285(1.52),296(1.53), 310(1.45),323(1.18)	22-2747-65
Cyclobutanone, 3-ethyl-2,2-dimethyl-	C_6H_{12}	283(1.42),292(1.42)	22-1968-64
	MeOH	289(1.46)	22-1968-64
Cyclohexanone, 2,5-dimethyl-	EtOH	286(1.34)	59-0529-65
	CCl_4	294(1.37)	59-0529-65
	HOAc	282(1.37)	59-0529-65
	CF_3COOH	266(1.56)	59-0529-65
	MeCN	288(1.30)	59-0529-65

Compound	Solvent	λ_{max}(log ϵ)	Ref.
Cyclohexanone, 2,6-dimethyl-	EtOH	285(1.34)	59-0529-65
	HOAc	283(1.37)	59-0529-65
	CF_3COOH	272(1.56)	59-0529-65
	CCl_4	292(1.37)	59-0529-65
	MeCN	287(1.34)	59-0529-65
cis	octane	286(1.27),313(0.79)	46-3225-65
trans	octane	295(1.44),313(1.20)	56-3225-65
Cyclohexanone, 3,3-dimethyl-	C_6H_{12}	287(1.30)	24-1858-65
	MeOH	282(1.32)	24-1858-65
	EtOH	284(1.28)	24-1858-65
	$CHCl_3$	288(1.34)	24-1858-65
	CCl_4	294(4.31)	24-1858-65
	MeCN	288(1.22)	24-1858-65
	EtOAc	287(1.23)	24-1858-65
Cyclohexanone, 3,5-dimethyl-	EtOH	286(1.23)	59-0529-65
	HOAc	282(1.30)	59-0529-65
	CF_3COOH	267(1.53)	59-0529-65
	MeCN	289(1.18)	59-0529-65
	CCl_4	295(1.28)	59-0529-65
Cyclooctanone	hexane	279(1.15),288(1.18), 300s(1.15),310s(1.00), 323s(0.65)	28-2374-65B
	C_6H_{12}	280s(1.16),289(1.20), 300s(1.16),310s(1.00), 323s(0.48)	28-2374-65B
	EtOH	282(1.23)	28-2374-65B
	Me_2SO	286(1.23)	22-2374-65B
Ethanol, 1-(2-methyl-1-cyclopenten-1-yl)-	EtOH	210(3.38)	22-0722-65
2-Hexenal, 2,4-dimethyl-	EtOH	229(4.15)	22-0161-64
2-Hexenal, 2-ethyl-	heptane	224(4.19),326(1.46)	33-0567-64
$C_8H_{14}O_2$			
4-Hexen-2-one, 3-hydroxy-3,5-dimethyl-	EtOH	287(2.20)	59-0529-65
	CCl_4	281(2.20)	59-0529-65
$C_8H_{14}O_2S_2$			
1,2-Dithiolane-4-valeric acid	$CHCl_3$	333.5(2.17)	1-2000-64
$C_8H_{14}O_3$			
Crotonic acid, 3-ethoxy-, ethyl ester	EtOH	235(4.26)	70-0110-64
2-Hexanone, 6-acetoxy-	EtOH	275(1.33)	44-0650-65
$C_8H_{14}O_5$			
α-D-threo-Hexopyranos-2-ene, 3-deoxy-2,4-di-O-methyl-	H_2O	198(3.97)	12-0837-65
$C_8H_{14}S$			
Thione, cyclohexyl methyl	C_6H_{12}	215(3.77),236(3.90), 508(1.00)	59-0299-64
	C_6H_{12}	508(1.00),516s(--), 570s(--)	89-0157-64
	EtOH	214(3.75),236(3.87), 492(0.96)	59-0299-64
$C_8H_{14}SSe$			
2-Thia-3-selenaspiro[4.5]decane	C_6H_{12}	410(2.0)	1-0815-64
$C_8H_{14}S_2$			
2,3-Dithiaspiro[4.5]decane	C_6H_{12}	330(2.2)	1-0815-64

Compound	Solvent	$\lambda_{max}(\log \epsilon)$	Ref.
$C_8H_{14}Se_2$			
2,3-Diselenaspiro[4.5]decane	C_6H_{12}	470(2.2)	1-0815-64
$C_8H_{15}Br_2N_4O_4Rh$			
Rhodium, dibromo(dimethylglyoximato)-(dimethylglyoxime)-	H_2O	264(4.28),335(3.76), 435s(1.90)	39-1951-65
$C_8H_{15}ClS$			
Sulfide, butyl 1-chloro-1-butenyl	EtOH	209(3.87),225s(3.50), 254(3.42)	44-2214-64
Sulfide, tert-butyl 1-chloro-1-butenyl	EtOH	212(4.01),258(3.16)	44-2214-64
$C_8H_{15}N$			
1-Pyrroline, 3,3,5,5-tetramethyl-	hexane	228(2.00)	39-2313-65
	EtOH	226(1.92)	39-2313-65
$C_8H_{15}NO$			
Cyclohexanol, 1-methyl-2-(methylimino)-	EtOH	305(2.21)	44-2967-65
Cyclohexanone, 2-methyl-2-(methylamino)-, hydrochloride	EtOH	283(1.35)	44-2967-65
Cyclopentanol, 1-(N-methylacetimidoyl)-	EtOH	233(2.22)	44-2967-65
Ketone, (1-methylamino)cyclopentyl methyl, hydrochloride	EtOH	279(1.34)	44-2967-65
3-Pyrrolidinone, 2,2,5,5-tetramethyl-	heptane	305(1.60)	22-3643-65
	MeOH	301(1.57)	22-3643-65
$C_8H_{15}NOS_2$			
Carbamic acid, dimethyldithio-, ester with 3-mercapto-2-pentanone	EtOH	248(4.06),276(4.08)	44-1703-64
1,2-Dithiolane-3-valeramide	MeOH	332(2.18)	95-0562-64
2-Pentanone, 4-methyl-4-(N-methylthiocarbamoylthio)-	H_2O	248(3.95),280(4.12)	39-4004-64
	EtOH	248(4.01),286(4.08)	39-4004-64
$C_8H_{15}NO_2$			
3-Pyrrolidinone, 2,2,5,5-tetramethyl-, 1-oxide	C_6H_{12}	233(3.26),455(0.83)	22-3643-65
	MeOH	230(3.28),311(1.98), 425(0.85)	22-3643-65
$C_8H_{15}NSi$			
Pyridine, 1,4-dihydro-1-(trimethylsilyl)-	n.s.g.	288(3.11)	35-3283-65
$C_8H_{15}N_2O_2$			
1-Pyrrolidinyloxy, 2,2,5,5-tetramethyl-3-oxo-, oxime	MeOH	237(3.45),420(0.84)	88-1781-64
	$CHCl_3$	430(0.89)	88-1781-64
$C_8H_{15}N_3O_4$			
Piperidine, 1-(2,2-dinitropropyl)-	hexane	268(2.60)	44-0354-65
	pH 1	270s(1.83)	44-0354-65
	H_2O	379(4.18)	44-0354-65
	pH 13	380(4.24)	44-0354-65
	MeOH	376.5(4.08)	44-0354-65
	EtOH	377(4.14)	44-0354-65
	CCl_4	270(2.62)	44-0354-65
$C_8H_{15}N_5$			
Pyrimidine, 4,5-diamino-6-(butylamino)-	pH 1	222(4.02),283(4.05)	44-2135-64
	pH 7	218(4.45),279(4.03)	44-2135-64
	pH 13	279(4.02)	44-2135-64
	EtOH	218(4.47),283(4.02)	44-2135-64

Compound	Solvent	$\lambda_{max}(\log \epsilon)$	Ref.
$C_8H_{15}N_5OS$			
Arginine, methylthiohydantoin deriv.	EtOH	235(3.94),265(4.20)	65-0988-65
$C_8H_{15}O_2P$			
2-Phospholene, 1-ethoxy-3,4-dimethyl-, 1-oxide	MeOH	200(3.89)	78-1593-64
3-Phospholene, 1-ethoxy-3,4-dimethyl-, 1-oxide	MeOH	200(3.72)	78-1593-64
$C_8H_{15}O_5P$			
Butyric acid, 4-hydroxy-2-phosphono-, γ-lactone, diethyl ester	MeOH	209(2.69)	5-0010-65E
$C_8H_{16}ClNO$			
Trimethyl(3-oxo-1-pentenyl)ammonium chloride	EtOH	208(4.84)	44-0385-64
$C_8H_{16}NO$			
1-Pyrrolidinyloxy, 2,2,5,5-tetramethyl-	C_6H_{12}	235(3.30),237(3.30), 435(0.70)	88-1781-64
	MeOH	233(3.40),410(0.78)	88-1781-64
$C_8H_{16}NO_2$			
1-Pyrrolidinyloxy, 3-hydroxy-2,2,5,5-tetramethyl-	C_6H_{12}	235(3.38),435(0.82)	88-1781-64
	MeOH	234(3.42),410(0.84)	88-1781-64
$C_8H_{16}N_2$			
2-Imidazoline, 2-pentyl-	EtOH	221(3.64)	4-0188-64
2-Pyrazoline, 3-isopropyl-4,4-dimethyl-	EtOH	228(3.61)	28-1343-65B
2-Pyrazoline, 5-isopropyl-4,4-dimethyl-	EtOH	232(3.48)	28-1343-65B
$C_8H_{16}N_2O$			
2-Imidazoline-1-ethanol, 2-isopropyl-	EtOH	231(3.75)	4-0188-64
$C_8H_{16}N_2O_2$			
2-Pentanone, 4-(N-ethyl-N-nitrosamino)-4-methyl-	H_2O	231.0(3.88)	28-5470-64A
$C_8H_{16}N_6O_6$			
D-arabino-Hexosulose, disemicarbazide	EtOH	292(4.31)	44-3074-64
L-Xylo-Hexosulose, disemicarbazide	EtOH	292(4.69)	44-3074-64
$C_8H_{16}O_2$			
Isobutyroin	EtOH	286(1.62)	59-0529-65
	HOAc	280(1.64)	59-0529-65
	CF_3COOH	271(1.79)	59-0529-65
	CCl_4	280(1.65)	59-0529-65
	MeCN	282(1.60)	59-0529-65
$C_8H_{16}S_2$			
m-Dithiane, 2-tert-butyl-	EtOH	229s(2.86),249(2.88)	78-0437-64
$C_8H_{17}ClN_2O$			
Trimethyl(3-oxo-1-pentenyl)ammonium chloride, oxime	EtOH	232(4.20)	44-1129-65
$C_8H_{17}NO$			
2-Hexanone, 6-(dimethylamino)-	EtOH	274(1.36)	44-0650-65

Compound	Solvent	$\lambda_{max}(\log \epsilon)$	Ref.
Piperidine, 2-(1-hydroxypropyl)- (or α-conhydrine)	EtOH	208(2.45),210(2.42)	24-2822-65
β-isomer	EtOH	210(2.78),260(2.63)	24-2822-65
$C_8H_{18}Si$			
Silane, trimethyl-4-pentenyl-	n.s.g.	175(4.2)	70-2203-64
$C_8H_{18}Si_2$			
Disilane, 1,1,2,2-tetramethyl-1,2-divinyl-	C_6H_{12}	225.0(3.90)	101-0369-64B
$C_8H_{20}Ge_2$			
Germane, vinylenebis[trimethyl-	n.s.g.	195(4.4)	70-1515-64
$C_8H_{20}N_4$			
Tetrazene, tetraethyl-	EtOH	285(3.90)	35-2395-64
$C_8H_{20}N_8O_8S$			
D-arabino-Hexosulose, bis(amidinohydrazone), sulfate	EtOH	289(4.56)	44-3074-64
L-xylo-Hexosulose, bis(amidinohydrazone), sulfate	EtOH	289(4.42)	44-3074-64
$C_8H_{24}N_6O_2P_2$			
Phosphonic diamide, P,P'-azobis-[N,N,N',N'-tetramethyl-	dioxan	300(3.4),561(1.11)	24-2273-65
$C_8H_{24}Si_3$			
Trisilane, octamethyl-	C_6H_{12}	215(3.96)	101-0163-65B
	C_6H_{12}	215.0(3.96)	101-0369-64B
$C_8H_{25}B_{10}Cl_9N_2O$			
Bis(tetramethylammonium)hydroxynonachlorodecaborate(2-)	MeCN	221(3.96),250(2.88)	35-3973-64
$C_8H_{26}B_{10}Cl_8N_2O_2$			
Bis(tetramethylammonium)dihydrooctachlorodecaborate(2-)	MeCN	216(3.57),248s(2.06)	35-3973-64
$C_8H_{26}B_{12}Br_{10}N_2O_2$			
Bis(tetramethylammonium)dihydroxydecabromododecaborate(2-)	MeCN	254(2.61)	35-3973-64
$C_8H_{30}ClN_6O_9Rh$			
Rhodium, bis(dimethylglyoximato)diammine-, chloride, pentahydrate, trans	H_2O	266(4.08),335(3.32)	39-1951-65

Compound	Solvent	$\lambda_{max}(\log \epsilon)$	Ref.
C_9F_{10}			
Styrene, heptafluoro-α-(trifluoro-methyl)-	n.s.g.	245(3.23)	78-1381-64
$C_9H_3F_9O$			
Anisole, 2,4,5-trifluoro-3,6-bis-(trifluoromethyl)-	EtOH	290.0(3.64)	39-2975-64
$C_9H_3F_9S$			
Sulfide, methyl 2,4,5-trifluoro-3,6-bis-(trifluoromethyl)phenyl	EtOH	236(3.86),305(3.79)	39-2975-64
$C_9H_3I_3O_3$			
Coumarin, 4-hydroxy-3,6,8-triiodo-	EtOH	232s(4.59),260s(3.90), 309(4.01)	7-0239-65
$C_9H_4ClNO_2$			
5,8-Isoquinolinedione, 6(or 7)-chloro-	MeCN	242(4.08),255(4.18), 262(4.17),300(3.64)	87-0801-64
$C_9H_4ClNO_2S$			
4,5-Thiazolidinedione, 2-(p-chloro-phenyl)-	dioxan	315(4.2),350s(3.6), 450(1.7)	24-2954-65
$C_9H_4Cl_4O_3$			
Benzoic acid, 2-acetyl-3,4,5,6-tetra-chloro-	EtOH	222(4.44),295(2.92), 305(2.95)	44-4293-65
$C_9H_4F_4O_4$			
Homoterephthalic acid, 2,3,5,6-tetrafluoro-	EtOH	272(3.18)	70-1798-65
$C_9H_4Fe_2O_5$			
Iron, (butatriene)pentacarbonyldi-	n.s.g.	346(3.73),448(3.36)	101-0007-65A
$C_9H_4I_2O_3$			
Coumarin, 4-hydroxy-3,6-diiodo-	EtOH	224(4.65),284s(3.72), 310s(4.00),323(4.08)	7-0239-65
Coumarin, 4-hydroxy-6,8-diiodo-	EtOH	232(4.52),255s(3.89), 289s(3.87),297s(3.95), 307(4.03),317s(3.95)	7-0239-65
$C_9H_4KN_3O_2$			
Malononitrile, (p-nitrophenyl)-, potassium derivative	50% EtOH-KOH	224(4.09),264(3.87), 475(4.30)	35-2174-64
$C_9H_4N_4$			
1,4-Cyclohexadiene, 3-diazo-6-di-cyanomethylene-	THF	290(3.54),456(4.43), 482(4.36)	35-2174-64
$C_9H_5BrN_2O_3$			
4-Bromo-1-hydroxy-2-nitroquinol-izinium hydroxide, inner salt	H_2O	207(4.55),268(4.33), 323(3.86),457(4.48)	39-3030-64
4H-Quinolizin-4-one, 3-bromo-1-nitro-	EtOH	258(3.99),380(3.80)	78-0945-65
C_9H_5BrO			
Indone, 3-bromo-	MeOH	235(4.52),241(4.50), 251s(3.95)	65-0448-64

Compound	Solvent	λ_{max}(log ε)	Ref.
$C_9H_5BrO_2S_2$			
[2,2'-Bithiophene]-5-carboxylic acid, 5'-bromo-	EtOH	239(3.74),335(4.25)	39-5134-65
$C_9H_5Br_2N$			
Cinnamonitrile, p,β-dibromo-	EtOH	282(4.32)	94-1446-64
$C_9H_5Br_2NO$			
4H-Quinolizin-4-one, 1,3-dibromo-	EtOH	255(4.42),415(4.39)	78-0945-65
$C_9H_5ClN_2O$			
2-Quinoxalinecarboxaldehyde, 6-chloro-	EtOH	240(4.46),321(3.84)	5-0146-65D
2-Quinoxalinecarboxaldehyde, 7-chloro-	EtOH	240(4.46),319(3.82)	5-0146-65D
$C_9H_5ClN_4O_4$			
3,4,5-Tricyano-N-methyl-pyridinium perchlorate	$HClO_4$	313(3.71)	89-0861-65
$C_9H_5ClO_2S_2$			
[2,2'-Bithiophene]-5-carboxylic acid, 5'-chloro-	EtOH	244(3.14),337(4.30)	39-5134-65
$C_9H_5Cl_3O_3$			
Phthalide, 3-hydroxy-3-(trichloromethyl)-	EtOH	275(2.97),283(2.92)	44-2784-65
$C_9H_5F_5O_2$			
Benzoic acid, pentafluoro-, ethyl ester	EtOH	270(3.07)	70-1798-65
$C_9H_5F_6NO$			
2(1H)-Pyridone, 1-[3,3,3-trifluoro-1-(trifluoromethyl)propenyl]-	EtOH	224(3.64),302(3.70)	44-2107-65
$C_9H_5IO_2S_2$			
[2,2'-Bithiophene]-5-carboxylic acid, 5-iodo-	EtOH	249(3.77),339(4.33)	39-5134-65
$C_9H_5IO_3$			
Coumarin, 4-hydroxy-3-iodo-	EtOH	238s(3.98),285s(3.77), 297s(3.93),316(4.05), 319s(4.03)	7-0239-65
Coumarin, 4-hydroxy-6-iodo-	EtOH	224(4.49),237s(4.22), 255s(3.55),263s(3.47), 275s(3.65),284s(3.80), 298(3.96),305s(3.93)	7-0239-65
$C_9H_5MnO_5$			
Manganese, tricarbonyl(carboxycyclopentadienyl)-	EtOH	338(3.10)	101-0269-65A
$C_9H_5NO_2$			
2-Benzofurancarbonitrile, 2,3-dihydro-3-oxo-	EtOH	209(4.03),232(4.10), 239(4.06),295s(3.97)	39-2361-65
5,8-Isoquinolinedione	MeCN	237(4.28),245(4.27), 254(4.13),325(3.52)	87-0801-64

Compound	Solvent	$\lambda_{max}(\log \epsilon)$	Ref.
$C_9H_5NO_2S$			
4,5-Thiazolinedione, 2-phenyl-	dioxan	300(4.3),360s(3.3), 450(1.6)	24-2954-65
	dioxan	209(4.1),304(4.3)	83-0124-65
$C_9H_5NO_3S_2$			
1,3-Dithiole, 2-oxo-4-(p-nitrophenyl)-	EtOH	243(4.03),335(4.14)	44-1708-64
$C_9H_5NO_5$			
2H-1,4-Benzoxazine-5-carboxylic acid, 3,4-dihydro-2,3-dioxo-	EtOH	318(3.79)	1-1051-65
$C_9H_5N_3O_2$			
1,2,4-Oxadiazolin-5-one, 3-cyano-4-phenyl-	EtOH	245(3.62)	33-0033-64
$C_9H_5N_3O_5$			
1-Hydroxy-2,4-dinitroquinolizinium hydroxide, inner salt	H_2O	208(4.25),247s(4.12), 254(3.98),420(4.33)	39-3030-64
4H-Quinolizin-4-one, 1,3-dinitro-	EtOH	255(3.89),340(3.86), 395(3.95),470(4.28)	78-1051-64
C_9H_6BrN			
Cinnamonitrile, β-bromo-, trans	EtOH	274(4.28)	94-1446-64
C_9H_6BrNO			
Isoxazole, 5-bromo-3-phenyl-	C_6H_{12}	210(4.19),240(4.17)	17-0203-65
	MeOH	210(4.14),241(4.21)	17-0203-65
4H-Quinolizin-4-one, 3-bromo-	EtOH	250(4.79),400(4.94)	78-0945-65
$C_9H_6BrN_3O_2$			
Pyrazole, 1-(p-bromophenyl)-4-nitro-	EtOH	238(4.13),297(4.08)	23-1605-64
$C_9H_6Br_2S_2$			
Thiophene, 2,2'-methylenebis[3-bromo-	EtOH	242(4.11)	44-2455-64
C_9H_6ClNO			
Isocarbostyril, 3-chloro-	EtOH	227(4.03),288(4.01), 328(3.70)	39-3856-64
Isoxazole, 5-chloro-3-phenyl-	C_6H_{12}	210(4.16),238(4.16)	17-0203-65
	MeOH	208(4.19),240(4.19)	17-0203-65
$C_9H_6ClNO_2$			
Isocarbostyril, 3-chloro-2-hydroxy-	EtOH	249(3.86),292(3.95), 328(3.71)	39-3856-64
$C_9H_6ClNO_4$			
Styrene, 2-chloro-4,5-(methylenedioxy)-β-nitro-	EtOH	252s(--),264(4.21), 316(4.06),368(4.26)	23-1901-64
$C_9H_6ClN_3O_4$			
1,3,4-Oxadiazole, 2-(chloromethyl)-5-[2-(5-nitro-2-furyl)vinyl]-	propylene glycol	266(4.08),364(4.37)	95-0255-64
$C_9H_6Cl_2N_2O_5$			
Anthranilic acid, N-(dichloronitroacetyl)-	MeOH	216(4.28),223(4.32), 256(4.05),301(3.81)	49-0415-64

Compound	Solvent	λ_{max}(log ε)	Ref.
$C_9H_6Cl_2OS_2$			
2,5-Cyclohexadien-1-one, 2,6-dichloro-4-(1,3-dithiolan-2-ylidene)-	MeOH	290(3.99),360(3.88) 428s(3.82),448(4.13), 476(4.28)	5-0037-65D
$C_9H_6Cl_2S_2$			
[2,2'-Bithiophene], 5-(dichloromethyl)-	EtOH	250(3.90),344(4.35)	39-7109-65
$C_9H_6Cl_4$			
Indan, 4,5,6,7-tetrachloro-	C_6H_6	282(2.7),293(2.6)	78-2217-64
$C_9H_6CrO_3$			
Chromium, (benzene)tricarbonyl-	n.s.g.	312.5(3.91)	28-2833-65A
$C_9H_6F_3N$			
2,4,6-Cycloheptatriene-1-carbonitrile, 1-(trifluoromethyl)-	C_6H_{12}	258(3.57)	35-1149-65
$C_9H_6N_2$			
Norcaradiene-7,7-dicarbonitrile	C_6H_{12}	271(3.47)	35-0652-65
$C_9H_6N_2OS$			
Thiazolo[5,4-e]benzoxazole, 2-methyl-	EtOH	240(4.5)	65-0278-64
$C_9H_6N_2O_2$			
2-Quinoxalinecarboxylic acid	MeOH	243(4.53),320(3.85)	78-2931-65
5,8-Quinoxalinedione, 6-methyl-	MeOH	264(4.03)	4-0171-64
$C_9H_6N_2O_3$			
1-Hydroxy-2-nitroquinolizinium hydroxide, inner salt	H_2O	212(4.21),229s(4.10), 257(4.03),429(4.24)	39-3030-64
1,3,4(2H)-Isoquinolinetrione, 4-oxime	EtOH	246(4.29),282s(3.81), 325s(3.56)	39-1423-64
5,8-Quinolinedione, 7-amino-6-hydroxy-	EtOH	234(4.21),280(4.03), 330(3.76),445(3.31)	95-0985-65
4H-Quinolizin-4-one, 3-nitro-	EtOH	265(3.96),340(3.69), 440(4.34)	78-1051-64
Sydnone, 4-formyl-3-phenyl-	EtOH	240(4.06),321(3.99)	44-2044-64
$C_9H_6N_2O_4$			
Indole-2,3-dione, 1-methyl-5-nitro-	EtOH	223s(4.06),249s(3.65), 285s(3.65),331(3.99)	44-3457-65
Sydnone, 3-[(3,4-methylenedioxy)phenyl]-	EtOH	308(3.98)	87-0531-65
$C_9H_6N_2O_6$			
1-Indanone, 7-hydroxy-4,6-dinitro-	EtOH	241(4.32),323(3.74)	39-2816-64
$C_9H_6N_4$			
2,3,5-Pyridinetricarbonitrile, 1,2-dihydro-1-methyl-	MeOH	246(3.92),350(3.95)	89-0861-65
3,4,5-Pyridinetricarbonitrile, 1,4-dihydro-1-methyl-	MeOH	359(3.87)	89-0861-65
$C_9H_6N_4O$			
3,4,5-Pyridinetricarbonitrile, 1,2-dihydro-2-hydroxy-1-methyl-	H_2O	261(3.76),384(3.93)	89-0861-65

Compound	Solvent	$\lambda_{max}(\log \epsilon)$	Ref.
$C_9H_6N_4O_4$			
Imidazole, 1-(2,4-dinitrophenyl)-	MeOH	259s(4.27)	65-0632-64
	HBr	250s(4.23)	65-0632-64
Pyrazole, 4-nitro-1-(p-nitrophenyl)-	EtOH	214(4.22),305(4.36)	23-1605-64
$C_9H_6N_4O_6$			
Indole, 2-methyl-3,5,6-trinitro-	EtOH	230(4.12),244(4.12), 317(4.05),386(3.92)	44-3457-65
$C_9H_6N_4S$			
3H-Imidazo[4,5-b]pyridine, 2-(4-thiazolyl)-	pH 1	321(4.47)	44-0259-65
$C_9H_6N_6S_3$			
Benzo[1,2-d:3,4-d':5,6-d'']tris-thiazole, 2,5,8-triamino-	EtOH	223(4.37),254(4.55), 295s(5.63),340s(3.64)	7-0080-64
C_9H_6O			
2,4-Nonadiene-6,8-diynal	Et_2O	233(4.06),243(4.28), 293s(4.44),307(4.63), 326(4.62)	24-1846-64
C_9H_6OS			
Benzothiophene-2-carboxaldehyde	EtOH	233(4.13),246s(3.82), 294s(4.12),299(4.26), 324s(3.63)	17-0151-65
	H_2SO_4	253(4.04),281(3.20), 367(4.61),452(3.35)	17-0151-65
Thiacoumarin	hexane	338(3.4)	73-3016-65
Thiocoumarin	hexane	373(4.1)	73-3016-65
C_9H_6OSe			
Benzoselenophene-2-carboxaldehyde	EtOH	241(4.13),262s(3.70), 303(4.27),349(3.52)	17-0151-65
	H_2SO_4	227(3.84),242s(3.87), 258(4.12),372(4.6), 442(3.33)	17-0151-65
$C_9H_6O_2$			
Coumarin	hexane	312(3.7)	73-3016-65
1,3-Indandione	MeOH	222(4.42),252(4.34), 260s(4.25)	24-1482-64
Phthalide, 3-methylene-	hexane	235(4.17),253(4.20), 299(3.54),309(3.43)	39-3811-65
$C_9H_6O_2S$			
Benzo[b]thiophene, 5,6-(methylenedioxy)-	EtOH	237(4.38),264(3.95), 273(3.96),303(3.53), 307(3.57),317(3.59)	4-0100-65
2,4-Pentadienoic acid, 4-hydroxy-5-(2-thienyl)-, γ-lactone, cis	Et_2O	290(3.74),361(4.35)	24-2236-65
trans	Et_2O	230(3.96),275(3.77), 359(4.42)	24-2236-65
$C_9H_6O_3$			
Coumarin, 4-hydroxy-	EtOH	232s(3.94),279s(3.93), 288(4.05),298(4.04)	7-0239-65
	MeOH-HCl	268(4.07),279(4.09), 303(3.92),317s(3.74)	49-0077-65
	MeOH-NaOH	287(4.14),298s(4.10)	49-0077-65

Compound	Solvent	$\lambda_{max}(\log \epsilon)$	Ref.
1,4-Isochromandione	EtOH	208(4.7),250(3.8)	54-0334-65
$C_9H_6O_4$			
Phthalide, 4,5-(methylenedioxy)-	EtOH	200(4.63),204s(4.61), 224(4.74),270(4.05), 294(3.97)	73-3479-65
$C_9H_6O_5$			
4-Cyclohexene-1,2,3-tricarboxylic acid, 6-hydroxy-, cyclic 1,2-anhydride, δ-lactone	MeCN	228(2.66)	35-1925-65
Glutaconic acid, 4-(1,3-dihydroxy-2-butenylidene)-3-hydroxy-, di-γ-lactone	EtOH	270(3.81),330(3.76)	35-3004-65
$C_9H_6S_2$			
Cyclopenta[1,2-b:4,3-b']dithiophene	C_6H_{12}	218(4.50),223(4.51), 259(3.38),264(3.34), 269(3.34),273(3.24), 279(3.18),298(3.04)	44-2455-64
Thiothiacoumarin	hexane	416(4.0)	73-3016-65
$C_9H_6S_3$			
3H-1,2-Dithiole-3-thione, 4-phenyl-	EtOH	238(4.14),281(3.68), 318(3.65),420(3.91)	24-0654-64
3H-1,2-Dithiole-3-thione, 5-phenyl-	EtOH	229(4.11),248s(4.01), 272(4.03),316(4.28), 433(4.02)	24-0654-64
$C_9H_7BrClNO$			
Acrylamide, 2-bromo-3-(2-chlorophenyl)-	MeOH	264(4.114)	23-2426-65
$C_9H_7BrN_2O$			
Cyclohepta[b]pyrrol-2(1H)-one, 1-amino-	EtOH	278(4.49),420(3.92)	94-0450-65
$C_9H_7BrN_4O$			
2H-1,2,3-Triazole-4-carboxamide, 2-(p-bromophenyl)-	EtOH	280(4.35)	39-2306-64
C_9H_7BrO			
Acrylophenone, p-bromo-	EtOH	310(2.377)	65-0190-64
$C_9H_7BrO_2$			
Bicyclo[4.2.0]octa-1,3,5-trien-7-one, 2-bromo-3-methoxy-	EtOH	235(4.32),275(4.05), 290(4.00)	39-5343-64
Styrene, β-bromo-3,4-(methylenedioxy)-	EtOH	276(4.26),308(4.17)	23-1901-64
$C_9H_7Br_2N$			
Indole, 2,6-dibromo-3-methyl-	EtOH	230(4.58),286(3.92), 297s(3.86)	44-1206-64
$C_9H_7Br_2NO$			
1-Bromo-2-hydroxyquinolizinium bromide	H_2O	236(4.49),305(4.13), 352(3.89)	39-2760-64
4-Bromo-3-hydroxyquinolizinium bromide	H_2O	246(4.40),265(4.30), 272(4.30),343(4.02)	44-0526-65

Compound	Solvent	$\lambda_{max}(\log \epsilon)$	Ref.
$C_9H_7Br_2N_3O$			
1-Diazonium-2-hydroxyquinolizinium dibromide	EtOH	231(4.16),248s(3.99), 278(3.70),360(3.75)	39-2763-64
C_9H_7Cl			
1-Propyne, 1-(p-chlorophenyl)-	n.s.g.	244(4.42),255(4.42)	22-1525-65
$C_9H_7ClN_2$			
Quinoxaline, 6-chloro-2-methyl-	EtOH	237(4.42),320(3.75)	5-0146-65D
Quinoxaline, 7-chloro-2-methyl-	EtOH	238(4.40),318(3.81)	5-0146-65D
$C_9H_7ClN_2O_5$			
Anthranilic acid, N-(chloronitroacetyl)-	MeOH	210(4.30),222(4.36), 252(4.08),301(3.78)	49-0415-64
C_9H_7ClO			
Acrylophenone, β-chloro-	EtOH	203(4.02),260(4.22)	44-0385-64
Acrylophenone, p-chloro-	EtOH	380(2.380)	65-0190-64
Cinnamaldehyde, α-chloro-	EtOH	221(3.85),227(3.84), 292(4.33)	28-1015-65B
2-Propyn-1-ol, 3-(p-chlorophenyl)-	Et_2O	247(4.31),257(4.27)	24-2118-64
$C_9H_7ClO_2$			
3(2H)-Benzofuranone, 7-chloro-5-methyl-	EtOH	252(3.92),258(3.96), 339(3.74)	39-4939-65
Cinnamic acid, o-chloro-	EtOH	270(4.27)	35-5548-64
Cinnamic acid, p-chloro-	EtOH	277(4.31)	35-5548-64
$C_9H_7ClO_4S$			
Benzothiopyrylium perchlorate	70% $HClO_4$	258(4.54),334(3.72), 386(3.54)	17-0021-65
	HOAC 1% $HClO_4$	384(3.5)	73-3016-65
2-Benzothiopyrylium perchlorate	HOAc-1% $HClO_4$	384(3.5)	73-3016-65
$C_9H_7ClO_4S_2$			
4-Phenyl-1,2-dithiolium perchlorate	pH 1	345(3.1)	73-3016-65
4-Phenyl-1,3-dithiolium perchlorate	EtOH-$HClO_4$	306(3.9)	73-3016-65
$C_9H_7ClO_4Se$			
Benzoseleninium perchlorate	70% $HClO_4$	260(4.51),348(3.76), 410(3.67)	17-0021-65
$C_9H_7ClO_5$			
Benzopyrylium perchlorate	70% $HClO_4$	242(4.34),324(4.14), 350s(3.50)	17-0021-65
$C_9H_7ClO_5S_2$			
4-Hydroxy-2-phenyl-1,3-dithiolium perchlorate	70% $HClO_4$	244(3.51),275(3.47), 367(4.18)	44-1708-64
$C_9H_7Cl_2NO_2$			
2-Indolinone, 5,7-dichloro-3-methoxy-	MeOH	214(4.40),258(3.99), 308(3.30)	44-3610-65
$C_9H_7FN_4O$			
2H-1,2,3-Triazole-4-carboxamide, 2-(p-fluorophenyl)-	EtOH	275(4.25)	39-2306-64

Compound	Solvent	λ_{max}(log ε)	Ref.
C_9H_7FO			
Cinnamaldehyde, α-fluoro-	C_6H_{12}	219(4.15),225s(4.12), 282(4.44)	22-2258-64
1-Indanone, 5-fluoro-	EtOH	244(4.05)	44-2165-65
$C_9H_7F_2N_3O$			
Benzimidazole, 4-acetamido-5,7-difluoro-	pH 1	258(3.81)	87-0737-65
	pH 7	249(3.89)	87-0737-65
	pH 13	267(3.91)	87-0737-65
$C_9H_7F_5N_2O_2S_2$			
Hydrazine, 1-(pentafluorothiopropionyl)-2-(phenylsulfonyl)-	iso-PrOH	267(3.79),273(3.79)	44-3739-65
ammonium salt	MeCN	273(3.96),283(3.97)	44-3739-65
C_9H_7I			
1,2-Propadiene, 1-iodo-3-phenyl-	EtOH	282(4.07)	39-4348-65
$C_9H_7KN_2O_5$			
Ketone, methyl 2,4-dinitrobenzyl, potassium derivative	H_2O	487(2.76)	25-2065-65
	MeOH	485(2.62)	25-2065-65
	EtOH	490(3.20)	25-2065-65
	iso-PrOH	495(3.60)	25-2065-65
	acetone	493(3.75)	25-2065-65
	methyl cellosolve	494(4.18)	25-2065-65
	THF	493(3.23)	25-2065-65
	DMF	495(4.45)	25-2065-65
	pyridine	500(3.60)	25-2065-65
	Me_2SO	498(4.30)	25-2065-65
C_9H_7N			
Pyridine, 4-(1-buten-3-ynyl)-	n.s.g.	276(4.18)	5-0062-65B
Quinoline	EtOH	230f(4.4),280(3.4), 315f(3.3)	61-0448-65
C_9H_7NO			
Acrylonitrile, 3-phenoxy-, cis	n.s.g.	240(4.19)	44-1800-64
Carbostyril	EtOH	229(4.54),268(3.86), 329(3.81)	1-1120-65
	EtOH	229(4.54),268(3.85), 328(3.80)	1-1389-64
	dioxan	232(3.92),276(3.28), 331(3.77)	1-1389-64
1-Hydroxyquinolizinium hydroxide, inner salt	EtOH	266(3.91),374s(3.91), 400(3.94)	44-0526-65
	$CHCl_3$	250(3.85),278(3.85), 382(3.85),431(3.85)	44-0526-65
3-Hydroxyquinolizinium hydroxide, inner salt	$CHCl_3$	250(4.38),269(4.36), 278(4.36),362(4.18)	44-0526-65
2-Indolinone, 3-methylene-	EtOH	248(4.37),254(4.37), 289(3.44)	44-2431-64
4-Isoquinolinol	EtOH	295(3.62),330(3.76)	28-1339-65B
Quinoline, 1-oxide	C_6H_{12}	351(4.05)	1-1120-65
	heptane	225(4.4),248(4.1), 350f(3.8)	61-0448-65
	N HCl	317(3.83)	1-1120-65
	H_2O	313(3.89)	1-1120-65
	EtOH	327(4.00)	1-1120-65

Compound	Solvent	$\lambda_{max}(\log \epsilon)$	Ref.
6-Quinolinol	pH 1	246(4.45),313(3.64), 341(3.57)	44-2860-64
	pH 7	226(4.49),273(3.345), 326(3.56)	44-2860-64
	pH 13	244(4.57),358(3.66)	44-2860-64
8-Quinolinol	C_6H_{12}	245(4.6),320(3.5)	101-0159-65B
2H-Quinolizin-2-one	H_2O	226(4.50),299(4.00), 325s(3.75)	39-2760-64
$C_9H_7NO_2$			
Acetonitrile, [3,4-(methylenedioxy)-phenyl]-	EtOH	286(--)	95-0857-65
3-Isoxazolol, 5-phenyl-	C_6H_{12}	262s(4.29)	78-2835-64
	0.01 N NaOH	261(4.30)	78-2835-64
	5% EtOH-20N H_2SO_4	277(4.30)	78-2835-64
	5% EtOH 0.01 N H_2SO_4	261(4.30)	78-2835-64
4H-Quinolizin-4-one, 3-hydroxy-	EtOH	222(4.25),253(4.15), 373(4.04)	44-0526-65
$C_9H_7NO_3$			
1,2-Benzisoxazole-3-acetic acid	EtOH	239(3.92),272(3.58), 279(3.59)	4-0385-65
4,1-Benzoxazepine-2,5(1H,3H)-dione	iso-PrOH	216(4.41),243(3.95), 302(3.49)	44-0582-64
Chroman, 4-hydroxyimino-2-oxo-	EtOH	238(3.86),244(3.67), 283(3.40)	4-0385-65
Inden-1-ol, 2-nitro-	EtOH	243(3.86),340(3.97)	44-3180-64
	H_2O	244(--),355(--)	44-3180-64
2H-Isoxazolo[2,3-a]pyridin-2-one, 3-acetyl-	EtOH	250(4.13),281(4.20), 349(3.84)	94-0595-64
Phthalimide, 4-methoxy-	EtOH	234(4.54),280(3.35), 321(3.47)	7-0128-64
Phthalimide, 5-methoxy-	EtOH	219(4.58),240s(3.80), 334(3.69)	7-0128-64
$C_9H_7NO_4$			
Cinnamic acid, o-nitro-	EtOH	307s(3.50)	35-5548-64
Cinnamic acid, p-nitro-	EtOH	296(4.20)	35-5548-64
1-Indanone, 7-hydroxy-4-nitro-	EtOH	232(4.16),273(3.63), 319(3.87)	39-2816-64
1-Indanone, 7-hydroxy-6-nitro-	EtOH	242(4.13),305(3.30), 334(3.53)	39-2816-64
Indole-2-carboxylic acid, 5,6-dihydroxy-	EtOH	209(4.35),320(4.23)	4-0387-65
Phthalide, 3-(nitromethyl)-	MeOH	273(3.19),281(3.18)	23-0190-65
	MeOH	273(3.18),280(3.18)	44-3180-64
Styrene, 3,4-(methylenedioxy)-β-nitro-	EtOH	245s(--),260(4.12), 310s(--),364(4.22)	23-1901-64
$C_9H_7NO_5$			
DL-Alanine, β-(6-carboxy-α-pyron-5-yl)-, lactam	pH 1.4	317(3.93)	88-3431-64
	pH 5.0	319(3.91)	88-3431-64
Benzoic acid, 2-acetyl-3-nitro-	EtOH	216(4.28),247(3.72)	44-4293-65
C_9H_7NS			
Isothiazole, 4-phenyl-	EtOH	210(4.21),242(4.03), 269(4.03)	39-0032-65

Compound	Solvent	$\lambda_{max}(\log \epsilon)$	Ref.
Isothiocyanic acid, p-vinylphenyl ester	MeCN	214(4.43),221(4.38), 238(4.19),274s(4.37), 281s(4.43),286(4.52), 299(4.46)	44-1926-65
$C_9H_7NS_2$			
1,3-Dithiol-2-one imine, N-phenyl-	EtOH-HI	300(3.91),485(2.76)	24-1298-64
$C_9H_7N_2NaO_2$			
Imidazo[1,2-a]pyridine-3-carboxylic acid, 2-methyl-, sodium salt	EtOH	237(4.64),243(4.7), 292(4.05)	44-2403-65
$C_9H_7N_3O$			
2-Indolinone, 3-diazo-1-methyl-	EtOH	262(4.39),265(4.38), 301(3.98)	44-3577-64
$C_9H_7N_3O_2$			
Quinoxaline, 2-methyl-6-nitro-	MeOH	255(6.04),297(5.67)	73-3102-65
$C_9H_7N_3O_4$			
Indole, 2-methyl-3,4-dinitro-	EtOH	233(4.08),300s(3.94), 338(4.00)	44-3457-65
Indole, 2-methyl-3,5-dinitro-	EtOH	251(4.37),314(4.04), 347(4.04)	44-3457-65
Indole, 2-methyl-3,6-dinitro-	EtOH	225(4.01),291s(4.14), 306(4.15),341(4.17)	44-3457-65
	EtOH	207(4.23),222s(4.00), 308(4.15),347(4.16)	78-1397-64
1,3,4-Oxadiazole, 2-methyl-5-[2-(5-nitro-2-furyl)vinyl]-	propylene glycol	268(4.00),365(4.28)	95-0225-64
1,3,4-Oxadiazole, 2-[1-methyl-2-(5-nitro-2-furyl)vinyl]-	propylene glycol	256(4.15),368(4.30)	95-0225-64
$C_9H_7N_3O_4S$			
1,3,4-Oxadiazole, 2-(methylthio)-5-[2-(5-nitro-2-furyl)vinyl]-	propylene glycol	227(4.10),296(4.08), 375(4.40)	95-0225-64
$C_9H_7N_3O_7$			
Anthranilic acid, N-acetyl-4,5-dinitro-	EtOH	234(4.20),326(3.95)	44-3457-65
$C_9H_7N_3S_2$			
Thiazolo[5,4-f]benzothiazole, 6-amino-2-methyl-	EtOH	240(4.47),244s(4.38)	7-0080-64
$C_9H_7N_5O_2S$			
1,2,4-Triazine, 3-amino-6-[2-(5-nitro-2-thienyl)vinyl]-	propylene glycol	287(4.12),410(4.33) 287(4.12),410(4.33)	95-0016-64 95-0109-64
$C_9H_7N_5O_3$			
1,2,4-Triazine, 3-amino-6-[2-(5-nitro-2-furyl)vinyl]-	propylene glycol	290(4.20),405(4.38)	95-0109-64
1,2,3-Triazole-4-carboxamide, 2-(p-nitrophenyl)-	EtOH	310(4.34)	39-2306-64
$C_9H_7N_5O_4$			
1,2,4-Triazin-5(4H)-one, 3-amino-6-[2-(5-nitro-2-furyl)vinyl]-	propylene glycol	290(4.05),390(4.46)	95-0001-64 95-0009-64
hydrochloride	propylene glycol	290(4.05),390(4.46)	95-0001-64

Compound	Solvent	λ_{max}(log ϵ)	Ref.
C_9H_8			
Indene	EtOH	249(4.0)	78-0231-64
C_9H_8BrN			
Indole, 2-bromo-3-methyl-	EtOH	223(4.55),277s(3.87), 282(3.89),291(3.82)	44-1206-64
Indole, 3-bromo-2-methyl-	EtOH	223(4.53),275s(3.84), 281(3.88),289(3.82)	44-1206-64
C_9H_8BrNO			
Cinnamamide, β-bromo-, trans	EtOH	262(4.13)	94-1446-64
2-Hydroxyquinolizinium bromide	H_2O	225(4.55),298(4.05), 324s(3.86)	39-2760-64
2-Indolinone, 3-bromo-3-methyl-	EtOH	217(4.23),229s(4.13), 310(3.97)	44-2431-64
2-Indolinone, 5-bromo-3-methyl-	EtOH	207(4.40),254(4.09), 288s(3.11)	44-2431-64
2-Indolinone, 6-bromo-3-methyl-	EtOH	213(4.57),254(3.85), 285(3.29),293s(3.19)	44-2431-64
$C_9H_8BrNO_2$			
2-Bromo-1-hydroxyquinolizinium hydroxide	H_2O	216(4.47),380(4.06)	39-2763-64
3,4-Dihydroxyquinolizinium bromide	EtOH	222(4.24),255(4.14), 373(4.03)	44-0526-65
2-Indolinone, 5-bromo-3-hydroxy-3-methyl-	EtOH	209(4.47),258(4.04), 296(3.15)	44-2431-64
2-Indolinone, 5-bromo-3-methoxy-	MeOH	259(4.03),302(3.17)	44-2431-64
$C_9H_8BrN_3$			
Pyrazole, 5-amino-4-bromo-3-phenyl-	EtOH	252(4.06)	95-1113-64
$C_9H_8BrN_3O_2$			
1,2,4-Triazolidine-3,5-dione, 1-(p-bromophenyl)-2-methyl-	EtOH	210(4.10),250(4.14)	22-0500-64
	NaOH	216(4.27),256(4.08)	22-0500-64
$C_9H_8BrN_5S$			
s-Triazine-2(1H)-thione, 1-(p-bromophenyl)hexahydro-4,6-diimino-	50% EtOH	218(4.51),286(4.33)	39-6296-65
$C_9H_8Br_2O_4$			
2,4-Hexadienedioic acid, 4-(dibromomethyl)-2-hydroxy-, δ-lactone, ethyl ester	50% EtOH	306(3.86)	88-3431-64
$C_9H_8ClNO_2$			
Allyl chloride, α-(p-nitrophenyl)-	MeOH	269(4.03)	78-1057-64
Benzene, 1-(1-chloropropenyl)-4-nitro-	MeOH	305(4.15)	78-1057-64
Benzene, 1-(3-chloropropenyl)-4-nitro-	MeOH	304(4.22)	78-1057-64
$C_9H_8ClNO_5$			
Piperonyl alcohol, 6-chloro-α-(nitromethyl)-	EtOH	244(3.93),265(3.93)	23-1901-64
$C_9H_8ClN_5$			
s-Triazine, 2,4-diamino-6-(m-chlorophenyl)-	MeOH	243(4.29)	4-0340-65
s-Triazine, 2,4-diamino-6-(p-chlorophenyl)-	MeOH	254(4.35)	4-0340-65

Compound	Solvent	λ max(log ε)	Ref.
$C_9H_8Cl_2$			
Bicyclo[6.1.0]nona-2,4,6-triene, 9,9-dichloro-	isooctane	243(3.49)	35-1941-65
	EtOH	243(3.46)	35-5194-64
$C_9H_8Cl_2S$			
Sulfide, 2,2-dichlorocyclopropyl phenyl	EtOH	240(3.88),250(3.86)	44-2211-64
$C_9H_8Cl_4O_2$			
1,3-Dioxolane-2-spirocyclopenta-2',4'-diene, 2',3',4',5'-tetrachloro-4,5-dimethyl-	iso-PrOH	312(3.27)	39-2305-65
$C_9H_8F_3N_3O_5S_2$			
1,2,4-Benzothiadiazine-7-sulfonamide, 3-(hydroxymethyl)-6-(trifluoromethyl)-, 1,1-dioxide	EtOH	276(4.03)	95-0095-65
$C_9H_8F_7NO$			
2-Cyclohexen-1-one, heptafluoro-3-(isopropylamino)-	EtOH	313.0(4.50)	39-5748-64
$C_9H_8HgO_4S$			
Benzoic acid, p-[[(carboxmethyl)thio]-mercuri]-	pH 9	259(4.3)	70-0111-64
$C_9H_8NNaO_4$			
1,3-Indandiol, 2-aci-nitro-, sodium derivative	H_2O	256(3.93),269(3.90), 275(3.89)	44-3180-64
	pH 12	256(4.09),269(4.05), 275(4.04)	44-3180-64
	EtOH-NaOH	261(3.99),269(4.00), 275(4.03)	44-3180-64
$C_9H_8N_2$			
9-Azabicyclo[4.2.1]nona-2,4,7-triene-9-carbonitrile	MeCN	255(3.65)	35-5512-65
Cinnamonitrile, β-amino-	EtOH	290(4.06)	94-1446-64
2,4,6-Cyclooctatrien-$\Delta^{1,N}$-carbamonitrile	MeCN	243(4.20),331(3.84)	35-5512-65
Imidazole, 1-phenyl-	octane	225s(3.79),243(3.84), 266s(3.27)	65-0632-64
	pH 1	218(3.94),230s(3.85), 261s(3.08)	65-0632-64
	MeOH	236(3.93),264s(3.27)	65-0632-64
2-Norbornene-2,3-dicarbonitrile	EtOH	249(4.00)	77-0391-65
Quinzoline, 4-methyl-	pH 0.3	234(4.52),270(3.47), 279(3.45),323(3.34)	39-5360-65
	pH 7.0	223(4.62),270(3.45), 305(3.45),314s(3.41)	39-5360-65
$C_9H_8N_2O$			
Benzimidazole, 2-acetyl-	pH 2	234(3.64),267(3.77), 275(3.83),300(3.76)	44-0476-64
	pH 12	237(3.74),320(3.97)	44-0476-64
	EtOH	235(3.65),300(3.92)	44-0476-64
Cinnoline, 4-methoxy-	EtOH	226(4.60),284s(3.72), 292(3.75),317(3.65)	39-5391-65

Compound	Solvent	λ max(log ε)	Ref.
4(1H)-Cinnolinone, 1-methyl-	pH -2.34	237(4.50),254s(3.95), 298s(3.31),308(3.52), 342(3.85)	39-2260-65
	pH 5.0	209(4.44),233(4.16), 240(4.18),256(3.92), 263s(3.85),286(3.33), 344(4.13),350s(4.11), 355s(4.07),359(4.03)	39-2260-65
	EtOH	209(4.47),234(4.01), 240(4.01),254(3.84), 263s(3.72),284(3.32), 296(3.36),341(4.10), 360(4.04)	39-5391-65
Cyclohepta[b]pyrrol-2(1H)-one, 1-amino-	EtOH	267(4.50),410(3.86)	94-0450-65
4-Hydroxy-2-methylcinnolinium hydroxide, anhydro base	EtOH	212(4.51),221s(4.21), 254(3.91),267s(3.66), 279s(3.48),352(4.09), 369(4.11)	39-5391-65
3H-Indol-3-one, 2-methyl-, oxime	EtOH	236s(4.20),240(4.25), 253(4.29),277(3.55), 314(3.68),355s(3.37)	44-3457-65
Isoxazole, 5-amino-3-phenyl-	C_6H_{12}	204(4.38),228(4.25), 265s(3.4)	17-0255-65
	MeOH	206(4.32),231(4.31), 270(3.69)	17-0255-65
	MeOH-HCl	244(3.96),290(4.24)	17-0255-65
Isoxazole, 5-amino-4-phenyl-	C_6H_{12}	205(4.10),218s(3.8), 258(4.14)	17-0255-65
	MeOH	204(4.12),218s(3.8), 262(4.24)	17-0255-65
	MeOH-HCl	247(3.92),275(3.82)	17-0255-65
1,2,4-Oxadiazole, 3-methyl-5-phenyl-	Et_2O	252(4.27)	33-0942-64
1,2,4-Oxadiazole, 5-methyl-3-phenyl-	Et_2O	238(4.12)	33-0942-64
$C_9H_8N_2O_2$			
Imidazo[1,2-a]pyridine-3-carboxylic acid, 2-methyl-, sodium salt	EtOH	237(4.64),243(4.7), 292(4.05)	44-2403-65
Indole, 1-methyl-3-nitro-	EtOH	210(4.1),260(4.0), 277s(3.8),359(4.0)	78-1397-64
Indole, 2-methyl-3-nitro-	EtOH	242s(3.97),248(4.00), 253(4.01),269(3.96), 276(3.99),352(4.05)	44-3457-65
	EtOH	210(4.1),255(3.8), 271(3.9),279(3.9), 358(4.0)	78-1397-64
Indole, 2-methyl-4-nitro-	EtOH	240(4.09),248s(4.05), 392(3.85)	44-3457-65
Indole, 2-methyl-5-nitro-	EtOH	204(4.29),264(4.31), 329(3.93)	35-3796-64
	H_2SO_4	212(4.02),284(4.09)	35-3796-64
Indole, 2-methyl-6-nitro-	EtOH	247(4.05),266s(3.76), 336s(3.89),376(4.00)	44-3457-65
Isosydnone, 4-methyl-5-phenyl-	EtOH	294(4.05)	23-1607-65
1,2,4-Oxadiazole, 5-methoxy-3-phenyl-	33N H_2SO_4	251(4.05)	78-1681-65
	H_2O	237(4.06)	78-1681-65
1,3,4-Oxadiazol-2-one, 2,3-dihydro-3-methyl-5-phenyl-	EtOH	266(3.97)	23-1607-65

Compound	Solvent	$\lambda_{max}(\log \epsilon)$	Ref.
4(3H)-Quinazolinone, 2-(hydroxymethyl)-	iso-PrOH	225(4.42),230s(4.36), 265(3.90),271(3.86), 304(3.65),315(3.54)	44-0582-64
Succinimide, N-2-pyridyl-	H_2O	258.0(3.51)	39-2763-64
$C_9H_8N_2O_2S$			
2-Thiazoline-4-carboxylic acid, 2-(4-pyridyl)-	EtOH	275(3.59)	44-2344-65
$C_9H_8N_2O_3$			
2,4,5-Benzimidazolinetrione, 1,3-dimethyl-	$CHCl_3$	595(3.44)	65-3788-64
1-Indanone, 2-nitro-, oxime	EtOH	256(4.11),284(3.62), 292(3.73),303(3.74)	95-0399-64
1,2,4-Oxadiazolinone, 3-methoxy-5-phenyl-	33N H_2SO_4	277(4.29)	78-1681-65
	H_2O	255(4.21)	78-1681-65
5,6-Phthalazinediol, 7-methoxy-	EtOH	220(4.16),256(4.58), 385(3.09)	39-4672-65
$C_9H_8N_2O_4$			
2-Indolinone, 3-methoxy-5-nitro-	MeOH	322(4.19)	44-3610-65
Uracil, 2,2'-anhydro-1-(3,5-epoxy-β-D-lyxosyl)-	H_2O	223(3.90),250(3.88)	44-0476-65
	EtOH	224(3.99),250(3.86)	44-0476-65
$C_9H_8N_2O_4S$			
Nicotinic acid, 5-cyano-1,2,3,6-tetrahydro-4-(methylthio)-2,6-dioxo-, methyl ester	MeOH	237(4.58),265(4.08), 291(4.46)	95-0387-65
$C_9H_8N_2O_5$			
Hydroxylamine, O-acetyl-N-(p-nitrobenzoyl)-	MeOH	261(4.07)	73-0940-65
	50% MeOH-HCl	264(4.13)	73-0940-65
	50% MeOH-borate	295(3.93)	73-0940-65
	50% MeOH-N NaOH	241s(3.91),260s(3.94), 346(3.73)	73-0940-65
$C_9H_8N_2S$			
4(1H)-Cinnolinethione, 1-methyl-	pH -3.29	211(4.11),239(4.36), 250s(4.23),300(3.42), 336(3.82),372(4.08), 386s(3.99)	39-2260-65
	pH 1.0	223(4.57),252(3.83), 264s(3.76),275(3.69), 308(3.30),427(4.36)	39-2260-65
	pH 9.5	221(4.52),247(3.82), 275(3.29),380(4.13)	39-2260-65
$C_9H_8N_3NaO_7$			
2-Propanone, (2,4,6-trinitrocyclohexadienyl)-, sodium derivative	EtOH	463(4.44),562(4.14)	25-2065-65
$C_9H_8N_4$			
1,2,4-Triazine, 3-amino-5-phenyl-	EtOH	272(4.04),333(3.90)	39-4157-64
1,2,4-Triazine, 3-amino-6-phenyl-	EtOH	266(4.43),343(3.62)	39-4157-64

Compound	Solvent	$\lambda_{max}(\log \epsilon)$	Ref.
$C_9H_8N_4O$			
Isoquinoline, 1,3-diamino-4-nitroso-	EtOH	228(4.20),268(4.15), 315(3.95),388(4.08)	39-1423-64
Pyrazolo[1,5-a]pyrimidine-6-carbonitrile, 4,7-dihydro-2,3-dimethyl-7-oxo-	EtOH	219(4.37),277(4.01), 314(3.92)	95-0442-65
Pyrazolo[1,5-a]pyrimidine-6-carbonitrile, 4,7-dihydro-2,5-dimethyl-7-oxo-	EtOH	219(4.49),265(4.03), 303(3.98)	95-0442-65
Pyrazolo[1,5-a]pyrimidine-6-carbonitrile, 4,7-dihydro-3,5-dimethyl-7-oxo-	EtOH	218(4.52),279(4.09), 309(4.12)	95-0442-65
Pyrazolo[1,5-a]pyrimidine-6-carbonitrile, 2-ethyl-4,7-dihydro-7-oxo-	EtOH	219(4.45),269(4.01), 308(3.94)	95-0442-65
Pyrazolo[1,5-a]pyrimidine-6-carbonitrile, 5-ethyl-4,7-dihydro-7-oxo-	EtOH	217(4.43),274(3.98), 304(4.09)	95-0442-65
$C_9H_8N_4O_2$			
Pyrazole, 1-(p-aminophenyl)-4-nitro-	EtOH	262.5(4.14)	23-1605-64
$C_9H_8N_4O_3$			
Pteridine-6-carboxylic acid, 7-hydroxy-, ethyl ester	pH 3.3	217(4.24),255(3.54), 321(3.97)	39-3357-64
	pH 7.7	231(4.32),262(3.78), 343(3.97)	39-3357-64
$C_9H_8N_4O_3S$			
1,2,4-Thiadiazole, 5-(methylamino)-3-[2-(5-nitro-2-furyl)vinyl]-	propylene glycol	291(4.08),376(4.32)	95-0948-65
1,2,4-Triazole, 3-(methylthio)-5-(5-nitro-2-furyl)vinyl-	propylene glycol	290(4.01),390(4.27)	95-0566-64
$C_9H_8N_4O_4$			
Bicarbamimide, 2-methyl-3-(p-nitrophenyl)-	NaOH	219(4.11),340(3.99)	22-0500-64
	EtOH	224(4.05),315(4.00)	22-0500-64
1,3,4-Oxadiazole, 2-amino-5-[1-methyl-2-(5-nitro-2-furyl)vinyl]-	propylene glycol	300(4.08),390(4.26)	95-0212-64
1,3,4-Oxadiazole, 2-(methylamino)-5-(5-nitro-2-furyl)vinyl-	propylene glycol	228(4.10),308(4.06), 395(4.35)	95-0219-64
$C_9H_8N_4O_5$			
Pyruvaldehyde, 1-(2,4-dinitrophenylhydrazone)	EtOH	230(4.03),270(3.92), 357(4.39)	24-0725-64
$C_9H_8N_4O_5S$			
Acetamide, 2-diazo-N-(methylsulfonyl)-2-(p-nitrophenyl)-	MeOH	220(4.18),281(3.86), 340(4.13)	88-1403-64
$C_9H_8N_4O_6$			
Hydantoin, 3-(hydroxymethyl)-1-[(5-nitrofurfurylidene)amino]-	2% DMF	265(4.11),368(4.25)	44-3416-64
$C_9H_8N_4O_7$			
Methanol, (methylazoxy)-, 3,5-dinitrobenzoate	aq. EtOH	215(4.6),285s(3.1)	10-0373-65B
$C_9H_8N_4S$			
Benzaldehyde, N-thiadiazolylhydrazone	EtOH	227(4.11),321(4.39)	1-0871-64
anion	EtOH	242(4.10),282(3.74), 378(4.34)	1-0871-64

Compound	Solvent	λ max(log ε)	Ref.
C_9H_8O			
Acrylophenone	EtOH	219(3.90),254(4.05), 345(1.83)	22-0704-65
	EtOH	360(2.455)	65-0190-64
Indanone	EtOH	242(4.01)	5-0021-64G
	EtOH	244(4.09)	44-2165-65
Indone, 3a,7a-dihydro-, cis	C_6H_{12}	267(3.50),323(2.07), 337(2.12),351(2.02)	24-3680-65
2,4-Nonadiene-6,8-diyn-1-ol	Et_2O	224(4.35),233(4.53), 286(4.49),297s(4.45)	24-1846-64
C_9H_8OS			
Benzothiophene-5-ol, 3-methyl-	aq. base	327.0(3.43)	22-1464-65
Benzothiophene-6-ol, 3-methyl-	aq. base	284.5(3.98)	22-1464-65
Benzothiophene-7-ol, 3-methyl-	aq. base	316.0(3.77)	22-1464-65
$C_9H_8OS_2$			
2,5-Cyclohexadien-1-one, 4-(1,3-di-thiolan-2-ylidene)-	MeOH	294(3.70),402(4.51), 420(4.53),440s(4.27)	5-0037-65D
Ketone, methyl 3-methylthieno[3,2-b]-thien-2-yl	n.s.g.	307(4.3),320s(4.3)	70-0510-65
C_9H_8OSe			
Selenochroman-4-one	n.s.g.	243(4.29),265(3.73), 295(1.8),356(3.40), 360(3.72)	20-0483-64
$C_9H_8O_2$			
Acrylic acid, α-phenyl-	EtOH	246.0(3.77)	1-0612-65
Benzocyclobutene-4-carboxylic acid	EtOH-HCl	238(4.09),279(3.32), 290(3.27)	78-2185-64
4-Benzofuranol, 2-methyl-	aq. base	294.0(3.75)	22-1464-65
4-Benzofuranol, 3-methyl-	aq. base	298.0(3.76)	22-1464-65
Benzoic acid, vinyl ester	nonane	233(4.15)	67-0375-65
Benzoquinone, allyl-	n.s.g.	247(4.21),251s(--)	39-5060-65
3-Chromanone	EtOH	274.5(3.28)	65-2699-64
Cinnamic acid	pet ether	204(4.16),216(4.24), 222(4.18),277(4.36)	54-0949-64
	EtOH	272(4.31)	35-5548-64
1-Indanone, 5-hydroxy-	pH 13	323(4.45)	44-2165-65
	EtOH	269(4.11),293(4.06)	44-2165-65
1-Indanone, 7-hydroxy-	EtOH	256(4.04),315(3.52)	39-2816-64
Isophthalaldehyde, 5-methyl-	EtOH	230(4.35),245s(3.36)	32-1221-64
Phthalide, 3-methyl-	hexane	224(4.00),230(3.90), 264(3.13),270(3.16), 277(3.25)	39-3811-65
	EtOH	228(3.98),273(3.24), 280(3.22)	39-3811-65
1,2-Propanedione, 1-phenyl-	EtOH	255(4.36)	23-0113-64
$C_9H_8O_2S_2$			
2,5-Cyclohexadien-1-one, 4-(1,3-dithio-lan-2-ylidene)-3-hydroxy-	MeOH	357(4.31),421(3.98)	5-0037-65D
	HOAc	350(3.89),433(3.46)	5-0037-65D
	$HCONH_2$	365(4.21),463(3.53)	5-0037-65D
$C_9H_8O_3$			
Benzaldehyde, 3-formyl-2-hydroxy-4-methyl-	pH 1	202(4.00),244(4.46), 346(3.81)	88-3439-65
	pH 13	220(4.36),242(4.33), 269(3.93),414(4.04)	88-3439-65

Compound	Solvent	λ max(log ε)	Ref.
Benzoic acid, o-acetyl-	MeOH	229(3.98),272(3.00), 279(3.00)	44-4293-65
Benzoic acid, p-acetyl-	EtOH	249(4.26)	44-2165-65
1-Oxaspiro[5,4]deca-6,9-diene-2,8-dione	20% MeCN	232(4.07)	39-3126-64
1,2-Propanedione, 1-(p-hydroxyphenyl)-	EtOH	225(3.82),296(4.06)	23-0113-64
$C_9H_8O_4$			
Acetic acid, salicyloyl-	EtOH	253(2.96),316(2.54)	28-6479-65A
Piperonylic acid, methyl ester	EtOH	298(--)	95-0857-65
4H-Pyran-3,5-dicarboxaldehyde, 4-acetyl-	Et_2O	282(3.66)	24-1959-64
$C_9H_8O_5$			
Isophthalaldehydic acid, 4-hydroxy-5-methoxy-	EtOH	230(4.26),290(4.03), 320(4.06)	23-2362-64
1,3-Pentadiene-1,3,4-tricarboxylic acid, 2-methyl-, cyclic 3,4-anhydride	EtOH	211(4.13)	44-0203-65
Phthalaldehyde, 3,4-dihydroxy-5-methoxy-	EtOH	219(3.95),264(4.41), 333(3.87)	39-4672-65
$C_9H_8O_6$			
2,3-Furandicarboxylic acid, 4-formyl-, dimethyl ester	MeOH	223(4.09),237(4.05)	24-1581-65
2,5-Heptadienedioic acid, 2,6-dihydroxy-3,5-bis(hydroxymethyl)-, di-γ-lactone	HCl	240(4.30)	39-0766-64
	H_2O	239(4.25)	39-0766-64
	NaOH	278(4.26)	39-0766-64
Isophthalic acid, 4-hydroxy-5-methoxy-	EtOH	227(4.44),267(3.87), 310(3.58)	23-2362-64
4H-Pyran-2,6-dicarboxylic acid, 4-oxo-, dimethyl ester	MeOH	271.2(4.02)	95-1227-64
α-Tetronic acid, β-methylenebis-	EtOH	241(4.31)	44-3560-64
C_9H_8S			
Benzo[b]thiophene, 3-methyl-	EtOH	231(4.53),262(3.64), 267s(3.62),290(3.42), 299(3.50)	4-0231-65
$C_9H_8S_3$			
1,3-Dithiolane-2-thione, 4-phenyl-	$CHCl_3$	320(4.17)	49-0631-65
$C_9H_9BrN_2O$			
1-Amino-2-hydroxyquinolizinium bromide, hydrobromide	H_2O	332(3.90),361(3.93)	39-2763-64
2-Amino-1-hydroxyquinolizinium bromide	H_2O	216(4.30),235(4.25), 323(3.88),357(3.89)	39-3030-64
$C_9H_9BrN_2O_6$			
D-erythro-Pentofuranuronic acid, 1-(5-bromo-3,4-dihydro-2,4-dioxo-1(2H)-pyrimidinyl)-1,2-dideoxy-	pH 2	280(4.03)	94-0007-65
	H_2O	280(3.99)	94-0007-65
	pH 12	276(3.86)	94-0007-65
$C_9H_9BrO_4$			
2,4-Hexadienedioic acid, 2-(bromomethyl)-5-hydroxy-, δ-lactone, ethyl ester	50% EtOH	303(4.03)	88-3431-64
2,4-Hexadienedioic acid, 4-(bromomethyl)-2-hydroxy-, δ-lactone, ethyl ester	50% EtOH	305(3.86)	88-3431-64

Compound	Solvent	λ_{max}(log ε)	Ref.
$C_9H_9Br_3O_2$			
Bicyclo[5.1.0]oct-4-en-3-one, 4,8,8-tribromo-1-methoxy-	EtOH	256(3.78)	39-5343-64
	n.s.g.	256(3.78)	25-1917-64
C_9H_9Cl			
Bicyclo[6.1.0]nona-2,4,6-triene, 9-chloro-, syn-anti mixture (3:1)	isooctane	248(3.49)	35-1941-65
	EtOH	248(3.57)	35-5194-65
syn isomer	EtOH	250(3.48)	35-4876-64
Indan, 1-chloro-	EtOH	262(3.52),270(3.44)	35-5194-64
$C_9H_9ClN_2$			
Malononitrile, (2-chlorocyclohexylidene)-	MeOH	240(4.23)	24-3571-65
$C_9H_9ClN_2O_3S$			
2H-1,2,4-Benzothiadiazine-3-ethanol, 7-chloro-, 1,1-dioxide	MeOH	269(4.03)	87-0269-64
$C_9H_9ClN_2S$			
2-Benzimidazoleethanethiol, 5-chloro-	EtOH	209(4.39),249(3.98), 254s(3.95),283(4.11), 290(4.10)	4-0306-65
$C_9H_9ClO_2$			
Propiophenone, 2'-chloro-4'-hydroxy-	C_6H_{12}	254(3.97)	28-5614-64A
	EtOH	269(4.01)	28-5614-64A
Propiophenone, 3'-chloro-2'-hydroxy-	C_6H_{12}	254(4.04),260(4.01), 333(3.63)	28-5614-64A
	EtOH	254(3.97),329(3.57)	28-5614-64A
Propiophenone, 3'-chloro-4'-hydroxy-	C_6H_{12}	259(4.13)	28-5614-64A
	EtOH	270(4.14)	28-5614-64A
Propiophenone, 4'-chloro-2'-hydroxy-	C_6H_{12}	257(4.16),264(4.15), 323(3.72)	28-5614-64A
	EtOH	257(4.14),318(3.67)	28-5614-64A
Propiophenone, 5'-chloro-2'-hydroxy-	C_6H_{12}	248(3.85),254(3.85), 341(3.61)	28-5614-64A
	EtOH	247(3.85),335(3.56)	28-5614-64A
$C_9H_9Cl_3O_3$			
3-Cyclohexene-1-carboxylic acid, 6-(trichloroacetyl)-	EtOH	287(1.93)	44-2784-65
$C_9H_9Cl_3O_4$			
1,4,8,11-Tetraoxadispiro[4.1.4.2]tridec-12-ene, 6,12,13-trichloro-	n.s.g.	215s(3.69)	39-4744-65
$C_9H_9FK_2N_2O_{11}S_2$			
Uridine, 5-fluorodeoxy-, 3',5'-sulfate	pH 5.9	268(3.94)	35-1636-64
C_9H_9FO			
Cinnamyl alcohol, α-fluoro-	C_6H_{12}	247(4.29)	22-2258-64
$C_9H_9F_3N_4S$			
s-Triazolo[2,3-c]pyrimidine-2-thiol, 5-propyl-7-(trifluoromethyl)-	MeOH	201(3.99),250(4.32), 333(3.56)	39-3369-65
$C_9H_9F_5Sn$			
Tin, trimethyl(pentafluorophenyl)-	C_6H_{12}	263(2.77)	39-4782-64
	MeOH	262.5(3.10)	39-4782-64

Compound	Solvent	$\lambda_{max}(\log \epsilon)$	Ref.
$C_9H_9IN_2O_5$			
3,6-Epoxy-2H,8H-pyrimid[6,1-b][1,3]oxazocine-8,10(9H)-dione, 3,4,5,6-tetrahydro-4-hydroxy-11-iodo-	pH 2	285(3.96)	44-3913-65
	H_2O	285(3.95)	44-3913-65
	pH 11	278(3.86)	44-3913-65
	EtOH	283(3.86)	44-3913-65
$C_9H_9IN_2O_6$			
D-erythro-Pentofuranuronic acid, 1,2-dideoxy-1-(3,4-dihydro-5-iodo-2,4-dioxo-1(2H)-pyrimidinyl)-	pH 2	287(3.88)	94-0007-65
	H_2O	287(3.9)	94-0007-65
	pH 12	279(3.77)	94-0007-65
$C_9H_9IO_4$			
Furan, 3-acetoxy-2-acetyl-4-iodo-5-methyl-	EtOH	244(3.86),287(4.01)	39-2543-65
$C_9H_9KN_2O_5$			
2-Propanone, (dinitrocyclohexadienyl)-, potassium derivative	H_2O	480(3.83)	25-2065-65
	MeOH	536(3.89)	25-2065-65
	EtOH	555(4.03)	25-2065-65
	iso-PrOH	565(3.28)	25-2065-65
	acetone	579(4.28)	25-2065-65
	THF	555(4.08)	25-2065-65
	methyl cellosolve	563(3.99)	25-2065-65
	DMF	582(4.26)	25-2065-65
	pyridine	580(3.75)	25-2065-65
	Me_2SO	587(4.29)	25-2065-65
C_9H_9Li			
Lithium, cyclononatetraenyl-	THF	251(5.0),319(3.7), 322(3.7)	35-5194-64
C_9H_9K			
Potassium, cyclononatetraenyl-	THF	249(5.0),314(4.0), 320(4.0)	35-5194-64
C_9H_9N			
Indole, 1-methyl-	EtOH	219(4.54),275(3.77), 282(3.78),287s(3.74), 293(3.66)	35-3796-64
	H_2SO_4	233(3.51),238(3.50), 282(3.67)	35-3796-64
Indole, 2-methyl-	EtOH	220(4.53),272(3.86), 277(3.85),281s(3.84), 289(3.73)	35-3796-64
	EtOH	220(4.53),272(3.85), 277(3.85),288(3.73)	44-1206-64
	EtOH	220(4.5),374(3.8)	56-1709-64
	H_2SO_4	230(3.71),236(3.66), 274(3.75)	35-3796-64
Indole, 3-methyl-	EtOH	222(4.51),284(3.73), 291(3.70)	39-5510-64
	EtOH	224(4.6),284(3.7)	56-1709-64
	EtOH	222(4.51),275s(3.73), 281(3.76),290(3.69)	35-3796-64
	EtOH	223(4.54),276s (3.74), 282(3.77),291(3.69)	44-1206-64
	H_2SO_4	236(3.59),241(3.56), 290(3.67)	39-5510-64
	H_2SO_4	236(3.60),240(3.58), 286(3.68)	35-3796-64

Compound	Solvent	$\lambda_{max}(\log \epsilon)$	Ref.
Indole, 5-methyl-	EtOH	219(4.48),269(3.75), 277s(3.72),284(3.69), 295(3.50)	35-3796-64
	H_2SO_4	239(3.61),246(3.69), 273(3.72)	35-3796-64
2,4-Nonadiynenitrile	hexane	200(4.83)	39-0543-64
C_9H_9NO			
Acetonitrile, (m-methoxyphenyl)-	hexane	223(3.79),274(3.18), 280(3.11)	78-0861-64
Acetonitrile, (o-methoxyphenyl)-	hexane	220(3.78),271(3.33), 277(3.29)	78-0861-64
Acetonitrile, (p-methoxyphenyl)-	hexane	226(3.90),277(3.19)	78-0861-64
2,1-Benzisoxazole, 3,4-dimethyl-	pH -4.0	266(3.64),275s(3.54), 342(3.51)	39-5360-65
	pH 3.0	215s(3.79),256s(3.06), 260s(3.06),267(3.11), 272s(3.05),279(3.11), 322(3.69)	39-5360-65
Carbostyril, 3,4-dihydro-	EtOH	210(4.33),248(4.14)	1-1389-64
	dioxan	252(4.00)	1-1389-64
Furo[2,3-b]pyridine, 2,6-dimethyl-	EtOH	217(4.18),250(3.84), 257s(3.80),290(3.98)	39-1632-64
1-Indanone, 5-amino-	EtOH	322(4.36)	44-2165-65
2-Indolinone, 3-methyl-	EtOH	207(4.44),249(3.94), 278s(3.16)	44-2431-64
1-Indolol, 2-methyl-	n.s.g.	223(4.54),280(3.81)	28-2851-65A
C_9H_9NOS			
Benzamide, N-acetyl-thio-	MeOH	218s(3.87),246s(3.84), 268(3.96),298(3.94), 476(2.19)	24-1556-65
	CCl_4	470(2.24)	24-1556-65
Benzamide, N-(thioacetyl)-	heptane	236(3.98),285(4.27), 435(1.72)	24-1556-65
	MeOH	240(3.93),290(4.18), 434(1.60)	24-1556-65
4,1-Benzothiazepin-2(3H)-one, 1,5-dihydro-	iso-PrOH	237(3.91)	44-3111-65
$C_9H_9NOS_2$			
Acrylamide, 3-(phenyldithio)-, cis	EtOH	240(4.11),260s(4.00)	44-2660-65
	base	271(4.21)	44-2660-65
Benzoic acid, thio-, anhydrosulfide with methyldithiocarbamic acid	hexane	250s(4.09),264(4.14), 302s(3.74),370s(2.53)	78-2865-65
2-Thiazolidinethione, 4-hydroxy-4-phenyl-	MeOH	245(4.16),273(3.88)	44-2146-64
$C_9H_9NOS_3$			
Disulfide, benzoyl methylthiocarbamoyl	hexane	252(4.51),280s(3.90)	78-2865-65
$C_9H_9NO_2$			
Butene, 1-nitro-2-propenyl-, trans	EtOH	239(4.16),323(3.33)	44-3604-65
Cinnamohydroxamic acid	MeCN	274(4.38)	35-4406-64
2-Indolinone, 3-hydroxy-1-methyl-	MeOH	210(4.43),258(3.79), 288(3.09)	87-0626-65
2-Indolinone, 3-hydroxy-3-methyl-	EtOH	208(4.51),252(3.81), 287(3.13)	44-2431-64
2-Indolinone, 3-methoxy-	MeOH	251(3.77),293(3.13)	44-3610-65

Compound	Solvent	λ_{max}(log ε)	Ref.
5-Indolol, 6-methoxy-	EtOH	208(4.38),219(4.36), 274(3.69),302(3.87)	4-0387-65
6-Indolol, 5-methoxy-	EtOH	224(4.11),274(3.67), 298(3.92),306(3.85)	4-0387-65
6,7-Isoquinolinediol, 3,4-dihydro-	EtOH-HCl	245(4.21),305(3.95), 353(4.02)	33-1945-65
	EtOH	224(3.98),270(3.94), 310(3.57),400(4.32)	33-1945-65
	EtOH-NaOH	245(4.08),331(4.26)	33-1945-65
hydrochloride or hydrobromide	EtOH-HCl	245(4.20),304(3.94), 354(4.02)	33-1945-65
	EtOH	246(4.17),305(3.91), 355(4.00)	33-1945-65
	EtOH-NaOH	245(4.11),332(4.26)	33-1945-65
$C_9H_9NO_2S$			
4,1-Benzothiazepin-2(3H)-one, 1,5-di-hydro-, 4-oxide	iso-PrOH	221(4.33),261(3.45)	44-3111-65
$C_9H_9NO_3$			
Acetanilide, 3',4'-(methylenedioxy)-	EtOH	296(--)	95-0857-65
2-Propen-1-ol, 3-(p-nitrophenyl)-	MeOH	309(4.19)	78-1057-64
3-Pyridinebutyric acid, γ-oxo-	EtOH	228(3.95),267(3.39)	37-3981-64
$C_9H_9NO_3S$			
4,1-Benzothiazepin-2(3H)-one, 1,5-di-hydro-, 4,4-dioxide	iso-PrOH	233(3.97)	44-3111-65
$C_9H_9NO_4$			
Acetic acid, (3-nitro-p-tolyl)-	H_2O	275(3.486)	65-1474-64
Benzaldehyde, 2-(1-hydroxy-2-nitro-ethyl)-, lactol	EtOH	254(2.78),261(2.73), 268(2.63)	44-3180-64
Benzoic acid, m-nitro-, ethyl ester	C_6H_{12}	250.5(3.92)	44-2088-64
	MeOH	255.5(3.86)	44-2088-64
	iso-PrOH	254(3.93)	44-2088-64
	dioxan	255.5(3.87)	44-2088-64
Benzoic acid, p-nitro-, ethyl ester	C_6H_{12}	254.5(4.25)	44-2088-64
	MeOH	259(4.31)	44-2088-64
	iso-PrOH	258.5(4.16)	44-2088-64
	dioxan	258.5(4.33)	44-2088-64
Glutaconic acid, 4-(1-cyanoethyli-dene)-, 5-methyl ester	EtOH	260(4.08)	25-1954-64
	EtOH	260(4.08)	44-0203-65
Glutaconic acid, 4-(1-cyanoethylidene)-3-methyl-	EtOH	210(4.10)	44-0203-65
1,3-Indandiol, 2-nitro-	EtOH	232(3.22),255(2.81), 262(2.78),271(2.65)	44-3180-64
$C_9H_9NO_4S$			
Acetic acid, [(2-nitrobenzyl)thio]-	iso-PrOH	247(3.72)	44-3111-65
$C_9H_9NO_5$			
Benzoic acid, 3-hydroxy-4-nitro-, ethyl ester	HCl	358(3.56)	35-4942-64
	NaOH	426(3.71)	35-4942-64
Benzyl alcohol, 3,4-(methylenedioxy)-α-nitromethyl-	EtOH	235s(--),285(3.93)	23-1901-64

Compound	Solvent	λ_{max}(log ε)	Ref.
$C_9H_9NO_6$			
2H-Pyran-3-alanine, 6-carboxy-2-oxo-	pH 1.4	234(3.38),304(4.02)	88-3437-64
	pH 5.0	234(3.42),306(4.01)	88-3437-64
	pH 12.4	330(4.11+)	88-3437-64
2H-Pyran-4-alanine, 6-carboxy-2-oxo-	pH 1.4	301(3.85)	88-3431-64
	pH 5.0	303(3.86)	88-3431-64
	pH 12.4	348(4.07)	88-3431-64
2H-Pyran-5-alanine, 6-carboxy-2-oxo-	aq. acid	303(3.82)	88-3431-64
Stizolobic acid	pH 5.0	303(3.89)	88-3431-64
C_9H_9NS			
1(2H)-Isoquinolinethione, 3,4-dihydro-	heptane	259(3.97),303(3.74), 329(3.84),412(2.06)	1-2432-65
	EtOH	260(3.97),303(3.79), 326(3.85),395s(2.2)	1-2432-65
$C_9H_9N_3$			
Isoquinoline, 1,3-diamino-	EtOH	234(4.32),307(4.02), 375(3.50)	39-1423-64
3,5-Pyridinedicarbonitrile, 1,2-dihydro-2,6-dimethyl-	EtOH	214(4.29),255(4.10), 375(3.83)	73-0143-64
3,5-Pyridinedicarbonitrile, 1,2-dihydro-4,6-dimethyl-	EtOH	216(4.35),258(4.08), 372(3.79)	73-3711-65
3,5-Pyridinedicarbonitrile, 1,4-dihydro-2,4-dimethyl-	EtOH	213(4.37),344(3.82)	73-0143-64
3,5-Pyridinedicarbonitrile, 1,4-dihydro-2,6-dimethyl-	EtOH	216(4.37),345(3.76)	73-1654-64
Quinoxaline, 6-amino-2-methyl-	n.s.g.	257(4.3),395(3.74)	73-3102-65
Quinoxaline, 7-amino-2-methyl-	n.s.g.	259(4.28),385(3.79)	73-3102-65
$C_9H_9N_3O$			
Formamidoxime, N-benzyl-1-cyano-	EtOH	255(3.95)	33-0033-64
Formamidoxime, 1-cyano-N-phenyl-, O-methyl derivative	EtOH	276(4.07)	33-0033-64
5-Hydroxy-4-methyl-2-phenyl-s-triazolium hydroxide, inner salt	n.s.g.	278(3.86)	4-0105-65
Isoquinoline, 1-amino-3-(hydroxyamino)-	EtOH	235(4.54),264(4.46), 305(3.84),375(3.71)	39-1423-64
1,3,4-Oxadiazole, 2-(methylamino)-5-phenyl-	EtOH	283(4.2)	28-4579-64A
1,3,4-Oxadiazole, 2-(N-methylanilino)-	n.s.g.	252(4.08)	28-4538-65A
1,3,4-Oxadiazoline, 5-imino-4-methyl-2-phenyl-	EtOH	293(4.1)	28-4579-64A
1,3,4-Oxadiazoline, 4-methyl-5-(phenylimino)-	n.s.g.	258(4.17)	28-4538-65A
Quinazoline, 2-amino-6-methoxy-	pH 1	241(4.21),251(4.21)	78-2059-65
	H_2O	236(4.66),260(4.06)	78-2059-65
8-Quinoxalinol, 5-amino-6-methyl-	MeOH	271(4.68),310(3.11), 420(2.99)	4-0171-64
1,2,4-Triazolinone, 3-methyl-4-phenyl-	H_2O	none	5-0123-65B
$C_9H_9N_3O_2$			
Benzimidazole, 1-ethyl-5-nitro-	MeOH	239(4.42),278s(3.86), 284s(3.92),302(4.05)	65-0632-64
1,4-Phthalazinedione, 2,3-dihydro-5-(N-methylamino)-	pH 12	223(3.89),303(3.91), 366(3.89)	46-0752-64
6-Quinazolinemethanol, 2-amino-8-hydroxy-, hydrochloride	pH 1	228(4.26),264(4.38)	78-2059-65
	H_2O	232(4.26),260(4.37)	78-2059-65
	pH 13	274(4.38)	78-2059-65

Compound	Solvent	λ_{max}(log ε)	Ref.
4(3H)-Quinazolinone, 3-amino-2-(hydroxymethyl)-	iso-PrOH	223(4.41),274(3.73), 305(3.48),316(3.40)	44-0582-64
1,2,4-Triazolidine-3,5-dione, 1-benzyl-	EtOH	211(4.01)	22-0500-64
	NaOH	218(3.89)	22-0500-64
1,2,4-Triazolidine-3,5-dione, 2-methyl-1-phenyl-	EtOH	210(4.01),243(3.98)	22-0500-64
	NaOH	220(3.84),250(3.92)	22-0500-64
$C_9H_9N_3O_3$			
Benzimidazole, 5-methoxy-2-methyl-4-nitro-	pH 2	247(3.67),290(3.76), 347(3.56)	44-0476-64
	pH 12	370(3.78)	44-0476-64
	EtOH	310(3.76)	44-0476-64
Benzimidazole, 5-methoxy-2-methyl-7-nitro-	pH 2	236(3.87),248(3.90), 340(3.40)	44-0476-64
	pH 12	256(3.79),374(3.94)	44-0476-64
	EtOH	237(3.94),253(3.90), 295(3.60),340(3.57)	44-0476-64
Imidazo[1,2-c]pyrimidin-5(1H)-one, 2-acetoxy-3-methyl-	n.s.g.	282(3.99),290(3.99), 304s(3.74)	44-1762-64
5-Pyrimidinecarbonitrile, 1,3-diacetyl-1,2,3,4-tetrahydro-2-oxo-	EtOH	235(4.11)	44-1740-64
Pyruvaldehyde, 1-(p-nitrophenylhydrazone)	EtOH	241(3.88),295(3.57), 379(4.55)	24-0725-64
$C_9H_9N_3S$			
5-Mercapto-4-methyl-2-phenyl-s-triazolium hydroxide, inner salt	n.s.g.	248(4.32),302(3.40)	4-0105-65
1,3,4-Thiadiazole, 2-(N-methylanilino)-	EtOH	239(3.79),279(3.98)	28-0766-65B
1,3,4-Thiadiazole, 3-methyl-2-(phenylimino)-	EtOH	246(3.88),282(3.90)	28-0766-65B
$C_9H_9N_5$			
Pyrazolo[1,5-a]pyrimidine-6-carbonitrile, 7-amino-2,5-dimethyl-	EtOH	232(4.59),300(4.16)	95-0442-65
Pyrazolo[1,5-a]pyrimidine-6-carbonitrile, 7-amino-3,5-dimethyl-	EtOH	230(4.47),307(4.16)	95-0442-65
$C_9H_9N_5O_4$			
1,2,4-Triazin-5-ol, 3-amino-4,5-dihydro-6-[2-(5-nitro-2-furyl)vinyl]-, hydrobromide	propylene glycol	292(4.16),383(4.27)	95-0009-64
$C_9H_9N_5S$			
Pyrrolo[2,3-d]pyrimidine-5-carbonitrile, 4-amino-6-(ethylthio)-	EtOH	229s(4.20),301(4.24)	35-1995-65
Pyrrolo[2,3-d]pyrimidine-5-carbonitrile, 4-amino-7-methyl-6-(methylthio)-	EtOH	235(4.24),269(4.09), 327(4.14)	35-1995-65
s-Triazine-2-thiol, 4-amino-6-anilino-	EtOH	268(4.29),303(4.25)	39-6296-65
s-Triazine-2(1H)-thione, tetrahydro-4,6-diimino-1-phenyl-	EtOH	290(4.31)	39-6296-65
C_9H_{10}			
Benzene, allyl-	hexane	263(2.49),269(2.49), 280s(1.35)	65-2088-65
	EtOH-HCl	268(2.49),280s(1.35)	65-2088-65
	EtOH	268(2.49),280s(1.35)	65-2088-65
	EtOH-NaOEt	268(2.55),280s(1.54)	65-2088-65

Compound	Solvent	λ max(log ε)	Ref.
Benzene, cyclopropyl-	EtOH	190(4.5),220(4.0), 250s(2.9),265(2.8), 275(2.8)	35-0908-64
Bicyclo[6.1.0]nona-2,4,6-triene	EtOH	247(3.69)	35-5194-64
Fulvene, 6-isopropenyl-	C_6H_{12}	277(4.31),288(4.48), 298(4.50),307(4.25), 395(2.31)	33-1022-64
	EtOH	289(4.44),296(4.46), 388(2.32)	33-1022-64
Fulvene, 6-methyl-6-vinyl-	C_6H_{12}	285(4.32),294(4.45), 303(4.47),313(4.26), 394(2.35)	33-1022-64
	EtOH	294(4.40),302(4.41), 392(2.33)	33-1022-64
Fulvene, 6-(1-propenyl)-	C_6H_{12}	288(4.42),296(4.55), 307(4.57),317(4.31), 392(2.39)	33-1022-64
	EtOH	389(2.39)	33-1022-64
Indene, 3,a,7a-dihydro-	hexane	258(3.43)	35-1941-65
	EtOH	262(3.57),270(3.51)	35-5194-64
trans	n.s.g.	260(3.57)	88-0391-65
3,5,7-Nonatrien-1-yne	n.s.g.	280(4.51),292(4.65), 321(4.60)	5-0062-65B
$C_9H_{10}BF_4NO$			
N-Ethyl-1,2-Benzisoxazolium tetra-fluoroborate	H_2O	258(4.12),297(3.46)	78-3019-65
$C_9H_{10}BrNO$			
1,2,3,4-Tetrahydro-1-oxoquinolizinium bromide	EtOH	274.3(3.78)	94-1338-64
$C_9H_{10}BrNS$			
N,2-Dimethylbenzothiazolium bromide	MeOH-HOAc	236(3.96),274(3.81)	22-2868-64
	MeOH-NaOMe	256(3.71),308(3.65)	22-2868-64
$C_9H_{10}BrN_5O_2$			
6-Amino-3-(1-carboxy-1-hydroxypropyl)-purinium bromide, γ-lactone	EtOH	261(4.35)	23-1599-64
$C_9H_{10}ClNO_2$			
p-Benzoquinone, [(2-chloroethyl)-methylamino]-	MeOH	220(4.2),280(4.0), 480(3.6)	44-4107-65
$C_9H_{10}ClNO_3$			
3-Pyrroline-3-carboxylic acid, 2-chloro-methylene-4-methyl-5-oxo-, ethyl ester	EtOH	274(3.85),336(3.96)	39-5999-64
$C_9H_{10}ClNS$			
N,2-Dimethylbenzothiazolium chloride	H_2O	274(3.80)	22-2868-64
	N NaOH	269(3.98)	22-2868-64
	2N NaOH	268(4.12)	22-2868-64
	MeOH-HOAc	238(3.92),274(3.80)	22-2868-64
	MeOH-NaOMe	256(3.70),308(3.66)	22-2868-64
$C_9H_{10}ClN_3$			
Pyrazolo[1,5-a]pyrimidine, 7-chloro-2,3,5-trimethyl-	EtOH	241(4.67),283s(3.26), 292(3.29),301s(3.18), 338(3.16)	94-1207-65

Compound	Solvent	λ_{max}(log ϵ)	Ref.
Pyrazolo[1,5-a]pyrimidine, 7-chloro-2,3,6-trimethyl-	EtOH	240(4.87),286s(3.18), 296(4.23),310(3.16), 344(3.24)	94-1207-65
$C_9H_{10}ClN_3O_5S_2$			
2H-1,2,4-Benzothiadiazine-7-sulfonamide, 6-chloro-3-(1-hydroxyethyl)-, 1,1-dioxide	EtOH	225(4.50),277(4.06)	95-0095-65
$C_9H_{10}Cl_2O$			
Phenol, 2,4-dichloro-3-ethyl-5-methyl-	C_6H_{12}	282(3.36),290(3.38)	2-0417-64
$C_9H_{10}FKN_2O_8S$			
Uridine, deoxy-5-fluoro-, 5'-sulfate	pH 5.9	268(3.94)	35-1636-64
$C_9H_{10}F_3N_3O_5$			
as-Triazine-3,5(2H,4H)-dione, 2-(2'-deoxy-β-D-ribofuranosyl)-6-(trifluoromethyl)-	pH 1	268(3.77)	4-0495-65
	pH 11	263(3.75)	4-0495-65
$C_9H_{10}F_3N_3O_5S_2$			
2H-1,2,4-Benzothiadiazine-7-sulfonamide, 3,4-dihydro-3-(hydroxymethyl)-6-(trifluoromethyl)-, 1,1-dioxide	EtOH	271.5(4.33)	95-0095-65
$C_9H_{10}INS$			
N,2-Dimethylbenzothiazolium iodide	H_2O	274(3.83)	22-2868-64
	N NaOH	269(3.99)	22-2868-64
	2N NaOH	268(4.13)	22-2868-64
	MeOH-HOAc	274(3.81)	22-2868-64
	MeOH-NaOMe	256(3.81),308(3.69)	22-2868-64
	EtOH	260(3.68)	22-2868-64
	EtOH-NaOEt	258(3.77),308(3.68)	22-2868-64
$C_9H_{10}IN_3$			
2-Amino-1-methylcycloheptimidazolium iodide	EtOH	256(4.46),345(3.96), 385(4.00)	94-0810-65
$C_9H_{10}IN_3O_5$			
D-erythro-Pentofuranuronic acid, 1-(4-amino-5-iodo-2-oxo-1(2H)-pyrimidinyl)-1,2-dideoxy-	pH 2	306(3.91)	94-0007-65
	H_2O	294(3.8)	94-0007-65
	pH 12	294(3.8)	94-0007-65
$C_9H_{10}N_2$			
Benzimidazole, 1-ethyl-	MeOH	249(3.78),254(3.78), 266(3.62),274(3.66), 281(3.64)	65-0632-64
Benzimidazole, 2-ethyl-	pH 2	235(3.61),268(3.92), 274(3.91)	44-0476-64
	pH 12	240(3.63),271(3.74), 277(3.75)	44-0476-64
	MeOH	281(3.91)	44-0259-65
	EtOH	243(3.80),272(3.91), 280(3.89)	44-0476-64
	EtOH	274(3.85),281(3.90)	94-0773-64
Benzonitrile, p-(dimethylamino)-	MeOH	290(4.42)	88-2729-64
2,2'-Bipyrrole, 3-methyl-	EtOH	276(4.21),281(4.23), 288(4.22)	44-2727-64
2,2'-Bipyrrole, 4-methyl-	EtOH	276(4.21),281(4.23), 288(4.23)	44-2727-64

Compound	Solvent	$\lambda_{max}(\log \epsilon)$	Ref.
2,2'-Bipyrrole, 5-methyl-	EtOH	278(4.19),283(4.21), 289(4.20)	44-2727-64
2-Imidazoline, 2-phenyl-	EtOH	230(4.06)	54-0314-65
Imidazo[1,2-a]pyridine, 2,3-dimethyl-	pH 1	283(3.96)	4-0053-65
	pH 11	283(3.79),308(3.78)	4-0053-65
Indazole, 3-ethyl-	EtOH	289(3.71)	32-0814-65
Quinoline, 2-amino-3,4-dihydro-	EtOH	252(4.07),267(4.13)	4-0330-65
$C_9H_{10}N_2O$			
Benzimidazole, 5-ethoxy-	pH 2	283(3.84)	44-0476-64
	pH 12	241(3.49),283(3.76)	44-0476-64
	EtOH	243(3.58),286(3.74), 290(3.71)	44-0476-64
Carbostyril, 3-amino-3,4-dihydro-	pH 2	254(4.415)	87-0632-64
	pH 7	255(4.362)	87-0632-64
	pH 10	255(4.245)	87-0632-64
3(2H)-Cinnolinone, 1,4-dihydro-2-methyl-	EtOH	206(4.19),253s(3.60)	39-5391-65
1H-Indazole, 3-methoxy-1-methyl-	EtOH	310(3.70)	7-0583-65
3-Indanzolinone, 1,2-dimethyl-	EtOH	239(4.05),312(3.70)	7-0583-65
1H-Pyrrolo[2,3-b]pyridine, 6-methoxy-4-methyl-	EtOH	222(4.34),298(4.00)	65-1454-64
$C_9H_{10}N_2O_2$			
Azetidine, 1-(p-nitrophenyl)-	n.s.g.	402(4.34)	5-0109-65I
Benzimidazole, 4,6-dimethoxy-	EtOH	254(3.80)	44-0476-64
	pH 2	266(3.79)	44-0476-64
	pH 12	254(3.47)	44-0476-64
Benzimidazole, 4,7-dimethoxy-	EtOH	251(3.77)	44-0476-64
	pH 2	265(3.79)	44-0476-64
	pH 12	249(3.58)	44-0476-64
Benzimidazole, 5,6-dimethoxy-	EtOH	243(3.48),289(3.84)	44-0476-64
Carbostyril, 3-amino-3,4-dihydro-1-hydroxy-	pH 2	257(4.111)	87-0632-64
	pH 7	267s(3.74),290(3.83)	87-0632-64
	pH 10	267s(3.77),290(3.86)	87-0632-64
$C_9H_{10}N_2O_3$			
Benzoic acid, 2-ureido-, methyl ester	n.s.g.	225(4.43),248(4.00), 316(3.64)	49-1068-64
Methanol, (methylazoxy)-, benzoate	aq. EtOH	235(4.4),275f(3.1)	10-0373-65B
1,7-Naphthyridin-2(1H)-one, 3,4-dihydro-3-hydroxy-6-methoxy-	EtOH	248(4.15)	35-3530-65
$C_9H_{10}N_2O_3S$			
Urea, 1-[(2,6-dimethyl-4-oxo-4H-pyran-3-yl)carbonyl]-2-thio-	EtOH	218(4.48),288(4.04)	70-0747-65
$C_9H_{10}N_2O_4$			
Benzohydroxamic acid, O-ethyl-p-nitro-	MeOH	222(4.00),304(3.97)	73-0940-65
	MeOH-NaOH	247(3.97),363(3.93)	73-0940-65
2,5-Piperazinedione, 3-(acetoxymethylene)-1-methyl-6-methylene-	MeOH	282(4.33)	39-0026-63
Uracil, 1-(2,3-dideoxy-β-D-glycero-pent-2-enofuranosyl)-	H_2O	231(3.36),261(4.01)	35-1896-64
$C_9H_{10}N_2O_5$			
Uridine, O^6,5',6-hydroxycyclodeoxy-	pH 2	262(4.08)	44-3913-65
	H_2O	262(4.09)	44-3913-65
	pH 11	263(3.99)	44-3913-65
	EtOH	261(4.07)	44-3913-65

Compound	Solvent	$\lambda_{max}(\log \epsilon)$	Ref.
Uracil, 1-(2,5-anhydro-β-D-arabino-	pH 6.9	264(4.03)	44-0467-65
furanosyl)-	N NaOH	264(3.93)	44-0476-65
Uracil, 1-(2,5-anhydro-β-D-lyxo- furanosyl)-	H_2O	263(4.04)	44-0476-65
Uracil, 1-(3,5-anhydro-β-D-lyxo- furanosyl)-	H_2O	205(3.93),260(3.97)	44-0476-65
Uracil, 1-(3,5-anhydro-β-D-xylo-	pH 6.9	260.5(3.98)	44-0467-65
furanosyl)-	N NaOH	262.5(3.86)	44-0467-65
$C_9H_{10}N_2O_6$			
D-erythro-Pentofuranuronic acid, 1,2-di-	pH 2	260(4.03)	94-0007-65
deoxy-1-(3,4-dihydro-2,4-dioxo-1(2H)-	H_2O	261(4.04)	94-0007-65
pyrimidinyl)-	pH 12	261(3.92)	94-0007-65
1(2H)-Pyrimidineacetic acid, 5-acetyl-	pH 2	283(4.14)	39-1642-65
3,4-dihydro-α-(hydroxymethyl)-2,4-	H_2O	231(4.07),287(4.15)	39-1642-65
dioxo-	pH 12	288(4.02)	39-1642-65
$C_9H_{10}N_2S$			
2-Benzimidazoleethanethiol	EtOH	208(4.36),246(4.05), 250s(4.05),272s(4.06), 276(4.13),282(4.17)	4-0306-65
Urea, 2-thio-1-(p-vinylphenyl)-	EtOH	255s(4.15),280(4.23)	44-1926-65
$C_9H_{10}N_4$			
Pyrazole, 3,5-diamino-1-phenyl-, hydrochloride	MeOH	249(4.26)	44-0942-64
$C_9H_{10}N_4O$			
Formamide, N-(2,3-dimethylpyrazolo- [1,5-a]pyrimidin-7-yl)-	EtOH	227(4.63),280s(3.24), 290(3.32),299s(3.21), 347(3.53)	94-0142-65
Tetrazole, 1-ethyl-2-(2-hydroxyphenyl)-	H_2O	283(3.51)	78-3019-65
	pH 3	288(3.45)	78-3019-65
1,2,4-Triazole, 3-allylamino-5-(2-	pH 2	268(4.22)	1-1191-65
furyl)-	pH 12	275(4.11)	1-1191-65
$C_9H_{10}N_4OS$			
s-Triazole, 3-anilino-5-(methyl- sulfinyl)-	EtOH	254(4.33)	39-3912-65
$C_9H_{10}N_4O_2$			
4,7-Diazatryptophan	pH 1	224(4.11),307(3.96)	44-3454-65
Formamidine, N,N'-bis(5-methyl- 3-isoxazolyl)-	EtOH	267(4.26)	78-0159-64
Imidazo[1,2-a]pyridine, 2-(dimethyl- amino)-3-nitro-	pH 1	260s(4.05),281(4.38), 366(4.17)	44-4085-65
	pH 11	231(4.21),285(4.27), 365(4.21)	44-4085-65
Purine-6-acetic acid, ethyl ester	pH 1	263(3.90)	44-1528-65
	pH 7	264(3.96)	44-1528-65
	pH 13	273(3.97)	44-1528-65
5H-Pyrrolo[3,2-d]pyrimidine-7-carboxylic	pH 1	234(4.47),270(4.19)	44-1528-65
acid, 4-amino-, ethyl ester	pH 7	233(4.58),277(4.01)	44-1528-65
	pH 13	248(4.65),294(3.82)	44-1528-65
$C_9H_{10}N_4O_2S$			
Carbonic acid, thio-, O-ethyl S-9-	pH 1	280(4.03)	87-0010-64
methyl-9H-purin-6-yl ester	pH 7	282(4.18)	87-0010-64
9H-Purine-9-carboxylic acid, 6-(methyl-	pH 1	288(4.31)	87-0010-64
thio)-, ethyl ester	pH 7	288(4.26)	87-0010-64

Compound	Solvent	$\lambda_{max}(\log \epsilon)$	Ref.
s-Triazole, 3-anilino-5-(methylsulfonyl)-	EtOH	252(4.32)	39-3912-65
$C_9H_{10}N_4O_3$			
Lumazine, 1,6,7-trimethyl-, 5-oxide	pH 6.0	238(4.40),287(3.71), 349(3.84)	89-1136-65
	pH 11.0	244(4.38),289(3.83), 353(3.88)	89-1136-65
Lumazine, 3,6,7-trimethyl-, 5-oxide	pH 3.0	238(4.44),281s(3.96), 343(3.88)	89-1136-65
	pH 9.0	239s(4.18),266(4.33), 382(3.90)	89-1136-65
Pyruvaldehyde, 1-(p-nitrophenylhydrazone), 2-oxime	EtOH	245(3.88),287(3.85), 323(3.70),396(4.49)	24-0725-64
$C_9H_{10}N_4O_4$			
Propionaldehyde, 2,4-dinitrophenylhydrazone	EtOH	358(4.32)	95-0158-65
$C_9H_{10}N_4O_6$			
Glycine, N,N-[(1,2,3,4-tetrahydro-2,4-dioxo-5-pyrimidinyl)carbonyl]glycyl-	pH 1	220(3.99),271(3.96)	87-0001-64
	pH 13	241(3.90),292(4.11)	87-0001-64
$C_9H_{10}N_4S$			
1,3,4-Thiadiazole, 2-(1-methylhydrazino)-5-phenyl-	EtOH	226(3.88),311(4.18)	1-0871-64
1,3,4-Thiadiazolin-5-one, 4-methyl-2-phenyl-, hydrazone	EtOH	244(4.12),333(3.87)	1-0871-64
$C_9H_{10}N_6$			
Melamine, phenyl-	EtOH	264(4.42)	39-3459-64
$C_9H_{10}O$			
Indan, 3a,7a-epoxy-3a,7a-dihydro-	isooctane	258(3.69)	88-0609-65
7-Nonene-3,5-diyn-1-ol	Et_2O	227(3.20),239(3.60), 251(3.92),265(4.09), 281(3.98)	24-2118-64
Phenol, 2-allyl-	pet ether	253(3.78),300s(1.35)	65-2088-65
	EtOH-HCl	274(3.60),305s(1.05)	65-2088-65
	EtOH	226(3.95),277(3.78), 300s(1.35)	65-2088-65
	EtOH-NaOEt	240(4.18),290(3.78), 335s(1.65)	65-2088-65
	10% H_2SO_4	282(3.38)	65-2088-65
Phenol, 2-propenyl-	hexane	248(3.87),292(3.88)	65-2088-65
	EtOH-HCl	245(3.45),281(3.55), 305s(3.28)	65-2088-65
	EtOH	296(3.95)	65-2088-65
	EtOH-NaOEt	325(4.03)	65-2088-65
	10% H_2SO_4	287(3.55),330s(2.08)	65-2088-65
Phenol, 4-propenyl-	EtOH-HCl	263(4.38),290(3.35)	65-2088-65
	EtOH	260(4.38),290(3.35)	65-2088-65
	EtOH-NaOEt	279(4.78),315(3.78)	65-2088-65
	10% H_2SO_4	265(3.38),330s(2.08)	65-2088-65
Propiophenone	EtOH	290(2.377)	65-0190-64
Tetracyclo[3.3.1.$0^{2,8}$.$0^{4,6}$]nonan-3-one	MeOH	212(3.73),258(1.60)	88-1477-65
$C_9H_{10}OS$			
Anisole, p-(epithioethyl)-	hexane	238(4.12)	35-4628-64

Compound	Solvent	λ_{max}(log ε)	Ref.
$C_9H_{10}O_2$			
Acetic acid, p-tolyl-	H_2O	265(2.383)	65-1474-64
Acetophenone, p-methoxy-	C_6H_{12}	219(3.72),245s(3.42), 265(4.96),270(2.78)	73-3462-65
	C_6H_{12}	263.5(4.25)	78-1555-64
	EtOH	216(4.08),272(4.21)	73-3462-65
Anisole, p-(epoxyethyl)-	hexane	230(4.09)	35-4628-64
Benzene, 1-methoxy-4-(vinyloxy)-	nonane	286(3.40)	67-0375-65
Cyclopenta[b]pyran-7(4H)-one, 5,6-dihydro-2-methyl-	EtOH	282.5(3.44)	44-2513-65
1,3-Indandione, 3a,4,7,7a-tetrahydro-	EtOH-HCl	242(4.09)	65-0808-64
	EtOH	244(4.10)	65-0808-64
	EtOH-NaOH	260(4.27)	65-0808-64
	dioxan	235(4.11)	65-0808-64
Propionic acid, 3-phenyl-	pet ether	207(3.99),248(2.11), 253(2.24),259(2.23), 264(2.22),268(2.15)	54-0949-64
Propiophenone, 2'-hydroxy-	C_6H_{12}	249(3.99),255(3.96), 327(3.62)	28-5614-64A
	EtOH	250(3.99),324(3.57)	28-5614-64A
Propiophenone, 3'-hydroxy-	C_6H_{12}	243(3.92),299(3.38)	28-5614-64A
	EtOH	250(3.90),308(3.41)	28-5614-64A
Propiophenone, 4'-hydroxy-	C_6H_{12}	258(4.13)	28-5614-64A
	EtOH	276(4.17)	28-5614-64A
Tricyclo[3.2.1.0^{2,7}]octane-3,6-dione, 1-methyl-	MeOH	206(3.58),290(1.73)	77-0149-65
$C_9H_{10}O_2Se$			
1,3-Pentanedione, 1-(selenophene-2-yl)-	n.s.g.	281(3.60),340(3.90)	67-0379-65
copper salt	n.s.g.	285(4.0),350(4.1)	67-0379-65
$C_9H_{10}O_3$			
Acetic acid, (p-methoxyphenyl)-	H_2O	270(3.017)	65-1474-64
Acetovanillone	EtOH	276(4.03),304(3.94)	25-0795-65
m-Anisic acid, methyl ester	EtOH	307(--)	95-0858-65
o-Anisic acid, methyl ester	EtOH	291(--)	95-0858-65
Benzaldehyde, 2,3-dimethoxy-	EtOH	219(4.32),258(3.99), 318(3.45)	78-1725-64
Carbonic acid, cyclopentadienylidenemethyl ethyl ester	hexane	261(4.36),268(4.30), 357(2.52)	5-0039-64H
2-Cyclohexene-$\Delta^{1,\alpha}$-glycolic acid, 2-hydroxy-3-methyl-, γ-lactone	n.s.g.	290(4.29)	39-0833-65
4-Cyclopentene-1,3-dione, 2-acetyl-4,5-dimethyl-	C_6H_{12}	238(4.37),262(4.26), 310(2.78)	1-0441-64
4,6-Nonadiyn-3-one, 8,9-dihydroxy-	EtOH	228(3.14),241(3.36), 254(3.62),267(3.75), 283(3.63)	39-1476-64
4,6-Octadiynoic acid, 8-hydroxy-, methyl ester	EtOH	230(2.62),242(2.60), 256(2.34)	70-1237-65
4H-Pyran-3,5-dicarboxaldehyde, 4-ethyl-	Et_2O	290(3.72)	24-1959-64
2H-Pyran-2-one, 5-acetyl-4,6-dimethyl-	EtOH	296(3.78)	44-2642-65
Resorcinol, 4-methyl-, 3-acetate	EtOH	215(3.89),278(3.38)	24-1926-64
	NaOH	297(3.64)	24-1926-64
$C_9H_{10}O_4$			
Acetic acid, (3,5-dihydroxyphenyl)-,	EtOH	277(3.43),282(3.43)	1-1677-65
methyl ester	EtOH	277(3.29),282(3.29)	1-1677-65
Benzoic acid, 2,5-dihydroxy-4-methyl-, methyl ester	EtOH	203(4.20),220(4.28), 249(3.90),335(3.66)	33-2234-64

Compound	Solvent	λ max(log ε)	Ref.
Benzoic acid, 2,3-dimethoxy-	EtOH	291(3.29)	78-1725-64
Benzoic acid, 2-hydroxy-4-methoxy-6-methyl-	EtOH	261(4.22),299(3.74)	1-1677-65
p-Benzoquinone, 2-hydroxy-3-methoxy-4,5-dimethyl-	EtOH	278(4.21),442(2.81)	1-2303-64
1,2,4-Cyclopentanetrione, 3-acetyl-	EtOH	282.5(4.01)	39-6543-65
5,5-dimethyl-	EtOH-base	250(4.14),323(4.01)	39-6543-65
1,2,4-Cyclopentanetrione, 5-methyl-	EtOH-HCl	255(4.38),275s(4.25)	39-2251-65
3-propionyl-	EtOH-NaOH	235(3.88),275(4.44), 300(4.39),430(2.73)	39-2251-65
Dehydroacetic acid, methyl ester	EtOH	225(3.98),315(3.94)	44-1255-65
Propionic acid, 3-(2,5-dihydroxyphenyl)-	EtOH	297(3.60)	39-3126-64
4H-Pyran-2-carboxylic acid, 6-ethyl-	EtOH-HCl	266(3.94)	39-2251-65
3-methyl-4-oxo-	EtOH-NaOH	259(4.19)	39-2251-65
$C_9H_{10}O_4S$			
2-Thiophenecarboxylic acid, 5-acetyl-	NaOH	292(4.12),365(3.84)	35-0107-64
4-hydroxy-, ethyl ester	EtOH	287(4.08),335(3.74)	35-0107-64
$C_9H_{10}O_5$			
Acetic acid, (3,4-dihydroxy-5-methoxyphenyl)-	EtOH	216(4.46),270(4.02)	78-1411-65
Acetic acid, (3,5-dihydroxy-4-methoxyphenyl)-	EtOH	275(3.20)	78-1411-65
Opuntiol, acetate	$CHCl_3$	280(3.72)	78-0093-65
4H-Pyran-4-one, 2-(hydroxymethyl)-3-methoxy-, acetate	C_6H_{12}	208(3.80),256(3.94)	78-0093-65
$C_9H_{10}O_6$			
Crotonic acid, 2,4,4-trihydroxy-3-methyl-, γ-lactone, diacetate	EtOH	212(4.08)	44-3560-64
$C_9H_{10}O_{11}$			
1,4-Pentadiene-1,3,5-tricarboxylic acid, 1,2,4,5-tetrahydroxy-3-(hydroxymethyl)-	pH 2	246(3.88)	28-2827-65A
$C_9H_{10}Se$			
Selenochroman	n.s.g.	255(3.91)	20-0483-64
$C_9H_{11}BrN_2$			
1,2-Dimethyl-1H-imidazo[1,2-a]pyridinium bromide	EtOH	210s(4.15),218(4.27), 223(4.25),285(3.92)	4-0331-65
1-Ethylbenzimidazolium bromide	MeOH	248(3.83),254(3.84), 267(3.74),275(3.74), 281(3.67)	65-0632-64
$C_9H_{11}BrN_2O$			
2,3,4,5-Tetrahydro-1-oxo-1H-pyrido-[1,2-a][1,4]diazepinium bromide	EtOH	278.0(3.56)	39-2763-64
2,3,4,5-Tetrahydro-2-oxo-1H-pyrido-[1,2-a][1,3]diazepinium bromide	H_2O	231(4.02),290(4.04)	39-2763-64
$C_9H_{11}BrN_2O_5$			
Uridine, 2'-bromodeoxy-	H_2O	205(3.95),260(4.02)	44-0558-64
$C_9H_{11}BrN_2O_6$			
Uridine, 5-bromo-	pH 1.9	279(3.95)	73-2956-64

Compound	Solvent	λ_{max}(log ϵ)	Ref.
$C_9H_{11}BrN_4$			
Pyrazolo[1,5-a]pyrimidine, 7-amino-3-bromo-2,5,6-trimethyl-	EtOH	232(4.57),280s(3.78), 290(3.88),308(3.87)	95-1113-64
s-Triazolo[1,5-c]pyrimidine, 2-bromo-7-methyl-5-propyl-	MeOH	213(4.49),254(3.79)	39-3357-65
$C_9H_{11}Br_3O_2$			
1,2-Cyclohexanedione, 3,3,6-tribromo-4,4,6-trimethyl-	EtOH	213s(3.61),287s(2.64)	22-3511-65
$C_9H_{11}ClF_4N_2$			
1-Cyclopenten-1-ylamine, 2-chloro-N-ethyl-3-(ethylimino)-4,4,5,5-tetrafluoro-	EtOH	291(4.47)	44-3698-65
$C_9H_{11}ClN_2$			
Quinoxaline, 6-chloro-1,2,3,4-tetrahydro-2-methyl-	EtOH	225(4.48),267(3.54), 321(3.54)	5-0146-65C
$C_9H_{11}ClN_2O_5$			
Uridine, 2'-chlorodeoxy-	H_2O	205(3.97),260(4.01)	44-0558-64
$C_9H_{11}ClN_2S$			
1-Phenylvinylisothiouronium chloride	EtOH	244(4.05)	95-0930-64
$C_9H_{11}ClO$			
m-Cresol, 4-chloro-5-ethyl-	C_6H_{12}	280(3.38),287(3.36)	2-0417-64
$C_9H_{11}Cl_2N_3O_3$			
5-Pyrimidinecarboxamide, N,N-bis(2-chloroethyl)-1,2,3,4-tetrahydro-2,4-dioxo-	pH 1	220(4.04),270(4.07)	65-1304-65
	pH 7	210(4.09),270(3.98)	65-1304-65
	pH 13	292(4.04)	65-1304-65
$C_9H_{11}Cl_3O_4$			
1,4-Dioxaspiro[4.4]non-8-en-7-one, 6,8,9-trichloro-, dimethyl acetal	n.s.g.	215s(3.78),250(2.58)	39-4744-65
$C_9H_{11}FN_2O_4S$			
Uridine, 2'-deoxy-5-fluoro-5'-thio-	pH 1	268(3.95)	44-0554-64
	pH 7	268(3.69)	44-0554-64
	pH 13	268(3.86)	44-0554-64
$C_9H_{11}FN_2O_5$			
Uridine, 2'-deoxy-2'-fluoro-	H_2O	206(3.93),260(3.99)	44-0558-64
$C_9H_{11}FN_2O_6$			
Uridine, 5-fluoro-	pH 6	269(3.95)	73-2956-64
$C_9H_{11}F_5N_2$			
1-Cyclopenten-1-ylamine, N-ethyl-3-(ethylimino)-2,4,4,5,5-pentafluoro-	EtOH	289(4.48)	44-3698-65
$C_9H_{11}I$			
1,2-Heptadien-4-yne, 1-iodo-6,6-dimethyl-	EtOH	215(3.62),240(3.17), 244(3.19)	39-4348-65
$C_9H_{11}IN_2O_4$			
Uridine, 2',3'-dideoxy-3'-iodo-	pH 1	263(4.05)	44-1508-64
	pH 13	263(3.94)	44-1508-64

Compound	Solvent	λ_{max}(log ε)	Ref.
$C_9H_{11}IN_2O_5$			
Uracil, 1-(2-deoxy-β-D-lyxofuranosyl)-5-iodo-	pH 1	255(3.30),288(3.64)	87-0385-64
	H_2O	289(3.81)	87-0385-64
	pH 13	278(3.73)	87-0385-64
$C_9H_{11}IN_2O_6$			
Uridine, 5-iodo-	pH 1.9	289(3.88)	73-2956-64
$C_9H_{11}N$			
Bicyclo[4.2.0]oct-1(8)-ene-8-carbonitrile	EtOH	223(4.08)	39-2165-64
1,6-Iminocyclodecapentaene, N-methyl-	C_6H_{12}	253(4.8),295(3.9), 370s(2.5)	88-3613-65
Indoline, 3-methyl-	MeOH	241(3.83),292(3.37)	33-0756-64
$C_9H_{11}NO$			
Benzaldehyde, p-(dimethylamino)-	MeOH	337(4.46)	88-2729-64
3-Buten-2-one, 4-(1-methylpyrrol-2-yl)-, trans	n.s.g.	214(3.70),357(4.33)	12-0875-65
Carbostyril, 5,6,7,8-tetrahydro-	EtOH	318(3.81)	44-1435-64
Isocarbostyril, 5,6,7,8-tetrahydro-	2N HCl	275(3.83)	39-3856-64
	2N NaOH	235(3.87),291(3.75)	39-3856-64
	EtOH	234(3.69),291(3.81)	39-3856-64
	EtOH	292(3.83)	32-0590-64
Phenol, o-(N-methylacetimidoyl)-	n.s.g.	396(3.77)	77-0464-65
6-Quinolinol, 1,2,3,4-tetrahydro-	pH 1	237(3.75),296(3.96)	44-2860-64
2H-Quinolizin-2-one, 6,7,8,9-tetrahydro-	MeOH-HCl	239.0(3.98)	39-2760-64
	MeOH	260.0(4.17)	39-2760-64
o-Toluidine, N-formyl-N-methyl-	MeOH	242(4.04)	56-0639-65
Tropone, 2-amino-6,7-dimethyl-	isooctane	250(4.4),340(4.0), 395(4.0),410(3.9)	49-0402-64
Tropone imine, 2-methoxy-N-methyl-	EtOH	246(4.46),335(3.92)	94-0457-65
$C_9H_{11}NOS$			
Salicylamide, N-ethylthio-	H_2O	264(3.97)	78-3019-65
	pH 11	267(3.89),286(3.74)	78-3019-65
$C_9H_{11}NO_2$			
Alanine, N-phenyl-	EtOH	242(4.01),291(3.28)	87-0147-65
Anthranilic acid, N,N-dimethyl-	10% MeOH	263s(2.74),270(2.81), 277(2.74)	24-1127-64
anion	10% MeOH	260(3.69),305s(3.1)	24-1127-64
Anthranilic acid, N-ethyl-	10% MeOH	248s(3.11),348(2.58)	24-1127-64
anion	10% MeOH	250(3.92),325(3.49)	24-1127-64
Anthranilic acid, N-methyl-, methyl ester	10% MeOH	254(3.83),350(3.67)	24-1127-64
1H-Azepine-1-carboxylic acid, ethyl ester	C_6H_{12}	216s(4.18),330(3.71)	44-0751-64
1H-Azepine-2,5-dione, 3,4,6-trimethyl-	n.s.g.	228(3.06),288(2.52)	88-1071-65
Cyclohexanepropionitrile, 2,6-dioxo-	MeOH	261(4.21),286s(3.87)	5-0084-65A
	0.05N HCl	261(4.23)	5-0084-65A
	0.05N NaOH	286.5(4.42)	5-0084-65A
	EtOH	258(4.17)	33-0799-65
Indol-2(4H)-one, 5,6-dihydro-3-methoxy-	EtOH	279(4.30)	39-0991-64
Isonicotinic acid, 2,4,6-trimethyl-	pH 1	279(3.89)	30-0939-65
	pH 13	272(3.63)	30-0939-65
Mesitylene, nitro-	heptane	196(4.75),215s(--), 255(3.32),335(2.60)	82-0174-64B

Compound	Solvent	$\lambda_{max}(\log \epsilon)$	Ref.
2-Pyridineacetic acid, ethyl ester	N H_2SO_4	203(3.58),261(3.84)	12-0455-64
	pH 10.0	256s(3.48),260(3.54), 266s(3.42)	12-0455-64
3-Pyridineacetic acid, ethyl ester	N H_2SO_4	259(3.71)	12-0455-64
	pH 10	255(3.38),261(3.45), 266(3.30)	12-0455-64
4-Pyridineacetic acid, ethyl ester	N H_2SO_4	216(3.67),250s(3.60), 254(3.68),259(3.60)	12-0455-64
	pH 10	251s(3.34),256(3.41), 262(3.32)	12-0455-64
Pyrrole-2-acrylic acid, ethyl ester, trans	n.s.g.	262(3.38),331(4.41)	12-0875-65
4H-Quinolizin-4-one, 6,7,8,9-tetra-hydro-3-hydroxy-	EtOH	247(3.64),316(3.92)	44-0526-65
	10% H_2SO_4	228(3.78),303(4.01)	44-0526-65
p-Toluhydroxamic acid, N-methyl-	MeOH	232(3.97)	73-0940-65
	50% MeOH-HCl	236(4.07)	73-0940-65
	50% MeOH-borate	226(4.00)	73-0940-65
$C_9H_{11}NO_3$			
Acetophenone, 2',6'-dihydroxy-4'-methyl-, oxime	EtOH	268(4.22)	39-2543-65
Anisole, 2,5-dimethyl-4-nitro-	MeOH	312(3.88)	44-2897-65
	pH 1	331(3.86)	44-2897-65
	pH 13	331(3.86)	44-2897-65
Benzoic acid, o-(2-aminoethoxy)-	pH 1	234(3.58),289(3.32)	39-1137-65
	pH 13	219(3.64),278(3.16)	39-1137-65
p-Benzoquinone, [(2-hydroxyethyl)-methylamino]-	aq. acid	255(4.2),368(3.3)	44-4107-65
	H_2O	210(4.33),368(4.05), 508(2.05)	44-4107-65
	at 55°	210(4.33),368(4.00), 508(2.44)	44-4107-65
	aq. NaOH	210(4.33),275(3.6), 368(3.3)	44-4107-65
	pH 7.7	220(4.2),370(4.0), 510(1.9)	44-4107-65
	50% MeOH-KOH	220(4.2),310(4.3), 365(3.4),500(1.2)	44-4107-65
Propionic acid, 3-(o-aminophenoxy)-	pH 1	211s(3.88),216(3.91), 268(3.35),274(3.30)	39-1137-65
	pH 13	213(4.50),232(3.82), 283(3.42)	39-1137-65
$C_9H_{11}NO_4$			
2,4,6-Heptatrienoic acid, 7-nitro-, ethyl ester	EtOH	327(4.32)	70-2093-64
Isoxazole-3-carboxylic acid, 4-acetyl-5-methyl-, ethyl ester	C_6H_{12}	229.1(3.83)	32-1478-65
	EtOH	226.2(3.88)	32-1478-65
Pyrrole-2,5-dicarboxylic acid, 1-methyl-, dimethyl ester	EtOH	277(4.34),287(4.29)	39-0459-65
$C_9H_{11}NO_5$			
Oxazole-4,5-dicarboxylic acid, diethyl ester	n.s.g.	241.5(3.95)	24-1414-64
$C_9H_{11}NS$			
4,1-Benzothiazepine, 1,2,3,5-tetrahydro-	iso-PrOH	242(3.81),285(3.24)	44-3111-65

Compound	Solvent	λ_{max}(log ε)	Ref.
$C_9H_{11}N_3$			
Imidazo[1,2-a]pyridine, 5-amino-2,3-dimethyl-	pH 1	231(4.58),271(3.57),323(4.06)	4-0053-65
	pH 11	235(4.52),313(3.96)	4-0053-65
Pyrazolo[1,5-a]pyrimidine, 2,3,5-trimethyl-	EtOH	239(4.65),281(3.32),290(3.23),299s(3.04),332(2.98)	94-1207-65
Pyrazolo[1,5-a]pyrimidine, 2,3,6-trimethyl-	EtOH	240(4.66),285s(3.14),294(3.15),309(3.06),345(3.06)	94-1207-65
$C_9H_{11}N_3O$			
Acetophenone, semicarbazone	EtOH	271(3.93)	33-1581-64
Carbonamide, 2,4-dimethylphenylazo-	n.s.g.	230(3.95),309(4.08),438(2.28)	49-1314-65
Pyrazolo[1,5a]pyrimidin-7(4H)-one, 2,3,4-trimethyl-	EtOH	216(4.38),267(3.86),313(3.71)	94-0142-65
$C_9H_{11}N_3O_2$			
Diacetamide, N-(4-methyl-2-pyrimidinyl)-	pH 1	246(3.54)	44-0115-65
	H_2O	246(3.54)	44-0115-65
$C_9H_{11}N_3O_4$			
2-Furanacrylic acid, α-ethyl-5-nitro-, hydrazide	propylene glycol	244(4.18),270(3.88),362(4.18)	95-0212-64
$C_9H_{11}N_3O_4Se$			
Piperidine, 1-(3,5-dinitroselenophene-2-yl)-	MeOH	228(3.92),265(3.96),400s(4.16),416(4.17)	7-1069-65
$C_9H_{11}N_3O_6$			
Arabinofuranuronic acid, 1-(4-amino-2-oxo-1(2H)-pyrimidinyl)-1-deoxy-	pH 2	279(4.13)	94-0007-65
	H_2O	271(3.99)	94-0007-65
	pH 12	271(3.99)	94-0007-65
1(2H)-Pyrimidinepropionic acid, 3,4-dihydro-5-nitro-2,4-dioxo-, ethyl ester	pH 1	241(3.85),308(4.01)	87-0187-65
	pH 7	239s(3.85),313(3.95)	87-0187-65
	pH 13	244s(3.82),324(3.99)	87-0187-65
$C_9H_{11}N_5$			
Pteridine, 3,4-dihydro-4-imino-3,6,7-trimethyl-	pH 7.0	237(4.12),313s(3.92),318(3.94),340s(3.55)	39-3770-65
$C_9H_{11}N_5O$			
Pteridine, 4-(dimethylamino)-7-methoxy-	MeOH	238(4.17),262(4.27),343(4.19),360(4.11)	4-0023-64
7(8H)-Pteridinone, 4-(dimethylamino)-8-methyl-	MeOH	210(4.22),230(4.24),245(4.10),262(4.05),302(3.74),355(4.01)	4-0023-64
$C_9H_{11}N_5O_2S$			
6-Azalumazine, 7-(ethylthio)-1,3-dimethyl-	H_2O	236(4.35),270(4.15),344(4.03)	24-0005-64
$C_9H_{11}N_5O_3$			
2-Furanmethacrolein, 5-nitro-, amidinohydrazone, hydrochloride	propylene glycol	290(4.16),385(4.10)	95-0001-64
4(3H)-Pteridinone, 2-amino-6-(L-erythro-1,2-dihydroxypropyl)-	pH 1	249(4.07),322(3.86)	37-0560-64
	pH 13	252(4.39),361(3.88)	37-0560-64

Compound	Solvent	λ_{max}(log ε)	Ref.
C_9H_{12}			
Bicyclo[5.2.0]nona-2,5-diene	C_6H_{12}	220(2.87)(end absorption)	5-0010-64A
Bicyclo[6.1.0]nona-2,4-diene	C_6H_{12}	234(3.81)	5-0010-64A
	EtOH	234(3.75)	35-3158-65
Bicyclo[6.1.0]nona-2,5-diene	C_6H_{12}	220(2.99)(end absorption)	5-0010-64A
Bicyclo[6.1.0]nona-3,5-diene	C_6H_{12}	240(3.48)(end absorption)	5-0010-64A
Cumene	pet ether	268(2.38)	65-2088-65
	EtOH-HCl	267(2.38)	65-2088-65
	EtOH	268(2.38)	65-2088-65
	10% H_2SO_4	265(2.55)	65-2088-65
Cycloheptene, 1-ethynyl-	n.s.g.	228(4.04)	22-1525-65
Cyclohexene, 1-propynyl-	n.s.g.	227(4.16)	22-1525-65
	n.s.g.	224(4.06)	44-3991-65
1,3,5-Cyclononatriene	n.s.g.	290(3.31)	88-0391-65
1,3,5-Cyclononatriene, all cis	C_6H_{12}	296(3.60)	88-0377-65
1,3,6-Cyclononatriene	MeOH	223(3.61)	5-0010-64A
1,4,7-Cyclononatriene	C_6H_{12}	220(3.34)(end absorption)	5-0010-64A
1,3,6,8-Nonatetraene	EtOH	237(4.51),261(3.91), 272(3.76)	44-2410-65
$C_9H_{12}Br_4$			
Propane, 1,3-bis(2,2-dibromocyclopropyl)-	hexane	199.5(3.67)	44-2951-64
$C_9H_{12}ClN$			
Pyridine, 3-chloro-2,4,5,6-tetramethyl-	EtOH	274(3.64)	32-0083-65
$C_9H_{12}ClN_3O$			
3(2H)-Pyridazinone, 4-chloro-5-piperidino-	EtOH	243(4.22),327(3.86)	22-2124-64
	NaOH	229(4.30),304(3.79)	22-2124-64
$C_9H_{12}ClN_3OS$			
3(2H)-Pyridazinethione, 6-chloro-2-(morpholinomethyl)-	THF	304(4.24)	49-0631-65
$C_9H_{12}ClN_3O_4$			
5-Pyrimidinecarboxylic acid, 1,2,3,4-tetrahydro-2,4-dioxo-, 2-[(2-chloroethyl)amino]ethyl ester	pH 1	220(4.05),270(4.08)	65-1304-65
	pH 7	223(4.07),275(4.08)	65-1304-65
	pH 13	240(3.99),292(4.20)	65-1304-65
$C_9H_{12}ClN_3O_5S_2$			
2H-1,2,4-Benzothiadiazine-7-sulfonamide, 6-chloro-3,4-dihydro-3-(1-hydroxyethyl)-, 1,1-dioxide	EtOH	226(4.59),270(4.31)	95-0095-65
$C_9H_{12}Cl_2O_2$			
2-Cyclohexen-1-one, 2,4-dichloro-3-methoxy-5,5-dimethyl-	MeOH	278(4.19)	20-0081-64
2-Cyclohexen-1-one, 2,6-dichloro-3-methoxy-5,5-dimethyl-	MeOH	280(4.13)	20-0081-64
$C_9H_{12}FN_3O_5$			
Cytidine, 5-fluoro-	pH 1.9	290(4.04)	73-2956-64
$C_9H_{12}F_3NO_4$			
Malonic acid, (1-amino-2,2,2-trifluoroethylidene)-, diethyl ester	EtOH	279(4.12)	44-0707-64

Compound	Solvent	$\lambda_{max}(\log \epsilon)$	Ref.
$C_9H_{12}IN_3O_3$			
Cytidine, 2',5'-dideoxy-5'-iodo-	MeOH	235(3.92),270(3.96)	44-3067-65
$C_9H_{12}NO_2$			
Pyridyl, 4-carboxy-1-ethyl-1,4-dihydro-, methyl ester	MeCN	304(4.05),395(3.67), 585(1.56),632(1.92), 690(1.86),775(1.56)	35-5515-64
$C_9H_{12}NO_5PS$			
Phosphorothioic acid, O,O-dimethyl S-3-nitro-o-tolyl ester	MeOH	200(4.41)	78-0177-64
$C_9H_{12}NO_6PS$			
Phosphorothioic acid, S-(2-methoxy-5-nitrophenyl) O,O-dimethyl ester	MeOH	200(4.19),233(3.98), 297(3.97)	78-0177-64
$C_9H_{12}N_2$			
Benzeneazoethane, 1'-methyl-	hexane	261(3.99),408(2.09)	39-2788-65
Benzimidazoline, 1,3-dimethyl-	EtOH	217(4.57),265(3.76), 310(3.76)	12-0877-64
Pyrroline, 2-(3-methylpyrrol-2-yl)-	EtOH-HCl	318(4.42)	44-2727-64
	EtOH-NaOH	281(4.33)	44-2727-64
Pyrroline, 2-(4-methylpyrrol-2-yl)-	EtOH-HCl	323(4.36)	44-2727-64
	EtOH-NaOH	285(4.19)	44-2727-64
Pyrroline, 2-(5-methylpyrrol-2-yl)-	EtOH-HCl	330(4.45)	44-2727-64
	EtOH-NaOH	290(4.19)	44-2727-64
Pyrroline, 5-methyl-2-(2-pyrrolyl)-	EtOH-HCl	318(4.36)	44-2727-64
	EtOH-NaOH	279(4.28)	44-2727-64
$C_9H_{12}N_2O$			
Aniline, N-isopropyl-N-nitroso-	C_6H_{12}	224(3.89),250(3.74)	35-4373-64
7-Azaindoline, 6-methoxy-4-methyl-	EtOH	247(3.76),306(3.92)	65-1454-64
Benzamide, N-(2-aminoethyl)-	EtOH-HCl	227(4.45)	54-0314-65
Benzamidine, N-ethyl-o-hydroxy-	pH 3	288(3.45)	78-3019-65
	pH 7.11	340(3.84)	78-3019-65
Cyclooct[c]isoxazole, 3-amino-4,5,8,9-tetrahydro-	EtOH	251(3.92)	88-2151-64
Indazole, 3-acetyl-4,5,6,7-tetrahydro-	EtOH	238(4.03),267s(3.56)	32-0814-65
Isonicotinonitrile, 5-acetyl-1,2,3,4-tetrahydro-1-methyl-	EtOH	298(4.48)	44-2534-65
Propiophenone, 2-amino-, oxime, anti	EtOH-HCl	233(3.84)	24-0567-65
	EtOH	228.5(3.82)	24-0567-65
Tropone, 2-(2,2-dimethylhydrazino)-	EtOH	240(4.43),336(4.13), 400(4.09)	94-0457-65
$C_9H_{12}N_2O_2$			
p-Benzoquinone, 3-methyl-2,5-bis-(methylamino)-	EtOH	217(4.36),343(4.39), 524(2.52)	39-5569-64
Nipecotonitrile, 3-acetyl-1-methyl-2-oxo-	MeOH	208(3.81)	24-1970-64
Picolinamide, N-(3-hydroxypropyl)-	H_2O	217(3.99),263(3.73)	39-2763-64
Propane, 2-hydroperoxy-2-phenylazo-	hexane	267(3.98),411(2.13)	39-2788-65
Pyridine, 1,4-diacetyl-1,4-dihydro-, oxime	EtOH	252(4.23)	44-2898-64
$C_9H_{12}N_2O_2S$			
2,4-Diazaspiro[5.5]undecane-1,3,5-trione, 3-thio-	70% EtOH	235(3.93),286(4.32)	87-0695-64

Compound	Solvent	λ_{max}(log ε)	Ref.
$C_9H_{12}N_2O_2S_2$			
p-Dithiin-2,3-dicarboximide, N-[(di-methylamino)methyl]-5,6-dihydro-	EtOH	250(4.00),407(3.52)	56-1713-65
p-Toluenesulfonamide, N-[(methylthio)-aminomethylene]-	EtOH	242.5(4.28)	95-0391-65
$C_9H_{12}N_2O_2S_3$			
Benzenesulfonamide, N-bis(methylthio)-methylene-p-amino-	EtOH	259(4.30),298(4.18)	95-0391-65
$C_9H_{12}N_2O_2Se$			
Piperidine, 1-(3-nitroselenophene-2-yl)-	MeOH	220(3.87),265(4.09), 295s(3.81),406(3.79)	7-1069-65
Piperidine, 1-(5-nitroselenophene-2-yl)-	MeOH	250(3.47),315(2.71), 458(4.54)	7-1069-65
$C_9H_{12}N_2O_3$			
γ-Butyrolactone, α-[(3-allylureido)-methylene]-	EtOH	272(4.45)	65-2599-64
2,4-Diazaspiro[5.5]undecane-1,3,5-trione	70% EtOH-NaOH	247.0(3.92)	87-0695-64
Piperidine, 1-(5-nitro-2-furyl)-	MeOH	228(3.99),440(4.31)	65-0710-64
Pyrazinemethanol, 3-ethoxy-, acetate	MeOH	216(4.04),294(3.88)	44-2623-64
4-Pyrimidinepropionic acid, α-methoxy-, methyl ester	EtOH	245(3.59),270s(2.71)	44-2398-65
2-Pyrroline-1-carboxylic acid, 2-cyano-3-methoxy-, ethyl ester	MeOH	268(3.98)	35-5293-64
$C_9H_{12}N_2O_4$			
4,5-Imidazoledicarboxylic acid, diethyl ester	n.s.g.	258(3.98)	24-1414-64
Orotic acid, butyl ester	pH 1	280(3.88)	65-1299-65
	pH 7	278(3.87)	65-1299-65
	pH 13	278(3.72)	65-1299-65
Uridine, 2',3'-dideoxy-	pH 2	263(4.00)	44-1508-64
	pH 12	263(3.86)	44-1508-64
$C_9H_{12}N_2O_5$			
3(2H)-Pyridazinone, 2-(2-deoxy-D-ribofuranosyl)-6-hydroxy-	H_2O	313(3.398)	73-3744-65
	Et.H	317(3.469)	73-3744-65
4(3H)-Pyrimidinone, 3-β-D-ribo-furanosyl-	MeOH	215s(3.80),274(3.58)	24-1511-65
Uracil, 1-(2-deoxy-β-D-lyxofuranosyl)-	pH 1	262(3.98)	87-0385-64
	H_2O	263(4.03)	87-0385-64
	pH 13	262(3.85)	87-0385-64
$C_9H_{12}N_2O_5S$			
Uridine, 5'-thio-	pH 1,7	262(3.86)	44-0554-64
	pH 13	262(3.81)	44-0554-64
$C_9H_{12}N_2O_6$			
3(2H)-Pyridazinone, 6-hydroxy-2-β-D-ribofuranosyl-	H_2O	316(3.505)	73-3744-65
	EtOH	317(3.487)	73-3744-65
Uracil, 1-β-D-arabinofuranosyl-	pH 7	263(4.05)	94-0803-65
	pH 9.6	263(3.97)	94-0803-65
Uracil, 1-α-D-lyxofuranosyl-	pH 3.2	263(4.04)	94-0803-65
	pH 7	263(4.04)	94-0803-65
	pH 9.6	264(4.00)	94-0803-65
	pH 14	266(3.94)	94-0803-65

Compound	Solvent	λ_{max}(log ε)	Ref.
Uracil, 1-β-D-lyxofuranosyl-	pH 3.2	263(4.02)	94-0803-65
	pH 7	263(4.02)	94-0803-65
	pH 9.5	264(4.01)	94-0803-65
	pH 14	266(3.95)	94-0803-65
Uracil, 1-α-D-ribofuranosyl-	pH 3-6	264(4.00)	94-1471-64
Uracil, 1-β-D-ribofuranosyl-	H_2O	262(4.01)	94-1471-64
$C_9H_{12}N_2O_7$			
Barbituric acid, 1-β-D-ribofuranosyl-, sodium salt	pH 1.0	257(3.07)	94-0459-64
	pH 6.0	259(4.18)	94-0459-64
	pH 13.0	263(4.05)	94-0459-64
$C_9H_{12}N_3O_6P$			
Cytidine, deoxy-, 3',5'-cyclic phosphate	pH 2.0	278(4.13)	35-1626-64
	pH 7.0	271(3.99)	35-1626-64
$C_9H_{12}N_4$			
4,7-Diazagramine	H_2O	215(4.18),307(3.97)	44-3454-65
Pyrazolo[1,5-a]pyrimidine, 7-amino-5-ethyl-6-methyl-	EtOH-HCl	228(4.58),263(3.59), 269(3.57),287(3.54), 319(3.85)	95-1113-64
Pyrazolo[1,5-a]pyrimidine, 7-amino-2,3,6-trimethyl-	EtOH	230(4.53),282s(3.71), 292(3.79),327(3.72)	94-0142-65
Pyrazolo[1,5-a]pyrimidine, 4,7-dihydro-7-imino-2,3,4-trimethyl-	EtOH	229(4.36),265(3.69), 322(3.61)	94-0142-65
Pyrazolo[1,5-a]pyrimidine, 4,7-dihydro-7-imino-3,4,6-trimethyl-	EtOH	231(4.44),272s(3.86), 320(3.85)	94-0142-65
Pyrazolo[1,5-a]pyrimidine, 2,3-dimethyl-7-(methylamino)-	EtOH	235(4.58),280(3.73), 290(3.83),321(3.79)	94-0142-65
4-Pyrimidinecarbamonitrile, 6-methyl-2-propyl-	MeOH	208(4.08),241(4.52)	39-3357-65
s-Triazolo[4,3-a]pyrazine, 3-ethyl-5,6-dimethyl-	C_6H_{12}	218(4.53),224(4.46), 260s(3.37),272s(3.41), 281(3.43),304s(3.56), 310(3.58),318s(3.53), 333s(3.22)	94-0204-64
s-Triazolo[1,5-a]pyrimidine, 2,5,6,7-tetramethyl-	EtOH	282(3.76)	94-0204-64
$C_9H_{12}N_4O$			
2-Penten-2-ol, 4-[(4-amino-5-pyrimidinyl)-imino]-	pH 1	280(4.03)	50-0970-64C
	pH 10	231(3.98),314(4.28)	50-0970-64C
	EtOH	232(4.01),314(4.21)	50-0970-64C
Pyrido[2,3-d]pyrimidin-7(6H)-one, 2-amino-5,8-dihydro-4,8-dimethyl-	pH 1	229(4.36),255(4.02), 302(3.94)	4-0263-64
	pH 8.4	224(4.50),302(4.02)	4-0263-64
	pH 13	290(3.97)	4-0263-64
5H-Pyrrolo[3,4-d]pyrimidine, 6-acetyl-4-amino-6,7-dihydro-2-methyl-	pH 1	248(4.02)	44-0194-65
	EtOH	235(4.00),268(3.74)	44-0194-65
5H-Pyrrolo[3,4-d]pyrimidine, 6-acetyl-4-amino-6,7-dihydro-7-methyl-	pH 1	249(3.99)	44-0194-65
	EtOH	236(4.12),269(3.44)	44-0194-65
4H-Pyrrolo[2,3-d]pyrimidin-4-one, 2-amino-3,7-dihydro-7-propyl-	pH 1	227(4.24),265(4.02)	4-0034-64
	pH 11	247(3.89),266(4.02)	4-0034-64
1,2,4-Triazole, 3-(2-furyl)-5-(propylamino)-	pH 2	268(4.22)	1-1191-65
	pH 12	275(4.12)	1-1191-65
1,2,4-Triazole, 3-(dimethylamino)-5-(2-furyl)-4-methyl-	pH 2	269(4.16)	1-1191-65
	EtOH	272(4.16)	1-1191-65
s-Triazolo[4,3-c]pyrimidin-3-ol, 7-methyl-5-propyl-	MeOH	207(4.00),247(4.02), 317(3.80)	39-3357-65

Compound	Solvent	λ_{max}(log ϵ)	Ref.
$C_9H_{12}N_4O_2$			
9H-Purine-9-ethanol, 6-methoxy-α-methyl-	pH 1,7,13	252(4.08)	87-0502-65
Pyrazolo[4,3-d]pyrimidine-dione, 1,4(or 6)-diethyl-	MeOH	290.5(3.76)	44-0199-65
$C_9H_{12}N_4O_2S$			
Sulfanilamide, N[1]-2-imidazolidin-ylidene-	EtOH	262.0(4.35)	95-0391-65
$C_9H_{12}N_4O_3$			
4-Pyrimidineacetic acid, 6-amino-5-formamido-, ethyl ester	pH 1	251(4.04)	44-1528-65
	pH 7	233(4.12),276(3.68)	44-1528-65
	pH 13	249(4.02),281(3.82)	44-1528-65
$C_9H_{12}N_4O_4$			
Acrylamide, N-carbamoyl-α-cyano-β-(ethoxycarbonylmethylamino)-	pH 2.5	290(4.39)	39-1642-65
	H_2O	290(4.39)	39-1642-65
	pH 11.5	292(4.29)	39-1642-65
$C_9H_{12}N_4O_5$			
4-Pyrimidinecarboxylic acid, 6-amino-2-ethoxy-5-nitro-, ethyl ester	pH -4.9	218(4.09),292(3.83)	39-3221-64
	pH 2.6	236(3.96),250s(3.76), 343(3.97)	39-3221-64
	pH 8.6	227(3.98),346(3.88)	39-3221-64
$C_9H_{12}N_4S$			
s-Triazolo[4,3-c]pyrimidine-3-thiol, 7-methyl-5-propyl-	MeOH	238(3.78),298(4.04), 347s(3.60)	39-3369-65
$C_9H_{12}N_5O_7P$			
1,2,3-Propanetriol, 1-(2-amino-4-hydroxy-6-pteridinyl)-, 3-(dihydrogen phosphate)	pH 1	240(4.35),326(3.87)	10-0444-65B
	pH 13	254(4.36),365(3.81)	10-0444-65B
$C_9H_{12}N_6$			
1,2,4-Triazole, 3-hydrazino-5-p-toluidino-	EtOH	258(4.32)	39-3912-65
$C_9H_{12}N_6O_3$			
Cytidine, 5'-azido-2',5'-dideoxy-	MeOH	235(3.89),270(3.93)	44-3067-65
3H-v-Triazolo[4,5-d]pyrimidine, 7-amino-3-(2-deoxy-α-D-erythro-pentofuranosyl)-	pH 1	263(4.10)	10-0076-65D
	pH 7	278(4.08)	10-0076-65D
	pH 13	278(4.08)	10-0076-65D
3H-v-Triazolo[4,5-d]pyrimidine, 7-amino-3-(2-deoxy-β-D-erythro-pentofuranosyl)-	pH 1	263(4.09)	10-0076-65D
	pH 7	278(4.07)	10-0076-65D
	pH 13	278(4.06)	10-0076-65D
$C_9H_{12}O$			
Anisole, 2,3-dimethyl-	C_6H_{12}	217s(3.88),270(3.13), 274(3.13),279(3.19)	23-2603-65
Anisole, 2,6-dimethyl-	C_6H_{12}	215s(3.95),268(2.59), 273(2.54)	23-2603-65
Benzaldehyde, dimethyl acetal	C_6H_{12}	251(2.11),257(2.24), 261(1.98),263(2.13), 267(1.64)	39-5957-64
1,3,5-Cycloheptatrien, 1-ethoxy-	EtOH	207(4.23),290(3.58)	35-0321-65
2-Cyclohexen-1-one, 4-propylidene	MeOH	286(4.18)	5-0028-64D
2-Cyclopenten-1-one, 2-allyl-3-methyl-	EtOH	236(4.08)	35-0936-64
Dispiro[2.1.2.1]nonan-4-one	n.s.g.	283(1.64)	88-3189-64

Compound	Solvent	$\lambda_{max}(\log \epsilon)$	Ref.
3,5-Hexadien-2-one, 6-cyclopropyl-	EtOH	294(4.34)	22-3218-64
4,5-Nonadien-2-yn-1-ol	EtOH	220(4.20)	39-4659-65
4,5-Octadien-2-yn-1-ol, 6-methyl-	EtOH	220(4.16)	39-4659-65
4,5-Octadien-2-yn-1-ol, 7-methyl-	EtOH	220(4.14)	39-4659-65
2,4,6-Octatrienal, 7-methyl-	isooctane	311(4.64),324(4.64)	35-5075-65
Phenol, 2-isopropyl-	hexane	258(3.38),268(3.38)	65-2088-65
	EtOH	275(3.38)	65-2088-65
	EtOH-HCl	258(3.78)	65-2088-65
	EtOH-NaOEt	242(4.18),286(3.78)	65-2088-65
	10% H_2SO_4	285(3.38)	65-2088-65
Phenol, 3-isopropyl-	hexane	279(3.78)	65-2088-65
	EtOH	268(3.78)	65-2088-65
	EtOH-HCl	276(3.49)	65-2088-65
	EtOH-NaOEt	239(4.09),291(3.88)	65-2088-65
Phenol, 4-isopropyl-	hexane	273(3.78)	65-2088-65
	EtOH	275(3.78)	65-2088-65
	EtOH-HCl	275(3.60)	65-2088-65
	EtOH-NaOEt	241(4.08),283(3.78)	65-2088-65
	10% H_2SO_4	288(3.60)	65-2088-65
$C_9H_{12}OS$			
Sulfoxide, ethyl p-tolyl	isooctane	210s(4.01),218s(3.97), 251s(3.70),270s(3.26), 279(2.88)	35-1958-65
	EtOH	218s(3.90),240(3.78), 265s(3.05),275s(2.63)	35-1958-65
(+)	isooctane	251(3.78)	35-5637-64
	EtOH-$HClO_4$	239(3.96),268s(3.16), 274s(2.99)	35-5637-64
	EtOH	215s(3.96),238(3.84), 262s(3.31),274s(3.01)	35-5637-64
$C_9H_{12}O_2$			
Bicyclo[3.2.1]oct-2-ene-2-carboxylic acid	EtOH	224(3.92)	35-4928-64
1,5-Cyclohexadiene-1-carboxylic acid, 3-methyl-, methyl ester	EtOH	282(3.42)	44-3679-65
2,4-Cyclohexadiene-1-carboxylic acid, 1-methyl-, methyl ester	n.s.g.	256(3.72)	44-3596-64
1,3-Cyclopentadiene-1-carboxylic acid,	EtOH	296(4.02)	70-1653-64
2,3,4-trimethyl-	EtOH	296(4.00)	70-1820-64
Cyclopentadienepropionic acid, methyl ester	C_6H_{12}	249(3.43)	44-3430-64
1,3-Cyclopentanedione, 4-allyl-5-methyl-	neutral	243(4.22)	39-3097-65
	basic	260(4.41)	39-3097-65
4-Cyclopentene-1,3-dione, 4-tert-butyl-	MeOH	222(3.90),303(2.78), 375(1.48)	24-2774-65
2-Cyclopenten-1-one, 2-(3-oxobutyl)-	EtOH	227(4.09)	44-1294-65
Fulvene, 6-carbethoxy-3,4-dihydro-	EtOH	268(4.22)	44-3430-64
4-Hepten-6-ynoic acid, 4,5-dimethyl-	EtOH	230(4.18)	70-1653-64
	EtOH	231(3.92),295(3.04)	70-1820-64
$\Delta^{2,\alpha}$-Norbornaneacetic acid	EtOH	228(4.05)	44-3241-64
2,5-Norbornanedione, 1,4-dimethyl-	MeOH	287.5(1.67)	23-2306-65
4H-Pyran-4-one, 2-ethyl-3,5-dimethyl-	EtOH	260(4.07)	24-1949-65
2H-Pyran-2-one, 4,5-trimethylene-6-methyl-	EtOH	304(3.31)	44-2642-65

Compound	Solvent	$\lambda_{max}(\log \epsilon)$	Ref.
$C_9H_{12}O_2S$			
2-Furaldehyde, 5-[(isopropylthio)-methyl]-	EtOH	285(4.25)	70-1281-65
3-Thiophenecarboxylic acid, 2,4-dimethyl-, ethyl ester	Et_2O	213(3.96),242(3.92)	24-2109-64
$C_9H_{12}O_3$			
Benzene, 3,4-dihydroxy-5-methoxy-1,2-dimethyl-	EtOH	213(4.18),270(2.86)	39-4672-65
1,3-Cycloheptanedione, 2-acetyl-	C_6H_{12}	236(3.92),280(3.99)	1-1368-64
1-Cyclohexene-1-carboxylic acid, 2,6-dimethyl-3-oxo-	EtOH	245(4.00)	44-3524-64
1,3-Cyclopentanedione, 2-acetyl-4,4-dimethyl-	C_6H_{12}	220(4.23),265(3.87)	1-1368-64
2-Cyclopentene-1-carboxylic acid, 3,4-di-methyl-2-oxo-, methyl ester	EtOH	239(4.11)	70-0164-64
2-Cyclopenten-1-one, 2-hydroxy-5-methyl-3-propionyl-	EtOH	268(3.83),305(2.53)	44-3520-64
	EtOH-base	355(4.04)	44-3520-64
2-Cyclopenten-1-one, 3-(3-oxobutyl)-2-hydroxy-	EtOH	260(4.11)	44-2513-65
3-Furoic acid, 2-ethyl-4,5-dimethyl-	MeOH	259(3.59)	39-5984-65
4-Hepten-2-ynoic acid, 5-ethoxy-	EtOH	279(4.16)	28-0594-64B
3,5-Nonadiyne-1,2,7-triol	EtOH	220(2.61),228(2.65), 242(2.56),256(2.36)	39-1476-64
2-Norbornene-2-carboxylic acid, 6-(hydroxymethyl)-, endo	EtOH	231(3.86)	44-3234-64
$C_9H_{12}O_4$			
Cycloheptanecarboxylic acid, 3,5-dioxo-, methyl ester	n.s.g.	249(3.70)	88-2711-64
Cyclohexanecarboxylic acid, 2,4-dioxo-, ethyl ester	EtOH	256(4.18)	44-0787-64
1,3-Cyclopentanedione, 2-acetyl-5-hydroxy-4,4-dimethyl-	EtOH	225(4.00),268(3.99)	39-6543-65
	EtOH-base	250(4.29),270(4.21)	39-6543-65
Genipic acid	EtOH	203(3.45)	78-1781-64
Malonic acid, isopropylidene-, cyclic isopropylidene ester	EtOH-HCl	237.5(3.94)	49-1283-64
	EtOH-NaOH	266(4.20)	49-1283-64
Meldrumic acid, isopropylidene-	MeOH	237(3.95)	88-0721-65
$C_9H_{12}O_6S$			
Maleic acid, α-acetoxy-α'-(methylthio)-, dimethyl ester	MeOH	281.5(4.13)	24-1581-65
$C_9H_{13}BrO_2$			
2-Cyclopenten-1-one, 3-bromo-2-hydroxy-4,4,5,5-tetramethyl-	EtOH	268(4.0)	5-0100-64I
$C_9H_{13}ClN_2O_6$			
m-Nitrophenyltrimethylammonium perchlorate	n.s.g.	253.5(3.86)	39-6851-65
p-Nitrophenyltrimethylammonium perchlorate	n.s.g.	252(3.99)	39-6851-65
$C_9H_{13}ClN_4O$			
3(2H)-Pyridazinone, 4-chloro-5-(4-methyl-1-piperazinyl)-	EtOH	239(4.23),321(3.85)	22-2124-64
	NaOH	228(4.32),303(3.78)	22-2124-64

Compound	Solvent	$\lambda_{max}(\log \epsilon)$	Ref.
$C_9H_{13}ClO$			
1,3-Cycloheptadiene, 3-chloro-2-ethoxy-	EtOH	203(4.06),243(3.6)	35-0321-65
1-Cyclooctene-1-carboxaldehyde, 2-chloro-	EtOH	258(4.05)	44-1126-65
$C_9H_{13}Cl_2N$			
Pyrrole, 2-(dichloromethyl)-2,3,4,5-tetramethyl-	EtOH	247(3.45)	32-0083-65
$C_9H_{13}Cl_2NO_2$			
3(2H)-Furanone, 2,2-dichloro-5-(diethylamino)-4-methyl-	EtOH	205s(3.89),310(4.31)	44-4303-65
$C_9H_{13}Cl_3O_4$			
4-Cyclopentene-1,3-dione, 2,4,5-trichloro-, bis(dimethyl acetal)	n.s.g.	215s(3.83),250(2.65)	39-4744-65
$C_9H_{13}N$			
2,3-Lutidine, 6-ethyl-	n.s.g.	255(2.9)	70-0322-65
Quinoline, 2,3,4,6,7,8-hexahydro-	0.05N HCl	265(3.94)	5-0084-65A
	0.05N NaOH	235(3.95)	5-0084-65A
	50% MeOH	230s(3.71),265(3.83)	5-0084-65A
$C_9H_{13}NO$			
m-Anisidine, N,N-dimethyl-	EtOH	294(--)	95-0858-65
Bicyclo[4.2.0]oct-1(8)-ene-8-carboxamide	EtOH	223(4.08)	39-2165-64
α,β-Butenolide, β-methyl-α-pyrrolidino-	EtOH	290(3.67)	44-3560-64
Carbostyril, 3,4,5,6,7,8-hexahydro-	EtOH	255(3.74)	32-0590-64
Carbostyril, 4a,5,6,7,8,8a-hexahydro-	EtOH	210(4.04)	32-0584-64
Ethylamine, 1-methyl-2-(p-hydroxyphenyl)-	EtOH-H_2SO_4	277(3.24)	54-0521-65
(+)	EtOH-H_2SO_4	277.5(3.23)	54-0521-65
Isocarbostyril, 3,4,5,6,7,8-hexahydro-	EtOH	216(4.06)	32-0584-64
Phenethylamine, p-hydroxy-N-methyl-	pH 1	222(3.84),275(3.20), 280s(3.42)	87-0368-65
	pH 13	239(4.02),294(3.42)	87-0368-65
Pyridine, 2-ethoxymethyl-5-methyl-	EtOH	267(3.49)	88-0153-65
2(1H)-Pyridone, 6-methyl-3-propyl-	EtOH	233(3.83),307(3.92)	39-1632-64
5(1H)-Quinolone, 2,3,4,6,7,8-hexahydro-	EtOH	300(4.41)	33-0799-65
$C_9H_{13}NO_2$			
Ethanolamine, 2-(4-hydroxy-3-methoxyphenyl)-	pH 1	208(3.84),229(3.81), 279(3.44),283s(3.39)	87-0368-65
	pH 13	218(3.84),247(4.00), 294(3.61)	87-0368-65
Ethanolamine, N-methyl-2-(4-hydroxyphenyl)-	pH 1	224(3.86),273(3.13), 279s(3.06)	87-0368-65
	pH 13	242(4.08),292(3.38)	87-0368-65
Ethylamine, 2-(p-hydroxyphenyl)-2-methoxy-	n.s.g.	227(4),277(3.23)	22-0817-64
2-Heptenoic acid, 2-cyano-3-methyl-	EtOH	232(4.07)	28-2859-64B
2-Hexenoic acid, 2-cyano-, ethyl ester	EtOH	223(4.00)	28-2859-64B
2-Pentenoic acid, 2-cyano-3-methyl-, ethyl ester	EtOH	231.5(4.07)	28-2859-64B
2-Pentenoic acid, 2-cyano-4-methyl-, ethyl ester	EtOH	222(4.01)	28-2859-64B
Pyrrole-2-carboxylic acid, 4,5-dimethyl-, ethyl ester	EtOH	238(3.34),286(3.97)	12-1977-65

Compound	Solvent	λ_{max}(log ε)	Ref.
Pyrrole-2-carboxylic acid, 4-isopropyl-, methyl ester	EtOH	233(3.95),274(4.20)	23-1279-64
Pyrrole-2-carboxylic acid, 5-isopropyl-, methyl ester	EtOH	233(3.46),278(4.31)	23-1279-64
3H-Pyrrolo[1,2-a]azepine-3,5(2H)-dione, hexahydro-	THF	233(3.23),279(1.89)	24-1548-64
2H-Quinolizine-4,6(1H,3H)-dione, tetrahydro-	THF	235(3.42),283(1.92)	24-1548-64
$C_9H_{13}NO_3$			
Ethanolamine, N-methyl-2-(3,4-dihydroxyphenyl)-	pH 1	208(3.84),224(3.78), 280(3.44),285s(3.40)	87-0368-65
2-Piperidine, 3-carbethoxy-1-methyl-	MeOH	208(3.81)	24-1970-64
$C_9H_{13}NO_3S$			
3-Thiophenecarboxylic acid, 2-acetamido-4,5-dihydro-, ethyl ester	MeOH	232(3.97),318(4.26)	24-1970-64
$C_9H_{13}NO_4$			
Crotonic acid, 3-[(dimethylamino)-methyl]-2,4-dihydroxy-, γ-lactone, acetate, hydrochloride	EtOH	210(3.96)	44-3560-64
$C_9H_{13}NS$			
2H-Azepine-2-thione, 1,3-dihydro-3,5,7-trimethyl-	EtOH	244(3.91),310(4.13), 364s(3.02)	35-4096-64
Piperidine, 1-(3-thienyl)-	C_6H_{12}	274(3.57)	54-1160-64
$C_9H_{13}N_3$			
Pyrazine, 2-methyl-3-(1-pyrrolidinyl)-, hydrochloride	H_2O	255(4.02),336(3.68)	44-0415-64
Pyrazine, 2-methyl-6-(1-pyrrolidinyl)-, hydrochloride	H_2O	251(4.15),350(3.80)	44-0415-64
$C_9H_{13}N_3O$			
Pyrazine, 3-morpholino-2-methyl-, hydrochloride	H_2O	235(3.75),248(3.77), 314(3.64)	44-0415-64
Pyrazine, 6-morpholino-2-methyl-, hydrochloride	H_2O	250(4.04),333(3.70)	44-0415-64
$C_9H_{13}N_3OS$			
3-Acetothienone, 2,5-dimethyl-, semicarbazone	EtOH	233(4.4),270(4.1)	70-2055-64
$C_9H_{13}N_3OS_2$			
3(2H)-Pyridazinethione, 6-mercapto-2-(morpholinomethyl)-	THF	310(4.37)	49-0631-65
$C_9H_{13}N_3O_2S$			
4-Thiazolecarboxamide, 5-acetamido-N,N,2-trimethyl-	EtOH	286(4.19)	94-1319-65
$C_9H_{13}N_3O_3$			
Cytidine, 2',5'-dideoxy-	MeOH	230(3.87),272(3.91)	44-3067-65
$C_9H_{13}N_3O_3S$			
Morpholine, 4-[6-(methylsulfonyl)-4-pyrimidinyl]-	pH 1	257(4.26),304(3.78)	4-0001-64
	pH 11	256(4.30),310(3.70)	4-0001-64

Compound	Solvent	$\lambda_{max}(\log \epsilon)$	Ref.
$C_9H_{13}N_3O_4$			
1(2H)-Pyrimidinepropionic acid, 5-amino-3,4-dihydro-2,4-dioxo-, ethyl ester	pH 1	267(3.92)	87-0187-65
	pH 7	224s(3.82),297(3.82)	87-0187-65
	pH 13	255s(3.94),292(3.75)	87-0187-65
1(2H)-Pyrimidinevaleric acid, α-amino-3,4-dihydro-2,4-dioxo-	pH 1	272(3.61)	65-2179-64
	H_2O	270(3.89)	65-2179-64
	pH 12	268(3.96)	65-2179-64
as-Triazine-6-butyric acid, 3,4,5,6-tetrahydro-3,5-dioxo-, ethyl ester	MeOH	210(3.83),252(3.77)	39-0868-64
as-Triazine-6-hexanoic acid, 3,4,5,6-tetrahydro-3,5-dioxo-	MeOH	211(3.81),254(3.78)	39-0868-64
$C_9H_{13}N_3O_5$			
Cytidine, 2'-deoxy-N-hydroxy-	pH 1	219(3.93),278(4.11)	39-0208-65
	pH 7	234(4.05),272(3.80)	39-0208-65
Cytosine, 1-β-D-arabinofuranosyl-	pH 2	213(4.01),281(4.12)	44-0835-65
	pH 12	227s(3.92),273(3.97)	44-0835-65
	pH 12	227s(3.88),272(3.93)	44-0835-65
1(2H)-Pyrimidinevaleric acid, α-amino-tetrahydro-2,4,6-trioxo-	pH 1	260(3.608)	65-2179-64
	H_2O	257(3.959)	65-2179-64
	pH 12	260(3.892)	65-2179-64
$C_9H_{13}N_3O_6$			
Cytidine, 3-oxide	pH 7	225(4.35),272(3.82)	44-2766-65
$C_9H_{13}N_3O_7$			
s-Triazine-2,4(1H,3H)-dione, 1-β-D-glucopyranosyl-	EtOH	240(3.25)	73-2060-64
$C_9H_{13}N_3S$			
Semicarbazide, 4-(α-methylbenzyl)-3-thio-, nickel complex	n.s.g.	435(1.69)	1-1239-65
$C_9H_{13}N_5O$			
9H-Purine-9-ethanol, α-methyl-6-(methylamino)-	pH 1	263(4.30)	87-0502-65
	pH 7,13	267(4.29)	87-0502-65
5H-Pyrrolo[3,4-d]pyrimidine, 6-acetyl-2,4-diamino-6,7-dihydro-7-methyl-	EtOH	281(3.98)	44-0194-65
	pH 1	272(3.91),302s(3.00)	44-0194-65
1,2,4-Triazolo[2,3-a]pyrimidine, 4-amino-2-butyl-6-oxo-	pH 7	266(4.2)	70-1475-64
	pH 12	224(4.5),257(3.9)	70-1475-64
1,2,4-Triazolo[2,3-a]pyrimidine, 4-amino-5-butyl-6-oxo-	pH 7	270(4.2)	70-1475-64
	pH 12	222(4.5),254(3.9)	70-1475-64
$C_9H_{13}N_5O_2$			
1,2-Propanediol, 3-[6-(methylamino)-9H-purin-9-yl]-	pH 1	263(4.24)	87-0502-65
	pH 7,13	266(4.22)	87-0502-65
$C_9H_{13}N_5O_2S$			
s-Triazolo[1,5-c]pyrimidine-2-sulfonamide, 7-methyl-5-propyl-	MeOH	208(4.38),257(3.77)	39-3369-65
$C_9H_{13}N_5O_3$			
Theobromine, 8-[(2-hydroxyethyl)amino]-, hydrobromide	pH 1	294(4.19)	65-2774-64
Theophylline, 8-[(2-hydroxyethyl)-amino]-, hydrobromide	pH 1	288(4.04)	65-2774-64

Compound	Solvent	λ_{max}(log ε)	Ref.
C_9H_{14}			
Cyclobutene, 1,2,3,3-tetramethyl-4-methylene-	EtOH	232(4.16)	35-1600-64
1,3-Cycloheptadiene, 2,3-dimethyl-	n.s.g.	233(3.66)	24-2949-64
Cyclohexene, 1-propenyl-, cis	n.s.g.	230(4.43)	44-3991-65
Cyclopropane, diisopropylidene-	EtOH	255(4.32),265(4.29)	35-5032-64
1,3,5-Heptatriene, 2,6-dimethyl-, trans	C_6H_{12}	262(4.44),272(4.56), 283(4.46)	23-2781-64
1,3,5-Heptatriene, 4,6-dimethyl-, cis	C_6H_{12}	207(4.12),246(4.19)	23-2781-64
trans	C_6H_{12}	267(4.33)	23-2781-64
1-Penten-3-yne, 2-tert-butyl-	n.s.g.	221(4.99),230(3.91)	22-1525-65
Spiro[2.5]octane, 1-methylene-	hexane	207(2.96)	28-1695-65B
	EtOH	203s(--),213(2.71)	28-1695-65B
Spiro[2.5]oct-1-ene, 1-methyl-	hexane	209(3.33)	28-1695-65B
	EtOH	215(3.18),218s(--)	28-1695-65B
$C_9H_{14}Br_2$			
Bicyclo[6.1.0]nonane, 9,9-dibromo-	heptane	206(3.49)	44-2951-64
$C_9H_{14}Br_2O$			
Cyclohexanone, 2,6-dibromo-3,3,5-trimethyl-	EtOH	305(2.16)	22-3511-65
$C_9H_{14}ClNO$			
1-Cyclooctene-1-carboxaldehyde, 2-chloro-, oxime	EtOH	252(4.24)	44-1126-65
$C_9H_{14}ClNO_2$			
Valeric acid, 5-chloro-2-cyano-2-methyl-, ethyl ester	MeOH	210(2.99)	24-1970-64
$C_9H_{14}ClN_3O$			
2-Cyclohexen-1-one, 3-chloro-5,5-dimethyl-, semicarbazone	EtOH	272(4.30)	44-0794-64
$C_9H_{14}ClO_3P$			
Tris(hydroxymethyl)phenylphosphonium chloride	EtOH	258(2.88),265(2.94), 272(2.84)	59-1143-64
$C_9H_{14}IP$			
Trimethylphenylphosphonium iodide	EtOH	218(3.50),260(2.96), 266(3.08),273(3.02)	59-1143-64
$C_9H_{14}N_2$			
Pyrrolidine, 1-methyl-2-pyrrol-2-yl-	EtOH	215(4.00)	39-0893-64
$C_9H_{14}N_2O$			
Cyclooct[c]isoxazole, 3-amino-4,5,6,7,8,9-hexahydro-	EtOH	252(3.99)	88-2151-64
Nipecotonitrile, 3-ethyl-1-methyl-2-oxo-	MeOH	208(3.77)	24-1970-64
2-Pyrrolidinemethanol, 5-pyrrol-2-yl-	EtOH	216(4.00)	39-0893-64
$C_9H_{14}N_2OS$			
4-Pyrimidinol, 2-butyl-6-(methylthio)-	pH -2.0	240(4.14),277(4.03)	39-3204-64
	pH 4.8	236(4.20),278(3.99)	39-3204-64
	pH 12.0	223(4.33),268(3.78)	39-3204-64
4-Pyrimidinol, 5-butyl-6-(methylthio)-	pH -2.0	238(3.94),286(3.99)	39-3204-64
	pH 4.8	238(4.14),292(3.94)	39-3204-64
	pH 12.0	224(4.29),279(3.82)	39-3204-64

Compound	Solvent	$\lambda_{max}(\log \epsilon)$	Ref.
$C_9H_{14}N_2O_2S$			
5-Pyrimidinecarboxylic acid, 1,2,3,4-tetrahydro-1,3-dimethyl-2-thioxo-, ethyl ester	EtOH	291(4.05),328(4.04)	44-2290-65
2H-1,3-Thiazine-5-carboxylic acid, 3,6-dihydro-3-methyl-2-(methylimino)-ethyl ester	EtOH	247(3.53),309(4.23)	44-2290-65
Uracil, 5-(pentylthio)-	pH 1	212(3.93),232s(3.74)	39-3987-65
	pH 7.2	212(3.92),232s(3.74)	39-3987-65
	pH 13	235(3.84),294(3.71)	39-3987-65
$C_9H_{14}N_2O_3$			
Pyridazine, 3,6-diethoxy-4-methyl-, 1-oxide	EtOH	253(3.95),335(3.90)	95-0344-65
Pyridazine, 3,6-diethoxy-4-methyl-, 2-oxide	EtOH	255(3.85),324(3.86)	95-0344-65
5-Pyrimidinecarboxylic acid, 1,2,3,4-tetrahydro-1,3-dimethyl-2-oxo-, ethyl ester	EtOH	221(3.82),300(3.84)	94-0804-64
Trimethylphenylammonium nitrate	n.s.g.	253.5(2.36)	39-6851-65
$C_9H_{14}N_2O_3S$			
Uracil, 5-(pentylsulfinyl)-	pH 1	214(4.03),268(3.86)	39-3987-65
	pH 7.2	217(3.96),281(3.88)	39-3987-65
	pH 13	235(3.99),286(3.95)	39-3987-65
$C_9H_{14}N_2O_4S$			
Uracil, 5-(pentylsulfonyl)-	pH 1	209(4.08),262(4.04)	39-3987-65
	pH 7.2	224(4.07),281(4.11)	39-3987-65
	pH 13	235(4.10),275(4.03)	39-3987-65
$C_9H_{14}N_2O_6$			
1,1-Hydrazinedicarboxylic acid, (carb-oxymethylene)-, 1,1-diethyl methyl ester	CH_2Cl_2	238(3.96)	5-0138-65J
$C_9H_{14}N_4$			
5H-Cyclohepta[d]pyrimidine, 2,4-diamino-6,7,8,9-tetrahydro-	pH 1	280(3.88)	44-1837-65
	pH 10	235(3.99),290(3.88)	44-1837-65
Pyrazine, 2-methyl-6-(1-piperazinyl)-, p-toluenesulfonate	H_2O	246(4.10),327(3.77)	44-0415-64
Quinazoline, 4-amino-5,6,7,8-tetra-hydro-2-(methylamino)-	pH 1	225(4.35),281(3.80)	44-1837-65
	pH 10	231(4.14),292(3.81)	44-1837-65
Quinazoline, 2,4-diamino-5,6,7,8-tetra-hydro-6-methyl-	pH 1	274(3.87)	44-1837-65
	pH 10	231(3.96),284(3.85)	44-1837-65
Quinazoline, 2,4-diamino-5,6,7,8-tetra-hydro-7(or 5)-methyl-	pH 1	274.5(3.89)	44-1837-65
	pH 10	229(4.00),285(3.88)	44-1837-65
$C_9H_{14}N_4O_4$			
1-Pyrimidinevaleric acid, α,6-diamino-2,4-dihydroxy-	pH 1.0	269(4.09)	65-0561-65
	H_2O	267(3.91)	65-0561-65
	pH 12	267(4.22)	65-0561-65
$C_9H_{14}N_4O_6$			
5-Azacytosine, 1-β-D-glucopyranosyl-	pH 5.0	243(3.81)	73-2060-64
$C_9H_{14}O$			
Cyclobutane, 1-acetyl-2,2-dimethyl-3-methylene-	C_6H_{12}	202(3.48),286(1.67)	28-0600-64A

Compound	Solvent	λ_{max}(log ε)	Ref.
Cyclohexanone, 2-isopropylidene-	EtOH	253(3.87)	44-3207-65
2-Cyclohexen-1-one, 3,5,5-trimethyl-	EtOH	226(4.14)	44-1129-65
Cyclopentene, 1-acetyl-2,3-dimethyl-	EtOH	251(3.95)	22-0722-65
Cyclopentene, 3-acetyl-1,2-dimethyl-	EtOH	275.5(1.46)	22-0722-65
Cyclopentenone, 4-methyl-5-propyl-	EtOH	222(3.9)	35-1326-65
2,4-Heptadienal, 2,4-dimethyl-	heptane	274(4.56)	33-0567-64
2-Hepten-4-yne, 2-ethoxy-	EtOH	238(4.17)	28-0594-64B
3,5-Hexadien-2-ol, 6-cyclopropyl-	EtOH	240.5(4.40)	22-3218-64
5-Hexen-2-one, 6-cyclopropyl-	EtOH	206.5(3.83)	22-3218-64
3,5,7-Nonatrien-1-ol	EtOH	256(4.62),265(4.73), 277(4.63)	22-3218-64
2-Nonen-4-yn-1-ol	Et_2O	226(4.07),235(4.00)	24-2118-64
3,5-Octadien-2-one, 7-methyl-	EtOH	276.5(4.38)	22-3218-64
4,5-Octadien-2-one, 7-methyl-	EtOH	240(4.22)	22-3218-64
5,7-Octadien-2-one, 6-methyl-	EtOH	230.1(4.34)	22-2533-64
cis	EtOH	232.5(4.18)	22-2533-64
2H-Pyran, 2,2,4,6-tetramethyl-	EtOH	282(3.59)	28-5895-64A
$C_9H_{14}OS$			
2-Cyclohexen-1-one, 5,5-dimethyl-3-(methylthio)-	EtOH	290(4.32)	87-0705-64
$C_9H_{14}O_2$			
Butyrolactone, α-isoamylidene-	EtOH	221(3.96)	65-3135-64
1,4-Cyclohexanedione, 2,2,6-trimethyl-	EtOH	289(1.65)	94-0043-65
2-Cyclohexen-1-one, 6-hydroxy-3,3,5-trimethyl-	EtOH	274.5(4.03)	22-3511-65
Cyclopentadienediethanol	EtOH	252(3.37)	44-3430-64
Cyclopentane, 1,2-diacetyl-, cis	n.s.g.	277(1.65)	24-2949-64
1,3-Cyclopentanedione, 2-butyl-	EtOH	251(4.23),271(4.40)	39-4472-64
1,3-Cyclopentanedione, 2-isobutyl-	EtOH	251(4.21),271(4.42)	39-4472-64
4-Cyclopentene-1,3-diol, 4-allyl-5-methyl-	EtOH	none	39-1854-64
3,5-Hexadien-1-ol, 4-methyl-, acetate	EtOH	228.6(4.35)	22-2533-64
2,4-Nonadienal, 9-hydroxy-, trans	EtOH	276(4.26)	70-2003-64
$C_9H_{14}O_2S$			
[(1-Cyclopenten-1-ylcarbonyl)methyl]dimethylsulfonium hydroxide, inner salt, oxide	EtOH	235(3.88),284(4.04)	35-1640-64
4H-Thiopyran-3-carboxylic acid, 5,6-dihydro-2-methyl-, ethyl ester	n.s.g.	283(4.18)	24-1963-64
$C_9H_{14}O_3$			
6-Ketonanolide	EtOH	278(1.26)	88-115-64
$C_9H_{14}O_3S$			
Crotonic acid, 3-(acetonylthio)-, ethyl ester	Et_2O	266(4.21)	24-2109-64
$C_9H_{14}O_4S$			
Butyric acid, 4-mercapto-3-methyl-2-oxo-, ethyl ester, acetate	EtOH	231(3.64)	39-0766-64
	EtOH-NaOH	248(3.85),305(3.04)	39-0766-64
Crotonic acid, 3-[(carboxymethyl)thio]-, 1-ethyl 3-methyl ester	Et_2O	267(4.16)	24-2109-64
$C_9H_{14}O_5S$			
Crotonic acid, 3-(acetonylsulfonyl)-, ethyl ester	Et_2O	211.5(4.11)	24-2109-64

Compound	Solvent	λ_{max}(log ε)	Ref.
$C_9H_{14}O_6S$			
Crotonic acid, 3-[(carboxymethyl)-sulfonyl]-, 1-ethyl 3-methyl ester	Et_2O	210(4.11)	24-2109-64
$C_9H_{15}BrO$			
Cyclohexanone, 2-bromo-2,6,6-trimethyl-	EtOH	320(2.00)	59-0529-65
	HOAc	320(2.02)	59-0529-65
	CF_3COOH	308(2.08)	59-0529-65
	MeCN	321(2.18)	59-0529-65
	CCl_4	323(2.00)	59-0529-65
Cyclohexanone, 2-bromo-3,3,5-trimethyl-	C_6H_{12}	309(2.1)	22-3511-65
Cyclohexanone, 2-bromo-3,5,5-trimethyl-	C_6H_{12}	310(2.1)	22-3511-65
$C_9H_{15}BrO_2$			
2-Hexenoic acid, 6-bromo-3-methyl-, ethyl ester	MeOH	217(4.06)	5-0079-65J
$C_9H_{15}ClS$			
Sulfide, butyl 1-chloro-2-methyl-1,3-butadienyl	EtOH	237(4.14),282(4.04)	44-0728-65
Sulfide, butyl 1-chloro-3-methyl-1,3-butadienyl	EtOH	236(4.00),278(3.93)	44-0728-65
$C_9H_{15}N$			
Pyridine, 1,2-dihydro-1,2,3,4-tetramethyl-	EtOH	323(3.60)	44-1647-64
Pyrrole, pentamethyl-	hexane	219(3.76)	24-0387-65
$C_9H_{15}NO$			
2H-Azepin-2-one, 1,3,4,5-tetrahydro-4,4,6-trimethyl-	MeOH	237(3.86)	5-0001-65B
2H-Azepin-2-one, 1,5,6,7-tetrahydro-3,5,5-trimethyl-	MeOH	218(4.07)	5-0001-65B
Caprolactam, N-allyl-	EtOH	none	70-0371-65
Caprolactam, N-propenyl-	EtOH	237(4.1)	70-0371-65
2-Cyclohexen-1-one, 3,5,5-trimethyl-, oxime, anti	EtOH	238.5(4.15)	44-1129-65
syn	EtOH	239(4.11)	44-1129-65
3-Furanamine, 2,5-dihydro-N,N,2,2-tetramethyl-5-methylene-	EtOH	251(3.92),336(4.35)	70-0729-65
3-Penten-2-one, 4-N-pyrrolidinyl-, perchlorate	EtOH	302(4.38)	44-1407-65
$C_9H_{15}NO_2S$			
2H-1,3-Thiazine-4-carboxylic acid, 3,6-dihydro-2,5-dimethyl-, ethyl ester	EtOH	286(3.49)	39-0766-64
4H-1,3-Thiazine-4-carboxylic acid, 5,6-dihydro-2,5-dimethyl-, ethyl ester	EtOH-HCl	250(3.93)	39-0766-64
	EtOH	236(3.68)	39-0766-64
Thiopyran-3-carboxylic acid, 2-amino-5,6-dihydro-5-methyl-, ethyl ester	MeOH	230(3.91),296(4.30)	24-1970-64
$C_9H_{15}NO_3$			
Nipecotic acid, 1-methyl-2-oxo-, ethyl ester	MeOH	208(3.81)	24-1970-64
$C_9H_{15}NO_3S$			
4H-1,3-Thiazine-4-carboxylic acid, 5,6-dihydro-4-hydroxy-2,5-dimethyl-, ethyl ester	EtOH-HCl	250(3.97)	39-0766-64
	EtOH	236(3.72)	39-0766-64

Compound	Solvent	λ_{max}(log ε)	Ref.
$C_9H_{15}NO_4$			
4-Morpholineacetic acid, α-acetonyl-	MeOH	219(3.91)	24-0363-64
$C_9H_{15}NS$			
Caprolactam, C-allylthio-	n.s.g.	280(4.1)	30-0734-64
Caprolactam, N-allylthio-	n.s.g.	276(4.16)	70-0702-65
	n.s.g.	278(4.0)	30-0734-64
	n.s.g.	275(4.1)	70-2064-64
Caprolactam, S-allylthio-	n.s.g.	232(4.9)	30-0734-64
	n.s.g.	231(3.90)	70-0702-65
Caprolactam, N-propenylthio-	n.s.g.	300(4.2)	70-2064-64
Piperidine, 1-(dihydro-3-thienyl)-	C_6H_{12}	218(3.83)	54-1160-64
$C_9H_{15}N_3$			
Pyrazolo[1,5-a]pyrimidine, 4,5,6,7-tetrahydro-2,3,6-trimethyl-	EtOH	244(3.82)	94-1207-65
$C_9H_{15}N_3O_2$			
Pyridazine, 3-[2-(dimethylamino)-ethoxy]-6-methoxy-	EtOH	286(3.31)	44-1751-64
$C_9H_{15}N_3O_2S$			
Pyridazine, 3-(butylamino)-6-(methylsulfonyl)-	pH 1	250(4.14)	4-0001-64
	pH 11	266(4.26)	4-0001-64
$C_9H_{15}N_3O_3$			
γ-Caprolactone, α-acetyl-, semicarbazone	n.s.g.	222(4.06),274(3.75)	39-01414-64
Cytosine, N,3-bis(2-hydroxyethyl)-1-methyl-	n.s.g.	285(3.9)	94-1240-64
	acid	291(4.10)	94-1240-64
Octanoic acid, 7-hydroxy-4-oxo-, lactone, semicarbazone	EtOH	228(4.10)	39-3239-64
$C_9H_{15}N_5O_2$			
4-Pyrimidinecarboxylic acid, 5,6-di-amino-2-(dimethylamino)-, ethyl ester	pH -4.4	235(4.30),265s(3.87), 375(3.39)	39-3221-64
	pH 3.9	209(4.16),252(4.25), 362(3.75)	39-3221-64
	pH 9.0	254(4.17),385(3.69)	39-3221-64
C_9H_{16}			
1-Cyclooctene, 1-methyl-, $PdCl_2$ olefin complex from	EtOH	232s(4.24),330(3.30)	24-2037-64
Cyclopentane, 1-isopropylidene-3-methyl-	isooctane	228(2.6)	77-0540-65
Cyclopentene, 1,2,3,5-tetramethyl-	96% H_2SO_4	294(4.12)	23-2768-64
Cyclopentene, 1,3,4,4-tetramethyl-	96% H_2SO_4	282.5(4.19)	23-2768-64
2,5-Heptadiene, 2,6-dimethyl-	96% H_2SO_4	396(4.70)	23-2768-64
1,3-Nonadiene	EtOH	226.5(4.44)	44-2410-65
1-Octene, 3-methylene-	EtOH	224.7(4.31)	44-2410-65
Unknown olefin	C_6H_{12}	192(4.04)	33-1385-64
$C_9H_{16}AlCl_4$			
Pentamethylcyclobutenylium chloro-aluminate	CH_2Cl_2	245(3.46)	35-1600-64
$C_9H_{16}NO_3P$			
Phosphonic acid, (1-methylpyrrol-2-yl)-, diethyl ester	EtOH	218(3.81)	44-0091-65

Compound	Solvent	$\lambda_{max}(\log \epsilon)$	Ref.
$C_9H_{16}N_2$			
2H-Indazole, 3,3a,4,5,6,7-hexahydro-3,3-dimethyl-	n.s.g.	230(2.35)	39-4636-64
$C_9H_{16}N_2O_2$			
2-Piperidone, 3-(aminoethoxymethylene)-1-methyl-	MeOH	206(3.46),274(2.77)	24-1970-64
$C_9H_{16}N_4O_6$			
Cytidine, 4,5-dihydro-N^6-hydroxy-4-hydroxyamino-2'-deoxy-	pH 7	222(4.08)	39-0208-65
$C_9H_{16}N_4O_7$			
Cytidine, 4,5-dihydro-N^6-hydroxy-4-hydroxyamino-	pH 7	224(4.05)	39-0208-65
$C_9H_{16}N_8O_2$			
Pyruvaldehyde, bis(1,3-diacetyl-guanylhydrazone)	pH 1	296(4.64)	44-0560-65
	pH 11	334(4.57)	44-0560-65
$C_9H_{16}O$			
Cyclobutane, 1-acetyl-2,2,3-trimethyl-	C_6H_{12}	289(1.56)	28-0600-64A
Cyclobutanone, 2,2,3,4,4-pentamethyl-	C_6H_{12}	299(1.32),311(1.34)	22-1968-64
	MeOH	307(1.40)	22-1968-64
Cyclohexanone, 2,2,6-trimethyl-	EtOH	292(1.43)	59-0529-65
	HOAc	290(1.45)	59-0529-65
	CF_3COOH	277(1.63)	59-0529-65
	MeCN	292(1.41)	59-0529-65
	CCl_4	295(1.45)	59-0529-65
Cyclohexanone, 3,3,5-trimethyl-	EtOH	272(1.53)	22-3511-65
Cyclopentane, 1-acetyl-2,3-dimethyl-	EtOH	283(1.42)	22-0722-65
3,5-Octadien-2-ol, 7-methyl-	EtOH	231(4.38)	22-3218-64
$C_9H_{16}O_2$			
2-Heptenal, 5-hydroxy-2,4-dimethyl-	heptane	225(4.26),318(1.64)	33-0567-64
2,4-Nonadiene-1,9-diol, trans-trans	EtOH	230(4.31)	70-2003-64
4,5-Nonadiene-3,7-diol	H_2O	200(3.31),220s(2.99)	28-0209-65A
	EtOH	209(3.04),220(2.98)	28-0209-65A
$C_9H_{16}O_3$			
Cyclohexanone, 2,4-dihydroxy-2,6,6-trimethyl-	EtOH	289(1.65)	94-1117-64
$C_9H_{16}O_5$			
α-D-threo-Hex-2-enopyranose, 3-deoxy-2,4,6-tri-O-methyl-	H_2O	200(3.93)	12-0837-65
$(C_9H_{16}S_3)_n$			
Polyoctamethylenetrithiocarbonate	$CHCl_3$	310(4.12)	49-0631-65
$C_9H_{17}BO_2$			
Diethylboryl acetylacetonate	n.s.g.	340(3.29)	5-0040-65I
$C_9H_{17}Cl_2N_5O_2S$			
2,6-Dihydroxy-4-methylpyridin-5-ylmethylaminoethylisothiuronium hydrochloride	pH 1	264(4.04)	65-2774-64

Compound	Solvent	λ_{max}(log ε)	Ref.
$C_9H_{17}N$			
1-Pyrroline, 2-isopropyl-3,3-dimethyl-	hexane	233(2.16)	39-2313-65
	EtOH	225(2.22)	39-2313-65
1-Pyrroline, 2,3,3,5,5-pentamethyl-	hexane	232(2.08)	39-2313-65
	EtOH	222(2.05)	39-2313-65
$C_9H_{17}NOS$			
Morpholine, 4-(methylthiobutyryl)-	MeOH	280(4.15),348(1.78)	35-0051-65
$C_9H_{17}NO_2S$			
Piperidine, 1-(2-methyl-3-thietanyl)-, 3,3-dioxide	hexane	218(2.52)	5-0092-65D
$C_9H_{17}NS$			
Pyrrolidine, 1-(2-methylthiobutyryl)-	MeOH	273(4.13),330(1.7)	35-0051-65
$C_9H_{17}N_4O_2$			
1-Pyrrolidinyloxy, 2,2,5,5-tetramethyl-3-oxo-, semicarbazone	$CHCl_3$	426(0.99)	22-3643-65
	MeOH	410(0.94)	22-3643-65
$C_9H_{17}N_5$			
Pyrimidine, 5-amino-4-(diethylamino)-6-(methylamino)-	pH 1	236(3.97),240s(3.97), 310(4.09)	44-3597-65
	pH 13	231(4.08),292(4.02)	44-3597-65
	EtOH	295(4.07)	44-3597-65
$C_9H_{17}O_4PS$			
Valeric acid, 5-mercapto-2-phosphono-, δ-thiolactone, diethyl ester	n.s.g.	242(3.60)	24-1963-64
C_9H_{18}			
Cyclopentane, sec-butyl-	C_6H_{12}	246.5(3.50)	33-1022-64
$C_9H_{18}ClNO$			
Trimethyl(4-methyl-3-oxo-1-pentenyl)-amonium chloride	EtOH	208(4.87)	44-0385-64
Trimethyl(3-oxo-1-hexenyl)ammonium chloride	EtOH	208(4.86)	44-0385-64
$C_9H_{18}CoN_3O_6$			
Cobalt, tris(L-alaninato)-, β (+)	n.s.g.	375(2.10),516(2.17)	39-6531-65
β (-)	n.s.g.	376(2.17),516(2.28)	39-6531-65
$C_9H_{18}NO$			
1-Piperidinoxy, 2,2,6,6-tetra-methyl-	heptane	470(1.02)	22-3273-65
	MeOH	245(3.28),450(1.01)	22-3273-65
$C_9H_{18}NO_2$			
1-Piperidinoxy, 4-hydroxy-2,2,6,6-tetramethyl-	C_6H_6	468(0.85)	22-3273-65
	C_6H_{12}	470(1.00)	22-3273-65
	heptane	475(0.95)	22-3273-65
	MeOH	239(3.24),445(1.04)	22-3273-65
	dioxan	240(3.10),465(0.90)	22-3273-65
$C_9H_{18}N_2$			
2-Imidazoline, 1-methyl-2-pentyl-	EtOH	230(3.75)	4-0188-64
$C_9H_{18}N_2O$			
2-Imidazoline-1-ethanol, 2-tert-butyl-	EtOH	232(3.71)	4-0188-64

Compound	Solvent	$\lambda_{max}(\log \epsilon)$	Ref.
$C_9H_{18}N_2O_2$			
2-Pentanone, 4-methyl-4-(N-nitroso-N-propylamino)-	H_2O	233.0(3.86)	28-5470-64A
$C_9H_{18}N_3O_6Rh$			
Rhodium, tris(L-α-alaninato)-, β (+)	n.s.g.	284(2.58),339(2.64)	39-6531-65
β (-)	n.s.g.	283(2.54),340(2.60)	39-6531-65
$C_9H_{18}N_4O$			
3-Pyrrolidinone, 2,2,5,5-tetra-methyl-, semicarbazone	MeOH	226.5(4.24)	22-3643-65
$C_9H_{18}O$			
3-Pentanone, 2,2,4,4-tetramethyl-	C_6H_{12}	297(1.31)	59-0529-65
	hexane	297(1.31)	
	H_2O	292(1.37)	
	MeOH	294(1.36)	
	EtOH	295(1.35)	
	PrOH	295(1.34)	
	BuOH	296(1.34)	
	pentanol	296(1.34)	
	hexanol	296(1.34)	
	HCOOH	291(1.40)	
	HOAc	294(1.35)	
	EtCOOH	295(1.34)	
	PrCOOH	295(1.34)	
	BuCOOH	295(1.34)	
	CF_3COOH	286(1.45)	
	$Cl_2CHCOOH$	290(1.45)	
	MeCN	295(1.33)	
	EtCN	295(1.32)	
	PrCN	295(1.32)	
	CH_2Cl_2	295(1.34)	
	$CHCl_3$	296(1.36)	
	CCl_4	297(1.37)	
	HCOOMe	295(1.32)	
	Et_2O	297(1.31)	
	dioxan	296(1.31)	
	Me_2SO	295(1.35)	
	$HCONH_2$	295(1.35)	
	HCONHMe	295(1.34)	
	DMF	295(1.34)	
	$MeCONMe_2$	296(1.34)	
$C_9H_{18}O_3$			
s-Trioxane, 2,4,6-triethyl-	heptane	226(1.82),276(1.22)	33-0567-64
$C_9H_{19}ClN_2O$			
Trimethyl(4-methyl-3-oxo-1-pentenyl)-ammonium chloride, oxime	EtOH	242(3.78)	44-1129-65
$C_9H_{19}N$			
Butylamin, N-neopentylidene-	hexane	244(1.94)	39-2313-65
	EtOH	235(2.03)	39-2313-65
sec-Butylamine, N-neopentylidene-	hexane	243(1.93)	39-2313-65
	EtOH	236(1.97)	39-2313-65
tert-Butylamine, N-neopentylidene-	hexane	250(1.90)	39-2313-65
	EtOH	242(1.92)	39-2313-65

Compound	Solvent	λ_{max}(log ε)	Ref.
$C_9H_{19}N_7$			
Tetrazoline, 5-cyanimino-1-methyl-, diisopropylammonium salt	EtOH	218(4.04)	44-0650-64
$C_9H_{20}CoN_3O_7$			
Cobalt, tris(L-α-alaninato)-, hydrate, α(+), α(-)	n.s.g.	375(2.18),545(2.02)	39-6531-65
	n.s.g.	375(1.98),540(1.78)	39-6531-65
$C_9H_{20}N_2O_4$			
L-Arabinose, butylhydrazone	MeOH	239(4.3)	24-1588-65
$C_9H_{27}Cl_6Fe_2O_9P_3S_3$			
Iron, hexachlorotris(O,O,S-trimethyl phosphorothioate)di-	$CHCl_3$	248(4.1),343(4.2)	24-0864-65
$C_9H_{28}Si_4$			
Trisilane, 1,1,1,3,3,3-hexamethyl-2-(trimethylsilyl)-	C_6H_{12}	none	101-0163-65B

Compound	Solvent	$\lambda_{max}(\log \epsilon)$	Ref.
$C_{10}Cl_2N_4$			
1,2,4,5-Benzenetetracarbonitrile, 3,6-dichloro-	MeCN	232(4.85),240(4.67), 253(4.12),260(4.06), 269(4.3),299(3.6), 354(3.75)	44-3250-65
$C_{10}Cl_8O_2$			
4,7-Methanoindene-1,8-dione, octachloro-3a,4,7,7a-tetrahydro-	C_6H_{12}	255(3.96)	23-1500-64
Tricyclo[5.3.0.0^{2,6}]deca-3,9-diene-5,8-dione, octachloro-, cis-trans-cis	EtOH	256(4.29),315(2.09), 328(2.19),342(2.23), 357(2.14),373(1.72)	44-3657-65
	dioxan	252(4.30),322(2.23), 337(2.25),351(2.14), 370(1.68)	44-3657-65
Tricyclo[5.3.0.0^{2,6}]deca-4,9-diene-3,8-dione, octachloro-, cis-trans-cis	EtOH	258(4.32),315(2.26), 328(2.14),342(2.34), 357(2.22),373(1.83)	44-3657-65
	dioxan	255(4.32),323(2.37), 337(2.40),352(2.29), 370(1.81)	44-3657-65
$C_{10}H_2ClN_5$			
1,2,4,5-Benzenetetracarbonitrile, amino-6-chloro-	MeCN	222(4.59),260(4.45), 275(3.97),432(3.88)	44-3250-65
$C_{10}H_2Cl_4N_2S$			
Naphtho[1,8-cd][1,2,6]thiadiazine, tetrachloro-	dioxan	252(4.57),361(4.00), 375(4.05),597(2.85)	24-3196-65
$C_{10}H_2N_4O_2$			
1,2,4,5-Benzenetetracarbonitrile, 3,6-dihydroxy-	EtOAc	216(4.57),254(4.40), 408(3.85)	44-3250-65
$C_{10}H_4BrF_3N_4$			
Quinoline, 4-azido-6-bromo-2-(trifluoromethyl)-	EtOH	239(4.72),306(4.02)	4-0113-65
$C_{10}H_4Cl_2N_2OS$			
4-Isothiazolecarbonitrile, 3-chloro-5-(p-chlorophenoxy)-	EtOH	252(4.00)	44-0660-64
$C_{10}H_4F_4O$			
1-Naphthol, 5,6,7,8-tetrafluoro-	EtOH	304(3.54),324(3.50), 336(3.44)	88-3575-64
$C_{10}H_4N_2S_3$			
Thiocyanic acid, penta[c]thiapyran-5,7-diyl ester	Et_2O	230(4.25),244(4.20), 288(4.40),320(3.64), 453(3.49)	35-0708-64
$C_{10}H_4N_6$			
1,2,4,5-Benzenetetracarbonitrile, 3,6-diamino-	MeCN	263(4.25),288(3.88), 498(3.84)	44-3250-65
$C_{10}H_5BrF_3NO$			
4-Quinolinol, 6-bromo-2-(trifluoromethyl)-	EtOH-HCl	239(4.62),294(3.76), 340(3.64)	4-0113-65
	EtOH	239(4.55),330(3.83), 342(3.81)	4-0113-65

Compound	Solvent	$\lambda_{max}(\log \epsilon)$	Ref.
$C_{10}H_5BrN_2O$			
2H-Cyclohepta[b]furan-3-carbonitrile, 8-bromo-2-imino-	EtOH	242(4.24),279(4.23), 287(4.17),405(4.25), 422(4.22),490(3.71)	94-0443-65
$C_{10}H_5BrO_5$			
2H-1-Benzopyran-3-carboxylic acid, 6-bromo-4-hydroxy-2-oxo-	MeOH	236(3.93),315(3.80)	44-4343-65
$C_{10}H_5Br_2NO_3$			
2H-Cyclohepta[b]furan-3-carboxamide, 6,8-dibromo-2-oxo-	EtOH	252(4.39),277(4.37), 289(4.35),413(4.36)	94-0443-65
$C_{10}H_5ClF_3NO$			
Acetoacetonitrile, 2-(p-chlorophenyl)-4,4,4-trifluoro-	EtOH	255(3.68),300(4.00)	4-0162-65
	10% EtOH-base	254(3.61),300(3.90)	4-0162-65
$C_{10}H_5ClN_2OS$			
4-Isothiazolecarbonitrile, 3-chloro-5-phenoxy-	EtOH	250(3.95)	44-0660-64
$C_{10}H_5ClN_2S_2$			
4-Isothiazolecarbonitrile, 3-chloro-5-(phenylthio)-	EtOH	274s(3.92),287(3.98)	44-0660-64
$C_{10}H_5ClO_5$			
2H-1-Benzopyran-3-carboxylic acid, 6-chloro-4-hydroxy-2-oxo-	MeOH	238(3.74),321(3.53)	44-4343-65
$C_{10}H_5Cl_2NO_3$			
4-Isoxazolecarboxylic acid, 3-(2,6-dichlorophenyl)-	MeOH	215s(4.06),220(4.07), 277(2.59)	39-5976-65
$C_{10}H_5Cl_2N_3S$			
4-Isothiazolecarbonitrile, 3-chloro-5-(p-chloroanilino)-	EtOH	227(4.30),301(4.20)	44-0660-64
$C_{10}H_5Cl_3F_2$			
Benzene, 1-chloro-4-(2,4-dichloro-3,3-difluoro-1-cyclobuten-1-yl)-	n.s.g.	273(4.42)	35-0449-64
Benzene, 1-chloro-4-(4,4-dichloro-3,3-difluoro-1-cyclobuten-1-yl)-	n.s.g.	266(4.37)	35-0449-64
$C_{10}H_5Cl_3N_2$			
Pyrimidine, 2,4,6-trichloro-5-phenyl-	MeOH	265(3.99)	87-0808-64
Pyrimidine, 4,5,6-trichloro-2-phenyl-	MeOH	275(4.52)	87-0808-64
$C_{10}H_5F_3N_4$			
Quinoline, 4-azido-2-(trifluoromethyl)-	EtOH	233(4.61),305(3.91)	4-0113-65
$C_{10}H_5F_{16}N_3O$			
Nonanal, 1H,9H-hexadecafluoro-, semicarbazone	EtOH	243(4.18)	44-0279-64
$C_{10}H_5MnO_3$			
Manganese, tricarbonyl(ethynylcyclopentadienyl)-	EtOH	333(3.03)	49-1520-65

Compound	Solvent	$\lambda_{max}(\log \epsilon)$	Ref.
$C_{10}H_5NO_2S_2$			
1H-Naphtho[1,2-d][1,2,3]dithiazole-4,5-dione	dioxan	243(4.28),249(4.39), 346(3.89),499(4.07)	5-0151-64E
$C_{10}H_5NO_7$			
2H-1-Benzopyran-3-carboxylic acid, 4-hydroxy-6-nitro-2-oxo-	MeOH	232(3.95),321(4.00)	44-4343-65
2H-1-Benzopyran-3-carboxylic acid, 4-hydroxy-8-nitro-2-oxo-	MeOH	239(3.65),316(3.50)	44-4343-65
$C_{10}H_5N_5$			
2,3,4,5-Pyridinetetracarbonitrile, 1,2-dihydro-1-methyl-	MeCN	263(3.60),418(3.84)	89-0861-65
$C_{10}H_6$			
Butadiyne, phenyl-	hexane	211(4.45),221(4.53), 233(3.59),244(3.90), 256(4.23),270(4.40), 285(4.28)	22-3518-65
$C_{10}H_6BrNO_2$			
5-Isoquinolinecarboxylic acid, 3-bromo-	EtOH	282(3.75),293(3.67), 325(3.56),334(3.57)	94-1296-64
$C_{10}H_6BrNO_3$			
2H-Cyclohepta[b]furan-3-carboxamide, 8-bromo-2-oxo-	EtOH	233(4.30),280(4.37), 407(4.37)	94-0443-65
$C_{10}H_6BrNO_5$			
Crotonic acid, 3-bromo-2,4-dihydroxy-4-(m-nitrophenyl)-, γ-lactone	MeOH	251(4.15)	44-1800-65
$C_{10}H_6Br_2$			
Naphthalene, 1,6-dibromo-	EtOH	285(4.0),295(4.1), 303(3.9)	38-1082-64B
Naphthalene, 2,3-dibromo-	EtOH	233(3.9),260(2.6), 278(2.6),294(2.5), 323(1.5)	38-1082-64B
$C_{10}H_6Br_2I_2N_2$			
Quinoxaline, 2,3-bis(bromoiodomethyl)-	EtOH	242(4.57),294(3.97), 320s(--)	44-1542-65
$C_{10}H_6Br_2N_2O$			
3(2H)-Pyridazinone, 4,5-dibromo-2-phenyl-	EtOH	269(3.74),322(3.80)	78-1323-65
Quinoxaline-2-carboxaldehyde, 3-(dibromomethyl)-	EtOH	244(4.50),325(3.83)	44-1542-75
$C_{10}H_6Br_4N_2$			
Quinoxaline, 2,3-bis(dibromomethyl)-	EtOH	252(4.65),330(3.90)	44-1542-65
$C_{10}H_6ClF_2NO_2$			
Benzene, (2-chloro-3,3-difluoro-4-nitro-1-cyclobuten-1-yl)-	n.s.g.	262(4.33)	35-5132-65
$C_{10}H_6ClN_3S$			
4-Isothiazolecarbonitrile, 5-anilino-3-chloro-	CH_2Cl_2	301(4.17)	44-0660-64
5H-Pyridazino[3,4-b][1,4]benzothiazine, 3-chloro-	EtOH	250(4.5)	94-0580-65

Compound	Solvent	$\lambda_{max}(\log \epsilon)$	Ref.
10H-Pyridazino[4,3-b][1,4]benzothiazine, 3-chloro-	EtOH	245(4.6),275s(4.1)	94-0580-65
10H-Pyridazino[4,3-b][1,4]benzothiazine, 4-chloro-	EtOH	250(4.5)	94-0580-65
10H-Pyridazino[4,5-b][1,4]benzothiazine, 1-chloro-	EtOH	270(4.5)	94-0580-65
$C_{10}H_6Cl_2N_2OS$			
4-Isothiazolecarboxanilide, 3,5-dichloro-	EtOH	254(4.20)	44-0660-64
$C_{10}H_6Cl_2N_2S$			
4-Pyrimidinethiol, 6-chloro-5-(2-chlorophenyl)-	EtOH	204(4.45),214s(4.34), 267(4.00),275(4.00)	73-3730-65
4-Pyrimidinethiol, 6-chloro-5-(4-chlorophenyl)-	EtOH	205(4.42),228s(4.17), 274(4.03)	73-3730-65
$C_{10}H_6Cl_2O_2$			
Cyclopentadienone, 2-chloro-, dimer	EtOH	233.8(3.96)	44-3503-64
$C_{10}H_6Cl_2O_3$			
Coumarin, 3,6-dichloro-4-methoxy-	EtOH	215(4.39),223(4.37), 280(4.08),291(4.04), 320(3.91)	44-4126-65
$C_{10}H_6Cl_3N_3O_2$			
Sydnone imine, 3-phenyl-N-(trichloroacetyl)-	pH 1.0	310(3.90)	65-2064-64
	pH 7.00	260(3.68),326(3.72)	65-2064-64
$C_{10}H_6Cl_4O_2$			
Benzene, p-bis[(2,2-dichlorovinyl)oxy]-	nonane	276(3.16)	67-0375-65
$C_{10}H_6Cl_4O_3$			
Phthalide, 4,5,6,7-tetrachloro-3-methoxy-3-methyl-	EtOH	224(4.59),295(3.06), 305(3.16)	44-4293-65
$C_{10}H_6F_3NO$			
4-Quinolinol, 2-(trifluoromethyl)-	EtOH-HCl	229(4.57),321(3.95), 335(3.89)	4-0113-65
	EtOH	225(4.45),319(3.84), 335(3.85)	4-0113-65
$C_{10}H_6F_8$			
Benzene, 1,4-difluoro-2,5-dimethyl-3,6-bis(trifluoromethyl)-	EtOH	283.0(3.56)	39-2975-64
$C_{10}H_6F_8O_2$			
Benzene, 2,5-difluoro-3,6-bis(trifluoromethyl)dimethoxy-	EtOH	294.0(3.74)	39-2975-64
$C_{10}H_6N_2$			
Heptafluvene, 8,8-dicyano-	hexane	250(3.91),255(3.87), 368(4.23),374(4.25), 381(4.25),400s(3.94), 445s(2.89),480s(2.71), 515s(2.44),555s(2.07)	24-2050-64
	EtOH	253(4.20),389(4.48), 405s(4.39)	24-2050-64
	50% EtOH	254(4.19),394(4.50), 405s(4.47)	24-2050-64

Compound	Solvent	$\lambda_{max}(\log \epsilon)$	Ref.
$C_{10}H_6N_2O$			
4-Isoquinolinecarbonitrile, 1,2-dihydro-1-oxo-	dioxan	230s(4.1),250s(4.0), 260s(4.0),295(4.0), 320s(3.6),335s(3.4)	83-0488-64
5-Isoquinolinecarbonitrile, 1,2-dihydro-1-oxo-	EtOH	226(4.27),240s(4.06), 248(4.02),276(4.01), 284(4.00),312s(3.63), 325(3.72),338s(3.57)	44-2534-64
5-Isoxazolecarbonitrile, 3-phenyl-	EtOH	225(4.36)	78-0817-65
$C_{10}H_6N_2O_2$			
2,3-Quinoxalinedicarboxaldehyde	EtOH	238(4.45),332(3.96)	44-1542-65
$C_{10}H_6N_2O_4SSe$			
Selenophene, 3,5-dinitro-2-(phenylthio)-	MeOH	268(3.94),292(3.91), 370(4.18)	7-1069-65
$C_{10}H_6N_2O_5$			
4H-Quinolizine-3-carboxylic acid, 1-nitro-4-oxo-	EtOH	258(3.86),380(4.24)	78-0945-65
$C_{10}H_6N_2O_6$			
2-Isoindolineacetic acid, 5-nitro-1,3-dioxo-	dil. HCl	233(4.43),289(3.58)	44-3151-64
	dil. NaOH	274(3.96)	44-3151-64
	EtOH	244s(4.45),278s(3.54)	44-3151-64
$C_{10}H_6N_2S$			
Naphtho[1,8-cd][1,2,6]thiadiazine	dioxan	234(4.63),331(3.91), 344(3.92),642(2.73),	24-3196-65
	n.s.g.	658(2.76)	77-0057-65
Naphtho[2,3-c][1,2,5]thiadiazole	dioxan	221(4.25),252(4.53), 327s(3.50),335s(2.70), 343(3.87),351(3.85), 359(4.01),436(3.31), 460s(3.21)	88-3815-64
$C_{10}H_6N_2Se$			
Naphtho[2,3-c][1,2,5]selenadiazole	THF	230(4.4),260(4.2), 360f(3.8),470(3.1)	46-1773-65
$C_{10}H_6N_4O_2$			
2H-Naphtho[2,3-d]triazole-4,9-dione, 2-amino-	EtOH	247(4.46),270(4.41), 331(3.47)	39-1003-65
$C_{10}H_6O_2$			
2,4-Hexadiyn-1-one, 1-(2-furyl)-	EtOH	310(4.33)	39-2983-65
1,4-Naphthoquinone	40% iso-PrOH	245(4.3),340(3.5)	61-0296-64
	MeCN	246(4.31),250(4.30), 330(3.48)	44-3819-65
$C_{10}H_6O_2S$			
Naphtho[1,8-bc]thiete, 1,1-dioxide	EtOH	225(4.6),274(3.6), 284(3.7),316(2.6)	89-0810-65
$C_{10}H_6O_3$			
1,4-Naphthoquinone, 2-hydroxy-	dioxan	243(4.22),249(4.27), 274(4.15),330(3.47)	5-0151-64E
	MeCN	244(4.22),249(4.27), 274(4.21),282(4.20), 332(3.52)	44-3819-65

Compound	Solvent	λ_{max}(log ε)	Ref.
$C_{10}H_6O_4$			
Coumarin, 6-formyl-7-hydroxy-	MeOH	229(4.11),256(4.22), 262(4.28),313(4.01), 343(4.09)	32-1073-64
1,3-Indandione, 2-hydroxy-, formate	MeOH	225(4.68),244(4.05)	24-1482-64
1,7-Naphthalenedione, 2,8-dihydroxy-, dianion	n.s.g.	247(4.20),333(3.85), 476(3.31),650(3.86)	24-1246-65
2,6-Naphthoquinone, 1,5-dihydroxy-, dianion	n.s.g.	263(4.36),271(4.35), 322(3.46),592(4.33)	24-1246-65
$C_{10}H_6O_5$			
2H-1-Benzopyran-3-carboxylic acid, 4-hydroxy-2-oxo-	MeOH	241(3.56),311(3.51)	44-4343-65
2H-1-Benzopyran-3-carboxylic acid, 8-hydroxy-2-oxo-	EtOH	252(3.92),306(4.18)	94-0443-65
2H-1-Benzopyran-6-carboxylic acid, 7-hydroxy-2-oxo-	MeOH	220(4.29),244(4.36), 335(4.18)	32-1073-64
$C_{10}H_6O_6$			
2H-1-Benzopyran-3-carboxylic acid, 4,7-dihydroxy-2-oxo-	MeOH	226(4.04),259(4.10), 298(3.98)	44-4343-65
1,4-Naphthoquinone, 2,5,6,7-tetra-hydroxy-	EtOH	274(4.11),317(3.85), 428(3.54)	94-1472-65
1,4-Naphthoquinone, 2,5,7,8-tetra-hydroxy-	EtOH	228(4.48),272(4.06), 318(3.93),486(3.78), 517(3.84),554(3.65)	94-0633-65
$C_{10}H_6O_7$			
1,4-Naphthoquinone, 2,3,5,6,8-penta-hydroxy-	EtOH	264(4.10),328(3.44), 463(3.41),485(3.41), 530(3.26)	39-2141-65
$C_{10}H_6S_2$			
2,2'-Bithiophene, 5-ethynyl-	hexane	248(--),326(--), 331(--)	39-7109-65
$C_{10}H_7Br$			
Naphthalene, 1-bromo-	EtOH	286(4.0),295s(3.9), 313(3.0)	38-1082-64B
Naphthalene, 2-bromo-	EtOH	227(4.1),267(2.8), 278(2.9),308(1.5)	38-1082-64B
Styrene, β-bromo-o-ethynyl-, trans	EtOH	237(4.27),245(4.17), 268s(4.06),274(4.03), 283s(3.92)	39-1151-64
$C_{10}H_7BrN_2$			
2,3'-Bipyridine, 5(?)-bromo-	MeOH	243(3.98),287(3.83)	65-4193-64
$C_{10}H_7BrN_2O$			
2-Quinoxalinecarboxaldehyde, 3-(bromomethyl)-	EtOH	241(4.47),321(3.81)	44-1542-65
$C_{10}H_7BrN_2O_3$			
4-Bromo-1-hydroxy-8-methyl-2-nitro-quinolizinium hydroxide, inner salt	H_2O	210(4.45),241(4.19), 325(3.66),379s(3.94) 444(4.31)	39-3030-64
$C_{10}H_7BrO_2$			
But-2-enolide, 4-bromo-3-phenyl-	C_6H_{12}	219(4.13),281(4.21)	39-3075-65

Compound	Solvent	λ_{max}(log ϵ)	Ref.
$C_{10}H_7BrO_2S_2$			
[2,2'-Bithiophene]-5-carboxylic acid, 5'-bromo-, methyl ester	EtOH	245(3.77),339(4.35)	39-5134-65
$C_{10}H_7Br_2NO$			
Pyrido[1,2-a]azepin-10-ol, 7,9-dibromo-	H_2O	261(3.87),316(3.67), 420(3.59)	44-1523-65
	EtOH	262(3.91),324(3.69), 455(3.62)	44-1523-65
	$CHCl_3$	518(2.93)	44-1523-65
$C_{10}H_7Cl$			
Naphthalene, 1-chloro-	EtOH	285(3.8),295s(3.6)	38-1082-64B
Naphthalene, 2-chloro-	EtOH	227(4.0),267(2.6), 310(1.6),323(1.6)	38-1082-64B
$C_{10}H_7ClN_2O$			
3(2H)-Pyridazinone, 6-(p-chlorophenyl)-	EtOH	257(4.44)	39-3342-65
$C_{10}H_7ClN_2O_2$			
6H-1,3-Oxazin-4-ol, 2-(p-chlorophenyl)-6-imino-	pH 2	245(4.25)	49-0212-65
	EtOH	245(4.25)	49-0212-65
$C_{10}H_7ClN_2O_3$			
Quinoline, 2-(chloromethyl)-4-nitro-, N-oxide	EtOH	260(4.34),376(4.04)	94-1495-64
Quinoline, 2-(chloromethyl)-x-nitro-, N-oxide	EtOH	233(4.49),261(4.11), 362(3.72)	94-1495-64
$C_{10}H_7ClN_2O_4$			
Sydnone, 4-chloro-3-piperonyl-	EtOH	292(4.03)	87-0531-65
$C_{10}H_7ClN_2S$			
4-Pyrimidinethiol, 6-chloro-5-phenyl-	EtOH	205(4.37),228s(4.02), 270(3.99),278(4.00)	73-3730-65
$C_{10}H_7ClN_4S$			
4-Isothiazolecarbonitrile, 3-chloro-5-(2-phenylhydrazino)-	CH_2Cl_2	268(4.03)	44-0660-64
$C_{10}H_7ClOS$			
1-Thiochromone, 3-chloro-2-methyl-	EtOH	216(4.04),227(4.05), 251(4.34),255s(4.33), 280(3.54),292(3.48), 344(4.03)	44-1575-64
$C_{10}H_7ClO_2S_2$			
[2,2'-Bithiophene]-5-carboxylic acid, 5'-chloro-, methyl ester	EtOH	245(3.89),339(4.27)	39-5134-65
$C_{10}H_7ClO_3$			
Succinic anhydride, (p-chlorophenyl)-	MeCN	220(4.04),252(2.20), 258(2.30),265(2.37), 275(2.20)	87-0259-65
$C_{10}H_7Cl_2NO_2S$			
Indole-2-carbonyl chloride, 3-(chlorosulfinyl)-1-methyl-	EtOH	295(4.09),312s(4.00), 324s(3.94),372s(3.40)	44-0178-64

Compound	Solvent	$\lambda_{max}(\log \epsilon)$	Ref.
$C_{10}H_7Cl_2N_3OS$			
4-Isothiazolecarboxanilide, 2'-amino-3,5-dichloro-	EtOH	260(3.96),298(3.58)	44-0660-64
$C_{10}H_7I$			
Naphthalene, 1-iodo-	EtOH	270(4.1),280(4.2), 313(3.0),317(2.9)	38-1082-64B
Naphthalene, 2-iodo-	EtOH	228(3.7),256(2.9), 310(1.7),315(1.6)	38-1082-64B
$C_{10}H_7IN_2O$			
2-Quinoxalinecarboxaldehyde, 3-(iodomethyl)-	EtOH	243(4.50),324(3.88)	44-1542-65
$C_{10}H_7IO_2S_2$			
[2,2'-Bithiophene]-5-carboxylic acid,5'-iodo-, methyl ester	EtOH	251(3.83),344(4.36)	39-5134-65
$C_{10}H_7NOS_2$			
3H-Naphtho[2,1-d]-1,2,3-dithiazole, 2-oxide	EtOH	218s(4.34),241(4.66), 262s(4.01),289(3.68), 302s(3.59),343(3.38)	44-2763-65
$C_{10}H_7NOS_3$			
3H-1,2-Dithiole-3-thione, 4-benzamido-	EtOH	227(4.30),241s(4.22), 295(3.95),414(4.07)	12-0447-64
$C_{10}H_7NO_2$			
Naphthalene, 2-nitro-	EtOH	220(4.68),264(4.38), 270(4.40),312(4.94)	39-5377-65
1,2-Naphthoquinone, 4-amino-	dioxan	268(3.96),432(3.04)	5-0151-64E
1,4-Naphthoquinone, 2-amino-	dioxan	264(4.24),328(3.15), 430(3.35)	5-0151-64E
$C_{10}H_7NO_2S$			
Phthalimide, N-(thioacetyl)-	heptane	219(4.48),239s(4.05), 273(4.15)	24-1556-65
	MeOH	205(4.27),288(4.16), 490(1.55)	24-1556-65
	dioxan	219(4.43),278(4.20), 495(1.60)	24-1556-65
4,5-Thiazolidinedione, 2-benzylidene-	MeOH	255(4.0),350(3.9)	24-2954-65
$C_{10}H_7NO_2SSe$			
Selenophene, 3-nitro-2-(phenylthio)-	MeOH	218(4.19),258(3.98), 298(3.84),370(3.79)	7-1069-65
Selenophene, 5-nitro-2-(phenylthio)-	MeOH	240(3.81),282s(3.56), 396(4.06)	7-1069-65
$C_{10}H_7NO_2S_2$			
3H-1,2-Dithiol-3-one, 4-benzamido-	EtOH	232(4.20),265s(3.90), 328(4.08)	12-0447-64
Thiophene, 3-nitro-2-(phenylthio)-	MeOH	366(3.82)	7-1252-65
	dioxan-HCl	374(3.82)	7-1252-65
Thiophene, 5-nitro-2-(phenylthio)-	MeOH	380(3.98)	7-1252-65
	dioxan-HCl	390(3.98)	7-1252-65

Compound	Solvent	$\lambda_{max}(\log \epsilon)$	Ref.
$C_{10}H_7NO_3$			
2H-Cyclohepta[b]furan-3-carboxamide, 2-oxo-	EtOH	224(4.26),258(4.37), 403(4.33)	94-0443-65
11-Oxabicyclo[4.4.1]undeca-1,3,5,7,9-pentaene, 2-nitro-	EtOH	238(4.39),283(4.19), 364(3.83)	35-3168-64
11-Oxabicyclo[4.4.1]undeca-1,3,5,7,9-pentaene, 3-nitro-	EtOH	242(4.33),279(4.42), 349(3.88)	35-3168-64
4H-Quinolizine-2-carboxylic acid, 4-oxo-	MeOH-acid	251(4.23),310(3.32), 376(4.04)	39-2633-65
	MeOH	225(4.39),251(4.09), 407(4.07)	39-2633-65
$C_{10}H_7NO_3S$			
4,5-Thiazoledione, 2-(p-methoxyphenyl)-	dioxan	245(3.9),350(4.2), 450s(2.0)	24-2954-65
$C_{10}H_7NO_4$			
1,2-Benzisoxazole-3-pyruvic acid	EtOH	316(4.18)	4-0385-65
3H-Indole-2-carboxylic acid, 3-oxo-, methyl ester, 1-oxide	MeOH	255(4.43),260s(4.40), 277s(4.03),335(3.44), 390(3.28)	17-0405-65
Phthalimidoacetic acid	EtOH	240(4.22),293(3.38)	44-3151-64
	dil. HCl	240(4.10),297(3.48)	44-3151-64
	dil. NaOH	269(3.04)	44-3151-64
$C_{10}H_7NO_5$			
3-Butenoic acid, 4-(m-nitrophenyl)-2-oxo-	pH 7.5	281(4.42)	44-1800-65
	MeOH	273(4.35)	44-1800-65
Coumarin, 6-hydroxy-4-methyl-5-nitro-	MeOH	272(4.34),345(4.00)	44-4344-65
Crotonic acid, 2,4-dihydroxy-4-(m-nitrophenyl)-, γ-lactone	MeOH	262(3.98)	44-1800-65
$C_{10}H_7NS$			
Benzo[b]thiophene-3-acetonitrile	EtOH	230(4.21),259(3.38), 265s(3.32),290(2.84), 298(3.05)	4-0231-65
Benzo[b]thiophene-2-carbonitrile, 3-methyl-	EtOH	233(4.02),242(3.98), 275(3.88),284(3.97), 308(3.23),319(3.21)	4-0231-65
$C_{10}H_7N_3$			
Malononitrile, (α-aminobenzylidene)-	EtOH	230(3.98),290(4.20)	94-0828-65
1H-Naphtho[1,8-de]triazine	EtOH	233(4.49),339(4.00), 453(2.87)	39-3005-64
$C_{10}H_7N_3O$			
Cyclohepta[b]pyrrole-3-carbonitrile, 1-amino-1,2-dihydro-2-oxo-	EtOH	282(4.45),425(4.06)	94-0450-65
$C_{10}H_7N_3O_2$			
Indole-3-acetonitrile, 4-nitro-	EtOH	234(3.98),332s(3.57), 374(3.68)	44-1158-64
$C_{10}H_7N_3O_5$			
1-Hydroxy-8-methyl-2,4-dinitroquinolizinium hydroxide, inner salt	75% EtOH	215(4.33),257(3.99), 432(4.42)	39-3030-64

Compound	Solvent	$\lambda_{max}(\log \epsilon)$	Ref.
$C_{10}H_7N_3S$			
Benzimidazole, 2-(4-thiazolyl)-	pH 1	241(4.12),301(4.41)	44-0259-65
	pH 1	243(4.14),301(4.42)	87-0399-64
	pH 13	235s(4.21),302(4.32)	87-0399-64
C^{14}-labelled	EtOH	243(4.18),299(4.37), 311(4.21)	87-0399-64
glucuronide	pH 1	248(4.03),308(4.25)	87-0399-64
	pH 13	237s(4.22),308(4.22)	87-0399-64
1H-Imidazo[4,5-b]pyridine, 2-(2-thienyl)-	MeOH	256(3.90),324(4.41), 340(4.24)	44-3403-64
$C_{10}H_7N_3S_2$			
1-[[(2,2-Dicyano-1-mercaptovinyl)thio]-methyl]pyridinium hydroxide, inner salt	MeCN	263(3.99),338(4.29)	44-0497-64
$C_{10}H_7N_5O_2$			
Alloxazine, 7-amino-	pH 13	262(4.57),428(4.30)	65-0675-65
$C_{10}H_7N_5O_3$			
as-Triazine, 3-(methyleneamino)-6-[2-(5-nitro-2-furyl)vinyl]-	DMF	300(4.22),420(4.44)	95-0016-64
$C_{10}H_8$			
Azulene	C_6H_{12}	582(2.5)	73-3016-65
1-Buten-3-yne, 1-phenyl-	n.s.g.	277(4.23)	5-0062-65B
trans	n.s.g.	288(4.36)	5-0062-65B
Naphthalene	EtOH	221(4.98),286(3.62), 312(3.40)	49-0285-65
	EtOH	220(4.1),265(2.6), 310(1.5)	38-1082-64B
bromanil complex	$CHCl_3$	478(2.92)	60-0465-64
chloranil complex	$CHCl_3$	470(2.85)	60-0465-64
iodoanil complex	$CHCl_3$	460(2.97)	60-0465-64
trinitrobenzene complex	heptane(50°)	360(3.12)	38-0166-64A
tetrabromophthalic anhydride complex	heptane(50°)	360(3.0)	38-0166-64A
	$CHCl_3$	365(2.96)	38-0166-64A
tetraiodophthalic anhydride complex (also other complexes)	$CHCl_3$	390(3.62)	38-0166-64A
$C_{10}H_8B_2O_3$			
1,8-Naphthalenediboronic anhydride	EtOH	229(4.66),291(3.90), 303(4.05),315(3.93)	44-0807-65
$C_{10}H_8BaO_6$			
Chorismic acid, barium salt	H_2O	272(3.43)	12-1227-65
$C_{10}H_8BrClO_4S_3$			
4-(p-Bromophenyl)-2-(methylthio)-1,3-dithiol-1-ium perchlorate	70% $HClO_4$	225(4.11),247(4.29), 287(4.01),375(4.16)	44-1711-64
$C_{10}H_8BrN$			
Cinnamonitrile, β-bromo-α-methyl-	EtOH	261(4.22)	94-1446-64
Isoquinoline, 3-bromo-5-methyl-	EtOH	278(3.59),326(3.40)	94-1296-64

Compound	Solvent	$\lambda_{max}(\log \epsilon)$	Ref.
$C_{10}H_8BrNO$			
Cinnamonitrile, β-bromo-4-methoxy-	EtOH	228(4.47),305(4.75)	94-1446-64
Isoxazole, 5-bromo-3-methyl-4-phenyl-	C_6H_{12}	208(4.03),233(3.88)	17-0203-65
	MeOH	210(4.05),231(3.98)	17-0203-65
Isoxazole, 5-bromo-4-methyl-3-phenyl-	C_6H_{12}	208(4.06),236(4.09)	17-0203-65
	MeOH	208(3.97),236(4.03)	17-0203-65
$C_{10}H_8BrNO_3$			
3-Indolineacetic acid, 3-bromo-2-oxo-	Et_2O	220(4.26),235s(4.12), 320(3.00)	44-2431-64
$C_{10}H_8BrN_3O$			
6(1H)-Pyridazinone, 4-amino-5-bromo-1-phenyl-	EtOH	227(4.40),310(4.01)	78-1323-65
$C_{10}H_8BrN_3O_2$			
Acetic acid, cyano-, 2-(6-bromo-7-oxo-1,3,5-cycloheptatrien-1-yl)hydrazide	EtOH	259(4.36),340(4.03), 405(4.09)	94-0450-65
2H-1,2,3-Triazole-4-carboxylic acid, 2-(p-bromophenyl)-, methyl ester	EtOH	276(4.37)	39-2306-64
$C_{10}H_8Br_2N_2Zn$			
Zinc, dibromo[2,2'-bipyridine)-	toluene	301(4.13),310(4.13)	101-0222-65A
$C_{10}H_8Br_2O_2S$			
1-Benzothiepin, 2,3-dibromo-2,3-dihydro-, 1,1-dioxide	EtOH	230(3.48),259(3.23), 296(3.13)	44-0366-64
$C_{10}H_8Br_3NO$			
7,9-Dibromo-10-hydroxy-6H-pyrido-[1,2-a]azepinium bromide	H_2O	258(3.98),311(3.78), 410(3.74)	44-1523-65
$C_{10}H_8Br_4O_2S$			
1-Benzothiepin, 2,3,4,5-tetrabromo-2,3,4,5-tetrahydro-, 1,1-dioxide	EtOH	278(3.54),284(3.51)	44-0366-64
	dioxan	277(3.34),284s(3.30)	44-0366-64
$C_{10}H_8ClNO$			
Isoxazole, 5-chloro-3-methyl-4-phenyl-	C_6H_{12}	210(3.97),234(3.96)	17-0203-65
	MeOH	208(4.07),231(3.97)	17-0203-65
Isoxazole, 5-chloro-4-methyl-3-phenyl-	C_6H_{12}	208(4.03),236(3.99)	17-0203-65
	MeOH	208(4.12),236(4.09)	17-0203-65
$C_{10}H_8ClNO_3$			
3-Buten-2-one, 1-chloro-4-(p-nitrophenyl)-	EtOH	306(4.28)	87-0035-65
$C_{10}H_8ClNO_4$			
Indole-2,3-dione, 4-chloro-5,7-dimethoxy-	EtOH	203(4.56),255(3.94), 304(3.45),499(2.52)	39-4310-64
Indole-2,3-dione, 7-chloro-4,5-dimethoxy-	EtOH	208(4.35),256(4.21), 329(3.79),481(2.56)	39-4310-64
$C_{10}H_8ClNO_6S_3$			
4-(p-Nitrophenyl)-2-methylthio-1,3-dithiolium perchlorate	70% $HClO_4$	232(4.19),268(3.98), 315(3.90),368(4.36)	44-1708-64
$C_{10}H_8ClN_3O$			
1H-Dipyrido[3,4-b:4',3'-d]pyrrol-1-one, 8-chloro-2,3,4,9-tetrahydro-	H_2O	229(4.41),292(4.07), 287s(4.05)	7-1223-65

Compound	Solvent	$\lambda_{max}(\log \epsilon)$	Ref.
$C_{10}H_8ClN_3OS$			
Benzaldehyde, p-chloro-, 2-azine with 2,4-thiazolidinedione	EtOH-NaOH	305(4.46)	7-0987-64
2-Isoxazoline-5-thione, 4-(2-chloro-phenylazo)-3-methyl-	EtOH	248(4.23),448(4.20)	39-3312-65
$C_{10}H_8ClN_3O_2$			
2-Isoxazolin-5-one, 4-(m-chloro-phenylazo)-3-methyl-	EtOH	251(4.02),392(4.33)	39-3312-65
2-Isoxazolin-5-one, 4-(o-chloro-phenylazo)-3-methyl-	EtOH	248(4.05),398(4.31)	25-1264-64
	EtOH	248(4.05),398(4.31)	39-3312-65
2H-1,2,3-Triazole-4-carboxylic acid, 2-(p-chlorophenyl)-, methyl ester	EtOH	275(4.41)	39-2306-64
1,2,3-Triazole-4-carboxylic acid, 2-(3-chlorophenyl)-5-methyl-	EtOH	278(4.30)	39-3312-65
$C_{10}H_8Cl_2N_2O_2$			
Benzoquinone, 2,5-bis(1-aziridinyl)-3,6-dichloro-	methyl cellosolve	226(4.18),348(4.23), 489(2.54)	78-1889-64
$C_{10}H_8Cl_2S_2$			
[2,2'-Bithiophene], 5-(1,1-dichloro-ethyl-	EtOH	249(4.07),345(4.35)	39-7109-65
$C_{10}H_8Cl_4$			
Tetralin, 1,2,3,4-tetrachloro-	C_6H_6	280(2.5),292s(2.39)	78-2217-64
$C_{10}H_8D_2O_2$			
2,4,6,8-Decatetraenoic-5,6-d_2 acid, 4-hydroxy-, γ-lactone, all trans	Et_2O	233(4.00),353(4.60)	24-2236-65
$C_{10}H_8F_2N_2O_2$			
Benzoquinone, 2,5-bis(1-aziridinyl)-3,6-difluoro-	methyl cellosolve	223(4.18),334(4.19), 467(2.55)	78-1889-64
$C_{10}H_8I_2N_2$			
Quinoxaline, 2,3-bis(iodomethyl)-	EtOH	252(4.42),334(3.83)	44-1542-65
$C_{10}H_8N_2$			
2,2'-Bipyridine	hexane	237(3.99),280(4.07)	65-0298-65
	octane	342(1.38)	65-0298-65
	H_2O	233(3.93),280(4.05)	65-0298-65
	n.s.g.	280(4.11)	39-6061-65
ferrous complex	n.s.g.	522(3.94)	39-6061-65
2,3'-Bipyridine	hexane	241(4.00),271(3.93)	65-0298-65
	heptane	363(0.61)	65-0298-65
	H_2O	236(4.06),275(4.03)	65-0298-65
4,4'-Bipyridine	hexane	237(4.11)	65-0298-65
	heptane	343(1.20)	65-0298-65
	H_2O	239(4.17)	65-0298-65
2,4,6,8-Decatetraenedinitrile	EtOH	314(4.78),330(4.95), 346(4.95)	70-0684-65
$C_{10}H_8N_2O$			
[2,3'-Bipyridin]-5-ol	H_2O	260(4.34)	65-0298-65
3(2H)-Pyridazinone, 6-phenyl-	EtOH	249(4.38)	39-3342-65
2-Quinoxalinecarboxaldehyde, 3-methyl-	EtOH	237(4.50),316(3.83)	44-1542-65

Compound	Solvent	λ_{max}(log ε)	Ref.
$C_{10}H_8N_2OS_2$			
Luciferin, decarboxy-	EtOH	268(3.86),328(4.24)	44-2344-65
3(2H)-Pyridazinone, 4,5-dimercapto-2-phenyl-	EtOH	268(4.38),306(3.86),362(3.92)	78-1323-65
$C_{10}H_8N_2O_2$			
Cinnamonitrile, α-methyl-p-nitro-	EtOH	302(4.17)	65-3150-64
Imidazo[1,2-a]pyridine-2-acrylic acid, hydrochloride	EtOH	248(4.24),256(4.32),300(3.83),324(3.91)	44-2403-65
2,6-Naphthoquinone, 1,5-diamino-	Me_2SO	320(3.7),595(4.2)	24-1625-64
6H-1,3-Oxazin-4-ol, 6-imino-2-phenyl-	pH 2	235(4.18),270s(3.20)	49-0212-65
	EtOH	235(4.18),270s(3.20)	49-0212-65
3H-Pyrrolo[1,2,3-de]quinoxalin-3-one, 5,6-dihydro-2-hydroxy-	MeOH	235(3.98),242(3.92),262(3.71),272(3.71),314(3.99),322(3.95),340s(3.59)	44-2589-65
5,8-Quinoxalinedione, 2,3-dimethyl-	MeOH	257(4.16)	4-0171-64
	MeCN	227(4.27),248(4.31),282(3.91),290(4.00)	44-2583-65
$C_{10}H_8N_2O_2S_2$			
Benzo[1,2-d:5,4-d']bisthiazole, 2,6-dimethoxy-	EtOH	240(4.68),260s(4.13),300(3.38),306(3.33),312(3.61)	7-0080-64
Benzo[1,2-d:5,4-d']bisthiazole-2,6(3H,5H)-dione, 3,5-dimethyl-	EtOH	230(4.76),264(3.79),314(4.11)	7-0080-64
$C_{10}H_8N_2O_3$			
Cyclohepta[b]pyrrole-3-carboxylic acid, 1-amino-1,2-dihydro-2-oxo-	EtOH	230(4.23),283(4.44),421(4.09)	94-0450-65
Furo[3,4-b]quinoxaline-1,3-diol, 1,3-dihydro-	EtOH	238(4.48),322(3.91)	44-1542-65
Glyoxylic acid, cyano-, benzyl ester, oxime	EtOH	237.0(4.02)	39-1262-65
Indole-3-carboxaldehyde, 1-methyl-4-nitro-	EtOH	227(4.31),292(3.89),362(3.61)	78-1397-64
Indole-3-carboxaldehyde, 1-methyl-5-nitro-	EtOH	267(4.46),318(3.93)	78-1923-65
Indole-3-carboxaldehyde, 1-methyl-6-nitro-	EtOH	284(4.47),330s(3.84)	78-1397-64
	EtOH	281(4.46)	78-1923-65
Indole-3-carboxaldehyde, 2-methyl-5-nitro-	EtOH	259(4.43),318(4.00)	78-1923-65
Indole-3-carboxaldehyde, 2-methyl-6-nitro-	EtOH	283(4.36),343(3.98)	78-1923-65
	EtOH	284(4.35),346s(3.97)	78-1397-64
Quinaldine, 4-nitro-, 1-oxide	EtOH	256(4.23),376(4.00)	94-1495-64
Sydnone, 3-benzyl-4-formyl-	MeOH	241(3.96),319(3.93)	44-2044-64
$C_{10}H_8N_2O_3S$			
[2,3'-Bipyridine]-5-sulfonic acid	H_2O	245(4.12),286(4.12)	65-0298-65
2H-Isothiazolo[4,5-b]indol-3(4H)-one, 4-methyl-, 1,1-dioxide	EtOH	229(4.45),252s(3.66),262(3.53),305(4.00)	44-0178-64
$C_{10}H_8N_2O_4$			
Acetic acid, [3,4-dihydro-4-oxo-1-phthalazinyl)oxy]-	2% EtOH	260(3.68),297(3.74)	20-0091-65
2(1H)-Phthalazineacetic acid, 4-hydroxy-1-oxo-	2% EtOH	262(3.47),311(3.87)	20-0091-65
Phthalimidoacetic acid, 3-amino-	EtOH	258s(3.90),388(3.71)	44-3151-64
	dil. HCl	295(2.87),387(3.65)	44-3151-64
	dil. NaOH	306(3.53)	44-3151-64

Compound	Solvent	λ_{max}(log ε)	Ref.
Phthalimidoacetic acid, 4-amino-	EtOH	246(4.44),275s(4.01), 307(3.72),372(3.65)	44-3151-64
	dil. HCl	232(4.12),240(4.16), 258(4.15),299(3.50), 368(3.42)	44-3151-64
	dil. NaOH	265(4.06)	44-3151-64
Sydnone, 3-piperonyl-	EtOH	288(4.06)	87-0531-65
$C_{10}H_8N_2O_6$			
1-Indanone, 7-methoxy-4,6-dinitro-	EtOH	226(4.30)	39-2816-64
1(2H)-Naphthalenone, 3,4-dihydro-8-hydroxy-5,7-dinitro-	EtOH	242(4.33),336(3.75)	39-2816-64
$C_{10}H_8N_2O_{12}S_4$			
[4,4'-Bipyridine]-2,2',6,6'-tetrasulfonic acid	H_2O	225(3.71),260(3.62)	65-0298-65
$C_{10}H_8N_2S$			
Imidazo[2,1-b]benzothiazole, 2-methyl-	EtOH	241(4.21),285(3.43), 293(3.36)	94-0813-64
Thieno[3,4-b]quinoxaline, 1,3-dihydro-	EtOH	231(3.79),239(4.38)	44-1542-65
$C_{10}H_8N_2S_2$			
Thiazolo[4,5-f]benzothiazole, 2,6-dimethyl-	EtOH	246(4.66),254(4.64), 278(3.97),304(3.30), 316(3.28)	7-0080-64
Thiazolo[5,4-f]benzothiazole, 2,6-dimethyl-	EtOH	238(4.50),274s(3.92), 282(3.98),294s(3.74), 313(3.23)	7-0080-64
Thiazolo[5,4-g]benzothiazole, 2,7-dimethyl-	EtOH	232(4.21),250(4.32), 270s(3.89),300(3.11), 306s(2.84),312(2.93)	7-0080-64
$C_{10}H_8N_2S_4$			
Benzo[1,2-d:5,4-d']bisthiazole, 2,6-bis(methylthio)-	EtOH	276(4.60),321(4.71), 334(4.25)	7-0080-64
Benzo[1,2-d:5,4-d']bisthiazole-2,6(3H,-5H)-dithione, 3,5-dimethyl-	EtOH	228(4.09),262(4.26), 294(4.22),355(4.51), 370(4.76)	7-0080-64
$C_{10}H_8N_4$			
1,4-Naphtho[1,8-de]triazine, 1-amino-	n.s.g.	233(4.42),339(4.01)	77-0193-65
1-Naphthylamine, 8-azido-	n.s.g.	232(4.20),253(4.19), 346(3.94)	77-0193-65
$C_{10}H_8N_4OS$			
Lysine, N^6-(methylthiocarbamoyl)-, methylthiohydantoin derivative	EtOH	235(4.31),265(4.25)	65-0988-65
Thiocyanic acid, 3-amino-5-oxo-1-phenyl-2-pyrazolin-4-yl ester	EtOH	247(4.14)	94-0023-64
$C_{10}H_8N_4O_3$			
Acetic acid, cyano-, (p-nitrobenzylidene)hydrazide	MeCN	294(4.14),346(4.07)	44-2862-65
Pyrimidine, 2-amino-4-[2-(5-nitro-2-furyl)vinyl]-	propylene glycol	234(4.10),310(3.74), 384(4.24)	95-0121-64

Compound	Solvent	λ_{max}(log ε)	Ref.
$C_{10}H_8N_4O_4$			
Pyrazole, 3-methyl-4-nitro-1-(p-nitrophenyl)-	EtOH	216(4.03),313(4.30)	23-1605-64
2H-1,2,3-Triazole-4-carboxylic acid, 2-(p-nitrophenyl)-, methyl ester	EtOH	300(4.47)	39-2306-64
$C_{10}H_8N_4O_6$			
Indole, 1,2-dimethyl-3,4,6-trinitro-	EtOH	294(4.20),347(4.08)	44-3457-65
Indole, 1,2-dimethyl-3,5,6-trinitro-	EtOH	230(4.09),272(4.30), 295s(4.19),353(4.03)	44-3457-65
Malealdehydic acid, 2,4-dinitrophenylhydrazone,cis	EtOH	374.5(4.48)	39-3075-65
$C_{10}H_8N_6O_3$			
7(6H)-Pteridinone, 6-(4,6-dihydroxypyrimidin-5-yl)-5,8-dihydro-	pH 1.0	223(4.39),254(4.01), 293(3.94),352(3.72)	39-6930-65
$C_{10}H_8N_6O_4$			
7(6H)-Pteridinone, 6-(2,4,6-trihydroxypyrimidin-5-yl)-5,8-dihydro-	pH 1.0	220(4.50),258(3.93), 288(3.89),346(3.51)	39-6930-65
$C_{10}H_8O$			
1-Benzoxepin	EtOH	211(4.17),231(4.03), 288(3.46)	35-3168-64
2-Cyclobuten-1-one, 3-phenyl-	EtOH	217(4.14),223(4.07), 286(4.36)	35-2645-64
1-Naphthol	C_6H_6	308(3.5),323(3.4)	60-2097-65
	heptane	310(3.6),322(3.4)	60-2097-65
	neutral	292(3.67)	59-1709-64
	base	332(3.93)	59-1709-64
2-Naphthol	C_6H_6	313(3.0),322(3.9), 329(3.1)	60-2097-65
	heptane	318(3.2),326(3.2), 329(4.3)	60-2097-65
	pH 0.09	225(4.8),260(3.5), 275(3.6),290(3.5), 310(3.2),327(3.2)	46-2044-65
	neutral	278(3.66),328(3.26)	59-1709-64
	base	282(3.79),346(3.47)	59-1709-64
	pH 12.28	240(4.7),280(3.8), 295(3.7),345(3.5)	46-2044-65
	EtOH	264(3.86),275(4.73), 286(3.30),318(1.76), 335(2.08)	44-1391-64
11-Oxabicyclo[4.4.1]undeca-1,3,5,7,9-pentaene	EtOH	255(4.86),299(3.84), 392(2.38)	35-3168-64
1,6-Oxacyclodecapentaene	n.s.g.	257(4.87),302(3.85), 345-435f(2.5)	89-0785-64
$C_{10}H_8OS$			
Benzo[b]cyclopropa[d]thiopyran-2(1H)-one, 1a,7b-dihydro-	heptane	255(3.65),280s(3.21)	44-1092-64
4-Hexen-2-yn-1-one, 1-(2-thienyl)-	EtOH	313(4.13)	39-2983-65
$C_{10}H_8O_2$			
Butenolide, 4-(2,4-hexadienylidene)-	Et_2O	246(3.98),333(4.41), 352s(4.36)	24-1616-65
2,4,8-Decatrien-6-ynoic acid, 4-hydroxy-, γ-lactone, cis-cis	Et_2O	232(3.95),343(4.29)	24-2236-65

Compound	Solvent	$\lambda_{max}(\log \epsilon)$	Ref.
4-Hexen-2-yn-1-one, 1-(2-furyl)-	EtOH	310(4.26)	39-2983-65
1-Indenecarboxylic acid	EtOH	262(3.51),270(3.49)	35-5194-64
1,8-Naphthalenediol	EtOH	227(4.74),305(3.83), 320(3.88),333(3.90)	39-2816-64
$C_{10}H_8O_2S$			
Benzo[b]thiophene, 2-acetyl-3-hydroxy-	EtOH	210(4.30),257(4.22), 265(4.27),298(4.10), 306(4.17)	39-4939-65
Benzo[b]thiophene-3-acetic acid	EtOH	229(4.47),262(3.72), 267s(3.71),290(3.40), 299(3.41)	4-0231-65
$C_{10}H_8O_3$			
Benzoic acid, p-(1-hydroxy-2-propynyl)-	EtOH	234(4.21)	39-2983-65
Chromone, 5-hydroxy-2-methyl-	MeOH	226(4.35),251(4.14), 324(3.69)	24-2774-65
	MeOH-KOH	235(4.38),258(4.14), 366(3.74)	24-2774-65
5-Indancarboxylic acid, 1-oxo-	EtOH	251(4.27)	44-2165-65
1-Indanone, 2-hydroxy-, formate	EtOH	248(4.12),292(3.42)	44-1723-64
Isocoumarin, 7-hydroxy-4-methyl-	MeOH	230(4.38),267(3.99), 345(3.61)	42-0821-64
Isocoumarin, 3-methoxy-	n.s.g.	270(3.9),277(3.9), 345(3.59)	83-0411-65
$C_{10}H_8O_3S$			
Benzo[b]thiophene, 2-(hydroxymethyl)-5,6-(methylenedioxy)-	EtOH	241(4.43),268(4.03), 277(4.19),305(3.67), 311(3.72),317(3.73)	4-0100-65
$C_{10}H_8O_4$			
Acetic acid, (2-acetyl-3,6-dihydroxyphenyl)-, γ-lactone	EtOH	234(4.12),250s(3.9), 351(3.54)	27-0358-64
Glutaconic acid, 4-(1,3-dihydroxy-2-butenylidene)-3-methyl-, di-γ-lactone	EtOH	280(3.90),335(3.90)	39-2283-65
Isocoumarin, 5,7-dihydroxy-4-methyl-	EtOH	237(4.34),365(3.71)	42-0821-64
Isocoumarin, 6,8-dihydroxy-3-methyl-	EtOH	237s(4.65),244(4.75), 260(4.49),276(4.47), 317(3.80)	39-5382-64
Maleic acid, di-2-propynyl ester	MeOH	none	70-1349-64
Pentacyclo[4.2.0.0^{2,5}.0^{3,8}.0^{4,7}]octane-1,4-dicarboxylic acid	EtOH	210(3.24)(end absorption)	35-0962-64
$C_{10}H_8O_5$			
Unknown lactone	EtOH	218(4.29),242(4.29), 327(4.18)	65-1013-64
$C_{10}H_9Br$			
Bullvalene, bromo-	n.s.g.	235s(3.42)	88-0773-64
	n.s.g.	235(3.42)	24-3385-65
Naphthalene, 4-bromo-1,2-dihydro-	EtOH	217(4.25),264(3.97)	54-0389-65
$C_{10}H_9BrN_2$			
1H-3-Benzazepine, 2-amino-4-bromo-	EtOH	296(4.05)	4-0026-65

Compound	Solvent	$\lambda_{max}(\log \epsilon)$	Ref.
$C_{10}H_9BrN_4$			
Butyronitrile, 3-imino-2-oxo-, (p-bromophenyl)hydrazone	EtOH	248(4.04),260s(3.99), 371(4.30)	5-0033-65J
	EtOH-NH	248(4.04),259s(4.00), 371(4.30)	5-0033-65J
$C_{10}H_9BrO$			
3-Buten-2-one, 4-(p-bromophenyl)-	hexane	288(5.00)	44-3000-65
	iso-PrOH	289(4.68)	44-3000-65
Indanone, 2-bromo-7-methyl-	EtOH	258(4.13),303(3.37)	44-0074-64
Indanone, 4-bromo-6-methyl-	EtOH	251(4.05),305(3.45)	22-3103-64
$C_{10}H_9BrO_2$			
1,3-Butanedione, 1-(p-bromophenyl)-	hexane	309(4.27)	44-3000-65
	iso-PrOH	314(4.26)	44-3000-65
Cinnamic acid, β-bromo-α-methyl-	EtOH	255(3.74)	94-1446-64
$C_{10}H_9Br_2NO$			
2-Bromo-1-hydroxy-6-methyl-quinolizinium bromide	H_2O	216(4.45),368(4.12)	39-3030-64
2-Bromo-1-hydroxy-8-methyl-quinolizinium bromide	H_2O	216(4.61),365(4.18)	39-3030-64
$C_{10}H_9ClN_2OS$			
Hydrouracil, 3-(o-chlorophenyl)-2-thio-	n.s.g.	280(4.11)	44-1623-64
6H-1,3-Thiazin-6-one, 2-[(o-chloro-phenyl)imino]tetrahydro-	EtOH	245(4.07),295(3.99)	44-1623-64
$C_{10}H_9ClN_2O_2$			
Malonamide, 2-(o-chlorobenzylidene)-	MeOH	269(4.072)	23-2426-65
$C_{10}H_9ClN_2O_3$			
D-Cycloserine, N-5-chlorosalicylidene-	MeOH	222(4.50),258(4.11), 332(3.66)	44-3436-65
$C_{10}H_9ClN_4$			
Butyronitrile, 3-imino-2-oxo-, (m-chlorophenylhydrazone)	EtOH	249(4.04),368(4.28)	5-0033-65J
	$EtOH-NH_3$	250(4.06),370(4.31)	5-0033-65J
Butyronitrile, 3-imino-2-oxo-, (o-chlorophenylhydrazone)	EtOH	247(4.13),378(4.27)	5-0033-65J
	$EtOH-NH_3$	247(4.13),377(4.28)	5-0033-65J
Butyronitrile, 3-imino-2-oxo-, (p-chlorophenylhydrazone)	EtOH	247(4.02),259(4.04), 371(4.37)	5-0033-65J
	$EtOH-NH_3$	247(4.08),259(4.03), 371(4.36)	5-0033-65J
$C_{10}H_9ClN_6$			
Pyrimidine, ,24-diamino-6-chloro-5-(phenylazo)-	pH 1	358(4.48)	87-0792-64
	pH 10	233(4.13),290s(3.88), 364(4.32)	87-0792-64
$C_{10}H_9ClO$			
Indanone, 5-chloro-4-methyl-	C_6H_{12}	251(4.27),259(4.02), 284(3.44),295(3.47)	22-3103-64
Indanone, 6-chloro-4-methyl-	C_6H_{12}	242(3.99),349(3.98), 296(3.36),307(3.41)	22-3103-64
$C_{10}H_9ClOS$			
1-Benzothiepin-5-ol, 2-chloro-4,5-dihydro-	heptane	217(3.94),240(3.56), 273(3.62)	44-1092-64

Compound	Solvent	$\lambda_{max}(\log \epsilon)$	Ref.
$C_{10}H_9ClO_4S_2$			
2-Methyl-4-phenyl-1,3-dithiolium perchlorate	EtOH-$HClO_4$	310(3.9)	73-3016-65
4-Methyl-2-phenyl-1,3-dithiolium perchlorate	EtOH-$HClO_4$	362(4.2)	73-3016-65
$C_{10}H_9ClO_4S_3$			
2-(Methylthio)-4-phenyl-1,3-dithiolium perchlorate	70% $HClO_4$	240(4.27),282(3.89), 373(4.14)	44-1708-64
$C_{10}H_9Cl_2N_3O$			
Indole-2-carboxylic acid, 3,x-dichloro-1-methyl-, hydrazide	EtOH	223(4.44),293(4.14)	44-0178-64
$C_{10}H_9FO$			
1(2H)-Naphthalenone, 6-fluoro-3,4-dihydro-	EtOH	248(4.06)	44-2165-65
$C_{10}H_9FO_2$			
Cinnamaldehyde, α-fluoro-p-methoxy-	EtOH	224(3.95),310(4.29)	22-2258-64
$C_{10}H_9F_2N_3O_4$			
Acetamide, N,N'-(4,6-difluoro-2-nitro-m-phenylene)bis-	pH 1,7	311.5(3.19)	87-0737-65
	pH 13	275(3.55),342(3.09)	87-0737-65
$C_{10}H_9FeNO_2$			
Ferrocene, nitro-	heptane	273(3.88),365(3.11), 473(2.89)	88-2881-64
$C_{10}H_9IO_4$			
Cinnamic acid, 4-hydroxy-5-iodo-3-methoxy-	EtOH	234(4.34),244(4.39), 322(4.31)	44-1812-64
$C_{10}H_9KO_6S$			
Potassium, [[1-carboxy-3-[[3-carboxy-2-(hydroxymethyl)allyl]thio]-2-(hydroxymethyl)propenyl]oxy]-, di-γ-lactone	EtOH	241(4.14),267(3.97)	44-3560-64
$C_{10}H_9MnO_5$			
Methylmanganese carbonyl, butadiene adduct	dioxan	215(4.34)	77-0370-64
$C_{10}H_9N$			
Cinnamonitrile, α-methyl-	EtOH	270(4.04)	65-3150-64
1-Naphthylamine	acid	278(3.73)	59-1709-64
	neutral	322(3.71)	59-1709-64
	EtOH	240(4.4),320(3.7)	44-1543-64
2-Naphthylamine	acid	275(3.62),303s(--)	59-1709-64
	neutral	280(3.78),340(3.30)	59-1709-64
$C_{10}H_9NO$			
Carbostyril, 1-methyl-	EtOH	229(4.59),270(3.87), 329(3.79)	1-1389-64
	dioxan	254(4.24),278(3.76), 334(3.79)	1-1389-64
Carbostyril, 3-methyl-	EtOH	220(4.25),269(3.91), 324(3.91)	1-1120-65
Carbostyril, 4-methyl-	EtOH	230(4.58),268(3.82), 327(3.84)	1-1120-65

Compound	Solvent	$\lambda_{max}(\log \epsilon)$	Ref.
Carbostyril, 5-methyl-	EtOH	234(4.51),281(3.97), 333(3.76)	1-1120-65
Carbostyril, 6-methyl-	EtOH	233(4.61),271(3.88), 337(3.81)	1-1120-65
Carbostyril, 7-methyl-	EtOH	232(4.61),276(3.95), 329(3.99)	1-1120-65
Carbostyril, 8-methyl-	EtOH	233(4.64),276(4.05), 332(3.88)	1-1120-65
1-Indanone, 3-imino-2-methyl-	MeOH	221(4.42),254(4.49), 261(4.51),302(2.95), 443(3.38)	65-0448-64
2-Indolinone, 3-ethylidene-	EtOH	246(4.43),249(4.43), 255(4.50),287(3.65)	44-2431-64
4-Isoquinolinol, 3-methyl-	EtOH	255(3.45),305(3.46), 385(3.75)	28-1339-65B
Ketone, 2-indolyl methyl	EtOH	225(4.10),305(4.33)	44-3604-65
Lepidine, 1-oxide	C_6H_{12}	360(3.83)	1-1120-65
	N HCl	316(3.88)	1-1120-65
	H_2O	315(3.88)	1-1120-65
	EtOH	330(3.89)	1-1120-65
1-Naphthol, 2-amino-	acid	288(3.67),322(3.34)	59-1709-64
	neutral	286(--),336(3.49)	59-1709-64
1-Naphthol, 4-amino-	acid	296(3.76)	59-1709-64
	neutral	340(3.54)	59-1709-64
1-Naphthol, 5-amino-	acid	298(3.74)	59-1709-64
	neutral	307(3.92)	59-1709-64
	base	342(4.07)	59-1709-64
2-Naphthol, 1-amino-	acid	274(3.66),330(3.28)	59-1709-64
	neutral	302(3.40),339(3.41)	59-1709-64
2-Naphthol, 3-amino-	acid	273(3.60),327(3.34)	59-1709-64
	neutral	285(3.63),332(3.58)	59-1709-64
	base	285s(--),344(3.82)	59-1709-64
2-Naphthol, 7-amino-	acid	275(3.63),328(3.30)	59-1709-64
	neutral	285(3.57),332(3.38)	59-1709-64
	base	288s(--),342(3.59)	59-1709-64
2-Naphthol, 8-amino-	acid	275(3.70),332(3.37)	59-1709-64
	neutral	298(3.62),335(3.43)	59-1709-64
	base	296(3.77),352(3.50)	59-1709-64
Quinaldine, 1-oxide	C_6H_{12}	344(3.78)	1-1120-65
	N HCl	319(3.71)	1-1120-65
	H_2O	314(3.82)	1-1120-65
	EtOH	319(3.93)	1-1120-65
Quinoline, 3-methyl-, 1-oxide	C_6H_{12}	351(3.96)	1-1120-65
	N HCl	320(3.84)	1-1120-65
	H_2O	313(3.88)	1-1120-65
	EtOH	317(3.87)	1-1120-65
Quinoline, 5-methyl-, 1-oxide	C_6H_{12}	351(3.73)	1-1120-65
	N HCl	319(3.78)	1-1120-65
	H_2O	328(4.00)	1-1120-65
	EtOH	332(3.99)	1-1120-65
Quinoline, 6-methyl-, 1-oxide	C_6H_{12}	349(4.07)	1-1120-65
	N HCl	323(3.88)	1-1120-65
	H_2O	319(3.98)	1-1120-65
	EtOH	323(3.97)	1-1120-65
Quinoline, 7-methyl-, 1-oxide	C_6H_{12}	354(4.05)	1-1120-65
	N HCl	328(3.81)	1-1120-65
	H_2O	314(3.86)	1-1120-65
	EtOH	318(3.92)	1-1120-65

Compound	Solvent	$\lambda_{max}(\log \epsilon)$	Ref.
Quinoline, 8-methyl-, 1-oxide	C_6H_{12}	350(4.09)	1-1120-65
	N HCl	319(3.64)	1-1120-65
	H_2O	330(3.92)	1-1120-65
	EtOH	334(4.12)	1-1120-65
$C_{10}H_9NOS_2$			
Ketone, 5-(5-amino-2-thienyl)-2-thienyl methyl-, hexachlorostannate	EtOH	263(4.06),412(4.60)	65-0674-64
	20% HCl	247(4.10),365(4.53)	65-0674-64
$C_{10}H_9NO_2$			
Benzamide, p-(1-hydroxyprop-2-ynyl)-	EtOH	233(4.15)	39-2983-65
Indole-3-acetic acid	EtOH	222(4.51),274s(3.75), 279(3.78),294(3.71)	35-3796-64
	H_2SO_4	234(3.65),239(3.63), 281(3.68)	35-3796-64
Isoxazole, 3-methoxy-5-phenyl-	C_6H_{12}	260f(4.23)	78-2835-64
	5% EtOH	260(4.32)	78-2835-64
	20N H_2SO_4-5% EtOH	279(4.33)	78-2835-64
Isoxazole, 5-(p-methoxyphenyl)-	$CHCl_3$	280(4.30)	24-3020-65
4-Isoxazolin-3-one, 2-methyl-5-phenyl-	C_6H_{12}	261(4.16)	78-2835-64
	5% EtOH	264(4.34)	78-2835-64
	20N H_2SO_4-5% EtOH	278(4.30)	78-2835-64
$C_{10}H_9NO_2S$			
2H-1,3-Benzoxazine-2,4(3H)-dione, 3-ethyl-4-thio-	H_2O	270(4.09),328(4.09)	78-3019-65
$C_{10}H_9NO_2S_3$			
1,2-Dithiole-3-thione, 4-carbobenzoxyamino-	EtOH	233(4.14),257(3.57), 298(3.59),417(4.02)	12-0447-64
$C_{10}H_9NO_2S_4$			
p-Toluenesulfonamide, N-(3-thioxo-3H-1,2-dithiole-4-yl)-	EtOH	225(4.30),293(3.57), 415(4.04)	12-0447-64
$C_{10}H_9NO_3$			
o-Anisic acid, α-cyano-, methyl ester	EtOH	208(4.15),230(3.85), 287(3.32)	39-2361-65
4,1-Benzoxazepine-2,5(1H,3H)-dione, 1-methyl-	iso-PrOH	216(4.38),297(3.36)	44-0582-64
4,1-Benzoxazepine-2,5(1H,3H)-dione, 3-methyl-	iso-PrOH	243(3.81),300(3.28)	44-0582-64
2H-1,3-Benzoxazine-2,4(3H)-dione, 3-ethyl-	MeOH	205(4.51),236(3.90), 290(3.00	4-0037-65
3-Buten-2-one, 4-(p-nitrophenyl)-	hexane	295(4.7)	44-3000-65
	EtOH	302(4.27)	7-0143-65
	iso-PrOH	298(4.2)	44-3000-65
Chroman, 4-hydroxyimino-6-methyl-2-oxo-	EtOH	239(3.87),246(3.77), 292(3.46)	4-0385-65
Chroman, 4-hydroxyimino-7-methyl-2-oxo-	EtOH	243(3.94),283(3.53)	4-0385-65
3-Indolineacetic acid, 2-oxo-	EtOH	205(4.44),247(3.94), 280s(3.15)	44-2431-64
1(2H)-Naphthalenone, 3,4-dihydro-7-nitro-	EtOH	234(4.34),269(3.92)	39-5377-65
2-Pyridinepropionic acid, α-(2-hydroxyethyl)-β-oxo-, γ-lactone	EtOH	234(3.93),270(3.69)	94-1338-64

Compound	Solvent	λ max(log ε)	Ref.
$C_{10}H_9NO_3S$			
4H-1,2-Benzothiazin-4-one, 3-ethylidene-2,3-dihydro-, 1,1-dioxide	EtOH	262(4.08),301s(3.65)	44-2241-65
$C_{10}H_9NO_3S_3$			
p-Toluenesulfonamide, N-(3-oxo-3H-1,2-dithiole-4-yl)-	EtOH	225(4.00),325(3.76)	12-0447-64
$C_{10}H_9NO_4$			
Acrylic acid, 2-amino-3-salicyloyl-	EtOH	365(4.59)	65-0541-64
1,3-Butanedione, 1-(p-nitrophenyl)-	iso-PrOH	327(4.70)	44-3000-65
1,3-Diox-4-ene, 4-p-nitrophenyl-	n.s.g.	320(3.98)	88-0499-64
1-Indanone, 7-methoxy-4-nitro-	EtOH	235(4.19),253(4.22), 315(4.05)	39-2816-64
1-Indanone, 7-methoxy-6-nitro-	EtOH	240(4.27)	39-2816-64
Indole-2-carboxylic acid, 5-hydroxy-6-methoxy-	EtOH	209(4.43),299(4.11), 315(4.17)	4-0387-65
Indole-2-carboxylic acid, 6-hydroxy-5-methoxy-	EtOH	210(4.40),321(4.26)	4-0387-65
Indole-2,3-dione, 5,6-dimethoxy-	EtOH	269(4.34),320(3.88), 470(3.15)	39-5551-65
1(2H)-Naphthalenone, 3,4-dihydro-8-hydroxy-5-nitro-	EtOH	249(4.20),274(3.67), 298(3.80)	39-2816-64
1(2H)-Naphthalenone, 3,4-dihydro-8-hydroxy-7-nitro-	EtOH	245(4.35),296(3.25), 342(3.90)	39-2816-64
$C_{10}H_9NO_4S$			
2H-1,2-Benzothiazin-4(3H)-one, 3-acetyl-, 1,1-dioxide	EtOH	244(3.86),324(4.04)	44-2241-65
	base	252(3.75),425(3.54)	44-2241-65
1-Benzothiepin, 2,3-dihydro-4-nitro-, 1,1-dioxide	EtOH	225(3.00),311(3.10)	44-0366-64
$C_{10}H_9NO_5$			
Benzoic acid, 2-acetyl-3-nitro-, methyl ether	EtOH	217(3.84),247(3.72)	44-4293-65
$C_{10}H_9NS$			
Acrylonitrile, 3-(p-tolylthio)-, cis	n.s.g.	281(4.22)	44-1800-64
trans	n.s.g.	273(4.46)	44-1800-64
$C_{10}H_9NS_2$			
2H-1,3-Thiazine-2-thione, 3,6-dihydro-2-thio-	H_2O	307(4.08)	39-4008-64
	EtOH	323(4.10)	39-4008-64
$C_{10}H_9N_3$			
Indole-3-acetonitrile, 4-amino-	EtOH	225(4.66),273(3.85), 288s(3.72),295(3.70)	44-1158-64
2-Pyrazoline-4-carbonitrile, 1-phenyl-	EtOH	239(3.86),282(4.02)	95-0158-65
$C_{10}H_9N_3O$			
Butyronitrile, 2,3-dioxo-, 2-(phenylhydrazone)	EtOH	241(4.06),285s(3.07), 362(4.32)	5-0033-65J
5-Aza-β-carboline, 3,4-dihydro-1-oxo-	H_2O	221(4.30),314(4.12)	7-1223-65
6-Aza-β-carboline, 3,4-dihydro-1-oxo-	H_2O	241(4.50),291(3.74), 309(3.75)	7-1223-65
$C_{10}H_9N_3OS$			
4-Thiazolecarboxanilide, 2'-amino-, C^{14} labelled	EtOH	232(4.28),302(3.71)	87-0399-64

Compound	Solvent	λ_{max}(log ε)	Ref.
$C_{10}H_9N_3O_2$			
2-Isoxazolin-5-one, 3-methyl-4-(phenylazo)-	EtOH	247(3.97),398(4.31)	39-3312-65
Pyrazole, 3-methyl-4-nitro-1-phenyl-	EtOH	300(3.95)	23-1605-64
Sydnone imine, N-acetyl-3-phenyl-	pH 1.0	296(4.18)	65-2064-64
	pH 7.00	250(4.18),330(4.08)	65-2064-64
$C_{10}H_9N_3O_3$			
Sydnone imine, N-carboxy-3-phenyl-, methyl ester	pH 1.0	292(4.18)	65-2064-64
	pH 7.00	240(4.20),342(4.04)	65-2064-64
Sydnone imine, 3-piperonyl-	EtOH-HCl	291(3.98)	87-0531-65
$C_{10}H_9N_3O_4$			
Acetic acid, diazo-p-nitrophenyl-, ethyl ester	MeOH	272(3.89),335(4.21)	88-1403-64
Indole, 1,2-dimethyl-3,5-dinitro-	EtOH	257(4.31),304s(3.97), 357(3.98)	44-3457-65
Indole, 1,2-dimethyl-3,6-dinitro-	EtOH	223s(3.95),273s(3.95), 303(4.18),339(4.20)	44-3457-65
	EtOH	212(4.17),220s(4.02), 308(4.18),344(4.19)	78-1397-64
Indole, 2-ethyl-3,5-dinitro-	EtOH	253(4.39),312(4.05), 349(4.02)	44-3457-65
1,3,4-Oxadiazole, 2-[1-ethyl-2-(5-nitro-2-furyl)vinyl]-	propylene glycol	244(4.04),368(4.28)	95-0225-64
1,3,4-Oxadiazole, 2-methyl-5-[1-methyl-2-(5-nitro-2-furyl)vinyl]-	propylene glycol	266(4.04),370(4.32)	95-0225-64
$C_{10}H_9N_3O_4S$			
1,3,4-Oxadiazole, 2-[1-methyl-2-(5-nitro-2-furyl)vinyl]-5-(methylthio)-	propylene glycol	230(4.07),294(3.98), 375(4.24)	95-0225-64
1,3,4-Oxadiazole-2-thiol, 5-[1-ethyl-2-(5-nitro-2-furyl)vinyl]-	propylene glycol	243(4.13),383(4.13)	95-0225-64
$C_{10}H_9N_3O_5$			
Indoline, 1-acetyl-5,7-dinitro-	EtOH	226(4.23),265s(3.75), 345(4.02)	44-0947-64
1,3,4-Oxadiazol-2-ol, 5-[1-ethyl-2-(5-nitro-2-furyl)vinyl]-	propylene glycol	284(4.05),378(4.25)	95-0225-64
$C_{10}H_9N_5O_2$			
Butyronitrile, 3-imino-2-oxo-, (m-nitrophenylhydrazone)	EtOH-NH_3	232(4.17),253(4.17), 355(4.08)	5-0033-65J
Butyronitrile, 3-imino-2-oxo-, (o-nitrophenylhydrazone)	EtOH-NH_3	232(4.09),382(4.23)	5-0033-65J
Butyronitrile, 3-imino-2-oxo-, (p-nitrophenylhydrazone)	EtOH	234s(3.83),260(4.03), 419(4.34)	5-0033-65J
	EtOH-NH_3	236s(3.91),263(3.99), 471(4.33)	5-0033-65J
$C_{10}H_9N_5O_2S$			
as-Triazine, 3-amino-6-methyl-5[2-(5-nitro-2-thienyl)vinyl]-	propylene glycol	236(4.08),394(4.30)	95-0109-64
$C_{10}H_9N_5O_3$			
Isoxazole, 5-methyl-3-[3-(m-nitrophenyl)-2-triazeno]-	EtOH	250(4.36)	78-0461-64
Isoxazole, 5-methyl-3-[3-(p-nitrophenyl)-2-triazeno]-	EtOH	286(4.3)	78-0461-64

Compound	Solvent	$\lambda_{max}(\log \epsilon)$	Ref.
1,2,4-Triazine, 3-amino-6-[1-methyl-2-(5-nitro-2-furyl)vinyl]-	propylene glycol	292(4.02),395(4.02)	95-0016-64
1,2,4-Triazine, 3-amino-6-methyl-5-[2-(5-nitro-2-furyl)vinyl]-	propylene glycol	237(4.25),380(4.41)	95-0109-64
$C_{10}H_9N_5O_4$			
Benzene, 1-(3-azido-2-methylpropenyl)-2,4-dinitro-	MeCN	245(4.03),290(3.94)	44-3049-64
Propionaldehyde, 2-cyano-, 2,4-dinitrophenylhydrazone	EtOH	222(4.16),348(4.28)	95-0158-65
Pyrazole, 5-amino-1-(2,4-dinitrophenyl)-4-methyl-	EtOH	317(3.89)	95-0158-65
$C_{10}H_9N_7O_8$			
2(1H)-Pyrimidinone, 4-hydrazino-, picrate	pH 1	276(4.04)	5-0134-65F
	H_2O	268(3.88)	5-0134-65F
$C_{10}H_{10}$			
Bullvalene	hexane	238s(3.23)	24-3140-64
Cycloprop[a]indene, 1,1a,6,6a-tetrahydro-	EtOH	200(4.5),218s(3.9), 230s(3.6),260s(2.9), 270(3.0),278(3.0)	35-0908-64
Indan, 1-methylene-	EtOH	250(3.9),280(3.0)	35-0908-64
Naphthalene, 1,2-dihydro-	EtOH	260(4.0)	78-0231-64
Nona-3,4-diene-5,8-diyne, 3-methyl-	EtOH	204(4.49),210(4.59), 238(4.09),250(4.13), 263(4.14),279(4.06), 291(3.86),310(3.81)	39-4659-65
Triquinacene	isooctane	187(4.11)	35-3162-64
Umbrellaene	n.s.g.	187(4.4),204(3.7)	35-2811-64
$C_{10}H_{10}BrN$			
Indole, 3-bromo-1,2-dimethyl-	EtOH	227(4.54),277s(3.82), 283(3.87),291s(3.85)	44-1206-64
1-Methylquinolizinium bromide	EtOH	229(4.01),289(3.52), 318(4.07),325(4.02), 332(4.27)	94-1338-64
2-Methylquinolizinium bromide	EtOH	227(4.32),274(3.49), 285(3.52),313(4.11), 320(4.04),328(4.30)	94-1338-64
3-Methylquinolizinium bromide	EtOH	232(4.32),236(4.31), 278(3.40),289(3.45), 303(3.68),316(4.07), 323(3.99),329(4.27)	94-1338-64
4-Methylquinolizinium bromide	EtOH	234(4.40),238(3.69), 289(3.69),318(4.08), 325(4.06),331(4.29)	94-1338-64
$C_{10}H_{10}BrNO$			
Cinnamamide, β-bromo-α-methyl-	EtOH	253s(3.65)	94-1446-64
1-Hydroxy-6-methylquinolizinium bromide	H_3O	205(4.52),240(4.07), 341(4.14)	39-3030-64
1-Hydroxy-8-methylquinolizinium bromide	H_3O	207(4.65),233(4.18), 337(4.19)	39-3030-64
$C_{10}H_{10}BrNO_3$			
Anthranilic acid, N-(2-bromopropionyl)-	iso-PrOH	212(4.28),225s(4.23), 258(4.03),303(3.75)	44-0582-64

Compound	Solvent	$\lambda_{max}(\log \epsilon)$	Ref.
$C_{10}H_{10}BrNO_4$			
m-Dioxane, 5-bromo-4-p-nitrophenyl-	n.s.g.	236(4.00)	88-0499-64
diastereoisomer	n.s.g.	269(4.00)	88-0499-64
$C_{10}H_{10}BrN_3O_3S$			
Acetimidic acid, 2-diazo-N-[(p-bromophenyl)sulfonyl]-, ethyl ester	n.s.g.	236(4.19),277(4.46)	24-0623-65
$C_{10}H_{10}Br_2O_4S$			
2,5-Thiophenedicarboxylic acid, 3,4-bis-(bromomethyl)-, dimethyl ester	EtOH	234(4.30),283(4.08)	44-1919-64
$C_{10}H_{10}Br_2Ti$			
Titanium, dibromodicyclopentadienyl-	acetone	425(3.38)	101-0271-65B
$C_{10}H_{10}Br_3NO$			
2,2-Dibromo-1,2,3,4-tetrahydro-6-methyl-1-oxoquinolizinium bromide	H_2O	276(3.83)	39-3030-64
2,2-Dibromo-1,2,3,4-tetrahydro-8-methyl-1-oxoquinolizinium bromide	H_2O	260(3.74)	39-3030-64
$C_{10}H_{10}ClI_2NO$			
2(3H)-Furanimine, dihydro-N-iodo-4-phenyl-, compound with ICl	THF	224(3.45)	39-0181-65
$C_{10}H_{10}ClNO$			
Cinnamamide, p-chloro-α-methyl-	EtOH	258(4.05)	65-3150-64
$C_{10}H_{10}ClNO_3$			
1-(2,5-Dihydro-4-hydroxy-5-oxo-3-furylmethyl)pyridinium chloride	aq. acid	231.0(4.51)	39-0783-64
	aq. base	257(4.44),262(4.44), 268(4.44)	
$C_{10}H_{10}Cl_2O_2$			
1-Norcaranepropiolic acid, 7,7-dichloro-	n.s.g.	224(3.86)	22-1518-65
$C_{10}H_{10}Cl_2S$			
Cyclopropane, 1,1-dichloro-3-methyl-2-(phenylthio)-	EtOH	240(3.90),251(3.92)	44-2211-64
$C_{10}H_{10}Cl_2Ti$			
Titanium, dichlorodicyclopentadienyl-	acetone	391(3.36)	101-0271-65B
$C_{10}H_{10}Co$			
Cobalt, dicyclopentadienyl-, chloranil complex	MeCN	260(4.54),289(3.84), 320(3.69),421(3.43), 448(3.58)	44-0975-64
trinitrobenzene complex	MeCN	262(4.95),427(3.61)	44-0975-64
$C_{10}H_{10}F_3N_3O_5S_2$			
2H-1,2,4-Benzothiadiazine-7-sulfonamide, 3-(1-hydroxyethyl)-6-(trifluoromethyl)-, 1,1-dioxide	EtOH	275(4.09)	95-0095-65
$C_{10}H_{10}F_4O_2$			
Bi-1-cyclobuten-1-yl, 3,3,3',3'-tetrafluoro-2,2'-dimethoxy-	n.s.g.	270(4.26)	44-1445-64

Compound	Solvent	λ max(log ε)	Ref.
$C_{10}H_{10}Fe$			
Ferrocene	heptane	260s(3.30),325(1.78), 440(1.96)	88-2881-64
	EtOH	260s(3.34),325(1.74), 440(1.98)	88-2881-64
$C_{10}H_{10}INS$			
2-(Methylthio)quinolizinium iodide	H_2O	213(4.55),258(4.25), 338(4.33)	39-2760-64
$C_{10}H_{10}N_2$			
1H-3-Benzazepine, 2-amino-	EtOH	295(4.15)	4-0026-65
Pyrazole, 5-methyl-4-phenyl-	EtOH	244(4.10)	44-1889-65
Pyrrolo[4,3,2-de]quinoline, 1,3,4,5-tetrahydro-	EtOH	227(4.53),277(3.82), 290(3.77),299(3.78)	44-1158-64
Quinazoline, 4,5-dimethyl-	pH 2.0	243(4.15),264(3.76), 289s(3.48),345(2.90)	39-5360-65
	pH 10.0	232(4.57),280(3.44), 317(3.50)	39-5360-65
Quinazoline, 4-ethyl-	pH 1.0	236(4.27),261(3.68), 323(3.11)	39-5360-65
	pH 7.0	224(4.62),271(3.43), 306(3.47),314s(3.43)	39-5360-65
Quinoxaline, 2,3-dimethyl-	EtOH	237(4.44),315(3.89)	44-1542-65
$C_{10}H_{10}N_2O$			
4(1H)-Cinnolinone, 1-ethyl-	EtOH	211(4.51),234(4.07), 240(4.08),255(3.90), 263s(3.79),284(3.46), 296(3.48),345(4.13), 358s(4.05)	39-5391-65
2-Ethyl-4-hydroxycinnolinium hydroxide, anhydrobase	EtOH	213(4.54),221s(4.25), 253(3.95),266s(3.68), 279s(3.48),351(4.10), 369(4.11)	39-5391-65
Indazole, 3-acetyl-6-methyl-	EtOH	234(4.08),242s(4.01) 297(4.11)	32-0814-65
Isoxazole, 5-amino-3-methyl-4-phenyl-	C_6H_{12}	205(4.04),226s(3.5), 249(3.96),268s(3.9)	17-0255-65
	MeOH-5N HCl	245s(4.0),272(4.13)	17-0255-65
	MeOH	205(4.02),225s(3.7), 256(4.16),280s(3.7)	17-0255-65
Isoxazole, 5-amino-4-methyl-3-phenyl-	C_6H_{12}	206(4.09),228(4.14), 270s(3.6)	17-0255-65
	MeOH-5N HCl	242(3.95),290(4.20)	17-0255-65
	MeOH	206(4.14),228(4.14), 268(3.77)	17-0255-65
3-Isoxazoline, 5-imino-2-methyl-3-phenyl-, hydrochloride	MeOH	208(4.05),247(4.04), 286(4.18)	17-0255-65
3-Isoxazoline, 5-imino-2-methyl-4-phenyl-, hydrochloride	MeOH	205(4.13),249(3.84), 280s(3.63)	17-0255-65
Pyrazolin-5-one, 1-methyl-3-phenyl-	C_6H_{12}	301(4.08)	78-0299-64
	20N H_2SO_4	255(4.30)	78-0299-64
	pH 5	255(4.21)	78-0299-64
	pH 13	252(4.24)	78-0299-64
Pyrazolin-5-one, 3-methyl-1-phenyl-	C_6H_{12}	245(4.26)	78-0299-64
	10N H_2SO_4	231(4.20)	78-0299-64
	pH 5	239(4.09)	78-0299-64
	pH 13	246(4.07)	78-0299-64

Compound	Solvent	λ max(log ε)	Ref.
3-Pyrazolol, 5-methyl-1-phenyl-	C_6H_{12}	264(4.12)	78-0315-64
	20N H_2SO_4	241(4.09)	78-0315-64
	pH 4.4	257(4.03)	78-0315-64
	pH 13	279(4.06)	78-0315-64
3(2H)-Pyridazinone, 4,5-dihydro-3-oxo-6-phenyl-	EtOH	286(4.19)	39-5302-64
$C_{10}H_{10}N_2OS$			
2H-1,3-Benzoxazine-2-thione, 4-ethylimino-3,4-dihydro-	pH 1-6	259(4.09),288(4.09), 297(4.18),325(3.92)	78-3019-65
	pH 11	250(4.03),290(4.10), 297(4.12)	78-3019-65
4H-1,3-Benzoxazine-4-thione, 3-ethyl-2,3-dihydro-2-imino-	pH 3-11	277(4.03),320(4.12)	78-3019-65
$C_{10}H_{10}N_2O_2$			
Benzonitrile, o-(3-nitropropyl)-	EtOH	224(3.95),275(3.05), 283(3.04)	88-2659-64
p-Benzoquinone, 2,5-bis(1-aziridinyl)-	methyl cellosolve	209(4.38),328(4.24), 417(2.79)	78-1889-64
2H-1,3-Benzoxazin-2-one, 4-ethylimino-3,4-dihydro-	pH 1-10	249(4.22),297(3.65)	78-3019-65
[2,2'-Bipyrrole]-5-carboxylic acid, methyl ester	MeOH	220(4.14),325(4.43)	44-0883-64
	EtOH	220(4.13),325(4.43)	44-2727-64
Butyric acid, 4-hydroxy-2-oxo-, γ-lactone, phenylhydrazone	n.s.g.	233(4.17),295(3.98), 333(4.49)	39-0141-64
Imidazo[1,2-a]pyridine-2-carboxylic acid, ethyl ester, picrate	EtOH	221s(4.65),225(4.67), 261s(3.38),272(3.46), 281(3.49),300(3.61), 311(3.59),325s(3.34)	44-2403-65
Indole, 1,2-dimethyl-4-nitro-	EtOH	212(4.41),243(4.08), 396(3.84)	44-3457-65
Indole, 1,2-dimethyl-5-nitro-	EtOH	260s(4.27),274(4.35), 335(3.98)	35-3796-64
	H_2SO_4	212s(4.08),281(4.12)	35-3796-64
Indole, 1,2-dimethyl-6-nitro-	EtOH	213(4.35),252(4.01), 267s(3.86),338(3.92), 384(3.95)	44-3457-65
Indole, 2-ethyl-5-nitro-	EtOH	208(4.30),266(4.37), 331(3.99)	44-3457-65
4(1H)-Quinazolinone, 2-(hydroxymethyl)-1-methyl-	iso-PrOH	230(4.22),269(3.63), 277(3.65),307(3.79), 317(3.90)	44-0582-64
4(3H)-Quinazolinone, 2-(1-hydroxyethyl)-	iso-PrOH	226(4.44),230s(4.41), 265(3.92),271(3.88), 305(3.62),316(3.54)	44-0582-64
4(3H)-Quinazolinone, 2-(hydroxymethyl)-3-methyl-	iso-PrOH	225(4.37),268(3.88), 275(3.90),304(3.59), 315(3.48)	44-0582-64
5,8-Quinoxalinediol, 2,3-dimethyl-	MeOH	266(4.62),310(3.11), 380(3.04)	4-0171-64
$C_{10}H_{10}N_2O_3$			
3-Benzimidazolineacetic acid, 1-methyl-2-oxo-, methyl ester	EtOH	230(3.90),282(3.95), 287s(3.89)	44-1118-65
Benzocyclobutene, 4-acetamido-5-nitro-	EtOH	246(3.97),281(3.32), 355(3.15)	78-2281-65
3-Buten-2-one, 4-(p-nitroanilino)-	MeOH	375(4.60)	83-0321-64
Cinnamamide, α-methyl-p-nitro-	EtOH	302(4.10)	65-3150-64
Indoline, 1-acetyl-5-nitro-	EtOH	231(4.01),340(4.12)	44-0947-64

Compound	Solvent	λ_{max}(log ϵ)	Ref.
Phenol, 2-(β-carbamino-β-amino-acroyl)-	EtOH	368(4.76)	65-0541-64
$C_{10}H_{10}N_2O_4$			
2-Indolinone, 3-ethoxy-5-nitro-	MeOH	323(4.07)	44-3610-65
2,4-Pentadienamide, 4-methyl-5-(5-nitro-2-furyl)-	propylene glycol	275(4.15),315(3.98), 390(4.30)	95-0646-64
$C_{10}H_{10}N_2O_9$			
5-Pyrimidinecarboxylic acid, 1,2,3,4-tetrahydro-2,4-dioxo-1-(β-D-ribo-furanosyl)-	pH 2	276(4.1)	94-0007-65
	H_2O	275(4.06)	94-0007-65
	pH 1	271(3.89)	94-0007-65
$C_{10}H_{10}N_2S_3$			
4-Isothiazolecarbonitrile, 3,5-bis-(allylthio)-	EtOH	229(4.11),286(4.06)	44-0665-64
$C_{10}H_{10}N_2S_4$			
Pyrrolidyl thiuram disulfide	hexane	230(4.31),285s(4.09)	78-2857-65
$C_{10}H_{10}N_4$			
Butyronitrile, 3-imino-2-oxo-, phenylhydrazone	EtOH	247(4.04),256(4.01), 363(4.30)	5-0033-65J
	$EtOH-NH_3$	246(4.04),255(4.00), 363(4.30)	5-0033-65J
5H-Tetrazolo[5,1-a][2]benzazepine, 6,7-dihydro-	EtOH	245(<u>3.9</u>),280s(<u>3.0</u>)	1-0191-64
$C_{10}H_{10}N_4O$			
Formamidine, N-(5-methyl-3-isoxazolyl)-N'-(2-pyridyl)-	EtOH	217(4.28),303(4.28)	78-0159-64
Pyrazolo[1,5-a]pyrimidine-6-carbo-nitrile, 4,7-dihydro-2,3,5-trimethyl-7-oxo-	EtOH	221(4.44),276(4.05), 310(3.96)	95-0442-65
Pyrazolo[1,5-a]pyrimidine-6-carbo-nitrile, 5-ethyl-4,7-dihydro-2-methyl-7-oxo-	EtOH	220(4.49),268(4.02), 303(3.96)	95-0442-65
$C_{10}H_{10}N_4OS$			
1,2,5-Thiadiazole-3-carboxamide, 4-amino-N-benzyl-	pH 1,7,13	326(3.86)	44-2141-64
$C_{10}H_{10}N_4O_2$			
Cycloheptimidazole, 2-(dimethylamino)-6-nitro-	EtOH	254(4.56),329(3.70), 435(4.52)	94-0465-65
Pyrazole, 5-amino-4-methyl-1-(o-nitro-phenyl)-	EtOH	236(4.16)	95-0158-65
Pyrazole, 1-(p-aminophenyl)-3-methyl-4-nitro-	EtOH	261.5(4.18)	23-1605-64
$C_{10}H_{10}N_4O_3$			
Isoxazole, 3,5-diamino-4-p-nitro-benzyl-, hydrochloride	MeOH	252(4.23)	44-2862-65
$C_{10}H_{10}N_4O_3S$			
1,2,4-Triazole, 4-methyl-3-(methyl-thio)-5-(5-nitro-2-furyl)vinyl-	propylene glycol	240(4.02),290(3.99), 388(4.27)	95-0566-64

Compound	Solvent	λ_{max}(log ε)	Ref.
$C_{10}H_{10}N_4O_4$			
4-Isoxazolemethanol, 3,5-diamino-α-p-nitrophenyl-	MeOH	232(4.03),269(3.92)	44-2862-65
1,3,4-Oxadiazole, 2-amino-5-[1-ethyl-2-(5-nitro-2-furyl)vinyl]-	propylene glycol	231(3.85),298(3.95), 385(4.24)	95-0212-64
1,3,4-Oxadiazole, 2-(dimethylamino)-5-[2-(5-nitro-2-furyl)vinyl]-	propylene glycol	230(4.12),310(4.04), 400(4.34)	95-0948-65
1,3,4-Oxadiazole, 2-(ethylamino)-5-(5-nitro-2-furyl)vinyl-	propylene glycol	227(4.13),304(4.10), 394(4.34)	94-0219-64
1,3,4-Oxadiazole, 2-(methylamino)-5-[1-methyl-2-(5-nitro-2-furyl)vinyl]-	propylene glycol	227(3.89),302(3.99), 398(4.26)	95-0219-64
$C_{10}H_{10}N_4O_4S$			
3(2H)-Thiophenone, dihydro-, 2,4-dinitrophenylhydrazone	EtOH	228(4.20),357(4.36)	54-0081-64
$C_{10}H_{10}N_4S$			
Benzaldehyde, azine with 3-methyl-2-thiadiazolinone	EtOH	234(4.13),286(3.86), 344(4.32)	1-0871-64
Benzaldehyde, N-methyl-N-thiadiazolyl-hydrazone	EtOH	226(4.16),323(4.43)	1-0871-64
$C_{10}H_{10}N_4S_2$			
Benzaldehyde, [5-(methylthio)thiadiazolyl]hydrazone	EtOH	226(4.14),254(3.92), 335(4.40)	1-0871-64
anion	EtOH	389(4.43)	1-0871-64
$C_{10}H_{10}Ni$			
Nickel, dicyclopentadienyl-, chloranil complex	MeCN	236(3.98),249(3.93), 256(3.97),269(3.98), 289(3.89),310(3.73), 449(3.11)	44-0975-64
$C_{10}H_{10}O$			
Acrylophenone, 4'-methyl-	EtOH	315(2.412)	65-0190-64
Benzocyclobutene, 4-aceto-	heptane	248(4.12),283(3.30), 292(3.22),320s(1.81)	78-2185-64
3-Buten-2-one, 4-phenyl-	hexane	281(4.38)	44-3000-65
	iso-PrOH	285(4.33)	44-3000-65
Crotonophenone	hexane	248(4.67)	44-3000-65
	iso-PrOH	254(4.23)	44-3000-65
2-Cyclobuten-1-ol, 3-phenyl-	EtOH	255(4.28)	35-2645-64
2,6,8-Decatrien-4-ynal, trans	Et_2O	232s(4.06),328(4.51), 342s(4.49)	24-0369-65
Indene, 5-methoxy-	EtOH	252(3.69)	44-3231-65
Indene, 6-methoxy-	EtOH	268(4.02)	44-3231-65
Methacrylophenone	EtOH	295(2.415)	65-0190-64
1(2H)-Naphthalenone, 3,4-dihydro-	EtOH	248(4.09)	5-0021-64G
	EtOH	247(4.05)	44-2165-65
$C_{10}H_{10}OS$			
Benzo[b]thiophene-3-ethanol	EtOH	231(4.34),262(3.43), 266s(3.42),292(3.11), 300(3.17)	4-0231-65
Benzo[b]thiophene-2-methanol, 3-methyl-	EtOH	234(4.38),266(3.64), 292(3.16),301(3.12)	4-0231-65
Benzothiophene-4-ol, 2-ethyl-	aq. base	324.0(3.85)	22-1464-65
Benzothiophene-4-ol, 7-ethyl-	aq. base	329.0(3.83)	22-1464-65
Benzothiophene-5-ol, 2,3-dimethyl-	aq. base	325.5(3.51)	22-1464-65
Benzothiophene-6-ol, 2,3-dimethyl-	aq. base	284.0(4.00)	22-1464-65

Compound	Solvent	$\lambda_{max}(\log \epsilon)$	Ref.
Benzothiophene-7-ol, 2,3-dimethyl-	aq. base	317.0(3.79)	22-1464-65
Cyclopentanone, 2-(2-thenylidene)-	hexane	320(4.36)	65-3645-64
	EtOH	332(4.29)	65-3645-64
	66% H_2SO_4	419(4.52)	65-3645-64
$C_{10}H_{10}O_2$			
Acrylophenone, p-methoxy-	EtOH	355(2.477)	65-0190-64
Benzene, m-bis(vinyloxy)-	nonane	274(3.33)	67-0375-65
Benzene, p-bis(vinyloxy)-	nonane	284(3.32)	67-0375-65
Benzocyclobutene-3-carboxylic acid, methyl ester	EtOH	241(3.91),290(3.41)	44-1333-65
4-Benzofuranol, 2,3-dimethyl-	aq. base	297.0(3.79)	22-1464-65
4-Benzofuranol, 2,7-dimethyl-	aq. base	300.5(3.79)	22-1464-65
5-Benzofuranol, 2,3-dihydro-2-vinyl-	EtOH	228(3.68),302(3.60)	39-5182-65
5-Benzofuranol, 2,3-dimethyl-	aq. base	313.0(3.63)	22-1464-65
6-Benzofuranol, 2,3-dimethyl-	aq. base	307.5(3.78)	22-1464-65
7-Benzofuranol, 2,3-dimethyl-	aq. base	286(3.48)	22-1464-65
p-Benzoquinone, 2-allyl-6-methyl-	n.s.g.	252s(--),255(4.27), 261s(--)	39-5060-65
1,3-Butanedione, 1-phenyl-	hexane	307(4.58)	44-3000-65
	iso-PrOH	307(4.20)	44-3000-65
3-Chromanone, 2-methyl-	EtOH	275(3.20)	65-2699-64
Cinnamic acid, o-methyl-	EtOH	270(4.23)	35-5548-64
Cinnamic acid, p-methyl-	EtOH	280(4.35)	35-5548-64
Crotonic acid, 2-phenyl-, cis	EtOH	247(3.99),291s(2.46)	1-0612-65
trans	EtOH	234s(3.74)	1-0612-65
2,8-Decadiene-4,6-diyne-1,10-diol, di-cis	EtOH	230(4.48),238(4.43), 247(4.32),261(3.86), 276(4.16),293(4.26), 312(4.35)	70-1460-65
2,4,6,8-Decatetraenoic acid, 4-hydroxy-, γ-lactone, all trans	Et_2O	235(3.97),353(4.59)	24-2236-65
1-Dicyclopentadienone, 8-hydroxy-, endo	EtOH	226(3.83)	78-0717-64
Δ^4-Dihydropyran-2-spirocyclohexa-2',5'-dien-4'-one	EtOH	223(4.07)	39-5182-65
1(2H)-Naphthalenone, 3,4-dihydro-6-hydroxy-	pH 13	330(4.41)	44-2165-65
	EtOH	278(4.16)	44-2165-65
2-Propyn-1-ol, 3-(p-methoxyphenyl)-	Et_2O	251(4.32)	24-2118-64
$C_{10}H_{10}O_3$			
Benzoic acid, 2-acetyl-, methyl ester	MeOH	230(4.18),272(2.88), 278(2.95)	44-4293-65
Benzoic acid, 2-acetyl-6-methyl-	EtOH	235(3.71),284(2.68)	44-4293-65
Cinnamic acid, o-methoxy-	EtOH	274(4.11)	35-5548-64
Cinnamic acid, p-methoxy-	EtOH	298(4.30)	35-5548-64
2-Decene-4,6-diynoic acid, 8-hydroxy-	Et_2O	217(4.49),223(4.60), 253(3.83),267(4.15), 283(4.38),301(4.36)	24-3469-64
9-Decene-4,6-diynoic acid, 8-hydroxy-	EtOH	228(2.76),242(2.71), 256(2.51),283(2.10), 302(2.00)	70-1237-65
1(2H)-Naphthalenone, 3,4-dihydro-4,8-dihydroxy-	EtOH	257(3.98),327(3.55)	70-0492-64
1(2H)-Naphthalenone, 3,4-dihydro-5,8-dihydroxy-	EtOH	236(4.12),266(3.93), 376(3.61)	39-2816-64
1,2-Propanedione, 1-(p-methoxyphenyl)-	EtOH	223(3.81),294(4.00)	23-0113-64
Propionic acid, 3-benzoyl-	EtOH	241(4.10),277(3.04)	32-0699-65
	EtOH	206(4.12),241(4.10)	88-2443-64

Compound	Solvent	$\lambda_{max}(\log \epsilon)$	Ref.
$C_{10}H_{10}O_3S$			
Acrylic acid, 3-[(o-hydroxyphenyl)-thio]-	Et_2O	208(4.02),273(4.12)	24-2109-64
1,3-Benzoxathiole-2-acetic acid, methyl ester	Et_2O	213(4.32),244(3.71), 292(3.64)	24-2109-64
Carbonic acid, thio-, cyclic O,O-ester with 3-phenoxy-1,2-propanediol	EtOH	224s(4.09),235(4.22), 269(3.34),276(3.24)	56-0931-65
$C_{10}H_{10}O_4$			
p-Benzoquinone, 2-acetonyl-5-methoxy-	EtOH	259(3.96),360(2.88)	1-0421-64
[Bicyclobutenyl]-3,3'-dione, 1,2'-dimethoxy-	n.s.g.	365(4.11)	44-1445-64
But-2-enolide, 4-(3-carboxy-2-methyl-allylidene)-3-methyl-	EtOH	320.5(4.41)	39-0594-64
Ketone, 2,3-dihydro-2,5-dihydroxy-4-benzofuranyl methyl	EtOH	232(4.12),256(3.85), 369(3.55)	27-0358-64
3-Norcarene-2,5-dione, 1-acetyl-4-methoxy-	EtOH	279(3.88)	39-0411-64
Phthalaldehyde, 3,6-dihydroxy-4,5-dimethyl-	EtOH	244(4.23),276(3.92), 400(3.84)	39-0042-64
1,2-Propanedione, 1-(4-hydroxy-3-methoxyphenyl)-	EtOH	235(4.86),291s(--), 320(4.93)	23-0113-64
Succinic acid, di-2-propynyl ester	MeOH	264(1.7),290(1.6)	70-1349-64
Terephthalic acid, dimethyl ester	EtOH	240(4.29),285(3.23), 294(3.15)	44-1431-65
$C_{10}H_{10}O_4S$			
Acrylic acid, 3-(phenylsulfonyl)-, methyl ester, trans	EtOH	205(4.09),234(3.97)	44-0043-65
1,3-Benzoxathiole-2-acetic acid, methyl ester, S-oxide	Et_2O	207(4.04),287(3.51)	24-2109-64
$C_{10}H_{10}O_4S_2$			
1H,3H-Thieno[3,4-c]thiophene-4,6-dicarboxylic acid, dimethyl ester	EtOH	205(4.15),278(4.32)	44-1919-64
$C_{10}H_{10}O_5$			
p-Benzoquinone-2-carboxylic acid, 5-methoxy-3-methyl-, methyl ester	EtOH	269(4.13),365s(2.9)	39-0411-64
2-Propanone, (2-carboxy-3,5-dihydroxy-phenyl)-	EtOH	270(4.11),305(3.79)	39-5382-64
$C_{10}H_{10}O_5S$			
1,3-Benzoxathiole-2-acetic acid, methyl ester, S,S-dioxide	Et_2O	216(3.74),282(3.48), 290(3.42)	24-2109-64
$C_{10}H_{10}O_6$			
Chorismic acid	H_2O	275(3.42)	12-1227-65
barium salt	H_2O	272(3.43)	12-1227-65
1,2,4-Cyclopentaentrione, 3-ethoxalyl-5-methyl-	EtOH-HCl	255(4.17),275s(4.12), 285s(4.09)	39-2251-65
	EtOH-NaOH	240(4.17),275(4.44), 302(4.35)	39-2251-65
$C_{10}H_{10}O_6S$			
Crotonic acid, 3,3'-(thiodimethylene)-bis[2,4-dihydroxy-, di-γ-lactone	EtOH	236(4.14)	44-3560-64
	NaOH	281(4.13)	44-3560-64

Compound	Solvent	$\lambda_{max}(\log \epsilon)$	Ref.
$C_{10}H_{10}O_6S_2$			
Crotonic acid, 3,3'-(dithiodimethylene)bis[2,4-dihydroxy-, di-γ-lactone	EtOH	244(4.35)	44-3560-64
$C_{10}H_{10}S_3$			
α-Pinenetrithione	isooctane	232(4.02),250(3.87), 283(3.72),330(3.81), 421(3.87),505s(2.3)	5-0062-64D
$C_{10}H_{11}BrN_2O_2$			
Benzimidazole, 5-bromo-4,7-dimethoxy-2-methyl-	EtOH	251(3.70)	44-0476-64
	pH 2	260(3.75)	44-0476-64
	pH 12	251(3.58)	44-0476-64
$C_{10}H_{11}BrN_4O_5$			
Inosine, 8-bromo-	pH 1	254(4.18)	35-1242-64
	pH 11	259(4.16)	35-1242-64
$C_{10}H_{11}BrN_4O_6$			
Xanthosine, 8-bromo-	pH 1	240(4.29),267(4.11)	35-1242-64
	pH 11	252(4.08),281(4.10)	35-1242-64
$C_{10}H_{11}Cl$			
Bicyclo[6.1.0]nonatriene, 9-chloro-9-methyl-	EtOH	249(3.61)	35-4876-64
Indan, 5-chloro-4-methyl-	C_6H_{12}	259(3.03),263(3.03), 271(3.00),275(2.85), 280(2.88)	22-3103-64
Indan, 6-chloro-4-methyl-	C_6H_{12}	260(2.94),265(2.97), 272(3.02),282(3.02)	22-3103-64
$C_{10}H_{11}ClN_2O_3$			
D-Cycloserine, N-(5-chloro-2-hydroxybenzyl)-	MeOH	205(3.94),227(3.93), 287(3.26)	44-3436-65
$C_{10}H_{11}ClN_4O$			
Pyrazolo-[1,5-a]pyrimidine, 7-(2-chloroacetamido)-2,3-dimethyl-	EtOH	237(4.61),280s(3.29), 290(3.35),299s(3.26), 346(3.61)	94-1207-65
$C_{10}H_{11}ClN_4O_5$			
Purine, 6-chloro-9-(2-deoxy-α-D-ribofuranosyl)-	H_2O	264(3.96)	35-4934-65
Purine, 6-chloro-9-(2-deoxy-β-D-ribofuranosyl)-	H_2O	264(4.00)	35-4934-65
$C_{10}H_{11}Cl_6Sb$			
Allylcycloheptatrienylium hexachloroantimonate (V)	MeCN	266(4.14),360(3.42)	35-3329-64
$C_{10}H_{11}FO_2$			
2-Propen-1-ol, 2-fluoro-3-(p-methoxyphenyl)-	EtOH	255(4.39)	22-2258-64
$C_{10}H_{11}F_3N_2O_5$			
Uridine, 2'-deoxy-5-(trifluoromethyl)-	pH 1	260(4.00)	87-0001-64
	pH 13	260(3.82)	87-0001-64

Compound	Solvent	λ_{max}(log ε)	Ref.
$C_{10}H_{11}F_3N_2O_6$			
Uracil, 1-β-D-arabinofuranosyl-5-(trifluoromethyl)-	pH 2	205(3.95),263(4.01)	44-0835-65
	pH 12	262.5(3.85) (changes with time)	44-0835-65
$C_{10}H_{11}N$			
Indole, 1,2-dimethyl-	EtOH	223(4.55),276(3.86), 282(3.88),291(3.81)	35-3796-64
	EtOH	223(4.54),276(3.86), 282(3.88),291(3.81)	44-1206-64
	H_2SO_4	230(3.66),237(3.63), 274(3.75)	35-3796-64
Indole, 1,3-dimethyl-	EtOH	225(4.50),278s(3.68), 288(3.72)	35-3796-64
	EtOH	226(4.50),292(3.78)	39-5510-64
	H_2SO_4	232(3.58),237(3.56), 274(3.69)	35-3796-64
	H_2SO_4	235(3.58),240(3.59), 286(3.67)	39-5510-64
Indole, 2,3-dimethyl-	MeOH	280(3.82)	65-3487-64
	EtOH	227(4.52),276s(3.82), 282(3.85),290(3.79)	35-3796-64
	EtOH	228(4.50),284(3.83), 293(3.79)	39-5510-64
	HCl	231(3.84),237s(3.76), 277(3.77)	35-3796-64
	H_2SO_4	235(3.70),241(3.68), 286(3.74)	39-5510-64
Indole, 2,5-dimethyl-	EtOH	221(4.47),272(3.86), 276s(3.85),282(3.83), 293(3.69)	35-3796-64
	H_2SO_4	234(3.71),241s(3.65), 284(3.80)	35-3796-64
Indole, 1-ethyl-	EtOH	219(4.54),274s(3.74), 281(3.76),292s(3.66)	35-3796-64
	H_2SO_4	233(3.54),238(3.52), 277(3.69)	35-3796-64
Indole, 2-ethyl-	EtOH	220(4.53),272(3.86), 277(3.86),280(3.85), 288(3.75)	35-3796-64
	H_2SO_4	224s(3.67),229(3.72), 235s(3.67),273(3.77)	35-3796-64
Indole, 3-ethyl-	EtOH	222(4.54),274s(3.73), 280(3.76),289(3.70)	35-3796-64
Indole, 6-ethyl-	EtOH	221(4.62),272(3.83), 275s(3.82),282(3.81), 285s(3.78),292(3.73)	39-7165-65
Indolenine, 3,3-dimethyl-	EtOH	256(4.24),298(3.70)	78-0989-65
	dil. HCl	232(3.72),236s(3.69), 280(3.64)	78-0989-65
$C_{10}H_{11}NO$			
Acetamide, N-styryl-	EtOH	221(4.14),287(4.33)	88-2473-65
3-Buten-2-one, 4-amino-3-phenyl-	EtOH	289(4.16)	44-1889-65
Carbostyril, 3,4-dihydro-N-methyl-	EtOH	210(4.32),250(4.01)	1-1389-64
	dioxan	253(4.14)	1-1389-64
1,6-Iminocyclodecapentaene, N-acetyl-	C_6H_{12}	253(4.8),300s(3.8), 385(2.3),400(2.5), 410(2.7)	88-3613-65

Compound	Solvent	$\lambda_{max}(\log \varepsilon)$	Ref.
Indole, 5-methoxy-6-methyl-	MeOH	217(4.38),271(3.84), 293(3.74),305(3.65)	44-2897-65
Indole, 6-methoxy-1-methyl-	EtOH	222(--),274(--), 291(--),299s(--)	39-3902-65
Indole, 6-methoxy-3-methyl-	MeOH	293(3.63)	24-1727-65
1(2H)-Naphthalenone, 6-amino-3,4-dihydro-	EtOH	323(4.34)	44-2165-65
5-Pseudoindolone, 2,2-dimethyl-	MeOH	272(4.3)	24-2939-65
5(6H)-Quinolone, 7,8-dihydro-7-methyl-	n.s.g.	275(3.57)	88-0171-65
$C_{10}H_{11}NOS$			
2H-5,1-Benzothiazocin-2-one, 1,3,4,6-tetrahydro-	iso-PrOH	233s(3.95),275(2.95)	44-3111-65
$C_{10}H_{11}NOS_2$			
Benzoic acid, thio-, anhydrosulfide with dimethyldithiocarbamic acid	hexane	240(4.38),285s(4.21), 400(2.39)	78-2865-65
Propionaldehyde, 3-phenyl-3-(thio-carbamoylthio)-	H_2O	241s(3.93),288(4.13)	39-4004-64
	EtOH	233(4.16),294(4.10)	39-4004-64
2-Thiazolidinethione, 4-hydroxy-3-methyl-4-phenyl-	MeOH	253s(4.00),275(4.21)	44-2146-64
$C_{10}H_{11}NOS_3$			
Carbamic acid, dimethyldithio-, anhydrosulfide with phenylxanthic acid	hexane	242(4.36),270s(4.09)	78-2865-65
$C_{10}H_{11}NO_2$			
Acetophenone, 2-acetamido-	EtOH	231(4.45),258(4.06), 266(3.99),324(3.66)	39-0928-64
Gentianine, dihydro-	EtOH	225s(3.65),272(3.15)	78-3721-65
2,4,6-Heptatrienoic acid, 7-cyano-, ethyl ester	EtOH	222(4.00),299(4.46), 311(4.39)	70-0684-65
2-Indolinone, 3-ethoxy-	MeOH	250(3.93),291(3.12)	44-3610-65
2-Indolinone, 3-methoxy-3-methyl-	EtOH	208(4.39),252(3.80), 289(3.14)	44-2431-64
2-Indolinone, 3-methoxy-5-methyl-	MeOH	257(3.95),301(3.14)	44-3610-65
6,7-Isoquinolinediol, 3,4-dihydro-1-methyl-	EtOH-HCl	249(4.20),312(3.93), 362(3.98)	33-1945-65
	EtOH	240(4.09),276(3.92), 312(3.66),404(4.11)	33-1945-65
	EtOH-NaOH	249(4.12),338(4.23)	33-1945-65
hydrochloride	EtOH-HCl	249(4.22),312(3.96), 362(4.00)	33-1945-65
	EtOH	238(3.99),249(3.99), 277(3.70),312(3.84), 395(3.89)	33-1945-65 33-1945-65
	EtOH-NaOH	249(4.14),338(4.24)	33-1945-65
6-Isoquinolinol, 3,4-dihydro-7-methoxy-	EtOH-HCl	248(4.19),307(3.96), 362(3.92)	33-1945-65
	EtOH	269(3.98),316(3.58), 402(4.32)	33-1945-65
	EtOH-NaOH	245(4.16),332(4.14)	33-1945-65
hydrobromide	EtOH-HCl	249(4.19),307(3.97), 361(3.96)	33-1945-65
	EtOH	248(4.18),307(3.96), 362(3.96)	33-1945-65
	EtOH-NaOH	246(4.18),333(4.16)	33-1945-65

Compound	Solvent	$\lambda_{max}(\log \epsilon)$	Ref.
7-Isoquinolinol, 3,4-dihydro-6-methoxy-	EtOH-HCl	247(4.27),303(4.00),360(3.90)	33-1945-65
	EtOH	238(4.36),270(3.82),310(3.82)	33-1945-65
	EtOH-NaOH	245(4.48),275s(3.74),347(3.73)	33-1945-65
hydrochloride	EtOH-HCl	246(4.32),303(4.06),358(3.96)	33-1945-65
	EtOH	246(4.32),302(4.06),360(3.96)	33-1945-65
	EtOH-NaOH	245(4.56),275s(3.88),345(3.84)	33-1945-65
1(2H)-Naphthalenone, 5-amino-3,4-dihydro-8-hydroxy-	EtOH	240(4.22),391(3.51)	39-2816-64
hydrochloride	EtOH	255(3.94),329(3.55)	39-2816-64
1(2H)-Naphthalenone, 7-amino-3,4-dihydro-8-hydroxy-	EtOH	240(4.23),277(3.94),381(3.41)	39-2816-64
	EtOH-HCl	256(3.98),325(3.59)	39-2816-64
1,4-Naphthoquinone, 4a,5,8,8a-tetra-hydro-, 1-oxime	EtOH	220(3.77),282(4.14)	44-3775-65
	base	235(3.83),346(4.23)	44-3775-65
2,4-Pentadienoic acid, 2?-tert-butyl-5-cyano-4-hydroxy-, γ-lactone	C_6H_{12}	281(4.46)	35-2819-64
2,4-Pentadienoic acid, 5-cyano-2,3-di-ethyl-4-hydroxy-, γ-lactone	C_6H_{12}	278(4.36)	35-2819-64
$C_{10}H_{11}NO_2S$			
2H-5,1-Benzothiazocin-2-one, 1,3,4,6-tetrahydro-, 5-oxide	iso-PrOH	238s(3.90)	44-3111-65
$C_{10}H_{11}NO_3$			
1,3-Isoindolediol, N-acetyl-	EtOH	222(3.13),243s(2.82),248s(2.88),255(2.96),260s(2.98),262(3.08),269(2.98)	44-2251-65
$C_{10}H_{11}NO_3S$			
2H-5,1-Benzothiazocin-2-one, 1,3,4,6-tetrahydro-, 5,5-dioxide	iso-PrOH	235s(3.81),270s(2.69),277s(2.51),300s(2.11)	44-3111-65
$C_{10}H_{11}NO_4$			
Aspartic acid, N-phenyl-	H_2O	240(3.81),286(2.93)	87-0821-65
barium salt	H_2O	242(4.00),288(3.10)	87-0821-65
Indole-7-carboxylic acid, 2,4,5,6-tetrahydro-3-hydroxy-2-oxo-, methyl ester	n.s.g.	312(4.39)	39-1648-65
α^5-Pyridoxylideneacetic acid	pH 1	233(4.37),305(4.06)	87-0112-65
$C_{10}H_{11}NO_4S$			
Propionic acid, 3-(2-nitrobenzyl-mercapto)-	iso-PrOH	248(3.72)	44-3111-65
$C_{10}H_{11}NO_5$			
Benzoic acid, 2-(1-methoxy-2-nitro-ethyl)-	pH 12	245(4.04)	23-0190-65
	MeOH	277.5(3.00)	23-0190-65
Serine, 3-[3,4-(methylenedioxy)-phenyl]-	H_2O	235(3.65),285(3.61)	23-1901-64

Compound	Solvent	$\lambda_{max}(\log \epsilon)$	Ref.
$C_{10}H_{11}NO_6S$			
Propionic acid, 3-[(2-nitrobenzyl)-sulfonyl]-	iso-PrOH	255(3.69)	44-3111-65
$C_{10}H_{11}NS$			
1(2H)-Isoquinolinethione, 3,4-di-hydro-N-methyl-	heptane	258(4.04),299(3.71), 323(3.81),425(2.09)	1-2432-65
	EtOH	261(4.04),300(3.87), 318(3.85),395(2.22)	1-2432-65
$C_{10}H_{11}N_3$			
Cycloheptimidazole, 2-(dimethylamino)-	EtOH	239(4.31),263(4.37), 298(3.71),370(4.23)	94-0465-65
Phenylazoisobutyronitrile	DMF	275(3.11),390(2.30)	60-1437-65
	$o-C_6H_4Cl_2$	395(2.25)	60-1437-65
	PhCN	395(2.28)	60-1437-65
3,5-Pyridinedicarbonitrile, 4-ethyl-1,2-dihydro-6-methyl-	EtOH	216(4.37),258(4.09), 376(3.80)	73-3711-65
3,5-Pyridinedicarbonitrile, 4-ethyl-1,4-dihydro-2-methyl-	EtOH	211(4.36),342(3.78)	73-3711-65
$C_{10}H_{11}N_3O$			
5-Hydroxy-2,4-dimethyl-3-phenyl-s-triazolium hydroxide, inner salt	n.s.g.	239(4.31)	4-0105-65
5-Hydroxy-3,4-dimethyl-2-phenyl-s-triazolium hydroxide, inner salt	n.s.g.	262(3.80)	4-0105-65
Indole-2-carboxylic acid, 1-methyl-, hydrazide	EtOH	218(4.45),234s(4.24), 293(4.20)	44-0178-64
1,3,4-Oxadiazole, 2-(dimethylamino)-5-phenyl-	EtOH	290(4.2)	28-4579-64A
Δ^2-1,3,4-Oxadiazoline, 4-methyl-5-(methylimino)-2-phenyl-	EtOH	302(4.0)	28-4579-64A
$C_{10}H_{11}N_3O_2$			
6,7-Benzimidazoledione, 2-(di-methylamino)-1-methyl-	$CHCl_3$	245(4.3),315(3.8), 537(3.2)	65-1642-64
1-Indanone, 2-hydroxy-, semicarbazone	EtOH	224(4.10),231s(3.96), 273(4.21),282(4.24), 299(4.24),311(4.21)	44-1723-64
Quinazoline, 2-amino-5,6-dimethoxy-	EtOH-HCl	252(4.19),296(3.52)	78-2059-65
	EtOH	238(4.51),257(4.30)	78-2059-65
Quinazoline-6-methanol, 2-amino-8-hydroxy-	EtOH	235(4.42),261(4.15)	78-2059-65
4(3H)-Quinazolinone, 3-amino-2-(1-hydroxyethyl)-	iso-PrOH	222(4.38),274(3.78), 306(3.54),317(3.23)	44-0582-64
$C_{10}H_{11}N_3O_3$			
Pyruvaldehyde, 1-[methyl(o-nitro-phenyl)hydrazone]	EtOH	237(4.02),305(4.32)	24-0725-64
$C_{10}H_{11}N_3O_3S$			
Acetimidic acid, 2-diazo-N-(phenyl-sulfonyl)-, ethyl ester	n.s.g.	277(4.45)	24-0623-65
$C_{10}H_{11}N_3S$			
5-Mercapto-2,4-dimethyl-3-phenyl-s-triazolium hydroxide, inner salt	n.s.g.	240(3.92)	4-0105-65
5-Mercapto-3,4-dimethyl-2-phenyl-s-triazolium hydroxide, inner salt	n.s.g.	242(4.15)	4-0105-65

Compound	Solvent	λ_{max}(log ε)	Ref.
$C_{10}H_{11}N_5$			
Benzimidazole, 2-(2-imidazolidinylideneamino)-, hydroiodide	MeOH-HCl	244(4.31),292(4.43)	4-0288-64
	MeOH-KOH	293(4.39),303(4.36)	4-0288-64
Pyrazolo[1,5-a]pyrimidine-6-carbonitrile, 7-amino-2,3,5-trimethyl-	EtOH	236(4.55),307(4.16)	95-0442-65
$C_{10}H_{11}N_5OS$			
s-Triazine-2(1H)-thione, tetrahydro-4,6-diimino-1-(p-methoxyphenyl)-	50% EtOH	285(4.37)	39-6296-65
$C_{10}H_{11}N_5O_2$			
Adenine, 9-(2,3-anhydro-5-deoxy-β-D-ribofuranosyl)-	pH 1	257(4.16)	44-3401-65
	pH 7	259(4.16)	44-3401-65
	pH 13	259(4.15)	44-3401-65
$C_{10}H_{11}N_5O_7$			
Allophanic acid, 4-(2,4-dinitroanilino)-, ethyl ester	EtOH	221(3.97),257(3.91), 331(4.08)	22-0500-64
	NaOH	216(4.25),239(4.14), 318(4.36)	22-0500-64
$C_{10}H_{11}N_5S$			
s-Triazinethione, tetrahydro-4,6-diimino-1-p-tolyl-	50% EtOH	286(4.33)	39-6296-65
$C_{10}H_{12}$			
Bicyclo[6.1.0]nonatriene, 9-methyl-, syn	EtOH	252(3.71)	35-4876-64
1-Buten-3-yne, 4-(1-cyclohexen-1-yl)-	n.s.g.	259(4.17),269s(4.09)	44-3991-65
Cycloheptatriene, 7-allyl-	heptane	257(3.52)	35-3329-64
1,3-Cyclohexadiene, 2,3-divinyl-	heptane	208(4.10),231(4.22)	35-4506-65
	n.s.g.	232.5(4.11)	88-1359-65
Indan, 1-methyl-	EtOH	215s(4.0),260s(3.0), 265(3.1),275(3.2), 285s(2.0)	35-0908-64
Naphthalene, tetrahydro-, trans	C_6H_{12}	260(3.6)	33-0558-64
Styrene, β,β-dimethyl-	isooctane	245(4.01)	95-0875-65
Tricyclo[3.3.2.0^{4,6}]deca-2,7-diene	hexane	230s(3.53)	24-3140-64
$C_{10}H_{12}BrNO$			
1,2,3,4-Tetrahydro-6-methyl-1-oxoquinolizinium bromide	EtOH	282.5(3.89)	94-1338-64
7,8,9,10-Tetrahydro-10-oxo-6H-pyrido-[1,2-a]azepinium bromide	H_2O	268(3.83)	44-1523-65
$C_{10}H_{12}BrN_5O_3$			
Adenosine, 8-bromo-2'-deoxy-	pH 1	263(4.25)	35-1242-64
	pH 11	264(4.23)	35-1242-64
$C_{10}H_{12}BrN_5O_4$			
Adenosine, 2-bromo-	pH 1	266(4.15)	4-0213-64
	pH 7,13	265(4.17)	4-0213-64
Adenosine, 8-bromo-	pH 1	263(4.28)	35-1242-64
	pH 11	264(4.25)	35-1242-64
$C_{10}H_{12}BrN_5O_5$			
Guanosine, 8-bromo-	pH 1	261(4.19)	35-1242-64
	pH 11	270(4.15)	35-1242-64

Compound	Solvent	λ_{max}(log ϵ)	Ref.
$C_{10}H_{12}Br_3NO_2$			
9,9-Dibromo-7,8,9,10-tetrahydro-10,10-dihydroxy-6H-pyrido[1,2-a]azepinium bromide	H_2O	267(3.88)	44-1523-65
$C_{10}H_{12}ClN$			
Isoquinoline, 3-chloro-5,6,7,8-tetrahydro-5-methyl-	EtOH	269(3.52)	94-1378-64
$C_{10}H_{12}ClNO$			
Propane, 2-[N-(5-chlorosalicylidene)-amino]-	MeOH	222(4.49),251(4.01), 275(3.45),326(3.57), 412(3.26)	59-0597-64
	EtOH	222(4.49),250(4.03), 275(3.36),327(3.57), 412(3.00)	59-0597-64
	$PhCH_2OH$	329(3.38),417(3.30)	59-0597-64
	dioxan	222(4.40),253(3.93), 327(3.59),410(2.60)	59-0597-64
	$CHCl_3$	255(3.95),327(3.59), 420(2.65)	59-0597-64
	THF	328(3.65),410(1.78)	59-0597-64
	Et_2O	225(4.40),252(4.01), 327(3.71)	59-0597-64
	pyridine	327(3.61),410(2.15)	59-0597-64
$C_{10}H_{12}ClNO_2$			
1,3-Dioxolane, 2-(p-aminophenyl)-2-chloromethyl-	EtOH	242(4.09)	87-0035-65
$C_{10}H_{12}ClNO_4$			
1,1-Dimethylindolium perchlorate	EtOH	215(4.20),250(3.94), 278(2.54),282(2.54), 288(2.36)	44-1449-64
$C_{10}H_{12}ClN_3O_4$			
Ketene, (6-chloro-5-nitro-4-pyrimidinyl)-, diethyl acetal	pH 1	245(3.53)	44-1528-65
	pH 7	327(4.33)	44-1528-65
	pH 13	270(3.92),320(4.01)	44-1528-65
	EtOH	322(4.32)	44-1528-65
$C_{10}H_{12}ClN_5O_3$			
Adenine, 2-chloro-9-(2'-deoxy-α-D-ribofuranosyl)-	pH 11	264(4.19)	35-4934-65
$C_{10}H_{12}ClN_5O_4$			
Adenosine, 2-chloro-	pH 1	265(4.16)	4-0213-64
	pH 7	264(4.18)	4-0213-64
	pH 13	265(4.18)	4-0213-64
$C_{10}H_{12}Cl_2N_2O_2$			
p-Benzoquinone, 2,5-dichloro-3,6-bis-(dimethylamino)-	methyl cellosolve	239(4.20),418(4.03), 550(2.53)	78-1889-64
p-Benzoquinone, 2,5-dichloro-3,6-bis-(ethylamino)-	methyl cellosolve	225(4.30),355(4.42), 527(2.42)	78-1889-64
$C_{10}H_{12}Cl_4O_4Te$			
3-Penten-2-one, 3-chloro-4-hydroxy-, dichlorotellurium derivative	$CHCl_3$	315(4.36)	39-0688-64

Compound	Solvent	λ_{max}(log ε)	Ref.
$C_{10}H_{12}FN_5O_4$			
Adenosine, 2-fluoro-	pH 7	262f(4.18)	10-0713-65C
$C_{10}H_{12}F_2N_2O_2$			
p-Benzoquinone, 2,5-bis(dimethylamino)-3,6-difluoro-	methyl cellosolve	234(4.21),399(4.14), 563(2.66)	78-1889-64
p-Benzoquinone, 2,5-bis(ethylamino)-3,6-difluoro-	methyl cellosolve	224(4.44),353(4.43), 543(2.35)	78-1889-64
$C_{10}H_{12}INO_4$			
Pyrrole-3-carboxylic acid, 4-acetoxy-5-iodo-2-methyl-, ethyl ester	EtOH	208(4.48),261(3.71)	23-1524-64
$C_{10}H_{12}INS$			
2-Ethyl-N-methylbenzothiazolium iodide	H_2O	274(3.84)	22-2868-64
	EtOH	258(3.72),286(3.65)	22-2868-64
	EtOH-NaOEt	306(3.70)	22-2868-64
$C_{10}H_{12}IN_5O_3$			
Adenosine, 2'-deoxy-8-iodo-	pH 1	271(4.28),279s(4.21)	35-1242-64
	pH 11	269(4.23),280s(4.12)	35-1242-64
$C_{10}H_{12}IN_5O_5$			
Guanosine, 8-iodo-	pH 1	261(4.25)	35-1242-64
	pH 11	271(4.23)	35-1242-64
$C_{10}H_{12}N_2$			
Benzimidazole, 1-isopropyl-	MeOH	249(3.77),254(3.78), 266(3.67),274(3.70), 281(3.67)	65-0632-64
Benzimidazole, 2-isopropyl-	pH 2	267(3.87),274(3.85)	44-0476-64
	pH 12	240(3.49),270(3.63), 278(3.65)	44-0476-64
	EtOH	241(3.73),272(3.86), 280(3.84)	44-0476-64
Benzimidazole, 1-propyl-	octane	250(3.78),254(3.78), 266s(3.56),278(3.60), 285(3.59)	65-0632-64
	MeOH	249(3.83),254(3.83), 266(3.66),274(3.70), 282(3.67)	65-0632-64
HBr salt	MeOH	248(3.90),254(3.88), 268(3.79),275(3.79), 281(3.60)	65-0632-64
2,2'-Bipyrrole, 4,4'-dimethyl-	EtOH	280(4.11),290(4.17), 295(4.17)	44-2727-64
1H-Indazole, 3-ethyl-6-methyl-	EtOH	289(3.62)	32-0814-65
Quinazoline, 4-ethyl-3,4-dihydro-	pH 2.0	213(4.22),217s(4.18), 223s(4.01),276(3.69)	39-5360-65
	pH 12.0	217(4.07),228s(4.05), 291(3.80)	39-5360-65
Tryptamine	H_2O	218(4.53),272s(3.71), 277(3.73),286(3.65)	35-3796-64
	EtOH	220(4.56),281(3.78), 290(3.71)	39-5510-64
	H_2SO_4	234(3.64),239(3.62), 288(3.68)	35-3796-64
	H_2SO_4	236(3.63),241(3.60), 295(3.68)	39-5510-64

Compound	Solvent	$\lambda_{max}(\log \epsilon)$	Ref.
$C_{10}H_{12}N_2O$			
Cotinine	EtOH	263(3.49)	35-3375-64
Indene-1-carboxylic acid, 3a,7a-dihydro-, hydrazide	n.s.g.	262(3.41),270(3.38)	35-0905-64
2-Indolinone, 3-(2-aminoethyl)- hydrobromide	EtOH	207(4.43),249(3.92), 280s(3.16)	44-2431-64
Pyrrolidine, 1-nitroso-2-phenyl-	EtOH	233(3.86)	44-3031-65
Pyruvaldehyde, (1-methylphenylhydrazone)	EtOH	235(4.03),285(3.53), 335(4.33)	24-0725-64
2(1H)-Quinoxalinone, 3,4-dihydro-3,3-dimethyl-	EtOH	224(4.59),265(3.45), 306(3.64)	12-0877-64
$C_{10}H_{12}N_2O_2$			
Pyrrolidine, N-(p-nitrophenyl)-	n.s.g.	417(4.39)	5-0109-65I
1-Pyrroline-5-carboxylic acid, 2-pyrrol-2-yl-, methyl ester	MeOH-HCl	270s(3.59),324(4.47)	44-0883-64
	MeOH-KOH	279(4.23)	44-0883-64
	EtOH-HCl	324(4.47)	44-2727-64
	EtOH-NaOH	279(4.23)	44-2727-64
$C_{10}H_{12}N_2O_3$			
Benzamide, o-(3-nitropropyl)-	EtOH	267(2.77)	88-2659-64
Benzoic acid, 2-ureido-, ethyl ester	n.s.g.	223(4.47),248(4.11), 316(3.72)	49-1068-64
Butyric acid, 4-hydroxy-2-oxo-, phenylhydrazone	n.s.g.	293(4.18),317(4.23)	39-0141-64
Indole-7-carboxylic acid, 3-amino-2,4,5,6-tetrahydro-2-oxo-, methyl ester	n.s.g.	318(4.06)	39-1648-65
Morpholine, N-(p-nitrophenyl)-	n.s.g.	397(4.21)	5-0109-65I
Nicotinamide, 1-acetyl-1,2-dihydro-5,6-dimethyl-2-oxo-	EtOH	240s(3.90),356(4.11)	94-0087-64
Propane, 2-[N-(5-nitrosalicylidene)-amino]-	C_6H_{12}	219(4.14),237(4.20), 260(4.04),308(3.99)	59-0597-64
	MeOH	257(4.26),265(3.78), 348(4.08),387(4.04)	59-0597-64
	EtOH	260(4.23),344(3.99), 390(3.93)	59-0597-64
	$PhCH_2OH$	355(4.07),393(4.04)	59-0597-64
	dioxan	220(4.15),239(4.20), 260(4.11),318(4.00), 400(3.40)	59-0597-64
	$CHCl_3$	260(4.15),327(4.00), 403(3.66)	59-0597-64
	pyridine	332(3.93),405(3.78)	59-0597-64
	$HCONH_2$	227(4.07),245(4.08), 355(4.13),382(4.13)	59-0597-64
4-Pyrimidineacrylic acid, α-methoxy-, ethyl ester	EtOH	290(4.52)	44-2398-65
5-Pyrimidinecarboxylic acid, 1-allyl-1,2-dihydro-2-oxo-, ethyl ester	EtOH	247(4.16),272(3.79), 310s(3.47)	94-1418-64
$C_{10}H_{12}N_2O_3S$			
Sulfanilamide, N^4-(3-oxo-1-butenyl)-	MeOH	338(4.60)	83-0321-64
$C_{10}H_{12}N_2O_4$			
Salicylic acid, 2-carboxyhydrazide, ethyl ester	MeOH	205(4.57),239(3.93), 305(3.57)	4-0037-65

Compound	Solvent	$\lambda_{max}(\log \epsilon)$	Ref.
$C_{10}H_{12}N_2O_4S$			
Crotonic acid, 4,4'-thiobis[2-amino-3-(hydroxymethyl)-, di-γ-lactone	EtOH	268(3.89)	44-3560-64
$C_{10}H_{12}N_2O_5$			
Phenol, 2-tert-butyl-4,6-dinitro-	hexane	260(4.21),344(3.56)	70-1666-64
Thymine, 2,2'-anhydro-1-β-D-arabinofuranosyl-	H_2O	224(3.70),254(3.85)	44-0558-64
$C_{10}H_{12}N_2O_7$			
Uridine-5-carboxaldehyde	pH 2	232(3.97),282(4.08)	94-0007-65
	H_2O	232(3.96),282(4.08)	94-0007-65
	pH 12	283(3.95)	94-0007-65
$C_{10}H_{12}N_2O_7S$			
Uracil, 1-(2,5-anhydro-β-D-lyxofuranosyl)-, 3'-methanesulfonate	H_2O	205(3.87),262(3.99)	44-0476-65
Uracil, 1-(3,5-anhydro-β-D-lyxofuranosyl)-, 2'-methanesulfonate	H_2O	204(3.87),259(3.93)	44-0476-65
$C_{10}H_{12}N_2O_8$			
Glucopyranuronic acid, 1-deoxy-1-(4-hydroxy-2-oxo-1(2H)-pyrimidinyl)-	H_2O	229(3.4),260(4.1)	94-1259-64
Uridine-5-carboxylic acid	pH 2	279(4.11)	94-0007-65
	H_2O	275(4.05)	94-0007-65
	pH 12	271(3.9)	94-0007-65
$C_{10}H_{12}N_2S$			
Benzimidazole, 2-[2-(methylthio)-ethyl]-	EtOH	208(4.37),243(4.09), 249s(4.07),271s(4.05), 274(4.13),280(4.17)	4-0306-65
2-Benzimidazoleethanethiol, 5-methyl-	EtOH	208(4.38),247(4.00), 253s(3.98),278s(4.07), 281(4.11),287(4.11)	4-0306-65
Urea, 1-methyl-2-thio-3-(p-vinylphenyl)-	EtOH	272(4.47)	44-1926-65
$C_{10}H_{12}N_4O$			
Formamide, N-(2,3,6-trimethylpyrazolo-[1,5-a]pyrimidin-7-yl)-	EtOH	239(4.68),284s(3.22), 294(3.29),311s(3.19), 348(3.43)	94-0142-65
Pyrazolo[1,5-a]pyrimidine, 7-acetamido-2,3-dimethyl-	EtOH	236(4.67),280s(3.33), 288(3.39),297s(3.34), 343(3.61)	94-0142-65
Pyrazolo[1,5-a]pyrimidine, 7-acetamido-3,6-dimethyl-	EtOH	236(4.73),282s(3.5), 290(3.53),300s(3.42), 343(3.60)	94-0142-65
$C_{10}H_{12}N_4O_2$			
Pyrazolo[1,5-a]pyrimidine-6-carboxylic acid, 7-amino-5-methyl-, ethyl ester	EtOH	226(4.43),307(4.16)	95-0442-65
$C_{10}H_{12}N_4O_2S_3$			
4-Isothiazolecarbonitrile, 3,5-bis(dimethylcarbamoylthio)-	EtOH	275(4.10)	44-0665-64
$C_{10}H_{12}N_4O_3$			
2-Isoxazolin-5-one, 4-[(3,4-dimethyl-5-isoxazolyl)azo]-3,4-dimethyl-	EtOH	298(4.10),414(2.23)	39-5414-65
Lumazine, 1,3,6,7-tetramethyl-, 5-oxide	pH 7.0	241(4.45),287(3.87), 351(3.83)	89-1136-65

Compound	Solvent	λ_{max}(log ϵ)	Ref.
Purine, 9-(2-deoxy-α-D-ribofuranosyl)-	H_2O	262.5(3.89)	35-4934-65
Purine, 9-(2-deoxy-β-D-ribofuranosyl)-	H_2O	262.5(3.86)	35-4934-65
Purine, 9-(3-deoxy-β-D-ribofuranosyl)-	pH 4	262.5(3.87)	87-0659-65
	pH 7	263(3.88)	87-0659-65
	pH 11	263(3.88)	87-0659-65
$C_{10}H_{12}N_4O_3S$			
6-Purinethiol, 9-(3-deoxy-β-D-ribo-furanosyl)-	pH 4	223(3.96),322(4.38)	87-0659-65
	pH 11	233(4.15),311(4.33)	87-0659-65
$C_{10}H_{12}N_4O_4$			
2-Butanone, 2,4-dinitrophenylhydrazone	$C_2H_4Cl_2$	344(4.36)	39-1761-65
Cytosinine	pH 1	274(4.14)	88-1405-65
	pH 13	267(3.87)	88-1405-65
$C_{10}H_{12}N_4O_4S$			
1,2,4-Triazolidine-3,5-dione, 1-[p-(N,N-dimethylsulfonamido)phenyl]-	EtOH	211(4.13),275(4.28)	22-0500-64
	NaOH	216(4.24),239(4.12),318(4.32)	22-0500-64
$C_{10}H_{12}N_4O_6$			
Inosine, N-oxide	pH 3.3	205(4.3),250(3.9)	33-0433-65
	pH 7.5	227(4.5),262(3.7)	33-0433-65
$C_{10}H_{12}N_4O_7$			
Uric acid, 9-β-D-ribofuranosyl-	pH 1	237(3.98),288(4.05)	35-1772-65
	pH 11	244(4.08),298(3.98)	35-1772-65
$C_{10}H_{12}N_5O_5P$			
Adenosine, deoxy-, cyclic 3',5'-phosphate	pH 2.0	257(4.16)	35-1626-64
	pH 7.0	259(4.16)	35-1626-64
$C_{10}H_{12}N_5O_6P$			
Guanosine, deoxy-, cyclic 3',5'-phosphate	pH 2.0	256(4.10)	35-1626-64
	pH 7.0	253(4.11)	35-1626-64
	pH 12.0	262(4.05)	35-1626-64
$C_{10}H_{12}N_6O_2$			
Pterin, 8-(4-morpholino)-	pH 1	226(4.02),290(4.29),405(3.65)	33-2195-64
	pH 13	266(4.20),283(4.18),404(3.77)	33-2195-64
$C_{10}H_{12}N_6O_6$			
Sydnone imine, 3,3'-ethylenebis[N-carboxy-, dimethyl ester	pH 1.0	286(4.36)	65-2064-64
	pH 7.00	228(4.22),318(4.22)	65-2064-64
$C_{10}H_{12}N_8O_4$			
Adenosine, 8-azido-	pH 1	281(4.23)	35-1772-65
	pH 11	228(4.13),281(4.13)	35-1772-65
$C_{10}H_{12}O$			
Anisole, 2-allyl-	hexane	270(3.08),300s(1.04)	65-2088-65
	EtOH-HCl	264(3.30),295s(2.42),300s(1.32)	65-2088-65
	EtOH	265(3.38),300s(1.33)	65-2088-65
	EtOH-NaOEt	262(3.38),335s(1.05)	65-2088-65
	10% H_2SO_4	279(3.38),335s(2.33),295s(3.03)	65-2088-65

Compound	Solvent	$\lambda_{max}(\log \epsilon)$	Ref.
Anisole, 4-allyl-	hexane	273(3.38),305s(1.85)	65-2088-65
	EtOH-HCl	276(3.49),305s(1.95)	65-2088-65
	EtOH	269(3.38),305s(2.02)	65-2088-65
	EtOH-NaOEt	270(3.38),305s(2.08)	65-2088-65
	10% H_2SO_4	280(3.49),310s(2.12)	65-2088-65
Anisole, 2-propenyl-	hexane	248(4.03),258(3.88), 280(3.35),308(3.35)	65-2088-65
	EtOH-HCl	292(3.60)	65-2088-65
	EtOH	246(4.03),280(3.35), 310(3.35)	65-2088-65
	EtOH-NaOEt	249(4.03),292(3.60), 335s(2.35)	65-2088-65
	10% H_2SO_4	295(3.08),325s(1.92)	65-2088-65
Anisole, 4-propenyl-	hexane	260(4.38),300(3.15)	65-2088-65
	EtOH-HCl	261(4.30),295(3.46)	65-2088-65
	EtOH	259(4.30),295(3.60), 300(3.60)	65-2088-65
	EtOH-NaOEt	260(4.30),300(3.85)	65-2088-65
	10% H_2SO_4	255(3.60),325s(2.33), 345s(2.05)	65-2088-65
Bicyclo[6.1.0]nona-2,4,6-triene, 9-methoxy-	EtOH	255(3.69)	35-5194-64
2,5-Cyclohexadien-1-one, 4-allyl-4-methyl-	EtOH	239(4.14)	33-0094-65
2,4,8-Decatrien-6-yn-1-ol	Et_2O	290(4.58),307(4.54)	24-0369-65
2,6,8-Decatrien-4-yn-1-ol	Et_2O	290(4.58),307(4.53)	24-0369-65
Indan, 1-methoxy-	EtOH	261(3.64),270(3.59)	35-5194-64
2,7-Octadien-4-ynal, 3,7-dimethyl-	MeOH	274s(4.14),333(2.7)	5-0026-64I
Tropone, 2,3,5-trimethyl-	hexane	236(4.4),295(3.7)	49-0402-64
$C_{10}H_{12}O_2$			
Acetophenone, 2'-hydroxy-3',5'-dimethyl-	n.s.g.	218(4.25),260(4.09), 348(3.87)	25-1686-64
Benzene, 1,2-(methylenedioxy)-4-propyl-	EtOH	288(--)	95-0857-65
Bicyclo[3.3.1]non-6-ene-2,4-dione, 6-methyl-	EtOH	240s(3.64),270(3.92)	44-0787-64
Butyrophenone, 2'-hydroxy-	EtOH	252(4.02),325(3.60)	28-5614-64A
Butyrophenone, 4'-hydroxy-	EtOH	277(4.21)	28-5614-64A
Chroman, 8-methoxy-	EtOH	210(4.19),277(3.26)	78-1185-64
2,5-Cyclohexadiene, 1,4-epoxy-2-methyl-2-propionyl-	EtOH	209(4.02),264(4.11)	28-0404-64B
5-Cyclohexene, 1,4-epoxy-2-methylene-3-propionyl-	EtOH	207(3.81),267(2.60), 315(2.00)	28-0404-64B
1,3-Naphthalenediol, 5,6,7,8-tetrahydro-,	EtOH	280s(3.31),284(3.32)	24-1926-64
	NaOH	225s(4.20),296(3.55)	24-1926-64
1,4-Naphthalenediol, 1,2,3,4-tetrahydro-, trans	EtOH	261(2.44),266(3.26), 271(2.25)	78-1931-65
1,4-Naphthalenediol, 5,6,7,8-tetrahydro-,	EtOH	220s(3.78),291(3.50)	24-1926-64
	NaOH	253(3.86),510(2.80)	24-1926-64
2,7(1H,3H)-Naphthalenedione, 4,4a,5,6-tetrahydro-	$CHCl_3$	242(4.12),300(3.32)	44-1292-65
2,7(1H,3H)-Naphthalenedione, 4,5,6,8-tetrahydro-	EtOH	318(2.39)	44-3209-65
	EtOH-NaOH	371s(4.53),386(4.80)	44-3209-65
2(1H)-Naphthalenone, 3,7,8,8a-tetrahydro-7-hydroxy-	H_2O	328(4.19)	44-3209-65
	aq. HCl	328(4.22)	44-3209-65
	aq. NaOH	380(5.11)	44-3209-65
	EtOH-HCl	323(4.44)	44-3209-65
	EtOH	323(4.39),385(4.12)	44-3209-65
	EtOH-NaOH	385(4.98)	44-3209-65

Compound	Solvent	λ_{max}(log ϵ)	Ref.
2(3H)-Naphthalenone, 4,4a,5,6-tetrahydro-7-hydroxy-	EtOH-acid	322(4.40)	44-1292-65
	EtOH	324(4.43),385(3.91)	44-1292-65
	EtOH-base	385(4.95)	44-1292-65
3-Oxetanol, 2-methyl-3-phenyl-	MeOH	258(2.46)	88-0485-65
Phenol, 4-allyl-2-methoxy- (eugenol)	heptane	228(3.85),270(3.58)	65-2094-65
	EtOH-HCl	279(4.08)	65-2094-65
	EtOH	231(4.03),281(3.78)	65-2094-65
	EtOH-NaOEt	249(4.03),294(3.77)	65-2094-65
	10% H_2SO_4	238(3.88),279(3.60)	65-2094-65
Phenol, 2-methoxy-4-(1-propenyl)-, cis	heptane	258(4.19),279(3.78)	65-2094-65
	EtOH	261(4.55),275(4.23), 325s(2.35)	65-2094-65
	EtOH-NaOEt	281(4.55),298(4.30), 340s(3.18)	65-2094-65
	10% H_2SO_4	280(3.78),313s(3.02)	65-2094-65
Phenol, 2-methoxy-4-(1-propenyl)-, trans	heptane	225(4.48),261(4.29), 290(3.45)	65-2094-64
	EtOH-HCl	280(3.95),305s(3.25), 325s(2.00)	65-2094-64
	EtOH	260(4.30),305(3.78), 327s(1.90)	65-2094-64
	EtOH-NaOEt	293(4.30),325(4.02), 343s(3.05),365s(2.08)	65-2094-64
	10% H_2SO_4	280(3.78)	65-2094-64
Propiophenone, 2'-hydroxy-3'-methyl-	C_6H_{12}	255(4.02),260(3.97), 333(3.60)	28-5614-64A
	EtOH	256(4.02),331(3.56)	28-5614-64A
Propiophenone, 2'-hydroxy-4'-methyl-	C_6H_{12}	256(4.10),263(4.10), 325(3.67)	28-5614-64A
	EtOH	260(4.11),324(3.63)	28-5614-64A
Propiophenone, 2'-hydroxy-5'-methyl-	C_6H_{12}	251(3.95),257(3.95), 340(3.61)	28-5614-64A
	EtOH	254(3.99),336(3.58)	28-5614-64A
Propiophenone, 4'-hydroxy-2'-methyl-	C_6H_{12}	259(4.06)	28-5614-64A
	EtOH	271(4.13)	28-5614-64A
Propiophenone, 4'-hydroxy-3'-methyl-	C_6H_{12}	261(4.16)	28-5614-64A
	EtOH	279(4.15)	28-5614-64A
2-Pyrone, 6-cyclopentyl-	MeOH	299(3.80)	54-0039-64
2-Pyrone, 4,5-tetramethylene-6-methyl-	EtOH	312(3.83)	44-2642-65
Spiro[3.5]nona-1,6-diene-2-carboxylic acid	EtOH	222(3.985)	28-1541-64A
$C_{10}H_{12}O_2S$			
1-Benzothiepin, 2,3,4,5-tetrahydro-, 1,1-dioxide	EtOH	262(3.51),269(3.56), 277(3.54)	44-0366-64
Benzothiophene-3-carboxylic acid, 4,5,6,7-tetrahydro-, methyl ester	Et_2O	207(4.38),244(3.93)	24-2109-64
Dimethylsulfonium phenacylide, oxide	EtOH	229(3.98),283(4.05)	35-1640-64
$C_{10}H_{12}O_3$			
p-Benzoquinone, 2-tert-butyl-6-hydroxy-	MeOH	267(4.05),383(3.95)	24-2774-65
	MeOH-KOH	222(4.18),272(3.91), 483(3.20)	24-2774-65
Chroman, 6-methoxy-	EtOH	206(4.15),230(3.79), 294(3.54)	78-1185-64
4,6-Decadiynoic acid, 9-hydroxy-	EtOH	225(2.79),240(2.70), 253(2.53),269(2.20), 285(2.21),303(2.14)	70-1237-65
Furan, 3,4-diacetyl-2,5-dimethyl-	n.s.g.	272(3.80)	32-0393-64
1,4,5-Naphthalenetriol, 1,2,3,4-tetrahydro-	EtOH	281(3.35),283(3.35)	70-0492-64

Compound	Solvent	λ_{max}(log ϵ)	Ref.
$C_{10}H_{12}O_4$			
Acetic acid, (3,5-dimethoxyphenyl)-	EtOH	274(3.30),281(3.30)	1-1677-65
	EtOH	273(3.29),280(3.29)	1-1677-65
Acetophenone, 3',6'-dihydroxy-4'-methoxy-2'-methyl-	EtOH	240(3.91),281(3.91), 344(3.58)	39-0411-64
Benzoic acid, 2-hydroxy-4-methoxy-6-methyl-, methyl ester	EtOH	263(4.20),303(3.68)	1-1677-65
Benzoic acid, 2,4-dimethoxy-6-methyl-	EtOH	280(3.38)	1-1677-65
p-Benzoquinone, 2,5-dihydroxy-3-isopropyl-6-methyl-	EtOH	293(4.31),435(2.36)	78-2319-64
	EtOH	293(4.31),435(2.36)	94-0511-65
Bicyclo[3.1.0]hex-2-ene-3,5-dicarboxylic acid, dimethyl ester	EtOH	237(3.81)	24-2201-65
Bicyclo[4.2.0]oct-7-ene-7,8-dicarboxylic acid, cis	EtOH	240(3.93)	39-2153-64
Butenolide, α-carboxy-γ-pentylidene-	EtOH	286(4.29)	78-2701-64
1,3-Cyclohexadiene-1,4-dicarboxylic acid, dimethyl ester	EtOH	309(4.12)	23-2852-64
	EtOH	309(4.11)	24-2201-65
	EtOH	309(4.13)	44-1431-65
1,4-Cyclohexadiene-1,4-dicarboxylic acid, dimethyl ester	EtOH	203(4.37),240s(3.15)	44-1431-65
Cyclohexadienone, 2-acetoxy-3-methoxy-2-methyl-	n.s.g.	360(3.6)	49-0649-64
Cyclohexadienone, 2-acetoxy-5-methoxy-2-methyl-	n.s.g.	300(3.7)	49-0649-64
3-Cyclopentene-1,1-dicarboxylic acid, 2-methylene-, ethyl ester	EtOH	231(4.13)	70-1460-65
Dehydroacetic acid, ethyl ester	EtOH	226(3.96),314(3.91)	44-1255-65
4H-Pyran-2-carboxylic acid, 6-ethyl-3-methyl-4-oxo-, methyl ester	EtOH-HCl	267(4.00)	39-2251-65
	EtOH-NaOH	260(4.15)	39-2251-65
$C_{10}H_{12}O_4S$			
2,5-Thiophenedicarboxylic acid, 3,4-dimethyl-, dimethyl ester	EtOH	209(4.06),280(4.22)	44-1919-64
2,5-Thiophenedicarboxylic acid, 3,4-dimethyl-, ethyl ester	EtOH	209(4.04),280(4.20)	44-1919-64
$C_{10}H_{12}O_4Se_2$			
1,3-Diselenacyclobutane, 2,2,4,4-tetraacetyl-	$CHCl_3$	286(4.21)	39-0688-64
$C_{10}H_{12}O_5$			
Cyclopentanecarboxylic acid, 3-acetyl-2,4-dioxo-, ethyl ester	MeOH-HCl	226(4.38),261(4.23)	20-0628-64
3-Cyclopentene-1-carboxylic acid, 4-formyl-3-hydroxy-1-methyl-2-oxo-, ethyl ester	EtOH	300(3.91),367(3.22)	44-3520-64
	EtOH-base	367(4.10)	44-3520-64
$C_{10}H_{12}O_6$			
Methyldegeranylmelicopol	EtOH	233s(3.96),285(4.20), 335(3.43)	12-2021-65
	EtOH-KOH	244(3.97),298(4.05), 327s(3.86)	12-2021-65
$C_{10}H_{12}S$			
Sulfide, 1-methylallyl phenyl	EtOH	256(3.64)	44-0728-65
Sulfide, 2-methylallyl phenyl	EtOH	255(3.81)	44-0728-65

Compound	Solvent	$\lambda_{max}(\log \epsilon)$	Ref.
$C_{10}H_{13}BrN_2O$			
7,8,9,10-Tetrahydro-10-oxo-6H-pyrido-[1,2-a]azepinium bromide, oxime	H_2O	284(3.96)	44-1523-65
$C_{10}H_{13}BrN_2O_5$			
Thymidine, 2'-bromo-	H_2O	205(3.95),265(3.97)	44-0558-64
$C_{10}H_{13}Br_2NO_2$			
9-Bromo-7,8,9,10-tetrahydro-10,10-di-hydroxy-6H-pyrido[1,2-a]azepinium bromide	H_2O	267(3.72)	44-1523-65
$C_{10}H_{13}Br_2NO_3$			
2-Pyrroline-3-carboxylic acid, 2-(di-bromomethyl)-4,4-dimethyl-5-oxo-, ethyl ester	EtOH	294(3.88)	39-5999-64
$C_{10}H_{13}Br_3O$			
2-Bornanone, 3,6,10-tribromo-, 3,6-endo	MeOH	299(1.70),306(1.70), 317(1.65)	78-0273-65
10α-Pinan-3-one, 2α,4α,10-tribromo-	MeOH	327(2.18),336(2.22), 347(2.13)	78-0273-65
$C_{10}H_{13}Cl$			
Cyclobutene, 4-(chloromethylene)-1-iso-propenyl-3,3-dimethyl-	EtOH	230(4.37),250(4.38)	28-0827-64B
$C_{10}H_{13}ClN_2O$			
Pyridazine, 3-chloro-6-(cyclohexyloxy)-	EtOH	283(3.28)	44-1751-64
$C_{10}H_{13}ClN_2O_2S$			
Acetic acid, [(5-butyl-6-chloro-4-pyri-midinyl)thio]-	EtOH	205(4.16),214s(4.06), 255(3.87),280(4.00)	73-3730-65
$C_{10}H_{13}ClN_2O_5$			
Thymidine, 2'-chloro-	H_2O	205(3.96),265(3.98)	44-0558-64
$C_{10}H_{13}ClN_2S$			
1-Phenylpropenylisothiouronium chloride	EtOH	252(4.09)	95-0930-64
$C_{10}H_{13}ClO$			
Phenol, 4-tert-butyl-2-chloro-	HCl	216(3.59)	44-2678-65
	NaOH	239(3.81)	44-2678-65
$C_{10}H_{13}Cl_2N$			
3-Chloro-1,1-dimethylindolinium chloride	H_2O	209s(3.87)	44-1449-64
	EtOH	216s(3.84),249s(2.23), 254(2.36),260(2.42), 267s(2.23)	44-1449-64
$C_{10}H_{13}FN_2O_5$			
Thymidine, 2'-fluoro-	H_2O	206(3.96),265(3.96)	44-0558-64
$C_{10}H_{13}IN_2$			
1,2,3-Trimethylimidazo[1,2-a]pyridinium iodide	EtOH	202(4.14),238(4.05), 326(3.65)	4-0331-65
1,2,4-Trimethylimidazo[1,2-a]pyridinium iodide	EtOH	218(4.53),288(3.97), 298s(3.91)	4-0331-65
1,2,6-Trimethylimidazo[1,2-a]pyridinium iodide	EtOH	219(4.54),224s(4.50), 283(3.96)	4-0331-65

Compound	Solvent	λ max(log ε)	Ref.
1,2,7-Trimethylimidazo[1,2-a]pyridinium iodide	EtOH	215s(4.45),221(4.56), 226s(4.51),287(3.99)	4-0331-65
$C_{10}H_{13}IN_2O_4$			
Thymine, 1-(2,5-dideoxy-5-iodo-β-D-lyxosyl)-	EtOH	266(4.04)	44-2076-64
$C_{10}H_{13}IN_2S$			
S-Methyl-N-(p-vinylphenyl)isothiuronium iodide	EtOH	274(4.53)	44-1926-65
$C_{10}H_{13}N$			
Propane, 2-N-benzylideneamino-	MeOH	212(3.93),245(3.92)	59-0597-64
	dioxan	215(--),238(4.00)	59-0597-64
$C_{10}H_{13}NO$			
Formanilide, 2'-ethyl-N-methyl-	MeOH	241(3.9)	56-0639-65
Propane, 2-[N-(p-hydroxybenzylidene)-amino]-	MeOH	215(4.15),269(4.23), 376(3.77)	59-0597-64
	dioxan	220(--),263(4.29)	59-0597-64
Propane, 2-(N-salicylideneamino)-	C_6H_{12}	217(4.53),254(4.10), 318(3.70)	59-0597-64
	H_2O	219(4.20),242(3.92), 274(3.98),389(3.78)	59-0597-64
	MeOH	218(4.32),252(3.60), 276(3.58),315(3.53), 400(3.20)	59-0597-64
	10% MeOH	274(4.1),385(3.8)	20-0518-65
	EtOH	215(4.45),252(4.06), 276(3.40),315(3.58), 400(2.95)	59-0597-64
	$PhCH_2OH$	314(3.45),404(3.28)	59-0597-64
	dioxan	225(4.23),254(4.08) 316(3.65),400(1.48)	59-0597-64
	$CHCl_3$	255(4.05),315(3.62), 409(2.18)	59-0597-64
	CCl_4	255(4.03),318(3.65)	59-0597-64
	MeCN	215(4.43),253(4.06), 313(3.61),404(2.18)	59-0597-64
	$HCONH_2$	275(3.91),315(3.08), 403(3.65)	59-0597-64
	DMF	315(3.62),405(2.15)	59-0597-64
	pyridine	316(3.60),405(1.90)	59-0597-64
$C_{10}H_{13}NOS$			
Benzothiazoline, 2-ethoxy-N-methyl-	EtOH	308(3.59)	22-2868-64
$C_{10}H_{13}NO_2$			
Alanine, N-phenyl-, methyl ester	EtOH	242(4.00),291(3.28)	87-0147-65
Anthranilic acid, N,N-dimethyl-, methyl ester	10% MeOH	255(3.62),337(3.35)	24-1127-64
Anthranilic acid, N-ethyl-, methyl ester	10% MeOH	252(3.88),349(3.65)	24-1127-64
Anthranilic acid, N-ethyl-N-methyl-	10% MeOH	265s(2.78),270(2.85), 277(2.77)	24-1127-64
anion	10% MeOH	260(3.66),300(3.1)	24-1127-64
1H-Azepine-2,5-dione, 6-isopropyl-4-methyl-	n.s.g.	228(3.16),288(2.47)	88-1071-65
1H-Azepine-2,5-dione, 3,4,6,7-tetra-methyl-	n.s.g.	233(3.07),298(2.43)	88-1071-65

Compound	Solvent	λ_{max}(log ε)	Ref.
Benzoic acid, p-dimethylamino-, methyl ester	MeOH	308(4.44)	88-2729-64
7-Isoquinolinol, 1,2,3,4-tetrahydro-6-methoxy-	EtOH-HCl	224(3.83),284(3.59)	33-1945-65
	EtOH	223s(3.85),284(3.59)	33-1945-65
	EtOH-NaOH	244(3.94),300(3.76)	
hydrochloride	EtOH-HCl	225(3.83),285(3.60)	33-1945-65
	EtOH	225(3.75),285(3.60)	33-1945-65
	EtOH-NaOH	245(3.88),302(3.71)	33-1945-65
Orcinol, 2-(2-cyanoethyl)dihydro-	MeOH	262(4.0)	5-0084-65A
	HCl	262(4.1)	5-0084-65A
	NaOH	285(4.3)	5-0084-65A
Oxaziridine, 2-ethyl-3-(p-methoxyphenyl)-	EtOH	233(4.15),275(3.3), 282(3.23)	44-3427-65
2-Pyridone, 1-methyl-, ethoxycarbonylmethide, anhydro base	EtOH	304s(4.21),313(4.27), 384(3.81)	12-0455-64
4-Pyridone, 1-methyl-, ethoxycarbonylmethide, anhydro base	EtOH	219(3.88),257(3.15), 265s(3.08),365(4.42), 378s(4.33)	12-0455-64
1-Pentanone, 4-hydroxy-1-(2-pyridyl)-	EtOH	229(3.84),267(3.58)	94-1338-64
Propane, 2-[N-(p-hydroxysalicylidene)-amino]-	MeOH	219(4.18),300(4.20), 310(3.60),372(3.89)	59-0597-64
	EtOH	220(4.20),275(3.30), 305(4.18),376(3.78)	59-0597-64
	$PhCH_2OH$	307(4.16),376(3.88)	59-0597-64
	dioxan	220(4.30),275(4.15), 307(3.93),382(2.48)	59-0597-64
	$CHCl_3$	275(3.98),305(4.05), 385(3.53)	59-0597-64
	DMF	305(3.96),382(3.04)	59-0597-64
	pyridine	310(3.99),383(2.85)	59-0597-64
Pyran-2-ol, tetrahydro-6-(2-pyridyl)-	EtOH	257(3.62),262(3.57), 268(3.44)	44-1523-65
2-Pyrone, 6-methyl-4-pyrrolidino-	MeOH	227(4.38),258(3.86), 265(3.86),296(4.04)	24-1266-64
Pyrrole-2-acrylic acid, 1-methyl-, ethyl ester, trans	n.s.g.	338(4.34)	12-0875-65
$C_{10}H_{13}NO_2S_3$			
Imidocarbonic acid, dithio(p-tolylsulfonyl)-, dimethyl ester	EtOH	229(3.98),263(4.30)	95-0391-65
$C_{10}H_{13}NO_3$			
Alanine, p-(hydroxymethyl)phenyl-	H_2O	252(2.34),257s(2.42), 262(2.47),265s(2.37), 270s(2.20)	87-0554-65
Glyoxal, (o-aminophenyl)-, dimethyl acetal	n.s.g.	378(3.76)	49-0889-65
6,7,8-Isoquinolinetriol, 1,2,3,4-tetrahydro-1-methyl-, hydrobromide	pH 2	272(2.92),283s(2.74)	49-0025-65
	pH 12	315s(3.73),367(3.44)	49-0025-65
	EtOH	272(2.92),283s(2.72)	49-0025-65
Propiophenone, 2-amino-3'-hydroxy-4'-methoxy-, hydrochloride	H_2O	230(4.10),275(3.96), 310(3.88)	78-1489-65
	pH 13	250(4.15),285(3.76), 355(3.65)	78-1489-65
2-Pyrone, 4-morpholino-6-methyl-	MeOH	228(4.31),260(3.87), 301(4.05)	24-1266-64

Compound	Solvent	$\lambda_{max}(\log \epsilon)$	Ref.
$C_{10}H_{13}NO_4$			
Benzamide, 3,4,5-trimethoxy-	H_2O	257(3.41)	23-1957-64
	H_2O	256(3.48)	23-2328-65
	HCl	283(3.92)	23-2328-65
	63.8% H_2SO_4	283(3.96)	23-1957-64
2-Butanol, 4-(p-nitrophenoxy)-	EtOH	228(3.92),309(4.13)	94-0987-64
$C_{10}H_{13}NO_4S$			
Propionic acid, 3-[(2-aminobenzyl)-sulfonyl]-	iso-PrOH	241(3.89),295(3.51)	44-3111-65
$C_{10}H_{13}NO_4S_2$			
N,2-Dimethylbenzothiazolium methyl sulfate	H_2O	273(3.81)	22-2868-64
	0.5N NaOH	271(3.88)	22-2868-64
	N NaOH	268(3.98)	22-2868-64
	2N NaOH	268(4.12)	22-2868-64
	EtOH	262(3.68)	22-2868-64
$C_{10}H_{13}NO_5$			
Indole-7-carboxylic acid, 2,4,5,6,7,7a-hexahydro-3,7a-dihydroxy-2-oxo-, methyl ester	n.s.g.	251(3.52)	39-1648-65
2(1H)-Pyridone, 1-D-ribofuranosyl-	EtOH	228(3.79),303(3.76)	94-0828-64
$C_{10}H_{13}NO_6$			
Pyrrole-2,5-dicarboxylic acid, 3,4-di-methoxy-, dimethyl ester	EtOH	272(4.49)	44-0859-65
$C_{10}H_{13}NS$			
2H-5,1-Benzothiazocine, 1,3,4,6-tetra-hydro-, hydrochloride	iso-PrOH	250(3.72),310(3.11)	44-3111-65
Benzothiazoline, 2,2,3-trimethyl-	C_6H_{12}	318(3.60)	22-2868-64
	EtOH	314(3.67)	22-2868-64
Hydratropamide, N-methylthio-	MeOH	265(4.03),335(1.82)	35-0051-65
$C_{10}H_{13}N_2NaO_8$			
Barbituric acid, 1-β-D-glucopyranosyl-, sodium derivative	pH 11.2	260(4.25)	94-0459-64
$C_{10}H_{13}N_2O_7P$			
Thymidine, 3',5'-phosphate,(cyclic)	pH 7.0	265(4.00)	35-1626-64
	pH 12.0	265(3.88)	35-1626-64
$C_{10}H_{13}N_3$			
Imidazo[1,2-a]pyridine, 3-(dimethyl-amino)-2-methyl-, hydrate	EtOH	228(4.44),234(4.44), 284(3.6),306(3.54)	44-2403-65
Imidazo[1,2-a]pyridine, 3-[(dimethyl-amino)methyl]-	pH 1	265s(3.86),273(3.91)	44-4085-65
	pH 11	269s(3.66),279(3.71), 297(3.61)	44-4085-65
	EtOH	224(4.50),228(4.49), 281(3.61),302(3.49)	44-2403-65
$C_{10}H_{13}N_3O$			
Pyruvaldehyde, 1-(methylphenylhydra-zone), 2-oxime	EtOH	250(3.65),320(4.42)	24-0725-64
$C_{10}H_{13}N_3O_2S$			
p-Toluenesulfonamide, N-2-imidazol-idinylidene-	EtOH	231.5(4.20)	95-0391-65

Compound	Solvent	λ max(log ε)	Ref.
$C_{10}H_{13}N_3O_3$			
5-Pyrimidinecarboxylic acid, 1-(2-cyano-ethyl)-1,2,3,4-tetrahydro-2-oxo-, ethyl ester	EtOH	214(3.93),290(3.98)	94-0681-65
Salicylaldehyde, 5-nitro-, ethyl-methylhydrazone	MeOH	295(4.47)	24-1631-64
	MeOH-NaOH	342(4.21),422(4.05)	24-1631-64
$C_{10}H_{13}N_3O_4$			
Aniline, N,N-diethyl-2,4-dinitro-	MeOH	375(4.23)	35-4018-64
$C_{10}H_{13}N_3O_4S$			
Malonic acid, [(1,3,4-thiadiazol-2-yl-amino)methylene]-, diethyl ester	EtOH	312(4.4)	70-1481-64
$C_{10}H_{13}N_3O_5$			
Isoguanosylriboside	H_2O	248(3.94),292(4.05)	88-3201-64
Uracil, 6-acetamido-5-(2-acetoxy-ethyl)-	pH 1	276(4.09)	44-2670-64
	pH 13	274.5(4.18)	44-2670-64
$C_{10}H_{13}N_3O_7$			
Nortetrodoic acid	H_2O	210(3.36)	78-2059-65
	pH 1	210(3.48)	78-2059-65
Uracil, 1-β-D-glucopyranosiduronamide	H_2O	229(3.42),259(4.02)	94-1259-64
$C_{10}H_{13}N_3O_8$			
Seconortetrododioic acid	pH 1	210(3.45)(end absorption)	78-2059-65
	H_2O	210(3.40)(end absorption)	78-2059-65
	pH 13	210(3.24)(end absorption)	78-2059-65
$C_{10}H_{13}N_4O_9P$			
Xanthosine, 5-phosphate	pH 7.5	249(4.00),278(3.95)	37-3407-64
$C_{10}H_{13}N_5O$			
7(8H)-Pteridinone, 4-(dimethylamino)-6,8-dimethyl-	MeOH	230(4.18),250(4.16), 304(3.84),347(4.05)	4-0023-64
$C_{10}H_{13}N_5O_2$			
Adenosine, 2',3'-dideoxy-	pH 1	260.5(4.14)	44-2854-65
	pH 7	260(4.18)	44-2854-65
	pH 13	260(4.17)	44-2854-65
$C_{10}H_{13}N_5O_3$			
Adenine, 9-(2-deoxy-α-D-ribofuranosyl)-	H_2O	259.5(4.19)	35-4934-65
Adenine, 9-(5-deoxy-β-D-xylofuranosyl)-	pH 1	257(4.17)	44-3401-65
	pH 7	259(4.17)	44-3401-65
	pH 13	260(4.17)	44-3401-65
Cordycepin (or 3'-deoxyadenosine)	pH 1	258(4.14)	87-0659-65
	H_2O	260(4.15)	87-0659-65
	pH 13	260(4.16)	87-0659-65
D-erythro-Pentose, 2,3-dideoxy-3-(6-amino-9-purinyl)-	pH 1	260(4.14)	35-0720-64
	H_2O	262(4.15)	35-0720-64
	pH 13	262(4.17)	35-0720-64
D-threo-Pentose, 2,3-dideoxy-3-(6-amino-9-purinyl)-	pH 1	260(4.14)	35-0720-64
	H_2O	262(4.16)	35-0720-64
	pH 13	262(4.16)	35-0720-64
$C_{10}H_{13}N_5O_3S$			
Adenine, 9-(4-thio-β-D-ribofuranosyl)-	pH 1	259(4.15)	25-1364-64
	pH 7, 13	261(4.17)	25-1364-64

Compound	Solvent	$\lambda_{max}(\log \epsilon)$	Ref.
Adenine, 9-(4-thio-β-L-ribofuranosyl)-	pH 1	259(4.16)	25-1364-64
	pH 7,13	261(4.17)	25-1364-64
Adenosine, 2'-deoxy-8-mercapto-	pH 1	222(4.11),242(4.07), 308(4.44)	35-1242-64
	pH 11	230(4.34),297(4.39)	35-1242-64
$C_{10}H_{13}N_5O_4$			
Adenine, 9-β-D-lyxofuranosyl-	pH 1	256(4.15)	25-1561-65
	pH 7	259(4.17)	25-1561-65
	pH 13	260(4.16)	25-1561-65
Adenine, 9-α-D-ribofuranosyl-	pH 1	257(--)	28-2453-64B
	pH 13	259(4.16)	28-2453-64B
Adenosine	pH 1	260(4.15)	28-2453-64B
	pH 13	260(4.16)	28-2453-64B
Guanine, 7-(3-deoxy-β-D-ribofuranosyl)-	pH 1	249(3.93),270s(3.7)	44-2851-65
	pH 4	240s(3.66),284(3.71)	44-2851-65
	pH 6	217(4.25),240s(3.71), 285(3.79)	44-2851-65
	pH 7	217(4.28),242s(3.75), 286(3.8)	44-2851-65
	pH 11	215(4.29),237s(3.78), 285(3.8)	44-2851-65
	pH 13	238s(3.85),282(3.75)	44-2851-65
Guanine, 9-(3-deoxy-β-D-ribofuranosyl)-	pH 1	255(4.05),275s(3.88)	44-2851-65
	pH 4	252(4.08),270s(3.94)	44-2851-65
	pH 6	252(4.08),272s(3.14)	44-2851-65
	pH 7	252(4.07),270s(3.93)	44-2851-65
	pH 11	253(4.05),270s(3.95)	44-2851-65
	pH 13	260s(3.98),267(3.99)	44-2851-65
Pyrazolo[3,4-d]pyrimidine, 4-amino-1-β-D-ribofuranosyl-	pH 1	258(4.02)	4-0215-64
	pH 7	260(3.99),274(4.09)	4-0215-64
	pH 13	260(3.99),275(4.04)	4-0215-64
Pyrazolo[3,4-d]pyrimidine, 4-amino-2-β-D-ribofuranosyl-	pH 1	268(4.03)	4-0215-64
	pH 7	268(3.91),290(4.04)	4-0215-64
	pH 13	268(3.95),290(4.02)	4-0215-64
Thymidine, 3'-azido-3'-deoxy-	H_2O	266.5(4.06)	44-2077-64
$C_{10}H_{13}N_5O_4S$			
Adenosine, 8-mercapto-	pH 1	222(4.16),240(4.08), 308(4.44)	35-1242-64
	pH 11	230(4.32),297(4.40)	35-1242-64
$C_{10}H_{13}N_5O_5$			
Adenosine, 1-N-oxide	H_2O	233(4.61),260(3.96)	88-3201-64
Adenosine, 8-hydroxy-	pH 1	264(3.95),284(3.93)	35-1772-65
Isoguanosylriboside	0.05N HCl	235(3.79),283(4.10)	88-3201-64
	0.05N NaOH	248(3.91),285(4.03)	88-3201-64
Purine, 6-N-hydroxylamino-9-β-D-ribofuranosyl-	pH 2	262.5(4.22)	87-0884-65
Urea, 4-cyano-1-β-D-ribofuranosylimidazol-5-yl)-	H_2O	229(3.93)	88-3201-64
$C_{10}H_{13}N_5O_5S$			
Guanosine, 8-mercapto-	pH 1	230(4.11),285(4.29), 302(4.29)	35-1242-64
	pH 11	290(4.32)	35-1242-64
$C_{10}H_{13}N_5O_6$			
Purine-6,8(1H,9H)-dione, 2-amino-9-β-D-ribofuranosyl-	pH 1	246(4.17),294(4.05)	35-1772-65
	pH 11	246(4.10),280(4.01)	35-1772-65

Compound	Solvent	λ max(log ε)	Ref.
$C_{10}H_{13}N_5S$			
[1,2,5]Thiadiazolo[3,4-d]pyrimidine, 7-(cyclohexylamino)-	pH 1	222(4.09),263(3.63), 344(4.23)	44-2135-64
	pH 7	226(4.14),272(3.66), 359(4.07)	44-2135-64
	pH 13	272(3.64),360(4.05)	44-2135-64
	EtOH	225(4.14),274(3.68), 283(3.60),363(4.04)	44-2135-64
$C_{10}H_{14}$			
Cycloheptene, 1-(1-propynyl)-	n.s.g.	234(3.93)	22-1525-65
Cyclohexene, 1-(1-butynyl)-	n.s.g.	225(4.1)	44-3991-65
p-Cymene	MeOH	258(2.6),263(2.7), 272(2.6)	70-0466-65
1,7-Octadiene, 3,6-dimethyl-	EtOH	221.8(4.55)	44-2410-65
Spiro[2.4]heptadiene, 1,1,2-trimethyl-	pentane	234(4.15),262(3.59)	77-0622-65
$C_{10}H_{14}BrNO$			
7,8,9,10-Tetrahydro-10-hydroxy-6H-pyrido-[1,2-a]azepinium bromide	H_2O	268(3.79)	44-1523-65
$C_{10}H_{14}BrNO_3$			
2-Pyrroline-3-carboxylic acid, 2-(bromo-methyl)-4,4-dimethyl-5-oxo-, ethyl ester	EtOH	291(4.00)	39-5999-64
$C_{10}H_{14}Br_2O$			
2-Bornanone, 3,6-dibromo-, endo	MeOH	300(1.81),309(1.88), 318(1.81)	78-0273-65
$C_{10}H_{14}Br_4$			
Bicyclopropyl, 2,2,2',2'-tetrabromo-3,3,3',3'-tetramethyl-	hexane	213(3.79)	44-2951-64
	MeOH	211.5(3.79)	44-2951-64
Butane, 1,4-bis(2,2-dibromocyclo-propyl)-	hexane	199.5(3.67)	44-2951-64
Ethane, 1,2-bis(2,2-dibromo-1-methyl-cyclopropyl)-	hexane	203(3.67)	44-2951-64
$C_{10}H_{14}ClNO$			
3-Hydroxy-1,1-dimethylindolinium chloride	EtOH	246s(2.18),252(2.32), 258(2.40),266(2.30)	44-1449-64
$C_{10}H_{14}ClNO_4$			
1-D-Ribofuranosylpyridinium chloride	pH 7.0	260(3.69)	39-0610-65
$C_{10}H_{14}ClN_3O_2$			
Pyrimidine, 2-amino-4-chloro-5-[2-(1,3-dioxolan-2-yl)ethyl]-6-methyl-	pH 1	229(4.24),314(3.78)	4-0079-64
	pH 7	233(4.22),302(3.67)	4-0079-64
	pH 13	302(3.68)	4-0079-64
$C_{10}H_{14}ClN_3O_3$			
Acetic acid, [(6-chloro-3-pyridazinyl)-oxy]-, 2-(dimethylamino)ethyl ester	1% EtOH-HCl	275(3.24)	20-0532-64
1(6H)-Pyridazineacetic acid, 3-chloro-6-oxo-, 2-(dimethylamino)ethyl ester	1% EtOH-HCl	298(3.45)	20-0532-64

Compound	Solvent	λ_{max}(log ϵ)	Ref.
$C_{10}H_{14}Cl_2$			
Cyclobutane, 2-chloro-4-chloromethylene-3-isopropylidene-1,1-dimethyl-	EtOH	206(3.86),262(4.02)	28-0827-64B
Cyclobutane, 1,2-dichloro-3,4-di-isopropylidene-	EtOH	206(3.98),273(4.12)	28-0827-64B
$C_{10}H_{14}Cl_2O_4Sn$			
Tin, dichlorobis(2,4-pentanedionato)-	$CHCl_3$	285(4.20),322(4.48)	101-0067-65B
$C_{10}H_{14}FN_3O_5$			
Cytosine, 1-(2-deoxy-D-glucopyranosyl)-5-fluoro-	pH 1	289(4.00)	87-0140-65
	pH 7	274(3.93)	87-0140-65
	pH 13	278(3.84)	87-0140-65
$C_{10}H_{14}INO_2$			
2-(Carboxymethyl)-1-methylpyridinium iodide, ethyl ester	N H_2SO_4	224(4.08),265(3.76)	12-0455-64
	N NaOH	222(4.23),308(4.08), 375(3.79)	12-0455-64
4-(Carboxymethyl)-1-methylpyridinium iodide, ethyl ester	N H_2SO_4	224(4.27),263(3.52), 267(3.60)	12-0455-64
	N NaOH	226(4.23),256s(3.18), 263s(3.04),367(4.35)	12-0455-64
$C_{10}H_{14}NNaO_3$			
2-Bornanone, 3-nitro-, sodium derivative	H_2O	227(3.66),317(4.11)	44-1687-65
$C_{10}H_{14}N_2$			
Benzeneazobutane	hexane	263(4.00),409(2.09)	39-2788-65
Benzeneazoethane, 1',1'-dimethyl-	hexane	260(4.01),413(2.04)	39-2788-65
Benzimidazoline, 1,2,3-trimethyl-	EtOH	218(4.52),266(3.76), 310(3.76)	12-0877-64
1-Pyrroline, 2-(3,4-dimethylpyrrol-1-yl)	EtOH	258(4.15)	23-1073-64
1-Pyrroline, 2-(3,5-dimethyl-pyrrol-2-yl)-	EtOH-HCl	280s(3.69),329(4.30)	39-0893-64
	EtOH	250s(3.53),293(4.25)	39-0893-64
1-Pyrroline, 5-methyl-2-(5-methyl-pyrrol-2-yl)-	EtOH-HCl	317(4.42)	44-2727-64
	EtOH-NaOH	281(4.31)	44-2727-64
Quinoxaline, 1,2,3,4-tetrahydro-2,6-dimethyl-	EtOH	256(3.69),314(3.58)	5-0146-65D
$C_{10}H_{14}N_2O$			
Anabasine, py-N-oxide	EtOH	268(4.11)	70-2241-64
2,4,6-Cycloheptatrien-1-one, 2-methoxy-, dimethylhydrazone	EtOH	230(4.24),335(3.89)	94-0457-65
Indazole, 3-acetyl-4,5,6,7-tetrahydro-6-methyl-	EtOH	238(4.01),267s(3.56)	32-0814-65
$C_{10}H_{14}N_2OS$			
5H-Thiopyrano[2,3-d]pyrimidine, 7-ethoxy-6,7-dihydro-4-methyl-	pH 1	228(3.92)	4-0079-64
	pH 7-13	249(3.82),282(3.97)	4-0079-64
$C_{10}H_{14}N_2O_2$			
Aniline, N,N-diethyl-p-nitro-	n.s.g.	415(4.38)	5-0109-65I
p-Benzoquinone, 2,5-dimethyl-3,6-bis(methylamino)-	EtOH	224(4.31),348(4.34), 545(2.26)	39-5569-64
4H-m-Dioxino[4,5-c]pyridine, 5-amino-2,2,8-trimethyl-	MeOH-acid	233(4.30),250s(3.53), 323(3.74)	44-2663-64
	MeOH-base	235s(3.92),302(3.73)	44-2663-64
Hydroperoxide, 1-(phenylazo)butyl-	hexane	269(3.97),411(2.15)	39-2788-65

Compound	Solvent	λ_{max}(log ε)	Ref.
Indazole-3-carboxylic acid, 4,5,6,7-tetrahydro-, ethyl ester	EtOH	226(3.95),250s(3.65)	32-0814-65
3-Pyridinebutyric acid, γ-(methylamino)-	EtOH-KOH	261(3.44)	35-3375-64
$C_{10}H_{14}N_2O_2S$			
5-Pyrimidinecarboxylic acid, 1,4-dihydro-1,2,6-trimethyl-4-thioxo-, ethyl ester	EtOH	228(3.75),337(4.32)	44-1115-64
$C_{10}H_{14}N_2O_3$			
Furan, 3,4-diacetyl-2,5-dimethyl-, dioxime	n.s.g.	246s(3.74)	32-0393-64
5-Pyrimidinecarboxylic acid, 1-allyl-1,2,3,4-tetrahydro-2-oxo-, ethyl ester	EtOH	215(3.94),297(4.01)	94-1418-64
5-Pyrimidinecarboxylic acid, 1-allyl-1,2,3,6-tetrahydro-2-oxo-, ethyl ester	EtOH	220(3.93),290(3.95)	94-1418-64
4-Pyrimidinol, 5-[2-(1,3-dioxolan-2-yl)-ethyl]-6-methyl-	pH 1	239(4.01)	4-0079-64
	pH 7	237(3.82),261(3.77)	4-0079-64
	pH 13	230s(4.03),268(3.73)	4-0079-64
$C_{10}H_{14}N_2O_3S$			
4-Pyrimidinol, 5-[2-(1,3-dioxolan-2-yl)-ethyl]-2-mercapto-6-methyl-	pH 1,7	280(4.26)	4-0079-64
	pH 13	263(4.16),309(3.90)	4-0079-64
6H-1,3-Thiazine-5-carboxylic acid, 2-(N-methylacetamido)-, ethyl ester	EtOH	207(3.9),277(3.87), 330(3.87)	44-2290-65
$C_{10}H_{14}N_2O_4$			
Ethanol, 2,2'-[(p-nitrophenyl)-imino]di-	n.s.g.	416(4.45)	5-0109-65I
8H-Oxazolo[3,2-a]pyrrolo[2,1-c]pyrazine-3,6(2H,5H)-dione, tetrahydro-10b-methoxy-	dioxan	220(3.06)	78-3537-65
$C_{10}H_{14}N_2O_5$			
Pyrrolidine-5-carboxylic acid, 3-acetamido-1-methyl-2,4-dioxo-, ethyl ester	H_2O	274(3.84)	44-2085-64
$C_{10}H_{14}N_2O_6$			
Thymine, 1-α-D-arabinofuranosyl-	pH 3.2	268(3.99)	94-0803-65
	pH 7	268(3.99)	94-0803-65
	pH 9.6	268(3.93)	94-0803-65
	pH 14.0	270(4.15)	94-0803-65
Thymine, 1-β-D-arabinofuranosyl-	pH 3.2	269(4.01)	94-0803-65
	pH 7	269(4.01)	94-0803-65
	pH 9.6	269(3.97)	94-0803-65
	pH 14	272(3.95)	94-0803-65
Thymine, 1-α-D-lyxofuranosyl-	pH 3.2	268(4.02)	94-0803-65
	pH 7	268(4.02)	94-0803-65
	pH 9.5	269(4.00)	94-0803-65
	pH 14	270(3.91)	94-0803-65
Thymine, 1-β-D-lyxofuranosyl-	pH 3.2	269(4.00)	94-0803-65
	pH 7	269(4.00)	94-0803-65
	pH 9.5	270(4.00)	94-0803-65
	pH 14	270.5(3.93)	94-0803-65
Thymine, 1-α-D-ribofuranosyl-	pH 4-6	267(4.00)	94-1471-64
Thymine, 1-β-D-ribofuranosyl-	pH 4-6	267(3.98)	94-1471-64
Thymine, N-β-D-ribofuranosyl-	pH 7.2	235(3.46),267(3.99)	94-0454-64
Uracil, 1-(2-deoxy-β-D-arabino-hexopyranosyl)-	MeOH	260(3.98)	44-3955-65

Compound	Solvent	$\lambda_{max}(\log \epsilon)$	Ref.
Uracil, 1-(2-deoxy-α,β-D-gluco-pyranosyl)-, α-form	H_2O	260(4.00)	24-1988-65
	pH 13	260(3.92)	24-1988-65
Uracil, 1-(2-deoxy-β-D-ribo-hexapyranosyl)-	MeOH	261(3.9)	44-3955-65
Uridine, 3-methyl-	pH 4.9	262(3.96)	87-0486-65
	pH 11.6	263(3.95)	87-0486-65
$C_{10}H_{14}N_2O_7$			
Uracil, 1-β-D-glucopyranosyl-	H_2O	258(4.03)	94-0357-64
$C_{10}H_{14}N_2O_8S$			
Uracil, 1-β-D-arabinofuranosyl-, 5'-methanesulfonate	pH 6.9	262(4.01)	44-0467-65
$C_{10}H_{14}N_4$			
4,7-Diazagramine, 1-methyl-	EtOH	220(4.24),294(3.94), 304(3.93)	44-3454-65
Pyrazolo[1,5-a]pyrimidine, 4,7-dihydro-7-imino-2,3,4,6-tetramethyl-	EtOH	234(4.39),268(3.86), 315(3.80)	94-0142-65
Pyrazolo[1,5-a]pyrimidine, 7-(di-methylamino)-2,3-dimethyl-	EtOH	231(4.55),287s(3.61), 297(3.67),338(3.83)	94-1207-65
Pyrazolo[1,5-c]pyrimidine, 7-(ethyl-amino)-2,3-dimethyl-	EtOH	225(4.78),290(3.84), 321(3.84)	94-0142-65
Pyrazolo[1,5-a]pyrimidine, 2,3,5-tri-methyl-7-(methylamino)-	EtOH	229(4.59),291(3.85), 316(3.80)	94-1207-65
Pyrazolo[1,5-a]pyrimidine, 2,3,6-tri-methyl-7-(methylamino)-	EtOH	233(4.52),285(3.65), 294(3.72),335(3.80)	94-0142-65
$C_{10}H_{14}N_4O$			
s-Triazole, 3-(diethylamino)-5-(2-furyl)-	pH 2	271(4.16)	1-1191-65
	pH 2	266(4.17)	1-1191-65
	pH 12	278(3.97)	1-1191-65
	EtOH	280(4.15)	1-1191-65
$C_{10}H_{14}N_4O_2S$			
s-Triazolo[1,5-c]pyrimidine, 7-methyl-2-(methylsulfonyl)-5-propyl-	MeOH	207(4.17),252(3.85)	39-3369-65
$C_{10}H_{14}N_4O_6$			
Glucopyranuronamide, 1-(4-amino-2-oxo-1(2H)-pyrimidinyl)-1-deoxy-	pH 1.2	277(4.08)	94-1259-64
	H_2O	239(3.9),269(3.92)	94-1259-64
	pH 12.6	236(3.90),269(3.91)	94-1259-64
$C_{10}H_{14}N_4S$			
s-Triazolo[1,5-c]pyrimidine, 7-methyl-2-(methylthio)-5-propyl-	MeOH	237(4.47),285s(3.28)	39-3369-65
$C_{10}H_{14}N_5O_8P$			
Guanosine, 5'-phosphate	pH 3.5	252(4.14)	10-0313-65D
dioxane and water adduct	pH 3.5	252(4.14)	10-0313-65D
$C_{10}H_{14}N_6O_3$			
9H-Purine, 2,6-diamino-9-(3-deoxy-D-ribofuranosyl)-	pH 1	215(4.70),253(4.03), 270(4.32),293(3.98)	87-0659-65
	H_2O	256(3.95),280(3.99)	87-0659-65
	pH 13	256(3.94),280(3.98)	87-0659-65

Compound	Solvent	λ max(log ε)	Ref.
$C_{10}H_{14}N_6O_4$			
Adenosine, 8-amino-	pH 1	270(4.13)	35-1772-65
	pH 11	273(4.21)	35-1772-65
Purine, 2,6-diamino-9-β-D-ribofuranosyl-	H_2O	256(4.03),280(4.05)	94-0951-64
$C_{10}H_{14}N_6O_5$			
Guanosine, 8-amino-	pH 1	250(4.23),289(3.98)	35-1772-65
	pH 11	258(4.13),271s(4.06)	35-1772-65
Purine, 2-amino-6-(hydroxyamino)-9-β-D-ribofuranosyl-	H_2O	264s(3.97),282(4.08)	94-0951-64
$C_{10}H_{14}N_6O_7$			
Inosine, 2-nitro-, ammonium salt	pH 1	222(4.08),335(3.62)	35-2948-64
	pH 7	233(4.16),343(3.58)	35-2948-64
$C_{10}H_{14}O$			
Anisole, 2-isopropyl-	C_6H_{12}	211(3.87),217(3.89), 221(3.89),265s(3.18), 272(3.33),278(3.31)	23-2603-65
Anisole, 3-isopropyl-	hexane	284(3.38)	65-2088-65
	EtOH-HCl	260(3.30)	65-2088-65
	EtOH	260(3.30)	65-2088-65
	EtOH-NaOEt	275(3.30)	65-2088-65
	10% H_2SO_4	280(3.55)	65-2088-65
Anisole, 4-isopropyl-	hexane	258(3.60)	65-2088-65
	EtOH-HCl	275(3.18)	65-2088-65
	EtOH	258(3.18)	65-2088-65
	EtOH-NaOEt	258(3.18)	65-2088-65
	10% H_2SO_4	277(3.55)	65-2088-65
Bicyclo[3.2.0]hept-3-en-6-one, 4,7,7-trimethyl-	isooctane	312(2.31),323(2.23)	44-4230-65
	MeOH	310(2.42)	44-4230-65
1-Butyn-4-ol, 1-cyclohexenyl-	n.s.g.	226(4.13)	44-3991-65
Carvacrol	pet ether	228(4.03),275(3.78)	65-2094-65
	EtOH	276(3.78)	65-2094-65
	EtOH-NaOEt	245(4.00),289(3.95)	65-2094-65
1-Cyclopentene-1-carboxaldehyde, 5-isopropenyl-2-methyl-	EtOH	250(4.07)	44-3740-64
Eucarvone	EtOH	303(3.81)	44-0669-65
3,5-Heptadien-2-one, 6-cyclopropyl-	EtOH	306(4.37)	22-3218-64
5(6H)-Indanone, 7,7a-dihydro-4-methyl-	EtOH	245(4.17)	35-0275-65
2(1H)-Naphthalenone, 4a,5,6,7,8,8a-hexahydro-	n.s.g.	229(4.05)	22-0111-65
5,6-Nonadien-3-yn-1-ol, 8-methyl-	EtOH	220(4.20)	39-4659-65
2-Norbornanone, 3,3-dimethyl-6-methylene-	MeOH	301(2.51)	44-4230-65
2,7-Octadien-4-yn-1-ol, 3,7-dimethyl-	MeOH	230(4.1)	5-0026-64I
5-Octen-7-yn-2-one, 5,6-dimethyl-	EtOH	230(4.10)	70-1820-64
Phenol, p-sec-butyl-	N NaOH	239(4.3),290(3.7)	65-2790-64
	$CHCl_3$	247(2.7),277(3.3)	65-2790-64
Phenol, 2-tert-butyl-	H_2O	270(3.23)	39-4603-65
anion	H_2O	240(3.92),293(3.53)	39-4603-65
	MeOH	274(3.35)	39-0676-65
	MeOH-NaOMe	244(3.94),291(3.56)	39-0676-65
Phenol, 4-tert-butyl-	MeOH	277(3.27)	39-0676-65
	MeOH-NaOMe	239(4.08),294(3.40)	39-0676-65
2-Propanol, 2-p-tolyl-	MeOH	220(3.86),233(3.24), 265(2.58),273(2.56)	70-0466-65

Compound	Solvent	$\lambda_{max}(\log \epsilon)$	Ref.
Thymol	pet ether	223(3.78),252(3.55), 266(3.55)	65-2094-65
	MeOH	220(3.77),278(3.37)	70-0466-65
	EtOH-NaOEt	240(3.95),284(3.78)	65-2094-65
Umbellulone	EtOH	219(3.82),265(3.49)	24-3045-65
$C_{10}H_{14}OS$			
Anisylthioethane	heptane	279(3.20)	46-1842-64
	H_2O	276(3.06)	46-1842-64
3,5-Heptadiyn-2-ol, 7-(ethylthio)-2-methyl-	n.s.g.	247?(3.20)	28-2847-65A
Sulfoxide, isopropyl p-tolyl	isooctane	211(3.99),218s(3.97), 255(3.68),265s(3.54), 270s(3.42)	35-1958-65
	isooctane-CF_3COOH	219s(3.94),242(3.79), 267s(3.11),276s(2.70)	35-1958-65
	EtOH	218s(3.94),244(3.77), 268s(3.17),275s(2.81)	35-1958-65
	EtOH-CF_3COOH	218s(3.94),246(3.77), 268s(3.18),277s(2.76)	35-1958-65
$C_{10}H_{14}O_2$			
d-Betuligenol	EtOH	224(3.90),279(3.28)	44-0988-64
Bicyclo[3.2.1]oct-2-ene-2-carboxylic acid, methyl ester	EtOH	228(3.89)	35-4928-64
Butyrolactone, α-cyclohexylidene-	EtOH	234(3.87)	65-3135-64
(+)-Cinerolone	EtOH	225(4.11)	39-5225-64
Cyclobutenedione, 1,2-dipropyl-	EtOH	218(4.25),355s(1.45)	35-1326-65
1,5-Cyclohexadiene-1-carboxylic acid, 3,3-dimethyl-, methyl ester	heptane	278(3.38)	44-3679-65
1,5-Cyclohexadiene-1-carboxylic acid, 3-ethyl-, methyl ester	EtOH	284(3.39)	44-3679-65
1,3-Cyclopentadiene-1-carboxylic acid, 2,3,4-trimethyl-, methyl ester	EtOH	297(4.00)	70-1653-64
Fulvene, 6,6-diethoxy-	hexane	293(4.26)	5-0039-64H
4-Hepten-6-ynoic acid, 2,4,5-trimethyl-	EtOH	230.5(4.12)	70-1653-64
Hydroquinone, ethyl-, dimethyl ether	95% THF	230(3.75),292(3.51)	44-2602-65
5-Indancarboxylic acid, tetrahydro-, cis	n.s.g.	219(3.26)	22-0301-64
trans	n.s.g.	218(3.99)	22-0301-64
Neonepetalactone	n.s.g.	241(4.0)	88-4097-65
3-Nonen-1-yn-5-one, 7-hydroxy-6-methyl-	n.s.g.	239(4.09)	70-2215-65
5-Norbornen-2-one, 7-methoxy-5,6-dimethyl-	C_6H_{12}	294(2.64),304(2.62), 316(2.38)	35-4211-64
Phenol, 4-(methoxymethyl)-2,6-dimethyl-	n.s.g.	275(3.30),280(3.32)	70-1675-65
Phenol, 2-methoxy-3,4,5-trimethyl-	EtOH-KOH	295(3.60)	35-0694-64
	EtOH	284(3.37)	35-0694-64
Propiophenone, 4'-methoxy-2',5'-dihydro-	MeOH	224.5(3.94)	5-0036-64D
2-Pyrone, 4,6-diethyl-5-methyl-	EtOH	304(3.82)	24-1949-65
4-Pyrone, 6-ethyl-2,3,5-trimethyl-	EtOH	259(4.07)	24-1949-65
Spiro[3.5]non-1-ene-2-carboxylic acid	EtOH	222(4.00)	28-1541-64A
$C_{10}H_{14}O_2S$			
2-Furaldehyde, 5-(tert-butylthio)methyl-	EtOH	285(4.25)	70-1281-65
$C_{10}H_{14}O_3$			
Benzofuran-2-one, 2,4,5,6,7,7a-hexahydro-7a-hydroxy-3,6-dimethyl-	MeOH	217(4.06)	39-4254-64
	MeOH-NaOH	264(3.61)	39-4254-64

Compound	Solvent	$\lambda_{max}(\log \epsilon)$	Ref.
Bicyclo[3.2.1]octan-8-one, 7-acetyl-1-hydroxy-	EtOH	293(1.73)	44-2513-65
Cyclohexaneacetoacetic acid, 1-hydroxy-, δ-lactone	MeOH-HCl	240(4.06)	5-0123-65A
	MeOH-NaOMe	269(4.36)	5-0123-65A
	H_2SO_4	265(4.08)	5-0123-65A
1-Cyclohexene-1-acetic acid, 2,5-dimethyl-6-oxo-	EtOH	242(4.15)	42-0242-64
3-Cyclohexene-1-carboxylic acid, 3-acetyl-4-methyl-	EtOH	248(3.85)	44-0787-64
3-Furoic acid, 4,5-diethyl-2-methyl-	MeOH	259(3.54)	39-5984-65
3-Furoic acid, 2-isopropyl-4,5-dimethyl-	MeOH	259(3.63)	39-5984-65
2(3H)-Naphthalenone, 4,4a,5,6,7,8-hexahydro-4a,5-dihydroxy-	EtOH	232.5(4.10)	39-3577-64
4-Octen-2-ynoic acid, 5-ethoxy-	EtOH	279(4.10)	28-0594-64B
Sarcomycin, methylethyl-	MeOH	254.5(4.05)	7-0206-64
$C_{10}H_{14}O_3S$			
Acrylic acid, 3-[(2-oxocyclohexyl)-thio]-, methyl ester	Et_2O	278(4.13)	24-2109-64
$C_{10}H_{14}O_4$			
1-Cyclobutene-1,2-dicarboxylic acid, 3,3-dimethyl-, dimethyl ester	MeOH	232(4.0)	24-3672-65
Cyclohexanecarboxylic acid, 3-(hydroxymethylene)-1-methyl-2-oxo-, methyl ester	EtOH	289(3.90)	44-0782-64
Genipic acid	EtOH	203(3.37)	78-1781-64
2,4-Hexadienedioic acid, tetramethyl-, trans-trans (or tetramethylmuconic acid)	MeOH	220(3.19)	24-1811-64
2-Hexenedioic acid, 4-hydroxy-2,3,4,5-tetramethyl-, γ-lactone	MeOH	215(3.71)	24-1811-64
1,7-Norcaranedicarboxylic acid, 6-methyl-, cis	EtOH	223(3.19),270s(2.65)	35-4053-64
Senecic acid lactone	EtOH	225(3.63)	39-2492-65
Succinic acid, isopropylidenemethylene-, dimethyl ester	C_6H_{12}	206(4.5)	24-3672-65
$C_{10}H_{14}O_4S$			
3-Penten-2-one, 3,3'-thiobis[4-hydroxy-	$CHCl_3$	288(4.13)	39-0688-64
$C_{10}H_{14}O_4S_2$			
3-Penten-2-one, 3,3'-dithiobis-[4-hydroxy-	$CHCl_3$	291(4.93),336(4.74)	39-0688-64
$C_{10}H_{14}O_4Se$			
2,4-Pentanedione, 3,3'-selenodi-	$CHCl_3$	281(3.92)	39-0688-64
$C_{10}H_{14}O_4Se_2$			
3-Penten-2-one, 3,3'-diselenobis-[4-hydroxy-	$CHCl_3$	287(4.45)	39-0688-64
$C_{10}H_{14}O_5$			
Hygrophyllinecic acid, lactone	EtOH	220(3.68)	39-5707-65

Compound	Solvent	$\lambda_{max}(\log \epsilon)$	Ref.
$C_{10}H_{14}O_5S$			
Acrylic acid, 3-[(2-oxocyclohexyl)-sulfonyl]-, methyl ester	Et_2O	211(3.99)	24-2109-64
$C_{10}H_{14}O_5S_3$			
β-L-Idofuranose, 1,2-O-isopropylidene-5,6-dithio-, cyclic 5,6-trithiocarbonate, thiono-oxide	MeOH	284(3.69),348(4.03)	44-3071-65
$C_{10}H_{14}O_6$			
Jacozinecic acid	n.s.g.	211(3.06)	12-0233-64
$C_{10}H_{14}O_6S$			
α-D-Glucofuranose, 1,2-O-isopropylidene-, cyclic 5,6-thiocarbonate	EtOH	234(4.20),287(1.64)	44-0162-65
	EtOH	238(4.09),274(3.48)	88-2531-64
	n.s.g.	234(4.18),288(1.64)	50-0186-64D
$C_{10}H_{15}Cl$			
Cyclobutane, 2-(chloromethylene)-3-isopropylidene-1,1-dimethyl-	EtOH	257(4.11)	28-0827-64B
$C_{10}H_{15}ClN_2O_6$			
Trimethyl(m-nitrobenzyl)ammonium perchlorate	n.s.g.	263(3.85)	39-6851-65
Trimethyl(p-nitrobenzyl)ammonium perchlorate	n.s.g.	263(4.01)	39-6851-65
$C_{10}H_{15}Cl_2N_3O$			
3(2H)-Pyridazinone, 4,5-dichloro-2-(diethylaminoethyl)-, hydrochloride	EtOH	220(3.98),301(3.63)	22-2124-64
	NaOH	219(4.24),299(3.71)	22-2124-64
$C_{10}H_{15}FO$			
2,6-Octadienal, 2-fluoro-3,7-dimethyl-	MeOH	248(4.14)	5-0020-64I
$C_{10}H_{15}IN_2$			
1,6-Dimethyl-2-phenylimidazo[1,2-a]-pyridinium iodide	EtOH	205(4.58),210s(4.56), 221(4.55),288(4.14)	4-0331-65
$C_{10}H_{15}N$			
Aniline, diethyl-, tetracyanoethylene complex	$CHCl_3$	835(3.48)	88-0189-64
Aniline, o-ethyl-N,N-dimethyl-	EtOH	208(4.18),247(3.51)	56-1437-65
2,4-Lutidine, 6-propyl-	hexane	215(3.57),264(3.51), 267(3.48),271(3.45)	1-1607-65
Pyridine, 2-pentyl-	EtOH	256(3.59),262(3.61), 268(3.47)	94-1344-64
Quinoline, 2,3,4,6,7,8-hexahydro-7-methyl-	MeOH	235(3.98)	5-0084-65A
Unknown alkylated pyridine	n.s.g.	265(3.1)	70-0322-65
$C_{10}H_{15}NO$			
2H-Azepin-2-one, 1,3-dihydro-1,3,5,7-tetramethyl-	EtOH	252(3.66)	35-0498-64
Cyclopentanecarbonitrile, 3,3-diethyl-2-oxo-	EtOH	235(3.42)	23-2512-65
	EtOH-base	264(4.05)	23-2512-65
Eucarvone, oxime	EtOH	291(4.03)	44-0669-65
Fulvene, 6-(dimethylamino)-6-ethoxy-	hexane	328(4.41)	5-0039-64H
Phenethylamine, 4-hydroxy-N,N-dimethyl-	pH 1	222(3.84),275(3.20), 280s(3.15)	87-0368-65
	pH 13	239(4.02),294(3.42)	87-0368-65

Compound	Solvent	$\lambda_{max}(\log \epsilon)$	Ref.
Pyrrole, 2-acetyl-3,4-diethyl-	EtOH	299(4.21)	39-1518-65
4(1H)-Quinolone, 2,3,5,6,7,8-hexahydro-1-methyl-	EtOH	199(3.54),242(3.68), 332(1.83)	33-0791-65
	EtOH	330(3.87)	94-1405-65
$C_{10}H_{15}NO_2$			
Ethanolamine, N,N-dimethyl-β-(4-hydroxyphenyl)-	pH 1	224(3.88),273(3.13), 279s(3.06)	87-0368-65
	pH 13	242(4.09),292(3.38)	87-0368-65
Ethanol, N-methyl-β-(4-hydroxy-3-methoxyphenyl)-	pH 1	208(3.84),229(3.80), 279(3.45),283s(3.40)	87-0368-65
	pH 13	218(3.84),247(4.03), 294(3.62)	87-0368-65
2-Heptenoic acid, 2-cyano-, ethyl ester	EtOH	224(4.02)	28-2859-64B
2-Hexenoic acid, 2-cyano-3-methyl-, ethyl ester	EtOH	233(4.08)	28-2859-64B
2-Hexenoic acid, 2-cyano-5-methyl-, ethyl ester	EtOH	224(4.01)	28-2859-64B
2-Pentenoic acid, 2-cyano-3,4-dimethyl-, ethyl ester	EtOH	231.5(4.08)	28-2859-64B
2-Pentenoic acid, 2-cyano-3-ethyl-, ethyl ester	EtOH	231(4.08)	28-2859-64B
Pyrido[1,2-a]azepine-4,6-dione, octahydro-	THF	228(3.42),275(1.90)	24-1548-64
$C_{10}H_{15}NO_3$			
Crotonic acid, 2-cyano-3-methoxy-, tert-butyl ester	dioxan	256(4.26)	33-1424-64
1-Piperidinecrotonic acid, α-hydroxy-β-(hydroxymethyl)-, γ-lactone	H_2O	235(3.73),267(3.72)	39-0766-64
	HCl	232.5(4.13)	39-0766-64
	NaOH	267(3.97)	39-0766-64
sulfate	H_2O	234(4.19),263(3.94)	39-0766-64
3-Pyridinepropanol, 5-hydroxy-4-(hydroxymethyl)-6-methyl-	pH 1	292(3.96)	87-0112-65
	pH 13	244(3.79),307(3.85)	87-0112-65
	EtOH	293(3.81)	87-0112-65
2-Pyrroline-3-carboxylic acid, 2,4,4-trimethyl-5-oxo-, ethyl ester	EtOH	216(3.59),279(4.10)	39-5999-64
3-Pyrrolin-2-one, 3-acetyl-5-sec-butyl-4-hydroxy- (or tenuazoic acid)	pH 13	239(3.98),279(4.08)	25-0419-64
	EtOH	217(3.71),277(4.11)	25-0419-64
$C_{10}H_{15}NO_3S$			
4H-Thiopyran-3-carboxylic acid, 2-acetamido-5,6-dihydro-, ethyl ester	MeOH	240(3.86),306(4.24)	24-1970-64
Valeric acid, 2-cyano-5-mercapto-, ethyl ester, acetate	MeOH	231(3.68)	24-1970-64
$C_{10}H_{15}NO_5$			
1,2-Pyrrolidinedicarboxylic acid, 3-oxo-, diethyl ester	MeOH	280(1.72)	35-5293-64
1,3-Pyrrolidinedicarboxylic acid, 4-oxo-, diethyl ester	MeOH	246(3.70)	35-5293-64
	MeOH-KOH	275(4.15)	35-5293-64
$C_{10}H_{15}NO_5S$			
Tropone imine, 2-methoxy-N-methyl-, methyl sulfate	EtOH	247(4.46),345(4.08)	94-0457-65
$C_{10}H_{15}N_3$			
Pyrazine, 2-methyl-3-piperidino-, hydrochloride	H_2O	239(3.74),257(3.79), 317(3.62)	44-0415-64

Compound	Solvent	λ max(log ε)	Ref.
Pyrazine, 2-methyl-6-piperidino-, hydrochloride	H_2O	227(3.73),256(4.18), 350(3.76)	44-0415-64
$C_{10}H_{15}N_3O$			
2-Cyclohexen-1-one, 4-propylidene-, semicarbazone	MeOH	293(4.50)	5-0028-64D
3,5-Hexadien-2-one, 6-cyclopropyl-, semicarbazone	EtOH	300(4.65)	22-3218-64
Pyrazine, 3-methyl-2-piperidino-, 1-oxide	N HCl	221(3.81),295(3.53), 325(3.53)	44-1645-64
$C_{10}H_{15}N_3OS_2$			
3(2H)-Pyridazinethione, 6-(methylthio)-2-(morpholinomethyl)-	THF	309(4.32)	49-0631-65
$C_{10}H_{15}N_3O_2$			
Pyrimidine, 2-amino-5-[2-(1,3-dioxolan-2-yl)ethyl]-6-methyl-	pH 1	227(4.22),310(3.60)	4-0079-64
	pH 7	230(4.17),300(3.53)	4-0079-64
	pH 13	300(3.52)	4-0079-64
Pyrimidine, 4-amino-5-[2-(1,3-dioxolan-2-yl)ethyl]-6-methyl-	pH 1	260(4.11)	4-0079-64
	pH 7	240(3.95),268(3.85)	4-0079-64
	pH 14	235(4.03),273(3.77)	4-0079-64
Sydnone imine, N-acetyl-3-cyclohexyl-	pH 1.0	280(4.10)	65-2064-64
	pH 7.00	236(3.80),310(4.10)	65-2064-64
$C_{10}H_{15}N_3O_3$			
3-Cyclopentene-1-carboxylic acid, 3,4-dimethyl-2-oxo-, methyl ester, semicarbazone	EtOH	267.5(4.37)	70-0164-64
$C_{10}H_{15}N_3O_4$			
6-Azauracilvaleric acid, ethyl ester	MeOH	210(3.85),252(3.77)	39-0868-64
Thymidine, 3'-amino-3'-deoxy-	pH 1	265(3.97)	44-1772-64
	pH 7.53	266.5(3.97)	44-1772-64
	pH 13	266.5(3.87)	44-1772-64
hydrochloride	H_2O	265.5(4.00)	44-2076-64
$C_{10}H_{15}N_3O_5$			
Cytosine, 1-(2-deoxy-β-D-arabino-hexopyranosyl)-	MeOH	271(3.98)	44-3955-65
Cytosine, 1-(2-deoxy-β-D-ribo-hexopyranosyl)-	MeOH	271(4.01)	44-3955-65
1(2H)-Pyrimidinenorleucine, tetrahydro-2,4,6-trioxo-	pH 1	259(3.638)	65-2179-64
	H_2O	260(3.966)	65-2179-64
	pH 12	263(4.017)	65-2179-64
$C_{10}H_{15}N_3S$			
Semicarbazide, 2-methyl-4-(α-methylbenzyl)-3-thio-, nickel complex	n.s.g.	428(2.11)	1-1239-65
$C_{10}H_{15}N_5$			
Adenine, N,N-diethyl-9-methyl-	pH 1	271(4.23)	44-3597-65
	pH 13	278(4.25)	44-3597-65
	MeOH	275(4.21)	44-3597-65
Biguanide, phenethyl-	MeOH-HCl	236(4.23)	44-0308-64
	MeOH-NaOMe	235s(--)	44-0308-64
	H_2O	233(3.15)	44-0308-64

Compound	Solvent	λ max(log ε)	Ref.
$C_{10}H_{15}N_5O$			
9H-Purine-9-ethanol, 6-(dimethyl-amino)-α-methyl-	pH 1	268(4.31)	87-0502-65
	pH 7	275(4.33)	87-0502-65
	pH 13	275(4.31)	87-0502-65
$C_{10}H_{15}N_5O_2$			
1,2-Propanediol, 3-[6-(dimethylamino)-9H-purin-9-yl]-	pH 1	269(4.19)	87-0502-65
	pH 7	275(4.20)	87-0502-65
	pH 13	276(4.20)	87-0502-65
$C_{10}H_{15}N_5O_3$			
Caffeine, 8-[(2-hydroxyethyl)amino]-, hydrobromide	pH 1	292(4.16)	65-2774-64
$C_{10}H_{15}N_5O_4$			
2-Furaldehyde, 5-nitro-, 2-[2-(di-methylamino)ethyl]semicarbazone	H_2O	375(4.21)	44-2582-64
$C_{10}H_{15}N_7O_4$			
Adenoxine, 8-hydrazino-	pH 1	264(4.11)	35-1772-65
	pH 11	262(4.13)	35-1772-65
9H-Purine, 2-amino-6-hydrazino-9-β-D-ribofuranosyl-	H_2O	259(3.99),283(4.11)	94-0951-64
$C_{10}H_{15}O_8P$			
Maleic acid, hydroxy(2-hydroxyethyl)-, γ-lactone, ethyl ester, dimethyl phosphate	MeOH	238(3.89)	5-0010-65E
Oxalacetic acid, (2-hydroxyethyl)phos-phono-, γ-lactone, dimethyl ester	MeOH	215(3.64)	5-0010-65E
$C_{10}H_{16}$			
Alloocimene	C_6H_{12}	265s(4.51),277(4.62), 287s(4.51)	23-2781-64
Bicyclo[7.1.0]dec-1-ene	hexane	205(3.65)	28-1695-65B
	EtOH	213(3.08)	28-1695-65B
Bicyclo[3.1.0]hex-2-ene, isopropyl-2-methyl-	C_6H_{12}	205(3.71)	35-3532-65
Bicyclopropylidene, 2,2,2',2'-tetra-methyl-, trans	hexane	207(3.49)	28-1695-65B
	EtOH	215(3.21)	28-1695-65B
1-Butene, 1-cyclohexenyl-, cis	n.s.g.	233(4.36)	44-3991-65
Cyclobutene, 1-isohex-3-enyl-	isooctane	191(4.20),200(4.02), 210(3.30)(end ab-sorptions)	78-1001-65
Cyclohexane, (2-cyclobuten-1-yl)-	isooctane	195(3.48),200(3.23), 205(3.00),210(2.85) (end absorptions)	78-1001-65
Cyclopentene, 2,3,3-trimethyl-1-vinyl-	EtOH	243(4.34)	44-2264-65
1-Cyclopropene, 1-(2,2-dimethyl-1-cyclo-propyl)-3,3-dimethyl-	hexane	214(3.74)	28-1695-65B
	EtOH	220(3.69)	28-1695-65B
2,5-Heptadiene, 2,6-dimethyl-4-methylene-	C_6H_{12}	218(4.19),241(4.06)	23-2781-64
1,3,5-Heptatriene, 2,4,6-trimethyl-, cis	C_6H_{12}	209(3.94),240(4.05)	23-2781-64
	isooctane	210s(--),240(4.05)	23-2744-65
trans	C_6H_{12}	258.5(4.00)	23-2781-64
	isooctane	258.5(4.00)	23-2744-65
1,4-Hexadiene, 2,5-dimethyl-3-vinyl-	EtOH	205.0(3.98)	88-3775-64
α-Myrcene	EtOH	225.5(4.25)	5-0083-64E
β-Myrcene	MeOH	225(4.20)	5-0083-64E
	EtOH	226(4.20)	5-0083-64E

Compound	Solvent	λ_{max}(log ε)	Ref.
α-Ocimene, cis	EtOH	234.5(4.33)	5-0083-64E
trans	EtOH	231(4.44)	5-0083-64E
β-Ocimene	MeOH	233(4.42)	5-0083-64E
cis	EtOH	237.5(4.32)	5-0083-64E
trans	EtOH	232(4.44)	5-0083-64E
1,3,7-Octatriene, 2,6-dimethyl-	EtOH	230(4.48)	24-2700-64
1,4,6-Octatriene, 2,6-dimethyl-	EtOH	236(4.36)	5-0083-64E
2,3,5-Octatriene, 2,6-dimethyl-	n.s.g.	225(4.39)	77-0017-64
2,4,6-Octatriene, 2,3-dimethyl-, trans	C_6H_{12}	276(4.62),287(4.51)	39-2465-64
β-Phellandrene	n.s.g.	232(5.26)	54-0317-65
$C_{10}H_{16}BrN_7O_2S$			
Theobromin-8-ylaminoethylisothiuronium hydrobromide	pH 1	297(4.27)	65-2774-64
$C_{10}H_{16}ClNO_4$			
Benzyltrimethylammonium perchlorate	n.s.g.	262(2.51)	39-6851-65
$C_{10}H_{16}Ge$			
Germane, benzyltrimethyl-	C_6H_{12}	261(2.59),268(2.65), 276(2.53)	39-4690-65
$C_{10}H_{16}NOP$			
Phosphine oxide, [p-(dimethylamino)-phenyl]dimethyl-	MeOH	274(4.39)	88-2729-64
$C_{10}H_{16}N_2$			
1-Cyclopentene-1-carbonitrile, 2-amino-3,3-diethyl-	EtOH	263(4.14)	23-2512-65
Fulvene, 6,6-bis(dimethylamino)-	hexane	251(3.87),342(4.44)	5-0039-64H
$C_{10}H_{16}N_2O$			
4(1H)-Pyridazinone, 3,5-dipropyl-	MeOH	275(4.16)	25-0839-64
	MeOH-base	257(4.07),275s(3.92)	
Pyrrole, 2,5-dimethyl-1-morpholino-	EtOH	210(2.7),220(2.7), 275s(2.0)	6-0073-64
$C_{10}H_{16}N_2OS$			
4,6(1H,5H)-Pyrimidinedione, 5,5-diethyl-1,2-dimethyl-4-thio-	C_6H_{12}	244(3.95),310(4.08)	39-3204-64
	pH -2.8	226(3.60),312(4.06)	39-3204-64
	pH 5.0	277(4.08)	39-3204-64
	EtOH	247(3.85),310(4.05)	39-3204-64
$C_{10}H_{16}N_2O_2$			
Acetaldehyde oxime, 2,2-dimethyl-3-carbamoylcyclopentylidene-	EtOH	226(3.81)	39-5819-64
4,6(1H,5H)-Pyrimidinedione, 5,5-diethyl-1,2-dimethyl-	pH 7.0	245(4.1+)	12-0567-64
	EtOH	244(4.18)	12-0567-64
4-Pyrimidinol, 5-(ethoxymethyl)-2-propyl-	EtOH	224(4.15),275(4.18)	94-0393-64
$C_{10}H_{16}N_2O_2SSi$			
2,4-Diaza-8-silaspiro[5.5]undecane-1,5-dione-3-thione, 8,8-dimethyl-	70% EtOH	237.0(3.89)	87-0695-64
$C_{10}H_{16}N_2O_2S_4$			
Disulfide, bis(morpholinothiocarbonyl)	hexane	221(4.31),250(4.18)	78-2857-65
$C_{10}H_{16}N_2O_2S_{15}$			
Tridecasulfide, bis(morpholinothiocarbonyl)	hexane	263(4.18),280(4.16)	78-2857-65

Compound	Solvent	$\lambda_{max}(\log \epsilon)$	Ref.
$C_{10}H_{16}N_2O_3Si$			
2,4-Diaza-8-silaspiro[5.5]undecane-1,3,5-trione, 8,8-dimethyl-	70% EtOH-NaOH	240.0(3.96)	87-0695-64
$C_{10}H_{16}N_2O_4$			
3-Pyrroline-1,3-dicarboxylic acid, 4-amino-, diethyl ester	MeOH	273(4.18)	35-5293-64
	MeOH-KOH	273(4.18)	35-5293-64
$C_{10}H_{16}N_2O_6S_2$			
Crotonic acid, 3,3'-(dithiodimethylene)-bis[2,4-dihydroxy-, di-γ-lactone, diammonium salt	EtOH	246(4.16)	44-3560-64
$C_{10}H_{16}N_2S_8$			
Hexasulfide, bis(1-pyrrolidinyl-thiocarbonyl)-	hexane	250(4.34),280(4.19)	78-2857-65
$C_{10}H_{16}N_4$			
Pyrimidine, 2,6-diamino-4,5-hexa-methylene-	pH 1	221(4.23),278(3.89)	44-1837-65
	pH 10	233(3.98),287(3.89)	44-1837-65
Quinazoline, 4-amino-2-(dimethyl-amino)-5,6,7,8-tetrahydro-	pH 1	231(4.45),283(3.72)	44-1837-65
	pH 10	236(4.27),298(3.74)	44-1837-65
$C_{10}H_{16}N_4O_3$			
Cytidine, 3'-amino-2',3'-dideoxy-5-methyl-	pH 1	285(4.08)	44-1772-64
	pH 7.53	240s(3.83),276(3.92)	44-1772-64
	pH 13	240s(3.80),278(3.91)	44-1772-64
$C_{10}H_{16}N_4S$			
7H-Thiopyrano[2,3-d]pyrimidine, 2-amino-5,6-dihydro-7-(dimethylamino)-4-methyl-	pH 1	217(4.21),270(3.96), 317(3.96)	4-0088-64
	pH 7	232(4.14),308(3.94)	4-0088-64
	pH 13	261s(3.87),317(4.16)	4-0088-64
$C_{10}H_{16}N_5O_{13}P_3$			
Adenosine 5'-phosphate	pH 2	257(4.16)	96-0803-64
$C_{10}H_{16}O$			
Bicyclo[3.2.0]heptan-6-one, 4,7,7-tri-methyl-	MeOH	309(1.72)	44-4230-65
2-Buten-1-one, 1-(2,2-dimethyl-cyclopropyl)-3-methyl-	isooctane	235(4.37),330(1.87)	88-3151-65
2-Caranone, cis	n.s.g.	210(3.63)	56-0007-65
2-Caranone, trans	n.s.g.	214(3.40)	56-0007-65
Cyclohexanone, 2-isopropylidene-4-methyl-	EtOH	250(3.85)	44-3207-65
2-Cyclohexen-1-one, 5-ethyl-3,5-dimethyl-	EtOH	235(4.00)	56-0591-64
2-Cyclohexen-1-one, 2,3,4,4-tetramethyl-	EtOH	239(3.51)	56-0385-64
2-Cyclohexen-1-one, 2,3,5,5-tetramethyl-	n.s.g.	251(3.84)	39-1511-64
2-Cyclohexen-1-one, 3,4,4,5-tetramethyl-	EtOH	235(3.94)	56-0599-64
Cyclononanone, 3-methylene-	n.s.g.	285(1.65),294(1.66), 303(1.62),313(1.46)	88-3189-64
Cyclopentanol, 2,3,3-trimethyl-4-vinyl-	EtOH	243(4.33)	44-2264-65
3,5-Heptadien-2-ol, 6-cyclopropyl-	EtOH	245.5(4.38)	22-3218-64
Ketone, ethyl 2-ethyl-1-cyclopentenyl	EtOH	252(3.90),307(1.88)	22-0722-65
p-Mentha-1,5-dien-8-ol	MeOH	262.5(3.55)	70-0466-65
2(1H)-Naphthalenone, octahydro-, trans	EtOH	239(2.23)	35-0275-65

Compound	Solvent	λ_{max}(log ε)	Ref.
2,7-Octadienal, 2,6-dimethyl-	EtOH	230(4.24)	24-2700-64
1,7-Octadien-3-one, 2,6-dimethyl-	EtOH	219(3.93)	24-2700-64
2,7-Octadien-4-one, 2,6-dimethyl-	EtOH	238(4.00)	24-2700-64
3,5-Octadien-2-one, 6,7-dimethyl-	EtOH	291.5(4.31)	22-3218-64
Sabinol, cis	EtOH	204(4.00)	24-3045-65
Spiro[2.7]decan-4-one	n.s.g.	285(1.30),294(1.23)	88-3189-64
α-Thujone	heptane	291(1.37),300(1.40), 311(1.28)	94-1439-64
β-Thujone	heptane	274(1.31)	94-1439-64
β-Thujen-4-ol, trans	EtOH	204(3.51)	24-3045-65
Umbellulol	EtOH	209(3.63)	24-3045-65
Umbellulone, dihydro-	EtOH	204(3.57),280(1.94)	24-3045-65
Unknown alcohol from 3-carene oxidation	MeOH	230(4.14)	70-0466-65
$C_{10}H_{16}OS$			
2-Cyclohexen-1-one, 3-(ethylthio)-5,5-dimethyl-	EtOH	292(4.27)	87-0705-64
4-Cyclooctene-1-thiol, acetate	C_6H_{12}	233(3.57)	77-0151-65
$C_{10}H_{16}O_2$			
1,3-Cyclohexanedione, 2,2,5,5-tetramethyl-	heptane	292(1.66)	88-0037-65
1,3-Cyclopentanedione, 2-isopentyl-	EtOH	250(4.17),271(4.34)	39-4472-64
	EtOH-acid	250(4.28)	39-4472-64
	EtOH-base	270(4.52)	39-4472-64
1,3-Cyclopentanedione, 4-isopentyl-	EtOH-acid	241.5(4.28)	39-1276-65
	EtOH-base	260(4.49)	39-1276-65
Isolauranolic acid, methyl ester	EtOH	229(3.99)	78-0115-65
p-Menth-1-en-6-one, 3-hydroxy-, trans	EtOH	232(3.90)	78-1889-65
p-Menth-3-en-2-one, 1-hydroxy-	EtOH	236(4.12),318(1.90)	44-0518-65
$C_{10}H_{16}O_2S$			
[(1-Cyclohexen-1-ylcarbonyl)methyl]-dimethylsulfonium hydroxide, inner salt, oxide	EtOH	228(4.05),279(4.21)	35-1640-64
Decahydro-2-methyl-4-oxo-2-benzothiopyrylium hydroxide, inner salt, 2-oxide	EtOH	245(4.16)	35-1640-64
$C_{10}H_{16}O_3$			
1,3-Cyclopentanedione, 4-hydroxy-2-isopentyl-	EtOH-acid	250(4.10)	39-1276-65
	EtOH-base	275(4.32)	39-1276-65
2,4-Nonadienoic acid, 9-hydroxy-, methyl ester, di-trans	EtOH	261(4.40)	70-2003-64
$C_{10}H_{16}O_3S$			
2-Hexanone, 6-acetylmercapto-3-acetyl-	n.s.g.	232(3.68),290(3.58)	24-1963-64
$C_{10}H_{16}O_4$			
Exogonic acid	C_6H_{12}	225(2.30),275(1.65)	24-2785-64
	hexane	230(2.35),275(1.69)	24-2785-64
	MeOH	210(2.32),271(1.79)	24-2785-64
	3% HCl	215(2.25),275(1.76)	24-2785-64
	3% KOH	216(2.04),270(1.78)	24-2785-64

Compound	Solvent	$\lambda_{max}(\log \epsilon)$	Ref.
$C_{10}H_{16}S$			
Thiocamphor	C_6H_{12}	214(3.62),244(4.06), 493(1.09)	59-0299-64
	EtOH	210(3.57),246(4.04), 481(1.02)	59-0299-64
	Et_2O	245(4.06),489(1.03)	59-0299-64
	MeCN	248(4.01),479(1.04)	59-0299-64
Thiofenchone	hexane	215(3.60),240(4.00), 488(1.04)	59-0299-64
	EtOH	212(3.54),243(4.09), 479(1.03)	59-0299-64
$C_{10}H_{16}S_3$			
Thiophene, 3,4-bis[(ethylthio)methyl]-	EtOH	246(3.75)	44-1919-64
$C_{10}H_{16}Si$			
Silane, benzyltrimethyl-	C_6H_{12}	255s(2.38),267(2.64), 274(2.60)	39-4690-65
$C_{10}H_{17}BO_2$			
Boron, (2,4-pentanedionato)(2-methyl-tetramethylene)-	n.s.g.	332(3.20)	5-0040-65I
$C_{10}H_{17}Br_2N_7O_2S$			
Theophyllin-8-ylaminoethylisothiuronium hydrobromide	pH 1	290(4.22)	65-2774-64
$C_{10}H_{17}N$			
Pyrrolidine, 1-(cyclopentylidene-methyl)-	C_6H_{12}	254(3.65)	44-3679-65
$C_{10}H_{17}NO_2$			
Nicotinic acid, 1,4,5,6-tetrahydro-, tert-butyl ester	EtOH	286(4.04)	35-5461-65
$C_{10}H_{17}NO_3$			
Acrolein, 3-cyano-2-ethoxy-, diethyl acetal	EtOH	232(4.08)	6-0583-65
Valeraldehyde, 3-cyano-2-oxo-, 1-(di-ethyl acetal)	EtOH	231(3.84)	6-0583-65
$C_{10}H_{17}NO_4$			
Maleic acid, (diethylamino)-, dimethyl ester	EtOH	287(4.55)	23-0700-65
$C_{10}H_{17}NS$			
Pyrrolidine, 1-(4,5-dihydro-4,4-di-methyl-3-thienyl)-	C_6H_{12}	262(3.86)	54-1160-64
$C_{10}H_{17}N_3O$			
3,5-Octadien-2-one, 7-methyl-, semicarbazone	EtOH	297(4.59)	22-3218-64
4,5-Octadien-2-one, 7-methyl-, semicarbazone	EtOH	243(4.45)	22-3218-64
5,7-Octadien-2-one, 6-methyl-, semicarbazone	EtOH	233(4.38)	22-2533-64
cis	EtOH	232.5(4.41)	22-2533-64
Pyrimidine, 4-amino-5-(ethoxymethyl)-2-propyl-	EtOH	235(3.97),273(3.66)	94-0393-64
s-Triazine, 2,4-diethyl-6-isopropoxy-	EtOH	238(3.39)	44-0702-65

Compound	Solvent	λ_{max}(log ε)	Ref.
$C_{10}H_{17}N_3O_2$			
1-Hydroxy-1-(3-methyl-2-pyrazinyl)-piperidinium hydroxide	N HCl	263(3.81),266(3.81), 291(3.06)	44-1645-64
$C_{10}H_{17}N_3O_3$			
2-Pyrazoline-3(or 4)-carboxylic acid, 4(or 3)-(N-butylcarbamoyl)-, methyl ester	EtOH	294(3.95)	35-4162-64
s-Triazine, 2-ethyl-4-isopropoxy-6-(2-hydroxyethoxy)-	n.s.g.	235(3.45)	44-0702-65
$C_{10}H_{17}N_4OP$			
Phosphonic diamide, N,N,N',N'-tetra-methyl-P-(phenylazo)-	dioxan	288(3.98),505(1.97)	24-2844-65
$C_{10}H_{17}O_5P$			
Acetoacetic acid, 2-(ethoxyethylphos-phinyl)-2-(2-hydroxyethyl)-, γ-lactone	MeOH	200(3.08),228(2.67)	5-0010-65E
$C_{10}H_{17}O_5PS$			
Crotonic acid, 3-hydroxy-2-(2-hydroxy-ethyl)-, γ-lactone, O-ester with O,O-diethyl phosphorothioate	MeOH	229(4.01)	5-0010-65E
$C_{10}H_{17}O_6P$			
Acetoacetic acid, 2-(2-hydroxyethyl)-2-phosphono-, γ-lactone, diethyl ester	MeOH	212(2.37)	5-0010-65E
Crotonic acid, 3-hydroxy-2-(2-hydroxy-ethyl)-, γ-lactone, diethyl phosphate	MeOH	225(4.18)	5-0010-65E
Hexanoic acid, 5-hydroxy-2-(1-hydroxy-ethylidene)-, δ-lactone, dimethyl phosphate	MeOH	220(3.85)	5-0010-65E
$C_{10}H_{18}$			
Cyclopentene, 1,2,3,3,5-pentamethyl-	96% H_2SO_4	297(4.15)	23-2768-64
2,5-Octadiene, 2,6-dimethyl-	96% H_2SO_4	399.5(4.58)	23-2768-64
1,3-Pentadiene, 2-tert-butyl-4-methyl-	C_6H_{12}	225(3.82)	44-3527-64
	EtOH	225(3.81)	44-3527-64
$C_{10}H_{18}N_2$			
Camphor hydrazone	$CHCl_3$	276(3.30)	44-0417-65
Cyclopentapyrazole, 3,6a-diethyl-1,3a,4,5,6,6a-hexahydro-	EtOH	238(3.55)	22-0722-65
Fenchone hydrazone	$CHCl_3$	276(3.54)	44-0417-65
$C_{10}H_{18}N_2O_2$			
1-Pyrazolin-3-ol, 3,5-diethyl-5-methyl-, acetate	EtOH	330(2.26)	44-1379-64
$C_{10}H_{18}N_4$			
Pyrazine, 2,5-bis(dimethylamino)-3,6-dimethyl-, dihydrochloride	EtOH	263(4.03),337(3.77)	44-1645-64
$C_{10}H_{18}O$			
1-Cyclobutanone, 4-isopropyl-2,2,3-trimethyl-	hexane	306(1.64)	28-0827-64B

Compound	Solvent	$\lambda_{max}(\log \epsilon)$	Ref.
Cyclohexanone, 4-tert-butyl-	EtOH	284(1.20)	59-0529-65
	HOAc	280(1.27)	59-0529-65
	CF_3COOH	266(1.42)	59-0529-65
	CCl_4	292(1.28)	59-0529-65
	MeCN	287(1.18)	59-0529-65
3,5-Heptadien-2-ol, 2,4,6-trimethyl-	EtOH	220(3.72),240(3.51)	28-5895-64A
Menthone	EtOH	290(1.44)	59-0529-65
	HOAc	286(1.48)	59-0529-65
	CF_3COOH	277(1.70)	59-0529-65
	CCl_4	295(1.47)	59-0529-65
	MeCN	292(1.38)	59-0529-65
Myrcenol	EtOH	225(4.23)	5-0083-64E
1-Nonen-4-one, 2-methyl-	EtOH	210(3.25),287(1.98)	44-2502-65
2-Nonen-4-one, 2-methyl-	EtOH	237(4.09)	44-2502-65
Ocimenol, cis	EtOH	234(4.30)	5-0083-64E
Ocimenol, trans	EtOH	233(4.32)	5-0083-64E
3,5-Octadien-2-ol, 6,7-dimethyl-	EtOH	239(4.36)	22-3218-64
1-Propanol, 1-(2-ethyl-1-cyclo-pentenyl)-	EtOH	210(3.46)	22-0722-65
Umbellulol, dihydro-	EtOH	201(2.41)	24-3045-65
$C_{10}H_{18}O_2$			
Acrylic acid, α-heptyl-	EtOH	214(3.86)	12-1260-64
1-Cyclobutene, 3,4-dimethoxy-1,2,3,4-tetramethyl-	MeOH	none	24-1811-64
3,8-Decanedione	EtOH	278.5(1.70)	22-0714-65
Neocarvomenthol, 4-hydroxy-	n.s.g.	203(2.35),281(1.56)	33-0010-65
3,5-Octadiene-2,7-diol, 2,7-dimethyl-, cis-cis	n.s.g.	235(3.95)	70-0174-64
trans-trans	n.s.g.	235(4.30)	70-0174-64
$C_{10}H_{18}O_2S$			
[(Cyclohexylcarbonyl)methyl]dimethyl-sulfonium hydroxide, inner salt, oxide	EtOH	252(4.13)	35-1640-64
$C_{10}H_{18}O_2S_4$			
Formic acid, dithiobis[thio-, O,O-di-butyl ester	hexane	360(1.9)	70-2175-64
	heptane	244(4.38),286(3.92), 365(1.91)	1-1113-65
	EtOH	243(4.37),287(3.91), 362(1.92)	1-1113-65
	MeCN	243(4.38),288(3.90), 360(1.92)	1-1113-65
$C_{10}H_{18}O_2S_5$			
Formic acid, trithiobis[thio-, O,O-dibutyl ester	hexane	none	70-2175-64
$C_{10}H_{18}O_5$			
Acetoacetic ester, 4,4-diethoxy-	EtOH	281.5(1.51)	6-0583-65
$(C_{10}H_{18}S_3)_n$			
Polynonamethylenetrithiocarbonate	$CHCl_3$	310(4.94)	49-0631-65
$C_{10}H_{19}N$			
Cyclopentane, 1-butylimino-3-methyl-	isooctane	257(2.2)	77-0540-65
Indolizine, octahydro-5,7-dimethyl-	hexane	none above 230 mμ	1-1607-65
$C_{10}H_{19}NOS_2$			
Conhydrin, N-dithiocarbomethoxy-	dioxan	280(4.06),345s(1.95)	78-0407-65

Compound	Solvent	λ_{max}(log ε)	Ref.
$C_{10}H_{19}NO_2$			
Acrylic acid, 3-(diisopropylamino)-, methyl ester, trans	EtOH	212(3.11),283(4.48)	5-0098-65H
L-Alanine, N-neopentylidene-, ethyl ester	hexane	217(2.46)	39-4503-65
Hexanoic acid, 2-(1-aminoethylidene)-, ethyl ester	EtOH	288(4.20)	39-2411-65
$C_{10}H_{19}NO_4$			
Acetoacetic ester imine, 4,4-diethoxy-	EtOH	277(3.98)	6-0583-65
$C_{10}H_{19}N_3O$			
s-Triazine, 2,2,6-triethyl-1,2-dihydro-4-methoxy-	EtOH	273(3.24)	44-0702-65
$C_{10}H_{19}N_5$			
Imidazole-4-carboxamidine, N,N-diethyl-1-methyl-5-(methylamino)-, HI salt	pH 1	288(3.70)	44-3597-65
	H_2O	295(3.65)	44-3597-65
	pH 13	290(3.27)	44-3597-65
$C_{10}H_{20}ClNO$			
(4,4-Dimethyl-3-oxo-1-pentenyl)trimethylammonium chloride	EtOH	208(4.93)	44-0385-64
Trimethyl(5-methyl-3-oxo-1-hexenyl)-ammonium chloride	EtOH	208(4.89)	44-0385-64
$C_{10}H_{20}NO_2$			
1-Piperidinooxy, 4-hydroxy-2,2,4,6,6-pentamethyl-	C_6H_{12}	241(3.70),445(0.89)	22-3273-65
	MeOH	241s(3.73),420(0.87)	22-3273-65
$C_{10}H_{20}N_2$			
2-Imidazoline, 1-ethyl-2-pentyl-	EtOH	232(3.77)	4-0188-64
$C_{10}H_{20}N_2O$			
2-Imidazoline-1-ethanol, 2-pentyl-	aq. HCl	231(3.72)	4-0229-64
	H_2O	231(3.72)	4-0229-64
	EtOH	231(3.73)	4-0229-64
2-Imidazoline-2-ethanol, 2-pentyl-	EtOH	231(3.72)	4-0188-64
$C_{10}H_{20}N_2O_2$			
2-Pentanone, 4-(N-butyl-N-nitrosamino)-4-methyl-	H_2O	233.5(3.84)	28-5470-64A
$C_{10}H_{20}N_2O_8$			
L-Arabinose, azine	MeOH	265(4.04)	24-1404-65
D-Ribose, azine	MeOH	260(4.02)	24-1404-65
D-Xylose, azine	MeOH	260(4.04)	24-1404-65
$C_{10}H_{20}N_2S_4$			
Disulfide, bis(diethylthiocarbamoyl)-	heptane	225(4.20),250(4.16), 278(4.03)	1-1113-65
	EtOH	220(4.10),252(3.91), 276(3.87)	1-1113-65
$C_{10}H_{20}N_4O$			
Pyrazolo[1,5-a]pyrimidine, 7-acetamido-2,5-dimethyl-	EtOH	238(4.53),295(3.67), 336(3.23)	94-1207-65

Compound	Solvent	$\lambda_{max}(\log \epsilon)$	Ref.
$C_{10}H_{20}O_2$			
Pivaloin	hexane	304(1.42)	59-0529-65
	isooctane	304(1.42)	59-0529-65
	MeOH	308(1.43)	59-0529-65
	EtOH	308(1.44)	59-0529-65
	50% EtOH	306(1.45)	59-0529-65
	HOAc	304(1.45)	59-0529-65
	50% HOAc	304(1.45)	59-0529-65
	CF_3COOH	296(1.62)	59-0529-65
	$ClCH_2CH_2OH$	306(1.43)	59-0529-65
	$Cl_2CHCOOH$	299(1.58)	59-0529-65
	MeCN	306(1.42)	59-0529-65
	50% MeCN	306(1.43)	59-0529-65
	CCl_4	304(1.46)	59-0529-65
$C_{10}H_{21}ClN_2O$			
(4,4-Dimethyl-3-oxo-1-pentenyl)-trimethylammonium chloride, oxime	EtOH	242.5(3.54)	44-1129-65
$C_{10}H_{21}NO_2$			
L-Alanine, N-neopentyl-, ethyl ester	hexane	220s(2.52)	39-4503-65
$C_{10}H_{24}N_2S_2$			
4,4'-Dithiobis(2-methyl-2-butylamine), dihydrochloride	n.s.g.	246(2.58)	44-3314-64
$C_{10}H_{30}Si_4$			
Tetrasilane, decamethyl-	C_6H_{12}	235.0(4.17)	101-0369-64B
$C_{10}N_6$			
1,3-Butadienehexacarbonitrile	MeCN	302(4.19)	35-2898-64
$C_{10}N_6Na$			
1,3-Butadienehexacarbonitrile, sodium derivative	MeCN	475(4.27),507(4.64), 613(3.38),667(3.57), 713(3.38)	35-2898-64
$C_{10}N_6Na_2$			
2-Butene-1,1,2,3,4,4-hexacarbonitrile, disodium derivative, cis	MeCN	325(3.65),392(3.73), 487(4.11)	35-2898-64
trans	MeCN	223(3.70),487(4.16)	35-2898-64

Compound	Solvent	λ_{max}(log ε)	Ref.
$C_{11}H_2Cl_6O_2$			
1,4-Methanonaphthalene-5,8-dione, 1,2,3,4,9,9-hexachloro-1,4-dihydro-	C_6H_{12}	257(4.09),336s(2.06), 382(4.00),476s(1.43)	39-3043-64
$C_{11}H_5F_3N_2OS$			
Nicotinonitrile, 1,2-dihydro-2-oxo-6-(2-thienyl)-4-(trifluoromethyl)-	EtOH	272(3.85),363(4.24)	44-3377-65
$C_{11}H_5N$			
2,4-Pentadiynenitrile, 4-phenyl-	EtOH	243(4.68),270(3.84), 287(4.12),303(4.27), 323(4.13)	39-0543-64
$C_{11}H_5NO_3$			
2-Naphthonitrile, 1,4-dihydro-3-hydroxy-1,4-dioxo-	MeCN	247(4.08),253(4.11), 279(4.18),287(4.16), 340(3.36)	44-3591-64
$C_{11}H_6Br_2N_2S$			
Imidazo[1,2-a]pyridine, 3-bromo-2-(5-bromo-2-thienyl)-	EtOH	211(4.4),267(4.3), 341(4.1)	70-1434-65
$C_{11}H_6ClMnO_4$			
Manganese, tricarbonyl[(1-chloro-2-formylvinyl)cyclopentadienyl]-	$CHCl_3$	276(4.28),371(3.60)	49-1520-65
$C_{11}H_6Cl_2N_2O_5$			
Pyoluteorin, 3'-nitro-	EtOH	245s(2.31),310(4.36)	39-2641-64
	EtOH-NaOH	325(4.33),400(4.36)	39-2641-64
$C_{11}H_6Cl_2N_4$			
Imidazo[4,5-d]pyridazine, 4,7-dichloro-2-phenyl-	EtOH	212(4.51),240(4.43), 310(4.35)	4-0182-64
$C_{11}H_6Cl_3NO_3$			
4-Isoxazolecarboxylic acid, 5-methyl-3-(2,3,6-trichlorophenyl)-	MeOH	217(4.32),291(2.66)	39-5976-65
4-Isoxazolecarboxylic acid, 5-methyl-3-(2,4,6-trichlorophenyl)-	MeOH	217(4.32),226s(4.27)	39-5976-65
$C_{11}H_6N_2O_6S$			
2-Propen-1-one, 1-(5-nitro-2-furyl)-3-(5-nitro-2-thienyl)-	$CHCl_3$	315(4.21),376(4.43)	65-2328-64
	95% H_2SO_4	272(3.81),336(3.98), 498(4.71)	65-2328-64
2-Propen-1-one, 3-(5-nitro-2-furyl)-1-(5-nitro-2-thineyl)-	EtOH	240(4.11),368(4.49)	65-2328-64
	95% H_2SO_4	270(3.94),335(4.03), 487(4.61)	65-2328-64
$C_{11}H_6O$			
2,4-Undecadiene-6,8,10-triynal	Et_2O	266(4.54),278(4.84), 294(4.09),312(4.39), 324s(4.35),333(4.58), 348(4.33),357(4.52)	24-1846-64
$C_{11}H_6O_3$			
Angelicin	EtOH	221(4.23),252(4.47), 303(4.11)	65-4171-64
	n.s.g.	245(4.38),297(3.96)	78-3591-65

Compound	Solvent	λ_{max}(log ε)	Ref.
$C_{11}H_6O_4$			
Bergaptol	EtOH	222(4.37),269(4.21), 313(4.09)	33-0408-64
Psoralen, 5-hydroxy-	EtOH	222(4.35),288(3.94), 325(4.12)	88-2559-65
$C_{11}H_7BO_3$			
1-Naphthoic acid, 8-borono-, cyclic anhydride	EtOH	230(4.39),314(3.85)	44-0807-65
$C_{11}H_7BrClNO_3$			
4-Isoxazolecarboxylic acid, 3-(2-chloro-6-bromophenyl)-5-methyl-	MeOH	217(4.25)	39-5976-65
$C_{11}H_7BrFNO_3$			
4-Isoxazolecarboxylic acid, 3-(6-bromo-2-fluorophenyl)-5-methyl-	MeOH	220(4.07),272(3.10)	39-5976-65
$C_{11}H_7BrN_2$			
Cinnamonitrile, β-(bromomethyl)-α-cyano-	MeOH	235(4.00),310(4.27)	24-3571-65
$C_{11}H_7BrN_2S$			
Imidazo[1,2-a]pyridine, 2-(5-bromo-2-thienyl)-	EtOH	211(4.4),262(4.3), 330(4.1)	70-1434-65
Imidazo[1,2-a]pyridine, 3-bromo-2-(2-thienyl)-	EtOH	208(4.3),258(4.3), 265s(4.3),332(4.0)	70-1434-65
Imidazo[1,2-a]pyridine, 5-bromo-2-(2-thienyl)-	EtOH	215(4.4),262(4.4), 336(4.0)	70-1434-65
$C_{11}H_7BrO_2$			
Furan, 2-benzoyl-5-bromo-	EtOH	263s(3.90),298(4.27)	39-6057-65
$C_{11}H_7ClFNO_3$			
4-Isoxazolecarboxylic acid, 3-(6-chloro-2-fluorophenyl)-5-methyl-	MeOH	220(4.11),270(3.18)	39-5976-65
$C_{11}H_7ClN_2OS$			
4-Isothiazolecarbonitrile, 5-(benzyloxy)-3-chloro-	EtOH	246(3.92)	44-0660-64
$C_{11}H_7ClN_2O_2$			
Pyridine, 3?-chloro-4-(o-nitrophenyl)-	EtOH	201(4.58),225s(4.32), 261(3.93)	39-2175-64
$C_{11}H_7ClN_2O_5$			
4-Isoxazolecarboxylic acid, 3-(2-chloro-4-nitrophenyl)-5-methyl-	MeOH	219(4.27),263(4.05)	39-5976-65
4-Isoxazolecarboxylic acid, 3-(2-chloro-5-nitrophenyl)-5-methyl-	MeOH	220(4.24),273(4.00)	39-5976-65
$C_{11}H_7ClN_2S$			
10H-Pyrido[3,2-d][1,4]benzothiazine, 3-chloro-	EtOH	253(4.56),320s(3.65)	95-0429-65
$C_{11}H_7ClN_2S_2$			
4-Isothiazolecarbonitrile, 5-(benzylthio)-3-chloro-	EtOH	218(4.23),288(4.00)	44-0660-64

Compound	Solvent	$\lambda_{max}(\log \epsilon)$	Ref.
$C_{11}H_7ClN_4O$			
1H-Imidazo[4,5-d]pyridazin-7(or 4)-ol, 4(or 7)-chloro-2-phenyl-	EtOH	207(4.36),249(4.34), 282(4.32)	4-0182-64
10H-Pyridazino[6,1-b]quinazolin-10-one, 3-amino-4-chloro-	EtOH	228(4.27),329(4.14), 343(4.18),389(3.76)	4-0042-64
$C_{11}H_7Cl_2NO_3$			
4-Isoxazolecarboxylic acid, 3-(2,3-dichlorophenyl)-5-methyl-	MeOH	217s(4.23)	39-5976-65
4-Isoxazolecarboxylic acid, 3-(2,4-dichlorophenyl)-5-methyl-	MeOH	215(4.25),223s(4.19)	39-5976-65
4-Isoxazolecarboxylic acid, 3-(2,5-dichlorophenyl)-5-methyl-	MeoH	215(4.22),225s(4.22)	39-5976-65
4-Isoxazolecarboxylic acid, 3-(2,6-dichlorophenyl)-5-methyl-	MeOH	218(4.22)	39-5976-65
4-Isoxazolecarboxylic acid, 3-(3,4-dichlorophenyl)-5-methyl-	MeOH	216(4.27),236s(4.14), 280s(2.61)	39-5976-65
4-Isoxazolecarboxylic acid, 3-(3,5-dichlorophenyl)-5-methyl-	MeOH	218(4.28),234s(3.97), 286(2.28)	39-5976-65
4-Isoxazolecarboxylic acid, 5-(2,3-dichlorophenyl)-3-methyl-	MeOH	218s(4.23),221(4.24), 240s(3.87)	39-5976-65
4-Isoxazolecarboxylic acid, 5-(2,4-dichlorophenyl)-3-methyl-	MeOH	217(4.21),249(3.97)	39-5976-65
4-Isoxazolecarboxylic acid, 5-(2,5-dichlorophenyl)-3-methyl-	MeOH	225(4.26),240s(3.83), 277s(3.29)	39-5976-65
4-Isoxazolecarboxylic acid, 5-(3,4-dichlorophenyl)-3-methyl-	MeOH	212s(4.15),218(4.18), 275(4.16)	39-5976-65
4-Isoxazolecarboxylic acid, 5-(3,5-dichlorophenyl)-3-methyl-	MeOH	216(4.25),222(4.27), 268(4.09)	39-5976-65
$C_{11}H_7Cl_2N_5S$			
v-Triazolo[4,5-c]pyridazine, 7-[(2,4-dichlorobenzyl)thio]-	EtOH	206(4.46),224(4.38), 281(3.88),329(4.21)	4-0247-64
v-Triazolo[4,5-c]pyridazine, 7-[(2,6-dichlorobenzyl)thio]-	EtOH	206(4.37),220s(4.34), 280(3.96),329(4.19)	4-0247-64
v-Triazolo[4,5-c]pyridazine, 7-[(3,4-dichlorobenzyl)thio]-	EtOH	205(4.46),220s(4.38), 280(3.84),328(4.19)	4-0247-64
$C_{11}H_7Cl_3O_2$			
Coumarin, 3,4,6-trichloro-5,7-dimethyl-	EtOH	220(4.36),304(4.11), 340(3.86)	44-4122-65
$C_{11}H_7Cl_4NO_2$			
Indole-2-carboxylic acid, 3,x,x,x-tetrachloro-1-methyl-, methyl ester	EtOH	216(4.29),245(4.50), 298s(4.14),307(4.12) 328s(3.81),344s(3.62)	44-0178-64
$C_{11}H_7Cl_6NO_2$			
2-Indolinecarboxylic acid, hexachloro-1-methyl-, methyl ester	isooctane	232(4.43),274(3.79), 304(3.33),333(3.42)	44-0178-64
$C_{11}H_7F_2NO_3$			
4-Isoxazolecarboxylic acid, 3-(2,6-difluorophenyl)-5-methyl-	MeOH	211s(3.92),222(3.98), 265s(3.11)	39-5976-65
$C_{11}H_7F_3N_4$			
Quinoline, 4-azido-6-methyl-2-(trifluoromethyl)-	EtOH	238(4.63),308(3.88)	4-0113-65
Quinoline, 4-azido-8-methyl-2-(trifluoromethyl)-	EtOH	237(4.55),314(3.85)	4-0113-65

Compound	Solvent	λ_{max}(log ϵ)	Ref.
$C_{11}H_7F_3N_4O$			
Quinoline, 4-azido-6-methoxy-2-(trifluoromethyl)-	EtOH	246(4.45),306(3.70)	4-0113-65
$C_{11}H_7F_5O_3$			
Acetic acid, (pentafluorobenzoyl)-, ethyl ester	heptane	264(3.95)	70-1798-65
$C_{11}H_7IO_2$			
Furan, 2-benzoyl-5-iodo-	EtOH	258(3.88),309(4.28)	39-6057-65
$C_{11}H_7NO_4$			
2,3-Dicarboxyquinolizinium hydroxide, inner salt	CF_3COOH	327s(4.04),337(4.15)	44-0452-64
$C_{11}H_7NO_4S$			
2-Propen-1-one, 1-(2-furyl)-3-(5-nitro-2-thienyl)-	EtOH	372(4.45)	65-2328-64
	95% H_2SO_4	268(3.81),333(4.00), 460(4.71)	65-2328-64
2-Propen-1-one, 3-(2-furyl)-1-(5-nitro-2-thienyl)-	EtOH	308(4.10),378(4.34)	65-2328-64
	95% H_2SO_4	277(3.42),347(3.86), 513(4.76)	65-2328-64
2-Propen-1-one, 1-(5-nitro-2-furyl)-3-(2-thienyl)-	EtOH	230(4.00),305(4.25), 374(4.37)	65-2328-64
	95% H_2SO_4	250(3.88),325(3.88), 515(4.72)	65-2328-64
2-Propen-1-one, 3-(5-nitro-2-furyl)-1-(2-thienyl)-	EtOH	362(4.43)	65-2328-64
	95% H_2SO_4	270(3.81),335(4.10), 456(4.63)	65-2328-64
$C_{11}H_7N_3$			
[2,2'-Bipyridine]-4-carbonitrile	octane	346(1.86),363(1.23)	65-0298-65
	EtOH	246(4.03),293(4.26)	65-0298-65
[2,3'-Bipyridine]-5-carbonitrile	H_2O	248(3.89),291(3.99)	65-0298-65
$C_{11}H_7N_3OS_2$			
Thiazolo[4,5-d]pyridazine-2,7(3H,6H)-dione, 6-phenyl-2-thio-	EtOH	236(4.16),302(4.40)	78-1323-65
$C_{11}H_7N_3O_2$			
Isoxazolo[5,4-d]pyrimidin-4(5H)-one, 3-phenyl-	EtOH	249(4.20)	44-2117-64
$C_{11}H_7N_3O_5S_3$			
3H-1,2-Dithiole-3-thione, 4-(N-methyl-3,5-dinitrobenzamido)-	EtOH	224(4.38),306(3.56), 416(3.83)	12-0061-65
$C_{11}H_7N_3O_6S_2$			
3H-1,2-Dithiole-3-one, 4-(N-methyl-3,5-dinitrobenzamido)-	EtOH	227(4.33),329(3.82)	12-0061-65
$C_{11}H_8$			
Pentadiyne, phenyl-	n.s.g.	211(4.50),221(4.64), 244(3.88),256(4.23), 270(4.42),287(4.30)	22-1525-65
$C_{11}H_8BrNO_5$			
Crotonic acid, 3-bromo-4-hydroxy-2-methoxy-4-(m-nitrophenyl)-, γ-lactone	MeOH	252(4.06)	44-1800-65

Compound	Solvent	λ max (log ε)	Ref.
$C_{11}H_8Br_2$			
1,6-Methanocyclodecapentaene, dibromo-	n.s.g.	234(4.13),277(4.79), 327(3.91),419(2.72)	89-0784B-64
$C_{11}H_8ClF_3N_4$			
Pyrimidine, 2,4-diamino-5-(p-chloro-phenyl)-6-(trifluoromethyl)-	3N HCl	274(3.76),316(3.45)	4-0162-65
	pH 7, 13	302(3.83)	4-0162-65
$C_{11}H_8ClNO$			
Pyrrole, 2-o-chlorobenzoyl-	MeOH	247(3.59),301(4.20)	39-2579-64
Pyrrole, 2-p-chlorobenzoyl-	MeOH	252(3.99),310(4.17)	39-2579-64
$C_{11}H_8ClNO_2S$			
2H-1,4-Thiazine-6-carboxylic acid, 5-(5-chloro-2-hydroxyphenyl)-3,4-dihydro-, δ-lactone	EtOH	222(4.24),243(4.01), 264(3.85),320(3.44), 365(3.89)	44-4122-65
$C_{11}H_8ClNO_4$			
4-Isoxazolecarboxylic acid, 3-(2-chloro-3-hydroxyphenyl)-5-methyl-	MeOH	216(4.21),287(3.47)	39-5976-65
4-Isoxazolecarboxylic acid, 3-(2-chloro-4-hydroxyphenyl)-5-methyl-	MeOH	216(4.22),278s(3.21), 286s(2.80)	39-5976-65
4-Isoxazolecarboxylic acid, 3-(2-chloro-5-hydroxyphenyl)-5-methyl-	MeOH	216(4.21),293(3.02)	39-5976-65
$C_{11}H_8ClN_3OS$			
Thiocyanic acid, 1-(p-chlorophenyl)-3-methyl-5-oxo-2-pyrazolin-4-yl ester	EtOH	252(4.31)	94-0023-64
$C_{11}H_8ClN_3O_2$			
Pyridine, 2-anilino-5-chloro-3-nitro-	EtOH	213(4.22),245(4.23), 302(4.18)	95-0429-65
$C_{11}H_8ClN_3O_2S$			
Pyridine, 2-[(o-aminophenyl)thio]-3-nitro-5-chloro-	EtOH	235(4.45),282(4.07), 310s(3.76)	95-0429-65
$C_{11}H_8ClN_3S$			
4-Isothiazolecarbonitrile, 3-chloro-5-(N-methylanilino)-	EtOH	222(4.29),282(4.03)	44-0660-64
5H-Pyridazino[3,4-b][1,4]benzothiazine, 3-chloro-5-methyl-	EtOH	255(4.6)	94-0580-65
10H-Pyridazino[4,3-b][1,4]benzothiazine, 3-chloro-10-methyl-	EtOH	245(4.55)	94-0580-65
10H-Pyridazino[4,5-b][1,4]benzothiazine, 4-chloro-10-methyl-	EtOH	265(4.45)	94-0580-65
$C_{11}H_8ClN_5$			
s-Triazolo[4,3-b]pyridazine, 7-amino-8-chloro-3-phenyl-	EtOH	221(4.36),294(4.29), 355(3.88)	4-0042-64
$C_{11}H_8ClN_5S$			
v-Triazolo[4,5-c]pyridazine, 7-[(p-chlorobenzyl)thio]-	EtOH	203(4.26),223(4.27), 280(3.89),328(4.27)	4-0247-64
$C_{11}H_8Cl_2N_2$			
Pyridazine, 3,6-dichloro-4-methyl-5-phenyl-	EtOH	260(3.60)	44-2128-64

Compound	Solvent	$\lambda_{max}(\log \epsilon)$	Ref.
$C_{11}H_8Cl_2N_2OS$			
4-Isothiazolecarboxy-o-toluidide, 3,5-dichloro-	EtOH	228(4.01),256(4.03)	44-0660-64
$C_{11}H_8Cl_2O_3$			
Coumarin, 3,6-dichloro-4-ethoxy-	EtOH	220s(4.54),267s(4.05), 278(4.17),284s(4.12), 290(4.12),322(3.96)	44-4126-65
$C_{11}H_8Cl_2O_4$			
Coumarin, 3,6-dichloro-4-(2-hydroxy-ethoxy)-	EtOH	218(4.28),225(4.31), 281(4.06),293(4.01), 323(3.83)	44-4126-65
$C_{11}H_8Cl_3NS_2$			
2H-1,3-Thiazine-2-thione, 3,6-dihydro-4-phenyl-6-(trichloromethyl)-	EtOH	247(4.16),328(4.01)	39-4008-64
$C_{11}H_8CrO_4$			
Chromium, (acetophenone)tricarbonyl-	n.s.g.	322.5(3.93)	28-2833-65A
$C_{11}H_8FNO_4S$			
2(1H)-Pyridone, 1-[[(p-fluorophenyl)-sulfonyl]oxy]-	EtOH	224(4.16),261s(3.13), 301(3.67)	35-5186-65
$C_{11}H_8F_3N$			
Quinoline, 6-methyl-2-(trifluoromethyl)-	EtOH	231(4.66),235(4.68), 293(3.51),306(3.51), 320(3.47)	4-0113-65
$C_{11}H_8F_3NO$			
Carbostyril, 6-methyl-4-(trifluoro-methyl)-	EtOH	233(4.59),257(3.90), 276(3.76),351(3.75)	4-0113-65
	EtOH-HCl	235(4.33),254(3.73), 274s(--)	4-0113-65
Carbostyril, 8-methyl-4-(trifluoro-methyl)-	EtOH	224(4.23),274(3.74), 346(3.65)	4-0113-65
	EtOH-HCl	233(4.12),255(3.90), 284(3.56),344(3.44)	4-0113-65
Quinoline, 6-methoxy-2-(trifluoro-methyl)-	EtOH	241(4.56),284(3.42), 330(3.60)	4-0113-65
4-Quinolinol, 6-methyl-2-(trifluoro-methyl)-	EtOH	231(4.55),293(3.69), 327(3.79),341(3.76)	4-0113-65
	EtOH-HCl	236(4.50),326(3.76)	4-0113-65
4-Quinolinol, 8-methyl-2-(trifluoro-methyl)-	EtOH	228(4.72),293(3.78), 319(3.63)	4-0113-65
	EtOH-HCl	219(3.88),292(3.90), 319(3.60)	4-0113-65
$C_{11}H_8F_3NO_2$			
4-Quinolinol, 6-methoxy-2-(trifluoro-methyl)-	EtOH	241(4.45),326(3.83)	4-0113-65
	EtOH-HCl	242(4.46),296(3.51)	4-0113-65
$C_{11}H_8N_2$			
Pyrrolo[1,2-a]quinoxaline	pH 1	225(4.52),240(4.44), 352(4.07)	25-1382-65
	pH 6.9	224(4.44),247(4.39), 334(3.93)	25-1382-65

Compound	Solvent	λ_{max}(log ε)	Ref.
$C_{11}H_8N_2O$			
Benzimidazole, 2-(2-furyl)-	EtOH	307(4.45),322(4.34)	94-0773-64
Cyclohepta[b]pyrrole-3-carbonitrile, 1,2-dihydro-5-methyl-2-oxo-	EtOH	276(4.59),420(4.21)	94-0450-65
$C_{11}H_8N_2O_2$			
[2,2'-Bipyridine]-4-carboxylic acid	n.s.g.	241(4.28),287(4.39)	65-0298-65
[2,3'-Bipyridine]-5-carboxylic acid	EtOH	253(4.30),292(4.36)	65-0298-65
5,8-Isoquinolinedione, 6(or 7)-(1-aziridinyl)-	MeCN	236(4.12),258(4.16), 318(3.77)	87-0801-64
Pyridine, 2-(m-nitrophenyl)-	EtOH	243(4.30),270(4.18)	39-5311-65
Pyridine, 2-(o-nitrophenyl)-	EtOH	202(4.40),223(4.34), 261(4.04)	39-2175-64
	EtOH	225(4.32),260(4.13)	39-5311-65
Pyridine, 2-(p-nitrophenyl)-	EtOH	300(4.28)	39-5311-65
Pyridine, 3-(m-nitrophenyl)-	EtOH	240(4.22),268s(4.18)	39-5311-65
Pyridine, 3-(o-nitrophenyl)-	EtOH	201(4.75),225(4.43), 261(4.08)	39-2175-64
	EtOH	220(4.21),252(4.18)	39-5311-65
Pyridine, 3-(p-nitrophenyl)-	EtOH	292(4.22)	39-5311-65
Pyridine, 4-(m-nitrophenyl)-	EtOH	243(4.29)	39-5311-65
Pyridine, 4-(p-nitrophenyl)-	EtOH	300(4.28)	39-5311-65
$C_{11}H_8N_2O_2S_2$			
2-Thiazoline-4-carboxylic acid, 2-(2-benzothiazolyl)-	EtOH	243(3.77),247(3.79), 251(3.78),295(4.22)	44-2344-65
$C_{11}H_8N_2O_3$			
Pyridine, 4-[2-(5-nitro-2-furyl)vinyl]-	MeOH	259(4.00),288(3.98), 368(4.33)	94-0389-65
Pyrrole, 2-o-nitrobenzoyl-	MeOH	255s(3.85),297(4.10)	39-2579-64
Pyrrole, 2-p-nitrobenzoyl-	MeOH	263(4.12),323(3.99)	39-2579-64
$C_{11}H_8N_2O_3S_2$			
2-Thiazoline-4-carboxylic acid, 2-(6-hydroxy-2-benzothiazolyl)-	EtOH	230(4.26),269(3.85)	44-2344-65
$C_{11}H_8N_2S$			
Imidazo[1,2-a]pyridine, 2-(2-thienyl)-	EtOH	206(4.4),254(4.4), 319s(4.0),328(4.1)	70-1434-65
Thiazolo[4,5-f]quinoline, 2-methyl-	EtOH	241(4.66),299(3.87), 325(3.67)	7-0530-64
Thiazolo[4,5-f]quinoline, 7-methyl-	EtOH	238(4.57),287(3.69), 311(3.52),318(3.46), 333(3.23)	4-0242-65
Thiazolo[4,5-f]quinoline, 9-methyl-	EtOH	241(4.46),292(3.63), 326(3.30)	4-0242-65
Thiazolo[5,4-f]quinoline, 7-methyl-	EtOH	258(4.50),318(3.29), 333(3.23)	4-0242-65
Thiazolo[5,4-f]quinoline, 9-methyl-	EtOH	248(4.46),285(3.56), 328(2.56)	4-0242-65
$C_{11}H_8N_4$			
Imidazo[4,5-d]pyridazine, 2-phenyl-	EtOH	208(4.57),231s(4.37), 236(4.38),245s(4.22), 287(4.35)	4-0182-64

Compound	Solvent	$\lambda_{max}(\log \epsilon)$	Ref.
Purine, 8-phenyl-	pH 1	236(4.13),303(4.38)	44-1916-65
	NaCl	232(4.08),298(4.40), 311s(4.23)	44-1916-65
	pH 13	233(4.26),304(4.42), 318s(4.23)	44-1916-65
$C_{11}H_8N_4O$			
Imidazo[4,5-d]pyridazin-4-ol, 2-phenyl-	EtOH	210(4.35),246(4.32), 283(4.30)	4-0182-64
Isoxazolo[5,4-d]pyrimidine, 4-amino-3-phenyl-	EtOH	244(4.09),276(3.91)	44-2117-64
$C_{11}H_8N_4O_2$			
5-Imidazolecarbonitrile, 1-benzyl-4-nitro-	EtOH-HCl	290(3.82)	4-0291-65
	EtOH	290(3.76)	4-0291-65
	EtOH-NaOH	292.5(3.83)	4-0291-65
Imidazo[4,5-d]pyridazine-4,7-diol, 2-phenyl-	EtOH	206(4.45),256(4.36)	4-0182-64
$C_{11}H_8N_4O_3S$			
Thiocyanic acid, 3-methyl-1-(p-nitrophenyl)-5-oxo-2-pyrazolin-4-yl ester	EtOH	236(4.08)	94-0023-64
$C_{11}H_8N_4O_4S$			
4H-Thiopyran-4-one, 2,4-dinitrophenylhydrazone	n.s.g.	420(4.45)	39-3037-65
$C_{11}H_8N_4S$			
Purine-6-thiol, 8-phenyl-	2N HCl	350(4.32)	44-1916-65
	pH 13	223s(4.19),258(4.27), 339(4.27)	44-1916-65
Thiazolo[5,4-d]pyrimidine, 7-amino-2-phenyl-	pH 1	220(4.29),232s(4.21), 302(4.21),318s(4.16), 336s(3.86)	44-1916-65
$C_{11}H_8N_4S_2$			
Imidazo[4,5-d]pyridazine-4,7-dithiol, 2-phenyl-	EtOH	206(4.54),279(4.44), 292s(4.38),349(4.19)	4-0182-64
Xanthine, 8-phenyl-2,6-dithio-	2N HCl	220(4.37),265(3.79), 333(4.42),352s(4.23)	44-1916-65
	pH 13	249s(4.09),344(4.24)	44-1916-65
$C_{11}H_8O$			
2,4-Pentadiyn-1-ol, 5-phenyl-	Et_2O	222(4.72),245(3.97), 256(4.27),271(4.44), 287(4.35)	24-2118-64
2,4-Undecadiene-6,8,10-triyn-1-ol	Et_2O	256(4.72),268(4.98), 282(4.11),289(4.13), 299s(4.27),305(4.37), 317s(4.32),324(4.55), 335(4.13),346(4.49)	24-1846-64
$C_{11}H_8OS$			
2-Heptene-4,6-diyn-1-ol, 7-(2-thienyl)-	Et_2O	208(4.41),250(4.38), 311(4.24)	24-2118-64
$C_{11}H_8OS_2$			
2-Propen-1-one, 1,3-di-2-thienyl-	n.s.g.	347(4.38)	101-0398-64B

Compound	Solvent	λ_{max}(log ϵ)	Ref.
$C_{11}H_8O_2$			
2-Naphthoic acid	EtOH	235(4.7),280(3.8), 320(3.0),335(3.1)	44-1543-64
1,4-Naphthoquinone, 2-methyl-	EtOH	245(4.25),251(4.27), 333(3.39)	54-1005-64
2-Pyrone, 6-phenyl-	EtOH	245(3.57),335(4.15)	54-0031-64
$C_{11}H_8O_2S$			
2H-Naphtho[2,3-b]thiete, 1,1-dioxide	EtOH	214s(4.49),233(4.84), 273s(3.74),284(3.67), 289(3.65),294(3.63), 313(3.21),319s(3.05), 327(3.19)	44-3883-65
2-Propen-1-one, 1-(2-furyl)-3-(2-thienyl)-	EtOH	352(4.39)	65-2328-64
	HOAc-H_2SO_4	352(3.88),470(4.60)	65-2328-64
2-Propen-1-one, 3-(2-furyl)-1-(2-thienyl)	EtOH	230(3.66),257(3.76), 350(4.41)	65-2328-64
	HOAc-30% H_2SO_4	290(3.51),356(3.93), 470(4.54)	65-2328-64
$C_{11}H_8O_2SSe$			
1,3-Propanedione, 1-selenophene-2-yl-3-(2-thienyl)-	n.s.g.	277(4.06),382(4.40)	67-0379-65
copper salt	n.s.g.	295(4.11),385(4.28)	67-0379-65
$C_{11}H_8O_2Se_2$			
1,3-Propanedione, 1,3-di-selenophene-2-yl-	n.s.g.	280(4.09),385(4.40)	67-0379-65
copper salt	n.s.g.	307(4.22),385(4.47)	67-0379-65
$C_{11}H_8O_3$			
1,2-Naphthoquinone, 3-hydroxy-4-methyl-	$CHCl_3$	270(4.14),358(3.19), 490(3.08)	24-0666-65
	dioxan	269(4.26),340(2.98), 470(3.04)	24-0666-65
	pyridine	343(3.02),482(3.12)	24-0666-65
1,2-Naphthoquinone, 8-methoxy-	EtOH	216(4.34),242(4.22), 422(3.82)	39-2355-65
$C_{11}H_8O_3Se$			
1,3-Propanedione, 1-(2-furyl)-3-selenophene-2-yl-	n.s.g.	277(4.33),372(4.77)	67-0379-65
copper salt	n.s.g.	307(4.33),380(4.57)	67-0379-65
$C_{11}H_8O_4$			
Coumarin, 6-acetyl-7-hydroxy-	EtOH	229(4.18),259(4.47), 316(4.04),349(4.09)	32-0513-65
Coumarin, 8-acetyl-7-hydroxy-	EtOH	214(4.20),236(3.95), 246(3.96),271(3.96), 323(4.09)	32-0513-65
2,4,6-Decatriynedioic acid, 10-methyl ester	EtOH	221(4.10),256(2.93), 271(3.12),287(3.36), 305(3.47),326(3.32)	70-0544-65
2-Furaldehyde, 5,5'-methylenedi-	EtOH	224(3.74),289(4.46)	95-0981-65
1,3-Indandione, 2-hydroxy-, acetate	MeOH	224(4.69),244(4.02)	24-1482-64
	MeOH-base	250(4.49),257(4.48)	24-1482-64
Phthalidylidene-3-acetic acid, methyl ester	EtOH	220(4.23),238(4.23), 271(4.25),279(4.19), 305(4.00),315(3.93)	39-3811-65

Compound	Solvent	$\lambda_{max}(\log \epsilon)$	Ref.
$C_{11}H_8O_6$			
2H-1-Benzopyran-6-carboxylic acid, 7-hydroxy-5-methoxy-2-oxo-	MeOH	250(4.26),331(4.18)	32-1054-64
Glauconin	C_6H_{12}	250(4.01)	39-1772-65
Glutaconic acid, 4-(2-acetyl-1,3-dihydroxy-2-butenylidene)-3-hydroxy-, di-γ-lactone	EtOH	230(3.84),240(3.81), 260(3.72),347(3.93)	35-3004-65
$C_{11}H_8S$			
Sulfide, 1,3-pentadiynyl phenyl	n.s.g.	248(4.14),261(4.14), 275(4.14),294(4.14)?	28-2847-65A
$C_{11}H_9BF_4FeO_3$			
Bicyclo[3.2.1]octadienyliumtricarbonyl iron, tetrafluoroborate	EtOH	218(4.13),250(3.86)	35-3269-65
$C_{11}H_9BN_2OS$			
4,5-Borazarothieno[2,3-c]pyridine, 4-hydroxy-5-phenyl-	EtOH	208(4.14),306(4.12)	1-1271-65
$C_{11}H_9Br$			
1,6-Methanocyclodecapentaene, bromo-	n.s.g.	244(4.25),266(4.69), 313(3.83),395(2.42), 405(2.45),415(2.35)	89-0784B-64
$C_{11}H_9BrN_2$			
Pyrrole, 2-[N-(m-bromophenyl)formimidoyl]-	EtOH	237(3.89),329(4.36)	12-0894-64
Pyrrole, 2-[N-(p-bromophenyl)formimidoyl]-	EtOH	227s(3.82),332(4.40)	12-0894-64
$C_{11}H_9BrN_4$			
3-Phenyltetrazolo[1,5-a]pyridinium bromide	EtOH	293(3.55)	89-0171-65
$C_{11}H_9BrO$			
2-Cyclopenten-1-one, 3-(p-bromophenyl)-	MeOH	290(4.47)	44-2205-65
$C_{11}H_9BrO_2$			
But-2-enolide, 4-bromo-2-methyl-3-phenyl-	C_6H_{12}	213(4.12),273(4.07)	39-3075-65
$C_{11}H_9BrO_4$			
Coumarin, 6-bromo-4-hydroxy-3-(2-hydroxyethyl)-	n.s.g.	274(3.9),282(3.9), 312(3.9)	95-0310-65
$C_{11}H_9ClN_2$			
Dipyrido[1,2-c:2',1'-e]imidazolium chloride	H_2O	241(4.53),285(3.36), 295(3.49),338(4.19), 353(4.17),390(3.45), 412(3.11)	12-1819-65
Pyridine, 2-anilino-5-chloro-	EtOH	275(4.98),312s(3.89)	95-0429-65
Pyrrole, 2-[N-(m-chlorophenyl)formimidoyl]-	EtOH	236(3.88),328(4.36)	12-0894-64
Pyrrole, 2-[N-(p-chlorophenyl)formimidoyl]-	EtOH	227s(3.85),330(4.39)	12-0894-64

Compound	Solvent	λ_{max}(log ε)	Ref.
$C_{11}H_9ClN_2O_3$			
4-Isoxazolecarboxylic acid, 3-(4-amino-2-chlorophenyl)-5-methyl-	MeOH	218(4.21),259(3.97), 296s(3.53)	39-5976-65
4-Isoxazolecarboxylic acid, 3-(5-amino-2-chlorophenyl)-5-methyl-	MeOH	218(4.17),245s(3.93), 311(3.23)	39-5976-65
$C_{11}H_9ClN_2O_6$			
Indole-2,3-dione, 5-chloro-6,7-dimethoxy-1-methyl-4-nitro-	MeOH	223(4.39),252s(4.02), 443(2.35)	39-0026-64
$C_{11}H_9ClN_4O_2$			
4-Pyrimidinecarboxylic acid, 5-(p-chlorophenyl)-2,6-diamino-	pH 1	288(4.03)	4-0162-65
	pH 7	290(4.11)	4-0162-65
	pH 13	256(4.25),293(4.20)	4-0162-65
$C_{11}H_9ClN_4O_2$			
1H-Tetrazole-1-acetic acid, α-(p-chlorobenzylidene)-5-methyl-	EtOH	273(4.29)	44-2222-65
$C_{11}H_9ClN_4O_4$			
2,4-Pentadienal, 2-chloro-, 2,4-dinitrophenylhydrazone	EtOH	246(4.20),265(4.22), 296(4.09),382(4.56)	78-2091-64
$C_{11}H_9ClO_5$			
2-Benzofurancarboxylic acid, 5-chloro-3-(2-hydroxyethoxy)-	EtOH	215(4.20),230s(4.14), 285(4.20)	44-4126-65
$C_{11}H_9FN_2$			
Pyrrole, 2-[N-(p-fluorophenyl)-formimidoyl]-	EtOH	327(4.29)	12-0894-64
$C_{11}H_9N$			
Pyridine, 2-phenyl-	pH 1	241(3.88),294(4.13)	32-0902-64
	EtOH	245(4.11),277(4.02)	32-0902-64
	EtOH	247(4.1),277(4.1)	56-1625-65
Pyridine, 3-phenyl-	pH 1	232(4.00),256(3.98)	32-0902-64
	EtOH	246(4.12)	32-0902-64
	EtOH	247(4.3),275s(4.0)	56-1625-65
Pyridine, 4-phenyl-	pH 1	290(4.24)	32-0902-64
	EtOH	255(4.21)	32-0902-64
	EtOH	257(4.2)	56-1625-65
$C_{11}H_9NO$			
1-Naphthamide	HOAc	282(3.75)	39-4399-65
	HOAc-H_2SO_4	304(3.66)	39-4399-65
2-Naphthamide	HOAc	280(3.73)	39-4399-65
	HOAc-H_2SO_4	290(3.88)	39-4399-65
Pyridine, 2-phenyl-, 1-oxide	N HCl	289(3.95)	32-0902-64
	pH 1	241(4.18)	32-0902-64
	pH 2	241(4.31)	32-0902-64
	EtOH	246(4.40),277(3.94)	32-0902-64
Pyridine, 3-phenyl-, 1-oxide	N HCl	234(4.16),256(4.10)	32-0902-64
	pH 1	250(4.34)	32-0902-64
	pH 2	250(4.41)	32-0902-64
	EtOH	254(4.41)	32-0902-64
Pyridine, 4-phenyl-, 1-oxide	N HCl	292(4.30)	32-0902-64
	EtOH	298(4.43)	32-0902-64
Pyrrole, 2-benzoyl-	MeOH	239(4.01),307(4.20)	39-2579-64
1H-Pyrrolo[1,2-a]indol-1-one, 2,3-dihydro-	MeOH	239(4.40),315(4.32)	87-0700-65

Compound	Solvent	$\lambda_{max}(\log \epsilon)$	Ref.
$C_{11}H_9NO_2$			
3,4-Benzazepine-2,5-dione, 6-methyl-	n.s.g.	207(3.19),233(3.08), 274(2.73)	88-1071-65
Ketone, 2-furyl phenyl, oxime, anti	EtOH	240(4.01),272(4.09)	22-2724-65
Ketone, 2-furyl phenyl, oxime, syn	EtOH	238(4.08),270(4.16)	22-2724-65
1,4-Naphthoquinone, 2-(methylamino)-	dioxan	232(4.15),270(4.34), 324(3.29),439(3.48)	5-0151-64E
4H-Quinolizin-4-one, 1-acetyl-	EtOH	258(3.89),288(3.90), 340(4.13),370(4.11)	78-0945-65
4H-Quinolizin-4-one, 3-acetyl-	EtOH	260(4.01),350(3.60), 420(4.33)	78-0945-65
$C_{11}H_9NO_2S$			
Succinimide, N-thiobenzoyl-	MeOH	227(3.87),323(4.27), 545(1.83)	24-1556-65
$C_{11}H_9NO_3$			
Doryanine	EtOH	231(4.39),248(4.44), 258(4.34),284(3.77), 294(3.83),325(3.58), 338(3.45)	100-0237-65
3-Indolinepropionic acid, 3-hydroxy-2-oxo-, γ-lactone	EtOH	209(4.53),253(3.70), 292(3.18)	44-2431-64
5-Isoxazoleacetic acid, 3-phenyl-	EtOH	227.5(4.24)	78-0817-65
2,4-Pentadienal, 5-(p-nitrophenyl)-	EtOH	339(4.14)	87-0035-65
$C_{11}H_9NO_3S$			
2H-1,4-Benzothiazine-$\Delta^{3(4H),\alpha}$-acetic acid, 2-oxo-, methyl ester	MeOH	232(2.64),258s(0.77), 363(0.73)	39-3200-65
$C_{11}H_9NO_3S_2$			
3H-1,2-Dithiole-4-carbamic acid, 3-oxo-, benzyl ester	EtOH	328(3.82)	12-0447-64
$C_{11}H_9NO_4$			
L-Alanine, N-phthalyl-	dioxan	220(4.58),235(4.08), 292(3.28),300s(3.23)	24-0533-64
2H-1,4-Benzoxazine-$\Delta^{2,\alpha}$-acetic acid, 3,4-dihydro-3-oxo-, methyl ester	MeOH	256(1.10),285s(0.42), 366(1.23)	39-3200-65
Coumarin, 3-carbamoyl-8-methoxy-	EtOH	251(4.03),310(4.24)	94-0443-65
Indene, 1-acetoxy-2-nitro-	EtOH	243(3.88),338(3.93)	44-3180-64
$C_{11}H_9NO_4S$			
2(1H)-Pyridone, 1-[(phenyl-sulfonyl)oxy]-	EtOH	223(4.18),261s(3.34), 270s(3.48),276(3.54), 301(3.69)	35-5186-65
$C_{11}H_9NO_5$			
Crotonic acid, 4-hydroxy-2-methoxy-4-(3-nitrophenyl)-, γ-lactone	MeOH	224s(4.11),258(3.90)	44-1800-65
Furo[2,3-b]pyridine-3,5-dicarboxylic acid, 2,6-dimethyl-	EtOH	218(4.47),249s(3.88), 293(3.85)	39-1632-64
$C_{11}H_9NO_6$			
Coumarin, 4-hydroxy-3-(2-hydroxy-ethyl)-6-nitro-	n.s.g.	263(4.3)	95-0310-65
$C_{11}H_9N_2NaO_3$			
Sodium, [2-(2-carbamoyl-2-cyanovinyl)-6-methoxyphenoxy]-	pH 1	256(3.81),316(4.06)	94-0443-65
	pH 13	235(4.28),280(3.81), 390(3.80)	94-0443-65

Compound	Solvent	λ_{max}(log ε)	Ref.
$C_{11}H_9N_3$			
Cyclohepta[b]pyrrole-3-carbonitrile, 1,2-dihydro-2-imino-1-methyl-	EtOH	228(4.16),281(4.45), 433(4.17)	94-0828-65
Malononitrile, (α-methylamino-benzylidene)-	EtOH	229(3.96),290(4.14)	94-0828-65
1H-Naphtho[1,8-de]triazine, 1-methyl-	EtOH	232(4.50),338(4.00), 451(2.97)	39-3005-64
2H-Naphtho[1,8-de]triazine, 2-methyl-	EtOH	232(4.51),355(4.11), 559(2.91),603(2.91), 655(2.75)	39-3005-64
	n.s.g.	655(2.76)	77-0057-65
Pyrazole-4-carbonitrile, 1-o-tolyl-	EtOH	236(4.01)	95-0158-65
Pyrazole-4-carbonitrile, 1-p-tolyl-	EtOH	261(4.26)	95-0158-65
$C_{11}H_9N_3O$			
2-Pyrazoline-3-carbonitrile, 5-oxo-1-p-tolyl-	EtOH	255(4.20)	44-2024-64
10H-Pyridazino[6,1-b]quinazolin-10-one, 1,2-dihydro-	EtOH	235(4.5),275(3.9), 285(3.9),317f(3.7)	95-0339-65
5-Pyrimidinecarbonitrile, 1,2,3,4-tetrahydro-2-oxo-1-phenyl-	EtOH	286.5(4.01)	44-1740-64
$C_{11}H_9N_3OS$			
Thiocyanic acid, 3-methyl-5-oxo-1-phenyl-2-pyrazolin-4-yl ester	EtOH	232(4.16)	94-0023-64
$C_{11}H_9N_3O_2S$			
Pyrazole-4-carbonitrile, 1-[p-(methyl-sulfonyl)phenyl]-	EtOH	271(4.41)	95-0158-65
$C_{11}H_9N_3O_3$			
Barbituric acid, 5-(p-tolylimino)-	EtOH	231(4.13),485(3.18)	39-5551-65
$C_{11}H_9N_3O_3S$			
Pyrimidine, 2-(methylthio)-4-(5-nitro-2-furyl)vinyl-	propylene glycol	252(4.19),302(3.87), 380(4.18)	95-0207-64
$C_{11}H_9N_5$			
Adenine, 8-phenyl-	pH 1	232(4.15),293(4.41), 300(4.40)	44-1916-65
	NaCl	234(4.29),294(4.33)	44-1916-65
	pH 13	238(4.31),303(4.34)	44-1916-65
$C_{11}H_9N_5O_2$			
Alloxazine, 7-amino-6-methyl-	pH 13	262(4.52),420(4.28)	65-0675-65
1,2,4-Triazine, 3-amino-6-(p-nitrostyryl)-	propylene glycol	272(4.02),355(4.28)	95-0016-64 95-0109-64
s-Triazine-2,4-dicarboxaldehyde, 6-phenyl-, dioxime	MeOH	262(4.62)	44-0678-64
3H-v-Triazolo[4,5-d]pyrimidine-5,7(4H,6H)-dione, 3-benzyl-	pH 2.0	233(3.80),256(4.02)	24-1060-65
	pH 8.0	248(3.94),279(3.94)	24-1060-65
$C_{11}H_9N_5O_3$			
1,2,4-Triazine, 3-amino-6-[4-(5-nitro-2-furyl)buta-1,3-dienyl]-	propylene glycol	316(4.28),420(4.46)	95-0016-64
$C_{11}H_9N_5O_3S$			
1,2,4-Triazine, 3-acetamido-6-[2-(5-nitro-2-thienyl)vinyl]-	propylene glycol	284(4.17),396(4.31)	95-0016-64

Compound	Solvent	$\lambda_{max}(\log \epsilon)$	Ref.
$C_{11}H_9N_5O_4$			
1H-Tetrazole-1-acetic acid, 5-methyl-α-(m-nitrobenzylidene)-	EtOH	256(4.47)	44-2222-65
1,2,4-Triazine, 3-acetamido-6-[2-(5-nitro-2-furyl)vinyl]-	propylene glycol	287(4.14),388(4.36)	95-0001-64
$C_{11}H_9N_5O_5$			
1,2,4-Triazin-5(4H)-one, 3-acetamido-6-[2-(5-nitro-2-furyl)vinyl]-	propylene glycol	230(4.42),290(3.93), 384(4.37)	95-0009-64
$C_{11}H_9N_5S$			
[1,2,5]Thiadiazolo[3,4-d]pyrimidine, 7-(benzylamino)- (shoulders not listed)	pH 1	219(4.20),264(3.70), 342(4.27)	44-2135-64
	pH 7	222(4.23),272(3.71), 353(4.09)	44-2135-64
	pH 13	358(3.99)	44-2135-64
	EtOH	223(4.23),274(3.73), 360(4.07)	44-2135-64
$C_{11}H_{10}$			
Azulene, 5-methyl-	hexane	566(2.49),588(2.61), 612(2.55),642(2.59), 672(2.30),710(2.23)	5-0031-64A
Azulene, 6-methyl-	hexane	525(2.45),544(2.55), 564(2.61),587(2.56), 615(2.59),644(2.41), 679(2.34)	5-0031-64A
Benzocycloheptene	EtOH	220(4.2),272(3.7)	88-1525-64
Benzonorbornene	heptane	259(2.82),265(2.99), 271(3.05)	35-4794-65
Benzonorcaradiene	EtOH	220(4.39),235s(--), 273(3.68),304s(--)	88-1525-64
Bicyclo[4.4.1]undeca-1,3,5,7,9-pentaene	C_6H_{12}	256(4.85),259s(4.82), 298(3.80),354s(2.23), 361(2.27),369(2.26), 378(2.20),385(2.13), 390s(2.06),395s(2.00), 400(1.98)	33-1494-65
	n.s.g.	256(4.83),259(4.80), 298(3.79)	89-0145-64
	n.s.g.	256(4.83),259(4.80), 298(3.79),375f(2.27)	89-0784B-64
$C_{11}H_{10}BrClN_4$			
Pyrimidine, 2,4-diamino-6-(bromomethyl)-5-(p-chlorophenyl)-	pH 1	290(3.79)	4-0021-65
	pH 13	307(3.83)	4-0021-65
	EtOH-HBr	291(3.78)	4-0021-65
$C_{11}H_{10}BrN$			
Quinoline, 3-bromo-2,4-dimethyl-	EtOH	279(3.64),294s(3.59), 307(3.57),321(3.62)	39-0938-64
$C_{11}H_{10}BrNO_3$			
4-Indolinecarboxylic acid, 5-bromo-1-methyl-2-oxo-, methyl ester	EtOH	262(4.18),312(3.25)	39-3229-64
$C_{11}H_{10}BrNS$			
Isothiazole, 5-benzyl-4-bromo-3-methyl-	EtOH	207(3.77),258(3.94)	39-0446-64

Compound	Solvent	λ_{max}(log ε)	Ref.
$C_{11}H_{10}BrN_3O_2$			
2H-1,2,3-Triazole-4-carboxylic acid, 2-(p-bromophenyl)-, ethyl ester	EtOH	278(4.35)	39-2306-64
$C_{11}H_{10}BrN_3O_2S$			
Pyrimidine, 4-(p-bromoanilino)-	pH 1	274(4.17)	4-0001-64
6-(methylsulfonyl)-	pH 11	276(4.21)	4-0001-64
$C_{11}H_{10}ClN$			
Quinoline, 3-chloro-2,4-dimethyl-	EtOH	278(3.63),293s(3.57), 307(3.55),321(3.62)	39-0928-64
$C_{11}H_{10}ClNO$			
Carbostyril, 3-chloro-1,4-dimethyl-	EtOH	277(3.91),286s(3.84), 321s(3.77),331(3.86)	39-0938-64
$C_{11}H_{10}ClNO_6$			
2-Carboxyquinolizinium perchlorate, methyl ester	MeOH	238(4.29),290(3.47), 315s(3.81),328(4.12), 343(4.23)	39-3225-64
$C_{11}H_{10}ClN_3O$			
3(2H)-Pyridazinone, 5-(benzylamino)-	EtOH	231(4.48),306(3.83)	22-2124-64
4-chloro-	NaOH	224(4.60),289(3.95)	22-2124-64
$C_{11}H_{10}ClN_3O_2$			
Isoxazole, 4-(o-chlorophenylazo)-	hexane	238(4.01),338(4.12)	25-1264-64
5-methoxy-3-methyl-	hexane	238(4.01),338(4.12)	39-3312-65
2-Isoxazolin-5-one, 4-(m-chlorophenylazo)-2,3-dimethyl-	EtOH	237(3.92),279(3.86), 351(4.24)	39-3312-65
2-Isoxazolin-5-one, 4-(o-chlorophenylazo)-2,3-dimethyl-	EtOH	242(4.01),272(3.89), 362(4.21)	39-3312-65 25-1264-64
2-Isoxazolin-5-one, 4-[methyl-(o-chlorophenyl)hydrazono]-3-methyl-	hexane	238(3.76),358(4.14)	25-1264-64
1,2,3-Triazole-4-carboxylic acid, 2-(p-chlorophenyl)-, ethyl ester	EtOH	275(4.25)	39-2306-64
1,2,3-Triazole-4-carboxylic acid, 2-(o-chlorophenyl)-5-methyl-, methyl ester	hexane	259(4.16)	25-1264-64
$C_{11}H_{10}Cl_2N_2O_3$			
1,2,4-Oxadiazolidin-5-one, 4-(dichloroacetyl)-2-methyl-3-phenyl-	n.s.g.	210(4.4),259(4.3)	83-0623-64
$C_{11}H_{10}Cl_3NOS_2$			
Butyrophenone, γ,γ,γ-trichloro-	H_2O	241(3.99),287(4.14)	39-4004-64
β-(thiocarbamoylthio)-	EtOH	243(4.17),292(4.22)	39-4004-64
$C_{11}H_{10}FeO_3$			
Iron, (bicyclo[3.2.1]octadiene)-tricarbonyl-	EtOH	215(4.35)	35-3269-65
$C_{11}H_{10}HgO_5S_2$			
Benzoic acid, p-[[(carboxymethyl)thio]-mercuri]thio-, S-ester with mercaptoacetic acid	pH 9	246(4.3),276(4.3)	70-0111-64
$C_{11}H_{10}INOS_3$			
4-Benzamido-3-(methylthio)-1,2-dithiolium iodide	n.s.g.	316(4.03),442(3.55)	12-1071-65

Compound	Solvent	λ_{max}(log ϵ)	Ref.
$C_{11}H_{10}N_2$			
2,4-Norcaradiene-7,7-dicarbonitrile, 1,4-dimethyl-	n.s.g.	235s(3.27),276(3.41)	35-0652-65
2,4-Norcaradiene-7,7-dicarbonitrile, 2,5-dimethyl-	n.s.g.	238(3.52),279(3.69)	35-0652-65
Pyridazine, 4-methyl-5-phenyl-	pH 1	297(3.30)	44-2128-64
Pyrrole, 2-(N-phenylformimidoyl)-	EtOH	325(4.34)	12-0894-64
$C_{11}H_{10}N_2O$			
3,4-Diazabicyclo[4.1.0]hept-4-en-2-one, 5-phenyl-	MeOH	287(4.29)	24-2438-65
3(2H)-Pyridazinone, 6-p-tolyl-	EtOH	256.5(4.43)	39-3342-65
Pyrimidine, 2-(benzyloxy)-	pH 7.0	206(4.31),263(4.26)	39-5542-65
2(1H)-Pyrimidinone, 1-benzyl-	pH 7.0	209(4.17),305(3.77)	39-5542-65
Pyrrolo[4,3,2-de]quinoline-5(1H)-carboxaldehyde, 3,4-dihydro-	EtOH	225(4.50),288s(3.98), 294(3.99)	44-1158-64
$C_{11}H_{10}N_2O_2$			
Hydantoin, 5-(α-methylbenzylidene)-	n.s.g.	222(4.03),299(4.18)	28-1987-65B
Indole-3-carboxaldehyde, 5-acetamido-	EtOH	246(4.37),304(4.06)	87-0389-64
3(2H)-Pyridazinone, 6-(p-methoxyphenyl)-	EtOH	268(4.45)	39-3342-65
Pyrido[1,2-b]indazole-7,10-dione, 1,2,3,4-tetrahydro-	EtOH	240(4.28),328(3.42)	39-5871-65
3H,5H-Pyrido[1,2,3-de]quinoxalin-3-one, 6,7-dihydro-2-hydroxy-	MeOH	236(3.98),242(3.93), 261(3.73),270(3.69), 312(3.97),325(3.91), 340s(3.55)	44-2589-65
5,8-Quinoxalinedione, 2,3,6-trimethyl-	MeOH	264(4.26)	4-0171-64
Uracil, 3-methyl-5-phenyl-	dioxan	242(4.2),274(4.2)	83-0367-64
$C_{11}H_{10}N_2O_2S$			
2H-Isothiazolo[4,5-b]indol-3(4H)-one, 2,4-dimethyl-, 1-oxide	EtOH	213(4.57),233(4.37), 262(3.56),305(3.92)	44-0178-64
2-Thiophenamine, 4-methyl-3-(p-nitrophenyl)-	MeOH	219(4.09),272(4.03), 381(3.51)	24-3571-65
$C_{11}H_{10}N_2O_3$			
2H-1-Benzopyran-3-carboxamide, 2-imino-8-methoxy-	EtOH	310(4.13)	94-0443-65
Indole, 3-acetyl-2-methyl-4-nitro-	EtOH	228(4.27),250s(4.11), 280(3.87),366(3.54)	44-3457-65
Indole, 3-acetyl-2-methyl-6-nitro-	EtOH	260s(4.11),283(4.33), 351(4.00)	44-3457-65
Indole-3-carboxaldehyde, 1,2-dimethyl-	EtOH	211(4.3),260(4.0), 270s(3.9),277(3.9), 360(4.04)	78-1397-64
Indole-3-carboxaldehyde, 1,2-dimethyl-5-nitro-	EtOH	269(4.45),320(3.97)	78-1923-65
Indole-3-carboxaldehyde, 1,2-dimethyl-6-nitro-	EtOH	286(4.42),348(3.95)	78-1923-65
$C_{11}H_{10}N_2O_4$			
Indole-3-acrylic acid, 5-nitro-	EtOH	276(4.28)	87-0389-64
Phthalimidoacetic acid, 3-amino-, methyl ester	EtOH	232s(4.53),258(3.89), 389(3.76)	44-3151-64
	dil. HCl	256s(3.77),294(2.47), 384(3.66)	44-3151-64
	dil. NaOH	306(3.47)	44-3151-64

Compound	Solvent	λ_{max}(log ε)	Ref.
Phthalimidoacetic acid, 4-amino-, methyl ester	EtOH	257(4.45),307(3.71), 375(3.67)	44-3151-64
	dil. HCl	232(4.13),240(4.18), 260(4.16),301(3.48)	44-3151-64
	dil. NaOH	266(4.03)	44-3151-64
2-Pyridineacrylic acid, 5-cyano-1,6-dihydro-α-hydroxy-6-oxo-, ethyl ester	EtOH	277(3.81),294s(3.73), 420(4.25)	35-5198-65
Sydnone, 4-methyl-3-piperonyl-	EtOH	291(4.10)	87-0531-65
Sydnone, 3-[(3,4-methylenedioxy)-phenethyl]-	EtOH	287(4.09)	87-0531-65
$C_{11}H_{10}N_4$			
5-Imidazolecarbonitrile, 4-amino-1-benzyl-	EtOH-HCl	236(4.00),264(3.81)	4-0291-65
	EtOH	231(3.65),267(3.90)	4-0291-65
	EtOH-NaOH	267(3.93)	4-0291-65
Triazene, 1-(α-pyridyl-3-phenyl)-	EtOH	339(4.32)	89-0171-65
$C_{11}H_{10}N_4O$			
Pyrimidine, 4-amino-5-benzamido-	pH 1	234(4.21),268s(4.02)	44-1916-65
	NaCl	228(4.25),280(3.85)	44-1916-65
	pH 13	230s(4.11),285(3.96)	44-1916-65
s-Triazine-2-carboxaldehyde, 4-methyl-6-phenyl-, oxime	MeOH	257(4.39)	44-0678-64
$C_{11}H_{10}N_4OS$			
4-Pyrimidinethiol, 6-amino-5-benzamido-	pH 13	225(4.38),245s(4.26), 292(4.11)	44-1916-65
$C_{11}H_{10}N_4OS_2$			
Uracil, 6-amino-5-benzamido-2,4-dithio-	pH 1	215(4.30),286(4.43), 329(4.32)	44-1916-65
	aq. NaCl	225(4.37),262(4.32), 320(4.19)	44-1916-65
	pH 13	222(4.43),270(4.42), 319(4.08)	44-1916-65
$C_{11}H_{10}N_4O_2$			
1H-Tetrazole-1-acetic acid, α-benzylidene-5-methyl-	EtOH	267(4.29)	44-2222-65
Urea, (3-anilino-2-cyanoacryloyl)-	H_2O	326(4.40)	39-1642-65
$C_{11}H_{10}N_4O_2S$			
Acetic acid, [benzylidene(1,3,4-thiadiazol-2-yl)hydrazino]-	EtOH	227(4.15),321(4.38)	1-0871-64
$C_{11}H_{10}N_4O_4$			
2-Isoxazolin-5-one, 3,4-dimethyl-4-(p-nitrophenylazo)-	EtOH	281(4.24),416(2.36)	39-5414-65
1,3,4-Oxadiazole, 2-amino-5-[3-methyl-4-(5-nitro-2-furyl)buta-1,3-dienyl]-	propylene glycol	246(4.03),324(4.23), 410(4.21)	95-0212-64
1,2,3-Triazole-4-carboxylic acid, 2-(p-nitrophenyl)-, ethyl ester	EtOH	305(4.28)	39-2306-64
$C_{11}H_{10}N_4O_6$			
Malealdehydic acid, 3-methyl-, 2,4-dinitrophenylhydrazone	EtOH	372(4.49)	39-0594-64
$C_{11}H_{10}N_4O_7$			
Hydantoin, 3-(acetoxymethyl)-1-[(5-nitrofurfurylidene)amino]-	2% DMF	263(4.12),365(4.27)	44-3416-64

Compound	Solvent	$\lambda_{max}(\log \epsilon)$	Ref.
$C_{11}H_{10}O$			
Azulene, 1-methoxy-	C_6H_{12}	236(4.15),285(4.71), 341(3.46),350(3.54), 358(3.76),367(3.48), 376(3.79),632s(2.40), 661(2.46),684(2.50), 725(2.44),815(2.07), 863(2.02)	44-2805-64
Azulene, 6-methoxy-	hexane	513(2.37),530(2.47), 548(2.39),573(2.32), 600(2.01),632(1.91)	5-0031-64A
2-Naphthol, 1-methyl-	$CHCl_3$	229(4.83),279(3.66), 299(3.57),337(3.57)	94-0112-64
1,10-Undecadiene-4,7-diyn-6-one	MeOH	244s(4.1),286(2.5)	5-0026-64I
$C_{11}H_{10}OS$			
4-Hexen-2-yn-1-one, 5-methyl- 1-(2-thienyl)-	EtOH	272(4.07),302(4.27), 314(4.27)	39-2983-65
$C_{11}H_{10}O_2$			
But-2-enolide, 2-methyl-3-phenyl-	C_6H_{12}	212(4.12),260(4.20)	39-3075-65
3-Chromen-6-ol, 5,8-dimethyl-	EtOH	233(4.21),273(3.83), 281s(3.83),340(3.51)	39-5060-65
3-Chromen-6-ol, 7,8-dimethyl-	EtOH	232(4.34),270(3.76), 278s(3.70),334(3.62)	39-5060-65
Cyclopentadiene, benzoquinone adduct	C_6H_{12}	226(4.11),238s(3.83), 278(2.39),385(1.76)	39-3043-64
	EtOH	226(4.11),239s(3.92), 284(2.41),374(1.80)	39-3043-64
1,2-Cyclopentanedione, 4-phenyl-	MeOH	253(3.97)	44-3573-65
1,3-Cyclopentanedione, 2-phenyl-	neutral	257(4.15)	94-1300-65
$\Delta^{1,\alpha}$-Indanacetic acid, trans	EtOH	226(4.15),233(4.06), 272(4.17),283(4.14), 304(4.12),315(4.04)	95-0975-65
Indene-2-acetic acid	EtOH	252(4.00)	95-0975-65
1-Naphthoic acid, 3,4-dihydro-	EtOH	222(4.08),270(3.79)	54-0389-65
1-Naphthol, 2-methoxy-	EtOH	237(4.53),292(3.54), 302(3.54),330(3.35)	39-2355-65
1-Naphthol, 8-methoxy-	EtOH	298(3.86),316(3.82), 330(3.83)	39-2355-65
2-Naphthol, 1-methoxy-	EtOH	230(4.49),280(3.52), 291(3.47),333(3.33)	39-2355-65
Pentacyclo[6.2.1.0^{2,7}.0^{4,10}.0^{5,9}]un- decane-3,6-dione	C_6H_{12}	219(2.09),301f(1.43), 318f(1.41)	39-3062-64
	EtOH	287f(1.38)	39-3062-64
4-Pentene-2,3-dione, 5-phenyl-	MeOH	223(3.9),228(3.92), 302(4.25)	44-3573-65
2H-Pyran-2-one, 5,6-dihydro- 6-phenyl-	MeOH	249(2.62),257(2.60), 261(2.51),267(2.33), 304(1.04)	78-2939-65
$C_{11}H_{10}O_2S$			
Benzo[b]thiophene-2-carboxylic acid, 3-methyl-, methyl ester	EtOH	235(4.06),246(4.11), 281(3.97),291(4.03), 315(3.29),326s(3.25)	4-0231-65
$C_{11}H_{10}O_3$			
Coumarin, 8-methoxy-4-methyl-	MeOH	218(4.18),283(4.11), 292s(4.10)	25-0383-65
1-Cyclopropanecarboxylic acid, 2-benzoyl-, cis	MeOH	245.5(4.18)	24-2438-65

Compound	Solvent	λ_{max}(log ε)	Ref.
Isocoumarin, 5-hydroxy-4,8-dimethyl-	EtOH	230(4.28),267(4.03), 353(3.76)	42-0821-64
Isocoumarin, 7-methoxy-4-methyl-	MeOH	230(4.40),268(4.08), 341(3.69)	42-0821-64
2-Naphthoic acid, 5,6,7,8-tetra-hydro-5-oxo-	EtOH	255(4.26)	44-2165-65
Psilotinin	MeOH	218(4.19),278(3.23), 280(3.17),320(1.63)	78-2939-65
$C_{11}H_{10}O_3S$			
Thiocoumarin, 7-acetoxy-3,4-dihydro-	EtOH	260(3.88)	39-3126-64
$C_{11}H_{10}O_4$			
Cinnamic acid, 3,4-methylenedioxy-, methyl ester	EtOH	325(--)	95-0857-65
Coumarin, 4-hydroxy-3-(2-hydroxy-ethyl)-	n.s.g.	268(3.9),278(3.9), 302(4.0)	95-0310-65
2-Decene-4,6-diynedioic acid, 10-methyl ester	EtOH	220(3.23),265(4.33), 280(4.01),298(4.23), 340(3.21)	70-1237-65
Limettine	EtOH	209(4.36),241(3.75), 255(3.74),325(4.09)	33-0408-64
Phthalide-3-acetic acid, methyl ester	EtOH	227(4.00),272(3.22), 279(3.21)	39-3811-65
Sepedonin, anhydro-	EtOH	249(4.20),286(4.47), 302s(4.27),354s(3.92), 375(3.86),386(3.91)	23-1835-65
Valeric acid, 3,5-dioxo-5-phenyl-	EtOH	248(2.77),312(3.19)	22-0525-65
$C_{11}H_{10}O_4S$			
2,4,8-Decatrien-6-ynoic acid, 5-hy-droxy-9-(methylsulfonyl)-, δ-lactone	ether	235(4.15),337(4.47)	24-1616-65
$C_{11}H_{10}O_7$			
Glyoxylic acid, (2-carboxy-3,5-di-methoxyphenyl)-	EtOH	259(4.16),295(3.85)	1-1677-65
$C_{11}H_{10}S$			
Azulene, 1-(methylthio)-	C_6H_{12}	200(4.24),235(4.23), 278(4.46),285(4.48), 289(4.48),314(3.86), 344(3.59),352(3.59), 360(3.62),581(2.43), 599(2.43),627(2.42), 695(2.15)	44-2715-65
$C_{11}H_{11}BN_2O_2S$			
2-Thiopheneboronic acid, 5-formyl-, phenylhydrazone	EtOH	253(4.26),294(3.86), 370(4.44)	1-1271-65
3-Thiopheneboronic acid, 4-formyl-, phenylhydrazone	EtOH	212(4.34),246(4.05), 303(4.22),334(4.42)	1-1271-65
$C_{11}H_{11}BrN_4O_2$			
Diacetamide, N-[3-bromo-6-methyl-pyrazolo[1,5-a]pyrimidin-7-yl)-	EtOH	238(4.57),282(3.20), 291(3.26),302(3.17), 345(3.32)	94-0142-65
$C_{11}H_{11}BrO_2$			
1,4-Pentanedione, 1-(p-bromophenyl)-	MeOH	255(4.15)	44-2205-65

Compound	Solvent	$\lambda_{max}(\log \epsilon)$	Ref.
$C_{11}H_{11}Br_2N$			
3H-Indole, 3-(dibromomethyl)-2,3-dimethyl-	EtOH	266(3.64)	39-0938-64
$C_{11}H_{11}ClN_2O_6S_2$			
2-(Dimethylamino)-4-(p-nitrophenyl)-1,3-dithiolium perchlorate	EtOH	223(4.32),255(3.96),323(4.32)	44-1703-64
$C_{11}H_{11}ClN_4O_5$			
2-Pentenal, 2-chloro-5-hydroxy-, 2,4-dinitrophenylhydrazone	EtOH	212(4.24),253(4.17),282s(3.96),371(4.45)	78-2091-64
$C_{11}H_{11}ClO$			
3-Penten-2-one, 5-chloro-1-phenyl-	heptane	212(3.26)	44-3327-64
$C_{11}H_{11}ClS$			
Sulfide, 1-chloro-3-methyl-1,3-butadienyl phenyl	EtOH	251(4.18),280s(3.92)	44-0728-65
$C_{11}H_{11}Cl_2N$			
3H-Indole, 3-(dichloromethyl)-2,3-dimethyl-	EtOH	262(3.71)	39-0928-64
$C_{11}H_{11}Cl_2NO_4S_2$			
4-(p-Chlorophenyl)-2-(dimethylamino)-1,3-dithiolium perchlorate	EtOH	230(4.23),237(4.22),305(4.16),315(4.15)	44-1703-64
$C_{11}H_{11}FO$			
5H-Benzocyclohepten-5-one, 2-fluoro-6,7,8,9-tetrahydro-	EtOH	247(3.96)	44-2165-65
$C_{11}H_{11}FO_2$			
Cinnamic acid, α-fluoro-, ethyl ester	C_6H_{12}	273(4.39)	22-2258-64
$C_{11}H_{11}F_2N$			
3H-Indole, 3-(difluoromethyl)-2,3-dimethyl-	EtOH	256(3.73)	39-0938-64
$C_{11}H_{11}IN_2O_6$			
Cyclodeoxyuridine, 5-iodo-3'-O-acetyl-O^6,5'-6-hydroxy-	pH 2	282(3.95)	44-3913-65
	pH 11	275(3.84)	44-3913-65
	H_2O	280(3.95)	44-3913-65
	90% EtOH	277(3.98)	44-3913-65
$C_{11}H_{11}N$			
Benz[c,d]indole, 1,3,4,5-tetrahydro-	EtOH	225(4.54),275(3.74),281(3.76),291(3.64)	44-0843-64
Pyridine, 2-cyclopentadienylidene-1,2-dihydro-1-methyl-	dioxan	240s(3.6),360(4.2),450s(3.9)	35-2908-65
Pyridine, 4-cyclopentadienylidene-1,4-dihydro-1-methyl-	dioxan	429(4.52)	35-2887-65
	dioxan	220(3.8),310s(3.3),420(4.5)	35-2908-65
$C_{11}H_{11}NO$			
1-Indanone, 2-ethyl-3-imino-	MeOH	221(4.38),254(4.57),263(4.60),303(3.03),440(3.42)	65-0448-64
2-Indolinone, 3-isopropylidene-	EtOH	252(4.44),260(4.51),291(3.82)	44-2431-64

Compound	Solvent	λ_{max}(log ϵ)	Ref.
1-Naphthol, 7-(methylamino)-	EtOH	250(4.57),291(3.95), 301(3.90),350(3.31)	44-1180-64
1-Pyridinol, 1,2-dihydro-2-phenyl-	EtOH	237(4.03),313(4.71)	44-0910-65
Skatole, 5-acetyl-	EtOH	257(4.53),298(3.88)	87-0141-64
$C_{11}H_{11}NOS_3$			
3H-1,2-Dithiol-3-one, 4-(methylamino)-5-(p-tolylthio)-	EtOH	367.5(3.83)	12-0061-65
$C_{11}H_{11}NO_2$			
2-Butenolide, 4-anilino-3-methyl-	n.s.g.	207(4.22),237(4.13), 280(3.43)	39-0594-64
Butyrolactam, N-benzoyl-	EtOH	206(4.10),229(3.99), 239s(3.94)	65-1394-65
Indole-3-acetic acid, 1-methyl-	EtOH	224(4.53),276(3.72), 287(3.76),297s(3.67)	35-3796-64
	H_2SO_4	232(3.63),237(3.62), 280(3.66)	35-3796-64
7-Isoquinolinol, 6-methoxy-1-methyl-	EtOH-HCl	251(4.80),310(3.85)	33-1945-65
	EtOH	235(4.78),268(3.67), 276s(3.62),317(3.55), 328(3.60)	33-1945-65
	EtOH-NaOH	254(4.70),289(3.90), 355(3.75)	33-1945-65
hydrochloride	EtOH-HCl	251(4.76),310(3.79)	33-1945-65
	EtOH	237(4.62),310(3.46), 327s(3.31)	33-1945-65
	EtOH-NaOH	254(4.64),289(3.89), 355(3.75)	33-1945-65
Isoquinolone, 7-methoxy-4-methyl-	MeOH	223(4.38),255(3.83), 278(4.00),345(3.76)	42-0821-64
$C_{11}H_{11}NO_2S_4$			
p-Toluenesulfonamide, N-methyl-N-(3-thioxo-3H-1,2-dithiol-4-yl)-	EtOH	226(4.29),323(3.57), 415(3.94)	12-0061-65
$C_{11}H_{11}NO_3$			
Anthranilic acid, N-(3-oxo-1-butenyl)-	MeOH	347(4.48)	83-0321-64
Benzoic acid, p-[(3-oxo-1-butenyl)-amino]-	MeOH	345(4.65)	83-0321-64
4,1-Benzoxazepine-2,5(1H,3H)-dione, 1,3-dimethyl-	iso-PrOH	217(4.58),296(3.56), 246s(4.17)	44-0582-64
Carbostyril, 3-hydroxy-6-methoxy-1-methyl-	pH 13	219(4.54),230(4.49), 338(4.17),352(4.13)	39-1080-65
Cinnamic acid, O-acetamido-	MeCN	286(4.36)	35-4406-64
Cinnamamide, N-acetoxy-	MeCN	274(4.46)	35-4406-64
Indole-2-carboxylic acid, 5-methoxy-6-methyl-	pH 1	298(4.29)	44-2897-65
	pH 13	291(4.24)	44-2897-65
	MeOH	294(4.26)	35-3877-64
	MeOH	294(4.26)	44-2897-65
4-Indolinecarboxylic acid, 1-methyl-2-oxo-, methyl ester	EtOH	255(4.06),315(3.43)	39-3229-64
3-Indolinepropionic acid, 2-oxo-	EtOH	250(3.93),280s(3.15)	44-2431-64
2-Pyridinepropionic acid, α-(2-hydroxyethyl)-α-methyl-β-oxo-, γ-lactone	EtOH	233(3.93),269(3.64)	94-1338-64
2-Pyridinepropionic acid, α-(2-hydroxyethyl)-3-methyl-β-oxo-, γ-lactone	EtOH	234(3.85),278(3.71)	94-1338-64
2-Pyridinepropionic acid, α-(2-hydroxyethyl)-5-methyl-β-oxo-, γ-lactone	EtOH	246(4.03),273(3.86)	94-1338-64

Compound	Solvent	$\lambda_{max}(\log \epsilon)$	Ref.
2-Pyridinepropionic acid, α-(2-hydroxypropyl)-β-oxo-, γ-lactone	EtOH	234(3.93),270(3.70), 302(3.42)	94-1338-64
2-Pyrrolidinone, 1-salicyloyl-	EtOH	210(4.27),225s(4.10), 302(3.36)	65-1394-65
	EtOH	302(3.36)	78-3537-65
	dioxan	210(4.27),225(4.10)	78-3537-65
$C_{11}H_{11}NO_3S$			
4H-1,2-Benzothiazin-4-one, 3-ethylidene-2,3-dihydro-2-methyl-, 1,1-dioxide	EtOH	263(4.11),301s(3.54)	44-2241-65
Indole-2-carboxylic acid, 1-methyl-3-sulfeno-, 2-methyl ester	EtOH	213(4.40),236s(4.10), 297(3.99)	44-0178-64
$C_{11}H_{11}NO_3S_3$			
p-Toluenesulfonamide, N-methyl-N-(3-oxo-3H-1,2-dithiol-4-yl)-	EtOH	230(4.16),320(3.79)	12-0061-65
$C_{11}H_{11}NO_4$			
Indole-2-carboxylic acid, 5-hydroxy-6-methoxy-, methyl ester	EtOH	255s(3.15),315(3.60)	33-0252-65
	EtOH-NaOH	310(3.69),363(3.49)	33-0252-65
3-Indolinepropionic acid, 3-hydroxy-2-oxo-	EtOH	209(4.42),252(3.78), 288(3.14)	44-2431-64
1(2H)-Naphthalenone, 3,4-dihydro-8-methoxy-5-nitro-	EtOH	243(4.12),305(3.82)	39-2816-64
1(2H)-Naphthalenone, 3,4-dihydro-8-methoxy-7-nitro-	EtOH	241(4.25),302(3.52)	39-2816-64
Salicylic acid, 4-[(3-oxo-1-butenyl)-amino]-	MeOH	347(4.66)	83-0321-64
$C_{11}H_{11}NO_6$			
Pyruvic acid, (5-methoxy-4-methyl-2-nitrophenyl)-	MeOH	285(3.87),308(3.86)	44-2897-65
	pH 1	331(3.84)	44-2897-65
	pH 13	318(4.1)	44-2897-65
$C_{11}H_{11}NS_2$			
2H-1,3-Thiazine-2-thione, 3,6-dihydro-3-methyl-6-phenyl-	H_2O	305(4.11)	39-4008-64
	EtOH	314(4.12)	39-4008-64
2H-1,3-Thiazine-2-thione, 3,6-dihydro-4-methyl-6-phenyl-	H_2O	305(4.06),325s(--)	39-4008-64
	EtOH	323(4.08)	39-4008-64
$C_{11}H_{11}N_3$			
Pyrimidine, 2-(benzylamino)-	pH 1.0	230(4.00),316(3.23)	39-5542-65
	pH 7.0	236(3.97),305(3.11)	39-5542-65
$C_{11}H_{11}N_3O$			
Cinnamamide, α-cyano-β-(methylamino)-	EtOH	220(4.01),291(4.25)	94-0828-65
Formamidine, N-(5-methyl-3-isoxazolyl)-N'-phenyl-	EtOH	278(4.29)	78-0159-64
$C_{11}H_{11}N_3O_2$			
3-Isoxazolin-5-one, 2,3-dimethyl-4-(phenylazo)-	EtOH	232(3.83),285(3.83), 342(4.21)	39-3312-65
2H-1,2,3-Triazole-4-carboxylic acid, 2-phenyl-, ethyl ester	EtOH	270(4.29)	39-2306-64
$C_{11}H_{11}N_3O_2S$			
3-Pyrazoline-4-carbonitrile, 1-[(o-methylsulfonyl)phenyl]-	EtOH	249(3.95),288(4.08)	95-0158-65

Compound	Solvent	$\lambda_{max}(\log \epsilon)$	Ref.
Pyridazine, 3-anilino-6-(methylsulfonyl)-	pH 1	272(4.03)	4-0001-64
	pH 11	288(4.17)	4-0001-64
	EtOH	294(4.30)	4-0001-64
$C_{11}H_{11}N_3O_3$			
Barbituric acid, 5-p-toluidino-, p-toluidine salt	EtOH	250(4.26)	39-5556-65
Sydnone imine, 4-methyl-3-piperonyl-	EtOH-HCl	294(4.06)	87-0531-65
$C_{11}H_{11}N_3O_4$			
Imidazo[1,2-c]pyrimidin-5(1H)-one, 2-acetoxy-1-acetyl-3-methyl-	EtOH	222(4.23),303(3.93)	44-1762-64
Indole, 2-ethyl-1-methyl-3,5-dinitro-	EtOH	256(4.38),308(4.02), 356(4.02)	44-3457-65
1,3,4-Oxadiazole, 5-[1-ethyl-2-(5-nitro-2-furyl)vinyl]-2-methyl-	propylene glycol	265(3.94),368(4.10)	95-0225-64
$C_{11}H_{11}N_3O_4S$			
Acetoacetamide, 2-diazo-N-(p-tolylsulfonyl)-	MeOH	233(4.48)	5-0101-64F
1,3,4-Oxadiazole, 5-[1-ethyl-2-(5-nitro-2-furyl)vinyl]-2-(methylthio)-	propylene glycol	222(4.06),282(3.98), 375(4.16)	95-0225-64
$C_{11}H_{11}N_3O_6$			
o-Diacetotoluidide, 3'',5''-dinitro-	EtOH	239(4.63),318(3.29)	44-3457-65
o-Diacetotoluidide, 5'',6''-dinitro-	EtOH	264(3.88),336s(2.85)	44-3457-65
$C_{11}H_{11}N_3O_7$			
Proline, 3-hydroxy-N-(2,4-dinitrophenyl)-, cis	MeOH	365(4.27)	35-5293-64
	N NaOH	385(4.24)	35-5293-64
trans	MeOH	364(4.26)	35-5293-64
	N NaOH	387(4.24)	35-5293-64
$C_{11}H_{11}N_5$			
Mesoxalonitrile, [o-(dimethylamino)-phenyl]hydrazone	MeCN	245s(3.92),277(3.49), 379(4.22)	44-4198-65
$C_{11}H_{11}N_5O$			
Pyrazolo[1,5-a]pyrimidine-6-carbonitrile, 7-acetamido-2,3-dimethyl-	EtOH	252(4.62),306(3.57), 318(3.59),346(3.31)	94-1207-65
Pyrimidine, 4,6-diamino-5-benzamido-	pH 1	218(4.52),264(4.12)	44-1916-65
	aq. NaCl	257s(4.02)	44-1916-65
	pH 13	270(3.98)	44-1916-65
$C_{11}H_{11}N_5O_2S$			
1,2,5-Thiadiazole-3-carboxamide, N-benzyl-4-ureido-	pH 1	233(4.20),305(3.94)	44-2141-64
	pH 7	233(4.22),304(3.94)	44-2141-64
$C_{11}H_{11}N_5O_3$			
Uracil, 6-amino-5-(benzylnitrosamino)-	pH 4.0	233s(3.98),259(4.23)	24-1060-65
	pH 10.0	249(4.16)	24-1060-65
$C_{11}H_{11}N_5O_4S$			
Adenosine, cyclic 2',3'-thiocarbonate	EtOH	225s(4.23),241(4.26)	44-2854-65
$C_{11}H_{11}N_5O_5$			
1,2,4-Triazine, 3-bis(hydroxymethyl)-amino-6-[2-(5-nitro-2-furyl)vinyl]-	DMF	292(4.24),405(4.44)	95-0016-64

Compound	Solvent	$\lambda_{max}(\log \epsilon)$	Ref.
$C_{11}H_{12}$			
1,2-Benzocyclohepta-1,3-diene	EtOH	254(4.2)	78-0231-64
1H-Cyclopropa[a]naphthalene, 1a,2,3,7b-tetrahydro-	EtOH	200(4.5),230s(3.8), 260s(2.5),270(2.7), 275(2.8)	35-0908-64
Methane, dicyclopentadienyl-	C_6H_{12}	249(3.86)	33-0955-65
1,3-Pentadiene, 1-phenyl-	n.s.g.	282(4.34)	5-0058-65B
Spiro[cyclopropane-1,1'-indan]	EtOH	190(4.6),220(4.0), 260s(3.0),270(3.2), 280(3.2)	35-0908-64
$C_{11}H_{12}BrN_3O_4$			
3H-Imidazo[4,5-b]pyridine, 6-bromo-3-β-D-ribofuranosyl-	pH 0.46	240(3.48),292(4.03)	44-4066-65
	pH 4.93	250(3.51),295(4.01)	44-4066-65
$C_{11}H_{12}ClNO$			
4H-1,3-Benzoxazine, 4-(chloromethyl)-2,4-dimethyl-	EtOH	259(3.87)	39-0928-64
$C_{11}H_{12}ClNOS_2$			
Carbamic acid, dimethyldithio-, ester with 4'-chloro-2-mercaptoacetophenone	EtOH	254(4.42)	44-1703-64
$C_{11}H_{12}ClNO_2$			
Benzamide, p-chloro-N-isobutyryl-	EtOH	244(4.20)	44-2222-65
$C_{11}H_{12}ClNO_4S_2$			
2-(Dimethylamino)-4-phenyl-1,3-dithiolium perchlorate	70% $HClO_4$	225(4.21),235(4.15), 300(4.03),315(4.08)	44-1703-64
$C_{11}H_{12}ClNO_5S_2$			
2-(Dimethylamino)-4-(p-hydroxyphenyl)-1,3-dithiolium perchlorate	EtOH	261(4.09),292(4.11), 330(3.96)	44-1703-64
$C_{11}H_{12}ClNO_8$			
1-(1,2-Dicarboxyvinyl)pyridinium perchlorate, dimethyl ester	H_2O	262(3.68)	39-2676-64
$C_{11}H_{12}ClN_3O_2$			
2,4-Dimethyl-7-nitro-1,5-benzodiazepinium chloride	MeOH-HCl	260(4.29),309(4.05), 350s(3.55),515(2.86)	39-3785-65
$C_{11}H_{12}ClN_3O_4$			
3H-Imidazo[4,5-c]pyridine, 4-chloro-3-β-D-ribofuranosyl-	EtOH	275(3.72)	44-2611-64
$C_{11}H_{12}Cl_2N_2O_4$			
7-Chloro-2,4-dimethyl-1,5-benzodiazepinium perchlorate	MeOH-HCl	237(4.24),268(4.54), 277(4.52),332(2.99), 343(2.96),516(2.97)	39-3785-65
$C_{11}H_{12}F_3N_5O_4$			
Adenosine, 2-(trifluoromethyl)-	pH 1	256(4.02)	87-0866-65
	pH 13	255(4.10)	87-0866-65
$C_{11}H_{12}N_2$			
Pyrido[1,2-a]benzimidazole, 6,7,8,9-tetrahydro-	n.s.g.	232(4.33),238(4.32), 278(3.59),284(3.63), 312(3.62)	32-0485-64

Compound	Solvent	λ_{max}(log ε)	Ref.
Pyrrolo[4,3,2-de]quinoline, 1,3,4,5-tetrahydro-1-methyl-, hydrochloride	EtOH	228(4.52),284(3.76), 303s(3.81),310(3.85)	44-1158-64
Pyrrolo[4,3,2-de]quinoline, 1,3,4,5-tetrahydro-5-methyl-	EtOH	227(4.53),282(3.86), 299(3.90)	44-1158-64
Quinazoline, 4-isopropyl-	pH 1.0	237(4.14),267(3.67), 322(3.01)	39-5360-65
	pH 7.0	224(4.57),271(3.42), 306(3.47),315s(3.42)	39-5360-65
Quinazoline, 2,4,5-trimethyl-	pH 2.0	244(4.15),261(3.81), 290s(3.43),347(2.83)	39-5360-65
	pH 10.0	233(4.59),274(3.42), 319(3.52)	39-5360-65
2-Quinolineacetonitrile, 1,2,3,4-tetrahydro-	EtOH	210(4.37),250(4.15), 300(3.44)	78-2961-65
$C_{11}H_{12}N_2O$			
1H-3-Benzazepine, 2-hydroxymethylamino-	EtOH	294(4.13)	4-0026-65
4(1H)-Cinnolinone, 1-propyl-	EtOH	211(4.48),234s(4.08), 240(4.09),254(3.90), 262s(3.80),285(3.46), 296(3.49),344(4.13), 360s(4.06)	39-5391-65
Imidazole, 2,5-dimethyl-4-phenyl-, 1-oxide	EtOH	270(4.07)	88-1565-65
Imidazole, 2,5-dimethyl-4-phenyl-, 3-oxide	EtOH	252(4.01)	88-1565-65
Isoxazole, 5-(dimethylamino)-3-phenyl-	C_6H_{12}	208(4.13),234(4.37), 272(3.63)	17-0255-65
	2N HCl	208(4.25),252(4.08), 306(3.97)	17-0255-65
	MeOH	208(4.06),234(4.38), 280(3.72)	17-0255-65
Isoxazoline, 5-imino-2,3-dimethyl-4-phenyl-, hydrochloride	MeOH	206(4.00),244(3.90), 280(4.16)	17-0255-65
Isoxazoline, 5-imino-2,4-dimethyl-3-phenyl-, hydrochloride	MeOH	204(4.03),242(3.84), 293(4.17)	17-0255-65
Pyrazole, 3-methoxy-5-methyl-1-phenyl-	C_6H_{12}	263(4.17)	78-0315-64
	20N H_2SO_4	241(4.05)	78-0315-64
	pH 7	250.5(4.05)	78-0315-64
3-Pyrazolinone, 2,5-dimethyl-1-phenyl-	C_6H_{12}	272(4.02)	78-0315-64
	20N H_2SO_4	234(4.10)	78-0315-64
	pH 7	258.5(4.07)	78-0315-64
5-Pyrazolinone, 2,3-dimethyl-1-phenyl-	C_6H_{12}	239(4.01),280(3.98)	78-0299-64
	10N H_2SO_4	230(4.10)	78-0299-64
	pH 5	241(3.96),253(3.94)	78-0299-64
5-Pyrazolinone, 3,4-dimethyl-1-phenyl-	C_6H_{12}	246(4.21)	78-0299-64
	10N H_2SO_4	238(4.16)	78-0299-64
	pH 5	244(4.13)	78-0299-64
	pH 13	250(4.10)	78-0299-64
3(2H)-Pyridazinone, 4,5-dihydro-6-p-tolyl-	EtOH	288(4.26)	39-5302-64
Pyrroloquinazolinone, tetrahydro-	iso-PrOH	222(4.39),257s(3.89), 335(3.62)	35-5793-65
$C_{11}H_{12}N_2OS$			
Hydantoin, 5-benzyl-3-methyl-2-thio-	EtOH	237(3.93),267(4.22)	65-0049-65
$C_{11}H_{12}N_2O_2$			
[2,2'-Bipyrrole]-5-carboxylic acid, 1'-methyl-, methyl ester	EtOH	314(4.23)	44-2727-64

Compound	Solvent	λ_{max}(log ϵ)	Ref.
3-(Carboxymethyl)-1,2-dimethylbenzimidazolium hydroxide, inner salt	EtOH	256(3.57),264(3.60), 271(3.71),278(3.74)	44-1118-65
4H-m-Dioxino[4,5-c]pyridine-5-carbonitrile, 2,2,8-trimethyl-	MeOH-acid	300(3.85)	44-2663-64
	MeOH-base	233(3.92),295(3.77)	44-2663-64
Imidazo[1,2-a]pyridine-3-carboxylic acid, 2-methyl-, ethyl ester	EtOH	240(4.8),246(4.9), 294(4.34),308s(4.28)	44-2403-65
1H-Indazole-3-carboxylic acid, 5-methyl-, ethyl ester	EtOH	301(3.93)	32-0814-65
1H-Indazole-3-carboxylic acid, 6-methyl-, ethyl ester	EtOH	290(4.01)	32-0814-65
3H-Indole, 2,3,3-trimethyl-5-nitro-	EtOH	222s(4.01),302(4.08)	44-3457-65
2,3,4-Pentanetrione, 3-phenylhydrazone	MeOH	242(4.01),362(4.30)	44-2959-64
3(2H)-Pyridazinone, 4,5-dihydro-6-(p-methoxyphenyl)-	EtOH	294(4.28)	39-5302-64
4(1H)-Quinazolinone, 2-(1-hydroxyethyl)-1-methyl-	iso-PrOH	230(4.22),267(3.62), 276(3.64),306(3.94), 315(3.89)	44-0582-64
4(3H)-Quinazolinone, 2-(1-hydroxyethyl)-3-methyl-	iso-PrOH	225(4.37),268(3.85), 275(3.83),304(3.54), 317(3.42)	44-0582-64
Tryptophan	EtOH	228(4.17),278(3.72)	39-5510-64
	90% EtOH	220(4.51),274(3.69), 282(3.71),290(3.64)	94-0088-65
	90% EtOH-HCl	219(4.52),273(3.73), 280(3.74),289(3.67)	94-0088-65
	H_2SO_4	234(3.73),240s(3.65), 290(3.64)	39-5510-64
γ-Valerolactone, α-oxo-, phenylhydrazone	n.s.g.	233(4.16),295(3.98), 333(4.50)	39-0141-64
$C_{11}H_{12}N_2O_2S$			
Hydantoin, 5-(p-hydroxybenzyl)-3-methyl-2-thio-	EtOH	228(4.13),268(4.22)	65-0049-65
$C_{11}H_{12}N_2O_3$			
5-Azaindole-2-carboxylic acid, 5-methoxy-, ethyl ester	EtOH	278(4.18),287(4.23), 344(3.55)	35-3530-65
Benzonitrile, 3-tert-butyl-4-hydroxy-5-nitro-	hexane	234(4.18),275(3.57), 353(3.39)	70-1666-64
4-Cinnolinol, 6,7-dimethoxy-3-methyl-	EtOH	226(4.26),256(4.20), 264(4.22),286s(3.40), 344(4.00),359(4.01)	39-6036-65
1-Indazolinecarboxylic acid, 2-methyl-3-oxo-, ethyl ester	EtOH	224(4.31),238(4.13), 242(4.12),304(3.81)	33-1986-64
$C_{11}H_{12}N_2O_3S$			
Indole-2-carboxylic acid, 3-(aminosulfinyl)-, ethyl ester	EtOH	301(4.19)	44-0178-64
Indole-2-carboxylic acid, 3-(aminosulfinyl)-1-methyl-, methyl ester	EtOH	216(4.38),234(4.38), 301(4.16)	44-0178-64
$C_{11}H_{12}N_2O_3S_2$			
Carbamic acid, dimethyldithio-, ester with 2-mercapto-4'-nitroacetophenone	EtOH	267(4.31)	44-1703-64
$C_{11}H_{12}N_2O_4$			
2,4-Pentadienamide, N,4-dimethyl-5-(5-nitro-2-furyl)-	propylene glycol	280(4.15),390(4.33)	95-0646-64
2,4-Pentadienamide, 4-ethyl-5-(5-nitro-2-furyl)	propylene glycol	275(4.08),315(3.90), 388(4.26)	95-0646-64

Compound	Solvent	$\lambda_{max}(\log \epsilon)$	Ref.
$C_{11}H_{12}N_2O_4S$			
Acetic anhydride, compound with 2-amino-benzothiazole-3-oxide	EtOH	310(4.24)	25-0368-64
Indole-2-carboxylic acid, 1-methyl-3-sulfamoyl-, methyl ester	EtOH	210(4.50),234s(4.15), 297(4.10)	44-0178-64
Indole-2-carboxylic acid, 3-sulfamoyl-, ethyl ester	EtOH	233(4.21),302(4.24)	44-0178-64
$C_{11}H_{12}N_2O_5$			
2,4-Pentadienamide, N-hydroxy-4-ethyl-5-(5-nitro-2-furyl)-	propylene glycol	275(4.10),315(4.06), 390(4.27)	95-0646-64
$C_{11}H_{12}N_2O_6$			
3'-O-Acetyl-O^6,5'-6-hydroxycyclo-deoxyuridine	pH 2	262(4.12)	44-3913-65
	H O	262(4.12)	44-3913-65
	pH 11	262.5(3.98)	44-3913-65
	90% EtOH	260(4.12)	44-3913-65
4-Pyridinepyruvic acid, 2-methoxy-4-nitro-, ethyl ester	EtOH	287(4.00)	35-3530-65
$C_{11}H_{12}N_4$			
Butyronitrile, 3-imino-2-oxo-, m-tolylhydrazone	EtOH	245(4.03),258s(4.00), 362(4.30)	5-0033-65J
	EtOH-NH_3	245(3.99),258(3.79), 362(4.29)	5-0033-65J
Butyronitrile, 3-imino-2-oxo-, o-tolylhydrazone	EtOH	248(4.05),260s(3.97), 364(4.29)	5-0033-65J
	EtOH-NH_3	248(4.05),261s(3.97), 366(4.30)	5-0033-65J
Butyronitrile, 3-imino-2-oxo-, p-tolylhydrazone	EtOH	247(4.02),262(3.99), 364(4.33)	5-0033-65J
	EtOH-NH_3	247(4.04),261(4.01), 364(4.31)	5-0033-65J
Pyrimidine, 2,4-diamino-5-methyl-6-phenyl-	EtOH	230s(4.13),289(3.85)	94-1446-64
$C_{11}H_{12}N_4O$			
Butyronitrile, 3-imino-2-oxo-, (p-methoxyphenyl)hydrazone	EtOH	248(4.01),255s(4.00), 268(3.98),305(3.92), 372(4.29),393s(4.22), 494(3.55)	5-0033-65J
	EtOH-NH_3	248(3.95),266(3.94), 305(3.91),371(4.27), 392s(4.22),487(3.56)	5-0033-65J
$C_{11}H_{12}N_4OS$			
Pyrrole-3,4-dicarbonitrile, 2-ethoxy-methyleneamino-5-ethylmercapto-	EtOH	282(4.15)	35-1995-65
$C_{11}H_{12}N_4O_2$			
Diformamide, N-(2,3,6-trimethyl-pyrazolo[1,5-a]pyrimidin-7-yl)-	EtOH	240(4.67),284s(3.18), 295(3.25),312(3.21), 348(3.36)	94-0142-65
$C_{11}H_{12}N_4O_3$			
2,4-Pentanedione, 3-(7,8-dihydro-6-hydroxy-7-pteridinyl)-	pH 2.0	210(4.41),292(4.06)	39-3357-64
	pH 6.0	209(4.49),292(4.05)	39-3357-64
	pH 9.5	209(4.54),296(4.43)	39-3357-64
2,4-Pentanedione, 3-(5,6,7,8-tetra-hydro-7-oxo-6-pteridinyl)-	pH 1.0	223(4.42),284(3.76), 349(3.70)	39-6930-65

Compound	Solvent	$\lambda_{max}(\log \epsilon)$	Ref.
$C_{11}H_{12}N_4O_3S$			
1,2,4-Triazole, 4-ethyl-3-(methylthio)-5-[(5-nitro-2-furyl)vinyl]-	propylene glycol	240(4.13),295(4.01), 390(4.25)	95-0566-64
$C_{11}H_{12}N_4O_4$			
Crotonaldehyde, 3-methyl-, 2,4-dinitrophenylhydrazone	EtOH	382(4.47)	24-0815-64
1,3,4-Oxadiazole, 2-(ethylamino)-5-[1-methyl-2-(5-nitro-2-furyl)vinyl]-	propylene glycol	227(4.23),302(4.13), 398(4.35)	95-0219-64
1,3,4-Oxadiazole, 2-(methylamino)-5-[1-ethyl-2-(5-nitro-2-furyl)vinyl]-	propylene glycol	228(4.08),310(4.01), 400(4.20)	95-0219-64
2-Pentenal, 2,4-dinitrophenylhydrazone, trans	EtOH	374(4.46)	39-2988-65
$C_{11}H_{12}N_4O_6S$			
Glycine, N-[2-(7-nitro-4H-1,2,4-benzothiadiazin-3-yl)ethyl]-, S,S-dioxide	2% $NaHCO_3$	237(4.04),358(4.10)	32-0786-65
$C_{11}H_{12}N_4S$			
Acetone, 5-phenylthiadiazolylhydrazone	EtOH	237(3.92),318(4.32)	1-0871-64
anion	EtOH	375(4.25)	1-0871-64
$C_{11}H_{12}N_4S_2$			
Benzaldehyde, azine with 3-methyl-5-(methylthio)-2-thiadiazolinone	EtOH	232(4.22),290(3.90), 355(4.37)	1-0871-64
Benzaldehyde, N-methyl-N-[5-(methylthio)thiadiazolyl]hydrazone	EtOH	227(4.13),260(3.95), 336(4.40)	1-0871-64
$C_{11}H_{12}N_6O$			
Imidazo[4,5-d]pyridazine, 4,7-diamino-2-phenyl-	EtOH	210(4.37),256(4.37), 287(4.32)	4-0182-64
$C_{11}H_{12}N_6O_2$			
4,5-Imidazoledicarboxylic acid dihydrazide, 2-phenyl-	EtOH	206(4.35),273(4.34)	4-0182-64
Pteridine, 2,4-bisacetamido-6-(dibromomethyl)-	pH -1	246(4.17),283(3.82), 331(4.07)	39-1530-65
	pH 4	251(4.29),339(3.94)	39-1530-65
$C_{11}H_{12}N_8$			
Imidazo[4,5-d]pyridazine, 4,7-dihydrazino-2-phenyl-, hydrochloride	EtOH	208(4.37),264(4.37), 292s(4.32)	4-0182-64
$C_{11}H_{12}O$			
5H-Benzocyclohepten-5-one, 6,7,8,9-tetrahydro- (or benzosuberone)	C_6H_{12}	241(3.92),280(2.95)	23-0579-64
	EtOH	245(3.96)	5-0021-64G
	EtOH	246(3.97)	44-2165-65
3-Buten-2-one, 4-p-tolyl-	hexane	285(4.50)	44-3000-65
	iso-PrOH	296(4.33)	44-3000-65
Crotonophenone, p-methyl-	hexane	254(4.57)	44-3000-65
	iso-PrOH	255(4.5)	44-3000-65
Ketone, methyl 2-phenylcyclopropyl	isooctane	260(2.57),267(2.57), 274(2.47)	88-3151-65
$C_{11}H_{12}OS$			
2H-1-Benzothiapyran, 4-ethoxy-	$CHCl_3$	256(4.02),323(3.16)	44-1575-64
Cyclohexanone, 2-(2-thenylidene)-	hexane	315(4.29)	65-3645-64
	EtOH	325(4.27)	65-3645-64
	66% H_2SO_4	422(4.61)	65-3645-64
4-Hexen-2-yn-1-ol, 5-methyl-1-(2-thienyl)-	EtOH	235(4.24)	39-2983-65

Compound	Solvent	λ max(log ε)	Ref.
$C_{11}H_{12}OS_2$			
2,5-Cyclohexadien-1-one, 4-(1,3-dithiolan-2-ylidene)-2,6-dimethyl-	MeOH	258(3.56),302(3.60), 420s(4.52),432(4.57), 450s(4.42)	5-0037-65D
$C_{11}H_{12}O_2$			
4-Benzofuranol, 2-ethyl-3-methyl-	aq. base	297.5(3.82)	22-1464-65
4-Benzofuranol, 7-ethyl-2-methyl-	aq. base	299.5(3.75)	22-1464-65
7-Benzofuranol, 2,3,4-trimethyl-	aq. base	295.0(3.57)	22-1464-65
7-Benzofuranol, 2,3,5-trimethyl-	aq. base	290.5(3.46)	22-1464-65
Benzoquinone, 3-allyl-2,5-dimethyl-	n.s.g.	256(4.32),260s(--)	39-5060-65
Benzoquinone, 5-allyl-2,3-dimethyl-	n.s.g.	256(4.24),260s(--)	39-5060-65
1-Benzosuberone, 7-hydroxy-	pH 13	327(4.30)	44-2165-65
	EtOH	276(4.09)	44-2165-65
Bicyclo[6.1.0]nona-2,4,6-triene-9-carboxylic acid, methyl ester	n.s.g.	242(3.63)	35-0905-64
1,3-Butanedione, 1-p-tolyl-	hexane	306(4.65)	44-3000-65
	iso-PrOH	312(4.55)	44-3000-65
3-Buten-2-one, 4-(p-methoxyphenyl)-	hexane	307(4.65)	44-3000-65
	iso-PrOH	317(4.38)	44-3000-65
3-Chromanone, 2,4-dimethyl-	EtOH	275(3.25)	65-2699-64
3-Chromanone, 4,4-dimethyl-	EtOH	278(3.20)	65-2699-64
Cinnamic acid, ethyl ester	n.s.g.	276(4.35)	5-0058-65B
Crotonophenone, p-methoxy-	hexane	283(4.29)	44-3000-65
	iso-PrOH	292(4.26)	44-3000-65
2-Decene-4,6-diynoic acid, methyl ester, trans	EtOH	224(4.77),257s(3.93), 271s(4.29),288(4.54), 306(4.53)	70-0851-65
8-Decene-4,6-diynoic acid, methyl ester	EtOH	225(3.60),248(3.93), 262(4.10),277(3.99), 304(3.46)	70-1237-65
Inden-1-one, 3a,7a-dihydro-, ethylene acetal, cis	C_6H_{12}	262(3.58),270(3.56)	24-3680-65
Lachnophyllum acid, methyl ester (or 2-decene-4,6-diynoic acid methyl ester)		218(4.52),225(4.56), 255(3.76),269(4.03), 285(4.23),303(4.18)	24-3469-64
2,4-Nonadien-6-ynoic acid, 4-hydroxy-8,8-dimethyl-, γ-lactone	ether	313(4.39)	24-2236-65
1,3,5-Norcaratriene-3-carboxylic acid, 7,7-dimethyl-, methyl ester	MeOH	238(3.95),279(3.41), 286s(3.41)	35-0525-64
1α(H),3β(H)-Pentacyclo[6.2.1.$0^{2,7}$.$0^{4,10}$.$0^{5,9}$]undecan-6-one, 3-hydroxy-	C_6H_{12}	287f(0.78)	39-3062-64
	EtOH	202s(2.69),283f(1.11)	39-3062-64
2-Pentenoic acid, 2-phenyl-, cis	EtOH	249(4.04),291s(2.57)	1-0612-65
2-Pentenoic acid, 2-phenyl-, trans	EtOH	235s(3.74)	1-0612-65
4-Pentenoic acid, 5-phenyl-, cis	n.s.g.	242(4.17)	28-3728-64A
Pyran-2-spirocyclohexa-2',5'-dien-4'-one, 2,3-dihydro-methyl-	EtOH	223(4.08)	39-5182-65
Spiro[2-cyclobutene-1,2-[5]norbornene]-3-carboxylic acid	EtOH	225(3.96)	28-1541-64A
$C_{11}H_{12}O_2S$			
Benzoic acid, thio-, S-ester with 4-mercapto-2-butanone	EtOH	237(4.32),265(4.19)	39-0788-64
1,4-Naphthoquinone, 4a,5,8,8a-tetrahydro-2-(methylthio)-, cis	EtOH	332(3.83)	44-0051-64
$C_{11}H_{12}O_3$			
Acetic acid, benzoyl-, ethyl ester	heptane	286(4.0)	70-1798-65
1,4-Benzodioxan, 5-allyloxy-	EtOH	270(2.84)	65-2994-64
1,4-Benzodioxan-5-ol, 6-allyl-	EtOH	270(3.13)	65-2994-64

Compound	Solvent	λ_{max}(log ε)	Ref.
Benzoic acid, 2-acetyl-6-methyl-, pseudomethyl ester	EtOH	234(3.57),284(3.06)	44-4293-65
1,3-Butanedione, 1-(p-methoxyphenyl)-	hexane	309(4.69)	44-3000-65
	iso-PrOH	320(4.75)	44-3000-65
2,6-Decadien-4-ynoic acid, 8-oxo-, methyl ester	ether	297(4.41),306s(4.36)	24-2608-65
2-Decene-4,6-diynoic acid, 8-hydroxy-, methyl ester	ether	217(4.63),223(4.68), 253(3.90),267(4.19), 284(4.42),301(4.40)	24-3469-64
	ether	272(3.99),287(4.20), 305(4.18)	24-2608-65
1,3-Indandione, 2-acetyl-3a,4,7,7a-tetrahydro-	n.s.g.	221(4.03),265(3.90)	1-0441-64
1-Indanone, 5,7-dimethoxy-	EtOH	273(4.25)	5-0021-64G
	HOAc-H_2SO_4	306(4.42)	5-0021-64G
4-Pyrone, 2-propenyl-6-propionyl-	EtOH	234(4.31),275(4.03), 310s(3.80)	39-3234-64
Vanillylideneacetone	buffer	240(4.4),280(4.0), 320(3.8),410(4.78)	83-0660-64
	acidic buffer	245(4.4),340(4.42)	83-0660-64
$C_{11}H_{12}O_4$			
p-Benzoquinone, 2-acetyl-3-ethyl-5-methoxy-	EtOH	264(4.18),360(2.90)	39-0411-64
p-Benzoquinone, 2,5-dihydroxy-3-methyl-6-(1-methylpropenyl)-	EtOH	287(4.24),436(2.40)	94-0511-65
Bicyclo[3.3.1]non-3-ene-1-carboxylic acid, 4-methyl-6,8-dioxo-	EtOH	242s(3.82),270(4.05)	44-0787-64
Butanoic acid, 3-(p-carboxyphenyl)-	EtOH	235(4.22),270s(3.02), 278s(2.83)	78-2593-65
Chroman-3-carboxylic acid, 7-methoxy-	n.s.g.	284(3.5),345(2.75)	39-5991-64
4,6-Decadiynedioic acid, methyl ester	EtOH	225(2.60),238(2.50), 253(2.25),283(1.30), 302(1.23)	70-1237-65
3-Norcarene-2,5-dione, 1-acetyl-4-methoxy-7-methyl-	EtOH	274(3.89)	39-0411-64
1,2-Propanedione, 1-(3,4-dimethoxybenzoyl)-	EtOH	235(3.55),284s(3.50), 317(3.64)	23-0113-64
Propionic acid, 3-(p-methoxybenzoyl)-	EtOH	270(4.29)	65-1985-64
Propionic acid, 3-(3,4-methylenedioxyphenyl)-, methyl ester	EtOH	288(--)	95-0857-65
Tropone, 4-acetyl-5-hydroxy-7-methoxy-2-methyl-	EtOH	247(4.38),344(3.98), 382(4.02)	39-0411-64
$C_{11}H_{12}O_5$			
Phthalide, 4,5,6-trimethoxy-	EtOH	257(4.12),288(3.45)	23-1595-64
1,2-Propanedione, 1-(4-hydroxy-3,5-dimethoxyphenyl)-	EtOH	326(4.02)	23-0113-64
Sepedonin	EtOH	253(4.57),327(3.81)	23-1835-65
$C_{11}H_{12}O_6$			
Acetic acid, (2-acetyl-3,5-dihydroxy-4-methoxyphenyl)-	EtOH	285(3.25)	78-1411-65
Acetic acid, (2-carboxy-3,5-dimethoxyphenyl)-	EtOH	250(3.78),284(3.53)	1-1677-65
3-Cyclopenten-1-ylideneacetic acid, 4-ethyl-α,3-dihydroxy-2,5-dioxo-, ethyl ester	neutral	260(4.1)	94-1300-65

Compound	Solvent	$\lambda_{max}(\log \epsilon)$	Ref.
$C_{11}H_{12}S$			
4H-Thiopyran, 5,6-dihydro-2-phenyl-	n.s.g.	210(4.00),224(4.16), 275(3.65)	24-1963-64
$C_{11}H_{12}S_2$			
Propane, 2,2-bis(2-thienyl)-	EtOH	235.0(4.25)	22-2635-65
$C_{11}H_{13}BrN_6$			
Acetone, [5-(p-bromoanilino)-s-triazol-3-yl]hydrazone	n.s.g.	265(4.48)	39-4448-65
$C_{11}H_{13}ClN_2O_4$			
2,4-Dimethyl-1,5-benzodiazepinium perchlorate	MeOH-HCl	224(4.01),260(4.21), 269(4.19),325(3.06), 339(3.04),510(2.93)	39-3785-65
$C_{11}H_{13}ClN_4O$			
Pyrazolo[1,5-a]pyrimidine, 7-(2-chloroacetamido)-2,3,6-trimethyl-	EtOH	240(4.64),295(3.22), 310(3.16),348(3.37)	94-1207-65
$C_{11}H_{13}ClN_4O_6$			
2,6-Pyridinedicarbamic acid, 4-chloro-3-nitro-, diethyl ester	pH 1, 7	230s(4.18),296(3.76), 343s(3.46)	44-0734-64
	pH 13	257(4.14),302(3.98)	44-0734-64
$C_{11}H_{13}Cl_3O_4$			
1,4,8,11-Tetraoxadispiro[4.1.4.2]tridec-12-ene, 6,12,13-trichloro-2,9-dimethyl-	n.s.g.	215s(3.79),298s(2.47)	39-4744-65
$C_{11}H_{13}F_3N_2O_6$			
2(1H)-Pyrimidinone, 1-β-D-arabinofuranosyl-4-methoxy-5-(trifluoromethyl)-	pH 2	205(4.26),270(3.77)	44-0835-65
	pH 12	217(4.11),225s(4.09), 278(3.81)	44-0835-65
$C_{11}H_{13}IN_4O_4$			
Valeraldehyde, 5-iodo-, 2,4-dinitrophenylhydrazone	MeOH	358(4.33)	35-1528-64
$C_{11}H_{13}N$			
Indole, 3-propyl-	EtOH	223(4.60),275s(3.78), 281(3.81),289(3.75)	35-3796-64
	H_2SO_4	233(3.68),238(3.64), 284(3.69)	35-3796-64
Indole, 1,2,3-trimethyl-	EtOH	230(4.54),279s(3.77), 286(3.82),293(3.80)	35-3796-64
	EtOH	231(4.52),287(3.83), 293s(3.81)	44-3457-65
	H_2SO_4	231(3.71),238(3.69), 275(3.74)	35-3796-64
3H-Indole, 2,3,3-trimethyl-	pH 1	229(4.00),235(3.95), 275(3.91)	39-5510-64
$C_{11}H_{13}NO$			
5H-Benzocyclohepten-5-one, 2-amino-6,7,8,9-tetrahydro-	EtOH	318(4.24)	44-2165-65
3-Buten-2-one, 4-p-toluidino-	MeOH	340(4.45)	83-0321-64
Cinnamamide, p,α-dimethyl-	EtOH	265(4.19)	65-3150-64
Cyclohexanone, 2-(2-pyridyl)-	EtOH	262(3.5)	94-0912-65
	N HCl	262(3.8)	94-0912-65

Compound	Solvent	λmax(log ε)	Ref.
Indole, 2-ethoxy-1-methyl-	n.s.g.	223(4.48),273(3.91)	77-0368-64
1-Naphthonitrile, 1,2,3,4,5,6,7,8-octahydro-6-oxo-	n.s.g.	235(4.21)	77-0377-65
isomer	n.s.g.	235(4.16)	77-0377-65
2-Pentenanilide	EtOH	270(4.16)	54-0581-65
$C_{11}H_{13}NOS$			
4H-1,3-Thiazin-4-ol, 5,6-dihydro-4-methyl-2-phenyl-	EtOH	238.0(4.16)	39-0788-64
$C_{11}H_{13}NOS_2$			
2-Butanone, 4-phenyl-4-(thiocarbamoylthio)-	H O	243s(3.87),289(4.11)	39-4004-64
	EtOH	238(4.15),295(4.08)	39-4004-64
Carbamic acid, dimethyldithio-, ester with 2-mercaptoacetophenone	MeOH	245(4.34),272(4.03)	44-2146-64
	EtOH	247(4.32),275(4.03)	44-1703-64
Propionaldehyde, β-phenyl-β-(N-methylthiocarbamoylthio)-	H_2O	243s(4.05),283(4.15)	39-4004-64
	EtOH	245(4.21),287(4.11)	39-4004-64
$C_{11}H_{13}NO_2$			
Acrylophenone, 3-(dimethylamino)-2'-hydroxy-	EtOH	214(4.11),227(4.10), 246(3.85),255(3.88), 261(3.88),360(4.45)	39-3610-65
4-Azabicyclo[5.2.0]nona-2,5,8-triene, N-carbethoxy-	isooctane	228(4.09)	89-0569-64
Benzamide, N-isobutyryl-	EtOH	234(4.06)	44-2222-65
2-Indolinone, 3-ethoxy-1-methyl-	MeOH	258(3.79),291(3.08)	44-3610-65
2-Indolinone, 3-propoxy-	MeOH	250(3.89),292(3.30)	44-3610-65
Isoquinoline, 3,4-dihydro-6,7-dimethoxy-	EtOH-HCl	243(4.29),301(3.97), 350(3.96)	33-1945-65
	EtOH	226(4.36),273(3.88), 306(3.84)	33-1945-65
	EtOH-NaOH	225(4.46),269(3.93), 306(3.86)	33-1945-65
hydrobromide	EtOH-HCl	237(4.25),301(3.95), 350(3.94)	33-1945-65
	EtOH	243(4.23),302(3.94), 352(3.91)	33-1945-65
	EtOH-NaOH	223(4.42),269(3.90), 305(3.83)	33-1945-65
6-Isoquinolinol, 3,4-dihydro-7-methoxy-1-methyl-	EtOH-HCl	249(4.14),311(3.91), 371(3.85)	33-1945-65
	EtOH	230(3.89),273(3.88), 317(3.58),408(4.15)	33-1945-65
	EtOH-NaOH	248(4.06),337(4.00)	33-1945-65
hydrobromide	EtOH-HCl	251(4.16),313(3.94), 376(3.92)	33-1945-65
	EtOH	250(4.14),313(3.94), 376(3.92)	33-1945-65
	EtOH-NaOH	250(4.14),338(4.07)	33-1945-65
7-Isoquinolinol, 3,4-dihydro-6-methoxy-1-methyl-	EtOH-HCl	250(4.30),310(4.05), 369(3.90)	33-1945-65
	EtOH	233(3.98),278(3.84), 314(3.81)	33-1945-65
	EtOH-NaOH	250(4.21),282s(3.76), 349(3.74)	33-1945-65
hydrobromide	EtOH-HCl	250(4.29),310(4.03), 368(3.89)	33-1945-65
	EtOH	250(4.24),311(4.00), 368(3.82)	33-1945-65
	EtOH-NaOH	250(4.01),282s(3.77), 350(3.74)	33-1945-65

Compound	Solvent	λ_{max}(log ε)	Ref.
Morpholine, N-benzoyl-	hexane	240(3.67)	78-2857-65
1-Pentene, 1-(o-nitrophenyl)-, trans	EtOH	239(4.18),323(3.35)	44-3604-65
2-Pyrrolidinone, 3-hydroxy-5-methyl-	EtOH	245(3.96)	94-0725-64
1-phenyl-	EtOH	240(3.95)	94-0725-64
$C_{11}H_{13}NO_2S_2$			
Carbamic acid, dimethyldithio-, ester with 4'-hydroxy-2-mercapto-acetophenone	EtOH	277(4.41)	44-1703-64
$C_{11}H_{13}NO_3$			
2H-1,4-Benzoxazine, 6-acetyl-3,4-di-hydro-7-hydroxy-3-methyl-	MeOH	253(4.3),283(3.9), 378(3.6)	39-7348-65
α[4]-3-O-Isopropylideneisopyridoxal	pH 1	292.5(3.87)	44-0574-64
	EtOH	282(3.74)	44-0574-64
Salicylamide, o-acetyl-N-ethyl-	pH 1-6	265(2.90)	78-3019-65
$C_{11}H_{13}NO_3S$			
Benzoic acid, p-nitrothio-, S-tert-butyl ester	C_6H_{12}	258(4.30),286s(4.14)	44-2471-64
$C_{11}H_{13}NO_4$			
Benzaldehyde, 3-tert-butyl-4-hydroxy-5-nitro-	hexane	260(4.20),350(3.40)	70-1666-64
4H-m-Dioxino[4,5-c]pyridine-5-carb-	MeOH-acid	235s(3.76),302(3.92)	44-2663-64
oxylic acid, 2,2,8-trimethyl-	MeOH-base	288(3.80)	44-2663-64
Glutaconic acid, 4-(1-cyanoethylidene)-	EtOH	210(4.19)	25-1954-64
3-methyl-, 5-ethyl ester, cis	EtOH	210(4.18)	44-0203-65
2-cis-4-trans	EtOH	210(4.19)	44-0203-65
Indole-7-carboxylic acid, 2,4,5,6-tetra-hydro-3-methoxy-2-oxo-, methyl ester	n.s.g.	312(4.40)	39-1648-65
Isocarbostyril, 3,4-dihydro-	pH 1	269(4.06),300s(3.52)	33-2089-64
8-hydroxy-6,7-dimethoxy-	pH 13	226(4.47),266(3.84), 326(3.76)	33-2089-64
	iso-PrOH	221(4.42),269(4.05), 303s(3.52)	33-2089-64
2,4,6,8-Nonatetraenoic acid, 9-nitro-, ethyl ester	EtOH	363(4.58)	70-2093-64
5-Pyridoxoic acid, α[4],3-O-isopro-	pH 1	299.5(3.93)	44-0574-64
pylidene-	EtOH	294(4.81)	44-0574-64
$C_{11}H_{13}NO_6$			
3-Carboxy-1-D-ribofuranosylpyridinium	H_2O	265(3.66)	39-0610-65
hydroxide, inner salt	M KCN	316(3.67)	39-0610-65
3-Carboxy-1-D-ribopyranosylpyridinium	H_2O	266(3.70)	39-0610-65
hydroxide, inner salt	M KCN	316(3.66)	39-0610-65
3-Carboxy-1-D-xylosylpyridinium	H_2O	265.5(3.69)	39-0610-65
hydroxide, inner salt	M KCN	314(3.72)	39-0610-65
$C_{11}H_{13}N_3$			
Acetamidine, 2-cyano-N-phenethyl-, hydrochloride	pH 13	258(4.04)	44-0308-64
Propionitrile, 2-methyl-2-(o-tolylazo)-	o-$C_6H_4Cl_2$	400(2.34)	60-1437-65
3,5-Pyridinedicarbonitrile, 4-ethyl-	EtOH	260(4.12),368(3.82)	73-1495-64
1,2-dihydro-2,6-dimethyl-	EtOH	217(4.35),260(4.12), 368(3.82)	73-1654-64
3,5-Pyridinedicarbonitrile, 4-ethyl-1,4-dihydro-2,6-dimethyl-	EtOH	336.6(3.75)	73-1495-64

Compound	Solvent	$\lambda_{max}(\log \epsilon)$	Ref.
$C_{11}H_{13}N_3O$			
Hydrazine, 1-acetyl-2-(3-indolyl-methyl)-	EtOH	281(3.81)	2-0423-64
Pyrazolo[1,5-a]pyrimidin-7(4H)-one, 4-allyl-3,6-dimethyl-	EtOH	222(4.33),269(4.01), 330(3.86)	94-0142-65
5-Quinoxalinol, 8-amino-2,3,7-trimethyl-	MeOH	272(4.55),310(3.46), 400(3.15)	4-0171-64
$C_{11}H_{13}N_3O_2$			
Gramine, 4-nitro-	EtOH	234s(3.96),350(3.60), 379(3.62)	44-1158-64
Gramine, 6-nitro-	EtOH	251(4.02),325(3.92), 366(3.86)	44-1158-64
$C_{11}H_{13}N_3O_3$			
Salicylaldehyde, 5-nitro-, N-amino-pyrrolidine derivative	MeOH	301(4.48)	24-1631-64
	MeOH-NaOH	346(4.25),423(4.04)	24-1631-64
Salicylaldehyde, 5-nitro-, methyl-allylhydrazone	MeOH	299(4.47)	24-1631-64
	MeOH-NaOH	350(4.22),420(4.07)	24-1631-64
$C_{11}H_{13}N_3O_3S$			
Acetimidic acid, α-diazo-N-(p-tolyl-sulfonyl)-, ethyl ester	n.s.g.	277.5(4.41)	24-0623-65
$C_{11}H_{13}N_3O_4$			
1H-Imidazo[4,5-b]pyridine, 1-β-D-ribofuranosyl-	pH 0.11	278s(3.95),284(4.00), 289s(3.98)	44-4066-65
	pH 5.6	259(3.56)	44-4066-65
1H-Imidazo[4,5-c]pyridine, 1-β-D-ribofuranosyl-	pH 1.72	264(3.67)	44-2611-64
Salicylic acid, 5-nitro-, N-amino-morpholine derivative	MeOH	290(4.44)	24-1631-64
	MeOH-NaOH	347(4.15),417(4.12)	24-1631-64
$C_{11}H_{13}N_3O_4S$			
7H-Pyrrolo[2,3-d]pyrimidine-4-thiol, 7-β-D-ribofuranosyl-	0.1N H_2SO_4	322(4.37)	4-0159-64
	H_2O	322(4.37)	4-0159-64
	pH 13	309(4.31)	4-0159-64
$C_{11}H_{13}N_3O_5$			
4H-Pyrrolo[2,3-d]pyrimidin-4-one, 3,7-dihydro-7-β-D-ribofuranosyl-	acid	259(3.98)	4-0159-64
	H_2O	259(3.98)	4-0159-64
	pH 12	263(4.03)	4-0159-64
$C_{11}H_{13}N_5O$			
Triazine, 2-amino-4-(2-hydroxyethyl-amino)-6-phenyl-	EtOH	219(4.39),242(4.36)	44-2766-64
$C_{11}H_{13}N_5O_2$			
Pyrido[3,4-b]pyrazine-7(or 5)-carbamic acid, 5(or 7)-amino-3-methyl-, ethyl ester	pH 1	238(4.22),307(4.26), 384(3.61)	44-0734-64
	pH 7	266(4.53),311(3.87), 396(3.56)	44-0734-64
	pH 13	266(4.51),311(3.86), 396(3.53)	44-0734-64
$C_{11}H_{13}N_5O_4$			
Adenosine, 2',3'-O-methylene-	pH 1	256(4.17),265s(4.1)	44-2854-65
	pH 7	259(4.18)	44-2854-65
	pH 13	259(4.18)	44-2854-65

Compound	Solvent	$\lambda_{max}(\log \epsilon)$	Ref.
$C_{11}H_{14}$			
5H-Benzocycloheptene, 6,7,8,9-tetrahydro-	C_6H_{12}	264(2.45),268(2.32), 272(2.42)	23-0579-64
Indan, 1,1-dimethyl-	EtOH	215s(4.0),255s(3.0), 270(3.1),275(3.1)	35-0908-64
Naphthalene, 1,2,3,4-tetrahydro-1-methyl-	C_6H_{12}	266(2.63),273(2.62)	23-0579-64
1-Pentene, 1-phenyl-	n.s.g.	250(4.20)	5-0058-65B
$C_{11}H_{14}BrN$			
3,4-Dihydro-4,6-dimethylquinolizinium bromide	n.s.g.	316(3.99)	78-0033-64
$C_{11}H_{14}BrNO$			
Spiro[5.5]undeca-1,4-dien-3-one, 2-syn-bromo-, oxime	MeOH-acid	255(4.20)	33-1988-65
	MeOH	255(4.22)	33-1988-65
$C_{11}H_{14}Br_2O$			
2,4-Cyclohexadien-1-one, 4,6-dibromo-6-tert-butyl-2-methyl-	n.s.g.	345(3.34)	70-0336-65
2,5-Cyclohexadien-1-one, 4,4-dibromo-6-tert-butyl-2-methyl-	n.s.g.	248(4.27)	70-0336-65
Phenol, 3,4-dibromo-6-tert-butyl-2-methyl-	n.s.g.	288(3.30)	70-0336-65
$C_{11}H_{14}Br_3NO_2$			
9,9-Dibromo-10-hydroxy-7,8,9,10-tetrahydro-10-methoxy-6H-pyrido-[1,2-a]azepinium bromide	H_2O	265(3.76)	44-1523-65
$C_{11}H_{14}ClNO$			
Benzamide, N-(2-chloro-tert-butyl)-	n.s.g.	226(4.01)	28-1440-65A
4,4-Dimethyl-2-phenyl-2-oxazolinium chloride	n.s.g.	242(4.10)	28-1440-65A
$C_{11}H_{14}ClNO_4$			
[2-(2,4,6-Cycloheptatrien-1-ylidene)-ethylidene]dimethylammonium perchlorate	MeCN	227(4.17),270(3.86), 348(3.97),465(4.27)	24-2050-64
$C_{11}H_{14}N_2$			
Benzimidazole, 2-ethyl-5,6-dimethyl-	pH 2	274(3.93),283(3.93)	44-0476-64
	pH 12	243(3.52),280(3.79), 286(3.71)	44-0476-64
	EtOH	244(3.65),281(3.85), 287(3.78)	44-0476-64
Quinazoline, 3,4-dihydro-4-isopropyl-	pH 2.0	213(4.23),217s(4.22), 224s(3.98),277(3.66)	39-5360-65
	pH 12.0	222(4.18),228s(4.08), 290(3.80)	39-5360-65
Tryptamine, N(a)-methyl-	EtOH	223(4.58),287(3.77)	39-5510-64
	H_2SO_4	233(3.63),240(3.62), 288(3.68)	39-5510-64
Tryptamine, N(b)-methyl-	EtOH	220(4.57),280(3.79), 290(3.72)	39-5510-64
	H_2SO_4	236(3.62),241(3.58), 294(3.64)	39-5510-64
Tryptamine, 2-methyl-	EtOH	224(4.57),281(3.89), 288(3.82)	39-5510-64
	H SO	233(3.77),241(3.76), 289(3.75)	39-5510-64

Compound	Solvent	$\lambda_{max}(\log \epsilon)$	Ref.
$C_{11}H_{14}N_2O$			
2,5-Cyclohexadien-1-one, 2-tert-butyl-	aq. EtOH	258(3.82),365(4.59)	30-1101-64
4-diazo-6-methyl-	EtOH-acid	325(4.28)	30-1101-64
2-Pyrrolidinone, 3-amino-	EtOH	239(3.95)	94-0725-64
5-methyl-1-phenyl-	EtOH	245(3.97)	94-0725-64
2-Quinolineacetamide, 1,2,3,4-tetra-hydro-	EtOH	209(4.28),249(3.91), 301(3.26)	78-2961-65
2(1H)-Quinoxalinone, 3,4-dihydro-1,3,3-trimethyl-	EtOH	226(4.56),271(3.51), 307(3.63)	12-0877-64
$C_{11}H_{14}N_2OS$			
2-Benzimidazoleethanethiol, 5-ethoxy-	EtOH	209(4.38),247(4.04), 291(4.16),296s(4.13)	4-0306-65
$C_{11}H_{14}N_2O_2$			
1H-Indazole-4,7-dione, 3a,7a-dihydro-3a,5,6,7a-tetramethyl-	EtOH	226(4.10),248(4.14)	88-1719-64
3H-Indazole-4,7-dione, 3a,7a-dihydro-3a,5,6,7a-tetramethyl-	EtOH	252(4.13)	88-1719-64
Piperidine, N-(p-nitrophenyl)-	n.s.g.	412(4.36)	5-0109-65I
4-Pyrimidineacrylic acid, butyl ester	EtOH	240(4.11),277(4.14)	44-2398-65
1-Pyrroline-5-carboxylic acid,	EtOH-HCl	324(4.43)	44-2727-64
2-(2-pyrrolyl)-, ethyl ester	EtOH-HCl	272s(3.58),324(4.47)	39-0893-64
	EtOH	245s(3.62),285(4.28)	39-0893-64
	EtOH-NaOH	279(4.28)	44-2727-64
Spiro[5.5]undecane-2,4-dione, 3-diazo-	n.s.g.	227(4.22),256(3.95)	5-0101-64F
$C_{11}H_{14}N_2O_2S_2$			
p-Dithiin-2,3-dicarboximide, 5,6-di-hydro-N-(1-pyrrolidinylmethyl)-	EtOH	250(4.00),407(3.51)	56-1713-65
$C_{11}H_{14}N_2O_3$			
sec-Butylamine, 5-nitrosalicylidene-	EtOH	230s(4.07),257(4.27), 348(4.11),390(4.08)	44-2265-64
4H-m-Dioxino[4,5-c]pyridine, 5-carb-	MeOH-acid	298(3.93)	44-2663-64
amyl-2,2,8-trimethyl-	MeOH-base	290(3.77)	44-2663-64
α^4,3-O-Isopropylidene-5-pyridoxamide	pH 1	295(3.87)	44-0574-64
	pH 13	288(3.72)	44-0574-64
	EtOH	290(3.71)	44-0574-64
Succinamic acid, N-2-pyridyl-, ethyl ester	EtOH	233(3.98),294(3.50)	39-2763-64
Valeric acid, 4-hydroxy-2-oxo-, phenylhydrazone	n.s.g.	293(4.16),317(4.23)	39-0141-64
$C_{11}H_{14}N_2O_3S_2$			
p-Dithiin-2,3-dicarboximide, 5,6-di-hydro-N-(morpholinomethyl)-	EtOH	252(4.00),405(3.52)	56-1713-65
$C_{11}H_{14}N_2O_3S_3$			
Imidocarbonic acid, (N-acetylsulfan-ilyl)dithio-, dimethyl ester	EtOH	272.0(4.41)	95-0391-65
$C_{11}H_{14}N_2O_4$			
Benzohydroxamic acid, N-tert-butyl-	MeOH	268(4.12)	73-0940-65
p-nitro-	50% MeOH-HCl	264(4.18)	73-0940-65
	50% MeOH-borate	272(4.13)	73-0940-65
	50% MeOH-N NaOH	271(4.15)	73-0940-65

Compound	Solvent	$\lambda_{max}(\log \epsilon)$	Ref.
$C_{11}H_{14}N_2O_4S$			
Malonic acid, [(2-thiazolylamino)-methylene]-, diethyl ester	EtOH	325(4.4)	70-1481-64
$C_{11}H_{14}N_2O_8$			
Glucopyranuronic acid, 1-deoxy-1-(4-hydroxy-2-oxo-1(2H)-pyrimidinyl)-, methyl ester	MeOH	229(3.37),259(3.92)	94-1259-64
$C_{11}H_{14}N_2O_9S_2$			
Uracil, 2,3'-anhydro-1-(2',5'-di-O-methylsulfonyl-β-D-lyxosyl)-	H_2O	230(3.97),245s(3.94)	44-0476-65
$C_{11}H_{14}N_2S$			
Urea, 1,1-dimethyl-2-thio-3-(p-vinylphenyl)-	EtOH	256(4.35),283s(4.08)	44-1926-65
$C_{11}H_{14}N_4$			
Pyrazole, 3,5-diamino-1-phenethyl-, hydrochloride	MeOH	237(4.12)	44-0942-64
Pyrazolo[1,5-a]pyrimidine, 4-allyl-7-imino-2,3-dimethyl-4,7-dihydro-	EtOH	234(4.42),267(3.79), 315(3.77)	94-0142-65
Pyrazolo[1,5-a]pyrimidine, 4-allyl-7-imino-3,6-dimethyl-4,7-dihydro-	EtOH	233(4.28),269(3.88), 313(3.92)	94-0142-65
$C_{11}H_{14}N_4O$			
Pyrazolo[1,5-a]pyrimidine, 7-acetamido-2,3,6-trimethyl-	EtOH	238(4.55),284s(3.27), 293(3.32),310s(3.24), 339(3.40)	94-0142-65
Pyrazolo[1,5-a]pyrimidine, 7-acetylimino-4,7-dihydro-2,3,4-trimethyl-	CF_3COOH	240(4.50),257(4.23), 365(3.62)	94-0142-65
$C_{11}H_{14}N_4O_2$			
Pyrazolo[1,5-a]pyrimidine-7-carbamic acid, 2,3-dimethyl-, ethyl ester	EtOH	233(4.63),287(3.44), 297(3.43),336(3.59)	94-1207-65
Pyrazolo[1,5-a]pyrimidine-6-carboxylic acid, 7-amino-2,5-dimethyl-, ethyl ester	EtOH	232(4.51),308(4.19)	95-0442-65
Pyrazolo[1,5-a]pyrimidine-6-carboxylic acid, 7-amino-3,5-dimethyl-, ethyl ester	EtOH	231(4.42),316(4.17)	95-0442-65
Pyrazolo[1,5-a]pyrimidine-6-carboxylic acid, 7-amino-2-ethyl-, ethyl ester	EtOH	232(4.53),307(4.22)	95-0442-65
Pyrazolo[1,5-a]pyrimidine-6-carboxylic acid, 7-amino-5-ethyl-, ethyl ester	EtOH	227(4.46),307(4.19)	95-0442-65
$C_{11}H_{14}N_4O_3$			
Purine, 6-methyl-9-(2'-deoxy-α-D-ribofuranosyl)-	pH 1	264(3.85)	35-4934-65
	pH 11	248s(3.91),260(3.91)	35-4934-65
$C_{11}H_{14}N_4O_3S$			
D-erythro-Pentose, 2,3-dideoxy-3-(6-methylthio-9-purinyl)-	pH 1	222(4.02),294(4.24)	35-0720-64
	H O	285(4.29),292s(4.29)	35-0720-64
	pH 13	285(4.28),292s(4.28)	35-0720-64
D-threo-Pentose, 2,3-dideoxy-3-(6-methylthio-9-purinyl)-	pH 1	222(4.03),293(4.24)	35-0720-64
	H O	285(4.28),292(4.28)	35-0720-64
	pH 13	285(4.28),292s(4.28)	35-0720-64
Sulfanilamide, N^4-acetyl-N^1-(2-imidazolidinylidene)-	EtOH	262.0(4.41)	95-0391-65

Compound	Solvent	$\lambda_{max}(\log \epsilon)$	Ref.
$C_{11}H_{14}N_4O_4$			
Tubercidin	pH 2	270(4.05)	4-0159-64
	H_2O	270(4.08)	4-0159-64
	pH 12	270(4.07)	4-0159-64
$C_{11}H_{14}N_4O_5$			
Inosine, 2'-O-methyl-	pH 1	248(4.15)	35-1145-65
	pH 11	251(4.18)	35-1145-65
$C_{11}H_{14}N_4O_6$			
Hypoxanthine, N^9-(β-D-glucopyranosyl)-	0.05N HCl	248.5(4.08)	88-3095-65
	H_2O	248.5(4.10)	88-3095-65
	0.05N NaOH	253.5(4.14)	88-3095-65
Lumazine, 8-(1'-D-ribityl)-	0.1N H_2SO_4	256(4.20),397(3.98)	37-3493-64
	pH 13	235(4.25),280(4.12), 310(3.91)	37-3493-64
$C_{11}H_{14}N_6$			
1,2,4-Triazole, 3-anilino-5-isopropylidenehydrazino-	EtOH	256(4.48)	39-3912-65
$C_{11}H_{14}O$			
Acetophenone, 2',4',6'-trimethyl-	C_6H_{12}	218(3.67),238(3.29), 243(3.22),248(3.03), 259(3.01),264(2.85)	73-3462-65
	EtOH	211(4.04),249(3.35), 268(2.35)	73-3462-65
5H-Benzocyclohepten-1-ol, 6,7,8,9-tetrahydro-	MeOH	276(3.30),280(3.28)	33-1988-65
Bicyclo[2.2.1]hept-5-ene, 2-methylene-3-propionyl-	EtOH	208(3.85),285(1.66)	28-0404-64B
2,4-Cyclohexadien-1-one, 6-allyl-2,6-dimethyl-	n.s.g.	303(3.64)	25-0366-64
2,4-Cyclohexadien-1-one, 6-allyl-4,6-dimethyl-	MeOH	318(3.66)	35-5115-65
2,5-Cyclohexadien-1-one, 4-allyl-2,4-dimethyl-	MeOH	242(4.14)	35-5115-65
3a,6-Etheno-3aH-inden-4(1H)-one, 2,3,5,6,7,7a-hexahydro-	EtOH	289(1.46)	44-2519-65
2H-1,2a-Methanocycloprop[f]inden-2-one, octahydro-	EtOH	277(2.04)	44-2519-65
Naphthalene, 1,2,3,4-tetrahydro-6-methoxy-	EtOH	280(3.31),288(2.25)	44-3209-65
1(2H)-Naphthalenone, 3,5,6,7-tetrahydro-4-methyl-	MeOH	228(4.06),298(3.52)	78-2297-65
2-Naphthol, 5,6,7,8-tetrahydro-1-methyl-	$CHCl_3$	282(3.48)	94-0112-64
2-Naphthol, 5,6,7,8-tetrahydro-3-methyl-	$CHCl_3$	283(3.48)	94-0112-64
Phenol, 2,6-dimethyl-4-propenyl-	hexane	261(4.09),292(3.18)	49-0512-64
	EtOH	262(4.22),294s(3.43)	44-3014-64
$C_{11}H_{14}OS$			
Acetophenone, p-(isopropylthio)-	hexane	230(3.9),237(3.9), 296(4.2),312(4.2)	39-4599-65
$C_{11}H_{14}O_2$			
Acetophenone, p-(isopropoxy)-	hexane	268(4.3)	39-4599-65
Benzene, 4-allyl-1,2-dimethoxy-	pet ether	228(4.03),278(3.78)	65-2094-65
	EtOH	279(3.60)	65-2094-65

Compound	Solvent	$\lambda_{max}(\log \epsilon)$	Ref.
Benzene, 4-allyl-1,2-dimethoxy- (contd.)	EtOH-HCl	277(3.78)	65-2094-65
	EtOH-NaOEt	279(3.60)	65-2094-65
	10% H_2SO_4	280(3.78)	65-2094-65
Benzene, 1,2-dimethoxy-4-propenyl-, cis	pet ether	260(4.18),283(3.28)	65-2094-65
	EtOH-HCl	250(3.58),285(3.60), 310s(3.22)	65-2094-65
	EtOH	260(4.18),283(3.58), 318s(3.08)	65-2094-65
	EtOH-NaOEt	257(4.38),285(3.53)	65-2094-65
	10% H_2SO_4	253(3.88),279(3.78), 308s(3.07)	65-2094-65
trans	pet ether	260(4.03),285(3.69)	65-2094-65
	EtOH-HCl	261(3.88),275(3.58)	65-2094-65
	EtOH	260(4.03),285(3.60)	65-2094-65
	EtOH-NaOEt	261(4.38),285(3.77)	65-2094-65
	10% H_2SO_4	268(4.18),295(3.85)	65-2094-65
Benzene, 2,4-dimethoxy-1-propenyl-	EtOH	259(4.22),301(3.72)	5-0021-64G
2,5-Cyclohexadiene, 2-butyryl-1,4-epoxy-3-methyl-	EtOH	208(4.04),265(4.03)	28-0404-64B
2,5-Cyclohexadiene, 2-isobutyryl-1,4-epoxy-3-methyl-	EtOH	208(4.13),265(4.11)	28-0404-64B
5-Cyclohexene, 3-butyryl-1,4-epoxy-2-methylene-	EtOH	210(3.80),267(2.66), 311(2.08)	28-0404-64B
5-Cyclohexene, 3-isobutyryl-1,4-epoxy-2-methylene-	EtOH	210(3.79),266(2.53), 306(2.04)	28-0404-64B
1,4-Cyclopentanedione, 3-methyl-2-(2,4-pentadien-1-yl)-, cis	n.s.g.	226(4.45),240s(4.33)	39-3097-65
2,4-Heptadien-6-ynoic acid, 4,5-dimethyl-, ethyl ester, 4-cis-2-trans	EtOH	295(4.36)	70-0546-65
trans-trans	EtOH	297.5(4.47)	70-0546-65
Indan-4-carboxylic acid, 7,7a-dihydro-, methyl ester	EtOH	235(4.12),292(3.42)	44-3679-65
2(4aH)-Naphthalenone, 3,4,5,6-tetrahydro-7-methoxy-	EtOH	312(4.45)	44-1292-65
	EtOH	312(4.50)	44-3209-65
1-Naphthol, 5,6,7,8-tetrahydro-4-methoxy-	EtOH	220s(3.83),280(3.50)	24-1926-64
	NaOH	241(3.87),302(3.55)	24-1926-64
2-Naphthol, 1,2,3,4,5,6,7,8-octahydro-3-hydroxymethylene-	EtOH	275(3.88)	88-1599-65
2,4-Nonadiynoic acid, ethyl ester	EtOH	235(3.24),247(3.60), 261(3.75),277(3.58)	39-0543-64
2,4,6-Nonatrienoic acid, 4-hydroxy-8,8-dimethyl-, γ-lactone	ether	321.5(4.41)	24-2236-65
Propiophenone, 2'-hydroxy-3',4'-dimethyl-	C_6H_{12}	261(4.13),269(4.11), 332(3.61)	28-5614-64A
	EtOH	264(4.14),332(3.58)	28-5614-64A
Propiophenone, 2'-hydroxy-3',5'-dimethyl-	C_6H_{12}	257(4.00),263(3.99), 345(3.61)	28-5614-64A
	EtOH	259(4.02),344(3.58)	28-5614-64A
Propiophenone, 2'-hydroxy-4',5'-dimethyl-	C_6H_{12}	257(4.09),264(4.10), 336(3.66)	28-5614-64A
	EtOH	261(4.11),334(3.63)	28-5614-64A
Propiophenone, 2'-hydroxy-4',6'-dimethyl-	C_6H_{12}	265(4.05),332(3.61)	28-5614-64A
	EtOH	257(3.62)	28-5614-64A
Propiophenone, 4'-hydroxy-2',3'-dimethyl-	C_6H_{12}	259(4.04)	28-5614-64A
	EtOH	274(4.03)	28-5614-64A
Propiophenone, 4'-hydroxy-2',5'-dimethyl-	C_6H_{12}	260(4.10)	28-5614-64A
	EtOH	274(4.09)	28-5614-64A
Propiophenone, 4'-hydroxy-3',5'-dimethyl-	C_6H_{12}	264(4.16)	28-5614-64A
	EtOH	280(4.13)	28-5614-64A
Pyrethrolone	EtOH	225(4.52)	39-5225-64

Compound	Solvent	$\lambda_{max}(\log \epsilon)$	Ref.
Spiro[bicyclo[2.2.1]heptane-2,1'-cyclobut-2'-ene]-3-carboxylic acid	EtOH	224.5(4.068)	28-1541-64A
Spiro[4.5]deca-6,8-diene-1-carboxylic acid	EtOH	283(3.49)	44-3679-65
$C_{11}H_{14}O_3$			
Acetophenone, 2',4'-dimethoxy-6'-methyl-	EtOH	267.0(3.74)	1-1677-65
Acetophenone, 3',4'-dimethoxy-2'-methyl-	MeOH	269(4.08)	44-3028-64
Acetophenone, 3',4'-dimethoxy-5'-methyl-	MeOH	269(4.05),299s(3.73)	44-3028-64
1,4-Benzodioxan-5-ol, 6-propyl-	EtOH	272(2.89)	65-2994-64
o-Benzoquinone, 5-tert-butyl-3-methoxy-	$CHCl_3$	246(3.70),467(3.16)	39-2914-65
p-Benzoquinone, 2-acetonyl-3,5-dimethyl-	EtOH	255(4.22),333(2.66)	1-0421-64
p-Benzoquinone, 5-tert-butyl-2-methoxy-	CCl_4	263(2.86),354(4.20)	39-2914-65
$\Delta^{1,\alpha}$-Cyclohexaneacetic acid, 2-hydroxy-	EtOH	212(4.19)	94-0043-65
2,6,6-trimethyl-4-oxo-, γ-lactone	EtOH	212(4.19)	94-0752-64
4,6-Decadiyn-1-oic acid, 8-hydroxy-, methyl ester	EtOH	228(3.39),240(3.39), 254(3.15)	70-1237-65
4,6-Decadiyn-1-oic acid, 9-hydroxy-, methyl ester	EtOH	226(2.67),240(2.65), 265(2.42)	70-1237-65
4,6-Decadiyn-1-oic acid, 8-hydroxy-8-methyl-	EtOH	228(2.44),240(2.22), 254(2.43)	70-1237-65
2,4,6-Nonatrienoic acid, 8-oxo-, ethyl ester	EtOH	310(4.63)	70-0684-65
Propiophenone, 2',4'-dimethoxy-	EtOH	266(4.18),302(3.88)	5-0021-64G
$C_{11}H_{14}O_4$			
Acetophenone, 2'-ethyl-3',6'-dihydroxy-4'-methoxy-	EtOH	238(3.95),275(3.62), 316(3.55)	39-0411-64
Benzoic acid, 2-hydroxy-4-methoxy-6-methyl-, ethyl ester	EtOH	262(4.24),300(3.74)	1-1677-65
Bicyclo[2.2.1]hept-2-ene-2,3-dicarboxylic acid, dimethyl ester	EtOH	236(3.92)	77-0391-65
1,3,5-Cyclohexanetrione, 4-acetyl-	EtOH-acid	275(3.77),330(3.93)	39-6543-65
2,2,6-trimethyl-	EtOH-base	355(4.22)	39-6543-65
1,2,4-Cyclopentanetrione, 5-methyl-3-pivaloyl-	EtOH-HCl	255(4.40),279(4.29), 340(2.93)	39-2251-65
	EtOH-NaOH	237(3.89),273(4.42), 300(4.40)	39-2251-65
Malonic acid, cyclopentylidene-, cyclic ester with acetone	EtOH-HCl	245.5(4.057)	49-1283-64
Malonic acid, (2,3-dimethyl-2-penten-4-ynyl)methyl-	EtOH	231(4.11)	70-1653-64
4H-Pyran-2-carboxylic acid, 6-ethyl-	EtOH-HCl	265(3.95)	39-2251-65
3-methyl-4-oxo-, ethyl ester	EtOH-NaOH	260(4.11)	39-2251-65
2-Pyrone, 6-butyl-3-carbomethoxy-	EtOH	321(4.10)	78-2701-64
o-Toluic acid, 4,6-dimethoxy-, methyl ester	EtOH	280(3.41)	1-1677-65
2,4,10-Trioxaadamantane, 3-(4-oxo-2-buten-1-yl)-	ether	215.5(4.37)	24-1839-64
$C_{11}H_{14}O_4S$			
Thiophene-2,5-dicarboxylic acid, 3,4-dimethyl-, ethyl methyl ester	EtOH	209(4.08),280(4.24)	44-1919-64

Compound	Solvent	λ_{max}(log ε)	Ref.
$C_{11}H_{14}O_5$			
Acetic acid, (3,4,5-trimethoxyphenyl)-	EtOH	270(3.57)	78-1411-65
Benzoic acid, 2-ethoxy-4,5-dimethoxy-	MeOH	225(4.35),258(3.97)	35-2177-64
1,3,5-Cyclohexanetrione, 6-acetyl-	EtOH	238(3.88),275(4.02)	39-6543-65
2-hydroxy-2,4,4-trimethyl-	EtOH-base	270(4.23)	39-6543-65
Cyclopentanecarboxylic acid, 3-acetyl-5-methyl-2,4-dioxo-, ethyl ester	MeOH-HCl	226(4.39),264(4.23)	20-0628-64
1,3-Cyclopentanedione, 2-acetyl-4-carbethoxy-5-methyl-	MeOH-NaOH	248(4.62),263s(4.55)	20-0628-64
3-Cyclopentene-1-carboxylic acid,	EtOH	293(4.12),360(3.08)	44-3520-64
4-acetyl-3-hydroxy-1-methyl-2-oxo-, ethyl ester	EtOH-base	360(4.14)	44-3520-64
$C_{11}H_{14}O_6$			
Dimethyldegeranylmelicopol	EtOH	218s(4.15),232(4.13), 281(4.17),333(3.62)	12-2021-65
	EtOH-KOH	239s(4.16),278(3.91), 360(3.73)	12-2021-65
Genipinic acid	EtOH	203(3.51)	78-1781-64
	EtOH-KOH	274(4.20)	78-1781-64
$C_{11}H_{14}S_3$			
Camphotrithione	isooctane	225(4.01),248(3.94), 280(3.79),343(3.88), 417(3.83),505s(2.3)	5-0062-64D
$C_{11}H_{14}Sn$			
Tin, trimethyl(phenylethynyl)-	EtOH	247(4.36),259(4.30)	22-0035-65
$C_{11}H_{15}Br$			
Toluene, p-(3-bromo-1-methylpropenyl)-	EtOH	251s(2.36),258(2.56), 264(2.70),272(2.55)	78-2593-65
$C_{11}H_{15}BrN_2$			
1,3-Diethylbenzimidazolium bromide	MeOH	256(3.73),263(3.74), 270(3.80),277(3.74)	65-0632-64
$C_{11}H_{15}BrO$			
o-Cresol, 4-bromo-6-tert-butyl-	n.s.g.	280(3.3)	70-0336-65
3a,6-Ethano-3aH-inden-4(1H)-one, 8-bromohexahydro-	EtOH	287(1.63)	44-2519-65
$C_{11}H_{15}ClN_2O_2$			
3-(2,2-Dihydroxyethyl)-1,2-dimethylbenzimidazolium chloride	EtOH	255(3.76),264(3.79), 270(3.91),277(3.95)	44-1118-65
$C_{11}H_{15}ClN_2O_4S$			
1-D-Ribofuranosyl-3-(thiocarbamoyl)-	pH 7.0	264.5(3.78)	39-0610-65
pyridinium chloride	M KCN	345(3.75)	39-0610-65
$C_{11}H_{15}ClN_2O_5$			
3-Carbamoyl-1-α-D-ribofuranosyl-	H_2O	267(3.68)	39-0610-65
pyridinium chloride	M KCN	329.5(3.79)	39-0610-65
$C_{11}H_{15}ClN_6OS$			
2-Acetamido-5-(2-cyanoethyl)-6-methyl-	pH 1, 8.4	242(4.28),292(3.91)	4-0263-64
4-pyrimidylisothiouronium chloride	pH 13	309(4.08)	4-0263-64
$C_{11}H_{15}ClO$			
2-Cyclohexen-1-one, 6-(3-chloro-2-butenyl)-6-methyl-	EtOH	227(3.98)	44-3642-65

Compound	Solvent	$\lambda_{max}(\log \epsilon)$	Ref.
Cyclopropenone, 2-propyl-3-(2-chloro-penten-1-yl)-, cis	ether	258.5(4.49)	88-2317-65
trans	ether	257(4.53)	88-2317-65
$C_{11}H_{15}Cl_2N_3O_4$			
1(2H)-Pyrimidinepropionic acid, 5-[bis(2-chloroethyl)amino]-3,4-dihydro-2,4-dioxo-	pH 1	268(3.80)	87-0187-65
	pH 7	260s(3.70),305s(3.48)	87-0187-65
	pH 13	257s(3.78)	87-0187-65
$C_{11}H_{15}Cl_3O_4$			
6,10-Dioxaspiro[4.5]dec-1-ene, 1,2,4-trichloro-3,3-dimethoxy-7-methyl-	n.s.g.	215s(3.91)	39-4744-65
$C_{11}H_{15}IN_2S$			
N-(p-Vinylphenyl)-N',S-dimethyl-isothiuronium iodide	EtOH	274(4.27)	44-1926-65
$C_{11}H_{15}N$			
Fulvene, 6-piperidino-	hexane	322(4.50)	5-0039-64H
$C_{11}H_{15}NO$			
5H-Benzocyclohepten-1-ol, 3-amino-6,7,8,9-tetrahydro-	MeOH	214(4.31),233(3.85), 289(3.36)	33-1988-65
sec-Butylamine, N-salicylidene-	EtOH	253(4.09),278s(3.40), 312(3.59),401(2.93)	44-2265-64
Oxaziridine, 2-tert-butyl-3-phenyl-	EtOH	210(3.95),249(2.97)	44-3427-65
Piperidine, 1-hydroxy-2-phenyl-	EtOH	260(2.30)	44-0910-65
Spiro[5.5]undeca-1,4-dien-3-one, oxime	MeOH-acid	276.5(4.20)	33-1988-65
	MeOH	246(4.22),259s(4.19)	33-1988-65
Spiro[5.5]undeca-1,8-dien-3-one, oxime	MeOH-acid	249(4.13)	33-1988-65
	MeOH	232(4.23)	33-1988-65
$C_{11}H_{15}NO_2$			
Anthranilic acid, N,N-diethyl-	10% MeOH	264s(2.77),270(2.85), 277(2.77)	24-1127-64
anion	10% MeOH	263(3.57),300s(2.95)	24-1127-64
Anthranilic acid, N-ethyl-N-methyl-, methyl ester	10% MeOH	257(3.54),332(3.25)	24-1127-64
Azepine-N-carboxylic acid, tert-butyl ester	hexane	210(3.95),332(2.53)	88-1733-64
7-Isoquinolinol, 1,2,3,4-tetrahydro-6-methoxy-1-methyl-	EtOH-HCl	224(3.81),284(3.59)	33-1945-65
	EtOH	223s(3.84),285(3.59)	33-1945-65
	EtOH-NaOH	245(3.85),301(3.71)	33-1945-65
hydrochloride	EtOH-HCl	224(3.82),284(3.60)	33-1945-65
	EtOH	224(3.82),285(3.59)	33-1945-65
	EtOH-NaOH	245(3.93),301(3.77)	33-1945-65
Propane, 2-[N-(3-methoxysalicylidene)amino]-	EtOH	222(4.26),258(3.96), 294(3.68),328(3.30), 420(3.23)	59-0597-64
	$PhCH_2OH$	421(3.42)	59-0597-64
	dioxan	222(4.30),260(4.08), 328(3.40)	59-0597-64
	$CHCl_3$	253(3.92),328(3.59), 425(2.40)	59-0597-64
	DMF	326(3.40),417(2.52)	59-0597-64
	pyridine	326(3.42),415(3.18)	59-0597-64
2-Pyrone, 6-methyl-4-piperidino-	MeOH	230(4.42),262(3.96), 299(4.10)	24-1266-64

Compound	Solvent	λ_{max}(log ε)	Ref.
$C_{11}H_{15}NO_2S$			
Benzothiophene-3-carboxylic acid, 2-amino-4,5,6,7-tetrahydro-, ethyl ester	MeOH	230(4.49),312(3.98)	24-3571-65
$C_{11}H_{15}NO_3$			
Anhalamine	pH 1	225s(3.92),270(2.88), 280s(2.70)	33-2089-64
	pH 13	240s(3.90),285(3.38)	33-2089-64
	iso-PrOH	227s(3.98),272(2.93), 280s(2.83)	33-2089-64
Phenol, 2-nitro-4-sec-pentyl-	N NaOH	227(4.6),292(4.1), 430(4.1)	65-2790-64
	$CHCl_3$	280(3.9),360(3.6)	65-2790-64
2-Pyrone, 5,6-dimethyl-4-morpholino-	MeOH	225(4.03),278(3.95), 300(3.91)	24-1266-64
$C_{11}H_{15}NO_4$			
Pyrrole-3-carboxylic acid, 4-acetoxy-1,2-dimethyl-, ethyl ester	EtOH	206(4.22),234(3.95), 259(3.72)	23-1524-64
$C_{11}H_{15}NO_4S$			
2,5-Pyridinedicarboxylic acid, 1,4,5,6-tetrahydro-3-methyl-6-thio-, 2-ethyl 5-methyl ester	EtOH	241(3.78),327(4.20)	39-1262-65
2H-1,3-Thiazinylideneacetic acid, 4-carboxy-3,6-dihydro-5-methyl-, 4-ethyl 2-methyl ester	EtOH	222s(3.90),280(3.67), 337(4.13)	39-1262-65
$C_{11}H_{15}NO_5$			
Anisole, 2-(3-hydroxybutoxy)-5-nitro-	EtOH	241(3.98),338(3.90)	94-0987-64
$C_{11}H_{15}NS$			
Butyramide, N-methyl-2-phenyl-thio-	MeOH	267(4.07),334(1.86)	35-0051-65
$C_{11}H_{15}N_2O_8P$			
3-Carbamoyl-1-α-D-ribofuranosylpyridinium hydroxide, 5'-phosphate, inner salt	n.s.g.	267(3.68)	5-0170-65J
$C_{11}H_{15}N_3O$			
Acetamide, N-(α-amidinophenethyl)-, hydrochloride	MeOH-NaOMe	276(4.00)	44-0308-64
Pyrazolo[1,5-a]pyrimidin-7(4H)-one, 4-ethyl-2,3,6-trimethyl-	EtOH	223(4.34),268(3.94), 318(3.78)	94-0142-65
$C_{11}H_{15}N_3O_2$			
Acetophenone, p-ethoxy-, semicarbazone	C_6H_{12}	264.5(4.27)	78-1555-64
α^4,3-O-Isopropylideneisopyridoxal, hydrazone	pH 1	310(3.93)	44-0574-64
$C_{11}H_{15}N_3O_3$			
α^4,3-O-Isopropylidene-5-pyridoxic acid hydrazide	pH 13	298(3.93)	44-0574-64
	EtOH	289(3.82)	44-0574-64
Salicylaldehyde, 5-nitro-, 1,1-diethylhydrazone	MeOH	300(4.46)	24-1631-64
	MeOH-NaOH	345(4.24),423(4.02)	24-1631-64
Salicylaldehyde, 5-nitro-, (1-methyl-1-isopropylhydrazone)	MeOH	294(4.46)	24-2713-64
	NaOH	337(4.24),431(4.01)	24-2713-64
Salicylaldehyde, 5-nitro-, (1-methyl-1-propylhydrazone)	MeOH	296(4.45)	24-2713-64
	NaOH	340(4.23),418(4.04)	24-2713-64

Compound	Solvent	$\lambda_{max}(\log \epsilon)$	Ref.
$C_{11}H_{15}N_3O_4S$			
Malonic acid, [[(5-methyl-1,3,4-thia-diazol-2-yl)amino]methylene]-, diethyl ester	EtOH	310(4.4)	70-1481-64
$C_{11}H_{15}N_4O_7P$			
7H-Pyrrolo[2,3-d]pyrimidine, 4-amino-7-β-D-ribofuranosyl-, 5'-phosphate	pH 1	228(4.32),271(4.02)	4-0159-64
$C_{11}H_{15}N_5$			
1,2,4-Triazolo[1,5-a]pyrimidine, 5-methyl-7-piperidino-	MeOH	304(4.29)	65-0204-64
$C_{11}H_{15}N_5O$			
Pyrazolo[1,5-a]pyrimidine, 2,3-di-methyl-7-(3,3-dimethylureido)-	EtOH	233(4.67),276s(3.25), 287(3.5),296(3.52), 332(3.69)	94-1207-65
$C_{11}H_{15}N_5OS$			
9H-Purine-9-carboxamide, N-butyl-6-(methylthio)-	pH 1	265(3.87)	87-0010-64
	pH 7	260(3.79)	87-0010-64
$C_{11}H_{15}N_5O_2$			
Glycine, N-(7-methyl-5-propyl-s-tria-zolo[1,5-c]pyrimidin-2-yl)-	MeOH	234(4.63),270s(3.54), 304(3.26)	39-3357-65
$C_{11}H_{15}N_5O_3$			
Orthoformic acid, ethyl ester, cyclic ester with 3-(6-amino-9H-purin-9-yl)-1,2-propanediol	pH 1	261(4.17)	87-0502-65
	pH 7,13	261(4.19)	87-0502-65
Purine, 9-(3-deoxy-β-D-ribofuranosyl)-6-(methylamino)-	pH 1	263(4.21)	87-0659-65
	pH 7	267(4.21)	87-0659-65
	pH 13	266(4.22)	87-0659-65
$C_{11}H_{15}N_5O_4$			
Adenine, 9-(2-deoxy-D-ribohexo-pyranosyl)-	H_2O	259.5(4.22)	44-2018-64
anomer	H_2O	260(4.1)	44-2018-64
$C_{11}H_{15}N_5O_4S$			
Adenosine, 8-(methylthio)-	pH 1	281(4.27)	35-1242-64
	pH 11	279(4.24)	35-1242-64
Guanosine, 7-methyl-6-thio-	pH 1	254(3.94),348(4.24)	4-0113-64
	MeOH	329(4.16)	4-0113-64
$C_{11}H_{15}N_5O_5$			
Adenine, 9-β-D-allopyranosyl-	H_2O	258(4.17)	87-0655-64
Adenine, 9-α-D-altropyranosyl-	H_2O	259(4.20)	87-0655-64
Adenine, 9-β-D-gulofuranosyl-	H_2O	259(4.16)	35-1457-64
Adenine, 9-β-D-gulopyranosyl-	H_2O	259(4.17)	87-0655-64
Adenine, 9-α-D-mannopyranosyl-	H_2O	260(4.16)	87-0655-64
Adenine, 9-α-D-talopyranosyl-	H_2O	259(4.18)	87-0655-64
Guanosine, N -methyl-	pH 1	258(4.16),281s(3.90)	35-3752-65
	pH 11	254(4.17),270s(4.06)	35-3752-65
$C_{11}H_{15}N_5O_5S$			
Guanosine, 8-(methylthio)-	pH 1	272(4.23),289s(4.13)	35-1242-64
	pH 11	284(4.23)	35-1242-64

Compound	Solvent	λ_{max}(log ε)	Ref.
$C_{11}H_{16}$			
Azulene, hexahydro-2-methyl-	EtOH	245(4.03)	78-2815-64
1-Cyclobutene, 3-allyl-1,2,3-trimethyl-4-methylene-	hexane	235(4.17)	24-2327-65
1,3,5-Cycloheptatriene, tetramethyl-	hexane	254(3.62)	24-2327-65
1,3,5,7-Nonatetraene, 2,8-dimethyl-	isooctane	288(4.60),301(4.77), 316(4.71)	35-5075-65
Spiro[2.4]heptadiene, 1-butyl-	pentane	228(3.75),257(3.40)	77-0622-65
Spiro[2.4]hepta-4,6-diene, 1-tert-butyl-	pentane	228(3.97),258(3.49)	77-0622-65
Spiro[2.4]hepta-4,6-diene, 1,1,2,2-tetramethyl-	pentane	237(3.98),273(3.26)	77-0622-65
Toluene, p-sec-butyl-	EtOH	252s(2.12),259(2.27), 264(2.36),272(2.26)	78-2921-64
$C_{11}H_{16}IN_3$			
Imidazo[1,2-a]pyridin-3-ylmethyltrimethylammonium iodide	EtOH	224(4.66),280(3.72)	44-2403-65
$C_{11}H_{16}NO_6PS$			
Phosphorothioic acid, O,O-diethyl S-(2-methoxy-5-nitrophenyl) ester	MeOH	200(4.30),211(4.20), 298(4.03)	78-0177-64
$C_{11}H_{16}N_2O$			
1-Benzimidazolineethanol, α,3-dimethyl-	EtOH	220(4.36),269(3.69), 315(3.65)	12-0877-64
1-Cyclopentene-1-carboxamide, 2-(cyclopentylideneamino)-	EtOH	306.5(3.80)	23-0010-64
Tropone, 2-(2-dimethylaminoethylamino)-	EtOH	246(4.99),338(4.60), 407(4.57)	94-0457-65
$C_{11}H_{16}N_2O_2$			
Indazole-3-carboxylic acid, 4,5,6,7-tetrahydro-6-methyl-, ethyl ester	EtOH	226(3.92),250s(3.67)	32-0814-65
$C_{11}H_{16}N_2O_7$			
Thymine, 1-β-D-glucopyranosyl-	H_2O	264(4.03)	94-0357-64
$C_{11}H_{16}N_2O_7S$			
Thymine, 1-(2-deoxy-β-D-lyxofuranosyl)-, 3'-methanesulfonate	EtOH	265(4.02)	44-2077-64
Thymine, 1-(2-deoxy-β-D-lyxofuranosyl)-, 5'-methanesulfonate	EtOH	266(3.98)	44-2077-64
$C_{11}H_{16}N_4$			
Pyrazolo[1,5-a]pyrimidine, 7-(dimethylamino)-2,3,5-trimethyl-	EtOH	234(4.62),288s(3.70), 298(3.78),330(3.85)	94-1207-65
Pyrazolo[1,5-a]pyrimidine, 7-(dimethylamino)-2,3,6-trimethyl-	EtOH	238(4.57),266s(3.67), 291(3.57),303s(3.47), 350(3.76)	94-1207-65
Pyrazolo[1,5-a]pyrimidine, 7-(ethylamino)-2,3,6-trimethyl-	EtOH	232(4.56),285s(3.68), 293(3.75),332(3.83)	94-0142-65
Pyrazolo[1,5-a]pyrimidine, 4-ethyl-4,7-dihydro-7-imino-2,3,6-trimethyl-	EtOH	234(4.43),265s(3.87), 318(3.76)	94-0142-65
hydrobromide	EtOH	235(4.52),267s(3.82), 345(3.76)	94-0142-65
$C_{11}H_{16}N_4O$			
s-Triazolo[1,5-c]pyrimidine, 2-ethoxy-7-methyl-5-propyl-	MeOH	214(4.54),258(3.70)	39-3357-65

Compound	Solvent	λ_{max}(log ϵ)	Ref.
$C_{11}H_{16}N_4O_2$			
1H-Pyrazolo[4,3-d]pyrimidine-5,7(4H,6H)-dione, 1,4,6-triethyl-	MeOH	291(3.76)	44-0199-65
2H-Pyrazolo[4,3-d]pyrimidine-5,7(4H,6H)-dione, 2,4,6-triethyl-	MeOH	289(3.71)	44-0199-65
Urea, [2-cyano-3-(cyclohexylamino)-acryloyl]-	H_2O	278(4.18)	39-1642-65
$C_{11}H_{16}N_4O_2S$			
s-Triazolo[2,3-c]pyrimidine, 2-(ethyl-sulfonyl)-7-methyl-5-propyl-	MeOH	209(4.34),255(3.88), 283s(3.55)	39-3369-65
$C_{11}H_{16}N_4S$			
s-Triazolo[2,3-c]pyrimidine, 2-(ethyl-thio)-7-methyl-5-propyl-	MeOH	238(4.51),285s(3.35)	39-3369-65
$C_{11}H_{16}N_6O_4$			
Adenine, 9-(2-amino-2-deoxy-α-D-glucopyranosyl)-	H_2O	262(4.29)	44-1556-65
Adenine, 9-(2-amino-2-deoxy-β-D-glucopyranosyl)-	H_2O	261(4.18)	44-1556-65
$C_{11}H_{16}O$			
Anisole, o-tert-butyl-	C_6H_{12}	218(3.88),222(3.87), 265s(3.18),270(3.30), 277(3.31)	23-2603-65
Anisole, 2,6-diethyl-	C_6H_{12}	215s(3.99),267(2.62), 273(2.58)	23-2603-65
1(2H)-Azulenone, 3,4,5,6,7,8-hexa-hydro-2-methyl-	EtOH	242(4.02)	78-2815-64
Bicyclo[3.3.]]hexane-$\Delta^{2,\alpha}$-acetaldehyde, 1,3,3-trimethyl-	n.s.g.	266(4.1)	27-0538-65
Bicyclo[3.3.1]non-2-en-9-one, 1,5-dimethyl-	EtOH	291(1.62)	39-0289-64
Bicyclo[5.1.0]oct-5-en-2-one, 1,4,4-trimethyl-	EtOH	207(3.57),272(2.42)	35-1353-65
Carvacrol, methyl ether	pet ether	275(3.55)	65-2094-65
2-Cyclohexen-1-one, 2-(3-butenyl)-3-methyl-	EtOH	243(4.08)	35-5148-65
Cyclopentanone, 2-allyl-2-isopropenyl-	isooctane	302(1.15)	88-2791-64
Cyclopentanone, 2-(1-methyl-4-penten-1-ylidene)-	isooctane	249(3.88),350(1.69), 354(1.70)	88-2791-64
isomer	isooctane	248(3.90),341(1.60), 356(1.62)	88-2791-64
Jasmone, cis	EtOH	237(4.04)	35-0936-64
4a-Naphthalenemethanol, 3,4,4a,5,6,7-hexahydro-	n.s.g.	228(--),238(4.02), 245(--)	28-3705-64A
2(3H)-Naphthalenone, 4,4a,5,6,7,8-hexa-hydro-3-methyl-	EtOH	238(4.08)	35-0275-65
4,5-Nonadien-2-yn-1-ol, 6,8-dimethyl-	EtOH	221(4.09)	39-4659-65
6,8-Nonadien-1-yn-3-ol, 3,7-dimethyl-	EtOH	231(4.31)	22-2533-64
1,6-Octadien-4-yne, 8-methoxy-2,6-dimethyl-	MeOH	230(4.2)	5-0026-64I
4,5-Octadien-2-yn-1-ol, 6,7,7-tri-methyl-	EtOH	221(4.23)	39-4659-65
Pentane, 3-(4-oxocyclohex-2-en-1-ylidene)-	MeOH	304(4.20)	5-0028-64D
Phenol, p-sec-pentyl-	N NaOH	239(4.2),290(3.6)	65-2790-64
	$CHCl_3$	279(3.7)	65-2790-64
Spiro[4.4]non-6-en-1-one, 6,9-dimethyl-	isooctane	309(1.72)	88-2791-64

Compound	Solvent	$\lambda_{max}(\log \epsilon)$	Ref.
Thymol, methyl ether	pet ether	223(4.03),250(3.48)	65-2094-65
$C_{11}H_{16}OS$			
Sulfoxide, butyl p-tolyl	EtOH	218s(3.91),241(3.78), 267s(3.04),275s(2.72)	35-1958-65
Sulfoxide, tert-butyl p-tolyl	isooctane	211(4.01),218(4.02), 261(3.70)	35-1958-65
	isooctane-CF_3COOH	217s(4.00),245(3.87), 276s(2.77)	35-1958-65
	EtOH	217s(4.01),250(3.81), 278s(3.00)	35-1958-65
	EtOH-CF_3COOH	217(4.03),250(3.81), 276(3.04)	35-1958-65
$C_{11}H_{16}O_2$			
Bicyclo[3.1.0]hexane-3α-acetic acid, 1α-methyl-2-methylene-, methyl ester	C_6H_{12}	200(4.05)	35-5218-65
	EtOH	200(4.02)	35-5218-65
1,5-Cyclohexadiene-1-carboxylic acid, 3-isopropyl-, methyl ester	EtOH	283(3.43)	44-3679-65
2-Cyclohexen-1-one, 2-acetyl-3,5,5-trimethyl-	n.s.g.	248(3.95)	39-1511-64
2-Cyclohexen-1-one, 6-methyl-6-(3-oxobutyl)-	EtOH	227(3.95)	44-3642-65
1,3-Cyclopentadiene-1-carboxylic acid, 2,3,4-trimethyl-, ethyl ester	EtOH	297(3.98)	70-1820-64
1,3-Cyclopentanedione, 4-methyl-5-(4-pentenyl)-	n.s.g.	244(4.13)	39-3097-65
4-Cyclopentene-1,3-diol, 4-methyl-5-(2,4-pentadienyl)-, cis	EtOH	230.5(4.36)	39-1854-64
1-Naphthoic acid, 2,3,4,4a,5,6,7,8-octahydro-	EtOH	224(3.85)	44-3679-65
1-Naphthol, 1,2,3,4,5,8-hexahydro-6-methoxy-	EtOH	238(3.98)	44-2942-65
2-Norbornylideneacetic acid, 3,3-dimethyl-	EtOH	227(4.17)	44-3241-64
8-Oxabicyclo[3.2.1]oct-6-en-3-one, 2,2,4,4-tetramethyl-	EtOH	301(1.56)	77-0144-64
2H-Pyran-2-one, 4,6-diethyl-3,5-dimethyl-	EtOH	304(3.86)	24-1949-65
4H-Pyran-4-one, 2,6-diethyl-3,5-dimethyl-	EtOH	259(4.09)	24-1949-65
Spiro[5.5]undecane-2,4-dione	MeOH-HCl	250(4.20)	5-0123-65A
	MeOH-NaOMe	280(4.44)	5-0123-65A
	H_2SO_4	277(4.29)	5-0123-65A
$C_{11}H_{16}O_2Pd$			
Palladium, (2,4-pentanedionato)-2-cyclohexen-1-yl-	hexane	222(4.31),301(3.87)	39-5002-64
$C_{11}H_{16}O_3$			
1-Cyclohexene-1-carboxylic acid, 2,6-dimethyl-3-oxo-, ethyl ester	EtOH	220(3.97)	44-3524-64
4-Cyclohexene-1,3-dione, 5-hydroxy-2,2,4,6,6-pentamethyl-	PrOH	260(4.10)	39-1028-65
1-Cyclohexene-1-propionic acid, 2,5-dimethyl-6-oxo-	EtOH	242(4.1)	42-0242-64
Cyclohexylideneacetic acid, 2,4-dihydroxy-2,6,6-trimethyl-, γ-lactone	EtOH	214(4.15)	94-0752-64
3-Cyclopentene-1-carboxylic acid, 4-isopropyl-1-methyl-2-oxo-, methyl ester	EtOH	232(4.11),310(2.34)	39-5640-64

Compound	Solvent	λ_{max}(log ϵ)	Ref.
Digiprolactone	EtOH	214(4.15)	94-0043-65
Ketone, 4-ethylenedioxycyclohex-1-en-1-yl ethyl	MeOH	229(4.03)	5-0036-64D
2-Naphthoic acid, decahydro-3-oxo-	EtOH	253(3.49)	28-1026-65B
2,4-Nonadienal, 9-acetoxy-, di-trans	EtOH	225(3.90),274(3.92)	70-2003-64
2,4-Pyrandione, tetrahydro-3-methyl-6,6-pentamethylene-	MeOH-HCl	249(4.00)	5-0123-65A
	MeOH-NaOMe	279(4.28)	5-0123-65A
	H_2SO_4	278(4.04)	5-0123-65A
2,4-Pyrandione, tetrahydro-6,6-pentamethylene-, enol methyl ether	MeOH	238(4.06)	5-0123-65A
Pyrethrolone, hydrate	H_2O	227(4.55)	39-5225-64
	EtOH	225(4.52)	39-5225-64
	ether	226(4.52)	39-5225-64
Sarcomycin, methylpropyl-	MeOH	255(4.05)	7-0206-64
$C_{11}H_{16}O_4$			
Cyclohexanecarboxylic acid, 3-(methoxymethylene)-1-methyl-2-oxo-, methyl ester	EtOH	280(3.88)	44-0782-64
3-Cyclohexene-1-carboxylic acid, 4-ethoxy-2-oxo-, ethyl ester	EtOH	252(4.20)	44-0787-64
$C_{11}H_{16}O_5$			
1-Cyclopentene-1-carboxylic acid, 2-hydroxy-, ethyl ester, ethyl carbonate	EtOH	228(3.87)	44-0087-64
$C_{11}H_{16}O_5S$			
Crotonic acid, 2-hydroxy-4-mercapto-3-methyl-, ethyl ester, diacetate	EtOH	226(4.18)	39-0766-64
$C_{11}H_{17}N$			
Pyrrolidine, N-(3-cyclohexen-1-ylidenemethyl)-	EtOH	245(3.77)	28-1541-64A
$C_{11}H_{17}NO$			
2H-Azepin-2-one, 1-ethyl-1,3-dihydro-3,5,7-trimethyl-	EtOH	253(3.68)	44-3447-64
4-Piperidone, 1-(1,1-dimethyl-2-propynyl)-3-methyl-	n.s.g.	none	70-0512-64
4(1H)-Quinolone, 2,3,5,6,7,8-hexahydro-1,3-dimethyl-	EtOH-HCl	329(3.94)	94-1405-64
$C_{11}H_{17}NO_2$			
Ethanolamine, N,N-dimethyl-β-(4-hydroxy-3-methoxyphenyl)-	pH 1	208(3.84),229(3.82),279(3.45),283s(3.40)	87-0368-65
	pH 13	218(3.84),247(4.03),294(3.62)	87-0368-65
2-Heptenoic acid, 2-cyano-3-methyl-, ethyl ester	EtOH	233(4.08)	28-2859-64B
2-Hexenoic acid, 2-cyano-3,5-dimethyl-, ethyl ester	EtOH	234(4.08)	28-2859-64B
Pyrrole-2-carboxylic acid, 4,5-diisopropyl-	EtOH	235(3.60),284(4.23)	23-1279-64
$C_{11}H_{17}NO_3$			
Phenethylamine, β,3,4-trimethoxy-	EtOH-HCl	247(4.32),308(3.99),363(3.91)	33-1945-65
	EtOH	230(4.47),278(3.87),308(3.77)	33-1945-65

Compound	Solvent	$\lambda_{max}(\log \epsilon)$	Ref.
Phenethylamine, β,3,4-trimethoxy- (contd.)	EtOH-NaOH	277(3.95),308(3.83)	33-1945-65
hydrochloride	EtOH-HCl	246(4.28),308(3.99), 362(3.86)	33-1945-65
	EtOH-NaOH	275(3.99),308(3.88)	33-1945-65
$C_{11}H_{17}NO_3S$			
Valeric acid, 2-cyano-5-mercapto-4-methyl-, ethyl ester, acetate	MeOH	230(3.73)	24-1970-64
$C_{11}H_{17}NO_4$			
Maleic acid, piperidino-, dimethyl ester	EtOH	286(4.35)	5-0098-65H
$C_{11}H_{17}NO_5S$			
2H-Pyran-6-carboxylic acid, 3,4-dihydro-2-hydroxy-2-methoxy-5-methyl-3-(thiocarbamoyl)-, ethyl ester	EtOH	264.0(4.41)	39-1262-65
$C_{11}H_{17}NS$			
3H-Azepine, 2-(ethylthio)-3,5,7-trimethyl-	EtOH	225(4.11),284(3.83)	35-4096-64
$C_{11}H_{17}N_3O$			
Eucarvone, semicarbazone	EtOH	310(4.29)	44-0669-65
3,5-Heptadien-2-one, 6-cyclopropyl-, semicarbazone	EtOH	307(4.65)	22-3218-64
$C_{11}H_{17}N_3O_2S$			
Pyridazine, 3-(cyclohexylamino)-6-(methylsulfonyl)-	pH 1	252(4.22)	4-0001-64
	pH 11	268(4.36)	4-0001-64
	EtOH	266(4.32)	4-0001-64
$C_{11}H_{17}N_3O_3$			
Cyclopentanecarboxylic acid, 2-(isobuten-2-yl)-3-oxo-, semicarbazone	MeOH	235(4.15)	7-0206-64
Sarcomycin, methylethyl-, semicarbazone	MeOH	278(4.18)	7-0206-64
$C_{11}H_{17}N_3O_4$			
3-Azabicyclo[3.3.1]non-6-ene, 3,8,9-trimethyl-1,5-dinitro-	MeOH	205(3.95)(end absorption)	24-0186-64
6-Azauracilhexanoic acid, ethyl ester	MeOH	210(3.81),252(3.77)	39-0868-64
Orotic acid, 2-(diethylamino)ethyl ester	pH 7	280(4.14)	65-1299-65
Valeric acid, 2-acetyl-5-(allyloxy)-4-hydroxy-, γ-lactone, semicarbazone	n.s.g.	222(4.07),274(3.73)	39-0141-64
$C_{11}H_{17}N_3O_8$			
Tetrodoic acid	pH 1	none	78-2059-65
Tetrodoic acid, anhydro-	pH 1	257(3.78)	78-2059-65
	H_2O	261(3.77)	78-2059-65
	pH 13	290(3.81)	78-2059-65
$C_{11}H_{17}N_3S$			
Semicarbazide, 1,2-dimethyl-4-(α-methylbenzyl)-3-thio-, nickel complex	n.s.g.	440(1.67)	1-1239-65
$C_{11}H_{17}N_3S_2$			
3(2H)-Pyridazinethione, 6-(methylthio)-2-(piperidinomethyl)-	THF	304(4.32)	49-0631-65

Compound	Solvent	λ max (log ε)	Ref.
$C_{11}H_{17}N_5$			
s-Triazolo[1,5-c]pyrimidine, 2-(ethyl-amino)-7-methyl-5-propyl-	MeOH	273(4.62),299(3.54)	39-3357-65
$C_{11}H_{17}N_5O$			
Isoxazolo[5,4-d]pyrimidine, 4-(3-di-methylaminopropylamino)-3-methyl-	EtOH	253(4.01),284(3.98)	44-2117-64
9H-Purine-9-propanol, 6-(iso-propylamino)-	pH 1	264(4.25)	87-0033-65
9H-Purine-9-propanol, 6-propylamino-	pH 1	266(4.26)	87-0033-65
	pH 7	270(4.24)	87-0033-65
	pH 13	270(4.22)	87-0033-65
Triazolo[1,5-c]pyrimidine, 2-(2-hy-droxyethylamino)-7-methyl-5-propyl-	MeOH	232(4.66),295(3.40)	39-3357-65
$C_{11}H_{17}N_7O$			
Pterin, 8-[(3-dimethylamino-1-propyl)-amino]-	pH 1	221(4.14),278(4.23), 384(3.64)	33-2195-64
	H_2O	242(3.96),285(4.20), 398(3.63)	33-2195-64
	pH 13	267(4.23),404(3.72)	33-2195-64
$C_{11}H_{17}N_7O_2S$			
Caffein-8-ylaminoethylisothiuronium hydrobromide	pH 1	295(4.27)	65-2774-64
$C_{11}H_{17}N_7O_4$			
Purine, 2-amino-6-(2-methylhydrazino)-9-β-D-ribofuranosyl-	H_2O	263(3.99),288(4.16)	94-0951-64
$C_{11}H_{17}O_7PS$			
Fumaric acid, hydroxy(2-hydroxyethyl)-, γ-lactone, methyl ester, O-ester with O,O-diethyl phosphorothioate	MeOH	240(4.20)	5-0010-65E
Fumaric acid, hydroxy(2-hydroxypropyl)-, γ-lactone, ethyl ester, O-ester with O,O-dimethyl phosphorothioate	MeOH	240(3.98)	5-0010-65E
$C_{11}H_{17}O_8P$			
Fumaric acid, hydroxy(3-hydroxy-propyl)-, δ-lactone, ethyl ester, dimethyl phosphate	MeOH	220(3.73)	5-0010-65E
Oxalacetic acid, (2-hydroxy-1-methyl-ethyl)phosphono-, γ-lactone, 1-ethyl dimethyl ester	MeOH	213(3.85)	5-0010-65E
Oxalacetic acid, (2-hydroxypropyl)-phosphono-, γ-lactone, 1-ethyl dimethyl ester	MeOH	213(3.75)	5-0010-65E
Oxalacetic acid, (3-hydroxypropyl)-phosphono-, δ-lactone, 1-ethyl dimethyl ester	MeOH	212(3.71)	5-0010-65E
$C_{11}H_{18}$			
Cyclopropane, diisopropylidene-1,1-dimethyl-	EtOH	256(4.42),266(4.37)	35-5032-64
$C_{11}H_{18}ClNO$			
Bicyclo[2.2.1]hexane-5-carboxamide, N-tert-butyl-6-chloro-, 5-endo-	EtOH	200(3.74)	44-3469-64
5-exo	EtOH	200(3.74)	44-3469-64

Compound	Solvent	λ max(log ε)	Ref.
$C_{11}H_{18}IN$			
Trimethylphenethylammonium iodide	EtOH	210(4.52),252(3.46), 259(3.44),265(3.36)	56-1437-65
$C_{11}H_{18}IN_5$			
6-(Diethylamino)-3,9-dimethyl-9H-purinium iodide	pH 1	288(4.26)	44-3597-65
	pH 13	287.5(3.54)	44-3597-65
	MeOH	289(4.26)	44-3597-65
$C_{11}H_{18}N_2$			
Pyrrole, 2,5-dimethyl-1-piperidino-	EtOH	210(2.7)	6-0073-64
$C_{11}H_{18}N_2O_2$			
2(1H)-Pyrazinone, 1-hydroxy-3-isobutyl-6-propyl-	EtOH	235(3.79),326(3.89)	44-3165-64
$C_{11}H_{18}N_2O_2SSi$			
2,4-Diaza-8-silaspiro[5.6]dodecane-1,3,5-trione, 8,8-dimethyl-3-thio-	70% EtOH	237(3.89),288(4.35)	87-0695-64
$C_{11}H_{18}N_2O_3$			
1,3(2H,4H)-Isoquinolinedione, 4-(dimethylamino)hexahydro-8a-hydroxy-	EtOH	none	65-0796-64
	EtOH-NaOH	237(4.24)	65-0796-64
Isoxazole, 3,5-di-tert-butyl-4-nitro-	isooctane	257(3.26),341(2.39)	32-0915-64
	MeOH	259(3.22),343(2.42)	32-0915-64
$C_{11}H_{18}N_2O_3Si$			
2,4-Diaza-8-silaspiro[5.6]dodecane-1,3,5-trione, 8,8-dimethyl-	70% EtOH-NaOH	241.5(3.96)	87-0695-64
2,4-Diaza-8-silaspiro[5.5]undecane-1,3,5-trione, 8,8,11-trimethyl-	70% EtOH-NaOH	241.0(3.96)	87-0695-64
$C_{11}H_{18}N_4$			
5H-Cyclononapyrimidine, 2,4-diamino-6,7,8,9,10,11-hexahydro-	pH 1	277(3.82)	44-1837-65
	pH 10	232(3.94),287(3.81)	44-1837-65
$C_{11}H_{18}N_4O_2$			
Pyrimidine, 6-(2-carbethoxyhydrazino)-4-methyl-2-propyl-	MeOH	206(3.95),233(4.03), 266(3.79)	39-3357-65
$C_{11}H_{18}O$			
Bicyclo[3.3.1]nonan-9-one, 1,5-dimethyl-	EtOH	290(1.34)	39-1243-65
Bicyclo[3.2.1]octan-2-one, 1,8,8-trimethyl-	n.s.g.	285(1.46)	24-2742-65
Cyclobutanone, 2-tert-butyl-4-isopropylidene-	C_6H_{12}	243(4.18),250(4.20), 339(1.75),354(1.90), 366(1.84),371(1.86), 386(1.51)	22-2747-65
Cyclobutanone, 4-isopropylidene-2,2,3,3-tetramethyl-	iso-PrOH	248(4.08),348(1.9)	44-4175-65
2-Cyclohexen-1-one, 2-butyl-3-methyl-	EtOH	242(4.13)	35-0935-64
2-Cyclohexen-1-one, 2-tert-butyl-5-methyl-	MeOH	234(4.00)	35-0078-64
2-Cyclohexen-1-one, 2,4-diethyl-5-methyl-	heptane	230(4.16),324(1.70)	33-0725-64
2-Cyclohexen-1-one, 2-ethyl-3,5,5-trimethyl-	n.s.g.	246(3.97)	39-1511-64
2-Cyclohexen-1-one, 5-ethyl-2,4,6-trimethyl-	heptane	231(3.84),331(1.30)	33-0725-64

Compound	Solvent	λ_{max}(log ϵ)	Ref.
Jasmone, dihydro-	EtOH	237(4.08)	35-0935-64
	EtOH	236(4.08)	44-1050-65
Ketone, bis(2,2-dimethylcyclopropyl)-	isooctane	291.5(2.19)	88-3151-65
1,6,8-Nonatrien-3-ol, 3,7-dimethyl-	EtOH	232.1(4.34)	22-2533-64
Pentane, 3-(4-hydroxycyclohex-5-en-1-ylidene)-	MeOH	248.5(4.31)	5-0028-64D
Pentane, 3-(4-oxocyclohex-5-en-1-yl)-	MeOH	228.5(3.95)	5-0028-64D
	ether	222(3.99)	5-0028-64D
2H-Pyran, 2,6-diethyl-2,4-dimethyl-	EtOH	283(3.61)	28-5895-64A
Spiro[4.4]nonan-1-one, 6,9-dimethyl-	isooctane	307(1.41)	88-2791-64
$C_{11}H_{18}O_2$			
Cyclopentadieneacetaldehyde, diethyl acetal	C_6H_{12}	248(3.59)	44-3430-64
2,4-Hexadien-3-ol, 2,5-dimethyl-, propionate	EtOH	220(3.65)	44-1038-65
3,4-Hexadien-2-ol, 2,5-dimethyl-, propionate	EtOH	230s(3.04)	44-1038-65
$C_{11}H_{18}O_3$			
4,5,6-Nonanetrione, 2,8-dimethyl-	n.s.g.	220(3.08),468(1.46)	35-5342-64
$C_{11}H_{18}O_4$			
Exogonic acid, methyl ester	C_6H_{12}	225(2.21),268(1.63)	24-2785-64
	MeOH	214(2.46),267(1.86)	24-2785-64
	MeOH-HCl	210(2.11),272(1.54)	24-2785-64
$C_{11}H_{18}O_4S$			
Valeric acid, 2-acetyl-5-mercapto-, ethyl ester, acetate	n.s.g.	230(3.74)	24-1963-64
$C_{11}H_{18}O_5S$			
Butyric acid, 2-hydroxy-4-mercapto-3-methyl-, ethyl ester, diacetate	EtOH	231(3.60)	39-0766-64
$C_{11}H_{19}INP$			
[p-(Dimethylamino)phenyl]trimethylphosphonium iodide	MeOH	282(4.45)	88-2729-64
$C_{11}H_{19}IN_2$			
[p-(Dimethylamino)phenyl]trimethylammonium iodide	H_2O	260(4.10)	46-1588-65
	8M LiCl	250(4.06)	46-1588-65
	MeOH	263(4.35)	88-2729-64
$C_{11}H_{19}N$			
Pyrrolidine, 1-(cyclohexylidenemethyl)-	EtOH	228.5(3.79)	28-1541-64A
$C_{11}H_{19}NO$			
1-Cyclohexene-1-carboxamide, N-tert-butyl-	EtOH	211(4.06)	35-0746-64
$C_{11}H_{19}NO_2$			
Nicotinic acid, 1,4,5,6-tetrahydro-1-methyl-, tert-butyl ester	EtOH	295(4.25)	35-5461-65
$C_{11}H_{19}NS$			
Thiophene, 2,3-dihydro-3,3-dimethyl-4-piperidino-	C_6H_{12}	257(3.86)	54-1160-64

Compound	Solvent	$\lambda_{max}(\log \epsilon)$	Ref.
$C_{11}H_{19}N_3$			
Pyrrole, 2,5-dimethyl-1-(1-methyl-4-piperazinyl)-	EtOH	210(2.6)	6-0073-64
$C_{11}H_{19}N_3O$			
6,8-Nonadien-2-one, 7-methyl-, semicarbazone	EtOH	233(4.41)	22-2533-64
3,5-Octadien-2-one, 6,7-dimethyl-, semicarbazone	EtOH	297.5(4.65)	22-3218-64
$C_{11}H_{19}N_3O_2$			
Pyridazine, 3-[2-(dimethylamino)-ethoxy]-6-isopropoxy-	EtOH	289(3.32)	44-1751-64
$C_{11}H_{19}N_3O_4$			
Valeric acid, 2-acetyl-4-hydroxy-5-isopropoxy-, γ-lactone, semicarbazone	n.s.g.	222(4.06),274(3.72)	39-0141-64
$C_{11}H_{19}N_3S$			
1-Cyclopentene-1-acetaldehyde, 2,3,3-trimethyl-, thiosemicarbazone	EtOH	233(3.84),273(4.33)	32-0546-65
$C_{11}H_{19}N_5O_5$			
Purine, 2-amino-6-benzyloxy-9-β-D-ribofuranosyl-	pH 1	243(4.01),288(4.03)	35-3752-65
	pH 11	246(4.06),282(4.03)	35-3752-65
$C_{11}H_{19}O_6P$			
Acetoacetic acid, 2-(2-hydroxy-1-methylethyl)-2-phosphono-, γ-lactone, diethyl ester	MeOH	210(2.81)	5-0010-65E
Hexanoic acid, 5-hydroxy-2-(hydroxymethylene)-, δ-lactone, diethyl phosphate	MeOH	228(3.82)	5-0010-65E
Malonaldehydic acid, (3-hydroxybutyl)-phosphono-, δ-lactone, diethyl ester	MeOH	210(2.40)	5-0010-65E
Valeric acid, 2-acetyl-4-hydroxy-2-phosphono-, γ-lactone, diethyl ester	MeOH	207(2.60)	5-0010-65E
Valeric acid, 2-(2-hydroxyethyl)-3-oxo-2-phosphono-, γ-lactone, diethyl ester	MeOH	210(2.86)	5-0010-65E
$C_{11}H_{20}FO_5P$			
Crotonic acid, 2-fluoro-3-methyl-4-phosphono-, triethyl ester	MeOH	224(4.09)	5-0020-64I
Crotonic acid, 3-(fluoromethyl)-4-phosphono-, triethyl ester	MeOH	218(4.16)	5-0021-65A
Crotonic acid, 4-fluoro-3-methyl-4-phosphono-, triethyl ester	MeOH	217(4.17)	5-0021-65A
$C_{11}H_{20}N_2O_3S$			
Butyramide, 2,4-dihydroxy-3,3-dimethyl-N-[2-(2-thiazolin-2-yl)-ethyl]-	N HCl	265(3.69)	94-0180-65
	EtOH	231(3.43),247(3.38)	94-0180-65
$C_{11}H_{20}O$			
Cyclohexanone, 4-tert-butyl-2-methyl-	C_6H_{12}	276(1.28)	88-2797-64
Cyclopentanone, 2-(1-methylpentyl)-	isooctane	299(1.43)	88-2791-64
2-Nonenal, 2,4-dimethyl-	EtOH	230(4.14)	22-0161-64

Compound	Solvent	λ_{max}(log ε)	Ref.
$C_{11}H_{20}O_2$			
4,5-Nonadiene-3,7-diol, 2,8-dimethyl-	H_2O	203(3.18),235s(2.79), 275s(2.51)	28-0209-65A
	EtOH	212(3.02),221s(2.98), 238s(2.89),271s(2.55)	28-0209-65A
5,6-Undecadiene-4,8-diol	H_2O	201(3.16),223s(2.69), 270(2.44)	28-0209-65A
	EtOH	212(2.96),266(2.63)	28-0209-65A
$C_{11}H_{20}O_3$			
2-Hexenoic acid, 3-ethoxy-5-methyl-, ethyl ester	EtOH	237.5(4.15)	70-0110-65
$C_{11}H_{20}S_3$			
2-Cyclohexene, 5,5-dimethyl-1,1,3-tris(methylthio)-	EtOH	236(4.00),252(4.03)	87-0705-64
$(C_{11}H_{20}S_3)_n$			
Polydecamethylenetrithiocarbonate	$CHCl_3$	310(4.04)	49-0631-65
$C_{11}H_{20}Si_2$			
Disilane, 1-(dimethylphenyl)-2,2,2-trimethyl-	C_6H_{12}	230.0(3.69)	39-4690-65
	C_6H_{12}	230.5(4.05)	101-0369-64B
	n.s.g.	230(3.69)	25-1492-64
$C_{11}H_{21}NO$			
2-Hexanone, 6-piperidino-	EtOH	275.5(1.39)	44-0650-65
$C_{11}H_{21}NO_4$			
Succinaldehydic acid, 3-(methylimino)-, ethyl ester, diethyl acetal	EtOH	281(2.84)	6-0583-65
$C_{11}H_{21}O_5P$			
Crotonic acid, 3-methyl-4-phosphono-, triethyl ester	MeOH	217(4.10)	5-0020-64I
$C_{11}H_{22}N_2$			
2-Imidazoline, 1-isopropyl-2-pentyl-	EtOH	231(3.74)	4-0188-64
$C_{11}H_{22}O$			
3-Nonanone, 5,7-dimethyl-	EtOH	280(1.50)	28-5228-64A
	EtOH	280(1.53)	28-5895-64A
2-Octanone, 6-ethyl-4-methyl-	EtOH	280(1.50)	28-5228-64A
$C_{11}H_{23}N_3S$			
Semicarbazide, 4-menth-3-yl-3-thio-, nickel complex	n.s.g.	440(1.92),480(1.90)	1-1239-65
$C_{11}H_{23}O_5PS$			
Valeric acid, 5-mercapto-2-phosphono-, triethyl ester	n.s.g.	210(3.16)	24-1963-64
$C_{11}H_{24}N_2O_5$			
D-Galactose, butylmethylhydrazone	MeOH	250(4.4)	24-1588-65

Compound	Solvent	λ_{max}(log ε)	Ref.
$C_{12}F_6O_2$			
Acenaphthene-1,2-quinone, hexafluoro-	EtOH	216(4.72),246(4.08), 304(3.67),327(3.53)	78-0927-65
$C_{12}HF_{10}N$			
Diphenylamine, decafluoro-	C_6H_{12}	251(4.08)	39-5017-64
$C_{12}H_3Cl_8N_3$			
Triazene, 1,3-bis(2,3,5,6-tetra- chlorophenyl)-	hexane	270s(4.08)	30-0031-65
$C_{12}H_4Co_2F_4K_6N_{10}O_2$			
Hexapotassium μ-tetrafluoroethylenebis- (pentacyano)dicobaltate, dihydrate	H_2O	276(4.18),320(3.32)	39-6629-65
$C_{12}H_4N_2O_2$			
2,3-Naphthalenedicarbonitrile, 1,4-dioxo-	MeOH	250(4.08),254(4.20), 282(4.07),335(3.23)	44-3591-64
	MeCN	267(4.15),277(4.09), 369(3.20)	44-3591-64
lithium salt	H_2O	226(4.49),267(4.13), 347(3.51),372(3.66), 390(3.72),447(3.42), 470(3.38),530(1.95), 680(3.04)	44-3591-64
$C_{12}H_4N_6O_{12}$			
Diphenylamine, 2,2',4,4',6,6'-hexa- nitro-, ammonium salt	H_2O	420(4.5)	28-2841-64B
$C_{12}H_5Br_6N_3$			
Triazene, 1,3-bis(2,4,6-tribromophenyl)-	hexane	270(4.12)	30-0031-65
$C_{12}H_5Cl_6N_3$			
Triazene, 1,3-bis(2,3,4-trichloro- phenyl)-	hexane	250(4.27),313(4.07), 365(4.36)	30-0031-65
	EtOH	250(4.26),360(4.37)	30-0031-65
Triazene, 1,3-bis(2,4,5-trichloro- phenyl)-	hexane	250(4.21),300(4.07), 365(4.37)	30-0031-65
	EtOH	245(4.15),305(4.05), 365(4.33)	30-0031-65
Triazene, 1,3-bis(2,4,6-trichloro- phenyl)-	hexane	278(4.01)	30-0031-65
	EtOH	255(4.23)	30-0031-65
$C_{12}H_5N_7O_{12}$			
Diphenylamine, 2,2',4,4',6,6'-hexa- nitro-	H_2O	390(4.20)	23-1681-64
anion	H_2O	425(4.36)	23-1681-64
$C_{12}H_6Cl_2O_2S_2$			
Thianthrene, 2,7-dichloro-, 5,10-di- oxide	EtOH	230(4.8)	35-2957-64
isomer	EtOH	220(4.7)	35-2957-64
$C_{12}H_6Cl_2S_2$			
Thianthrene, 2,7-dichloro-	96% H_2SO_4 (six days)	276(4.42),297(4.68), 585(4.08)	44-0021-64
	(ten days)	276(4.44),297(4.70), 585(4.13)	44-0021-64

Compound	Solvent	λ max(log ε)	Ref.
$C_{12}H_6Cl_4N_2$			
Azobenzene, 3,3',5,5'-tetrachloro-	EtOH	216(4.51),317(4.29), 430(2.83)	78-0189-64
$C_{12}H_6F_3N_3O$			
Benzamide, N-[2,2-dicyano-1-tri- fluoromethyl)vinyl]-	CH_2Cl_2	270(3.83),338(4.22)	44-0707-64
$C_{12}H_6F_5N$			
Diphenylamine, 2,3,4,5,6-pentafluoro-	C_6H_{12}	266(4.07)	39-5017-64
$C_{12}H_6N_2O_2$			
2,3-Naphthalenedicarbonitrile, 1,4-dihydroxy-	MeOH	238(4.90),260(4.15), 278(3.70),335(3.74), 351(3.92),367(3.98), 395(3.63)	44-3591-64
	MeCN	240(4.94),261(4.19), 280(3.83),330(4.01), 346(4.16),363(4.19)	44-3591-64
$C_{12}H_6N_2S_2$			
Malononitrile, 4-phenyl-1,3-dithiol- 2-ylidene-	CF_3COOH	383(4.37)	44-1711-64
	THF	243(3.99),380(4.18)	44-1711-64
$C_{12}H_6N_4$			
1,3,6,8-Tetrazapyrene	EtOH	233(4.58),246s(4.24), 284s(3.86),295(4.09), 307(4.16),330(3.50), 339(3.52),346(3.68), 356(3.60),364(3.78), 452(1.70),481(1.81)	33-1484-64
4,5,9,10-Tetrazapyrene	EtOH	227s(4.5),235(4.68), 254s(4.13),263(4.21), 340(4.22),349s(4.21), 373(3.79),380s(3.58), 391(3.87)	39-6090-64
$C_{12}H_6N_4O_2$			
4,5,9,10-Tetrazapyrene, dioxide	EtOH	222(4.52),267s(4.16), 281(4.17),314(3.92), 327(3.95),358(3.93), 398(3.98),420(3.97)	39-6090-64
$C_{12}H_6N_6O_{10}$			
Diphenylamine, 2,2',4,4',6-pentanitro-	EtOH	378(4.28)	23-1681-64
anion	H_2O	480(4.41)	23-1681-64
$C_{12}H_6N_8O_{12}$			
3,3'-Biphenyldiamine, 2,2',4,4',6,6'-hexanitro-	dioxan	326(4.31),394(4.02)	77-0232-64
$C_{12}H_6OS$			
2-Thiophenepropynal, 5-(1,3-penta- diynyl)-	ether	342(4.51),364(4.56)	24-0155-65
$C_{12}H_6O_3$			
1,2-Naphthalenedicarboxylic anhydride	EtOH	260(4.8),325(3.6), 310(3.6),345(3.4), 355(3.4),360(3.4)	44-1543-64

Compound	Solvent	λ_{max}(log ϵ)	Ref.
$C_{12}H_6O_5$			
1H-2-Benzopyran-4,5-dicarboxylic anhydride, 3-methyl-1-oxo-	dioxan	246(4.29),296(3.98), 335(3.84)	44-2534-64
$C_{12}H_6O_7$			
2H,5H,10H-Dipyrano[4,3-b:2',3'-d]pyran-2,5,10-trione, 4-hydroxy-8-methyl-	EtOH	253(3.83),280(3.94), 373(4.04)	35-3004-65
$C_{12}H_7BrN_2$			
Benzo[c]cinnoline, 4-bromo-	EtOH	207(4.61),250(4.66), 320(4.26),362(3.53)	39-1265-64
$C_{12}H_7BrN_2O$			
Benzo[c]cinnoline, 4-bromo-, 6-oxide	EtOH	244(4.59),250(4.59), 258(4.59),289(4.26), 298s(--),350(4.18)	39-1265-64
$C_{12}H_7BrO_2S_2$			
Thianthrene, 2-bromo-, 5,10-dioxide	EtOH	225(4.7)	35-2957-64
isomer	EtOH	218(4.7)	35-2957-64
$C_{12}H_7BrS_2$			
Thianthrene, 2-bromo-	96% H_2SO_4, 5 days	275(4.51),295(4.72), 584(4.07)	44-0021-64
	10 days	275(4.51),295(4.72), 584(4.07)	44-0021-64
$C_{12}H_7Br_4N_3$			
Triazene, 1,3-bis(2,6-dibromophenyl)-	hexane	268s(3.94)	30-0031-65
$C_{12}H_7ClN_2O$			
Phenazine, 2-chloro-, 5-oxide	benzene	371(3.75),389(4.02), 407(3.88),429(3.86)	95-1080-64
Phenazine, 2-chloro-, 10-oxide	benzene	371(3.78),389(4.05), 405(4.04),429(4.11)	95-1080-64
$C_{12}H_7ClO$			
2,4-Hexadiyn-1-one, 1-(p-chlorophenyl)-	EtOH	283(4.32),300(4.27)	39-2983-65
$C_{12}H_7ClS_2$			
Thianthrene, 1-chloro-	96% H_2SO_4, 5 days	274(4.49),291(4.45), 299s(4.40),539(3.93)	44-0021-64
Thianthrene, 2-chloro-	96% H_2SO_4, 5 days	274(4.50),294(4.69), 572(4.04)	44-0021-64
$C_{12}H_7Cl_3$			
Benzocyclooctene, 6,7,8-trichloro-	EtOH	217(4.65),235s(4.51)	39-2549-65
$C_{12}H_7Cl_4N_3$			
Triazene, 1,3-bis(2,5-dichlorophenyl)-	hexane	245(4.20),295(3.94), 363(4.24)	30-0031-65
	EtOH	245(4.26),300(4.00), 363(4.30)	30-0031-65
$C_{12}H_7FN_4$			
1-Propene-1,1,3-tricarbonitrile, 2-amino-3-fluoro-3-phenyl-	EtOH	285(4.07)	44-0707-64
$C_{12}H_7N_3$			
Malononitrile, 2(1H)-quinolylidene-	$CHCl_3$	290(4.40),391(4.16), 411(4.22),434(4.00)	44-0243-65

Compound	Solvent	λ_{max}(log ϵ)	Ref.
$C_{12}H_7N_3O_3$			
1H-Naphtho[2,3-d]triazole-4,9-dione, 1-acetyl-	EtOH	245(4.57),265s(4.16), 327(3.50)	39-1003-65
$C_{12}H_7N_3O_4$			
Carbazole, 2,4-dinitro-	C_6H_{12}	252(4.25),295(4.42), 312s(4.20),386(3.81)	23-2674-64
$C_{12}H_7N_5O_8$			
Diphenylamine, 2,2',4,4'-tetranitro-	H_2O	413(4.33)	23-1681-64
anion	H_2O	510(4.46)	23-1681-64
Diphenylamine, 2,3',4,6-tetranitro-	benzene	337(4.32)	23-1681-64
anion	pyridine	441(4.41)	23-1681-64
Diphenylamine, 2,4,4',6-tetranitro-	benzene	377(4.24)	23-1681-64
anion	H_2O	465(4.38)	23-1681-64
$C_{12}H_8$			
Biphenylene	pentane	239(4.78),248(5.06), 325(3.44),330(3.47), 338(3.77),343(3.73), 348s(3.53),358(3.95)	35-5720-65
Cyclododecatetraenediyne, isomer A	isooctane	244(4.72),249(4.74), 285s(3.21),465(2.26), 479s(2.26),500s(2.21), 517s(2.16),536s(2.06), 580s(1.74),605s(1.40)	35-5720-65
isomer B	isooctane	238(4.56),247(4.73), 276s(3.24),282s(3.18), 293(3.13),311(2.93), 332(2.61),395s(1.91), 408s(2.01),417(2.08), 431s(2.14),442(2.19), 457(2.22),472(2.19), 487(2.22),505(2.14), 522s(1.97),544(1.93), 563s(1.62)	35-5720-65
$C_{12}H_8BrNO_3$			
2(1H)-Pyridone, 1-[(m-bromo-benzoyl)oxy]-	EtOH	226(4.19),295(3.75)	35-5186-65
$C_{12}H_8BrN_3$			
2H-Benzotriazole, 2-(p-bromophenyl)-	n.s.g.	220(4.18),240(4.07), 245(4.08),253(3.83), 309(4.45),313(4.44)	39-4831-65
$C_{12}H_8BrN_3O_4$			
Diphenylamine, 2'-bromo-2,4-dinitro-	C_6H_{12}	260s(4.02),332(4.22)	23-2674-64
$C_{12}H_8Br_2$			
Benzocycloctene, 5,10-dibromo-	EtOH	214(3.85)	39-1622-64
$C_{12}H_8Br_4$			
Benzocyclooctene, 5,7,8,10-tetra-bromo-7,8-dihydro-	EtOH	238(4.19)	39-1622-64
$C_{12}H_8ClHgN_3O_3$			
Phenol, 2-(chloromercuri)-4-(p-nitrophenylazo)-	15% EtOH-HCl	480(4.45)	50-1065-64C
	15% EtOH	385(4.37)	50-1065-64C

Compound	Solvent	λ_{max}(log ε)	Ref.
$C_{12}H_8ClNO_3$			
2(1H)-Pyridone, 1-[(m-chloro-benzoyl)oxy]-	EtOH	229(3.92),295(3.45)	35-5186-65
2(1H)-Pyridone, 1-[(p-chloro-benzoyl)oxy]-	EtOH	244(4.28),276s(3.61), 288s(3.70),299(3.75)	35-5186-65
$C_{12}H_8ClN_3$			
2H-Benzotriazole, 2-(p-chlorophenyl)-	n.s.g.	219(4.03),240(3.94), 243(3.91),253(3.67), 308(4.49),315(4.43)	39-4831-65
$C_{12}H_8ClN_3O_4$			
Diphenylamine, 2'-chloro-2,4-dinitro-	C_6H_{12}	250s(4.05),331(4.18)	23-2674-64
Diphenylamine, 4'-chloro-2,4-dinitro-	C_6H_{12}	260s(4.04),333(4.22)	23-2674-64
$C_{12}H_8Cl_2N_2O_2$			
3-Furonitrile, 5,5-dichloro-4,5-di-hydro-2-(N-methylanilino)-4-oxo-	EtOH	223s(3.97),288(4.31)	44-4303-65
$C_{12}H_8CrO_4$			
Chromium, tricarbonyl(1-indanone)	n.s.g.	321.2(3.94)	28-2833-65A
$C_{12}H_8FN_3O_4$			
Diphenylamine, 2'-fluoro-2,4-dinitro-	C_6H_{12}	260s(4.03),330(4.22)	23-2674-64
$C_{12}H_8F_2N_2$			
Azobenzene, 2,2'-difluoro-	EtOH	227(4.22),325(4.09), 465(2.84)	78-0189-64
$C_{12}H_8I_2N_2$			
Azobenzene, 2,2'-diiodo-	EtOH	221(4.37),250(4.06), 326(4.10),475(2.53)	39-6972-65
$C_{12}H_8I_2N_2O$			
Azoxybenzene, 2,2'-diiodo-	EtOH	228(4.39),304(3.83)	39-6972-65
$C_{12}H_8N_2$			
3,4-Benzocinnoline	EtOH	307(3.97)	39-2319-64
Heptafulvene, 8-(2,2-dicyanovinyl)-	hexane	239(3.69),244(3.69), 250(3.71),256s(3.76), 262(3.79),270(3.83), 280s(3.72),430(4.49), 443s(4.47)	24-2050-64
	benzene	273(3.95),283s(3.86), 464(4.55)	24-2050-64
	EtOH	247(3.72),252(3.73), 260(3.78),270(3.94), 280s(3.90),465(4.61)	24-2050-64
	50% EtOH	255s(4.20),272(4.32), 278s(4.32),477(4.97), 494s(4.96)	24-2050-64
1,4,5-Methenoindene-2,3-dicarbonitrile, 3a,4,5,7a-tetrahydro-	EtOH	251(3.94),285(3.58)	35-1434-64
Phenazine	C_6H_{12}	210(4.73),250(5.08), 288(4.11),364(4.22)	65-1969-64
	EtOH	248(5.08)	39-2319-64
	50% MeOH-acid	251(5.0),257(4.9), 366s(4.2),375s(4.2), 383(4.2),430s(3.2)	59-1665-64
	MeOH	249(5.2),352s(4.1), 365(4.2),390s(3.8)	59-1665-64

Compound	Solvent	λ_{max}(log ϵ)	Ref.
$C_{12}H_8N_2O$			
3,4-Benzocinnoline, 5-oxide	EtOH	330(3.97),363(3.78), 382(3.64)	39-2319-64
Phenazine, 5-oxide	EtOH	265(4.96)	39-2319-64
1-Phenazinol	50% MeOH-HCl	274(4.6),375s(4.0), 385(4.2),505(3.2)	59-1665-64
	MeOH	265(4.7),343s(3.6), 353s(3.7),362s(3.8), 369(3.9),425(3.36)	59-1665-64
	50% MeOH-base	295(4.65),364(3.5), 375(3.5),530(3.35)	59-1665-64
2-Phenazinol	50% MeOH-acid	263(4.77),390(4.20), 453(3.68)	59-1665-64
	MeOH	257(4.82),362(3.86), 406(3.70)	59-1665-64
	50% MeOH-base	277(4.62),369(3.74), 478(3.73)	59-1665-64
$C_{12}H_8N_2O_2$			
1H-Benz[f]indazole-4,9-dione, 1-methyl-	EtOH	246(4.3),273(4.1), 318(3.8)	23-2665-64
1H-Benz[f]indazole-4,9-dione, O-methyl derivative	EtOH	242(4.3),262s(4.1), 315(3.5)	23-2665-64
Benzo[c]cinnoline, 5,6-dioxide	EtOH	292(4.12),302(4.10), 345(4.04)	39-2319-64
Carbazole, 3-nitro-	MeOH	279(4.39),305(4.62), 364(4.02)	23-1681-64
anion	MeOH	488(4.28)	23-1681-64
Phenazine, 5,10-dioxide	C_6H_{12}	210(4.67),284(4.47)	65-1969-64
	EtOH	285(4.90)	39-2319-64
$C_{12}H_8N_2O_4S$			
Sulfide, bis(p-nitrophenyl)	dioxan	230(4.15),250(4.11), 338(4.24)	65-2073-65
	$C_2H_4Cl_2$	252(3.90),348(4.07)	65-2073-65
$C_{12}H_8N_2O_5$			
2(1H)-Pyridone, 1-[(m-nitrobenzoyl)oxy]-	EtOH	220(4.50),255(3.90), 297(3.81)	35-5186-65
2(1H)-Pyridone, 1-[(p-nitrobenzoyl)oxy]-	EtOH	231(4.05),258(4.13), 296(3.85)	35-5186-65
$C_{12}H_8N_2S$			
6,7-Ethenonaphtho[1,8-cd][1,2,6]thiadiazine	dioxan	233(4.66),345(3.89), 357(3.89),690(2.72)	24-3196-65
1,2,5-Thiadiazole, 3-(2-naphthyl)-	isooctane	212(4.39),236(4.58), 257(4.11),269(4.21), 278(4.29),293(4.28), 304(4.32),324(3.71), 340(3.32)	89-0262-65
$C_{12}H_8N_4OS_2$			
Thiocyanic acid, diester with 4-mercapto-1-(p-mercaptophenyl)-3-methyl-2-pyrazolin-5-one	EtOH	267(4.28)	94-0023-64
$C_{12}H_8N_4O_3$			
2H-Naphtho[2,3-d]triazole-4,9-dione, 2-acetamido-	$CHCl_3$	252(4.25),328(3.19)	39-1003-65

Compound	Solvent	λ_{max}(log ϵ)	Ref.
$C_{12}H_8N_4O_4$			
Azobenzene, p,p'-dinitro-	EtOH	330(4.80)	31-0128-64
$C_{12}H_8N_4O_5$			
Azoxybenzene, p,p'-dinitro-	EtOH	268(4.30),343(4.08)	31-0128-64
$C_{12}H_8N_4O_6$			
Diphenylamine, 2,3',4-trinitro-	C_6H_{12}	243(4.31),328(4.27)	23-2674-64
	sulfolane	360(4.25)	23-1681-64
anion	sulfolane	450(4.27)	23-1681-64
Diphenylamine, 2,4,4'-trinitro-	CCl_4	348(4.38)	23-2674-64
	pyridine	385(4.34)	23-1681-64
anion	pyridine	520(4.44)	23-1681-64
Diphenylamine, 2,4,6-trinitro-	H_2O	372(4.12)	23-1681-64
anion	H_2O	450(4.28)	23-1681-64
$C_{12}H_8O$			
2,4-Dodecadiene-6,8,10-triynal	ether	269(4.55),282(4.77), 304(4.16),317(4.34), 339(4.52),362(4.46)	24-0809-64
2,4-Hexadiyn-1-one, 1-phenyl-	EtOH	266(4.20),279(4.26), 295(4.16)	39-2983-65
$C_{12}H_8OS_2$			
1-Thianthrenol	96% H_2SO_4, 30 min.	528(3.87)	44-0021-64
	5 days	515(3.89)	44-0021-64
	10 days	515(3.89)	44-0021-64
2-Thianthrenol	96% H_2SO_4, 30 min.	597(4.03)	44-0021-64
	2 hours	291(4.84)	44-0021-64
	1 day	291(4.80),318(4.42)	44-0021-64
	5 days	291(4.74),318(4.42), 592(4.00)	44-0021-64
$C_{12}H_8O_2S_2$			
Thianthrene, 5,10-dioxide	EtOH	220(4.7)	35-2957-64
isomer	EtOH	210(4.7)	35-2957-64
2,7-Thianthrenediol	96% H_2SO_4, 5 min.	604(4.06)	44-0021-64
	1 hour	296(4.56),333(4.41)	44-0021-64
	1 day	296(4.51),333(4.58)	44-0021-64
	5 days	540(4.04)	44-0021-64
	10 days	296(4.41),333(4.73), 540(4.04)	44-0021-64
$C_{12}H_8O_3$			
5-Benzofuranacrylic acid, 6-hydroxy- β-methyl-, δ-lactone	EtOH	244(4.39),289(3.99), 329(3.85)	44-2467-64
$C_{12}H_8O_4$			
Furano[2',3':7,8]-5-methoxycoumarin	MeOH	223(4.2),250(4.2), 308(3.9)	2-0354-65
Furo[3,2-g]coumarin, 4',5'-dihydro- 4'-oxo-5-methoxy-	MeOH	215(4.17),275(4.41), 325(4.07)	32-1054-64
3-Furoic acid, 2,5-dihydro- 2-oxo-5-benzylidene-	EtOH	237(3.93),344(4.44)	54-0031-64
Malonic acid, (3-phenyl-2-propyn- ylidene)-	EtOH	225(3.99),320(4.14)	54-0031-64
Psoralen, 5-methoxy-	EtOH	237(4.43),287(4.03), 321(3.97)	88-2559-65

Compound	Solvent	$\lambda_{max}(\log \epsilon)$	Ref.
2-Pyrone, 3-carboxy-6-phenyl-	EtOH	244(3.91),361(4.29)	54-0031-64
2-Pyrone, 6-carboxy-3-phenyl-	EtOH	237(3.85),331(4.15)	54-0031-64
$C_{12}H_8O_4S_5$			
Thiophene, 2,5-bis(2-thienylsulfonyl)-	EtOH	255(4.04),295(4.11)	39-7018-65
Thiophene, 2,5-bis(3-thienylsulfonyl)-	EtOH	206(3.98),231(3.95), 277(4.14)	39-7018-65
Thiophene, 3,4-bis(2-thienylsulfonyl)-	EtOH	209(4.03),242(4.15), 268(4.10)	39-7018-65
Thiophene, 3,4-bis(3-thienylsulfonyl)-	EtOH	209(4.40),245(4.28)	39-7018-65
$C_{12}H_8O_5$			
2-Naphthaleneacetic acid, 1,4-dihydro-3-hydroxy-1,4-dioxo-	MeCN	243(4.28),248(4.33), 275(4.12),332(3.53)	44-3819-65
Psoralen, 5-hydroxy-5-methoxy-	EtOH	225(4.30),242(3.90), 249(3.89),275(4.24), 317(3.99)	83-0672-65
$C_{12}H_8O_7$			
Spinochrome M	MeOH	251(4.16),270s(4.14), 317(4.10),514(3.67)	35-2959-64
	MeOH-NaOH	287s(4.21),328(4.28), 443(3.49),469(3.49), 569(3.80)	35-2959-64
$C_{12}H_8O_8$			
1,4-Naphthoquinone, 6-acetyl-2,3,5,7,8-pentahydroxy-	MeOH	239(4.13),293(4.15), 348(3.98),460(3.77)	88-3557-64
	MeOH-KOH	254(4.07),307(4.17), 365(3.98),466(3.90)	88-3557-64
$C_{12}H_8S$			
Benzo[b]thialene	dioxan	522(2.2)	73-3016-65
Naphtho[1,2-b]thiophene	C_6H_{12}	304(3.5)	73-0195-65
Naphtho[2,1-b]thiophene	C_6H_{12}	304(4.1)	73-0195-65
Naphtho[2,3-b]thiophene	C_6H_{12}	352(3.8)	73-0195-65
Thiaphenalene	EtOH	420(3.2)	73-3016-65
$C_{12}H_8S_2$			
1,3-Benzodithiole, 2-(2,4-cyclopentadien-1-ylidene)-	$CHCl_3$	268s(3.72),319(3.51), 392(4.47),405(4.47)	24-2825-65
2,2'-Bithiophene, 5-(3-buten-1-ynyl)-	EtOH	249(3.95),341(4.36)	39-7109-65
Thianthrene	96% H_2SO_4	270(4.50),290(4.58), 546(3.95)	44-0021-64
$C_{12}H_8S_3$			
2,2':5',2''-Terthiophene	EtOH	252(3.96),350(4.35)	39-7109-65
$C_{12}H_8S_5$			
Thiophene, 2,5-bis(2-thienylthio)-	hexane	205(4.55),240(4.27), 274(4.07)	39-7018-65
Thiophene, 2,5-bis(3-thienylthio)-	hexane	206(4.57)	39-7018-65
Thiophene, 3,4-bis(2-thienylthio)-	hexane	206(4.55),237(4.30), 276(3.91)	39-7018-65
Thiophene, 3,4-bis(3-thienylthio)-	hexane	208(4.65),274(4.00)	39-7018-65
$C_{12}H_9BrN_2$			
Azobenzene, p-bromo-, trans	EtOH	329(4.43),446(2.80)	39-1045-64

Compound	Solvent	λ max(log ε)	Ref.
$C_{12}H_9BrN_2O$			
Azoxybenzene, p-bromo-, cis	EtOH	242(4.17),332(3.71)	35-2419-64
12-Oxo-12H-dipyrido[1,2-a:1',2'-d]pyrazin-5-ium bromide	H_2O	213(4.42),261(3.94), 308(3.07),318(3.03), 458(4.44)	39-3366-64
$C_{12}H_9BrN_4O_3$			
7-Bromo-1,3,4,6-tetrahydro-4'-methyl-2,4,6-trioxospiro[pyrimidine-5(2H)-2'(1'H)-quinoxalinium]hydroxide, inner salt	EtOH	210(4.46),247(4.32), 278(3.90),286(3.85)	12-0877-64
$C_{12}H_9ClHgN_2O$			
Phenol, 2-(chloromercuri)-4-(phenylazo)-	15% EtOH-6N HCl	470(4.61)	50-1065-64C
	15% EtOH	350(4.41)	50-1065-64C
	15% EtOH-NaOH	410(4.36)	50-1065-64C
$C_{12}H_9ClN_2$			
Azobenzene, p-chloro-, trans	EtOH	328(4.37),443(2.78)	39-1045-64
Pyrrolo[1,2-a]quinoxaline, 1-chloro-4-methyl-	H_2O	214(4.36),227(4.44), 258s(4.12),329(3.96)	39-3678-65
	EtOH	232s(4.41),257s(4.16), 277(4.45),333(3.96), 347s(3.80)	35-1830-64
	EtOH	215(4.40),227(4.48), 258s(4.17),329(3.98), 347s(3.83)	39-3678-65
$C_{12}H_9ClN_2O$			
Azoxybenzene, p-chloro-, cis	EtOH	241(4.31),330(3.82)	35-2419-64
Pyrrolo[1,2-a]quinoxaline, 4-chloro-1-methoxy-	EtOH	230(4.53),234s(4.52), 269(4.11),275s(4.06), 314(3.63),326(3.76), 360(3.92)	35-1830-64
$C_{12}H_9ClO$			
Cyclopropenone, 1-(1-chloro-1-propenyl)-2-phenyl-	ether	282s(4.51),291(4.55), 305(4.39)	88-4003-65
Phenol, 2-chloro-4-phenyl-	HCl	256(4.26)	44-2678-65
	NaOH	285(4.29)	44-2678-65
$C_{12}H_9ClOS_2$			
3-Butyn-2-ol, 1-chloro-4-[5-(2-thienyl)-2-thienyl]-	EtOH	245(--),331(--), 336(--)	39-7109-65
$C_{12}H_9ClO_3$			
3-Furoic acid, 2-(p-chlorophenyl)-4-methyl-	MeOH	298(4.23)	39-5984-65
3-Furoic acid, 4-(o-chlorophenyl)-2-methyl-	MeOH	245(3.79)	39-5984-65
3-Furoic acid, 4-(p-chlorophenyl)-2-methyl-	MeOH	214(4.25),249(3.98)	39-5984-65
3-Furoic acid, 5-(o-chlorophenyl)-2-methyl-	MeOH	281(4.36)	39-5984-65
$C_{12}H_9ClO_4$			
1,4-Naphthoquinone, 2-chloro-5-hydroxy-3-methoxy-6-methyl-	EtOH	210(4.3),215s(4.2), 250s(4.0),297(4.0), 430(3.5)	33-0538-65

Compound	Solvent	λ max(log ε)	Ref.
1,4-Naphthoquinone, 5-chloro-8-hydroxy-2-methoxy-7-methyl-	EtOH	210(4.4),250(4.1),295(4.1),435(3.6)	33-0538-65
$C_{12}H_9FN_2$			
Azobenzene, p-fluoro-, trans	EtOH	323(4.31),441(2.71)	39-1045-64
$C_{12}H_9F_2N_3$			
Aniline, 2,6-difluoro-4-(phenylazo)-, cis	EtOH-HCl	231(3.77),262(3.80),324(4.07),517(4.55)	44-3878-65
	EtOH	245(4.08),366(4.20)	44-3878-65
Aniline, 2,6-difluoro-4-(phenylazo)-, trans	EtOH-HCl	231(3.74),262(3.78),324(4.06),517(4.55)	44-3878-65
	EtOH	242(4.06),377(4.46)	44-3878-65
$C_{12}H_9N$			
Carbazole	C_6H_{12}	232(4.63),291(4.31)	23-2674-64
$C_{12}H_9NO$			
Naphth[1,2-d]oxazole, 2-methyl-	heptane	225(4.71),280(3.75),290(3.76),310(3.46),325(3.26)	65-0427-64
Naphth[2,1-d]oxazole, 2-methyl-	heptane	245(4.92),270(3.80),320(3.32)	65-0427-64
$C_{12}H_9NOS$			
Phenothiazine, 5-oxide	benzene	300(3.81),340(3.65)	44-2130-65
	C_6H_{12}	273(3.98),305(3.72),340(3.48)	44-2130-65
	EtOH	229(4.52),270(4.21),303(3.94),337(3.78)	44-2130-65
	$CHCl_3$	272(4.15),300(3.85),335(3.69)	44-2130-65
	CCl_4	275(4.06),310(3.88),340(3.67)	44-2130-65
	HOAc	270(4.30),304(4.01),340(3.81)	44-2130-65
$C_{12}H_9NO_2S$			
Thieno[3,2-c]quinoline-6,9-dione, 2,3-dihydro-4-methyl-	EtOH	232(4.32),253(4.22),300(3.99),351(2.96),425(3.40)	95-0271-65
$C_{12}H_9NO_2Se$			
1,3-Propanedione, 1-(2-pyridyl)-3-selenophene-2-yl-	n.s.g.	285(3.91),365(4.37)	67-0379-65
1,3-Propanedione, 1-(3-pyridyl)-3-selenophene-2-yl-	n.s.g.	282(3.85),360(4.42)	67-0379-65
1,3-Propanedione, 1-(4-pyridyl)-3-selenophene-2-yl-	n.s.g.	287(3.97),360(4.39)	67-0379-65
$C_{12}H_9NO_3$			
4-Benzoxazoleacrylic acid, 5-hydroxy-β,2-dimethyl-, δ-lactone	EtOH	244s(3.57),292(4.26),336s(3.41)	44-4344-65
Furo[3,2-c]quinoline-6,9-dione, 2,3-dihydro-4-methyl-	EtOH	247(4.35),347(4.18)	95-0271-65
5-Hexen-2-ynophenone, 4'-nitro-	MeOH	270(4.69)	5-0026-64I
Indole-3-crotonic acid, γ-oxo-	NaOH	273(4.11),350(4.11)	65-3025-64
Phenol, 2-nitro-5-phenyl-	HCl	328(4.15)	35-4942-64
	NaOH	422(3.82)	35-4942-64
2(1H)-Pyridone, 1-(benzoyloxy)-	EtOH	231(4.28),287s(3.74),297(3.76)	35-5186-65

Compound	Solvent	λ_{max}(log ε)	Ref.
1H-Pyrrolo[1,2-a]indole-9-carboxaldehyde, 2,3,7,8-tetrahydro-7,8-dioxo-	MeOH	228(4.87),279(4.36), 345(4.04),515(3.62)	44-4381-65
$C_{12}H_9NO_4$			
Furo[2,3-b]quinolin-3(2H)-one, 4-hydroxy-6-methoxy-	EtOH	242(4.57),268(4.38), 300s(3.63),340(3.54)	2-0491-64
4H-Quinolizine-1-carboxylic acid, 3-acetyl-4-oxo-	EtOH	260(4.10),345(3.65), 420(4.11)	78-0945-65
4H-Quinolizine-3-carboxylic acid, 1-acetyl-4-oxo-	EtOH	265(3.94),350(4.02), 395(4.04)	78-0945-65
$C_{12}H_9NO_5S$			
Benzenesulfonic acid, p-nitrophenyl ester	EtOH	216(4.18),265(4.00), 340s(2.20)	65-2080-65
	ether	214(4.17),262(4.07), 335s(2.35)	65-2080-65
	dioxan	266(4.06),338(2.47)	65-2080-65
	H_2SO_4	220(4.12),285(3.90), 410(3.22)	65-2080-65
$C_{12}H_9NS_2$			
1-Thianthrenamine	96% H_2SO_4, 30 min.	539(3.72)	44-0021-64
	70 min.	271(4.39),290(4.42)	44-0021-64
	10 days	271(4.39),290(4.42), 539(3.72)	44-0021-64
2-Thianthrenamine	96% H_2SO_4, 10 min.	531(3.79)	44-0021-64
	25 min.	274(4.48),290(4.59)	44-0021-64
	1 day	290(4.68)	44-0021-64
	5 days	524(3.99)	44-0021-64
	10 days	290(4.70),524(3.99)	44-0021-64
$C_{12}H_9N_3$			
Benzimidazole, 2-(3-pyridyl)-	EtOH	310(4.34)	94-0773-64
2H-Benzotriazole, 2-phenyl-	MeOH	306(4.32)	39-0751-64
	n.s.g.	219(4.16),237(4.01), 241(3.94),251(3.65), 305(4.45),310(4.39)	39-4831-65
Glutacononitrile, 4-(anilinomethylene)-	n.s.g.	351(4.64)	44-1800-64
Imidazo[1,2-a]pyridazine, 2-phenyl-	C_6H_{12}	250(4.6),285(3.8), 300(3.7),320(3.9), 340(4.0),350(3.8)	94-1351-64
	EtOH-HCl	240(4.3),300(4.3)	94-1351-64
	EtOH	245(4.5),280s(3.7), 300s(3.8),320(3.9)	94-1351-64
Imidazo[1,2-b]pyridazine, 2-phenyl-	C_6H_{12}	240(4.5),280(3.8), 290(3.9),300s(3.8), 350s(4.2),365(4.2), 385(4.1)	94-1351-64
	EtOH-HCl	245(4.2),325(4.2)	94-1351-64
	EtOH	238(4.5),280(3.7), 295(3.7),350(4.1)	94-1351-64
1H-Imidazo[4,5-b]pyridine, 2-phenyl-	MeOH	234(3.96),307(4.47), 319(4.43)	44-3403-64
Normacrorine	EtOH	214(4.63),236(4.20), 264(4.44),325(4.02), 336(4.04)	39-3001-65
Pyrido[2,1-c]-3-phenyl-s-triazole	EtOH	209(4.42),241(4.13), 282(4.04)	2-0162-65

Compound	Solvent	$\lambda_{max}(\log \epsilon)$	Ref.
$C_{12}H_9N_3O$			
Benzo[c]cinnoline, 4-amino-, 6-oxide	EtOH	230(4.49),255(4.52), 329(4.35),364(4.02), 384(3.96),463(3.84)	39-1265-64
2H-Benzotriazole, 2-(p-hydroxyphenyl)-	n.s.g.	220(4.11),248(4.00), 252(4.00),259(4.01), 316(4.44)	39-4831-65
Imidazo[1,2-b]pyridazin-6-ol, 2-phenyl-	EtOH	235s(4.4),260s(4.2), 290(3.7),345(4.3)	94-1351-64
1H-Pyrazolo[4,3-c]pyridin-3-ol, 1-phenyl-	N NaOH	231(4.12),250(4.17), 306(4.08)	12-0379-65
2(1H)-Quinolylideneacetamide, α-cyano-	$CHCl_3$	288(4.38),386s(4.14), 402(4.25),426(4.10)	44-0243-65
$C_{12}H_9N_3OS_2$			
Thiazolo[4,5-d]pyridazin-7(6H)-one, 2-(methylmercapto)-6-phenyl-	EtOH	268(4.23),305(4.07)	78-1323-65
$C_{12}H_9N_3O_2$			
Maleimide, N-(2-benzimidazolylmethyl)-	EtOH	300(2.7)	94-0127-64
1H-Pyrazolo[4,3-c]pyridine, 3-hydroxy-1-phenyl-, 5-oxide	N NaOH	277(4.35),335(4.02)	12-0379-65
$C_{12}H_9N_3O_4$			
Diphenylamine, 2,4-dinitro-	C_6H_{12}	259s(4.03),336(4.24)	23-2674-64
	ether	342.5(4.27)	23-2674-64
	pyridine	362(4.23)	23-1681-64
anion	H_2O	495(4.20)	23-1681-64
Diphenylamine, 4,4'-dinitro-	pyridine	416(4.55)	23-1681-64
anion	H_2O	580(4.57)	23-1681-64
$C_{12}H_9N_3S_3$			
Benzo[1,2-d:3,4-d':5,6-d'']tristhiazole, 2,5,8-trimethyl-	EtOH	219(3.52),259(4.56), 275s(4.02),312(3.50)	7-0080-64
$C_{12}H_9N_5O_4S$			
Thiocyanic acid, 1-(2,4-dinitrophenyl)-3,5-dimethylpyrazol-4-yl ester	EtOH	290(3.86)	94-0023-64
$C_{12}H_9N_5O_6$			
p-Phenylenediamine, N-picryl-	sulfolane	420(4.07)	23-1681-64
anion	sulfolane	470(4.29)	23-1681-64
$C_{12}H_9N_7$			
1,2,4-Triazolo[1,5-a]pyrimidine, 7-(1H-benzotriazol-1-yl)-5-methyl-	MeOH	245s(3.8),270(4.0), 305(4.0)	65-0204-64
$C_{12}H_9OPS_3$			
Phosphine oxide, tri-2-thienyl-	EtOH	238(4.52),250s(4.45)	44-0097-65
$C_{12}H_9O_3P$			
Phenoxaphosphinic acid	EtOH	215(4.53),241(4.11), 275s(3.36),287s(3.61), 294(3.71)	44-2382-64
$C_{12}H_9O_4P$			
Phosphine oxide, tri-2-furyl-	EtOH	238(4.53)	44-0097-65
$C_{12}H_{10}$			
1,2-Benzoheptafulvene	ether	250(4.34),300(3.63)	35-3719-65

Compound	Solvent	λ_{max}(log ϵ)	Ref.
3,4-Benzoheptafulvene	ether	237s(3.96),245s(4.06), 252(4.22),261(4.29), 271(4.09),281(3.97), 296s(3.17),328(3.15), 345(2.98)	35-3719-65
Biphenyl	EtOH	250(4.25)	32-0902-64
bromanil complex	$CHCl_3$	420(2.99)	60-0465-64
chloranil complex	$CHCl_3$	410(2.78)	60-0465-64
iodoanil complex	$CHCl_3$	395(2.92)	60-0465-64
1,3-Hexadien-5-yne, 1-phenyl-, cis-trans	n.s.g.	306(4.48),318s(4.40)	5-0062-65B
all trans	n.s.g.	298(4.48),308(4.56), 321(4.46)	5-0062-65B
$C_{12}H_{10}BrNO_3$			
4H-Quinolizine-1-carboxylic acid, 3-bromo-4-oxo-, ethyl ester	EtOH	263(4.13),283(3.91), 385(4.24)	78-0945-65
$C_{12}H_{10}BrN_3O_2$			
3(2H)-Pyridazinone, 5-acetamido-4-bromo-2-phenyl-	EtOH	226(4.30),284(4.06)	78-1323-65
$C_{12}H_{10}BrN_3S$			
Thiocyanic acid, 1-(p-bromophenyl)-3,5-dimethylpyrazol-4-yl ester	EtOH	252(4.20)	94-0023-64
$C_{12}H_{10}BrN_5O_3$			
1,2,4-Triazine, 3-amino-6-methyl-5-[3-bromo-4-(5-nitro-2-furyl)-1,3-butadienyl]-	propylene glycol	292(4.09),410(4.33)	95-0109-64
$C_{12}H_{10}Br_2$			
Naphthalene, 7-bromo-2-(bromomethyl)-1-methyl-	MeOH	243(4.78),284(3.76)	44-2109-64
$C_{12}H_{10}Br_2N_2$			
Dipyrido[1,2-a:1',2'-d]pyrazinediium dibromide	H_2O	199(4.03),262(3.94), 448(2.97)	39-3885-65
Dipyrido[1,2-a:2',1'-c]pyrazinediium dibromide	H_2O	215(4.13),224(4.11), 239(4.19),265(4.44), 272(4.58),310s(4.05), 322(4.15)	25-0847-65
	H_2O	224(4.33),239(4.37), 264(4.61),272(4.73), 323(4.29)	88-1465-65
$C_{12}H_{10}Br_2O_4$			
3-Benzofurancarboxylic acid, 4,6-dibromo-5-hydroxy-2-methyl-, ethyl ester	EtOH	217(4.53),265(4.05), 301(3.75),309s(3.74)	44-1453-64
$C_{12}H_{10}Br_4$			
Benzene, o-bis(2,2-dibromocyclopropyl)-	heptane	200.5(3.51)	44-2951-64
Butadiyne, bis(2,2-dibromo-1-methylcyclopropyl)-	n.s.g.	203(4.68),233(3.96), 245(3.78),259(3.71), 225(3.52)[sic]	22-1518-65
$C_{12}H_{10}ClIO_4$			
3-Benzofurancarboxylic acid, 4-chloro-5-hydroxy-6-iodo-2-methyl-, ethyl ester	EtOH	215(4.59),265(4.12), 302(3.83),309s(3.80)	44-1453-64

Compound	Solvent	$\lambda_{max}(\log \epsilon)$	Ref.
$C_{12}H_{10}ClNO_3$			
Cinnamic acid, p-chloro-β-cyano-α-hydroxy-, ethyl ester	pH 1	305(3.84)	4-0162-65
	pH 7	319(4.01)	4-0162-65
	pH 13	313(4.07)	4-0162-65
3,5-Hexadien-2-one, 1-chloro-6-(p-nitrophenyl)-	EtOH	224(4.47)	87-0035-65
	EtOH	224(3.86),345(4.48)	87-0035-65
Isoxazole-3-carboxylic acid, 5-chloro-4-phenyl-, ethyl ester	C_6H_{12}	206(4.07),232(3.96)	17-0203-65
	MeOH	208(3.97),230(3.94)	17-0203-65
Isoxazole-4-carboxylic acid, 5-chloro-3-phenyl-, ethyl ester	C_6H_{12}	226(4.19)	17-0203-65
	MeOH	224(4.19)	17-0203-65
$C_{12}H_{10}ClNO_4$			
Isoxazole-4-carboxylic acid, 3-(2-chloro-3-methoxyphenyl)-5-methyl-	MeOH	216(4.24),286(3.41)	39-5976-65
$C_{12}H_{10}ClN_3O_2S$			
Pyridine, 5-chloro-2-[(o-methylthio)-anilino]-3-nitro-	EtOH	251(4.24),308(4.16)	95-0429-65
$C_{12}H_{10}ClN_3S$			
Thiocyanic acid, 1-(p-chlorophenyl)-3,5-dimethylpyrazol-4-yl ester	EtOH	252(4.17)	94-0023-64
Thionine	n.s.g.	598(4.78)	46-0641-65
$C_{12}H_{10}ClN_5$			
s-Triazolo[1,5-a]pyrimidine, 7-(o-chloroanilino)-5-methyl-	MeOH	294(4.30)	65-0204-64
$C_{12}H_{10}ClN_5O$			
Ethenetricarbonitrile, [2-chloro-3-(2-oxo-1-pyrrolidinyl)-2-azetidinyl]-	MeCN	265(4.19)	44-1194-64
$C_{12}H_{10}ClN_5O_3$			
1,2,4-Triazine, 3-amino-6-methyl-5-[3-chloro-4-(5-nitro-2-furyl)-1,3-butadienyl]-	propylene glycol	293(4.12),410(4.37)	95-0109-64
$C_{12}H_{10}Cl_2Ge$			
Germane, dichlorodiphenyl-	heptane	248(2.68),253(2.93), 258(2.92),263(2.98), 270(2.86)	28-3931-65A
$C_{12}H_{10}Cl_2N_4O_3$			
Spiro[pyrimidine-5(2H),2'(1'H)-quinoxaline]-2,4,6(1H,3H)-trione, 6',7'-dichloro-3',4'-dihydro-4'-methyl-	EtOH	230(4.33),260s(3.74), 312(3.62)	12-0877-64
$C_{12}H_{10}Cl_2O$			
Cyclobutenone, 3,3-dichloro-1-ethyl-2-phenyl-	ether	286(4.44)	88-4003-65
Cyclobutenone, 4,4-dichloro-1-ethyl-2-phenyl-	ether	272(4.22)	88-4003-65
$C_{12}H_{10}Cl_2O_4$			
3-Benzofurancarboxylic acid, 4,6-dichloro-5-hydroxy-2-methyl-, ethyl ester	EtOH	212(4.54),262(4.00), 300(3.71)	44-1453-64

Compound	Solvent	λ max(log ε)	Ref.
$C_{12}H_{10}Cl_2Sn$			
Tin, dichlorodiphenyl-	heptane	247(2.57),252(2.91), 259(2.81),263(2.84), 270(2.72)	28-3931-65A
$C_{12}H_{10}Cl_3NO_4$			
2-Indolinecarboxylic acid, trichloro-2-methoxy-1-methyl-3-oxo-, methyl ester	EtOH	256(4.49),276s(3.89)	44-0178-64
$C_{12}H_{10}Cl_3NS_2$			
2H-1,3-Thiazine-2-thione, 3,6-dihydro-3-methyl-4-phenyl-6-(trichloromethyl)-	EtOH	240(4.12),321(4.04)	39-4008-64
$C_{12}H_{10}Cl_4$			
Butadiyne, bis(2,2-dichloro-1-methylcyclopropyl)-	n.s.g.	209(4.42),230(3.28), 242(3.34),255(3.42), 270(3.24),293(3.42)	22-1518-65
$C_{12}H_{10}Fe$			
Ferrocene, ethynyl-	MeOH	264(3.72),440(2.26)	49-1750-64
$C_{12}H_{10}I_2O_4$			
3-Benzofurancarboxylic acid, 5-hydroxy-4,6-diiodo-2-methyl-, ethyl ester	EtOH	218(4.32),230(4.38), 273(4.20),304(3.83), 312s(3.84)	44-1453-64
$C_{12}H_{10}NO$			
Nitroxide, diphenyl-	n.s.g.	280s(4.4),300(4.5), 310(4.6)	70-0800-65
$C_{12}H_{10}N_2$			
Azobenzene	hexane	227(4.1),322(4.2)	61-0973-64
trans	EtOH	320(4.33),444(2.71)	39-1045-64
	MeCN	320(4.3)	49-1173-65
antimony trichloride complex	MeCN	410(4.46)	49-1173-65
Benzocyclobutene, 1-cyano-1-(2-cyanoethyl)-	EtOH	253s(2.76),258(3.01), 264(3.18),270(3.17)	87-0732-65
2,4,6,8,10-Dodecapentaenedinitrile	EtOH	265(3.62),342(4.74), 359(4.96),379(4.97)	70-0684-65
2-Naphthonitrile, 1-amino-3-methyl-	EtOH	230s(4.4),265(4.4), 320(3.6),360(3.8)	44-1543-64
	50% H_2SO_4	240(5.0),280(3.8), 290(3.8),335(3.5), 350(3.5)	44-1543-64
2-Naphthonitrile, 1-amino-4-methyl-	EtOH	245s(3.8),265(4.4), 330(3.8),360(3.8)	44-1543-64
	50% H_2SO_4	240(4.7),275s(3.6), 290(3.8),300(3.8)	44-1543-64
1H,5H-Pyrazolo[1,2-a]pyrazole, 1,5-didehydro-2-phenyl-	EtOH	254(4.1),480(3.5)	35-5256-65
Pyrrolo[1,2-a]quinoxaline, 4-methyl-	pH 1.5	224(4.48),239(4.44), 246s(4.39),260s(4.12), 280s(3.83),345(4.13), 360s(4.09)	39-3678-65
	pH 7.8	224(4.44),242(4.38), 245(4.38),259s(4.08), 333(3.94)	39-3678-65

Compound	Solvent	$\lambda_{max}(\log \epsilon)$	Ref.
Pyrrolo[1,2-a]quinoxaline, 4-methyl- (contd.)	EtOH	224(4.42),242s(4.35), 246(4.37),259s(4.08), 315s(3.79),329(3.92), 340(3.93),352s(3.74)	39-3678-65
$C_{12}H_{10}N_2O$			
Azoxybenzene	90% H_2SO_4	230(3.7),390(4.2)	23-0862-65
Azoxybenzene, cis	EtOH	239(4.05),327(3.59)	35-2419-64
Indole-2-acetonitrile, 5-acetyl-	EtOH	248(4.57),285(3.99), 294s(3.95)	87-0141-64
Norharmine	MeOH	237(3.73),240(4.63), 302(4.24),325(3.78), 337(3.73)	73-0447-64
Phenol, (p-phenylazo)-	90% H_2SO_4	463(4.6)	23-0862-65
2-Propen-1-one, 1-imidazol-1-yl-3-phenyl-	pH 6.87	307(4.40)	35-4406-64
Pyridine, 4-amino-3-benzoyl-	MeOH	228(4.38),244(4.12), 312(3.74)	87-0722-65
$C_{12}H_{10}N_2OS_2$			
p-Dithiino[2,3-d]pyridazin-5(6H)-one, 2,3-dihydro-6-phenyl-	EtOH	256(4.32),312(3.91)	78-1323-65
$C_{12}H_{10}N_2O_2$			
Diphenylamine, 2-nitro-	sulfolane	435(3.80)	23-1681-64
anion	sulfolane	545(3.96)	23-1681-64
Diphenylamine, 4-nitro-	C_6H_{12}	260s(3.87),343(4.27)	23-2674-64
	pyridine	400(4.29)	23-1681-64
anion	pyridine	508(4.54)	23-1681-64
3-Pyridazinecarboxylic acid,	0.2N KOH	246(3.87)	44-2128-64
4-methyl-5-phenyl-	EtOH	255(3.85)	44-2128-64
1-(2-Pyridylmethyl)pyridinium-	pH 1	262(4.0)	39-3366-64
2-carboxylate	pH 13	262(3.8),268(3.83), 278s(3.75)	39-3366-64
Quinaldamide, N-formyl-N-methyl-	EtOH	207(4.53),237(4.69), 290(3.67)	39-5969-64
$C_{12}H_{10}N_2O_2S$			
Acetic acid, [(5-phenyl-4-pyrimidinyl)-thio]-	EtOH	204(4.35),227s(4.11), 274(4.01)	73-3730-65
Aniline, p-[(p-nitrophenyl)thio]-	EtOH	245(4.10),266(4.20), 344(4.11)	65-2073-65
	ether	245(4.11),264(4.23), 340(4.13)	65-2073-65
	dioxan	247(4.20),264(4.30), 340(4.18)	65-2073-65
	$C_2H_4Cl_2$	250(--),260(4.27), 350(4.20)	65-2073-65
$C_{12}H_{10}N_2O_3$			
Pyridine, 4-[1-methyl-2-(5-nitro-2-furyl)vinyl]-	MeOH	259(3.94),285(3.88), 368(4.29)	94-0389-65
$C_{12}H_{10}N_2O_3S_2$			
2-Thiazoline-4-carboxylic acid, 2-(6-methoxy-2-benzothiazolyl)-	EtOH	268(3.87),326(4.27)	44-2344-65
$C_{12}H_{10}N_2O_4S$			
Benzenesulfonic acid, p-(p-hydroxy-	pH 5.0	none above 350 nm	10-0111-64A
phenylazo)-, sodium salt	pH 9.0	425(4.24)	10-0111-64A

Compound	Solvent	λ_{max}(log ε)	Ref.
$C_{12}H_{10}N_2O_5$			
4H-Quinolizine-1-carboxylic acid, 3-nitro-4-oxo-, ethyl ester	EtOH	265(4.08),340(4.03), 430(4.3)	78-1051-64
4H-Quinolizine-3-carboxylic acid, 1-nitro-4-oxo-, ethyl ester	EtOH	385(4.309)	78-0945-65
$C_{12}H_{10}N_2O_6$			
3,3'-Azo-2-pyrone, 4,4'-dihydroxy-6,6'-dimethyl-	DMF	277(3.95),383(3.42), 476(4.39),498(4.40)	5-0214-65G
2(1H)-Phthalazineacetic acid, 4-(carboxymethoxy)-1-oxo-	2% EtOH	262(3.73),300(3.92)	20-0091-65
$C_{12}H_{10}N_2S$			
Thiazolo[4,5-f]quinoline, 2,7-dimethyl-	EtOH	241(4.46),300(3.61), 314(3.46),328(3.33)	4-0242-65
Thiazolo[4,5-f]quinoline, 2,9-dimethyl-	EtOH	245(4.74),300(3.93), 326(3.61)	4-0242-65
	EtOH	245(4.74),300(3.93), 326(3.61)	7-0530-64
Thiazolo[5,4-f]quinoline, 2,7-dimethyl-	EtOH	253(4.54),319(3.21), 334(3.09)	4-0242-65
Thiazolo[5,4-f]quinoline, 2,9-dimethyl-	EtOH	252(4.57),286(3.53), 329(2.78)	4-0242-65
$C_{12}H_{10}N_2S_2Ti$			
Titanium, dicyclopentadienyldithiocyanato-	acetone	438(4.09)	101-0271-65B
$C_{12}H_{10}N_4$			
Formamidine, N-(3-cyano-4-phenylpyrrol-2-yl)-	pH 1	275(4.00)	35-1995-65
	EtOH	229(4.28),305(4.11)	35-1995-65
1H-Imidazo[4,5-b]pyridine, 2-(o-aminophenyl)-	MeOH	250(4.11),294(4.09), 305(4.16),355(4.10)	44-3403-64
Purine, 6-methyl-8-phenyl-	pH 1	236(4.23),301(4.43)	44-1916-65
	aq. NaCl	231(4.14),293(4.43), 298(4.43),312s(4.11)	44-1916-65
	pH 13	233(4.28),304(4.44), 318s(4.21)	44-1916-65
Pyrazole, 1,1'-p-phenylenedi-	EtOH	283(4.37)	23-1605-64
7H-Pyrrolo[2,3-d]pyrimidine, 4-amino-5-phenyl-	pH 1	240(4.16),286(4.04)	35-1995-65
	pH 13	239(4.06),283(4.05)	35-1995-65
	EtOH	230(4.02),255(4.00), 282(4.13)	35-1995-65
$C_{12}H_{10}N_4O$			
Hypoxanthine, 2-methyl-8-phenyl-	pH 1	233(4.09),288(4.33)	44-1916-65
	aq. NaCl	236(4.23),295(4.33)	44-1916-65
	pH 13	237(4.30),304(4.31)	44-1916-65
Isoxazolo[5,4-d]pyrimidine, 4-(methylamino)-3-phenyl-	EtOH	240(4.04),290(3.91)	44-2117-64
Purine-8-methanol, α-phenyl-	pH 1	268(4.04)	87-0797-65
	pH 11	223(3.93),275(4.11)	87-0797-65
$C_{12}H_{10}N_4OS$			
Purine-8-methanol, 6-mercapto-α-phenyl-	pH 1	228(4.03),326(4.24)	87-0797-65
	pH 11	234(4.26),312(4.37)	87-0797-65
$C_{12}H_{10}N_4OS_2$			
Imidazo[4,5-d]pyridazin-7(4)-ol, 1-benzyl-2,4(7)-dimercapto-	0.5N NaOH	221(4.34),268(4.38)	4-0247-65

Compound	Solvent	λ_{max}(log ϵ)	Ref.
$C_{12}H_{10}N_4O_2$			
1,2,4-Oxadiazole, 3-methyl-5-[p-(3-methyl-1,2,4-oxadiazol-5-yl)phenyl]-	ether	287(4.45)	33-0942-64
1,2,4-Oxadiazole, 5-methyl-3-[p-(5-methyl-1,2,4-oxadiazol-3-yl)phenyl]-	ether	264(4.48)	33-0942-64
Purine-8-methanol, 6-hydroxy-α-phenyl-	pH 1	253(4.21)	87-0797-65
	pH 11	262(4.17)	87-0797-65
Pyrimidine, 2-amino-4-(p-nitrostyryl)-	propylene glycol	232(4.20),320(4.16)	95-0121-64
$C_{12}H_{10}N_4O_2S$			
1H-Imidazo[4,5-d]pyridazine-4,7-diol, 1-benzyl-2-mercapto-	EtOH	206(4.35),234s(4.18), 270(4.36)	4-0247-65
Thiocyanic acid, 3,5-dimethyl-1-(m-nitrophenyl)pyrazol-4-yl ester	EtOH	251(4.33)	94-0023-64
Thiocyanic acid, 3,5-dimethyl-1-(o-nitrophenyl)pyrazol-4-yl ester	EtOH	233(4.10)	94-0023-64
Thiocyanic acid, 3,5-dimethyl-1-(p-nitrophenyl)pyrazol-4-yl ester	EtOH	299(4.16)	94-0023-64
$C_{12}H_{10}N_4O_2S_2$			
Thiazolo[4,5-f]benzothiazole, 2,6-diacetamido-	EtOH	212(4.28),267(4.58), 274(4.60),294s(4.26), 316(4.17),322s(4.00), 331(4.32)	7-0080-64
$C_{12}H_{10}N_4O_3$			
Pyrimidine, 2-amino-4-[4-(5-nitro-2-furyl)butadienyl]-	propylene glycol	292(4.04),340(4.11), 408(4.33)	95-0121-64
Xanthine, 7-benzyl-1-hydroxy-	pH 12	287(4.29)	4-0275-64
	EtOH	268(4.22)	4-0275-64
$C_{12}H_{10}N_4O_4$			
Diphenylamine, 2'-amino-2,4-dinitro-	C_6H_{12}	257(4.04),329(4.18)	23-2674-64
Diphenylamine, 4'-amino-2,4-dinitro-	sulfolane	375(4.26)	23-1681-64
anion	sulfolane	495(4.26)	23-1681-64
1H-Tetrazole-1-acetic acid, 5-methyl-α-piperonylidene-	EtOH	235(4.48),288(4.15), 323(4.30)	44-2222-65
$C_{12}H_{10}N_4O_5S$			
Xanthine, 7-methyl-1-phenylsulfonoxy-	H O	271(3.92)	4-0275-64
	pH 12	292(3.96)	4-0275-64
	EtOH	269(3.94)	4-0275-64
	90% EtOH-NaOH	288(4.02)	4-0275-64
$C_{12}H_{10}N_4O_6$			
Maleimide, N-[4-(dimethylamino)-3,5-dinitrophenyl]-	$C_2H_4Cl_2$	425(3.38)	50-1065-64C
$C_{12}H_{10}N_4S$			
Purine, 6-phenylthiomethyl-	N HCl	249(3.87)	87-0667-65
	pH 7.65	250(3.93),268(3.97)	87-0667-65
	N NaOH	251(3.52),277(3.81)	87-0667-65
$C_{12}H_{10}N_4S_2$			
1H-Imidazo[4,5-d]pyridazine-7(4)-thiol, 4(7)-(methylthio)-2-phenyl-	EtOH	206(4.41),273s(4.35), 284(4.39),325s(4.04), 344(4.11)	4-0182-64

Compound	Solvent	$\lambda_{max}(\log \epsilon)$	Ref.
$C_{12}H_{10}N_4S_3$			
1H-Imidazo[4,5-d]pyridazine-2,4,7-trithiol, 1-benzyl-	0.5N NaOH	230(4.18),282(4.49), 310(4.25)	4-0247-65
$C_{12}H_{10}N_4S_4$			
Disulfide, bis(5-amino-4-cyano-3-thenyl)	MeOH	220(4.28),284(3.79)	24-3571-65
$C_{12}H_{10}O$			
1'-Acetonaphthone	EtOH	213(4.67),298(3.80)	39-2072-65
2'-Acetonaphthone	EtOH	240(4.74),291(3.99)	39-2072-65
1H-Cyclopentacycloocten-1-one, 3-methyl-	heptane	262(4)	88-0365-65
2,4-Dodecadiene-6,8,10-triyn-1-ol	ether	248(4.52),258(4.80), 269(4.08),289(4.28), 306(4.49),326(4.65), 348(4.57)	24-0809-64
Furan, 2-styryl-	MeOH	225(3.90),232(3.91), 315(4.51)	4-0318-65
2,4-Hexadiyn-1-ol, 1-phenyl-	EtOH	242(2.93),256(2.79)	39-2983-65
3,5-Hexadiyn-1-ol, 6-phenyl-	ether	244(3.95),257(4.31), 271(4.51),288(4.41)	24-2118-64
4-Hexen-2-ynophenone	EtOH	270(4.23),282(3.23)	39-2983-65
5-Hexen-2-ynophenone	MeOH	263(4.13)	5-0026-64I
4-Penten-2-ynophenone, 4-methyl-	EtOH	267(3.86)	39-2983-65
Phenyl ether	nonane	272(3.26)	67-0375-65
$C_{12}H_{10}OS_2$			
2,2'-Bithienyl, 5-(4-hydroxy-but-2-ynyl)-	EtOH	242(3.82),328(4.34), 334(4.35)	39-7109-65
	n.s.g.	242(3.82),328(4.34), 334(4.35)	88-3159-64
$C_{12}H_{10}O_2$			
Benzene, m-bis(2-propynyloxy)-	MeOH	274(3.3),280(3.3)	70-1349-64
1,4-Ethanonaphthalene-5,8-dione, 1,4-dihydro-	C_6H_{12}	243s(4.06),252(4.20), 260s(4.14),302s(2.54), 356(2.84),435(1.39), 455(1.38),472s(1.20)	39-3043-64
	EtOH	252(4.16),261s(4.11), 302s(2.54),363(2.70), 450s(1.23)	39-3043-64
$C_{12}H_{10}O_2S$			
Cyclopenta[c]thiopyran, 5,7-diacetyl-	hexane	240(4.25),284(4.67), 294(4.43),308(4.38), 358(3.46),366(3.53), 470(3.64),508s(3.39), 517s(3.29),527s(3.09)	35-0708-64
2-Furaldehyde, 5-[(phenylthio)methyl]-	EtOH	285(4.25)	70-1281-65
$C_{12}H_{10}O_2S_2$			
2,2'-Bithienyl, 5-(3,4-dihydroxy-1-butynyl)-	ether	238(3.81),325(4.34), 332(4.34)	24-0155-65
$C_{12}H_{10}O_3$			
2'-Acetonaphthone, 1',3'-dihydroxy-	EtOH	222(4.41),263s(4.27), 275(4.31),286s(4.20), 300s(4.00),312(3.70), 403(3.43)	94-0316-64

Compound	Solvent	λ_{max}(log ϵ)	Ref.
2'-Acetonaphthone, 1',8'-dihydroxy-	C_6H_{12}	219(4.33),262(4.68), 396(3.95)	54-1005-64
5-Benzofuranacrylic acid, 2,3-dihydro-6-hydroxy-β-methyl-, δ-lactone	EtOH	225(4.21),254(3.57), 294(3.78),332(4.26)	44-2467-64
3(2H)-Dibenzofuranone, 1,4-dihydro-8-hydroxy-	EtOH	254(4.09),296(3.68)	39-2932-64
3-Furoic acid, 2-methyl-4-phenyl-	MeOH	246(3.83)	39-5984-65
3-Furoic acid, 4-methyl-2-phenyl-	MeOH	291(4.11)	39-5984-65
4-Indancarboxylic acid, 5-(1-hydroxyethyl)-1-oxo-, γ-lactone	EtOH	238(4.02),289(3.55), 296(3.58)	39-3811-65
2-Naphthaldehyde, 3-hydroxy-8-methoxy-	n.s.g.	233(4.48),257(4.48), 270(4.32),315(3.84), 323(3.88),398(3.40)	88-0041-65
2-Naphthoic acid, 3-hydroxy-4-methyl-	$CHCl_3$	239(4.65),276(3.94), 285(3.61),299(3.32)	94-0112-64
1,2-Naphthoquinone, 5-methoxy-7-methyl-	EtOH	216(4.32),262(4.26), 376(3.24),472(3.64)	39-2355-65
1,2-Naphthoquinone, 8-methoxy-6-methyl-	EtOH	218(4.44),244(4.27), 420(3.87)	39-2355-65
$C_{12}H_{10}O_4$			
Coumarin, 6-formyl-7-hydroxy-4,8-dimethyl-	EtOH	267(4.67),312(3.98), 348(3.95)	32-0513-65
2-Furaldehyde, 5,5'-ethylenedi-	EtOH	224(3.81),289(4.54)	95-0981-65
4-Indancarboxylic acid, 5-(1,1-dihydroxyethyl)-1-oxo-, γ-lactone	H_2O	267(4.13),303(3.52)	39-3811-65
2-Naphthoic acid, 1-hydroxy-8-methoxy-	EtOH	242(4.63),287(3.49), 298(3.60),311(3.62), 356(3.96),369(3.89)	33-0769-64
2-Naphthoic acid, 3-hydroxy-8-methoxy-	n.s.g.	227(4.53),236(4.61), 260(4.34),292(3.62), 301(3.74),310(3.65), 368(3.45)	88-0041-65
2-Naphthoic acid, 4-hydroxy-5-methoxy-	EtOH	235(4.62),302(3.67), 311(3.69),348(3.69), 362(3.70)	33-0769-64
1,4-Naphthoquinone, 5,7-dimethoxy-	EtOH	258(4.19),410(3.59)	39-2941-64
1,4-Naphthoquinone, 8-hydroxy-2-methoxy-1-methyl-	EtOH	212(4.5),250(4.2), 290(4.3),415(3.7)	33-0538-65
$C_{12}H_{10}O_5$			
2H-1-Benzopyran-6-acetic acid, 7-hydroxy-2-oxo-, methyl ester	MeOH	233(4.29),328(4.33)	32-1073-64
2H-1-Benzopyran-6-carboxylic acid, 7-hydroxy-4,8-dimethyl-2-oxo-	EtOH	244(4.37),268(3.86), 278(3.84),338(4.20)	32-0513-65
2H-1-Benzopyran-6-carboxylic acid, 7-methoxy-2-oxo-, methyl ester	MeOH	219(4.47),244(4.34), 329(4.22)	32-1073-64
	EtOH	219(4.47),244(4.34), 329(4.22)	32-0513-65
2-Furaldehyde, 5,5'-(oxydimethylene)di-	EtOH	226(3.88),281(4.48)	95-0981-65
$C_{12}H_{10}O_6$			
2H-1-Benzopyran-6-acetic acid, 7-hydroxy-5-methoxy-2-oxo-	MeOH	264(3.99),278(3.99), 330(4.09)	32-1054-64
1,4-Naphthoquinone, 2,5-dihydroxy-6,7-dimethoxy-	EtOH	261(4.25),315(4.07), 415(3.66)	94-1472-65
1,4-Naphthoquinone, 5,8-dihydroxy-2,6-dimethoxy-	EtOH	227(4.46),277(3.82), 308(3.83),475(3.73), 507(3.79),544(3.61)	94-0633-65

Compound	Solvent	λ_{max}(log ε)	Ref.
1,4-Naphthoquinone, 5,8-dihydroxy-2,7-dimethoxy-	MeOH	279(3.60),307(3.61), 477(3.46),505(3.53), 542(3.32)	35-2959-64
	MeOH-KOH	301(3.60),319s(3.51), 497s(3.78),532(3.86), 565(3.84)	35-2959-64
$C_{12}H_{10}S$			
Thiophene, 2-styryl-	MeOH	229(4.00),323(4.44)	4-0318-65
$C_{12}H_{10}S_3$			
[1,2]Dithiolo[1,5-b][1,2]dithiole-7-S^{IV}, 2-methyl-5-phenyl-	dioxan	240(4.7),275(4.7), 330(4.0),495(4.0)	5-0188-65B
$C_{12}H_{11}BF_4$			
Bicyclo[5.4.1]dodecapentaenylium tetrafluoroborate	60% H_2SO_4	272(4.50),302(4.71), 320(4.15),385(3.36), 423(3.60)	89-0348-65
$C_{12}H_{11}BrN_2$			
Pyrrole, 2-[N-(m-bromophenyl)formimidoyl]-1-methyl-	EtOH	241(3.94),330(4.35)	12-0894-64
Pyrrole, 2-[N-(p-bromophenyl)formimidoyl]-1-methyl-	EtOH	221s(3.98),331(4.38)	12-0894-64
$C_{12}H_{11}BrN_2O$			
1H-3-Benzazepine, 2-acetamido-4-bromo-	EtOH	276(3.92)	4-0026-65
$C_{12}H_{11}BrN_4O_3$			
Urea, [(7-bromo-4-ethyl-3,4-dihydro-3-oxo-2-quinoxalinyl)carbonyl]-	EtOH	245(4.46),312(3.84), 395(3.65)	12-0877-64
$C_{12}H_{11}BrO$			
2-Naphthalenemethanol, 7-bromo-1-methyl-	MeOH	236(4.91),285(3.72)	44-2109-64
$C_{12}H_{11}BrO_2$			
2-Naphthoic acid, 7-bromo-3,4-dihydro-1-methyl-	MeOH	228(4.51),275(4.23), 296s(3.77)	44-2109-64
$C_{12}H_{11}BrO_4$			
3-Benzofurancarboxylic acid, 6-bromo-5-hydroxy-2-methyl-, ethyl ester	EtOH	208(4.60),227(4.14), 258(3.96),301(3.86)	44-1453-64
$C_{12}H_{11}Br_2NO_2$			
10-Acetoxy-9-bromo-6H-pyrido[1,2-a]azepinium bromide	H_2O	244(4.02),317(4.07)	44-1523-65
Indole-3-butyric acid, 2,6-dibromo-	EtOH	229(4.62),286(3.94)	44-1206-64
$C_{12}H_{11}ClN_2$			
1H-Azepino[5,4,3-cd]indole, 9-chloro-3,4-dihydro-6-methyl-	EtOH	245(4.39),339(3.81)	87-0200-65
Pyridine Red	EtOH	220(4.5),250(3.9), 330(4.0),400(3.5), 500(4.2)	27-0325-65
Pyrrole, 2-[N-(m-chlorophenyl)formimidoyl]-1-methyl-	EtOH	236(3.88),326(4.32)	12-0894-64
Pyrrole, 2-[N-(p-chlorophenyl)formimidoyl]-1-methyl-	EtOH	225s(3.89),330(4.36)	12-0894-64

Compound	Solvent	$\lambda_{max}(\log \epsilon)$	Ref.
$C_{12}H_{11}ClN_4O_4$			
3,5-Hexadien-2-one, 3-chloro-, 2,4-dinitrophenylhydrazone	n.s.g.	242(4.17),270(4.14), 298(4.05)	88-0821-65
$C_{12}H_{11}ClO$			
2-Naphthalenemethanol, 7-chloro-1-methyl-	MeOH	233(4.93),278(3.71)	44-2109-64
$C_{12}H_{11}ClO_2$			
2-Naphthoic acid, 7-chloro-3,4-dihydro-1-methyl-	EtOH	228(4.31),278(4.05), 304s(3.77)	44-2109-64
$C_{12}H_{11}ClO_2S$			
Benzo[b]thiepin, 5-acetoxy-2-chloro-4,5-dihydro-	heptane	240(3.55),274(3.72)	44-1092-64
$C_{12}H_{11}ClO_4$			
3-Benzofurancarboxylic acid, 4-chloro-5-hydroxy-2-methyl-, ethyl ester	EtOH	211(4.48),238(3.98), 256(3.98),298(3.60)	44-1453-64
3-Benzofurancarboxylic acid, 6-chloro-5-hydroxy-2-methyl-, ethyl ester	EtOH	209(4.59),228(4.11), 257(3.95),301(3.82)	44-1453-64
$C_{12}H_{11}ClO_5$			
2-Benzofurancarboxylic acid, 5-chloro-3-(2-hydroxyethoxy)-, methyl ester	EtOH	213(4.15),230s(4.05), 287(4.18)	44-4126-65
$C_{12}H_{11}Cl_2NO_3$			
Coumarin, 3,6-dichloro-4-(N-methyl-2-hydroxyethylamino)-	EtOH	215s(4.08),224(4.39), 228s(3.46),258s(3.96), 272(4.00),330(4.00)	44-4122-65
3(2H)-Furanone, 2,2-dichloro-4-methoxy-5-(N-methylanilino)-	EtOH	230s(3.71),325(4.25)	44-4303-65
$C_{12}H_{11}Cl_6Sb$			
1-Methyl-2,3-benzotropenium hexachloroantimonate	H_2SO_4	246(4.39),278(4.68), 338(3.53),425(3.35)	35-3719-65
$C_{12}H_{11}FN_2$			
Pyrrole, 2-[N-(p-fluorophenyl)formimidoyl]-1-methyl-	EtOH	327(4.26)	12-0894-64
$C_{12}H_{11}FN_2O_7$			
Glutamic acid, N-(3-fluoro-4-nitrobenzoyl)-	pH 1	258(4.03)	87-0727-65
	pH 7	262(4.01)	87-0727-65
	pH 13	262(3.99)	87-0727-65
$C_{12}H_{11}IO_4$			
3-Benzofurancarboxylic acid, 5-hydroxy-4-iodo-2-methyl-, ethyl ester	EtOH	215(4.49),244s(3.93), 263(4.06),302(3.66), 309s(3.61)	44-1453-64
3-Benzofurancarboxylic acid, 5-hydroxy-6-iodo-2-methyl-, ethyl ester	EtOH	210(4.61),232(4.22), 260(4.03),305(3.93)	44-1453-64
$C_{12}H_{11}I_3N_2$			
6-Methyldipyrido[1,2-c:2',1'-e]imidazolium triiodide	H_2O	225s(4.62),230(4.70), 236s(4.68),282(3.84), 291(3.83),334s(4.10), 345(4.28),362(4.25), 380s(3.72),382(3.72), 403(3.62),415(3.52), 427(3.30)	12-1819-65

Compound	Solvent	λ_{max}(log ε)	Ref.
$C_{12}H_{11}KO_4$			
Malonic acid, phenyl-, cyclic isopropylidene ester, potassium deriv.	$(MeOCH_2)_2$	272.5(4.27)	35-1857-65
$C_{12}H_{11}LiO_4$			
Malonic acid, phenyl-, cyclic isopropylidene ester, lithium deriv.	$(MeOCH_2)_2$	271(4.29)	35-1857-65
	DMF	273.5(4.29)	35-1857-65
$C_{12}H_{11}N$			
1H-Cyclopenta[c]quinoline, 2,3-dihydro-	n.s.g.	237(3.91),279(3.40), 306(3.18),320(3.28)	39-2421-64
Diphenylamine	C_6H_{12}	240s(3.68),282(4.24)	23-2674-64
	ether	284(4.33)	23-2674-64
3-Picoline, 2-phenyl-	hexane	238(4.02),274(3.80)	32-1287-64
	MeOH	234(3.95),274(3.81)	32-1287-64
Pyridine, 2-benzyl-	EtOH	258(3.7),264(3.7), 270(3.6)	56-1625-65
Pyridine, 2-o-tolyl-	hexane	240(4.00),268(3.79)	32-1287-64
	MeOH	236(4.01),268(3.90)	32-1287-64
Pyrrole, 2-styryl-, cis	n.s.g.	229(4.05),320(4.09)	12-0875-65
trans	n.s.g.	236(3.92),242(3.90), 330(4.44)	12-0875-65
$C_{12}H_{11}NO$			
1-Azabicyclo[4.4.1]undeca-1,3,5,7,9-pentaene, 11-acetyl-	n.s.g.	252(4.76),299(3.74), 387(2.48),398(2.70), 408(2.76)	89-0785-64
Carbazol-1(2H)-one, 3,4-dihydro-	EtOH	309(4.37)	12-0246-64
Cyclopentanecarbonitrile,	EtOH	238(3.77)	23-2512-65
2-oxo-3-phenyl-	EtOH-base	264(4.11)	23-2512-65
Furo[3,2-c]quinoline, 2,3-dihydro-2-methyl-	EtOH	214s(4.44),230(4.69), 295(3.78),306(3.71), 320(3.58)	94-1424-64
Hydroxylamine, N,N-diphenyl-	n.s.g.	240(4.1),280(4.0)	70-0800-65
Isoxazole, 4,5-trimethylene-3-phenyl-	C_6H_{12}	240(4.07)	44-1582-64
1-Naphthamide, N-methyl-	HOAc	281(3.78)	39-4399-65
	HOAc-H_2SO_4	292(3.71)	39-4399-65
2-Naphthamide, N-methyl-	HOAc	280(3.82)	39-4399-65
	HOAc-H_2SO_4	286(3.92)	39-4399-65
Pyrrole, 2-o-toluoyl-	MeOH	251(3.75),299(4.25)	39-2579-64
Pyrrole, 2-p-toluoyl-	MeOH	254(3.97),307(4.24)	39-2579-64
1H-Pyrrolo[1,2-a]indole-9-carboxaldehyde, 2,3-dihydro-	MeOH	248(4.30),265(4.18), 307(4.28)	87-0700-65
Quinoline, 4-(allyloxy)-	EtOH	224(4.81),278s(3.85), 286(3.87),300s(3.59)	94-1424-64
4-Quinolinol, 3-allyl-	EtOH	212(4.46),238s(4.42), 242(4.43),291(3.49), 323(4.05),337(4.08)	94-1424-64
4(1H)-Quinolone, 1-allyl-	EtOH	212(4.40),238(4.33), 242s(4.30),289(3.47), 323(4.15),337(4.21)	94-1424-64
$C_{12}H_{11}NOS$			
Sulfoximine, S,S-diphenyl-	N HCl	240(4.22),262(3.50), 268(3.53),276s(3.34)	25-1261-64
	EtOH	231(4.10),260(3.42), 266(3.37),273(3.15)	25-1261-64

Compound	Solvent	λ_{max}(log ε)	Ref.
$C_{12}H_{11}NOS_3$			
2-Oxazoline-5-thione, 4-carbonyl-2-phenyl-, 4-(dimethyl mercaptole)	EtOH	221(4.35),311(4.37), 387(4.15)	12-0061-65
2-Thiazolin-5-one, 4-carbonyl-2-phenyl-, 4-(dimethyl mercaptole)	EtOH	284(4.21),410(4.33)	12-0061-65
$C_{12}H_{11}NO_2$			
Benz[cd]indole-3-carboxylic acid, 1,3,4,5-tetrahydro-	EtOH	223(4.52),274(3.76), 280(3.77),291(3.66)	44-0843-64
1-Naphthonitrile, 1,2,3,4-tetrahydro-5-methoxy-2-oxo-	EtOH	272(3.92),278(3.88)	88-0103-64
Pyrrole, 2-o-anisoyl-	MeOH	255s(3.63),301(4.22)	39-2579-64
Pyrrole, 2-p-anisoyl-	MeOH	260s(3.88),311(4.27)	39-2579-64
1H-Pyrrolo[1,2-a]indole-9-carboxaldehyde, 2,3-dihydro-7-hydroxy-	MeOH	214(4.21),255(4.24), 279(4.14),309(4.07)	44-4381-65
Spiro[bicyclo[4.2.0]octa-1,3,5-triene-7,3'-piperidine]-2',6'-dione	EtOH	238(3.05),245(3.05), 253(3.05),259(3.15), 265(3.26),272(3.20)	87-0732-65
$C_{12}H_{11}NO_2S$			
2(5H)-Furanone, 4-methyl-3-thiobenzamido-	EtOH	254(4.13),305s(3.78)	39-0783-64
4-Isothiazolecarboxylic acid, 3-methyl-5-benzyl-	EtOH	207(4.33),256(3.85)	39-0446-64
Pyridine, 2-[(phenylsulfonyl)methyl]-	cation	218s(4.04),265(4.03)	39-3090-65
	neutral	217(4.12),262(3.72), 268(2.62)	39-3090-65
Pyridine, 3-[(phenylsulfonyl)methyl]-	cation	219(4.13),262(3.83)	39-3090-65
	neutral	218(4.15),262(3.60)	39-3090-65
Pyridine, 4-[(phenylsulfonyl)methyl]-	cation	223(4.17),260(3.77)	39-3090-65
	neutral	219(4.13),264(3.53)	39-3090-65
Thieno[3,2-c]quinoline-6,9-diol, 2,3-dihydro-4-methyl-	EtOH-HBr	272(4.60),327(3.98), 423(3.53)	95-0271-65
$C_{12}H_{11}NO_2S_2$			
2-Oxazolin-5-one, 4-carbonyl-2-phenyl-, 4-(dimethyl mercaptole)	EtOH	226(3.84),238(3.89), 246(3.92),263(4.05), 284s(3.88),290(4.43)	12-0061-65
$C_{12}H_{11}NO_3$			
Furo[3,2-c]quinoline-6,9-diol, 2,3-dihydro-4-methyl-	EtOH-HBr	227(4.28),255(4.54), 338(3.65),390(4.08)	95-0271-65
Indole-3-acetic acid, 5-acetyl-	EtOH	253(4.53),297(3.91)	87-0141-64
Indole-3-butyric acid, γ-oxo-	NaOH	262(4.29),330(4.33)	65-3025-64
Indole-2-carboxylic acid, 5-acetyl-3-methyl-	EtOH	269(4.73),302(3.94)	87-0141-64
3-Indolinebutyric acid, 3-hydroxy-2-oxo-, δ-lactone	EtOH	208(4.49),253(3.75), 292(3.18)	44-2431-64
$C_{12}H_{11}NO_3S_4$			
3H-1,2-Dithiole-3-thione, 4-(N-acetyl-p-toluenesulfonamido)-	n.s.g.	229(4.34),268(3.65), 319(3.56),413(3.95)	12-1071-65
$C_{12}H_{11}NO_4$			
Carbostyril, 4-ethoxy-7,8-methylenedioxy-	EtOH	228(4.49),232s(4.39), 239s(4.35),251(4.35), 258(4.33),305(3.86), 315s(--)	78-1725-64
Carbostyril, 4-methoxy-1-methyl-7,8-methylenedioxy-	EtOH	227(4.55),237(4.49), 252(4.38),260(4.38), 301(3.87)	78-1725-64

Compound	Solvent	λ_{max}(log ε)	Ref.
Coumarin, 8-methoxy-3-(methylcarbamoyl)-	EtOH	254(4.18),310(4.42)	94-0443-65
Indole-3-succinic acid	MeOH	274s(3.83),281(3.86), 289(3.79)	39-0526-64
3-Isoxazoline-4-carboxylic acid, 5-oxo-2-phenyl-, ethyl ester	EtOH	211(4.33),311(4.55)	78-2735-65
$C_{12}H_{11}NO_4S$			
2(1H)-Pyridone, 1-((p-tolyl-sulfonyl)oxy]-	EtOH	229(4.24),270s(3.43), 278s(3.51),301(3.69), 324s(3.40)	35-5186-65
$C_{12}H_{11}NO_6S$			
2H-1,2-Benzothiazine-2-acetic acid, 3-acetyl-3,4-dihydro-4-oxo-, 1,1-dioxide	EtOH	252(3.78),320(3.99)	44-2241-65
$C_{12}H_{11}NS$			
Aniline, p-(phenylthio)-	EtOH	262(4.17),350(2.80)	65-2073-65
	ether	264(4.20)	65-2073-65
	dioxan	262(4.20)	65-2073-65
	$C_2H_4Cl_2$	260(4.38)	65-2073-65
$C_{12}H_{11}N_3$			
Benzaldehyde, 2-pyridylhydrazone	EtOH	204(4.19),239(4.07), 330(4.39)	2-0162-65
Isonicotinaldehyde, phenylhydrazone	EtOH	244(4.25),300(3.70), 366(4.41)	7-0180-64
Nicotinaldehyde, phenylhydrazone	EtOH	242(4.08),300(3.99), 360(4.34)	7-0180-64
Picolinaldehyde, phenylhydrazone	EtOH	240(4.11),300(3.85), 356(4.43)	7-0180-64
2,2':5',2''-Terpyrrole	MeOH	319(4.42),327(4.41), 345s(4.11)	44-0883-64
$C_{12}H_{11}N_3O$			
p-Benzoquinone, azine with 1-methyl-4(1H)-pyridone	95% C_5H_5N	550(4.83)	5-0009-65J
	90% C_5H_5N	553(4.84)	
	80% C_5H_5N	551(4.86)	
	70% C_5H_5N	550(4.87)	
	60% C_5H_5N	549(4.87)	
	50% C_5H_5N	546(4.88)	
	40% C_5H_5N	548(4.87)	
	30% C_5H_5N	545(4.86)	
	20% C_5H_5N	542(4.85)	
	10% C_5H_5N	540(4.84)	
	5% C_5H_5N	536(4.82)	
	2% C_5H_5N	535(4.80)	
	1% C_5H_5N	534(4.80)	
Isonicotinaldehyde, phenylhydrazone, 1-oxide	EtOH	224(4.12),252(4.10), 296(3.99),400(4.51)	7-0180-64
Picolinaldehyde, phenylhydrazone, 1-oxide	EtOH	248(4.25),288(3.94), 388(4.47)	7-0180-64
$C_{12}H_{11}N_3O_2$			
p-Phenylenediamine, N-(p-nitrophenyl)-	EtOH	227(3.98),258(4.05), 300(3.55),307(--), 405(4.28),530s(2.80)	65-2073-65
	ether	228(3.90),255(3.95), 295(3.49),305(--), 382(4.20),480s(2.60)	65-2073-65

Compound	Solvent	λ_{max}(log ϵ)	Ref.
p-Phenylenediamine, N-(p-nitrophenyl)- (contd.)	dioxan	230(3.96),256(4.04), 300(3.55),390(4.23)	65-2073-65
Urea, 1-methyl-3-quinaldoyl-	EtOH	208(4.43),243(4.61), 292(3.75)	39-5969-64
$C_{12}H_{11}N_3O_3S_2$			
Benzenesulfonic acid, p-(3,5-dimethyl-4-thiocyanatopyrazol-1-yl)-	EtOH	253(4.16)	94-0023-64
$C_{12}H_{11}N_3O_4$			
Pyrazole-1-acetic acid, 5-methyl-4-p-nitrophenyl-	EtOH	222(4.08),325(4.12)	44-1889-65
$C_{12}H_{11}N_3S$			
Thiocyanic acid, 3,5-dimethyl-1-phenylpyrazol-4-yl ester	EtOH	246(4.06)	94-0023-64
$C_{12}H_{11}N_5$			
Adenine, 3-benzyl-	pH 1	275(4.24)	4-0115-64
	pH 7	272(4.09)	4-0115-64
	pH 13	272(4.08)	4-0115-64
	EtOH-HCl	278(4.24)	4-0115-64
	EtOH	274(4.10)	23-1599-64
Adenine, 9-benzyl-	pH 1	259(4.17)	4-0115-64
	pH 7	261(4.17)	4-0115-64
	pH 13	261(4.18)	4-0115-64
1,2,4-Triazolo[1,5-a]pyrimidine, 7-anilino-5-methyl-	MeOH	300(4.26)	65-0204-64
$C_{12}H_{11}N_5O$			
Guanine, 8-p-tolyl-	pH 1	254(4.20),308(4.29)	44-1916-65
	pH 7	231(4.07),254(4.12), 307(4.26)	44-1916-65
	pH 13	241(4.24),317(4.24)	44-1916-65
Purine-8-methanol, 2-amino-α-phenyl-	pH 1	236s(4.07),315(3.76)	87-0797-65
	pH 11	303(3.97)	87-0797-65
Purine-8-methanol, 6-amino-α-phenyl-	pH 1	266(4.22)	87-0797-65
	pH 11	271(4.19)	87-0797-65
$C_{12}H_{11}N_5O_2$			
3-Benzyltetrahydro-1-methyl-5,7-dioxo-v-triazolo[4,5-d]pyrimidinium hydroxide, inner salt	pH -0.89	275(3.89)	24-1060-65
	pH 4.0	228s(3.96),245(3.80), 311(3.84)	24-1060-65
1,2,4-Triazine, 3-amino-6-methyl-5-(p-nitrostyryl)-	propylene glycol	230(4.29),312(4.12)	95-0109-64
3H-v-Triazolo[4,5-d]pyrimidine-5,7(4H,6H)-dione, 3-benzyl-6-methyl-	pH 2.0	231(3.72),256(4.02)	24-1060-65
	pH 8.0	251(3.77),280(3.92)	24-1060-65
$C_{12}H_{11}N_5O_3$			
1,2,4-Triazine, 3-amino-6-methyl-5-[4-(5-nitro-2-furyl)-1,3-butadienyl]-	propylene glycol	295(4.01),415(4.56)	95-0109-64
1,2,4-Triazine, 3-amino-6-[4-(5-nitro-2-furyl)-3-methyl-1,3-butadienyl]-	propylene glycol	316(4.26),420(4.38)	95-0016-64
$C_{12}H_{11}N_5S$			
7H-Pyrimido[4,5-b][1,4]thiazine, 4-hydrazino-6-phenyl-	EtOH	276(4.32),378(3.88)	44-2121-64

Compound	Solvent	λ_{max}(log ε)	Ref.
$C_{12}H_{11}NaO_4$			
Malonic acid, phenyl-, cyclic isopropylidene ester, sodium deriv.	$(MeOCH_2)_2$	272(4.30)	35-1857-65
	DMF	274(4.30)	35-1857-65
$C_{12}H_{11}O_4P$			
Phosphonic acid, (m-phenoxyphenyl)-	EtOH	208(4.41),273(3.29), 279(3.35)	87-0891-65
Phosphonic acid, (p-phenoxyphenyl)-	EtOH	236(4.20),278(3.08)	87-0891-65
$C_{12}H_{12}$			
7,8-Benzotricyclo[4.2.0^{1,3}.0^{4,6}]octene	EtOH	219(4.2),223s(--), 274(3.18),304s(--)	88-1525-64
Bicyclo[5.4.1]dodeca-2,5,7,9,11-pentaene	C_6H_{12}	215(4.24),250(4.12), 335(3.54)	89-0348-65
Naphthalene, 2,6-dimethyl-	C_6H_{12}	227(5.06),274(3.67), 284(3.52),303(2.70), 309(2.93),315(2.74), 324(3.09)	89-0348-65
Naphthalene, 1-ethyl-	EtOH	225(4.93),282(4.11)	39-2072-65
Naphthalene, 2-ethyl-	EtOH	225(5.14),276(3.69)	39-2072-65
1,4-Pentadiene, 3-benzylidene-	EtOH	285(4.81)	28-0237-64A
$C_{12}H_{12}B_3N_3$			
Tris[1,2]azaborino[1,2-a:1',2'-c:-1'',2''-e]-s-triazatriborine	EtOH	280(4.6),285(4.7), 310s(3.8),320(3.7)	35-1125-64
$C_{12}H_{12}BrNO_2$			
Indole-3-butyric acid, 2-bromo-	EtOH	224(4.55),276s(3.89), 283(3.91),291(3.84)	44-1206-64
$C_{12}H_{12}BrNO_3$			
3-Indolinebutyric acid, 3-bromo-2-oxo-	EtOH	217(4.22),227s(4.15), 310(3.00)	44-2431-64
3-Indolinebutyric acid, 5-bromo-2-oxo-	EtOH	207(4.45),255(4.11), 290s(3.12)	44-2431-64
$C_{12}H_{12}Br_2N_2$			
6,7-Dihydrodipyrido[1,2-a:2',1'-c]pyrazinediium dibromide	H_2O	310(4.29)	25-0782-65
$C_{12}H_{12}Br_2N_2O$			
6,7-Dihydro-6-hydroxydipyrido-[1,2-a:2',1'-c]pyrazinediium dibromide	10% H_2SO_4	312(4.30),317s(4.28)	88-1465-65
$C_{12}H_{12}ClNO_4$			
Coumarin, 6-chloro-4-(2-hydroxyethylamino)-3-methoxy-	EtOH	212(4.42),235(4.36), 263(4.12),305(3.99), 314(4.04),347(4.03)	44-4122-65
1,3-Dioxolane, 2-(chloromethyl)-2-(p-nitrostyryl)-	EtOH	222(3.97),302(4.12)	87-0035-65
$C_{12}H_{12}ClNO_{10}$			
1-(Tricarboxyvinyl)pyridinium perchlorate, 1,2-dimethyl ester, cis	H_2O	260(3.79)	39-2676-64
trans	H_2O	262(3.76)	39-2676-64
$C_{12}H_{12}ClN_3O$			
Cytosine, 5-(p-chlorophenyl)-6-ethyl-	pH 1	284(4.08)	44-2674-64
	pH 11	275(3.90)	44-2674-64

Compound	Solvent	$\lambda_{max}(\log \epsilon)$	Ref.
3(2H)-Pyridazinone, 4-chloro-5-(phenethylamino)-	EtOH	231(4.47),306(3.82)	22-2124-64
	NaOH	225(4.54),290(3.88)	22-2124-64
4-Pyrimidinol, 2-amino-5-(p-chlorophenyl)-6-ethyl-	pH 1	278(3.98)	44-2674-64
	pH 11	223(4.19),276(3.98)	44-2674-64
$C_{12}H_{12}ClN_3O_2$			
Isoxazole, 4-(3-chlorophenylazo)-5-ethoxy-3-methyl-	hexane	242(4.00),334(4.10)	39-3312-65
4,5-Isoxazoledione, 3-methyl-, 4-[(3-chlorophenyl)ethylhydrazone]	EtOH	249(3.86),384(4.24)	39-3312-65
1,2,3-Triazole-4-carboxylic acid, 2-(3-chlorophenyl)-5-methyl-, ethyl ester	hexane	278(4.23)	39-3312-65
$C_{12}H_{12}Cl_2N_2O_2$			
p-Benzoquinone, 2,5-bis(1-azetidinyl)-3,6-dichloro-	methyl cellosolve	241(4.19),255(4.18), 378(4.42),545(2.57)	78-1889-64
$C_{12}H_{12}Cl_2N_2O_3$			
Benzimidazole, 5,6-dichloro-1-(2-deoxy-α-D-ribofuranosyl)-	pH 1	246(3.73),278s(3.72), 284(3.88),293(3.87)	35-4940-65
	pH 11	253(3.84),281s(3.63), 287(3.75),295(3.74)	35-4940-65
	MeOH	254(3.84),281s(3.64), 287(3.76),296(3.77)	35-4940-65
$C_{12}H_{12}Cl_2N_6O$			
Pyrimidine, 2-amino-4-chloro-5-(p-chlorophenylazo)-6-[(2-hydroxyethyl)-amino]-	pH 1	237(4.23),367(4.43)	87-0182-65
	pH 11	304(3.77),396(4.16)	87-0182-65
$C_{12}H_{12}Cl_2OS$			
Benzo[b]cyclopropa[d]thiopyran, 1,1-dichloro-7b-ethoxy-1,1a,2,7b-tetrahydro-	EtOH	223(4.13),262(3.83), 295(3.09)	44-1575-64
$C_{12}H_{12}Cl_3NOS_2$			
Butyrophenone, γ,γ,γ-trichloro-β-(N-methylthiocarbamoylthio)-	EtOH	245(4.30),284(4.12)	39-4004-64
$C_{12}H_{12}Cl_3NO_2$			
p-Benzoquinone, 2,3,5-trichloro-6-(2-diethylaminovinyl)-	MeOH	244(3.58),322(4.25), 680(3.72)	54-0711-64
$C_{12}H_{12}F_2N_2O_2$			
p-Benzoquinone, 2,5-bis(1-azetidinyl)-3,6-difluoro-	methyl cellosolve	239(4.36),381(4.47), 577(2.32)	78-1889-64
$C_{12}H_{12}N_2$			
2,2'-Bi-3-picoline	hexane	224(3.61),271(3.83)	32-1287-64
	MeOH	266(3.80)	32-1287-64
2,2'-Bi-4-picoline	n.s.g.	281(4.15)	39-6061-65
ferrous complex	n.s.g.	529(3.93)	39-6061-65
6,6'-Bi-3-picoline	n.s.g.	289(4.26)	39-6061-65
ferrous complex	n.s.g.	510(3.92)	39-6061-65
1-Cyclopentene-1-carbonitrile, 2-amino-3-phenyl-	EtOH	263(4.15)	23-2512-65
Diphenylamine, 2-amino-	EtOH	233(4.00),286(3.85)	12-1241-65
Diphenylamine, 4-amino-	EtOH	245(3.93),288(4.33), 348(2.91),356(2.54)	12-1241-65

Compound	Solvent	λ_{max}(log ε)	Ref.
Diphenylamine, 4-amino- (contd.)	ether	245(4.02),288(4.34)	12-1241-65
	dioxan	228(3.62),253(3.73), 289(4.08)	12-1241-65
Indole, 3-(1-pyrrolin-2-yl)-	EtOH	217(4.37),255(4.07), 271s(3.97),289(4.08), 332s(3.08)	87-0415-64
Indole-3-propionitrile, 2-methyl-	MeOH	225(4.53),280(3.85)	65-3487-64
Pyridine, 2-[2-(1-methylpyrrol-2-yl)vinyl]-	EtOH	362(4.35)	54-0441-65
Pyrrole, 2-(N-benzylformimidoyl)-	EtOH	244s(3.63),288(4.25)	12-0894-64
Pyrrole, 1-methyl-2-(N-phenylformimidoyl)-	EtOH	219s(3.86),324(4.27)	12-0894-64
Pyrrole, 2-(N-m-tolylformimidoyl)-	EtOH	234(3.82),241(3.79), 327(4.30)	12-0894-64
Pyrrole, 2-(N-p-tolylformimidoyl)-	EtOH	222s(3.81),331(4.35)	12-0894-64
$C_{12}H_{12}N_2O$			
2-Naphthamide, 1-amino-3-methyl-	EtOH	240s(4.3),330(3.8)	44-1543-64
	50% H_2SO_4	220(4.8),280(3.6), 328(3.0)	44-1543-64
2-Naphthamide, 1-amino-4-methyl-	EtOH	270(4.3),340(3.8), 370(3.8)	44-1543-64
	50% H_2SO_4	235(4.6),275(3.8), 300(3.7),325(3.4), 340(3.4)	44-1543-64
3(2H)-Pyridazinone, 6-(2,4-xylyl)-	EtOH	254(4.34)	39-7005-65
3(2H)-Pyridazinone, 6-(2,5-xylyl)-	EtOH	248(4.28)	39-7005-65
3(2H)-Pyridazinone, 6-(3,4-xylyl)-	EtOH	258(4.44)	39-7005-65
3-Pyridinol, 2-amino-4-methyl-5-phenyl-	EtOH-HCl	322(3.86)	44-2124-64
	EtOH-NaOH	324(3.81)	44-2124-64
Pyrrole, 2-[N-(p-methoxyphenyl)-formimidoyl]-	EtOH	227s(3.88),337(4.37)	12-0894-64
$C_{12}H_{12}N_2OS$			
Sulfoximine, S-(p-aminophenyl)-S-phenyl-	N HCl	296(4.44)	25-1261-64
	EtOH	288(4.20)	25-1261-64
$C_{12}H_{12}N_2OS_2$			
3(2H)-Pyridazinone, 4,5-bis(methylthio)-2-phenyl-	EtOH	250(4.08),318(4.01)	78-1323-65
$C_{12}H_{12}N_2OS_3$			
4-Isothiazolecarboxanilide, 3,5-bis(methylthio)-	EtOH	255(4.22),287(4.12)	44-0665-64
$C_{12}H_{12}N_2O_2$			
1H-3-Benzazepin-4-one, 2-acetamido-4,5-dihydro-	EtOH	235(3.98),294(4.09)	4-0026-65
Carbamic acid, (α-cyano-β-methylstyryl)-, methyl ester	EtOH	260.5(4.16)	28-1987-65B
Carbamic acid, (β-cyanostyryl)-, ethyl ester	MeOH	220(4.2),280(4.3)	83-0488-64
Indole, 2-methyl-3-(2-nitropropenyl)-	n.s.g.	281(4.00),287(3.98), 418(4.10)	28-0609-64A
3-Indolinepropionitrile, 3-methoxy-2-oxo-	MeOH	248(3.82),292(3.16)	44-3610-65
2-Piperazinone, 3-phenacylidene-	EtOH	258(3.81),370(4.24)	65-0541-64
Pyrazole-1-acetic acid, 5-methyl-4-phenyl-	EtOH	244(4.16)	44-1889-65

Compound	Solvent	λ_{max}(log ϵ)	Ref.
$C_{12}H_{12}N_2O_2S$			
2,4-Oxazolidinedione, 5-(anilino-methylene)-3-ethyl-2-thio-	EtOH	383(2.85)	70-0576-64
$C_{12}H_{12}N_2O_3$			
1-Benzimidazoleacetic acid, α-formyl-2-methyl-, methyl ester	EtOH	258(4.36),267s(4.26), 276(4.08)	44-1118-65
1-Butanone, 1-indol-3-yl-4-nitro-	EtOH	209(4.51),242(4.14), 257s(3.98),298(4.12)	87-0415-64
2-Cyclobuten-1-one, 2-(dimethyl-amino)-3-nitro-4-phenyl-	n.s.g.	403(4.04)	35-5132-65
4H-Quinolizine-1-carboxylic acid, 3-amino-4-oxo-, ethyl ester	EtOH	270(4.17),300(3.51), 380(4.12)	78-1051-64
$C_{12}H_{12}N_2O_3S$			
4-Imidazolidinecarboxylic acid, 5-oxo-1-phenyl-2-thioxo-, ethyl ester, sodium deriv.	H_2O	305(4.26)	44-2003-64
$C_{12}H_{12}N_2O_4$			
4-Imidazolidinecarboxylic acid, 2,5-di-oxo-1-phenyl-, ethyl ester, sodium deriv.	H_2O	293(4.18)	44-2003-64
$C_{12}H_{12}N_2O_5$			
3-Cinnolineacetic acid, 4-hydroxy-6,7-dimethoxy-	EtOH	227(4.54),257(4.43), 265(4.45),285s(3.34), 343(4.26),357(4.25)	39-6036-65
Quinazoline, 7-methoxy-4-methyl-, oxalate	MeOH	236(4.68),311(3.61)	39-5911-64
$C_{12}H_{12}N_2O_7P_2$			
Phosphonic acid, (azoxydi-p-phen-ylene)di-	EtOH	232(3.94),269(3.99), 330(4.28)	87-0891-65
$C_{12}H_{12}N_3OP$			
Phosphine oxide, tripyrrol-2-yl-	EtOH	238(4.06),240s(4.06)	44-0097-65
$C_{12}H_{12}N_4$			
1,4;5,8-Bistrimethylenepyrid-azo[4,5-d]pyridazine	EtOH	263(3.81),272(3.76), 285(3.69),297(3.76), 308(3.72)	35-0661-64
$C_{12}H_{12}N_4O$			
Malononitrile, [anilino[(2-hydroxy-ethyl)amino]methylene]-	EtOH	249(4.09),286(4.20)	95-0387-65
Pyrimidine, 4-amino-5-benzamido-6-methyl-	pH 1	235(4.24)	44-1916-65
	aq. NaCl	229(4.26),275(3.85)	44-1916-65
	pH 13	231(4.11),291(4.02)	44-1916-65
2(1H)-Pyrimidone, 4-hydrazino-, acetophenone reaction product	MeOH	312(4.35)	5-0134-65F
$C_{12}H_{12}N_4O_2$			
4-Pyrimidinol, 6-amino-5-benzamido-2-methyl-	2N HCl	231(4.03),258(4.09)	44-1916-65
	pH 13	259s(3.89)	44-1916-65
Urea, [3-(benzylamino)-2-cyano-acryloyl]-	H_2O	294(4.45)	39-1642-65
$C_{12}H_{12}N_4O_3$			
Benzamide, N-(2-imidazol-1-ylethyl)-p-nitro-	MeOH	261(4.09)	87-0107-65

Compound	Solvent	λ_{max}(log ε)	Ref.
1H-Tetrazole-1-acetic acid, α-(p-methoxybenzylidene)-5-methyl-	EtOH	222(4.20),295(4.39)	44-2222-65
$C_{12}H_{12}N_4O_3S_2$			
Sulfanilamide, N[1]-(1-methylthio-2,2-dicyanoethylene)-N[4]-acetyl-	EtOH	262(4.45),311(4.41)	95-0391-65
$C_{12}H_{12}N_4O_4$			
1,3,4-Oxadiazole, 2-amino-5-[3-ethyl-4-(5-nitro-2-furyl)-1,3-butadienyl]-	propylene glycol	260(3.90),324(4.11), 405(4.06)	95-0212-64
1,3,4-Oxadiazole, 2-(methylamino)-5-[3-methyl-4-(5-nitro-2-furyl)-1,3-butadienyl]-	propylene glycol	246(4.06),324(4.16), 414(4.14)	95-0219-64
$C_{12}H_{12}N_4O_5$			
3,6-Anhydrogalactose, p-nitrophenylosotriazole	EtOH	310(4.45)	39-2306-64
3,6-Anhydroglucose, p-nitrophenylosotriazole	EtOH	310(4.25)	39-2306-64
$C_{12}H_{12}N_4O_5S$			
1,3,4-Thiadiazole, 2-[N-(acetoxymethyl)-methylamino]-5-[2-(5-nitro-2-furyl)-vinyl]-	propylene glycol	234(3.93),319(4.00), 404(4.33)	95-0948-65
$C_{12}H_{12}N_4O_6$			
1,3,4-Oxadiazole, 2-[(N-acetoxymethyl)-methylamino]-5-[2-(5-nitro-2-furyl)-vinyl]-	propylene glycol	225(4.05),300(4.02), 389(4.27)	95-0948-65
$C_{12}H_{12}N_6$			
s-Triazole, 3-(4-amino-1-benzyl-imidazol-5-yl)-	pH 1	244(3.92),265s(3.87)	44-3601-65
	pH 7	265(3.96)	44-3601-65
	pH 13	260(3.94)	44-3601-65
s-Triazole, 3-(5-amino-1-benzyl-imidazol-4-yl)-	pH 1	246(4.09),261(4.07)	44-3601-65
	pH 7	263(4.19)	44-3601-65
$C_{12}H_{12}N_6O$			
5-Pyrazolol, 1-(5-anilino-s-triazol-3-yl)-3-methyl-	EtOH	257(4.50)	39-3912-65
$C_{12}H_{12}O$			
Azulene, 1-ethoxy-	C_6H_{12}	236(4.16),286(4.72), 343(3.47),351(3.55), 359(3.75),368(3.49), 378(3.77),637s(2.40), 662(2.47),686(2.50), 725(2.43),765(2.41), 814(2.06),867(1.99)	44-2805-64
Cyclopentanone, 2-benzylidene-, trans	MeOH	298(4.23)	78-2201-64
2-Cyclopenten-1-one, 2-methyl-3-phenyl-	EtOH	275(4.23)	23-2408-65
1-Hexen-5-yn-3-ol, 1-phenyl-	n.s.g.	250(4.23)	5-0062-65B
4a,8a-(Methanoxymethano)naphthalene	n.s.g.	242(3.53)	35-2738-64
Naphthalene, 7-methoxy-1-methyl-	EtOH	232(5.02),266(3.39), 277(3.62),288(3.53), 317(3.13),332(3.26)	87-0559-65
2-Naphthol, 7,8-dimethyl-	EtOH	233(4.38),265s(3.62), 274(3.65),283s(3.60), 336(2.81)	39-0361-65

Compound	Solvent	λ_{max}(log ε)	Ref.
$C_{12}H_{12}OS_2$			
2,2'-Bithienyl, 5-(4-hydroxy-1-butenyl)-, cis	EtOH	243(3.97),335(4.32)	39-7109-65
trans	EtOH	245(3.94),340(4.31)	39-7109-65
$C_{12}H_{12}O_2$			
1,3-Cyclopentanedione, 2-benzyl-	neutral	246(4.20)	94-1300-65
2-Cyclopenten-1-one, 3-(p-methoxyphenyl)-	MeOH	230(4.14),311(4.44)	44-2205-65
2-Dibenzofuranol, 6,7,8,9-tetrahydro-	EtOH	256(3.93),295(3.52)	70-2086-64
1,4-Ethanonaphthalene-5,8-dione, 1,4,4a,8a-tetrahydro-	C_6H_{12}	223(4.11),267s(2.51), 388(1.81)	39-3043-64
	EtOH	227(4.04),278s(2.42), 376(1.78)	39-3043-64
photoisomer	EtOH	302(1.68)	44-2593-65
1-Hexene-3,5-dione, 1-phenyl-	EtOH	229(2.84),340(3.50)	22-0525-65
$\Delta^{1,\alpha}$-Indanacetic acid, 2-methyl-	EtOH	225(4.18),231(4.07), 272(4.18),303(4.07)	95-0975-65
4-Indancarboxylic acid, 5-(1-hydroxyethyl)-, γ-lactone	EtOH	237(3.95),290(3.58), 295(3.58)	39-3811-65
5-Indancarboxylic acid, 4-(1-hydroxyethyl)-, γ-lactone	EtOH	244(4.05),275(3.19), 284(3.13)	39-3811-65
3-Indeneacetic acid, 2-methyl-	EtOH	259(4.10)	95-0975-65
Indenone, 7-methyl-, ethylene ketal	EtOH	219(4.46),226(4.40), 278(3.42),287s(3.37), 300(3.27),311s(3.14)	44-0074-64
Naphthalene, 2,7-dimethoxy-	EtOH	235(4.99)	44-3209-65
$\Delta^{1,\alpha}$-Naphthaleneacetic acid, 1,2,3,4-tetrahydro-, trans	EtOH	278(4.09)	95-0975-65
1-Naphthaleneacetic acid, 3,4-dihydro-	EtOH	262(3.85)	95-0975-65
1,8-Naphthalenediol, 3,6-dimethyl-	EtOH	235(4.73),308(3.74), 322(3.82),337(3.92)	54-1005-64
Pentacyclo[6.2.2.$0^{2,7}$.$0^{5,10}$.$0^{5,9}$]dodecane-3,6-dione	C_6H_{12}	219(2.15),304f(1.67)	39-3062-64
	EtOH	289.5f(1.65)	39-3062-64
4-Pentynoic acid, 3-methyl-5-phenyl-	EtOH	241(4.30),252(4.28)	22-2541-64
2-Tetralone, 3-acetyl-	EtOH	251(3.80),258s(3.79), 291(3.91),385(2.40)	44-1391-64
Tricyclo[6.2.2.$0^{2,7}$]dodeca-4,9-diene-3,6-dione, syn	MeOH	284.5(4.44)	35-5202-64
$C_{12}H_{12}O_2S$			
2,8-Decadiene-4,6-diyn-1-oic acid, 9-(methylthio)-, methyl ester, cis-cis	ether	259(4.17),288(4.27), 312(4.22),339(4.21), 356(4.24)	24-1736-65
cis-trans	ether	260(4.16),289(4.29), 306(4.26),327(4.23), 355(4.27)	24-1736-65
trans-cis	ether	256(4.09),292(4.14), 342(4.08),363(4.10)	24-1736-65
trans-trans	ether	269(4.11),291(4.19), 308(4.17),329(4.13), 357(4.17)	24-1963-64
4H-Thiopyran-3-carboxylic acid, 5,6-dihydro-2-phenyl-	n.s.g.	210(3.84),227(3.97), 283(3.97)	24-1963-64
Valeric acid, 2-benzoyl-5-mercapto-, thiolactone	n.s.g.	208(3.95),247(4.26), 315(3.62)	24-1963-64
$C_{12}H_{12}O_3$			
1-Benzosuberone, 7-carboxy-	EtOH	253(4.18)	44-2165-65
Butyric acid, α-oxo-β-methylene-, benzyl ester	EtOH	229.0(3.81)	39-0788-64

Compound	Solvent	$\lambda_{max}(\log \epsilon)$	Ref.
Chroman, 6-carbomethoxy-5-hydroxy-8-methoxy-2,2-dimethyl-	EtOH	213(4.33),230s(4.16),270(4.13),320(3.84)	78-1331-64
3(2H)-Dibenzofuranone, 1,4,4a,9b-tetrahydro-8-hydroxy-	EtOH	231(3.69),304(3.64)	39-2932-64
1,8-Dicyclopentadienedione, 8-ethylene ketal	EtOH	226(3.79)	78-0717-64
endo	C_6H_{12}	222(3.88)	24-3680-65
2-Indanacetic acid, α-methyl-1-oxo-	EtOH	244(4.13)	42-0479-64
Isocoumarin, 7-methoxy-4,6-dimethyl-	EtOH	240(4.44),262(3.89),272(3.93),340(3.69)	42-0821-64
1(2H)-Naphthalenone, 8-acetoxy-3,4-dihydro-	EtOH	248(3.95)	39-2816-64
2,4-Pentadienoic acid, 3-methoxy-5-phenyl-	EtOH	229(4.12),235(4.08),308(4.35)	44-3161-64
Spiro[chroman-2,5'-tetrahydrofuran-2'-one]	EtOH	272(3.22),278(3.25)	49-0220-65
$C_{12}H_{12}O_3S$			
Benzo[b]thiophene, 2-ethoxymethyl-5,6-methylenedioxy-	EtOH	241(4.43),267(4.05),277(4.09),304(3.68),310(3.72),317(3.76)	4-0100-65
2(5H)-Furanone, 4-benzylthiomethyl-3-hydroxy-	EtOH	232(4.30)	39-0766-64
	EtOH-NaOH	277(4.02)	39-0766-64
$C_{12}H_{12}O_4$			
Benzocyclobutene, 4,5-dihydroxy-	EtOH	269s(3.40),273(3.42),278s(3.36)	78-2281-65
3-Benzofurancarboxylic acid, 5-hydroxy-2-methyl-, ethyl ester	EtOH	225(4.05),255(3.90),296(3.67)	44-1453-64
5-Benzofuranol, 4-acetyl-2-ethoxy-	EtOH	230(4.15),302(4.02),368(3.82)	27-0358-64
4-Chromancarboxylic acid, 3-oxo-, ethyl ester	EtOH	294(3.64)	65-2699-64
Coumarin, 3-(2-hydroxyethyl)-4-methoxy-	n.s.g.	270(4.0),276(4.0),295(3.9)	95-0310-65
Isocoumarin, 5,7-dimethoxy-4-methyl-	EtOH	242(4.33),357(3.71)	42-0821-64
Ketone, methyl 2,3,3a,8a-tetrahydro-5-hydroxyfuro[2,3-b]benzofuran-4-yl	EtOH	232(3.96),255s(3.6),358(3.46)	27-0358-64
4,7-Methanoindene-2,5-dicarboxylic acid, 3a,4,7,7a-tetrahydro-	EtOH	222(4.23)	44-3234-64
1(2H)-Naphthalenone, 4-acetoxy-3,4-dihydro-8-hydroxy-	EtOH	259(3.92),325(3.49)	70-0492-64
1(2H)-Naphthalenone, 8-acetoxy-3,4-dihydro-4-hydroxy-	EtOH	247(4.17),297(3.26)	70-0492-64
Sepedonin, anhydro-O-methyl-	EtOH	249(4.24),285(4.46),383(3.73)	23-1835-65
$C_{12}H_{12}O_4S$			
2,6-Decadiene-4,8-diynoic acid, 7-(methylsulfonyl)-, methyl ester, cis-cis	ether	231(4.07),313(4.24)	24-1616-65
trans-trans	ether	311(4.48),333s(4.45)	24-1616-65
$C_{12}H_{12}O_5$			
Succinic acid, phenacyl-	EtOH	242(4.05)	44-3312-65
$C_{12}H_{12}O_5S$			
Safrole, sulfone autooxidation product	EtOH	256(4.22),305(4.14),352(4.44)	39-3563-65

Compound	Solvent	λ_{max}(log ε)	Ref.
$C_{12}H_{12}S_2$			
Azulene, 1,3-bis(methylthio)-	C_6H_{12}	204(4.27),235(4.27), 283(4.35),289(4.35), 320(4.03),370(3.69), 606s(2.43),627(2.45)	44-2715-65
$C_{12}H_{13}ClN_2$			
1H-Azepino[5,4,3-cd]indole, 9-chloro-3,4,5,6-tetrahydro-6-methyl-	EtOH	228(4.52),292(3.85), 301(3.82)	87-0200-65
$C_{12}H_{13}ClN_2O_2S$			
Pyrido[2,1-c][1,2,4]benzothiadiazine, 3-chloro-7,8,9,10-tetrahydro-2-methyl-, 5,5-dioxide	EtOH	275(3.99)	44-3206-64
$C_{12}H_{13}ClN_2O_3$			
3-[(Carboxycarbonyl)methyl]-1,2-dimethylbenzimidazolium chloride	EtOH	255(3.80),264(3.82), 271(3.93),278(3.96)	44-1118-65
$C_{12}H_{13}ClN_2O_6$			
7-Carboxy-2,4-dimethyl-1,5-benzodiazepinium perchlorate	MeOH-HCl	246(4.37),280(4.49), 333(3.07),345s(3.04), 526(2.96)	39-3785-65
$C_{12}H_{13}ClN_2O_6S$			
Cephalosporanic acid, 7-(chloroacetamido)-	pH 6.0	260(3.97)	39-5015-65
$C_{12}H_{13}ClN_6OS$			
4-Pyrimidinethiol, 2-amino-5-(p-chlorophenylazo)-6-(2-hydroxyethyl)amino-	pH 1	292(4.24),326(3.91), 450(4.32)	87-0182-65
	pH 11	292(4.31),415(4.24)	87-0182-65
$C_{12}H_{13}ClO_5$			
2,4-Cyclohexadien-1-one, 2,4-diacetoxy-3-chloro-6,6-dimethyl-	MeOH	220(3.76),310(3.71)	20-0081-64
$C_{12}H_{13}Cl_3N_2O_6$			
Thymidine, 5'-trichloroacetate	EtOH	264.5(4.09)	31-0432-65
$C_{12}H_{13}FO$			
5(6H)-Benzocyclooctenone, 2-fluoro-7,8,9,10-tetrahydro-	EtOH	249(3.89)	44-2165-65
$C_{12}H_{13}FO_3$			
Cinnamic acid, α-fluoro-p-methoxy-, ethyl ester	C_6H_{12}	224(4.01),297(4.33)	22-2258-64
$C_{12}H_{13}IN_2O_2$			
Cyclobutene, 1-(dimethylamino)-4-iodo-2-nitro-3-phenyl-	n.s.g.	384(4.27)	35-5132-65
$C_{12}H_{13}N$			
Benz[cd]indole, 1,3,4,5-tetrahydro-2-methyl-	EtOH	230(4.52),275(3.81), 282(3.83),291(3.70)	44-0843-64
Carbazole, tetrahydro-	C_6H_{12}	225(4.50),273(3.88)	23-2674-64
Carbazole, 1,2,3,4-tetrahydro-	MeOH	227(4.54),282(3.83)	65-3487-64
Indene-$\Delta^{1,\alpha}$-methylamine, N,N-dimethyl-	MeCN	360(4.43)	73-2783-65
Indole, 1-allyl-3-methyl-	EtOH	226(4.53),289(3.75)	78-0989-65

Compound	Solvent	λmax(log ε)	Ref.
3H-Indole, 3-allyl-3-methyl-	pH 1	225(3.72),236s(3.54), 270(3.42)(changing)	78-0989-65
	EtOH	253(3.83)	78-0989-65
2-Naphthylamine, 1,4-dimethyl-	EtOH	215(4.34),243(4.74), 279s(3.65),289(3.78), 300(3.74),343(3.39)	44-0190-65
Pyrrole, 2,5-dimethyl-1-phenyl-	EtOH	238s(3.83),268(3.14)	44-0190-65
$C_{12}H_{13}NO$			
Azepino[3,2,1-hi]indol-6(7H)-one, 1,2,4,5-tetrahydro-	MeOH	257(3.81),300s(3.18)	35-1397-65
1-Butanone, 1-indol-3-yl-	EtOH	241(4.11),256f(3.96), 296(4.10)	87-0094-64
	base	241(4.08),260(3.98), 296(4.07),332f(3.42)	87-0094-64
2-Butanone, 1-indol-3-yl-	EtOH	220(4.52),275(3.78), 281(3.80),289(3.72)	87-0094-64
Indole, 2-acetyl-1,3-dimethyl-	EtOH	239(4.22),310(4.30)	12-0246-64
Indole, 3-acetyl-1,2-dimethyl-	EtOH	217(4.43),245(4.13), 266(3.93),305(4.07)	87-0094-64
1-Naphthol, 7-(dimethylamino)-	EtOH	256(4.51),296(3.97), 306(3.93),350(3.28)	44-1180-64
1-Naphthol, 7-(ethylamino)-	EtOH	251(4.52),292(3.89), 303(3.84),347(3.26)	44-1180-64
5-Pseudoindolone, 2,2-tetramethylene-	MeOH	278(4.38)	24-2939-65
Pyridine, 1,2-dihydro-1-hydroxy-6-methyl-2-phenyl-	EtOH	238(4.02),314(4.70)	44-0910-65
6(2H)-Quinolone, 2,2,4-trimethyl-	MeOH	242(4.0),252(4.0), 280(4.1),362(3.7)	24-2939-65
Sorbanilide	EtOH	272(4.48)	54-0581-65
$C_{12}H_{13}NOS_2$			
1-Pyrrolidinecarbodithioic acid, anhydrosulfide with BzSH	hexane	238(4.38),283(4.19), 400(2.35)	78-2857-65
$C_{12}H_{13}NO_2$			
Indole, 2-acetyl-5-methoxy-6-methyl-	MeOH	315(4.49)	35-4612-64
Isoquinoline, 6,7-dimethoxy-1-methyl-	EtOH-HCl	246(4.29),307(3.98), 356(3.95)	33-1945-65
	EtOH	225(4.51),273(3.91), 306(3.82)	33-1945-65
	EtOH-NaOH	271(4.11),305(3.98)	33-1945-65
hydrochloride	EtOH-HCl	245(4.28),307(3.97), 358(3.94)	33-1945-65
	EtOH	248(4.22),308(3.95), 358(3.88)	33-1945-65
	EtOH-NaOH	272(3.99),306(3.87)	33-1945-65
Norsecurinine	EtOH	256(4.34),257(4.34)	94-0786-65
	EtOH	256.5(4.34)	94-1520-64
hydrochloride	EtOH	254(4.21)	94-0786-65
Valerolactam, N-benzoyl-	EtOH	205(4.11),230(3.91), 248(3.87)	65-1394-65
	EtOH	205(4.11),230(3.91), 248(3.87)	70-0774-64
$C_{12}H_{13}NO_2S$			
2(5H)-Furanone, 3-amino-4-benzylthiomethyl-	EtOH	266(4.03)	39-0766-64
	0.5N HCl	267(3.73)	39-0766-64

Compound	Solvent	$\lambda_{max}(\log \epsilon)$	Ref.
$C_{12}H_{13}NO_2S_2$			
4-Morpholinecarbodithioic acid, anhydrosulfide with thiobenzoic acid	hexane	240(4.29),292(4.12), 412(2.30)	78-2857-65
$C_{12}H_{13}NO_3$			
Benzocyclobutene, 4-acetamido-5-acetoxy-	EtOH	237(3.68),279(3.29)	78-2281-65
2,3-Benz-1-oxa-10-azabicyclo[0.4.4]-decan-4-one, 9-hydroxy-	EtOH	208(4.53),238(3.97), 298(3.46)	65-1394-65
Butyrolactam, N-(o-methoxybenzoyl)-	EtOH	214(4.18),290(3.28)	65-1394-65
	EtOH	214(4.18),290(3.08)	78-3537-65
Indole-2-carboxylic acid, 5-methoxy-6-methyl-, methyl ester	MeOH	298(4.30)	35-3877-64
	MeOH	298(4.3)	44-2897-65
3-Indolinebutyric acid, 2-oxo-	EtOH	206(4.42),250(3.93), 281s(3.15)	44-2431-64
7-Isoquinolinol, 6,8-dimethoxy-1-methyl-, sulfate	pH 2	226(4.21),260(4.68), 315(3.81),355(3.63)	49-0025-65
	pH 12	229(4.44),262(4.63), 296(3.91),357(3.72)	49-0025-65
	EtOH	248(4.49),260(4.48), 317(3.89),362s(3.46)	49-0025-65
7H,11H-Pyrido[2,1-b][1,3]benzoxazin-11-one, 5a,6,8,9-tetrahydro-5a-hydroxy-	EtOH	208(4.53),238(3.97), 298(3.46)	78-3537-65
$C_{12}H_{13}NO_4$			
3-Indolinebutyric acid, 3-hydroxy-2-oxo-	EtOH	208(4.42),253(3.77), 288(3.13)	44-2431-64
$C_{12}H_{13}NO_4S$			
Indole-2-carboxylic acid, 1-methyl-3-sulfino-, dimethyl ester	EtOH	215(4.41),237(4.38), 304(4.19)	44-0178-64
$C_{12}H_{13}NO_5$			
Indole-7-carboxylic acid, 3-acetoxy-2,4,5,6-tetrahydro-2-oxo-, methyl ester	EtOH	300(4.26)	39-1648-65
$C_{12}H_{13}NO_5S$			
Benzoic acid, p-nitro-, anhydride with S-tert-butyl thiocarbonate	C_6H_{12}	254(4.43)	44-2471-64
4H-1,2-Benzothiazin-4-one, 3-acetyl-2,3-dihydro-3-methoxy-2-methyl-, 1,1-dioxide	EtOH	250(3.89)	44-2241-65
$C_{12}H_{13}NS_2$			
2H-1,3-Thiazine-2-thione, 3,6-dihydro-3,4-dimethyl-6-phenyl-	H_2O	303(4.06),320s(--)	39-4008-64
	EtOH	310(4.02)	39-4008-64
2H-1,3-Thiazine-2-thione, 3,6-dihydro-4,5-dimethyl-6-phenyl-	EtOH	325(4.03)	39-4008-64
$C_{12}H_{13}N_3$			
Pyrimidine, 1-benzyl-1,2-dihydro-2-(methylimino)-	pH 7.0	229(4.54),318(3.63)	39-5542-65
	pH 13.1	242(4.15),370(3.39)	39-5542-65
Pyrimidine, 2-(benzylimino)-1,2-dihydro-1-methyl-	pH 7.0	230(4.27),315(3.63)	39-5542-65
	pH 13.1	245(4.29),364(3.41)	39-5542-65
s-Triazine, 2-ethyl-4-methyl-6-phenyl-	EtOH	263(4.33)	44-0707-65
$C_{12}H_{13}N_3O$			
s-Triazine, 2-ethyl-4-methoxy-6-phenyl-	EtOH	263(4.31)	44-0707-65

Compound	Solvent	$\lambda_{max}(\log \epsilon)$	Ref.
$C_{12}H_{13}N_3OS_3$			
4-Isothiazolecarboxanilide, 2'-amino-3,5-bis(methylthio)-	EtOH	233(4.00),287(4.09)	44-0665-64
$C_{12}H_{13}N_3O_2$			
Formamidine, N'-(p-methoxyphenyl)-N-(5-methyl-3-isoxazolyl)-	EtOH	287(4.35)	78-0159-64
$C_{12}H_{13}N_3O_4$			
Indole, 3-(2-nitrobutyl)-6-nitro-	EtOH	251(4.00),264s(3.99), 324(3.92),366(3.85)	87-0274-64
$C_{12}H_{13}N_3O_5$			
2-Cyclobuten-1-one, 2-(dimethylamino)-3-nitro-4-phenyl-	n.s.g.	372(4.26)	35-5132-65
$C_{12}H_{13}N_3O_5S$			
Malonamic acid, 2-diazo-N-(p-tolylsulfonyl)-, ethyl ester	MeOH	230(4.37),250s(--)	5-0101-64F
$C_{12}H_{13}N_3S$			
s-Triazine, 2-ethyl-4-(methylthio)-6-phenyl-	EtOH	263(4.54)	44-0707-65
$C_{12}H_{13}N_5O_2$			
p-Toluamide, N-(2,4-diamino-6-hydroxy-5-pyrimidinyl)-	pH 1	247s(4.31),263(4.36)	44-1916-65
	aq. NaCl	242(4.23),266(4.26)	44-1916-65
	pH 13	241(4.26),262s(4.13)	44-1916-65
$C_{12}H_{13}N_5O_3$			
Uracil, 6-amino-5-(benzylnitrosamino)-3-methyl-	pH 5.0	228(4.00),259(4.20)	24-1060-65
	pH 11.0	255(4.08)	24-1060-65
$C_{12}H_{13}N_5O_4$			
1H-Pyrrolo[2,3-d]pyrimidine-5-carbonitrile, 4-amino-1-D-ribosyl-	pH 1	234(4.27),274(4.16)	35-1995-65
	pH 13	229(4.14),277(4.24)	35-1995-65
$C_{12}H_{14}$			
Benzocyclobutene, 1-isopropenyl-1-methyl-	EtOH	260(2.10),266(2.45), 272(2.50)	5-0055-64B
5H-Benzocycloheptene, 6,7-dihydro-9-methyl-	EtOH	224(4.56)	35-3719-65
1,2-Benzocycloocta-1,3-diene	EtOH	239(3.98)	78-0231-64
Naphthalene, 3,4-dihydro-1,3-dimethyl-	EtOH	259(3.97)	28-1259-64A
1,3-Pentadiene, 2-methyl-4-phenyl-	C_6H_{12}	221(2.85),229(2.81), 249(2.59)	35-5244-64
1,3-Pentadiene, 4-methyl-2-phenyl-	C_6H_{12}	262(3.67)	35-5244-64
$C_{12}H_{14}BrNO_2$			
10-Acetoxy-7,8-dihydro-6H-pyrido-[1,2-a]azepinium bromide	H_2O	242(3.88),300(4.07)	44-1523-65
$C_{12}H_{14}ClN$			
Quinoline, 7-chloro-1,2-dihydro-2,2,4-trimethyl-	EtOH-acid	264(4.16)	44-1832-65
	EtOH-base	231(4.60),273(3.50), 330(3.59)	44-1832-65
$C_{12}H_{14}ClNO_2$			
1,3-Dioxolane, 2-(p-aminostyryl)-2-(chloromethyl)-	pH 13	280(4.53)	87-0035-65
	EtOH	286(4.53)	87-0035-65

Compound	Solvent	λ_{max}(log ϵ)	Ref.
Norsecurinine, hydrochloride	EtOH	254(4.24)	94-1520-64
$C_{12}H_{14}ClNO_4$			
1-Ethyllepidinium perchlorate	EtOH	230(4.57),315(3.91)	65-0506-65
$C_{12}H_{14}ClNO_4S_2$			
2-(Dimethylamino)-4-p-tolyl-1,3-dithiol-1-ium perchlorate	EtOH	228(4.14),242(4.13), 302(4.03),320(4.04)	44-1703-64
$C_{12}H_{14}ClNO_5S_2$			
2-(Dimethylamino)-4-(p-methoxyphenyl)-1,3-dithiol-1-ium perchlorate	EtOH	256(4.1),292(4.12), 326(4.01)	44-1703-64
$C_{12}H_{14}INO_2$			
Indancarbamic acid, 2-iodo-, ethyl ester, trans	MeOH	261(3.20),268(3.31), 273(3.31)	78-1037-64
1-Naphthalenecarbamic acid, 1,2,3,4-tetrahydro-2-iodo-, methyl ester, trans	MeOH	274(3.04)	44-3640-64
	MeOH	267(3.04),274(2.95)	78-1037-64
$C_{12}H_{14}I_2N_2$			
1,1'-Dimethyl-2,2'-bipyridinium diiodide	H_2O	270(4.13)	25-0782-65
$C_{12}H_{14}N_2$			
1H-Azepino[5,4,3-cd]indole, 3,4,5,6-tetrahydro-6-methyl-	EtOH	226(4.51),284(3.83)	87-0200-65
Benzeneazo-1'-cyclohexene	n.s.g.	222(3.97),227(4.02), 233(3.86),300(4.33), 310s(--),325s(--)	88-4545-65
2-Carboline, 1,2,3,4-tetrahydro-1-methyl-	pH 1	271(3.84),277(3.83), 287(3.71)	25-0622-64
	EtOH	275s(3.84),281(3.86), 289(3.78)	25-0622-64
Cyclobutane, 3,4-bis(cyanomethylene)-1,1,2,2-tetramethyl-	n.s.g.	281(4.06)	77-0321-65
Indole, 3-(2-pyrrolidinyl)-	EtOH	216(4.57),274s(3.76), 280(3.78),287(3.69)	87-0415-64
$C_{12}H_{14}N_2O$			
1,2-Cyclohexanedione, phenylhydrazone	MeOH	237(4.00),290(3.72), 295(3.74),355(4.18)	24-2111-65
Dinor-5-eserolone, 9-methyl-	MeOH	266(4.3)	24-2939-65
Isoxazole, 5-(dimethylamino)-3-methyl-4-phenyl-	C_6H_{12}	206(4.06),254(3.94), 274(3.84)	17-0255-65
	MeOH-HCl	207(3.98),245s(3.65), 282(4.14)	17-0255-65
	MeOH	207(4.03),260(4.08), 270s(4.05)	17-0255-65
Isoxazole, 5-(dimethylamino)-4-methyl-3-phenyl-	C_6H_{12}	208(3.98),230(4.18), 270(3.67)	17-0255-65
	MeOH-HCl	208(4.07),245(3.96), 282(4.05)	17-0255-65
	MeOH	204(4.11),230(4.19), 280(3.84)	17-0255-65
Pyrazole, 5-ethoxy-3-methyl-1-phenyl-	C_6H_{12}	252(4.29)	78-0299-64
	10N H_2SO_4	233.5(4.18)	78-0299-64
	pH 5	233.5(4.11)	78-0299-64
Pyrazole-1-ethanol, 5-methyl-4-phenyl-	EtOH	243(4.16)	44-1889-65

Compound	Solvent	λ_{max}(log ε)	Ref.
5-Pyrazolinone, 2,3,4-trimethyl-1-phenyl-	C_6H_{12}	241(4.02),278(4.02)	78-0299-64
	10N H_2SO_4	236(4.02)	78-0299-64
	pH 5	246(3.95),265(4.00)	78-0299-64
5-Pyrazolinone, 3,4,4-trimethyl-1-phenyl-	C_6H_{12}	246(4.28)	78-0299-64
	29N H_2SO_4	223.5(4.02)	78-0299-64
	pH 5	235(4.10)	78-0299-64
3(2H)-Pyridazinone, 4,5-dihydro-6-(2,4-xylyl)-	EtOH	270(4.16)	39-7005-65
3(2H)-Pyridazinone, 4,5-dihydro-6-(2,5-xylyl)-	EtOH	266(4.11)	39-7005-65
3(2H)-Pyridazinone, 4,5-dihydro-6-(3,4-xylyl)-	EtOH	287(4.26)	39-7005-65
11H-Pyrido[2,1-b]quinazolin-11-one, 5,5a,6,7,8,9-hexahydro-	iso-PrOH	229(4.51),257s(3.60), 350(3.46)	35-5793-65
$C_{12}H_{14}N_2O_2$			
Aniline, N,N-dipropenyl-p-nitro-	n.s.g.	402(4.38)	5-0109-65I
p-Benzoquinone, 2,5-bis(1-azetidinyl)-	methyl cellosolve	224(4.44),367(4.48), 507(2.65)	78-1889-64
γ-Caprolactone, α-oxo-, phenylhydrazone	n.s.g.	233(4.17),295(3.97), 333(4.50)	39-0141-64
1-Cyclobuten-1-ylamine, N,N-dimethyl-2-nitro-3-phenyl-	n.s.g.	379(4.32)	35-5132-65
Cyclohepta[b]pyrrole-3-carboxylic acid, 1,2-dihydro-2-imino-1-methyl-, methyl ester	EtOH	240(4.18),283(4.47), 425(4.10)	94-0828-65
Indole-3-acrylic acid, 5-amino-, ethyl ester	EtOH	232(4.34),282(4.03), 339(4.11)	87-0389-64
2-Pyrazoline-4-carboxylic acid, 1-phenyl-, ethyl ester	EtOH	250(3.86),284(3.98)	95-0158-65
Pyrrole-3-carboxylic acid, 4-methyl-5-(pyrrol-2-yl)-, ethyl ester	EtOH	231(4.32),267(4.22)	44-2727-64
Quinoxaline, 5,8-dimethoxy-2,3-dimethyl-	MeOH	265(4.58),308(3.48), 360(3.23)	4-0171-64
Tryptophan, 1-methyl-	90% EtOH	223(4.52),277s(3.72), 286(3.76),296s(3.68)	94-0088-65
	90% EtOH-HCl	222(4.54),277s(3.77), 285(3.81),296s(3.71)	94-0088-65
Vasicinol, O-methyl ether	EtOH	296(4.03)	2-0524-65
$C_{12}H_{14}N_2O_2S$			
2H-1,2,6-Thiadiazine, 2-benzyl-3,5-dimethyl-, 1,1-dioxide	EtOH	253f(3.1),323(3.9)	25-0182-65
$C_{12}H_{14}N_2O_3$			
Benzimidazole, 1-(2-deoxy-D-ribofuranosyl)-	pH 1	249(3.65),262(3.68), 269(3.79),275(3.76)	35-4940-65
	pH 11	246(3.82),250s(3.80), 265(3.54),273(3.59), 280(3.57)	35-4940-65
	MeOH	246(3.83),251s(3.82), 265(3.53),273(3.61), 281(3.61)	35-4940-65
Cinnoline, 4,6,7-trimethoxy-3-methyl-	EtOH	243(4.65),304(3.87)	39-6036-65
4(1H)-Cinnolinone, 6,7-dimethoxy-1,3-dimethyl-	EtOH	229(4.25),260(4.26), 268(4.28),283s(3.45), 292s(3.32),350(4.04), 365(4.05)	39-6036-65
4-Hydroxy-6,7-dimethoxy-2,3-dimethylcinnolinium hydroxide, inner salt	EtOH	239(4.37),271(4.06), 279(4.07),361(4.00), 378(4.05)	39-6036-65

Compound	Solvent	$\lambda_{max}(\log \epsilon)$	Ref.
Quinoline, 7-amino-5,6,8-trimethoxy-	EtOH-HCl	228(4.42),279(4.62), 348(3.66),388(3.88)	95-0985-65
	EtOH	216(4.45),260(4.64), 297(3.47),358(3.68)	95-0985-65
Urea, 1-(3-methoxy-2-methylacryloyl)-3-phenyl-	MeOH	225(3.8),275(4.3)	83-0367-64
$C_{12}H_{14}N_2O_3S$			
Indole-2-carboxylic acid, 1-methyl-3-[(methylamino)sulfinyl]-, methyl ester	EtOH	216(4.38),234(4.42), 302(4.18),328s(3.79), 336s(3.62)	44-0178-64
$C_{12}H_{14}N_2O_4$			
1,3-Cyclohexanedione, 2,2'-azodi-	CH_2Cl_2	261(4.03),424(4.25)	5-0214-65G
Indole-7-carboxylic acid, 3-acetamido-2,4,5,6-tetrahydro-2-oxo-, methyl ester	n.s.g.	319(3.91)	39-1648-65
2,4-Pentadienamide, 4-ethyl-N-methyl-5-(5-nitro-2-furyl)-	propylene glycol	280(4.20),390(4.39)	95-0646-64
2,4-Pentadienamide, N,N,4-trimethyl-5-(5-nitro-2-furyl)-	propylene glycol	280(4.11),390(4.37)	95-0646-64
Sydnone, 3-(4-ethoxy-3-methoxybenzyl)-	EtOH	285(4.00)	87-0531-65
$C_{12}H_{14}N_2O_4S$			
Indole-2-carboxylic acid, 1-methyl-3-(methylsulfamoyl)-, methyl ester	EtOH	211(4.52),236s(4.01), 291(4.04)	44-0178-64
$C_{12}H_{14}N_2O_5$			
Acetanilide, 3'-(2-acetoxyethyl)-4'-nitro-	EtOH	229(3.93),309(3.86)	78-2281-65
Quinazoline, 3,4-dihydro-7-methoxy-4-methyl-, oxalate	MeOH	229(4.37),278s(3.49), 284(3.50),302(3.44)	39-5911-64
$C_{12}H_{14}N_4$			
2,2'-Bipyrimidine, 4,4',6,6'-tetramethyl-	n.s.g.	248(4.17)	44-0943-64
$C_{12}H_{14}N_4O$			
Butyronitrile, 3-imino-2-oxo-, (p-ethoxyphenyl)hydrazone	EtOH	240(4.03),268(3.95), 305(3.70),371(4.28), 393s(4.22),500(2.35)	5-0033-65J
	EtOH-NH_3	246(4.03),268(3.97), 305(3.71),372(4.34), 392(4.20),505s(2.27)	5-0033-65J
$C_{12}H_{14}N_4O_2$			
Acetamide, N-(4-acetyl-3,6-dimethyl-pyrazolo[1,5-a]pyrimidin-7(4H)-ylidene-	EtOH	237(4.89),283s(3.43), 292(3.49),303(3.41), 341(3.58)	94-0142-65
Formamide, N-acetyl-N-(2,3,6-trimethyl-pyrazolo[1,5-a]pyrimidin-7-yl)-	EtOH	240(4.67),287(3.11), 298(3.19),312(3.17), 345(3.18)	94-0142-65
$C_{12}H_{14}N_4O_3$			
Pyrazolo[1,5-a]pyrimidine-6-carboxylic acid, 7-acetamido-2-methyl-, ethyl ester	EtOH	245(4.62),298(3.54), 334(3.44)	94-1207-65
$C_{12}H_{14}N_4O_3S$			
4H-1,2,4-Triazole, 3-(methylthio)-5-(5-nitro-2-furyl)vinyl-4-propyl-	propylene glycol	235(4.02),295(3.97), 390(4.22)	95-0566-64

Compound	Solvent	λ_{max}(log ε)	Ref.
$C_{12}H_{14}N_4O_4$			
Cyclobutanone, 2,3-dimethyl-, 2,4-dinitrophenylhydrazone	$CHCl_3$	364.1(4.32)	22-2755-65
Cyclopropanecarboxaldehyde, 1,2-dimethyl-, 2,4-dinitrophenylhydrazone, cis	$CHCl_3$	371.0(4.41)	22-2755-65 88-0979-65
trans	$CHCl_3$	370.3(4.18)	22-2755-65 88-0979-65
1,3,4-Oxadiazole, 2-(butylamino)-5-[2-(5-nitro-2-furyl)vinyl]-	propylene glycol	230(4.04),306(4.06), 398(4.34)	95-0219-64
1,3,4-Oxadiazole, 2-(diethylamino)-5-[2-(5-nitro-2-furyl)vinyl]-	propylene glycol	233(4.15),300(4.05), 395(4.34)	95-0948-65
1,3,4-Oxadiazole, 2-(ethylamino)-5-[1-ethyl-2-(5-nitro-2-furyl)vinyl]-	propylene glycol	310(3.93),400(4.14)	95-0219-64
2-Pentenal, 2-methyl-, 2,4-dinitrophenylhydrazone	n.s.g.	375(4.45)	39-5815-64
3-Penten-2-one, 4-methyl-, 2,4-dinitrophenylhydrazone	$C_2H_4Cl_2$	351(4.33)	39-1761-65
6-Pteridineacetic acid, α-acetyl-5,6,7,8-tetrahydro-7-oxo-, ethyl ester	pH 1.0	223(4.48),285(3.75), 350(3.72)	39-6930-65
	EtOH	214(4.55),275(3.67), 325(3.77)	39-6930-65
$C_{12}H_{14}N_4O_5$			
2,6-Pyridinedicarboxylic acid, 4-(formylmethyl)-, dimethyl ester, semicarbazone	MeOH-HCl	255s(3.81),416(4.01)	33-1922-65
	MeOH	230s(4.21),271(3.67), 279(3.53)	33-1922-65
$C_{12}H_{14}N_4O_6S$			
Alanine, N-[2-(7-nitro-4H-1,2,4-benzothiadiazin-3-yl)ethyl]-, S,S-dioxide	2% $NaHCO_3$	240(4.04),359(4.11)	32-0786-65
β-Alanine, N-[2-(7-nitro-4H-1,2,4-benzothiadiazin-3-yl)ethyl]-, S,S-dioxide	2% $NaHCO_3$	242(4.06),357(4.14)	32-0786-65
$C_{12}H_{14}N_4O_7S$			
Serine, N-[2-(7-nitro-4H-1,2,4-benzothiadiazin-3-yl)ethyl]-, S,S-dioxide	2% $NaHCO_3$	241(3.95),358(4.08)	32-0786-65
$C_{12}H_{14}N_4S$			
Acetone, N-methyl-N-(5-phenylthiadiazolylhydrazone)	EtOH	319(4.22)	1-0871-64
2-Thiadiazolinone, 3-methyl-5-phenyl-, azine with acetone	EtOH	241(4.10),345(4.08)	1-0871-64
$C_{12}H_{14}N_6O_3$			
1,2,4-Triazine, 3-(N,N-dimethylaminomethylamino)-6-[2-(5-nitro-2-furyl)vinyl]-, hydrochloride	propylene glycol	293(--),402(--)	95-0016-64
$C_{12}H_{14}O$			
5(6H)-Benzocyclooctenone, 7,8,9,10-tetrahydro-	EtOH	244(3.86)	5-0021-64G
	EtOH	248(3.76)	44-2165-65
Bullvalene, ethoxy-	n.s.g.	235(3.48)(end absorption)	24-3385-65
1(2H)-Naphthalenone, 3,4-dihydro-4,4-dimethyl-	C_6H_{12}	245(4.08),285(3.28), 295(3.17)	23-0579-64
2-Pentenal, 2-methyl-4-phenyl-	EtOH	230(4.11),240s(--)	22-0161-64
$C_{12}H_{14}OS$			
Benzothiophene-4-ol, 2,7-diethyl-	aq. base	330.0(3.87)	22-1464-65

Compound	Solvent	$\lambda_{max}(\log \epsilon)$	Ref.
Benzothiophene-7-ol, 4-ethyl-2,3-dimethyl-	aq. base	319.5(3.81)	22-1464-65
$C_{12}H_{14}O_2$			
1-Benzocyclooctenone, tetrahydro-8-hydroxy-	pH 13	331(4.28)	44-2165-65
	EtOH	282(4.11)	44-2165-65
5-Benzofuranol, 2,3-dihydro-2-isopropenyl-2-methyl-	EtOH	228(3.69),303(3.60)	39-5182-65
1,3-Butanedione, 2,2-dimethyl-1-phenyl-	hexane	253(4.29)	44-3000-65
	iso-PrOH	242(4.19)	44-3000-65
3-Buten-2-one, 4-ethoxy-3-phenyl-	EtOH	227(3.99),260(4.00)	44-1889-65
3-Chromen-6-ol, 5,7,8-trimethyl-	EtOH	233(4.24),273(3.90), 281s(3.83),333(3.45)	39-5060-65
2,6,8-Decatrien-4-yn-1-ol, acetate	ether	290(4.58),307(4.53)	24-0369-65
8-Decene-4,6-diynoic acid, 8-methyl-, methyl ester	EtOH	238(3.75),250(4.03), 264(4.20),279(4.10)	70-1237-65
3,5,7,9-Dodecatetraene-2,11-dione	EtOH	258(3.58),351(4.80)	70-0684-65
4a,6-Ethanonaphthalen-2-one, 5,6,7,8-tetrahydro-10-hydroxy-	MeOH	245(4.26)	35-0288-64
Indan, 1-acetyl-5-methoxy-	EtOH	281(3.43)	22-3718-65
2-Naphthaldehyde, 3,5,6,7,8,8a-hexahydro-8a-methyl-3-oxo-	EtOH	220(4.08),244(4.04)	88-4585-65
	EtOH-NaOH	242(4.08),347(3.98)	88-4585-65
Naphthalene, 1,2-dihydro-5,7-dimethoxy-	EtOH	223(4.23),269(4.13), 278(4.09),301(3.70)	5-0021-64G
1,4-Naphthalenediol, 5,6-dihydro-2,7-dimethyl-	EtOH	267(3.94),273(3.95), 318(3.55)	94-0533-64
1,6-Naphthalenediol, 1,2,3,4-tetrahydro-1-vinyl-	EtOH	218(3.94),276(3.22), 342(1.69),347(1.72)	70-1809-65
1-Naphthoic acid, 5,6,7,8-tetrahydro-4-methyl-	n.s.g.	242(3.95),284(3.10)	39-0833-65
Δ^4-Dihydro-4,5-dimethylpyran-2-spirocyclohexa-2',5'-dien-4'-one	EtOH	225(4.13)	39-5182-65
	n.s.g.	225(4.13)	77-0195-64
$C_{12}H_{14}O_2S$			
Benzoic acid, p-[2-(methylthio)propenyl]-, methyl ester	ether	228(4.10),310(4.34)	24-3087-65
Cinnamic acid, β-methyl-p-(methylthio)-, methyl ester	ether	229(4.07),305(4.30)	24-3087-65
1,4-Naphthoquinone, 4a,5,8,8a-tetrahydro-4a-methyl-3-(methylthio)-	EtOH	332(3.95)	44-0051-64
$C_{12}H_{14}O_3$			
p-Benzoquinone, 2-acetonyl-3,5,6-trimethyl-	EtOH	259(4.28),335(2.50)	1-0421-64
Butanoic acid, 3-(p-acetylphenyl)-	EtOH	254(4.23)	78-2593-65
1,3-Naphthalenediol, 5,6,7,8-tetrahydro-, 1-acetate	EtOH	220s(3.92),280(3.35), 282s(3.35)	24-1926-64
	NaOH	296(3.56)	24-1926-64
1(2H)-Naphthalenone, 3,4-dihydro-6,8-dimethoxy-	EtOH	275(4.24)	5-0021-64G
	$HOAc-H_2SO_4$	319(4.43)	5-0021-64G
2-Naphthoic acid; 5,6,7,8-tetrahydro-3-hydroxy-4-methyl-	$CHCl_3$	250(4.00)	94-0112-64
3,5-Nonadiyn-7-one, 1,2-dihydroxy-, isopropylidene derivative	EtOH	227(3.20),240(3.39), 253(3.67),267(3.81), 283(3.67)	39-1476-64
1,4-Pentanedione, 1-(p-methoxyphenyl)-	MeOH	264(4.37)	44-2205-65
3-Pentenoic acid, 4-(m-methoxyphenyl)-	n.s.g.	286(3.36)	35-1884-64
Phenol, 4-allyl-2-methoxy-, acetate	pet ether	275(3.49),295s(1.95)	65-2094-65
	EtOH-HCl	280(3.78),340(1.15)	65-2094-65
	EtOH	275(3.38),300s(1.37)	65-2094-65

Compound	Solvent	$\lambda_{max}(\log \epsilon)$	Ref.
Phenol, 2-methoxy-4-propenyl-, acetate, trans	EtOH-NaOEt	284(4.78),325(4.30)	65-2094-65
	10% H_2SO_4	278(4.18)	65-2094-65
$C_{12}H_{14}O_4$			
p-Benzoquinone, 2-acetyl-5-methoxy-3-propyl-	EtOH	268(4.16),367(2.89)	39-0411-64
[Bi-1-cyclobuten-1-yl]-3,3'-dione, 2,2'-diethoxy-	n.s.g.	365(4.10)	44-1445-64
2-Chromanpropionic acid, 2-hydroxy-	EtOH	275(3.36),281(3.32)	49-0220-65
3-Norcarene-2,5-dione, 1-acetyl-7-ethyl-4-methoxy-	EtOH	271(3.90)	39-0411-64
Propionic acid, 3-(p-methoxybenzoyl)-, methyl ester	EtOH	270(4.31)	65-1985-64
Propionic acid, 3-(3,4-methylenedioxyphenyl)-, ethyl ester	EtOH	287(--)	95-0857-65
Terephthalic acid, diethyl ester	EtOH	240(4.31)	23-2852-64
Terephthalic acid, 2,3-dimethyl-, dimethyl ester	hexane	241(4.01)	24-2953-64
Tropone, 4-acetyl-5-hydroxy-7-methoxy-2-ethyl-	EtOH	246(4.40),344(4.00), 375(4.03)	39-0411-64
$C_{12}H_{14}O_4S$			
Benzoic acid, o-[2-(methylsulfonyl)-propenyl]-, methyl ester, cis	ether	213(4.39),236s(4.04), 282(3.36)	24-3087-65
trans	ether	211(4.27),237s(3.98), 284(3.29)	24-3087-65
Benzoic acid, p-[2-(methylsulfonyl)-propenyl]-, methyl ester	ether	263(4.38)	24-3087-65
$C_{12}H_{14}O_5$			
6-Chromancarboxylic acid, 5,8-dihydroxy-2,2-dimethyl-	EtOH	213(4.28),225s(4.08), 267(3.96),323(3.73)	78-1331-64
1,2-Propanedione, 1-(3,4,5-trimethoxyphenyl)-	EtOH	295(3.74)	23-0113-64
Radicinin, dihydro-	EtOH	250s(3.46),310(4.07)	39-3234-64
$C_{12}H_{14}O_6S$			
Crotonic acid, 4,4'-thiobis[3-(hydroxymethyl)-2-methoxy-, di-γ-lactone	EtOH	229(4.3)	44-3560-64
$C_{12}H_{14}S_2$			
Butane, 2,2-bis(2-thienyl)-	EtOH	235.0(4.19)	22-2635-65
2,8-Decadiene-4,6-diyne, 2,9-bis-(methylthio)-, cis-cis	ether	294(4.40),320(4.44), 337(4.50),361(4.43)	24-1736-65
trans-trans	ether	301(4.23),322(4.26), 344(4.30),368(4.22)	24-1736-65
$C_{12}H_{15}Br$			
Dodeca-3,5,11-trien-1-yne, 1-bromo-, trans-trans	ether	272(4.51)	24-3010-65
$C_{12}H_{15}BrO_3S$			
Benzenesulfonic acid, p-bromo-, 2,2-dimethyl-3-buten-1-yl ester	EtOH	234(4.28)	44-0010-65
$C_{12}H_{15}BrO_7S$			
Glucopyranose, 1-[(p-bromophenyl)-sulfonyl]-1-deoxy-	MeOH	234(4.24)	44-1782-64

Compound	Solvent	λ_{max}(log ϵ)	Ref.
$C_{12}H_{15}ClN_2O_4$			
2,4,7-Trimethyl-1,5-benzodiazepinium perchlorate	MeOH-HCl	230(3.89),267(4.39), 276(4.36),335(3.03), 343(3.05),525(2.82)	39-3785-65
$C_{12}H_{15}ClN_2O_5$			
7-Methoxy-2,4-dimethyl-1,5-benzodiazepinium perchlorate	MeOH-HCl	229(3.87),270(4.18), 286(4.18),340(3.30), 347(3.31),497(3.04)	39-3785-65
$C_{12}H_{15}ClN_2S$			
2,4-Dimethyl-7-(methylthio)-1,5-benzodiazepinium chloride	MeOH-HCl	227(3.97),270(4.08), 278(4.08),309(3.86), 364(3.36),541(2.57)	39-3785-65
$C_{12}H_{15}ClO$			
Phenol, 2-chloro-4-cyclohexyl-	HCl	217(4.73)	44-2678-65
	NaOH	239(3.87)	44-2678-65
$C_{12}H_{15}ClO_4S$			
2,3-Benzo-7,7-dimethylbicyclo[2.2.1]-heptene-1-sulfonium perchlorate	n.s.g.	263s(2.84),268(2.95), 276(2.89)	88-3569-65
$C_{12}H_{15}FN_2$			
Indole, 3-(2-aminobutyl)-6-fluoro-	EtOH-HCl	218(4.48),271s(3.69), 278(3.62),283(3.74), 288s(3.71)	87-0274-64
$C_{12}H_{15}FN_4O_6$			
Lysine, N-(1-fluoro-2,4-dinitro-5-phenyl)-	H_2O	346(4.15)	37-3264-65
$C_{12}H_{15}F_7N_2$			
Cyclohexene, heptafluoro-1-(isopropylamino)-3-(isopropylimino)-	EtOH	305.5(4.42)	39-5748-64
$C_{12}H_{15}N$			
Azacyclohept-3(or 4)-ene, 4-phenyl-, hydrochloride	EtOH	242(4.10)	39-5130-64
3H-Azepine, 4,5,6,7-tetrahydro-2-phenyl-	EtOH-HCl	270(4.23)	95-0671-64
	EtOH	240(4.10)	95-0671-64
Butane, 1-cyano-3-p-tolyl-	EtOH	251s(2.29),258(2.49), 263(2.62),271(2.51)	78-2593-65
Indole, 3-tert-butyl-	EtOH	222(4.56),272s(3.71), 281(3.75),289(3.68)	35-3796-64
	H_2SO_4	235(3.75),240(3.73), 289(3.68)	35-3796-64
Indole, 1,3-diethyl-	EtOH	227(4.51),280s(3.68), 290(3.73)	35-3796-64
Indole, 2-methyl-3-propyl-	EtOH	225(4.58),274s(3.85), 282(3.88),289(3.84)	35-3796-64
	H_2SO_4	230(3.78),237s(3.72), 280(3.73)	35-3796-64
Indole, 3-methyl-1-propyl-	EtOH	228(4.49),291(3.72)	78-0989-65
Indole, 3-methyl-2-propyl-	EtOH	229(4.42),284(3.85), 292(3.82)	78-0989-65
Indole, 5-methyl-2-propyl-	EtOH	223(4.46),274(3.90), 284(3.88),295(3.73)	87-0204-65
3H-Indole, 3-methyl-3-propyl-	EtOH	253(4.11)	78-0989-65
	dil. HCl	233(3.70),238s(3.68), 282(3.61)	78-0989-65

Compound	Solvent	λ_{max}(log ε)	Ref.
Indolizine, 1,2,6,7-tetramethyl-	EtOH-HCl	225(4.38),250(3.96), 326(3.79)	39-1518-65
	EtOH	248(4.50),283s(3.28), 294(3.40),308(3.32), 358(3.16)	39-1518-65
Indolizine, 1,3,5,7-tetramethyl-	EtOH-HCl	217(4.38),240(3.78), 310(3.81)	39-0893-64
	EtOH	242(4.46),285(3.58), 294(3.74),307(3.78), 370(3.36)	39-0893-64
1-Pyrroline, 3,3-dimethyl-2-phenyl-	hexane	239(4.06)	39-2313-65
	EtOH	239(3.95)	39-2313-65
$C_{12}H_{15}NO$			
5(6H)-Benzocyclooctenone, 2-amino-7,8,9,10-tetrahydro-	EtOH	321(4.28)	44-2165-65
2-Hexenanilide	EtOH	270(4.16)	54-0581-65
3H-Indole, 2-ethoxy-3,3-dimethyl-	n.s.g.	253(3.89)	77-0368-64
5-Indolol, 1-ethyl-2,6-dimethyl-	MeOH	280(3.93),297(3.85), 309(3.68)	35-3878-64
Piperidine, 1-benzoyl-	C_6H_{12}	240s(3.66)	78-0035-65
Piperidine, 3,4-epoxy-1-methyl-4-phenyl-	iso-PrOH	253(2.20),258(2.27), 264(2.13)	44-0394-65
Propiophenone, 2-(1-aziridinyl)-2-methyl-	EtOH	243(3.91)	44-3146-64
3-Pyridinol, 1,2,3,6-tetrahydro-1-methyl-4-phenyl-	iso-PrOH	244(4.06)	44-0394-65
6(2H)-Quinolone, 3,4-dihydro-2,2,4-trimethyl-	MeOH	274(4.5),358(3.1)	24-2939-65
$C_{12}H_{15}NOS$			
Indole, 2-(ethylthio)-6-methoxy-3-methyl-	MeOH	303(4.06)	24-1727-65
$C_{12}H_{15}NOS_2$			
2-Butanone, 3-methyl-4-phenyl-4-(thiocarbamoylthio)-	EtOH	235(4.12),294(4.14)	39-4004-64
2-Butanone, 4-(N-methylthiocarbamoylthio)-4-phenyl-	H_2O	252(3.99),273(3.98)	39-4004-64
	EtOH	252(4.05),281(4.06)	39-4004-64
Carbamic acid, dimethyldithio-, ester with 2-mercapto-4'-methylacetophenone	EtOH	254(4.03)	44-1703-64
2-Thiazolidinethione, 4-hydroxy-3,5,5-trimethyl-4-phenyl-	MeOH	250s(3.93),274(4.22)	44-2146-64
$C_{12}H_{15}NO_2$			
3H-2-Benzazepine, 4,5-dihydro-7,8-dimethoxy-	EtOH-HCl	246(4.05),326(4.14), 363(4.09)	88-2419-64
	EtOH	236(4.19),273(4.16), 312(3.91)	88-2419-64
3-Buten-2-one, 4-p-phenetidino-	MeOH	345(4.41)	83-0321-64
Hydroperoxide, 2-methyl-3-propyl-3H-indol-3-yl-	EtOH	220(4.18),257(3.48)	78-0989-65
Hydroperoxide, 3-methyl-2-propyl-3H-indol-3-yl-	EtOH	218(4.19),269(3.48)	78-0989-65
Isoquinoline, 3,4-dihydro-6,7-dimethoxy-1-methyl-	EtOH-HCl	247(4.29),308(3.96), 362(3.90)	33-1945-65
	EtOH	230(4.42),278(3.84), 308(3.79)	33-1945-65
	EtOH-NaOH	229(4.45),278(3.87), 310(3.81)	33-1945-65

Compound	Solvent	λ_{max}(log ϵ)	Ref.
Isoquinoline, 3,4-dihydro-6,7-dimethoxy-1-methyl-, hydrobromide	EtOH-HCl	246(4.34),307(4.01), 363(3.93)	33-1945-65
	EtOH	242(4.24),307(3.96), 363(3.84)	33-1945-65
	EtOH-NaOH	228(4.49),276(3.91), 308(3.84)	33-1945-65
Norsecurinine, dihydro-	EtOH	214(4.16)	94-1520-64
Pyrrolidine, 3,4-epoxy-2-p-methoxyphenylmethyl-	MeOH	225(4.05),275(3.21), 282(3.15)	44-2334-65
$C_{12}H_{15}NO_2S_2$			
Carbamic acid, dimethyldithio-, ester with 2-mercapto-4'-methoxyacetophenone	EtOH	275(4.41)	44-1703-64
Ketene, (p-nitrophenyl)-, diethyl mercaptole	EtOH	254(4.06),371(4.21)	88-0245-64
$C_{12}H_{15}NO_3$			
Alanine, N-salicylidene-, ethyl ester	EtOH	214(4.35),256(4.09), 318(3.62),407(2.21)	35-1757-65
7-Isoquinolinol, 3,4-dihydro-6,8-dimethoxy-1-methyl-	pH 2	247(4.10),317(4.15), 364s(3.64)	49-0025-65
	pH 12	251(4.38),291s(3.72), 347(3.59)	49-0025-65
	EtOH	247(4.08),317(4.12), 370s(3.57)	49-0025-65
Methacrylamide, N,N-dimethacryloyl-	MeOH	221(4.34)	88-0023-65
Salicylamide, N-ethyl-o-methoxyacetyl-	pH 1-7	265(2.98)	78-3019-65
$C_{12}H_{15}NO_4$			
Acetanilide, 2'-glyoxyloyl-, aldehydo-, dimethyl acetal	n.s.g.	331(3.62)	49-0889-65
3H-Azepine-3,6-dicarboxylic acid, 2,7-dimethyl-, dimethyl ester	hexane	243(3.66),280(3.80)	39-2411-65
4H-Azepine-3,6-dicarboxylic acid, 2,7-dimethyl-, dimethyl ester	EtOH	216(4.32),275(3.56), 328(3.46)	39-2411-65
4H-m-Dioxino[4,5-c]pyridine-5-carboxylic acid, 2,2,8-trimethyl-, methyl ester	MeOH-acid	235s(3.77),302(3.92)	44-2663-64
	MeOH-base	234s(3.87),296(3.81)	44-2663-64
Isocarbostyril, 3,4-dihydro-6,7,8-trimethoxy-	pH 1	264(4.02),297s(3.38)	33-2089-64
	pH 13	264(4.01),296s(3.34)	33-2089-64
	iso-PrOH	219(4.51),259(3.99), 295s(3.26)	33-2089-64
3,4-Pyridinedicarboxylic acid, 6-sec-butyl-5-methyl-	EtOH	274(3.72)	25-0186-65
Pyrrole-1-acrylic acid, 3-carboxy-β,2-dimethyl-, 3-ethyl ester	EtOH	248(3.01)	39-2411-65
$C_{12}H_{15}NO_5$			
Pyrrole-3-carboxylic acid, 4-acetoxy-1-acetyl-2-methyl-, ethyl ester	EtOH	223(4.33),241(4.02)	23-1524-64
$C_{12}H_{15}NO_5S$			
Benzenesulfonamide, N-acetonyl-o-carbethoxy-	EtOH	225(3.85),276(3.15)	44-2241-65
Benzenesulfonic acid, p-nitro-, 5-hexenyl ester	EtOH	250(4.07)	35-1959-64
Benzenesulfonic acid, p-nitro-, 4-methyl-4-pentenyl ester	EtOH	250(4.07)	35-5593-64

Compound	Solvent	$\lambda_{max}(\log \epsilon)$	Ref.
$C_{12}H_{15}NO_7$			
3-Carboxy-1-D-galactosylpyridinium hydroxide, inner salt	H_2O	265(3.68)	39-0610-65
	M KCN	315(3.68)	39-0610-65
3-Carboxy-1-D-glucosylpyridinium hydroxide, inner salt	H_2O	265.5(3.67)	39-0610-65
	M KCN	315(3.72)	39-0610-65
$C_{12}H_{15}N_3$			
3,5-Pyridinedicarbonitrile, 2-butyl-1,2-dihydro-4-methyl-	EtOH	217(4.28),258(3.90), 372(3.60)	73-2609-65
$C_{12}H_{15}N_3O_2$			
Pyruvaldehyde, 2-O-acetyloxime, 1-(methylphenylhydrazone)	EtOH	236(4.06),330(4.46)	24-0725-64
$C_{12}H_{15}N_3O_2S$			
Indole-2-carboxamide, N,1-dimethyl-3-(methylaminosulfinyl)-	EtOH	217(4.48),225s(4.46), 291(4.22)	44-0178-64
$C_{12}H_{15}N_3O_3$			
Salicylic acid, 5-nitro-, N-amino-piperidine derivative	MeOH	294(4.44)	24-1631-64
	MeOH-NaOH	346(4.17),423(4.11)	24-1631-64
$C_{12}H_{15}N_3O_3S$			
Indole-2-carboxamide, N,1-dimethyl-3-(methylsulfamoyl)-	EtOH	215(4.61),282(4.04)	44-0178-64
$C_{12}H_{15}N_3O_4$			
1,2,3,4-Butanetetrol, 1-(6-amino-2-quinoxalinyl)-	EtOH	258(4.48),333s(2.84), 388(3.75)	5-0146-65D
1,2,3,4-Butanetetrol, 1-(7-amino-2-quinoxalinyl)-	EtOH	259(4.42),333s(3.15), 385(3.81)	5-0146-65D
$C_{12}H_{15}N_5$			
Guanidine, (4,6,7-trimethyl-2-quinazolinyl)-	pH 1	249(4.82),328(3.59)	44-0285-65
	pH 10	250(4.59),264(4.48), 282s(4.14),341(3.60)	44-0285-65
	EtOH	249(4.76),264s(4.27), 331(3.53),340(3.59)	44-0285-65
$C_{12}H_{15}N_5O_6$			
Purin-8(9H)-one, 6-acetamido-9-β-D-ribofuranosyl-	pH 1	239(3.76),289(4.11)	35-1772-65
	pH 11	265(3.85),301(4.13)	35-1772-65
$C_{12}H_{15}N_5O_7$			
Glycine, N-(1,6-dihydro-6-oxo-9-β-D-ribofuranosyl-9H-purin-2-yl)-	pH 1	257(4.17),277s(3.98)	35-3752-65
	pH 11	257(4.14),270s(4.06)	35-3752-65
$C_{12}H_{15}P$			
Phosphine, diethyl(phenylethynyl)-	EtOH	220(4.08),237(4.08), 247(4.11),258(4.08), 282(3.88)	28-1537-64A
$C_{12}H_{16}$			
Naphthalene, 1,2,3,4-tetrahydro-1,3-dimethyl-	EtOH	265.5(2.74)	28-1259-64A
$C_{12}H_{16}BrNO$			
4-Piperidinol, 3-bromo-1-methyl-4-phenyl-, hydrobromide	H_2O	250(2.17),256(2.22), 261(2.07),268(1.77)	44-0394-65

Compound	Solvent	$\lambda_{max}(\log \epsilon)$	Ref.
$C_{12}H_{16}Br_2O$			
2,4-Cyclohexadien-1-one, 4,6-dibromo-2,6-diisopropyl-	n.s.g.	345(3.32)	70-0336-65
2,5-Cyclohexadien-1-one, 4,4-dibromo-2,6-diisopropyl-	n.s.g.	248(4.26)	70-0336-65
$C_{12}H_{16}Br_3NO_2$			
9,9-Dibromo-10-ethoxy-7,8,9,10-tetrahydro-10-hydroxy-6H-pyrido[1,2-a]azepinium bromide	H_2O	270(3.90)	44-1523-65
$C_{12}H_{16}ClNO$			
(2-Benzoylvinyl)trimethylammonium chloride	EtOH	207(4.95),216s(4.86),267(4.89)	44-0385-64
$C_{12}H_{16}ClNO_2$			
1,3-Dioxolane, 2-(p-aminophenethyl)-2-(chloromethyl)-	EtOH	238(4.00),276(3.92)	87-0035-65
$C_{12}H_{16}ClNO_4$			
8-Dimethylimoniomethyl-8-methylheptafulvene perchlorate	MeCN	241s(3.90),263(3.96),434(4.41)	24-2050-64
7-Pentamethyleneimoniotropylidene perchlorate	MeCN	243(4.40),337(4.32)	24-2050-64
3,5-Pyridinedicarboxylic acid, 4-(chloromethyl)-1,4-dihydro-2,6-dimethyl-, dimethyl ester	EtOH	232(4.26),349(3.87)	39-2411-65
$C_{12}H_{16}ClNO_7S$			
Glucosamine, N-p-chlorophenylsulfonyl-	n.s.g.	234(4.17),267(2.79),278(2.53)	95-0399-65
$C_{12}H_{16}ClNO_9$			
1-(1,2-Dicarboxy-2-methoxyethyl)pyridinium perchlorate, dimethyl ester	H_2O	256s(3.53),261(3.57),267(3.48)	39-2676-64
$C_{12}H_{16}ClN_5$			
Pteridine, 6-chloro-3,4-dihydro-4-(cyclohexylamino)-	$CHCl_3$	274(3.63),311(3.83),324(3.75)	39-4920-64
$C_{12}H_{16}Cl_2N_2O_2$			
p-Benzoquinone, 2,5-dichloro-3,6-bis(isopropylamino)-	methyl cellosolve	224(4.33),356(4.43),529(4.33)	78-1889-64
$C_{12}H_{16}FNO_7S$			
Glucosamine, N-p-fluorophenylsulfonyl-	n.s.g.	224(3.93),254(2.60),260(2.58)	95-0399-65
$C_{12}H_{16}F_2N_2O_2$			
p-Benzoquinone, 2,5-difluoro-3,6-bis(isopropylamino)-	methyl cellosolve	223(4.47),353(4.44),544(2.35)	78-1889-64
$C_{12}H_{16}N_2$			
Benzimidazole, 2-isopropyl-5,6-dimethyl-	EtOH	246(3.71),282(3.88),286(3.81)	44-0476-64
	pH 2	274(3.92),283(3.91)	44-0476-64
	pH 12	243(3.41),281(3.75),286(3.68)	44-0476-64
Cyclohexaneazobenzene	hexane	263(4.01),410(2.13)	39-2788-65

Compound	Solvent	$\lambda_{max}(\log \epsilon)$	Ref.
2,6-Methanobenzo[d][1,3]diazocine, 1,2,3,4,5,6-hexahydro-3-methyl-	EtOH-HCl	241(4.07),291(3.32)	95-0871-65
	EtOH	250(4.01),302(3.34)	95-0871-65
Noreseroline, deoxy-	EtOH-6N HCl	234(3.67),240(3.63), 284(3.58)	39-5510-64
	EtOH-0.01N HCl	246(3.97),298(3.40)	39-5510-64
	EtOH	252(4.02),302(3.44)	39-5510-64
Pyrazine, 3,4,5,6-tetrahydro-3,3-dimethyl-2-phenyl-	5N HCl	254(3.81)	44-3574-64
Quinazoline, 4-tert-butyl-3,4-dihydro-	pH 7.0	213(4.22),218s(4.19), 276(3.67)	39-5360-65
	pH 11.5	218s(4.00),223(4.06), 287(3.74)	39-5360-65
Tryptamine, N(a),N(b)-dimethyl-	EtOH	223(4.55),287(3.74)	39-5510-64
	H_2SO_4	235(3.50),241(3.49), 290(3.64)	39-5510-64
Tryptamine, N(b)-dimethyl-	EtOH	223(4.55),282(3.76), 291(3.69)	39-5510-64
	H_2SO_4	236(3.59),242(3.57), 295(3.64)	39-5510-64
$C_{12}H_{16}N_2O$			
2,5-Cyclohexadien-1-one, 4-diazo-2,6-diisopropyl-	aq. EtOH	260(3.97),365(4.57)	30-1101-64
	EtOH-acid	325(4.27)	30-1101-64
2(1H)-Quinoxalinone, 3,4-dihydro-1,3,3,4-tetramethyl-	EtOH	225(4.57),265(3.53), 307(3.62)	12-0877-64
$C_{12}H_{16}N_2O_2$			
Hexamethyleneimine, N-(p-nitrophenyl)-	n.s.g.	417(4.41)	5-0109-65I
Hydroperoxide, 1-(phenylazo)-cyclohexyl-	hexane	268(4.02),413(2.15)	39-2788-65
Piperidine, 2-methyl-1-(p-nitrophenyl)-	n.s.g.	413(4.40)	5-0109-65I
Piperidine, 3-methyl-1-(p-nitrophenyl)-	n.s.g.	412(4.33)	5-0109-65I
Piperidine, 4-methyl-1-(p-nitrophenyl)-	n.s.g.	413(4.35)	5-0109-65I
Pyrrole-3-carboxylic acid, 4-methyl-5-(2-pyrrolin-2-yl)-, ethyl ester	EtOH-HCl	313(4.45)	44-2727-64
	EtOH-NaOH	281(4.13)	44-2727-64
$C_{12}H_{16}N_2O_2S_2$			
p-Dithiin-2,3-dicarboximide, 5,6-dihydro-N-(piperidinomethyl)-	EtOH	252(4.00),407(3.51)	56-1713-65
$C_{12}H_{16}N_2O_3$			
Barbituric acid, 5-cyclohexenyl-1,5-dimethyl-	EtOH	224(3.8)	83-0885-65
Hexanoic acid, 4-hydroxy-2-oxo-, phenylhydrazone	n.s.g.	293(4.18),317(4.23)	39-0141-64
$C_{12}H_{16}N_2O_4$			
Tryptamine, 6-hydroxy-5-methoxy-, formate	EtOH	220(4.40),280(3.72), 304s(3.92)	4-0387-65
$C_{12}H_{16}N_2O_4S$			
Malonic acid, [[(4-methyl-2-thiazolyl)-amino]methylene]-, diethyl ester	EtOH	333(4.4)	70-1481-64
$C_{12}H_{16}N_2O_5$			
Uridine, 5-allyl-2'-deoxy-	pH 1	267(3.99)	4-0313-65
	pH 7	266.5(3.99)	4-0313-65
	pH 13	266(3.86)	4-0313-65

Compound	Solvent	λ_{max}(log ε)	Ref.
Uridine, 5-allyl-2'-deoxy-,			
α-anomer	pH 7	268(4.00)	87-0567-64
	pH 13	267(3.87)	87-0567-64
	pH 14	267(3.88)	87-0567-64
β-anomer	pH 7	268(3.98)	87-0567-64
	pH 13	267(3.84)	87-0567-64
	pH 14	267(3.86)	87-0567-64
$C_{12}H_{16}N_2O_6$			
Ribofuranosylamine, N-(3-nitro-p-tolyl)-	EtOH	240(4.36),372(3.54)	65-0675-65
Uracil, 1-(2',3'-isopropylidene-α-D-ribofuranosyl)-	H_2O	262(4.01)	94-1471-64
$C_{12}H_{16}N_2O_7$			
1-L-Arabinosyl-3-carbamoyl-pyridinium formate	pH 7	266(3.67)	39-0610-65
	M KCN	325(3.75)	39-0610-65
3-Carbamoyl-1-D-ribopyranosyl-pyridinium formate	pH 7	266(3.67)	39-0610-65
	M KCN	324(3.82)	39-0610-65
3-Carbamoyl-1-D-xylosylpyridinium formate	pH 7	266(3.67)	39-0610-65
	M KCN	323.5(3.81)	39-0610-65
$C_{12}H_{16}N_4$			
Pyrazolo[1,5-a]pyrimidine, 4-allyl-4,7-dihydro-7-imino-2,3,6-trimethyl-	EtOH	235(4.29),265s(3.93), 314(3.85)	94-0142-65
$C_{12}H_{16}N_4O$			
Pyrazolo[1,5-a]pyrimidine, 7-(N-ethyl-acetamido)-2,3-dimethyl-	EtOH	239(4.64),285s(3.20), 295(3.26),306(3.16), 349(3.24)	94-0142-65
$C_{12}H_{16}N_4O_2$			
Pyrazolo[1,5-a]pyrimidine-7-carbamic acid, 2,3,6-trimethyl-, ethyl ester	EtOH	238(4.66),283s(3.21), 293(3.28),308s(3.21), 342(3.39)	94-1207-65
Pyrazolo[1,5-a]pyrimidine-6-carboxylic acid, 7-amino-5-ethyl-2-methyl-, ethyl ester	EtOH	234(4.51),309(4.19)	95-0442-65
Pyrazolo[1,5-a]pyrimidine-6-carboxylic acid, 7-amino-2,3,5-trimethyl-, ethyl ester	EtOH	236(4.47),316(4.19)	95-0442-65
$C_{12}H_{16}N_4O_3$			
Salicylic acid, 5-nitro-, 4-amino-1-methylpiperazine derivative	MeOH	290(4.45)	24-1631-64
	MeOH-NaOH	347(4.15),416(4.13)	24-1631-64
$C_{12}H_{16}N_4O_7$			
Lysine, N^6-(2,4-dinitrophenyl)-5-hydroxy-	0.48N HCl	345(4.15)	37-3264-65
	0.3N NaOH	333(4.18),415(3.94)	37-3264-65
$C_{12}H_{16}N_6$			
Acetone, (5-p-toluidino-s-triazol-3-yl)hydrazone	EtOH	257(4.48)	39-3912-65
$C_{12}H_{16}N_6O_8$			
o-Phenylenediamine, N,N'-bis-(2,2-dinitropropyl)-	hexane-2% dioxan	240(3.96),292(3.46)	44-0354-65
p-Phenylenediamine, N,N'-bis-(2,2-dinitropropyl)-	MeCN	252(4.22),320(3.42)	44-0354-65

Compound	Solvent	$\lambda_{max}(\log \epsilon)$	Ref.
$C_{12}H_{16}O$			
Acetophenone, p-tert-butyl-	C_6H_{12}	256(4.15)	78-1555-64
Anisole, 4-(3-butenyl)-3-methyl-	EtOH	226(3.95),277(3.22), 284(3.18)	35-5148-65
Bicyclo[2.2.1]hept-5-ene, 3-butyryl-2-methylene-	EtOH	209(3.85),287(1.72)	28-0404-64B
Bicyclo[2.2.1]hept-5-ene, 3-isobutyryl-2-methylene-	EtOH	208(3.86),293(1.77)	28-0404-64B
2,4-Cyclohexadien-1-one, 6-allyl-2,4,6-trimethyl-	MeOH	319(3.68)	35-5115-65
2,4-Cyclohexadien-1-one, 6-(2-butenyl)-2,6-dimethyl-, trans	MeOH	309(3.62)	35-5115-65
2,5-Cyclohexadien-1-one, 4-allyl-2,4,6-trimethyl-	MeOH	244(4.15)	35-5115-65
2,5-Cyclohexadien-1-one, 4-(2-butenyl)-2,4-dimethyl-, trans	MeOH	242(4.17)	35-5115-65
2,5-Cyclohexadien-1-one, 4-(3-methyl-2-butenyl)-4-methyl-	EtOH	235(4.12)	33-0094-65
1,4a-Cyclo-2(1H)-naphthalenone, 4a,5,6,7,8,8a-hexahydro-3,8a-dimethyl-	EtOH	234(3.72),270(3.28)	35-4053-64
as-Indacen-4(1H)-one, 2,3,5,5a,6,7,8,8a-octahydro-	n.s.g.	253(4.00)	28-1844-64A
Isobenzofuran, 1,3-dihydro-4,5,6,7-tetramethyl-	MeOH	283(2.58)	24-2331-65
Naphthalene, 1,2,3,4-tetrahydro-7-methoxy-1-methyl-	n.s.g.	225(4.04),280(3.38), 287(3.34)	39-4521-64
1(5H)-Naphthalenone, 6,7,8,8a-tetra-hydro-3,8a-dimethyl-	EtOH	316(3.63)	35-4053-64
2(3H)-Naphthalenone, 4,4a,5,6-tetra-hydro-4a,8-dimethyl-	EtOH	287(4.38)	44-3642-65
2(4aH)-Naphthalenone, 5,6,7,8-tetra-hydro-1,4a-dimethyl-	EtOH	238(4.04)	44-3110-64
2(4aH)-Naphthalenone, 5,6,7,8-tetra-hydro-3,4a-dimethyl-	EtOH	242(4.23)	35-4053-64
Spiro[cyclopentane-1,6'-(1',4'-di-methylbicyclo[3.1.0]hex-3'-en-2'-one]	EtOH	232(3.62),265(3.35), 320(2.60)	35-4053-64
Spiro[4.5]deca-3,6-dien-2-one, 3,6-dimethyl-	EtOH	246(4.19)	78-2183-65
Tricyclo[5.3.0.0^{3,7}]decan-4-one, 3-methyl-2-methylene-	EtOH	304(2.23)	44-3647-65
Tricyclo[3.3.2.0^{1,5}]dec-9-en-2-one, 9,10-dimethyl-	EtOH	225s(3.08),307(2.32)	44-3647-65
$C_{12}H_{16}OS$			
Acetophenone, p-(tert-butylthio)-	hexane	248(4.2),305(3.5)	39-4599-65
$C_{12}H_{16}O_2$			
Carvacrol, acetate	pet ether	265(2.88)	65-2094-65
Isoprethrolone, methyl ether	EtOH	260(4.27),270(4.31), 279s(4.24)	39-0888-64
Naphthalene, 1,4,5,8-tetrahydro-2,7-dimethoxy-	EtOH	278(1.64)	44-3209-65
2(3H)-Naphthalenone, 4,4a,5,6,7,8-hexa-hydro-3-(hydroxymethylene)-4a-methyl-	EtOH	250(4.02),306(3.75)	88-4585-65
1-Naphthoic acid, 4,4a,5,6,7,8-hexa-hydro-, methyl ester	EtOH	232s(3.48),286(3.45)	44-3679-65
1-Naphthol, 4-ethoxy-5,6,7,8-tetra-hydro-	EtOH	220s(3.84),288(3.48)	24-1926-64
	NaOH	241(3.89),301(3.54)	24-1926-64

Compound	Solvent	$\lambda_{max}(\log \epsilon)$	Ref.
1,6-Octadien-4-yne, 8,8-ethylenedioxy-2,6-dimethyl-	MeOH	230(4.12)	5-0026-64I
4-Pentynoic acid, 5-(1-cyclohexen-1-yl)-3-methyl-	EtOH	228(4.14)	22-2541-64
Pyrethrolone, methyl ether	EtOH	225(4.33)	39-0888-64
Spiro[4.5]deca-6,8-diene-7-carboxylic acid, methyl ester	EtOH	284(3.52)	44-3679-65
Thymol, acetate	pet ether	268(2.78)	65-2094-65
Thymol, 8(9)-dehydro-4-hydroxy-, dimethyl ether	pet ether	296(3.4)	83-0182-64
$C_{12}H_{16}O_2Pd$			
Palladium, (2,4-pentanedionato)-2,4-cycloheptadien-1-yl-	hexane	211(4.12),265(4.04), 291s(3.93)	39-5002-64
$C_{12}H_{16}O_2S$			
Benzothiophene-3-carboxylic acid, 4,5,6,7-tetrahydro-2-methyl-, ethyl ester	ether	215(4.20),243(3.98)	24-2109-64
$C_{12}H_{16}O_3$			
Bicyclo[3.3.1]non-3-ene-1-acetic acid, 5-methyl-2-oxo-	EtOH	232(3.91)	39-1344-65
Bicyclo[3.3.1]non-3-ene-1-carboxylic acid, 5-methyl-2-oxo-, methyl ester	EtOH	229(3.91)	39-1344-65
5-Chromanol, 8-methoxy-2,2-dimethyl-	EtOH	286(3.57)	78-1317-64
	EtOH-KOH	297(3.64)	78-1317-64
4,6-Decadiyn-1-oic acid, 8-hydroxy-8-methyl-, methyl ester	EtOH	228(3.56),241(3.58), 254(3.39)	70-1237-65
Propiophenone, 3',4'-dimethoxy-2'-methyl-	MeOH	267(4.05)	44-3028-64
Propiophenone, 3',4'-dimethoxy-5'-methyl-	MeOH	268(4.04),300s(3.67)	44-3028-64
$C_{12}H_{16}O_4$			
Benzoic acid, 2,4-dihydroxy-6-pentyl-	EtOH	262(4.11),301(3.65)	1-1677-65
Bicyclo[2.2.0]hex-2-ene-1,4-dicarboxylic acid, 2,3-dimethyl-, dimethyl ester	hexane	none	24-2953-64
1,3-Cyclohexadiene-1,4-dicarboxylic acid, 2,3-dimethyl-, dimethyl ester	hexane	301(4.06)	24-2953-64
3-Cyclohexene-1-acrylic acid, 2-carboxy-α,2-dimethyl-	n.s.g.	220(4.14)	39-5815-64
3-Cyclopentene-1,1-dicarboxylic acid, 2-methylene-, diethyl ester	EtOH	230(4.17)	70-1460-65
2,4-Hexadienedioic acid, 2,3-diethyl-4-hydroxy-, γ-lactone, ethyl ester	C_6H_{12}	274(4.45)	35-2819-64
Malonic acid, (2-heptynylidene)-, dimethyl ester	EtOH	268(4.26)	78-2701-64
Malonic acid, cyclohexylidene-, cyclic isopropylidene ester	EtOH-HCl	243(4.007)	49-1283-64
	EtOH-NaOH	262(4.22)	49-1283-64
1,4-Naphthoquinone, octahydro-5-hydroxy-, acetate	EtOH	286(1.60)	44-1341-64
$C_{12}H_{16}O_5$			
Acetic acid, 3,4,5-trimethoxyphenyl-, methyl ester	EtOH	277(3.82)	78-1411-65
3-Cyclopentene-1-carboxylic acid, 3-hydroxy-1-methyl-2-oxo-4-propionyl, ethyl ester	EtOH	293(4.06),360(2.71)	44-3520-64
	EtOH-base	360(4.14)	44-3520-64

Compound	Solvent	λ_{max}(log ϵ)	Ref.
$C_{12}H_{16}O_5S_3$			
β-L-Idofuranose, 1,2-O-isopropylidene-5,6-dithio-, cyclic 5,6-trithiocarbonate, 3-acetate	MeOH	316(4.04)	44-3071-65
$C_{12}H_{16}O_6$			
Trimethyldegeranylmelicopol	EtOH	280(3.80)	12-2021-65
$C_{12}H_{16}O_6S_3$			
β-L-Idofuranose, 1,2-O-isopropylidene-5,6-dithio-, cyclic 5,6-trithiocarbonate, 3-acetate, thiono-oxide	MeOH	285(3.64),347(4.01)	44-3071-65
$C_{12}H_{16}S$			
Benzo[b]thiophene, 3-ethyl-2,3-dihydro-2,2-dimethyl-	n.s.g.	250(3.97),265s(3.69), 288s(3.19),300s(3.04)	88-3569-65
$C_{12}H_{16}S_3$			
Decalin-2β,3α-trithiocarbonate, 9-methyl-, trans	dioxan	227(3.46),305s(4.2), 317(4.24),454(1.94)	78-0583-65
6a-Thiathiophthene, 2,5-diethyl-3,4-trimethylene-	C_6H_{12}	235(4.31),261(4.73), 481(3.90)	24-1732-64
$C_{12}H_{17}$			
Pentamethylbenzyl (carbonium ion)	H_2SO_4	220(3.93),240(3.95), 265(3.80),330(4.16), 520(3.28)	88-2987-64
$C_{12}H_{17}BrOS$			
Diethylphenacylsulfonium bromide	acid	252.5(4.21)	28-0585-64A
$C_{12}H_{17}Br_2NO_2$			
9-Bromo-10-ethoxy-7,8,9,10-tetrahydro-10-hydroxy-6H-pyrido[1,2-a]azepinium bromide	H_2O	267(3.86)	44-1523-65
$C_{12}H_{17}ClO$			
2-Cyclohexen-1-one, 6-(3-chloro-2-butenyl)-2,6-dimethyl-	EtOH	235(3.94)	44-3642-65
$C_{12}H_{17}ClO_4S$			
2,3,4,5-Tetrahydro-2,2-dimethyl-1-phenylthiophenium perchlorate	n.s.g.	255s(3.06),258(3.04), 261(3.03),263(3.0), 270(2.95),277(2.72)	88-3569-65
$C_{12}H_{17}Cl_2N_3O_4$			
1(2H)-Pyrimidinebutyric acid, 5-[bis-(2-chloroethyl)amino]-3,4-dihydro-2,4-dioxo-	pH 1	266(3.81),304s(3.49)	87-0187-65
	pH 7	262s(3.74),304s(3.51)	87-0187-65
	pH 13	253s(3.80)	87-0187-65
$C_{12}H_{17}Cl_2N_5S$			
9H-Purine, 9-[2-[bis(2-chloroethyl)-amino]ethyl]-6-(methylthio)-	pH 1	224(4.03),293(4.22)	87-0182-65
	pH 11	230(3.79),290(4.24)	87-0182-65
$C_{12}H_{17}Cl_3O_4$			
6,10-Dioxaspiro[4.5]dec-3-en-2-one, 1,3,4-trichloro-8,8-dimethyl-, dimethyl acetal	n.s.g.	215s(3.84),251(2.94)	39-4744-65

Compound	Solvent	$\lambda_{max}(\log \epsilon)$	Ref.
$C_{12}H_{17}IN_2S$			
N-(p-Vinylphenyl)-N',N',S-trimethyl-isothiuronium iodide	EtOH	274s(3.05),303(3.18)	44-1926-65
$C_{12}H_{17}N$			
Carbazole, 1,2,3,4,5,6,7,8-octahydro-	EtOH-HCl	211(3.63),274(3.77)	78-0515-64
	EtOH	214(3.81)	78-0515-64
Indoline, 3-methyl-3-propyl-	EtOH	243(3.88),288(3.50)	78-0989-65
1-Naphthonitrile, 2,3,4,4a,5,6,7,8-octahydro-4a-methyl-	EtOH	224(3.84)	78-2641-65
Pyrrolidine, 1-(5-norbornen-2-ylidene-methyl)-	EtOH	262.5(3.85)	28-1541-64A
$C_{12}H_{17}NO$			
1H,8H-Benzo[ij]quinolizin-8-one, 2,3,5,6,7,9,10,10a-octahydro-, hydrochloride	H_2O	312(4.38)	24-1354-64
Propylamine, N-(o-methoxybenzyli-dene)-1-methyl-	EtOH	250(4.40),304(4.00)	44-2265-64
$C_{12}H_{17}NO_2$			
Azepine-N-carboxylic acid, 2,4,6-tri-methyl-, ethyl ester	hexane	215(4.44),293(3.16)	88-1733-64
1-Pentanone, 5-ethyl-1-(2-pyridyl)-	EtOH	230(3.94),270(3.68)	44-1523-65
2-Pyrone, 5,6-dimethyl-4-piperidino-	MeOH	227(4.08),284(4.03), 300s(4.00)	24-1266-64
$C_{12}H_{17}NO_2S$			
2-Benzothiazolinone, 3-methyl-, diethyl acetal	EtOH	300(3.66)	22-2868-64
$C_{12}H_{17}NO_3$			
Anhalidine	pH 1	225s(4.03),270(2.95), 277(2.84)	33-2089-64
	pH 13	240s(3.97),285(3.47)	33-2089-64
	iso-PrOH	227s(4.06),272(2.98), 280s(3.85)	33-2089-64
Anhalonidine	pH 1	270(2.79),278s(2.72)	33-2089-64
	pH 13	245s(3.84),286(3.38)	33-2089-64
	iso-PrOH	270(2.87),278s(2.81)	33-2089-64
2,5-Cyclohexadien-1-one, 6-tert-butyl-2,4-dimethyl-4-nitro-	hexane	233(4.05)	70-1666-64
6-Isoquinolinol, 1,2,3,4-tetrahydro-7,8-dimethoxy-1-methyl-	pH 1	225s(3.95),274s(3.11), 280(3.15)	49-1409-65
	pH 2	230s(3.87),275s(3.22), 282(3.26)	49-0025-65
	pH 12	250(3.86),297(3.62)	49-0025-65
	pH 13	237s(3.95),295(3.54)	49-1409-65
	EtOH	230s(3.87),275s(3.28), 282(3.30)	49-0025-65
	iso-PrOH	230s(3.95),275s(3.11), 282(3.15)	49-1409-65
7-Isoquinolinol, 1,2,3,4-tetrahydro-6,8-dimethoxy-1-methyl-, hydro-chloride	pH 2	228s(3.9),282(3.30)	49-0025-65
	pH 12	249(3.83),297(3.60)	49-0025-65
	EtOH	228s(3.90),281(3.29)	49-0025-65
2-Pyrone, 5-ethyl-6-methyl-4-morpholino-	MeOH	207(3.67),289(4.40)	24-1266-64
$C_{12}H_{17}NO_4$			
Alanine, 3-(3,4-dimethoxyphenyl)-α-methyl-	MeOH	233(3.98),280(3.51)	44-1424-64

Compound	Solvent	λ_{max}(log ϵ)	Ref.
Alanine, 3-(3,4-dimethoxyphenyl)- -methyl-, hydrochloride	MeOH	231(3.91),278(3.43)	44-1424-64
1H-Azepine-3,6-dicarboxylic acid, 4,5-dihydro-2,7-dimethyl-, dimethyl ester	hexane	226(4.14),252s(3.58), 321(4.13)	39-2411-65
$C_{12}H_{17}NO_7S$			
Glucosamine, N-phenylsulfonyl-	n.s.g.	224(4.01),259(2.85), 266(2.96),272(2.88)	95-0399-65
$C_{12}H_{17}NS$			
2H-1,3-Thiazine, tetrahydro- 4,4-dimethyl-2-phenyl-	EtOH	214(3.72)	44-3314-64
	EtOH-acid	214(4.09),257(3.80)	44-3314-64
$C_{12}H_{17}N_3$			
Indole, 6-amino-3-(2-aminobutyl)-	EtOH-HCl	225(4.51),276(3.66), 291(3.62),304s(3.48)	87-0274-64
$C_{12}H_{17}N_3O_2$			
Acetophenone, p-isopropoxy-, semicarbazone	C_6H_{12}	266(4.26)	78-1555-64
Pyrethrolone, semicarbazone	EtOH	229(4.43),265(4.33)	39-5225-64
$C_{12}H_{17}N_3O_3$			
Salicylaldehyde, 5-nitro-, butylmethyl- hydrazone	MeOH	297(4.47)	24-1631-64
	MeOH-NaOH	342(4.24),426(4.03)	24-1631-64
Salicylaldehyde, 5-nitro-, 1-sec-butyl- 1-methylhydrazone)	MeOH	297(4.48)	24-2713-64
	MeOH-NaOH	338(4.27),427(4.02)	24-2713-64
Salicylaldehyde, 5-nitro-, 1-tert- butyl-1-methylhydrazone)	MeOH	295(4.48)	24-2713-64
	MeOH-NaOH	337(4.24),426(4.02)	24-2713-64
Salicylaldehyde, 5-nitro-, (1-isobutyl- 1-methylhydrazone)	MeOH	297(4.46)	24-2713-64
	MeOH-NaOH	340(4.26),427(4.01)	24-2713-64
$C_{12}H_{17}N_3O_3S_2$			
4-Morpholinecarboximidic acid, N-sulfa- nilylthio-, methyl ester	EtOH	260.0(4.27)	95-0391-65
$C_{12}H_{17}N_3O_4$			
Agaritine	pH 7.0	238(4.06),278(3.1)	37-2267-64
$C_{12}H_{17}N_3O_4S_2$			
Glycine, N-[1-(methylthio)-N-sulfa- nilylformimidoyl]-, ethyl ester	EtOH	275.5(4.32)	95-0391-65
$C_{12}H_{17}N_5$			
Formamidine, N,N-dimethyl-N'-(2,3,6- trimethylpyrazolo[1,5-a]pyrimidin- 7-yl)-	EtOH	242(4.44),274(4.05), 371(3.85)	94-1207-65
1,2,4-Triazolo[1,5-a]pyrimidine, 7-cyclohexylamino-5-methyl-	MeOH	304(4.35)	65-0204-64
1,2,4-Triazolo[1,5-c]pyrimidine, 2-allylamino-7-methyl-5-propyl-	MeOH	234(4.66),297s(3.38)	39-3357-65
$C_{12}H_{17}N_5O$			
Acetamide, 2-[2,3-dimethylpyrazolo- [1,5-a]pyrimidin-7-yl)amino]- N,N-dimethyl-	EtOH	236(4.64),280s(3.32), 289(3.38),298s(3.29), 346(3.63)	94-1207-65

Compound	Solvent	$\lambda_{max}(\log \epsilon)$	Ref.
$C_{12}H_{17}N_5O_2$			
s-Triazolo[2,3-c]pyrimidine-2-carbamic acid, 7-methyl-5-propyl-, methyl ester	MeOH	230(4.57),258s(3.61), 294s(3.34)	39-3357-65
$C_{12}H_{17}N_5O_2S$			
Adenine, 9-(2',3'-dideoxy-2'-(ethylthio)-β-D-arabinofuranosyl)-	pH 1	259(4.16)	44-2854-65
	pH 13	261(4.19)	44-2854-65
	EtOH	261(4.18)	44-2854-65
1-Pyrrolidinecarboxamide, N-(4-pyrrolidinylcarbonyl-1,2,5-thiadiazolyl)-	pH 1, 7	242(4.16),298(3.84)	44-2141-64
$C_{12}H_{17}N_5O_3$			
Adenosine, 3'-deoxy-N,N-dimethyl-	pH 1	210(4.24),268(4.26)	87-0659-65
	H_2O	214(4.20),276(4.26)	87-0659-65
	pH 13	276(4.25)	87-0659-65
erythro	pH 1	210(4.21),268(4.26)	35-0720-64
	H_2O	213(4.22),275(4.28)	35-0720-64
	pH 13	213(4.18),277(4.28)	35-0720-64
threo	pH 1	210(4.25),268(4.27)	35-0720-64
	H_2O	213(4.23),275(4.28)	35-0720-64
	pH 13	213(4.19),277(4.28)	35-0720-64
Adenosine, 3'-deoxy-N-ethyl-	pH 1	263(4.23)	87-0659-65
	pH 7	268(4.22)	87-0659-65
	pH 13	268(4.23)	87-0659-65
$C_{12}H_{17}N_5O_4$			
Adenine, N,N-dimethyl-3-β-D-ribofuranosyl-	pH 1	291(4.36)	35-5320-64
	pH 12	298(4.24)	35-5320-64
	MeOH	226(4.18),298(4.26)	35-5320-64
Adenine, N,N-dimethyl-9-β-D-ribofuranosyl-	pH 1	268(4.27)	35-5320-64
	pH 7	215(4.19),275(4.27)	35-5320-64
	pH 12	275(4.28)	35-5320-64
$C_{12}H_{17}N_5O_5$			
2-Furaldehyde, 5-nitro-, 2-(2-morpholinoethyl)semicarbazone, hydrochloride	H_2O	378(4.22)	44-2582-64
Guanosine, N,N-dimethyl-	pH 1	264(4.11),293s(3.77)	35-3752-65
	pH 11	262(4.09),273s(4.03)	35-3752-65
$C_{12}H_{17}N_5O_6S$			
Guanosine, 8-(2-hydroxyethylthio)-	pH 1	272(4.20),289s(4.11)	35-1242-64
	pH 11	285(4.20)	35-1242-64
$C_{12}H_{17}O_4P$			
Phosphonic acid, 1-phenylepoxyethyl-, diethyl ester	EtOH	214(3.91),252(2.60), 257(2.57),264(2.49)	12-0168-65
$C_{12}H_{18}$			
Benzene, 1-tert-butyl-2,6-dimethyl-	isooctane	270(2.3)	88-1625-65
Benzene, hexamethyl- (protonated)	H_2SO_4	280(3.75),395(3.95)	88-2987-64
Biphenylene, decahydro-	isooctane	190(3.93),204(4.02), 214(3.72),220(3.00), 230(1.53)	78-1001-65
1-Cyclobutene, 1,2,3-trimethyl-3-methallyl-4-methylene-	hexane	236(4.14)	24-2327-65
Cyclopropane, 2-isopropylidene-1,1-dimethyl-3-(2-methylpropenylidene)-	EtOH	217(4.35),233(4.37), 251(4.34)	35-5032-64
1,3,9,11-Dodecatetraene	EtOH	228(4.62)	44-2410-65
3,4-Undecadien-6-yne, 2-methyl-	EtOH	221(4.02),268(3.45), 282(3.34)	39-4659-65

Compound	Solvent	λ max (log ε)	Ref.
$C_{12}H_{18}Cl_2Pd_2$			
Palladium, di-μ-chlorodi(2-cyclohexenyl)di-	$CHCl_3$	244(4.00),253s(3.92), 320(3.04)	39-5002-64
$C_{12}H_{18}F_2O_2$			
2,6-Octadienoic acid, 2,4-difluoro-3,7-dimethyl-, ethyl ester	MeOH	216(4.14)	5-0001-64D
$C_{12}H_{18}INO$			
(2-Benzoylethyl)trimethylammonium iodide	H_2O	230(4.25),242s(4.19)	7-0968-65
$C_{12}H_{18}N_2$			
Benzimidazoline, 1,3-diethyl-2-methyl-	EtOH	220(4.42),268(3.70), 315(3.66)	12-0877-64
1-Pyrroline, 2-(4-ethyl-3,5-dimethyl-2-pyrrolyl)-	EtOH-HCl	292s(3.77),344(4.44)	39-0893-64
	EtOH	261s(3.66),304(4.23)	39-0893-64
Quinoxaline, 1,2,3,4-tetrahydro-1,2,2,4-tetramethyl-	EtOH	227(4.50),270(3.81), 313(3.72)	12-0877-64
$C_{12}H_{18}N_2O$			
Acetamidoxime, N-butyl-2-phenyl-, hydrochloride	EtOH	251(2.34),258(2.31), 264(2.18)	39-1147-64
Propiophenone, 2-(2-aminoethylamino)-2-methyl-, dihydrochloride	H_2O	224(3.54)	44-3574-64
	EtOH	254s(2.70),256s(2.70), 260s(2.61),263s(2.59), 267s(2.44)	44-3574-64
$C_{12}H_{18}N_2O_2$			
Aniline, p-nitro-N,N-dipropyl-	n.s.g.	417(4.43)	5-0109-65I
3H-Cyclopenta[c]pyridazin-3-one, 2,5,6,7-tetrahydro-4-hydroxy-7-pentyl-	EtOH	222(4.01),252(4.01)?	54-0648-65
$C_{12}H_{18}N_2O_4$			
1,5-Diazecine-6,10(1H,7H)-dione, 1,5-diacetylhexahydro-	H_2O	221(4.23)	35-2003-65
5-Pyrimidinecarboxylic acid, 1,2,3,6-tetrahydro-2-oxo-1-(tetrahydro-2-pyranyl)-, ethyl ester	EtOH	287(3.99)	94-0681-65
$C_{12}H_{18}N_2O_5$			
1(2H)-Pyrimidinepropionic acid, 5-carboxy-3,4-dihydro-2-oxo-, diethyl ester	EtOH	214(3.98),293(4.02)	94-0681-65
$C_{12}H_{18}N_2O_7S$			
Glucosamine, N-p-aminophenylsulfonyl-	n.s.g.	262.5(4.47)	95-0399-65
$C_{12}H_{18}N_4$			
Pyrazolo[1,5-a]pyrimidine, 4-ethyl-7-(ethylimino)-4,7-dihydro-2,3-dimethyl-	EtOH	230(4.50),254(4.13), 333(3.83)	94-0142-65
$C_{12}H_{18}N_4O_2S$			
Sulfanilamide, N^1-(aminopiperidinomethylene)-	EtOH	266.5(4.34)	95-0391-65

Compound	Solvent	$\lambda_{max}(\log \epsilon)$	Ref.
$C_{12}H_{18}N_6O_4$			
[$\Delta^{2,2'(1H,1'H)}$-Bi-s-triazine-4,4',6,6'(3H,3'H,5H,5'H)-tetrone, 1,1',3,3',5,5'-hexamethyl-	H_2O	244(4.40)	88-2587-64
9H-Purine, 2-amino-6-(dimethylamino)-9-β-D-ribofuranosyl-	H_2O	228(4.28),268s(4.06), 284(4.19)	94-0951-64
$C_{12}H_{18}O$			
2H-Benzocyclohepten-2-one, 1,4a,5,6,7,8,9,9a-octahydro-4a-methyl-	EtOH	240(4.17)	35-1755-64
2H-Benzocyclohepten-2-one, 3,4,4a,5,6,7,8,9-octahydro-4a-methyl-	EtOH	227(3.99)	35-1755-64
2,4-Cyclohexadien-1-one, 2,3,4,5,6,6-hexamethyl-	EtOH	330(3.62)	35-1454-64
Cyclohexanone, 2-cyclohexylidene-	EtOH	254(3.83)	39-5617-64
2-Cyclohexen-1-one, 2-allyl-3,5,5-trimethyl-	n.s.g.	249(3.99)	39-1511-64
1-Indanone, 4,5,6,7-tetrahydro-6-isopropyl-	EtOH	238(4.16)	28-1983-65B
2(3H)-Naphthalenone, 4,4a,5,6,7,8-hexahydro-1,4a-dimethyl-	EtOH	246(4.20)	44-3110-64
2(3H)-Naphthalenone, 4,4a,5,6,7,8-hexahydro-4a,8-dimethyl-	EtOH	240(4.13)	44-2501-64
	EtOH	240(4.19)	44-3642-65
2-Norbornaneacrolein, 3,3-dimethyl-	EtOH	232(4.20)	28-4783-65A
4,5-Octadien-2-yn-1-ol, 6-isopropyl-7-methyl-	EtOH	222(4.26)	39-4659-65
Spiro[bicyclo[3.1.0]hexane-6,1'-cyclopentan]-2-one, 1,4-dimethyl-	EtOH	216(3.57)	35-4053-64
$C_{12}H_{18}OS$			
2,4-Cyclohexadien-1-one, 2,3,5,6-tetramethyl-6-[(methylthio)methyl]-	MeOH	325(3.66)	35-4656-65
$C_{12}H_{18}O_2$			
2(1H)-Azulenone, 4,5,6,7,8,8a-hexahydro-8-hydroxy-3,8-dimethyl-	EtOH	242(4.16)	44-3110-64
isomer	EtOH	242(4.16)	44-3110-64
Benzene, 1,2-bis(hydroxymethyl)-3,4,5,6-tetramethyl-	MeOH	275(2.73)	24-2331-65
Bicyclo[3.3.1]non-3-en-2-one, 1-(hydroxyethyl)-5-methyl-	EtOH	232(3.91)	39-1344-65
1-Butanol, 4-(4-hydroxy-2,5-xylyl)-	EtOH	217s(3.86),282(3.30)	78-2183-65
1-Butanol, 4-(4-methoxy-o-tolyl)-	EtOH	226(3.97),277(3.30), 284(3.27)	35-5148-65
2-Cyclohexen-1-one, 2,6-dimethyl-6-(3-oxobutyl)-	EtOH	237(3.95)	44-3642-65
2-Cyclopenten-1-ol, 4-methoxy-3-methyl-2-(2,4-pentadienyl)-, cis	EtOH	230.5(4.36)	39-0888-64
trans-1	EtOH	268(4.46),278(4.54), 288(4.45)	39-0888-64
Isocnidilide	EtOH	216(3.87)	78-1971-64
p-Menta-1,5-dien-8-ol, acetate	MeOH	263(3.55),310s(2.0)	70-0466-65
2(1H)-Naphthalenone, 4a,5,6,7,8,8a-hexahydro-8a-hydroxy-3,8a-dimethyl-, 1 (8a-4aβ)-abeo-	EtOH	234(3.98)	35-4053-64
2(3H)-Naphthalenone, 4,4a,5,6,7,8-hexahydro-4a-(hydroxymethyl)-1-methyl-	EtOH	251(4.28)	35-0275-65

Compound	Solvent	λ_{max}(log ε)	Ref.
Phthalide, 3-butyltetrahydro-	EtOH	220.0(4.01)	77-0479-65
cis	EtOH	218(3.63)	78-1433-65
trans	EtOH	216(3.52)	78-1433-65
$C_{12}H_{18}O_3$			
1,3-Cyclohexanedione, 2-(4-methyl-3-oxopentyl)-	EtOH	264(4.17)	39-2340-65
1,3,5-Cyclohexanetrione, 2,2,4,4,6,6-hexamethyl-	C_6H_{12}	300(1.90)	39-1028-65
(±)-Cyclohex-2-ene-1-acetic acid, 2,6,6-trimethyl-4-oxo-, methyl ester	EtOH	236(4.14)	94-0043-65
(-)-isomer	EtOH	236(4.12)	94-0043-65
3-Cyclohexene-1-propionic acid, 1,4-dimethyl-2-oxo-, methyl ester	EtOH	235(4.11)	35-0465-64
2-Cyclohexen-1-one, 6-acetoxy-3-isopropyl-6-methyl-	EtOH	237(4.16),317(1.95)	44-0518-65
Cyclopentaneacetaldehyde, 2-(1,3-dioxolan-2-yl)-3-methyl-α-methylene-	EtOH	223(3.95)	12-1260-64
1,3-Cyclopentanedione, 2-acetyl-4-isopentyl-	MeOH-HCl	226(4.10),262(3.98)	20-0628-64
	MeOH-NaOH	248(4.40),263s(4.33)	20-0628-64
3-Furoic acid, 2-methyl-4,5-dipropyl-	MeOH	259(3.55)	39-5984-65
2-Naphthoic acid, decahydro-3-oxo-, methyl ester, 2-cis 3-trans	EtOH	256(3.87)	28-1026-65B
$C_{12}H_{18}O_3S$			
Crotonic acid, 3-[(2-oxocyclohexyl)-thio]-, ethyl ester	ether	269(4.13)	24-2109-64
$C_{12}H_{18}O_5$			
Carbonic acid, ethyl ester, ester with ethyl 2-hydroxy-1-cyclohexene-1-carboxylate	EtOH	222(3.93)	44-0087-64
$C_{12}H_{18}O_5S$			
Crotonic acid, 3-[(2-oxocyclohexyl)-sulfonyl]-, ethyl ester	ether	212.5(4.11)	24-2109-64
$C_{12}H_{18}O_7$			
α-D-erythro-Hex-2-enopyranose, 3-deoxy-2,4-di-O-methyl-, diacetate	H_2O	201(4.03)	12-0837-65
$C_{12}H_{18}S_3$			
Carbonic acid, trithio-, cyclic ester with decahydro-4a-methyl-2,3-naphthalenedithiol, trans	dioxan	227(3.46),305s(4.2),317(4.24),454(1.94)	78-1581-65
$C_{12}H_{19}ClO$			
6(1H)-Azulenone, 7-chlorooctahydro-1,4-dimethyl-	EtOH	286(1.39)	94-0942-65
$C_{12}H_{19}Cl_2N$			
N-(3-Chloro-5,5-dimethyl-2-cyclohexen-1-ylidene)pyrrolidinium chloride	EtOH	274(4.40)	44-0794-64
$C_{12}H_{19}Cl_2NO$			
N-(3-Chloro-5,5-dimethyl-2-cyclohexen-1-ylidene)morpholinium chloride	EtOH	283(4.23)	44-0794-64

Compound	Solvent	λ_{max}(log ε)	Ref.
$C_{12}H_{19}Cl_2NO_4$			
N-(3-Chloro-5,5-dimethyl-2-cyclohexen-1-ylidene)pyrrolidinium perchlorate	$CHCl_3$	272(4.43)	44-0794-64
$C_{12}H_{19}Cl_2NO_5$			
N-(3-Chloro-5,5-dimethyl-2-cyclohexen-1-ylidene)morpholinium perchlorate	EtOH	283(4.30)	44-0794-64
$C_{12}H_{19}FO_2$			
2,6-Octadienoic acid, 2-fluoro-3,7-dimethyl-, ethyl ester	MeOH	224(4.06)	5-0001-64D
$C_{12}H_{19}N$			
Isoquinoline, 3,4,5,6,7,8-hexahydro-1,3,3-trimethyl-	EtOH	255(3.62)	4-0013-64
Pyrrolidine, 1-(2-norbornylidene-methyl)-	EtOH	245(3.86)	28-1541-64A
$C_{12}H_{19}NO$			
1H-Benzocycloheptene-2-carboxamide, 4,4a,5,6,7,8,9,9a-octahydro-, cis	n.s.g.	214(3.49)	28-0597-64A
low melting form	n.s.g.	217.5(4.02)	28-0597-64A
4-Hexen-3-one, 1-[(1,1-dimethyl-2-propynyl)amino]-4-methyl-, hydrochloride	EtOH	231.5(4.05)	70-0512-64
4-Piperidone, N-(1,1-dimethyl-2-propynyl)-2,5-dimethyl-	n.s.g.	none	70-0512-64
$C_{12}H_{19}NO_2$			
2-Cyclohexen-1-one, 5,5-dimethyl-3-morpholino-	EtOH	303(4.51)	44-0794-64
hydrochloride	EtOH	301(4.45)	44-0794-64
perchlorate	EtOH	302(4.48)	44-1407-65
2-Nonenoic acid, 2-cyano-, ethyl ester	EtOH	228(3.93)	12-1260-64
Pyrrole-2-carboxylic acid, 4,5-diisopropyl-, methyl ester	EtOH	236(3.72),286(4.31)	23-1279-64
$C_{12}H_{19}NO_3$			
1,4-Hexadien-3-ol, 1-(1-nitrocyclohexyl)-	n.s.g.	none	70-1245-64
1,3(2H,4H)-Naphthalenedione, 4-(dimethylamino)-hexahydro-8a-hydroxy-, trans-trans	EtOH-HCl	262(3.66)	65-0796-64
	EtOH	285(3.79)	65-0796-64
	EtOH-KOH	288(3.86)	65-0796-64
sodium salt	EtOH-HCl	262(3.67)	65-0796-64
	EtOH-KOH	288(3.86)	65-0796-64
$C_{12}H_{19}NO_4$			
Maleic acid, (isopropenylisopropylamino)-, dimethyl ester	EtOH	286(4.25)	5-0098-65H
$C_{12}H_{19}NO_5$			
3-Furoic acid, 2,5-dihydro-4-hydroxy-5-oxo-, ethyl ester, piperidine salt	EtOH	302(4.17)	39-0766-64
	EtOH-NaOH	303(4.20)	39-0766-64
$C_{12}H_{19}N_3O$			
2-Cyclohexen-1-one, 2-(3-butenyl)-3-methyl-, semicarbazone	EtOH	267(4.40)	35-5148-65
$C_{12}H_{19}N_3O_3$			
Cyclopentanecarboxylic acid, 2-(1-isopenten-2-yl)-3-oxo-, semicarbazone	MeOH	235(4.15)	7-0206-64

Compound	Solvent	$\lambda_{max}(\log \epsilon)$	Ref.
Cyclopentanecarboxylic acid, 2-(2-iso-penten-3-yl)-3-oxo-, semicarbazone	MeOH	235(4.17)	7-0206-64
Sarcomycin, methylpropyl-, semicarbazone	MeOH	278(4.22)	7-0206-64
$C_{12}H_{19}N_5$			
Pyrazolo[1,5-a]pyrimidine, 7-(2-dimethylaminoethylamino)-2,3-dimethyl-	EtOH	225(4.55),282s(3.77), 291(3.81),323(3.83)	94-1207-65
3H-Pyrrolo[2,3-d]pyrimidine, 3-(3-dimethylaminopropyl)-4,7-dihydro-4-imino-5-methyl-	pH 1	234(4.33),288(3.91)	35-1995-65
	pH 13	273(3.98)	35-1995-65
	EtOH	230(4.09),274(3.97)	35-1995-65
7H-Pyrrolo[2,3-d]pyrimidine, 4-(3-dimethylaminopropylamino)-5-methyl-	pH 1	237(4.20),285(3.98)	35-1995-65
	pH 13	281(4.04)	35-1995-65
	EtOH	283(4.05)	35-1995-65
Triazolo[1,5-c]pyrimidine, 2-(isopropylamino)-7-methyl-5-propyl-	MeOH	233(4.63),301(3.40)	39-3357-65
$C_{12}H_{19}N_5O$			
9H-Purine-9-propanol, 6-(tert-butylamino)-	pH 1	267(4.27)	87-0033-65
	pH 7	270(4.27)	87-0033-65
	pH 13	270(4.27)	87-0033-65
$C_{12}H_{19}N_5O_2S$			
Purine, 9-[2-[N,N-bis(2-hydroxyethyl)-amino]ethyl]-6-(methylthio)-	pH 1	229(4.09),306(4.28)	87-0182-65
	pH 11	228(3.95),286(4.28)	87-0182-65
Sulfanilamide, N^1-(hydrazinopiperidinomethylene)-	EtOH	268.0(4.28)	95-0391-65
$C_{12}H_{19}N_5S$			
[1,2,5]Thiadiazolo[3,4-d]pyrimidine, 7-(dibutylamino)-	pH 1	229(4.11),265(3.67), 355(4.23)	44-2135-64
	pH 7	235(4.20),275(3.59), 285s(--),374(4.08)	44-2135-64
	pH 13	235(4.20),276(3.60), 374(4.08)	44-2135-64
	EtOH	234(4.21),277(3.65), 287(3.51),378(4.07)	44-2135-64
$C_{12}H_{19}O_6P$			
Cyclopropanepropionic acid, α-(2-hydroxyethyl)-β-oxo-α-phosphono-, γ-lactone	MeOH	210(3.55)	5-0010-65E
$C_{12}H_{19}O_7PS$			
Maleic acid, hydroxy(2-hydroxyethyl)-, γ-lactone, ethyl ester, O-ester with O,O-diethyl phosphorothioate	MeOH	240(3.98)	5-0010-65E
$C_{12}H_{19}O_8P$			
Maleic acid, hydroxy(3-hydroxybutyl)-, δ-lactone, ethyl ester, dimethyl phosphate	MeOH	214(3.87)	5-0010-65E
Maleic acid, hydroxy(2-hydroxyethyl)-, γ-lactone, ethyl ester, diethyl phosphate	MeOH	233(4.55)	5-0010-65E
Oxalacetic acid, (3-hydroxybutyl)-phosphono-, δ-lactone, 1-ethyl dimethyl ester	MeOH	213(3.87)	5-0010-65E
Oxalacetic acid, (2-hydroxyethyl)-phosphono-, γ-lactone, triethyl ester	MeOH	216(3.85)	5-0010-65E

Compound	Solvent	λ_{max}(log ε)	Ref.
Oxalacetic acid, (2-hydroxy-2-methyl-propyl)phosphono-, γ-lactone, 1-ethyl dimethyl ester	MeOH	212(3.83)	5-0010-65E
$C_{12}H_{19}P$			
Phosphine, (1-cyclohexen-1-ylethynyl)-diethyl-	EtOH	219(4.08),229(4.04), 240(4.00),252(3.95)	28-1537-64A
$C_{12}H_{20}Br_2O$			
Cyclobutanone, 2,4-dibromo-2,4-di-tert-butyl-, cis	C_6H_{12}	366(2.14)	22-2751-65
	dioxan	366(2.17)	22-2751-65
	CCl_4	365(2.26)	22-2751-65
	MeCN	366(2.13)	22-2751-65
$C_{12}H_{20}ClNO_4$			
2,3,4,5,6,6a,7,8,9,9a-Decahydro-1H-pyrido[2,1,6-de]quinolizinium perchlorate	EtOH	266(2.38)	94-0747-64
$C_{12}H_{20}ClO_2P$			
Butylbis(hydroxymethyl)phenyl-phosphonium chloride	EtOH	259(2.99),266(3.01), 272(2.89)	59-1143-64
$C_{12}H_{20}Cl_2N_5OP$			
Phosphorane, [(3,5-dichloro-6-oxo-2,4-cyclohexadien-1-ylidene)hydrazono]tris(dimethylamino)-	MeOH	516(3.081)	5-0056-64I
$C_{12}H_{20}F_2N_2O_2$			
p-Benzoquinone, 2,5-bis(diethylamino)-3,6-difluoro-	methyl cellosolve	235(4.36),399)4.26), 572(2.53)	78-1889-64
$C_{12}H_{20}N_2$			
1,2-Propanediamine, N,N',2-trimethyl-1-phenyl-, dihydrochloride	H_2O	251(2.33),256(2.47), 261(2.54),267(2.43)	44-3146-64
$C_{12}H_{20}N_2O$			
4H-Cyclopenta[c]pyridazin-3(2H)-one, 2,4,4a,5,6,7-hexahydro-7-pentyl-	EtOH	210(--),248(4.00)	54-0648-65
Ethylenediamine, N-(1,1-dimethyl-2-hydroxy-2-phenylethyl)-, dihydrochloride	EtOH	251(2.19),257(2.31), 263(2.21),267(1.98)	44-3574-64
2-Pyrazinol, 3,6-diisobutyl-	EtOH	229(3.92),325(3.92)	39-1507-64
$C_{12}H_{20}N_2O_2$			
2,2'-Bi-1-pyrroline, 5,5,5',5'-tetramethyl-, N,N'-dioxide	n.s.g.	331(4.27)	77-0524-65
Neoaspergillic acid	EtOH	234(3.86),328(3.97)	88-4837-65
2-Pyrazinol, 3,6-diisobutyl-, 1-oxide	EtOH	236(3.96),328(4.02)	39-1507-64
$C_{12}H_{20}N_2O_3$			
5-Pyrimidinepropionaldehyde, 4-hydroxy-6-methyl-, diethyl acetal	pH 1	238(3.95)	4-0079-64
	pH 7	236(3.79),263(3.75)	4-0079-64
	pH 13	231(3.96),271(3.72)	4-0079-64
$C_{12}H_{20}N_2O_3Si$			
2,4-Diaza-8-silaspiro[5.5]undecane-1,3,5-trione, 11-ethyl-8,8-dimethyl-	70% EtOH-NaOH	240.5(3.94)	87-0695-64
$C_{12}H_{20}N_2O_4$			
Hydrazine, N,N'-dimethyl-N,N'-bis-(2-carbethoxyvinyl)-	ligroin	260(4.42),286(4.49)	5-0134-65F

Compound	Solvent	$\lambda_{max}(\log \epsilon)$	Ref.
$C_{12}H_{20}N_2S_4$			
Disulfide, bis(piperidinothiocarbonyl)	C_6H_{12}	219(4.37),258s(4.15)	78-0035-65
$C_{12}H_{20}N_4$			
Cyclodecapyrimidine, 2,4-diamino- 5,6,7,8,9,10,11,12-octahydro-	pH 1	277(3.89)	44-1837-65
	pH 10	230(3.99),287(3.89)	44-1837-65
$C_{12}H_{20}N_4O$			
Butyramide, N-[(4-amino-2-propyl- 5-pyrimidinyl)methyl]-	EtOH	236(3.93),278(3.75)	94-0393-64
$C_{12}H_{20}N_4O_2$			
1-Pyrroline, 2,2'-azobis[5,5-dimethyl-, 1,1'-dioxide	EtOH	240(3.68),435(4.50), 590(3.22)	39-1224-65
$C_{12}H_{20}O$			
5(1H)-Azulenone, octahydro- 3,8-dimethyl-	EtOH	287(1.46)	94-0942-65
6(1H)-Azulenone, octahydro- 1,4-dimethyl-	EtOH	283(1.46)	94-0942-65
Cyclohexaneacetaldehyde, 1,3,3-tri- methyl-2-methylene-	isooctane	297(1.20)	35-2884-64
Cyclohexaneacrolein, 3,3,5-trimethyl-	EtOH	226(4.09)	28-4712-64B
2-Cyclohexen-1-one, 2,3,4,5,6,6-hexa- methyl-	MeOH	246(3.93)	35-4656-65
2-Cyclohexen-1-one, 3,5,5-trimethyl- 2-propyl-	n.s.g.	244(3.97)	39-1511-64
Furan, 2,4-di-tert-butyl-	EtOH	215(3.80)	88-2831-65
Furan, 2,5-di-tert-butyl-	hexane	218(4.02)	44-1058-65
Ketone, (2,3-diethyl-1-cyclo- penten-1-yl) ethyl	EtOH	251.5(3.90)	22-0722-65
Ketone, (2,3-diethyl-2-cyclo- penten-1-yl) ethyl	EtOH	276(1.47)	22-0722-65
Ketone, propyl 2-propyl-1-cyclo- penten-1-yl	EtOH	251(3.85),306(1.85)	22-0722-65
Ketone, propyl 2-propyl-2-cyclo- penten-1-yl	EtOH	209(3.45),276(1.47)	22-0722-65
5,7-Octadien-2-ol, 2-cyclopropyl- 6-methyl-	EtOH	230(4.28)	22-2533-64
$C_{12}H_{20}O_2$			
1,3-Cyclohexanedione, 2-methyl- 5,5-pentamethylene-	MeOH-HCl	251(4.21)	5-0123-65A
	MeOH-NaOMe	280(4.47)	5-0123-65A
	H_2SO_4	291(4.20)	5-0123-65A
1,3-Cyclohexanedione, 5,5-penta- methylene-, enol methyl ether	MeOH	250(4.21)	5-0123-65A
Myrcenyl acetate	MeOH	224.7(4.23)	5-0083-64E
2,4-Nonadienal, 7-hydroxy- 2,4,6-trimethyl-	heptane	273(3.82)	33-0567-64
$C_{12}H_{20}O_3$			
Carvomenthone, 4-acetoxy-	n.s.g.	204(2.03),278(0.97)	33-0010-65
1-Cyclohexene-1-carboxylic acid, 6-(1-hydroxypentyl)-, erythro	EtOH	215(3.97)	78-1433-65
threo	EtOH	210(3.41)	78-1433-65
3,5,11-Trioxadec-9-en-7-yne, 4,6,6-trimethyl-	n.s.g.	239(4.19)	70-1318-64
$C_{12}H_{20}O_5S$			
Valeric acid, δ-acetylmercapto-α-carb- ethoxy-, ethyl ester	n.s.g.	230(3.71)	24-1963-64

Compound	Solvent	λ max(log ε)	Ref.
$C_{12}H_{20}S$			
Thiophene, 2,4-di-tert-butyl-	EtOH	234(3.89)	44-1058-65
Thiophene, 2,5-di-tert-butyl-	C_6H_{12}	237(3.97)	22-3136-65
	MeOH	237(3.98)	22-3136-65
	EtOH	237.5(4.26)	22-2635-65
	EtOH	239(3.95)	44-1058-65
$C_{12}H_{21}BrO$			
Cyclobutanone, 2-bromo-2,4-di-tert-butyl-, cis	C_6H_{12}	336(2.20)	22-2751-65
	MeCN	334(2.21)	22-2751-65
trans	C_6H_{12}	334(2.42)	22-2751-65
	MeCN	332(2.53)	22-2751-65
$C_{12}H_{21}ClN_5OP$			
Phosphorane, [(3-chloro-6-oxo-2,4-cyclohexadien-1-ylidene)hydrazono]tris-(dimethylamino)-	MeOH	504(3.082)	5-0056-64I
$C_{12}H_{21}N$			
Pyrrole, 2,5-di-tert-butyl-	C_6H_{12}	218(4.00),289s(1.30)	22-3136-65
	C_6H_{12}	218(4.00),289s(1.30)	77-0453-65
	MeOH	217(3.99),320(1.0)	22-3136-65
$C_{12}H_{21}NO_4$			
Maleic acid, diisopropylamino-, dimethyl ester	EtOH	287(4.37)	5-0098-65H
$C_{12}H_{21}N_3O_2$			
Pyridazine, 3-butoxy-6-[2-(dimethylamino)ethoxy]-	EtOH	287(3.34)	44-1751-64
5-Pyrimidinepropionaldehyde, 2-amino-4-methyl-, diethyl acetal	pH 1	227(4.25),310(3.57)	4-0079-64
$C_{12}H_{21}N_3O_4$			
Valeric acid, 2-acetyl-5-butoxy-4-hydroxy-, γ-lactone, semicarbazone	n.s.g.	222(4.06),274(3.72)	39-0141-64
$C_{12}H_{21}O_5P$			
Crotonic acid, 3-hydroxy-2-(2-hydroxyethyl)-, γ-lactone, ethyl butylphosphonate	MeOH	230(4.14)	5-0010-65E
$C_{12}H_{21}O_6P$			
Hexanoic acid, 2-acetyl-5-hydroxy-2-phosphono-, δ-lactone, diethyl ester	MeOH	210(2.83)	5-0010-65E
Hexanoic acid, 5-hydroxy-2-(1-hydroxyethylidene)-, δ-lactone, diethyl phosphate	MeOH	229(3.91)	5-0010-65E
$C_{12}H_{22}$			
1,3-Dodecadiene	EtOH	225.3(4.47)	44-2410-65
$C_{12}H_{22}N_2$			
Cyclopentapyrazole, 1,3a,4,5,6,6a-hexahydro-3,6a-dipropyl-	EtOH	237(3.53)	22-0722-65
$C_{12}H_{22}N_2S$			
3,4-Thiophenebis(methylamine), N,N'-diethyl-2,5-dimethyl-	n.s.g.	240(3.9),246s(3.8)	70-2182-64

Compound	Solvent	$\lambda_{max}(\log \epsilon)$	Ref.
$C_{12}H_{22}N_4O$			
4-Pyrimidinol, 2-amino-5-[3-(butyl-amino)propyl]-6-methyl-, dihydro-chloride	pH 1	229(4.02),267(3.98)	4-0088-64
	pH 6	272(3.75)	4-0088-64
	pH 13	233s(3.95),280(3.88)	4-0088-64
$C_{12}H_{22}N_5OP$			
Phosphorane, tris(dimethylamino)-[(6-oxo-2,4-cyclohexadien-1-yli-dene)hydrazono]-	MeOH	494(3.009)	5-0056-64I
$C_{12}H_{22}O$			
Cyclobutanone, 2,4-di-tert-butyl-, cis	C_6H_{12}	310(1.89),322(1.71)	22-2747-65
	dioxan	309(1.93)	22-2751-65
	CCl_4	309.5(2.09)	22-2751-65
	MeCN	307.5(1.91)	22-2751-65
trans	C_6H_{12}	310(1.62)	22-2747-65
	MeCN	311(1.66)	22-2751-65
Cyclopentane, 1-propionyl-2,3-diethyl-	EtOH	283.5(1.42)	22-0722-65
1-Cyclopentene-1-methanol, α,2-di-propyl-	EtOH	210(3.48)	22-0722-65
2-Dodecenal, trans	EtOH	223(4.00),315(1.28)	57-0540-64B
$C_{12}H_{22}O_2$			
3,8-Decanedione, 2,9-dimethyl-	EtOH	278.5(1.89)	22-0714-65
4,9-Dodecanedione	EtOH	280(1.71)	22-0714-65
2-Octenal, 2,4-diethyl-5-hydroxy-	heptane	227(4.26),320(1.80)	33-0567-64
$C_{12}H_{22}O_2S_4$			
Diisopentyl xanthogen disulfide	hexane	360s(1.9)	70-2175-64
Dipentyl xanthogen disulfide	hexane	360s(1.8)	70-2175-64
$C_{12}H_{22}O_2S_5$			
Diisopentyl xanthogen trisulfide	hexane	none from 320 to 480 nm	70-2175-64
Dipentyl xanthogen trisulfide	hexane	none from 320 to 480 nm	70-2175-64
$C_{12}H_{23}NO$			
2-Heptanone, 7-piperidino-	EtOH	277(1.39)	44-0650-65
$C_{12}H_{24}N_2O$			
2-Imidazoline, 1-(2-ethoxyethyl)-2-pentyl-	EtOH	231(3.79)	4-0188-64
$C_{12}H_{24}N_2O_4$			
Propionic acid, 2,2'-azobis[2-methyl-, diethyl ester	C_6H_{12}	363(1.32)	35-1918-64
$C_{12}H_{24}N_2O_8$			
L-Rhamnose, azine	MeOH	260(4.04)	24-1404-65
$C_{12}H_{24}N_2O_{10}$			
D-Fructose, azine	MeOH	260(4.04)	24-1404-65
D-Galactose, azine	MeOH	260(4.00)	24-1404-65
D-Glucose, azine	MeOH	265(4.06)	24-1404-65
D-Mannose, azine	MeOH	260(4.06)	24-1404-65
L-Sorbose, azine	MeOH	260(4.03)	24-1404-65
$C_{12}H_{24}O_3$			
s-Trioxane, 2,4,6-tripropyl-	heptane	225(1.68),280(1.42)	33-0567-64

Compound	Solvent	$\lambda_{max}(\log \epsilon)$	Ref.
$C_{12}H_{25}N_3S$			
Semicarbazide, 4-menth-3-yl-2-methyl-3-thio-, nickel complex	n.s.g.	438(2.11)	1-1239-65
$C_{12}H_{26}N_2O_5$			
D-Galactose, N,N-dipropylhydrazone	MeOH	252(4.4)	24-1588-65
$C_{12}H_{28}N_4$			
Tetrazene, tetraisopropyl-	heptane	291(4.05)	35-2395-64
$C_{12}H_{30}Si_4$			
Tetrasilane, octamethyl-1,2-divinyl-	C_6H_{12}	243.5(4.15)	101-0369-64B
$C_{12}H_{36}Cl_2Si_6$			
Hexasilane, 1,6-dichlorododecamethyl-	C_6H_{12}	222s(--),259(4.35)	101-0369-64B
$C_{12}H_{36}Si_5$			
Pentasilane, dodecamethyl-	C_6H_{12}	250.0(4.27)	101-0369-64B
Trisilane, hexamethyl-2,2-bis-(trimethylsilyl)-	C_6H_{12}	none	101-0163-65B
$C_{12}H_{36}Si_6$			
Cyclohexasilane, dodecamethyl-	n.s.g.	232(3.76),255s(3.30)	101-0176-65B

Compound	Solvent	$\lambda_{max}(\log \epsilon)$	Ref.
$C_{13}F_{10}O$			
Benzophenone, decafluoro-	EtOH	250(4.12)	30-1135-64
$C_{13}HF_{10}N$			
Aniline, 2,3,4,5,6-pentafluoro-N-(2,3,4,5,6-pentafluorobenzylidene)-	EtOH	254(4.22)	30-1135-64
$C_{13}H_4BrN_3O_7$			
Fluorenone, 4-bromo-2,5,7-trinitro-	benzene	272(4.30),282(4.31), 385s(3.62)	45-5600-64
	EtOH	243(4.29),249(4.27), 255(4.25),261(4.24), 283(4.42),334(3.97), 352s(3.83),376s(3.60)	35-5600-64
	$CHCl_3$	244(4.34),250(4.36), 256(4.37),262(4.32), 283(4.45),334(3.98), 353s(3.68)	35-5600-64
$C_{13}H_4Cl_6N_2$			
Methane, diazobis-(2,4,6-trichlorophenyl)-	C_6H_{12}	237s(4.17),255s(4.09), 283(4.15),315(4.12), 323(4.11),450(2.45), 470(2.47)	35-2149-64
$C_{13}H_4IN_3O_7$			
Fluorenone, 4-iodo-2,5,7-trinitro-	benzene	273(4.26),282(4.26), 396s(3.54)	35-5600-64
	EtOH	227(4.46),243(4.40), 249(4.39),255(4.38), 261(4.35),285(4.35), 346(3.99),390s(3.73)	35-5600-64
	$CHCl_3$	239(4.45),244(4.48), 250(4.53),256(4.54), 262(4.47),287(4.39), 348(3.98),390s(3.66)	35-5600-64
$C_{13}H_6BrNO_3$			
Fluorenone, 6-bromo-2-nitro-	n.s.g.	245(4.2),289(4.5)	78-0803-64
$C_{13}H_6ClNO_4$			
2-Propyn-1-one, 3-(p-chlorophenyl)-1-(5-nitro-2-furyl)-	EtOH	236(4.21),326(4.36)	30-0833-65
$C_{13}H_6Cl_3NOS$			
Phenothiazine-1-carbonyl chloride, dichloro-	THF	248(4.53),300(3.70), 458(3.76)	4-0272-65
$C_{13}H_6Cl_4O$			
1,4-Methanobiphenylen-9-one, 1,2,3,4-tetrachloro-1,4,4a,8b-tetrahydro-	EtOH	254s(--),259(3.04), 265(3.24),271(3.25)	39-2549-65
$C_{13}H_6I_4O_3$			
Benzophenone, 4,4'-dihydroxy-3,3',5,5'-tetraiodo-	EtOH	209(4.53),247(4.5), 372(4.1)	44-1812-64
$C_{13}H_6N_2O_2$			
Phenalene-1,3(2H)-dione, 2-diazo-	MeOH	234(4.77),293(4.11), 333(4.06)	5-0101-64F

Compound	Solvent	λ max(log ε)	Ref.
$C_{13}H_7BrO$			
Fluorenone, 4-bromo-	iso-PrOH	384(2.60)	39-5518-65
$C_{13}H_7BrO_2$			
2-Propyn-1-one, 3-(m-bromophenyl)-1-(2-furyl)-	EtOH	312(4.33)	30-0833-65
2-Propyn-1-one, 3-(p-bromophenyl)-1-(2-furyl)-	EtOH	230(4.30),316(4.56)	30-0833-65
$C_{13}H_7Br_2ClN_2$			
Benzimidazole, 5,6-dibromo-2-(o-chlorophenyl)-	EtOH	231(4.41),253(4.49),264(4.45),274(4.29),303(4.33),350s(4.10)	49-0614-65
$C_{13}H_7Br_4N_3O_2$			
Benzoyl bromide, m-nitro-, (2,4,6-tribromophenyl)hydrazone	EtOH	314(4.16)	39-1500-64
$C_{13}H_7ClO$			
Fluorenone, 4-chloro-	iso-PrOH	380(2.62)	39-5518-65
$C_{13}H_7ClO_2$			
2-Propyn-1-one, 3-(p-chlorophenyl)-1-(2-furyl)-	EtOH	228(4.15),320(4.39)	30-0833-65
$C_{13}H_7FO$			
Fluorenone, 4-fluoro-	iso-PrOH	395(2.72)	39-5518-65
$C_{13}H_7F_3N_2O$			
Nicotinonitrile, 1,2-dihydro-2-oxo-6-phenyl-4-(trifluoromethyl)-	EtOH	264(4.04),342(4.01),376(4.12)	44-3377-65
1,7-Phenanthrolin-4-ol, 2-(trifluoromethyl)-	EtOH-HCl	230(4.70),236(4.69),269(4.48)	4-0120-65
	EtOH	244(4.38),325(2.74)	4-0120-65
$C_{13}H_7IO$			
Fluorenone, 4-iodo-	iso-PrOH	384(2.62)	39-5518-65
$C_{13}H_7NO_2$			
Benzo[g]quinoline-5,10-dione	EtOH	249(4.53),324(3.51)	44-0432-65
$C_{13}H_7NO_2S$			
4,5-Thiazoledione, 2-(2-naphthyl)-	dioxan	230(4.5),278(4.2),331(4.3)	83-0124-65
$C_{13}H_7NO_4$			
2-Propyn-1-one, 1-(5-nitro-2-furyl)-3-phenyl-	EtOH	230(4.15),329(4.32)	30-0833-65
$C_{13}H_7N_2NaO_2$			
Sodium, [(2,3-dicyano-4-methoxy-1-naphthyl)oxy]-	MeCN	240(3.81),272(4.15),282(4.11),338(3.93),355(4.07),395(4.11),412(4.16)	44-3591-64
$C_{13}H_7N_3$			
Pyridazino[3,4,5,6-lmn]phenanthridine	EtOH	235(4.73),251(4.19),269(4.07),347(4.10),367s(3.85),380(3.76)	39-6090-64

Compound	Solvent	λ_{max}(log ε)	Ref.
$C_{13}H_7N_3O$			
Pyridazino[3,4,5,6-lmn]phenanthridine, oxide	EtOH	223(4.45),236(4.47), 280(4.13),288(4.21), 358s(3.97),363(4.00), 368s(3.96),382(4.00), 402(3.97)	39-6090-64
$C_{13}H_8BrClN_2$			
Benzimidazole, 4-bromo-2-(o-chloro-phenyl)-	EtOH	244(4.12),267(4.20), 290(4.22),295(4.22)	49-0614-65
Benzimidazole, 5-bromo-2-(o-chloro-phenyl)-	EtOH	248s(3.03),258(3.96), 265s(2.93),303(3.05)	49-0614-65
$C_{13}H_8BrN$			
Acridine, 1-bromo-	EtOH	248s(4.84),254(5.05), 324s(5.37),332s(3.52), 340(3.71),349(3.75), 357(3.90),369s(3.65), 382s(3.45)	12-0108-65
	H_2SO_4	256(4.79),261(4.94), 327s(3.45),343(3.92), 352s(3.95),358(4.33), 400s(3.46),417(3.50), 440s(3.29)	12-0108-65
Acridine, 2-bromo-	EtOH	246s(4.69),253(5.26), 332s(3.62),343s(3.78), 350(3.81),359(3.92), 378s(3.44)	12-0108-65
	H_2SO_4	244s(4.32),255s(4.94), 261(5.19),330s(3.59), 346(4.03),362(4.33), 389(3.59),407(3.59), 429s(3.36)	12-0108-65
Acridine, 3-bromo-	EtOH	224(4.16),250s(4.94), 256(5.18),328s(3.57), 335s(3.71),343(3.89), 351(3.91),359(4.03)	12-0108-65
	H_2SO_4	255s(4.72),260(5.03), 331s(3.65),348(4.31), 364(4.48),406s(3.53), 429s(3.23)	12-0108-65
Acridine, 4-bromo-	EtOH	247(4.85),252(3.81), 332s(3.65),338(3.81), 348s(3.86),356(4.00), 366s(3.79),384s(3.58)	12-0108-65
	H_2SO_4	256s(4.82),262(4.98), 326(3.50),340(3.98), 350s(4.03),357(4.33), 405s(3.45),423(3.48), 441s(3.30)	12-0108-65
Phenanthridine, 4-bromo-	EtOH	252(4.37),274s(4.08), 297(3.86),310s(3.76), 334(3.08),350(2.90)	12-0206-65
Phenanthridine, 7-bromo-	n.s.g.	253(4.40),276s(4.04), 297(3.89),309(3.83), 337(3.34),353(3.30)	12-1406-64
Phenanthridine, 9-bromo-	n.s.g.	251(4.30),275s(4.08), 296(3.80),307s(3.69), 332(3.15),347(3.04)	12-1406-64

Compound	Solvent	$\lambda_{max}(\log \epsilon)$	Ref.
$C_{13}H_8BrNOS$			
2H-Thieno[3,2-b]pyrrol-3(4H)-one, 2-benzylidene-6-bromo-	EtOH	263(3.88),347(4.32)	44-1012-65
$C_{13}H_8Br_2N_2O$			
Benzimidazole, 4,6-dibromo-2-(o-hydroxyphenyl)-	EtOH	249(3.29),253(3.09), 259(3.18),265(3.07)	49-0614-65
$C_{13}H_8ClHgNO_4$			
Benzoic acid, (p-chloromercuri)-, p-nitrophenyl ester	THF	250(4.4),274s(4.3)	70-0111-64
$C_{13}H_8ClNOS$			
Benzothiazole, 5-chloro-2-(p-hydroxyphenyl)-	EtOH	263(3.86),315(4.37), 327(4.44)	39-0954-65
$C_{13}H_8ClNS$			
Benzothiazole, 5-chloro-2-phenyl-	EtOH	222(4.29),238(4.41), 260(3.97),289(4.26), 299s(4.23)	39-0954-65
$C_{13}H_8F_3NO_3$			
2(1H)-Pyridone, 1-[(α,α,α-trifluoro-m-toluoyl)oxy]-	EtOH	226(4.26),271s(3.35), 279s(3.48),304(3.70)	35-5186-65
2(1H)-Pyridone, 1-[(α,α,α-trifluoro-p-toluoyl)oxy]-	EtOH	226(4.30),275s(3.47), 284s(3.59),304(3.71)	35-5186-65
$C_{13}H_8N_2$			
5H-Benzocycloheptene-5,5-dicarbonitrile	n.s.g.	277(3.90)	35-0652-65
7H-Benzocycloheptene-7,7-dicarbonitrile	n.s.g.	228(4.66),257(3.77)	35-0652-65
1H-Cyclopropa[a]naphthalene, 1,1-dicarbonitrile, 1a,7b-dihydro-	n.s.g.	227(4.40),272(3.90)	35-0652-65
$C_{13}H_8N_2O$			
2-Phenazinecarboxaldehyde	C_6H_{12}	266(4.63),365(4.02)	65-1969-64
$C_{13}H_8N_2O_2$			
Benzimidazole-4,5-quinone, 2-phenyl-	$CHCl_3$	285(4.59),480(3.31)	65-0949-64
$C_{13}H_8N_2O_3$			
Benzo[f]quinazoline-5,6-dione, 3-hydroxy-1-methyl-	pH 1	224(4.40),231(4.43), 240s(4.49),246s(4.54), 251(4.56),297s(3.74), 307(3.81),336(3.63), 350(3.60)	44-2881-64
	pH 10	252(4.64),259s(4.61), 281s(3.95),288s(3.86), 302(3.71),338(3.60), 351(3.71),371(3.52)	44-2881-64
	EtOH	225s(4.45),245(4.63), 252(4.65),262(4.36), 284s(3.61),295(3.74), 307(3.81),325s(3.46), 338(3.64),352(3.62)	44-2881-64
2(1H)-Pyridone, 1-[(p-cyanobenzoyl)-oxy]-	EtOH	237(4.39),246s(4.27), 284s(3.72),293(3.80), 304s(3.74)	35-5186-65
$C_{13}H_8N_2O_4S$			
Thiobenzophenone, 4,4'-dinitro-	EtOH	298(4.31),631(2.21)	54-0289-65

Compound	Solvent	λ_{max}(log ε)	Ref.
$C_{13}H_8N_2O_5S_2$			
1,4-Pentadien-3-one, 1,5-bis-(5-nitro-2-thienyl)-	propylene glycol	392(4.462)	95-0001-64
$C_{13}H_8N_2O_6$			
Benzophenone, 2-hydroxy-4',5-dinitro-	EtOH	271(4.30),305s(4.12)	78-1015-65
	EtOH-NaOH	268(4.20),397(4.30)	78-1015-65
2-Propen-1-one, 3-(5-nitrofuryl)-1-(p-nitrophenyl)-	EtOH	271(4.07),357(4.35)	65-0489-64
	95% H_2SO_4	295(3.95),440(4.58)	65-0489-64
$C_{13}H_8N_2O_7$			
1,4-Pentadien-3-one, 1,5-bis-(5-nitro-2-furyl)-	EtOH	237(4.24),376(4.58)	65-0489-64
	95% H_2SO_4	278(4.09),365(4.18), 512(4.74)	65-0489-64
	propylene glycol	350(4.185)	95-0001-64
$C_{13}H_8N_4O_4$			
Benzimidazole, 1-(2,4-dinitrophenyl)-	MeOH	243(4.31),270s(3.96), 279s(3.82)	65-0632-64
$C_{13}H_8N_4O_8$			
Methane, bis(2,4-dinitrophenyl)-	EtOH	242(4.47),260s(4.41)	44-0636-64
$C_{13}H_8N_8O_2$			
6-Pteridinol, 3,7-dihydro-7-(7-hydroxy-pteridin-6-ylmethylene)-	pH 4.5	213(4.58),319(4.09), 433(4.47),457(4.44)	39-3357-64
	pH 8.1	219(4.50),330(4.05), 426(4.48),453(4.46)	39-3357-64
	pH 11.5	222(4.56),338(4.09), 428(4.41),452(4.48)	39-3357-64
$C_{13}H_8O$			
Fluorenone	EtOH	248(4.81),257(5.04), 284(3.49),294(3.59), 307(3.35),314(3.25), 322(3.14),329s(3.04), 360s(2.37),380(2.47), 393(2.44)	22-2953-64
	iso-PrOH	383(2.40)	39-5518-65
Furan, 2-(1-nonene-3,5,7-triynyl)-	ether	270(4.28),282(4.62), 313s(4.35),319s(4.36), 333(4.59),339s(4.56), 361(4.57)	24-1411-65
2,4,6-Heptatriyn-1-ol, 7-phenyl-	ether	238(4.85),251(5.17), 274(4.07),291(4.33), 310(4.45),332(4.30)	24-2135-64
$C_{13}H_8OS$			
Fluorene-9-thione, S-oxide	dioxan	232(4.45),238(4.45), 265(4.42),273(4.48), 362(4.18)	35-1891-64
Thiophene, 2-(3,4-epoxy-1-butynyl)-5-(1,3-pentadiynyl)-	ether	209(4.42),234(3.89), 246(4.05),320(4.40), 341(4.47)	24-0155-65
$C_{13}H_8O_2$			
1-Acenaphthenecarboxaldehyde, 2-oxo-	DMF	293s(4.30),305s(4.36), 315(4.48),336(3.75), 353(3.54),460(3.32)	32-1130-65

Compound	Solvent	λ_{max}(log ε)	Ref.
1-Acenaphthenecarboxaldehyde, 2-oxo-, cobalt complex	DMF	283(4.49),308s(4.70), 320(4.76),340s(4.18), 355(3.96),448(3.54)	32-1130-65
copper complex	DMF	283(4.35),308s(4.65), 320(4.83),340s(4.40), 355s(3.85),435(3.45), 680(2.08)	32-1130-65
nickel complex	DMF	283(4.48),308s(4.70), 320(4.80),340(4.12), 355(3.75),448(3.48), 632(1.20)	32-1130-65
palladium complex	DMF	283(4.35),315s(4.70), 325(4.77),342s(4.21), 365s(4.08),448(3.38)	32-1130-65
zinc complex	DMF	283(4.41),308s(4.65), 320(4.76),340(4.03), 355s(3.70),444(3.46)	32-1130-65
2-Propyn-1-one, 1-(2-furyl)-3-phenyl-	EtOH	226(4.07),315(4.35)	30-0833-65
$C_{13}H_8O_2S_2$			
1-Thianthrenecarboxylic acid	96% H_2SO_4 10 min.	538(4.04)	44-0021-64
	1 hour	291(4.63)	44-0021-64
	1 day	291(4.65)	44-0021-64
	5 days	291(4.65),537(4.05)	44-0021-64
	10 days	537(4.03)	44-0021-64
2-Thianthrenecarboxylic acid	96% H_2SO_4 15 min.	539(3.86)	44-0021-64
	1 hour	300(4.60)	44-0021-64
	1 day	300(4.63)	44-0021-64
	5 days	534(4.67)	44-0021-64
	10 days	300(4.67),534(3.98)	44-0021-64
$C_{13}H_8O_3$			
Naphtho[2,3-b]furan-4,9-dione, 2-methyl-	MeCN	243(4.56),293(3.80), 350(3.46)	44-3819-65
$C_{13}H_8O_4$			
Xanthen-9-one, 2,6-dihydroxy-	EtOH	238(4.70),316(4.21)	39-5074-64
Xanthen-9-one, 3,5-dihydroxy-	EtOH	237(4.56),268(4.33), 333(3.80)	39-5074-64
$C_{13}H_8O_6$			
Xanthen-9-one, 1,3,5,8-tetrahydroxy-	EtOH	231(4.3),250(4.3), 274(4.2),350(4.1)	78-1449-65
$C_{13}H_8S$			
Thiophene, 2-(3-butyn-1-yl)- 5-(1,3-pentadiynyl)-	ether	253(3.98),332(4.48), 357(4.42)	24-0155-65
Thiophene, 2-(5-hexene-1,3-diyn-1-yl)- 5-(1-propynyl)-	ether	258(4.17),264(4.16), 274(4.22),324s(4.43), 338(4.50),358(4.35)	24-2125-64
$C_{13}H_9BrN_2O$			
Benzimidazole, 4-bromo- 2-(o-hydroxyphenyl)-	EtOH	245(4.08),257(3.98), 263(3.44),298(4.02), 327(4.10),340(4.17)	49-0614-65
Benzimidazole, 5-bromo- 2-(o-hydroxyphenyl)-	EtOH	265(3.00),275(3.05), 330(3.18)	49-0614-65

Compound	Solvent	$\lambda_{max}(\log \epsilon)$	Ref.
$C_{13}H_9BrS$			
Thiobenzophenone, 4-bromo-	C_6H_{12}	230(4.08),321(3.96), 611(1.96)	54-0289-65
	EtOH	231(4.02),323(4.04), 602(2.00)	54-0289-65
$C_{13}H_9Br_4N$			
Diphenylamine, 2,2',4,4'-tetra-bromo-N-methyl-	$CHCl_3$	291(4.16)	44-3248-65
$C_{13}H_9ClN_2$			
Phenazine, 2-(chloromethyl)-	C_6H_{12}	210(4.31),254(5.04), 280(3.90),367(4.18)	65-1969-64
$C_{13}H_9ClN_2O$			
8-Nitrobenzo[c]quinolizinium chloride	n.s.g.	255(4.54),288(4.10), 343(3.66),357(4.02), 374(4.06)	77-0288-65
$C_{13}H_9ClN_2O_2S$			
1H-2,1,3-Benzothiadiazine, 6-chloro-4-phenyl-, 2,2-dioxide	EtOH	231(4.44),268(4.01), 295s(3.72),372(3.5)	44-3960-65
$C_{13}H_9ClN_2O_6$			
7-Nitroacridizinium perchlorate	EtOH	215(4.25),239(4.28), 268s(3.89),276s(3.86), 329(3.40),344(3.64), 358(3.92),390(3.79), 401s(3.78)	4-0030-64
9-Nitroacridizinium perchlorate	EtOH	214s(4.02),241(4.26), 257(4.28),290s(3.83), 324(3.56),339(3.62), 356(3.72),385s(3.60), 403(3.81),424(3.81)	4-0030-64
$C_{13}H_9ClO$			
Benzophenone, o-chloro-	MeOH	251(4.27),293s(3.34)	39-2579-64
Benzophenone, p-chloro-	MeOH	256(4.32)	39-2579-64
2H-Pyran, 5-chloro-5,6-dihydro-2-(2,4,6-octatriynylidene)-	ether	236s(4.31),245s(4.29), 269(4.40),280(4.43), 341(4.33),360s(4.31)	24-1416-65
$C_{13}H_9ClO_4S$			
Dibenzo[b,d]thiopyrylium perchlorate	H_2SO_4	260(4.6),265(4.3), 385(4.0),440(3.4)	17-0165-65
1-Thiaphenanthrenium perchlorate	HOAc	246(4.58),291(4.10), 305(4.11),428(3.84)	5-0136-64C
4-Thiaphenanthrenium perchlorate	HOAc	247(3.89),306(4.59), 382(3.70)	5-0136-64C
Thioxanthylium perchlorate	70% $HClO_4$	242(3.95),280(4.48), 382(4.40)	17-0021-65
$C_{13}H_9ClO_4Se$			
Dibenzo[b,d]seleninium perchlorate	H_2SO_4	264(4.51),282(4.27), 318(3.93),404(4.07), 440s(3.69)	17-0165-65
Selenoxanthylium perchlorate	70% $HClO_4$	254(3.96),284(4.85), 390(4.44)	17-0021-65

Compound	Solvent	λ_{max}(log ϵ)	Ref.
$C_{13}H_9ClO_5$			
Dibenzo[b,d]pyrylium perchlorate	H_2SO_4	249(4.57),260s(4.18), 286s(3.56),354(3.75), 417(3.79)	17-0165-65
Xanthylium perchlorate	70% $HClO_4$	258(4.70),375(4.63)	17-0021-65
$C_{13}H_9Cl_2N$			
Aniline, p-chloro-N-(p-chloro-benzylidene)-	isooctane	271(4.33),320(4.05)	9-0091-65
	EtOH	269(4.29),320(4.07)	9-0091-65
$C_{13}H_9Co_3O_9Sn$			
Cobalt, (butylstannylidyne)-tris[tricarbonyl-	n.s.g.	207(4.62),364(3.85)	77-0114-65
$C_{13}H_9Cs$			
Cesium, fluoren-9-yl-	$C_6H_{11}NH_2$	447(2.99),472(3.08), 504(2.94)	35-0384-65
$C_{13}H_9IO_2$			
(o-Carboxyphenyl)phenyliodonium hydroxide, inner salt	H_2O	205(4.43),266(4.02)	44-0445-64
$C_{13}H_9Li$			
Lithium, fluorenyl-	$C_6H_{11}NH_2$	452(3.03),477(3.11), 510(2.92)	35-0384-65
$C_{13}H_9N$			
Acridine	EtOH	243s(4.87),246s(5.18), 249(5.26),324s(3.47), 332s(3.62),339(3.79), 347(3.83),356(3.98), 377s(3.39)	12-0108-65
	H_2SO_4	250s(4.88),255(5.08), 323s(3.44),332s(3.72), 338(3.97),348s(4.03), 354(4.31),385(3.48), 400(3.46),424s(3.21)	12-0108-65
Phenanthridine	pH 1.00	248(4.54),316(3.83), 355(3.53)	39-4426-65
	pH 9.17	248(4.61),290(3.75), 330(3.22),346(3.18)	39-4426-65
$C_{13}H_9NO$			
Benzoxazole, 2-phenyl-	EtOH	298(3.16)	88-2365-65
Carbazole-3-carboxaldehyde	n.s.g.	238(4.40),244s(4.30), 273(4.52),288(4.49), 327(4.09)	78-0681-65
Cyanic acid, biphenylyl ester	$CHCl_3$	251(4.28)	28-2839-65A
Fluorenone, 4-amino-	iso-PrOH	405(2.86)	39-5518-65
Isocyanic acid, biphenylyl ester	$CHCl_3$	263(4.36)	28-2839-65A
$C_{13}H_9NO_2$			
1-Acenaphthenecarboxamide, 2-oxo-	EtOH	228(4.09),244(3.87), 252(3.87),283(3.43), 303(3.46),315(3.39), 335(3.26)	44-2610-65
2H-Isoxazolo[2,3-a]pyridin-2-one, 3-phenyl-	EtOH	266(3.24),305(3.40), 388(2.80)	94-0595-64

Compound	Solvent	$\lambda_{max}(\log \epsilon)$	Ref.
$C_{13}H_9NO_2S$			
5H-Cyclopenta[d,e]-1-benzothiopyran-4-carboxamide, 6-methyl-5-oxo-	EtOH	212(4.59),244(4.28), 300(4.29),350s(3.43)	88-2489-64
	H_2SO_4	207(4.17),222(4.26), 240(4.25),301(4.41), 371(4.16)	88-2489-64
Thiobenzophenone, 4-nitro-	C_6H_{12}	299(4.20),320s(--), 625(2.23)	54-0289-65
	EtOH	302(4.19),320s(--), 615(2.22)	54-0289-65
$C_{13}H_9NO_3$			
Benzophenone, o-nitro-	MeOH	249(4.21)	39-2579-64
Benzophenone, p-nitro-	MeOH	266(4.33)	39-2579-64
$C_{13}H_9NO_3S$			
7-Sulfoacridizinium hydroxide, inner salt	H_2O	208(4.26),248s(3.58), 270(3.58),358(2.82), 375(2.82),396(2.78)	4-0228-65
9-Sulfoacridizinium hydroxide, inner salt	EtOH	215s(3.70),245(4.62), 253(4.58),267s(4.31), 347s(3.68),360(3.92), 380(4.02),399(4.01)	4-0030-64
$C_{13}H_9NO_4$			
Kokusagine	EtOH	255(4.83),312s(3.62), 326(3.71),348(3.80)	2-0449-64
$C_{13}H_9NS$			
Benzo[b]cyclohepta[e][1,4]thiazine	MeOH-HCl	470(3.8)	73-3016-65
	MeOH	412(3.8)	73-3016-65
Benzothiazole, 2-phenyl-	EtOH	248(2.90),290(3.18), 296(3.19)	88-2365-65
$C_{13}H_9N_3O_2$			
Benzimidazole, 2-p-nitrophenyl-	EtOH	340(4.31)	94-0773-64
Benzimidazole, 5-nitro-1-phenyl-	MeOH	245(4.52),274(4.07), 281(4.07),306(4.01)	65-0632-64
$C_{13}H_9N_3O_6$			
Methane, (2,4-dinitrophenyl)-(p-nitrophenyl)-	EtOH	248s(4.26),262(4.30)	44-0636-64
$C_{13}H_9N_3S$			
2-Thia-3,6,10c-triazaaceanthrene, 4-methyl-	n.s.g.	277(4.51),347(4.02)	44-2064-64
$C_{13}H_9N_5$			
Benzenediazocyanide, p-phenylazo-, anti	EtOH	226(4.10),341(4.21), 354s(4.16),448(3.32)	39-0751-64
1-(Cyanoamino)-2-phenyl-2H-benzotriazolium hydroxide, inner salt	EtOH	219(4.30),303(4.16), 375(3.75)	39-0751-64
$C_{13}H_{10}$			
Fluorene	EtOH-DMSO	455(3.58),483(3.69), 517(3.56)	78-0261-65
$C_{13}H_{10}BrClO_4$			
7-(m-Bromophenyl)tropylium perchlorate	MeCN	269(4.15),358(4.14)	24-0029-64
7-(p-Bromophenyl)tropylium perchlorate	MeCN	274(4.11),379(4.26)	24-0029-64

Compound	Solvent	λ max(log ε)	Ref.
$C_{13}H_{10}BrN$			
Aniline, N-benzylidene-p-bromo-	isooctane	222(4.14),265(4.28), 320(3.98)	9-0091-65
	EtOH	220(4.16),265(4.24), 315(4.04)	9-0091-65
Aniline, N-(p-bromobenzylidene)-	isooctane	271(4.34),320(3.95)	9-0091-65
	EtOH	271(4.32),315(4.03)	9-0091-65
Pyridine, 3-(p-bromostyryl)-, cis	hexane	224s(4.25),273(4.09)	32-1322-65
trans	hexane	225s(4.02),312(4.43)	32-1322-65
$C_{13}H_{10}BrNO$			
Aniline, m-bromo-N-salicylidene-	EtOH	220(4.4),272(4.11), 342(4.03)	65-3837-64
Aniline, p-bromo-N-salicylidene-	EtOH	230(4.3),272(4.17), 340(4.16)	65-3837-64
$C_{13}H_{10}BrNO_2$			
7,10-Dihydroxybenzo[b]quinolizinium bromide	MeOH	248(4.41),375(3.73)	44-0252-65
$C_{13}H_{10}Br_2N_2O_4$			
3-Indolinealanine, N-acetyl-5,7-dibromo-3-hydroxy-2-oxo-, γ-lactone	H_2O	266(4.00),315(3.48)	37-1165-65
$C_{13}H_{10}Br_3NO$			
2,5-Cyclohexadien-1-one, 2,6-dibromo-4-(o-bromoanilino)-4-methyl-	MeOH	240(4.08),256(4.09)	35-1127-64
$C_{13}H_{10}Br_3NO_5$			
3,4-Indolinedicarboxylic acid, 3,5,6-tribromo-1-methyl-2-oxo-, dimethyl ester	EtOH	285s(4.38),330(3.69)	39-3229-64
$C_{13}H_{10}ClFO_4$			
7-(m-Fluorophenyl)tropylium perchlorate	MeCN	269(4.16),357(4.11)	24-0029-64
7-(p-Fluorophenyl)tropylium perchlorate	MeCN	271(4.14),373(4.19)	24-0029-64
$C_{13}H_{10}ClN$			
Aniline, N-benzylidene-p-chloro-	EtOH	221(4.16),265(4.25), 315(4.02)	9-0091-65
Aniline, N-(p-chlorobenzylidene)-	isooctane	269(4.32),310(3.94)	9-0091-65
	EtOH	270(4.28),310(3.98)	9-0091-65
Pyridine, 2-(o-chlorostyryl)-, cis	n.s.g.	285(4.03)	77-0288-65
Pyridine, 3-(p-chlorostyryl)-, cis	hexane	224s(4.26),273(4.09)	32-1322-65
Pyridine, 3-(p-chlorostyryl)-, trans	hexane	227s(4.09),305(4.43)	32-1322-65
Pyridine, 4-(p-chlorostyryl)-, cis	hexane	225s(4.23),282(4.08)	32-1322-65
$C_{13}H_{10}ClNO$			
Aniline, m-chloro-N-salicylidene-	EtOH	220(4.3),272(4.11), 340(4.02)	65-3837-64
Aniline, p-chloro-N-salicylidene-	EtOH	230(4.3),272(4.13), 340(4.11)	65-3837-64
Benzophenone, 2-amino-5-chloro-	MeOH	238(4.45)	10-0334-64E
$C_{13}H_{10}ClN_3O_3S$			
Acetanilide, 2'-[(5-chloro-3-nitro-2-pyridyl)thio]-	EtOH	236(4.34),284(4.10), 371(3.64)	95-0429-65

Compound	Solvent	λ_{max}(log ϵ)	Ref.
$C_{13}H_{10}ClN_3O_5S$			
Acetanilide, 2'-[(5-chloro-3-nitro-2-pyridyl)sulfonyl]-	EtOH	237s(4.25),276s(3.80), 317s(3.54)	95-0429-65
$C_{13}H_{10}Cl_2N_4$			
Formazan, N,N'-bis(o-chlorophenyl)-	EtOH	244(3.99),296(3.85), 418(4.45)	5-0099-65B
Formazan, N,N'-bis(p-chlorophenyl)-	EtOH	262(4.08),304(3.79), 432(4.30)	5-0099-65B
$C_{13}H_{10}Cl_2N_4O_4$			
Methanediamine, N,N'-bis-(2-chloro-4-nitrophenyl)-	DMF	423(4.0)	7-0969-64
$C_{13}H_{10}Cl_2O_4$			
7-(m-Chlorophenyl)tropylium perchlorate	MeCN	269(4.17),356(4.13)	24-0029-64
7-(p-Chlorophenyl)tropylium perchlorate	MeCN	273(4.13),374(4.24)	24-0029-64
$C_{13}H_{10}Cl_4N_2$			
Methanediamine, N,N'-bis(2,5-dichloro-phenyl)-	EtOH	260(4.4),301(3.8)	7-0969-64
$C_{13}H_{10}CrO_4$			
Chromium, tricarbonyl(1-tetralone)	n.s.g.	322.5(3.96)	28-2833-65A
$C_{13}H_{10}FeO$			
Ferrocenylpropynal	$CHCl_3$	302(4.13),366(3.34), 474(3.15)	49-1750-64
$C_{13}H_{10}IN$			
Pyridine, 3-(p-iodostyryl)-, cis	hexane	225s(4.15),275(4.01)	32-1322-65
Pyridine, 3-(p-iodostyryl)-, trans	hexane	229(4.07),317(4.49)	32-1322-65
$C_{13}H_{10}N_2$			
Acridine, 1-amino-	5N HCl	258(5.01),341(3.97), 358(4.29),389(3.51), 403(3.52)	39-4653-65
	pH 2.5	236(4.50),287(4.55), 343(3.68),358(3.95), 524(3.45)	39-4653-65
	pH 11.0	238(4.56),262(4.64), 325(3.14),343(3.37), 349(3.32),358(3.57), 413(3.49)	39-4653-65
Acridine, 2-amino-	5N HCl	257(5.03),339(3.96), 355(4.27),384(3.50)	39-4653-65
	pH 2.5	256(4.50),273(4.69), 356(3.81),371(3.92), 462(3.55)	39-4653-65
	pH 11.0	242s(4.42),260(4.89), 325s(3.32),339(3.56), 355(3.76)	39-4653-65
Acridine, 3-amino-	pH 2.5	233(4.62),274(4.65), 349(4.03),365(4.15), 454(4.10)	39-4653-65
	pH 11.0	237(4.46),262(4.83), 321(3.35),337(3.65), 353(3.92),410(3.79)	39-4653-65

Compound	Solvent	λ_{max}(log ϵ)	Ref.
Acridine, 4-amino-	5N HCl	259(4.99),338(3.97), 355(4.30),396(3.51), 412(3.53)	39-4653-65
	pH 2.5	243(4.48),250(4.49), 277(4.53),363(3.92), 463(3.20)	39-4653-65
	pH 11.0	239(4.27),262(4.33), 327(3.13),342(3.33), 359(3.47),405(3.46)	39-4653-65
Acridine, 9-amino-	pH 6.0	217(4.36),260(4.92), 311(3.21),326(3.24), 381(3.81),401(4.00), 423(3.90)	39-4653-65
	pH 12-14	218(4.33),260(4.86), 389(3.83),406(3.89), 425(3.66)	39-4653-65
Benzimidazole, 1-phenyl-	MeOH	244(4.19),274(3.72), 281s(3.64)	65-0632-64
hydrobromide	MeOH	245(4.11),275(3.72), 282(3.59)	65-0632-64
Benzimidazole, 2-phenyl-	MeOH	302(3.38)	44-0259-65
	EtOH	242(3.07),297(3.15), 301(3.16)	88-2365-65
	EtOH	304(4.41)	94-0773-64
Fluorenone, hydrazone	$CHCl_3$	297(4.22),306s(4.20), 332(4.27)	44-0417-65
Heptafulvene, 1-methyl- 8-(2,2-dicyanovinyl)-	benzene	273(4.08),287s(4.01), 340s(3.09),360(3.24), 464(4.57)	24-2050-64
	EtOH	258s(3.94),276(4.07), 283s(4.07),352(3.21), 465(4.62)	24-2050-64
	50% EtOH	258(4.38),280(4.51), 323(3.70),328(3.68), 350(3.73),489(5.05)	24-2050-64
Imidazo[1,2-a]pyridine, 2-phenyl-	EtOH	204(4.5),245(4.6), 280s(3.7),310(3.9), 322(3.9)	70-1434-65
Indole, 1-(2-pyridyl)-	EtOH	210(4.00),255(3.95), 309(3.67)	44-2534-65
Indole, 3-(2-pyridyl)-	EtOH	223(4.44),267(4.07), 311(4.09)	44-2534-65
Phenanthridine, 6-amino-	pH 2.00	240(4.62),294(3.68), 332(3.94),348(3.93)	39-4426-65
	pH 10.0	238(4.66),302(3.76), 316(3.77),332(3.80), 348(3.79)	39-4426-65
Phenazine, 2-methyl-	C_6H_{12}	212(4.30),252(5.05), 292(3.70),362(4.10)	65-1969-64
$C_{13}H_{10}N_2O$			
1-Azulenecarbonitrile, 3-acetyl-2-amino-	EtOH	233(4.52),260(4.14), 291(4.67),303(4.74)	94-0443-65
1-Benzimidazolol, 2-phenyl-	EtOH	298(4.33)	44-1537-64
Benzo[f]quinazolin-1-ol, 3-methyl-	pH 1	240s(4.05),270(4.49), 324(3.67),340(3.46)	44-2881-64
	pH 10	262(4.61),290(3.86), 316(3.50),331(3.57), 346(3.58)	44-2881-64

Compound	Solvent	λ_{max}(log ε)	Ref.
Benzo[f]quinazolin-1-ol, 3-methyl- (contd.)	50% EtOH	261(4.63),270(4.58), 298(3.90),316(3.61), 331(3.67),340(3.46), 346(3.70)	44-2881-64
Benzo[f]quinazolin-3-ol, 1-methyl-	pH 1	232(4.72),257(4.23), 277(3.80),288(3.90), 360(3.98)	44-2881-64
	pH 10	235(4.58),248(4.44), 258(4.47),264(4.47), 282(4.09),305s(3.69), 356(3.63),367(3.62)	44-2881-64
	50% EtOH	235(4.72),250(4.39), 259(4.32),284(3.92), 318(3.85),358(3.72), 371(3.72)	44-2881-64
Benzo[g]quinazolin-4-ol, 2-methyl-	pH 1	249s(4.47),266(4.66), 274(4.64),311(3.52), 352(3.50)	44-2881-64
	pH 10	240(4.59),253s(4.59), 260(4.65),269(4.66), 299(3.55),311(3.60), 324(3.48),360(3.56), 372s(3.47)	44-2881-64
	50% EtOH	235(4.72),250(4.39), 259(4.32),284(3.92), 318(3.85),358(3.72), 371(3.72)	44-2881-64
Benzo[g]quinazolin-4-ol, 2-methyl-	pH 1	249s(4.47),266(4.66), 274(4.64),311(3.52), 352(3.50)	44-2881-64
	pH 10	240(4.59),253s(4.59), 260(4.65),269(4.66), 299(3.55),311(3.60), 324(3.48),360(3.56), 372s(3.47)	44-2881-64
	EtOH	235(4.58),252s(4.62), 260(4.72),296(3.62), 308(3.69),322(3.59), 345s(3.39),357(3.57), 373(3.44)	44-2881-64
Phenazine, 1-methoxy-	EtOH	260(4.74),360(3.87), 402(3.45)	12-1241-65
	50% MeOH-acid	268(4.5),372(4.0), 383(4.0),437(3.2), 481s(3.17)	59-1665-64
	MeOH	261(4.8),350s(3.8), 360(3.9),366(3.9), 410(3.4)	59-1665-64
Phenazine, 2-methoxy-	EtOH	255(4.9),360(4.0), 395s(3.9)	12-1241-65
	50% MeOH-acid	263(4.8),390(3.8), 453(3.4)	59-1665-64
	MeOH	257(4.9),361(4.0), 392(3.9)	59-1665-64
2-Phenazinemethanol	C_6H_{12}	210(4.35),254(5.01), 290(3.87),365(4.15)	65-1969-64
2-Phenazinol, 3-methyl-	MeOH	259(4.11),372(3.33), 395(3.15)	59-1665-64
	50% MeOH-HCl	263(4.09),396(3.59), 452(3.08)	59-1665-64

Compound	Solvent	λ_{max}(log ϵ)	Ref.
2-Phenazinol, 3-methyl- (contd.)	50% MeOH-NaOH	278(4.01),377(3.29), 479(3.20)	59-1665-64
2-Phenazinol, 4-methyl-	MeOH	262(4.30),362(3.34), 415(3.20)	59-1665-64
	50% MeOH-HCl	268(4.25),389(3.70), 470(3.22)	59-1665-64
	50% MeOH-NaOH	282(4.19),372(3.24), 485(3.25)	59-1665-64
Pyocyanine (as perchlorate)	KCl disc	290(4.1),402(3.7), 552(3.2)	59-1665-64
$C_{13}H_{10}N_2O_2$			
Aniline, N-benzylidene-p-nitro-	isooctane	295(4.25),315(4.19)	9-0091-65
	EtOH	328(4.23)	9-0091-65
	$C_2H_4Cl_2$	226(4.21),332(4.25)	30-1017-64
Aniline, N-(o-nitrobenzylidene)-	isooctane	223(4.27),263(4.16), 330(3.82)	9-0091-65
	EtOH	260(4.15),325(3.79)	9-0091-65
Aniline, N-(p-nitrobenzylidene)-	isooctane	230(4.09),288(4.09), 344(4.01)	9-0091-65
	EtOH	240(4.03),290(4.08), 340(3.92)	9-0091-65
	$C_2H_4Cl_2$	292(4.21),347(4.06)	30-1017-64
4,5'-Biisoxazole, 5-methyl-3'-phenyl-	EtOH	233(4.33)	7-1223-65
Pyridine, 3-(p-nitrostyryl)-, cis	hexane	307(4.04)	32-1322-65
trans	hexane	326s(4.47),335(4.48)	32-1322-65
Pyridine, 4-(p-nitrostyryl)-, cis	hexane	222s(4.12),307(4.13)	32-1322-65
$C_{13}H_{10}N_2O_2S$			
1H-2,1,3-Benzothiadiazine, 4-phenyl-, 2,2-dioxide	EtOH	223(4.31),269(4.1), 359(3.51)	44-3960-65
$C_{13}H_{10}N_2O_3$			
Aniline, m-nitro-N-salicylidene-	EtOH	270(4.27),340(3.97)	65-3837-64
Aniline, p-nitro-N-salicylidene-	EtOH	320(4.18),356(4.24)	65-3837-64
Aniline, N-(5-nitrosalicylidene)-	MeOH	232(4.19),307(4.07)	24-1631-64
	MeOH-NaOH	400(4.28)	24-1631-64
Benzoic acid, o-[(p-hydroxyphenyl)azo]-	6N HCl	476(4.60)	10-0375-64E
	2.4N HCl	360(4.1),470(4.1)	10-0375-64E
	0.24N HCl	355(4.3)	10-0375-64E
	pH 6.2	348(4.31)	10-0375-64E
	pH 8.5	375(4.1)	10-0375-64E
	pH 13	400(4.4)	10-0375-64E
	EtOH	355(4.3),480(3.7)	10-0375-64E
	PrOH	360(4.3),480(3.9)	10-0375-64E
	acetone	370(4.2),495(3.9)	10-0375-64E
	DMSO	348(4.1),500(4.2)	10-0375-64E
	DMF	345(3.9),502(4.5)	10-0375-64E
Formanilide, N-(p-nitrophenyl)-	EtOH	228(4.14),312(4.05)	44-3427-65
3-Isoxazolecarboxylic acid, 4-cyano-5-phenyl-, ethyl ester	C_6H_{12}	276.5(4.27)	32-1478-65
	EtOH	277.5(4.35)	32-1478-65
3-Isoquinolinecarboxylic acid, 5-cyano-1,2-dihydro-2-methyl-1-oxo-, methyl ester	EtOH	256(3.98),322(4.02)	44-2534-64
$C_{13}H_{10}N_2O_4$			
Azepine-N-carboxylic acid, p-nitro-, phenyl ester	hexane	208(4.56),265(4.08)	88-1733-64
Benzohydroxamic acid, p-nitro-N-phenyl-	MeOH	253(4.21)	73-0940-65
	50% MeOH-HCl	252s(4.10)	73-0940-65

Compound	Solvent	$\lambda_{max}(\log \epsilon)$	Ref.
Benzohydroxamic acid, p-nitro-N-phenyl- (contd.)	50% MeOH-borate	272s(4.04)	73-0940-65
	50% MeOH-NaOH	280(4.10)	73-0940-65
1,2-Indolizinedicarboxylic acid, 3-cyano-, dimethyl ester	MeOH	216(4.28),241(4.58), 273(3.97),317(4.10)	35-3651-65
Methane, bis(p-nitrophenyl)-	EtOH-DMSO	704(4.61)	78-0261-65
Methane, (m-nitrophenyl)-(p-nitrophenyl)-	EtOH-DMSO	550(4.50)	78-0261-65
Methane, (o-nitrophenyl)-(p-nitrophenyl)-	EtOH-DMSO	555(4.19)	78-0261-65
$C_{13}H_{10}N_2O_4S_2$			
Methane, diazobis(phenylsulfonyl)-	MeCN	232(4.24),375(1.86)	44-2272-65
$C_{13}H_{10}N_2S$			
Thiochroman-4-ylidenemalononitrile, 6-methyl-	EtOH	245(4.11),267(4.04), 273(4.04),310(4.02)	88-2489-64
$C_{13}H_{10}N_4$			
Aniline, N,N-dimethyl-4-(tricyanovinyl)-	$CHCl_3$	515(4.68)	39-1334-64
Benzonitrile, o-[(p-aminophenyl)azo]-	EtOH	255(4.12),422(4.34)	39-3663-64
Benzonitrile, o-(3-phenyl-1-triazeno)-	EtOH	237(4.08),300s(3.65), 368(4.31)	39-3663-64
1,2,3-Benzotriazine, 4-anilino-	EtOH	241s(4.11),272(3.85), 334(4.11)	39-3663-64
1,2,3-Benzotriazine, 3,4-dihydro-4-imino-3-phenyl-	EtOH	260(3.97),268(3.95), 307(3.76),318(3.77)	39-3663-64
1H-Imidazo[4,5-b]pyridine, 2-[2-(3-pyridyl)vinyl]-	MeOH	255(4.02),335(4.58)	44-3403-64
1H-Imidazo[4,5-b]pyridine, 2-[2-(4-pyridyl)vinyl]-	MeOH	255(3.86),336(4.50)	44-3403-64
$C_{13}H_{10}N_4O$			
Tetrazolone, 1,3-diphenyl-	EtOH	225(4.16),261(3.97), 325(4.12)	39-0906-64
$C_{13}H_{10}N_4O_2S$			
2,1,3-Benzothiadiazol-4-ol, 7-[(o-methoxyphenyl)azo]-	EtOH	467(4.39)	17-0041-64
2,1,3-Benzothiadiazol-4-ol, 7-[(p-methoxyphenyl)azo]-	EtOH	465(4.35)	17-0041-64
2,1,3-Benzothiadiazol-5-ol, 4-[(o-methoxyphenyl)azo]-	EtOH	490(4.32)	17-0041-64
2,1,3-Benzothiadiazol-5-ol, 4-[(p-methoxyphenyl)azo]-	EtOH	492(4.25)	17-0041-64
$C_{13}H_{10}N_4O_3$			
Benzamide, p-[(p-nitrophenyl)azo]-	EtOH	337(4.47)	24-0172-64
$C_{13}H_{10}N_4O_4$			
Formamidine, N,N'-bis(p-nitrophenyl)-	pyridine	387(4.65)	22-0393-65
$C_{13}H_{10}N_4S_2$			
Thiocyanic acid, diester with 1-(p-mercaptophenyl)-3,5-dimethylpyrazole-4-thiol	EtOH	263(4.27)	94-0023-64

Compound	Solvent	$\lambda_{max}(\log \epsilon)$	Ref.
$C_{13}H_{10}N_6$			
1H-s-Triazolo[5,1-i]purine, 1-benzyl-	pH 1	276(3.89)	44-3601-65
	pH 7	280(3.91)	44-3601-65
3H-s-Triazolo[5,1-i]purine, 3-benzyl-	pH 7	264(3.90),278(3.89)	44-3601-65
7H-s-Triazolo[3,4-i]purine, 7-benzyl-	pH 1	258(3.96),265(3.99)	44-3601-65
	pH 7	248(3.74),256(3.85), 265(3.88),288(3.72)	44-3601-65
9H-s-Triazolo[3,4-i]purine, 9-benzyl-	pH 7	260(3.99),268(3.98), 285(3.67)	44-3601-65
1,2,4-Triazolo[1,5-a]pyrimidine, 7-(benzimidazol-1-yl)-5-methyl-	MeOH	242(3.9),275(3.9), 305(4.0)	65-0204-64
$C_{13}H_{10}N_8O_2$			
6(5H)-Pteridinone, 7-[(7,8-dihydro-7-oxo-6-pteridinyl)methyl]-7,8-dihydro-	pH 2.6	297(4.30)	39-3357-64
	pH 5.7	298(4.25)	39-3357-64
	pH 9.0	264s(3.92),305(4.12)	39-3357-64
$C_{13}H_{10}N_{10}O_4$			
6-Pteridinecarboxaldehyde, 2,4-diamino-, 2,4-dinitrophenylhydrazone	pH 1	244(4.16),262s(3.86), 334(4.04)	87-0713-65
	pH 13	225(4.02),257(4.27), 275s(4.00),372(3.93)	87-0713-65
$C_{13}H_{10}O$			
2'-Acrylonaphthone	EtOH	290(2.42),330(2.37)	65-0190-64
2-Biphenylcarboxaldehyde	hexane	213(4.25),231(4.29), 234s(4.28),245s(4.08), 296(3.52)	22-2953-64
	EtOH	214(4.26),233(4.32), 249s(4.09),300(3.52)	22-2953-64
3-Biphenylcarboxaldehyde	hexane	210(4.27),238(4.47), 239(4.46),299(3.12)	22-2953-64
	EtOH	211(4.16),242(4.48), 303(3.17)	22-2953-64
$C_{13}H_{10}OS$			
Benzophenone, o-mercapto-	EtOH	242(4.16),332(3.48)	44-1859-65
Benzophenone, p-mercapto-	EtOH	240(4.06),300(3.88), 380(4.01)	44-1859-65
$C_{13}H_{10}OS_2$			
1(4H)-Naphthalenone, 4-(1,3-dithiolan-2-ylidene)-	MeOH	310(4.19),322(4.20), 434(4.11),457(4.23), 482(4.08)	5-0037-65D
	dioxan	434(4.10)	5-0037-65D
	80% dioxan-20% $HCONH_2$	445(4.09)	5-0037-65D
	20% dioxan-80% $HCONH_2$	459(4.08)	5-0037-65D
2(1H)-Naphthalenone, 1-(1,3-dithiolan-2-ylidene)-	MeOH	301(4.14),450(4.06), 528(4.08)	5-0037-65D
	$HCONH_2$	430(4.05),450(4.03)	5-0037-65D
	HOAc	423(4.03),446(3.92)	5-0037-65D
1,4-Pentadien-3-one, 1,5-di(2-thienyl)-	hexane	275(3.94),355(4.51)	65-3645-64
	EtOH	278(3.89),370(4.51)	65-3645-64
	66% H_2SO_4	245(3.80),315(3.89), 385(4.00),520(4.73)	65-3645-64
$C_{13}H_{10}O_2$			
Fluorene-2,7-diol	EtOH	276(4.42),288s(4.32), 321(3.70)	88-3505-65

Compound	Solvent	λ max(log ε)	Ref.
2-Furanacrolein, 5-(2,4-hexadiyn-1-yl)-, trans	ether	317.5(4.51)	24-2596-65
2,4-Hexadiynophenone, 4'-methoxy-	EtOH	315(4.25)	39-2983-65
$C_{13}H_{10}O_2S$			
1,4-Pentadien-3-one, 1-(2-furyl)-5-(2-thienyl)-	EtOH	257(3.79),374(4.47)	65-2328-64
	HOAc-30% H_2SO_4	315(3.90),383(4.02),518(4.83)	65-2328-64
2,4-Pentadien-1-one, 1-(2-furyl)-5-(2-thienyl)-	EtOH	220(4.00),378(4.50)	65-2328-64
	HOAc-30% H_2SO_4	305(3.57),351(3.81),525(4.69)	65-2328-64
2,4-Pentadien-1-one, 5-(2-furyl)-1-(2-thienyl)-	EtOH	278(4.00),380(4.53)	65-2328-64
	HOAc-30% H_2SO_4	315(3.96),367(4.01),523(4.61)	65-2328-64
Thiophene, 2-(3,4-dihydroxy-1-butynyl)-5-(1,3-pentadiynyl)-	ether	208(4.45),233(3.92),246(4.07),319(4.50),340(4.49)	24-0155-65
$C_{13}H_{10}O_2Se$			
1,3-Propanedione, 1-phenyl-3-selenophene-2-yl-	n.s.g.	285(4.20),360(4.76)	67-0379-65
copper salt	n.s.g.	297(4.47),355(4.06)	67-0379-65
$C_{13}H_{10}O_3$			
5-Benzofuranacrylic acid, 6-hydroxy-α,β-dimethyl-, δ-lactone	EtOH	245(4.38),290(4.02),329(3.95)	44-2467-64
Benzoic acid, p-(1-hydroxy-2,4-hexadiynyl)-	EtOH	235(4.39)	39-2983-65
Benzophenone, 2,2'-dihydroxy-	EtOH	259(4.04),336(3.72)	39-5074-64
Benzophenone, 4,4'-dihydroxy-	EtOH	227(4.17),300(4.36)	44-1812-64
1,4-Pentadien-3-one, 1,5-di-2-furyl-	hexane	243(3.79),350(4.61)	65-3645-64
	EtOH	241(3.91),370(4.55)	65-3645-64
	HOAc-H_2SO_4	517(4.78)	65-3645-64
2-Pyrone, 4-hydroxy-6-styryl-	EtOH	288(3.90)	65-2766-64
3,9(2H)-Xanthenedione, 1,4-dihydro-	MeOH	226(4.33),243(3.99),263(3.79),299(3.91)	94-0214-64
$C_{13}H_{10}O_4$			
1,4-Naphthoquinone, 6-acetyl-5-hydroxy-7-methyl-	EtOH	212(4.40),226s(4.30),249s(4.19),425(3.64)	12-0218-65
1,4-Naphthoquinone, 2-acetyl-8-methoxy-	EtOH	247(4.15),327(3.27),402(3.53)	33-0769-64
1,4-Naphthoquinone, 5-acetyl-8-methoxy-	EtOH	248(4.26),357(3.34),398(3.48)	54-1005-64
2H-Pyran-4-carboxylic acid, 6-methyl-2-oxo-, methyl ester	EtOH	253(4.02),360(4.06)	44-3312-65
$C_{13}H_{10}O_6$			
1H-2-Benzopyran-3,5-dicarboxylic acid, 1-oxo-, dimethyl ester	EtOH	243(4.12),292(3.98),335s(3.32)	44-2534-64
$C_{13}H_{10}S$			
Naphtho[2,1-b]thiophene, 4-methyl-	EtOH	235s(4.62),244(4.67),257(4.46),290s(3.92),300(4.07),313(4.02),333s(2.73),339(2.71)	39-6221-65
Naphtho[2,1-b]thiophene, 5-methyl-	EtOH	233s(4.59),246(4.75),256(4.39),286(3.92),294(4.03),306(4.01),320(3.43),327(3.26),335(3.39)	39-6221-65

Compound	Solvent	λ max(log ε)	Ref.
Naphtho[2,1-b]thiophene, 6-methyl-	EtOH	225(4.28),233(4.31), 247(4.51),255(4.31), 296(3.95),306(3.90), 322(3.42),337(3.25)	39-6221-65
Naphtho[2,1-b]thiophene, 7-methyl-	EtOH	238s(4.59),248(4.63), 258(4.48),289(3.98), 301(4.19),313(4.11), 337(3.05)	39-6221-65
Thiobenzophenone	C_6H_{12}	236(3.95),315(4.19), 609(2.25)	54-0289-65
	EtOH	235(3.95),317(4.20), 599(2.26)	54-0289-65
Thiophene, 2-(3-buten-1-yn-1-yl)- 5-(3-penten-1-yn-1-yl)-	ether	334(4.51),355s(4.31)	24-2125-64
	n.s.g.	334(4.51),354s(4.31)	88-0297-65
Thiophene, 2-(3,5-hexadien-1-ynyl)- 5-(1-propynyl)-	ether	253(4.05),334(4.48), 358(4.42)	24-2125-64
	n.s.g.	334(4.42),357(4.47)	88-0297-65
$C_{13}H_{10}S_2$			
2,2'-Bithiophene, 5-(3-buten-1-yn-1-yl)-5'-methyl-	ether	250(3.97),349(4.46)	24-0883-65
$C_{13}H_{11}BrN_4$			
Pyrazolo[1,5-a]pyrimidine, 7-amino-3-bromo-1-methyl-5-phenyl-	EtOH-HCl	257(4.51),288s(3.84), 336(3.74)	95-1113-64
$C_{13}H_{11}BrOS_2$			
4,5-Dihydro-2-(2-hydroxy-1-naphthyl)-1,3-dithiolium bromide	MeCN	300(4.13),426(4.00), 446(3.98)	24-1374-65
$C_{13}H_{11}BrO_2$			
2-Naphthoic acid, 7-bromo-1-methyl-, methyl ester	MeOH	235(4.70),248(3.78), 285(3.80),295(3.60)	44-2109-64
$C_{13}H_{11}BrO_5$			
2-Naphthoic acid, 8-bromo-4-hydroxy-5,6-dimethoxy-	EtOH	252(4.44),332(3.82), 345(3.85),355(3.84)	39-4292-65
$C_{13}H_{11}Br_2NO$			
2,5-Cyclohexadien-1-one, 4-anilino-2,6-dibromo-4-methyl-	MeOH	242(4.14),256(4.09)	35-1127-64
$C_{13}H_{11}Br_2NO_2$			
1H-Pyrrolo[1,2-a]indol-1-one, 2,9-di-bromo-2,3-dihydro-7-methoxy-6-methyl-	MeOH	219(4.46),249s(3.99), 343(4.3)	44-2904-65
$C_{13}H_{11}Br_2NO_4$			
Indole-2,4-dicarboxylic acid, 3,6-di-bromo-1-methyl-, dimethyl ester	MeOH	243(4.39),316(4.14)	39-3229-64
$C_{13}H_{11}ClN_2$			
Pyrrolo[1,2-a]quinoxaline, 1-chloro-2,4-dimethyl-	pH 1.15	234s(4.49),240(4.50), 267(3.97),364(4.17)	39-3678-65
	pH 7.1	216(4.27),236(4.48), 260s(4.07),337(4.01)	39-3678-65
	EtOH	232(4.46),240s(4.43), 260s(4.10),336(3.99), 349s(3.85)	35-1830-64

Compound	Solvent	$\lambda_{max}(\log \epsilon)$	Ref.
$C_{13}H_{11}ClN_2O$			
Benzanilide, 4-amino-2'-chloro-	EtOH	214(4.3),301(4.4)	28-4295-64B
Benzanilide, 4-amino-3'-chloro-	EtOH	216(4.3),304(4.5)	28-4295-64B
Benzanilide, 4-amino-4'-chloro-	EtOH	227(4.4),304(4.6)	28-4295-64B
Pyrrolo[1,2-a]quinoxaline, 4-chloro-1-ethoxy-	EtOH	231(4.57),234s(4.57), 270(4.15),276s(4.09), 304(3.48),314(3.67), 327(3.78),362(3.95)	35-1830-64
$C_{13}H_{11}ClN_2O_2$			
1H-Azepino[5,4,3-cd]indole-2-carboxylic acid, 9-chloro-3,4-dihydro-6-methyl-	EtOH-HCl	221(4.35),248(4.09), 267(4.19),346(3.90), 397(3.92)	87-0200-65
β-Carboline, 5-acetyl-8-chloro-1,2,3,4-tetrahydro-1-oxo-	EtOH	226(4.34),255(4.24), 322(4.08)	87-0200-65
$C_{13}H_{11}ClN_4$			
5-Pyrimidinepropionitrile, 2-amino-4-chloro-6-phenyl-	pH 1	235(4.21),325(4.00)	4-0263-64
	pH 8.4	237(4.43),310(3.97)	4-0263-64
	pH 13	237(4.44),310(3.91)	4-0263-64
$C_{13}H_{11}ClN_4O_4$			
5-Norbornen-2-one, 3-chloro-, 2,4-dinitrophenylhydrazone	EtOH	359(4.36)	35-4074-64
$C_{13}H_{11}ClN_4S$			
Pteridine, 4-(benzylthio)-6-chloro-3,4-dihydro-	$CHCl_3$	281(3.73),327(3.63)	39-4920-64
Pteridine, 4-(benzylthio)-7-chloro-3,4-dihydro-	$CHCl_3$	341(3.96)	39-4920-64
$C_{13}H_{11}ClO_2$			
2-Naphthoic acid, 7-chloro-1-methyl-, methyl ester	EtOH	233(4.74),243s(4.69), 274(3.90),284(3.91), 295(3.72)	44-2109-64
$C_{13}H_{11}ClO_4$			
7-Phenyltropylium perchlorate	MeCN	271(4.15),369(4.18)	24-0029-64
	MeCN	271(4.15),369(4.18)	24-1590-64
	$CHCl_3$	272(4.03),385(4.16)	24-0029-64
$C_{13}H_{11}ClO_5$			
Coumarilic acid, 5-chloro-3-(1-carboxypropyl)-	EtOH	217(4.24),265s(4.05), 274(4.11),282s(4.05), 303s(3.64)	44-4126-65
7-(p-Hydroxyphenyl)tropylium perchlorate	MeCN	274(4.03),435(4.37)	24-0029-64
$C_{13}H_{11}Cs$			
Cesium, (diphenylmethyl)-	$C_6H_{11}NH_2$	443(4.57)	35-0384-65
$C_{13}H_{11}F_2N_3$			
Aniline, 2,6-difluoro-N-methyl-4-(phenylazo)-, cis	EtOH-HCl	228(3.76),322(4.14), 539(4.26)	44-3878-65
	EtOH	251(4.02),379(4.15)	44-3878-65
trans	EtOH-HCl	229(3.83),322(4.19), 538(4.31)	44-3878-65
	EtOH	244(3.96),290s(3.63), 397(4.42)	44-3878-65

Compound	Solvent	$\lambda_{max}(\log \epsilon)$	Ref.
$C_{13}H_{11}N$			
Aniline, N-benzylidene-	isooctane	262(4.24),310(3.83)	9-0091-65
	EtOH	262(4.23),310(3.93)	9-0091-65
	$C_2H_4Cl_2$	237s(3.94),265(4.21), 312s(3.91)	30-1017-64
Macrorine	EtOH-HCl	216(4.42),260(4.58), 306(3.82),320(3.89), 335(3.98),365(3.74)	39-5969-64
Pyridine, 2-styryl-, cis	hexane	285(3.99)	32-1322-65
	aq. HCl	317(4.10)	23-1345-65
	H_2O	288(3.98)	23-1345-65
	50% MeOH-HCl	318(3.99)	23-1345-65
	50% MeOH	288(4.01)	23-1345-65
iodine complex	hexane	242(4.75),308(4.34), 426(3.22)	60-1406-65
Pyridine, 2-styryl-, trans	aq. HCl	334(4.45)	23-1345-65
	H_2O	309(4.42)	23-1345-65
	50% MeOH	310(4.42)	23-1345-65
iodine complex	hexane	237(4.69),317(4.37), 432(3.17)	60-1406-65
Pyridine, 3-styryl-, cis	hexane	270(4.03)	32-1322-65
iodine complex	hexane	242(4.75),422(3.28)	60-1406-65
Pyridine, 3-styryl-, trans	hexane	227s(4.05),300(4.38)	32-1322-65
iodine complex	hexane	238(4.74),295(4.52), 420(3.24)	60-1406-65
Pyridine, 4-styryl-, cis	hexane	270(3.90)	32-1322-65
iodine complex	hexane	240(4.76),290s(--), 420(3.28)	60-1406-65
Pyridine, 4-styryl-, trans, iodine complex	hexane	239(4.59),315(4.63), 421(3.22)	60-1406-65
$C_{13}H_{11}NO$			
Aniline, N-benzylidene-o-hydroxy-	isooctane	273(4.08),355(4.03)	9-0091-65
	EtOH	265(4.11),346(3.95)	9-0091-65
Aniline, N-(p-hydroxybenzylidene)-	EtOH	224(4.24),295(4.26), 315(4.29),417(1.54)	9-0091-65
Aniline, N-salicylidene-	isooctane	225(4.31),268(4.13), 340(4.05)	9-0091-65
	EtOH	225(4.29),270(4.11), 339(4.06),435(1.79)	9-0091-65
	EtOH	425s(2.0)	60-2177-64
	EtOH	220(4.3),270(4.09), 337(4.06)	65-3837-64
	EtOH	340(4.04)	59-1625-65
	EtOH-acid	395(4.08)	59-1625-65
	EtOH-base	390(4.14)	59-1625-65
Benzanilide	EtOH	266(4.11)	54-0949-64
4-Biphenylcarboxamide	HOAc	272(4.34)	39-4399-65
	HOAc-H_2SO_4	294(4.31)	39-4399-65
Carbazole, 1-methoxy-	n.s.g.	240(4.59),252s(4.39), 286(3.84),320(3.46)	78-0681-65
Nitrone, N,α-diphenyl-	n.s.g.	322(4.30)	46-1205-64
Oxaziridine, diphenyl-	n.s.g.	223(4.15)	46-1205-64
2,4-Pentadienal, 5-indol-3-yl-	MeOH	226(3.98),270(3.76), 278(3.75),390(4.09)	44-2534-65
Pyridine, 2-phenacyl-	N H_2SO_4	262(4.21)	39-3093-65
	pH 6.8	255(4.21),400(2.15)	39-3093-65
	N NaOH	259(4.08),350(3.98)	39-3093-65

Compound	Solvent	λ_{max}(log ε)	Ref.
Pyridine, 3-phenacyl-	N H_2SO_4	255(4.23)	39-3093-65
	pH 6.8	251(4.19)	39-3093-65
	N NaOH	232(4.02),343(4.18)	39-3093-65
Pyridine, 4-phenacyl-	N H_2SO_4	250(4.25)	39-3093-65
	pH 6.8	247(4.23),403(2.00)	39-3093-65
	N NaOH	235(3.97),353(4.34)	39-3093-65
$C_{13}H_{11}NO_2$			
Aniline, m-hydroxy-N-salicylidene-	EtOH	220(4.4),266(4.09), 340(4.08)	65-3837-64
Aniline, o-hydroxy-N-salicylidene-	EtOH	269(4.03),350(4.11), 448(3.14)	9-0091-65
Aniline, p-hydroxy-N-salicylidene-	EtOH	230(4.3),270(4.04), 352(4.33)	65-3837-64
1-Azaanthraquinone, 1,2,3,4-tetrahydro-	EtOH	236(4.27),241(4.27), 268s(4.32),276(?), 290s(4.43),330s(4.19), 470(3.70)	39-2941-64
1H-Azepine-1-carboxylic acid, phenyl ester	C_6H_{12}	325(2.81)	44-0751-64
	EtOH	220(4.31),242s(3.70), 303(2.96)	44-0751-64
$C_{13}H_{11}NO_2S$			
2H-Cyclopenta]d,e]-1-benzothiopyran-4-carboxamide, 3,5-dihydro-6-methyl-5-oxo-	EtOH	243s(4.35),249(4.40), 286(3.90)	88-2489-64
$C_{13}H_{11}NO_3$			
Dictamnine, 6-methoxy-	EtOH	237(4.66),250(4.56), 295(3.91),306(3.99), 334(3.68),348(3.67)	2-0491-64
Ketone, 2-furyl phenyl, O-acetyloxime, anti	EtOH	230(3.92),281(4.21)	22-2724-65
syn	EtOH	238(4.12),275(4.29)	22-2724-65
Phthalide, 3-(1-methyl-2-oxo-3-pyrrolidinylidene)-	EtOH	220(4.05),226(4.07), 248(4.11),256(4.07), 276(4.17),284(4.18), 297(4.09),326(4.21)	95-0839-65
2(1H)-Pyridone, 1-(p-toluoyloxy)-	EtOH	245(4.27),287s(3.77), 297(3.78),304s(3.77)	35-5186-65
1H-Pyrrolo[1,2-a]indole-9-carboxaldehyde, 2,3,7,8-tetrahydro-6-methyl-7,8-dioxo-	MeOH	225(4.94),280(4.35), 345(4.06)	35-3877-64
	MeOH	225(4.34),280(3.74), 345(3.47),520(3.20)	44-2897-65
4H-Quinolizin-4-one, 1,3-diacetyl-	EtOH	270(4.06),350(4.16), 410(4.33)	78-0945-65
$C_{13}H_{11}NO_4$			
Furo[2,3-b]quinoline-3,4(2H,9H)-dione, 6-methoxy-9-methyl-	EtOH	220s(4.37),236s(4.53), 244(4.57),270(4.42), 334(3.67),344s(3.64)	2-0491-64
	EtOH	248(4.66),294s(4.19), 304(4.28),360(3.44), 394(3.62)	2-0491-64
3-Indolinepropionic acid, 1-acetyl-3-hydroxy-2-oxo-, γ-lactone	EtOH	219(4.20),231s(4.06), 275s(2.70)	44-2431-64
2(1H)-Pyridone, 1-(m-anisoyloxy)-	EtOH	213(4.52),244(3.95), 301(3.92)	35-5186-65
2(1H)-Pyridone, 1-(p-anisoyloxy)-	EtOH	213(4.24),232s(3.76), 266(4.33),304s(3.80)	35-5186-65

Compound	Solvent	λ_{max}(log ε)	Ref.
1H-Pyrrolo[1,2-a]indole-9-carboxaldehyde, 2,3,5,8-tetrahydro-7-hydroxy-6-methyl-5,8-dioxo-	pH 13	236(4.38),299s(4.11), 325(4.14)	35-3877-64
	MeOH	219(4.33),299(4.17), 330(3.91)	35-3877-64 44-2897-65
$C_{13}H_{11}NO_5$			
2-Cyclohexen-1-one, 5-[3,4-(methylenedioxy)phenyl]-4-nitro-	EtOH	287(3.72),337(3.68), 410(3.67)	44-1416-65
2H-Quinolizine-3,4-dicarboxylic acid, 2-oxo-, dimethyl ester	MeOH-ether	230(4.48),327(3.99)	24-3537-65
4H-Quinolizine-1,2-dicarboxylic acid, 4-oxo-, dimethyl ester	MeOH-ether	248(4.04),275(3.79), 386(4.12)	24-3537-65
4H-Quinolizine-1,3-dicarboxylic acid, 4-oxo-, ethyl ester	EtOH	265(4.22),340(4.05), 390(4.29)	78-1051-64
$C_{13}H_{11}NO_5S$			
p-Toluenesulfonic acid, m-nitrophenyl ester	EtOH	228(4.30),256(3.98), 288(3.25),345s(2.10)	65-2080-65
	ether	228(4.12),255(3.85), 285(3.17),333s(2.10)	65-2080-65
	dioxan	232(4.10),258(3.85), 288(3.25)	65-2080-65
	H_2SO_4	230(4.00),270(3.92), 332(3.15)	65-2080-65
p-Toluenesulfonic acid, o-nitrophenyl ester	EtOH	226(4.20),252(3.65), 280(3.20),340(2.40)	65-2080-65
	ether	230(4.25),250(3.75), 285(3.22),310s(2.50)	65-2080-65
	dioxan	234(4.20),250(3.85), 282(3.38),335s(2.70)	65-2080-65
	H_2SO_4	228(4.37),295(3.90), 370(3.40),370(3.40)	65-2080-65
p-Toluenesulfonic acid, p-nitrophenyl ester	EtOH	218(4.23),265(4.10)	65-2080-65
	ether	227(4.00),262(3.95)	65-2080-65
	dioxan	230(4.15),264(4.00)	65-2080-65
	H_2SO_4	226(4.28),320(3.88), 405(3.50)	65-2080-65
$C_{13}H_{11}NS$			
6H-Thieno[2,3-b]pyrrole, 6-benzyl-	EtOH	250s(3.99)	44-0184-65
$C_{13}H_{11}N_3$			
Acridine, 3,6-diamino- (or proflavine)	pH 7.0	211(4.32),261(4.73), 277s(4.38),444(4.59)	39-4653-65
	pH 12-14	262(4.76),282s(4.52), 292s(4.40),395(4.25)	39-4653-65
Acridine, 4,5-diamino-	pH 0.0	213(4.11),249(5.23), 323s(3.47),337(3.79), 347s(3.84),353(3.99), 366s(3.70),385s(3.44)	39-4653-65
	pH 7.0	270(4.84),362s(3.22), 425(3.57)	39-4653-65
Acridine, 9-hydrazino-	pH 5.0	223(4.41),264(4.69), 394(3.83),409(4.03), 431(3.99)	39-4653-65
	pH 9.5	230(4.72),288(3.92), 384(3.91)	39-4653-65
2H-Benzotriazole, 2-p-tolyl-	n.s.g.	221(4.08),238(3.81), 242(3.81),253(3.64), 309(4.45),314(4.38)	39-4831-65

Compound	Solvent	$\lambda_{max}(\log \epsilon)$	Ref.
5H-Dibenzo[d,f][1,3]diazepine, 6-amino-	EtOH	240(4.54),278(3.49)	87-0310-64
1H-Imidazo[4,5-b]pyridine, 2-m-tolyl-	MeOH	237(3.92),307(4.47), 319(4.32)	44-3403-64
1H-Imidazo[4,5-b]pyridine, 2-p-tolyl-	MeOH	238(4.02),307(4.52), 321(4.37)	44-3403-64
Isomacrorine	EtOH-HCl	218(4.33),252(4.57), 292(3.85),319(3.84), 334(3.81)	39-5969-64
	EtOH	214(4.44),235(4.11), 269(4.45),328(4.04), 342(4.05)	39-5969-64
Macrorine	EtOH	217(4.59),238(4.18), 265(4.36),341(3.97)	39-5969-64
Proflavine, as sulfate	pH 4	265(4.7),450(4.6)	60-0386-64
$C_{13}H_{11}N_3O$			
2H-Benzotriazole, 2-(p-methoxyphenyl)-	n.s.g.	220(4.00),250(3.73), 252(3.73),259(3.75), 317(4.39)	39-4831-65
Imidazo[1,2-b]pyridazine, 6-methoxy-2-phenyl-	EtOH	295(3.7),340(4.2), 375s(3.7)	94-1351-64
Imidazo[1,2-b]pyridazin-6(5H)-one, 5-methyl-2-phenyl-	EtOH	255(4.2),290s(3.7), 350(4.3)	94-1351-64
Macrorungine	EtOH-HCl	203(4.35),225(4.46), 253(3.96),300(4.04), 413(4.58)	39-5969-64
	EtOH	220(4.60),284(4.24), 301(4.22),354(4.26)	39-5969-64
	EtOH-KOH	227(4.45),283(4.07), 323(3.91),382(4.04)	39-5969-64
Triazolo[4,3-a]pyridin-3(2H)-one, 1-benzyl-	EtOH	234(4.08),283(3.52), 344(3.58)	7-0935-65
$C_{13}H_{11}N_3OS$			
7H-Pyrimido[4,5-b][1,4]thiazine, 4-methoxy-6-phenyl-	EtOH	233(4.13),268(4.30), 295(3.81),344(3.92)	44-2121-64
$C_{13}H_{11}N_3O_2$			
5-Pyrimidinecarbonitrile, 3-acetyl-1,2,3,4-tetrahydro-2-oxo-1-phenyl-	EtOH	283(3.99)	44-1740-64
$C_{13}H_{11}N_3O_2S$			
Diimide, [(p-nitrophenyl)thio]-p-tolyl-	n.s.g.	360(4.24)	77-0313-65
$C_{13}H_{11}N_3O_3$			
Acetophenone, o-[(5-nitro-2-pyridyl)-amino]-	EtOH	206(4.19),238(4.39), 262s(3.86),387(4.46)	44-1539-65
Maleimide, N-[(4 or 7-methoxy-2-benzimidazolyl)methyl]-	EtOH	251(3.95),280(3.44), 300(2.69)	94-0127-64
Salicylaldehyde, 5-nitro-, phenylhydrazone	MeOH	231(4.27),301(4.31), 354(4.45)	24-1631-64
	MeOH-NaOH	370(4.42),428(3.99)	24-1631-64
Urea, 1-formyl-1-methyl-3-quinaldoyl-	EtOH	206(4.67),231(4.68), 278(3.69),303(3.64), 309(3.53),317(3.69)	39-5969-64
$C_{13}H_{11}N_3O_3S$			
Pyrimidine, 2-(methylthio)-4-[4-(5-nitro-2-furyl)-1,3-butadienyl]-	propylene glycol	252(4.17),330(4.25)	95-0207-64

Compound	Solvent	λ_{max}(log ϵ)	Ref.
$C_{13}H_{11}N_3O_4$			
Diphenylamine, N-methyl-2,4-dinitro-	C_6H_{12}	358(4.15)	23-2674-64
	ether	367.5(4.08)	23-2674-64
Diphenylamine, 2'-methyl-2,4-dinitro-	C_6H_{12}	260s(4.02),332(4.24)	23-2674-64
Diphenylamine, 3'-methyl-2,4-dinitro-	C_6H_{12}	260s(4.05),338(4.25)	23-2674-64
Diphenylamine, 4'-methyl-2,4-dinitro-	C_6H_{12}	260s(4.04),336(4.25)	23-2674-64
	ether	338(4.29)	23-2674-64
$C_{13}H_{11}N_3O_5$			
p-Anisidine, N-(2,4-dinitrophenyl)-	C_6H_{12}	258s(4.10),338(4.22)	23-2674-64
$C_{13}H_{11}N_5O$			
Alanine, N-benzoyl-N-methyl-	EtOH	282.5(4.07)	87-0659-65
$C_{13}H_{11}N_5OS$			
9H-Purine-9-carboxanilide, 6-(methylthio)-	pH 1	275(4.04)	87-0010-64
$C_{13}H_{11}N_5O_3$			
as-Triazine, 3-acetamido-6-(p-nitrostyryl)-	propylene glycol	270(4.03),342(4.28)	95-0016-64
$C_{13}H_{11}N_5O_5$			
Diacetamide, N-[6-[2-(5-nitro-2-furyl)-vinyl]-as-triazin-3-yl]-	propylene glycol	288(4.15),386(4.40)	95-0009-64
$C_{13}H_{11}N_5S$			
Pteridine, 2-amino-4-(benzylthio)-	EtOH	211(4.41),232(4.28), 268(4.13),310(3.57), 383(3.93)	44-3370-64
$C_{13}H_{11}O_2P$			
Dibenz[b,e]phosphorin, 5,10-dihydro-5-hydroxy-, 5-oxide	EtOH	205(4.56),227s(3.96), 264s(3.10),269(3.21), 276(3.21)	44-2382-64
$C_{13}H_{12}$			
Cycloheptatriene, 1-phenyl-	isooctane	200(4.40),234(4.18), 295(3.90)	54-0245-65
Cycloheptatriene, 2-phenyl-	isooctane	200(4.46),238(4.24)	54-0245-65
Cycloheptatriene, 3-phenyl-	isooctane	203(4.40),231(4.15), 283(3.94)	54-0245-65
Cycloheptatriene, 7-phenyl-	isooctane	201(4.60),258(3.53)	54-0245-65
$C_{13}H_{12}BN$			
Dibenz[c,e][1,2]azaborine, 5,6-dihydro-6-methyl-	$CHCl_3$	252(4.48),272(4.08), 300(3.81),313(4.00), 324(4.08)	44-1757-64
$C_{13}H_{12}BNO$			
Dibenz[c,e][1,2]azaborine, 5,6-dihydro-6-methoxy-	$CHCl_3$	262(4.23),272(4.23), 300s(3.79),312(3.99), 325(4.03)	44-1757-64
$C_{13}H_{12}BrNO$			
1,2,3,4-Tetrahydro-4-oxobenzo[c]-quinolizinium bromide	H_2O	202(4.47),253(4.53), 325(3.92)	78-2529-65
	10% NaOH	233(4.45),301(3.93), 436(3.23)	78-2529-65

Compound	Solvent	$\lambda_{max}(\log \epsilon)$	Ref.
$C_{13}H_{12}BrNO_2$			
1H-Pyrrolo[1,2-a]indol-1-one, 9-bromo-2,3-dihydro-7-methoxy-6-methyl-	MeOH	221(4.5),330(4.37)	44-2904-65
$C_{13}H_{12}Br_2N_2$			
6-Methyldipyrido[1,2-a:2',1'-c]pyrazinediium dibromide	H_2O	238(4.21),244(4.58), 267(4.72),276(4.06), 301s(3.91),312(4.14), 325(3.51),351(--)	88-3871-64
$C_{13}H_{12}ClF_3O_2$			
Acetic acid, trifluoro-, 4-(p-chlorophenyl)-3-pentenyl ester	n.s.g.	250(4.18)	88-1053-65
$C_{13}H_{12}ClNO_2$			
Cinnamic acid, p-chloro-α-cyano-β-methyl-, ethyl ester	EtOH	287.5(4.09)	28-3102-65A
$C_{13}H_{12}ClNO_2S$			
p-Toluenesulfonanilide, 2'-chloro-	hexane	218(4.19),275(3.17)	65-2080-65
	EtOH	220(4.20),273(3.13)	65-2080-65
	EtOH-NaOEt	248(4.31),280(3.50)	65-2080-65
	ether	220(4.27),230(4.14), 275(3.23)	65-2080-65
	dioxan	220(4.22),273(3.17)	65-2080-65
	H_2SO_4	230(3.96),250(4.12), 270(3.81),310s(2.50)	65-2080-65
p-Toluenesulfonanilide, 4'-chloro-	hexane	224(4.27),232(4.25), 275(3.40)	65-2080-65
	EtOH	225(4.30),238(4.27), 275(3.47)	65-2080-65
	EtOH-NaOEt	252(4.33),290(3.65)	65-2080-65
	ether	225(4.30),238(4.22), 275(3.46)	65-2080-65
	dioxan	226(4.20),235(4.15), 275(3.46)	65-2080-65
	H_2SO_4	227(4.02),246(4.23), 270(3.85),308s(2.33)	65-2080-65
$C_{13}H_{12}ClNO_3$			
2H-1,4-Oxazine-6-carboxylic acid, 5-(3-chloro-6-hydroxy-2,4-xylyl)-3,4-dihydro-, δ-lactone	EtOH	212(4.22),238(4.22), 298(4.02),308(4.14), 338(3.99)	44-4122-65
$C_{13}H_{12}ClNO_4$			
4-Isoxazolecarboxylic acid, 3-(2-chloro-3-methoxyphenyl)-5-methyl-, methyl ester	MeOH	217(4.27),286(3.43)	39-5976-65
$C_{13}H_{12}ClN_5$			
Pteridine, 4-(benzylamino)-6-chloro-3,4-dihydro-	$CHCl_3$	274(3.63),311(3.79), 325(3.74)	39-4920-64
$C_{13}H_{12}F_4O_4$			
p-Toluic acid, α-carboxy-2,3,5,6-tetrafluoro-, diethyl ester	EtOH	272(3.63),326(2.76)	70-1798-65
$C_{13}H_{12}Fe$			
Ferrocene, dehydro-1,1'-trimethylene-	EtOH	260s(3.5),320s(1.8), 440(2.1)	44-2452-64

Compound	Solvent	λ max(log ε)	Ref.
$C_{13}H_{12}FeO$			
Ferrocenylacrolein	$CHCl_3$	304(4.26),488(3.26)	5-0088-64F
$C_{13}H_{12}FeO_5$			
Iron, tricarbonyl(duroquinone)-	CH_2Cl_2	260s(3.8),290s(3.4), 350(3.5)	101-0336-64B
$C_{13}H_{12}N_2$			
Azobenzene, p-methyl-, trans	EtOH	333(4.37),451(2.83)	39-1045-64
Benzaldehyde, phenylhydrazone	EtOH	236(4.15),304(4.05), 344(4.37)	7-0180-64
Benzophenone, hydrazone	$CHCl_3$	275(3.98)	44-0417-65
2,4-Heptadien-6-ynal, phenyl-hydrazone	ether	262(4.14),297(4.10), 380(4.69)	24-0809-64
p-Phenylenediamine, N-benzylidene-	EtOH	247(4.17),355(4.14)	9-0091-65
Pyridine, 2-(p-aminostyryl)-, cis	aq. HCl	317(4.10)	23-1345-65
	H_2O	313(4.11)	23-1345-65
	50% MeOH	319(4.11)	23-1345-65
	50% MeOH-HCl	318(4.10)	23-1345-65
Pyridine, 2-(p-aminostyryl)-, trans	aq. HCl	328(4.49)	23-1345-65
	H_2O	337(4.48)	23-1345-65
	50% MeOH	339(4.47)	23-1345-65
	50% MeOH-HCl	328(4.48)	23-1345-65
$C_{13}H_{12}N_2O$			
Azobenzene, p-methoxy-	MeCN	322(3.3),348(4.4)	49-1173-65
$SbCl_5$ complex	MeCN	460(4.6)	49-1173-65
Azoxybenzene, p-methyl-, cis	EtOH	240(4.11),339(3.58)	35-2419-64
Benzaldehyde, phenylhydrazone, N-oxide	EtOH	232(3.94),252(3.97)	7-0180-64
Benzamide, o-anilino-	EtOH	220s(4.31),288(4.20), 345(3.81)	44-1020-65
Benzanilide, p-amino-	EtOH	223(4.1),297(4.5)	28-4295-64B
Benzo[f]quinazolin-3(4H)-one, 1,2-dihydro-1-methyl-	pH 1	243(4.66),248(4.66), 275(3.63),286(3.72), 297(3.61),324(3.18), 336(3.18)	44-2881-64
	pH 10	243(4.65),248(4.65), 276(3.62),287(3.72), 298(3.61),325(3.17), 337(3.18)	44-2881-64
	50% EtOH	244(4.68),249(4.70), 276(3.65),286(3.74), 297(3.63),326(3.22), 338(3.23)	44-2881-64
Indole-3-butyronitrile, β-methyl-γ-oxo-	EtOH	243(4.12),258(3.98), 300(4.11)	87-0415-64
Pyrrolo[1,2-a]quinoxaline, 4-ethoxy-	EtOH	230(4.49),236s(4.44), 266s(3.98),272(4.04), 280(3.98),306s(3.37), 318(3.58),330(3.73), 367(3.93)	35-1830-64
Tropone, 2-(2-phenylhydrazino)-	EtOH	244(4.48),334(4.07), 395(4.08)	94-0457-65
$C_{13}H_{12}N_2O_2$			
Aniline, p-(p-nitrobenzyl)-	EtOH	241(4.13),275(4.10), 328(3.30),335(3.25), 357s(2.94)	65-2073-65

Compound	Solvent	$\lambda_{max}(\log \epsilon)$	Ref.
Aniline, p-(p-nitrobenzyl)- (contd.)	dioxan	248(4.15),265(4.05), 322(3.30)	65-2073-65
	H_2SO_4	310(4.00)	65-2073-65
2-Isoxazoline, 5-(5-methyl-4-isoxazolyl)-3-phenyl-	EtOH	263(4.15)	7-1223-65
Pyridine, 1,2-dihydro-1-methyl-2-(p-nitrobenzylidene)-	EtOH	267(4.19)	35-4917-64
	MeCN	547(4.03)	35-4917-64
$C_{13}H_{12}N_2O_2S$			
Benzenesulfonic acid, benzylidene-hydrazide	EtOH	274(4.3)	39-1020-65
1H-2,1,3-Benzothiadiazine, 3,4-dihydro-4-phenyl-, 2,2-dioxide	EtOH	228(3.96),278(3.50), 355(2.85)	44-3960-65
$C_{13}H_{12}N_2O_3$			
5-Pyrimidinecarboxylic acid, 1,2-di-hydro-2-oxo-1-phenyl-, ethyl ester	EtOH	253(4.05),273(4.02), 315s(3.29)	94-1418-64
$C_{13}H_{12}N_2O_3S_2$			
2-Thiazoline-4-carboxylic acid, 2-(6-hydroxy-2-benzothiazolyl)-5,5-dimethyl-	EtOH	269(3.84),330(4.24)	44-2344-65
$C_{13}H_{12}N_2O_4$			
Cinnamic acid, α-cyano-β-methyl-p-nitro-, ethyl ester, trans	EtOH	279.5(4.18)	28-3102-65A
3-Indolinealanine, N-acetyl-3-hydroxy-2-oxo-, γ-lactone	EtOH	254(3.68),298(3.16)	44-2431-64
$C_{13}H_{12}N_2O_4S$			
p-Toluenesulfonanilide, 2'-nitro-	hexane	218(4.34),270(3.87), 348(3.57)	65-2080-65
	EtOH	220(4.20),270(3.72), 344(3.37)	65-2080-65
	EtOH-NaOEt	240(4.08),275(3.55), 396(3.38)	65-2080-65
	ether	222(4.26),269(3.76), 348(3.43)	65-2080-65
	dioxan	222(4.34),270(3.83), 344(3.55)	65-2080-65
	H_2SO_4	226(4.14),268(3.84), 355s(2.60)	65-2080-65
p-Toluenesulfonanilide, 3'-nitro-	hexane	222(4.35),270(3.80), 318(3.20)	65-2080-65
	EtOH	220(4.33),268(3.79), 325(3.30)	65-2080-65
	EtOH-NaOEt	250(4.30),280(3.90), 375(3.08)	65-2080-65
	ether	226(4.41),236(4.37), 268(3.95),318(3.37)	65-2080-65
	dioxan	222(4.38),268(3.83), 319(3.32)	65-2080-65
	H_2SO_4	228(4.20),265(4.00), 350s(2.40)	65-2080-65
p-Toluenesulfonanilide, 4'-nitro-	hexane	224(4.29),302(4.16), 320s(2.63)	65-2080-65
	EtOH	224(4.27),310(4.05)	65-2080-65
	EtOH-NaOEt	380(4.15)	65-2080-65
	ether	226(4.28),308(4.21)	65-2080-65
	dioxan	228(4.23),308(4.12), 325s(2.75)	65-2080-65

Compound	Solvent	λ max (log ε)	Ref.
p-Toluenesulfonanilide, 4'-nitro- (contd.)	H_2SO_4	228(4.10),264(3.93), 355s(2.40)	65-2080-65
$C_{13}H_{12}N_2O_4S_2$			
Gliotoxin, dehydro-	EtOH	214(4.34),272(3.73), 300(3.67)	39-4315-64
$C_{13}H_{12}N_2O_5$			
6-Azaindole-3-propionic acid, 2-carboxy-5-methoxy-	EtOH	283(4.11),292(4.20), 350(3.64)	35-3530-65
4H-Quinolizine-3-carboxylic acid, 6-methyl-1-nitro-4-oxo-, ethyl ester	EtOH	415(4.34)	78-3305-65
$C_{13}H_{12}N_4$			
Pyrazolo[1,5-a]pyrimidine, 7-amino-2-methyl-5-phenyl-	EtOH	209(4.48),258(4.68), 282s(4.01),332(3.72)	95-1113-64
Pyrazolo[1,5-a]pyrimidine, 7-amino-6-methyl-2-phenyl-	EtOH	249s(4.40),261(4.51), 270(4.44),308(3.95)	95-1113-64
5H-Pyrrolo[2,3-d]pyrimidine, 4,7-dihydro-4-imino-3-methyl-5-phenyl-	pH 1	240(4.15),288(4.03)	35-1995-65
	pH 13	273(4.09)	35-1995-65
	EtOH	248(4.01),282(4.10)	35-1995-65
7H-Pyrrolo[2,3-d]pyrimidine, 4-(methylamino)-5-phenyl-	pH 1	250(4.13),285(4.11)	35-1995-65
	pH 13	282(4.15)	35-1995-65
	EtOH	262s(4.05),283(4.19)	35-1995-65
s-Triazolo[1,5-a]pyrazine, 5,6-dimethyl-2-phenyl-	C_6H_{12}	249(4.65),258(4.62), 281s(3.91),294(3.82), 301s(3.78),315(3.55)	44-2542-64
s-Triazolo[1,5-a]pyrazine, 6,8-dimethyl-2-phenyl-	C_6H_{12}	247(4.61),255s(4.59), 278s(4.01),286s(3.91), 303s(3.54)	44-2542-64
$C_{13}H_{12}N_4O$			
Purine-8-methanol, 6-methyl-α-phenyl-	pH 1	270(4.06)	87-0797-65
	pH 11	275(3.99)	87-0797-65
4H-Pyrrolo[2,3-d]pyrimidin-4-one, 2-amino-7-benzyl-3,7-dihydro-	pH 1	224(4.28),263(4.04)	4-0034-64
	pH 11	266(4.06)	4-0034-64
$C_{13}H_{12}N_4OS$			
Purine-8-methanol, 2-(methylthio)-α-phenyl-	pH 1	240(4.37),312(3.78)	87-0797-65
	pH 11	240(4.33),305(3.90)	87-0797-65
Purine-8-methanol, 6-(methylthio)-α-phenyl-	pH 1	227(4.10),299(4.20)	87-0797-65
	pH 11	230(4.15),293(4.23)	87-0797-65
$C_{13}H_{12}N_4O_2$			
Formazan, N,N'-bis(2-hydroxyphenyl)-	benzene	470(4.33)	65-0502-65
	DMF	475(4.45)	65-0502-65
$C_{13}H_{12}N_4O_3$			
Pyrimidine, 2-amino-4-[3-methyl-4-(5-nitro-2-furyl)-1,3-butadienyl]-	propylene glycol	290(3.81),340(3.90), 408(3.95)	95-0121-64
$C_{13}H_{12}N_4O_3S$			
Xanthine, 7-benzyl-1-hydroxy-8-(methylthio)-	EtOH	290(4.44)	4-0275-64
	EtOH-NaOH	304(4.23)	4-0275-64
$C_{13}H_{12}N_4O_4$			
2-Norbornen-7-one, 2,4-dinitrophenylhydrazone	EtOH	224(4.27),358(4.36)	44-0160-64
	$CHCl_3$	362(4.38)	44-0160-64

Compound	Solvent	$\lambda_{max}(\log \epsilon)$	Ref.
$C_{13}H_{12}N_4O_5S_2$			
Xanthine, 1-[(phenylsulfonyl)oxy]-7-methyl-8-(methylthio)-	EtOH	292(4.31)	4-0275-64
	EtOH-NaOH	302(4.35)	4-0275-64
$C_{13}H_{12}N_4S$			
Purine, 6-benzylthiomethyl-	N HCl	265(3.61)	87-0667-65
	pH 7.65	268(3.79)	87-0667-65
	N NaOH	280(3.84)	87-0667-65
$C_{13}H_{12}N_4S_2$			
Imidazo[4,5-d]pyridazine, 4,7-bis-(methylthio)-2-phenyl-	EtOH	205(4.39),266(4.43), 296(4.35),307s(4.29)	4-0182-64
1H-Imidazo[4,5-d]pyridazine-7(4)-thiol, 4(7)-(ethylthio)-2-phenyl-	EtOH	206(4.46),274s(4.41), 284(4.44),324(4.11), 342(4.14)	4-0182-64
6(5H)-Pteridinethione, 7-(benzyl-thio)-7,8-dihydro-	pH 1.0	333(4.17)	39-0027-65
	pH 4.4	236s(4.08),331(4.22)	39-0027-65
	pH 11	237(4.20),299(4.16), 411(3.80)	39-0027-65
$C_{13}H_{12}N_6O$			
Formamide, N-(1-benzyl-5-s-triazol-3-ylimidazol-4-yl)-	pH 1	247(4.14)	44-3601-65
	pH 7	244(3.99)	44-3601-65
	pH 13	245(4.01)	44-3601-65
$C_{13}H_{12}O$			
Benzhydrol, carbonium ion from	H_2SO_4	441(--)	60-0264-64
	$HClO_4$	292(3.38),304s(3.32), 493(4.71)	60-0264-64
carbanion	ether	255(3.95),300(3.00), 434(4.34)	60-0264-64
2,4-Heptadien-6-yn-1-ol, 7-phenyl-	ether	223(4.28),298(4.63), 317(4.55)	24-2118-64
4,6-Heptadiyn-1-ol, 7-phenyl-	ether	221(4.73),244(3.97), 257(4.29),272(4.48), 288(4.38)	24-2118-64
6-Hepten-4-yn-3-one, 7-phenyl-	EtOH	297(4.22)	39-2983-65
4-Hexen-2-ynophenone, 5-methyl-	EtOH	261(4.12),289(4.18), 305(4.16)	39-2983-65
Ketone, ethyl 1-naphthyl	EtOH	213(4.70),299(3.79)	39-2072-65
Ketone, ethyl 2-naphthyl	EtOH	240(4.71),292(3.97)	39-2072-65
$C_{13}H_{12}OS$			
Sulfoxide, ethyl 5-phenyl-2,4-pentadiynyl	n.s.g.	247(4.28),260(4.28), 274(4.28),291(4.28)	28-2847-65A
Sulfoxide, phenyl p-tolyl	isooctane	232(4.24),265s(3.52), 273s(3.32)	35-1958-65
	EtOH	237(4.21),266s(3.41), 274s(3.23)	35-1958-65
R(+)	EtOH	236(4.24),265s(3.62), 272s(3.51)	35-1958-65
$C_{13}H_{12}O_2$			
Benzaldehyde, di-2-propynyl acetal	MeOH	241(3.0),248(3.0), 254(2.9),278(2.3)	70-1349-64
Fluoren-2(1H)-one, 3,4-dihydro-7-hydroxy-	EtOH	267(4.08)	88-3505-65
	MeOH-KOH	299(4.28)	88-3505-65
1-Naphthaleneacetic acid, methyl ester	EtOH	225(4.94),281(3.85)	39-2072-65
2-Naphthaleneacetic acid, methyl ester	EtOH	224(5.03),276(3.73)	39-2072-65
1-Naphthoic acid, ethyl ester	EtOH	220(5.62),297(3.81)	39-2072-65

Compound	Solvent	$\lambda_{max}(\log \epsilon)$	Ref.
2-Naphthoic acid, ethyl ester	EtOH	237(4.89),279(3.90)	39-2072-65
2H-Pyran-2-one, 4,6-dimethyl-5-phenyl-	EtOH	231(3.87),301(3.83)	44-2642-65
Toluene, 2,4-dipropargyloxy-	MeOH	256(3.1),271(3.2), 278(3.2)	70-1349-64
$C_{13}H_{12}O_2S$			
2-Furaldehyde, 5-(benzylthiomethyl)-	EtOH	285(4.25)	70-1281-65
$C_{13}H_{12}O_3$			
2'-Acetonaphthone, 1'-hydroxy-3'-methoxy-	EtOH	221(4.60),261(4.46), 271(4.44),285s(4.00), 296(3.98),305s(3.82), 389(3.49)	94-0316-64
2'-Acetonaphthone, 3'-hydroxy-1'-methoxy-	EtOH	224(4.54),256(4.39), 264s(4.33),285s(3.79), 294(3.73),304s(3.62), 336(3.09),398(3.09)	94-0316-64
5-Benzofuranacrylic acid, 2,3-dihydro-6-hydroxy-α,β-dimethyl-, γ-lactone	EtOH	225(4.10),296(3.75), 332(4.22)	44-2467-64
2-Dibenzofuranol, 8,9-dihydro-7-methoxy-	EtOH	238s(3.93),325(4.25)	39-2932-64
1,4-Ethanonaphthalene-5,8-dione, 1,4-dihydro-1-methoxy-	EtOH	251(4.28)	39-2941-64
3-Furoic acid, 4,5-dimethyl-2-phenyl-	MeOH	297(4.25)	39-5984-65
3-Furoic acid, 2-methyl-5-p-tolyl-	CH_2Cl_2	280(4.36)	44-2205-65
Grisan-3,4'-dione	MeOH	251(4.02),326(3.71)	94-0214-64
Spiro[3,6-dioxabicyclo[3.1.0]hexane-2,2'(3'H)-furan], 4-(2,4-hexadiyn-ylidene)-4',5'-dihydro-	ether	224(4.52),267(4.15), 279(4.29),293(4.22)	24-2596-65
$C_{13}H_{12}O_3S$			
3-Thiophenecarboxylic acid, 4-hydroxy-2-phenyl-, ethyl ester	ether	211(4.13),260(4.01), 319(3.66)	24-2109-64
p-Toluenesulfonic acid, phenyl ester	EtOH	227(4.08),266(2.83)	65-2080-65
	ether	226(4.10),274(2.93)	65-2080-65
	dioxan	230(4.09),257(2.98), 274(2.98)	65-2080-65
	H_2SO_4	232(4.00),266(3.10), 324s(2.40)	65-2080-65
$C_{13}H_{12}O_4$			
6-Heptenoic acid, 3,5-dioxo-7-phenyl-	EtOH	230(2.85),342(3.50)	22-0525-65
4-Indanecarboxylic acid, 5-acetyl-1-oxo-, methyl ester	EtOH	254(3.96),302(3.58), 304(3.59)	39-3811-65
4-Indanecarboxylic acid, 5-(1-hydroxy-1-methoxyethyl)-1-oxo-, γ-lactone	EtOH	218(4.68),238(3.99), 289(3.57),298(3.66)	39-3811-65
Malonic acid, benzylidene-, cyclic isopropylidene ester	EtOH-HCl	320(4.28)	49-1283-64
	EtOH-NaOH	262.5(4.25)	49-1283-64
2-Naphthaldehyde, 6-hydroxy-4,5-dimethoxy-	EtOH	255(4.21),261(4.23), 330(3.76),365(3.85)	39-4292-65
1,4-Naphthalenediol, 8-acetyl-5-meth-oxy-	EtOH	246(4.14),354(3.81)	54-1005-64
1,4-Naphthoquinone, 5,6-dimethoxy-2-methyl-	EtOH	262(4.20),393(3.48)	39-4292-65
1,4-Naphthoquinone, 6,7-dimethoxy-2-methyl-	EtOH	270(4.44),276(4.42), 349(3.36)	73-0197-64
$C_{13}H_{12}O_5$			
2H-1-Benzopyran-3-carboxylic acid, 8-methoxy-2-oxo-, ethyl ester	EtOH	252(3.90),306(4.22)	94-0443-65

Compound	Solvent	$\lambda_{max}(\log \epsilon)$	Ref.
Coumarin-6α-propionic acid, 7-hydroxy-, methyl ester	EtOH	222(4.17),248(3.64), 258(3.56),334(4.28)	32-0513-65
2-Indanacetic acid, 3-carboxy-4-methyl-1-oxo-	MeOH	251(4.04),292(3.45)	39-0990-65
4,5-Indandicarboxylic acid, 1-oxo-, dimethyl ester	EtOH	252(4.07),258(4.03), 300(3.64),304(3.65)	39-3811-65
4,5-Indandicarboxylic acid, 3-oxo-, dimethyl ester	EtOH	224(4.64),286(3.13), 296(3.15)	39-3811-65
2-Naphthoic acid, 4-hydroxy-5,6-dimethoxy-	EtOH	252(4.42),325(3.74), 340(3.74),355(3.73)	39-4292-65
$C_{13}H_{12}O_6$			
2H-1-Benzopyran-6-acetic acid, 7-hydroxy-5-methoxy-2-oxo-, methyl ester	MeOH	226(4.10),253(3.82), 262(3.85),331(4.33)	32-1054-64
1,4-Naphthoquinone, 2-hydroxy-5,6,7-trimethoxy-	EtOH	267(4.33),307(4.12), 355(3.56)	94-1472-65
1,4-Naphthoquinone, 2-hydroxy-6,7,8-trimethoxy-	EtOH	269(4.38),305(4.21), 354(3.54)	94-1472-65
$C_{13}H_{12}S$			
Sulfide, ethyl 5-phenyl-2,4-pentadiynyl	n.s.g.	247(4.34),260(4.34), 275(4.34),292(4.34)	28-2847-65A
Thiophene, 2-(m-methylstyryl)-, cis	EtOH	236(3.85),294(3.50)	39-6221-65
trans	EtOH	238(4.01),244s(3.91), 330(4.45)	39-6221-65
Thiophene, 2-(o-methylstyryl)-, cis	EtOH	238(3.79),288(3.94)	39-6221-65
trans	EtOH	237(3.78),254s(3.16), 262(3.25),329(4.30)	39-6221-65
Thiophene, 2-(p-methylstyryl)-, trans	EtOH	233(3.95),242s(3.83), 333(4.37),345(4.17)	39-6221-65
$C_{13}H_{12}S_2$			
2,2'-Bithiophene, 5-(1-propenyl)-5'-vinyl-	ether	254(3.99),262(4.00), 366(4.46)	24-1228-65
$C_{13}H_{13}BrN_2O$			
5,11-Dihydro-11-hydroxy-11-methyl-pyrido[2,1-b]quinazolinium bromide	EtOH	258(4.16),288(3.42), 298(3.43),349(3.86)	44-1539-65
1,2,3,4-Tetrahydro-4-oxobenzo[c]-quinolizinium bromide, oxime	EtOH	220(4.37),243(4.48), 280(--),372(4.05)?	78-2529-65
$C_{13}H_{13}BrO$			
Naphthalene, 6-bromo-7-methoxy-1,2-dimethyl-	EtOH	239(4.88),261(3.56), 272(3.67),283(3.70), 293(3.57),324(3.18), 340(3.27)	39-0361-65
$C_{13}H_{13}BrO_2$			
2-Naphthoic acid, 7-bromo-3,4-dihydro-1-methyl-, methyl ester	MeOH	231(4.45),248s(--), 280(4.06),310(3.67)	44-2109-64
$C_{13}H_{13}ClN_2$			
1H-Azepino[5,4,3-cd]indole, 9-chloro-3,4-dihydro-1,6-dimethyl-	EtOH-HCl	234(3.99),260(4.16), 343(3.60),411(3.79)	87-0200-65
2,10-Dimethyldipyrido[1,2-c:2',1'-e]-imidazol-5-ium chloride	n.s.g.	208(4.34),224(4.60), 231(4.41),245(4.55), 273(3.69),308(3.95), 330s(4.06),344(4.20), 362(4.17),387(3.67), 407(3.63),432(3.43)	12-1819-65

Compound	Solvent	λ_{max}(log ε)	Ref.
3,9-Dimethyldipyrido[1,2-c:2',1'-e]-imidazol-5-ium chloride	n.s.g.	246(4.65),253s(4.63), 304(3.74),330s(3.96), 343(4.13),356(4.12), 385(3.67),406(3.45)	12-1819-65
$C_{13}H_{13}ClN_2O_3$			
Tryptamine, 4-acetyl-2-carboxy-7-chloro-	EtOH	219(4.36),258(4.21), 347(3.86),408(3.85)	87-0200-65
$C_{13}H_{13}ClN_4O_2$			
4-Pyrimidinecarboxylic acid, 2,6-diamino-5-(p-chlorophenyl)-, ethyl ester	pH 1	299(4.22)	4-0162-65
	pH 7	308(4.26)	4-0162-65
	pH 13	297(4.34)	4-0162-65
$C_{13}H_{13}ClN_4O_4$			
2-Cyclohepten-1-one, 2-chloro-, 2,4-dinitrophenylhydrazone	EtOH	365(4.42)	35-0321-65
$C_{13}H_{13}ClO_2$			
2,4-Pentadienoic acid, 5-(p-chloro-phenyl)-2,3-dimethyl-, di-trans	MeOH-acid	228(4.04),306(4.51)	44-2986-64
$C_{13}H_{13}ClO_6$			
2-Benzofurancarboxylic acid, 5-chloro-3-(2-hydroxyethoxy)-, 2-hydroxy-ethyl ester	EtOH	215(4.28),230(4.18), 288(4.28)	44-4126-65
$C_{13}H_{13}Cl_2NO_3$			
Coumarin, 3,6-dichloro-4-(2-hydroxy-ethylamino)-5,7-dimethyl-	EtOH	217(4.37),249(4.22), 325(4.18)	44-4122-65
$C_{13}H_{13}F_3O_2$			
Acetic acid, trifluoro-, 4-phenyl-2-pentenyl ester	n.s.g.	243(4.06)	88-1053-65
$C_{13}H_{13}F_6NO$			
2H-Azepin-2-one, 1,3-dihydro-3,5-di-methyl-7-[4,4,4-trifluoro-2-(tri-fluoromethyl)-1-butenyl]-	EtOH	215(4.17),296(3.97)	44-3447-64
$C_{13}H_{13}N$			
2,6-Lutidine, 3-phenyl-	n.s.g.	272(3.9)	70-0322-65
Phenanthridine, 7,8,9,10-tetrahydro-	n.s.g.	230(4.21),284(3.66), 310(3.47),323(3.53)	39-2421-64
Pyridine, 2-benzylidene-1,2-dihydro-1-methyl-	C_6H_{12}	336(4.20),423(3.44)	35-4917-64
Pyrrole, 1-methyl-2-styryl-	MeOH	237(3.99),337(4.32)	4-0318-65
Pyrrole, 1-methyl-2-styryl-, cis	n.s.g.	233(4.14),322(4.01)	12-0875-65
Pyrrole, 1-methyl-2-styryl-, trans	n.s.g.	237(4.01),338(4.29)	12-0875-65
Pyrrole, 2-(α-methylstyryl)-, cis	n.s.g.	232(4.18),303(4.05)	12-0875-65
trans	n.s.g.	228(4.11),311(4.29)	12-0875-65
p-Toluidine, α-(2,4-cyclopentadien-1-ylidene)-N-methyl-	MeOH	400(4.5)	87-0390-65
p-Toluidine, α-phenyl-	hexane	240(--),244(4.40), 270(3.30),290(--), 294(--),298(3.56)	65-2073-65
	EtOH	240(4.03),269(--), 312(3.21),380s(2.49)	65-2073-65
	EtOH	240(4.12)	78-0861-64

Compound	Solvent	λ_{max}(log ε)	Ref.
p-Toluidine, α-phenyl- (contd.)	ether	246(4.10),298(3.30), 321s(2.42),326s(2.12)	65-2073-65
	dioxan	245(4.15),275(3.07), 297(3.33)	65-2073-65
$C_{13}H_{13}NO$			
Acetamide, N-1-azulenyl-N-methyl-	C_6H_{12}	237(4.28),277(4.61), 282(4.61),286s(4.51), 333(3.51),342s(3.61), 344(3.63),357(3.40), 572s(2.46),590(2.53), 613(2.50),642(2.49), 669s(2.25),710(2.09)	35-3137-64
1,2-Benzisoxazole, 4,5,6,7-tetra-hydro-3-phenyl-	C_6H_{12}	236(4.18)	44-1582-64
2,1-Benzisoxazole, 4,5,6,7-tetra-hydro-3-phenyl-	C_6H_{12}	226(4.32)	44-1582-64
Carbazol-1(2H)-one, 3,4-dihydro-9-methyl-	EtOH	309(4.36)	12-0246-64
Furo[3,2-c]quinoline, 2,3-dihydro-2,2-dimethyl-	EtOH	214s(4.42),230(4.64), 295(3.77),306(3.71), 320(3.60)	94-1424-64
Furo[3,2-c]quinoline, 2,3-dihydro-2,3-dimethyl-	EtOH	214s(4.46),230(4.70), 294(3.78),306(3.71), 319(3.54)	94-1424-64
Furo[3,2-c]quinoline, 2,3-dihydro-2,4-dimethyl-	EtOH	214(4.29),230(4.69), 286(3.72),292(3.75), 303(3.66),310(3.50), 316(3.36)	94-0789-64
2-Indolinone, 3-cyclopentylidene-	EtOH	217(3.95),251(4.42), 255(4.43),260(4.51), 293(3.89),349(3.27)	87-0626-65
1-Naphthamide, N,N-dimethyl-	HOAc	282(3.81)	39-4399-65
	HOAc-H_2SO_4	288(3.68)	39-4399-65
Quinoline, 4-(2-butenyloxy)-	EtOH	224(4.79),277s(3.86), 286(3.88),299s(3.62), 304s(3.43)	94-1424-64
Quinoline, 4-[(2-methylallyl)oxy]-	EtOH	224(4.78),278s(3.91), 286(3.92),299s(3.66), 304s(3.44)	94-1424-64
4-Quinolinol, 3-(1-methylallyl)-	EtOH	212(4.49),238s(4.44), 243(4.45),291(3.53), 323(4.06),336(4.09)	94-1424-64
4-Quinolinol, 3-(2-methylallyl)-	EtOH	212(4.46),238s(4.37), 243(4.38),292(3.55), 323(4.04),337(4.08)	94-1424-64
4(1H)-Quinolone, 1-(2-butenyl)-	EtOH	211(4.41),238(4.42), 243s(4.30),290(3.51), 323(4.12),337(4.21)	94-1424-64
4(1H)-Quinolone, 1-(2-methylallyl)-	EtOH	211(4.40),239(4.32), 243s(4.29),289(3.47), 324(4.15),337(4.21)	94-1424-64
$C_{13}H_{13}NO_2$			
1-Acenaphthenecarboxamide, 6,7,8,8a-tetrahydro-2-oxo-	EtOH	257(4.05),304(3.45)	44-2610-65
1-Acenaphthylenecarboxamide, 2,6,7,8-tetrahydro-2-hydroxy-	EtOH	237(4.23),245(4.20), 304(4.02)	44-2610-65
Allosecurinine	hexane	342(3.17)	94-1118-64
	H_2O	259(4.24),299s(3.39)	44-3441-64

Compound	Solvent	$\lambda_{max}(\log \varepsilon)$	Ref.
Allosecurinine (contd.)	aq. acid	256.5(4.26)	44-3441-64
	MeOH	300(3.26)	94-1118-64
	EtOH	302(3.27)	94-1118-64
	99.5% EtOH	256(4.04),306(3.11)	44-3441-64
	95% EtOH	257(4.19),304s(3.36)	44-3441-64
	50% EtOH	259(4.20),299s(3.37)	44-3441-64
	dioxan	257(4.19),345(3.22)	94-1118-64
	CCl_4	342(3.17)	44-3441-64
Benz[cd]indole-3-carboxylic acid, 1,3,4,5-tetrahydro-2-methyl-	EtOH	226(4.53),275(3.86), 279(3.87),290(3.73)	44-0843-64
Carbazol-1(2H)-one, 3,4-dihydro-6-methoxy-	EtOH	314(4.40)	12-0246-64
Carbazol-1(2H)-one, 3,4-dihydro-7-methoxy-	EtOH	333(4.42)	12-0246-64
Carbazol-1(2H)-one, 3,4-dihydro-8-methoxy-	EtOH	306(4.33)	12-0246-64
Crotonic acid, 2-cyano-3-methyl-4-phenyl-, methyl ester, trans	EtOH	282(4.03)	28-3102-65A
1H-Pyrrolo[1,2-a]indole-9-carboxaldehyde, 2,3-dihydro-7-hydroxy-6-methyl-	MeOH	256(4.20),283(4.17), 311(4.11)	35-3877-64
	MeOH	215(4.47),256(4.20), 283(4.17),311(4.11)	44-2897-65
1H-Pyrrolo[1,2-a]indol-1-one, 2,3-dihydro-7-methoxy-6-methyl-	MeOH	331(4.33)	35-3877-64
	MeOH	218(4.46),331(4.33)	44-2897-65
Securinine	hexane	328(3.11)	94-1118-64
	MeOH	325(3.23)	94-1118-64
	EtOH	256(4.15),325(3.23)	28-0337-65A
	EtOH	330(3.24)	94-1118-64
	dioxan	255(4.20),333(3.35)	94-1118-64
Virosecurinine	H_2O	258(4.25),300s(3.46)	44-3441-64
	aq. HCl	256.5(4.26)	44-3441-64
	99.5% EtOH	255(4.21),331(3.26)	44-3441-64
	95% EtOH	255(4.22),325(3.35)	44-3441-64
	50% EtOH	258(4.23),300s(3.44)	44-3441-64
	CCl_4	332(3.30)	44-3441-64
$C_{13}H_{13}NO_2S$			
5H-Cyclopenta[de]-1-benzothiopyran-4-carboxamide, 2,3,3a,4-tetrahydro-6-methyl-5-oxo-	EtOH	243s(4.28),249(4.32), 269s(4.01),340(3.55)	88-2489-64
Pyridine, 1,2-dihydro-1-methyl-2-[(phenylsulfonyl)methylene]-, cation	n.s.g.	218(4.06),270(4.03)	39-3090-65
neutral	n.s.g.	219(3.99),293(4.02), 365(3.97)	39-3090-65
Pyridine, 1,4-dihydro-1-methyl-4-[(phenylsulfonyl)methylene]-, cation	n.s.g.	226(4.18),264(3.74)	39-3060-65
neutral	n.s.g.	220(4.10),353(4.51)	39-3060-65
p-Toluenesulfonanilide	hexane	224(4.17),270(3.35)	65-2080-65
	EtOH	224(4.26),272(3.35)	65-2080-65
	EtOH-NaOEt	246(4.07),280(3.43)	65-2080-65
	ether	224(4.16),270(3.40), 300s(2.75)	65-2080-65
	dioxan	224(4.23),270(3.24)	65-2080-65
	H_2SO_4	246(4.20),270(3.72)	65-2080-65
$C_{13}H_{13}NO_3$			
Cyclohepta[b]pyrrole-3-carboxylic acid, 1,2-dihydro-1-methyl-2-oxo-, ethyl ester	EtOH	230(4.22),280(4.56), 428(4.27)	94-0828-65

Compound	Solvent	λ_{max}(log ε)	Ref.
3-Indolevaleric acid, δ-oxo-	NaOH	267(4.21),332(4.33)	65-3025-64
Pyoluteorin, didechloro-O,O-dimethyl-	EtOH	255s(3.70),297(4.19)	39-2641-64
$C_{13}H_{13}NO_4$			
Benzoic acid, 1,2,3,6-tetrahydro-2-(o-nitrophenyl)-, cis	EtOH	254(3.62)	44-0681-64
Benzoic acid, 1,2,5,6-tetrahydro-2-(o-nitrophenyl)-	EtOH	251(3.69)	44-0681-64
Cyclohepta[b]pyrrole-3-carboxylic acid, 1,2-dihydro-8-methoxy-2-oxo-, ethyl ester	EtOH	237(4.31),280(4.07), 314(4.29),408(4.10)	94-0828-65
2-Cyclohexen-1-one, 2-(m-nitrophenyl)-3-nitro-	EtOH	220(4.23),273(3.55), 337(3.70),409(3.57)	44-1416-65
Furo[2,3-b]quinolin-7-ol, 2,3-dihydro-4,8-dimethoxy-	EtOH	225(4.59),240(4.63), 310(3.82),322(3.85)	95-0731-65
2-Indolecarboxylic acid, 3-formyl-5-methoxy-6-methyl-, methyl ester	MeOH	218(4.38),251(4.23), 335(4.09)	35-4612-64
L-Valine, N-phthalyl-	MeOH	292(3.25),300s(3.21)	24-0533-64
$C_{13}H_{13}NO_5S$			
2H-1,2-Benzothiazine-2-acetic acid, 3-ethylidene-3,4-dihydro-4-oxo-, methyl ester, 1,1-dioxide	EtOH	262(4.09),300s(3.48)	44-2241-65
Ketone, 4-hydroxy-2-methyl-2H-1,2-benzothiazin-3-yl methyl, acetate, 1,1-dioxide	EtOH	293(4.04),305s(4.02)	44-2241-65
$C_{13}H_{13}NO_6$			
Isophthalic acid, 5-cyano-2,4-dihydroxy-, diethyl ester	EtOH	207(4.1),227(4.64), 244(4.30),261s(4.1), 322(3.81)	39-1629-65
$C_{13}H_{13}NO_9$			
1,3-Butadiene-1,2,3,4-tetracarboxylic acid, 1-cyano-4-hydroxy-, tetramethyl ester	MeOH	277(4.20),331(3.81)	24-1581-65
$C_{13}H_{13}NS_2$			
2,1-Benzisothiazoline-3-thione, 4,5,6,7-tetrahydro-1-phenyl-	EtOH	262(3.86),377(4.27)	24-0654-64
$C_{13}H_{13}N_3O$			
Benzanilide, p,p'-diamino-	EtOH	222(4.4),300(4.5)	28-4295-64B
p-Benzoquinone, 2-methyl-, 4-azine with 1-methyl-4(1H)-pyridone	C_5H_5N	531(4.78)	5-0009-65J
	99% C_5H_5N	535(4.78)	5-0009-65J
	98% C_5H_5N	535(4.77)	5-0009-65J
	95% C_5H_5N	541(4.77)	5-0009-65J
	90% C_5H_5N	544(4.77)	5-0009-65J
	80% C_5H_5N	549(4.80)	5-0009-65J
	70% C_5H_5N	549(4.81)	5-0009-65J
	60% C_5H_5N	549(4.82)	5-0009-65J
	50% C_5H_5N	549(4.83)	5-0009-65J
	40% C_5H_5N	549(4.83)	5-0009-65J
	30% C_5H_5N	549(4.83)	5-0009-65J
	20% C_5H_5N	548(4.84)	5-0009-65J
	10% C_5H_5N	548(4.84)	5-0009-65J
	5% C_5H_5N	545(4.83)	5-0009-65J
	2% C_5H_5N	540(4.82)	5-0009-65J
	1% C_5H_5N	540(4.82)	5-0009-65J

Compound	Solvent	λ_{max}(log ε)	Ref.
$C_{13}H_{13}N_3OS$			
Hydantoin, 5-(indol-3-ylmethyl)-3-methyl-2-thio-	97% HOAc	265(4.28)	65-0554-65
$C_{13}H_{13}N_3O_2$			
p-Benzoquinone, 2-(hydroxymethyl)-, 4-azine with 1-methyl-4(1H)-pyridone	C_5H_5N	550(4.77)	5-0009-65J
	99% C_5H_5N	550(4.78)	5-0009-65J
	98% C_5H_5N	545(4.79)	5-0009-65J
	95% C_5H_5N	547(4.79)	5-0009-65J
	90% C_5H_5N	550(4.80)	5-0009-65J
	80% C_5H_5N	554(4.83)	5-0009-65J
	70% C_5H_5N	554(4.84)	5-0009-65J
	60% C_5H_5N	550(4.86)	5-0009-65J
	50% C_5H_5N	550(4.85)	5-0009-65J
	40% C_5H_5N	550(4.85)	5-0009-65J
	30% C_5H_5N	549(4.84)	5-0009-65J
	20% C_5H_5N	546(4.84)	5-0009-65J
	10% C_5H_5N	545(4.84)	5-0009-65J
	5% C_5H_5N	541(4.83)	5-0009-65J
	2% C_5H_5N	540(4.82)	5-0009-65J
	1% C_5H_5N	539(4.82)	5-0009-65J
"free" merocyanine	C_5H_5N	539(4.78)	5-0009-65J
	99% C_5H_5N	541(4.79)	5-0009-65J
	98% C_5H_5N	545(4.79)	5-0009-65J
	95% C_5H_5N	547(4.79)	5-0009-65J
$C_{13}H_{13}N_3O_2S$			
5-Pyrimidinepropionic acid, 2-amino-4-mercapto-6-phenyl-	pH 1	262(3.92),347(4.26)	4-0263-64
	pH 8.4	260(4.09),360(4.27)	4-0263-64
	pH 13	327(4.18)	4-0263-64
$C_{13}H_{13}N_3O_3$			
4-Indancarboxylic acid, 5-(1-hydroxyethyl)-1-oxo-, γ-lactone, semicarbazone	EtOH	279(4.22),319(3.89)	39-3811-65
5-Pyrimidinepropionic acid, 2-amino-4-hydroxy-6-phenyl-	pH 1	232(4.24),276(4.07)	4-0263-64
	pH 8.4	232(4.30),290(3.96)	4-0263-64
	pH 13	289(3.95)	4-0263-64
$C_{13}H_{13}N_3O_3S_2$			
Sulfanilamide, N^4-(3-oxo-1-butenyl)-N^1-2-thiazolyl-	MeOH	342(4.61)	83-0321-64
$C_{13}H_{13}N_3S$			
Thiocyanic acid, 3,5-dimethyl-1-o-tolylpyrazol-4-yl ester	EtOH	226(3.92)	94-0023-64
Thiocyanic acid, 3,5-dimethyl-1-p-tolylpyrazol-4-yl ester	EtOH	245(4.09)	94-0023-64
$C_{13}H_{13}N_5$			
Adenine, 3-(p-methylbenzyl)-	EtOH-HCl	278(4.29)	23-1599-64
	EtOH	273(4.13)	23-1599-64
s-Triazolo[1,5-a]pyrimidine, 7-(benzylamino)-5-methyl-	MeOH	292(4.23)	65-0204-64
s-Triazolo[1,5-a]pyrimidine, 5-methyl-7-(N-methylanilino)-	MeOH	308(4.26)	65-0204-64
s-Triazolo[1,5-a]pyrimidine, 5-methyl-7-(o-toluidino)-	MeOH	294(4.29)	65-0204-64

Compound	Solvent	λ_{max}(log ε)	Ref.
$C_{13}H_{13}N_5O$			
Adenine, 3-(benzyloxymethyl)-	pH 1	276(4.18)	35-5442-65
	pH 7	276(4.05)	35-5442-65
	pH 13	276(4.04)	35-5442-65
Purine-8-methanol, 6-amino-2-methyl-α-phenyl-	pH 1	269(4.25)	87-0797-65
	pH 11	273(4.20)	87-0797-65
s-Triazolo[1,5-a]pyrimidine, 7-o-anisidino-5-methyl-	MeOH	296(4.26)	65-0204-64
s-Triazolo[1,5-a]pyrimidine, 7-p-anisidino-5-methyl-	MeOH	296(4.27)	65-0204-64
$C_{13}H_{13}N_5OS$			
Purine-8-methanol, 2-amino-6-(methylthio)-α-phenyl-	pH 1	274(3.89),324(4.09)	87-0797-65
	pH 11	316(4.11)	87-0797-65
$C_{13}H_{13}N_5O_3$			
as-Triazine, 3-amino-6-methyl-5-[3-methyl-4-(5-nitro-2-furyl)-1,3-butadienyl]-	propylene glycol	290(4.02),417(4.46)	95-0109-64
$C_{13}H_{13}N_5O_4S$			
Theophylline, 7-phenyl-8-sulfamoyl-	MeOH	287(4.01)	54-1215-64
$C_{13}H_{13}N_5O_5$			
4-Oxazolecarboxylic acid, 2-methyl-5-[3-(p-nitrophenyl)-2-triazeno]-, ethyl ester	EtOH	282(4.18)	7-0576-65
$C_{13}H_{13}O_3P$			
Phosphonic acid, o-benzylphenyl-	EtOH	218s(4.18),263s(3.00), 269(3.14),276(3.10)	44-2382-64
$C_{13}H_{14}$			
1,3,5-Heptatriene, 1-phenyl-, 3-cis-1,5-trans	n.s.g.	307(3.9)	39-2898-65
all trans	n.s.g.	310(4.58),324(4.60), 340(4.45)	39-2898-65
Naphthalene, 1-propyl-	EtOH	225(4.93),283(4.11)	39-2072-65
Naphthalene, 2-propyl-	EtOH	225(5.08),276(3.70)	39-2072-65
$C_{13}H_{14}BrNO$			
1,2,3,4-Tetrahydro-4-hydroxybenzo-[c]quinolizinium bromide	EtOH	204(4.55),239(4.51), 320(4.09)	78-2529-65
1H-Pyrrolo[1,2-a]indole, 9-bromo-2,3-dihydro-7-methoxy-6-methyl-	MeOH	289(3.95),300(3.95), 309s(3.83)	44-2904-65
$C_{13}H_{14}Br_2N_2$			
7,8-Dihydro-6H-dipyrido[1,2-a:2',1'-c]-[1,4]diazepinediium dibromide	H_2O	287(4.19)	25-0782-65
$C_{13}H_{14}Br_2N_2O$			
6,7-Dihydro-6-hydroxy-6-methyldipyrido-[1,2-a:2',1'-c]pyrazinediium dibromide	H_2O	224s(3.97),286(4.05)	88-3871-64
$C_{13}H_{14}Br_2O_2$			
Naphthalene, 1,4-dibromo-5,6-dihydro-2,3-dimethoxy-8-methyl-	EtOH	206(4.34),239(4.36), 312(3.94)	39-0361-65

Compound	Solvent	λ_{max}(log ε)	Ref.
$C_{13}H_{14}ClIN_2$			
9-Chloro-3,4-dihydro-5,6-dimethyl-1H-azepino[5,4,3-cd]indolium iodide	EtOH	215(4.51),340(3.72), 405(3.79)	87-0200-65
$C_{13}H_{14}ClNO_2$			
Indole-3,5-dicarboxaldehyde, 4-chloro-6,7-dihydro-	MeOH	228(4.45),306(3.95), 375(4.17)	35-5262-65
$C_{13}H_{14}ClNO_2S$			
Morpholine, 4-[(p-chlorobenzoyl)-thioacetyl]-	n.s.g.	257(4.19),279(4.11), 329(3.92)	32-0693-65
$C_{13}H_{14}ClNO_4$			
(1-Azulenylmethylene)dimethyl-ammonium perchlorate	MeCN	276(4.19),323(4.62), 402(4.55),485s(3.16)	24-2050-64
Coumarin, 6-chloro-3-ethoxy-4-(2-hydroxyethylamino)-	EtOH	217(4.16),235(4.11), 262(4.04),305(3.97), 312(4.00),345(4.00)	44-4122-65
$C_{13}H_{14}ClN_3O_2$			
2,3-Piperidinedione, 3-[(5-acetyl-2-chlorophenyl)hydrazone]	EtOH	240(4.33),258(4.00), 334(4.33)	87-0200-65
$C_{13}H_{14}Cl_2N_2O_2$			
Nicotinamide, N-(2,6-dichlorobenzyl)-1,4,5,6-tetrahydro-6-hydroxy-	MeOH	287(4.16)	5-0079-64D
$C_{13}H_{14}Cl_2N_2O_4S$			
2-Pyridinesulfonic acid, 5-carbamoyl-N-(2,6-dichlorobenzyl)-1,2,3,4-tetrahydro-	MeOH	294.8(4.29)	5-0079-64D
4-Pyridinesulfonic acid, 5-carbamoyl-N-(2,6-dichlorobenzyl)-1,2,3,4-tetrahydro-	MeOH	295.7(4.35)	5-0079-64D
$C_{13}H_{14}Fe$			
Ferrocene, 1,1'-trimethylene-	EtOH	260s(3.5),320s(1.8), 440(2.3)	44-2452-64
$C_{13}H_{14}N_2$			
Acridine, 9-amino-1,2,3,4-tetrahydro-	pH 7	218(4.21),240(4.58), 323(4.04),336(3.99)	39-4653-65
	pH 12	218(4.40),237(4.54), 313(3.85),317(3.85)	39-4653-65
Indole, 3-(3-methyl-1-pyrrolin-2-yl)-	EtOH	219(4.39),250s(4.02), 256(4.07),273s(3.98), 280s(4.02),291(4.07), 330s(3.18)	87-0415-64
3H-Indole, 3-(1-methyl-2-pyrrolidin-ylidene)-	EtOH-HI	225s(4.27),245s(4.06), 251(4.12),265(3.98), 271(4.00),332(4.25)	87-0415-64
	EtOH	244s(3.87),251(3.95), 271(4.07),276s(3.96), 339(4.18)	87-0415-64
3H-Indole-3-propionitrile, 2,3-dimethyl-	MeOH	257(3.79),345(2.37)	65-3487-64
Pyridine, 4-(p-dimethylaminophenyl)-	N HCl	272(4.32)	32-0902-64
	pH 1	272(4.32),412(2.97)	32-0902-64
	pH 2	272(4.32),412(4.01)	32-0902-64
	EtOH	232(4.08),334(4.39)	32-0902-64

Compound	Solvent	λ_{max}(log ε)	Ref.
Pyrrole, 2-(N-benzylformimidoyl)-1-methyl-	EtOH	287(4.24)	12-0894-64
Pyrrole, 1-methyl-2-(N-m-tolylformimidoyl)-	EtOH	231(3.83),240(3.78), 327(4.25)	12-0894-64
Pyrrole, 1-methyl-2-(N-p-tolylformimidoyl)-	EtOH	220s(3.77),330(4.30)	12-0894-64
$C_{13}H_{14}N_2O$			
Benz[cd]indole-3-carboxamide, 1,3,4,5-tetrahydro-2-methyl-	EtOH	225(4.53),274(3.86), 278(3.86),289(3.71)	44-0843-64
3H-Indazol-3-one, 2,3a,4,5,6,7-hexahydro-2-phenyl-	C_6H_{12}	247(4.31)	78-0299-64
	10N H_2SO_4	239(4.14)	78-0299-64
	pH 5	246(4.12)	78-0299-64
	pH 13	252(4.12)	78-0299-64
3-Pyridinol, 4-methyl-2-(methylamino)-5-phenyl-	MeOH-HCl	313(4.04)	44-2124-64
	MeOH	310(4.00)	44-2124-64
3-Pyridinol, 4-methyl-6-(methylamino)-5-phenyl-	MeOH-HCl	219(4.30),332(3.86)	44-1887-65
	MeOH	327(3.71)	44-1887-65
	MeOH-KOH	228(4.16),339(3.69)	44-1887-65
1H-Pyrido[3,4-b]indol-1-one, 2,3,4,9-tetrahydro-2,3-dimethyl-	EtOH	228(4.36),304(4.20)	39-7165-65
2(1H)-Pyridone, 1-(2-anilinoethyl)-	EtOH	237(3.63),247(3.66), 298(3.37)	44-3687-64
Pyrrole, 2-[N-(p-methoxyphenyl)-formimidoyl]-1-methyl-	EtOH	219s(3.97),337(4.31)	12-0894-64
$C_{13}H_{14}N_2O_2$			
Carbamic acid, (α-cyano-β-ethylstyryl)-, methyl ester	EtOH	256(4.05)	28-1987-65B
Cinnamic acid, α-cyano-β-(methylamino)-, ethyl ester	EtOH	230(3.86),293(4.25)	94-0828-65
Pyrazole-1-acetic acid, 5-methyl-4-phenyl-, methyl ester	EtOH	243(4.17)	44-1889-65
4-Pyrimidinol, 5-(ethoxymethyl)-2-phenyl-	EtOH	242(4.10),294(3.95)	94-0393-64
$C_{13}H_{14}N_2O_2S$			
5-Pyrimidinecarboxylic acid, 1,2,3,4-tetrahydro-1-phenyl-2-thio-, ethyl ester	EtOH	315(4.13)	44-2290-65
5-Pyrimidinecarboxylic acid, 1,2,3,6-tetrahydro-1-phenyl-2-thio-, ethyl ester	EtOH	316(4.17)	44-2290-65
6H-1,3-Thiazine-5-carboxylic acid, 2-anilino-, ethyl ester	EtOH	311(4.34)	44-2290-65
p-Toluenesulfonanilide, 2'-amino-	hexane	228(4.26),296(3.57)	65-2080-65
	EtOH	230(4.27),296(3.53)	65-2080-65
	EtOH-NaOEt	294(3.49)	65-2080-65
	ether	230(4.32),298(3.62)	65-2080-65
	dioxan	230(4.20),298(3.57)	65-2080-65
	H_2SO_4	236(4.20),265(3.26), 274(3.08),307s(2.56)	65-2080-65
p-Toluenesulfonanilide, 3'-amino-	hexane	292(3.49),330s(2.43)	65-2080-65
	EtOH	236(4.22),294(3.42)	65-2080-65
	EtOH-NaOEt	295(3.48),330s(2.43)	65-2080-65
	ether	214(4.56),236(4.23), 298(3.54)	65-2080-65
	dioxan	234(4.20),298(3.50), 330s(2.40)	65-2080-65
	H_2SO_4	248(4.12),270(3.10), 310s(2.75)	65-2080-65

Compound	Solvent	λ max(log ε)	Ref.
p-Toluenesulfonanilide, 4'-amino-	EtOH	226(4.09),248(4.02), 300(3.35)	65-2080-65
	EtOH-NaOEt	252(4.02),300(3.32), 340s(2.57)	65-2080-65
	ether	226(4.34),248(4.22), 294(3.68)	65-2080-65
	dioxan	220(4.16),252(4.02), 296(3.67)	65-2080-65
	H_2SO_4	248(4.43),270(3.96), 310s(2.75)	65-2080-65
$C_{13}H_{14}N_2O_3$			
3-(1-Carboxy-2-hydroxyvinyl)-2,3-dimethylbenzimidazolium hydroxide, inner salt, methyl ester	EtOH	257(4.42),277(4.07)	44-1118-65
5-Pyrimidinecarboxylic acid, 1,2,3,4-tetrahydro-2-oxo-1-phenyl-, ethyl ester	EtOH	236s(3.96),295(4.04)	94-0804-64
Tryptophan, 5-acetyl-	EtOH-HCl	251(4.52),295(3.90)	87-0141-64
$C_{13}H_{14}N_2O_3S$			
6H-1,3-Thiazin-6-one, 3-acetyltetrahydro-2-[(p-methoxyphenyl)imino]-	EtOH	241(4.22),295(4.06)	44-1623-64
$C_{13}H_{14}N_2O_5$			
3-(Carboxymethyl)-4-hydroxy-6,7-dimethoxy-2-methylcinnolinium hydroxide, inner salt	EtOH	239(4.48),272(4.13), 281(4.10),362(4.12), 379(4.16)	39-6036-65
3-Cinnolineacetic acid, 1,4-dihydro-6,7-dimethoxy-1-methyl-4-oxo-	EtOH	235(4.42),260(4.29), 269(4.31),280s(3.90), 349(4.08),363(4.14)	39-6036-65
3-Cinnolineacetic acid, 4-hydroxy-6,7-dimethoxy-, methyl ester	EtOH	228(4.42),257(4.28), 266(4.30),290s(3.00), 342(4.08),357(4.07)	39-6036-65
Isophthalic acid, 4-amino-5-cyano-2-hydroxy-, diethyl ester	EtOH	210(4.1),234(4.52), 260(4.28),285(4.20), 335(3.85)	39-1629-65
$C_{13}H_{14}N_4$			
m-Phenylenediamine, 4-methyl-6-(phenylazo)-	EtOH	284(3.95),440(4.28)	65-0675-65
s-Triazine, 2-amino-4-ethyl-6-styryl-	EtOH	300(4.46)	44-0707-65
$C_{13}H_{14}N_4O_4$			
7-Norbornanone, 2,4-dinitrophenylhydrazone	EtOH	227(4.22),358(4.31)	44-0160-64
	$CHCl_3$	362(4.35)	44-0160-64
1,3,4-Oxadiazole, 2-(ethylamino)-5-[3-methyl-4-(5-nitro-2-furyl)-1,3-butadienyl]-	propylene glycol	246(3.94),328(4.07), 412(4.19)	95-0219-64
Sorbaldehyde, 2-methyl-, 2,4-dinitrophenylhydrazone	$CHCl_3$	395(4.45)	22-0161-64
$C_{13}H_{14}N_4O_6$			
3-Butenoic acid, 3-methyl-2-oxo-, ethyl ester, 2,4-dinitrophenylhydrazone	EtOH	249s(4.10),365(4.46)	39-0766-64

Compound	Solvent	λ max(log ε)	Ref.
$C_{13}H_{14}N_4O_{10}$			
Crotonic acid, 3-[(dimethylamino)-methyl]-2,4-dihydroxy-, γ-lactone, picrate	50% EtOH	231(4.20),358(4.19)	39-0766-64
$C_{13}H_{14}N_6$			
1H-Imidazo[4,5-d]pyridazine, 4,7-bis-(methylamino)-2-phenyl-	EtOH	211(4.58),262(4.48), 288s(4.36),293(4.36)	4-0182-64
s-Triazole, 3-anilino-5-(3,5-di-methylpyrazol-1-yl)-	EtOH	256(4.48)	39-3912-65
$C_{13}H_{14}O$			
Cyclohexanone, 2-benzylidene-, trans	MeOH	290(4.05)	78-2201-64
Cyclopentanone, 2-benzylidene-5-methyl-, trans	MeOH	297(4.34)	78-2201-64
2-Cyclopentenone, 2-benzyl-5-methyl-	MeOH	227(3.90)	78-2201-64
6-Hepten-4-yn-3-ol, 7-phenyl-	EtOH	284(4.45)	39-2983-65
Indanone, 2-isopropylidene-6-methyl-	C_6H_{12}	265(4.32),275(4.29)	22-3103-64
1(4H)-Naphthalenone, 3,4,4-trimethyl-	EtOH	252(4.09),271(3.99)	44-2956-65
$C_{13}H_{14}OS$			
5,7-Nonadiyn-2-ol, 9-(2-thienyl)-	ether	234(3.62),290(4.21)	24-0801-64
6-Nonen-8-yn-3-one, 9-(2-thienyl)-	ether	290(4.21)	24-0801-64
$C_{13}H_{14}O_2$			
2-Cyclopentene-1-carboxylic acid, 3-methyl-2-phenyl-	EtOH	243(3.58)	23-2408-65
Fluoren-2(1H)-one, 3,4,4a,9a-tetra-hydro-7-hydroxy-	MeOH-KOH	242(3.93),302(3.52)	88-3505-65
	EtOH	283(3.42)	88-3505-65
Fluoren-2(1H)-one, 3,4,4a,9a-tetra-hydro-7-hydroxy-, trans	MeOH-KOH	242(4.00),301(3.53)	88-3505-65
	EtOH	282.5(3.45)	88-3505-65
3-Furanmethanol, 2-methyl-5-p-tolyl-	MeOH	288(4.37)	44-2205-65
Naphthalene, 1-acetyl-3,4-dihydro-6-methoxy-	EtOH	243(4.09),286(3.72)	44-2527-64
Naphthalene, 6,7-dimethoxy-1-methyl-	EtOH	236(4.78),243(4.61), 264(3.59),273(3.66), 281(3.67),292(3.51), 311(3.31),318(3.16), 325(3.49)	39-0361-65
1(4H)-Naphthalenone, 6-methoxy-4,4-dimethyl-	EtOH	226(4.05),237(4.02), 337(3.99)	39-0361-65
2,4-Pentadienoic acid, 2,3-dimethyl-5-phenyl-, 2-cis-4-trans	MeOH-acid	230(4.14),300(4.46)	44-2986-64
all trans	MeOH-acid	230(4.00),302(4.50)	44-2986-64
2,4-Pentadienoic acid, 5-phenyl-, ethyl ester	n.s.g.	311(4.58)	5-0058-65B
$C_{13}H_{14}O_2S$			
1,6-Dioxaspiro[4.4]non-3-ene, 2-[5-(methylthio)-4-penten-2-ynylidene]-	ether	269(3.99),339(4.35)	24-1179-64
$C_{13}H_{14}O_3$			
2-Benzocyclooctenecarboxylic acid, 5,6,7,8,9,10-hexahydro-5-oxo-	EtOH	246(3.98)	44-2165-65
2-Benzofuranvaleric acid	MeOH	247(4.18),276(3.65), 283(3.65)	94-0214-64
1-Cyclopropanecarboxylic acid, 2-benzoyl-1,2-dimethyl-, cis	MeOH	245(3.87)	24-2438-65
3(2H)-Dibenzofuranone, 1,4,4a,9b-tetra-hydro-8-hydroxy-9b-methyl-	EtOH	230(3.63),303(3.63)	39-2932-64

Compound	Solvent	λ_{max}(log ε)	Ref.
4,6,8-Dodecatriynoic acid, 10-hydroxy-, methyl ester	EtOH	239(2.79),274(2.47), 291(2.65),310(2.75), 328(2.62)	70-0544-65
1,4-Ethanonaphthalene-5,8-dione, 1,4,4a,8a-tetrahydro-1-methoxy-	EtOH	221(4.06)	39-2932-64
Indone, 4-hydroxy-6-(2-hydroxyethyl)-2,5-dimethyl-	EtOH	247(4.26),369(3.43), 440(3.24)	78-2671-65
1-Naphthol, 2,8-dimethoxy-6-methyl-	EtOH	235(4.23),291(3.32), 305(3.32),340(3.29)	39-2355-65
2-Naphthol, 1,8-dimethoxy-6-methyl-	EtOH	230(4.59),235(4.54), 285(3.74),297(3.71), 335(3.46),347(3.51)	39-2355-65
2,4-Pentadienoic acid, 3-ethoxy-5-phenyl-	EtOH	229(4.26),235(4.22), 310(4.47)	44-3161-64
$C_{13}H_{14}O_4$			
1,4-Benzodioxan-5-ol, 6-allyl-, acetate	EtOH	279(3.31)	65-2994-64
Benzofuro[2,3-b]pyran, 5-acetyl-tetrahydro-6-hydroxy-	EtOH	234(4.10),257(3.81), 374(3.52)	27-0358-64
Bicyclo[3.2.2]nona-3,6,8-triene, 2,3-dicarbomethoxy-	MeCN	245(3.61)	88-4413-65
2-Chromancarboxylic acid, 4-methyl-3-oxo-, ethyl ester	EtOH	283(3.61)	65-2699-64
4-Chromancarboxylic acid, 2-methyl-3-oxo-, ethyl ester	EtOH	295(3.63)	65-2699-64
m-Cresol, 2,4,6-triacetyl-	EtOH	241(4.43),317(3.90)	39-5646-64
5-Hexene-2,4-dione, 6-(4-hydroxy-3-methoxyphenyl)-	iso-PrOH-HOAc	240(4.5),300s(4.1), 345(4.47)	83-0660-64
	iso-PrOH-NH_3	235(4.4),290(4.2), 350(4.2),425(4.44)	83-0660-64
Indan-4,5-dicarboxylic acid, dimethyl ester	EtOH	243(4.02),284(3.47), 290(3.50)	39-3811-65
Malonic acid, benzyl-, cyclic isopropylidene ester	EtOH-NaOH	272(4.32)	49-1283-64
2-Naphthoic acid, 1,2,3,4-tetrahydro-8-methoxy-3-oxo-, methyl ester	EtOH	214(4.11),262(3.81), 278(3.59)	88-0103-64
	EtOH-base	225(3.88),279(4.05), 288(4.06)	88-0103-64
2-Naphthol, 3-(1,2-dihydroxyethyl)-5-methoxy-	n.s.g.	222(4.60),244(4.43), 248(4.42),279(4.36), 287(3.58),296(3.51), 318(3.18),332(3.20)	88-0041-65
2,4-Pentadienoic acid, 5-(3,4-dimethoxyphenyl)-, cis-trans	EtOH	206(4.24),241(4.09), 334(4.44)	39-4274-64
4-Phthalanpropionic acid, 1-methyl-3-oxo-, methyl ester	EtOH	231(3.96),277(3.37), 284(3.39)	39-3811-65
Sepedonin, anhydro-di-O-methyl-	EtOH	251(4.19),271s(4.27), 281(4.37),370(4.23)	23-1835-65
$C_{13}H_{14}O_4S$			
1,6-Dioxaspiro[4.4]non-3-ene, 2-[5-(methylsulfonyl)-4-penten-2-ynylidene]-	ether	227(3.90),345(4.37)	24-1179-64
$C_{13}H_{14}O_5$			
2-Hexenedioic acid, 3-(p-methoxyphenyl)-	EtOH	284(4.19)	65-1985-64
$C_{13}H_{14}O_6$			
Propionic acid, 3-(2,4-diacetoxyphenyl)-	EtOH	273(3.36)	39-3126-64

Compound	Solvent	$\lambda_{max}(\log \varepsilon)$	Ref.
$C_{13}H_{14}O_7S_2$			
2,7-Naphthalenedisulfonic acid, 1,4-dihydro-3,4,4-trimethyl-1-oxo-	EtOH	262(3.81),280(3.79)	44-2956-65
$C_{13}H_{14}S$			
6H-Benzo[b]cyclohepta[d]thiophene, 7,8,9,10-tetrahydro-	n.s.g.	236(4.49),267(3.82), 292(3.57),302(3.42)	22-0989-64
$C_{13}H_{15}BrO$			
2,5-Cyclohexadien-1-one, 2,4-diallyl-6-bromo-4-methyl-	EtOH	254(4.02)	33-0094-65
$C_{13}H_{15}BrO_3$			
1,4-Benzodioxan, 6-allyl-5-(2-bromoethoxy)-	EtOH	275(3.11)	65-2994-64
$C_{13}H_{15}BrO_6$			
o-Toluic acid, 5-bromo-α-carboxy-4,6-dimethoxy-, 1-ethyl ester	EtOH	291.0(3.57)	1-1677-65
$C_{13}H_{15}Br_2IO_4$			
Bicyclo[2.2.0]hexane-2,3-dicarboxylic acid, 2,3-dibromo-5-iodo-6-hydroxy-1,4,5,6-tetramethyl-, 3-methyl ester, lactone	hexane	273(2.85)	24-3838-65
$C_{13}H_{15}Cl$			
Norcarane, 7-chloro-7-phenyl-	C_6H_{12}	207(3.94),221(3.92)	35-4080-64
$C_{13}H_{15}ClN_2$			
1H-Azepino[5,4,3-cd]indole, 9-chloro-3,4,5,6-tetrahydro-5,6-dimethyl-	EtOH	227(4.53),291(3.85), 301(3.82)	87-0200-65
Indole, 5-chloro-3-(1-methyl-2-pyrrolidinyl)-	EtOH	226(4.55),283s(3.72), 289(3.74),297s(3.63)	87-0415-64
$C_{13}H_{15}ClN_2O$			
1H-Azepino[5,4,3-cd]indole-2-methanol, 9-chloro-3,4,5,6-tetrahydro-6-methyl-	EtOH	232(4.48),290(3.86)	87-0200-65
$C_{13}H_{15}ClN_2O_6S_2$			
2-(Diethylamino)-4-(p-nitrophenyl)-1,3-dithiol-1-ium perchlorate	EtOH	225(4.21),256(3.95), 326(4.33)	44-1703-64
$C_{13}H_{15}ClN_4O$			
Pyrimidine, 2,4-diamino-5-(p-chlorophenyl)-6-(ethoxymethyl)-	pH 1	273(3.80)	4-0021-65
	pH 13	246(3.85),295(3.90)	4-0021-65
	EtOH	255(3.85),296(3.93)	4-0021-65
$C_{13}H_{15}ClO_2$			
6-Heptenoic acid, 7-(p-chlorophenyl)-, trans	EtOH	257(4.29)	22-2988-65
$C_{13}H_{15}Cl_2IO_4$			
Bicyclo[2.2.0]hexane-2,3-dicarboxylic acid, 2,3-dichloro-5-iodo-6-hydroxy-1,4,5,6-tetramethyl-, 3-methyl ester, lactone	MeOH	271(2.75)	24-3838-65

Compound	Solvent	λ_{max}(log ϵ)	Ref.
$C_{13}H_{15}Cl_2NO_4S_2$			
4-(p-Chlorophenyl)-2-(diethylamino)-1,3-dithiol-1-ium perchlorate	EtOH	230(4.20),238(4.21),305(4.11),320(4.13)	44-1703-64
$C_{13}H_{15}Cl_2N_2OP$			
Phosphonic dichloride, (1,3,4,9-tetrahydro-1,1-dimethyl-2H-pyrido[3,4-b]indol-2-yl)-	EtOH	223(4.60),273(3.88),280(3.89),289(3.80)	44-2864-64
$C_{13}H_{15}FN_2O_6S$			
Uridine, 2'-deoxy-5-fluoro-5'-thio-, 3',5'-diacetate	pH 1	267.5(3.96)	44-0554-64
	pH 7	267.5(3.92)	44-0554-64
	pH 13	268(3.86)	44-0554-64
$C_{13}H_{15}IN_2$			
Pyridine, 2-[2-(1-methylpyrrol-2-yl)-vinyl]-, methiodide	H_2O	414(4.43)	54-0441-65
$C_{13}H_{15}I_3O_4$			
Bicyclo[2.2.0]hexane-2,3-dicarboxylic acid, 2,3,5-triiodo-6-hydroxy-1,4,5,6-tetramethyl-, 3-methyl ester, lactone	MeOH	271(3.20)	24-3838-65
$C_{13}H_{15}N$			
1H-Carbazole, 2,3,4,4a-tetrahydro-4a-methyl-	MeOH-HCl	230(3.9),239s(3.8),275(3.8)	78-1737-64
	MeOH	252(3.7)	78-1737-64
Cyclopenta[b]pyrrole, 3,3a,4,5,6,6a-hexahydro-2-phenyl-, cis	EtOH	243(4.29)	95-0674-64
Macrorine, tetrahydro-	EtOH-HCl	207(4.25),247(3.66),345(3.60)	39-5969-64
	EtOH	207(4.43),250(3.93),301(3.34)	39-5969-64
Phenanthridine, 1,2,3,4,4a,10b-hexahydro-, cis	EtOH	258(4.02)	44-1419-64
trans	EtOH	254(4.19)	44-1419-64
Pyrrole, 1-benzyl-2,5-dimethyl-	EtOH	209s(4.22),263s(2.73),269s(2.58)	44-0190-65
Quinoline, 5,7-diethyl-	EtOH	238(4.78),298(3.59),310(3.58),323(3.51)	95-0314-65
$C_{13}H_{15}NO$			
Benz[cd]indole-3-methanol, 1,3,4,5-tetrahydro-2-methyl-	EtOH	227(4.54),274(3.85),279(3.85),290f(3.72)	44-0843-64
Benzoquinolizinone, hexahydro-	EtOH	264(2.68),274(2.58)	42-0163-64
Carbazole, 1,2,3,4-tetrahydro-6-methoxy-	MeOH	229(4.39),288(3.95),297(3.88)	88-0973-64
Carbazole, 1,2,3,4-tetrahydro-7-methoxy-	MeOH	229(4.51),270(3.65),300(3.68)	88-0973-64
Cyclopropenone, (diethylamino)phenyl-	EtOH	280(4.43)	35-1326-65
1-Indanone, 6-(1-pyrrolidinyl)-	EtOH	246(4.49),287(3.45)	44-2513-65
Indole, 3-isobutyryl-1-methyl-	EtOH	244(4.16),248s(--),301(4.16)	87-0094-64
Isoxazole, 5-butyl-3-phenyl-	EtOH	241(4.10)	78-0817-65
1-Naphthol, 7-(isopropylamino)-	EtOH	253(4.66),293(4.06),304(4.03),349(3.43)	44-1180-64
6(5H)-Phenanthridinone, 6a,7,8,9,10,10a-hexahydro-, cis	EtOH	250(4.11)	44-0681-64
trans	EtOH	249(4.11)	44-0681-64

Compound	Solvent	λ_{max}(log ε)	Ref.
5-Pseudoindolone, 2,2-pentamethylene-	MeOH	279(4.3)	24-2939-65
1H-Pyrrolo[1,2-a]indole, 2,3-dihydro-7-methoxy-6-methyl-	MeOH	279(3.90),295(3.84), 308(3.66)	35-3877-64
	MeOH	220(4.5),279(3.9), 295(3.8),308(3.66)	44-2897-65
6(2H)-Quinolone, 2,2,4,7-tetramethyl-	MeOH	242(3.9),294(4.2), 370(3.6)	24-2939-65
8(2H)-Quinolone, 2,2,4,6-tetramethyl-	MeOH	243(4.2),252(4.2), 455(3.6)	24-2939-65
Valerolactam, N-styryl-, trans	MeOH	286(4.39)	83-0474-64
$C_{13}H_{15}NOS_2$			
1-Piperidinecarbodithioic acid, anhydrosulfide with thiobenzoic acid	C_6H_{12}	239(4.35),292(4.16), 400(2.41)	78-0035-65
$C_{13}H_{15}NOS_3$			
1-Piperidinecarbodithioic acid, anhydrosulfide with phenylxanthic acid	C_6H_{12}	243(4.34),273s(4.11)	78-0035-65
$C_{13}H_{15}NO_2$			
2H-Azepin-2-one, 1-benzoylhexahydro-	EtOH	205(4.10),232(3.92), 247(3.86)	65-1394-65
3H-Indole-3-propionic acid, 2,3-dimethyl-	MeOH	255(3.75)	65-3487-64
Ochropine dehydrogenation product	EtOH	233(4.09),258(3.88), 335(4.37)	12-0246-64
1H-Pyrrolo[1,2-a]indol-1-ol, 2,3-dihydro-7-methoxy-6-methyl-	MeOH	280(3.94),298s(3.81), 308s(3.64)	44-2904-65
8(2H)-Quinolone, 6-methoxy-2,2,4-trimethyl-	MeOH	253(4.3),312(3.7), 445(3.6)	24-2939-65
$C_{13}H_{15}NO_3$			
12H-Azepino[2,1-b][1,3]benzoxazin-12-one, 5a,6,7,8,9,10-hexahydro-5a-hydroxy-	EtOH	207(4.51),236(3.94), 295(3.38)	65-1394-65
	EtOH	207(4.51),236(3.99), 295(3.38)	78-3537-65
Benzoic acid, p-[(3-oxo-1-butenyl)-amino]-, ethyl ester	EtOH	234(3.9),261(3.56), 348(4.65)	44-3404-65
Butyrolactam, N-(α-benzyloxy)-acetyl-	dioxan	215(4.20)	78-3537-65
Isoquinoline, 6,7,8-trimethoxy-1-methyl-, hydrochloride	EtOH	238(4.75),270s(3.50), 286(3.54),310s(3.38), 323(3.34)	49-0025-65
Phyllantidine	EtOH	258(4.20)	28-0337-65A
Valerolactam, N-(o-methoxybenzoyl)-	EtOH	213(4.22),223(4.18), 290(3.35)	65-1394-65
$C_{13}H_{15}NO_4$			
Benzoic acid, p-[N-(formylmethyl)-acetamido]-, ethyl ester	EtOH	256(3.95)	35-5664-64
Carbostyril, 3-(2-hydroxyethoxy)-6-methoxy-1-methyl-	pH 13	223(4.63),281(3.89), 337(3.95)	39-1080-65
Maleic acid, (N-methylanilino)-, dimethyl ester	EtOH	238(3.70),289(4.29)	5-0098-65H
α^5-Pyridoxylideneacetic acid, α^4,3-O-isopropylidene-	pH 1	237(4.30),308(3.99)	87-0112-65
$C_{13}H_{15}NO_5$			
β-Alanine, N-formyl-N-(p-carbethoxyphenyl)-	EtOH	264(4.18)	44-3404-65

Compound	Solvent	$\lambda_{max}(\log \epsilon)$	Ref.
$C_{13}H_{15}NO_5S_2$			
Benzothiazole, 2-(β-D-galacto-pyranosylthio)-	EtOH	223(4.3),275(4.1), 300s(3.9)	24-3508-65
Benzothiazole, 2-(β-D-gluco-pyranosylthio)-	EtOH	224(4.3),275(4.1), 300s(3.9)	24-3508-65
$C_{13}H_{15}NO_7$			
2H-Pyran-5-pyruvic acid, 6-carboxy-2-oxo-, diethyl ester, oxime	EtOH	309(3.84)	88-3431-64
$C_{13}H_{15}NS_2$			
2H-1,3-Thiazine-2-thione, 3,6-dihydro-3,4,5-trimethyl-6-phenyl-	EtOH	265(3.65),314(4.02)	39-4008-64
$C_{13}H_{15}N_3$			
Pyrrole, 2-[N-[p-(dimethylamino)-phenyl]formimidoyl]-	EtOH	245(3.90),297(4.05), 366(4.31)	12-0894-64
3-Pyrroline-3-carbonitrile, 4-amino-1-benzyl-5-methyl-	pH 1	208(3.95),260(4.06)	39-4546-65
	pH 13	217(4.18),264(3.98)	39-4546-65
$C_{13}H_{15}N_3O$			
Pyrimidine, 4-amino-5-(ethoxy-methyl)-2-phenyl-	EtOH	240(3.75),282(3.94)	94-0393-64
$C_{13}H_{15}N_3O_2$			
Sydnone imine, N-benzoyl-3-butyl-	pH 1.0	240(3.92),288(4.36)	65-2064-64
	pH 7.00	254(4.00),324(4.40)	65-2064-64
$C_{13}H_{15}N_3O_3$			
γ-Butyrolactone, α-acetyl-γ-phenyl-, semicarbazone	n.s.g.	222(4.24),274(3.72)	39-0141-64
Salicylaldehyde, 5-nitro-, diallylhydrazone	MeOH	296(4.47)	24-1631-64
	MeOH-NaOH	343(4.24),425(4.08)	24-1631-64
$C_{13}H_{16}$			
1H-Benz[e]indene, 2,3,6,7,8,9-hexa-hydro-	n.s.g.	269(2.75),273(2.67), 278(2.75)	88-1717-65
5H-Benzocyclononene, 6,7,8,9-tetra-hydro-	EtOH	240s(3.4)	78-0231-64
Cyclohexane, benzylidene-	C_6H_{12}	209(4.21),246(4.15)	35-4080-64
	n.s.g.	247(4.16)	5-0058-65B
Naphthalene, 1,2-dihydro-1,1,6-trimethyl-	n.s.g.	265(3.72)	39-2892-65
$C_{13}H_{16}Br_2O$			
Naphthalene, 2,3-dibromo-1,2,3,4-tetra-hydro-7-methoxy-1,1-dimethyl-	EtOH	220s(3.95),281(3.37), 288(3.35)	39-0361-65
$C_{13}H_{16}ClNOS_2$			
Carbamic acid, diethyldithio-, ester with 4'-chloro-2-mercaptoaceto-phenone	EtOH	254(4.40)	44-1703-64
$C_{13}H_{16}ClNO_3$			
Isoquinoline, 6,7,8-trimethoxy-1-methyl-, hydrochloride	EtOH	238(4.75),270s(3.50), 286(3.54),310s(3.38), 323(3.34)	49-0025-65
$C_{13}H_{16}ClNO_4$			
[(2,3-Dihydro-1-azulenyl)methylene]-dimethylammonium perchlorate	MeCN	265(4.01),322(3.10), 473(4.48)	24-2050-64

Compound	Solvent	λ_{max}(log ε)	Ref.
1-Ethyl-2,4-dimethylquinolinium perchlorate	EtOH	314(3.90)	65-0506-65
$C_{13}H_{16}ClNO_4S_2$			
2-(Diethylamino)-4-phenyl-1,3-dithiol-1-ium perchlorate	EtOH	225(4.18),236(4.12), 300(3.95),320(4.06)	44-1703-64
$C_{13}H_{16}ClNO_5S_2$			
2-(Diethylamino)-4-(p-hydroxyphenyl)-1,3-dithiol-1-ium perchlorate	EtOH	264(4.13),288(4.12), 335(3.96)	44-1703-64
$C_{13}H_{16}ClNO_8$			
1-(1,2-Dicarboxyvinyl)-3,5-dimethylpyridinium perchlorate, dimethyl ester	H_2O	273(3.77),281s(3.73)	39-2676-64
$C_{13}H_{16}INO_2$			
1-Naphthalenecarbamic acid, 1,2,3,4-tetrahydro-2-iodo-, ethyl ester, trans	MeOH	267(3.04),274(2.94)	44-3640-64
$C_{13}H_{16}I_2O_4$			
Bicyclo[2.2.0]hexane-2,3-dicarboxylic acid, 2,5(or 3,5)-diiodo-6-hydroxy-1,4,5,6-tetramethyl-, 3-methyl ester, lactone	MeOH	264(3.11)	24-3838-65
$C_{13}H_{16}N_2$			
1H-Azepino[5,4,3-cd]indole, 3,4,5,6-tetrahydro-5,6-dimethyl-	EtOH	225(4.49),284(3.81)	87-0200-65
Benzocyclobutene, 1-cyano-1-(2-dimethylaminoethyl)-	EtOH	253s(3.77),258(3.02), 264(3.20),270(3.18)	87-0732-65
Dipyrromethene, 3,3',5,5'-tetramethyl-	EtOH-HCl	226(3.94),463(5.03)	12-0363-65
	EtOH	289(3.66),435(4.52)	12-0363-65
	CCl_4	437(4.47)	12-0363-65
hydrochloride	EtOH	286(3.15),303(3.25), 346(3.64),469(5.01)	39-0893-64
Dipyrromethene, 3,4',5,5'-tetramethyl-	EtOH-HCl	238(3.72),476(4.80)	12-0363-65
	EtOH	295(3.57),439(4.40)	12-0363-65
	CCl_4	439(4.40)	12-0363-65
Dipyrromethene, 4,4',5,5'-tetramethyl-	EtOH-HCl	357(3.73),472(4.77)	12-0363-65
	EtOH	233(3.81),441(4.36)	12-0363-65
	CCl_4	442(4.38)	12-0363-65
Indole, 3-(2-isopropylideneaminoethyl)-	EtOH	221(4.53),275(3.75), 281(3.78),289(3.71)	44-2864-64
Indole, 3-(1-methyl-2-pyrrolidinyl)-	EtOH	219(4.57),274s(3.77), 281(3.80),289(3.73)	87-0415-64
Indole, 3-(5-methyl-2-pyrrolidinyl)-	EtOH-HCl	215(4.62),270(3.81), 276(3.80),279(3.80), 286(3.75),326(2.62)	87-0415-64
1H-Pyrido[3,4-b]indole, 2,3,4,9-tetrahydro-1,1-dimethyl-	EtOH	224(4.49),281(3.79), 289(3.72)	44-2864-64
$C_{13}H_{16}N_2O$			
Acetimidic acid, 2-cyano-N-phenethyl-, ethyl ester	MeOH	256(4.35)	44-0308-64
2-Carboline, 6-methoxy-1-methyl-1,2,3,4-tetrahydro-	pH 1	220(4.40),273(3.86), 293s(3.69),304s(3.52)	25-0622-64
	EtOH	225(4.34),280(3.86), 294s(3.80),307s(3.56)	25-0622-64

Compound	Solvent	$\lambda_{max}(\log \epsilon)$	Ref.
2,2'-Dipyrryl ketone, 3,4',5,5'-tetramethyl-	EtOH	269(3.20),297(3.44), 361(4.15)	12-1977-65
	EtOH	269(3.44),299(3.20), 361(4.15)	12-1977-65
	EtOH-HBr	302(4.25),354(4.34), 433(4.86)	12-1977-65
Gramine, 5-acetyl-	EtOH	251(4.53),296(3.92)	87-0141-64
Indol-5-ol, 3-(1-methyl-2-pyrrolidinyl)-	EtOH	275(3.79),299(3.67), 310s(3.56)	87-0415-64
Indol-6-ol, 3-(1-methyl-2-pyrrolidinyl)-	EtOH	215(4.48),254s(3.44), 262s(3.57),270(3.63), 293(3.67),302s(3.55)	87-0415-64
Pyrazole, 5-ethoxy-3,4-dimethyl-1-phenyl-	C_6H_{12}	259(4.21)	78-0299-64
	20N H_2SO_4	238.5(4.12)	78-0299-64
	pH 5	249(4.05)	78-0299-64
1H-Pyrido[3,4-b]indole, 3-ethyl-2,3,4,9-tetrahydro-7-hydroxy-, hydrochloride	EtOH	222(4.51),269(3.70), 295(3.76)	44-2864-64
$C_{13}H_{16}N_2O_2$			
Tryptophan, 1-ethyl-	90% EtOH-HCl	223(4.53),246(3.22), 277s(3.74),286(3.78), 296s(3.70)	94-0088-65
	90% EtOH	223(4.53),278s(3.71), 287(3.76),297s(3.67)	94-0088-65
$C_{13}H_{16}N_2O_3$			
Benzimidazole, 1-(2'-deoxy-β-D-ribofuranosyl)-2-methyl-	MeOH	245(3.85),266(3.51), 274(3.53),281(3.72)	35-4940-65
	pH 1	243(3.73),263s(3.75), 269(3.87),276(3.90)	35-4940-65
	pH 11	245(3.88),266s(3.64), 273(3.72),280(3.75)	35-4940-65
Carbamic acid, [2-(2-oxo-3-indolinyl)-ethyl]-, ethyl ester	EtOH	249(3.85),281s(3.22)	78-0565-64
Tryptamine, N-acetyl-6-hydroxy-5-methoxy-	EtOH	208(4.38),220(4.40), 303(3.89)	4-0387-65
$C_{13}H_{16}N_2O_3S$			
Indole-2-carboxylic acid, 1-methyl-3-(dimethylaminosulfinyl)-, methyl ester	EtOH	215(4.38),235(4.42), 303(4.19),326s(3.67)	44-0178-64
$C_{13}H_{16}N_2O_3S_2$			
Carbamic acid, diethyldithio-, ester with 2-mercapto-4'-nitroacetophenone	EtOH	267(4.41)	44-1703-64
$C_{13}H_{16}N_2O_4$			
3,6-Azepinedicarboxylic acid, 4-cyano-4,5-dihydro-2,7-dimethyl-, dimethyl ester	EtOH	229(4.14),326(4.15)	39-2411-65
Benzohydroxamic acid, N-cyclohexyl-p-nitro-	MeOH	270(4.02)	73-0940-65
	50% MeOH-HCl	272(4.02)	73-0940-65
	50% MeOH-borate	270(4.02)	73-0940-65
	50% MeOH-NaOH	267(4.01)	73-0940-65
2-Hexenamide, N,N-dimethyl-4-(5-nitrofurfurylidene)-	propylene glycol	280(4.15),390(4.41)	95-0646-64

Compound	Solvent	$\lambda_{max}(\log \epsilon)$	Ref.
2-Hexenamide, N-ethyl-4-(5-nitro-furfurylidene)-	propylene glycol	275(4.11),390(4.26)	95-0646-64
$C_{13}H_{16}N_2O_4S$			
Indole-2-carboxylic acid, 1-methyl-3-(dimethylsulfamoyl)-, methyl ester	EtOH	211(4.56),286(4.03)	44-0178-64
Pyrazinecarboxylic acid, 1,4,5,6-tetra-hydro-4-(p-tolylsulfonyl)-, methyl ester	THF	238(3.83),270(3.83), 308(3.20)	44-0043-65
$C_{13}H_{16}N_2O_5$			
L-Glutamic acid, N-[p-(methylamino)-benzoyl]-, hydrobromide	H_2O	291(4.18)	44-1277-65
$C_{13}H_{16}N_2O_6S_2$			
5-Thia-1-azabicyclo[4.2.0]oct-2-ene-2-carboxylic acid, 3-(hydroxy-methyl)-7-[2-(methylthio)acet-amido]-8-oxo-, acetate	pH 6.0	258.5(3.95)	39-5015-65
$C_{13}H_{16}N_2S_2$			
2-Benzothiazolesulfenamide, N-cyclohexyl-	EtOH	225(4.36),278(4.07), 288(4.03),298(3.93)	65-1588-64
3-Pyrroline-3-thiocarboxamide, 1-ben-zyl-4-mercapto-5-methyl-	EtOH	208(4.33),304(3.55), 375(4.31)	39-4546-65
	pH 13	249(3.66),291(3.60), 365(4.31)	39-4546-65
$C_{13}H_{16}N_4O_2$			
Diacetamide, N-(2,3,6-trimethyl-pyrazolo[1,5-a]pyrimidin-7-yl)-	EtOH	241(4.66),287s(3.16), 298(3.23),311(3.20), 347(3.20)	94-0142-65
$C_{13}H_{16}N_4O_3$			
Pyrazolo[1,5-a]pyrimidine-6-carboxylic acid, 7-acetamido-2,3-dimethyl-, ethyl ester	EtOH	246(4.63),299(3.53), 340(3.29)	94-1207-65
$C_{13}H_{16}N_4O_3S$			
4H-1,2,4-Triazole, 4-butyl-3-(methyl-thio)-5-(5-nitro-2-furyl)vinyl-	propylene glycol	240(4.17),295(4.03), 390(4.27)	95-0566-64
$C_{13}H_{16}N_4O_4$			
2-Pentenal, 2,4-dimethyl-, 2,4-di-nitrophenylhydrazone	$CHCl_3$	381.5(4.48)	22-0161-64
$C_{13}H_{16}N_4O_5$			
6-Pteridinemalonic acid, 5,6,7,8-tetra-hydro-7-oxo-, diethyl ester	pH 1.0	223(4.61),285(3.86), 351(3.85)	39-6930-65
	EtOH	214(4.66),232s(3.81), 275(3.79),322(3.95)	39-6930-65
$C_{13}H_{16}N_4O_7$			
Glycine, N-[N-(dinitrophenyl)-L-valyl]-	HOAc	338(4.21)	35-1848-64
$C_{13}H_{16}N_4O_7S$			
Threonine, N-[2-(7-nitro-4H-1,2,4-benzothiadiazin-3-yl)ethyl]-, S,S-dioxide	2% $NaHCO_3$	241(4.00),357(4.10)	32-0786-65

Compound	Solvent	λ_{max}(log ϵ)	Ref.
$C_{13}H_{16}N_6O_2S_2$			
Adenine, 9-(3-deoxy-2-(ethylthio)-3-thiocyanato-β-D-arabinofuranosyl)-	pH 1	258(4.20)	44-2854-65
	pH 13	261(4.23)	44-2854-65
	EtOH	261(4.21)	44-2854-65
$C_{13}H_{16}O$			
Anisole, m-1-cyclohexen-1-yl-	EtOH	215(4.3),248(3.9), 288(3.25)	39-3872-65
5H-Benzocyclononen-5-one, 6,7,8,9,10,11-hexahydro-	EtOH	245(3.70)	5-0021-64G
Bullvalene, isopropoxy-	n.s.g.	235(3.48)(end absorption)	24-3385-65
2,5-Cyclohexadien-1-one, 2,4-di-allyl-4-methyl-	EtOH	241(4.07)	33-0094-65
2,4,6,8,10-Dodecapentaenal, 11-methyl-	isooctane	374(4.80),395(4.76)	35-5075-65
Furan, 3-cinnamyltetrahydro-	EtOH	251(4.27),284(3.18)	39-4972-64
as-Indacene, 1,2,3,6,7,8-hexahydro-4-methoxy-	EtOH	277(3.39),287(3.39)	28-1844-64A
1-Indanol, 2-isopropylidene-6-methyl-	C_6H_{12}	265(3.32),270(3.35)	22-3103-64
Indanone, 2-isopropyl-6-methyl-	C_6H_{12}	242(4.08),291(3.50)	22-3103-64
Indanone, 5-isopropyl-6-methyl-	EtOH	251(4.15),302(3.57)	22-3103-64
	EtOH	265(3.91),301(3.35)	22-3103-64
Indanone, 6-isopropyl-4-methyl-	EtOH	252(4.13),304(4.18)	22-3103-64
Indanone, 6-isopropyl-7-methyl-	EtOH	251(4.08),306(3.35)	22-3103-64
Indanone, 7-isopropyl-4-methyl-	EtOH	254(4.06),301(3.37)	22-3103-64
Indanone, 7-isopropyl-6-methyl-	EtOH	253(3.87),306(3.28)	22-3103-64
Ketone, cyclohexyl phenyl	C_6H_{12}	215(4.27),237(3.78), 245(3.60),255(3.81), 265(2.88),275(2.91), 288(2.90)	73-3462-65
	EtOH	200(3.91),242(4.04), 280(2.62)	73-3462-65
Naphthalene, 1,4-dihydro-7-methoxy-1,1-dimethyl-	EtOH	227(3.95),275(3.49)	39-0361-65
Naphthalene, 2-ethyl-1,2-dihydro-5-methoxy-	EtOH	368(3.72)	94-0651-65
1(2H)-Naphthalenone, 3,4-dihydro-3,4,4-trimethyl-	EtOH	249(4.08),292(3.23)	44-2956-65
2(4aH)-Naphthalenone, 4a-allyl-5,6,7,8-tetrahydro-	EtOH	245(4.14)	33-0094-65
$C_{13}H_{16}OS_2$			
2,5-Cyclohexadien-1-one, 4-(1,3-di-thiolan-2-ylidene)-2,6-diethyl-	MeOH	260(3.56),303(3.62), 417s(4.48),432(4.53), 450s(4.38)	5-0037-65D
$C_{13}H_{16}O_2$			
1'-Acetonaphthone, 1',2',6',7',8',8'a-hexahydro-8'a-methyl-6'-oxo-	EtOH	279(4.41)	25-1767-65
5H-Benzocycloheptene, 6,7-dihydro-1,3-dimethoxy-	EtOH	210(4.35),263(4.14), 295(3.63)	5-0021-64G
4-Benzofuranol, 3,7-diethyl-2-methyl-	aq. base	306.5(3.82)	22-1464-65
p-Benzoquinone, 2,3-dimethyl-5-(3-methyl-2-butenyl)-	EtOH	254(4.22),318(2.82)	37-1374-65
1,3-Butanedione, 2,2-dimethyl-1-p-tolyl-	hexane	253(4.45)	44-3000-65
	iso-PrOH	254.5(4.13)	44-3000-65
1,4-Ethanonaphthalen-6(4H)-one, 1,7,8,8a-tetrahydro-1-hydroxy-8a-methyl-	EtOH	257(4.04)	78-1119-64
6-Heptenoic acid, 7-phenyl-, cis	EtOH	246(4.05)	22-2988-65
6-Heptenoic acid, 7-phenyl-, trans	EtOH	250(4.25)	22-2988-65

Compound	Solvent	λ_{max}(log ε)	Ref.
2(1H)-Naphthalenone, 3,4-dihydro-7-methoxy-1,1-dimethyl-	EtOH	206(4.17),220(3.86), 282(3.32),287(3.29)	39-0361-65
1-Naphthol, tetrahydro-6-methoxy-1-vinyl-	dioxan	277(3.26),284(--)	25-1729-65
4-Pentenoic acid, 3,3-dimethyl-5-phenyl-, trans	n.s.g.	252(4.23)	28-3728-64A
Renifolin, methyl-, hydrolysis product	EtOH	250(3.85),269(3.77), 320(3.39)	94-0533-64
$C_{13}H_{16}O_2S$			
Acetic acid, thio-, S-ester with 5-mercaptovalerophenone	n.s.g.	209(4.00),237(4.32)	24-1963-64
$C_{13}H_{16}O_3$			
5H-Benzocyclohepten-9-one,	EtOH	272(3.86)	5-0021-64G
6,7,8,9-tetrahydro-1,3-dimethoxy-	$HOAc-H_2SO_4$	320(4.18)	5-0021-64G
2(1H)-Naphthalenone, 3,4-dihydro-6,7-dimethoxy-1-methyl-	EtOH	230s(3.85),285(3.58)	39-0361-65
4-Pentene-2,3-dione, 5-phenyl-, 2-(dimethyl acetal)	MeOH	223(3.94),229(3.9), 299(4.32)	44-3573-65
$C_{13}H_{16}O_4$			
Bicyclo[3.3.1]nona-2,6-diene-1-carboxylic acid, 2-methoxy-6-methyl-4-oxo-, methyl ester	EtOH	240s(3.90),259(3.96)	44-0787-64
Butanoic acid, 3-(p-methoxycarbonylphenyl)-, methyl ester	EtOH	236(4.26),271s(3.02), 279s(2.87)	78-2593-65
1,4-Hexadiene-1,3-dione, 1-(methyl 3-oxo-1-cyclohexen-1-yl acetal)	MeOH	250(4.57)	35-0471-64
Malonic acid, [3-(1-cyclohexen-1-yl)-1-methyl-2-propynyl]-	EtOH	228.2(4.14)	22-2541-64
1,6(2H,7H)-Naphthalenedione, 3,4,8,8a-tetrahydro-8a-hydroxy-5-methyl-, acetate	EtOH	247(4.11)	44-1626-65
2,4,6-Octatrienedioic acid, 2,3-diethyl-4-hydroxy-, γ-lactone, methyl ester	C_6H_{12}	220(3.94),318(4.57), 330s(4.52)	35-2819-64
Spiro[5.5]undec-7-ene-1,5,9-trione, 7-methoxy-11-methyl-	MeOH	249(4.18)	35-0471-64
2,4,10-Trioxaadamantane, 3-(6-oxo-2,4-hexadien-1-yl)-	ether	262.5(4.64)	24-1839-64
Valeric acid, 5-m-anisoyl-	EtOH	217(4.34),249(3.9), 306(3.4)	39-3872-65
$C_{13}H_{16}O_5$			
Hexanedioic acid, 3-(p-methoxyphenyl)-	EtOH	275(3.24)	65-1985-64
Sepedonin, di-O-methyl-	EtOH	252(4.36),322(3.78)	23-1835-65
$C_{13}H_{16}O_6$			
Benzoic acid, 2-(3-carboxypropyl)-4,6-dimethoxy-	EtOH	251(3.90),286(3.54)	1-1677-65
Curvulic acid, O,O-dimethyl-	EtOH	215(4.02),270(3.17)	78-1411-65
Curvulic acid, O-methyl-, methyl ester	EtOH	273(3.21)	78-1411-65
$C_{13}H_{16}O_7$			
Isophthalic acid, 2,4,6-trimethoxy-, dimethyl ester	EtOH	250s(3.86),280(3.27)	12-2021-65

Compound	Solvent	λ_{max}(log ε)	Ref.
$C_{13}H_{16}Pd$			
Palladium, cyclopentadienyl-(cycloocta-2,4-dienyl)-	hexane	194(4.67),203(4.22), 233(4.13),289(4.45), 348(4.01)	39-5002-64
$C_{13}H_{16}S_2$			
Propane, 2,2-bis(5-methyl-2-thienyl)-	EtOH	239.0(4.18)	22-2635-65
$C_{13}H_{17}BrN_2O_5$			
L-Glutamic acid, N-[p-(methylamino)-benzoyl]-, hydrobromide	H_2O	291(4.18)	44-1277-65
$C_{13}H_{17}ClN_2O$			
1H-Pyrido[3,4-b]indole, 3-ethyl-2,3,4,9-tetrahydro-7-hydroxy-, hydrochloride	EtOH	222(4.51),269(3.70), 295(3.76)	44-2864-64
$C_{13}H_{17}ClO_4S$			
3,4-Dihydro-4,9,9-trimethyl-2H-1,4-methano-1-benzothio-pyrylium perchlorate	n.s.g.	262s(2.99),269(3.05), 276(2.96)	88-3569-65
$C_{13}H_{17}N$			
Acridine, 1,2,3,4,5,6,7,8-octahydro-	EtOH	216(3.80),222(3.65), 272(3.63),277(3.81), 281(3.89),285(3.85), 290(3.75)	23-0010-64
Benzo[f]quinoline, 1,2,3,4,7,8,9,10-octahydro-	EtOH	251(3.93),300(3.30)	44-1419-64
2H-Benzo[a]quinolizine, 1,3,4,6,7,11b-hexahydro-, hydrochloride	EtOH	265(2.64),273(2.60)	42-0163-64
Indole, 1-butyl-3-methyl-	EtOH	228(4.44),291(3.72)	78-0989-65
Indole, 2-butyl-3-methyl-	EtOH	227(4.44),284(3.80), 291(3.75)	78-0989-65
Indole, 2-tert-butyl-3-methyl-	MeOH	227(4.58),283(3.90), 291(3.86)	22-0101-64
Indole, 2,5-dimethyl-3-propyl-	EtOH	228(4.47),279s(3.83), 286(3.85),297(3.78)	35-3796-64
	H_2SO_4	232(3.82),243(3.77), 291(3.80)	35-3796-64
Indolizine, 1,3,5,6,7-pentamethyl-	EtOH	251(4.36),285(3.42), 298(3.54),310(3.59), 345(3.23)	39-0893-64
	EtOH-HCl	221(4.34),249(3.85), 322(3.78)	39-0893-64
Phenanthridine, 1,2,3,4,4a,5,6,10b-octahydro-, cis	EtOH	266(2.62),273(2.60)	44-1419-64
trans	EtOH	266(2.58),273(2.57)	44-1419-64
Phenanthridine, 5,6,6a,7,8,9,10,10a-octahydro-, cis	EtOH	251(3.92),304(3.42)	44-0681-64
trans	EtOH	253(3.90),304(3.36)	44-0681-64
Pyridine, 4-benzyl-1,2,3,6-tetrahydro-1-methyl-	EtOH	252.5(2.72)	44-1840-65
Quinoline, 1,2-dihydro-2,2,4,6-tetramethyl-	EtOH-acid	260(4.05)	44-1832-65
	EtOH-base	228(4.48),270s(3.51), 333(3.46)	44-1832-65
Quinoline, 1,2-dihydro-2,2,4,7-tetramethyl-	EtOH-acid	262(4.17)	44-1832-65
	EtOH-base	229(4.59),262(3.51), 325(3.58)	44-1832-65
Spiro[cyclopentane-1,3'-indoline], 1'-methyl-	EtOH	249(3.89),295(3.42)	78-1327-65

Compound	Solvent	$\lambda_{max}(\log \epsilon)$	Ref.
$C_{13}H_{17}NO$			
Butyrophenone, 2-(1-aziridinyl)-2-methyl-	EtOH	242(3.94)	44-3146-64
Cyclohexanol, 1-methyl-2-(phenylimino)-	EtOH	276(3.28)	44-2967-65
Cyclohexanone, 2-anilino-2-methyl-	EtOH	248(4.22),294(3.40)	44-2967-65
Cyclopentanol, 1-(N-phenylacetimidoyl)-	EtOH	274(3.31)	44-2967-65
6-Heptenamide, 7-phenyl-, cis	EtOH	244.5(4.10)	22-2988-65
6-Heptenamide, 7-phenyl-, trans	EtOH	249.5(4.32)	22-2988-65
2-Heptenanilide	EtOH	270(4.18)	54-0581-65
Indole, 1-ethoxy-2-propyl-	EtOH	274(3.8),282(3.8), 292s(3.7)	44-3604-65
Indole, 1-ethyl-5-methoxy-2,6-dimethyl-	MeOH	217(4.50),278(3.92), 297(3.85),307(3.66)	35-3878-64
2-Isoxazoline, 5-butyl-3-phenyl-	EtOH	264(4.11)	78-0817-65
Ketone, 1-anilinocyclopentyl methyl	EtOH	248(4.16),293(3.31)	44-2967-65
4-Pentenamide, 3,3-dimethyl-5-phenyl-, cis	n.s.g.	223(3.78)	28-3728-64A
trans	n.s.g.	251.5(4.25)	28-3728-64A
Quinoline, 1,2-dihydro-7-methoxy-2,2,4-trimethyl-	EtOH-acid	267(4.16)	44-1832-65
	EtOH-base	229(4.54),276(3.52), 321(3.66)	44-1832-65
$C_{13}H_{17}NOS$			
Morpholine, 4-(thiohydratropoyl)-	MeOH	283(4.3),355(1.86)	35-0051-65
$C_{13}H_{17}NOS_2$			
2-Butanone, 3-methyl-4-phenyl-4-(N-methylthiocarbamoylthio)-	H_2O	253(4.00),275(3.98)	39-4004-64
	EtOH	252(4.02),276(3.90)	39-4004-64
Carbamic acid, diethyldithio-, ester with 2-mercaptoacetophenone	EtOH	248(4.36),278(4.1)	44-1703-64
Carbamic acid, dimethyldithio-, ester with 2-mercapto-2-methylpropiophenone	MeOH	246(4.23),277(4.02)	44-2146-64
$C_{13}H_{17}NO_2$			
Acrylophenone, 3-(diethylamino)-2'-hydroxy-	EtOH	214(4.10),226(4.05), 248(3.84),256(3.88), 262(3.88),362(4.48)	39-3610-65
Allosecurinine, dihydro-	EtOH	213(4.33)	42-0163-64
3H-2-Benzazepine, 4,5-dihydro-7,8-dimethoxy-1-methyl-	EtOH-HCl	301(3.77),346(3.81)	88-2419-64
	EtOH	263(3.78),297(3.68)	88-2419-64
Benzosuberane, 1-hydroxy-3-N-acetamido-	MeOH	250(4.21),280(3.74), 290(3.74)	33-1988-65
Diphenyl-2-carboxylic acid, 1,2,3,4,5,6-hexahydro-2'-amino-	EtOH	257(4.08)	44-0681-64
$C_{13}H_{17}NO_2S_2$			
Carbamic acid, diethyldithio-, S-ester with 4'-hydroxy-2-mercaptoacetophenone	EtOH	278(4.42)	44-1703-64
$C_{13}H_{17}NO_3$			
Isocarbostyril, 7-ethoxy-3,4-dihydro-6-methoxy-2-methyl-	EtOH	261(4.08),296(3.93)	5-0200-65E
Securinol A	EtOH	214(4.15)	94-1307-65
Securinol B	EtOH	215(4.10)	94-1307-65
$C_{13}H_{17}NO_4$			
2,4-Cyclohexadiene-1-carboxylic acid, 1-carbamoyl-2,3,4-trimethyl-5-oxo-	n.s.g.	348(4.08)	25-1313-64
α^5-Pyridoxylacetic acid, α^4,3-O-isopropylidene-	pH 1	225s(3.52),290(3.95)	87-0112-65

Compound	Solvent	λ max(log ε)	Ref.
$C_{13}H_{17}NO_5$			
Fumaramic acid, N-(4,4-dimethyl-2,6-dioxocyclohexyl)-, methyl ester	MeOH-HCl	260(4.0),320s(3.6)	32-0948-65
	MeOH	260(4.0),320s(3.6)	32-0948-65
	MeOH-NaOH	280(4.3),330s(3.3)	32-0948-65
$C_{13}H_{17}NO_5S$			
Benzenesulfonic acid, p-nitro-, 6-heptenyl ester	EtOH	250(4.07)	35-1959-64
Benzenesulfonic acid, p-nitro-, 5-methyl-5-hexenyl ester	EtOH	250(4.06)	35-5593-64
$C_{13}H_{17}N_3$			
Hexanenitrile, 4-methyl-5-oxo-, phenylhydrazone	MeOH	275(4.20),411(4.61)	65-3487-64
$C_{13}H_{17}N_3O$			
2,3-Piperidinedione, 1,6-dimethyl-, 3-phenylhydrazone	EtOH	230(4.01),295(4.01), 328(4.33)	39-7165-65
$C_{13}H_{17}N_3O_3$			
Benzamide, p-nitro-N-(2-pyrrolidinoethyl)-	MeOH	262(4.05)	87-0107-65
$C_{13}H_{17}N_3O_4$			
Acetic acid, cyano[(4,4-dimethyl-2,6-dioxocyclohexyl)azo]-, ethyl ester	CH_2Cl_2	290(3.84),384(4.27)	89-0920-64
Benzamide, N-(2-morpholinoethyl)-p-nitro-	MeOH	262(4.05)	87-0107-65
1H-Pyrrolo[3,2-d]pyrimidine-6-carboxylic acid, 1,3-diethyl-2,3,4,5-tetrahydro-2,4-dioxo-, ethyl ester	N NaOH	280(4.2),310s(3.8)	94-1030-64
	MeOH	220(4.3),275(4.2), 320(3.7)	94-1030-64
5H-Pyrrolo[3,2-d]pyrimidine-6-carboxylic acid, 2,4-diethoxy-, ethyl ester	N HCl	300(4.3)	94-1030-64
	0.5N NaOH	300(4.2)	94-1030-64
	MeOH	220(4.4),280(4.2), 310s(3.7)	94-1030-64
$C_{13}H_{17}N_3O_7$			
Glycine, N-[2-(5-acetyl-3,4-dihydro-2,4-dioxo-1(2H)-pyrimidinyl)hydracryloyl]-, ethyl ester	pH 2	284(4.15)	39-1642-65
	H_2O	230(4.11),284(4.15)	39-1642-65
	pH 12	285(4.06)	39-1642-65
Malonic acid, acetamido[(2,4-dihydroxy-6-methyl-5-pyrimidinyl)methyl]-, dimethyl ester	EtOH	266(3.786)	65-2171-64
$C_{13}H_{17}N_5OS$			
Pyrimidine, 2,5-diamino-4-(benzylthio)-	pH 1	238(4.19),318(3.95)	87-0182-65
	pH 11	225(4.26),316(4.01)	87-0182-65
$C_{13}H_{17}N_5O_7$			
L-Alanine, N-(1,6-dihydro-2-oxo-9-β-D-ribofuranosyl-9H-purin-2-yl)-	pH 1	258(4.17),277s(3.96)	35-3752-65
	pH 11	258(4.11),270s(4.04)	35-3752-65
D-isomer	pH 1	258(4.16),277s(3.97)	35-3752-65
	pH 11	257(4.11),270s(4.03)	35-3752-65
$C_{13}H_{18}$			
Indan, 1-isopropyl-4-methyl-	C_6H_{12}	260s(2.47),266(2.60), 269(2.58),274(2.63)	22-3103-64
Indan, 1-isopropyl-5-methyl-	C_6H_{12}	248(2.99),271(2.99), 279(3.03),287(2.70), 300s(2.12)	22-3103-64

Compound	Solvent	λ_{max}(log ε)	Ref.
Indan, 2-isopropyl-4-methyl-	C_6H_{12}	260(2.64),265(2.75), 266(2.75),274(2.78)	22-3103-64
Indan, 4-isopropyl-5-methyl-	C_6H_{12}	255(2.78),268(3.18), 269(2.95),280(2.03)	22-3103-64
Indan, 4-isopropyl-6-methyl-	C_6H_{12}	270(3.02),279(3.11)	22-3103-64
Indan, 5-isopropyl-5-methyl-	C_6H_{12}	260(2.67),264(2.70), 268(2.79),273(2.80), 278(2.79)	22-3103-64
Indan, 6-isopropyl-4-methyl-	C_6H_{12}	264(3.18),268(3.20), 277(3.21)	22-3103-64
Indan, 6-isopropyl-5-methyl-	C_6H_{12}	264(2.94),269(2.97), 273(3.10),276(3.08), 282(3.14)	22-3103-64
Indan, 7-isopropyl-4-methyl-	C_6H_{12}	246(2.92),254(2.93), 271(2.60),293(2.27), 303(2.15)	22-3103-64
Naphthalene, 1,2,3,4-tetrahydro- 1,1,6-trimethyl-	n.s.g.	272(2.74),279(2.75)	39-2892-65
Naphthalene, 1,2,3,4-tetrahydro- 1,2,2-trimethyl-	EtOH	264(2.92),272(2.90)	28-3780-64B
1,3,5,7,9-Undecapentaene, 2,10-dimethyl-	isooctane	302(4.46),315(4.79), 331(5.02),349(5.02)	35-5075-65
$C_{13}H_{18}ClNO_8$			
1-(1,2-Dicarboxyethyl)-3,5-dimethyl- pyridinium perchlorate, dimethyl ester	H_2O	273(3.77),280s(3.69)	39-2676-64
$C_{13}H_{18}Cl_2O_6$			
3-Butene-2,2,3-tricarboxylic acid, 4,4-dichloro-, triethyl ester	EtOH	229(3.93)	39-3841-64
$C_{13}H_{18}F_3N_5$			
s-Triazolo[1,5-c]pyrimidine, 2-(butyl- amino)-5-propyl-7-(trifluoromethyl)-	MeOH	236(4.42),271(3.46), 316(3.45)	39-3357-65
$C_{13}H_{18}N_2$			
Indole, 3-(2-aminobutyl)-1-methyl-	EtOH-HCl	223(4.56),276s(3.73), 286(3.77),296s(3.68)	87-0274-64
Indole, 3-(2-aminobutyl)-2-methyl-, maleate	EtOH	221(4.65),289s(3.85), 298(3.87),304(3.81)	87-0274-64
Indole, 3-(2-aminobutyl)-7-methyl-	EtOH-HCl	218(4.64),271(3.84), 279(3.84),289(3.71)	87-0274-64
$C_{13}H_{18}N_2O$			
Cyclododeca[c]isoxazole, 3-amino- 4,5,8,9,12,13-hexahydro-	EtOH	244(3.90)	88-2151-64
all trans	EtOH	256(3.90)	88-2151-64
Indole, 3-(2-aminobutyl)-5-methoxy-, acetic acid salt	EtOH	221(4.41),275(3.79), 295(3.70),306s(3.56)	87-0274-64
Indole, 3-(2-aminobutyl)-6-methoxy-, acetic acid salt	EtOH	223(4.54),263s(3.57), 273s(3.64),293(3.73), 302s(3.59)	87-0274-64
Indole, 3-[2-(dimethylamino)ethyl]- 7-methoxy-	EtOH	221(4.75),270(3.87), 291(3.75)	5-0168-64F
Lespedamine	MeOH	223(4.49),278(3.66), 291(3.68)	5-0212-65B
Pyrrolo[2,3-b]indole, 1,2,3,3a,8,8a- hexahydro-5-methoxy-3a,8-dimethyl-	EtOH-HCl	245(4.02),318(3.50)	39-3336-65
	EtOH	250(4.00),324(3.48)	39-3336-65

Compound	Solvent	λ_{max}(log ϵ)	Ref.
$C_{13}H_{18}N_2O_2$			
Ethyl N-phenethylamidinoacetate, hydrochloride	MeOH-NaOMe	271(4.61)	44-0308-64
Piperidine, 2,6-dimethyl-1-(p-nitrophenyl)-	n.s.g.	405(4.34)	5-0109-65I
$C_{13}H_{18}N_2O_4$			
Octanamide, N-(1,2,5,6-tetrahydro-2,5,6-trioxo-3-pyridyl)-	EtOH	208(4.02),261(4.14), 348(3.80)	77-0640-65
$C_{13}H_{18}N_2O_5$			
D-Fructosone, 1-methylphenylhydrazone	EtOH	234(4.07),343(4.30)	24-0725-64
4-Pyrimidinepyruvic acid,	pH 1	259(4.01),319(3.45)	4-0049-65
2,6-diethoxy-, ethyl ester	pH 11	244(3.62),285(3.63), 348(4.33)	4-0049-65
$C_{13}H_{18}N_2O_6$			
Thymine, 1-(2,3-isopropylidene-α-D-ribofuranosyl)-	H_2O	268(4.00)	94-1471-64
Uridine, 2',3'-isopropylidene-	pH 5.4	261(3.98)	87-0486-65
3-methyl-	pH 11.5	259(3.98)	87-0486-65
$C_{13}H_{18}N_2O_7$			
Uridine, 5-(hydroxymethyl)-2',3'-O-isopropylidene-	MeOH	263(4.01)	88-1031-65
$C_{13}H_{18}N_2O_8$			
3-Carbamoyl-1-D-galactosyl-	pH 7	266(3.69)	39-0610-65
pyridinium formate	M KCN	325(3.80)	39-0610-65
3-Carbamoyl-1-D-glucosyl-	pH 7	266(3.68)	39-0610-65
pyridinium formate	M KCN	324(3.82)	39-0610-65
$C_{13}H_{18}N_4O$			
Acetamide, N-(4-ethyl-2,3,6-trimethyl-pyrazolo[1,5-a]pyrimidin-7(4H)-yl-idene)-	EtOH	226(4.25),269(4.03), 334(3.93)	94-0142-65
Pyrazolo[1,5-a]pyrimidine, 7-(N-ethyl-acetamido)-2,3,6-trimethyl-	EtOH	241(4.66),288s(3.16), 299(3.26),312(3.22), 350(3.26)	94-0142-65
$C_{13}H_{18}N_4O_4S_3$			
5-Thia-1-azabicyclo[4.2.0]oct-2-ene-2-carboxylic acid, 3-[(amidinothio)-methyl]-7-[2-(ethylthio)acetamido]-8-oxo-	pH 6.0	262(3.97)	39-5015-65
$C_{13}H_{18}O$			
3-Buten-2-one, 4-(5-isopropenyl-2-methyl-1-cyclopenten-1-yl)-	n.s.g.	297(4.08)	44-3740-64
2-Hexene, 3-anisyl-, cis	n.s.g.	250(4.07)	20-0703-64
trans	n.s.g.	240(3.80)	20-0703-64
3-Hexene, 3-anisyl-, cis	n.s.g.	252(4.10)	20-0703-64
trans	n.s.g.	238(3.95)	20-0703-64
β-Ionone, 4,5-dehydro-	n.s.g.	226(3.8),346(4.0)	27-0538-65
Naphthalene, 1,2,3,4-tetrahydro-7-methoxy-1,6-dimethyl-	EtOH	283.0(3.6)	39-3319-65
Tricyclo[5.4.0.0^{3,7}]undec-1-en-4-one, 2,3-dimethyl-	EtOH	228s(3.29),310(2.66)	44-3647-65
Tricyclo[4.3.2.0^{1,6}]undec-10-en-7-one, 10,11-dimethyl-	EtOH	228s(3.19),307(2.53)	44-3647-65

Compound	Solvent	λ_{max}(log ε)	Ref.
$C_{13}H_{18}O_2$			
Butanoic acid, p-tolyl-, ethyl ester	EtOH	251s(2.47),257(2.57), 263(2.63),272(2.55)	78-2921-64
Ethanol, 2-[(5,6,7,8-tetrahydro-4-methyl-1-naphthyl)oxy]-	MeOH	220s(4.09),275(3.33)	78-2297-65
2(1H)-Naphthalenone, 3,4-dihydro-7-methoxy-1,1-dimethyl-	EtOH	221(3.88),282(3.29), 288(3.25)	39-0361-65
$C_{13}H_{18}O_2Pd$			
Palladium, (2,4-pentanedionato)-2,4-cyclooctadien-1-yl-	hexane	221(4.24),250(4.20), 286(4.18)	39-5002-64
$C_{13}H_{18}O_3$			
1'-Acetonaphthone, octahydro-1'α-hydroxy-8'a-methyl-6'-oxo-	EtOH	239(4.19)	78-1119-64
Bicyclo[2.2.2]octane-2,3-dione,	EtOH	249(2.85),434(1.49)	35-5646-64
6-acetyl-1,5,5-trimethyl-	EtOH-KOH	249(2.84),432(1.48)	35-5646-64
	$CHCl_3$	437(1.52)	35-5646-64
isomer	$CHCl_3$	462(1.68)	35-5646-64
Chroman, 5,8-dimethoxy-2,2-dimethyl-	EtOH	285(3.50)	78-1317-64
Chroman, 7,8-dimethoxy-2,2-dimethyl-	EtOH	279(3.04)	78-1317-64
Cyclohexadienone, 2-acetoxy-2,6-dimethyl-4-propyl-	hexane	304.5(3.54)	49-0512-64
3-Cyclopentene-1-carboxylic acid, 1-acetyl-3,4-dimethyl-2-methylene-, ethyl ester	EtOH	240(4.25)	70-1820-64
Hexanoic acid, 6-(m-methoxyphenyl)-	EtOH	216(3.8),275(3.2)	39-3872-65
1,6(2H,7H)-Naphthalenedione, 3,4,8,8a-tetrahydro-8a-methyl-, 1-ethylene ketal	n.s.g.	241(4.0)	35-0478-64
2-Naphthoic acid, 1,2,3,4,6,7,8,8a-octahydro-6-oxo-, ethyl ester	MeOH	238(4.22)	44-4145-65
$C_{13}H_{18}O_4$			
Agglomerone	EtOH	235(4.07),264(3.79), 322(4.06)	12-1418-64
Cyclohexylideneacetic acid, 2,4-dihydroxy-2,6,6-trimethyl-, γ-lactone, acetate	EtOH	212(4.23)	94-0752-64
Cyclopentadienedipropionic acid, dimethyl ester	C_6H_{12}	246(2.63)	44-3430-64
Digiprolactone, acetate	EtOH	212(4.23)	94-0043-65
Fulvene, carbethoxymethyl-6-carbethoxy-3,4-dihydro-	EtOH	268(4.17)	44-3430-64
2(3H)-Naphthalenone, 4,4a,5,6,7,8-hexahydro-4a,5-dihydroxy-1-methyl-, 5-acetate	EtOH	244(4.12)	44-1626-65
Spiro[5.5]undecane-2,4-dione,	MeOH-HCl	259(4.17)	5-0123-65A
3-acetoxy-	MeOH-NaOMe	286(4.37)	5-0123-65A
$C_{13}H_{18}O_7$			
Acetophenone, 2'-hydroxy-2,3',4',5',6'-pentamethoxy-	EtOH	220s(4.22),284(4.09), 345(3.46)	12-2021-65
	EtOH-KOH	236(4.17),276s(3.68), 365(3.52)	12-2021-65
β-D-Glucoside, o-hydroxybenzyl-	acid	273(4.28)	44-0603-65
	base	294(3.6)	44-0603-65
β-D-Glucoside, p-hydroxybenzyl-	acid	222(3.9)	44-0603-65
	base	244(4.1)	44-0603-65
Isohomoarbutin	EtOH	287(3.43)	95-0337-64

Compound	Solvent	λ max(log ε)	Ref.
$C_{13}H_{18}O_7S$			
Galactopyranose, 1-deoxy-1-(p-tolyl-sulfonyl)-	MeOH	225(4.15)	44-1782-64
Glucopyranose, 1-deoxy-1-(p-tolyl-sulfonyl)-	MeOH	225(4.19)	44-1782-64
$C_{13}H_{18}S$			
Benzo[b]thiophene, 3-ethyl-2,3-dihydro-2,2,3-trimethyl-	n.s.g.	252(3.78),265s(3.45),290s(3.25),300s(3.05)	88-3569-65
$C_{13}H_{19}BrN_2$			
1,3-Dipropylbenzimidazolium bromide	MeOH	256(3.83),262(3.82),270(3.87),277(3.80)	65-0632-64
$C_{13}H_{19}BrO$			
β-Ionone, 5-bromo-	n.s.g.	220(3.9),291(4.0)	27-0538-65
$C_{13}H_{19}ClO_4$			
Coumarilic acid, 5-chloro-3-(1-carboxypropyl)-, anhydride	EtOH	217(4.36),270(4.19),277(4.27),288s(4.13),303s(3.75),312s(3.60)	44-4126-65
$C_{13}H_{19}N$			
Carbazole, 1,2,3,4,5,6,7,8-octahydro-9-methyl-	EtOH-HCl	274(3.78)	78-0515-64
	EtOH	247(3.62)	78-0515-64
Cinnamylamine, N,N-diethyl-	n.s.g.	256(4.26)	39-0531-64
4,8,10-Undecatrienenitrile, 5,9-dimethyl-	EtOH	235(4.25)	22-2533-64
$C_{13}H_{19}NO$			
1-Aziridineethanol, β-ethyl-β-methyl-α-phenyl-	EtOH	247(2.03),252(2.19),258(2.28),264(2.16)	44-3146-64
Hydroxylamine, N-methyl-N-(6-phenyl-5-hexenyl)-	EtOH	250(4.02),284(3.31),294(2.75)	39-1653-65
Morpholine, 3-ethyl-3-methyl-2-phenyl-, hydrochloride	EtOH	251(2.21),257(2.34),261(2.19),263(2.27)	44-3146-64
2H-1,2-Oxazocine, hexahydro-2-methyl-8-phenyl-	EtOH	250s(1.88),253(2.09),259(2.20),265(2.03)	39-1653-65
4H-1-Pyrindin-4-one, 1-(1,1-dimethyl-2-propynyl)octahydro-	n.s.g.	none	70-0512-64
$C_{13}H_{19}NO_2$			
Benzaldehyde, p-[3-(dimethylamino)-2-methylpropoxy]-	EtOH	284(4.26)	87-0511-64
2-Pyrone, 5-ethyl-6-methyl-4-piperidino-	MeOH	226(4.05),289(4.04)	24-1266-64
$C_{13}H_{19}NO_3$			
Acetamide, N-(2,4-dioxospiro[5.5]undec-3-yl)-	MeOH-HCl	256(4.04)	5-0123-65A
	MeOH-NaOMe	281(4.34)	5-0123-65A
7-Isoquinolinol, 1,2,3,4-tetrahydro-	pH 2	230s(3.88),280(3.29)	49-0025-65
	pH 12	250(3.87),295(3.64)	49-0025-65
	EtOH	230s(3.89),282(3.34)	49-0025-65
Pellotine	pH 1	230s(3.90),271(2.90),280s(2.74)	33-2089-64
	pH 13	245s(3.87),286(3.42)	33-2089-64
	iso-PrOH	230s(3.97),272(2.91),280s(2.81)	33-2089-64
α^5-Pyridoxyl-1-ethanol, α^4,3-O-isopropylidene-	pH 1	232s(3.43),290(3.94)	87-0112-65
	pH 13	281(3.81)	87-0112-65
	EtOH	281(3.76)	87-0112-65

Compound	Solvent	λ_{max}(log ε)	Ref.
$C_{13}H_{19}NO_7S$			
Glucosamine, N-p-toluenesulfonyl-	n.s.g.	230(4.23),261(2.91), 272(2.68)	95-0399-65
$C_{13}H_{19}NS$			
Cyclohexene, 6-dimethylaminomethyl-1-(2-thienyl)-, hydrochloride	EtOH	273(4.05)	39-3635-64
$C_{13}H_{19}N_3O$			
Acetimidic acid, N-phenethylamidino-, ethyl ester, dihydrochloride	MeOH	281(3.89)	44-0308-64
Benzamide, p-amino-N-(2-pyrrolidino-ethyl)-	MeOH	280(4.20)	87-0107-65
$C_{13}H_{19}N_3O_2$			
Benzamide, p-amino-N-(2-morpholino-ethyl)-	MeOH	280(4.23)	87-0107-65
Isopyrethrolone, methyl ether, semicarbazone	EtOH	220(4.23),264s(4.45), 273(4.50),296s(4.37)	39-0888-64
$C_{13}H_{19}N_3O_2S_2$			
Sulfanilamide, N[1]-[(methylthio)-piperidinomethylene]-	EtOH	259.0(4.29)	95-0391-65
$C_{13}H_{19}N_3O_3$			
Benzamide, N-(2-diethylaminoethyl)-p-nitro-	MeOH	262(4.06)	87-0107-65
Salicylaldehyde, 5-nitro-,	MeOH	305(4.48)	24-1631-64
1,1-diisopropylhydrazone	MeOH-NaOH	342(4.36),438(3.95)	24-1631-64
Salicylaldehyde, 5-nitro-,	MeOH	302(4.49)	24-1631-64
1,1-dipropylhydrazone	MeOH-NaOH	344(4.30),427(4.01)	24-1631-64
Salicylaldehyde, 5-nitro-,	MeOH	297(4.38)	24-2713-64
(1-isopentyl-1-methylhydrazone)	NaOH	343(4.22),415(4.04)	24-2713-64
Salicylaldehyde, 5-nitro-,	MeOH	298(4.48)	24-1631-64
methylpentylhydrazone	MeOH-NaOH	341(4.29),431(4.02)	24-1631-64
$C_{13}H_{19}N_3O_5$			
D-Fructosone, 1-(methylphenyl-hydrazone), 2-oxime	EtOH	233(3.90),333(4.34)	24-0725-64
4-Pyrimidinepyruvic acid, 2,6-di-	pH 1	259(4.10)	4-0049-65
ethoxy-, ethyl ester, oxime	pH 11	259(4.27)	4-0049-65
$C_{13}H_{19}N_3O_6S$			
Gramine, 4-nitro-, methyl sulfate	EtOH	223(3.93),328s(3.59), 368(3.69)	44-1158-64
$C_{13}H_{19}N_5$			
s-Triazine, 4,6-diamino-1,2-dihydro-	pH 1	235(3.77),271s(2.86)	44-0285-65
2,2-dimethyl-1-(3,4-xylyl)-	pH 10	241(4.04),271(2.96)	44-0285-65
	EtOH	244(4.10)	44-0285-65
$C_{13}H_{19}N_5O$			
Acetamide, N,N-dimethyl-2-[(2,3,6-tri-methylpyrazolo[1,5-a]pyrimidin-7-yl)amino]-	EtOH	233(4.63),285s(3.62), 294(3.70),336(3.80)	94-1207-65
$C_{13}H_{19}N_5O_4$			
2-Furaldehyde, 5-nitro-, 2-(2-piperi-dinoethyl)semicarbazone, hydrochloride	H_2O	375(4.23)	44-2582-64

Compound	Solvent	λ max(log ε)	Ref.
$C_{13}H_{19}N_5O_5$			
Adenine, 3-β-D-glucopyranosyl-N,N-dimethyl-	pH 1	292(4.29)	35-5320-64
	pH 7	222(4.07),298(4.18)	35-5320-64
	pH 12	299(4.18)	35-5320-64
$C_{13}H_{19}O_2P$			
Benzoic acid, p-(diethylphosphino)-, ethyl ester	n.s.g.	380(3.7)	65-3905-64
$C_{13}H_{19}O_5P$			
Diethyl 1-phenylacetonyl phosphate	EtOH	220(3.6),240(3.4), 280s(2.8),330s(1.9)	65-0542-65
$C_{13}H_{20}$			
Benzene, 1-tert-butyl-2,4,6-trimethyl-	isooctane	270(2.3)	88-1625-65
1,3,5-Heptatriene, 2,6-dimethyl-4-(2-methyl-1-propenyl)-	isooctane	225(4.02),281(4.20)	23-2744-65
cis	isooctane	210s(--),242(4.09)	23-2744-65
trans	isooctane	261.0(4.06)	23-2744-65
Toluene, 2,4-diisopropyl-	EtOH	257(2.58),264(2.63), 273(2.53)	78-2911-64
$C_{13}H_{20}BrNO$			
β-Ionone, 5-bromo-, oxime	n.s.g.	233s(4.0),266(4.1)	27-0538-65
$C_{13}H_{20}ClNO_5$			
1-Hexanone, 6-(methylamino)-1-phenyl-, perchlorate	n.s.g.	242(5.09),280(4.19)	73-3111-65
$C_{13}H_{20}ClNO_7S$			
Benzenesulfonamide, p-chloro-N-methyl-N-(D-gluco-2,3,4,5,6-pentahydroxyhexyl)-	n.s.g.	235(4.21),277(2.67)	95-0399-65
$C_{13}H_{20}ClN_3O_4$			
3-Azabicyclo[3.3.1]non-6-ene, 6-chloro-8,9-diethyl-3-methyl-1,5-dinitro-	MeOH	203(4.08)	24-0186-64
$C_{13}H_{20}NO_5$			
Piperidinooxy, 4-(1,2-dicarboxyethylidene)-2,2,6,6-tetramethyl-	MeOH	211(3.48),228s(3.4)	70-0716-65
$C_{13}H_{20}N_2$			
Pyrrolidine, 2-(4-ethyl-3,5-dimethyl-2H-pyrrol-2-ylidene)-1-methyl-	EtOH	278s(3.57),343(4.39)	39-2614-65
$C_{13}H_{20}N_2O$			
1-Cyclohexene-1-carboxamide, 2-(cyclohexylideneamino)-	EtOH	305(3.83)	23-0010-64
2(1H)-Pyrimidinone, 4-(2,4-dimethyl-1,3-pentadienyl)-5,6-dihydro-6,6-dimethyl-	MeOH	280(4.03)	77-0439-65
$C_{13}H_{20}N_2S$			
1-Cyclohexene-1-carboxamide, 2-(cyclohexylideneamino)thio-	EtOH	280(3.69),377(3.26)	23-0010-64
$C_{13}H_{20}N_4$			
Pyrazolo[1,5-a]pyrimidine, 4-ethyl-7-(ethylimino)-4,7-dihydro-2,3,6-trimethyl-	EtOH	234(4.59),251s(4.13), 348(3.90)	94-0142-65

Compound	Solvent	$\lambda_{max}(\log \epsilon)$	Ref.
$C_{13}H_{20}O$			
Anisole, 2,6-diisopropyl-	C_6H_{12}	212s(4.02),265(2.63), 273(2.59)	23-2603-65
7H-Benzocyclohepten-7-one, 1,2,3,4,4a,8,9,9a-octahydro-5,6-dimethyl-	EtOH	258(3.94)	44-3647-65
Bicycloionone	EtOH	278(1.74),284(1.73)	35-5646-64
Bicyclo[3.3.1]non-2(3)-en-9-one, 2,4,4,8-tetramethyl-	EtOH	230(3.44),286(2.15)	88-1337-64
3-Buten-2-one, 4-(3,3-dimethyl-2-norbornyl)-	EtOH	234(4.00)	28-4783-65A
Epibicycloionone	EtOH	293(1.52)	35-5646-64
α-Ionone	$C_2H_4Cl_2$-acid	375(4.19)	39-1761-65
β-Ionone	$C_2H_4Cl_2$-acid	380(4.48)	39-1761-65
2(3H)-Naphthalenone, 4,4a,5,6,7,8-hexahydro-4a,7,7-trimethyl-	EtOH	241(4.26)	35-0465-64
2-Norbornaneacrolein, 1,3,3-trimethyl-	EtOH	233(4.12)	28-4712-64B
4-Penten-3-one, 4-methyl-1-(2-methylenecyclohexyl)-	EtOH	220(3.92)	44-2420-65
Tricycloionone	EtOH	286(1.60)	35-5646-64
$C_{13}H_{20}OS$			
2,4-Cyclohexadien-1-one, 2,3,4,5,6-pentamethyl-6-[(methylthio)methyl]-	MeOH	336(3.6)	35-4656-65
$C_{13}H_{20}O_2$			
2(1H)-Naphthalenone, 4a,5,6,7,8,8a-hexahydro-3,8aα-dimethyl-8aβ-methoxy-, 1-(8a-4aβ)-abeo-	EtOH	230(3.85)	35-4053-64
Resorcinol, 5-heptyl-	EtOH	273(4.20),280(4.20)	1-1677-65
	EtOH	273(3.14),280(3.13)	1-1677-65
$C_{13}H_{20}O_3$			
1-Cyclohexene-1-butyric acid, 2-methyl-6-oxo-, ethyl ester	EtOH	243(4.08)	39-3554-64
2-Cyclohexene-1-carboxylic acid, 2,3,6,6-tetramethyl-4-oxo-, ethyl ester	n.s.g.	250(3.90)	39-1511-64
1-Cyclohexene-1-propionic acid, 2-isopropyl-4-methyl-6-oxo-	n.s.g.	245(4.12)	39-2340-65
Cyclopentanone, 2,5-bis(3-oxobutyl)-	EtOH	281(1.81)	44-2513-65
4a(2H)-Naphthalenecarboxylic acid, octahydro-1,1-dimethyl-2-oxo-, trans	EtOH	287(1.08)	35-1573-65
5,7-Octadienoic acid, 2-acetyl-6-methyl-, ethyl ester	EtOH	230(4.36)	22-2533-64
4,5-Octadien-2-yne, 1-[2-(2-hydroxyethoxy)ethoxy]-7-methyl-	EtOH	218(4.06)	39-4659-65
$C_{13}H_{20}O_3Pd$			
Palladium, (4-methoxycyclohept-2-enyl)(2,4-pentanedionato)-	hexane	213s(4.25),229(4.27), 258s(4.01),292(3.91)	39-5002-64
$C_{13}H_{20}O_4$			
Cyclopentaneacetic acid, 2-(1,3-dioxolan-2-yl)-3-methyl-α-methylene-, methyl ester	EtOH	212(4.11)	12-1260-64

Compound	Solvent	$\lambda_{max}(\log \epsilon)$	Ref.
$C_{13}H_{20}O_5$			
Carbonic acid, ethyl ester, ester with ethyl 2-hydroxy-1-cyclo-heptene-1-carboxylate	EtOH	223(3.56)	44-0087-64
$C_{13}H_{21}BO_2$			
9-Borabicyclo[3.3.1]nonyl acetyl-acetonate	n.s.g.	348(3.13)	5-0040-65I
$C_{13}H_{21}BrO$			
3-Buten-2-ol, 4-(4-bromo-2,6,6-tri-methyl-1-cyclohexen-1-yl)-	n.s.g.	225(3.8)	27-0538-65
$C_{13}H_{21}IN_2$			
2-(4-Ethyl-3,5-dimethyl-2-pyrrolyl)-1-methyl-1-pyrrolinium iodide	EtOH	295(3.77),344(4.43)	39-0893-64
$C_{13}H_{21}NO$			
Cyclohexanone, 2-cyclohexylamino-methylene-	EtOH	327(4.26)	70-1054-64
5(1H)-Quinolone, 2,3,6,7,8,8a-hexa-hydro-7-methyl-1-propyl-	n.s.g.	236(3.54)	88-0171-65
$C_{13}H_{21}NOS$			
Cyclohexanol, 2-dimethylaminomethyl-1-(2-thienyl)-	EtOH	234(3.99)	39-3635-64
$C_{13}H_{21}NO_2$			
Benzyl alcohol, α-[1-[(2-hydroxyethyl)-amino]-1-methylpropyl]-, hydro-chloride	EtOH	247(2.00),252(2.17), 258(2.29),264(2.19)	44-3146-64
2-Butanone, 3-hydroxy-4-(1,3,3-tri-methylbicyclo[3.1.0]hex-2-yli-dene)-, oxime	n.s.g.	217(4.0)	27-0538-65
Cyclohexanecarboxamide, N-cyclo-hexyl-2-oxo-	EtOH	256(3.08)	70-1054-64
	$CHCl_3$	255(3.92)	70-1054-64
hydrogen peroxide adduct	EtOH	257(2.99)	70-1054-64
	$CHCl_3$	255(3.76)	70-1054-64
$C_{13}H_{21}NO_3$			
3,5-Heptadienamide, 2-carbethoxy-3,4,5-trimethyl-	n.s.g.	227(4.04)	25-1313-64
$C_{13}H_{21}N_3O_2$			
2(3H)-Naphthalenone, 4,4a,5,6,7,8-hexahydro-6-methoxy-4a-methyl-	EtOH	239(4.21)	35-1761-64
$C_{13}H_{21}N_3O_4$			
3-Azabicyclo[3.3.1]non-6-ene, 8,9-di-ethyl-3-methyl-1,5-dinitro-	MeOH	205(3.94) (end absorption)	24-0186-64
$C_{13}H_{21}N_3O_6$			
1(2H)-Pyrimidinepropionic acid, 5-bis(2-hydroxyethyl)amino-3,4-di-hydro-2,4-dioxo-, ethyl ester	pH 1	268(3.90)	87-0187-65
	pH 7	255s(3.71),302s(3.47)	87-0187-65
	pH 13	279s(3.57)	87-0187-65
$C_{13}H_{21}N_5$			
s-Triazolo[1,5-c]pyrimidine, 2-(butyl-amino)-7-methyl-5-propyl-	MeOH	233(4.63),301(3.45)	39-3357-65

Compound	Solvent	$\lambda_{max}(\log \epsilon)$	Ref.
$C_{13}H_{21}N_5O$			
3-Cyclohexyl-5-(cyclohexylidene-hydrazino)-1,2,3,4-oxatriazolium hydroxide, inner salt	n.s.g.	240(4.37),357(3.85)	44-0567-65
Isoxazolo[5,4-d]pyrimidine, 4-(3-di-ethylaminopropylamino)-3-methyl-	EtOH	253(4.03),284(4.00)	44-2116-64
$C_{13}H_{21}N_5O_2$			
Triazolo[1,5-c]pyrimidine, 2-bis(hy-droxyethyl)amino-7-methyl-5-propyl-	MeOH	239(4.56),307(3.42)	39-3357-65
$C_{13}H_{21}N_5O_4$			
s-Triazine, 2,4-bis(2-hydroxyethyl-amino)-6-phenyl-, dihydrate	EtOH	224(4.51),241(4.45)	44-2766-64
$C_{13}H_{21}N_5O_4S$			
Theophylline, 7-hexyl-8-sulfamoyl-	MeOH	285(3.92)	54-1215-64
$C_{13}H_{21}O_7PS$			
Maleic acid, hydroxy(2-hydroxy-1-meth-ylethyl)-, γ-lactone, ethyl ester, O-ester with O,O-diethyl phosphorothioate	MeOH	240(3.98)	5-0010-65E
Maleic acid, hydroxy(2-hydroxypropyl)-, γ-lactone, ethyl ester, O-ester with O,O-diethyl phosphorothioate	MeOH	233(3.81)	5-0010-65E
$C_{13}H_{21}O_8P$			
Maleic acid, hydroxy(2-hydroxy-1-meth-ylethyl)-, γ-lactone, ethyl ester, diethyl phosphate	MeOH	237(3.89)	5-0010-65E
Maleic acid, hydroxy(2-hydroxypropyl)-, γ-lactone, ethyl ester, diethyl phosphate	MeOH	238(4.01)	5-0010-65E
Oxalacetic acid, (2-hydroxy-1-methyl-ethyl)phosphono-, γ-lactone, triethyl ester	MeOH	212(3.78)	5-0010-65E
Oxalacetic acid, (2-hydroxypropyl)-phosphono-, γ-lactone, triethyl ester	MeOH	211(3.84)	5-0010-65E
Oxalacetic acid, (3-hydroxypropyl)-phosphono-, δ-lactone, triethyl ester	MeOH	212(3.92)	5-0010-65E
$C_{13}H_{22}INO_2$			
4-(5,5-Dimethyl-3-oxo-1-cyclohexen-1-yl)-4-methylmorpholinium iodide	EtOH	294(4.38)	44-0794-64
$C_{13}H_{22}N_2O$			
Cyclododeca[c]isoxazole, 3-amino-4,5,6,7,8,9,10,11,12,13-decahydro-	EtOH	253(3.90)	88-2151-64
$C_{13}H_{22}O$			
p-Menthane-3-acrolein	EtOH	229(4.06)	28-4712-64B
6,8-Nonadien-2-one, 5-isopropyl-8-methyl- (or solanone)	hexane	230(4.07)	44-2918-65
	EtOH	230(4.07)	44-2918-65
$C_{13}H_{22}O_6$			
Chalcomycin derivative	MeOH	222(4.14)	35-2726-64
$C_{13}H_{22}Pb$			
Lead, (1-cyclopenten-1-ylethynyl)-triethyl-	hexane	207(4.27)	22-3518-65

Compound	Solvent	$\lambda_{max}(\log \epsilon)$	Ref.
$C_{13}H_{23}NO_2$			
Acetaldehyde, 4-(dimethylamino)-1,1-dimethyl-4-penten-2-ynyl ethyl acetal	n.s.g.	280(4.12)	70-0729-65
$C_{13}H_{23}N_3O$			
Macrorungine, dodecahydro-	EtOH	207(3.64)	39-5969-64
$C_{13}H_{23}N_3O_2$			
Pyridazine, 3-(isopentyloxy)-6-(dimethylaminoethoxy)-	EtOH	286(3.32)	87-0129-65
$C_{13}H_{25}N_3O_{10}S$			
Cellobiose, thiosemicarbazone	H_2O	237(4.09)	25-0546-64
Lactose, thiosemicarbazone	H_2O	237(4.05)	25-0546-64
$C_{13}H_{25}O_6PS$			
Valeric acid, 5-mercapto-2-phosphono-, triethyl ester, acetate	n.s.g.	232(3.70)	24-1963-64
$C_{13}H_{26}N_2$			
2-Imidazoline, 1-isopropyl-4,4-dimethyl-2-pentyl-	EtOH	230(3.81)	4-0188-64
$C_{13}H_{26}Si_3$			
Trisilane, 1,1,1,2,3,3,3-heptamethyl-2-phenyl-	C_6H_{12}	243(4.05)	101-0163-65B
$C_{13}H_{27}As_3I_2O_2W$			
Tungsten, diiododicarbonyl[[2-[(dimethylarsino)methyl]-2-methyltrimethylene]bis[dimethylarsino]]-	DMSO	262(3.40),332(2.79)	39-6570-65

Compound	Solvent	λ_{max}(log ε)	Ref.
$C_{14}H_4O_{10}S_2$			
1,4-Anthracenedione, 5,8,9,10-tetra-hydroxy-, cyclic 5,10:8,9-disulfate	dioxan	240(4.76),249(4.60), 255(4.58),261(4.44), 285(3.92),325(3.73), 404(3.78)	33-0119-65
$C_{14}H_5F_{12}ORh$			
Rhodium, cyclopentadienyl[tetrakis-(trifluoromethyl)cyclopentadienone]-	C_6H_{12}	246(4.51),283s(--), 332(3.67)	39-2699-64
	MeOH	246(4.51),280s(--), 334(3.74)	39-2699-64
$C_{14}H_6Cl_4O_3$			
Benzoic acid, 2-benzoyl-3,4,5,6-tetrachloro-	EtOH	244(4.01)	44-4293-65
$C_{14}H_6Cl_4S_2$			
1,3-Dithiole, 4-phenyl-2-(tetrachloro-2,4-cyclopentadien-1-ylidene)-	$CHCl_3$	445(4.62)	88-2121-65
	CF_3COOH	470(4.35)	88-2121-65
$C_{14}H_6F_{10}Sn$			
Tin, dimethylbis(pentafluorophenyl)-	C_6H_{12}	265(3.18)	39-4782-64
	MeOH	261.5(3.05)	39-4782-64
$C_{14}H_6N_6$			
1,3-Cyclohexadiene-1,4-dicarbonitrile, tetracyanoethylene adduct	CH_2Cl_2	247(4.17),266(4.25), 277(4.27),295(4.06), 307(4.06),320(3.82)	23-2852-64
4-Cyclohexene-1,1,2-tricarbonitrile, 2-(tricyanovinyl)-	MeCN	252(4.13)	35-2898-64
$C_{14}H_6O_8S_2$			
1,4,9,10-Anthracenetetrayl sulfate	hexane	260(4.99),367(3.90), 375(3.90),387(4.08), 410(4.01)	33-0119-65
1,5,9,10-Anthracenetetrayl sulfate	hexane	260(4.26),367(3.11), 387(3.25),410(3.23)	33-0119-65
$C_{14}H_7BrO_5S$			
2-Anthracenesulfonic acid, 3-bromo-9,10-dihydro-9,10-dioxo-, sodium salt	0.02N NaOH	510(3.90),786(3.15)	27-0333-65
leuco form	0.02N NaOH	439(4.18),532(3.59)	27-0333-65
$C_{14}H_7ClN_2O_4$			
3H-Indol-3-one, 5-chloro-2-(m-nitro-phenyl)-, 1-oxide	MeOH	242(4.01),279(4.41), 330s(3.20),434(3.16)	17-0405-65
$C_{14}H_7ClO_5S$			
2-Anthracenesulfonic acid, 3-chloro-9,10-dihydro-9,10-dioxo-, sodium salt	0.02N NaOH	508(3.88),786(3.18)	27-0333-65
leuco form	0.02N NaOH	437(4.18),532(3.59)	27-0333-65
$C_{14}H_7Cl_2NO_2$			
Phenanthrene, 9,10-dichloro-3-nitro-	THF	244(4.73),251(4.79), 303(3.59),341(3.92)	39-2587-65
$C_{14}H_7F_3N_4$			
Benzo[h]quinoline, 4-azido-2-(trifluoromethyl)-	EtOH	258(4.47),317(3.98)	4-0120-65

Compound	Solvent	$\lambda_{max}(\log \epsilon)$	Ref.
$C_{14}H_7NO$			
Fluorene-4-carbonitrile, 9-oxo-	iso-PrOH	368(2.43)	39-5518-65
$C_{14}H_7NO_4$			
Benzo[g]quinoline-4-carboxylic acid, 5,10-dihydro-5,10-dioxo-	EtOH	252(4.54),317(3.73)	44-0432-65
Phenanthrenequinone, 2-nitro-	n.s.g.	273(4.1),289(4.1), 370(3.4),392(3.3), 498s(1.7)	78-0803-64
Phenanthrenequinone, 4-nitro-	n.s.g.	248(4.4),319(3.63), 325(3.6),366(3.3), 382(3.3),498(1.7)	78-0803-64
$C_{14}H_7N_2O_{11}P$			
Phenoxaphosphine-2,8-dicarboxylic acid, 10-hydroxy-4,6-dinitro-	EtOH	255(4.21),271(4.20), 305s(3.69)	44-1983-64
$C_{14}H_8ClNO_2$			
Phenanthrene, 9-chloro-10-nitro-	THF	224(4.65),248(4.79), 254(4.83),288(4.05), 300(3.9)	39-2587-65
$C_{14}H_8ClNO_3$			
Lambertellin, 8-chloro-, lactam	EtOH	244(4.78),270(3.87), 405(4.10)	100-0359-65
$C_{14}H_8Cl_2CuO_4$			
Copper m-chlorobenzoate, hydrate	dioxan	422s(--),676(2.46)	39-6464-65
Copper o-chlorobenzoate	dioxan	414s(--),681(2.42)	39-6464-65
Copper o-chlorobenzoate, hydrate	dioxan	412s(--),681(2.42)	39-6464-65
Copper p-chlorobenzoate	dioxan	410(--),675(2.44)	39-6464-65
$C_{14}H_8CuN_2O_8$			
Copper m-nitrobenzoate	MeCN	704(2.35)	39-6464-65
Copper o-nitrobenzoate	dioxan	687(2.40)	39-6464-65
hydrated form A	dioxan	684(2.40)	39-6464-65
hydrated form B	dioxan	683(2.41)	39-6464-65
Copper p-nitrobenzoate, hydrate	dioxan	680(2.42)	39-6464-65
$C_{14}H_8F_3NO$			
Benzo[h]quinolin-2-ol, 4-(trifluoromethyl)-	EtOH	230(4.44),275(4.18), 287(4.17),304(3.73)	4-0120-65
	EtOH-HCl	229(4.45),275(4.15), 303(3.68)	4-0120-65
Benzo[h]quinolin-4-ol, 2-(trifluoromethyl)-	EtOH	240(4.60),287(3.83), 298(3.96)	4-0120-65
	EtOH-HCl	253(4.62),296(3.99), 308(3.99)	4-0120-65
$C_{14}H_8INO_2$			
3H-Indol-3-one, 2-(p-iodophenyl)-, 1-oxide	C_6H_{12}	288(4.40),299(4.43), 315s(4.01),452(3.39)	17-0405-65
	MeOH	316(4.55),325(4.54), 443(3.45)	17-0405-65
$C_{14}H_8N_2$			
4,10-Diazapyrene	EtOH	234(4.71),257(4.43), 267(4.36),304s(3.92), 319(4.11),332(4.17)	39-3379-65

Compound	Solvent	λ_{max}(log ε)	Ref.
Malononitrile, (1-azulenyl-methylene)-	hexane	242s(3.96),250(3.92), 256(3.93),261(3.93), 290(3.95),310s(3.98), 334(4.22),342(4.21), 358(4.07),420(4.45), 440(4.39)	24-2050-64
	benzene	292(4.06),310(3.99), 340(4.22),356(4.19), 430(4.52),449(4.50)	24-2050-64
	EtOH	232(4.14),260(3.90), 289(4.04),335(4.22), 351(--),430(4.52), 445(4.53)	24-2050-64
$C_{14}H_8N_2O$			
Anthrone, 10-diazo-	C_6H_{12}	238(4.47),253(4.28), 268(4.21),304(4.04), 403(4.33)	28-2259-65A
4H-4,9-Diazapyren-5-one	MeOH	243(4.53),253(4.53), 273(4.34),306(4.00), 317(3.95),344(3.72), 358(3.97),376(4.05)	39-3379-65
	n.s.g.	243(2.87),253(3.19), 273(3.20),306(3.51), 317(3.55),344(3.66), 358(3.53),376(3.63)	39-5135-64
4H-4,10-Diazapyren-5-one	EtOH	246(4.29),254(4.27), 280(4.16),308(3.69), 320(3.69),336(3.56), 360(3.62),376(3.72)	39-3379-65
11H-Isoindolo[2,1-a]benz-imidazol-11-one	$C_{10}H_7Cl$	none (in visible)	33-1999-65
$C_{14}H_8N_2O_2S$			
Benzoxazole, 2,2'-thiobis-	$CHCl_3$	285(4.40),312(4.27)	44-3618-65
$C_{14}H_8N_2O_4$			
3H-Indol-3-one, 2-(p-nitrophenyl)-, 1-oxide	MeOH	228(4.11),285(4.32), 311s(4.23),427(3.53)	17-0405-65
3H-Indol-3-one, 6-nitro-2-phenyl-, 1-oxide	C_6H_{12}	240(4.21),276s(4.30), 283(4.32),303(4.28), 310s(4.25),339(4.00), 356(3.85),457(3.17)	17-0405-65
	MeOH	247(4.27),284(4.27), 298s(4.23),335(3.82), 351s(3.67),445(3.17)	17-0405-65
$C_{14}H_8O_2$			
Anthraquinone	iso-PrOH	250(4.8),270(4.3), 330(3.8)	60-1981-65
Phenanthrenequinone	EtOH	258(4.53),265(4.56), 322(3.7),422(3.2)	78-0803-64
	$CHCl_3$	258(4.1),267(4.1), 298(3.3),420(3.1)	78-0803-64
	MeCN	256(4.6),264(4.6), 316(3.7),412(3.2)	78-0803-64
$C_{14}H_8O_4$			
Anthraquinone, 1,4-dihydroxy-	o-chloro-phenol	305(3.73),480(4.00)	56-1251-64

Compound	Solvent	λ_{max}(log ϵ)	Ref.
4H-Naphtho[1,2-b]pyran-4,5,6-trione, 2-methyl-	EtOH	216(4.25),260(4.24), 310(3.63)	94-0316-64
4H-Naphtho[2,3-b]pyran-4,5,10-trione, 3-methyl-	EtOH	233s(4.32),241(4.37), 253s(4.31),270s(3.92), 305(3.78)	94-0316-64
$C_{14}H_8O_5$			
Lambertellin	MeOH	284s(4.08),290(4.10), 430(3.68)	39-5927-65
Naphtho[2,3-b]furan-3-carboxylic acid, 4,9-dihydro-2-methyl-4,9-dioxo-	MeCN	250(4.54),301(3.88), 345(3.54)	44-3819-65
$C_{14}H_8O_5S$			
2-Anthracenesulfonic acid, 9,10-dihydro-9,10-dioxo-, sodium salt	0.02N NaOH	496(3.94),770(3.33)	27-0333-65
leuco form	0.02N NaOH	430(4.18),518(3.61)	27-0333-65
$C_{14}H_8O_6$			
Anthraquinone, 1,4,5,8-tetrahydroxy-	$CHCl_3$	488(4.2),511(4.3), 523(4.4),548(4.4), 562(4.5)	24-3145-65
$C_{14}H_9BClN$			
Phenanthro[4,5-cde][1,2]azaborine, 5-chloro-4,5-dihydro-	$CHCl_3$	252(4.60),270(4.41), 278(4.39),295(3.91), 307(4.05),331s(3.54), 340s(3.38),348(3.76), 359s(3.43),365(3.92)	44-1757-64
$C_{14}H_9BrN_4O_5$			
Benzocyclobutenone, 3-bromo-4-hydroxy-, 2,4-dinitrophenylhydrazone	$CHCl_3$	383(4.52)	39-1390-65
$C_{14}H_9Br_2N_3O_2$			
1,2,4-Triazolidine-3,5-dione, 1,2-bis(p-bromophenyl)-	EtOH	216(4.17),250(4.35)	22-0500-64
	NaOH	218(4.40),257(4.32)	22-0500-64
$C_{14}H_9ClN_2$			
3-Pyridineacrylonitrile, α-(p-chlorophenyl)-	MeOH	230(4.06),315(4.33)	87-0583-65
$C_{14}H_9ClN_2O$			
Ketone, 2-chloro-4-pyridyl 3-indolyl	MeOH	207(4.68),268(4.11), 320(4.08)	44-2534-65
$C_{14}H_9ClN_2O_2$			
3,1,4-Benzoxadiazepin-2(1H)-one, 7-chloro-5-phenyl-	EtOH	320(3.6)	44-2368-64
$C_{14}H_9Cl_3S$			
Thiobenzophenone, 4-(trichloromethyl)-	C_6H_{12}	236(4.11),300(4.02), 315s(--),617(2.01)	54-0289-65
	EtOH	235(4.10),300(3.95), 315s(--),607(2.00)	54-0289-65
$C_{14}H_9Cl_4N_3O_2$			
Acetophenone, 3'-nitro-, chloro-(2,4,6-trichlorophenyl)hydrazone	EtOH	251(4.10),417(2.34)	39-1500-64
Benzoyl chloride, m-nitro-, (4,6-dichloro-o-tolyl)hydrazone	EtOH	295(3.99),418(2.73)	39-1500-64

Compound	Solvent	$\lambda_{max}(\log \epsilon)$	Ref.
$C_{14}H_9CsO_2$			
Cesium, [(3-phenyl-2-benzo-furanyl)oxy]-	$(MeOCH_2)_2$	345(4.35)	35-1857-65
	DMF	354(4.39)	35-1857-65
$C_{14}H_9F_3N_2O$			
Benzo[h]-1,6-naphthyridin-4-ol, 5-methyl-2-(trifluoromethyl)-	EtOH-HCl	222(4.28),248(4.40)	4-0120-65
	EtOH	246(4.47),320(2.79)	4-0120-65
$C_{14}H_9F_3N_2O_2$			
1,10-Phenanthrolin-4-ol, 5-methoxy-2-(trifluoromethyl)-	EtOH-HCl	230(4.70),236(4.69), 269(4.48)	4-0120-65
	EtOH	231(4.81),254(4.69), 303(4.13),345(3.90)	4-0120-65
$C_{14}H_9F_3S$			
Thiobenzophenone, 4-(trifluoromethyl)-	C_6H_{12}	300s(--),320(3.89), 618(1.92)	54-0289-65
	EtOH	300s(--),323(3.98), 608(2.01)	54-0289-65
$C_{14}H_9KO_2$			
Potassium, [(3-phenyl-2-benzo-furanyl)oxy]-	$(MeOCH_2)_2$	342.5(4.35)	35-1857-65
$C_{14}H_9LiO_2$			
Lithium, [(3-phenyl-2-benzo-furanyl)oxy]-	$(MeOCH_2)_2$	326(4.29)	35-1857-65
	DMF	354(4.35)	35-1857-65
$C_{14}H_9N$			
Fluorene-9-carbonitrile	EtOH	410(3.35)	78-0261-65
	EtOH-base	410(3.35)	23-1225-65
$C_{14}H_9NO$			
Acenaphth[1,2-d]isoxazole, 9-methyl-	EtOH	231(4.69),266(3.96), 276(3.98),318(3.98), 340(3.76)	88-2157-64
$C_{14}H_9NO_2$			
[14]Annulene, 1,8-bisdehydro-3-nitro-	isooctane	327(4.93),467(4.28), 605(3.89)	35-0521-64
Anthracene, 2-nitro-	EtOH	236(4.67),265(4.31), 302(4.33)	39-5377-65
Anthranil, 3-benzoyl-	EtOH	218s(3.85),261(3.95), 293s(3.78),308s(3.69), 354(4.00)	44-1104-65
	EtOH	258s(4.07),352(4.03)	44-1104-65
4H-3,1-Benzoxazin-4-one, 2-phenyl-	EtOH	241(4.35),275s(4.21), 285(4.33),298(4.29), 321(4.06)	88-2597-65
3H-Indol-3-one, 2-phenyl-, 1-oxide	C_6H_{12}	281(4.65),290(4.69), 313s(3.74),445(3.38)	17-0405-65
	MeOH	277(4.60),285(4.61), 440(3.36)	17-0405-65
$C_{14}H_9NO_3$			
2H-1,3-Benzoxazine-2,4(3H)-dione, 3-phenyl-	MeOH	206(4.59),240(3.98), 290(3.18)	4-0037-65
Phthalimide, N-(2-hydroxyphenyl)-	EtOH	219(4.65),234s(4.20), 280(3.73),294s(3.45)	7-0128-64

Compound	Solvent	λ_{max}(log ε)	Ref.
Phthalimide, N-(3-hydroxyphenyl)-	EtOH	220(4.64),234s(4.35), 282(3.68),297s(3.33)	7-0128-64
Phthalimide, N-(4-hydroxyphenyl)-	EtOH	217(4.63),240s(4.24), 247s(4.15),296s(3.28)	7-0128-64
Phthalimide, 4-hydroxy-N-phenyl-	EtOH	241(4.55),264s(4.15), 314s(3.44),333s(3.23)	7-0128-64
Phthalimide, 5-hydroxy-N-phenyl-	EtOH	226(4.56),278s(4.18), 341(3.71)	7-0128-64
$C_{14}H_9NO_4$			
2-Propyn-1-one, 1-(5-nitro-2-furyl)-3-p-tolyl-	EtOH	231(4.21),338(4.36)	30-0833-65
$C_{14}H_9NO_5$			
Benzoic acid, 2-benzoyl-3-nitro-	EtOH	246(3.96)	44-4293-65
Benzoic acid, 2-benzoyl-6-nitro-	EtOH	248(4.28)	44-4293-65
$C_{14}H_9N_3$			
Nicotinonitrile, 2-indol-3-yl-	EtOH	217(4.49),258(3.91), 267(3.93),316(4.16)	44-2534-65
$C_{14}H_9N_3O_2$			
2-Pyridineacrylonitrile, α-(p-nitrophenyl)-	MeOH	219(4.11),269(3.80), 325(3.38)	87-0583-65
4-Pyridineacrylonitrile, α-(p-nitrophenyl)-	MeOH	256(3.90),309(4.32)	87-0583-65
$C_{14}H_9N_3O_4$			
1,3,4-Oxadiazole, 2-[2-(5-nitro-2-furyl)vinyl]-5-phenyl-	propylene glycol	238(4.12),297(4.14), 376(4.42)	95-0225-64
$C_{14}H_9N_3O_5$			
1,3,4-Oxadiazol-2-ol, 5-[1-phenyl-2-(5-nitro-2-furyl)vinyl]-	propylene glycol	287(3.87),382(4.00)	95-0225-64
$C_{14}H_9Na$			
Anthracene, sodium derivative	THF	550(--),595(--), 650s(--),720(4.00)	60-1424-64
Phenanthrene, sodium derivative	THF	415(2.6),445(2.7)	60-0891-65
$C_{14}H_9NaO_2$			
Sodium, [(3-phenyl-2-benzofuranyl)oxy]-	$(MeOCH_2)_2$	337(4.36)	35-1857-65
	DMF	354(4.39)	35-1857-65
$C_{14}H_9O_7P$			
Phenoxaphosphine-2,8-dicarboxylic acid, 10-hydroxy-, 10-oxide	EtOH	218(4.58),271(4.33), 282s(4.33),285(4.34)	44-1983-64
$C_{14}H_{10}$			
Anthracene	C_6H_{12}	255(5.25),380(3.88)	60-0274-64
	n.s.g.	223(4.1),250s(4.9), 254(5.2),299s(2.5), 313(2.9),327(3.2), 342(3.6),357(3.7), 374(3.7)	28-4795-64A
	n.s.g.	252(5.15),322(3.47), 338(3.77),355(3.90), 374(3.85)	70-1260-64
bromanil complex	$CHCl_3$	640(2.71)	60-0465-64
chloranil complex	$CHCl_3$	630(2.70)	60-0465-64

Compound	Solvent	$\lambda_{max}(\log \epsilon)$	Ref.
Anthracene, iodoanil complex	$CHCl_3$	627(2.58)	60-0465-64
tetrachlorophthalic anhydride complex	$CHCl_3$	420(4.38)	60-0465-64
1,3,5-trinitrobenzene complex	$CHCl_3$	440(4.36)	38-0166-64A
2,4,7-trinitrofluorenone complex	$CHCl_3$	500(4.30)	38-0166-64A
Phenanthrene	n.s.g.	224(4.3),247(4.7), 252(4.8),274(4.2), 282(4.0),293(4.2), 309(2.2),316(2.3), 325(2.3),331(2.3), 338(2.3),346(2.3)	28-4795-64A
bromanil complex	$CHCl_3$	486(3.16)	60-0465-64
chloranil complex	$CHCl_3$	480(3.16)	60-0465-64
iodoanil complex	$CHCl_3$	478(3.02)	60-0465-64
1,3,5-trinitrobenzene complex	$CHCl_3$	365(4.44)	38-0166-64A
$C_{14}H_{10}BN$			
Phenanthro[4,5-cde]azaborine, 4,5-dihydro-	$C_6H_{11}Me$	238(4.53),250(4.57), 267(4.36),295(3.88), 307(4.09),317(4.09), 320(4.18),331(3.54), 351s(3.38),358(3.54), 365(4.14)	44-1757-64
$C_{14}H_{10}BrN$			
10-Phenanthrylamine, 1-bromo-	EtOH	258(4.53),333(3.91)	44-3933-65
$C_{14}H_{10}BrNO$			
Acenaphth[1,2-d]isoxazole, 9a-bromo-6b,9a-dihydro-9-methyl-	EtOH	224(4.75),292(3.81), 325(3.16)	88-2157-64
Phthalimidine, 2-(m-bromophenyl)-	MeOH-acid	280(4.17)	95-1042-65
	MeOH	279(4.17)	95-1042-65
Phthalimidine, 2-(p-bromophenyl)-	MeOH-acid	282(4.24)	95-1042-65
	MeOH	283(4.24)	95-1042-65
$C_{14}H_{10}BrNO_2S_2$			
1,3-Dithiol-2-ylideneacetic acid, 4-(p-bromophenyl)-α-cyano-, ethyl ester	CF_3COOH	378(4.56)	44-1711-64
	THF	247(4.31),382(4.43)	44-1711-64
$C_{14}H_{10}BrN_3O_2$			
1H-Indazole, 1-benzyl-3-bromo-6-nitro-	MeOH	264(4.29),352(3.46)	65-2799-64
2H-Indazole, 2-benzyl-3-bromo-6-nitro-	MeOH	268(4.38),350(3.37)	65-2799-64
$C_{14}H_{10}Br_4N_2$			
Benzoyl bromide, p-bromo-, (4,6-dibromo-o-tolyl)hydrazone	EtOH	234(4.21),319(4.14)	39-1500-64
$C_{14}H_{10}ClF_3O_4$			
(α,α,α-Trifluoro-m-tolyl)cycloheptatrienylium perchlorate	MeCN	267(4.25),347(4.17)	24-0029-64
$C_{14}H_{10}ClNO$			
Phthalimidine, 2-(m-chlorophenyl)-	MeOH-acid	281(4.13)	95-1042-65
	MeOH	279(4.12)	95-1042-65
Phthalimidine, 2-(o-chlorophenyl)-	MeOH-acid	250(3.91)	95-1042-65
	MeOH	280s(--)	95-1042-65
Phthalimidine, 2-(p-chlorophenyl)-	MeOH-acid	281(4.16)	95-1042-65
	MeOH	281(4.14)	95-1042-65
$C_{14}H_{10}ClNO_2$			
Stilbene, α-chloro-4'-nitro-	EtOH	320(4.25)	65-3393-64

Compound	Solvent	$\lambda_{max}(\log \epsilon)$	Ref.
$C_{14}H_{10}ClNO_6$			
7-Carboxybenzo[b]quinolizinium perchlorate	EtOH	209s(3.88),242(4.62), 383s(3.76),343s(3.71), 361(4.01),381(4.06), 402(4.03)	4-0030-64
9-Carboxybenzo[b]quinolizinium perchlorate	EtOH	246(4.62),277s(4.05), 287s(4.02),324(3.44), 341(3.63),357(3.89), 368(3.81),389(4.05), 410(4.10)	4-0030-64
$C_{14}H_{10}Cl_2$			
Stilbene, 2,4-dichloro-, cis	EtOH	290(3.88)	44-1473-65
Stilbene, 2,4-dichloro-, trans	EtOH	330(4.26)	44-1473-65
Stilbene, 4,4'-dichloro-, trans	hexane	228(4.21),302(--), 328(4.56)	65-2073-65
	EtOH	305(4.53),315(4.52)	44-1473-65
	EtOH	229(4.18),304(--), 328(4.53)	65-2073-65
	ether	227(4.33),303(--), 328(4.54)	65-2073-65
	$C_2H_4Cl_2$	310(4.47)	65-2073-65
$C_{14}H_{10}Cl_2N_4$			
1,3,4-Triazole, 4-amino-3,5-bis-(2-chlorophenyl)-	EtOH	246s(4.0),283s(2.9)	28-1262-64A
1,3,4-Triazole, 4-amino-3,5-bis-(4-chlorophenyl)-	EtOH	267(4.6)	28-1262-64A
$C_{14}H_{10}Cl_2O_2S$			
Dibenzo[c,e]thiepin, 5,5-dichloro-5,7-dihydro-, 1,1-dioxide	EtOH	213(4.56),250s(3.91), 279(3.61)	35-4089-64
Dibenzo[c,e]thiepin, 5,7-dichloro-5,7-dihydro-, 1,1-dioxide	EtOH	225(4.55),270(3.53)	35-4089-64
$C_{14}H_{10}Cl_3N_3O_2$			
Acetophenone, m-nitro-, (2,4,6-trichlorophenyl)hydrazone	EtOH	312(4.27)	39-1500-64
Benzoyl chloride, m-nitro-, (4,6-dichloro-o-tolyl)hydrazone	EtOH	315(4.25)	39-1500-64
$C_{14}H_{10}CuO_4$			
Copper benzoate	dioxan	384s(--),669(2.43)	39-6464-65
	MeCN	386s(--),688(2.42)	39-6464-65
$C_{14}H_{10}INO$			
Phthalimidine, 2-(m-iodophenyl)-	MeOH-acid	281(4.15)	95-1042-65
	MeOH	281(4.11)	95-1042-65
Phthalimidine, 2-(p-iodophenyl)-	MeOH-acid	284(4.06)	95-1042-65
	MeOH	284(4.05)	95-1042-65
$C_{14}H_{10}INO_2$			
Stilbene, 2'-iodo-α-nitro-, cis	EtOH	314(3.88)	44-3792-65
	EtOH	314(3.88)	88-0359-65
Stilbene, 3'-iodo-α-nitro-, cis	EtOH	310(4.08)	44-3792-65
$C_{14}H_{10}N_2$			
Copyrine, 3-phenyl-	EtOH	230(4.05),254(4.34), 295(3.96)	44-2298-64

Compound	Solvent	λ_{max}(log ϵ)	Ref.
Cycloheptimidazole, 2-phenyl-	EtOH	252(4.44),272(4.39), 342(4.34),354(4.35)	94-0465-65
Heptafulvene, 8-(4,4-dicyano- 1,3-butadienyl)-	hexane	248s(3.79),255(3.63), 275s(3.78),284(3.84), 295s(3.74),330(3.42), 350(3.45),380s(3.58), 472(4.79),493s(4.71)	24-2050-64
	EtOH	236(4.24),251s(3.81), 256(3.64),286(3.89), 295(3.92),305s(3.80), 330s(3.47),345(3.45), 371(3.40),505(4.82)	24-2050-64
$C_{14}H_{10}N_2O$			
Acetophenone, α-diazo-α-phenyl-	MeOH	257(4.20),312(3.80)	88-1403-64
2-Indolinone, 3-(2-pyridyl- methylene)-	EtOH	256(4.19),333(4.17)	87-0626-65
2-Indolinone, 3-(3-pyridyl- methylene)-	EtOH	256(4.15),316(4.04), 390(3.39)	87-0626-65
2-Indolinone, 3-(4-pyridyl- methylene)-	EtOH	258(4.19),312(3.93), 398(3.31)	87-0626-65
Ketone, 3-indolyl 4-pyridyl	MeOH	208(4.55),258(4.36), 310(4.04)	44-2534-65
Nitrone, α-cyano-N,α-diphenyl-, cis	C_6H_{12}	240(5.0),325(5.0)	46-2545-65
	EtOH	240(5.0),320(4.9)	46-2545-65
trans	C_6H_{12}	230(5.2),245s(4.9), 320(5.3)	46-2545-65
	EtOH	225(5.1),245s(4.7), 325(5.0)	46-2545-65
Oxadiazole, 2,5-diphenyl-	DMF	281.4(4.38)	24-2966-65
1,2,4-Oxadiazole, 3,5-diphenyl-	ether	245(4.57)	33-0942-64
$C_{14}H_{10}N_2O_2$			
6,7-Benzimidazoledione, 1-methyl-2-phenyl-	$CHCl_3$	272(4.2),325s(3.6), 467(2.8)	65-0949-64
4H-3,1-Benzoxazin-4-one, 1,2-dihydro- 2-phenylimino-	EtOH	244(3.91),282(3.96), 292(3.93),332(3.42)	88-2597-65
Halfordinol	EtOH	230(4.03),250(4.09), 330(4.40)	4-0310-65
	pH 13	261(4.16),348(4.38)	4-0310-65
Nicotinonitrile, 1,2-dihydro- 2-oxo-6-phenacyl-	EtOH	242(4.23),338(4.06), 435(3.64)	35-5198-65
Uretidinedione, 1,3-diphenyl-	n.s.g.	209(4.3),256(4.5)	83-0623-64
$C_{14}H_{10}N_2O_3$			
Phthalimidine, 2-(m-nitrophenyl)-	MeOH-acid	273(4.34)	95-1042-65
	MeOH	273(4.34)	95-1042-65
Phthalimidine, 2-(o-nitrophenyl)-	MeOH-acid	320s(--)	95-1042-65
	MeOH	320s(--)	95-1042-65
Phthalimidine, 2-(p-nitrophenyl)-	MeOH-acid	325(4.32)	95-1042-65
	MeOH	326(4.36)	95-1042-65
$C_{14}H_{10}N_2O_4$			
Fluorene, 9-methyl-2,9-dinitro-	ether	214(4.63),241(4.25), 322(4.57)	44-1961-64
Stilbene, 4,4'-dinitro-	$C_2H_4Cl_2$	342(4.30)	65-2073-65
Stilbene, 4,4'-dinitro-, trans	EtOH	303(4.27)	44-1473-65

Compound	Solvent	λ_{max}(log ε)	Ref.
$C_{14}H_{10}N_2O_4S$			
Naphtho[2,3-c][1,2,5]thiadiazole-4,9-diol, diacetate	$CHCl_3$	259(4.83),329(3.63),344(3.91),360(4.06)	39-1003-65
$C_{14}H_{10}N_2O_5S_2$			
1,4-Pentadien-3-one, 2-methyl-1,5-bis(5-nitro-2-thienyl)-	propylene glycol	380(4.394)	95-0001-64
$C_{14}H_{10}N_2O_7$			
1,4-Pentadien-3-one, 2-methyl-1,5-bis(5-nitro-2-furyl)-	propylene glycol	350(4.308)	95-0001-64
$C_{14}H_{10}N_2O_{10}$			
1,2-Ethenediol, bis(5-nitro-2-furyl)-, diacetate	MeOH-dioxan	287(4.09),379(4.20)	94-0389-65
$C_{14}H_{10}N_2S$			
Cyclopenta[c]thiapyran, 5(or 7)-(phenylazo)-	hexane	234(4.16),290(4.19),297(4.19),396(4.21),525(3.83)	35-0708-64
1,2,5-Thiadiazole, 3,4-diphenyl-	isooctane	214(4.38),250(4.03),295(4.12)	89-0262-65
$C_{14}H_{10}N_4$			
Unknown compound (PhN:NN:C(Ph)CN ?)	heptane	343(4.36)	35-2025-64
	$C_2H_4Cl_2$	351(4.45)	35-2025-64
	DMF	350(4.30)	35-2025-64
$C_{14}H_{10}N_4O$			
Isobenzofuran, 1,1,3,3-tetracyano-1,3,3a,7a-tetrahydro-4,7-dimethyl-	MeCN	267(3.85)	35-3657-65
$C_{14}H_{10}N_4O_4$			
2H-Naphtho[2,3-d]triazole-4,9-dione, 2-diacetamido-	$CHCl_3$	257(4.58),328(3.53)	39-1003-65
1,3,4-Oxadiazole, 2-amino-5-[1-phenyl-2-(5-nitro-2-furyl)vinyl]-	propylene glycol	232(4.06),308(4.13),395(4.39)	95-0212-64
1,3,4-Oxadiazole, 2-anilino-5-(5-nitro-2-furyl)vinyl-	propylene glycol	242(4.14),305(4.03),403(4.27)	95-0219-64
$C_{14}H_{10}N_4O_5$			
Benzocyclobutenone, 4-hydroxy-, 2,4-dinitrophenylhydrazone	$CHCl_3$	385(4.46)	39-1390-65
$C_{14}H_{10}N_8O_2$			
Sydnone, 4-formyl-3-phenyl-, purin-6-ylhydrazone	dioxan	274(4.65),385(4.11)	44-2044-64
$C_{14}H_{10}N_8O_8$			
Glyoxal, bis(2,4-dinitrophenylhydrazone)	EtOH-KOH	550(4.84)	37-2189-64
$C_{14}H_{10}O$			
Benzofuran, 2-phenyl-	n.s.g.	209(4.36),214s(4.20),226(4.03),230s(4.01),241s(3.78),278s(4.24),288s(4.37),297s(4.42),303(4.48),316(4.35)	12-0379-64

Compound	Solvent	λ_{max}(log ε)	Ref.
Fluorenone, 2-methyl-	EtOH	252(5.04),260(4.97), 285s(3.49),296(3.56), 308(3.36),315s(3.23), 323s(3.05),330(3.01), 383(2.42),400(2.44)	22-2953-64
Fluorenone, 4-methyl-	iso-PrOH	383(2.55)	39-5518-65
Furan, 2-(inden-1-ylidenemethyl)-	dioxan	248(4.02),292(3.93), 300(3.91),357(4.50), 365(4.47)	5-0018-64D
$C_{14}H_{10}OS$			
Fluoren-9-one, 4-(methylthio)-	iso-PrOH	421(2.59)	39-5518-65
$C_{14}H_{10}O_2$			
1,4-Anthracenedione, 2,3-dihydro-	n.s.g.	265(4.77),293(3.72), 347(3.39),362(3.55)	77-0015-65
Benzil	EtOH	259(4.31),370(1.89)	23-0113-64
Benzofuran, 2-(p-hydroxyphenyl)-	EtOH	247s(--),309(4.44)	4-0158-64
	EtOH-NaOEt	332(4.49)	4-0158-64
Fluoren-9-one, 4-methoxy-	iso-PrOH	418(3.00)	39-5518-65
Naphtho[1,2-b]pyran-2-one, 4-methyl-	MeOH	221(4.40),266(4.51), 274(4.40)	44-0502-64
4H-Naphtho[2,3-b]pyran-4-one, 2-methyl-	EtOH	225(4.25),251(4.66), 267s(4.40),280s(4.25), 300s(3.40),314(3.40), 328(3.25),356s(3.50), 380(3.61)	94-0316-64
1,7-Phenanthrenediol	EtOH	236(4.36),254s(4.42), 267(4.52),312(3.79), 328(3.18),343(3.30), 360(3.33)	88-3671-64
2-Propyn-1-one, 1-(2-furyl)-3-p-tolyl-	EtOH	231(4.13),319(4.38)	30-0833-65
$C_{14}H_{10}O_2S$			
Sulfone, methyl 1-(1,3-pentadiynyl)-1,7-octadiene-3,5-diynyl, cis	ether	227(4.39),239(4.35), 250s(4.20),267(4.00), 342(4.36)	24-3015-65
trans	ether	227(4.51),239(4.46), 253s(4.26),268(4.11), 350(4.57)	24-3015-65
Thiirene, diphenyl-, 1,1-dioxide	EtOH	223(4.26),296(4.34), 307(4.41),322(4.27)	35-5804-65
$C_{14}H_{10}O_2S_2$			
Benzoyl disulfide	C_6H_{12}	240(4.46),262s(4.26)	78-0035-65
$C_{14}H_{10}O_3$			
Benzoic acid, o-benzoyl-	EtOH	244(3.69)	44-4293-65
1-Cyclopentene-1-carboxylic acid, 2-(6-hydroxy-5-benzofuranyl)-, δ-lactone	EtOH	246(4.35),298(4.01), 326(4.00)	44-2467-64
Naphtho[2,1-b]pyran-1-one, 5-hydroxy-3-methyl-	MeOH	217(4.23),287(4.24)	44-0502-64
Naphtho[2,1-b]pyran-1-one, 6-hydroxy-3-methyl-	MeOH	230(4.15),300(3.71), 330(3.15)	44-0502-64
1H-Naphtho[1,2-b]pyran-4-one, 5-hydroxy-2-methyl-	EtOH	224(4.67),265(4.51), 375(3.36)	94-0312-64
	EtOH	224(4.67),265(4.51), 280s(4.21),310s(4.62), 375(3.36)	94-0316-64

Compound	Solvent	λ_{max}(log ϵ)	Ref.
1H-Naphtho[1,2-b]pyran-4-one, 6-hydroxy-2-methyl-	EtOH	221(4.64),263(4.42), 283s(4.02),304(3.66), 365(3.81)	94-0316-64
Naphtho[2,3-b]pyran-4-one, 10-hydroxy-2-methyl-	EtOH	225(4.16),251(4.63), 270s(4.36),279s(4.25), 301(3.38),314(3.43), 328(3.26),356s(3.55), 382(3.64)	94-0316-64
Oroselone	EtOH	220(4.06), 270(4.14), 293(4.18),320(3.97)	65-4171-64
	EtOH	286(4.33),299(4.26)	78-2605-64
$C_{14}H_{10}O_4$			
Isophthalic acid, di-2-propynyl ester	MeOH	276(3.8),284(3.8)	70-1349-64
4H-Naphtho[2,3-b]pyran-4-one, 5,10-dihydroxy-2-methyl-	EtOH	220s(4.19),255s(4.32), 269(4.46),275(4.46), 312s(3.47),323(3.43), 341(3.40),427(3.41)	94-0316-64
Terephthalic acid, di-2-propynyl ester	MeOH	288(3.3),294(3.3)	70-1349-64
$C_{14}H_{10}O_4S_2$			
2-Thianthrenecarboxylic acid, methyl ester, 5,10-dioxide	EtOH	220(4.6)	35-2957-64
isomer	EtOH	228(4.6)	35-2957-64
$C_{14}H_{10}O_5$			
Norrubrofusarin	EtOH	224(4.48),249s(4.38), 270(4.66),325(3.55), 340(3.62),373(3.78)	44-0112-65
$C_{14}H_{10}O_6$			
Bellidifolin	EtOH	234(4.1),255(4.2), 279(4.1),302s(3.7), 334(3.9)	78-0991-64
2,2'-Bi-p-benzoquinone, 5,5'-dihydroxy-3,3'-dimethyl-	MeOH	271(4.40),387(3.11)	24-2774-65
	MeOH-KOH	222(4.55),280(4.26), 496(3.60)	24-2774-65
Isobellidifolin	EtOH	231(4.35),251(4.3), 276(4.2),342(4.1)	78-1449-65
2-Naphthaleneacrylic acid, 1,4-dihydro-3,8-dihydroxy-α-methyl-1,4-dioxo-	MeOH	232(4.10),270(4.16), 415(3.63)	39-5927-65
Xanthone, 1,3,5-trihydroxy-6-methoxy-	MeOH	248(4.70),266(4.46), 338(4.24)	78-2653-65
$C_{14}H_{10}O_8$			
2,2'-Bi-p-benzoquinone, 3,3',6,6'-tetrahydroxy-5,5'-dimethyl-	EtOH	290(4.60),437(2.85)	57-0152-65A
$C_{14}H_{10}S$			
Sulfide, methyl 1-(1,3-pentadiynyl)-1,7-octadiene-3,5-diynyl-, cis	ether	216(4.51),238s(4.27), 256(4.01),271(4.05), 276(4.07),307s(3.97), 358(4.39)	24-3015-65
trans	ether	238s(4.37),256(4.13), 271(4.17),276s(4.16), 292s(4.05),307(4.12), 360(4.48),381s(4.41)	24-3015-65
Thiophene, 2-(inden-1-ylidenemethyl)-	dioxan	242(4.04),286(4.01), 368(4.11)	5-0018-64D

Compound	Solvent	λ_{max}(log ϵ)	Ref.
$C_{14}H_{11}BrN_6O_6$			
1,4-Pentadien-3-one, 2-bromo-1,5-bis-(5-nitro-2-furyl)-, amidinohydrazone, hydrochloride	propylene glycol	374(4.574)	95-0001-64
$C_{14}H_{11}BrO_2S_2$			
2,4-Pentanedione, 3-[4-(p-bromophenyl)-1,3-dithiol-2-ylidene]-	CF_3COOH	388(4.28)	44-1711-64
	THF	389(4.42)	44-1711-64
$C_{14}H_{11}Cl$			
Stilbene,α-chloro-	EtOH	287(4.35)	65-3393-64
Stilbene, 2-chloro-, cis	EtOH	276(4.05)	44-1473-65
Stilbene, 2-chloro-, trans	EtOH	295(4.29)	44-1473-65
Stilbene, 4-chloro-, trans	EtOH	348(4.44)	44-1473-65
$C_{14}H_{11}ClN_2O_2$			
4-Pyridinepropionanilide, 2'-chloro-β-oxo-	EtOH	318(4.16)	65-2400-64
4-Pyridinepropionanilide, 3'-chloro-β-oxo-	EtOH	243(4.08),326(4.22)	65-2400-64
4-Pyridinepropionanilide, 4'-chloro-β-oxo-	EtOH	246(4.14),326(4.28)	65-2400-64
$C_{14}H_{11}ClN_2O_3$			
Acetanilide, 2-chloro-2'-nitro-N-phenyl-	EtOH	232(4.21),300(3.20)	44-1279-65
$C_{14}H_{11}ClN_6O_6$			
1,4-Pentadien-3-one, 2-chloro-1,5-bis-(5-nitro-2-furyl)-, amidinohydrazone, hydrochloride	propylene glycol	376(4.415)	95-0001-64
$C_{14}H_{11}ClO_2$			
2,5-Norbornadiene-2-carboxylic acid, 3-(o-chlorophenyl)-	EtOH	276(3.58)	35-5548-64
2,5-Norbornadiene-2-carboxylic acid, 3-(p-chlorophenyl)-	EtOH	294(3.90)	35-5548-64
$C_{14}H_{11}ClO_3$			
1,2-Benzoquinone, 4-[(3-chloro-p-tolyl)oxy]-5-methyl-	MeOH	267(3.85),328s(3.45), 406(3.23)	24-2643-65
1,2-Benzoquinone, 4-[(4-chloro-m-tolyl)oxy]-5-methyl-	MeOH	270(3.74),321s(3.49), 405(3.11)	24-2643-65
$C_{14}H_{11}ClS$			
1-Heptene-3,5-diyne, 2-chloro-2-(methylthio)-1-phenyl-	ether	215(4.53),241s(3.93), 327(4.15)	24-3015-65
$C_{14}H_{11}Cl_2I$			
(α-Chlorostyryl)phenyliodonium chloride	MeOH	268(4.03)	44-1930-65
$C_{14}H_{11}Cl_2N_3O_5S_2$			
2H-1,2,4-Benzothiadiazine-7-sulfonamide, 6-chloro-3-(p-chlorobenzoyl)-3,4-dihydro-, 1,1-dioxide	EtOH	225(4.60),269(4.54)	95-0095-65
2H-1,2,4-Benzothiadiazine-7-sulfonamide, 6-chloro-3-(o-chloro-α-hydroxybenzyl)-, 1,1-dioxide	EtOH	222(4.56),279(4.08)	95-0095-65

Compound	Solvent	$\lambda_{max}(\log \epsilon)$	Ref.
2H-1,2,4-Benzothiadiazine-7-sulfonamide, 6-chloro-3-(p-chloro-α-hydroxybenzyl)-, 1,1-dioxide	EtOH	226(4.61),277(4.12)	95-0095-65
$C_{14}H_{11}Cl_3O_2$			
Hydroquinone, trichloro-p-xylyl-	EtOH	269(3.02),275(3.14), 305(3.76)	5-0152-64D
$C_{14}H_{11}F$			
Stilbene, 3-fluoro-, trans	EtOH	227(4.19),234s(4.08), 295(4.43),307(4.44), 321s(4.22)	35-3665-65
Stilbene, 4-fluoro-, cis	EtOH	280(4.03)	44-1473-65
$C_{14}H_{11}F_5O_5$			
Malonic acid, (pentafluorobenzoyl)-, diethyl ester	EtOH	278(4.15)	70-1798-65
$C_{14}H_{11}I$			
Stilbene, 2-iodo-, cis	EtOH	271.5(4.23)	44-3792-65
	EtOH	271.5(4.23)	88-0359-65
Stilbene, 2-iodo-, trans	EtOH	297(4.40)	44-3792-65
$C_{14}H_{11}IN_6O_6$			
1,4-Pentadien-3-one, 2-iodo-1,5-bis-(5-nitro-2-furyl)-, amidinohydrazone, hydrochloride	propylene glycol	383(4.454)	95-0001-64
$C_{14}H_{11}IO$			
4-Stilbenol, 4'-iodo-	EtOH	234(3.92),240s(3.87), 315s(4.35),331(4.42)	7-0805-64
$C_{14}H_{11}MnO_4$			
Manganese, (pentenophenonato)-tricarbonyl-	dioxan	231(3.78),266(3.97), 307(3.76)	77-0370-64
$C_{14}H_{11}N$			
Indole, 2-phenyl-	n.s.g.	241(4.23),310(4.36)	88-2285-65
Indole, 3-phenyl-	n.s.g.	225(4.49),270(4.18)	88-2285-65
Isoindole, 1-phenyl-	n.s.g.	270(3.8),280(3.9), 325(4.0),360(4.1)	35-4152-64
Phenanthridine, 1-methyl-	pH 1.00	242(4.62),328(3.89)	39-4426-65
	pH 9.17	246(4.69),294(3.83), 332(3.16),348(3.11)	39-4426-65
	n.s.g.	220(4.26),246(4.69), 295(3.85),302(3.80), 334(3.14),348(3.07)	39-3032-65
Phenanthridine, 3-methyl-	pH 1.00	252(4.61),316(3.85), 370(3.56)	39-4426-65
	pH 9.17	250(4.68),292(3.71) 320(3.06),336(3.26), 350(3.25)	39-4426-65
Phenanthridine, 8-methyl-	pH 1.00	252(4.64),318(3.86), 372(3.57)	39-4426-65
	pH 9.17	250(4.70),290(3.73), 320(3.06),336(3.27), 350(3.27)	39-4426-65
Phenanthridine, 10-methyl-	pH 1.00	244(4.64),324(3.86)	39-4426-65
	pH 9.12	246(4.72),294(3.82), 334(3.24),348(3.24)	39-4426-65

Compound	Solvent	$\lambda_{max}(\log \epsilon)$	Ref.
Phenanthridine, 10-methyl- (contd.)	n.s.g.	221(4.20),246(4.69), 294(3.85),304s(3.78), 334(3.24),349(3.26)	39-3032-65
2,3-Phenylenedihydroindole	MeOH	250(3.1),290(2.8)	44-1270-64
Pyrrole, 2-(inden-1-ylidenemethyl)-	dioxan	248(4.01),289(3.59), 292(3.56),386(4.13)	5-0018-64D
$C_{14}H_{11}NO$			
Acenaphth[1,2-d]isoxazole, 6b,9a-dihydro-9-methyl-	EtOH	224(4.74),276(3.44), 286(3.54),298(3.37)	88-2157-64
Acetophenone, (phenylimino)-	EtOH	206(4.26),220(4.15), 254(4.18)	44-0043-65
Benzo[f]quinolin-3(4H)-one, N-methyl-	EtOH	243(4.89),250s(4.46), 272(4.01),287(4.04), 298(4.11),311(4.01), 338(3.72),346(3.80)	1-1806-64
Benzo[h]quinolinone, N-methyl-	EtOH	219(4.31),233(4.54), 270(4.45),281(4.48), 307(3.60),346(3.43)	1-1806-64
Phthalimidine, 2-phenyl-	$MeOH-H_2SO_4$	281(4.15)	95-1035-65
	MeOH	281(4.11)	95-1035-65
$C_{14}H_{11}NOS$			
Dibenzamide, thio-	heptane	243(4.14),285(4.11), 320s(3.92)	24-1556-65
	MeOH	246(4.13),280(4.09), 320(3.92),485(2.20)	24-1556-65
	dioxan	235(4.13),288(4.06), 325(3.91),495(1.67)	24-1556-65
$C_{14}H_{11}NO_2$			
Carbazole-3-carboxaldehyde, 1-methoxy-	n.s.g.	238(4.47),247(4.30), 274(4.56),289(4.56)	78-0681-65
Fluorene, 9-methyl-2-nitro-	ether	237(4.25),327(4.67)	44-1960-64
Nicotinic acid, 6-styryl-, cis	n.s.g.	509(4.11)	87-0150-64
Nicotinic acid, 6-styryl-, trans	n.s.g.	323(4.51)	87-0150-64
Phthalimidine, 2-(m-hydroxyphenyl)-	MeOH-acid	282(4.02)	95-1042-65
	MeOH	282(4.02)	95-1042-65
Phthalimidine, 2-(o-hydroxyphenyl)-	MeOH-acid	279(3.85)	95-1042-65
	MeOH	279(3.85)	95-1042-65
Phthalimidine, 2-(p-hydroxyphenyl)-	MeOH-acid	283(3.95)	95-1042-65
	MeOH	283(3.96)	95-1042-65
Stilbene, 4-nitro-	EtOH	348(4.44)	44-1473-65
$C_{14}H_{11}NO_2S$			
Benzophenone, 2-methyl-4'-nitrothio-	C_6H_{12}	308(4.17)	54-0289-65
	EtOH	308(4.15)	54-0289-65
6H-Thieno[2,3-b]pyrrole-4-carboxylic acid, 6-benzyl-	EtOH	232(4.00),267(4.16)	44-0184-65
$C_{14}H_{11}NO_2S_2$			
1,3-Dithiol-2-ylideneacetic acid, α-cyano-4-phenyl-, ethyl ester	CF_3COOH	383(4.44)	44-1711-64
	THF	380(4.39)	44-1711-64
$C_{14}H_{11}NO_3$			
Benzoic acid, m-(salicylideneamino)-	EtOH	220(4.4),260(4.00), 336(3.93)	65-3837-64
Benzoic acid, p-(salicylideneamino)-	EtOH	230(4.3),278(4.26), 340(4.09)	65-3837-64

Compound	Solvent	λ_{max}(log ε)	Ref.
2-Naphthonitrile, 3-acetoxy-8-methoxy-	n.s.g.	215(4.58),252(4.61), 290(3.45),299(3.53), 310(3.48),352(3.72)	88-0041-65
$C_{14}H_{11}NO_3S$			
Benzophenone, 4-methoxy-4'-nitrothio-	C_6H_{12}	253(4.07),297(4.15), 368(4.19),612(2.36)	54-0289-65
	EtOH	253(4.06),299(4.16), 373(4.20),600(2.37)	54-0289-65
11-Methyl-7-sulfobenzo[b]quinolizinium hydroxide, inner salt	EtOH	241(3.78),250(3.75), 367(3.28),384(3.34), 404(3.30)	4-0228-65
2H-Thieno[2,3-b]pyrrole-4-carboxylic acid, 6-benzyl-3,6-dihydro-3-oxo-	EtOH	238(3.91),322(3.79)	44-0184-65
$C_{14}H_{11}NO_4$			
Benz[cd]indole-3-carboxylic acid, 1-acetyl-1,3,4,5-tetrahydro-5-oxo-	EtOH	226(4.18),257(4.22), 292s(3.94),303(4.03), 326(3.58)	44-0843-64
Indole-3-succinic anhydride, 1-acetyl-	EtOH	239(4.29),260(3.94), 270s(3.89),280s(3.67), 290(3.86),298(3.89)	44-0843-64
2,5-Norbornadiene-2-carboxylic acid, 3-(o-nitrophenyl)-	EtOH	306s(3.47)	35-5548-64
2,5-Norbornadiene-2-carboxylic acid, 3-(p-nitrophenyl)-	EtOH	327(4.03)	35-5548-64
$C_{14}H_{11}NO_4S$			
1,2,3-Benzoxathiazine, 7-methoxy-4-phenyl-, 2,2-dioxide	EtOH	241s(3.9),246(3.9), 302(4.18),320(4.17)	44-3960-65
$C_{14}H_{11}N_2O_7P$			
Phenoxaphosphine, 10-hydroxy-2,8-dimethyl-4,6-dinitro-, 10-oxide	EtOH	235s(4.24),330(3.60)	44-1983-64
$C_{14}H_{11}N_3$			
1H-Imidazo[4,5-b]pyridine, 2-styryl-	MeOH	265(3.90),335(4.52)	44-3403-64
1(4H)-Pyridinecarbonitrile, 2-indol-3-yl-	EtOH	220(4.59),273(3.80), 280(3.81),290(3.72)	44-2534-65
3,5-Pyridinedicarbonitrile, 1,2-dihydro-4-methyl-2-phenyl-	EtOH	217(4.35),241s(4.16), 380(3.63)	73-2609-65
$C_{14}H_{11}N_3O$			
5-Hydroxy-2,4-diphenyl-s-triazolium hydroxide, inner salt	n.s.g.	223(4.22),292(4.01)	4-0105-65
1,3,4-Oxadiazole, 2-anilino-5-phenyl-	n.s.g.	226s(4.1),245(4.2), 296(4.4)	28-2868-64B
2H-Pyrido[4,3-e]-1,4-diazepin-2-one, 1,3-dihydro-5-phenyl-	MeOH	222(4.49),248s(4.26)	87-0722-65
$C_{14}H_{11}N_3O_2$			
1,2,4-Triazolidine-3,5-dione, 1,2-diphenyl-	EtOH	214(3.88),242(4.09)	22-0500-64
	NaOH	216(4.11),250(4.06)	22-0500-64
$C_{14}H_{11}N_3O_3S$			
Sydnone imine, 3-phenyl-N-(phenylsulfonyl)-	pH 7.00	236(4.12),322(4.00)	65-2064-64
$C_{14}H_{11}N_3O_4$			
3-Pyridinepropionanilide, 2'-nitro-β-oxo-	EtOH	227(4.27),280(4.12), 324(4.12)	65-2400-64

Compound	Solvent	λ_{max}(log ε)	Ref.
3-Pyridinepropionanilide, 3'-nitro-β-oxo-	EtOH	240(4.18),270(4.15), 318(4.24)	65-2400-64
3-Pyridinepropionanilide, 4'-nitro-β-oxo-	EtOH	226(4.24),248(4.40)	65-2400-64
$C_{14}H_{11}N_3O_7$			
Toluene, α-benzoyldihydro-2,4,6-tri-nitro-, sodium salt	EtOH	464(4.36),565(4.03)	25-2065-65
$C_{14}H_{11}N_3S$			
5-Mercapto-2,4-diphenyl-s-triazolium hydroxide, inner salt	n.s.g.	245(4.31),314(3.44)	4-0105-65
$C_{14}H_{11}N_5$			
Benzimidazole, 2,2'-iminobis-, hydrochloride	MeOH-HCl	214(4.46),319(4.69)	4-0288-64
	MeOH-KOH	226(4.47),335(4.61)	4-0288-64
$C_{14}H_{11}N_5O_{10}S$			
Hypotaurine, N,S-bis-(2,4-dinitrophenyl)-	acetone	347.0(4.31)	39-0826-65
$C_{14}H_{12}$			
Benzocalicene	hexane	335(3.61)	89-0258-65
	ether	267(3.92),336(3.64)	89-0258-65
	$CHCl_3$	343(3.70)	89-0258-65
Ethylene, 1,1-diphenyl-	C_6H_{12}	250(4.06)	22-0693-64
	C_6H_{12}	229(4.20),251(4.04)	54-0806-65
	EtOH	231(4.19),250(4.03)	54-0806-65
	98% H_2SO_4	315(3.99),428(4.55)	54-0806-65
Fluorene, 1-methyl-	EtOH	264(4.29)	42-0479-64
Phenanthrene, 9,10-dihydro-	isooctane	264(4.24)	35-1710-64
	EtOH	265(4.24),296(3.70)	24-0593-65
	EtOH	209(4.64),220s(4.01), 265(4.25),290s(3.63), 298(3.62)	39-5544-64
Stilbene	n.s.g.	295(4.45)	5-0058-65B
cis	EtOH	278(4.01)	44-1473-65
trans	EtOH	296(4.45),305(4.43)	44-1473-65
	$CHCl_3$	295(4.43)	83-0715-65
Stilbene, bromanil complex	$CHCl_3$	526(2.85)	60-0465-64
Stilbene, chloranil complex	$CHCl_3$	520(2.80)	60-0465-64
Stilbene, iodoanil complex	$CHCl_3$	510(2.80)	60-0465-64
$C_{14}H_{12}BrN$			
6-Methylacridizinium bromide	EtOH	246s(4.54),251(4.61), 364(3.88),383(3.89), 404(3.83)	4-0121-64
$C_{14}H_{12}ClNO$			
Benzophenone, 5-chloro-2-(methylamino)-	MeOH	237(4.40)	10-0334-64E
$C_{14}H_{12}ClNO_3$			
3,5,7-Octatrien-2-one, 1-chloro-8-(p-nitrophenyl)-	EtOH	281(3.86),379(4.67)	87-0035-65
$C_{14}H_{12}ClNS$			
N-Methyl-2-phenylbenzothiazolium chloride	EtOH-H_2SO_4	258(3.93),305(4.22)	22-2868-64
	EtOH	256(3.88),305(4.06)	22-2868-64
	EtOH-NaOEt	308(3.70)	22-2868-64

Compound	Solvent	$\lambda_{max}(\log \epsilon)$	Ref.
$C_{14}H_{12}ClN_3O_3S$			
Acetanilide, N-(5-chloro-3-nitro-2-pyridyl)-2'-(methylthio)-	EtOH	222(4.36),254(4.09), 330s(3.11)	95-0429-65
$C_{14}H_{12}ClN_3O_5S_2$			
2H-1,2,4-Benzothiadiazine-7-sulfonamide, 3-benzoyl-6-chloro-3,4-dihydro-, 1,1-dioxide	EtOH	225(4.43),267(4.37)	95-0095-65
$C_{14}H_{12}Cl_2N_2O_2S$			
p-Toluenesulfonic acid, (2,4-dichlorobenzylidene)hydrazide	EtOH	285(4.29)	44-1242-65
anion	H_2O	313(4.27)	44-1242-65
	MeOH	329(4.29)	44-1242-65
	EtOH	334.5(4.25)	44-1242-65
	MeCN	355(4.28)	44-1242-65
	DMF	368.5(4.13)	44-1242-65
$C_{14}H_{12}Cl_2N_4S$			
9H-Purine, 6-[(m-chlorobenzyl)thio]-9-(2-chloroethyl)-	EtOH	285(4.31)	87-0182-65
$C_{14}H_{12}N_2$			
Benzimidazole, 1-benzyl-	MeOH	249(3.68),253(3.68), 266(3.64),274(3.62), 281(3.55)	65-0632-64
Benzimidazole, 2-benzyl-	EtOH	276(3.89),282(3.93)	94-0773-64
Indazole, 3-benzyl-	EtOH	290(3.68)	32-0814-65
1H-Indeno[6,7,1-def]isoquinoline, 2-amino-2,3-dihydro-	EtOH	268(3.52),279(3.52), 323(4.00),333(3.91), 348(3.74)	44-2824-64
Indole, 1-methyl-2-(2-pyridyl)-	EtOH	204(4.55),219(4.53), 314(4.43)	44-3584-64
Indole, 3-methyl-2-(2-pyridyl)-	EtOH	207(4.31),232(4.20), 326(4.31)	44-3584-64
Phenanthridine, 6-amino-1-methyl-	pH 2.00	240(4.58),290s(3.82), 334(3.89),350(3.88)	39-4426-65
	pH 10.0	238(4.59),250s(4.40), 334(3.71),350(3.68)	39-4426-65
	n.s.g.	240(4.61),266(4.30), 286(3.81),320(3.78), 338(3.74),354(3.70)	39-3032-65
Phenanthridine, 6-amino-10-methyl-	pH 2.00	242(4.64),290s(3.75), 336(3.90),350(3.90)	39-4426-65
	pH 10.0	240(4.65),284s(3.80), 304(3.80),334(3.75), 352(3.74)	39-4426-65
	n.s.g.	240(4.64),286(3.75), 310(3.77),320(3.80), 338(3.77),354(3.75)	39-3032-65
1H-Pyrrolo[2,3-b]pyridine, 4-methyl-1-phenyl-	n.s.g.	280(4.2),320(4.1)	65-0495-64
Pyrrolo[1,2-a]quinoxaline, 2,4-di-2,4-dimethyl-	pH 2.0	235s(4.44),242(4.48), 248s(4.43),263s(3.88), 356(4.14)	39-3678-65
	pH 8.0	228(4.43),245s(4.43), 248(4.44),259s(4.03), 336(3.98)	39-3678-65

Compound	Solvent	λ_{max}(log ε)	Ref.
Pyrrolo[1,2-a]quinoxaline, 2,4-dimethyl- (contd.)	EtOH	228(4.45),244s(4.43), 250(4.46),262s(4.07), 332s(3.98),340(4.01), 358s(3.79)	39-3678-65
$C_{14}H_{12}N_2O$			
Benzimidazole, 2-methyl-1-phenyl-, 3-oxide	EtOH	283(3.90)	44-1279-65
2-Benzimidazolemethanol, 1-phenyl-	EtOH	217(4.13),248(3.82), 275(3.93),283(3.93)	50-0378-64A
Imidazo[1,2-a]pyridine, 5-(benzyloxy)-	pH 1	250(3.63),295(4.12)	4-0053-65
	pH 11	283s(4.00),296(4.03), 302(3.89)	4-0053-65
Imidazo[1,5-a]pyridin-3(2H)-one, 1-benzyl-	EtOH	270(3.90),380(3.30)	95-0812-65
Indazole, 3-benzyloxy-	EtOH	217(4.48),252(3.27), 300(3.72)	33-1986-64
	EtOH-NaOH	212(4.62),251(3.33), 300(3.73)	33-1986-64
Indazolinone, 1-benzyl-	EtOH	222(4.36),313(3.69)	33-1986-64
	EtOH-NaOH	216(4.59),240(4.37), 333(3.62)	33-1986-64
Indazolinone, 2-benzyl-	EtOH	218(4.40),240(4.15), 260(3.40),313(3.69)	33-1986-64
3-Indazolinone, 2-methyl-1-phenyl-	EtOH	312(3.75)	7-0583-65
2-Indolinone, 3-(2-pyridylmethyl)-	EtOH	252(4.01),260s(3.94), 284s(3.14)	87-0626-65
2-Indolinone, 3-(3-pyridylmethyl)-	EtOH	252(3.98),261s(3.89), 279s(3.15)	87-0626-65
2-Indolinone, 3-(4-pyridylmethyl)-	EtOH	251(3.97),263(3.80), 279(3.12)	87-0626-65
Nicotinonitrile, 1,2-dihydro-4,6-dimethyl-2-oxo-5-phenyl-	EtOH	242(4.01),337(4.01)	44-3593-65
Nicotinonitrile, 1,2-dihydro-2-oxo-6-phenethyl-	EtOH	237(3.89),337(4.12)	44-3593-65
2-Phenazinol, 1,3-dimethyl-	50% MeOH-HCl	270(4.01),399(3.57), 474(2.84)	59-1665-64
	MeOH	263(4.11),372(3.31), 408(2.93)	59-1665-64
	50% MeOH-NaOH	288(3.99),383(3.25), 505(3.05)	59-1665-64
2-Phenazinol, 1,4-dimethyl-	50% MeOH-HCl	275(4.75),391(4.22), 496(3.70)	59-1665-64
	MeOH	267(4.83),362(3.91), 425(3.63)	59-1665-64
	50% MeOH-NaOH	289(4.78),376(3.84), 508(3.74)	59-1665-64
2-Phenazinol, 3,4-dimethyl-	50% MeOH-HCl	268(4.00),397(3.56), 469(2.82)	59-1665-64
	MeOH	263(4.05),374(3.21), 405s(2.98)	59-1665-64
	50% MeOH-NaOH	281(3.90),382(3.18), 485(3.05)	59-1665-64
$C_{14}H_{12}N_2O_2$			
1-Acenaphthenecarboxamide, 8a-cyano-6,7,8,8a-tetrahydro-2-oxo-	EtOH	256(4.00),296(3.43)	44-2610-65
1H-Benz[f]indazole-4,9-dione, 3-methyl-, O-ethyl derivative	EtOH	248(4.5),266(4.2), 275(4.2),318(3.8)	23-2665-64

Compound	Solvent	λ_{max}(log ϵ)	Ref.
Glyoxal bis(o-hydroxyanil), nickel complex	MeOH	350(4.0),590(4.3)	89-0076-64
Glyoxal bis(o-hydroxyanil), zinc complex	MeOH	290(4.0),350(3.8), 560(3.8)	89-0076-64
Nicotinonitrile, 1,2-dihydro-6-(β-hydroxyphenethyl)-2-oxo-	EtOH	236(3.92),338(4.13)	35-5198-65
Pyridine, 4-acetamido-3-benzoyl-	MeOH	229(4.33),257(4.20)	87-0722-65
Pyridine, 3-methyl-6-(p-nitrostyryl)-	EtOH	330(4.17)	65-4106-64
4-Pyridinepropionanilide, β-oxo-	EtOH	240(4.16),326(4.26)	65-2400-64
1H-Pyrrolo[1,2-a]indole-2-carbonitrile, 2,3-dihydro-7-methoxy-6-methyl-1-oxo-	pH 1	215(4.24),343(4.34)	44-2897-65
	pH 13	221(4.57),345(4.31)	44-2897-65
	MeOH	218(4.52),336(4.34)	44-2897-65
1H-Pyrrolo[1,2-a]indole-9-carbonitrile, 2,3-dihydro-7-methoxy-6-methyl-1-oxo-	MeOH	218(4.57),252s(3.87), 339(4.28)	44-2904-65
2(1H)-Quinolylideneacetic acid, α-cyano-, ethyl ester	EtOH	287(4.43),382s(4.16), 398(4.25),414(4.11)	44-0243-65
	$CHCl_3$	290(4.40),386s(4.15), 403(4.25),426(4.09)	44-0243-65
4-Stilbenamine, 4'-nitro-	EtOH	287(4.10),410(4.45)	65-2073-65
	ether	285(4.21),410(4.53)	65-2073-65
	$C_2H_4Cl_2$	280(4.13),390(4.33)	65-2073-65
$C_{14}H_{12}N_2O_2S$			
1H-2,1,3-Benzothiadiazine, 1-methyl-4-phenyl-, 2,2-dioxide	EtOH	228(4.34),276(4.1), 365(3.6)	44-3960-65
$C_{14}H_{12}N_2O_3S$			
Benzenesulfonamide, p-(1-oxo-2-isoindolinyl)-	MeOH-acid	284(4.23)	95-1035-65
	MeOH	284(4.22)	95-1035-65
$C_{14}H_{12}N_2O_4$			
Benzohydroxamic acid, p-nitro-, benzyl ester	MeOH	263(4.08)	73-0940-65
	50% MeOH-HCl	266(4.11)	73-0940-65
	50% MeOH-borate	264(4.07)	73-0940-65
	50% MeOH-NaOH	240s(3.97),346(3.77)	73-0940-65
2-Pyridinemethanol, α-(p-nitrophenyl)-, acetate	MeCN	266(4.13)	35-4917-64
$C_{14}H_{12}N_2O_4S$			
Sulfide, 2,4-dinitrophenyl o-ethylphenyl	MeOH	275(4.05)	56-0681-65
$C_{14}H_{12}N_2O_5$			
2-Pyridineacetic acid, α-(5-nitrofurfurylidene)-, ethyl ester	EtOH	242(3.99),360(4.05)	95-0565-65
isomer	EtOH	240(4.01),360(4.22)	95-0565-65
$C_{14}H_{12}N_2O_6S$			
Sulfone, 2,4-dinitrophenyl o-ethylphenyl	MeOH	273(3.89)	56-0681-65
$C_{14}H_{12}N_2S$			
5H-Dibenzo[d,f][1,3]diazepine, 6-(methylthio)-	EtOH	250(4.63),285(3.66)	87-0310-64
$C_{14}H_{12}N_2S_2$			
2,2'-Bibenzothiazoline	MeOH	255s(3.81),315(3.93)	35-3056-64

Compound	Solvent	λ_{max}(log ε)	Ref.
2,2'-Bibenzothiazoline, cadmium chelate	pyridine	376(4.10),575(3.64)	35-3056-64
2,2'-Bibenzothiazoline, zinc chelate	pyridine	380(4.12),590(3.56)	35-5056-64
$C_{14}H_{12}N_4$			
1,2,3-Benzotriazine, 3-benzyl-3,4-dihydro-4-imino-	EtOH	261(4.03),269(3.10),307(3.72),318(3.73)	39-3663-64
1,2,3-Benzotriazine, 3,4-dihydro-4-(methylimino)-3-phenyl-	EtOH	262(4.15),270(4.14),310(3.83)	39-3663-64
$C_{14}H_{12}N_4O_2$			
Phenol, 2,2'-(4-amino-4H-1,2,4-triazole-3,5-diyl)di-	EtOH	257(4.3),302(4.2)	28-1262-64A
$C_{14}H_{12}N_4O_2S$			
2,1,3-Benzothiadiazol-4-ol, 5-[(o-methoxyphenyl)azo]-7-methyl-	EtOH	527(4.27)	17-0041-64
2,1,3-Benzothiadiazol-4-ol, 5-[(p-methoxyphenyl)azo]-7-methyl-	EtOH	531(4.22)	17-0041-64
$C_{14}H_{12}N_4O_3S$			
2,1,3-Benzothiadiazol-4-ol, 7-methoxy-5-[(o-methoxyphenyl)azo]-	EtOH	552(4.18)	17-0041-64
2,1,3-Benzothiadiazol-4-ol, 7-methoxy-5-[(p-methoxyphenyl)azo]-	EtOH	552(4.11)	17-0041-64
Sulfanilamide, N^1-(3-phenyl-1,2,4-oxadiazol-5-yl)-	EtOH	269(4.27)	95-1061-64
Sulfanilamide, N^1-(5-phenyl-1,2,4-oxadiazol-3-yl)-	EtOH	266(4.45)	95-1061-64
$C_{14}H_{12}N_4O_4$			
Acetophenone, 2,4-dinitrophenylhydrazone	EtOH	378(4.42)	24-0815-64
p-Tolualdehyde, 2,4-dinitrophenylhydrazone	$CHCl_3$	386(4.48)	39-5963-64
$C_{14}H_{12}N_4O_5S$			
Anisole, 2-[(2,4-dinitrophenyl)azo]-5-(methylthio)-	EtOH	242s(4.14),278(3.98),440(4.29)	32-1137-64
Anisole, 4-[(2,4-dinitrophenyl)azo]-5-(methylthio)-	EtOH	237s(4.26),275(4.08),376(4.17),460(4.07)	32-1137-64
$C_{14}H_{12}N_6O_4S_2$			
1,4-Pentadien-3-one, 1,5-bis(5-nitro-2-thienyl)-, amidinohydrazone, hydrochloride	propylene glycol	387(4.58)	95-0001-64
$C_{14}H_{12}N_6O_6$			
1,4-Pentadien-3-one, 1,5-bis(5-nitro-2-furyl)-, amidinohydrazone, hydrochloride	propylene glycol	384(4.568)	95-0001-64
1,2,4-Triazine, 3-amino-4,5-dihydro-5-(5-nitrofurfuryl)-6-[2-(5-nitro-2-furyl)vinyl]-, α-form, hydrochloride	propylene glycol	234(4.24),317(4.37),380(4.41)	95-0001-64
β-form, hydrochloride	propylene glycol	235(4.23),317(4.35),380(4.39)	95-0001-64
$C_{14}H_{12}N_8O_2$			
6-Pteridinol, 7,8-dihydro-7-(7-hydroxypteridin-6-ylmethyl)-7-methyl-	pH 2.0	294(4.30)	39-3357-64
	pH 5.63	298(4.23)	39-3357-64

Compound	Solvent	$\lambda_{max}(\log \epsilon)$	Ref.
6-Pteridinol, 7,8-dihydro-7-(7-hydroxy-pteridin-6-ylmethyl)-7-methyl- (contd.)	pH 9.1	305(4.11)	39-3357-64
	pH 13.0	313(4.25)	39-3357-64
$C_{14}H_{12}O$			
Benzophenone, o-methyl-	MeOH	250(4.28),292s(3.40)	39-2579-64
2-Biphenylcarboxaldehyde, 4-methyl-	EtOH	218(4.23),235(4.36), 256(4.11),309(3.49)	22-2953-64
3-Biphenylcarboxaldehyde, 4-methyl-	EtOH	243(4.47),312(3.17)	22-2953-64
Dibenz[c,e]oxepin, 5,7-dihydro-	isooctane	250(4.23)	35-1710-64
Ether, 4-biphenyl vinyl	nonane	256(4.31)	67-0375-65
Fluoren-2-ol, 1-methyl-	EtOH	275(4.31)	42-0479-64
Naphtho[2,3-b]furan, 3,5-dimethyl-	EtOH	240(4.79),244(4.78), 303(3.84),315(4.02), 330(3.92),336(3.75)	78-2655-64
Ujacazulene	heptane	226(4.17),285(4.63), 291(4.61),296(4.61), 304(4.31),318(4.25), 349(3.75),358(3.84), 363(3.77),366(3.79), 376(3.70),534(2.71), 545(2.69),575(2.64), 600(2.45),632(2.34)	39-3577-64
$C_{14}H_{12}OS$			
Benzophenone, 2-(methylthio)-	EtOH	245(4.11),340(3.08)	44-1859-65
Thiobenzophenone, 4-methoxy-	C_6H_{12}	240(4.02),320(4.03), 354(4.18),598(2.31)	54-0289-65
	EtOH	240(4.04),322(4.04), 362(4.20),587(2.34)	54-0289-65
$C_{14}H_{12}O_2$			
Benzophenone, 2-hydroxy-4-methyl-	EtOH	272(4.17),336(3.72)	28-5614-64A
Benzophenone, 4-hydroxy-2-methyl-	EtOH	293(4.00)	28-5614-64A
Benzophenone, 2-methoxy-	MeOH	248(4.26),293s(3.54)	39-2579-64
Benzophenone, 4-methoxy-	MeOH	250(3.96),290(4.21)	39-2579-64
Ether, p-phenoxyphenyl vinyl	nonane	271(3.39),278(3.38)	67-0375-65
Mycomycin, methyl ester, trans-trans	ether	254(4.49),266(4.70), 280(4.70)	24-1846-64
2,5-Norbornadiene-2-carboxylic acid, 3-phenyl-	EtOH	284(3.94)	35-5548-64
2-Oxaphenanthrene, 2,3,5,6-tetra-hydro-3-oxo-1-methyl-	EtOH	250(3.34),318(3.61)	44-2642-65
9,10-Phenanthrenediol, 9,10-dihydro-, cis	EtOH	208(4.60),225s(4.00), 270(4.22),282s(4.04)	39-5544-64
trans	EtOH	211(4.60),225s(4.00), 269(4.18),298s(3.42)	39-5544-64
4H-Pyran-4-one, 2-methyl-6-styryl-	EtOH	226(4.21),320(4.47)	65-2766-64
4,4'-Stilbenediol	EtOH	228(4.04),325(4.50)	65-2073-65
	EtOH-NaOEt	324(--),355(4.55)	65-2073-65
	ether	227(4.07),325(4.50)	65-2073-65
2,4,6-Tridecatriene-8,10,12-triynoic acid, methyl ester, 2-cis	ether	281(4.46),294(4.76), 327(4.60),346(4.85), 365(4.85)	24-1846-64
trans-trans	ether	280(4.47),293(4.80), 326(4.58),345(4.82), 365(4.83)	24-1846-64
$C_{14}H_{12}O_2S$			
Sulfone, 1-benzylidene-2,4-hexadiynyl-methyl, cis	ether	251(4.04),321(4.29)	24-3015-65

Compound	Solvent	λ_{max}(log ϵ)	Ref.
Sulfone, 1-benzylidene-2,4-hexadiynyl methyl, trans	ether	249(3.85),314(4.37)	24-3015-65
$C_{14}H_{12}O_2S_2$			
2,4-Pentanedione, 3-(4-phenyl-1,3-dithiol-2-ylidene)-	CF_3COOH	275(4.23),388(4.36)	44-1711-64
	THF	243(4.20),390(4.42)	44-1711-64
Thianthrene, 2,7-dimethyl-, 5,10-dioxide	EtOH	225(4.8)	35-2957-64
isomer	EtOH	215(4.7)	35-2957-64
$C_{14}H_{12}O_2Se$			
1,3-Propanedione, 1-selenophene-2-yl-3-p-tolyl-	n.s.g.	277(3.95),366(4.45)	67-0379-65
copper salt	n.s.g.	292(4.45),370(4.67)	67-0379-65
$C_{14}H_{12}O_3$			
1(2H)-Anthracenone, 3,4-dihydro-8,9-dihydroxy-	EtOH	267(4.67),295(3.65), 304(3.55),410(3.84)	88-2355-64
1H-Benz[e]indene-6,9-dione, 2,3-dihydro-7-methoxy-	EtOH	249(4.10),256(4.11), 283(4.18),350(3.62)	39-3811-65
5-Benzofuranacrylic acid, α-ethyl-6-hydroxy-β-methyl-, δ-lactone	EtOH	245(4.42),290(4.06), 326(3.99),329(3.99)	44-2467-64
Cyclohexanone, 2-phthalyl-	hexane	248(3.92),288(4.14), 318(4.06)	1-2139-65
	EtOH	250(3.89)	1-2139-65
1-Cyclopentene-1-carboxylic acid, 2-(2,3-dihydro-6-hydroxy-5-benzofuranyl)-, δ-lactone	EtOH	227(4.16),255(3.60), 295(3.82),332(4.30)	44-2467-64
Resorcinol, 4-methyl-, 3-benzoate	EtOH	225(4.29),276(3.60)	24-1926-64
	NaOH	230(4.34),285s(3.49), 295(3.56)	24-1926-64
Seselin	EtOH	218(4.42),284(4.04), 294(4.05),330(4.07)	78-2605-64
$C_{14}H_{12}O_3Se$			
1,3-Propanedione, 1-(p-methoxyphenyl)-3-(selenophene-2-yl)-	n.s.g.	288(4.10),370(4.53)	67-0379-65
copper salt	n.s.g.	305(4.51),375(4.79)	67-0379-65
$C_{14}H_{12}O_4$			
Benzophenone, 2,4,4'-trihydroxy-2'-methyl-	EtOH	285(4.39),325(4.30)	39-6960-65
2'-Grisen-3,4'-dione, 2'-methoxy-	MeOH	254(4.47),328(3.72)	94-0214-64
3'-Grisen-2',3-dione, 4'-methoxy-	MeOH	250(4.45),326(3.72)	94-0214-64
Itaconitin, anhydro-	EtOH	257(3.95),300(4.1), 353(4.22),396(4.23)	94-0058-65
Jatamansinone	EtOH	218(4.15),256(3.5), 321(4.12)	78-2605-64
1,4-Naphthoquinone, 6-acetyl-5-hydroxy-2,7-dimethyl-	EtOH	260(4.09),418(3.67)	54-1005-64
1,4-Naphthoquinone, 3-acetyl-5-methoxy-2-methyl-	EtOH	212(4.57),252(4.23), 405(3.66)	12-0218-65
1,4-Naphthoquinone, 6-acetyl-5-methoxy-7-methyl-	EtOH	215(4.57),240s(4.48), 356(3.47)	12-0218-65
Oroselol	EtOH	218(4.18),250(4.40), 300(4.02)	65-4171-64
	EtOH	252(4.40),301(4.02)	78-2605-64

Compound	Solvent	$\lambda_{max}(\log \epsilon)$	Ref.
$C_{14}H_{12}O_5$			
2'-Acetonaphthone, 1',3',4'-tri-hydroxy-, 4'-acetate	EtOH	223(4.38),265(4.21), 275s(4.19),292(4.19), 325s(3.53)	94-0316-64
p-Benzoquinone, 2-(4,6-dihydroxy-o-tolyl)-5-hydroxy-3-methyl-	MeOH	272(4.26),371(3.08)	24-2774-65
	MeOH-KOH	222(4.16),283(4.11), 468(3.40)	24-2774-65
1-Cyclopentene-1-carboxylic acid, 2-(2-hydroxy-4,6-dimethoxyphenyl)-3-oxo-, δ-lactone	EtOH	227s(4.34),238s(4.26), 257(4.11),355(4.57)	25-1865-64
Cyclopentenone[2,3-c]coumarin, 5,7-dimethoxy-	EtOH	215(4.35),237s(4.17), 257(3.98),345s(4.41), 355(4.43)	35-0882-65
Cyclopentenone[3,2-c]coumarin, 5,7-dimethoxy-	EtOH	245(4.12),268(3.94), 356(3.95)	35-0882-65
5H-Furo[3,2-g][1]benzopyran-5-one, 4,9-dimethoxy-7-methyl-	EtOH	220s(4.21),248(4.54), 260s(4.30),282(3.66), 332(3.67)	94-0316-64
2-Indanacrylic acid, 2,3-dihydroxy-4-methoxy-α-methyl-1-oxo-, γ-lactone	EtOH	230(4.81),260s(4.28), 315(3.76)	35-4507-64
4-Indanhydracrylic acid, 5-carboxy-3-oxo-, γ-lactone, methyl ester	EtOH	240(4.11),287(3.14), 296(3.20)	39-3811-65
$C_{14}H_{12}O_5S$			
p-Benzoquinone, 2-acetyl-3-(3,4-di-methoxy-2-thienyl)-	EtOH	227(4.06),270(4.10), 500(3.65)	33-0769-64
$C_{14}H_{12}O_7$			
1,2-Naphthoquinone, 3-acetyl-4,8-di-hydroxy-5,7-dimethoxy-	MeOH	270s(3.76),305(3.86), 490(3.69)	35-2959-64
	MeOH	294(3.89),320s(3.82), 558(3.98),588s(3.77)	35-2959-64
1,4-Naphthoquinone, 6-acetyl-5,8-di-hydroxy-2,7-dimethoxy-	MeOH	263(4.07),302(4.18), 450(3.43)	35-2959-64
	MeOH-KOH	260(4.28),302(4.34), 377(3.70),460(3.58)	35-2959-64
Purpurogallin, 9-carbethoxy-	$CHCl_3$	290(4.3),310(4.4), 360(3.6),380s(3.6), 440(3.4)	24-0307-64
$C_{14}H_{12}O_8$			
Phthalide, 4,5,6-triacetoxy-	EtOH	236(4.02),277s(3.36), 283(3.40)	23-1595-64
$C_{14}H_{12}S$			
Dibenzo[c,e]thiepin, 5,7-dihydro-	isooctane	244(4.01)	35-1710-64
Sulfide, 1-benzylidene-2,4-hex-adiynyl methyl, trans	ether	326(4.43),343(4.36)	24-3015-65
Thiobenzophenone, 2-methyl-	C_6H_{12}	228(4.09),317(4.18), 612(2.06)	54-0289-65
	EtOH	227(4.03),320(4.16), 600(2.06)	54-0289-65
Thiobenzophenone, 4-methyl-	C_6H_{12}	236(3.98),319(4.16), 605(2.20)	54-0289-65
	EtOH	235(3.99),322(4.16), 592(2.20)	54-0289-65
$C_{14}H_{12}S_2$			
Thianthrene, 2,7-dimethyl-	96% H_2SO_4- 20 min.	273(4.41),295(4.64)	44-0021-64

Compound	Solvent	λ_{max}(log ε)	Ref.
Thianthrene, 2,7-dimethyl- (contd.)	96% H_2SO_4-30 min.	580(4.09)	44-0021-64
	4 days	273(4.41),295(4.64), 580(4.09)	44-0021-64
$C_{14}H_{13}BrN_2$			
1-Benzylbenzimidazolium bromide	MeOH	248(3.97),252(3.96), 264s(3.75),274(3.75), 281(3.72)	65-0632-64
1-Methyl-2-phenyl-1H-imidazo[1,2-a]-pyridinium bromide	EtOH	205(4.39),238(4.15), 328(3.66)	4-0331-65
2-Methyl-1-phenyl-1H-imidazo[1,2-a]-pyridinium bromide	EtOH	204(4.47),287(4.06)	4-0331-65
$C_{14}H_{13}BrN_2O_2$			
3H-3-Benzazepine, 2-acetamido-3-acetyl-4-bromo-	EtOH	250(4.59),280(4.03)	4-0026-65
$C_{14}H_{13}BrN_2O_2S$			
5-Pyrimidinecarboxylic acid, 2-(p-bromophenyl)-1,4-dihydro-6-methyl-4-thioxo-, ethyl ester	EtOH	267(4.20),308(4.14), 356(3.68)	44-1115-64
$C_{14}H_{13}BrN_4O_3$			
5-(6-Bromo-3-ethyl-2-methylbenzimidazolium)barbiturate	EtOH	209(4.61),247(4.37), 279(4.00),287(3.96)	12-0877-64
Pyrimidine, 5-bromo-2-(methylamino)-4-[3-methyl-4-(5-nitro-2-furyl)-butadienyl]-	propylene glycol	230(4.27),260(4.11), 293(4.08),410(4.28)	95-0131-64
$C_{14}H_{13}BrO$			
Anisole, p-(α-bromo-p-tolyl)-	hexane	276.7(4.37)	78-0861-64
$C_{14}H_{13}BrO_2$			
2-Naphthoic acid, 7-bromo-1-methyl-, ethyl ester	MeOH	235(4.69),275(3.76), 285(3.77),295(3.57)	78-0861-64
$C_{14}H_{13}Br_2NO$			
2,5-Cyclohexadien-1-one, 2,6-dibromo-4-methyl-4-m-toluidino-	MeOH	242(4.11),258(4.08), 290s(--)	35-1127-64
2,5-Cyclohexadien-1-one, 2,6-dibromo-4-methyl-4-o-toluidino-	MeOH	239(4.11),257(4.08)	35-1127-64
$C_{14}H_{13}ClN_2$			
Pyrrolo[1,2-a]quinoxaline, 1-chloro-2-ethyl-4-methyl-	pH 1	234s(4.50),240(4.52), 267(3.94),364(4.17)	39-3678-65
	pH 7	216(4.26),239(4.47), 261s(4.05),337(4.00)	39-3678-65
	EtOH	216(4.28),233(4.50), 239s(4.48),260s(4.11), 329s(4.02),337(4.04), 352s(3.89)	39-3678-65
$C_{14}H_{13}ClN_2O$			
1H-Azepino[5,4,3-cd]indole, 5-acetyl-9-chloro-3,4,5,6-tetrahydro-6-methylene-	EtOH	224(4.35),242(4.22), 321(3.96)	87-0200-65
2(3H)-Phenazinone, 1-chloro-4,10-dihydro-4,4-dimethyl-	MeOH	227(4.43),313(4.13), 408(4.10)	20-0081-64

Compound	Solvent	λ_{max}(log ε)	Ref.
$C_{14}H_{13}ClN_2O_5$			
D-Cycloserine, diacetyl-N-(5-chloro-salicylidene)-	MeOH	200(4.26),252(4.18), 297(3.39)	44-3436-65
$C_{14}H_{13}ClN_4S$			
9H-Purine, 6-(benzylthio)-9-(2-chloroethyl)-	EtOH	285(4.31)	87-0182-65
$C_{14}H_{13}ClO$			
Anisole, p-(α-chloro-p-tolyl)-	hexane	271.2(4.36)	78-0861-64
$C_{14}H_{13}ClO_2$			
2,10(1H,3H)-Anthracenedione, 3-chloro-4,4aβ,9,9aα-tetrahydro-	EtOH	244(3.11)	70-0310-64
2-Norbornene-2-carboxylic acid, 3-(o-chlorophenyl)-	EtOH	244s(3.88)	35-5548-64
2-Norbornene-2-carboxylic acid, 3-(p-chlorophenyl)-	EtOH	281(4.09)	35-5548-64
$C_{14}H_{13}ClO_4$			
7-m-Tolyltropylium perchlorate	MeCN	271(4.09),376(4.16)	24-0029-64
7-p-Tolyltropylium perchlorate	MeCN	273(4.08),393(4.27)	24-0029-64
$C_{14}H_{13}ClO_4S$			
7-[p-(Methylthio)phenyl]tropylium perchlorate	MeCN	275s(4.11),467(4.33)	24-0029-64
$C_{14}H_{13}ClO_5$			
7-(m-Methoxyphenyl)tropylium perchlorate	MeCN	264(4.14),356(4.06)	24-0029-64
7-(p-Methoxyphenyl)tropylium perchlorate	MeCN	275(3.96),435(4.35)	24-0029-64
1,4-Naphthoquinone, 2-chloro-5,6,8-trimethoxy-3-methyl-	EtOH	220(4.5),275(4.0), 430(3.6)	33-0538-65
$C_{14}H_{13}Cl_2NO$			
2,5-Cyclohexadien-1-one, 2,6-dichloro-4-methyl-4-o-toluidino-	EtOH	243(4.14),275s(--)	35-1127-64
$C_{14}H_{13}Cl_2N_3O_5S_2$			
2H-1,2,4-Benzothiadiazine-7-sulfonamide, 6-chloro-3-(p-chloro-α-hydroxybenzyl)-3,4-dihydro-, 1,1-dioxide	EtOH	228(4.60),272(4.27)	95-0095-65
$C_{14}H_{13}Cl_3N_4O_5$			
9H-Purine, 2,6,8-trichloro-9-(3,5-di-O-acetyl-2-deoxy-β-D-ribofuranosyl)-	pH 1	246(3.9),278(4.10)	35-4934-65
	pH 11	278(4.18)	35-4934-65
$C_{14}H_{13}Cl_4NO_2$			
3(2H)-Furanone, 2,2-dichloro-4-(3,4-dichlorophenyl)-5-(diethylamino)-	EtOH	304(4.25)	44-4303-65
$C_{14}H_{13}F_2N_3$			
Aniline, 2,6-difluoro-N,N-dimethyl-4-(phenylazo)-, cis	EtOH-HCl	230(3.94),323(4.37), 445(2.83)	44-3878-65
	EtOH	239(4.02),292(3.76), 381(4.00)	44-3878-65
trans	EtOH-HCl	231(3.95),325(4.38), 438(3.00)	44-3878-65

Compound	Solvent	λ_{max}(log ε)	Ref.
Aniline, 2,6-difluoro-N,N-dimethyl-4-(phenylazo)-, trans (contd.)	EtOH	235(4.12),297(4.00), 387(4.27)	44-3878-65
$C_{14}H_{13}IN_2$			
2-Indol-3-yl-1-methylpyridinium iodide	MeOH	216(4.46),268(3.85), 365(3.82)	44-2534-65
$C_{14}H_{13}IO$			
Anisole, p-(α-iodo-p-tolyl)-	hexane	291(4.37)	78-0861-64
$C_{14}H_{13}N$			
Aniline, N-(p-methylbenzylidene)-	isooctane	267(4.29),310(3.94)	9-0091-65
	EtOH	220(4.20),268(4.26), 310(4.04)	9-0091-65
2-Picoline, 6-styryl-, cis	hexane	292(4.03)	32-1322-65
3-Picoline, 2-styryl-, cis	hexane	257(3.99)	32-1322-65
3-Picoline, 4-styryl-, cis	hexane	265(4.06)	32-1322-65
3-Picoline, 6-styryl-	EtOH	308(4.15)	65-4106-64
cis	hexane	292(4.03)	32-1322-65
4-Picoline, 2-styryl-, cis	hexane	285(4.02)	32-1322-65
Pyridine, 2-(p-methylstyryl)-, cis	hexane	222s(4.18),292(4.10)	32-1322-65
Pyridine, 3-(p-methylstyryl)-, cis	hexane	225s(4.24),273(4.02)	32-1322-65
trans	hexane	228s(4.05),310(4.39)	32-1322-65
Pyridine, 4-(m-methylstyryl)-, cis	hexane	283(4.05)	32-1322-65
Pyridine, 4-(o-methylstyryl)-, cis	hexane	271(3.97)	32-1322-65
Pyridine, 4-(p-methylstyryl)-, cis	hexane	222s(4.19),295(4.12)	32-1322-65
Pyridine, 2-(2-phenylcyclopropyl)-	n.s.g.	201(4.26),209s(4.15), 229(4.15),270(3.83)	87-0716-64
p-Toluidine, N-benzylidene-	isooctane	262(4.20),320(3.92)	9-0091-65
	EtOH	264(4.19),318(4.00)	9-0091-65
Vinylamine, 2,2-diphenyl-	EtOH	283(4.18)	35-0863-65
$C_{14}H_{13}NO$			
Acetanilide, phenyl-	EtOH	245(4.20)	54-0949-64
9-Acridanmethanol	EtOH	283(4.20)	33-1395-65
Aniline, N-p-methoxybenzylidene-	isooctane	221(4.28),280(4.32), 315(4.13)	9-0091-65
	EtOH	223(4.24),290(4.26), 314(4.24)	9-0091-65
p-Anisidine, N-benzylidene-	isooctane	235(4.16),269(4.13), 335(4.07)	9-0091-65
	EtOH	235(4.09),266(4.10), 332(4.09)	9-0091-65
Benzamide, N-benzyl-	C_6H_{12}	220(4.1)	56-0789-64
	MeOH	227(4.1),267s(3.3)	56-0789-64
	20% C_6H_{12} in MeOH	226(4.1)	56-0789-64
	20% C_6H_{12} in EtOH	225(4.1)	56-0789-64
4-Biphenylcarboxamide, N-methyl-	HOAc	269(4.32)	39-4399-65
	HOAc-H_2SO_4	287(4.32)	39-4399-65
2,4-Hexadiynophenone, 4'-(dimethylamino)-	EtOH	255(4.14),289(3.71), 310(3.29)	39-2983-65
N-Methyl-2-phenacylpyridine, anhydro base	pH 2	263(4.30)	39-3093-65
	pH 10	250(4.02),325(4.02), 410(4.40)	39-3093-65
Phenol, o-(N-benzylformimidoyl)-	EtOH	220(4.4),256(4.13), 316(3.62)	65-3837-64
Pyridine, 3-(p-methoxystyryl)-, cis	hexane	228s(4.19),287(4.06)	32-1322-65
Pyridine, 3-(p-methoxystyryl)-, trans	hexane	229s(4.07),320(4.42)	32-1322-65

Compound	Solvent	$\lambda_{max}(\log \epsilon)$	Ref.
Pyridine, 4-(p-methoxystyryl)-, cis	hexane	229s(4.25),293(4.04)	32-1322-65
m-Toluidine, N-salicylidene-	EtOH	225s(4.3),270(4.12), 340(4.10)	65-3837-64
p-Toluidine, N-salicylidene-	EtOH	220(4.4),270(4.15), 340(4.19)	65-3837-64
Tropone, 2-(benzylamino)-	EtOH	243(4.64),336(4.24), 403(4.25)	94-0457-65
$C_{14}H_{13}NO_2$			
m-Anisidine, N-salicylidene-	EtOH	220(4.4),268(4.10), 340(4.08)	65-3837-64
p-Anisidine, N-salicylidene-	EtOH	220(4.3),270(4.03), 348(4.27)	65-3837-64
Anthracene, 1,2,3,4-tetrahydro-7-nitro-	EtOH	217(4.41),263(4.16), 271(4.16),320(3.89)	39-5377-65
Carbazole-3-methanol, 1-methoxy-	n.s.g.	225(4.42),242(4.58), 252(4.49),258s(4.26), 290(3.94),330(3.55)	78-0681-65
Furo[3,2-c]quinolin-4(5H)-one, 2-isopropyl-	n.s.g.	230(4.56),238s(4.47), 246s(4.36),277(3.39), 287(3.97),318(4.08), 333(4.11)	44-2598-64
Phenol, o-[(p-methoxybenzylidene)-amino]-	EtOH	350(4.13)	59-1625-65
	acid	380(3.95)	59-1625-65
	base	380(3.90)	59-1625-65
Pyridineacetic acid, α-phenyl-, methyl ester	C_6H_{12}	257(3.52)	35-4917-64
Pyrrolido[3,4-b]-3a,4,5,6-tetrahydro-acenaphthen-10-one, 1'-oxo-	EtOH	260(4.05),304(3.40)	44-2610-65
$C_{14}H_{13}NO_3$			
4-Biphenylmethanol, 2-methyl-4'-nitro-	EtOH	300(4.07)	88-2655-65
Furo[3,2-c]quinoline-3,4(2H,5H)-dione, 2-isopropyl-	n.s.g.	235(4.30),248s(4.21), 287(3.81),298(3.81)	44-2598-64
2-Naphthonitrile, 3-acetoxy-1,4-dihydro-8-methoxy-	n.s.g.	202(4.46),275(3.32), 281(3.38)	88-0041-65
1H-Pyrrolo[1,2-a]indole-2-carboxylic acid, 2,3-dihydro-1-oxo-, ethyl ester	MeOH	242(4.41),320(4.32)	87-0700-65
$C_{14}H_{13}NO_3S$			
3-Pyrrolin-2-one, 3-acetyl-4-hydroxy-5-(S-benzylthio)methylene-	EtOH	288(4.27),350(4.07)	35-5654-64
$C_{14}H_{13}NO_4$			
3-Indolinebutyric acid, 1-acetyl-3-hydroxy-2-oxo-, δ-lactone	EtOH	218(4.18),232s(4.06), 275s(2.70)	44-2431-64
2-Norbornene-2-carboxylic acid, 3-(o-nitrophenyl)-	EtOH	307s(3.45)	35-5548-64
2-Norbornene-2-carboxylic acid, 3-(p-nitrophenyl)-	EtOH	314(4.06)	35-5548-64
1H-Pyrrolo[1,2-a]indole-9-carboxalde-hyde, 2,3,5,8-tetrahydro-7-methoxy-6-methyl-5,8-dioxo-	MeOH	216(4.40),243(4.17), 272(4.15),289(4.14), 332(3.85)	35-3877-64
	MeOH	216(4.40),243(4.17), 272(4.15),289(4.14), 332(3.85)	44-2897-65
4H-Quinolizine-1-carboxylic acid, 3-acetyl-4-oxo-, ethyl ester	EtOH	270(4.21),345(3.95), 410(4.23)	78-0945-65

Compound	Solvent	λ_{max}(log ε)	Ref.
4H-Quinolizine-3-carboxylic acid, 1-acetyl-4-oxo-, ethyl ester	EtOH	240(3.67),265(4.00), 280(3.97),345(4.20), 400(4.21)	78-0945-65
$C_{14}H_{13}NO_4S$			
3-Thiophenecarboxylic acid, 4-methyl-2-(p-nitrophenyl)-, ethyl ester	ether	207(4.11),253(4.12), 282(4.20)	24-2109-64
$C_{14}H_{13}NO_5$			
1-[1-(4-Acetoxy-2,5-dihydro-5-oxo-3-furyl)-2-hydroxypropenyl]pyridinium betaine	EtOH	216(3.67),402(4.65)	39-0783-64
2H,6H-Chromeno[7,6-b][1,4]oxazine, 3,4-dihydro-8-methoxycarbonyl-3-methyl-6-oxo-	MeOH	256(4.4),293(3.9), 403(3.9)	39-7348-65
Indene, 3-acetonyl-1-methyl-5,6-methylene-2-nitro- ?	EtOH	255(4.14),352(4.07)	44-1416-65
4H-Quinolizine-1,3-dicarboxylic acid, 4-oxo-, 1-ethyl 3-methyl ester	EtOH	264(4.18),341(3.96), 393(4.21)	78-1051-64
$C_{14}H_{13}N_3$			
Benzotriazole, 5-methyl-2-p-tolyl-	EtOH	313(4.14)	39-0751-64
$C_{14}H_{13}N_3O$			
Acetanilide, p-(phenylazo)-	EtOH-NH_3	349(4.46)	35-2315-64
2H-Benzotriazole, 2-(p-ethoxyphenyl)-	n.s.g.	220(4.10),249(3.97), 251(3.98),257(4.00), 316(4.38)	39-4831-65
$C_{14}H_{13}N_3O_2$			
[2,2':5',2''-Terpyrrole]-5-carboxylic acid, methyl ester	MeOH	358(4.46)	44-0883-64
$C_{14}H_{13}N_3O_3$			
Salicylaldehyde, 5-nitro-, methylphenylhydrazone	MeOH	303(4.35),343(4.47)	24-1631-64
	MeOH-NaOH	363(4.44),438(4.00)	24-1631-64
$C_{14}H_{13}N_3O_3S$			
Pyrimidine, 2-(ethylthio)-4-[4-(5-nitro-2-furyl)-1,3-butadienyl]-	propylene glycol	256(4.30),335(4.28), 404(4.35)	95-0207-64
Pyrimidine, 4-[3-methyl-4-(5-nitro-2-furyl)-1,3-butadienyl]-2-(methylthio)-	propylene glycol	253(4.02),330(4.07)	95-0207-64
$C_{14}H_{13}N_3O_4$			
2,6-Xylidine, N-(2,4-dinitrophenyl)-	C_6H_{12}	256s(4.05),329(4.27)	23-2674-64
$C_{14}H_{13}N_3O_4S$			
Benzaldehyde, m-nitro-, p-toluenesulfonhydrazone	EtOH	265(4.34)	44-1242-65
anion	H_2O	297(4.29)	44-1242-65
	MeOH	315(4.25)	44-1242-65
	EtOH	321(4.25)	44-1242-65
	MeCN	342(4.30)	44-1242-65
	DMF	351(4.18)	44-1242-65
Benzaldehyde, p-nitro-, p-toluenesulfonhydrazone	EtOH	317(4.20)	44-1242-65
anion	benzene	403(4.02)	44-1242-65
	H_2O	373(4.22)	44-1242-65
	MeOH	387(4.24)	44-1242-65

Compound	Solvent	$\lambda_{max}(\log \epsilon)$	Ref.
Benzaldehyde, p-nitro-, p-toluenesulfonhydrazone, anion (contd.)	90% MeOH	386(4.25)	44-1242-65
	EtOH	403(4.23)	44-1242-65
	70% EtOH	395(4.22)	44-1242-65
	iso-PrOH	411(4.20)	44-1242-65
	BuOH	405(4.18)	44-1242-65
	acetone	473(4.18)	44-1242-65
	90% acetone	427(4.25)	44-1242-65
	$HCONH_2$	403(4.25)	44-1242-65
	DMF	475(4.15)	44-1242-65
	MeCN	450(4.02)	44-1242-65
	pyridine	475(4.14)	44-1242-65
	$CHCl_3$	400(3.97)	44-1242-65
	CCl_4	395(3.97)	44-1242-65
$C_{14}H_{13}N_4$			
Verdazyl, 1,5-diphenyl-	benzene	340(4.0),670(3.72)	49-0457-64
	MeOH-HCl	480(3.9)	49-0457-64
leuco form	benzene	290(4.1)	49-0457-64
$C_{14}H_{13}N_5$			
Adenine, 3-cinnamyl-	EtOH-HCl	256(4.34),276(4.35)	23-1599-64
	EtOH	256(4.39),276s(--)	23-1599-64
Guanidine, (1-methylbenzo[f]quinazolin-3-yl)-	pH 1	233(4.48),269(4.69), 307(3.77),335(3.51), 351(3.46)	44-2881-64
	pH 10	233(4.26),280(4.68), 352(3.46)	44-2881-64
	EtOH	233(4.48),269(4.72), 305s(3.83),336(3.50), 352(3.50)	44-2881-64
$C_{14}H_{13}N_5O$			
Adenine, N-acetyl-3-benzyl-	pH 1	275(4.27)	4-0115-64
	pH 7	272(4.13)	4-0115-64
	pH 13	272(4.00)	4-0115-64
$C_{14}H_{13}O_2PS$			
Phenothiaphosphine, 10-hydroxy-2,8-dimethyl-, 10-oxide	EtOH	221(4.43),267(4.16), 287(3.76),310s(3.54)	44-1983-64
$C_{14}H_{13}O_3P$			
Phenoxaphosphine, 10-hydroxy-2,8-dimethyl-, 10-oxide	EtOH	218(4.60),243(4.31), 281s(3.45),296s(3.71), 301(3.74)	44-1983-64
$C_{14}H_{13}O_4PS$			
Phenothiaphosphine, 10-hydroxy-2,8-dimethyl-, 5,5,10-trioxide	EtOH	222(4.63),275(3.41), 284(3.43)	44-1983-64
$C_{14}H_{14}$			
Acenaphthene, 5,6-dimethyl-	EtOH	297(3.89),308(3.73), 316(3.66),323(3.29), 331(3.63)	44-2824-64
7,7'-Bi-2,5-norbornadiene	heptane	177(4.16),220s(3.59), 240(2.59)	44-2259-65
Ethane, 1,1-diphenyl-	C_6H_{12}	254(2.74),256(2.74), 259(2.76),262(2.81), 269(2.69)	22-0587-65
	isooctane	254(2.89),255(2.88), 258(2.89),262(2.89), 269(2.78)	22-0587-65

Compound	Solvent	$\lambda_{max}(\log \epsilon)$	Ref.
$C_{14}H_{14}BrNO$			
2,5-Cyclohexadien-1-one, 4-anilino-2-bromo-4,6-dimethyl-	MeOH	250(4.26)	35-1127-64
1,2-Dihydro-4-hydroxy-3-methylbenzo-[c]quinolizinium bromide	H_2O	201(4.57),246(4.29), 373(3.99)	78-2529-65
$C_{14}H_{14}ClNOS$			
Morpholine, 4-[5-(p-chlorophenyl)-2-thienyl]-	$CHCl_3$	340(4.27)	4-0315-65
$C_{14}H_{14}ClNO_4$			
1,3-Dioxoloane, 2-chloromethyl-2-[4-(p-nitrophenyl)-1,3-butadienyl]-	EtOH	242(4.21),337(4.27)	87-0035-65
$C_{14}H_{14}ClN_3O_2$			
1-Pyrazoline-3-carboxylic acid, 4-(p-chlorophenyl)-3-cyano-4-methyl-, ethyl ester, cis	$CHCl_3$	327.5(1.85)	28-1332-65B
$C_{14}H_{14}ClN_3O_5S_2$			
2H-1,2,4-Benzothiadiazine-7-sulfon-amide, 6-chloro-3,4-dihydro-3-(α-hydroxybenzyl)-, 1,1-dioxide	EtOH	226(4.55),270(4.23)	95-0095-65
$C_{14}H_{14}ClN_5O$			
9H-Purine-9-propanol, 6-p-chloro-anilino-	pH 1	277(4.27)	87-0033-65
	pH 7	291(4.39)	87-0033-65
	pH 13	291(4.39)	87-0033-65
$C_{14}H_{14}ClN_5O_4S$			
Theophylline, 7-(o-chlorobenzyl)-8-sulfamoyl-	MeOH	286(4.03)	54-0193-65
Theophylline, 7-(p-chlorobenzyl)-8-sulfamoyl-	MeOH	287(4.00)	54-0193-65
$C_{14}H_{14}ClN_5S$			
9H-Purine, 2-amino-6-(benzylthio)-9-(2-chloroethyl)-	EtOH	245(4.10),310(4.05)	87-0182-65
$C_{14}H_{14}Cl_2N_4O_5$			
9H-Purine, 2,6-dichloro-9-(3,5-di-O-acetyl-2-deoxy-α-D-ribofuranosyl)-	H_2O	275(4.05)	35-4934-65
$C_{14}H_{14}Cl_4O_2$			
Spirooxetane from photoaddition of cyclooctene to chloranil	n.s.g.	262(3.81)	88-3471-64
$C_{14}H_{14}Cl_6Sn$			
Dicycloheptatrienylium hexachloro-stannate	96% H_2SO_4	274(3.94)	35-5511-64
$C_{14}H_{14}Fe$			
Ferroceno[1,2]cyclohexa-1,3-diene	EtOH	300(3.8),450(2.4)	49-0576-64
$C_{14}H_{14}FeO$			
Ferrocene, 1,1'-(α-oxo- α'-methyl-trimethylene)-	n.s.g.	273(3.46),338(2.66)	44-3941-65
Ferroceno[1,2]cyclohex-1-en-3-one	EtOH	240(4.2),275(3.9), 345(3.0),475(2.7)	49-0576-64

Compound	Solvent	λ_{max}(log ε)	Ref.
$C_{14}H_{14}FeO_2$			
Ferrocenylacrylic acid, methyl ester	$CHCl_3$	296(4.20),471(3.08)	5-0088-64F
$C_{14}H_{14}IN$			
1-Methyl-2-styrylpyridinium iodide, cis	H_2O	325(4.16)	23-1345-65
1-Methyl-2-styrylpyridinium iodide, trans	H_2O	340(4.45)	23-1345-65
$C_{14}H_{14}NO_3$			
Nitroxide, bis(o-methoxyphenyl)-	n.s.g.	265(2.8),320(3.1), 330(3.2)	70-0800-65
$C_{14}H_{14}N_2$			
1H-Indeno[6,7,1-def]isoquinoline, 2-amino-2,3,6,7-tetrahydro-	EtOH	285(3.96),294(4.04), 297(4.05),306(3.91), 312(3.86),326(3.60)	44-2824-64
3-Picoline, 6,6'-vinylenedi-	n.s.g.	271(4.16),319(4.41)	77-0217-64
1H-Pyrrolo[2,3-b]pyridine, 2,3-dihydro-4-methyl-1-phenyl-	n.s.g.	255(4.2),286(3.9)	65-0495-64
Pyrrolo[1,2-a]quinoxaline, 2-ethyl-4-methyl-	pH 2.0	236(4.46),243(4.49), 249s(4.43),263s(3.87), 357(4.13)	39-3678-65
	pH 7.9	229(4.43),245s(4.44), 249(4.46),262s(4.01), 338(4.00)	39-3678-65
4,4'-Stilbenediamine	EtOH	226(4.28),335(4.43)	65-2073-65
	ether	224(4.30),340(4.48)	65-2073-65
	$C_2H_4Cl_2$	340(4.18)	65-2073-65
$C_{14}H_{14}N_2O$			
Acetanilide, 2'-amino-N-phenyl-	EtOH	296(3.52)	44-1279-65
p,p'-Azoxytoluene, cis	EtOH	241(4.17),338(3.61)	35-2419-64
Benzanilide, 4-amino-N-methyl-	EtOH	220(4.2),296(4.3)	28-4295-64B
1H-Benzo[c]quinolizine-3-carbonitrile, 2,3,4,4a,5,6-hexahydro-4-oxo-	EtOH	209(4.19),246(4.02), 293(3.22)	78-2961-65
1H-Benzo[c]quinolizine-4-carbonitrile, 2,3,4,4a,5,6-hexahydro-3-oxo-	EtOH	210(4.33),248(4.11), 294(3.20)	78-2961-65
o-Benzotoluidide, 4-amino-	EtOH	290(4.5)	28-4295-64B
p-Benzotoluidide, 4-amino-	EtOH	297(4.6)	28-4295-64B
Indazole, 3-benzoyl-4,5,6,7-tetrahydro-	EtOH	257(4.23)	32-0814-65
Indole, 1-acetyl-3-(1-pyrrolin-2-yl)-	dioxan	236(4.04),256s(3.75), 268(3.69),276(3.69), 295(3.84),305(3.88)	87-0415-64
Phenetole, p-(phenylazo)-, trans	EtOH	348(4.42),433?[sic]	39-1045-64
1H-Pyrrolo[1,2-a]indole-9-carbonitrile, 2,3-dihydro-7-methoxy-6-methyl-	MeOH	278(3.97),290(3.96), 305(3.82)	44-2904-65
Pyrrolo[1,2-a]quinoxalin-1(5H)-one, 4-methyl-	EtOH	234(4.46),256s(3.87), 291s(4.01),298(4.02), 321s(3.42),417(4.15)	39-3678-65
$C_{14}H_{14}N_2OS$			
Diimide, [(p-methoxyphenyl)thio]-p-tolyl-	n.s.g.	334(4.12)	77-0313-65
$C_{14}H_{14}N_2O_2$			
1,8a(6H)-Acenaphthylenedicarboxamide, 7,8-dihydro-	EtOH	234(3.90),298(4.02)	44-2610-65
Acetanilide, 2'-anilino-2-hydroxy-	EtOH	282(4.12)	44-1279-65
o-Benzanisidide, 4-amino-	EtOH	219(4.2),305(4.5)	28-4295-64B

Compound	Solvent	λ_{max}(log ε)	Ref.
p-Benzanisidide, 4-amino-	EtOH	220s(4.1),303f(4.5)	28-4295-64B
3H-3-Benzazepine, 2-acetamido-3-acetyl-	EtOH	247(4.46),286(3.87)	4-0026-65
Benzylamine, N-methyl-N-(p-nitrophenyl)-	n.s.g.	403(4.38)	5-0109-65I
2-Propen-1-ol, 3-(3-methyl-2-quinoxalinyl)-, acetate	EtOH	253(4.47),331s(3.99), 338(4.01)	39-3678-65
Pyrrolido[3,4-b]-3a,4,5,6-tetrahydro-acenaphthen-10-ol, 1'-oxo-3'-imino-	EtOH	237s(4.26)	44-2610-65
2-Pyrrylmethylenimine, N-(p-carbethoxyphenyl)-	EtOH	232(3.92),342(4.43)	12-0894-64
$C_{14}H_{14}N_2O_2S$			
5-Pyrimidinecarboxylic acid, 1,4-dihydro-6-methyl-2-phenyl-4-thioxo-, ethyl ester	EtOH	240(4.16),308(4.14), 352(3.68)	44-1115-64
p-Toluenesulfonic acid,	EtOH	272(4.28)	39-1020-65
benzylidenehydrazide	EtOH	273(4.24)	44-1242-65
anion	H_2O	303(4.24)	44-1242-65
	MeOH	310.5(4.26)	44-1242-65
	EtOH	313(4.12)	44-1242-65
	MeCN	334(4.15)	44-1242-65
	DMF	345(4.23)	44-1242-65
$C_{14}H_{14}N_2O_3S$			
Hydrazine, N-benzoyl-N'-(p-tolylsulfonyl)-	EtOH	226(4.17)	39-1020-65
3-Pyrrolin-2-one, 3-acetamido-4-hydroxy-5-(benzylthio)methylene-	EtOH	235(4.00),330(4.38)	35-5654-64
3-Pyrrolin-2-one, 3-acetyl-5-[(benzylthio)methylene]-4-hydroxy-, 3-oxime	EtOH	297(4.19),334(4.27)	35-5654-64
p-Toluenesulfonic acid,	EtOH	275(4.15)	39-1020-65
salicylidenehydrazide	EtOH	273(4.15)	44-1242-65
anion	H_2O	322(4.19)	44-1242-65
	MeOH	322(4.23)	44-1242-65
	EtOH	325(4.23)	44-1242-65
	MeCN	337(4.37)	44-1242-65
	DMF	342(4.34)	44-1242-65
$C_{14}H_{14}N_2O_4$			
1-[1-(4-Acetamido-2,5-dihydro-5-oxo-3-furyl)-2-hydroxypropenyl]pyridinium betaine	EtOH	218(3.87),412(4.48)	39-0783-64
$C_{14}H_{14}N_2O_4S$			
N-Methylphenazinium methyl sulfate	H_2O	387.5(4.42)	37-3964-64
p-Toluenesulfonanilide, N-methyl-2'-nitro-	hexane	225(4.12),280(3.15), 310(2.80),327s(2.60)	65-2080-65
	EtOH	228(4.15),280(3.19), 320(2.90),338s(2.60)	65-2080-65
	ether	228(4.23),279(3.28), 306(3.08),315(2.90), 336s(2.55)	65-2080-65
	dioxan	226(4.26),280(3.26), 310(3.00)	65-2080-65
	H_2SO_4	227(4.20),242(4.20), 270(3.90),355s(2.58), 380s(2.28)	65-2080-65
p-Toluenesulfonanilide, N-methyl-3'-nitro-	hexane	224(4.22),307(2.90)	65-2080-65

Compound	Solvent	λ_{max}(log ϵ)	Ref.
p-Toluenesulfonanilide, N-methyl-3'-nitro- (contd.)	EtOH	224(4.20),240(4.22), 304(3.12)	65-2080-65
	ether	224(4.20),240(4.22), 306(3.20),335s(2.48)	65-2080-65
	dioxan	225(4.35),244(4.33), 310(3.07)	65-2080-65
	H_2SO_4	225(4.36),265(4.18), 385s(1.94)	65-2080-65
p-Toluenesulfonanilide, N-methyl-4'-nitro-	ether	224(4.16),298(3.96)	65-2080-65
	dioxan	225(4.40),250(4.08), 296(4.30)	65-2080-65
	H_2SO_4	228(4.28),263(4.07), 355(2.25),382(1.70)	65-2080-65
$C_{14}H_{14}N_2O_5$			
1-Benzimidazoleoxalacetic acid, 2-methyl-, dimethyl ester	EtOH	263(4.34),267s(4.33), 276(4.21)	44-1118-65
$C_{14}H_{14}N_2O_5S_2$			
Cephalosporadesic acid, 7-(2-thienylacetamido)-, as K salt	n.s.g.	235(4.08),260(3.85)	87-0022-65
$C_{14}H_{14}N_4$			
Pyrazolo[1,5-a]pyrimidine, 7-amino-2,3-dimethyl-5-phenyl-	EtOH	208(4.46),262(4.68), 296s(3.72),343(3.66)	95-1113-64
Pyrazolo[1,5-a]pyrimidine, 7-amino-5,6-dimethyl-2-phenyl-	EtOH-HCl	248s(4.30),262(4.48), 294s(3.89),322(3.86)	95-1113-64
$C_{14}H_{14}N_4O$			
Imidazole, 1-benzyl-5-cyano-4-ethoxymethyleneamino-	EtOH-HCl	236(4.01),264(3.86)	4-0291-65
	EtOH	209(4.31),237(4.14)	4-0291-65
	EtOH-NaOH	280.5(4.16)	4-0291-65
5H-Pyrrolo[3,4-d]pyrimidine, 4-amino-6-benzoyl-6,7-dihydro-7-methyl-	pH 1	249(3.99)	44-0194-65
	EtOH	236(4.20),265(3.76)	44-0194-65
2H-1,2,4-Triazole, 3-(methylamino)-5-(2-furyl)-2-benzyl-	pH 2	270(4.22)	1-1191-65
	EtOH	265(4.18)	1-1191-65
$C_{14}H_{14}N_4OS$			
9H-Purine-9-ethanol, 6-(benzylthio)-	pH 1	293(4.30)	87-0182-65
	pH 11	289(4.32)	87-0182-65
$C_{14}H_{14}N_4O_2$			
1,4-Pentadien-3-one, 1,5-di-2-furyl-, amidinohydrazone, hydrochloride	propylene glycol	335(4.568)	95-0001-64
$C_{14}H_{14}N_4O_3$			
5-(3-Ethyl-2-methylbenzimidazolium) barbiturate	H_2O	247(4.34),270(3.98), 276(3.95)	12-0877-64
Mandelamide, N-(4-amino-5-pyrimidinyl)-, acetate	pH 1	251(4.08)	87-0797-65
	pH 11	236(3.93),281(3.87)	87-0797-65
Pyrimidine, 2-amino-5-methyl-4-[3-methyl-4-(5-nitro-2-furyl)-butadienyl]-	propylene glycol	230(3.85),340(4.04), 414(4.10)	95-0121-64
Pyrimidine, 2-(methylamino)-4-[3-methyl-4-(5-nitro-2-furyl)-butadienyl]-	propylene glycol	294(4.17),410(4.31)	95-0131-64
$C_{14}H_{14}N_4O_4$			
Malonamide, 2-[α-(4,6-dihydroxy-5-pyrimidinyl)benzyl]-	pH 4.0	260(3.97),290s(3.43)	12-0567-64
	pH 9.0	259(3.91),283s(3.56)	12-0567-64

Compound	Solvent	λ_{max}(log ϵ)	Ref.
Mandelamide, N-(4-amino-6-hydroxy-5-pyrimidinyl)-, α-acetate	pH 1	258(3.94)	87-0797-65
	pH 11	255(3.80)	87-0797-65
1(3aH)-Pentalenone, 4,5,6,6a-tetrahydro-, 2,4-dinitrophenylhydrazone, cis	$CHCl_3$	381(4.41)	35-0321-65
1-Pyrazoline-3-carboxylic acid, 3-cyano-4-methyl-4-p-nitrophenyl-, ethyl ester, cis	$CHCl_3$	324s(2.65)	28-1332-65B
trans	$CHCl_3$	327s(2.65)	28-1332-65B
$C_{14}H_{14}N_4O_5S$			
Theophylline-8-sulfonic acid, 7-benzyl-	MeOH	281(4.00)	54-0193-65
$C_{14}H_{14}N_6$			
4H-1,2,4-Triazole, 4-amino-3,5-bis(2-aminophenyl)-	EtOH	220(4.6),314(3.9)	28-1262-64A
4H-1,2,4-Triazole, 4-amino-3,5-bis(3-aminophenyl)-	EtOH	241(4.6),306(3.8)	28-1262-64A
4H-1,2,4-Triazole, 4-amino-3,5-bis(4-aminophenyl)-	EtOH	294(4.7)	28-1262-64A
$C_{14}H_{14}N_6O_3$			
Pteroic acid, dihydro-	pH 1	220(4.0),255(3.9), 290(4.0)	37-2259-64
	pH 7	230(4.3),277(4.30), 330s(3.9),390(3.6)	37-2259-64
	pH 13	225(4.2),277(4.33), 320s(4.0)	37-2259-64
$C_{14}H_{14}N_6O_6S$			
Theophylline, 7-(p-nitrobenzyl)-8-sulfamoyl-	MeOH	278(4.22)	54-0193-65
$C_{14}H_{14}O$			
Anisole, m-benzyl-	hexane	274(3.34),281(3.38)	78-0861-64
Anisole, o-benzyl-	hexane	272(3.26),279(3.23)	78-0861-64
Anisole, p-benzyl-	hexane	227(3.79),276(3.04), 280(3.01)	78-0861-64
Anisole, p-(p-tolyl)-	hexane	260.2(4.37)	78-0861-64
Anthrone, 1,4,4a,9a-tetrahydro-	EtOH	248(3.07)	70-0310-64
2,5-Cyclohexadien-1-one, 4-benzyl-4-methyl-	EtOH	235(4.12)	33-0094-65
2-Cyclohexen-1-one, 6-benzylidene-3-methyl-	MeOH	238(3.84),263(3.86), 303(4.13)	5-0079-65J
1,2-Cyclopentenonaphthalene, 6-methoxy-	EtOH	227(4.78),264(3.64), 274(3.70),285(3.48), 323(3.36),337(3.46)	39-3811-65
Fluoren-9-one, 1,4,4a,9a-tetrahydro-8-methyl-	EtOH	250(4.08),298(3.40)	44-0074-64
Fluoren-2(3H)-one, 4,4a-dihydro-1-methyl-	EtOH	252(3.97)	42-0479-64
$C_{14}H_{14}OS$			
3,5-Heptadiyn-2-ol, 2-methyl-7-(phenylthio)-	n.s.g.	236(3.82),247(3.82), 262(3.82)	28-2847-65A
Sulfoxide, m-tolyl p-tolyl	isooctane	234(4.14),271s(3.49), 279s(3.35)	35-1958-65
	EtOH	238(4.22),265s(3.45), 271(3.37),278(3.20)	35-1958-65
S-(+)	EtOH	238(4.23),268s(3.83), 276s(3.78)	35-5637-64

Compound	Solvent	$\lambda_{max}(\log \epsilon)$	Ref.
Sulfoxide, o-tolyl p-tolyl	isooctane	235(4.14),273s(3.40), 280s(3.26)	35-1958-65
	EtOH	239(4.25),265s(3.46), 276s(3.32)	35-1958-65
S-(-)	EtOH	239(4.31),264s(3.78), 276s(3.73)	35-5637-64
$C_{14}H_{14}OSi$			
Phenoxasilin, 10,10-dimethyl-	$CHCl_3$	252(3.21),298(2.92)	25-0794-65
$C_{14}H_{14}OSn$			
Phenoxastannin, 10,10-dimethyl-	$CHCl_3$	250(3.61),282(3.38)	25-0794-65
$C_{14}H_{14}O_2$			
5,6-Acenaphthenedimethanol	EtOH	299(4.07)	44-2824-64
Anthrone, 2,3α-epoxy-1,2,3,4,4aβ,9aα-hexahydro-	EtOH	248(3.28)	70-0310-64
4-Biphenylmethanol, 4'-methoxy-	hexane	263.8(4.39)	78-0861-64
Fluoren-2-ol, 5,8-dihydro-7-methoxy-	EtOH	264(4.11),272s(4.01)	88-3505-65
	MeOH-KOH	284(4.15)	88-3505-65
Fluoren-2(1H)-one, 3,4-dihydro-7-methoxy-	EtOH	266.5(4.18)	88-3505-65
2α(H),12α(H)-Hexacyclo[$6.4.2.0^{2,7}.0^{4,14}.0^{5,13}.0^{9,12}$]tetradecane-3,6-dione	C_6H_{12}	219(2.19),303f(1.77)	39-3062-64
1-Naphthaleneacetic acid, ethyl ester	EtOH	225(4.99),281(3.84)	39-2072-65
2-Naphthaleneacetic acid, ethyl ester	EtOH	224(5.02),275(3.72)	39-2072-65
1-Naphthoic acid, propyl ester	EtOH	220(5.67),295(3.81)	39-2072-65
2-Naphthoic acid, propyl ester	EtOH	237(4.92),280(3.88)	39-2072-65
2-Norbornene-2-carboxylic acid, 3-phenyl-	EtOH	276(4.04)	35-5548-64
4,6-Tetradecadiene-8,10,12-triyne-1,3-diol	ether	287(4.14),304(4.42), 324(4.60),346(4.51)	24-2605-65
Xanthorrhoein	EtOH	225(4.70),244(4.44), 287(3.67),298(3.64), 323(3.40),338(3.49)	88-1623-64
$C_{14}H_{14}O_2S$			
Sulfoxide, p-anisyl p-tolyl, R-(+)	EtOH	246(4.34)	35-5637-64
S-(-)	EtOH	246(4.34)	35-5637-64
3-Thiophenecarboxylic acid, 4-methyl-2-phenyl-, ethyl ester	ether	211(4.20),254(3.96)	24-2109-64
$C_{14}H_{14}O_3$			
2'-Acetonaphthone, 1',8'-dihydroxy-3',6'-dimethyl-	C_6H_{12}	225(4.38),273(4.65), 410(3.92)	54-1005-64
5-Benzofuranacrylic acid, α-ethyl-2,3-dihydro-6-hydroxy-β-methyl-, δ-lactone	EtOH	226(4.11),256(3.47), 295(3.76),332(4.26)	44-2467-64
p-Benzoquinone, 2-methyl-5-(1-methyl-4-oxocyclohex-2-en-1-yl)-	EtOH	257(4.26)	39-2932-64
3-Cyclohexene-1-carboxylic acid, 6-benzoyl-, cis	EtOH	240(4.00)	70-0310-64
trans	EtOH	241(4.35)	70-0310-64
2,4,6-Heptatrienoic acid, 3-methoxy-7-phenyl-	EtOH	244(4.17),251(4.18), 332(4.72)	44-3161-64
Kavaic acid	EtOH	244(3.92),251(3.98), 332(4.62)	44-3161-64
Seselin, dihydro-	EtOH	213(4.17),255(3.37), 329(4.09)	78-2605-64

Compound	Solvent	$\lambda_{max}(\log \epsilon)$	Ref.
$C_{14}H_{14}O_3S$			
Sulfoxide, o-anisyl p-anisyl, S-(-)	EtOH	244(4.29),278(3.84)	35-5637-64
p-Toluenesulfonic acid, m-tolyl ester	hexane	218(4.20),255(3.03)	65-2080-65
	EtOH	228(4.10),256(2.95), 274(2.95)	65-2080-65
	ether	217(4.23),256(3.00)	65-2080-65
	dioxan	228(4.12),256(3.11), 274(3.11)	65-2080-65
	H_2SO_4	230(4.22),274(3.43), 286(3.43),320s(2.90), 360s(2.59)	65-2080-65
p-Toluenesulfonic acid, o-tolyl ester	hexane	225(4.20),275(3.00)	65-2080-65
	EtOH	227(4.10),275(2.95)	65-2080-65
	ether	224(4.15),255(3.02)	65-2080-65
	dioxan	228(4.07),273(3.03)	65-2080-65
	H_2SO_4	232(4.06),275(3.28), 315s(2.88)	65-2080-65
p-Toluenesulfonic acid, p-tolyl ester	hexane	224(4.12),275(2.86)	65-2080-65
	EtOH	224(4.18),256(3.00), 273(3.00)	65-2080-65
	ether	225(4.13),256(3.03), 274(3.03)	65-2080-65
	dioxan	227(4.08),256(3.00), 274(3.00)	65-2080-65
	H_2SO_4	233(4.05),273(3.22), 324s(2.90),373s(2.33)	65-2080-65
$C_{14}H_{14}O_4$			
2'-Acetonaphthone, 1'-hydroxy- 3',4'-dimethoxy-	EtOH	223(4.57),265(4.45), 274(4.44),290s(3.90), 300s(3.75),314s(3.30), 396(3.64)	94-0316-64
2'-Acetonaphthone, 3'-hydroxy- 1',4'-dimethoxy-	EtOH	226(4.63),237s(4.56), 258(4.49),299(3.77), 314s(3.51),344(3.26), 396s(3.15)	94-0316-64
2'-Acetonaphthone, 4'-hydroxy- 1',3'-dimethoxy-	EtOH	212(4.40),240(4.47), 260s(3.78),302(3.63), 332(3.49)	94-0316-64
1-Cyclopentene-1-carboxylic acid, 2-(2-hydroxy-4,6-dimethoxyphenyl)-, δ-lactone	EtOH	248(3.89),257(3.85), 325(4.21)	35-0882-65
4,5-Cyclopentenophthalide-3-acetic acid, methyl ester	EtOH	243(4.09),274(3.12), 283(3.06)	39-3811-65
Itaconitin, anhydrodihydro-	EtOH	252(3.86),275(3.73), 283(3.64)	94-0069-65
Jatamansinol	EtOH	216(4.01),246(3.41), 256(3.36),329(4.06)	78-2605-64
1,4,5-Methenoindene-2,3-dicarboxylic acid, 3a,4,5,7a-tetrahydro-, dimethyl ester	EtOH	245(3.84)	35-1434-64
2-Naphthaldehyde, 4,5,6-trimethoxy-	EtOH	262(4.33),330(3.77), 367(3.88)	39-4292-65
4a,8a-Naphthalenedicarboxylic acid, dimethyl ester, cis	n.s.g.	258(3.85)	89-0786-64
2-Naphthoic acid, 1-hydroxy- 8-methoxy-, ethyl ester	EtOH	216(4.25),246(4.62), 286(3.52),297(3.62), 310(3.62),358(3.97), 373(3.94)	33-0769-64
1,4-Naphthoquinone, 6,7-dimethoxy-	n.s.g.	276(4.48),345(3.32), 404(3.05)	39-4011-64

Compound	Solvent	$\lambda_{max}(\log \epsilon)$	Ref.
Xanthogalol	EtOH	222(4.1),245(3.5), 258(3.5),325(4.2)	30-1164-64
Zosimol	EtOH	220s(4.1),252(3.5), 264(3.5),330(4.2)	30-1164-64
$C_{14}H_{14}O_5$			
Cinnamic acid, 3-acetyl-2-carboxy-, dimethyl ester	EtOH	235(4.25),275(3.6)	39-3811-65
5H-Furo[3,2-g][1]benzopyran-5-one, 6,7-dihydro-4,9-dimethoxy-7-methyl-	MeOH	250(4.51)	5-0167-65E
1,2-Indandicarboxylic acid, 7-methyl-3-oxo-, dimethyl ester	MeOH	255(4.04),297(3.94)	39-0990-65
1,4-Naphthoquinone, 5,6,8-trimethoxy-3-methyl-	EtOH	215(4.4),265(4.0), 410(3.5)	33-0538-65
1,4-Naphthoquinone, 5,7,8-trimethoxy-3-methyl-	n.s.g.	220(4.32),272(4.0), 430(3.51)	33-2017-64
$C_{14}H_{14}O_5S_3$			
2-(2-Hydroxy-1-naphthyl)-1,3-dithiolanium methyl sulfate	MeCN	318(4.16),370(3.30), 460(3.41)	24-1374-65
$C_{14}H_{14}O_6$			
Coumarin-6-acetic acid, 5,7-dimethoxy-, methyl ester	MeOH	224(4.16),245(3.70), 255(3.61),331(4.09)	32-1054-64
1,4-Naphthoquinone, 2,5,6,7-tetramethoxy-	EtOH	267(4.33),301(4.09), 361(3.47)	94-1472-65
1,4-Naphthoquinone, 2,6,7,8-tetramethoxy-	EtOH	269(4.35),299(4.17), 359(3.49)	94-1472-65
$C_{14}H_{14}O_6S$			
Maleic acid, hydroxy(methylthio)-, dimethyl ester, benzoate	MeOH	231(4.29),284(4.25)	24-1581-65
$C_{14}H_{14}O_8$			
Aphloiol	EtOH	241(4.31),259(4.39), 317(4.08),367(3.97)	22-0376-64
$C_{14}H_{14}O_8S_2$			
Disulfide, bis(4-acetoxy-2,5-dihydro-3-methyl-5-oxo-2-furyl)-	EtOH	242(4.13),290(3.51)	39-0766-64
$C_{14}H_{15}BrN_4O$			
1,2,3,4-Tetrahydro-4-oxobenzo[c]quinolizinium bromide, semicarbazone	EtOH	202(4.48),252(4.15), 288(4.09),368(4.47)	78-2529-65
$C_{14}H_{15}BrO$			
Cyclohexanone, 2-(α-bromobenzylidene)-6-methyl-, trans	MeOH	296(4.19)	78-2201-64
$C_{14}H_{15}BrO_4$			
Acetoacetic acid, 2-(p-bromophenacyl)-, ethyl ester	MeOH	256(4.34)	44-2205-65
$C_{14}H_{15}ClN_2O_2$			
Tryptamine, N,4-diacetyl-7-chloro-	EtOH	242(4.38),314(3.87)	87-0200-65
$C_{14}H_{15}ClN_4O_5$			
Purine, 6-chloro-9-(3,4-di-O-acetyl-2-deoxy-α-D-ribopyranosyl)-	pH 1	263(3.95)	35-1251-64
	pH 11	264(4.03)	35-1251-64

Compound	Solvent	$\lambda_{max}(\log \epsilon)$	Ref.
$C_{14}H_{15}ClO$			
Anthrone, 2(or 3)-chloro-1,2,3,4,4aβ,-9aα-hexahydro-	EtOH	250(3.22)	70-0310-64
Cyclohexanone, 2-(p-chlorobenzylidene)-6-methyl-, trans	MeOH	286(4.22)	78-2201-64
Cyclohexanone, 2-(α-chlorobenzylidene)-6-methyl-, trans	MeOH	292(4.12)	78-2201-64
$C_{14}H_{15}ClO_2$			
Anthrone, 2β-chloro-1,2,3,4,4aα,9aβ-hexahydro-3α-hydroxy-	EtOH	245(4.00)	70-0310-64
$C_{14}H_{15}Cl_2N$			
Carbazole, 4a-(dichloromethyl)-2,3,4,4a-tetrahydro-9-methyl-	EtOH	278(4.15),310s(3.36)	39-0938-64
$C_{14}H_{15}Cl_2NO_2$			
Quinaldine, 4-chloro-3-(2-chloroethyl)-5,8-dimethoxy-	EtOH	259(4.63),340(3.69)	95-0271-65
$C_{14}H_{15}Cl_2NO_2S$			
3(2H)-Furanone, 2,2-dichloro-5-(diethylamino)-4-(phenylthio)-	EtOH	240(4.09),248(4.10), 291(4.16)	44-4303-65
$C_{14}H_{15}Cl_2NO_3$			
3(2H)-Furanone, 2,2-dichloro-5-(diethylamino)-4-phenoxy-	EtOH	214s(4.13),311(4.25)	44-4303-65
$C_{14}H_{15}Cl_2NO_4$			
Carbostyril, 3-(2-chloroethyl)-5,7,8-trimethoxy-	n.s.g.	220(4.51),262(4.36), 268(4.38),326(4.12)	2-0071-65
$C_{14}H_{15}F_3O_3$			
Acetic acid, trifluoro-, 4-(p-methoxyphenyl)-3-pentenyl ester	n.s.g.	226(3.98),251(3.71)	88-1053-65
$C_{14}H_{15}FeN$			
Propane, 2-cyano-2-ferrocenyl-	EtOH	322(2.03),436(1.89)	78-0791-64
$C_{14}H_{15}IN_2$			
2-(p-Aminostyryl)-1-methylpyridinium iodide, cis	H_2O	374(3.96)	23-1345-65
	aq. HCl	313(4.04)	23-1345-65
trans	H_2O	392(4.40)	23-1345-65
	aq. HCl	327(4.41)	23-1345-65
$C_{14}H_{15}N$			
4-Biphenylamine, N,N-dimethyl-	pH 1	250(4.29)	32-0902-64
	EtOH	295(4.34)	32-0902-64
7H-Cyclohepta[c]quinoline, 8,9,10,11-tetrahydro-	n.s.g.	227(4.59),288(3.49), 306(3.36),320(3.37)	39-2421-64
Cyclopentadiene, 1-(p-dimethylaminobenzylidene)-	MeOH	405(4.5)	87-0390-65
	HOAc	405(3.8)	87-0390-65
Phenanthridine, 7,8,9,10-tetrahydro-6-methyl-	n.s.g.	233(4.37),281(4.05), 307(3.66),315(3.19), 321(3.36)	39-2421-64
$C_{14}H_{15}NO$			
Benz[cd]indole, 3-acetyl-1,3,4,5-tetrahydro-2-methyl-	EtOH	225(4.55),274(3.87), 280(3.87),289(3.72)	44-0843-64

Compound	Solvent	λ_{max}(log ε)	Ref.
Benzo[f]quinolin-1(2H)-one, 3,4,5,6-tetrahydro-4-methyl-	EtOH-HCl	253(4.27),353(3.80)	94-1405-64
	EtOH-HCl	253(4.26),353(3.80)	95-1220-64
	EtOH	280(4.19),356(4.02)	94-1405-64
	EtOH	280(4.18),356(4.02)	95-1220-64
1(2H)-Carbazoleninone, 3,4-dihydro-2,11-dimethyl-	MeOH-HCl	246(3.97),252(3.99), 350(4.31)	24-2111-65
	MeOH	228(4.12),235(4.16), 242(4.08),315(4.20)	24-2111-65
	MeOH-NaOH	232(4.16),240s(4.08), 336(4.06)	24-2111-65
3(2H)-Carbazolenine, 1,4-dihydro-1,11-dimethyl-	MeOH-HCl	231(4.07),279(3.98)	24-2111-65
	MeOH	260(3.98)	24-2111-65
2(3H)-Carbazolone, 4,11-dihydro-3,11-dimethyl-	MeOH-HCl	248(4.07),252(4.08), 340(4.34)	24-2111-65
	MeOH	233(4.27),297(3.98), 336(4.47)	24-2111-65
	MeOH-NaOH	239(4.10),348(4.39)	24-2111-65
4H-Cyclohept[d]isoxazole, 5,6,7,8-tetrahydro-3-phenyl-	C_6H_{12}	229(4.09)	44-1582-64
Cyclopentanecarbonitrile, 3-ethyl-2-oxo-3-phenyl-	EtOH	297(2.16)	23-2512-65
	EtOH-base	270(4.06)	23-2512-65
Furo[3,2-c]quinoline, 2,3-dihydro-2,2,4-trimethyl-	EtOH	214(4.51),230(4.73), 285(3.76),292(3.80), 303(3.71),310(3.59), 316(3.42)	94-0789-64
Furo[3,2-c]quinoline, 2,3-dihydro-2,3,4-trimethyl-	EtOH	214(4.48),230(4.75), 286(3.76),292(3.79), 302(3.69),308(3.57), 315(3.37)	94-0789-64
2-Indancarbonitrile, 1-tert-butyl-3-oxo-	EtOH	234(3.96),242(3.94), 293(3.96)	44-0396-65
2-Indolinone, 3-cyclohexylidene-	EtOH	255(4.44),262(4.51), 294(3.91),352(3.30)	87-0626-65
2-Indolinone, 3-cyclopentylidene-1-methyl-	MeOH	257(4.48),262(4.55), 294(3.87),350(3.12)	87-0626-65
4H-Indol-4-one, 2,3,3a,5,6,7-hexahydro-3a-phenyl-	EtOH	284(2.63)	88-1389-65
2-Naphthol, 1-(N-isopropylformimidoyl)-	MeOH	224(4.61),310(3.90), 398(3.84),415(3.85)	59-0597-64
	dioxan	226(4.61),309(3.99), 358(3.70),400(3.56), 420(3.56)	59-0597-64
	$HCONH_2$	302(3.90),395(3.85), 415(3.85)	59-0597-64
	DMF	306(3.66),360(3.32), 400(3.48),420(3.49)	59-0597-64
Phenanthridone, 7,8,9,10-tetrahydro-N-methyl-	EtOH	228(4.58),271(3.88), 281(3.83),324(3.92)	44-0681-64
Pyrrolo[1,2-a]quinolin-5(1H)-one, 2,3-dihydro-1,4-dimethyl-	EtOH	213(4.45),243(4.43), 249(4.44),284s(3.30), 295s(3.44),329(4.10), 341(4.13)	44-1989-65
4-Quinolinol, 2-methyl-3-(1-methylallyl)-	EtOH	213(4.48),241(4.47), 248(4.47),281(3.36), 292(3.50),322(4.05), 335(4.06)	94-0789-64
4-Quinolinol, 2-methyl-3-(methallyl)-	EtOH	213(4.48),240(4.46), 247(4.44),281(3.38), 293(3.53),321(4.05), 336(4.06)	94-0789-64

Compound	Solvent	$\lambda_{max}(\log \epsilon)$	Ref.
4(1H)-Quinolone, 1-allyl-2,3-dimethyl-	EtOH	214(4.40),243(4.44), 249(4.42),282s(3.23), 294s(3.33),329(3.12), 342(4.16)	44-1989-65
4(1H)-Quinolone, 2-(3-butenyl)-3-methyl-	EtOH	213(4.45),240(4.51), 247(4.50),283s(3.31), 294s(3.47),323(4.09), 336(4.11)	44-1989-65
$C_{14}H_{15}NOS$			
Morpholine, 4-(5-phenyl-2-thienyl)-	EtOH	254(3.32),330(4.26)	32-0699-65
	$CHCl_3$	334(4.28)	4-0315-65
$C_{14}H_{15}NO_2$			
Benz[cd]indole-3-carboxylic acid, 1,3,4,5-tetrahydro-2-methyl-, methyl ester	EtOH	225(4.55),274(3.86), 279(3.87),289(3.72)	44-0843-64
Cinnamic acid, α-cyano-p,β -dimethyl-, ethyl ester	EtOH	294.5(4.04)	28-3102-65A
Pyrrole-3-carboxylic acid, N-ethyl-4-phenyl-, methyl ester	ether	218(4.43),245(3.98)	24-1952-64
1H-Pyrrolo[1,2-a]indole-9-carboxaldehyde, 2,3-dihydro-7-methoxy-6-methyl-	MeOH	256(4.26),282(4.23), 309(4.13)	35-3877-64
	MeOH	213(4.63),256(4.26), 282(4.22),309(4.13)	44-2897-65
Pyrrolo[3,2,1-hi]indole-5-carboxylic acid, 1,2-dihydro-4-methyl-, ethyl ester	MeOH	220(4.58),240(4.35), 297(4.15)	35-1397-65
2-Quinolineacetic acid, α-methyl-, ethyl ester	EtOH	213(4.35),229(4.49), 232(4.49)	78-2961-65
$C_{14}H_{15}NO_2S$			
Thieno[3,2-c]quinoline, 2,3-dihydro-6,9-dimethoxy-4-methyl-	EtOH	250(4.51),343(3.90)	95-0271-65
p-Toluenesulfonanilide, N-methyl-	hexane	222(4.03),235(3.97), 270(3.34),300s(1.95)	65-2080-65
	EtOH	223(4.06),235(4.06), 270(3.35),310s(2.10)	65-2080-65
	ether	222(4.05),237(3.90), 275(3.15)	65-2080-65
	dioxan	223(4.10),270(3.36)	65-2080-65
	H_2SO_4	248(4.15),285(3.38), 315(2.55),325(2.50)	65-2080-65
$C_{14}H_{15}NO_3$			
Anthrone, 1,2,3,4,4a,9a-hexahydro-7-nitro-	EtOH	234(4.43),269(4.34)	39-5377-65
Azepino[3,2,1-hi]indole-7-carboxylic acid, 1,2,4,5,6,7-hexahydro-6-oxo-, methyl ester	EtOH	218(4.20),253(4.03)	35-1397-65
	EtOH-NaOH	245(4.20),285(4.07)	35-1397-65
Carbazol-1(2H)-one, 3,4-dihydro-5,7-dimethoxy-	EtOH	327(4.33)	12-0246-64
Carbazol-1(2H)-one, 3,4-dihydro-6,7-dimethoxy-	EtOH	339(4.41)	12-0246-64
Carbostyril, 4-hydroxy-3-isovaleryl-	n.s.g.	234(4.19),300(3.81)	44-2598-64
	base	230(4.21),265(3.73), 310(3.73)	44-2598-64
Cinnamic acid, α-cyano-p-methoxy-β-methyl-, ethyl ester, trans	EtOH	320.5(4.09)	28-3102-65A

Compound	Solvent	$\lambda_{max}(\log \epsilon)$	Ref.
Furo[3,2-c]quinoline, 2,3-dihydro-6,9-dimethoxy-4-methyl-	EtOH	225(4.55),333(3.92)	95-0271-65
Indole-2-carboxylic acid, 5-acetyl-3-methyl-, ethyl ester	EtOH	270(4.79),310(3.97)	87-0141-64
3-Indolinylideneacetic acid, 2-oxo-, tert-butyl ester	EtOH	252(4.36),257(4.32),312(3.86)	44-2431-64
3-Indolizinecarboxylic acid, 1-acetyl-5-methyl-, ethyl ester	EtOH	227(4.16),252(4.44),287(3.97),340(4.30)	78-3305-65
$C_{14}H_{15}NO_3S$			
3-Pyrrolin-2-one, 3-acetyl-4-hydroxy-5-(benzylthio)methyl-	EtOH	244(3.71),277(4.10)	35-5654-64
$C_{14}H_{15}NO_4$			
3-Cyclohexene-1-carboxylic acid, 2-(o-nitrophenyl)-, methyl ester	EtOH	252(3.62)	44-0681-64
L-Leucine, N-phthalyl-	MeOH	292(3.30),300s(3.27)	24-0533-64
1H-Pyrrolo[1,2-a]indole-5,8-dione, 2,3-dihydro-9-(hydroxymethyl)-7-methoxy-6-methyl-	MeOH	230(4.25),287(4.14),350(3.52),460(3.30)	35-3877-64
	MeOH	230(4.25),287(4.14),350(3.52),460(3.12)	44-2897-65
$C_{14}H_{15}NO_5$			
2H,6H-Chromeno[7,6-b][1,4]oxazine, 3,4,7,8-tetrahydro-8-methoxycarbonyl-3-methyl-6-oxo-	MeOH	256(4.2),284(3.8),378(3.4)	39-7348-65
$C_{14}H_{15}NO_5S$			
Cinnamic acid, β-(acetonylthio)-p-nitro-, ethyl ester	ether	209(4.15),262(4.37)	24-2109-64
$C_{14}H_{15}NO_7S$			
Cinnamic acid, β-(acetonylsulfonyl)-p-nitro-, ethyl ester	ether	209(4.20),265(4.04)	24-2109-64
$C_{14}H_{15}NO_8$			
2,3,4,5-Pyridinetetracarboxylic acid, 6-methyl-, tetramethyl ester	MeOH	283(3.52)	39-0948-65
$C_{14}H_{15}NS_2$			
2(1H)-Pyridinethione, 1-[2-(p-tolylthio)ethyl]-	pH 1	247(4.04),254(4.06),347(3.84)	4-0097-65
	pH 7	254(4.06),274(4.04),347(3.01)	4-0097-65
	pH 13	253(4.06),274(4.04),347(3.83)	4-0097-65
	MeOH	252(4.01),286(4.06),363(3.77)	4-0097-65
$C_{14}H_{15}N_3$			
Azobenzene, 4-(dimethylamino)-	N HCl	320(4.00),520(4.53)	78-1547-64
	EtOH	254(4.03),305s(3.70),408(4.44)	65-3602-64
	EtOH	420(4.38)	78-1547-64
	MeCN	325(3.6),410(4.4)	49-1173-65
$SbCl_5$ complex	MeCN	515(4.7)	49-1173-65
$(SbCl_5)_2$ complex	MeCN	405(4.45)	49-1173-65
s-Triazine, 2-ethyl-4-methyl-6-styryl-	EtOH	224(4.35),311(4.53)	44-0707-65

Compound	Solvent	λ_{max}(log ε)	Ref.
$C_{14}H_{15}N_3O$			
p-Benzoquinone, azine with 1,2,6-trimethyl-4(1H)-pyridone	C_5H_5N	554(4.92)	5-0009-65J
	99% C_5H_5N	554(4.93)	5-0009-65J
	98% C_5H_5N	555(4.93)	5-0009-65J
	90% C_5H_5N	555(4.94)	5-0009-65J
	80% C_5H_5N	551(4.95)	5-0009-65J
	70% C_5H_5N	550(4.93)	5-0009-65J
	60% C_5H_5N	545(4.92)	5-0009-65J
	50% C_5H_5N	545(4.90)	5-0009-65J
	40% C_5H_5N	542(4.88)	5-0009-65J
	30% C_5H_5N	542(4.87)	5-0009-65J
	10% C_5H_5N	534(4.82)	5-0009-65J
	5% C_5H_5N	530(4.81)	5-0009-65J
	1% C_5H_5N	525(4.76)	5-0009-65J
p-Benzoquinone, 2,6-dimethyl-, 4-azine with 1-methyl-4(1H)-pyridone	C_5H_5N	530(4.76)	5-0009-65J
	99% C_5H_5N	531(4.76)	5-0009-65J
	98% C_5H_5N	533(4.76)	5-0009-65J
	95% C_5H_5N	535(4.76)	5-0009-65J
	90% C_5H_5N	539(4.76)	5-0009-65J
	80% C_5H_5N	541(4.76)	5-0009-65J
	70% C_5H_5N	542(4.76)	5-0009-65J
	60% C_5H_5N	545(4.76)	5-0009-65J
	50% C_5H_5N	547(4.77)	5-0009-65J
	40% C_5H_5N	548(4.77)	5-0009-65J
	30% C_5H_5N	547(4.77)	5-0009-65J
	20% C_5H_5N	547(4.78)	5-0009-65J
	10% C_5H_5N	547(4.78)	5-0009-65J
	5% C_5H_5N	545(4.78)	5-0009-65J
	2% C_5H_5N	545(4.79)	5-0009-65J
	1% C_5H_5N	540(4.78)	5-0009-65J
Benzo[f]pyrazolo[3,4-a]quinolizin-1(3bH)-one, 2,4,5,11,12,12a-hexahydro-	2N NaOH	250(4.19)	78-2529-65
	EtOH	208(4.46),255(4.08)	78-2529-65
Formamide, N-methyl-N-[2-(methylamino)-1-(2-quinolyl)vinyl]-	EtOH	218(4.69),287(4.16), 312(4.08),323(4.03), 380(4.11)	39-5969-64
s-Triazine, 2-ethyl-4-methoxy-6-styryl-	EtOH	226(4.08),310(4.42)	44-0707-65
$C_{14}H_{15}N_3O_2$			
[2,2'-Bipyrrole]-5-carboxylic acid, 5'-(1-pyrrolin-2-yl)-, methyl ester	MeOH-HCl	258(4.02),384s(4.58), 389(4.63)	44-0883-64
	MeOH-KOH	300s(4.05),350s(4.45), 363(4.49)	44-0883-64
1-Pyrazoline-3-carboxylic acid, 3-cyano-4-methyl-4-phenyl-, ethyl ester, trans	$CHCl_3$	326s(2.20)	28-1332-65B
$C_{14}H_{15}N_3O_2S$			
5-Pyrimidinepropionanilide, 4-hydroxy-2-mercapto-6-methyl-	pH 1, 8.4	280(4.25)	4-0263-64
	pH 13	240(4.31),260s(4.24), 310s(3.82)	4-0263-64
$C_{14}H_{15}N_3O_3$			
p-Benzoquinone, 2,6-bis(hydroxymethyl)-, 4-azine with 1-methyl-4(1H)-pyridone	C_5H_5N	539(4.76)	5-0009-65J
	99% C_5H_5N	540(4.76)	5-0009-65J
	98% C_5H_5N	541(4.76)	5-0009-65J
	95% C_5H_5N	530(4.76)	5-0009-65J
	90% C_5H_5N	549(4.77)	5-0009-65J
	80% C_5H_5N	550(4.78)	5-0009-65J

Compound	Solvent	$\lambda_{max}(\log \epsilon)$	Ref.
p-Benzoquinone, 2,6-bis(hydroxymethyl)-, 4-azine with 1-methyl-4(1H)-pyridone (contd.)	70% C_5H_5N	551(4.80)	5-0009-65J
	60% C_5H_5N	554(4.81)	5-0009-65J
	50% C_5H_5N	550(4.81)	5-0009-65J
	40% C_5H_5N	550(4.82)	5-0009-65J
	30% C_5H_5N	550(4.83)	5-0009-65J
	20% C_5H_5N	549(4.83)	5-0009-65J
	10% C_5H_5N	547(4.82)	5-0009-65J
	5% C_5H_5N	545(4.82)	5-0009-65J
	2% C_5H_5N	541(4.82)	5-0009-65J
	1% C_5H_5N	541(4.82)	5-0009-65J
$C_{14}H_{15}N_3O_3S$			
Benzenesulfonic acid, 2-[[p-(dimethylamino)phenyl]azo]-	N HCl	321(3.64),518(4.72)	78-1547-64
	EtOH	446(4.32)	78-1547-64
Benzenesulfonic acid, 3-[[p-(dimethylamino)phenyl]azo]-	N HCl	318(3.90),513(4.60)	78-1547-64
	EtOH	467(4.30)	78-1547-64
Benzenesulfonic acid, 4-[[p-(dimethylamino)phenyl]azo]-	N HCl	321(4.11),500(4.48)	78-1547-64
	EtOH	472(4.34)	78-1547-64
$C_{14}H_{15}N_5$			
Adenine, 3-(3-phenylpropyl)-	EtOH-HCl	277(4.21)	23-1599-64
	EtOH	275(4.09)	23-1599-64
s-Triazolo[1,5-a]pyrimidine, 7-(N-ethylanilino)-5-methyl-	MeOH	308(4.30)	65-0204-64
$C_{14}H_{15}N_5O$			
9H-Purine-9-propanol, 6-anilino-	pH 1	275(4.26)	87-0033-65
	pH 7	287(4.33)	87-0033-65
	pH 13	287(4.33)	87-0033-65
5H-Pyrrolo[3,4-d]pyrimidine, 2,4-diamino-6-benzoyl-6,7-dihydro-7-methyl-	pH 1	273(3.90),300s(3.43)	44-0194-65
	EtOH	282(3.69)	44-0194-65
s-Triazolo[1,5-a]pyrimidine, 5-methyl-7-p-phenetidino-	MeOH	298(4.31)	65-0204-64
$C_{14}H_{15}N_5OS$			
9H-Purine-9-ethanol, 2-amino-6-(benzylthio)-	pH 1	247(3.98),319(4.08)	87-0182-65
	pH 11	245(4.06),310(4.09)	87-0182-65
$C_{14}H_{15}N_5O_3$			
Mandelamide, N-(2,4-diamino-5-pyrimidinyl)-, acetate	pH 1	265s(3.76)	87-0797-65
	pH 11	227(4.13),287(3.87)	87-0797-65
Mandelamide, N-(4,6-diamino-5-pyrimidinyl)-, acetate	pH 1	263(4.08)	87-0797-65
	pH 11	257(3.78)	87-0797-65
as-Triazine, 3-amino-5-[3-ethyl-4-(5-nitro-2-furyl)-1,3-butadienyl]-6-methyl-	propylene glycol	295(3.94),420(4.48)	95-0109-64
$C_{14}H_{15}N_5O_4S$			
Isoxanthine, 9-benzyl-1,3-dimethyl-8-sulfamoyl-	MeOH	255(4.11),264(4.09)	54-0193-65
Theobromine, 1-benzyl-8-sulfamoyl-	MeOH	288(4.06)	54-0193-65
Theophylline, 7-benzyl-8-sulfamoyl-	MeOH	286(4.01)	54-1215-64
$C_{14}H_{15}OP$			
Phosphine oxide, ethyldiphenyl-	MeOH	220(4.35),255(3.0),260(3.1),265(3.28),272(3.2)	70-0669-65
Phosphinous acid, diphenyl-, ethyl ester	MeOH	285s(3.2)	70-0669-65

Compound	Solvent	λmax(log ε)	Ref.
$C_{14}H_{15}O_2P$			
Phosphinic acid, diphenyl-, ethyl ester	n.s.g.	224(4.18),255(3.6), 260(3.1),265(3.30), 272(3.2)	70-0669-65
$C_{14}H_{15}O_5Na$			
Malonic acid, benzylidene, cyclic isopropylidene ester, NaOMe adduct	EtOH-NaOH	262(4.30)	49-1283-64
$C_{14}H_{16}$			
Azulene, 6-ethyl-1,4-dimethyl-	hexane	243(4.29),286(4.71), 292(4.72),343(3.52), 352(3.6),369(3.11), 541(2.35),563(2.44), 585(2.47),614(2.43), 638(2.36),675(2.05), 708(1.83)	94-0717-65
Azulene, 1,4,6,8-tetramethyl-	C_6H_{12}	247(4.07),287(4.36), 294(4.40),330s(3.23), 340(3.32),348s(3.29), 355s(3.36),357(3.37), 369(2.62),550s(2.43), 576(2.50),600s(2.44), 627(2.38),664s(2.04), 692(1.92)	44-0131-65
	hexane	563(2.62),576(2.65), 600(2.58),627(2.53), 689(2.03)	5-0194-64B
Azulene, 2,4,6,8-tetramethyl-	C_6H_{12}	247(4.24),287(4.56), 296(4.72),314(4.02), 330s(3.53),337s(3.58), 339(3.65),354s(3.77), 355(3.77),535(2.57), 564(2.54),580s(2.48), 613s(2.16),634s(1.99)	44-0131-65
1-Butene, 2-methyl-1-(phenylcyclopropylidene)-	EtOH	204(4.38),216s(3.15), 275s(2.88)	39-5625-65
1,3-Cyclohexadiene, 1,6-dimethyl-5-phenyl-	n.s.g.	265s(3.45),270(3.46), 275s(3.45)	39-2898-65
1,3-Cyclohexadiene, 5,6-dimethyl-1-phenyl-	n.s.g.	309(4.1)	39-2898-65
1,3,5-Heptatriene, 5-methyl-1-phenyl-, trans	n.s.g.	320(4.42),325(4.45), 342(4.29)	39-2898-65
Naphthalene, 1-butyl-	EtOH	225(4.99),283(3.97)	39-2072-65
Naphthalene, 2-butyl-	EtOH	225(5.04),276(3.68)	39-2072-65
$C_{14}H_{16}BrNO_3$			
3-Indolineacetic acid, 3-bromo-2-oxo-, tert-butyl ester	EtOH	218(4.20),231s(4.13), 317(2.94)	44-2431-64
$C_{14}H_{16}BrN_5O_4$			
Pyrido[3,4-b]pyrazine-5,7-dicarbamic acid, 8-bromo-3-methyl-, diethyl ester	pH 1	264(4.41),310(3.92), 385(3.72)	44-0734-64
	pH 7	263(4.50),297(3.58), 368(3.70)	44-0734-64
	pH 13	269(4.46),354(3.48)	44-0734-64
$C_{14}H_{16}Br_2N_2$			
1,1'-Tetramethylene-2,2'-bipyridylium dibromide	H_2O	275(4.10)	25-0782-65

Compound	Solvent	λ_{max}(log ε)	Ref.
$C_{14}H_{16}Br_2O_4$			
2,4,6,8-Decatetraenedioic acid, 2,9-dibromo-, diethyl ester	EtOH	362(4.69),379(4.62)	70-0684-65
$C_{14}H_{16}Br_4ClNO_4$			
3,5-Pyridinedicarboxylic acid, 4-(chloromethyl)-2,6-bis(dibromomethyl)-1,4-dihydro-, diethyl ester	EtOH	239(4.23),350(3.68)	39-2411-65
$C_{14}H_{16}Br_5NO_4$			
3,5-Pyridinedicarboxylic acid, 4-(bromomethyl)-2,6-bis(dibromomethyl)-1,4-dihydro-, diethyl ester	EtOH	238(4.19),348(3.63)	39-2411-65
$C_{14}H_{16}ClFN_2O_2$			
Benzylamine, α-(1-chloro-2-fluoro-3-nitroallylidene)-N,N-diethyl-	n.s.g.	276(4.04),377(4.06)	35-5132-65
$C_{14}H_{16}ClNO$			
6(2H)-Quinolone, 5-chloro-2,2,4,7,8-pentamethyl-	MeOH	247(4.15),301(4.30), 369(3.65)	24-2939-65
$C_{14}H_{16}ClNO_2S_2$			
3-(p-Methoxyphenyl)-5-morpholino-1,2-dithiol-3-ylium chloride	n.s.g.	244(3.69),362(4.38)	32-1078-65
$C_{14}H_{16}ClNO_3$			
Coumarin, 6-chloro-3-ethoxy-4-(1-propylamino)-	EtOH	208(4.41),234(4.19), 262(4.02),303(3.88), 311(3.95),347(3.98)	44-4126-65
$C_{14}H_{16}ClO_2P$			
Bis(hydroxymethyl)diphenylphosphonium chloride	EtOH	260(3.16),267(3.29), 273(3.18)	59-1143-64
$C_{14}H_{16}Cl_2N_2O_4$			
p-Benzoquinone, 2,5-dichloro-3,6-dimorpholino-	methyl cellosolve	240(4.26),436(4.05), 562s(2.73)	78-1889-64
$C_{14}H_{16}F_2N_2O_2$			
p-Benzoquinone, 2,5-difluoro-3,6-di-1-pyrrolidinyl-	methyl cellosolve	237(4.40),419(4.09), 569(2.66)	78-1889-64
$C_{14}H_{16}F_2N_2O_4$			
p-Benzoquinone, 2,5-difluoro-3,6-dimorpholino-	methyl cellosolve	233(4.36),411(4.04), 550(2.67)	78-1889-64
$C_{14}H_{16}IP$			
Dimethyldiphenylphosphonium iodide	EtOH	223(3.86),261(3.16), 267(3.30),274(3.24)	59-1143-64
$C_{14}H_{16}N_2$			
1-Cyclopentene-1-carbonitrile, 2-amino-3-ethyl-3-phenyl-	EtOH	264(4.12)	23-2512-65
Indole, 3-(5-ethyl-1-pyrrolin-2-yl)-	EtOH-HCl	244s(4.05),250(4.11), 264(4.03),270(4.05), 329(4.28)	87-0415-64

Compound	Solvent	$\lambda_{max}(\log \epsilon)$	Ref.
3H-Indole, 3-(1,5-dimethyl-2-pyrrolidinylidene)-	EtOH-HI	246s(4.05),251(4.13),265(4.00),271(4.02),332(4.27)	87-0415-64
3H-Indole, 3-(5,5-dimethyl-2-pyrrolidinylidene)-	EtOH-HI	250(4.10),264(4.00),270(4.02),330(4.25)	87-0415-64
Indole-3-acetonitrile, 5-methyl-2-propyl-	MeOH	226(4.52),277(3.90),286(3.89),296(3.76)	87-0204-65
$C_{14}H_{16}N_2O$			
1H-3-Benzazepine, 2-morpholino-	EtOH	307(4.06)	4-0026-65
Benz[cd]indole, 3-acetyl-1,3,4,5-tetrahydro-2-methyl-, oxime	EtOH	227(4.56),274(3.85),280(3.86),290(3.71)	44-0843-64
Indole, 3-(2-aminoethyl)-1-(3-oxo-1-butenyl)-, hydrochloride	MeOH	211(4.43),273(4.18),335(4.38)	24-0557-64
	H_2SO_4	216(4.47),272(3.94),350(3.61)	24-0557-64
	2N H_2SO_4	216(4.26),272(4.06),340(4.26)	24-0557-64
	dioxan	232(3.93),276(3.91),324(4.01)	24-0557-64
2-Indolinone, 3-(1-methyl-4-piperidylidene)-	EtOH	254(4.44),260(4.45),294(3.89),355(3.28)	87-0626-65
$C_{14}H_{16}N_2O_2$			
1H-Pyrido[3,4-b]indol-1-one, 2,3,4,9-tetrahydro-6-methoxy-2,3-dimethyl-	EtOH	310(4.30)	39-7165-65
4-Pyrimidinol, 2-benzyl-5-(ethoxymethyl)-	EtOH	223s(3.92),256(3.83)	94-0393-64
Tryptophan, 1-allyl-	90% EtOH-HCl	221(4.61),275s(3.81),284(3.84),295s(3.77)	94-0088-65
	90% EtOH	222(4.60),276s(3.78),284(3.82),295s(3.76)	94-0088-65
$C_{14}H_{16}N_2O_3$			
Gramine, 5-acetyl-2-carboxy-	EtOH	261(4.71),303(3.94)	87-0141-64
5-Pyrimidinecarboxylic acid, 1-benzyl-1,2,3,4-tetrahydro-2-oxo-, ethyl ester	EtOH	208(3.89),297(4.01)	94-1418-64
5-Pyrimidinecarboxylic acid, 1-benzyl-1,2,3,6-tetrahydro-2-oxo-, ethyl ester	EtOH	208(4.14),290(4.18)	94-1418-64
2-Quinoxalinelactic acid, 3-methyl-, ethyl ester	EtOH	236(4.45),240s(4.40),318(3.19)	39-3678-65
Valeric acid, 5-(allyloxy)-4-hydroxy-2-oxo-, γ-lactone, phenylhydrazone	n.s.g.	233(4.08),295(3.91),333(4.35)	39-0141-64
$C_{14}H_{16}N_2O_3S$			
6,7-Dihydro-5H-pyrrolo[1,2-c]pyrimidin-8-ium p-toluenesulfonate	EtOH	220(4.09),249(3.55),297(3.42)	44-2398-65
$C_{14}H_{16}N_2O_4$			
Indole-3-carboxaldehyde, 1-ethyl-5-methoxy-2,6-dimethyl-4-nitro-	MeOH	218(4.60),247(4.21),295(4.08)	35-3878-64
Pyrrolidine, 1-[4-methyl-5-(5-nitro-2-furyl)-2,4-pentadienoyl]-	propylene glycol	280(4.11),390(4.34)	95-0646-64

Compound	Solvent	$\lambda_{max}(\log \epsilon)$	Ref.
$C_{14}H_{16}N_2O_5$			
3-Cinnolineacetic acid, 1,4-dihydro-6,7-dimethoxy-1-methyl-4-oxo-, methyl ester	EtOH	235(4.40),261(4.31), 269(4.31),283s(3.64), 296s(3.38),348(4.15), 363(4.16)	39-6036-65
Morpholine, 4-[4-methyl-5-(5-nitro-2-furyl)-2,4-pentadienoyl]-	propylene glycol	280(4.16),390(4.41)	95-0646-64
$C_{14}H_{16}N_2O_6$			
Indicaxanthine	H_2O	260(3.88),297(3.42), 483(4.79)	33-0361-65
$C_{14}H_{16}N_2O_6S$			
1-Methyl-2-(p-nitrobenzyl)pyridinium methyl sulfate	MeCN	268(4.24)	35-4917-64
$C_{14}H_{16}N_2S_2$			
2-Pyrroline-3-thiocarboxamide, 1-benzyl-5,5-dimethyl-4-thioxo-	pH 1	265(3.97),336(4.06), 382(4.29)	39-4546-65
	pH 13	265(3.92),333(4.15), 382(4.32)	39-4546-65
$C_{14}H_{16}N_4OS$			
5-Pyrimidinepropionanilide, 2-amino-4-mercapto-6-methyl-	pH 1	247(4.24),337(4.25)	4-0263-64
	pH 13	240s(4.36),317(4.13)	4-0263-64
$C_{14}H_{16}N_4O_2$			
p-Benzoquinone, tetrakis-(1-aziridinyl)-	methyl cellosolve	221(4.63),246s(4.61), 359(4.08),559(2.47)	78-1889-64
$C_{14}H_{16}N_4O_3$			
1,3-Cyclohexanedione, 5,5-dimethyl-2-(5,6,7,8-tetrahydro-7-oxo-6-pteridinyl)-	pH 1.0	223(4.48),250s(4.15), 286(4.02),359(3.68)	39-6930-65
	EtOH	215(4.46),286(4.19), 334(3.75)	39-6930-65
$C_{14}H_{16}N_4O_4$			
2,4-Heptadienal, 2-methyl-, 2,4-dinitrophenylhydrazone	$CHCl_3$	393.5(4.55)	22-0161-64
Pyrazolo[1,5-a]pyrimidine-6-carboxylic acid, 7-(diacetylamino)-2-methyl-, ethyl ester	EtOH	250(4.62),297(3.50), 330(3.24)	94-1207-65
$C_{14}H_{16}N_4O_6S$			
L-Proline, 1-[2-(7-nitro-4H-1,2,4-benzothiadiazin-3-yl)ethyl]-, S,S-dioxide	2% $NaHCO_3$	242(4.03),359(4.13)	32-0786-65
$C_{14}H_{16}N_6O_4S$			
Theophylline, 7-(p-aminobenzyl)-8-sulfamoyl-, sulfate	MeOH	284(4.08)	54-0193-65
$C_{14}H_{16}N_6O_6S_2$			
Theophylline, 7-(p-sulfamoylbenzyl)-8-sulfamoyl-	MeOH	287(4.01)	54-0193-65
$C_{14}H_{16}O$			
Cyclohexanone, 2-benzylidene-6-methyl-, cis	MeOH	270(3.83)	78-2201-64
trans	MeOH	284(4.20)	78-2201-64

Compound	Solvent	$\lambda_{max}(\log \epsilon)$	Ref.
2-Cyclohexen-1-one, 2-benzyl-3-methyl-	EtOH	242(4.0)	88-3145-65
2-Cyclohexen-1-one, 2-benzyl-6-methyl-	MeOH	234(3.95)	78-2201-64
Ketone, endo-2-norbornyl phenyl	EtOH	242(4.15),278(3.16), 319(2.26)	101-0261-65B
4,6,12-Tetradecatriene-8,10-diyn-1-ol	ether	249(4.50),265(4.44), 278s(4.21),295(4.46), 313(4.60),335(4.46)	24-0872-65
6-Tetradecene-8,10,12-triyn-3-ol	ether	231(4.83),242(5.05), 257(3.63),271(3.93), 288(4.18),307(4.31), 328(4.15)	24-1411-65
$C_{14}H_{16}OPd$			
Palladium, cyclopentadienyl(8-methoxy-2,4,6-cyclooctatrien-1-yl)-	hexane	200(4.33),223s(4.13), 312(4.49),383s(3.49)	39-5002-64
$C_{14}H_{16}OS_2$			
2(1H)-Naphthalenone, 2,2-bis[(methyl-thio)methyl]-	MeOH	234(4.51)	35-4656-65
$C_{14}H_{16}O_2$			
2(1H)-Anthracenone, 3,4,4aβ,9,9aα,10-hexahydro-9-hydroxy-	EtOH	203(3.91),266(2.33), 273(2.33)	70-1039-65
Fluoren-2-ol, 4b,5,8,8a-tetrahydro-7-methoxy-	EtOH	282(3.46)	88-3505-65
	MeOH-KOH	241(3.89),298(3.38)	88-3505-65
Naphthalene, 1,2-dihydro-7-methoxy-4-(1-methoxyvinyl)-	EtOH	275(4.01)	44-2527-64
Naphthalene, 6,7-dimethoxy-1,2-dimethyl-	EtOH	236(4.82),266(3.62), 275(3.68),284(3.67), 294(3.52),314(3.39), 321(3.26),329(3.55)	39-0361-65
2,6-Octadiene, 1,8-(1,4-phenylenedioxy)-	EtOH	227(3.57),279(3.00)	39-5182-65
2,4-Pentadienoic acid, 2,3-dimethyl-5-p-tolyl-, 2-cis-4-trans	MeOH-acid	305(4.50)	44-2986-64
2-trans-4-trans	MeOH-acid	230(3.95),307(4.49)	44-2986-64
3,5,7,9,11-Tetradecapentaene-2,13-dione	EtOH	284(3.79),381(4.40), 398(3.39)	70-0684-65
$C_{14}H_{16}O_2S$			
4H-Thiopyran-3-carboxylic acid, 5,6-di-hydro-2-phenyl-, ethyl ester	n.s.g.	208(3.87),232(3.93), 287(4.05)	24-1963-64
$C_{14}H_{16}O_3$			
1,4-Benzodioxan-5-ol, 6,8-diallyl-	EtOH	273(2.81)	65-2994-64
2-Cyclohexene-$\Delta^{1,\alpha}$-glycolic acid, 3-(1-cyclohexen-1-yl)-2-hydroxy-, γ-lactone	EtOH	232(3.82),235(4.38)	39-5617-64
	EtOH-NaOH	245(3.94),352(4.37)	39-5617-64
1-Cyclopropanecarboxylic acid, 2-ben-zoyl-1,2-dimethyl-, methyl ester, cis	MeOH	245(4.14)	24-2438-65
3(2H)-Dibenzofuranone, 1,4,4a,9b-tetra-hydro-8-hydroxy-6,9-dimethyl-	EtOH	231(3.69),305(3.65)	39-2932-64
3(2H)-Dibenzofuranone, 1,4,4a,9b-tetra-hydro-8-hydroxy-7,9b-dimethyl-	EtOH	231(3.65),303(3.62)	39-2932-64
1,4-Ethanonaphthalene-5,8-dione, 1,4,4a,8a-tetrahydro-1-methoxy-4-methyl-	EtOH	221(4.04)	39-2932-64
Fraxinellon	MeOH	215(4.1),315s(1.2)	49-1324-65
4H-Inden-4-one, 1-hydroxy-6-(2-hy-droxyethyl)-2,5,7-trimethyl-	EtOH	249(4.30),374(3.43), 440(3.20)	78-1231-65
	EtOH	249(4.30),374(3.43), 440(3.20)	94-0853-64

Compound	Solvent	$\lambda_{max}(\log \epsilon)$	Ref.
$C_{14}H_{16}O_3S$			
Cinnamic acid, β-(acetonylthio)-, ethyl ester	ether	209(4.14),274(4.07)	24-2109-64
$C_{14}H_{16}O_4$			
1,3-Cyclopentadiene-1-carboxylic acid, 2-methyl-, dimer	EtOH	231(3.96)	70-1460-65
Cyclopentanecarboxylic acid, 4-(p-methoxyphenyl)-2-oxo-, methyl ester	EtOH	275(3.39)	65-1985-64
3(2H)-Dibenzofuranone, 1,4-dihydro-8-hydroxy-, dimethyl acetal	EtOH	225(4.04),294(3.76)	39-2932-64
1,4-Ethanonaphthalene-5,8-dione, 1,4,4a,8a-tetrahydro-1,3-dimethoxy-	EtOH	221(4.06)	39-2932-64
$C_{14}H_{16}O_4S$			
Cinnamic acid, β-[(carboxymethyl)thio]-, ethyl methyl ester	ether	211(4.02),275(4.02)	24-2109-64
$C_{14}H_{16}O_5$			
Furo[3,2-g][1]benzopyran-5-one, 2,3,6,7-tetrahydro-4,9-dimethoxy-7-methyl-	MeOH	227(4.23),290(4.13)	5-0167-65E
4-Chromone, 6-ethyl-7-hydroxy-5,8-dimethoxy-2-methyl-	MeOH	303(3.95)	5-0167-65E
4-Cyclohexene-1,2-dicarboxylic anhydride, 3-(4-carboxy-3-butenyl)-, methyl ester	heptane	204(3.20)	44-1061-65
Hexanedioic acid, 3-hydroxy-3-(p-methoxyphenyl)-, γ-lactone, methyl ester	EtOH	274(3.19)	65-1985-64
2-Hexenedioic acid, 3-(p-methoxyphenyl)-, monomethyl ester	EtOH	291(4.24)	65-1985-64
$C_{14}H_{16}O_5S$			
Cinnamic acid, β-(acetonylsulfonyl)-, ethyl ester	ether	211(4.23),260(3.62)	24-2109-64
$C_{14}H_{16}O_5S_3$			
2-Hydroxy-1-naphthoylium methyl sulfate, dimethyl mercaptole	MeOH	318(4.16),370(3.30), 460(3.41)	5-0037-65D
$C_{14}H_{16}O_6$			
Acetoacetic acid, 2-(2-acetyl-3,6-dihydroxyphenyl)-, ethyl ester	EtOH	227(4.09),354(3.62)	33-0769-64
Coriamyrtinisonorketone	EtOH	232(3.93),314(1.65)	88-4191-65
$C_{14}H_{16}O_6S$			
Cinnamic acid, β-[(carboxymethyl)sulfonyl]-, ethyl methyl ester	ether	212(4.14),258(3.49)	24-2109-64
Malic acid, 3-(methylthio)-, dimethyl ester, benzoate	MeOH	229(4.21)	24-1581-65
$C_{14}H_{16}O_7$			
Acetic acid, 3,4-diacetoxy-5-methoxyphenyl)-, methyl ester	EtOH	212(4.47),246(3.94), 294(3.41)	78-1411-65
Tutinisonorketone	EtOH	232(3.99),320(1.71)	88-4191-65
$C_{14}H_{16}O_8$			
Isophthalic acid, 4-(carboxymethyl)-2,6-dihydroxy-, 3,4-diethyl ester	EtOH	315.5(3.51)	1-1677-65

Compound	Solvent	$\lambda_{max}(\log \varepsilon)$	Ref.
$C_{14}H_{16}S$			
Azulene, 4,6,8-trimethyl-1-(methylthio)-	C_6H_{12}	246(4.44),302(4.48), 320(4.46),360(3.70), 372(3.70),392(3.67), 589(2.58)	44-2715-65
6H-Benzo[b]cyclohepta[d]thiophene, 7,8,9,10-tetrahydro-4-methyl-	n.s.g.	234(4.53),266(3.90), 294(3.57),304(3.58)	22-0989-64
$C_{14}H_{17}BO_2$			
Boron, (2,4-pentanedionato)(o-phenyl-enetrimethylene)-	n.s.g.	331(3.18)	5-0040-65I
$C_{14}H_{17}BrN_2$			
2,2'-Dipyrrylmethine, 5-bromo-3,3',4,4',5'-pentamethyl-	EtOH	232(3.83),472(4.40)	12-1835-65
$C_{14}H_{17}BrN_4O$			
2-Amino-1-[3-(2-pyridylcarbamoyl)-propyl]pyridinium bromide	H_2O	230(4.15),275(3.93), 297(3.86)	39-2763-64
$C_{14}H_{17}BrO_6$			
Acetic acid, (4-bromo-2-carbethoxy-3,5-dimethoxyphenyl)-, methyl ester	EtOH	293.5(3.59)	1-1677-65
$C_{14}H_{17}BrO_8S$			
Glucopyranose, 1-[(p-bromophenyl)sul-fonyl]-1-deoxy-, 6-acetate	MeOH	234(4.29)	44-1782-64
$C_{14}H_{17}ClN_2O$			
1H-Azepino[5,4,3-cd]indole-2-methanol, 9-chloro-3,4,5,6-tetrahydro-5,6-dimethyl-	EtOH	230(3.59),290(3.90)	87-0200-65
$C_{14}H_{17}ClN_2O_6S$			
5-Thia-1-azabicyclo[4.2.0]oct-2-ene-2-carboxylic acid, 7-(4-chlorobutyr-amido)-3-(hydroxymethyl)-8-oxo-, acetate	pH 6.0	261(3.95)	39-5015-65
$C_{14}H_{17}ClN_4S$			
4-Pyrimidinethiol, 2-amino-5-[3-(p-chloroanilino)propyl]-6-methyl-	pH 1	260(3.75),338(4.19)	4-0088-64
	pH 7	251(4.26),350(4.19)	4-0088-64
	pH 13	248(4.35),318(4.17)	4-0088-64
$C_{14}H_{17}ClO$			
Inden-7-ol, 5-(2-chloroethyl)-2,4,6-trimethyl-	EtOH	266(3.99),302(3.48), 312(3.46)	35-1594-65
$C_{14}H_{17}Cl_2N_2O_2P$			
Phosphonic dichloride, (1,3,4,9-tetra-hydro-6-methoxy-1,1-dimethyl-2H-pyrido[3,4-b]indol-2-yl)-	EtOH	226(4.47),274(3.95), 293s(3.80),306s(3.58)	44-2864-64
$C_{14}H_{17}IN_2$			
1-Methyl-3-(1-methyl-2-pyrrolidinyl-idene)-3H-indolium iodide	EtOH	212(4.58),254(4.17), 262s(--),270(3.95), 335(4.30)	87-0415-64

Compound	Solvent	$\lambda_{max}(\log \epsilon)$	Ref.
$C_{14}H_{17}N$			
3H-Indole, 3a,4,5,6,7,7a-hexahydro-2-phenyl-, trans	EtOH-HCl	273(4.24)	95-0674-64
	EtOH	245(4.20)	95-0674-64
Lepidine, 5,7-diethyl-	EtOH	238(4.70),298(3.60),310(3.52),323(3.36)	95-0314-65
Phenanthridine, hexahydro-N-methyl-	EtOH	234(4.42),324(3.53)	44-0681-64
Pyridine, 2-benzyl-1,2-dihydro-1,4-dimethyl-	EtOH	330(3.54)	44-1647-64
Quinaldine, 5,7-diethyl-	EtOH	240(4.70),296(3.64),309(3.70),322(3.71)	95-0314-65
$C_{14}H_{17}NO$			
3H-Benzo[c]quinolizin-3-one, 1,2,4,4a,5,6-hexahydro-2-methyl-	EtOH	210(4.25),254(3.98),294(3.27)	78-2961-65
Carbazole, 1,2,3,4-tetrahydro-7-methoxy-9-methyl-	EtOH	233(4.54),281(3.71),296(3.73)	33-1349-65
1H-Carbazol-3-ol, 2,3,4,4a-tetrahydro-1,4a-dimethyl-	MeOH-HCl	230(3.86),275(3.83)	24-2111-65
	MeOH	254(3.83)	24-2111-65
9a,4a-(Epoxyethano)carbazole, 1,2,3,4-tet ahydro-	hexane	245(3.9),290(3.4)	78-2047-64
	MeOH-HCl	230(3.9),270(3.7)	78-2047-64
	MeOH	247(3.8),282(3.3)	78-2047-64
	MeOH-NaOH	244(3.9),284(3.3)	78-2047-64
Indole, 1-ethyl-3-isobutyryl-	EtOH	212(4.42),245(4.13),250s(4.11),302(4.17)	87-0094-64
2,4-Octadienanilide	EtOH	276(4.46)	54-0581-65
6(2H)-Quinolone, 2,2,4,7,8-pentamethyl-	MeOH	241(3.9),303(4.2),370(3.5)	24-2939-65
8(2H)-Quinolone, 2,2,4,6,7-pentamethyl-	MeOH	256(4.25),485(3.52)	24-2939-65
Spiro[cyclohexane-1,4''(3'H)-isoquinolin]-1'-one, 2',3'-dihydro-	EtOH	230(4.01)	94-1084-65
Spiro[cyclohexane-1,4'(1'H)-quinolin]-2'(3'H)-one	EtOH	251(4.07)	94-1084-65
$C_{14}H_{17}NO_2$			
1-Butanone, 2-ethyl-2-hydroxy-1-indol-3-yl-	EtOH	243(4.06),258(3.91),301(4.06)	87-0094-64
Carbostyril, 4-hydroxy-3-isopentyl-	n.s.g.	227(4.44),276(3.72),312(3.67),325s(3.60)	44-2598-64
	base	225(4.4),252(3.86),305(3.90)	44-2598-64
Cyclohexanecarboxamide, N-benzoyl-	EtOH	233(4.13),275(3.23)	44-2222-65
2-Indancarboxamide, 1-tert-butyl-3-oxo-	EtOH	250(4.19),298(3.37)	44-0396-65
Indole-3-acetic acid, 2,5-dimethyl-, ethyl ester	MeOH	225(4.46),279(3.85),286(3.86),297(3.75)	87-0204-65
Indole-3-carboxaldehyde, 1-ethyl-5-methoxy-2,6-dimethyl-	MeOH	216(4.55),258(4.26),283(4.18),310(4.10)	35-3878-64
3H-Indole-3-propionic acid, α,2,3-trimethyl-	MeOH-HCl	231(4.01),278(3.86)	24-2111-65
	MeOH	218(4.37),258(3.91)	24-2111-65
Ketone, o-hydroxyphenyl 2-piperidinovinyl	EtOH	213(4.15),227(4.09),249(3.88),256(3.93),262(3.93),362(4.50)	39-3610-65
1-Propanone, 1-(1-ethyl-3-indolyl)-2-hydroxy-2-methyl-	EtOH	211(4.41),247(4.10),306(4.15)	87-0094-64
$C_{14}H_{17}NO_2S$			
2H-1,3-Thiazine-4-carboxylic acid, 3,6-dihydro-2,5-dimethyl-, benzyl ester	EtOH	285.0(3.51)	39-0788-64

Compound	Solvent	λ_{max}(log ε)	Ref.
4H-1,3-Thiazine-4-carboxylic acid, 5,6-dihydro-2,5-dimethyl-, benzyl ester	EtOH-HCl	264(4.13)	39-0766-64
	EtOH	239(4.06)	39-0766-64
	EtOH-NaOH	236(4.08)	39-0766-64
$C_{14}H_{17}NO_3$			
12H-Azepino[2,1-b][1,3]benzoxazin-12-one, 5a,6,7,8,9,10-hexahydro-5a-methoxy-	EtOH	209(4.55),236(3.99), 294(3.42)	78-3537-65
7H-Benzo[de]quinolin-7-one, 1,2,3,8,9,9a-hexahydro-6-hydroxy-5-methoxy-1-methyl-	EtOH	227(4.20),272(3.91), 364(3.55)	44-2486-64
	N NaOH	239(4.23),279(3.73), 390(3.76)	44-2486-64
Caprolactam, N-(o-methoxybenzoyl)-	EtOH	212(4.22),223(4.18), 290(3.35)	78-3537-65
Cyclopent[ij]insoquinolin-7(1H)-one, 2,3,8,8a-tetrahydro-5,6-dimethoxy-1-methyl-	EtOH	259(3.90),334(3.56)	31-0380-64
Indole-2-carboxylic acid, 5-(1-hydroxyethyl)-3-methyl-, ethyl ester	EtOH	232(4.42),297(4.31)	87-0141-64
Phyllantine	EtOH	256(4.32),327(3.24)	28-0337-65A
2-Piperidone, 1-[(benzyloxy)acetyl]-	dioxan	216(4.05)	78-3537-65
2-Pyrrolidinone, 1-[2-(benzyloxy)-propionyl]-	dioxan	216(4.13)	78-3537-65
$C_{14}H_{17}NO_3S$			
4H-1,3-Thiazine-4-carboxylic acid, 5,6-dihydro-4-hydroxy-2,5-dimethyl-, benzyl ester	EtOH	238.0(3.70)	39-0788-64
$C_{14}H_{17}NO_4$			
Indole-4,7-dione, 1-ethyl-3-(hydroxymethyl)-5-methoxy-2,6-dimethyl-	MeOH	231(4.23),287(4.14), 350(3.50),460(3.11)	35-3878-64
3-Indolineacetic acid, 3-hydroxy-2-oxo-, tert-butyl ester	EtOH	209(4.44),253(3.79), 291(3.15)	44-2431-64
$C_{14}H_{17}NO_4S$			
2H-1,2-Benzothiazine, 3-acetyl-4-isopropoxy-2-methyl-, 1,1-dioxide	EtOH	235s(3.67),296(4.00), 312(4.03)	44-2241-65
2-Benzyl-1-methylpyridinium methyl sulfate	MeCN	267(3.93)	35-4917-64
$C_{14}H_{17}NO_5$			
β-Alanine, N-acetyl-N-(p-carbethoxyphenyl)-	EtOH	241(4.13)	44-3404-65
2H,6H-Chromeno[7,6-b]-1,4-oxazine, 3,4,7,8-tetrahydro-6-hydroxy-8-methoxycarbonyl-3-methyl-	MeOH	241(3.6),304(3.4)	39-7348-65
Malonic acid, [(N-hydroxyanilino)-methylene]-, diethyl ester	EtOH	222(4.04),317(4.51)	78-2735-65
$C_{14}H_{17}NO_6$			
Carbostyril, 4-hydroxy-3-(2-hydroxyethyl)-5,7,8-trimethoxy-	n.s.g.	220(4.50),232s(4.45), 252(4.35),260(4.36), 308(4.19),318s(4.10)	2-0071-65
Pyrrole-3-carboxylic acid, 1,5-diacetyl-4-acetoxy-2-methyl-, ethyl ester	EtOH	228(4.29),287(4.13)	23-1524-64

Compound	Solvent	λ_{max}(log ε)	Ref.
$C_{14}H_{17}NO_7$			
L-Mandelonitrile, p-hydroxy-,	acid	230(4.03)	44-0603-65
β-D-glucopyranoside	base	255(4.18)	44-0603-65
$C_{14}H_{17}N_3$			
1H-Azepine-3-carbonitrile, 4-amino-1-benzyl-2,5,6,7-tetrahydro- ?	EtOH	272(4.05)	39-5130-64
Pyrimidine, 1-benzyl-1,2-dihydro-	pH 1.0	232(4.21),319(3.63)	39-5542-65
2-(isopropylimino)-	pH 13.7	245(4.14),373(3.38)	39-5542-65
Pyrimidine, 2-(benzylimino)-	pH 1.0	232(4.26),317(3.65)	39-5542-65
1,2-dihydro-1-isopropyl-	pH 13.2	247(4.27),371(3.44)	39-5542-65
Pyrrole, 2-[N-[p-(dimethylamino)-phenyl]formimidoyl]-1-methyl-	EtOH	239(3.92),297(4.07), 366(4.29)	12-0894-64
3-Pyrroline-3-carbonitrile, 4-amino-	pH 1	207(3.78),259(4.10)	39-4546-65
1-benzyl-5,5-dimethyl-	pH 13	216(4.20),262(4.07)	39-4546-65
$C_{14}H_{17}N_3O$			
4-Pyrimidinol, 5-(3-anilinopropyl)-	pH 1	238(3.99)	4-0079-64
6-methyl-	pH 7	240(4.21),273s(--)	4-0079-64
	pH 13	238(4.27),273s(--)	4-0079-64
$C_{14}H_{17}N_3O_2$			
Indoline, 3-isonitroso-2-(α-oximino-hexyl)-	EtOH	258(4.40)	32-1248-64
4-Pyrimidinol, 2-amino-6-methyl-	pH 1	267(4.01)	4-0088-64
5-(3-phenoxypropyl)-	pH 7	271(3.88)	4-0088-64
	pH 13	277(3.98)	4-0088-64
$C_{14}H_{17}N_3O_3$			
Acetic acid, 2-carboxy-1-(indol-3-yl-methyl)hydrazide, ethyl ester	EtOH	280(4.01)	2-0423-64
Acrylic acid, 3-anilino-2-cyano-3-(2-hydroxyethylamino)-, ethyl ester	EtOH	287.0(4.32)	95-0387-65
$C_{14}H_{17}N_3O_4$			
Valeric acid, 2-acetyl-4-hydroxy-5-phenoxy-, γ-lactone, semicarbazone	n.s.g.	222(4.24),274(3.85)	39-0141-64
$C_{14}H_{17}N_3S$			
4-Pyrimidinethiol, 5-(3-anilino-	pH 1	322(4.14)	4-0079-64
propyl)-6-methyl-	pH 7	240s(4.12),288(4.01), 333(4.05)	4-0079-64
	pH 13	240s(4.23),302(4.15)	4-0079-64
$C_{14}H_{17}N_5O_4$			
Pyrido[3,4-b]pyrazine-5,7-dicarbamic acid, 3-methyl-, diethyl ester	pH 1	242(4.49),301(4.19), 385(3.75)	44-0734-64
	pH 7	257(4.54),290(3.82), 376(3.69)	44-0734-64
	pH 13	259(3.47),345(3.50), 392(3.67)	44-0734-64
$C_{14}H_{17}N_5O_5$			
Triazene, 1,3-bis(2'-methyl-4-carbethoxyoxazolo)-	EtOH	311(3.36)	7-0576-65
$C_{14}H_{18}$			
Benzene, o-bis(2-methylpropenyl)-	n.s.g.	229(4.34)	35-0136-65
Cyclooctene, 1-phenyl-	n.s.g.	248(4.07)	44-3467-64

Compound	Solvent	λ_{max}(log ϵ)	Ref.
$C_{14}H_{18}ClNO_3$			
Pyrrolidine, 3-acetoxy-4-chloro-2-(p-methoxybenzyl)-, trans, hydrochloride	MeOH	225(4.06),276(3.21), 282(3.16)	44-3467-64
$C_{14}H_{18}ClNO_4S_2$			
2-(Diethylamino)-4-p-tolyl-1,3-dithiol-1-ium perchlorate	EtOH	228(4.12),244(4.12), 302(4.00),325(4.04)	44-1703-64
$C_{14}H_{18}ClNO_5S_2$			
2-(Diethylamino)-4-(p-methoxyphenyl)-1,3-dithiol-1-ium perchlorate	EtOH	260(4.06),289(4.05), 330(3.94)	44-1703-64
$C_{14}H_{18}Cl_2Pd_2$			
Palladium, dichloro-π-cycloheptadienyldi-	$CHCl_3$	264(4.05),291(3.58)	39-5002-64
$C_{14}H_{18}INO_2$			
1-Naphthalenecarbamic acid, 1,2,3,4-tetrahydro-2-iodo-, isopropyl ester	MeOH	267(3.01),274(2.94)	44-3640-64
$C_{14}H_{18}N_2$			
Benz[cd]indole, 3-(1-aminoethyl)-1,3,4,5-tetrahydro-2-methyl-	EtOH	227(4.55),274(3.84), 279(3.85),289f(3.70)	44-0843-64
isomer	EtOH	228(4.54),275f(3.85), 280(3.85),290f(3.72)	44-0843-64
Benz[c]cinnoline, 1,2,3,4,5,6-hexahydro-2,4-dimethyl-	MeOH-HCl	230(3.89),313(3.53)	24-2111-65
	MeOH	300(3.69)	24-2111-65
Cyclobutane, 1,2-diethyl-1,2-dimethyl-3,4-di(cyanomethylene)-	n.s.g.	282(4.26)	77-0321-65
Dipyrromethene, 3,3',4,5,5'-pentamethyl-	EtOH-N HCl	232(3.84),357(3.59), 474(4.84)	12-0363-65
	EtOH	222(3.97),270(3.41), 433(4.39)	12-0363-65
	CCl_4	434(4.41)	12-0363-65
9a,4a-(Iminoethano)carbazole, 1,2,3,4-tetrahydro-	MeOH-acid	233(3.8),238(3.7), 283(3.7)	78-1737-64
	MeOH	239(3.8),289(3.3)	87-0415-64
Indole, 3-(1,5-dimethyl-2-pyrrolidinyl)-	EtOH	217(4.58),274s(3.77), 280(3.80),289(3.72)	87-0415-64
Indole, 1-methyl-3-(1-methyl-2-pyrrolidinyl)-	EtOH	221(4.54),287(3.77), 328s(2.26)	87-0415-64
Indole, 5-methyl-3-(1-methyl-2-pyrrolidinyl)-	EtOH	218(4.52),277(3.76), 285(3.76),295(3.61)	87-0415-64
Indole, 3-(1-methyl-2-piperidyl)-	EtOH	218(4.58),274(3.81), 280(3.82),288(3.74)	44-2534-65
$C_{14}H_{18}N_2O$			
Benz[cd]indole, 1,3,4,5-tetrahydro-3-[1-(hydroxyamino)ethyl]-2-methyl-	EtOH	225(4.48),275f(3.78), 280(3.79),289f(3.65)	44-0843-64
1,2-Cyclohexanedione, 3,5-dimethyl-, 2-(phenylhydrazone)	MeOH	233(4.04),290(3.31), 300(3.31),370(4.23)	24-2111-65
1,2-Cyclohexanedione, 3,6-dimethyl-, 1-(phenylhydrazone)	MeOH	234(3.93),291(3.21), 301(3.21),370(4.18)	24-2111-65
Indene-2-carboxamide, 3-amino-1-tert-butyl-	EtOH	243(3.99),330(4.11)	44-0396-65
Indole, 3-(5-methylaminovaleryl)-	EtOH	207(4.49),242(4.04), 256(3.94),297(3.96)	44-2534-65

Compound	Solvent	λ_{max}(log ε)	Ref.
Indoline, 1-acetyl-3-(2-pyrrolidinyl)-	EtOH-HCl	252(4.19),278s(3.49), 287s(3.34)	87-0415-64
2-Indoline, 3-methyl-3-piperidino-	EtOH	207(4.50),249(3.89), 282s(3.21)	44-2431-64
2-Indolinone, 3-(1-methyl-4-piperidyl)-	EtOH	250(3.82),277(3.15)	87-0626-65
2-Indolinone, 3-(2-piperidylmethyl)-	EtOH	235(3.83),287(3.38)	87-0626-65
2-Indolinone, 3-(3-piperidylmethyl)-	EtOH	249(3.93),277(3.15)	87-0626-65
2-Indolinone, 3-(4-piperidylmethyl)-	EtOH	249(3.94),280s(3.16)	87-0626-65
3H-Indol-3-one, 2-hexyl-, oxime	EtOH	241(4.28),253(4.28), 314(3.66)	32-1248-64
Ketone, 5-methyl-3-indazyl tert-pentyl	EtOH	239(3.95),244(3.96), 307(3.99)	32-1248-64
1H-Pyrido[3,4-b]indole, 2,3,4,9-tetra- hydro-6-methoxy-1,1-dimethyl-	EtOH	226(4.44),278(3.91), 295s(3.85),308s(3.59)	44-2864-64
1H-Pyrido[3,4-b]indole, 2,3,4,9-tetra- hydro-7-methoxy-1,1-dimethyl-	EtOH	226(4.58),269(3.66), 296(3.76)	44-2864-64
Tryptamine, 5-acetyl-N,N-dimethyl-	EtOH-HCl	253(4.52),297(3.88)	87-0141-64
$C_{14}H_{18}N_2O_2$			
p-Benzoquinone, 2,5-dipyrrolidino-	methyl cellosolve	226(4.45),373(4.42), 507(2.63)	78-1889-64
2-Indolinone, 3-methoxy-3-(1-pyrro- lidinylmethyl)-	MeOH	250(3.76),292(3.16)	44-3610-65
2-Pentanone, 1-(3,4-dihydro-4-quin- azolinyl)-4-hydroxy-4-methyl-	EtOH	227(4.19),233(4.12), 303(3.92)	94-0291-65
$C_{14}H_{18}N_2O_3$			
Benzimidazole, 5,6-dimethyl-1-(2-deoxy- D-ribofuranosyl)-	pH 1	255s(3.68),271s(3.82), 277(3.91),285(3.89)	35-4940-65
	pH 11	249(3.86),278(3.72), 288(3.72)	35-4940-65
	MeOH	249(3.87),279(3.72), 282(3.72),289(3.73)	35-4940-65
Quinoline, 7-(dimethylamino)- 5,6,8-trimethoxy-	EtOH	240(4.05),281(4.08), 356(3.39)	95-0985-65
Valeric acid, 4-hydroxy-5-isopropoxy- 2-oxo-, γ-lactone, phenylhydrazone	n.s.g.	233(4.16),295(4.00), 333(4.50)	39-0141-64
$C_{14}H_{18}N_2O_4$			
Valeric acid, 5-allyloxy-4-hydroxy- 2-oxo-, phenylhydrazone	n.s.g.	293(4.18),317(4.24)	39-0141-64
$C_{14}H_{18}N_2O_5$			
Benzimidazole, 5,6-dimethoxy- 1-(2-deoxy-D-ribofuranosyl)-	pH 1	293(4.05),301(3.99)	35-4940-65
	pH 11	249(3.77),257s(3.69), 291(3.90),299s(3.89)	35-4940-65
	MeOH	244(3.69),249(3.74), 255s(3.66),292(3.94), 296s(3.92),300(3.87)	35-4940-65
$C_{14}H_{18}N_2O_6S$			
Thymidine, 5'-thio-, 3',5'-diacetate	pH 1, 7	264(4.03)	44-0554-64
	pH 13	267(3.93)	44-0554-64
$C_{14}H_{18}N_2O_6S_2$			
5-Thia-1-azabicyclo[4.2.0]oct-2-ene- 2-carboxylic acid, 7-[2-(ethylthio)- acetamido]-3-(hydroxymethyl)-8-oxo-, acetate, sodium salt	pH 6.0	259(3.96)	39-5015-65

Compound	Solvent	$\lambda_{max}(\log \varepsilon)$	Ref.
$C_{14}H_{18}N_2S_2$			
2-Benzothiazolinethione, 3-hexamethyleneiminomethyl-	EtOH	240(4.24),319(4.38)	65-1588-64
3-Pyrroline-3-thiocarboxamide, 1-benzyl-4-mercapto-5,5-dimethyl-	pH 13	256s(3.49),300(3.52), 370(4.27)	39-4546-65
	EtOH	208(4.32),298(3.47), 376(4.32)	39-4546-65
$C_{14}H_{18}N_2Zn$			
Zinc, (2,2'-bipyridine)diethyl-	toluene	420(2.77)	101-0222-65A
$C_{14}H_{18}N_4$			
Pyrimidine, 2-amino-5-(3-anilinopropyl)-4-methyl-	pH 1	228(4.22),312(3.67)	4-0079-64
	pH 7	235(4.35),298(3.67)	4-0079-64
	pH 13	297(3.73)	4-0079-64
$C_{14}H_{18}N_4O$			
4-Pyrimidinol, 2-amino-5-(3-anilinopropyl)-6-methyl-	pH 1	225(4.05),264(4.00)	87-0024-64
	pH 13	239(4.24),280(3.97)	87-0024-64
$C_{14}H_{18}N_4O_4$			
Cyclobutanone, 2-tert-butyl-, 2,4-dinitrophenylhydrazone	$CHCl_3$	365(4.34)	22-2747-65
Cyclobutanone, 3-ethyl-2,2-dimethyl-, 2,4-dinitrophenylhydrazone	$CHCl_3$	364(4.34)	22-1968-64
2-Hexenal, 2,4-dimethyl-, 2,4-dinitrophenylhydrazone	$CHCl_3$	382.5(4.47)	22-0161-64
$C_{14}H_{18}N_4O_6S$			
Valine, N-[2-(7-nitro-4H-1,2,4-benzothiadiazin-3-yl)ethyl]-, S,S-dioxide	2% $NaHCO_3$	241(4.00),358(4.11)	32-0786-65
$C_{14}H_{18}N_4O_6S_2$			
Methionine, N-[2-(7-nitro-4H-1,2,4-benzothiadiazin-3-yl)ethyl]-, dioxide	2% $NaHCO_3$	241(4.02),359(4.11)	32-0786-65
$C_{14}H_{18}N_4O_7$			
Indolizine, octahydro-, picrate	MeOH	238(4.22),269(3.65), 280(3.63),293(3.64), 312(3.01),344(3.19), 370(2.97)	70-2216-65
$C_{14}H_{18}O$			
Bullvalene, tert-butoxy-	n.s.g.	235(3.40)(end absorption)	24-3385-65
	n.s.g.	237(3.34)	24-3385-65
2,5-Cyclohexadien-1-one, 2,4-diallyl-4,6-dimethyl-	EtOH	246(4.05)	33-0094-65
Cyclohexylideneethanol, α-phenyl-	EtOH	284(3.54)	39-3484-64
Inden-7-ol, 5-ethyl-2,4,6-trimethyl-	EtOH	266(3.99),300(3.45), 311(3.42)	35-1594-65
Khusilal	EtOH	232(4.09)	78-2617-64
1(2H)-Naphthalenone, 5,7-diethyl-3,4-dihydro-	EtOH	216(4.45),257(4.04), 308(3.37)	95-0555-64
1(2H)-Naphthalenone, 3,4-dihydro-7-isopropyl-5-methyl-	EtOH	249(4.23),296(3.17), 304(3.16)	23-3103-64
$C_{14}H_{18}O_2$			
p-Benzoquinone, trimethyl-(3-methyl-2-butenyl)-	EtOH	266(4.23),310(2.74)	37-1374-65

Compound	Solvent	λ_{max}(log ε)	Ref.
Naphthalene, 1,4-dihydro-6,7-dimethoxy-1,1-dimethyl-	EtOH	225(4.04),282(3.62), 285(3.63)	39-0361-65
2(1H)-Naphthalenone, 7-hydroxy-1,1,3,3-tetramethyl-	EtOH	220(3.85),283(3.35)	39-0361-65
2-Naphthoic acid, 3,4,4a,5,6,7,8,8a-octahydro-5-methylene-8-vinyl-	EtOH	215(3.98)	78-2617-64
$C_{14}H_{18}O_3$			
5(6H)-Benzocyclooctenone, 7,8,9,10-tetrahydro-2,4-dimethoxy-	EtOH	282(3.81)	5-0021-64G
	$HOAc-H_2SO_4$	325(3.83)	5-0021-64G
4,6-Heptadiynoic acid, 7-(1-hydroxy-1-cyclohexyl)-, methyl ester	EtOH	229(3.53),241(3.54), 255(3.31)	70-1237-65
Mexicanin E, dihydro-	EtOH	222(4.02)	44-2101-64
2-Naphthaleneacetic acid, 1,2,3,4,-4a,5,8,8a-octahydro-3-hydroxy-α,4a-dimethyl-8-oxo-, γ-lactone	EtOH	227(3.96)	78-2655-64
isomer	EtOH	230(3.93)	78-2655-64
2-Naphthaleneacetic acid, 1,2,3,4,-4a,7,8,8a-octahydro-3-hydroxy-α,4a-dimethyl-8-oxo-, γ-lactone	EtOH	281(1.45)	78-2655-64
2(1H)-Naphthalenone, 3,4-dihydro-6,7-dimethoxy-1,1-dimethyl-	EtOH	205(4.48),227s(3.85), 286(3.51)	39-0361-65
1-Naphthoic acid, 1,2,3,4,4a,7-hexahydro-1,4a-dimethyl-7-oxo-, methyl ester	EtOH	243(4.14)	44-0782-64
$C_{14}H_{18}O_4$			
5-Azuleneacetic acid, 2,3,3a,4,5,6,7,8-octahydro-2,6-dihydroxy-α,3a-dimethyl-8-oxo-, γ-lactone	EtOH	235(3.91)	78-0341-64
4-Cyclopentene-1,3-dione, 4-hydroxy-2-(1-hydroxy-2-methylpropylidene)-5-(3-methyl-2-butenyl)-	EtOH-HCl	255(4.21),280(4.08)	39-2251-65
	EtOH-NaOH	275(4.42),300(4.39)	39-2251-65
$C_{14}H_{18}O_5$			
Chroman-6-carboxylic acid, 5-hydroxy-8-methoxy-2,2-dimethyl-, methyl ester	EtOH	212(4.40),228(4.17), 270(4.11),320(3.85)	78-1317-64
Succinic acid, 2-(2-hydroxy-4-methylphenethyl)-3-methyl-	EtOH	219(3.78),276(3.23), 283(3.20)	94-0064-65
$C_{14}H_{18}O_9S$			
Dimethyl(1,2,3,4-tetracarboxy-4-hydroxy-1,3-butadienyl)sulfonium hydroxide, inner salt, tetramethyl ester	MeOH	220(4.02),264(3.90), 408(4.09)	24-1581-65
$C_{14}H_{19}BrN_6S$			
Urea, 1-[bis(isopropylidenehydrazino)-methylene]-3-(p-bromophenyl)-2-thio-	n.s.g.	255(4.52),300(4.40)	39-4448-65
$C_{14}H_{19}ClO_3$			
Crotonic acid, 3-[(8a-chlorodecahydro-1-hydroxy-1-naphthyl)oxy]-, δ-lactone	EtOH	248(3.80)	39-4456-65
isomer	EtOH	251(3.87)	39-4456-65
$C_{14}H_{19}N$			
Acridine, octahydro-N-methyl-, cis	EtOH	256(4.18),305(3.57)	44-0681-64
Acridine, octahydro-N-methyl-, trans	EtOH	260(4.15),301(3.59)	44-0681-64
Indole, 1-isopentyl-3-methyl-	EtOH	228(4.53),291(3.69)	78-0989-65
Indole, 2-isopentyl-3-methyl-	EtOH	228(4.57),284(3.85), 292(3.81)	78-0989-65

Compound	Solvent	$\lambda_{max}(\log \epsilon)$	Ref.
Indolizine, 1,2-diethyl-5,8-dimethyl-	EtOH	241(4.56),281s(3.41), 292(3.62),305(3.73), 341(3.45)	39-1518-65
Indolizine, 1,2,3,5,6,7-hexamethyl-	EtOH-HCl	228(4.28),252(3.75), 331(3.66)	39-0893-64
	EtOH	246(4.32),294(3.46), 301(3.85),309(3.86), 357(3.57)	39-0893-64
Phenanthridine, octahydro-N-methyl-, cis	EtOH	257(4.08),304(3.57)	44-0681-64
trans	EtOH	257(4.08),302(3.60)	44-0681-64
Quinoline, 1,2-dihydro-2,2,4,6,7-pentamethyl-	pH 1	222s(4.24),264(4.07), 294s(2.99)	44-0285-65
	pH 10	228(4.48),265s(3.44), 330(3.49)	44-0285-65
	EtOH	231(4.51),265s(3.44), 339(3.49)	44-0285-65
Spiro[cyclohexane-1,4'(1'H)-isoquinoline], 2',3'-dihydro-	EtOH	260(2.65),266(2.69), 273(2.62)	94-1084-65
hydrochloride	EtOH	264(2.53),269(2.48), 272(2.50)	94-1084-65
Spiro[cyclohexane-1,4'(1'H)-quinoline], 2',3'-dihydro-	EtOH	248(3.95),300(3.39)	94-1084-65
hydrochloride	EtOH	257s(2.5),261(2.55), 269(2.37)	94-1084-65
$C_{14}H_{19}NO$			
Carbazol-2-ol, 1,2,3,4,4a,9a-hexahydro-3,4a-dimethyl-	MeOH	242(3.81),291(3.38)	24-2111-65
Carbazol-3-ol, 1,2,3,4,4a,9a-hexahydro-1,4a-dimethyl-	MeOH-HCl	255(2.90),261(3.05), 267(3.04)	24-2111-65
	MeOH	243(4.12),292(3.67)	24-2111-65
Khusilal, oxime	EtOH	235(4.26)	78-2617-64
2-Octenanilide	EtOH	270(4.18)	54-0581-65
2-Oxa-1-azabicyclo[3.3.0]octane, 6,6-dimethyl-3-phenyl-	EtOH	258(2.45)	23-2717-65
2-Oxa-1-azabicyclo[3.3.0]octane, 8,8-dimethyl-3-phenyl-	EtOH	258(2.44)	23-2717-65
Pyridine, 2-[[2-(methoxymethyl)-1-cyclohexen-1-yl]methyl]-	EtOH	262(3.60)	65-2632-64
$C_{14}H_{19}NOS$			
Morpholine, N-(thion-α-phenylbutyryl)-	MeOH	284(4.15),359(1.82)	35-0051-65
$C_{14}H_{19}NOS_2$			
Carbamic acid, diethyldithio-, ester with 2-mercapto-4'-methylacetophenone	EtOH	255(4.42)	44-1703-64
$C_{14}H_{19}NO_2$			
Acrylophenone, 3-(diethylamino)-2'-methoxy-	EtOH	256(3.64),308(4.25), 339(4.23)	39-3610-65
Benzaldehyde, p-(2-piperidinoethoxy)-	EtOH	284(4.25)	87-0511-64
Furo[3,2-c]quinolin-4(2H)-one, 3,5,6,7,8,9-hexahydro-2-isopropyl-	n.s.g.	288(3.76)	44-2598-64
2-Quinolineacetic acid, 1,2,3,4-tetrahydro-α-methyl-, ethyl ester	EtOH	209(4.30),250(3.92), 303(3.27)	78-2961-65
$C_{14}H_{19}NO_2S_2$			
Carbamic acid, diethyldithio-, ester with 2-mercapto-4'-methoxyacetophenone	EtOH	276(4.45)	44-1703-64

Compound	Solvent	λ_{max}(log ϵ)	Ref.
$C_{14}H_{19}NO_3$			
Ethylamine, 2-[(6-allyl-1,4-benzodioxan-5-yl)oxy]-N-methyl-	EtOH	273(3.01)	65-2994-64
$C_{14}H_{19}NO_3S$			
L-Methionine, N-salicylidene-, ethyl ester	EtOH	214(4.35),221s(4.32), 258(4.13),320(3.65), 408(2.05)	35-1757-65
$C_{14}H_{19}NO_4$			
3H-Azepine-3,6-dicarboxylic acid, 2,7-dimethyl-, diethyl ester	EtOH	235s(3.70),282(3.79)	39-2411-65
4H-Azepine-3,6-dicarboxylic acid, 2,7-dimethyl-, diethyl ester	hexane	216(4.43),269(3.60), 326(3.41)	39-2411-65
4-Morpholineacetic acid, α-(2-hydroxy-3,5-xylyl)-	pH 13	228s(3.81),290(3.39), 308s(2.99)	24-0363-64
	EtOH	220s(3.88),288(3.47)	24-0363-64
Nordehydrocycloheximide	EtOH-NaOH	315(4.27)	44-2595-64
Pyrrole-1-acrylic acid, 3-carboxy-β,2-dimethyl-, diethyl ester	EtOH	248(3.97)	39-2411-65
3,4-Pyrrolidinediol, 2-(p-methoxybenzyl)-, 3-acetate	MeOH	224(4.03),277(3.2)	44-2334-65
3,4-Pyrrolidinediol, 2-(p-methoxybenzyl)-, 4-acetate	MeOH	225(4.04),276(3.14), 283(3.07)	44-2334-65
$C_{14}H_{19}NO_4S$			
2H-1,2-Benzothiazine-3-methanol, 4-isopropoxy-α,2-dimethyl-, 1,1-dioxide	EtOH	272(3.88),298(3.76)	44-2241-65
$C_{14}H_{19}NO_5$			
Alanine, N-acetyl-3-(3,4-dimethoxyphenyl)-2-methyl-	MeOH	230(3.95),279(3.47)	44-2053-64
	MeCN	232(3.95),288(3.47)	44-1424-64
$C_{14}H_{19}NO_5S$			
Benzenesulfonic acid, p-nitro-, 6-methyl-5-heptenyl ester	EtOH	250(4.06)	35-5593-64
Benzenesulfonic acid, p-nitro-, 6-methyl-6-heptenyl ester	EtOH	250(4.06)	35-5593-64
$C_{14}H_{19}N_3O$			
Indazole, 3-(2-piperidinoethoxy)-	EtOH	216(4.41),252(3.15), 301(3.65)	33-1986-64
	EtOH	216(4.53),252(3.18), 301(3.68)	33-1986-64
3-Indazolinone, 2-(2-piperidinoethyl)-	EtOH-HCl	223(4.38),310(3.68)	33-1986-64
	EtOH	218(4.37),235(4.17), 311(3.70)	33-1986-64
	EtOH-NaOH	216(4.29),236(4.46), 354(3.68)	33-1986-64
$C_{14}H_{19}N_3O_2$			
2,3-Piperidinedione, 1,6-dimethyl-, 3-(p-methoxyphenylhydrazone)	EtOH	202(4.12),231(4.06), 308(4.19),340(4.31)	39-7165-65
isomer m. 68-70°	EtOH	202(4.08),246(3.99), 319(3.96),366(4.27)	39-7165-65
Pyrido[2,3-d]pyrimidine-2,4-diol, 7-butyl-6-propyl-	pH 1	246(3.88),320(4.04)	44-2674-64
	pH 11	265(3.92),318(3.95)	44-2674-64
	EtOH	248(3.95),314(3.93)	44-2674-64

Compound	Solvent	$\lambda_{max}(\log \epsilon)$	Ref.
$C_{14}H_{19}N_3O_3$			
Benzamide, p-nitro-N-(2-piperidino-ethyl)-	MeOH	262(4.06)	87-0107-65
1H-Pyrrolo[2,3-c]pyridine-2-carboxylic acid, 3-[(dimethylamino)methyl]-5-methoxy-, ethyl ester, dihydro-chloride	EtOH	283(4.08),292(4.14), 347(3.57)	35-3530-65
$C_{14}H_{19}N_3O_4S_2$			
4-Morpholinecarboximidic acid, N-(N-acetylsulfanilyl)thio-, methyl ester	EtOH	276.0(4.33)	95-0391-65
$C_{14}H_{19}N_3O_7$			
Malonic acid, acetamido[(2,4-dihydroxy-5-pyrimidinyl)methyl]-, diethyl ester	pH 1	261(3.11)	4-0049-65
	pH 11	286(4.06)	4-0049-65
$C_{14}H_{19}N_3O_8S$			
β-D-Glucopyranoside, ethyl 2-deoxy-2-(2,4-dinitroanilino)-1-thio-	EtOH	265(3.88),349(4.00)	44-1556-65
$C_{14}H_{19}N_5O$			
1-Piperidinecarboxamide, N-(2,3-di-methylpyrazolo[1,5-a]pyrimidin-7-yl)-	EtOH	233(4.63),286(3.54), 294(3.57),329(3.82)	94-1207-65
hydrochloride	EtOH	233(4.66),276s(3.38), 286(3.54),294(3.57), 329(3.74)	94-1207-65
$C_{14}H_{19}N_5O_4$			
Pyrido[3,4-b]pyrazine-5,7-dicarbamic acid, 1,2-dihydro-3-methyl-, diethyl ester	pH 1	255(4.76),312(3.95)	44-0734-64
	pH 7	250(4.58),299(3.87)	44-0734-64
	pH 13	248(4.53),289(3.89)	44-0734-64
$C_{14}H_{19}N_5O_6$			
Pteridine, 4-(dimethylamino)-7-(β-D-glucopyranosyloxy)-	MeOH	240(4.15),256(4.03), 350(4.03)	4-0023-64
$C_{14}H_{19}N_5O_7$			
2,6-Pyridinedicarbamic acid, 4-(acet-onylamino)-3-nitro-, diethyl ester	pH 1	224(4.37),252(4.38), 275(4.20),331(4.00)	44-0734-64
	pH 7	225(4.46),246(4.33), 351(4.07)	44-0734-64
	pH 13	234(4.34),293(3.92), 350(3.96)	44-0734-64
$C_{14}H_{20}$			
1,3-Cyclohexadiene, 2,6,6-trimethyl-1-(3-methyl-1,3-butadienyl)-	n.s.g.	235(4.04),313(4.09)	24-0549-64
2,4,6,8,10-Dodecapentaene, 2,3-dimethyl-	C_6H_{12}	321(4.65),337(4.84), 355(4.83)	39-2465-64
s-Indacene, 1,2,3,3a,4,4a,5,6-octa-hydro-3a,4a-dimethyl-	C_6H_{12}	248(4.30)	35-3510-65
1H,4H-3a,8a-Methano-s-indacene, 2,3,4a,5,6,7-hexahydro-4a-methyl-	C_6H_{12}	215(3.88)	35-3510-65
Naphthalene, decahydro-1,6-di-methylene-4-vinyl-	EtOH	246(2.81),252(2.95), 257(2.93),263(2.78)	78-2617-64
Naphthalene, 4-ethyl-1,2,3,4-tetra-hydro-1,1-dimethyl-	C_6H_{12}	265(2.56),272(2.52)	23-0579-64
Naphthalene, 1,2,3,4-tetrahydro-1,1,4,4-tetramethyl-	C_6H_{12}	257(2.51),264(2.66), 271(2.60)	23-0579-64

Compound	Solvent	$\lambda_{max}(\log \epsilon)$	Ref.
$C_{14}H_{20}BrClO$			
2,5-Cyclohexadien-1-one, 4-bromo-2,6-di-tert-butyl-4-chloro-	n.s.g.	248(4.16)	70-0336-65
$C_{14}H_{20}BrNO_3$			
2,5-Cyclohexadien-1-one, 4-bromo-2,6-di-tert-butyl-4-nitro-	hexane	245(3.97)	70-0371-64
	hexane	243(4.1)	70-1666-64
$C_{14}H_{20}Br_2O$			
2,5-Cyclohexadien-1-one, 4,4-dibromo-2,6-di-tert-butyl-	n.s.g.	248(4.18)	70-0336-65
$C_{14}H_{20}ClNO_3$			
2,4-Cyclohexadien-1-one, 2,6-di-tert-butyl-4-chloro-6-nitro-	hexane	315(3.68)	70-1666-64
$C_{14}H_{20}ClNO_4$			
3,5-Pyridinedicarboxylic acid, 4-(chloromethyl)-1,4-dihydro-2,6-dimethyl-, diethyl ester	C_6H_{12}	227(4.22),339(3.79)	39-4226-65
$C_{14}H_{20}Cl_2N_2O_2$			
p-Benzoquinone, 2,5-bis(tert-butylamino)-3,6-dichloro-	methyl cellosolve	226(4.29),355(4.40), 523(2.41)	78-1889-64
p-Benzoquinone, 2,5-dichloro-3,6-bis(diethylamino)-	MeOH	245(3.88)	54-0711-64
$C_{14}H_{20}F_2N_2O_2$			
p-Benzoquinone, 2,5-bis(tert-butylamino)-3,6-difluoro-	methyl cellosolve	223(4.49),351(4.43), 531(2.31)	78-1889-64
$C_{14}H_{20}IN$			
1,2,3,4,4a,9a-Hexahydro-9,9-dimethylcarbazolium iodide	EtOH	207(3.99),250(3.54), 293(2.91)	56-1437-65
$C_{14}H_{20}INO_2$			
4-(2-Benzoylethyl)-4-methylmorpholinium iodide	H_2O	228(4.23),245s(4.13)	7-0968-65
$C_{14}H_{20}INO_4$			
3,5-Pyridinedicarboxylic acid, 1,4-dihydro-4-(iodomethyl)-2,6-dimethyl-, diethyl ester	C_6H_{12}	227(4.21),339(3.77)	39-4226-65
$C_{14}H_{20}N_2$			
Indole, 3-(2-aminobutyl)-1,2-dimethyl-, hydrochloride	EtOH	227(4.57),280s(3.83), 285(3.72),294(3.84)	87-0274-64
Indole, 3-(2-aminobutyl)-2,7-dimethyl-, acetic acid salt	EtOH	222(4.58),274s(3.87), 279(3.88),288s(3.79)	87-0274-64
Indole, 3-(2-ethylaminobutyl)-, acetic acid salt	EtOH	220(4.58),274(3.76), 280(3.80),289(3.74)	87-0274-64
$C_{14}H_{20}N_2O$			
2,4-Cyclohexadien-1-one, 2,6-di-tert-butyl-4-diazo-	aq. EtOH	265(3.81),364(4.46)	30-1101-64
	acid	323(4.37)	30-1101-64
Piperazine, 1-acetyl-2,2-dimethyl-3-phenyl-, hydrochloride	EtOH	251(2.42),257(2.45), 262(2.41),268(2.26)	44-3574-64
Pyrrolo[2,3-b]indole, 1,2,3,3a,8,8a-hexahydro-5-methoxy-1,3a,8-trimethyl- (or esermethole)	EtOH-6N HCl	252(3.66),331(3.74)	39-5510-64
	EtOH-0.01N HCl	244(4.02),313(3.53)	39-5510-64

Compound	Solvent	$\lambda_{max}(\log \epsilon)$	Ref.
Esermethole (contd.)	EtOH	248(4.02),320(3.52)	39-5510-64
$C_{14}H_{20}N_2O_4$			
Benzene, 1,2-di-tert-butyl-4,5-dinitro-	EtOH	220(4.15),276(3.89)	35-5281-64
	EtOH	220(4.15),276(3.89)	88-0061-64
Valeric acid, 4-hydroxy-5-isopropoxy-2-oxo-, phenylhydrazone	n.s.g.	293(4.18),317(4.24)	39-0141-64
$C_{14}H_{20}N_2O_8S$			
Glucosamine, N-p-acetamidophenyl-	n.s.g.	261.5(4.39)	95-0399-65
$C_{14}H_{20}N_4$			
Pyrazolo[1,5-a]pyrimidine, 2,3,5-trimethyl-7-piperidino-	EtOH	236(4.64),288s(3.60), 299(3.68),332(3.78)	94-1207-65
$C_{14}H_{20}N_4O$			
Pyrido[2,3-d]pyrimidine, 2-amino-7-butyl-4-hydroxy-6-propyl-	pH 1	278(4.12),352(4.05)	44-2674-64
	pH 11	268(3.95),333(3.93)	44-2674-64
$C_{14}H_{20}N_4O_3$			
Benzamide, N-[2-(4-methyl-1-piperazinyl)ethyl]-p-nitro-	MeOH	262(4.11)	87-0107-65
$C_{14}H_{20}N_4O_4$			
Hydrazine, 1,2-dicarbethoxy-2-(α-phenylazo)ethyl-	MeOH	404(2.20)	89-0684-64
$C_{14}H_{20}N_4O_4S_2$			
5-Thia-1-azabicyclo[4.2.0]oct-2-ene-2-carboxylic acid, 3-[(amidinothio)-methyl]-8-oxo-7-valeramido-	pH 6.0	262(3.97)	39-5015-65
$C_{14}H_{20}N_6S$			
Urea, 1-[bis(isopropylidenehydrazino)-methylene]-3-phenyl-2-thio-	n.s.g.	254(4.47),296(4.38)	39-4448-65
$C_{14}H_{20}O$			
4-Cloven-3-one, 4-demethyl-	EtOH	244(4.11)	39-1344-65
2H-Cyclobuta[cd]benzofuran, 2a,3,3a-6,6a,6b-hexahydro-6-isopropylidene-2,2-dimethyl-	n.s.g.	244(4.19)	35-0136-65
7H-2,4a-Ethanonaphthalen-7-one, 1,2,3,4,5,6-hexahydro-1,1-dimethyl-	EtOH	247(4.16)	35-4438-64
4-Hexen-2-one, 5-methyl-, dehydrodimer	C_6H_{12}	216(4.40)	54-1233-65
4,7-Methanoazulen-1(2H)-one, 3,4,5,6,7,8-hexahydro-4,9,9-trimethyl-	EtOH	246(4.11)	35-4438-64
Naphtho[2,3-b]furan, 4,4a,5,6,7,8,8a,9-octahydro-3,8a-dimethyl-	heptane	222.5(3.84)	25-0899-65
Phenanthren-9(1H)-one, 2,3,4,4a,5,-6,7,8,10,10a-decahydro-	EtOH	248(4.07)	28-1844-64A
Phenanthren-9(1H)-one, 2,3,4,5,6,-7,8,8a,10,10a-decahydro-	EtOH	210(4.12),244(3.18)	28-1844-64A
$C_{14}H_{20}OPd$			
Palladium, cyclopentadienyl(4-methoxy-1-cycloocten-1-yl)-	hexane	199(4.23),229(4.01), 270(4.33),328(3.96)	39-5002-64
Palladium, cyclopentadienyl(8-methoxy-1-cycloocten-1-yl)-	hexane	202(4.35),256(4.47), 329(3.84),410(2.40)	39-5002-64

Compound	Solvent	$\lambda_{max}(\log \epsilon)$	Ref.
$C_{14}H_{20}OS$			
15-Thiabicyclo[10.2.1]penta-deca-12,14-dien-2-one	EtOH	263s(3.8),300(4.1)	70-2055-64
$C_{14}H_{20}O_2$			
o-Benzoquinone, 3,5-di-tert-butyl-	C_6H_{12}	385(3.45)	44-2493-64
p-Benzoquinone, 2,6-di-tert-butyl-	C_6H_{12}	255(4.38),262s(4.30)	44-2493-64
Butanonaphthalenedione, hexahydro-	EtOH	243(1.51),287(1.65)	44-2513-65
1-Naphthol, 4-butoxy-5,6,7,8-tetra-hydro-	EtOH	220s(3.86),230s(3.73), 288(3.50)	24-1926-64
	NaOH	241(3.93),301(3.56)	24-1926-64
Tricyclo[4.4.4.0]tetradecane-2,3-dione	MeOH	271(3.74)	5-0062-65D
Valeric acid, α-ethyl-β-(m-tolyl)-	C_6H_{12}	260(2.61),273(2.59)	39-1772-65
Valeric acid, 4-p-tolyl-, ethyl ester	EtOH	251s(2.29),258(2.46), 264(2.56),272(2.45)	78-2593-65
$C_{14}H_{20}O_3$			
1,4-Benzodioxan-5-ol, 6,8-dipropyl-	EtOH	275(3.04)	65-2994-64
Bicyclo[3.3.1]non-2(3)-ene-3-carboxylic acid, 2,4,4,8-tetramethyl-9-oxo-	EtOH	270(3.42)	88-1337-64
Carbonic acid, ethyl 3,4,4a,5,6,7-hexahydro-4a-methyl-2-naphthyl ester	EtOH	235(4.30)	78-2641-65
4a(2H)-Naphthalenecarboxylic acid, 3,4,5,6,7,8-hexahydro-1-methyl-2-oxo-, ethyl ester	EtOH	247(4.09)	35-0275-65
1-Naphthoic acid, 1,2,3,4,4a,5,6,7-octahydro-1,4a-dimethyl-7-oxo-, methyl ester	EtOH	241(4.10)	44-0782-64
2-Naphthoic acid, 2,3,4,4a,5,6,7,8-octahydro-2-methyl-5-oxo-, ethyl ester	EtOH	245(4.1)	44-2754-65
2-Naphthol, 1,2,3,4-tetrahydro-6,7-dimethoxy-1,1-dimethyl-	EtOH	209(4.27),224(3.90), 283(3.55),286(3.55)	39-0361-65
$C_{14}H_{20}O_3S$			
α-Tetronic acid, β-triphenyl-methylmercapto-	EtOH	230(4.24)	44-3560-64
$C_{14}H_{20}O_4$			
Benzoic acid, 2-heptyl-4,6-dihydroxy-	EtOH	263(4.19),302(3.81)	1-1677-65
Carbonic acid, 4-(4-methoxy-o-tolyl)butyl methyl ester	EtOH	226(3.92),277(3.22), 284(3.18)	35-5148-65
Cyclobutaneacrylic acid, 3-carboxy-2-methyl-4-propenyl-, dimethyl ester	hexane	209(3.79)	5-0039-65A
1,2-Cyclobutanediacrylic acid, 3,4-dimethyl-, dimethyl ester	hexane	213(4.63)	5-0039-65A
2-Cyclohexene-1-carboxylic acid, 3-acetyl-2,6,6-trimethyl-4-oxo-, ethyl ester	n.s.g.	259(3.95)	39-1511-64
1,2,4-Cyclopentanetrione, 3-isobutyl-5-(α-hydroxybutylidene)-	EtOH-HCl	255(4.36),280(4.24)	39-2251-65
	EtOH-NaOH	235(4.09),275(4.48), 300(4.44)	39-2251-65
3-Cyclopentene-1,1-dicarboxylic acid, 3,4-dimethyl-5-methylene-, diethyl ester	EtOH	240(4.19)	70-1653-64
Flavesone	C_6H_{12}	234(4.6),280(4.5)	39-3690-65
Malonic acid, (2,2-dimethylpenta-3,4-dienylidene)-, diethyl ester	EtOH	204(3.93)	39-6784-65
1-Naphthoic acid, 2,3,4,4a,5,6,7,8-octahydro-5-hydroxy-4a-methyl-2-oxo-, ethyl ester	EtOH	244(4.08)	44-3333-65

Compound	Solvent	$\lambda_{max}(\log \epsilon)$	Ref.
$C_{14}H_{20}O_4S_3$			
2,5-Thiophenedicarboxylic acid, 3,4-bis[(ethylthio)methyl]-, dimethyl ester	EtOH	218(4.34),277(4.23)	44-1919-64
$C_{14}H_{20}O_5$			
2-Cyclohexene-1,2-dicarboxylic acid, 1,3-dimethyl-4-oxo-, diethyl ester	EtOH	244(4.00)	44-3524-64
Itaconitin, hexahydro-	EtOH	210(3.91),250s(3.50)	94-0058-65
$C_{14}H_{20}O_8Si$			
Silicon, bis(acetato)bis(2,4-pentane-dionato)-	$CHCl_3$	279(4.30)	35-1403-65
$C_{14}H_{20}Pb$			
Lead, triethyl(phenylethynyl)-	hexane	239(4.46),250(4.46), 272(3.10),279(2.92)	22-3518-65
$C_{14}H_{21}BrO$			
Phenol, 4-bromo-2,6-di-tert-butyl-	hexane	280(3.4)	70-1666-64
$C_{14}H_{21}ClO$			
Phenol, 2,6-di-tert-butyl-4-chloro-	hexane	280(3.5)	70-1666-64
$C_{14}H_{21}FO_2$			
1-Cyclohexene-1-acrylic acid, α-fluoro-2,6,6-trimethyl-, ethyl ester	MeOH	203(3.95),272(3.36)	5-0001-64D
$C_{14}H_{21}IN_2O$			
Bufotenine, O-methyl ether, methiodide	EtOH	224(4.46),277(3.84), 295(3.76)	25-1800-64
$C_{14}H_{21}NO$			
Ketone, 1-cyclohexen-1-yl β-(3-methyl-1-butyn-3-yl)aminoethyl, hydrochloride	EtOH	235.5(4.13)	70-0512-64
Pyridine, 2-[[2-(methoxymethyl)-cyclohexyl]methyl]-	EtOH	262(3.43)	65-2632-64
$C_{14}H_{21}NO_2$			
Benzene, 1,2-di-tert-butyl-4-nitro-	EtOH	280(4.00)	35-5281-64
Cyclohexanol, 2-(methoxymethyl)-1-(2-pyridylmethyl)-	EtOH	262(3.47)	65-2632-64
2-Cyclohexen-1-one, 2-acetyl-5,5-di-methyl-3-(1-pyrrolidinyl)-	EtOH	274(4.12),310(4.20)	44-0798-64
perchlorate	EtOH	274(4.10),310(4.23)	44-0798-64
$C_{14}H_{21}NO_3$			
Acetophenone, 4'-(2-diethylamino-ethoxy)-2'-hydroxy-	EtOH	227(4.00),234(3.93), 274(4.20),313(3.87)	87-0446-64
	EtOH-KOH	228(4.30),238(4.25), 274(4.00),353(3.94)	87-0446-64
1,3-Cyclohexanedione, 2-acetamido-5,5-pentamethylene-, enol methyl ether	MeOH	262(4.11)	5-0123-65A
4-Pentene-2,3-diol, 1-(dimethylamino)-5-(p-methoxyphenyl)-	MeOH	262(4.26),292s(3.64), 303s(3.37)	44-2334-65
Phenol, 2,6-di-tert-butyl-4-nitro-	hexane	300(4.0)	70-1666-64
Phenol, 4,5-di-tert-butyl-2-nitro-	EtOH-HCl	219(4.03),294(3.94), 358(3.39)	35-5281-64
	EtOH	218(4.26),287(3.97), 356(3.54)	35-5281-64

Compound	Solvent	$\lambda_{max}(\log \epsilon)$	Ref.
Phenol, 4,5-di-tert-butyl-2-nitro- (contd.)	EtOH-KOH	302(3.89),435(3.58)	35-5281-64
$C_{14}H_{21}NO_4S$			
6H-1,3-Thiazin-2-ylacetic acid, 4-carbethoxy-5-methyl-, tert-butyl ester	EtOH	224s(3.88),282(3.72), 338(4.09)	39-1262-65
2-Thiophenedecanoic acid, 5-nitro-	EtOH	200(3.9),325(4.0)	70-2055-64
$C_{14}H_{21}NO_5$			
Aniline, N-carbethoxy-3,4,5-trimethoxy-	iso-PrOH	225s(3.98),269(2.83), 280s(2.60)	33-2089-64
Azepine-3,6-dicarboxylic acid, 4-ethoxy-4,5-dihydro-2,7-dimethyl-, dimethyl ester	hexane	223(4.13),247(3.68), 310(4.07)	39-2411-65
$C_{14}H_{21}NO_7S$			
Glucosamine, N-(3,4-xylylsulfonyl)-	n.s.g.	233(4.07),270(2.95), 278(2.88)	95-0399-65
$C_{14}H_{21}N_3O$			
Benzamide, p-amino-N-(2-piperidinoethyl)-	MeOH	280(4.24)	87-0107-65
$C_{14}H_{21}N_3O_2$			
3-Camphorylideneacetone, semicarbazone	EtOH	216(3.89),310(4.35)	39-6072-64
$C_{14}H_{21}N_3O_2S_2$			
Sulfanilamide, N^1-[cyclohexylamino-(methylthio)methylene]-	EtOH	276.0(4.30)	95-0391-65
$C_{14}H_{21}N_3O_5$			
Glycine, N-[N-(5,5-dimethyl-2,3-dioxocyclohexylidene)glycyl]-, ethyl ester, 2-oxime	EtOH	250(4.01),330(3.61)	25-1183-65
$C_{14}H_{21}N_3O_9$			
Pseudourea, 2-methyl-1-(β-D-ribofuranosylcarbamoyl)-, triacetate	EtOH	221(4.27)	73-2060-64
Pseudourea, 2-methyl-1-(β-D-ribopyranosylcarbamoyl)-, triacetate	EtOH	221(4.28)	73-2060-64
$C_{14}H_{21}N_5$			
s-Triazolo[1,5-c]pyrimidine, 2-(allylamino)-7-methyl-5-pentyl-	MeOH	234(4.66),297(3.34)	39-3357-65
s-Triazolo[1,5-c]pyrimidine, 7-methyl-2-piperidino-5-propyl-	MeOH	237(4.60),307(3.49)	39-3357-65
$C_{14}H_{21}N_5O$			
s-Triazolo[1,5-c]pyrimidine, 2-(N-methylacetamido)-5,7-dipropyl-	MeOH	234(4.65),261s(3.62), 293(3.30)	39-3357-65
$C_{14}H_{21}N_5O_4S$			
Theophylline, 7-(cyclohexylmethyl)-8-sulfamoyl-	MeOH	286(4.03)	54-1215-64
$C_{14}H_{21}N_7O_6$			
2,6-Pyridinedicarbamic acid, 4-(acetonylamino)-3-nitro-, diethyl ester, hydrazone	pH 1	225(4.43),251(4.43), 274(4.24),332(4.00)	44-0734-64
	pH 7	225(4.54),245(4.38), 352(4.11)	44-0734-64

Compound	Solvent	$\lambda_{max}(\log \epsilon)$	Ref.
2,6-Pyridinedicarbamic acid, 4-(acetonylamino)-3-nitro-, diethyl ester, hydrazone (contd.)	pH 13	234(4.49),374(3.91)	44-0734-64
$C_{14}H_{21}O_5P$			
6-Chromanol, 2,2,5,7,8-pentamethyl-, phosphate	EtOH	280s(3.27),285(3.33)	37-1374-65
$C_{14}H_{22}$			
Toluene, 4-ethyl-2-pentyl-	EtOH	261(2.82),267(2.85), 276(2.77)	39-5503-64
3,4-Undecadien-6-yne, 2,2,3-trimethyl-	EtOH	222(4.18)	39-4659-65
$C_{14}H_{22}ClNO$			
Ketone, 1-cyclohexen-1-yl β-(3-methyl-1-butyn-3-yl)aminoethyl, hydrochloride	EtOH	235.5(4.13)	70-0512-64
$C_{14}H_{22}Cl_2O_2Pd$			
Palladium, dichlorobis(4-methoxy-2-cyclohexen-1-yl)di-	$CHCl_3$	245(3.94),252s(3.88), 302(3.04),338s(2.95)	39-5002-64
$C_{14}H_{22}IN$			
1,2,3,4,5,6,7,8-Octahydro-9,9-dimethylcarbazolium iodide	EtOH	276(3.83)	78-0515-64
	EtOH-NaOH	295(3.88)	78-0515-64
$C_{14}H_{22}INO_7$			
α-D-Galactopyranoside, ethyl 2-acetamido-2,6-dideoxy-6-iodo-, diacetate	$CHCl_3$	256(2.70)	44-3654-64
$C_{14}H_{22}NO_5$			
Piperidinooxy, 4-(1,2-dicarboxyethylidene)-2,2,6,6-tetramethyl-, 1-methyl ester	MeOH	211(3.47),228s(3.4)	70-0716-65
$C_{14}H_{22}N_2$			
Piperazine, 1-ethyl-2,2-dimethyl-3-phenyl-, dihydrochloride	EtOH	251(2.27),257(2.41), 262(2.43),267(2.31)	44-3574-64
$C_{14}H_{22}N_2O_2$			
Aniline, N,N-dibutyl-p-nitro-	n.s.g.	415(4.40)	5-0109-65I
Aniline, N,N-diisobutyl-p-nitro-	n.s.g.	417(4.41)	5-0109-65I
$C_{14}H_{22}N_3O_4P$			
Phosphoric acid, 2-pyridyl ester, ester with hydracrylonitrile, compound with cyclohexylamine	n.s.g.	261(3.58)	24-1031-65
$C_{14}H_{22}N_4O_2S_2$			
3(2H)-Pyridazinethione, 2-(morpholinomethyl)-6-[(morpholinomethyl)thio]-	THF	310(4.33),390(3.38)	49-0631-65
$C_{14}H_{22}N_6O_5$			
2-Furaldehyde, 5-nitro-, 2-[2-[4-(2-hydroxyethyl)-1-piperazinyl]ethyl]-, semicarbazone, dihydrochloride	H_2O	383(4.22)	44-2582-64

Compound	Solvent	λ_{max}(log ε)	Ref.
$C_{14}H_{22}O$			
1'-Acetonaphthone, 3',4',4'a,5',6',7',-8',8'a-octahydro-2',8'a-dimethyl-, cis	EtOH	244(3.19)	35-1966-64
trans	EtOH	242(3.17)	35-1966-64
2-Butenal, 2-methyl-4-(2,6,6-trimethyl-1-cyclohexen-1-yl)-	n.s.g.	231(4.25)	24-0549-64
3-Butenal, 2-methyl-4-(2,6,6-trimethyl-1-cyclohexen-1-yl)-	n.s.g.	232(3.73)	24-0549-64
2-Cyclohexen-1-one, 4-(1,5-dimethyl-4-hexenyl)-	EtOH	227(4.07)	2-0091-65
5,9,11-Dodecatrien-2-one, 6,10-dimethyl-	EtOH	231.5(4.34)	22-2533-64
Inden-2(4H)-one, 6-tert-butyl-5,6,7,7a-tetrahydro-7a-methyl-	C_6H_{12}	226(4.14),316(1.52), 327(1.56),339(1.45)	88-2797-64
Isobenzofuran, 1,3,3a,7a-tetrahydro-3a,4,5,6,7,7a-hexamethyl-	MeOH	265(3.79)	24-2331-65
3,5-Octadien-2-one, 8-(2,2-dimethylcyclopropyl)-6-methyl-, trans	EtOH	294(4.39)	27-0174-64
Phenol, 2,4-di-tert-butyl-	MeOH	278(3.36)	39-0676-65
	MeOH-NaOMe	245(4.03),296(3.53)	39-0676-65
Phenol, 2,6-di-tert-butyl-	MeOH	272(3.20)	39-0676-65
	MeOH-NaOMe	251(3.95),298(3.72)	39-0676-65
Phenol, 3,4-di-tert-butyl-	EtOH	277(3.21)	35-5281-64
	EtOH	277(3.21)	88-0061-64
	EtOH-KOH	294(3.50)	35-5281-64
	EtOH-KOH	294(3.50)	88-0061-64
Phenol, 3,5-di-tert-butyl-	EtOH	218(3.84),273(3.28), 278(3.27)	35-5281-64
	EtOH-KOH	290(3.57)	35-5281-64
$C_{14}H_{22}OS$			
Sulfoxide, 2-octyl phenyl	EtOH	205s(4.09),217s(3.82), 243(3.68),265s(3.18), 273(2.88)	35-1958-65
$C_{14}H_{22}O_2$			
[Bicyclohexyl]-2,2'-dione, 1,1'-dimethyl-	EtOH	290(1.60)	35-5208-64
Bicyclo[3.3.1]non-2(3)-en-9-one, 3-(hydroxymethyl)-2,4,4,8-tetramethyl-	EtOH	237(3.43),285(1.30)	88-1337-64
Bicyclo[7.2.0]undecan-2-one, 5-hydroxy-10,10-dimethyl-6-methylene-	EtOH	287(1.43)	39-3154-65
Cyclohexaneacetic acid, 3-cyclohexyl-2-hydroxy-, lactone	EtOH	258(1.54)	39-5617-64
Furan, 3-acetyl-2,5-di-tert-butyl-	C_6H_{12}	201(4.34),273(3.64)	22-3136-65
	MeOH	205s(4.30),280(3.60)	22-3136-65
2,7-Octadien-4-ynal, 3,7-dimethyl-, diethyl acetal	MeOH	230(4.12)	5-0026-64I
Tricyclo[6.3.1.0^{2,5}]dodecan-9-one, 1-hydroxy-4,4-dimethyl-	EtOH	286(1.36)	39-3154-65
2,5-Xylenol, 4-(4-ethoxybutyl)-	EtOH	282(3.35),287(3.32)	78-2183-65
$C_{14}H_{22}O_3$			
2,5-Cyclohexadien-1-one, 2,4-di-tert-butyl-4,5-dihydroxy-	MeOH	246(3.92),293(3.49)	5-0093-65I
	MeOH-KOH	273(3.51),347(3.74)	5-0093-65I
2-Cyclohexene-1-carboxylic acid, 3-ethyl-2,6,6-trimethyl-4-oxo-, ethyl ester	n.s.g.	250(3.94)	39-1511-64

Compound	Solvent	λ_{max}(log ε)	Ref.
2-Cyclohexene-1,4-dione, 5,6-di-tert-butyl-2-hydroxy-	MeOH	283(3.98)	5-0093-65I
	MeOH	282(4.00)	24-2774-65
	MeOH-KOH	238(3.94),335(4.00)	5-0093-65I
	MeOH-KOH	239(3.90),334(4.00)	24-2774-65
Resorcinol, 4-tert-butyl-6-tert-butoxy-	MeOH	221(3.85),286(3.62)	5-0093-65I
	MeOH-KOH	294(3.64)	5-0093-65I
$C_{14}H_{22}O_3Pd$			
Palladium, (4-methoxy-2-cycloocten-1-yl)(2,4-pentanedionato)-	hexane	213s(4.25),229(4.27), 258s(4.01),292(3.91)	39-5002-64
Palladium, (8-methoxy-4-cycloocten-1-yl)(2,4-pentanedionato)-	hexane	205(4.38),228s(4.27), 306(3.93)	39-5002-64
$C_{14}H_{22}O_4$			
Hydroperoxide, 1,5-di-tert-butyl-2-hydroxy-4-oxo-2,5-cyclohexadien-1-yl-	MeOH	242(4.02),301(3.54)	5-0093-65I
	MeOH-KOH	288(3.20),350(3.80)	5-0093-65I
$C_{14}H_{23}N$			
Aniline, dibutyl-, tetracyanoethylene complex	$CHCl_3$	845(3.45)	88-0189-64
Aniline, 3,4-di-tert-butyl-	EtOH	292(3.24)	35-5281-64
$C_{14}H_{23}NO$			
2-Cyclohexen-1-one, 3-(cyclohexylamino)-5,5-dimethyl-	EtOH	292.5(4.51)	44-1407-65
hydrochloride	EtOH	291(4.43)	44-1407-65
perchlorate	EtOH	290(4.46)	44-1407-65
$C_{14}H_{23}NO_2$			
2-Cyclohexene-1-acrylic acid, 2-(diethylamino)-, methyl ester	heptane	205(4.21)	44-3679-65
$C_{14}H_{23}NO_3$			
2-Pyridinevaleraldehyde, δ-hydroxy-, diethyl acetal	EtOH	257(3.46),262(3.51), 268(3.4)	44-1523-65
$C_{14}H_{23}NO_7S$			
p-Toluenesulfonamide, N-methyl-N-(D-gluco-2,3,4,5,6-pentahydroxyhexyl)-	n.s.g.	231(4.13),263(2.95), 274(2.65)	95-0399-65
$C_{14}H_{23}N_3O_2$			
Pyridazine, 3-(cyclohexyloxy)-6-[2-(dimethylamino)ethoxy]-	EtOH	289(3.32)	44-1751-64
$C_{14}H_{23}N_5$			
s-Triazolo[1,5-c]pyrimidine, 7-methyl-5-pentyl-2-(propylamino)-	MeOH	234(4.64),300(3.40)	39-3357-65
$C_{14}H_{23}O_7PS$			
Fumaric acid, hydroxy(3-hydroxybutyl)-, δ-lactone, ethyl ester, O-ester with O,O-diethyl phosphorothioate	MeOH	211(3.85)	5-0010-65E
$C_{14}H_{23}O_8P$			
Fumaric acid, hydroxy(3-hydroxybutyl)-, δ-lactone, ethyl ester, diethyl phosphate	MeOH	200(4.21)	5-0010-65E
Oxalacetic acid, (3-hydroxybutyl)phosphono-, δ-lactone, triethyl ester	MeOH	215(3.91)	5-0010-65E

Compound	Solvent	λ_{max}(log ε)	Ref.
$C_{14}H_{24}$			
Cyclobutane, 3,4-diisopropylidene-1,1,2,2-tetramethyl-	EtOH	250(4.15)	44-1038-65
1,3-Cyclohexadiene, 1,6-di-tert-butyl-	EtOH	274(3.58)	35-5281-64
$C_{14}H_{24}N_2O_2$			
Pyridazine, 3,6-bis(isopentyloxy)-	EtOH	287(3.32)	87-0129-65
$C_{14}H_{24}N_4O_2$			
p-Benzoquinone, tetrakis-(dimethylamino)-	methyl cellosolve	236(4.20),388(3.94), 595s(2.59)	78-1889-64
2(1H)-Pyrimidinone, 3,4,5,5',6,6'-hexahydro-4,6,6,6',6'-pentamethyl-4,4'-methylenedi-	MeOH	242(3.48)	77-0439-65
$C_{14}H_{24}O$			
3-Buten-2-one, 4-(3-p-menthyl)-	n.s.g.	231(4.16)	28-4054-64B
Cyclohexanone, 4-tert-butyl-2-(methylallyl)-	C_6H_{12}	282(1.54)	88-2797-64
5,7-Decadien-2-one, 5-methyl-4-propyl-	n.s.g.	238.5(4.45)	22-0225-64
6-Decen-2-one, 5-methylene-4-propyl-	n.s.g.	232.5(4.31)	22-0225-64
Ketone, butyl 2-butyl-1-cyclopenten-1-yl	EtOH	254(3.98),313(1.93)	22-0722-65
Ketone, butyl 2-butyl-2-cyclopenten-1-yl	EtOH	209(3.44),278(1.57)	22-0722-65
5,7-Nonadien-2-one, 4-isopropyl-5,8-dimethyl-	n.s.g.	243(4.00)	22-0225-64
6-Nonen-2-one, 4-isopropyl-8-methyl-5-methylene-	n.s.g.	231.3(3.90)	22-0225-64
β-Thujadicarboxylic acid, dimethyl ester	EtOH	219(4.10)	39-5640-64
$C_{14}H_{24}O_2$			
1,2-Butanediol, 2-methyl-4-(1,3,3-trimethylbicyclo[3.1.0]hex-2-ylidene)-	n.s.g.	220(3.6)	27-0538-65
Cyclododecanone, 2-acetyl-	dioxan	279(3.96)	28-0759-65B
copper derivative	$CHCl_3$	311(4.52)	28-0759-65B
1,3-Cyclotetradecanedione	dioxan	275(4.14)	28-0759-65B
copper derivative	$CHCl_3$	250(4.52),299(4.63)	28-0759-65B
$C_{14}H_{24}S_2$			
Spiro[azulene-5(1H),2'-[1,3]dithiolane], octahydro-3,8-dimethyl-	EtOH	242(2.63)	94-0942-65
$C_{14}H_{25}N_3S_2$			
5-Pyrimidinepropionaldehyde, 2-amino-4-methyl-, dipropyl mercaptal, hydrochloride	pH 1	228(4.31),313(3.75)	4-0079-64
	pH 7	230(4.26),300(3.61)	4-0079-64
	pH 13	300(3.61)	4-0079-64
Tetraethylammonium salt with [(ethylthio)mercaptomethylene]malononitrile	EtOH	285(4.00),342(4.34)	44-0497-64
$C_{14}H_{25}O_5P$			
Acetoacetic acid, 2-(butoxybutylphosphinyl)-2-(2-hydroxyethyl)-, γ-lactone	MeOH	229(3.05)	5-0010-65E
$C_{14}H_{26}N_2$			
Cyclopentapyrazole, 3,6a-dibutyl-1,3a,4,5,6,6a-hexahydro-	EtOH	240.5(3.58)	22-0722-65

Compound	Solvent	λ_{max}(log ϵ)	Ref.
$C_{14}H_{26}O$			
Cyclohexanone, 4-tert-butyl-2-isobutyl-	C_6H_{12}	276(1.53),283(1.52)	88-2797-64
1-Cyclopentene-1-methanol, α,2-dibutyl-	EtOH	210.5(3.59)	22-0722-65
$C_{14}H_{26}O_2$			
3,8-Decanedione, 2,2,9,9-tetramethyl-	EtOH	288(1.84)	22-0714-65
$C_{14}H_{27}As_3Br_2MoO_3$			
Bromotricarbonyl[[2-[(dimethylarsino)-methyl]-2-methyltrimethylene]bis-[dimethylarsine]molybdenum bromide	MeOH	289(3.27),376(2.81)	39-6570-65
$C_{14}H_{27}As_3Br_2O_3W$			
Bromotricarbonyl[[2-[(dimethylarsino)-methyl]-2-methyltrimethylene]bis-[dimethylarsine]]tungsten bromide	MeOH	258(4.16),372(2.92)	39-6570-65
$C_{14}H_{27}As_3I_2MoO_3$			
Iodotricarbonyl[[2-[(dimethylarsino)-methyl]-2-methyltrimethylene]bis-[dimethylarsine]]molybdenum iodide	MeOH	305(3.51),384(2.93)	39-6570-65
$C_{14}H_{27}As_3I_2O_3W$			
Iodotricarbonyl[[2-[(dimethylarsino)-methyl]-2-methyltrimethylene]bis-[dimethylarsine]]tungsten iodide	MeOH	290(3.65),377(3.15)	39-6570-65
$C_{14}H_{28}N_2S_4$			
Disulfide, bis(diisopropyl-thiocarbamoyl)-	heptane	221(4.25),254(4.23), 279(4.21)	1-1113-65
	EtOH	220(4.23),259(4.21), 272(4.21)	1-1113-65
$C_{14}H_{28}O_2S$			
1-Butanesulfinic acid, p-menth-3-yl ester	heptane	224(3.38)	35-1958-65
	EtOH	215.5(3.35)	35-1958-65
$C_{14}H_{42}Si_6$			
Hexasilane, tetradecamethyl-	C_6H_{12}	220s(4.15),260(4.32)	101-0369-64B

Compound	Solvent	$\lambda_{max}(\log \epsilon)$	Ref.
$C_{15}H_7NO_3$			
3H,7H-Anthra[9,1-cd][1,2]oxazine-3,7-dione	$C_6H_3Cl_3$	309(3.63),325(3.62), 342(3.63),355(3.58)	65-3819-64
$C_{15}H_8BrNO_4$			
2H-1,3-Benzoxazine-2,4(3H)-dione, 3-(m-bromobenzoyl)-	MeOH	207(4.65),248(4.18), 295s(3.43)	4-0037-65
$C_{15}H_8ClNO_3S$			
Benzo[b]thiophen-3(2H)-one, 2-(o-chlorobenzylidene)-5-nitro-	n.s.g.	426(4.13),434(4.13)	65-0519-65
Benzo[b]thiophen-3(2H)-one, 2-(p-chlorobenzylidene)-5-nitro-	n.s.g.	424(4.24),435(4.31)	65-0519-65
$C_{15}H_8Cl_2N_2O_2$			
Acrylonitrile, 3-(2,4-dichlorophenyl)-2-(p-nitrophenyl)-	MeOH	264(4.06),317(4.19)	87-0583-65
$C_{15}H_8Cl_2O_3$			
Coumarin, 3,6-dichloro-4-phenoxy-	EtOH	213(4.44),220(4.46), 280(4.09),289(4.01), 328(3.84)	44-4126-65
$C_{15}H_8Cl_2O_5$			
Flavone, 5,6-dichloro-2',4',7-trihydroxy-	MeOH	291.5(3.33)	44-3445-64
$C_{15}H_8Cl_2O_7$			
Pyrano[3,2-b]pyran-4,8-dione, 6-chloromethyl-2-[6-(chloromethyl)-3-hydroxy-4-oxo-4H-pyran-2-yl]-	MeOH	286.7(3.69)	44-3445-64
$C_{15}H_8Cl_4OS_2$			
1,3-Dithiole, 4-(p-methoxyphenyl)-2-(tetrachloro-2,4-cyclopentadien-1-ylidene)-	$CHCl_3$	450(4.54)	88-2121-65
	CF_3COOH	496(4.19)	88-2121-65
$C_{15}H_8Cl_4O_3$			
Benzoic acid, 2-benzoyl-3,4,5,6-tetrachloro-, methyl ester	EtOH	247(3.19)	44-4293-65
$C_{15}H_8Cl_6N_4O_3$			
2-Propanone, 1-[(2,3,5-trichloro-6-hydroxyphenyl)azo]-1-[2,3,5-trichloro-6-hydroxyphenyl)hydrazono]-	benzene	530(4.18)	65-0502-65
	DMF	400(3.89),578(4.39)	65-0502-65
$C_{15}H_8I_4O_4$			
Cinnamic acid, 4-(4-hydroxy-3,5-diiodophenoxy)-3,5-diiodo-	EtOH	213(4.58),247(4.61), 279(4.37)	44-1812-64
$C_{15}H_8N_2O_2$			
3H-Dibenzo[de,h]cinnoline-3,7(2H)-dione	$C_6H_3Cl_3$	309(3.82),378(3.92)	65-3819-64
$C_{15}H_8N_2O_3$			
3H,7H-Anthra[9,1-cd][1,2]oxazine-3,7-dione, 4-amino-	$C_6H_3Cl_3$	333(3.64),408(3.71)	65-3819-64
3H,7H-Anthra[9,1-cd][1,2]oxazine-3,7-dione, 5-amino-	$C_6H_3Cl_3$	478(3.75)	65-3819-64

Compound	Solvent	λ max(log ε)	Ref.
$C_{15}H_8N_2O_3S_2$			
Carbonic acid, dithio-, S,S-bis-(2-benzoxazolyl) ester	$CHCl_3$	293(4.45),333(3.98)	44-3618-65
$C_{15}H_8N_2O_5S$			
Benzo[b]thiophen-3(2H)-one, 5-nitro-2-(o-nitrobenzylidene)-	n.s.g.	429(4.02)	65-0519-65
Benzo[b]thiophen-3(2H)-one, 5-nitro-2-(p-nitrobenzylidene)-	n.s.g.	432(4.22),442(4.24)	65-0519-65
$C_{15}H_8N_2O_6$			
2H-1,3-Benzoxazine-2,4(3H)-dione, 3-(p-nitrobenzoyl)-	MeOH	205(4.72),250s(4.27), 263(4.32),290(--)	4-0037-65
$C_{15}H_8O_4$			
3H-Benzofuro[3,2-b][1]benzopyran-3-one, 8-hydroxy-	n.s.g.	258(4.04),309(3.71), 363(4.17),409(4.44)	2-0319-64
11H-Benzofuro[3,2-b][1]benzopyran-11-one, 3-hydroxy-	$CHCl_3$	252(4.17),271(3.87), 316(4.47)	39-5140-65
$C_{15}H_8O_5$			
3H-Benzofuro[3,2-b][1]benzopyran-3-one, 1,8-dihydroxy-	MeOH	500(3.58)	2-0319-64
$C_{15}H_8S_3$			
Phenanthreno[1,2-d]trithione	$CHCl_3$	272s(4.36),289(4.56), 306s(4.39),315(4.43), 340s(4.23),380(3.71), 433s(4.00),453(4.12)	5-0062-64D
$C_{15}H_9BrCl_2O_2$			
Chalcone, 5'-bromo-3,5-dichloro-2'-hydroxy-	n.s.g.	239(4.31),349(3.95)	49-0450-65
$C_{15}H_9BrO$			
Indone, 3-bromo-2-phenyl-	MeOH	245(4.54),296(3.87), 425(3.08)	65-0448-64
$C_{15}H_9Br_3O_2$			
Chalcone, 2,3',4-tribromo-2'-hydroxy-	n.s.g.	325(4.54)	49-0450-65
$C_{15}H_9ClN_2O_6S$			
2-Nitro-11H-[1]benzothiopyrano-[4,3-b]indol-5-ium perchlorate	MeCN-1% $HClO_4$	230(4.40),263(4.68), 307(4.41),324s(4.29), 403(4.08)	44-3613-65
$C_{15}H_9ClO$			
5H-Dibenzo[a,d]cyclohepten-5-one, 10-chloro-	isooctane	252(3.13),303(2.73)	87-0886-65
$C_{15}H_9ClOS$			
Benzo[b]thiophen-3(2H)-one, 2-(o-chlorobenzylidene)-	n.s.g.	425(4.06),435(4.07)	65-0519-65
Benzo[b]thiophen-3(2H)-one, 2-(p-chlorobenzylidene)-	n.s.g.	423(4.08),436(4.13)	65-0519-65
$C_{15}H_9ClO_2S$			
Anthraquinone, 1-chloro-4-(methylthio)-	benzene	312(3.98),440(3.68)	22-1648-65

Compound	Solvent	$\lambda_{max}(\log \varepsilon)$	Ref.
$C_{15}H_9Cl_2N$			
Acrylonitrile, 3-(o-chlorophenyl)-2-(p-chlorophenyl)-	MeOH	230(4.13),308(4.28)	87-0583-65
$C_{15}H_9Cl_2NO_2$			
Coumarin, 4-anilino-3,6-dichloro-	EtOH	220(4.27),228(4.30), 254(4.14),278(3.89), 333(4.08)	44-4126-65
$C_{15}H_9Cl_2NO_4S$			
2-Chloro-11H-[1]benzothiopyran-[4,3-b]indol-5-ium perchlorate	MeCN-1% $HClO_4$	239(4.28),272(4.56), 295(4.55),406(4.12)	44-3613-65
$C_{15}H_9Cs$			
Cesium, 4H-cyclopenta[def]phen-anthren-4-yl)-	$C_6H_{11}NH_2$	505(3.84)	35-0384-65
$C_{15}H_9I_2NO_5$			
Acrylic acid, 3-(4-mercapto-3,5-di-iodophenyl)-2-(p-nitrophenyl)-	EtOH	237(4.47),268(4.29), 367(4.32)	7-0805-64
$C_{15}H_9I_3O_3$			
Acrylic acid, 2-(4-hydroxy-3,5-diiodo-phenyl)-3-(p-iodophenyl)-	EtOH	251(4.04),295(4.02)	7-0805-64
Acrylic acid, 3-(4-hydroxy-3,5-diiodo-phenyl)-2-(p-iodophenyl)-	EtOH	237(4.50),275s(4.14), 385(3.99)	7-0805-64
$C_{15}H_9Li$			
Lithium, 4H-cyclopenta[def]phen-anthren-4-yl-	$C_6H_{11}NH_2$	505(3.87)	35-0384-65
$C_{15}H_9NO$			
1-Phenanthrenecarboxylic acid, 10-amino-, lactam	EtOH	222(4.71),230(4.45), 244(4.44),284(4.32), 295(4.34)	44-3933-65
$C_{15}H_9NO_3S$			
Benzo[b]thiophen-3(2H)-one, 2-benzylidene-5-nitro-	n.s.g.	422(4.22),433(4.28)	65-0519-65
Benzo[b]thiophen-3(2H)-one, 2-(o-nitrobenzylidene)-	n.s.g.	428(3.85)	65-0519-65
Benzo[b]thiophen-3(2H)-one, 2-(p-nitrobenzylidene)-	n.s.g.	432(4.17),448(4.21)	65-0519-65
$C_{15}H_9NO_4$			
2H-1,3-Benzoxazine-2,4(3H)-dione, 3-benzoyl-	MeOH	205(4.59),246(4.21), 290(--)	4-0037-65
Coumarin, 3-(p-nitrophenyl)-	MeOH	338(4.16)	2-0182-64
Coumarin, 6-nitro-3-phenyl-	MeOH	283(4.48),325(4.48)	2-0182-64
1-Phenanthrenecarboxylic acid,	EtOH	256.5(4.57)	44-3792-65
10-nitro-	EtOH	256(4.57)	44-3933-65
Phthalimide, 4-hydroxy-N-(2-hydroxyphenyl)-	EtOH	237(4.51),322s(3.39)	7-0128-64
Phthalimide, 4-hydroxy-N-(4-hydroxyphenyl)-	EtOH	240(4.50),275s(4.04), 300s(3.67),336s(3.20)	7-0128-64
Phthalimide, 5-hydroxy-N-(4-hydroxyphenyl)-	EtOH	220(4.50),234s(4.35), 341(3.69)	7-0128-64
$C_{15}H_9N_3O$			
Benzonitrile, m-(5-phenyl-1,3,4-oxadia-	DMF	284(4.37)	24-2966-65
zol-2-yl)-	DMF	284.2(4.35)	24-2966-65

Compound	Solvent	$\lambda_{max}(\log \epsilon)$	Ref.
Dibenzo[3,4:6,7]cyclohepta-[1,2-d]triazol-8(1H)-one	EtOH	255.5(4.46)	24-1318-64
	EtOH-NaOH	261(4.49),370(3.63)	24-1318-64
$C_{15}H_{10}BrClO_3$			
Chalcone, 3'-bromo-5-chloro-2,2'-dihydroxy-	n.s.g.	222(4.42),259(4.11),265(4.10),313(3.21),395(3.29)	49-0450-65
Chalcone, 5'-bromo-5-chloro-2,2'-dihydroxy-	n.s.g.	226(4.19),300(3.29),399(3.25)	49-0450-65
$C_{15}H_{10}BrN$			
Acrylonitrile, 3-bromo-2,3-diphenyl-	EtOH	224(3.84),294(3.70)	94-1446-64
$C_{15}H_{10}BrNO_4$			
Acrylic acid, 2-(o-bromophenyl)-3-(o-nitrophenyl)-, cis	EtOH	303(3.60)	44-3933-65
$C_{15}H_{10}BrN_3O_5$			
Isocarbostyril, 5-bromo-2-(2,4-dinitrophenyl)-	EtOH	240(4.28),291(4.10),380(3.96)	24-1023-65
Isocarbostyril, 8-bromo-2-(2,4-dinitrophenyl)-	EtOH	241(4.24),289(4.08),388(3.90)	24-1023-65
$C_{15}H_{10}Br_2I_2O_4$			
Hydrocinnamic acid, 3,5-dibromo-4-(4-hydroxy-3,5-diiodophenoxy)-	EtOH	211(4.77),240(3.58),287(3.58),302(3.65)	44-1812-64
$C_{15}H_{10}Br_2N_4$			
6-Bromo-3-phenyltetrazolo[5,1-a]quinolinium bromide	EtOH	293(3.95)	89-0171-65
$C_{15}H_{10}Br_2O_2$			
Chalcone, 4,5'-dibromo-2'-hydroxy-	n.s.g.	229(4.41),344(3.97)	49-0450-65
$C_{15}H_{10}ClIO_2$			
Acrylic acid, 3-(p-chlorophenyl)-2-(p-iodophenyl)-	EtOH	236(4.54),265(4.12)	7-0805-64
$C_{15}H_{10}ClNO$			
Indole-2-carboxaldehyde, 5-chloro-3-phenyl-	iso-PrOH	251(4.49),320(4.38)	44-1621-64
2-Indolinone, 3-(p-chlorobenzylidene)-	EtOH	254(4.11),329(4.21),396(3.56)	87-0626-65
$C_{15}H_{10}ClNO_3$			
Phthalimidine, 2-(m-chlorophenyl)-5,6-(methylenedioxy)-	MeOH	310(4.30)	95-1049-65
Phthalimidine, 2-(o-chlorophenyl)-5,6-(methylenedioxy)-	MeOH	303(4.01)	95-1049-65
Phthalimidine, 2-(p-chlorophenyl)-5,6-(methylenedioxy)-	MeOH	312(4.33)	95-1049-65
$C_{15}H_{10}ClNO_4S$			
11H-[1]Benzothiopyrano[4,3-b]indol-5-ium perchlorate	$MeCN-H_2SO_4$	231s(4.21),234(4.23),269(4.52),292(4.47),400(4.07)	44-3613-65
$C_{15}H_{10}ClN_3O_5$			
Isocarbostyril, 5-chloro-2-(2,4-dinitrophenyl)-	EtOH	239(4.23),291(4.08),392(3.94)	24-1023-65

Compound	Solvent	λ_{max}(log ε)	Ref.
Isocarbostyril, 8-chloro-2-(2,4-dinitrophenyl)-	EtOH	241(4.28),289(4.12), 383(3.99)	24-1023-65
$C_{15}H_{10}Cl_2I_2O_4$			
Hydrocinnamic acid, 3,5-dichloro-4-(4-hydroxy-3,5-diiodophenoxy)-	EtOH	213(4.65),221(4.63), 237(4.24),285(3.59), 303(3.69)	44-1812-64
$C_{15}H_{10}Cl_2O_2$			
Acrylic acid, 2,3-bis(p-chlorophenyl)-	EtOH	286(4.28)	44-1473-65
Acrylic acid, 2-(2,4-dichlorophenyl)-3-phenyl-	EtOH	280(4.14)	44-1473-65
$C_{15}H_{10}Cl_4N_2O$			
1H-4,1,2-Benzoxadiazine, 5,6,7,8-tetrachloro-3-(3,5-xylyl)-	EtOH	208(4.5),260(4.4), 335(3.8)	24-2884-64
$C_{15}H_{10}FNO$			
2-Indolinone, 3-(p-fluorobenzylidene)-	MeOH	224s(4.13),252(4.11), 272s(3.94),322(4.14), 388s(3.52)	87-0626-65
$C_{15}H_{10}INO_4$			
Acrylic acid, 2-(p-iodophenyl)-3-(p-nitrophenyl)-	EtOH	235(4.49),248(4.46), 348(4.39)	7-0805-64
$C_{15}H_{10}I_2O_3$			
Acrylic acid, 3-(4-hydroxy-3,5-diiodophenyl)-2-phenyl-	EtOH	243(4.57),277s(4.26), 380(4.21)	7-0805-64
$C_{15}H_{10}I_4O_4$			
Hydrocinnamic acid, 4-(4-hydroxy-3,5-diiodophenoxy)-3,5-diiodo-	EtOH	216(4.68),225(4.69), 238(4.52),295(3.62), 303(3.63)	44-1812-64
$C_{15}H_{10}I_4O_5$			
Lactic acid, 3-[4-(4-hydroxy-3,5-diiodophenoxy)-3,5-diiodophenyl]-	EtOH	214(4.70),225(4.70), 238(4.45),294(3.64), 302(3.65)	44-1812-64
$C_{15}H_{10}N_2$			
4,10-Diazapyrene, 5-methyl-	EtOH	235(3.67),267(3.11), 304(2.99),332(2.85)	39-3379-65
12H-Indolo[2,3-a]quinolizin-5-ium hydroxide, inner salt	EtOH-acid	219(4.00),245(3.93), 292(3.64),343(3.76), 387(3.58)	49-0909-65
	EtOH-base	212(4.23),240s(3.86), 254s(3.73),287(3.92), 315(3.61),360(3.80), 444(3.10)	49-0909-65
Malononitrile, (biphenylyl)-	MeCN	253(4.30)	35-2174-64
$C_{15}H_{10}N_2O$			
4H-4,10-Diazapyren-5-one, 9-methyl-	EtOH	234(3.86),254(4.31), 274(4.21),291s(4.06), 302s(3.86),308s(3.79), 318(3.79),338(3.72), 357(3.72),373(3.83)	39-3379-65

Compound	Solvent	$\lambda_{max}(\log \epsilon)$	Ref.
$C_{15}H_{10}N_2O_2$			
6-Canthinone, 4-methoxy-	MeOH-acid	265(4.1),320(4.1), 360(4.0)	100-0095-65
	MeOH	240(4.3),265(4.3), 288(3.9),296(3.9), 350(4.0),370(4.1)	100-0095-65
	MeOH-base	268(4.1),288(3.8), 298(3.8),350(3.8), 368(3.9)	100-0095-65
Indazolo[2,3-a]quinoline-7,10-dione, 5,6-dihydro-	MeOH	270(4.32),325(3.42)	39-5871-65
$C_{15}H_{10}N_2O_3$			
1-Indanone, 3-imino-2-(p-nitrophenyl)-	MeOH	224(4.41),261(4.46), 311(3.60),320(3.66), 393(4.28)	65-0448-64
Indone, 3-amino-2-(p-nitrophenyl)-	MeOH	224(4.41),261(4.47), 311(3.60),393(4.28)	65-0448-64
	MeOH-base	229(4.41),260(4.39), 320s(3.60),333s(3.48), 486(4.40)	65-0448-64
4-Quinolinol, 3-(4-nitrophenyl)-	EtOH	238(4.41),290(3.85), 328s(4.26),336(4.28), 353s(4.15)	35-2086-64
	EtOH-NaOH	243s(4.28),285(4.03), 315(4.07),415(4.08)	35-2086-64
$C_{15}H_{10}N_2O_5$			
2H-1,3-Benzoxazine-2,4(3H)-dione, 3-(p-nitrobenzyl)-	MeOH	206(4.71),243(4.14), 270(4.09)	4-0037-65
3H-Indol-3-one, 5-methoxy-2-(m-nitrophenyl)-, 1-oxide	MeOH	250s(3.80),283(4.31), 341s(3.32),477(2.06)	17-0405-65
3H-Indol-3-one, 6-methoxy-2-(p-nitrophenyl)-, 1-oxide	MeOH	220(3.99),267s(4.21), 288(4.27),327(4.37), 428(3.16)	17-0405-65
Phthalimidine, 5,6-(methylenedioxy)-2-(m-nitrophenyl)-	MeOH	311(4.26)	95-1049-65
Phthalimidine, 5,6-(methylenedioxy)-2-(p-nitrophenyl)-	MeOH	339(4.44)	95-1049-65
$C_{15}H_{10}N_2O_6$			
Acrylic acid, 2,3-bis(p-nitrophenyl)-, cis	EtOH	358(4.34)	87-0614-64
trans	EtOH	286(4.17)	87-0614-64
1,4-Pentadien-3-one, 1-(5-nitro-2-furyl)-5-(p-nitrophenyl)-	EtOH	224(4.13),360(4.55)	65-0489-64
	95% H_2SO_4	273(3.92),345(4.15), 478(4.67)	65-0489-64
2,4-Pentadien-1-one, 1-(5-nitro-2-furyl)-5-(p-nitrophenyl)-	EtOH	312s(--),366(4.60)	65-0489-64
	95% H_2SO_4	335(4.09),508(4.79)	65-0489-64
2,4-Pentadien-1-one, 5-(5-nitro-2-furyl)-1-(p-nitrophenyl)-	EtOH	276(4.25),389(4.43)	65-0489-64
	95% H_2SO_4	325(4.05),500(4.66)	65-0489-64
$C_{15}H_{10}N_4$			
Bicyclo[6.1.0]nonatriene, tetracyanoethylene adduct	EtOH	230(3.41)	35-5194-64
Mesoxalonitrile, diphenylhydrazone	MeCN	232(3.95),250s(3.82), 348(4.26)	44-4198-65
$C_{15}H_{10}N_4O_4$			
Pyrazole, 1,3-bis(p-nitrophenyl)-	EtOH	340(4.34)	23-1605-64

Compound	Solvent	$\lambda_{max}(\log \epsilon)$	Ref.
Pyrazole, 4-nitro-1-(p-nitrophenyl)-5-phenyl-	EtOH	285(4.29)	23-1605-64
$C_{15}H_{10}N_4O_{10}$			
Acetic acid, bis(2,4-dinitrophenyl)-, methyl ester	EtOH	241(4.50),262s(4.38)	44-0636-64
$C_{15}H_{10}OS$			
Benzo[b]thiophen-3(2H)-one, 2-benzylidene-	n.s.g.	422(4.08),434(4.12)	65-0519-65
$C_{15}H_{10}O_2$			
6H-Benzofuro[3,2-c][1]benzopyran	n.s.g.	240(4.28),295(3.97), 300(4.09),333(4.28)	39-4212-64
Coumarin, 3-phenyl-	MeOH	325(4.18)	2-0182-64
	EtOH	324(4.17)	42-0093-64
Indandione, 3-phenyl-	MeOH	252(4.51)	35-2510-65
1,3-Indandione, 2-phenyl-	H_2O	229(4.71),440s(1.0)	30-0184-64A
	MeOH-HCl	226(4.62),440(2.60)	30-0184-64A
	EtOH-HCl	225(4.48),440(2.88)	30-0184-64A
	iso-BuOH-HCl	226(4.52),440(2.80)	30-0184-64A
	acetone	432(2.15)	30-0184-64A
	acetone-HCl	225(4.63),435(2.49)	30-0184-64A
	dioxan	420s(1.90)	30-0184-64A
	dioxan-HCl	226(4.64),428(2.00)	30-0184-64A
Isocoumarin, 3-phenyl-	$CHCl_3$	297(4.35),309(4.27)	83-0715-65
9-Phenanthrenecarboxylic acid	EtOH	211(4.45),231(4.36), 254(4.62),300(4.00), 339(2.79),356(2.59)	39-5544-64
$C_{15}H_{10}O_2S$			
Anthraquinone, 1-(methylthio)-	EtOH	244(4.52),304(3.83), 440(3.65)	22-1648-65
$C_{15}H_{10}O_3$			
3(2H)-Benzofuranone, 2-benzoyl-	EtOH	214(4.10),242(4.03), 343(4.21)	39-2361-65
$C_{15}H_{10}O_3S$			
Anthraquinone, 1-(methylsulfinyl)-	EtOH	253(4.60),302(3.67)	22-1648-65
$C_{15}H_{10}O_4$			
Chrysin	n.s.g.	268(4.5),312(4.1)	88-1987-65
$C_{15}H_{10}O_4S$			
Anthraquinone, 1-(methylsulfonyl)-	EtOH	254(4.55),320(3.54)	22-1648-65
$C_{15}H_{10}O_5$			
3(2H)-Benzofuranone, 2-(2,4-dihydroxybenzoyl)-	$CHCl_3$	297(3.90),367(4.41)	39-5140-65
Flavone, 2',4',7-trihydroxy-	MeOH	281.5(2.92)	44-3445-64
Genistein	EtOH	262(4.51)	22-1038-64
$C_{15}H_{10}O_6$			
Flavone, 2',3,5,6-tetrahydroxy-	MeOH	216(4.22),247(4.12), 270(4.10),371(3.96), 433(3.96)	24-0164-65

Compound	Solvent	λ_{max}(log ϵ)	Ref.
Flavone, 2',3,5,6-tetrahydroxy- (contd.)	MeOH-NaOMe	218(4.22),244(4.10), 295(4.03),371(4.04), 462s(3.59)	24-0164-65
Flavone, 2',5,6,7-tetrahydroxy-	MeOH	214(4.21),281(4.11), 338(4.06)	24-0164-65
	MeOH-NaOMe	214(4.27),285(4.05), 305s(3.95),397(3.95)	24-0164-65
Flavone, 2',5,7,8-tetrahydroxy-	MeOH	214(4.27),268(4.20), 334(4.03),376(3.95)	24-0164-65
	MeOH-NaOMe	230(4.27),268(4.20), 385(4.14)	24-0164-65
Flavonol, 4',7,8-trihydroxy-	EtOH	268(4.28),315(4.10), 370(4.27)	12-1170-64
Kaempferol	EtOH	266(4.26),328s(3.85), 368(4.05)	22-0779-65
	EtOH-$AlCl_3$	270(4.30),304(3.82), 347(3.92),421(4.15)	22-0779-65
Luteolin	EtOH	258(4.31),268s(4.29), 296(4.05),354(4.36)	94-0841-64
	EtOH-NaOEt	268(4.39),340(4.04), 405(4.47)	94-0841-64
	EtOH-NaOAc	237(4.29),270(4.33), 296(4.04),381(4.27)	94-0841-64
	EtOH-$AlCl_3$	270(4.25),292(4.05), 361(4.20),389(4.21)	94-0841-64
	EtOH-H_3BO_3-NaOAc	262(4.43),296(3.99), 377(4.43)	94-0841-64
$C_{15}H_{10}O_7$			
Flavone, 2',4',5,6',7-pentahydroxy-	MeOH	291(3.04)	44-3445-64
Quercetin	EtOH	257(4.30),295s(3.83), 373(4.29)	22-0779-65
	EtOH	257(4.34),374(4.34)	95-0047-64
	EtOH-$AlCl_3$	267(4.33),303s(3.86), 361(3.86),426(4.27)	22-0779-65
	EtOH-$AlCl_3$	270(4.39),438(4.38)	95-0047-64
$C_{15}H_{10}O_8$			
Myricetin	EtOH	255(4.21),378(4.29)	22-0779-65
$C_{15}H_{10}O_9$			
Pyrano[3,2-b]pyran-4,8-dione, 2-[3-hydroxy-6-(hydroxymethyl)-4-oxo-4H-pyran-2-yl]-6-(hydroxymethyl)-	MeOH	287(3.63)	44-3445-64
$C_{15}H_{10}S_3$			
2,3-Benzo-6a-thiathiophthene, 5-phenyl-	$CHCl_3$	258(4.30),270(4.63), 331(4.04),367(3.78), 515(4.11)	24-1732-64
Phenanthreno[1,2-d]trithione, 10,11-dihydro-	$CHCl_3$	242(4.48),303(4.26), 346(4.22),386(4.12), 452(4.09)	5-0062-64D
$C_{15}H_{11}BrClNO_2$			
Acetanilide, 4'-bromo-2-(p-chlorobenzoyl)-	0.04N NaOH	338(4.44)	20-0782-64
	50% EtOH	256(4.49)	20-0782-64
$C_{15}H_{11}BrN_2O_2$			
4(3H)-Quinazolinone, 3-(p-bromophenyl)-2-methyl-, 1-oxide	EtOH	315(4.02)	95-0507-65

Compound	Solvent	λ_{max}(log ε)	Ref.
$C_{15}H_{11}BrN_2O_4$			
Acetanilide, 4'-bromo-2-(m-nitro-	0.04N NaOH	341(4.38)	20-0782-64
benzoyl)-	60% EtOH	250(4.40)	20-0782-64
$C_{15}H_{11}BrO$			
Chalcone, 4-bromo-	n.s.g.	242(4.47),294(4.63), 351(4.58)	49-0450-65
$C_{15}H_{11}BrO_2$			
Chalcone, 5'-bromo-2'-hydroxy-	n.s.g.	225(4.74),345(3.99)	49-0450-65
$C_{15}H_{11}BrO_3$			
Chalcone, 3'-bromo-2,2'-dihydroxy-	n.s.g.	226(4.25),334(4.32)	49-0450-65
Chalcone, 5'-bromo-2,2'-dihydroxy-	n.s.g.	223(4.49),268(3.91), 322(4.11),396(4.38)	49-0450-65
$C_{15}H_{11}Cl$			
Anthracene, 1-chloro-9-methyl-	EtOH	260(5.1),325s(3.1), 343s(3.4),359(3.6), 375(3.7),395(3.6)	28-6922-65A
Anthracene, 1-chloro-10-methyl-	EtOH	258(5.2),321(3.2), 337(3.5),353(3.6), 370(3.8),390(3.8)	28-6922-65A
$C_{15}H_{11}ClF_3N_3O_5S_2$			
2H-1,2,4-Benzothiadiazine-7-sulfonamide, 6-chloro-3-[α-hydroxy-m-(trifluoromethyl)benzyl]-, 1,1-dioxide	EtOH	227(4.54),279(4.11)	95-0095-65
$C_{15}H_{11}ClN_2$			
Benzimidazole, 5(6)-chloro-2-styryl-	EtOH	263(4.03),324(4.43), 336(4.44),352s(4.22)	2-0169-64
$C_{15}H_{11}ClN_2O$			
Carbostyril, 3-amino-6-chloro-4-phenyl-	iso-PrOH	235(4.6),330(4.1), 345(4.1)	39-3097-64
Phenol, p-[(7-chloro-4-quinolyl)-	pH 1	342(4.18)	4-0006-64
amino]-	EtOH	339(4.10)	4-0006-64
$C_{15}H_{11}ClN_2O_4$			
Acetanilide, 2-(4-chlorobenzoyl)-	0.04N NaOH	336(4.43)	20-0782-64
3'-nitro-	60% EtOH	253(4.45)	20-0782-64
Acetanilide, 3'-chloro-2-(3-nitro-	0.04N NaOH	333(4.35)	20-0782-64
benzoyl)-	60% EtOH	239(4.41)	20-0782-64
$C_{15}H_{11}ClN_4$			
2-Quinoxalinecarboxaldehyde, 6-chloro-, phenylhydrazone	EtOH	229(4.41),292(3.91), 348s(4.12),410(4.47)	5-0146-65D
2-Quinoxalinecarboxaldehyde, 7-chloro-, phenylhydrazone	EtOH	237(4.47),292(4.02), 361s(4.09),413(4.49)	5-0146-65D
$C_{15}H_{11}ClO$			
5H-Dibenzo[a,d]cyclohepten-5-ol, 2-chloro-	EtOH	223(4.52),283(4.15)	87-0829-65
$C_{15}H_{11}ClO_2$			
Acrylic acid, 2-(o-chlorophenyl)-3-phenyl-	EtOH	275(4.28)	44-1473-65
Acrylic acid, 2-(p-chlorophenyl)-3-phenyl-	EtOH	282(4.19)	44-1473-65

Compound	Solvent	$\lambda_{max}(\log \epsilon)$	Ref.
$C_{15}H_{11}ClO_4$			
7-(Phenylethynyl)tropylium perchlorate	MeCN	236(4.32),256(4.31), 275s(4.07),338(3.84), 417(4.48)	24-1337-64
2',4',7-Trihydroxyflavylium chloride	n.s.g.	515(3.86)	2-0319-64
$C_{15}H_{11}ClO_4S_2$			
2,4-Diphenyl-1,3-dithiol-1-ium perchlorate	70% $HClO_4$	243(4.17),277(3.82), 310(3.72),394(4.18)	44-1703-64
3,5-Diphenyl-1,2-dithiol-1-ium perchlorate	EtOH-$HClO_4$	381(4.4)	73-3016-65
$C_{15}H_{11}ClO_5$			
Flavylium perchlorate	MeCN	250(4.09),260(4.10), 390(4.12)	35-3142-64
$C_{15}H_{11}Cl_2NO_2$			
Acetanilide, 3'-chloro-2-(4-chlorobenzoyl)-	0.04N NaOH	334(4.42)	20-0782-64
	60% EtOH	253(4.42)	20-0782-64
$C_{15}H_{11}Cl_2NO_3$			
9H-Pyrrolo[1,2-a]indole-3-glyoxyloyl chloride, 1-chloro-7-methoxy-6-methyl-	$CHCl_3$	247(4.11),305(4.00), 330(4.00)	35-4608-64
$C_{15}H_{11}FO_2$			
Acrylic acid, 2-(p-fluorophenyl)-3-phenyl-	EtOH	277(4.14)	44-1473-65
$C_{15}H_{11}HgNO_6S$			
Benzoic acid, p-[[(carboxymethyl)thio]-mercuri]-, p-(p-nitrophenyl) ester	pH 9	235(4.3)	70-0111-64
$C_{15}H_{11}IO_3$			
Acrylic acid, 3-(p-hydroxyphenyl)-2-(p-iodophenyl)-	EtOH	232(4.05),303(3.96)	7-0805-64
$C_{15}H_{11}MnO_5$			
Phenylmanganese carbonyl, butadiene adduct	dioxan	231(3.76),268(4.03)	77-0370-64
$C_{15}H_{11}N$			
Pyridine, 2-(inden-1-ylidenemethyl)-	EtOH-HCl	247(4.23),275(4.17), 358(4.23)	5-0018-64D
	dioxan	244s(4.09),274(4.24), 277(4.25),347(4.39), 360(4.34)	5-0018-64D
Pyridine, 3-(inden-1-ylidenemethyl)-	EtOH-HCl	242(4.31),275(4.40), 345(4.20)	5-0018-64D
	dioxan	248(4.08),278(4.33), 342(4.28)	5-0018-64D
Pyridine, 4-(inden-1-ylidenemethyl)-	EtOH-HCl	240(4.20),289(4.21), 354(4.24)	5-0018-64D
	dioxan	245(4.02),275(4.36), 338(4.22)	5-0018-64D
Pyridine, 2-(4-phenyl-1-buten-3-ynyl)-, trans	EtOH	221(4.00),320(4.57)	94-1344-64
Quinoline, 2-phenyl-	EtOH	253(4.6),316(4.0)	28-0954-64A

Compound	Solvent	$\lambda_{max}(\log \epsilon)$	Ref.
$C_{15}H_{11}NO$			
Acenaphth[1,2-d]isoxazole, 9-ethyl-	EtOH	231(4.71),266(3.97), 276(3.99),318(3.98), 340(3.76)	88-2157-64
1-Indanone, 3-imino-2-phenyl-	MeOH	222(4.42),276(4.49), 282s(4.48),324(3.74), 441(3.50)	65-0448-64
Indole-2-carboxaldehyde, 3-phenyl-	iso-PrOH	230(4.22),250(4.35), 320(4.32)	44-1621-64
2-Indolinone, 3-benzylidene-	EtOH	243(4.05),253(4.08), 273s(3.92),321(4.25), 392s(3.48)	87-0626-65
	EtOH	256(4.17),327(4.26), 385s(3.61)	39-5762-65
Indone, 3-amino-2-phenyl-	MeOH	222(4.42),276(4.49), 282s(4.48),441(3.50)	65-0448-64
	EtOH	275(4.56),445(3.29)	35-0874-65
Indone, 3-anilino-	MeOH	225(4.31),253(4.32), 262s(4.27),283(4.08), 326(3.72),427(3.71)	65-0448-64
Isoxazole, 3,5-diphenyl-	$CHCl_3$	247(4.29),268(4.33)	24-3020-65
	n.s.g.	244(4.3),267(4.35)	32-0393-64
Ketone, 2-indolyl phenyl	EtOH	245(4.12),322(4.33)	44-3604-65
Phenanthrene-9-carboxamide	EtOH	212(4.45),221(4.33), 248(4.64),254(4.71), 275s(4.35),285(4.26), 296(4.34),332(2.49), 342(2.43),348(1.69)	39-5544-64
4-Quinolinol, 3-phenyl-	EtOH	263(4.42),309(3.99), 329(4.01),334s(4.00)	35-2086-64
$C_{15}H_{11}NO_2$			
2-Indolinone, 3-(hydroxymethylene)-1-phenyl-	MeOH	219s(4.37),262(4.38), 294s(3.99),301(4.00), 328s(3.33)	87-0626-65
Phenanthrene, 9-methyl-10-nitro-	THF	222(4.7),245(4.84), 253(5.47),274(4.25), 283(4.12),294(4.08)	39-2587-65
$C_{15}H_{11}NO_3$			
4H-Benzo[a]quinolizin-4-one, 3-acetyl-2-hydroxy-	EtOH	234(4.42),295(4.36), 400(4.45),420(4.50)	95-0874-64
4,1-Benzoxazepine-2,5(1H,3H)-dione, 1-phenyl-	iso-PrOH	244s(4.11),300(3.57)	4-0323-65
2H-1,3-Benzoxazine-2,4(3H)-dione, 3-benzyl-	MeOH	207(4.59),240(3.98), 290s(3.26)	4-0037-65
3H-Indol-3-one, 5-methoxy-2-phenyl-, 1-oxide	C_6H_{12}	281s(4.57),287(4.64), 310s(4.23),353s(3.38), 480(3.32)	17-0405-65
	MeOH	284(4.64),338s(3.60), 485(3.36)	17-0405-65
Phthalimide, N-(m-methoxyphenyl)-	EtOH	220(4.65),234s(4.30), 281(3.64),296s(3.55)	7-0128-64
Phthalimide, N-(o-methoxyphenyl)-	EtOH	219(4.66),226s(4.22), 278(3.77),294s(3.41)	7-0128-64
Phthalimide, N-(p-methoxyphenyl)-	EtOH	216(4.60),238s(4.30), 249s(4.20),295s(3.29)	7-0128-64
Phthalimide, 4-methoxy-N-phenyl-	EtOH	240(4.59),262s(4.13), 314s(3.32),324s(3.25)	7-0128-64
Phthalimide, 5-methoxy-N-phenyl-	EtOH	226(4.56),336(3.76)	7-0128-64

Compound	Solvent	$\lambda_{max}(\log \epsilon)$	Ref.
Phthalimidine, 2-(m-carboxyphenyl)-	MeOH-acid	276(4.13)	95-1042-65
	MeOH	281(4.09)	95-1042-65
Phthalimidine, 2-(o-carboxyphenyl)-	MeOH-acid	270s(--)	95-1042-65
	MeOH	270s(--)	95-1042-65
Phthalimidine, 2-(p-carboxyphenyl)-	MeOH-acid	294(4.39)	95-1042-65
	MeOH	294(4.39)	95-1042-65
Phthalimidine, 5,6-(methylenedioxy)- 2-phenyl-	MeOH	309(4.24)	95-1049-65
$C_{15}H_{11}NO_3S$			
Acrylophenone, 3-[(m-nitrophenyl)thio]-	MeOH	255(4.34),329(4.34)	5-0136-64C
4H-1,2-Benzothiazin-4-one, 3-benzylidene-2,3-dihydro-, 1,1-dioxide	EtOH	235(4.03),262(4.00), 335(4.22)	44-2241-65
$C_{15}H_{11}NO_4$			
Acrylic acid, 2-(p-nitrophenyl)- 3-phenyl-, cis	EtOH	332(4.28)	87-0614-64
trans	EtOH	269(4.31)	87-0614-64
Acrylic acid, 3-(p-nitrophenyl)- 2-phenyl-, cis	EtOH	333(4.27)	87-0614-64
trans	EtOH	314(4.15)	87-0614-64
Chalcone, 2'-hydroxy-4-nitro-	MeOH	318(4.43)	24-1910-64
Phthalimidine, 2-(m-hydroxyphenyl)- 5,6-(methylenedioxy)-	MeOH	311(4.28)	95-1049-65
Phthalimidine, 2-(o-hydroxyphenyl)- 5,6-(methylenedioxy)-	MeOH	304(4.31)	95-1049-65
Phthalimidine, 2-(p-hydroxyphenyl)- 5,6-(methylenedioxy)-	MeOH	308(4.23)	95-1049-65
$C_{15}H_{11}NO_5$			
Benzoic acid, 2-benzoyl-3-nitro-, methyl ester	EtOH	246(4.07)	44-4293-65
Benzoic acid, 2-benzoyl-6-nitro-, methyl ester	EtOH	247(4.03)	44-4293-65
4H,5H-Pyrano[2,3-b]quinolizine- 9-carboxylic acid, 2-methyl- 4,5-dioxo-, methyl ester	EtOH	345(3.64),350(3.73), 434(3.99)	95-0874-64
$C_{15}H_{11}NO_8$			
4H-Quinolizine-1,2-dicarboxylic acid, 3-(carboxycarbonyl)-4-oxo-, 1,2-dimethyl ester	MeOH	252s(4.32),410(4.26)	39-2633-65
$C_{15}H_{11}NS$			
Isothiazole, 3,5-diphenyl-	EtOH	207(4.39),251(4.43), 277(4.34)	39-0032-65
$C_{15}H_{11}N_3O$			
10H-Dibenzo[b,e][1,4]diazepine-10-acetonitrile, 5,11-dihydro-11-oxo-	EtOH	223(4.52)	33-1590-65
Formic acid, cyano-, (diphenylmethylene)hydrazide	C_6H_{12}	225s(4.14),289(4.28)	44-4198-65
2-Naphthol, 1-(2-pyridylazo)-	pH 1	280(4.08)	96-0161-65
$C_{15}H_{11}N_3O_2$			
Pyrazole, 1-(p-biphenylyl)-4-nitro-	EtOH	265(4.06),315(4.08)	23-1605-64
Pyrazole, 4-nitro-1,3-diphenyl-	EtOH	256(4.24),300s(3.95)	23-1605-64
Pyrazole, 4-nitro-1,5-diphenyl-	EtOH	245(4.24),280(3.85)	23-1605-64
Pyrazole, 1-(p-nitrophenyl)-3-phenyl-	EtOH	262(4.16),337(4.24)	23-1605-64

Compound	Solvent	λ_{max}(log ϵ)	Ref.
as-Triazin-3(2H)-one, 5,6-diphenyl-, 1-oxide, hydrate	EtOH	236(4.37),275(--), 310s(--)	94-1168-65
$C_{15}H_{11}N_3O_3$			
Cyclohepta[b]pyrrol-2(1H)-one, 1-amino-3-(p-nitrophenyl)-	EtOH	235(4.26),287(4.41), 440(4.21)	94-0450-65
$C_{15}H_{11}N_3O_4$			
Indole, 2-methyl-5-nitro-3-(p-nitrophenyl)-	EtOH	270(4.34),344(4.10)	78-0823-65
1,3,4-Oxadiazole, 2-[1-methyl-2-(5-nitro-2-furyl)vinyl]-5-phenyl-	propylene glycol	238(4.18),295(4.12), 378(4.36)	95-0225-64
$C_{15}H_{11}N_3O_6$			
Acetanilide, 2-(3,5-dinitrobenzoyl)-	0.04N NaOH	314(4.19)	20-0782-64
	60% EtOH	234(4.45)	20-0782-64
$C_{15}H_{11}N_3O_8$			
Acetic acid, (2,4-dinitrophenyl)-(p-nitrophenyl)-, methyl ester	EtOH	247s(4.31),261(4.34)	44-0636-64
$C_{15}H_{12}$			
Anthracene, 9-methyl-	EtOH	257(5.28)	87-0088-64
	n.s.g.	none above 390 nm	10-0133-65B
Cyclopentadienoheptalene, 2-methyl-	hexane	370(4.06),385(4.13), 389(4.16),424(2.64), 745(1.92),813(2.03), 887(1.93),928(1.99), 954(1.94),1070(1.62), 1120(1.64)	5-0194-64B
Cyclopentadienoheptalene, 4-methyl-	hexane	372(3.96),388(4.02), 393(4.08),458(2.36), 753(1.81),822(1.92), 838(1.91),941(1.87), 1085(1.41),1140(1.43)	5-0194-64B
Dibenzonorcaradiene	EtOH	223(4.5),253(4.74), 276s(--),284s(--), 296(4.03)	88-1525-64
	n.s.g.	271(4.14),283(4.05), 302(3.81),313(3.84)	88-0423-65
Indene, 3-phenyl-	n.s.g.	215(4.5),260(3.9)	88-4569-65
Phenanthrene, 9-methyl-	n.s.g.	253(4.75),277(4.13), 284(4.00),297(4.07), 325(4.07)	88-0423-65
$C_{15}H_{12}BN$			
Phenanthro[4,5-cde][1,2]azaborine, 4,5-dihydro-5-methyl-	$CHCl_3$	252(4.61),270(4.42), 279s(4.28),296(3.90), 308(4.10),332s(3.60), 341s(3.38),348(3.83), 359s(3.48),366(4.02)	44-1757-64
$C_{15}H_{12}BrClO_4$			
7-(4-Bromostyryl)tropylium perchlorate	MeCN	254(4.23),280(4.19), 443(4.55)	24-1337-64
$C_{15}H_{12}BrNO$			
Acenaphth[1,2-d]isoxazole, 9-(1-bromoethyl)-6b,9a-dihydro-	EtOH	223(4.83),276(3.78), 286(3.87),298(3.70)	88-2157-64

Compound	Solvent	$\lambda_{max}(\log \varepsilon)$	Ref.
Acenaphth[1,2-d]isoxazole, 9a-bromo-9-ethyl-6b,9a-dihydro-	EtOH	224(4.73),292(3.80), 325(3.17)	88-2157-64
Acrylophenone, 3-anilino-4'-bromo-	heptane	239s(3.99),261(4.23), 371(4.48)	70-1382-65
$C_{15}H_{12}BrNO_2$			
Acetanilide, 2-benzoyl-4'-bromo-	0.04N NaOH	340(4.45)	20-0782-64
	60% EtOH	252(4.46)	20-0782-64
Acrylic acid, 3-(o-aminophenyl)-2-(o-bromophenyl)-	EtOH	284(3.95),350(3.73)	44-3933-65
$C_{15}H_{12}BrNO_5$			
4H-m-Dioxino[5,4-c]pyridine-4,5(6H)-dione, 6-(m-bromophenyl)-7-hydroxy-2,2-dimethyl-	EtOH	347(4.72)	78-1917-65
4H-m-Dioxino[5,4-c]pyridine-4,5(6H)-dione, 6-(p-bromophenyl)-7-hydroxy-2,2-dimethyl-	EtOH	338(4.20)	78-1917-65
Malonic acid, [3-(m-bromoanilino)-3-hydroxy-1-(1-hydroxy-1-methylethoxy)allylidene]-, di-δ-lactone	EtOH	313(4.43)	78-1917-65
Malonic acid, [3-(p-bromoanilino)-3-hydroxy-1-(1-hydroxy-1-methylethoxy)allylidene]-, di-δ-lactone	EtOH	313(4.39)	78-1917-65
$C_{15}H_{12}BrN_3O$			
Isocarbostyril, 5-bromo-2-(2,4-diaminophenyl)-	EtOH	216(4.55),240(4.34), 294(4.15),336s(3.74)	24-1023-65
$C_{15}H_{12}Br_2$			
Bicyclo[4.2.0]octa-1,3,5-triene, 7,8-dibromo-7-methyl-8-phenyl-	isooctane	271s(3.60),277(3.65), 283(3.60)	35-0086-65
$C_{15}H_{12}ClNO$			
2-Indolinone, 3-(p-chlorophenyl)-1-methyl-	EtOH	220s(--),255(3.93), 265(--),283(--)	87-0626-65
$C_{15}H_{12}ClNO_2$			
Acetanilide, 2-benzoyl-3'-chloro-	0.04N NaOH	336(4.40)	20-0782-64
	60% EtOH	247(4.42)	20-0782-64
Acetanilide, 2-benzoyl-4'-chloro-	0.04N HaOH	330(4.41)	20-0782-64
	60% EtOH	241(4.46)	20-0782-64
Acetanilide, 2-(4-chlorobenzoyl)-	0.04N NaOH	338(4.36)	20-0782-64
	60% EtOH	253(4.37)	20-0782-64
$C_{15}H_{12}ClNO_3$			
Anthranilic acid, N-(chloroacetyl)-N-phenyl-	iso-PrOH	222s(4.24),250(3.92), 290(3.26)	4-0323-65
$C_{15}H_{12}ClNO_5$			
4H-m-Dioxino[5,4-c]pyridine-4,5(6H)-dione, 6-(m-chlorophenyl)-7-hydroxy-2,2-dimethyl-	EtOH	312(4.68)	78-1917-65
4H-m-Dioxino[5,4-c]pyridine-4,5(6H)-dione, 6-(p-chlorophenyl)-7-hydroxy-2,2-dimethyl-	EtOH	312(4.67)	78-1917-65
Malonic acid, [3-(m-chloroanilino)-3-hydroxy-1-(1-hydroxy-1-methylethoxy)a-lylidene]-, di-δ-lactone	EtOH	338(4.56)	78-1917-65

Compound	Solvent	λ_{max}(log ϵ)	Ref.
Malonic acid, [3-(p-chloroanilino)-3-hydroxy-1-(1-hydroxy-1-methylethoxy)allylidene]-, di-δ-lactone	EtOH	347(4.73)	78-1917-65
$C_{15}H_{12}ClNO_6$			
7-Carboxybenzo[b]quinolizinium perchlorate, methyl ester	EtOH	205(4.00),239(4.62), 263(4.35),343s(3.74), 367(4.02),382(4.08), 403(4.04)	4-0030-64
9-Carboxybenzo[b]quinolizinium perchlorate, methyl ester	EtOH	245(4.60),274s(4.28), 323(3.44),339(3.67), 357(3.91),370(3.78), 390(4.08),412(4.16)	4-0030-64
$C_{15}H_{12}ClN_3O$			
Isocarbostyril, 5-chloro-2-(2,4-diaminophenyl)-	EtOH	214(4.51),238s(4.35), 284(4.16),330s(3.78)	24-1023-65
$C_{15}H_{12}Cl_2N_2$			
1-Pyrazoline, 3,5-bis(p-chlorophenyl)-	EtOH	327(2.83)	35-0658-64
$C_{15}H_{12}Cl_2N_4O$			
2-Propanone, 1-[(o-chlorophenyl)azo]-1-[(o-chlorophenyl)hydrazono]-	EtOH	260(4.05),298(4.15), 440(4.34)	5-0099-65B
$C_{15}H_{12}Cl_2O_4$			
1,4-Naphthoquinone, 8-dichloroacetyl-5-methoxy-2,7-dimethyl-	EtOH	207(4.39),261(4.38), 394(3.45)	54-0995-64
$C_{15}H_{12}Cl_4O_2$			
1,4-Methanobiphenylen-9-one, 1,2,3,4-tetrachloro-1,4,4a,8b-tetrahydro-, dimethyl acetal	EtOH	254s(--),260(2.98), 265(3.22),272(3.24)	39-2549-65
$C_{15}H_{12}FNO_2$			
Acetanilide, 2-benzoyl-4'-fluoro-	0.04N NaOH	332(4.33)	20-0782-64
	60% EtOH	242(4.36)	20-0782-64
$C_{15}H_{12}INO_2$			
Acetanilide, 2-benzoyl-4'-iodo-	0.04N NaOH	338(4.45)	20-0782-64
	60% EtOH	251(4.48)	20-0782-64
$C_{15}H_{12}IN_3$			
3-Cyano-2-indol-3-yl-1-methylpyridinium iodide	EtOH	212(4.79),275(4.02), 284(3.98),370(3.76)	44-2534-65
$C_{15}H_{12}N_2$			
Benzimidazole, 2-styryl-	EtOH	262(4.05),322(4.43), 334(4.43),350s(4.20)	2-0169-64
5H-Benzo[g]pyrido[2,3-b]indole, 6,11-dihydro-	EtOH-acid	235(3.67),292(3.33), 348(3.45),389(3.28)	49-0909-65
	EtOH-base	240s(3.76),255(3.61), 287(3.71),316(3.37), 357(3.51)	49-0909-65
$C_{15}H_{12}N_2O$			
Benzimidazole, 2-benzoyl-1-methyl-	MeOH	265(3.98),306(4.18)	78-0017-64
4(1H)-Cinnolinone, 1-benzyl-	EtOH	209(4.56),238(4.07), 254s(3.95),265s(3.79), 283(3.48),294(3.46), 343(4.16),358s(4.09)	39-5391-65

Compound	Solvent	λ_{max}(log ε)	Ref.
Furo[3,4-a]phenazine, 1,3-dihydro-5-methoxy-	EtOH	237(3.41),273(3.79),370(3.15)	39-4672-65
Indazole, 3-benzoyl-6-methyl-	EtOH	254(4.24),314(4.08)	32-0814-65
2-Indolinone, 1-methyl-3-(2-pyridylmethylene)-	EtOH	260(4.21),334(4.23),397(3.25)	87-0626-65
2-Pyrazolin-5-one, 1,3-diphenyl-	C_6H_{12}	263(4.25)	78-0299-64
	10N H_2SO_4	266(4.35)	78-0299-64
	pH 5	265.5(4.32)	78-0299-64
	pH 13	261(4.32)	78-0299-64
4-Quinazolinone, 3-benzyl-	MeOH	267(3.91),274s(3.87),301(3.56),313(3.47)	4-0475-65
4(1H)-Quinazolinone, 2-methyl-1-phenyl-	EtOH	229(4.33),234s(4.29),265(3.72),274(3.75)	44-1020-65
$C_{15}H_{12}N_2O_2$			
1H-1,4-Benzodiazepine-2,5-dione, 3,4-dihydro-1-phenyl-	iso-PrOH	214(4.45),242s(4.17),293(3.62)	4-0323-65
Cycloheptimidazole-2,6-dione, 1-benzyl-1,3-dihydro-	EtOH	251(4.31),364(4.31)	94-0473-65
Indole, 2-methyl-3-(p-nitrophenyl)-	EtOH	223(4.50),268(4.05),396(4.05)	78-0823-65
6,9-Methanobenzo[g]quinoxaline-5,10-dione, 6,9-dihydro-2,3-dimethyl-	MeCN	220(4.96),251(4.60),282(4.52),292(4.37)	44-2583-65
2H-1,3,4-Oxadiazin-2-one, 3,4-dihydro-5,6-diphenyl-	EtOH	217(4.11),297(4.08)	35-5716-65
Pyridine, 3-[5-(p-methoxyphenyl)-2-oxazolyl]-	iso-PrOH-HCl	247(3.95),325(4.34)	4-0310-65
Pyrido[1,2-b]benzo[f]indazole, 1,2,3,4,7,12-hexahydro-7,12-dioxo-	EtOH	248(4.54),318(3.75)	39-5871-65
4(1H)-Quinazolinone, 2-(hydroxymethyl)-1-phenyl-	iso-PrOH	230(4.31),275(3.70),304(4.00),315(3.93)	4-0323-65
4(3H)-Quinazolinone, 2-methyl-3-phenyl-, 1-oxide	EtOH	315(3.98)	95-0507-65
$C_{15}H_{12}N_2O_2S$			
Indole-3-carbonitrile, 5,6-dimethoxy-2-(2-thienyl)-	MeOH	234(4.18),344(4.33)	44-2253-65
$C_{15}H_{12}N_2O_3$			
Acrylophenone, 3-anilino-4'-nitro-	MeCN	238(3.95),270(4.15),398(4.34)	70-1382-65
Indole-3-carbonitrile, 2-(2-furyl)-5,6-(methylenedioxy)-	MeOH	237(4.28),336(4.47)	44-2253-65
$C_{15}H_{12}N_2O_4$			
Acetanilide, 2-benzoyl-3'-nitro-	0.04N NaOH	334(4.42)	20-0782-64
	60% EtOH	246(4.52)	20-0782-64
Acetanilide, 2-benzoyl-4'-nitro-	0.04N NaOH	384(4.38)	20-0782-64
	60% EtOH	323(4.18)	20-0782-64
Acetanilide, 2-(3-nitrobenzoyl)-	0.04N NaOH	333(4.29)	20-0782-64
	60% EtOH	236(4.39)	20-0782-64
Acetanilide, 2-(4-nitrobenzoyl)-	0.04N NaOH	394(4.04)	20-0782-64
	60% EtOH	258(4.26)	20-0782-64
$C_{15}H_{12}N_2O_4S$			
2-Thiopheneacrylonitrile, α-(4,5-dimethoxy-2-nitrophenyl)-	MeOH	245(4.09),322(4.34)	44-2253-65
$C_{15}H_{12}N_2O_5$			
2-Furanacrylonitrile, α-(4,5-dimethoxy-2-nitrophenyl)-	MeOH	244(4.11),316(4.41)	44-2253-65

Compound	Solvent	$\lambda_{max}(\log \epsilon)$	Ref.
$C_{15}H_{12}N_2S_2$			
Thiocyanic acid, 4,6,8-trimethyl-1,3-azulenylene ester	$CHCl_3$	247(4.45),308(4.58),346(3.90),511(2.9)	44-2715-65
$C_{15}H_{12}N_4$			
Isoquinoline, 1-(3-phenyl-2-triazeno)-	EtOH	397(4.28)	89-0171-65
$C_{15}H_{12}N_4O$			
as-Triazine, 3-amino-5,6-diphenyl-, 1-oxide	EtOH	231(4.44),298(4.2)	94-1168-65
$C_{15}H_{12}N_4O_2$			
Sydnone, 4-formyl-3-phenyl-, phenylhydrazone	EtOH	245(3.94),297(4.11),424(4.08)	44-2044-64
$C_{15}H_{12}N_4O_4$			
1-Indanone, 2,4-dinitrophenylhydrazone	$CHCl_3$	388(4.48)	5-0021-64G
	diglyme	298s(3.94),312(3.83),387(4.50)	44-1723-64
2-Indanone, 2,4-dinitrophenylhydrazone	diglyme	256(4.08),267(4.08),275s(4.02),362(4.37)	44-1723-64
1,3,4-Oxadiazole, 2-anilino-5-[1-methyl-2-(5-nitro-2-furyl)vinyl]-	propylene glycol	246(4.29),306(4.03),400(4.27)	95-0219-64
1,3,4-Oxadiazole, 2-(methylamino)-5-[1-phenyl-2-(5-nitro-2-furyl)vinyl]-	propylene glycol	298(3.96),378(4.18)	95-0219-64
$C_{15}H_{12}N_4O_5S$			
Acetamide, 2-diazo-2-(p-nitrophenyl)-N-(p-tolylsulfonyl)-	MeOH	227(4.37),280(3.91),347(4.12)	88-1403-64
$C_{15}H_{12}N_4S$			
Benzaldehyde, (5-phenylthiadiazolylhydrazone)	EtOH	229(4.27),347(4.43)	1-0871-64
anion	EtOH	325(3.90),408(4.45)	1-0871-64
$C_{15}H_{12}N_6O$			
6(5H)-Pteridinone, 7,8-dihydro-7-(4-quinazolinylmethyl)-	pH 2.0	224(4.56),295(4.09)	39-3357-64
	pH 7.0	224(4.53),293(3.96)	39-3357-64
	pH 13.0	304(4.04)	39-3357-64
7(8H)-Pteridinone, 6-[(3,4-dihydro-4-quinazolinyl)methyl]-	pH 3.0	216s(4.46),299(4.12)	39-3357-64
	pH 7.85	214(4.55),264(3.90),326(4.05)	39-3357-64
	pH 11.5	214(4.54),229s(4.34),265(3.89),327(4.06)	39-3357-64
$C_{15}H_{12}N_6O_7$			
2-Propanone, 1-[(2-hydroxy-4-nitrophenyl)azo]-1-[(2-hydroxy-4-nitrophenyl)hydrazono]-	dioxan	500(4.28)	65-0502-65
	DMF	446(4.19),680(4.27)	65-0502-65
$C_{15}H_{12}O$			
Benzocyclobutene, 4-benzoyl-	EtOH	253(4.28),344(2.18)	78-0245-65
Biphenylene, 2-propionyl-	EtOH	261(4.58),348(3.60),362(3.64)	39-4930-65
Chalcone	benzene	308(4.45)	65-0094-65
	EtOH	226(4.02),310(4.43)	23-2580-64
	EtOH	225(4.08),310(4.41)	65-0094-65
	HOAc	310(4.42)	65-0094-65
6H-Dibenzo[a,c]cyclohepten-6-one, 5,7-dihydro-	isooctane	250(4.18)	35-1710-64

Compound	Solvent	$\lambda_{max}(\log \epsilon)$	Ref.
Fluorenone, 2,4-dimethyl-	EtOH	252(4.74),261(4.92), 289(3.40),299(3.45), 301(3.48),328s(3.26), 388(2.53),407(2.58)	22-2953-64
$C_{15}H_{12}OS$			
Acrylophenone, 3-(phenylthio)-	MeOH	245(4.16),285s(3.54), 320(2.70)	5-0136-64C
12H-Dibenzo[b,e]thiocin-12-one, 6,7-dihydro-	MeOH	255(4.12)	24-0685-65
1,4-Pentadien-3-one, 1-phenyl-5-(2-thienyl)-	hexane	230(4.00),335(4.47)	65-3645-64
	EtOH	350(4.54)	65-3645-64
	66% H_2SO_4	230(3.85),307(3.74), 362(3.95),495(4.68)	65-3645-64
$C_{15}H_{12}OS_2$			
Benzoic acid, dithio-, ester with 2-mercaptoacetophenone	EtOH	243(4.16),297(4.22)	44-1703-64
Cyclopentanone, 2,5-di-2-thenylidene-	hexane	252(3.96),385(4.64)	65-3645-64
	EtOH	250(3.90),402(4.57)	65-3645-64
	66% H_2SO_4	244(3.39),330(3.57), 400(4.46),557(5.00)	65-3645-64
$C_{15}H_{12}O_2$			
Acrylic acid, 2,3-diphenyl-	EtOH	277(4.04)	44-1473-65
Acrylic acid, 2,3-diphenyl-, cis	MeOH	224(4.25),289(4.33)	2-0182-64
	EtOH	288(4.34)	87-0614-64
Acrylic acid, 2,3-diphenyl-, trans	MeOH	280(4.19)	2-0182-64
	EtOH	282(4.18)	87-0614-64
Benzoic acid, 2-(o-hydroxy-phenethyl)-, lactone	EtOH	283(3.98),294(3.95)	42-0093-64
Chalcone, 2'-hydroxy-	MeOH	317(4.36)	24-1910-64
Chalcone, 3-hydroxy-	EtOH	246(4.05),312(4.20)	23-2580-64
Chalcone, 3'-hydroxy-	EtOH	227(4.04),308(4.41)	23-2580-64
Dibenzonorcaradienecarboxylic acid	n.s.g.	271(4.18),280(4.15), 298(3.72),310(--)	88-0423-65
Fluorene-9-carboxylic acid, methyl ester	EtOH	389(3.67)	78-0261-65
2-Naphthaleneacetic acid, 4-cyclo-propyl-1-hydroxy-, lactone	n.s.g.	230(4.80),277(3.82)	25-0467-65
Naphtho[1,2-b]pyran-2-one, 3,4-dimethyl-	MeOH	221(4.45),267(4.42), 276(4.49)	44-0502-64
9-Phenanthrenecarboxylic acid, 9,10-dihydro-	EtOH	265(4.27),297(3.73)	54-0364-64
2-Propyn-1-one, 3-(p-ethylphenyl)-1-(2-furyl)-	EtOH	230(4.26),326(4.48)	30-0833-65
$C_{15}H_{12}O_2S$			
Benzo[b]thiophene, 2-methyl-3-phenyl-, 1,1-dioxide	EtOH	228(4.54),254s(3.79), 312(3.38)	44-2840-65
1,4,6-Heptatrien-3-one, 1-(2-furyl)-7-(2-thienyl)-	EtOH	230(3.94),275(3.92), 387(4.60)	65-2328-64
	HOAc-30% H_2SO_4	257(3.79),340(3.91), 562(4.83)	65-2328-64
1,4,6-Heptatrien-3-one, 7-(2-furyl)-1-(2-thienyl)-	EtOH	276(4.02),387(4.60)	65-2328-64
	HOAc-30% H_2SO_4	355(3.85),410(3.90), 563(4.72)	65-2328-64
2,4,6-Heptatrien-1-one, 7-(2-furyl)-1-(2-thienyl)-	EtOH	302(4.04),400(4.62)	65-2328-64
	HOAc-30% H_2SO_4	265(4.07),307(4.04), 370(4.18),592(4.72)	65-2328-64

Compound	Solvent	λ_{max}(log ε)	Ref.
$C_{15}H_{12}O_3$			
6,7-Benzochromone, 5-methoxy-2-methyl-	EtOH	223(4.57),258(4.56), 355(3.44)	94-0312-64
Benzoic acid, o-benzoyl-, methyl ester	EtOH	244(4.15)	44-4293-65
Benzoic acid, o-benzoyl-, pseudo methyl ester	EtOH	244(4.15),280(3.18)	44-4293-65
Chalcone, 2',4-dihydroxy-	EtOH	222(4.18),243(3.94), 285(4.29),335(4.55)	23-2580-64
Chalcone, 2',5'-dihydroxy-	EtOH	234(3.75),326(3.88), 405(3.62)	24-1453-64
Chalcone, 3,3'-dihydroxy-	EtOH	252(4.07),308(4.31)	23-2580-64
Cyclopentanone, 2,5-difurfurylidene-	EtOH	250(3.78),402(4.44)	65-3645-64
	$HOAc-H_2SO_4$	340(3.42),416(3.68), 560(4.80)	65-3645-64
Flavanone, 3-hydroxy-	EtOH	254(3.91),320(3.46)	22-3350-65
Flavanone, 6-hydroxy-	EtOH	262(3.72),360(3.95)	24-1453-64
Fluorene-3-carboxylic acid, 2-hydroxy-1-methyl-	EtOH	279(4.30),330(4.15)	42-0479-64
Mansonone F	EtOH	234(--),555(--)	88-4857-65
4H-Naphtho[1,2-b]pyran-1-one, 5-methoxy-3-methyl-	EtOH	213(4.57),258(4.56), 273s(4.26),286s(4.01), 299s(3.60),355(3.38)	94-0316-64
4H-Naphtho[2,3-b]pyran-4-one, 5-methoxy-2-methyl-	EtOH	218(4.29),248(4.80), 260s(4.45),278s(3.84), 293(3.51),304(3.68), 316(3.56),366(3.69)	94-0316-64
2H-Pyran, 5-acetoxy-2-(2,4,6-octa-triynylidene)-5,6-dihydro-	ether	269(4.39),279(4.43), 338(4.33),356s(4.30)	24-1416-65
Stilbene-2-carboxylic acid, 2'-hydroxy-, cis	EtOH	280(3.63)	42-0093-64
$C_{15}H_{12}O_3S$			
[14]Annulene-3-sulfonic acid, 1,8-bis-dehydro-, methyl ester	isooctane	315(5.25),442(4.31), 591(3.89)	35-0521-64
$C_{15}H_{12}O_4$			
3(2H)-Benzofuranone, 2-benzyl-2,6-dihydroxy-	EtOH	236(3.85),280(4.03), 326(3.84)	22-3572-65
	EtOH-NaOH	256(3.98),339(4.32)	22-3572-65
1,3-Bis(p-benzoquinonyl)propane	95% THF	248(4.62),315(3.22), 444(1.76)	44-2602-65
Isoliquiritigenin	n.s.g.	239(4.10),311s(4.05), 371(4.49)	12-0379-64
Liquiritigenin	n.s.g.	234(4.26),275(4.18), 310(3.86)	2-0422-65
4H-Naphtho[1,2-b]pyran-4-one, 5-hydroxy-6-methoxy-2-methyl-	EtOH	225(4.61),268(4.47), 290s(4.03),315s(3.65), 389(3.40)	94-0316-64
4H-Naphtho[1,2-b]pyran-4-one, 6-hydroxy-5-methoxy-2-methyl-	EtOH	221(4.56),260(4.50), 285s(3.93),320(3.73), 374(3.55)	94-0316-64
2-Pyrone, 4-acetoxy-6-styryl-	EtOH	248(4.11),300(4.02)	65-2766-64
2-Pyrone, 3-cinnamoyl-4-hydroxy-6-methyl-, cis	EtOH	310(4.17),350(4.20)	65-2766-64
trans	EtOH	356(4.39)	65-2766-64
Thyroacrylic acid	EtOH	207(4.32),225(4.21), 297(4.32),308s(4.31)	44-1812-64
Xanthone, 2,6-dimethoxy-	EtOH	240(4.51),310(4.41)	39-5074-64
Xanthone, 3,5-dimethoxy-	EtOH	237(4.54),268(4.24), 304(4.05)	39-5074-64

Compound	Solvent	λ_{max}(log ε)	Ref.
$C_{15}H_{12}O_5$			
p-Benzoquinone, 2-(o-anisoyl)-5-methoxy-	EtOH	255(4.18),292(3.51)	78-1495-65
Spiro[benzofuran-2(3H),1'-[2,5]cyclohexadiene]-3,4'-dione, 4,6-dimethoxy-	EtOH	249(4.18),277(4.05)	39-6960-65
$C_{15}H_{12}O_6$			
o-Anisic acid, α-(β-resorcyloyl)-	EtOH	225(4.22),289(4.07)	39-2361-65
Bellidifolium, methyl-	EtOH	230(4.3),254(4.4), 278(4.2),300s(3.8), 335(4.0)	78-0991-64
Xanthone, 1,5-dihydroxy-3,6-dimethoxy-	MeOH	247(4.64),287(4.05), 326(4.27)	78-2653-65
$C_{15}H_{12}O_7$			
o-Anisic acid, α-(2,4,6-trihydroxybenzoyl)-	EtOH	208(4.26),229(4.02), 294(4.13)	39-2361-65
$C_{15}H_{13}Br$			
Bicyclo[4.2.0]octa-1,3,5-triene, 7-bromo-8-methyl-7-phenyl-	isooctane	268s(3.55),274(3.61), 279(3.55)	35-0086-65
$C_{15}H_{13}BrN_2O_3$			
Propiophenone, 2-anilino-4'-bromo-3-nitro-	heptane	240(4.18),263(4.30)	70-1382-65
$C_{15}H_{13}BrO$			
2-Propen-1-ol, 1-(p-bromophenyl)-3-phenyl-	n.s.g.	255(4.38)	39-4978-64
$C_{15}H_{13}BrO_3S_2$			
1,3-Dithiol-2-ylideneacetic acid, α-acetyl-4-(p-bromophenyl)-, ethyl ester	THF	245(4.40),384(4.49)	44-1711-64
	CF_3COOH	390(4.31)	44-1711-64
$C_{15}H_{13}Br_2NO_2$			
Acetophenone, 3'-[(3,5-dibromo-1-methyl-4-oxo-2,5-cyclohexadien-1-ylamino]-	MeOH	233(4.35),257(4.23), 338(3.30)	35-1127-64
$C_{15}H_{13}Cl$			
5H-Dibenzo[a,c]cycloheptene, 3-chloro-6,7-dihydro-	pet ether	253(4.29),282s(3.20)	39-5317-64
Stilbene, 4-chloro-4'-methyl-, trans	EtOH	230(4.15),303(4.49), 315(4.5)	35-3665-65
$C_{15}H_{13}ClN_2$			
1H-1,4-Benzodiazepine, 7-chloro-2,3-dihydro-5-phenyl-	EtOH	225(4.4),370(2.6)	44-2368-64
$C_{15}H_{13}ClN_2O_2$			
Benzanilide, 4-acetamido-2'-chloro-	EtOH	277(4.4),316s(3.3)	28-4295-64B
Benzanilide, 4-acetamido-3'-chloro-	EtOH	214(4.4),283(4.6)	28-4295-64B
Benzanilide, 4-acetamido-4'-chloro-	EtOH	226s(4.0),286(4.6)	28-4295-64B
$C_{15}H_{13}ClO_2S_2$			
4,5-Dihydro-2-(2-acetoxy-1-naphthyl)-1,3-dithiol-1-ium chloride	MeCN	299(4.24),423(4.09), 442(4.02)	24-1374-65

Compound	Solvent	λ_{max}(log ε)	Ref.
$C_{15}H_{13}ClO_4$			
Spiro[benzofuran-2(3H),1'-[2]cyclohexene]-3,4'-dione, 7-chloro-2'-methoxy-6'-methyl-	EtOH	216(4.38),255(4.37), 335(3.68)	39-4939-65
7-Styryltropylium perchlorate	MeCN	244(4.17),278(4.19), 440(4.53)	24-1337-64
	MeCN	244(4.17),278(4.19), 440(4.53)	24-1590-64
$C_{15}H_{13}Cl_3O$			
Azulene, 1-(trichloroacetyl)-4,6,8-trimethyl-	C_6H_{12}	243(4.38),289(4.29), 324(4.30),388(3.94), 515(2.92)	44-0131-65
$C_{15}H_{13}Cl_3O_2$			
1,4-Methanobiphenylene, 1,2,3-trichloro-1,4,4a,8b-tetrahydro-9,9-dimethoxy-	EtOH	256s(--),261(3.10), 267(3.32),273(3.34)	39-2549-65
$C_{15}H_{13}F$			
Benzene, 1-fluoro-3-(2-phenylcyclopropyl)-, cis	n.s.g.	267(3.09)	65-0244-65
trans	n.s.g.	232(4.26),269(3.34)	65-0244-65
Benzene, 1-fluoro-4-(2-phenylcyclopropyl)-, cis	n.s.g.	267(3.05),274(3.03)	65-0244-65
trans	n.s.g.	227(4.19),268(3.03)	65-0244-65
$C_{15}H_{13}I$			
Stilbene, 2-iodo-2'-methyl-, cis	EtOH	265(4.37)	44-3792-65
	EtOH	265(4.37)	88-0359-65
$C_{15}H_{13}N$			
Phenanthridine, 1,10-dimethyl-	pH 1.00	250(4.56),346(3.90)	39-4426-65
	pH 9.17	250(4.58),310(3.87)	39-4426-65
	n.s.g.	250(4.58),312(3.67)	39-3032-65
Pyridine, 1,2-dihydro-2-inden-1-ylidene-1-methyl-	dioxan	290(3.8),400(4.2)	35-2908-65
Pyridine, 1,4-dihydro-4-inden-1-ylidene-1-methyl-	dioxan	225(4.4),290(3.8), 445(4.5)	35-2908-65
Pyridine, 2-(4-phenyl-1,3-butadienyl)-, 1-cis-3-trans	EtOH	240(4.18),247(4.13), 330(4.49)	94-1344-64
3-cis-1-trans	EtOH	238(4.15),244(4.18), 325(4.45)	94-1344-64
1,3-ditrans	EtOH	231(3.95),332(4.72)	94-1344-64
$C_{15}H_{13}NO$			
Acenaphth[1,2-d]isoxazole, 9-ethyl-6b,9a-dihydro-	EtOH	225(4.73),276(3.75), 286(3.85),298(3.67)	88-2157-64
Acetonitrile, (4'-methoxy-4-biphenylyl)-	hexane	265.7(4.36)	78-0861-64
Acrylophenone, 2-anilino-	heptane	224(4.16),252(4.00)	70-1382-65
Acrylophenone, 3-anilino-	heptane	239(4.04),252(4.15), 369(4.41)	70-1382-65
Carbazole, 3-carboxaldehyde, 1,4-dimethyl-	n.s.g.	237(4.39),242s(4.34), 275(4.51),287(4.42), 330(4.00)	78-0681-65
Carbostyril, 3,4-dihydro-4-phenyl-	EtOH	254(4.03)	32-1115-65
Cinnamanilide	EtOH	296(4.46)	54-0949-64
3-Indolinone, 2-methyl-2-phenyl-	EtOH	234(4.53),255(4.12), 402(3.67)	39-4320-64
3H-Indol-3-ol, 3-methyl-2-phenyl-	EtOH	239(4.29),247(4.27), 317(4.22)	39-4320-64

Compound	Solvent	$\lambda_{max}(\log \epsilon)$	Ref.
Isoindole, 1-(p-methoxyphenyl)-	n.s.g.	215(4.3),270(4.1), 280(4.3),310(3.9), 355(3.9)	35-4152-64
Phthalimidine, 2-m-tolyl-	MeOH-acid	277(4.24)	95-1042-65
	MeOH	275(4.24)	95-1042-65
Phthalimidine, 2-p-tolyl-	MeOH-acid	281(4.04)	95-1042-65
	MeOH	280(4.03)	95-1042-65
2-Propen-1-one, 1-(p-amino-phenyl)-3-phenyl-	H_2SO_4	257(3.30),298(3.53), 424(4.51)	65-1724-65
2-Propen-1-one, 3-(p-amino-phenyl)-1-phenyl-	H_2SO_4	240(3.43),322(3.84), 395(4.54)	65-1724-65
$C_{15}H_{13}NOS$			
2-Indolinone, 3-methyl-3-(phenylthio)-	EtOH	207(4.37),230s(4.24), 258s(4.21),287s(3.20)	44-2431-64
2-Indolinone, 3-(phenylthiomethyl)-	dioxan	213(4.45),252(4.17), 285s(3.41)	44-2431-64
1-Thioflavanone, 3-amino-	EtOH-HCl	239(4.27),348(3.29)	88-0137-64
$C_{15}H_{13}NO_2$			
Acetanilide, 2-benzoyl-	pH 13	250(4.18),330(4.39)	20-0843-64
	0.04N NaOH	330(4.39)	20-0782-64
	60% EtOH	247(4.41)	20-0782-64
Acetophenone, m-salicylideneamino-	EtOH	225(4.4),255(4.21), 338(4.03)	65-3837-64
Acetophenone, p-salicylideneamino-	EtOH	275(4.20),348(4.18)	65-3837-64
Azepino[3,2,1-hi]indole-$\Delta^{7(4H),\alpha}$-acetic acid, 1,2-dihydro-6-(hydroxymethyl)-, δ-lactone	EtOH	260(4.00),300(4.04), 428(3.7)	35-1397-65
Phthalimidine, 2-(m-methoxyphenyl)-	MeOH-acid	281.5(4.08)	95-1042-65
	MeOH	282(4.08)	95-1042-65
Phthalimidine, 2-(p-methoxyphenyl)-	MeOH-acid	283(4.09)	95-1042-65
	MeOH	283(4.09)	95-1042-65
$C_{15}H_{13}NO_2S$			
4H-1,3-Benzothiazin-4-one, 2,3-dihydro-3-methyl-2-phenyl-, 1-oxide	EtOH	272(3.45),280(3.44), 289s(3.38)	44-2068-64
Thiobenzophenone, 2,6-dimethyl-4'-nitro-	C_6H_{12}	312(4.12)	54-0289-65
$C_{15}H_{13}NO_3$			
Benzoic acid, m-(salicylideneamino)-, methyl ester	EtOH	220(4.4),272(4.16), 340(4.06)	65-3837-64
Benzoic acid, p-(salicylideneamino)-,	EtOH	220(4.4),295(4.43), 340s(3.75)	65-3837-64
4-Benzoxazoleacrylic acid, 7-allyl-5-hydroxy-β,2-dimethyl-, δ-lactone	EtOH	240s(3.55),298(4.27), 324(3.96),340(3.58)	44-4344-65
Hydrochalcone, 4-nitro-	n.s.g.	245(4.20),275(4.10)	39-4978-64
$C_{15}H_{13}NO_3S$			
Benzoic acid, p-nitrothio-, S-o-ethylphenyl ester	MeOH	272(4.49)	56-0681-65
4H-1,3-Benzothiazin-4-one, 2,3-dihydro-3-methyl-2-phenyl-, 1,1-dioxide	EtOH	270(3.51),277(3.50), 286(3.43)	44-2068-64
$C_{15}H_{13}NO_4$			
Benz[cd]indole-3-carboxylic acid, 1-acetyl-1,3,4,5-tetrahydro-2-methyl-5-oxo-	EtOH	227(4.25),259(4.18), 292(3.94),303(4.03), 341(3.62)	44-0843-64

Compound	Solvent	λ_{max}(log ε)	Ref.
Benz[cd]indole-3-carboxylic acid, 1-acetyl-1,3,4,5-tetrahydro-2-methyl-5-oxo- (contd.)	EtOH-base	220(4.22),251(4.24), 272(3.52),309(3.65), 360(3.61)	44-0843-64
Benz[cd]indole-3-carboxylic acid, 1-acetyl-1,3,4,5-tetrahydro-5-oxo-, methyl ester	EtOH	226(4.19),256(4.24), 292s(3.97),302(4.05), 328s(3.59)	44-0843-64
3-Indolesuccinic anhydride, 1-acetyl-2-methyl-	EtOH	244(4.19),264(3.98), 272(3.94),276(3.90), 290(3.72),299(3.72)	44-0843-64
$C_{15}H_{13}NO_5$			
4H-m-Dioxino[5,4-c]pyridine-4,5(6H)-dione, 7-hydroxy-2,2-dimethyl-6-phenyl-	EtOH	316(4.69)	78-1917-65
Malonic acid, [3-anilino-3-hydroxy-1-(1-hydroxy-1-methylethoxy)allylidene]-, di-δ-lactone	EtOH	350(4.69)	78-1917-65
4H,5H-Pyrano[2,3-b]quinolizine-9-carboxylic acid, 10a,11-dihydro-2-methyl-4,5-dioxo-, methyl ester	EtOH	245s(4.05),350(3.64)	95-0874-64
$C_{15}H_{13}NS$			
Acrylophenone, 3-anilinothio-	EtOH	225(4.03),236(4.04), 250(4.03),324(4.07), 440(4.39)	39-0032-65
$C_{15}H_{13}N_3$			
3H-Dibenz[b,f]imidazo[1,2-d][1,4]diazepine, 2,9-dihydro-	EtOH	234(4.54),293(4.04)	33-1590-65
3,5-Pyridinedicarbonitrile, 2-benzyl-1,2-dihydro-4-methyl-	EtOH	217(4.38),255(4.03), 372(3.66)	73-2609-65
$C_{15}H_{13}N_3O$			
Acrylanilide, 4'-phenylazo-	$EtOH-NH_3$	353(4.51)	35-2315-64
Benzamide, N-(2-benzimidazolylmethyl)-	EtOH	275(3.94),282(3.94)	94-0773-64
3-Hydroxy-4-methyl-1,5-diphenyl-4H-1,2,4-triazolium hydroxide, inner salt	n.s.g.	286(3.82)	4-0105-65
Isocarbostyril, 2-(2,4-diaminophenyl)-	EtOH	212(4.65),284(4.08), 327s(3.75)	24-1013-65
1,3,4-Oxadiazole, 2-(N-methylanilino)-5-phenyl-	n.s.g.	219(4.1),245s(3.9), 293(4.2)	28-2868-64B
Δ^2-1,3,4-Oxadiazoline, 4-methyl-2-phenyl-5-(phenylimino)-	n.s.g.	230(4.2),260(4.1), 311(3.9)	28-2868-64B
2H-Pyrido[4,3-e]-1,4-diazepin-2-one, 1,3-dihydro-1-methyl-5-phenyl-	MeOH	228(4.40),250s(4.23)	87-0722-65
$C_{15}H_{13}N_3O_2S$			
7H-Thiopyrano[2,3-d]pyrimidine, 2-acetamido-5,6-dihydro-7-oxo-4-phenyl-	pH 1	238(4.29),297s(3.94), 347s(3.71)	4-0263-64
	pH 8.4	242(4.43),302(3.97)	4-0263-64
	pH 13	320(4.03)	4-0263-64
$C_{15}H_{13}N_3O_3$			
2-Indolinone, 3-[2-nitro-1-(2-pyridylethyl]-	EtOH	251(3.97),259s(3.91), 283s(3.16)	87-0626-65
$C_{15}H_{13}N_3O_3S$			
p-Toluenesulfonamide, N-(3-phenyl-1,2,4-oxadiazol-5-yl)-	EtOH	233.5(4.47)	95-1061-64

Compound	Solvent	λ_{max}(log ε)	Ref.
$C_{15}H_{13}N_3O_4$			
Pyrrole-2-acrylonitrile, α-(4,5-dimethoxy-2-nitrophenyl)-	MeOH	247(4.05),337(4.39)	44-2253-65
$C_{15}H_{13}N_3O_5$			
Benzyl alcohol, o-[2-(2,4-dinitroanilino)vinyl]-	EtOH	224(4.22),261(4.15), 287s(3.97),395(4.29)	24-1013-65
$C_{15}H_{13}N_3S$			
3-Mercapto-4-methyl-1,5-diphenyl-1H-1,2,4-triazolium hydroxide, inner salt	n.s.g.	240(3.92)	4-0105-65
$C_{15}H_{13}N_5$			
1-(Cyanoamino)-5-methyl-2-p-tolyl-2H-benzotriazolium hydroxide, inner salt	EtOH	227(4.39),312(4.19), 367(3.76)	39-0751-64
$C_{15}H_{14}$			
Anthracene, 1,4-dihydro-9-methyl-	EtOH	230(4.93),279(3.70), 289(3.73)	70-1024-64
5H-Dibenzo[a,c]cycloheptene, 6,7-dihydro-	isooctane	247.5(4.18)	35-1710-64
	pet ether	249(4.20),276(3.18)	39-5317-64
Ethylene, 1-phenyl-1-o-tolyl-	C_6H_{12}	231(4.10),249(4.05)	54-0806-65
	EtOH	232(4.08),248(4.05)	54-0806-65
	98% H_2SO_4	321(3.97),432(4.36)	54-0806-65
Ethylene, 1-phenyl-1-p-tolyl-	C_6H_{12}	235(4.21),253(4.05)	54-0806-65
	EtOH	235(4.21),253(4.06)	54-0806-65
	98% H_2SO_4	319(3.98),442(4.67)	54-0806-65
Fluorene, 2,3-dimethyl-	C_6H_{12}	265(4.35),295(3.95), 300(3.91),307(4.14)	33-1800-65
1-Propene, 1,1-diphenyl-	C_6H_{12}	248(4.25)	22-0693-64
Stilbene, α-methyl-	n.s.g.	276(4.32)	5-0058-65B
Stilbene, α-methyl-, cis	isooctane	243(4.14)	35-1410-65
Stilbene, 4-methyl-, cis	EtOH	278(4.11)	44-1473-65
Stilbene, 4-methyl-, trans	EtOH	300(4.46),310(4.44)	44-1473-65
$C_{15}H_{14}BN$			
Phenanthro[4,5-cde][1,2]azaborine, 4,5,9,10-tetrahydro-5-methyl-	$CHCl_3$	257s(4.32),268s(4.05), 285s(3.91),304(3.60), 316(3.89),330(4.00)	44-1757-64
$C_{15}H_{14}BNO$			
Phenanthro[4,5-cde][1,2]azaborine, 4,5,9,10-tetrahydro-5-methoxy-	$CHCl_3$	275(3.97),284(3.97), 300(3.69),314(3.87), 327(3.97)	44-1757-64
$C_{15}H_{14}BrClN_2$			
2-Chloro-1-(2-indol-3-ylethyl)-pyridinium bromide	EtOH	275(4.03),289s(3.79)	94-0931-65
$C_{15}H_{14}BrN$			
6,11-Dimethylbenzo[b]quinolizinium bromide	EtOH	210s(3.86),247s(4.52), 254(4.62),371(4.00), 389(4.00),411(3.90)	4-0121-64
$C_{15}H_{14}BrNO$			
Ethylamine, α-phenyl-N-(5-bromosalicylidene)-	MeOH	325(3.58),414(2.91)	65-3055-64
	EtOH	253(4.04),282s(3.12), 328(3.60),415(2.68)	44-2265-64

Compound	Solvent	$\lambda_{max}(\log \epsilon)$	Ref.
$C_{15}H_{14}BrNO_4S$			
1,3,6-Cycloheptatriene-1-sulfonic acid, 6-bromo-5-oxo-4-p-toluidino-, methyl ester	EtOH	258(4.34),370(4.19), 418(4.33)	94-0457-65
$C_{15}H_{14}BrNO_7$			
Indoline-3,3,4-tricarboxylic acid, 5-bromo-1-methyl-2-oxo-, trimethyl ester	MeOH	215(4.32),267(4.13), 312s(3.39)	39-3229-64
$C_{15}H_{14}Br_2N_2$			
2-Bromo-1-(2-indol-3-ylethyl)-pyridinium bromide	EtOH	280(4.06),289s(3.90)	94-0931-65
$C_{15}H_{14}ClNO$			
Ethylamine, α-phenyl-N-(5-chloro-salicylidene)-	MeOH	255(4.00),325(3.66), 415(2.85)	65-3055-64
(R)-(-)	EtOH	254(3.96),280s(3.21), 328(3.55),415(2.80)	44-2265-64
$C_{15}H_{14}ClNO_3$			
3-Carboxy-6,7-dihydro-9-methoxy-benzo[a]quinolizinium chloride	EtOH	220(3.98),254s(3.74), 283(3.87),355(4.38)	23-1527-65
$C_{15}H_{14}ClNO_7$			
3-Carboxy-6,7-dihydro-9-methoxy-benzo[a]quinolizinium perchlorate	EtOH	220(4.38),254s(3.69), 284(3.82),355(4.37)	23-1527-65
$C_{15}H_{14}ClNO_8$			
2-(1,2-Dicarboxyvinyl)isoquinolinium perchlorate, dimethyl ester	H_2O	237(4.69),270s(3.48), 280(3.53),292s(3.25), 344(3.56)	39-2676-64
$C_{15}H_{14}Cl_2N_2O$			
Ethanol, 2-[[2-amino-5-chloro-α-(o-chlorophenyl)benzylidene]amino]-	EtOH	231(4.47),363(3.71)	44-2368-64
$C_{15}H_{14}FeO$			
2,4-Pentadienal, 5-ferrocenyl-	C_6H_{12}	324(4.43),482(3.45)	5-0088-64F
	MeOH	337(4.43),505(3.57)	5-0088-64F
	$CHCl_3$	338(4.43),502(3.58)	5-0088-64F
$C_{15}H_{14}FeOS$			
Thiophene, 2-ferrocenyl-5-(hydroxymethyl)-	$CHCl_3$	298(4.09),446(1.68)	49-1750-64
$C_{15}H_{14}FeO_2$			
Propiolic acid, ferrocenyl-, ethyl ester	$CHCl_3$	285(4.02),335(3.25), 452(2.81)	49-1750-64
$C_{15}H_{14}FeS$			
Thiophene, 2-ferrocenyl-5-methyl-	$CHCl_3$	299(4.05),446(1.61)	49-1750-64
$C_{15}H_{14}N_2$			
Benzimidazole, 1-(α-methylbenzyl)-	MeOH	249(4.02),253(4.01), 265(3.82),274(3.81), 281s(3.64)	65-0632-64
Benzimidazole, 2-phenethyl-	EtOH	248(3.68),274(4.02), 280(3.98)	2-0169-64

Compound	Solvent	λ_{max}(log ε)	Ref.
Benzo[c]cinnoline, 1,2,4-trimethyl-	C_6H_{12}	253(4.62),274s(4.22), 317(4.00),327(4.02), 357(3.26),374s(3.16), 403s(2.61)	12-1036-64
	2N HCl	259(4.52),269s(4.43), 292s(3.88),383(4.04), 445(3.69)	12-1036-64
Cyclohexylidenemalononitrile, 2-phenyl-	aq. MeOH	240(4.15)	95-0381-64
Indazole, 3-benzyl-6-methyl-	EtOH	290(3.73)	32-0814-65
Isoquinoline, 1-(o-aminophenyl)- 3,4-dihydro-	EtOH	245(4.3),340(3.3)	44-2368-64
9-Phenanthrenecarbonitrile, 10-amino- 5,6,7,8-tetrahydro-	EtOH	265(4.4),330(3.7), 360(3.7),370(3.7)	44-1543-64
	50% H_2SO_4	245(4.8),290(3.8), 300(3.8),335(3.4), 350(3.45)	44-1543-64
Phenanthridine, 6-amino- 1,10-dimethyl-	pH 2.00	228(4.28),250(4.52), 346(3.84)	39-4426-65
	pH 10.0	246(4.55),316(3.83), 360(3.67)	39-4426-65
	n.s.g.	246(4.51),263(4.34), 322(3.80),348(3.74), 370(3.67)	39-3032-65
1-Pyrazoline, 3,5-diphenyl-	EtOH	329(2.46)	35-0658-64
$C_{15}H_{14}N_2O$			
Acetanilide, 4'-(benzylideneamino)-	EtOH	225(4.22),305(4.35)	9-0091-65
2-Benzimidazolemethanol, 1-methyl- α-phenyl-	EtOH	217(4.18),254(3.89), 277(3.86),285(3.86)	50-0378-64A
1H-Indazole, 1-benzyl-3-methoxy-	EtOH	307(3.73)	7-0583-65
3-Indazolinone, 1-benzyl-2-methyl-	EtOH	241(4.04),316(3.69)	7-0583-65
2-Indolinone, 1-methyl- 3-(2-pyridylmethyl)-	EtOH-HCl	255(4.08),262s(4.03)	87-0626-65
Indolo[2,3-a]quinolizin-4(3H)-one, 2,6,7,12-tetrahydro-	EtOH	220s(4.43),232(4.48), 308(4.34),319(4.31)	44-2771-64
Nicotinonitrile, 1,2-dihydro-4(or 6)- methyl-2-oxo-6(or 4)-phenethyl-	EtOH	233(3.91),332(3.97)	44-3593-65
2-Phenazinol, 1,3,4-trimethyl-	50% MeOH- HCl	275(4.19),402(3.74), 495(3.02)	59-1665-64
	MeOH	267(4.27),371(3.44), 415s(3.05)	59-1665-64
	50% MeOH- NaOH	292(4.16),387(3.39), 513(3.18)	59-1665-64
2(10H)-Phenazinone, 3,4,10-trimethyl-	50% MeOH- acid	271(4.5),385s(4.0), 403(4.1),474(3.6)	59-1665-64
	MeOH	235(4.2),287(4.4), 372(3.9),519(3.8)	59-1665-64
2(1H)-Pyridone, 1-(2-indol-3-ylethyl)-	EtOH	285(3.97),292(3.99), 308s(3.79)	94-0931-65
4(1H)-Pyridone, 1-(2-indol-3-ylethyl)-	MeOH	264(4.36)	24-2463-64
$C_{15}H_{14}N_2O_2$			
Benzanilide, p-acetamido-	EtOH	201(4.6),282(4.4)	28-4295-64B
Benzophenone, p-(dimethylamino)- p'-nitro-	benzene	314(4.22),360(4.10)	65-0094-65
	EtOH	244(4.16),312(4.15), 367(4.15)	65-0094-65
	HOAc	260(4.12),313(4.11), 374(4.17)	65-0094-65
Malondialdehyde, bis(2-hydroxyanil), nickel complex	MeOH	435(4.0),450(4.0), 490(4.1)	89-0076-64

Compound	Solvent	$\lambda_{max}(\log \epsilon)$	Ref.
Tropone, 2-(2-phenylacetylhydrazino)-	EtOH	240(4.31),282(3.74), 333(4.00),387(4.09)	94-0450-65
$C_{15}H_{14}N_2O_3$			
Ethylamine, α-phenyl-N-(5-nitro-salicylidene)-	MeOH	245(4.35),330(3.98), 390(3.93)	65-3055-64
(R)-(-)	EtOH	233s(4.11),253(4.17), 348(4.02),392(3.99)	44-2265-64
Indole-2-carboxylic acid, 5-acetyl-3-(cyanomethyl)-, ethyl ester	EtOH	266(4.77),310(4.01)	87-0141-64
Propiophenone, 2-anilino-3-nitro-	heptane	242(4.33),285(3.54)	70-1382-65
4-Pyridinepropion-m-anisidide, β-oxo-	EtOH	248(4.16),326(4.21)	65-2400-64
4-Pyridinepropion-o-anisidide, β-oxo-	EtOH	334(4.14)	65-2400-64
4-Pyridinepropion-p-anisidide, β-oxo-	EtOH	248(4.18),338(4.21)	65-2400-64
Pyrrolo[1,2-a]quinoxaline-2-acetic acid, 1,5-dihydro-4-methyl-1-oxo-, methyl ester	EtOH	234(4.44),256s(3.93), 289s(3.97),298(4.00), 322s(3.52),420(4.10)	39-3678-65
$C_{15}H_{14}N_2O_4$			
Uracil, 5-acetoacetyl-6-methyl-1-phenyl-	EtOH	260.5(3.88)	70-0747-65
Urea, 1-[2,6-dimethyl-4-oxo-4H-pyran-3-yl)carbonyl]-3-phenyl-	EtOH	210(4.60)	70-0747-65
$C_{15}H_{14}N_4$			
Aniline, N,N-diethyl-4-(tricyanovinyl)-	$CHCl_3$	524(4.71)	39-1334-64
1,2,3-Benzotriazine, 4-(ethylimino)-3,4-dihydro-3-phenyl-	EtOH	263(4.15),271(4.15), 312(3.86)	39-3663-64
$C_{15}H_{14}N_4O$			
2-Propanone, 1-(phenylazo)-1-(phenylhydrazono)-	MeOH	254(4.01),300(4.19), 440(4.34)	44-2959-64
Pyrazolo[1,5-a]pyrimidine, 7-benzamido-2,5-dimethyl-	EtOH	238(4.60),285(3.63), 331(3.82)	94-1207-65
$C_{15}H_{14}N_4O_3$			
2-Propanone, 1-[(o-hydroxyphenyl)azo]-1-[(o-hydroxyphenyl)hydrazono]-	benzene	500(4.15)	65-0502-65
	DMF	520(4.22)	65-0502-65
$C_{15}H_{14}N_4O_3S$			
Sulfanilamide, N^1-(3-benzyl-1,2,4-oxadiazol-5-yl)-	EtOH	232(4.24),259(4.22)	95-1061-64
$C_{15}H_{14}N_4O_5$			
Acetophenone, o-methoxy-, 2,4-dinitrophenylhydrazone	$CHCl_3$	372.5(4.39)	5-0021-64G
Acetophenone, p-methoxy-, 2,4-dinitrophenylhydrazone	$CHCl_3$	391(4.43)	5-0021-64G
Formanilide, 2'-(2,4-dinitroanilino)-N-ethyl-	EtOH	348(4.40)	65-0269-64
	$CHCl_3$	352(4.23)	65-0269-64
5-Pyrimidinepropionaldehyde, 2-acetamido-4-hydroxy-6-(p-nitrophenyl)-	pH 1, 7	248(4.24),276(4.29)	87-0283-65
	pH 13	275(4.26)	87-0283-65
$C_{15}H_{14}N_6$			
Melamine, 2,4-diphenyl-	EtOH	270(4.64)	39-3459-64
$C_{15}H_{14}N_6O_3$			
Homopteroic acid	pH 1	224(4.35),310(4.16)	44-3404-65
	pH 13	256(4.43),277(4.34), 365(3.88)	

Compound	Solvent	$\lambda_{max}(\log \epsilon)$	Ref.
$C_{15}H_{14}N_6O_6$			
1,4-Pentadien-3-one, 2-methyl-1,5-bis-(5-nitro-2-furyl)-, amidinohydrazone, hydrochloride	propylene glycol	378(4.455)	95-0001-64
$C_{15}H_{14}O$			
Anisole, o-styryl-, cis	EtOH	280(4.02)	44-1473-65
Anisole, p-styryl-, cis	EtOH	276(4.07)	44-1473-65
Anisole, p-styryl-, trans	EtOH	306(4.46),310(4.44)	44-1473-65
Anthracene, 2,3-epoxy-1,2,3,4-tetra-hydro-9-methyl-	EtOH	230(5.00),287(3.75)	70-1024-64
2(1H)-Anthracenone, 3,4-di-hydro-9-methyl-	EtOH	231(4.99),280(3.70), 290(3.74)	70-1013-64
2-Biphenylcarboxaldehyde, 4,6-di-methyl-	hexane	218(4.33),253(4.05), 260s(3.96),302(3.42), 307s(3.41)	22-2953-64
	EtOH	219(4.30),257(4.01), 310(3.40)	22-2953-64
1,4-Pentadien-3-one, 1,5-diphenyl-	hexane	230(4.17),320(4.51)	65-3645-64
	EtOH	329(4.50)	65-3645-64
	66% H_2SO_4	282(3.68),352(4.03), 455(4.52)	65-3645-64
2-Propen-1-ol, 2,3-diphenyl-, cis	EtOH	222(4.20),257(4.10)	44-0105-64
2-Propen-1-ol, 2,3-diphenyl-, trans	MeOH	257(4.10)	88-2181-65
	EtOH	273(4.30)	44-0105-64
$C_{15}H_{14}OS$			
6H-Dibenzo[b,e]thiocin-12-ol, 7,12-dihydro-	MeOH	235(4.00),262(3.49), 268(3.47),273s(3.38)	24-0685-65
Methylphenylsulfonium phenacylide	EtOH	310(4.16)	88-0251-65
$C_{15}H_{14}O_2$			
2(1H)-Anthracenone, 3,4-dihydro-9-hydroxy-10-methyl-	EtOH	239(4.73),303(3.73)	70-1013-64
2(1H)-Anthracenone, 3,4-dihydro-10-hydroxy-9-methyl-	EtOH	238(4.87),294(3.81)	70-1013-64
1,3-Indandione, 4,5,6,7-tetrahydro-2-phenyl-	EtOH-HCl	248(4.06)	65-0818-64
2,5-Norbornadiene-2-carboxylic acid, 3-o-tolyl-	EtOH	282(3.59)	35-5548-64
2,5-Norbornadiene-2-carboxylic acid, 3-p-tolyl-	EtOH	301(3.89)	35-5548-64
$C_{15}H_{14}O_2S$			
Thiobenzophenone, 4,4'-dimethoxy-	EtOH	352.5(4.30)	54-0289-65
$C_{15}H_{14}O_3$			
1(2H)-Anthracenone, 3,4-dihydro-8,9-dihydroxy-10-methyl-	EtOH	220(4.15),270(3.88), 356(4.22)	65-2570-64
5-Benzofuranacrylic acid, 6-hydroxy-α-isopropyl-β-methyl-, δ-lactone	EtOH	245(4.30),290(3.96), 329(3.94)	44-2467-64
Mansonone D	EtOH	219(4.3),243(4.1), 278(4.11),405(3.88)	88-4857-65
Mansonone E	EtOH	219(4.25),264(4.31), 370(3.2),445(3.38)	88-4857-65
2-Naphthaleneacetic acid, 4-cyclopropyl-1-hydroxy-	EtOH	234(4.61),281(3.62)	25-0467-65
4,6-Nonadiynoic acid, 8-hydroxy-9-phenyl-	EtOH	243(2.90),255(2.78), 283(2.06),302(2.95)	70-1237-65

Compound	Solvent	λ_{max}(log ϵ)	Ref.
2,5-Norbornadiene-2-carboxylic acid, 3-(o-methoxyphenyl)-	EtOH	307(3.70)	35-5548-64
2,5-Norbornadiene-2-carboxylic acid, 3-(p-methoxyphenyl)-	EtOH	317(3.99)	35-5548-64
Xanthorrhoeol	C_6H_{12}	228(4.51),265(4.52), 318(3.72),331(3.74), 415(3.38)	12-0575-65
	EtOH	226(4.59),257(4.36), 290s(3.73),350(3.47)	12-0575-65
	EtOH-NaOH	240(4.59),272(4.39), 380(3.53)	12-0575-65
$C_{15}H_{14}O_4$			
Catalpalactone	EtOH	275(3.21),282(3.20)	88-1261-65
2-Furanacrylic acid, 5-(4-hepten-2-ynoyl)-, methyl ester	EtOH	351(4.43)	77-0422-65
Furo[3,2-g][1]benzopyran-5-one, 4-hydroxy-2-isopropyl-7-methyl-	EtOH	216(4.14),255(4.49), 342(3.38)	95-0055-65
1,8-Naphthalenediol, 2,4-diacetyl-6-methyl-	EtOH	258(4.48),279s(3.73), 382(4.27)	39-5646-64
1,4-Naphthoquinone, 6-acetyl-5-methoxy-2,7-dimethyl-	EtOH	255(4.26),359(3.56)	54-1005-64
1,4-Naphthoquinone, 8-acetyl-5-methoxy-2,7-dimethyl-	EtOH	213(4.46),254(4.25), 402(3.56)	54-1005-64
	EtOH	212(4.45),253(4.25), 401(3.56)	54-0995-64
Oroselol, methyl ether	EtOH	230(4.21),252(4.58), 295(4.27)	65-4171-64
Spiro[benzofuran-2(3H),1'-[2]cyclohexene-3,4'-dione, 2'-methoxy-6'-methyl-	EtOH	251(4.43),326(3.73)	39-4939-65
$C_{15}H_{14}O_5$			
Acetophenone, 2',4',5'-trihydroxy-2-(p-methoxyphenyl)-	EtOH	219(4.31),244(4.08), 283(4.09),354(3.90)	78-1331-64
Benzophenone, 2,4'-dihydroxy-4,6-dimethoxy-	EtOH	290(4.14)	39-6960-65
Cyclohexanecarboxylic acid, 4-hydroxy-1-[2-formyl-4,5-(methylenedioxy)-phenyl]-, lactone	EtOH	232(4.24),260(4.00), 341(3.55)	94-0489-64
4-Indanacrylic acid, 5-carboxy-3-oxo-, dimethyl ester	EtOH	229(4.53),270(3.9)	39-3811-65
1,4-Naphthalenediol, 5-methoxy-, diacetate	EtOH	224(4.72),300(4.02), 315(3.89),328(3.75)	33-0769-64
Scleroin	EtOH	255(4.01),293(4.00), 373(3.67)	78-2697-65
$C_{15}H_{14}O_6$			
5H-Benzocycloheptene-6,8-dicarboxylic acid, 6,7,8,9-tetrahydro-5,9-dioxo-, dimethyl ester	MeOH	209(4.1),252(4.3), 274(4.2),316(4.0)	56-1423-65
Fonsecin	EtOH	233(4.42),275(4.49), 321(3.86),332(3.90), 400(3.94)	44-0112-65
4-Indanglycidic acid, 5-carboxy-3-oxo-, dimethyl ester	EtOH	212(4.46),242(3.95), 278(4.11),324(3.83)	39-3811-65
Teracacidin	EtOH	275(3.40)	12-1170-64
$C_{15}H_{14}O_7$			
Canescin	n.s.g.	247(4.68),278(3.84), 290(3.65),327(3.77)	88-0029-65

Compound	Solvent	λ_{max}(log ε)	Ref.
$C_{15}H_{14}O_8$			
1,4-Naphthoquinone, 6-acetyl-5,8-dihydroxy-2,3,7-trimethoxy-	MeOH	314(3.69),469s(3.59),494(3.62),529s(3.48),591s(2.74)	88-3557-64
	MeOH-KOH	307(3.88),564(3.95),596(3.93)	88-3557-64
$C_{15}H_{14}S$			
Dibenzothiophene, 2,4,8-trimethyl-	EtOH	233(4.68),240(4.60),258(4.18),266(4.08),281(3.84),291(4.12),305s(3.16),321(3.44),333(3.54)	39-4077-64
Dibenzothiophene, 2,6,7-trimethyl-	EtOH	234(4.59),240s(4.57),258(4.22),268(4.15),280(3.90),290(4.14),314(3.38),327(3.45)	39-4077-64
Thiobenzophenone, 2,2'-dimethyl-	C_6H_{12}	324(4.03),607(2.23)	54-0289-65
	EtOH	325(4.05),594(2.24)	54-0289-65
Thiobenzophenone, 2,4-dimethyl-	C_6H_{12}	227s(--),317(4.16),611(2.16)	54-0289-65
	EtOH	227s(--),320(4.17),596(2.17)	54-0289-65
Thiobenzophenone, 2,4'-dimethyl-	C_6H_{12}	234(4.12),329(4.20),608(2.01)	54-0289-65
	EtOH	234(3.99),333(4.19),597(2.03)	54-0289-65
Thiobenzophenone, 2,6-dimethyl-	C_6H_{12}	228(4.07),317(4.13),621(1.67)	54-0289-65
	EtOH	228(4.05),320(4.13),607(1.68)	54-0289-65
Thiobenzophenone, 4,4'-dimethyl-	C_6H_{12}	237(4.21),325(4.32),605(2.33)	54-0289-65
	EtOH	240(4.06),328(4.32),587(2.34)	54-0289-65
$C_{15}H_{15}BrN_2$			
1-Benzyl-2-methylimidazo[1,2-a]pyridinium bromide	EtOH	204(4.46),218s(4.37),285(4.02)	4-0331-65
$C_{15}H_{15}BrN_2O_2S$			
5-Pyrimidinecarboxylic acid, 2-(p-bromophenyl)-1,4-dihydro-1,6-dimethyl-4-thioxo-, ethyl ester	EtOH	259(4.09),340(4.28)	44-1115-64
$C_{15}H_{15}BrN_4O_3$			
Pyrimidine, 5-bromo-4-[3-ethyl-4-(5-nitro-2-furyl)butadienyl]-2-(methylamino)-	propylene glycol	230(4.33),260(4.15),294(4.15),415(4.31)	95-0131-64
$C_{15}H_{15}BrOS$			
Methylphenacylphenylsulfonium bromide	acid	253(4.21)	28-0585-64A
	base	221(4.20),295(4.05)	28-0585-64A
$C_{15}H_{15}BrO_2$			
Anthrone, 3β-bromo-2,3,4,4aβ-tetrahydro-10α-hydroxy-10-methyl-	EtOH	249(4.20),287(2.90)	65-2576-64

Compound	Solvent	λ_{max}(log ε)	Ref.
$C_{15}H_{15}BrO_3$			
2,9-Anthracenedione, 3-bromo-1,3,4,4aβ,9aβ,10-hexahydro-10α-hydroxy-10-methyl-	EtOH	250(4.02),290(3.10)	70-1013-64
2,10(1H,3H)-Anthracenedione, 3-bromo-4,4aα,9,9aβ-tetrahydro-9β-hydroxy-9α-methyl-	EtOH	251(4.03),291(3.18)	70-1013-64
$C_{15}H_{15}Br_2ClO_2$			
Anthrone, 2α,3β-dibromo-5-chloro-8-methoxy-1,2,3,4,4aβ,9aα-hexahydro-	EtOH	254(3.90),326(3.65)	70-0806-65
$C_{15}H_{15}Br_2NO$			
2,5-Cyclohexadien-1-one, 2,6-dibromo-4-(o-ethylanilino)-4-methyl-	MeOH	240(4.10),258(4.05), 290s(--)	35-1127-64
$C_{15}H_{15}ClN_2$			
Aniline, 4-chloro-N',N'-dimethyl-N,4'-methylidynedi-	isooctane	240(4.20),353(4.53)	9-0091-65
	EtOH	240(4.13),360(4.56)	9-0091-65
$C_{15}H_{15}ClN_2O$			
Ethanol, 2-[(2-amino-5-chloro-α-phenylbenzylidene)amino]-	EtOH	233(4.38),360(3.67)	44-2368-64
$C_{15}H_{15}ClN_2O_2$			
1H-Azepino[5,4,3-cd]indole-2-carboxylic acid, 9-chloro-3,4-dihydro-6-methyl-, ethyl ester	EtOH	226(4.32),249(4.12), 263(4.15),343(3.97), 387(3.90)	87-0200-65
$C_{15}H_{15}ClN_2O_2S$			
5-Pyrimidinecarboxylic acid, 2-(chloromethyl)-1,4-dihydro-6-methyl-1-phenyl-4-thioxo-, ethyl ester	EtOH	234(4.07),326(4.18)	44-1115-64
$C_{15}H_{15}ClN_2S$			
1,2-Diphenylvinylisothiouronium chloride	EtOH	299(4.26)	95-0930-64
$C_{15}H_{15}ClO$			
2-Anthrol, 3α-chloro-1,2α,3,4-tetrahydro-9-methyl-	EtOH	232(5.03),280(3.73), 290(3.77)	70-1024-64
2-Anthrol, 3β-chloro-1,2β,3,4-tetrahydro-10-methyl-	EtOH	232(5.06),280(3.74), 290(3.78)	70-1024-64
$C_{15}H_{15}ClO_2$			
2,10-Epoxyanthracen-9(2βH)-one, 3β-chloro-1,3,4,4aβ,9aβ,10-hexahydro-10β-methyl-	EtOH	248(3.98),289(3.12)	70-1024-64
$C_{15}H_{15}ClO_3$			
2,9-Anthracenedione, 3-chloro-1,3,4,4aβ,9aβ,10-hexahydro-10α-hydroxy-10-methyl-	EtOH	252(4.04),292(3.22)	70-1013-64
2,9-Anthracenedione, 5-chloro-1,3,4,4a,9a,10-hexahydro-8-methoxy-	EtOH	223(4.34),255(3.86), 323(3.58)	70-0806-65
2,10(1H,3H)-Anthracenedione, 3-chloro-4,4aα,9,9aα-tetrahydro-9β-hydroxy-9-methyl-	EtOH	249(4.01),290(3.17)	70-1013-64
3-Cyclohexene-1-carboxylic acid, 6-(2-chloro-α-hydroxy-5-methoxybenzyl)-, γ-lactone	EtOH	207(4.37),230(3.99), 283(3.26)	70-0806-65

Compound	Solvent	$\lambda_{max}(\log \epsilon)$	Ref.
$C_{15}H_{15}ClO_4$			
3-Cyclohexene-1-carboxylic acid, 6-(6-chloro-4-anisoyl)-, cis	EtOH	214(4.23),291(3.26)	70-0806-65
trans	EtOH	216(4.21),300(3.21)	70-0806-65
$C_{15}H_{15}Cl_3N_3Rh$			
Rhodium, 1,2,3-trichlorotris(pyridine)-	n.s.g.	407(2.14)	39-1951-65
$C_{15}H_{15}F_3OS$			
p-Toluic acid, α,α,α-trifluorothio-, 2-norbornyl ester	HOAc	420(2.04)	44-2882-65
	HOAc	420(2.04)	88-3363-64
$C_{15}H_{15}IN_2$			
1,7-Dimethyl-2-phenyl-1H-imidazo[1,2-a]pyridinium iodide	EtOH	212(4.50),222(4.53), 289(4.14)	4-0331-65
$C_{15}H_{15}N$			
Benzylamine, N-benzylidene-α-methyl-	EtOH	249(4.31)	44-2265-64
5H-Dibenz[c,e]azepine, 6,7-dihydro-6-methyl-	isooctane	241.5(4.23)	35-1710-64
	6N HCl	248.5(3.59)	35-1710-64
3-Picoline, 4-(o-methylstyryl)-, cis	hexane	269(4.01)	32-1322-65
3-Picoline, 4-(p-methylstyryl)-, cis	hexane	221s(4.20),262(4.10)	32-1322-65
Pyridine, 3,6-dimethyl-2-styryl-, cis	hexane	258(3.99)	32-1322-65
Pyridine, 4,6-dimethyl-2-styryl-, cis	hexane	222s(4.19),225(4.00)	32-1322-65
Quinoline, 1,2,3,4-tetrahydro-3-phenyl-	EtOH	208(4.53),251(3.95), 304(3.44)	39-0249-64
$C_{15}H_{15}NO$			
Acetanilide, p-tolyl-	EtOH	244(4.22)	54-0949-64
9-Acridanmethanol, α-methyl-	EtOH	280(4.12),283(4.13)	33-1395-65
9-Acridanmethanol, 9-methyl-	EtOH	284(4.20)	33-1395-65
9-Acridanmethanol, 10-methyl-	EtOH	284(4.21)	33-1395-65
10b-Azabenz[a]acenaphthylene, 1,2,3,3a,4,5,6,10b-octahydro-1-oxo-	MeOH	242(4.21),267(4.03), 297(3.63)	65-3487-64
Benzamide, N-benzyl-N-methyl-	MeOH	235(3.9),260s(3.1)	56-0789-64
Benzamide, N-(α-methylbenzyl)-	C_6H_{12}	223(4.1)	56-0789-64
	MeOH	225(4.1)	56-0789-64
	20% C_6H_{12} in MeOH	227(4.1)	56-0789-64
	20% C_6H_{12} in EtOH	227(4.1)	56-0789-64
Benzophenone, p-(dimethylamino)-	benzene	342(4.37)	65-0094-65
	EtOH	248(4.17),357(4.37)	65-0094-65
	HOAc	361(4.34)	65-0094-65
Cyclohexanone, 2-(2-quinolyl)-	EtOH	217(4.57),278(3.95), 440(3.89)	94-0912-65
	N HCl	237(4.57),317(4.06)	94-0912-65
Ethylamine, α-phenyl-N-salicylidene-	MeOH	257(4.17),316(3.60), 403(2.85)	65-3055-64
	EtOH	256(4.14),283s(3.35), 315(3.61),404(2.78)	65-3055-64
3-Picoline, 6-(p-methoxystyryl)-	EtOH	325(4.34)	65-4106-64
Propionanilide, β-phenyl-	EtOH	244(4.20)	54-0949-64
4(1H)-Quinolone, 1,3-diallyl-	EtOH	213(4.39),242s(4.39), 246(4.40),293(3.51), 330(4.11),344(4.18)	94-1424-64
p-Toluidine, N-(p-methoxybenzylidene)-	isooctane	221(4.28),280(4.33), 320(4.17)	9-0091-65
	EtOH	222(4.19),285(4.26), 320(4.25)	9-0091-65

Compound	Solvent	$\lambda_{max}(\log \epsilon)$	Ref.
$C_{15}H_{15}NO_2$			
o-Anisidine, N-(p-methoxybenzylidene)-	isooctane	276(4.35),330(4.97)	9-0091-65
	EtOH	280(4.23),325(4.99)	9-0091-65
p-Anisidine, N-(p-methoxybenzylidene)-	isooctane	280(4.29),330(4.18)	9-0091-65
	EtOH	282(4.26),331(4.28)	9-0091-65
Apo-β-erythroidine	EtOH	240(4.39),345(3.54)	35-1397-65
Heptafulvene, 8-(2-carbethoxy-2-cyanovinyl)-1-methyl-	benzene	272s(4.06),360s(3.37), 454(4.56)	24-2050-64
Isoapo-β-erythroidine	EtOH	253(4.23),288(4.03), 379(3.81)	35-1397-65
Propionanilide, 3-hydroxy-3-phenyl-	EtOH	244(4.22)	32-1115-65
$C_{15}H_{15}NO_3$			
Azepino[3,2,1-hi]indole-$\Delta^{7(4H),\alpha}$-acetic acid, 1,2,5,6-tetrahydro-6-hydroxy-6-methoxy-, γ-lactone	MeOH	255(4.21),290(3.95), 402(3.74)	35-1397-65
Morpholine, 4-(3-hydroxy-2-naphthoyl)-	$CHCl_3$	227(4.84),277(3.59), 288(3.30)	94-0112-64
$C_{15}H_{15}NO_4$			
Nicotinic acid, 1,6-dihydro-1-(m-methoxyphenethyl)-6-oxo-	EtOH	261(4.21),303(3.73)	23-1527-65
2,4-Pentadienoic acid, 2,3-diethyl-4-hydroxy-5-(p-nitrophenyl)-, γ-lactone	C_6H_{12}	332(4.43),347(4.50), 366(4.35)	35-2819-64
1H-Pyrrolo[1,2-a]indole-2-carboxylic acid, 2,3-dihydro-7-methoxy-6-methyl-1-oxo-, methyl ester	MeOH	336(4.34)	35-3877-64
	MeOH	216(4.48),336(4.34)	44-2897-65
4H-Quinolizine-3-carboxylic acid, 1-acetyl-6-methyl-4-oxo-, ethyl ester	EtOH	260(3.89),353(3.98), 420(4.03)	78-3305-65
$C_{15}H_{15}NO_4S$			
6H-1,3-Thiazin-2-ylacetic acid, 4-carboxy-5-methyl-, benzyl ester	EtOH	225s(3.53),280(3.89), 338(4.40)	39-1262-65
sodium salt	EtOH	225s(3.58),282(3.36), 350(4.10)	39-1262-65
$C_{15}H_{15}NO_5$			
Aureothinic acid	EtOH	249(4.14),345(4.20)	88-2655-65
2H,6H-Chromeno[7,6-b][1,4]oxazine, 8-carbethoxy-3,4-dihydro-3-methyl-6-oxo-	MeOH	256(4.3),294(3.8), 404(3.8)	39-7348-65
$C_{15}H_{15}NO_6$			
1H-3-Benzazepine-2,4-dicarboxylic acid, 2,3,4,5-tetrahydro-3-methyl-1,5-dioxo-, dimethyl ester	MeOH	204(4.1),243(4.40), 272(4.11),302(4.13)	56-1423-65
2(1H)-Pyridone, 1-[(3,4,5-trimethoxybenzoyl)oxy]-	EtOH	214(4.54),277(4.10), 302s(3.99)	35-5186-65
$C_{15}H_{15}NO_7$			
3,3,4-Indolinetricarboxylic acid, 1-methyl-2-oxo-, trimethyl ester	EtOH	213(4.31),262(3.90), 313(3.30)	39-3229-64
$C_{15}H_{15}NO_8$			
8aH-Thiazolo[3,2-a]pyridine-5,6,7,8-tetracarboxylic acid, tetramethyl ester	MeOH	227(1.11),283(1.84), 435(0.42)	39-3200-65

Compound	Solvent	λ_{max}(log ε)	Ref.
$C_{15}H_{15}NS$			
Thiazole, dimethyl acetylene-dicarboxylate adduct	MeOH	225(4.16),282(4.33), 430(3.63)	88-1797-64
$C_{15}H_{15}N_3$			
Nicotinonitrile, 1,4,5,6-tetrahydro-2-indol-3-yl-1-methyl-	EtOH	222(4.04),273(3.95), 300(3.81)	44-2534-65
$C_{15}H_{15}N_3O$			
5H-Dibenzo[b,e]-1,4-diazepine, 10-(2-aminoethyl)-10,11-dihydro-11-oxo-, oxalate	EtOH	222(4.55)	33-1590-65
2-Indolinone, 3-[2-amino-1-(2-pyridyl)ethyl]-	EtOH	253(4.00),262s(3.91), 283s(3.19)	87-0626-65
$C_{15}H_{15}N_3O_2$			
Methyl Red	EtOH	494(4.50)	46-0821-65
p-Phenylenediamine, N,N-dimethyl-N'-(p-nitrobenzylidene)-	EtOH	275(4.25),442(4.24)	9-0091-65
	$C_2H_4Cl_2$	282(4.28),450(4.27)	30-1017-64
1H-Pyrrolo[2,1-c][1,4]benzodiazepine-2-acrylamide, 5,10,11,11a-tetra-hydro-5-oxo-	pH 1	230(4.05),328(4.51)	35-5791-65
	iso-PrOH	238(4.26),253s(4.18), 324(4.52),335(4.52), 361(4.20)	35-5791-65
$C_{15}H_{15}N_3O_3$			
Benzamide, N-(2-anilinoethyl)-p-nitro-	MeOH	248(4.32)	87-0107-65
Nitrone, α-[p-(dimethylamino)-phenyl]-N-(m-nitrophenyl)-	benzene	390(4.44)	44-3427-65
	EtOH	402(4.52)	44-3427-65
	ether	386(4.44)	44-3427-65
	acetone	390(4.46)	44-3427-65
	MeCN	388(--)	44-3427-65
Salicylaldehyde, 5-nitro-benzylmethylhydrazone	MeOH	296(4.48)	24-1631-64
	MeOH-NaOH	342(4.25),424(4.07)	24-1631-64
Salicylaldehyde, 5-nitro-, ethylphenylhydrazone	MeOH	302(4.31),348(4.46),	24-1631-64
	MeOH-NaOH	368(4.44),438(3.96)	24-1631-64
$C_{15}H_{15}N_3O_3S$			
Pyrimidine, 2-(ethylthio)-4-[3-methyl-4-(5-nitro-2-furyl)-1,3-butadienyl]-	propylene glycol	256(4.26),337(4.25), 410(4.27)	95-0207-64
$C_{15}H_{15}N_3O_4$			
5-Pyrimidinealanine, N-carboxy-, N-benzyl ester	pH 1	248(3.49)	4-0001-65
	pH 11	249(3.48)	4-0001-65
$C_{15}H_{15}N_3O_4S$			
Benzaldehyde, p-nitro-, N-methyl-(p-tolylsulfonyl)hydrazone	EtOH	323(4.24)	44-1242-65
$C_{15}H_{15}N_3O_5$			
6-Quinazolinemethanol, 2-acetamido-8-hydroxy-, diacetate	EtOH	255(3.95),294(3.63)	78-2059-65
$C_{15}H_{15}N_5$			
Quinoline, 1,2,3,4-tetrahydro-5-methyl-1-s-triazolo[1,5-a]pyrimidin-7-yl-	MeOH	331(4.24)	65-0204-64
$C_{15}H_{15}N_5O_6S$			
p-Toluic acid, α-(1,2,3,6-tetrahydro-1,3-dimethyl-2,6-dioxo-8-sulfamoyl-purin-7-yl)-	MeOH	285(4.03)	54-0193-65

Compound	Solvent	λ max(log ε)	Ref.
$C_{15}H_{15}O_2P$			
2-Propanone, (diphenylphosphinyl)-	MeOH	226(4.30),255(2.9), 260(3.0),265(3.22), 278(3.2),283(2.18)	70-0669-65
	MeOH-NaOMe	260(3.2),265(3.2), 273(3.2),290s(2.9)	70-0669-65
Phosphinic acid, diphenyl-, isopropenyl ester	MeOH	224(4.12),255(2.9), 260(3.0),265(3.16), 272(3.0)	70-0669-65
$C_{15}H_{16}$			
Biphenyl, 2,4,6-trimethyl-	EtOH	240s(3.76)	25-1837-65
Biphenyl, 3,3',4-trimethyl-	EtOH	256(4.16)	39-4077-64
Propane, 1,1-diphenyl-	C_6H_{12}	255(2.73),260(2.71), 262(2.72),269(2.61)	22-0587-65
Tricyclo[2.2.1.0^{2,6}]heptane, 1-(1-phenylvinyl)-	EtOH	243(3.9)	44-3569-65
Unknown hydrocarbon	EtOH	244(3.94)	25-1837-65
$C_{15}H_{16}BF_4N$			
4-(Dimethylamino)phenyltropylium tetrafluoroborate	EtOH	306(4.08),570(3.38)	44-4180-65
$C_{15}H_{16}BrClO_3$			
Cyclohexanecarboxylic acid, 4α-bromo-2β-(2-chloro-5-methoxybenzyl)-5α-hydroxy-, γ-lactone	EtOH	207(4.34),230(4.00), 282(3.22),290(3.19)	70-0806-65
$C_{15}H_{16}BrNO$			
2,5-Cyclohexadien-1-one, 2-bromo-4,6-dimethyl-4-m-toluidino-	MeOH	250(4.26)	35-1127-64
2,5-Cyclohexadien-1-one, 2-bromo-4,6-dimethyl-4-o-toluidino-	MeOH	245(4.27)	35-1127-64
$C_{15}H_{16}Br_2O_2$			
Anthrone, 2,3-dibromo-1,2,3,4,4aβ,9aβ-hexahydro-10α-hydroxy-10-methyl-	EtOH	250(3.68),290(2.88)	70-1013-64
$C_{15}H_{16}ClNO_4$			
[2-(7H-Benzocyclohepten-7-ylidene)-ethylidene]dimethylammonium perchlorate	MeCN	238(4.47),256s(4.15), 311(4.34),318s(4.30), 444(4.67)	24-2050-64
7-(p-Dimethylaminophenyl)tropylium perchlorate	MeCN	275(4.13),569(4.56)	24-0029-64
$C_{15}H_{16}ClN_3O_3$			
Pyrrolo[2,3-b]indole-2-carboxamide, 5-chloro-1,8-dihydro-6,7-dimethoxy-N,8-dimethyl-	MeOH	235(4.42),281(4.39), 328(4.44)	39-0026-64
$C_{15}H_{16}F_3N_3O$			
4-Pyrimidinol, 2-amino-5-(4-phenylbutyl)-6-(trifluoromethyl)-	N HCl	268(3.89)	4-0162-65
	pH 7	302(3.94)	4-0162-65
	pH 13	291(3.82)	4-0162-65
$C_{15}H_{16}FeO$			
Ferrocene, 1,1'-(α-oxo-α',β-dimethyltrimethylene)-	n.s.g.	272(3.5),337(3.00)	44-3941-65

Compound	Solvent	λ_{max}(log ϵ)	Ref.
$C_{15}H_{16}N_2$			
Aniline, N,N-dimethyl-p-(N-phenyl-formimidoyl)-	isooctane	234(4.17),340(4.49)	9-0091-65
	EtOH	239(4.09),355(4.51)	9-0091-65
	$C_2H_4Cl_2$	240(4.14),320s(4.25), 355(4.55)	30-1017-64
Carbazole-1-propionitrile, 1,2,3,4-tetrahydro-	MeOH	227(4.55),282(3.92)	65-3487-64
4aH-Carbazole-4a-propionitrile, 1,2,3,4-tetrahydro-	MeOH	260(3.82)	65-3487-64
p-Phenylenediamine, N-benzylidene-N,N-dimethyl-	isooctane	250(4.29),370(4.25)	9-0091-65
	EtOH	255(4.24),375(4.21)	9-0091-65
Pyridine, 2-(p-dimethylaminostyryl)-	EtOH	370(4.52)	54-0441-65
Pyridine, 2-(p-dimethylaminostyryl)-, cis	aq. HCl	310(4.00)	23-1345-65
	H_2O	323(4.04)	23-1345-65
	50% MeOH	323(4.11)	23-1345-65
	50% MeOH-HCl	308(3.97)	23-1345-65
trans	aq. HCl	324(4.50)	23-1345-65
	H_2O	348(4.42)	23-1345-65
	50% MeOH	359(4.44)	23-1345-65
	50% MeOH-HCl	325(4.50)	23-1345-65
$C_{15}H_{16}N_2O$			
Benzanilide, p-amino-2',6'-dimethyl-	EtOH	215(4.2),284(4.4)	28-4295-64B
Dipyrrolo[1,2-c:2',1'-f]pyrimidin-10-one, 1,3,7,8-tetramethyl-5-methylene-	EtOH	258(3.62),307(3.80), 315s(3.78),395(3.63)	12-1977-65
Formanilide, N-(p-dimethylaminophenyl)-	EtOH	268(4.34)	44-3427-65
Indazole, 3-benzoyl-4,5,6,7-tetra-hydro-6-methyl-	EtOH	256(4.15)	32-0814-65
Indolo[2,3-a]quinolizin-2(1H)-one, 3,4,6,7,12,12b-hexahydro-	MeOH	283s(3.90),290(3.82)	44-3407-64
9-Phenanthrenecarboxamide, 10-amino-5,6,7,8-tetrahydro-	50% H_2SO_4	235(4.7),285(3.7), 295s(3.4),325(3.3)	44-1543-64
Phenol, o-[[p-(dimethylamino)benzyli-dene]amino]-	isooctane	240(4.08),323(4.39), 368(4.26)	9-0091-65
	EtOH	242(4.08),346(4.42)	9-0091-65
m-Phenylenediamine, N,N-dimethyl-N'-salicylidene-	EtOH	220(4.5),266(4.31), 338(4.02)	65-3837-64
p-Phenylenediamine, N,N-dimethyl-N'-salicylidene-	EtOH	215(4.2),244(4.20), 385(4.36)	65-3837-64
4(1H)-Pyridone, 2,3-dihydro-1-[2-(in-dol-3-yl)ethyl]-	MeOH	282(3.88),290(3.91)	24-2463-64
$C_{15}H_{16}N_2OS$			
4H-1,3,4-Oxadiazine, 5,6-dihydro-4,5-dimethyl-6-phenyl-2-(2-thienyl)-	$CHCl_3$	242(3.87),308(3.97)	44-0668-64
2-Thiazolin-4-one, 5-(cyclohexyl-imino)-2-phenyl-	dioxan	209(4.1),300(4.4)	83-0124-65
$C_{15}H_{16}N_2O_2$			
Cyclopenta[c]pyrrole, 2,4,5,6-tetrahy-dro-1,3-dimethyl-N-(p-nitrophenyl)-	n.s.g.	227(4.35),337(4.0)	24-2949-64
4H-1,3,4-Oxadiazine, 2-(2-furyl)-5,6-dihydro-4,5-dimethyl-6-phenyl-	$CHCl_3$	240(3.78),290(4.06)	44-0668-64
2,5-Piperazinedione, 3-benzylidene-6-isobutylidene-	EtOH	234(3.9),318(4.4)	44-0277-65
Spiro[indoline-3,1'(5'H)-indolizine]-2,5'-dione, 2',3',6',7',8',8'a-hexahydro-	EtOH	251(3.90),281(3.18)	78-1327-65

Compound	Solvent	$\lambda_{max}(\log \epsilon)$	Ref.
$C_{15}H_{16}N_2O_2S$			
5-Pyrimidinecarboxylic acid, 2-benzyl-1,4-dihydro-6-methyl-4-thioxo-, ethyl ester	EtOH	300(4.11),345(3.77)	44-1115-64
5-Pyrimidinecarboxylic acid, 1,4-dihydro-1,6-dimethyl-2-phenyl-4-thioxo-, ethyl ester	EtOH	250(4.01),340(4.31)	44-1115-64
5-Pyrimidinecarboxylic acid, 1,4-dihydro-2,6-dimethyl-1-phenyl-4-thioxo-, ethyl ester	EtOH	228(3.80),337(4.36)	44-1115-64
$C_{15}H_{16}N_2O_3$			
2,2'-Dipyrrylketone, 5,5'-diformyl-3,3',4,4'-tetramethyl-	EtOH	234(4.05),277(4.30), 354(4.41)	12-1835-65
2-Indolecarboxylic acid, 1-(2-cyanoethyl)-5-methoxy-6-methyl-, methyl ester	MeOH	210(4.44),301(4.34)	44-2897-65
$C_{15}H_{16}N_2O_4$			
2-Isoindolineacetic acid, α-(aminomethylene)-1,3-dioxo-, tert-butyl ester	EtOH	237(4.07),266(4.14)	44-3560-64
$C_{15}H_{16}N_2O_5$			
Carbamic acid, ester with 2,3-dihydro-9-(hydroxymethyl)-7-methoxy-6-methyl-1H-pyrrolo[1,2-a]indole-5,8-dione	MeOH	230(4.28),287(4.16), 345(3.6),460(3.14)	44-2897-65
3-(1,2-Dicarboxy-2-hydroxyvinyl)-1,2-dimethylbenzimidazolium hydroxide, inner salt, dimethyl ester	EtOH	264(4.31),277(4.16)	44-1118-65
Mitosene, 7-methoxy-	MeOH	230(4.28),287(4.17), 345(3.59),460(3.14)	35-3877-64
$C_{15}H_{16}N_4$			
Formaldehyde, (m-tolylazo)-, m-tolylhydrazone	EtOH	258(4.02),430(4.43)	5-0099-65B
Formaldehyde, (o-tolylazo)-, o-tolylhydrazone	EtOH	254(4.06),298(3.80), 424(4.43)	5-0099-65B
Formaldehyde, (p-tolylazo)-, p-tolylhydrazone	EtOH	260(4.03),305(3.83), 438(4.41)	5-0099-65B
$C_{15}H_{16}N_4O$			
5H-Pyrrolo[3,4-d]pyrimidine, 4-amino-6-benzoyl-6,7-dihydro-2,7-dimethyl-	pH 1	244(4.12)	44-0194-65
	EtOH	232(4.08),266(3.65)	44-0194-65
$C_{15}H_{16}N_4O_3$			
Lumiflavine, 5-acetyl-1,5-dihydro-	MeOH	220(4.47),262(4.35), 302(3.99)	33-1354-64
Mandelamide, N-(4-amino-6-methyl-5-pyrimidinyl)-, acetate	pH 1	249(4.05)	87-0797-65
	pH 11	234(3.89),274(3.60)	87-0797-65
Pyrimidine, 2-(ethylamino)-4-[3-methyl-4-(5-nitro-2-furyl)butadienyl]-	propylene glycol	282(4.06),410(4.38)	95-0131-64
$C_{15}H_{16}N_4O_3S$			
Mandelamide, N-(4-amino-2-(methylthio)-5-pyrimidinyl]-, acetate	pH 1	243(4.40)	87-0797-65
	pH 11	226(4.19),253(4.06), 291(3.90)	87-0797-65
Mandelamide, N-(4-amino-6-(methylthio)-5-pyrimidinyl]-, acetate	pH 1	242(4.24),294(4.10)	87-0797-65
	pH 11	230(4.27),286(3.90)	87-0797-65

Compound	Solvent	$\lambda_{max}(\log \epsilon)$	Ref.
$C_{15}H_{16}N_4O_4$			
1,3-Cycloheptadiene, 1-acetyl-, 2,4-dinitrophenylhydrazone	$CHCl_3$	391.5(4.47)	35-0321-65
2-Cyclohexen-1-one, 4-propylidene-, 2,4-dinitrophenylhydrazone	MeOH	229(4.21),260(4.22), 383(4.51)	5-0028-64D
3,5-Hexadien-2-one, 6-cyclopropyl-, 2,4-dinitrophenylhydrazone	$CHCl_3$	399(4.54)	22-3218-64
$C_{15}H_{16}N_4O_4S$			
Theophylline, 7-benzyl-8-(methylsulfonyl)-	MeOH	291(3.99)	54-0193-65
$C_{15}H_{16}N_4S_2$			
1H-Imidazo[4,5-d]pyridazine, 4,7-bis-(ethylthio)-2-phenyl-	EtOH	207(4.49),268(4.45), 298(4.38),309(4.35)	4-0182-64
$C_{15}H_{16}N_6O_2S$			
p-Toluenesulfonic acid, 2-(5-anilino-s-triazol-3-yl)hydrazide	EtOH	226(4.28),258(4.33)	39-3912-65
$C_{15}H_{16}O$			
2-Anthrol, 1,2,3,4-tetrahydro-9-methyl-	EtOH	233(4.93),280(3.73), 290(3.76)	70-1024-64
2-Anthrol, 1,2,3,4-tetrahydro-10-methyl-	EtOH	233(5.02),280(3.73), 290(3.76)	70-1024-64
Anthrone, 1,4,4aα,9aβ-tetrahydro-10-methyl-	EtOH	249(4.00)	70-0310-64
Azulene, 1-acetyl-4,6,8-trimethyl-	C_6H_{12}	229(4.27),247(4.37), 310(4.52),528(2.81)	44-0131-65
1-Azulenecarboxaldehyde, 2,4,6,8-tetramethyl-	hexane	508(2.78)	5-0194-64B
2(1H)-Phenanthrone, 3,4,9,10-tetrahydro-1-methyl-	EtOH	257(4.00)	35-2038-64
1'-Valeronaphthone	EtOH	215(4.73),298(3.78)	39-2072-65
2'-Valeronaphthone	EtOH	240(4.71),291(3.97)	39-2072-65
$C_{15}H_{16}O_2$			
2,9-Epoxyanthracen-10(1H)-one, 2α,3,4,4aα,9,9aα-hexahydro-9α-methyl-	EtOH	249(4.06),287(3.01)	70-1024-64
Fluorene, 1,4-dihydro-2,7-dimethoxy-	EtOH	264(4.16)	88-3505-65
1,3-Indandione, hexahydro-2-phenyl-	EtOH-HCl	251(4.22)	65-0808-64
Mansonone C	EtOH	206(4.14),258(4.24), 432(3.39)	88-4857-65
1,4-Methanobiphenylene, 1,4,4a,8b-tetrahydro-9,9-dimethoxy-	EtOH	257s(--),262(3.14), 268(3.33),274(3.32)	39-2549-65
1-Naphthaleneacetic acid, propyl ester	EtOH	225(5.00),281(3.86)	39-2072-65
2-Naphthaleneacetic acid, propyl ester	EtOH	224(5.01),276(3.73)	39-2072-65
1-Naphthoic acid, butyl ester	EtOH	220(5.65),296(3.80)	39-2072-65
2-Naphthoic acid, butyl ester	EtOH	237(4.91),280(3.88)	39-2072-65
2-Norbornene-2-carboxylic acid, 3-o-tolyl-	EtOH	256s(3.74)	39-5548-64
2-Norbornene-2-carboxylic acid, 3-p-tolyl-	EtOH	283(4.03)	35-5548-64
4-Phenanthrol, 1,2,3,4-tetrahydro-7-methoxy-	EtOH	234(4.85),255(3.62), 264(3.74),274(3.76), 283s(3.56),309s(3.00), 322(3.31),327(3.41)	88-3671-64

Compound	Solvent	λ_{max}(log ε)	Ref.
$C_{15}H_{16}O_3$			
5-Benzofuranacrylic acid, 2,3-dihydro-6-hydroxy-α-isopropyl-β-methyl-, δ-lactone	EtOH	226(4.14),255(3.57), 295(3.80),332(4.28)	44-2467-64
6H,11H-Benzo[b]naphtho[1,8-cd]furan-11-one, hexahydro-8-hydroxy-	EtOH	238(4.30),288(3.90), 316(3.70)	94-1012-64
Cacalone	EtOH	212(3.82),250(4.05), 320(3.93)	78-2331-64
2-Cyclopentene-1-carboxylic acid, 3-methyl-4-oxo-2-phenyl-, ethyl ester	EtOH	236(4.00),243(4.02),	23-2408-65
2,3α-Epoxyanthracen-10(1H)-one, 2,3,4,4aα,9,9aα-hexahydro-9β-hydroxy-9α-methyl-	EtOH	249(4.03),289(3.22)	70-1013-64
2,3β-Epoxyanthracen-10(1H)-one, 2,3,4,4aα,9,9aα-hexahydro-9β-hydroxy-9α-methyl-	EtOH	250(4.03),289(3.17)	70-1013-64
2,9β-Epoxyanthracen-10(1H)-one, 2α,3,4,4aα,9,9aα-hexahydro-3α-hydroxy-9α-methyl-	EtOH	247(3.44),287(2.95)	70-1024-64
2,9β-Epoxyanthracene-3,10-dione, 1,2,3,4,4aα,9,9aα,10-octahydro-9α-methyl-	EtOH	248(4.00),288(3.09)	70-1024-64
2(1H)-Fluorenone, 7-acetoxy-3,4,4a,9a-tetrahydro-	EtOH	262s(3.00),270(3.18), 276(3.15)	88-3505-65
trans	EtOH	262s(3.00),269(3.15), 275(3.13)	88-3505-65
3-Furoic acid, 4,5-diethyl-2-phenyl-	MeOH	297(4.26)	39-5984-65
3-Furoic acid, 2-methyl-5-p-tolyl-, ethyl ester	MeOH	278(4.26)	44-2205-65
Isolinderalactone	EtOH	210(4.22)	39-4578-64
Linderalactone	EtOH	208(4.18)	39-4578-64
Naphtho[1,8-bc]pyran-4-ethanol, 2,3-dihydro-5-hydroxy-α,2-dimethyl-	EtOH	227(4.61),247(4.40), 281s(3.50),291(3.61), 326(3.33),340(3.41)	12-0575-65
2-Norbornene-2-carboxylic acid, 3-(o-methoxyphenyl)-	EtOH	268(3.69)	35-5548-64
2-Norbornene-2-carboxylic acid, 3-(p-methoxyphenyl)-	EtOH	298(4.02)	35-5548-64
$C_{15}H_{16}O_4$			
Anthrone, 1,4,4aα,9aα-tetrahydro-4β,10β-dihydroxy-5-methoxy-	EtOH	223(4.21),256(3.81), 318(3.37)	65-2558-64
Auraptenol	EtOH	246(3.6),256(3.6), 322(4.17)	78-0089-65
2-Benzofuranvaleric acid, δ-oxo-, ethyl ester	MeOH	225(3.82),293(4.32)	94-0214-64
Catalpalactone, dihydro-	EtOH	228(4.00),274(3.22), 281(3.22)	88-1261-65
4-Indanacrylic acid, 5-carboxy-, dimethyl ester	EtOH	235(4.42),268(4.11)	39-3811-65
2H-Indeno[5,6-b]furan-5,7(3H,6H)-dione, 6-(hydroxymethyl)-4,6,8-trimethyl-	EtOH	256(4.14),296(3.44), 330(3.14)	78-1231-65
	EtOH	256(4.14),296(3.44), 330(3.14)	94-0853-64
Indene-3-propionic acid, 2-(1-carboxyethyl)-	EtOH	262(4.10)	42-0479-64
Naphthoquinone, 6-ethoxy-7-methoxy-2,3-dimethyl-	n.s.g.	277(4.51),346(3.35), 408(3.10)	39-4011-64
Propane, 1,3-bis(2,5-dihydroxyphenyl)-	95% THF	299(3.94)	44-2602-65
Spiro[chroman-2,5'-tetrahydrofuran-2'-one-4,1''-propan-2'-one]	EtOH	272(3.28),279(3.25)	49-0220-65

Compound	Solvent	λ_{max}(log ε)	Ref.
$C_{15}H_{16}O_5$			
1,3-Cyclohexanedione, 4-[o-(methoxy-methoxy)benzoyl]-	MeOH	258(4.08),281(4.26)	94-0214-64
Hamaudol	EtOH	229(4.22),252(4.29), 258(4.29),299(4.02)	95-0055-65
Lactulin	EtOH	256(4.16)	83-0703-64
Lactulin decomposition product	EtOH	223(4.24)	83-0703-64
Phenol, 4-acetoacetyl-2,6-di-acetyl-3-methyl-	EtOH	242(4.27),303(4.23)	39-5646-64
2-Propanone, (2,4,6-triacetyl-3-hydroxyphenyl)-	EtOH	245(4.40),321(4.02)	39-5646-64
$C_{15}H_{16}O_5S_3$			
4,5-Dihydro-2-(2-methoxy-1-naphthyl)-1,3-dithiol-1-ium methyl sulfate	MeCN	309(3.71),376(3.89), 464(4.19)	24-1374-65
$C_{15}H_{16}O_{10}$			
1,3-Cyclopentadiene, 1,2,3,4,5-penta-carbomethoxy-	MeOH	262(4.69),295(4.16)	44-0423-64
$C_{15}H_{17}BrO_3$			
Acetic acid, bromo-, ester with 1,7,8,8a-tetrahydro-1-hydroxy-8a-methyl-1,4-ethanonaphthalen-6(4H)-one	EtOH	254(4.24)	78-1119-64
Anthrone, 2α-bromohexahydro-3β,10α-di-hydroxy-10-methyl-	EtOH	252(4.12),292(3.24)	70-1013-64
Anthrone, 3β-bromohexahydro-2α,10α-di-hydroxy-10-methyl-	EtOH	250(3.92),290(3.28)	70-1013-64
Lumisantonin, 2-bromo-	EtOH	243(3.65),283(3.40)	39-2518-64
$C_{15}H_{17}Br_2ClO_3$			
Cyclohexanecarboxylic acid, 4α,5β-di-bromo-2β-(2-chloro-5-methoxybenzyl)-	EtOH	282(3.32),289(3.28)	70-0806-65
$C_{15}H_{17}ClN_2O_2$			
1H-Azepino[5,4,3-cd]indole-2-carboxylic acid, 9-chloro-3,4,5,6-tetrahydro-6-methyl-, ethyl ester	EtOH	240(4.44),297(4.27)	87-0200-65
$C_{15}H_{17}ClO_3$			
Anthrone, 2α-chloro-3β,10α-dihydroxy-hexahydro-10-methyl-	EtOH	252(4.01),291(3.19)	70-1013-64
Anthrone, 3β-chloro-2α,10α-dihydroxy-hexahydro-10-methyl-	EtOH	250(3.99),291(3.21)	70-1013-64
3-Cyclohexene-1-carboxylic acid, 6-(2-chloro-5-methoxybenzyl)-, trans	EtOH	204(4.39),273(3.39), 280(3.06)	70-0806-65
$C_{15}H_{17}CoO_2$			
Cobalt, cyclopentadienyl(tetra-methyl-p-benzoquinone)	C_6H_{12}	285(4.32),362(3.35), 415(2.97)	101-0336-64B
	H_2O	286(4.20),356(3.36), 400(3.00)	101-0336-64B
	EtOH	287(4.36),362(3.37), 410(2.97)	101-0336-64B
	$HClO_4$	283(4.38),335(3.05), 390(2.73)	101-0336-64B
	CCl_4	287(4.38),364(3.42), 415(3.08)	101-0336-64B

Compound	Solvent	$\lambda_{max}(\log \epsilon)$	Ref.
$C_{15}H_{17}IN_2O$			
2-(1-Acetyl-3-indolyl)-1-methyl-1-pyrrolinium iodide	dioxan-1% DMF	258s(3.99),270s(3.93),291(3.93),299(3.96)	87-0415-64
$C_{15}H_{17}IrO_2$			
Iridium, cyclopentadienyl(tetramethyl-p-benzoquinone)	H_2O	215s(4.2),260s(3.9),315(3.9)	35-3265-64
	$HClO_4$	225(4.1),235(4.0),275s(3.7)	35-3265-64
	$CHCl_3$	215s(4.2),250s(3.8),315(3.9)	35-3265-64
$C_{15}H_{17}N$			
1,3,5-Cycloheptatriene, 7-[p-(dimethylamino)phenyl]-	EtOH	256(4.45),300s(3.76)	44-4180-65
Pyridine, 2-(4-phenylbutyl)-	EtOH	257(3.56),262(3.62),268(3.49)	94-1344-64
$C_{15}H_{17}NO$			
Acetamide, N-(4,6,8-trimethyl-1-azulenyl)-	C_6H_{12}	563(2.72)	35-3137-64
Benzo[f]quinolin-1(2H)-one, 3,4,5,6-tetrahydro-2,4-dimethyl-	EtOH-HCl	251(4.27),352(3.88)	94-1405-64
	EtOH	278(4.18),353(4.03)	94-1405-64
Benzo[g]quinolin-4(1H)-one, 2,3,5,10-tetrahydro-1,3-dimethyl-	EtOH-HCl	268(4.00),322(3.94)	94-1405-64
2-Indolinone, 3-cycloheptylidene-	EtOH	254(4.44),262(4.52),295(3.91),304(3.81),352(3.29)	87-0626-65
4(1H)-Quinolone, 1-allyl-3-propyl-	EtOH	213(4.34),246(4.35),282s(3.31),293(3.44),332(4.10),346(4.18)	44-1986-65
$C_{15}H_{17}NOS$			
Aniline, N,N-dimethyl-p-(tolylsulfinyl)-	isooctane	280(4.44)	35-5637-64
	EtOH	220s(4.23),293(4.36)	35-5637-64
hydrochloride	EtOH	238(4.39),265s(3.82)	35-5637-64
$C_{15}H_{17}NO_2$			
Carbazole-1-propionic acid, 1,2,3,4-tetrahydro-	MeOH	227(4.53),282(3.88)	65-3487-64
4aH-Carbazole-4a-propionic acid, 1,2,3,4-tetrahydro-	MeOH	257(3.86)	65-3487-64
Furo[2,3-b]quinoline, 2,3-dihydro-	n.s.g.	251(4.21),272(4.15),282(4.07),306(3.69),320(3.69)	44-2598-64
	acid	288(4.18),315s(3.81)	44-2598-64
Lunacrine, demethoxy-	n.s.g.	235(4.40),249s(4.14),296s(3.94),307(4.03),318(3.98)	44-2598-64
	acid	233(4.72),239s(4.58),300s(4.09),313(3.86)	44-2598-64
2-Naphthol, 3-(morpholinomethyl)-	$CHCl_3$	228(5.08),267(3.61),278(3.65),289(3.46),335(3.54)	94-0112-64
7-Phenanthridinol, 7,8,9,10-tetrahydro-3-methoxy-6-methyl-	MeOH	215(4.46),240(4.65)	39-5916-64
Pyrrolo[2,1-a]isoquinoline, 5,6-dihydro-8,9-dimethoxy-2-methyl-	EtOH	298(4.22),313(4.18)	94-0775-65

Compound	Solvent	$\lambda_{max}(\log \epsilon)$	Ref.
$C_{15}H_{17}NO_2S$			
p-Toluenesulfono-m-toluidide, N-methyl-	hexane	236(4.00)	65-2080-65
	EtOH	236(4.00),262(3.45)	65-2080-65
	ether	235(3.95),264(3.27)	65-2080-65
	dioxan	236(4.06)	65-2080-65
	H_2SO_4	248(4.23),270(3.83), 310s(2.82)	65-2080-65
p-Toluenesulfono-o-toluidide, N-methyl-	hexane	234(4.06),263(3.27)	65-2080-65
	EtOH	233(4.11),263(3.28)	65-2080-65
	ether	233(4.08),265(3.32)	65-2080-65
	dioxan	234(4.08),264(3.38)	65-2080-65
	H_2SO_4	248(4.23),270(3.80), 325s(2.72)	65-2080-65
p-Toluenesulfono-p-toluidide, N-methyl-	hexane	222(4.15)	65-2080-65
	EtOH	222(4.15)	65-2080-65
	ether	223(4.09),265(3.28)	65-2080-65
	dioxan	232(4.00),260(3.53)	65-2080-65
	H_2SO_4	226(4.02),245(4.26), 270(3.80),310s(2.60)	65-2080-65
$C_{15}H_{17}NO_3$			
2-Azabicyclo[2.2.2]octan-3-one, 6-p-anisoyl-	EtOH	280(4.11)	28-3425-65A
Carbostyril, 4-hydroxy-8-methoxy-3-(3-methyl-2-butenyl)-	MeOH	243(4.45),250(4.42), 280(3.93),288(3.94), 320(3.61)	39-0438-64
3-Indolizinecarboxylic acid, 5-methyl-1-propionyl-, ethyl ester	EtOH	225(4.2),252(4.45), 285(3.92),338(4.25)	78-3305-65
2H-Pyrano[2,3-b]quinolin-5-ol, 3,4-dihydro-9-methoxy-2,2-dimethyl-	MeOH	239(4.69),298(3.91), 313(3.98),325(3.88)	39-4190-64
5H-Pyrano[3,2-c]quinolin-5-one, 2,3,4,6-tetrahydro-7-methoxy-2,2-dimethyl-	MeOH	245(4.45),277(3.90), 287(3.89),319(3.50), 330(3.36)	39-4190-64
$C_{15}H_{17}NO_3S$			
2,3-Dihydro-1H-indolizinium p-toluenesulfonate	EtOH	220(4.10),264(3.72), 271s(3.59)	44-2398-65
3-Pyrrolin-2-one, 3-acetyl-5-[(benzylthio)methyl]-4-hydroxy-1-methyl-	EtOH	283(4.03)	35-5654-64
p-Toluenesulfonamide, N-methyl-N-(o-methoxyphenyl)-	hexane	218(4.22),280(3.45)	65-2080-65
	EtOH	221(4.18),277(3.48)	65-2080-65
	ether	276(3.40)	65-2080-65
	dioxan	276(3.50)	65-2080-65
	H_2SO_4	227(4.02),246(4.27), 275(3.73),330s(2.35)	65-2080-65
p-Toluenesulfonamide, N-methyl-N-(p-methoxyphenyl)-	hexane	223(4.18),260(3.70)	65-2080-65
	EtOH	228(4.30),260(3.72)	65-2080-65
	ether	228(4.00),270(3.15)	65-2080-65
	dioxan	228(4.12),265(3.55)	65-2080-65
	H_2SO_4	230(4.12),270(4.02), 327s(2.63)	65-2080-65
$C_{15}H_{17}NO_4$			
Carbazol-1(2H)-one, 3,4-dihydro-5,6,7-trimethoxy-	EtOH	330(4.44)	12-0246-64
Glutarimide, 3-(2-hydroxy-3,5-dimethylphenacyl)-	n.s.g.	216(4.40),260(4.08), 342(3.80)	25-1686-64
$C_{15}H_{17}NO_5$			
Acronycidine, dihydro-	n.s.g.	236s(4.25),256(4.63), 310(3.77),328s(3.70)	2-0071-65

Compound	Solvent	λ max(log ε)	Ref.
2H,6H-Chromeno[7,6-b][1,4]oxazine, 8-carbethoxy-3,4,7,8-tetrahydro-3-methyl-6-oxo-	MeOH	255(4.2),284s(--), 377(3.5)	39-7348-65
Furo[2,3-b]pyridine-3,5-dicarboxylic acid, 2,6-dimethyl-, diethyl ester	EtOH	219(4.57),249s(4.08), 296(3.99)	39-1632-64
$C_{15}H_{17}NO_6$			
Malonic acid, (α-methyl-p-nitrobenzylidene)-, diethyl ester	EtOH	276(4.16)	28-4776-65B
$C_{15}H_{17}NO_8$			
2,3,4,5-Pyridinetetracarboxylic acid, 6-ethyl-, tetramethyl ester	MeOH	284(3.56)	39-0948-65
$C_{15}H_{17}N_3$			
Azobenzene, 4-(dimethylamino)-2-methyl-	N HCl	326(3.60),520(4.70)	78-1547-64
	EtOH	420(4.43)	78-1547-64
Azobenzene, 4-(dimethylamino)-2'-methyl-	N HCl	326(4.20),520(4.00)	78-1547-64
	EtOH	410(4.43)	78-1547-64
Azobenzene, 4-(dimethylamino)-3'-methyl-	N HCl	320(3.95),520(4.58)	78-1547-64
	EtOH	410(4.42)	78-1547-64
$C_{15}H_{17}N_3O$			
Benzamide, p-amino-N-(2-anilinoethyl)-	MeOH	248(4.14),258(4.11), 282(4.27)	87-0107-65
$C_{15}H_{17}N_3O_2$			
1-Pyrazoline-3-carboxylic acid, 3-cyano-4-methyl-4-p-tolyl-, ethyl ester, trans	$CHCl_3$	301s(2.28)	28-1332-65B
Sydnone imine, N-benzoyl-3-cyclohexyl-	pH 1.00	220(3.90),288(4.34)	65-2064-64
	pH 7.00	254(4.04),328(4.36)	65-2064-64
$C_{15}H_{17}N_3O_3$			
4-Pyrimidinecarboxylic acid, 2-amino-6-hydroxy-5-(4-phenylbutyl)-	pH 1	283(3.73)	4-0162-65
	pH 7	286(3.93)	4-0162-65
	pH 13	284(3.92)	4-0162-65
$C_{15}H_{17}N_3O_3S$			
Benzenesulfonic acid, m-[[4-(dimethylamino)-o-tolyl]azo]-	N HCl	321(3.46),513(4.67)	78-1547-64
	EtOH	467(4.32)	78-1547-64
Benzenesulfonic acid, o-[[4-(dimethylamino)-o-tolyl]azo]-	N HCl	333(3.46),513(4.70)	78-1547-64
	EtOH	446(4.32)	78-1547-64
Benzenesulfonic acid, p-[[4-(dimethylamino)-o-tolyl]azo]-	EtOH	481(4.40)	78-1547-64
m-Toluenesulfonic acid, 4-[[p-(dimethylamino)phenyl]azo]-	N HCl	316(4.24),513(4.36)	78-1547-64
	EtOH	455(4.46)	78-1547-64
o-Toluenesulfonic acid, 4-[[p-(dimethylamino)phenyl]azo]-	N HCl	325(3.87),521(4.68)	78-1547-64
	EtOH	470(4.40)	78-1547-64
$C_{15}H_{17}N_3O_6$			
1-Pyrazoline-3,3-dicarboxylic acid, 4-p-nitrophenyl-, diethyl ester	hexane	327(2.70)	28-4776-65B
$C_{15}H_{17}N_3S$			
1-Pyrroline, 2-(2-pyrrolyl)-, thiocarbanilide	EtOH	204(4.38),249(4.21)	39-0888-64
$C_{15}H_{17}N_5$			
s-Triazolo[1,5-c]pyrimidine, 2-anilino-7-methyl-5-propyl-	MeOH	209(--),245(4.63), 301s(3.93)	39-3357-65

Compound	Solvent	λ_{max}(log ϵ)	Ref.
$C_{15}H_{17}N_5O$			
Adenine, 3-benzyl-7-(3-hydroxypropyl)-	pH 1	277(4.19)	87-0710-65
	pH 7	279(4.22)	87-0710-65
	pH 13	282(4.17)	87-0710-65
9H-Purine-9-propanol, 6-(benzylamino)-	pH 1	268(4.29)	87-0033-65
	pH 7	269(4.31)	87-0033-65
	pH 13	269(4.30)	87-0033-65
5H-Pyrrolo[3,4-d]pyrimidine, 6-acetyl-2,4-diamino-7-benzyl-6,7-dihydro-	pH 1	273(3.76)	44-0194-65
	EtOH	284(3.87)	44-0194-65
$C_{15}H_{17}N_5O_3$			
Mandelamide, N-(4,6-diamino-2-methyl-5-pyrimidinyl)-, acetate	pH 1	264(4.03)	87-0797-65
	pH 11	258(4.00)	87-0797-65
$C_{15}H_{17}N_5O_4S$			
Theophylline, 7-benzyl-8-methylsulfamoyl-	MeOH	288(4.02)	54-0193-65
Theophylline, 7-phenethyl-8-sulfamoyl-	MeOH	286(4.02)	54-1215-64
$C_{15}H_{17}O_2Rh$			
Rhodium, cyclopentadienyl(tetramethyl-p-benzoquinone)-	H_2O	222(4.26),262(4.42), 362(3.98),422(2.48)	35-3265-64
	$HClO_4$	270(4.3),340(3.6), 460s(1.5)	35-3265-64
	$CHCl_3$	270(4.3),362(3.9), 450s(2.4)	35-3265-64
$C_{15}H_{18}$			
Azulene, 6-isopropyl-1,4-dimethyl-	hexane	243(4.27),285(4.69), 291(4.69),343(3.51), 351(3.59),368(3.30), 539(2.34),563(2.41), 585(2.46),613(2.39), 638(2.34),675(2.03), 706(1.86)	94-0717-65
Azulene, 1,2,4,6,8-pentamethyl-	hexane	562(2.59),593(2.53), 608(2.47),650(2.10), 670(1.92)	5-0194-64B
Naphthalene, 1-pentyl-	EtOH	225(5.05),283(3.89)	39-2072-65
Naphthalene, 2-pentyl-	EtOH	225(5.08),276(3.72)	39-2072-65
$C_{15}H_{18}BrN_3$			
Quinoxaline[2,3-c]pyrroline-1'-spiro-1''-piperidinium bromide	EtOH	241(4.53),322(3.97)	44-1542-65
$C_{15}H_{18}Br_2N_2O_3$			
7,10-Dimethoxybenzo[b]quinolizinium bromide, compound with hydroxylamine hydrobromide	MeOH	251(4.36),396(4.08)	44-0061-64
$C_{15}H_{18}ClNO_4$			
Coumarin, 6-chloro-3-ethoxy-4-(2-hydroxyethylamino)-5,7-dimethyl-	EtOH	211(4.23),240(4.22), 263s(3.98),295s(4.17), 302(4.23),306(4.23), 330s(3.91)	44-4122-65
Coumarin, 6-chloro-3-isopropoxy-4-(N-methyl-2-hydroxyethylamino)-	EtOH	213(4.33),249(4.19), 304(3.92),311(3.94), 353(3.81)	44-4122-65

Compound	Solvent	λ_{max}(log ε)	Ref.
$C_{15}H_{18}Cl_2N_2O_4$			
[[2-Chloro-1-(dimethylamino)methylene]-inden-3-yl]methylene]dimethyl-ammonium perchlorate	MeCN	442(4.62)	73-2783-65
$C_{15}H_{18}INO_4$			
7-Acetoxy-6,8-dimethoxy-1,2-dimethyl-isoquinolinium iodide	pH 2	219(4.48),253(4.72), 300s(3.67),314(3.74), 335(3.74)	49-0025-65
	pH 12	250s(4.01),288(4.69), 333(3.78)	49-0025-65
	EtOH	219(4.47),254(4.72), 302s(3.70),316(3.74), 335(3.74)	49-0025-65
$C_{15}H_{18}N_2$			
Cyclopentadiene, tetramethyl-5-(phenylazo)-	n.s.g.	284(4.29),408(4.16)	25-1313-64
$C_{15}H_{18}N_2O$			
Benz[cd]indole, 1,3,4,5-tetrahydro-5-morpholino-	EtOH	225(4.41),284(3.78), 293(3.74)	39-5573-64
Benz[cd]indole-3-carboxamide, 1,3,4,5-tetrahydro-N,N,2-trimethyl-	EtOH	227(4.80),274(3.83), 280(3.84),290(3.71)	44-0843-64
4-Piperidone, 1-(2-indol-3-ylethyl)-	MeOH	281(3.79),290(3.73)	24-2579-65
	MeOH	275s(3.66),282(3.69), 290(3.62)	44-3407-64
1H-Pyrrolo[1,2-a]indol-7-ol, 2,3-dihydro-1-pyrrolidinyl-	MeOH	278(3.95),302s(3.64), 312s(3.56)	44-2904-65
$C_{15}H_{18}N_2O_2$			
1H-Azepino[5,4,3-cd]indole-2-carboxylic acid, 3,4,5,6-tetrahydro-6-methyl-, ethyl ester	EtOH	233(4.33),300(4.26)	87-0200-65
3H-Indol-3-one, 2-tert-butyl-5-methyl-, O-acetyloxime	EtOH	248(4.61),316(3.80)	32-1248-64
Ketone, methyl 3-(morpholinomethyl)-indol-5-yl	EtOH	251(4.55),296(3.94)	87-0141-64
$C_{15}H_{18}N_2O_3$			
2-Indolinone, 3-[1-(nitromethyl)-cyclohexyl]-	EtOH	251(3.89),283(3.19)	87-0626-65
3-Pyrroline-3-carboxylic acid, 2-(3,4-dimethyl-2-pyrrolylmethylene)-4-methyl-5-oxo-, ethyl ester	EtOH	282(4.10),434(4.40), 450(4.33)	39-5999-64
L-Tryptophan, N-acetyl-, ethyl ester	pH 4.1	275s(3.7),280(3.74), 290(3.7)	37-3580-65
$C_{15}H_{18}N_2O_3S$			
3-Pyrrolin-2-one, 3-acetamido-5-[(ben-zylthio)methyl]-4-hydroxy-1-methyl-	EtOH	210(4.30),270(3.73)	35-5654-64
3-Pyrrolin-2-one, 3-acetyl-5-[(benzyl-thio)methyl]-4-hydroxy-1-methyl-, 3-oxime	EtOH	248(3.80),311(4.15)	35-5654-64
$C_{15}H_{18}N_2O_4$			
Piperidine, 1-[4-methyl-5-(5-nitro-2-furyl)-2,4-pentadienoyl]-	propylene glycol	280(4.15),390(4.34)	95-0646-64
1-Pyrazoline-3,3-dicarboxylic acid, 4-phenyl-, diethyl ester	hexane	329(2.28)	28-4776-65B

Compound	Solvent	λ_{max}(log ε)	Ref.
Pyrrolo[2,3-b]pyridine-3,5-dicarboxylic acid, 2,6-dimethyl-, diethyl ester	EtOH-HCl	235(4.53),253s(4.31), 301(3.93)	39-1632-64
	EtOH	235(4.60),253s(4.26), 299(3.88)	39-1632-64
	EtOH-NaOH	241(4.41),248(4.41), 274(4.36),311(3.99)	39-1632-64
$C_{15}H_{18}N_2O_5$			
Carbamic acid ester with 1-ethyl-3-(hydroxymethyl)-5-methoxy-2,6-dimethylindole-4,7-dione	MeOH	231(4.26),286(4.14), 345(3.55),455(3.11)	35-3878-64
$C_{15}H_{18}N_2O_8$			
4(3H)-Pyrimidinone, 3-(2,3,5-tri-O-acetyl-β-D-ribofuranosyl)-	MeOH	220(3.75),274(3.58)	24-1511-65
$C_{15}H_{18}N_3OP$			
Phosphine oxide, tris(1-methylpyrrol-2-yl)-	EtOH	218(3.92),243(4.11)	44-0097-65
$C_{15}H_{18}N_4O_3$			
Benzoic acid, p-[[3-(2-amino-4-hydroxy-6-methyl-5-pyrimidinyl)propyl]amino]-	pH 1	227(4.33),267(4.04), 303(--)	87-0024-64
	pH 13	282(4.39)	87-0024-64
7-Pteridinol, 5,6-dihydro-6-(4,4-dimethyl-2,6-dioxocyclohexyl)-2-methyl-	pH 1.0	224(4.51),250s(4.20), 285(4.08),359(3.68)	39-6930-65
7-Pteridinol, 5,6-dihydro-6-(4,4-dimethyl-2,6-dioxocyclohexyl)-4-methyl-	pH 1.0	223(4.53),250s(4.16), 288(4.05),358(3.75)	39-6930-65
$C_{15}H_{18}N_4O_4$			
2-Norbornanone, 3,3-dimethyl-, 2,4-dinitrophenylhydrazone	$CHCl_3$	369(4.36)	28-4783-65A
3,5-Octadien-2-one, 7-methyl-, 2,4-dinitrophenylhydrazone	$CHCl_3$	394(4.49)	22-3218-64
4,5-Octadien-2-one, 7-methyl-, 2,4-dinitrophenylhydrazone	$CHCl_3$	369(4.29)	22-3218-64
5,7-Octadien-2-one, 6-methyl-, 2,4-dinitrophenylhydrazone	$CHCl_3$	229(4.02),362(4.57)	22-2533-64
cis	$CHCl_3$	365(4.35)	22-2533-64
Pyrazolo[1,5-a]pyrimidine-6-carboxylic acid, 7-(diacetylamino)-2,3-dimethyl-, ethyl ester	EtOH	256(4.64),300(3.52), 320s(3.39),348(3.15)	94-1207-65
$C_{15}H_{18}N_4O_5$			
Mitomycin C	H_2O	367(4.34)	87-0001-65
	MeOH	357(4.36),550(2.37)	87-0001-65
$C_{15}H_{18}N_4O_7S$			
Butyric acid, 3-methyl-2-oxo-4-thioacetoxy-, ethyl ester, 2,4-dinitrophenylhydrazone	EtOH	227(4.19),356(4.45)	39-0766-64
$C_{15}H_{18}O$			
Cacalol, 9-deoxy-	EtOH	216(4.54),258(4.07), 284(3.45),294(3.45)	78-2331-64
Lindestrene	EtOH	218(3.91)	78-2655-64
2-Naphthol, 4-isopropyl-1,6-dimethyl-	isooctane	234(4.83),264s(3.45), 281(3.74),292(3.69), 325s(3.39),329(3.43), 340(3.50),343s(3.45)	12-1111-65

Compound	Solvent	$\lambda_{max}(\log \epsilon)$	Ref.
2-Naphthol, 5-isopropyl-3,8-dimethyl-	isooctane	222(4.56),238(4.56), 266s(3.32),275s(3.53), 285(3.68),293(3.64), 297(3.65),304s(3.30), 317(3.26),331(3.34)	12-1111-65
4H-Pyran, 4-benzyl-2,4,6-trimethyl-	EtOH	230(4.26)	5-0183-64H
$C_{15}H_{18}OS$			
4-Cyclooctene-1-thiol, benzoate	C_6H_{12}	238(4.06),244(--), 270(3.86)	77-0151-65
$C_{15}H_{18}O_2$			
9β,10α-Anthracenediol, 1,4,4aα,9,9aα,10-hexahydro-9α-methyl-	EtOH	262(2.32)	70-1024-64
10β-isomer	EtOH	262(2.35)	70-1024-64
Cacalol	EtOH	218(4.48),256(4.02), 264(4.0),286(3.26)	78-2331-64
Coumarin, octahydro-8a-phenyl-	EtOH	254(3.24),259(3.20), 265(3.20)	39-5189-65
1,3-Cyclohexanedione, 2-benzyl-5,5-dimethyl-	EtOH	265(4.18)	78-0515-65
2-Cyclopentene-1-carboxylic acid, 3-methyl-2-phenyl-, ethyl ester	EtOH	248.5(3.57)	23-2408-65
$\Delta^{2(1H),\alpha}$-Naphthaleneacetic acid, 4a,5,6,7,8,8a-hexahydro-3-hydroxy-α,4a-dimethyl-8-methylene-, γ-lactone	EtOH	275(4.32)	94-0755-64
2,4-Pentadienoic acid, 2,3-dimethyl-5-phenyl-, ethyl ester	MeOH	227(4.04),235s(3.98), 300(4.42)	44-2986-64
Spiro[cyclohexane-1,1'-indan]-3'-one, 4'-methoxy-	EtOH	255(4.03),313(3.68)	94-1084-65
Spiro[cyclohexane-1,1'-indan]-3'-one, 6'-methoxy-	EtOH	225(4.19),270(4.17), 288(4.04),295s(4.0)	94-1084-65
$C_{15}H_{18}O_3$			
3α,10α-Anthracenediol, 2,9-epoxy-1,2,3,4,4aα,9,9aα,10-octahydro-9α-methyl-	EtOH	260(2.26)	70-1024-64
9β,10α-Anthracenediol, 2,3-epoxy-1,2,3,4,4aα,9,9aα,10-octahydro-9α-methyl-	EtOH	261(2.29)	70-1024-64
2(1H)-Anthracenone, 3,4,4aα,9,9aα,10-hexahydro-9β,10-dihydroxy-9α-methyl-	EtOH	261(2.46)	70-1013-64
Aromaticin	EtOH	215(4.19),320(1.70)	78-0079-64
Aromatin	EtOH	215(4.18),320(1.70)	78-0079-64
Cacalone, dihydro-	EtOH	278(4.12)	78-2331-64
Coronopilic acid	n.s.g.	203(4.10),310(4.21)	44-2553-64
Cyclohexanepropionic acid, 5-oxo-2-phenyl-	EtOH	255(2.20),260(2.29), 265(2.15)	39-5189-65
3-Cyclohexene-1-carboxylic acid, 6-(m-methoxybenzyl)-, trans	EtOH	208(4.32),280(3.50), 286(3.45)	70-0806-65
11-Epigeigerin, anhydro-	EtOH	299(4.19)	39-2518-64
1,4-Ethanonaphthalene-5,8-dione, 1,4,4a,8a-tetrahydro-1-methoxy-4,7-dimethyl-	EtOH	223(4.00),350(2.50)	39-2932-64
1,4-Ethanonaphthalene-5,8-dione, 1,4,4a,8a-tetrahydro-1-methoxy-4a,7-dimethyl-	EtOH	233.5(3.88)	39-2932-64
1,4-Ethanonaphthalene-5,8-dione, 1,4,4a,8a-tetrahydro-1-methoxy-6,8a-dimethyl-	EtOH	233(3.88),350(2.52)	39-2932-64

Compound	Solvent	λ_{max}(log ϵ)	Ref.
Isolinderalactone, dihydro-	EtOH	216(3.90)	39-4578-64
Linderalactone, dihydro-	EtOH	221(3.83)	39-4578-64
$\Delta^{2(1H),\alpha}$-Naphthaleneacetic acid, 3,4,4a,7,8,8a-hexahydro-3,3-dihydroxy-α,4a-dimethyl-8-methylene-, γ-lactone	EtOH	216(4.07)	78-2655-64
$C_{15}H_{18}O_4$			
3aH-Cyclopenta[b]benzofuran-3aα-carboxylic acid, 1,4aβ,7,8,8a,8b-hexahydro-6,8aβ,8bα-trimethyl-1-oxo-	EtOH	208(3.90)	1-1088-65
Malonic acid, (α-methylbenzylidene)-, diethyl ester	EtOH	254(3.94)	28-4776-65B
Santonin, α-hydroxy-	EtOH	240(3.97)	12-0543-65
Unknown acid, m. 206-7°	EtOH	225(3.89)	1-1088-65
$C_{15}H_{18}O_5$			
4-Cyclohexene-1,2-dicarboxylic anhydride, 4-(5-carboxy-4-pentenyl)-, methyl ester	heptane	200(2.95)	44-1061-65
$C_{15}H_{18}O_6$			
Lactulin decomposition product	EtOH	221.5(4.16)	83-0703-64
$C_{15}H_{18}O_8$			
1,3,5-Benzenetricarboxylic acid, 2,4-dihydroxy-, triethyl ester	EtOH	203(3.8),235(4.71), 256s(4.1),308(3.74)	39-1629-65
2H-Pyran-3,5-dicarboxylic acid, 6-ethoxycarbonylmethyl-2-oxo-, diethyl ester	C_6H_{12}	208(4.0),249(3.87), 318(3.75)	39-1629-65
	EtOH-NaOEt	216(3.8),262(4.00), 311s(4.0),320(4.13), 353(3.88),396(4.11)	39-1629-65
$C_{15}H_{18}S$			
6H-Benzo[b]cyclohepta[d]thiophene, 7,8,9,10-tetrahydro-1,4-dimethyl-	n.s.g.	234(4.47),268(3.87), 298(3.61),308(3.61)	22-0989-64
$C_{15}H_{18}S_2$			
Azulene, 4,6,8-trimethyl-1,3-bis(methylthio)-	C_6H_{12}	206(4.22),254(4.34), 326(4.34),344s(4.16), 393(3.88),628(2.61)	44-2715-65
$C_{15}H_{19}ClN_2O_4$			
[[1-[(Dimethylamino)methylene]inden-2-yl]methylene]dimethylammonium perchlorate	MeCN	389(4.34),470(4.00)	73-2783-65
[[1-[(Dimethylamino)methylene]inden-3-yl]methylene]dimethylammonium perchlorate	MeCN	454(4.66)	73-2783-65
$C_{15}H_{19}ClN_2O_6S$			
5-Thia-1-azabicyclo[4.2.0]oct-2-ene-2-carboxylic acid, 7-(5-chlorovaleramido)-3-(hydroxymethyl)-8-oxo-, acetate	pH 6.0	260.5(3.94)	39-5015-65
$C_{15}H_{19}FN_2O_6S$			
Uridine, 2'-deoxy-5-fluoro-5'-thio-, 3',5'-dipropionate	pH 1	266(3.93)	44-0554-64
	pH 7	266(3.89)	44-0554-64
	pH 13	266(3.84)	44-0554-64

Compound	Solvent	$\lambda_{max}(\log \epsilon)$	Ref.
$C_{15}H_{19}FeN$			
Pyrrolidine, N-(ferrocenylmethyl)-	MeCN	328(1.95),440(2.07)	44-2028-64
$C_{15}H_{19}IN_2$			
1,1-Dimethyl-2-(1-methylindol-3-yl)-2-pyrrolinium iodide	EtOH	212(4.53),243(4.11), 260s(3.85),303(4.11)	87-0415-64
$C_{15}H_{19}N$			
Pyrrole, N-benzyl-2,3,4,5-tetramethyl-	MeOH	208(4.07),226s(3.79)	44-3225-65
Quinoline, 3,4,4a,5,6,7,8,8a-octahydro-2-phenyl-, cis	EtOH	236(4.20)	95-0674-64
trans	EtOH	236.5(4.04)	95-0674-64
$C_{15}H_{19}NO$			
9a,4a-(Epoxyethano)carbazole, 1,2,3,4-tetrahydro-9-methyl-	hexane	255(3.9),315(3.3)	78-2047-64
	MeOH-HCl	225(3.9),275(3.9)	78-2047-64
	MeOH	250(4.2),300(3.4)	78-2047-64
	MeOH-NaOH	250(4.2),300(3.4)	78-2047-64
2,4-Nonadienanilide	EtOH	265(4.48)	54-0581-65
Pyrrolidine, 1-(3,4-dihydro-5-methoxy-2-naphthyl)-	EtOH	245(4.11),322(4.26)	88-3935-64
Spiro[cyclohexane-1,4'(3'H)-isoquinoline]-1'(2'H)-one, 2'-methyl-	EtOH	229(3.95),250s(3.74), 262s(3.6)	94-1084-65
Spiro[cyclohexane-1,4'(1'H)-quinoline]-2'(3'H)-one, 1'-methyl-	EtOH	254(4.06)	94-1084-65
$C_{15}H_{19}NO_2$			
1-Butanone, 2-ethyl-2-hydroxy-1-(1-methylindol-3-yl)-	EtOH	210(4.44),246(4.14), 306(4.16)	87-0094-64
Carbostyril, 3-isopentyl-4-methoxy-	n.s.g.	269(4.19),279(4.15), 321(3.95),334(3.80)	44-2598-64
Indole-3-acetic acid, 5-methyl-2-propyl-, methyl ester	MeOH	225(4.50),280(3.88), 287(3.89),297(3.79)	87-0204-65
Isoquinoline, 1-tert-butyl-6,7-dimethoxy-	MeOH	233(4.83),311(3.53), 323(3.60)	83-0561-65
1-Penten-3-one, 5-morpholino-1-phenyl-, hydrochloride	H_2O	293(4.17)	7-0652-65
Spiro[cyclohexane-1,4'(3'H)-isoquinolin]-1'(2'H)-one, 6'-methoxy-	EtOH	261(4.12)	94-1084-65
Spiro[cyclohexane-1,4'(3'H)-isoquinolin]-1'(2'H)-one, 8'-methoxy-	EtOH	241(3.81),301(3.65)	94-1084-65
Spiro[cyclohexane-1,4'(1'H)-quinolin]-2'(3'H)-one, 6'-methoxy-	EtOH	261(4.12)	94-1084-65
$C_{15}H_{19}NO_3$			
2H-Azepin-2-one, 1-[(benzyloxy)acetyl]-hexahydro-	dioxan	221(4.07)	78-3537-65
2H-Benzo[a]quinolizine-3-carboxylic acid, 1,3,4,6,7,11b-hexahydro-9-methoxy-, cis	EtOH	227(3.96),277(3.24), 285(3.23)	23-1527-65
trans	EtOH	227(3.96),277(3.25), 285(3.24)	23-1527-65
Benzosuberane, 1-acetoxy-3-acetamido-	MeOH	248(4.32),280(3.81)	33-1988-65
Carbostyril, 4-hydroxy-3-isopentyl-8-methoxy-	MeOH	243(4.43),280(3.93), 288(3.94),319(3.62), 331(3.44)	39-0438-64
Geigerin, anhydro-, oxime	EtOH	288(4.36)	39-2518-64
Indole-3-carboxylic acid, 1-ethyl-5-hydroxy-2,6-dimethyl-, ethyl ester	MeOH	218(4.59),245(4.17), 298(4.10)	35-3878-64
Morpholine, 4-(5,6,7,8-tetrahydro-3-hydroxy-2-naphthoyl)-	$CHCl_3$	295(3.54)	94-0112-64

Compound	Solvent	λ max(log ε)	Ref.
2-Piperidone, 1-[2-(benzyloxy)-propionyl]-	dioxan	219(4.03)	78-3537-65
2-Piperidone, 1-[3-(benzyloxy)-propionyl]-	dioxan	214(4.05)	78-3537-65
$C_{15}H_{19}NO_4$			
3-Indolineacetic acid, 3-methoxy-2-oxo-, tert-butyl ester	EtOH	208(4.44),253(3.75), 293(3.14)	44-2431-64
$C_{15}H_{19}NO_5$			
2-Azabicyclo[3.2.2]nona-6,8-diene-6,7-dicarboxylic acid, 1,4,9-tri-methyl-3-oxo-, dimethyl ester	EtOH	212(2.40),264s(3.11)	44-3447-64
$C_{15}H_{19}NO_6$			
Carbostyril, 3-(2-hydroxyethyl)-4,5,7,8-tetramethoxy-	n.s.g.	235s(4.23),258s(4.28), 266(4.36),315(4.11), 330s(4.03)	2-0071-65
$C_{15}H_{19}NO_7$			
2H-Pyran-3,5-dicarboxylic acid, 6-amino-2-(ethoxycarbonylmethylene)-, diethyl ester	EtOH	203(4.0),221(4.00), 255(4.12),312(4.33), 322(4.32),393(4.18)	39-1629-65
3,5-Pyridinedicarboxylic acid, 6-(ethoxycarbonylmethyl)-1,2-di-hydro-2-oxo-, diethyl ester	EtOH	208(4.3),262(4.19), 332(3.98)	39-1629-65
$C_{15}H_{19}N_3O$			
4-Pyrimidinol, 2-amino-6-methyl-5-(4-phenylbutyl)-	pH 1	266(4.02)	4-0088-64
	pH 13	280(3.97)	4-0088-64
$C_{15}H_{19}N_3O_3$			
Pyrimidine, 2,4,6-trimethoxy-5-(N-methyl-p-toluidino)-	EtOH	249(4.28),285(3.90)	39-5556-65
$C_{15}H_{19}N_3O_4$			
Valeric acid, 2-acetyl-4-hydroxy-5-(o-tolyloxy)-, γ-lactone, semicarbazone	n.s.g.	222(4.24),274(3.82)	39-0141-64
$C_{15}H_{19}N_3O_5$			
Valeric acid, 2-acetyl-4-hydroxy-5-(p-methoxyphenoxy)-, γ-lactone, semicarbazone	n.s.g.	222(4.36),274(3.83)	39-0141-64
$C_{15}H_{19}N_3O_5S$			
Indole-2-carboxylic acid, 3-[(butyl-carbamoyl)sulfamoyl]-1-methyl-	EtOH	212(4.53),222s(4.47), 282s(4.00),286(4.03), 300s(3.77)	44-0178-64
$C_{15}H_{19}N_3O_9$			
s-Triazin-2(1H)-one, 1-(2,3,4-tri-O-acetyl-β-D-ribofuranosyl)-4-methoxy-	MeCN	253(3.36)	73-2060-64
s-Triazin-2(1H)-one, 1-(2,3,4-tri-O-acetyl-β-D-ribopyranosyl)-4-methoxy-	MeCN	253(3.39)	73-2060-64
$C_{15}H_{19}N_5O_2$			
4-Pyrimidinecarboxylic acid, 2-amino-6-hydroxy-5-(4-phenylbutyl)-, hydrazide	pH 1	284(3.73)	4-0162-65
	pH 7	302(3.83)	4-0162-65
	pH 13	290(3.69)	4-0162-65

Compound	Solvent	$\lambda_{max}(\log \epsilon)$	Ref.
$C_{15}H_{19}O_6P$			
Butyric acid, 2-benzoyl-4-hydroxy-2-phosphono-, γ-lactone, diethyl ester	MeOH	217(3.90),253(3.29)	5-0010-65E
Cinnamic acid, β-hydroxy-α-(2-hydroxyethyl)-, γ-lactone, diethyl phosphate	MeOH	207(4.05),267(4.16)	5-0010-65E
$C_{15}H_{20}$			
Laurene	EtOH	253(2.45),259(2.45), 265(2.45),274(2.38)	88-3619-65
1,3,5,7,9,11-Tridecahexaene, 2,12-dimethyl-	isooctane	322(4.50),338(4.85), 356(5.09),376(5.11)	35-5075-65
$C_{15}H_{20}N_2$			
Benz[cd]indole, 3-[(dimethylamino)-methyl]-1,3,4,5-tetrahydro-2-methyl-	EtOH	227(4.51),274(3.84), 280(3.84),290(3.71)	44-0843-64
1,4-Diazaspiro[5.5]undec-4-ene, 5-phenyl-	EtOH	257(2.63),263(2.54)	44-3146-64
Dipyrromethene, 3,3',4,4',5,5'-hexa-methyl-	EtOH-HCl	229(3.89),362(3.59), 480(4.80)	12-0363-65
	EtOH	224(4.08),280s(3.39), 325(3.63),442(4.48)	12-0363-65
	CCl_4	444(4.42)	12-0363-65
Indole, 3-(1,5,5-trimethyl-2-pyrrolidinyl)-	EtOH	219(4.41),274s(3.75), 279(3.77),287s(3.69)	87-0415-64
$C_{15}H_{20}N_2O$			
2-Indolinone, 3-[1-(aminomethyl)-cyclohexyl]-	EtOH	238(3.86),291(3.41)	87-0626-65
Indol-5-ol, 3-[(1-methyl-2-piperidyl)-methyl]-	EtOH	223(4.43),280(3.86), 302(3.76)	44-2860-64
Ketone, bis(3,4,5-trimethylpyrrol-2-yl)	EtOH	302(3.89),363(4.31)	12-1977-65
Ketone, 1,3,5-trimethylpyrrol-2-yl 1,4,5-trimethylpyrrol-2-yl	EtOH	274(3.42),300(3.10), 369(4.04)	12-1977-65
1H-Pyrido[3,4-b]indol-7-ol, 3-ethyl-2,3,4,9-tetrahydro-1,1-dimethyl-	EtOH	225(4.55),269(3.65), 298(3.75)	44-2864-64
3-Pyrrolin-2-one, 4-ethyl-3-methyl-5-[(3,4,5-trimethyl-2-pyrrolyl)-methylene]-	EtOH	233(3.94),272(3.58), 418(4.52)	39-5999-64
$C_{15}H_{20}N_2O_2$			
Benz[cd]indole, 3-(aminomethyl)-1,3,4,5-tetrahydro-2-methyl-, acetic acid salt	EtOH	226(4.54),274(3.86), 279(3.86),290(3.71)	44-0843-64
$C_{15}H_{20}N_2O_3$			
Valeric acid, 5-butoxy-4-hydroxy-2-oxo-, γ-lactone, phenylhydrazone	n.s.g.	233(4.17),295(3.97), 333(4.35)	39-0141-64
$C_{15}H_{20}N_2O_6S$			
5-Thia-1-azabicyclo[4.2.0]oct-2-ene-2-carboxylic acid, 3-(hydroxymethyl)-8-oxo-7-valeramido-, acetate	pH 6.0	260(3.97)	39-5015-65
$C_{15}H_{20}N_2O_7S$			
5-Thia-1-azabicyclo[4.2.0]oct-2-ene-2-carboxylic acid, 3-(hydroxymethyl)-8-oxo-7-valeramido-, acetate, 5-oxide	pH 6.0	258(4.03)	39-5015-65

Compound	Solvent	λ max(log ε)	Ref.
$C_{15}H_{20}N_4O_4$			
Cyclobutanone, 2,2,3,4,4-pentamethyl-, 2,4-dinitrophenylhydrazone	$CHCl_3$	367(4.45)	22-1968-64
$C_{15}H_{20}N_4O_6S$			
Isoleucine, N-[2-(7-nitro-4H-1,2,4-benzothiadiazin-3-yl)ethyl]-, S,S-dioxide	2% $NaHCO_3$	241(4.01),358(4.11)	32-0786-65
Leucine, N-[2-(7-nitro-4H-1,2,4-benzo-thiadiazin-3-yl)ethyl]-, S,S-dioxide	2% $NaHCO_3$	242(4.00),358(4.12)	32-0786-65
$C_{15}H_{20}N_4O_7$			
Glycine, N-[N-(dinitrophenyl)-L-isoleucyl]-, methyl ester	EtOH	341(4.22)	35-1848-64
	HOAc	339(4.17)	35-1848-64
$C_{15}H_{20}O$			
Atractylon	EtOH	220(3.89)	94-0755-64
2(3H)-Naphthalenone, 4,4a,5,6-tetra-hydro-5-isopropenyl-3,8-dimethyl-	isooctane	272(4.37)	12-1111-65
$C_{15}H_{20}OS_2$			
2,5-Cyclohexadien-1-one, 4-(1,3-di-thiolan-2-ylidene)-2,6-diisopropyl-	MeOH	258(3.78),300(3.65), 414s(4.50),430(4.55), 450s(4.38)	5-0037-65D
$C_{15}H_{20}O_2$			
Cacalol, dihydro-	EtOH	288(3.28)	78-2331-64
Cinnamic acid, α-butyl-, ethyl ester	EtOH	267(4.24)	95-0757-65
α-Cyclocostunolide	EtOH	210(4.0)(end absorption)	78-2639-64
2,4,6,8,10-Dodecapentaenoic acid, 11-methyl-, ethyl ester	isooctane	347(4.76),363(4.92), 384(4.85)	35-5075-65
10-Epi-5αH,6,11βH-eudesma-1,3-dien-6,13-olide	MeOH	265(3.68)	35-5736-65
6-Heptenoic acid, 7-phenyl-, ethyl ester, cis	EtOH	247(4.04)	22-2988-65
Inden-7-ol, 5-(2-methoxyethyl)-2,4,6-trimethyl-	EtOH	266(3.97),300(3.43), 311(3.42)	35-1594-65
Ligularone	EtOH	269(3.51)	78-2605-65
Mansonone A	EtOH	209(4.32),430(2.93)	88-4857-65
$\Delta^{2(1H),\alpha}$-Naphthaleneacetic acid, octa-hydro-3-hydroxy-α,4a-dimethyl-8-methylene-, γ-lactone	EtOH	220(4.21),276(1.41)	94-0755-64
	EtOH-NaOH	231(3.74)	94-0755-64
2-Naphthaleneacetic acid, 1,2,3,4,4a,8a-hexahydro-1α-hydroxy-α,4aβ,8-tri-methyl-, γ-lactone	MeOH	262(3.69)	35-5736-65
2-Naphthoic acid, 3,4,4a,5,6,7,8,8a-octahydro-5-methylene-8-vinyl-, methyl ester	EtOH	217(4.02)	78-2617-64
1,15-Pentadeca-10,12-diynolide	EtOH	228(2.91),238(2.83), 254(2.46)	78-1773-64
2,4-Pentadienoic acid, 3-methyl-5-(2,6,6-trimethyl-1,3-cyclo-hexadien-1-yl)-	n.s.g.	255(4.11),334(4.26)	24-0549-64
2-Pentenoic acid, 4-hydroxy-3-methyl-5-(2,6,6-trimethyl-2-cyclohexen-1-ylidene)-, γ-lactone	n.s.g.	243(4.16)	24-0549-64
$C_{15}H_{20}O_3$			
Asperilin	EtOH	211(3.94)	44-1022-64
Carabrone	EtOH	213(3.91)	39-5503-64

Compound	Solvent	λ_{max}(log ϵ)	Ref.
Carabrone (contd.)	n.s.g.	213(3.91)	77-0120-64
1-Epigeigerin, deoxy-	EtOH	238(4.15)	39-2518-64
11-Epigeigerin, deoxy-	EtOH	237(4.23)	39-2518-64
Eremophilenolide, oxo-	n.s.g.	240(4.03),328(1.90)	73-2189-64
Eremophilenolide, 3-oxo-	n.s.g.	220(4.199)	73-2182-64
Geigerin, ahydrodihydro-	EtOH	238(4.22)	39-2518-64
Illudin M	EtOH	228(4.14),318(3.56)	35-1594-65
Isoaromaticin, dihydro-	EtOH	220(4.04),291(1.65)	78-0079-64
Isoaromatin, dihydro-	EtOH	218(4.20),290(1.69)	78-0079-64
Isotenulin, anhydrodeacetyldihydro-	EtOH	219(4.17)	44-1549-64
Mansonone B	EtOH	226(3.68),272(3.69), 408(2.4)	88-4857-65
Naphthalene, 3,4-dihydro-1-(1,1-dimethoxyethyl)-6-methoxy-	EtOH	270(4.10),300s(3.26)	44-2527-64
$\Delta^{2(1H),\alpha}$-Naphthaleneacetic acid, octahydro-3,3-dihydroxy-α,4a-dimethyl-8-methylene-, γ-lactone	EtOH	220(4.19)	94-0755-64
	EtOH-NaOH	264(3.73)	94-0755-64
Neotrichodermone	EtOH	204(4.01),215s(--)	1-1088-65
Parthenolide	EtOH	214(4.22)	78-1509-65
	n.s.g.	214(4.22)	88-3927-64
Pentadeca-10,12-diyn-1,15-olide, 3-hydroxy-	n.s.g.	226(2.56),240(2.52), 254(2.37)	70-0860-64
Santamarin	EtOH	208(3.91)	78-1741-65
2,4,6,8,10-Tetradecapentaenal, 12,13-dihydroxy-2-methyl-	$CHCl_3$	374(4.89),390(4.88)	39-0842-64
Trichodermone	EtOH	205(3.45)	1-1088-65
$C_{15}H_{20}O_4$			
Acetophenone, 2',4'-dimethoxy-6'-[(3-methyl-2-butenyl)oxy]-	MeOH	224(4.11),279(3.69)	78-2653-65
Acetophenone, 2'-hydroxy-4',6'-dimethoxy-3'-(1,1-dimethylallyl)-	MeOH	294(4.30),334s(3.49)	78-2653-65
Acetophenone, 6'-hydroxy-2',4'-dimethoxy-3'-(3-methyl-2-butenyl)-	MeOH	294(4.30),326s(3.61)	78-2653-65
5-Azuleneacetic acid, 3,3a,4,5,6,7,8,8a-octahydro-2,6-dihydroxy-α,3a,8-trimethyl-3-oxo-, γ-lactone	EtOH	261(3.85)	44-1549-64
	acid	260(3.75)	44-1549-64
	base	300(3.78)	44-1549-64
p-Benzoquinone, 2,5-dihydroxy-3-methyl-6-(1,2,2-trimethylcyclopentyl)-	EtOH	210(4.13),297(4.15), 377(2.61),430(2.47)	94-0511-65
p-Benzoquinone, 2-(1,5-dimethyl-4-hexenyl)-3,6-dihydroxy-5-methyl-	EtOH	297(4.15),430(2.47)	78-2319-64
1,2,4-Cyclopentanetrione, 3-(3-methyl-2-butenyl)-5-pivaloyl-	EtOH-HCl	255(4.40),280(4.26)	39-2251-65
	EtOH-NaOH	275(4.41),303(4.42)	39-2251-65
1-Epigeigerin	EtOH	238(4.13)	39-2518-64
Helicobasidin	EtOH	297(4.15),377(2.61), 430(2.47)	94-0236-64
Hulupinic acid	EtOH-acid	301(4.01)	39-0952-64
	EtOH-base	261(4.16),393(4.09)	39-0952-64
Illudin S	MeOH	235(4.10),320(3.54)	78-1231-65
	MeOH	235(4.10),320(3.54)	94-0853-64
	EtOH	233(4.12),319(3.56)	35-1594-65
Isoilludin S	EtOH	252(4.31)	78-1231-65
	EtOH	252(4.31)	94-0853-64
Isolampterol	EtOH	252(4.3)	78-2671-65
Ivasperin	EtOH	210(3.89)	44-1022-64
2-Naphthaleneacetic acid, 1,2,3,4,4a,-5,8,8a-octahydro-1,8-dihydroxy-α,4a,8-trimethyl-5-oxo-, γ-lactone	EtOH	215(3.95),330(1.76)	78-1741-65
Pseudoivalin, 1,10-epoxy-	EtOH	211(3.87)	44-0118-65
Pulchellin C, didehydrotetrahydro-	EtOH-base	328(3.32)	78-0341-64

Compound	Solvent	λ_{max}(log ε)	Ref.
Santamarin, epoxide	EtOH	211(3.93)	78-1741-65
$C_{15}H_{20}O_5$			
3,5-Cyclohexadien-1-one, 2,2-diacetoxy-5-tert-butyl-	C_6H_{12}	244(4.12),330(3.46)	39-2904-65
2(3H)-Naphthalenone, 4,4a,5,6,7,8-hexahydro-4a,5-dihydroxy-1-methyl-, diacetate, cis	EtOH	247(4.14)	44-1626-65
Psilostachyin	n.s.g.	212(4.10)	88-3397-65
$C_{15}H_{20}O_6$			
3,5-Cyclohexadien-1-one, 2,2-diacetoxy-6-tert-butyl-4-methoxy-	C_6H_{12}	242(3.94),330(3.44)	39-2904-65
2H-Pyran-2-one, 3,3'-methylenebis-[5,6-dihydro-4-hydroxy-5,5-dimethyl-	n.s.g.	247(4.19)	22-0651-64
Scirpen-8-one, 3a,4β,15-trihydroxy-	EtOH	224(3.95),319(1.38)	33-0962-65
$C_{15}H_{20}O_7$			
3-Octene-1,2,3,6-tetracarboxylic acid, 5-ethyl-2-methyl-, 2,3-anhydride	EtOH	224(3.88)	39-1779-65
isomer	EtOH	222(3.93)	39-1779-65
$C_{15}H_{20}O_8S$			
Glucopyranose, 1-deoxy-1-(p-tolylsulfonyl)-, 6-acetate	MeOH	225(4.18)	44-1782-64
$C_{15}H_{21}BrO$			
Cyclocolor-4-en-3-one, 2α-bromo-	EtOH	271(4.16)	44-1744-65
$C_{15}H_{21}FO$			
2,4-Pentadienal, 2-fluoro-3-methyl-5-(2,6,6-trimethyl-1-cyclohexen-1-yl)-	MeOH	276(3.29),329(3.29)	5-0020-64I
$C_{15}H_{21}N$			
Indole, 2-tert-butyl-3,5,7-trimethyl-	MeOH	227(3.85),283(3.85), 291(3.74)	22-0101-64
1-Naphthalenepropylamine, 3,4-dihydro-N,N-dimethyl-	EtOH	261(3.95)	44-1419-64
$C_{15}H_{21}NO$			
Spiro[cyclohexane-1,4'(1'H)-isoquinoline], 2',3'-dihydro-6'-methoxy-	EtOH	280(3.48),286s(3.4)	94-1084-65
hydrochloride	EtOH	224(3.96),278(3.37), 286(3.34)	94-1084-65
Spiro[cyclohexane-1,4'(1'H)-isoquinoline], 2',3'-dihydro-8'-methoxy-	EtOH	272(3.23),279(3.24)	94-1084-65
hydrochloride	EtOH	218(3.88),273(3.27), 280(3.27)	94-1084-65
$C_{15}H_{21}NO_2$			
2H-Benzo[a]quinolizine, 1,3,4,6,7,11b-hexahydro-9,10-dimethoxy-	EtOH	287(3.50)	95-0412-64
Isoquinoline, 3,4-dihydro-6,7-dimethoxy-1-tert-butyl-, hydroiodide	MeOH	305(3.87),354(3.69)	83-0561-65
2-Naphthol, 5,6,7,8-tetrahydro-3-(morpholinomethyl)-	$CHCl_3$	285(3.51)	94-0112-64

Compound	Solvent	$\lambda_{max}(\log \epsilon)$	Ref.
$C_{15}H_{21}NO_3$			
D-Alloisoleucine, N-salicylidene-, ethyl ester	EtOH	216(4.34),221s(4.31), 257(4.13),319(3.63), 406(2.40)	35-1757-65
11-Epigeigerin, deoxy-, oxime	EtOH	211(4.28)	39-2518-64
Isoleucine, N-salicylidene-, ethyl ester	EtOH	216(4.33),221s(4.30), 257(4.12),319(3.63), 407(2.36)	35-1757-65
Leucine, N-salicylidene-, ethyl ester	EtOH	216(4.33),223s(4.31), 257(4.12),321(3.63), 408(2.27)	35-1757-65
1-Penten-3-one, 5-(dimethylamino)-4-methoxy-1-(p-methoxyphenyl)-, hydrochloride	MeOH	237(3.93),331(4.39)	44-2334-65
Pipitzol, oxime	n.s.g.	278(3.87)	88-1577-65
$C_{15}H_{21}NO_4$			
Dehydrocycloheximide	EtOH-NaOH	315(4.25)	44-2595-64
$C_{15}H_{21}NO_5$			
Alanine, N-acetyl-3-(3,4-dimethoxyphenyl)-2-methyl-, methyl ester	MeCN	233(3.95),280(3.47)	44-1424-64
$C_{15}H_{21}N_3O$			
1H-Indazole, 3-methoxy-1-(2-piperidinoethyl)-	EtOH-HCl	222(4.35),250(3.17), 307(3.70)	33-1986-64
1H-Indazole, 1-methyl-3-(2-piperidinoethoxy)-	EtOH-HCl	220(4.36),250(3.11), 308(3.68)	33-1986-64
1H-Indazole, 3-(3-piperidinopropoxy)-	EtOH	217(4.45),253(3.17), 301(3.69)	33-1986-64
	EtOH-NaOH	215(4.41),247(3.10), 302(3.69)	33-1986-64
3-Indazolinone, 1-methyl-2-(2-piperidinoethyl)-	EtOH	221(4.30),242(4.08), 317(3.64)	33-1986-64
Indazolinone, 1-(3-piperidinopropyl)-	EtOH	223(4.34),315(3.61)	33-1986-64
3-Indazolinone, 2-(3-piperidinopropyl)-	EtOH	218(4.32),235(4.15), 312(3.63)	33-1986-64
	EtOH-NaOH	216(4.57),233(4.48), 350(3.68)	33-1986-64
Khusilal, semicarbazone	EtOH	265(4.46)	78-2617-64
$C_{15}H_{21}N_3O_2$			
Physostigmine (or eserine)	EtOH-6N HCl	237(3.78),294(3.76)	39-5510-64
	EtOH-0.1N HCl	246(4.05),303(3.47)	39-5510-64
	EtOH	254(4.07),312(3.47)	39-5510-64
$C_{15}H_{21}N_5O_6$			
7(8H)-Pteridinone, 4-(dimethylamino)-8-β-D-glucopyranosyl-6-methyl-	MeOH	210(4.20),220s(4.17), 252(4.20),297(3.78), 352(3.96)	4-0023-64
$C_{15}H_{21}N_5O_6S_2$			
5-Thia-1-azabicyclo[4.2.0]oct-2-ene-2-carboxylic acid, 3-[(amidinothio)-methyl]-7-(5-amino-5-carboxyvaleramido]-8-oxo-	pH 6.0	261(3.91)	39-5015-65
$C_{15}H_{21}N_5O_8$			
L-Lysine, N^6-(2-amino-2-carboxyethyl)-N^6-(2,4-dinitrophenyl)-	n.s.g.	264(3.88),350(4.16)	24-1164-65

Compound	Solvent	λ_{max}(log ε)	Ref.
$C_{15}H_{21}N_7O_6S$			
Arginine, N^2-[2-(7-nitro-4H-1,2,4-benzothiadiazin-3-yl)ethyl]-, S,S-dioxide	2% $NaHCO_3$	241(4.01),358(4.08)	32-0786-65
$C_{15}H_{22}$			
Benzsuberane, 1,1,4,4-tetramethyl-	C_6H_{12}	262(2.38),268(2.27), 271s(2.14)	23-0579-64
Naphthalene, 1,2,3,4-tetrahydro-1-isopropyl-3,3-dimethyl-	n.s.g.	260(2.97),267(3.01), 274(2.97)	35-3135-65
Naphthalene, 1,2,3,4-tetrahydro-4-isopropyl-1,1-dimethyl-	C_6H_{12}	265(2.52),273(2.43)	23-0579-64
Naphthalene, 1,2,3,4-tetrahydro-4-isopropyl-1,6-dimethyl-	EtOH	269(2.80),278(2.71)	78-2911-64
$C_{15}H_{22}BrNO_3$			
2,5-Cyclohexadien-1-one, 4-bromo-2-tert-butyl-4-nitro-6-(tert-pentyl)-	hexane	245(3.97)	70-0371-64
$C_{15}H_{22}INO$			
1-(2-Benzoylethyl)-1-methylpiperidinium iodide	H_2O	228(4.21),246s(4.15)	7-0968-65
1,2,3,4,4a,9a-Hexahydro-8-methoxy-9,9-dimethylcarbazolium iodide	EtOH	221(4.35),276(3.65)	56-1437-65
$C_{15}H_{22}N_2$			
Gramine, 5-methyl-2-propyl-	MeOH	276(3.88),284(3.88), 295(3.76)	87-0204-65
$C_{15}H_{22}N_2O$			
Aphyllidine	EtOH	240(4.08)	23-0764-64
Monospessulanine	MeOH	241.5(4.05)	39-4613-64
	EtOH	241(4.06)	88-2433-65
α-Obscurine	MeOH	255(3.75)	23-2456-64
$C_{15}H_{22}N_2O_2$			
Argyrolobine	EtOH	239(3.97)	23-0764-64
Cinnamylamine, N,N-diisopropyl-p-nitro-	MeOH	305(4.21)	78-1057-64
$C_{15}H_{22}N_2O_4$			
Valeric acid, 5-butoxy-4-hydroxy-2-oxo-, phenylhydrazone	n.s.g.	293(4.18),317(4.24)	39-0141-64
$C_{15}H_{22}N_2O_5S$			
Gramine, 5-acetyl-, methyl sulfate	EtOH	243(4.60),286(4.01)	87-0141-64
$C_{15}H_{22}N_2O_6$			
Malonic acid, [(4,4-dimethyl-2,6-dioxocyclohexyl)azo]-, diethyl ester	CH_2Cl_2	293(3.89),376(4.23)	89-0920-64
$C_{15}H_{22}N_6S$			
Urea, 1-[bis(isopropylidenehydrazino)-methylene]-2-thio-3-p-tolyl-	n.s.g.	255(4.46),295(4.39)	39-4448-65
$C_{15}H_{22}N_8O_7$			
2,6-Pyridin dicarbamic acid, 4-(acetonylamino)-3-nitro-, diethyl ester, semicarbazone	pH 1	224(4.45),253(4.42), 275(4.23),331(4.03)	44-0734-64
	pH 13	378(3.87)	44-0734-64

Compound	Solvent	λ_{max}(log ε)	Ref.
$C_{15}H_{22}O$			
Atractylon, dihydro-	EtOH	221(3.82)	94-0755-64
2(1H)-Azulenone, 6,7,8,8a-tetra-hydro-5-isopropyl-3,8-dimethyl-	EtOH	296(4.3)	39-1119-64
Costal	EtOH	216(4.24)	78-1521-65
3-Cyclohexen-1-ol, 3-[2-(1-cyclohex-en-1-yl)vinyl]-4-methyl-, cis	n.s.g.	240(4.06)	54-1173-64
trans	n.s.g.	266(--),277(4.36), 288(--)	54-1173-64
Cyperotundone	EtOH	245(3.96)	94-0628-65
7βH-Eudesma-4,11-dien-3-one	EtOH	252.0(4.15)	39-1029-64
3,5-Hexadien-2-one, 6-(3,3-di-methyl-2-norbornyl)-	EtOH	283(4.49)	28-4783-65A
$\Delta^{9(11)}$-Hinesene, 4-oxo-	EtOH	240(4.11)	94-1430-65
β-Ionylideneacetaldehyde	MeOH	273(4.29),326(4.34)	5-0020-64I
	iso-PrOH	285(4.00),325(4.05)	28-2466-64B
Isopatchoulenone	n.s.g.	243(4.15)	88-4053-65
Lanceal	EtOH	230(3.73)	78-0333-64
Mustakone	EtOH	255(3.72)	78-0607-65
1-Naphthaleneethanol, 5,6,7,8-tetra-hydro-α,4,7-trimethyl-	EtOH	264(2.30)	78-2911-64
2(1H)-Naphthalenone, 4a,5,6,7-tetra-hydro-5-isopropyl-3,8-dimethyl-	EtOH	244(4.02),278(3.87)	78-2911-64
2-Naphthol, 5,6,7,8-tetrahydro-	EtOH	283.0(3.6)	39-3319-65
5-isopropyl-3,8-dimethyl-	EtOH	274(3.29),281(3.26)	78-2911-64
$C_{15}H_{22}O_2$			
Benzaldehyde, 3,5-di-tert-butyl-	pH 7.00	290(4.2)	70-0293-64
4-hydroxy-	pH 7.60	290(4.0),367(4.0)	70-0293-64
	pH 8.2	290(3.7),367(4.4)	70-0293-64
	pH 10.0	260(3.9),367(4.5)	70-0293-64
Bicyclo[5.3.0]deca-1(2),7(8)-dien-9-one, 5-(1-hydroxy-1-methylethyl)-2,8-dimethyl-	EtOH	298(3.9)	39-1119-64
Costic acid	EtOH	210(3.70)	78-1521-65
Eudesm-4-ene-3,6-dione	EtOH	254(4.09)	78-1455-64
Furopelargone A	EtOH	219(3.87)	78-1789-64
Furopelargone B	EtOH	219(3.73)	78-1789-64
Lanceolic acid	hexane	219(4.11)	78-0333-64
Ligularol	EtOH	220(3.82)	78-2605-65
$\Delta^{2(1H),\alpha}$-Naphthaleneacetic acid, octahydro-3-hydroxy-α,4a,8-tri-methyl-, γ-lactone	EtOH	221(3.80)	94-0755-64
2-Naphthol, 1,2,3,4-tetrahydro-7-meth-oxy-1,1,3,3-tetramethyl-	EtOH	220(3.87),281(3.32), 288(3.29)	39-0361-65
4-Pentynoic acid, 3-methyl-5-(2,6,6-trimethylcyclohex-1-enyl)-	EtOH	235(4.08),245(3.88)	22-2541-64
Petasalbin	n.s.g.	220(3.836)	73-2189-64
Spiro[1,3-dioxolan-2,1'(2'H)-naphtha-lene], 6'-ethylidene-3',4',6',7',-8',8'a-hexahydro-8'a-methyl-	EtOH	235(4.34),243(4.36)	35-0478-64
$C_{15}H_{22}O_3$			
p-Benzoquinone, 5-(1,5-dimethylhexyl)-3-hydroxy-2-methyl-	EtOH	270(4.08),406(2.99)	94-0236-64
p-Benzoquinone, 3-hydroxy-2-methyl-5-octyl-	EtOH	270(4.10),406(2.99)	94-0236-64
Carabrone, dihydro-	EtOH	221(4.00)	39-5503-64
Carbonic acid, ethyl 1,4a-dimethyl-3,4,4a,5,6,7-hexahydro-2-naphthyl ester	EtOH	237(4.25)	78-2641-65

Compound	Solvent	λ_{max}(log ε)	Ref.
2-Cyclohexene-1-carboxylic acid, 3-allyl-2,6,6-trimethyl-4-oxo-, ethyl ester	n.s.g.	252(3.94)	39-1511-64
1,3-Cyclopentanedione, 4-isopentylidene-2-isovaleryl-	EtOH-acid	238(4.14),305(4.01)	39-1276-65
	EtOH-base	240(4.11),270(4.23)	39-1276-65
Eremophilenolide, hydroxy-	n.s.g.	220(4.120)	73-2189-64
2-Hexenoic acid, 2-methyl-6-(4-methyl-3-cyclohexen-1-yl)-6-oxo-, methyl ester	hexane	217(4.06)	78-0333-64
$\Delta^{2(1H),\alpha}$-Naphthaleneacetic acid, octahydro-3,3-dihydroxy-α,4a,8-trimethyl-, γ-lactone	EtOH	221(4.10)	94-0755-64
3-Pentenoic acid, 2-hydroxy-3-methyl-5-(2,6,6-trimethyl-2-cyclohexen-1-ylidene)-	n.s.g.	287(4.52)	24-0549-64
2(3H)-Phenanthrone, 4,4a,4b,5,6,7,8,8a-9,10-decahydro-5,8-dihydroxy-4a-methyl-, 4,4a-cis, (-)-(1S,10aS)	EtOH	237(4.24)	33-1725-65
(+)-(1S,10aR)	EtOH	241(4.21)	33-1725-65
Spiro[azulene-5(8H),1'-cyclopropane]-8-one, 1,2,3,3a,4,8a-hexahydro-3-hydroxy-2-(hydroxymethyl)-2,4-dimethyl-	EtOH	253.5(4.21)	78-2671-65
Spiro[cyclopropane-1,5'-[5H]indene]-3',6',7'-triol, 2',3',6',7'-tetrahydro-2',2',4',6'-tetramethyl-	EtOH	256(4.35)	35-1594-65
$C_{15}H_{22}O_4$			
o-Anisic acid, 6-heptyl-4-hydroxy-	EtOH	245s(3.60),282(3.40)	1-1677-65
p-Anisic acid, 2-heptyl-6-hydroxy-	EtOH	262(4.20),302(3.70)	1-1677-65
Carbonic acid, ethyl 4-(4-methoxy-o-tolyl)butyl ester	EtOH	226(3.93),277(3.28), 284(3.25)	35-5148-65
3,5-Cyclooctadiene-1-malonic acid, diethyl ester	MeOH	229(3.69)	35-3273-65
Humulinic acid C	EtOH-acid	225(4.15),265(4.00)	39-1276-65
	EtOH-base	250(4.26),269(4.18)	39-1276-65
Malonic acid, (2,3-dimethyl-2-penten-4-ynyl)methyl-, diethyl ester	EtOH	230(4.13)	70-1653-64
Microcephalin	EtOH	212(3.86)	44-1700-64
11-Oxabicyclo[8.1.0]undecane-4-acetic acid, 3-hydroxy-α,1,7-trimethyl-2-oxo-, γ-lactone	EtOH	202(3.12),302(1.51)	78-1509-65
β-Resorcylic acid, 6-heptyl-, methyl ester	EtOH	265(4.22),301(3.78)	1-1677-65
Verrucarol	EtOH	195(3.9)	33-2234-64
$C_{15}H_{22}O_6$			
2-Cyclohexene-1,1-dicarboxylic acid, 2-ethoxy-6-methyl-4-oxo-, diethyl ester	EtOH	250(4.23)	35-0471-64
$C_{15}H_{23}BrO$			
Longihomocamphenylone, 3-bromo-	C_6H_{12}	320(1.95)	22-0729-64
$C_{15}H_{23}BrO_2$			
Daucone, bromo-	n.s.g.	340(2.08)	22-2020-64
$C_{15}H_{23}FO$			
2,4-Pentadien-1-ol, 2-fluoro-3-methyl-5-(2,6,6-trimethyl-1-cyclohexen-1-yl)-	MeOH	235(4.10),264(4.10)	5-0001-64D

Compound	Solvent	$\lambda_{max}(\log \epsilon)$	Ref.
$C_{15}H_{23}NO$			
o-Anisidine, 6-cyclohexyl-N,N-dimethyl-	EtOH	275(3.28)	56-1437-65
Cyclohexylamine, 2-(m-methoxyphenyl)-N,N-dimethyl-	EtOH	279(3.25),286(3.19)	56-1437-65
$C_{15}H_{23}NO_2$			
Isoquinoline, 1-tert-butyl-1,2,3,4-tetrahydro-6,7-dimethoxy-	MeOH	283(3.53)	83-0561-65
Nuphamine	EtOH	210.5(3.89)	94-1247-65
Propiophenone, 4'-(2-diethylamino-ethoxy)-	EtOH	271(4.22)	87-0511-64
dihydrogen citrate	EtOH	267(4.21)	87-0511-64
$C_{15}H_{23}N_3O_3$			
Benzamide, N-(2-diisopropylamino-ethyl)-p-nitro-	MeOH	262(4.07)	87-0107-65
Salicylaldehyde, 5-nitro-, dibutylhydrazone	MeOH	303(4.46)	24-1631-64
	MeOH-NaOH	344(4.28),430(4.01)	24-1631-64
Salicylaldehyde, 5-nitro-, diisobutylhydrazone	MeOH	304(4.47)	24-1631-64
	MeOH-NaOH	344(4.32),436(4.00)	24-1631-64
$C_{15}H_{23}O_9PS$			
1-Propene-1,2,3-tricarboxylic acid, 1-hydroxy-3-(hydroxymethyl)-, γ-lactone, diethyl ester, O-ester with O,O-diethyl phosphorothioate	MeOH	238(4.18)	5-0010-65E
$C_{15}H_{23}O_{10}P$			
1-Propene-1,2,3-tricarboxylic acid, 1-hydroxy-3-(hydroxymethyl)-, γ-lactone, diethyl ester, diethyl phosphate	MeOH	235(4.21)	5-0010-65E
$C_{15}H_{24}$			
ar-Curcumene, dihydro-	EtOH	247(2.80),250(2.79), 256(2.78),264(2.74), 272(2.57)	78-2911-64
Se-guaiazulene, hexahydro-	EtOH	249(3.84)	78-2821-64
2-Heptene, 2-methyl-6-(4-methylene-2-cyclohexen-1-yl)-	n.s.g.	232(3.73),265(3.44)	2-0091-65
Naphthalene, octahydro-2-isopropenyl-4a,8-dimethyl-	EtOH	238(4.29),243(4.34), 249(4.32),254(4.18)	78-2927-64
Spiro[4.5]dec-6-ene, 2-isopropenyl-6,10-dimethyl-	EtOH	210(3.59) (end absorption)	78-0115-65
$C_{15}H_{24}IN$			
Trimethyl(2-phenylcyclohexyl)ammonium iodide	EtOH	217(4.49),258(2.59), 264(2.23)	56-1437-65
$C_{15}H_{24}NO_5$			
Piperidinooxy, 4-(1,2-dicarboxyethylidene)-2,2,6,6-tetramethyl-, 1-ethyl ester	heptane	216(3.5),228s(3.4)	70-0716-65
	MeOH	211(3.51),228s(3.4)	70-0716-65
$C_{15}H_{24}N_2Na_2O_{17}P_2$			
Hydrouracil, 1-β-D-ribofuranosyl-, 5'-pyrophosphate, D-glucosyl ester, disodium salt	n.s.g.	198(4.02),206s(3.88)	70-0914-65

Compound	Solvent	λ max(log ε)	Ref.
$C_{15}H_{24}N_2O$			
1-Cyclohexene-1-carboxamide, 5-methyl-2-[(4-methylcyclohexylidene)amino]-	EtOH	305(3.80)	23-0010-64
Propiophenone, 2-[[2-(dimethylamino)-ethyl]methylamino]-2-methyl-, dihydrochloride	EtOH	243(3.98)	44-3574-64
$C_{15}H_{24}N_2O_2$			
Virgiline	n.s.g.	210(4)(end absorption)	39-5243-64
$C_{15}H_{24}N_2O_3$			
2H,6H-Pyrimido[2,1-b][1,3]oxazine-2,4(3H)-dione, 3,3-dibutyl-7,8-dihydro-	hexane	217(4.11)	33-0066-64
$C_{15}H_{24}O$			
Anisole, 2,6-di-tert-butyl-	C_6H_{12}	270(2.53),275s(2.51), 300s(1.54)	23-2603-65
Eudesm-4-en-6-one	EtOH	253(3.87)	78-2593-64
7βH-Eudesm-4-en-6-one	EtOH	252.5(3.84)	78-2593-64
α-Ionol, vinyl-	EtOH	220(4.32)	39-2019-65
Ψ-Ionol, vinyl-	EtOH	243(4.38)	39-2019-65
β-Ionone, enol ethyl ether	heptane	287(4.43)	70-0934-64
β-Ionylideneethanol	MeOH	235(4.13),265(4.12)	5-0020-64I
Phenol, 2,6-di-tert-butyl-4-methyl-	MeOH	279(3.26)	39-0676-65
	MeOH-NaOMe	254(3.91),307(3.69)	39-0676-65
Pseudoionone, bismethylene-	EtOH	231(3.81),280s(2.79)	27-0174-64
$C_{15}H_{24}OS$			
Sulfoxide, 1-methylheptyl p-tolyl (R)	EtOH	219(3.98),246(3.70), 266s(3.25),276s(2.88)	35-1958-65
(S)	EtOH	208s(4.12),220s(3.91), 244(3.87),267s(3.19), 276s(2.76)	35-1958-65
$C_{15}H_{24}O_2$			
Benzene, 1-heptyl-3,5-dimethoxy-	EtOH	272(3.23),278(3.23)	1-1677-65
Daucone	n.s.g.	319(1.56)	22-2020-64
4βH-Eudesmane-3,6-dione	EtOH	298(1.95)	78-1455-64
4βH,5β,7βH-Eudesmane-3,6-dione	EtOH	296(2.08)	78-1455-64
Hinesol, 4-oxo-	EtOH	242(4.28)	94-1430-65
1,15-Pentadeca-10,12-dienolide	EtOH	234(4.13)	78-1773-64
$C_{15}H_{24}O_3$			
2,5-Cyclohexadien-1-one, 2,4-di-tert-butyl-4-hydroxy-5-methoxy-	MeOH	242(4.05),290(3.59)	5-0093-65I
2-Cyclohexene-1-carboxylic acid, 2,6,6-trimethyl-4-oxo-3-propyl-, ethyl ester	n.s.g.	251(4.00)	39-1511-64
2-Cyclohexene-1,4-dione, 5,6-di-tert-butyl-2-methoxy-	MeOH	237(--)	5-0093-65I
1,3-Cyclopentanedione, 4-isopentyl-4-isovaleryl-	EtOH-acid	225(4.05),265(3.93)	39-1276-65
	EtOH-base	245(4.38),265(4.23)	39-1276-65
1-Cyclopentene-1-propionic acid, 2-acetyl-α-isopropyl-5-methyl-, methyl ester	EtOH	257(3.67)	94-1484-65
Humulinic acid C, deoxo-	pH 1	253(3.91)	20-0629-65
	pH 13	275(4.09)	20-0629-65
4a(2H)-Naphthalenecarboxylic acid, octahydro-1,1-dimethyl-2-oxo-, ethyl ester, trans	EtOH	280(1.74)	35-1573-65

Compound	Solvent	λ_{max}(log ϵ)	Ref.
10,12-Pentadecadien-1,15-olide, 3-hydroxy-	n.s.g.	234(4.26)	70-0860-64
$C_{15}H_{24}O_4$			
Hulupinic acid, tetrahydro-	EtOH-acid	300(4.04)	39-0952-64
	EtOH-base	260(4.15),395(4.08)	39-0952-64
Hydroperoxide, 1,5-di-tert-butyl-2-methoxy-4-oxo-2,5-cyclohexa-dien-1-yl-	MeOH	239(4.08),292(3.52)	5-0093-65I
$C_{15}H_{25}NO$			
2-Cyclohexen-1-one, 5,5-dimethyl-3-(N-methylcyclohexylamino)-	EtOH	303(4.51)	44-1407-65
perchlorate	EtOH	302(4.45)	44-1407-65
$C_{15}H_{25}NO_2$			
Quinoline, 1,2,6,7,8,8a-hexahydro-5-(2-hydroxyethoxy)-7-methyl-1-propyl-	n.s.g.	247(4.31)	88-0171-65
$C_{15}H_{25}N_3O$			
Benzamide, p-amino-N-[2-(diisopropyl-amino)ethyl]-	MeOH	283(4.23)	87-0107-65
5,9,11-Dodecatrien-2-one, 6,10-di-methyl-, semicarbazone	EtOH	228.5(4.45)	22-2533-64
$C_{15}H_{25}N_5$			
s-Triazolo[1,5-c]pyrimidine, 2-(hexyl-amino)-7-methyl-5-propyl-	MeOH	233(4.62),301(3.42)	39-3357-65
$C_{15}H_{27}O_7PS$			
Fumaric acid, hydroxy(3-hydroxy-1-methylbutyl)-, δ-lactone, ethyl ester, O-ester with O,O-diethyl phosphorothioate	MeOH	212(3.84)	5-0010-65E
$C_{15}H_{25}O_8P$			
Fumaric acid, hydroxy(3-hydroxy-1-methylbutyl)-, δ-lactone, ethyl ester, diethyl phosphate	MeOH	203(4.33)	5-0010-65E
$C_{15}H_{26}$			
Spiro[5.4]dec-1-ene, 8-isopropyl-1,5-dimethyl-	EtOH	210(3.42)(end absorption)	78-0115-65
$C_{15}H_{26}ClNO_5$			
N-(5,5-Dimethyl-3-methoxycyclohex-2-en-1-ylidene)cyclohexyliminium per-chlorate	EtOH	282(4.37)	44-1407-65
$C_{15}H_{26}O$			
Cyclohexanone, 2-methallyl-2-methyl-4-tert-butyl-, cis	C_6H_{12}	296(1.53)	88-2797-64
trans	C_6H_{12}	294(1.48)	88-2797-64
2-Cyclohexen-1-one, 2,4,6-triethyl-5-propyl-	heptane	234(3.82),335(1.50)	33-0725-64
	EtOH	241(3.88)	44-3650-65
Cyclopentene, 1-butyryl-2,3-dipropyl-	EtOH	250.5(3.85)	22-0722-65
Cyclopentene, 3-butyryl-1,2-dipropyl-	EtOH	276(1.47)	22-0722-65

Compound	Solvent	λ_{max}(log ε)	Ref.
$C_{15}H_{26}O_2$			
1,3-Cyclotetradecanedione, 2-methyl-	dioxan	294(2.91)	28-0759-65B
copper derivative	$CHCl_3$	286(4.64)	28-0759-65B
Valeranone, hydroxy-	n.s.g.	290(1.70)	78-1289-64
$C_{15}H_{26}O_3$			
4-Cyclohexene-1-carboxylic acid, 2-[hydroxy-(3-methoxyphenyl)-methyl]-, lactone	EtOH	206(4.08),275(3.35), 281(3.32)	70-0806-65
1,3-Cyclopentanedione, 4-hydroxy-2,5-diisopentyl-	EtOH-acid	250(4.19)	39-3816-64
	EtOH-base	273(4.37)	39-3816-64
Humulinic acid A, dehydrodeoxo-	EtOH-HCl	252(4.21)	20-0275-64
	EtOH-NaOH	275(4.40)	20-0275-64
Humulinic acid B, dehydrodeoxo-	EtOH-HCl	251(4.21)	20-0275-64
	EtOH-NaOH	274(4.39)	20-0275-64
$C_{15}H_{27}NO_5S$			
Cysteamine, S-(2-hydroxy-2-methyl-succinyl)-N-capryloyl-	n.s.g.	235(3.68)	15-0040-65
$C_{15}H_{28}O$			
Cyclohexanone, 4-tert-butyl-2-iso-butyl-2-methyl-, cis	C_6H_{12}	295(1.43)	88-2797-64
trans	C_6H_{12}	295(1.48)	88-2797-64
Cyclohexanone, 2,4,6-triethyl-3-propyl-	heptane	284(1.40)	33-0725-64
Cyclopentane, 1-butyryl-2,3-dipropyl-	EtOH	286(1.42)	22-0722-65
$C_{15}H_{29}B_{10}Cl_9N_2O$			
Bis(tetramethylammonium)nona-chlorobenzoyldecaborate(2-)	MeCN	244(3.97),278(3.50)	35-3973-64
$C_{15}H_{29}B_{10}Cl_9N_2O_2$			
Bis(tetramethylammonium)nona-chloro(benzoato)decaborate(2-)	MeCN	225(4.28),270(3.48), 279(3.40)	35-3973-64
$C_{15}H_{29}OSi_3$			
Phenoxy, 2,4,6-tris(trimethylsilyl)-	n.s.g.	421(4.00)	70-0776-64
$C_{15}H_{32}Si_4$			
Trisilane, 1,1,1,3,3,3-hexamethyl-2-phenyl-2-(trimethylsilyl)-	C_6H_{12}	241(4.12)	101-0163-65B
$C_{15}H_{38}B_{10}N_2O$			
Bis(tetramethylammonium)nonahydro-benzoyldecaborate(2-)	MeCN	232(4.04),319(3.55)	35-3973-64
$C_{15}H_{38}B_{10}N_2O_2$			
Bis(tetramethylammonium)nonahydro-(benzoato)decaborate(2-)	MeCN	270(3.56),288(4.08)	35-3973-64

Compound	Solvent	$\lambda_{max}(\log \epsilon)$	Ref.
$C_{16}F_{18}S_2$			
Disulfide, bis(nonafluoro-2,5-xylyl)-	EtOH	233(4.53),257(4.34), 290(4.26)	39-2975-64
$C_{16}H_5LiN_5O_6$			
Fluorene-$\Delta^{9,\alpha}$-malononitrile, 2,4,7-trinitro-, lithium complex	MeCN	425(3.20),525(3.24), 562(3.29),740(3.21)	44-0644-65
$C_{16}H_5N_5O_6$			
Fluorene-$\Delta^{9,\alpha}$-malononitrile, 2,4,7-trinitro-	CH_2Cl_2	365(4.38)	44-0644-65
$C_{16}H_6Br_2N_4O$			
Naphtho[1,2-c]furan-1,1,3,3-tetracarbonitrile, 6,9-dibromo-3a,9b-dihydro-	MeCN	230(3.37),259(3.97), 267(3.95),278s(3.75), 298s(3.09),308(3.28), 320(3.25)	35-3657-65
$C_{16}H_6N_2O_4$			
Anthra[9,1-cd:10,4-c'd']bis[1,2]oxazine-3,6-dione	$C_6H_3Cl_3$	325(3.95)	65-3819-64
$C_{16}H_8Cl_4$			
7H-Benzocycloheptene, 7-(tetrachloro-2,4-cyclopentadien-1-ylidene)-	MeOH	241(4.38),320(3.95), 480(4.44)	89-0345-65
$C_{16}H_8Cl_{10}$			
1,4:9,10-Dimethanoanthracene, 1,2,3,4,-5,6,7,8,12,12-decachloro-1,2,3,4,4a,-9,9a,10-octahydro-, endo-endo	EtOH	215s(4.67),220(4.08), 242s(--),287(2.20), 297(2.20)?	39-4646-65
1,4:9,10-Dimethanoanthracene, 1,2,3,4,-5,6,7,8,12,12-decachloro-1,4,4a,8a,-9,9a,10,10a-octahydro-, endo-endo-exo	EtOH	216(4.31),265(3.04), 276(3.20),287(3.35), 299(3.39),313(3.18)	39-4646-65
$C_{16}H_8F_8$			
[2,2]Paracyclophane, octafluoro-	C_6H_{12}	222(4.08),242s(3.62), 291(3.21)	25-0767-65
$C_{16}H_8N_2$			
Phenalene-$\Delta^{1,\alpha}$-malononitrile	EtOH	476(4.28)	89-0346-65
	MeCN	268(4.42),275(4.44), 331(3.56),374(3.60), 392(3.76),473(4.34)	44-3166-65
	MeCN	472(4.27)	89-0346-65
$C_{16}H_8N_2O_4S_2$			
Oxalic acid, 1,2-dithio-, S,S'-bis-(2-benzoxazolyl) ester	$CHCl_3$	283(4.31),318(4.10)	44-3618-65
$C_{16}H_8N_4$			
Hydrazine, 1-dicyanomethylene-2-fluorene-9-ylidene-	MeCN	240(4.46),270(4.50), 358(4.20),400(3.72), 500(2.36)	44-4198-65
$C_{16}H_8N_4O$			
Naphtho[1,2-c]furan-1,1,3,3-tetracarbonitrile, 3a,9b-dihydro-	MeCN	211s(4.35),216(4.41), 222(4.33),261(3.90), 286s(3.35),298(3.50), 328(1.27)	35-3657-65

Compound	Solvent	λ_{max}(log ε)	Ref.
$C_{16}H_8O_3$			
1'-Oxoindeno[2',3':4,3]coumarin	EtOH	263(4.31),281(4.41)	42-0093-64
9,10-Phenanthrenedicarboxylic anhydride	EtOH	213(4.40),229(4.37), 258(4.63),303(3.99), 338(3.89),355(3.77)	39-5544-64
$C_{16}H_8O_4$			
Biphthalide, cis	MeCN	215(4.7),235s(4.4), 250(4.2),300s(4.2), 315(4.3),340(4.2), 355(4.2),370(4.0)	44-3070-64
Biphthalide, trans	MeCN	220(4.6),225(4.6), 255(4.2),290(4.1), 355(4.5),375s(4.4)	44-3070-64
$C_{16}H_8O_4S$			
Carbonic acid, thio-, cyclic O,O-ester with 2-benzylidene-6,7-dihydroxy-3(2H)-benzofuranone	EtOH	208(4.28),227(4.22), 321(4.05),382(4.08)	56-0931-65
$C_{16}H_8O_6$			
Coumestan, 7-hydroxy-11,12-(methylenedioxy)-	EtOH	245(4.22),309(4.01), 347(4.43)	44-2353-65
$C_{16}H_9Br$			
Pyrene, 1-bromo-	isooctane	233(4.59),242(4.80), 265(4.39),276(4.67), 313(4.06),326(4.46), 343(4.67)	44-1470-65
Pyrene, 2-bromo-	isooctane	239(4.75),248(5.01), 266(4.44),277(4.64), 306(4.08),320(4.51), 335(4.75)	44-1470-65
Pyrene, 4-bromo-	isooctane	232(4.53),241(4.80), 264(4.29),275(4.52), 308(3.97),322(4.36), 338(4.58)	44-1470-65
$C_{16}H_9ClN_2O$			
Benzo[a]phenazin-6-ol, 5-chloro-	dioxan	235(4.72),307(4.80), 360(3.99),377(4.05), 418(3.88)	24-0666-65
$C_{16}H_9NO_2$			
5H-Benzo[a]phenoxazin-5-one	$CHCl_3$	238(4.33),246(4.34), 260(4.17),283(3.93), 358(4.09),373(4.07), 439(4.13)	39-0480-65
9H-Benzo[a]phenoxazin-9-one	$CHCl_3$	255(4.43),261(4.43), 291(4.22),312(3.97), 407(3.79),496(4.25)	39-0480-65
$C_{16}H_9N_3O_3$			
2-Indolinone, 3-(α-cyano-p-nitrobenzylidene)-	MeOH	262(4.27),339(4.13), 450(3.24)	87-0583-65
$C_{16}H_9N_3O_6$			
1,3-Indandione, 2-formyloxy-2-[(p-nitrophenyl)azo]-	MeOH	226(4.54),283(4.22)	24-1482-64

Compound	Solvent	$\lambda_{max}(\log \epsilon)$	Ref.
$C_{16}H_9Na$			
Sodium, pyrenyl-	THF	460(--),491(4.69), 560s(--),730(3.58)	60-1424-64
$C_{16}H_{10}$			
Fluoranthene	EtOH	277(4.51),283(4.39), 289(4.68),310(3.64), 326(3.88),344(3.98), 362(3.99)	44-3092-64
	EtOH	236(4.66),253(4.17), 262(4.16),272(4.17), 276(4.40),282(4.26), 287(4.66),309(3.56), 323(3.76),342(3.90), 359(3.95)	78-1559-64
Pyrene, bromanil complex	$CHCl_3$	617(2.92)	60-0465-64
Pyrene, chloranil complex	$CHCl_3$	610(2.94)	60-0465-64
Pyrene, iodoanil complex	$CHCl_3$	605(2.92)	60-0465-64
Pyrene, tetracyanoethylene complex	n.s.g.	410(--),440(--), 500(--),750(--)	38-3749-64A
$C_{16}H_{10}BrFO$			
Furan, 2-bromo-4-fluoro-3,5-diphenyl-	EtOH	218(4.29),240(4.32), 292(4.40)	78-3325-65
	H_2SO_4	340(3.97),423(4.48), 565(3.02)	78-3325-65
$C_{16}H_{10}BrNO_2$			
Ketone, p-bromophenyl 3-phenyl-5-isoxazolyl	EtOH	233(4.28),281(4.31)	7-1165-64
$C_{16}H_{10}BrN_3$			
9H-Pyridazino[3,4-b]indole, 3-(p-bromophenyl)-	EtOH	277(4.71),375(3.51)	94-1129-64
$C_{16}H_{10}ClNO_2$			
Cinchoninic acid, 2-(2-chlorophenyl)-	EtOH	248(4.5),322(3.9)	28-0954-64A
Cinchoninic acid, 2-(3-chlorophenyl)-	EtOH	261(4.7),329(3.9)	28-0954-64A
Cinchoninic acid, 2-(4-chlorophenyl)-	EtOH	269(4.7),333(4.0)	28-0954-64A
Ketone, p-chlorophenyl 3-phenyl-5-isoxazolyl	EtOH	233(4.24),276(4.31)	7-1165-64
$C_{16}H_{10}ClN_3$			
9H-Pyridazino[3,4-b]indole, 3-(p-chlorophenyl)-	EtOH	276(4.69),376(3.47)	94-1129-64
$C_{16}H_{10}CrO_4$			
Chromium, (benzophenone)tricarbonyl-	n.s.g.	325.0(3.91)	28-2833-65A
$C_{16}H_{10}I_4O_3$			
3-Butenal, 2,4-bis(4-hydroxy-3,5-diiodophenyl)-	EtOH	220(4.57),245(4.42), 260(4.38),348(4.45), 415s(3.8)	44-3061-64
$C_{16}H_{10}N_2$			
Fumaronitrile, diphenyl-	EtOH	236(4.08),324(4.28)	39-2222-64
Maleonitrile, diphenyl-	EtOH	227(4.11),273(3.83), 320(4.05)	39-2222-64

Compound	Solvent	$\lambda_{max}(\log \epsilon)$	Ref.
Malononitrile, [3-(1-azulenyl)-allylidene]-	hexane	255(4.14),262(4.12), 275s(4.12),285(4.19), 308s(4.03),324s(4.14), 340(4.23),468(4.81)	24-2050-64
	benzene	272(4.22),285(4.21), 326s(4.07),345(4.15), 483(4.79)	24-2050-64
	EtOH	275s(3.75),285(4.19), 305(3.94),323(4.02), 343(4.13),483(4.79)	24-2050-64
Phenalene-$\Delta^{1,\alpha}$-malononitrile, 2,3-dihydro-	$CHCl_3$	274(4.4),380(4.14)	44-3166-65
9,10-Phenanthrenedicarbonitrile, 9,10-dihydro-, trans	EtOH	206(4.59),211(4.61), 268(4.24)	39-5544-64
$C_{16}H_{10}N_2O$			
Benzo[a]phenazin-6-ol	dioxan	240(4.68),313(4.70), 360(3.86),380(3.90), 430(3.68)	24-0666-65
$C_{16}H_{10}N_2O_2$			
2-Quinoxalinecarboxylic acid, 3-(α-hydroxybenzyl)-, lactone	EtOH	248(4.64),328(3.95)	33-1860-64
$C_{16}H_{10}N_2O_4$			
Ketone, p-nitrophenyl 3-phenyl-5-isoxazolyl	EtOH	242s(4.22),270(4.38)	7-1165-64
Naphthalene-2,3-dicarbonitrile, 1,4-diacetoxy-	MeCN	246(4.91),285(3.64), 296(3.68),309(3.59), 333(3.54),349(3.70)	44-3591-64
Uretidinedione, 1,3-dibenzoyl-	n.s.g.	210(4.4),262(4.4)	83-0623-64
$C_{16}H_{10}N_4$			
2,2'-Biquinazoline	pH 0	253(4.54),317(4.11)	39-1258-65
	pH 8	230(4.31),259(4.81), 309(4.09)	39-1258-65
	C_6H_{12}-$CHCl_3$	227s(4.20),256(4.78), 301(4.11)	39-1258-65
4,4'-Biquinazoline	C_6H_{12}	223(4.56),245s(4.17), 284s(3.61),319(3.88)	39-1258-65
	pH -0.9	321(3.89)	39-1258-65
	after hydration	271(3.94),292s(3.78)	39-1258-65
	pH 8	218(4.70),230s(4.59), 253s(3.95),321(3.83)	39-1258-65
Mesoxalonitrile, azine with benzophenone	C_6H_{12}	230(4.11),275s(3.97), 335(4.12)	44-4198-65
$C_{16}H_{10}N_4O$			
Terephthalic acid, dimethyl ester, tetracyanoethylene adduct	CH_2Cl_2	242(4.48),266(4.24), 276(4.16),295(3.30)	23-2852-64
$C_{16}H_{10}N_4OS$			
2,1,3-Benzothiadiazol-4-ol, 7-(1-naphthylazo)-	EtOH	469(4.29)	17-0041-64
2,1,3-Benzothiadiazol-4-ol, 7-(2-naphthylazo)-	EtOH	462(4.40)	17-0041-64
2,1,3-Benzothiadiazol-5-ol, 4-(1-naphthylazo)-	EtOH	500(4.27)	17-0041-64
2,1,3-Benzothiadiazol-5-ol, 4-(2-naphthylazo)-	EtOH	486(4.39)	17-0041-64

Compound	Solvent	$\lambda_{max}(\log \epsilon)$	Ref.
$C_{16}H_{10}N_4O_4$			
2-Decene-4,6,8-triynal, 2,4-dinitrophenylhydrazone, trans	MeOH	243(4.44),270s(4.16), 285s(3.96),312(3.92), 334s(4.03),406(4.60)	44-2359-65
$C_{16}H_{10}O_3$			
Homophthalic acid anhydride, 4-benzylidene-	$CHCl_3$	240(3.8),262(3.7), 345(4.11)	83-0411-65
Maleic anhydride, diphenyl-	EtOH	253(3.86),275(3.75), 347(3.79)	39-2222-64
9,10-Phenanthrenecarboxylic anhydride, 9,10-dihydro-, cis	EtOH	208(4.56),262s(4.23), 269(4.28),280s(4.13)	39-5544-64
$C_{16}H_{10}O_4$			
11H-Benzofuro[3,2-b][1]benzopyran-11-one, 3-methoxy-	$CHCl_3$	255(4.20),268(3.70), 318(4.51)	39-5140-65
Benzoic acid, o-(2-oxo-2H-1-benzopyran-3-yl)-	EtOH	289(4.10),294(4.11), 316(4.06)	42-0093-64
1,3-Indandione, 2-hydroxy-, benzoate	MeOH	226(4.75)	24-1482-64
	MeOH-base	248(4.46),257(4.45)	24-1482-64
4-Isocoumarincarboxylic acid, 3-phenyl-	$CHCl_3$	297(4.21),325(3.84)	83-0715-65
$C_{16}H_{10}O_5$			
3H-Benzofuro[3,2-b][1]benzopyran-3-one, 1-hydroxy-8-methoxy-	MeOH	425(3.78),480(3.81)	2-0319-64
4H-Naphtho[1,2-b]pyran-4,5,6-trione, 3-acetyl-2-methyl-	EtOH	220(4.12),258(4.22), 320s(3.65)	94-0316-64
$C_{16}H_{10}O_6$			
Trifoliol	EtOH	350(4.35)	78-1963-64
$C_{16}H_{10}S$			
Dibenzo[b,f]thialene	dioxan	505(3.2)	73-3016-65
Phenanthro[9,10-c]thiophene	$CHCl_3$	321(3.9)	73-0195-65
9-Thia-1,2-benzofluorene	EtOH	245(4.73),253(4.76), 267(4.47),275(4.64), 292(4.18),303(4.24), 318s(3.75),332(3.40), 337s(3.18),341s(3.01), 349(3.40)	39-1637-64
$C_{16}H_{11}BrClNO_2$			
Ketone, p-bromophenyl 4-chloro-3-phenyl-2-isoxazolin-5-yl	EtOH	273(4.33)	7-1165-64
$C_{16}H_{11}BrO_2$			
9-Phenanthrenecarboxylic acid, 8-bromo-, methyl ester	EtOH	260(4.71),294(4.06), 306(4.07)	44-3933-65
$C_{16}H_{11}Br_3O_2$			
Chalcone, α,β,2-tribromo-2'-hydroxy-4-methyl-	n.s.g.	268(4.10),353(3.62)	49-0450-65
$C_{16}H_{11}Cl$			
Indene, 1-(m-chlorobenzylidene)-	dioxan	242(4.18),280(4.30), 340(4.28)	5-0018-64D
Indene, 1-(p-chlorobenzylidene)-	dioxan	241(4.12),286(4.29), 345(4.32)	5-0018-64D

Compound	Solvent	λ_{max}(log ε)	Ref.
$C_{16}H_{11}ClN_2$			
Pyridazine, 4-chloro-3,5-diphenyl-	$(MeOCH_2)_2$	250(4.33)	25-0839-64
Pyridazine, 3-(p-chlorophenyl)-6-phenyl-	EtOH	284(4.57)	39-3342-65
$C_{16}H_{11}ClN_4O_2$			
1H-Tetrazole-1-acetic acid, α-(m-chlorobenzylidene)-5-phenyl-	EtOH	242(4.24),251(4.23), 258(4.22)	44-2222-65
1H-Tetrazole-1-acetic acid, α-(o-chlorobenzylidene)-5-phenyl-	EtOH	223(4.25),240(4.18), 252(4.16)	44-2222-65
1H-Tetrazole-1-acetic acid, α-(p-chlorobenzylidene)-5-phenyl-	EtOH	218(4.34),225(4.33), 273(4.38)	44-2222-65
$C_{16}H_{11}ClO$			
1-Indanone, 2-(α-chlorobenzylidene)-, cis	MeOH	226(3.94),278s(4.17), 301(4.25)	44-0382-64
trans	MeOH	227(4.16),314(4.54)	44-0382-64
$C_{16}H_{11}ClO_5$			
Fragilin	$CHCl_3$	272(4.56),313(4.15), 435(4.18)	1-0839-65
$C_{16}H_{11}Cl_2NO_2$			
Ketone, p-chlorophenyl 4-chloro-3-phenyl-2-isoxazolin-5-yl	EtOH	268(4.37)	7-1165-64
$C_{16}H_{11}FO$			
Furan, 3-fluoro-2,4-diphenyl-	EtOH	217(4.21),223(4.18), 243(4.29),274(4.34)	78-3325-65
	H_2SO_4	241(3.41),310(3.92), 315(3.92),420(4.79)	78-3325-65
$C_{16}H_{11}N$			
Fluoranthene, 9-amino-	EtOH	262(4.70),283(3.99), 288(4.30),294(3.80), 326(4.17),342(3.63), 358(3.77)	44-3092-64
Pyrene, 2-amino-	C_6H_{12}	264(4.81),283(4.23), 296(4.00),309(3.91), 323(4.26),338(4.61), 372s(--),377(3.53), 391s(--),399(3.54)	44-1470-65
$C_{16}H_{11}NO_2$			
Cinchoninic acid, 2-phenyl-	EtOH	264(4.6),333(4.0)	28-0954-64A
Indene, 1-(m-nitrobenzylidene)-	dioxan	221(4.34),275(4.43), 338(4.26)	5-0018-64D
Indene, 1-(p-nitrobenzylidene)-	dioxan	248(4.17),285(4.13), 360(4.30)	5-0018-64D
Isoxazole, 5-benzoyl-3-phenyl-	EtOH	245s(4.21),266(4.25)	7-1165-64
	EtOH	232s(4.15),243s(4.17), 251s(4.18),266(4.21)	32-0206-65
Maleimide, diphenyl-	EtOH	224(4.20),250s(4.08), 263(3.93),307(3.53), 319s(3.56),353(3.71)	39-2222-64
1,2-Naphthoquinone, 4-anilino-	dioxan	248(3.95),292(4.15), 342(3.78),442(3.50)	5-0151-64E
1,4-Naphthoquinone, 2-anilino-	dioxan	273(4.45),455(3.64)	5-0151-64E
6H-1,2-Oxazin-6-one, 2,5-diphenyl-	MeOH	345(4.2)	83-0367-64

Compound	Solvent	$\lambda_{max}(\log \epsilon)$	Ref.
9,10-Phenanthrenedicarboximide, 9,10-dihydro-	EtOH	208(4.55),230s(4.00), 240(3.76),265s(4.15), 275(4.20),284s(4.06), 303(3.08)	39-5544-64
$C_{16}H_{11}NO_2S$			
Phthalimide, N-(thio-p-toluyl)-	heptane	213(4.51),228(4.40), 285(3.95),336(4.23)	24-1556-65
	MeOH	213(4.46),229(4.36), 288(4.06),340(4.13), 535(2.04)	24-1556-65
	dioxan	210(4.48),225(4.45), 288(4.09),340(4.27), 360(3.94),540(2.07)	24-1556-65
$C_{16}H_{11}NO_3$			
Benzoic acid, o-(3-indolylcarbonyl)-	NaOH	270(4.16),340(4.13)	65-3025-64
Conchoninic acid, 2-(2-hydroxyphenyl)-	EtOH	217(4.7),268(4.6), 356(4.2)	28-0954-64A
Cinchoninic acid, 2-(4-hydroxyphenyl)-	EtOH	250s(4.3),279(4.5), 352(4.2)	28-0954-64A
Cinchoninic acid, 2-phenyl-, N-oxide	EtOH	232(4.4),276(4.6), 349(4.1)	28-0954-64A
$C_{16}H_{11}NO_4$			
1-Phenanthrenecarboxylic acid, 10-nitro-, methyl ester	EtOH	257.5(4.57)	44-3792-65
	EtOH	257.5(4.57)	88-0359-65
$C_{16}H_{11}NO_4S$			
Benzo[b]thiophen-3(2H)-one, 2-(o-methoxybenzylidene)-5-nitro-	n.s.g.	425(4.30),445(4.38)	65-0519-65
Benzo[b]thiophen-3(2H)-one, 2-(p-methoxybenzylidene)-5-nitro-	n.s.g.	423(4.35),445(4.46)	65-0519-65
$C_{16}H_{11}NO_5$			
2H-1,3-Benzoxazine-2,4(3H)-dione, 3-p-anisoyl-	MeOH	206(4.57),238s(4.15), 243s(3.94),293(4.29)	4-0037-65
Coumarin, 8-methoxy-3-(p-nitrophenyl)-	EtOH	255(4.00),328(4.36)	94-0443-65
Lambertellin lactam acetyl ester	EtOH	256(4.06),292(3.81)	100-0359-65
Phthalimidine, 2-(m-carboxyphenyl)-5,6-(methylenedioxy)-	MeOH	309(4.28)	95-1049-65
Phthalimidine, 2-(o-carboxyphenyl)-5,6-(methylenedioxy)-	MeOH	304(4.11)	95-1049-65
Phthalimidine, 2-(p-carboxyphenyl)-5,6-(methylenedioxy)-	MeOH	315(4.45)	95-1049-65
$C_{16}H_{11}N_3$			
9H-Pyridazino[3,4-b]indole, 3-phenyl-	EtOH	270(4.66),374(3.47)	94-1129-64
s-Triazolo[4,3-a]quinoline, 1-phenyl-	EtOH	219(4.50),243(4.19), 294(3.94),318(3.73)	2-0162-65
s-Triazolo[3,4-a]isoquinoline, 3-phenyl-	EtOH	208(4.37),249(4.53), 310(3.54),312(3.55)	2-0162-65
$C_{16}H_{11}N_3O$			
Cyclohepta[b]pyrrole-3-carbonitrile, 1-anilino-1,2-dihydro-2-oxo-	EtOH	230(4.40),275(4.51), 423(4.14)	94-0450-65
s-Triazine-2-carboxaldehyde, 4,6-diphenyl-	MeOH	267(4.54)	44-0678-64

Compound	Solvent	λ_{max}(log ε)	Ref.
$C_{16}H_{11}N_3O_2$			
Quinoxaline, 6-nitro-2-styryl-	dioxan	255(5.78),365(6.0)[sic]	73-3102-65
$C_{16}H_{11}N_5O_4$			
s-Tetrazole-1-acetic acid, α-(m-nitrobenzylidene)-5-phenyl-	EtOH	248(4.23)	44-2222-65
s-Tetrazole-1-acetic acid, α-(o-nitrobenzylidene)-5-phenyl-	EtOH	236(4.41),305(3.40)	44-2222-65
s-Tetrazole-1-acetic acid, α-(p-nitrobenzylidene)-5-phenyl-	EtOH	230(4.06),238(3.99), 294(4.03)	44-2222-65
$C_{16}H_{12}$			
Indene, 1-benzylidene-	dioxan	239(4.18),282(4.33), 339(4.31)	5-0018-64D
Naphthalene, 2-phenyl-	C_6H_{12}	250(4.70),285(4.06)	24-0593-65
Pentaleno[2,1,6-def]heptalene, 2-methyl-	hexane	255(4.55),262(4.57), 284(4.75),291(4.83), 317(4.24),367(3.55), 382(3.61),396(3.38), 405(3.35),421(3.58), 449(3.76),504(2.74), 539(3.00),582(3.11), 710(2.08),813(1.70)	89-0042-65
$C_{16}H_{12}BN$			
Benzo[c]naphth[2,1-e][1,2]azaborine, 5,6-dihydro-	$CHCl_3$	268(4.64),276(4.67), 309(3.98),322(3.95), 337(3.98),353(4.04)	44-1757-64
14,13-Borazarotriphenylene	EtOH	258(4.71),269(4.69), 295(3.62),316(3.93), 330(3.97)	35-1125-64
$C_{16}H_{12}Br_2$			
Cyclobutene, 3,4-dibromo-1,2-diphenyl-	MeCN-10% EtOH	230s(4.38),238(4.42), 303(4.31)	44-2331-64
$C_{16}H_{12}Br_2O_2$			
Benzofuro[3,2-b]benzofuran, 4b,9b-dihydro-3,8-dibromo-4b,9b-dimethyl-	EtOH	206s(4.45),226(4.32), 292(3.76)	78-2289-65
2-Butyne, 1,4-bis(p-bromophenoxy)-	EtOH	278(3.3042)	78-2289-65
$C_{16}H_{12}ClNO$			
Indole-2-carboxaldehyde, 5-chloro-1-methyl-3-phenyl-	iso-PrOH	252(4.48),318(4.35)	44-1621-64
2-Indolinone, 3-(p-chlorobenzylidene)-1-methyl-	EtOH	256(4.07),265(4.06), 278s(--),330(4.23), 390s(--)	87-0626-65
$C_{16}H_{12}ClNO_2$			
Ketone, 4-chloro-3-phenyl-2-isoxazolin-5-yl phenyl	EtOH	263(4.18)	7-1165-64
$C_{16}H_{12}ClNO_4$			
2-Butyne, 1-(p-chlorophenoxy)-4-(p-nitrophenoxy)-	EtOH	301(4.00)	78-2289-65
2-Stilbenecarboxylic acid, 2'-chloro-α-nitro-, methyl ester	EtOH	308(4.01)	88-0359-65
cis	EtOH	308(4.02)	44-3792-65

Compound	Solvent	λ_{max}(log ε)	Ref.
$C_{16}H_{12}ClNO_4S$			
2-Methyl-11H-[1]benzothiopyrano-[4,3-b]indol-5-ium perchlorate	MeCN-1% $HClO_4$	237(4.26),273(4.55), 293(4.52),404(4.17)	44-3613-65
8-Methyl-11H-[1]benzothiopyrano-[4,3-b]indol-5-ium perchlorate	MeCN-1% $HClO_4$	231(4.24),236s(4.16), 272(4.59),300s(4.39), 322s(3.91),403(4.07)	44-3613-65
11-Methyl-11H-[1]benzothiopyrano-[4,3-b]indol-5-ium perchlorate	MeCN-1% $HClO_4$	237(4.19),269(4.47), 266(4.51),323s(3.41), 407(4.14)	44-3613-65
$C_{16}H_{12}ClNO_5$			
2,6-Diphenyl-1,3-oxazin-1-ium perchlorate	n.s.g.	394(4.27)	24-3892-65
$C_{16}H_{12}Cl_2O_2$			
Benzofuro[3,2-b]benzofuran, 3,8-di-chloro-4b,9b-dihydro-4b,9b-dimethyl-	EtOH	203(4.59),226(4.25), 293(3.79)	78-2289-65
2-Butyne, 1,4-bis(o-chlorophenoxy)-	EtOH	204(4.50),274(3.50), 281(3.48)	78-2289-65
$C_{16}H_{12}INO_4$			
2-Stilbenecarboxylic acid, 2'-iodo-α-nitro-, methyl ester	EtOH	306(3.95)	88-0359-65
cis	EtOH	306(3.95)	44-3792-65
2-Stilbenecarboxylic acid, 3'-iodo-α-nitro-, methyl ester, cis	EtOH	309(3.91)	44-3792-65
2-Stilbenecarboxylic acid, 4'-iodo-α-nitro-, methyl ester, cis	EtOH	230(4.34),328(4.25)	44-3792-65
$C_{16}H_{12}I_4O_4$			
Butyric acid, 4-[4-(4-hydroxy-3,5-di-iodophenoxy)-3,5-diiodophenyl]-	EtOH	213(4.67),226(4.69), 238(4.43),295(3.64), 303(3.63)	44-1812-64
Hydrocinnamic acid, 4-(4-hydroxy-3,5-diiodophenoxy)-3,5-diiodo-α-methyl-	EtOH	213(4.7),226(4.69), 237(4.45),295(3.65), 303(3.66)	44-1812-64
$C_{16}H_{12}N_2$			
3,3'-Biindole	EtOH	233(4.60),282(3.91), 291(3.92)	23-2900-64
1,4-Diazafulvene, 6,6-diphenyl-	C_6H_{12}	241(3.95),361(4.27)	89-1077-65
	H_2SO_4	438(4.56)	89-1077-65
	MeCN	356(4.24)	89-1077-65
1,6-Pyrenediamine	benzene	375(4.4),400(4.2), 425(4.2)	38-0890-64A
Pyridazine, 3,5-diphenyl-	$(MeOCH_2)_2$	256(4.58)	25-0839-64
Pyridazine, 3,6-diphenyl-	EtOH	279(4.44)	39-3342-65
$C_{16}H_{12}N_2O$			
Benzamide, N-(α-cyanostyryl)-	MeOH	222(4.19),226(3.20), 228(4.23),231(4.70), 296(4.28),298(4.33), 301(4.35),303(4.35)	49-1214-65
Cyclohepta[b]pyrrol-2(1H)-one, 1-(benzylideneamino)-	EtOH	251(4.31),310(4.60), 415(3.84)	94-0450-65
Imidazole, 2-benzoyl-4(5)-phenyl-	EtOH	256(3.98),286(3.41), 336(3.88)	7-1364-64
11H-Indolo[3,2-c]quinoline, 3-methoxy-	MeOH	236(4.52),245s(4.43), 270s(4.50),277(4.75), 292s(4.30),302(4.14), 312(3.93)	39-5919-64

Compound	Solvent	λ_{max}(log ε)	Ref.
2-Naphthol, (phenylazo)-	EtOH	260(4.0),308(3.8), 417s(4.0),477(3.9)	70-2206-65
Phenalene-Δ$^{1,\alpha}$-acetamide, α-cyano-2,3-dihydro-	MeCN	266(4.25),360(3.98)	44-3166-65
4(1H)-Pyridazinone, 3,5-diphenyl-	$(MeOCH_2)_2$	330(4.24)	25-0839-64
4(3H)-Pyrimidinone, 2,5-diphenyl-	dioxan	340(4.2)	83-0367-64
3-Pyrrolin-2-one, 5-imino-3,4-diphenyl-	EtOH	225(4.15),239(4.11), 251(4.09),271(4.14), 330(3.66)	39-2222-64
	EtOH-KOH	210(4.33),229(4.27), 287(4.25)	39-2222-64
$C_{16}H_{12}N_2O_2$			
Indazolo[2,3-a]quinoline-7,10-dione, 5,6-dihydro-8-methyl-	MeOH	273(4.51),330(3.47)	39-5871-65
Indazolo[2,3-a]quinoline-7,10-dione, 5,6-dihydro-9-methyl-	MeOH	260(4.15),320s(3.36)	39-5871-65
Indole, 3-(2-nitrovinyl)-2-phenyl-	n.s.g.	242(4.43),298(4.30), 418(4.29)	28-0609-64A
Isoxazole, 5-benzoyl-3-phenyl-, oxime	EtOH	232(4.43)	32-0206-65
$C_{16}H_{12}N_2O_3$			
Benzoic acid, m-(5-phenyl-1,3,4-oxadiazol-2-yl)-, methyl ester	DMF	286.0(4.49)	24-2966-65
$C_{16}H_{12}N_2O_4$			
Acrylonitrile, 3-(4-hydroxy-3-methoxyphenyl)-2-(p-nitrophenyl)-	MeOH	276(3.81),385(4.36)	87-0583-65
$C_{16}H_{12}N_4O$			
s-Triazine-2-carboxaldehyde, 4,6-diphenyl-, oxime	MeOH	267(4.70)	44-0678-64
$C_{16}H_{12}N_4O_2$			
1H-Tetrazole-1-acetic acid, α-benzylidene-5-phenyl-	EtOH	247(4.28),263(4.28)	44-2222-65
$C_{16}H_{12}N_4O_4$			
2-Isoxazolin-5-one, 4-methyl-4-(4-nitrophenylazo)-3-phenyl-	EtOH	270(4.36),420(2.43)	39-5414-65
$C_{16}H_{12}N_4O_6$			
1-Indanone, 2-hydroxy-, formate, 2,4-dinitrophenylhydrazone	diglyme	267s(4.06),302s(3.69), 316(3.65),385(4.47)	44-1723-64
$C_{16}H_{12}N_4O_{10}$			
Acetic acid, bis(2,4-dinitrophenyl)-, ethyl ester	EtOH	241(4.50),262s(4.38)	44-0636-64
$C_{16}H_{12}O$			
Anthracene, 1-acetyl-	EtOH	241(4.79),255(4.84), 349s(3.78),364(3.74), 385(3.63)	39-5744-65
Anthracene, 9-acetyl-	EtOH	253(5.11),330(3.42), 348(3.77),363(3.90), 380(3.83)	39-5744-65
5H-Dibenzo[a,d]cycloheptene-5-carboxaldehyde	EtOH	291(4.15)	88-1059-65
1-Indanone, 2-benzylidene, cis	MeOH	228(3.97),316(4.45)	44-1276-64

Compound	Solvent	λ_{max}(log ϵ)	Ref.
$C_{16}H_{12}O_2$			
Biphenylene, 2,6-diacetyl-	EtOH	219(4.08),226s(--), 261(4.51),284s(4.60), 287(4.61),344s(3.64), 360(3.93),377(4.10)	39-4930-65
Biphenylene, 2,7-diacetyl-	EtOH	220s(4.18),229(4.19), 237s(4.13),267s(4.74), 275(4.81),308s(3.31), 325(3.37),343s(3.45), 369s(--),393(3.71)	39-4930-65
1H-Cyclopropa[l]phenanthrene-1-carboxylic acid, 1a,9b-dihydro-	EtOH	247(3.64),271(4.18), 280(4.15),298(3.72), 310(3.76)	88-2919-64
Dibenzo[a,e]cycloheptatriene-5-carboxylic acid	EtOH	288(4.16)	87-0088-64
1,4:5,8-Dimethanoanthracene-9,10-dione, 1α,4α,5β,8β-tetrahydro-	C_6H_{12}	278s(4.12),285(4.17), 345(2.77),420(2.11), 470s(1.77),500s(1.36)	39-3043-64
	EtOH	288(4.10),329(2.71), 423(2.15)	39-3043-64
3(2H)-Furanone, 2,4-diphenyl-	EtOH	410(3.10)	78-3325-65
(inflections not listed)	H_2SO_4	447(3.68)	78-3325-65
Indone, 3-methoxy-2-phenyl-	MeOH	245(4.59),410(2.97)	65-0448-64
1,4-Naphthoquinone, 5,8-dihydro-5-phenyl-	EtOH	247(4.21),340(3.06)	39-5110-64
Phthalide, (1-phenylethylidene)-, cis	ether	317(4.01)	22-3047-65
trans	ether	327(3.97)	22-3047-65
$C_{16}H_{12}O_2S$			
Anthraquinone, 1-(ethylthio)-	EtOH	245(4.58),305(3.83), 440(3.66)	22-1648-65
Benzo[b]thiophen-3(2H)-one, 2-(o-methoxybenzylidene)-	n.s.g.	424(4.11),442(4.22)	65-0519-65
Benzo[b]thiophen-3(2H)-one, 2-(p-methoxybenzylidene)-	n.s.g.	421(4.20),440(4.34)	65-0519-65
$C_{16}H_{12}O_3$			
Benzofuran, 3-acetyl-2-(4-hydroxyphenyl)-	EtOH	233(4.21),314(4.15)	44-2602-64
	EtOH-NaOEt	249(4.24),267(4.20), 363(4.29)	44-2602-64
Chalcone, 2'-benzyloxy-4-methoxy-, epoxide	EtOH	252(4.2),307(3.6)	78-0963-65
Coumarin, 7-methoxy-3-phenyl-	MeOH	244(4.11),340(4.18)	2-0182-64
Coumarin, 8-methoxy-3-phenyl-	MeOH	249(3.90),310(4.03)	2-0182-64
$C_{16}H_{12}O_3S$			
Anthraquinone, 1-(ethylsulfinyl)-	EtOH	255(4.60),335(3.69)	22-1648-65
$C_{16}H_{12}O_3S_2$			
Anthraquinone, 1-(methylsulfinyl)-4-(methylthio)-	EtOH	246(4.63),316(4.10), 451(3.60)	22-1648-65
$C_{16}H_{12}O_4$			
5,6-Acenaphthylenedicarboxylic acid, dimethyl ester	EtOH	335(4.07)	44-2824-64
Chalcone, 2'-hydroxy-3,4-(methylenedioxy)-	EtOH	248(4.00),284(3.76), 322(3.68),373(3.96)	23-2580-64
Chalcone, 3'-hydroxy-3,4-(methylenedioxy)-	EtOH	258(4.14),310(3.99), 357(4.38)	23-2580-64
Flavone, 3'-hydroxy-4'-methoxy-	EtOH	243(4.34),340(4.36)	2-0351-65

Compound	Solvent	λ_{max}(log ε)	Ref.
Flavone, 3-hydroxy-7-methoxy-	EtOH	228(4.22),253(4.14), 321(4.18),340(4.20)	78-1141-64
Flavone, 4'-hydroxy-3'-methoxy-	EtOH	243(4.91),295(4.26), 341(4.57)	2-0351-65
Isoflavone, 7-hydroxy-4'-methoxy-	EtOH	249(4.39),302(3.96)	78-1141-64
9,10-Phenanthrenedicarboxylic acid, 9,10-dihydro-, trans	EtOH	208(4.56),211(4.56), 266(4.20),298s(3.20)	39-5544-64
$C_{16}H_{12}O_4S$			
Anthraquinone, 1-(ethylsulfonyl)-	EtOH	255(4.67),314(3.67)	22-1648-65
Anthraquinone, 1-methoxy-4-(methylsulfinyl)-	EtOH	255(4.35),327(3.44), 378(3.48)	22-1648-65
$C_{16}H_{12}O_4S_2$			
Anthraquinone, 1,4-bis(methylsulfinyl)-	EtOH	255(4.57),329(3.69), 409(3.10)	22-1648-65
$C_{16}H_{12}O_5$			
Echioidinin	EtOH	267(4.50),340(4.32)	78-2633-65
	EtOH-NaOEt	265(4.35),413(4.22)	78-2633-65
	EtOH-$AlCl_3$	255(4.15),280(4.43), 350(4.24),383s(4.05)	78-2633-65
Flavone, 4',5'-dihydroxy-7-methoxy-	EtOH	268(4.33),339(4.37), 300s(4.18)	12-1871-65
	EtOH-NaOH	275(4.28),304(4.03), 360s(4.41),387(4.47)	12-1871-65
Genisteine, 5-methyl-	EtOH	257(4.49)	22-1038-64
Isoflavone, 6,7-dihydroxy-4'-methoxy-	EtOH	231(4.39),258(4.52), 326(4.13)	78-1331-64
Naphtho[2,3-b]furan-3-carboxylic acid, 4,9-dihydro-2-methyl-4,9-dioxo-, ethyl ester	MeCN	250(4.36),290(3.90), 333(3.60)	44-3819-65
4H-Naphtho[1,2-b]pyran-4-one, 3-acetyl-5,6-dihydroxy-2-methyl-	EtOH	233(4.40),274(4.33), 336(3.69),407(3.41)	94-0316-64
$C_{16}H_{12}O_5S$			
Anthraquinone, 1-methoxy-4-(methylsulfonyl)-	EtOH	255(4.62),358(3.60)	22-1648-65
$C_{16}H_{12}O_6$			
2,5'-Biphenyldicarboxylic acid, 2'-hydroxy-4,5-dimethoxy-, δ-lactone	MeOH	252(4.74),300(4.05)	100-0090-65
$\Delta^{1,\alpha:2,\alpha}$-Heptalenediacetic acid, 3,4,5,7-tetrahydro-α,8-dihydroxy-7-oxo-, δ-lactone	EtOH	233(3.99),365(3.85)	78-3605-65
	base	239(4.26),404(4.12)	78-3605-65
Kaempferol, 3-methyl-	EtOH	268(4.29),299(4.08), 353(4.26)	22-0779-65
	EtOH-NaOH	274(4.36),325(4.12), 398(4.42)	22-0779-65
	EtOH-$AlCl_3$	277(4.28),304(4.10), 345(4.22),397(4.13)	22-0779-65
$C_{16}H_{12}O_6S_2$			
Anthraquinone, 1,4-bis(methylsulfonyl)-	EtOH	257(4.48),300(3.50)	22-1648-65
$C_{16}H_{12}O_7$			
Flavone, 3,3',4',5,6-pentahydroxy-8-methyl-	EtOH	255(4.18),295s(--), 375(4.19)	78-3727-65
Isorhamnetin	EtOH	255(4.30),305(3.86), 375(4.32)	22-0779-65

Compound	Solvent	$\lambda_{max}(\log \epsilon)$	Ref.
Isorhamnetin (contd.)	EtOH-AlCl3	264(4.35),303(3.78), 359(3.99),428(4.33)	22-0779-65
Quercetin, 3-methyl-	EtOH	257(4.33),298(3.96), 360(4.30)	22-0779-65
	EtOH-NaOH	271(4.38),326(4.03), 405(4.49)	22-0779-65
	EtOH-AlCl3	269(4.31),298(3.96), 360(4.15),403(4.23)	22-0779-65
Quercetin, 5-methyl-	EtOH	254(4.34),298(3.87), 368(4.35)	22-0779-65
	EtOH-NaOH	243(4.19),325(4.48)	22-0779-65
	EtOH-AlCl3	263(4.40),333(3.50), 428(4.49)	22-0779-65
$C_{16}H_{12}O_8$			
Myricetin, 3-methyl-	EtOH	255(4.22),305(3.91), 365(4.27)	22-0779-65
	EtOH-AlCl3	272(4.23),309(3.85), 355(4.00),410(4.17)	22-0779-65
$C_{16}H_{12}S$			
Azulene, 1-(phenylthio)-	C_6H_{12}	236(4.46),281(4.68), 291(4.57),334(3.73), 343(3.75),359(3.76), 570(2.50),588(2.49), 616(2.46),675(2.11)	44-2715-65
Benzo[b]thiophene, 3-styryl-, trans	EtOH	240(4.20),263s(4.02), 323(4.19)	39-6221-65
Thiophene, 2,4-diphenyl-	EtOH	256(4.33),302(4.00)	23-1928-64
Thiophene, 2,5-diphenyl-	$CHCl_3$	325(4.41)	49-1750-64
$C_{16}H_{13}BrN_2$			
7-Methylpyrido[2',1':3,4]pyrazino-[1,2-a]indol-5-ium bromide	EtOH	212(4.41),253(4.00), 294(3.72),385(4.31)	44-3584-64
$C_{16}H_{13}BrN_2O_4$			
2,4-Pentadienanilide, p-bromo-4-methyl-5-(5-nitro-2-furyl)-	propylene glycol	315(4.30),398(4.52)	95-0646-64
$C_{16}H_{13}BrO$			
Chalcone, α-(bromomethyl)-, trans	EtOH	257(4.14),290(4.14)	88-4833-65
Chalcone, 4'-bromo-4-methyl-	EtOH	229(4.02),320(4.45)	23-2580-64
$C_{16}H_{13}BrO_2$			
Chalcone, 2-bromo-2'-hydroxy-4-methyl-	n.s.g.	248(4.14),333(4.40)	49-0450-65
Chalcone, 4'-bromo-2-methoxy-	EtOH	280(4.10),293(4.14), 350(4.08)	23-2580-64
Chalcone, 4'-bromo-4-methoxy-	EtOH	240(4.06),276(4.06), 291(4.01),344(4.45)	23-2580-64
$C_{16}H_{13}BrO_3$			
Chalcone, 2-bromo-2'-hydroxy-4'-methoxy-	n.s.g.	335(4.1)	49-0450-65
Chalcone, 5'-bromo-2'-hydroxy-4-methoxy-	n.s.g.	232(4.41),354(3.08)	49-0450-65
$C_{16}H_{13}BrO_6$			
2,3-Biphenylenedione, 4-bromo-1,6,7,8-tetramethoxy-	EtOH	289(4.26),318(4.31), 436(4.04)	39-1067-64

Compound	Solvent	$\lambda_{max}(\log \epsilon)$	Ref.
$C_{16}H_{13}ClN_2O$			
Carbostyril, 6-chloro-3-(methylamino)-4-phenyl-	iso-PrOH	240(4.5),330(4.2), 345(4.3)	39-3097-64
Phenol, p-[(7-chloro-4-quinolyl)-methylamino]-	pH 1	352(4.13)	4-0006-64
	EtOH	345(3.94)	4-0006-64
3(2H)-Pyridazinone, 6-(p-chloro-phenyl)-4,5-dihydro-4-phenyl-	EtOH	289(4.23)	39-3342-65
$C_{16}H_{13}ClN_2O_4$			
2,4-Pentadienanilide, p-chloro-4-methyl-5-(5-nitro-2-furyl)-	propylene glycol	315(4.22),395(4.45)	95-0646-64
$C_{16}H_{13}ClO_2$			
Chalcone, 2-chloro-4'-methoxy-	EtOH	227(4.14),299(4.18)	23-2580-64
4'-Hydroxy-3-methylflavylium chloride	EtOH-HCl	251(--),269(--), 306s(--),448(--)	44-2602-64
$C_{16}H_{13}ClO_4S$			
7-Methyl-2-phenyl-1-benzothio-pyrylium perchlorate	HOAc	269(4.46),293(3.86), 406(4.40)	5-0136-64C
$C_{16}H_{13}ClO_5$			
7-[(p-Methoxyphenyl)ethynyl]tropylium perchlorate	MeCN	255(4.26),262s(4.39), 334(3.73),472(4.53)	24-1337-64
2-Naphthaleneacetic acid, α-acetyl-3-chloro-1,4-dihydro-1,4-dioxo-, ethyl ester	MeCN	249(4.38),254(4.48), 267(4.28),278(4.25), 338(3.54)	44-3819-65
2',4',7-Trihydroxy-3-methoxy-flavylium chloride	EtOH-HCl	240(4.35),263(4.02), 282(4.04),508(4.66)	44-3036-64
$C_{16}H_{13}FO_2$			
Butyrophenone, 2,3-epoxy-4-fluoro-3-phenyl-	EtOH	246(4.11)	78-3325-65
	H_2SO_4	261(4.32),298(3.82)	78-3325-65
$C_{16}H_{13}IN_2$			
5-Methylindolizino[1,2-c]quinolinium iodide	MeOH	245(4.67),275(4.38), 320(4.23),425(4.30)	5-0196-65H
$C_{16}H_{13}IO_2$			
Acrylic acid, 3-(o-iodophenyl)-2-phenyl-, methyl ester, cis	EtOH	286(4.04)	44-3792-65
	EtOH	286(4.04)	88-0359-65
$C_{16}H_{13}IO_3$			
Acrylic acid, 2-(p-iodophenyl)-3-(p-methoxyphenyl)-	EtOH	231(4.40),245s(4.32), 301(4.39)	7-0805-64
$C_{16}H_{13}N$			
Bicyclo[4.2.0]octa-1,3,5-triene, 7-benzyl-7-cyano-	EtOH	254s(2.92),259(3.14), 265(3.29),271(3.25), 293(2.32)	87-0732-65
Cyclopropane-1-carbonitrile, 1-(4-biphenylyl)-	EtOH	256(4.41)	87-0504-64
Isoquinoline, 1-methyl-3-phenyl-	pH 2	294(3.92),345(3.86)	56-0893-64
	pH 12	290(4.10),323s(3.59)	56-0893-64
Isoquinoline, 1-methyl-4-phenyl-	pH 2	284(3.62),346(3.96)	56-0515-64
	pH 12	283(3.85),325(3.83)	56-0515-64
Pyrrole, 3,4-diphenyl-	EtOH	238(4.18),271(4.04)	44-0859-65
Quinaldine, 3-phenyl-	pH 2	321(3.97)	56-0893-64
	pH 12	309(3.62),317(3.65)	56-0893-64

Compound	Solvent	λ max(log ε)	Ref.
Quinaldine, 4-phenyl-	pH 2	316.5(4.14)	56-0515-64
	pH 12	291(3.91),315s(3.72)	56-0515-64
p-Toluidine, α-inden-1-ylidene-	MeOH	385(4.4)	87-0390-65
	HOAc	370(4.2)	87-0390-65
$C_{16}H_{13}NO$			
1,3-Indandione imine, 2-benzyl-	MeOH	256(4.55),263(4.56), 300(3.38),427(3.69)	65-0448-64
Indone, 3-anilino-2-methyl-	MeOH	221(4.33),263(4.43), 453(3.63)	65-0448-64
Indone, 3-(methylamino)-2-phenyl-	MeOH	218(4.47),267(4.50), 315(3.51),435(3.47)	65-0448-64
$C_{16}H_{13}NOS$			
2,5-Cyclohexadien-1-one, 4-[2-(3-methyl-2-benzothiazolinylidene)-ethylidene]-	C_5H_5N	530(4.76),566(4.99)	5-0009-65J
	99% C_5H_5N	531(4.71),566(5.04)	5-0009-65J
	98% C_5H_5N	533(4.75),567(5.07)	5-0009-65J
	95% C_5H_5N	531(4.72),566(5.10)	5-0009-65J
	50% C_5H_5N	544(4.85)	5-0009-65J
	30% C_5H_5N	533(4.77)	5-0009-65J
	20% C_5H_5N	526(4.73)	5-0009-65J
	10% C_5H_5N	515(4.69)	5-0009-65J
	5% C_5H_5N	497(4.64)	5-0009-65J
	2% C_5H_5N	494(4.61)	5-0009-65J
	1% C_5H_5N	494(4.61)	5-0009-65J
$C_{16}H_{13}NO_2$			
Cinnamaldehyde, 2-benzamido-	EtOH	225(4.29),256(4.22), 285(4.28)	44-0305-64
Isoxazole, 3-(p-methoxyphenyl)-5-phenyl-	$CHCl_3$	268(4.47)	24-3020-65
Isoxazole, 5-(p-methoxyphenyl)-3-phenyl-	$CHCl_3$	246(4.20),285(4.36)	24-3020-65
Phthalimide, N-(α-methylbenzyl)-	MeOH	293(3.29),300s(3.26)	24-0533-64
4H-Pyran-4-one, 2-methyl-6-)2-quinolylmethyl)-	EtOH	285(3.8),300(3.7), 315(3.7)	94-0018-64
Viridicatin, 3-O-methyl-	EtOH	223(4.61),281(3.91), 313(3.86),325(3.96), 337(3.81)	39-1197-64
$C_{16}H_{13}NO_3$			
4,1-Benzoxazepine-2,5(1H,3H)-dione, 3-methyl-1-phenyl-	iso-PrOH	220s(4.36),236(4.11), 321(3.51)	4-0323-65
Carbazole-3-acrylic acid, 1-methoxy-	n.s.g.	218(4.36),245(4.51), 256(4.42),281(4.58), 320(4.17)	78-0681-65
Chalcone, 4'-methyl-4-nitro-	EtOH	255(3.83),318(4.50)	23-2580-64
Cinchoninic acid, 1,2,3,4-tetrahydro-2-oxo-1-phenyl-	MeOH	206(4.53),214(4.53), 230(4.45),284(3.77), 340(2.53)	44-2973-65
Cinnamic acid, 2-benzamido-	EtOH	227(4.31),265(4.30)	44-0305-64
$C_{16}H_{13}NO_3S$			
4H-1,2-Benzothiazin-4-one, 3-benzylidene-2,3-dihydro-2-methyl-, 1,1-dioxide	EtOH	234(4.10),264(3.94), 331(4.31)	44-2241-65
2H-Thieno[3,2-b]pyrrol-3(4H)-one, 6-(acetoxymethyl)-2-benzylidene-	EtOH	268(3.89),345(4.41)	44-1012-65

Compound	Solvent	$\lambda_{max}(\log \epsilon)$	Ref.
$C_{16}H_{13}NO_4$			
Chalcone, 4-methoxy-3'-nitro-	EtOH	244(4.37),352(4.36)	23-2580-64
Phthalimidine, 2-(m-methoxyphenyl)-5,6-(methylenedioxy)-	MeOH	310(4.25)	95-1049-65
Phthalimidine, 2-(p-methoxyphenyl)-5,6-(methylenedioxy)-	MeOH	310(4.23)	95-1049-65
2-Stilbenecarboxylic acid, α-nitro-,	EtOH	315(4.16)	44-3792-65
methyl ester, cis	EtOH	315(4.16)	88-0359-65
$C_{16}H_{13}NO_5$			
Acrylic acid, 3-(2-methoxy-5-nitrophenyl)-2-phenyl-, cis	MeOH	290(4.36)	2-0182-64
Acrylic acid, 2-(p-methoxyphenyl)-3-(p-nitrophenyl)-, cis	EtOH	357(4.37)	87-0614-64
trans	EtOH	288(4.36)	87-0614-64
Acrylic acid, 3-(o-methoxyphenyl)-2-(p-nitrophenyl)-, cis	MeOH	270(3.57),355(3.74)	2-0182-64
trans	MeOH	270(4.02)	2-0182-64
Acrylic acid, 3-(p-methoxyphenyl)-2-(p-nitrophenyl)-, cis	EtOH	340(4.41)	87-0614-64
trans	EtOH	300(4.27)	87-0614-64
1,3-Dioxolo[4,5-g]pyrrolo[1,2-b]isoquinoline-9-acetaldehyde, 10-formyl-5,7,8,9-tetrahydro-5-oxo-	EtOH	248(4.50),290(4.32), 333(3.86),346(3.96)	94-0253-64
$C_{16}H_{13}NS_2$			
2H-1,3-Thiazine-2-thione, 3,6-dihydro-4,6-diphenyl-	H_2O	251(4.22),310s(4.00), 325s(3.97)	39-4008-64
	EtOH	248(4.29)	39-4008-64
$C_{16}H_{13}N_3$			
Benzaldehyde, 1-isoquinolylhydrazone	EtOH	207(4.49),248(4.41)	2-0162-65
Benzaldehyde, 2-quinolylhydrazone	EtOH	209(4.46),236(4.40), 321(4.40)	2-0162-65
$C_{16}H_{13}N_3O$			
p-Benzoquinone, azine with 1-methyl-4(1H)-quinolone	benzene	532(4.60)	5-0009-65J
	C_6H_5Cl	543(4.62)	5-0009-65J
	$C_6H_4Cl_2$	550(4.51)	5-0009-65J
	$CHCl_3$	549(4.65)	5-0009-65J
	C_5H_5N	555(4.74)	5-0009-65J
	99% C_5H_5N	558(4.73)	5-0009-65J
	50% C_5H_5N	595(4.85)	5-0009-65J
	40% C_5H_5N	594(4.85)	5-0009-65J
	30% C_5H_5N	593(4.85)	5-0009-65J
	20% C_5H_5N	590(4.84)	5-0009-65J
	10% C_5H_5N	586(4.83)	5-0009-65J
10H-Dibenzo[b,e][1,4]diazepine-10-acetonitrile, 5,11-dihydro-11-oxo-	EtOH	224(4.50)	33-1590-65
$C_{16}H_{13}N_3O_2$			
Indole-3-carbonitrile, 5,6-dimethoxy-2-(3-pyridyl)-	MeOH	222(4.54),340(4.34)	44-2253-65
Indole-3-carbonitrile, 5,6-dimethoxy-2-(4-pyridyl)-	MeOH	224(4.61),354(4.38)	44-2253-65
as-Triazin-3(2H)-one, 2-methyl-5,6-diphenyl-, 1-oxide, hydrate	EtOH	232(4.41),272(3.44)	94-1168-65

Compound	Solvent	$\lambda_{max}(\log \epsilon)$	Ref.
$C_{16}H_{13}N_3O_4$			
3-Butenoic acid, 4-(m-nitrophenyl)-2-oxo-, phenylhydrazone	MeOH	217(4.27),266(4.34), 383(4.48)	44-1800-65
2-Pyrazoline-3-carboxylic acid, 5-(m-nitrophenyl)-1-phenyl-	MeOH	264(4.42),384(4.32)	44-1800-65
3-Pyridineacrylonitrile, α-(4,5-dimethoxy-2-nitrophenyl)-	MeOH	247(4.25),291(4.26)	44-2253-65
4-Pyridineacrylonitrile, α-(4,5-dimethoxy-2-nitrophenyl)-	MeOH	254(4.30)	44-2253-65
$C_{16}H_{13}N_3O_4S$			
Acetamide, 2-benzoyl-2-diazo-N-(p-tolylsulfonyl)-	MeOH	240(4.46)	5-0101-64F
$C_{16}H_{13}N_3O_5$			
1H-Naphtho[2,3-d]triazole-4,9-diol, 1-acetyl-, diacetate	$CHCl_3$	245(4.44),272(4.33), 282(4.29),298(3.52), 312(3.69),326(3.65), 374(3.85)	39-1003-65
$C_{16}H_{13}N_3O_6$			
Isocarbostyril, 5-methoxy-2-(2,4-dinitrophenyl)-	EtOH	240(4.24),262(4.27), 398(3.78)	24-1023-65
Isocarbostyril, 8-methoxy-2-(2,4-dinitrophenyl)-	EtOH	254(4.32),405(3.93)	24-1023-65
$C_{16}H_{13}N_3O_8$			
Acetic acid, (2,4-dinitrophenyl)-(p-nitrophenyl)-, ethyl ester	EtOH	240s(4.30),261(4.34)	44-0636-64
$C_{16}H_{13}N_5$			
s-Triazolo[1,5-a]pyrimidine, 5-methyl-7-(2-naphthylamino)-	MeOH	304(4.36)	65-0204-64
$C_{16}H_{13}Na$			
Sodium, (9,10-dimethylanthracenyl)-	THF	580(--),625(--), 695(--),725(3.97)	60-1424-64
$C_{16}H_{14}$			
Anthracene, dimethyl-	C_6H_{12}	257(5.28),330(3.47), 342(3.63),358(3.73), 378(3.64)	33-0437-65
Anthracene, 1,9-dimethyl-	C_6H_{12}	346s(3.4),354(3.7), 372(3.8),391(3.8)	28-3688-65A
Anthracene, 1,10-dimethyl-	C_6H_{12}	351(3.8),365s(3.8), 370(4.0),385s(3.7), 387s(3.9),388(4.0)	28-3688-65A
	EtOH	257s(3.4),350(3.7), 268(3.9),389(3.8)	44-4384-65
Anthracene, 2,9-dimethyl-	EtOH	251(5.01),258(5.30), 352(3.75),371(3.87), 392(3.84)	70-0482-64
	EtOH	251(5.01),258(5.30), 352(3.74),371(3.87), 392(3.84)	70-1013-64
Anthracene, 3,9-dimethyl-	EtOH	251(5.13),258(5.44), 351(3.88),370(4.00), 390(3.95)	70-0482-64
	EtOH	251(5.13),258(5.43), 351(3.88),370(3.99), 391(3.96)	70-1013-64

Compound	Solvent	$\lambda_{max}(\log \epsilon)$	Ref.
1,3-Butadiene, 1,3-diphenyl-, cis	n.s.g.	223(4.24),276(3.96)	12-0353-64
1,3-Butadiene, 1,3-diphenyl-, trans	n.s.g.	212(4.32),215(4.34), 221(4.30),265(4.40), 273s(4.41),280(4.53), 288(4.38),297(4.48)	12-0353-64
Butadiene, 1,4-diphenyl-	CHCl	328(4.61),345(4.55)	83-0715-65
1,3-Butadiene, 1,4-diphenyl-	n.s.g.	316(4.64),330(4.72), 347(4.53)	5-0058-65B
1,3-Butadiene, 2,3-diphenyl-	EtOH	242(4.33)	94-0663-65
Cyclobutene, 1,2-diphenyl-	hexane	223(4.35),227(4.38), 297(4.26),307s(4.23)	88-3937-65
	EtOH	297(4.28),308(4.27)	44-3295-65
	EtOH	228(4.36),298(4.25), 309s(4.23)	77-0353-65
Cyclobutene, 1,3-diphenyl-	MeOH	254(4.26)	88-0945-65
Phenanthrene, 1,7-dimethyl-	hexane	252s(4.85),257(4.90), 280(4.28),288(4.18), 301(4.24),319(2.60), 327(2.46),335(2.55), 352(2.22)	1-1875-65
Phenanthrene, 9,10-dimethyl-	EtOH	214(4.47),223(4.33), 248(4.66),254(4.75), 271(4.25),278(4.13), 296(3.99),298(4.00), 320(2.46),328(2.41), 336(2.62),353(2.59)	39-5544-64
Pyrene, 4,5,9,10-tetrahydro-	isooctane	260s(3.93),270(4.16), 280(4.25),293(4.07)	35-1710-64
Unknown compound	n.s.g.	314(4.60),334(4.41)	35-0453-64
$C_{16}H_{14}BF_4N_2PS_2$			
3-Methyl-2-[(3-methyl-2-benzothiazolinylidene)phosphino]benzothiazolium tetrafluoroborate	n.s.g.	472(4.64)	89-0433-64
$C_{16}H_{14}BaO_6S_2$			
Benzocyclobutene-4-sulfonic acid, barium salt	H_2O	220(3.86),262(3.03), 268(3.19),275(3.19)	78-0245-65
$C_{16}H_{14}BrNO$			
1-(Benzyloxy)quinolizinium bromide	H_2O	246(4.33),350(4.27)	44-0526-65
$C_{16}H_{14}BrNO_2$			
Acetanilide, 4'-bromo-2-(p-toluoyl)-	0.04N NaOH	336(4.47)	20-0782-64
	60% EtOH	258(4.50)	20-0782-64
$C_{16}H_{14}BrNO_3$			
Acetanilide, 2-p-anisoyl-4'-bromo-	0.04N NaOH	337(4.52)	20-0782-64
	60% EtOH	275(4.35)	20-0782-64
$C_{16}H_{14}ClNO$			
2-Indolinone, 3-(p-chlorobenzyl)-1-methyl-	EtOH	222s(--),254(3.91), 265s(--),276s(--)	87-0626-65
$C_{16}H_{14}ClNO_2$			
Acetanilide, 3'-chloro-2-(p-toluoyl)-	0.04N NaOH	338(4.46)	20-0782-64
	60% EtOH	257(4.40)	20-0782-64
p-Acetotoluidide, 2-(p-chlorobenzoyl)-	0.04N NaOH	338(4.52)	20-0782-64
	60% EtOH	255(4.41)	20-0782-64

Compound	Solvent	$\lambda_{max}(\log \epsilon)$	Ref.
$C_{16}H_{14}ClNO_3$			
Acetanilide, 2-p-anisoyl-4'-chloro-	0.04N NaOH	333(4.48)	20-0782-64
	60% EtOH	278(4.28)	20-0782-64
p-Acetanisidide, 2-(p-chlorobenzoyl)-	0.04N NaOH	337(4.36)	20-0782-64
	60% EtOH	256(4.43)	20-0782-64
$C_{16}H_{14}ClNO_4$			
1-Phenyllepidinium perchlorate	EtOH	236(4.62),315(3.97)	65-0506-65
1-Phenylquinaldinium perchlorate	EtOH	240(4.46),320(4.01)	65-0506-65
	EtOH	240(4.5),320(4.0)	65-3360-64
9H-Pyrrolo[1,2-a]indole-3-glyoxylic acid, 1-chloro-7-methoxy-6-methyl-, methyl ester	MeOH	247(4.11),305(4.00), 330(4.00)	35-4608-64
$C_{16}H_{14}ClNO_4S_2$			
2-(N-Methylanilino)-4-phenyl-1,3-dithiol-1-ium perchlorate	EtOH	234(4.13),330(4.03)	44-2877-64
$C_{16}H_{14}ClNO_5$			
6-Hydroxy-1-phenylquinaldinium perchlorate	EtOH	245(4.5),320(4.0)	65-3360-64
$C_{16}H_{14}ClNO_6$			
6-Hydroxy-1-(p-hydroxyphenyl)-lepidinium perchlorate	EtOH	251(4.43),320(3.67)	65-0506-65
dianion	EtOH	281(4.45),427(3.75)	65-0506-65
$C_{16}H_{14}Cl_4CuN_4O_2$			
Mandelamidine, 2,4-dichloro-, copper complex	MeOH	263(3.85),590(1.61)	39-4004-65
$C_{16}H_{14}CuO_4$			
Copper m-toluate, form A	dioxan	397s(--),663(2.42)	39-6464-65
	MeCN	400s(--),689(2.40)	39-6464-65
Copper m-toluate, form B	dioxan	402s(--),669(2.43)	39-6464-65
	MeCN	400s(--),682(2.43)	39-6464-65
Copper o-toluate	dioxan	416s(--),669(2.45)	39-6464-65
	MeCN	417s(--),684(2.44)	39-6464-65
Copper p-toluate	dioxan	415s(--),668(2.48)	39-6464-65
$C_{16}H_{14}FNO$			
2-Indolinone, 3-(p-fluorobenzyl)-1-methyl-	MeOH	206(4.50),253(3.92), 281s(3.18)	87-0626-65
$C_{16}H_{14}FeN_2$			
Ferrocene, (phenylazo)-	isooctane	267(3.92),275(3.94), 320(4.27),515(3.39)	70-0197-64
$C_{16}H_{14}FeN_2O$			
Ferrocene, (phenylazoxy)-	isooctane	328(4.05),540(3.20)	70-0197-64
$C_{16}H_{14}FeS$			
Ethylene, 1-ferrocenyl-2-(2-thienyl)-, cis	$CHCl_3$	304(4.01),448(2.76)	5-0088-64F
trans	$CHCl_3$	325(4.28),462(3.11)	5-0088-64F
$C_{16}H_{14}N_2O$			
Acetophenone, 2-(3,4-dihydro-4-quinazolinyl)-	EtOH	245(4.12),285(3.80)	94-1111-64
Propiophenone, 3-(1-benzimidazolyl)-	EtOH	249(4.25)	31-0617-65

Compound	Solvent	$\lambda_{max}(\log \epsilon)$	Ref.
3(2H)-Pyridazinone, 4,5-dihydro-4,6-diphenyl-	EtOH	286(4.17)	39-3342-65
3(2H)-Pyridazinone, 4,5-dihydro-5,6-diphenyl-	EtOH	288(4.20)	39-5302-64
$(C_{16}H_{14}N_2O)_n$			
Polyazine of 4,4'-diacetyl-diphenyl oxide	n.s.g.	283(4.4)	70-1703-64
$C_{16}H_{14}N_2OS$			
4H-1,3-Thiazin-4-one, 2,3,5,6-tetra-hydro-3-phenyl-2-(phenylimino)-	MeOH	227(4.35),275(3.60)	44-1720-64
$C_{16}H_{14}N_2O_2$			
1,2,3-Butanetrione, 1-phenyl-, 2-phenylhydrazone	MeOH	240(3.99),280(3.79), 365(4.24)	44-2959-64
Butyric acid, 4-hydroxy-2-oxo-4 phenyl-, γ-lactone, phenylhydrazone	n.s.g.	233(4.27),295(3.98), 333(4.50)	39-0141-64
6,9-Ethanobenzo[g]quinoxaline-5,10-di-one, 6,9-dihydro-2,3-dimethyl-	MeCN	220(4.76),252(4.45), 276(4.46),292(4.24)	44-2583-65
Fluorene, 9-dimethylaminomethylene-2-nitro-	MeCN	230(4.65),284(4.32), 364(4.56),445(3.76)	24-3331-64
4(1H)-Quinazolinone, 2-(1-hydroxy-ethyl)-1-phenyl-	iso-PrOH	230(4.30),268(3.68), 277(3.72),303(4.00), 315(3.95)	4-0323-65
4(3H)-Quinazolinone, 2-methyl-3-o-tolyl-, 1-oxide	EtOH	315(4.13)	95-0507-65
$C_{16}H_{14}N_2O_3$			
2-Indolinone, 3-[α-(nitromethyl)-benzyl]-	EtOH	251(3.85),280(3.15)	87-0626-65
Nicotinonitrile, 1,2-dihydro-6-(β-hy-droxyphenethyl)-2-oxo-, acetate	EtOH	235(3.95),339(4.07), 380(3.38)	35-5198-65
1,2,4-Oxadiazolidin-5-one, 4-benzoyl-2-methyl-3-phenyl-	n.s.g.	210(4.5),258(4.4)	83-0623-64
$C_{16}H_{14}N_2O_4$			
Acetanilide, 3'-nitro-2-(p-toluoyl)-	0.04N NaOH	333(4.47)	20-0782-64
	60% EtOH	253(4.47)	20-0782-64
Acetotoluidide, 2-benzoyl-3'-nitro-	0.04N NaOH	339(4.46)	20-0782-64
	60% EtOH	248(4.52)	20-0782-64
Acetotoluidide, 2-(m-nitrobenzoyl)-	0.04N NaOH	334(4.30)	20-0782-64
	60% EtOH	240(4.38)	20-0782-64
Fluorene, 2,9-dinitro-9-propyl-	ether	216(4.71),241(4.35), 324(4.62)	44-1961-64
$C_{16}H_{14}N_2O_4S$			
5-Thia-1-azabicyclo[4.2.0]oct-2-ene-2-carboxylic acid, 3-(hydroxymethyl)-8-oxo-7-(2-phenylacetamido)-, γ-lactone	$CHCl_3$	258(3.88)	39-5015-65
$C_{16}H_{14}N_2O_4S_2$			
5-Thia-1-azabicyclo[4.2.0]oct-2-ene-2-carboxylic acid, 3-(hydroxymethyl)-8-oxo-7-[2-(phenylthio)acetamido]-, γ-lactone	n.s.g.	250(4.08)	87-0022-65
$C_{16}H_{14}N_2O_5$			
Acetanilide, 2-p-anisoyl-3'-nitro-	0.04N NaOH	336(4.53)	20-0782-64

Compound	Solvent	$\lambda_{max}(\log \epsilon)$	Ref.
Acetanilide, 2-p-anisoyl-3'-nitro- (contd.)	60% EtOH	279(4.36)	20-0782-64
2,4-Pentadienanilide, p-hydroxy-4-methyl-5-(5-nitro-2-furyl)-	propylene glycol	285(4.15),315(4.18), 405(4.46)	95-0646-64
Uracil, 1-(2,3-dideoxy-β-D-glycero-pent-2-enofuranosyl)-, 5'-benzoate	EtOH	230(4.16),261(3.97)	35-1896-64
$(C_{16}H_{14}N_2S)_n$			
Polyazine of 4,4'-diacetyldiphenyl sulfide	n.s.g.	247(4.3),300(4.3), 327(4.3)	70-1703-64
$C_{16}H_{14}N_4$			
Quinazoline, 3,4-dihydro-, dimer	pH 1	282(3.9)	1-1984-64
	pH 12	292(4.0)	1-1984-64
$C_{16}H_{14}N_4O_2$			
s-Tetrazine, 3,6-bis(p-methoxyphenyl)-	$CHCl_3$	336(4.6),390s(3.7), 564(2.8)	28-1262-64A
$C_{16}H_{14}N_4O_4$			
1(2H)-Naphthalenone, 3,4-dihydro-, 2,4-dinitrophenylhydrazone	$CHCl_3$	388(4.45)	5-0021-64G
1,3,4-Oxadiazole, 2-(ethylamino)-5-[1-phenyl-2-(5-nitro-2-furyl)vinyl]-	propylene glycol	300(3.95),384(4.20)	95-0219-64
$C_{16}H_{14}N_4O_4S$			
Sulfanilamide, N^1-(3-phenyl-1,2,4-oxadiazol-5-yl)-N^4-acetyl-	EtOH	233(4.49)	95-1061-64
$C_{16}H_{14}N_4O_5$			
2H-Naphtho[2,3-d]triazole-4,9-diol, 2-acetamido-, diacetate	$CHCl_3$	246(4.75),313(3.22), 327(3.50),343(3.62)	39-1003-65
$C_{16}H_{14}N_4O_7$			
Isoquinoline, 3,4-dihydro-1-methyl-, picrate	EtOH-HCl	275(3.97)	94-0249-64
	EtOH	248(3.89)	94-0249-64
$C_{16}H_{14}N_4S$			
Benzaldehyde, azine with 4-methyl-2-phenyl-2-thiadiazolin-5-one	EtOH	237(4.38),316(4.04), 369(4.38)	1-0871-64
Benzaldehyde, N-benzyl-N-thiadiazolylhydrazone	EtOH	227(4.18),322(4.41)	1-0871-64
Benzaldehyde, N-methyl-N-(5-phenylthiadiazolyl)hydrazone	EtOH	228(4.28),278(3.95), 349(4.49)	1-0871-64
$C_{16}H_{14}N_6O$			
6(5H)-Pteridinone, 7,8-dihydro-7-(quinazolin-4-ylmethyl)-7-methyl-	pH 2.0	225(4.56),297(4.07)	39-3357-64
	pH 7.8	224(4.50),292(3.91)	39-3357-64
	pH 13.0	220(4.64),305(4.05)	39-3357-64
$C_{16}H_{14}O$			
Anthracene, 1-methoxy-10-methyl-	EtOH	239(4.60),256(5.02), 356(3.80),364(3.75), 375(3.94),396(3.84)	65-2563-64
Crotonaldehyde, 2,3-diphenyl-	n.s.g.	277(4.00)	88-1049-65

Compound	Solvent	λ max(log ε)	Ref.
Cyclopenta[e,f]heptalene, 5-methoxy-3-methyl-	hexane	373(4.13),386(4.07), 390(4.10),420(3.00), 438(2.59),448(3.03), 677(2.04),733(2.14), 749(2.15),780(2.06), 823(2.09),847(2.06), 930(1.70),973(1.68)	5-0194-64B
Cyclopropane, 1-benzoyl-2-phenyl-	C_6H_{12}	256(4.30)	35-1353-65
cis	hexane	240(4.15)	35-1410-65
isomeric mixture	hexane	242(4.26)	35-1353-65
Fluorene, 9-acetyl-9-methyl-	EtOH	212(4.46),268(4.33), 292(3.88),305(3.88)	22-2345-65
Fluoren-9-one, 2,4,5-trimethyl-	EtOH	259(4.69),267(4.78), 300s(3.36),311(3.46), 385(2.57),408(2.64)	22-2953-64
Fluoren-9-one, 2,4,7-trimethyl-	EtOH	255(4.66),264(4.84), 291s(3.23),302(3.51), 309(3.41),328s(3.23), 405s(2.54),422(2.60)	22-2953-64
1(2H)-Naphthalenone, 3,4-dihydro-7-phenyl-	EtOH	245(4.40),314(3.16)	78-0195-64
Phenanthrene, 7-methoxy-1-methyl-	n.s.g.	223(4.3),236(4.3), 257(4.7),277(4.0), 300(3.9),323(2.8), 337(3.0),354(3.0)	70-2021-64
Phenanthrene, 7-methoxy-2-methyl-	n.s.g.	226(4.3),236(4.3), 256(4.8),280(4.3), 293(4.0),323(2.8), 331s(2.6),337(3.0), 356(3.0)	70-2021-64
$C_{16}H_{14}OS$			
Acrylophenone, 3-(m-tolylthio)-	MeOH	257(4.11),337(4.29)	5-0136-64C
4H,6H-5-Oxa-11-thiodibenzo[ef,kl]heptalene, 10,12-dihydro-	isooctane	250(4.01),255s(4.00), 292s(2.81)	35-1710-64
	EtOH	253.5(4.04)	35-1710-64
$C_{16}H_{14}OS_2$			
Cyclohexanone, 2,6-di-2-thenylidene-	hexane	266(4.00),362(4.37)	65-3645-64
	EtOH	266(4.00),380(4.46)	65-3645-64
	66% H_2SO_4	244(3.84),326(3.42), 398(3.65),548(4.94)	65-3645-64
$C_{16}H_{14}O_2$			
Acrylic acid, 3-phenyl-2-p-tolyl-	EtOH	276(4.15)	44-1473-65
Benzofuro[2,3-b]benzofuran, 5a,10b-dihydro-5a,10b-dimethyl-	EtOH	203(4.24),283(3.80)	78-2289-65
4H,6H-[2]Benzoxepino[6,5,4-def][2]benzoxepin, 10,12-dihydro-	isooctane	256(4.16)	35-1710-64
2-Butyne, 1,4-diphenoxy-	EtOH	270(3.40)	78-2289-65
Chalcone, 3-hydroxy-4'-methyl-	EtOH	260(3.90),308(4.40)	23-2580-64
Chalcone, 4'-hydroxy-4-methyl-	EtOH	230(3.89),320(4.21), 386(3.81)	23-2580-64
Cyclopropanecarboxylic acid, 1-(4-biphenylyl)-	EtOH	255(4.39)	87-0504-64
1,4:5,8-Dimethanoanthracene-9,10-dione, 1α,2,3,4α,5β,8β-hexahydro-	C_6H_{12}	278s(4.15),285(4.21), 406(2.16),441s(1.89), 471s(1.59),505s(1.20)	39-3043-64
	EtOH	283(4.14),321s(2.85), 395s(2.23)	39-3043-64

Compound	Solvent	λ_{max}(log ε)	Ref.
Diphenacyl	n.s.g.	244(4.42)	23-2822-64
Fluorene-2-carboxylic acid, ethyl ester	C_6H_{12}	272(4.33),283(4.43), 294(4.34),299(4.36), 307(4.54)	35-2088-64
1α(H),5α(H)-Heptacyclo[10.2.1.1^{5,8}.-0^{2,11}.0^{2,6}.0^{7,11}]hexadec-13-ene-3,10-dione	EtOH	310(1.62)	39-3043-64
Naphtho[1,2-b]pyran-2-one, 4-propyl-	MeOH	221(4.36),265(4.26), 275(4.35)	44-0502-64
2,4,6-Tetradecatriene-8,10,12-triyn-oic acid, ethyl ester	ether	280(4.50),293(4.79), 327(4.57),348(4.80), 371(4.76)	24-0809-64
o-Tolil	EtOH	260(4.29)	44-0859-65
$C_{16}H_{14}O_2S$			
Acetophenone, 4',4'''-thiodi-	n.s.g.	247(4.3),300(4.3), 327(4.3)	70-1703-64
$C_{16}H_{14}O_3$			
Acetophenone, 4',4'''-oxydi-	n.s.g.	283(4.4)	70-1703-64
Acrylic acid, 2-(o-methoxyphenyl)-3-phenyl-	EtOH	273(4.20)	44-1473-65
Acrylic acid, 2-(p-methoxyphenyl)-3-phenyl-	EtOH	270(4.11)	44-1473-65
cis	EtOH	298(4.37)	87-0614-64
trans	EtOH	274(4.16)	87-0614-64
Acrylic acid, 3-(o-methoxyphenyl)-2-phenyl-, cis	MeOH	280(3.97),309(3.87)	2-0182-64
trans	MeOH	275(3.97),305(3.87)	2-0182-64
Acrylic acid, 3-(p-methoxyphenyl)-2-phenyl-, cis	EtOH	309(4.31)	87-0614-64
trans	EtOH	302(4.36)	87-0614-64
1H-Benz[e]indene-2-acetic acid, 2,3-dihydro-α-methyl-3-oxo-	EtOH	251(4.72),273(3.90), 282(3.97)	42-0643-64
Benzofuran, 5-methoxy-2-(m-methoxy-phenyl)-	EtOH	214(4.59),290(4.32), 303s(4.35),316(4.41)	35-5277-64
Benzofuran, 6-methoxy-2-(p-methoxy-phenyl)-	EtOH	286s(4.31),317(4.65), 332(4.55)	35-5277-64
Benzofuran, 7-methoxy-2-(m-methoxy-phenyl)-	EtOH	211(4.52),241(4.19), 296(4.43),304s(4.39), 318(4.26)	35-5277-64
3-Butenoic acid, 2-hydroxy-4,4-diphenyl-	n.s.g.	253(4.16)	28-1649-64B
Chalcone, 2'-hydroxy-4-methoxy-	MeOH	364(4.48)	24-1910-64
Chalcone, 3-hydroxy-4'-methoxy-	EtOH	282(4.07),327(4.43)	23-2580-64
9H-Cyclohepta[a]naphthalen-9-one, 7,8,10,11-tetrahydro-2,3-(meth-ylenedioxy)-	EtOH	230(4.70),250s(4.32), 273(3.77),283(3.80), 294(3.67),310(3.21), 317(3.49),324(3.42), 330(3.73)	95-0437-65
Cyclohexanone, 2,6-difurfurylidene-	EtOH	248(3.86),377(4.38)	65-3645-64
	HOAc-H_2SO_4	408(3.80),550(4.88)	65-3645-64
Fluorene-3-carboxylic acid, 2-hydroxy-1-methyl-, methyl ester	EtOH	247(4.46),273(4.21), 284(4.13)	42-0479-64
Stilbene-2-carboxylic acid, 2'-methoxy-, trans	EtOH	290(4.15),326(4.24)	42-0093-64
4,6-Tetradecadiene-8,10,12-triynoic acid, 3-oxo-, ethyl ester	ether	356(4.74),378(4.78)	24-2605-65

Compound	Solvent	λ max(log ε)	Ref.
$C_{16}H_{14}O_4$			
5,6-Acenaphthenedicarboxylic acid, dimethyl ester	EtOH	315(3.88)	44-2824-64
Acrylic acid, 3-(2-hydroxy-3-methoxy-phenyl)-2-phenyl-, trans	MeOH	286(4.11)	2-0182-64
Acrylic acid, 3-(2-hydroxy-4-methoxy-phenyl)-2-phenyl-, trans	MeOH	290(3.98),325(4.12)	2-0182-64
Anthracene-1,4-peroxide, 1,4-dihydro-1,4-dimethoxy-, endo	ether	234(3.7),259(3.9), 281s(3.8),292s(3.5), 309(3.0),330s(2.5)	28-5031-65A
Anthracene-9,10-peroxide, 9,10-dihydro-9,10-dimethoxy-, endo	ether	259(3.0),312s(3.0), 321(3.0),330s(2.9)	28-5031-65A
3(2H)-Benzofuranone, 2-benzyl-2-hydroxy-6-methoxy-	EtOH	235(3.91),276(4.04), 326(3.83)	22-3572-65
	EtOH-NaOH	380(3.75)	22-3572-65
3(2H)-Benzofuranone, 2-hydroxy-2-(p-methoxybenzyl)-	EtOH	256(3.93),336(3.47)	22-3572-65
	EtOH-NaOH	398(3.88)	22-3572-65
Flavanone, 5,7-dihydroxy-6-methyl-	n.s.g.	295(4.24),336(3.57)	2-0399-64
Flavanone, 3-hydroxy-7-methoxy-	EtOH	230s(--),275(4.12), 310(3.80)	22-3350-65
	EtOH	274(4.12),311(3.82)	78-1141-64
Imperatorin	EtOH	299(4.07)	2-0464-64
1H-Naphtho[2,1-b]pyran-1-one, 5,6-dimethoxy-2-methyl-	EtOH	222(4.57),259(4.53), 277s(4.17),293s(3.80), 306(3.76),353(3.61)	94-0316-64
4H-Naphtho[2,3-b]pyran-4-one, 5,10-dimethoxy-2-methyl-	EtOH	221(4.10),251(4.72), 270s(4.30),278s(3.95), 297(3.41),308(3.53), 321(3.36),380(3.72)	94-0316-64
1,3-Propanedione, 1-(o-hydroxyphenyl)-3-(p-methoxyphenyl)-	EtOH	249(4.11),255(4.13), 375(4.56)	2-0351-65
Pterocarpan, 7-hydroxy-4'-methoxy-	n.s.g.	282(3.84),287(3.89)	25-0562-65
4,6,8,10-Tetradecatetraynedioic acid, dimethyl ester	EtOH	227(5.23),238(5.43), 267(3.23),279(3.42), 302(2.85),316(3.05), 335(3.22),359(3.15)	70-0544-65
$C_{16}H_{14}O_4S_2$			
3-Butyne-1,2-diol, 4-[5-(2-thienyl)-2-thienyl]-, diacetate	ether	239(3.81),325(4.36), 332(4.36)	24-0155-65
$C_{16}H_{14}O_5$			
Benzo[a]heptalen-9(5H)-one, 6,7-di-hydro-1,2,3,10-tetrahydroxy-	EtOH	242(4.4),363(4.2)	78-3605-65
	base	260(4.5),402(4.3)	78-3605-65
p-Benzoquinone, 5-(2-ethoxybenzoyl)-2-methoxy-	EtOH	260(4.16),315(3.52)	78-1495-65
Heraclenin	EtOH	250(4.31),305(4.02)	78-0087-64
Heraclinin	EtOH	263(4.11),298(4.07)	2-0464-64
Itaconitin, anhydro-, acetate	EtOH	236s(4.27),316(4.44), 374s(3.50)	94-0058-65
Oroselol, acetate	EtOH	253(4.37),301(3.9)	78-2605-64
Pterofuran	n.s.g.	220(4.47),236s(4.22), 286s(4.21),295s(4.26), 305s(4.39),317(4.58), 332(4.56)	12-0379-64
Rubrofusarin, methyl ether	EtOH	225(4.47),274(4.75), 330(3.52),345(3.54), 385(3.73)	44-0112-65
Spiro[benzofuran-2(3H),1'-[2,5]cyclo-hexadiene]-3,4'-dione, 4,6-dimethoxy-2'-methyl-	EtOH	290(4.47)	39-6960-65

Compound	Solvent	λ_{max}(log ε)	Ref.
$C_{16}H_{14}O_6$			
Acetic acid, (5-benzyloxy-2-carboxy-3-hydroxyphenyl)-	EtOH	262(4.16),303(3.80)	39-5382-64
2'-Acetonaphthone, 1',3',4'-trihydroxy-, 3',4'-diacetate	EtOH	219(4.51),244s(4.27), 258(4.38),268s(4.31), 284s(3.86),296s(3.74), 308s(3.46),378(3.70)	94-0316-64
Acetophenone, 2',4'-dihydroxy-2-(2-carbomethoxyphenoxy)-	EtOH	219(4.21),233s(4.16), 286(4.10)	39-2361-65
2,3-Biphenylenedione, 1,6,7,8-tetramethoxy-	EtOH	264(4.23),306(4.36), 423(4.02)	39-1067-64
2,7-Biphenylenedione, 1,3,6,8-tetramethoxy-	EtOH	258(4.17),298(4.32), 401(4.15)	39-1067-64
6-Chromanacrylic acid, 4,5-dihydroxy-2,2-dimethyl-3-oxo-, δ-lactone	n.s.g.	261(3.55),324(4.11)	78-3591-65
1,2,3-Naphthalenetriol, triacetate	dioxan	232(4.59),292(3.80)	24-0666-65
$C_{16}H_{14}O_7$			
Acetophenone, 2,2',4',6'-tetrahydroxy-3'-methoxy-, 2-benzoate	EtOH	229(4.32),290(4.27), 332(3.54)	78-2977-64
o-Anisic acid, α-(2,4,6-trihydroxybenzoyl)-, methyl ester	EtOH	213(4.37),230(4.31)	39-2361-65
Deodarin	EtOH	290(4.09),330s(--)	78-3727-65
	EtOH-NaOAc	290s(--),330(4.14)	78-3727-65
	EtOH-$AlCl_3$	300(3.84)	78-3727-65
$C_{16}H_{14}O_8$			
3,9-Dodecadiene-5,7-diyne-1,1,12,12-tetracarboxylic acid, di-cis	EtOH	221(4.45),231(4.48), 238(4.48),249(4.36), 263(3.94),277(4.20), 294(4.37),314(4.27)	70-1460-65
Hydroquinone, 2-acetyl-3-(4,5-dioxotetrahydrofuran-2-yl)-, diacetate	EtOH	296(3.50)	33-0769-64
$C_{16}H_{14}S_2$			
4H,6H-[2]Benzothiepino[6,5,4-def]-[2]benzothiepin, 10,12-dihydro-	isooctane	245.5(4.06)	35-1710-64
$C_{16}H_{15}Br$			
Ethylene, 1-(p-bromophenyl)-1-(2,6-xylyl)-	EtOH	257(4.27)	54-0806-65
$C_{16}H_{15}BrN_2O_2$			
1-[(Indol-3-ylcarbonyl)methyl]-4-methoxypyridinium bromide	MeOH	245(4.32),257s(4.30), 303(4.12)	44-3407-64
$C_{16}H_{15}BrO$			
Isobutyrophenone, β-bromo-p-phenyl-	MeOH	285(4.37)	44-0499-64
$C_{16}H_{15}BrO_3$			
Fluorene-8a(4bH)-acetic acid, 7-bromo-5,6,7,8-tetrahydro-8-hydroxy-1-methyl-9-oxo-, γ-lactone	EtOH	252(4.13),300(3.36)	44-0074-64
$C_{16}H_{15}BrS$			
Thiobenzophenone, 3'-bromo-2,4,6-trimethyl-	C_6H_{12}	307(4.02),627(1.67)	54-0289-65
	EtOH	309(3.95),617(1.68)	54-0289-65
Thiobenzophenone, 4'-bromo-2,4,6-trimethyl-	C_6H_{12}	328(4.31),627(1.83)	54-0289-65
	EtOH	238s(--),330(4.26), 614(1.82)	54-0289-65

Compound	Solvent	$\lambda_{max}(\log \epsilon)$	Ref.
$C_{16}H_{15}ClN_2O$			
1H-1,4-Benzodiazepine, 7-chloro-2,3-dihydro-1-methyl-5-phenyl-, 4-oxide	iso-PrOH	241(4.36),268(4.20), 306(4.04)	44-1621-64
1H-1,4-Benzodiazepin-3-ol, 7-chloro-2,3-dihydro-1-methyl-5-phenyl-	iso-PrOH	228(4.28),249(4.23)	44-1621-64
$C_{16}H_{15}ClO_4$			
7-(m-Methylstyryl)tropylium perchlorate	MeCN	247(4.12),278(4.15), 448(4.52)	24-1337-64
7-(o-Methylstyryl)tropylium perchlorate	MeCN	249(4.15),279(4.13), 453(4.51)	24-1337-64
7-(p-Methylstyryl)tropylium perchlorate	MeCN	251(4.16),281(4.15), 461(4.67)	24-1337-64
Spiro[benzofuran-2(3H),1'-[2]cyclohexene]-3,4'-dione, 5-chloro-2'-ethoxy-6'-methyl-	EtOH	223(4.38),248(4.39), 252(4.40),340(3.69)	39-4939-65
Spiro[benzofuran-2(3H),1'-[2]cyclohexene]-3,4'-dione, 6-chloro-2'-ethoxy-6'-methyl-	EtOH	223(4.31),229(4.31), 257(4.52),327(3.88)	39-4939-65
Spiro[benzofuran-2(3H),1'-[2]cyclohexene]-3,4'-dione, 7-chloro-2'-ethoxy-6'-methyl-	EtOH	216(4.38),254(4.41), 331(3.71)	39-4939-65
$C_{16}H_{15}ClO_5$			
6-Chromanacrylic acid, 3-chloro-4,5-dihydroxy-2,2-dimethyl-, δ-lactone, acetate	n.s.g.	261(3.62),324(4.16)	78-3591-65
7-(m-Methoxystyryl)tropylium perchlorate	MeCN	222(4.38),245s(4.12), 278(4.22),448(4.44)	24-1337-64
7-(o-Methoxystyryl)tropylium perchlorate	MeCN	224(4.36),281(4.18), 386s(3.91),481(4.44)	24-1337-64
7-(p-Methoxystyryl)tropylium perchlorate	MeCN	259(4.24),287s(4.11), 503(4.62)	24-1337-64
$C_{16}H_{15}Cl_2O_3P$			
(5,6-Dichloro-3-hydroxy-p-benzoquinonyl)diethylphenylphosphonium hydroxide, inner salt	CH_2Cl_2	266(4.07),289(4.13), 441(3.05)	78-1941-65
$C_{16}H_{15}Cl_3O$			
Azulene, 1-(trichloroacetyl)-3,4,6,8-tetramethyl-	C_6H_{12}	245(4.14),295(3.39), 330(4.02),408(3.92), 540(2.93)	44-0131-65
$C_{16}H_{15}F_3O$			
Azulene, 1-(trifluoroacetyl)-3,4,6,8-tetramethyl- (several inflections not listed)	C_6H_{12}	245(4.46),288(4.04), 332(4.41),413(3.02), 533(2.96)	44-0131-65
$C_{16}H_{15}IN_2O_6$			
Uracil, 1-(3-deoxy-3-iodo-β-D-arabinofuranosyl)-, 5'-benzoate	EtOH	229(4.29),261(3.94)	35-1896-64
$C_{16}H_{15}N$			
2-Picoline, 6-(4-phenyl-1,3-butadienyl)-, trans-trans	EtOH	231(4.01),235(4.73)	94-0503-65
3-Picoline, 2-(4-phenyl-1,3-butadienyl)-, trans-trans	EtOH	234(4.03),289(4.29), 340(4.68)	94-0503-65
3-Picoline, 6-(4-phenyl-1,3-butadienyl)-, 1-cis-3-trans	EtOH	242(4.15),249(4.11), 330(4.43)	94-0503-65

Compound	Solvent	λ_{max}(log ϵ)	Ref.
3-Picoline, 6-(4-phenyl-1,3-butadienyl)-, di-trans	EtOH	229(3.94),333(4.74)	94-0503-65
4-Picoline, 2-(4-phenyl-1,3-butadienyl)-, trans-trans	EtOH	232(4.00),332(4.72)	94-0503-65
Pyridine, 2-(2-methyl-4-phenyl-1,3-butadienyl)-, trans-trans	EtOH	325(4.65)	94-0503-65
$C_{16}H_{15}NO$			
Acrylophenone, 3-anilino-4'-methyl-	heptane	237(3.97),257(4.11), 367(4.48)	70-1382-65
8-Azagona-1,3,5(10),9(11),13-pentaen-12-one	EtOH	256(4.57),286(4.23)	44-3667-65
2-Indolinone, 3-benzyl-1-methyl-	EtOH	253(3.91),263s(--), 281s(--)	87-0626-65
$C_{16}H_{15}NOS_2$			
Carbanilic acid, N-methyldithio-, ester with 2-mercaptoacetophenone	H_2O	246s(4.04),288(4.16)	39-4004-64
	EtOH	240(4.19),293(4.17)	39-4004-64
$C_{16}H_{15}NO_2$			
Acetanilide, 2-m-toluoyl-	0.04N NaOH	336(4.35)	20-0782-64
	60% EtOH	249(4.34)	20-0782-64
p-Acetotoluidide, 2-benzoyl-	0.04N NaOH	332(4.35)	20-0782-64
	60% EtOH	245(4.40)	20-0782-64
Cinnamyl alcohol, 2-benzamido-	EtOH	234(4.36)	44-0305-64
Indole, 5-(benzyloxy)-6-methoxy-	EtOH	217(4.52),273(3.72), 298(3.89)	4-0387-65
Indole, 6-(benzyloxy)-5-methoxy-	EtOH	217(4.49),273(3.76), 298(3.89)	4-0387-65
$C_{16}H_{15}NO_2S$			
Alanine, 3-(fluoren-2-ylthio)-	EtOH	292(4.23),304(4.13)	39-0515-64
Thiobenzophenone, 2,4,6-trimethyl-3'-nitro-	C_6H_{12}	299(4.09),632(1.63)	54-0289-65
	EtOH	300(4.00),620(1.66)	54-0289-65
Thiobenzophenone, 2,4,6-trimethyl-4'-nitro-	C_6H_{12}	312(4.16),638(3.61)	54-0289-65
	EtOH	312(4.11),625(3.60)	54-0289-65
$C_{16}H_{15}NO_3$			
Acetanilide, 2-p-anisoyl-	0.04N NaOH	333(4.44)	20-0782-64
	60% EtOH	278(4.27)	20-0782-64
p-Acetanisidide, 2-benzoyl-	0.04N NaOH	335(4.37)	20-0782-64
	60% EtOH	250(4.42)	20-0782-64
3-Cyclohexene-1-carboxylic acid, 3-(indol-3-ylcarbonyl)-, cis	NaOH	265(4.18),330(4.22)	65-3025-64
$C_{16}H_{15}NO_4$			
Biphenyl, 4-acetoxymethyl-2-methyl-4'-nitro-	EtOH	295(4.07)	88-2655-65
Indole-7-carboxylic acid, 2,4,5,6-tetrahydro-3-hydroxy-2-oxo-1-phenyl-, methyl ester	n.s.g.	300(3.87)	39-1648-65
Propiophenone, 2-hydroxy-α-methyl-4'-phenyl-, nitrate	MeOH	288(4.38)	44-0499-64
$C_{16}H_{15}NO_5$			
4H-m-Dioxino[5,4-c]pyridine-4,5(6H)-dione, 7-hydroxy-2,2-dimethyl-6-m-tolyl-	EtOH	313(4.51)	78-1917-65

Compound	Solvent	$\lambda_{max}(\log \epsilon)$	Ref.
4H-m-Dioxino[5,4-c]pyridine-4,5(6H)-dione, 7-hydroxy-2,2-dimethyl-6-o-tolyl-	EtOH	316(4.57)	78-1917-65
4H-m-Dioxino[5,4-c]pyridine-4,5(6H)-dione, 7-hydroxy-2,2-dimethyl-6-p-tolyl-	EtOH	312(4.39)	78-1917-65
Malonic acid, [3-hydroxy-1-(1-hydroxy-1-methylethoxy)-3-m-toluidino-allylidene]-, di-δ-lactone	EtOH	340(4.92)	78-1917-65
Malonic acid, [3-hydroxy-1-(1-hydroxy-1-methylethoxy)-3-o-toluidino-allylidene]-, di-δ-lactone	EtOH	340(4.52)	78-1917-65
Malonic acid, [3-hydroxy-1-(1-hydroxy-1-methylethoxy)-3-p-toluidino-allylidene]-, di-δ-lactone	EtOH	336(4.48)	78-1917-65
$C_{16}H_{15}NO_5S$			
o-Toluic acid, 6-acetyl-α-benzene-sulfonamido-	n.s.g.	284(3.20)	44-0722-65
$C_{16}H_{15}NO_6$			
2,3-Biphenylenedione, 1,6,7,8-tetra-methoxy-, 3-oxime	EtOH	235(4.09),301(4.42), 419(4.07)	39-1067-64
$C_{16}H_{15}NO_8$			
Coumarin-6-carboxylic acid, 3-acetyl-5-hydroxy-4,7-dimethyl-8-nitro-, ethyl ester	EtOH	255(4.24),293s(4.04)	39-2543-65
$C_{16}H_{15}NS$			
Acrylophenone, 3-N-methyl-anilinothio-	EtOH	211(4.24),255(4.04), 319(4.07),440(4.24)	39-0032-65
$C_{16}H_{15}N_3$			
3H-Dibenz[b,f]imidazo[1,2-d][1,4]di-azepine, 2,9-dihydro-9-methyl-	EtOH	233(4.50),255s(4.23), 294(4.00)	33-1590-65
$C_{16}H_{15}N_3O$			
Benzamide, N-[1-(2-benzimidazolyl)-ethyl]-	EtOH	275(3.93),282(3.92)	94-0773-64
$C_{16}H_{15}N_3O_2$			
10H-Dibenzo[b,e][1,4]diazepine-10-pro-pionamide, 5,11-dihydro-11-oxo-	EtOH	212(4.35),224(4.49)	33-1590-65
3-Pyridineacrylonitrile, α-(2-amino-4,5-dimethoxyphenyl)-	MeOH	247(4.17),284(4.18)	44-2253-65
4-Pyridineacrylonitrile, α-(2-amino-4,5-dimethoxyphenyl)-	MeOH	249(4.17),284(4.19)	44-2253-65
$C_{16}H_{15}N_3O_3$			
Anthramycin, anhydro-	MeCN	231(4.34),334(4.57), 360s(4.35)	35-5791-65
$C_{16}H_{15}N_3O_3S$			
p-Toluenesulfonamide, N-(3-benzyl-1,2,4-oxadiazol-5-yl)-	EtOH	232.7(4.17)	95-1061-64
$C_{16}H_{15}N_3O_4$			
Pyrrole-2-acrylonitrile, α-(4,5-di-methoxy-2-nitrophenyl)-1-methyl-	MeOH	249(4.05),342(4.33)	44-2253-65

Compound	Solvent	$\lambda_{max}(\log \epsilon)$	Ref.
[2,2':5',2''-Terpyrrole]-5,5''-dicarboxylic acid, dimethyl ester	MeOH	226(4.29),367(4.64), 385s(4.53)	44-0883-64
$C_{16}H_{15}N_3O_5$			
L-Tyrosine, mono(p-azobenzenecarboxylic acid)	pH 6.0	330(4.25)	37-0699-65
	pH 13	330(4.17),485(3.99)	37-0699-65
$C_{16}H_{16}$			
Anthracene, 9,10-dihydro-9,9-dimethyl-	EtOH	255(2.87),262(2.95), 269(2.94)	87-0088-64
1-Butene, 1,1-diphenyl-	C_6H_{12}	250(3.90)	22-0693-64
Cyclooctatetraene dimer, m. 41°	n.s.g.	232(3.38)	35-3398-64
5H-Dibenzo[a,c]cycloheptene, 6,7-dihydro-5-methyl-	pet ether	249(4.18),276(3.14)	39-5317-64
Dibenzocyclooctene, tetrahydro-	EtOH	225s(4.20),286(3.85)	23-2183-65
Dibenzo[a,c]cyclooctene, 5,6,7,8-tetrahydro-	pet ether	236(4.08),264(3.02), 273(2.77)	39-5317-64
Dibenzo[a,e]cyclooctene, 5,6,11,12-tetrahydro-	hexane	257(2.59),265(2.70), 268(2.62),272(2.69)	88-3937-65
1,4:9,10-Dimethanoanthracene, 1,4,4a,9,9a,10-hexahydro-, endo-endo	EtOH	209(4.06),259(2.68), 266(2.80),272(2.87), 280(2.58)	39-4646-65
Ethylene, 1,1-di-o-tolyl-	C_6H_{12}	231(4.21)	54-0806-65
	EtOH	231.5(4.20)	54-0806-65
	98% H_2SO_4	329(3.89),440(4.26)	54-0806-65
Ethylene, 1,1-di-p-tolyl-	C_6H_{12}	239(4.33),255(4.12)	54-0806-65
	EtOH	238(4.33),253(4.12)	54-0806-65
	98% H_2SO_4	323(4.03),457(4.80)	54-0806-65
Ethylene, 1-phenyl-1-(2,6-xylyl)-	C_6H_{12}	249(4.09)	54-0806-65
	EtOH	249(4.07)	54-0806-65
	98% H_2SO_4	333(4.29),570(3.90)	54-0806-65
Ethylene, 1-o-tolyl-1-p-tolyl-	C_6H_{12}	236(4.11),254(4.13)	54-0806-65
	EtOH	236(4.10),254(4.13)	54-0806-65
	98% H_2SO_4	329(4.10),452(4.42)	54-0806-65
Fluorene, 2,3,6-trimethyl-	C_6H_{12}	265(4.35),289(3.79), 296(4.03),301(4.00), 306(4.25)	33-1800-65
[2.2]Paracyclophane	C_6H_{12}	225(4.38),244s(3.52), 286(2.41),302(2.19)	25-0767-65
Pentacyclo[9.3.2.0^{2,9}.0^{3,8}.0^{10,12}]hexadeca-4,6,13,15-tetraene	hexane	232s(3.66),282(3.36)	24-3131-64
Phenanthrene, 9,10-dihydro-4,5-dimethyl-	isooctane	261(4.18)	35-1710-64
Tricyclo[8.4.2.0^{2,9}]hexadeca-3,5,7,11,13,15-hexaene	hexane	217(4.35),262(3.51)	24-3131-64
$C_{16}H_{16}BrN$			
4-Stilbenamine, 3'-bromo-N,N-dimethyl-, trans	EtOH	360(4.50)	33-0517-65
conjugate acid	EtOH	298(4.50)	33-0517-65
4-Stilbenamine, 4'-bromo-N,N-dimethyl-, trans	80% EtOH	358(4.53)	33-0517-65
conjugate acid	EtOH	315(4.59)	33-0517-65
$C_{16}H_{16}BrNO$			
Ethylamine, N-(5-bromosalicylidene)-α-benzyl-	EtOH	254(4.01),278s(3.36), 327(3.55),413(3.06)	44-2265-64
$C_{16}H_{16}Br_2CuN_4O_2$			
Mandelamidine, o-bromo-, copper complex	MeOH	266(3.79),580(1.58)	39-4004-65

Compound	Solvent	$\lambda_{max}(\log \epsilon)$	Ref.
$C_{16}H_{16}ClNO$			
Ethylamine, N-(5-chlorosalicylidene)-α-benzyl-	EtOH	254(3.98),278s(3.28), 327(3.53),414(3.01)	44-2265-64
$C_{16}H_{16}ClNO_4$			
1,3-Dioxolane , 2-(chloromethyl)-2-[6-(p-nitrophenyl)-1,3,5-hexatrienyl]-	EtOH	280(4.27),372(4.47)	87-0035-65
$C_{16}H_{16}ClO_4P$			
(5-Chloro-3,6-dihydroxy-p-benzoquinonyl)diethylphenylphosphonium hydroxide, inner salt	CH_2Cl_2	240(4.11),294(4.25), 423(2.95)	78-1941-65
$C_{16}H_{16}Cl_2CuN_4O_2$			
Mandelamidine, o-chloro-, copper complex	MeOH	266(3.77),585(1.62)	39-4004-65
$C_{16}H_{16}Cl_2N_2O_4$			
Azobenzene, 4,4'-dichloro-2,2',5,5'-tetramethoxy-	EtOH	215(4.15),278(4.12), 318(3.84),434(2.88)	78-0189-64
Azobenzene, 5,5'-dichloro-2,2',4,4'-tetramethoxy-	EtOH	215(4.47),274(4.40), 314(3.66),415(3.02)	78-0189-64
$C_{16}H_{16}Cl_4$			
1,4:9,10-Dimethanoanthracene, 5,6,7,8-tetrachloro-1,2,3,4,4a,8a,-9,9a,10,10a-decahydro-, exo-endo-endo	EtOH	267(3.37),278(3.57), 289(3.62),300(3.75), 315(3.53)	39-4646-65
$C_{16}H_{16}D_4N_4O_4$			
Hydrindan-2-one, 8-methyl-4,4,6,6-d_4-, 2,4-dinitrophenylhydrazone, trans	$CHCl_3$	365(4.41)	35-0580-65
$C_{16}H_{16}FeO_2$			
2,4-Pentadienoic acid, 5-ferrocenyl-, methyl ester	$CHCl_3$	324(4.41),482(3.42)	5-0088-64F
$C_{16}H_{16}FeO_4$			
Malonic acid, ferrocenylmethylene-, dimethyl ester	MeOH	299(4.14),482(3.20)	5-0088-64F
	$CHCl_3$	301(4.17),482(3.20)	5-0088-64F
$C_{16}H_{16}N_2$			
Benzimidazole, 2-benzyl-5,6-dimethyl-	pH 2	277(3.97),285(3.96)	44-0476-64
	pH 12	286(3.40)	44-0476-64
	EtOH	246(3.73),283(3.89), 287(3.83)	44-0476-64
$\Delta^{1,\alpha}$-Cyclohexanemalononitrile, 2-methyl-2-phenyl-	aq. MeOH	240(4.04)	95-0381-64
$C_{16}H_{16}N_2O$			
1H-Indazole, 3-methoxy-1-phenethyl-	EtOH	308(3.68)	7-0583-65
3-Indazolinone, 2-methyl-1-phenethyl-	EtOH	242(4.13)	7-0583-65
2-Indolinone, 3-[α-(aminomethyl)-benzyl]-	EtOH	251(3.87),279(3.14)	87-0626-65
2-Indolinone, 1-(dimethylamino)-3-phenyl-	MeOH	253(3.84)	44-3451-65
4(1H)-Pyridone, 1-[2-(2-methylindol-3-yl)ethyl]-	MeOH	264(4.32)	24-2463-64

Compound	Solvent	$\lambda_{max}(\log \epsilon)$	Ref.
$C_{16}H_{16}N_2OS$			
2H-Thieno[3,2-b]pyrrol-3(4H)-one, 2-benzylidene-6-[(dimethylamino)-methyl]-	EtOH	268(3.89),347(4.35)	44-1012-65
$C_{16}H_{16}N_2OS_2$			
Spiro[cyclohexane-1,2'-[1,3]dithiolo-[4,5-d]pyridazin-4'(5'H)-one, 5'-phenyl-	EtOH	256(4.42),312(3.92), 361(3.67)	78-1323-65
$C_{16}H_{16}N_2O_2$			
Benzanilide, 4-acetamido-N-methyl-	EtOH	205(4.5),278(4.3)	28-4295-64B
Benzoic acid, methyl ester, azine	n.s.g.	270.0(4.09)	23-0356-65
Benzo[g]quinoxaline-5,10-dione, 6,9-dihydro-2,3,7,8-tetramethyl-	MeCN	285(4.48)	44-2583-65
o-Benzotoluidide, p-acetamido-	EtOH	204(4.5),271(4.4)	28-4295-64B
p-Benzotoluidide, p-acetamido-	EtOH	203(4.50),282(4.4)	28-4295-64B
Ergol-8-ene-8-carboxylic acid, 6-methyl-	pH 1	218(4.60),278(3.78)	33-1052-64
Ethylenediamine, N,N'-dibenzoyl-	EtOH	230(4.33)	54-0314-65
Indole, 1-acetyl-3-(1-acetyl-2-pyrrolin-2-yl)-	dioxan	238(4.41),292s(3.86), 301(3.94)	87-0415-64
2-Indolinone, 1-(dimethylamino)-3-hydroxy-3-phenyl-	MeOH	259(3.75)	44-3451-65
Lysergic acid	pH 13	240(4.3),310(4.0)	33-1052-64
Pyrazine, 2,5-diethyl-3,6-di-2-furyl-	$CHCl_3$	293(4.22),361(4.43)	22-3476-65
1(2H)-Pyridinecarboxylic acid, 2-indol-3-yl-, ethyl ester	EtOH	223(4.74),275(4.00), 290(3.92)	44-2534-65
2(1H)-Quinolylideneacetic acid, α-cyano-, tert-butyl ester	$CHCl_3$	289(4.38),390s(4.16), 405(4.26),427(4.14)	44-0243-65
	EtOH	285(4.42),382s(4.15), 396(4.24),410s(4.09)	44-0243-65
4-Stilbenamine, N,N-dimethyl-3'-nitro-, trans	80% EtOH	359(4.62)	33-0517-65
conjugate acid	80% EtOH	308(4.69)	33-0517-65
4-Stilbenamine, N,N-dimethyl-4'-nitro-, trans	80% EtOH	435(4.48)	33-0517-65
conjugate acid	80% EtOH	344(4.55)	33-0517-65
$C_{16}H_{16}N_2O_3$			
o-Benzanisidide, p-acetamido-	EtOH	285(4.5)	28-4295-64B
p-Benzanisidide, p-acetamido-	EtOH	284(4.4)	28-4295-64B
Butyric acid, 4-hydroxy-2-oxo-4-phenyl-, phenylhydrazone	n.s.g.	293(4.18),317(4.27)	39-0141-64
Ethylamine, N-(5-nitrosalicylidene)-α-benzyl-	EtOH	228(4.08),248(4.13), 354(4.12),391(4.01)	44-2265-64
2-Indolinone, 5,6-dimethoxy-3-(2-pyridylmethyl)-	MeOH	262(3.96),267s(3.94), 300s(3.57)	87-0626-65
Propiophenone, 2-anilino-4'-methyl-3-nitro-	heptane	242(4.23),256(4.24)	70-1382-65
$C_{16}H_{16}N_2O_4$			
Indole-2-carboxylic acid, 1-(2-cyano-ethyl)-3-formyl-5-methoxy-6-methyl-, methyl ester	MeOH	251(4.29),352(4.12)	35-4612-64
3-Pyridinealanine, N-carboxy-, N-benzyl ester	n.s.g.	260(3.40)	44-2658-64
$C_{16}H_{16}N_2O_4S$			
Acetanilide, 4',4'''-sulfonylbis-	MeOH	256(4.41),284(4.56)	50-0630-65B

Compound	Solvent	$\lambda_{max}(\log \epsilon)$	Ref.
$C_{16}H_{16}N_2O_5S$			
1,3-Cyclohexanedione, 2-[2,3-dihydro-3-oxobenzo[b]thien-2-yl)-azo]-5,5-dimethyl-, S,S-dioxide	MeOH	452(4.15)	24-0036-65
	CH_2Cl_2	453(4.31)	24-0036-65
dipotassium salt	MeOH	508(4.28)	24-0036-65
$C_{16}H_{16}N_2O_6$			
Benzo[1,2-b:4,5-b']difuran-3,7-dicarboxylic acid, 2,6-diamino-, diethyl ester	THF	274(4.45),320(4.36)	39-0974-65
$C_{16}H_{16}N_2O_6S_2$			
5-Thia-1-azabicyclo[4.2.0]oct-3-ene-2-carboxylic acid, 3-(hydroxymethyl)-8-oxo-7-[2-(2-thienyl)acetamido]-, acetate, potassium salt	n.s.g.	236(4.11),260(3.90)	87-0022-65
sodium salt	pH 6.0	237(4.17)	39-5015-65
$C_{16}H_{16}N_2O_7$			
Uracil, 1-β-D-arabinofuranosyl-, 5'-benzoate	H_2O	232(4.11),262(3.97)	44-0467-65
$C_{16}H_{16}N_2S_2$			
2,2'-Bibenzothiazoline, 2,2'-dimethyl-	MeOH	225s(3.89),317(3.98)	35-3056-64
cadmium chelate	pyridine	335(3.69),463(3.31)	35-3056-64
zinc chelate	pyridine	350(3.88),510(3.33)	35-3056-64
$C_{16}H_{16}N_3OP$			
Phosphonic diamide, N,N'-diphenyl-P-pyrrol-2-yl-	EtOH	228(4.52),268s(3.21), 273(3.27),280s(3.14)	44-0091-65
$C_{16}H_{16}N_4$			
1,2,3-Benzotriazine, 3,4-dihydro-4-(isopropylimino)-3-phenyl-	EtOH	263(4.13),271(4.13), 317(3.86)	39-3663-64
4H-1,2,4-Triazole, 4-amino-3,5-di-o-tolyl-	EtOH	233(4.3)	28-1262-64A
$C_{16}H_{16}N_4O_2$			
s-Tetrazine, 3,6-bis(p-methoxyphenyl)-1,2-dihydro-	EtOH	270(4.6),330(3.9)	28-1262-64A
	$CHCl_3$	270(4.6),333s(2.8)	28-1262-64A
4H-1,2,4-Triazole, 4-amino-3,5-bis-(m-methoxyphenyl)-	EtOH	260(4.4),300s(4.1)	28-1262-64A
hydrochloride	EtOH	250(4.6),285(3.2)	28-1262-64A
4H-1,2,4-Triazole, 4-amino-3,5-bis-(p-methoxyphenyl)-	EtOH	269(4.6)	28-1262-64A
$C_{16}H_{16}N_4O_4$			
2,5-Cyclohexadien-1-one, 4-allyl-4-methyl-, 2,4-dinitrophenylhydrazone	EtOH	388(4.52)	33-0094-65
$C_{16}H_{16}N_4O_5$			
p-Benzoquinone, 2-methyl-5-propyl-, 1-(2,4-dinitrophenylhydrazone)	EtOH	425(4.31),520(3.69)	5-0134-65E
$C_{16}H_{16}N_4O_6$			
Acetophenone, 2',4'-dimethoxy-, 2,4-dinitrophenylhydrazone	$CHCl_3$	382.5(4.39)	5-0021-64G
$C_{16}H_{16}N_4O_9S$			
Thymidine, 3'-(2,4-dinitrobenzenesulfenate)	dioxan	264(4.26),328(4.02)	44-2615-64

Compound	Solvent	$\lambda_{max}(\log \epsilon)$	Ref.
$C_{16}H_{16}N_6$			
Melamine, 2-phenyl-4-p-tolyl-	EtOH	270(4.63)	39-3459-64
$C_{16}H_{16}N_6O_8S_2$			
Aniline, N,N'-(dithiodiethylene)-bis[2,4-dinitro-	acetone	349(4.47)	96-0788-64
Cystamine, bis(2,4-dinitrophenyl)-	acetone	352(4.55)	96-0788-64
$C_{16}H_{16}O$			
Anthracene, 9-acetyl-	C_6H_{12}	219(3.66),235(3.95), 239(4.02),253s(2.85), 258(3.95),262(3.53), 312(2.88),344(3.37), 361(3.54),380(3.50)	73-3462-65
	EtOH	215(4.07),245(4.94), 254(5.17),310(3.04), 346(3.66),363(3.85), 382(3.79)	73-3462-65
Bicyclo[4.2.0]octa-1,3,5-triene, 7-methoxy-8-methyl-7-phenyl-	isooctane	250s(2.81),255s(3.00), 261(3.20),267(3.35), 273(3.33),290(1.88), 300(1.86)	35-0086-65
Biphenyl-2-carboxaldehyde, 2',4,6-trimethyl-	hexane	217(4.37),253(4.04), 302(3.41),307s(3.36)	22-2953-64
	EtOH	218(4.30),255(4.07), 309(3.39)	22-2953-64
	$CHCl_3$	309(3.43)	22-2953-64
	CCl_4	303(3.40),310s(3.38)	22-2953-64
Biphenyl-2-carboxaldehyde, 4,4',6-trimethyl-	hexane	218(4.34),226(4.00), 230s(4.18),302(3.44), 312s(3.39)	22-2953-64
	EtOH	220(4.27),259(4.06), 311(3.42)	22-2953-64
Chalcol, 4-methyl-	n.s.g.	258(4.39)	39-4978-64
1-Cyclohexanol, 2-(phenylbutadiynyl)-	ether	245(3.87),258(4.30), 272(4.51),288(4.43)	24-2118-64
Dibenz[c,e]oxepin, 5,7-dihydro-1,11-dimethyl-	isooctane	243(4.06)	35-1710-64
Dibenz[c,e]oxepin, 5,7-dihydro-5,5-dimethyl-	EtOH	215(4.27),254(4.15)	78-3401-65
2(3H)-Naphthalenone, 4,6,7,8-tetrahydro-5-phenyl-	EtOH	232(3.94),307(4.27)	44-2942-65
$C_{16}H_{16}OS$			
Sulfoxide, 2,3-diphenylallyl methyl, cis	EtOH	227(4.28),272(4.16)	94-0663-65
trans	EtOH	276(4.25)	94-0663-65
$C_{16}H_{16}O_2$			
Azulene, 1,3-dipropionyl-	hexane	518(2.80),556(2.73), 609(2.27)	65-0894-64
	EtOH	502(2.84)	65-0894-64
Benzo[b]biphenylene-5,10-diol, 4a,4b,5,10,10a,10b-hexahydro-	n.s.g.	273(3.51)	35-0136-65
Benzophenone, 4'-hydroxy-2,4,6-trimethyl-	EtOH	222s(4.25),282(4.25)	78-1015-65
	EtOH-NaOH	214(4.19),246(3.23), 346(4.50)	78-1015-65
1α(H),5α(H)-Heptacyclo[10.2.1.1^{5,8}.-0^{4,9}.0^{2,6}.0^{7,11}]hexadeca-3,10-dione	EtOH	310(1.50)	39-3043-64
Stilbene, p,p'-dimethoxy-	EtOH	233(4.31),294(4.37)	65-2073-65

Compound	Solvent	λ_{max}(log ε)	Ref.
Stilbene, p,p'-dimethoxy- (contd.)	$C_2H_4Cl_2$	325(4.28)	65-2073-65
tetracyanoethylene complex	CH_2Cl_2	425(3.78?),775(3.78?)	54-1478-65
$C_{16}H_{16}O_2S$			
Dibenzo[c,e]thiepin-1,11-dimethanol, 5,7-dihydro-	dioxan	240s(4.03),284(2.86)	35-1710-64
$C_{16}H_{16}O_3$			
5-Benzofuranacrylic acid, α-butyl-6-hydroxy-β-methyl-, δ-lactone	EtOH	244(4.40),290(4.04), 328(3.98)	44-2467-64
Cyclohexanedione, 2-acetoxy-2,6-dimethyl-4-phenyl-	n.s.g.	242(4.30),318(3.10), 366s(2.63)	49-0512-64
Dalbergione, dihydro-4-methoxy-	EtOH	264(4.17)	78-2683-65
Dibenz[c,e]oxepin-1,11-dimethanol, 5,7-dihydro-	dioxan	246(4.18),285s(3.78)	35-1710-64
Fluorene-8a(4bH)-acetic acid, 5,6-dihydro-1-methyl-9-oxo-	EtOH	252(4.11),300(3.37)	44-0074-64
Fluorene-3-carboxylic acid, 1,2,3,4-tetrahydro-1-methyl-2-oxo-, methyl ester	EtOH	249(4.17)	42-0479-64
Fluoren-2-ol, 5,8-dihydro-7-methoxy-, acetate	EtOH	261(4.11)	88-3505-65
Ichthyothereol, acetate	n.s.g.	210(4.53),232(4.72), 245(4.78),257(3.79), 273(3.95),289(4.23), 308(4.35),325(4.18)	35-5237-65
4,6-Nonadiynoic acid, 8-hydroxy-9-phenyl-, methyl ester	EtOH	243(2.69),258(2.57), 280(2.43),295(2.31)	70-1237-65
2-Phenanthrenecarboxylic acid, 1,2,3,4-tetrahydro-7-methoxy-	EtOH	228(4.78),255(3.59), 265(3.69),275(3.71), 285(3.52),321(3.28), 330(3.24),335(3.38)	44-1213-65
Xanthorrhoeol, methyl ether	EtOH	226(4.59),254(4.36), 295(3.62),345(3.35)	12-0575-65
$C_{16}H_{16}O_4$			
Angolensin	n.s.g.	215(4.35),219s(4.34), 228s(4.18),239s(3.98), 279(4.16),316(3.97)	12-0379-64
10(9H)-Anthracenone, 1,9aα-dihydro-4,9β-dihydroxy-5-methoxy-9α-methyl-	EtOH	244(3.93),389(4.14)	65-2570-64
Benzophenone, 2'-ethoxy-2-hydroxy-4-methoxy-	MeOH	282(4.20),322(4.10)	78-1495-65
Benzophenone, 1,2',3-trimethoxy-	EtOH	250s(2.95),308(4.00)	39-5074-64
2,9β-Epoxyanthracene-3,10-dione, 1,2,3,4,4aα,9,9aα,10-octahydro-5-methoxy-9α-methyl-	EtOH	216(4.32),252(3.88), 316(3.68)	70-1024-64
3,9β-Epoxyanthracene-2,10-dione, 1,2,3,4,4aα,9,9aα,10-octahydro-8-methoxy-9α-methyl-	EtOH	226(4.11),255(3.75), 319(3.50)	70-1024-64
Fluorene-8a(4bH)-acetic acid, 5,6,7,8-tetrahydro-1-methyl-7,9-dioxo-	n.s.g.	251(4.11),298(3.36)	44-2528-65
Hamaudol, dehydration product methyl ether	EtOH	213(4.28),253(4.45), 285(3.97)	95-0055-65
Spiro[benzofuran-2(3H),1'-[2]cyclohexene]-3,4'-dione, 2'-ethoxy-6'-methyl-	EtOH	212(4.34),252(4.47), 329(3.73)	39-4939-65

Compound	Solvent	λ max(log ε)	Ref.
$C_{16}H_{16}O_5$			
2'-Acetonaphthone, 3'-hydroxy-1',4'-dimethoxy-, acetate	EtOH	229(4.73),250s(4.06), 285s(3.71),295s(3.63), 326(3.14)	94-0316-64
Benzophenone, 2,4'-dihydroxy-4,6-dimethoxy-2'-methyl-	EtOH	297(4.22)	39-6960-65
2-Butanone, 3,3-bis(3,4-dihydroxyphenyl)-	EtOH	285.0(3.933)	39-3040-65
2,4,6-Cycloheptatrien-1-one, 2-hydroxy-6-[3-(3,4,5-trihydroxyphenyl)-propyl]-	EtOH	238(4.4),325(3.6), 350(3.7)	78-3605-65
Selinetin	MeOH	247(3.63),257(3.52), 329(4.19)	88-3367-64
Spiro[benzofuran-2(3H),1'-[2]cyclohexene]-3,4'-dione, 4,6-dimethoxy-2'-methyl-	EtOH	285(4.37)	39-6960-65
$C_{16}H_{16}O_6$			
Avipirin	EtOH	310(4.17)	2-0464-64
Benzoic acid, 2-(p-hydroxyphenoxy)-4,6-dimethoxy-, methyl ester	EtOH	282(3.49)	39-6960-65
Benzoic acid, 2-[(4-hydroxy-o-tolyl)-oxy]-4,6-dimethoxy-	EtOH	280(3.52)	39-6960-65
Coumarin, 3-acetyl-6-carbethoxy-5-hydroxy-4,7-dimethyl-	EtOH	257(4.31),312(4.10), 338s(4.02)	39-2543-65
Fonsecin, O-methyl-	EtOH	232(4.45),277(4.61), 317(3.96),330(4.00), 395(3.92)	44-0112-65
Heraclenin, hydrate	EtOH	249(4.35),263(4.15), 300(4.06)	2-0464-64
Jatamansinol, acetate	EtOH	216(4.15),246(3.56), 256(3.51),327(4.17)	78-2605-64
Khellactone, monoacetyl-	n.s.g.	249(3.57),261(3.52), 324(4.14)	78-3591-65
Komalin	EtOH	249(4.36),263(4.15), 300(4.07)	2-0464-64
Psoralen, 5-(2,3-dihydroxy-3-methyl-butoxy)-	EtOH	222(4.00),250(3.90), 268(3.87),310(3.83)	65-1353-64
$C_{16}H_{16}O_9$			
Fomecin A, tetraacetate	EtOH	264(2.92),270s(2.90)	23-1595-64
$C_{16}H_{16}S$			
Dibenzo[c,e]thiepin, 5,7-dihydro-1,11-dimethyl-	isooctane	236s(4.01)	35-1710-64
Dibenzothiophene, 2,4,6,7-tetramethyl-	EtOH	244(4.86),260s(4.24), 270(4.14),278(4.10), 287(4.17),312(3.32), 325(3.45)	39-4077-64
Dibenzothiophene, 3,4,6,7-tetramethyl-	EtOH	236s(4.66),243(4.75), 270(4.10),281(3.90), 290(4.07),310(3.36), 323(3.42)	39-4077-64
Thiobenzophenone, 2,4,4'-trimethyl-	C_6H_{12}	232(4.11),329(4.24), 608(2.12)	54-0289-65
	EtOH	232(4.05),333(4.25), 595(2.15)	54-0289-65
Thiobenzophenone, 2,4,6-trimethyl-	C_6H_{12}	317(4.18),621(2.81)	54-0289-65
	EtOH	228(4.18),320(4.18), 607(2.83)	54-0289-65

Compound	Solvent	λ max(log ε)	Ref.
$C_{16}H_{17}BrN_2$			
1-Phenyl-3-propylbenzimidazolium bromide	MeOH	263(4.05),270(4.06), 277(3.99)	65-0632-64
$C_{16}H_{17}BrN_2O$			
1-(2-Indol-3-ylethyl)-4-methoxy-pyridinium bromide	MeOH	242(4.08),279(3.74), 288(3.66)	24-2463-64
$C_{16}H_{17}BrN_4O_8$			
Inosine, 8-bromo-, 2',3',5'-triacetate	pH 1	254(4.17)	35-1242-64
	pH 11	259(4.14)	35-1242-64
$C_{16}H_{17}BrO_4$			
2,10-Anthracenedione, 3-bromo-1,2,3,4,4aα,9,9aα,10-octa-hydro-9β-hydroxy-5-methoxy-9α-methyl-	EtOH	255(3.86),317(3.64)	70-1013-64
3,10-Anthracenedione, 2-bromo-1,2,3,4,4aα,9,9aα,10-octa-hydro-9β-hydroxy-5-methoxy-9α-methyl-	EtOH	258(3.93),318(3.68)	70-1013-64
$C_{16}H_{17}Br_2NSi$			
Phenazasiline, 2,8-dibromo-5-ethyl-5,10-dihydro-10,10-dimethyl-	$CHCl_3$	295(4.31)	44-3248-65
$C_{16}H_{17}ClN_2O$			
1-Propanol, 3-[(2-amino-5-chloro-α-phenylbenzylidene)amino]-	EtOH	234(4.40),358(3.71)	44-2368-64
$C_{16}H_{17}ClO_4$			
2,10-Anthracenedione, 3-chloro-1,2,3,4,4aα,9,9aα,10-octahydro-9β-hydroxy-8-methoxy-9α-methyl-	EtOH	222(4.26),256(3.86), 312(3.25)	70-1013-64
3,10-Anthracenedione, 2-chloro-1,2,3,4,4aα,9,9aα,10-octahydro-9β-hydroxy-5-methoxy-9α-methyl-	EtOH	223(4.43),257(4.22), 319(3.95)	70-1013-64
Cyclohexane-1α-carboxylic acid, 2β-(2-chloro-5-methoxybenzyl)-5-oxo-	EtOH	229(4.02),282(3.31), 289(3.26)	70-0806-65
$C_{16}H_{17}ClO_5$			
1,8-Naphthalenediol, 5-chloro-3,4-di-hydro-7-methoxy-3-methyl-, diacetate	EtOH	224(4.14),253(3.56), 323(3.22)	39-2355-65
$C_{16}H_{17}Cl_4O_2P$			
Diethylphenyl(2,4,5-trichloro-3,6-di-hydroxyphenyl)phosphonium chloride	EtOH	273(3.14),337(3.82), 368(3.49)	78-1941-65
$C_{16}H_{17}N$			
Cyclohepta[d,e]-1-pyrindine, 1,2,5,7-tetramethyl-	C_6H_{12}	236(4.52),243s(4.47), 270(4.51),290(4.26), 358(3.99),387(3.74), 399(3.81),423(3.56)	35-3137-64
Phenethylamine, N-benzylidene-α-methyl-	EtOH	248(3.84)	44-2265-64
4-Stilbenamine, N,N-dimethyl-, trans	80% EtOH	347(4.49)	33-0517-65
conjugate acid	80% EtOH	308(4.49)	33-0517-65

Compound	Solvent	$\lambda_{max}(\log \epsilon)$	Ref.
$C_{16}H_{17}NO$			
9-Acridanmethanol, α,10-dimethyl-	EtOH	284(4.19)	33-1395-65
9-Acridanmethanol, 9,10-dimethyl-	EtOH	282(4.21)	33-1395-65
Aniline, 2,4,6-trimethyl- N-salicylidene-	EtOH	220(4.4),260(4.25), 328(3.88)	65-3837-64
8-Aza-9-gona-1,3,5(10),13-tetra- en-12-one	C_6H_{12}	310(4.08)	44-3667-65
	EtOH	333(4.17)	44-3667-65
perchlorate	EtOH	333(4.16)	44-3667-65
Benzanilide, 2-propyl-	EtOH	225(4.09),270(3.64)	44-0305-64
Benzylamine, N-(o-methoxybenzyli- dene)-α-methyl-	EtOH	251(4.21),304(3.80)	44-2265-64
Butyranilide, γ-phenyl-	EtOH	243(4.20)	54-0949-64
Ethylamine, α-benzyl-N-salicylidene-	EtOH	253(4.09),280s(3.44), 315(3.58),402(3.01)	44-2265-64
2,4-Heptadiynophenone, 6-(di- methylamino)-6-methyl-	EtOH	269(4.23),281(4.28), 305(4.11)	39-2983-65
Propylamine, α-phenyl-N-salicylidene-	hexane	216(4.50),228s(4.35), 256(4.18),261s(4.13), 320(3.70)	35-1757-65
	EtOH	213(4.43),254(4.12), 317(3.60),405(2.57)	35-1757-65
	dioxan	212(4.36),227s(4.09), 250s(4.16),255(4.28), 262(4.22),317(3.73), 410s(1.31)	35-1757-65
Pyrrolo[1,5-a]quinolin-5-one, 4-allyl- 1,2,3,5-tetrahydro-1-methyl-	EtOH	214s(4.45),231(4.74), 286s(3.74),293(3.79), 303s(3.70),310s(3.57), 317s(3.41)	44-1989-65
4(1H)-Quinolone, 3-allyl- 1-(2-butenyl)-	EtOH	213(4.43),246(4.41), 282s(3.39),293(3.51), 332(4.13),345(4.19)	44-1986-65
4(1H)-Quinolone, 3-allyl- 2-(3-butenyl)-	EtOH	213(4.43),241(4.49), 248(4.49),282s(3.36), 294s(3.50),323(4.06), 336(4.08)	44-1989-65
4(1H)-Quinolone, 1-allyl- 3-(1-methylallyl)-	EtOH	213(4.42),246(4.45), 282s(3.43),292(3.57), 330(4.17),345(4.24)	44-1986-65
4(1H)-Quinolone, 1-allyl- 3-(2-methylallyl)-	EtOH	214(4.41),246(4.40), 283s(3.40),293(3.52), 331(4.13),345(4.18)	44-1986-65
4(1H)-Quinolone, 3-allyl- 1-(2-methylallyl)-	EtOH	213(4.40),246(4.40), 282s(3.43),292(3.54), 331(4.14),345(4.22)	44-1986-65
4(1H)-Quinolone, 1,3-diallyl-2-methyl-	EtOH	213(4.38),242(4.44), 250(4.40),281(3.23), 293s(3.31),329(4.14), 343(4.18)	44-1989-65
$C_{16}H_{17}NOS$			
Benzothiazoline, 2-ethoxy- 3-methyl-2-phenyl-	EtOH	308(3.72)	22-2868-64
$C_{16}H_{17}NO_2$			
2(3H)-Benzofuranone, 3-phenyl-,	$(MeOCH_2)_2$	323(4.24)	35-1857-65
anion, salt with dimethylamine	DMF	354(4.37)	35-1857-65
Benzophenone, p-(dimethylamino)- p'-methoxy-	benzene	340(4.58)	65-0094-65
	EtOH	253(4.01),280(4.03), 360(4.42)	65-0094-65
	HOAc	284(4.02),361(4.36)	65-0094-65

Compound	Solvent	λ_{max}(log ε)	Ref.
$C_{16}H_{17}NO_3$			
Apogalanthamine, hydroxy-, HI salt	EtOH	294(3.60)	94-0696-64
2(1H)-Pyridone, 1-hydroxy-, p-tert-butylbenzoate	EtOH	245(4.30),288(3.76), 298(3.77)	35-5186-65
$C_{16}H_{17}NO_3S_2$			
2,3-Dimethylbenzothiazolium p-toluenesulfonate	H_2O	274(3.81)	22-2868-64
	0.5N NaOH	270(3.89)	22-2868-64
	N NaOH	268(3.99)	22-2868-64
	1.5N NaOH	268(4.07)	22-2868-64
	2N NaOH	267(4.13)	22-2868-64
	EtOH-HOAc	274(3.81)	22-2868-64
	EtOH	260(3.69)	22-2868-64
	EtOH-NaOEt	256(3.76),308(3.68)	22-2868-64
	MeOH-HOAc	235(3.95),275(3.81)	22-2868-64
	MeOH	274(3.79)	22-2868-64
	MeOH-NaOMe	256(3.73),308(3.66)	22-2868-64
$C_{16}H_{17}NO_4$			
Lycorine	EtOH	292(3.70)	95-1194-64
4-Morpholineacetic acid, α-(2-hydroxy-1-naphthyl)-	EtOH	230(4.81),258s(3.42), 268(3.61),278(3.72), 290(3.64),322(3.39), 335(3.46)	24-0363-64
2-Naphthoic acid, 3-hydroxy-4-(morpholinomethyl)-	$CHCl_3$	238(4.71),276(3.73), 287(3.75)	94-0112-64
Pyrrole-2,3-dicarboxylic acid, 1-ethyl-4-phenyl-, dimethyl ester	ether	230(4.33),274(4.06)	24-1952-64
4H-Quinolizine-3-carboxylic acid, 6-methyl-4-oxo-1-propionyl-, ethyl ester	EtOH	265(4.06),350(4.1), 425(4.28)	78-3305-65
$C_{16}H_{17}NO_5$			
1H-Pyrrolo[1,2-a]indole-5,8-dione, 2,3-dihydro-9-(hydroxymethyl)-7-methoxy-6-methyl-, acetate	MeOH	230(4.27),286(4.15), 346(3.58),450(3.08)	44-2897-65
$C_{16}H_{17}NO_8S$			
8aH-Thiazolo[3,2-a]pyridine-5,6,7,8-tetracarboxylic acid, 3-methyl-, tetramethyl ester	MeOH	228(1.43),287(1.99), 445(0.43)	39-3200-65
$C_{16}H_{17}NS$			
Thiazole, 2-methyl-, dimethyl acetylenedicarboxylate adduct	MeOH	224(4.11),283(4.32), 428(3.64)	88-1797-64
Thiazole, 4-methyl-, dimethyl acetylenedicarboxylate adduct	MeOH	227(4.18),285(4.33), 441(3.65)	88-1797-64
Thiobenzophenone, 4'-amino-2,4,6-trimethyl-	C_6H_{12}	383(4.46),595(1.96)	54-0289-65
	EtOH	264(3.79),424(4.43)	54-0289-65
Thiocyanic acid, 5-isopropyl-3,8-dimethyl-1-azulenyl ester	C_6H_{12}	245(4.39),296(4.53), 307(4.42),357(3.77), 375(3.85),578(2.68), 616(1.61),683(2.28)	44-2715-65
$C_{16}H_{17}N_3$			
5H-Dibenzo[d,f][1,3]diazepine, 6-(isopropylamino)-	EtOH	241(4.63),275(3.66)	87-0310-64

Compound	Solvent	$\lambda_{max}(\log \epsilon)$	Ref.
$C_{16}H_{17}N_3O$			
11H-Dibenzo[b,e][1,4]diazepin-11-one, 10-(2-aminoethyl)-5,10-dihydro-5-methyl-	EtOH-HCl	215s(4.36),226(4.44), 285s(3.43)	33-1590-65
11H-Dibenzo[b,e][1,4]diazepin-11-one, 5,10-dihydro-10-[2-(methylamino)-ethyl]-	EtOH-HCl	223(4.48)	33-1590-65
$C_{16}H_{17}N_3O_2S$			
Diimide, [(p-tert-butylphenyl)thio]-(p-nitrophenyl)-	n.s.g.	355(4.10)	77-0313-65
$C_{16}H_{17}N_3O_3$			
Benzamide, N-[2-(N-methylanilino)-ethyl]-p-nitro-	MeOH	254(4.39)	87-0107-65
$C_{16}H_{17}N_3O_3S$			
Pyrimidine, 4-[3-methyl-4-(5-nitro-2-furyl)buta-1,3-dienyl]-2-propyl-thio)-	propylene glycol	256(4.12),330(4.19)	95-0207-64
$C_{16}H_{17}N_3O_4$			
Anthramycin	MeCN	235(4.26),333(4.50)	35-5791-65
[2,2'-Bipyrrole]-5-carboxylic acid, 5'-(5-carboxy-1-pyrrolin-2-yl)-, dimethyl ester	MeOH-HCl	402(4.66)	44-0883-64
	MeOH-KOH	345(4.56),360s(4.48)	44-0883-64
$C_{16}H_{17}N_3O_5$			
Cytidine, N-benzoyl-2'-deoxy-	H_2O	258(4.31),301(4.04)	44-3067-65
$C_{16}H_{18}$			
Bibenzyl, 4,4'-dimethyl-	EtOH	213(4.26),219(4.26), 260(3.85),265(2.97), 268(2.95),274(3.00)	5-0152-64D
Butane, 1,1-diphenyl-	C_6H_{12}	255(2.65),260(2.78), 262(2.79),269(2.68)	22-0587-65
Butane, 1,3-diphenyl-	n.s.g.	253(2.51)	12-0353-64
2,4,8,10-Dodecatetrayne, 6,6,7,7-tetramethyl-	EtOH	218(3.08),230(3.16), 242(3.21),255(3.00)	78-1357-65
Methane, p-tolyl-2,5-xylyl-	EtOH	215(4.30),268(3.00), 276(2.97)	5-0152-64D
$C_{16}H_{18}BrNO$			
2,5-Cyclohexadien-1-one, 2-bromo-4-(o-ethylanilino)-4,6-dimethyl-	MeOH	247(4.25)	35-1127-64
$C_{16}H_{18}BrN_5O_7$			
Adenosine, 8-bromo-, 2',3',5'-triacetate	pH 1	263(4.28)	35-1242-64
	pH 11	264(4.24)	35-1242-64
$C_{16}H_{18}Br_2O_3$			
Anthrone, 2α,3β-dibromo-1,2,3,4,4aβ,-9aβ-hexahydro-10α-hydroxy-8-methoxy-10-methyl-	EtOH	260(3.92),319(3.66)	70-1013-64
$C_{16}H_{18}Br_2O_4$			
2,4,6,8,10-Decapentaenedioic acid 2,11-dibromo-, diethyl ester	EtOH	386(4.90),408(4.84)	70-0684-65

Compound	Solvent	λ_{max}(log ϵ)	Ref.
$C_{16}H_{18}ClIN_2O_2$			
1H-Azepino[5,4,3-cd]indole-2-carboxylic acid, 9-chloro-3,4-dihydro-6-methyl-, ethyl ester, methiodide	EtOH	222(4.54),250(4.09), 271(4.17),346(3.95), 390(4.00)	87-0200-65
$C_{16}H_{18}ClN_3S$			
Methylene blue	pH 4.1	295(4.6),605(4.7), 665(4.9)	60-1787-65
	pH 6.3	610(4.7),660(4.9)	60-1787-65
	pH 6.5	660(4.7)	43-0817-64
	H_2O	620(4.7),660(4.9)	60-1787-65
	0.9M Na_2SO_4	605(4.7),660(4.5)	60-1800-65
	0.9N NaCl	615(4.7),660(4.8)	60-1800-65
	BuOH	650(5.0)	60-1800-65
	n.s.g.	662(4.81)	46-0647-65
$C_{16}H_{18}CuN_4O_2$			
Copper, bis(mandelamidinato)-	MeOH	261(3.61),580(1.50)	39-4004-65
$C_{16}H_{18}D_2N_4O_4$			
Hydrindan-2-one-5,6-d_2, 8-methyl-, 2,4-dinitrophenylhydrazone	$CHCl_3$	365(4.33)	35-0580-65
$C_{16}H_{18}NO_5$			
Nitroxide, bis(2,4-dimethoxyphenyl)-	n.s.g.	275(3.8),315(4.1), 328(4.2)	70-0800-65
$C_{16}H_{18}N_2$			
Agroclavine	EtOH	224(4.42),282(3.81), 293(3.74)	39-1858-64
Azobenzene, 2,2'-diethyl-	EtOH	235(4.05),330(4.17), 460(2.75)	23-0836-64
	EtOH	240(4.06),336(4.15), 446(2.92)	78-0189-64
Azobenzene, 4,4'-diethyl-	EtOH	236(4.14),331(4.32), 440(3.00)	23-0836-64
Indoline, 1-(dimethylamino)-	MeOH-HCl	280(3.51)	44-3451-65
3-phenyl-	MeOH	253(3.89),298(3.56)	44-3451-65
3-Picoline, 6-(p-dimethylamino-styryl)-	EtOH	360(4.35)	65-4106-64
$C_{16}H_{18}N_2O$			
Azoxybenzene, 2,2'-diethyl-	EtOH	234(4.03),309(3.97)	23-0836-64
Azoxybenzene, 4,4'-diethyl-	EtOH	236(4.13),334(4.32)	23-0836-64
Benzomesidide, p-amino-	EtOH	282(4.4)	28-4295-64B
Dasycarpidone, N-demethyl-	EtOH	238(4.15),317(4.23)	78-1717-65
4H-Indeno[1,2-d]pyrimidin-4-one, 5-tert-butyl-3,5-dihydro-2-methyl-	EtOH	245(4.42),290s(3.90), 298(3.95),312s(3.89)	44-0396-65
Naphthalene-1-carbonitrile, 3,4-dihydro-5-methoxy-2-(N-pyrrolidinyl)-	EtOH	214(4.07),280(4.10), 289(4.12)	88-0103-64
1H,4H-Naphthalene-1,4-dione, 2,3,4a,5,8,8a-hexahydro-, phenylhydrazone	EtOH	277(4.29)	88-2443-64
2-Pyridinol, 4-methyl-2-(methylamino)-	EtOH-HCl	313(4.13)	44-2128-64
5-phenyl- (solvated)	EtOH	310(4.09)	44-2128-64
	EtOH-NaOH	319(4.19)	44-2128-64
4(1H)-Pyridone, 2,3-dihydro-1-[2-(2-methyl-3-indolyl)ethyl]-	MeOH	283(3.95),290(3.96), 327(4.23)	24-2463-64

Compound	Solvent	λ_{max}(log ϵ)	Ref.
$C_{16}H_{18}N_2O_2$			
2H-Benzo[a]quinolizine-1-carbonitrile, 3,4,6,7-tetrahydro-9,10-dimethoxy-	EtOH-HCl	251(4.18),315(4.02), 369(4.04)	95-0412-64
	EtOH	276(4.0),314(4.05)	95-0412-64
$C_{16}H_{18}N_2O_4$			
Azobenzene, 2,2',6,6'-tetramethoxy-	EtOH	208(4.55),236(4.26), 345(4.33),444(3.01)	78-0189-64
3-Pyrroline-3-carboxylic acid, 2-[(5-formyl-3,4-dimethyl-2-pyrrolyl)methylene]-4-methyl-5-oxo-, ethyl ester	EtOH	282(4.10),434(4.40), 450(4.33)	39-5999-64
$C_{16}H_{18}N_2O_5$			
1H-Benzo[c]quinolizine-3-carboxylic acid, 2,3,4,4a,5,6-hexahydro-8-nitro-4-oxo-, ethyl ester	EtOH	208(3.99),246(4.00), 395(4.15)	78-2961-65
$C_{16}H_{18}N_2S$			
Diimide, [(p-tert-butylphenyl)-thio]phenyl-	n.s.g.	336(4.03)	77-0313-65
$C_{16}H_{18}N_2Zn$			
Zinc, diethyl(1,10-phenanthroline)-	toluene	427(2.95)	101-0222-65A
$C_{16}H_{18}N_4$			
Pyrazolo[1,5-a]pyrimidine, 7-amino-2-ethyl-3,5-dimethyl-6-phenyl-	EtOH	231(4.59),255(3.80), 297(4.02)	95-1113-64
Pyrazolo[1,5-a]pyrimidine, 7-amino-2-methyl-5-phenyl-6-propyl-	EtOH-HCl	239(4.49),284s(3.64), 328(3.74)	95-1113-64
$C_{16}H_{18}N_4O$			
Azobenzene, 4-acetamido-4'-(dimethylamino)-	EtOH	233(4.06),258(4.08), 316(3.90),415(4.53)	65-3602-64
$C_{16}H_{18}N_4O_3$			
Lumiflavin, 5-acetyl-1,5-dihydro-3-methyl-	MeOH	218(4.59),260(4.45), 302(4.08)	33-1354-64
$C_{16}H_{18}N_4O_4$			
1-Cyclopentene-1-carboxaldehyde, 5-isopropenyl-2-methyl-, 2,4-dinitrophenylhydrazone	EtOH	386(4.48)	44-3740-64
Eucarvone, 2,4-dinitrophenylhydrazone	EtOH	384(4.68)	44-0669-65
3,5-Heptadien-2-one, 6-cyclopropyl-, 2,4-dinitrophenylhydrazone	$CHCl_3$	394(4.55)	22-3218-64
Leucoflavin, 5-acetyl-10-(hydroxyethyl)-7,8-dimethyl-	MeOH	220(4.52),262(4.39), 304(4.00)	33-1354-64
p-Mentha-1,3-diene-7-carboxaldehyde, 2,4-dinitrophenylhydrazone	$CHCl_3$	409(4.46)	44-0518-65
2(1H)-Naphthalenone, 3,4,4a,5,6,8a-hexahydro-, 2,4-dinitrophenylhydrazone, cis-anti	$CHCl_3$	367(4.41)	35-1972-64
2(1H)-Naphthalenone, 3,4,6,7,8,8a-hexahydro-, 2,4-dinitrophenylhydrazone	$CHCl_3$	367(4.42)	35-1972-64
2(1H)-Naphthalenone, 4a,5,6,7,8,8a-hexahydro-, 2,4-dinitrophenylhydrazone	n.s.g.	256(4.21),378(4.45)	22-0111-65

Compound	Solvent	λ_{max}(log ε)	Ref.
$C_{16}H_{18}N_4O_5$			
Cinerolone, 2,4-dinitro-phenylhydrazone	EtOH	215(4.21),250(4.23),283s(3.97),380(4.44)	39-5225-64
$C_{16}H_{18}N_4O_7S$			
9H-Purine-6-thiol, 9-(2,3,5-tri-O-acetyl-β-D-ribofuranosyl)-	pH 1	320(4.41)	87-0200-64
	pH 11	236(4.15),309(4.38)	87-0200-64
$C_{16}H_{18}O$			
Ketone, 1-naphthyl pentyl	EtOH	215(4.69),299(3.80)	39-2072-65
Ketone, 2-naphthyl pentyl	EtOH	240(4.71),292(3.97)	39-2072-65
1(4H)-Naphthalenone, 4a,5,6,7,8,8a-hexahydro-8-phenyl-	EtOH	226(3.96)	44-2942-65
2(3H)-Naphthalenone, 4,4a,5,6,7,8-hexahydro-6-phenyl-	EtOH	238(4.31)	87-0519-64
2(1H)-Phenanthrone, 3,4,9,10-tetra-hydro-1,1-dimethyl-	EtOH	260(4.11),292s(3.40)	35-2038-64
$C_{16}H_{18}OS$			
Sulfoxide, mesityl p-tolyl	isooctane	202(4.72),238s(4.08),251s(3.95),277s(3.57),285s(3.48)	35-1958-65
	isooctane-CF_3COOH	203(4.67),244(4.19),269(3.61),278s(3.57),288s(3.42)	35-1958-65
	EtOH	245(4.22),270(3.58),277(3.57),287s(3.47)	35-1958-65
	EtOH-CF_3COOH	245(4.20),270(3.56),278(3.54),287s(3.44)	35-1958-65
S-(-)	EtOH	244(4.26),269(3.68),277(3.66),285(3.57)	35-5637-64
	EtOH-$HClO_4$	229s(4.38),256(4.40)	35-5637-64
$C_{16}H_{18}O_2$			
1-Azulenecarboxylic acid, 3,4,6,8-tetramethyl-, methyl ester	C_6H_{12}	250(3.43),308(4.47),355(3.58),376(3.56),557(2.64),562s(2.63)	44-0131-65
7,8-Benzochroman-6-ol, 2,2,5-trimethyl-	C_6H_{12}	249(4.31),324(--),339(--)	22-2513-65
Benzo[b]naphtho[1,2-d]furan-2(3H)-one, 4,4a,5,6,8,9,10,11-octahydro-	EtOH	230(4.77),247(4.54),272(5.19),322(5.32)	24-2738-65
2,2'-Biphenyldimethanol, α,α-di-methyl-	EtOH	212(4.29)	78-3401-65
3,5,7,9,11,15-Hexadecahexaene-2,15-dione	EtOH	392(4.59),406(4.72),428(4.71)	70-0684-65
2(3H)-Naphthalenone, 4,4a,5,6,7,8-hexahydro-6-(p-hydroxyphenyl)-	EtOH	230(4.32)	87-0519-64
1-Naphthoic acid, pentyl ester	EtOH	220(5.63),297(3.81)	39-2072-65
2-Naphthoic acid, pentyl ester	EtOH	238(4.89),281(3.89)	39-2072-65
$C_{16}H_{18}O_3$			
3-Anthracenone, 10-acetoxy-1,2,3,4,4aβ,9,9aα,10-octahydro-	EtOH	203(3.67),266(2.69),273(2.69)	70-1039-65
5-Benzofuranacrylic acid, α-butyl-2,3-dihydro-6-hydroxy-β-methyl-, δ-lactone	EtOH	226(4.08),256(3.45),296(3.73),332(4.23)	44-2467-64
4,6,8-Decatriynoic acid, 10-(1-hydroxy-1-cyclopentyl)-, methyl ester	EtOH	241(2.93),258(2.58),277(2.69),293(2.75),328(2.62)	70-0544-65

Compound	Solvent	λ_{max}(log ϵ)	Ref.
3,9β-Epoxyanthracen-10-one, 1,2,3,4,4aα,9,9aα,10-octahydro-3α-methoxy-9 -methyl-	EtOH	249(4.14),290(3.35)	70-1024-64
Fluoren-2-ol, 4b,5,8,8a-tetrahydro-7-methoxy-, acetate	EtOH	263s(3.15),269(3.24), 275(3.19)	88-3505-65
1,4-Naphthoquinone, 2-(3-hydroxy-3-methylbutyl)-3-methyl-	C_6H_{12}	243(4.20),249(4.22), 264(4.18),273(4.2), 326(3.3)	22-2513-65
2α-Phenanthrenecarboxylic acid, 1,2,3,9,10,10aβ-hexahydro-7-methoxy-	EtOH	263(4.31),298(3.52)	44-1213-65
2α-Phenanthrenecarboxylic acid, 1,2,3,4,4aα,9,10,10aβ-octahydro-4α-hydroxy-7-methoxy-, lactone	EtOH	222(3.94),278(3.27)	44-1213-65
2α-Phenanthrenecarboxylic acid, 1,2,3,4,4aβ,9,10,10aβ-octahydro-4α-hydroxy-7-methoxy-, lactone	EtOH	223(3.91),278(3.29), 286(3.26)	44-1213-65
2β-Phenanthrenecarboxylic acid, 1,2,3,4,4aβ,9,10,10aβ-octahydro-4β-hydroxy-7-methoxy-, lactone	EtOH	280(3.34),288(3.32)	44-1213-65
$C_{16}H_{18}O_4$			
10(9H)-Anthracenone, 1,2,3,9aα-tetrahydro-4,9β-dihydroxy-5-methoxy-9α-methyl-	EtOH	266(3.83),341(4.20)	65-2570-64
Biphenyl, 2,2',4,4'-tetramethoxy-	EtOH	208(4.78),251(4.12), 288(3.97)	12-0901-64
Biphenyl, 2,2',4,6'-tetramethoxy-	EtOH	222(4.31),248(3.91), 280(3.79)	94-1232-64
6-Chromanacrylic acid, 5-hydroxy-8-methoxy-β,2,2-trimethyl-, δ-lactone	EtOH	212(4.47),231(4.23), 310s(3.58),347(4.04)	78-1331-64
Coumarin, 7-hydroxy-6-methoxy-4-methyl-8-(3-methyl-2-butenyl)-	EtOH	216(4.20),232(4.18), 300s(3.66),344(4.10)	78-1331-64
2,3α-Epoxyanthracen-10-one, 1,2,3,4,-4aα,9,9aα,10-octahydro-9β-hydroxy-5-methoxy-9α-methyl-	EtOH	255(4.12),317(3.86)	70-1013-64
3β-isomer	EtCH	228(3.37),257(3.43), 317(3.36)	70-1013-64
2,9β-Epoxyanthracen-10-one, 1,2,3,4,-4aα,9,9aα,10-octahydro-3α-hydroxy-5-methoxy-9α-methyl-	EtOH	216(4.32),252(3.87), 317(3.68)	70-1024-64
2H-2,4a-Ethenonaphthalene-3,4-dicarboxylic anhydride, 1,3,4,5,6,7,8,8a-octahydro-8a,10-dimethyl-1-oxo-	EtOH	299(2.11)	35-4053-64
Indone, 4-hydroxy-6-(2-hydroxyethyl)-2,5,7-trimethyl-, acetate	EtOH	249(4.43),340(3.18), 400(2.70)	78-2671-65
Lachnophyllic acid, 8-hydroxy-, tiglic acid ester	ether	223(4.72),253(3.91), 267(4.21),282(4.44), 300(4.41)	24-3469-64
1α-Phenanthrenecarboxylic acid, 1,2,3,4,4aβ,9,10,10aβ-octahydro-7-methoxy-4-oxo-	EtOH	220s(3.97),276(3.25), 285(3.19)	44-2527-64
1β-Phenanthrenecarboxylic acid, 1,2,3,4,4aα,9,10,10aβ-octahydro-7-methoxy-4-oxo-	EtOH	220s(3.89),278(3.20), 285(3.19)	44-2527-64
4aβ-isomer	EtOH	220s(3.94),280(3.28), 287s(3.23)	44-2527-64
2α-Phenanthrenecarboxylic acid, 1,2,3,4,4aα,9,10,10aβ-octahydro-7-methoxy-4-oxo-	EtOH	220(3.92),277(3.23), 284(3.23)	44-1213-65

Compound	Solvent	λ_{max}(log ε)	Ref.
2α-Phenanthrenecarboxylic acid, 1,2,3,4,4aβ,9,10,10aβ-octahydro-7-methoxy-4-oxo-	EtOH	225(3.86),278(3.26), 287s(3.18)	44-1213-65
2β-Phenanthrenecarboxylic acid, 1,2,3,4,4aβ,9,10,10aβ-octahydro-7-methoxy-4-oxo-	EtOH	278(3.26),284(3.20)	44-1213-65
$C_{16}H_{18}O_5$			
Anthrone, 4β-acetoxy-1,2,3,4,4aα,9aα-hexahydro-5,10β-dihydroxy-	EtOH	224(4.15),260(3.87), 315(3.44)	65-2563-64
1,3-Cyclohexanedione, 4-[o-(methoxymethoxy)benzoyl]-5-methyl-	MeOH	254(4.23),279(4.12)	94-0214-64
1,8-Naphthalenediol, 3,4-dihydro-7-methoxy-3-methyl-, diacetate	EtOH	224(4.30),252(3.89), 323(3.08)	39-2355-65
$C_{16}H_{18}O_6$			
2H-1-Benzopyran-4-propionic acid, 5,7-dimethoxy-2-oxo-, ethyl ester	EtOH	207(3.66),247(3.90), 254(3.82),322(4.19)	25-1865-64
α-D-Glucoside, 1-naphthyl-	EtOH	219(4.65),283s(3.76), 287(3.76),304s(3.43), 319(3.11)	39-4363-65
$C_{16}H_{18}O_7$			
β-D-Glucoside, 8-hydroxy-1-naphthyl	EtOH	225(4.69),293s(3.75), 302(3.81),308s(3.69), 316(3.75),324s(3.55), 331(3.75)	39-4363-65
$C_{16}H_{18}O_9$			
Isobenzofuran-1,1,3,3-tetracarboxylic acid, 1,3,3a,7a-tetrahydro-, tetramethyl ester	MeOH	246(3.62),265(3.60)	35-3657-65
$C_{16}H_{18}O_{10}$			
1,3-Cyclopentadiene-1,2,3,4,5-pentacarboxylic acid, 5-methyl-, pentamethyl ester	MeOH	220(3.95),292(3.80)	44-0423-64
$C_{16}H_{18}S$			
2,3-Xylyl sulfide	EtOH	252(4.18),276(3.73)	39-4077-64
$C_{16}H_{19}BrO_4$			
10-Anthracenone, 2α-bromo-1,2,3,4,4aα,-9,9aα,10-octahydro-3β,9β-dihydroxy-5-methoxy-9α-methyl-	EtOH	228(3.47),258(3.53), 317(3.41)	70-1013-64
10-Anthracenone, 3β-bromo-1,2,3,4,4aα,-9,9aα,10-octahydro-2α,3β-dihydroxy-5-methoxy-9-methyl-	EtOH	238(3.95),260(3.84), 318(3.65)	70-1013-64
$C_{16}H_{19}ClN_2O_2$			
1H-Azepino[5,4,3-cd]indole-2-carboxylic acid, 9-chloro-3,4,5,6-tetrahydro-5,6-dimethyl-, ethyl ester	EtOH	238(4.48),297(4.30)	87-0200-65
$C_{16}H_{19}ClN_2O_5$			
[[1-[(Dimethylamino)methylene]-2-formyl-3-indenyl]methylene]dimethylammonium perchlorate	MeCN	483(4.45)	73-2783-65

Compound	Solvent	$\lambda_{max}(\log \epsilon)$	Ref.
$C_{16}H_{19}ClN_4O_2$			
Acetophenone, 4'-[[3-(2-amino-4-hydroxy-6-methyl-5-pyrimidyl)propyl]amino]-2-chloro-	pH 0.5	249(4.14)	87-0035-65
	pH 13	282(3.98),348(4.35)	87-0035-65
$C_{16}H_{19}ClO_4$			
10-Anthracenone, 2α-chloro-1,2,3,4,-4a,9,9aα,10-octahydro-3β,9β-dihydroxy-5-methoxy-9α-methyl-	EtOH	255(4.08),316(3.63)	70-1013-64
10-Anthracenone, 3β-chloro-1,2,3,4,-4aα,9,9aα,10-octahydro-2α,9β-dihydroxy-5-methoxy-9α-methyl-	EtOH	257(3.92),317(3.65)	70-1013-64
$C_{16}H_{19}IN_2$			
2-[p-(Dimethylamino)styryl]-1-methylpyridinium iodide, cis	aq. HCl	308(4.04)	23-1345-65
	H_2O	403(3.92)	23-1345-65
	50% MeOH	403(3.92)	23-1345-65
trans	aq. HCl	326(4.44)	23-1345-65
	H_2O	437(4.42)	23-1345-65
	50% MeOH	453(4.52)	23-1345-65
	50% MeOH-HCl	327(4.45)	23-1345-65
$C_{16}H_{19}N$			
Aniline, N-methyl-N-(1-phenylpropyl)-	EtOH	256(4.17),296(3.40)	39-0249-64
	HCl	250(2.97),259(2.97),263(2.95),269(2.74)	39-0249-64
Camphorindole	EtOH	239(4.33),304(4.31)	32-0546-65
	n.s.g.	205(4.28),240(4.36),303(4.32)	88-2285-65
$C_{16}H_{19}NO$			
Acetamide, N-methyl-N-(4,6,8-trimethyl-1-azulenyl)-	C_6H_{12}	246(4.46),291(4.65),337(3.63),347(3.70),352(3.75),553(2.72),592s(2.65),650s(2.20)	35-3137-64
2(3H)-Carbazolone, 4,11-dihydro-4,4,9,11-tetramethyl-	MeOH-HCl	247(4.27),252(4.25),341(4.45)	24-2111-65
	MeOH	235(4.34),299(4.00),338(4.55)	24-2111-65
2-Indolinone, 3-(1-cyclohexylethylidene)-	EtOH	254(4.48),261(4.55),293(3.91),351(3.27)	87-0626-65
4(1H)-Quinolone, 1-(2-butenyl)-3-propyl-	EtOH	213(4.39),246(4.39),283s(3.35),293(3.47),332(4.14),346(4.19)	44-1986-65
$C_{16}H_{19}NO_2$			
Phenol, p-[2-[(2-phenoxyethyl)amino]ethyl]-, hydrochloride	EtOH-acid	269(3.43),276(3.47)	54-0521-65
Quinoline, 2-(4-ethoxy-2-methylbutyryl)-	EtOH	212(4.34),243(4.53),289(3.86)	78-2529-65
$C_{16}H_{19}NO_3$			
1H-Benzo[c]quinolizine-2-carboxylic acid, 2,3,4,4a,5,6-hexahydro-3-oxo-, ethyl ester	EtOH	210(4.24),252(4.14),293(3.32)	78-2961-65
Carbostyril, 4,8-dimethoxy-3-(3-methyl-2-buten-1-yl)-	MeOH	239(4.32),254(4.45),282(3.95),290(3.91),321(3.55),331(3.62),344(3.46)	39-0438-64

Compound	Solvent	λ_{max}(log ε)	Ref.
Carbostyril, 4-hydroxy-8-methoxy-1-methyl-3-(3-methyl-2-buten-1-yl)-	MeOH	218(4.35),240(4.46),246(4.43),255(4.42),284(3.91),294(3.93),319(3.59)	39-0438-64
Ethanol, 1-(p-hydroxyphenyl)-2-(2-phenoxyethylamino)-	EtOH-acid	276(3.47)	54-0521-65
Furo[3,2-c]quinolin-4(2H)-one, 3,5-dihydro-2-isopropyl-6-methoxy-5-methyl-	MeOH	220(4.32),239(4.42),248(4.35),292(3.79),301(3.84),326(3.36)	39-0438-64
5H-Pyrano[3,2-c]quinolin-5-one, 2,3,4,6-tetrahydro-7-methoxy-2,2,6-trimethyl-	MeOH	237(4.43),248(4.38),253(4.38),281(3.81),291(3.81),323(3.39)	39-4190-64
5H-Pyrano[2,3-b]quinolin-5-one, 2,3,4,10-tetrahydro-9-methoxy-2,2,10-trimethyl-	MeOH	220(4.19),244(4.59),302(3.79),322(3.98),333(3.90)	39-4190-64
Zephyranthine, deoxy-	EtOH	291(3.68)	95-0200-65
$C_{16}H_{19}NO_4$			
2α-Phenanthrenecarboxylic acid, 1,2,3,4,4aα,9,10,10aβ-octahydro-7-methoxy-4-oxo-, oxime	EtOH	225(3.91),278(3.23),285(3.20)	44-1213-65
2β-Phenanthrenecarboxylic acid, 1,2,3,4,4aβ,9,10,10aβ-octahydro-7-methoxy-4-oxo-, oxime	EtOH	278(3.25),286(3.22)	44-1213-65
Zephyranthine	EtOH	291(3.7)	94-0253-64
	EtOH	291(3.71)	95-1194-64
$C_{16}H_{19}NO_5$			
Macranthine	EtOH	236(3.53),292(3.67)	33-0185-64
$C_{16}H_{19}NO_7$			
Alanine, N-acetyl-α-methyl-(3,4-diacetoxyphenyl)-	MeOH	263(2.62),268(2.57)	44-1424-64
Alanine, N-acetyl-2-methyl-3-(3,4-diacetoxyphenyl)-	MeOH	265(2.73),271(2.70)	44-2053-64
$C_{16}H_{19}N_3$			
Azobenzene, 4-(dimethylamino)-2,2'-dimethyl-	N HCl	332(4.15),510(4.18)	78-1547-64
$C_{16}H_{19}N_3O$			
Benzamide, p-amino-N-[2-(N-methylanilino)ethyl]-	MeOH	255(4.31),266(4.25),280(4.27)	87-0107-65
d-Lysergamide, 2,3-dihydro-	MeOH	243(4.42),318(3.31)	33-0756-64
$C_{16}H_{19}N_3O_3$			
2-Cyclopentene-1-carboxylic acid, 3-methyl-4-oxo-2-phenyl-, ethyl ester, semicarbazone	EtOH	290(4.39)	23-2408-65
$C_{16}H_{19}N_3O_3S$			
Azobenzene-4'-sulfonic acid, 4-(dimethylamino)-2,2'-dimethyl-	N HCl	333(3.87),513(4.60)	78-1547-64
	EtOH	455(4.35)	78-1547-64
Azobenzene-4'-sulfonic acid, 4-(dimethylamino)-2,3'-dimethyl-	N HCl	333(3.53),521(4.71)	78-1547-64
	EtOH	476(4.35)	78-1547-64
Azobenzene-4'-sulfonic acid, 4-(dimethylamino)-2',6'-dimethyl-	N HCl	305(3.86),472(4.20)	78-1547-64
	EtOH	420(4.34)	78-1547-64
$C_{16}H_{19}N_5$			
s-Triazolo[1,5-c]pyrimidine, 2-(benzylamino)-7-methyl-5-propyl-	MeOH	213(4.10),234(4.67),298(3.42)	39-3357-65

Compound	Solvent	λ_{max}(log ε)	Ref.
$C_{16}H_{19}N_5O$			
Isoxazolo[5,4-d]pyrimidine, 4-[[3-(dimethylamino)propyl]amino]-3-phenyl-	EtOH	238(4.09),291(3.95)	44-2116-64
$C_{16}H_{19}N_5O_4S$			
Theophylline, 7-(3-phenylpropyl)-8-sulfamoyl-	MeOH	287(3.98)	54-1215-64
$C_{16}H_{19}N_5O_5S$			
Theophylline, 7-(benzyloxyethyl)-8-sulfamoyl-	MeOH	287(4.01)	54-1215-64
$C_{16}H_{19}N_5O_6$			
Purine, 6-acetamido-9-(3,5-di-O-acetyl-2-deoxy-β-D-ribofuranosyl)-	pH 11	273(4.15),282(4.30)	35-4934-65
	N NaOH	291(4.09)	35-4934-65
Riboflavin, 7-aminodemethyl-	H_2O	252(4.69),296(3.94), 490(4.62)	65-0675-65
$C_{16}H_{19}N_5O_7S$			
6-Purinethiol, 2-amino-9-(2,3,5-tri-acetyl-β-D-ribofuranosyl)-	pH 1	264(3.58),344(4.18)	87-0200-64
	pH 11	251(3.92),317(4.13)	87-0200-64
$C_{16}H_{19}N_5S_2$			
5H-Thiopyrano[2,3-d]pyrimidine, 7,7'-iminobis[6,7-dihydro-4-methyl-	pH 1	227(3.89),298(4.18)	4-0079-64
	pH 7	253(3.75),284(3.95)	4-0079-64
	pH 13	290(3.98)	4-0079-64
$C_{16}H_{20}$			
Naphthalene, 2,3-diethyl-1,4-dimethyl-	EtOH	214(4.44),234(4.91), 264s(3.26),274s(3.52), 286(3.70),295(3.74), 306s(3.59),325(2.75)	44-0963-64
Naphthalene, 6-(5-hexyn-1-yl)-1,2,3,4-tetrahydro-	hexane	202(4.69),218(4.00), 222s(3.96),264(2.66), 270(2.81),273s(2.76), 278(2.87)	39-3160-65
$C_{16}H_{20}BrN$			
Dimethylphenethylammonium bromide	EtOH	256(2.85),260(2.87), 263(2.82),270(2.60)	39-0249-64
$C_{16}H_{20}ClFN_2O_2$			
Benzylamine, α-(1-chloro-2-fluoro-3-nitroallylidene)-N,N-dipropyl-	n.s.g.	277(4.06),377(4.08)	35-5132-65
$C_{16}H_{20}ClNO_4$			
Ribalinium chloride	50% EtOH	221(4.42),247(4.35), 293s(3.83),299(3.88), 334(3.65)	78-0909-65
$C_{16}H_{20}Cl_2N_2O_2$			
p-Benzoquinone, 2,5-dichloro-3,6-dipiperidino-	methyl cellosolve	241(4.23),441(4.07), 565(2.72)	78-1889-64
$C_{16}H_{20}F_2N_2O_2$			
p-Benzoquinone, 2,5-bis(2,2-dimethyl-1-azetidinyl)-3,6-difluoro-	methyl cellosolve	236(4.43),380(4.44), 563(2.45)	78-1889-64
p-Benzoquinone, 2,5-difluoro-3,6-dipiperidino-	methyl cellosolve	237(4.40),419(4.09), 569(2.65)	78-1889-64

Compound	Solvent	λ max(log ε)	Ref.
$C_{16}H_{20}Ge$			
Germane, dibenzyldimethyl-	C_6H_{12}	253(2.81),259(2.86), 262s(2.85),265(2.77), 268(2.72)	39-4690-65
$C_{16}H_{20}N_2$			
Benz[cd]indole, 1,3,4,5-tetrahydro-5-piperidino-	EtOH	225(4.35),283(3.77), 293(3.73)	39-5573-64
Hydrazobenzene, 2,2'-diethyl- (changing spectrum)	EtOH	245(4.27),287(3.51)	23-0836-64
Hydrazobenzene, 4,4'-diethyl- (changing spectrum)	EtOH	248(4.34),298(3.65), 437(1.93)	23-0836-64
$C_{16}H_{20}N_2O$			
Chanoclavine I	MeOH	222(4.50),275s(3.86), 281(3.89),291(3.82)	33-2186-64
Chanoclavine II	MeOH	222(4.50),275s(3.86), 281(3.89),291(3.82)	33-2186-64
racemic forms	MeOH	222(4.50),275s(3.86), 281(3.89),291(3.82)	33-2186-64
Indole, 3-(1-acetyl-5,5-dimethyl-2-pyrrolidinyl)-	EtOH	220(4.61),274s(3.75), 281(3.78),289(3.72)	87-0415-64
Indole, 5-acetyl-3-piperidinomethyl-	EtOH	251(4.53),296(3.92)	87-0141-64
Indole, 3-[2-(3,6-dihydro-4-methoxy-1(2H)-pyridyl)ethyl]-	MeOH	281(3.87),290(3.71)	24-2463-64
Isochanoclavine I	MeOH	222(4.50),275s(3.86), 281(3.89),291(3.82)	33-2186-64
$C_{16}H_{20}N_2O_2$			
Hexanoic acid, δ-oxo-, tryptamide	MeOH	277(3.94)	24-2463-64
Indene-2-carboxamide, 3-acetamido-1-tert-butyl-	EtOH	225(4.15),282(3.89)	44-0396-65
Indole, 3-(3-carbomethoxy-1-methyl-2-piperidyl)-, hydrochloride	EtOH	217(4.61),274(3.81), 280(3.83),287(3.75)	44-2534-65
Quinaldic acid, 1-(3-cyanopropyl)-1,2,3,4-tetrahydro-, ethyl ester	EtOH	209(4.35),250(3.92), 301(3.37)	78-2529-65
1(2H)-Quinolinepropionic acid, 2-(cyanomethyl)-3,4-dihydro-, ethyl ester	EtOH	210(4.35),255(4.17), 302(3.41)	78-2961-65
Tryptophan, α-(3-methyl-2-butenyl)-	MeOH	220(4.79),273(4.00), 280(4.02),289(3.96)	22-2306-65
$C_{16}H_{20}N_2O_3$			
[2,2'-Bipyrrole]-5-carboxylic acid, 3,3'-diethyl-5'-formyl-4,4'-dimethyl-	$CHCl_3$	221(4.01),267(4.18), 331(4.29)	39-1460-65
[2,2'-Bipyrrole]-5-carboxylic acid, 5'-formyl-3,3',4,4'-tetramethyl-, ethyl ester	$CHCl_3$	221(4.06),264(4.15), 344(4.38)	39-1460-65
Gramine, 5-acetyl-2-carbethoxy-	EtOH-HCl	264(4.70),313(4.02), 322(3.97)	87-0141-64
1H-Pyrrolo[2,1-c][1,4]benzodiazepine-2-propionic acid, 2,3,5,10,11,11a-hexahydro-5-oxo-, methyl ester	iso-PrOH	222(4.43),257s(3.93), 335(3.68)	35-5793-65
$C_{16}H_{20}N_2O_3S$			
Indole-2-carboxylic acid, 1-methyl-3-piperidinosulfinyl)-, methyl ester	EtOH	215(4.39),235(4.43), 303(4.17),324s(3.88), 336s(3.66)	44-0178-64

Compound	Solvent	λ_{max}(log ϵ)	Ref.
$C_{16}H_{20}N_2O_4$			
Piperidine, 1-[4-(5-nitrofurfurylidene)-2-hexenoyl]-	propylene glycol	280(4.10),390(4.29)	95-0646-64
$C_{16}H_{20}N_2O_4S$			
Indole-2-carboxylic acid, 1-methyl-3-(piperidinosulfonyl)-, methyl ester	EtOH	287(4.03)	44-0178-64
$C_{16}H_{20}N_2O_5$			
2-Pyrroline-3-carboxylic acid, 2-methyl-4-oxo-5-(3-carbethoxy-2-methyl-4-pyrrolyl)-, ethyl ester	EtOH	203(4.24),241(4.38), 274(3.79),311(3.81)	23-1524-64
$C_{16}H_{20}N_3$			
Amidogen, bis[p-(dimethylamino)phenyl]-	benzene	380(4.31),784(3.94)	24-0844-65
$C_{16}H_{20}N_4$			
Cinnolino[5,4,3-cde]cinnoline, 1,2,3,6,7,8-hexahydro-2,2,7,7-tetramethyl-	EtOH	264(3.81),273(3.76), 283(3.67),297(3.70), 308(3.64)	35-0661-64
$C_{16}H_{20}N_4OS$			
Acetanilide, N-[3-(2-amino-4-mercapto-6-methyl-5-pyrimidinyl)propyl]-	pH 1	261s(3.78),340(4.21)	4-0088-64
	pH 7	262s(3.88),350(4.23)	4-0088-64
	pH 13	264s(3.88),319(4.16)	4-0088-64
$C_{16}H_{20}N_4O_4$			
2-Cyclohexen-1-one, 2,3,4,4-tetramethyl-, 2,4-dinitrophenylhydrazone	EtOH	222(4.0),255(4.0), 385(4.78)	56-0385-64
2-Cyclohexen-1-one, 3,4,4,5-tetramethyl-, 2,4-dinitrophenylhydrazone	EtOH	382(4.25)	56-0599-64
Hydrindan-2-one, 8-methyl-, 2,4-dinitrophenylhydrazone, trans	$CHCl_3$	364(4.35)	35-0580-65
6,8-Nonadien-2-one, 7-methyl-, 2,4-dinitrophenylhydrazone	$CHCl_3$	364(4.33)	22-2533-64
2-Norbornanecarboxaldehyde, 3,3-dimethyl-, 2,4-dinitrophenylhydrazone	EtOH	362(4.38)	28-4783-65A
3,5-Octadien-2-one, 6,7-dimethyl-, 2,4-dinitrophenylhydrazone	$CHCl_3$	397(4.51)	22-3218-64
(+)-Pulegone, 2,4-dinitrophenylhydrazone	EtOH	379(4.40)	33-0051-64
(±)-Pulegone, 2,4-dinitrophenylhydrazone	EtOH	378.5(4.35)	33-0051-64
$C_{16}H_{20}N_4O_5$			
p-Menth-1-en-3-one, 4-hydroxy-, 2,4-dinitrophenylhydrazone	$CHCl_3$	389(4.46)	44-0518-65
p-Menth-3-en-2-one, 1-hydroxy-, 2,4-dinitrophenylhydrazone	$CHCl_3$	392(4.42)	44-0518-65
p-Menth-6-en-2-one, 5-hydroxy-, trans, 2,4-dinitrophenylhydrazone	EtOH	247(4.20),260(4.22), 375(4.46)	78-1889-65
Mitomycin C, N-methyl-	MeOH	215(4.44),363(4.36)	87-0001-65
$C_{16}H_{20}O$			
2,5-Cyclohexadien-1-one, 2,4,6-triallyl-4-methyl-	EtOH	246(4.03)	33-0094-65
2(4aH)-Naphthalenone, 3,4a-diallyl-5,6,7,8-tetrahydro-	EtOH	247(4.09)	33-0094-65

Compound	Solvent	λ max(log ε)	Ref.
$C_{16}H_{20}O_2$			
Benzo[b]naphtho[1,2-d]furan-2(3H)-one, 4,4a,5,6,7a,8,9,10,11,11a-decahydro-	EtOH	269(4.54),321(4.52)	24-2738-65
Cacalol, methyl-	EtOH	218(4.49),256(4.05)	78-2331-64
2,4-Pentadienoic acid, 2,3-dimethyl-5-p-tolyl-, ethyl ester	MeOH	229(4.12),238(4.05), 307(4.41)	44-2986-64
Spiro[cyclohexane-1,1'(4'H)-naphthalen]-4'-one, 2',3'-dihydro-5'-methoxy-	EtOH	257(3.93),314(3.66)	94-1084-65
Spiro[cyclohexane-1,1'(4'H)-naphthalen]-4'-one, 2',3'-dihydro-7'-methoxy-	EtOH	227(4.15),278(4.19)	94-1084-65
$C_{16}H_{20}O_3$			
1α-Phenanthrenecarboxylic acid, 1,2,3,4,4aβ,9,10,10aβ-octahydro-7-methoxy-	EtOH	220(3.90),279(3.30), 287(3.28)	44-2527-64
1β-Phenanthrenecarboxylic acid, 1,2,3,4,4aα,9,10,10aβ-octahydro-7-methoxy-	EtOH	221(3.91),278(3.30), 287(3.27)	44-2527-64
4aβ-isomer	EtOH	220(3.84),280(3.24), 287(3.23)	44-2527-64
$C_{16}H_{20}O_4$			
4β,9β,10β-Anthracenetriol, 1,4,4aα,9,9aα,10-hexahydro-5-methoxy-9α-methyl-	EtOH	275(3.36),281(3.36)	65-2570-64
4H,8H-Benzo[1,2-b:3,4-b']dipyran-4-one, 2,3,9,10-tetrahydro-5-hydroxy-	EtOH	216(4.36),293(4.32), 330s(3.49)	44-3604-64
2,3,8,8-tetramethyl-	EtOH-base	240s(4.09),292(4.25), 305(3.61)	44-3604-64
Clov-5-ene-3-carboxylic acid, 7,9-dioxo-	MeOH	206(3.63),244(3.95)	39-1338-65
2α-Phenanthrenecarboxylic acid, 1,2,3,4,4aα,9,10,10aβ-octahydro-4α-hydroxy-7-methoxy-	EtOH	220(3.93),277(3.24), 285s(3.19)	44-1213-65
2α-Phenanthrenecarboxylic acid, 1,2,3,4,4aβ,9,10,10aβ-octahydro-4α-hydroxy-7-methoxy-	EtOH	220(3.91),278(3.27), 286s(3.19)	44-1213-65
2α-Phenanthrenecarboxylic acid, 1,2,3,4,4aβ,9,10,10aβ-octahydro-4β-hydroxy-7-methoxy-	EtOH	221(3.92),279(3.29), 287(3.24)	44-1213-65
4aα-isomer	EtOH	219(3.99),278(3.31), 286(3.28)	44-1213-65
2β-Phenanthrenecarboxylic acid, 1,2,3,4,4aβ,9,10,10aβ-octahydro-4 -hydroxy-7-methoxy-	EtOH	220(3.98),278(3.29), 285(3.26)	44-1213-65
$C_{16}H_{20}O_4S$			
Valeric acid, 2-benzoyl-5-mercapto-, ethyl ester, acetate	n.s.g.	208(4.23),242(4.27)	24-1963-64
$C_{16}H_{20}O_5$			
Phthalide, 3-acetyl-4-methoxy-6-(2-methoxyethyl)-5,7-dimethyl-	EtOH	248(3.98),300(3.53)	35-1594-65
	EtOH-base	251(4.14),323(4.08), 410(4.08)	35-1594-65
$C_{16}H_{20}O_6$			
Pyrenophorin	n.s.g.	220(4.37)	88-4675-65
$C_{16}H_{20}O_8$			
1,2-Cyclobutanediacrylic acid, 3,4-dicarboxy-, tetramethyl ester, all trans	hexane	212(4.45)	5-0039-65A

Compound	Solvent	$\lambda_{max}(\log \epsilon)$	Ref.
1,3-Cyclobutanediacrylic acid, 2,4-dicarboxy-, tetramethyl ester, all trans	hexane	213(4.61)	5-0039-65A
Isophthalic acid, 4-(carboxymethyl)-2,6-dihydroxy-, triethyl ester	EtOH	312.5(3.78)	1-1677-65
$C_{16}H_{20}S$			
Azulene, 7-isopropyl-1,4-dimethyl-3-(methylthio)-	C_6H_{12}	249(3.3),294(3.38),309(3.33),373(3.85),392(3.79),654(2.55)	44-2715-65
$C_{16}H_{21}N$			
Camphorindole, dihydro-	n.s.g.	223(4.56),283(3.91),292(3.83)	88-2285-65
Quinoline, 2-benzyl-3,4,4a,5,6,7,8,8a-octahydro-, cis	EtOH	240s(3.5)	95-0674-64
$C_{16}H_{21}NO$			
2,4-Decadienanilide	EtOH	276(4.45)	54-0581-65
Quinoline, N-benzoyldecahydro-, cis	H_2SO_4	241(3.92)	70-2006-65
Quinoline, N-benzoyldecahydro-, trans	H_2SO_4	241(3.98)	70-2006-65
$C_{16}H_{21}NO_2$			
1-Butanone, 2-ethyl-1-(1-ethylindol-3-yl)-2-hydroxy-	EtOH	213(4.42),246(4.12),306(4.17)	87-0094-64
2,6-Methano-11bH-benzo[a]quinolizine, 1,2,3,4,6,7-hexahydro-9,10-dimethoxy-	EtOH	284(3.59),288(3.59)	94-1478-64
Spiro[5H-1-benzazepine-5,1'-cyclohexan]-2-one, 1,2,3,4-tetrahydro-7-methoxy-	EtOH	248(4.15),280s(3.5)	94-1084-65
Spiro[5H-2-benzazepine-5,1'-cyclohexan]-1-one, 1,2,3,4-tetrahydro-7-methoxy-	EtOH	252(4.03)	94-1084-65
Spiro[cyclohexane-1,4'(3'H)-isoquinolin]-1'(2'H)-one, 6'-methoxy-2'-methyl-	EtOH	262(4.11)	94-1084-65
Spiro[cyclohexane-1,4'(3'H)-isoquinolin]-1'(2'H)-one, 8'-methoxy-2'-methyl-	EtOH	241(3.78),300(3.67)	94-1084-65
$C_{16}H_{21}NO_3$			
2H-Azepin-2-one, 1-[3-(benzyloxy)-propionyl]-hexahydro-	dioxan	221(4.07)	78-3537-65
Carbostyril, 4-hydroxy-8-methoxy-1-methyl-3-isopentyl-	MeOH	217(4.30),239(4.45),246(4.43),254(4.40),284(3.89),293(3.91),322(3.51)	39-0438-64
Carbostyril, 3-isopentyl-4,8-dimethoxy-	MeOH	238(4.28),254(4.40),282(3.90),290(3.88),330(3.59)	39-0438-64
1-Cyclohexene, 1-morpholino-6-(2,5-dihydroxyphenyl)-	EtOH	218(3.65),231(3.63),304(3.59)	70-2086-64
$C_{16}H_{21}NO_4$			
Carbostyril, 4-hydroxy-3-(2-hydroxy-3-methylbutyl)-8-methoxy-1-methyl-	MeOH	216(4.35),237(4.49),245(4.45),254(4.45),285(3.95),295(3.96),325(3.50)	39-0438-64
Carbostyril, 3-(2-hydroxy-3-methylbutyl)-4,8-dimethoxy-	MeOH	239(4.25),255(4.39),283(3.88),292(3.85),319(3.49),331(3.53),342(3.44)	39-0438-64

Compound	Solvent	λ_{max}(log ε)	Ref.
2-Naphthoic acid, 5,6,7,8-tetrahydro-3-hydroxy-4-(morpholinomethyl)-	$CHCl_3$	238(4.17)	94-0112-64
$C_{16}H_{21}NO_5S$			
Benzenesulfonic acid, p-nitro-, 5,9-decadienyl ester, trans	EtOH	251(4.07)	35-1959-64
$C_{16}H_{21}N_3O$			
d-Lysergamide, 2,3,9,10-tetrahydro-	MeOH	245(3.77),291(3.29)	33-0756-64
$C_{16}H_{21}N_3O_2$			
1H-Indazole, 1-acetyl-3-(2-piperidino-ethoxy)-	EtOH	243(4.20),287(3.79), 293(4.06),297(3.97), 305(4.25)	33-1986-64
$C_{16}H_{21}N_3O_5S$			
Indole-2-carboxylic acid, 3-[(butyl-carbamoyl)sulfamoyl]-1-methyl-, methyl ester	EtOH	210(4.51),292(4.04)	44-0178-64
$C_{16}H_{21}N_3O_6$			
Cytidine, N-acetyl-2',3'-O-cyclo-pentylidene-	pH 2	213(--),244(4.13), 297(--)	73-0214-64
$C_{16}H_{21}N_5S$			
5H-Thiopyrano[2,3-d]pyrimidine, 2-amino-7-[p-(dimethylamino)-anilino]-6,7-dihydro-4-methyl-	pH 1	220(4.34),309(4.04)	4-0088-64
	pH 7	237(4.34),309(4.04)	4-0088-64
	pH 13	316(4.19)	4-0088-64
$C_{16}H_{22}$			
Toluene, m-(2,6,6-trimethyl-1-cyclohexen-1-yl)-	n.s.g.	266s(2.93),272(2.84)	39-2892-65
$C_{16}H_{22}BrNO$			
1,5-Cyclohexadiene-1-acetonitrile, 3-bromo-3,5-di-tert-butyl-4-oxo-	n.s.g.	325(3.40)	70-0336-65
2,5-Cyclohexadiene-1-acetonitrile, 1-bromo-3,5-di-tert-butyl-4-oxo-	n.s.g.	248(4.22)	70-0336-65
$C_{16}H_{22}ClN_3O$			
Oxychloroquine	pH 1	328(4.24),340(4.25)	4-0006-64
	EtOH	328(4.07)	4-0006-64
$C_{16}H_{22}ClN_3O_2$			
Oxychloroquine, 1-oxide	pH 1	343(4.24),353(4.23)	4-0006-64
	EtOH	390(4.10)	4-0006-64
$C_{16}H_{22}Cl_2Pd_2$			
Palladium, dichlorodi-2,4-cyclo-octadien-1-yldi-	$CHCl_3$	232(3.93),259(4.09), 285s(3.87),353s(3.04)	39-5002-64
$C_{16}H_{22}F_2N_4O_2$			
p-Benzoquinone, 2,5-difluoro-3,6-bis-(4-methyl-1-piperazinyl)-	methyl cellosolve	234(4.40),407(4.07), 552(2.68)	78-1889-64
$C_{16}H_{22}N_2$			
Indole, 3-(1-ethyl-5,5-dimethyl-2-pyrrolidinyl)-	EtOH	220(4.52),275s(3.73), 281(3.76),288s(3.70)	87-0415-64

Compound	Solvent	$\lambda_{max}(\log \epsilon)$	Ref.
$C_{16}H_{22}N_2O$			
1H-Azecino[5,4-b]indol-6-ol, 2,3,4,5,6,7,8,9-octahydro-3-methyl-	EtOH	229(4.51),285(3.85), 292(3.83)	44-1550-65
2-Indolinone, 3-[1-(aminomethyl)-cycloheptyl]-	EtOH	237(3.87),292(3.43)	87-0626-65
1-Pentanone, 5-(ethylmethylamino)-1-indol-3-yl)-, hydriodide	EtOH	210(4.60),240(4.18), 258(3.98),297(4.10)	44-2534-65
$C_{16}H_{22}N_2O_2$			
p-Benzoquinone, 2,5-dipiperidino-	methyl cellosolve	229(4.43),379(4.30), 528(2.76)	78-1889-64
2-Hexanone, 1-(3,4-dihydro-4-quinazolinyl)-4-hydroxy-3,4-dimethyl-	EtOH	228(4.16),234(4.09), 305(3.90)	94-0291-65
Hydrazine, 1-benzoyl-1-(2-hydroxy-cyclohexyl)-2-isopropylidene-, trans	EtOH	227(3.91)	44-1097-64
$C_{16}H_{22}N_2O_3$			
2,5-Cyclohexadiene-1-acetonitrile, 3,5-di-tert-butyl-1-nitro-4-oxo-	hexane	232(4.05)	70-1666-64
2-Indolinone, 3-(4-piperidylmethyl)-	EtOH	249(3.93),277(3.15)	87-0626-65
$C_{16}H_{22}N_2O_4$			
2H-Benzo[a]quinolizine-3-carboxamide, 1,3,4,6,7,11b-hexahydro-2-hydroxy-9,10-dimethoxy-	n.s.g.	285(3.54)	87-0635-64
1,3-Cyclohexanedione, 2,2'-azobis-[5,5-dimethyl-	MeOH	295(4.02),440(4.22)	5-0101-64F
	MeOH-base	295(4.2),425(4.3)	5-0101-64F
	CH_2Cl_2	265(3.98),425(4.27)	5-0214-65G
$C_{16}H_{22}N_2O_6S$			
5-Thia-1-azabicyclo[4.2.0]oct-2-ene-2-carboxylic acid, 3-(hydroxymethyl)-8-oxo-7-valeramido-, methyl ester, acetate	EtOH	262(3.89)	39-5015-65
Thymidine, 5'-thio-, 3',5'-di-propionate	pH 1	263.5(3.94)	44-0554-64
	pH 7	264.5(4.01)	44-0554-64
	pH 13	264.5(3.94)	44-0554-64
$C_{16}H_{22}N_2Zn$			
Zinc, (2,2'-bipyridine)diisopropyl-	toluene	365(2.70),480(2.63)	101-0222-65A
$C_{16}H_{22}N_4O_3$			
Eseramine	n.s.g.	256(4.12),311(3.43)	25-0459-64
$C_{16}H_{22}N_4O_4$			
Cyclobutanone, hexamethyl-, 2,4-dinitrophenylhydrazone	n.s.g.	361(4.34)	77-0098-65
Cyclobutanone, 4-isopropyl-2,2,3-trimethyl-, 2,4-dinitrophenylhydrazone	hexane	368(4.36)	28-0827-64B
$C_{16}H_{22}N_4O_9S$			
Ammonium sulfisomezole-N^1-glucosiduronate	MeOH	273(4.34)	95-1104-64
$C_{16}H_{22}O$			
6,8,12,14-Hexadecatetraen-10-yn-1-ol	ether	303s(4.50),315(4.66), 337(4.67)	24-0872-65
Phenanthrene, 1,2,3,4,4aβ,9,10,10aβ-octahydro-7-methoxy-1α-methyl-	EtOH	278(3.32),286(3.30)	44-2527-64

Compound	Solvent	$\lambda_{max}(\log \epsilon)$	Ref.
$C_{16}H_{22}O_3$			
7H-Benz[e]inden-7-one, 1,2,3,3a,-4,5,8,9,9a,9b-decahydro-3-hydroxy-3a-methyl-, acetate, anti-trans	EtOH	238(4.24)	13-0185s-65
Cyclohexanepropionic acid, 1-(m-methoxyphenyl)-	EtOH	278(3.33),281(3.28)	94-1084-65
Dispiro[5.2.5.2]hexadecane-3,7,16-trione	EtOH	286.5(1.98)	78-2553-64
2α-Phenanthrenemethanol, 1,2,3,4,4aα,9-10,10aβ-octahydro-4α-hydroxy-7-methoxy-	EtOH	224(3.95),276(3.27), 283(3.22)	44-1213-65
2α-Phenanthrenemethanol, 1,2,3,4,4aβ,9-10,10aβ-octahydro-4β-hydroxy-7-methoxy-	EtOH	221(3.93),278(3.29), 285(3.27)	44-1213-65
$C_{16}H_{22}O_3Si_3$			
Cyclotrisiloxane, 2,2,4,4-tetramethyl-6,6-diphenyl-	$CHCl_3$	254(2.7),259(2.9), 265(2.91),272(2.8)	46-1066-65
$C_{16}H_{22}O_4$			
Hexanoic acid, 4-(p-carboxyphenyl)-2,2,3-trimethyl-	EtOH	240(4.15)	39-5704-64
4-Pentynoic acid, 5-(2,6,6-trimethyl-1-cyclohexen-1-yl)-3-methyl-2-carboxy-	EtOH	235(4.18),245s(3.97)	22-2541-64
$C_{16}H_{22}O_5$			
1,4-Epoxy-1H-2,3-benzodioxepin, 4,5-dihydro-6-methoxy-9-(2-methoxyethyl)-4,7,9-trimethyl-	EtOH	274(3.04),283(3.04)	35-1594-65
3a(4H)-Indanpropionic acid, 2-carboxy-5,6-dihydro-1,1,5-trimethyl-6-oxo-	MeOH	243(4.09)	39-1338-65
1-Naphthoic acid, 2,3,4,4a,5,6,7,8-octahydro-5β-hydroxy-4aβ-methyl-2-oxo-, ethyl ester	EtOH	240(4.06)	44-3333-65
$C_{16}H_{22}O_6$			
Acid, m. 139.5-140°	EtOH	204(3.60)	1-1088-65
$C_{16}H_{22}O_6S$			
1-Epigeigerin, methanesulfonate	EtOH	236(4.11)	39-2518-64
Trichothecolone, methanesulfonate	MeOH	226(3.94),325(1.51)	39-2234-64
$C_{16}H_{22}O_7S$			
Isophotoartemisic lactone, methanesulfonate	EtOH	238(4.18)	39-2518-64
$C_{16}H_{22}O_{11}$			
Monotropein	EtOH	235(3.83)	25-0931-64
	EtOH	235(3.98)	94-0888-64
$C_{16}H_{22}Si_2$			
Disilane, 1,1,1,2-tetramethyl-2,2-diphenyl-	C_6H_{12}	230.5(4.22)	101-0369-64B
Disilane, 1,1,2,2-tetramethyl-1,2-diphenyl-	C_6H_{12}	236.0(4.26)	25-1063-64
	C_6H_{12}	236.0(4.26)	101-0369-64B
	C_6H_{12}	237.5(4.22)	39-4690-65
	n.s.g.	237.5(4.22)	25-1492-64

Compound	Solvent	$\lambda_{max}(\log \epsilon)$	Ref.
$C_{16}H_{23}ClO_6S$			
Tricothecolone, O-methanesulfonyl-, chlorohydrin	MeOH	227(3.90),326(1.40)	33-2234-64
$C_{16}H_{23}IO_2S$			
Benzenesulfinic acid, p-iodo-, p-menth-3-yl ester	EtOH	215s(4.09),247(4.25), 265s(4.06),273s(3.83)	35-1958-65
$C_{16}H_{23}N$			
Indolizine, 2,6-diethyl-1,3,5,7-tetramethyl-	EtOH	248(4.42),286(3.59), 297(3.63),310(3.62), 353(3.36)	39-0893-64
	EtOH-HCl	229(4.28),256(3.77), 333(3.70)	39-0893-64
Indolizine, 1,2,6,7-tetraethyl-	EtOH	248(4.50),287(3.34), 297(3.49),309(3.48), 360(3.30)	39-1518-65
$C_{16}H_{23}NO$			
4-Eudesmen-7α-carbonitrile, 3-oxo-	EtOH	251.5(4.14)	78-0791-65
Spiro[5H-1-benzazepine-5,1'-cyclohexane], 1,2,3,4-tetrahydro-7-methoxy-	EtOH	246(3.89),300(3.41)	94-1084-65
hydrochloride	EtOH	228(3.68),273(2.98), 280(2.98)	94-1084-65
Spiro[5H-2-benzazepine-5,1'-cyclohexane], 1,2,3,4-tetrahydro-7-methoxy-	EtOH	230(3.91),276(3.20), 282(3.17)	94-1084-65
hydrochloride	EtOH	231(3.99),275(3.13), 280(3.10)	94-1084-65
Spiro[cyclohexane-1,4'(1'H)-isoquinoline], 2',3'-dihydro-8'-methoxy-2'-methyl-	EtOH	272(3.22),280(3.24)	94-1084-65
hydrochloride	EtOH	219(3.80),274(3.32), 281(3.33)	94-1084-65
$C_{16}H_{23}NO_2$			
Condensation product of 1-cyclopenten-1-yl vinyl ketone with ammonia	n.s.g.	240(4.33)	70-2155-65
$C_{16}H_{23}NO_3$			
1H,5H-Benzo[ij]quinolzine-8-acetic acid, 2,3,6,7,7a,8,9,10-octahydro-10-oxo-, ethyl ester	ether	302(4.37)	24-1354-64
11-Epigeigerin, deoxy-, O-methyloxime	EtOH	248(4.30)	39-2518-64
$C_{16}H_{23}N_3$			
Valeronitrile, 2-isobutyl-4-methyl-2-(phenylazo)-	C_6H_5Cl	390(2.3)	60-1437-65
	o-$C_6H_4Cl_2$	395(2.30)	60-1437-65
$C_{16}H_{23}N_3O$			
11H-Dibenzo[b,e]-1,4-diazepin-11-one, 1,2,3,4,4a,5,10,11a-octahydro-10-(2-methylaminoethyl)-	EtOH	224(4.45),290(3.25)	33-1590-65
1H-Indazole, 3-methoxy-1-(3-piperidinopropyl)-	EtOH-HCl	222(4.42),309(3.69)	33-1986-64
1H-Indazole, 1-methyl-3-(3-piperidinopropoxy)-	EtOH-HCl	222(4.43),252(3.12), 311(3.72)	33-1986-64
3-Indazolone, 1-methyl-2-(3-piperidinopropyl)-	EtOH-HCl	219(4.28),240(4.03), 316(3.64)	33-1986-64

Compound	Solvent	λ_{max}(log ε)	Ref.
$C_{16}H_{23}N_3O_3$			
Butyric acid, 4-(dimethylamino)-2-oxo-, ethyl ester, (p-acetylphenyl)-hydrazone	EtOH	236(4.00),351(4.61)	87-0141-64
$C_{16}H_{23}N_5S$			
4-Pyrimidinethiol, 2-amino-5-[3-(p-dimethylaminoanilino)propyl]-6-methyl-	pH 1	258s(4.07),328(4.07)	4-0088-64
	pH 7	252(4.27),317(3.93), 350(3.94)	4-0088-64
	pH 13	248s(4.33),317(4.14)	4-0088-64
$C_{16}H_{24}$			
1,3,9,11-Cyclohexadecatetraene	hexane	229(4.35)	39-6674-65
1,3-Cyclohexadiene, 2-methyl-6-(2,6,6-trimethylcyclohex-1-enyl)-	n.s.g.	272(3.74)	39-2892-65
Toluene, 3-(2,2,6-trimethylcyclohexyl)-	n.s.g.	261(2.39),267(2.46), 275(2.39)	39-2892-65
$C_{16}H_{24}BrNO_3$			
2,5-Cyclohexadien-1-one, 4-bromo-4-nitro-2,6-di-tert-pentyl-	hexane	250(3.95)	70-0371-64
$C_{16}H_{24}IN_3O_2$			
Physostigmine, methiodide	EtOH	245(4.08),303(3.40)	39-5510-64
$C_{16}H_{24}N_2$			
Spiro[indoline-3,3'-pyrrolidine], 1'-ethyl-2'-propyl-	EtOH	242(3.80),290(3.43)	78-1327-65
$C_{16}H_{24}N_2O$			
α-Obscurine, N-demethyl-	MeOH	255(3.81)	23-2456-64
$C_{16}H_{24}N_2O_2$			
Lycocernuine, dehydro-	n.s.g.	318(1.81)	88-2201-64
α-Obscurine, N-demethylhydroxy-	MeOH	255(3.78)	23-2456-64
Pyridazine, 3,6-bis(cyclohexyloxy)-	EtOH	291(3.32)	44-1751-64
$C_{16}H_{24}N_2O_5$			
Pyrrole-3,4-dicarboxylic acid, 2,5-dimethyl-1-morpholino-, diethyl ester	EtOH	210(2.8),263(2.4)	6-0073-64
$C_{16}H_{24}N_4O_4$			
Bicarbamic acid, [2-methyl-1-(phenylazo)propyl]-, diethyl ester	EtOH	270(4.08),408(2.09)	7-0441-65
Bicarbamic acid, [1-(phenylazo)-butyl]-, diethyl ester	EtOH	270(4.08),408(2.09)	7-0441-65
$C_{16}H_{24}N_4O_8S$			
Cephalosporin C	H_2O	260(3.97)	87-0689-64
$C_{16}H_{24}O$			
Acetophenone, 3',4'-di-tert-butyl-	EtOH	257(4.23)	35-5281-64
2-Cyclohexen-1-one, 3-(2-bornyl)-	EtOH	245(4.07)	33-1766-64
2-Cyclohexen-1-one, 3-(1,5,5-trimethyl-2-exo-norbornyl)-	EtOH	243(4.19)	33-1766-64
2-Cyclohexen-1-one, 3-(5,5,6-exo-trimethyl-2-exo-norbornyl)-	EtOH	240(4.21)	33-1766-64
	EtOH	240(4.152)	33-0319-64
$C_{16}H_{24}OS$			
Ketone, methyl 15-thiabicyclo[10.2.1]-pentadeca-12,14-dien-13-yl	EtOH	258(3.9),285(3.4)	70-2055-64

Compound	Solvent	$\lambda_{max}(\log \epsilon)$	Ref.
$C_{16}H_{24}O_2$			
Acetophenone, 2'-hydroxy-	EtOH	286(3.39)	25-1837-65
4',6'-di-tert-butyl-	EtOH-NaOH	240(4.00),304(3.35), 335s(--)	25-1837-65
Bicyclo[2.2.0]hex-2-ene, 5,6-diacetyl-	C_6H_{12}	239(3.24),301(2.44)	39-0194-65
1,2,3,4,5,6-hexamethyl-	EtOH	239(3.16),300(2.62)	39-0194-65
Costic acid, methyl ester	EtOH	210(3.75)	78-1521-65
Lanceolic acid, methyl ester	hexane	217(4.06)	78-0333-64
2(3H)-Phenanthrone, 4,4a,4b,5,6,7,8,-8a,9,10-decahydro-8-hydroxy-1,8a-dimethyl-, (1S,10aR)	EtOH	250(4.16)	33-1725-65
(1S,10aS)	EtOH	252(4.16)	33-1725-65
$C_{16}H_{24}O_3$			
Acetophenone, 2'-heptyl-6'-hydroxy-4'-methoxy-	EtOH	270(3.76)	1-1677-65
Bicyclo[3.3.1]non-2-ene-3-carboxylic acid, 2,4,4,8-tetramethyl-9-oxo-, ethyl ester	EtOH	206(3.77),272(3.40)	88-1337-64
11βH-Eudesm-4-en-13-oic acid, 3-oxo-, methyl ester	EtOH	248(3.97)	78-1167-65
1,4-Indandiol, 6-(2-methoxyethyl)-2,2,5,7-tetramethyl-	EtOH	280s(3.18),286(3.23)	35-1594-65
$C_{16}H_{24}O_4$			
Benzoic acid, 2-heptyl-4,6-dimethoxy-	EtOH	239s(3.66),280(3.34)	1-1677-65
Benzoic acid, 2-heptyl-4-hydroxy-6-methoxy-, methyl ester	EtOH	244s(3.62),281(3.43)	1-1677-65
3-Cyclohexene-1-acrylic acid, 2-carboxy-α,2-dimethyl-, diethyl ester	n.s.g.	219(4.15)	39-5815-64
Humulinic acid D, methyl ester, allo	pH 1	242(3.87),303(4.02)	20-0629-65
	pH 13	263(4.03),297(4.00)	
$C_{16}H_{24}O_5$			
1-Naphthoic acid, decahydro-5-hydroxy-4a-methyl-2-oxo-, ethyl ester, cis	EtOH	258(4.01)	44-3333-65
trans	EtOH	258(3.70)	44-3333-65
$C_{16}H_{24}O_7$			
2-Cyclopentene-1-propionic acid, 3-carboxy-2-hydroxy-, diethyl ester, ethyl carbonate	EtOH	229(3.80)	44-0087-64
$C_{16}H_{25}NO_2$			
Benzene, 2,4-diisopentyl-1-nitro-	EtOH	272(3.51)	32-1221-64
$C_{16}H_{25}NO_4$			
2-Azabicyclo[3.2.2]nona-6,8-diene-6,7-dicarboxylic acid, 1,2,4,9-tetramethyl-, dimethyl ester	EtOH	221s(3.59),292s(2.81)	35-4092-64
2,5-Cyclohexadien-1-one, 2,6-di-tert-butyl-4-(methoxymethyl)-	hexane	235(4.06)	70-1666-64
	n.s.g.	236(4.06)	70-1675-65
$C_{16}H_{25}N_3O$			
Costal, semicarbazone	EtOH	262(4.42)	78-1521-65
$C_{16}H_{25}N_3O_3$			
Salicylaldehyde- 5-nitro-, ethyl-heptylhydrazone	MeOH	302(4.46)	24-1631-64
	MeOH-NaOH	344(4.27),428(4.01)	24-1631-64

Compound	Solvent	λ_{max}(log ε)	Ref.
$C_{16}H_{25}N_3O_{11}$			
Cytosine, 1-(β-cellobiosyl)-, hydrochloride	pH 1	275(4.13)	44-2723-65
Cytosine, 1-(β-lactosyl)-, hydrochloride	50% MeOH	277(4.12)	44-2723-65
$C_{16}H_{25}O_7PS$			
Fumaric acid, hydroxy(2-hydroxycyclohexyl)-, γ-lactone, ethyl ester, O-ester with O,O-diethyl phosphorothioate	MeOH	236(3.78)	5-0010-65E
$C_{16}H_{25}O_8P$			
Fumaric acid, hydroxy(2-hydroxycyclohexyl)-, γ-lactone, ethyl ester, diethyl phosphate	MeOH	235(3.90)	5-0010-65E
$C_{16}H_{26}$			
Benzene, m-diisopentyl-	EtOH	258s(2.38),264(2.45), 271(2.35)	32-1221-64
$C_{16}H_{26}Cl_2O_2Pd_2$			
Palladium, dichlorobis(4-methoxy-2-cyclohepten-1-yl)di-	$CHCl_3$	249(3.90),297(3.11), 315s(3.00)	39-5002-64
$C_{16}H_{26}INO_2$			
N-Methyllophocerinium iodide	MeOH	286(3.88)	5-0207-65E
$C_{16}H_{26}N_2$			
Acetamidine, N,N'-dibutyl-2-phenyl-, hydrochloride	EtOH	252(2.05),258(2.19), 264(2.07)	39-1147-64
Aziridine, 1-[β-(butylamino)-α,α-dimethylphenethyl]-	EtOH	252(2.32),259(2.33), 264(2.14)	44-3146-64
$C_{16}H_{26}N_4S_2$			
3(2H)-Pyridazinethione, 2-(piperidinomethyl)-6-[(piperidinomethyl)thio]-	hexane	310(4.34)	49-0631-65
$C_{16}H_{26}O$			
3,5-Hexadien-2-one, 6-p-menth-3-yl-	EtOH	279(4.51)	28-4712-64B
$C_{16}H_{26}O_2$			
β-Erythroidinol, deazaanhydro-octahydro-	n.s.g.	240(3.91)	44-1307-64
Spiro[4.4]non-1-ene-2-carboxylic acid, 7-isopropyl-1,4-dimethyl-, methyl ester	EtOH	229(3.89)	78-0115-65
$C_{16}H_{26}O_3$			
2-Cyclohexen-1-one, 3-isobutoxy-2-(4-methyl-3-oxopentyl)-	EtOH	271(3.91)	39-2340-65
$C_{16}H_{26}O_4$			
4-Pentenoic acid, 2-hydroxy-3-methoxy-3-methyl-5-(2,6,6-trimethyl-1-cyclohexen-1-yl)-	n.s.g.	239(3.76)	24-0549-64
$C_{16}H_{26}O_5$			
8-Tridecynoic acid, 10,11-dihydroxy-10-methyl-7-oxo-, ethyl ester	EtOH	225(3.80)	70-0818-65

Compound	Solvent	λ_{max}(log ϵ)	Ref.
$C_{16}H_{27}ClN_2O_4$			
N-(5,5-Dimethyl-3,N'-pyrrolidyl-2-cyclohexen-1-ylidene)pyrrolidinium perchlorate	EtOH	325s(4.64),331(4.66)	44-0794-64
$C_{16}H_{27}ClN_2O_6$			
N-(5,5-Dimethyl-3,N'-morpholinyl-cyclohex-2-en-1-ylidene)morpholinium perchlorate	EtOH	332s(4.63),342(4.66)	44-0794-64
$C_{16}H_{27}N$			
Aniline, 2,4-diisopentyl-	EtOH	236(3.92),290(3.33)	32-1221-64
$C_{16}H_{27}N_5O_7S$			
6-Indolol, 3-(2-aminobutyl)-, creatinine sulfate salt	EtOH	222(4.52),264s(3.56), 273(3.61),295(3.66), 300s(3.53)	87-0274-64
$C_{16}H_{27}O_7PS$			
Fumaric acid, hydroxy(3-hydroxy-1,3-dimethylbutyl)-, δ-lactone, ethyl ester, O-ester with O,O-diethyl phosphorothioate	MeOH	209(3.95)	5-0010-65E
$C_{16}H_{27}O_8P$			
Fumaric acid, hydroxy(3-hydroxy-1,3-dimethylbutyl)-, δ-lactone, ethyl ester, diethyl phosphate	MeOH	220(3.83)	5-0010-65E
$C_{16}H_{28}N_4O_2$			
1-Pyrrolidinyloxy, 2,2,5,5-tetramethyl-3-oxo-, azine	C_6H_{12}	209(4.28),228(4.19)	22-3290-65
	EtOH	228(4.23),420(1.21)	22-3290-65
	$CHCl_3$	427.5(1.25)	22-3290-65
$C_{16}H_{28}O$			
Furan, 2,3,5-tri-tert-butyl-	hexane	221(3.95)	77-0001-65
$C_{16}H_{28}O_2$			
Costic acid, tetrahydro-, methyl ester	EtOH	210(2.86)	78-1521-65
Cyclohexaneacetic acid, 3-cyclohexyl-, ethyl ester	EtOH	256(1.30)	39-5617-64
2-Cyclohexen-1-one, 6-(hydroxymethyl)-5-propyl-2,4,6-triethyl-	EtOH	239.5(3.95)	44-3650-65
isomer	EtOH	242.5(3.76)	44-3650-65
$C_{16}H_{28}O_3$			
Cyclohexaneacetic acid, 3-cyclohexyl-2-hydroxy-, ethyl ester	EtOH	259(2.36)	39-5617-64
$C_{16}H_{29}ClN_2O_4$			
N-(3-Diethylamino-5,5-dimethylcyclohex-2-en-1-ylidene)pyrrolidinium perchlorate	EtOH	325s(4.64),335(4.68)	44-0794-64
$C_{16}H_{29}N$			
Pyrrole, 2,3,5-tri-tert-butyl-	C_6H_{12}	218(3.98),283s(1.52)	77-0453-65
$C_{16}H_{29}NO$			
Morpholine, 4-(1-cyclododecen-1-yl)-	C_6H_{12}	223(3.93)	28-0759-65B

Compound	Solvent	λ_{max}(log ϵ)	Ref.
$C_{16}H_{29}NO_4$			
O-Methylribalinium indole base	50% EtOH	234(4.55),243(4.51), 301(4.07),319(3.98), 333(3.90)	78-0909-65
$C_{16}H_{29}N_4O$			
1-Pyrrolidyloxy, 2,2,5,5-tetramethyl-3-oxo-, azine with 2,2,5,5-tetramethyl-3-pyrrolidinone	C_6H_{12}	208(4.46),228(4.22)	22-3290-65
	EtOH	228(4.15),420(0.91)	22-3290-65
	$CHCl_3$	427.5(0.95)	22-3290-65
$C_{16}H_{30}N_3O_9P$			
Uridine, 3-methyl-, 3'-phosphate, compound with triethylamine	pH 5.6	261(3.98)	87-0486-65
	pH 11.6	260(3.97)	87-0486-65
$C_{16}H_{30}N_4$			
3-Pyrrolidone, 2,2,5,5-tetramethyl-, azine	MeOH	208(4.31),229(3.67)	22-3290-65
$C_{16}H_{32}O_2$			
Palmitic acid, lead salt	C_6H_{12}	230(4.33)	88-2847-65
$C_{16}H_{32}Sn$			
Tin, tributyl(1-methyleneallyl)-	C_6H_{12}	220(4.06)	44-1994-64
$C_{16}H_{36}B_4Cl_4N_4$			
1,3,5,7,2,4,6,8-Tetrazatetraborocine, 1,3,5,7-tetra-tert-butyl-2,4,6,8-tetrachlorooctahydro-	n.s.g.	204.6(3.40)	39-6421-65
$C_{16}H_{36}B_4N_{16}$			
1,3,5,7,2,4,6,8-Tetrazatetraborocine, 2,4,6,8-tetraazido-1,3,5,7-tetra-tert-butyloctahydro-	n.s.g.	204(3.78),229(3.72)	39-6421-65

Compound	Solvent	$\lambda_{max}(\log \epsilon)$	Ref.
$C_{17}H_9NO_3$			
Liriodenine	EtOH	248(4.42),269(4.33), 312(3.82),415(4.07)	95-0077-65
Spermatheridine	pH 1	257(4.33),280(4.25), 334(3.70)	88-1629-64
	EtOH	248(4.23),269(4.16), 302(3.70)	88-1629-64
$C_{17}H_{10}ClN$			
Benzo[j]phenanthridine, 6-chloro-	EtOH	218(4.59),232(4.50), 272s(4.76),281(4.80), 303(3.93),322(3.77), 337(3.87),352(3.77), 370(3.45),387(3.41)	4-0015-65
$C_{17}H_{10}ClNO$			
Naphtho[1,2-d]oxazole, 2-(p-chlorophenyl)-	heptane	230(4.42),250(3.91), 265s(3.70),295(4.04), 305(4.09),337(4.34), 350(4.11)	65-0427-64
Naphtho[2,1-d]oxazole, 2-(p-chlorophenyl)-	heptane	222(4.47),267(4.27), 277(4.31),302(4.10), 315(4.14),335(4.13), 350(4.18)	65-0427-64
$C_{17}H_{10}Cl_2S_3$			
[1,2]Dithiolo[1,5-b][1,2]dithiole-7-S^{IV}, 2,5-bis(p-chlorophenyl)-	dioxan	259(4.66),316(4.42), 512(4.18)	24-1732-64
$C_{17}H_{10}FNO$			
Naphth[1,2-d]oxazole, 2-(p-fluorophenyl)-	heptane	225(4.68),290(4.31), 305(4.33),330(4.47), 345(4.47)	65-0427-64
$C_{17}H_{10}N_4O_5$			
Sydnone, 4,4'-carbonylbis[3-phenyl-	THF	225(3.99)	78-1369-65
$C_{17}H_{10}O$			
Cyclopropenone, 2-phenyl-3-(phenylethynyl)-	ether	216(4.37),243(4.27), 318(4.44),339(4.26)	88-2317-65
2,4-Pentadiynophenone, 5-phenyl-	EtOH	269(3.73),315(3.58), 336(3.53)	39-2983-65
$C_{17}H_{10}O_3$			
5-Benzofuranacrylic acid, 6-hydroxy-β-phenyl-, δ-lactone	EtOH	225(4.43),248(4.41), 298(4.06),331(3.83)	44-2467-64
6H-Benzo[d]naphtho[1,2-b]pyran-6-one, 11-hydroxy-	MeOH	243(4.7),259(4.6), 268(4.7),293(4.0), 319(4.1),331(4.1)	83-0529-64
12H-Benzo[b]xanthen-12-one, 11-hydroxy-	C_6H_{12}	226(4.44),269(4.85), 284s(4.42),309(4.08), 323(3.72),424(3.78)	35-5424-65
$C_{17}H_{10}O_4$			
Benzo[b]naphtho[2,3-d]furan-6,11-dione, 10-hydroxy-2-methyl-	CH_2Cl_2	259(4.59),308(3.92), 413(3.95)	33-1459-64
$C_{17}H_{10}O_5S$			
Carbonic acid, thio-, cyclic O,O-ester with 6,7-dihydroxy-2-(p-methoxybenzylidene)-3(2H)-benzofuranone	EtOH	206(4.26),228(4.30), 353s(4.23),405(4.35)	56-0931-65

Compound	Solvent	λ_{max}(log ϵ)	Ref.
$C_{17}H_{10}O_9$			
Flavone, 5',6-dicarboxy-2',4',7-trihydroxy-	MeOH	268(3.22),300(4.01)	44-3445-64
$C_{17}H_{11}BrN_2O_2$			
2,3'-Biindolyl, 5'-bromo-5,6-(methylenedioxy)-	EtOH	248(4.31),281s(4.03), 291s(4.00),333(4.37)	44-2030-64
$C_{17}H_{11}BrO_3$			
Chromone, 3-benzoyl-2-(bromomethyl)-	MeOH	235(4.40),246(4.40), 301(3.91)	35-5417-65
$C_{17}H_{11}ClN_2$			
Pyrrolo[1,2-a]quinoxaline, 1-chloro-4-phenyl-	EtOH	233s(4.41),240(4.43), 243s(4.42),272(4.40), 323s(3.70),336(3.92), 346s(3.91)	35-1830-64
$C_{17}H_{11}ClS_3$			
6a-Thiathiophthene, 5-(p-chlorophenyl)-2-phenyl-	dioxan	256(4.56),310(4.41), 509(4.19)	24-1732-64
$C_{17}H_{11}Cl_2N_3OS$			
Benzaldehyde, p-chloro-, 2-azine with 5-(p-chlorobenzylidene)-2,4-thiazolidinedione	EtOH-HOAc	292(4.47),304(4.48), 360(4.52)	7-0987-64
$C_{17}H_{11}Cl_9O$			
1,4:9,10-Dimethanoanthracene, 1,3,4,5,-6,7,8,11,11-nonachloro-1,2,3,4,4a,9,-9a,10-octahydro-2-methoxy-, endo-endo	EtOH	215(4.68),220s(4.64), 242s(4.06),287(2.18), 297(2.18)	39-4646-65
1,4:9,10-Dimethanoanthracene, 1,3,4,5,-6,7,8,11,11-nonachloro-1,4,4a,8a,9,-9a,10,10a-octahydro-2-methoxy-, endo-endo-exo	EtOH	216(3.92),265(3.35), 276(3.51),287(3.68), 299(3.73),313(3.52)	39-4646-65
$C_{17}H_{11}Cs$			
Cesium, 7H-benzo[c]fluoren-7-yl-	$C_6H_{11}NH_2$	418(4.29),599(3.19), 648(3.09)	35-0384-65
$C_{17}H_{11}Li$			
Lithium, 7H-benzo[c]fluoren-7-yl-	$C_6H_{11}NH_2$	397(3.68),465(3.32), 487(3.42),519(3.31)	35-0384-65
Lithium, 11H-benzo[a]fluoren-11-yl-	$C_6H_{11}NH_2$	425(3.91)	35-0384-65
Lithium, 11H-benzo[b]fluoren-11-yl-	$C_6H_{11}NH_2$	420(4.34),605(3.23), 657(3.11)	35-0384-65
$C_{17}H_{11}N$			
Benzo[a]phenanthridine	pH 1.00	238(4.32),248(4.33), 282(4.49),400(3.66)	39-4426-65
	pH 9.20	276(4.50),312(3.88), 340(3.13),358(3.30), 376(3.36)	39-4426-65
Benzo[b]phenanthridine	EtOH	218(4.49),256(4.62), 264s(4.57),274(4.61), 284(4.69),297(4.56), 324(3.72),338(3.80), 352(3.72)	4-0015-65

Compound	Solvent	$\lambda_{max}(\log \epsilon)$	Ref.
Benzo[b]phenanthridine (contd.)	EtOH-HCl	219s(4.39),252(4.64), 284s(4.31),297(4.46), 311(4.55),341s(3.60), 356(3.85),372(3.98)	4-0015-65
Benzo[j]phenanthridine	EtOH	219(4.47),231(4.49), 253(4.51),271s(4.63), 280(4.80),289s(4.67), 321(3.77),336(3.87), 352(3.76),368(3.41), 388(3.36)	4-0015-65
	EtOH-HCl	220(4.38),246(4.42), 285(4.54),293(4.57), 307(4.41),334s(3.70), 346(3.85),363(3.88)	4-0015-65
Benzo[k]phenanthridine	pH 1.00	254(4.30),282(4.61), 330(3.79),344(3.87), 380(3.46),410(3.57)	39-4426-65
	pH 9.22	272(4.66),310(3.89), 340s(3.12),358(3.34), 376(3.38)	39-4426-65
$C_{17}H_{11}NO$			
Benzo[b]phenanthridin-5(6H)-one	EtOH	227(4.48),245(4.57), 252(4.62),261(4.69), 270(4.69),317s(3.92), 358(3.62),374(3.75)	4-0015-65
Benzo[j]phenanthridin-6(5H)-one	EtOH	221(4.56),236(4.47), 262(4.81),271(4.88), 288(4.20),327(3.84), 341(3.94),358(3.60), 375(3.48)	4-0015-65
Naphth[1,2-d]oxazole, 2-phenyl-	heptane	225(4.68),290(4.29), 305(4.33),330(4.45), 346(4.41)	65-0427-64
Naphth[2,1-d]oxazole, 2-phenyl-	heptane	220(4.62),265(4.52), 277(4.59),300(4.29), 315(4.28),330(4.26), 345(4.31)	65-0427-64
Naphth[2,3-d]oxazole, 2-phenyl-	heptane	262(4.73),325(4.50)	65-0427-64
$C_{17}H_{11}NO_2$			
5H-Benzo[a]phenoxazin-5-one, 10-methyl-	$CHCl_3$	240(4.31),248(4.35), 262(4.17),286(4.00), 361(4.11),456(4.16)	39-0480-65
9H-Benzo[a]phenoxazin-9-one, 8-methyl-	$CHCl_3$	291(4.16),313(3.95), 402(3.86),494(4.20)	39-0480-65
$C_{17}H_{11}NO_3$			
8H,9H-Benzo[a]pyrano[3,2-g]quinolizine-8,9-dione, 11-methyl-	EtOH	249(4.35),301(4.40), 349(3.58),392(4.27), 414(4.36)	95-0874-64
$C_{17}H_{11}NO_3S$			
9-Sulfobenzo[a]acridizinium hydroxide, inner salt	EtOH	211s(3.73),224(3.87), 251s(3.80),259(3.91), 269(3.92),288(3.96), 295(4.02),307(4.10), 317s(3.91),348(3.51), 365(3.63),383(3.84), 404(3.97)	4-0228-65

Compound	Solvent	λ_{max}(log ε)	Ref.
$C_{17}H_{11}NO_4$			
2H-1,3-Benzoxazine-2,4(3H)-dione, 3-cinnamoyl-, trans	MeOH	206(4.53),230s(4.15), 303(4.34)	4-0037-65
2,3-Dicarboxy-1-phenylquinolizinium hydroxide, inner salt	MeOH	227(4.26),255(4.22), 332s(3.96),342(4.12)	44-0452-64
$C_{17}H_{11}NO_6$			
1-Phenanthrenecarboxylic acid, 6,7-(methylenedioxy)-10-nitro-, methyl ester	EtOH	234(4.26),268(4.31), 282(4.25),357(3.69)	44-3792-65
$C_{17}H_{11}NO_7$			
Aristolochic acid	EtOH	250(4.43),318(4.08), 390(3.81)	88-0359-65
$C_{17}H_{11}N_3O_2$			
Isoxazolo[3,4-d]pyridazin-7-one, 3,4-diphenyl-	C_6H_{12}	240.1(4.13)	32-1478-65
	EtOH	243.3(4.18)	32-1478-65
	70% $HClO_4$	254(3.88),289(3.92)	32-1478-65
$C_{17}H_{11}N_3O_6$			
1,3-Indandione, 2-(p-nitrophenylazo)-2-acetoxy-	MeOH	226(4.65),282(4.31)	24-1482-64
$C_{17}H_{12}$			
9H-Cyclopropa[e]pyrene, 8b,9a-dihydro-	EtOH	260(4.56),290(4.15), 303(4.08)	88-2673-65
$C_{17}H_{12}BrN$			
Benzo[a]phenanthridizinium bromide	EtOH	211s(4.48),216(4.54), 240(4.30),274s(4.44), 283(4.60),340s(3.68), 355(3.92),377(4.12), 399(4.23)	44-1846-65
$C_{17}H_{12}BrN_3$			
9H-Pyridazino[3,4-b]indole, 3-(p-bromophenyl)-9-methyl-	EtOH	279(4.70),385(3.45)	94-1129-64
$C_{17}H_{12}BrN_3O$			
9H-Pyridazino[3,4-b]indole, 3-(p-bromophenyl)-6-methoxy-	EtOH	285(4.62),405(3.51)	94-1129-64
$C_{17}H_{12}BrN_3S$			
Thiocyanic acid, 1-(p-bromophenyl)-3-methyl-5-phenylpyrazol-4-yl ester	EtOH	258(4.26)	94-0023-64
$C_{17}H_{12}ClN$			
Quinoline, 4-(p-chlorostyryl)-	pH 1	240(4.47),327(3.97), 380(4.15)	7-1154-65
	pH 3	238(4.43),325(4.04), 375(4.02)	7-1154-65
	EtOH	234(4.38),267s(4.01), 332(4.40)	7-1154-65
Quinoline, 6-chloro-4-styryl-	pH 1	246(4.37),337(4.03), 390(4.26)	7-1154-65
	pH 3	240(4.43),335(4.28), 390s(3.80)	7-1154-65
	EtOH	240(4.47),262s(4.05), 335(4.35)	7-1154-65

Compound	Solvent	$\lambda_{max}(\log \epsilon)$	Ref.
$C_{17}H_{12}ClNO_3$			
[1]Benzopyrano[3,4-b][1,4]oxazin-5(1H)-one, 2,3-dihydro-9-chloro-2-phenyl-	EtOH	210(4.38),230(4.19), 251(4.05),290s(3.66), 298(3.87),302(3.90), 309(4.02),341(3.91)	44-4126-65
$C_{17}H_{12}ClN_3$			
9H-Pyridazino[3,4-b]indole, 3-(p-chlorophenyl)-9-methyl-	EtOH	278(4.68),385(3.46)	94-1129-64
$C_{17}H_{12}ClN_3O$			
9H-Pyridazino[3,4-b]indole, 3-(p-chlorophenyl)-6-methoxy-	EtOH	284(4.62),405(3.49)	94-1129-64
$C_{17}H_{12}ClN_3S$			
Thiocyanic acid, 1-(p-chlorophenyl)-3-methyl-5-phenylpyrazol-4-yl ester	EtOH	256(4.23)	94-0023-64
$C_{17}H_{12}CrO_4$			
Chromium, tricarbonyl(2-phenylacetophenone)-	n.s.g.	313.2(3.96)	28-2833-65A
$C_{17}H_{12}FeO$			
2,3-Ferrocoindenone	EtOH	235(4.15),276(4.11), 407(3.74),512(3.79)	35-5607-65
$C_{17}H_{12}INO_6$			
Stilbene-2-carboxylic acid, 2'-iodo-4',5'-(methylenedioxy)-α-nitro-, methyl ester, cis	EtOH	370(3.94)	44-3792-65
$C_{17}H_{12}N_2$			
Benzo[a]phenanthridine, 5-amino-	pH 2.00	272(4.68),310(4.00), 364(3.93),382(3.94)	39-4426-65
	pH 10.0	258(4.53),272(4.50), 344(3.62),360(3.74), 378(3.75)	39-4426-65
Benzo[k]phenanthridine, 6-amino-	pH 2.00	272(4.70),310(3.86), 320(3.93),348(3.64), 364(3.89),382(3.93)	39-4426-65
	pH 10.0	258(4.54),318(4.02), 362(3.67),380(3.72)	39-4426-65
1,2-Benzophenanthridine, 6-amino-	n.s.g.	258(4.53),274(4.50), 316(3.97),345(3.62), 360(3.74),379(3.76)	39-3032-65
9,10-Benzophenanthridine, 6-amino-	n.s.g.	258(4.57),275(4.36), 292(4.19),322(3.95), 368(3.68),386(3.71)	39-3032-65
Pyrrolo[1,2-a]quinoxaline, 4-phenyl-	EtOH	229(4.37),247(4.47), 271(4.37),339(3.87), 351(3.87)	25-1382-65
$C_{17}H_{12}N_2O$			
Benzo[a]phenazin-6-ol, 5-methyl-	dioxan	236(4.64),309(4.72), 360(3.76),377(3.83), 430(3.63)	24-0666-65
Cyclohepta[b]pyrrole-3-carbonitrile, 1-benzyl-1,2-dihydro-2-oxo-	EtOH	279(4.37),425(4.05)	94-0473-65

Compound	Solvent	λ_{max}(log ϵ)	Ref.
$C_{17}H_{12}N_2OS_2$			
1,3-Dithiolo[4,5-d]pyridazin-4(5H)-one, 2,5-diphenyl-	EtOH	258(4.35),314(3.90), 372(3.58)	78-1323-65
$C_{17}H_{12}N_2O_2$			
2,3'-Biindolyl, 5,6-(methylenedioxy)-	EtOH	240(4.44),279(4.04), 289s(4.00),335(4.40)	44-2030-64
Quinoline, 4-(p-nitrostyryl)-	pH 1	236(4.41),317(4.16), 350s(4.04)	7-1154-65
	pH 3	226(4.54),317(4.17), 345s(4.05)	7-1154-65
	EtOH	225(4.60),305s(4.15), 318(4.16)	7-1154-65
Spiro[5H-dibenzo[a,d]cycloheptene-5,4'-imidazolidine]-2',5'-dione	EtOH	297(4.13)	87-0439-64
$C_{17}H_{12}N_2O_3$			
1,7-Dioxa-2,8-diazaspiro[4.4]nona-2,8-dien-6-one, 3,9-diphenyl-	EtOH	261(4.29)	32-0206-65
2-Quinoxalinecarboxylic acid, 3-(α-hydroxy-p-methoxybenzyl)-, lactone	EtOH	245(4.63),265(3.58), 326(3.97)	33-1860-64
$C_{17}H_{12}N_2O_4$			
Pyrrolo[2,1-a]isoquinoline-1,2-dicarboxylic acid, 3-cyano-, dimethyl ester	n.s.g.	242s(4.33),263(4.74), 340(3.95)	35-3651-65
$C_{17}H_{12}N_2O_5$			
1,4-Pentadien-3-one, 1,5-bis-(p-nitrophenyl)-	propylene glycol	325(4.311)	95-0001-64
$C_{17}H_{12}N_2O_7$			
1,3,6,8-Nonatetraen-5-one, 1,9-bis-(5-nitro-2-furyl)-	propylene glycol	414(4.29)	95-0001-64
$C_{17}H_{12}N_2O_8$			
3-Butenoic acid, 3-(α-hydroxy-m-nitrobenzyl)-4-(m-nitrophenyl)-2-oxo-	MeOH	210(4.35),262(4.25)	44-1800-65
$C_{17}H_{12}N_4$			
s-Triazolo[4,3-a]pyrazine, 5,6-diphenyl-	C_6H_{12}	205(4.56),229s(3.58), 253(4.33),321(3.71)	44-2542-64
$C_{17}H_{12}N_4OS$			
2,1,3-Benzothiadiazol-4-ol, 7-methyl-5-(1-naphthylazo)-	EtOH	536(4.24)	17-0041-64
2,1,3-Benzothiadiazol-4-ol, 7-methyl-5-(2-naphthylazo)-	EtOH	525(4.27)	17-0041-64
$C_{17}H_{12}N_4O_2$			
1,2,4-Oxadiazole, 3-phenyl-5-(5-phenyl-1,2,4-oxadiazol-3-yl)methyl-	ether	246(4.49)	33-0942-64
$C_{17}H_{12}N_4O_2S$			
2,1,3-Benzothiadiazol-4-ol, 7-methoxy-5-(1-naphthylazo)-	EtOH	562(4.15)	17-0041-64
2,1,3-Benzothiadiazol-4-ol, 7-methoxy-5-(2-naphthylazo)-	EtOH	552(4.23)	17-0041-64
Thiocyanic acid, 3-methyl-1-(o-nitrophenyl)-5-phenylpyrazol-4-yl ester	EtOH	256(4.12)	94-0023-64

Compound	Solvent	λ_{max}(log ε)	Ref.
Thiocyanic acid, 3-methyl-1-(p-nitrophenyl)-5-phenylpyrazol-4-yl ester	EtOH	302(4.17)	94-0023-64
$C_{17}H_{12}N_4O_4$			
1H-Tetrazole-1-acetic acid, 5-phenyl-α-piperonylidene-	EtOH	235(4.26),288(3.90), 321(4.08)	44-2222-65
$C_{17}H_{12}N_4O_5$			
Sydnone, 4,4'-(hydroxymethylene)-bis[3-phenyl-	THF	245(--),325(4.11)	78-1369-65
$C_{17}H_{12}O$			
17H-Cyclopenta[a]phenanthren-17-one, 15,16-dihydro-	EtOH	265(4.89),284(4.52), 297(4.38)	25-0270-65
Fluoranthene, 3-methoxy-	EtOH	218(4.46),239(4.65), 283(4.22),295(4.55), 321(3.66),361(3.94)	39-5110-64
4-Penten-2-yn-1-one, 1,5-diphenyl-	EtOH	264(4.25),321(4.43)	39-2983-65
$C_{17}H_{12}OS$			
2H-Thiopyran-2-one, 4,6-diphenyl-	EtOH	212(4.38),221s(4.3), 265(4.36),290s(4.0), 369(3.92)	39-0032-65
$C_{17}H_{12}OS_2$			
Anthrone, 10-(1,3-dithiolan-2-ylidene)-	MeOH	317(3.95),408(4.05)	5-0037-65D
	dioxan	406(4.04)	5-0037-65D
	80% dioxan-20% $HCONH_2$	415(4.05)	5-0037-65D
	50% dioxan-50% $HCONH_2$	421(4.02)	5-0037-65D
	20% dioxan-80% $HCONH_2$	425(4.00)	5-0037-65D
	HCONH	308(4.00),425(3.99)	5-0037-65D
	HOAc	317(3.95),408(4.05)	5-0037-65D
$C_{17}H_{12}O_2$			
2-Cyclopenten-1-one, 4,5-epoxy-3,4-diphenyl-	EtOH	307(4.01)	88-0813-64
2-Furaldehyde, 4,5-diphenyl-	EtOH	234(4.32),256(4.14), 329(4.29)	88-1049-65
2-Pyrone, 3,6-diphenyl-	EtOH	245(4.25),360(4.46)	54-0031-64
2-Pyrone, 4,5-diphenyl-	EtOH	237(4.24),282(3.85)	88-0813-64
4-Pyrone, 2,5-diphenyl-	EtOH	259(3.72)	35-3186-65
$C_{17}H_{12}O_3$			
5-Benzofuranacrylic acid, 2,3-dihydro-6-hydroxy-β-phenyl-, δ-lactone	EtOH	235(4.13),261(4.00), 337(4.19)	44-2467-64
Chromone, 3-benzoyl-2-methyl-	MeOH	221(4.45),250(4.40), 293(3.91),302(3.86)	35-5417-65
1,3-Indandione, 2-acetyl-2-phenyl-	EtOH	227(4.54)	35-0874-65
1,2-Naphthoquinone, 4-benzyl-3-hydroxy-	$CHCl_3$	270(4.20),352(3.36), 488(3.16)	24-0666-65
	dioxan	269(4.38),334(3.33), 470(3.15)	24-0666-65
	pyridine	482(3.16)	24-0666-65
$C_{17}H_{12}O_4$			
Isocoumarin-4-carboxylic acid, 3-phenyl-, methyl ester	$CHCl_3$	289(4.31),322(4.02)	83-0715-65

Compound	Solvent	λ_{max}(log ε)	Ref.
3,5,7,9,11-Tridecapentayne-1,2-diol, diacetate	ether	237(4.89),249(5.31), 262(5.49),311(2.22), 337(2.26),362(2.22), 391(1.93)	24-1228-65
$C_{17}H_{12}O_5$			
8H-Benzofuro[3,2-b][1]benzopyran-8-one, 1,3-dimethoxy-	MeOH	260(4.08),294(3.91), 485(3.99),522(4.08)	2-0319-64
11H-Benzofuro[3,2-b][1]benzopyran-11-one, 1,3-dimethoxy-	$CHCl_3$	259(4.41),309(4.31), 330(4.24)	39-5140-65
1,4-Naphthoquinone, 2-acetyl-3-(2-furyl)-5-methoxy-	EtOH	241(4.45),320(3.85), 399(4.12)	33-0769-64
1,4-Naphthoquinone, 2-acetyl-3-(2-furyl)-8-methoxy-	EtOH	254(4.43),308(3.98), 412(4.17)	33-0769-64
$C_{17}H_{12}O_6$			
Keto acid, m. above 300°	H_2O	247(4.44),316(4.08)	39-3811-65
$C_{17}H_{12}O_8$			
2,3-Anthracenedicarboxylic acid, 9,10-dihydro-1,8,10-trihydroxy-10-methyl-9-oxo-	MeOH-HCl	255s(3.80),265(3.88), 274(3.89),304(3.97), 376(3.97)	35-1794-65
Clavorubin, methyl ester	diisopropyl ether	260(4.37),305(3.85), 380(3.28),471(3.85), 500(3.99),536(3.86)	24-1514-65
$C_{17}H_{12}S$			
Benzo[b]naphtho[1,8-de]thiopyran, 9-methyl-	EtOH	216(4.69),234s(4.61), 244(4.71),255s(4.34), 268s(4.23),278s(4.01), 300s(3.82),309(3.91), 326(3.80)	39-1637-64
9-Thia-1,2-benzofluorene, 5-methyl-	EtOH	246(4.72),255(4.72), 268s(4.51),276(4.73), 300(4.17),318(3.82), 334(3.54),343(3.10), 351(3.83)	39-1637-64
9-Thia-1,2-benzofluorene, 6-methyl-	EtOH	246(4.68),255(4.69), 268s(4.28),277(4.46), 296(4.08),307(4.12), 318s(3.70),332(3.36), 338(3.17),342(3.06), 350(3.31)	39-1637-64
9-Thia-1,2-benzofluorene, 7-methyl-	EtOH	247(4.69),255(4.70), 267(4.46),277(4.63), 294(4.20),305(4.24), 318s(3.78),332(3.32), 340(2.99),349(3.17)	39-1637-64
11-Thiabenzo[a]fluorene, 1-methyl-	EtOH	250s(4.82),258(4.89), 263s(4.84),268s(4.77), 278(4.74),296(4.35), 307(4.50),321s(3.89), 342(3.64),360(3.80)	39-6221-65
11-Thiabenzo[a]fluorene, 2-methyl-	EtOH	248(4.36),256(4.40), 270s(4.19),278(4.40), 294(3.99),306(4.09), 321s(3.74),337(3.22), 353(3.29)	39-6221-65

Compound	Solvent	$\lambda_{max}(\log \epsilon)$	Ref.
11-Thiabenzo[a]fluorene, 3-methyl-	EtOH	248(4.71),256(4.70), 270(4.41),280(4.55), 294(4.17),306(4.21), 320s(3.78),337(3.29), 354(3.26)	39-6221-65
11-Thiabenzo[a]fluorene, 4-methyl-	EtOH	250(4.56),258(4.74), 270(4.39),278(4.44), 296(4.06),308(4.12), 320s(3.74),337(3.32), 355(3.40)	39-6221-65
11-Thiabenzo[a]fluorene, 5-methyl-	EtOH	248(4.90),256(4.96), 263(4.79),270(4.75), 280(4.89),294(4.38), 306(4.47),321(4.06), 336(3.78),356(3.79)	39-6221-65
11-Thiabenzo[a]fluorene, 6-methyl-	EtOH	248(4.79),256(4.79), 268(4.70),278(4.78), 292(4.13),303(4.19), 321s(3.73),339(3.47), 356(3.53)	39-6221-65
$C_{17}H_{12}S_2$			
2H-Thiopyran-2-thione, 4,6-diphenyl-	EtOH	242(4.27),309(4.44), 480(3.85)	39-0032-65
	EtOH	245(4.16),308(4.17), 470(3.64)	89-0261-65
$C_{17}H_{12}S_3$			
6a-Thiathiophthene, 2,5-diphenyl-	dioxan	252(4.68),305(4.36), 509(4.14)	24-1732-64
$C_{17}H_{13}BF_4Fe$			
2,3-Ferrocoindenol, tetrafluoroborate	H_2SO_4	255(3.90),305(3.77), 356(3.63),495(3.43), 544(3.41),585(3.35), 754(2.92)	35-5607-65
$C_{17}H_{13}BrN_2O$			
Fluoline, bromo-	EtOH	255(4.6),340(4.5), 375s(3.8)	5-0089-65A
$C_{17}H_{13}ClN_2$			
6-Phenyldipyrido[1,2-c:2',1'-e]imidazol-5-ium chloride	MeOH	245(4.43),345(4.18), 363(4.16),402(3.59), 426(3.20)	12-1819-65
$C_{17}H_{13}ClN_2O$			
Pyridazine, 3-(p-chlorophenyl)-6-(p-methoxyphenyl)-	EtOH	302(4.59)	39-3342-65
$C_{17}H_{13}ClN_2O_2$			
Indole-3-carbonitrile, 2-(p-chlorophenyl)-5,6-dimethoxy-	MeOH	236(4.39),331(4.34)	44-2253-65
$C_{17}H_{13}ClN_2O_4$			
Acrylonitrile, 3-(p-chlorophenyl)-2-(4,5-dimethoxy-2-nitrophenyl)-	MeOH	223(4.27),295(4.41)	44-2253-65
$C_{17}H_{13}ClO_4$			
7-(2-Naphthyl)tropylium perchlorate	MeCN	269(4.27),375(4.18), 430s(4.01)	24-0029-64

Compound	Solvent	λ_{max}(log ϵ)	Ref.
$C_{17}H_{13}Cl_2NO_3$			
Coumarin, 3,6-dichloro-4-(2-phenyl-2-hydroxyethylamino)-	EtOH	220(4.57),250s(4.09),312(4.09),325(4.09),337s(4.00)	44-4126-65
3(2H)-Furanone, 2,2-dichloro-5-(N-methylanilino)-4-phenoxy-	EtOH	218(4.13),317(4.28)	44-4303-65
$C_{17}H_{13}F_3N_6O_4$			
Homopteroic acid, N''-trifluoroacetyl-	pH 11	256(4.41),277(4.32),365(3.88)	44-3404-65
$C_{17}H_{13}N$			
Benzo[b]phenanthridine, 5,6-dihydro-	EtOH	234(4.38),268(4.70),302s(4.01),316(3.99),375(3.51)	4-0015-65
	EtOH-HCl	254(4.61),264(4.68),290s(4.11),298(4.20),311(4.17)	4-0015-65
Benzo[j]phenanthridine, 5,6-dihydro-	EtOH	220(4.60),245(4.70),332(3.87),356(3.86)	4-0015-65
	EtOH-HCl	255(4.66),264(4.74),290s(4.14),298(4.21),309(4.17)	4-0015-65
Pyridine, 2-(6-phenyl-1,3-hexadien-5-ynyl)-, 1-cis 3-trans	EtOH	248(4.05),340(4.55),356(4.10)	94-1344-64
di-trans	EtOH	341(4.73)	94-1344-64
$C_{17}H_{13}NO$			
2(1H)-Pyridone, 3,4-diphenyl-	MeOH	236(4.26),320(3.88)	88-4647-65
$C_{17}H_{13}NO_2$			
Acetamide, N-(1-oxo-2-phenylinden-3-yl)-	EtOH	430(3.28)	35-0874-65
Cinchoninic acid, 2-m-tolyl-	EtOH	263(4.6),334(4.0)	28-0954-64A
Cinchoninic acid, 2-p-tolyl-	EtOH	264(4.6),334(4.0)	28-0954-64A
Isoxazole, 5-acetyl-3,4-diphenyl-	EtOH	225(4.12),262(3.84)	32-0127-65
1,4-Naphthoquinone, 2-p-toluidino-	dioxan	273(4.42),464(3.61)	5-0151-64E
Spiro[indan-1,3'-indoline]-2',3-dione, 1'-methyl-	EtOH	245(4.24),286(3.58),293s(3.56)	44-2973-65
Spiro[indoline-3,1'(4'H)-naphthalene]-2,4'-dione, 2',3'-dihydro-	EtOH	248(4.24),289(3.54)	44-2973-65
$C_{17}H_{13}NO_3$			
8H,9H-Benzo[a]pyrano[3,2-g]quinolizine-8,9-dione, 13,13a-dihydro-11-methyl-	EtOH	236(4.28),308s(3.84),330(3.85)	95-0874-64
Chromene-2-carboxylic acid, 4-(phenylimino)-, methyl ester	EtOH	322(3.97)	65-1924-64
Indone, 3-(carboxymethylamino)-2-phenyl-	MeOH	221(4.42),271(4.52),313(3.51),435(3.45)	65-0448-64
sodium salt	MeOH	221(4.41),272(4.47),315(3.53),440(3.47)	65-0448-64
1,4-Naphthoquinone, 2-p-anisidino-	dioxan	273(4.42),470(3.60)	5-0151-64E
$C_{17}H_{13}NO_3S_4$			
3H-1,2-Dithiole-3-thione, 4-(N-benzoyl-p-toluenesulfonamido)-	n.s.g.	229(4.45),276(3.75),320(3.48),419(3.83)	12-1071-65
$C_{17}H_{13}NO_4$			
1,3-Butanedione, 1-(2-hydroxy-4-oxo-4H-benzo[a]quinolizin-3-yl)-	EtOH	240(4.60),300(4.46),355(3.68),395(4.26),424(4.31)	95-0874-64

Compound	Solvent	λ_{max}(log ϵ)	Ref.
3-Indolineglyoxylic acid, 2-oxo-1-phenyl-, methyl ester	MeOH	248(4.18),265(4.23), 309(3.92),328s(3.84)	44-2973-65
$C_{17}H_{13}NO_5$			
Chalcone, 2'-acetoxy-4-nitro-	MeOH	304(4.27)	24-1910-64
1-Phenanthrenecarboxylic acid, 8-methoxy-10-nitro-, methyl ester	EtOH	230(4.54),254(4.39), 279(4.11)	44-3792-65
Styrylamine, 3,4-(methylenedioxy)-N-piperonylidene-, N-oxide	EtOH	235(4.36),292s(--), 344(4.47)	23-1901-64
L-Tyrosine, N-phthalyl-	MeOH	280(3.44),284s(3.43), 297s(3.24)	24-0533-64
$C_{17}H_{13}N_3$			
9H-Pyridazino[3,4-b]indole, 9-methyl-3-phenyl-	EtOH	272(4.66),384(3.45)	94-1129-64
$C_{17}H_{13}N_3O$			
9H-Pyridazino[3,4-b]indole, 6-methoxy-3-phenyl-	EtOH	280(4.58),402(3.49)	94-1129-64
9H-Pyridazino[3,4-b]indole, 3-(p-methoxyphenyl)-	EtOH	277(4.69),380(3.48)	94-1129-64
$C_{17}H_{13}N_3S$			
Thiocyanic acid, 3-methyl-1,5-diphenylpyrazol-4-yl ester	EtOH	251(4.18)	94-0023-64
$C_{17}H_{13}N_5OS$			
9H-Purine-9-carboxamide, 6-(methylthio)-N-1-naphthyl-	pH 1	280(3.74)	87-0010-64
	pH 7	280(4.05)	87-0010-64
$C_{17}H_{14}$			
Cyclopentadiene, 1,3-diphenyl-	EtOH-HOAc	250(4.08)	35-4533-65
Cyclopropane, 2-ethynyl-1,1-diphenyl-	C_6H_{12}	222(4.17)	32-0091-64
Fluorene, 9-(2-butenylidene)-	ether	257(4.63),267(4.61), 302(4.31),321(4.38), 334(4.58),350(4.54)	5-0050-65J
Naphthalene, 1-methyl-3-phenyl-	EtOH	250(4.74),290(4.06)	28-3780-64B
$C_{17}H_{14}BN$			
Benzo[c]naphtho[2,1-e][1,2]azaborine, 5,6-dihydro-6-methyl-	$CHCl_3$	258s(4.51),267(4.71), 277(4.77),309(4.01), 322(4.01),338(4.04), 354(4.11)	44-1757-64
$C_{17}H_{14}BNO$			
Benzo[c]naphtho[2,1-e][1,2]azaborine, 5,6-dihydro-6-methoxy-	$CHCl_3$	251(4.31),259s(4.41), 268(4.62),278(4.83), 311(3.97),324(3.96), 339(4.00),356(4.05)	44-1757-64
$C_{17}H_{14}ClNO$			
Cinnamonitrile, p-chloro-β-ethyl-α-(p-hydroxyphenyl)-, cis	EtOH	301(3.98)	87-0511-64
trans	EtOH	289(4.12)	87-0511-64
Cinnamonitrile, α-(p-chlorophenyl)-β-ethyl-p-hydroxy-, trans	EtOH	301(4.15)	87-0511-64

Compound	Solvent	λ_{max}(log ϵ)	Ref.
$C_{17}H_{14}ClNO_3$			
Benzo[f]quinoline-3-carboxylic acid, 1-chloro-8-methoxy-, ethyl ester	MeOH	218(4.37),236(4.39), 242s(4.34),284(4.62), 325(4.01),353(3.73), 373(3.63)	39-5907-64
Coumarin, 4-anilino-6-chloro-3-ethoxy-	EtOH	208(4.42),220s(4.34), 260(4.36),304(3.97), 313(4.04),350(4.05)	44-4126-65
3-Indolineacetic acid, 3-(p-chlorophenyl)-1-methyl-2-oxo-	EtOH	230s(4.16),256(3.9), 280s(3.35)	44-2973-65
$C_{17}H_{14}ClNO_4$			
4,5-Dimethylpyrrolo[3,2,1-de]phenanthridinium perchlorate	EtOH	204(4.59),238(4.19), 247(4.17),257(4.21), 267(4.18),312s(4.14), 325(4.08),334(4.07)	4-0208-64
$C_{17}H_{14}ClNO_8$			
7,8-Dihydroxybenzo[b]quinolizinium perchlorate, diacetate	MeCN	248(4.79),283s(4.09), 357(4.01),382(3.97), 402(3.95)	44-0252-65
7,10-Dihydroxybenzo[b]quinolzinium perchlorate, diacetate	MeCN	250(4.78),364(4.04)	44-0252-65
$C_{17}H_{14}ClN_3O_5$			
2H-Pyrazino[1',2':1,5]pyrrolo[2,3-b]indole-1,3,4(6H)-trione, 9-chloro-7,8-dimethoxy-2,6-dimethyl-	MeOH	219(4.30),257(4.45), 293(4.40),398(3.81)	39-0026-64
$C_{17}H_{14}Cl_2O$			
1,4-Pentadien-3-ol, 1,5-bis-(p-chlorophenyl)-	MeOH	268(4.60)	5-0064-65F
	20% dioxan	269(4.56)	5-0064-65F
2,4-Pentadien-1-ol, 1,5-bis-(p-chlorophenyl)-	MeOH	292(4.62)	5-0064-65F
	20% dioxan	293(4.58)	5-0064-65F
$C_{17}H_{14}Cl_2O_7$			
Lecanoric acid, 3,5-dichloro-, methyl ester	EtOH	259(4.2),318(4.1)	1-1188-65
$C_{17}H_{14}FNO$			
Cinnamonitrile, β-ethyl-α-(p-fluorophenyl)-p-hydroxy-, trans	EtOH	298(4.11)	87-0511-64
$C_{17}H_{14}Fe$			
2,3-Ferrocoindene	EtOH	212(4.53),241(4.2), 286(4.02),345(2.8), 456(2.56)	35-5607-65
$C_{17}H_{14}FeO$			
Benzaldehyde, 4-ferrocenyl-	$CHCl_3$	306(4.24),383(3.48), 472(3.23)	49-1750-64
2,3-Ferrocoindenol	EtOH	212(4.52),243(4.11), 289(4.09),345(2.99), 459(2.57)	35-5607-65
	H_2SO_4	238(4.06),260(4.05), 304(4.00),360(3.59), 412(3.37),502(2.79), 543(2.84),586(2.49), 754(2.52)	35-5607-65

Compound	Solvent	$\lambda_{max}(\log \epsilon)$	Ref.
2,3-Ferrocoindenol (contd.)	70% $HClO_4$	206(4.40),252(4.28), 265(4.10),305(4.08), 354(3.69),410(3.40), 496(2.73),592(2.59)	35-5607-65
$C_{17}H_{14}FeOS$			
Acrolein, 2-ferrocenyl-3-(2-thienyl)-	n.s.g.	341(4.18),503(3.28)	101-0398-64B
isomer	n.s.g.	348(4.08),515(3.32)	101-0398-64B
2-Propen-1-one, 1-ferrocenyl-3-(2-thienyl)-	n.s.g.	333(4.27),500(3.28)	101-0398-64B
2-Propen-1-one, 3-ferrocenyl-1-(2-theinyl)-	n.s.g.	326(4.32),390s(3.45), 506(3.53)	101-0398-64B
$C_{17}H_{14}INO_5$			
Stilbene-2-carboxylic acid, 2'-iodo-6'-methoxy-α-nitro-, methyl ester, cis	EtOH	314(4.11)	44-3792-65
$C_{17}H_{14}N_2$			
3,4-Diazabicyclo[4.1.0]hepta-2,4-diene, 2,5-diphenyl-	MeOH	252(4.02),320(4.28)	24-2438-65
Fluorene-2-carbonitrile, 9-(dimethylaminomethylene)-	MeCN	251s(4.65),255(4.67), 279(4.44),304(4.29), 366(4.34),405(4.24)	24-3331-64
Pyridazine, 4-methyl-3,6-diphenyl-	MeOH	262.5(4.42)	24-2438-65
Pyridazine, 3-phenyl-6-p-tolyl-	EtOH	284(4.52)	39-3342-65
$C_{17}H_{14}N_2O$			
Fluoline	EtOH	252(3.8),355(4.0)	5-0089-65A
	50% EtOH-NaOH	349(4.1)	5-0089-65A
3-Indolineacetonitrile, 1-methyl-2-oxo-3-phenyl-	EtOH	256(3.87),268s(3.73), 287s(3.2)	44-2973-65
Olivacine, N-oxide	EtOH	236(4.21),252(4.15), 300(4.71),311(4.82), 330(3.73),345(3.72)	78-1141-65
	EtOH-HCl	242(4.40),308(4.88), 354(3.64)	78-1141-65
Pyridazine, 3-(p-methoxyphenyl)-6-phenyl-	EtOH	298(4.58)	39-3342-65
3(2H)-Pyridazinone, 6-(2-fluorenyl)-4,5-dihydro-	EtOH	320.5(4.56)	39-7005-65
2(1H)-Pyrimidinone, 1-methyl-5,6-diphenyl-	MeOH	250(4.3),335(3.8)	83-0367-64
$C_{17}H_{14}N_2O_2$			
Indole, 3-(2-nitro-1-propenyl)-2-phenyl-	n.s.g.	237(4.30),305(4.30), 416(3.85)	28-0609-64A
Indole-3-carbonitrile, 5,6-dimethoxy-2-phenyl-	MeOH	225(4.44),330(4.32)	44-2253-65
11H-Indolo[3,2-c]quinoline, 3,9-dimethoxy-, hydrochloride	MeOH	234(4.59),253(4.56), 280(4.67),294(4.54), 308(4.30),317s(4.17), 340(3.75)	39-5919-64
11H-Indolo[3,2-c]quinoline, 3,10-dimethoxy-	MeOH	241(4.71),270s(4.57), 278(4.75),307s(3.98)	39-5919-64
3(2H)-Pyridazinone, 4,5-dihydro-6-(2-xanthenyl)-	EtOH	298(4.34)	39-7005-65

Compound	Solvent	λ max(log ε)	Ref.
4(1H)-Quinazolinone, 2-acetonylidene-2,3-dihydro-1-phenyl-	C_6H_{12}	216(4.30),223s(4.29), 278(3.87),312(4.50)	44-1020-65
	MeOH-HCl	220s(4.24),268s(3.80), 276(3.83),318(4.46)	44-1020-65
	MeOH	220s(4.27),268(3.71), 318(4.55)	44-1020-65
	MeOH-KOH	280(3.76),317(4.52)	44-1020-65
Spiro[5H-dibenzo[a,d]cycloheptene-5,4'-imidazolidine]-2',5'-dione, 10,11-dihydro-	EtOH	242s(3.33),267s(2.99), 277s(2.84),297s(1.57)	87-0439-64
$C_{17}H_{14}N_2O_3S$			
Benzo[b]thiophen-3(2H)-one, 2-(p-dimethylaminobenzylidene)-5-nitro-	n.s.g.	472(4.57),498(4.69)	65-0519-65
$C_{17}H_{14}N_2O_4$			
Acrylonitrile, 3-(3,4-dimethoxyphenyl)-2-(p-nitrophenyl)-	MeOH	278(4.07),374(4.23)	87-0583-65
4H-Quinolizine-1-carboxylic acid, 9-hydroxy-4-oxo-3-(2-pyridyl)-, ethyl ester	ether	254(4.03),274(4.05), 365(3.97),390(4.19)	24-0653-65
$C_{17}H_{14}N_4O$			
as-Triazine, 3-acetamido-5,6-diphenyl-	EtOH	231(4.45),267(4.42), 330(4.0)	94-1168-65
$C_{17}H_{14}N_4O_2$			
as-Triazine, 3-acetamido-5,6-diphenyl-, 1-oxide	EtOH	237(4.35),309(4.15), 390(4.1)	94-1168-65
$C_{17}H_{14}N_4O_3$			
1H-Tetrazole-1-acetic acid, α-(p-methoxybenzylidene)-5-phenyl-	EtOH	225(4.29),298(4.31)	44-2222-65
$C_{17}H_{14}N_8O_8$			
Glutaconaldehyde, bis(2,4-dinitrophenylhydrazone)	$CHCl_3$	244(4.33),363(4.43)	88-0821-65
$C_{17}H_{14}O$			
Anisole, m-(inden-1-ylidenemethyl)-	dioxan	252(4.10),278(4.07), 341(4.07)	5-0018-64D
Cyclopenta[ef]heptalene-1-carboxaldehyde, dimethyl-	benzene	399(4.02),445(3.61), 464(3.68),483(3.61), 686(2.16)	5-0194-64B
isomer	benzene	364(3.92),387(4.00), 402(4.05),439(3.70), 466(3.79),712(2.39), 769(2.53),865(2.51), 978(2.16)	5-0194-64B
Cyclopentaheptalene-2'-carboxaldehyde, 2,4-dimethyl-	benzene	384(4.08),400(4.22), 448(3.56),473(3.44), 710(2.38),771(2.50), 861(2.47),972(2.14)	5-0194-64B
1-Cyclopropene, 3-acetyl-1,2-diphenyl-	EtOH	227(4.37),234(4.32), 283(4.24),294(4.35), 318(4.38),330(4.34)	88-2817-65
Indene, 1-(p-methoxybenzylidene)-	MeOH	360(4.5)	87-0390-65
	HOAc	360(4.4)	87-0390-65
	dioxan	247(4.23),294(4.04), 356(4.40)	5-0018-64D

Compound	Solvent	$\lambda_{max}(\log \epsilon)$	Ref.
4-Penten-2-yn-1-ol, 1,5-diphenyl-	EtOH	222(4.18),229(4.05), 282(4.28),305(4.16)	39-2983-65
$C_{17}H_{14}OS$			
Cyclopentanone, 2-benzylidene- 5-(2-thenylidene)-	hexane	372(4.49)	65-3645-64
	EtOH	236(4.00),380(4.54)	65-3645-64
	66% H_2SO_4	236(4.00),318(3.54), 380(3.84),525(4.92)	65-3645-64
Sulfoxide, 1-naphthyl o-tolyl, R-(+)	EtOH	223(4.61),291(3.92)	35-5637-64
S-(-)	EtOH	223(4.61),291(3.92)	35-5637-64
$C_{17}H_{14}OS$			
Anthrone, 10-carbonyl-, 10-(dimethyl mercaptole)	MeOH	317(4.00),360(3.81), 380(4.00),402(3.95)	5-0037-65D
$C_{17}H_{14}O_2$			
Biphenylene, 2-acetyl-6-propionyl-	EtOH	219(3.94),227s(--), 262(4.49),284(4.59), 287(4.59),344s(3.63), 359(3.92),376(4.09)	39-4930-65
Cyclopenta[ef]heptalene-3-acetic acid, 5-methyl-	EtOAc	373(4.08),391(4.17), 449(2.39),733(2.03), 800(2.12),905(2.05), 1035(1.64)	5-0194-64B
9-Phenanthrol, 10-methyl-, acetate	EtOH	211(4.59),220(4.43), 246(4.74),253(4.86), 276(4.18),284(4.06), 296(4.10),317(2.57), 325s(2.53),333(2.67), 340s(2.47),349(2.64)	5-0109-64E
4H-Pyran-4-one, 2,3-dihydro- 2,2-diphenyl-	EtOH	263(4.06)	35-3186-65
$C_{17}H_{14}O_2S$			
Propionic acid, 3-(9-anthrylthio)-	EtOH	219(4.06),261(4.57), 340(3.37),356(3.68), 373(3.87),392(3.84)	12-0109-64
$C_{17}H_{14}O_3$			
Chalcone, 2'-acetoxy-	MeOH	306(4.35)	24-1910-64
Coumarin, 7-methoxy-4-methyl-3-phenyl-	MeOH	326(4.26)	2-0182-64
1,4-Ethanoanthracene-9,10-dione, 1,4-dihydro-1-methoxy-	EtOH	243(4.30),268(4.15), 334(3.47)	39-2941-64
$C_{17}H_{14}O_4$			
Chromenocoumarone, 4',7-dimethoxy-	EtOH	335(4.45),350(4.39)	39-5991-64
Chromeno[3',4':3,2]coumarone, 6,7'-dimethoxy-	n.s.g.	230(4.19),242(4.16), 335(4.45),352(4.39)	39-4212-64
Coumarin, 7,8-dimethoxy-3-phenyl-	MeOH	244(3.83),258(3.84), 340(4.26)	2-0182-64
Coumarin, 7-methoxy-3-(p-methoxyphenyl)-	n.s.g.	264(4.25),298(4.06)	39-4212-64
Homopterocarpin, 2,3-dehydro-	n.s.g.	227(4.21),303(4.07)	39-4212-64
Indanone, 3-acetoxy-2-hydroxy- 2-phenyl-, cis	EtOH	246(4.24)	35-0874-65
Isocoumarin, 6-(benzyloxy)-8-hydroxy- 3-methyl-	EtOH	235s(4.66),244(4.77), 259s(4.18),277(4.05), 326(3.80)	39-5382-64
4H-Naphtho[1,2-b]pyran-4-one, 3-acetyl- 6-methoxy-2-methyl-	EtOH	221(4.62),260(4.49), 285s(3.99),313(3.81), 343(3.83)	94-0316-64

Compound	Solvent	λ_{max}(log ε)	Ref.
Pterocarp-3-ene, 4',7-dimethoxy-	n.s.g.	230(4.23),242(4.21), 335(4.48),352(4.42)	77-0309-65
$C_{17}H_{14}O_5$			
Benzofuran, 2-(2-hydroxy-4-methoxy-phenyl)-3-methyl-6,7-(methylene-dioxy)-	n.s.g.	271(4.13),322(4.39)	39-5991-64
3(2H)-Benzofuranone, 2-(2,4-di-methoxybenzoyl)-	EtOH	211(4.32),236(4.19), 348(4.31)	39-2361-65
3(2H)-Benzofuranone, 6-hydroxy-4-meth-oxy-2-(p-methoxybenzylidene)-	EtOH	391(4.44)	28-5582-65A
3(2H)-Benzofuranone, 6-methoxy-2-(p-methoxysalicylidene)-	MeOH	257(4.12),285(3.79), 402(4.49)	2-0319-64
Flavone, 2'-hydroxy-5,7-dimethoxy-	EtOH	262(4.46),333(4.28)	78-2633-65
Flavone, 5-hydroxy-4',7-dimethoxy-	EtOH	271(4.34),328(4.38)	12-1871-65
	EtOH-NaOH	282(4.47),378(3.86)	12-1871-65
Flavone, 7-hydroxy-5,8-dimethoxy-	EtOH-HCl	272(4.53),320s(3.96)	39-2743-65
	EtOH	272(4.53),320s(3.94)	39-2743-65
	EtOH-NaOEt	243(4.30),283(4.55), 373(3.91)	39-2743-65
Isoflav-3-ene, 2'-hydroxy-7-methoxy-4',5'-(methylenedioxy)-	n.s.g.	340(4.34)	39-5991-64
4H-Naphtho[1,2-b]pyran-4-one, 3-acetyl-5-hydroxy-6-methoxy-2-methyl-	EtOH	220(4.55),228s(4.56), 233(4.57),270(4.48), 295s(4.07),323(3.69), 390(3.44)	94-0316-64
4H-Naphtho[1,2-b]pyran-4-one, 3-acetyl-6-hydroxy-5-methoxy-2-methyl-	EtOH	223(4.55),265(4.47), 280s(4.06),320(3.85), 372(3.65)	94-0316-64
4H-Naphtho[1,2-b]pyran-4-one, 5-hydroxy-6-methoxy-2-methyl-, acetate	EtOH	221(4.60),259(4.52), 278s(4.14),290s(3.83), 302(3.73),343(3.65), 353(3.66)	94-0316-64
4H-Naphtho[1,2-b]pyran-4-one, 6-hydroxy-5-methoxy-2-methyl-, acetate	EtOH	222(4.53),257(4.52), 278s(4.09),288s(3.85), 301(3.70),344(3.48), 352(3.49)	94-0316-64
Pterocarpin	n.s.g.	280(3.67),286(3.72), 310(3.90)	12-0379-64
$C_{17}H_{14}O_6$			
3(2H)-Benzofuranone, 2-(4,6-di-methoxysalicyloyl)-	$CHCl_3$	304(3.58),363(3.60)	39-5140-65
2-Biphenylcarboxylic acid, 5-carbo-methoxy-2'-hydroxy-4,5-dimethoxy-, δ-lactone	MeOH	253(4.80),248s(4.78), 305(4.09)	100-0090-65
Flavone, 5,7-dihydroxy-6,8-dimethoxy-	EtOH-HCl	256s(4.12),281(4.49), 324(4.05)	39-2743-65
	EtOH	252s(4.18),280(4.48), 323(3.88)	39-2743-65
	EtOH-NaOEt	250(4.16),270(4.37), 286(4.34),384(3.98)	39-2743-65
Irisolidone	EtOH	270(4.65)	44-3561-65
Kaempferol, 3,5-dimethyl-	EtOH	265(4.31),302(4.12), 340(4.32)	22-0779-65
	EtOH-NaOH	271(4.39),319(4.10), 386(4.43)	22-0779-65
	EtOH-$AlCl_3$	264(4.31),305(4.15), 340(4.32)	22-0779-65
Pisatin	EtOH	286(3.56),309(3.94)	39-5991-64

Compound	Solvent	λ_{max}(log ϵ)	Ref.
$C_{17}H_{14}O_7$			
Flavone, 4',5,7-trihydroxy-6,8-dimethoxy-	EtOH-HCl	283(4.38),338(4.47)	39-6255-64
	EtOH	283(4.33),337(4.33)	39-6255-64
Quercetin, 3,5-dimethyl-	EtOH	252(4.33),300(3.99), 346(4.32)	22-0779-65
	EtOH	253(4.39),347(4.37)	95-0047-64
	EtOH	253(4.38),347(4.37)	95-0195-64
	EtOH-NaOH	265(4.40),316(4.03), 385(4.15)	22-0779-65
	EtOH-$AlCl_3$	253(4.29),300(3.97), 346(4.29)	22-0779-65
Xanthone, 5-acetoxy-1-hydroxy-3,6-dimethoxy-	MeOH	241(4.62),278(3.77), 311(4.37),340(3.77)	78-2653-65
$C_{17}H_{14}O_8$			
Flavone, 3',4',5,7-tetrahydroxy-3,8-dimethoxy-	EtOH	267s(4.35),276(4.40), 338s(4.15),373(4.21)	78-3219-65
	EtOH-NaOAc	284(4.48),329(4.11), 410(4.24)	78-3219-65
	EtOH-NaOH	276s(4.47),282(4.48), 343s(4.05),420(4.44)	78-3219-65
	EtOH-NaOAc-H_3BO_3	265(4.45),385(4.32)	78-3219-65
Myricetin, 3,5-dimethyl-	EtOH	252(4.18),300(3.84), 346(4.15)	22-0779-65
	EtOH-NaOH	265(4.27),315(3.86), 390(4.07)	22-0779-65
	EtOH-$AlCl_3$	252(4.18),300(3.84), 346(4.15)	22-0779-65
$C_{17}H_{14}O_9$			
Griseorhodin A, HCl product from	EtOH	232(4.7),255(4.2), 315(3.8),364(3.7), 508(3.7)	24-0024-65
$C_{17}H_{14}S$			
Benzo[b]thiophene, 3-(m-methylstyryl)-, cis	EtOH	242(4.47),263s(4.23), 294(4.12),323(4.12)	39-6221-65
trans	EtOH	242(4.32),263s(4.09), 323(4.30)	39-6221-65
Benzo[b]thiophene, 3-(o-methylstyryl)-, trans	EtOH	244(4.12),263s(4.19), 330(4.31)	39-6221-65
Benzo[b]thiophene, 3-(p-methylstyryl)-, trans	EtOH	245(4.42),260s(4.26), 323(4.44)	39-6221-65
Benzo[b]thiophene, 3-(α-methylstyryl)-, cis	EtOH	236(4.48),256s(4.25), 294(3.96),303(3.94)	39-6221-65
trans	EtOH	240(4.31),282(4.06)	39-6221-65
Benzo[b]thiophene, 3-(β-methylstyryl)-, trans	EtOH	240(4.38),294(4.10), 311(4.13)	39-6221-65
$C_{17}H_{15}Br$			
5H-Dibenzo[a,d]cycloheptene, 5-(2-bromoethylidene)-10,11-dihydro-	MeOH	242.5(4.12)	87-0392-64
$C_{17}H_{15}BrClNO_4S_2$			
2-(p-Dimethylaminophenyl)-4-(p-bromophenyl)-1,3-dithiolium perchlorate	70% $HClO_4$	253(4.24),284(4.15), 395(4.11)	44-1711-64
	CF_3COOH	262(4.18),285(4.22), 305(4.15),405(4.06), 537(3.91)	44-1711-64

Compound	Solvent	$\lambda_{max}(\log \epsilon)$	Ref.
$C_{17}H_{15}BrN_2$			
7,12-Dimethyl-12H-indolo[2,3-a]quinolizin-5-ium bromide	EtOH	201(4.74),226(4.82), 240s(4.8),248(4.85), 272s(4.52),290(4.50), 327(4.70),395(4.51)	44-3584-64
7,13-Dimethylpyrido[2',1':3,4]pyrazino-[1,2-a]indol-5-ium bromide	EtOH	215(4.42),253(4.05), 283s(3.82),298(3.83), 330(3.93),387(4.24)	44-3584-64
$C_{17}H_{15}BrN_2O_4$			
2-Hexenanilide, p-bromo-4-(5-nitro-2-furfurylidene)-	propylene glycol	315(4.30),398(4.52)	95-0646-64
$C_{17}H_{15}BrO_3$			
Chalcone, 2-bromo-2'-hydroxy-4'-methoxy-4-methyl-	n.s.g.	249(4.10),359(4.49)	49-0450-65
$C_{17}H_{15}ClN_2O$			
Carbostyril, 6-chloro-3-(dimethylamino)-4-phenyl-	iso-PrOH	240(4.6),355(3.9)	39-3097-64
Carbostyril, 6-chloro-1-methyl-3-(methylamino)-4-phenyl-	iso-PrOH	240(4.5),260(4.1), 340(4.2),360(4.2)	39-3097-64
7,12-Dihydro-12-hydroxy-12-methylquino[2,1-b]quinazolin-13-ium chloride	EtOH	245(4.80),272(4.71), 372(4.62)	44-1539-65
$C_{17}H_{15}ClN_2O_2$			
Acrylonitrile, 2-(2-amino-4,5-dimethoxyphenyl)-3-(p-chlorophenyl)-	MeOH	235(4.19),330(4.11)	44-2253-65
1H-1,4-Benzodiazepin-3-ol, 1-acetyl-7-chloro-2,3-dihydro-5-phenyl-	iso-PrOH	220s(4.58),253(4.11)	44-1621-64
$C_{17}H_{15}ClN_2O_3$			
Coumarin, 4-(o-aminoanilino)-6-chloro-3-ethoxy-	EtOH	210(4.19),240(4.03), 255(3.98),299(4.12), 303(4.12),310(4.11), 345(4.04)	44-4126-65
$C_{17}H_{15}ClN_2O_4$			
2-Hexenanilide, p-chloro-4-(5-nitrofurfurylidene)-	propylene glycol	315(4.20),395(4.42)	95-0646-64
$C_{17}H_{15}ClN_2O_5$			
2-(Formylmethyl)-1-phenylisoquinolinium perchlorate, oxime	EtOH	222(4.53),236(4.48), 280(3.72),333(3.70), 345(3.70)	44-1846-65
$C_{17}H_{15}ClN_2O_6S_2$			
2-(p-Dimethylaminophenyl)-4-(p-nitrophenyl)-1,3-dithiolium perchlorate	70% $HClO_4$	225(4.11),262(4.04), 295(4.13),375(4.30)	44-1711-64
	CF_3COOH	265(4.22),283(4.25), 370(4.21),538(4.44)	44-1711-64
$C_{17}H_{15}ClO_2$			
Benzofuro[3,2-b]benzofuran, 3-chloro-4b,9b-dihydro-3,8,9b-trimethyl-	EtOH	206(4.28),226(4.22), 290(3.81)	78-2289-65
2-Butyne, 1-(p-chlorophenoxy)-4-(p-tolyloxy)-	EtOH	276(3.4808)	78-2289-65

Compound	Solvent	λ_{max}(log ϵ)	Ref.
$C_{17}H_{15}ClO_3$			
4'-Hydroxy-8-methoxy-3-methylflavylium chloride	EtOH-HCl	272(4.15),443(4.28)	78-1471-65
$C_{17}H_{15}ClO_4$			
7-(4-Phenyl-1,3-butadien-1-yl)-tropylium perchlorate	MeCN	227(4.30),266s(4.07), 294(4.29),340(3.64), 493(4.69)	24-1590-64
$C_{17}H_{15}ClO_6S$			
6,7-Dimethoxy-2-phenyl-1-benzothio-pyrylium perchlorate	HOAc	249(4.19),288(4.27), 320(3.41),442(4.29)	5-0136-64C
6,7-Dimethoxy-4-phenyl-1-benzothio-pyrylium perchlorate	HOAc	250(4.23),276(4.28), 325(3.77),413(4.17)	5-0136-64C
$C_{17}H_{15}F_3S$			
Thiobenzophenone, 2,4,6-trimethyl-4'-(trifluoromethyl)-	C_6H_{12}	305(4.00),631(1.63)	54-0289-65
	EtOH	305(3.97),622(1.62)	54-0289-65
$C_{17}H_{15}IO_2$			
Acrylic acid, 3-(o-iodophenyl)-2-o-tolyl-, methyl ester, cis	EtOH	273(3.87)	44-3792-65
	EtOH	273(3.81)	88-0359-65
$C_{17}H_{15}N$			
Cyclobutanecarbonitrile, 1-(4-biphenylyl)-	EtOH	254(4.35)	87-0504-64
15H-Cyclopenta[a]phenanthren-12-amine, 16,17-dihydro-	EtOH	256(4.64),314(3.98), 360(3.11)	5-0200-65D
	EtOH-HCl	252(4.68),258(4.74), 279(4.10),288(4.00), 300(4.06),319(2.89), 334(2.92),350(2.85)	5-0200-65D
Indene, 1-[p-(methylamino)benzylidene]-	MeOH	405(4.5)	87-0390-65
	HOAc	400(4.2)	87-0390-65
Pyridine, 1-benzyl-4-cyclopentadien-ylidene-1,4-dihydro-	dioxan	430(4.55)	35-2887-65
Pyridine, 2-(6-phenyl-1,3,5-hexatrien-yl)-, 1-cis-5-cis-3-trans	EtOH	339(4.64)	94-1344-64
1-cis-3-trans-5-trans	EtOH	254(4.07),263(4.08), 356(4.70)	94-1344-64
all trans	EtOH	356(4.85),371(4.74)	94-1344-64
Pyrrole, 1-methyl-2,5-diphenyl-	EtOH	230(3.97),307(4.35)	44-0190-65
$C_{17}H_{15}NO$			
Inden-1-imine, 3-ethoxy-2-phenyl-	MeOH	251(4.37),348(3.53)	65-0448-64
2-Indolinone, 3-(1-phenylethylidene)-	EtOH	253(4.50),260(4.51), 295(3.91),353(3.24)	87-0626-65
Indone, 3-(benzylamino)-2-methyl-	MeOH	219(4.47),264(4.59), 309(3.25),442(3.59)	65-0448-64
Indone, 3-(N-ethylanilino)-	MeOH	216(4.38),258(4.35), 265s(4.33),280s(4.15), 331(3.34),432(3.65)	65-0448-64
Ketone, 1-ethyl-2-indolyl phenyl	EtOH	247(4.13),319(4.33)	44-3604-65
2-Naphthol, 1-(α-aminobenzyl)-	90% EtOH-10% THF	226s(4.69),231(4.74), 256(3.54),267(3.63), 277(3.72),288(3.64), 322s(3.42),334(3.49)	7-0696-64
	base	241s(4.74),246(4.76), 265s(3.86),277(3.81), 287(3.89),298(3.79), 345s(3.51),359(3.65)	7-0696-64

Compound	Solvent	λ_{max}(log ε)	Ref.
2-Pentenenitrile, 2-(p-hydroxyphenyl)-3-phenyl-, cis	EtOH	298(3.99)	87-0511-64
trans	EtOH	285(4.08)	87-0511-64
2-Pentenenitrile, 3-(p-hydroxyphenyl)-2-phenyl-, cis	EtOH	307.5(4.07)	87-0511-64
trans	EtOH	297.5(4.11)	87-0511-64
Spiro[indan-1,3'-indolin]-2'-one, 1'-methyl-	MeOH	257(3.90),266s(3.82), 280s(3.27)	44-2973-65
Spiro[indoline-3,1'(2'H)-naphthalen]-2-one, 3',4'-dihydro-	EtOH	251(3.88),273s(3.31), 282s(3.21)	44-2973-65
$C_{17}H_{15}NOS$			
Benzo[b]thiophen-3(2H)-one, 2-(p-dimethylaminobenzylidene)-	n.s.g.	456(4.48),478(4.60)	65-0519-65
$C_{17}H_{15}NO_2$			
Acrylonitrile, 2-(3,4-dimethoxyphenyl)-3-phenyl-	EtOH	236(4.12),273(4.01), 338(4.21)	87-0583-65
Anonaine	EtOH	272(4.28),316(3.58)	33-2119-64
Aziridine, 2,3-dibenzoyl-1-methyl-, cis	EtOH	250.5(4.37)	35-1050-65
trans	EtOH	252(4.42)	35-1050-65
Homoveratrylamine, β-benzyl-	MeOH	236(4.20),274(4.12), 338(4.43)	83-0879-65
Isoquinoline, 6,7-dimethoxy-1-phenyl-	MeOH	241(4.81),328(3.75)	83-0561-65
Ketone, 1-ethoxy-2-indolyl phenyl	EtOH	250(4.12),314(4.30)	44-3604-65
$C_{17}H_{15}NO_3$			
Acetanilide, 4'-acetyl-2-benzoyl-	0.04N NaOH	360(4.60)	20-0782-64
	60% EtOH	291(4.35)	20-0782-64
Acrylic acid, 3-benzamido-2-phenyl-, methyl ester	MeOH	312(4.3)	83-0367-64
Anolobine	EtOH	216(4.47),282(4.29), 320(3.60)	95-0077-65
Benzo[f]quinoline, 3-carbethoxy-8-methoxy-	MeOH	219(4.35),234(4.45), 245s(4.40),277(4.52), 321(4.04),350(3.79), 368(3.74)	39-5907-64
Cinchoninic acid, 1,2,3,4-tetrahydro-2-oxo-1-phenyl-, methyl ester	MeOH	205(4.65),231(4.38), 286(3.77),344(3.70)	44-2973-65
3-Indolineacetic acid, 1-methyl-2-oxo-3-phenyl-	EtOH	256(3.9),284s(3.27)	44-2973-65
3-Indolineacetic acid, 3-methyl-2-oxo-1-phenyl-	MeOH	246(4.11),276s(3.21)	44-2973-65
3-Indolineacetic acid, 2-oxo-1-phenyl-, methyl ester	MeOH	247(3.99),313s(2.99)	44-2973-65
3-Indolinepropionic acid, 2-oxo-3-phenyl-	EtOH	225(3.76),265s(3.71), 280(3.26)	44-2973-65
2-Indolinone, 3-benzylidene-5,6-dimethoxy-	MeOH	270(4.22),325(4.08), 414(3.59)	87-0626-65
2-Indolinone, 3-veratrylidene-	EtOH	251(4.21),371(4.25)	87-0626-65
Michelalbine	EtOH	274(4.28),320(3.64)	95-0077-65
5-Norbornene-2-carboxylic acid, 6-(3-indolylcarbonyl)-, cis	NaOH	265(4.51),332(4.48)	65-3025-64
$C_{17}H_{15}NO_4$			
Indole-2-carboxylic acid, 5-(benzyloxy)-6-methoxy-	EtOH	215(4.58),294(4.11), 313(4.23)	4-0387-65
Indole-2-carboxylic acid, 6-(benzyloxy)-5-methoxy-	EtOH	212(4.57),298(4.14), 313(4.23)	4-0387-65

Compound	Solvent	$\lambda_{max}(\log \epsilon)$	Ref.
$C_{17}H_{15}NO_5$			
4H-1,3-Benzoxazin-4-one, 2-(3,4,5-trimethoxyphenyl)-	EtOH	220s(4.52),298(4.22), 317(4.22)	44-1020-65
$C_{17}H_{15}NO_5S$			
2H-1,3-Benzothiazine-2-carboxylic acid, 3,4-dihydro-4-oxo-2-phenyl-, ethyl ester, 1,1-dioxide	EtOH	265s(3.46),271s(3.42), 278(3.41),286s(3.31)	44-2068-64
$C_{17}H_{15}N_3O$			
10H-Dibenzo[b,e][1,4]diazepine-10-propionitrile, 5,11-dihydro-5-methyl-11-oxo-	EtOH	224(4.43)	33-1590-65
$C_{17}H_{15}N_3O_2$			
2-Stilbenecarbonitrile, p'-(dimethylamino)-p-nitro-	MeOH	261(4.00),448(4.44)	87-0583-65
$C_{17}H_{16}$			
Anthracene, 10-ethyl-1-methyl-	EtOH	257(5.28),333(3.53), 350(3.68),367(3.81), 387(3.80)	95-1072-64
Anthracene, 1,5,9-trimethyl-	EtOH	255(5.20),328(3.52), 348(3.71),362(3.80), 380(3.76)	95-1072-64
Anthracene, 1,7,10-trimethyl-	EtOH	258(5.18),334(3.54), 350(3.70),368(3.80), 390(3.75)	95-1072-64
Cyclobutane, diphenylmethylene-	C_6H_{12}	257(4.23)	44-3215-65
	EtOH	255(4.21)	88-0913-65
Cyclopenta[ef]heptalene, 2-ethyl-4-methyl-	hexane	374(4.11),392(4.17), 423(2.80),454(2.28), 742(2.01),803(2.12), 910(2.06),1060(1.66)	5-0194-64B
Cyclopenta[ef]heptalene, 2,2',4-trimethyl-	hexane	380(4.06),395(4.11), 435(2.78),468(2.56), 796(2.05),880(1.97), 1050(1.52)	5-0194-64B
Cyclopenta[ef]heptalene, 2,3',4-trimethyl-	hexane	381(4.10),393(4.15), 432(2.75),455(2.30), 464(2.41),732(1.87), 897(1.96),906(1.88), 1038(1.45)	5-0194-64B
Cyclopenta[ef]heptalene, 2,4,9-trimethyl-	hexane	373(4.06),392(4.14), 417(2.96),444(1.93), 700(2.17),770(2.27), 850(2.21),868(2.20), 967(1.88),1004(1.85)	5-0194-64B
Dibenzonorbornadiene, 7,7-dimethyl-	EtOH	274(3.30),281(3.41)	35-3896-64
Naphthalene, 1,2-dihydro-4-methyl-1-phenyl-	EtOH	260(3.93)	28-3780-64B
Naphthalene, 1,2-dihydro-4-methyl-2-phenyl-	EtOH	259(3.95)	28-3780-64B
Naphthalene, 1,2-dihydro-4-methyl-3-phenyl-	EtOH	279(4.15)	28-3780-64B
$C_{17}H_{16}BrNO$			
Chalcone, 4'-bromo-4-(dimethylamino)-	EtOH	275(4.30),326(3.49), 420(4.56)	23-2580-64

Compound	Solvent	λ max(log ε)	Ref.
$C_{17}H_{16}BrNO_2$			
Chalcone, 3'-bromo-2'-hydroxy-4-(dimethylamino)-	n.s.g.	224(4.13),280(4.22)	49-0450-65
Chalcone, 5'-bromo-2'-hydroxy-4-(dimethylamino)-	n.s.g.	223(4.16),258(4.09), 279(4.08)	49-0450-65
$C_{17}H_{16}BrNO_3$			
7-Isoquinolinol, 8-bromo-3,4-dihydro-1-(p-hydroxybenzyl)-6-methoxy-	EtOH-HCl	225(4.16),286(3.89)	65-0550-64
	EtOH	223(4.07),287(3.77)	65-0550-64
$C_{17}H_{16}Br_2N_2O_2$			
1,2-Propanediamine, N,N'-bis(5-bromosalicylidene)-	MeOH	328(3.82),415(3.11)	65-3049-64
$C_{17}H_{16}ClNO$			
7-(Dimethylamino)flavylium chloride	50% MeOH-pH 2	290(4.14),335(4.03), 520(4.36)	35-3142-64
	pH 4	340(3.73),445(4.17)	35-3142-64
	pH 7	273(3.96),440(4.38)	35-3142-64
	pH 9	270(3.98),455(4.33)	35-3142-64
	pH 11	268(3.94),485(4.42)	35-3142-64
	pH 13	268(3.94),485(4.42)	35-3142-64
$C_{17}H_{16}ClNO_2$			
Indole-2-carboxaldehyde, 5-chloro-3-phenyl-, dimethyl acetal	iso-PrOH	229(4.54),267(4.04)	44-1621-64
6,7-Isoquinolinediol, 1-(p-chlorophenethyl)-3,4-dihydro-	EtOH-HCl	216s(4.28),247(4.17), 308(3.92),360(4.02)	33-1945-65
	EtOH	217s(4.41),270(3.92), 310(3.65)	33-1945-65
	EtOH-NaOH	244(4.07),334(4.25)	33-1945-65
hydrochloride	EtOH-HCl	216(4.30),248(4.20), 310(3.93),362(4.04)	33-1945-65
	EtOH	217(4.30),248(4.18), 310(3.92),363(4.03)	33-1945-65
	EtOH-NaOH	248(4.06),336(4.28)	33-1945-65
$C_{17}H_{16}ClNO_4$			
6-Methyl-1-phenylquinaldinium perchlorate	EtOH	243(4.5),320(4.0)	65-3360-64
$C_{17}H_{16}ClNO_4S_2$			
2-Dimethylamino-4,5-diphenyl-1,3-dithiolium perchlorate	EtOH	210(4.35),230(4.29), 315(4.00),320(3.98)	44-1711-64
2-(p-Dimethylaminophenyl)-4-phenyl-1,3-dithiolium perchlorate	70% $HClO_4$	243(4.19),278(4.01), 390(4.08)	44-1711-64
	CF_3COOH	275(4.09),300(4.03), 398(4.08)	44-1711-64
$C_{17}H_{16}ClNO_5$			
7-(Dimethylamino)flavylium perchlorate	MeCN	290(4.28),335(4.17), 520(4.41)	35-3142-64
$C_{17}H_{16}ClN_3$			
3H-1,4-Benzodiazepine, 7-chloro-2-(dimethylamino)-5-phenyl-	EtOH-acid	251(3.51)	87-0235-64
$C_{17}H_{16}ClN_3O$			
3H-1,4-Benzodiazepine, 7-chloro-2-(dimethylamino)-5-phenyl-, 4-oxide	EtOH-acid	251(4.56),305s(4.00)	87-0235-64

Compound	Solvent	λ_{max}(log ε)	Ref.
Quinazoline, 6-chloro-2-(dimethyl-aminomethyl)-4-phenyl-, 3-oxide	EtOH-acid	233(4.38),265(4.54)	87-0235-64
$C_{17}H_{16}Cl_2N_2O_2$			
1,2-Propanediamine, N,N'-bis-(5-chlorosalicylidene)-	MeOH	253(4.19),325(3.81), 420(3.11)	65-3049-64
$C_{17}H_{16}CoN_2O_2$			
1,2-Propanediamine, N,N'-di-salicylidene-, cobalt chelate	MeOH	253(4.72),387(3.74)	65-3049-64
$C_{17}H_{16}CuN_2O_2$			
1,2-Propanediamine, N,N'-di-salicylidene-, copper chelate	MeOH	270(4.35),357(3.93), 560(2.42)	65-2049-64
$C_{17}H_{16}FeO$			
2,4,6-Heptatrienal, 7-ferrocenyl-	$CHCl_3$	370(4.56),508(3.77)	5-0088-64F
$C_{17}H_{16}NNaO_2$			
Sodium, [2-(2-benzoylvinyl)-5-(di-methylamino)phenoxy]-	50% MeOH	268(3.97),485(4.42)	35-3142-64
$C_{17}H_{16}N_2$			
1H-1,4-Diazepine, 2,3-dihydro-2,3-diphenyl-	THF	305(3.78)	88-0051-65
3-Stilbenecarbonitrile, 4'-(dimethyl-amino)-, trans	80% EtOH	363(4.46)	33-0517-65
conjugate acid	80% EtOH	295(4.51)	33-0517-65
4-Stilbenecarbonitrile, 4'-(dimethyl-amino)-, trans	80% EtOH	387(4.57)	33-0517-65
conjugate acid	80% EtOH	319(4.67)	33-0517-65
$C_{17}H_{16}N_2NiO_2$			
1,2-Propanediamine, N,N'-disalicyli-dene-, nickel chelate	MeOH	243(4.79),327(3.93), 400(3.80)	65-3049-64
$C_{17}H_{16}N_2O$			
Acetophenone, 2-diazo-2',4',6'-tri-methyl-2-phenyl-	MeOH	256(4.24),300(3.64)	88-1403-64
3(2H)-Pyridazinone, 4,5-dihydro-4-phenyl-6-p-tolyl-	EtOH	289(4.22)	39-3342-65
3-Quinolinecarbonitrile, 8-benzyl-1,2,5,6,7,8-hexahydro-2-oxo-	EtOH	238(3.97),348(4.12)	44-3593-65
$C_{17}H_{16}N_2OS$			
4H-1,3-Thiazin-4-one, 2,3,5,6-tetra-hydro-5-methyl-3-phenyl-2-(phenyl-imino)-	MeOH	228(4.36),275(3.62)	44-1720-64
$C_{17}H_{16}N_2O_2$			
Benzamide, N-[2-(2-oxo-3-indolinyl)-ethyl]-	EtOH	230(4.17),280s(3.26)	44-2431-64
Nitrone, α-phenyl-N-(1,2,3,4-tetra-hydro-1-oxo-2-naphthyl)-, oxime	n.s.g.	260s(4.1),295(4.2)	65-0149-64
3(2H)-Pyridazinone, 4,5-dihydro-6-(p-methoxyphenyl)-4-phenyl-	EtOH	296(4.28)	39-3342-65
$C_{17}H_{16}N_2O_3$			
Chalcone, 4-(dimethylamino)-4'-nitro-	benzene	441(4.38)	65-0094-65
	EtOH	272(4.24),450(4.37)	65-0094-65

Compound	Solvent	λ_{max}(log ϵ)	Ref.
Chalcone, 4-(dimethylamino)-4'-nitro- (contd.)	HOAc	271(4.27),452(4.36)	65-0094-65
	+$FeCl_3$	620(3.49)	65-0094-65
Chalcone, 4'-(dimethylamino)-4-nitro-	benzene	314(4.37)	65-0094-65
	EtOH	316(4.44),412(4.24)	65-0094-65
	HOAc	314(4.43),418(4.29)	65-0094-65
	+$FeCl_3$	507(4.08)	65-0094-65
1-Indazolinecarboxylic acid, 2-benzyl-3-oxo-, ethyl ester	EtOH	224(4.39),307(3.81)	33-1986-64
3-Indolineacetic acid, 1-methyl-2-oxo-3-(2-pyridylmethyl)-	EtOH	255(4.03),262s(4.01), 282s(3.26)	44-2973-65
Valeric acid, 4-hydroxy-2-oxo-5-phenoxy-, γ-lactone, phenylhydrazone	n.s.g.	233(4.26),295(3.98), 333(4.49)	39-0141-64
$C_{17}H_{16}N_2O_4$			
1,3-Indandione, 2-(4,4-dimethyl-2,6-dioxocyclohexylazo)-	CH_2Cl_2	246(4.40),446(4.42)	89-0920-64
$C_{17}H_{16}N_2O_4S$			
4-Quinolinesulfonic acid, 1-benzyl-3-carbamoyl-1,4-dihydro-, sodium salt	EtOH	240(3.93),332(3.90)	77-0607-65
$C_{17}H_{16}N_2O_4S_2$			
5-Thia-1-azabicyclo[4.2.0]oct-2-ene-2-carboxylic acid, 7-[2-(benzylthio)-acetamido]-3-(hydroxymethyl)-8-oxo-, γ-lactone	EtOH	254(3.90)	39-5015-65
$C_{17}H_{16}N_2O_5$			
2,4-Pentadienanilide, 4-ethyl-p-hydroxy-5-(5-nitro-2-furyl)-	propylene glycol	285(4.18),315(4.20), 405(4.51)	95-0646-64
$C_{17}H_{16}N_2O_6$			
8H-[1,3]Dioxolo[6,7][2]benzopyrano-[3,4-c]indol-8-one, 3,4,4a,5,6,6a-hexahydro-3-methoxy-5-nitroso-	EtOH	231(4.54),266(3.83), 308(3.86)	35-4912-65
$C_{17}H_{16}N_2O_7$			
Acetic acid, [3,5-dinitro-4-(3,5-xylyl-oxy)phenyl]-, methyl ester	EtOH	234.5(4.16)	87-0474-65
$C_{17}H_{16}N_2S_2$			
Benzothiazoline, 2,2'-methylidyne-bis[3-methyl-	hexane	308(4.29)	22-2879-64
	EtOH	306(3.95)	22-2879-64
	HOAc	278(4.11)	22-2879-64
$C_{17}H_{16}N_4O_2$			
Nicotinonitrile, 6,6'-pentamethylene-bis[1,2-dihydro-2-oxo-	EtOH	237(3.83),338(4.03)	44-3593-65
$C_{17}H_{16}N_4O_4$			
5H-Benzocyclohepten-5-one, 6,7,8,9-tetrahydro-, 2,4-dinitro-phenylhydrazone	$CHCl_3$	375(4.43)	5-0021-64G
$C_{17}H_{16}N_4O_4S$			
Acetanilide, 4'-[(3-benzyl-1,2,4-oxa-diazol-5-yl)sulfamoyl]-	EtOH	230(4.21),261(4.29)	95-1061-64
Benzenesulfonic acid, m-(3-methyl-5-oxo-4-(o-tolylazo)-2-pyrazolin-1-yl]-	MeOH	400(5.1)	27-0244-64

Compound	Solvent	λ_{max}(log ε)	Ref.
$C_{17}H_{16}N_4O_6$			
1-Indanone, 5,7-dimethoxy-, 2,4-dinitrophenylhydrazone	$CHCl_3$	402.5(4.47)	5-0021-64G
$C_{17}H_{16}N_4O_8$			
2,4-Cyclopentadiene-1,2-dicarboxylic acid, 5-formyl-3-methyl-, dimethyl ester, 2,4-dinitrophenylhydrazone	n.s.g.	415(4.45)	65-0058-65
$C_{17}H_{16}N_6O_4$			
Homopteroic acid, N''-acetyl-	pH 13	253(4.46),365(3.85)	44-3404-65
$C_{17}H_{16}O$			
Cyclopenta[ef]heptalene-1-carboxaldehyde, 3,6-dihydro-8,10-dimethyl-	hexane	548(2.81)	5-0194-64B
isomer	hexane	580(2.73)	5-0194-64B
5H-Dibenzo[a,d]cycloheptene-$\Delta^{5,\beta}$-ethanol, 10,11-dihydro-	MeOH	240(4.14)	87-0392-64
5H-Dibenzo[a,d]cyclohepten-5-ol, 10,11-dihydro-5-vinyl-	MeOH	263(2.84),266s(2.81), 270s(2.75),273s(2.67)	87-0392-64
6H-Dibenzo[a,c]cyclohepten-6-one, 5,7-dihydro-1,11-dimethyl-	isooctane	246(3.99)	35-1710-64
Fluoren-9-one, 2,4-diethyl-	EtOH	254(4.73),262(4.40), 292(3.43),301(3.50), 313(3.32),384(2.54), 405(2.60)	22-2953-64
Naphthalene, 1,2-dihydro-7-methoxy-3-phenyl-	EtOH	233(4.18),313(4.35)	87-0519-64
Naphthalene, 1,2-dihydro-7-methoxy-4-phenyl-	EtOH	273(4.07)	44-2942-65
1,4-Pentadien-3-ol, 1,5-diphenyl-	MeOH	262(4.49)	5-0064-65F
	20% dioxan	264(4.47)	5-0064-65F
2,4-Pentadien-1-ol, 1,5-diphenyl-	MeOH	288(4.57)	5-0064-65F
	20% dioxan	289(4.54)	5-0064-65F
$C_{17}H_{16}OS$			
1-Penten-3-one, 1-phenyl-5-(phenylthio)-	EtOH	260s(4.11),290(4.39)	7-1093-65
$C_{17}H_{16}OS_2$			
Cycloheptanone, 2,7-di-2-thenylidene-	hexane	274(4.00),326(4.35)	65-3645-64
	EtOH	274(4.16),337(4.37)	65-3645-64
	66% H_2SO_4	258(3.88),519(4.76)	65-3645-64
$C_{17}H_{16}O_2$			
2α-Anthraceneacetic acid, 1,2,3,4-tetrahydro-3β-hydroxy-9-methyl-, lactone	EtOH	231(5.02),280(3.66), 290(3.68)	65-2576-64
9-Anthroic acid, 9,10-dihydro-10,10-dimethyl-	EtOH	257s(2.61),263(2.72), 271(2.63)	87-0088-64
Chalcone, 4-methoxy-4'-methyl-	EtOH	236(4.05),350(4.46)	23-2580-64
Chalcone, 4'-methoxy-4-methyl-	EtOH	235(4.02),298s(4.45), 316(4.50)	23-2580-64
Cyclobutanecarboxylic acid, 1-(4-biphenylyl)-	EtOH	246(4.34)	87-0504-64
1(2H)-Naphthalenone, 3,4-dihydro-7-(p-methoxyphenyl)-	EtOH	209(4.54),253(4.53), 320(3.19)	39-5110-64
$C_{17}H_{16}O_3$			
1H-Benz[e]indene-2-acetic acid, 2,3-dihydro-α-methyl-3-oxo-, methyl ester	EtOH	251(4.69),273(3.88), 282(3.96),293(3.46), 341(3.17)	42-0643-64

Compound	Solvent	λ_{max}(log ϵ)	Ref.
Benzofuran, 2-(p-hydroxyphenyl)-3-α-methoxyethyl-	EtOH	298(4.31)	4-0158-64
	EtOH-NaOEt	321(4.40)	4-0158-64
Chalcone, 4,4'-dimethoxy-	EtOH	236(4.11),350(4.51)	23-2580-64
Chalcone, 2'-hydroxy-4-methoxy-5'-methyl-	MeOH	359(4.43)	24-1910-64
Cinnamic acid, 2-methoxy-α-phenyl-, methyl ester, trans	MeOH	282(3.81),320(3.81)	2-0182-64
Cycloheptanone, 2,7-difurfurylidene-	EtOH	266(4.00),340(4.38)	65-3645-64
	HOAc-H_2SO_4	304(3.41),515(4.64)	65-3645-64
1,4-Ethanoanthracene-9,10-diol, 1,4-dihydro-1-methoxy-	EtOH	243(4.40),272s(3.69), 337(3.73)	39-2941-64
1,4-Ethanoanthracene-9,10-dione, 1,4,4a,9a-tetrahydro-1-methoxy-	EtOH	225(4.62),254(3.16), 295(3.11),304(3.10), 345(2.48)	39-2941-64
2-Flavene, 4'-hydroxy-8-methoxy-3-methyl-	EtOH	222(4.33),243(4.30), 270(3.90)	78-1471-65
1,3-Naphthalenediol, 5,6,7,8-tetrahydro-, 1-benzoate	EtOH	228(4.31),277(3.56), 281(3.56)	24-1926-64
	NaOH	296(3.52)	24-1926-64
2-Pentenoic acid, 3-(p-hydroxyphenyl)-2-phenyl-	EtOH	260(4.07)	87-0511-65
Phenanthrene, 1,5,6-trimethoxy-	EtOH	219(4.44),255(4.77), 289(4.07),301(4.18), 315(4.24),331(3.51), 348(3.64),367(3.67)	94-0695-65
Phenol, 4-allyl-2-methoxy-, benzoate	pet ether	230(4.49),275(3.88), 303s(2.03)	65-2094-65
	EtOH	275(3.55),310s(1.70)	65-2094-65
	EtOH-HCl	273(3.88),320s(1.50)	65-2094-65
	EtOH-NaOEt	240(4.03),295(3.78)	65-2094-65
	10% H_2SO_4	245(4.03),273(3.78)	65-2094-65
Phenol, 2-methoxy-4-propenyl-, benzoate, cis	pet ether	246(4.30),280(3.80)	65-2094-65
	EtOH	280(3.78)	65-2094-65
	EtOH-HCl	240(4.47),276(3.80)	65-2094-65
	EtOH-NaOEt	281(4.30)	65-2094-65
	10% H_2SO_4	272(3.78)	65-2094-65
Phenol, 2-methoxy-4-propenyl-, benzoate, trans	pet ether	249(4.18),282(3.72)	65-2094-65
	EtOH	250(3.83),285(3.26)	65-2094-65
	EtOH-NaOEt	282(4.38),340(3.05)	65-2094-65
	10% H_2SO_4	285(3.57)	65-2094-65
4-Pyrone, tetrahydro-2-hydroxy-6,6-diphenyl-	EtOH	258(2.88)	35-3186-65
$C_{17}H_{16}O_3S$			
Acrylophenone, 3-[(3,4-dimethoxyphenyl)thio]-	MeOH	254(4.15),339(4.19)	5-0136-64C
$C_{17}H_{16}O_4$			
4,9-Anthracenedione, 1,4a,9a,10-tetrahydro-5,10β-(isopropylidenedioxy)-	EtOH	269(3.78),319(3.45)	65-2563-64
Benzofuran, 2-(2-hydroxy-4-methoxyphenyl)-5-methoxy-3-methyl-	n.s.g.	271(4.12),311(4.27)	39-5991-64
Cinnamic acid, 2,3-dimethoxy-α-phenyl-, cis	MeOH	296(4.46)	2-0182-64
trans	MeOH	281(4.18)	2-0182-64
Cinnamic acid, 2,4-dimethoxy-α-phenyl-, cis	MeOH	236(3.80),320(4.24)	2-0182-64
trans	MeOH	289(4.04),319(4.12)	2-0182-64
Cinnamic acid, p-methoxy-α-(p-methoxyphenyl)-, cis	EtOH	302(4.33)	87-0614-64

Compound	Solvent	λ_{max}(log ε)	Ref.
Cinnamic acid, p-methoxy-α-(p-methoxyphenyl)-, trans	EtOH	300(4.27)	87-0614-64
Dalbergione, 3,4-dimethoxy-	EtOH	260(4.06),405(3.00)	78-2697-65
Dalbergione, 4,4'-dimethoxy-	EtOH	228(4.16),258(4.14), 333(3.23)	78-2683-65
Flavanone, 7-hydroxy-5-methoxy-6-methyl-	MeOH	284(4.17),315(3.73)	2-0399-64
Isoflav-3-ene, 2'-hydroxy-4',7-dimethoxy-	n.s.g.	242(3.94),328(4.02)	39-5991-64
1(2H)-Naphthalenone, 2-furfurylidene-3,4-dihydro-6,8-dimethoxy-	EtOH	352.5(4.52)	5-0021-64G
Pentane, 1,5-bis(p-benzoquinonyl)-	95% THF	249(4.50),315(3.20)	44-2602-65
Xanthorrhoeol, acetate	EtOH	220(4.61),245(4.45), 300(3.86)	12-0575-65
$C_{17}H_{16}O_5$			
1,10-Anthracenedione, 9α-acetoxy-4,4aα,9,9aβ-tetrahydro-8-methoxy-	EtOH	223(4.30),225(3.67), 317(3.34)	65-2558-64
Anthraquinone, 1α-acetoxy-1,4,4aβ,9aβ-tetrahydro-8-methoxy-	EtOH	229(4.51),341(3.81)	65-2558-64
Chromanocoumaran, 3-hydroxy-4',7-dimethoxy-	n.s.g.	285(3.93),301(2.33)	39-5991-64
Cinnamic acid, 3,4-dimethoxy-2-hydroxy-α-phenyl-, trans	MeOH	299(4.23)	2-0182-64
3-Coumaranone, 2-benzyl-2-hydroxy-4',6-dimethoxy-	EtOH	227(4.13),275(4.08), 326(3.81)	22-3572-65
	EtOH-NaOH	384(3.84)	22-3572-65
Flavanone, 3-hydroxy-3',4'-dimethoxy-	EtOH	254(3.95),282s(--), 321(3.50)	22-3572-65
Itaconitin, anhydro-, propionate	EtOH	236(4.22),315(4.61), 370s(4.07)	94-0058-65
Phellopterin	EtOH	225(4.30),242(4.01), 249(4.02),270(4.18), 313(4.00)	83-0672-65
1,3-Propanedione, 1-(3,4-dimethoxyphenyl)-3-(o-hydroxyphenyl)-	EtOH	243(4.04),249(4.07), 381(4.38)	2-0351-65
2-Propanone, (5-benzyloxy-2-carboxy-3-hydroxyphenyl)-	EtOH	263(4.11),302(3.79)	39-5382-64
$C_{17}H_{16}O_6$			
2'-Acetonaphthone, 3',4'-dihydroxy-1-methoxy-, diacetate	EtOH	223(4.63),243s(4.35), 253s(4.24),275s(3.74), 286(3.78),296(3.67), 330(3.27)	94-0316-64
Acetophenone, 2-(2-carboxyphenoxy)-2',4'-dimethoxy-	EtOH	212(4.16),231(4.05), 274(3.92),305(3.85)	39-2361-65
2-Biphenylcarboxylic acid, 2'-hydroxy-4,4',5,6-tetramethoxy-, δ-lactone	dioxan	281(4.23),301(4.19), 340(3.94)	88-1743-64
Methane, bis(2,4-dihydroxy-3-acetophenonyl)-	MeOH	284(4.43),320(4.34)	78-0555-64
Olivin	n.s.g.	250(4.25),268(4.23), 344(4.36)	73-1484-64
Xanthone, 1,3,5,6-tetramethoxy-	EtOH	244(4.69),287s(4.09), 304(4.34)	44-0692-64
$C_{17}H_{16}O_7$			
Acetophenone, 2-(2-carboxyphenoxy)-2'-hydroxy-4',6'-dimethoxy-	$CHCl_3$	295(4.39)	39-5140-65

Compound	Solvent	λ_{max}(log ε)	Ref.
$C_{17}H_{16}O_8$			
Acetophenone, 2,2',4',6'-tetra-hydroxy-3',5'-dimethoxy-, 2-benzoate	EtOH	228(4.42),293(4.35), 345(3.43)	78-2977-64
$C_{17}H_{16}S$			
Cyclohepta[b]naphthothiophene, tetrahydro-	n.s.g.	234(4.45),268(3.87), 294(3.44),306(3.44)	22-0989-64
isomer	n.s.g.	245(4.49),263(4.27), 308(4.07),335(3.27)	22-0989-64
$C_{17}H_{17}BO_2$			
3-Oxa-1-oxonia-2-boratacyclohexa-4,6-diene, 4,6-dimethyl-2,2-diphenyl-	n.s.g.	322(3.67)	5-0040-65I
$C_{17}H_{17}Br$			
Cyclopentane, 2-bromo-1,1-diphenyl-	C_6H_{12}	255(2.71),262(2.74), 270(2.57)	44-3215-65
$C_{17}H_{17}BrN_2O$			
3-Acetyl-1-(2-indol-3-ylethyl)pyrid-inium bromide	EtOH-HCl	219(4.55),268(3.88), 288(3.67),315s(3.01)	35-1580-65
	EtOH-NaOH	272(4.15),279s(4.12), 288s(3.97),325(4.04)	35-1580-65
$C_{17}H_{17}BrN_2O_2$			
4-Ethoxy-1-[2-(3-indolyl)-2-oxo-ethyl]pyridinium bromide	MeOH	245(4.32),257s(4.26), 303(4.13)	44-3407-64
$C_{17}H_{17}BrO$			
Benzocyclobutene, 1-bromo-2-ethoxy-1-methyl-2-phenyl-	isooctane	256s(3.02),264(3.26), 270(3.40),277(3.36)	35-0086-65
$C_{17}H_{17}BrO_4$			
2,10(1H,3H)-Anthracenedione, 3-bromo-4,4aα,9,9aα-tetrahydro-9β-hydroxy-9-methyl-, acetate	EtOH	251(3.99),294(3.29)	70-1013-64
$C_{17}H_{17}ClN_2O$			
4H-1,3,4-Oxadiazine, 2-(o-chlorophenyl)-5,6-dihydro-4,5-dimethyl-6-phenyl-	$CHCl_3$	244(3.68),278(3.66)	44-0668-64
4H-1,3,4-Oxadiazine, 2-(p-chlorophenyl)-5,6-dihydro-4,5-dimethyl-6-phenyl-	$CHCl_3$	240(3.87),300(4.05)	44-0668-64
Quinazoline, 6-chloro-1,2-dihydro-1,2,2-trimethyl-4-phenyl-, 3-oxide	iso-PrOH	237(4.3),255s(4.23), 311(3.8),390(3.74)	44-3957-65
$C_{17}H_{17}ClN_2O_2$			
Acetanilide, 2-(p-chlorobenzoyl)-4'-(dimethylamino)-	0.04N NaOH	347(4.37)	20-0782-64
	60% EtOH	261(4.43)	20-0782-64
Acetanilide, 4'-chloro-2'-[N-(2-hy-droxyethyl)benzimidoyl]-	EtOH	237(4.45),326(3.46)	44-2368-64
$C_{17}H_{17}ClN_2O_6$			
4-Ethoxy-1-[2-(3-indolyl)-2-oxoethyl]-pyridinium perchlorate	MeOH	246(4.23),257s(4.18), 303(4.06)	44-3407-64
$C_{17}H_{17}ClN_2O_8S$			
Uridine, 2'-chloro-2'-deoxy-, 5'-benzoate 3'-methanesulfonate	EtOH	229(4.11),255(3.96)	44-0558-64

Compound	Solvent	λ_{max}(log ϵ)	Ref.
$C_{17}H_{17}ClN_2S$			
4H-1,3,4-Thiadiazine, 2-(p-chloro-phenyl)-5,6-dihydro-4,5-dimethyl-6-phenyl-	$CHCl_3$	246(4.27),324(3.99)	44-2228-65
isomer	$CHCl_3$	245(4.25),325(3.97)	44-2228-65
$C_{17}H_{17}ClO_2$			
2-Anthrol, 3β-chloro-1,2β,3,4-tetra-hydro-10-methyl-, acetate	EtOH	231(4.98),279(3.68), 290(3.71)	70-1024-64
$C_{17}H_{17}ClO_4$			
2,9-Anthracenedione, 3-chloro-1,3,4,4aβ,9aβ,10-hexahydro-10α-hydroxy-10-methyl-, acetate	EtOH	250(4.14),288(3.15)	70-1013-64
2,10(1H,3H)-Anthracenedione, 3-chloro-4,4aα,9,9aα-tetrahydro-9β-hydroxy-9-methyl-, acetate	EtOH	244(4.28),287(3.23)	70-1013-64
7-(2,4-Dimethylstyryl)tropylium perchlorate	MeCN	252(4.16),283(4.12), 473(4.55)	24-1337-64
7-(2,5-Dimethylstyryl)tropylium perchlorate	MeCN	249(4.14),280(4.13), 460(4.47)	24-1337-64
7-(3,4-Dimethylstyryl)tropylium perchlorate	MeCN	254(4.17),281(4.14), 470(4.57)	24-1337-64
Spiro[benzofuran-2(3H),1'-[2]cyclo-hexene]-3,4'-dione, 2'-ethoxy-5,6'-dimethyl-	EtOH	218(4.41),255(4.40), 262(4.35),343(3.70)	39-4939-65
$C_{17}H_{17}ClO_5$			
7-(2-Methoxy-5-methylstyryl)tropylium perchlorate	MeCN	226(4.39),283(4.15), 410s(4.05),498(4.41)	24-1337-64
7-(4-Methoxy-2-methylstyryl)tropylium perchlorate	MeCN	255(4.05),285(4.09), 508(4.55)	24-1337-64
7-(4-Methoxy-3-methylstyryl)tropylium perchlorate	MeCN	260(4.32),287(4.05), 514(4.63)	24-1337-64
$C_{17}H_{17}ClO_5S$			
Spiro[benzofuran-2(3H),1'-[2]cyclo-hexene]-3,4'-dione, 7-chloro-4,6-di-methoxy-6'-methyl-2'-(methylthio)-	EtOH	230s(4.18),290(4.61)	87-0705-64
Spiro[benzofuran-2(3H),1'-[3]cyclo-hexene]-2',3-dione, 7-chloro-4,6-di-methoxy-6'-methyl-4'-(methylthio)-	EtOH	230s(4.14),296(4.60)	87-0705-64
$C_{17}H_{17}ClO_6$			
7-(2,4-Dimethoxystyryl)tropylium perchlorate	MeCN	225(4.35),258(4.04), 290(4.17),380s(3.55), 539(4.65)	24-1337-64
7-(2,5-Dimethoxystyryl)tropylium perchlorate	MeCN	229(4.47),282(4.18), 305s(3.70),409(4.26), 519(4.31)	24-1337-64
7-(2,6-Dimethoxystyryl)tropylium perchlorate	MeCN	222s(4.38),242s(4.24), 285(4.08),501(4.57)	24-1337-64
7-(3,4-Dimethoxystyryl)tropylium perchlorate	MeCN	220s(4.36),258(4.17), 290(4.10),390s(3.75), 524(4.57)	24-1337-64
$C_{17}H_{17}Cl_2N_3$			
Benzimidazole, 6-chloro-2-(p-chloro-phenyl)-1-(2-dimethylaminoethyl)-	EtOH	256(3.99),294(4.27)	4-0453-65

Compound	Solvent	λ_{max}(log ε)	Ref.
$C_{17}H_{17}Cl_3O$			
Azulene, 3-(trichloroacetyl)-7-isopropyl-1,4-dimethyl-	C_6H_{12}	238(4.37),288(4.32),325(4.21),462(4.03),582(2.80)	44-0131-65
$C_{17}H_{17}FN_4O_4S$			
9H-Purine, 6-(benzylthio)-2-fluoro-9-β-D-ribofuranosyl-	pH 1	297(4.41)	35-3752-65
	pH 11	297(4.42)	35-3752-65
	MeOH	291s(4.40),296(4.44)	35-3752-65
$C_{17}H_{17}FN_4O_5$			
9H-Purine, 6-(benzyloxy)-2-fluoro-9-β-D-ribofuranosyl-	MeOH	256(4.14)	35-3752-65
$C_{17}H_{17}IN_2O_2$			
4-Ethoxy-1-[2-(3-indolyl)-2-oxoethyl]-pyridinium iodide	MeOH	244(4.38),257s(4.32),302(4.19)	44-3407-64
$C_{17}H_{17}IN_2O_8S$			
Uracil, 1-(3-deoxy-3-iodo-β-D-arabino-furanosyl)-, 5'-benzoate 2'-methane-sulfonate	EtOH	232(4.16),260(4.00)	35-1896-64
$C_{17}H_{17}N$			
Gona-1,3,5(10),8,11,13-hexaen-12-amine	EtOH-HCl	215(4.57),272(4.25),300s(3.58)	5-0200-65D
	EtOH-NaOH	224(4.29),273(4.11),315(3.65)	5-0200-65D
Indole, 4,5,7-trimethyl-2-phenyl-	EtOH	252(4.34),313(4.51)	44-0563-65
$C_{17}H_{17}NO$			
Acridan, 10-isobutyryl-	n.s.g.	238s(3.95),245(3.97)	39-5877-65
Chalcone, 4-(dimethylamino)-	benzene	402(4.37)	65-0094-65
	EtOH	269(4.20),418(4.45)	65-0094-65
	HOAc	270(4.16),424(4.38)	65-0094-65
Chalcone, 4'-(dimethylamino)-	benzene	300(4.26),372(4.38)	65-0094-65
	EtOH	304(4.26),385(4.36)	65-0094-65
	HOAc	304(4.26),394(4.39)	65-0094-65
1-Penten-3-one, 5-anilino-1-phenyl-	EtOH	226(4.16),248(4.14),290(4.40)	7-0652-65
2-Propen-1-one, 1-(p-dimethylamino-phenyl)-3-phenyl-	H_2SO_4	259(3.49),295(3.78),425(4.59)	65-1724-65
2-Propen-1-one, 3-(p-dimethylamino-phenyl)-1-phenyl-	H_2SO_4	240(3.61),395(4.57)	65-1724-65
L-Tyrosine, N-salicylidene-, methyl ester	MeOH	220s(4.53),256(4.07),286s(3.46),319(3.58),406(2.41)	35-1757-65
	dioxan	216(4.48),222s(4.45),258(4.14),285s(3.50),319(3.71),420s(0.96)	35-1757-65
$C_{17}H_{17}NOS_2$			
Carbamic acid, dimethyldithio-, ester with 2-mercapto-2-phenylacetophenone	EtOH	248(4.35),272s(4.01),310s(--)	39-4004-64
	EtOH	248(4.41)	44-1711-64
$C_{17}H_{17}NO_2$			
Asimilobine	EtOH	274(4.21),308(3.51)	95-0077-65
13-Aza-18-norequilenin methyl ether	EtOH	265(4.01),275(4.01),285(3.98)	39-3007-65

Compound	Solvent	λ_{max}(log ε)	Ref.
Caaverine	MeOH	272(4.25),310s(3.7)	78-1435-64
Chalcone, 4-(dimethylamino)-3'-hydroxy-	EtOH	268(4.43),337(3.66), 420(4.50)	23-2580-64
$C_{17}H_{17}NO_3$			
p-Acetanisidide, 2-p-toluoyl-	0.04N NaOH	334(4.40)	20-0782-64
	60% EtOH	256(4.44)	20-0782-64
p-Acetotoluidide, 2-p-anisoyl-	0.04N NaOH	334(4.45)	20-0782-64
	60% EtOH	279(4.29)	20-0782-64
2,3-Indolinediol, 1-benzoyl-2,3-dimethyl-, cis	EtOH	272(4.14)	39-4320-64
Propionanilide, 3-acetoxy-3-phenyl-	EtOH	243(4.30)	32-1115-65
Sparsiflorine	EtOH-HCl	226(4.54),266(4.09), 275(4.23),310(3.95)	88-1539-65
Succinimide, N-[2-(6-methoxy-1-naphthyl)ethyl]-	EtOH	262(4.01),271(4.09), 284(3.99)	39-3007-65
$C_{17}H_{17}NO_4$			
Acrylic acid, 3-(2-amino-3-methoxyphenyl)-2-(p-methoxyphenyl)-	EtOH	240(4.4),255(4.7), 280(4.0),290(4.0), 310(4.1),330(3.3), 342(3.7),362(3.8)	49-1512-65
Powelline, 3-dehydro-	MeOH	280(3.29)	83-0704-65
$C_{17}H_{17}NO_8$			
Coumarin-6-carboxylic acid, 3-acetyl-5-methoxy-4,7-dimethyl-8-nitro-, ethyl ester	EtOH	294(4.15)	39-2543-65
$C_{17}H_{17}N_3O_2$			
10H-Dibenzo[b,e][1,4]diazepine-10-propionamide, 5,11-dihydro-5-methyl-11-oxo-	EtOH	212(4.37),227(4.44), 285s(3.48)	33-1590-65
$C_{17}H_{17}N_3O_3$			
Benzamide, N-[2-(1-indolinyl)ethyl]-p-nitro-	MeOH	257(4.31)	87-0107-65
Benzanilide, p,p'-diacetamido-	EtOH	216(4.3),283(4.5)	28-4295-64B
$C_{17}H_{17}N_3O_4$			
Acetanilide, 4'-(dimethylamino)-2-(3-nitrobenzoyl)-	0.04N NaOH	341(4.32)	20-0782-64
	60% EtOH	267(4.37)	20-0782-64
Acetanilide, 4'-(dimethylamino)-2-(4-nitrobenzoyl)-	0.04N NaOH	403(4.00)	20-0782-64
	60% EtOH	268(4.35)	20-0782-64
$C_{17}H_{17}N_3O_5S$			
5-Pyrimidinecarboxylic acid, 4-[3-methyl-4-(5-nitro-2-furyl)buta-1,3-dienyl]-2-(methylthio)-, ethyl ester	propylene glycol	288(4.38),335(4.22), 410(4.19)	95-0207-64
$C_{17}H_{17}N_3O_8$			
Thymidine, 3'-p-nitrobenzoate	MeOH	263(4.36)	35-5661-65
	MeOH	208(4.15),264(4.33)	22-2489-65
$C_{17}H_{17}N_5O_6$			
Guanosine, N-benzoyl-	MeOH	238(--),260(4.17), 295(--)	73-0214-64
$C_{17}H_{18}$			
5H-Dibenzo[a,c]cycloheptene, 6,7-dihydro-1,11-dimethyl-	n.s.g.	240.5(4.07)	35-1710-64

Compound	Solvent	λ_{max}(log ϵ)	Ref.
Gona-5,7,9,11,13-pentaene, 5,6,7,8-tetrahydro-	EtOH	229(4.82),235(4.98), 283(3.79),291(3.72), 311(3.19),317(3.41), 326(3.34),332(3.62)	5-0200-65D
Naphthalene, 1,2,3,4-tetrahydro-1-methyl-3-phenyl-	EtOH	258(2.91),265(2.91), 273(2.81)	28-3780-64B
1-Pentene, 1,1-diphenyl-	C_6H_{12}	250.5(4.18)	22-0693-64
$C_{17}H_{18}BrNO_4$			
Acetamide, N-(3-bromo-4-hydroxy-5-methoxyphenethyl)-2-(p-hydroxyphenyl)-	EtOH	225(4.22),285(3.74)	65-0550-64
$C_{17}H_{18}Br_2N_6$			
Biguanide, 1,2-bis(p-bromophenyl)-5-isopropylideneamino-	EtOH	233(4.32),275s(4.52)	39-0932-65
$C_{17}H_{18}ClNO_4$			
8-(Methylphenylimoniomethyl)-8-methylheptafulvene perchlorate	MeCN	266(3.98),475(4.49)	24-2050-64
$C_{17}H_{18}ClNO_5S$			
Spiro[benzofuran-2(3H),1'-[2]cyclohexene]-3,4'-dione, 7-chloro-4,6-dimethoxy-6'-methyl-2'-(methylthio)-, 4'-oxime	EtOH	240s(4.13),288(4.57), 321s(3.81)	87-0705-64
$C_{17}H_{18}ClN_3$			
Benzimidazole, 5-chloro-1-(2-dimethylaminoethyl)-2-phenyl-	EtOH	253(3.87),295(4.10)	4-0453-65
Benzimidazole, 6-chloro-1-(2-dimethylaminoethyl)-2-phenyl-	EtOH	254(3.89),293(4.21)	4-0453-65
$C_{17}H_{18}INO_3$			
3-Carboxy-6,7-dihydro-9-methoxybenzo-[a]quinolizinium iodide, ethyl ester	EtOH	220(4.38),258s(3.71), 291(3.82),364(4.44)	23-1527-65
$C_{17}H_{18}N_2$			
Indole, 1-(dimethylamino)-2-methyl-3-phenyl-	MeOH	229(4.41),277(4.12), 282(4.12)	44-3451-65
Uleine, N-demethyldehydro-	EtOH	303(4.21),310(4.21)	78-1717-65
$C_{17}H_{18}N_2O$			
2-Indolinone, 1-(dimethylamino)-3-methyl-3-phenyl-	MeOH	253(3.85)	44-3451-65
4H-1,3,4-Oxadiazine, 5,6-dihydro-4,5-dimethyl-2,6-diphenyl-	$CHCl_3$	240(3.67),294(3.96)	44-0668-64
Phenazine, 1-tert-butyl-3-methoxy-	EtOH	218(4.28),258(4.81), 357(3.90)	12-1241-65
Phenazine, 3-tert-butyl-1-methoxy-	EtOH	215(4.45),267(5.05), 368(4.15),402s(3.66)	12-1241-65
	EtOH	215(4.45),267(5.05), 368(4.15),402(3.66)	39-2914-65
Phenazine, 3-tert-butyl-2-methoxy-	EtOH	219(4.41),259(4.88), 370(4.03)	12-1241-65
$C_{17}H_{18}N_2O_2$			
Acetanilide, 2-benzoyl-4'-(dimethylamino)-	0.04N NaOH	341(4.38)	20-0782-64
	60% EtOH	254(4.30)	20-0782-64
2',6'-Benzoxylidide, 4-acetamido-	EtOH	270(4.5)	28-4295-64B
2-Indolinone, 1-(dimethylamino)-3-methoxy-3-phenyl-	MeOH	259(3.73)	44-3451-65

Compound	Solvent	λ_{max}(log ε)	Ref.
1,2-Propanediamine, N,N'-di-salicylidene-	MeOH	243(4.42),253s(4.37), 317(3.89),400(3.17)	65-3049-64
1-Pyrazoline, 3,5-bis(p-methoxyphenyl)-, cis	EtOH	329(2.52)	35-5364-64
trans	EtOH	332(2.73)	35-5364-64
11bH-Quino[1,2-c]quinazolin-1-one, 1,2,3,4,12,13-hexahydro-9-methoxy-	MeOH-HCl	228(4.39),263(4.20), 300s(3.50)	39-5911-64
$C_{17}H_{18}N_2O_3$			
1,3-Cyclohexanedione, 2-[2-(7-methoxy-quinazolin-4-yl)ethyl]-	MeOH	214s(4.30),235(4.63), 262(4.19),287s(3.99), 311s(3.76)	39-5911-64
$C_{17}H_{18}N_2O_4S$			
4-Isothiazoline-2-acetic acid, α-iso-propenyl-3-oxo-4-(2-phenylacetamido)-, methyl ester	EtOH	223s(4.00),297(4.03)	35-5307-64
	EtOH	223s(4.01),297(4.03)	88-1471-64
4-Isothiazoline-2-acetic acid, α-iso-propylidene-3-oxo-4-(2-phenyl-acetamido)-, methyl ester	EtOH	296(4.00),307s(3.90)	35-5307-64
	EtOH	296(4.00),307s(3.90)	88-1471-64
Malonic acid, [[(4-phenyl-2-thiazolyl)-amino]methylene]-, diethyl ester	EtOH	263(4.2),340(4.2)	70-1481-64
$C_{17}H_{18}N_2O_5S_2$			
5-Thia-1-azabicyclo[4.2.0]oct-2-ene-2-carboxylic acid, 3-(hydroxymethyl)-8-oxo-7-[2-(phenylthio)acetamido]-, methyl ester	n.s.g.	250(4.05)	87-0022-65
$C_{17}H_{18}N_2O_5S_4$			
Xanthic acid, ethyl-, ester with 3-(mercaptomethyl)-8-oxo-7-[2-(2-thienyl)acetamido]-5-thia-1-azabicyclo[4.2.0]oct-2-ene-2-carboxylic acid, sodium salt	n.s.g.	223(4.28),284(4.31)	87-0174-65
$C_{17}H_{18}N_2O_6$			
Crinamine, 6-hydroxy-, product from nitrous acid reaction of	EtOH	239(4.15),290(3.68)	35-4912-65
$C_{17}H_{18}N_2O_6S$			
3-[5-(p-Methoxyphenyl)-2-oxazolyl]-1-methylpyridinium methyl sulfate	iso-PrOH	262(4.18),352(4.20)	4-0310-65
$C_{17}H_{18}N_2O_6S_2$			
5-Thia-1-azabicyclo[4.2.0]oct-3-ene-2-carboxylic acid, 3-(hydroxymethyl)-8-oxo-7-[2-(2-thienyl)acetamido]-, methyl ester, acetate	EtOH	237(4.11),260s(3.89)	39-5015-65
$C_{17}H_{18}N_2S$			
4H-1,3,4-Thiadiazine, 5,6-dihydro-4,5-dimethyl-2,6-diphenyl-	$CHCl_3$	245(4.18),319(3.91)	44-2228-65
$C_{17}H_{18}N_3OP$			
Phosphonic diamide, P-(1-methyl-pyrrol-2-yl)-N,N'-diphenyl-	EtOH	232(4.26),276(3.19), 270s(3.15),283s(3.08)	44-0091-65
$C_{17}H_{18}N_4$			
Aniline, N,N-dipropyl-4-tricyanovinyl-	$CHCl_3$	526(4.74)	39-1334-64

Compound	Solvent	$\lambda_{max}(\log \epsilon)$	Ref.
$C_{17}H_{18}N_4O$			
Formazane, C-acetyl-N,N'-di-m-tolyl-	EtOH	250(3.96),266(3.94), 305(4.19),444(4.39)	5-0099-65B
Formazane, C-acetyl-N,N'-di-o-tolyl-	EtOH	256(3.93),304(4.15), 432(4.34)	5-0099-65B
Pyrazolo[1,5-a]pyrimidine, 7-acetamido-2,3,5-trimethyl-6-phenyl-	EtOH	244(4.72),296(3.62), 339(3.27)	94-1207-65
$C_{17}H_{18}N_4O_2$			
4-Pyrimidinol, 2-amino-5-(3-anilino-propyl)-6-(α-furyl)-	pH 1	235(3.97),315(4.23)	87-0283-65
	pH 7	240(4.30),285(4.15), 320(4.05)	87-0283-65
	pH 13	310(3.96)	87-0283-65
$C_{17}H_{18}N_4O_3$			
4-Pyrimidinecarboxylic acid, 2-amino-6-hydroxy-5-(4-phenylbutyl)-, glycolonitrile ester	pH 1	290(3.81)	4-0162-65
	pH 7	304(3.79)	4-0162-65
	pH 13	295(3.64)	4-0162-65
$C_{17}H_{18}N_4O_4$			
Benzaldehyde, p-tert-butyl-, 2,4-dinitrophenylhydrazone	$CHCl_3$	385(4.49)	39-5963-64
$C_{17}H_{18}N_4O_7$			
Acetophenone, 2',4',6'-trimethoxy-, 2,4-dinitrophenylhydrazone	$CHCl_3$	373(4.40)	5-0021-64G
Xanthosine, 8-(benzyloxy)-	pH 1	235(4.05),284(4.01)	35-1772-65
	pH 11	243(4.18),286(3.93)	35-1772-65
$C_{17}H_{18}N_4O_9$			
3-Cyclopentene-1,2-dicarboxylic acid, 3-acetyl-5-hydroxy-, methyl ester, 2,4-dinitrophenylhydrazone	n.s.g.	365(4.42)	65-0058-65
$C_{17}H_{18}N_8O_8$			
Adenine, 9-[2-deoxy-2-(2,4-dinitro-anilino)-β-D-glucopyranosyl]-	EtOH	265(4.10),347(4.09)	44-1556-65
$C_{17}H_{18}O$			
Anisole, p-[1-(2,6-xylyl)vinyl]-	EtOH	263(4.22)	54-0806-65
Benzocyclobutene, 2-ethoxy-1-methyl-2-phenyl-	isooctane	248s(2.83),254s(3.01), 2 1(3.20),267(3.34), 274(3.30),291(1.97), 300(2.00)	35-0086-65
2-Biphenylcarboxaldehyde, 4,6-diethyl-	EtOH	219(4.35),258(4.02), 308(3.41)	22-2953-64
2,8,10,16-Heptadecatetraene-4,6-di-ynal, 2-cis	hexane	292(--),314(--), 346(--),373(--)	24-1225-65
all trans	hexane	257(4.38),265(4.41), 283(4.45),316(4.35), 343(4.43),371(4.40)	24-1225-65
$C_{17}H_{18}OS$			
2-Butanone, 4-phenyl-4-(p-tolylthio)-	EtOH	262(3.66)	7-0652-65
Thiobenzophenone, 3-methoxy-2',4',6'-trimethyl-	C_6H_{12}	313(4.17),620(1.79)	54-0289-65
	EtOH	316(4.17),609(1.79)	54-0289-65
Thiobenzophenone, 4-methoxy-2',4',6'-trimethyl-	C_6H_{12}	242(4.08),352(4.34), 604(1.86)	54-0289-65
	EtOH	239(4.03),362(4.33), 591(1.88)	54-0289-65

Compound	Solvent	λ_{max}(log ε)	Ref.
$C_{17}H_{18}O_2$			
3-Anthracenone, 1,2,3,4,4aβ,9,9aα,10-octahydro-10-hydroxy-2β-(2-propynyl)-	EtOH	204(4.07),266(2.57), 273(2.57)	70-1039-65
Azulene, 1,3-diacetyl-4,6,8-trimethyl-	CH_2Cl_2	257(4.44),300(4.34), 319(4.40),387(3.93), 503(2.99)	44-0131-65
Mycomycin, tert-butyl ester, trans-trans	ether	256(4.26),267(4.46), 281(4.48)	24-1846-64
Thymol, benzoate	pet ether	227(4.49),267(3.55)	65-2094-65
$C_{17}H_{18}O_2S$			
Benzothiophene-3-carboxylic acid, 4,5,6,7-tetrahydro-2-phenyl-, ethyl ester	ether	211(4.25),249(3.97), 290(3.92)	24-2109-64
$C_{17}H_{18}O_3$			
2α-Anthraceneacetic acid, 1,2,3,4-tetrahydro-3β-hydroxy-9-methyl-	EtOH	233(4.99),281(3.72), 291(3.75)	65-2576-64
Anthrone, 10α-acetoxy-1,4,4aβ,9aβ-tetrahydro-10-methyl-	EtOH	242(3.96),287(3.16)	70-1013-64
Butyric acid, γ-(2-methoxy-5-phenyl)-phenyl-	EtOH	262(4.30)	39-5110-64
Fluorene-8a(4bH)-acetic acid, 5,6-dihydro-1-methyl-9-oxo-, methyl ester	EtOH	251(4.08),300(3.36)	44-2528-65
Fluorene-8a(4bH)-acetic acid, 5,8-dihydro-1-methyl-9-oxo-, methyl ester	EtOH	250(4.09),299(3.41)	44-0074-64
1-Phenanthrenecarboxylic acid, 1,2,3,4,9,10-hexahydro-1-methyl-2-oxo-, methyl ester	EtOH	262(4.08),292s(3.43)	35-2038-64
1(or 2)-Phenanthrenecarboxylic acid, 1,2,9,10-tetrahydro-7-methoxy-2-methyl-	EtOH	233(3.85),239(3.84), 317(4.22),335(4.28)	70-2021-64
$C_{17}H_{18}O_4$			
Acetophenone, 2'-benzyloxy-2,4'-dimethoxy-	EtOH	272(4.07),305(3.90)	22-3572-65
4,9-Anthracenedione, 1,2,3,4,4aα,9aα,10-hexahydro-5,10β-(isopropylidene-dioxy)-	EtOH	260(3.92),318(3.54)	65-2563-64
Anthracen-10-one, 2,3α-epoxy-1,2,3,4,4aα,9,9aα,10-octahydro-9β-hydroxy-9α-methyl-, acetate	EtOH	247(4.04),286(3.16)	70-1013-64
Anthracen-10-one, 2,3β-epoxy-1,2,3,4,4aα,9,9aα,10-octahydro-9β-hydroxy-9α-methyl-, acetate	EtOH	246(4.16),286(3.16)	70-1013-64
Anthrone, 1,4,4aα,9aα-tetrahydro-4β-hydroxy-5,10β-(isopropylidenedioxy)-	EtOH	226(4.05),263(3.85), 318(3.36)	65-2563-64
Benzophenone, 2'-ethoxy-2,4-dimethoxy-	MeOH	278(3.90),310(4.01)	78-1495-65
Dalbergione, dihydro-4,4'-dimethoxy-	EtOH	223(4.12),264(4.21), 350(3.15)	78-2683-65
Fluorene-8aβ(4bβH)-acetic acid, 5,6,7,8-tetrahydro-1-methyl-7,9-dioxo-, methyl ester	EtOH	252(4.11),300(3.37)	44-0074-64
$C_{17}H_{18}O_5$			
Anthrone, 4β-acetoxy-1,4,4aα,9aα-tetrahydro-10β-hydroxy-5-methoxy-	EtOH	223(4.22),253(3.88), 310(3.41)	65-2558-64
Anthrone, 10β-acetoxy-1,4,4aα,9aβ-tetrahydro-4-hydroxy-5-methoxy-	EtOH	224(4.09),257(3.66), 319(3.20)	65-2558-64

Compound	Solvent	λ_{max}(log ε)	Ref.
Propionic acid, 2-hydroxy-2-(2-methoxyphenyl)-3-(4-methoxyphenyl)-	EtOH	275(3.50)	22-3572-65
$C_{17}H_{18}O_6$			
Coumarin-6-carboxylic acid, 3-acetyl-5-methoxy-4,7-dimethyl-, ethyl ester	EtOH	294(4.20)	39-2543-65
$C_{17}H_{18}O_7$			
Glauconic acid	EtOH	223(4.02),260s(--)	39-1772-65
2-Naphthaleneacetic acid, 1,4-dihydro-5,7,8-trimethoxy-3-methyl-1,4-dioxo-, methyl ester	n.s.g.	220(4.26),268(3.91), 430(3.20)	33-2017-64
$C_{17}H_{18}O_8$			
Xanthophanic acid, ethyl methyl ester	EtOH	252(4.36),352(4.26), 430(3.29)	88-2313-64
$C_{17}H_{18}S$			
Thiobenzophenone, 2,2',3,3'-tetramethyl-	C_6H_{12}	327(3.97),606(2.19)	54-0289-65
	EtOH	329(3.96),592(2.19)	54-0289-65
Thiobenzophenone, 2,3',4,6-tetramethyl-	C_6H_{12}	319(4.07),617(1.68)	54-0289-65
	EtOH	321(4.03),602(1.67)	54-0289-65
Thiobenzophenone, 2,4,4',6-tetramethyl-	C_6H_{12}	234(4.13),329(4.20), 616(1.76)	54-0289-65
	EtOH	233(4.10),333(4.20), 601(1.76)	54-0289-65
$C_{17}H_{19}BrN_2O$			
2-Ethoxy-1-(2-indol-3-ylethyl)pyridinium bromide	EtOH	283(4.10),289s(4.04)	94-0931-65
4-Ethoxy-1-(2-indol-3-ylethyl)pyridinium bromide	MeOH	246(4.18),281(3.76), 290(3.70)	44-3407-64
4-Methoxy-1-[2-(2-methyl-3-indolyl)-ethyl]pyridinium bromide	MeOH	243(4.23),280(3.90)	24-2463-64
$C_{17}H_{19}BrO_3$			
Cacalol, bromo-, acetate	EtOH	215(4.47),260(4.21)	78-2331-64
$C_{17}H_{19}BrO_4$			
Anthracen-10-one, 2α-bromo-3β-hydroxy-9β-acetoxy-1,2,3,4,4aα,9,9a,10-octahydro-9α-methyl-	EtOH	249(4.06),288(3.17)	70-1013-64
Anthracen-10-one, 3β-bromo-2α-hydroxy-9β-acetoxy-1,2,3,4,4aα,9,9a,10-octahydro-9-methyl-	EtOH	251(4.11),288(3.21)	70-1013-64
$C_{17}H_{19}ClN_2O_5$			
4-Ethoxy-1-(2-indol-3-ylethyl)-pyridinium perchlorate	MeOH	246(4.26),280(3.86), 290(3.79)	44-3407-64
$C_{17}H_{19}ClO_3$			
2,9β-Epoxyanthracene, 10β-acetoxy-3α-chloro-1,2,3,4,4aα,9,9aα,10-octahydro-9α-methyl-	EtOH	260(2.23)	70-1024-64
$C_{17}H_{19}ClO_4$			
Anthracen-10-one, 3β-acetoxy-2α-chloro-1,2,3,4,4aα,9,9aα,10-octahydro-9β-hydroxy-9α-methyl-	EtOH	251(4.08),291(3.22)	70-1024-64

Compound	Solvent	λ_{max}(log ε)	Ref.
Anthracen-10-one, 9β-acetoxy-2α-chloro-1,2,3,4,4aα,9,9aα,10-octahydro-3β-hydroxy-9α-methyl-	EtOH	250(4.01),288(3.15)	70-1013-64
7-(p-tert-Butylphenyl)tropylium perchlorate	MeCN	274(4.12),393(4.32)	24-0029-64
$C_{17}H_{19}ClO_5S$			
Spiro[benzofuran-2(3H),1'-[2]cyclohexen]-3-one, 7-chloro-4'-hydroxy-4,6-dimethoxy-6'-methyl-2'-(methylthio)-	EtOH	240(4.25),288(4.28), 322s(3.71)	87-0705-64
Spiro[benzofuran-2(3H),1'-[3]cyclohexen]-3-one, 7-chloro-2'-hydroxy-4,6-dimethoxy-6'-methyl-4'-(methylthio)-	EtOH	228s(4.39),235s(4.34), 289(4.35),322(3.72)	87-0705-64
$C_{17}H_{19}ClO_6$			
1,3-Diacetyl-4,6,8-trimethyl-azulenylium perchlorate	MeCN	254(4.47),318(4.35), 388(3.88)	65-0894-64
$C_{17}H_{19}N$			
Aniline, o-(1-benzylvinyl)-N,N-dimethyl-	EtOH	206(4.48),244(4.14)	39-0249-64
5H-Dibenz[c,e]azepine, 6,7-dihydro-1,6,11-trimethyl-	isooctane	236.5(4.13)	35-1710-64
	6N HCl	242(4.01),272(3.58)	35-1710-64
Stilbene, 4-(dimethylamino)-3'-methyl-, trans	EtOH	346(4.49)	33-0517-65
conjugate acid	EtOH	312(4.49)	33-0517-65
Stilbene, 4-(dimethylamino)-4'-methyl-, trans	EtOH	347(4.52)	33-0517-65
conjugate acid	EtOH	314(4.53)	33-0517-65
$C_{17}H_{19}NO$			
13H-Dibenzo[a,f]quinolizin-13-one, 1,2,3,4,6,7,11b,12-octahydro-	C_6H_{12}	312(4.05)	44-3667-65
	EtOH	336(4.11)	44-3667-65
perchlorate	EtOH	337(4.08)	44-3667-65
Mesidine, 3-phenacyl-	EtOH	238(4.32),287(3.48)	44-0563-65
	EtOH-HCl	242(4.20)	44-0563-65
Phenethylamine, N-(o-methoxybenzylidene)-α-methyl-	EtOH	252(4.20),305(3.75)	44-2265-64
Pyrrolo[1,2-a]quinolin-5(1H)-one, 3-allyl-2,3-dihydro-1,4-dimethyl-	EtOH	214(4.46),243(4.46), 250(4.47),284s(3.29), 296s(3.43),329(4.10), 342(4.13)	44-1989-65
3H-Pseudoindol-4-one, 3a,4,7,7a-tetrahydro-3a,6,7-trimethyl-2-phenyl- (or isomer)	EtOH	242(4.52)	44-0563-65
	EtOH-HCl	240(4.28),273(4.36)	44-0563-65
4(1H)-Quinolone, 1-allyl-2-(3-butenyl)-3-methyl-	EtOH	216(4.35),244(4.47), 251(4.45),283s(3.22), 294s(3.37),331(4.13), 346(4.20)	44-1989-65
4(1H)-Quinolone, 2-(1-allyl-3-butenyl)-3-methyl-	EtOH	213(4.47),241(4.53), 247(4.53),283s(3.30), 294s(3.42),324(4.10), 338(4.14)	44-1989-65
4(1H)-Quinolone, 3-allyl-1-crotyl-2-methyl-	EtOH	244(4.45),251(4.42), 328(4.12),342(4.16)	88-1635-64
4(1H)-Quinolone, 1,3-bis(2-butenyl)-	EtOH	212(4.45),243s(4.41), 246(4.42),293(3.55), 332(4.13),345(4.18)	94-1424-64
4(1H)-Quinolone, 1,3-bis(2-methylallyl)-	EtOH	212(4.40),243s(4.36), 246(4.37),293(3.51), 330(4.11),344(4.17)	94-1424-64

Compound	Solvent	$\lambda_{max}(\log \epsilon)$	Ref.
Stilbene, 4-(dimethylamino)-3'-methoxy-, trans	EtOH	348(4.48)	33-0517-65
conjugate acid	EtOH	298(4.44)	33-0517-65
Stilbene, 4-(dimethylamino)-4'-methoxy-, trans	EtOH	345(4.51)	33-0517-65
conjugate acid	EtOH	323(4.54)	33-0517-65
$C_{17}H_{19}NO_2$			
8-Aza-9ξ-gona-1,3,5(10),13-tetraen-12-one, 3-methoxy-	C_6H_{12}	310(4.09)	44-3667-65
	EtOH	333(4.17)	44-3667-65
perchlorate	EtOH	332(4.17)	44-3667-65
8-Aza-9ξ-gona-1,3,5(10),13-tetraen-12-one, 4-methoxy-	C_6H_{12}	310(4.08)	44-3667-65
	EtOH	333(4.18)	44-3667-65
perchlorate	EtOH	333(4.16)	44-3667-65
Isoquinoline, 1,2,3,4-tetrahydro-6,7-dimethoxy-1-phenyl-	MeOH	286(3.52)	83-0561-65
1H,3H-[1,3]Oxazino[3,4-a]quinolin-3-one, 4,4a-dihydro-1-isopropylidene-4,4-dimethyl-	EtOH	237(4.35),285(3.44), 342(3.60)	77-0574-65
$C_{17}H_{19}NO_3$			
Acetophenone, 4'-hydroxy-2-[(1-methyl-2-phenoxyethyl)amino]-	$EtOH-H_2SO_4$	286.5(4.20)	54-0521-65
hydrochloride	$EtOH-H_2SO_4$	286(4.23)	54-0521-65
Benzo[f]quinoline-3-carboxylic acid, 1,2,3,4-tetrahydro-8-methoxy-, ethyl ester	MeOH	226(4.45),246(4.75), 281(3.99),367(3.49)	39-5907-64
9a(5aH)-Dibenzofuranpropionitrile, 4-ethoxy-6,7,8,9-tetrahydro-6-oxo-	EtOH	278(3.20)	94-1012-64
2-Propen-1-one, 1-(3,4-dimethyl-3-cyclohexen-1-yl)-3-(p-nitrophenyl)-	EtOH	304(4.28)	7-0143-65
$C_{17}H_{19}NO_3S_2$			
2-Ethyl-N-methylbenzothiazolium p-toluenesulfonate	EtOH	258(3.73),286(3.64)	22-2868-64
$C_{17}H_{19}NO_4$			
Aureothinic ketone	EtOH	248(4.20),342(4.28)	88-2655-65
9-Azabicyclo[3.3.1]nonan-3-carboxylic acid, 9-benzoyl-2-oxo-, methyl ester	C_6H_{12}	250(4.11)	39-5378-64
	EtOH	251(4.09)	39-5378-64
Powelline, 3-dehydrodihydro-	MeOH	287(3.22)	83-0704-65
$C_{17}H_{19}NO_5$			
Dubinine	EtOH	225(5.2),270(4.3), 280(4.3),305(4.1)	65-0345-64
Hemanthidine	EtOH	292(3.73)	95-1194-64
$C_{17}H_{19}NO_6$			
3H-3-Benzazepine-2,4-dicarboxylic acid, 1,5-dimethoxy-3-methyl-, dimethyl ester	MeOH	238(4.4),286(4.1)	56-1423-65
$C_{17}H_{19}NO_8S$			
Thiazolo[3,2-a]azepine-5,6,7,8-tetracarboxylic acid, 5,6-dihydro-3-methyl-, tetramethyl ester	MeOH	446.0(2.80)	39-3200-65
	MeOH-HCl	246(0.84),324(1.76), 335s(1.58)	39-3200-65
$C_{17}H_{19}NS$			
Thiazole, 2,4-dimethyl-, dimethyl acetylenedicarboxylate adduct	MeOH	215(4.09),239(4.13), 295(4.24),446(3.72)	88-1797-64

Compound	Solvent	$\lambda_{max}(\log \epsilon)$	Ref.
Thiazole, 2-ethyl-, dimethyl acetylenedicarboxylate adduct	MeOH	215(3.94),450(4.49)	88-1797-64
$C_{17}H_{19}N_3O$			
11H-Dibenzo[b,e][1,4]diazepin-11-one, 5,10-dihydro-5-methyl-10-(2-methylaminoethyl)-	EtOH-HCl	226(4.43)	33-1590-65
$C_{17}H_{19}N_3O_3$			
Benzamide, p-nitro-N-[2-(2,6-xylidino)ethyl]-	MeOH	249(4.13)	87-0107-65
$C_{17}H_{19}N_3O_3S$			
Pyrimidine, 2-(butylthio)-4-[3-methyl-4-(5-nitro-2-furyl)buta-1,3-dienyl]-	propylene glycol	256(4.15),330(4.22)	95-0207-64
$C_{17}H_{19}N_5$			
Acridine orange	H_2O	492(4.78)	46-3872-65
$C_{17}H_{19}N_5O_6$			
Guanosine, 8-(benzyloxy)-	pH 1	246(4.07),294(3.94)	35-1772-65
	pH 11	248(4.09),268s(3.95)	35-1772-65
$C_{17}H_{19}N_5O_6S$			
Theophylline, 7-(p-carbethoxybenzyl)-8-sulfamoyl-	MeOH	285(4.03)	54-0193-65
$C_{17}H_{19}N_5O_9$			
2-Amino-1-(3-carboxypropyl)pyridinium picrate, ethyl ester	H_2O	232(4.36),314s(4.03), 354(4.20)	39-2763-64
$C_{17}H_{20}$			
Pentane, 1,1-diphenyl-	C_6H_{12}	255(2.73),260(2.78), 262(2.79),269(2.68)	22-0587-65
$C_{17}H_{20}FeO$			
Cyclohexanone, 2-(ferrocenylmethyl)-	MeCN	328s(2.13),436(2.13)	44-2028-64
$C_{17}H_{20}IN$			
1,2,3,4-Tetrahydro-1,1-dimethyl-3-phenylquinolinium iodide	EtOH	258(2.74),263(2.76), 270s(2.59)	39-0249-64
$C_{17}H_{20}N_2$			
Isoquinoline, 4-(N-benzylaminomethyl)-1,2,3,4-tetrahydro-	EtOH	258(2.76),262(2.80), 268(2.72),290(5.59)	94-1225-65
2,6-Methanoindolo[2,3-a]quinolizine, 1,2,3,4,6,7,12,12b-octahydro-12-methyl-	EtOH	230(4.60),278s(3.86), 285(3.90),292s(3.86)	94-0622-65
	EtOH	230(4.60),278s(3.86), 285(3.90),292s(3.86)	94-1231-65
Pyridine, 2-[p-(diethylamino)styryl]-, cis	aq. HCl	307(4.00)	23-1345-65
	H_2O	328(4.04)	23-1345-65
	50% MeOH	333(4.06),353(4.06)	23-1345-65
	50% MeOH-HCl	308(4.02)	23-1345-65
trans	aq. HCl	326(4.51)	23-1345-65
	H_2O	368(4.36)	23-1345-65
	50% MeOH	376(4.51)	23-1345-65
	50% MeOH-HCl	327(4.51)	23-1345-65
5,6-Secoagroclavine, 5(10)-dehydro-6-methyl-	EtOH	211(4.42),234(4.02), 342(3.77)	39-1858-64

Compound	Solvent	$\lambda_{max}(\log \epsilon)$	Ref.
Uleine, N-demethyl-	EtOH	305(4.24),312(4.23)	78-1717-65
$C_{17}H_{20}N_2O$			
Benzophenone, p,p'-bis(dimethylamino)-	benzene	348(4.39)	65-0094-65
	EtOH	245(4.18),365(4.51)	65-0094-65
	HOAc	373(4.47)	65-0094-65
Dasycarpidone	EtOH	237(4.15)	31-0363-64
	EtOH	237(4.15),316(4.29)	78-1717-65
Isocyanic acid, 1-naphthyl ester, adduct with N,N-dimethylisobutenyl amine	EtOH	224(4.72),303(3.94)	44-0812-65
Ketone, methyl 1,4,5,6-tetrahydro-1-(2-indol-3-ylethyl)-3-pyridyl-	EtOH	221(4.50),314(4.41)	35-1580-65
$C_{17}H_{20}N_2OS$			
Diimide, [(p-butylphenyl)thio]-(p-methoxyphenyl)-	n.s.g.	348(4.20)	77-0313-65
$C_{17}H_{20}N_2O_2$			
Indole-3-carbonitrile, 2-cyclohexyl-5,6-dimethoxy-	MeOH	214(4.6),296(4.1), 302(4.07)	44-2253-65
1,4-Naphthoquinone, 2-(methylamino)-3-(piperidinomethyl)-	EtOH	236(4.14),277(4.33), 330s(3.37),470(3.47)	39-5569-64
Nicotinic acid, 1,4,5,6-tetrahydro-1-(2-indol-3-ylethyl)-, methyl ester	EtOH	228(4.53),295(4.41)	35-5461-65
4-Piperidone, 1-[2-(2-acetylindol-3-yl)ethyl]-	MeOH	233(4.24),309(4.29)	24-2579-65
$C_{17}H_{20}N_2O_4$			
Cyclohexaneacrylonitrile, α-(4,5-dimethoxy-2-nitrophenyl)-	MeOH	248(4.09)	44-2253-65
Dipyrromethene, 4,4'-diacetyl-3,3',5,5'-tetramethyl-	EtOH	248(--),280(--), 424(--)	12-0363-65
	EtOH-HCl	232(--),284(--), 472(--)	12-0363-65
Dipyrromethene-4,4'-dicarboxylic acid, 3,3'-dimethyl-, diethyl ester	EtOH	219(4.44),263(3.84), 434(4.38)	12-0363-65
	EtOH-HCl	211(4.43),286(3.70), 353(3.47),444(4.81)	12-0363-65
	CCl_4	437(4.51)	12-0363-65
Glycine, N-(5,5-dimethyl-2,3-dioxocyclohexylidene)-, benzyl ester, 2-oxime	$CHCl_3$	332(3.83)	25-1183-65
Indole-2-carboxylic acid, 1-(2-cyanoethyl)-3-(hydroxymethyl)-5-methoxy-6-methyl-, ethyl ester	MeOH	213(4.48),305(4.31)	35-4612-64
$C_{17}H_{20}N_2O_4S$			
1,4-Thiazepine-3-carboxylic acid, 2,3,4,5-tetrahydro-2,2-dimethyl-5-oxo-6-(phenylacetamido)-, methyl ester	EtOH	235(3.98),305(3.72)	35-5307-64
	EtOH	235(3.98),305(3.72)	88-1465-64
$C_{17}H_{20}N_2O_5$			
Radiatine, oxime	EtOH	271(3.94),310(3.70)	95-0615-65
$C_{17}H_{20}N_2O_6S$			
Penicillenic acid, 2,6-dimethoxyphenyl-	MeOH	333(4.37)	44-1826-64

Compound	Solvent	$\lambda_{max}(\log \epsilon)$	Ref.
1,4-Thiazepine-3-carboxylic acid, 2,3,4,5-tetrahydro-2,2-dimethyl-5-oxo-6-(phenylacetamido)-, 1,1-dioxide, methyl ester	EtOH	261(4.12)	35-5307-64
$C_{17}H_{20}N_2S$			
Benzophenone, 4,4'-bis(dimethylamino)thio-	benzene	311(3.55),323(3.55), 430(4.71),573(3.08)	88-0073-65
Diimide, [(p-butylphenyl)thio]-p-tolyl-	n.s.g.	339(4.10)	77-0313-65
$C_{17}H_{20}N_4O_2$			
Alanine, 3-[p-(p-dimethylamino-phenylazo)phenyl]-	EtOH	250(4.1),300s(3.8), 400(4.4)	94-1375-64
Hydrazine, 1-phenyl-2-carbethoxy-2-(α-phenylazo)ethyl]-	MeOH	406(2.19)	89-0684-64
$C_{17}H_{20}N_4O_3$			
Leucoflavin, 5-acetyl-1,3,7,8,10-pentamethyl-	MeOH	214(4.36),250s(4.18), 308(3.89)	33-1354-64
Lumiflavin, 5-acetyl-3,4-dimethyl-1,5-dihydro-	MeOH	216(4.34),242(4.31), 308(3.81)	33-1354-64
$C_{17}H_{20}N_4O_4$			
2-Cyclohexenone, 3-methyl-4-(3-butenyl)-, 2,4-dinitrophenylhydrazone	EtOH	254(4.25),287s(4.05), 384(4.40)	35-5148-65
Cyclopentanone, 2-(1-methylpent-4-en-1-ylidene)-, 2,4-dinitrophenylhydrazone	$CHCl_3$	293(4.37)	88-2791-64
isomer	$CHCl_3$	291(4.37)	88-2791-64
4a,8a-Methanonaphthalen-2(1H)-one, hexahydro-, 2,4-dinitrophenylhydrazone, cis	$CHCl_3$	366(4.31)	35-0275-65
Pentane, 3-(4-oxocyclohex-2-en-1-ylidene)-, 2,4-dinitrophenylhydrazone	CH_2Cl_2	265(4.22),316(4.14), 401(4.57)	5-0028-64D
$C_{17}H_{20}N_4O_6$			
Ketone, 4-ethylenedioxycyclohex-1-en-1-yl ethyl, 2,4-dinitrophenylhydrazone	MeOH	254(4.21),379(4.44)	5-0036-64D
Mitomycin C, acetate	MeOH	218(4.34),358(4.34)	87-0001-65
Riboflavine	98% dioxan	224(4.4),269(4.5), 352(3.9),445(4.1)	31-0189-65
$C_{17}H_{20}N_4O_7$			
Theophylline, 7-(2,3-didehydro-2,3-dideoxy-D-glucopyranosyl)-, diacetate	pH 1	273(3.89)	35-1252-64
	pH 11	231(3.63),273(3.95)	35-1252-64
$C_{17}H_{20}N_6$			
Biguanide, 5-(isopropylidene-amino)-1,2-diphenyl-	EtOH	274s(4.42)	39-0932-65
$C_{17}H_{20}N_6O$			
Tetraethylammonium (1,2,3,4,4-pentacyano-1,3-butadienyl)oxide	MeCN	266(3.60),424(4.40), 442(4.46)	35-2898-64
$C_{17}H_{20}O$			
Azulene, 3-acetyl-1-ethyl-4,6,8-trimethyl-	hexane	556(2.78)	65-0894-64
	EtOH	249(4.38),321(4.42), 395(3.87),536(2.89)	65-0894-64

Compound	Solvent	$\lambda_{max}(\log \epsilon)$	Ref.
Azulene, 1-isobutyryl-4,6,8-trimethyl-	hexane	531(2.78)	65-0894-64
	EtOH	519(2.88)	65-0894-64
Benzhydrol, o-ethyl-, ethyl ether	isooctane	250s(2.50),254s(2.61), 260(2.71),266(2.69), 269s(2.55),273(2.47), 308(1.10),320(0.90)	35-0086-65
Cyclohexene, 4-benzyloxy-1-phenyl-	EtOH	247(4.3)	39-5189-65
2,8,10,16-Heptadecatetraene-4,6-diyn-1-ol	ether	245(4.48),266(4.45), 280(4.21),296(4.44), 315(4.57),337(4.43)	24-3010-65
1,9,16-Heptadecatriene-4,6-diyn-3-one, cis	ether	209(4.25),246(3.58), 259(3.79),274(4.02), 291(3.96)	24-3010-65
Ketone, hexyl 1-naphthyl	EtOH	216(4.67),298(3.81)	39-2072-65
Ketone, hexyl 2-naphthyl	EtOH	240(4.71),291(3.96)	39-2072-65
2-Propen-1-one, 1-(3,4-dimethyl-3-cyclohexenyl)-3-phenyl-	EtOH	292(4.33)	7-0143-65
$C_{17}H_{20}O_2$			
1-Naphthaleneacetic acid, pentyl ester	EtOH	225(4.96),281(3.86)	39-2072-65
2-Naphthaleneacetic acid, pentyl ester	EtOH	224(5.03),276(3.73)	39-2072-65
2(3H)-Naphthalenone, 4,4a,5,6,7,8-hexahydro-6-(p-methoxyphenyl)-	EtOH	229(4.36)	87-0519-64
1-Naphthoic acid, hexyl ester	EtOH	220(5.65),296(3.81)	39-2072-65
2-Naphthoic acid, hexyl ester	EtOH	238(4.88),280(3.90)	39-2072-65
Phenanthrene, 1-acetyl-1,2,3,4,9,10-hexahydro-7-methoxy-	MeOH	274(4.19)	44-2527-64
Phenanthrene, 2α-acetyl-1,2,3,9,10,10aβ-hexahydro-7-methoxy-	EtOH	263(4.33),296(3.56)	44-1213-65
2(3H)-Phenanthrone, 4,4a,9,10-tetrahydro-6-methoxy-1,4a-dimethyl-	n.s.g.	229(4.12),247(4.23)	39-4521-64
$C_{17}H_{20}O_3$			
Anthracene-9β,10α-diol, 1,4,4aα,9,-9aα,10-hexahydro-9α-methyl-, 10-acetate	EtOH	261(2.37)	70-1024-64
Anthracene-9β,10β-diol, 1,4,4aα,9,-9aα,10-hexahydro-9α-methyl-, 10-acetate	EtOH	261(2.39)	70-1024-64
Cacalol, acetate	EtOH	218(4.43),255(4.08), 280(3.32),292(3.12)	78-2331-64
Fluorene-8a(4bH)-acetic acid, 5,6,7,8-tetrahydro-1-methyl-7-oxo-, methyl ester	EtOH	265(2.69),273(2.66)	44-0074-64
1H-9,9b-(1-Oxopropano)dibenzofuran, 6-ethoxy-2,3,4,4a-tetrahydro-	EtOH	238(4.32),286(4.02), 318(3.85)	94-1012-64
2α-Phenanthrenecarboxylic acid, 1,2,3,9,10,10aβ-hexahydro-7-methoxy-, methyl ester	EtOH	263(4.30),296(3.51)	44-1213-65
$C_{17}H_{20}O_3S$			
Cinnamic acid, β-[(2-oxocyclohexyl)-thio]-, ethyl ester	ether	210(4.13),278(3.96)	24-2109-64
$C_{17}H_{20}O_4$			
Anthracene-9β,10β-diol, 2,3α-epoxy-1,2,3,4,4aα,9,9aα,10-octahydro-9α-methyl-, 10-acetate	EtOH	260(2.32)	70-1024-64
Anthracene-4β,9β-diol, 1,4,4aα,9,9aα,-10-hexahydro-5,10β-(isopropylidene-dioxy)-	EtOH	278(3.38),286(3.38)	65-2563-64

Compound	Solvent	$\lambda_{max}(\log \epsilon)$	Ref.
Indene-3-propionic acid, 2-(1-carboxyethyl)-, dimethyl ester	EtOH	258(4.10)	42-0479-64
Methane, bis(3,4-dimethoxyphenyl)-	hexane	204(4.78),230(4.20), 282(3.79)	39-1685-65
Pentane, 1,5-bis(2,5-dihydroxyphenyl)-	95% THF	299(3.89)	44-2602-65
1α-Phenanthrenecarboxylic acid, 1,2,3,-4,4aβ,9,10,10aβ-octahydro-7-methoxy-4-oxo-, methyl ester	EtOH	219(3.96),279(3.26), 285(3.21)	44-2527-64
1β-Phenanthrenecarboxylic acid, 1,2,3,-4,4aα,9,10,10aβ-octahydro-7-methoxy-4-oxo-, methyl ester	EtOH	220s(3.99),278(3.29), 284(3.26)	44-2527-64
1β-Phenanthrenecarboxylic acid, 1,2,3,-4,4aβ,9,10,10aβ-octahydro-7-methoxy-4-oxo-, methyl ester	EtOH	220(3.99),279(3.31), 287s(3.26)	44-2527-64
$C_{17}H_{20}O_5$			
Anthrone, 4β-acetoxy-1,2,3,4,4aα,9aα-hexahydro-10β-hydroxy-5-methoxy-	EtOH	224(4.40),255(4.02), 310(3.57)	65-2563-64
6-Epiartemisin, acetate	EtOH	243(4.09)	12-0543-65
Santonin, acetate	EtOH	241(4.08)	12-0543-65
$C_{17}H_{20}O_5S$			
Cinnamic acid, β-[(2-oxocyclohexyl)-sulfonyl]-, ethyl ester	ether	211(4.18)	24-2109-64
$C_{17}H_{20}O_6S_4$			
β-L-Idofuranose, 1,2-O-isopropylidene-5,6-dithio-, cyclic 5,6-trithiocarbonate, 3-p-toluenesulfonate	MeOH	316(4.14)	44-3071-65
$C_{17}H_{20}O_7$			
Glauconic acid, dihydro-	EtOH	230(3.90)	39-1772-65
$C_{17}H_{21}BrO_5$			
Geigerin acetate, 1-bromo-	EtOH	234(4.04)	39-2518-64
$C_{17}H_{21}ClN_2O_4$			
1-Ethyl-2,3,4,6,7,12-hexahydro-1H-indolo[2,3-a]quinolizin-5-ium perchlorate	EtOH-HCl	247(4.02),251s(4.00), 353(4.39)	35-1580-65
$C_{17}H_{21}ClO_4$			
3β,9β,10β-Anthracenetriol, 2α-chloro-1,2,3,4,4aα,9,9aα,10-octahydro-9α-methyl-, 3-acetate	EtOH	260(2.35)	70-1024-64
$C_{17}H_{21}ClO_5$			
1-Ethyldihydro-3-(1-hydroxyethylidene)-4,6,8-trimethylazulenylium perchlorate	MeCN	254(4.50),318(4.35), 386(3.94)	65-0894-64
$C_{17}H_{21}N$			
Aniline, N,N-dimethyl-o-(2-phenylpropyl)-	EtOH	207(4.35),251(3.64)	39-0249-64
4a-Azachrysene, 1,2,3,4,4a,5,6,11,12,12a-decahydro-	EtOH	236(3.93),313(4.01)	44-1834-64
perchlorate	EtOH	251(3.91)	44-1834-64
4a-Azachrysene, 1,2,3,4,5,5a,6,11,11a,12-decahydro-	EtOH	265(2.83),272(2.85)	44-1834-64
perchlorate	EtOH	264(3.08),272(3.08)	44-1834-64

Compound	Solvent	$\lambda_{max}(\log \epsilon)$	Ref.
Pyridine, 2-(6-phenylhexyl)-	EtOH	257(3.59),262(3.65), 268(3.52)	94-1344-64
$C_{17}H_{21}NO$			
1-Hexanone, 6-(methylamino)- 1-α-naphthyl-	n.s.g.	216(4.65),238s(4.12), 302(3.74)	73-3111-65
perchlorate	n.s.g.	216(4.55),238s(4.10), 304(3.74)	73-3111-65
$C_{17}H_{21}NO_2$			
Benzo[f]quinoline-2-carboxylic acid, 1,2,3,4,5,6-hexahydro-4-methyl-, ethyl ester	EtOH	224.5(4.16)	94-0420-65
Cyclohexene, 2-benzoyl-1-morpholino-	C_6H_{12}	240(4.10),285(3.30), 345(3.19)	1-2139-65
	EtOH	246(4.07),385(3.60)	1-2139-65
Diethylamine, 2-(p-hydroxyphenyl)-1-methyl-2'-phenoxy-, hydrochloride	EtOH-H_2SO_4	269(3.42),276(3.45)	54-0521-65
(+)	EtOH-H_2SO_4	269(3.41),276(3.45)	54-0521-65
(-)	EtOH-H_2SO_4	269(3.42),276(3.46)	54-0521-65
Diethylamine, 2-(p-hydroxyphenyl)-1'-methyl-2'-phenoxy-, hydrochloride	EtOH-H_2SO_4	269(3.41),276(3.45)	54-0521-65
(+)	EtOH-H_2SO_4	269(3.42),276(3.46)	54-0521-65
	EtOH-H_2SO_4	269(3.43),276(3.47)	54-0521-65
$C_{17}H_{21}NO_3$			
1H-Benzo[c]quinolizine-2-carboxylic acid, 2,3,4,4a,5,6-hexahydro-2-methyl-3-oxo-, ethyl ester	EtOH	210(4.37),254(4.21), 294(3.47)	78-2961-65
Benzyl alcohol, p-hydroxy-α-[[1-(methyl-2-phenoxyethyl)amino]methyl]-	EtOH-H_2SO_4	269(3.44),276(3.46)	54-0521-65
Carbostyril, 4,8-dimethoxy-1-methyl-3-(3-methylbut-2-enyl)-	MeOH	239(4.35),258(4.33), 286(3.86),294(3.85), 332(3.53)	39-0438-64
Crinine methine, tetrahydrooxo-	EtOH	242(3.62),290(3.58)	94-0427-65
Lycoramine, deoxyoxo-	EtOH	260(4.05),297(3.71)	94-0696-64
Phenethylamine, 4-(benzyloxy)-β,3-dimethoxy-	EtOH-HCl	244(4.33),302(3.98), 348(3.96)	33-1945-65
	EtOH	226(4.44),270(3.87), 304(3.82)	33-1945-65
	EtOH-NaOH	269(4.00),304(3.87)	33-1945-65
hydrochloride	EtOH-HCl	243(4.33),303(3.96), 352(3.91)	33-1945-65
	EtOH	243(4.27),303(3.94), 352(3.86)	33-1945-65
	EtOH-NaOH	268(3.96),306(3.83)	33-1945-65
Propanol, 1-(p-hydroxyphenyl)-2-(2-phenoxyethylamino)-, hydrochloride	EtOH-H_2SO_4	275.5(3.47)	54-0521-65
(+)	EtOH-H_2SO_4	276(3.47)	54-0521-65
(-)	EtOH-H_2SO_4	276(3.48)	54-0521-65
$C_{17}H_{21}NO_4$			
Amaryllisine	EtOH	283(3.77)	35-4976-64
	EtOH-NaOH	252(3.82),297(3.61)	35-4976-64
$C_{17}H_{21}NO_5$			
Nerbowdine	MeOH	287(3.21)	83-0704-65
$C_{17}H_{21}N_3$			
Azobenzene, 4-(dimethylamino)-2,2',6'-trimethyl-	EtOH	385(4.32)	78-1547-64
	N HCl	320(3.90),470(4.30)	78-1547-64

Compound	Solvent	λ_{max}(log ε)	Ref.
p-Phenylenediamine, N'-[p-(dimethylamino)benzylidene]-N,N-dimethyl-	isooctane	233(4.05),319(4.16), 370(4.28)	9-0091-65
	EtOH	239(4.11),325(4.13), 378(4.42)	9-0091-65
$C_{17}H_{21}N_3O_2$			
1-Pyrroline-4-carboxylic acid, 3,5-dimethyl-2,5-di-2-pyrrolyl-, ethyl ester	EtOH-HCl	209(4.05),326(4.37)	39-5999-64
	EtOH	208(3.98),286(4.22)	39-5999-64
$C_{17}H_{21}N_3O_3$			
2-Imidazoline-$\Delta^{5,\alpha}$-acetic acid, 2-amino-4-oxo-α-(4-phenylbutyl)-, ethyl ester	pH 1	242(3.83),309(4.17)	4-0162-65
	H_2O	279(4.18),325(4.03)	4-0162-65
	pH 13	279(4.07),330(4.02)	4-0162-65
4-Pyrimidinecarboxylic acid, 2-amino-6-hydroxy-5-(4-phenylbutyl)-, ethyl ester	pH 1	288(3.84)	4-0162-65
	pH 7	303(3.77)	4-0162-65
	pH 13	294(3.67)	4-0162-65
$C_{17}H_{21}N_3O_3S$			
Azobenzene, 4-(dimethylamino)-2,2',6'-trimethyl-4'-sulfo-	EtOH	424(4.28)	78-1547-64
	N HCl	316(3.64),470(4.49)	78-1547-64
$C_{17}H_{21}N_3O_7$			
Atropine, 4-nitro-, nitrate	MeOH	278(5.19)	83-0826-65
	acetone-MeOH-KOH	561(4.54)	83-0826-65
	acetone-MeOH-MeONa	565(4.48)	83-0826-65
$C_{17}H_{21}N_3S$			
1-Pyrrolidinethiocarboxanilide, 2-(3,5-dimethyl-2-pyrrolyl)-	EtOH	207(4.40),248(4.25), 265(4.14)	39-0893-64
$C_{17}H_{21}N_4O_9P$			
Riboflavine, 5'-phosphate (also other spectra)	pH 5.4	375(4.0),450(4.1)	11-0281-64
$C_{17}H_{21}N_5O_8$			
6-Purinone, 2-(methylamino)-9-(2,3,5-tri-O-acetyl-β-D-ribofuranosyl)-	pH 1	259(4.15),282s(3.93)	35-3752-65
	pH 11	257(4.14),268s(4.04)	35-3752-65
	MeOH	252(4.25),275(4.01)	35-3752-65
$C_{17}H_{21}O_7PS$			
$\Delta^{3,\alpha}$-Chromanglycolic acid, 2-oxo-, ethyl ester, O-ester with O,O-diethyl phosphorothioate	MeOH	209(4.13)	5-0010-65E
$C_{17}H_{21}O_8P$			
$\Delta^{3,\alpha}$-Chromanglycolic acid, 2-oxo-, ethyl ester, diethyl phosphate	MeOH	200(4.35),272(3.31)	5-0010-65E
$\Delta^{4,\alpha}$-Isochromanglycolic acid, 3-oxo-, ethyl ester, diethyl phsophate	MeOH	200(4.33),300(3.60)	5-0010-65E
$C_{17}H_{22}$			
Gona-1,3,5(10)-triene	EtOH	214(3.83),217s(3.77), 259s(2.43),266(2.62), 273(2.62)	5-0200-65D
1,3,5,7,9,11,13-Pentadecaheptaene, 2,14-dimethyl-	isooctane	343(4.56),361(4.88), 379(5.13),402(5.17)	35-5075-65

Compound	Solvent	$\lambda_{max}(\log \epsilon)$	Ref.
$C_{17}H_{22}ClNO_7$			
3-Carboxy-1,2,3,4,6,7-hexahydro-9-methoxybenzo[a]quinolizinium perchlorate, ethyl ester	EtOH	230(3.77),237(3.87), 244s(3.75),323(4.28)	23-1527-65
$C_{17}H_{22}ClN_2O_4P$			
Phosphoric acid, p-chlorophenyl 2-pyridyl ester, compound with cyclohexylamine	n.s.g.	262(3.64)	24-1031-65
$C_{17}H_{22}INO_4$			
6-Methoxyribalinium iodide	50% EtOH	222(4.67),246(4.48), 298(3.96),330(3.76)	78-0909-65
$C_{17}H_{22}N_2O$			
Dasycarpidol	EtOH	220(4.54),282(3.89), 290(3.81)	31-0363-64
	EtOH	220(4.54),282(3.89), 290(3.81)	78-1717-65
$C_{17}H_{22}N_2O_2$			
Dipyrromethene-3-carboxylic acid, 3',4,4',5,5'-pentamethyl-,	EtOH-HCl	263(3.83),370s(3.89), 470(4.91)	12-0363-65
ethyl ester	EtOH	212(4.32),226s(4.23), 260s(3.99),407(4.48)	12-0363-65
Dipyrromethene-3'-carboxylic acid,	EtOH-HCl	470(4.91)	12-0363-65
3,4,4',5,5'-pentamethyl-, ethyl	EtOH	407(4.91)	12-0363-65
ester	CCl_4	416(4.46)	12-0363-65
Dipyrromethene-4'-carboxylic acid, 3,3',4,5,5'-pentamethyl-, ethyl	EtOH-HCl	226(4.04),263(3.68), 470(4.86)	12-0363-65
ester	EtOH	225(4.01),262(3.73), 409(4.47)	12-0363-65
	CCl_4	420(4.52)	12-0363-65
Dipyrromethene-5'-carboxylic acid,	EtOH-HCl	251(4.12),467(4.60)	12-0363-65
3,3',4,4',5-pentamethyl-, ethyl	EtOH	253(3.93),410(4.35)	12-0363-65
ester	CCl_4	417(4.40)	12-0363-65
$C_{17}H_{22}N_2O_3$			
Coronopilic acid, methyl ester, pyrazoline derivative	n.s.g.	309(4.19)	44-2553-64
Tryptamine, 5-acetyl-2-carbethoxy-N,N-dimethyl-	EtOH-HCl	268(4.72),314(3.93)	87-0141-64
$C_{17}H_{22}N_2O_4$			
3-Azabicyclo[3.3.1]nonan-9-ol, 3,9-dimethyl-, p-nitrobenzoate	EtOH	259(4.13)	44-3634-65
isomer	EtOH	259(4.15)	44-3634-65
$C_{17}H_{22}N_2O_4S$			
1,4-Thiazepine-3-carboxylic acid, hexahydro-2,2-dimethyl-5-oxo-6-(2-phenyl-	EtOH	252(2.33),258(2.34), 264(2.22),268(2.00)	35-5307-64
acetamido)-, methyl ester	EtOH	252(2.33),258(2.34), 264(2.22),268(2.01)	88-1465-64
$C_{17}H_{22}N_4O_4$			
Bicyclo[5.3.0]decan-9-one, 1β-methyl-, trans, 2,4-dinitrophenylhydrazone	$CHCl_3$	363(4.24)	35-1755-64
Cyclobutanone, 2-tert-butyl-4-isopropylidene-, 2,4-dinitrophenylhydrazone	$CHCl_3$	394(4.34)	22-2747-65

Compound	Solvent	λ_{max}(log ε)	Ref.
2-Cyclohexen-1-one, 2-tert-butyl-5-methyl-, 2,4-dinitrophenylhydrazone	$CHCl_3$	383(4.37)	35-0078-64
2-Cyclohexen-1-one, 4-(1-ethylpropyl)-, 2,4-dinitrophenylhydrazone	MeOH	252(4.20),285s(--), 378(4.43)	5-0028-64D
	CH_2Cl_2	255(4.20),287s(--), 383(4.44)	5-0028-64D
2(1H)-Naphthalenone, octahydro-8a-methyl-, 2,4-dinitrophenylhydrazone	$CHCl_3$	368(4.35)	35-1755-64
$C_{17}H_{22}N_4O_5$			
5-Pyrimidinepropionaldehyde, 2-amino-4-hydroxy-6-(p-nitrophenyl)-, diethyl acetal	pH 1	225(4.21),255(4.19), 300s(3.98)	87-0283-65
	pH 7	275(4.17)	87-0283-65
	pH 13	275(4.21)	87-0283-65
$C_{17}H_{22}N_4O_{10}$			
Atropine, 4-nitro-, nitric acid ester (also other solvent mixtures)	acetone-MeOH-KOH	561(3.65)	83-0826-65
	acetone MeOH-KCN	541(3.88)	83-0826-65
$C_{17}H_{22}O_2$			
3-Cyclohexene-1-carboxylic acid, 6-ethyl-3,4-dimethyl-1-phenyl-, cis	EtOH	247s(2.33),253(2.32), 259(2.37),266(2.24), 269s(1.96)	1-1951-65
trans	EtOH	220s(3.73),253(2.33), 259(2.36),266(2.22), 269s(1.85)	1-1951-65
Phenanthrene, 1α-acetyl-1,2,3,4,4aβ,-9,10,10aβ-octahydro-7-methoxy-	EtOH	220(3.94),278(3.35), 286(3.34)	44-2527-64
1β-acetyl isomer	EtOH	220(3.95),279(3.36), 288(3.33)	44-2527-64
$C_{17}H_{22}O_3$			
1H-Benz[e]indene-3,7-diol, hexahydro-3a,6-dimethyl-, 3-acetate. trans	EtOH	280(3.15)	78-2487-64
1H-Benz[e]inden-3-ol, hexahydro-7-methoxy-3a-methyl-, acetate, trans	MeOH	279(3.36)	13-0185s-65
3-Cyclohexene-1-carboxylic acid, 3-ethyl-4-(p-methoxyphenyl)-2-methyl-	EtOH	230(4.00)	70-2246-64
3(2H)-Dibenzofuranone, 1,4,4a,9b-tetrahydro-8-methoxy-4,4,7,9b-tetramethyl-	EtOH	231(3.69),304(3.67)	39-2932-64
1α-Phenanthrenecarboxylic acid, 1,2,3,4,4aβ,9,10,10aβ-octahydro-7-methoxy-, methyl ester	EtOH	220(3.90),279(3.29), 287(3.27)	44-2527-64
1β-Phenanthrenecarboxylic acid, 1,2,3,4,4aα,9,10,10aβ-octahydro-7-methoxy-, methyl ester	EtOH	220(3.90),278(3.29), 286(3.26)	44-2527-64
$C_{17}H_{22}O_4$			
Illudin M, acetate	EtOH	226(4.12),243s(4.07), 312(3.56)	35-1594-65
Mansonone B, acetate	EtOH	263(4.1),339(2.53), 435(1.8)	88-4857-65
2α-Phenanthrenecarboxylic acid, 1,2,3,4,4aβ,9,10,10aβ-octahydro-7-methoxy-, methyl ester	EtOH	221(3.91),278(3.26), 287(3.22)	44-1213-65

Compound	Solvent	$\lambda_{max}(\log \epsilon)$	Ref.
Spiro[benzofuran-2(3H),1'-cyclohexane]-2',4,6'(5H)-trione, 6,7-dihydro-4',4',6,6-tetramethyl-	EtOH	267(4.07)	44-1251-65
$C_{17}H_{22}O_5$			
5-Azuleneacetic acid, 3,3a,4,5,6,7,8,8a-octahydro-2,6-dihydroxy-α,3a,8-trimethyl-3-oxo-, γ-lactone, acetate	EtOH	231(3.94),315(1.78)	44-1549-64
	HCl	246(--)	44-1549-64
	KOH	300(3.80)	44-1549-64
5H-Benzocycloheptene-9-propionic acid, 6,7-dihydro-1,2,3-trimethoxy-	EtOH	218(4.43),250(3.98), 282s(3.08)	44-1752-65
1-Epigeigerin, acetate	EtOH	235(4.15)	39-2518-64
Geigerin, deoxy-, 2-acetoxy-	EtOH	240(4.15)	39-2518-64
Isotenulin	EtOH	225(3.94)	44-1549-64
Pulchellin B	EtOH	210(3.95),325(1.72)	78-0341-64
Tenulin	EtOH	226(3.93)	44-1549-64
$C_{17}H_{22}O_6$			
5H-Benzocycloheptene-5-propionic acid, 6,7,8,9-tetrahydro-5,6-dihydroxy-2,3,4-trimethoxy-, γ-lactone	EtOH	229s(4.04),273(2.85), 280(2.78)	44-1752-65
9-Cyclononene-1,2,4-tricarboxylic acid, 7,8-diethyl-2-methyl-6-oxo-, 1,2-anhydride	EtOH	227(3.94)	39-1772-65
Isoflexuosin A, dehydro-	EtOH	222.5(4.15)	78-0979-64
Isotenulin, oxide	EtOH	225(3.29)	44-1549-64
$C_{17}H_{23}BrO_3$			
Laurencin	EtOH	224(4.22),232s(4.04)	88-1091-65
$C_{17}H_{23}Cl_2O_7Rh$			
Rhodium, bis(3-chloro-2,4-pentanedionato)[3-(methoxymethyl)-2,4-pentanedionato]-	n.s.g.	265(4.84),333(4.88)	44-3216-64
$C_{17}H_{23}N$			
4a-Azachrysene, 1,2,3,4,5,6,11,12,-12a,4a,11a,5a-dodecahydro-	EtOH	267(2.70),274(2.70)	44-1834-64
$C_{17}H_{23}NO$			
Cinnamonitrile, 3,5-di-tert-butyl-4-hydroxy-	MeOH	230(4.2),315(4.3)	5-0141-65A
Isoxazole, 5-octyl-3-phenyl-	EtOH	240(4.17)	78-0817-65
$C_{17}H_{23}NO_2$			
Spiro[5H-1-benzazepine-5,1'-cyclohexan]-2(1H)-one, 3,4-dihydro-7-methoxy-1-methyl-	EtOH	246(4.12),282s(3.4)	94-1084-65
Spiro[5H-2-benzazepine-5,1'-cyclohexan]-1(2H)-one, 3,4-dihydro-7-methoxy-2-methyl-	EtOH	252(4.03)	94-1084-65
$C_{17}H_{23}NO_3$			
2H-Benzo[a]quinolizine-3-carboxylic acid, 1,3,4,6,7,11b-hexahydro-9-methoxy-, ethyl ester, cis	EtOH	220(3.99),278(3.28), 286(3.27)	23-1527-65
trans	EtOH-HCl	227(3.94),277(3.24), 284(3.24)	23-1527-65
	EtOH	220(3.97),278(3.28), 286(3.27)	23-1527-65

Compound	Solvent	λ_{max}(log ϵ)	Ref.
Carbostyril, 3-isopentyl-4,8-dimethoxy-1-methyl-	MeOH	239(4.36),258(4.34), 285(3.86),292(3.85), 333(3.53)	39-0438-64
$C_{17}H_{23}NO_4$			
Carbostyril, 3-(3-hydroxy-1-methylbutyl)-4,8-dimethoxy-1-methyl-	MeOH	240(4.39),258(4.39), 285(3.90),292(3.89), 332(3.51)	39-4190-64
Lunacridine	MeOH	239(4.38),258(4.42), 285(3.91),292(3.89), 333(3.50)	39-0438-64
$C_{17}H_{23}N_2O_4P$			
Phenyl 2-pyridyl phosphate, compound with cyclohexylamine	n.s.g.	261(3.63)	24-1031-65
$C_{17}H_{23}N_3O_2$			
1H-Indazole, 1-acetyl-3-(3-piperidinopropoxy)-	EtOH	242(4.27),287(3.79), 293(4.12),297(3.96), 305(4.27)	33-1986-64
$C_{17}H_{23}N_3O_3$			
1H-Indazole-1-carboxylic acid, 3-(2-piperidinoethoxy)-, ethyl ester	EtOH-HCl	228(4.43),233(4.42), 284(3.72),289(3.92), 293(3.89),301(4.01)	33-1986-64
1-Indazolinecarboxylic acid, 3-oxo-2-(2-piperidinoethyl)-, ethyl ester	EtOH-HCl	223(4.34),239(4.11), 306(3.76)	33-1986-64
$C_{17}H_{23}N_3O_4$			
Succinamic acid, N-amidino-3-oxo-2-(4-phenylbutyl)-, ethyl ester	H_2O	234(4.48)	4-0162-65
	pH 13	233(4.44)	4-0162-65
$C_{17}H_{23}N_3O_5$			
Cyclopentanecarboxylic acid, 1-[[2-(p-nitrobenzamido)ethyl]amino]-, ethyl ester	MeOH	262(4.09)	87-0107-65
$C_{17}H_{23}N_5O_{12}$			
4-Pteridinol, 2-amino-6(7)-(1,2,3,4,5-pentahydroxypentyl)-, 2'-β-glucuronide	pH 1	255(4.3),320(4.0)	5-0221-65I
	pH 13	260(4.6),370(4.3)	5-0221-65I
$C_{17}H_{23}N_7O_5$			
Cytomycin	pH 1	274(4.08)	88-1411-65
	pH 13	266(3.84)	88-1411-65
$C_{17}H_{24}ClNO_3$			
N-(2-Ethyl-3-oxobutyl)-3,4-dihydro-6,7-dimethoxyisoquinolinium chloride	n.s.g.	247(4.27),311(4.06), 364(4.04)	24-0557-65
$C_{17}H_{24}N_2$			
Dipyrromethene, 3,3'-diethyl-4,4',5,5'-tetramethyl-	EtOH	226(3.80),320(3.52), 443(4.59)	12-0363-65
	EtOH-N HCl	230(3.92),280(3.40), 363(3.79),478(4.90)	12-0363-65
	CCl_4	454(4.50)	12-0363-65
Dipyrromethene, 4,4'-diethyl-3,3',5,5'-tetramethyl-	EtOH	225(3.80),286s(3.28), 322(3.50),447(4.53)	12-0363-65
	EtOH-N HCl	229(4.04),280(3.33), 360(3.82),483(4.86)	12-0363-65
	CCl_4	451(4.40)	12-0363-65

Compound	Solvent	λ_{max}(log ε)	Ref.
Dipyrromethene, 5,5'-diethyl-3,3',4,4'-tetramethyl-	EtOH	225(3.94),448(4.43)	12-0363-65
	EtOH-N HCl	227s(4.10),273(3.56),366(3.81),484(4.80)	12-0363-65
	CCl_4	449(4.42)	12-0363-65
6,7-Secoagroclavine, 8,9-dihydro-6-methyl-	EtOH	224(4.36),283(3.83)	39-1858-64
$C_{17}H_{24}N_2O_2$			
2,3-Diaza-19-norandrosta-3,5(10)-dien-1-one, 17-hydroxy-4-methyl-	MeOH	230s(3.49),286(3.58)	87-0590-64
2,3-Diaza-19-norandrost-3-ene-1,17-dione, 4-methyl-	MeOH	244(3.78)	87-0590-64
$C_{17}H_{24}N_2O_4$			
11bH-Benzo[a]quinolizine-2,9(?)-diol, 10(?)-methoxy-3-diethylcarbamoyl-1,2,3,4,6,7-hexahydro-	MeOH-HCl	287(3.55)	87-0635-64
	MeOH-NaOH	294(3.59)	87-0635-64
$C_{17}H_{24}N_4O_2S_2$			
4-Thiazoline-2-thione, 3-[(4-amino-2-methyl-5-pyrimidinyl)methyl]-4-methyl-5-[2-[(tetrahydropyran-2-yl)-oxy]ethyl]-	EtOH	232(4.09),278(3.76),322(4.17)	94-0558-64
$C_{17}H_{24}N_4O_4$			
Bicarbamic acid, [1-(phenylazo)cyclopentyl]-, diethyl ester	EtOH	268(4.02),400(2.20)	7-0441-65
Cyclohexanone, 2-tert-butyl-5-methyl-, 2,4-dinitrophenylhydrazone	$CHCl_3$	368(4.40)	35-0078-64
Cyclohexanone, 4-(1-ethylpropyl)-, 2,4-dinitrophenylhydrazone	MeOH	230(4.15),364(4.32)	5-0028-64D
	CH_2Cl_2	368(4.36)	5-0028-64D
2-Nonenal, 2,4-dimethyl-, 2,4-dinitrophenylhydrazone	$CHCl_3$	382.5(4.49)	22-0161-64
$C_{17}H_{24}O$			
2,5-Cyclohexadien-1-one, 4-methyl-2,4-bis(3-methyl-2-butenyl)-	EtOH	236(4.04)	33-0094-65
Phenanthrene-1α-methanol, 1,2,3,4,9,-10,11,12-octahydro-1β,12β-dimethyl-	n.s.g.	267(2.86)	39-4521-64
$C_{17}H_{24}OS_2$			
2,5-Cyclohexadien-1-one, 2,6-di-tert-butyl-4-(1,3-dithiolan-2-ylidene)-	MeOH	257(3.92),448s(4.34),462(4.41),480s(4.28)	5-0037-65D
$C_{17}H_{24}O_2$			
Cinnamaldehyde, 3,5-di-tert-butyl-4-hydroxy-	MeOH	230(4.2),340(4.5)	5-0141-65A
Hexanoic acid, 2,2-dimethyl-3-methylene-4-phenyl-, ethyl ester	EtOH	259(2.56)	39-5704-64
Phenanthrene-1α-methanol, 1,2,3,4,9,-10,11a,12-octahydro-2β-hydroxy-1β,12β-dimethyl-	n.s.g.	265(2.8)	39-4521-64
$C_{17}H_{24}O_3$			
3-Buten-2-one, 4-(2,6-dimethoxy-4-pentylphenyl)-	MeOH	210(4.06),325(4.55)	5-0122-65E
19-Nor-6-oxatestosterone	n.s.g.	259(4.34)	31-0418-64
$C_{17}H_{24}O_4$			
Clov-5-ene-3-carboxylic acid, 9-hydroxy-7-oxo-, methyl ester	MeOH	244(4.00)	39-1338-65

Compound	Solvent	λ max(log ε)	Ref.
Helicobasidin, O,O'-dimethyl-	EtOH	284(3.99),381(2.61)	94-0236-64
4-Pentynoic acid, 5-(1-cyclohexenyl)-3-methyl-2-carbethoxy-, ethyl ester	EtOH	228(4.14)	22-2541-64
Trichodermin	EtOH	205(3.38)	1-1088-65
$C_{17}H_{24}O_5$			
Flexuosin A, didehydrodeoxy-	EtOH	218(4.18),320(1.51)	78-0979-64
$C_{17}H_{24}O_6$			
Isoflexuosin A	EtOH	223(4.17)	78-0979-64
$C_{17}H_{25}ClN_2$			
Trimethyl[2-(1,2,3,4-tetrahydro-4aH-carbazol-4a-yl)ethyl]ammonium chloride	MeOH-HCl	232(3.9),237s(3.9),281(3.8)	78-1737-64
	MeOH	258(3.8)	78-1737-64
	MeOH-NaOH	258(3.86)	78-1737-64
$C_{17}H_{25}FO_2$			
2,4-Pentadienoic acid, 2-fluoro-3-methyl-5-(2,6,6-trimethyl-1-cyclohexenyl)-, ethyl ester	MeOH	263(4.04),302(4.07)	5-0001-64D
$C_{17}H_{25}NO$			
2-Isoxazoline, 5-octyl-3-phenyl-	EtOH	265(4.12)	78-0817-65
Spiro[5H-1-benzazepine-5,1'-cyclohexane], 1,2,3,4-tetrahydro-7-methoxy-1-methyl-	EtOH	254(3.87),299(3.44)	94-1084-65
$C_{17}H_{25}NO_3$			
1,4-Benzodioxan, 6-allyl-5-(2-diethylaminoethoxy)-	EtOH	273(3.13)	65-2994-64
$C_{17}H_{25}NO_4$			
1-Isoquinolinepropionic acid, 1,2,3,4-tetrahydro-6,7-dimethoxy-2-methyl-, ethyl ester	EtOH	227(3.92),279(3.42)	44-2486-64
$C_{17}H_{25}NO_5$			
Succinamic acid, N-(3,4-dimethoxyphenethyl)-N-methyl-, ethyl ester	EtOH	229(3.95),279(3.45)	44-2486-64
$C_{17}H_{25}NO_8S$			
α-D-Galactopyranoside, 2-acetamido-2-deoxy-, ethyl ester, 6-p-toluenesulfonate	$CHCl_3$	241(2.67),257(2.66),263(2.78),266(2.75),268(2.75),274(2.71)	44-3654-64
$C_{17}H_{25}N_3O$			
Benzamide, p-amino-N-[2-(1-carbethoxycyclopentylamino)ethyl]-	MeOH	280(4.23)	87-0107-65
$C_{17}H_{25}N_3O_3$			
Valeric acid, 5-(dimethylamino)-2-oxo-, ethyl ester, p-acetylphenylhydrazone	EtOH-HCl	236(3.98),348(4.60)	87-0141-64
$C_{17}H_{25}N_3O_4$			
Barbituric acid, 5-(1-cyclohexen-1-yl)-1,5-dimethyl-3-(morpholinomethyl)-	EtOH	224(3.8)	83-0885-65
$C_{17}H_{25}N_3O_{11}$			
Isobiuret, 1-(2,3,4,6-tetra-O-acetyl-β-D-glucopyranosyl)-4-methyl-	EtOH	221(4.26)	73-2060-64

Compound	Solvent	λ_{max}(log ε)	Ref.
$C_{17}H_{26}N_2O_2$			
2,3-Diaza-19-norandrost-3-en-1-one, 17-hydroxy-4-methyl-	MeOH	244(3.72)	87-0590-64
$C_{17}H_{26}N_2O_4$			
Pyrrole-3,4-dicarboxylic acid, 2,5-dimethyl-1-piperidino-, diethyl ester	EtOH	210(2.7),263(2.3)	6-0073-64
$C_{17}H_{26}N_4O_3S$			
Sulfanilamide, N^1-[(cyclohexylamino)-morpholinomethylene]-	EtOH	269.0(4.30)	95-0391-65
$C_{17}H_{26}N_8O_5$			
Blasticidin S	pH 1	274(4.13)	88-1411-65
	pH 13	266(3.95)	88-1411-65
$C_{17}H_{26}O$			
Longifolene, ω-acetyl-	EtOH	256(4.2)	44-2838-65
$C_{17}H_{26}OS$			
15-Thiabicyclo[10.2.1]pentadeca-12,14-dien-2-one, 13-isopropyl-	EtOH	260s(3.8),313(4.1)	70-2055-64
$C_{17}H_{26}OS_2$			
7βH-Eudesm-4-ene-3,6-dione, 3-ethylene thioketal	EtOH	253(3.86)	78-2593-64
epimer	EtOH	253(3.83)	78-2593-64
$C_{17}H_{26}O_2$			
7βH-Eudesma-3,5-diene, 3-acetoxy-	EtOH	240(4.21)	78-1455-64
β-Ionylideneacetic acid, ethyl ester	MeOH	260(4.13),302(4.14)	5-0020-64I
Phenol, 2-acetonyl-4,6-di-tert-butyl-	EtOH	283(3.44)	1-0421-64
	EtOH-NaOH	291(3.54)	1-0421-64
$C_{17}H_{26}O_2S$			
p-Toluenesulfinic acid, menthyl ester	isooctane	223(4.06),256(3.64)	35-1958-65
	isooctane-CF_3COOH	226(3.92),241(3.89), 264s(3.05),275s(2.62)	35-1958-65
	EtOH	225(4.00),247(3.76), 263s(3.30),275(2.70)	35-1958-65
(-)-S-(-)	isooctane	223(4.09),256(3.66),	35-5637-64
	EtOH	225(4.06),246(3.82), 273s(3.05)	35-5637-64
$C_{17}H_{26}O_3$			
Acetophenone, 2'-heptyl-4',6'-dimethoxy-	EtOH	266.5(3.54)	1-1677-65
7βH-Eudesm-4-en-3-one, 6β-acetoxy-	EtOH	251(4.15)	78-2593-64
$C_{17}H_{26}O_3S$			
Benzenesulfinic acid, p-methoxy-, methyl ester	EtOH	247(4.05),281s(2.94)	35-5637-64
$C_{17}H_{26}O_4$			
Benzoic acid, 2-heptyl-4,6-dimethoxy-, methyl ester	EtOH	280.0(3.40)	1-1677-65
p-Benzoquinone, 2,5-dihydroxy-3-undecyl-	EtOH	291(4.25),427(2.45)	78-2319-64
	EtOH	202(4.16),291(4.25), 427(2.45)	94-0236-64

Compound	Solvent	λ_{max}(log ε)	Ref.
$C_{17}H_{26}S_2$			
Spiro[1,3-dithiolane-2,2'(3'H)-naphthalene]-4',4'a,5',6'-tetrahydro-7'-isopropyl-1',4'aβ-dimethyl-	EtOH	244(4.07)	78-2593-64
$C_{17}H_{27}FO_2$			
2,6,10-Dodecatrienoic acid, 2-fluoro-3,7,11-trimethyl-, ethyl ester	MeOH	224(4.10)	5-0001-64D
$C_{17}H_{27}N$			
Morphinan, 1,2,3,4-tetrahydro-N-methyl-	EtOH	210(3.89)(end absorption)	44-1769-65
$C_{17}H_{27}NO_2$			
2-Cyclohexen-1-one, 5,5-dimethyl-2-pivaloyl-3-(N-pyrrolidinyl)-	EtOH	305(4.36)	44-0798-64
$C_{17}H_{27}NO_3$			
1,4-Benzodioxan, 5-(2-diethylamino-ethoxy)-6-propyl-	EtOH	275(3.03)	65-2994-64
$C_{17}H_{27}N_3OS$			
Ketone, methyl 15-thiabicyclo-[10.2.1]pentadeca-12,14-dien-13-yl, semicarbazone	EtOH	240(4.5),288(4.0)	70-2055-64
$C_{17}H_{27}N_3O_4$			
Pyrrole-3,4-dicarboxylic acid, 2,5-dimethyl-1-(4-methyl-1-piperazinyl)-, diethyl ester	EtOH	210(2.7),263(2.4)	6-0073-64
$C_{17}H_{27}O_7PS$			
Fumaric acid, hydroxy[2-(hydroxymethyl)-cyclohexyl]-, δ-lactone, ethyl ester, O-ester with O,O-diethyl phosphorothioate	MeOH	225(3.90)	5-0010-65E
$C_{17}H_{27}O_8P$			
Fumaric acid, hydroxy[2-(hydroxymethyl)-cyclohexyl]-, δ-lactone, ethyl ester, diethyl phosphate	MeOH	220(3.77)	5-0010-65E
$C_{17}H_{28}NO_5$			
2,2,6,6-Tetramethyl-1-oxylpiperidin-alsuccinic acid, diethyl ester	heptane	216(3.5),228s(3.4)	70-0716-65
$C_{17}H_{28}O$			
2,4-Cyclohexadien-1-one, 2,5-di-tert-butyl-6-ethyl-6-methyl-	C_6H_{12}	315(3.63)	77-0314-65
	MeOH	322(3.59)	77-0314-65
$C_{17}H_{28}O_3$			
4-Pentynal, 3-hydroxy-3-methyl-5-(2,6,6-trimethyl-1-cyclohexen-1-yl)-, dimethyl acetal	iso-PrOH	235(4.13)	28-2466-64B
$C_{17}H_{28}O_4$			
4-Cyclopentene-1,3-dione, 2,2-di-isopentyl-4,5-dimethoxy-	EtOH	293(3.82)	39-0952-64

Compound	Solvent	$\lambda_{max}(\log \epsilon)$	Ref.
$C_{17}H_{29}N$			
Tetraethylammonium cyclononatetra-enide	MeCN	250(4.82),317(3.82), 322(3.83)	35-1941-65
o-Toluidine, 4,6-diisopentyl-	EtOH	237(3.89),289(3.33)	32-1221-64
$C_{17}H_{29}NO$			
4,6-Octadienamide, N-cyclohexyl-3,3,7-trimethyl-	EtOH	232s(4.41),238(4.45)	39-4035-64
$C_{17}H_{29}N_3O_3$			
4,11βH,5αH-Eudesman-13-oic acid, 3-oxo-, methyl ester, semicarbazone	EtOH	228(4.14)	78-1167-65
$C_{17}H_{29}N_5$			
s-Triazolo[1,5-c]pyrimidine, 7-methyl-2-tert-octylamino-5-propyl-	MeOH	235(4.63),303(3.52)	39-3357-65
$C_{17}H_{30}N_4$			
5H-Cyclopentadecapyrimidine, 2,4-diaminododecahydro-	pH 1	278(3.92)	44-1837-65
	pH 10	232(4.02),289(3.91)	44-1837-65
$C_{17}H_{30}O_2$			
1,3-Cycloheptadecanedione	dioxan	275(4.17)	28-0759-65B
copper derivative	$CHCl_3$	250(4.58),299(4.66)	28-0759-65B
$C_{17}H_{30}O_3$			
1-Cyclohexene, 1-(3-hydroxy-5,5-di-methoxy-3-methyl-1-penten-1-yl)-2,6,6-trimethyl-, cis	iso-PrOH	212(3.64)	28-2466-64B
trans	iso-PrOH	235(3.76)	28-2466-64B
$C_{17}H_{31}NO_2S$			
2,4-Thiazolidinedione, 5-tetradecyl-	n.s.g.	240(3.74),300s(2.70)	65-0886-65
$C_{17}H_{32}N_2OS$			
4-Thiazolidinone, 2-imino-5-tetradecyl-	n.s.g.	248(3.92),300s(1.98)	65-0886-65

Compound	Solvent	$\lambda_{max}(\log \epsilon)$	Ref.
$C_{18}F_{15}N$			
Triphenylamine, pentadecafluoro-	C_6H_{12}	255(4.12)	39-5017-64
$C_{18}H_2F_{32}N_2O_2$			
1,2,5-Oxadiazole, 3,4-bis(8H-hexadeca-fluorooctyl)-, N-oxide	EtOH	266(3.59)	44-0279-64
$C_{18}H_5F_{10}N$			
Triphenylamine, 2,2',3,3',4,4',-5,5',6,6'-decafluoro-	C_6H_{12}	265(4.12)	39-5017-64
$C_{18}H_8Cl_4N_2O_2Zn$			
8-Quinolinol, 5,7-dichloro-, zinc chelate	$CHCl_3$	268(4.83),346(3.74), 406(3.74)	59-1229-65
$C_{18}H_8N_2O_3$			
1-Indanone, 2-diazo-3-(1,3-dioxo-indan-2-ylidene)-	MeOH	242(4.58),309(4.28), 350(4.15),430(4.09)	5-0101-64F
$C_{18}H_8N_4$			
1,4-Pentadien-3-one, 1,5-diphenyl-, amidinohydrazone, hydrochloride	propylene glycol	313(4.549)	95-0001-64
$C_{18}H_8O_6$			
Benzo[1,2-b:4,5-b']bisbenzofuran-6,12-dione, 3,9-dihydroxy-	dioxan	262(4.60),292(4.46), 407(4.11),475s(2.81)	1-1063-65
Benzo[1,2-b:5,4-b']bisbenzofuran-6,12-dione, 3,9-dihydroxy-	dioxan	257(4.67),360(3.95), 538(3.65)	1-1063-65
Erosnin	EtOH	239(4.58),285(3.95), 354(4.30),366(4.26)	88-2559-65
$C_{18}H_8O_8$			
Benzo[1,2-b:4,5-b']bisbenzofuran-6,12-dione, 1,3,7,9-tetrahydroxy-	dioxan	270s(4.36),285s(4.43), 294s(4.51),299(4.52), 446(4.02),580s(2.85)	1-1063-65
$C_{18}H_{10}$			
[18]Annulene, tetradehydro-	isooctane	228(4.21),297(4.40), 316(4.82),327(5.00), 338(4.73),354(4.05), 363(3.94),374(3.91), 383(4.10),395(4.16), 414(3.41)	35-3638-65
	benzene	324s(4.80),334(4.98), 367(3.95),378(3.91), 387(4.07),399(4.15), 418(3.38)	35-3638-65
$C_{18}H_{10}F_5N$			
Triphenylamine, 2,3,4,5,6-pentafluoro-	C_6H_{12}	282(4.16)	39-5017-64
$C_{18}H_{10}N_2O$			
7H-Benzimidazo[2,1-a]benz[de]iso-quinolin-7-one	$1-C_{10}H_7Cl$	400(3.84)	33-1999-65
12H-Isoindolo[2,1-a]perimidine, 12-oxo-	$1-C_{10}H_7Cl$	452(3.77),474(3.76)	33-1999-65
$C_{18}H_{10}N_2O_2$			
Indeno[2,1-e]perimidine-2,12-(1H,3H)-dione	MeOH	262(4.35),275(4.30), 400(4.23),410s(4.20), 460(3.60)	24-1282-65

Compound	Solvent	λ_{max}(log ε)	Ref.
$C_{18}H_{10}N_2O_2S_2$			
Isothiazolecarbonitrile, 3,5-bis(benzoylthio)-	EtOH	249(4.31),280(4.16), 288(4.16)	44-0665-64
$C_{18}H_{10}O_2S_2$			
1,3-Indandione, 2-(4-phenyl-1,3-dithiol-2-ylidene)-	CF_3COOH	258(4.37),300(4.21), 420(4.60)	44-1711-64
	THF	425(4.43)	44-1711-64
$C_{18}H_{10}O_6$			
Naphthacenequinone, 1,4,6,11-tetrahydroxy-	$CHCl_3$	487(4.20),521(4.53), 561(4.63)	24-3145-65
	H_2SO_4	575(4.56),617(4.61)	24-3145-65
Naphthacenequinone, 1,6,7,11-tetrahydroxy-	$CHCl_3$	479(4.11),512(4.52), 550(4.70)	24-2797-65
	H_2SO_4	552(4.38),595(4.73)	24-2797-65
	DMF	582(4.15),626(4.04)	24-2797-65
Naphthacenequinone, 1,6,10,11-tetrahydroxy-	$CHCl_3$	482(4.20),515(4.53), 554(4.60)	24-2797-65
	H_2SO_4	565(4.45),618(4.70)	24-2797-65
	DMF	568(4.30),609(4.34)	24-2797-65
$C_{18}H_{10}O_6S$			
Coumarin, 3,3'-thiobis[4-hydroxy-	MeOH-HCl	280s(3.96),291(4.02), 314(4.09),325(4.04)	49-0077-65
	MeOH-NaOH	315(4.05)	49-0077-65
$C_{18}H_{10}O_7$			
Coumestan, 7-acetoxy-11,12-(methylenedioxy)-	EtOH	239(4.35),298(4.13), 333(4.52),349(4.45)	44-2353-65
Naphthacenequinone, 1,4,6,7,11-pentahydroxy-	$CHCl_3$	504(4.65),540(4.56), 584(4.68)	24-3145-65
	H_2SO_4	601(4.38),653(4.69)	24-3145-65
Naphthacenequinone, 1,4,6,8,11-pentahydroxy-	$CHCl_3$	490(4.08),525(4.48), 565(4.58)	24-3145-65
	H_2SO_4	580(4.38),625(4.64)	24-3145-65
$C_{18}H_{11}BrO_3$			
3,4;7,8-Dibenzocoumarin, 6-bromo-5-methoxy-	MeOH	243(3.8),259(3.7), 268(3.8)	83-0529-64
$C_{18}H_{11}ClN_4$			
Pyrazino[2,3-c]pyridazine, 3-chloro-6,7-diphenyl-	EtOH	208(4.46),230s(4.39), 242(4.42),260s(4.24), 385(4.01)	4-0247-64
Pyrazino[2,3-c]pyridazine, 4-chloro-6,7-diphenyl-	EtOH	207(4.33),233(4.29), 274s(4.19),372(4.07)	4-0247-64
$C_{18}H_{11}NO_4$			
Atherospermidine	1% HCl	263(4.46),283(4.36), 410(3.78),505(3.58)	44-0432-65
	pH 1	262(4.24),283(4.16)	88-1629-64
	EtOH	247(4.39),281(4.53), 316s(3.80),383(3.71), 440(3.92)	44-0432-65
	EtOH	247(4.38),281(4.52), 312s(3.95)	88-1629-64
8H-Benzo[g]-1,3-benzodioxolo[6,5,4-de]quinolin-8-one, 4-methoxy-	EtOH	247(4.33),281(4.46), 310s(3.95),380(3.72), 437(3.93)	12-1997-65

Compound	Solvent	λ max(log ε)	Ref.
8H-Benzo[g]-1,3-benzodioxolo[6,5,4-de]-quinolin-8-one, 4-methoxy- (contd.)	EtOH	247(4.37),281(4.49), 312s(3.89)	78-2579-65
$C_{18}H_{11}Na$			
Chrysene, sodium derivative	THF	380(3.1),395(3.1), 470(2.8),480(2.9)	60-0891-65
$C_{18}H_{12}$			
Benz[a]anthracene, 1,3,5-trinitro-benzene complex	$CHCl_3$	425(4.37)	38-0166-64A
Chrysene	EtOH	225(4.5),240s(4.2), 260s(4.9),275(5.2), 280s(4.0),292(4.0), 315(4.1),325(4.2)	12-0190-65
	EtOH	220(4.56),241(4.36), 259(5.00),267(5.20), 283(4.14),295(4.13), 306(4.19),319(4.19), 344(2.88),351(2.62), 360(3.00)	78-2107-64
	dioxan	260(4.98),270(5.20), 292(4.29),307(4.26), 323(4.24),346(2.80), 353(2.58),362(2.85)	24-2860-64
bromanil complex	$CHCl_3$	500(--)	60-0465-64
chloranil complex	$CHCl_3$	495(--)	60-0465-64
iodoanil complex	$CHCl_3$	493(--)	60-0465-64
Tridehydro[18]annulene, isomer I	benzene	317(4.90),328(5.04), 343(5.25),390(4.10), 406(4.21),436(3.02), 441(3.06)	35-3638-65
	isooctane	233(4.44),237(4.52), 244(4.53),252(4.38), 295(4.34),313(4.78), 322(5.05),335(5.28), 364(3.96),368(3.95), 378(3.95),385(4.11), 395s(4.05),399(4.23), 433(3.28)	35-3638-65
isomer II	benzene	326(4.95),338(5.16), 347s(3.98),374(3.95), 392(4.06),407(4.20)	35-3638-65
	isooctane	234(4.49),244(4.42), 253(4.36),295(4.35), 310(4.77),318(4.97), 331(5.22),340(4.96), 358(3.91),367(3.91), 370(3.91),380(3.93), 387(4.10),402(4.23), 420(3.46)	35-3638-65
Triphenylene	EtOH	240(4.9),255(5.0), 275(4.4),285s(4.2), 320(2.9),330(2.9), 335(2.9)	35-1125-64
	EtOH	249(4.97),257(5.18), 273(4.40),284(4.26), 302(3.50),312(3.00), 321(2.90),327(2.88), 334(2.88)	78-3289-65

Compound	Solvent	$\lambda_{max}(\log \epsilon)$	Ref.
$C_{18}H_{12}Cl_2N_2O_2$			
p-Benzoquinone, 2,5-dianilino-3,6-dichloro-	methyl cellosolve	268(4.20),391(4.23), 548(2.85)	78-1889-64
$C_{18}H_{12}F_2N_2O_2$			
p-Benzoquinone, 2,5-dianilino-3,6-difluoro-	methyl cellosolve	268(4.46),402(4.18), 555(2.79)	78-1889-64
$C_{18}H_{12}MgN_2O_2$			
8-Quinolinol, magnesium chelate	$CHCl_3$	318(3.52),333(3.56), 366(3.51)	59-1229-65
$C_{18}H_{12}MgN_2O_8S_2$			
5-Quinolinesulfonic acid, 8-hydroxy-, magnesium chelate	H_2O	326(3.72),334(3.72), 355(3.76)	59-1229-65
$C_{18}H_{12}N_2$			
2,2'-Biquinoline	C_6H_{12}	227s(4.23),252s(4.72), 285s(4.88),302s(4.22), 312(4.28),324(4.34), 339(4.26)	39-1258-65
	pH 0	264(4.60),290(3.84), 300(3.78),354(4.39)	39-1258-65
	pH 8.3	256(4.77),329(4.27)	39-1258-65
$C_{18}H_{12}N_2O$			
2'-Acetonaphthone, 2-diazo-2-phenyl-	MeOH	252(4.46),319(3.92)	88-1403-64
1-Azulenecarbonitrile, 2-amino-3-benzoyl-	EtOH	230(4.34),270(4.28), 319(4.72)	94-0443-65
$C_{18}H_{12}N_2OS$			
Thiazole, 4-(2-benzofuranyl)-2-(benzylideneamino)-	EtOH	300(4.63),372(3.97)	7-0987-64
$C_{18}H_{12}N_2O_2$			
Perimidone, 4-benzoyl-	MeOH	262(4.28),338(3.25), 415(3.83)	24-1282-65
Perimidone, 5(6)-benzoyl-	MeOH	228(4.53),257(4.22), 351(3.88),390(3.90)	24-1282-65
9H-Pyrido[3,4-b]indole, 1-(3,4-methylenedioxyphenyl)-	EtOH	202(4.48),214(4.47), 224(4.46),235s(4.44), 250s(4.23),265(4.17), 297(4.15),357s(3.92), 362(3.93)	4-0168-64
$C_{18}H_{12}N_2O_2Zn$			
8-Quinolinol, zinc chelate	$CHCl_3$	257(4.65),340(3.50), 379(3.61)	59-1229-65
$C_{18}H_{12}N_2O_4$			
Atherospermidine, oxime	EtOH	280(4.47),376s(3.78), 410(3.92)	44-0432-65
$C_{18}H_{12}N_2O_6$			
Isoquinoline, 5-methoxy-6,7-(methylenedioxy)-1-(2-nitrobenzoyl)-	EtOH	235(4.68),270(4.80), 353(3.98)	78-2579-65
$C_{18}H_{12}N_2O_8S_2Zn$			
5-Quinolinesulfonic acid, 8-hydroxy-, zinc chelate	H_2O	255(4.68),362(3.88)	59-1229-65

Compound	Solvent	λ_{max}(log ϵ)	Ref.
$C_{18}H_{12}N_2S_4$			
Thiazole, 2,2'-dithiobis[4-phenyl-	MeOH	235(4.53),258(4.49)	44-2146-64
$C_{18}H_{12}N_3O_6P_3$			
Trispiro[1,3,5,2,4,6-triazatriphosphorine-2,2':4,2'':6,2'''-tris[1,3,2]-benzodioxaphosphole]	MeCN	209(4.38),266s(3.68), 271(3.79),276(3.72)	35-2591-64
$C_{18}H_{12}N_4$			
Cycloheptimidazole, 6-dicyanomethylene-1,6-dihydro-1-methyl-2-phenyl-	EtOH	257(4.42),437(4.62)	94-0810-65
$C_{18}H_{12}N_4S_2$			
Thiocyanic acid, diester with 1-(p-mercaptophenyl)-3-methyl-5-phenylpyrazole-4-thiol	EtOH	266(4.30)	94-0023-64
$C_{18}H_{12}N_5O_6$			
Hydrazyl, diphenylpicryl-	ether	260(4.6),460(3.8)	29-0153-63B
	Ph_2O	260(4.7),520(4.0)	29-0153-63B
$C_{18}H_{12}O$			
Benzo[c]phenanthren-1-ol	EtOH	205s(4.07),221(4.43), 282(4.57),306(4.17), 326s(3.85),345s(3.61), 364(3.35),383(3.27)	35-0503-64
Benzo[c]phenanthren-2-ol	EtOH	219(4.51),232(4.37), 246s(4.22),252(4.23), 277(4.58),287(4.55), 307(4.19),316s(4.04), 330s(3.65),346(3.32), 364(3.44),383(3.46)	35-0503-64
Benzo[c]phenanthren-3-ol	EtOH	219(4.59),249s(4.04), 278s(4.64),284(4.72), 308(3.92),321(3.95), 333s(3.76),363(3.44), 380(3.40)	35-0503-64
Benzo[c]phenanthren-4-ol	EtOH	220(4.59),241s(4.31), 268(4.50),280(4.62), 292(4.68),304(4.38), 322s(4.06),341s(3.94), 362s(3.69),381(3.62)	35-0503-64
Benzo[c]phenanthren-5-ol	EtOH	219(4.57),229s(4.41), 242s(4.15),257s(4.22), 269s(4.48),279(4.72), 289(4.81),311(4.10), 322s(4.08),338(3.99), 365(3.71),383(3.69)	35-0503-64
Benzo[c]phenanthren-6-ol	EtOH	219(4.62),236(4.40), 253(4.35),268s(4.52), 279(4.69),289(4.62), 307(4.35),319(4.24), 350(3.65),369(3.76), 387(3.79)	35-0503-64
Tribenz[b,d,f]oxepin	EtOH	241(4.65),263(4.22)	78-1299-65
$C_{18}H_{12}OS$			
Acetic acid, thio-, S-1-pyrenyl ester	EtOH	237(4.54),245(4.71), 259(4.09),269(4.34), 280(4.55),306s(3.79),	39-2571-64

Compound	Solvent	λ_{max}(log ε)	Ref.
Acetic acid, thio-, S-1-pyrenyl ester (contd.)		319(4.11),334(4.42), 349(4.57),369(3.22), 376(3.35)	
$C_{18}H_{12}OS_2$			
[18]Annulene 1,4-oxide 7,10;13,16-disulfide	EtOH	229(4.19),281(4.46), 292(4.47),405(3.95), 425(3.98)	77-0269-65
$C_{18}H_{12}O_2S$			
[18]Annulene 1(4),7(10)-dioxide 13(16)-sulfide	EtOH	224(4.16),244(4.07), 320(4.26),333(4.63), 343(4.98),376(3.71), 391(3.85),413(4.22), 431(3.32)	77-0492-65
$C_{18}H_{12}O_3$			
5-Benzofuranacrylic acid, 6-hydroxy-α-methyl-β-phenyl-, δ-lactone	EtOH	225(4.35),247(4.38), 296(4.05),328(3.92)	44-2467-64
3,4;7,8-Dibenzocoumarin, 5-methoxy-	MeOH	243(4.5),268(4.5), 293(3.8),330(3.8)	83-0529-64
1-Naphthoic acid, 2-benzoyl-	EtOH	234(4.51),292(3.78)	44-4293-65
2-Naphthoic acid, 1-benzoyl-	EtOH	228(4.56),291(3.72)	44-4293-65
1,4;7,10;13,16-Triepoxy[18]annulene	EtOH	220(3.89),239(3.87), 309(4.26),323(4.81), 332(5.46),370(3.51), 391(3.74),405(4.24)	77-0269-65
$C_{18}H_{12}O_4$			
1H-2-Benzopyran-4-carboxylic acid, 1-oxo-3-styryl-	$CHCl_3$	320(4.37),335(4.39), 355(4.39)	83-0715-65
$C_{18}H_{12}O_6$			
Atromentin	MeOH	276(4.46),360(3.68)	100-0203-65
	dioxan	268(4.53),385(3.74)	100-0203-65
Spiro[naphthalene-2(1H),1'-phthalan]-1,3',4(3H)-trione, 7',8-dihydroxy-4'-methyl-	CH_2Cl_2	236(4.47),272(3.85), 309(3.71),358(3.82)	33-1459-64
dimorphic form m. 252-253°	CH_2Cl_2	236(4.46),272(3.84), 309(3.71),357(3.80)	33-1459-64
$C_{18}H_{12}S$			
Tribenzo[b,d,f]thiepin	EtOH	236(4.66),251(4.39)	78-1299-65
$C_{18}H_{12}S_3$			
[18]Annulene 1,4;7,10;13,16-trisulfide	EtOH	204(4.26),224(4.23), 288(4.24)	12-0070-65
	n.s.g.	204(4.26),224(4.23), 288(4.24)	77-0082-64
$C_{18}H_{12}S_4$			
$\Delta^{2,2'}$-Bi-1,3-dithiole, 4,4'-diphenyl-	THF	259(4.08),329(3.95), 403(3.43)	89-0453-65
$C_{18}H_{13}ClN_2$			
Pyrrolo[1,2-a]quinoxaline, 1-chloro-2-methyl-4-phenyl-	EtOH	235s(4.44),244s(4.50), 248(4.51),276(4.36), 328s(3.90),342(3.99)	35-1830-64

Compound	Solvent	λ_{max}(log ϵ)	Ref.
$C_{18}H_{13}ClN_4S_2$			
Imidazo[4,5-d]pyridazine-7(4)-thiol, 4(7)-p-chlorobenzylthio-2-phenyl- (as dihydrate)	EtOH	205(4.38),222s(4.36), 283(4.37),323s(4.20), 346(4.25)	4-0182-64
$C_{18}H_{13}ClS_3$			
6a-Thiathiophthene, 2-(p-chlorophenyl)-5-p-tolyl-	dioxan	258(4.67),318(4.41), 513(4.20)	24-1732-64
	$CHCl_3$	260(4.45),319(4.25), 510(4.07)	24-1732-64
$C_{18}H_{13}Cl_2NO_6$			
3H-Phenoxazine-1,9-dicarboxylic acid, 2,7-dichloro-4,6-dimethyl-3-oxo-, dimethyl ester	MeOH	230(4.40),264(4.2), 370(4.2),480(3.99)	44-3185-65
$C_{18}H_{13}Cl_9O$			
1,4:9,10-Dimethanoanthracene, 1,3,4,5,-6,7,8,11,11-nonachloro-2-ethoxy-1,4,4a,8a,9,9a,10,10a-octahydro-, endo-endo-exo	EtOH	216(3.90),265(3.39), 276(3.54),287(3.71), 299(3.77),313(3.55)	39-4646-65
1,4:9,10-Dimethanoanthracene, 1,3,4,5,-6,7,8,11,11-nonachloro-2-ethoxy-1,2,3,4,4a,9,9a,10-octahydro-, endo-endo	EtOH	215(4.66),220s(4.63), 242s(4.05),287(2.20), 297(2.20)	39-4646-65
$C_{18}H_{13}F$			
Biphenyl, 2-fluoro-2'-phenyl-	hexane	230(3.1)	44-1270-64
$C_{18}H_{13}N$			
Benzo[c]phenanthrene, 1-amino-	EtOH	213(4.64),272(4.52), 287(4.40),318(4.04), 353s(3.52),371(3.46), 391s(3.32)	35-1835-64
hydrochloride	80% EtOH-HCl	220(4.64),231s(4.25), 247s(4.02),267s(4.39), 277(4.70),286(4.83), 298s(4.20),308(4.02), 321(3.99),333s(3.67), 362(2.75),384(2.86)	35-1835-64
Benzo[c]phenanthrene, 2-amino-	EtOH	221(4.44),237(4.48), 248s(4.48),256(4.51), 280(4.52),292(4.51), 317s(4.19),372s(3.32), 388(3.40)	35-1835-64
hydrochloride	80% EtOH-HCl	218(4.67),223s(4.50), 231(4.41),244s(3.99), 254s(4.15),264s(4.46), 273(4.72),283(4.88), 296s(4.18),303(4.06), 317(4.01),327s(3.75), 355(2.76),373(2.59)	35-1835-64
Benzo[c]phenanthrene, 3-amino-	EtOH	222(4.69),257s(4.30), 287(4.76),325(3.92), 337s(3.88),384s(3.27)	35-1835-64
	80% EtOH-HCl	218(4.61),223s(4.47), 229(4.32),245s(3.91), 254s(4.12),264s(4.39), 273(4.71),283(4.86), 295s(4.11),303(4.05),	35-1835-64

Compound	Solvent	$\lambda_{max}(\log \epsilon)$	Ref.
Benzo[c]phenanthrene, 3-amino- (contd.)		316(4.09),327s(3.59), 355(2.55),373(2.31)	35-1835-64
Benzo[c]phenanthrene, 4-amino-	EtOH	210(4.64),226(4.51), 265(4.56),298(4.53), 348(3.77),368s(3.70), 385s(3.48)	35-1835-64
hydrochloride	EtOH-HCl	219(4.65),223s(4.53), 230(4.39),245s(4.05), 254s(4.16),265s(4.41), 275(4.75),285(4.88), 297s(4.25),305s(4.11), 318(4.04),328s(3.77), 355(2.68),375(2.46)	35-1835-64
Benzo[c]phenanthrene, 5-amino-	EtOH	218(4.53),232s(4.35), 248s(4.14),263s(4.27), 293(4.71),346(3.97), 388s(3.28)	35-1835-64
hydrochloride	80% EtOH-HCl	218(4.67),224s(4.50), 230(4.38),245s(3.91), 254s(4.10),264s(4.46), 273(4.75),283(4.90), 297(4.20),304(4.10), 317(4.08),328s(3.75), 357(2.78),375(2.45)	35-1835-64
Benzo[c]phenanthrene, 6-amino-	EtOH	222(4.55),240(4.40), 248(4.40),256(4.44), 274s(4.43),284(4.48), 292s(4.47),321s(4.15), 373s(3.23),395(3.35)	35-1835-64
hydrochloride	80% EtOH-HCl	219(4.64),223s(4.50), 229(4.32),245s(3.97), 254s(4.21),265s(4.53), 273(4.47),283(4.84), 297(4.20),303s(4.11), 316(4.03),327s(3.72), 358(2.67),377(2.39)	35-1835-64
$C_{18}H_{13}NO$			
9-Anthraceneacetonitrile, 10-acetyl-	n.s.g.	274(4.40),328(3.36), 345(3.60),364(3.67), 385(3.54),405(3.58)	44-2037-65
Acrylophenone, 3-(2-quinolyl)-	95% H_2SO_4	235(4.19),300(4.00), 402(4.40)	65-1724-65
	EtOH	275(4.4),327(4.28)	65-2914-64
Naphth[1,2-d]oxazole, 2-p-tolyl-	heptane	230(4.74),250(4.17), 290(4.34),307(4.44), 330(4.54),345(4.46)	65-0427-64
Phthalimidine, 2-(1-naphthyl)-	MeOH-acid	280(4.03)	95-1042-65
	MeOH	280(4.03)	95-1042-65
Phthalimidine, 2-(2-naphthyl)-	MeOH-acid	266(4.32)	95-1042-65
	MeOH	266(4.31)	95-1042-65
2-Propen-1-one, 3-phenyl-1-(2-quinolyl)-	95% H_2SO_4	228(4.22),258(4.03), 350(3.74),492(4.56)	65-1724-65
	EtOH	260(4.3),326(4.41)	65-2914-64
$C_{18}H_{13}NO_2$			
9H-Benzo[a]phenoxazin-9-one, 8,10-dimethyl-	$CHCl_3$	264(4.38),271(4.37), 293(4.23),314(3.94), 389(3.95),405(3.95), 495(4.17)	39-0480-65

Compound	Solvent	$\lambda_{max}(\log \varepsilon)$	Ref.
3H-Naphth[1,2-e][1,3]oxazin-3-one, 1,2-dihydro-1-phenyl-	EtOH	229(4.51),278(3.73), 289(3.62),309(3.02), 316(2.93),324(3.16)	70-1107-65
$\Delta^{1,\alpha}$-Perinaphtheneacetic acid, α-cyano-, ethyl ester	MeCN	270(4.31),330(3.36), 390(3.7),462(4.2)	44-3166-65
Phenalene, ethoxycarbonyl- cyanomethylene-	EtOH	492(4.06)	89-0346-65
	MeCN	482(4.08)	89-0346-65
Quinoline, 4-(3,4-methylene- dioxystyryl)-	MeOH	355(4.2)	87-0137-65
	HOAc	315(3.9),418(4.0)	87-0137-65
$C_{18}H_{13}NO_3$			
11H,12H-Benzo[f]pyrano[2,3-b]quino- lizine-11,12-dione, 5,9-dimethyl-	EtOH	250s(4.35),292(4.12), 405(4.36),425(4.32)	95-0874-64
Isoindole-3-succinic anhydride, 1-phenyl-	n.s.g.	270(3.8),280(3.9), 325(4.0),360(4.1)	35-4152-64
$C_{18}H_{13}NO_4$			
Dibenzo[de,g]quinolin-7-one, 2-hydroxy-1,3-dimethoxy-	EtOH-HCl	246(4.37),281(4.40), 390(3.63),496(3.36)	88-4655-65
	EtOH	237(4.47),272(4.41), 315s(4.10),374(3.55), 440(3.67)	88-4655-65
	EtOH-NaOH	247(4.42),283(4.31), 310(4.25),407(3.99), 517(3.33)	88-4655-65
Fumaric acid, (β-hydroxy-α-2-pyridyl- styryl)-, δ-lactone, methyl ester	MeOH-ether	250(4.32),408(4.07)	24-3537-65
Pyrrole-2,5-dicarboxylic acid, 3,4-diphenyl-	EtOH	245(4.21),284(4.07)	44-0859-65
$C_{18}H_{13}NO_7$			
Aristolochic acid C, O-methyl-, methyl ester	EtOH	255(4.62),281(4.2), 302(4.12)	44-3935-65
$C_{18}H_{13}N_3$			
Benzo[c]cinnoline, 2-(p-aminophenyl)-	22N H_2SO_4	255(4.5),395(4.2)	12-0190-65
2-Styryl-1H-naphtho[1,8-de]triazinium hydroxide, inner salt	EtOH	655(2.90)	39-3005-64
$C_{18}H_{13}N_3O_2$			
Indole-3-carbonitrile, 2-(p-cyano- phenyl)-5,6-dimethoxy-	MeOH	219(4.45),242(4.21), 359(4.30)	44-2253-65
$C_{18}H_{13}N_3O_4$			
Benzonitrile, p-(β-cyano-4,5-di- methoxy-2-nitrostyryl)-	MeOH	246(4.15),289(4.43)	44-2253-65
$C_{18}H_{13}N_3O_6$			
Azulene, 1-vinyl-, compound with 1,3,5-trinitrobenzene	C_6H_{12}	234(5.61),292(5.98), 580(3.78),603(3.84), 628(3.87),660(3.8), 691(3.76),735(3.4)	44-0270-65
$C_{18}H_{13}N_3O_6S_3$			
3H-1,2-Dithiol-3-one, 4-(methylamino)- 5-(p-tolylthio)-, 3,5-dinitro- benzoate	EtOH	293(4.08),335(3.95)	12-0061-65
$C_{18}H_{14}$			
Azulene, 1-styryl-, cis	C_6H_{12}	259(5.25),313(5.47), 625(2.46)	44-0270-65

Compound	Solvent	λ_{max}(log ε)	Ref.
Azulene, 1-styryl-, trans	C_6H_{12}	255(4.18),318(4.48), 647(2.52),717(2.34)	44-0270-65
1,2-Benzoheptafulvene, 8-phenyl-	ether	267(4.04),321(3.88)	35-3719-65
2,2'-Biindene	C_6H_{12}	190(4.53),211(4.23), 249(4.15),257(4.08), 331(4.45),348(4.58), 367(4.44)	24-0140-65
Chrysene, 1,2-dihydro-	EtOH	256(4.73),265(4.86), 282(4.07),294(4.13), 305(4.16),318(4.02), 342(2.89)	24-0593-65
Pleiadene, 7,12-dihydro-	dioxan	215(5.02),226(5.06), 267s(3.89),271s(3.94), 278(4.08),288(4.16), 299(4.02),307s(3.63), 317s(3.18),322(3.26)	78-3073-65
p-Terphenyl-	C_6H_{12}	276(4.54)	39-3342-65
	$CHCl_3$	280(4.40)	39-3342-65
$C_{18}H_{14}BrN$			
13-Methylbenz[h]acridizinium bromide	EtOH	229(4.39),233s(4.38), 273(4.45),308(4.22), 321(4.28),364s(3.84), 380(4.10),400(4.22), 437(2.87)	4-0121-64
7-Methylbenz[j]acridizinium bromide	EtOH	217s(4.32),230(4.40), 268(4.48),278(4.60), 310s(4.36),321(4.57), 368(3.95),386(3.92), 408(3.91)	4-0121-64
$C_{18}H_{14}BrNO_2$			
1H-Pyrrolo[1,2-a]indol-1-one, 7-(benzyloxy)-2-bromo-2,3-dihydro-	MeOH	215(4.52),255(3.95), 335(4.27)	44-2904-65
1H-Pyrrolo[1,2-a]indol-1-one, 7-(benzyloxy)-9-bromo-2,3-dihydro-	MeOH	218(4.52),241(4.36), 321(4.36)	44-2904-65
$C_{18}H_{14}ClNO$			
Quinoline, 4-(p-chlorostyryl)-6-methoxy-	EtOH	223(4.50),238(4.52), 322(4.14),335s(4.10), 350s(4.05)	7-1154-65
	0.1N HCl	254(4.52),325s(3.84), 340s(3.88),365(3.91), 380s(3.90)	7-1154-65
	0.001N HCl	226(4.43),243(4.45), 325s(3.86),340(3.89), 380s(3.71)	7-1154-65
$C_{18}H_{14}ClNO_2$			
Oxazole, 2-(4-chlorostyryl)-5-(4-methoxyphenyl)-	C_6H_{12}	270(4.30),348(4.53), 365(4.52)	4-0310-65
$C_{18}H_{14}ClNO_4$			
3-Methylbenzo[a]phenanthridizinium perchlorate	EtOH	213s(4.47),217(4.52), 243(4.26),286(4.61), 340s(3.62),355(3.94), 372(4.16),390(4.27)	44-1846-65

Compound	Solvent	λ_{max}(log ε)	Ref.
5-Methylbenzo[a]phenanthridizinium perchlorate	EtOH	210s(4.37),217(4.48), 241(4.28),248s(4.24), 275s(4.41),282(4.58), 340s(3.71),355(3.92), 391(4.21)	44-1846-65
6-Methylbenzo[a]phenanthridizinium perchlorate	EtOH	214s(4.40),218(4.54), 242(4.43),277s(4.45), 286(4.62),350s(3.68), 365(3.90),381(4.06), 400(4.20)	44-1846-65
$C_{18}H_{14}ClNO_6$			
3H-Phenoxazine-1,9-dicarboxylic acid, 2-chloro-4,6-dimethyl-3-oxo-, dimethyl ester	MeOH	224(4.20),262(4.05), 358(3.97),500(3.67)	44-3185-65
$C_{18}H_{14}F_6O_2$			
Azulene, 2,4,6,8-tetramethyl-1,3-bis(trifluoroacetyl)-	C_6H_{12}	233(4.54),237s(4.55), 244(4.59),286(4.58), 318(4.55),334s(4.53), 387(2.98),500(2.88)	44-0131-65
$C_{18}H_{14}Fe$			
Iron, diindenyl-	hexane	260(4.33),415(2.77), 563(2.49)	101-0107-65A
$C_{18}H_{14}INO_7$			
2-Stilbenecarboxylic acid, 2'-iodo-4'-methoxy-4,5-(methylenedioxy)-α-nitro-, methyl ester, cis	EtOH	260(4.1),308(3.98), 345(4.08)	44-3935-65
2-Stilbenecarboxylic acid, 2'-iodo-6'-methoxy-4,5-(methylenedioxy)-α-nitro-, methyl ester, cis	EtOH	304(3.97)	44-3792-65
	EtOH	304(3.97)	88-0359-65
$C_{18}H_{14}N_2$			
2,4-Dodecadiene-6,8,10-triynal, phenylhydrazone	ether	420(4.81)	24-0809-64
11H-Imidazo[2,1-b][3]benzazepine, 2-phenyl-	EtOH	262(4.46)	4-0026-65
Isoquinoline, 3-(1-methylindol-2-yl)-	EtOH	228(4.64),241s(4.48), 279(4.04),319(4.33)	44-3584-64
Pyrrolo[1,2-a]quinoxaline, 2-methyl-4-phenyl-	EtOH	236s(4.41),247s(4.50), 252(4.52),275(4.31), 346(3.97)	35-1830-64
Pyrrolo[1,2-a]quinoxaline, 4-methyl-2-phenyl-	pH 1.4	265(4.59),282s(4.39), 368(4.18)	39-3678-65
	pH 7.3	267s(4.53),273(4.54), 279s(4.49),342(4.01)	39-3678-65
	EtOH	265s(4.56),273(4.58), 281s(4.51),331s(4.00), 342(4.05),356s(3.95)	39-3678-65
$C_{18}H_{14}N_2O$			
Acrylophenone, 4'-amino-3-(2-quinolyl)-	H_2SO_4	236(4.14),279(4.19), 353(4.30)	65-1724-65
2-Propen-1-one, 3-(p-aminophenyl)-1-(2-quinolyl)-	H_2SO_4	268(4.26),379(4.34)	65-1724-65
Pyrrolo[1,2-a]quinoxalin-1(5H)-one, 2-methyl-4-phenyl-	EtOH	238(4.45),272(4.28), 292s(4.16),330s(3.66), 340s(3.59)	35-1830-64

Compound	Solvent	$\lambda_{max}(\log \epsilon)$	Ref.
$C_{18}H_{14}N_2OS$			
Thiazole, 4-(2-benzofuranyl)- 2-(benzylamino)-	EtOH	266(4.41),310(4.31)	7-0987-64
$C_{18}H_{14}N_2O_2$			
Carbostyril, dimer	dioxan	224(4.33),259(4.23), 290(3.66)	1-1389-64
2,5-Piperazinedione, 3,6-dibenzylidene-	EtOH	234(3.9),338(4.5)	44-0277-65
4(1H)-Pyridazinone, 1-acetyl- 3,5-diphenyl-	$(MeOCH_2)_2$	313(4.37)	25-0839-64
3H-Pyrrolo[1,2,3-de]quinoxaline, 2,3-di-2-furyl-5,6-dihydro-	MeOH	250s(4.03),279(4.15), 318(4.18),425(3.70)	44-2589-65
$C_{18}H_{14}N_2O_3$			
1,7-Dioxa-2,8-diazaspiro[4.4]nona- 2,8-dien-6-one, 9-methyl-3,4-diphenyl-	EtOH	232(4.00),247s(4.07), 256(4.10)	32-0127-65
$C_{18}H_{14}N_2O_4S_3$			
4-Isothiazolecarbonitrile, 3,5-bis(benzylsulfonyl)-	EtOH	268(3.80)	44-0665-64
$C_{18}H_{14}N_2O_5$			
4H-1-Benzopyran-2-carboxylic acid, 4-(p-nitrophenylimino)-, ethyl ester	EtOH	336(4.26),370s(4.11)	65-1924-64
$C_{18}H_{14}N_2O_8$			
Butyric acid, 4-hydroxy-3-(m-nitro- benzylidene)-4-(m-nitrophenyl)- 2-oxo-, methyl ester	MeOH	215(4.69),244(4.55)	44-1800-65
$C_{18}H_{14}N_2S_3$			
4-Isothiazolecarbonitrile, 3,5-bis(benzylthio)-	EtOH	287(4.14)	44-0665-64
$C_{18}H_{14}N_4$			
2,2'-Biquinazoline, 4,4'-dimethyl-	pH 0	238s(4.26),267(4.62), 320(4.10)	39-1258-65
	pH 8	224s(4.23),257(4.70), 308(4.02)	39-1258-65
	C_6H_{12}-$CHCl_3$	250(4.73),294s(4.18)	39-1258-65
4,4'-Biquinazoline, 2,2'-dimethyl-	C_6H_{12}	216(4.81),227s(4.67), 250s(4.05),322(3.88)	39-1258-65
	pH 0	225(4.51),317(3.85)	39-1258-65
	pH 0, after hydration	271(3.98),286s(3.83)	39-1258-65
	pH 8	221(4.74),259s(3.84), 280s(3.67),327(3.85)	39-1258-65
Pyrimido[4,5-d]pyrimidine, 5,6-dihydro-2,7-diphenyl-	EtOH	254(4.42),315(3.78)	94-0393-64
s-Triazolo[4,3-a]pyrazine, 3-methyl-5,6-diphenyl-	C_6H_{12}	205(4.59),251(4.39), 316(3.63),325s(3.61), 336s(3.37)	44-2542-64
$C_{18}H_{14}N_4O$			
Acetanilide, 3'-(9H-pyridazino- [3,4-b]indol-3-yl)-	EtOH	271(4.73),376(3.51)	94-1129-64
$C_{18}H_{14}N_4O_2$			
1H-Tetrazole-1-acetic acid, α-cinnam- ylidene-5-phenyl-	EtOH	232(4.06),312(4.12)	44-2222-65

Compound	Solvent	$\lambda_{max}(\log \epsilon)$	Ref.
$C_{18}H_{14}N_4O_2S$			
1-Benzenesulfonamide, 2-phenyl-2H-benzotriazolium hydroxide, inner salt	EtOH	213s(4.39),301(4.15)	39-0751-64
$C_{18}H_{14}N_4O_5$			
Formamide, N-[2-(2,4-dinitroanilino)-1-naphthyl]-N-methyl-	EtOH	348(4.25)	65-0269-64
	$CHCl_3$	352(4.21)	65-0269-64
$C_{18}H_{14}N_4O_5S$			
Xanthine, 7-benzyl-1-hydroxy-, benzenesulfonate	EtOH	266(4.12)	4-0275-64
	EtOH-NaOH	283(4.25)	4-0275-64
$C_{18}H_{14}O$			
Naphthalene, 1-benzoyl-2-methyl-	MeOH	224(4.84),251(4.14), 277(3.78),321(3.01)	35-5417-65
Naphtho[2,3-c]furan, 1,3-dihydro-4-phenyl-	EtOH	228(4.75),275s(3.80), 284(3.93),294s(3.91), 321s(2.96)	94-1094-64
Tricyclo[2.1.1.0^{5,6}]hexan-2-one, 5,6-diphenyl-	MeOH	255(4.08)	88-0945-65
$C_{18}H_{14}OS_2$			
2H-1-Benzothiopyran, 2,2'-oxybis-	EtOH	242(4.74),272(4.03), 311(3.45)	17-0151-65
$C_{18}H_{14}O_2$			
Bicyclo[2.2.2]oct-5-ene-2,3-dione, 7-(1-naphthyl)-	EtOH	226(4.5),274(3.5), 283(3.6),295s(3.5), 314(3.3)	12-1775-65
Cinnamic acid, 3-phenyl-2-propynyl ester, trans	EtOH	218(4.28),224(4.31), 243(4.31),279(4.43)	78-0871-64
Dibenzo[ef,kl]heptalene-5,11(4H,6H)-dione, 10,12-dihydro-	isooctane	256.5(4.09)	35-1710-64
4H-Pyran-4-one, 5-benzyl-2-phenyl-	EtOH	273(3.64)	35-3186-65
$C_{18}H_{14}O_3$			
5-Benzofuranacrylic acid, 2,3-dihydro-6-hydroxy-α-methyl-β-phenyl-, δ-lactone	EtOH	335(4.25)	44-2467-64
2H-1-Benzopyran, 2,2'-oxybis-	EtOH	254(4.21),296(3.72)	17-0151-65
3-Furoic acid, 2-methyl-4,5-diphenyl-	MeOH	284(4.25)	39-5984-65
$C_{18}H_{14}O_4$			
Anthracene-2,6-diol, 1,5-diacetyl-	$CHCl_3$	276(4.67),345(4.11), 363(4.29),403(3.72), 425(3.95),451(3.45)	39-4565-64
Anthrone, 10-ethynyl-10-hydroxy-2,6-dimethoxy-	EtOH	229(4.34),246(4.31), 276s(3.85),304(4.21), 332s(3.80)	39-4565-64
2-Butyne, 1,4-bis(p-formylphenoxy)-	EtOH	273(4.48)	78-2289-65
Dibenzo[a,e]cyclooctatriene, bis(methylenedioxy)-	EtOH	303.7s(4.10)	23-2183-65
1,7-Phenanthrenediol, diacetate	EtOH	256(4.81),278(4.19), 286(4.07),298(4.08), 318(2.66),327(2.56), 334(2.62),340(2.47), 349(2.38)	88-3671-64

Compound	Solvent	λ_{max}(log ϵ)	Ref.
$C_{18}H_{14}O_5$			
Osajaxanthone	EtOH	240(3.87),285(4.67), 339(3.90),382(3.68)	44-0689-64
$C_{18}H_{14}O_6$			
Coumestan, 7,11,12-trimethoxy-	EtOH	247(3.02),306(3.84), 347(3.42)	44-2353-65
11H-Furo[2,3-a]xanthen-11-one, 4,7,8-trimethoxy-	MeOH	254(4.58),263(4.60), 281(4.14),315(4.25)	78-2653-65
2-Naphthoic acid, 3-(2-furyl)-1,4-dihydro-8-methoxy-1,4-dioxo-, ethyl ester	EtOH	258(4.32),311(3.75), 430(3.92)	33-0769-64
4H-Naphtho[1,2-b]pyran-4-one, 3-acetyl-5,6-dihydroxy-2-methyl-, 6-acetate	EtOH	218s(4.40),230(4.50), 261(4.41),270s(4.38), 285s(4.18),315s(3.64), 376(3.35)	94-0316-64
4H-Naphtho[1,2-b]pyran-4-one, 5,6-dihydroxy-2-methyl-, diacetate	EtOH	220(4.62),248s(4.50), 256(4.55),277s(4.13), 287s(3.86),299(3.74), 320s(3.36),335(3.57), 348(3.57)	94-0316-64
Succinic acid, (o-carboxy-α-phenylbenzylidene)-	pH 1	265s(3.93)	35-3814-64
	pH 13	230s(4.16),263(4.03)	35-3814-64
Unknown keto acid, methyl ester	EtOH	241(4.4),300(4.05)	39-3811-65
$C_{18}H_{14}O_7$			
Quercetin, 7-O-allyl-	EtOH	258(4.4),372(4.4)	24-0114-65
	EtOH-NaOAc	258(4.3),384(4.1)	24-0114-65
	EtOH-EtONa	294(4.2),358(4.0)	24-0114-65
	EtOH-$AlCl_3$	268(4.4),433(4.4)	24-0114-65
$C_{18}H_{14}O_8$			
Flavone, 5,7-dihydroxy-6,8-dimethoxy-3',4'-(methylenedioxy)-	EtOH-HCl	243s(4.20),284(4.24), 344(4.31)	39-2743-65
	EtOH	285(4.29),343(4.26)	39-2743-65
	EtOH-NaOEt	240(4.35),285(4.32), 317(4.07),384(3.94)	39-2743-65
$C_{18}H_{14}O_{10}$			
Mompain, tetraacetate	EtOH	248(4.18),266s(4.02), 352(3.53)	94-0633-65
$C_{18}H_{14}Ru$			
Ruthenium, diindenyl-	hexane	230(4.29),292(3.73), 340(3.21),420(2.87)	101-0107-65A
$C_{18}H_{14}S$			
Naphtho[2,3-c]thiophene, 1,3-dihydro-4-phenyl-	EtOH	229(4.94),277s(3.84), 287(3.93),296s(3.85), 322(2.83)	94-1094-64
Sulfide, bis(3-phenyl-2-propynyl)-	EtOH	245(4.63),262(4.59)	94-1094-64
$C_{18}H_{14}S_2$			
Benzene, 1,4-bis(2-vinylthienyl)-	dioxan	253(4.65),347(4.72)	38-2839-64A
$C_{18}H_{14}S_3$			
6a-Thiathiophthene, 2-methyl-3,5-diphenyl-	dioxan	250(4.7),285(4.7), 330s(4.3),490(4.4)	5-0188-65B
6a-Thiathiophthene, 2-phenyl-5-p-tolyl-	dioxan	256(4.69),313(4.39), 511(4.21)	24-1732-64

Compound	Solvent	λ_{max}(log ε)	Ref.
$C_{18}H_{15}AsBrI_3$ Arsenic, bromotriphenyl(triiodo)-	MeCN	289(4.61),360(4.15)	39-6076-64
$C_{18}H_{15}AsBr_2I_2$ Arsenic, (dibromoiodo)iodotriphenyl-	MeCN	275(4.64),351(3.98)	39-6076-64
$C_{18}H_{15}AsBr_3I$ Arsenic, iodotriphenyl(tribromo)-	MeCN	257(4.72)	39-6076-64
$C_{18}H_{15}AsBr_4$ Arsenic, bromotriphenyl(tribromo)-	MeCN	269(4.73)	39-6076-64
$C_{18}H_{15}AsI_4$ Arsenic, iodotriphenyl(triiodo)-	MeCN	292(4.76),362(4.49)	39-6076-64
$C_{18}H_{15}BF_4$ 1,2;3,4-Dibenzo-5,6-dimethylcalicenium tetrafluoroborate	70% $HClO_4$	261(4.17)	89-0621-65
$C_{18}H_{15}BrN_2O_2$ 2,3'-Biindole, 5'-bromo-5,6-dimethoxy-	EtOH	247(4.37),293s(4.14), 323(4.40)	44-2030-64
$C_{18}H_{15}BrN_2S_2$ 2,3-Dihydro-3-(3-methyl-2-benzothiazolinylidene)-1H-pyrrolo[2,1-b]benzothiazolium bromide	EtOH	453(4.88)	65-2455-64
$C_{18}H_{15}BrO_3$ Chalcone, 2'-acetoxy-2-bromo-4-methyl-	n.s.g.	243(4.13),317(4.42)	49-0450-65
$C_{18}H_{15}BrO_4$ Chalcone, 2'-acetoxy-2-bromo-4'-methoxy-	n.s.g.	239(4.21),317(4.40)	49-0450-65
$C_{18}H_{15}BrO_6$ Anthraquinone, 1-bromo-2,4,5,7-tetramethoxy-	EtOH	226(4.67),286(4.51), 411(3.85)	12-0182-65
$C_{18}H_{15}Cl$ Cyclobuta[1,2-a:4,3-a']diindene, 9a-chloro-4b,4c,9,9a,9b,10-hexahydro-	EtOH	261(3.10),268(3.29), 274(3.36),328(2.38), 344(2.51),366(2.28)	24-2762-65
$C_{18}H_{15}ClGe$ Germane, chlorotriphenyl-	heptane	247(2.57),253(2.81), 259(2.95),264(2.94), 269(2.83)	28-3931-65A
$C_{18}H_{15}ClO_5$ Fragilin, di-O-methyl-	EtOH	273(4.62),385(3.74)	1-0839-65
$C_{18}H_{15}ClO_6$ 2-Naphthaleneacetic acid, 3-chloro-1,4-dihydro-α-(1-hydroxyethylidene)-1,4-dioxo-, ethyl ester, acetate	MeCN	247(4.18),253(4.19), 275(4.11)338(3.42)	44-3819-65
$C_{18}H_{15}ClSn$ Tin, chlorotriphenyl-	heptane	247(2.79),253(2.92), 259(3.00),264(2.93), 269(2.75)	28-3931-65A

Compound	Solvent	λ_{max}(log ε)	Ref.
$C_{18}H_{15}Cl_2NO_2$			
3(2H)-Furanone, 2,2-dichloro-5-(N-methylanilino)-4-p-tolyl-	EtOH	247(4.07),318(4.18)	44-4303-65
$C_{18}H_{15}Cl_3O_4$			
Hydroquinone, trichloro(p-methylbenzyl)-, diacetate	EtOH	266(2.88),274(2.86), 285s(2.51)	5-0152-64D
$C_{18}H_{15}Cl_6Sb$			
5-Benzylbenzocycloheptenylium hexachloroantimonate (V)	H_2SO_4	227(4.48),246(4.54), 284(4.67),342(3.63), 435(3.55)	35-3719-65
$C_{18}H_{15}LiSi$			
Lithium, triphenylsilyl-	n.s.g.	335(4.0)	25-0563-65
$C_{18}H_{15}N$			
Quinoline, 3,4-cyclopenteno-2-phenyl-	n.s.g.	252(4.32),287(3.73), 315(3.66),330(3.61)	39-2421-64
Quinoline, 6-methyl-4-styryl-	0.1N HCl	251(4.37),337s(4.07), 385(4.31)	7-1154-65
	0.001N HCl	251(4.37),337(4.12), 385(4.29)	7-1154-65
	EtOH	235(4.41),264s(3.97), 330(4.29)	7-1154-65
Quinoline, 8-methyl-4-styryl-	0.1N HCl	244(4.58),320(3.86), 362(3.85)	7-1154-65
	0.001N HCl	240(4.53),320(3.90), 370s(3.61)	7-1154-65
	EtOH	224(4.40),330(4.28)	7-1154-65
$C_{18}H_{15}NO$			
9-Anthraceneacetonitrile, 10-(1-hydroxyethyl)-	n.s.g.	260(5.25),317(3.00), 333(3.8),349(3.63), 366(3.79),387(3.74)	44-2037-65
Cyclopentanecarbonitrile, 2-oxo-3,3-diphenyl-	EtOH	296(2.39)	23-2512-65
	EtOH-base	278(4.14)	23-2512-65
Indene, 1-(4-acetamidobenzylidene)-	MeOH	360(4.5)	87-0390-65
	HOAc	360(4.6)	87-0390-65
Phenol, 4-amino-2,5-diphenyl-	EtOH-HCl	259(4.57),301(4.28)	4-0140-65
	EtOH	240(4.68),333(3.88)	4-0140-65
4(1H)-Pyridone, 3-benzyl-2-phenyl-	EtOH	268(4.01)	35-3186-65
9H-Pyrrolo[1,2-a]indole, 7-(benzyloxy)-	MeOH	264(4.3),301(3.51)	44-2904-65
Quinoline, 4-(2-methoxystyryl)-	MeOH	345(4.3)	87-0137-65
	HOAc	320(4.0),390(3.85)	87-0137-65
Quinoline, 4-(3-methoxystyryl)-	MeOH	330(4.3)	87-0137-65
	HOAc	320(4.0),365(3.85)	87-0137-65
Quinoline, 4-(4-methoxystyryl)-	MeOH	350(4.3)	87-0137-65
	HOAc	315(3.8),405(3.8)	87-0137-65
Quinoline, 6-methoxy-4-styryl-	0.1N HCl	254(4.40),350(4.02), 390(4.10)	7-1154-65
	0.001N HCl	242(4.36),254s(4.35), 345(4.05),390(4.06)	7-1154-65
	EtOH	238(4.46),320(4.23), 352(4.16)	7-1154-65
Quinoline, 8-methoxy-4-styryl-	0.1N HCl	233(4.04),262(4.48), 365(4.29),397s(4.18)	7-1154-65
	0.001N HCl	233(4.09),262(4.47), 365(4.29),397s(4.17)	7-1154-65
	EtOH	232s(4.32),248(4.40), 314(4.17),350s(4.07)	7-1154-65

Compound	Solvent	$\lambda_{max}(\log \epsilon)$	Ref.
$C_{18}H_{15}NO_2$			
Cinchoninic acid, 2-(2,4-xylyl)-	EtOH	240(4.4),323(3.9)	28-0954-64A
Cinchoninic acid, 2-(2,5-xylyl)-	EtOH	237(4.6),322(3.9)	28-0954-64A
Cinchoninic acid, 2-(3,4-xylyl)-	EtOH	268(4.5),341(4.0)	28-0954-64A
Isoxazole, 5-methyl-3-phenyl-4-phenacyl-	EtOH	240(4.29)	32-0206-65
$\Delta^{1,\alpha}$-Perinaphtheneacetic acid, α-cyano-2,3-dihydro-, ethyl ester	$CHCl_3$	267(4.26),365(4.01)	44-3166-65
1H-Pyrrolo[1,2-a]indol-1-one, 2,3-dihydro-7-(benzyloxy)-	MeOH	320(4.31)	44-2904-65
Spiro[indoline-3,1'(4'H)-naphthalene]-2,4'-dione, 2',3'-dihydro-1-methyl-	EtOH	250(4.24),289(3.59)	44-2973-65
Spiro[indoline-3,2'(1'H)-naphthalene]-2,4'(3'H)-dione, 1-methyl-	EtOH	251(4.27),295(3.37), 300s(3.32)	44-2973-65
$C_{18}H_{15}NO_3$			
11H,12H-Benzo[f]pyrano[2,3-b]quinolizine-11,12-dione, 6a,7-dihydro-5,9-dimethyl-	EtOH	230(4.44),310(3.70)	95-0874-64
$C_{18}H_{15}NO_4$			
4H-1-Benzopyran-2-carboxylic acid, 4-(p-hydroxyphenylimino)-, ethyl ester	EtOH	318(3.04),405(3.69)	65-1924-64
$C_{18}H_{15}NO_4S$			
2H-Thieno[3,2-b]pyrrol-3(4H)-one, 4-acetyl-6-acetoxymethyl-2-benzylidene-	EtOH	272(3.90),345(4.40)	44-1012-65
$C_{18}H_{15}N_2OP$			
Phosphine oxide, diphenyl(phenylazo)-	dioxan	296(4.14),502(2.00)	24-2844-65
$C_{18}H_{15}N_2PS$			
Phosphine sulfide, diphenyl(phenylazo)-	dioxan	294(4.13),493(2.16)	24-2844-65
$C_{18}H_{15}N_3O$			
9H-Pyridazino[3,4-b]indole, 3-(p-methoxyphenyl)-9-methyl-	EtOH	278(4.73),390(3.48)	94-1129-64
s-Triazine-2-carboxaldehyde, 4,6-di-p-tolyl-	MeOH	283(4.68)	44-0678-64
$C_{18}H_{15}N_3O_2$			
Hydrazine, 1-(p-nitrophenyl)-1,2-diphenyl-	n.s.g.	240(4.21),385(4.22)	33-1047-64
9H-Pyridazino[3,4-b]indole, 6-methoxy-3-(p-methoxyphenyl)-	EtOH	280(4.69),407(3.52)	94-1129-64
$C_{18}H_{15}N_3O_4S$			
Phthalimidine, 2-[p-(5-methyl-3-isoxazolsulfamoyl)phenyl]-	MeOH-acid	293(4.36)	95-1035-65
	MeOH	293(4.36)	95-1035-65
$C_{18}H_{15}N_5$			
Adenine, N^6-(diphenylmethyl)-	pH 1	280(4.33)	4-0115-64
	pH 7	271(4.33)	4-0115-64
	pH 13	276(4.31),284s(4.20)	4-0115-64
Adenine, 3-(diphenylmethyl)-	pH 1	274(4.22)	4-0115-64
	pH 7	269(4.11)	4-0115-64
	pH 13	270(4.09)	4-0115-64

Compound	Solvent	λ max(log ε)	Ref.
Adenine, 9-(diphenylmethyl)-	pH 1	258(4.24)	4-0115-64
	pH 7	263(4.23)	4-0115-64
	pH 13	263(4.24)	4-0115-64
$C_{18}H_{15}N_5O$			
Pterin, 5,6-dihydro-6,7-diphenyl-	MeOH	266(4.40),454s(3.67)	33-0764-65
Pterin, 7,8-dihydro-6,7-diphenyl-	MeOH	248(4.36),288(4.02), 376(4.16)	33-0764-65
$C_{18}H_{15}OP$			
Phosphine oxide, triphenyl-	EtOH	223(4.38),254(2.99), 259(3.15),265(3.28), 271(3.21),295(1.00)	39-2184-65
	EtOH	225(4.30),259(3.13)	49-0285-65
$C_{18}H_{15}O_3P$			
Phenyl phosphite	C_6H_{12}	264(3.30)	12-1579-65
$C_{18}H_{15}O_4P$			
Phenyl phosphate	C_6H_{12}	262(3.10)	12-1579-65
$C_{18}H_{15}P$			
Phosphine, triphenyl-	n.s.g.	262(4.02)	49-0285-65
$C_{18}H_{15}PS$			
Phosphine sulfide, triphenyl-	EtOH	220(4.12),249(3.56)	39-2184-65
$C_{18}H_{15}PSe$			
Phosphine selenide, triphenyl-	EtOH	220(4.54),266(3.73)	39-2184-65
$C_{18}H_{16}$			
7H-Benzocycloheptene, 5-benzyl-	EtOH	223(4.61)	35-3719-65
Chrysene, 1,2,3,4-tetrahydro-	EtOH	258(4.75),280(4.16), 289(4.04),301(4.09), 319(2.56),328(2.47), 335(2.67),351(2.50)	24-0593-65
1,2;3,4-Dibenzo-5,6-dimethylcalicene	C_6H_{12}	235(4.69),241(4.67), 248s(4.41),254s(4.18), 259(4.13),265(4.08), 284s(4.09),295(4.24), 336(4.27),353(4.24)	89-0621-65
	EtOH	233(4.66),237s(4.64), 247(4.45),263(4.31), 271s(4.15),295(4.22), 325(4.34),356s(4.15)	89-0621-65
$C_{18}H_{16}BrNO_2$			
1H-Pyrrolo[1,2-a]indol-1-ol, 7-(benzyl-oxy)-2-bromo-2,3-dihydro-	MeOH	278(4.00),297(3.67), 305s(3.51)	44-2904-65
$C_{18}H_{16}ClNO$			
Cinnamonitrile, α-(p-chlorophenyl)-β-ethyl-p-methoxy-, trans	EtOH	296(4.15)	87-0511-64
$C_{18}H_{16}ClNO_2$			
Indole-3-acetic acid, 1-(p-chloro-benzyl)-5-methyl-	MeOH	223(4.59),278(3.82), 292(3.78),302(3.68)	87-0204-65
Oxazole, 2-(p-chlorophenethyl)-5-(p-methoxyphenyl)-	C_6H_{12}	274(4.41)	4-0310-65

Compound	Solvent	$\lambda_{max}(\log \epsilon)$	Ref.
$C_{18}H_{16}ClNO_3$			
3-Indolineacetic acid, 3-(p-chloro-benzyl)-1-methyl-2-oxo-	EtOH	221s(4.23),255(3.90), 263s(3.69),282s(3.24)	44-2973-65
3-Indolineacetic acid, 3-(p-chloro-phenyl)-1-methyl-2-oxo-, methyl ester	EtOH	230s(4.15),257(3.9), 285s(3.24)	44-2973-65
$C_{18}H_{16}ClNO_4$			
(1-Azulenylmethylene)methylphenyl-ammonium perchlorate	MeCN	278(4.07),327(4.38), 409(4.52),485s(3.50)	24-2050-64
5H-Dibenzo[d,f]azonine-5,6(7H)-dione, 3-chloro-8,9-dihydro-11,12-dimethoxy-	MeOH	217(4.47),250(4.19), 340(3.55)	24-0046-65
$C_{18}H_{16}ClN_3O_4$			
2H-Pyrazino[1',2':1,5]pyrrolo[2,3-b]in-dole-1,4(3H,6H)-dione, 9-chloro-7,8-dimethoxy-2,6-dimethyl-3-methylene-	MeOH	258(4.42),297(4.50), 394(3.82)	39-0026-64
$C_{18}H_{16}Fe$			
Ferrocene, styryl-, cis	$CHCl_3$	292(4.06),446(2.70)	5-0088-64F
Ferrocene, styryl-, trans	C_6H_{12}	303(4.41),450(3.04)	5-0088-64F
	$CHCl_3$	307(4.31),458(2.96)	5-0088-64F
$C_{18}H_{16}FeS$			
Butadiene, 1-ferrocenyl-2-(2-thienyl)-	$CHCl_3$	348(4.53),360s(--), 467(3.36)	5-0088-64F
$C_{18}H_{16}N_2$			
3,3'-Biindole, 2,2'-dimethyl-	EtOH	232(4.64),283(4.10), 291(4.08)	23-2900-64
Cyclopentapyrazole, 1,4,5,6-tetra-hydro-1,3-diphenyl-	C_6H_{12}	287(4.32)	44-1582-64
1-Cyclopentene-1-carbonitrile, 2-amino-3,3-diphenyl-	EtOH	269(4.14)	23-2512-65
p-Phenylenediamine, 2,5-diphenyl-	EtOH-HCl	238(4.68),318(3.90)	4-0140-65
	EtOH	235(4.66),338(3.65)	4-0140-65
Pyridazine, 3-phenyl-6-(2,4-xylyl)-	EtOH	270(4.42)	39-7005-65
Pyridazine, 3-phenyl-6-(2,5-xylyl)-	EtOH	264.5(4.38)	39-7005-65
Pyridazine, 3-phenyl-6-(3,4-xylyl)-	EtOH	285(4.47)	39-7005-65
$C_{18}H_{16}N_2O$			
Fluoline, methyl-	EtOH	352(4.2)	5-0089-65A
3-Indolineacetonitrile, 3-benzyl-1-methyl-2-oxo-	EtOH	255(3.84),266s(3.69), 286s(3.07)	44-2973-65
3-Indolinepropionitrile, 1-methyl-2-oxo-3-phenyl-	EtOH	256(3.88),286s(3.20)	44-2973-65
Pyridazine, 3-(p-methoxyphenyl)-6-p-tolyl-	EtOH	299(4.58)	39-3342-65
$C_{18}H_{16}N_2O_2$			
Fluorene, 2-nitro-9-(3-dimethylamino-2-propen-1-ylidene)-	MeCN	243(4.69),281(4.32), 320(4.25),422(4.74), 475(4.34)	24-3331-64
2,5-Piperazinedione, 3-benzyl-6-benzylidene-	EtOH	296(4.1)	44-0277-65
2,5-Piperazinedione, 3,6-dibenzyli-dene-dihydro-	EtOH	296(4.1)	44-0277-65
3,5-Pyrazolidinedione, 1,2-diphenyl-4-propylidene-	MeOH	247(4.17),402(4.35)	32-0320-65
Pyridazine, 3,6-bis(p-methoxyphenyl)-	EtOH	305(4.60)	39-3342-65
1H-Pyrrolo[1,2-a]indole, 7-benzyloxy-2,3-dihydro-1-oxo-, oxime	MeOH	215(4.51),313(4.38)	44-2910-65

Compound	Solvent	λ_{max}(log ε)	Ref.
$C_{18}H_{16}N_2O_3$			
1,7-Dioxa-2,8-diazaspiro[4.4]nona-2,8-dien-6-ol, 6-methyl-3,9-diphenyl-	EtOH	258(4.32)	32-0206-65
Indole-3-carbonitrile, 5,6-dimethoxy-2-(p-methoxyphenyl)-	MeOH	238(4.36),332(4.44)	44-2253-65
$C_{18}H_{16}N_2O_4$			
Indole-3-glyoxylamide, 6-(benzyloxy)-5-methoxy-	EtOH	213(4.51),252(4.00), 259(3.99),292(4.21), 332(3.99)	4-0387-65
$C_{18}H_{16}N_2O_5$			
Acrylonitrile, 2-(4,5-dimethoxy-2-nitrophenyl)-3-(p-methoxyphenyl)-	MeOH	237(4.22),319(4.38)	44-2253-65
Acrylonitrile, 2-(p-nitrophenyl)-3-(3,4,5-trimethoxyphenyl)-	MeOH	280(3.94),364(4.29)	87-0583-65
Isoquinoline, 3,4-dihydro-5-methoxy-6,7-(methylenedioxy)-1-(o-nitrobenzyl)-	EtOH	280(3.99),305s(3.82)	12-1997-65
$C_{18}H_{16}N_2O_5S$			
4H-1-Benzopyran-2-carboxylic acid, 4-[(p-sulfamoylphenyl)imino]-, ethyl ester	EtOH	245s(4.19),324(4.05), 355s(3.89)	65-1924-64
$C_{18}H_{16}N_4$			
2,4'-Biphenyldiamine, 5-(phenylazo)-	22N H_2SO_4	220s(4.0),410(4.5)	12-0190-65
$C_{18}H_{16}N_4O$			
Benzamide, N-[(4-amino-2-phenyl-5-pyrimidinyl)methyl]-	EtOH	238(4.49),284(3.99)	94-0393-64
$C_{18}H_{16}N_4O_2$			
1H-Pyrrolo[1,2-a]indol-1-ol, 2-azido-7-(benzyloxy)-2,3-dihydro-, trans	MeOH	217(4.51),276(3.89), 310s(3.5)	44-2910-65
$C_{18}H_{16}N_4O_5$			
2-Benzimidazolone-4,5-quinone, 1,3-dimethyl-, dimer	$CHCl_3$	665(4.095)	65-3788-64
$C_{18}H_{16}N_4O_6S$			
Sydnone imine, N-(N-acetylsulfanilyl)-3-piperonyl-	EtOH	262(4.37),317(4.07)	87-0531-65
$C_{18}H_{16}N_4S_2$			
2H-Pyrimido[1,2-a]-s-triazine-2,4(3H)-dithione, 1,6,7,8-tetrahydro-1,3-diphenyl-	n.s.g.	270(4.43),295(4.45)	32-0735-65
$C_{18}H_{16}N_6O_4$			
1,4-Pentadien-3-one, 1,5-bis(p-nitrophenyl)-, amidinohydrazone, hydrochloride	propylene glycol	340(4.516)	95-0001-64
$C_{18}H_{16}N_6O_6$			
1,3,6,8-Nonatetraen-5-one, 1,9-bis-(5-nitro-2-furyl)-, amidinohydrazone, hydrochloride	propylene glycol	411(4.20)	95-0001-64

Compound	Solvent	λ max(log ε)	Ref.
$C_{18}H_{16}N_6O_8$			
1,2,4-Triazine, 3-acetamido-4-acetyl-4,5-dihydro-5-(5-nitrofurfuryl)-6-[2-(5-nitro-2-furyl)vinyl]-	propylene glycol	240(4.18),317(4.21), 393(4.27)	95-0001-64
$C_{18}H_{16}N_8O_8$			
2-Hexenal, 4-oxo-, bis(2,4-dinitrophenylhydrazone), trans	$CHCl_3$	435(4.62),478(4.58)	39-2955-65
4-Hexene-2,3-dione, bis(2,4-dinitrophenylhydrazone)	EtOH	342(4.30),397(4.31)	39-2187-64
$C_{18}H_{16}O$			
Bicyclo[3.1.0]hexan-2-one, 5,6-diphenyl-, cis (end absorptions)	EtOH	215(4.23),220(4.18), 230(4.03)	35-4036-64
trans (end absorptions)	EtOH	215(4.23),220(4.17), 230(3.88)	35-4036-64
Butadiene, p-monomethoxydiphenyl-	MeOH	325(4.69)	2-0017-64
Cyclopentadienoheptalene, 2-acetonyl-4-methyl-	EtOAc	375(4.08),392(4.16), 452(2.26),735(2.00), 805(2.11),910(2.05), 1040(1.63)	5-0194-64B
Cyclopentadienoheptalene, acetyl-2,4-dimethyl-	benzene	378(3.96),395(4.02), 499(2.87),790(2.08), 890(1.97),1025(1.54)	5-0194-64B
isomer	benzene	364(3.96),386(4.03), 400(4.08),438(3.67), 462(3.70),710(2.36), 768(2.50),865(2.47), 983(2.10)	5-0194-64B
1-Cyclopentene, 1-benzoyl-2-phenyl-	EtOH	211(4.22),247(4.33), 258(4.27),287(3.28), 315(3.19)	22-0722-65
1-Cyclopentene, 3-benzoyl-2-phenyl-	EtOH	211(4.20),218(4.10), 247(4.42),271(3.88), 326(2.04)	22-0722-65
1-Indanone, 2-benzylidene-3,3-dimethyl-, cis	MeOH	230(3.89),280s(3.85), 322(4.25)	44-1276-64
$C_{18}H_{16}OS$			
Cyclohexanone, 2-benzylidene-6-(2-thenylidene)-	hexane	347(4.33)	65-3645-64
	EtOH	236(4.08),366(4.27)	65-3645-64
	66% H_2SO_4	238(3.88),378(3.73), 511(4.83)	65-3645-64
Sulfoxide, 2,5-diphenyl-2-penten-4-ynyl methyl, cis	EtOH	243(4.28),307(4.35)	94-0663-65
trans	EtOH	317(4.50)	94-0663-65
1H-Thiopyran, 1-methyl-3,5-diphenyl-, 1-oxide	MeOH	240(4.42),364(4.00)	35-4972-65
$C_{18}H_{16}O_2$			
Bicyclo[1.1.0]butane-2-carboxylic acid, 1,3-diphenyl-, methyl ester	MeOH	255(4.08)	88-0945-65
[1,2'-Biindan]-2-one, 1'-hydroxy-	C_6H_{12}	214(4.08),261(3.07), 268(3.23),274(3.22), 276(3.16),280(2.46), 310(2.40),324(2.39)	24-0140-65
Biphenylene, 2,6-dipropionyl-	EtOH	219(4.01),227s(--), 260(4.51),285(4.59), 344s(3.63),359(3.92), 376(4.08)	39-4930-65

Compound	Solvent	λ max(log ε)	Ref.
Biphenylene, 2,7-dipropionyl-	EtOH	219s(4.17),228(4.18), 236s(4.14),266s(4.75), 274(4.82),307s(3.25), 326(3.32),344s(3.45), 367s(3.64),376(3.70), 388s(3.69)	39-4930-65
Cinnamic acid, cinnamyl ester, trans-trans	EtOH	216(3.45),223(3.25), 256s(3.42),273(3.46)	78-0871-64
Indene, 1-(2,5-dimethoxybenzylidene)-	MeOH	370(4.0)	87-0390-65
	HOAc	370(4.3)	87-0390-65
Indene, 1-(3,4-dimethoxybenzylidene)-	MeOH	365(4.5)	87-0390-65
	HOAc	365(4.4)	87-0390-65
Phenanthrene, 1-acetyl-7-methoxy-2-methyl-	EtOH	256(4.28),291(3.67)	70-1051-65
Phenanthrene, 3,6-dimethoxy-1-vinyl-	MeOH	224(4.57),259(4.61), 289(4.30),312(4.19), 349(3.34),365(3.36)	35-2177-64
$C_{18}H_{16}O_2S_2$			
Anthraquinone, 1,4-bis(ethylthio)-	EtOH	257(4.50),340(3.82), 504(3.67)	22-1648-65
$C_{18}H_{16}O_3$			
Crotonolactone, 3-(p-methoxybenzyl)-1-phenyl-	MeOH	261(4.01)	2-0017-64
Crotonolactone, 1-(p-methoxyphenyl)-3-benzyl-	MeOH	282(4.05)	2-0017-64
2,4-Pentadienoic acid, 2-(p-methoxyphenyl)-5-phenyl-	MeOH	227(4.19),325(4.45)	2-0017-64
2,4-Pentadienoic acid, 5-(p-methoxyphenyl)-2-phenyl-	MeOH	229(4.09),335(4.58)	2-0017-64
$C_{18}H_{16}O_4$			
Anthracene, 9-acetoxy-2,6-dimethoxy-	EtOH	232s(4.24),236(4.25), 254(4.94),263(5.32), 295(3.28),310(3.53), 325(3.69),342(3.59), 367s(3.32),384(3.63), 406(3.67)	39-4565-64
2-Anthracenemalonic acid, 1,2-dihydro-9-methyl-	EtOH	232(5.16),275(3.65), 285(3.70),295(3.56)	65-2576-64
2α-Anthracenemalonic acid, 1,2,3,4-tetrahydro-3β-hydroxy-9-methyl-, lactone	EtOH	232(5.03),280(3.79), 290(3.82)	65-2576-64
Chalcone, 2'-acetoxy-4-methoxy-	MeOH	340(4.29)	24-1910-64
Cyclopropaneacetic acid, 2-carboxy-1,3-diphenyl-	n.s.g.	220(4.20(end absorption)	35-4036-64
9,10-Phenanthrenediol, 9,10-dihydro-, diacetate, cis	EtOH	211(4.65),222s(4.02), 230s(3.76),268(4.26), 281s(4.04)	39-5544-64
trans	EtOH	211(4.71),225(4.08), 232s(3.83),270(4.26), 301(3.08)	39-5544-64
1,2-Phenanthrenedicarboxylic anhydride, 1,2,9,10-tetrahydro-7-methoxy-2-methyl-	EtOH	252(4.38),300(3.93)	70-2021-64
isomer	EtOH	245(4.35),251(4.34), 300(3.86)	70-2021-64
Pterocarpan, 7-acetoxy-4'-methoxy-	n.s.g.	285(3.80)	25-0562-65
Pyrano[2',3':3,4]xanthone, 5',6'-dihydro-1,5,6-trihydroxy-6',6'-dimethyl-	MeOH	251(4.63),285(4.05), 326(4.34)	78-2653-65

Compound	Solvent	λ_{max}(log ϵ)	Ref.
$C_{18}H_{16}O_4S$			
Dibenzo[c,e]thiepin-1,11-dicarboxylic acid, 5,7-dihydro-, dimethyl ester	dioxan	215(4.65),295s(3.45)	35-1710-64
$C_{18}H_{16}O_4S_2$			
Anthraquinone, 1,4-bis(ethylsulfinyl)-	EtOH	255(4.58),336(3.66), 414(3.09)	22-1648-65
$C_{18}H_{16}O_4Si$			
Phenyl 2-propynyl silicate	EtOH	265(3.5),272(3.6), 280(3.5)	70-1349-64
$C_{18}H_{16}O_5$			
Angelic acid, 4-(2,4-hexadiynylidene)-spiro[3,6-dioxabicyclo[3.1.0]hexane-2,2'(3'H)-furan]-3'-yl ester	ether	222(4.56),264(4.11), 277(4.26),292(4.18)	24-2596-65
Dibenzofuran, 2,7-diacetoxy-4,5-dimethyl-	MeOH	226(4.56),252(4.00), 260(4.13),287(4.28)	5-0010-64F
Dibenz[c,e]oxepin-1,11-dicarboxylic acid, 5,7-dihydro-, dimethyl ester	dioxan	258(3.97),282(3.74)	35-1710-64
Flavone, 5-hydroxy-4',7-dimethoxy-6-methyl-	n.s.g.	213(4.60),276(4.37), 327(4.42)	12-0692-64
Isoflavone, 3',4',7-trimethoxy-	EtOH	255(4.51),261(4.52)	78-0963-65
4H-Naphtho[1,2-b]pyran-4-one, 3-acetyl-5,6-dimethoxy-2-methyl-	EtOH	224(4.58),258(4.53), 285s(4.03),309(3.83), 356(3.57)	94-0316-64
Osajaxanthone, dihydro-	EtOH	237(4.49),262(4.55), 316(4.21),380(3.78)	44-0144-65
9-Phenanthrenecarboxylic acid, 3,5,6-trimethoxy-	EtOH	250(4.61),257(4.51), 272(4.24),316(4.08), 372(3.50)	78-2573-65
$C_{18}H_{16}O_6$			
1,4-Anthracenedione, 9,10-diacetoxy-1,2,3,4-tetrahydro-	n.s.g.	264(4.68),305(3.57), 348(3.53),362(3.61)	77-0015-65
Anthraquinone, 1,3,6,8-tetramethoxy-	EtOH	223(4.62),280(4.47), 300s(4.01),342(3.57), 411(3.74)	12-0182-65
Benzo[b]furan-6-ol, 2,3-dihydro-5-(2-methoxy-4,5-(methylenedioxy)-phenylacetyl)-	EtOH	239(4.21),284(4.14), 330(3.70)	31-0668-64
3(2H)-Benzofuranone, 2-(4,6-dimethoxy-salicylidene)-6-methoxy-	MeOH	256(3.88),294(3.68), 345(3.77),400(3.90)	2-0319-64
3(2H)-Benzofuranone, 2-(2,4,6-tri-methoxybenzoyl)-	EtOH	212(4.37),230s(4.02), 320(4.11)	39-2361-65
1,3-Cyclohexadiene, 2,3-divinyl-, maleic anhydride adduct	MeCN	257(3.5)	35-4506-65
Flavone, 3-hydroxy-2',5,6-trimethoxy-	MeOH	218(4.18),245(4.24), 351(3.83),395(3.70)	24-0164-65
	MeOH-NaOMe	218(4.18),250(4.26), 337(3.68),394(3.97)	24-0164-65
Flavone, 4'-hydroxy-3',5,7-trimethoxy-	EtOH	242(4.39),265(4.23), 338(4.37)	94-0841-64
	EtOH-NaOEt	256(4.28),289(3.93), 400(4.49)	94-0841-64
Flavone, 5-hydroxy-2',3,6-trimethoxy-	MeOH	214(4.15),256(4.16), 313s(3.68),367(3.51)	24-0164-65
	MeOH-NaOMe	214(4.16),253(4.14), 269(4.14),398(3.49)	24-0164-65

Compound	Solvent	λ_{max}(log ϵ)	Ref.
Flavone, 5-hydroxy-2',6,7-trimethoxy-	MeOH	214(4.33),269(4.17), 329(4.06)	24-0164-65
	MeOH-NaOMe	216(4.33),274(4.14), 328(4.02)	24-0164-65
Flavone, 5-hydroxy-6,7,8-trimethoxy-	EtOH	282(4.52),316s(4.05)	39-2743-65
Flavone, 7-hydroxy-5,6,8-trimethoxy-	EtOH	246s(4.30),271(4.45), 313(4.18)	39-2743-65
	EtOH-HCl	270(4.45),313(4.21)	39-2743-65
	EtOH-NaOEt	242(4.26),277(4.41), 373(3.94)	39-2743-65
Fonsecin, O-methyl-, triacetate	EtOH	225(4.35),247s(4.48), 268(4.55),355(3.8)	44-0112-65
$C_{18}H_{16}O_6S_2$			
Anthraquinone, 1,4-bis(ethylsulfonyl)-	EtOH	236(4.35),258(4.44), 300(3.48)	22-1648-65
2,7-Thianthrenediacetic acid, dimethyl ester, 5,10-dioxide	EtOH	227(4.8)	35-2957-64
isomer	EtOH	217(4.7)	35-2957-64
$C_{18}H_{16}O_7$			
2'-Acetonaphthone, 1',3',4'-triacetoxy-	EtOH	223(4.54),240s(4.38), 250s(4.13),277s(3.77), 284(3.82),295s(3.71), 330(3.25)	94-0316-64
Flavone, 4',5-dihydroxy-3,7,8-trimethoxy-	EtOH	274(4.34),331(4.14), 370(4.17)	12-0934-64
	EtOH-NaOH	250(4.25),276(4.26), 360(4.26),404(4.37)	12-0934-64
Gibbs reagent product	EtOH-borate	692(2.68)	12-0934-64
Flavone, 5,7-dihydroxy-3,4',8-trimethoxy-	EtOH	278(4.37),308(4.22), 358(4.09)	12-1871-65
	EtOH-NaOAc	286(4.49),305(4.30), 393(3.96)	12-1871-65
	EtOH-$AlCl_3$	286(4.28),313(4.24), 348(4.24),410(3.96)	12-1871-65
Trypacidin	n.s.g.	286(4.37),325s(3.84)	39-6658-65
Wightin	EtOH	275(4.39),335s(3.86)	78-3237-65
	EtOH-$AlCl_3$	293(4.40),330(4.10), 400(3.81)	78-3237-65
$C_{18}H_{16}O_8$			
Centaureidine	MeOH	256(4.25),350(4.31)	24-1666-64
Flavone, 3',4',5-trihydroxy-3,7,8-trimethoxy-	EtOH	262(4.34),271s(4.32), 374(4.19)	78-3219-65
	EtOH-NaOH	269(4.33),408(4.29)	78-3219-65
	EtOH-NaOAc	267(4.33),424(4.25)	78-3219-65
	EtOH-NaOAc-H_3BO_3	264(4.39),387(4.26)	78-3219-65
Flavone, 3',5,5'-trihydroxy-3,4',7-trimethoxy-	EtOH	264(4.27),350(4.24)	12-0934-64
	EtOH-NaOH	271(4.38),378(4.11)	12-0934-64
Gibbs reagent product	EtOH-borate	648(3.97)	12-0934-64
Flavone, 4',5,7-trihydroxy-3',6,8-trimethoxy-	EtOH	283(4.15),349(4.08)	39-6255-64
	EtOH-HCl	250s(3.99),282(4.06), 349(4.19)	39-6255-64
Jaceidin	EtOH	255(4.25),270(4.19), 354(4.35)	39-5651-65
	EtOH-NaOEt	275(4.11),330(4.17), 400(4.34)	39-5651-65

Compound	Solvent	λ max(log ε)	Ref.
Jaceidin (contd.)	EtOH-NaOAc-H_3BO_3	273(4.33),316(4.08),374(4.22)	39-5651-65
	EtOH-AlCl	245(4.00),265(4.21),385(4.31)	39-5651-65
$C_{18}H_{16}Si$			
Silane, triphenyl-	C_6H_{12}	254(2.92),260(3.06),264(3.08),266s(3.06),271(3.02)	39-4690-65
$C_{18}H_{16}Sn$			
Tin, dimethylbis(phenylethynyl)-	EtOH	247(4.53),260(4.49)	22-0035-65
$C_{18}H_{17}BrO$			
Ketone, 4-biphenylyl 1-bromocyclopentyl	EtOH	295(4.37)	87-0504-64
$C_{18}H_{17}ClN_2O$			
Carbostyril, 6-chloro-3-(dimethylamino)-1-methyl-4-phenyl-	iso-PrOH	240(4.6),355(4.0)	39-3097-64
3-Ethyl-10-methoxy-12H-indolo[2,3-a]quinolizinium chloride	EtOH	262(4.46),312(4.38),384(4.47)	94-1381-64
	EtOH-KOH	293(4.59),335(4.34),398(4.37)	94-1381-64
2-(Formylmethyl)-1-p-tolylisoquinolinium chloride, oxime	EtOH	221(4.59),280(3.79),318s(3.66),325(3.71),350s(3.20)	44-1846-65
$C_{18}H_{17}ClN_2O_6S_2$			
Cephalosporanic acid, 7-(p-chlorophenylthio)acetamido-	pH 6.0	255.5(4.19)	39-5015-65
$C_{18}H_{17}ClO_4$			
7-(4-p-Tolyl-1,3-butadien-1-yl)-tropylium perchlorate	MeCN	228(4.29),269(4.11),299(4.20),343(3.69),513(4.70)	24-1590-64
$C_{18}H_{17}ClO_5$			
7-[4-(p-Methoxyphenyl)-1,3-butadien-1-yl]tropylium perchlorate	MeCN	231(4.35),273(4.26),300(4.09),350(3.65),550(4.76)	24-1590-64
$C_{18}H_{17}ClO_6$			
Radicicol	EtOH	265(4.25)	88-0365-64
	EtOH-base	254(4.35),274(4.35),319(4.18)	88-0365-64
$C_{18}H_{17}ClS$			
1-Benzothiepin, 5-[2-(2-chloroethyl)-phenyl]-2,3-dihydro-	MeOH	240s(4.16)	24-0685-65
$C_{18}H_{17}N$			
Benzofulvene, ω-(p-dimethylaminophenyl)-	dioxan	253(4.37),400(4.60)	5-0018-64D
Cyclopentanecarbonitrile, 1-(4-biphenylyl)-	EtOH	252(4.46),258(4.47)	87-0504-64
Fluorene, 9-(3-dimethylamino-2-propen-1-ylidene)-	MeCN	246(4.82),249(4.82),292(4.02),304(3.91),415(4.79)	24-3331-64

Compound	Solvent	λ_{max}(log ε)	Ref.
Indene, 1-[4-(dimethylamino)-benzylidene]-	MeOH	420(4.6)	87-0390-65
	HOAc	340(4.3),420(3.9)	87-0390-65
2-Naphthylamine, 1,4-dimethyl-N-phenyl-	EtOH	211(4.49),226(4.65), 246s(4.19),273s(4.35), 278(4.37),303(4.15), 312(4.14),345s(3.45)	44-0190-65
Pyrrole, 3,4-di-o-tolyl-	EtOH	235(4.19),257(3.96)	44-0859-65
Pyrrole, 3,4-di-p-tolyl-	EtOH	243(4.13),264(4.10)	44-0859-65
$C_{18}H_{17}NO$			
Indene, 1-(4-methylamino-3-methoxy-benzylidene)-	MeOH	415(4.5)	87-0390-65
	HOAc	405(4.3)	87-0390-65
2-Indolinone, 3-(α-ethyl-phenethylidene)-	EtOH	254(4.47),261(4.51), 294(3.89),354(3.23)	87-0626-65
2-Indolinone, 3-(1-methyl-3-phenyl-propylidene)-	EtOH	255(4.47),261(4.47), 296(3.91),357(3.26)	87-0626-65
2-Pentenenitrile, 3-(p-methoxyphenyl)-2-phenyl-, cis	EtOH	304(4.06)	87-0511-64
trans	EtOH	292(4.13)	87-0511-64
Tetralin, 1-carboxy-1-(o-N-methyl-aminophenyl)-, lactam	EtOH	255(3.91),280s(3.32)	44-2973-65
Tetralin, 3-carboxy-3-(o-N-methyl-aminophenyl)-, lactam	MeOH	253(3.94),273(3.44), 284s(3.17)	44-2973-65
$C_{18}H_{17}NO_2$			
Anthracene-3-carbonitrile, 1,2,3,4,-4aβ,9,9aα,10-octahydro-3α-hydroxy-10-oxo-2-(2-propynyl)-	EtOH	248(3.94),292(3.12)	70-1039-65
13-Aza-D-homo-18-norequilenin methyl ether, 14(15)-dehydro-	EtOH	245(4.25),254(4.31), 305(3.52)	39-3007-65
Carbamic acid, N,N-dimethyl-2,3-di-phenyl-2-cyclopropenyl ester	CH_2Cl_2	227(4.37),232s(4.34), 289s(4.44),304(4.56), 320(4.44)	88-3459-64
4bH-Cyclopropa[1]phenanthrene-5-carbox-amide, 5,5a-dihydro-N,N-dimethyl-	EtOH	250(3.72),272(4.14), 281(4.09),299(3.69), 311(3.73)	88-2919-64
Isoquinoline, 7-(benzyloxy)-6-methoxy-1-methyl-	EtOH-HCl	251(4.84),307(3.93)	33-1945-65
	EtOH	236(4.81),265(3.70), 275s(3.63),311(3.49), 324(3.53)	33-1945-65
	EtOH-NaOH	235(4.84),265(3.72), 274s(3.66),311(3.51), 324(3.54)	33-1945-65
Pyrrole, 3,4-bis(p-methoxyphenyl)-	EtOH	235(3.95),260(4.07)	44-0859-65
1H-Pyrrolo[1,2-a]indol-1-ol, 7-(benzyloxy)-2,3-dihydro-	MeOH	278(4.03),298s(3.8), 310s(3.6)	44-2904-65
Remerine	EtOH	271(4.27),314(3.54)	33-2119-64
Vinylamine, benzoyl-α-acetyl-β-methyl-β-phenyl-	EtOH	268(4.20)	32-1115-65
$C_{18}H_{17}NO_3$			
Apoerysopin-5-one, dimethyl ether	MeOH	217(4.47),256(4.29), 272(4.18),306(3.97)	78-1729-64
Butyrolactam, N-(o-benzyloxybenzoyl)-	EtOH	213(4.49),294(3.34)	65-1394-65
	dioxan	212(4.49),294(3.34)	78-3537-65
3-Indolineacetic acid, 3-benzyl-1-methyl-2-oxo-	EtOH	255(3.9),265s(3.80), 285s(3.24)	44-2973-65
3-Indolineacetic acid, 1-methyl-2-oxo-3-phenyl-, methyl ester	EtOH	256(3.88),286s(3.18)	44-2973-65

Compound	Solvent	$\lambda_{max}(\log \epsilon)$	Ref.
3-Indolinepropionic acid, 1-methyl-2-oxo-3-phenyl-	EtOH	255(3.88),282s(3.26)	44-2973-65
2-Indolinone, 3-(3,4-dimethoxy-benzylidene)-1-methyl-	MeOH	263(4.21),368(4.37)	87-0626-65
Isofugapavine	n.s.g.	222(4.50),266(4.20), 274(4.23),307(4.00)	70-0502-65
4-Quinolinepropionic acid, 1,2,3,4-tetrahydro-2-oxo-4-phenyl-	EtOH	251(4.04),263s(3.89), 280s(3.36)	44-2973-65
$C_{18}H_{17}NO_3S$			
L-Cysteine, S-(9-anthryl)-N-acetyl-	EtOH	222(4.13),249s(4.63), 259(4.89),339(3.25), 357(3.56),375(3.73), 393(3.71)	12-0109-64
$C_{18}H_{17}NO_4$			
2-Indolinone, 3-(3,4,5-trimethoxy-benzylidene)-	EtOH	253(4.18),354(4.22)	87-0626-65
$C_{18}H_{17}NO_5$			
Indole-7-carboxylic acid, 3-acetoxy-2,4,5,6-tetrahydro-2-oxo-1-phenyl-, methyl ester	n.s.g.	288(4.02)	39-1648-65
$C_{18}H_{17}NO_5S$			
2H-1,3-Benzothiazine-2-carboxylic acid, 3,4-dihydro-3-methyl-4-oxo-2-phenyl-, ethyl ester, 1,1-dioxide	EtOH	266(3.46),272(3.49), 279(3.49),287(3.42)	44-2068-64
$C_{18}H_{17}NO_7$			
1H-Pyrrolo[1,2-a]indole-9-carboxaldehyde, 5,7,8-triacetoxy-2,3-dihydro-	MeOH	219(4.44),247(4.16), 304(4.06)	44-4381-65
$C_{18}H_{17}N_3$			
Pyrimidine, 1-benzyl-2-(benzyl-imino)-1,2-dihydro-	pH 7.0	232(4.24),317(3.66)	39-5542-65
	pH 13.1	246(4.24),370(3.45)	39-5542-65
$C_{18}H_{17}N_3O_2$			
Urea, N,N-dimethyl-N'-nitroso-N'-(2,3-diphenyl-2-cyclopropen-1-yl)-	n.s.g.	305(4.4),318(4.3)	88-3459-64
$C_{18}H_{17}N_3O_3S_3$			
1-[[[2,2-Dicyano-1-(methylthio)vinyl]-thio]methyl]pyridinium p-toluene-sulfonate	MeCN	219(3.99),263(3.76), 332(4.06)	44-0497-64
$C_{18}H_{17}N_3O_5S$			
Carbanilic acid, p-[(5-methyl-3-isoxazol)sulfamoyl]-, benzyl ester	MeOH	257(4.38)	95-1104-64
$C_{18}H_{17}N_5O_2S$			
1,2,5-Thiadiazole-3-carboxamide, N-benzyl-4-(3-benzylureido)-	EtOH	232(4.31),311(3.92)	44-2141-64
$C_{18}H_{18}$			
Anthracene, 1,2,3,5-tetramethyl-	EtOH	260(5.10),330(3.33), 346(3.60),365(3.75), 385(3.65)	95-1072-64
Anthracene, 1,2,3,6-tetramethyl-	EtOH	260(5.25),343(3.76), 365(3.81),385(3.79)	95-1072-64

Compound	Solvent	$\lambda_{max}(\log \epsilon)$	Ref.
Anthracene, 1,2,3,8-tetramethyl-	EtOH	260(5.10),330(3.33), 347(3.60),365(3.73), 385(3.67)	95-1072-64
Anthracene, 1,4,5,8-tetramethyl-	C_6H_{12}	257(5.50),311(3.05), 327(3.39),342(3.61), 358(3.72),378(3.62)	33-0437-65
Chrysene, 3,4,5,6,7,8-hexahydro-	EtOH	274(4.28)	24-0593-65
Cyclopenta[ef]heptalene, 9-isopropyl-1-methyl-	hexane	381(4.13),394(4.24), 466(2.24),777(1.71), 859(1.79),992(1.70), 1175(1.18)	5-0194-64B
Cyclopenta[ef]heptalene, 1,2,3,5-tetramethyl-	hexane	384(4.05),397(4.10), 445(2.77),477(2.64), 790(1.97)	5-0194-64B
Dibenzo[ef,kl]heptalene, 4,5,6,10,11,12-hexahydro-	isooctane	252(4.15)	35-1710-64
9,10-Ethanoanthracene, 9,10-dihydro-9,10-dimethyl-	n.s.g.	258s(--),265(3.06), 272(3.14)	23-1754-65
Phenanthrene, 9,10-diethyl-	EtOH	225(4.34),249s(4.68), 255(4.77),271(4.25), 279(4.09),290(4.02), 302(4.07),336(2.70), 344(2.48),352(2.74)	5-0036-64D
Phenanthrene, 7-isopropyl-1-methyl-	EtOH	215(4.39),259(4.69), 280(4.04),300(4.04), 320(2.62)	65-0280-65
$C_{18}H_{18}BF_4N_2O_2PS_2$			
6-Methoxy-2-[(6-methoxy-3-methyl-2-benzothiazolinylidene)phosphino]-3-methylbenzothiazolium tetrafluoroborate	n.s.g.	485(4.64)	89-0433-64
$C_{18}H_{18}BF_4N_2PS_2$			
3-Ethyl-2-[(3-ethyl-2-benzothiazolinylidene)phosphino]benzothiazolium tetrafluoroborate	n.s.g.	472(4.64)	89-0433-64
$C_{18}H_{18}BrNO_3$			
Isoquinoline, 8-bromo-3,4-dihydro-1-(4-hydroxybenzyl)-6,7-dimethoxy-	EtOH	275(3.916)	65-0550-64
$C_{18}H_{18}Br_2N_8O_4$			
Theobromine, 1,1'-(2-butenylene)-bis[8-bromo-	n.s.g.	240(3.7),282(4.3)	83-0146-64
$C_{18}H_{18}ClNO_2$			
6-Isoquinolinol, 1-(p-chlorophenethyl)-3,4-dihydro-7-methoxy-	EtOH-HCl	249(4.19),312(3.97), 368(3.99)	33-1945-65
	EtOH	268(3.95),318(3.58), 410(4.34)	33-1945-65
	EtOH-NaOH	247(4.40),336(4.16)	33-1945-65
hydrobromide	EtOH-HCl	252(4.19),312(3.97), 367(3.98)	33-1945-65
	EtOH	252(4.16),312(3.94), 370(3.96)	33-1945-65
	EtOH-NaOH	248(4.21),337(4.16)	33-1945-65
7-Isoquinolinol, 1-(p-chlorophenethyl)-3,4-dihydro-6-methoxy-	EtOH-HCl	217(4.31),248(4.25), 308(3.99),368(3.91)	33-1945-65

Compound	Solvent	λ max(log ε)	Ref.
7-Isoquinolinol, 1-(p-chlorophenethyl)-3,4-dihydro-6-methoxy- (contd.)	EtOH	224(4.46),272(3.83), 313(3.81)	33-1945-65
	EtOH-NaOH	247(4.43),280s(3.70), 352(3.69)	33-1945-65
hydrobromide	EtOH-HCl	217(4.32),248(4.25), 308(3.99),368(3.91)	33-1945-65
	EtOH	217(4.35),248(4.20), 308(3.97),370(3.86)	33-1945-65
	EtOH-NaOH	247(4.45),280s(3.70), 350(3.72)	33-1945-65
$C_{18}H_{18}ClNO_3$			
6H-Dibenzo[d,f]azonin-6-one, 5-chloro-5,7,8,9-tetrahydro-11,12-dimethoxy-	MeOH	216(4.58),255s(3.90), 294(3.81)	24-0046-65
$C_{18}H_{18}ClNO_4$			
Erythrinane-4,6-dien-8-one, 4-chloro-7-hydroxy-15,16-dimethoxy-	MeOH	210(4.33),233(4.26), 285(4.17)	24-0046-65
1-Ethyl-2-phenyllepidinium perchlorate	EtOH	235(4.48),316(4.09)	65-0506-65
6-Methyl-1-p-tolyllepidinium perchlorate	EtOH	243(4.56),320(3.98)	65-0506-65
6-Methyl-1-p-tolylquinaldinium perchlorate	EtOH	245(4.52),325(3.99)	65-0506-65
	EtOH	245(4.5),325(4.0)	65-3360-64
$C_{18}H_{18}ClNO_6$			
6-Methoxy-1-(p-methoxyphenyl)-lepidinium perchlorate	EtOH	252(4.48),320(3.78)	65-0506-65
6-Methoxy-1-(p-methoxyphenyl)-quinaldinium perchlorate	EtOH	240(4.67),320(4.10)	65-0506-65
$C_{18}H_{18}ClN_3O_2$			
3H-1,4-Benzodiazepine, 7-chloro-2-[(methoxymethyl)methylamino]-5-phenyl-, 4-oxide	EtOH-acid	250(4.52),307s(3.98)	87-0235-64
$C_{18}H_{18}ClN_3O_4$			
3H-Pyrazino[1',2':1,5]pyrrolo[2,3-b]indole, 7-chloro-1,2,4,10-tetrahydro-8,9-dimethoxy-2,3,6-trimethyl-1,4-dioxo-	MeOH	243(4.37),282(4.31), 335(4.37)	39-0026-64
$C_{18}H_{18}ClN_3O_4S_2$			
Sporidesmin B, anhydro-	MeOH	232(4.42),321(4.08)	39-4315-64
$C_{18}H_{18}Cl_2O_6$			
2H-1-Benzopyran-4-malonic acid, 3,6-dichloro-α-ethyl-2-oxo-, diethyl ester	EtOH	223(4.21),286(3.91), 333(3.68)	44-4126-65
$C_{18}H_{18}Cl_4$			
Butadiene, bis(7,7-dichloro-1-norcaryl)-	n.s.g.	245(3.73),260(3.81), 275(3.67)	22-1518-65
$C_{18}H_{18}F_3N_3$			
Benzimidazole, 1-(2-dimethylaminoethyl)-2-phenyl-5-(trifluoromethyl)-	EtOH-HCl	251(3.85),286(4.12)	4-0453-65
$C_{18}H_{18}FeO_2$			
2,4,6-Heptatrienoic acid, 7-ferrocenyl-, methyl ester	$CHCl_3$	353(4.56),492(3.64)	5-0088-64F

Compound	Solvent	λ_{max}(log ε)	Ref.
$C_{18}H_{18}FeO_4$			
Malonic acid, 3-ferrocenylprop-2-en-1-ylidene-, dimethyl ester	$CHCl_3$	342(4.38),510(3.60)	5-0088-64F
$C_{18}H_{18}IN_3O_5$			
Cytidine, N-benzoyl-2',5'-dideoxy-5'-iodo-, 3'-acetate	MeOH	260(4.36),303(4.02)	44-3067-65
$C_{18}H_{18}N_2$			
1,2-Diazocine, 4,5,6,7-tetrahydro-3,8-diphenyl-	EtOH	212(4.29),270(4.45)	22-0722-65
Quinoline, 1,2-dihydro-, dimer	EtOH	208(4.67),243(4.23), 300(3.78)	1-1389-64
$C_{18}H_{18}N_2O$			
1H-Cyclopent[f]indolo[2,3-a]quinolizin-13(5H)-one, 2,3,6,11,11b,12-hexahydro-	EtOH	224(4.53),280(3.94), 292(3.92),334(4.19)	4-0329-65
Pyrrolo[1,2-a]quinoxalin-1(5H)-one, tetrahydro-2-methyl-4-phenyl-	EtOH	231(4.35),262s(3.92), 319(3.76)	35-1830-64
$C_{18}H_{18}N_2O_2$			
3-Indolineacetamide, N,1-dimethyl-2-oxo-3-phenyl-	EtOH	256(3.91),262s(3.86), 283(3.29)	44-2973-65
1H-Pyrrolo[1,2-a]indol-1-ol, 2-amino-7-(benzyloxy)-2,3-dihydro-, trans	MeOH	278(4.60)	44-2910-65
Tryptophan, 1-benzyl-	90% EtOH	223(4.55),277s(3.76), 286(3.80),296s(3.73)	94-0088-65
	90% EtOH-HCl	221(4.56),276s(3.80), 284(3.83),295s(3.75)	94-0088-65
$C_{18}H_{18}N_2O_3$			
2-Indolinone, 3-(1-methyl-1-phenyl-2-nitroethyl)-	EtOH	251(3.91),282(3.17), 293s(3.06)	87-0626-65
Valeric acid, 4-hydroxy-2-oxo-5-(o-tolyloxy)-, γ-lactone, phenylhydrazone	n.s.g.	295(3.92),333(4.39)	39-0141-64
$C_{18}H_{18}N_2O_3S$			
Malonamic acid, 2-(2-phenylacetamido)-3-thio-, benzyl ester	EtOH	272.0(4.04)	39-1262-65
$C_{18}H_{18}N_2O_4$			
Benzoic acid, 4,4'-azodi-, diethyl ester	EtOH	223s(4.23),327(4.46), 466(2.90)	78-0189-64
Valeric acid, 4-hydroxy-5-(p-methoxyphenoxy)-2-oxo-, γ-lactone, phenylhydrazone	n.s.g.	295(4.01),333(4.39)	39-0141-64
$C_{18}H_{18}N_2O_6S$			
Cephalosporanic acid, 7-phenylacetamido-	pH 6.0	259(3.97)	39-5015-65
$C_{18}H_{18}N_2O_6S_2$			
Cephalosporanic acid, 7-phenylmercaptoacetamido-, potassium salt	n.s.g.	250(4.08)	87-0022-65
$C_{18}H_{18}N_2S$			
1-Cyclopentene-1-carboxanilide, 2-anilinothio-	$CHCl_3$	314(3.95),386(4.38)	35-4509-64

Compound	Solvent	λ_{max}(log ε)	Ref.
$C_{18}H_{18}N_4$			
4,4'-Bicinnoline, 1,1',4,4'-tetrahydro-1,1'-dimethyl-	EtOH	306(4.10)	39-5391-65
$C_{18}H_{18}N_4O_4$			
5(6H)-Benzocyclooctenone, 7,8,9,10-tetrahydro-, 2,4-dinitrophenylhydrazone	$CHCl_3$	367(4.40)	5-0021-64G
1(2H)-Naphthalenone, 3,4-dihydro-2,2-dimethyl-, 2,4-dinitrophenylhydrazone	EtOH	380(4.43)	44-2956-65
1,3,4-Oxadiazole, 2-(butylamino)-5-[1-phenyl-2-(5-nitro-2-furyl)vinyl]-	propylene glycol	298(3.93),382(4.16)	95-0219-64
2-Pentenal, 2-methyl-4-phenyl-, 2,4-dinitrophenylhydrazone	$CHCl_3$	362(4.40)	22-0161-64
red isomer	$CHCl_3$	381(4.50)	22-0161-64
2-Quinoxalinecarboxanilide, 2'-amino-3-(1,2,3-trihydroxypropyl)-	EtOH	241(4.54),300(3.89)	33-1860-64
$C_{18}H_{18}N_4O_4S$			
Benzenesulfonic acid, m-[3-methyl-5-oxo-4-(2,6-xylylazo)-2-pyrazolin-1-yl]-	MeOH	380(5.0)	27-0244-64
$C_{18}H_{18}N_4O_5$			
p-Benzoquinone, 2-methyl-5-propyl-, 1-(N-ethyl-2,4-dinitrophenylhydrazone)	EtOH	410(4.31)	5-0134-65E
Indan, 1-acetyl-5-methoxy-, 2,4-dinitrophenylhydrazone	$CHCl_3$	364(4.35)	22-3718-65
$C_{18}H_{18}N_4O_6$			
1-Tetralone, 6,7-dimethoxy-, 2,4-dinitrophenylhydrazone	$CHCl_3$	250(4.19),309(3.93),407(4.41)	39-0361-65
1-Tetralone, 6,8-dimethoxy-, 2,4-dinitrophenylhydrazone	$CHCl_3$	402.5(4.42)	5-0021-64G
$C_{18}H_{18}N_4O_6S$			
Alanine, N-[2-(7-nitro-4H-1,2,4-benzothiadiazin-3-yl)ethyl]-3-phenyl-, S,S-dioxide	2% $NaHCO_3$	241(4.04),359(4.10)	32-0786-65
$C_{18}H_{18}N_4O_7$			
Theophylline, N^7-(β-D-glucopyranosyl)-	H_2O	275(3.94)	88-3095-65
$C_{18}H_{18}N_4O_7S$			
Tyrosine, N-[2-(7-nitro-4H-1,2,4-benzothiadiazin-3-yl)ethyl]-, S,S-dioxide	2% $NaHCO_3$	242s(--),360(4.10)	32-0786-65
$C_{18}H_{18}N_4O_8$			
2,4-Cyclopentadiene-1,2-dicarboxylic acid, 5-acetyl-3-methyl-, dimethyl ester, 2,4-dinitrophenylhydrazone	n.s.g.	400(4.46)	65-0058-65
$C_{18}H_{18}N_6O_5$			
Cytidine, 5'-azido-N-benzoyl-2',5'-dideoxy-, 3'-acetate	MeOH	260(4.29),302(3.98)	44-3067-65

Compound	Solvent	λ_{max}(log ϵ)	Ref.
$C_{18}H_{18}N_6O_9$			
Thymine, 1-(2-deoxy-β-D-erythro-pento-dialdo-1,4-furanosyl)-, 2,4-dinitro-phenylhydrazone, 3'-acetate	MeOH	261(4.29),350(4.33)	35-5661-65
$C_{18}H_{18}O$			
1'-Acetonaphthone, 2-cyclohexylidene-	EtOH	304(3.90)	39-3484-64
Chrysofluorene, 1,1a,2,3,5,6-hexa-hydro-1a-methyl-3-oxo-	EtOH	226(4.26),241(4.26), 251(4.26),257(4.22)	42-0479-64
Cyclopenta[ef]heptalene-1-methanol, α,3,5-trimethyl-	benzene	379(4.11),396(4.17), 460(2.44),731(2.05), 800(2.17),898(2.10), 1035(1.69)	5-0194-64B
Fluorene, 9-ethyl-9-propionyl-	EtOH	212(4.48),268(4.33), 292(3.92),305(3.92)	22-2345-65
Ketone, 4-biphenylyl cyclopentyl	EtOH	283(4.40)	87-0504-64
Phenanthrene, 1-ethyl-7-methoxy-2-methyl-	EtOH	225(4.26),258(4.77), 277(4.22),290(4.00), 300(3.92),322(2.86), 337(3.09)	70-1051-65
Phenanthrene, 2-ethyl-7-methoxy-1-methyl-	EtOH	225(4.3),258(4.8), 277(4.2),290(4.0), 300(3.9),322(2.9), 337(3.1)	70-1051-65
3-Phenanthrol, 2-isopropyl-8-methyl-	EtOH	219(4.40),254(4.68), 279(4.21),300s(3.96), 308(4.03),320s(3.53), 356(3.41)	33-1234-64
$C_{18}H_{18}OS$			
6H-Dibenzo[b,e]thiocine, 12-(3-hydroxy-propylidene)-7,12-dihydro-	MeOH	229(4.26),264(3.80)	24-0685-65
Ketone, 2-methyl-5-phenyl-3-cyclohex-en-1-yl 2-thienyl	EtOH	262(3.91),285(3.82)	65-2258-64
1-Penten-3-one, 1-phenyl-5-(p-tolyl-thio)-	EtOH	260s(4.17),289(4.41)	7-0652-65
$C_{18}H_{18}O_2$			
Benzofuro[2,3-b]benzofuran, 5a,10b-di-hydro-2,5a,9,10b-tetramethyl-	EtOH	288(3.84)	78-2289-65
Benzofuro[3,2-b]benzofuran, 4b,9b-di-hydro-1,4b,6,9b-tetramethyl-	EtOH	280(3.76)	78-2289-65
Biphenyl, 3,4'-dipropionyl-	EtOH	216(4.39),255(4.29), 280(4.38)	39-4930-65
2-Butyne, 1,4-bis(o-tolyloxy)-	EtOH	273(3.38)	78-2289-65
Cyclopentanecarboxylic acid, 1-(4-biphenylyl)-	EtOH	256(4.38)	87-0504-64
Equilenin	MeOH	229(4.80),269(3.66), 280(3.72),291(3.57), 326(3.30),340(3.38)	44-0316-65
1,3,5(10),8,14-Estrapentaen-17-one, 3-hydroxy-	EtOH	312(4.53)	70-1809-65
Naphthalene, 1,2-dihydro-7-methoxy-3-(p-methoxyphenyl)-	EtOH	233(4.30),314(4.38)	87-0519-64
1,4-Pentadien-3-ol, 1-(p-methoxy-phenyl)-5-phenyl-	MeOH	270(4.51)	5-0064-65F
	20% dioxan	272(4.47)	5-0064-65F
2,4-Pentadien-1-ol, 1-(p-methoxy-phenyl)-5-phenyl-	MeOH	289(4.61)	5-0064-65F
	20% dioxan	290(4.57)	5-0064-65F
2,4-Pentadien-1-ol, 5-(p-methoxy-phenyl)-1-phenyl-	MeOH	295(4.59)	5-0064-65F
	20% dioxan	296(4.56)	5-0064-65F

Compound	Solvent	λ_{max}(log ε)	Ref.
Phenanthrene, 1-acetyl-9,10-dihydro-7-methoxy-2-methyl-	EtOH	278(4.32)	70-1051-65
Phenanthrene, 1-ethyl-3,6-dimethoxy-	MeOH	238(4.51),251(4.62),259(4.66),278s(4.18),286(4.26),297(4.02),308(4.11),331(2.95),347(3.18),363(3.23)	35-2177-64
$C_{18}H_{18}O_2S$			
1-Penten-3-one, 5-(p-methoxyphenylthio)-1-phenyl-	EtOH	264s(4.19),290(4.38)	7-1093-65
$C_{18}H_{18}O_3$			
Anthrone, 1,8-dihydroxy-4,5,10,10-tetramethyl-	MeOH-HCl	252s(3.80),260(3.95),268(3.97),300(4.13),371(4.00)	35-1794-65
Benzofuran, 2-(p-methoxyphenyl)-3-α-methoxyethyl-	EtOH	298(4.36)	4-0158-64
2-Butyn-1-ol, 1,1-diphenyl-	H_2SO_4	518(4.63)	35-1381-65
Cinnamic acid, o-methoxy-α-phenyl-, ethyl ester, trans	MeOH	284(4.04),320(4.06)	2-0182-64
2-Flavene, 4',8-dimethoxy-3-methyl-	EtOH	222(4.36),242(4.33),270(3.91)	78-1471-65
Gibberic acid, dehydro-	MeOH	260(4.19),270(4.19),290(3.61),301(3.61)	39-0990-65
Hexanoic acid, 6-oxo-6-(4-biphenylyl)-	EtOH	286(4.31)	87-0504-64
18-Norestr-9(11)-ene-15,17-dione, 3-methyl-, cis	n.s.g.	259(4.60)	70-0843-65
3-Pentenoic acid, 2-(p-methoxyphenyl)-5-phenyl-	MeOH	258(3.36)	2-0017-64
3-Pentenoic acid, 5-(p-methoxyphenyl)-2-phenyl-	MeOH	270(3.41)	2-0017-64
Phenanthrene, 2,3,4-trimethoxy-7-methyl-	EtOH	258(4.97),282(4.25),302(4.04)	39-4257-64
Propene, 3-(2-acetoxy-4-methoxyphenyl)-3-phenyl-	EtOH	227(4.13),275(3.33)	78-2707-65
Propene, 3-(2-acetoxy-6-methoxyphenyl)-3-phenyl-	EtOH	272(3.33)	78-2707-65
$C_{18}H_{18}O_4$			
Acetophenone, 4',4'''-(ethylenedioxy)di-	n.s.g.	280(4.6)	70-1703-64
5H-Benzocyclohepten-5-one, 6-furfurylidene-6,7,8,9-tetrahydro-2,4-dimethoxy-	EtOH	342.5(4.42)	5-0021-64G
2-Butyne, 1,4-bis(o-methoxyphenoxy)-	EtOH	225(3.96),272(3.59)	78-2289-65
2-Butyne, 1,4-bis(p-methoxyphenoxy)-	EtOH	286(3.77)	78-2289-65
Cinnamic acid, 2,4-dimethoxy-β-methyl-α-phenyl-, trans	MeOH	275(3.87)	2-0182-64
Cinnamic acid, 2,3-dimethoxy-α-phenyl-, methyl ester, trans	MeOH	284(4.19)	2-0182-64
Cinnamic acid, 2,4-dimethoxy-α-phenyl-, methyl ester, trans	MeOH	280(4.07),321(4.03)	2-0182-64
9β-Estra-1,4-diene-3,11,17-trione, 9,10-epoxy-	MeOH	261.5(4.28)	94-0473-64
Naphthacene, 1,4,4a,5,5aα,6,11,11aβ-12,12a-decahydro-3,12aα- dihydroxy-1,11-dioxo-	EtOH	249(4.00),290(4.06)	70-1039-65
4,6-Tetradecadiene-8,10,12-triyne-1,3-diol, diacetate	ether	256(4.79),267(5.09),286(4.15),304(4.48),324(4.66),347(4.58)	24-2605-65

Compound	Solvent	λ_{max}(log ε)	Ref.
$C_{18}H_{18}O_5$			
Benzoin, 3,3'-dimethoxy-, acetate	EtOH	220(4.46),256(3.99), 287(3.51),312(3.48)	35-5277-64
Chalcone, 2'-hydroxy-2,5',6'-trimethoxy-	MeOH	212(4.20),292(4.06), 344(4.03)	24-0164-65
Cinnamic acid, 2,3,4-trimethoxy-α-phenyl-, cis	MeOH	310(4.24)	2-0182-64
trans	MeOH	224(3.83),240(3.94), 300(4.09)	2-0182-64
Flavanone, 2',5,6-trimethoxy-	MeOH	218(4.17),265(3.82), 360(3.34)	24-0164-65
7,8-Phenanthrenedicarboxylic acid, 7,8,9,10-tetrahydro-2-methoxy-7-methyl-	EtOH	245(4.36),298(3.91)	70-2021-64
$C_{18}H_{18}O_6$			
Acetic acid, (5-benzyloxy-2-carbethoxy-3-hydroxyphenyl)-	EtOH	263(4.13),303(3.85)	39-5382-64
Acetophenone, 2-(2-carbomethoxyphenoxy)-2',4'-dimethoxy-	EtOH	213(4.33),231(4.27), 272(4.14),305(4.10)	39-2361-65
Matsuzake lactone, oxidation product	EtOH	240(4.45),276(4.43)	94-1232-64
$C_{18}H_{18}O_7$			
Acetophenone, 2-(2-carbomethoxyphenoxy)-2'-hydroxy-4',6'-dimethoxy-	$CHCl_3$	293(4.37)	39-5140-65
Acetophenone, 2-(2-carboxyphenoxy)-2',4',6'-trimethoxy-	EtOH	223(4.20),282(3.83)	39-2361-65
Glauconic acid ketone	EtOH	220(4.00)	39-1772-65
3,4-Isoflavandiol, 2',7-dimethoxy-4',5'-(methylenedioxy)-	n.s.g.	290(3.85)	39-5991-64
Propiophenone, 2,2',3,4'-tetrahydroxy-3-(p-methoxyphenyl)-, 3-acetate	EtOH	282(3.99),320(3.80)	78-1141-64
Sulochrin, monomethyl-	n.s.g.	285(4.17),325s(3.87)	39-6658-65
$C_{18}H_{18}O_8$			
Matsuzake lactone, oxidation product	EtOH	226(4.47),285(3.85)	94-1232-64
$C_{18}H_{19}BrO_6$			
Biphenylene, 4-bromo-1,2,3,6,7,8-hexamethoxy-	EtOH	286(4.64),350(4.23), 368(4.26),382(4.31)	39-1067-64
Isobyssochlamic acid, bromo-	C_6H_{12}	240(3.93)	39-1787-65
$C_{18}H_{19}Cl$			
Dibenzo[a,c]cyclooctene, 3-chloro-5,6,7,8-tetrahydro-6,7-dimethyl-	pet ether	243.5(4.17)	39-5317-64
$C_{18}H_{19}ClN_2O$			
3-Ethyl-6,7-dihydro-10-methoxy-12H-indolo[2,3-a]quinolizinium chloride	EtOH	268(3.80),335(4.18), 433(4.24)	94-1381-64
$C_{18}H_{19}ClO_4$			
7-(2,4,6-Trimethylstyryl)tropylium perchlorate	MeCN	258(4.31),280s(3.89), 324s(3.40),463(4.38)	24-1337-64
$C_{18}H_{19}ClO_5$			
Propiophenone, 3-chloro-2-hydroxy-2',4'-dimethoxy-3-(p-methoxyphenyl)-	MeOH	271(4.21),305(4.04)	78-1141-64

Compound	Solvent	$\lambda_{max}(\log \epsilon)$	Ref.
$C_{18}H_{19}ClO_5S$			
Spiro[benzofuran-2(3H),1'-[2]cyclohexene]-3,4'-dione, 7-chloro-2'-(ethylthio)-4,6-dimethoxy-6'-methyl-	EtOH	290(4.61)	87-0705-64
Spiro[benzofuran-2(3H),1'-[3]cyclohexene]-2',3-dione, 7-chloro-4'-(ethylthio)-4,6-dimethoxy-6'-methyl-	EtOH	296(4.62)	87-0705-64
$C_{18}H_{19}ClO_6$			
Spiro[benzofuran-2(3H),1'-[2]cyclohexene]-3,4'-dione, 7-chloro-2'-ethoxy-4,6-dimethoxy-6'-methyl-	EtOH	237(4.42),253s(4.25), 290(4.37),320s(3.74)	39-4939-65
Spiro[benzofuran-2(3H),1'-[3]cyclohexene]-2',3-dione, 7-chloro-4'-ethoxy-4,6-dimethoxy-6'-methyl-	EtOH	215s(4.29),225s(4.29), 262(4.30),289(4.33), 325(3.73)	39-4939-65
$C_{18}H_{19}ClO_7$			
7-(2,3,4-Trimethoxystyryl)tropylium perchlorate	MeCN	221(4.40),258(4.18), 286s(4.01),513(4.59)	24-1337-64
7-(2,4,5-Trimethoxystyryl)tropylium perchlorate	MeCN	234(4.25),296(3.88), 415(3.89),587(4.52)	24-1337-64
7-(2,4,6-Trimethoxystyryl)tropylium perchlorate	MeCN	229s(4.31),293(4.10), 557(4.79)	24-1337-64
7-(3,4,5-Trimethoxystyryl)tropylium perchlorate	MeCN	220s(4.50),265(4.33), 282s(4.07),332(4.27), 495(4.51)	24-1337-64
$C_{18}H_{19}N$			
4bH-Cyclopropa[l]phenanthrene, 5-dimethylaminomethyl-5,5a-dihydro-	EtOH-HCl	248(3.56),271(4.14), 281(4.09),300(3.69), 310(3.74)	88-2919-64
$C_{18}H_{19}NO$			
Acridan, 10-(1-methoxy-2-methylpropenyl)-	n.s.g.	287(4.16)	39-5877-65
Chalcone, 4-(dimethylamino)-4'-methyl-	EtOH	266(4.21),418(4.45)	23-2580-64
Cyclopentanecarboxamide, 1-(4-biphenylyl)-	EtOH	257(4.37)	87-0504-64
Dibenzo[b,e]oxepine, 11-(3-methylaminopropylidene)-6,11-dihydro-, hydrochloride	n.s.g.	250s(3.98),296(3.71)	49-0485-64
Morpholine, 4-(1,2-diphenylvinyl)-	C_6H_{12}	224(4.17),306(4.06)	44-3705-65
isomer	C_6H_{12}	240(4.17),320(4.06)	44-3705-65
$C_{18}H_{19}NO_2$			
Apoerysopine, dimethyl ether	MeOH	246(4.31),336(3.69)	78-1729-64
	0.1N HCl	229(4.33),272(4.01), 296(3.95)	78-1729-64
Aporphine, 5-hydroxy-6-methoxy-	EtOH	271(4.14),310(3.59)	35-0694-64
	EtOH-KOH	270(3.85),345(3.69)	35-0694-64
Aporphine, 6-hydroxy-5-methoxy-	EtOH	370(4.15)	33-2119-64
6-Azaequilenin, methyl ether	EtOH	236(4.65),258(3.67), 290(3.47),328(3.72), 340(3.74)	78-2517-65
6-Aza-14β-equilenin, methyl ether	EtOH	236(4.74),276(3.56), 325(3.66),337(3.69)	78-2517-65
8-Azaestra-1,3,5(10),9(11),14-pentaen-17-one, 3-methoxy-	pH 1	246(3.83),367(4.34)	39-4900-65
	MeOH	260(4.43),306(3.77)	39-4900-65
13-Aza-D-homo-18-norequilenin, methyl ether	EtOH	236(4.70),264(3.97), 274(3.97),280(3.80)	39-3007-65

Compound	Solvent	λ_{max}(log ϵ)	Ref.
5H-2-Benzazepine, 3,4-dihydro-	EtOH-HCl	270(4.10),380(3.80)	88-2419-64
7,8-dimethoxy-1-phenyl-	EtOH	239(4.27),305(3.67)	88-2419-64
1,2-Benzopyran, 7-(dimethylamino)-2-methoxy-2-phenyl-	MeOH	325(3.96),430(4.14)	35-3142-64
Chalcone, 4-(dimethylamino)-	benzene	286(4.15),400(4.41)	65-0094-65
4'-methoxy-	EtOH	234(4.03),300(4.07), 416(4.52)	65-0094-65
	HOAc	312(4.19),422(4.40)	65-0094-65
	+$FeCl_3$	585(4.20)	65-0094-65
Chalcone, 4'-(dimethylamino)-	benzene	370(4.31)	65-0094-65
4-methoxy-	EtOH	238(4.13),387(4.47)	65-0094-65
	HOAc	396(4.50)	65-0094-65
	+$FeCl_3$	510(4.27)	65-0094-65
$C_{18}H_{19}NO_3$			
Anthracen-10-one, 3α-hydroxy-3β-cyano-2β-(2-oxopropyl)-1,2,3,4,4aβ,9,9aα,10-octahydro-	EtOH	248(4.19),291(3.18)	70-1039-65
Apoerysopine, 1,2,3,3a-tetrahydro-, dimethyl ether	MeOH	212(4.10),253(3.99), 365(4.30)	78-1729-64
6H-Dibenz[d,f]azonin-6-one, 5,7,8,9-tetrahydro-11,12-dimethoxy-	MeOH	215(4.51),252s(3.74), 288(3.61)	24-0046-65
Glaziovine	EtOH	288(3.57)	35-0694-64
	EtOH-KOH	308(3.74)	35-0694-64
Glutarimide, N-[2-(6-methoxy-1-naphthyl)ethyl]-	EtOH	230(4.53),265(3.85), 275(3.91),287(3.83)	39-3007-65
Hydrocinnamic acid, (β-salicylidene-amino)-, ethyl ester	EtOH	214(4.48),256(4.17), 317(3.66),404(2.11)	35-1757-65
	dioxan	255(4.19),318(3.70), 420s(0.82)	35-1757-65
2-Indolinone, 3-benzyl-5,6-dimethoxy-1-methyl-	MeOH	208(4.50),274(3.80), 297s(3.67)	87-0626-65
2-Indolinone, 1-methyl-3-veratryl-	MeOH	231(4.10),253(4.00), 279(3.73)	87-0626-65
Isoquinoline, 8-(benzyloxy)-3,4-di-	pH 1	243(4.15),323(4.13)	33-2089-64
hydro-6,7-dimethoxy-	pH 13	232(4.34),279(3.96)	33-2089-64
	iso-PrOH	241(4.16),330(4.15)	33-2089-64
Phthalide, 3-(2-morpholino-2-cyclohexen-1-ylidene)-	hexane	228(4.08),237s(4.04), 290(4.15),335(4.01)	1-2139-65
$C_{18}H_{19}NO_4$			
6H-Dibenz[d,f]azonin-6-one, 5,7,8,9-tetrahydro-5-hydroxy-11,12-dimethoxy-	MeOH	214(4.48),254s(3.78), 290(3.70)	24-0046-65
Galanthaminone, N-acetyl-N-demethyl-	EtOH	263(3.69),300s(3.40)	35-4434-64
2-Indolinone, 3-(3,4,5-tri-methoxybenzyl)-	EtOH	232s(4.06),249s(3.94), 279s(3.29)	87-0626-65
Isocarbostyril, 8-(benzyloxy)-3,4-dihydro-6,7-dimethoxy-	iso-PrOH	218(4.44),259(3.99), 295s(3.28)	33-2089-64
$C_{18}H_{19}NO_5$			
Macronine	EtOH	229(4.49),268(3.74), 308(3.78)	33-0185-64
$C_{18}H_{19}NO_8$			
9aH-Quinolizine-1,2,3,4-tetracarboxylic acid, 6-methyl-, tetramethyl ester	MeOH	236(4.21),294(4.08), 433(3.72)	39-0948-65

Compound	Solvent	$\lambda_{max}(\log \epsilon)$	Ref.
$C_{18}H_{19}N_2NaO_4S_2$			
5-Thia-1-azabicyclo[4.2.0]oct-2-ene-2-carboxylic acid, 3-[(ethylthio)-methyl]-8-oxo-7-(2-phenylacetamido)-, sodium salt	pH 6.0	264(4.02)	39-5015-65
$C_{18}H_{19}N_3O$			
Fluorene, 9-acetyl-9-ethyl-, semicarbazone	EtOH	212(4.52),269(4.35), 292(3.92),304(3.92)	22-2345-65
$C_{18}H_{19}N_3O_3$			
Azobenzene, 4-methacrylamido-2',3-dimethoxy-	90% EtOH-NH_3	378(4.41)	35-2315-64
$C_{18}H_{19}N_3O_5S$			
5-Pyrimidinecarboxylic acid, 2-(ethyl-thio)-4-[3-methyl-4-(5-nitro-2-furyl)-buta-1,3-dienyl]-, ethyl ester	propylene glycol	284(4.23),328(4.16)	95-0207-64
$C_{18}H_{20}$			
Chrysene, 1,2,3,4,7,8,9,10-octahydro-	EtOH	236(4.91),284(3.80), 290(3.74),315(3.24), 322(3.07),329(3.40)	24-0593-65
Dibenzo[a,c]cyclooctane, 5,6,7,8-tetra-hydro-6,7-dimethyl-	pet ether	239(4.10),266(3.07), 275(2.81)	39-5317-64
1-Hexene, 1,1-diphenyl-	C_6H_{12}	250(4.03)	22-0693-64
$C_{18}H_{20}BrNO_3$			
Indol-2(4H)-one, 7-bromo-1-(3,4-di-methoxyphenethyl)-5,6-dihydro-	MeOH	230(3.96),280(4.16)	78-1729-64
$C_{18}H_{20}Br_2N_8O_4$			
Theobromine, 1,1'-(2,3-dibromo-tetramethylene)di-	n.s.g.	240(3.8),282(4.4)	83-0146-64
$C_{18}H_{20}ClNO_2$			
7-Isoquinolinol, 1-(p-chlorophenethyl)-1,2,3,4-tetrahydro-6-methoxy-	EtOH-HCl	219(4.29),285(3.62)	33-1945-65
	EtOH	252(4.08),279s(3.78)	33-1945-65
	EtOH-NaOH	245(3.89),278s(3.36), 302(3.73)	33-1945-65
hydrochloride	EtOH-HCl	219(4.34),285(3.67)	33-1945-65
	EtOH	249(4.06),278s(3.73), 283(3.74)	33-1945-65
	EtOH-NaOH	242(4.00),278s(3.71), 301(3.91)	33-1945-65
$C_{18}H_{20}ClNO_3$			
Erythrinan-4-en-8-one, 4-chloro-15,16-dimethoxy-, cis	MeOH	212(4.28),232s(3.92), 282(3.58)	24-0046-65
$C_{18}H_{20}ClN_3O_5$			
Alanine, N-[(5-chloro-1,8-dihydro-6,7-dimethoxy-8-methylpyrrolo[2,3-b]-indol-2-yl)carbonyl]-N-methyl-	MeOH	236(4.43),282(4.31), 330(4.37)	39-0026-64
$C_{18}H_{20}Cl_3N_5S$			
Purine, 6-(m-chlorobenzylthio)-9-[2-[N,N-bis(2-chloroethyl)amino]ethyl]-	pH 1	292(4.24)	87-0182-65
	pH 11	297(4.30)	87-0182-65

Compound	Solvent	$\lambda_{max}(\log \epsilon)$	Ref.
$C_{18}H_{20}NO_3P$			
Phosphonic acid, (2-phenylindol-3-yl)-, diethyl ester	EtOH	230s(4.36),293(4.25)	44-3604-65
$C_{18}H_{20}N_2$			
Appariciné	EtOH	230s(4.70),303(4.67), 312s(4.61)	39-4773-65
(+)	EtOH	230s(4.70),303(4.67), 312s(4.61)	78-1717-65
[2]Benzazepino[6,5,4-def][2]benzazepine, 4,5,6,10,11,12-hexahydro-5,11-dimethyl-	isooctane	241(4.25)	35-1710-64
	6N HCl	255.5(4.16)	35-1710-64
1,2-Diazocine, 3,4,5,6,7,8-hexahydro-3,8-diphenyl-	EtOH	215(4.05),252(2.93), 259(2.90),276(1.79)	22-0722-65
Indoline, 1-(dimethylamino)-3-methyl-2-methylene-3-phenyl-	MeOH	273(4.20)	44-3451-65
1H-2,6-Methanobenzodiazocine, 4-benzyl-3,4,5,6-tetrahydro-	pH 1	252(2.64),257(2.69), 262(2.56),268(2.54)	94-1225-65
	EtOH	254(2.96),260(2.99), 266(3.00),274(2.87)	94-1225-65
$C_{18}H_{20}N_2O$			
2-Indolinone, 3-(1-methyl-1-phenyl-2-aminoethyl)-	EtOH	237(3.86),292(3.44)	87-0626-65
Nicotinonitrile, 6-(α-butylphenethyl)-1,2-dihydro-2-oxo-	EtOH	238(3.82),339(4.10)	44-3593-65
4H-1,3,4-Oxadiazine, 5,6-dihydro-4,5-dimethyl-6-phenyl-2-m-tolyl-	$CHCl_3$	242(3.66),280(3.81)	44-0668-64
4H-1,3,4-Oxadiazine, 5,6-dihydro-4,5-dimethyl-6-phenyl-2-o-tolyl-	$CHCl_3$	244(3.70),290(3.98)	44-0668-64
4H-1,3,4-Oxadiazine, 5,6-dihydro-4,5-dimethyl-6-phenyl-2-p-tolyl-	$CHCl_3$	242(3.85),290(3.99)	44-0668-64
Vincanidine, deformyl-	n.s.g.	222(4.30),262(3.66)	70-1992-65
Vinervine, decarbomethoxy-	n.s.g.	226(4.32),266(3.74)	70-2152-65
$C_{18}H_{20}N_2OS$			
4H-1,3,4-Thiadiazine, 5,6-dihydro-2-(p-methoxyphenyl)-4,5-dimethyl-6-phenyl-, cis	$CHCl_3$	254(4.17),308(3.82)	44-2228-65
trans	$CHCl_3$	253(4.20),308(4.01)	44-2228-65
$C_{18}H_{20}N_2O_2$			
Benzomesidine, p-acetamido-	EtOH	269(4.5)	28-4295-64B
4H-1,3,4-Oxadiazine, 5,6-dihydro-2-(o-methoxyphenyl)-4,5-dimethyl-6-phenyl-	$CHCl_3$	244(3.75),282(3.77)	44-0668-64
4H-1,3,4-Oxadiazine, 5,6-dihydro-2-(p-methoxyphenyl)-4,5-dimethyl-6-phenyl-	$CHCl_3$	240(3.96),289(4.13)	44-0668-64
$C_{18}H_{20}N_2O_3$			
3-Pyrrolin-2-one, 5-(5-carbethoxy-3-ethyl-4-methylpyrrolyl-2-methyl)-4-ethyl-3-methyl-	EtOH	209(4.24),283(4.28)	39-5999-64
$C_{18}H_{20}N_2O_5$			
D-Fructosone-1-diphenylhydrazone	EtOH	233(4.14),347(4.26)	24-0725-64

Compound	Solvent	$\lambda_{max}(\log \epsilon)$	Ref.
$C_{18}H_{20}N_2O_5S_4$			
Cephalosporanic acid, 7-(2-thiophene-acetamido)-, isopropyl xanthate derivative, sodium salt	n.s.g.	227(4.24),285(4.24)	87-0174-65
Cephalosporanic acid, 7-(2-thiophene-acetamido)-, propyl xanthate derivative, sodium salt	n.s.g.	226(4.24),284(4.25)	87-0174-65
$C_{18}H_{20}N_2O_7$			
Uridine, 2',3'-(2,4-dimethoxy-benzylidene)-	MeOH	232(4.09),262(4.06)	5-0156-64I
$C_{18}H_{20}N_2S$			
4H-1,3,4-Thiadiazine, 5,6-dihydro-4,5-dimethyl-6-phenyl-2-p-tolyl-	$CHCl_3$	243(4.25),315(3.96)	44-2228-65
isomer	$CHCl_3$	246(4.30),316(3.74)	44-2228-65
$C_{18}H_{20}N_4O_4S_3$			
5-Thia-1-azabicyclo[4.2.0]oct-2-ene-2-carboxylic acid, 3-[(amidinothio)-methyl]-7-[2-(benzylthio)acetamido]-8-oxo-	pH 6.0	260(3.98)	39-5015-65
$C_{18}H_{20}N_4O_5$			
Isopyrethrolone, methyl ether, 2,4-dinitrophenylhydrazone	EtOH	232(4.40),262s(4.37), 282(4.43),308s(4.10), 380(4.36)	39-0888-64
Pyrethrolone, methyl ether, 2,4-dinitrophenylhydrazone	EtOH	225(4.54),282s(3.97), 380(4.44)	39-1854-64
$C_{18}H_{20}N_4O_6$			
Bicyclo[3.3.1]non-3-ene-1-carboxylic acid, 5-methyl-2-oxo-, methyl ester, 2,4-dinitrophenylhydrazone	$CHCl_3$	260(3.98),375(4.43)	39-1344-65
isomer	$CHCl_3$	260(3.97),380(4.38)	39-1344-65
Propiophenone, 4'-ethoxy-3'-methoxy-, 2,4-dinitrophenylhydrazone	n.s.g.	298(4.04),395(4.44)	39-1572-65
$C_{18}H_{20}N_4O_7S$			
2H-1,3-Thiazine, tetrahydro-4,4-di-methyl-2-phenyl-, picrate	n.s.g.	232(3.87),337(3.90)	44-3314-64
$C_{18}H_{20}O$			
Azulene, 1-ethyl-3-propionyl-4,6,8-trimethyl-	hexane	557(2.77)	65-0894-64
	EtOH	543(2.86)	65-0894-64
Chrysofluorene, 1,1a,1b,2,3,5,6,6a-octahydro-1a-methyl-3-oxo-	EtOH	238(4.29)	42-0479-64
$C_{18}H_{20}O_2$			
Azulene, 1,3-diisobutyryl-	hexane	516(2.78),554(2.71), 607(2.23)	65-0894-64
	EtOH	501(2.86)	65-0894-64
Estra-1,3,5(10),8-tetraen-17-one, 3-hydroxy-	EtOH	276(4.15)	13-0031-64
	EtOH	215(4.16),276(4.14)	88-0171-64
Estra-1,3,5(10),9-tetraen-17-one, 3-hydroxy-	EtOH	264(4.2)	13-0031-64
	EtOH	264(4.18)	88-0171-64
Estra-1,3,5,(10),14-tetraen-17-one, 3-hydroxy-	MeOH	222(3.93),280(3.35)	44-0214-64
Estra-1,3,5(10),15-tetraen-17-one, 3-hydroxy-	MeOH	222(4.14),280(3.37)	44-0214-64

Compound	Solvent	λ_{max}(log ε)	Ref.
14β-Estra-1,3,5(10),15-tetraen-17-one, 3-hydroxy-	MeOH	222(4.19),280(3.41)	44-0214-64
3-Hexene-1,5-diyne, 3,4-bis-(4-oxocyclohexyl)-	ether	262(4.33),271(4.24)	5-0024-65D
1(2H)-Naphthalenone, 3,7,8,8a-tetrahydro-6-(p-methoxyphenyl)-	EtOH	289(4.47)	70-1355-64
Spiro[2-cyclohexene-1,1'-inden]-4-one, 2',3'-diethyl-6'-hydroxy-	MeOH	267(4.15),324(3.46)	5-0036-64D
	MeOH-KOH	290(4.22),364(3.27)	5-0036-64D
$C_{18}H_{20}O_3$			
Carbonic acid, ethyl 3,4,9,10-tetrahydro-1-methyl-2-phenanthryl ester	EtOH	310(4.15),323(4.21), 335(4.00)	35-2038-64
Cyclohexadienone, 2-acetoxy-2,4-dimethyl-6-(2-phenylethyl)-	n.s.g.	308(3.56)	49-0512-64
Estra-2,5(10)-diene-1,4,17-trione	EtOH	249(4.18),338(2.99)	24-1940-64
Estra-1,3,5(10),9(11)-tetraen-17-one, 3,14-dihydroxy-	EtOH	275(4.18)	70-1809-65
14β-Estra-1,3,5(10)-triene-15,17-dione, 3-hydroxy-	EtOH	279(3.44)	44-1333-64
	MeOH-KOH	240(3.92),277(4.18)	44-1333-64
Fluorene-8aβ(4bβH)-propionic acid, 5,6-dihydro-1-methyl-9-oxo-, methyl ester	EtOH	250(4.11),299(3.42)	44-2528-65
6-Oxaestra-1,3,5(10),8-tetraen-17-one, 3-methoxy-	n.s.g.	287(3.95),307(4.02)	31-0418-64
8,14-Secoestra-1,3,5(10),9(11)-tetraene-14,17-dione, 3-hydroxy-	EtOH	269(4.15)	13-0031-64
	EtOH	215(4.14),269(4.15)	88-0171-64
$C_{18}H_{20}O_4$			
10-Anthrone, 9β-acetoxy-1,4,4aα,9aα-tetrahydro-5-methoxy-9α-methyl-	EtOH	258(4.03),333(3.69)	70-1013-64
Anthrone, 1,4,4aα,9aα-tetrahydro-4β,10β-(isopropylidenedioxy)-5-methoxy-	EtOH	222(4.54),255(4.07), 316(3.72)	65-2563-64
2,2'-Diphenoquinone, 5,5'-dimethoxy-3,3',4,4'-tetramethyl-	C_6H_{12}	240(4.4),380(3.5), 615(4.0)	5-0092-64A
Estra-1,3,5(10)-triene-11,17-dione, 3,9-dihydroxy-	MeOH	276(3.13),282(3.10)	94-0473-64
Estra-1,3,5(10)-triene-11,17-dione, 3,9β-dihydroxy-	MeOH	278.2(3.16)	94-0473-64
$C_{18}H_{20}O_5$			
Agatharesinol	EtOH	266(4.41)	1-0913-65
4(1H)-Anthracenone, 10β-acetoxy-4aα,9,9aα,10-tetrahydro-9β-hydroxy-5-methoxy-9α-methyl-	EtOH	276(3.49),281(3.49)	65-2570-64
10(9H)-Anthracenone, 4β-hydroxy-1,4,4aα,9aα-tetrahydro-9β-hydroxy-5-methoxy-9α-methyl-	EtOH	258(3.90),315(3.72)	65-2570-64
2,4,6-Cycloheptatrien-1-one, 2-hydroxy-4-[3-(3-hydroxy-4,5-dimethoxyphenyl)propyl]-	EtOH	322(3.70),349(3.68)	78-3605-65
	base	335(3.99),391(3.99)	78-3605-65
Indone, 4-hydroxy-6-(2-hydroxyethyl)-2,5,7-trimethyl-, diacetate	EtOH	247(4.57),344(3.36), 400(2.08)	78-1231-65
	EtOH	248(4.57),344(3.36), 400(2.08)	94-0853-64
Lactic acid, 2-(o-methoxyphenyl)-3-(p-methoxyphenyl)-, methyl ester	EtOH	275(3.50)	22-3572-65
Spiro[1,3-dioxolane-2,2'(1'H)-fluorene]-9'a(3'H)-acetic acid, 4',4'a-dihydro-8'-methyl-9'-oxo-	EtOH	249(4.11),297(3.38)	44-2528-65

Compound	Solvent	λ_{max}(log ε)	Ref.
Unknown product from 9,10-epoxy-9-estra-1,4-diene-9,11,17-trione and dioxan	MeOH	294(4.1)	94-0473-64
$C_{18}H_{20}O_6$			
Benzoic acid, 2,4-dimethoxy-6-[(4-methoxy-o-tolyl)oxy]-, methyl ester	EtOH	280(3.54)	39-6960-65
Hamaudol, monomethyl ether, monoacetate	EtOH	244(4.26),291(4.13)	95-0055-65
Isobyssochlamic acid	EtOH	207(3.78),236(3.80)	39-1787-65
hydrate	EtOH	234(3.52)	39-1787-65
Lactic acid, 2-(2-methoxyphenyl)-3-(3,4-dimethoxyphenyl)-	EtOH	278(3.61)	22-3572-65
$C_{18}H_{20}O_7$			
2α-Anthracenemalonic acid, 1,2,3,4,4a,-9,9aα,10-octahydro-3β,9β-dihydroxy-9α-methyl-10-oxo-	EtOH	247(4.07),287(3.15)	65-2576-64
$C_{18}H_{20}O_8$			
Xanthophanic acid, diethyl ester	EtOH	298(4.12),440(4.35),520(3.60)	88-2313-64
$C_{18}H_{20}S_3$			
Thiophene, 2,5-bis(α,α-dimethyl-3-thenyl)-	EtOH	237.5(4.34)	22-2635-65
$C_{18}H_{21}BrO_4$			
Podocarpic acid, 6α-bromo-7-oxo-, methyl ester	MeOH	232(4.18),297(4.26)	44-0501-65
$C_{18}H_{21}Br_2NO_3$			
Erythrinan-8-one, 4,5-dibromo-15,16-dimethoxy-, cis	MeOH	211(4.25),285(3.56)	24-0046-65
$C_{18}H_{21}ClN_4O_2$			
3-Buten-2-one, 1-chloro-4-[p-[(2-amino-4-hydroxy-6-methyl-5-pyrimidyl)-3-propyl]amino]phenyl-	pH 1	222(4.20),282(4.31)	87-0035-65
	pH 7	259(3.99),402(4.28)	87-0035-65
	pH 13	278(4.01),402(4.27)	87-0035-65
$C_{18}H_{21}ClO_5S_2$			
Grisan-2',3-dione, 7-chloro-4,6-dimethoxy-6'-methyl-4',4'-bis(methylthio)-	EtOH	294.5(4.33)	87-0705-64
$C_{18}H_{21}ClO_6$			
Radicicol, tetrahydro-	EtOH	215(4.40),265(3.90),310(3.68)	88-0365-64
	EtOH-base	251(4.10),320(4.30)	88-0365-64
$C_{18}H_{21}Cl_2N_5S$			
Purine, 6-(benzylthio)-9-[2-[N,N-bis-(2-chloroethyl)amino]ethyl]-	pH 1	293(4.28)	87-0182-65
	pH 11	289(4.32)	87-0182-65
$C_{18}H_{21}N$			
Cyclohepta[d,e]-1-pyrindine, 6-isopropyl-1,2,8-trimethyl-	C_6H_{12}	238(4.47),253(4.43),270(4.49),300s(3.84),361(4.12),398(3.69),422(3.62),449(3.43),690s(2.38),763(2.50),862(2.50),996(2.23)	35-3137-64
Pyridine, 3-methyl-6-(p-tert-butylstyryl)-	EtOH	315(4.69)	65-4106-64

Compound	Solvent	λ_{max}(log ϵ)	Ref.
$C_{18}H_{21}NO$			
5-Azachrysene, 1,2,3,4,7,8,9,10-octahydro-7-hydroxy-6-methyl-	MeOH	215(4.46),240(4.62)	39-5916-64
Benzo[a]cyclohepta[f]quinolizin-14(1H)-one, 2,3,4,5,7,8,12b,13-octahydro-	C_6H_{12}	318(4.00)	44-3667-65
	EtOH	341(4.13)	44-3667-65
perchlorate	EtOH	338(4.14)	44-3667-65
Neopentylamine, N-salicylidene-α-phenyl-	hexane	223s(4.44),256(4.18), 263(4.15),321(3.72)	35-1757-65
	EtOH	215(4.46),256(4.13), 263s(4.08),317(3.61), 403(2.60)	35-1757-65
	dioxan	215(4.45),257(4.12), 262s(4.06),318(3.64), 410s(1.26)	35-1757-65
$C_{18}H_{21}NO_2$			
6-Azaequilenin, dihydro-	n.s.g.	234(4.72),329(3.08), 341(3.71)	31-0418-64
6-Azaestra-1,3,5(10),6,8-pentaen-17β-ol, 3-methoxy-	EtOH	215(4.42),235(4.65), 256(3.64),327(3.67), 339(3.71)	78-2517-65
13H-Dibenzo[a,f]quinolizin-13-one, 1,2,3,4,6,7,11b,12-octahydro-2-methoxy-	C_6H_{12}	313(4.07)	44-3667-65
	EtOH	335(4.11)	44-3667-65
perchlorate	EtOH	335(4.10)	44-3667-65
13H-Dibenzo[a,f]quinolizin-13-one, 1,2,3,4,6,7,11b,12-octahydro-8-methoxy-	C_6H_{12}	313(4.06)	44-3667-65
	EtOH	336(4.12)	44-3667-65
perchlorate	EtOH	336(4.11)	44-3667-65
1H,3H-[1,3]Oxazino[3,4-a]quinolin-3-one, 4,4a-dihydro-1-isopropylidene-4,4,4a-trimethyl-	EtOH	239(4.44),284(3.51), 342(3.74)	77-0574-65
$C_{18}H_{21}NO_3$			
Azepino[3,2,1-hi]indole, 7-tert-butoxycarbonylmethylene-1,2,4,5,6,7-hexahydro-6-oxo-	EtOH	251(4.18),378(3.72)	35-1397-65
Benzo[a]cyclopenta[i]quinolizin-12(1H)-one, 2,3,5,6,10b,11-hexahydro-8,9-dimethoxy-	C_6H_{12}	312(4.08)	44-3667-65
	EtOH	332(4.17)	44-3667-65
perchlorate	EtOH	339(4.10)	44-3667-65
Glaziovine, tetrahydro-	EtOH	286(3.42)	35-0694-64
	EtOH-KOH	253(3.79),299(3.65)	35-0694-64
Isoquinoline, 8-(benzyloxy)-1,2,3,4-tetrahydro-6,7-dimethoxy-	pH 1	280(3.08)	33-2089-64
	pH 13	280(3.21)	33-2089-64
	iso-PrOH	225s(4.11),259s(2.83), 265s(2.93),275(3.04), 280(3.05)	33-2089-64
N-Norarmepavine	EtOH	282(3.74)	95-0362-64
$C_{18}H_{21}NO_4$			
1-Dibenzofurancarboxylic acid, 9α-cyanomethyl-4-ethoxy-5a,6,7,8,9,9a-hexahydro-, methyl ester	EtOH	257(3.85),299(3.73)	94-1012-64
Erythrinane-4,8-dione, 15,16-dimethoxy-, cis	MeOH	210(4.33),232s(3.90), 282(3.60)	24-0046-65
Maleic acid, (N-1-cyclohexen-1-ylanilino)-, dimethyl ester	EtOH	319(3.89)	5-0098-65H
Pluviine, β-dihydro-7-oxo-	EtOH	223(4.46),249(3.84), 263(3.89),298(3.79)	94-0408-64

Compound	Solvent	λ_{max}(log ε)	Ref.
Pyrrolo[2,1-a]isoquinoline-3-carboxylic acid, 5,6-dihydro-8,9-dimethoxy-2-methyl-, ethyl ester	EtOH	324(4.49),336(4.47)	94-0775-65
$C_{18}H_{21}NO_5$			
Criwelline	EtOH	239(3.71),291(3.65)	33-0185-64
Homolycorine, α-dihydro-2-oxo-	EtOH	226(3.89),267(3.48), 302(3.23)	94-0408-64
Norhomolycorine, N-formyl-β-dihydro-	EtOH	227(4.37),268(3.96), 302(3.77)	94-0408-64
Spiro[cyclohexane-1,9'-[9H-1,3]dioxolo-[4,5-h][1]benzazepin]-6'(5'H)-one, 7',8'-dihydro-4-hydroxy-, acetate	EtOH	256(3.82),297(3.76)	94-0427-65
Spiro[cyclohexane-1,9'-[9H-1,3]dioxolo-[4,5-h][2]benzazepin]-5'(6'H)-one, 7',8'-dihydro-4-hydroxy-, acetate	EtOH	262(3.47),294(3.49)	94-0427-65
Tazettine	EtOH	240(3.74),291(3.68)	95-1194-64
$C_{18}H_{21}NO_6$			
Macranthine, acetate	EtOH	238(3.56),293(3.69)	33-0185-64
$C_{18}H_{21}NO_8$			
9aH-Quinolizine-1,2,3,4-tetracarboxylic acid, dihydro-6-methyl-, tetramethyl ester	MeOH	236(3.98),278(4.17), 417(3.84)	39-0948-65
$C_{18}H_{21}N_3O_2$			
5H-Dibenzo[b,e][1,4]diazepin-11-one, 10,11-dihydro-10-(2-dimethylamino-ethyl)-8-hydroxy-5-methyl-	EtOH-HCl	210(4.52),302(3.59)	33-1590-65
$C_{18}H_{21}N_3O_2S$			
5-Pyrrolidinecarboxylic acid, 2-(2-pyrrolyl)-, ethyl ester, thiocarbanilide	EtOH	206(4.34),220(4.27), 252(4.14)	39-0893-64
$C_{18}H_{21}N_3O_3$			
o-Propionanisidide, 4'-[(o-methoxy-phenyl)azo]-2-methyl-	EtOH-NH_3	376(4.38)	35-2315-64
$C_{18}H_{21}N_3O_4$			
D-Xylosone-1-benzylphenylhydrazone 2-oxime	EtOH	235(4.00),330(4.37)	24-0725-64
$C_{18}H_{21}N_3O_5$			
Benzoin, 2-(2-aminoethyl)-2'-nitro-4,5-dimethoxydeoxy-, oxime	EtOH	270(4.25)	39-4014-65
D-Fructosone-1-diphenylhydrazone 2-oxime	EtOH	231(4.03),334(4.30)	24-0725-64
Indole-2-carboxylic acid, 1-(2-cyano-ethyl)-3-(hydroxymethyl)-5-methoxy-6-methyl-, ethyl ester, carbamate	MeOH	211(4.47),303(4.3)	44-2897-65
$C_{18}H_{21}N_3O_7$			
Cytidine, 2',3'-O-(2,4-dimethoxy-benzylidene)-	MeOH	274(3.03)	5-0156-64I
$C_{18}H_{21}N_5O_4S$			
9H-Purine, 6-(benzylthio)-2-(methyl-amino)-9-β-D-ribofuranosyl-	pH 1	256(4.11),330(3.96)	35-3752-65
	pH 11	256(4.27),330(4.00)	35-3752-65
	MeOH	253(4.30),320(4.02)	35-3752-65

Compound	Solvent	λ_{max}(log ε)	Ref.
$C_{18}H_{21}N_5O_5$			
9H-Purine, 6-(benzyloxy)-2-(methyl-amino)-9-β-D-ribofuranosyl-	pH 1	247(4.02),300(3.93)	35-3752-65
	pH 11	252(4.10),291(3.93)	35-3752-65
$C_{18}H_{22}$			
Hexane, 1,1-diphenyl-	C_6H_{12}	255(2.85),256(2.84), 260(2.88),262(2.88), 269(2.76)	22-0587-65
$C_{18}H_{22}B_{10}N_2O$			
2-Naphthol, 1-[(m-1,2-dicarbadodeca-boran(12)-1-ylphenyl)azo]-	EtOH	262(4.0),282s(3.9), 303(3.9),417s(4.0), 477(4.2)	70-2206-65
2-Naphthol, 1-[(p-1,2-dicarbadodeca-boran(12)-1-ylphenyl)azo]-	EtOH	264(4.2),304(4.0), 417s(4.0),477(4.3)	70-2206-65
2-Naphthol, 1-[(p-1,7-dicarbadodeca-boran(12)-1-ylphenyl)azo]-	EtOH	264(4.1),305(4.0), 417s(4.0),477(4.3)	70-2206-65
$C_{18}H_{22}BrNO_2$			
3-Ethyl-6,7-dihydro-9,10-dimethoxy-2-methylbenzo[a]quinolizinium bromide	EtOH	230s(4.17),267s(4.06), 281(4.18),362(4.19)	33-1117-64
$C_{18}H_{22}BrNO_4$			
Erythrinan-8-one, 6α-bromo-7α-hydroxy-15,16-dimethoxy-, cis	MeOH	283(3.48)	24-0046-65
2-Indolinone, 7-bromo-1-(3,4-dimethoxy-phenethyl)-3a,4,5,6-tetrahydro-3a-hydroxy-	MeOH	231(4.27),278(3.46)	78-1729-64
$C_{18}H_{22}ClN_5O_2S$			
Purine, 6-(m-chlorobenzylthio)-9-[2-[N,N-bis(2-hydroxyethyl)amino]ethyl]-	pH 1	291(4.28)	87-0182-65
	pH 11	289(4.28)	87-0182-65
$C_{18}H_{22}CuN_4O_2$			
Atrolactamidine, copper complex	MeOH	252(3.70),580(1.47)	39-4004-65
$C_{18}H_{22}Fe$			
Iron, bis(tetrahydroindenyl)-	hexane	215(4.04),325(1.96), 435(2.09)	101-0107-65A
$C_{18}H_{22}N_2$			
Bis(propylideneaniline)	MeOH	250(4.35),303(3.70)	65-0356-64
Tubifoline	2N H_2SO_4	235(4.00),295(3.77)	33-1497-64
	ether	216(4.33),250(3.95)	33-1497-64
$C_{18}H_{22}N_2O$			
Vincanidine derivative	n.s.g.	226(4.48),276(3.66), 294(3.48)	70-1992-65
Vincanidine derivative	n.s.g.	245(3.91),292(3.49)	70-1992-65
$C_{18}H_{22}N_2O_2$			
Compactinervine, $\Delta^{1,2}$-decarbomethoxy-	EtOH-HCl	220(4.25),260(3.45), 280(3.44)	78-1141-65
	EtOH	215(4.30),250(3.80)	78-1141-65
Compactinervine, $\Delta^{2,16}$-decarbomethoxy-	EtOH	225s(3.75),272(4.15), 312(3.70)	78-1141-65
Indole, 1-acetyl-3-(1-acetyl-5,5-di-methyl-2-pyrrolidinyl)-	EtOH	239(4.25),260(3.91), 270s(3.86),290(3.83), 299(3.87)	87-0415-64

Compound	Solvent	λ_{max}(log ε)	Ref.
$C_{18}H_{22}N_2O_3S$			
2-Cyclohexen-1-one, 5,5-dimethyl-	EtOH	242(4.06),306(4.29)	44-0798-64
2-(o-nitrophenylthio)-3-N-pyrrolidyl- perchlorate	EtOH	242(4.20),309(4.34)	44-0798-64
$C_{18}H_{22}N_2O_4$			
Dipyrromethene, 4,4'-dicarbethoxy- 3,3',5'-trimethyl-	EtOH-HCl	242(3.97),351(3.65), 459(5.04)	12-0363-65
	EtOH	219(4.43),338s(3.82), 433(4.44)	12-0363-65
	CCl_4	440(4.52)	12-0363-65
Indole-3-propionic acid, α-(3-methyl- 2-butenyl)-α-nitro-, ethyl ester	MeOH	272(3.66),280(3.70), 289(4.05)	22-2306-65
$C_{18}H_{22}N_2O_5$			
1-Anilinium-4-ethyl-2-anilinomalate	pH 3.63	200(4.1),222s(3.5)	12-0337-65
	pH 4.26	188(4.2),225(3.7)	12-0337-65
	pH 4.98	185(4.3),225(4.1), 280(3.3)	12-0337-65
	pH 7.36	280(3.6)	12-0337-65
	pH 10.1	270(4.1)	12-0337-65
	pH 11.8	220(4.3),270(4.3)	12-0337-65
	EtOH	242(4.0),321(4.18)	12-0337-65
$C_{18}H_{22}N_2O_7$			
Estriol, 2,4-dinitro-	EtOH	275(3.82),346(3.52), 420(3.07)	1-0483-64
$C_{18}H_{22}N_2O_{11}$			
Glucopyranuronic acid, 1-deoxy-1- (4-methoxy-2-oxo-1(2H)-pyrimidinyl)-, methyl ester, 2,3,4-triacetate	EtOH	277(3.75)	94-1259-64
Uracil, 1-(2,3,4,6-tetra-O-acetyl- β-D-glucopyranosyl)-	EtOH	254(4.04)	94-0357-64
$C_{18}H_{22}N_4O_2$			
Azobenzene, 4-acetamido-4'-(dimethyl- amino)-5-methoxy-2-methyl-	EtOH	235(4.06),256(4.09), 316(3.93),413(4.59)	65-3602-64
$C_{18}H_{22}N_4O_3$			
5-Deoxy-D-ribophenyl osazone	pH 1	252(4.12),400(4.28)	44-3436-64
Leucoflavin, 5-acetyl-10-ethyl- 1,3,7,8-tetramethyl-	MeOH	212(4.38),250s(4.13), 303(3.84)	33-1354-64
$C_{18}H_{22}N_4O_4$			
Bicyclo[5.4.0]undec-7-en-9-one, 1β- methyl-, 2,4-dinitrophenylhydrazone	$CHCl_3$	391(4.42)	35-1755-64
Bicyclo[5.4.0]undec-10-en-9-one, 1β- methyl-, 2,4-dinitrophenylhydrazone	$CHCl_3$	382(4.40)	35-1755-64
Cyclohexanone, 2-(2-methylcyclopentyli- dene)-, 2,4-dinitrophenylhydrazone	n.s.g.	370(4.31)	70-0121-64
Cyclohexanone, 2-cyclohexylidene, 2,4-dinitrophenylhydrazone	n.s.g.	359(4.3)	70-0121-64
solid isomer	n.s.g.	373(4.3)	70-0121-64
4a,8a-Methanonaphthalen-2(1H)-one, hexahydro-1-methyl-, 2,4-dinitro- phenylhydrazone	$CHCl_3$	363(4.36)	35-0275-65
2(1H)-Naphthalenone, 3,4,4a,5,6,7-hexa- hydro-1,1-dimethyl-, 2,4-dinitro- phenylhydrazone	$CHCl_3$	368(4.34)	39-2340-65

Compound	Solvent	$\lambda_{max}(\log \epsilon)$	Ref.
2(1H)-Naphthalenone, 3,4,5,6,7,8-hexahydro-1,1-dimethyl-, 2,4-dinitrophenylhydrazone	$CHCl_3$	368(4.41)	39-2340-65
$C_{18}H_{22}N_4O_{10}$			
Glucopyranuronamide, 1-(4-acetamido-2-oxo-1(2H)-pyrimidinyl)-1-deoxy-, 2,3,4-triacetate	50% EtOH	251(4.2),300(3.85)	94-1259-64
$C_{18}H_{22}N_8O_9$			
Glucitol, 1-(6-amino-9H-purin-9-yl)-2-deoxy-2-(2,4-dinitroanilino)-1-O-methyl-	EtOH	263(4.09),351(4.02)	44-1096-65
$C_{18}H_{22}Na_2O_8S_2$			
Estradiol, bis(hydrogen sulfate)-, disodium salt	MeOH	270(3.00),276(2.99)	13-0845-65
$C_{18}H_{22}O$			
Chrysofluorene, 1,1a,1b,2,3,4,4a,5,-6,6a-decahydro-1a-methyl-3-oxo-	EtOH	266(3.01)	42-0479-64
1,4:9,10-Dimethanoanthracene, 2-ethoxy-1,2,3,4,4a,9,9a,10-octahydro-	EtOH	205(4.17),270(2.45), 278(2.43)	39-4646-65
Estra-1,3,5(10)-trien-17-one	C_6H_{12}	213(4.04),274(2.68), 298(1.62)	60-0285-64
	MeOH	266(2.73),273(2.70)	87-0409-65
	EtOH	214(3.91),267(2.67), 274(2.69)	44-0295-65
	EtOH	213(3.85),274(2.70), 294(1.75)	60-0285-64
3(2H)-Phenanthrone, 9,10-diethyl-1,4,5,8-tetrahydro-	MeOH	273(2.46),279s(--)	5-0036-64D
$C_{18}H_{22}OS$			
3-Thioestrone	EtOH	213(4.37),242(4.01), 285(3.03)	44-0295-65
	pH 13	265(4.20)	44-0295-65
$C_{18}H_{22}O_2$			
Estra-4,9-diene-3,17-dione	EtOH	301(4.34)	44-2195-64
Estrone	EtOH	201(4.49),219s(2.78), 283(3.32)	60-0285-64
2(3H)-Naphthalenone, 4,4a,5,6,7,8-hexahydro-6-(4-methoxy-m-tolyl)-	EtOH	231(4.34)	87-0519-64
1-Naphthoic acid, heptyl ester	EtOH	220(5.64),296(3.80)	39-2072-65
2-Naphthoic acid, heptyl ester	EtOH	237(4.85),281(3.90)	39-2072-65
1-Phenanthrenecarboxylic acid, 1,2,3,4,5,9-hexahydro-1,4a-dimethyl-, methyl ester	EtOH	260(2.76),267(2.83), 273(2.78)	78-2133-65
	EtOH	260(2.76),267(2.83), 273(2.78)	94-0984-64
isomer	EtOH	264(3.93)	78-2133-65
	EtOH	264(3.93)	94-0984-64
Spiro[cyclohexane-1,1'-inden]-4-one, 2',3'-diethyl-6'-hydroxy-	MeOH	270(4.15)	5-0036-64D
	MeOH-NaOH	292(4.20)	5-0036-64D
Spiro[2-cyclohexene-1,1'-indene]-4,6'(2'H)-dione, 2',3'-diethyl-4',5'-dihydro-	MeOH	224(4.16),303(4.28)	5-0036-64D
	ether	220(4.15),290(4.31)	5-0036-64D
$C_{18}H_{22}O_3$			
1,3-Cyclohexanedione, 2-methyl-2-(γ-methyl-p-methoxycinnamyl)-, ethyl ester	EtOH	257(4.18)	70-1355-64

Compound	Solvent	λ_{max}(log ϵ)	Ref.
4-Cyclopentene-1,3-dione, 4-tert-butyl-3-(3-tert-butyl-4-oxocyclopent-2-en-1-ylidene)-	MeOH	233(4.20),304(4.49)	24-2774-65
	MeOH-KOH	227(4.18),330(4.10), 555(3.66)	24-2774-65
Estra-1,3,5(10)-trien-17-one, 1,3-dihydroxy-	EtOH	225s(3.96),281s(3.35), 286(3.35)	24-1940-64
	NaOH	296(3.57),325s(2.49)	24-1940-64
Estra-1,3,5(10)-trien-17-one, 1,4-dihydroxy-	EtOH	220s(3.81),294(3.58)	24-1940-64
Estra-1,3,5(10)-trien-17-one, 2α,3-dihydroxy-	MeOH	290(2.45)	7-0310-65
Estra-1,3,5(10)-trien-17-one, 3,7α-dihydroxy-	EtOH	281(3.38)	44-1333-64
	MeOH-KOH	297(3.47)	44-1333-64
Estra-1,3,5(10)-trien-17-one, 3,15-dihydroxy-	EtOH	281(3.36)	44-1333-64
	MeOH-KOH	297(3.48)	44-1333-64
Estra-1,3,5(10)-trien-17-one, 3,15α-dihydroxy-	EtOH	282(3.30)	44-2731-64
Estra-1,3,5(10)-trien-17-one, 3,15β-dihydroxy-	MeOH	222(3.86),280(3.30)	44-0214-64
Estr-4-ene-3,16,17-trione	MeOH	240(2.70)	7-0310-65
8,14-Seco-19-norandrosta-4,9-diene-3,14,17-trione	EtOH	304(4.20)	13-0729-64
$C_{18}H_{22}O_4$			
2-Cyclohexene-1-carboxylic acid, 5-acetyl-4-(p-methoxyphenyl)-6-methyl-, methyl ester	EtOH	227(4.15),275(3.28), 284(3.21)	70-1911-64
Hydroperoxide, 3,17-dioxoestra-4,9-dien-11β-yl-	EtOH	293(4.30)	28-5669-64A
1β-Phenanthrenecarboxylic acid, 1,2,3,9,10,10aβ-hexahydro-4,7-dimethoxy-, methyl ester	EtOH	265(4.26),308s(3.20)	44-2527-64
Podocarpic acid, 6,7-dehydro-13-hydroxy-, methyl ester	EtOH	224(4.34),281(3.85), 305(3.67),315(3.62)	39-0361-65
Podocarpic acid, 7-oxo-, methyl ester	EtOH	282(4.54),282(4.34)[sic]	78-0409-64
$C_{18}H_{22}O_4S_2$			
3,5-Xylenesulfonic acid, 4-methoxythio-, S-4-methoxy-3,5-xylyl ester	EtOH	242(4.17)	44-0759-64
$C_{18}H_{22}O_5$			
4β,9β-Anthracenediol, 10β-acetoxy-1,4,4aα,9,9aα,10-hexahydro-5-methoxy-9α-methyl-	EtOH	277(3.54),283(3.53)	65-2570-64
Podocarpic acid, 13-hydroxy-7-oxo-, methyl ester	EtOH	211(4.20),237(4.22), 283(4.02),320(3.87)	39-0361-65
$C_{18}H_{22}O_6$			
Biphenyl, 2,2',4,4',5,5'-hexamethoxy-	EtOH	245s(4.01),298(4.01)	39-3040-65
1,4-Naphthoquinone, 2-hydroxy-3-(5-hydroxyhexyl)-5,7-dimethoxy-	EtOH	212(4.49),261(4.27), 305(4.03),370(3.56), 423(3.22)	39-2289-64
$C_{18}H_{22}O_7$			
5H-Benzocycloheptene-6-butyric acid, 6,7,8,9-tetrahydro-2,3,4-trimethoxy-γ,5-dioxo-	EtOH	240s(3.77),319(4.21)	44-1752-65
5H-Benzocycloheptene-9-propionic acid, 8-carboxy-6,7-dihydro-1,2,3-trimethoxy-	EtOH	222(4.29),271(3.97)	44-1752-65
Glauconic acid, dihydro-, methyl ester	EtOH	232(3.93)	39-1772-65

Compound	Solvent	λ_{max}(log ε)	Ref.
$C_{18}H_{22}Ru$			
Ruthenium, bis(4,5,6,7-tetrahydro-indenyl)-	hexane	215(3.60),315(2.40)	101-0107-65A
$C_{18}H_{23}BrO_2$			
Estra-1,3,5(10)-triene-3,17β-diol, 4-bromo-	EtOH	284(3.26)	44-2195-64
	EtOH-base	244(3.88),314(3.51)	44-2195-64
$C_{18}H_{23}ClN_4O_2$			
2-Butanone, 1-chloro-4-[p-[(2-amino-4-hydroxy-6-methyl-5-pyrimidyl)-3-propyl]amino]phenyl-	pH 1	265(3.85)	87-0035-65
	pH 13	240(4.11),280(3.83)	87-0035-65
$C_{18}H_{23}ClN_4O_3$			
Acetophenone, N-[1-(2-amino-4-hydroxy-6-methyl-5-pyrimidyl)-3-propyl]-p-amino-α-chloro-, ethylene ketal	EtOH	253(4.28)	87-0035-65
$C_{18}H_{23}IN_2$			
2-[p-(Diethylamino)styryl]-1-methyl-pyridinium iodide, cis	aq. HCl	309(3.83)	23-1345-65
	H_2O	434(4.00)	23-1345-65
trans	aq. HCl	325(4.45)	23-1345-65
	H_2O	456(4.52)	23-1345-65
	50% MeOH	469(4.60)	23-1345-65
	50% MeOH-HCl	328(4.45)	23-1345-65
$C_{18}H_{23}NO$			
Acetamide, N-(5-isopropyl-3,8-di-methyl-1-azulenyl)-N-methyl-	C_6H_{12}	246(4.38),288(4.57), 304(4.25),351(3.76), 368(3.73),610(2.69), 660s(2.61),731s(2.16)	35-3137-64
Morphinan, 5-dehydro-3-methoxy-N-methyl-	EtOH	228(3.86),280(3.35), 288(3.32)	44-1769-65
$C_{18}H_{23}NO_2$			
6-Azaestrone, methyl ether	EtOH	214(4.43),251(3.86), 299(3.60)	78-2517-65
	EtOH	214(4.4),253(3.84), 301(3.62)	88-1275-64
6-Aza-14β-estrone, methyl ether	EtOH	213(4.41),250(3.77), 301(3.59)	78-2517-65
Estra-1,3,5(10)-trien-17-one, 1-amino-3-hydroxy-	MeOH	212(4.60),240s(3.90), 292(3.49)	35-2943-64
$C_{18}H_{23}NO_2S$			
p-Toluenesulfonamide, N-[2-(1,3,3-tri-methylbicyclo[3.1.0]hex-2-ylidene)-ethylidene]-	n.s.g.	222(4.0),304(4.3)	27-0538-65
$C_{18}H_{23}NO_3$			
Apoerysopin-5-one, 1,2,3,3a,3b,12b-hexahydro-, dimethyl ether	MeOH	230(3.89),282(3.48)	78-1729-64
Diethylamine, 1,1'-dimethyl-2-phenoxy-2'-(p-hydroxyphenyl)-, (racemic)	EtOH-H_2SO_4	270(3.43),276(3.47)	54-0521-65
(+) (also allo forms, same spectra)	EtOH-H_2SO_4	270(3.43),276(3.47)	54-0521-65
Erythrinan-7-one, 15,16-dimethoxy-, cis	MeOH	210(4.16),282(3.52)	24-0046-65

Compound	Solvent	$\lambda_{max}(\log \epsilon)$	Ref.
$C_{18}H_{23}NO_4$			
5H-Benzo[a]carbazole, 6,6a,6b,7,8,9,-10,10a,11,11a-decahydro-, oxalate	MeOH	209(4.21),265f(3.00)	39-5919-64
Lycorenine, α-deoxydihydro-2-oxo-	EtOH	284(3.64)	94-0408-64
$C_{18}H_{23}NO_5$			
Erythrinan-8-one, 6,7-dihydroxy-15,16-dimethoxy-	MeOH	212(4.29),232s(3.93), 282(3.61)	24-0046-65
Estra-1,3,5(10)-triene-3,16α,17β-triol, 2-nitro-	EtOH	290(3.86),363(3.53)	1-0483-64
$C_{18}H_{23}NO_6$			
Jacozine	n.s.g.	233(3.24)	12-0233-64
$C_{18}H_{23}N_3O$			
1H-Azepino[5,4,3-cd]indole, 3,4,5,6-tetrahydro-6-methyl-2-piperidinocarbonyl-	EtOH	224(4.47),294(4.09)	87-0200-65
Bicyclo[3.1.0]hexane-$\Delta^{2,\alpha}$-acetaldehyde, 1,3,3-trimethyl-, 4-phenylsemicarbazone	n.s.g.	237(4.2)	27-0538-65
$C_{18}H_{23}N_3O_2S$			
p-Toluenesulfonamide, N-[p-(diethylamino)benzylidene]-	EtOH	341(4.37)	44-1242-65
anion	H_2O	325(4.36)	44-1242-65
	MeOH	325(4.41)	44-1242-65
	EtOH	325(4.53)	44-1242-65
	MeCN	332(4.42)	44-1242-65
	DMF	337(4.40)	44-1242-65
$C_{18}H_{23}N_3O_{10}$			
Cytosine-O-glucoside, tetraacetate	EtOH	271(3.84)	39-6125-65
$C_{18}H_{23}N_3O_{11}$			
1,3,5-Triazin-2(1H)-one, 1-(2,3,4,6-tetra-O-acetyl-β-D-glucopyranosyl)-4-methoxy-	MeCN	254(3.40)	73-2060-64
$C_{18}H_{23}N_5O$			
Isoxazolo[5,4-d]pyrimidine, 4-(3-diethylaminopropylamino)-3-phenyl-	EtOH	238(4.08),291(3.96)	44-2117-64
$C_{18}H_{23}N_5O_2S$			
Purine, 6-(benzylthio)-9-[2-[N,N-bis-(2-hydroxyethyl)amino]ethyl]-	pH 1	291(4.29)	87-0182-65
	pH 11	292(4.27)	87-0182-65
$C_{18}H_{23}NaO_5S$			
Estradiol, 3-(hydrogen sulfate)-, sodium salt	MeOH	270(2.43),276(2.88)	13-0845-65
Estradiol, 17-(hydrogen sulfate)-, sodium salt	MeOH	281(3.35)	13-0845-65
$C_{18}H_{23}O_5P$			
Estrone, 3-phosphate	MeOH	270(3.09),276(3.04)	22-0018-65
$C_{18}H_{24}$			
Chrysene, dodecahydro-	n.s.g.	244(2.89),253(2.77), 263(2.51),277(2.37)	12-0055-64
Elliotin	EtOH	231(4.97)	44-0429-65

Compound	Solvent	λ_{max}(log ε)	Ref.
Estra-1,3,5(10)-triene	EtOH	260s(2.54),266(2.69), 273(2.69)	24-0140-64
Mesitylene, 2,4,6-tris(1-propen-1-yl)-, cis	hexane	220(4.54),275(2.52)	39-3160-65
Naphthalene, 1-n-octyl-	EtOH	225(5.14),283(3.88)	39-2072-65
Naphthalene, 2-n-octyl-	EtOH	225(5.09),275(3.73)	39-2072-65
Tricyclo[9.7.0.0^{2,10}]octadecatetraene	heptane	252(3.96)	35-4506-65
$C_{18}H_{24}ClFN_2O_2$			
Benzylamine, N,N-dibutyl-α-(1-chloro-2-fluoro-3-nitroallylidene)-	n.s.g.	278(4.07),378(4.08)	35-5132-65
$C_{18}H_{24}F_2N_2O_2$			
p-Benzoquinone, 2,5-bis(cyclohexylamino)-3,6-difluoro-	methyl cellosolve	223(4.49),356(4.46), 546(2.36)	78-1889-64
p-Benzoquinone, 2,5-difluoro-3,6-bis-(hexahydro-1H-azepin-1-yl)-	methyl cellosolve	237(4.34),403(4.19), 582(2.47)	78-1889-64
$C_{18}H_{24}IN$			
1,3,4,6,7,12,13,13a-Octahydro-5-methyl-2H-naphtho[1,2-c]quinolizinium iodide	EtOH	215(4.59),221(4.51), 259(4.07)	44-1834-64
$C_{18}H_{24}INO_2$			
3-Ethoxy-4a,5,7,8,9,10-hexahydro-11-methyl-6H-benzofuro[3a,3,2-ef]-[2]benzazepinium iodide	EtOH	256(3.72),318(4.04)	94-1012-64
$C_{18}H_{24}N_2$			
Tubifolidine	EtOH	207(4.47),244(3.93), 298(3.61)	33-1497-64
$C_{18}H_{24}N_2O$			
Carbazol-3-ol, 5,6,7,8-tetrahydro-4-(piperidinomethyl)-	EtOH	232(4.36),285(3.96), 295s(3.92)	35-4631-64
Uleine, 1,13-dihydro-13-hydroxy-	EtOH	221(4.56),282(3.91), 289(3.87)	31-0363-64
	EtOH	211(4.56),282(3.91), 289(3.87)	78-1717-65
$C_{18}H_{24}N_2O_2$			
Dipyrromethene, 4'-carbethoxy-4-ethyl-3,3',5,5'-tetramethyl-	EtOH-HCl	223s(4.07),262(3.70), 471(4.88)	12-0363-65
	EtOH	214(4.28),259(3.85), 410(4.44)	12-0363-65
	CCl_4	420(4.48)	12-0363-65
$C_{18}H_{24}N_2O_3S$			
2,4-Diazaestr-4-en-1-one-3-thione, 17β-acetoxy-	MeOH-HCl	220(4.13),282(4.33)	13-0361-64
	MeOH	219(4.08),280(4.32)	13-0631-64
	MeOH-NaOH	264(4.16),315(4.11)	13-0631-64
$C_{18}H_{24}N_2O_4$			
8-Azabicyclo[4.3.1]decan-10-ol, 8,10-dimethyl-, p-nitrobenzoate	EtOH	259(4.15)	44-3634-64
3,3'-Bipyrrole-5,5'-dicarboxylic acid, 2,2',4,4'-tetramethyl-, diethyl ester	$CHCl_3$	255(4.19),285(4.53)	39-3315-64
2-Quinolone, 3-(3-acetamido-3-methylbutyl)-4-hydroxy-8-methoxy-1-methyl-	MeOH	216(4.35),238(4.48), 253(4.43),284(3.90), 293(3.93),319(3.60)	39-4190-64

Compound	Solvent	$\lambda_{max}(\log \epsilon)$	Ref.
$C_{18}H_{24}N_2O_5$			
11bH-Benzo[a]quinolizine-3-carboxamide, 2-acetoxy-1,2,3,4,6,7-hexahydro-9,10-dimethoxy-	n.s.g.	284.5(3.53)	87-0635-64
2-Pyrroline-3-carboxylic acid, 1,2-dimethyl-4-oxo-5-(3-carbethoxy-1,2-dimethyl-4-pyrrolyl)-, ethyl ester	EtOH	205(4.45),246(4.39), 297s(3.64)	23-1524-64
$C_{18}H_{24}N_2O_6$			
Quinaldinic acid, 6-nitro-N-(3-carbethoxypropyl)-1,2,3,4-tetrahydro-, ethyl ester	EtOH	209(3.96),229(3.73), 393(3.97)	78-2961-65
$C_{18}H_{24}N_2O_{10}$			
2(1H)-Pyrimidone, 1-(2,3,4-tri-O-acetyl-β-D-glucopyranosyl)-4-ethoxy-	EtOH	276(3.81)	44-2723-65
$C_{18}H_{24}N_4$			
2-Pyrroline, 1-methyl-2-(2-pyrrolyl)-3-[1-methyl-2-(2-pyrrolyl)-2-pyrrolidinyl]-	EtOH	216(4.11),263(3.25), 321(4.27)	39-2614-65
$C_{18}H_{24}N_4O_4$			
5(1H)-Azulenone, octahydro-3,8-dimethyl-, 2,4-dinitrophenylhydrazone	EtOH	230(4.32),362(4.36)	94-0942-65
isomer	EtOH	229(4.34),362(4.38)	94-0942-65
2H-Benzocyclohepten-2-one, decahydro-4a-methyl-, 2,4-dinitrophenylhydrazone	$CHCl_3$	368(4.29)	35-1755-64
2H-Benzocyclohepten-2-one, decahydro-9a-methyl-, 2,4-dinitrophenylhydrazone	$CHCl_3$	368(4.40)	35-1755-64
Bicyclo[6.3.0]undecanone, 1β-methyl-, 2,4-dinitrophenylhydrazone, trans	$CHCl_3$	363(4.38)	35-1755-64
Cyclohexaneacrolein, 3,3,5-trimethyl-, 2,4-dinitrophenylhydrazone	$CHCl_3$	379(4.48)	28-4712-64B
$C_{18}H_{24}N_4O_6$			
p-Menthan-2-one, 3-hydroxy-, 2,4-dinitrophenylhydrazone, acetate	$CHCl_3$	381(4.47)	44-0518-65
$C_{18}H_{24}O$			
Estra-1,3,5(10)-trien-2-ol	EtOH	217s(3.86),282(3.40)	24-0140-64
	pH 13	243(3.98),298(3.56)	24-0140-64
Estra-1,3,5(10)-trien-3-ol	EtOH	281(3.31),287(3.27)	24-0140-64
	pH 13	241(3.93),298(3.43)	24-0140-64
Estra-1,3,5(10)-trien-17β-ol	MeOH	266(2.70),274(2.71)	87-0409-65
$C_{18}H_{24}OS$			
3-Thioestradiol	EtOH	214(4.34),242(3.99), 286(2.99)	44-0295-65
$C_{18}H_{24}O_2$			
Abietic acid, dehydrodeisopropyl-, methyl ester	EtOH	260(2.64),266(2.74)	44-2754-65
p-Benzoquinone, 2,3-dimethyl-5,6-bis-(3-methyl-2-butenyl)-	EtOH	266(4.21),304(2.96)	37-1374-65
Estra-1,3,5(10)-triene-2,17β-diol	EtOH	218(3.86),282(3.42)	24-0140-64
	pH 13	243(3.94),299(3.55)	24-0140-64
5α-Estr-1(10)-ene-3,17-dione	EtOH-KOH	231(3.8)	33-1961-64

Compound	Solvent	λ max(log ε)	Ref.
Estr-4-ene-3,16-dione	MeOH	238(4.24)	44-2351-64
Estr-4-ene-3,17-dione	MeOH	239(2.79)	7-0310-65
1,3,5-Hexatriene, 3,4-bis(4-oxocyclohexyl)-	MeOH	239.5(3.81)	5-0024-65D
$C_{18}H_{24}O_3$			
2,9-Anthracenedione, 1,3,4,4a,9a,10-hexahydro-, 2-(diethyl acetal)	EtOH	248(4.08),294(3.24)	70-1039-65
Chromone, 3,7-di-tert-butyl-5-hydroxy-2-methyl-	MeOH	216(4.51),247(4.15), 278(3.54)	24-2774-65
	MeOH-KOH	225(4.43),257(4.04), 303(3.56)	24-2774-65
Cyclohexanecarboxylic acid, 2-benzyl-5-oxo-, tert-butyl ester	EtOH	209(3.70),248(2.21), 253(2.24),259(2.27), 265(2.18),269(2.10)	70-1039-65
Estra-1,3,5(10)-triene-1,3,17β-triol	EtOH	282s(3.33),287(3.34)	24-1940-64
	NaOH	297(3.53),333s(3.03)	24-1940-64
Estra-1,3,5(10)-triene-3,7β,17-triol	EtOH	220s(3.79),281(3.31)	44-1325-64
Estra-1,3,5(10)-triene-3,15α,17β-triol	EtOH	281(3.32)	44-2731-64
4-Estrene-3,17-dione, 1β-hydroxy-	EtOH	242(4.19)	54-0626-65
4-Estrene-3,17-dione, 2α-hydroxy-	MeOH	236(2.68)	7-0310-65
4-Estrene-3,17-dione, 9α-hydroxy-	n.s.g.	245(4.19)	39-4492-64
4-Estrene-3,17-dione, 16α-hydroxy-	MeOH	238(2.78)	7-0310-65
Estr-4-en-3-one, 9α,10α-epoxy-17β-hydroxy-	n.s.g.	245(4.13)	39-4492-64
A-Norandrost-3(5)-ene-1,2-dione, 17β-hydroxy-	EtOH	284(3.77)	94-0050-65
19-Nortestolactone	MeOH	238(4.23)	35-1528-64
6-Oxaestra-1,3,5(10)-trien-17-ol, 3-methoxy-	n.s.g.	282(3.56),288(3.44)	31-0418-64
Podocarpic acid, 7,13-dihydroxy-, methyl ester	EtOH	225(3.98),286(3.59)	39-0361-65
$C_{18}H_{24}O_5$			
5H-Benzocycloheptene-9-propionic acid, 6,7-dihydro-1,2,3-trimethoxy-, methyl ester	EtOH	218(4.45),250(3.98), 282s(3.16)	44-1752-65
1,3-Naphthalenediol, 2-(5-hydroxyhexyl)-6,8-dimethoxy-	EtOH	245(4.85),296(3.59), 302(3.63),313(3.56), 330(3.37)	39-2289-64
1β-Phenanthrenecarboxylic acid, 1,2,3,4,4aβ,9,10,10aβ-octahydro-4,4,7-trimethoxy-	EtOH	222(3.93),278(3.21), 286s(3.12)	44-2527-64
$C_{18}H_{24}O_6$			
9-Cyclononene-1,2,4-tricarboxylic acid, 7,8-diethyl-2-methyl-6-oxo-, 1,2-anhydride, methyl ester	EtOH	225(3.92)	39-1772-65
Hyptolide	EtOH	212(3.93)	39-4167-64
$C_{18}H_{24}O_7$			
Carbonic acid, ethyl ester, ester with isophotoartemisic lactone	EtOH	238(4.16)	39-2518-64
Renifolin	EtOH	283(3.30)	94-0533-64
$C_{18}H_{25}Cl_2N_3O_8$			
[[1-[(Dimethylamino)methylene]inden-2,3-ylene]dimethylidyne]bis[dimethylammonium perchlorate]	MeCN	485(4.30)	73-2783-65

Compound	Solvent	λ_{max}(log ϵ)	Ref.
$C_{18}H_{25}Cl_2O_7Rh$			
Rhodium, bis(3-chloro-2,4-pentanedionato)[3-(ethoxymethyl)-2,4-pentanedionato]-	n.s.g.	265(4.84),334(4.89)	44-3216-64
$C_{18}H_{25}IN_2O$			
[2-(1-Formyl-2,3,4,9-tetrahydro-4aH-carbazol-4a-yl)ethyl]trimethylammonium iodide	MeOH	302(3.5),360(4.3)	78-1737-64
	N NaOH	248(3.9),370(4.2)	78-1737-64
$C_{18}H_{25}N$			
Estra-1,3,5(10)-trien-2-amine	pH 1	212s(4.00),260s(2.88), 265(2.93),274(2.89)	24-0140-64
	EtOH	239(3.96),292(3.32)	24-0140-64
$C_{18}H_{25}NO_2$			
6-Azaestradiol, methyl ether	EtOH	214(4.43),251(3.86), 299(3.60)	78-2517-65
17a-Aza-D-homo-19-norandrost-4-ene-3,17-dione	MeOH	239(4.20)	78-0743-65
8aH-Dibenzofuro[1,9b:9a,cd]azepine, 7-ethoxy-1,2,3,4,9,10,11,12-octahydro-3-methyl-, perchlorate	EtOH	232s(3.91),288(3.50)	94-1012-64
$C_{18}H_{25}NO_3$			
17a-Aza-D-homo-19-norandrost-4-en-17-one, 17a-hydroxy-	MeOH	237(4.27)	78-0743-65
1,4-Benzodioxan, 6-allyl-5-(β-piperidinoethoxy)-	EtOH	274(2.90)	65-2994-64
Lycorenine, deoxy-α-dihydro-	EtOH	285(3.50)	95-0699-65
β-isomer	EtOH	285(3.48)	95-0699-65
A-Norandrost-3(5)-ene-1,2-dione, 17β-hydroxy-, 2-oxime	EtOH	233(3.89),291(4.02)	94-0156-65
$C_{18}H_{25}NO_4$			
Lycorenine, α-dihydro-	EtOH	231(3.90),282(3.50)	94-0408-64
Lycorenine, β-dihydro-	EtOH	233(3.97),282(3.52)	94-0408-64
1(2H)-Quinolinebutyric acid, 2-carboxy-3,4-dihydro-, diethyl ester	EtOH	209(4.35),250(3.93), 301(3.44)	78-2529-65
1(2H)-Quinolinepropionic acid, 2-(carboxymethyl)-3,4-dihydro-, diethyl ester	EtOH	210(4.35),257(4.14), 306(3.40)	78-2961-65
$C_{18}H_{25}NO_4S$			
N-Methyl-N-(1-phenylpropyl)anilinium methyl sulfate	EtOH	253(3.04),260(3.06), 264(3.01),271(2.84)	39-0249-64
$C_{18}H_{25}N_2O_4P$			
Benzyl 2-pyridyl phosphate, compound with cyclohexylamine	n.s.g.	262(3.60)	24-1031-65
$C_{18}H_{25}N_3$			
1H-Azepino[5,4,3-cd]indole, 3,4,5,6-tetrahydro-6-methyl-2-(piperidinomethyl)-	EtOH	229(4.53),286(3.94)	87-0200-65
$C_{18}H_{25}N_3O_5$			
Benzamide, N-[2-(1-carbethoxycyclohexylamino)ethyl]-p-nitro-	MeOH	262(4.09)	87-0107-65

Compound	Solvent	λ_{max}(log ε)	Ref.
$C_{18}H_{25}O_5P$			
Estradiol, 3-(dihydrogen phosphate)	MeOH	270(3.04),276(2.99)	22-0018-65
$C_{18}H_{26}Cl_2N_5O_4P$			
Phosphorane, [3,5-dichloro-6-oxo-2,4-cyclohexadien-1-ylidene)hydrazono]-trimorpholino-	MeOH	519(3.099)	5-0056-64I
$C_{18}H_{26}N_2OS$			
2,4-Diazaandrosta-1,5-diene-3-thione, 17β-hydroxy-17α-methyl-	MeOH	240(3.89),272(4.34)	13-0639-64
$C_{18}H_{26}N_2O_2$			
p-Benzoquinone, 2,5-dimethyl-3,6-dipiperidino-	EtOH	235(4.16),300(3.44), 443(3.56)	39-5569-64
2,3-Diazaestra-3,5(10)-dien-1-one, 17β-hydroxy-2,4-dimethyl-	MeOH	230s(3.45),290(3.61)	87-0590-64
$C_{18}H_{26}N_2O_3$			
Acrylamide, N-butyl-2-(butylamino)-3-salicyloyl-	EtOH	368(4.37)	65-0541-64
$C_{18}H_{26}N_2O_4$			
2H-Benzo[a]quinolizine-3-carboxamide, N-ethyl-1,3,4,6,7,11b-hexahydro-2-hydroxy-9,10-dimethoxy-	n.s.g.	284.5(3.56)	87-0635-64
Lycorenine,β-dihydro-, oxime	EtOH	271(4.04),310(3.62)	94-0408-64
1(2H)-Quinolinebutyric acid, 6-amino-2-carboxy-3,4-dihydro-, diethyl ester	EtOH	209(4.01),260(3.72)	78-2961-65
$C_{18}H_{26}N_2O_5S$			
Pantetheine, S-benzoyl-	EtOH	239(4.03),267(3.91)	94-0180-65
$C_{18}H_{26}N_2Zn$			
Zinc, (2,2'-bipyridine)dibutyl-	toluene	425(2.56)	101-0222-65A
$C_{18}H_{26}N_4O$			
4-Pyrimidinol, 2-amino-5-(p-butyl-anilinopropyl)-6-methyl-	pH 1	265(3.97)	87-0035-65
	pH 7	240(4.19),291(3.80)	87-0035-65
	pH 13	240(4.24),280(3.90)	87-0035-65
$C_{18}H_{26}N_4O_3$			
Tryptamine, N-acetyl-5,6-dihydroxy-, salt with 1,4-diazabicyclo[2.2.2]-octane	EtOH	210(4.39),220(4.39), 305(3.93)	4-0387-65
$C_{18}H_{26}N_4O_4$			
Bicarbamic acid, [1-(phenylazo)cyclo-hexyl]-, diethyl ester	EtOH	274(4.49),400(2.17)	7-0441-65
$C_{18}H_{26}O$			
Estr-4-en-3-one	MeOH	240(4.20)	44-2351-64
	EtOH	240(4.23)	35-0085-64
1H-Phenalene, 2,3,3a,4,5,6-hexahydro-1-isopropyl-8-methoxy-3a-methyl-	EtOH	280(3.38)	39-2146-64
$C_{18}H_{26}O_2$			
5βH-Estr-1-en-3-one, 17 -hydroxy-	EtOH	231(3.98)	54-0626-65
Estr-4-en-3-one, 16α-hydroxy-	MeOH	240(4.21)	44-2351-64
Estr-4-en-3-one, 16β-hydroxy-	MeOH	239(4.24)	44-2351-64

Compound	Solvent	λ_{max}(log ε)	Ref.
Estr-4-en-3-one, 17β-hydroxy-	EtOH	243(4.21)	39-5488-64
2,4-Hexadiene, 3,4-bis(4-oxocyclohexyl)-, 2-trans-4-cis	hexane	182(4.39)	5-0024-65D
A-Nortestosterone	EtOH	235(4.20)	94-0156-65
19-Nortestosterone	EtOH	241(4.26),306(2.58)	94-0905-64
Phenanthrene, 1,2,3,4,9,10,11,12-octahydro-1α-hydroxymethyl-6-methoxy-1β,12β-dimethyl-	n.s.g.	222s(3.85),280(3.40), 286(3.36)	39-4521-64
Spiro[cyclohexane-1,1'-indan]-4,6'(3'aH)-dione, 2',3'-diethyl-4',5'-dihydro-	MeOH	240(4.15)	5-0036-64D
Spiro[2-cyclohexene-1,1'-indene]-4,6'-diol, 2',3'-diethyl-2',4',5',6'-tetrahydro-	MeOH	249(4.22)	5-0036-64D
5,7,11-Tridecatrien-1-yn-4-ol, 4,8,12-trimethyl-, acetate	EtOH	242(4.28)	5-0014-65D
$C_{18}H_{26}O_3$			
Anthracene, 3,3-diethoxy-1,2,3,4,4aβ,-9,9aα,10-octahydro-10-hydroxy-	EtOH	204(4.06),266(2.60), 273(2.60)	70-1039-65
Estr-4-en-3-one, 10β,17β-dihydroxy-	MeOH	234(4.16)	88-0663-64
A-Norandrost-3(5)-en-2-one, 1β,17β-dihydroxy-	EtOH	236(4.16)	94-0156-65
A-Norandrost-3-en-2-one, 3,17β-dihydroxy-	EtOH	265(4.10)	44-1325-65
	NaOH	303(3.99)	44-1325-65
A-Nor-5β-androst-2-en-1-one, 2,17β-dihydroxy-	EtOH	262(3.78)	78-0759-65
19-Nortestosterone, 1β-hydroxy-	EtOH	243(4.18)	54-0626-65
19-Nortestosterone, 16α-hydroxy-	MeOH	242(2.75)	7-0310-65
$C_{18}H_{26}O_4$			
Estr-4-en-3-one, 17β-hydroxy-10β-hydroperoxy-	MeOH	234(4.18)	88-0663-64
$C_{18}H_{26}O_5$			
3,5-Seco-4-norestran-3-oic acid, 17β-hydroxy-6-(hydroxymethylene)-5-oxo-	MeOH	288(3.97)	39-1356-65
$C_{18}H_{27}BrN_2$			
Acetamidine, 2-[o-(2-bromovinyl)-phenyl]-N,N'-dibutyl-	EtOH	250(4.13)	39-1147-64
$C_{18}H_{27}ClN_5O_4P$			
Phosphorane, [(3-chloro-6-oxo-2,4-cyclohexadien-1-ylidene)hydrazono]-trimorpholino-	MeOH	507(3.097)	5-0056-64I
$C_{18}H_{27}F_3$			
Benzene, 1,2,3-tri-tert-butyl-trifluoro-	n.s.g.	209(4.17),228(4.28), 266(3.61),319(2.90)	89-0768-65
Benzene, 1,2,4-tri-tert-butyl-3,5,6-trifluoro-	hexane	201(4.40),238(3.82), 297(3.18)	89-0922-64
	n.s.g.	201(4.40),238(3.82), 297(3.18)	89-0768-65
Tricyclo[2.1.1.0^{5,6}]hex-2-ene, tri-tert-butyl-trifluoro-	hexane	232(3.40),300(2.48)	89-0922-64
	n.s.g.	232(3.40),300(2.60)	89-0768-65

Compound	Solvent	$\lambda_{max}(\log \epsilon)$	Ref.
$C_{18}H_{27}NO$			
Benzo[g]quinoline, 4-ethyl-1,2,3,4,-5,10,11,12-octahydro-7-methoxy-1,5-dimethyl-	EtOH	278(3.33),286s(--)	87-0559-65
$C_{18}H_{27}NO_3$			
1,4-Benzodioxan, 5-(β-piperidino-ethoxy)-6-propyl-	EtOH	275(3.05)	65-2994-64
hydrochloride	EtOH	276(3.10),280(3.10)	65-2994-64
A-Norandrost-3(5)-en-2-one, 1β,17β-dihydroxy-, oxime	EtOH	252(4.13)	94-0156-65
$C_{18}H_{27}O_5P$			
Estr-4-en-3-one, 17β-hydroxy-, dihydrogen phosphate	MeOH	240(4.26)	22-0018-65
$C_{18}H_{28}$			
1,3,10,12-Cyclooctadecatetraene	hexane	229(4.74)	39-6674-65
Tricyclo[9.7.0.0^{2,10}]octadeca-9,11-diene	heptane	254(4.04),263(4.17),274(4.04)	35-4506-65
isomer m. 80-83°	heptane	250(4.08),259(4.2),269(4.07)	35-4506-65
$C_{18}H_{28}Cl_2N_2O_2$			
p-Benzoquinone, 2,5-dichloro-3,6-bis(hexylamino)-	methyl cellosolve	224(4.52),355(4.45),529(2.46)	78-1889-64
$C_{18}H_{28}F_2N_2O_2$			
p-Benzoquinone, 2,5-difluoro-3,6-bis(hexylamino)-	methyl cellosolve	224(4.49),355(4.45),547(2.36)	78-1889-64
$C_{18}H_{28}INO_3$			
2-[(6-Allyl-1,4-benzodioxan-5-yl)oxy]-ethyl]diethylmethylammonium iodide	EtOH	275(3.26)	65-2994-64
$C_{18}H_{28}N_2O$			
Flabellidine	MeOH-perchlorate	237(3.84)	23-2456-64
$C_{18}H_{28}N_2O_3$			
2-Pyrrolidone, 5-(5-carbethoxy-3-ethyl-4-methylpyrrolyl-2-methyl)-4-ethyl-3-methyl-	EtOH	241s(3.67),281(4.26)	39-5999-64
$C_{18}H_{28}N_2O_4$			
Hydroquinone, 2,3-dimethyl-5m6-bis(morpholinomethyl)-	dioxan	303(3.71)	39-0042-64
$C_{18}H_{28}N_5O_4P$			
Phosphorane, trimorpholino[(6-oxo-2,4-cyclohexadien-1-ylidene)-hydrazono]-	MeOH	498(3.066)	5-0056-64I
$C_{18}H_{28}N_6O_8$			
Lysine, N^6,N$^{6'}$-(4,6-dimethyl-m phenylene)di-	H_2O	214(4.27),343(4.48),428(4.06)	37-3264-65
$C_{18}H_{28}O$			
2,4-Cyclohexadien-1-one, 6-allyl-2,6-di-tert-butyl-4-methyl-	MeOH	320(3.57)	44-3895-65

Compound	Solvent	λ_{max}(log ε)	Ref.
$C_{18}H_{28}O_3$			
p-Benzoquinone, 2-methoxy-6-undecyl-	EtOH	268(4.09),364(2.90)	94-0236-64
Bicyclo[3.1.0]hexane-$\Delta^{2,\gamma}$-butyric acid, α,β-epoxy-α,1,3,3-tetramethyl-, tert-butyl ester	n.s.g.	220(4.1)	27-0538-65
Hydroquinone, 3-heptyl-2-hydroxy-6-isopentenyl-	EtOH	274(4.26)	32-0311-65
	0.5N NaOH	270(--)	32-0311-65
$C_{18}H_{28}O_4$			
1,3-Cyclohexanedione, 2,2-bis-(4-methyl-3-oxopentyl)-	EtOH	288(2.05)	39-2340-65
$C_{18}H_{28}O_4Si_4$			
Cyclotetrasiloxane, 2,2,4,4,6,6-hexamethyl-8,8-diphenyl-	$CHCl_3$	265f(3.6)	46-1066-65
$C_{18}H_{28}O_7$			
2-Cyclohexene-1-pyruvic acid, 1-carboxy-4-oxo-, diethyl ester, diethyl acetal	n.s.g.	222(4.08)	24-1774-65
$C_{18}H_{28}Si_3$			
Trisilane, 1,1,2,2,3,3-hexamethyl-1,3-diphenyl-	C_6H_{12}	243.0(4.28)	25-1063-64
	C_6H_{12}	243(4.29)	101-0163-65B
$C_{18}H_{29}N$			
Morphinan methine, 1,2,3,4-tetrahydro-N-methyl-	EtOH	268(3.60)	44-1769-65
$C_{18}H_{29}NO_3$			
2,5-Cyclohexadien-1-one, 2,4,6-tri-tert-butyl-4-nitro-	n.s.g.	239(4.02)	70-1675-65
$C_{18}H_{29}NO_4$			
Maleic acid, (dicyclohexylamino)-, dimethyl ester	EtOH	241s(3.48),290(4.39)	5-0098-65H
$C_{18}H_{29}N_3O$			
Longifolene, ω-acetyl-, semicarbazone	EtOH	272(4.33)	44-2838-65
$C_{18}H_{30}Cl_2O_2Pd_2$			
Palladium, dichlorobis(4-methoxy-2-cycloocten-1-yl)di-	$CHCl_3$	244(4.11),256(4.10), 298s(3.38),339(3.10)	39-5002-64
$C_{18}H_{30}O$			
2,4-Cyclohexadien-1-one, 2,5-di-tert-butyl-6-methyl-6-propyl-	n.s.g.	310(3.59)	35-5515-65
Phenol, 2,4,6-tri-tert-butyl-	MeOH	274(3.20)	39-0676-65
	MeOH-NaOMe	252(4.01),302(3.64)	
$C_{18}H_{30}O_3$			
α-Campholenic acid	EtOH	226(4.04),261(3.91), 271(4.03),282(3.92)	70-2003-64
β-Campholenic acid	EtOH	258(4.62),268(4.76), 279(4.65)	70-2003-64
4-Cyclohexene-1,3-dione, 5-ethoxy-2,2,4,6,6-pentamethyl-	C_6H_{12}	245(3.95),320(2.18)	39-1028-65
Nonanoic acid, 9-(5-methyl-2-oxocyclohexylidene)-, ethyl ester	EtOH	244(3.76)	39-4154-64
11-Octadecen-9-ynoic acid, 8-hydroxy-, trans	C_6H_{12}	229(4.24),239(4.17)	44-0610-65

Compound	Solvent	λ_{max}(log ϵ)	Ref.
$C_{18}H_{32}$			
Des-AB-cholest-8(9)-ene	EtOH	235(4.05)	39-4907-64
$C_{18}H_{32}N_4O_2$			
Piperidinooxy, 2,2,6,6-tetra-methyl-4-oxo-, azine	C_6H_{12}	210(4.43),233s(4.18), 465(1.26)	22-3290-65
	MeOH	208(4.54),230s(4.21), 430(1.27)	22-3290-65
$C_{18}H_{32}O$			
Cyclopentene, 2,3-dibutyl-1-valeryl-	EtOH	254(3.98)	22-0722-65
Cyclopentene, 1,2-dibutyl-3-valeryl-	EtOH	277.5(1.57)	22-0722-65
$C_{18}H_{32}O_2$			
9,13-Hexadecadienoic acid, ethyl ester, cis-cis	n.s.g.	none	70-1453-64
$C_{18}H_{33}N_4O$			
Piperidinooxy, 2,2,6,6-tetramethyl-4-[(2,2,6,6-tetramethyl-4-piperi-dylidene)hydrazino]-	C_6H_{12}	211(4.20),240(3.95), 445(0.97)	22-3290-65
	MeOH	209(4.24),235(3.98), 420(1.04)	22-3290-65
	DMF	445(0.92)	22-3290-65
$C_{18}H_{34}N_2O_2$			
L-Lysine, N^2,N^6-dineopentylidene-, ethyl ester	hexane	215s(2.59),245s(2.36)	39-4503-65
$C_{18}H_{34}N_4$			
4-Piperidone, 2,2,6,6-tetramethyl-, azine	C_6H_{12}	210(4.35),248(4.00), 308f(2.62)	22-3290-65
$C_{18}H_{34}O$			
Cyclopentane, 2,3-dibutyl-1-valeryl-	EtOH	286(1.47)	22-0722-65
$C_{18}H_{34}O_2S_5$			
Formic acid, trithiobis[thio-, O,O-dioctl ester	C_6H_{12}	none 320-470nm	70-2175-64
$C_{18}H_{36}N_2S_4$			
Disulfide, bis(dibutylthiocarbamoyl)	heptane	220(4.35),255(4.19), 278(4.20)	1-1113-65
	EtOH	218(4.37),264(4.20), 272(4.21)	1-1113-65
$C_{18}H_{54}Si_8$			
Octasilane, octadecamethyl-	C_6H_{12}	215s(4.46),241(4.26), 273(4.59)	101-0369-64B

Compound	Solvent	λ max(log ε)	Ref.
$C_{19}H_3F_{15}Sn$			
Tin, methyltris(pentafluorophenyl)-	C_6H_{12}	266.5(3.45)	39-4782-64
	MeOH	263(3.29)	39-4782-64
$C_{19}H_9NO$			
Lambertellin lactam	EtOH	255(4.24),302(4.15), 420(3.66)	100-0359-65
$C_{19}H_{10}N_2O$			
2,5-Cyclopentadiene-1,2-dicarbonitrile, 4-oxo-3,5-diphenyl-	hexane	252(4.68),305(3.63), 540(3.45)	39-2009-65
$C_{19}H_{10}O$			
5H-Benzo[cd]pyren-5-one	MeOH	209(4.70),230(4.69), 256(4.35),263(4.42), 284(4.35),296(4.42), 309(4.58),390(3.59), 396(3.73),418(3.95), 479(4.08)	39-5920-65
$C_{19}H_{10}O_4$			
Sesquixanthydrol	EtOH	280(3.86),288(3.95)	35-2252-64
12cH-4,8,12-Trioxa-4H,8H-dibenzo- [cd,mn]pyren-12c-ol	HOAc	240(4.68),282(4.47), 330(4.52),452(3.95), 475s(3.93)	35-2252-64
$C_{19}H_{10}O_5$			
2-Naphthoic acid, 3-(2-carboxy-6-hydroxy-m-tolyl)-1,4-dihydroxy-, di-δ-lactone	EtOH	234(4.70),268(4.59), 277(4.66),310s(3.70), 325(3.76),370(3.99), 389(4.15),410(4.20)	33-1459-64
$C_{19}H_{11}ClO_4$			
Benzo[cd]pyrenylium perchlorate	MeCN	201(4.80),218(4.70), 236(4.52),245(4.52), 269s(4.27),279(4.35), 290(4.40),299s(4.22), 318s(3.99),434(4.31), 475s(3.94),616(3.60)	39-5920-65
$C_{19}H_{11}NO$			
Acenaphtho[1,2-d]isoxazole, 9-phenyl-	EtOH	230(4.65),277(4.18), 287(4.18),319(3.97), 344(3.68)	88-2157-64
$C_{19}H_{12}$			
Benzpyrene	EtOH	285(4.68),297(4.79), 332(3.72),347(4.10), 365(4.41),385(4.46), 404(3.59)	35-1444-64
$C_{19}H_{12}BrIO_7$			
Flavone, 6-bromo-4',5,7-trihydroxy-8-iodo-, 4',7-diacetate	EtOH	288(4.51),315(4.28), 344(3.99)	44-0897-65
$C_{19}H_{12}BrNO$			
Acenaphth[1,2-d]isoxazole, 9a-bromo-6b,9a-dihydro-9-phenyl-	EtOH	222(4.97),285(3.76)	88-2157-64

Compound	Solvent	λ_{max}(log ε)	Ref.
$C_{19}H_{12}Br_2N_2O$			
Pyrazolo[1,2-a]pyrazole, 1-(p-bromo-benzoyl)-2-(p-bromophenyl)-	EtOH	235(4.22),405(3.95)	35-0528-65
$C_{19}H_{12}Br_2O_7$			
Flavone, 6,8-dibromo-4',5,7-trihydroxy-, 4',7-diacetate	EtOH	284(4.43),312(4.23), 342(3.99)	44-0897-65
$C_{19}H_{12}ClNO$			
Naphth[1,2-d]oxazole, 2-(p-chloro-styryl)-	heptane	225(4.55),285(4.10), 305(4.19),350(4.56), 365(4.58)	65-0427-64
$C_{19}H_{12}ClNO_4$			
Fluoreno[1,9-bc]quinolizinium perchlorate	EtOH	225(4.39),252(4.58), 272s(4.30),295s(3.94), 313(3.80),335(3.87), 352(3.85),370(3.77), 437(3.97),461(4.00)	4-0121-64
$C_{19}H_{12}Cl_2N_2O$			
Pyrazolo[1,2-a]pyrazole, 1-(p-chloro-benzoyl)-2-(p-chlorophenyl)-	EtOH	230(4.22),275s(3.90), 400(4.10)	35-0528-65
$C_{19}H_{12}I_2O_7$			
Flavone, 4',5,7-trihydroxy-6,8-di-iodo-, 4',7-diacetate	EtOH	288(4.49),320(4.18), 340(3.87)	44-0897-65
$C_{19}H_{12}N_2OS$			
2-Thiazolin-4-one, 2-(2-naphthyl)-5-(phenylimino)-	dioxan	230(4.5),279(4.4), 328(4.4)	83-0124-65
$C_{19}H_{12}N_2O_2$			
Benz[5,6]indazolo[2,3-a]quinoline-7,12-dione, 5,6-dihydro-	EtOH	238(4.44),278(4.85), 330(3.93)	39-5871-65
$C_{19}H_{12}N_4O_5$			
Pyrazolo[1,2-a]pyrazole, 1-(m-nitro-benzoyl)-2-(m-nitrophenyl)-	EtOH	220(4.45),260s(4.32), 400(4.16)	35-0528-65
$C_{19}H_{12}O$			
Fluoren-9-one, 4-phenyl-	iso-PrOH	385(2.58)	39-5518-65
$C_{19}H_{12}O_4$			
3,4:7,8-Dibenzocoumarin, 5-acetoxy-	MeOH	243(4.3),259(4.2), 270(4.3),290(3.6), 305(3.7),342(3.2)	83-0529-64
$C_{19}H_{12}O_6$			
Dicoumarol	MeOH-HCl	275s(3.59),286(3.68), 307(3.73),320(3.58)	49-0077-65
	MeOH-NaOH	316(4.06)	49-0077-65
5H-Furo[3,2-g][1]benzopyran-5-one, 6-[2-methoxy-4,5-(methylenedi-oxy)phenyl]-	EtOH	238(4.59),309(4.20)	31-0668-64
Methane, bis(1,4-dioxoisochroman-3-yl)-	EtOH	220(4.6),250(3.91)	54-0334-65
$C_{19}H_{12}O_8$			
Coumestan, 7,11,12-triacetyl-	$CHCl_3$	236(3.34),298(3.12), 328(3.42),344(3.32)	44-2353-65

Compound	Solvent	λ_{max}(log ε)	Ref.
$C_{19}H_{13}BrO_7$			
Flavone, 8-bromo-4',5,7-trihydroxy-, 4',7-diacetate	EtOH	275(4.43),303(4.23), 335(4.01)	44-0897-65
$C_{19}H_{13}ClN_2O_2$			
4,5'-Biisoxazole, 3'-(m-chlorophenyl)-5-methyl-3-phenyl-	EtOH	235(4.39)	7-1223-65
$C_{19}H_{13}Cl_3N_4O_4$			
5-(3,4-Dichlorophenyl)-2,3-diphenyl-2H-tetrazolium perchlorate	n.s.g.	258.5(4.43)	5-0197-65I
isomer	n.s.g.	258.5(4.49)	5-0197-65I
$C_{19}H_{13}Li$			
Lithium, (9-phenylfluoren-9-yl)-	$C_6H_{11}NH_2$	452(3.29),487(3.36), 520(3.23)	35-0384-65
$C_{19}H_{13}N$			
Acridine, 2-phenyl-	pH 12	368(3.9)	20-0591-65
ion	pH 2	368(4.0),412(3.6)	20-0591-65
Azatriptycene	EtOH	263s(3.34),270(3.58), 278(3.69)	5-0021-64F
5,6-Benzoquinoline, 2-phenyl-	EtOH	279(4.43),326(3.76), 340(3.95),357(3.99)	88-0659-64
Isoquinoline, 1-(2-naphthyl)-	EtOH	217(4.97),238s(4.48), 245s(4.38),266s(4.11), 275(4.14),285s(4.09), 324(3.98)	44-1846-65
$C_{19}H_{13}NO$			
Acenaphth[1,2-d]isoxazole, 6b,9a-dihydro-9-phenyl-	EtOH	225(4.86),266(4.06), 275(4.08),286(4.03), 298(3.78)	88-2157-64
Isoxazole, 5-(1-naphthyl)-3-phenyl-	EtOH	225(4.72),302(4.10)	78-0817-65
Isoxazole, 5-(2-naphthyl)-3-phenyl-	EtOH	236(4.54),256(4.62), 298(4.33)	78-0817-65
$C_{19}H_{13}NO_3$			
Phthalimidine, 5,6-(methylenedioxy)-2-(1-naphthyl)-	MeOH	302(4.21)	95-1049-65
Phthalimidine, 5,6-(methylenedioxy)-2-(2-naphthyl)-	MeOH	315(4.36)	95-1049-65
$C_{19}H_{13}NO_3S$			
11-(3-Sulfophenyl)acridizinium hydroxide, inner salt	EtOH	244(4.39),366(3.92), 384(3.90),405(3.84)	4-0228-65
$C_{19}H_{13}NS_3$			
6a-Thiathiophthene, 4-cyano-5-phenyl-2-p-tolyl-	dioxan	259(4.67),307(4.34), 346(4.11),489(4.16)	24-1732-64
$C_{19}H_{13}N_3O_6$			
Methane, tris(p-nitrophenyl)-	EtOH-DMSO	707(4.29)	78-0261-65
$C_{19}H_{14}$			
Fluorene, 9-phenyl-	EtOH-DMSO	375(3.99),409(4.05)	78-0261-65
Methane, inden-1-yl-inden-1-ylidene-	DMSO-KO-tert-Bu	365(3.49),515s(4.64), 543(5.15)	5-0050-65J

Compound	Solvent	λ_{max}(log ε)	Ref.
Naphthalene, 1-(3-phenyl-2-propynyl)-	MeOH	222(4.89),238(4.36), 250(4.32),270(3.89), 280(3.96),288(3.76), 291(3.77),313(2.56)	39-6710-65
Triphenylene, 1-methyl-	EtOH	254(4.74),261(5.05), 279(4.16),288(4.14)	78-3289-65
Triphenylene, 2-methyl-	EtOH	251(4.84),259(5.51), 275(4.18),286(4.16)	78-3289-65
$C_{19}H_{14}BrClN_4O_4$			
3-(p-Bromophenyl)-2,5-diphenyl-2H-tetrazolium perchlorate	n.s.g.	247.5(4.42)	5-0197-65I
isomer	n.s.g.	247.5(4.43)	5-0197-65I
$C_{19}H_{14}BrIO_7$			
Flavanone, 6-bromo-4',5,7-trihydroxy-8-iodo-, 4',7-diacetate	EtOH	282(4.13),358(3.69)	44-0897-65
$C_{19}H_{14}BrNO_2$			
7,8-Dihydroxy-10-phenylbenzo[b]quinolizinium bromide	MeOH	340s(3.93),356(4.08), 372(4.10),400(3.79)	44-0252-65
$C_{19}H_{14}ClNO_9$			
Tetracycline, 7-chloro-6-demethyl-	MeOH-HCl	258(4.40),350(3.66)	35-3874-64
$C_{19}H_{14}I_2O_7$			
Flavanone, 4',5,7-trihydroxy-6,8-diiodo-, 4',7-diacetate	EtOH	285(3.96),365(3.67)	44-0897-65
$C_{19}H_{14}N_2$			
Acridine, 9-amino-2-phenyl-	pH 12	320(2.1),344s(2.2), 368s(2.5),395s(2.6), 438s(2.5)	20-0591-65
monocation	acid	395s(2.7),428(2.8), 219(2.9)	20-0591-65
dication	H_o= -9	356s(3.8),377(3.9), 422s(3.6)	20-0591-65
Isoquinoline, 1-(2-indol-3-ylvinyl)-	MeOH	385(3.6)	87-0397-65
	HOAc	465(4.0)	87-0397-65
Phenanthridine, 3-amino-2-phenyl-	EtOH-HCl	261(4.59),324(3.81)	4-0140-65
	EtOH	253(4.58),336s(3.89), 347(3.93),368(3.92)	4-0140-65
Quinoline, 2-(2-indol-3-ylvinyl)-	MeOH	375(4.4)	87-0397-65
	HOAc	465(4.6)	87-0397-65
Quinoline, 4-(2-indol-3-ylvinyl)-	MeOH	380(4.3),495(--)	87-0397-65
	HOAc	480(4.5)	87-0397-65
$C_{19}H_{14}N_2O_2$			
4,5'-Biisoxazole, 5-methyl-3,3'-diphenyl-	EtOH	230(4.38)	7-1223-65
1H-Pyrrolo[1,2-a]indole-2-carbonitrile, 7-(benzyloxy)-2,3-dihydro-1-oxo-	MeOH	328(4.32)	44-2910-65
1H-Pyrrolo[1,2-a]indole-9-carbonitrile, 7-(benzyloxy)-2,3-dihydro-1-oxo-	MeOH	219(4.7),255(4.13), 328(4.29)	44-2904-65
$C_{19}H_{14}N_2O_3$			
Pyrrolo[1,2-a]quinoxaline-2-acetic acid, 1,5-dihydro-1-oxo-4-phenyl-	EtOH	238(4.43),273(4.28), 292s(4.18),330s(3.68), 340s(3.61)	35-1830-64

Compound	Solvent	$\lambda_{max}(\log \epsilon)$	Ref.
$C_{19}H_{14}O$			
2'-Acrylonaphthone, 3-phenyl-	H_2SO_4	255(4.09),350(3.89), 435(4.59)	65-1724-65
	dioxan	313(4.28)	23-2580-64
Acrylophenone, 3-(1-naphthyl)-	EtOH	352(4.17)	23-2580-64
Acrylophenone, 3-(2-naphthyl)-	EtOH	328(4.39)	23-2580-64
Benzo[c]phenanthrene, 1-methoxy-	EtOH	221(4.57),248s(4.10), 281(4.72),304(4.27), 321s(3.95),336s(3.79), 362(3.35),381(3.29)	35-0503-64
Benzo[c]phenanthrene, 2-methoxy-	EtOH	218(4.51),233(4.52), 251(4.22),277(4.62), 287(4.61),306(4.22), 316s(4.08),346(3.48), 359(3.56),381(3.57)	35-0503-64
Benzo[c]phenanthrene, 3-methoxy-	EtOH	220(4.40),248s(3.87), 278(4.46),283(4.60), 306(3.79),318(3.80), 332s(3.63),360(3.26), 378(3.24)	35-0503-64
Benzophenone, 2-phenyl-	EtOH	240(4.30)	78-0195-64
Tropone, 2,7-diphenyl-	EtOH	229(4.33),276(4.06), 339(4.02)	35-1404-65
$C_{19}H_{14}O_2$			
9H-Cyclopropa[e]pyrene-9-carboxylic acid, 8b,9a-dihydro-, methyl ester	EtOH	260(4.60),290(4.11), 302(4.10)	88-2673-65
7(12H)-Pleiadenone, 1-methoxy-	EtOH	225(4.48),260(4.16), 376(3.81)	12-1865-65
$C_{19}H_{14}O_3$			
5-Benzofuranacrylic acid, α-ethyl-6-hydroxy-β-phenyl-, δ-lactone	EtOH	225(4.28),247(4.32), 296(3.99),328(3.84)	44-2467-64
1-Naphthoic acid, 2-benzoyl-, methyl pseudo ester	EtOH	241(4.50),306(3.80)	44-4293-65
2-Naphthoic acid, 1-benzoyl-, methyl ester	EtOH	227(4.41),240(4.50)	44-4293-65
2-Naphthoic acid, 1-benzoyl-, methyl pseudo ester	EtOH	225(4.59),312(3.80)	44-4293-65
Xanthene, 9-(2,4-dihydroxyphenyl)-	EtOH	240(4.2),282(3.7)	95-0057-64
$C_{19}H_{14}O_5$			
2-Naphthoic acid, 3-(2-carboxy-6-hydroxy-m-tolyl)-5,6,7,8-tetrahydro-1,4-dihydroxy-, di-δ-lactone	EtOH	231(4.64),259s(4.32), 280(3.74),369(4.14), 388(4.13)	33-1459-64
$C_{19}H_{14}O_5S$			
Phenol red	pH 7.8	560(4.40)	43-0817-64
$C_{19}H_{14}O_6$			
5H-Furo[3,2-g][1]benzopyran-5-one, 2,3-dihydro-6-[2-methoxy-4,5-(methylenedioxy)phenyl]-	EtOH	241(4.22),251(4.19), 310(4.23)	31-0668-64
$C_{19}H_{14}O_7$			
Flavone, 4',5,7-trihydroxy-, 4',7-diacetate	EtOH	271(4.42),301(4.21), 331(4.02)	44-0897-65
Veratric acid, 6-(3-hydroxy-7-methoxy-4-oxo-4H-1-benzopyran-2-yl)-, δ-lactone	EtOH	266(4.44),324(4.32), 342(4.42),358(4.42)	39-4941-64

Compound	Solvent	λ_{max}(log ε)	Ref.
$C_{19}H_{14}O_9$			
Xanthone, 3,5,6-triacetoxy-1-hydroxy-	EtOH	237(4.32),254(4.40), 292s(4.11),297(4.16), 351(3.74)	44-0692-64
$C_{19}H_{15}$			
Trityl radical	EtOH	337(--),514(--)	38-0979-64B
carbanion	ether	245(4.08),300s(3.00), 410(3.85),475(4.18)	60-0264-64
$C_{19}H_{15}BCl_4$			
Tritylium tetrachloroborate	H_2SO_4	432(4.58)	35-5511-64
$C_{19}H_{15}BI_4$			
Tritylium tetraiodoborate	H_2SO_4	432(4.55)	35-0539-65
$C_{19}H_{15}Cl$			
Methane, chlorotriphenyl-	SO_2	410(4.5),430(4.5)	49-0678-64
$C_{19}H_{15}ClN_2$			
Pyrazolo[1,2-a]quinoxaline, 2-benzyl-1-chloro-4-methyl-	EtOH	233(4.56),240s(4.53), 246s(4.52),259s(4.17), 327s(4.03),337(4.05), 351s(3.91)	39-3678-65
$C_{19}H_{15}ClN_2O_2$			
2-Isoxazoline, 3-(m-chlorophenyl)-5-(5-methyl-3-phenyl-4-isoxazolyl)-	EtOH	265(4.14)	7-1223-65
$C_{19}H_{15}ClO_4$			
7-m-Biphenylyltropylium perchlorate	MeCN	250(4.38),263(4.40), 363(4.15)	24-0029-64
7-p-Biphenylyltropylium perchlorate	MeCN	248(4.45),275(4.11), 415(4.33)	24-0029-64
7-[2-(1-Naphthyl)vinyl]tropylium perchlorate	MeCN	226(4.76),270(4.29), 290s(3.86),346(3.90), 495(4.38)	24-1349-64
7-[2-(2-Naphthyl)vinyl]tropylium perchlorate	MeCN	224(4.56),264(4.44), 285s(4.14),311s(3.55), 477(4.55)	24-1349-64
$C_{19}H_{15}Cs$			
Cesium, trityl-	$C_6H_{11}NH_2$	488(4.46)	35-0384-65
$C_{19}H_{15}F_3N_6O_5$			
Benzoic acid, p-[N-[2-(2-acetamido-4-hydroxy-6-pteridinyl)ethyl-2,2,2-trifluoroacetamido]-	pH 11	255(4.47),277(4.33), 353(3.85)	44-3404-65
$C_{19}H_{15}I$			
Methane, iodotriphenyl-	96% H_2SO_4	432(4.58)	35-0539-65
$C_{19}H_{15}IO_7$			
Flavanone, 4',5,7-trihydroxy-8-iodo-, 4',7-diacetate	EtOH	278(4.00),355(3.59)	44-0897-65
$C_{19}H_{15}N$			
Pyridine, 2-fluoren-9-ylidene-1,2-dihydro-1-methyl-	dioxan	240(4.25),260(4.2), 300s(3.7),410(4.0), 530(3.8)	35-2908-65

Compound	Solvent	$\lambda_{max}(\log \epsilon)$	Ref.
Pyridine, 4-fluoren-9-ylidene-1,4-dihydro-1-methyl-	dioxan	260(4.55),290s(3.8), 310s(3.8),440(4.5)	35-2908-65
$C_{19}H_{15}NO$			
Benzo[j]phenanthridine, 6-ethoxy- (two bands not listed)	EtOH	220(4.61),229s(4.51), 272(4.88),280s(4.63), 289(4.46),304(3.69), 334(3.78),346s(3.71)	4-0015-65
2-Isoxazoline, 5-(1-naphthyl)-3-phenyl-	EtOH	223(4.91),271(4.21), 281(4.17)	78-0817-65
2-Isoxazoline, 5-(2-naphthyl)-3-phenyl-	EtOH	266(4.33)	78-0817-65
Nitrone, α,α,N-triphenyl-	EtOH	232(4.14),308(4.01)	78-2735-65
2(1H)-Pyridone, 6-(2,2-diphenylvinyl)-	EtOH	245(4.48),354(4.18)	35-5198-65
$C_{19}H_{15}NO_2$			
Acrylophenone, 4'-methoxy-3-(2-quinolyl)-	H_2SO_4	236(4.46),278(4.12), 312(3.99),446(4.55)	65-1724-65
2-Propen-1-one, 3-(p-methoxyphenyl)-1-(2-quinolyl)-	H_2SO_4	230(4.20),275(3.85), 368(3.58),538(4.81)	65-1724-65
$C_{19}H_{15}NO_2S$			
Dibenzo[d,f][1,2]thiazepine, 6,7-dihydro-6-phenyl-, 5,5-dioxide	EtOH	240s(4.11),280s(3.80)	12-0206-65
$C_{19}H_{15}NO_3$			
Diacetamide, N-(1-oxo-2-phenyl-inden-3-yl)-	EtOH	415(3.39)	35-0874-65
1H-Pyrrolo[1,2-a]indole-9-carboxaldehyde, 6-benzyl-2,3,7,8-tetrahydro-7,8-dioxo-	MeOH	225(4.87),281(4.36), 345(4.04),525(3.61)	44-4381-65
1H-Pyrrolo[1,2-a]indole-9-carboxaldehyde, 7-(benzyloxy)-2,3-dihydro-1-oxo-	MeOH	245(4.30),253(4.32), 262(4.18),280(3.86), 343(4.20)	35-4612-64
$C_{19}H_{15}NO_4$			
1H-Pyrrolo[1,2-a]indole-9-carboxaldehyde, 6-benzyl-2,3,5,8-tetrahydro-7-hydroxy-5,8-dioxo-	pH 13	240(4.37),301(4.10), 335(4.13),550(3.10)	44-4381-65
	MeOH	222(4.37),299(4.10), 330(3.93),475(2.69)	44-4381-65
$C_{19}H_{15}NO_5$			
Atheroline	EtOH-HCl	257(4.12),282(4.12), 385(4.05),500(3.38)	88-2399-65
	EtOH	244(4.09),273(4.17), 292s(3.96),355(3.90), 380s(3.83),435(3.62)	88-2399-65
	EtOH-NaOH	252(4.04),294(3.99), 320(3.98),390(3.74), 535(3.46)	88-2399-65
4H-m-Dioxino[5,4-c]pyridine-4,5(6H)-dione, 7-hydroxy-2,2-dimethyl-6-(2-naphthyl)-	EtOH	313(4.49)	78-1917-65
2-Naphthacenecarboxamide, 4,4a,5,12-tetrahydro-1,3,10-trihydroxy-12-oxo-	n.s.g.	246(4.41),261(4.46), 320(3.99),383(4.54)	70-0945-64
Pyrano[4,3-d][1,3]dioxin, 7-(2-naphthylamino)-2,2-dimethyl-4,5-dioxo-	EtOH	347(4.53)	78-1917-65

Compound	Solvent	λ_{max}(log ε)	Ref.
$C_{19}H_{15}NO_9$			
1,11-Epoxynaphthacene-3-carboxamide, 1,4,4a,5,5a,6,11,11a,12,12a-decahydro-1,2,4a,7-tetrahydroxy-4,5,6-trioxo-	MeOH-HCl	258(4.64),338(3.67)	35-3874-64
$C_{19}H_{15}N_2O_2P$			
Diimide, benzoyl(diphenylphosphinyl)-	dioxan	509(1.58)	24-2273-65
$C_{19}H_{15}N_3$			
Acridine, 9-hydrazino-2-phenyl-	EtOH-base	400(3.5),488f(2.2)	20-0591-65
monocation	pH 8	400(3.5),500s(1.6)	20-0591-65
dication	pH 3	422(3.63),449(3.59)	20-0591-65
$C_{19}H_{15}N_3O_3$			
Salicylaldehyde, 5-nitro-, diphenylhydrazone	MeOH	232s(4.35),302(4.29), 350(4.42)	24-1631-64
	MeOH-NaOH	372(4.42),438(4.06)	24-1631-64
$C_{19}H_{15}N_3O_4$			
Carbazic acid, 2,3-diphenyl-, p-nitrophenyl ester	n.s.g.	238(4.21),271(4.09), 375(2.94)	33-1047-64
$C_{19}H_{15}N_3O_5$			
4-Pyridone-3,5-dicarboxylic acid, 2,6-bis(2-pyridyl)-, dimethyl ester	EtOH	246(4.49),277(4.52)	88-3175-65
	$CHCl_3$	247(4.34),281(4.39)	88-3175-65
$C_{19}H_{15}N_3O_7$			
1,3-Cyclopentadiene-1,2,3-tricarboxylic acid, 4-(dicyanomethyl)-5-oxo-, trimethyl ester, compound with pyridine	EtOH	227s(3.93),257(4.35), 262s(4.33),313(4.40), 401s(3.68),668(3.68)	44-0423-64
$C_{19}H_{15}N_5O$			
Adenine, N-benzoyl-3-benzyl-	pH 1	300(4.45)	4-0115-64
	pH 7	240s(4.23),301(4.23)	4-0115-64
	pH 13	231s(4.23),333(4.25)	4-0115-64
Adenine, N-benzoyl-7-benzyl-	pH 1	240(4.01),280(4.10), 330(4.29)	4-0115-64
	pH 7	237(4.05),279(4.07), 333(3.45)	4-0115-64
	pH 13	298(4.05)	4-0115-64
Adenine, N-benzoyl-9-benzyl-	pH 1	290(4.34)	88-2059-65
	pH 7	281(4.26)	88-2059-65
	pH 13	299(4.12)	88-2059-65
$C_{19}H_{15}OP$			
Benzaldehyde, p-(diphenylphosphino)-	EtOH	265(4.08),377(2.39), 390(2.43),404(2.24)	23-1175-65
$C_{19}H_{16}$			
Azuleno[2,1,8-kla]heptalene, 1,4-dihydro-11-methyl-	hexane	273(4.74),324(4.11), 340(3.82),389(3.99), 409(4.09),446(3.08), 477(3.07),514(3.01), 802(2.17),890(2.11), 1040(1.69)	5-0194-64B
1H-Benz[f]indene, 2,3-dihydro-4-phenyl-	EtOH	229(4.51),277s(3.45), 287(3.71),296s(3.60), 323s(3.13)	94-1094-64

Compound	Solvent	λ_{max}(log ε)	Ref.
Biphenyl, 2-benzyl-	hexane	250s(3.81)	70-0124-65
Biphenyl, 3-benzyl-	hexane	250(4.02)	70-0124-65
Biphenyl, 4-benzyl-	hexane	252(4.30)	70-0124-65
1,6-Heptadiyne, 1,7-diphenyl-	EtOH	242(4.53),251(4.45)	94-1094-64
1H-Indeno[6,7,1-mna]anthracene, 2,10,11,11a-tetrahydro-	n.s.g.	226s(4.1),256(5.1), 301(2.6),308(2.5), 327(2.9),340(3.2), 353(3.3),369(3.2), 384(3.0)	28-4795-64A
Naphthalene, 1-cinnamyl-	MeOH	223(4.88),254(4.30), 270(4.11),282(4.06), 292(3.92),313(2.70), 317(2.40)	39-6710-65
cis isomer	MeOH	223(4.87),271(3.96), 282(3.99),288(3.82), 291(3.81),313(2.70), 317(2.48)	39-6710-65
Naphthalene, 2-cinnamyl-	MeOH	255(4.51),305(2.84)	39-6710-65
Naphthalene, 1-(3-phenylpropenyl)-	MeOH	227(4.73),298(4.10)	39-6710-65
Naphthalene, 2-(3-phenylpropenyl)-	MeOH	250(4.70),276(4.18), 286(4.28),294(4.19)	39-6710-65
Perinaphthan, 1-phenyl-	EtOH	212(4.5),235(4.6), 244(4.6),285(3.9)	5-0129-64D
$C_{19}H_{16}ClNO_4$			
7,13-Dimethylbenz[j]acridizinium perchlorate	EtOH	215(4.19),233(4.27), 274(4.55),315s(4.35), 325(4.51),378(4.01), 393(4.01),416(3.93)	4-0121-64
5,6-Dimethylbenzo[a]phenanthridizinium perchlorate	EtOH	215s(4.43),219(4.52), 242(4.30),250s(4.22), 288(4.63),350(3.75), 366(3.94),382(4.08), 402(4.22)	44-1846-65
Thalifendine, chloride	EtOH	231(4.17),269(4.15), 348(4.10)	88-3595-65
$C_{19}H_{16}FeO$			
Cinnamaldehyde, α-ferrocenyl-	n.s.g.	313(4.14),488(3.20)	101-0398-64B
isomer	n.s.g.	311(4.09),494(3.20)	101-0398-64B
Ferrocene, 1,1'-(α-oxo-α'-phenyltrimethylene)-	n.s.g.	269(3.7)	44-3941-65
2-Propenal, 3-ferrocenyl-2-phenyl-	n.s.g.	310(4.14),488(3.30)	101-0398-64B
2-Propen-1-one, 1-ferrocenyl-3-phenyl-	n.s.g.	300(4.42),384(3.28), 493(3.26)	101-0398-64B
2-Propen-1-one, 3-ferrocenyl-1-phenyl-	n.s.g.	321(4.28),385s(3.40), 502(3.45)	101-0398-64B
$C_{19}H_{16}IN_3$			
11,12-Dihydro-5,6-dimethylpyrido-[3,2-b:4,5-b']diindolium iodide	EtOH	280(4.81),333(4.36), 348(4.40)	7-0452-65
$C_{19}H_{16}N_2$			
Fluorene-2-carbonitrile, 9-(3-dimethylamino-2-propen-1-ylidene)-	MeCN	257(4.82),287(4.20), 300(4.22),315(4.25), 435(4.69),453(4.73)	24-3331-64
Methane, α-(benzeneazo)diphenyl-	EtOH	274(4.1),403(2.4)	28-2113-64B
Pyrrolo[1,2-a]quinoxaline, 2-benzyl-4-methyl-	pH 1.5	233s(4.47),244(4.54), 249s(4.50),265s(3.87), 355(4.13)	39-3678-65

Compound	Solvent	$\lambda_{max}(\log \epsilon)$	Ref.
Pyrrolo[1,2-a]quinoxaline, 2-benzyl-4-methyl- (cont.)	pH 7.5	231(4.48),245s(4.48), 249(4.50),337(4.03)	39-3678-65
	EtOH	230(4.47),246s(4.46), 251(4.51),260s(4.12), 334s(3.98),340(3.99), 357s(3.78)	39-3678-65
$C_{19}H_{16}N_2O$			
Benzanilide, 4-amino-2'-phenyl-	EtOH	223(4.4),303(4.5)	28-4295-64B
Benzanilide, 4-amino-N-phenyl-	EtOH	224(4.2),253(4.2), 308(4.3)	28-4295-64B
Ketone, benzyl 4-methyl-5-phenyl-3-pyridazinyl	0.2N NaOH	291(4.16),345(3.79)	44-2128-64
	EtOH	250(3.95),310(2.95)	44-2128-64
Naphth[1,2-d]oxazole, 2-(p-dimethylaminophenyl)-	heptane	235(4.34),300(4.17), 310(4.23),350(4.55), 368s(4.41)	65-0427-64
Naphth[2,1-d]oxazole, 2-(p-dimethylaminophenyl)-	heptane	245(4.52),290(4.42), 345(4.73),362(4.63)	65-0427-64
Naphth[2,3-d]oxazole, 2-(p-dimethylaminophenyl)-	heptane	242(4.30),306s(4.02), 367(4.62)	65-0427-64
Pyrrolo[1,2-a]quinoxalin-1(5H)-one, 2-benzyl-4-methyl-	EtOH	236(4.52),259s(3.88), 294s(4.02),300(4.04), 328s(3.48),424(4.20)	39-3678-65
Pyrrolo[1,2-a]quinoxalin-1(5H)-one, 2,5-dimethyl-4-phenyl-	EtOH	238(4.45),264(4.12), 297(4.13),303(4.12), 330s(3.61),343s(3.49)	35-1830-64
$C_{19}H_{16}N_2O_2$			
5,5'-Biisoxazole, 4,5-dihydro-5-methyl-3,3'-diphenyl-	EtOH	235(4.13),262(4.17)	7-1223-65
3,5-Pyrazolidinedione, 4-(2-butenylidene)-1,2-diphenyl-	MeOH	242(4.09),310(4.42)	32-0320-65
9H-Pyrido[3,4-b]indole, 1-(3,4-dimethoxyphenyl)-	EtOH	212(4.57),234(4.55), 250s(4.35),265(4.29), 296(4.25),354s(4.02), 361(4.03)	4-0168-64
$C_{19}H_{16}N_2O_3$			
Carbazole, 9-benzoyl-1,2,3,4-tetrahydro-7-nitro-	EtOH	239(4.32),324(4.05), 362(4.12)	39-4320-64
1H-Pyrrolo[1,2-a]indole-2-carboxamide, 7-(benzyloxy)-2,3-dihydro-1-oxo-	MeOH	315(4.31)	44-2910-65
$C_{19}H_{16}N_2O_4$			
1H-Pyrrolo[1,2-a]indole-2-carboxaldehyde, 7-(benzyloxy)-2,3-dihydro-8-nitro-	MeOH	217(4.56),254(4.26), 295(4.03)	44-4381-65
$C_{19}H_{16}N_2O_4$			
Indole-3-glyoxylic acid, 5-(benzyloxy)-4-nitro-, ethyl ester	MeOH	260(4.09),280(4.08), 310s(4.00)	44-4381-65
Indole-3-glyoxylic acid, 5-(benzyloxy)-6-nitro-, ethyl ester	MeOH	260s(3.96),300(4.12)	44-4381-65
5H-Oxazolo[4,5-b]phenoxazine-4,6-dicarboxylic acid, 9,11-dimethyl-, dimethyl ester	MeOH	226(4.61),252(4.57), 419(4.22)	44-3185-65

Compound	Solvent	λ_{max}(log ε)	Ref.
$C_{19}H_{16}N_2O_6S_3$			
5-Thia-1-azabicyclo[4.2.0]oct-2-ene-2-carboxylic acid, 3-(hydroxymethyl)-8-oxo-7-[2-(2-thienyl)acetamido]-, 2-thiophenecarboxylate, sodium salt	n.s.g.	240(4.27)	87-0022-65
$C_{19}H_{16}N_2O_7$			
1,3,6,8-Nonatetraen-5-one, 2,8-dimethyl-1,9-bis(5-nitro-2-furyl)-	propylene glycol	415(4.29)	95-0001-64
$C_{19}H_{16}N_4$			
s-Triazolo[4,3-a]pyrazine, 3-ethyl-5,6-diphenyl-	C_6H_{12}	251(4.33),316(3.59), 324s(3.58),341s(3.37)	44-2542-64
$C_{19}H_{16}N_4O$			
Acetanilide, 3'-(9-methyl-9H-pyridazino[3,4-b]indol-3-yl)-	EtOH	273(4.69),384(3.47)	94-1129-64
$C_{19}H_{16}N_4OS$			
Purine-8-methanol, 6-(benzylthio)-α-phenyl-	pH 1	298(4.22)	87-0797-65
	pH 11	297(4.27)	87-0797-65
$C_{19}H_{16}N_4O_2$			
Diacetamide, N-(5,6-diphenyl-as-triazin-3-yl)-	EtOH	225(4.36),268(4.26), 322(3.99)	94-1168-65
$C_{19}H_{16}N_4O_5S_2$			
Xanthine, 7-benzyl-1-hydroxy-8-(methylthio)-, benzenesulfonate	EtOH	292(4.26)	4-0275-64
	EtOH-NaOH	301(4.33)	4-0275-64
$C_{19}H_{16}N_4O_8$			
1,2,3,4-Tetrahydro-4-hydroxybenzo-[c]quinolizinium picrate	EtOH	205(4.60),239(4.56), 320(4.20),357(4.19)	78-2529-65
$C_{19}H_{16}O$			
Biphenyl, 2-(p-methoxyphenyl)-	n.s.g.	245(4.39),262(4.08), 270(3.95)	39-2898-65
Cyclopentanone, 2,5-dibenzylidene-	hexane	338(4.54)	65-3645-64
	EtOH	232(4.15),352(4.50)	65-3645-64
	66% H_2SO_4	232(3.52),300(3.00), 370(3.44),492(4.45)	65-3645-64
15H-Cyclopenta[a]phenanthrene, 12-acetyl-16,17-dihydro-	EtOH	242(4.50),262(4.66), 291(4.03),321(4.16), 350(3.50),369(3.43)	5-0200-65D
11H-Dicyclopenta[a,c]fluoren-11-one, 1,2,3,4,5,6-hexahydro-	EtOH	256(4.82),266(4.87), 294(3.56),305(3.55), 346(3.62)	54-1094-65
Methanol, triphenyl-	$HClO_4$	266s(3.20),291(3.08), 408(4.63),434(4.64)	60-0264-64
	HCOOH	437(--)	60-0264-64
	H_2SO_4	431(--)	60-0264-64
	$POCl_3$	435(--)	60-0264-64
$C_{19}H_{16}O_2$			
2-Cyclopropene-$\Delta^{1,\alpha}$-acetic acid, 2,3-diphenyl-, ethyl ester	isooctane	244(4.42),252(4.43), 268(4.33),286(4.24), 299(4.26),312(4.12), 382(3.89),400(3.87), 423s(3.71),452s(3.23)	35-0942-64

Compound	Solvent	λ_{max}(log ε)	Ref.
2-Cyclopropene-$\Delta^{1,\alpha}$-acetic acid, 2,3-diphenyl-, ethyl ester (cont.)	MeCN	242(4.39),250(4.39), 267(4.34),286s(4.30), 295(4.35),308(4.24), 371(3.95)	35-0942-64
	MeCN-HBF_4	247(4.22),293(4.51), 307(4.53)	35-0942-64
2-Naphthol, 7,8-dimethyl-, benzoate	EtOH	232(4.85),266(3.96), 275(3.94),282s(3.87), 298s(3.60),318s(2.76)	39-0361-65
7-Pleiadenol, 7,12-dihydro-1-methoxy-	EtOH	212(4.32),233(4.54), 287(3.79),299(3.81), 327(3.40),339(3.44)	12-1865-65
$C_{19}H_{16}O_3$			
5-Benzofuranacrylic acid, α-ethyl-2,3-dihydro-6-hydroxy-β-phenyl-, δ-lactone	EtOH	335(4.16)	44-2467-64
Cyclopentanone, 2-benzoyl-, enol benzoate, exo-trans	EtOH	231(4.42),286(4.39)	1-2139-65
$C_{19}H_{16}O_4$			
2H-Benzo[d]naphtho[1,2-b]pyran-11-carboxylic acid, 1,3,4,6-tetrahydro-10,12-dihydroxy-7-methyl-, δ-lactone	EtOH	242(4.57),276(4.13), 373(4.01),386(3.98)	33-1459-64
$C_{19}H_{16}O_5$			
Indanone, 2,3-diacetoxy-2-phenyl-, cis	EtOH	247(4.12)	35-0874-65
Osajaxanthone, methyl ether	EtOH	240(4.33),248(4.30), 286(4.64),339(3.96), 375s(3.65)	44-0144-65
$C_{19}H_{16}O_6$			
Flavone, 5-acetoxy-4',7-dimethoxy-	EtOH	258(4.19),323(4.47)	12-1871-65
4H-Naphtho[1,2-b]pyran-4-one, 3-acetyl-5-hydroxy-6-methoxy-2-methyl-, acetate	EtOH	222(4.59),255(4.52), 280s(4.12),305(3.82), 342(3.61),355(3.60)	94-0316-64
4H-Naphtho[1,2-b]pyran-4-one, 3-acetyl-6-hydroxy-5-methoxy-2-methyl-, acetate	EtOH	223(4.55),254(4.55), 280s(4.12),304(3.86), 343(3.50),354(3.49)	94-0316-64
$C_{19}H_{16}O_7$			
Flavanone, 4',5,7-trihydroxy-, 4',7-diacetate	EtOH	276(4.10),340(3.59)	44-0897-65
Fonsecin, diacetate	EtOH	240s(4.43),260(4.61), 315(3.59),358(3.81)	44-0112-65
Isoflavone, 2-carbethoxy-6,7-dihydroxy-4'-methoxy-	EtOH	215(4.10),240(3.96), 341(4.58)	78-1331-64
$C_{19}H_{16}O_8$			
Flavone, 5-hydroxy-6,7,8-trimethoxy-3',4'-(methylenedioxy)-	EtOH	256(4.18),284(4.23), 343(4.33)	39-2743-65
Flavone, 7-hydroxy-5,6,8-trimethoxy-3',4'-(methylenedioxy)-	EtOH-HCl	245(4.32),271(4.18), 330(4.39)	39-2743-65
	EtOH	238s(4.33),274(4.19), 330(4.29)	39-2743-65
	EtOH-NaOEt	279(4.34),322(4.12), 375(4.15)	39-2743-65

Compound	Solvent	λ_{max}(log ε)	Ref.
$C_{19}H_{16}O_9$			
Flavone, 5,8-dihydroxy-3,6,7-trimethoxy-3',4'-(methylenedioxy)-	n.s.g.	257(4.07),288(4.17), 342(4.12)	12-0461-64
$C_{19}H_{16}S_3$			
6a-Thiathiophthene, 2,5-bis(p-tolyl)-	dioxan	257(4.65),321(4.40), 512(4.19)	24-1732-64
$C_{19}H_{16}Si$			
7-Silabicyclo[4.2.0]octa-1,3,5-triene, 7,7-diphenyl-	EtOH	264(3.23),270(3.34), 277(3.24)	35-5589-64
$C_{19}H_{17}BrN_2O_3$			
Butadiene, 1-benzoyl-1-bromo-2-(dimethylamino)-3-nitro-4-phenyl-	EtOH	250(4.0),325(4.11)	35-5132-65
Cyclobutene, 4-benzoyl-4-bromo-1-(dimethylamino)-2-nitro-3-phenyl-	EtOH	250(4.00),380(4.18)	35-5132-65
$C_{19}H_{17}BrO_4$			
Chalcone, 2'-acetoxy-2-bromo-4'-methoxy-4-methyl-	n.s.g.	241(4.19),328(4.43)	49-0450-65
$C_{19}H_{17}ClN_2O_3$			
1H-1,4-Benzodiazepine, 3-acetoxy-1-acetyl-7-chloro-2,3-dihydro-5-phenyl-	iso-PrOH	223s(4.48),254(4.11)	44-1621-64
$C_{19}H_{17}ClN_2O_4S_2$			
1,2,3,4-Tetrahydro-4-(3-methyl-2-benzothiazolinylidene)pyrido[2,1-b]benzothiazolium perchlorate	EtOH	460(4.74)	65-2455-64
$C_{19}H_{17}ClN_2O_6$			
2-Naphthacenecarboxamide, 4-amino-7-chloro-1,4,4a,5,5a,6,11,11a-octahydro-3,10,12-trihydroxy-1,11-dioxo-	MeOH-borate	272(3.93),287(3.88), 471(4.67),497(4.53)	35-0933-65
$C_{19}H_{17}ClO_4$			
7-(6-Phenyl-1,3,5-hexatrien-1-yl)-tropylium perchlorate	MeCN	262(4.20),271(4.20), 313(4.23),539(4.76)	24-1590-64
$C_{19}H_{17}ClO_5$			
Tri-O-methylpeltogynidin chloride	EtOH	530(4.40)	39-4941-64
$C_{19}H_{17}N$			
Benz[f]isoindoline, 2-methyl-4-phenyl-	EtOH	266(4.87),275s(3.86), 285(3.96),293s(3.94), 323s(2.88)	94-1094-64
Di-2-propynylamine, N-methyl-3,3'-diphenyl-	EtOH	243(4.57),253(4.53)	94-1094-64
Phenanthridine, 7,8,9,10-tetrahydro-6-phenyl-	n.s.g.	246(3.89),287(3.63), 312(3.54),324(3.54)	39-2421-64
$C_{19}H_{17}NO$			
Acetamide, N-(16,17-dihydro-15H-cyclopenta[a]phenanthren-6-yl)-	EtOH	262(4.68),291s(3.97), 304(4.02),338(2.96), 350(2.81)	5-0200-65D
Acetamide, N-(16,17-dihydro-15H-cyclopenta[a]phenanthren-12-yl)-	EtOH	223(4.48),243s(4.49), 263(4.81),303(4.17), 312s(4.03)	5-0200-65D

Compound	Solvent	λ_{max}(log ε)	Ref.
Ethylamine, N-salicylidene-α-(1-naphthyl)-	hexane	224(5.03),257(4.24), 270(4.07),281(4.05), 288(3.94),294(3.94), 314(3.80)	35-1757-65
	EtOH	225(5.09),256(4.32), 272s(4.03),283(4.02), 290s(3.90),314(3.71), 406(2.65)	35-1757-65
	dioxan	226(4.99),258(4.23), 272s(4.03),285(4.01), 292s(3.89),315(3.73), 410s(1.32)	35-1757-65
Indone, 2-phenyl-3-(1-pyrrolidinyl)-	EtOH	276(4.38)	35-0874-65
2-Pyridone, 3-benzyl-1-methyl-5-phenyl-	EtOH	316(4.09)	78-3255-65
$C_{19}H_{17}NO_2$			
Cinchoninic acid, 2-(2,4,5-trimethylphenyl)-	EtOH	238(4.5),328(4.0)	28-0954-64A
Cyclopropanecarboxylic acid, 1-cyano-2,2-diphenyl-, ethyl ester	EtOH	282(3.99)	28-1332-65B
1H-Pyrrolo[1,2-a]indole-9-carboxaldehyde, 6-benzyl-2,3-dihydro-7-hydroxy-	MeOH	256(4.43),278(4.32), 309(4.26)	44-4381-65
1H-Pyrrolo[1,2-a]indole-9-carboxaldehyde, 7-(benzyloxy)-2,3-dihydro-	MeOH	257(4.30),276(4.15), 308(4.08)	87-0700-65
Pyrrolo[3,2,1-hi]indole-5-carboxylic acid, 1,2-dihydro-4-phenyl-, ethyl ester	MeOH	251(4.25),309(3.97)	35-1397-65
Quinoline, 4-(2,3-dimethoxystyryl)-	MeOH	330(4.4)(changing)	87-0137-65
	HOAc	380(4.3)	87-0137-65
Quinoline, 4-(2,4-dimethoxystyryl)-	MeOH	360(4.1)(changing)	87-0137-65
	HOAc	315(3.8),430(3.7)	87-0137-65
Quinoline, 4-(2,5-dimethoxystyryl)-	MeOH	325(4.2),360(4.2) (changing)	87-0137-65
	HOAc	330(4.0),370(4.1), 420(4.1)	87-0137-65
Quinoline, 4-(3,4-dimethoxystyryl)-	MeOH	355(4.2)(changing)	87-0137-65
	HOAc	310(3.8),420(4.0)	87-0137-65
$C_{19}H_{17}NO_2S$			
Benzenesulfonanilide, N-benzyl-	EtOH	230s(4.00),265s(3.33)	12-0206-65
$C_{19}H_{17}NO_3$			
Chromene-2-carboxylic acid, 4-(p-tolylimino)-, ethyl ester	EtOH	322(2.95),375(3.65)	65-1924-64
Indolizine-3-carboxylic acid, 1-benzoyl-5-methyl-, ethyl ester	EtOH	250(4.46),288(3.82), 358(4.29)	78-3305-65
2-Pyridinol, 1-benzoyl-1,4,5,6-tetrahydro-, benzoate	dioxan	228(4.34),260s(3.88)	70-0774-64
1H-Pyrrolo[1,2-a]indole-9-carboxaldehyde, 7-(benzyloxy)-2,3-dihydro-1-hydroxy-	MeOH	257(4.43),275(4.15), 308(4.11)	35-4612-64
1H-Pyrrolo[1,2-a]indol-1-one, 7-(benzyloxy)-2,3-dihydro-9-(hydroxymethyl)-	MeOH	322(4.34)	35-4612-64

Compound	Solvent	$\lambda_{max}(\log \epsilon)$	Ref.
$C_{19}H_{17}NO_3S$			
Alanine, N-acetyl-3-(9-anthrylthio)-	EtOH	222(4.14),249s(4.63), 259(4.89),339(3.26), 357(3.56),375(3.74), 393(3.71)	12-0109-64
$C_{19}H_{17}NO_4$			
12H-Benz[6,7]oxepin[2,3,4-ij]isoquinoline, 6,9,10-trimethoxy-	EtOH-HCl	251(4.51),284(3.76), 396(3.73)	95-0667-65
	EtOH	228(4.61),284(3.92), 348(3.74)	95-0667-65
Chromene-2-carboxylic acid, 4-(p-methoxyphenylimino)-, ethyl ester	EtOH	318(4.03),398(3.68)	65-1924-64
Escholtzine	EtOH	295(4.04)	23-2180-65
Neolitsine	EtOH	284(3.9),310(4.11)	39-2285-65
$C_{19}H_{17}NO_6$			
2-Naphthaleneacetic acid, α-cyano-1,4-diacetoxy-, ethyl ester	EtOH	226(4.84),287(3.85), 323(2.95)	39-0974-65
Naphtho[1,2-b]furan-3-carboxylic acid, 2-acetamido-5-acetoxy-, ethyl ester	EtOH	245(4.69),337(3.66)	39-0974-65
$C_{19}H_{17}NO_8S$			
4aH-Pyrido[2,1-b]benzothiazole-1,2,3,4-tetracarboxylic acid, tetramethyl ester	MeOH	223(1.97),270(1.95), 295(0.87),427(0.73)	39-3200-65
$C_{19}H_{17}NO_9$			
4aH-Pyrido[2,1-b]benzoxazole-1,2,3,4-tetracarboxylic acid, tetramethyl ester	MeOH	246(1.27),275s(0.96), 296(2.62),391(1.42)	39-3200-65
4H-Quinolizine-3-fumaric acid, 1,2-dicarboxy-4-oxo-, tetramethyl ester	MeOH-ether	260(4.10),396(4.23)	24-3537-65
$C_{19}H_{17}N_3O_2$			
1,10-Phenanthroline-$\Delta^{2(1H),\alpha}$-acetic acid, α-cyano-, tert-butyl ester	$CHCl_3$	251(4.45),310(4.18), 323(4.46),368(4.01), 405s(3.91),426(3.98), 450s(3.80)	44-0243-65
$C_{19}H_{17}N_3O_2S$			
5-Pyrimidinecarboxylic acid, 1,4-dihydro-6-methyl-1-phenyl-2-(2-pyridyl)-4-thioxo-, ethyl ester	EtOH	240(4.04),272s(3.81), 341(4.38)	44-1115-64
$C_{19}H_{17}N_3O_4S$			
Benzenesulfonamide, N-(3,4-dimethyl-5-isoxazolyl)-p-(1-oxo-2-isoindolinyl)-	MeOH-acid	293(4.41)	95-1035-65
	MeOH	271(4.32)	95-1035-65
$C_{19}H_{17}N_5$			
Adenine, N,N-dibenzyl-	EtOH-HCl	287(4.34)	4-0291-65
	EtOH	278(4.38)	4-0291-65
	EtOH-NaOH	284(4.33),292s(4.27)	4-0291-65
Adenine, N,1-dibenzyl-	EtOH-HCl	267(4.11)	4-0291-65
	EtOH	233(4.27),280(4.11)	4-0291-65
	EtOH-NaOH	275(4.18),280(4.18)	4-0291-65
Adenine, N,3-dibenzyl-	pH 1	287(4.39)	4-0115-64
	pH 7	290(4.25)	4-0115-64
	pH 13	290(4.24)	4-0115-64
	EtOH-HCl	288(4.39)	4-0291-65

Compound	Solvent	λ_{max}(log ε)	Ref.
Adenine, N,3-dibenzyl- (cont.)	EtOH	218(4.35),293(4.24)	4-0291-65
	EtOH-NaOH	294(4.24)	4-0291-65
Adenine, N,7-dibenzyl-	EtOH-HCl	285(4.30)	4-0291-65
	EtOH	279(4.18)	4-0291-65
	EtOH-NaOH	279(4.20)	4-0291-65
Adenine, N,9-dibenzyl-	EtOH-HCl	266(4.31)	4-0291-65
	EtOH	271(4.28)	4-0291-65
	EtOH-NaOH	271(4.28)	4-0291-65
Adenine, 1,7-dibenzyl-	EtOH-HCl	277(3.92)	4-0291-65
	EtOH	266(3.99),275s(3.94)	4-0291-65
	EtOH-NaOH	266(4.04),275s(3.98)	4-0291-65
Adenine, 1,9-dibenzyl-	EtOH-HCl	261.5(4.16)	4-0291-65
	EtOH	262(4.13),269s(4.05), 290s(3.66)	4-0291-65
	EtOH-NaOH	262(4.16),269s(4.09)	4-0291-65
Adenine, 3,7-dibenzyl-	pH 1	278(4.18)	4-0115-64
	pH 7	278(4.18)	4-0115-64
	pH 13	281(4.09)	4-0115-64
	EtOH-HCl	224s(4.17),281(4.23)	4-0291-65
	EtOH	281(4.20)	4-0291-65
	EtOH-NaOH	281(4.13)	4-0291-65
$C_{19}H_{17}N_5O$			
4(3H)-Pteridinone, 2-amino-5,6-dihydro-5-methyl-6,7-diphenyl-	MeOH	269(4.40),423s(3.61)	33-0764-65
$C_{19}H_{17}N_5O_3S$			
1-[(Aminomethylene)amino]-2-phenyl-2H-benzotriazolium benzenesulfonate	EtOH	220(4.38),317(4.11), 382(3.81)	39-0751-64
$C_{19}H_{17}N_5O_7$			
L-Tryptophan, dinitrophenylglycyl-	HOAc	340(4.17)	35-1848-64
$C_{19}H_{18}$			
Cyclopentane, 1,2-dibenzylidene-	EtOH	236(3.96),332(4.48)	44-2335-64
$C_{19}H_{18}BF_4N$			
[4-(Dimethylamino)-1-naphthyl]tropylium tetrafluoroborate	EtOH	333(3.22),628(3.36)	44-4180-65
$C_{19}H_{18}BrNO$			
2-(1-Methylacetonyl)-1-phenylisoquinolinium bromide	EtOH	236(4.69),272(3.26), 280(3.26),344(3.38)	44-1846-65
$C_{19}H_{18}ClN$			
4H-Benzo[a]quinolizine, 2-(p-chlorophenyl)-1,6,7,11b-tetrahydro-	EtOH	253(4.18)	33-1852-64
$C_{19}H_{18}ClNO_2$			
Indole-3-acetic acid, 1-(p-chlorobenzyl)-2,5-dimethyl-	MeOH	226(4.57),278(3.89), 290(3.88),300s(3.78)	87-0204-65
sodium salt	EtOH	226(4.46),284(3.77), 292(3.79),300(3.72)	87-0204-65
Indole-3-acetic acid, 1-(p-chlorobenzyl)-5-methyl-, methyl ester	MeOH	223(4.63),278(3.84), 292(3.75),303(3.68)	87-0204-65
$C_{19}H_{18}ClN_3O$			
3H-1,4-Benzodiazepine, 2-(allylmethylamino)-7-chloro-5-phenyl-, 4-oxide	EtOH-acid	253(4.52),307s(3.96)	87-0235-64

Compound	Solvent	λ max(log ε)	Ref.
$C_{19}H_{18}ClOP$			
(Hydroxymethyl)triphenyl-phosphonium chloride	EtOH	261(3.40),267(3.41), 274(3.34)	59-1143-64
$C_{19}H_{18}FeO$			
2,4,6,8-Nonatetraenal, 9-ferrocenyl-	$CHCl_3$	400(4.68),512(3.97)	5-0088-64F
$C_{19}H_{18}INS$			
2-Benzo[b]thien-2-yl-1,3,3-tri-methyl-3H-indolium iodide	EtOH-HCl	248(3.99),378(4.45)	44-2875-65
	EtOH	248(4.06),378(4.21)	44-2875-65
$C_{19}H_{18}IN_7O_6$			
Pteroylglutamic acid, 3'-iodo-	pH 1	223(4.48),301(4.28)	44-2837-65
	pH 13	223(4.77),255(4.48), 282(4.38),366(4.12)	44-2837-65
$C_{19}H_{18}IP$			
Methyltriphenylphosphonium iodide	EtOH	227(4.08),262(3.35), 268(3.49),275(3.42)	59-1143-64
$C_{19}H_{18}N_2$			
3,4-Diazabicyclo[4.1.0]hepta-2,4-diene, 1,6-dimethyl-2,5-diphenyl-	MeOH	255(3.94),290(4.07)	24-2438-65
1H-Indazole, 4,5,6,7-tetrahydro-1,3-diphenyl-	C_6H_{12}	276(4.36)	44-1582-64
$C_{19}H_{18}N_2O$			
3-Indolinepropionitrile, 3-benzyl-1-methyl-2-oxo-	MeOH	206(4.48),257(3.85), 278s(3.24)	44-2973-65
Pyridazine, 3-(p-methoxyphenyl)-6-(2,4-xylyl)-	EtOH	289(4.42)	39-7005-65
Pyridazine, 3-(p-methoxyphenyl)-6-(2,5-xylyl)-	EtOH	289(4.42)	39-7005-65
Pyridazine, 3-(p-methoxyphenyl)-6-(3,4-xylyl)-	EtOH	301(4.57)	39-7005-65
3-Pyridazinemethanol, α-benzyl-4-methyl-5-phenyl-	pH 1	251s(3.81),301(3.70)	44-2128-64
	EtOH	251(3.91)	44-2128-64
$C_{19}H_{18}N_2OS$			
2-Thiazolin-4-one, 5-(cyclohexyl-imino)-2-(2-naphthyl)-	dioxan	225(4.5),277(4.4), 325(4.4)	83-0124-65
$C_{19}H_{18}N_2O_2$			
Cyclobutene, 4-benzylidene-1-(di-methylamino)-2-nitro-3-phenyl-	n.s.g.	333(3.93),422(4.22)	35-5132-65
4H-1,2-Diazepin-4-one, 2,3-dihydro-3-(α-hydroxybenzyl)-5-methyl-6-phenyl-	pH 13	243(4.42),344(3.43), 412(3.73)	44-2128-64
	EtOH	309(3.72),395(3.43)	44-2128-64
1-Penten-3-one, 5-(1-benzimidazolyl)-1-(p-methoxyphenyl)-	EtOH	236(4.09),325(4.26)	31-0617-65
3,5-Pyrazolidinedione, 4-butylidene-1,2-diphenyl-	MeOH	246(4.31),400(4.40)	32-0320-65
$C_{19}H_{18}N_2O_3$			
Cyclobutene, 4-benzoyl-1-(dimethyl-amino)-2-nitro-3-phenyl-	n.s.g.	370(4.27)	35-5132-65
Indone, 3-butylamino-2-(p-nitrophenyl)-	MeOH	218(4.39),265(4.47), 322(3.62),390(4.08)	65-0448-64
	MeOH-base	228(4.62),264(4.54), 325s(4.18),500(4.39)	65-0448-64

Compound	Solvent	λ_{max}(log ϵ)	Ref.
$C_{19}H_{18}N_2O_6$			
3H-Phenoxazine-1,9-dicarboxylic acid, 4,6-dimethyl-2-(methylamino)-3-oxo-, dimethyl ester	MeOH	220(4.7),253(4.69), 425(4.61),445(4.64)	44-3185-65
$C_{19}H_{18}N_2S$			
Quinoline, 1,4-dihydro-1-methyl-2-[(3-methyl-2-benzothiazolinyl)-methylene]-	hexane	272(4.24),316(3.77), 386(4.21)	22-2879-64
	EtOH	308(4.06)	22-2879-64
	HOAc	302(3.97),314(3.97)	22-2879-64
$C_{19}H_{18}N_4O_5$			
1(4H)-Naphthalenone, 6-methoxy-4,4-dimethyl-, 2,4-dinitrophenylhydrazone	EtOH	238(4.37),255s(4.31), 408(4.56)	39-0361-65
$C_{19}H_{18}N_6O_5$			
Homopteroic acid, N^2,N"-diacetyl-	pH 13	254(4.50),353(3.84)	44-3404-65
$C_{19}H_{18}O$			
Anisole, p-(6-phenyl-1,3,5-hexatrienyl)-	n.s.g.	357(4.70),370(4.75), 389(4.58)	39-2898-65
1(2H)-Naphthalenone, 2-benzylidene-3,4-dihydro-4,4-dimethyl-, cis	MeOH	234(3.99),269(3.97), 311(4.05)	44-1276-64
2H-Pyran, 2,4-dimethyl-2,6-diphenyl-	EtOH	227(3.88),250(3.87), 323(3.95)	28-5228-64A
$C_{19}H_{18}O_2$			
Cyclopenta[ef]heptalene-3-acetic acid, 5-methyl-, ethyl ester	EtOAc	374(4.07),391(4.16), 451(2.24),737(2.00), 801(2.11),908(2.05), 1038(1.64)	5-0194-64B
Cyclopropeneacetic acid, diphenyl-, ethyl ester	C_6H_{12}	228(4.22),237(4.13), 302s(4.26),307(4.31), 316(4.39),323s(4.30), 355(4.28)	88-0479-65
$C_{19}H_{18}O_2S$			
Acetophenone, 4'-[(3-oxo-5-phenyl-4-pentenyl)thio]-	EtOH	226(4.27),238s(3.96), 290s(4.52),300(4.56)	7-1093-65
$C_{19}H_{18}O_3$			
Indene, 1-(2,4,5-trimethoxybenzylidene)-	MeOH	385(4.3)	87-0390-65
	HOAc	385(4.3)	87-0390-65
Indene, 1-(3,4,5-trimethoxybenzylidene)-	MeOH	355(4.4)	87-0390-65
	HOAc	360(4.4)	87-0390-65
1-Naphthoic acid, 1,2,3,4-tetrahydro-4-oxo-1-phenyl-, ethyl ester	EtOH	248(4.08),292(3.31)	44-2973-65
2,4-Pentadienoic acid, 2-(p-methoxyphenyl)-5-phenyl-, methyl ester	MeOH	230(4.17),331(4.41)	2-0017-64
2,4-Pentadienoic acid, 5-(p-methoxyphenyl)-2-phenyl-, methyl ester	MeOH	241(4.03),335(4.46)	2-0017-64
Phenanthrene, 3,4,6-trimethoxy-1-vinyl-	EtOH	222(4.18),245s(4.20), 263(4.32),318(3.71), 328(3.75)	88-1539-65
$C_{19}H_{18}O_3S_2$			
Spiro[5.5]undecane-1,5,9-trione, 7,11-di-2-thienyl-	EtOH	236(4.22)	78-0515-65

Compound	Solvent	$\lambda_{max}(\log \epsilon)$	Ref.
$C_{19}H_{18}O_4$			
Chalcone, 2'-acetoxy-4-methoxy-5'-methyl-	MeOH	338(4.20)	24-1910-64
p-Dioxane, [(10-methoxy-9-phenanthryl)oxy]-	dioxan	272s(4.28),281s(4.08), 293(4.04),304(4.08), 326s(2.70),341(2.94), 357(2.93)	44-2362-64
2-Flavene, 4'-acetoxy-8-methoxy-3-methyl-	EtOH	220(4.41),240s(4.19), 275(3.82)	78-1471-65
Phenanthrene-10-carboxaldehyde, 2,3,4-trimethoxy-7-methyl-	EtOH	261(4.71),332(3.96)	39-4257-64
$C_{19}H_{18}O_5$			
Eucalyptin	EtOH	281(4.35),287(4.34), 323(4.41)	12-0464-64
Isobenzopyranol, 1,6-anhydro-1-(3-methoxy-4-hydroxyphenyl)-3-methyl-4-ethyl-6,7-dihydroxy-	EtOH	250(4.5),400(4.5)	49-0369-65
3-Isoflavene, 2'-acetyl-4',7-dimethoxy-	n.s.g.	242(3.91),321(4.03)	39-5991-64
Osajaxanthone, dihydro-, methyl ether	EtOH	233(4.50),263(4.55), 316(4.22),375(3.79)	44-0144-65
Phenanthrene, 8-acetoxy-3,4,7-trimethoxy-	EtOH	224(4.33),268(4.67), 287(4.18),305(4.01), 316(4.00),345(3.19), 364(3.08)	88-3617-64
Phenanthrene-9-carboxylic acid, 3,5,6-trimethoxy-, methyl ester	EtOH	250(4.78),276s(4.36), 332(4.17),354s(3.95), 372(3.64)	78-2573-65
Spiro[5.5]undecane-1,5,9-trione, 7,11-di-2-furyl-	EtOH	216(4.17)	78-0515-65
$C_{19}H_{18}O_6$			
Flavone, 2',3,5,6-tetramethoxy-	MeOH	216(4.14),243(4.20), 304(3.76),336(3.79)	24-0164-65
Flavone, 2',5,6,7-tetramethoxy-	MeOH	213(4.31),234(4.11), 261(4.11),320(4.12)	24-0164-65
	EtOH	234s(4.20),262(4.20), 322(4.21)	78-3573-65
Flavone, 2',5,7,8-tetramethoxy-	MeOH	213(4.33),267(4.31), 333(4.04)	24-0164-65
Flavone, 3,4',5,8-tetramethoxy-	n.s.g.	272(4.11),299(4.13), 321(4.16)	23-0306-65
Flavone, 3',4',5',7-tetramethoxy-	EtOH	267(3.72),312(4.40)	28-6930-65A
Flavone, 3,4',7,8-tetramethoxy-	EtOH	260(4.20),320(4.13), 352(4.21)	12-1170-64
Flavone, 4',5,6,8-tetramethoxy-	EtOH	222(4.43),298(4.48), 322(4.39)	2-0351-65
Flavone, 5,6,7,8-tetramethoxy-	EtOH	271(4.52),305(4.26)	39-2743-65
Xanthone, 2-allyl-1-hydroxy-3,5,6-trimethoxy-	MeOH	251(4.47),284(4.06), 323(4.40)	78-2653-65
Xanthone, 1-(allyloxy)-3,5,6-trimethoxy-	MeOH	246(4.69),291s(3.93), 305(4.28)	78-2653-65
$C_{19}H_{18}O_7$			
Chalcone-2-carboxylic acid, 2'-hydroxy-4,4',5-trimethoxy-	EtOH	257(4.21),364(4.39)	39-3822-64
	EtOH	260(4.21),370(4.44)	39-4941-64
Flavone, 5-hydroxy-3,4',7,8-tetramethoxy-	EtOH	275(4.36),324(4.19), 362(4.12)	12-0934-64
	EtOH-NaOH	290(4.42),398(3.80)	12-0934-64
Gibbs reagent product	EtOH-borate	702(3.39)	12-0934-64

Compound	Solvent	$\lambda_{max}(\log \epsilon)$	Ref.
Quercetin, 3',4',5,7-tetramethyl ether	EtOH	253(4.37),362(4.34)	95-0047-64
	EtOH-$AlCl_3$	262(4.41),423(4.43)	95-0047-64
Wightin, monomethyl ether	EtOH	270(4.48),335s(3.85)	78-3237-65
$C_{19}H_{18}O_8$			
Flavone, 3',5-dihydroxy-3,4',6,7-tetramethoxy-	MeOH	258(4.30),272(4.24), 349(4.33)	24-2857-64
Flavone, 3',5-dihydroxy-3,4',7,8-tetramethoxy-	EtOH	260(4.32),275(4.30), 366(4.20)	12-0934-64
	EtOH-NaOH	274(4.45),298s(4.15), 395(4.01)	12-0934-64
Gibbs reagent product	EtOH-borate	694(3.73)	12-0934-64
Flavone, 4',5-dihydroxy-3,3',6,7-tetramethoxy-	MeOH	257(4.24),271(4.19), 351(4.33)	24-0548-65
$C_{19}H_{18}O_{11}$			
Mangiferin	EtOH	242(4.0),259(4.13), 317(3.82),370(3.68)	95-0374-65
$C_{19}H_{19}BrN_2O$			
13a-Ethyl-1,2,3,12,13,13a-hexahydro-12-oxoindolo[3,2,1-de]pyrido[3,2,1-ij][1,5]naphthyridin-4-ium bromide	EtOH	233(4.30),256(4.26), 305(4.28),353(3.65)	35-1580-65
	EtOH-NaOH	223(4.24),251(4.07), 282s(4.59),287(4.69), 322(3.96),332(4.05), 429(3.46)	35-1580-65
$C_{19}H_{19}ClN_2O_3$			
Acetamide, N-[[2-(p-chlorophenyl)-5,6-dimethoxyindol-3-yl]methyl]-	MeOH	224(4.57),323(4.52)	44-2253-65
Acetanilide, 4'-chloro-2'-[N-(2-hydroxyethyl)benzimidoyl]-, acetate	EtOH	237(4.48),329(3.53)	44-2368-64
$C_{19}H_{19}ClN_2O_4S$			
3-[p-(Dimethylamino)benzylidene]-2,3-dihydro-1H-pyrrolo[2,1-b]benzothiazolium perchlorate	EtOH	500(4.76)	65-2447-64
$C_{19}H_{19}IN_4OS$			
7-[p-(Dimethylamino)styryl]-4,6-dimethyloxazolo[5,4-e]-2,1,3-benzothiadiazolium iodide	EtOH	506(4.29)	17-0048-64
7-[p-(Dimethylamino)styryl]-8-ethyloxazolo[4,5-e]-2,1,3-benzothiadiazolium iodide	EtOH	507(4.54)	17-0048-64
$C_{19}H_{19}IN_8O_5$			
Aminopterin, 3'-iodo-	pH 1	220(4.53),287(4.31), 336(4.04)	87-0713-65
	pH 13	225(4.54),260(4.42), 279s(4.35),371(3.90)	87-0713-65
$C_{19}H_{19}N$			
1,3,5-Cycloheptatriene, 7-(1-dimethylamino-4-naphthyl)-	EtOH	313(2.98)	44-4180-65
Indene, 1-(4-dimethylamino-3-methylbenzylidene)-	MeOH	375(4.4)	87-0390-65
	HOAc	345(4.4)	87-0390-65
2-Naphthylamine, N-benzyl-1,4-dimethyl-	EtOH	212(4.45),254(4.75), 285s(3.77),296(3.94), 309(3.92),356(3.52)	44-0190-65

Compound	Solvent	λ_{max}(log ε)	Ref.
$C_{19}H_{19}NO$			
2,5-Cyclohexadien-1-one, 4-[2-(1,3,3-trimethyl-2-indolinylidene)ethylidene]-	C_5H_5N	513(4.81),541(4.91)	5-0009-65J
	99% C_5H_5N	515(4.79),545(4.91)	5-0009-65J
	98% C_5H_5N	516(4.77),548(4.92)	5-0009-65J
(other mixtures of solvents also listed)	95% C_5H_5N	519(4.76),550(4.97)	5-0009-65J
	90% C_5H_5N	520(4.75),551(5.00)	5-0009-65J
	80% C_5H_5N	521(4.73),551(5.08)	5-0009-65J
1,2-Cyclopentenophenanthrene, 3-acetamido-9,10-dihydro-	EtOH	214(4.51),244(4.27), 273(4.28)	5-0200-65D
Indene, 1-(4-dimethylamino-3-methoxybenzylidene)-	MeOH	380(4.3)	87-0390-65
	HOAc	345(4.3)	87-0390-65
Indone, 3-(butylamino)-2-phenyl-	MeOH	220(4.45),267(4.50), 315(3.48),425(3.51)	65-0448-64
$C_{19}H_{19}NOS$			
2-Indolinol, 2-benzo[b]thien-2-yl-1,3,3-trimethyl-	EtOH-HCl	248(4.03),378(4.44)	44-2875-65
	EtOH	232(4.51),240(4.45), 258(4.13),266(4.05), 291(3.78),300(3.75)	44-2875-65
$C_{19}H_{19}NO_3$			
1H-1-Benzazepine-5-carboxylic acid, 2,3,4,5-tetrahydro-2-oxo-5-phenyl-, ethyl ester	EtOH	230(4.05)	44-2973-65
3-Cyclohexene-$\Delta^{1,\alpha}$-acetic acid, 6-(1-benzyl-2-pyrrolidinyl)-6-hydroxy-, γ-lactone	EtOH	210(3.68)	94-0786-65
3-Indolineacetic acid, 3-benzyl-1-methyl-2-oxo-, methyl ester	EtOH	254(3.90),262s(3.83), 284s(3.20)	44-2973-65
3-Indolinepropionic acid, 2-oxo-3-phenyl-, ethyl ester	EtOH	252(3.88),264s(3.72), 282s(3.21)	44-2973-65
1-Naphthoic acid, 1,2,3,4-tetrahydro-4-oxo-1-phenyl-, ethyl ester, oxime	EtOH	220s(4.35),256(4.13), 298(2.81)	44-2973-65
2-Piperidone, N-[o-(benzyloxy)-benzoyl]-	EtOH	212(4.45),300(3.49)	65-1394-65
	EtOH	212(4.45),300(3.39)	70-0774-64
	dioxan	212(4.45),300(3.39)	78-3537-65
4-Quinolinepropionic acid, 1,2,3,4-tetrahydro-1-methyl-2-oxo-4-phenyl-	EtOH	254(4.02),262s(3.94)	44-2973-65
$C_{19}H_{19}NO_4$			
Amurine	EtOH	245(4.2),294(3.9)	83-0209-65
12H-[1]Benzoxepino[2,3,4-ij]isoquinoline, 2,3-dihydro-6,9,10-trimethoxy-	EtOH-HCl	226(4.34),285(3.95), 368(3.53),394(3.49)	95-0667-65
	EtOH	284(3.88),351(3.61)	95-0667-65
Guatterine	EtOH	242(4.27),281(4.26)	44-0432-65
Isoquinoline, 3,4-dihydro-1-(p-methoxybenzoyl)-6,7-(methylenedioxy)-	MeOH	214(4.6),229s(4.3), 290(4.2)	83-0362-64
$C_{19}H_{19}NO_5$			
1H-[1]Benzoxepino[2,3,4-ij]isoquinolin-2(3H)-one, 12,12a-dihydro-6,9,10-trimethoxy-	EtOH-HCl	225(4.53),269(4.43), 416(3.52)	95-0667-65
	EtOH	213(4.60),255(4.46), 301(3.64),404(3.72)	95-0667-65
Cassyfiline	EtOH	220(4.54),285(4.28), 305(4.26)	95-0827-65
$C_{19}H_{19}NO_5S$			
6H-Thieno[2,3-b]pyrrole-2,4-dicarboxylic acid, 6-benzyl-3-hydroxy-, diethyl ester	EtOH	262(3.92),298(4.24)	44-0184-65

Compound	Solvent	λ_{max}(log ε)	Ref.
$C_{19}H_{19}NO_7$			
1H-Pyrrolo[1,2-a]indole-9-carboxaldehyde, 2,3-dihydro-5,7,8-trihydroxy-6-methyl-, triacetate	MeOH	218(4.45),248(4.26), 305(4.05)	35-3877-64
	MeOH	218(4.45),248(4.26), 305(4.05)	44-2897-65
$C_{19}H_{19}NO_8$			
2,3,4,5-Pyridinetetracarboxylic acid, 1,2-dihydro-1-phenyl-, tetramethyl ester	MeOH-acid	297(2.88),393(0.12)	39-3200-65
	MeOH	262(2.16),278s(1.55), 385(1.17)	39-3200-65
$C_{19}H_{19}NO_9$			
3H-Pyrido[2,1-b]benzoxazole-1,2,3,4-tetracarboxylic acid, 4,4a-dihydro-, tetramethyl ester	MeOH	252s(0.52),268(0.65), 273(0.65),317(3.91)	39-3200-65
$C_{19}H_{19}N_2NaO_6S_2$			
5-Thia-1-azabicyclo[4.2.0]oct-2-ene-2-carboxylic acid, 7-[2-(benzylthio)acetamido]-3-(hydroxymethyl)-8-oxo-, acetate, sodium salt	pH 6.0	259(3.98)	39-5015-65
$C_{19}H_{19}N_3O_2$			
Indole-3-carbonitrile, 2-[p-(dimethylamino)phenyl]-5,6-dimethoxy-	MeOH	240(4.31),354(4.56)	44-2253-65
$C_{19}H_{19}N_3O_9$			
Uridine, 2',3'-O-isopropylidene-, 5'-p-nitrobenzoate	MeOH	208(4.20),259(4.38)	22-2489-65
$C_{19}H_{19}N_5O_3$			
4-Pyrimidinol, 2-amino-5-(3-anilinopropyl)-6-(p-nitrophenyl)-	pH 1	225(4.20),256(4.17), 300s(3.93)	87-0283-65
	pH 7 or 13	243(4.31),277(4.18)	87-0283-65
$C_{19}H_{20}$			
Indeno[1,2-a]indene, 4b,9,9a,10-tetrahydro-4b,9,9-trimethyl-	C_6H_{12}	260(3.06),266(3.25), 273(3.22)	23-0025-64
$C_{19}H_{20}BrNO_2$			
5,6-Dihydroxy-4a-azoniaanthracene bromide	MeOH	245(4.07),347(4.05), 363(4.08)	44-0252-65
$C_{19}H_{20}ClNO$			
2H-Benzo[a]quinolizin-2-ol, 2-(p-chlorophenyl)-1,3,4,6,7,11b-hexahydro-, m. 141°	EtOH	265(2.81),272(2.73)	33-1852-64
form m. 181°	EtOH	258(2.84),265(2.90), 272(2.79)	33-1852-64
$C_{19}H_{20}ClNO_2$			
Isoquinoline, 1-(p-chlorophenethyl)-3,4-dihydro-6,7-dimethoxy-	EtOH-HCl	245(4.25),304(3.96), 358(3.98)	33-1945-65
	EtOH	223(4.48),272(3.89), 308(3.85)	33-1945-65
	EtOH-NaOH	223(4.48),272(3.93), 307(3.87)	33-1945-65
hydrobromide	EtOH-HCl	244(4.24),304(3.97), 358(3.94)	33-1945-65

Compound	Solvent	λ max(log ε)	Ref.
hydrobromide (cont.)	EtOH	244(4.18),305(3.95), 360(3.91)	33-1945-65
	EtOH-NaOH	225(4.45),269(4.07), 307(3.98)	33-1945-65
$C_{19}H_{20}ClNO_4$			
Δ6-Codeine, 6-chloro-6-deoxy-7-formyl-14-hydroxy-	EtOH	252(3.92)	78-1407-64
2-(2,4,6-Cycloheptatrien-1-ylidene-methyl)-1,3,3-trimethyl-3H-indolium perchlorate	MeCN	243(3.90),256s(3.71), 383(4.51)	24-2050-64
$C_{19}H_{20}ClNO_{12}$			
1,2,3,4-Tetracarboxy-7,9-dimethyl-quinolizinium perchlorate, tetramethyl ester	MeOH-HClO4	252(4.24),342(3.90), 356(3.93)	39-3225-64
$C_{19}H_{20}ClN_3O$			
Quinazoline, 6-chloro-2-(diethyl-amino)methyl-4-phenyl-, 3-oxide	EtOH-acid	238(4.03),263(4.53)	87-0235-64
$C_{19}H_{20}CoN_2O_4$			
Cobalt, [α,α'-(propylenedinitrilo)bis-[6-methoxy-o-cresol]ato(2-)]-	MeOH	243(4.68),390(3.67)	65-3049-64
$C_{19}H_{20}CuN_2O_4$			
Copper, [α,α'-(propylenedinitrilo)bis-[6-methoxy-o-cresol]ato(2-)]-	MeOH	233(4.70),280(4.40), 370(3.89),590(2.53)	65-3049-64
$C_{19}H_{20}INO_3$			
Aporphine, 1,2-(methylenedioxy)-3-methoxy-	EtOH-HI	240(4.46),279(4.41), 312s(3.59)	78-2579-65
$C_{19}H_{20}N_2$			
1H-1,4-Diazepine, 2,3-dihydro-5,7-di-methyl-2,3-diphenyl-, cis	THF	302(3.82)	88-0051-65
Indole, 2-(piperidinophenyl)-	MeOH	230(4.52),320(4.20)	44-1270-64
$C_{19}H_{20}N_2NiO_4$			
Nickel, [α,α'-(propylenedinitrilo)bis-[6-methoxy-o-cresol]ato(2-)]-	MeOH	245(4.67),345(3.93), 405(3.71)	65-3049-64
$C_{19}H_{20}N_2O$			
Benz[f]indolo[2,3-a]quinolizin-14(6H)-one, 1,2,3,4,6,7,12,12b-octahydro-	EtOH	224(4.57),282(3.90), 292(3.86),336(4.14)	4-0329-65
Unknown alkaloid from Melodinus scandeus	EtOH	229(3.82),253(4.07), 285s(3.4)	31-0374-65
$C_{19}H_{20}N_2OS$			
2H-Thieno[3,2-b]pyrrol-3(4H)-one, 2-benzylidene-6-(piperidinomethyl)-	EtOH	227(3.88),265(3.62), 348(4.40)	44-1012-65
$C_{19}H_{20}N_2O_2$			
Imidazole, 1-[(1,2,3,9,10,10a -hexahy-dro-7-methoxy-2α-phenanthryl)-carbonyl]-	EtOH	214(4.47),264(4.31), 297(3.55)	44-1213-65
3-Indolineacetamide, N,N,1-trimethyl-2-oxo-3-phenyl-	EtOH	257(3.9),265s(3.82), 281s(3.34)	44-2973-65
Vincanidine	n.s.g.	242(3.95),291(3.26), 375(4.14)	70-1992-65

Compound	Solvent	$\lambda_{max}(\log \epsilon)$	Ref.
$C_{19}H_{20}N_2O_3$			
Cyclobutene, 1-dimethylamino-4-(α-hydroxybenzyl)-2-nitro-3-phenyl-	n.s.g.	376(4.26)	35-5132-65
Indole, 6-(benzyloxy)-3-(2-nitrobutyl)-	EtOH	222(4.62),264s(3.66), 274s(3.69),292(3.76)	87-0274-64
$C_{19}H_{20}N_2O_4$			
Acrylic acid, 2-(p-nitrophenyl)-3-phenyl-, 2-(dimethylamino)-ethyl ester, cis	EtOH-HCl	318(4.31)	87-0614-64
trans	EtOH-HCl	267(4.31)	87-0614-64
Acrylic acid, 3-(p-nitrophenyl)-2-phenyl-, 2-(dimethylamino)-ethyl ester, cis	EtOH-HCl	318(4.28)	87-0614-64
trans	EtOH-HCl	306(4.22)	87-0614-64
$C_{19}H_{20}N_2O_5$			
Isoquinoline, 3,4-dihydro-5,6,7-trimethoxy-1-(o-nitrobenzyl)-	EtOH	267(4.06),305s(3.70)	12-1997-65
$C_{19}H_{20}N_2O_6$			
2-Indolinone, 3-[3,4,5-trimethoxy-α-(nitromethyl)benzyl]-	EtOH	248s(3.85),279s(3.35)	87-0626-65
$C_{19}H_{20}N_2O_6S$			
Cephalosporanic acid, 7-phenylacetamido-, methyl ester	EtOH	260(3.89)	39-5015-65
$C_{19}H_{20}N_2O_6S_2$			
Cephalosporanic acid, 7-phenylmercaptoacetamido-, methyl ester	n.s.g.	252(4.07)	87-0022-65
$C_{19}H_{20}N_2O_7$			
L-Threonine, carbobenzoxy-, p-nitrobenzyl ester	EtOH	270(3.97)	44-2272-64
$C_{19}H_{20}N_4O_2$			
Diacetamide, N-(2,3,5-trimethyl-6-phenylpyrazolo[1,5-a]pyrimidin-7-yl)-	EtOH	247(4.75),297(3.46), 341(3.19)	94-1207-65
$C_{19}H_{20}N_4O_4$			
6H-Benzocyclononen-6-one, 5,7,8,9,10,11-hexahydro-, 2,4-dinitrophenylhydrazone	CHCl	366(4.32)	78-0231-64
1-Benzosuberone, 3,4-dimethyl-, 2,4-dinitrophenylhydrazone	EtOH	374(4.47)	44-2956-65
2,5-Cyclohexadien-1-one, 2,4-diallyl-4-methyl-, 2,4-dinitrophenylhydrazone	EtOH	391(4.51)	33-0094-65
1-Indanone, 3-isopropyl-3-methyl-, 2,4-dinitrophenylhydrazone	EtOH	384(4.50)	44-2956-65
1(2H)-Naphthalenone, 3,4-dihydro-3,3,4-trimethyl-, 2,4-dinitrophenylhydrazone	EtOH	385(4.36)	44-2956-65
1(2H)-Naphthalenone, 3,4-dihydro-3,4,4-trimethyl-, 2,4-dinitrophenylhydrazone	EtOH	380(4.46)	44-2956-65
2(4aH)-Naphthalenone, 4a-allyl-5,6,7,8-tetrahydro-, 2,4-dinitrophenylhydrazone	EtOH	395(4.52)	33-0094-65

Compound	Solvent	λ_{max}(log ε)	Ref.
$C_{19}H_{20}N_4O_6$			
5H-Benzocyclohepten-9-one, 6,7,8,9-tetrahydro-1,3-dimethoxy-, 2,4-dinitrophenylhydrazone	$CHCl_3$	375(4.41)	5-0021-64G
$C_{19}H_{20}N_4S$			
4-Pyrimidinethiol, 2-amino-5-(3-anilinopropyl)-6-phenyl-	pH 1	265(3.89),350(4.09),	87-0283-65
	pH 7	245(4.29),363(4.12)	87-0283-65
	pH 13	328(4.05)	87-0283-65
$C_{19}H_{20}N_6O_4$			
Homopteroic acid, N"-acetyl-, ethyl ester	pH 13	253(4.4),366(3.8)	44-3404-65
$C_{19}H_{20}N_8O_5$			
Aminopterin	pH 13	261(4.41),282(4.39), 373(3.91)	87-0139-65
$C_{19}H_{20}O$			
1,2-Cyclopentenophenanthrene, 5(or 8)-acetyl-1,2,3,4-tetrahydro-	EtOH	244(4.31),319(3.89)	5-0200-65D
1,2-Cyclopentenophenanthrene, 9-acetyl-1,2,3,4-tetrahydro-	EtOH	228(4.64),312(3.84), 333s(3.78)	5-0200-65D
Phenanthrene, 9,10-diethyl-3-methoxy-	ether	227s(4.32),233(4.42), 252(4.62),254s(4.62), 279(4.21),298(3.95), 309(3.96),328(2.85), 345(3.08),360(3.08)	5-0036-64D
$C_{19}H_{20}O_2$			
Equilenin, methyl ether, cis	EtOH	230(4.71),265(3.72), 275(3.67),320(3.17), 330(3.22)	70-1051-65
dl-Equilenin, 3-methyl ether	EtOH	230(4.53),268(3.53), 278(3.53),289(3.38), 323(3.15),338(3.23)	94-1285-65
Estra-1,3,5(10),8,14-pentaen-17-one, 3-methoxy-	EtOH	310(4.36)	94-1285-65
Gona-1,3,5(10),8,14-pentaen-17-one, 13-ethyl-3-hydroxy-	EtOH	311(4.41)	39-4472-64
Gona-1,3,5(10),8,13-pentaen-17-one, 3-methoxy-15-methyl-	EtOH	260(4.18),282(4.22)	70-0843-65
Ketone, 4-biphenylyl 1-hydroxy-cyclohexyl	EtOH	286(4.40)	87-0504-64
2-Phenanthrol, 9,10-diethyl-6-methoxy-	MeOH	232(4.48),251s(4.55), 258(4.68),276(4.45), 288(4.28),305s(3.98), 315s(3.74),342(2.82), 360(2.82)	5-0036-64D
Phenol, p-(1-ethyl-3,4-dihydro-6-methoxy-2-naphthyl)-	EtOH	229(4.09),255(4.08)	87-0519-64
Spiro[2,5-cyclohexadiene-1,1'-inden]-4-one, 2',3'-diethyl-6'-methoxy-	MeOH	236(4.39),271(4.19), 301s(--),312s(--)	5-0036-64D
$C_{19}H_{20}O_3$			
Dibenzo[a,c]cyclohepta-1,3,5-triene, 1,2,3-trimethoxy-9-methyl-	EtOH	239(4.60),262(4.14)	39-4257-64
Gibberic acid, dehydro-, methyl ester	MeOH	213(4.40),260(4.23), 270(4.16),288(3.51), 298(3.45)	39-0990-65

Compound	Solvent	λ_{max}(log ε)	Ref.
Hexanoic acid, 6-oxo-6-(p-biphenylyl)-, methyl ester	EtOH	285(4.38)	87-0504-64
1,4-Pentadien-3-ol, 1,5-bis-(p-methoxyphenyl)-	MeOH	274(4.57)	5-0064-65F
	dioxan	276(4.56)	5-0064-65F
2,4-Pentadien-1-ol, 1,5-bis-(p-methoxyphenyl)-	MeOH	296(4.57)	5-0064-65F
	dioxan	299(4.56)	5-0064-65F
18,14-Seco-18-nor-D-homoestra-1,3,5(10),9(11)-tetraene-14,17a-dione, 3-hydroxy-	EtOH	230(3.9),265(4.32)	88-0171-64
$C_{19}H_{20}O_4$			
2H-Benzo[d]naphtho[1,2-b]pyran-11-carboxylic acid, 1,3,4,6,6a,7,8,9-octahydro-10,12-dihydroxy-7-methyl-, δ-lactone	EtOH	236(4.52),240(4.51), 277(4.03),288s(3.88), 372(3.90)	33-1459-64
Phenol, 4-cinnamyl-2,3-dimethoxy-, acetate	EtOH	254(4.35),292(3.32)	78-2707-65
Phenol, 2,3-dimethoxy-6-(1-phenylallyl)-, acetate	EtOH	273(3.22)	78-2707-65
3,3'-Spirobichroman, 8,8'-dimethoxy-	PrOH	280(3.73)	88-0457-65
$C_{19}H_{20}O_5$			
Anthrone, 4β-acetoxy-1,4,4aα,9aα-tetrahydro-5,10β-(isopropylidenedioxy)-	EtOH	228(4.05),263(3.78), 317(3.31)	65-2563-64
[2]Benzopyrano[4,3-b][1]benzopyran, 5,6a,7,12a-tetrahydro-2,3,10-trimethoxy-	EtOH	282(3.76)	39-2844-65
Colchiceine, deacetamido-	EtOH	244(4.54),350(4.27)	28-0243-64A
2,4,6-Cycloheptatrien-1-one, 2-hydroxy-4-(3,4,5-trimethoxy)-	EtOH	273(4.35),374(3.76)	78-3605-65
Fluoren-9aα-ylacetic acid, 1,2,3,9a-tetrahydro-2,8-dimethyl-9β-carbomethoxy-3-oxo-	MeOH	231(3.98),236(3.98), 296(4.34)	39-0990-65
2H-Furo[2,3-h][1]benzopyran-2-one, 8-(1-angeloyloxy-1-methylethyl)-8,9-dihydro-	EtOH	218s(4.31),251(3.41), 261(3.46),328(4.20)	1-1379-64
Jatamansin	EtOH	220(4.30),246(3.47), 256(3.38),326(4.13)	78-2605-64
1,2-Phenanthrenedicarboxylic acid, 1,2,9,10-tetrahydro-7-methoxy-2-methyl-, 2-methyl ester	EtOH	245(4.38),301(3.88)	70-2021-64
Salicylic acid, 6-acetonyl-4-(benzyloxy)-, ethyl ester	EtOH	263(4.16),303(3.85)	39-5382-64
Selinidin	MeOH	256(3.52),325(4.17)	88-3367-64
Zozimin	n.s.g.	252(3.60),268(3.61), 327(4.27)	65-3912-64
$C_{19}H_{20}O_6$			
Gibb-3-ene-1α,10β-dicarboxylic acid, 4aα,7-dihydroxy-1β-methyl-8-methylene-2-oxo-, 1→4a-lactone	EtOH	228(3.82)	39-1605-65
$C_{19}H_{20}O_7$			
Acetophenone, 2-(2-carbomethoxyphenoxy)-2',4',6'-trimethoxy-	EtOH	215(4.25),287(4.23), 323s(3.95)	39-2361-65
Dibenzoylmethane, 2-hydroxy-2',3,4,6-tetramethoxy-	MeOH	214(4.27),236s(4.05), 294(4.09),380(4.00)	24-0164-65
Dibenzoylmethane, 2-hydroxy-2',4,5,6-tetramethoxy-	MeOH	214(4.24),236(3.95), 284(3.95),318(3.90), 379(4.10)	24-0164-65

Compound	Solvent	$\lambda_{max}(\log \epsilon)$	Ref.
$C_{19}H_{20}O_9$			
Cervicarcin	EtOH	227(4.17),264(3.90), 323(3.57)	35-4507-64
Coumarin, 7-(2,3-O-carbonyl-α-novio-syloxy)-4-hydroxy-8-methyl-	EtOH	285(4.11),306(4.18)	33-0390-64
$C_{19}H_{21}BO_2$			
3-Oxa-1-oxonia-2-boratacyclohexa-4,6-diene, 2,2-dibenzyl-4,6-dimethyl-	n.s.g.	342(3.26)	5-0040-65I
3-Oxa-1-oxonia-2-boratacyclohexa-4,6-diene, 4,6-dimethyl-2,2-bis(p-tolyl)-	n.s.g.	320(3.68)	5-0040-65I
$C_{19}H_{21}BrO_3$			
Androsta-1,4-diene-3,11,17-trione, 18-bromo-	MeOH	239(4.24)	35-3394-64
$C_{19}H_{21}ClN_2O$			
2-Indolinone, 3-(p-chlorophenyl)-3-(2-dimethylaminoethyl)-1-methyl-	MeOH-HCl	256(3.91),282s(3.27)	87-0626-65
$C_{19}H_{21}ClN_2OS$			
Benzothiazole, 5-chloro-2-(p-diethyl-aminoethoxyphenyl)-, hydrochloride	EtOH	263(3.90),312(4.39), 325(4.50)	39-0954-65
$C_{19}H_{21}ClN_2O_2$			
Benzoxazole, 5-chloro-2-(p-diethyl-aminoethoxyphenyl)-	EtOH	276(4.25),312(4.58)	39-0954-65
$C_{19}H_{21}ClN_2O_5$			
13a-Ethyl-1,2,3,5,6,12,13,13a-octahy-dro-12-oxoindolo[3,2,1-de]pyrido-[3,2,1-ij][1,5]naphthyridin-4-ium perchlorate	EtOH-HCl	219(4.28),231(4.25), 260(3.84),354(4.33)	35-1580-65
$C_{19}H_{21}ClO_5S$			
Spiro[benzofuran-2(3H),1'-[2]cyclohex-ene]-3,4'-dione, 7-chloro-4,6-di-methoxy-6'-methyl-2'-(isopropylthio)-	EtOH	291(4.58)	87-0705-64
Spiro[benzofuran-2(3H),1'-[2]cyclohex-ene]-3,4'-dione, 7-chloro-4,6-di-methoxy-6'-methyl-2'-(propylthio)-	EtOH	290(4.60)	87-0705-64
Spiro[benzofuran-2(3H),1'-[3]cyclohex-ene]-2',3-dione, 7-chloro-4,6-di-methoxy-6'-methyl-4'-(isopropylthio)-	EtOH	296(4.61)	87-0705-64
Spiro[benzofuran-2(3H),1'-[3]cyclohex-ene]-2',3-dione, 7-chloro-4,6-di-methoxy-6'-methyl-4'-(propylthio)-	EtOH	296(4.64)	87-0705-64
$C_{19}H_{21}ClO_7$			
Spiro[benzofuran-2(3H),1'-[2]cyclohex-ene]-3,4'-dione, 7-chloro-2'-(2-meth-oxyethoxy)-4,6-dimethoxy-6'-methyl-	EtOH	236(4.43),253(4.23), 289(4.36),320(2.76)	39-4939-65
$C_{19}H_{21}Cl_2NO_2$			
2-Cyclohexen-1-one, 2-(3,4-dichloro-benzoyl)-5,5-dimethyl-3-(N-pyrrol-idinyl)-	EtOH	298(4.45)	44-0798-64

Compound	Solvent	λ_{max}(log ε)	Ref.
$C_{19}H_{21}Cl_2NO_3$			
2-Cyclohexen-1-one, 2-(3,4-dichloro-benzoyl)-5,5-dimethyl-3-(N-mor-pholinyl)-	EtOH	255(4.11),302(4.42)	44-0798-64
$C_{19}H_{21}FN_2O_8S$			
Uridine, 2'-deoxy-5-fluoro-, 3'-pro-pionate 5'-p-toluenesulfonate	pH 1	265.5(3.96)	44-0554-64
	pH 7	265.5(3.93)	44-0554-64
	pH 13	265.5(3.85)	44-0554-64
$C_{19}H_{21}IN_2S_2$			
3-Methyl-2-[2-[N-methyl-o-(methylthio)-anilino]propenyl]benzothiazolium iodide	EtOH	385(4.77)	22-2879-64
$C_{19}H_{21}N$			
Cyclopenta[ef]heptalene-1-methyl-amine, N,N,3,5-tetramethyl-	benzene	376(4.02),391(4.08), 450(2.48),735(2.11), 780(2.12)	5-0194-64B
isomer	benzene	383(4.03),397(4.06), 456(2.48),730(2.00), 795(2.06),894(1.99), 1030(1.59)	5-0194-64B
$C_{19}H_{21}NO$			
Estra-1,3,5(10)-triene-3-carbo-nitrile, 17-oxo-	n.s.g.	265(3.82)	39-5889-64
Gona-1,3,5(10),6,8-pentaene, 6-acet-amido-	EtOH	233(4.84),294(3.85), 323s(3.39)	5-0200-65D
$C_{19}H_{21}NO_2$			
8-Aza-D-homoestra-1,3,5(10),9(11),14-pentaen-17a-one, 3-methoxy-	pH 1	245(3.80),358(4.30)	39-4900-65
	MeOH	261(4.38),306(3.68)	39-4900-65
Benzo[def]cyclopenta[b]carbazol-6(4H)-one, 5,5a,7,8,8a,8b,9,10-octahydro-2-methoxy-5a-methyl-	MeOH	229(4.53),272(3.74), 298(3.57)	35-2943-64
Cinnamic acid, 2-dimethylamino-ethyl ester, cis	EtOH-HCl	284(4.29)	87-0614-64
trans	EtOH-HCl	284(4.23)	87-0614-64
Estra-1,3,5(10)-triene-15β-carbo-nitrile, 3-hydroxy-17-oxo-	MeOH	222(3.92),280(3.32)	94-0214-64
Gona-1,3,5(10),9(11)-tetraen-12-one, O-acetyloxime	EtOH	217(4.14),225(4.12), 231(4.16),238(4.01), 300(4.41)	5-0200-65D
Nuciferine	EtOH	271(4.23)	33-2119-64
$C_{19}H_{21}NO_3$			
Aporphine, 3-hydroxy-5,6-dimethoxy-	EtOH	215(4.58),265(4.15), 273(4.15),301(3.87)	33-2122-64
Berbin, 3,13-dimethoxy-, hydrochloride hydrate	EtOH	278(3.74),286(3.71)	73-0400-64
Chalcone, 4-(dimethylamino)-2',4'-dimethoxy-	benzene	323(3.97),395(4.46)	65-0094-65
	EtOH	252(4.07),324(3.94), 407(4.48)	65-0094-65
	HOAc	417(4.34)	65-0094-65
	HOAc-$FeCl_3$	585(4.38)	65-0094-65
Chalcone, 4'-(dimethylamino)-2,4-dimethoxy-	benzene	330(4.36)	65-0094-65
	EtOH	246(4.31),344(4.37), 392(4.42)	65-0094-65
	HOAc	348(4.37),402(4.43)	65-0094-65
	HOAc-$FeCl_3$	540(4.53)	65-0094-65

Compound	Solvent	λ max(log ε)	Ref.
Isoquinoline, 7-(benzyloxy)-3,4-dihydro-6,8-dimethoxy-1-methyl-, hydrochloride	pH 2	244(4.19),316(4.22)	49-0025-65
	pH 12	238(4.42),273(4.05)	49-0025-65
	EtOH	241(4.16),316(4.17)	49-0025-65
Pronuciferine	EtOH-HCl	229(4.40),278(3.42)	33-2122-64
	EtOH	230(4.46),282(3.47)	33-2119-64
$C_{19}H_{21}NO_4$			
Canadine	MeOH	286(3.73)	95-0955-64
Coreximine	EtOH	289(3.95)	95-0077-65
Salutaridine	EtOH	242(4.25),277(3.76)	39-2423-65
$C_{19}H_{21}NO_5$			
1H-[1]Benzoxepino[2,3,4-ij]isoquinolin-2-ol, 2,3,12,12a-tetrahydro-6,9,10-trimethoxy-	EtOH-HCl	284(3.85)	95-0667-65
	EtOH	284(3.72)	95-0667-65
$C_{19}H_{21}NO_6$			
Crinamine, 6-hydroxy-, 11-acetate	EtOH	240(3.55),290(3.65)	35-4912-65
Pyrrolo[2,1-a]isoquinoline-2-acetic acid, 3-carboxy-5,6-dihydro-8,9-dimethoxy-, ethyl ester	EtOH	321(4.45),331(4.42)	94-0775-65
$C_{19}H_{21}NO_8$			
4H-Quinolizine-1,2,3,4-tetracarboxylic acid, 6-ethyl-, tetramethyl ester	MeOH	266(3.95),307(4.02), 350(4.06),445(4.06)	39-0948-65
	MeOH-HClO4	322(4.09),329s(4.05)	39-0948-65
9aH-Quinolizine-1,2,3,4-tetracarboxylic acid, 6,9a-dimethyl-, tetramethyl ester	MeOH	294(4.08),330(3.46), 445(3.64)	39-0948-65
9aH-Quinolizine-1,2,3,4-tetracarboxylic acid, 6-ethyl-, tetramethyl ester	MeOH	236(4.19),294(3.96), 430(3.74)	39-0948-65
9aH-Quinolizine-1,2,3,4-tetracarboxylic acid, 9a-ethyl-, tetramethyl ester	MeOH	281(4.06),334(3.65), 418(3.81)	39-0948-65
$C_{19}H_{21}NO_9$			
Coumarin, 3-amino-7-[(5,5-di-C-methyl-4-O-methyl-α-L-lyxopyranosyl)oxy]-4-hydroxy-8-methyl-, cyclic carbonate	EtOH	232(4.16),283(4.11), 298(4.16)	33-0390-64
$C_{19}H_{21}NS$			
6H-Dibenzo[b,e]thiocine, 12-(3-methylaminopropylidene)-7,12-dihydro-	MeOH-HCl	230(4.23),263s(3.77)	24-0685-65
isomer	MeOH-HCl	227(4.26),268(3.78)	24-0685-65
$C_{19}H_{21}N_3O_2$			
3-Indolinepropionic acid, 3-benzyl-1-methyl-2-oxo-, hydrazide	MeOH	255(3.85),284s(3.15)	44-2973-65
$C_{19}H_{21}N_3O_7$			
Uridine, 2',3'-O-isopropylidene-, 5'-p-aminobenzoate	MeOH	205(4.03),272s(4.22), 295(4.29)	22-2489-65
$C_{19}H_{21}N_5$			
Pyrimidine, 2,4-diamino-5-(3-anilinopropyl)-6-phenyl-	pH 1	286(3.92)	87-0283-65
	pH 7	296(3.92)	87-0283-65
	pH 13	298(3.89)	87-0283-65
$C_{19}H_{21}N_5O_6$			
Adenosine, 2',3'-O-(2,4-dimethoxybenzylidene)-	MeOH	233(4.10),260(4.24)	5-0156-64I

Compound	Solvent	$\lambda_{max}(\log \epsilon)$	Ref.
$C_{19}H_{21}N_5O_6S$			
Glycine, N-[6-(benzylthio)-9-β-D-ribofuranosyl-9H-purin-2-yl]-	pH 1	252(4.16),322(4.08)	35-3752-65
	pH 11	253(4.27),320(4.05)	35-3752-65
$C_{19}H_{21}N_5O_7$			
Glycine, N-[6-(benzyloxy)-9-β-D-ribofuranosyl-9H-purin-2-yl]-	pH 1	246(4.04),295(3.96)	35-3752-65
	pH 11	250(4.10),290(3.96)	35-3752-65
Gramine, 2,7-dimethyl-, picrate	EtOH	219(4.77),264s(4.06), 276s(4.00),287(3.89), 358(4.20),410(3.99)	87-0274-64
$C_{19}H_{22}$			
Bicyclo[4.2.0]octa-1,3,5-triene, 8-mesityl-2,4-dimethyl-	C_6H_{12}	271(3.83),276(3.78), 281(3.84)	35-2149-64
Biphenyl, 2-(cyclohexylmethyl)-	hexane	235(3.8)	70-0124-65
Biphenyl, 3-(cyclohexylmethyl)-	hexane	235(3.62)	70-0124-65
Biphenyl, 4-(cyclohexylmethyl)-	hexane	252(4.18)	70-0124-65
1-Heptene, 1,1-diphenyl-	C_6H_{12}	250(4.07)	22-0693-64
$C_{19}H_{22}BrNO_3$			
8-(Benzyloxy)-3,4-dihydro-6,7-dimethoxy-2-methylisoquinolinium bromide	pH 1	247(4.19),326(4.16)	33-2089-64
	iso-PrOH	248(4.17),333(4.19), 265s(3.89)	33-2089-64
$C_{19}H_{22}ClN$			
(2,4-Diphenyl-2-cyclobuten-1-yl)-trimethylammonium chloride	n.s.g.	243(4.23)	35-0453-64
$C_{19}H_{22}ClNO_3$			
Doryafranine	EtOH	225(4.20)	100-0237-65
$C_{19}H_{22}ClN_3O$			
Morpholine, 4-[2-[(2-amino-5-chloro-α-phenylbenzylidene)amino]ethyl]-	EtOH	233(4.38),362(3.68)	44-2368-64
isomer	EtOH	248(4.40)	44-2368-64
$C_{19}H_{22}Cl_2N_4O_2$			
Alanine, 3-[p-[[p-[bis(2-chloroethyl)-amino]phenyl]azo]phenyl-	EtOH	250(4.1),300s(3.8), 400(4.4)	94-1375-64
$C_{19}H_{22}N_2$			
Methane, diazodimesityl-	C_6H_{12}	236(4.05),278(4.26), 500(2.32)	35-2149-64
Spiro[indoline-3,3'-pyrrolidine], 1'-ethyl-2'-phenyl-	EtOH	242(3.81),290(3.38)	78-1327-65
Yohimban, 3-dehydro-	EtOH	230(4.39),258s(3.93), 295s(4.24),306(4.35), 318(4.30)	44-0105-65
	MeCN	227(4.51),274(3.86), 280(3.87),289(3.81), 320(3.43)	44-0105-65
	CH_2Cl_2	234(4.3),305(4.18), 312s(4.17)	44-0105-65
$C_{19}H_{22}N_2O$			
Chalcone, 4,4'-bis(dimethylamino)-	benzene	400(4.65)	65-0094-65
	EtOH	256(4.16),422(4.59)	65-0094-65
	HOAc	262(4.08),426(4.47)	65-0094-65
	HOAc-$FeCl_3$	597(4.82)	65-0094-65

Compound	Solvent	$\lambda_{max}(\log \epsilon)$	Ref.
Cinchonine (as sulfate)	pH 1.2	316(3.88)	2-0251-65
	pH 1.8	316(3.88)	2-0251-65
	pH 3.2	316(3.84)	2-0251-65
	pH 3.4	315.5(3.83)	2-0251-65
	pH 3.6	308(3.76),316(3.80)	2-0251-65
	pH 3.7	306(3.73),316(3.76)	2-0251-65
	pH 4.0	304(3.82),316(3.73)	2-0251-65
	pH 4.4	289(3.68),302(3.61), 315(3.53)	2-0251-65
	pH 6.1	289(3.67),302(3.60), 315(3.53)	2-0251-65
Eburnamonine	MeOH	242(4.28),268(4.00), 294(3.68),301(3.68)	73-0433-64
Epieburnamonine	EtOH	242(4.29),266(3.98), 295(3.59),302(3.60)	35-1580-65
16-Epipleiocarpaminol	EtOH	233(4.43),288(3.85)	33-0878-64
Indole, 3-(2-aminobutyl)-6-(benzyloxy)-	EtOH-HCl	223(4.61),258s(3.59), 264s(3.64),273(3.67), 292(3.74),302s(3.57)	87-0274-64
2-Indolinone, 3-(1-benzyl-1-methyl-2-aminoethyl)-	EtOH	238(3.88),268s(2.93), 293(3.41)	87-0626-65
2-Indolinone, 3-(2-dimethylamino-ethyl)-3-methyl-1-phenyl-	MeOH-HCl	244(4.08),292s(2.79)	87-0626-65
Morpholine, 4-[2-(diphenylmethylene)-amino]ethyl]-	EtOH	247(4.15)	44-2368-64
Normacusine B	EtOH	227(4.49),281(3.98), 290(3.88)	25-0319-64
	EtOH	226(4.58),275(3.86), 282(3.88),290(3.79)	39-4419-64
4H-1,3,4-Oxadiazine, 5,6-dihydro-4,5-dimethyl-6-phenyl-2-(3,5-xylyl)-	$CHCl_3$	242(3.78),294(4.00)	44-0668-64
Pleiocarpaminol	EtOH	234(4.53),287(3.92)	33-0878-64
Tombozine	EtOH	230(4.58),279(3.88)	100-0220-64
Uleine, N-acetyl-N-demethyl-	EtOH	237s(4.10),308(4.29), 313(4.29)	78-1717-65
Yohimban-17β-ol, 3-dehydro-	EtOH	230(4.38),260s(3.90), 296s(4.23),307(4.33), 318(4.29)	44-0105-65
$C_{19}H_{22}N_2OS$			
4H-1,3,4-Thiadiazine, 2-(p-ethoxy-phenyl)-5,6-dihydro-4,5-dimethyl-6-phenyl-	$CHCl_3$	254(4.19),309(3.84)	44-2228-65
$C_{19}H_{22}N_2O_2$			
Alkaloid RP-5	EtOH	225(4.57),278(3.86)	100-0220-64
Methane, diazobis(2,6-dimethyl-4-methoxyphenyl)-	C_6H_{12}	274(4.34),503(2.31)	35-2149-64
4H-1,3,4-Oxadiazine, 2-(o-ethoxyphenyl)-5,6-dihydro-4,5-dimethyl-6-phenyl-	$CHCl_3$	244(3.75),282(3.77)	44-0668-64
4H-1,3,4-Oxadiazine, 2-(p-ethoxyphenyl)-5,6-dihydro-4,5-dimethyl-6-phenyl-	$CHCl_3$	242(3.96),288(4.13)	44-0668-64
$C_{19}H_{22}N_2O_3$			
4H-1,3,4-Oxadiazine, 2-(3,4-dimethoxy-phenyl)-5,6-dihydro-4,5-dimethyl-6-phenyl-	$CHCl_3$	244(3.90),300(4.12)	44-0668-64
$C_{19}H_{22}N_2O_4$			
2-Cyclohexen-1-one, 2-(p-nitrobenzoyl)-5,5-dimethyl-3-(N-pyrrolidinyl)-	EtOH	300(4.23)	44-0798-64

Compound	Solvent	$\lambda_{max}(\log \epsilon)$	Ref.
1,2-Propanediamine, N,N'-bis(3-methoxysalicylidene)-	MeOH	263(4.42),330(3.63), 425(3.45)	65-3049-64
Tryptamine, 6-(benzyloxy)-5-methoxy-, formate	EtOH	222(4.47),279s(3.72), 297(3.90)	4-0387-65
$C_{19}H_{22}N_2O_5$			
Acetic acid, benzoyl[(4,4-dimethyl-2,6-dioxocyclohexyl)azo]-, ethyl ester	CH_2Cl_2	287(4.03),392(4.18)	89-0920-64
D-Fructosone-1-benzylphenylhydrazone	EtOH	236(4.04),341(4.22)	24-0725-64
$C_{19}H_{22}N_2O_5S_4$			
Cephalosporanic acid, 7-(2-thiopheneacetamido)-, butyl xanthate derivative, sodium salt	n.s.g.	226(4.23),285(4.24)	87-0174-65
$C_{19}H_{22}N_2S$			
4H-1,3,4-Thiadiazine, 5,6-dihydro-4,5-dimethyl-2-phenethyl-6-phenyl-	$CHCl_3$	262(3.48)	44-2228-65
$C_{19}H_{22}N_4$			
Aniline, N,N-dibutyl-4-tricyanovinyl-	$CHCl_3$	530(4.75)	39-1334-64
$C_{19}H_{22}N_4O_4$			
Bicarbamic acid, [α-(phenylazo)-benzyl]-, diethyl ester	MeOH	403(2.27)	89-0684-64
$C_{19}H_{22}N_4O_5$			
Hydrazine, 1,2-dicarbethoxy-2-(α-phenylazo)-p-hydroxybenzyl-	MeOH	400(2.27)	89-0684-64
$C_{19}H_{22}N_4O_{10}$			
Hypoxanthine, 9-β-D-glucopyranosyl-, 2',3',4',6'-tetraacetate	EtOH	244(4.12)	88-3095-65
$C_{19}H_{22}N_6O_4$			
Benzoic acid, p-[N-[2-(2-amino-7,8-dihydro-4-hydroxy-6-pteridinyl)ethyl]-acetamido]-, ethyl ester	pH 13	234(4.30),277(3.88), 330(3.72)	44-3404-65
$C_{19}H_{22}O$			
5,7,9,14-Anthrastatetraen-17β-ol	EtOH	221(4.38),227(4.41), 233s(4.22),266(4.23), 298(3.4),309(3.31)	44-4384-65
Bicyclo[4.2.0]octa-1,3,5-triene, 7-tert-butoxy-3-methyl-7-phenyl-	isooctane	255s(2.90),262(3.17), 268(3.32),274(3.29)	35-0086-65
$C_{19}H_{22}O_2$			
Azulene, 4,6,8-trimethyl-1,3-dipropionyl-	hexane	517(2.91)	65-0894-64
	EtOH	501(3.03)	65-0894-64
Bicyclo[4.2.0]octa-1,3,5-triene, 4-methoxy-8-(4-methoxy-2,6-xylyl)-2-methyl-	C_6H_{12}	237(4.22),279(3.45), 283(3.55),287(3.58)	35-2149-64
Chalcol, 4-methoxy-, isopropyl ether	n.s.g.	258(4.30)	39-4978-64
Estra-1,3,5(10),6,8-pentaen-17β-ol, 3-methoxy-	EtOH	233(4.30),255(3.83), 265(3.85),279(3.6), 291(3.6),306(3.5)	94-1285-65
Estra-1,3,5(10),8,14-pentaen-17β-ol, 3-methoxy-	EtOH	310(4.35)	94-1285-65
Estra-2,5,7,9-tetraen-17-one, 3-methoxy-	dioxan	269(2.48),276(2.36)	87-0536-65
Estrone, 9-dehydro-, methyl ether	EtOH	263(4.3),296(3.5)	13-0031-64

Compound	Solvent	λ_{max}(log ε)	Ref.
Gona-1,3,5(10),8,13-pentaen-17-ol, 3-methoxy-15-methyl-	EtOH	280(4.22)	70-0843-65
Gona-1,3,5(10),8-tetraen-17-one, 13β-ethyl-3-hydroxy-	EtOH-NaOH	279(4.19)	39-4472-64
Gona-1,3,5(10),9(11)-tetraen-17-one, 13β-ethyl-3-hydroxy-	EtOH	266(4.18)	39-4472-64
D-Homoestra-1,3,5(10),8,14-pentaene-3,17aβ-diol	EtOH	306(4.44)	70-1814-64
D-Homoestra-1,3,5(10),8-tetraen-17a-one, 3-hydroxy-	EtOH	225(3.96),277(4.05)	70-1814-64
A-Homo-8α-estra-2,4a,10-triene-4,17-dione	n.s.g.	235(4.40),312(4.02)	39-5137-65
Octadeca-3,5,7,9,11,13,15-heptaene-2,17-dione, 5-methyl-	benzene	409(4.88),431(5.08), 458(5.08)	39-0842-64
Spiro[2-cyclohexene-1,1'-inden]-4-one, 2',3'-diethyl-6'-methoxy-	MeOH	268(4.18),321(3.48)	5-0036-64D
$C_{19}H_{22}O_3$			
Androsta-1,4-diene-3,17-dione, 6,19-epoxy-	EtOH	244(4.14)	39-3621-64
Androsta-1,4-diene-3,17-dione, 11β,18-epoxy-	MeOH	242(4.18)	35-3394-64
1-Butanol, 4-(4-methoxy-o-tolyl)-, benzoate	EtOH	228(4.36),277(3.44)	35-5148-65
Estra-1,3,5(10),9-tetraen-17-one, 14-hydroxy-3-methoxy-	EtOH	267(4.26)	94-1285-65
Estra-1,3,5(10)-triene-3-carboxylic acid, 17-oxo-	n.s.g.	242(4.02),270(3.82)	39-5889-64
D-Homo-8β-estra-1,3,5(10),9(11)-tetraen-17a-one, 3,14α-dihydroxy-	EtOH	262(4.13),299(3.47)	70-1809-65
8,14-Secoestra-1,3,5(10),9-tetraene-14, 17-dione, 3-methoxy-	EtOH	266(4.11)	94-1285-65
8,14-Seco-D-homoestra-1,3,5(10),9(11)-tetraene-14,17a-dione, 3-hydroxy-	EtOH	267(4.21)	70-1809-65
$C_{19}H_{22}O_4$			
Androst-4-en-19-oic acid, 6β-hydroxy-3,17-dioxo-, γ-lactone	MeOH	235(4.08)	35-1528-64
2H-Benzo[d]naphtho[1,2-b]pyran-11-carboxylic acid, 1,3,4,6,6a,7,8,9,10,10a-decahydro-10,12-dihydroxy-7-methyl-, δ-lactone	EtOH	235(4.10),267(3.93), 352(3.75)	33-1459-64
Podocarpic acid, O-methyl-7-oxo-5,6-dehydro-, methyl ester	EtOH	202(4.38),242(4.18), 302(4.05)	39-0361-65
Propane, 1-(2-acetoxy-3,4-dimethoxyphenyl)-1-phenyl-	EtOH	272(3.18)	78-2707-65
Propane, 3-(4-acetoxy-2,3-dimethoxyphenyl)-1-phenyl-	EtOH	261(2.92),267(2.89)	78-2707-65
$C_{19}H_{22}O_5$			
2-Anthroic acid, 3α-acetonyl-1,2,3,4,-4aβ,9,9aα,10-octahydro-2β-hydroxy-9-oxo-, methyl ester	EtOH	248(4.11),290(3.16)	70-1039-65
Anthrone, 4β-acetoxy-1,2,3,4,4aα,9aα-hexahydro-5,10β-(isopropylidenedioxy)-	EtOH	228(4.16),263(3.91), 310(3.44)	65-2563-64
Benzo[a]heptalen-10(5H)-one, 6,7,11,12-tetrahydro-9-hydroxy-1,2,3-trimethoxy-	EtOH	220s(4.26),258(3.89), 370(4.02)	44-1752-65
Benzophenone, 4,4',5,5'-tetramethoxy-2,2'-dimethyl-	EtOH	204(4.55),235(4.37), 284(3.98),322(4.01)	39-1685-65

Compound	Solvent	$\lambda_{max}(\log \epsilon)$	Ref.
p-Benzoquinone, 3-tert-butyl-2-(3-tert-butyl-2,5-dioxocyclopent-3-en-1-yl)-5-hydroxy-	MeOH	232(4.14),270(4.12), 394(3.00)	24-2774-65
	MeOH-KOH	224(4.32),270s(3.96), 492(3.26)	24-2774-65
Colchiceine, 7-deacetamido-11,12-dihydro-	EtOH	370(4.02)	28-0243-64A
2,4,6-Cycloheptatrien-1-one, 2-hydroxy-4-[3-(3,4,5-trimethoxyphenyl)propyl]-	EtOH	250(4.4),334(3.7)	78-3605-65
9β-Estra-1,4-diene-3,11,17-trione, 9,10β-epoxy-, methanol adduct	MeOH	226(3.83),283(3.39)	94-0473-64
Jatamansin, dihydro-	EtOH	212(4.24),224(4.16), 246(3.49),256(3.47), 326(4.13)	78-2605-64
17α-Oxa-D-homoandrost-4-en-19-oic acid, 6β-hydroxy-3,17-dioxo-, 6β,19-lactone	MeOH	233(4.21)	35-1528-64
$C_{19}H_{22}O_6$			
$\Delta^{1,\alpha}$-Indansuccinic acid, 2-(1-carboxyethyl)-, trimethyl ester	EtOH	272(4.10)	42-0479-64
Lactic acid, 3-(3,4-dimethoxyphenyl)-2-(o-methoxyphenyl)-, methyl ester	EtOH	278(3.62)	22-3572-65
Propionic acid, 2,2-bis(3,4-dimethoxyphenyl)-	EtOH	280.5(3.750)	39-3040-65
$C_{19}H_{22}O_7$			
Peucedanin irradiation product	n.s.g.	285(3.83),330(3.98)	65-2848-64
$C_{19}H_{23}BrO$			
Androsta-1,3,5-trien-17-one, 3-bromo-	MeOH	307(3.77),318(3.79)	44-2495-64
Estra-1,3,5(10)-trien-17-one, 1-bromo-4-methyl-	MeOH	271(2.44)	44-1893-64
Estra-1,3,5(10)-trien-17-one, 16α-bromo-4-methyl-	EtOH	263(2.51),270(2.38), 316(2.08)	94-0687-65
$C_{19}H_{23}BrO_2$			
Estrone, 4-bromo-, methyl ether	MeOH	222(4.01),278(3.31), 288(3.31)	44-2234-65
$C_{19}H_{23}BrO_4$			
Podocarpic acid, 6α-bromo-7-oxo-O-methyl-, methyl ester	MeOH	235(3.93),295(4.07)	44-0501-65
$C_{19}H_{23}ClN_2$			
Yohimban, 3-dehydro-, chloride	EtOH	246(4.00),352(4.34)	44-0105-65
$C_{19}H_{23}ClO$			
Androsta-1,3,5-trien-17-one, 3-chloro-	MeOH	306(3.75),316(3.76)	44-2495-64
Estra-1,3,5(10)-trien-17-one, 1-chloro-4-methyl-	MeOH	271(2.45)	44-2495-64
$C_{19}H_{23}ClO_4S_3$			
Spiro[benzofuran-2(3H),1'-[2]cyclohexen]-3-one, 7-chloro-4,6-dimethoxy-6'-methyl-2',4',4'-tris(methylthio)-	EtOH	240(4.24),288(4.35), 322s(3.72)	87-0705-64
$C_{19}H_{23}ClO_5$			
Anthracene-3β,9β,10β-triol, 2α-chloro-1,2,3,4,4aα,9,9aα,10-octahydro-9α-methyl-, 3,10-diacetate	EtOH	262(2.36)	70-1024-64

Compound	Solvent	λ_{max}(log ϵ)	Ref.
$C_{19}H_{23}N$			
Hexamethylenimine, 1-methyl-2,2-diphenyl-	EtOH	250(3.45)	39-1653-65
$C_{19}H_{23}NO$			
Estra-3,5(10)-diene-3-carbonitrile, 17-oxo-	n.s.g.	305(3.96)	39-5889-64
Pyridine, 4-(2-furyl)-1,2,3,6-tetrahydro-3,5-dimethyl-N-phenethyl-	EtOH-HCl	257(4.15)	87-0726-64
Stilbene, 4-(dimethylaminopropoxy)-	EtOH	304(4.5),319(4.5)	87-0511-64
$C_{19}H_{23}NO_2$			
Benzo[def]cyclopenta[b]carbazol-6(4H)-one, 4a,5,5a,7,8,8a,8b,9,10,10c-decahydro-2-methoxy-5a-methyl-	MeOH	212(4.45),235s(3.76), 293(3.55)	35-2943-64
2-Cyclohexen-1-one, 2-benzoyl-5,5-dimethyl-3-N-pyrrolidinyl-	EtOH	299(4.36)	44-0798-64
Estra-1,3,5(10)-triene-2-carbonitrile, 3,17β-dihydroxy-	EtOH	236(4.04),305(3.57)	88-4603-65
Estra-1,3,5(10)-triene-15β-carbonitrile, 3,17β-dihydroxy-	MeOH	222(3.88),280(3.34)	44-0214-64
$C_{19}H_{23}NO_3$			
13H-Dibenzo[a,f]quinolizin-13-one, 1,2,3,4,6,7,11b,12-octahydro-2,8-dimethoxy-	C_6H_{12}	312(4.08)	44-3667-65
	EtOH	335(4.14)	44-3667-65
perchlorate	EtOH	335(4.15)	44-3667-65
13H-Dibenzo[a,f]quinolizin-13-one, 1,2,3,4,6,7,11b,12-octahydro-9,10-dimethoxy-	C H	313(4.04)	44-3667-65
	EtOH	334(4.11)	44-3667-65
perchlorate	EtOH	335(4.12)	44-3667-65
Isoquinoline, 1,2,3,4-tetrahydro-1-(m-methoxybenzyl)-6,7-dimethoxy-, hydrochloride hydrate	EtOH	281(3.77)	73-0400-64
Pronuciferine, dihydro-	EtOH	287(3.37)	33-2122-64
$C_{19}H_{23}NO_4$			
Benzo[f]quinoline-3-hydracrylic acid, 1,2,3,4-tetrahydro-8-methoxy-, ethyl ester	MeOH	221(4.60),244(4.83), 280(4.27),370(3.15)	39-5907-64
Picolinic acid, 6-[2-(hexahydro-7a-methyl-1,5-dioxo-4-indanyl)ethyl]-5-methyl-	EtOH	273(3.67)	35-1386-65
Reticuline (as perchlorate)	EtOH	285(3.83)	95-0362-64
Zephyranthine, O,O'-isopropylidene-	EtOH	235(3.52),290(3.64)	95-0200-65
$C_{19}H_{23}NO_5$			
Radiatine, O-ethyl-	EtOH	239(3.55),290(3.55)	95-0615-65
$C_{19}H_{23}NO_6$			
Crinamine, dihydro-6-hydroxy-, 11-acetate	EtOH	237(3.64),288(3.72)	35-4912-65
$C_{19}H_{23}N_3O_2$			
5H-Dibenzo[b,e][1,4]diazepin-11-one, 5-methyl-8-methoxy-10-(2-dimethylaminoethyl)-10,11-dihydro-	EtOH-HCl	209(4.54),225(4.47), 302(3.58)	33-1590-65
Estra-1,3,5(10)-trien-17-one, 1-azido-3-methoxy-	MeOH	218(4.36),255(3.81), 293(3.56),303s(3.48)	35-2943-64

Compound	Solvent	λ_{max}(log ε)	Ref.
$C_{19}H_{23}N_3O_5$			
L-Sorbosone-1-(benzylphenylhydrazone) 2-oxime	EtOH	234(4.30),329(4.37)	24-0725-64
$C_{19}H_{23}N_5O_4S$			
9H-Purine, 6-(benzylthio)-2-(dimethyl-amino)-9-β-D-ribofuranosyl-	pH 1	222(4.12),263(4.12), 340(3.75)	35-3752-65
	pH 11	228(4.04),260(4.20), 330(3.78)	35-3752-65
	MeOH	258(4.20),327(3.80)	35-3752-65
$C_{19}H_{23}N_5O_5$			
Purine, 6-(benzyloxy)-2-(dimethyl-amino)-9-β-D-ribofuranosyl-	pH 1	257(4.16),312(3.91)	35-3752-65
	pH 11	258(4.17),298(3.92)	35-3752-65
$C_{19}H_{24}$			
Gona-1,3,5(10),13(17)-tetraene, 4,17-dimethyl-	EtOH	263(2.51)	78-1611-65
Heptane, 1,1-diphenyl-	C_6H_{12}	255(2.76),260(2.80), 262(2.80),269(2.67)	22-0587-65
$C_{19}H_{24}Br_2O$			
Androsta-3,5-dien-17-one, 1,3-dibromo-	MeOH	237(4.30),245(4.30), 253(4.12)	44-2495-64
$C_{19}H_{24}ClNO_3$			
Lotusine chloride	EtOH	255(3.79)	95-0472-65
$C_{19}H_{24}ClNO_8$			
Reticuline perchlorate	EtOH	225s(4.27),285(3.95)	39-2244-64
	EtOH-KOH	246(4.22),300(4.07)	39-2244-64
$C_{19}H_{24}Cl_2N_2O_4$			
Pyrrole-2-carboxylic acid, 5-[2-carb-oxy-2-(dichloromethyl)-3,4-dimethyl-2H-pyrrol-5-yl]-3,4-dimethyl-, diethyl ester	EtOH-HCl	248(4.07),254(4.06), 310(3.70),385(4.32)	39-1460-65
	$CHCl_3$	258(4.27),320(4.17)	39-1460-65
$C_{19}H_{24}Cl_2O$			
Androsta-3,5-dien-17-one, 1,3-di-chloro-	MeOH	235(4.32),242(4.38), 251(4.21)	44-2495-64
$C_{19}H_{24}N_2$			
Apparicine, N^b-methyldihydro-	EtOH	227(3.60),295(3.15)	39-4773-65
Aspidofractinine	EtOH	241(3.83),291(3.46)	33-1147-64
Pyridine, 2-[p-(dipropylamino)-styryl]-, cis	aq. HCl	309(4.04)	23-1345-65
	50% MeOH	335(4.10),356(4.11)	23-1345-65
	50% MeOH-HCl	309(4.04)	23-1345-65
trans	50% MeOH	380(4.54)	23-1345-65
	50% MeOH-HCl	326(4.51)	23-1345-65
Tuboxenine	EtOH	206(4.39),244(3.81), 295(3.44)	33-0358-64
$C_{19}H_{24}N_2O$			
Fendleridine	EtOH	243(4.86),293(3.48)	25-0033-64
Normacusine B, dihydro-	EtOH	226(--),275(--), 282(--),290(--)	39-4419-64
Normavacurine, 19,20-dihydro-	EtOH	233(4.50),289(3.92)	33-0878-64

Compound	Solvent	λ_{max}(log ϵ)	Ref.
Pleiocarpaminol, 2,7-dihydro-	EtOH-HCl	253(4.09),294(3.45)	33-0878-64
	EtOH	258(4.06),298(3.46)	33-0878-64
Tubifolidin, N-formyl-	EtOH	251(4.08),291(3.66)	33-1497-64
$C_{19}H_{24}N_2O_2$			
Androsta-4,6-dieno[2,3-c]furazan, 17β-hydroxy-	EtOH	286(4.37)	94-1445-65
Androst-4-eno[2,3-c]furazan, 17-oxo-	EtOH	254(4.03)	94-1445-65
Strychnan, 19β,20α-dihydroxy-16-methylene-	EtOH	246(3.75),300(3.56)	78-1141-65
$C_{19}H_{24}N_2O_3$			
Podocarpic acid, O-methyl-7-oxo-5,6-dehydro-, methyl ester, hydrazone	EtOH	244(4.16),319(3.90)	39-0361-65
$C_{19}H_{24}N_2O_4$			
Dipyrromethene-3,3'-dicarboxylic acid, 4,4',5,5'-tetramethyl-, diethyl ester	EtOH-HCl	227(3.93),362(3.78), 526(4.92)	12-0363-65
	EtOH	209(4.34),244s(3.98), 479(4.41)	12-0363-65
	CCl_4	485(4.50)	12-0363-65
Dipyrromethene-3',4-dicarboxylic acid, 3,4',5,5'-tetramethyl-, diethyl ester	EtOH-HCl	210(4.49),243(3.99), 480(4.50)	12-0363-65
	EtOH	222(4.37),256(3.91), 453(4.45)	12-0363-65
	CCl_4	459(4.57)	12-0363-65
Dipyrromethene-4,4'-dicarboxylic acid, 3,3',5,5'-tetramethyl-, diethyl ester	EtOH-HCl	212(4.43),247(3.89), 344(3.54),465(5.06)	12-0363-65
	EtOH	220(4.38),261(3.89), 452(4.55)	12-0363-65
	CCl_4	451(4.63)	12-0363-65
Dipyrromethene-5,5'-dicarboxylic acid, 3,3'-diethyl-4,4'-dimethyl-, dimethyl ester	CCl_4	462(4.30)	12-0363-65
Dipyrromethene-5,5'-dicarboxylic acid, 3,3',4,4'-tetramethyl-, diethyl ester	CCl_4	458(4.40)	12-0363-65
$C_{19}H_{24}N_2O_5$			
2,2'-Dipyrryl ketone, 5,5'-dicarbethoxy-3,3',4,4'-tetramethyl-	EtOH	248(4.14),309(3.99), 343(4.20)	12-1977-65
$C_{19}H_{24}N_2O_{11}$			
Glucopyranuronic acid, 1-deoxy-1-(4-ethoxy-2-oxo-1(2H)-pyrimidinyl)-, methyl ester, 2,3,4-triacetate	EtOH	277(3.74)	94-1259-64
Thymine, 1-β-D-glucopyranosyl-, 2',3',4',6'-tetraacetate	EtOH	261(4.02)	94-0357-64
$C_{19}H_{24}N_4O_4$			
3-Buten-2-one, 4-(3,3-dimethyl-2-norbornyl)-, 2,4-dinitrophenylhydrazone	$CHCl_3$	387(4.46)	28-4783-65A
β-Ionone, 2,4-dinitrophenylhydrazone	$C_2H_4Cl_2$	357(4.46)	39-1761-65
2-Norbornaneacrolein, 1,3,3-trimethyl-, 2,4-dinitrophenylhydrazone	$CHCl_3$	380(4.53)	28-4712-64B
$C_{19}H_{24}N_4O_6$			
5-Pyrimidinepropionaldehyde, 2-acetamido-4-hydroxy-6-(p-nitrophenyl)-, diethyl acetal	pH 1 or 7	250(4.24),276(4.27)	87-0283-65
	pH 13	275(4.27)	87-0283-65

Compound	Solvent	λ_{max}(log ε)	Ref.
$C_{19}H_{24}N_6$			
Guanidine, 1-(N,N'-diphenylamidino)-3-diethylmethyleneamino-	EtOH	274s(4.42)	39-0932-65
Guanidine, 1-(N,N'-di-p-tolylamidino)-3-isopropylideneamino-	EtOH	275s(4.43)	39-0932-65
$C_{19}H_{24}N_8O_9$			
Glucitol, 1-(6-amino-9H-purin-9-yl)-2-deoxy-2-(2,4-dinitroanilino)-1-O-ethyl-	EtOH	260(4.47),351(4.35)	44-1096-65
$C_{19}H_{24}O$			
Estra-1,3,5(10)-trien-17-one, 2-methyl-	MeOH	269(2.87),278(2.92)	87-0409-65
Estra-1,3,5(10)-trien-17-one, 3-methyl-	MeOH	269(2.88),278(2.92)	87-0409-65
Estra-1,3,5(10)-trien-17-one, 4-methyl-	EtOH	263(2.46),270(2.34), 290(1.61)	24-0140-64
	EtOH	264(2.48),270(2.43)	94-0687-65
Gona-1,3,5(10),13-tetraene, 3-methoxy-17β-methyl-	EtOH	280(3.30),287(3.20)	44-2187-64
$C_{19}H_{24}O_2$			
Androsta-3,5-diene-11,17-dione	n.s.g.	227(4.26),237(4.34), 243(4.11)	13-0391-64
Androsta-4,14-diene-3,17-dione	MeOH	239(4.23)	44-2776-65
	MeOH	239(4.23)	88-2955-64
8α,10α-Androsta-4,6-diene-3,17-dione	MeOH	287(4.40)	54-0841-65
9β-Androsta-4,6-diene-3,17-dione	MeOH	282(4.41)	54-0889-65
8α,10α-Androsta-1,4,6-trien-3-one, 17β-hydroxy-	MeOH	225(4.05),256(3.98), 303(4.06)	54-0841-65
5β,19-Cycloandrost-1-ene-3,17-dione	EtOH	272(3.80)	35-3727-65
5β,19-Cycloandrost-6-ene-3,17-dione	EtOH	213(3.72)	13-0001-64
6β,19-Cycloandrost-4-ene-3,17-dione	EtOH	243(4.19),298(2.12)	13-0001-64
	EtOH-NaOH	242(4.19),299(2.12)	13-0001-64
Estra-1,3,5(10),8-tetraen-17β-ol, 3-methoxy-	EtOH	278(4.15)	94-1285-65
Estra-2,5,7,9-tetraen-17β-ol, 3-methoxy-	dioxan	269(2.48),276(2.32)	87-0536-65
Estra-1,3,5(10)-trien-17-one, 1-hydroxy-4-methyl-	EtOH	220s(3.92),283(3.38)	24-1926-64
	NaOH	245(3.86),300(3.56)	24-1926-64
Estra-1,3,5(10)-trien-17-one, 3-hydroxy-1-methyl-	EtOH	223s(3.95),282s(3.25), 285(3.25)	24-1926-64
	NaOH	242(3.98),298(3.43)	24-1926-64
Estra-1,3,5(10)-trien-17-one, 3-hydroxy-2-methyl-	EtOH	283(3.42)	94-0196-64
Estra-1,3,5(10)-trien-17-one, 3-hydroxy-4-methyl-	EtOH	284(3.27)	94-0196-64
Estr-9-ene-3,17-dione, 5β,6β-methylene-	EtOH	217(3.86)	44-4160-65
Gona-4,9-diene-3,17-dione, 13β-ethyl-	n.s.g.	303(4.30)	39-4472-64
9β-Gona-1,3,5(10)-trien-17-one, 13β-ethyl-3-hydroxy-	n.s.g.	279(3.27),285(3.23)	39-4472-64
A-Homo-17β-estra-1(10),2,4a-trien-4-one, 17β-hydroxy-	n.s.g.	230(4.43),234(4.46), 238(4.47),314(4.05)	39-5137-65
D-Homoestradiol, 8-dehydro-	EtOH	274(4.20)	70-1814-64
D-Homoestrone	EtOH	278(3.33)	70-1814-64
14-Isoandrosta-4,15-diene-3,17-dione	MeOH	217(4.1),239(4.33)	44-2776-65
8-Iso-D-homoestrone	EtOH	227(4.13),280(3.63), 286(3.56)	70-1814-64
1-Naphthaleneacetic acid, heptyl ester	EtOH	224(5.00),281(3.84)	39-2072-65
2-Naphthaleneacetic acid, heptyl ester	EtOH	224(5.01),276(3.73)	39-2072-65

Compound	Solvent	λ_{max}(log ε)	Ref.
Spiro[cyclohexane-1,1'-inden]-4-one, 2',3'-diethyl-6'-methoxy-	MeOH	270(4.20),299s(--), 311s(--)	5-0036-64D
$C_{19}H_{24}O_3$			
Androsta-4,15-diene-3,17-dione, 14β-hydroxy-	MeOH	217(4.09),239(4.25)	44-2776-65
	MeOH	217(4.09),239(4.26)	88-2955-64
Androst-4-en-19-al, 3,17-dioxo-	MeOH	244(4.10)	35-5670-65
Androst-4-ene-3,17-dione, 14α,15α-epoxy-	MeOH	239(4.16)	44-2776-65
	MeOH	239(4.16)	88-2955-64
5α-Androst-14-ene-3,17-dione, 8β,19-epoxy-	EtOH-KOH	227(3.44)	33-1961-64
5β-Androst-4-ene-3,17-dione, 8β,19-epoxy-	EtOH-KOH	227(3.83)	33-1961-64
Androst-4-ene-3,6,17-trione	MeOH	242(4.26)	13-0713-64
1,3-Cyclohexanedione, 2-[2-(2,3,4,6,7,8-hexahydro-6-oxo-1-naphthyl)ethyl]-2-methyl-	EtOH	302(4.23)	70-1131-64
Estra-1,3,5(10)-trien-6-one, 17β-hydroxy-3-methoxy-	EtOH	222(4.34),255(3.94), 324(3.48)	44-2731-64
Estra-1,3,5(10)-trien-17-one, 1-hydroxy-4-methoxy-	EtOH	223s(3.86),292(3.58)	24-1940-64
	NaOH	242(3.88),306(3.61)	24-1940-64
Estra-1,3,5(10)-trien-17-one, 3-hydroxy-15β-methoxy-	MeOH	222(3.92),278(3.32)	44-0214-64
Estra-1,3,5(10)-trien-17-one, 15α-hydroxy-3-methoxy-	EtOH	278(3.32),287(3.29)	44-2731-64
Estra-1,3,5(10)-trien-17-one, 15β-hydroxy-3-methoxy-	MeOH	220(3.95),278(3.34), 288(3.30)	44-0064-64
Etiojerv-4-ene-3,11,17-trione	EtOH	240(4.20)	35-0701-64
A-Homo-3,17β-dihydroxy-estra-1(10),2,4a-trien-4-one	n.s.g.	235(4.23),320(3.83)	39-5137-65
17a-Oxa-D-homo-9β-androsta-4,6-diene-3,17-dione	MeOH	280(4.40)	54-0889-65
6-Oxa-19-nor-17α-pregn-4-en-20-yn-3-one, 17-hydroxy-	n.s.g.	259(4.34)	31-0418-64
1-Phenanthrenebutyric acid, 2,3,4,5,-6,7,9,10-octahydro-2-methyl-7-oxo-	EtOH	326(3.98)	13-0729-64
2α-Phenanthrenecarboxylic acid, 1,2,3,9,10,10aβ-hexahydro-7-methoxy-, isopropyl ester	EtOH	264(4.31),298(3.53)	44-1213-65
8,14-Seco-19-nor-D-homoandrosta-4,9-diene-3,14,17a-trione	EtOH	301(4.26)	13-0729-64
$C_{19}H_{24}O_3S_2$			
1α-Phenanthrenecarboxylic acid, 1,2,3,4,4aβ,9,10,10aβ-octahydro-7-methoxy-4-oxo-, methyl ester, 4-ethylenethioketal	EtOH	278(3.27),285(3.26)	44-2527-64
1β-Phenanthrenecarboxylic acid, 1,2,3,4,4aα,9,10,10aβ-octahydro-7-methoxy-4-oxo-, methyl ester, 4-ethylenethioketal	EtOH	225s(3.97),276(3.15), 285(3.08)	44-2527-64
4aβ-isomer	EtOH	277(3.22),285(3.19)	44-2527-64
$C_{19}H_{24}O_4$			
Androst-4-en-18-al, 9-hydroxy-3,17-dioxo-	EtOH	240.5(4.20)	35-0736-64
Androst-4-ene-3,17-dione, 6,19-epoxy-9α-hydroxy-	EtOH	240(4.07)	35-1385-65
Androst-4-ene-3,17-dione, 11β,18-epoxy-9-hydroxy-	EtOH	241(4.2)	35-4655-65

Compound	Solvent	λ_{max}(log ε)	Ref.
Androst-4-ene-3,12,17-trione, 9-hydroxy-	MeOH	240.5(4.21)	35-0736-64
Androst-4-ene-3,15,17-trione, 9-hydroxy-	MeOH	242.5(4.22)	35-0736-64
9-Anthrol, 1,4,4aα,9,9aβ,10-hexahydro-4β,10β-(isopropylidenedioxy)-5-methoxy-9β-methyl-	EtOH	275(3.29),281(3.29)	65-2563-64
4-Cyclohexene-1,2-dicarboxylic anhydride, 4-(8-formyl-4-methyl-3,7-nonadienyl)-	EtOH	229(4.18)	44-1690-65
1,5(4H)-Indandione, tetrahydro-7a-(hydroxymethyl)-4-(5-hydroxy-2-methylphenethyl)-	EtOH	279.5(3.37)	35-0736-64
Podocarpic acid, O-methyl-7-oxo-, methyl ester	EtOH	226(4.11),277(4.16)	39-0361-65
Propane, 1,3-bis(2,5-dimethoxyphenyl)-	95% THF	229(4.06),292(3.89)	44-2602-65
$C_{19}H_{24}O_5$			
9-Anthrol, 2,3-epoxy-1,4,4aα,9,9aβ,10-hexahydro-4β,10β-(isopropylidenedioxy)-5-methoxy-9β-methyl-	EtOH	276(3.42),281(3.41)	65-2563-64
2-Cyclohexene-1,4-dione, 6-tert-butyl-5-(2,5-dioxo-3-tert-butyl-cyclopent-3-en-1-yl)-2-hydroxy-	MeOH	229(4.13),281(3.95)	24-2774-65
	MeOH-KOH	234(4.28),330(3.95), 477(3.00)	24-2774-65
1β,8-Epidimethylgibb-2-ene-1α,10β-dicarboxylic acid, 4aα,7-dihydroxy-, 1→4a-lactone	EtOH	220(3.14)	39-1605-65
1,2-Phenanthrenedicarboxylic acid, 1,2,3,4b,5,6,7,8,8a,9-decahydro-4b-methyl-7-oxo-, dimethyl ester	EtOH	239(3.94)	70-1241-64
$C_{19}H_{24}O_6$			
5H-Benzocycloheptene-9-propionic acid, 8-formyl-6,7-dihydro-1,2,3-trimethoxy-, methyl ester	EtOH	223s(4.02),252(4.03), 309(4.03)	44-1752-65
Helicobasidin, diacetate	EtOH	267(4.23),337(2.53)	94-0236-64
Illudin S, diacetate	isooctane	226(4.10),245(4.05), 314(3.54)	78-1231-65
	MeOH	226(4.08),245s(4.02), 313(3.54)	78-1231-65
	EtOH	227(4.11),243s(4.05), 313(3.53)	35-1594-65
Lampterol, diacetate	isooctane	223(4.20),249s(4.04), 308(3.60)	78-2671-65
	EtOH	227(4.1),317(3.5)	78-2671-65
$C_{19}H_{25}BrO_4$			
5α-Androstan-19-oic acid, 5-bromo-3β,6β-dihydroxy-17-oxo-, γ-lactone	MeOH	241(4.23)	35-1528-64
$C_{19}H_{25}Cl$			
Azulene, 6-benzylidene-7-chloro-decahydro-1,4-dimethyl-	EtOH	279(4.22)	94-0942-65
$C_{19}H_{25}ClO_2$			
Androsta-3,5-dien-7-one, 6-chloro-17β-hydroxy-	MeOH	292(4.27)	31-0249-65
19-Norandrost-1-ene-3,17-dione, 5β-chloromethyl-	EtOH	233(3.93)	35-3727-65

Compound	Solvent	λ max(log ε)	Ref.
$C_{19}H_{25}FO$			
Estra-1,3,5(10)-triene, 17α-fluoro-3-methoxy-	EtOH	279(3.31),287(3.27)	44-2187-64
$C_{19}H_{25}FO_2$			
9β,10α-Androsta-4,6-dien-3-one, 6-fluoro-17β-hydroxy-	MeOH	286(4.37)	54-0918-65
B-Homoestr-4-ene-3,17-dione, 7ξ-fluoro-	EtOH	238(4.19)	44-4160-65
5,10-Seco-5,19-cycloandrost-4-ene-3,17-dione, 10β-fluoro-	EtOH	241(4.09)	35-3727-65
$C_{19}H_{25}NO$			
Butylamine, N,N-dimethyl-4-(2-hydroxyphenyl)-4-(2-methylphenyl)-	n.s.g.	275(3.48),282(3.43)	49-0485-64
Estra-1,3,5(10)-trien-17-one, 1-amino-4-methyl-	MeOH	241(3.87),293(3.34)	44-1893-64
$C_{19}H_{25}NO_2$			
8-Aza-D-homoestra-1,3,5(10)-trien-17a-one, 3-methoxy-	MeOH	280(3.33),287(3.31)	39-4900-65
17-Aza-D-homo-13α-estra-1,3,5(10)-trien-17-one, 3-methoxy-	MeOH	276(3.30),285(3.11)	78-0743-65
1H-Cyclopenta[7,8]phenanthro[3,2-d]isoxazol-1-ol, 2,3,3a,3b,4,5,10,10a,10b,-11,12,12a-dodecahydro-12a-methyl-	EtOH	285(4.01)	88-4603-65
Estra-1,3,5(10)-trien-17-one, 1-amino-3-methoxy-	MeOH	245s(3.83),291(3.48)	35-2943-64
Estr-4-ene-2α-carbonitrile, 17β-hydroxy-3-oxo-	EtOH	241(4.20)	88-4603-65
Estr-5(10)-ene-2α-carbonitrile, 17β-hydroxy-3-oxo-	EtOH	240(3.77)	88-4603-65
Metaphanine A, deoxo-	EtOH-$HClO_4$	217(4.32),274(4.13), 280(4.11),296(3.73), 310(3.66)	88-3605-64
	EtOH-$HClO_4$	217(4.32),274(4.13), 280(4.11),296(3.73), 310(3.66)	94-0695-65
Unknown structure	n.s.g.	239.5(4.19)	35-2451-65
$C_{19}H_{25}NO_3$			
Androsta-4,6-diene-2,3-dione, 17β-hydroxy-, 2-oxime	EtOH	215(3.81),310(4.23)	94-1445-65
Metaphanine B, dehydrodeoxo-	EtOH	230(4.24),275(3.93), 298s(3.6)	88-3605-64
	EtOH	230(4.24),275(3.93)	94-0695-65
Naphtho[2,1-f]quinolin-2(1H)-one, 3,4,4a,4b,5,6,10b,11,12,12a-decahydro-1-hydroxy-8-methoxy-12a-methyl-	MeOH	276(3.30),285(3.26)	78-0743-65
isomer	MeOH	276(3.30),285(3.26)	78-0743-65
$C_{19}H_{25}NO_4$			
Amaryllisine, O-methyl-	EtOH	275(3.08),282(3.08)	35-4976-64
Estra-1,3,5(10)-trien-17β-ol, 3-methoxy-16ξ-nitro-	MeOH	279(3.51),287(3.33)	78-0373-64
Nerinine, deoxy-	EtOH	210(4.14),279(3.20)	95-0699-65
$C_{19}H_{25}NO_5$			
Nerinine	EtOH	272(3.53)	95-1194-64

Compound	Solvent	λ_{max}(log ε)	Ref.
$C_{19}H_{25}NO_5S$			
2H-Pyran-6-carboxylic acid, 2-tert-butoxy-3,4-dihydro-2-hydroxy-5-methyl-3-(thiocarbamoyl)-, benzyl ester	EtOH	265.0(4.07)	39-1262-65
$C_{19}H_{25}N_3O_3$			
3(2H)-Phenanthrone, 10a-ethyl-1,9,10,10a-tetrahydro-5,6-dimethoxy-, semicarbazone	EtOH	215(4.23),308(4.48)	87-0200-64
$C_{19}H_{25}N_5O_7S$			
9H-Purine-6-thiol, 2-amino-9-(2,3,5-tripropionyl-β-D-ribofuranosyl)-	pH 1	264(3.84),343(4.34)	87-0200-64
	pH 11	251(4.22),317(4.30)	87-0200-64
$C_{19}H_{25}O_4P$			
Estrone, hydrogen methylphosphonate	MeOH	270(3.05),276(3.03)	22-0933-65
$C_{19}H_{26}$			
Estra-1,3,5(10)-triene, 4-methyl-	EtOH	263(2.39)	24-0140-64
$C_{19}H_{26}ClN_3$			
Chloroquine	pH 5.9	329(4.2),343(4.29)	37-3123-65
	pH 7.0	343(4.28)	
$C_{19}H_{26}N_2$			
Apparicine, N^b-methyltetrahydro-	EtOH	226(4.59),284(3.91),292(3.84)	39-4773-65
(-)-Quebrachamine	EtOH	230(4.51),286(3.83)	33-1822-65
$C_{19}H_{26}N_2O_2$			
Rhazidigenin	MeOH	235(3.99),292(3.40)	49-1228-64
$C_{19}H_{26}N_2O_2$			
Androst-4-eno[2,3-c]furazan, 17β-hydroxy-	EtOH	255(4.02)	94-1445-65
2,3-Diaza-19-norandrosta-3,5(10)-dien-1-one, 17-acetyl-4-methyl-	MeOH	230s(3.46),286(3.56)	87-0590-64
$C_{19}H_{26}N_2O_4$			
Acetamide, N-[3-(1,2-dihydro-4,8-dimethoxy-1-methyl-2-oxo-3-quinolyl)-1,1-dimethylpropyl]-	MeOH	239(4.39),257(4.42),283(3.92),292(3.89),332(3.56)	39-4190-64
Androst-5-ene-2,3,4-trione, 17β-hydroxy-, 2,4-dioxime	EtOH	263(3.99)	94-1445-65
$C_{19}H_{26}N_2O_8$			
Guaiacol, bis(morpholinocarboxymethyl)-	H_2O	233s(3.94),288(3.60)	24-0363-64
	pH 13	257(4.06),302(3.80)	24-0363-64
$C_{19}H_{26}N_4O_4$			
2(1H)-Benzocyclooctenone, decahydro-10a-methyl-, 2,4-dinitrophenylhydrazone	$CHCl_3$	368(4.34)	35-1755-64
p-Menthane-3-acrolein, 2,4-dinitrophenylhydrazone	$CHCl_3$	375(4.53)	28-4712-64B
$C_{19}H_{26}O$			
Androsta-3,5-dien-7-one	EtOH	279(4.47)	35-2832-64

Compound	Solvent	λ_{max}(log ε)	Ref.
Androsta-3,5-dien-17-one	EtOH	236(4.35),290s(1.95)	60-0285-64
	EtOH	234(4.21)	94-1435-65
Androsta-4,6-dien-3-one	EtOH	285(4.41)	44-0068-64
10α-Androsta-3,5,7-trien-17β-ol	MeOH	302(4.20),315(4.28), 330(4.13)	54-0841-65
5-Azulenol, 6-benzylidenedecahydro-3,8-dimethyl-	EtOH	250(4.18)	94-0942-65
11H-Benzo[a]fluoren-3β-ol, 1,2,3,4,4aα-5,6,6aβ,11aα,11b-decahydro-10,11bβ-dimethyl-	EtOH	265(2.76),273(2.74)	13-0463-64
Estra-1,3,5(10)-trien-17β-ol, 2-methyl-	MeOH	273(2.97),278(3.11), 286s(--)	87-0409-65
Estra-1,3,5(10)-trien-17β-ol, 3-methyl-	MeOH	269(2.86),278(2.91)	87-0409-65
19-Norandrost-4-en-17-one, 3-methylene-	EtOH	238(4.40)	87-0345-64
$C_{19}H_{26}O_2$			
Androsta-1,4-dien-3-one, 17β-hydroxy-	EtOH	246(4.18)	94-0050-65
Androsta-1,5-dien-3-one, 17β-hydroxy-	MeOH	226(4.02)	13-0183-64
8α,10α-Androsta-4,6-dien-3-one, 17β-hydroxy-	MeOH	287.5(4.40)	54-0841-65
9β-Androsta-4,6-dien-3-one, 17β-hydroxy-	MeOH	284(4.42)	54-0889-65
5α-Androst-1-ene-3,17-dione	EtOH	230(4.04)	44-2925-65
Androst-4-ene-3,17-dione	EtOH	240(4.21)	35-3727-65
p-hydroxybenzaldehyde complex	HOAc-$HClO_4$	600(3.85)	96-0134-65
vanillin complex	HOAc-$HClO_4$	550(3.81)	96-0134-65
9β-Androst-4-ene-3,17-dione	MeOH	245.5(4.16)	54-1069-64
Androst-5-ene-3,17-dione	MeOH	268(2.38)	35-5670-65
p-hydroxybenzaldehyde complex	HOAc-$HClO_4$	560(3.81)	96-0134-65
vanillin complex	HOAc-$HClO_4$	550(3.78)	96-0134-65
8α-Androst-5-ene-3,17-dione	EtOH	241(4.25)	39-5139-65
Androst-5-ene-7,17-dione	EtOH	238(3.73)	44-0068-64
2H-Benzopyran-6-ol, 2,5,7,8-tetra-methyl-2-(4-methyl-3-penten-1-yl)-	hexane	236(4.29),264s(--), 274(3.95),284(3.91), 340(3.54)	39-5060-65
p-Benzoquinone, (3,7-dimethyl-2,6-octa-dienyl)trimethyl-	EtOH	259(4.23),267(4.25)	39-5060-65
2,19-Cyclo-5α-androstane-3,17-dione	EtOH	282(1.87)	44-2198-65
5β,19-Cycloandrost-1-en-3-one, 17β-hydroxy-	EtOH	272(3.76)	35-3727-65
5β,19-Cycloandrost-6-en-3-one. 17β-hydroxy-	EtOH	213(3.69)	13-0001-64
5β,19-Cycloandrost-6-en-17-one, 3β-hydroxy-	EtOH	213(3.75)	13-0001-64
Estra-1,3,5(10)-triene-1, 17β-diol, 4-methyl-	EtOH	284(3.25)	39-3621-64
Estra-1,3,5(10)-triene-3,17β-diol, 2-methyl-	EtOH	285(3.38)	94-0196-64
Estra-1,3,5(10)-triene-3,17β-diol, 4-methyl-	EtOH	284(4.19)	94-0196-64
Estr-1-ene-3,17-dione, 5α-methyl-	EtOH	234(3.92),300(1.85)	35-4629-65
Estr-4-ene-3,17-dione, 1β-methyl-	EtOH	244(4.19)	44-3781-65
Estr-9-en-17-one, 3β-hydroxy-5β,6β-methylene-	EtOH	218(3.95)	44-4160-65
Etiojerva-1,4-dien-3-one, 17β-hydroxy-	MeOH	242(4.23)	44-0123-65
Etiojerv-4-ene-3,17-dione	EtOH	240(4.19)	35-0701-64
Gona-4,9-dien-3-one, 13β-ethyl-17β-hydroxy-	n.s.g.	303(4.28)	39-4472-64

Compound	Solvent	$\lambda_{max}(\log \epsilon)$	Ref.
D-Homoestradiol	EtOH	280(3.35)	70-1814-64
8-Iso-D-homoestradiol	EtOH	218(3.84),281(3.35)	70-1814-64
Phenalen-1-one, 2,3,3a,4,5,6-hexahydro-6-isopropyl-7-methoxy-3a,8-dimethyl-	EtOH	275(4.08)	39-2146-64
13,14-Secoandrosta-4,13(17)-diene-3,14-dione, cis	MeOH	238(4.17)	44-2776-65
	MeOH	238(4.17)	88-2955-64
5,10-Seco-5,19-cycloandrost-4-ene-3,17-dione	EtOH	246(4.16)	35-3727-65
Spiro[cyclohexane-1,1'-indan]-4-one, 2',3'-diethyl-6'-methoxy-	MeOH	221s(--),226(3.85), 279(3.47),286s(3.42)	5-0036-64D
$C_{19}H_{26}O_3$			
13α-Androsta-5,15-dien-17-one, 3β,16-dihydroxy-	MeOH	258(3.79)	44-3304-64
	MeOH-KOH	292(3.69)	44-3304-64
5α-Androstane-1,3,17-trione	EtOH	254(4.08)	44-3552-65
Androst-4-ene-3,17-dione, 14β-hydroxy-	MeOH	238(4.21)	44-2776-65
6β,19-Cycloandrostane-3,17-dione, 5α-hydroxy-	EtOH	289(2.03)	13-0001-64
Estra-1,3,5(10)-triene-3,17β-diol, 15β-methoxy-	MeOH	222(3.88),280(3.36)	44-0214-64
Estra-1,3,5(10)-triene-6α,17β-diol, 3-methoxy-	EtOH	279(3.36),287(3.28)	44-2731-64
Estra-1,3,5(10)-triene-6β,17β-diol, 3-methoxy-	EtOH	278(3.76)	44-2731-64
Estra-1,3,5(10)-triene-15α,17β-diol, 3-methoxy-	EtOH	279(3.31),287(3.27)	44-2731-64
Estra-1,3,5(10)-triene-15β,17β-diol, 3-methoxy-	MeOH	220(3.94),278(3.36), 288(3.32)	44-0064-64
Estr-4-en-3-one, 17β-hydroxy-2-(hydroxymethylene)-	EtOH	248(4.06),303(3.72)	88-4603-65
Etiojerv-4-ene-3,11-dione, 17α-hydroxy-	EtOH	240(4.19)	35-0701-64
17a-Oxa-D-homoandrosta-1,4-dien-3-one, 17α-hydroxy-	MeOH	243(4.20)	44-3564-65
Taxininol, amhydrodeformyldehydro-	EtOH	228(3.97)	94-0386-64
Testosterone, 1-dehydro-7α-hydroxy-	EtOH	244.5(4.19)	24-3363-64
$C_{19}H_{26}O_4$			
Androst-4-ene-3,17-dione, 2ξ,9α-dihydroxy-	EtOH	243(4.08)	35-1385-65
Androst-4-ene-3,17-dione, 9,12-dihydroxy-	EtOH	241(4.21)	35-0736-64
Androst-4-ene-3,17-dione, 9,14-dihydroxy-	EtOH	242(4.20)	35-0736-64
Androst-4-ene-3,17-dione, 9,15-dihydroxy-	EtOH	242.5(4.22)	35-0736-64
Androst-4-ene-3,17-dione, 9,18-dihydroxy-	EtOH	242(4.19)	35-0736-64
Androst-4-ene-3,17-dione, 9α,19-dihydroxy-	EtOH	245(4.17)	35-1385-65
9-Anthrol, 1,2,3,4,4aα,9,9aβ,10-octahydro-4β,10β-(isopropylidenedioxy)-5-methoxy-9β-methyl-	EtOH	275(3.36),281(3.35)	65-2563-64
18,19-Dinorcass-8-en-16-oic acid, 7,13-dioxo-, methyl ester	EtOH	243(4.0)	39-0403-65
1-Naphthalenebutyric acid, α-acetyl-2,3,4,6,7,8-hexahydro-α-methyl-6-oxo-, ethyl ester	EtOH	302(4.19)	70-1131-64

Compound	Solvent	$\lambda_{max}(\log \epsilon)$	Ref.
$C_{19}H_{26}O_5$			
1,2,4-Cyclohexanetrione, 6-tert-butyl-5-(3-tert-butyl-2,5-dihydroxycyclopenta-2,4-dien-1-yl)-	MeOH	227(3.90),272(3.30)	24-2774-65
	MeOH-KOH	224(3.85),234s(3.81), 283(4.08)	24-2774-65
1-Naphthol, 2-(5-hydroxyhexyl)-3,6,8-trimethoxy-	EtOH	246(4.87),289(3.63), 299(3.61),314(3.46), 330(3.34)	39-2289-64
1β-Phenanthrenecarboxylic acid, 1,2,3,4,4aβ,9,10,10aβ-octahydro-4,4,7-trimethoxy-, methyl ester	EtOH	224(3.93),278(3.21), 286s(3.14)	44-2527-64
$C_{19}H_{26}O_6$			
Acetic acid, [3,5-dimethoxy-2-(7-oxooctanoyl)phenyl]-, methyl ester	EtOH	266(3.81),290(3.64)	39-2289-64
Lampterol, dihydro-, diacetate	EtOH	257(4.37)	78-2671-65
Melicopol	EtOH	235s(4.19),292(4.26), 328s(3.61)	12-2021-65
	EtOH-KOH	245(3.90),325(4.46)	12-2021-65
$C_{19}H_{27}BrO_2$			
5β-Androstane-3,11-dione, 9α-bromo-	EtOH	316.0(2.04)	39-1161-64
5β-Androstane-3,11-dione, 12α-bromo-	EtOH	319.0(2.14)	39-1161-64
Testosterone, 2α-bromo-	MeOH	244(4.15)	94-1217-64
$C_{19}H_{27}ClO_2$			
Androst-5-en-7-one, 6-chloro-17β-hydroxy-	MeOH	254(4.03)	31-0249-64
Testosterone, 2α-chloro-	MeOH	243(4.16)	94-1217-64
$C_{19}H_{27}FO$			
Androst-4-en-3-one, 17α-fluoro-	EtOH	241(4.23)	44-2187-64
$C_{19}H_{27}FO_2$			
B-Homoestr-4-en-3-one, 7-fluoro-17β-hydroxy-	EtOH	240(4.18)	44-4160-65
5,10-Seco-5,19-cycloandrost-4-en-3-one, 10β-fluoro-17β-hydroxy-	EtOH	242(4.14)	35-3727-65
$C_{19}H_{27}NO$			
11H-Benzo[a]fluoren-3β-ol, 9-amino-1,2,3,4,4aα,5,6,6aβ,11aα,11b-decahydro-10,11bβ-dimethyl-	EtOH	236(3.92),290(3.29)	13-0463-64
$C_{19}H_{27}NO_2$			
Androsta-1,4-dien-3-one, 17β-hydroxy-, oxime	MeOH-acid	286(4.29)	33-1988-65
	MeOH	253s(4.20),264(4.21)	33-1988-65
$C_{19}H_{27}NO_3$			
Androst-4-ene-2,3-dione, 17β-hydroxy-, 2-oxime	EtOH	264(4.16)	94-1445-65
Androst-4-en-3-one, 17β-hydroxy-, 17-nitrite	MeOH	239(4.25)	78-0743-65
17α-Aza-D-homoandrost-4-ene-3,17-dione, 17α-hydroxy-	MeOH	238(4.25)	78-0743-65
2,5-Cyclohexadien-1-one, 2,6-dicyclohexyl-4-methyl-4-nitro-	n.s.g.	239(4.07)	70-1675-65
Pronuciferine, hexahydro-	EtOH-HCl	225s(3.95),280(3.29)	33-2122-64
$C_{19}H_{27}NO_4$			
Nerinine, α-deoxydihydro-	EtOH	277(3.33)	95-0699-65

Compound	Solvent	λ_{max}(log ε)	Ref.
Nerinine, β-deoxydihydro-	EtOH	277(3.32)	95-0699-65
1(2H)-Quinolinebutyric acid, 2-carboxy-3,4-dihydro-α-methyl-, diethyl ester	EtOH	209(4.38),250(3.97), 300(3.41)	78-2961-65
1(2H)-Quinolinepropionic acid, 2-(1-carboxyethyl)-3,4-dihydro-, diethyl ester	EtOH	210(4.34),258(4.18), 306(3.39)	78-2961-65
Testosterone, 2α-nitro-	pH 1	252(4.17)	78-0373-64
	pH 13	242(3.86),268(3.89), 355(4.08)	78-0373-64
	MeOH	244(4.16),370(2.96)	78-0373-64
$C_{19}H_{27}NO_5$			
Nerinine, dihydro-	EtOH	272(3.51)	95-0206-65
Nerininediol	EtOH	210(4.20),230(3.75), 274(3.08)	95-0699-65
$C_{19}H_{27}NO_6$			
Senkirkine	EtOH	219(4.02)	39-2492-65
$C_{19}H_{27}O_4P$			
Estradiol, 3-(hydrogen methylphosphonate)	MeOH	280.5(3.29)	22-0933-65
Estradiol, 17-(hydrogen methylphosphonate)	MeOH	270(3.03),276(3.00)	22-0933-65
$C_{19}H_{28}INO_3$			
1-[2-[(6-Allyl-1,4-benzodioxan-5-yl)-oxy]ethyl]-1-methylpiperidinium iodide	EtOH	275(3.11)	65-2994-64
$C_{19}H_{28}N_2O_2$			
Androstano[2,3-c]furazan, 17β-hydroxy-	EtOH	217(3.7)	94-1445-65
2,3-Diaza-19-norandrosta-3,5(10)-dien-1-one, 17-hydroxyethyl-4-methyl-	MeOH	230s(3.57),286(3.61)	87-0590-64
2,3-Diaza-19-norandrost-3-en-1-one, 17-acetyl-4-methyl-	MeOH	245(3.74)	87-0590-64
$C_{19}H_{28}N_2O_3$			
Androstano[2,3-c]furazan, 17β-hydroxy-, N-oxide	EtOH	264(3.84)	94-1445-65
$C_{19}H_{28}O$			
Androst-2-en-17-one	EtOH	296(1.60)	60-0285-64
5α-Androst-1-en-3-one	EtOH	229(4.04)	35-2825-64
5α-Androst-2-en-1-one	EtOH	225(3.88)	35-2623-64
5α-Androst-2-en-17-one, p-hydroxybenzaldehyde complex	HOAc-$HClO_4$	580(4.15)	96-0134-65
resorcylic aldehyde complex	HOAc-$HClO_4$	540(3.48)	96-0134-65
vanillin complex	HOAc-$HClO_4$	600(4.0)	96-0134-65
5α-Androst-8(14)-en-15-one	EtOH	259(3.98)	13-0239-65
5α-Androst-15-en-17-one	EtOH	233(3.90)	35-0817-65
5α-Androst-16-en-15-one	EtOH	222.5(3.83)	35-0817-65
19-Norandrost-4-en-17β-ol, 3-methylene-	EtOH	237(4.40)	87-0345-64
1H-Phenalene, 2,3,3a,4,5,6-hexahydro-1-isopropyl-9-methoxy-3a,8-dimethyl-	EtOH	280(3.06)	39-2146-64
$C_{19}H_{28}OS$			
5α-Androstan-17-one, 2α,3α-epithio-	EtOH	265(1.77),283(1.70)	78-0329-65

Compound	Solvent	$\lambda_{max}(\log \epsilon)$	Ref.
$C_{19}H_{28}O_2$			
Androsta-3,5-diene-11β,17β-diol	n.s.g.	228(4.22),236(4.27), 243(4.11)	13-0391-64
5α-Androstane-3,17-dione, p-hydroxybenzaldehyde complex	$HOAc-HClO_4$	600(3.85)	96-0134-65
resorcylic aldehyde complex	$HOAc-HClO_4$	540(3.40)	96-0134-65
vanillin complex	$HOAc-HClO_4$	550(4.38)	96-0134-65
Androst-5-en-17-one, 3β-hydroxy-	C_6H_{12}	197(3.82),296(1.51)	60-0285-64
	EtOH	198(3.88),295(1.64)	60-0285-65
p-hydroxybenzaldehyde complex	$HOAc-HClO_4$	550(4.20)	96-0134-65
resorcylic aldehyde complex	$HOAc-HClO_4$	540(3.70)	96-0134-65
vanillin complex	$HOAc-HClO_4$	550(4.20)	96-0134-65
Androst-9(11)-en-17-one, 3β-hydroxy-	C_6H_{12}	193(3.89),299(1.52)	60-0285-64
	EtOH	198(3.81),295(1.63)	60-0285-64
5α-Androst-1-en-3-one, 17β-hydroxy-	EtOH	230(4.0)	44-2925-65
tetrahydropyranyl ether	MeOH	229.5(4.01)	87-0048-65
5α-Androst-2-en-1-one, 17β-hydroxy-	MeOH	224.5(3.88)	87-0048-65
5α-Androst-7-en-6-one, 3β-hydroxy-	EtOH	244(4.08)	35-2837-64
5β-Androst-9(11)-en-12-one, 3α-hydroxy-	EtOH	241(4.10)	35-2825-64
9β-Androst-4-en-3-one, 17β-hydroxy-	MeOH	246.5(4.14)	54-1069-64
Androsterone, resorcylic aldehyde complex [wrong formula--see $C_{19}H_{30}O_2$]	$HOAc-HClO_4$	540(3.60)	96-0134-65
6-Chromanol, 2,5,7,8-tetramethyl-2-(4-methylpent-3-enyl)-	EtOH	293(3.60)	77-0040-65
3α,5-Cyclo-5α-androstan-17-one, 6β-hydroxy-, p-hydroxybenzaldehyde complex	$HOAc-HClO_4$	560(4.08)	96-0134-65
resorcylic acid complex	$HOAc-HClO_4$	540(3.65)	96-0134-65
vanillin complex	$HOAc-HClO_4$	550(4.22)	96-0134-65
5β,19-Cycloandrost-6-ene-3β,17β-diol	EtOH	213(3.73)	13-0001-64
8H-Cyclopent[a]anthracen-8-one, tetradecahydro-3α-hydroxy-3aα,6α-dimethyl-	EtOH	241(4.17)	78-2487-64
Estr-1-en-3-one, 17β-hydroxy-5α-methyl-	EtOH	234(3.98)	35-4629-65
Estr-4-en-3-one, 17β-hydroxy-2β-methyl-	EtOH	240.5(4.23)	78-2501-65
Etiojerv-12-en-17-one, 3β-hydroxy-	MeOH	247(4.17)	44-0123-65
Gon-4-en-3-one, 13β-ethyl-17β-hydroxy-	n.s.g.	238(4.18)	39-4472-64
8-Isotestosterone	EtOH	242(4.20)	78-2487-64
5,10-Seco-5,19-cycloandrost-4-en-3-one, 17β-hydroxy-	EtOH	246(4.16)	35-3727-65
Testosterone	MeOH	241(4.18)	94-1217-64
$C_{19}H_{28}O_3$			
5α-Androstane-11,17-dione, 3α-hydroxy-, p-hydroxybenzaldehyde complex	$HOAc-HClO_4$	570(3.88)	96-0134-65
vanillin complex	$HOAc-HClO_4$	600(3.70)	96-0134-65
Androst-4-en-3-one, 14β,17α-dihydroxy-	MeOH	240(4.17)	44-2776-65
	MeOH	240(4.17)	88-2955-64
Cyclohexanecarboxylic acid, 4-hydroxy-1,3-dimethyl-2-(2,2,4-trimethyl-5-oxo-7-norcaryl)-, δ-lactone	EtOH	205(3.77)	94-0386-64
A-Norandrost-3-en-2-one, 17β-hydroxy-3-methoxy-	EtOH	253(4.05),306(2.53)	44-1325-65
A-Nor-5β-androst-2-en-1-one, 17β-hydroxy-2-methoxy-	EtOH	258(3.77)	78-0759-65
19-Nortestosterone, 16α-hydroxy-17α-methyl-	MeOH	241(2.72)	7-0221-65
Testosterone, 4-hydroxy-	EtOH	278(4.08)	44-1348-64
Testosterone, 7α-hydroxy-	EtOH	242.5(4.24)	24-3363-64
10α-Testosterone, 19-hydroxy-	EtOH	246(4.14)	78-2473-64

Compound	Solvent	λ_{max}(log ϵ)	Ref.
$C_{19}H_{28}O_4$			
Androst-1-en-3-one, 4α,5α,17β-tri-hydroxy-	EtOH	231(3.96)	94-0050-65
Androst-4-en-3-one, 1α,2α,17β-tri-hydroxy-	EtOH	241(4.14)	94-0050-65
Androst-4-en-3-one, 4,9α,17β-tri-hydroxy-	EtOH	278(4.10)	35-1385-65
$C_{19}H_{29}BrO$			
5α-Androstan-2-one, 3α-bromo-	EtOH	310(2.12)	35-5542-64
$C_{19}H_{29}BrO_2$			
5β-Androstan-11-one, 12α-bromo-3α-hydroxy-	EtOH	319.0(2.15)	39-1161-64
$C_{19}H_{29}NO_2$			
17α-Aza-D-homoandrost-4-en-3-one, N-hydroxy-	MeOH	237(4.08)	44-0639-65
$C_{19}H_{29}NO_3$			
Androstane-2,3-dione, 17β-hydroxy-, 2-oxime	EtOH	243(3.84)	94-1445-65
	EtOH-NaOEt	301(4.12)	94-1445-65
$C_{19}H_{29}NO_4$			
5α-Androstan-3-one, 17β-hydroxy-4-nitro-	pH 13	232(3.99)	78-0373-64
	NaOMe	239(3.79)	78-0373-64
	KO-tert-Bu	224(3.66)	78-0373-64
5β-Androstan-3-one, 17β-hydroxy-4β-nitro-	pH 13	235(4.08),338(2.70)	78-0373-64
	NaOMe	239(3.95),347(2.77)	78-0373-64
	KO-tert-Bu	348(3.78)	78-0373-64
$C_{19}H_{29}N_2O_{11}P$			
Uridine, 2',3'-bis-O-(tetrahydropyran-2-yl)-, 5'-phosphate, ammonium salt	pH 4	261(3.98)	94-1503-64
	pH 12	261(3.84)	94-1503-64
$C_{19}H_{29}N_3$			
Indole, 1-(2-diethylaminoethyl)-3-(1-methyl-2-pyrrolidinyl)-	EtOH-HCl	216(4.60),272(3.83), 280(3.82),284s(3.81), 341(3.71)	87-0415-64
$C_{19}H_{29}N_3O_3S$			
1H-Pyrido[3,4-b]indole, 2,3,4,9-tetra-hydro-1,1-dimethyl-, cyclohexane-sulfamic acid salt	EtOH	221(4.56),270(3.84), 278(3.83),280s(3.82), 288(3.72)	44-2864-64
$C_{19}H_{29}N_5O_6S$			
1H-Pyrido[3,4-b]indole, 3-ethyl-2,3,4,9-tetrahydro-1,1-dimethyl-, compound with creatinine sulfate	pH 2	222(4.59),262s(3.67), 267(3.68),293(3.71)	44-2864-64
$C_{19}H_{29}O_4P$			
Estr-4-en-3-one, 17β-hydroxy-, hydrogen methylphosphonate	MeOH	240(4.21)	22-0933-65
$C_{19}H_{29}O_5P$			
Testosterone, dihydrogen phosphate	MeOH	241(4.22)	22-0018-65
$C_{19}H_{30}N_2O_2$			
4-Azaandrost-5-en-3-one, 4-amino-17β-hydroxy-17α-methyl-	EtOH	244(4.12)	4-0212-65

Compound	Solvent	$\lambda_{max}(\log \epsilon)$	Ref.
2,3-Diaza-19-norandrost-3-en-1-one, 17-hydroxyethyl-4-methyl-	MeOH	245(3.74)	87-0590-64
$C_{19}H_{30}N_2O_3$			
Androstane-2,3-dione, 17β-hydroxy-, dioxime	EtOH	240(3.8)	94-1445-65
$C_{19}H_{30}O$			
5α-Androstan-2-one	EtOH	287(1.43)	35-5542-64
Androstan-17-one	C_6H_{12}	298(1.52)	60-0285-64
	EtOH	294(1.63)	60-0285-64
16-Norpimar-7-en-3-one	EtOH	273(2.23)	1-1875-65
$C_{19}H_{30}O_2$			
5β-Androstan-3-one, 17β-hydroxy-, p-hydroxybenzaldehyde complex	$HOAc-HClO_4$	550(3.88)	96-0134-65
resorcylic aldehyde complex	$HOAc-HClO_4$	540(3.0)	96-0134-65
vanillin complex	$HOAc-HClO_4$	600(4.00)	96-0134-65
Androstan-17-one, 3β-hydroxy-	C_6H_{12}	297(1.51)	60-0285-64
	EtOH	293.5(1.63)	60-0285-64
5α-Androstan-17-one, 3β-hydroxy-, vanillin complex	$HOAc-HClO_4$	600(4.20)	96-0134-65
5α-Androstan-17-one, 5α-hydroxy-, p-hydroxybenzaldehyde complex	$HOAc-HClO_4$	550(4.36)	96-0134-65
resorcylic aldehyde complex	$HOAc-HClO_4$	540(3.85)	96-0134-65
5β-Androstan-17-one, 3α-hydroxy-, resorcylic aldehyde complex	$HOAc-HClO_4$	540(3.74)	96-0134-65
vanillin complex	$HOAc-HClO_4$	600(4.15)	96-0134-65
Androst-5-ene-3β,17β-diol, p-hydroxybenzaldehyde complex	$HOAc-HClO_4$	560(4.20)	96-0134-65
o-phthalaldehyde complex	$HOAc-HClO_4$	560(3.81)	96-0134-65
vanillin complex	$HOAc-HClO_4$	600(3.93)	96-0134-65
Androsterone, p-hydroxybenzaldehye complex	$HOAc-HClO_4$	550(4.26)	96-0134-65
vanillin complex	$HOAc-HClO_4$	600(4.04)	96-0134-65
Etiocholanolane, p-hydroxybenzaldehyde complex	$HOAc-HClO_4$	550(4.28)	96-0134-65
16-Norros-5(6)-en-7-one, 10-hydroxy-	n.s.g.	238(4.03)	39-7246-65
Testosterone, p-hydroxybenzaldehyde complex	$HOAc-HClO_4$	560(3.81)	96-0134-65
vanillin complex	$HOAc-HClO_4$	600(3.70)	96-0134-65
$C_{19}H_{30}O_3$			
5α-Androstan-17-one, 3β,16α-dihydroxy-, p-hydroxybenzaldehyde complex	$HOAc-HClO_4$	550(4.34)	96-0134-65
resorcylic aldehyde complex	$HOAc-HClO_4$	600(4.20)	96-0134-65
vanillin complex	$HOAc-HClO_4$	580(4.36)	96-0134-65
5β-Androstan-11-one, 3α,17β-dihydroxy-, p-hydroxybenzaldehyde complex	$HOAc-HClO_4$	560(4.04)	96-0134-65
resorcylic aldehyde complex	$HOAc-HClO_4$	540(3.95)	96-0134-65
vanillin complex	$HOAc-HClO_4$	600(3.78)	96-0134-65
5β-Androstan-17-one, 3α,11β-dihydroxy-, p-hydroxybenzaldehyde complex	$HOAc-HClO_4$	550(4.28)	96-0134-65
resorcylic aldehyde complex	$HOAc-HClO_4$	540(3.98)	96-0134-65
vanillin complex	$HOAc-HClO_4$	550(4.26)	96-0134-65
Androst-5-ene-3β,17β-diol, resorcylic aldehyde complex	$HOAc-HClO_4$	540(3.85)	96-0134-65
p-Benzoquinone, 2-dodecyl-6-methoxy-	EtOH	268(4.03),363(2.91)	94-0236-64

Compound	Solvent	λ_{max}(log ε)	Ref.
$C_{19}H_{30}O_4$			
p-Benzoquinone, 2,5-dihydroxy-3-tridecyl-	EtOH	291(4.26),425(2.46)	78-2319-64
	EtOH	203(4.18),291(4.26), 425(2.46)	94-0236-64
$C_{19}H_{30}O_5$			
8-Nonynoic acid, 9-(5-ethyl-2,2,4-trimethyl-1,3-dioxolan-4-yl)-7-oxo-, ethyl ester	EtOH	222(3.87)	70-0818-65
$C_{19}H_{30}O_6$			
Melicopol, tetrahydro-	EtOH	235s(4.15),290(4.25), 330(3.55)	12-2021-65
	EtOH-KOH	240(3.89),322(4.40)	12-2021-65
7,9a-Methano-9aH-cyclopenta[b]heptalen-4(1H)-one, dodecahydro-2,8,11,11a,12-pentahydroxy-1,1,8-trimethyl-	EtOH	280(1.54)	44-2756-64
$C_{19}H_{30}O_7$			
Pyran-2-crotonic acid, 6-(carboxymethyl)-tetrahydro-4-hydroxy-,α,γ,3,5-tetramethyl-, dimethyl ester, acetate	n.s.g.	219(4.13)	31-0336-64
$C_{19}H_{31}BF_4$			
Dodecyltropylium tetrafluoroborate	96% H_2SO_4	225(4.46),298(3.74)	46-1466-65
$C_{19}H_{31}NO$			
Naphthalene, 1-(2-dimethylaminoethyl)-1,2-diethyl-1,2,3,4-tetrahydro-7-methoxy-	EtOH	281(3.58),288s(3.52)	87-0559-65
$C_{19}H_{31}NO_2$			
Cassine, dehydro-	EtOH-acid	230(3.87),301(4.01)	44-0471-64
	EtOH	224(3.96),288(3.81)	44-0471-64
	EtOH-base	245(4.10),311(3.93)	44-0471-64
$C_{19}H_{31}N_3O_3$			
Salicylaldehyde, 5-nitro-, 1,1-dihexylhydrazone	MeOH	303(4.47)	24-1631-64
	MeOH-NaOH	344(4.29),428(4.01)	24-1631-64
$C_{19}H_{32}$			
Toluene, 2,4,6-tri-tert-butyl-	$C_6H_{11}Me$	259(2.34)	88-0043-64
$C_{19}H_{32}IN$			
Morphinan methine, 1,2,3,4-tetrahydro-N-methyl-, methiodide	EtOH	219(4.16),267(3.64)	44-1769-65
$C_{19}H_{32}O_2$			
5α-Androstane-5β,17β-diol, p-hydroxybenzaldehyde complex	HOAc-$HClO_4$	550(4.13)	96-0134-65
resorcylic aldehyde complex	HOAc-$HClO_4$	540(3.30)	96-0134-65
vanillin complex	HOAc-$HClO_4$	600(4.0)	96-0134-65
5β-Androstane-3α,17β-diol, resorcylic aldehyde complex	HOAc-$HClO_4$	540(3.78)	96-0134-65
vanillin complex	HOAc-$HClO_4$	600(4.0)	96-0134-65
Colensan-1-one	EtOH	297(1.65)	12-0066-64
Des-AB-cholest-8(9)-ene-8-carboxylic acid	EtOH	224(3.95)	39-4907-64
Grevillol	EtOH	274(3.2),281(3.19)	12-2015-65
	EtOH-KOH	290(3.48)	12-2015-65

Compound	Solvent	λ_{max}(log ϵ)	Ref.
$C_{19}H_{32}O_3$			
β-Campholenic acid, methyl ester	EtOH	258(4.63),268(4.76), 279(4.64)	70-2003-64
Helenynolic acid, methyl ester	isooctane	228(4.24),238(4.15)	44-0610-65
$C_{19}H_{32}O_5$			
8-Nonenoic acid, 9-(5-ethyl-2,2,4-trimethyl-1,3-dioxolan-4-yl)-7-oxo-, ethyl ester	n.s.g.	229(3.91)	70-0818-65
$C_{19}H_{33}N_5$			
s-Triazolo[2,3-c]pyrimidine, 2-(butylamino)-7-heptyl-5-propyl-	MeOH	235(4.56),303(3.40)	39-3357-65
$C_{19}H_{34}$			
Bicyclohexyl, 2-(cyclohexylmethyl)-	hexane	none	70-0124-65
Bicyclohexyl, 3-(cyclohexylmethyl)-	hexane	none	70-0124-65
Bicyclohexyl, 4-(cyclohexylmethyl)-	hexane	none	70-0124-65
$C_{19}H_{34}N_2O_5$			
Piperidinooxy, 4-hydroxy-2,2,6,6-tetramethyl-, carbonate	C_6H_{12}	241(3.55),446(1.37)	22-3390-65
	MeOH	240(3.39)	22-3390-65
$C_{19}H_{35}NO$			
Morpholine, 4-(1-cyclopentadecen-1-yl)-	C_6H_{12}	221(3.91)	28-0759-65B
$C_{19}H_{37}NO_2$			
Cassine (as hydrochloride)	EtOH	276(1.52)	44-0471-64
$C_{19}H_{38}O_2$			
Unknown acid, methyl ester	isooctane	181.5(4.30)	25-1861-64

Compound	Solvent	λ_{max}(log ε)	Ref.
$C_{20}H_6Cl_4N_2O_4$			
Quino[2,3-b]acridine-6,7,13,14(5H,12H)-tetrone, 1,3,8,10-tetrachloro-	H_2SO_4	413(3.97)	27-0242-65
$C_{20}H_6N_6O_{12}$			
Quino[2,3-b]acridine-6,7,13,14(5H,12H)-tetrone, 1,3,8,10-tetranitro-	H_2SO_4	366(4.09),390s(4.07)	27-0242-65
$C_{20}H_8Br_4O_5$			
Eosin	pH 7.1	516(4.9)	28-5087-64A
$C_{20}H_8Cl_2N_2O_4$			
Quino[2,3-b]acridine-6,7,13,14(5H,12H)-tetrone, 2,9-dichloro-	H_2SO_4	444(4.14)	27-0242-65
Quino[2,3-b]acridine-6,7,13,14(5H,12H)-tetrone, 4,11-dichloro-	H_2SO_4	446(3.97)	27-0242-65
$C_{20}H_8Cl_4$			
Benzo[e]pyrene, 1,3,6,8-tetrachloro-	benzene	296(4.52),308(4.59), 329(4.22),343(4.35), 360(4.41),404(2.88)	24-0218-64
	EtOH	223(4.64),243(4.56)	24-0218-64
$C_{20}H_8Mn_2O_6$			
Manganese. (butadiynylenedicyclopentadienylene)bis[tricarbonyl-	EtOH	250s(4.40),301(4.09), 344(3.83)	49-1520-65
$C_{20}H_8N_4O_8$			
Quino[2,3-b]acridine-6,7,13,14(5H,12H)-tetrone, 2,9-dinitro-	H_2SO_4	395(4.12)	27-0242-65
$C_{20}H_9Br_3$			
Benzo[a]pyrene, 1,3,6-tribromo-	benzene	292(4.32),304(4.56), 316(4.67),356(3.70), 375(4.16),392(4.50), 400(4.40),415(4.58), 424(4.49)	24-0218-64
$C_{20}H_9Cl_7$			
Benzo[a]pyrene, 1,3,4,5,6,11,12-heptachloro-4,5,11,12-tetrahydro-	dioxan	288(4.60),322(4.04), 332(4.00),338(3.98), 363(3.38)	24-0218-64
$C_{20}H_{10}$			
Butadiyne, bis(o-ethynylphenyl)-	C_6H_{12}	231(4.75),242(4.70), 256(4.40),269(4.28), 276(4.26),284(4.29), 304(4.42),324(4.37), 347?(--)	39-1151-64
$C_{20}H_{10}Br_2$			
Benzo[e]pyrene, 3,6-dibromo-	benzene	292(4.60),304(4.76), 322(4.18),335(4.42), 352(4.54),378(2.79), 389(2.59)	24-0218-64
	EtOH	226(4.74),238(4.58), 268(4.38),278(4.44)	24-0218-64

Compound	Solvent	λ max(log ε)	Ref.
$C_{20}H_{10}Br_2N_2$			
Benzo[f]naphtho[2,1-c]cinnoline, 1,6-dibromo-	$CHCl_3$	279(4.15),313(4.38), 339(4.19),368(3.68), 407(3.12),430(3.26), 455(3.24)	39-6095-64
Benzo[f]naphtho[2,1-c]cinnoline, 7,12-dibromo-	$CHCl_3$	270(4.20),279(4.25), 317(4.29),334(4.01), 383(3.22),406s(3.29), 429(3.30)	39-6095-64
Benzo[f]naphtho[2,1-c]cinnoline, 7,14-dibromo-	$CHCl_3$	269(4.29),277(4.30), 313(4.41),333(4.14), 383s(3.21),405s(3.27), 429s(3.33)	39-6095-64
Benzo[f]naphtho[2,1-c]cinnoline, 8,13-dibromo-	$CHCl_3$	261(4.35),282(4.39), 312s(4.47),326(4.15), 350(3.89),396s(3.35), 420(3.31),445(2.61)	39-6095-64
$C_{20}H_{10}Cl_4$			
Calicene, tetrachlorodiphenyl-	C_6H_{12}	243s(4.10),249(4.13), 254(4.16),267(4.20), 347(4.52)	88-2967-65
	MeOH	338(4.26)	88-2967-65
$C_{20}H_{10}F_8$			
p-Terphenyl, 2',5'-difluoro-3',6'-bis-(trifluoromethyl)-	EtOH	240(3.96),292(4.18)	39-2975-64
$C_{20}H_{10}N_2O_4$			
Quino[2,3-b]acridine-6,7,13,14(5H,12H)-tetrone	H_2SO_4	430(4.02)	27-0242-65
$C_{20}H_{10}N_4O_4$			
Benzo[f]naphtho[2,1-c]cinnoline, 7,12-dinitro-	$CHCl_3$	261(4.41),285(4.41), 301(4.43),322(4.31), 397(3.49),420s(3.39)	39-6095-64
Benzo[f]naphtho[2,1-c]cinnoline, 7,14-dinitro-	$CHCl_3$	287s(4.08),300(4.37), 347(4.05),397(3.42), 421(3.22)	39-6095-64
$C_{20}H_{10}N_4O_5$			
Benzo[f]naphtho[2,1-c]cinnoline, 7,14-dinitro-, N-oxide	$CHCl_3$	286(4.22),309(4.19), 302(4.21),335(4.17), 356(4.01),374(3.75), 396(3.35),419s(3.53), 444(3.64)	39-6095-64
$C_{20}H_{10}O_2$			
peri-Xanthenoxanthene	n.s.g.	250(4.4),290f(4.2), 320f(3.9),410(4.1), 450(4.2)	27-0208-65
$C_{20}H_{10}O_6S$			
1,4-Naphthoquinone, 2,2'-thio-bis[3-hydroxy-	dioxan	242(4.44),280(4.53), 330(3.73)	5-0151-64E
$C_{20}H_{10}O_7$			
Unknown compound, m. 260°	dioxan	251(4.22),321(4.10), 373(4.13)	24-0666-65

Compound	Solvent	$\lambda_{max}(\log \epsilon)$	Ref.
$C_{20}H_{11}Br$			
Benzo[a]pyrene, 6-bromo-	benzene	294(4.62),306(4.75), 345(3.70),361(4.14), 380(4.45),402(4.56), 410(4.14)	24-0218-64
$C_{20}H_{11}I_2NO_5$			
Fluorescein, 4-amino-4',5'-diiodo-	EtOH	205(4.63),229(4.67), 290(4.35)	44-0490-64
$C_{20}H_{11}NO_7$			
Fluorescein, 4-nitro-	EtOH	225(4.77)	44-0490-64
Fluorescein, 5-nitro-	EtOH	222(4.83)	44-0490-64
$C_{20}H_{11}N_3O$			
2-Cyclopentene-1,1,2-tricarbonitrile, 4-oxo-3,5-diphenyl-	$CHCl_3$	325(3.95)	39-2009-65
$C_{20}H_{12}$			
Benzo[j]fluoranthene	EtOH	223(4.69),240(4.65), 279(4.12),291(4.28), 307(4.45),318(4.50), 332(4.11),347(3.62), 365(3.91),384(4.10)	39-2380-64
11,12-Benzofluoranthene	benzene	360(3.9),380(4.1), 400(4.2)	12-1138-64
Benzo[a]pyrene	C_6H_{12}	368(--),382(--), 388(--),406(--)	96-0735-64
1,2-Benzopyrene	dioxan	238(4.44),258(4.35), 268(4.43),278(4.56), 290(4.70),306(4.10), 318(4.34),332(4.52), 348(2.87),358(2.72), 366(2.74),376(2.42)	24-0218-64
tetracyanoquinodimethan complex	CH_2Cl_2	560(--),720(--)	60-0408-65
3,4-Benzopyrene	benzene	284(4.70),296(4.81), 336(3.67),350(4.12), 368(4.39),389(4.48), 405(3.64)	24-0218-64
tetracyanoquinodimethan complex	CH_2Cl_2	530(--),840(--)	60-0408-65
Dibenzo[a,c]biphenylene	EtOH	252(4.33),263(4.25), 268(4.22),293(3.57), 304(4.75),316(4.73), 337s(3.19),359(2.97), 400(3.34),420(3.40)	39-5537-65
$C_{20}H_{12}Br_2$			
Butadiyne, bis[o-trans-(2-bromovinyl)phenyl]-	EtOH	249(4.70),262s(4.65), 273s(4.57),308(4.36), 328(4.41),351(4.31)	39-1151-64
Dibenzo[a,c]biphenylene dibromide	EtOH	225(4.44),273(4.19)	39-5537-65
$C_{20}H_{12}Br_2N_2$			
1,1'-Azonaphthalene, 2,2'-dibromo-	$CHCl_3$	279(4.18),356(3.85)	39-5245-65
1,1'-Azonaphthalene, 4,4'-dibromo-	$CHCl_3$	274(4.32),415(4.26)	39-5245-65
1,1'-Azonaphthalene, 5,5'-dibromo-	$CHCl_3$	274(4.26),405(4.11)	39-5245-65
2,2'-Azonaphthalene, 1,1'-dibromo-	$CHCl_3$	272(4.26),279(4.27), 289(4.28),301s(4.18), 350(4.39),397(4.28)	39-6095-64

Compound	Solvent	λ_{max}(log ϵ)	Ref.
2,2'-Azonaphthalene, 4,4'-dibromo-	$CHCl_3$	279(4.30),289(4.32), 298(4.29),336(4.19), 382s(4.03),394(4.04)	39-6095-64
2,2'-Azonaphthalene, 5,5'-dibromo-	$CHCl_3$	280(4.31),288(4.38), 298(4.29),338(4.28), 378(4.01)	39-6095-64
2,2'-Azonaphthalene, 6,6'-dibromo-	$CHCl_3$	260(4.22),267(4.21), 291(4.20),305(4.30), 324(4.05),355(4.21), 395(4.08)	39-6095-64
$C_{20}H_{12}Br_2N_2O_2$			
4a,7a-Diazoniapentacene-6,13-dione dibromide	H_2O	253(4.78),383(4.36)	44-0252-65
4a,11a-Diazoniapentacene-6,13-dione dibromide	pH 3	251(4.70),378s(4.40), 385(4.41)	44-0061-64
	MeOH	227(4.64),241s(4.57), 359s(4.34),372(4.38)	44-0061-64
	MeOH	224(4.66),251(4.70), 377(4.41)	44-0252-65
$C_{20}H_{12}N_2$			
Benzo[f]naphtho[2,1-c]cinnoline	EtOH	229(4.57),236s(4.51), 257(4.34),272(4.34), 295s(4.40),308(4.55), 326(4.24),346(3.88), 375(3.43),400(3.37), 423(3.37)	39-6095-64
	2N HCl	221(4.56),241s(4.58), 268(4.06),282(4.08), 297s(4.15),327(4.43), 397(3.83)	39-6095-64
	$CHCl_3$	272s(4.32),297(4.36), 309s(4.49),328(4.20), 345(3.87),377(3.32), 399s(3.37),421(3.40)	39-6095-64
6,13-Diazadibenz[a,h]anthracene	EtOH	218(4.66),228(4.45), 282(4.86),287(4.95), 293(4.94),318(4.18), 332(4.15),347(4.11), 369(3.81),387(3.83)	4-0140-65
	EtOH-HCl	218(4.48),293(4.80), 323(4.00),352s(4.14), 368(4.18)	4-0140-65
4,10-Diazapyrene, 5-phenyl-	EtOH	237(4.86),264s(4.49), 308(4.18),323(4.26), 336(4.32)	39-3379-65
$C_{20}H_{12}N_2O$			
4H-4,10-Diazapyren-5-one, 9-phenyl- [extra log ε listed]	EtOH	257(4.32),280(4.33), 327(4.33),340s(3.87), 344(3.88),362(3.79), 378(3.80),(3.90)	39-3379-65
$C_{20}H_{12}N_2O_2$			
6,7-Phthaloylcarbazole, 3-amino-	C_6H_{12}	222(4.6),250(4.6), 300(4.7)	65-0282-64
$C_{20}H_{12}N_2O_4$			
Pyrazine, tetra-2-furyl-	$CHCl_3$	305(4.49),358(4.35)	22-3476-65

Compound	Solvent	λ_{max}(log ϵ)	Ref.
$C_{20}H_{12}N_2O_4S$			
1,4-Naphthoquinone, 2,2'-thio-bis[3-amino-	dioxan	247(4.45),278(4.53), 332(3.57),450(3.60)	5-0151-64E
$C_{20}H_{12}N_2O_4S_2$			
1,4-Naphthoquinone, 2,2'-dithio-bis[3-amino-	dioxan	222(3.94),270(4.47), 464(3.72)	5-0151-64E
$C_{20}H_{12}N_2S_2$			
Thiazolo[5,4-f]benzothiazole, 2,6-diphenyl-	EtOH	236(4.50),276(4.27), 332(4.54),345(4.56), 356s(4.37)	7-0080-64
$C_{20}H_{12}N_4O_4$			
1,1'-Azonaphthalene, 2,2'-dinitro-	$CHCl_3$	273(4.30),418(4.11)	39-5245-65
1,1'-Azonaphthalene, 4,4'-dinitro-	$CHCl_3$	276(4.20),431(4.04)	39-5245-65
1,1'-Azonaphthalene, 5,5'-dinitro-	$CHCl_3$	255(4.23),398(4.00)	39-5245-65
1,1'-Azonaphthalene, 7,7'-dinitro-	$CHCl_3$	307(4.30),410(4.26)	39-5245-65
$C_{20}H_{12}N_4S_2$			
Naphtho[2,3-c]-1,2,5-thiadiazole, dimer	dioxan	218(4.26),280(4.40)	88-3815-64
$C_{20}H_{12}O$			
Dinaphtho[1,2-b:1',2'-d]furan	EtOH	214(4.78),226(4.65), 240s(4.62),246(4.67), 260(4.71),268(4.72), 292(4.42),305(4.45), 325(4.50),339(4.64), 350(4.43)	39-5161-64
$C_{20}H_{12}O_3S$			
Anthraquinone, 1-(phenylsulfinyl)-	EtOH	252(4.59),332(3.69)	22-1648-65
$C_{20}H_{12}O_4$			
Anthraquinone, 1,4-bis(2-propynyloxy)-	MeOH	244(4.5),274s(4.1), 320(3.4),445(3.7)	70-1349-64
$C_{20}H_{12}O_4S$			
Anthraquinone, 1-(phenylsulfonyl)-	EtOH	253(4.56),322(3.60)	22-1648-65
$C_{20}H_{12}O_4S_2$			
Benzo[1",2":4,5;4",5":4',5']dithieno-[2,3-f:2',3'-f]bis[1,3]benzodioxole, 6,13-dihydro-	DMF	267(4.32),280(4.30), 286(4.32),310(3.96), 315(3.96),320(3.99)	4-0100-65
Dinaphtho[1,8-bc:1',8'-fg][1,5]di-thiocin, S,S,S',S'-tetraoxide	THF	295(3.1)	89-0810-65
$C_{20}H_{12}O_5$			
[1,1'-Binaphthalene]-3,4-dione, 2,2',3'-trihydroxy-	$CHCl_3$	269(4.40),311(3.60), 317(3.52),326(3.68), 358(3.35),488(3.25)	24-0666-65
$C_{20}H_{12}O_6$			
1,2-Naphthoquinone, 3-hydroxy-, dimer	dioxan	250(4.32),294(3.50)	24-0666-65
$C_{20}H_{13}Cl$			
Triptycene, 9-chloro-	heptane	270(3.40),277(3.51)	24-0428-65

Compound	Solvent	$\lambda_{max}(\log \epsilon)$	Ref.
$C_{20}H_{13}ClN_2O$			
4(1H)-Quinazolinone, 1-(p-chloro-phenyl)-2-phenyl-	MeOH	233(4.46),281(3.88)	44-1020-65
$C_{20}H_{13}FN_2O$			
4(1H)-Quinazolinone, 1-(p-fluoro-phenyl)-2-phenyl-	EtOH	234(4.39),265s(3.94)	44-1020-65
$C_{20}H_{13}FN_2S$			
4(1H)-Quinazolinethione, 1-(p-fluoro-phenyl)-2-phenyl-	EtOH	218(4.55),262(4.10)	44-1020-65
$C_{20}H_{13}N$			
5H-Naphtho[2,3-b]carbazole	benzene	292(4.93),312(4.55), 325(4.48)	24-0212-64
	$C_6H_3Cl_3$	362(3.70),382(3.80), 412(3.38),435(3.64), 463(3.65)	24-0212-64
5H-Naphtho[2,3-c]carbazole	EtOH	258(3.62),283(3.70), 306(4.54),317(4.50), 332(3.67),348(3.82), 367(3.93),397(3.71), 420(3.64)	24-0212-64
$C_{20}H_{13}NO_2$			
Triptycene, 9-nitro-	heptane	270(3.22),278(3.26)	24-0428-65
$C_{20}H_{13}NO_5$			
Fluorescein, 4-amino-	EtOH	222(4.83),285(4.32)	44-0490-64
Fluorescein, 5-amino-	EtOH	222(4.78)	44-0490-64
Oxysanguinarine	EtOH	256(4.33),286(3.82), 310(4.0),330(3.8)	23-0679-65
$C_{20}H_{13}N_3$			
Acetonitrile, 2-quinolyl-2(1H)-quino-lylidene-	$CHCl_3$	276(4.37),322s(4.17), 363(3.84),382(3.90), 432(4.23),456(4.43), 485(4.33)	44-0243-65
$C_{20}H_{13}N_3O_4$			
Indole, 5-nitro-3-(p-nitrophenyl)-2-phenyl-	EtOH	292(4.52)	78-0823-65
$C_{20}H_{14}$			
Anthracene, 1-phenyl-	CH_2Cl_2	258(5.14),315(3.08), 333(3.45),349(3.74), 367(3.92),387(3.84)	44-1981-65
Anthracene, 2-phenyl-	CH_2Cl_2	261(4.72),281(4.85), 319(3.15),334(3.45), 350(3.70),368(3.82), 389(3.67)	44-1981-65
Anthracene, 9-phenyl-	CH_2Cl_2	259(5.13),318(3.09), 332(3.46),350(3.78), 368(3.97),387(3.93)	44-1981-65
Phenanthrene, 9-phenyl-	EtOH	209(4.53),212(4.53), 255(4.75),296(4.13), 333(2.59),341(2.44), 350(2.50)	39-5544-64
Triptycene	heptane	271(3.55),278(3.68)	24-0428-65
	EtOH	264s(3.3),270(3.54), 278(3.66)	5-0021-64F

Compound	Solvent	λ_{max}(log ε)	Ref.
$C_{20}H_{14}Br_2N_2$			
4a,8a-Diazoniapentaphene dibromide	EtOH	232s(4.32),255(4.44), 274s(4.18),286(4.29), 298(4.35),325s(4.30), 341(4.67),355(4.68), 362(4.51),384(4.59), 422(3.73),447(3.72)	44-0856-64
4a,12a-Diazoniapentaphene dibromide	EtOH	235(4.51),258(4.59), 286(4.39),296(4.36), 347s(4.65),357(4.79), 375(4.51),404(3.82), 427(3.70)	44-0856-64
12a,14a-Diazoniapentaphene dibromide	EtOH	231(4.48),242(4.47), 263(4.63),296(4.26), 323s(4.20),340(4.47), 348s(4.31),369(4.49), 389(4.54),395s(4.45), 415(3.60)	44-0856-64
x-Bromo-5-methyl-14H-indolo[2,3-a]phenanthridizinium bromide	EtOH	222(4.50),237(4.32), 255(4.25),277(4.22), 297(4.33),318s(4.00), 379(4.29),400(4.23)	4-0168-64
$C_{20}H_{14}Br_2N_2O_2$			
4a,8a-Diazoniapentaphene dibromide, 6,7-dihydroxy-	MeOH	229(4.51),262(4.42), 363(4.61),380(4.60)	44-0252-65
$C_{20}H_{14}Br_2O_8$			
Flavone, 6,8-dibromo-4',5,7-trihydroxy-3'-methoxy-, 4',7-diacetate	EtOH	235(4.16),281(4.32), 330(4.15)	44-0897-65
$C_{20}H_{14}Cl_2N_4O$			
Acetophenone, 2-[(m-chlorophenyl)azo]-2-[(m-chlorophenyl)hydrazono]-	EtOH	252(4.29),308(4.17), 436(4.38)	5-0099-65B
Acetophenone, 2-[(p-chlorophenyl)azo]-2-[(p-chlorophenyl)hydrazono]-	EtOH	249(4.29),321(4.26), 438(4.38)	5-0099-65B
$C_{20}H_{14}Cl_2N_4O_6$			
3-(p-Chlorophenyl)-5-[3,4-(methylenedioxy)phenyl]-2-phenyl-2H-tetrazolium perchlorate	n.s.g.	227(4.48),276(4.29), 304(4.28)	5-0197-65I
isomer	n.s.g.	227(4.55),276(4.35), 304(4.35)	5-0197-65I
$C_{20}H_{14}FNO_3$			
Anthranilic acid, N-benzoyl-N-(p-fluorophenyl)-	EtOH	227s(4.27),278s(3.83)	44-1020-65
$C_{20}H_{14}Fe$			
Butadiyne, 1-ferrocenyl-4-phenyl-	MeOH	275(4.28),292(4.26), 310(4.25),446(3.04)	49-1750-64
1,2-Perinaphthaleneferrocene	EtOH	229(4.64),290(4.12), 321(3.83),420(3.20), 485(3.26)	25-0195-64
$C_{20}H_{14}I_2O_8$			
Flavone, 3',5,7-trihydroxy-6,8-diiodo-4'-methoxy-, 3',7-diacetate	EtOH	229(4.48),282(4.27), 307(4.30),332(4.44)	44-0897-65

Compound	Solvent	λ_{max}(log ϵ)	Ref.
$C_{20}H_{14}N_2$			
1,1'-Azonaphthalene	$CHCl_3$	271(4.23),403(4.08)	39-5245-65
$C_{20}H_{14}N_2O$			
Nicotinonitrile, 6-(2,2-diphenylvinyl)-1,2-dihydro-2-oxo-	EtOH	237(4.30),246(4.32), 338(4.28)	35-5198-65
4(1H)-Quinazolinone, 1,2-diphenyl-	EtOH	236(4.40)	44-1020-65
$C_{20}H_{14}N_2O_2$			
Indole, 2-(2-naphthyl)-3-(2-nitrovinyl)-	n.s.g.	276(3.57),316(3.57), 424(3.54)	28-0609-64A
Indole, 3-(p-nitrophenyl)-2-phenyl-	EtOH	239(4.43),304(4.24), 405(3.94)	78-0823-65
Nicotinonitrile, 1,2-dihydro-2-oxo-4-phenacyl-6-phenyl-	EtOH	270(4.25),359(4.22), 405s(3.76)	35-5198-65
Nicotinonitrile, 1,2-dihydro-2-oxo-6-phenacyl-5-phenyl-	EtOH	245(4.38),345(4.04), 448(3.69)	35-5198-65
$C_{20}H_{14}N_2O_2S$			
3,5-Pyrazolidinedione, 1,2-diphenyl-4-(2-thenylidene)-	MeOH	243(4.24),375(4.49)	32-0320-65
$C_{20}H_{14}N_2O_3$			
3,5-Pyrazolidinedione, 4-furfurylidene-1,2-diphenyl-	MeOH	243(4.19),375(4.51)	32-0320-65
$C_{20}H_{14}N_2S$			
4(1H)-Quinazolinethione, 1,2-diphenyl-	EtOH	219(4.56),241s(4.25), 266s(4.11)	44-1020-65
$C_{20}H_{14}N_4$			
Benzo[f]naphtho[2,1-c]cinnoline, 7,12-diamino-	pH 1	223(4.58),234s(4.50), 275(4.20),310(4.30), 323s(4.20)	39-6095-64
	EtOH	257(4.58),339s(4.19), 460(3.46)	39-6095-64
	80% H_2SO_4	222(4.64),237s(4.54), 267(4.18),280(4.16), 327(4.39),390(3.85)	39-6095-64
Benzo[f]naphtho[2,1-c]cinnoline, 7,14-diamino-	pH 1	222(4.50),230s(4.49), 276(4.14),310(4.27), 323s(4.20)	39-6095-64
	EtOH	260(4.47),330(4.20), 461(3.56)	39-6095-64
	80% H_2SO_4	222(4.57),238s(4.52), 269(4.19),281(4.19), 327(4.41),390(3.85)	39-6095-64
$C_{20}H_{14}N_4O_2$			
9-Acridinecarboxaldehyde, p-nitrophenylhydrazone	pH 1.25-4	515(4.49)	65-2371-64
	pH 5.35	495(4.32)	65-2371-64
	pH 5.85	465(4.37)	65-2371-64
$C_{20}H_{14}N_4O_4$			
Ethylene, 1-(2,4-dinitrophenylazo)-2,2-diphenyl-	EtOH	285(3.83),400(4.42)	35-0863-65
1,3,4-Oxadiazole, 2-anilino-5-[1-phenyl-2-(5-nitro-2-furyl)vinyl]-	propylene glycol	248(4.34),392(4.23)	95-0219-64

Compound	Solvent	λ_{max}(log ε)	Ref.
$C_{20}H_{14}N_8O_8$			
Phthalaldehyde, bis(2,4-dinitro-phenylhydrazone)	$CHCl_3$	370(4.48)	35-0874-65
$C_{20}H_{14}O$			
Fluorene, 1-benzoyl-	MeOH	255(4.53),263(4.54), 287(3.81),320(3.34)	35-5417-65
Fluorenone, 3-methyl-4-phenyl-	EtOH	261(4.69),300(3.50), 375s(2.48)	24-1329-64
$C_{20}H_{14}OS$			
1,2-Benzanthracene, 10-acetylthio-	EtOH	223(4.50),232(4.43), 248s(4.18),256s(4.22), 264s(4.24),274(4.37), 284(4.62),295(4.70), 329(3.48),343(3.70), 360(3.78),373(3.66), 391(3.37)	12-0109-64
$C_{20}H_{14}O_2$			
1,1'-Binaphthol	EtOH	268(4.08),278(4.73), 288(4.17),324s(2.81), 335(3.65)	44-1391-64
2,5-Cyclohexadien-1-one, 4-(α-phenyl-phenacylidene)-	EtOH	262(4.32),365(4.45)	88-2131-64
	n.s.g.	207(4.15),245(4.16), 301s(4.14),332(4.22)	88-2137-64
$C_{20}H_{14}O_3$			
Phthalide, 3-(p-hydroxyphenyl)-3-phenyl-	80% H_2SO_4	376(3.8),462(4.4)	56-0245-65
	2N NaOH	240(4.2),295(3.0)	56-0245-65
	EtOH	283(3.2)	56-0245-65
$C_{20}H_{14}O_4$			
Phenolphthalein	80% H_2SO_4	380(4.0),495(4.51)	56-1533-64
	pH 11.5	295(4.1),375(3.9), 552(4.3)	56-1533-64
	2.5N NaOH	243(4.5),295(3.7)	56-1533-64
	EtOH	228(4.5),278(3.7)	56-1533-64
Phthalide, 3-(o-hydroxyphenyl)-3-(p-hydroxyphenyl)-	80% H_2SO_4	449(3.9),496(3.9)	56-1767-64
	pH 11.5	410(3.0),520(2.9)	56-1767-64
	2N NaOH	236(4.3),293(3.7)	56-1767-64
	EtOH	226(4.5),281(3.8)	56-1767-64
Xanthene, 9-(3-carboxy-2-hydroxyphenyl)-	EtOH	293(3.8),305(3.8)	95-0052-64
Xanthene, 9-(3-carboxy-4-hydroxyphenyl)-	EtOH	288(3.7),293(3.7), 310(3.5)	95-0052-64
Xanthene, 9-(4-carboxy-2-hydroxyphenyl)-	EtOH	243(4.1),292(3.7)	95-0052-64
Xanthene, 9-(5-carboxy-2-hydroxyphenyl)-	EtOH	253(4.2),282(3.6), 290(3.5)	95-0052-64
$C_{20}H_{14}O_4Se_2$			
1,3-Propanedione, 1,1'-p-phenylene-bis[3-selenophene-2-yl-	n.s.g.	255(4.17),305(4.08), 365(4.60)	67-0379-65
$C_{20}H_{14}O_5$			
γ-Rhodomycinone, bisanhydro-	EtOH	496(4.25),532(4.30)	39-3927-64
$C_{20}H_{14}O_6$			
Matsukaze lactone	EtOH	220(4.37),257(4.02), 326(4.44)	94-1232-64
	EtOH-NaOH	221(4.48),337(4.13)	94-1232-64

Compound	Solvent	λ_{max}(log ε)	Ref.
β-Rhodomycinone, bisanhydro-	$CHCl_3$	483(4.15),516(4.58), 556(4.77)	24-2797-65
	H_2SO_4	554(4.43),602(4.78)	24-2797-65
	DMF	582(4.23),628(4.11)	24-2797-65
5,12-Tetracenequinone, 8-ethyl-1,4,6,11-tetrahydroxy-	$CHCl_3$	488(4.20),523(4.57), 562(4.68)	24-3145-65
	H_2SO_4	576(4.51),618(4.70)	24-3145-65
$C_{20}H_{14}O_7$			
5,12-Tetracenequinone, 8-ethyl-1,4,6,7,11-pentahydroxy-	$CHCl_3$	508(4.20),545(4.60), 589(4.73)	24-3145-65
	H_2SO_4	607(4.51),660(4.86)	24-3145-65
$C_{20}H_{15}ClN_2$			
Quinoline Red	EtOH	230(4.8),260(4.2), 350f(3.8),400(3.3), 500(4.3),520(4.5), 580(4.4)	27-0325-65
$C_{20}H_{15}ClN_2O_4$			
5-Methyl-14H-indolo[2,3-a]phenanthridizinium perchlorate	EtOH	215(4.48),248s(4.12), 277(4.28),300(4.26), 379(4.23),405(4.22)	4-0168-64
$C_{20}H_{15}ClN_4O$			
Acetophenone, 4'-chloro-2-(phenylazo)-2-(phenylhydrazono)-	EtOH	255(4.25),318(4.21), 440(4.34)	5-0099-65B
Acetophenone, 2-[(o-chlorophenyl)-hydrazono]-2-(phenylazo)-	EtOH	248(4.18),312(4.24), 448(4.34)	5-0099-65B
Acetophenone, 2-[(p-chlorophenyl)-hydrazono]-2-(phenylazo)-	EtOH	246(4.26),317(4.23), 444(4.37)	5-0099-65B
$C_{20}H_{15}ClO_6$			
7-(p-Benzoyloxyphenyl)tropylium perchlorate	MeCN	272(4.28),378(4.30)	24-0029-64
$C_{20}H_{15}ClO_7$			
Fragilin, diacetate	EtOH	275(4.47),343(3.57)	1-0839-65
$C_{20}H_{15}FN_2O_2$			
Benzamide, N-(o-carbamoylphenyl)-N-(p-fluorophenyl)-	EtOH	276(3.84)	44-1020-65
$C_{20}H_{15}N$			
Isoindole, 1,3-diphenyl-	n.s.g.	228s(4.24),237(4.27), 268s(4.22),273(4.27), 322(4.13),335(4.16), 387(4.37)	77-0272-65
Triptycene, 9-amino-	heptane	270(3.51),279(3.62)	24-0428-65
$C_{20}H_{15}NO$			
5,6-Benzoquinoline, 2-(p-methoxyphenyl)-	EtOH	262(4.44),283(4.55), 305s(4.37),345(4.09), 363(5.10)	88-0659-64
2,4-Pentadien-1-one, 5-phenyl-1-(2-quinolyl)-	H_2SO_4	595(2.75)	65-1724-65
$C_{20}H_{15}NO_2$			
3-Anthranilyldiphenylcarbinol	EtOH	283s(3.31),316(3.78)	44-1104-65

Compound	Solvent	λ_{max}(log ε)	Ref.
Phthalimide, N-[1-(1-naphthyl)-ethyl]-	MeOH	262s(3.76),271(3.93), 281(4.02),288(3.90), 292(3.68),302(2.82), 312(2.52)	24-0533-64
$C_{20}H_{15}NO_4$			
Sanguinarine, dihydro-	pH 1	238(4.53),250s(4.49), 265(4.40),274(4.40), 307(4.20),321(4.36), 338(4.20),355(4.30)	100-0199-65
	EtOH	237(4.55),284(4.57), 322(4.21),335s(4.12), 350s(3.72)	100-0191-65
	EtOH-KOH	235(4.53),282(4.56), 322(4.23),335s(4.12), 350s(3.72)	100-0199-65
$C_{20}H_{15}NO_5$			
Sanguinarine	EtOH	259(4.34),312(4.61), 338(4.18)	23-0679-65
$C_{20}H_{15}NO_6$			
Pretetramide, 6-methyl-	H_2SO_4-boric acid	263(4.36),278(4.34), 341(4.15),400(4.17), 512(4.14)	35-1794-65
$C_{20}H_{15}NO_7$			
Pretetramide, 4-hydroxy-6-methyl-	H_2SO_4-boric acid	280(4.34),316(4.62), 467(4.19),520(4.22)	35-1794-65
$C_{20}H_{15}N_3O$			
p-Benzoquinone, azine with 10-methyl-9(10H)-acridone	C_5H_5N	584(4.42)	5-0009-65J
	99% C_5H_5N	585(4.42)	
	98% C_5H_5N	588(4.41)	
	95% C_5H_5N	594(4.40)	
	90% C_5H_5N	600(4.40)	
	80% C_5H_5N	610(4.42)	
	70% C_5H_5N	619(4.43)	
	60% C_5H_5N	636(4.44)	
	50% C_5H_5N	633(4.45)	
	40% C_5H_5N	635(4.46)	
	30% C_5H_5N	645(4.47)	
	20% C_5H_5N	650(4.48)	
	10% C_5H_5N	646(4.38)	
3-Hydroxy-1,4,5-triphenyl-1H-1,2,4-triazolium hydroxide, inner salt	n.s.g.	290(3.84)	4-0105-65
$C_{20}H_{15}N_3O_2$			
3,5-Pyrazolidinedione, 1,2-diphenyl-4-(pyrrol-2-ylmethylene)-	MeOH	244(4.21),305(3.52), 410(4.74)	32-0320-65
$C_{20}H_{15}N_3S$			
3-Mercapto-1,4,5-triphenyl-1H-1,2,4-triazolium hydroxide, inner salt	n.s.g.	241(4.34),316(3.44)	4-0105-65
$C_{20}H_{15}OP$			
Phosphine oxide, diphenyl-(phenylethynyl)-	EtOH	230(4.26),244(4.26), 253(4.36),264(4.28), 286(3.26)	28-1537-64A

Compound	Solvent	$\lambda_{max}(\log \epsilon)$	Ref.
$C_{20}H_{15}P$			
Phosphine, diphenyl-(phenylethynyl)-	EtOH	230(4.20),246(4.20), 254(4.28),265(4.20), 287(3.56)	28-1537-64A
$C_{20}H_{15}PS$			
Phosphine sulfide, diphenyl-(phenylethynyl)-	EtOH	229(4.08),245(4.32), 252(4.36),264(4.26), 285(3.9)	28-1537-64A
$C_{20}H_{16}$			
Chrysene, 3,x-dimethyl-	n.s.g.	225(4.4),245(4.2), 258s(4.8),267(5.0), 289(4.0),299(4.0), 309(4.0),347(4.0)	70-2021-64
Triphenylene, 2,7-dimethyl-	EtOH	252(4.96),261(5.21), 276(4.28),288(4.22), 317(2.92),325(2.74), 332(2.94),340(2.68), 348(2.88)	78-3289-65
$C_{20}H_{16}BrIO_8$			
Flavanone, 6-bromo-3',5,7-trihydroxy-8-iodo-4'-methoxy-, 3',7-diacetate	EtOH	280(4.04),357(3.73)	44-0897-65
$C_{20}H_{16}BrNO_2S$			
Biphenyl-2'-sulfonic acid, N-(o-bromobenzyl)-2-amino-5'-methyl-, sultam	n.s.g.	220(4.34),268(4.09), 307(3.74)	12-1406-64
Biphenyl-2'-sulfonic acid, N-(p-bromobenzyl)-2-amino-5'-methyl-, sultam	n.s.g.	228(4.36),263s(3.89), 310(3.64)	12-1406-64
$C_{20}H_{16}Br_2O_3$			
Cyclohexanone, 2,6-bis(o-bromobenzoyl)-	n.s.g.	242(4.13)	44-3711-65
$C_{20}H_{16}Br_2O_8$			
Flavanone, 6,8-dibromo-4',5,7-trihydroxy-3'-methoxy-, 4',7-diacetate	EtOH	278(4.16),357(3.72)	44-0897-65
$C_{20}H_{16}ClNO_5$			
7-(p-Benzamidophenyl)tropylium perchlorate	MeCN	277(4.25),444(4.37)	24-0029-64
$C_{20}H_{16}ClNO_8$			
2-Naphthoic acid, 3-(4-carbamoyl-2,3,5-trihydroxybenzyl)-5-chloro-1,2,3,4-tetrahydro-4,8-dihydroxy-4-methyl-1-oxo-, γ-lactone	pH 1	260(4.09),352(3.68)	35-1795-65
Tetracycline, 4a,12a-anhydro-7-chloro-4-dedimethylamino-4-hydroxy-	pH 1	250(4.32),362(4.01)	35-1795-65
$C_{20}H_{16}ClNO_9$			
Tetracycline-4,6-hemiketal, 11a-chloro-4-oxo-4-dedimethylamino-	MeOH-HCl	258(4.42),343(3.68)	23-1382-65
	MeOH-HCl	258(4.42),343(3.68)	35-2736-64
$C_{20}H_{16}Cl_2N_2O_2$			
p-Benozquinone, 2,5-bis(benzylamino)-3,6-dichloro-	methyl cellosolve	234s(4.27),356(4.40), 529(2.49)	78-1889-64

Compound	Solvent	λ_{max}(log ϵ)	Ref.
$C_{20}H_{16}Cl_2N_2O_2S$			
5-Pyrimidinecarboxylic acid, 2-(2,4-dichlorophenyl)-1,4-dihydro-6-methyl-1-phenyl-4-thioxo-, ethyl ester	EtOH	230(4.22),340(4.39)	44-1115-64
$C_{20}H_{16}Cl_2O_3$			
Cyclohexanone, 2,6-bis(o-chlorobenzoyl)-	n.s.g.	250(4.14)	44-3711-65
$C_{20}H_{16}F_2N_2O_2$			
p-Benzoquinone, 2,5-bis(benzylamino)-3,6-difluoro-	methyl cellosolve	207(4.43),355(4.45), 545(2.32)	78-1889-64
$C_{20}H_{16}F_2O_3$			
Cyclohexanone, 2,6-bis(o-fluorobenzoyl)-	n.s.g.	232(4.21)	44-3711-65
$C_{20}H_{16}Fe$			
Ferrocene, 1-naphthyl-	EtOH	225(4.70),279(3.81), 302(3.87),356(3.06), 446(2.52)	25-0195-64
$C_{20}H_{16}FeS$			
Thiophene, 2-ferrocenyl-5-phenyl-	$CHCl_3$	327(4.35),446(3.18)	49-1750-64
$C_{20}H_{16}I_2O_3$			
Cyclohexanone, 2,6-bis(o-iodobenzoyl)-	n.s.g.	248(4.14)	44-3711-65
$C_{20}H_{16}I_2O_8$			
Flavanone, 3',5,7-trihydroxy-6,8-diiodo-4'-methoxy-, 3',7-diacetate	EtOH	280(4.03),364(3.68)	44-0897-65
$C_{20}H_{16}KNO_9$			
1,11-Epoxynaphthacene-3-carboxamide, 1,4,4a,5,5a,6,11,11a,12,12a-decahydro-1,2,4a,7-tetrahydroxy-11-methyl-4,5,6-trioxo-, potassium derivative	MeOH-HCl	268(4.40),346(3.72)	35-2736-64
	MeOH-NaOH	370(4.4)	35-2736-64
$C_{20}H_{16}MgN_2O_2$			
Magnesium, bis(2-methyl-8-quinolinolato)-	$CHCl_3$	310(3.61),370(3.28)	59-1229-65
$C_{20}H_{16}N_2$			
Benzimidazole, 1-benzhydryl-	n.s.g.	249(3.93),253(3.93), 266(3.71),274(3.69), 282(3.64)	65-1582-64
Pyridine, 2,2'-(p-phenylenedivinylene)di-	dioxan	244(4.08),356(4.80)	38-2839-64A
Pyridine, 3,3'-(p-phenylenedivinylene)di-	dioxan	243(3.96),353(4.35)	38-2839-64A
Quinoline, 4-[2-(1-methylindol-3-yl)vinyl]-	MeOH	380(3.9),495(3.8)	87-0397-65
	HOAc	485(4.5)	87-0397-65
$C_{20}H_{16}N_2O$			
Nicotinonitrile, 1,2-dihydro-2-oxo-4-phenethyl-6-phenyl-	EtOH	253(4.07),354(4.23)	44-3593-65
Nicotinonitrile, 1,2-dihydro-2-oxo-6-phenethyl-4-phenyl-	EtOH	246(4.23),349(4.10)	44-3593-65
Nicotinonitrile, 4-(p-methoxyphenyl)-2-methyl-6-phenyl-	EtOH	270(4.46),300(4.35)	2-0561-65

Compound	Solvent	$\lambda_{max}(\log \epsilon)$	Ref.
$C_{20}H_{16}N_2O_2$			
Benzamide, N-(o-carbamoylphenyl)-N-phenyl-	EtOH	244s(4.08),276s(3.86)	44-1020-65
Benzamide, α-(N-phenylbenzimidoyloxy)-	EtOH	233s(4.38),276s(3.76)	44-1020-65
2-Indolinone, 1-benzyl-3-hydroxy-3-(2-pyridyl)-	MeOH	211(4.42),258(3.81), 290(3.02)	44-2534-65
Nicotinonitrile, 1,2-dihydro-6-(2-hydroxy-2,2-diphenylethyl)-2-oxo-	EtOH	235(4.04),338(4.12)	35-5198-65
$C_{20}H_{16}N_2O_2S$			
1H-2,1,3-Benzothiadiazine, 1-benzyl-4-phenyl-, 2,2-dioxide	EtOH	226(4.34),274(4.13), 293s(4.02),360(3.7)	44-3960-65
$C_{20}H_{16}N_2O_2Zn$			
Zinc, bis(2-methyl-8-quinolinolato)-	$CHCl_3$	269(4.85),339(3.36), 386(3.63)	59-1229-65
1-Indolinecrotonic acid, 2-indol-3-yl-γ-oxo-, cis	MeOH	257(4.12),280(4.09), 290(4.04)	39-0526-64
Pyrrolo[1,2-a]quinoxaline-2-acetic acid, 1,5-dihydro-4-methyl-1-oxo-α-phenyl-	EtOH	236(4.32),255s(4.07), 291s(3.92),300(3.94), 328s(3.73),429(3.80)	39-3678-65
Pyrrolo[1,2-a]quinoxaline-2-acetic acid, 1,5-dihydro-1-oxo-4-phenyl-, methyl ester	EtOH	238(4.43),273(4.30), 292s(4.19),330s(3.67), 340(3.58)	35-1830-64
$C_{20}H_{16}N_2O_7$			
Cyclohexanone, 2,6-bis(p-nitrobenzoyl)-	n.s.g.	264(4.42)	44-3711-65
$C_{20}H_{16}N_2O_8$			
Tetracycline, 5a,6-anhydro-4-hydroxyimino-4-dedimethylamino-	MeOH-HCl	225(4.48),263(4.64), 302(4.26),422(4.07)	23-1382-65
$C_{20}H_{16}N_4$			
Quinoxaline, 2,2'-vinylenebis-[3-methyl-, trans	EtOH	235(4.23),278(4.38), 382(4.39)	44-1542-65
$C_{20}H_{16}N_4O$			
Acetophenone, 2-(phenylazo)-2-(phenylhydrazono)-	MeOH	248(4.23),315(4.20), 440(4.34)	44-2959-64
	EtOH	246(4.23),314(4.31), 440(4.45)	5-0099-65B
$C_{20}H_{16}N_4O_8$			
1,2,3,4-Tetrahydro-3-methyl-4-oxo-benzo[c]quinolizinium picrate	H_2O	202(4.49),253(4.56), 326(3.93)	78-2529-65
$C_{20}H_{16}N_5O_2$			
Verdazyl, 1,5-diphenyl-3-(p-nitrophenyl)-	CCl_4	730(3.55)	49-0457-64
	DMF	740(3.54)	49-0457-64
$C_{20}H_{16}O$			
Furan, 2,5-distyryl-	EtOH	234(4.16),273(4.34), 281(4.34),379(4.59)	4-0318-65
$C_{20}H_{16}O_2$			
Xanthene, 9-(2-hydroxy-3-methylphenyl)-	EtOH	250(3.9),282(3.7)	95-0052-64
Xanthene, 9-(2-hydroxy-4-methylphenyl)-	EtOH	282(3.7)	95-0052-64
Xanthene, 9-(2-hydroxy-5-methylphenyl)-	EtOH	254(4.0),287(3.7), 292(3.7)	95-0052-64

Compound	Solvent	λ_{max}(log ε)	Ref.
Xanthene, 9-(4-hydroxy-2-methylphenyl)-	EtOH	235s(4.2),280(3.6)	95-0052-64
Xanthene, 9-(4-hydroxy-3-methylphenyl)-	EtOH	239(4.2),281(3.6)	95-0052-64
$C_{20}H_{16}O_2S_4$			
$\Delta^{2,2'}$-Bi-1,3-dithiole, 4,4'-bis-(p-methoxyphenyl)-	THF	270(4.36),332(4.08), 392(3.71)	89-0453-65
$C_{20}H_{16}O_4S$			
2,5-Thiophenedicarboxylic acid, 3,4-diphenyl-, ethyl ester	EtOH	226(4.32),282(4.06)	35-1739-65
$C_{20}H_{16}O_6$			
Cinnamic acid, 3,4-(methylenedioxy)-, 3,4-(methylenedioxy)cinnamyl ester, cis-trans	EtOH	270(4.21),295(4.22), 321(4.23)	78-0871-64
trans-trans	EtOH	270(4.22),291(4.24), 322(4.30)	78-0871-64
Isopinastric acid	MeOH	231(4.19),270(4.09), 321(4.25)	2-0017-64
Leprapinic acid	MeOH	270(4.18),316s(3.92)	78-3205-65
Pinastric acid	MeOH	226(3.94),295(4.29), 388(3.85)	2-0017-64
Viridin	EtOH	242(4.49),300(4.22)	39-3803-65
β-Viridin	EtOH	243(4.45),300(4.25)	39-3803-65
$C_{20}H_{16}O_7$			
Flavone, 4',5-diacetoxy-7-methoxy-	EtOH	258(4.31),307(4.40)	12-1871-65
4H-Naphtho[1,2-b]pyran-4-one, 3-acetyl-5,6-dihydroxy-2-methyl-, diacetate	EtOH	222(4.57),253(4.55), 260s(4.49),279s(4.17), 288(3.92),302(3.83), 320s(3.51),335(3.56), 350(3.56)	94-0316-64
$C_{20}H_{16}O_8$			
Flavone, 3',5,7-trihydroxy-4'-methoxy-, 3',7-diacetate	EtOH	271(4.27),324(4.35)	44-0897-65
Flavone, 4',5,7-trihydroxy-3'-methoxy-, 4',7-diacetate	EtOH	243(4.45),272(4.36), 324(4.16)	44-0897-65
$C_{20}H_{16}O_9$			
Xanthen-9-one, 1,3,5-trihydroxy-6-methoxy-, triacetate	MeOH	235(4.56),276(4.19), 345(4.32)	78-2653-65
$C_{20}H_{16}O_{12}$			
1,4-Naphthoquinone, 2,3,5,6,8-penta-acetoxy-	EtOH	252(3.12),274(3.13), 357(2.36)	39-2141-65
$C_{20}H_{16}S$			
Thiophene, 2,5-distyryl-	MeOH	239(4.21),266(4.07), 274(4.08),380(4.51)	4-0318-65
$C_{20}H_{17}BrN_2$			
1-Benzyl-2-phenyl-1H-imidazo[1,2-a]-pyridinium bromide	EtOH	238(4.18),327(3.75)	4-0331-65
$C_{20}H_{17}BrN_2O$			
2-Acetonyl-1-phenyl-9H-pyrido[3,4-b]-indolium bromide	EtOH	205(4.48),263(4.42), 316(4.34),386(3.68)	4-0168-64

Compound	Solvent	$\lambda_{max}(\log \epsilon)$	Ref.
$C_{20}H_{17}BrN_2O_2S$			
5-Pyrimidinecarboxylic acid, 2-(p-bromophenyl)-1,4-dihydro-6-methyl-1-phenyl-4-thioxo-, ethyl ester	EtOH	254(4.06),341(4.40)	44-1115-64
$C_{20}H_{17}ClO_5$			
7-[2-(2-Methoxy-1-naphthyl)vinyl]-tropylium perchlorate	MeCN	235(4.70),270(4.22), 349(3.91),552(4.45)	24-1349-64
7-[2-(4-Methoxy-1-naphthyl)vinyl]-tropylium perchlorate	MeCN	238(4.45),260(4.17), 295s(3.89),347(3.83), 568(4.55)	24-1349-64
$C_{20}H_{17}FeNO_2$			
Butadiene, 1-ferrocenyl-4-(p-nitrophenyl)-	C_6H_{12}	375(4.43),488(3.81)	5-0088-64F
	MeOH	384(4.29),495(3.71)	5-0088-64F
	$CHCl_3$	388(4.39),508(3.86)	5-0088-64F
$C_{20}H_{17}IN_4OSSe$			
4,6-Dimethyl-7-[3-(3-methyl-2-benzoselenazolinylidene)propenyl]oxazolo-[5,4-e]-2,1,3-benzothiadiazolium iodide	EtOH	546(4.69)	17-0048-64
$C_{20}H_{17}IN_4OS_2$			
4,6-Dimethyl-7-[3-(3-methyl-2-benzothiazolinylidene)propenyl]oxazolo[5,4-e]-2,1,3-benzothiadiazolium iodide	EtOH	540(4.92)	17-0048-64
$C_{20}H_{17}IN_4O_2S$			
4,6-Dimethyl-7-[3-(3-methyl-2-benzoxazolinylidene)propenyl]oxazolo[5,4-e]-2,1,3-benzothiadiazolium iodide	EtOH	503(4.88)	17-0048-64
$C_{20}H_{17}IO_8$			
Flavanone, 3',5,7-trihydroxy-8-iodo-4'-methoxy-, 3',7-diacetate	EtOH	277(4.07),352(3.68)	44-0897-65
$C_{20}H_{17}NO$			
Benzamide, N-(diphenylmethyl)-	C_6H_{12}	220(4.3),265s(3.05)	56-0789-64
	MeOH	220(4.3),265s(3.2)	56-0789-64
$C_{20}H_{17}NO_3$			
Glyoxal, (o-aminophenyl)-, 1-(diphenyl acetal)	n.s.g.	380(3.82)	49-0889-65
Nicotinic acid, 5-benzyl-1,6-dihydro-1-methyl-6-oxo-2-phenyl-	EtOH	267(4.03),307(3.94)	78-3255-65
2-Propen-1-one, 1-(2,4-dimethoxyphenyl)-3-(2-quinolyl)-	H_2SO_4	236(4.47),274(4.08), 310(4.03),430(4.56)	65-1724-65
2-Propen-1-one, 3-(2,4-dimethoxyphenyl)-1-(2-quinolyl)-	H_2SO_4	230(4.23),269(3.97), 510(4.62)	65-1724-65
$C_{20}H_{17}NO_4$			
5H-Benzo[g]-1,3-benzodioxolo[6,5,4-de]-quinoline, 7-acetyl-6,7-dihydro-4-methoxy-	EtOH	262(4.85),286(4.58), 320(4.08),330(4.08), 368(3.23)	44-0432-65
Pyrrole-2,5-dicarboxylic acid, 3,4-di-p-tolyl-	EtOH	247(4.24),283(4.04)	44-0859-65
1H-Pyrrolo[1,2-a]indole-2-carboxylic acid, 7-(benzyloxy)-2,3-dihydro-1-oxo-, methyl ester	MeOH	325(4.36)	44-2904-65
	NaOH	356(4.42)	44-2904-65

Compound	Solvent	λmax(log ε)	Ref.
1H-Pyrrolo[1,2-a]indol-1-one, 2-acetoxy-7-(benzyloxy)-2,3-dihydro-	MeOH	216(4.57),248s(4.12), 327(4.34)	44-2910-65
4H-Quinolizine-3-carboxylic acid, 1-benzoyl-6-methyl-4-oxo-, ethyl ester	EtOH	258(4.21),350(4.11), 430(4.26)	78-3305-65
Tetracycline, 5a,6-anhydro-4-hydroxy-4-dedimethylamino-	MeOH-HCl	222(4.46),262(4.70), 310(3.68),423(4.01)	23-1382-65
$C_{20}H_{17}NO_5$			
2-Naphthacenecarboxamide, 4,4a,5,12-tetrahydro-1,3,10-trihydroxy-6-methyl-12-oxo-	EtOH	246(4.35),261(4.37), 320(3.91),382(4.46)	65-0670-65
	EtOH	246(4.35),261(4.37), 320(3.91),382(4.46)	70-0945-64
$C_{20}H_{17}NO_6$			
3H-3-Benzazepine-2,4-dicarboxylic acid, 1,5-dihydroxy-3-phenyl-, dimethyl ester	MeOH	203(4.4),248(4.5), 311(4.1)	56-1423-65
Pyrrole-2,5-dicarboxylic acid, 3,4-bis(p-methoxyphenyl)-	EtOH	253(4.14),283(3.95)	44-0859-65
$C_{20}H_{17}NO_8$			
2-Naphthacenecarboxamide, 1,4,4a,5,-12,12a-hexahydro-3,4,10,11,12a-pentahydroxy-6-methyl-1,12-dioxo-	MeOH-HCl	222(4.46),262(4.70), 310(3.68),423(4.01)	35-2736-64
2-Naphthoic acid, 3-(4-carbamoyl-2,3,5-trihydroxybenzyl)-1,2,3,4-tetrahydro-4,8-dihydroxy-4-methyl-1-oxo-, γ-lactone	pH 1	260(4.14),335(3.79)	35-1795-65
2-Naphthoic acid, 3-(4-carbamoyl-2,3,5-trihydroxybenzyl)-1,2,3,4-tetrahydro-4,8-dihydroxy-4-methyl-1-oxo-, ε-lactone	pH 1	261(4.12),335(3.79)	35-1795-65
Tetracycline, 4a,12a-anhydro-4-dedimethylamino-4-hydroxy-	pH 1	370(3.75),485(4.17)	35-1795-65
Tetracycline, 12a-epidedimethylamino-anhydrooxy-	MeOH-HCl	224(4.49),268(4.63), 413(3.92)	35-0134-65
	MeOH-NaOH	232(4.45),266(4.53), 340(3.77),418(4.08)	35-0134-65
$C_{20}H_{17}NO_9$			
Tetracycline-4,6-hemiketal, 4-dedimethylamino-4-oxo-	MeOH-HCl	268(4.40),346(3.72)	23-1382-65
	MeOH-NaOH	370(4.40)(changing)	23-1382-65
$C_{20}H_{17}N_3O_4S$			
5-Pyrimidinecarboxylic acid, 1,4-dihydro-6-methyl-2-(m-nitrophenyl)-1-phenyl-4-thioxo-, ethyl ester	EtOH	243(4.25),342(4.35)	44-1115-64
5-Pyrimidinecarboxylic acid, 1,4-dihydro-6-methyl-2-(p-nitrophenyl)-1-phenyl-4-thioxo-, ethyl ester	EtOH	262(4.12),340(4.31)	44-1115-64
$C_{20}H_{17}N_4$			
Verdazyl, 1,3,5-triphenyl-	benzene	400(3.9),720(3.64)	49-0457-64
	MeOH-HCl	280(4.2),340(4.3), 545(4.1)	49-0457-64
$C_{20}H_{17}N_5$			
4H-1,2,4-Triazole, 3,5-dianilino-4-phenyl-	EtOH	260(4.36)	39-3912-65

Compound	Solvent	$\lambda_{max}(\log \epsilon)$	Ref.
$C_{20}H_{17}N_5O$			
Adenine, N-acetyl-3-benzhydryl-	pH 1	282(4.22)	35-5442-65
	pH 7	274(4.23)	35-5442-65
	pH 13	267(4.04),292(4.00)	35-5442-65
$C_{20}H_{17}O_2P$			
Phosphine oxide, diphenylphenacyl-	MeOH	250(4.20),280(3.18), 320(2.24)	70-0669-65
	MeOH-NaOMe	250(4.20),320(3.3)	70-0669-65
$C_{20}H_{18}$			
Benzene, p-di-1,3,5-cycloheptatrien-1-yl-	EtOH	259(3.82),272s(--)	88-2903-65
Benzo[a]fluorene, 9-ethyl-10-methyl-	C_6H_{12}	222(4.58),241(4.29), 249(4.46),258(4.85), 268(5.08),305(4.31), 317(4.30),333(3.47)	42-0643-64
Fluorene, 9-(7-norcarylidene)-	EtOH	230(4.74),246(4.46), 256(4.65),275(4.23), 285(4.33),301(4.24), 314(4.33)	35-0863-65
$C_{20}H_{18}Cl_2N_4O_4$			
Piperazine-2,5-dione, 3,6-bis[N-(5-chlorosalicylidene)aminoxymethyl]-	MeOH	220(4.60),258(4.35), 323(4.01)	44-3436-65
$C_{20}H_{18}Cl_2O_4Sn$			
Tin, dichlorobis(1-phenyl-1,3-butanedionato)-	$CHCl_3$	258(4.28),348(4.65)	101-0067-65B
$C_{20}H_{18}Fe$			
Butadiene, 1-ferrocenyl-4-phenyl-	$CHCl_3$	333(4.52),465(3.28), 345s(--)	5-0088-64F
$C_{20}H_{18}FeS$			
Hexatriene, 1-ferrocenyl-6-(2-thienyl)-	$CHCl_3$	372(4.68),382s(--), 470(3.61)	5-0088-64F
$C_{20}H_{18}INO_4$			
Berberine, iodide	EtOH	225(4.60),270(4.45), 350(4.45)	95-0721-64
$C_{20}H_{18}I_2N_2O_5$			
L-Tyrosine, N-(α-acetamido-cinnamoyl)diiodo-	pH 2	282(4.31),315(3.44)	35-1627-65
$C_{20}H_{18}N_2O$			
2-Propen-1-one, 1-(p-dimethylamino-phenyl)-3-(2-quinolyl)-	H_2SO_4	235(4.08),279(4.20), 354(4.29)	65-1724-65
2-Propen-1-one, 3-(p-dimethylamino-phenyl)-1-(2-quinolyl)-	H_2SO_4	267(4.30),377(4.37)	65-1724-65
Yobyrine, 5-methoxy-	EtOH	219(4.58),247(4.74), 289(4.17),335(3.84), 349(3.89)	78-2957-65
Yobyrine, 8-methoxy-	EtOH	213(4.56),243(4.69), 277(3.94),287(4.04), 342(3.74),352s(3.69)	78-2957-65
$C_{20}H_{18}N_2O_2$			
Carbostyril, N-methyl-, photodimer	EtOH	225s(4.15),260(3.83)	44-0305-64

Compound	Solvent	λ_{max}(log ε)	Ref.
Carbostyril, N-methyl-, photodimer (cont.)	EtOH	211(4.62),260(4.26)	1-1389-64
	dioxan	259(4.29)	1-1389-64
1,3-Pentadienylamine, N,N-dimethyl-5-(2-nitrofluoren-9-ylidene)-	MeCN	250(4.42),277(4.24), 341(4.13),497(4.64)	24-3331-64
3-Pyridinemethanol, 4-acetamido-α,α-diphenyl-	MeOH	250(4.18)	87-0722-65
3-Pyridinol, 4-methyl-6-(N-methylbenzamido)-5-phenyl-	MeOH	290(3.72)	44-1881-65
$C_{20}H_{18}N_2O_2S$			
5-Pyrimidinecarboxylic acid, 1,4-dihydro-6-methyl-1,2-diphenyl-4-thioxo-, ethyl ester	EtOH	248(3.95),341(4.40)	44-1115-64
$C_{20}H_{18}N_2O_3$			
1-Indolinebutyric acid, 2-indol-3-yl-γ-oxo-	MeOH	254(4.25),279(4.05), 289(3.97)	39-0526-64
5-Pyrimidinecarboxylic acid, 1,4-dihydro-6-methyl-4-oxo-1,2-diphenyl-, ethyl ester	EtOH	244(4.35),274s(3.91)	44-1115-64
$C_{20}H_{18}N_2O_4$			
2-Naphthoic acid, 1-hydroxy-8-methoxy-4-(phenylazo)-, ethyl ester	EtOH	386(4.26)	33-0769-64
1H-Pyrrolo[1,2-a]indol-1-one, 2-acetoxy-7-(benzyloxy)-2,3-dihydro-, oxime	MeOH	319(4.30)	44-2910-65
$C_{20}H_{18}N_2O_6$			
5H-Oxazolo[4,5-b]phenoxazine-4,6-dicarboxylic acid, 2-9,11-trimethyl-, dimethyl ester	MeOH	226(4.59),250(4.52), 416(4.2)	44-3185-65
$C_{20}H_{18}N_2O_7$			
Phenoxazine-1,9-dicarboxylic acid, 2-acetamido-4,6-dimethyl-3-oxo-, dimethyl ester	MeOH	237(4.43),401(4.32)	44-3185-65
Tetracycline, 5a,6-anhydro-4-epi-N^4,N^4-didemethyl-	MeOH-HCl	223(4.45),272(4.71), 423(3.93)	35-2736-64
	MeOH-HCl	223(4.45),272(4.79), 423(3.93)	23-1382-65
	MeOH-NaOH	229(4.39),271(4.57), 338(3.73),428(4.09)	23-1382-65
$C_{20}H_{18}N_2O_9$			
Tetracycline, 4-dedimethylamino-4-hydroxyimino-	MeOH-HCl	222(4.25),237s(4.17), 268s(4.06),276(4.07), 308(4.26),359(4.19)	23-1382-65
	MeOH-NaOH	235(4.25),245(4.24), 256s(4.21),325s(4.16), 373(4.28)	23-1382-65
$C_{20}H_{18}N_2S$			
2-Indolinecarbonitrile, 2-benzo[b]thien-2-yl-1,3,3-trimethyl-	EtOH-HCl	248(4.06),378(4.43)	44-2875-65
	EtOH	229(4.54),258(4.23), 266(4.17),291(3.80), 301(3.68)	44-2875-65
$C_{20}H_{18}N_4$			
s-Tetrazine, 1,2,3,4-tetrahydro-2,4,6-triphenyl-	benzene	365(4.0)	49-0457-64

Compound	Solvent	$\lambda_{max}(\log \epsilon)$	Ref.
$C_{20}H_{18}N_4O$			
5H-Pyrrolo[3,4-d]pyrimidine, 4-amino-6-benzoyl-7-benzyl-6,7-dihydro-	pH 1	250(4.10)	44-0194-65
	EtOH	237(4.11),264s(3.68)	44-0194-65
$C_{20}H_{18}N_4O_2S$			
Benzotriazole 1-benzenesulfonylimide, 5-methyl-2-p-tolyl-	EtOH	220(4.43),306(4.15)	39-0751-64
$C_{20}H_{18}N_4O_4$			
2,5-Cyclohexadien-1-one, 4-benzyl-4-methyl-, 2,4-dinitrophenylhydrazone	EtOH	390(4.52)	33-0094-65
$C_{20}H_{18}N_4O_5S$			
Phthalimidine, 2-[p-(2,6-dimethoxy-4-pyrimidinylsulfamoyl)phenyl]-	MeOH-acid	293(4.43)	95-1035-65
	MeOH	289(4.37)	95-1035-65
$C_{20}H_{18}N_6O_8$			
3,6-Tetrazinedicarboxylic acid, 1,4-dihydro-1,4-bis(p-nitrophenyl)-, diethyl ester	$CHCl_3$	250(4.20),379(4.49)	24-1476-65
$C_{20}H_{18}N_8O_{14}$			
1-(3-Aminopropyl)pyridinium dipicrate	H_2O	210(4.47),248(4.32), 353(4.41)	39-2763-64
$C_{20}H_{18}O$			
Biphenyl, 4-benzyl-4'-methoxy-	hexane	263.2(4.45)	78-0861-64
Cyclohexanone, 2,6-dibenzylidene-	hexane	313(4.34)	65-3645-64
	EtOH	232(3.20),320(4.27)	65-3645-64
	66% H_2SO_4	467(4.41)	65-3645-64
2-Norbornanone, 3-benzhydrylidene-	EtOH	230(3.93),305(3.91)	101-0261-65B
o-Terphenyl, 2-hydroxy-3',6'-dimethyl-	EtOH	235(4.74),294(3.19)	24-1329-64
$C_{20}H_{18}O_2S$			
Anthraquinone, 1-(cyclohexylthio)-	EtOH	245(4.58),265(4.24), 307(3.88),443(3.65)	22-1648-65
$C_{20}H_{18}O_3$			
Cyclohexanone, 2-benzoyl-, exo-cis-enol benzoate	EtOH	236(4.28)	1-2139-65
Cyclohexanone, 2-benzoyl-, exo-trans-enol benzoate	EtOH	231(4.33),265(3.92)	1-2139-65
$C_{20}H_{18}O_3S$			
Anthraquinone, 1-(cyclohexylsulfinyl)-	EtOH	255(4.64),336(3.64), 400(2.96)	22-1648-65
$C_{20}H_{18}O_4$			
Anthracene, 1,5-diacetyl-2,6-dimethoxy-	$CHCl_3$	269(5.09),322(3.43), 338(3.60),357(3.68), 408(3.74),427s(3.68)	39-4565-64
2,4-Hexadienedioic acid, 2,5-diphenyl-, dimethyl ester	n.s.g.	321(4.27)	44-2101-64
2-Naphthol, 6,7-dimethoxy-1-methyl-, benzoate	EtOH	272(4.00),315(3.47), 322(3.35),329(3.56)	39-0361-65
Phaseolin (as anion)	EtOH	250(4.17)	88-0029-64
$C_{20}H_{18}O_4S$			
Anthraquinone, 1-(cyclohexylsulfonyl)-	EtOH	255(4.37),317(3.53)	22-1648-65

Compound	Solvent	λ_{max}(log ϵ)	Ref.
$C_{20}H_{18}O_5$			
Cyclohept[f]indene-6,8-dicarboxylic acid, 1,7-dihydro-7-oxo-, diethyl ester	EtOH	248(4.44),255(4.44), 286(4.58)	44-3032-64
$C_{20}H_{18}O_6$			
Averythrin	EtOH	223(4.47),255s(4.12), 266(4.18),294(4.45), 324(4.02),453(3.95)	39-3666-65
Phenanthrene, 1,2-diacetoxy-5,6-dimethoxy-	EtOH	232(4.41),257(4.77), 281(4.19),306(4.11), 312(4.12),343(3.39), 361(3.32)	94-0695-65
Succinic acid, 2,3-dibenzoyl-, dimethyl ester	MeOH	252(4.44)	35-2392-64
$C_{20}H_{18}O_7$			
Echioidinin, di-O-acetyl-	EtOH	280(4.32),302(4.28)	78-2633-65
Pterofuran, acetate	n.s.g.	213(4.46),224s(4.32), 280s(4.27),289s(4.35), 296s(4.44),309(4.61), 323(4.52)	12-0379-64
$C_{20}H_{18}O_8$			
Flavanone, 3',5,7-trihydroxy-4'-methoxy-, 3',7-diacetate	EtOH	275(4.15),339(3.54)	44-0897-65
Flavanone, 4',5,7-trihydroxy-3'-methoxy-, 4',7-diacetate	EtOH	276(4.17),339(3.54)	44-0897-65
Flavone, 5,6,7,8-tetramethoxy-3',4'-(methylenedioxy)-	EtOH	251(4.37),271s(4.24), 335(4.42)	39-2743-65
$C_{20}H_{18}O_9$			
β-Isorhodomycinone	$CHCl_3$	491(4.0),514(4.2), 524(4.3),552(4.3), 564(4.3)	24-3145-65
Melibentin, 5-demethyl-	EtOH	260(4.27),289(4.24), 345(4.25)	12-2021-65
	EtOH-KOH	294(4.32),315(4.33)	12-2021-65
$C_{20}H_{19}BrN_2O_2S$			
7H-Dibenzo[d,f][1,2]thiazepine, 6-(o-bromophenyl)-2-methyl-, 5,5-dioxide	EtOH	222s(4.57),260s(3.96)	12-0206-65
$C_{20}H_{19}BrO$			
Ketone, 1-bromocyclohexyl 2-fluorenyl	EtOH	242(3.93),318(4.40)	87-0504-64
Ketone, p-bromophenyl 2-methyl-5-phenyl-3-cyclohexen-1-yl	EtOH	257(4.243)	65-2258-64
$C_{20}H_{19}BrO_5$			
Cyclohept[f]indene-6,8-dicarboxylic acid, 1-bromo-1,2,3,7-tetrahydro-7-oxo-, diethyl ester	EtOH	249(4.42),283(4.64)	44-3032-64
$C_{20}H_{19}ClN_2O_4S_2$			
7,8,9,10-Tetrahydro-6-(3-methyl-2-benzothiazolinylidene)-6H-azepino[2,1-b]benzothiazolinium perchlorate	EtOH	475(4.67)	65-2455-64

Compound	Solvent	λ_{max}(log ε)	Ref.
$C_{20}H_{19}ClN_2O_7$			
Tetracycline, 5a,6-anhydro-4-dedimethyl-amino-4-epiamino-	MeOH	224(4.39),269(4.67), 308s(3.65),324s(3.34), 424(3.88)	31-0162-65
$C_{20}H_{19}ClO$			
Ketone, p-chlorophenyl 2-methyl-5-phenyl-3-cyclohexen-1-yl	EtOH	254(4.188)	65-2258-64
$C_{20}H_{19}CuN_2O_4S_3$			
Bis[8-(methylthio)quinoline]copper hydrogen sulfate	H_2O	248s(4.20),294(3.94), 312s(3.89),351(3.56)	5-0202-65I
$C_{20}H_{19}FO_3$			
2H-1-Benzopyran-7-ol, 3-(p-fluoro-phenyl)-2,2,4-trimethyl-, acetate	MeOH	269(3.96),309(3.93)	44-4114-65
$C_{20}H_{19}N$			
2,4-Cyclopentadiene-$\Delta^{1,\alpha}$-methylamine, N,N-dimethyl-2,5-diphenyl-	MeCN	240(4.22),307(4.42), 361(4.43)	24-3331-64
1,3-Pentadienylamine, 5-fluoren-9-ylidene-N,N-dimethyl-	MeCN	251s(4.71),254(4.72), 278s(4.18),463(4.82)	24-3331-64
$C_{20}H_{19}NO$			
Benzamide, N-(4,6,8-trimethyl-1-azulenyl)-	$CHCl_3$	244(4.45),296(4.56), 350s(3.90),385(3.68), 563(2.66)	35-3137-64
Indone, 3-phenyl-2-piperidino-	n.s.g.	271(4.49),521(3.23)	35-2510-65
$C_{20}H_{19}NO_3$			
4(1H)-Pyridone, 2-(β-hydroxy-p-meth-oxyphenethyl)-6-phenyl-	MeOH	238(4.42)	44-4263-65
1H-Pyrrolo[1,2-a]indole, 1-acetoxy-7-(benzyloxy)-2,3-dihydro-	MeOH	220(4.64),278(3.97), 302(3.63),315(3.46)	35-4612-64
Quinoline, 4-(2,4,5-trimethoxystyryl)-	MeOH	375(4.2)(changing)	87-0137-65
	HOAc	460(4.3)	87-0137-65
Quinoline, 4-(3,4,5-trimethoxystyryl)-	MeOH	345(4.5)(changing)	87-0137-65
	HOAc	315(4.0),410(4.2)	87-0137-65
	HOAc	410(4.3)	87-0137-65
$C_{20}H_{19}NO_4$			
Corysamine, tetrahydro-	MeOH	289(4.0)	73-3697-65
Eschscholtzine methine	EtOH-HCl	297.5(4.05)	23-2183-65
	EtOH	298.4(4.04)	23-2183-65
Ficine	n.s.g.	275(4.5),329(4.0)	88-1987-65
Maleic acid, (2,3-diphenyl-1-aziri-dinyl)-, dimethyl ester	EtOH	262(4.34)	88-4363-65
Spiro[indoline-3,2'(1'H)-naphthalene]-2,4'(3'H)-dione, 6',7'-dimethoxy-1-methyl-	MeOH	238(4.41),279(4.05), 318(3.86)	44-2973-65
$C_{20}H_{19}NO_5$			
Protopine	pH 1	238(3.93),286(3.88)	100-0199-65
	EtOH	289(3.92)	100-0199-65
	EtOH-KOH	236(3.94),286(3.89)	100-0199-65
1,2,4-Trioxa-7-azaspiro[4.5]decane-6,8-dione, 9,9-dimethyl-3,3-diphenyl-	n.s.g.	252(3.15),257(3.15), 263(3.08)	12-0154-64
$C_{20}H_{19}NO_6$			
Papaverrubine A	MeOH	236(3.98),288(3.91)	83-0385-65

Compound	Solvent	$\lambda_{max}(\log \epsilon)$	Ref.
Papaverrubine E	MeOH	239(3.97),289(3.93)	83-0385-65
$C_{20}H_{19}NO_8S$			
Azepino[2,1-b]benzothiazole -7,8,9,10-tetracarboxylic acid, 9,10-dihydro-, tetramethyl ester	MeOH	269s(0.64),320s(0.30), 429(4.11)	39-3200-65
	MeOH-acid	254s(1.84),259(1.95), 344(2.76)	39-3200-65
$C_{20}H_{19}NO_8Se$			
Azepino[2,1-b]benzoselenazole-7,8,9,10-tetracarboxylic acid, 9,10-dihydro-, tetramethyl ester	MeOH	280(0.80),321(0.32), 429(4.43)	39-3200-65
	MeOH-acid	263(1.82),350(2.59)	39-3200-65
$C_{20}H_{19}NO_9$			
2-Naphthacenecarboxamide, 1,4,4a,5,5a,-6,11,12a-octahydro-3,4,6,10,12,12a-hexahydroxy-6-methyl-1,11-dioxo-	MeOH-HCl	257(4.23),362(4.20)	35-2736-64
Tetracycline, dedimethylamino-12a-epioxy-	MeOH-HCl	262(4.37),335(3.67)	35-0134-65
	MeOH-NaOH	249(4.21),260(4.23), 379(4.16)	35-0134-65
(after reacidification)	MeOH-acid	262(4.22),357(4.20)	35-0134-65
Tetracycline, 4-hydroxy-4-de-dimethylamino-	MeOH-HCl	257(4.42),362(3.68)	23-1382-65
	MeOH-NaOH	246(4.26),264(4.23), 378(4.28)	23-1382-65
$C_{20}H_{19}N_3O$			
Benzo[f]pyrazolo[3,4-a]quinolizin-1(3bH)-one, 2-4,5,11,12,12a-hexahydro-2-phenyl-	2N NaOH	248(4.31)	78-2529-65
	EtOH	207(4.64),250(4.43)	78-2529-65
$C_{20}H_{19}N_3O_8$			
Tetracycline, 4-hydrazono-4-dedimethylamino-	MeOH-HCl	264(4.16),335(4.53)	23-1382-65
	MeOH-HCl	264(4.16),335(4.53)	35-2736-64
	MeOH-NaOH	261(4.33),323(4.25), 373(4.29)	23-1382-65
	MeOH-NaOH	261(4.33),323(4.25), 373(4.29)	35-2736-64
$C_{20}H_{19}N_5O$			
5H-Pyrrolo[3,4-d]pyrimidine, 2,4-di-amino-6-benzoyl-7-benzyl-6,7-dihydro-	pH 1	273(3.77),306s(2.85)	44-0194-65
	EtOH	286(3.88)	44-0194-65
$C_{20}H_{19}N_5O_4S$			
Theophylline, 7-(diphenylmethyl)-8-sulfamoyl-	MeOH	286(3.98)	54-0193-65
$C_{20}H_{19}N_5O_6S$			
Tryptophan, N-[2-(7-nitro-4H-1,2,4-ben-zothiadiazin-3-yl)ethyl]-, S,S-dioxide	2% $NaHCO_3$	242s(--),360(4.07)	32-0786-65
$C_{20}H_{19}OP$			
Phosphine oxide, (1-cyclohexen-1-yl-ethynyl)diphenyl-	EtOH	229(4.00),242(3.95), 252(3.95)	28-1537-64A
$C_{20}H_{19}P$			
Phosphine, (1-cyclohexen-1-ylethynyl)-diphenyl-	EtOH	231(4.04),243(3.95), 255(3.9)	28-1537-64A
$C_{20}H_{19}PS$			
Phosphine sulfide, (1-cyclohexen-1-yl-ethynyl)diphenyl-	EtOH	228(4.08),241(4.00), 252(3.9)	28-1537-64A

Compound	Solvent	$\lambda_{max}(\log \epsilon)$	Ref.
$C_{20}H_{20}$			
Cyclohexane, 1,2-dibenzylidene-	EtOH	285(4.28)	44-2335-64
Fluorene, 9-(7-norcaryl)-	EtOH	264(4.22),289(3.77), 301(3.96)	35-0863-65
Fluorene, 9-(tetramethylcyclopropylidene)-	EtOH	229(4.72),246(4.46), 254(4.64),275(4.23), 285(4.29),300(4.22), 313(4.33)	35-0863-65
Perylene, octahydro-	n.s.g.	234(4.64),241(4.66), 252(4.23),262(4.11), 304(4.17),317(4.29), 332(4.34),348(4.27), 357(2.04)	12-0055-64
$C_{20}H_{20}ClNO_2$			
Indole-3-acetic acid, 1-(p-chlorobenzyl)-2-ethyl-5-methyl-	MeOH	225(4.48),240(3.82), 282(3.81),300(3.74)	87-0204-65
$C_{20}H_{20}ClNO_3$			
Oxindole-3-acetic acid, 3-(p-chlorobenzyl)-1-methyl-, methyl ester	EtOH	255(3.93),283s(3.24)	44-2973-65
$C_{20}H_{20}ClNO_9$			
Protopine perchlorate	EtOH	240(3.81),290(3.75)	95-0721-64
$C_{20}H_{20}ClN_3$			
Pyrimidine, 2-amino-4-chloro-6-phenyl-5-(4-phenylbutyl)-	pH 1, 7	238(4.29),315(3.75)	87-0283-65
	pH 13	238(4.32),315(3.77)	87-0283-65
$C_{20}H_{20}ClN_3O$			
Quinazoline, 6-chloro-4-phenyl-2-(piperidinomethyl)-, 3-oxide	EtOH-acid	238(4.28),262(4.54)	87-0235-64
$C_{20}H_{20}ClOP$			
(1-Hydroxyethyl)triphenylphosphonium chloride	EtOH	260(3.43),267(5.40), 273(3.35)	59-1143-64
$C_{20}H_{20}FeO_2$			
2,4,6,8-Nonatetraenoic acid, 9-ferrocenyl-, methyl ester	$CHCl_3$	380(4.67),492(3.83)	5-0088-64F
$C_{20}H_{20}FeO_4$			
Malonic acid, (5-ferrocenyl-2,4-pentadien-1-ylidene)-, dimethyl ester	$CHCl_3$	372(4.52),514(3.85)	5-0088-64F
$C_{20}H_{20}INO_4$			
Columbamine iodide	neutral	220(4.5),265(4.4), 355(4.4),440(3.8)	100-0073-65
	weak base	220(4.5),265(4.4), 355(4.4),440(3.8)	100-0073-65
	strong base	220(4.5),270(4.5), 330(4.2),390(4.2)	100-0073-65
Jatrorrhizine iodide	neutral	215(4.5),280(4.4), 345(4.5),440(3.8)	100-0073-65
	weak base	215(4.5),260(4.4), 360(4.3),400(4.3)	100-0073-65
	strong base	225(4.5),245(4.3), 260(4.3),400(4.5)	100-0073-65

Compound	Solvent	λ_{max}(log ε)	Ref.
$C_{20}H_{20}NOP$			
Phosphine oxide, (p-dimethylamino-phenyl)diphenyl-	C_6H_{12}	277(4.44)	88-2729-64
	MeOH	284(4.44)	88-2729-64
$C_{20}H_{20}NP$			
Phosphine, (p-dimethylamino-phenyl)diphenyl-	C_6H_{12}	279(4.44)	88-2729-64
	MeOH	282(4.44)	88-2729-64
$C_{20}H_{20}NPS$			
Phosphine sulfide, (p-dimethylamino-phenyl)diphenyl-	MeOH	286(4.34)	88-2729-64
$C_{20}H_{20}N_2$			
Cycloheptapyrazole, 1,4,5,6,7,8-hexa-hydro-1,3-diphenyl-	C_6H_{12}	268(4.22)	44-1582-64
$C_{20}H_{20}N_2O$			
4,7-Methanobenzopyrazolidin-3-one, perhydro-1,2-diphenyl-	EtOH	245(4.4)	35-4509-64
$C_{20}H_{20}N_2O_2$			
Alkaloid C	MeOH	242(3.90),294(3.53)	35-4944-65
3,5-Pyrazolidinedione, 4-pentyli-dene-1,2-diphenyl-	MeOH	241(4.31)	32-0320-65
	$CHCl_3$	400(2.95)	32-0320-65
Pyrrolo[1,2-a]quinoxalin-1(5H)-one, acetyltetrahydro-2-methyl-4-phenyl-	EtOH	235(4.31),267(4.21)	35-1830-64
Schizozygine, deoxy-	MeOH	277(3.94),334(3.95)	33-0308-65
$C_{20}H_{20}N_2O_4$			
Mesoxalic acid, diethyl ester, azine with benzophenone	C_6H_{12}	267(4.24),310s(3.86)	44-4198-65
$C_{20}H_{20}N_2O_5$			
1H-2,4-Benzoxazocine, 5,6-dihydro-8,9-dimethoxy-1-(2-nitrobenzylidene)-3-methyl-, cis	EtOH	270(4.20)	39-4014-65
Isoquinoline, 2-acetyl-1,2,3,4-tetra-hydro-6,7-dimethoxy-1-(2-nitrobenzyl-idene)-, cis	EtOH	270(4.18)	39-4014-65
trans	EtOH	290(4.09)	39-4014-65
$C_{20}H_{20}N_2O_6$			
Phenoxazine-1,9-dicarboxylic acid, 2-(ethylamino)-4,6-dimethyl-3-oxo-, dimethyl ester	MeOH	224(4.43),252(4.53), 425(4.49),446(4.62)	44-3185-65
$C_{20}H_{20}N_2O_8$			
Tetracycline, 4-epi-N^4,N^4-dedimethyl-	MeOH-HCl	260(4.25),360(4.17)	35-2736-64
	MeOH-HCl-DMF	260(4.25),360(4.17)	23-1382-65
	MeOH-NaOH-DMF	246(4.23),266(4.22), 375(4.25)	23-1382-65
$C_{20}H_{20}N_4O_4$			
s-Tetrazine-3,6-dicarboxylic acid, 1,4-dihydro-1,4-diphenyl-, diethyl ester	$CHCl_3$	256(3.92),316(4.14)	24-1476-65
$C_{20}H_{20}N_6O_2S_2$			
Indole-2-carboxylic acid, 3,3'-dithio-bis[1-methyl-, dihydrazide	EtOH	219(4.67),280(4.31), 296(4.29),344s(3.82)	44-0178-64

Compound	Solvent	λ_{max}(log ε)	Ref.
$C_{20}H_{20}N_8$			
Tetraethylammonium 1,1,2,3,4,5,5-hepta-cyanopenta-2,4-dienide	MeCN	528(4.52)	35-2898-64
$C_{20}H_{20}O$			
Ketone, cyclohexyl 2-fluorenyl	EtOH	297(4.38),317(4.42)	87-0504-64
Ketone, 2-methyl-5-phenyl-3-cyclo-hexen-1-yl phenyl	EtOH	245(4.033)	65-2258-64
$C_{20}H_{20}O_2$			
Chrysene, 5,6,8,9,10,10a-hexahydro-2-methoxy-10a-methyl-8-oxo-	EtOH	230s(4.83),236(4.86), 262s(3.94),274(3.84), 284s(3.65),306(3.05), 314s(3.12),319(3.28), 327(3.22),334(3.20)	23-0591-64
Cyclohexanecarboxylic acid, 1-(2-fluorenyl)-	EtOH	270(4.47),295(3.97), 307(4.12)	87-0504-64
Ketone, 2-fluorenyl 1-hydroxycyclohexyl	EtOH	314(4.41)	87-0504-64
7,11-Methanocycloocta[a]naphthalen-13-one, 7,8,11,12-tetrahydro-3-methoxy-7,10-dimethyl-	EtOH	231(4.76),266(3.77), 276(3.76),286(3.57), 307(3.02),315s(3.06), 320(3.72),329(3.16), 335(3.34)	23-0591-64
3-Phenanthrol, 2-isopropyl-8-methyl-, acetate	EtOH	213(4.51),220s(4.38), 250s(4.66),257(4.76), 278(4.16),288(4.02), 300(4.10),318(2.78), 328s(2.62),334(2.86), 343(2.57),351(2.85)	33-1234-64
$C_{20}H_{20}O_3$			
Estra-1,3,5(10),8(9),11-pentaene-16,17-dione, 3-methoxy-15-methyl-	EtOH	235(4.42),300(3.91)	70-1051-65
$C_{20}H_{20}O_5$			
Cyclohepta[f]indene-6,8-dicarboxylic acid, 1,2,3,7-tetrahydro-7-oxo-, diethyl ester	EtOH	246(4.45),282(4.68)	44-3032-64
Dalbergione, 4-methoxy-, quinol diacetate	EtOH	278(3.45)	78-2683-65
Homostephanol, O-ethyl-, acetate	EtOH	257(4.92),294(4.15), 307(4.30),320(4.36), 353(3.60),370(3.92)	95-0584-65
Isocoleon A, anhydro-	ether	234(4.39),270(4.59), 322(3.82),450(3.82)	33-0471-65
$C_{20}H_{20}O_6$			
Averythrin, dihydro-	EtOH	223(4.45),255(4.08), 265s(4.11),293(4.49), 325(3.97),453(3.95)	39-3666-65
Coleon A lactone	ether	245(4.21),304(4.04), 420(3.75),440s(3.65)	33-0471-65
Flavone, 7-ethoxy-3',4',5'-trimethoxy-	n.s.g.	268(3.75),310(4.40)	28-6930-65A
9-Phenanthrenecarboxylic acid, 3-ethoxy-4,6,8-trimethoxy-	EtOH	264(4.70),320(4.11), 356(3.48),376(3.56)	95-0584-65
Pyrano[2',3':3,4]xanthone, 5',6'-dihy-dro-1-hydroxy-5,6-dimethoxy-6',6'-dimethyl-	MeOH	246(4.60),284(3.93), 321(4.32)	78-2653-65
β-D-Ribofuranoside, methyl 2,5-di-O-benzoyl-3-deoxy-	MeOH	229(4.41),273(3.27), 281(3.18)	87-0659-65

Compound	Solvent	λ_{max}(log ε)	Ref.
$C_{20}H_{20}O_7$			
Benzophenone, 2,5-dihydroxy-4-methoxy-2'-ethoxy-, diacetate	EtOH	243(4.19),286(4.07), 355(4.30)	78-1495-65
1,3,5-Cycloheptatriene-1-carboxylic acid, 6-hydroxy-2-[2-hydroxy-3-(3,4,5-trimethoxyphenyl)propyl-7-oxo-, δ-lactone	EtOH	255(4.20),332(3.72), 377(3.83),387(3.83)	78-3605-65
Flavone, 2',3',5,7,8-pentamethoxy-	EtOH	266(4.54),328s(4.07)	78-3237-65
Flavone, 3,3',4',7,8-pentamethoxy-	EtOH	250(4.35),347(4.34), 318s(4.15)	12-1164-64
Tangeretin	EtOH	270(4.23)	78-1441-65
$C_{20}H_{20}O_8$			
Avipirin, acetate	EtOH	222(4.37),249(4.28), 258(4.22),267(4.23), 308(4.16)	2-0464-64
Biphenylene, 2,3-diacetoxy-1,6,7,8-tetramethoxy-	EtOH	269(4.70),346(4.59), 365(4.57)	39-1067-64
Biphenylene, 2,7-diacetoxy-1,3,6,8-tetramethoxy-	EtOH	271(4.59),345(4.36), 364(4.71)	39-1067-64
Flavone, 3'-hydroxy-3,4',5,6,7-penta-methoxy-	MeOH	252(4.35),335(4.37)	24-2857-64
Flavone, 4'-hydroxy-3,3',5,6,7-penta-methoxy-	MeOH	240(4.28),339(4.35)	24-0548-65
Flavone, 5-hydroxy-3,3',4',7,8-penta-methoxy-	EtOH	258(4.26),276(4.28), 340s(4.15),364(4.16)	12-0934-64
	EtOH-NaOH	288(4.34),312(4.27), 400(3.78)	12-0934-64
Gibbs reagent product	EtOH-borate	695(3.53)	12-0934-64
Myricetin, 3,3',4',5',7-pentamethyl ether	EtOH	268(4.38),347(4.36)	12-0934-64
	EtOH-NaOH	289(4.56),386(3.92)	12-0934-64
Gibbs reagent product	EtOH-borate	683(3.21)	12-0934-64
$C_{20}H_{20}O_9$			
Anthraquinone, 4a,9a-epoxy-2-(2,3-epoxy-butyryl)-1,2,3,4,4a,9a-hexahydro-1,2,4-trihydroxy-8-methoxy-3-methyl-	EtOH	237(4.10),349(3.66)	35-4507-64
$C_{20}H_{21}BrN_2O_5$			
α-D-Glucoside, methyl 4,6-O-benzyli-dene-3-p-bromophenylazo-3-deoxy-	EtOH	283(4.17),398(2.24)	39-1045-64
$C_{20}H_{21}BrN_2S$			
4-[p-(Dimethylamino)benzylidene]-1,2,3,4-tetrahydropyrido[2,1-b]-benzothiazolium bromide	EtOH	515(3.78)	65-2447-64
$C_{20}H_{21}BrO_2$			
Indene-3-carboxylic acid, 2-bromo-1-(dipropyl-2-cyclopropen-1-yli-dene)-, methyl ester, cation	MeOH	234(4.57),281(4.06), 365(4.51)	35-1609-65
$C_{20}H_{21}ClN_2$			
1,2,3,4,7,8-Hexahydro-1-methyl-13H-benz[g]indolo[2,3-a]quinolizinium chloride	EtOH	223(4.35),317(4.23), 386(4.10)	94-1378-64
$C_{20}H_{21}ClN_2O_3$			
Acetanilide, 4'-chloro-2'-[N-(3-hy-droxypropyl)benzimidoyl]-, acetate	EtOH	237(4.48),330(3.48)	44-2368-64

Compound	Solvent	λ_{max}(log ε)	Ref.
$C_{20}H_{21}ClO_2$			
Estra-5,7,9-trien-3-one, 17α-chloroethynyl-17β-hydroxy-	dioxan	270(2.72),307(2.34)	87-0536-65
$C_{20}H_{21}ClO_6$			
Radicicol, O,O'-dimethyl-	EtOH	279(4.28)	88-0365-64
$C_{20}H_{21}FN_2O_5$			
α-D-Glucoside, methyl 4,6-O-benzylidene-3-(p-fluorophenylazo)-3-deoxy-	EtOH	266(3.99),388(2.28)	39-1045-64
$C_{20}H_{21}N$			
Indene, 1-[4-(butylamino)benzylidene]-	MeOH	410(4.4)	87-0390-65
	HOAc	400(4.4)	87-0390-65
$C_{20}H_{21}NO_2$			
Dibenzamide, N-cyclohexyl-	EtOH	253(4.04)	35-4365-65
Indene, 1-(4-dimethylamino-2,5-dimethoxybenzylidene)-	MeOH	400(4.5)	87-0390-65
	HOAc	336(4.3)	87-0390-65
Ketone, 2-cyclohexyl-3-phenyl-3-oxaziridinyl phenyl	EtOH	253(4.00)	35-4365-65
Quinoline, 1-benzoyl-1,2,3,4-tetrahydro-2-(2-tetrahydrofuryl)-	EtOH	264.5(3.88)	7-0763-65
diastereoisomer	EtOH	260(3.94)	7-0763-65
$C_{20}H_{21}NO_3$			
2H-Azepin-2-one, N-[o-(benzoyloxy)-benzoyl]hexahydro-	EtOH	209(4.45),283s(3.35)	65-1394-65
3-Indolinepropionic acid, 3-benzyl-1-methyl-2-oxo-, methyl ester	EtOH	205(4.52),256(3.88), 284s(3.18)	44-2973-65
1-Naphthaleneacetic acid, 1,2,3,4-tetrahydro-4-oxo-1-phenyl-, ethyl ester, oxime	EtOH	218s(4.4),256(4.10), 278s(2.93)	44-2973-65
Noraporphine, N-acetyl-5,6-dimethoxy-	EtOH	274(4.53)	39-4014-65
$C_{20}H_{21}NO_3S$			
Thiocyanic acid, 3,11,17-trioxoandrosta-1,4-dien-9-yl ester	EtOH	238(4.13)	44-0339-64
$C_{20}H_{21}NO_4$			
(-)-Canadine	EtOH	286.5(3.78)	94-1072-64
Eschscholtzine methine, dihydro-	EtOH-HCl	239s(3.92),294(3.93)	23-2183-65
Isoquinoline, 5,6,7-trimethoxy-1-(p-methoxybenzyl)-	MeOH-HCl	242(4.67),278(3.70), 327(3.52),337(3.60)	88-3599-65
Oxazole, 2-(3,4-dimethoxyphenethyl)-5-(p-methoxyphenyl)-	C_6H_{12}	278(4.42),281(4.40)	4-0310-65
$C_{20}H_{21}NO_5$			
13-Epiophiocarpine	EtOH	286(3.76)	94-1072-64
3-Indolineacetic acid, 3-benzyl-5,6-dimethoxy-1-methyl-2-oxo-	MeOH	209(4.51),275(3.77), 297s(3.71)	44-2973-65
3-Indolineacetic acid, 1-methyl-2-oxo-3-veratryl-	MeOH	232(4.03),256(3.88), 276s(3.66),285s(3.57)	44-2973-65
Ophiocarpine	EtOH	291(3.74)	94-1072-64
Papaveraldine, 3,4-dihydro-	MeOH	212(4.2),231(4.2), 276(4.0),316(4.0)	83-0362-64
$C_{20}H_{21}NO_6$			
Rheageninediol	EtOH	204(4.87),241(3.87), 292(3.93)	73-3479-65

Compound	Solvent	λ max(log ε)	Ref.
$C_{20}H_{21}NO_7$			
2-Anthraceneacetic acid, 1,2,3,4-tetrahydro-10-hydroxy-5-methoxy-9-methyl-α-nitro-4-oxo-	EtOH	222(4.51),268(4.66), 295(3.91),308(3.91), 319(3.79),407(4.00)	65-0655-65
$C_{20}H_{21}NO_8$			
Acrylic acid, 2-(2,4-dimethoxyphenyl)-3-(4-ethoxy-3-methoxy-2-nitrophenyl)-, trans	EtOH	289(4.14)	95-0584-65
1H-Azepine-2,3,4,5-tetracarboxylic acid, 2,3-dihydro-1-phenyl-, tetramethyl ester	MeOH	247(0.85),378(1.65)	39-3200-65
$C_{20}H_{21}N_3O$			
4-Pyrimidinol, 2-amino-6-phenyl-5-(4-phenylbutyl)-	pH 1	233(4.06),278(3.88)	87-0283-65
	pH 7	233(4.16),300(3.81)	87-0283-65
	pH 13	290(3.71)	87-0283-65
$C_{20}H_{21}N_3O_6S$			
Cytidine, N-benzoyl-2'-deoxy-5'-thio-	MeOH	260(4.36),303(4.00)	44-3067-65
$C_{20}H_{21}N_5O_5S$			
3-Benzyltetrahydro-1,6-dimethyl-5,7-dioxo-v-triazolo[4,5-d]pyrimidinium p-toluenesulfonate	pH -0.89	272(3.90)	24-1060-65
	pH 4.0	224(4.34),250s(3.60), 313(3.85)	24-1060-65
$C_{20}H_{21}N_5O_7$			
Indole, 1-methyl-3-(1-methyl-2-pyrrolidinyl)-, picrate	EtOH	218(4.77),244s(4.09), 270(3.94),282(3.94), 286s(--),293(3.89), 358(4.21),400s(3.00)	87-0415-64
$C_{20}H_{21}N_7O_6$			
Homofolic acid	pH 13	255(4.4),281(4.3), 365(3.9)	44-3404-65
$C_{20}H_{22}$			
Benz[a]anthracene, 5,6,8,9,10,11-hexahydro-10,10-dimethyl-	EtOH	266(4.28),272s(4.25), 294s(3.83),303(3.87)	39-0724-64
Benzo[e]pyrene, 1,2,3,6,7,8,9,10,11,12-decahydro-	benzene	283(3.82),306(3.93), 320(3.77),335(3.22)	24-0218-64
	EtOH	240(4.80)	24-0218-64
Cyclooctene, 1,5-diphenyl-	n.s.g.	249(4.13)	44-3467-64
$C_{20}H_{22}BF_4N_2O_2PS_2$			
3-Ethyl-2-[(3-ethyl-6-methoxy-2-benzothiazolinylidene)phosphino]-6-methoxybenzothiazolium tetrafluoroborate	n.s.g.	485(4.64)	89-0433-64
$C_{20}H_{22}BrClN_2$			
3-Chloro-5,6,7,8-tetrahydro-2-(2-indol-3-ylethyl)-5-methylisoquinolinium bromide	EtOH	280(4.01)	94-1378-64
$C_{20}H_{22}ClN_3O$			
Amodiaquine	pH 1	342(4.24)	4-0006-64
	EtOH	339(4.10)	4-0006-64
$C_{20}H_{22}ClN_3O_2$			
Amodiaquine, 1-oxide	pH 1	356(4.24)	4-0006-64

Compound	Solvent	$\lambda_{max}(\log \epsilon)$	Ref.
Amodiaquine, 1-oxide (cont.)	EtOH	400(4.14)	4-0006-64
Amodiaquine, N-α-oxide	pH 1	341(4.23)	4-0006-64
	EtOH	338(4.10)	4-0006-64
$C_{20}H_{22}ClN_3O_3$			
Amodiaquine, N-α,1-dioxide	pH 1	357(4.23)	4-0006-64
	EtOH	400(4.14)	4-0006-64
$C_{20}H_{22}FNO_2$			
Estra-1,3,5(10)-triene-16β-carbonitrile, 16α-fluoro-3-methoxy-17-oxo-	MeOH	222(3.95),279(3.29), 288(3.29)	44-2775-64
$C_{20}H_{22}F_3N_3$			
Benzimidazole, 1-(2-diethylaminoethyl)-2-phenyl-5-(trifluoromethyl)-	EtOH-HCl	252(3.91),287(4.18)	4-0453-65
$C_{20}H_{22}N_2$			
Indole, 1-benzyl-3-(1-methyl-2-pyrrolidinyl)-, cyclohexanesulfamate	EtOH	218(4.62),260s(3.69), 268s(3.83),273s(3.85), 280(3.87),291s(3.77)	87-0415-64
Quinoline, 1,2-dihydro-1-methyl-, dimer	EtOH	209(4.55),252(4.27) 300(3.77)	1-1389-64
$C_{20}H_{22}N_2O$			
1H-Pyrido[3,4-b]indole, 7-(benzyloxy)-3-ethyl-2,3,4,9-tetrahydro-	EtOH	228(4.59),258s(3.69), 268(3.74),296(3.73)	44-2864-64
$C_{20}H_{22}N_2O_2$			
Akuammiline, deacetyldeformo-	n.s.g.	218(4.17),262(3.71)	28-5538-65B
Benz[f]indolo[2,3-a]quinolizin-14(6H)-one, 1,2,3,4,6,7,12,12b-octahydro-2-methoxy-	EtOH	224(4.52),278(4.05), 292(3.93),336(4.23)	4-0329-65
Condylocarpine	EtOH	230(4.01),297(3.99), 329(4.15)	33-1822-65
16-Epipleiocarpamine	EtOH	230(4.42),286(3.90)	33-0878-64
6H-Indolo[3,4:1',2']imidazo[1,2-a]indole, 6a,7,12a,13-tetrahydro-7,13-dihydroxy-7,12a,13-trimethyl-	EtOH	204(4.05),244(3.54), 297(3.26)	39-4320-64
Mitoridine	EtOH	215(4.55),249(3.89), 293(3.42)	22-2683-64
	base	225(4.35),252(3.60), 305(3.85)	22-2683-64
Normacusine A, deformo-	EtOH	230(4.34),275(3.87), 280(3.87),290(3.79)	39-4419-64
Pleiocarpamine	EtOH	230(4.47),285(3.91)	33-0689-65
	EtOH	230(4.47),285(3.91)	33-0878-64
Schizozygine, dihydrodeoxo-	MeOH	258(3.77),335(3.66)	33-0308-65
$C_{20}H_{22}N_2O_3$			
Acetamide, N-(2-indol-2-yl-4,5-dimethoxyphenethyl)-	EtOH	295(4.29)	39-4014-65
Akuammicine, 19,20-epoxy-19,20-dihydro-	EtOH	235(4.00),300(3.98), 330(4.20)	78-1141-65
Fluorocarpamine	EtOH	235(4.05),257(3.64), 398(3.13)	49-0909-65
Isoquinoline, 2-acetyl-1-(2-aminobenzylidene)-1,2,3,4-tetrahydro-6,7-dimethoxy-, trans	EtOH	224(4.66),290(4.28), 341(4.28)	39-4014-65
	EtOH-HCl	227(4.47),330(4.37)	39-4014-65
Isoquinoline, 2-acetyl-1-(4-aminobenzylidene)-1,2,3,4-tetrahydro-6,7-dimethoxy-	EtOH	228(4.42),345(4.47)	39-4014-65

Compound	Solvent	$\lambda_{max}(\log \epsilon)$	Ref.
Lochrovicine	EtOH	225(5.05),298(5.09), 326(5.25)	100-0203-64
Perivine	n.s.g.	226(4.28)	100-0374-64
	n.s.g.	227(4.14),314(4.20)	100-0470-64
Schizozygine	MeOH	269(3.99),313(3.97)	33-0308-65
Tryptamine, N-acetyl-6-(benzyloxy)-5-methoxy-	EtOH	228(4.34),298(3.72)	4-0387-65
Vinervine	n.s.g.	234(4.22),290(3.86), 336(4.22)	70-2152-65
$C_{20}H_{22}N_2O_3S$			
Spiro[indoline-3,3'-pyrrolidin]-2-one, 2'-ethyl-1'-(p-tolylsulfonyl)-	EtOH	233(4.13),282(3.20)	78-1327-65
$C_{20}H_{22}N_2O_4$			
Perividine	n.s.g.	240(4.25),286(4.47), 315(4.19)	100-0374-64
$C_{20}H_{22}N_2O_5$			
Cinnamic acid, p-methoxy-α-(p-nitrophenyl)-, β-dimethylaminoethyl ester, cis	EtOH-HCl	349(4.33)	87-0614-64
trans	EtOH-HCl	300(3.81)	87-0614-64
Cinnamic acid, α-(p-methoxyphenyl)-p-nitro-, β-dimethylaminoethyl ester, cis	EtOH-HCl	350(4.27)	87-0614-64
trans	EtOH-HCl	292(4.20)	87-0614-64
α-D-Glucoside, methyl 4,6-O-benzylidene-3-(phenylazo)-3-deoxy-	EtOH	266(4.01),390(2.36)	39-1045-64
Isoquinoline, 2-acetyl-1,2,3,4-tetrahydro-6,7-dimethoxy-1-(o-nitrobenzyl)-	EtOH	285(3.92)	39-4014-65
$C_{20}H_{22}N_2O_5S_4$			
Cephalosporanic acid, 7-(2-thiopheneacetamido)-, cyclopentyl xanthate derivative, sodium salt	n.s.g.	228(4.26),285(4.23)	87-0174-65
$C_{20}H_{22}N_2O_6$			
Benzoin, 2-(2-acetamidoethyl)-2'-nitro-4,5-dimethoxydeoxy-	EtOH	230(4.33),274(4.04)	39-4014-65
Isoquinoline, 3,4-dihydro-6,7-dimethoxy-1-(6-nitroveratryl)-, (hydrochloride)	EtOH	245(4.41),306(4.21), 353(4.21)	39-4014-65
$C_{20}H_{22}N_2O_8$			
1,2,3,4-Butanetetrol, 1-(2-quinoxalinyl)-, tetraacetate	EtOH	211(4.38),237(4.51), 317(3.88)	44-2457-65
$C_{20}H_{22}N_4$			
Pyrimidine, 2,4-diamino-6-phenyl-5-(4-phenylbutyl)-	pH 1	288(4.03)	87-0283-65
	pH 7	290(3.93)	87-0283-65
	pH 13	298(3.88)	87-0283-65
$C_{20}H_{22}N_4O$			
2-Pyrazolin-5-one, 4-[[4-(diethylamino)-o-tolyl]imino]-1-phenyl-	n.s.g.	450(4.2),520(4.4)	46-1501-64
4-Pyrimidinol, 2-amino-5-(3-anilinopropyl)-6-p-tolyl-	pH 1	230(4.13),280(4.00)	87-0283-65
	pH 7	242(4.37),295(3.94)	87-0283-65
	pH 13	290(3.92)	87-0283-65

Compound	Solvent	λ max(log ε)	Ref.
$C_{20}H_{22}N_4O_4$			
Khusilal, 2,4-dinitrophenylhydrazone	$CHCl_3$	378(4.30)	78-2617-64
$C_{20}H_{22}N_4O_6$			
Benzocycloocten-10-one, 5,6,7,8,9,10-hexahydro-1,3-dimethoxy-, 2,4-dinitrophenylhydrazone	$CHCl_3$	371(4.39)	5-0021-64G
2(1H)-Naphthalenone, 3,4-dihydro-6,7-dimethoxy-1,1-dimethyl-, 2,4-dinitrophenylhydrazone	EtOH	232(4.39),269(4.06), 365(4.38)	39-0361-65
$C_{20}H_{22}N_4O_7S$			
Glutamic acid, N-[p-[3-(4-hydroxy-2-mercapto-6-methyl-5-pyrimidinyl)-propionamido]benzoyl]-	pH 1	275(4.56)	4-0263-64
	pH 8.4	275(4.63)	4-0263-64
	pH 13	267(4.63),315s(4.08)	4-0263-64
$C_{20}H_{22}N_8O_8$			
Heptanal, 2-methyl-6-oxo-, bis(2,4-dinitrophenylhydrazone)	EtOH	355(4.59)	39-2165-64
$C_{20}H_{22}O_2$			
2-Butyne, 1,4-bis(2,6-xylyloxy)-	EtOH	262(2.81)	78-2289-65
Chrysene, 5,6,6a,7,8,9,10,10a-octahydro-2-methoxy-10a-methyl-8-oxo-, cis	EtOH	229(4.82),255(3.69), 266(3.76),276(3.77), 286(3.59),307(3.03), 315s(3.11),320(3.28), 328(3.21),335(3.38)	23-0591-64
Chrysofluorene, 3-ethylenedioxy-1,1a,2,3,4,6-hexahydro-1a-methyl-	EtOH	259(4.07)	42-0479-64
Gona-1,3,5(10),8,14-pentaen-17-one, 13-ethyl-3-methoxy-	EtOH	311(4.44)	39-4472-64
Gona-1,3,5(10),8,14-pentaen-17-one, 13β-ethyl-3-methoxy-	EtOH	314(4.40)	94-1289-65
D-Homogona-1,3,5(10),8,14-pentaen-17a-one, 13-ethyl-3-hydroxy-	n.s.g.	311(4.4)	70-0760-65
Indene-3-carboxylic acid, 1-(dipropyl-2-cyclopropen-1-ylidene)-, methyl ester	MeOH	234(4.55),248s(4.41), 275(4.09),363(4.56)	35-1609-65
Naphthalene, 1-ethyl-3,4-dihydro-6-methoxy-2-(p-methoxyphenyl)-	EtOH	230(4.14),254(4.13)	87-0519-64
Naphtho[2',3':1,2]dibenzofuran, 2,3,4,6,7,8,9,4',4'a,5',6',7'-dodecahydro-7'-oxo-	EtOH	230(4.67),363(5.40)	24-2738-65
Phenanthrene, 9,10-diethyl-3,7-dimethoxy-	MeOH	232(4.50),256(4.74), 274(4.50),287(4.34), 313s(--),342(3.15), 360(3.15)	5-0036-64D
2-Phenanthrenebutyric acid, 9,10-dihydro-α,α-dimethyl-	EtOH	270(4.20),299(3.58), 323(2.68)	39-0724-64
$C_{20}H_{22}O_3$			
2H-1-Benzopyran-7-ol, 4-ethyl-3-(p-methoxyphenyl)-2,2-dimethyl-	MeOH	277(4.07),306(4.00)	44-4114-65
Estra-1,3,5(10),8(9),15-pentaen-17-one, 16-hydroxy-3-methoxy-15-methyl-	EtOH	270(4.17)	70-1051-65
Gona-1,3,5(10),8(9)-tetraene-15,16-dione, 3-methoxy-14,17-dimethyl-	n.s.g.	265(4.25)	70-1058-65
Gona-1,3,5(10),9(11)-tetraene-15,16-dione, 3-methoxy-14,17-dimethyl-	n.s.g.	270(4.29)	70-1058-65

Compound	Solvent	λ_{max}(log ε)	Ref.
Inden-4-ol, 6-methoxy-2-(p-methoxyphenyl)-1,1,3-trimethyl-	MeOH	276(4.25)	44-4120-65
$C_{20}H_{22}O_4$			
2H-Benzo[d]naphtho[1,2-b]pyran-11-carboxylic acid, 1,3,4,6,6a,7,8,9-octahydro-10-hydroxy-12-methoxy-7-methyl-, δ-lactone	EtOH	237s(4.48),245(4.58), 272(4.11),284s(3.92)	33-1459-64
$C_{20}H_{22}O_5$			
Chalcone, 2,4,4',5'-tetramethoxy-2'-methyl-	EtOH	211(4.42),247(4.17), 353(4.24)	39-2844-65
Estra-1,3,5(10)-triene-11,17-dione, 3,9β-dihydroxy-, 3-acetate	MeOH	266(2.63),273(2.59), 292(2.21)	94-0473-64
	MeOH	264(2.60),272(2.53), 297(2.02)	94-0473-64
Isocoleon A, anhydrodihydro-	EtOH	232(4.33),271(4.54), 320(3.82),450(3.78)	33-0471-65
2-Phenanthreneglycolic acid, 1-carboxy-1,2,3,9,10,10a-hexahydro-7-methoxy-α,1-dimethyl-, γ-lactone	n.s.g.	275(4.06),278(4.07)	70-1058-65
$C_{20}H_{22}O_6$			
Acetic acid, (5-benzyloxy-2-carbethoxy-3-hydroxyphenyl)-, ethyl ester	EtOH	265(4.18),306(3.84)	39-5382-64
Biphenyl-2,5:2',5'-diquinone, 4',6-di-tert-butyl-4,6'-dihydroxy-	MeOH	269(4.43),387(3.20)	24-2774-65
	MeOH-KOH	223(4.42),273(4.18), 487(3.46)	24-2774-65
1-Butanone, 2,3-epoxy-1-(1,2,3,4-tetrahydro-2,4,10-trihydroxy-8-methoxy-3-methylanthracen-2-yl)-	EtOH	235(4.87),290(3.74), 303(3.85),318(3.81)	35-4507-64
Coleon A	ether	213(4.33),230(4.30), 251(4.22),273(4.07), 315(4.00),432(3.77), 454s(3.73)	33-0471-65
Coleon, dihydro-, γ-lactone	EtOH	217(4.49),248(4.36), 305(4.18),418(3.91), 440s(3.81)	33-0471-65
Isocoleon A	ether	234(4.35),271(4.57), 322(3.82),450(3.77)	33-0471-65
Shikonin, isobutyl-	EtOH	273(4.25)	88-4737-65
$C_{20}H_{22}O_7$			
Propiophenone, 2,3-dihydroxy-2',4'-dimethoxy-3-(p-methoxyphenyl)-, 3-acetate	MeOH	272(4.09),305(3.91)	78-1141-64
$C_{20}H_{22}O_8$			
Isoglauconic acid, acetate	EtOH	220(4.02)	39-1772-65
1,3-Propanedione, 1-(2,3-dimethoxyphenyl)-3-(2-hydroxy-3,4,6-trimethoxyphenyl)-	EtOH	295(4.52),370(4.61)	78-3237-65
$C_{20}H_{23}BrO_3$			
Estra-1,3,5(10)-trien-17-one, 4-bromo-3-methoxy-16-(hydroxymethylene)-	pH 1	222(4.03),268(4.09)	44-2234-65
	pH 13	222(4.03),303(4.34)	44-2234-65
	MeOH	222(4.03),268(3.99), 308(3.29)	44-2234-65

Compound	Solvent	$\lambda_{max}(\log \epsilon)$	Ref.
$C_{20}H_{23}BrO_4$			
Estra-1,3,5(10)-trien-6-one, 7α-bromo-3,17β-dihydroxy-, 17-acetate	EtOH	216(4.23),261(3.95),316s(3.22),336(3.23)	44-1325-64
$C_{20}H_{23}BrO_5$			
Podocarpic acid, O-acetyl-6α-bromo-7-oxo-, methyl ester	MeOH	263(3.94),294s(3.62),337s(3.02)	44-0501-65
$C_{20}H_{23}Br_2N_7$			
Isobiguanide, 3,4-bis(p-bromophenyl)-1,2-bis(isopropylideneamino)-	EtOH	235-260(4.51),275(4.59)	39-3912-65
$C_{20}H_{23}ClN_2$			
Voachalotane, 16β-(chloromethyl)-	MeOH	229(4.04),285(3.74)	20-0253-65
$C_{20}H_{23}ClN_2O$			
2-Indolinone, 3-(p-chlorobenzyl)-3-(2-dimethylaminoethyl)-1-methyl-	EtOH-HCl	256(3.89),280s(3.29)	87-0626-65
	EtOH	255(3.86),266s(3.72),281s(3.24)	87-0626-65
$C_{20}H_{23}ClN_2OS$			
Benzothiazole, 5-chloro-2-(p-diethylaminopropoxyphenyl)-	EtOH	262(3.91),312(4.40),325(4.44)	39-0954-65
$C_{20}H_{23}ClN_2O_4$			
N-Methylhalfordinium chloride	EtOH	265(4.2),360(4.1)	12-0119-64
$C_{20}H_{23}ClO_5S$			
Spiro[benzofuran-2(3H),1'-[2]cyclohexene]-3,4'-dione, 2'-(butylthio)-7-chloro-4,6-dimethoxy-6'-methyl-	EtOH	291(4.58)	87-0705-64
Spiro[benzofuran-2(3H),1'-[3]cyclohexene]-2',3-dione, 4'-(butylthio)-7-chloro-4,6-dimethoxy-6'-methyl-	EtOH	296(4.61)	87-0705-64
$C_{20}H_{23}Cl_3N_2O_{11}$			
2(1H)-Pyrimidinone, 1-(2,3,4-tri-O-acetyl-6-O-trichloroacetyl-β-D-glucopyranosyl)-4-ethoxy-	EtOH	276(3.81)	44-2723-65
$C_{20}H_{23}FN_2O$			
2-Indolinone, 3-(p-fluorobenzyl)-3-(2-dimethylaminoethyl)-1-methyl-	MeOH-HCl	207(4.45),255(3.84),284s(3.13)	87-0626-65
$C_{20}H_{23}N$			
1H-Azepine, 2,3,6,7-tetrahydro-1-phenethyl-4-phenyl-, hydrochloride	EtOH	242(4.11)	39-5130-64
$C_{20}H_{23}NO$			
Acrylophenone, 2-[α-(tert-butylamino)-benzyl]-	EtOH	243(4.08)	88-4833-65
Chalcone, α-[(tert-butylamino)methyl]-	isooctane	255(4.06),283(4.21)	88-4833-65
Indene, 2-(2-dimethylaminoethyl)-3-(2-methoxyphenyl)-	MeOH	239(4.10),262(4.08)	49-0485-64
$C_{20}H_{23}NO_2$			
1H-3-Benzazepine, 7,8-diethoxy-2,3-dihydro-5-phenyl-	CHCl	248(4.14),300(4.21)	6-0213-65
Estra-1,3,5(10)-triene-15β-carbonitrile, 17-oxo-	MeOH	222(3.93),278(3.30),288(3.27)	44-0064-64

Compound	Solvent	λ_{max}(log ϵ)	Ref.
Noraporphine, N-ethyl-5,6-dimethoxy-, (as hydrochloride)	EtOH	274(4.27)	39-4014-65
1-Phenanthreneethylamine, 3,6-dimethoxy-N,N-dimethyl-	MeOH	236(4.57),251(4.64), 259(4.64),277(4.20), 286(4.23),298(4.02), 309(4.11),330s(3.38), 348(3.15),364(3.18)	35-2177-64
$C_{20}H_{23}NO_3$			
Berbine, 2,3,13-trimethoxy-, hydrochloride, hydrate	EtOH	282(3.91),286(3.90)	73-0400-64
Cinnamic acid, p-methoxy-, 2-dimethylaminoethyl ester, cis	EtOH-HCl	298(4.34)	87-0614-64
Cinnamic acid, α-(p-methoxyphenyl)-, 2-dimethylaminoethyl ester, cis	EtOH-HCl	296(4.29)	87-0614-64
trans	EtOH-HCl	274(4.23)	87-0614-64
Isoquinoline, 2-acetyl-1-benzyl-1,2,3,4-tetrahydro-6,7-dimethoxy-	EtOH	285(4.18)	39-4014-65
$C_{20}H_{23}NO_3S$			
Thiocyanic acid, 11β-hydroxy-3,17-dioxoandrosta-1,4-dien-9α-yl-	EtOH	242(4.10)	44-0339-64
$C_{20}H_{23}NO_4$			
Aporphine, 1-hydroxy-2,9,10-trimethoxy-	EtOH	220(4.52),280(4.12), 305(4.12)	88-1509-65
Aporphine, 10-hydroxy-1,2,9-trimethoxy-, HI salt	EtOH	222(4.56),280(4.14), 303(4.13)	78-2155-65
	EtOH-KOH	232(4.51),258(4.35), 290(3.93),327(3.92)	78-2155-65
Chalcone, 4-(dimethylamino)-2',4',6'-trimethoxy-	benzene	377(4.54)	65-0094-65
	EtOH	268(3.69),392(4.47)	65-0094-65
	HOAc	402(4.42)	65-0094-65
	HOAc-$FeCl_3$	566(4.21)	65-0094-65
Chalcone, 4'-(dimethylamino)-2,4,6-trimethoxy-	benzene	375(4.55)	65-0094-65
	EtOH	246(3.91),392(4.46)	65-0094-65
	HOAc	403(4.56)	65-0094-65
	HOAc-$FeCl_3$	547(4.54)	65-0094-65
Isocorydine	EtOH	221(4.45),266(4.10), 303(3.73)	39-2244-64
	EtOH-KOH	225(4.51),273s(3.92), 345(3.78)	39-2244-64
Isoquinoline, 3,4-dihydro-5,6,7-trimethoxy-1-(p-methoxybenzyl)-	MeOH	272(3.96),315(3.43)	88-3599-65
Rogersine	EtOH	219(4.81),282(4.40), 304(4.40)	39-2244-64
	EtOH-KOH	223(4.70),329(4.59)	39-2244-64
Spiro[13H-dibenzo[a,f]quinolizine-2(1H),2'-[1,3]dioxolan]-13-one, 3,4,6,7,11bβ,12-hexahydro-8-methoxy-	C_6H_{12}	311(4.06)	44-3667-65
$C_{20}H_{23}NO_5$			
Δ^6-Codeine, 6-deoxy-7-formyl-14-hydroxy-6-methoxy-	EtOH	277(4.05)	78-1407-64
Estra-1,3,5(10)-trien-17-one, 16-(hydroxymethylene)-3-methoxy-4-nitro-	pH 1	222(4.08),268(3.78),	44-2234-65
	pH 13	222(4.04),302(4.33)	44-2234-65
	MeOH	222(4.08),265(3.96)	44-2234-65

Compound	Solvent	$\lambda_{max}(\log \epsilon)$	Ref.
$C_{20}H_{23}NO_6$			
Acrylic acid, 3-(2-amino-4-ethoxy-3-methoxyphenyl)-2-(2,4-dimethoxyphenyl)-, trans	EtOH	292(4.03)	95-0584-65
$C_{20}H_{23}NO_7$			
Macranthine, diacetate	EtOH	235(3.58),292(3.69)	33-0185-64
$C_{20}H_{23}NO_8$			
Anthracene, 2-[(ethoxycarbonyl)nitromethyl]-1,2,3,9,9aα,10-hexahydro-4,9β-dihydroxy-5-methoxy-9α-methyl-10-oxo-	EtOH	268(3.73),341(4.15)	65-0655-65
9aH-Quinolizine-1,2,3,4-tetracarboxylic acid, 9a-ethyl-6-methyl-, tetramethyl ester	MeOH	294(3.97),335s(3.40),450(3.60)	39-0948-65
$C_{20}H_{23}NS$			
Phenethylamine, o-(2,3-dihydro-1-benzothiepin-5-yl)-N,N-dimethyl-	MeOH	247(4.2)	24-0685-65
$C_{20}H_{23}N_3$			
Yohimban-3ξ-carbonitrile	MeCN	223(4.47),282(3.87),290(3.86),309(3.67),318(3.69)	44-0105-65
$C_{20}H_{23}N_3O_2$			
Hydrouracil, 6-(dimethylamino)-5,5-dimethyl-1,3-diphenyl-	EtOH	240(3.89)	44-0812-65
2-Indolinone, 1-(dimethylamino)-3-morpholino-3-phenyl-	MeOH	259(3.78)	44-3451-65
	MeOH-HCl	259(3.67),302(3.06)	44-3451-65
$C_{20}H_{23}N_3O_4$			
Hydrocornicularic acid, 4-methoxy-, methyl ester, semicarbazone	MeOH	230(4.38)	2-0017-64
Levulinic acid, 2-(p-methoxyphenyl)-5-phenyl-, methyl ester, semicarbazone	MeOH	232(4.36)	2-0017-64
Levulinic acid, 5-(p-methoxyphenyl)-2-phenyl-, methyl ester, semicarbazone	MeOH	225(4.34),280(3.27),310(2.66)	2-0017-64
$C_{20}H_{23}N_3O_4S_3$			
5-Thia-1-azabicyclo[4.2.0]oct-2-ene-2-carboxylic acid, 3-(mercaptomethyl)-8-oxo-7-(2-phenylacetamido)-, methyl ester, dimethyldithiocarbamate	EtOH	275(4.27)	39-5015-65
$C_{20}H_{23}N_3O_{13}$			
α-D-Glucopyranose, 1,3,4,6-tetra-O-acetyl-2-deoxy-2-(2,4-dinitroanilino)-	EtOH	208(3.81),264(3.70),333(3.89)	44-1776-64
β-isomer	EtOH	208(3.86),261(3.72),336(3.92)	44-1776-64
$C_{20}H_{23}N_5O_6S$			
L-Glutamic acid, N-[p-[3-(2-amino-4-mercapto-6-methyl-5-pyrimidinyl)propionamido]benzoyl]-	pH 1	267(4.44),337(4.18)	4-0263-64
	pH 8.4	267(4.46),347(4.23)	4-0263-64
	pH 13	267(4.49),317(4.18)	4-0263-64

Compound	Solvent	λ_{max}(log ε)	Ref.
$C_{20}H_{23}N_5O_7$			
D-Alanine, N-[6-(benzyloxy)-9-β-D-ribofuranosyl-9H-purin-2-yl]-	pH 1	247(4.07),295(3.96)	35-3752-65
	pH 11	251(4.14),290(3.96)	35-3752-65
L-isomer	pH 1	247(4.09),294(4.00)	35-3752-65
	pH 11	251(4.13),290(4.00)	35-3752-65
L-Glutamic acid, N-[p-[3-(2-amino-4-hydroxy-5-methyl-5-pyrimidinyl)-propionamido]benzoyl]-	pH 1	266(4.43)	87-0024-64
	pH 8.4	268(4.40)	87-0024-64
	pH 13	270(4.41)	87-0024-64
$C_{20}H_{23}N_8O_{12}P$			
5'-Inosinic acid, 5'→5'-ester with inosine, barium salt	pH 1	250.4(4.32)	44-3211-65
	pH 13	255.0(4.41)	44-3211-65
$C_{20}H_{24}$			
1-Octene, 1,1-diphenyl-	C_6H_{12}	251(4.04)	22-0693-64
Phenanthrene, 4a,4b-dihydro-1,3,4a,4b,6,8-hexamethyl-	$C_6H_{11}Me$	245(4.20),310(3.85), 321(3.86),475(3.49)	77-0447-65
$C_{20}H_{24}Br_2O_2$			
9β,10α-Androst-4-ene-3,17-dione, 6-(dibromomethylene)-	MeOH	249(3.96),285(3.79)	54-0904-65
$C_{20}H_{24}ClN_3O$			
Morpholine, 4-[3-[(2-amino-5-chloro-α-phenylbenzylidene)amino]propyl]-	EtOH	230(4.47),359(3.70)	44-2368-64
isomer	EtOH	248(4.41)	44-2368-64
Morpholine, 4-[2-[[5-chloro-2-(methyl-amino)-2-phenylbenzylidene]amino]-ethyl]-	EtOH	230(4.48),378(3.78)	44-2368-64
isomer	EtOH	253(4.45)	44-2368-64
$C_{20}H_{24}IN$			
[(3,5-Dimethylcyclopenta[ef]hepta-len-1-yl)methyl]trimethylammonium iodide	MeOH	371(4.01),389(4.06), 440(2.84),705(2.34), 765(2.44),855(2.39), 975(2.01)	5-0194-64B
$C_{20}H_{24}INO_2$			
3,4-Dihydro-6,7-dimethoxy-2-methyl-1-phenethylisoquinolinium iodide	MeOH	249(4.19),308(3.90), 360(3.94)	24-3452-64
5,6,6a,7-Tetrahydro-2,10-dimethoxy-6,6-dimethyl-4H-dibenzo[de,g]quin-olinium iodide	MeOH	268(4.03),273(4.04), 300(3.66),311(3.70), 319(3.72)	35-2177-64
$C_{20}H_{24}INO_9$			
Galactopyranose, 2-(carboxyamino)-2,6-dideoxy-6-iodo-, benzyl ester, triacetate	$CHCl_3$	254(2.87),259(2.82), 264(2.77),269(2.64)	44-3654-64
$C_{20}H_{24}N_2O$			
Affinisine	MeOH	227(4.57),284(3.83)	20-0253-65
Tuboxenine, N-formyl-	EtOH	251(4.12),277(3.59), 287(3.53)	33-0358-64
Unnamed indole base	n.s.g.	217(4.37),265(3.71)	77-0197-65
$C_{20}H_{24}N_2O_2$			
Affinine	n.s.g.	238(4.18)	100-0374-64
Indole, 2-[2-[2-(ethylamino)ethyl]-4,5-dimethoxyphenyl]-	EtOH	296(4.25)	39-4014-65
Pleiocarpamine, 2,7-dihydro-	EtOH	254(4.06),295(3.46)	33-0689-65

Compound	Solvent	λ_{max}(log ϵ)	Ref.
Pleiocarpamine, 2,7-dihydro- (cont.)	EtOH	254(4.06),295(3.46)	33-0878-64
Pleiocarpamine, 19,20-dihydro-	EtOH	230(4.49),279s(3.88), 284(3.89)	33-0689-65
Quinine, sulfate	pH 0.8	318(3.63),346(3.69)	2-0251-65
	pH 2.0	318(3.63),346(3.69)	2-0251-65
	pH 3.1	318(3.61),343(3.65)	2-0251-65
	pH 3,7	318(3.60),336(3.63)	2-0251-65
	pH 4.0	334(3.64)	2-0251-65
	pH 4.6	282(3.44),331(3.64)	2-0251-65
	pH 5.2	281(3.48),330(3.64)	2-0251-65
	pH 6.0	281(3.49),330(3.64)	2-0251-65
	pH 6.3	281(3.49),330(3.64)	2-0251-65
	pH 7.1	281(3.50),330(3.64)	2-0251-65
Tubotaiwin	EtOH	204(4.12),232(4.01), 298(3.96),328(4.12)	33-1497-64
$C_{20}H_{24}N_2O_3$			
Aspidodasycarpine, deformo-	EtOH	240(3.94),296(3.77)	78-1717-65
	EtOH	240(3.94),296(3.77)	88-3899-64
Isoquinoline, 2-acetyl-1-(2-aminobenzyl)-1,2,3,4-tetrahydro-6,7-dimethoxy-	EtOH	286(3.78)	39-4014-65
Schizozygine, tetrahydro-	MeOH	267(4.03),313(3.98)	33-0308-65
Vallesamine	EtOH	222(4.64),283(3.94), 292(3.86)	33-2072-64
Vincaminic acid	MeOH-HCl	220(4.44),270(3.91)	73-0433-64
	MeOH-NaOH	230(4.44),281(3.86)	73-0433-64
Vinervine, dihydro-	n.s.g.	245(3.90),294(3.46)	70-2152-65
$C_{20}H_{24}N_2O_4$			
Compactinervine	EtOH	237(3.97),297(3.95), 331(4.15)	78-1141-65
$C_{20}H_{24}N_2O_5$			
Δ^6-Codeine, 6-deoxy-7-formyl-14-hydroxy-6-methoxy-, oxime	EtOH	255(4.17)	78-1407-64
D-Fructosone-1-dibenzylhydrazone	EtOH	230(3.65),305(4.38)	24-0725-64
$C_{20}H_{24}N_2O_8$			
[2,2'-Bipyrrole]-4,4'-dicarboxylic acid, 3,3'-diacetoxy-5,5'-dimethyl-, diethyl ester	EtOH	222(4.49),263(4.20)	23-1524-64
$C_{20}H_{24}N_4O_4$			
Bicarbamic acid, benzoyl-, diethyl ester, m-tolylhydrazone	MeOH	405(2.28)	89-0684-64
Clov-4-en-3-one, 4-demethyl-, 2,4-dinitrophenylhydrazone	$CHCl_3$	395(4.49)	39-1344-65
$C_{20}H_{24}N_4O_6$			
Bicyclo[3.3.1]non-2(3)-ene-3-carboxylic acid, 2,4,4,8-tetramethyl-9-oxo-, ethyl ester, 2,4-dinitrophenylhydrazide	n.s.g.	326(4.25)	88-1337-64
$C_{20}H_{24}N_4O_7$			
Aniline, o-cyclohexyl-N,N-dimethyl-, picrate	EtOH	208(4.15),258(3.53), 300(2.99)	56-1437-65

Compound	Solvent	λ_{max}(log ε)	Ref.
$C_{20}H_{24}N_4O_8$			
Itaconitin, hexahydro-, 2,4-dinitrophenylhydrazide	EtOH	222(4.58),237s(4.36), 257(4.17),322(4.30)	94-0058-65
	EtOH	222(4.67),237s(4.51), 255(4.35),325(4.47)	94-0058-65
$C_{20}H_{24}O$			
Estra-1,3,5(10)-trien-17β-ol, 17α-ethynyl-	MeOH	266(2.71),273(2.71)	87-0409-65
2-Stilbenol, α-ethyl-5-methyl-α'-propyl-	n.s.g.	208(4.49),284(3.60)	39-3887-65
$C_{20}H_{24}O_2$			
Chrysene, 5,6,6a,7,8,9,10,10a-octahydro-8α-hydroxy-2-methoxy-10a-methyl-, cis	EtOH	228(4.82),255(3.67), 266(3.75),276(3.76), 287(3.60),307(3.08), 315s(3.15),321(3.31), 329(3.24),336(3.40)	23-0591-64
Chrysofluorene, 3-ethylenedioxy-1,1a,-2,3,4,6,6a,?-octahydro-la-methyl-	EtOH	266(2.87)	42-0479-64
Gona-1,3,5(10),8,14-pentaen-17β-ol, 13β-ethyl-3-methoxy-	EtOH	311(4.50)	94-1289-65
Gona-1,3,5(10),8-tetraen-17-one, 13β-ethyl-3-methoxy-	n.s.g.	280(4.20)	39-4472-64
Gona-1,3,5(10),9(11)-tetraen-17-one, 13β-ethyl-3-methoxy-	EtOH	264(4.24)	39-4472-64
2-Hexene, 3,4-bis(4-hydroxy-3-methylphenyl)-	EtOH	213(4.34),236(4.12), 277s(3.75)	39-6509-65
2-Hexene, 3-(3,5-dimethyl-4-hydroxyphenyl)-4-(4-hydroxyphenyl)-	EtOH	226s(4.22),275(3.55), 280(3.51)	39-6509-65
Naphtho[1,2-b:4,3-b']dipyran, 2,3,4,5,-6,7-hexahydro-2,2,7,7-tetramethyl-	n.s.g.	252(4.32),325(3.65), 340(3.68)	24-0588-64
4,4'-Stilbenediol, α,α'-diethyl-3,3'-dimethyl-	EtOH	214(4.31),242(4.15), 277s(3.72)	39-6509-65
$C_{20}H_{24}O_3$			
o-Benzoquinone, 5-tert-butyl-3-(5-tert-butyl-2-hydroxyphenyl)-	MeOH	260(4.00)	70-1717-64
1,3-Cyclohexanedione, 2-[2-(3,4-dihydro-6-hydroxy-1(2H)-naphthylidene)-ethyl]-2-ethyl-	n.s.g.	266(4.4)	70-0760-65
Estra-1,3,5(10),6-tetraen-3-ol, 17,17-ethylenedioxy-	EtOH	262(3.83),303(3.37)	87-0755-64
Estra-1,3,5(10),7-tetraen-3-ol, 17,17-ethylenedioxy-	EtOH	280(3.28)	87-0755-64
Estra-1,3,5-triene-2-carboxaldehyde, 3-methoxy-17-oxo-	MeOH	224(4.29),266(4.12), 332(3.69)	44-3993-65
Estra-1,3,5(10)-trien-17-one, 16-(hydroxymethylene)-3-methoxy-	EtOH	270(4.03),304(3.82)	39-1184-64
D-Homoestra-1,3,5(10),9(11)-tetraen-17a-one, 14α-hydroxy-3-methoxy-	EtOH	265(4.34),300(3.58)	70-1413-65
isomer	EtOH	262(4.25),299(3.51)	70-1413-65
5β-Podocarpa-8,11,13-trien-15-oic acid, 6-hydroxy-13-isopropyl-7-oxo-, γ-lactone	EtOH	255(3.92),300(3.23)	44-4387-65
8,14-Secogona-1,3,5(10),9(11)-tetraene-14,17-dione, 13-ethyl-3-methoxy-	EtOH	266(4.23)	39-4472-64
	EtOH	266(4.23)	94-1289-65

Compound	Solvent	λ max(log ε)	Ref.
$C_{20}H_{24}O_4$			
2H-Benzo[d]naphtho[1,2-b]pyran-11-carboxylic acid, 1,3,4,6,6a,7,8,9,10,10a-decahydro-10-hydroxy-12-methoxy-7-methyl-, δ-lactone	EtOH	232(4.25),262(3.90), 326(3.65)	33-1459-64
Dibenzo[a,e]cyclooctene, 5,6,11,12-tetrahydro-2,3,8,9-tetramethoxy-	EtOH	225s(4.20),286(3.85)	23-2183-65
1,12-Dioxacyclodocosa-14,20-diene-16,18-diyne-2,11-dione	ether	265(3.90),280(4.15), 297(4.30),317(4.23)	24-0794-64
Dithymoquinone	$CHCl_3$	241(3.3)	5-0134-65E
Estra-1,3,5(10)-triene-3,17β-diol, 6β,7β-epoxy-, 17-acetate	EtOH	244(3.81),286(3.36)	44-1325-64
Estra-1,3,5(10)-trien-17-one, 1-acetoxy-3-hydroxy-	NaOH	297(3.52)	24-1940-64
	EtOH	280(3.30)	24-1940-64
Estra-1,3,5-trien-6-one, 3,17β-dihydroxy-, 17-acetate	EtOH	221(4.32),256(3.95), 327(3.49)	44-1325-64
Gibb-2-ene-10β-carboxylic acid, 1α-carboxy-4aα-hydroxy-1β-methyl-8-methylene-, 1→4a-lactone, methyl ester	EtOH	223s(3.22)	39-3550-65
Guaiaretic acid	n.s.g.	207(4.64),260(4.28)	39-4011-64
$C_{20}H_{24}O_5$			
Badkhysin	n.s.g.	230(4.03),252(4.17), 335(1.66)	65-2843-64
p-Benzoquinone, 5-tert-butyl-2-(2-tert-butyl-4,6-dihydroxyphenyl)-3-hydroxy-	MeOH	270(4.05),398(2.93)	24-2774-65
	MeOH-KOH	273(4.00),515(3.23)	24-2774-65
2-Butanone, 3,3-bis(2,4-dimethoxyphenyl)-	EtOH	283.5(3.781)	39-3040-65
2-Butanone, 3,3-bis(3,4-dimethoxyphenyl)-	EtOH	230s(4.21),282(3.81)	39-3040-65
Chaparrol	EtOH	268(2.60)	23-2996-65
Chroman, 3-(2,4-dimethoxybenzyl)-6,7-dimethoxy-	EtOH	282(3.83)	39-2844-65
Estra-1,4-diene-3,11,17-trione, 9,10β-epoxy-, ethanol adduct	MeOH	226(3.77),284(3.41)	94-0473-64
Estra-1,4-diene-9,11,17-trione, 9,10β-epoxy, ethanol adduct	MeOH	226(3.94),281(3.34)	94-0473-64
Ovatodiolide, epoxide	EtOH	216(4.11)	78-2117-65
$C_{20}H_{24}O_6$			
Chaparrin, anhydrodehydro-	dioxan	276(4.00)	23-2996-65
Glaucanol	EtOH	270(2.58)	22-1818-64
Isoglaucanol C	EtOH	267(2.70),278(2.66)	22-1818-64
$C_{20}H_{24}O_7$			
Ailanthone	n.s.g.	240(4.00)	88-3983-64
Angelol	EtOH	223(4.29),243s(3.82), 254s(3.72),296s(3.93), 330(4.20)	88-4557-65
$C_{20}H_{24}O_8$			
Isoglauconic acid, dihydro-, acetate	EtOH	210(2.48)(end absorption)	39-1772-65
$C_{20}H_{24}S_2$			
Thianthrene, 2,7-di-tert-butyl-	96% H_2SO_4, 1 day	273(4.45),296(4.66), 585(4.11)	44-0021-64
	3 days	273(4.45),296(4.66), 585(4.11)	44-0021-64

Compound	Solvent	λ_{max}(log ε)	Ref.
$C_{20}H_{24}S_3$			
Thiophene, 2,5-bis(α-ethyl-α-methyl-2-thenyl)-	EtOH	237.5(4.34)	22-2635-65
$C_{20}H_{25}BrO_2$			
Estra-1,3,5-trien-17-one, 16α-bromo-3-methoxy-1-methyl-	EtOH	280(3.21),287(3.23),312(2.18)	94-0687-65
$C_{20}H_{25}ClN_2O_5$			
Norvenenatic acid, hydrochloride	EtOH	223(4.58),270(3.84),285(3.69),295(3.66)	78-2951-65
$C_{20}H_{25}ClN_4O_3$			
3-Buten-2-one, 1-chloro-4-[p-[(2-amino-4-hydroxy-6-methyl-5-pyrimidinyl)-3-propyl]amino]phenyl-, ethylene ketal	pH 1	256(4.45)	87-0035-65
	pH 7	292(4.44)	87-0035-65
	pH 13	288(4.44)	87-0035-65
$C_{20}H_{25}ClO_6$			
Radicicol, tetrahydro-O,O'-dimethyl-	EtOH	245s(3.72),292(3.60)	88-0365-64
$C_{20}H_{25}FO_4$			
Androsta-1,4-diene-3,11-dione, 9α-fluoro-2,17β-dihydroxy-17α-methyl-	MeOH	248(4.12)	13-0345-65
$C_{20}H_{25}IN_2O$			
Pleiocarpaminol, methiodide	EtOH	222(4.67),279(3.89)	33-0878-64
$C_{20}H_{25}NO$			
4-Stilbenamine, α,α'-diethyl-4'-methoxy-2'-methyl-, trans	n.s.g.	236(4.22),278(3.81)	39-3887-65
Stilbene, 4-diethylaminoethoxy-	EtOH	304(4.5),319(4.5)	87-0511-64
Stilbene, 4-(3-dimethylamino-2-methylpropyloxy)-	EtOH	304(4.5),319(4.5)	87-0511-64
$C_{20}H_{25}NOS$			
6H-Dibenzo[b,e]thiocin-12-ol, 12-(3-dimethylaminopropyl)-7,12-dihydro-	MeOH	264(2.95),268(2.96),278s(2.82)	24-0689-65
$C_{20}H_{25}NO_3$			
Benzo[a]cyclohepta[f]quinolizin-14(1H)-one, 2,3,4,5,7,8,12b,13-octahydro-9,10-dimethoxy-	C_6H_{12}	318(4.06)	44-3667-65
	EtOH	340(4.13)	44-3667-65
perchlorate	EtOH	338(4.17)	44-3667-65
Isohypognavinolone	EtOH	227(3.83)	94-1124-64
Petaline methine	n.s.g.	299(4.32)	88-3841-64
3-Stilbenol, 2-(2-dimethylamino-ethyl)-4,4'-dimethoxy-	n.s.g.	317(4.37)	88-3841-64
$C_{20}H_{25}NO_3S$			
Thiocyanic acid, 11β-hydroxy-3,20-dioxoandrost-4-en-9α-yl ester	EtOH	242(4.15)	13-0305-64
$C_{20}H_{25}NO_4$			
17a-Aza-D-homoestra-1,3,5(10)-trien-17-one, 3,17a-dihydroxy-, 3-acetate	MeOH	266(2.88),275(2.88)	78-0743-65
Isoquinoline, 1,2,3,4-tetrahydro-1-(p-methoxybenzyl)-5,6,7-trimethoxy-	MeOH-HCl	278(3.54),284(3.51)	88-3599-65
Nerinine methine	EtOH	270(3.90)	95-0206-65
Tembetarine methine	EtOH	214(4.39),335(4.41)	5-0200-65E
Thalifendlerine	EtOH	282(3.49)	88-3595-65

Compound	Solvent	$\lambda_{max}(\log \epsilon)$	Ref.
$C_{20}H_{25}N_3O_5$			
D-Fructosone-1-dibenzylhydrazone, 2-oxime	EtOH	230(3.64),300(4.32)	24-0725-64
D-Fructosone-1-di-p-tolylhydrazone, 2-oxime	EtOH	230(4.12),341(4.33)	24-0725-64
$C_{20}H_{25}N_3O_{11}$			
Cytosine, N-acetyl-1-(2,3,4,6-tetra-O-acetyl-β-D-glucopyranosyl)-	EtOH	249(4.26),297(3.83)	94-0357-64
$C_{20}H_{25}N_3O_{11}S$			
β-D-Glucopyranoside, ethyl 3,4,6-tri-O-acetyl-2-deoxy-2-(2,4-dinitro-anilino)-1-thio-	EtOH	267(3.80),343(3.96)	44-1556-65
$C_{20}H_{25}N_7$			
Isobiguanide, 1,2-bis(isopropylidene-amino)-3,4-diphenyl-	EtOH	240(4.46),272(4.50)	39-3912-65
$C_{20}H_{26}$			
Biphenyl, 3,5-di-tert-butyl-	EtOH	252(4.32)	25-1837-65
1,3-Butadiene, 2-methyl-1-phenyl-4-(2,6,6-trimethyl-1-cyclohexen-1-yl)-	n.s.g.	286(4.32)	5-0058-65B
Octane, 1,1-diphenyl-	C_6H_{12}	255(2.81),259(2.83), 262(2.82),269(2.66)	22-0587-65
$C_{20}H_{26}Br_2O_2$			
9β,10α-Androsta-4,6-dien-3-one, 6-(dibromomethyl)-17β-hydroxy-	MeOH	259(3.98),292(3.87)	54-0904-65
9β,10α-Androst-4-en-3-one, 6-(dibromo-methylene)-17β-hydroxy-	MeOH	250(3.99),284(3.76)	54-0904-65
$C_{20}H_{26}ClNO_4$			
Tembetarine, chloride	EtOH	284(3.83)	5-0200-65E
	EtOH	284(3.83)	25-1580-64
$C_{20}H_{26}ClNO_5$			
Cissamine, chloride	pH 1	284(3.81)	24-2732-64
	pH 13	218(4.52),248(4.15), 299(4.04)	24-2732-64
	EtOH	235(4.11),290(3.84)	24-2732-64
$C_{20}H_{26}Cl_2O_2$			
9β,10α-Androst-4-en-3-one, 6-(dichloro-methylene)-17β-hydroxy-	MeOH	243(3.97),270(3.83)	54-0904-65
$C_{20}H_{26}CuN_4O_4$			
Mandelamidine, 2-ethoxy-, copper complex	MeOH	270(3.81),585(1.61)	39-4004-65
Mandelamidine, 3-ethoxy-, copper complex	MeOH	277(3.91),585(1.61)	39-4004-65
Mandelamidine, 4-ethoxy-, copper complex	MeOH	270(3.72),590(1.69)	39-4004-65
$C_{20}H_{26}N_2$			
Bis(butylideneaniline)	MeOH	246(4.15),274(4.10), 298(3.66)	65-0356-64
isomer m. 102°	MeOH	246(4.20),274(4.11), 298(3.70)	65-0356-64
Yohimban, 3ξ-methyl-	EtOH	225(4.56),272s(3.86), 281(3.87),289(3.81)	44-0105-65

Compound	Solvent	λ_{max}(log ϵ)	Ref.
$C_{20}H_{26}N_2O$			
Apparicine, N[b]-methyl-6-methoxydihydro-	EtOH	223(3.56),292(3.16)	39-4773-65
Astrocasine	EtOH	263(4.09)	88-1761-65
[17,16-c]Pyrazoloandrost-4-en-3-one	EtOH	237(4.27)	32-0338-65
$C_{20}H_{26}N_2O_2$			
Androsta-4,6-dieno[2,3-c]furazan, 17β-hydroxy-17α-methyl-	EtOH	286(4.40)	94-1445-65
Burnamicine	n.s.g.	311(4.16)	100-0374-64
Corynantheol, 19,20-dehydrodihydro-10-methoxy-	EtOH	225(4.44),280(3.93), 297s(3.84),308s(3.54)	78-1141-65
Cyclohexanone, 2-(3,4-dihydro-4-quinazolinyl)-6-(1-hydroxycyclohexyl)-	EtOH	228(4.25),235(4.20), 300(3.95)	94-0291-65
21-Deoxyajmalol A oxidation product	5N HCl	236(3.79),241(3.76), 307(3.57)	35-0729-64
Isoiboxygaine	EtOH	231(4.40),295(3.76)	25-2064-65
2,3-Seco-A-norandrost-5-ene-2,3-dicarbonitrile, 17β-acetoxy-	EtOH	210(4.08)	94-1445-65
$C_{20}H_{26}N_2O_3$			
β-Indolylhex-4-ene, 2-acetamido-2-carbethoxy-5-methyl-	MeOH	221(4.95),275(4.03), 282(4.08)	22-2306-65
$C_{20}H_{26}N_2O_4$			
Compactinervine, 2,16-dihydro-	EtOH	245(3.86),298(3.55)	78-1141-65
$C_{20}H_{26}N_2O_6$			
Cyclohexaneacetoacetic acid, α,α'-azobis[1-hydroxy-, di-δ-lactone	CH_2Cl_2	244(4.00),410(4.27)	5-0214-65G
$C_{20}H_{26}N_2Zn$			
Zinc, dibutyl(1,10-phenanthroline)-	toluene	440(2.79)	101-0222-65A
$C_{20}H_{26}N_4$			
[3,2-c][17,16-c]Dipyrazolo-5α-estrane	MeOH	225(4.04)	32-0257-65
$C_{20}H_{26}N_4O_4$			
2(3H)-Naphthalenone, 4,4a,5,6,7,8-hexahydro-7α-isopropyl-4aβ-methyl-, 2,4-dinitrophenylhydrazone	$CHCl_3$	390(4.53)	35-0465-64
$C_{20}H_{26}N_4O_8S_2$			
Thymidine, 3',3'''-dithiobis-[3'-methoxy-	pH 1-7	267(4.04)	44-1772-64
	pH 13	261.5(3.91)	44-1772-64
$C_{20}H_{26}N_6O_6$			
Glutamic acid, N-[p-[[2-[(2-amino-4-hydroxy-5-pyrimidinyl)amino]ethyl]-amino]benzoyl]-, dimethyl ester	pH 1	216s(4.16),265(4.11), 296(4.10)	35-5664-64
	pH 13	296(4.41)	35-5664-64
	EtOH	215s(4.28),301(4.47)	35-5664-64
$C_{20}H_{26}O$			
p-Cresol, 2-(α-ethyl-β-propylphenethyl)-	n.s.g.	284(3.40)	39-3887-65
Estra-1,3,5(10)-triene, 2-acetyl-	EtOH	210(4.35),260(4.18)	24-0140-64
Estra-1,3,5(10)-trien-3-ol, 17-ethylidene-	EtOH	280(3.34)	39-5772-65
Estra-1,3,5(10)-trien-17β-ol, 17α-vinyl-	MeOH	266(2.72),284(2.66)	87-0409-65

Compound	Solvent	$\lambda_{max}(\log \epsilon)$	Ref.
$C_{20}H_{26}OS$			
Estra-1,3,5(10)-trien-17-one, 3-(ethylthio)-	EtOH	258(3.94)	44-0295-65
$C_{20}H_{26}OS_2$			
Estra-1,3,5(10)-trien-3-ol, 17,17-ethylenedithio-	EtOH	280(3.27)	87-0755-64
$C_{20}H_{26}O_2$			
Androsta-4,7-diene-3,17-dione, 6α-methyl-	EtOH	237.5(4.17)	78-1753-65
Androst-4-ene-3,17-dione, 6-methylene-	EtOH	261(4.05)	78-0597-65
Clovan-3-one, 2-furfurylidene-	EtOH	334(4.34)	39-1344-65
2-Cyclopenten-1-one, 4-benzoyl-2,5-di-tert-butyl-	MeOH	243(4.35),277(3.16), 328(2.40),318(2.40)	44-3087-64
Estra-1,3,5(10)-triene, 17β-acetoxy-	MeOH	266(2.72),274(2.71)	87-0409-65
Estra-1,3,5-trien-17-one, 3-ethoxy-	MeOH	222(3.95),230(3.88), 278(3.32),288(3.26)	44-2234-65
	EtOH	279(3.34),287(3.29)	13-0013-64
Estra-1,3,5-trien-17-one, 3-methoxy-1-methyl-	EtOH	280(3.23),287(3.24)	94-0687-65
Estra-1,3,5-trien-17-one, 3-methoxy-2-methyl-	EtOH	285(3.45)	94-0196-64
9β,10α-Estr-4-en-3-one, 17α-ethynyl-17β-hydroxy-	EtOH	239(4.20)	39-5488-64
Gona-1,3,5(10),8-tetraen-17β-ol,	EtOH	280(4.15)	94-1289-65
13β-ethyl-3-methoxy-	EtOH-NaOH	278(4.13)	39-4472-64
Gona-1,3,5-trien-17-one, 13β-ethyl-3-methoxy-	EtOH	279(3.27)	39-4472-64
19-Norpregna-1,3,5(10),16-tetraene-3,20α-diol	EtOH	280(3.38)	44-1120-64
19-Norpregna-4,9,11-trien-3-one, 20-hydroxy-	EtOH	238(3.79),345(4.47)	28-4545-65A
5,10-Seco-5,19-cycloandrosta-1(10),2,4-trien-17-one, 3-methoxy-	EtOH	257(3.76)	35-3727-65
$C_{20}H_{26}O_2S$			
Estra-1,3,5(10)-trien-3-ol, 17,17-ethylenoxythio-	EtOH	280(3.27)	87-0755-64
1-Naphthalenesulfinic acid, menthyl ester	EtOH	223(4.75),291(3.95), 304s(3.75),314(3.34), 319(3.22)	35-5637-64
Thieno[4',3',2'-4,5,6]-19-norandrost-5-en-3-one, 17 -hydroxy-5'-methyl-	EtOH	222(4.12),268(4.08), 303(3.35)	94-1433-64
$C_{20}H_{26}O_3$			
Androsta-1,4,6-trien-3-one, 4,17β-dihydroxy-17α-methyl-	EtOH	232(4.07),348(3.92)	7-0288-65
Estra-1,3,5(10)-trien-3-ol, 17,17-ethylenedioxy-	EtOH	281(4.36),289(4.32)	87-0755-64
Estra-1,3,5-trien-17-one, 2-(hydroxymethyl)-3-methoxy-	MeOH	280(3.45),286(3.45)	44-3993-65
Estra-1,3,5-trien-17-one, 4-(hydroxymethyl)-3-methoxy-	MeOH	282(3.37),288(3.36)	44-3993-65
Estra-1,3,5-trien-17-one, 2-(methoxymethyl)-3-hydroxy-	EtOH	285(3.46)	94-0196-64
Estra-1,3,5-trien-17-one, 4-(methoxymethyl)-3-hydroxy-	EtOH	284(3.43)	94-0196-64
Estr-4-en-3-one, 17α-ethynyl-10β,17β-dihydroxy-	MeOH	234(4.14)	88-0663-64

Compound	Solvent	λ_{max}(log ε)	Ref.
A-Norpregn-3(5)-ene-1,2,20-trione	EtOH	283(3.8)	94-0050-65
8,14-Seco-19-nor-D-homoandrosta-2,5(10),9(11)-triene-14,17a-dione, 3-methoxy-	EtOH	243(4.20)	13-0729-64
$C_{20}H_{26}O_4$			
Androst-4-en-19-oic acid, 3,17-dioxo-, methyl ester	EtOH	242(4.15)	78-2473-64
Atractyligenin, dioxo-, methyl ester	n.s.g.	234(3.89)	88-1829-65
2,2',3,3'-Biphenyltetrol, 5,5'-di-tert-butyl-	EtOH	217(4.54),283(3.51)	39-2914-65
Chromone, 5-acetoxy-3,7-di-tert-butyl-2-methyl-	MeOH	251(4.22),278(3.69)	24-2774-65
Cyclopenta[7,8]phenanthro[10,1-bc]-furan-1,3-diol, 1,2,3,3a,4,5a,6,-7,8,9,11b,11c-dodecahydro-2-methoxy-11b-methyl-	n.s.g.	269(2.90),278(2.92)	77-0343-65
5α-Estrane-3,17-dione, 2,16-bis-(hydroxymethylene)-	EtOH	268(4.23)	32-0257-65
Estr-4-ene-3,17-dione, 2α-acetoxy-	MeOH	240(2.83)	7-0310-65
Estr-4-ene-3,17-dione, 16α-acetoxy-	MeOH	238(2.72)	7-0310-65
Estr-4-en-3-one, 17α-ethynyl-10β-hydro-peroxy-17β-hydroxy-	MeOH	235(4.16)	88-0663-64
Ovatodiolide, dihydro-	EtOH	216(4.70)	78-2117-65
Saunderolide, dehydro-	EtOH	199(3.93),236(3.90), 319(1.82)	33-2330-64
Viridin, dideoxydecahydro-	EtOH	269(2.90),278(2.92)	39-3803-65
$C_{20}H_{26}O_4S$			
5α-Androstane-5α-carboxylic acid, 11β-hydroxy-3,17-dioxo-9α-mercapto-, γ-(thio lactone)	EtOH	240(3.56)	13-0305-64
$C_{20}H_{26}O_5$			
Badkhysin, dihydro-	n.s.g.	247(3.91),333(1.88)	65-2843-64
Estra-1,3,5-triene-3,6α,7β,17β-tetrol, 17-acetate	EtOH	282(3.33)	44-1325-64
Gibb-2-ene-10β-carboxylic acid, 1α-carboxy-4aα,7-dihydroxy-1β,8-dimethyl-, methyl ester, 1→4a-lactone	EtOH	220(3.19)	39-1605-65
Thurberilin	EtOH	223(4.19),320(1.73)	78-1711-65
$C_{20}H_{26}O_6$			
Chaparrin A, anhydro-	EtOH	230(4.24)	23-2996-65
Ovatodiolide, dihydro-, diepoxide	EtOH	218(3.82)	78-2117-65
$C_{20}H_{26}O_7$			
Chaparrinone	EtOH	241(4.08)	22-2793-65
$C_{20}H_{26}O_8$			
Glaucarubolone	n.s.g.	240(3.95)	22-2793-65
$C_{20}H_{27}BrO_2$			
Testosterone, 6-(exo-bromomethylene)-	EtOH	250(3.94),270(3.99)	88-4487-65
$C_{20}H_{27}ClN_2O_2$			
Nα-Methylhunterburninium chloride	H_2O	274(4.08),297s(3.82), 308s(3.66)	78-3731-65
	0.25N NaOH	269(4.00),323(3.74)	78-3731-65

Compound	Solvent	$\lambda_{max}(\log \epsilon)$	Ref.
Nα-Methylhunterburninium chloride (cont.)	MeOH	268s(4.05),274(4.08), 302(3.78),311s(3.72)	78-3731-65
	EtOH	275(3.99),303(3.72), 313s(3.63)	33-1957-65
	EtOH-NaOH	274(3.98),325(3.65)	33-1957-65
Nβ-Methylhunterburninium chloride	EtOH	276(4.00),300s(3.66)	33-1957-65
	EtOH-NaOH	272(3.96),328(3.74)	33-1957-65
N-Methylhuntrabrinium chloride	EtOH	271(3.99),302(3.68), 310s(3.58)	33-1957-65
	EtOH-NaOH	267(3.94),325(3.70)	33-1957-65
$C_{20}H_{27}ClN_4O_2$			
2-Hexanone, 1-chloro-6-[p-[(2-amino-4-hydroxy-6-methyl-5-pyrimidinyl)-3-propyl]amino]phenyl-	pH 1	265(3.94)	87-0035-65
	pH 7	241(4.22)	87-0035-65
	pH 13	241(4.26),278(3.97)	87-0035-65
$C_{20}H_{27}ClN_4O_3$			
4-Pyrimidinol, 2-amino-5-[3-[p-[2-[2-(chloromethyl)-1,3-dioxolan-2-yl]-ethyl]anilino]propyl]-6-methyl-	pH 1	265(3.90)	87-0035-65
	pH 7	238(4.21),292(3.77)	87-0035-65
	pH 13	240(4.26),281(3.92)	87-0035-65
$C_{20}H_{27}ClO_2$			
Androsta-1,4-dien-3-one, 2-chloro-17β-hydroxy-17α-methyl-	EtOH	254(4.19)	32-0138-65
9β,10α-Androsta-4,6-dien-3-one, 6-chloro-17β-hydroxy-17α-methyl-	EtOH	287(4.32)	33-0989-65
19-Norandrosta-3,5-dien-17β-ol, 3-chloro-, acetate	EtOH	237s(4.32),243(4.33), 249s(4.20)	88-0137-65
Testosterone, 6-(exo-chloromethylene)-	EtOH	250(3.97),270(3.91)	88-4487-65
$C_{20}H_{27}ClO_3$			
Androsta-1,4-dien-3-one, 2-chloro-11α,17β-dihydroxy-17α-methyl-	EtOH	253(4.18)	32-0138-65
A-Nortestosterone, 3-chloro-, acetate	EtOH	246(4.13)	88-2233-65
$C_{20}H_{27}FO_2$			
Androst-4-ene-3,17-dione, 16α-fluoro-6α-methyl-	EtOH	241(4.22)	87-0108-64
Androst-4-ene-3,17-dione, 16α-fluoro-16β-methyl-	EtOH	241(4.21)	87-0108-64
$C_{20}H_{27}FO_3$			
Androsta-1,4-dien-3-one, 9α-fluoro-11β,17β-dihydroxy-17α-methyl-	MeOH	240(4.19)	13-0345-65
$C_{20}H_{27}FO_5$			
Androst-4-ene-3,11-dione, 9α-fluoro-1α,2α,17β-trihydroxy-17α-methyl-	MeOH	234(4.11)	13-0345-65
$C_{20}H_{27}IN_2$			
2-[p-(Dipropylamino)styryl]-1-methyl-pyridinium iodide, cis	aq. HCl	310(4.03)	23-1345-65
	H_2O	440(4.05)	23-1345-65
trans	aq. HCl	327(4.46)	23-1345-65
	H_2O	459(4.57)	23-1345-65
	50% MeOH	470(4.62)	23-1345-65
	50% MeOH-HCl	328(4.47)	23-1345-65
$C_{20}H_{27}IN_2O$			
Pleiocarpaminol, 2,7-dihydro-, methiodide	MeOH	252(4.05),293(3.39)	33-0878-64

Compound	Solvent	λ_{max}(log ε)	Ref.
$C_{20}H_{27}NO$			
Estra-1,3,5(10)-triene, 2-acetamido-	EtOH	207(4.49),248(4.19), 288(3.16)	24-0140-64
Estra-1,3,5(10)-triene, 3-acetamido-	EtOH	208(4.48),248(4.19), 288(3.09)	24-0140-64
Estra-1,3,5(10)-triene, 2-acetyl-, oxime	EtOH	211(4.45),253(4.14), 298s(2.67)	24-0140-64
Estra-1,3,5(10)-triene, 3-acetyl-, oxime	EtOH	212(4.48),253(4.17)	24-0140-64
$C_{20}H_{27}NO_2$			
1H,7H-6a,9-Ethano-4,10b-propanobenz[h]-isoquinoline-3,7(2H)-dione, octa-hydro-4-methyl-8-methylene-	EtOH	228.5(3.89)	35-0777-65
3H,7H-6a,9-Ethano-4,10b-propanobenz[h]-isoquinolin-3-one, decahydro-7-hy-droxy-4-methyl-8-methylene-	EtOH	205(3.81)	35-0777-65
$C_{20}H_{27}NO_2S$			
Podocarpamide, N-methylthion-O-acetyl-	MeOH	268(4.05),340(1.65)	35-0051-65
$C_{20}H_{27}NO_3$			
Androsta-4,6-diene-2,3-dione, 17β-hy-droxy-17α-methyl-, 2-oxime	EtOH	215(3.92),310(4.33)	94-1445-65
17a-Aza-D-homo-19-norandrost-4-en-17-one, 17a-acetoxy-	MeOH	238(4.26)	78-0743-65
$C_{20}H_{27}NO_3S$			
5α-Androstane-5α-carboximidic acid, 11β-hydroxy-3,17-dioxo-9α-mercapto-, γ-(thio lactone)	pH 1	258(3.82)	13-0305-64
$C_{20}H_{27}NO_5$			
Lycorenine, α-dihydro-, acetate	EtOH	234(4.12),281(3.97), 311(3.83)	94-0408-64
$C_{20}H_{27}N_3O$			
p-Benzoquinone, 2,6-di-tert-butyl-, 4-azine with 1-methyl-4(1H)-pyridone	C_5H_5N	525(4.76)	5-0009-65J
	99% C_5H_5N	525(4.76)	
	98% C_5H_5N	525(4.76)	
	95% C_5H_5N	525(4.74)	
	90% C_5H_5N	525(4.74)	
	80% C_5H_5N	526(4.74)	
	70% C_5H_5N	526(4.74)	
	50% C_5H_5N	525(4.74)	
	30% C_5H_5N	524(4.73)	
	10% C_5H_5N	522(4.72)	
	5% C_5H_5N	519(4.72)	
	2% C_5H_5N	516(4.71)	
	1% C_5H_5N	516(4.69)	
$C_{20}H_{27}N_3O_3$			
Virgiline, O-(2-pyrrolylcarbonyl)-	MeOH	267(4.23)	39-5243-64
$C_{20}H_{28}$			
Estra-1,3,5(10)-triene, 2-ethyl-	EtOH	215s(3.99),265s(2.77), 270(2.91),278s(2.96), 300s(1.87)	24-0140-64

Compound	Solvent	$\lambda_{max}(\log \epsilon)$	Ref.
$C_{20}H_{28}BrClO_3$			
Testosterone, 2α-bromo-6α-chloro-4-hydroxy-17α-methyl-	EtOH	293(3.98)	7-0288-65
Testosterone, 2α-bromo-6β-chloro-4-hydroxy-17α-methyl-	EtOH	293(3.97)	7-0288-65
$C_{20}H_{28}Br_4O_6$			
Melicopol, methyl-, tetrabromide	EtOH	233s(4.14),289(4.33), 330(3.71)	12-2021-65
$C_{20}H_{28}N_2$			
Astrocasine, deoxy-	EtOH-HCl	244(3.94)	88-1761-65
$C_{20}H_{28}N_2O$			
Astrocasine, dihydro-	EtOH	262(2.44),271(2.29)	88-1761-65
Astrophylline, N-methyl-, cis	n.s.g.	254(4.05)	88-4537-65
Astrophylline, N-methyl-, trans	n.s.g.	281(4.30)	88-4537-65
$C_{20}H_{28}N_2O_2$			
Androst-4-eno[2,3-c]furazan, 17β-hydroxy-17α-methyl-	EtOH	255(4.06)	94-1445-65
Burnamicine, dihydro-	EtOH	220(4.59),274(3.91), 282(3.92),291(3.86), 310s(3.31)	35-5362-64
	ether	307(4.14)	35-5362-64
2,3-Diaza-19-norandrosta-3,5(10)-dien-1-one, 17-acetyl-2,4-dimethyl-	MeOH	230s(3.43),290(3.58)	87-0590-64
$C_{20}H_{28}N_2O_3$			
Androsta-4,6-diene-2,3-dione, 17β-hydroxy-17α-methyl-, dioxime	EtOH	251s(4.04),260(4.06), 297(4.43)	94-1445-65
$C_{20}H_{28}N_2O_3S$			
p-Toluenesulfonamide, N-[1-(1,3,3-trimethylbicyclo[3.1.0]hex-2-ylidene)-methyl]acetonyl]-, oxime	n.s.g.	226(4.2),265(2.7)	27-0538-65
$C_{20}H_{28}N_2O_4$			
19-Epicompactinervine	EtOH	235(4.00),295(3.95), 325(4.18)	78-1141-65
$C_{20}H_{28}N_4O_2$			
p-Benzoquinone, 2,5-bis(1-aziridinyl)-3,6-dipiperidino-	methyl cellosolve	233(4.23),464(3.75), 604(2.64)	78-1889-64
$C_{20}H_{28}O$			
Androsta-4,13-dien-3-one, 17β-methyl-	EtOH	239(4.23)	44-2187-64
5α-Androst-1-en-3-one, 4-methylene-	EtOH	245(3.98),345(1.79)	44-2925-65
5α-Androst-1-en-17-one, 3-methylene-	EtOH	234.5(4.36)	87-0345-64
Estra-1,3,5(10)-trien-17β-ol, 4,17α-dimethyl-	MeOH	263.5(2.39)	87-0409-65
Estra-1,3,5(10)-trien-17β-ol, 2-ethyl-	EtOH	215s(4.03),265s(2.77), 270(2.91),278(2.94)	24-0140-64
Estra-1,3,5(10)-trien-17β-ol, 17α-ethyl-	MeOH	266(2.69),273(2.69)	87-0409-65
Gona-4,13-dien-3-one, 17α-ethyl-17β-methyl-	EtOH	239(4.20)	87-0268-65
Gona-1,3,5(10)-trien-3-ol, 17α-ethyl-17-methyl-	MeOH	280(3.35),286(3.32)	44-0257-65

Compound	Solvent	$\lambda_{max}(\log \epsilon)$	Ref.
18-Norandrosta-1,13-dien-3-one, 17,17-dimethyl-	MeOH	228(4.03)	73-1029-64
γ-Retinal	EtOH	408(4.76)	39-2019-65
5,10-Seco-5,19-cycloandrosta-1(10),2,4-trien-17β-ol, 3-methyl-	EtOH	216(4.32),256(3.75)	35-3727-65
Vitamin A aldehyde, 6-cis	EtOH	254(3.95),376(4.56)	54-1113-65
Vitamin A aldehyde, all trans	EtOH	254(3.90),381(4.63)	54-1113-65
$C_{20}H_{28}O_2$			
Abietic acid, dehydro-	EtOH	268(3.78),272(3.79)	78-2911-64
Androsta-1,4-dien-3-one, 17β-(hydroxymethyl)-	n.s.g.	245(4.2)	31-0688-65
Androsta-1,5-dien-3-one, 17β-hydroxy-17α-methyl-	MeOH	226(4.05)	13-0183-64
Androst-2-ene-1,17-dione, 3-methyl-	ether	233(4.16)	24-1770-64
Androst-4-en-3-one, 17β-hydroxy-6-methylene-	EtOH	261(4.06)	44-2925-65
5α-Androst-1-en-3-one, 17β-hydroxy-4-methylene-	EtOH	244(3.98),344(1.83)	44-2925-65
Chrysene, 4b,5,6,6a,7,8,9,10,10a,10b,-11,12-dodecahydro-8α-hydroxy-2-methoxy-10a-methyl-, cis-syn-cis	EtOH	221(3.88),279(3.31), 287(3.31)	23-0591-64
5β,19-Cycloandrost-6-en-3-one, 17β-hydroxy-17α-methyl-	EtOH	210(3.81)	13-0001-64
5,19-Cyclo-5β-androst-6-en-17-one, 3β-methoxy-	MeOH	215(3.76)	31-0563-65
Estra-3,5-dien-17-one, 3-methoxy-6-methyl-	EtOH	247(4.29)	78-0569-65
Estra-1,3,5-triene-3,17β-diol, 2,17α-dimethyl-	EtOH	285(3.46)	94-0196-64
Estra-1,3,5-trien-17β-ol, 3-methoxy-1-methyl-	EtOH	280(3.34),286(3.34)	94-0687-65
Estra-1,3,5-trien-17β-ol, 3-methoxy-2-methyl-	EtOH	284(3.42)	94-0196-64
Etiojerva-12,15,17-trien-3-ol, 17-methoxy-	MeOH	281(2.98),285(2.98), 290(2.98)	44-0123-65
Gona-1,3,5(10)-triene-3,17β-diol, 13β-propyl-	EtOH	281(3.36),288(3.32)	39-4472-64
Gona-1,3,5(10)-trien-17β-ol, 13β-ethyl-3-methoxy-	EtOH	278(3.28)	39-4472-64
D-Homogona-4,9-dien-3-one, 13β-ethyl-17aβ-hydroxy-	n.s.g.	306(4.34)	39-4472-64
19-Norprogesterone	MeOH	240(4.23)	35-1528-64
5,10-Seco-5,19-cycloandrosta-1(10),2,4-trien-17β-ol, 3-methoxy-	EtOH	257(3.74)	35-3727-65
Sugiol	EtOH	233(4.22),288(4.15)	78-0409-64
Testosterone, 6-methylene-	EtOH	259(4.04)	78-0597-64
Vitamin A acid, 2-cis	EtOH	352(4.59)	95-0998-65
6-cis	EtOH	342(4.61)	95-0998-65
2,4-di-cis	EtOH	346(4.41)	77-0347-65
all trans	EtOH	348(4.67)	95-0998-65
$C_{20}H_{28}O_3$			
Androsta-1,4-dien-3-one, 4,17β-dihydroxy-17α-methyl-	EtOH	244(3.90),305(3.75)	7-0288-65
Androsta-4,6-dien-3-one, 4,17β-dihydroxy-17α-methyl-	EtOH	319(4.33)	7-0288-65
9β,10α-Androsta-4,6-dien-3-one, 17β-hydroxy-6-methoxy-	MeOH	248(3.88),308(4.19)	54-0918-65

Compound	Solvent	λ max(log ε)	Ref.
5α-Androst-1-ene-2-carboxaldehyde, 17β-hydroxy-3-oxo-	EtOH-HCl	236(3.92)	44-3481-64
	EtOH	241(3.90)	44-3481-64
	EtOH-NaOH	305(4.22)	44-3481-64
	dioxan	245(3.98)	44-3481-64
Androst-4-ene-3,16-dione, 17β-hydroxy-17α-methyl-	MeOH	240(4.21)	7-0205-65
5β,19-Cycloandrostane-6,17-dione, 3β-methoxy-	EtOH	210(3.63)	88-1345-64
6β,19-Cycloandrostane-3,17-dione, 5α-methoxy-	EtOH-HCl	243(3.39),289(1.75)	13-0001-64
	EtOH	291(1.96)	13-0001-64
	EtOH-NaOH	291(1.94)	13-0001-64
Dicyclopenta[a,g]naphthalen-8(1H)-one, dodecahydro-3α-hydroxy-3aα,6-dimethyl-, acetate	EtOH	236(4.22)	35-2073-64
Estradiol, 2-(methoxymethyl)-	EtOH	285(3.46)	94-0196-64
Estradiol, 4-(methoxymethyl)-	EtOH	285(3.42)	94-0196-64
Estr-4-en-3-one, 16β-acetoxy-	MeOH	238(4.18)	44-2351-64
B-Homoestr-5(10)-ene-7,17-dione, 3β-methoxy-	EtOH	287(2.59)	88-1345-64
19-Nor-16α-kaur-1-en-17-oic acid, 3-oxo-, methyl ester	EtOH	230(4.04)	12-0578-64
19-Norprogesterone, 11β-hydroxy-	MeOH	242(4.20)	44-1723-65
19-Nor-17α-progesterone, 8-hydroxy-	EtOH	243(4.20)	44-0351-64
A-Nortestosterone, acetate	EtOH	233(4.23)	94-0156-65
Ros-5(6)-en-16-oic acid, 10-hydroxy-7-oxo-, γ-lactone	n.s.g.	235(4.04)	39-7246-65
Unknown compound from Leonotis leonurus	n.s.g.	245(4.06),318(1.70)	39-1857-64
$C_{20}H_{28}O_3Se$			
Dihydroisorosenonolactone SeO oxidation product	n.s.g.	269(3.81)	39-7246-65
$C_{20}H_{28}O_4$			
Araucarenolone	EtOH	263(3.89)	1-1875-65
19-Norcortexone, 8-hydroxy-	EtOH	243(4.27)	44-0345-64
16-Norisopimara-1,7-dien-15-oic acid, 2-hydroxy-3-oxo-, methyl ester	EtOH	267(3.97)	1-1875-65
A-Norpregn-3-ene-2,20-dione, 16α,17α-dihydroxy-	EtOH	234(4.20)	13-0493-64
Plectrin	EtOH	232(3.93),295(1.37), 347(1.47)	65-0280-65
Saunderolide	EtOH	286(1.57)	33-2330-64
Unknown diosphenol from dihydroisorosenonolactone	n.s.g.	287(4.02)	39-7246-65
$C_{20}H_{28}O_5$			
Badkhysin, tetrahydro-	n.s.g.	248(3.8),337(1.74)	65-2843-64
4-Etiocholenic acid, 14α,17α-dihydroxy-3-oxo-	MeOH	240(4.22)	44-2776-65
Flexuosin B, anhydrodihydro-	EtOH	227(4.29),318(1.66)	78-0979-64
16α-Kauran-17,19-dioic acid, 1-oxo-	EtOH	295(1.49)	12-2005-65
Unknown compound from Leonotis leonurus	n.s.g.	214(2.29)	39-1857-64
$C_{20}H_{28}O_6$			
Flexuosin B	EtOH	220(4.08),300(1.48)	78-0979-64
Klaineanone	n.s.g.	243(4.04)	22-2793-65
Melicopol, methyl-	EtOH	234s(4.06),289(4.25), 333(3.59)	12-2021-65
	EtOH-KOH	233s(4.18),292(4.01), 348(3.59)	12-2021-65

Compound	Solvent	λ_{max}(log ε)	Ref.
$C_{20}H_{28}O_7$			
Chaparrin	DMF	203(3.75)	23-2996-65
Isooxoryanodol anhydride	EtOH	270(3.9)	31-0425-65
$C_{20}H_{28}O_9$			
Taxininol, anhydro-	n.s.g.	215(3.72),260s(2.90)	95-0762-64
$C_{20}H_{28}O_{14}$			
Phloroacetophenone, 2,4-β-D-diglucosyl-	EtOH	280(4.19)	22-2937-65
$C_{20}H_{29}BrO_3$			
Rosenonolactone, 6-bromodihydro-	n.s.g.	278-296(1.52)	39-7246-65
Testosterone, 2α-bromo-4-hydroxy-17α-methyl-	EtOH	286(4.03)	7-0288-65
$C_{20}H_{29}ClN_2O_2$			
Ochrosandwine, chloride	H_2O	275(3.92),295s(3.75),306s(3.58)	78-3731-65
	0.25N NaOH	268(3.71),321(3.47)	78-3731-65
	MeOH	267s(3.87),275(3.92),299(3.65),309s(3.58)	78-3731-65
$C_{20}H_{29}ClO_2$			
Androst-1-en-3-one, 2-chloro-17β-hydroxy-17α-methyl-	EtOH	249(3.96)	32-0138-65
Testosterone, 2α-chloro-17α-methyl-	MeOH	244(4.14)	94-1217-64
$C_{20}H_{29}ClO_3$			
Testosterone, 6α-chloro-4-hydroxy-17α-methyl-	EtOH	280.5(4.06)	7-0288-65
Testosterone, 6β-chloro-4-hydroxy-17α-methyl-	EtOH	282(4.05)	7-0288-65
$C_{20}H_{29}FO$			
Androst-4-en-3-one, 13α-fluoro-17β-methyl-	EtOH	241(4.34)	44-2187-64
$C_{20}H_{29}FO_2$			
Androst-4-en-3-one, 16α-fluoro-17β-hydroxy-16β-methyl-	EtOH	241(4.06)	87-0108-64
$C_{20}H_{29}FO_5$			
Androst-4-en-3-one, 9α-fluoro-1α,2α,11β,17β-tetrahydroxy-17α-methyl-	MeOH	236(4.24)	13-0345-65
$C_{20}H_{29}N$			
Tuberostemonan	EtOH	231(4.50),278(4.05)	95-0548-64
$C_{20}H_{29}NO$			
5α-Androst-2-ene-2-carbonitrile, 17β-hydroxy-	EtOH	210(3.96)	44-3300-64
$C_{20}H_{29}NO_2$			
Androst-4-en-3-one, 2-(aminomethylene)-17β-hydroxy-	EtOH	245(4.19),354(4.03)	94-0077-64
17-Azapregn-4-ene-3,20-dione	MeOH	240(4.22)	13-0291-64
	MeOH	240(4.22)	88-0223-64

Compound	Solvent	$\lambda_{max}(\log \epsilon)$	Ref.
$C_{20}H_{29}NO_3$			
Androst-4-ene-2,3-dione, 17β-hydroxy-17α-methyl-, 2-oxime	EtOH	264(4.14)	94-1445-65
1,4-Benzodioxan, 6,8-diallyl-5-(2-diethylaminoethoxy)-	EtOH	275(3.02)	65-2994-64
$C_{20}H_{29}NO_4$			
Testosterone, 17-methyl-2α-nitro-	MeOH	241(3.97),360(2.24)	78-0373-64
$C_{20}H_{29}O_4P$			
Estra-1,3,5(10)-trien-17β-ol, 3-methoxy-, hydrogen methylphosphonate	MeOH	279(3.10),288(3.08)	22-0933-65
$C_{20}H_{30}$			
p-Xylene, α,α'-dicyclohexyl-	hexane	254(2.68),260(2.68), 264(2.68),266(2.68)	70-0124-65
$C_{20}H_{30}N_2$			
5α-Androstano[3,2-c]pyrazole	EtOH	222(3.54)	32-0455-65
5α-Androstano[17,16-c]pyrazole	EtOH	224(3.65)	32-0455-65
$C_{20}H_{30}N_2O_2$			
5α-Androstano[2,3-c]furazan, 17β-hydroxy-17α-methyl-	EtOH-HCl	217(3.59)	94-1445-65
	EtOH	217(3.63)	94-0895-65
	EtOH	217(3.63)	94-1445-65
1,3-Cyclohexadiene-1,4-dicarboxamide, N,N'-dicyclohexyl-	EtOH	308(4.15)	23-2852-64
Flabellidine, acetyl-	MeOH	237(3.84)	23-2456-64
$C_{20}H_{30}N_2O_3$			
Androstano[2,3-c]furazan, 17β-hydroxy-17α-methyl-, N-oxide	EtOH	263(3.87)	94-1445-65
Androst-4-ene-2,3-dione, 17β-hydroxy-17α-methyl-, dioxime	EtOH	230(4.15),261(4.13)	94-1445-65
	EtOH-NaOEt	253(4.1),306(3.96)	94-1445-65
$C_{20}H_{30}N_2O_4$			
11bH-Benzo[a]quinolizine-3-carboxamide, N,N-diethyl-1,2,3,4,6,7-hexahydro-9,10-dimethoxy-	n.s.g.	285(3.59)	87-0635-64
$C_{20}H_{30}N_2O_7$			
L-Glutamic acid, N-[p-(2,2-diethoxyethylamino)benzoyl]-, dimethyl ester	EtOH	215s(4.05),298(4.32)	35-5664-64
$C_{20}H_{30}N_2S$			
4H,10H-Dithieno[3,4-c:3',4'-h][1,6]diazecine, 5,11-diethyl-5,6,11,12-tetrahydro-1,3,7,9-tetramethyl-	n.s.g.	220(4.2),238(4.2), 244s(4.1)	70-2182-64
$C_{20}H_{30}O$			
Androsta-3,5-diene, 3-methoxy-	EtOH	239(4.30)	35-2837-64
5α-Androst-1-en-17β-ol, 3-methylene-	EtOH	234.5(4.32)	87-0345-64
5,8-Epoxy-1(15),3,9(17),13-duvatetraene	C_6H_{12}	237(4.28)	44-0016-64
1,4-Epoxyphenanthrene, 1,2,3,4,7,8,8a-9,10,10a-decahydro-8a-isopropyl-4,6,10a-trimethyl-	C_6H_{12}	251(3.99)	44-0016-64
19-Norandrost-4-en-17β-ol, 17α-methyl-3-methylene-	EtOH	238(4.42)	87-0345-64
15-Oxabicyclo[10.2.1]pentadeca-2,6,10-triene, 8-isopropyl-1,11-dimethyl-5-methylene-	C_6H_{12}	236(4.4)	44-0016-64

Compound	Solvent	λ_{max}(log ϵ)	Ref.
γ-Vitamin A	EtOH	324(4.75),341(4.93), 359(4.89)	39-2019-65
$C_{20}H_{30}O_2$			
Abieta-7,9(11)-dienoic acid	EtOH	241(4.01)	44-1881-65
(±)-Abietic acid	EtOH	235(4.29),242(4.34), 250(4.16)	35-0096-64
(-)-Abietic acid	EtOH	235(4.33),242(4.36), 250(4.19)	35-0096-64
Androsta-3,5-dien-17β-ol, 3-methoxy-	EtOH	240(4.24)	78-0597-64
5α-Androstan-17-one, 16-(hydroxymethylene)-	EtOH	265(4.02)	32-0338-65
Androst-2-en-1-one, 17β-hydroxy-3-methyl-	ether	233(4.08)	24-1770-64
Androst-4-en-3-one, 17β-(hydroxymethyl)-	MeOH	242(4.2)	31-0499-65
	n.s.g.	240(4.2)	31-0688-65
Androst-4-en-3-one, 17β-hydroxy-1α-methyl-	MeOH	245(4.15)	73-3575-65
	EtOH	243(4.18)	35-3727-65
5α-Androst-2-en-1-one, 17β-hydroxy-17α-methyl-	MeOH	224.5(3.84)	87-0048-65
5α-Androst-3-en-2-one, 17β-hydroxy-17α-methyl-	MeOH	232(3.80)	87-0048-65
9β,10α-Androst-4-en-3-one, 17β-hydroxy-17α-methyl-	EtOH	241(4.24)	33-0989-65
Epitestosterone, 17β-methyl-	MeOH	242(4.2)	31-0499-65
Estr-4-en-3-one, 17β-hydroxy-2β,17α-dimethyl-	EtOH	240.5(4.28)	78-2501-65
Etiojerv-1-en-3-one, 17α-hydroxy-17-methyl-	MeOH	229.5(3.95)	44-0123-65
Gon-4-en-3-one, 13β-ethyl-17β-hydroxy-17α-methyl-	n.s.g.	240(4.20)	39-4472-64
8α-Gon-4-en-3-one, 17β-hydroxy-13β-isopropyl-	EtOH	240(4.08)	94-1294-65
A-Homo-5α-androstane-4,17-dione	MeOH	288(1.9)	78-2937-64
Manoyl oxide, 3,18-cyclopropyl-2-oxo-	EtOH	215(2.45),284(1.85)	78-3599-65
19-Nortestosterone, 18,18-dimethyl-	EtOH	240(4.21)	94-1294-65
Podocarpa-8,11,13-triene-13,16-diol, 14-isopropyl-	EtOH	279(3.33),283s(3.31)	39-3001-64
Testosterone, 17α-methyl-	EtOH	241(4.2)	94-1217-64
$C_{20}H_{30}O_3$			
13αH-Abiet-7-enoic acid, 12-oxo-	EtOH	288(2.15)	44-3190-65
Androstane-2,3-dione, 17β-hydroxy-17α-methyl-	EtOH	270(3.91)	94-1445-65
5α-Androstan-2-one, 17β-hydroxy-3-(hydroxymethylene)-	n.s.g.	282(3.95)	44-3786-65
5β-Androstan-3-one, 17β-hydroxy-4-(hydroxymethylene)-	EtOH	280(4.06)	88-1839-64
5α-Androst-2-ene-2-carboxylic acid, 17β-hydroxy-	EtOH	218(3.94)	44-3300-64
Androst-4-en-3-one, 11β-hydroxy-17β-(hydroxymethyl)-	MeOH	242(4.19)	39-0586-64
Araucarone	EtOH	280(1.83)	1-1875-65
Estr-4-en-3-one, 17β-(2-hydroxyethoxy)-	EtOH	240(4.19)	13-0557-65
Isorosenolic acid	n.s.g.	203(3.78)	88-0849-64
16-Norisopimar-7-en-15-oic acid, 2-oxo-, methyl ester	EtOH	285(1.58)	1-1875-65
16-Norisopimar-7-en-15-oic acid, 3-oxo-, methyl ester	EtOH	273(2.23)	1-1875-65
Ros-5(10)-en-16-oic acid, 7-oxo-	n.s.g.	284(1.58)	39-7246-65

Compound	Solvent	λ_{max}(log ε)	Ref.
Testosterone, 7α-hydroxy-17α-methyl-	EtOH	242(4.17)	24-3363-64
	MeOH-KOH	244(3.91),285(3.92)	24-3363-64
Testosterone, 16α-hydroxy-17α-methyl-	MeOH	243(4.20)	7-0205-65
Testosterone, 16β-hydroxy-17α-methyl-	MeOH	243(4.27)	7-0205-65
$C_{20}H_{30}O_4$			
Agathic acid	EtOH	218(4.0)	12-0393-64
Araucarolone	EtOH	280(1.92)	1-1875-65
Kolavic acid	n.s.g.	216(4.32)	88-3751-64
4-Pentynoic acid, 5-(2,4,6-trimethyl-1-cyclohexenyl)-3-methyl-2-carbethoxy-, ethyl ester	EtOH	235(4.11),245s(4.12)	22-2541-64
Testosterone, 4,16α-dihydroxy-17α-methyl-	MeOH	277(4.12)	7-0205-65
$C_{20}H_{30}O_5$			
Flexuosin B, anhydrotetrahydro-	EtOH	285(1.59)	78-0979-64
Isoandrographolide	EtOH	214(3.92),220(3.73), 230(3.23)	78-2617-65
$C_{20}H_{30}O_6$			
Bicyclo[3.2.1]octane-1-acetic acid, 6,8-dihydroxy-2-[1-(4-hydroxy-3,3-dimethyl-2-oxocyclopentylidene)-ethyl]-6-methyl-	EtOH	260(4.08)	78-3091-65
$C_{20}H_{30}O_8$			
Secoryanodol	EtOH	250s(2.5),332(2.14)	31-0425-65
$C_{20}H_{31}NO$			
17-Aza-N-ethylandrost-4-en-3-one	MeOH	240(4.20)	13-0291-64
$C_{20}H_{31}NO_2$			
5α-Androstan-3-one, 2-aminomethylene-17β-hydroxy-	EtOH	315(4.18)	87-0238-64
	EtOH	316(4.20)	94-0077-64
5α-Androst-2-ene-2-carboxaldehyde, 17β-hydroxy-, oxime	EtOH	234(4.23)	44-3300-64
$C_{20}H_{31}NO_3$			
Androstan-3-one, 17β-hydroxy-2-hydroxyimino-17α-methyl-	EtOH	243(3.85)	94-0895-65
	EtOH	243(3.84)	94-1445-65
$C_{20}H_{31}N_3$			
Ormojanine	EtOH	212.5(3.47)	39-2657-64
$C_{20}H_{31}O_4P$			
Androst-5-en-17-one, 3β-hydroxy-, hydrogen methylphosphonate	MeOH	293(1.63)	22-0933-65
Testosterone, hydrogen methylphosphonate	MeOH	241(4.27)	22-0933-65
$C_{20}H_{32}$			
Cembrene	EtOH	245(4.23)	44-1693-65
1,3,11,13-Cycloeicosatetraene	hexane	229(4.65)	39-6674-65
Rimua-1(10),5-diene	EtOH	232(--),239(--), 247s(4.21)	88-1903-64
$C_{20}H_{32}N_2O_2$			
Hydroquinone, 2,3-dimethyl-5,6-bis-(piperidinomethyl)-	3N HCl	297(3.65)	39-0042-64
	ether	304(3.70)	39-0042-64

Compound	Solvent	λ_{max}(log ε)	Ref.
Hydroquinone, 2,5-dimethyl-3,6-bis(piperidinomethyl)-	dioxan	263(3.35),307(3.65)	39-0042-64
$C_{20}H_{32}N_2O_3$			
Androstane-2,3-dione, 17β-hydroxy-17α-methyl-, dioxime	EtOH	240(3.82)	94-0895-65
	EtOH	240(3.82)	94-1445-65
$C_{20}H_{32}N_4O_2$			
p-Benzoquinone, 2,5-bis(methylamino)-3,6-bis(piperidinomethyl)-	EtOH	240s(3.96),343(4.30)	39-5569-64
$C_{20}H_{32}N_4O_8$			
p-Benzoquinone, 2,3-dimethyl-5,6-bis-(piperidinomethyl)-, dinitrate	H_2O	260(4.16),305(2.67), 358(2.73)	39-0042-64
$C_{20}H_{32}O$			
Communol	EtOH	233(4.34)	12-0390-64
Rimu-5(10)-en-6-one	EtOH	249.5(4.06)	88-1903-64
$C_{20}H_{32}O_2$			
A-Homo-5α-androstan-4-one, 17β-hydroxy-	MeOH	282(1.6)	5-0047-64G
Manoyl oxide, 18-hydroxy-2-oxo-	EtOH	291(1.54)	78-3599-65
$C_{20}H_{32}O_3$			
2,6,10-Cyclotetradecatrien-1-one, 8,12-dihydroxy-5-isopropyl-2,8,12-trimethyl-	EtOH	235(4.0)	44-0016-64
10-Norabiet-5(10)-enoic acid, 8α-hydroxy-9β-methyl-	EtOH	215(3.38)	44-1881-65
Quinone from grevillol dimethyl ether	EtOH	266(3.87)	12-2015-65
	EtOH-KOH	297(3.75)	12-2015-65
$C_{20}H_{32}O_5$			
Andromed-10(18)enol	EtOH	200(3.76)	44-2756-64
Wightionolide	EtOH	214(3.96)	78-3237-65
$C_{20}H_{33}NO_3$			
1,4-Benzodioxan, 5-(2-diethyl-aminoethyl)-6,8-dipropyl-	EtOH	275(2.93)	65-2994-64
$C_{20}H_{33}N_3$			
Ormojine	EtOH	211(3.42)	39-2657-64
Ormosajine	EtOH	212.5(3.42)	39-2657-64
$C_{20}H_{34}$			
Cembrene, dihydro-	EtOH	200(4.16)(end absorption)	44-1693-65
$C_{20}H_{34}O$			
Abienol	EtOH	238(4.29)	39-5822-64
Verticillol	EtOH	205(4.15)	94-1510-64
$C_{20}H_{34}O_2$			
Rimuan-6-one, 5-hydroxy-	EtOH	305(1.68)	88-1903-64
$C_{20}H_{34}Si_4$			
Tetrasilane, 1,1,2,2,3,3,4,4-octa-methyl-1,4-diphenyl-	C_6H_{12}	250.5(4.33)	25-1063-64

Compound	Solvent	$\lambda_{max}(\log \epsilon)$	Ref.
$C_{20}H_{35}N_3$			
Piptanthine	EtOH	211(3.32)	39-2657-64
$C_{20}H_{35}N_3O$			
4-Indancarboxaldehyde, 1-(1,5-dimethylhexyl)-3a,6,7,7a-tetrahydro-7a-methyl-, semicarbazone	EtOH	268(4.42)	39-4907-64
$C_{20}H_{35}N_3O_3S_2$			
Thieno[3,4-d]imidazoline-4-valeric acid, thio-, S-ester with N-(2-mercaptoethyl)octanamide	n.s.g.	232(3.65)	37-2865-64
$C_{20}H_{36}$			
Cyclohexane, 1,2-bis(cyclohexylmethyl)-	hexane	none	70-0124-65
Cyclohexane, 1,4-bis(cyclohexylmethyl)-	hexane	none	70-0124-65
$C_{20}H_{36}B_4N_8O_4$			
1,3,5,7,2,4,6,8-Tetrazatetraborocine, 1,3,5,7-tetra-tert-butyloctahydro-2,4,6,8-tetraisocyanato-	n.s.g.	206.1(3.67)	39-6421-65
$C_{20}H_{36}B_4N_8S_4$			
1,3,5,7,2,4,6,8-Tetrazatetraborocine, 1,3,5,7-tetra-tert-butyloctahydro-2,4,6,8-tetraisothiocyanato-	n.s.g.	209(4.44),264(3.92)	39-6421-65
$C_{20}H_{36}B_4N_8Se_4$			
1,3,5,7,2,4,6,8-Tetrazatetraborocine, 1,3,5,7-tetra-tert-butyloctahydro-2,4,6,8-tetraisoselenocyanato-	n.s.g.	229(4.60),287(3.90)	39-6421-65
$C_{20}H_{36}N_4$			
Quinazoline, 4-amino-2-(dodecylamino)-5,6,7,8-tetrahydro-	pH 1	227(4.31),283(3.74)	44-1837-65
	pH 10	235(4.14),294(3.80)	44-1837-65
$C_{20}H_{36}O_2$			
Labd-8-ene-3β,15-diol	EtOH	210.0(3.74)	39-3648-64
Labd-8(20)-ene-3β,15-diol	EtOH	206.0(3.70)	39-3648-64
$C_{20}H_{36}O_5$			
Pentadecanedioic acid, 4-methyl-6-oxo-, diethyl ester	EtOH	280(2.11)	39-4154-64
$C_{20}H_{36}O_6$			
Mogoltavinin	EtOH	215(4.41),323(4.27)	65-1013-64
$C_{20}H_{38}ClOP$			
Tributyl(5,5-dimethyl-3-oxocyclohex-2-en-1-yl)phosphonium chloride	EtOH	224(3.88)	44-1407-65
$C_{20}H_{38}O_2S_4$			
Dinonyl xanthogen disulfide	hexane	360s(2.0)	70-2175-64
$C_{20}H_{38}O_2S_5$			
Dinonyl xanthogen trisulfide	hexane	none above 320 nm	70-2175-64

Compound	Solvent	λ_{max}(log ϵ)	Ref.
$C_{20}H_{40}N_4S_2$ Bis(tetraethylammonium) salt with (dimercaptomethylene)malono-nitrile	EtOH	272(4.16),340(4.33)	44-0497-64

Compound	Solvent	λ_{max}(log ε)	Ref.
$C_{21}H_{12}N_2O_2$			
3H-Dibenzo[de,h]cinnoline-3,7(2H)-dione, 2-phenyl-	$C_6H_3Cl_3$	316(3.85),398(3.99)	65-3819-64
$C_{21}H_{12}N_2O_3$			
Naphth[2,3-c]acridan-5,8,14-trione, 6-amino-	$C_6H_3Cl_3$	665(4.15)	65-1575-64
$C_{21}H_{12}N_4O$			
1-Tricycloquinazolinol	$CHCl_3$	254(4.58),287(4.33), 299(4.37),333(4.24), 382(4.36),402(4.34), 430(3.90),457(3.43)	39-3670-64
$C_{21}H_{12}O_6S_3$			
19,20,21-Trithiatetracyclo-[14.2.1.1^{4,7}.1^{10,13}]heneicosa-2,4,6,8,10,12,14,16,18-nonaene-2,8,15-tricarboxylic acid	EtOH	204(4.36),248(4.31), 323(4.19),389(4.14)	12-0070-65
$C_{21}H_{13}ClO_2$			
Anthraquinone, 1-(p-chlorophenyl)-2-methyl-	EtOH	258(4.48),331(3.72)	65-2740-64
$C_{21}H_{13}ClO_2S$			
Anthraquinone, 1-(benzylthio)-4-chloro-	benzene	313(4.04),437(3.70)	22-1648-65
$C_{21}H_{13}N$			
Dibenzo[c,i]phenanthridine	dioxan	230(4.74),252(4.92), 273(4.93),282(4.96), 315(4.15),330(4.11), 355(3.78),374(3.91)	88-2109-64
Triptycene-9-carbonitrile	heptane	267(2.99),274(3.03)	24-0428-65
$C_{21}H_{13}NO$			
Naphth[1,2-d]oxazole, 2-(1-naphthyl)-	heptane	230(4.84),310(4.08), 340(4.41),355(4.45), 372(4.10)	65-0427-64
Naphth[2,1-d]oxazole, 2-(1-naphthyl)-	heptane	225(4.78),240(4.66), 255(4.58),280s(4.09), 327(4.32),335(4.34), 352(4.43),370(4.15)	65-0427-64
Naphth[2,3-d]oxazole, 2-(1-naphthyl)-	heptane	248(4.98),341(4.96)	65-0427-64
$C_{21}H_{13}NO_4$			
Anthraquinone, 2-methyl-1-(o-nitrophenyl)-	EtOH	257(4.64),329(3.81)	65-2740-64
Anthraquinone, 2-methyl-1-(p-nitrophenyl)-	EtOH	259(4.52),329(3.78)	65-2740-64
$C_{21}H_{13}N_5O_4$			
2H-Benzo[a]quinolizine-3,4-dicarboxylic acid, 1,1,2,2-tetracyano-1,11b-dihydro-, dimethyl ester	MeOH	245(4.35),298(3.83), 312(3.83),363(4.26)	39-2676-64

Compound	Solvent	λ_{max}(log ϵ)	Ref.
$C_{21}H_{14}$			
1,2;6,7-Dibenzofluorene	EtOH	252(4.80),261(5.11), 278(4.46),299(4.30), 312(4.53),328(4.57), 338(3.87),344(3.48), 355(3.91)	24-4033-65
2,3;7,8-Dibenzophenalene	EtOH	269(4.29),276(4.28), 285(4.26),302(3.87), 315(4.02),327(4.05), 339(4.03),372(3.06)	24-0743-65
Fluorene, 9-(phenylethynyl)-	heptane	255f(4.5),285(3.7), 302(3.9)	24-2611-65
Fluorene, 9-styrylidene-	heptane	253f(4.7),272f(4.5), 298(4.3),309(4.4)	24-2611-65
1-Propyne, 3-biphenylene-1-phenyl-	DMSO-tert-BuOK	360(4.14),441(4.60), 469s(4.37),495s(3.92), 549(3.23)	5-0050-65J
$C_{21}H_{14}Br_4O_5S$			
m-Cresolsulfonphthalein, tetrabromo-	pH 6.95	615(4.63)	10-0510-64E
$C_{21}H_{14}ClNO_4$			
Naphtho[2,1-a]phenanthridizinium perchlorate	EtOH	206s(4.46),227(4.64), 258s(4.33),278(4.43), 294s(4.47),305(4.53), 310s(4.52),325(4.40), 370(3.80),391(4.03), 409(4.16),455s(3.34)	44-1846-65
$C_{21}H_{14}F_6O_2$			
Propane, hexafluoro-2,2-bis[p-(2-propynyloxy)phenyl]-	EtOH	270(3.4),276(3.4)	70-1349-64
$C_{21}H_{14}INO$			
Pyridine, 4-(5-iodo-2-furyl)-2,6-diphenyl-	MeOH	234(4.38),272(4.37), 313(4.41)	65-3979-64
$C_{21}H_{14}N_2O$			
2-Propen-1-one, 1,3-di-2-quinolyl-	n.s.g.	250s(4.5),331(4.393)	65-2914-64
$C_{21}H_{14}N_2O_3$			
Indone, 3-anilino-2-(p-nitrophenyl)-	MeOH	222(4.43),261(4.44), 325(3.66),378(4.08)	65-0448-64
	MeOH-base	239(4.46),246(4.45), 259(4.46),473(4.36)	65-0448-64
$C_{21}H_{14}N_4$			
7H-5,6a,12-Triazabenz[a]anthracene, 7-(phenylimino)-	EtOH	243(4.58),321(4.07), 364(4.09)	39-3670-64
$C_{21}H_{14}N_4O_8$			
1,2-Indandione, bis(2,4-dinitrophenylhydrazone)	diglyme	257s(4.21),375(4.68)	44-1723-64
$C_{21}H_{14}O$			
Anthracene, 1-benzoyl-	MeOH	220(4.45),253(4.95) 312(3.48),328(3.49), 344(3.60),361(3.70), 379(3.71)	39-5666-64

Compound	Solvent	$\lambda_{max}(\log \epsilon)$	Ref.
Anthracene, 2-benzoyl-	MeOH	197(4.08),214(4.29), 256(4.78),340(3.59), 355(3.45),374(3.30), 395(3.15)	39-5666-64
Anthracene, 9-benzoyl-	MeOH	254(5.14),332s(3.49), 345(3.75),364(3.89), 383(3.84)	39-5666-64
Indone, 2,3-diphenyl-	MeOH	234(4.48),256(4.50), 440(3.24)	65-0448-64
9(10H)-Phenanthrone, 10-benzylidene-	n.s.g.	229(4.33),248(4.4), 280(4.11),320(3.94), 352(3.42),370(3.51)	23-0190-64
$C_{21}H_{14}O_2$			
Anthraquinone, 2-methyl-1-phenyl-	EtOH	255(4.37),333(3.68)	65-2740-64
Phthalide, 3-(diphenylmethylene)-	ether	337(4.30)	22-3047-65
$C_{21}H_{14}O_2S$			
Anthraquinone, 1-(benzylthio)-	benzene	294(3.96),430(3.63)	22-1648-65
$C_{21}H_{14}O_2S_2$			
Anthraquinone, 1-(methylthio)- 4-(phenylthio)-	benzene	342(4.14),503(3.93)	22-1648-65
$C_{21}H_{14}O_3S$			
Anthraquinone, 1-(benzylsulfinyl)-	EtOH	255(4.45),324(3.49), 378(3.55)	22-1648-65
$C_{21}H_{14}O_4S$			
Anthraquinone, 1-(benzylsulfonyl)-	EtOH	254(4.39),321(3.47)	22-1648-65
$C_{21}H_{15}Cl$			
Indene, 3-chloro-1,2-diphenyl-	EtOH	304(4.31)	44-3105-65
$C_{21}H_{15}ClN_2O_6$			
5-Methyl-2,3-(methylenedioxy)-14H-indo-10[2,3-a]phenanthridizinium perchlorate	EtOH	215(4.55),241(4.43), 260(4.38),292s(4.43), 296(4.45),325s(3.90), 380s(4.18),397(4.40), 413(4.62)	4-0168-64
$C_{21}H_{15}Cl_3N_2O$			
Benzamide, o-chloro-N-[o-chloro-α-[o-chlorobenzylidene)amino]benzyl]-	EtOH	252(4.20)	35-1701-64
Benzamide, p-chloro-N-[p-chloro-α-[p-chlorobenzylidene)amino]benzyl]-	EtOH	255(4.48)	35-1701-64
$C_{21}H_{15}Cl_4FeO$			
3,4-Diphenyl-2-benzopyrylium tetrachloroferrate	EtOH	231(4.28),306(4.30)	6-0359-64
$C_{21}H_{15}F_3N_2O$			
Benzamide, p-fluoro-N-[p-fluoro-α-[p-fluorobenzylidene)amino]benzyl]-	EtOH	246(4.23)	35-1701-64
$C_{21}H_{15}N$			
2,3;5,6-Dibenzocarbazole, 9-methyl-	EtOH	233(4.62),282(4.90), 291(4.96),322(4.00), 338(4.06),346(3.94), 354(4.44),378(3.50), 396(3.76),418(4.65)	24-2814-65

Compound	Solvent	$\lambda_{max}(\log \epsilon)$	Ref.
2,3;6,7-Dibenzocarbazole, N-methyl-	EtOH	262(4.76),295s(4.65), 304(4.88),324(4.56), 342(4.43),375s(2.72), 395s(3.04),416(3.33), 440(3.41)	24-2814-65
Pyridine, 2-(5H-dibenzo[a,d]cyclohepten-5-ylidenemethyl)-	EtOH-HCl	225(4.41),235(4.40), 290(4.28)	87-0457-64
$C_{21}H_{15}NO$			
Indone, 3-anilino-2-phenyl-	MeOH	224(4.31),275(4.45), 310s(4.00),455(3.59)	65-0448-64
$C_{21}H_{15}NO_4$			
9(10H)-Phenanthrone, 10-hydroxy-10-(p-nitrobenzyl)-	MeOH	244(4.47),250(4.46), 279(4.26),325s(3.60)	44-2362-64
$C_{21}H_{15}N_3$			
Glyoxylonitrile, phenyl-, azine with benzophenone	C_6H_{12}	233s(4.11),280s(4.04), 338(4.36)	44-4198-65
$C_{21}H_{15}N_5O_7$			
Benzamide, p-nitro-N-[p-nitro-α-[p-nitrobenzylidene)amino]benzyl]-	EtOH	269(4.56)	35-1701-64
$C_{21}H_{16}$			
Anthracene, 9-benzyl-	EtOH	257(5.05),332(3.44), 348(3.73),366(3.96), 385(3.94)	39-4396-65
Cyclopenta[ef]heptalene, 3-methyl-5-phenyl-	hexane	376(4.15),388(4.14), 394(4.16),435(2.95), 465(2.85),749(2.04), 813(2.16),829(2.15), 892(2.08),927(2.14), 1068(1.76),1118(1.76)	5-0194-64B
2,3;5,6-Dibenzonorborna-2,5-diene, 7-phenyl-	dioxan	266(3.08),270s(2.98), 274(2.89)	78-1657-65
2,3;4,5-Dibenzonorcaradiene, 7-phenyl-	EtOH	245s(4.00),276(3.98), 286s(3.90),306(3.60), 317(3.64)	78-1657-65
Naphtho[2,3-b]norcara-2,4-diene, 7-exo-phenyl-	dioxan	264(4.66),303(4.29), 315(4.23),342(2.83), 359(2.83)	78-1657-65
Propadiene, triphenyl-	hexane	234(4.35),240(4.36), 260(4.50),289s(3.95)	78-2177-64
1-Propene, 1-biphenylene-3-phenyl-	DMSO-tert-BuOK	375(4.19),488(4.67), 508s(4.57),530s(4.28)	5-0050-65J
Propyne, triphenyl-	hexane	242(4.39),253(4.34), 271s(2.95),279(2.63)	78-2177-64
Triptycene, 9-methyl-	heptane	270(3.56),278(3.68)	24-0428-65
$C_{21}H_{16}ClNO_3$			
3H-Pyrrolo[1,2-a]indole-$\Delta^{3,\alpha}$-acetic acid, 7-(benzyloxy)-1-chloro-, methyl ester	MeOH	261(4.08),280(4.04), 407(4.36)	35-4608-64
$C_{21}H_{16}ClNO_4$			
9H-Pyrrolo[1,2-a]indole-3-glyoxylic acid, 7-(benzyloxy)-1-chloro-, methyl ester	MeOH	240(4.2),285(4.0), 315(4.0)	35-4608-64

Compound	Solvent	λ_{max}(log ε)	Ref.
$C_{21}H_{16}ClNO_4S_2$			
2-Diphenylamino-4-phenyl-1,3-dithiol-1-ium perchlorate	EtOH	238(4.21),343(3.97)	44-2877-64
$C_{21}H_{16}FNO_3$			
Anthranilic acid, N-benzoyl-N-(p-fluorophenyl)-, methyl ester	EtOH	230s(4.29),278s(3.83)	44-1020-65
Salicylic acid, methyl ester, N-(p-fluorophenyl)benzimidate	EtOH	230s(4.39),278(3.74)	44-1020-65
$C_{21}H_{16}N_2O_2$			
Halfordinol, O-benzyl-	EtOH	228(4.02),249(4.08), 260(4.07),326(4.42)	4-0310-65
Indole, 2-(2-naphthyl)-3-(2-nitro-1-propenyl)-	n.s.g.	238(4.67),322(4.38), 420(3.96)	28-0609-64A
1,2,3-Propanetrione, 1,3-diphenyl-, 2-phenylhydrazone	MeOH	246(4.25),378(4.29)	44-2959-64
1-Propanone, 3-hydroxy-1,3-di-2-quinolyl-	n.s.g.	235(4.7),250(4.6), 290(4.05),320(3.9)	65-2914-64
4(1H)-Quinazolinone, 1-(p-methoxyphenyl)-2-phenyl-	EtOH	236(4.53),284(3.92)	44-1020-65
$C_{21}H_{16}N_4$			
Benzaldehyde, (3-phenyl-2-quinoxalinyl)hydrazone	EtOH	239(4.31),313(4.33), 399(4.06)	95-1085-64
$C_{21}H_{16}N_4O_6$			
1,4-Epoxy-2,3-benzoxazepine, 4-(2,4-dinitroanilino)-1,3,4,5-tetrahydro-3-phenyl-	EtOH	205(4.29),233s(4.18), 336(4.14)	24-1013-65
$C_{21}H_{16}N_4O_7$			
Isoquinoline, 3,4-dihydro-1-phenyl-, picrate	EtOH-HCl	282(4.14)	94-0249-64
	EtOH	256(4.04)	94-0249-64
$C_{21}H_{16}O$			
Anisole, m-9-anthryl-	EtOH	254(5.16),346(3.99), 364(4.03),384(3.83)	44-3695-64
Anisole, o-9-anthryl-	EtOH	255(4.93),346(3.83), 364(4.03),384(4.00)	44-3695-64
Fluorene, 9-benzoyl-9-methyl-	EtOH	216(4.43),256(4.35), 290(3.81),300(3.78)	22-2345-65
Fluorenone, 1,3-dimethyl-4-phenyl-	EtOH	263(4.85),406(3.03)	24-1329-64
Indenol, 2,3-diphenyl-	EtOH	246(4.25),318(4.04)	44-3105-65
Ketone methyl 9-phenyl-9-fluorenyl	EtOH	217(4.42),269(4.19), 303(3.54)	22-2345-65
Triptycene, 9-methoxy-	heptane	270(3.50),278(3.61)	24-0428-65
$C_{21}H_{16}OS$			
Chalcone,β-(phenylthio)-	MeOH	261(4.12),335(4.24)	5-0136-64C
$C_{21}H_{16}O_2$			
Naphtho[1,2-b]pyran-2-one, 3-benzyl-4-methyl-	MeOH	225(4.13),268s(4.36), 277(3.61)	44-0502-64
9(10H)-Phenanthrone, 10-benzyl-10-hydroxy-	MeOH	242(4.44),248s(4.40), 275s(3.89),327(3.45)	44-2362-64
4H-Pyran-4-one, 2,6-distyryl-	EtOH	315(4.11)	65-2766-64

Compound	Solvent	λ_{max}(log ϵ)	Ref.
$C_{21}H_{16}O_3$			
Phthalide, 3-(4-hydroxy-m-tolyl)-3-phenyl-	80% H_2SO_4	372(4.0),470(4.5)	56-0245-65
	2N NaOH	248(3.9),295(3.4)	56-0245-65
	EtOH	283(3.3)	56-0245-65
$C_{21}H_{16}O_4$			
4,9-Anthracenedione, 5,10β-(benzylidenedioxy)-1,4aα,9aα,10α-tetrahydro-	EtOH	263(3.66),318(3.23)	65-2563-64
Phenolphthalein, 3'-methyl-	80% H_2SO_4	394(4.1),504(4.8)	56-0237-65
	80% H_2SO_4	395(4.04),505(4.5)	56-1533-64
	pH 11.5	558(4.7)	56-0237-65
	pH 11.5	295(4.1),375(4.0), 558(4.5)	56-1533-64
	2.5N NaOH	240(4.3),295(3.8)	56-0237-65
	2.5N NaOH	245(4.3),295(3.8)	56-1533-64
	EtOH	280(3.6)	56-0237-65
	EtOH	228(4.4),278(3.7)	56-1533-64
Phthalide, 3-(4-hydroxy-3-methoxyphenyl)-3-phenyl-	80% H_2SO_4	392(4.0),495(4.2)	56-0245-65
	2N NaOH	248(4.0),259(3.5)	56-0245-65
	EtOH	283(3.6)	56-0245-65
$C_{21}H_{16}O_5$			
Naphthacenequinone, 1,4,6-trimethoxy-	EtOH	242(4.90),295(4.42), 425(4.38)	94-0797-65
Phenolphthalein, 3'-methoxy-	80% H_2SO_4	398(4.0),526(4.6)	56-0237-65
	pH 11.5	574(4.5)	56-0237-65
	2.5N NaOH	244(4.4),296(3.9)	56-0237-65
	EtOH	280(3.58)	56-0237-65
Phthalide, 3-(o-hydroxyphenyl)-3-(4-hydroxy-3-methoxyphenyl)-	H_2SO_4	400(3.9),528(3.9)	56-0375-65
	80% H_2SO_4	400(3.9),539(3.88)	56-1767-64
	pH 11.5	410(3.4),432(3.0)	56-1767-64
	2N NaOH	292(3.9)	56-0375-65
	2N NaOH	238(4.17),293(3.9)	56-1767-64
	EtOH	284(3.9)	56-0375-65
	EtOH	283(4.0)	56-1767-64
Phthalide, 3-(p-hydroxyphenyl)-3-(4-hydroxy-3-methoxyphenyl)-	80% H_2SO_4	397(3.0),528(4.57)	56-1767-64
	pH 11.5	374(3.5),569(4.5)	56-1767-64
	2N NaOH	243(4.5),297(3.8)	56-1767-64
	EtOH	229(4.4),281(3.8)	56-1767-64
$C_{21}H_{16}O_6$			
Justicidin B	$CHCl_3$	260(4.52),295(4.13), 310(4.13),350(3.41)	88-4167-65
$C_{21}H_{16}O_8$			
11H-Benzofuro[3,2-b][1]benzopyran-3,8,11-triol, triacetate	MeOH	248(4.44),321(4.35), 370(4.44)	2-0319-64
$C_{21}H_{16}S$			
1,1'-Binaphthyl, 7-(methylthio)-	EtOH	222(4.70),282(3.96)	39-2380-64
$C_{21}H_{17}BrN_2$			
8,13-Dimethyl-13H-indolo[2,3-a]acridizinium bromide	EtOH	202(4.41),243s(4.46), 255(4.63),273(4.40), 285s(4.31),332s(4.37), 343(4.45),410(4.11), 440s(3.93)	44-3584-64

Compound	Solvent	λ max(log ε)	Ref.
$C_{21}H_{17}ClN_2O_4S_2$			
3-[(1,2-Dihydropyrrolo[2,1-b]benzothiazol-3-yl)methylene]-2,3-dihydro-1H-pyrrolo[2,1-b]benzothiazolium perchlorate	EtOH	578(5.26)	65-2455-64
$C_{21}H_{17}ClO_4$			
7-(2,2-Diphenylvinyl)tropylium perchlorate	MeCN	221(4.54),276(4.09), 343(3.80),461(4.32)	24-1590-64
7-(4-Phenylstyryl)tropylium perchlorate	MeCN	274(4.26),284s(4.23), 475(4.62)	24-1337-64
$C_{21}H_{17}ClO_5S$			
Ketone, 2-(phenylthio)-2-phenylvinyl phenyl, perchlorate	HOAc	250(4.02),363(3.97), 428(3.87),480(4.17)	5-0136-64C
$C_{21}H_{17}IN_2O$			
1,4-Dihydro-3-methyl-4-oxo-1,2-diphenylquinazolinium iodide	EtOH	220(4.71),285s(3.94), 296(3.96),306s(3.88), 348s(2.68)	44-1020-65
$C_{21}H_{17}IN_4O_2$			
9-Formyl-10-methylacridinium iodide, (p-nitrophenyl)hydrazone	pH 2.15	525(4.44)	65-2371-64
	pH 4.95	530(4.43)	65-2371-64
	pH 5.35	530(4.42)	65-2371-64
$C_{21}H_{17}IN_6O_2S_2$			
7-[3-(4,6-Dimethyloxazolo[5,4-e]-2,1,3-benzothiadiazol-7(6H)-ylidene)propenyl]-4,6-dimethyloxazolo[5,4-e]-2,1,3-benzothiadiazolium iodide	EtOH	520(4.58)	17-0048-64
5-Ethyl-7-[3-(8-ethyloxazolo[4,5-e]-2,1,3-benzothiadiazol-7(8H)-ylidene)-propenyl]thiazolo[4,5-e][1,2,3]benzothiadiazolium iodide	EtOH	516(4.82)	17-0048-64
$C_{21}H_{17}Li$			
Lithium, (1,3,3-triphenylpropenyl)-	$C_6H_{11}NH_2$	470(4.33),556(4.66)	35-0384-65
$C_{21}H_{17}N$			
Isoindole, 2-methyl-1,3-diphenyl-	EtOH	228(4.47),268s(4.13), 276(4.20),332(4.01), 371(4.27)	77-0272-65
Pyridine, 2-[(10,11-dihydro-5H-dibenzo-[a,d]cyclohepten-5-ylidene)methyl]-	EtOH	295(4.24)	87-0457-64
$C_{21}H_{17}NO$			
2-Indolinone, 1-benzyl-3-phenyl-	EtOH	253(3.89)	87-0626-65
Vinylamine, benzoyl-β,β-diphenyl-	EtOH	230(4.27),310(4.25)	32-1115-65
$C_{21}H_{17}NO_2$			
Benzamide, N-(p-methoxy-α-phenylbenzylidene)-	EtOH	229(4.25),250(4.19), 285(4.19),297(4.19)	35-4365-65
$C_{21}H_{17}NO_3S$			
Indol-3-ol, 2-phenyl-, p-toluenesulfonate	EtOH	225(4.50),303(4.31)	44-3604-65

Compound	Solvent	λ_{max}(log ε)	Ref.
$C_{21}H_{17}NO_4$			
9H-Pyrrolo[1,2-a]indole-3-glyoxylic acid, 7-(benzyloxy)-, methyl ester	MeOH	240(4.2),280(4.0), 320s(4.0)	35-4608-64
$C_{21}H_{17}NS$			
Chalcone, β-anilinothio-	EtOH	232(4.13),313(4.10), 435(4.34)	39-0032-65
$C_{21}H_{17}N_3O$			
Benzpyrene-cytosine adduct	EtOH	288(4.57),301(4.65), 338(3.45),354(4.10), 373(4.37),393(4.44), 405(3.99)	35-1444-64
$C_{21}H_{17}N_5O_7$			
Benz[cd]indole-3-carboxylic acid, 1-acetyl-1,3,4,5-tetrahydro-2-methyl-, 2,4-dinitrophenylhydrazone	$CHCl_3$	265(4.32),307(4.03), 394(4.48)	44-0843-64
$C_{21}H_{18}$			
3,4-Aceperinaphthan, 1-phenyl-	EtOH	238(4.7),253(4.7), 285(4.0),300(4.0)	5-0129-64D
Cyclopropane, 1,1,2-triphenyl-	n.s.g.	225(4.25),232s(4.21)	35-5046-64
Triphenylene, 1,6,11-trimethyl-	EtOH	256(4.89),264(5.06), 280(4.39),291(4.46), 309(3.63),335(2.97), 344(2.64),351(2.91)	78-3289-65
Triphenylene, 2,6,10-trimethyl-	EtOH	254(4.87),263(5.50), 278(4.20),289(4.17)	78-3289-65
Triphenylene, 2,6,11-trimethyl-	EtOH	253(4.99),262(5.21), 278(4.31),289(4.25), 318(2.98),325(2.98), 334(2.76),341(2.64), 349(2.82)	78-3289-65
$C_{21}H_{18}ClNO_4$			
9H-Pyrrolo[1,2-a]indole-3-glycolic acid, 7-(benzyloxy)-1-chloro-, methyl ester	MeOH	271(4.18)	35-4608-64
$C_{21}H_{18}ClNO_6$			
7,10-Dimethoxy-11-phenylacridizinium perchlorate	MeOH	250(5.08),323s(3.31), 409(4.02),455s(3.84)	44-0452-64
$C_{21}H_{18}N_2$			
Fluorene-2-carbonitrile, 9-(5-dimethylamino)-2,4-pentadien-1-ylidene)-	MeCN	264(4.77),306(4.18), 317(4.15),496(4.80)	24-3331-64
Malondialdehyde, phenyl-, dianil, cis	EtOH	211(4.13),247(4.10), 287(4.23),404(4.30)	39-0032-65
trans	EtOH	225s(4.0),250(3.86), 310(4.43),370s(3.8)	39-0032-65
Methanediamine, N,N'-di-1-naphthyl-	$HCONH_2$	331(4.0)	7-0969-64
$C_{21}H_{18}N_2O$			
Benzamide, N-[α-(benzylideneamino)-benzyl]-	EtOH	248(4.35)	35-1701-64
$C_{21}H_{18}N_2O_2$			
Acetanilide, 4'-(diphenylcarbamoyl)-	EtOH	245(4.3),303(4.3)	28-4295-64B
Benzanilide, 4-acetamido-2'-phenyl-	EtOH	275(4.5)	28-4295-64B

Compound	Solvent	λ_{max}(log ϵ)	Ref.
4H-1,2-Diazepin-4-one, 2-acetyl-3-benzylidene-2,3-dihydro-5-methyl-6-phenyl-	EtOH	227(4.26),304(4.37), 335s(--)	44-2128-64
5H-Dibenzo[b,e]-1,4-diazepine, 8-(benzyloxy)-10,11-dihydro-5-methyl-11-oxo-	EtOH	209(4.61),295s(3.64)	33-1590-65
$C_{21}H_{18}N_2O_3$			
1-Indolinecrotonic acid, 2-indol-3-yl-γ-oxo-, methyl ester, trans	MeOH	280(4.03),288(4.01), 323(3.91)	39-0526-64
Pyrrolo[1,2-a]quinoxaline-2-acetic acid, 1,5-dihydro-4-methyl-1-oxo-α-phenyl-, methyl ester	EtOH	235(4.44),259s(3.81), 292s(3.92),300(3.96), 329s(3.40),427(4.12)	39-3678-65
$C_{21}H_{18}N_2O_6S_2$			
Cephalosporadesic acid, O-benzoyl-7-(2-thiopheneacetamido)-, sodium salt	n.s.g.	233(4.34)	87-0022-65
$C_{21}H_{18}N_2S_2$			
Naphtho[1,2-d]thiazoline, 1-methyl-2-[(3-methyl-2-benzothiazolinyl)-methylene]-	hexane	357(3.91)	22-2879-64
	HOAc	265(4.14),328(3.70)	22-2879-64
$C_{21}H_{18}O$			
2-Propanone, 1,1,3-triphenyl-	EtOH	255(2.95),292(2.62)	23-0298-64
Triphenylene, 2-methoxy-6,11-dimethyl-	EtOH	256(4.91),264(5.09), 282(4.26),337(3.21), 353(3.21)	78-3289-65
$C_{21}H_{18}O_2S$			
Sulfone, 2,2-diphenylvinyl p-tolyl	EtOH	275(4.26)	35-4898-64
$C_{21}H_{18}O_4$			
1,10-Anthracenedione, 8-(benzyloxy)-4,4aβ,9,9aα-tetrahydro-9β-hydroxy-	EtOH	220(4.59),257(3.94), 316(3.44)	65-0662-65
2-Cyclopropene-$\Delta^{1,\alpha}$-succinic acid, 2,3-diphenyl-, dimethyl ester	MeCN	244(4.34),252(4.34), 268(4.26),298(4.30), 378(3.84)	35-0944-64
cation	MeCN-HBF_4	249(4.22),294(4.47), 308(4.49)	35-0944-64
5(12H)-Naphthacenone, 1,4,11-tri-methoxy-	EtOH	225(4.74),264(4.56), 304(3.94),316(3.97), 362(3.94)	94-0797-65
Spiropent-4-ene-1,2-dicarboxylic acid, 4,5-diphenyl-, dimethyl ester	MeCN	234(4.55),240s(4.50), 328(4.48)	88-3459-64
$C_{21}H_{18}O_6$			
Pulvinic acid, 2-methoxy-O-methyl-, methyl ester	MeOH	229s(4.27),261(4.14), 336(4.44)	78-3205-65
$C_{21}H_{18}O_7$			
2'-Acetonaphthone, 1'.4'-diacetoxy-3'-(2-furyl)-5'-methoxy-	EtOH	218(4.21),274(4.23), 322(3.87)	33-0769-64
2'-Acetonaphthone, 1',4'-diacetoxy-3'-(2-furyl)-8'-methoxy-	EtOH	215(4.42),236(4.39), 275(4.55),355(--)	33-0769-64
Mitorubrin	EtOH	216(4.26),266(4.26), 292(4.00),346(4.21)	35-3484-65
	NaOH	246(4.31),320(4.37), 346(4.46),485(3.75)	35-3484-65

Compound	Solvent	λ_{max}(log ε)	Ref.
$C_{21}H_{18}O_9$			
Hemiergoflavin, 2-(2,4-dihydroxy-phenyl)-	n.s.g.	255(4.18),284(4.20), 394(3.41)	39-4144-65
$C_{21}H_{18}S$			
Sulfide, 2,2-diphenylvinyl p-tolyl	EtOH	310(4.38)	35-4898-64
$C_{21}H_{19}AsO$			
Acetone, triphenylarsenic derivative	n.s.g.	270(3.56)	30-0424-64B
$C_{21}H_{19}AsO_2$			
Acetic acid, methyl ester, triphenyl-arsenic derivative	n.s.g.	265(3.73)	30-0424-64B
$C_{21}H_{19}ClN_4O_4$			
2,5-Diphenyl-3-(3,4-xylyl)-2H-tetra-zolium perchlorate	n.s.g.	245.5(4.40)	5-0197-65I
isomer	n.s.g.	245.5(4.42)	5-0197-65I
$C_{21}H_{19}ClN_4O_5$			
3-(p-Methoxyphenyl)-2-phenyl-5-p-tolyl-2H-tetrazolium perchlorate	MeOH	254(4.45)	5-0197-65I
isomer	MeOH	254(4.48)	5-0197-65I
$C_{21}H_{19}ClO_4$			
7-(8-Phenyl-1,3,5,7-octatetraen-1-yl)tropylium perchlorate	MeCN	232s(4.26),287s(4.29), 297(4.30),338(4.28), 582(4.84)	24-1590-64
$C_{21}H_{19}ClO_6$			
2,4-Dimethoxytriphenylcarbonium perchlorate	HOAc-$HClO_4$	502(4.23)	27-0339-65
2,5-Dimethoxytriphenylcarbonium perchlorate	HOAc-$HClO_4$	680(2.98)	27-0339-65
3,4-Dimethoxytriphenylcarbonium perchlorate	HOAc-$HClO_4$	517(4.32)	27-0339-65
$C_{21}H_{19}Cl_2NO_2$			
2-(p-Chlorophenyl)-6,7-dihydro-9,10-di-methoxybenzo[a]quinolizinium chloride	EtOH	301(4.49),385(4.09)	33-1852-64
$C_{21}H_{19}Cl_2NO_6$			
2-(p-Chlorophenyl)-6,7-dihydro-9,10-dimethoxybenzo[a]quinolizinium perchlorate	EtOH	301(4.5),384(4.10)	33-1852-64
$C_{21}H_{19}IN_4OSSe$			
8-Ethyl-7-[3-(3-ethyl-2-benzoselenazol-inylidene)propenyl]oxazolo[4,5-e]-2,1,3-benzothiadiazolium iodide	EtOH	543(4.87)	17-0048-64
$C_{21}H_{19}IN_4OS_2$			
8-Ethyl-7-[3-(3-ethyl-2-benzothiazolin-ylidene)propenyl]oxazolo[4,5-e]-2,1,3-benzothiadiazolium iodide	EtOH	540(5.07)	17-0048-64
$C_{21}H_{19}IN_4O_2S$			
8-Ethyl-7-[3-(3-ethyl-2-benzoxazolinyl-idene)propenyl]oxazolo[4,5-e]-2,1,3-benzothiadiazolium iodide	EtOH	503(5.07)	17-0048-64

Compound	Solvent	$\lambda_{max}(\log \epsilon)$	Ref.
$C_{21}H_{19}N$			
1H-Azuleno[1,8-bc]pyridine, 1,5,7-trimethyl-2-phenyl-	n.s.g.	230(4.40),244(4.45), 274(4.56),285s(4.45), 350s(3.95),361(4.09), 387(3.66),404(3.74), 425(3.54),645s(2.54), 702(2.67),783(2.67), 887(2.37)	35-3137-64
Pyrrole, 1-methyl-2,5-distyryl-	MeOH	244(4.18),379(4.57)	4-0318-65
$C_{21}H_{19}NO$			
Ethylamine, N-salicylidene-α,β-diphenyl-	hexane	213(4.57),257(4.18), 263s(4.12),320(3.72)	35-1757-65
	EtOH	213(4.59),258(4.19), 264s(4.12),317(3.66), 407(2.45)	35-1757-65
	dioxan	257(4.21),262s(4.17), 317(3.70),410s(1.29)	35-1757-65
$C_{21}H_{19}NO_4$			
Chelerythrine, dihydro-	pH 1	231(4.48),255s(4.47), 265(4.58),273(4.62), 291s(4.16),304(4.31), 317(4.39),335(4.13), 352(4.25)	100-0199-65
	EtOH	227(4.53),282(4.65), 318(4.18),350s(3.52)	100-0199-65
	EtOH-KOH	281(4.65),318(4.21), 350s(3.54)	100-0199-65
2-Propen-1-one, 1-(2-quinolyl)-3-(2,4,6-trimethoxyphenyl)-	H_2SO_4	484(4.51)	65-1724-65
2-Propen-1-one, 3-(2-quinolyl)-1-(2,4,6-trimethoxyphenyl)-	H_2SO_4	234(4.19),270(4.05), 305(3.78),412(4.56)	65-1724-65
1H-Pyrrolo[1,2-a]indole, 1-acetoxy-7-(benzyloxy)-9-formyl-2,3-dihydro-	MeOH	257(4.32),276(3.99), 308(4.04)	35-4612-64
1H-Pyrrolo[1,2-a]indole-2-carboxylic acid, 7-(benzyloxy)-2,3-dihydro-1-oxo-, ethyl ester	HCl	330(4.38)	44-2904-65
	NaOH	356(4.42)	44-2904-65
	MeOH	325(4.36)	44-2904-65
$C_{21}H_{19}NO_5$			
Chelerythrine	EtOH	255(4.6),298(4.68)	23-0679-65
$C_{21}H_{19}NO_6$			
Carbonic acid, phenyl ester, ester with 2,3-dihydro-9-(hydroxymethyl)-7-methoxy-6-methyl-1H-pyrrolo[1,2-a]indole-5,8-dione	MeOH	230(4.28),285(4.14), 345(3.58),450(2.98)	35-3877-64
	MeOH	230(4.3),285(4.14), 345(3.6),450(2.97)	44-2897-65
Mitorubramine	EtOH	212(4.41),275(4.51), 365(4.27)	35-3484-65
	10% NaOH	240(4.13),297(4.64), 335(4.27)	35-3484-65
$C_{21}H_{19}N_3O_3$			
Benzamide, N-(2-diphenylaminoethyl)-p-nitro-	MeOH	248(4.27),275(4.24)	87-0107-65
$C_{21}H_{19}N_3O_5$			
3,5-Pyridinedicarboxylic acid, 1,4-dihydro-4-oxo-2,6-bis-(2-pyridyl)-, diethyl ester	$CHCl_3$	246(4.27),281(4.31)	88-3175-65

Compound	Solvent	$\lambda_{max}(\log \epsilon)$	Ref.
$C_{21}H_{20}$			
6H-Cyclohepta[b]naphthalene, 7,8,9,10-tetrahydro-7-phenyl-	dioxan	277(3.34),287(3.33), 309(2.98),319(2.67), 324(2.94)	78-1657-65
$C_{21}H_{20}Cl_4O_3$			
11H-Benz[b]indeno[1,2-e]-p-dioxin, 6,7,8,9-tetrachloro-3-ethyl-4b,10a-dihydro-1-methoxy-2,4,10a-trimethyl-	EtOH	295s(3.51),303(3.60)	35-1594-65
$C_{21}H_{20}N$			
Methyl, [p-(dimethylamino)phenyl]-diphenyl-	EtOH-acid	460(5.00)	28-6933-65A
$C_{21}H_{20}N_2$			
Cycloheptatriene, 1-(benzylamino)-7-(benzylimino)-	EtOH	259(4.33),344(4.05), 357(4.08),413(3.98)	94-0810-65
hydrochloride	EtOH	260(4.44),346(4.04), 412(4.15)	94-0810-65
$C_{21}H_{20}N_2O$			
4H-Indeno[1,2-d]pyrimidin-4-one, 5-tert-butyl-3,5-dihydro-2-phenyl-	EtOH	265(4.72),289(4.23), 300(4.21),322(4.11)	44-0396-65
1,3-Propano-β-carboline, 2-benzoyl-1,2,3,4-tetrahydro-	EtOH	226(4.68),278(3.90), 290(3.81)	39-5378-64
$C_{21}H_{20}N_2O_2$			
Alstonilinine, 11-demethoxytetrahydro-	EtOH	223(4.70),282(4.02)	35-2083-64
$C_{21}H_{20}N_2O_2S$			
5-Pyrimidinecarboxylic acid, 2-benzyl-1,4-dihydro-6-methyl-1-phenyl-4-thioxo-, ethyl ester	EtOH	234(3.90),339(4.34)	44-1115-64
$C_{21}H_{20}N_2O_2S_2$			
3-Pyrrolin-2-one, 3-acetamido-4-(benzylthio)-5-[(benzylthio)methylene]-	EtOH	360(4.37)	35-5654-64
$C_{21}H_{20}N_2O_5S_2$			
3-Pyrrolin-2-one, 3-acetamido-5-[(benzylthio)methylene]-4-hydroxy-, p-toluenesulfonate	EtOH	218(4.26),356(4.32)	35-5654-64
$C_{21}H_{20}N_2O_9$			
1,3-Cyclopentadiene-1,2,3-tricarboxylic acid, 4-(carboxycyanomethyl)-5-oxo-,	EtOH	257(4.25),317(4.31), 625(3.22)	39-4355-65
4-ethyl trimethyl ester, compound with pyridine	EtOH	252s(4.29),257(4.29), 262s(4.25),317(4.36), 625(3.28)	44-0423-64
$C_{21}H_{20}N_4O$			
Azobenzene, 4-benzamido-4'-(dimethylamino)-	EtOH	233(4.19),315(4.01), 422s(4.52)	65-3602-64
5H-Pyrrolo[3,4-d]pyrimidine, 4-amino-6-benzoyl-7-benzyl-6,7-dihydro-2-methyl-	pH 1	240(4.11)	44-0194-65
	EtOH	238(4.19),265(3.78)	44-0194-65
$C_{21}H_{20}N_4O_3S$			
Pyrimidine, 4-amino-5-(acetylmandelamido)-6-(benzylthio)-	pH 1	240(4.21),297(4.06)	87-0797-65
	pH 11	233(4.31),280(4.04)	87-0797-65

Compound	Solvent	λ_{max}(log ε)	Ref.
$C_{21}H_{20}N_4O_6$			
2-Cyclopentene-1-carboxylic acid, 3-methyl-4-oxo-2-phenyl-, ethyl ester, 2,4-dinitrophenylhydrazone	$CHCl_3$	389(4.46)	23-2408-65
$C_{21}H_{20}O$			
Cycloheptanone, 2,7-dibenzylidene-	hexane	290(4.41)	65-3645-64
	EtOH	230(4.16),298(4.36)	65-3645-64
	66% H_2SO_4	344(3.52),434(4.61)	65-3645-64
2,5-Cyclohexadien-1-one, 2,4-dibenzyl-4-methyl-	EtOH	238(4.00)	33-0094-65
Fluorenone, 1,2;3,4-bis(tetramethylene)-	EtOH	235(4.18),258(4.60), 266(4.73),296(3.42), 308(3.39),350(3.50)	54-1094-65
$C_{21}H_{20}O_2$			
Propane, 2,2-bis[p-(2-propynyloxy)-phenyl]-	MeOH	278(3.4),280(3.4)	70-1349-64
$C_{21}H_{20}O_4$			
4β,9β-Anthracenediol, 5,10β-(benzylidenedioxy)-1,4,4aα,9,9aα,10-hexahydro-	EtOH	277(3.25),286(3.27)	65-2563-64
Anthrone, 5-(benzyloxy)-1,4,4aα,9aβ-tetrahydro-4β,10β-dihydroxy-	EtOH	225(4.45),257(3.95), 315(3.47)	65-2558-64
Bicyclo[2.2.2]oct-5-ene-2,3-dicarboxylic anhydride, 7-allyl-4,7-dimethyl-γ-oxo-6-phenyl-	EtOH	260(4.14),305(3.38)	44-3014-64
2,3-Naphthalenedicarboxylic anhydride, 4-ethoxy-1,2,3,4-tetrahydro-1-methyl-4-phenyl-	isooctane	244s(2.55),250s(2.58), 255(2.64),261(2.69), 264(2.67),271(2.51)	35-0086-65
$C_{21}H_{20}O_6$			
Cinnamic acid, 3,4-dimethoxy-, cis, 3,4-(methylenedioxy)cinnamyl ester, trans	EtOH	270(4.23),315(4.34)	78-0871-64
trans isomer	EtOH	272(4.23),295s(4.34), 320(4.37)	78-0871-64
Curcumin	iso-PrOH-HOAc	425(4.57)	83-0660-64
	iso-PrOH-NH_3	505(4.50)	83-0660-64
Peucedanin irradiation product	n.s.g.	223(4.32),325(4.12)	65-2848-64
$C_{21}H_{20}O_7$			
Podophyllotoxin, 4'-demethyldeoxy-	MeOH	288(3.66)	33-1203-64
$C_{21}H_{20}O_9$			
Chrysophanol mono-β-D-1-glucoside	n.s.g.	220(4.30),257(4.12), 410(3.70)	24-2859-65
Melibentin	EtOH	254(4.49),268s(4.20), 342(4.34)	12-2021-65
$C_{21}H_{20}O_{10}$			
Afzelin	EtOH	270(4.36),354(4.13)	95-0360-64
Aloe-emodin glucoside	MeOH	223(4.42),255(4.32), 410(3.87)	24-1662-64
$C_{21}H_{20}O_{11}$			
Luteolin, 4'-β-D-glucoside	EtOH	245(4.21),271(4.29), 340(4.27)	94-0841-64

Compound	Solvent	λ_{max}(log ε)	Ref.
Luteolin, 4'-β-D-glucoside (cont.)	EtOH	269(4.38),340(4.27)	95-0895-64
	EtOH-NaOEt	239(4.33),277(4.43), 320(4.07),372(4.14)	94-0841-64
	EtOH-NaOEt	274(4.48),380(4.20)	95-0895-64
	EtOH-NaOAc	274(4.42),322(4.15), 356(4.16)	94-0841-64
	EtOH-NaOAc	274(4.38),323(4.11)	95-0895-64
	EtOH-$AlCl_3$	259(4.12),282(4.27), 290s(4.25),343(4.09), 382(4.09)	94-0841-64
	EtOH-$AlCl_3$	282(4.28),290s(4.26), 344(4.24),380(4.07)	95-0895-64
	EtOH-H_3BO_3- NaOAc	271(4.25),340(4.21)	94-0841-64
		270(4.30),340(4.25)	95-0895-64
Luteolin, 7-glucoside	n.s.g.	254(4.30),345(4.33)	73-1484-64
Quercitrin	EtOH	260(4.34),359(4.20)	95-0360-64
Trifolin	EtOH	267(4.34),354(4.17)	95-0360-64
$C_{21}H_{21}B_3O_{11}$			
Curcumin, boric acid complex	iso-PrOH- HOAc	425(4.57)	83-0660-64
	iso-PrOH- NH_3	505(4.50)	83-0660-64
$C_{21}H_{21}ClN_2$			
Cycloheptatriene, 1-(benzylamino)- 7-(benzylimino)-, hydrochloride	EtOH	260(4.44),346(4.04), 412(4.15)	94-0810-65
$C_{21}H_{21}ClN_2S_2$			
3,3'-Diethylthiacarbocyanine chloride	H_2O	525s(4.8),550(5.2)	46-1894-65
	MeOH	530s(4.9),553(5.2)	46-1894-65
$C_{21}H_{21}N$			
Fluorene, 9-(5-dimethylamino-1-methyl- 2,4-pentadien-1-ylidene)-	MeCN	256(4.67),463(4.69)	24-3331-64
Phenanthro[9,10-b]quinolizidine	EtOH-HBr	213(4.45),221(4.36), 247(4.67),254(4.78), 271(4.14),277(4.12), 285(3.98),297(4.00)	54-0593-64
Piperidine, 4-(5H-dibenzo[a,d]cyclo- hepten-5-ylidene)-1-methyl-	aq. HCl	224(4.54),240s(--), 285(4.04)	87-0829-65
$C_{21}H_{21}NO$			
Benzamide, N-methyl-N-(4,6,8-tri- methyl-1-azulenyl)-	C_6H_{12}	246(4.44),299(4.56), 340s(3.66),354(3.76), 560(2.68)	35-3137-64
Indone, 2-cyclohexylamino-3-phenyl-	n.s.g.	265(4.76),525(4.14)	35-2510-65
$C_{21}H_{21}NO_2$			
Acridine, 9,10-dihydro-10-(1-meth- acryloyloxy-2-methylpropenyl)-	n.s.g.	206(4.69),278(4.11)	39-5877-65
1,2-Cyclopentenophenanthrene, N,N-di- acetyl-3-amino-9,10-dihydro-	EtOH	211(4.77),273(4.31), 284s(4.18),301s(3.55)	5-0200-65D
Phenanthridine, dimethylketene adduct	EtOH	232(4.23),278(3.83), 340(3.76)	77-0574-65
Quinoline, 4-(3,4-diethoxystyryl)-	MeOH	360(4.3)(changing)	87-0137-65
	HOAc	315(3.9),425(4.1)	87-0137-65

Compound	Solvent	λ max(log ε)	Ref.
$C_{21}H_{21}NO_6$			
Carbonic acid, phenyl ester, ester with 1-ethyl-3-(hydroxymethyl)-5-methoxy-2,6-dimethylindole-4,7-dione	MeOH	231(4.25),285(4.14), 345(3.51),455(3.06)	35-3874-64
α-Hydrastine	EtOH	297.5(3.86)	94-1072-64
β-Hydrastine	EtOH	297.5(3.86)	94-1072-64
Rheadine	EtOH	205(4.91),240(3.96), 292(3.94)	73-3479-65
$C_{21}H_{21}NO_8$			
Glucuronic acid, 1-O-(N-fluoren-2-yl-acetamido)-	MeOH	290(4.50)	37-1011-65
$C_{21}H_{21}NO_8S$			
Azepino[2,1-b]benzothiazole-7,8,9,10-tetracarboxylic acid, 9,10-dihydro-6-methyl-, tetramethyl ester	MeOH	265s(0.70),318(0.27), 432(3.15)	39-3200-65
$C_{21}H_{21}OP$			
Phosphine oxide, tri-m-tolyl-	EtOH	271(3.45),279(3.42)	49-1793-65
Phosphine oxide, tri-o-tolyl-	EtOH	272(3.43),279(3.47)	49-1793-65
Phosphine oxide, tri-p-tolyl-	EtOH	231(4.54),257(3.20), 264(3.23),269(3.11), 275(2.97)	49-1793-65
$C_{21}H_{21}P$			
Phosphine, tri-m-tolyl-	EtOH	263(4.04)	49-1793-65
Phosphine, tri-o-tolyl-	EtOH	276(4.05)	49-1793-65
Phosphine, tri-p-tolyl-	EtOH	261(4.06)	49-1793-65
$C_{21}H_{21}PS$			
Phosphine sulfide, tri-m-tolyl-	EtOH	273(3.66),280(3.55)	49-1793-65
Phosphine sulfide, tri-o-tolyl-	EtOH	268(3.82),273(3.84), 281(3.76)	49-1793-65
Phosphine sulfide, tri-p-tolyl-	EtOH	227(4.49)	49-1793-65
$C_{21}H_{22}$			
1H-Cyclopent[a]anthracene, 3-ethyl-2,3-dihydro-3,7-dimethyl-	EtOH	260(5.20),319(3.12), 334(3.42),350(3.65), 367(3.78),386(3.75)	35-4414-64
1H-Cyclopenta[a]phenanthrene, 17-ethyl-16,17-dihydro-1,17-dimethyl-	EtOH	261(4.79),284(4.09), 295(4.05),307(4.15), 323(2.91),338(2.95), 354(2.91)	35-4414-64
$C_{21}H_{22}BrNO$			
Ketone, p-bromophenyl 1-cyclohexyl-3-phenyl-2-aziridinyl, cis	MeOH	264(4.36)	88-4369-65
trans	MeOH	265(4.52)	88-4369-65
$C_{21}H_{22}BrNO_6$			
Acetamide, N-(3-bromo-4-hydroxy-5-methoxyphenethyl)-2-(p-hydroxy-phenyl)-, diacetate	EtOH	222(4.15),282(3.40)	65-0550-64
$C_{21}H_{22}ClNO_2$			
11bH-Benzo[a]quinolizine, 2-(p-chloro-phenyl)-1,4,6,7-tetrahydro-9,10-di-methoxy-	EtOH	238(4.2),252(4.23)	33-1852-64

Compound	Solvent	λ_{max}(log ϵ)	Ref.
11bH-Benzo[a]quinolizine, 2-(p-chlorophenyl)-3,4,6,7-tetrahydro-9,10-dimethoxy-	EtOH	252(4.3)	33-1852-64
Indole-3-acetic acid, 1-(p-chlorobenzyl)-5-methyl-2-propyl-	MeOH	226(4.57),273(3.88), 290(3.89),300s(3.78)	87-0204-65
$C_{21}H_{22}ClN_{3}$			
Pyrimidine, 2-amino-6-benzyl-4-chloro-5-(4-phenylbutyl)-	pH 1, 7	236(4.19),305(3.64)	87-0283-65
	pH 13	235(4.21),308(3.66)	87-0283-65
$C_{21}H_{22}N_{2}O_{2}$			
Alkaloid B	MeOH	250(3.96),298(3.56)	35-4944-65
Benzamide, N-[β-(piperidinocarbonyl)styryl]-	MeOH	304(4.3)	83-0367-64
Voachalotinic acid, de(hydroxymethyl)-	MeOH	227(4.54),284(3.83)	20-0253-65
$C_{21}H_{2}$ $N_{2}O_{3}$			
Polyneuridine	EtOH	226(4.50),282(3.67), 290(3.65)	78-1717-65
Schizozygine isomethine	MeOH	230(4.32),333(3.98), 346(3.96)	33-0308-65
Schizozygine methine	MeOH	215(4.23),269(3.99), 312(3.94)	33-0308-65
$C_{21}H_{22}N_{2}O_{4}$			
Perimivine	EtOH	232(4.04),302(4.01), 340(4.26)	100-0203-64
Spiro[indoline-3,3'-pyrrolidin]-2-one, 1'-acetyl-2'-(3,4-dimethoxyphenyl)-	EtOH	231(4.12),279(3.64), 283(3.62)	78-1327-65
$C_{21}H_{22}N_{2}O_{6}$			
2-Naphthacenecarboxamide, 4-(dimethylamino)-1,4,4a,5,5a,6,11,11a-octahydro-3,10,12-trihydroxy-1,11-dioxo-	MeOH-NaOH	248(4.20),380(4.10), 471(4.31),495(4.24)	35-0933-65
	MeOH-borax	312(3.79),470(4.67), 496(4.56)	35-0933-65
Phenoxazine-1,9-dicarboxylic acid, 2-(isopropylamino)-4,6-dimethyl-3-oxo-, dimethyl ester	MeOH	226(4.35),254(4.47), 424(4.44),444(4.5)	44-3185-65
$C_{21}H_{22}N_{2}O_{7}$			
Oxazolo[4,5-b]phenoxazine-4,6-dicarboxylic acid, 2,3-dihydro-11a-hydroxy-2,2,9,11-tetramethyl-, dimethyl ester	MeOH	222(4.22),244(4.41), 268(4.24),310(4.07), 376(4.4)	44-3185-65
$C_{21}H_{22}O$			
13H-Indeno[1,2-l]phenanthren-13-one, 1,2,3,4,5,6,7,8,8a,13b-decahydro-	EtOH	232(3.72),265(2.48), 308(2.48),338(2.49), 346(2.49)	54-1094-65
Ketone, 2-methyl-5-phenyl-3-cyclohexen-1-yl p-tolyl	EtOH	253(4.190)	65-2258-64
$C_{21}H_{22}O_{2}$			
Ketone, 2-methyl-5-phenyl-3-cyclohexen-1-yl p-anisyl	EtOH	275(4.115)	65-2258-64
$C_{21}H_{22}O_{3}$			
2-Phenanthrol, 9,10-diethyl-6-methoxy-	MeOH	233(4.47),252(4.68), 269s(4.42),280(4.31), 301(3.97),310(3.98), 343(3.26),360(3.26)	5-0036-64D

Compound	Solvent	$\lambda_{max}(\log \epsilon)$	Ref.
$C_{21}H_{22}O_3S_2$			
Spiro[5.5]undecane-1,5,9-trione, 3,3-dimethyl-7,11-di-2-thienyl-	EtOH	238(4.19)	78-0515-65
$C_{21}H_{22}O_4$			
Archangelin	EtOH	222(4.38),251(4.19), 310(4.08)	88-1961-64
1H-Benz[e]indene-3-propionic acid, 2-(1-carboxyethyl)-, dimethyl ester	EtOH	252(4.61),262(4.58), 285(3.69),295(3.72)	42-0643-64
Bicyclo[2.2.2]oct-5-ene-2,3-dicarboxylic anhydride, 4,7-dimethyl-8-oxo-6-phenyl-7-propyl-	EtOH	260(4.11),305(3.30)	44-3014-64
Hydrocinnamic acid, β-acetonyl-α-benzoyl-, ethyl ester	EtOH	248(4.25),285(3.22)	78-0195-64
Isoflav-3-ene, 7-hydroxy-4'-methoxy-2,2,4-trimethyl-	MeOH	271(4.04),308(3.98)	44-4114-65
Naphthalene, 1,4,8-trimethoxy-2-(6-methoxy-m-tolyl)-	EtOH	228(4.86),243s(4.38), 291(4.05),308(4.03), 322(3.99),337(3.99)	33-1459-64
Pregna-1,4,17(20)-trien-21-oic acid, 16β-hydroxy-3,11-dioxo-, γ-lactone	EtOH	228(4.35)	87-0348-64
$C_{21}H_{22}O_5$			
Gibba-A,4b-tetraene-10β-carboxylic acid, 8,8-(ethylenedioxy)-1,7-dimethyl-6-oxo-, methyl ester	MeOH	229(3.99),235(3.98), 298(4.31)	39-0990-65
Spiro[5.5]undecane-1,5,9-trione, 7,11-di-2-furyl-3,3-dimethyl-	EtOH	220(4.17)	78-0515-65
$C_{21}H_{22}O_6$			
Dalbergione, 3,4-dimethoxy-, quinol diacetate	EtOH	270(3.00)	78-2697-65
Dalbergione, 4,4'-dimethoxy-, quinol diacetate	EtOH	230(4.31),278(3.64)	78-2683-65
11H-Furo[2,3-a]xanthen-11-one, 2,3-dihydro-4,7,8-trimethoxy-2,3,3-trimethyl-	MeOH	248(4.66),285s(4.04), 299(4.39)	78-2653-65
Shikonin, β,β-dimethylacryl-	EtOH	273(3.87)	88-4737-65
Xanthone, 2-(1,1-dimethylallyl)-1-hydroxy-3,5,6-trimethoxy-	MeOH	246(4.61),282(4.01), 317(4.25)	78-2653-65
Xanthone, 1-hydroxy-3,5,6-trimethoxy-4-(3-methyl-2-butenyl)-	MeOH	244(4.58),283(3.89), 316(4.27),400(3.80)	78-2653-65
Xanthone, 3,5,6-trimethoxy-1-(3-methyl-2-butenyloxy)-	MeOH	246(4.66),291s(3.91), 305(4.28)	78-2653-65
$C_{21}H_{22}O_7$			
Peucenidine	EtOH	220(4.45),250(3.77), 323(4.19)	65-4171-64
Podophyllotoxin, 6,7-O-demethylenedeoxy-	MeOH	288(3.58)	33-1529-64
$C_{21}H_{22}O_8$			
Epipodophyllotoxin, 6,7-O-demethylene-	MeOH	286(3.55)	33-1529-64
Flavone, 3,3',4',5,7,8-hexamethoxy-	EtOH	253(4.33),272(4.31), 353(4.30)	12-0934-64
Myricetin, hexamethyl ether	EtOH	239(4.25),263(4.19), 336(4.24)	12-0934-64
Podophyllotoxin, 6,7-O-demethylene-	MeOH	286(3.55)	33-1529-64

Compound	Solvent	λ_{max}(log ε)	Ref.
$C_{21}H_{22}O_9$			
Aloin	MeOH	260(3.84),269(3.92), 296(3.98),359(4.08)	83-0262-65
Isoliquiritin	EtOH	260(4.00),360(4.56)	30-0277-64B
Neoisoliquiritin	EtOH	240(4.01),370(4.44)	30-0277-64B
$C_{21}H_{22}O_{10}$			
Naringenin, 7-β-D-glucosyl-	EtOH	284(4.19)	22-2937-65
$C_{21}H_{22}O_{11}$			
Chalcone, 2',3,4,4',6-pentahydroxy-, 4'-glucoside	EtOH	265(3.56),380(4.14)	65-3340-64
Chalcone, 2',3,4,4',6'-pentahydroxy-, 2'-glucoside	n.s.g.	209(4.56),248(3.83), 261(3.80),383(4.80)	83-0838-65
Flavanone, 3',4',5,7-tetrahydroxy-, 7-glucoside	EtOH	279(4.12)	65-3340-64
$C_{21}H_{22}O_{12}$			
Chalcone, 2',3,4,4',5,6'-hexahydroxy-, 2'-glucoside	n.s.g.	213(4.63),246(3.94), 255(3.95),388(4.47)	83-0838-65
Homoorientin	EtOH	259(4.26),270(4.28), 351(4.38)	95-0553-65
$C_{21}H_{22}S$			
Azulene, 7-isopropyl-1,4-dimethyl-3-(phenylthio)-	C_6H_{12}	197(4.43),248(4.52), 294(4.58),307(4.32), 359(3.89),593(2.68), 639(2.61),795(2.21)	44-2715-65
$C_{21}H_{23}BrN_2S$			
6-[p-(Dimethylamino)benzylidene]-7,8,9,10-tetrahydro-6H-azepino-[2,1-b]benzothiazolium bromide	EtOH	485(4.53)	65-2447-64
$C_{21}H_{23}ClN_2O$			
Acrylonitrile, 2-(p-chlorophenyl)-3-[p-(diethylaminoethoxy)phenyl]-	EtOH-HCl	339.5(4.50)	87-0511-64
Acrylonitrile, 3-(o-chlorophenyl)-2-[p-(diethylaminoethoxy)phenyl]-	EtOH-HCl	328(4.30)	87-0511-64
Acrylonitrile, 3-(p-chlorophenyl)-2-[p-(diethylaminoethoxy)phenyl]-	EtOH-HCl	337(4.41)	87-0511-64
$C_{21}H_{23}ClN_2O_3$			
Yohimbine, py-tetradehydro-, chloride	MeOH-HCl	280(4.63),324(4.03)	73-0433-64
$C_{21}H_{23}ClO_2$			
Estra-2,5,7,9-tetraen-17β-ol, 17α-chloroethynyl-3-methoxy-	dioxan	269(2.65),275(2.60)	87-0536-65
$C_{21}H_{23}ClO_3$			
9β,10α-Androsta-1,4,6-trien-3-one, 6-chloro-17α-ethynyl-17β-hydroxy-	EtOH	227(4.04),252(4.01), 300(4.03)	33-0989-65
$C_{21}H_{23}Cl_2NO_5$			
2-(p-Chlorophenyl)-1,2,3,4,6,7-hexahydro-9,10-dimethoxybenzo[a]quinolizinium perchlorate	EtOH-HCl	244(4.24),303(4.00), 357(4.05)	33-1852-64
	EtOH-NaOH	240(4.22),290(4.00)	33-1852-64

Compound	Solvent	λ max(log ε)	Ref.
$C_{21}H_{23}FN_2O$			
Acrylonitrile, 3-[p-(diethylamino-ethoxy)phenyl]-2-(p-fluorophenyl)-	EtOH-HCl	337(4.45)	87-0511-64
$C_{21}H_{23}INP$			
p-Dimethylaminophenylmethyl-diphenylphosphonium iodide	MeOH	300(4.42)	88-2729-64
$C_{21}H_{23}N$			
2,3:5,6-Dibenzocarbazole, 1',2',3',4'-1",2",3",4"-octahydro-9-methyl-	EtOH	246(4.68),274(4.36), 295s(4.12),303(4.33), 340(3.61),356(3.67)	24-2814-65
Naphthalen-1,4-imine, N-benzyl-1,4-dihydro-1,2,3,4-tetramethyl-	MeOH	208(4.53),253(2.95), 260(2.97),266(2.99), 274(3.00),282(2.92)	44-3225-65
Piperidine, 2-(10,11-dihydro-5H-dibenzo[a,d]cyclohepten-5-ylidene-methyl)-	EtOH	240(4.01)	87-0457-64
$C_{21}H_{23}NO$			
Chalcone, α-(piperidinomethyl)-	EtOH	250(4.15),285(4.14)	88-4833-65
$C_{21}H_{23}NO_2$			
Isobutyric acid, 1-(10-acridanyl)-2-methylpropenyl ester	n.s.g.	280(4.11)	39-5877-65
$C_{21}H_{23}NO_4$			
Carbazole-3,6-dibutyric acid, 9-methyl-, as potassium salt	50% EtOH	240(4.64),250s(4.40), 267(4.34),295s(4.13), 299(4.28),340(3.56), 354(3.56)	24-2814-65
Thalictricavine	MeOH	290(3.58)	95-0955-64
Thalictrifoline	MeOH	285(3.80)	95-0955-64
$C_{21}H_{23}NO_5$			
α-Allocryptopine	EtOH	236(4.2),288(3.6)	23-0679-65
	EtOH	285(3.65)	95-0721-64
Chelamine	MeOH	287(3.8)	73-3697-65
8H-Dibenzo[a,g]quinolizine, 5,6,13,13a-tetrahydro-1,9,10-trimethoxy-2,3-(methylenedioxy)-	EtOH	283.5(3.47)	94-1080-64
Protoberberine, tetrahydro-1,9,10-trimethoxy-2,3-(methylenedioxy)-	50% EtOH	282(3.52)	39-1087-65
$C_{21}H_{23}NO_6$			
Alkaloid K-4	MeOH	228(4.33),296(4.30)	65-1686-64
Carbostyril, 3-[2-(benzyloxy)ethyl]-4-hydroxy-5,7,8-trimethoxy-	n.s.g.	254(4.33),262(4.35), 310(4.20),325s(4.10)	2-0071-65
Chelamidine	MeOH	287(3.9)	73-3697-65
8H-Dibenzo[a,g]quinolizine, 5,6,13,13a-tetrahydro-13-hydroxy-1,9,10-trimethoxy-2,3-(methylenedioxy)-, 1-α	EtOH	282(3.51)	94-1080-64
1-β	EtOH	281.5(3.47)	94-1080-64
Isoquinoline-2-carboxaldehyde, 1,2,3,4-tetrahydro-6,7-dimethoxy-1-(3,4-dimethoxybenzylidene)-	EtOH	220(4.44),335(4.34)	39-4014-65
Protoberberine, tetrahydro-13α-hydroxy-1,9,10-trimethoxy-2,3-(methylene-dioxy)-	50% EtOH	282(3.60)	39-1087-65
13β-isomer	50% EtOH	280(3.48)	39-1087-65

Compound	Solvent	$\lambda_{max}(\log \epsilon)$	Ref.
$C_{21}H_{23}NO_7$			
Crinamine, 6-hydroxy-, diacetate	EtOH	241(3.62),291(3.68)	35-4912-65
$C_{21}H_{23}NO_9S_2$			
Benzothiazolethione, S-(tetraacetyl-β-D-galactopyranosyl)-	EtOH	222(4.2),275(4.1)	24-3508-65
Benzothiazolethione, S-(tetraacetyl-β-D-glucopyranosyl)-	EtOH	220(4.3),275(4.1)	24-3508-65
$C_{21}H_{23}NO_{10}$			
[5,6'-Bi-2H-pyran]-3,3',5-tricarboxylic acid, 6-ethoxy-2-imino-2'-oxo-, triethyl ester	EtOH	207(4.4),240(4.27), 301(4.03),365(4.17)	39-1629-65
4H-Quinolizine-1,3,7,9-tetracarboxylic acid, 6-hydroxy-4-oxo-, tetraethyl ester	EtOH	209(4.4),235s(4.2), 273(4.02),360(4.04), 427(4.35),450(4.37), 472s(4.2)	39-1629-65
$C_{21}H_{23}NO_{10}S$			
Benzoxazolinethione, S-(tetraacetyl-α-D-glucopyranosyl)-	EtOH	250(4.1),275(4.1), 282(4.1)	24-3515-65
$C_{21}H_{23}N_3O$			
4-Pyrimidinol, 2-amino-6-benzyl-5-(4-phenylbutyl)-	pH 1	228(4.12),266(3.95)	87-0283-65
	pH 7	228(4.07),295(3.86)	87-0283-65
	pH 13	283(3.86)	87-0283-65
$C_{21}H_{23}N_3O_2$			
Indazole, 1-benzoyl-3-piperidinoethoxy-	EtOH	242(4.23),309(4.18)	33-1986-64
Indole-3-carbonitrile, 2-[p-(diethylamino)phenyl]-5,6-dimethoxy-	MeOH	245(4.15),360(4.50)	44-2253-65
$C_{21}H_{23}N_3O_4$			
Acrylonitrile, 3-[p-(diethylamino)-phenyl]-2-(4,5-dimethoxy-2-nitrophenyl)-	MeOH	223(4.25),244(4.22), 328(4.07)	44-2253-65
$C_{21}H_{23}N_3O_9$			
Uridine, 2',3'-O-isopropylidene-, 5'-(2,6-dimethyl-4-nitrobenzoate)	MeOH	208(4.30),262(4.25)	22-2489-65
$C_{21}H_{23}N_5O_4S$			
Hydrazine, 1-[[2-amino-6-hydroxy-5-(4-phenylbutyl)-4-pyrimidinyl]carbonyl]-2-(phenylsulfonyl)-	pH 1	271(3.85)	4-0162-65
	pH 13	287(3.85)	4-0162-65
	EtOH	302(3.89)	4-0162-65
$C_{21}H_{23}N_5O_8S$			
Dimethylbis[(4-nitroindol-3-yl)methyl]-ammonium methyl sulfate	EtOH	234(4.25),328s(3.86), 370(3.97)	44-1158-64
$C_{21}H_{24}$			
1,1'-Spirobiindan, 3,3,3',3'-tetramethyl-	C_6H_{12}	260(2.99),266(3.05), 273(3.12)	23-0025-64
$C_{21}H_{24}ClNO_2$			
11bH-Benzo[a]quinolizine, 2-(p-chlorophenyl)-1,2,3,4,6,7-hexanhydro-9,10-dimethoxy-, cis	EtOH	282(3.57),286(3.57)	33-1852-64
trans	EtOH	282(3.58),285(3.58)	33-1852-64

Compound	Solvent	λ_{max}(log ε)	Ref.
$C_{21}H_{24}ClNO_3$			
11bH-Benzo[a]quinolizin-2-ol, 2-(p-chlorophenyl)-1,2,3,4,6,7-hexahydro-9,10-dimethoxy-, m. 173°	EtOH	282(3.58),286(3.58)	33-1852-64
isomer m. 116°	EtOH	282(3.52),285(3.52)	33-1852-64
$C_{21}H_{24}ClNO_5$			
13-Epihydroxy-7-methyl-2,3-(methylenedioxy)-9,10-dimethoxy-5,6,13,13a-tetrahydro-8H-dibenzo[a,g]quinolizinium chloride	EtOH	236(4.25),287(3.89)	94-1072-64
Isoquinoline, 1-β-1-(α-hydroxy-2'-chloromethyl-3',4'-dimethoxybenzyl)-2-methyl-6,7-(methylenedioxy)-1,2,3,4-tetrahydro-	EtOH	288(3.84)	94-1072-64
N-Methylophiocarpinium chloride	EtOH	235(4.11),287(3.80)	94-1072-64
Succinamic acid, 2-(p-chlorophenyl)-N-(3,4-dimethoxyphenethyl)-N-methyl-	EtOH	221(4.29),275(3.48)	87-0259-65
$C_{21}H_{24}ClN_3O$			
Amodiaquine, N^4-methyl-	pH 1	355(4.11)	4-0006-64
	EtOH	348(3.93)	4-0006-64
Amodiaquine, O-methyl-	pH 1	342(4.22)	4-0006-64
	EtOH	340(4.09)	4-0006-64
$C_{21}H_{24}ClN_3O_2$			
Acetanilide, 4'-chloro-2'-[N-(2-morpholinoethyl)benzimidoyl]-	EtOH	236(4.46),327(3.50)	44-2368-64
Amodiaquine, O-methyl-, 1-oxide	pH 1	355(4.25)	4-0006-64
	EtOH	400(4.16)	4-0006-64
$C_{21}H_{24}ClN_3O_3$			
Amodiaquine, N^4-methyl-, Nα,1-dioxide	pH 1	374(4.12)	4-0006-64
	EtOH	404(4.07)	4-0006-64
$C_{21}H_{24}INO_4$			
1-(p-Methoxybenzyl)-5,6,7-trimethoxy-2-methylisoquinolinium iodide	MeOH	265(4.61),318(3.69)	88-3599-65
Pseudoberberine α-methiodide	EtOH	290(3.91)	78-2971-64
Pseudoberberine β-methiodide	EtOH	290(3.90)	78-2971-64
$C_{21}H_{24}N_2O$			
Acrylonitrile, 2-[p-(2-diethylaminoethoxy)phenyl]-3-phenyl-	EtOH-HCl	330.5(4.39)	87-0511-64
Acrylonitrile, 3-[m-(2-diethylaminoethoxy)phenyl]-2-phenyl-	EtOH-HCl	313.5(4.31)	87-0511-64
Acrylonitrile, 3-[o-(2-diethylaminoethoxy)phenyl]-2-phenyl-	EtOH-HCl	338(4.14)	87-0511-64
Acrylonitrile, 3-[p-(2-diethylaminoethoxy)phenyl]-2-phenyl-	EtOH-HCl	335(4.37)	87-0511-64
Acrylonitrile, 3-[p-(3-dimethylamino-2-methylpropoxy)phenyl]-2-phenyl-	EtOH-HCl	336(4.46)	87-0511-64
Kopsan-22-one, N-methyl-	MeOH	253(3.93),298(3.57)	35-4944-65
$C_{21}H_{24}N_2O_2$			
Catharanthine	EtOH	202(4.42)	35-0093-65
Purpeline	EtOH	251(3.95),290(3.42)	22-2683-64
Venalstonine	EtOH	243(3.86),293(3.45)	33-1822-65
	EtOH	245(3.87),292(3.46)	88-2239-65
Voachalotine, de(hydroxymethyl)-	MeOH	228(4.61),285(3.89)	20-0253-65

Compound	Solvent	$\lambda_{max}(\log \epsilon)$	Ref.
$C_{21}H_{24}N_2O_2S$			
p-Toluenesulfonic acid, tricyclo-[2.2.1.0^{2,6}]hept-3-yltricyclo-[2.2.1.0^{2,6}]hept-3-ylidenehydrazide	MeCN	222(4.06),262s(3.40), 274(3.15)	44-1673-64
$C_{21}H_{24}N_2O_3$			
Akuammidine	EtOH	225(4.54),275(3.90), 281(3.91),291(3.81)	33-1822-65
invert-Akuammigine	EtOH	225(4.75),289(3.89)	35-2229-65
Alloyohimb-16-ene-16-carboxylic acid, 17-hydroxy-, methyl ester	pH 13	215(4.76),283(4.37)	44-1107-65
	MeOH	224(4.58),264(4.10), 282s(3.96),289(3.83)	44-1107-65
invert-Alstonine, tetrahydro-	EtOH	225(4.59),289(3.75)	35-2229-65
Venalstonidine	EtOH	243(3.85),292(3.45)	33-1822-65
	EtOH	244(3.85),292(3.45)	88-2239-65
Vobasine	n.s.g.	239(4.19),315(4.27)	100-0374-64
	n.s.g.	239(4.20),316(4.28)	100-0456-64
Yohimbine, 19-dehydro-	EtOH	226(4.48),283(3.85), 293(3.77)	31-0566-65
$C_{21}H_{24}N_2O_4$			
Ajmalicine pseudoindoxyl	EtOH	234(4.37),398(3.61)	35-2229-65
Burnamine	n.s.g.	234(3.83),288(3.49)	22-0392-64
Isovoacarpine	MeOH	225(4.53),282(3.83), 313(3.61)	20-0170-65
Vincoline	EtOH	244(3.87),300(3.50)	100-0203-64
Voacarpine	MeOH-HCl	225(4.50),282(3.80), 292s(3.71)	20-0170-65
	MeOH	225(4.50),282(3.80), 292s(3.71)	20-0170-65
	MeOH-KOH	225(4.50),282(3.80), 292s(3.71)	20-0170-65
$C_{21}H_{24}N_2O_5$			
α-D-Glucoside, methyl 4,6-O-benzylidene-3-o-tolylazo-3-deoxy-	EtOH	270(2.93),395(2.44)	39-1045-64
α-D-Glucoside, methyl 4,6-O-benzylidene-3-p-tolylazo-3-deoxy-	EtOH	274(4.08),394(2.41)	39-1045-64
Voamonine	MeOH-HCl	225(4.62),282(3.85), 292s(3.76)	20-0170-65
	MeOH	225(4.62),282(3.85), 292s(3.76)	20-0170-65
	MeOH-KOH	225(4.62),282(3.85), 292s(3.76)	20-0170-65
$C_{21}H_{24}N_2O_5S_4$			
Cephalosporanic acid, 7-(2-thiopheneacetamido)-, cyclohexl xanthate derivative, sodium salt	n.s.g.	230(4.27),286(4.25)	87-0174-65
$C_{21}H_{24}N_2O_6$			
α-D-Glucoside, methyl 4,6-O-benzylidene-p-methoxyphenylazo-3-deoxy-	EtOH	298(4.03),392(2.64)	39-1045-64
Glycine, deacetylcolchicidyl-	EtOH	250(4.48),358(4.31), 416(4.02)	65-0620-64
$C_{21}H_{24}N_2O_9$			
2H-Pyran-3,5-dicarboxylic acid, 6-amino-2-(3-cyano-1,3-diethoxycarbonylallylidene)-, diethyl ester	EtOH	205(4.20),255(4.08), 305(4.14),350(4.07), 511(4.45)	39-1629-65

Compound	Solvent	$\lambda_{max}(\log \epsilon)$	Ref.
$C_{21}H_{24}N_4$			
Pyrimidine, 2,4-diamino-6-benzyl-5-(4-phenylbutyl)-	pH 1	280(3.92)	87-0283-65
	pH 7	287(3.92)	87-0283-65
	pH 13	292(3.90)	87-0283-65
$C_{21}H_{24}N_4O$			
2-Pyrazolin-5-one, 4-[[4-(diethylamino)-o-tolyl]imino]-3-methyl-1-phenyl-	n.s.g.	455(4.3),495(4.4)	46-1501-64
$C_{21}H_{24}O$			
13H-Indeno[1,2-l]phenanthren-13-one, 1,2,3,4,4a,4b,5,6,7,8,8a,8b-dodecahydro-	EtOH	287(4.33),355(2.38)	54-1094-65
$C_{21}H_{24}O_2$			
Etiojerva-4,12,14,16-tetraen-3-one, 17-acetyl-	MeOH	246(4.35)	44-4220-65
Gona-1,3,5(10),8,14-pentaen-17-one, 13β-isopropyl-3-methoxy-	EtOH	314(4.45)	94-1294-65
2-Phenanthrenebutyric acid, 9,10-dihydro-α,α-dimethyl-, methyl ester	EtOH	276(4.24),308(3.69)	39-0724-64
$C_{21}H_{24}O_3$			
Estra-1,3,5(10),8,14-pentaene, 17,17-ethylenedioxy-3-methoxy-	EtOH	311(4.49)	87-0755-64
Gona-1,3,5(10),8-tetraen-17-one, 3-acetoxy-13β-ethyl-3-hydroxy-	EtOH	271s(4.09),277(4.10), 288s(3.94)	39-4472-64
D-Homoestrone, 8-dehydro-, 3-acetate	EtOH	220(4.23),272(4.18)	70-1814-64
18,14-Seco-18-nor-D-homoestra-1,3,5(10),9(11)-tetraene-14,17a-dione, 3-hydroxy-16,16-dimethyl-	EtOH	227(3.9),264(4.1)	88-0171-64
$C_{21}H_{24}O_4$			
1H-Benz[e]indene-3-propionic acid, 2-(1-carboxyethyl)-2,3-dihydro-, dimethyl ester	EtOH	251(3.77),275(3.62), 280(3.65),291(3.55)	42-0643-64
Pregna-4,17(20)-dien-21-oic acid, 16β-hydroxy-3,11-dioxo-, γ-lactone	EtOH	230(4.35)	87-0348-64
$C_{21}H_{24}O_5$			
Anthrone, 10-(3-ethoxypropyl)-10-hydroxy-2,6-dimethoxy-	EtOH	218s(4.36),224(4.33), 304(4.22),337s(3.77)	39-4565-64
2H-Benzo[d]naphtho[1,2-b]pyren-11-carboxylic acid, 1,3,4,6,6a,7,8,9,10,10a-decahydro-10,12-dihydroxy-7-methyl-, δ-lactone, 12-acetate	EtOH	230(4.27),262(3.91), 320(3.61)	23-1459-64
1-Naphthoic acid, 2,3,4,4a,5,6,7,8-octahydro-5-hydroxy-4a-methyl-2-oxo-, ethyl ester, benzoate	EtOH	231(4.40)	44-3333-65
2-Phenanthreneglycolic acid, 1-carboxy-1,2,3,9,10,10a-hexahydro-7-methoxy-α,1-dimethyl-, γ-lactone, methyl ester	n.s.g.	276(4.21),278(4.22)	70-1058-65
$C_{21}H_{24}O_6$			
Acetophenone, 3',3'''-methylenebis-[2',4'-dimethoxy-	MeOH	274(4.42)	78-0555-64
Tinophyllone	EtOH	238(3.88)	25-1074-65

Compound	Solvent	λ_{max}(log ϵ)	Ref.
$C_{21}H_{24}O_7$			
Samidin, dihydro-	n.s.g.	249(3.65),260(3.57), 324(4.17)	78-3591-65
$C_{21}H_{24}O_9$			
Chromomycinone	EtOH	232(4.38),282(4.60), 326(3.85),340(3.85), 412(4.01)	88-2355-64
$C_{21}H_{24}Si_2$			
Disilane, 1,1,1-trimethyl-2,2,2-triphenyl-	C_6H_{12}	234.0(4.29)	101-0369-64B
	C_6H_{12}	236.0(4.27)	39-4690-65
	n.s.g.	236.0(4.27)	25-1492-64
Disilane, 1,1,2-trimethyl-1,2,2-triphenyl-	C_6H_{12}	237.0(4.33)	101-0369-64B
$C_{21}H_{25}BrN_4O_4$			
Cyclocolor-4-en-3-one, 2α-bromo-, 2,4-dinitrophenylhydrazone	EtOH	398(4.49)	44-1744-65
$C_{21}H_{25}BrO_4$			
B-Homoestra-1(10),5-diene-7,17-dione, 3β-acetoxy-6-bromo-	EtOH	228(3.69),310(3.97)	44-4154-65
$C_{21}H_{25}ClN_2O_2$			
N-Methylpleiocarpaminium chloride	EtOH	231(4.58),286(3.95), 294s(3.91)	33-1957-65
$C_{21}H_{25}ClO_2$			
9β,10α-Androsta-4,6-dien-3-one, 6-chloro-17β-hydroxy-17α-ethynyl-	EtOH	286(4.33)	33-0989-65
$C_{21}H_{25}ClO_3$			
9β,10α-Androsta-1,4,6-trien-3-one, 17β-acetoxy-6-chloro-	MeOH	254(4.02),301(4.02)	54-0918-65
$C_{21}H_{25}FO_2$			
9β,10α-Pregna-1,4,6-triene-3,20-dione, 6-fluoro-	MeOH	254(4.02),302(4.06)	54-0918-65
$C_{21}H_{25}FO_3$			
9β,10α-Androsta-1,4,6-trien-3-one, 17β-acetoxy-6-fluoro-	MeOH	255(4.02),301(4.05)	54-0918-65
$C_{21}H_{25}IO_4$			
Pregn-4-en-21-oic acid, 16β-hydroxy-17α-iodo-3,11-dioxo-, γ-lactone	EtOH	237.5(4.22)	87-0348-64
$C_{21}H_{25}NO$			
19-Norpregna-1,3,5(10),17(20)-tetraene-21-nitrile, 3-methoxy-	EtOH	219(4.05)	44-0505-65
Stilbene, 4-[2-(1-pyrrolidinyl)ethoxy]-	EtOH	304(4.5),319(4.5)	87-0511-64
$C_{21}H_{25}NO_2$			
Estra-1,3,5(10)-triene-3-carbonitrile, 17,17-ethylenedioxy-	n.s.g.	270(3.08),277(3.17), 286(3.13)	39-5889-64
Estra-1,3,5(10)-triene-16β-carbonitrile, 3-methoxy-16α-methyl-	MeOH	221(3.97),279(3.32), 288(3.29)	44-2775-64

Compound	Solvent	$\lambda_{max}(\log \epsilon)$	Ref.
$C_{21}H_{25}NO_2S$			
Estra-1,3,5(10)-triene-16β-carbonitrile, 3-methoxy-16α-(methylthio)-17-oxo-	MeOH	225(3.87),278(3.33), 288(3.29)	44-2775-64
$C_{21}H_{25}NO_3S$			
Estra-1,3,5(10)-triene-16β-carbonitrile, 16α-(methylsulfinyl)-	MeOH	221(4.03),279(3.31), 288(3.31)	44-2775-64
$C_{21}H_{25}NO_4$			
Acrylic acid, 2,3-bis(p-methoxyphenyl)-, 2-dimethylaminoethyl ester, cis	EtOH-HCl	312(4.36)	87-0614-64
trans	EtOH-HCl	312(4.31)	87-0614-64
Pavine, N-methyl-	EtOH	287(4.00)	23-2183-65
$C_{21}H_{25}NO_4S$			
Cyclopenta[7,8]phenanthro[4,4b-de][1,3]-thiazine-2,7,9(3H)-trione, 3a,8,8a,10-11,11a,11b,12,13,13b-decahydro-5-methoxy-8a,13b-dimethyl-	EtOH	238(4.11)	44-0339-64
Cyclopenta[1,2]phenanthro[4,4a-d]thia-zole-2,9(4aH,7aH)-dione, 8,8a,10,11-11a,11b,12,13-octahydro-7a-hydroxy-6-methoxy-4a,8a-dimethyl-	EtOH	244(4.13)	44-0339-64
9,5-(Epithiomethenonitrilo)-10H-cyclo-penta[a]phenanthrene-3,11,17(4H)-trione, octahydro-7a-hydroxy-6-methoxy-4a,8a-dimethyl-	EtOH	223(4.03)	44-0339-64
Estra-1,3,5(10)-triene-16β-carbonitrile, 3-methoxy-16α-(methylsulfonyl)-17-oxo-	MeOH	222(3.63),279(3.33), 288(3.33)	44-2775-64
$C_{21}H_{25}NO_5$			
Bechuanine	EtOH	260(4.1),292(3.8)	73-1689-64
1-Demecolcine	MeOH	244(4.49),350(4.25)	5-0122-64D
Isoquinoline, 1,2,3,4-tetrahydro-1-(α-hydroxy-2-methyl-3,4-dimethoxy-benzyl)-6,7-(methylenedioxy)-, 1-α-	EtOH	290(3.66)	94-1072-64
1-β-isomer	EtOH	287(3.67)	94-1072-64
Lumidemecolcine	MeOH	224(4.40),266(4.34), 294s(4.01),334(3.30), 348s(3.26)	5-0122-64D
$C_{21}H_{25}NO_6$			
Pyrrolo[2,1-a]isoquinoline-2-acetic acid, 3-carbethoxy-5,6-dihydro-8,9-dimethoxy-, ethyl ester	EtOH	324(4.47),333(4.45)	94-0775-65
$C_{21}H_{25}NO_7$			
Crinamine, 6-hydroxy-, diacetate	EtOH	239(3.72),290(3.75)	35-4912-65
Hemanthidine, dihydro-, diacetate	EtOH	238(3.59),289(3.61)	35-4912-65
$C_{21}H_{25}NO_9S$			
Galactopyranose, 2-(carboxyamino)-2-deoxy-, benzyl ester, 6-p-toluene-sulfonate	dioxan	259(2.79),263(2.84), 268(2.80),274(2.68)	44-3654-64
$C_{21}H_{25}N_2O_4P$			
2-Naphthyl 2-pyridyl phosphate, compound with cyclohexylamine	n.s.g.	264(3.87)	24-1031-65

Compound	Solvent	$\lambda_{max}(\log \epsilon)$	Ref.
$C_{21}H_{25}N_3O$			
Acrylonitrile, 2-(p-aminophenyl)-3-[p-(diethylaminoethoxy)phenyl]-	EtOH-HCl	365(4.38)	87-0511-64
Indazole, 3-(benzyloxy)-1-(2-piperidinoethyl)-	EtOH-HCl	218(4.46),247(3.26),306(3.74)	33-1986-64
	EtOH-NaOH	250(3.28),309(3.80)	33-1986-64
Indazole, 1-benzyl-3-(2-piperidinoethoxy)-	EtOH-HCl	220(4.42),308(3.75)	33-1986-64
Indazolinone, 1-benzyl-2-(2-piperidinoethyl)-	EtOH-HCl	220(4.32),241(4.07),318(3.68)	33-1986-64
Indazolinone, 2-benzyl-1-(2-piperidinoethyl)-	EtOH-HCl	222(4.38),240(4.16),317(3.69)	33-1986-64
	EtOH-NaOH	220(4.52),240(4.19),321(3.71)	33-1986-64
$C_{21}H_{25}N_3O_2$			
Acrylonitrile, 2-(2-amino-4,5-dimethoxyphenyl)-3-[p-(diethylamino)-phenyl]-	MeOH	239(4.23),328(3.88),360(4.28)	44-2253-65
$C_{21}H_{25}N_3O_3$			
Acetamide, N-[[2-[p-(dimethylamino)-phenyl]-5,6-dimethoxyindol-3-yl]-methyl]-	MeOH	216(4.43),270(4.00),328(4.44)	44-2253-65
$C_{21}H_{26}$			
1-Nonene, 1,1-diphenyl-	C_6H_{12}	250(4.16)	22-0693-64
$C_{21}H_{26}ClFO_2$			
Androst-4-en-3-one, 17α-chloroethynyl-16α-fluoro-17β-hydroxy-	MeOH	241(4.23)	44-3739-64
Androst-4-en-3-one, 17β-chloroethynyl-16α-fluoro-17α-hydroxy-	MeOH	242(4.23)	44-3739-64
$C_{21}H_{26}ClFO_3$			
Estra-1,3,5(10)-trien-17β-ol, 3-methoxy-, chlorofluoroacetate	EtOH	278(3.35),287(3.29)	44-2187-64
$C_{21}H_{26}ClNO_4$			
Acrylic acid, 2,3-bis(p-methoxyphenyl)-, 2-dimethylaminoethyl ester, hydrochloride, cis	EtOH	312(4.36)	87-0614-64
trans	EtOH	312(4.31)	87-0614-64
$C_{21}H_{26}INO$			
[2-[3-(o-Methoxyphenyl)inden-2-yl]-ethyl]trimethylammonium iodide	MeOH	237s(4.16),262(4.08)	49-0485-64
$C_{21}H_{26}INO_2$			
Atherosperminine methiodide	EtOH	251s(4.69),257(4.73),279(4.08),302(4.08),312(4.08)	12-1997-65
[2-(3,6-Dimethoxy-1-phenanthryl)ethyl-trimethylammonium iodide	MeOH	236(4.60),251(4.67),259(4.71),278(4.16),288(4.27),300(4.04),311(4.12),333(2.90),349(3.15),366(3.20)	35-2177-64
$C_{21}H_{26}INO_3$			
1,2,3-Trimethoxy-N-methylaporphinium iodide	EtOH	277(4.26)	12-1997-65

Compound	Solvent	λ_{max}(log ε)	Ref.
$C_{21}H_{26}N_2O$			
Aspidofractinine, N-acetyl-	EtOH	229(3.65),277(3.58)	33-1147-64
Kopsan, 22-hydroxy-N-methyl-	MeOH	256(4.02),303(3.57)	35-4944-65
[17,16-d]Pyrimidinoandrost-4-en-3-one	EtOH	242(4.28)	32-0338-65
Tuboxenine, N-acetyl-	EtOH	251(4.15),281(3.60)	33-0358-64
$C_{21}H_{26}N_2O_2$			
Androst-4-eno[2,3-c]furazan, 17α-ethynyl-17β-hydroxy-	EtOH	255(4.05)	94-1445-65
Catharanthine, dihydro-	EtOH	202(4.26)	35-0093-65
Kopsinine	EtOH	204(4.35),246(3.83), 295(3.45)	78-2951-65
Limatine	EtOH	220(4.66),259(4.17), 290(3.83)	33-0822-65
	EtOH-KOH	230(4.69),258s(4.12), 310(4.09)	33-0822-65
Macroline	EtOH	231(4.56),282(3.67), 288(3.66)	33-0689-65
1,4-Naphthoquinone, 2-piperidino-3-(piperidinomethyl)-	EtOH	247(4.16),283(4.08), 501(3.51)	39-0042-64
Seredamine	EtOH	210(4.51),253(3.93), 290(3.41)	22-2683-64
Voachalotine, de(hydroxymethyl)-dihydro-	MeOH	227(4.33),285(3.63)	20-0534-65
$C_{21}H_{26}N_2O_3$			
Dregamine	n.s.g.	239(4.18),316(4.27)	100-0374-64
Heyneanine	n.s.g.	225(4.50),286(3.91), 293(3.86)	88-3873-65
Pseudoyohimbine	EtOH	225(4.53),290(3.79)	23-1760-64
Rauwolscine, nitrate	EtOH	225(4.57),279(3.38)	100-0220-64
Schizozygine, tetrahydro-N-methyl-	MeOH	266(4.02),313(3.98)	33-0308-65
Stemmadenine	EtOH	228(4.54),285(3.85), 293(3.83)	33-1822-65
Tabernamontanine	n.s.g.	237(4.16),312(4.24)	100-0374-64
Vincarine	n.s.g.	242(3.84),292(3.50)	30-0644-65
Vobasinol	n.s.g.	224(4.48),287f(3.9)	100-0374-64
Yohimbine	n.s.g.	223(4.70),274(3.91)	100-0470-64
invert-Yohimbine	EtOH	224(4.57),282(3.85), 289(3.78)	35-2229-65
$C_{21}H_{26}N_2O_4$			
Aspidodasycarpine	EtOH	205(4.16),240(3.96), 297(3.63)	78-1717-65
Aspidodasycarpine, deformo-N-(hydroxymethyl)-	EtOH	242(3.83),298(3.45)	78-1717-65
Δ^6-Codeine, 6-deoxy-6-(dimethylamino)-7-formyl-14-hydroxy-	EtOH	345(4.02)	78-1407-64
7H-Yohimbine, 7-hydroxy-	EtOH	218(4.33),262(3.62)	35-2229-65
Yohimbine pseudoindoxyl	EtOH	234(4.38),400(3.58)	35-2229-65
$C_{21}H_{26}N_2O_5S$			
2,4-Diazapregna-1,5-diene-11,20-dione-3-thione, 21-acetoxy-17α-hydroxy-	MeOH	240(3.93),272(4.36)	13-0639-64
$C_{21}H_{26}N_2O_5S_4$			
Cephalosporanic acid, 7-(2-thiopheneacetamido)-, hexyl xanthate derivative, sodium salt	n.s.g.	229(4.25),284(4.16)	87-0174-65

Compound	Solvent	λ_{max}(log ε)	Ref.
$C_{21}H_{26}N_2O_6$			
Dipyrromethene, 3,4',5-tricarbethoxy-3',4,5'-trimethyl-	EtOH	208(4.59),270(4.31), 442(4.30)	12-0363-65
	CCl_4	438(4.40)	12-0363-65
$C_{21}H_{26}N_2O_9$			
D-Fructosone, tetraacetyl-, 1-(methylphenylhydrazone)	EtOH	236(4.03),345(4.32)	24-0725-64
$C_{21}H_{26}N_4$			
Pyrazolo[3",4":3',4']cyclopenta[1',2'-5,6]naphtho[1,2-g]quinazoline, 2,4,4a,4b,5,6,6a,7,12,12a,12b,13,14-14a-tetradecahydro-14a-methyl-	EtOH	257(4.00)	32-0257-65
Pyrimido[4",5":3',4']cyclopenta[1',2':-5,6]naphth[1,2-f]indazole, tetradecahydro-6a-methyl-	EtOH	222(3.75),252(3.67)	32-0338-65
$C_{21}H_{26}N_4O_4$			
Cypertundone, 2,4-dinitrophenylhydrazone	EtOH	396(4.76)	94-0628-65
3,5-Hexadien-2-one, 6-(3,3-dimethylbicyclo[2.2.1]-2-heptyl)-, 2,4-dinitrophenylhydrazone	$CHCl_3$	400(4.66)	28-4783-65A
α-Ionylideneacetaldehyde, 2,4-dinitrophenylhydrazone	n.s.g.	394(4.57)	70-1197-65
isomer	n.s.g.	400(4.61)	70-1197-65
Retroionylideneacetaldehyde, 2,4-dinitrophenylhydrazone, cis	n.s.g.	373(4.49)	70-1197-65
trans	n.s.g.	210(4.28),254(4.26), 378(4.51)	70-1197-65
Sinensal, 2,4-dinitrophenylhydrazone	EtOH	227(4.49)	44-1690-65
$C_{21}H_{26}N_4O_6$			
4a(2H)-Naphthalenecarboxylic acid, 1,3,4,5,6,7-hexahydro-1,1-dimethyl-2-oxo-, ethyl ester, 2,4-dinitrophenylhydrazone	$CHCl_3$	368(4.33)	39-2340-65
Unknown 2,4-dinitrophenylhydrazone, m. 139-140°	$CHCl_3$	366(4.41)	39-2340-65
$C_{21}H_{26}N_6O$			
Tetraethylammonium[2,2-dicyano-1-(1,6,6-tricyano-3-cyclohexen-1-yl)vinyl]oxide	MeCN	268(4.08)	35-2898-64
$C_{21}H_{26}O$			
Anisole, 2-(α-ethyl-β-propylstyryl)-4-methyl-, trans	n.s.g.	211(4.41),281(3.69), 321(3.28)	39-3887-65
Estra-1,3,5(10)-trien-17β-ol, 17α-ethynyl-3-methyl-	MeOH	269(2.85),278(2.91)	87-0409-65
Estra-1,3,5(10)-trien-17β-ol, 17α-ethynyl-4-methyl-	MeOH	263(2.37)	87-0409-65
4-Heptanone, 2,6-dimethyl-2,6-diphenyl-	C_6H_{12}	254(2.71),259(2.77), 264(2.70)	23-0025-64
$C_{21}H_{26}O_2$			
Androsta-1,4-diene-3,17-dione, 19-vinyl-	EtOH	244(4.15)	33-0094-65

Compound	Solvent	$\lambda_{max}(\log \epsilon)$	Ref.
Androsta-1,5-dien-3-one, 17α-ethynyl-17β-hydroxy-	MeOH	226(4.01)	13-0183-64
Androst-1-ene-3,17-dione, 4,16-dimethylene-	EtOH	233(4.22),344(.08)	44-2925-65
Androst-4-ene-3,17-dione, 4-ethynyl-	EtOH	268(4.08)	22-0321-64
11H-Benzo[a]fluoren-3-one, 9-acetyl-1,2,3,4,4a,5,6,6aβ,11a,11b-decahydro-10,11bβ-dimethyl-	EtOH	259(4.11)	13-0463-64
Benzophenone, 2,4-di-tert-butyl-6-hydroxy-	EtOH	251(4.18),285s(3.36)	25-1837-65
	EtOH-NaOH	247(4.30),296(3.65)	25-1837-65
Estra-1,3,5(10),16-tetraen-17-ol, 4-methyl-, acetate	EtOH	263(2.53),270(2.43)	94-0687-65
Estra-1,3,5(10)-trien-17-one, 2-acetyl-4-methyl-	EtOH	214(4.43),262(4.17)	24-0140-64
Etiojerva-4,12,17(20)-triene-3,11-dione, 17-ethyl-	EtOH	238(4.20),301(4.43)	44-0755-64
Etiojerva-12,14,16-trien-3-one, 17-acetyl-	MeOH	258(4.09)	44-4220-65
Gona-1,3,5(10),8,14-pentaen-17β-ol, 13β-isopropyl-3-methoxy-	EtOH	312(4.50)	94-1294-65
Gona-1,3,5(10),8-tetraen-17-one, 13β-isopropyl-3-methoxy-	EtOH	282(4.21)	94-1294-65
Norethindrone	EtOH	240(4.22)	13-0441-65
19-Nor-17α-pregna-1,3,5(10)-trien-20-yn-17β-ol, 3-methoxy-	MeOH	219(3.98),278(3.32), 287(3.30)	4-0207-65
8α,10α-Pregna-1,4,6-triene-3,20-dione	MeOH	223(4.05),257(3.99), 303(4.10)	54-0841-65
9β-Pregna-1,4,6-triene-3,20-dione	MeOH	221(4.02),253(3.99), 299(4.09)	54-0889-65
9β,10α-Pregna-1,4,6-triene-3,20-dione	MeOH	255(3.98),305(4.10)	54-0918-65
$C_{21}H_{26}O_2S$			
Estra-1,3,5(10)-trien-17-one, 16β-mercapto-4-methyl-, acetate	EtOH	237(3.33),264(2.64), 300(2.23)	94-0687-65
$C_{21}H_{26}O_3$			
Androsta-1,4-dien-3-one, 17α-ethynyl-2,17β-dihydroxy-	MeOH	255(4.13)	13-0345-65
1,3-Cyclohexanedione, 2-[3,4-dihydro-6-hydroxy-1(2H)-naphthylidene)-ethyl]-2-propyl-	n.s.g.	267(4.2)	70-0760-65
Estra-1,3,5(10),6-tetraene, 17,17-ethylenedioxy-3-methoxy-	EtOH	261(3.83),301(3.37)	87-0755-64
Estra-1,3,5(10),7-tetraene, 17,17-ethylenedioxy-3-methoxy-	EtOH	280(3.23)	87-0755-64
Estra-1,3,5(10),9(11)-tetraene, 17,17-ethylenedioxy-3-methoxy-	EtOH	264(4.26)	87-0755-64
Estra-1,3,5(10),14-tetraene, 17,17-ethylenedioxy-3-methoxy-	EtOH	277(3.32)	87-0755-64
Estra-1,3,5(10),15-tetraene, 17,17-ethylenedioxy-3-methoxy-	EtOH	277(3.27)	87-0755-64
Estra-1,3,5(10)-trien-17-one, 15β-allyloxy-3-hydroxy-	MeOH	222(4.01),282(3.45)	44-0214-64
Estra-1,3,5(10)-trien-17-one, 3-ethoxy-16-(hydroxymethylene)-	pH 1	222(4.10),265(4.17)	44-2234-65
	pH 13	222(3.95),302(4.36)	44-2234-65
	MeOH	222(3.98),265(4.06), 302(3.23)	44-2234-65
Estra-1,3,5(10)-trien-17-one, 16-(hydroxymethylene)-3-methoxy-1-methyl-	EtOH	268(4.02),303(3.79)	39-1184-64

Compound	Solvent	$\lambda_{max}(\log \epsilon)$	Ref.
Etiojerva-4,12-diene-3,11-dione, 17,20-epoxy-17-ethyl-	EtOH	252(4.39)	44-0755-64
A-Homoestra-2,4a,10-trien-4-one, 17β-acetoxy-	EtOH	238(4.43),308s(3.99), 315(4.01)	22-0906-64
hydrochloride	$CHCl_3$	307(3.92),315(3.95)	22-0906-64
Pregna-4,6-dien-18-oic acid, 20-hydroxy-3-oxo-, 18→20 lactone	EtOH	286(4.45)	88-1671-64
5,10-Seco-5,19-cycloandrosta-1(10),2,4-trien-17-one, 3-acetoxy-	EtOH	257(3.78)	35-3727-65
8,14-Secogona-1,3,5(10),9-tetraene-14,17-dione, 13-isopropyl-3-methoxy-	EtOH	266(4.26)	94-1294-65
$C_{21}H_{26}O_3S$			
Estra-1,3,5(10)-trien-17-one, 16β-mercapto-3-methoxy-, acetate	EtOH	224(4.05),278(3.34), 287(3.32)	94-0905-64
$C_{21}H_{26}O_4$			
Androsta-5,15-diene-7,17-dione, 3β-acetoxy-	EtOH	235(4.27)	13-0651-65
Androst-4-ene-3,17-dione, 2,16-bis-(hydroxymethylene)-	EtOH	256(4.21),302(3.85)	32-0257-65
Estra-1,3,5(10)-trien-15-one, 17β-acetoxy-3-methoxy-	MeOH	222(4.01),278(3.40), 288(3.38)	44-0064-64
Estra-1,3,5(10)-trien-16-one, 17,17-ethylenedioxy-3-methoxy-	EtOH	281(3.20),288(3.24)	87-0755-64
Etiojerva-5,12-diene-3,11,17-trione, 3-ethylene ketal	EtOH	267(4.15)	35-0701-64
Isolatifolin, diethyl ether	EtOH	290(3.77)	78-1495-65
Latifolin, diethyl ether	EtOH	280(3.74),290(3.70)	78-1495-65
Pregna-4,17(20)-dien-21-al,	MeOH	240(4.20),285(4.09)	39-0586-64
20-hydroxy-3,11-dioxo-	MeOH	238(4.19),282(4.12)	44-0513-64
Pregna-4,16-diene-3,11,20-trione, 21-hydroxy-	MeOH	238(4.38)	44-0513-64
Pregn-4-ene-3,12,15,20-tetrone	MeOH	238.5(4.20)	33-1933-65
Pregn-4-en-19-oic acid, 6β-hydroxy-3,20-dioxo-, γ-lactone	MeOH	237(4.10)	35-1528-64
Pregn-4-en-19-oic acid, 11β-hydroxy-3,20-dioxo-, γ-lactone	MeOH	242(3.48)	44-1723-65
$C_{21}H_{26}O_5$			
Phenanthrene-12-carboxylic acid, 1,2,3,4,9,12-hexahydro-5,6-dihydroxy-7-isopropyl-1,1-dimethyl-9-oxo-, methyl ester	EtOH	222s(4.24),251(4.15), 313(3.82)	33-1234-64
17α-Pregn-4-ene-1,3,15,20-tetrone,	EtOH	243(3.8),310(3.15)	39-3611-64
17β-hydroxy-	EtOH-NaOH	256(3.78),368(4.03)	39-3611-64
$C_{21}H_{26}O_6$			
Pregna-1,4-diene-3,11,20-trione, 16α,17,21-trihydroxy-	EtOH	237.5(4.19)	39-0130-65
$C_{21}H_{26}O_7$			
Chaparrinone, 12-dehydromethoxy-	n.s.g.	241(3.70)	22-2793-65
Isoilludin S, triacetate	EtOH	252(4.30)	78-1231-65
Isolampterol, triacetate	isooctane	250(4.18)	78-2671-65
	EtOH	253(4.3)	78-2671-65
$C_{21}H_{26}O_9$			
Scirpen-8-one, 3α,4β,15-triacetoxy-	EtOH	227(3.90),324(1.31)	33-0962-65

Compound	Solvent	λ_{max}(log ϵ)	Ref.
$C_{21}H_{27}BrO_2$			
9β,10α-Pregna-4,6-diene-3,20-dione, 4-bromo-	MeOH	301(4.34)	54-0918-65
9β,10α-Pregna-4,6-diene-3,20-dione, 6-bromo-	MeOH	292(4.28)	54-0918-65
$C_{21}H_{27}BrO_3$			
Abietic acid, 6α-bromo-7-oxodehydro-, methyl ester	MeOH	262(3.72),310(2.95)	44-0501-65
Estra-1,3,5(10)-triene, 16α-bromo-17,17-ethylenedioxy-3-methoxy-	EtOH	278(3.24)	87-0755-64
$C_{21}H_{27}ClN_2O_2$			
N,N'-Dimethylsarpaginium chloride	EtOH	275(3.96),295(3.68), 305(3.68)	33-1957-65
	EtOH-NaOH	272(4.03),330(3.84)	33-1957-65
Kopsinic acid, methochloride	EtOH	248(3.85),295(3.45)	33-1957-65
$C_{21}H_{27}ClN_2O_6$			
1-(Carboxymethyl)-1-ethyl-2,3,4,6,7,12-hexahydro-1H-indolo[2,3-a]quinolizin-5-ium perchlorate, ethyl ester	EtOH-HCl	247(3.98),251s(3.95), 352(4.33)	35-1580-65
$C_{21}H_{27}ClO_2$			
9β,10α-Pregna-4,6-diene-3,20-dione, 4-chloro-	MeOH	298(4.35)	54-0918-65
9β,10α-Pregna-4,6-diene-3,20-dione, 6-chloro-	MeOH	287.5(4.33)	54-0918-65
$C_{21}H_{27}ClO_3$			
Androsta-1,4-diene-3,11-dione, 4-chloro-17β-hydroxy-2,17α-dimethyl-	EtOH	250(4.13)	32-0159-65
Androsta-3,5-dien-7-one, 17β-acetoxy-6-chloro-	MeOH	292(4.29)	31-0249-64
9β,10α-Androsta-4,6-dien-3-one, 17β-acetoxy-6-chloro-	MeOH	286(4.33)	54-0918-65
Estra-1,3,5(10)-triene, 16α-chloro-17,17-ethylenedioxy-3-methoxy-	EtOH	279(3.24),287(3.21)	87-0755-64
19-Norandrosta-3,5-diene-4-carboxaldehyde, 3-chloro-17β-hydroxy-, acetate	EtOH	228(4.11),302(3.63)	88-0137-65
19-Norandrosta-3,5-diene-6-carboxaldehyde, 3-chloro-17β-hydroxy-, acetate	EtOH	292(4.21)	88-0137-65
$C_{21}H_{27}Cl_3O_3$			
Testosterone, 2,2,6β-trichloro-, acetate	MeOH	249(4.10)	94-1217-64
$C_{21}H_{27}FO_2$			
9β,10α-Pregna-1,4-diene-3,20-dione, 6β-fluoro-	MeOH	242(4.22)	54-0863-65
9β,10α-Pregna-4,6-diene-3,20-dione, 6-fluoro-	MeOH	285(4.37)	54-0918-65
$C_{21}H_{27}FO_3$			
9β,10α-Androsta-1,4-dien-3-one, 6β-fluoro-17β-hydroxy-, acetate	MeOH	241(4.20)	54-0863-65
9β,10α-Androsta-4,6-dien-3-one, 6-fluoro-17β-hydroxy-, acetate	MeOH	285(4.37)	54-0918-65

Compound	Solvent	$\lambda_{max}(\log \epsilon)$	Ref.
$C_{21}H_{27}NO$			
3-Heptanone, 6-(dimethylamino)-4,4-diphenyl-, hydrochloride	H_2O	292(2.67)	35-5277-64
$C_{21}H_{27}NO_2$			
Androsta-1,4-diene-2-carbonitrile, 17β-hydroxy-17α-methyl-3-oxo-	EtOH	250(4.02)	44-3300-64
Estra-3,5(10)-diene-3-carbonitrile, 17,17-ethylenedioxy-	n.s.g.	305(3.95)	39-5889-64
Estra-1,3,5(10)-triene-16β-carbonitrile, 17β-hydroxy-3-methoxy-16α-methyl-	MeOH	220(3.90),279(3.27), 288(3.25)	44-2775-64
Estra-1,3,5(10)-trien-17-one, 2-acetamido-4-methyl-	EtOH	249(4.13)	24-0140-64
Estra-1,3,5(10)-trien-17-one, 2-(α-oximinoethyl)-4-methyl-	EtOH	214(4.45),255(4.09)	24-0140-64
$C_{21}H_{27}NO_2S$			
Estra-1,3,5(10)-triene-16-carbonitrile, 17β-hydroxy-3-methoxy-16α-(methylthio)-	MeOH	220(3.93),279(3.29), 288(3.27)	44-2775-64
$C_{21}H_{27}NO_4$			
Isoquinoline, 1,2,3,4-tetrahydro-1-(p-methoxybenzyl)-5,6,7-trimethoxy-2-methyl-	MeOH-HCl	281(3.51)	88-3599-65
$C_{21}H_{27}NO_5$			
5H-Benzocycloheptene-9-propionic acid, 8-(2-cyanoethyl)-6,7-dihydro-1,2,3-trimethoxy-, methyl ester	EtOH	220(4.46),252(4.09), 283s(3.15)	44-1752-65
Isoquinoline, 1,2,3,4-tetrahydro-6,7-dimethoxy-1-(2-hydroxy-4,5-dimethoxyphenyl)-2-methyl-	MeOH	289(3.89)	35-2177-64
HI salt	MeOH	289(3.76)	35-2177-64
$C_{21}H_{27}N_3O_3$			
d-Lysergic acid, 2'-butanolamide	MeOH	227s(4.29),245(4.16), 322(4.03)	33-0756-64
$C_{21}H_{27}N_3O_9$			
D-Fructosone, tetraacetyl-, 1-(methylphenylhydrazone) 2-oxime	EtOH	234(3.97),338(4.38)	24-0725-64
$C_{21}H_{27}N_7O_{14}P_2$			
3-Carbamoyl-1-α-D-ribofuranosylpyridinium hydroxide, 5'-ester with adenosine 5'-pyrophosphate, inner salt	n.s.g.	260(4.25)	5-0170-65J
1-β- isomer	n.s.g.	260(4.26)	5-0170-65J
$C_{21}H_{28}$			
Nonane, 1,1-diphenyl-	C_6H_{12}	255(2.73),260(2.80), 263(2.83),270(2.82)	22-0587-65
Phenanthrene, 1,2-diethyl-1,2,3,4-tetrahydro-1,2,8-trimethyl-	EtOH	234(5.05),281(3.77), 285s(3.78),291(3.82), 311(3.31),318(3.04), 326(3.18)	35-4414-64
$C_{21}H_{28}BrClO_3$			
Testosterone, 2α-bromo-4-chloro-, acetate	MeOH	261(4.09)	94-1217-64

Compound	Solvent	λ_{max}(log ϵ)	Ref.
$C_{21}H_{28}ClFO_3$			
Testosterone, chlorofluoroacetate	EtOH	241(4.21)	44-2187-64
$C_{21}H_{28}ClNO$			
3-Heptanone, 6-(dimethylamino)-4,4-diphenyl-, hydrochloride	H_2O	292(2.67)	35-5277-64
$C_{21}H_{28}ClNO_4$			
Pareirine, chloride	pH 1	281(3.83)	24-2732-64
	pH 13	239(4.20),289(3.98)	24-2732-64
	EtOH	283(3.91)	24-2732-64
$C_{21}H_{28}Cl_2O_3$			
Testosterone, 2α,4-dichloro-, acetate	MeOH	260(4.05)	94-1217-64
Testosterone, 2α,6α-dichloro-, acetate	MeOH	239(4.03)	94-1217-64
Testosterone, 2α,6β-dichloro-, acetate	MeOH	242(4.05)	94-1217-64
$C_{21}H_{28}F_2O_2$			
9β,10α-Pregn-4-ene-3,20-dione, 6,6-difluoro-	MeOH	229(4.08)	54-0918-65
$C_{21}H_{28}F_2O_3$			
9β,10α-Androst-4-en-3-one, 17β-acetoxy-6,6-difluoro-	MeOH	229(4.08)	54-0918-65
$C_{21}H_{28}INO_4$			
Menismine, iodide	pH 13	242(4.26),295(3.87)	24-2732-64
	EtOH	281(3.65)	24-2732-64
$C_{21}H_{28}N_2O$			
1H-Indolizino[8,1-cd]carbazole, 6-acetyl-2-ethyl-2,3,3a,4,5,5a,6,11,12,13a-decahydro-	MeOH	253(4.15),280(3.61), 289(3.53)	35-0953-64
1'-Methyl[17,16-c]pyrazoloandrost-4-en-3-one	MeOH	238(4.38)	32-0338-65
2'-Methyl[3,2-c]pyrazoloandrost-4-en-17-one	EtOH	219(4.01),276(4.15)	32-0257-65
[17,16-d]Pyrimidinoandrost-5-en-3β-ol	MeOH-HCl	219(3.54),256(3.74)	32-0338-65
	MeOH	218(3.45),253(3.71)	32-0338-65
Vindolinol, dihydro-N-methyl-	EtOH	257(3.99),305(3.49)	33-0827-64
$C_{21}H_{28}N_2O_2$			
Androstano[2,3-c]furazan, 17α-ethynyl-17β-hydroxy-	EtOH	218(3.63)	94-1445-65
Estrone, 3-methyl ether, acetylhydrazone	MeOH	231(4.33),278(3.32), 287(3.26)	44-3242-65
Macrolinol	EtOH	230(4.81),284(4.02), 294(3.92)	33-0689-65
$C_{21}H_{28}N_2O_3$			
Corynantheine, demethyltetrahydro-	MeOH	225(4.6),280(3.9), 288(3.8)	54-0154-64
Echitinolide, α,β-dihydro-	EtOH	248(3.99),307(3.57)	33-1598-65
$C_{21}H_{28}N_2O_4$			
Dipyrromethene, 4,4'-bis(2-methoxycarbonylethyl)-3,3',5,5'-tetramethyl-	EtOH-HCl	228(3.99),288(3.13), 362(3.77),482(4.98)	12-0363-65
	EtOH	300(3.40),320(3.51), 445(4.58)	12-0363-65
	CCl_4	450(4.53)	12-0363-65

Compound	Solvent	λ_{max}(log ε)	Ref.
Dipyrromethene, 5,5'-dicarbethoxy-3,3'-diethyl-4,4'-dimethyl-	CCl_4	462(4.30)	12-0363-65
Yohimbine, pseudoindoxyl, dihydro-	EtOH	246(3.98),301(3.35)	35-2229-65
$C_{21}H_{28}N_2O_5$			
7H-Pseudoyohimbine, 7-hydroxy-	EtOH	245(3.69),297(3.45)	35-2229-65
Pyrrole-2-carboxylic acid, 5,5'-carbonylbis[4-ethyl-3-methyl-, diethyl ester	EtOH	251(4.32),309(4.17), 341(4.31)	12-1977-65
$C_{21}H_{28}N_4$			
2H-Pyrazolo[3",4":3',4']cyclopenta-[1',2':5,6]naphth[1,2-f]indazole, tetradecahydro-4a,6a-dimethyl-	MeOH	225(4.01)	32-0257-65
$C_{21}H_{28}N_4O_4$			
Carophyllene oxide, 2,4-dinitrophenylhydrazone	$CHCl_3$	362(4.35)	23-1664-64
isomer	$CHCl_3$	369(4.31)	23-1664-64
Cyclocoloranone, 2,4-dinitrophenylhydrazone	EtOH	363(4.33)	44-1744-65
$C_{21}H_{28}O$			
Biphenyl, 3,5-di-tert-butyl-4'-methoxy-	EtOH	285(4.18)	25-1837-65
Estra-1,3,5(10)-triene, 2-acetyl-4-methyl-	EtOH	215(4.43),263(4.17)	24-0140-64
Estra-1,3,5(10)-trien-17β-ol, 3-methyl-17α-vinyl-	MeOH	269(2.86),278(2.91)	87-0409-65
Gona-1,3,5(10),13-tetraene, 17α-ethyl-3-methoxy-17β-methyl-	EtOH	278(3.34),287(3.30)	87-0268-65
$C_{21}H_{28}OS_2$			
Estra-1,3,5(10)-trien-17-one, 3-methoxy-, cyclic ethylene mercaptole	EtOH	279(2.90),287(2.86)	13-0557-65
	EtOH	280(3.28)	87-0755-64
$C_{21}H_{28}O_2$			
Androsta-1,4-diene-3,17-dione, 19-ethyl-	EtOH	244(4.10)	33-0094-65
Androsta-1,4-dien-3-one, 17β-hydroxy-19-vinyl-	EtOH	246(4.16)	33-0094-65
Androst-5-ene-3,17-dione, 4-ethylidene-	EtOH	242(4.24)	22-0321-64
Androst-4-ene-3,17-dione, 4-vinyl-	EtOH	259(3.99)	22-0321-64
Androst-4-en-3-one, 17α-ethynyl-17β-hydroxy-, vanillin complex	$HOAc-HClO_4$	600(3.48)	96-0134-65
11H-Benzo[a]fluorene, 9-acetyl-1,2,3,4-4aα,5,6,6aβ,11aα,11b-decahydro-10,11b-dimethyl-	EtOH	212(4.29),258(4.05)	13-0463-64
Estra-1,3,5(10)-triene, 17β-acetoxy-2-methyl-	MeOH	269(2.88),278(2.94)	87-0409-65
Estra-1,3,5(10)-triene, 17β-acetoxy-3-methyl-	MeOH	269(2.86),278(2.90)	87-0409-65
Estra-1,3,5(10)-trien-17β-ol, 2-acetyl-4-methyl-	EtOH	214(4.45),263(4.17)	24-0140-64
Etiojerva-12,14,16-trien-3β-ol, 17-acetyl-	MeOH	258(4.10)	44-4220-65
Gona-1,3,5(10),8-tetraen-17β-ol, 13β-isopropyl-3-methoxy-	EtOH	281(4.13)	94-1294-65
Gona-1,3,5(10)-trien-17-one, 13β-isopropyl-3-methoxy-	EtOH	280(3.30),288(3.28)	94-1294-65

Compound	Solvent	λ_{max}(log ε)	Ref.
Gon-4-en-3-one, 13β-ethyl-17α-ethynyl-17β-hydroxy-	n.s.g.	241(4.22)	39-4472-64
19-Norpregna-1,3,5(10)-trien-20-one, 3-methoxy-	EtOH	278(3.34),287(3.29)	13-0013-64
19-Norpregn-9-ene-3,20-dione, 5β,6β-methylene-	EtOH	215(3.82)	44-4160-65
Pregna-1,4-diene-3,20-dione	EtOH	246(4.21)	94-0050-65
Pregna-4,17(20)-diene-3,16-dione, cis	MeOH	241(4.44)	44-1142-64
trans	MeOH	241(4.40)	44-1142-64
8α,10α-Pregna-4,6-diene-3,20-dione	MeOH	288.5(4.41)	54-0841-65
9β-Pregna-4,6-diene-3,20-dione	MeOH	284(4.42)	54-0889-65
9β,10α-Pregna-1,4-diene-3,20-dione	MeOH	242.5(4.19)	54-0863-65
Pregna-3,5-dien-20-one, 16α,17α-epoxy-	EtOH	234(4.29)	73-1178-64
$C_{21}H_{28}O_2S$			
Estra-1,3,5(10)-trien-17-one, 16β-(ethylthio)-3-methoxy-	EtOH	221(3.93),279(3.29), 287(3.26),322(2.23)	94-0905-64
Estra-1,3,5(10)-trien-17-one, 3-methoxy-, cyclic ethylenemonothioacetal	EtOH	279(3.29),287(3.25)	13-0557-65
	EtOH	280(3.31)	87-0755-64
$C_{21}H_{28}O_3$			
Abietic acid, dehydro-7-oxo-, methyl ester	isooctane	248(4.06),255s(3.97), 292(3.28),303(3.23)	78-0409-64
Acetaldehyde, [(3-methoxyestra-1,3,5(10)-trien-17β-yl)oxy]-	EtOH	278(3.22),287(3.11)	13-0557-65
Androsta-1,4-diene-2-carboxaldehyde, 17β-hydroxy-17α-methyl-3-oxo-	EtOH-HCl	248(4.14)	44-3481-64
	EtOH	222(4.17),247(4.07)	44-3481-64
	EtOH-NaOH	242(4.17),348(4.00)	44-3481-64
Androsta-3,5-diene-6-carboxaldehyde, 3-methoxy-17-oxo-	EtOH	220(4.02),322(4.18)	78-0597-64
Androsta-1,4-diene-3,11-dione, 17β-hydroxy-2,17α-dimethyl-	EtOH	244(4.19)	32-0159-65
Androsta-1,4-dien-3-one, 16α,17β-dihydroxy-19-vinyl-	EtOH	249(4.07)	33-0094-65
Androsta-1,4-dien-3-one, 1,11α-epoxy-17β-hydroxy-2,17α-dimethyl-	EtOH	243(4.13),292(3.81)	32-0138-65
Androsta-1,4-dien-3-one, 17β-hydroxy-, acetate	EtOH	244(4.18)	44-1348-64
8α,10α-Androsta-4,6-dien-3-one, 17β-hydroxy-, acetate	MeOH	287(4.41)	54-0841-65
9β,10α-Androsta-4,6-dien-3-one, 17β-hydroxy-, acetate	MeOH	285(4.39)	54-0863-65
10α-Androsta-4,6-dien-3-one, 17β-hydroxy-, acetate	MeOH	285(4.28)	54-0841-65
Androsta-1,4,6-trien-3-one, 11α,17β-dihydroxy-2,17α-dimethyl-	EtOH	266(4.14),301(3.88)	32-0138-65
5β,19-Cycloandrost-1-en-3-one, 17β-acetoxy-	EtOH	272(3.76)	35-3727-65
5β,19-Cycloandrost-6-en-17-one, 3β-acetoxy-	EtOH	213(3.77)	13-0001-64
Estra-1,3,5(10)-trien-17-one, 2-(ethoxymethyl)-3-hydroxy-	EtOH	286(3.45)	94-0196-64
Estra-1,3,5(10)-trien-17-one, 1-hydroxy-4-methyl-, cyclic ethylene acetal	EtOH	282(3.36)	87-0755-64
Estra-1,3,5(10)-trien-17-one, 3-hydroxy-1-methyl-, cyclic ethylene acetal	EtOH	285(3.33)	87-0755-64
Estra-1,3,5(10)-trien-17-one, 3-methoxy-, cyclic ethylene acetal	EtOH	280(4.38),288(3.33)	87-0755-64

Compound	Solvent	λ_{max}(log ε)	Ref.
Estra-1,3,5(10)-trien-17-one,	MeOH	281(3.44),286(3.44)	44-3993-65
3-methoxy-2-(methoxymethyl)-	EtOH	282(3.44),287(3.43)	94-0196-64
Estra-1,3,5(10)-trien-17-one,	EtOH	282(3.39),287(3.39)	94-0196-64
3-methoxy-4-(methoxymethyl)-			
Estr-9-en-17-one, 3β-acetoxy-	EtOH	216(3.95)	44-4154-65
5β,6β-methylene-	EtOH	217(3.93)	44-4160-65
Etiojerva-12,14,16-triene-3β,16-diol,	MeOH	263(3.62),305(3.38)	44-4220-65
17-acetyl-			
Etiojerva-12,15,17-triene-3,17-diol,	MeOH	281(3.38)	44-0123-65
3-acetate			
19-Nortestosterone, 6-methylene-,	EtOH	265(4.05)	78-0597-64
acetate			
1,3-Phenanthrenedione, 1,2,3,4,4a,5,6-	EtOH-acid	256(4.16)	88-2643-64
10b-octahydro-7-isopropyl-8-methoxy-	EtOH-base	288(4.42)	88-2643-64
4,4,10b-trimethyl-			
Pregna-4,17(20)-dien-21-al,	MeOH	242(4.24),285(4.06)	39-0586-64
20-hydroxy-3-oxo-			
Pregna-1,4-diene-3,20-dione,	MeOH	247(4.25)	24-1974-65
11α-hydroxy-			
Pregna-1,4,17(20)-trien-3-one,	EtOH	243(4.17)	13-0189-64
11β,21-dihydroxy-, cis			
Pregn-4-ene-3,20-dione, 6β,19-epoxy-	MeOH	239(4.09)	35-1528-64
9β,10α-Pregn-4-ene-3,20-dione,	MeOH	241(4.15)	54-0918-65
6β,7β-epoxy-			
Pregn-4-ene-3,11,20-trione	MeOH	237(4.21)	35-5670-65
	EtOH	240(4.22)	70-2008-64
Pregn-4-en-18-oic acid, 20-hydroxy-	EtOH	240(4.21)	88-1671-64
3-oxo-. lactone			
Δ^6-Testosterone, acetate	EtOH	284(4.40)	50-0523-64C
$C_{21}H_{28}O_4$			
Acetic acid, [(3-methoxyestra-	EtOH	275(3.41),278(3.30)	13-0557-65
1,3,5(10)-trien-17β-yl)oxy]-			
Androsta-1,4-diene-2-carboxaldehyde,	5% NaOH	242(4.11),344(4.01)	32-0138-65
11α,17β-dihydroxy-17α-methyl-3-oxo-	EtOH	251(4.14)	32-0138-65
Androsta-5,15-dien-17-one,	EtOH	232(3.9)	13-0651-65
3β,7α-dihydroxy-, 3-acetate			
5α-Androstane-3,17-dione, 2,16-bis-	EtOH	270(4.18)	32-0257-65
(hydroxymethylene)-			
Androst-4-ene-3,11-dione, 1α,2α-epoxy-	EtOH	230(4.01),250s(3.91)	32-0159-65
17β-hydroxy-2β,17α-dimethyl-			
Androst-4-ene-3,17-dione, 19-acetoxy-	EtOH	238(4.21)	44-2198-65
5α-Androst-1-ene-3,11-dione,	EtOH	228(4.03)	87-0555-64
17β-acetoxy-			
9β,10α-Androst-4-ene-3,6-dione,	MeOH	248.5(4.01)	54-0863-65
17β-acetoxy-			
Androst-4-en-3-one, 6β,19-epoxy-	MeOH	237(4.11)	35-1528-64
17β-hydroxy-, acetate			
Androst-4-en-3-one, 17α-ethynyl-	MeOH	239(4.17)	13-0345-65
1α,2α,17β-trihydroxy-			
14β-Digacetigenin, anhydro-	EtOH	242(4.1)	39-3611-64
Estra-1,3,5(10)-trien-15β-ol,	MeOH	222(3.94),278(3.31),	44-0064-64
17β-acetoxy-3-methoxy-		288(3.28)	
Estra-1,3,5(10)-trien-16α-ol,	EtOH	280(3.18),288(3.18)	87-0755-64
17,17-ethylenedioxy-3-methoxy-			
Etiojerva-5,12-diene-3,11-dione,	$CHCl_3$	250(4.15)	35-0701-64
17ζ-hydroxy-, 3-ethylene ketal			
Etiojerva-5,12(14)-diene-3,11-dione,	$CHCl_3$	243(4.00)	35-0701-64
17ζ-hydroxy-, 3-ethylene ketal			

Compound	Solvent	λ_{max}(log ε)	Ref.
B-Homoestr-5(10)-ene-7,17-dione, 3β-acetoxy-	EtOH	289(2.17)	44-4154-65
A-Norandrost-3(5)-ene-1,2-dione, 17β-hydroxy-, propionate	EtOH	283(3.79)	94-0050-65
19-Norpregna-1(10),5-diene-15,20-dione, 3β,17β-dihydroxy-1-methyl-	EtOH	252(4.05)	39-3611-64
Pentane, 1,5-bis(2,5-dimethoxyphenyl)-	95% THF	232(3.95),292(3.87)	44-2602-65
Pregna-1,4-dien-3-one, 11α,20-dihydroxy-18,20-epoxy-	MeOH	247.5(4.26)	24-1974-65
17α-Pregn-4-ene-3,20-dione, 8,19-epoxy-19-hydroxy-	EtOH	243(4.11)	44-0351-64
Pregn-4-ene-3,11,20-trione, 12α-hydroxy-	EtOH	238(4.19)	44-2169-65
Pregn-4-ene-3,11,20-trione, 12β-hydroxy-	EtOH	238(4.21),288(1.60)	44-2169-65
Pregn-4-ene-3,12,20-trione, 11α-hydroxy-	EtOH	240(4.21),280(2.30)	44-2169-65
Pregn-4-ene-3,19,20-trione, 11β-hydroxy-, 11,19-hemiacetal	MeOH	246(3.89)	44-1723-65
Pregn-4-en-18-oic acid, 16α,20-dihydroxy-3-oxo-, 18→20-lactone	EtOH	240(4.20)	88-1671-64
5α-Pregn-17(20)-en-20-oic acid, 3β,16β-dihydroxy-11-oxo-, γ-lactone	EtOH	216(4.12)	87-0348-64
Propionic acid, 3-(3-oxo-10β,17β-dihydroxyestr-4-en-17α-yl)-, lactone	MeOH	234.5(4.20)	94-0859-64
Propionic acid, 3-(3-oxo-15β,17β-dihydroxyestr-4-en-17α-yl)-, lactone	MeOH	239(4.23)	94-0859-64
$C_{21}H_{28}O_5$			
Androst-4-ene-3,17-dione, 9,12-dihydroxy-, 12-acetate	EtOH	241(4.20)	35-0736-64
Androst-4-ene-3,17-dione, 9,15-dihydroxy-, 15-acetate	EtOH	242(4.21)	35-0736-64
Cassamic acid, dehydro-	EtOH	220(4.1),244s(4.0)	39-0403-65
Cortexone, 8,19-epoxy-19-hydroxy-	EtOH	243(4.18)	44-0345-64
4,6-Etiadienoic acid, 8β,14β-dihydroxy-3-oxo-, methyl ester	EtOH	280(4.44)	33-1113-65
5β-Eti-1-enic acid, 7,8β-epoxy-14β-hydroxy-3-oxo-, methyl ester	EtOH	231(3.85),322(1.46)	33-1113-65
Pregn-4-ene-3,20-dione, 11β,19-epoxy-17α,21-dihydroxy-	n.s.g.	243(4.22)	13-0371-65
Pregn-4-ene-3,11,20-trione, 21,21-dihydroxy-	MeOH	238(4.19)	44-0513-64
14α,17α-Pregn-5-ene-1,15,20-trione, 3β,17β-dihydroxy-	EtOH	283(1.7)	39-3611-64
Propane, 1-(2,4-dimethoxyphenyl)-3-(4,5-dimethoxy-o-tolyl)-2-methoxy-	EtOH	282(3.94)	39-2844-65
$C_{21}H_{28}O_7$			
Chaparrinone, methoxy-	n.s.g.	241(4.02)	22-2793-65
$C_{21}H_{28}O_8$			
Flexuosin, diacetyl-	EtOH	211(4.00)	78-0979-64
$C_{21}H_{29}BrO_2$			
Androsta-3,5-dien-17β-ol, 3-bromo-, acetate	EtOH	240(4.4)	44-2784-64
Pregna-3,5-dien-20-one, 16β-bromo-17α-hydroxy-	EtOH	234(4.3)	73-1178-64
9β,10α-Pregn-4-ene-3,20-dione, 6α-bromo-	MeOH	247(4.09)	54-0863-65

Compound	Solvent	λ_{max}(log ε)	Ref.
9β,10α-Pregn-4-ene-3,20-dione, 6β-bromo-	MeOH	237.5(4.11)	54-0863-65
Sugiol, 6β-bromo-, methyl ether	isooctane	232(4.07),238s(4.00), 282(4.07),307s(3.82), 341(2.32)	44-0501-65
	MeOH	242(2.98),304(3.61)	44-0501-65
Totarol, 6α-bromo-7-oxo-, methyl ether	MeOH	263(3.72),321(3.39)	44-0501-65
$C_{21}H_{29}BrO_3$			
Androst-4-en-3-one, 2α-bromo-1α,11α-epoxy-17β-hydroxy-2α,17α-dimethyl-	EtOH	250(3.88)	32-0138-65
9β,10α-Androst-4-en-3-one, 6β-bromo-17β-hydroxy-, acetate	MeOH	237.5(4.12)	54-0863-65
Testosterone, 2α-bromo-, acetate	MeOH	244(4.18)	94-1217-64
$C_{21}H_{29}ClOS$			
Estra-1,3,5(10)-triene, 17β-[(2-chloroethyl)thio]-3-methoxy-	EtOH	278(3.35),287(3.29)	13-0557-65
$C_{21}H_{29}ClO_2$			
Androsta-3,5-dien-17-one, 19-chloro-3-ethoxy-	EtOH	244(4.28)	44-2198-65
9β,10α-Androsta-4,6-dien-3-one, 6-chloro-17α-ethyl-17β-hydroxy-	EtOH	288(4.31)	33-0989-65
Estra-1,3,5(10)-triene, 17β-(2-chloroethoxy)-3-methoxy-	EtOH	278(3.28),288(3.23)	13-0557-65
9β,10α-Pregn-4-ene-3,20-dione, 6β-chloro-	MeOH	236(4.18)	54-0863-65
$C_{21}H_{29}ClO_3$			
Androst-5-en-7-one, 3β-chloro-17β-hydroxy-, acetate	MeOH	238(4.13)	31-0249-64
9β,10α-Androst-4-en-3-one, 6β-chloro-17β-hydroxy-, acetate	MeOH	237(4.16)	54-0863-65
Androst-4-en-3-one, 2β-chloro-17β-hydroxy-2α,17α-dimethyl-1α,11α-oxido-	EtOH	246(3.99)	32-0138-65
19-Norandrosta-3,5-dien-17β-ol, 3-chloro-4-(hydroxymethyl)-, acetate	EtOH	245(4.23)	88-0137-65
9β,10α-Pregn-4-ene-3,20-dione, 6β-chloro-7β-hydroxy-	MeOH	237.5(4.14)	54-0918-65
Testosterone, chloroacetate	MeOH	241(4.21)	13-0195-65
Testosterone, 2α-chloro-, acetate	MeOH	243(4.16)	94-1217-64
	iso-PrOH	242(4.14)	94-1217-64
Testosterone, 4-chloro-, acetate	MeOH	255(4.17)	94-1217-64
Testosterone, 6α-chloro-, acetate	MeOH	236(4.16)	94-1217-64
	EtOH	236(4.14)	94-1217-64
Testosterone, 6β-chloro-, acetate	MeOH	240(4.19)	94-1217-64
	EtOH	240(4.16)	94-1217-64
$C_{21}H_{29}ClO_4$			
Androst-4-ene-3,11-dione, 2β-chloro-1α,17β-dihydroxy-2α,17α-dimethyl-	EtOH	240(4.08)	32-0159-65
Pregn-4-ene-3,20-dione, 19-chloro-17α,21-dihydroxy-	n.s.g.	240(4.21)	13-0371-65
$C_{21}H_{29}FOS$			
Estra-1,3,5(10)-triene, 17β-[(2-fluoroethyl)thio]-3-methoxy-	EtOH	278(3.32),287(3.27)	13-0557-65

Compound	Solvent	$\lambda_{max}(\log \epsilon)$	Ref.
$C_{21}H_{29}FO_2$			
Estra-1,3,5(10)-triene, 17β-(2-fluoroethoxy)-3-methoxy-	EtOH	279(3.32),287(3.27)	13-0557-65
B-Homo-19-norpregn-4-ene-3,20-dione, 7ξ-fluoro-	EtOH	239(4.19)	44-4160-65
9β,10α-Pregn-4-ene-3,20-dione, 6β-fluoro-	MeOH	237.5(4.20)	54-0863-65
$C_{21}H_{29}FO_3$			
9β,10α-Androst-4-en-3-one, 6α-fluoro-17β-hydroxy-, acetate	MeOH	234(4.10)	54-0863-65
6β-fluoro isomer	MeOH	236(4.21)	54-0863-64
Pregna-4,17(20)-dien-3-one, 4-fluoro-11β,21-dihydroxy-, cis	EtOH	249(4.17)	44-2982-64
5,10-Seco-5,19-cycloandrost-4-en-3-one, 17β-acetoxy-10β-fluoro-	EtOH	242(4.13)	35-3727-65
$C_{21}H_{29}FO_4$			
B-Homo-19-norpregnene-3,20-dione, 7ξ-fluoro-17α,21-dihydroxy-	EtOH	238(4.19)	44-4160-65
$C_{21}H_{29}IO_2$			
Estra-1,3,5(10)-triene, 17β-(2-iodoethoxy)-3-methoxy-	EtOH	279(3.36),287(3.31)	13-0557-65
$C_{21}H_{29}NO$			
Estra-1,3,5(10)-triene, 2-acetamido-4-methyl-	EtOH	211(4.51),250(4.17), 288(2.81)	24-0140-64
Estra-1,3,5(10)-triene, 2-acetyl-4-methyl-, oxime	EtOH	214(4.47),255(4.14)	24-0140-64
2-Indolinone, 3-[α-(2,2,6-trimethylcyclohexyl)ethylidene]-	EtOH	253(4.46),261(4.54), 293(3.89),352(3.24)	87-0626-65
$C_{21}H_{29}NO_2$			
5α-Androst-1-ene-2-carbonitrile, 17β-hydroxy-17α-methyl-3-oxo-	EtOH	238(4.00),328(1.18)	44-3300-64
Estra-1,3,5(10)-trien-17β-ol, 2-(α-oximinoethyl)-4-methyl-	EtOH	214(4.47),254(4.11)	24-0140-64
Estrone, 2-(dimethylaminomethyl)-	EtOH	286(3.49)	94-0196-64
Estrone, 4-(dimethylaminomethyl)-	EtOH	284(3.36)	94-0196-64
$C_{21}H_{29}NO_3$			
1,4-Benzodioxan, 6,8-diallyl-5-(β-piperidinoethoxy)-	EtOH	276(3.13)	65-2994-64
Pregn-4-ene-3,16-dione, 17,20-epoxy-, 16-oxime	n.s.g.	239.5(4.25)	35-2451-65
Pregn-4-ene-3,16,20-trione, 16-oxime	n.s.g.	239.5(4.24)	35-2451-65
$C_{21}H_{29}NO_4$			
17α-Aza-D-homoandrost-4-ene-3,17-dione, 17α-acetoxy-	MeOH	239(4.23)	78-0743-65
A-Norandrost-3(5)-ene-1,2-dione, 17β-hydroxy-, propionate, 2-oxime	EtOH	233(3.90),291(4.03)	94-0156-65
Pregn-4-ene-3,20-dione, 2α-nitro-	pH 1	252(4.20)	78-0373-64
	pH 13	242(3.93),265(3.93), 356(4.11)	78-0373-64
	MeOH	243(4.22),378(2.86)	78-0373-64
$C_{21}H_{29}NO_5$			
Deoxycorticosterone, 2α-nitro-	pH 1	253(4.16)	78-0373-64

Compound	Solvent	λ_{max}(log ε)	Ref.
Deoxycorticosterone, 2α-nitro- (cont.)	pH 13	243(3.83),268(3.85), 355(4.04)	78-0373-64
	MeOH	245(4.14),370(2.88)	78-0373-64
$C_{21}H_{29}NO_7$			
Senkirkine, acetate	EtOH	218(3.81)	39-2492-65
$C_{21}H_{29}NO_{10}S$			
α-D-Galactopyranoside, 2-acetamido-2-deoxy-, ethyl ester, 3,4-diacetate, 6-p-toluenesulfonate	$CHCl_3$	258(2.68),263(2.78), 266(2.76),269(2.76), 274(2.72)	44-3654-64
$C_{21}H_{29}N_3O_2S$			
Naphth[2',1':4,5]indeno[1,2-c]pyrazole-7(2H)-carboxamide, tetradecahydro-6b-hydroxy-4a,6a-dimethyl-2-oxothio-	pH 1	243(4.36),265(4.30)	44-2234-65
	pH 13	248(4.29),362(4.44)	44-2234-65
	MeOH	240(4.36),275(4.26)	44-2234-65
$C_{21}H_{29}N_5O_{10}$			
Riboflavin, 7-(ribitylamino)-demethyl-	H_2O	256(4.63),306(3.97), 498(4.58)	65-0675-65
$C_{21}H_{30}$			
1H-Benzotriscycloheptene, tetradecahydro-	EtOH	274(2.42)	35-1326-65
$C_{21}H_{30}BrFO_3$			
Androstan-17-one, 5α-bromo-6β-fluoro-3β-hydroxy-, acetate	EtOH	280(1.63)	44-2187-64
$C_{21}H_{30}Cl_2O_2$			
Testosterone, 6α-(dichloromethyl)-6β-methyl-	EtOH	242(4.13)	88-4487-65
Testosterone, 6β-(dichloromethyl)-6α-methyl-	EtOH	243(4.06)	88-4487-65
$C_{21}H_{30}N_2$			
5α-Androstano[3,2-d]pyrimidine	EtOH	253(3.62)	32-0455-65
5α-Androstano[17,16-d]pyrimidine	EtOH-HCl	252(3.68)	32-0338-65
	EtOH	251(3.67)	32-0338-65
$C_{21}H_{30}N_2O$			
5α-Androstano[17,16-d]pyrimidine, 3β-hydroxy-	MeOH-HCl	219(3.54),256(3.74)	32-0338-65
	MeOH	218(3.45),252(3.70)	32-0338-65
3β-Hydroxyandrost-5-eno[16,17-c]-5'-methylpyrazole	MeOH	225(3.79)	44-1142-64
2'-Methylpyrazolo[3,2-c]androst-4-en-17β-ol	EtOH	273(4.15)	32-0257-65
$C_{21}H_{30}N_2O_2$			
Androstano[2,3-c]furazan, 17β-hydroxy-17α-vinyl-	EtOH	217(3.64)	94-1445-65
$C_{21}H_{30}N_2O_3$			
Androstano[2,3-c]furazan, 17β-acetoxy-	EtOH	217(3.70)	94-1445-65
A-Norandrost-3(5)-en-1-one, 17β-hydroxy-, 2-hydrazone, 17-propionate	EtOH	253(4.17),345(3.80)	94-0156-65
$C_{21}H_{30}N_2O_4$			
Androstano[2,3-c]furazan, 17β-acetoxy-, N-oxide	EtOH	264(3.83)	94-1445-65

Compound	Solvent	λ_{max}(log ε)	Ref.
$C_{21}H_{30}N_2O_8$			
L-Glutamic acid, N-[p-[N-(2,2-diethoxyethyl)formamido]benzoyl]-, dimethyl ester	EtOH	260(4.18)	35-5664-64
$C_{21}H_{30}N_4O_4$			
Caryophyllene oxide, dihydro-, 2,4-dinitrophenylhydrazone	$CHCl_3$	363(4.35)	23-1664-64
2-Cyclohexen-1-one, 2-(2-carbethoxyethyl)-3-isopropyl-5-methyl-, 2,4-dinitrophenylhydrazone	$CHCl_3$	392(4.44)	39-2340-65
$C_{21}H_{30}O$			
Estra-1,3,5(10)-trien-17β-ol, 17α-ethyl-4-methyl-	MeOH	263(2.39)	87-0409-65
Gona-1,3,5(10)-triene, 17α-ethyl-3-methoxy-17-methyl-	MeOH	277(3.30),287(3.27)	44-0257-65
$C_{21}H_{30}OS$			
Estra-1,3,5(10)-triene, 17β-(ethylthio)-3-methoxy-	EtOH	278(3.36),287(3.29)	13-0557-65
$C_{21}H_{30}O_2$			
Androsta-5,7-dien-3β-ol, acetate	EtOH	270(4.14),280(4.15), 292(3.82)	35-2837-64
Androsta-1,4-dien-3-one, 19-ethyl-17β-hydroxy-	EtOH	246(4.20)	33-0094-65
Androsta-1,5-dien-3-one, 17α-ethyl-17β-hydroxy-	MeOH	225(4.03)	13-0183-64
Androsta-4,13-dien-3-one, 7α-hydroxy-17,17-dimethyl-	EtOH	241(4.21)	24-3363-64
Androsta-3,5-dien-17-one, 3-methoxy-6-methyl-	EtOH	246.5(4.29)	78-0569-65
Androst-4-ene-3,17-dione, 19-ethyl-	EtOH	243(4.22)	33-0094-65
Cannabinol, tetrahydro-	EtOH	278(3.31),282(3.32), 300s(2.92)	35-1646-64
Estradiol, 2,17α-dimethyl-, 3-methyl ether	EtOH	284(3.45)	94-0196-64
Etiojerva-12,14,16-triene-3β,16-diol, 17-ethyl-	MeOH	285(3.70)	44-4220-65
Etiojerv-16-en-3-one, 17-acetyl-	MeOH	235(4.05)	44-2545-64
Gona-4,9-dien-3-one, 13β-ethyl-17α-ethynyl-17β-hydroxy-	n.s.g.	307(4.32)	39-4472-64
Gona-4,9-dien-3-one, 17β-hydroxy-17α-methyl-13β-propyl-	n.s.g.	307(4.32)	39-4472-64
Gona-1,3,5(10)-trien-17β-ol, 13β-ethyl-3-methoxy-17α-methyl-	n.s.g.	279(3.29)	39-4472-64
Gona-1,3,5(10)-trien-17-ol, 13β-isopropyl-3-methoxy-	EtOH	279(3.32),287(3.3)	94-1294-65
8α-Gona-1,3,5(10)-trien-17β-ol, 13β-isopropyl-3-methoxy-	EtOH	278(3.23),287(3.20)	94-1294-65
Gon-4-en-3-one, 13β-ethyl-17β-hydroxy-17α-vinyl-	n.s.g.	240(4.18)	39-4472-64
1,3,8-Nonatrien-5-yne, 7-hydroxy-1-methoxy-3,7-dimethyl-9-(2,2,6-trimethylcyclohex-6-enyl)-	EtOH	280(4.45)	54-1113-65
19-Norpregn-4-ene-3,20-dione, 17-methyl-	MeOH	240(4.27)	78-0357-64
19-Norpregn-9-en-20-one, 3β-hydroxy-5β,6β-methylene-	EtOH	218(3.96)	44-4160-65

Compound	Solvent	λ_{max}(log ε)	Ref.
19-Nortestosterone, 17α-cyclopropyl-	MeOH	240(4.23)	24-1470-65
Pregna-4,17(20)-dien-3-one, 16α-hydroxy-, trans	MeOH	240(4.22)	44-1142-64
Pregna-4,17(20)-dien-3-one, 16β-hydroxy-, trans	MeOH	241(4.23)	44-1142-64
Pregna-5,16-dien-20-one, 3β-hydroxy-	n.s.g.	239(3.92)	78-0387-64
5α-Pregna-14,16-dien-20-one, 3β-hydroxy-	MeOH	310(4.12)	24-1188-65
9β,10α-Pregna-4,6-dien-3-one, 20α-hydroxy-	MeOH	287(4.42)	54-0853-65
9β,10α-Pregna-4,6-dien-3-one, 20β-hydroxy-	MeOH	287(4.42)	54-0853-65
5α-Pregn-16-ene-3,20-dione	EtOH	239(3.96),318(1.76)	1-0750-64
9β-Pregn-4-ene-3,20-dione	MeOH	247(4.16)	54-1069-64
Testosterone, 2,2-ethylene-	MeOH	241(4.21)	24-1470-65
Testosterone, 17α-methyl-6-methylene-	EtOH	260(4.04)	78-0597-64
$C_{21}H_{30}O_2S$			
Estra-1,3,5(10)-triene, 17β-[(2-hydroxyethyl)thio]-3-methoxy-	EtOH	278(3.35),288(3.29)	13-0557-65
Estra-1,3,5(10)-trien-17β-ol, 16β-(ethylthio)-3-methoxy-	EtOH	279(3.32),287(3.29)	94-0905-64
$C_{21}H_{30}O_3$			
Androsta-3,5-diene-6-carboxaldehyde, 17β-hydroxy-3-methoxy-	EtOH	220(4.00),322(4.18)	78-0597-64
Androsta-4,6-dien-17β-ol, 3-ethylenedioxy-	MeOH	232s(4.37),238(4.41), 248s(4.19)	35-2183-64
Androsta-1,4-dien-3-one, 11α,17β-dihydroxy-2,17α-dimethyl-	EtOH	250(4.18)	32-0138-65
Androsta-3,5-dien-17-one, 6-(hydroxymethyl)-3-methoxy-	EtOH	249(4.28)	78-0597-64
Androst-4-en-3-one, 6α-acetyl-17-hydroxy-	EtOH	246.5(4.13)	35-5213-64
isomer	EtOH	238(4.15)	35-5213-64
Androst-5-en-17-one, 3β-hydroxy-,	C_6H_{12}	196(3.84),296(1.53)	60-0285-64
acetate	EtOH	198(3.89),295(1.64)	60-0285-64
p-hydroxybenzaldehyde complex	$HOAc-HClO_4$	600(4.26)	96-0134-65
resorcylic aldehyde complex	$HOAc-HClO_4$	540(3.70)	96-0134-65
vanillin complex	$HOAc-HClO_4$	600(4.13)	96-0134-65
5β-Androst-9(11)-en-12-one, 3α-hydroxy-, acetate	EtOH	240(4.11)	35-2825-64
5β,19-Cycloandrost-6-ene-3β,17β-diol, 3-acetate	EtOH	213(3.76)	13-0001-64
Estradiol, 2-(methoxymethyl)-, 3-methyl ether	EtOH	282(3.42),286(3.41)	94-0196-64
Estradiol, 4-(methoxymethyl)-, 3-methyl ether	EtOH	282(3.39),286(3.38)	94-0196-64
Estr-4-en-3-one, 17β-acetoxy-1β-methyl-	EtOH	246(4.13)	44-3781-65
Ethanol, 2-[(3-methoxyestra-1,3,5(10)-trien-17β-yl)oxy]-	EtOH	279(3.32),287(3.27)	13-0557-65
Etiojerv-4-en-3-one, 17α-acetoxy-	EtOH	240(4.20)	35-0701-64
Etiojerv-4-en-3-one, 17β-acetoxy-	EtOH	240(4.19)	35-0701-64
Gon-4-en-3-one, 13β-ethyl-17β-hydroxy-, acetate	n.s.g.	240(4.22)	39-4472-64
Hardwickiic acid, methyl ester	n.s.g.	213(4.10)	88-3751-64
A-Nor-5α-pregn-2-ene-2-carboxylic acid, 20-oxo-	EtOH	226(4.00)	44-1677-64
Podocarpa-8,11,13-trien-16-oic acid, 13-hydroxy-14-isopropyl-, methyl ester	EtOH	279(3.37),283s(3.32)	39-3001-64

Compound	Solvent	λ_{max}(log ϵ)	Ref.
5ξ,9β,10α-Pregnane-3,6,20-trione	MeOH	none	54-0863-65
Pregn-4-ene-3,11-dione, 20β-hydroxy-	EtOH	238(4.20)	70-2008-64
Pregn-4-ene-3,20-dione, 2β-hydroxy-	EtOH	240(4.20)	13-0057-65
Pregn-4-ene-3,20-dione, 11β-hydroxy-	EtOH	240(4.20)	70-2008-64
Pregn-4-ene-3,20-dione, 19-hydroxy-	MeOH	244(4.15)	35-1528-64
Pregn-4-ene-3,20-dione, 21-hydroxy-, vanillin complex	HOAc-$HClO_4$	600(4.00)	96-0134-65
5α-Pregn-1-ene-3,20-dione, 2-hydroxy-	EtOH	270(3.89)	44-1677-64
9β,10α-Pregn-4-ene-3,20-dione, 6α-hydroxy-	MeOH	240(4.10)	54-0863-65
9β,10α-Pregn-4-ene-3,20-dione, 6β-hydroxy-	MeOH	241.5(4.20)	54-0863-65
Testosterone, acetate	MeOH	241(4.2)	94-1217-64
	EtOH	241(4.21)	50-0523-64C
10α-Testosterone, acetate	EtOH	244(4.19)	35-4629-65
Testosterone, 1-dehydro-7α-methoxy-17α-methyl-	EtOH	245(4.17)	24-3363-64
$C_{21}H_{30}O_4$			
Androst-4-en-3-one, 11α,17β-dihydroxy-2-(hydroxymethylene)-17α-methyl-	EtOH	254(4.01),312(3.74)	32-0138-65
9β,10α-Androst-4-en-3-one, 6α,17β-dihydroxy-, 17-acetate	MeOH	240(4.09)	54-0863-65
6β,17β-isomer	MeOH	241(4.19)	54-0863-65
Cortexone, 2β-hydroxy-	EtOH	242(4.16)	13-0057-65
	MeOH-KOH	233(4.23)	13-0057-65
Corticosterone, vanillin complex	HOAc-$HClO_4$	560(3.88)	96-0134-65
Estrane-2-glyoxylic acid, 3-oxo-, methyl ester	EtOH	302(3.94)	35-0085-64
A-Norandrost-3(5)-en-2-one, 1β,17β-dihydroxy-, 17-propionate	EtOH	235(4.18)	94-0156-65
19-Nortestosterone, 16α-hydroxy-17α-methyl-, acetate	MeOH	240(2.67)	7-0221-65
A-Nortestosterone, 3-methoxy-, acetate	EtOH	250(4.07)	88-2233-65
Pacholide	EtOH	293(1.69)	33-2330-64
Pregn-1-ene-3,20-dione, 4α,5α-dihydroxy-	EtOH	231(3.98)	94-0050-65
Pregn-4-ene-3,20-dione, 1α,2α-dihydroxy-	EtOH	241(4.14)	94-0050-65
Pregn-4-ene-3,20-dione, 2β,16α-dihydroxy-	EtOH	243(4.16)	13-0057-65
Pregn-4-ene-3,20-dione, 11α,12α-dihydroxy-	EtOH	242(4.21)	44-2169-65
Pregn-4-ene-3,20-dione, 11α,12β-dihydroxy-	EtOH	241(4.16)	44-2169-65
Pregn-4-ene-3,20-dione, 11β,12α-dihydroxy-	EtOH	242(4.17)	44-2169-65
Pregn-4-ene-3,20-dione, 11β,12β-dihydroxy-	EtOH	240(4.25)	44-2169-65
Pregn-4-ene-3,20-dione, 11β,19-dihydroxy-	MeOH	243.4(4.16)	44-1723-65
Pregn-4-ene-3,20-dione, 16α,17α-dihydroxy-	MeOH	241(4.26)	13-0493-64
Pregn-4-ene-3,20-dione, 17α,21-dihydroxy-, vanillin complex	HOAc-$HClO_4$	600(3.70)	96-0134-65
Pregn-5-ene-7,20-dione, 3β,11α-dihydroxy-	EtOH	240(4.00)	70-2016-64
5α-Pregn-15-ene-12,20-dione, 3β,17α-dihydroxy-	MeOH	290(1.85)	24-1188-65
Pregn-4-en-21-oic acid, 20α-hydroxy-3-oxo-	MeOH	242(4.22)	44-2559-64

Compound	Solvent	$\lambda_{max}(\log \epsilon)$	Ref.
Pregn-4-en-21-oic acid, 20β-hydroxy-3-oxo-	MeOH	242(4.23)	44-2559-64
$C_{21}H_{30}O_5$			
Cortisol, 11-deoxy-1β-hydroxy-	EtOH	241(4.18)	13-0645-64
14α-Digacetigenin, deacetyl-	EtOH	285(1.71)	39-3611-64
14β-Digacetigenin, deacetyl-	EtOH	280(1.85)	39-3611-64
4-Etienic acid, 8β,14β-dihydroxy-3-oxo-, methyl ester	EtOH	242(4.31),310(1.93)	33-1113-65
19-Nortestosterone, 4,16α-dihydroxy-17α-methyl-	MeOH	278(2.59)	7-0221-65
Pregna-1,4-dien-3-one, 11β,17α,20α-tetrahydroxy-?	EtOH	244(4.14)	78-0179-65
Pregn-4-ene-3,20-dione, 6,11β,21-trihydroxy-	n.s.g.	237(4.11)	10-0015-64C
Pregn-4-ene-3,20-dione, 15α,17α,21-trihydroxy-	EtOH	242(4.23)	13-0713-64
$C_{21}H_{30}O_6$			
4-Etienic acid, 6β,8β,14β-trihydroxy-3-oxo-, methyl ester	EtOH	234(4.04),322(1.71)	33-1113-65
Melicopol, dimethyl-	EtOH	230s(3.90),271(3.39)	12-2021-65
	EtOH-base	230s(3.90),271(3.39)	12-2021-65
$C_{21}H_{31}BrO_3$			
5α-Androstan-11-one, 3β-acetoxy-9α-bromo-	EtOH	319.0(1.97)	39-1161-64
5α-Androstan-11-one, 3β-acetoxy-12α-bromo-	EtOH	317.5(2.20)	39-2933-65
5β-Androstan-11-one, 3α-acetoxy-9α-bromo-	EtOH	320.0(1.95)	39-1161-64
5β-Androstan-11-one, 3α-acetoxy-12α-bromo-	EtOH	317.0(2.15)	39-1161-64
5β-Androstan-11-one, 9α-bromo-3,3-ethylenedioxy-	EtOH	319.0(1.98)	39-1161-64
5β-Androstan-11-one, 12α-bromo-3,3-ethylenedioxy-	EtOH	320.0(2.15)	39-1161-64
$C_{21}H_{31}FO_2$			
Androst-4-ene-11β,17β-diol, 9α-fluoro-17α-methyl-3-methylene-	EtOH	238(4.37)	87-0345-64
$C_{21}H_{31}NO$			
5α-Androstane-$\Delta^{3,\alpha}$-acetonitrile, 17β-hydroxy-	EtOH	219(4.19)	44-0505-65
$C_{21}H_{31}NO_2$			
3-Aza-A-homopregn-4a-ene-4,20-dione	EtOH	224(4.20)	39-3388-64
Estradiol, 2-(dimethylaminomethyl)-	EtOH	285(3.48)	94-0196-64
Estradiol, 4-(dimethylaminomethyl)-	EtOH	286(3.33)	94-0196-64
Pregn-4-ene-3,20-dione, 17β-amino-	MeOH	240(4.19)	44-0579-65
hydrochloride	MeOH	239(4.23)	44-0579-65
Pregn-4-ene-3,20-dione, 3-oxime	EtOH	240(4.34)	39-3388-64
$C_{21}H_{31}NO_2S$			
Cyclopenta[7,8]phenanthro[3,2-b][1,4]-thiazin-8(9H)-one, hexadecahydro-1-hydroxy-11a,13a-dimethyl-	EtOH	230(3.59),298(3.38)	87-0555-64

Compound	Solvent	$\lambda_{max}(\log \epsilon)$	Ref.
$C_{21}H_{31}NO_3$			
Azacyclotridecan-2-one, 1-[(benzyl-oxy)acetyl]-	dioxan	219(3.99)	78-3537-65
$C_{21}H_{31}NO_4$			
Androstane-2,3-dione, 17β-acetoxy-, 2-oxime	EtOH	243(3.92)	94-1445-65
$C_{21}H_{31}NS$			
Abietamide, N-methylthiodehydro-	MeOH	267(4.06),335(1.68)	35-0051-65
$C_{21}H_{31}N_2O_4P$			
2-Pyridyl thymyl phosphate, compound with cyclohexylamine	n.s.g.	262(3.65)	24-1031-65
$C_{21}H_{31}N_3$			
1H-Cyclopenta[5,6]naphtho[1,2-g]quin-azoline, 8-aminotetradecahydro-11a,13a-dimethyl-	EtOH-HCl	228(4.20),289(3.66)	32-0455-65
$C_{21}H_{32}N_2$			
5α-Androstano[3,2-c]-1'-methylpyrazole	EtOH	230(3.77)	32-0455-65
5α-Androstano[17,16-c]-1'-methyl-pyrazole	EtOH	227(3.62)	32-0455-65
$C_{21}H_{32}N_2O$			
5ζ-Androstan-3β-olo[16,17-c]-3'-methyl-pyrazole	n.s.g.	225(3.82)	65-3565-64
$C_{21}H_{32}N_2O_2$			
Androstano[2,3-c]furazan, 17α-ethyl-17β-hydroxy-	EtOH	217(3.64)	94-1445-65
$C_{21}H_{32}N_2O_4$			
Androstan-17β-ol, 2,3-dihydroxy-imino-, 17-acetate	EtOH	239(3.84)	94-1445-65
$C_{21}H_{32}O$			
Androsta-3,5-dien-17β-ol, 3,17α-di-methyl-	EtOH	232(4.34),239(4.36), 247(4.18)	87-0345-64
Androst-4-en-17β-ol, 17α-methyl-3-methylene-	EtOH	238(4.44)	87-0345-64
5α-Androst-1-en-17β-ol, 17α-methyl-3-methylene-	EtOH	235(4.32)	87-0345-64
5α-Androst-1-en-3-one, 4,4-dimethyl-	EtOH	229(4.35)	78-1987-64
19-Norandrost-4-en-17β-ol, 17α-ethyl-3-methylene-	EtOH	238(4.42)	87-0345-64
Pregn-4-en-3-one	EtOH	241(4.20)	39-3388-64
5α-Pregn-2-en-20-one, p-hydroxybenzaldehyde complex	$HOAc-HClO_4$	510(4.18)	96-0134-65
resorcylic aldehyde complex	$HOAc-HClO_4$	540(3.18)	96-0134-65
vanillin complex	$HOAc-HClO_4$	600(4.11)	96-0134-65
5α-Pregn-17(20)-en-16-one, cis	C_6H_{12}	235(3.95),240(3.95)	35-0269-64
5α-Pregn-17(20)-en-16-one, trans	C_6H_{12}	236(3.98)	35-0269-64
$C_{21}H_{32}O_2$			
Androst-4-en-3-one, 1α-ethyl-17β-hydroxy-	MeOH	245(4.14)	73-3575-65
Androst-4-en-3-one, 19-ethyl-17β-hydroxy-	EtOH	244(4.20)	33-0094-65

Compound	Solvent	λ_{max}(log ε)	Ref.
9β,10α-Androst-4-en-3-one, 17α-ethyl-17β-hydroxy-	EtOH	241(4.17)	33-0989-65
Communic acid, methyl ester	EtOH	233(4.41)	12-0390-64
	n.s.g.	232(4.45)	25-2059-64
Elliotinoic acid, methyl ester	EtOH	232(4.44)	44-0429-65
Estr-4-en-3-one, 17β-hydroxy-2,2,17α-trimethyl-	EtOH	239(4.20)	78-2501-65
13β-Etiojerv-16-en-3β-ol, 17-acetyl-	MeOH	237(4.05)	44-2545-64
Gon-4-en-3-one, 13β,17α-diethyl-17β-hydroxy-	EtOH	240(4.20)	94-1289-65
	n.s.g.	241(4.21)	39-4472-64
D-Homogon-4-en-3-one, 13β-ethyl-17aβ-hydroxy-17aα-methyl-	n.s.g.	242(4.23)	39-4472-64
1-Naphthoic acid, decahydro-6-(2-isopropylallylidene)-1,4a-dimethyl-5-methylene-, methyl ester	n.s.g.	232(4.41)	32-1108-64
1,3,5,8-Nonatetraen-7-ol, 1-methoxy-3,7-dimethyl-9-(2,6,6-trimethylcyclohex-1-en-1-yl)-	EtOH	284(4.47)	54-1113-65
5α-Pregnane-3,20-dione, p-hydroxybenzaldehyde complex	HOAc-$HClO_4$	395(4.23)	96-0134-65
vanillin complex	HOAc-$HClO_4$	420(4.13)	96-0134-65
5β-Pregnane-3,20-dione, vanillin complex	HOAc-$HClO_4$	420(3.65)	96-0134-65
Pregn-5-en-20-one, 3β-hydroxy-, p-hydroxybenzaldehyde complex	HOAc-$HClO_4$	560(4.00)	96-0134-65
o-phthalaldehyde complex	HOAc-$HClO_4$	560(3.85)	96-0134-65
resorcylic aldehyde complex	HOAc-$HClO_4$	540(3.70)	96-0134-65
vanillin complex	HOAc-$HClO_4$	600(4.04)	96-0134-65
9β,10α-Pregn-4-en-3-one, 20α-hydroxy-	MeOH	242(4.23)	54-0853-65
20β-isomer	MeOH	242(4.22)	54-0853-65
Testosterone, 6,6-dimethyl-	EtOH	242(4.05)	88-4487-65
$C_{21}H_{32}O_2S$			
5α-Androstan-17β-ol, 2α,3α-epithio-, acetate	EtOH	264(1.72)	78-0329-65
5α-Androstan-17β-ol, 2β,3β-epithio-, acetate	EtOH	263(1.69)	78-0329-65
Testosterone, 17-O-(thiomethoxymethyl)-	MeOH	241(4.20)	35-5670-65
$C_{21}H_{32}O_3$			
Androsta-3,5-diene-6-methanol, 17β-hydroxy-3-methoxy-	EtOH	250(4.27)	78-0597-64
5α-Androstan-11-one, 3β-acetoxy-	EtOH	298(1.43)	39-1161-64
5α-Androstan-17-one, 3α-acetoxy-p-hydroxybenzaldehyde complex	HOAc-$HClO_4$	580(4.36)	96-0134-65
resorcylic aldehyde complex	HOAc-$HClO_4$	540(3.60)	96-0134-65
vanillin complex	HOAc-$HClO_4$	600(4.06)	96-0134-65
5α-Androst-2-ene-2-carboxylic acid, 17β-hydroxy-17α-methyl-	EtOH	218(3.94)	44-3300-64
Androst-4-en-3-one, 2β,17β-dihydroxy-2α,17α-dimethyl-	EtOH	246(4.19)	7-0277-65
Podocarpa-7,13-dien-15-oic acid, 12-(hydroxymethyl)-13-isopropyl-	EtOH	243(4.37)	44-2356-65
Podocarp-8(14)-en-16-oic acid, 14-isopropyl-13-oxo-, methyl ester	EtOH	250(4.14)	39-3001-64
Pregn-4-en-3-one, 11α,20β-dihydroxy-	EtOH	243(4.14)	32-0151-65
Pregn-4-en-3-one, 20α,21-dihydroxy-	MeOH	241(4.24)	44-2559-64
Pregn-5-en-20-one, 3β,16α-dihydroxy-, p-hydroxybenzaldehyde complex	HOAc-$HClO_4$	480(3.85)	96-0134-65
o-phthalaldehyde complex	HOAc-$HClO_4$	540(3.48)	96-0134-65

Compound	Solvent	λ_{max}(log ε)	Ref.
Pregn-5-en-20-one, 3β,16α-dihydroxy-, resorcylic aldehyde complex	HOAc-$HClO_4$	540(3.60)	96-0134-65
vanillin complex	HOAc-$HClO_4$	560(3.74)	96-0134-65
Pregn-5-en-20-one, 3β,17α-dihydroxy-, p-hydroxybenzaldehyde complex	HOAc-$HClO_4$	580(4.00)	96-0134-65
o-phthalaldehyde complex	HOAc-$HClO_4$	580(3.48)	96-0134-65
resorcylic aldehyde complex	HOAc-$HClO_4$	540(3.65)	96-0134-65
vanillin complex	HOAc-$HClO_4$	600(3.85)	96-0134-65
5α-Pregn-16-en-20-one, 3β,15β-dihydroxy-	MeOH	231.5(3.85)	24-1188-65
Testosterone, 16β-hydroxy-16α,17α-dimethyl-	MeOH	241(4.22)	7-0205-65
Testosterone, 7α-methoxy-17α-methyl-	EtOH	243(4.20)	24-3363-64
$C_{21}H_{32}O_4$			
17a-Oxa-5α-D-homoandrostan-17-one. 3β-hydroxy-16-(hydroxymethylene)-	MeOH	250(4.05)	4-0229-64
5α-Pregnane-11,20-dione, 3α,17α-dihydroxy-, p-hydroxybenzaldehyde complex	HOAc-$HClO_4$	480(3.78)	96-0134-65
vanillin complex	HOAc-$HClO_4$	560(3.70)	96-0134-65
Pregn-4-en-3-one, 11α,20β,21-trihydroxy-	MeOH	242(4.19)	44-2559-64
Pregn-5-en-20-one, 3β,7α,11α-trihydroxy-	EtOH	240(2.57)	70-2016-64
$C_{21}H_{32}O_5$			
5α-Pregnane-3,20-dione, 11β,17α,21-trihydroxy-, vanillin complex	HOAc-$HClO_4$	560(3.88)	96-0134-65
5β-isomer, vanillin complex	HOAc-$HClO_4$	600(3.85)	96-0134-65
Pregn-4-en-3-one, 11β,17α,20α,21-tetrahydroxy-	EtOH	243(4.16)	78-0179-65
$C_{21}H_{32}O_7$			
8-Nonynoic acid, 9-(5-ethyl-2,2,4-trimethyl-1,3-dioxolan-4-yl)-3-hydroxy-2-methyl-7-oxo-, methyl ester, acetate	EtOH	222(3.82)	70-0818-65
$C_{21}H_{33}NO$			
17a-Aza-D-homoandrost-4-en-3-one, N-ethyl-	MeOH	237(4.09)	44-0639-65
3-Aza-A-homopregn-4a-en-4-one	EtOH	223(4.18)	39-3388-64
Pregn-4-en-3-one, oxime	EtOH	240(4.28)	39-3388-64
$C_{21}H_{33}NO_2$			
5α-Androstan-3-one, 2-aminomethylene-17β-hydroxy-17α-methyl-	EtOH	315(4.16)	87-0238-64
	EtOH	316(4.20)	94-0077-64
5α-Androst-2-ene-2-carboxaldehyde, 17β-hydroxy-17α-methyl-, oxime	EtOH	234(4.26)	44-3300-64
$C_{21}H_{33}NO_3$			
1,4-Benzodioxan, 5-(2-piperidinoethoxy)-6,8-dipropyl-	EtOH	275(3.00)	65-2994-64
$C_{21}H_{34}O_2$			
5α-Pregnan-20-one, 3α-hydroxy-, vanillin complex	HOAc-$HClO_4$	600(4.15)	96-0134-65
5α-Pregnan-20-one, 3β-hydroxy-, p-hydroxybenzaldehyde complex	HOAc-$HClO_4$	520(4.26)	96-0134-65
vanillin complex	HOAc-$HClO_4$	600(4.15)	96-0134-65

Compound	Solvent	λ_{max}(log ϵ)	Ref.
5β-Pregnan-20-one, 3α-hydroxy-, p-hydroxybenzaldehyde complex	HOAc-$HClO_4$	520(4.28)	96-0134-65
vanillin complex	HOAc-$HClO_4$	600(4.23)	96-0134-65
Pregn-5-ene-3β,20β-diol, p-hydroxybenzaldehyde complex	HOAc-$HClO_4$	560(4.28)	96-0134-65
o-phthalaldehyde complex	HOAc-$HClO_4$	560(3.78)	96-0134-65
resorcylic aldehyde complex	HOAc-$HClO_4$	540(4.18)	96-0134-65
$C_{21}H_{34}O_3$			
Hardwickiic acid, tetrahydro-, methyl ester	n.s.g.	215(3.84)	88-3751-64
A-Homo-5α-androstan-4-one, 17β-hydroxy-2α-methoxy-	MeOH	284(1.8)	78-2937-64
Ketone, 3,5-dimethoxyphenyl dodecyl	EtOH	262(3.80),316(3.37)	12-2015-65
5α-Pregnan-20-one, 3α,17α-dihydroxy-, p-hydroxybenzaldehyde complex	HOAc-$HClO_4$	500(4.15)	96-0134-65
resorcylic aldehyde complex	HOAc-$HClO_4$	540(3.30)	96-0134-65
vanillin complex	HOAc-$HClO_4$	560(4.00)	96-0134-65
5β-Pregnan-20-one, 3α,6α-dihydroxy-, p-hydroxybenzaldehyde complex	HOAc-$HClO_4$	560(4.04)	96-0134-65
vanillin complex	HOAc-$HClO_4$	600(4.00)	96-0134-65
5β-Pregnan-20-one, 3α,16α-dihydroxy-, p-hydroxybenzaldehyde complex	HOAc-$HClO_4$	510(3.54)	96-0134-65
vanillin complex	HOAc-$HClO_4$	540(3.30)	96-0134-65
5β-Pregnan-20-one, 3α,17α-dihydroxy-, p-hydroxybenzaldehyde complex	HOAc-$HClO_4$	500(4.20)	96-0134-65
resorcylic aldehyde complex	HOAc-$HClO_4$	425(3.78)	96-0134-65
vanillin complex	HOAc-$HClO_4$	560(4.04)	96-0134-65
Pregn-5-ene-3β,16α,20β-triol, p-hydroxybenzaldehyde complex	HOAc-$HClO_4$	580(4.11)	96-0134-65
o-phthalaldehyde complex	HOAc-$HClO_4$	580(3.78)	96-0134-65
resorcylic aldehyde complex	HOAc-$HClO_4$	540(4.00)	96-0134-65
vanillin complex	HOAc-$HClO_4$	550(4.23)	96-0134-65
Pregn-5-ene-3β,17α,20α-triol, p-hydroxybenzaldehyde complex	HOAc-$HClO_4$	580(4.08)	96-0134-65
o-phthalaldehyde complex	HOAc-$HClO_4$	620(4.26)	96-0134-65
resorcylic aldehyde complex	HOAc-$HClO_4$	540(4.08)	96-0134-65
vanillin complex	HOAc-$HClO_4$	600(4.08)	96-0134-65
Pregn-5-ene-3β,17α,20β-triol, p-hydroxybenzaldehyde complex	HOAc-$HClO_4$	580(4.15)	96-0134-65
o-phthaladehyde complex	HOAc-$HClO_4$	620(4.20)	96-0134-65
resorcylic aldehyde complex	HOAc-$HClO_4$	540(4.00)	96-0134-65
vanillin complex	HOAc-$HClO_4$	600(4.15)	96-0134-65
$C_{21}H_{34}O_4$			
5β-Pregnan-11-one, 3α,17α,20β-trihydroxy-, p-hydroxybenzaldehyde complex	HOAc-$HClO_4$	530(3.70)	96-0134-65
resorcylic aldehyde complex	HOAc-$HClO_4$	540(3.18)	96-0134-65
vanillin complex	HOAc-$HClO_4$	600(3.65)	96-0134-65
Sciadopic acid, methyl ester	EtOH	202(4.20)	94-0744-64
$C_{21}H_{34}S_2$			
5α-Androstan-3-one, cyclic ethylene mercaptole	dioxan	243(2.47)	78-1581-65
5α-Androstan-16-one, cyclic ethylene mercaptole	dioxan	240s(1.91)	78-1581-65

Compound	Solvent	$\lambda_{max}(\log \epsilon)$	Ref.
$C_{21}H_{35}Cl_3N_2$			
Pyrimidine, 4,5,6-trichloro-2-heptadecyl-	MeOH	270(3.68)	87-0808-64
$C_{21}H_{36}N_2O$			
1-Cyclohexene-1-carboxamide, 5-tert-butyl-2-[(4-tert-butylcyclohexylidene)amino]-	EtOH	306(3.30)	23-0010-64
5β-Pregnan-20-one, 3β-amino-, oxime	EtOH	204(3.64)	22-0761-64
isomer	EtOH	202(3.62)	22-0761-64
$C_{21}H_{36}O_2$			
Grevillol dimethyl ether	EtOH	272(3.21),279(3.22)	12-2015-65
5α-Pregnane-3α,20α-diol, p-hydroxybenzaldehyde complex	$HOAc-HClO_4$	560(4.30)	96-0134-65
resorcylic aldehyde complex	$HOAc-HClO_4$	540(4.48)	96-0134-65
vanillin complex	$HOAc-HClO_4$	600(4.26)	96-0134-65
(also other isomers)			
$C_{21}H_{36}O_3$			
Grevillol, 6-hydroxy-, dimethyl ether	EtOH	287(3.59)	12-2015-65
	EtOH-KOH	302(3.67)	12-2015-65
5α-Pregnane-3α,17α,20α-triol, p-hydroxybenzaldehyde complex	$HOAc-HClO_4$	600(4.23)	96-0134-65
o-phthalaldehyde complex	$HOAc-HClO_4$	460(3.74)	96-0134-65
resorcylic aldehyde complex	$HOAc-HClO_4$	540(4.34)	96-0134-65
vanillin complex	$HOAc-HClO_4$	600(4.18)	96-0134-65
(also other isomers)			
Tridecanol, 1-(3,5-dimethoxyphenyl)-	EtOH	273(3.26),279(3.27)	12-2015-65
$C_{21}H_{38}O_2$			
1,3-Cyclopentanedione, 2-hexadecyl-	n.s.g.	252(4.18),267(4.43)	39-4472-64
$C_{21}H_{39}NO_2S$			
5α-Androstano[3,2-e]-1',4'-thiaz-5'-en-3'-one, 17-oxo-	DMF	295(3.36)	87-0555-64
$C_{21}H_{40}O_3$			
11-Eicosenic acid, 14-hydroxy-, methyl ester, cis	C_6H_{12}	183.5(4.14)	88-3391-65
$C_{21}H_{42}FeN_3S_6$			
Iron, tris(N,N-dipropyldithiocarbamato)-	$CHCl_3$	267(4.52),282s(4.49), 345(4.02),385s(3.93), 513(3.51),596s(3.40), 1540(0.64),1720(0.69)	12-0294-64

Compound	Solvent	λmax(log ε)	Ref.
$C_{22}H_8F_{10}N_2Zn$			
Zinc, (2,2'-bipyridine)bis-(pentafluorophenyl)-	toluene	298(4.14),309(4.15)	101-0222-65A
$C_{22}H_{10}N_2$			
Benzo[e]pyrene, 3,6-dicyano-	benzene	286(4.28),298(4.52)	24-0218-64
	$C_6H_3Cl_3$	314(4.68),333(4.14), 348(4.40),367(4.52), 397(3.20)	24-0218-64
$C_{22}H_{10}N_4O_2$			
Terephthaloylenebisbenzimidazole	$CHCl_3$	287(4.62),322(4.53), 332(4.59),390(4.02)	33-2211-64
	$C_{10}H_7Cl$	389(3.93),465s(3.47)	33-2211-64
$C_{22}H_{10}O_4$			
5,8,13,14-Pentaphenetetrone	dioxan	245(4.38),275(4.58)	65-3484-64
$C_{22}H_{12}$			
Acenaphth[1,2-a]acenaphthylene	EtOH	225(4.72),232(4.61), 245(4.36),291(4.09), 385(4.19),404(4.17)	44-0243-64
$C_{22}H_{12}Cl_2N_2O_6$			
Quino[2,3-b]acridine-6,7,13,14-(15H,12H)-tetrone, 1,8-dichloro-4,11-dimethoxy-	H_2SO_4	445(3.89),575(4.01)	27-0242-65
$C_{22}H_{12}N_2O$			
14H-Benz[4,5]isoquino[2,1-a]perimidin-14-one	$C_{10}H_7Cl$	480(4.00)	33-1999-65
$C_{22}H_{12}O_2$			
1,8-Naphthalylnaphthalene	EtOH	220(3.86),280(3.05)	44-0243-64
	H_2SO_4	220(4.07),340(3.78)	44-0243-64
$C_{22}H_{12}O_3$			
Phthaloyleneanthrone	$CHCl_3$	246(4.1),260s(4.05), 307(3.6),387(3.42)	6-0359-64
$C_{22}H_{12}O_4$			
Benzo[e]pyrene-3,6-dicarboxylic acid (as potassium salt)	50% EtOH	226(4.56),240(4.54), 288(4.54),299(4.63), 332(4.32),346(4.38), 388(2.98)	24-0218-64
$C_{22}H_{13}N_3O_6$			
1,3-Indandione, 2-benzoyloxy-2-(p-nitrophenylazo)-	MeOH	229(4.51),283(4.14)	24-1482-64
$C_{22}H_{14}$			
1,2-Benzotetracene	benzene	306(5.01),320(4.98), 361(3.16),381(3.49), 402(3.80),425(3.97), 453(4.00)	24-0212-64
Fluoranthene, 1-phenyl-	EtOH	242(4.61),279(4.43), 290(4.41),338(4.14)	39-5760-65
Fluoranthene, 2-phenyl-	EtOH	227(4.52),260(4.67), 277(4.69),331(4.10), 350(4.15),365(4.16)	39-5760-65

Compound	Solvent	$\lambda_{max}(\log \epsilon)$	Ref.
Fluoranthene, 3-phenyl-	EtOH	240(4.65),294(4.43), 314(3.65),358(4.10), 370(4.12)	39-5760-65
Fluoranthene, 7-phenyl-	EtOH	234(4.67),282(4.26), 289(4.32),312(3.60), 327(3.76),360(3.86)	39-5760-65
Fluoranthene, 8-phenyl-	EtOH	227(4.55),236(4.55), 263(4.23),319(4.89), 356(3.74),373(3.83)	39-5760-65
$C_{22}H_{14}Cl_2N_4O_4S_2$			
Pyridine, 2-(o-aminophenylthio)- 3-nitro-5-chloro-, disulfide from oxidation of	EtOH	246(4.58),302(4.52)	95-0429-65
$C_{22}H_{14}Cl_4$			
Benzene, p-bis(2,4-dichlorostyryl)-	dioxan	245(4.15),280(4.15), 358(4.66)	38-2839-64A
Benzene, p-bis(2,6-dichlorostyryl)-	dioxan	242(4.16),330(4.60)	38-2839-64A
Benzene, p-bis(3,4-dichlorostyryl)-	dioxan	242(4.61),278(4.23), 362(4.87)	38-2839-64A
$C_{22}H_{14}N_2O_6$			
Quino[2,3-b]acridine-6,7,13,14- (5H,12H)-tetrone, 2,9-dimethoxy-	H_2SO_4	412(3.78),533(4.01)	27-0242-65
$C_{22}H_{14}N_4$			
Benzo[g]quinoxaline, 2,3-di-2-pyridyl-	pH 2.70	227(4.56),304(4.63), 387(3.87)	4-0295-64
Cyclopentadienoheptalene, tetracyanoethylene adduct	benzene	572(2.65),590(2.68), 610(2.64),643(2.53), 708(1.99)	5-0194-64B
$C_{22}H_{14}N_4O_2$			
1,3,4-Oxadiazole, 2,2'-m-phenylene- bis[5-phenyl-	DMF	286.0(4.70)	24-2966-65
$C_{22}H_{14}O$			
Dibenz[a,h]anthracen-7(14H)-one	$CHCl_3$	298(3.92),313(3.93), 342(3.89)	44-0686-64
$C_{22}H_{14}OS$			
3,4-Benzopyrene, 5-acetylthio-	$CHCl_3$	261(4.62),271(4.69), 284(4.51),295(4.68), 308(4.77),349s(4.13), 369(4.31),386(4.50), 406(4.51),411(4.53)	12-0109-64
$C_{22}H_{14}O_2$			
Acenaphth[1,2-a]acenaphthylene-6b,12b- diol, 6b,12b-dihydro-, cis	EtOH	224(4.78),270(3.67), 281(3.79),306(4.07), 323(4.08)	44-0243-64
$C_{22}H_{14}O_3$			
Coumarone, phenanthrenequinone adduct	dioxan	244(4.50),280(3.88), 330(3.42)	24-3102-65
Indone, 2-(o-carboxyphenyl)-3-phenyl-	EtOH	246(4.34),280s(3.93), 412(2.99),480s(2.58)	6-0359-64

Compound	Solvent	λ max(log ε)	Ref.
$C_{22}H_{14}O_4$			
1,3-Indandione, 2-(o-carboxyphenyl)-2-phenyl-	EtOH	228(4.52),290s(3.27), 340(2.43)	6-0359-64
$C_{22}H_{14}O_4S$			
Thionaphthene-1,1-dioxide, phenanthrenequinone adduct	dioxan	246(4.49),281(3.82), 332(3.45)	24-3102-65
$C_{22}H_{14}O_6$			
2,3;6,7-Anthracenetetracarboxylic dianhydride, 1,4,5,8-tetramethyl-	$CHCl_3$	249(4.83),324(4.03), 395(2.48)	33-0437-65
$C_{22}H_{15}ClO_2$			
Anthraquinone, 1-(p-chlorophenyl)-2,3-dimethyl-	EtOH	255(4.38),333(3.75)	65-2740-64
$C_{22}H_{15}Cl_2NO_5$			
2-(p-Chlorophenyl)-4,6-diphenyl-1,3-oxazin-1-ium perchlorate	n.s.g.	362(4.59),409(4.54)	24-3892-65
4-(p-Chlorophenyl)-2,6-diphenyl-1,3-oxazin-1-ium perchlorate	n.s.g.	364(--),411(--)	24-3892-65
6-(p-Chlorophenyl)-2,4-diphenyl-1,3-oxazin-1-ium perchlorate	n.s.g.	360(4.23),413(4.26)	24-3892-65
$C_{22}H_{15}NO_2$			
1-Indanone, 3-imino-2-(9-xanthenyl)-	MeOH	255(4.62),264(4.60), 290s(3.18),313(3.10), 329(2.77),430(3.31)	65-0448-64
Isoxazole, 5-benzoyl-3,4-diphenyl-	EtOH	262(4.01),288s(3.80)	32-0127-65
4-Quinolinol, 2-benzoyl-3-phenyl-	EtOH	255(4.50),288s(3.98), 328s(3.89),335(3.90)	35-2086-64
$C_{22}H_{15}NO_4$			
Anthraquinone, 2,3-dimethyl-1-(p-nitrophenyl)-	EtOH	260(4.65),310(3.99), 330(3.86)	65-2740-64
2H-1,3-Benzoxazine-2,4(3H)-dione, 3-(p-phenylphenacyl)-	MeOH	204(4.76),235s(4.05), 290(4.34)	4-0037-65
$C_{22}H_{15}N_3$			
Indole, 3-(3H-indol-3-ylidene-4-pyridylmethyl)-	N HCl	282(4.10),540(2.32)	50-0077-64D
	EtOH	282(4.08),450(1.52)	50-0077-64D
$C_{22}H_{16}$			
Benzo[j]fluoranthene, 6,11-dimethyl-	EtOH	226(4.68),243(4.68), 294(4.28),304(4.32), 317(4.38),339(4.10), 352(3.66),372(3.92), 390(4.03)	39-2380-64
	EtOH	244(4.78),268s(--), 280(4.23),291(3.58), 317(4.25),339(4.00), 370(3.58),390(3.58)	44-3092-64
Bicyclo[4.2.0]octa-1,3,5-triene, 7,8-dibenzylidene-	EtOH	233(4.39),239(4.35), 282(4.26),307(4.01), 333(4.22),335(4.21)	35-5041-64
Ethylene, 1,1-di-1-azulenyl-	hexane	234(4.64),277(4.68), 296(4.82),367(4.12), 608(2.79),660(2.70), 735(2.20)	65-0894-64
	EtOH	604(2.80)	65-0894-64

Compound	Solvent	λ max(log ε)	Ref.
Ethylene, 1,1-di-1-azulenyl- (cont.)	HOAc	632(4.20)	65-0894-64
Naphthalene, 1,4-diphenyl-	C_6H_{12}	225(4.7),301(4.2)	28-5447-64A
Naphthalene, 1,6-diphenyl-	EtOH	238(4.42),258(4.67)	44-2942-65
Naphthalene, 1,7-diphenyl-	EtOH	253(4.80),300(4.12)	39-5110-64
	$CHCl_3$	256(4.54),302(4.05)	78-0195-64
Naphthalene, 1,8-diphenyl-	EtOH	214(4.60),235(4.76), 299(4.09)	39-5110-64
$C_{22}H_{16}BrNO$			
6-Methoxyphenanthro[9,10-b]quinolizinium bromide	EtOH	215(4.78),245(4.72), 257s(4.66),267(4.67), 275(4.68),295(4.51), 330(4.90),415s(3.91)	54-0593-64
$C_{22}H_{16}ClNO_4$			
5-Methylbenzo[a]naphtho[1,2-h]quinolizinium perchlorate	EtOH	206(4.47),228(4.56), 248s(4.30),278(4.43), 298(4.52),308(4.54), 317(4.38),345s(3.91), 366(3.76),388s(3.92), 406(4.07),415(4.02), 454s(3.32)	44-1846-65
4-Methyl-7-phenylpyrrolo[3,2,1-de]-phenanthridinium perchlorate	EtOH	206(4.58),233(4.34), 253s(4.15),265(3.99), 294s(3.60),313s(3.79), 327(4.08),340(4.09)	4-0208-64
$C_{22}H_{16}ClNO_5$			
2,4,6-Triphenyl-1,3-oxazin-1-ium perchlorate	n.s.g.	356(4.46),406(4.45)	24-3892-65
$C_{22}H_{16}Cl_2$			
Benzene, p-bis(m-chlorostyryl)-	dioxan	242(4.56),356(4.75)	38-2839-64A
Benzene, p-bis(o-chlorostyryl)-	dioxan	244(4.23),280s(4.22), 353(4.74)	38-2839-64A
Benzene, p-bis(p-chlorostyryl)-	dioxan	243(4.32),361(4.89)	38-2839-64A
$C_{22}H_{16}F_2$			
Cyclobutene, 3,3-difluoro-1,2,4-triphenyl-	n.s.g.	220(4.46),294(4.23)	35-0449-64
$C_{22}H_{16}N_2$			
6,13-Diazadibenz[a,h]anthracene, 5,12-dimethyl-	EtOH-HCl	224(4.45),282(4.71), 293(4.81),347(3.97), 363(3.94),384(3.89)	4-0140-65
	EtOH	221(4.55),231(4.40), 291(4.83),333(3.95), 348(3.97),366(3.82), 384(3.83)	4-0140-65
Fluorene-2-carbonitrile, 9-[(N-methylanilino)methylene]-	MeCN	251(4.77),301(4.20), 312(4.24),411(4.50)	24-3331-64
3H-Pyrrolo[1,2,3-de]quinoxaline, 2,3-diphenyl-	MeOH	262(4.24),377(3.97)	44-2589-65
$C_{22}H_{16}N_2O_2$			
1H-1,3-Benzodiazepine-2,4(3H,5H)-dione, 5-benzylidene-3-phenyl-	EtOH	250(4.44),286s(3.89), 410(3.74)	39-5762-65
Hydantoin, 4-benzylidene-1,3-diphenyl-	EtOH	319(4.18)	39-5762-65
3,5-Pyrazolidinedione, 4-benzylidene-1,2-diphenyl-	MeOH	235(4.23),334(4.46)	32-0320-65

Compound	Solvent	λ_{max}(log ε)	Ref.
$C_{22}H_{16}N_2O_4$			
2-Indolinone, 3,3'-ethanediylidenebis[1-acetyl-	benzene	395(4.19),414(4.23), 470(4.31)	39-1455-65
isomer	benzene	390(4.15),411(4.18), 465(4.28)	39-1455-65
isomer	benzene	390(4.13),411(4.16), 461(4.22)	39-1455-65
$C_{22}H_{16}N_2O_4S$			
1,4-Naphthoquinone, 2,2'-thiobis-[3-(methylamino)-	dioxan	223(4.39),258(4.57), 284(4.50),467(3.76)	5-0151-64E
$C_{22}H_{16}N_2S$			
3(2H)-Pyridazinethione, 4,5,6-triphenyl-	EtOH	308(4.38)	88-0253-64
$C_{22}H_{16}O$			
1-Cyclopropene, 3-benzoyl-1,2-diphenyl-	EtOH	228(4.39),235(4.41), 244(4.28),294(4.31), 308(4.40),321(4.24)	88-2817-65
5H-Dibenzo[a,c]cyclohepten-5-one, 6-p-tolyl-	MeOH	247(4.56),320(4.02)	44-3333-64
$C_{22}H_{16}O_2$			
Anthraquinone, 2,3-dimethyl-1-phenyl-	EtOH	257(4.37),333(3.69)	65-2740-64
Anthraquinone, 2-methyl-1-p-tolyl-	EtOH	257(4.23),333(3.42)	65-2740-64
Benzo[j]fluoranthene, 6,11-dimethoxy-	EtOH	249(4.65),300(4.38), 313(4.19),333(3.73), 349(3.99),359(3.71), 389(4.09),405(4.14)	39-2380-64
Benzofuran, 2-benzoyl-3-benzyl-	MeOH	232(3.99),260(3.91), 318(4.27)	35-5417-65
Benzofuran, 3-benzoyl-2-benzyl-	MeOH	243(4.36),276(3.93), 283(3.91)	35-5417-65
Phthalide, 3-(1,2-diphenylethylidene)-, cis	ether	317(4.10)	22-3047-65
trans	ether	327(4.02)	22-3047-65
$C_{22}H_{16}O_3$			
Anthraquinone, 1-(p-methoxyphenyl)-2-methyl-	EtOH	258(4.57),330(3.70)	65-2740-64
Flavanone, 3-benzylidene-6-hydroxy-	EtOH	240(3.99),312(3.83), 398(3.73)	24-1453-64
Indanone, 2-(o-carboxyphenyl)-3-phenyl-	EtOH	227(3.95),283(3.20), 345s(2.42)	6-0359-64
$C_{22}H_{16}O_4$			
2-Pyrone, 3-cinnamoyl-4-hydroxy-6-styryl-	EtOH	294(4.08)	65-2766-64
$C_{22}H_{16}O_5$			
Desoxybenzoin-2,2'-dicarboxylic acid, α-phenyl-	EtOH	230s(3.79),280(3.06)	6-0359-64
$C_{22}H_{16}O_6$			
1,2-Naphthoquinone, 3-hydroxy-4-methyl-, dimer	dioxan	255(4.33),296(3.48)	24-0666-65

Compound	Solvent	λ_{max}(log ϵ)	Ref.
$C_{22}H_{16}O_8$			
Benzo[1,2-b:4,5-b']bisbenzofuran-6,12-dione, 1,3,7,9-tetramethoxy-	$CHCl_3$	284(4.53),294(4.59), 407(3.93),496s(2.91)	1-1063-65
Benzo[1,2-b:5,4-b']bisbenzofuran-6,12-dione, 1,3,9,11-tetramethoxy-	dioxan	260(4.47),270(4.66), 388(3.91),538(3.61)	1-1063-65
Benzo[1,2-b:5,4-b']bisbenzofuran-6,12-dione, 2,3,9,10-tetramethoxy-	dioxan	261(4.47),283(4.31), 367(3.62),416(3.72), 554(3.89)	1-1063-65
η-Isopyrromycinone	EtOH	271(4.56),498(4.35), 526(4.38),534(4.37), 552(4.34),564(4.38)	39-3927-64
$C_{22}H_{16}Pb$			
Lead, butadiynyltriphenyl-	hexane	215(3.47),258(2.68), 264(3.90)	22-3518-65
$C_{22}H_{16}S_2$			
Azulene, 1,3-bis(phenylthio)-	C_6H_{12}	203(4.69),237(4.51), 253(4.52),284(4.52), 295(4.43),361(3.84), 578(2.54)	44-2715-65
$C_{22}H_{16}Sn$			
Tin, butadiynyltriphenyl-	EtOH	203(4.78),250(3.18), 258(3.04),264(3.08)	22-0035-65
$C_{22}H_{17}Cl$			
5H-Dibenzo[a,d]cycloheptene, 5-(p-chlorobenzylidene)-10,11-dihydro-	EtOH	285(4.34)	87-0457-64
$C_{22}H_{17}ClN_2O$			
7,12-Dihydro-12-hydroxy-12-phenylquino[2,1-b]quinazolin-13-ium chloride	EtOH	205(4.74),254(4.58), 348(4.06)	44-1539-65
$C_{22}H_{17}ClO_4$			
1-[1-(1-Azulenyl)ethylidene]-1,5-dihydroazulenylium perchlorate	HOAc	634(4.32)	65-0894-64
$C_{22}H_{17}IN_2$			
5-Methyl-x-phenylindolizino[1,2-c]quinolinium iodide	MeOH	259(4.60),315(4.26), 343(3.78),448(4.28)	5-0196-65H
$C_{22}H_{17}N$			
Pyrrole, 2,3,4-triphenyl-	EtOH	249(4.35),300(4.22)	88-0253-64
$C_{22}H_{17}NO$			
1-Indanone, 2-(diphenylmethyl)-3-imino-	MeOH	256(4.53),265s(4.46), 301(3.07),312(3.10), 431(3.32)	65-0448-64
Indone, 3-(benzylamino)-2-phenyl-	MeOH	220(4.48),268(4.51), 315(3.45),437(3.53)	65-0448-64
$C_{22}H_{17}NO_2$			
2-Propen-1-one, 3-(3-methyl-1-ureido)-1,2-diphenyl-	MeOH	240(4.1),298(4.2)	83-0367-64
$C_{22}H_{17}N_3O_2$			
Nicotinonitrile, 6-(m-cyanophenyl)-4-(2,3-dimethoxyphenyl)-2-methyl-	EtOH	223(4.6),260(4.43), 295(4.32)	2-0561-65

Compound	Solvent	λ_{max}(log ε)	Ref.
$C_{22}H_{17}OP$			
Phosphine oxide, 1-naphthyldiphenyl-	EtOH	221(4.75),270s(3.68), 275(3.78),292(3.89), 318s(3.04)	49-0285-65
Phosphole, 1,2,5-triphenyl-, oxide	EtOH	221(4.32),243(4.22), 394(4.16)	39-2184-65
$C_{22}H_{17}P$			
Phosphine, 1-naphthyldiphenyl-	EtOH	221(4.8),285(4.00)	49-0285-65
Phosphine, 2-naphthyldiphenyl-	EtOH	228(4.9),262(4.34), 326(3.27)	49-0285-65
Phosphole, 1,2,5-triphenyl-	EtOH	222(4.38),369(4.25)	39-2184-65
$C_{22}H_{17}PSe$			
Phosphole, 1,2,5-triphenyl-, 1-selenide	EtOH	222(4.41),380(4.14)	39-2184-65
$C_{22}H_{18}$			
Benzene, distyryl-	dioxan	233(4.43),246(4.30), 251(4.23),357(4.76)	38-2839-64A
1,1'-Binaphthyl, 7,7'-dimethyl-	EtOH	225(5.03),287(4.11), 293(4.11)	39-2380-64
	EtOH	287(4.09),296(4.09)	44-3092-64
5H-Dibenzo[a,c]cycloheptene, 6-p-tolyl-	MeOH	229(4.39),250(4.47), 294(4.28)	44-3333-64
Naphthalene, 3,4-dihydro-1,7-diphenyl-	EtOH	250(4.68)	39-5110-64
	EtOH	251(4.40)	78-0195-64
Phenanthrene, 9-(p-methylbenzyl)-	MeOH	252(4.78),275(4.15), 285(4.04),296(4.11), 316(2.48),325(2.48), 331(2.60),339(2.48), 347(2.64)	44-3333-64
$C_{22}H_{18}BrNO$			
2-Acetonyl-1-(2-naphthyl)isoquinolinium bromide	EtOH	221(4.90),239(4.79), 270(4.00),280s(3.98), 320s(3.64),345(3.86)	44-1846-65
$C_{22}H_{18}ClNO_4$			
1,2-Diphenyllepidinium perchlorate	EtOH	240(4.71),317(4.08)	65-0506-65
$C_{22}H_{18}ClNO_6$			
6-Hydroxy-1-(p-hydroxyphenyl)-2-phenyllepidinium perchlorate	EtOH	270(4.53),395(4.00)	65-0506-65
dianion	EtOH	274(4.52),412(4.04)	65-0506-65
$C_{22}H_{18}N_2$			
3H-Pyrrolo[1,2,3-de]quinoxaline, 5,6-dihydro-2,3-diphenyl-	MeOH	258s(4.26),276(4.34), 318s(3.83),428(3.55)	44-2589-65
$C_{22}H_{18}N_2O_2$			
Cycloheptimidazole-2,4(1H,3H)-dione, 1,3-dibenzyl-	EtOH	231(4.33),261(4.36), 271(4.33),325(3.93), 380(3.79)	94-0473-65
Cycloheptimidazole-2,6(1H,3H)-dione, 1,3-dibenzyl-	EtOH	256(4.32),360(4.31)	94-0473-65
Nicotinonitrile, 1,2-dihydro-2-oxo-6-(β-phenacylphenethyl)-	EtOH	242(4.23),339(4.08)	35-5198-65

Compound	Solvent	$\lambda_{max}(\log \epsilon)$	Ref.
$C_{22}H_{18}N_2O_2S$			
2-Isoxazolin-5-one, 4-[4-(3-ethyl-2-benzothiazolinylidene)-2-butylidene]-3-phenyl-	benzene	558(4.79),592(4.89)	5-0009-65J
	PhCl	562s(4.85),598(5.10)	5-0009-65J
	$o-C_6H_4Cl_2$	562s(4.78),599(5.09)	5-0009-65J
	$CHCl_3$	558s(4.22),596(5.15)	5-0009-65J
(addditional spectra also given)	pyridine	560s(4.74),600(5.18)	5-0009-65J
$C_{22}H_{18}N_2O_2S_2$			
Glycolic acid, di-2-thienyl-, 2,2-diphenylhydrazide	n.s.g.	238(4.3),282(4.1)	65-1786-64
$C_{22}H_{18}N_2Zn$			
Zinc, (2,2'-bipyridine)diphenyl-	toluene	350(2.91)	101-0222-65A
$C_{22}H_{18}N_4$			
Acetophenone, (3-phenyl-2-quinoxalinyl)hydrazone	EtOH	305(4.37),408(3.99)	95-1085-64
Benzaldehyde, methyl(3-phenyl-2-quinoxalinyl)hydrazone	EtOH	318(4.33),389(4.13)	95-1085-64
$C_{22}H_{18}N_4O_2$			
2-Quinoxalinecarboxanilide, 2'-amino-3-(α-hydroxybenzyl)-	EtOH	242(4.62),325(3.8)	33-1860-64
$C_{22}H_{18}N_4O_7$			
Isoquinoline, 1-benzyl-3,4-dihydro-, picrate	EtOH-HCl	276(3.96)	94-0249-64
	EtOH	252(3.89)	94-0249-64
$C_{22}H_{18}N_4O_{12}$			
1-Cyclohexene-1,2-dimethanol, bis(3,5-dinitrobenzoate)	EtOH	209(4.72),224(4.61)	39-1787-65
$C_{22}H_{18}N_6O_2$			
Phenazine, 1-tert-butyl-3-(5-nitro-1H-benzotriazol-1-yl)-	EtOH	255(4.70),372(4.25)	12-1241-65
$C_{22}H_{18}O$			
Anthracene, 1-(benzyloxy)-10-methyl-	EtOH	239(4.47),256(4.96), 355(3.74),363(3.68), 374(3.88),394(3.80)	65-2563-64
Dibenzo[a,c]cycloheptan-5-one, 6-p-tolyl-	MeOH	232(4.45),255(4.01), 299(3.29)	44-3333-64
Fluorene, 9-benzoyl-9-ethyl-	EtOH	218(4.41),256(4.32), 290(3.80),300(3.73)	22-2345-65
Ketone, ethyl 9-phenyl-9-fluorenyl	EtOH	217(4.41),269(4.18), 303(3.56)	22-2345-65
$C_{22}H_{18}O_2$			
1-Anthrol, 8-(benzyloxy)-10-methyl-	EtOH	261(5.04),350(3.67), 367(3.94),386(3.93), 407(3.78)	65-0655-65
1,1'-Binaphthyl, 7,7'-dimethoxy-	EtOH	229(4.95),278(3.96), 289(3.97),319(3.69), 334(3.81)	39-2380-64
Chalcone, 2'-(benzyloxy)-	EtOH	308(4.33)	22-3350-65
Dibenzo[a,c]cycloheptan-5-one, 6-p-tolyl-, cis	MeOH	235(4.39),256s(4.00), 301(3.34)	44-3333-64
	MeOH-KOH	245(4.57),320(4.04)	44-3333-64
trans	MeOH	230(4.45),295(3.23)	44-3333-64
	MeOH-KOH	248(4.58)	44-3333-64

Compound	Solvent	$\lambda_{max}(\log \epsilon)$	Ref.
9,10-Phenanthrenediol, p-methylbenzyl ether	dioxan	272s(4.10),299(3.86), 309(3.86),330s(2.85), 347(3.00),363(3.04)	44-2362-64
9(10H)-Phenanthrone, 10-hydroxy-10-(m-methylbenzyl)-	MeOH	242(4.41),248s(4.36), 275s(3.90),327(3.45)	44-2362-64
9(10H)-Phenanthrone, 10-hydroxy-10-(o-methylbenzyl)-	MeOH	242(4.43),248s(4.38), 275s(3.95),327(3.46)	44-2362-64
$C_{22}H_{18}O_2S$			
Chalcone, β-[(p-methoxyphenyl)thio]-	MeOH	259(4.15),338(4.22)	5-0136-64C
$C_{22}H_{18}O_3$			
Biphenyl-2-carboxylic acid, 2'-(2,5-dimethylbenzoyl)-	MeOH	205(4.72),240s(4.26), 290s(3.69)	44-3333-64
Biphenyl-2-carboxylic acid, 2'-(p-tolylacetyl)-	MeOH	286(3.49)	44-3333-64
Chalcone, 2'-(benzyloxy)-, epoxide	EtOH	258(4.02),317(3.60)	22-3350-65
1,2-Propanedione, 1-[o-(benzyloxy)-phenyl]-3-phenyl-	EtOH	259(3.77),324(3.47)	22-3350-65
	EtOH-NaOH	378(3.96)	22-3350-65
$C_{22}H_{18}O_4$			
Anthracene, 1,4-dihydro-1,4-dimethoxy-9-phenyl-, 1,4-endoperoxide	ether	234(4.6),243(4.6), 270(4.1),280(4.0), 288s(3.9),298(2.9), 303s(2.8),312(2.7), 319(2.9)	28-5031-65A
[1,1'-Binaphthalene]-3,3'-dimethanol, 2,2'-dihydroxy-	EtOH	267(4.59),278(4.59), 288(2.77),320s(2.03), 332(2.49)	44-1391-64
Phenolphthalein, 3',3"-dimethyl-	pH 11.5	568(4.7)	56-0237-65
	pH 11.5	295(4.07),375(3.97), 564(4.48)	56-1533-64
	2.5N NaOH	240(4.4),292(3.9)	56-0237-65
	2.5N NaOH	243(4.4),295(3.8)	56-1533-64
	EtOH	278(3.9)	56-0237-65
	EtOH	228(4.4),278(3.9)	56-1533-64
	80% H_2SO_4	398(4.0),512(4.7)	56-0237-65
	80% H_2SO_4	395(4.0),515(4.50)	56-1533-64
$C_{22}H_{18}O_5$			
Phenophthalein, 3'-methoxy-3"-methyl-	pH 11.5	581(4.7)	56-0237-65
	2 N NaOH	298(3.9)	56-0375-65
	2.5N NaOH	244(4.4),295(3.9)	56-0237-65
	EtOH	282(3.9)	56-0237-65
	EtOH	283(3.9)	56-0375-65
	80% H_2SO_4	396(4.1),528(4.6)	56-0237-65
	H_2SO_4	396(4.1),568(4.6)	56-0375-65
Phthalide, 3-(p-hydroxy-3-methoxyphenyl)-3-(2-hydroxy-m-tolyl)-	2N NaOH	295(3.9)	56-0375-65
	2N NaOH	290(3.9)	56-1019-65
	EtOH	283(3.92)	56-0375-65
	80% H_2SO_4	370(4.1),520(3.6)	56-0375-65
Phthalide, 3-(p-hydroxy-3-methoxyphenyl)-3-(2-hydroxy-p-tolyl)-	2N NaOH	296(3.8)	56-1019-65
	EtOH	282(3.9)	56-1019-65
	80% H_2SO_4	374(4.1),530(3.6)	56-1019-65
Phthalide, 3-(p-hydroxy-3-methoxyphenyl)-3-(6-hydroxy-m-tolyl)-	2N NaOH	300(3.9)	56-1019-65
	EtOH	285(3.9)	56-1019-65
	80% H_2SO_4	386(4.1),516(3.9)	56-1019-65

Compound	Solvent	λ_{max}(log ε)	Ref.
$C_{22}H_{18}O_6$			
Anthracene-2,6-diol, 1,5-diacetyl-, diacetate	$CHCl_3$	261(5.06),350s(3.68), 363(3.76),383(3.68)	39-4565-64
Phenolphthalein, 3',3"-dimethoxy-	pH 11.5	591(4.5)	56-0237-65
	2.5N NaOH	243(4.4),300(4.1), 356(3.6)	56-0237-65
	EtOH	283(3.9)	56-0237-65
	80% H_2SO_4	396(4.1),547(4.6)	56-0237-65
$C_{22}H_{18}O_7$			
Justicidin A	$CHCl_3$	265(4.35),295(4.13), 315(4.13),355(3.33)	88-4167-65
Viridin, acetate	EtOH	240(4.54),300(4.14)	39-3803-65
	n.s.g.	300(4.1)	77-0343-65
$C_{22}H_{18}O_9$			
Flavone, 4',5,7-triacetoxy-6-methoxy-	EtOH	265(4.29),304(4.33)	44-3438-64
$C_{22}H_{18}O_{12}$			
2,2'-Bi-p-benzoquinone, 3,3',6,6'-tetraacetoxy-5,5'-dimethyl-	EtOH	262(4.39),325s(3)	57-0152-65A
$C_{22}H_{18}Sn$			
Tin, 3-buten-1-ynyltriphenyl-	EtOH	243(4.15)	22-0035-65
$C_{22}H_{19}ClN_2O$			
1,3-Dibenzyl-2,3-dihydro-2-oxo-cycloheptimidazolinium chloride	EtOH	260(4.45),325(3.90), 345(3.90),370(3.8)	94-0810-65
$C_{22}H_{19}ClN_2O_2$			
Propionanilide, 3-anilino-3-benzoyl-4'-chloro-	EtOH	247(4.47)	39-2040-65
$C_{22}H_{19}ClN_2O_6$			
2,3-Dimethoxy-5-methyl-14H-indolo-[2,3-a]phenanthridizinium perchlorate	EtOH	215(4.52),241(4.46), 261(4.38),297(4.47), 325s(3.95),378s(4.16), 395(4.40),412(4.61)	4-0168-64
$C_{22}H_{19}ClN_4O_6$			
5-(3,4-Methylenedioxyphenyl)-2,3-di-p-tolyl-2H-tetrazolium perchlorate	n.s.g.	225(4.50),277(4.29), 305(4.32)	5-0197-65I
isomer	n.s.g.	225(4.52),277(4.32), 305(4.35)	5-0197-65I
$C_{22}H_{19}ClO_3$			
Propiophenone, 2'-(benzyloxy)-3-chloro-2-hydroxy-3-phenyl-	EtOH	253(3.88),309(3.49)	22-3350-65
$C_{22}H_{19}ClO_6S$			
1-Hydroxy-3-[(p-methoxyphenyl)thio]-1,3-diphenylallylium perchlorate	HOAc	250(4.88),334(3.92), 385(3.88),485(4.23)	5-0136-64C
$C_{22}H_{19}Cs$			
Cesium, (9,10-dihydro-10,10-dimethyl-9-phenyl-9-anthryl)-	$C_6H_{11}NH_2$	445(4.53)	35-0384-65
$C_{22}H_{19}Li$			
Lithium, (9,10-dihydro-10,10-dimethyl-9-phenyl-9-anthryl)-	$C_6H_{11}NH_2$	454(4.49)	35-0384-65

Compound	Solvent	$\lambda_{max}(\log \epsilon)$	Ref.
$C_{22}H_{19}N$			
1,2-Benzofluorene, 9-(3-dimethylamino-2-propen-1-ylidene)-	MeCN	263(4.86),275s(4.71), 285s(4.66),310s(4.04), 353(3.93),448(4.73), 465s(4.68)	24-3331-64
2,3-Benzofluorene, 9-(3-dimethylamino-2-propen-1-ylidene)-	MeCN	266(4.93),311s(4.22), 325(4.10),365s(4.21), 427(4.77),445s(4.71)	24-3331-64
3,4-Benzofluorene, 9-(3-dimethylamino-2-propen-1-ylidene)-	MeCN	268(4.67),335(4.12), 369s(4.17),441(4.84), 455s(4.81)	24-3331-64
$C_{22}H_{19}NO$			
2-Indolinone, 3-benzyl-1-methyl-3-phenyl-	MeOH	257(3.83),284s(3.23)	87-0626-65
$C_{22}H_{19}NO_2$			
Indole, 5,6-bis(benzyloxy)-	EtOH	212(4.63),216(4.71), 298(3.93)	4-0387-65
$C_{22}H_{19}NO_4$			
Benzamide, N-(o-methoxycarbonyl-phenyl)-N-(p-methoxyphenyl)-	EtOH	230s(4.33),250s(4.13), 281(3.87)	44-1020-65
Benzoic acid, o-(N-p-methoxyphenyl-benzimidoyloxy)-, methyl ester	EtOH	230(4.44),288(3.81), 328s(3.57)	44-1020-65
$C_{22}H_{19}NO_6S$			
Benzenesulfonamide, N-(2,2a-dihydroxy-5(2aH)-acenaphthenylidene)-, diacetate	MeOH	270(3.83),310s(3.43)	44-4078-65
$C_{22}H_{19}N_5$			
Phenazine, 3-(1H-benzotriazol-1-yl)-1-tert-butyl-	EtOH	232s(4.37),258(4.70), 302(4.15),372(4.20)	12-1241-65
$C_{22}H_{20}$			
Benzo[j]fluoranthene, 2,3,7,8-tetra-hydro-6,11-dimethyl-	EtOH	318(4.38),334(4.28)	44-3092-64
5H-Dibenzo[a,c]cycloheptene, 6,7-dihydro-6-p-tolyl-	MeOH	249(4.18)	44-3333-64
Triphenylene, 1,3,6,11-tetramethyl-	EtOH	258(4.82),266(5.01), 280(5.25),293(4.15), 331(2.64),338(2.84), 354(2.77)	78-3289-65
Triphenylene, 2,3,6,11-tetramethyl-	EtOH	255(4.81),264(5.06), 279(4.17),290(4.12), 326(2.66),334(2.60), 341(2.45),349(2.30)	78-3289-65
$C_{22}H_{20}Br_2N_2S_2$			
2,2'-Vinylenebis[3-methyl-4-phenyl-thiazolium bromide]	n.s.g.	385(4.0)	65-0075-64
$C_{22}H_{20}ClNO_4$			
(β-2,4,6-Cycloheptatrien-1-ylidene-phenethylidene)dimethylphenyl-ammonium perchlorate	MeCN	259(4.18),486(4.55)	24-2050-64
$C_{22}H_{20}Fe$			
1,3,5-Hexatriene, 1-ferrocenyl-6-phenyl-	$CHCl_3$	356(4.70),370s(--), 469(3.58)	5-0088-64F

Compound	Solvent	$\lambda_{max}(\log \epsilon)$	Ref.
$C_{22}H_{20}Fe_2$			
Ethylene, 1,2-diferrocenyl-	$CHCl_3$	314(4.19),460(3.15)	5-0088-64F
$C_{22}H_{20}Fe_2N_2$			
Iron, cyclopentadienyl(formyl-cyclopentadienyl)-, azine	$CHCl_3$	314(4.26),482(3.51)	5-0088-64F
$C_{22}H_{20}Fe_2O$			
Biferrocene, 1'-acetyl-	EtOH	217(4.66),260s(4.10), 292(4.00)	44-0323-64
	EtOH	220(4.7),260s(4.1), 300s(4.0),450(3.2)	44-0996-64
Biferrocene, 2-acetyl-	EtOH	225s(4.4),275s(4.0), 350s(3.4),450(3.0)	44-0996-64
Biferrocene, 3-acetyl-	EtOH	240(4.6),275(4.3), 340s(3.6),375s(3.5), 465(3.2)	44-0996-64
$C_{22}H_{20}N_2O$			
Nicotinonitrile, 1,2-dihydro-2-oxo-4,6-diphenethyl-	EtOH	236(3.98),334(4.30)	44-3593-65
2,4-Pentadien-1-one, 5-(p-dimethyl-aminophenyl)-1-(2-quinolyl)-	H_2SO_4	558(4.74)	65-1724-65
$C_{22}H_{20}N_2O_2$			
Acetamide, N,N'-p-terphenyl-2',5'-ylenebis-	EtOH	245(4.52)	4-0140-65
Phthalimide, N-[1-(1,3,4,5-tetrahydro-2-methylbenz[cd]indol-3-yl)ethyl]-	EtOH	222(4.86),242f(4.06), 275(3.96),279(3.97), 290(3.86)	44-0843-64
isomer	EtOH	225(4.85),240f(4.11), 275(3.91),280(3.91), 289f(3.82)	44-0843-64
Propionanilide, 3-anilino-3-benzoyl-	EtOH	245(4.48)	39-2040-65
$C_{22}H_{20}N_2O_4S$			
Indole-2-carboxylic acid, 3,3'-thio-bis[1-methyl-, dimethyl ester	EtOH	237(4.61),300(4.38), 352(3.98)	44-0178-64
$C_{22}H_{20}N_2O_4S_2$			
Indole-2-carboxylic acid, 3,3'-dithio-bis[1-methyl-, dimethyl ester	EtOH	216(4.61),234(4.52), 279(4.33),302(4.35), 356s(3.77)	44-0178-64
$C_{22}H_{20}N_2O_4S_3$			
Indole-2-carboxylic acid, 3,3'-trithio-bis[1-methyl-, dimethyl ester	EtOH	222(4.58),276s(4.23), 299(4.33),344s(3.97)	44-0178-64
$C_{22}H_{20}N_2O_5S$			
Indole-2-carboxylic acid, 3,3'-sul-finylbis[1-methyl-, dimethyl ester	EtOH	215(4.64),233(4.63), 299(4.41),328s(4.13), 344s(3.94)	44-0178-64
Indole-2-carboxylic acid, 3,3'-sul-finyldi-, diethyl ester	EtOH	214(4.68),232(4.61), 301(4.46),335s(3.98)	44-0178-64
$C_{22}H_{20}N_2O_9$			
Illudin S, dioxo-, 3,5-dinitrobenzoate	EtOH	294(4.09)	78-1231-65

Compound	Solvent	λ_{max}(log ε)	Ref.
$C_{22}H_{20}N_4O$			
Acetophenone, 2-(m-tolylazo)-2-(m-tolylhydrazono)-	EtOH	240(4.21),318(4.20), 447(4.36)	5-0099-65B
Acetophenone, 2-(o-tolylazo)-2-(o-tolylhydrazono)-	EtOH	242(4.20),319(4.21), 440(4.35)	5-0099-65B
Acetophenone, 2-(p-tolylazo)-2-(p-tolylhydrazono)-	EtOH	244(4.21),327(4.25), 452(4.47)	5-0099-65B
$C_{22}H_{20}N_4O_3$			
Acetophenone, 2-[(p-methoxyphenyl)-azo]-2-[(p-methoxyphenyl)hydrazono]-	EtOH	242(4.29),351(4.30), 468(4.41)	5-0099-65B
$C_{22}H_{20}N_4O_6$			
Xanthorrhoeol methyl ether, 2,4-dinitrophenylhydrazone	EtOH	225(4.60),243(4.40), 275s(4.13),303(3.76), 361(4.28)	12-0575-65
$C_{22}H_{20}N_4O_7S$			
N'-(2,4-Dinitrophenyl)-N-ethylbenzimidazolium p-toluenesulfonate	EtOH	348(4.09)	65-0269-64
$C_{22}H_{20}N_6S$			
Urea, 1-[bis(benzylidenehydrazino)-methylene]-3-phenyl-2-thio-	n.s.g.	273(4.42)	39-4448-65
$C_{22}H_{20}O$			
Dibenzo[a,c]cyclohepten-5-ol, 6,7-dihydro-6-p-tolyl-, cis	MeOH	250(4.19)	44-3333-64
trans	MeOH	250(4.31)	44-3333-64
1-Naphthol, 1,2,3,4-tetrahydro-1,7-diphenyl-	EtOH	250(4.23)	78-0195-64
$C_{22}H_{20}O_2$			
9,10-Phenanthrenediol, 9,10-dihydro-9-(p-methylbenzyl)-, cis	MeOH	273(4.21)	44-3333-64
trans	MeOH	271(4.16),300s(3.49)	44-3333-64
$C_{22}H_{20}O_4$			
1,4-Cyclohexadiene-1,2-dicarboxylic acid, 3,6-diphenyl-, dimethyl ester	MeOH	240(3.52),262(3.28), 270(3.15)	35-2392-64
3,5-Cyclohexadiene-1,2-dicarboxylic acid, 3,6-diphenyl-, dimethyl ester	MeOH	230s(4.00),335(4.35)	35-2392-64
$C_{22}H_{20}O_8$			
Neoepipicropodophyllin, 3-O-methyl-	EtOH	293(3.58)	44-1594-64
$C_{22}H_{20}O_9$			
Flavone, 4',5-diacetoxy-3,7,8-trimethoxy-	EtOH	263(4.36),308(4.17), 334(4.17)	12-0934-64
Flavone, 5,7-diacetoxy-3,4',8-trimethoxy-	EtOH	260(4.22),349(4.27)	12-1871-65
θ-Rhodomycinone	EtOH	234(4.59),258(4.35), 296(3.82),480(4.10), 493(4.13),513(4.00), 527(3.94)	39-3927-64
Wightin, di-O-acetyl-	EtOH	260(4.43),306(4.30)	78-3237-65
$C_{22}H_{20}O_{10}$			
Melibentin, 5-demethyl-, acetate	EtOH	253(4.49),268s(4.20), 342(4.34)	12-2021-65

Compound	Solvent	$\lambda_{max}(\log \epsilon)$	Ref.
$C_{22}H_{20}O_{12}$			
1,2,4,5,7,8-Naphthalenehexol, hexaacetate	EtOH	231(4.83),295(3.92), 328s(3.23)	94-0633-65
$C_{22}H_{21}BBr_4O_3$			
Trianisylmethylium tetrabromoborate	MeCN	483(4.87)	35-5511-64
$C_{22}H_{21}BrN_2O_3$			
2-Acetonyl-1-(3,4-dimethoxyphenyl)-9H-pyrido[3,4-b]indolium bromide	EtOH	207(4.67),237(4.30), 264(4.44),314(4.28), 384(3.75)	4-0168-64
$C_{22}H_{21}BrO_3$			
Methane, bromotris(p-methoxyphenyl)-	MeCN	483(4.87)	35-5511-64
$C_{22}H_{21}ClN_2O_4S$			
3-(1-Ethyl-6-methyl-2(1H)-quinolylidene)-2,3-dihydro-1H-pyrrolo[2,1-b]-benzothiazolium perchlorate	EtOH	517(4.72)	65-2455-64
$C_{22}H_{21}ClN_2O_9$			
Tetracycline, 7-chloro-6-deoxy-6-peroxydehydro-	MeOH-HCl	249(4.31),370(3.57)	44-2746-64
	MeOH-NaOH	243(4.37),265(4.31), 407(3.90)	44-2746-64
$C_{22}H_{21}ClN_4O_5$			
3-(p-Methoxyphenyl)-2,5-di-p-tolyl-2H-tetrazolium perchlorate	n.s.g.	253.5(4.35)	5-0197-65I
isomer	n.s.g.	253.5(4.45)	5-0197-65I
$C_{22}H_{21}Cl_2NO_4$			
Atropic acid, β,β'-iminobis-[o-chloro-, diethyl ester	MeOH	331(4.61)	32-0831-65
isomer	MeOH	339(4.62)	32-0831-65
Atropic acid, β,β'-iminobis-[p-chloro-, diethyl ester	MeOH	343(4.60)	32-0831-65
isomer	MeOH	353(4.62)	32-0831-65
$C_{22}H_{21}N$			
Cyclopentadiene, 5-(3-dimethylamino-2-propen-1-ylidene)-1,4-diphenyl-	MeCN	278(4.54),450(4.81)	24-3331-64
Fluorene, 9-(7-dimethylaminohepta-2,4,6-trien-1-ylidene)-	MeCN	255(4.70),300(4.00), 341(3.68),365s(3.84), 490(4.84)	24-3331-64
Quinoline, 1,2,3,4-tetrahydro-1-methyl-2,3-diphenyl-, cis	HCl	258(2.97),263(2.97), 270(2.94)	39-0249-64
	EtOH	209(4.69),250(4.06), 306(3.60)	39-0249-64
$C_{22}H_{21}NO$			
Tropinone, 2,4-dibenzylidene, cis-trans	EtOH	233(4.16),327(4.31)	39-2723-65
trans-trans	EtOH	233(4.18),328(4.42)	39-2723-65
$C_{22}H_{21}NO_2S$			
Nicotinic acid, 5-benzylidene-1,2,5,6-tetrahydro-1-methyl-2-phenyl-6-thioxo-, ethyl ester	EtOH	320(4.32)	78-3255-65

Compound	Solvent	λ_{max}(log ε)	Ref.
$C_{22}H_{21}NO_3$			
Nicotinic acid, 5-benzyl-1,6-dihydro-1-methyl-6-oxo-2-phenyl-, ethyl ester	EtOH	270(4.15),306(4.03)	78-3255-65
Nicotinic acid, 5-benzylidene-1,2,5,6-tetrahydro-1-methyl-6-oxo-2-phenyl-, ethyl ester	EtOH	322(4.39)	78-3255-65
$C_{22}H_{21}NO_4S$			
2,5-Pyridinedicarboxylic acid, 1,4,5,6-tetrahydro-3-methyl-6-thioxo-, dibenzyl ester	EtOH	245(3.75),327(4.07)	39-1262-65
6H-1,3-Thiazin-2-ylacetic acid, 4-benzyloxycarbonyl-5-methyl-, benzyl ester	EtOH	225s(4.02),282(3.82), 338(4.15)	39-1262-65
$C_{22}H_{21}NO_6$			
3H-2-Benzazepine-2,4-dicarboxylic acid, 1,5-dimethoxy-3-phenyl-, dimethyl ester	MeOH	204(4.5),249(4.5)	56-1423-65
$C_{22}H_{21}N_3$			
1,2,4-Triazole, 4,5-dihydro-5,5-dimethyl-1,3,4-triphenyl-	EtOH	260s(4.0),345(4.0)	24-2174-65
$C_{22}H_{21}N_3O_3$			
Benzamide, N-[2-(N-benzylanilino)-ethyl]-p-nitro-	MeOH	254(4.42)	87-0107-65
$C_{22}H_{21}N_3O_5$			
4-Pyridone-3,5-dicarboxylic acid, 2,6-bis(2-pyridyl)-, diethyl ester, O-methyl derivative	EtOH	244(4.39),278(4.41)	88-3175-65
$C_{22}H_{21}N_4$			
Verdazyl, 6-ethyl-1,3,5-triphenyl-	benzene	720(3.64)	49-0457-64
$C_{22}H_{21}N_4O_2$			
Verdazyl, 1,3-bis(p-methoxyphenyl)-5-phenyl-	benzene	740(3.60)	49-0457-64
$C_{22}H_{22}$			
Benzene, o-diphenethyl-	EtOH	262(2.34),266(2.31), 269(2.29),275(1.96)	44-0886-64
Naphthalene, 4,4a,5,6,7,8-hexahydro-2,5-diphenyl-	EtOH	286(4.30)	78-0195-64
$C_{22}H_{22}BF_4N_2P$			
1-Ethyl-2-[1-ethyl-2(1H)-quinolylidene)phosphino]quinolinium tetrafluoroborate	n.s.g.	592(4.63)	89-0433-64
$C_{22}H_{22}Cl_2CrN_3$			
Chromium, dichlorobenzyltris(pyridine)-	pyridine	378(2.28),430s(--)	77-0509-65
$C_{22}H_{22}FeO_2$			
2,4,6,8,10-Undecapentaenoic acid, 11-ferrocenyl-, methyl ester	$CHCl_3$	400(4.79)	5-0088-64F

Compound	Solvent	$\lambda_{max}(\log \epsilon)$	Ref.
$C_{22}H_{22}FeO_4$			
Malonic acid, 7-ferrocenyl-2,4,6-hepta-trien-1-ylidene, dimethyl ester	$CHCl_3$	402(4.62),520(4.23)	5-0088-64F
$C_{22}H_{22}N_2O$			
3H-Pyrrolo[1,2-a]indole, 7-(benzyloxy)-1-(1-pyrrolidinyl)-	MeOH	315(4.15),342(4.07)	44-2904-65
$C_{22}H_{22}N_2O_4$			
Kopsane, N_a-carbomethoxy-10,22-dioxo-	MeOH	240(4.09),279(3.33)	35-4944-65
$C_{22}H_{22}N_4$			
Benzene, [2-(o-tolylazo)-2-(o-tolyl-hydrazono)ethyl]-	EtOH	254(4.21),296(3.83), 415(4.26)	5-0099-65B
Phenazine, 3-(o-aminoanilino)-1-tert-butyl-	EtOH	235(4.55),285(4.63), 362(3.76),475(4.00)	12-1241-65
$C_{22}H_{22}N_4O_2$			
Hydrazine, 1-carbethoxy-1-(α-phenyl-azo)benzyl-2-phenyl-	MeOH	406(2.29)	89-0684-64
$C_{22}H_{22}N_4O_2S$			
Indole-2-carboxamide, 3,3'-thio-bis[N,1-dimethyl-	EtOH	230(4.69),294(4.31), 328s(3.95)	44-0178-64
$C_{22}H_{22}N_4O_3$			
Leucoflavin, 5-acetyl-3-benzoyl-1,7,8-trimethyl-	MeOH	212(4.44),238(4.31), 308(3.84)	33-1354-64
$C_{22}H_{22}N_8O_8$			
Acraldehyde, α-(2-formyl-3-methyl-cyclopentyl)-, bis(2,4-dinitro-phenylhydrazone)	$CHCl_3$	362(4.68)	12-1260-64
$C_{22}H_{22}O_2$			
Bicyclo[5.2.1]decane-4,10-dione, 2,6-diphenyl-	EtOH	249(2.77),253(2.70), 259(2.71),265(2.59), 268(2.42),314(2.42)	44-1409-65
Bicyclo[3.3.1]non-2-en-9-one, 2-(p-methoxyphenyl)-4-phenyl-	EtOH	232(4.2)	2-0561-65
Indone, 3-(benzyloxy)-3a,4,5,6,7,7a-hexahydro-2-phenyl-	EtOH	250(4.13)	65-0808-64
1-Penten-3-one, 5-(2-oxocyclo-pentyl)-1,5-diphenyl-	EtOH	292(4.38)	44-1409-65
$C_{22}H_{22}O_4$			
4,10(3H,9H)-Anthracenedione, 5-(benzyl-oxy)-1,2,4aα,9aα-tetrahydro-9β-hy-droxy-9α-methyl-	CH_2Cl_2	260(4.04),320(3.63)	65-2570-64
4(1H)-Anthracenone, 5-(benzyloxy)-4aα,9,9aα,10-tetrahydro-9β,10β-di-hydroxy-9α-methyl-	EtOH	275(3.32),282(3.33)	65-2570-64
10(9H)-Anthracenone, 5-(benzyloxy)-1,2,3,9aα-tetrahydro-4,9β-di-hydroxy-9α-methyl-	EtOH	266(3.76),342(4.12)	65-2570-64
$C_{22}H_{22}O_5$			
2H-1-Benzopyran-7-ol, 3-(p-hydroxy-phenyl)-2,2,4-trimethyl-, diacetate	MeOH	270(3.98),308(3.94)	44-4114-65

Compound	Solvent	$\lambda_{max}(\log \epsilon)$	Ref.
$C_{22}H_{22}O_7$			
Cinnamic acid, 3,4,5-trimethoxy-, 3,4-(methylenedioxy)cinnamyl ester, cis-trans	EtOH	270(4.23),308(4.35)	78-0871-64
trans-trans	EtOH	270(4.21),310(4.42)	78-0871-64
Dalbergione, 4'-hydroxy-4-methoxy-, quinol triacetate	EtOH	279(3.45)	78-2683-65
$C_{22}H_{22}O_8$			
Matsuzake lactone, O-dimethyl ether	EtOH	229(4.44),263(4.47)	94-1232-64
$C_{22}H_{22}O_9$			
Flavone, 5-acetoxy-3,3',4',7,8-penta-methoxy-	EtOH	251(4.33),353(4.23)	12-0934-64
$C_{22}H_{22}O_{10}$			
Trifolirhizin	MeOH	285(3.64),311(3.87)	94-0093-65
$C_{22}H_{23}BrN_2O_6$			
α-D-Glucoside, methyl 4,6-O-benzyli-dene-3-(p-bromophenylazo)-3-deoxy-, 2-acetate	EtOH	281(4.11),392(2.46)	39-1045-64
$C_{22}H_{23}F_3N_2O$			
Acrylonitrile, 2-[p-(diethylamino-ethoxy)phenyl]-3-(p-trifluoromethyl-phenyl)-	EtOH	337(4.36)	87-0511-64
$C_{22}H_{23}FeN$			
Butadiene, 1-ferrocenyl-4-(p-di-methylaminophenyl)-	C_6H_{12}	367(4.67),446(3.40)	5-0088-64F
	MeOH	364(4.67),446(3.43)	5-0088-64F
	$CHCl_3$	373(4.66),446(3.48)	5-0088-64F
$C_{22}H_{23}IN_2S$			
3-Ethylbenzothiazol-2-yl 3-ethyl-6-methylquinolin-2-yl monomethine cyanine iodide	EtOH	491(4.68)	65-2455-64
$C_{22}H_{23}NO_2$			
9-Acridanacetic acid, 10-(1-hydroxy-2-methylpropenyl)-α,α,9-trimethyl-, ζ-lactone	EtOH	239(3.92),273(3.44)	77-0569-65
Aziridine, 2,3-dibenzoyl-1-cyclo-hexyl-, cis	EtOH	251(4.33)	35-1050-65
trans	EtOH	253.5(4.45)	35-1050-65
Methacrylic acid, 2-methyl-1-(9-methyl-10-acridanyl)propenyl ester	EtOH	207(4.69),277(4.20)	77-0569-65
$C_{22}H_{23}NO_4$			
Atropic acid, β,β'-iminodi-, diethyl ester	MeOH	344(4.58)	32-0831-65
isomer	MeOH	351(4.60)	32-0831-65
Ochotensimine	n.s.g.	226(4.41),287(4.12)	88-3819-64
$C_{22}H_{23}NO_5$			
Isoquinoline, 2-acetyl-1,2,3,4-tetra-hydro-6,7-dimethoxy-1-(3,4-dimeth-oxybenzylidene)-	EtOH	220(4.45),332(4.44)	39-4014-65

Compound	Solvent	λ_{max}(log ϵ)	Ref.
$C_{22}H_{23}NO_6$			
13-Epiophiocarpine, acetate	EtOH	290(3.75)	94-1072-64
Ophiocarpine, acetate	EtOH	290(3.72)	94-1072-64
$C_{22}H_{23}NO_7$			
α-Narcotine	EtOH	291(3.62),310(3.70)	94-1080-64
β-Narcotine	EtOH	290(3.58),312(3.68)	94-1080-64
$C_{22}H_{23}N_3O$			
Acrylonitrile, 2-[p-(diethylamino-ethoxy)phenyl]-3-(p-cyanophenyl)-	EtOH-HCl	349(4.47)	87-0511-64
$C_{22}H_{24}$			
3,4-Benzotetraphene, 1,3,5-trinitro-benzene complex	$CHCl_3$	400(4.40)	38-0166-64A
Naphthalene, 1,2,3,4,4a,5,6,8a-octa-hydro-1,8-diphenyl-	EtOH	215(4.14),245(3.81)	44-2942-65
$C_{22}H_{24}BrP$			
Tribenzylmethylphosphonium bromide	EtOH	254(2.86),260(2.92), 266(2.84)	59-1143-64
$C_{22}H_{24}Br_2O_2$			
Anthracene, 9,10-bis(3-bromopropyl)-2,6-dimethoxy-	EtOH	233(4.26),260(4.86), 269(5.24),303(3.14), 318(3.43),333(3.69), 350(3.74),374(3.43), 394(3.76),417(3.83)	39-4565-64
$C_{22}H_{24}Cl_2N_2$			
1,4-Cyclohexanebis(methylamine), N,N'-bis(2-chlorobenzylidene)-, trans	n.s.g.	250(4.50)	87-0826-64
$C_{22}H_{24}FeO$			
Ferrocene, 4',5'-(α-oxotrimethylene)-1,1';3,3';4,5-tris(trimethylene)-	n.s.g.	232(4.19),271(3.83)	88-0827-64
$C_{22}H_{24}IP$			
Butyltriphenylphosphonium iodide	EtOH	262(3.36),268(3.49), 275(3.39)	59-1143-64
Isobutyltriphenylphosphonium iodide	EtOH	262(3.36),268(3.49), 275(3.39)	59-1143-64
Methyltri-m-tolylphosphonium iodide	EtOH	274(3.59),281(3.56)	49-1793-65
Methyltri-o-tolylphosphonium iodide	EtOH	218(4.56),274(3.65), 281(3.61)	49-1793-65
Methyltri-p-tolylphosphonium iodide	EtOH	234(4.62),266(3.64), 276(3.44)	49-1793-65
$C_{22}H_{24}N_2$			
3,3'-Biindole, 2,2'-dipropyl-	EtOH	230(4.71),283(4.17), 290(4.15)	44-3604-65
$C_{22}H_{24}N_2O$			
Acrylonitrile, 2-phenyl-3-[p-N-piperi-dinylethoxy)phenyl]-	EtOH-HCl	335(4.48)	87-0511-64
1H-Pyrrolo[1,2-a]indole, 7-(benzyloxy)-2,3-dihydro-1-(1-pyrrolidinyl)-	MeOH	278(3.97),297s(3.72), 310s(3.52)	44-2904-65
$C_{22}H_{24}N_2O_2$			
Alkaloid RP-7, nitrate	H_2O	215(4.37),263(3.73)	100-0220-64

Compound	Solvent	λ max(log ε)	Ref.
$C_{22}H_{24}N_2O_3$			
Icajine	n.s.g.	254(4.14),291(3.51)	28-5237-65B
	n.s.g.	253(4.23),291(3.52)	28-5237-65B
$C_{22}H_{24}N_2O_4$			
3-Azabicyclo[3.3.1]nonan-9-ol, 3-methyl-19-phenyl-, p-nitrobenzoate	EtOH	260(4.18)	44-3634-65
Erinine	acid	240(4.13)	33-1598-65
	EtOH	256(4.16)	33-1598-65
Voacafricine	n.s.g.	238(4.27),315(4.35)	100-0374-64
Vomicine	n.s.g.	223(3.57),265(3.70), 299(4.25)	28-5237-65B
$C_{22}H_{24}N_2O_6$			
α-D-Glucoside, methyl 4,6-O-benzylidene-3-phenylazo-3-deoxy-, 2-acetate	EtOH	267(4.02),400(2.40)	39-1045-64
$C_{22}H_{24}N_2O_8$			
Tetracycline	EtOH	268(4.2),310s(3.8), 366(4.2)	83-0034-65
$C_{22}H_{24}N_2O_{10}$			
Pentadiene-1,2,3,4,5-pentacarboxylic acid, 1-cyano-, 1-ethyl 2,3,4,5-tetramethyl ester, compound with pyridine	EtOH	229(4.11),250s(4.03), 370(4.24)	39-4355-65
$C_{22}H_{24}N_4O$			
p-Benzoquinone 4-imine, 2-(2-aminoanilino)-6-tert-butyl-N-(o-aminophenyl)-	EtOH	228(4.34),298(4.10), 540(3.70)	12-1241-65
$C_{22}H_{24}O$			
Estra-1,3,5(10)-trien-17β-ol, 17α-butadiynyl-	EtOH	242(2.58),256(2.63), 267(2.66),274(2.62)	78-1197-65
Spiro[4.5]decan-1-one, 6,10-diphenyl-	EtOH	249(2.42),254(2.54), 259(2.64),265(2.52), 303(1.99)	44-1409-65
$C_{22}H_{24}O_2$			
4(3H)-Anthracenone, 1,2,4aα,9,9aβ,10-hexahydro-5-hydroxy-9α-methyl-, benzyl ether	EtOH	249(3.21),278(3.46)	65-2563-64
Estra-1,3,5(10)-triene-3,17β-diol, 17α-butadiynyl-	MeOH	287(3.34),294(3.30)	13-0801-64
	EtOH	250(2.60),255(2.71), 262s(2.75),281s(3.27)	78-1197-65
$C_{22}H_{24}O_4$			
5,9-Anthracenediol, 4β,10β-(benzylidenedioxy)-1,2,3,4,4aα,9,9aβ,10-octahydro-9β-methyl-	EtOH	280(3.68)	65-2563-64
2H-1-Benzopyran-7-ol, 4-ethyl-3-(p-methoxyphenyl)-2,2-dimethyl-	MeOH	270(4.07),307(3.97)	44-4114-65
5(6H)-Chrysenone, 8α-acetoxy-6a,7,8,9,10,10a-hexahydro-2-methoxy-10a-methyl-, cis	EtOH	220(4.71),247(4.45), 316(3.87),348(3.61)	23-0591-64
Guaiaretic acid, dehydrodimethyl-	n.s.g.	237(5.01),284(4.14), 314(3.62),329(3.74)	39-4011-64
	n.s.g.	237(4.85),283(4.10), 314(3.47),329(3.57)	39-4011-64

Compound	Solvent	$\lambda_{max}(\log \epsilon)$	Ref.
Inden-7-ol, 5-methoxy-2-(p-methoxyphenyl)-1,3,3-trimethyl-, acetate	MeOH	278(4.31)	44-4120-65
7,11-Methanocycloocta[a]naphthalen-13-one, 7,8,9,10,11,12-hexahydro-10-hydroxy-3-methoxy-7,10-dimethyl-, acetate	EtOH	231(4.75),266(3.76), 277(3.74),287(3.57), 307(3.03),315s(3.09), 321(3.25),330(3.18), 336(3.36)	23-0591-64
19-Norcarda-4,14,20(22)-trienolide, 3,11-dioxo-	MeOH	219(4.34),240(4.28)	44-0527-64
$C_{22}H_{24}O_5$			
Furoguaiacin, dimethyl-	n.s.g.	252(4.11),326(4.47)	39-4011-64
19-Norpregna-1,3,5(10),6,8-pentaen-20-one, 21-acetoxy-3,17α-dihydroxy-	MeOH	230(4.82),270(3.68), 280(3.74),292(3.60), 327(3.36),341(3.42)	35-2309-64
$C_{22}H_{24}O_6$			
Naphthalene, 1,2,3-trimethoxy-7-(3,4,5-trimethoxyphenyl)-	EtOH	230(4.59),261(4.66), 299(4.18)	78-3605-65
Viridin, deoxyhexahydro-, acetate	EtOH	216(4.34),257(4.01), 309(3.53)	39-3803-65
$C_{22}H_{24}O_7$			
Flavone, 3',5-diethoxy-2',7,8-trimethoxy-	EtOH	267(4.48),325s(4.00)	78-3237-65
Ichthynol, dihydro-	EtOH	216s(4.35),237(4.17), 294(4.20),348(3.94)	78-1317-64
	EtOH-KOH	226s(4.94),289(4.23), 371(4.06)	78-1317-64
$C_{22}H_{24}O_8$			
2-Naphthoic acid, 1,2,3,4-tetrahydro-6,7-dimethoxy-4-oxo-1-(3,4,5-trimethoxyphenyl)-	EtOH	276(4.24),313(4.02)	2-0190-64
$C_{22}H_{24}O_9$			
Succinic acid, [α-(3,4,5-trimethoxyphenyl)veratrylidene]-	EtOH	215(4.74),282(4.24)	2-0190-64
	EtOH	214(4.7),282(4.22)	2-0190-64
$C_{22}H_{24}O_{10}$			
Isosakuranetin, 7-β-D-glucosyl-	EtOH	284(4.23)	22-2937-65
$C_{22}H_{24}O_{11}$			
Chalcone, 2',3,4,6-tetrahydroxy-4-methoxy-, 4'-glucoside	EtOH	260(3.57),370(4.14)	65-3340-64
Flavanone, 3',5,7-trihydroxy-4'-methoxy-, 7-glucoside	EtOH	280(4.08)	65-3340-64
Hesperetin, 7-β-D-glucosyl-	EtOH	284(4.20)	22-2937-65
$C_{22}H_{25}BrO_5$			
Estra-1,3,5(10)-trien-6-one, 7α-bromo-3,17β-dihydroxy-, diacetate	EtOH	212(4.31),258(4.03), 307(3.36)	44-1325-64
$C_{22}H_{25}ClN_2O_4$			
Vincine, py-tetradehydro-, chloride	MeOH	255(4.29),334(4.18)	73-0447-64
$C_{22}H_{25}ClO_5$			
4-Ethyl-6,7-dimethoxy-1-(3,4-dimethoxyhydroxyphenyl)-3-methyl-2-benzopyryl-chloride	EtOH-HCl	250(4.4),270(4.3), 366(4.0),435(4.2)	49-0369-65
	EtOH	250(4.5),283(4.3), 366(4.0),444(4.2)	49-0369-65

Compound	Solvent	λ_{max}(log ϵ)	Ref.
$C_{22}H_{25}Cl_2N_3O_2$			
Acetanilide, 2,4'-dichloro-2'-[N-(3-morpholinopropyl)benzimidoyl]-	EtOH	237(4.48),323(3.59)	44-2368-64
$C_{22}H_{25}F_3O_5$			
Estra-1,3,5(10)-triene-1,3,17β-triol, 1-acetate 17α-trifluoroacetate	EtOH	280(3.31)	24-1940-64
	NaOH	297(3.51)	24-1940-64
$C_{22}H_{25}NO_2S$			
6H-Dibenzo[b,e]thiocine, 12-[3-(methylcarbethoxyamino)propylidene]-7,12-dihydro-	MeOH	231(4.22),263(3.79)	24-0685-65
$C_{22}H_{25}NO_4$			
Ochotensimine, dihydro-	n.s.g.	233s(4.09),282(3.77)	88-3819-64
$C_{22}H_{25}NO_5$			
Papaveraldine, 3,4-dihydro-4,4-dimethyl-	MeOH	214(4.41),233(4.5), 276(4.3),316(4.3)	83-0362-64
Prednisone, 2-cyano-	EtOH	239(4.00)	44-3300-64
$C_{22}H_{25}NO_6$			
Alkaloid K-3	MeOH	228(4.32),290(4.28)	65-1686-64
Carbostyril, 3-[2-(benzyloxy)-ethyl]-4,5,7,8-tetramethoxy-	n.s.g.	235s(4.25),295s(4.32), 265(4.37),316(4.13), 330s(4.04)	2-0071-65
$C_{22}H_{25}NO_7$			
Oxycolchicine	EtOH	281(2.51)	78-1449-64
$C_{22}H_{25}N_2OP$			
Phosphine oxide, bis[p-(dimethylamino)phenyl]phenyl-	MeOH	284(4.61)	88-2729-64
$C_{22}H_{25}N_2P$			
Phosphine, bis[p-(dimethylamino)-phenyl]phenyl-	C_6H_{12}	281(4.58)	88-2729-64
	MeOH	285(4.59)	88-2729-64
$C_{22}H_{25}N_2PS$			
Phosphine sulfide, bis[p-(dimethylamino)phenyl]phenyl-	MeOH	285(4.60)	88-2729-64
$C_{22}H_{25}N_3$			
Cyclopentapyrazole, 1,3a,4,5,6,6a-hexahydro-1,3-diphenyl-6a-(1-methylpyrrolidinyl)-	C_6H_{12}	251(4.45),299(4.02)	44-1582-64
$C_{22}H_{25}N_3O$			
3-Pyridine Green	0.5N HCl	425(4.08),655(4.89)	1-0447-64
	acetate	419(4.20),632(4.95)	1-0447-64
$C_{22}H_{25}N_3O_3$			
Yohimbine, 3ξ-cyano-	EtOH	228(4.38),257s(3.93), 295s(4.25),307(4.36), 318(4.30)	44-0105-65
	CH_2Cl_2	283(3.94),291(3.95), 305(3.94),316(3.92)	44-0105-65
	MeCN	223(4.46),282(3.89), 289(3.90),307(3.78), 318(3.78)	44-0105-65

Compound	Solvent	λ_{max}(log ε)	Ref.
$C_{22}H_{25}N_3O_7$			
Griseoviridin, dehydro-	EtOH	209(4.40),277(4.30)	23-0371-64
$C_{22}H_{26}$			
Biphenyl, 4-methyl-2-(2,6,6-trimethyl-cyclohex-1-enyl)-	n.s.g.	248(4.28)	39-2892-65
1,1'-Spirobiindan, 2,3,3,3',3'-pentamethyl-	C_6H_{12}	259(3.08),266(3.29), 273(3.33)	23-0025-64
$C_{22}H_{26}Br_2O_2$			
9β,10α-Pregna-1,4-diene-3,20-dione, 6-(dibromomethylene)-	MeOH	247(4.18),286(3.66)	54-0904-65
$C_{22}H_{26}ClNO_6$			
1-α-13-Hydroxy-2,3-(methylenedioxy)-1,9,10-trimethoxy-5,6,13,13a-tetra-hydro-8H-dibenzo[a,g]quinolizinium chloride	EtOH 50% EtOH	282.5(3.57) 282(3.49)	94-1080-64 39-1087-65
1-β-isomer	EtOH	282.5(3.54)	94-1080-64
Isoquinoline, 1-α-1-(α-hydroxy-2'-chloromethyl-3',4'-dimethoxybenzyl)-8-methoxy-2-methyl-6,7-(methylene-dioxy)-1,2,3,4-tetrahydro-	EtOH	285(3.55)	94-1080-64
1-β-isomer	EtOH	285(3.47)	94-1080-64
$C_{22}H_{26}FN_5O_{15}$			
2(1H)-Pyrimidone, 1-(3,4,6-tri-O-p-nitrobenzoyl-2-deoxy-D-arabinohexo-pyranosyl)-4-ethoxy-5-fluoro-	CH_2Cl_2	262(5.66)	87-0140-65
$C_{22}H_{26}F_2O_4$			
Pregna-5(6),17(20)-dien-21-oic acid, 4,4-difluoro-3,11-dioxo-, methyl ester, cis	EtOH	224(4.08)	44-2982-64
$C_{22}H_{26}N_2O$			
Acrylonitrile, 2-[p-(diethylamino-ethoxy)phenyl]-3-p-tolyl-	EtOH-HCl	335(4.44)	87-0511-64
1H-Pyrido[3,4-b]indole, 7-(benzyloxy)-3-ethyl-2,3,4,9-tetrahydro-1,1-dimethyl-	EtOH	228(4.60),267(3.71), 294(3.74)	44-2864-64
$C_{22}H_{26}N_2OS$			
Acrylonitrile, 2-[p-(diethylamino-ethoxy)phenyl]-3-(p-methylthio-phenyl)-	EtOH-HCl	360(4.53)	87-0511-64
$C_{22}H_{26}N_2O_2$			
Acrylonitrile, 2-[p-(diethylaminoeth-oxy)phenyl]-3-(o-methoxyphenyl)-	EtOH	345(4.29)	87-0511-64
Acrylonitrile, 2-[p-(diethylaminoeth-oxy)phenyl]-3-(p-methoxyphenyl)-	EtOH	346(4.49)	87-0511-64
Acrylonitrile, 3-[p-(diethylaminoeth-oxy)phenyl]-2-(p-methoxyphenyl)-, (as citrate)	EtOH	344(4.41)	87-0511-64
2,4-Diazaestra-3,5(10)-dien-1-one, 17β-hydroxy-3-phenyl-	MeOH-HCl MeOH MeOH-NaOH	245(4.22),282(4.11) 239(4.10),295(4.02) 258s(4.00),286(3.94)	13-0631-64 13-0631-64 13-0631-64

Compound	Solvent	$\lambda_{max}(\log \epsilon)$	Ref.
$C_{22}H_{26}N_2O_3$			
Alstophylline	EtOH	234(4.69),258(4.23), 294(3.93)	33-1349-65
Ochropamine	EtOH	243(4.28),315(4.25)	12-0246-64
	n.s.g.	243(4.28),315(4.25)	100-0374-64
Powerchrine	EtOH	232(4.37),279(3.75), 297(3.77)	12-0246-64
Unknown isoschizogamine derivative	MeOH	214(4.33),261(4.09), 291(3.82)	100-0406-64
Vincamajine, 2α-hydroxy-17-deoxy-	MeOH-HCl	243(3.85),248(3.95)	20-0253-65
	MeOH	238(3.85),243(3.85), 295(3.81)	20-0253-65
	MeOH-NaOH	249(3.95),300(3.54)	20-0253-65
$C_{22}H_{26}N_2O_4$			
Aspidodasycarpine, N-acetyldeformo-	EtOH	240(3.98),295(3.59)	78-1717-65
Aspidodasycarpine, N^b-acetyldeformo-	EtOH	240(3.98),295(3.59)	88-3899-64
Erinicine	acid	242(3.93)	33-1598-65
	EtOH	255(4.03)	33-1598-65
Lochrovidine	EtOH	226(5.02),297(5.07), 328(5.25)	100-0203-64
Picraphylline	n.s.g.	238(4.14),313(3.98)	100-0456-64
$\Delta^{2,16}$-Strychnene, 19β,20α-diacetoxy-	EtOH	225(3.76),284(3.39), 290(3.38),325(3.55)	78-1141-65
Vallesamine, acetate	EtOH	222(4.63),284(3.93), 292(3.83)	33-2072-64
Voacafrine	n.s.g.	240(4.23),315(4.32)	100-0374-64
Voacangine lactam	EtOH	222(4.33),282(4.00)	12-1279-65
Voacaline	MeOH-acid	230(4.54),285(4.86)	20-0534-65
	MeOH-base	230(4.54),285(4.85)	20-0534-65
$C_{22}H_{26}N_2O_5$			
Africanine	n.s.g.	255(4.0),310s(3.1)	28-3872-64B
Aricine pseudoindoxyl	EtOH	228(4.55),428(3.60)	35-2229-65
Henningsoline	EtOH	223(4.42),256(3.84), 285s(2.9)	39-2812-65
$C_{22}H_{26}O_2$			
1-Anthrol, 8-(benzyloxy)-1β,2,3,4,-4aα,9,9aβ,10-octahydro-10β-methyl-	EtOH	281(3.57)	65-2563-64
Estr-4-en-3-one, 17α-butadiynyl-17β-hydroxy-	MeOH	239(4.24)	13-0801-64
Estr-5(10)-en-3-one, 17α-butadiynyl-17β-hydroxy-	MeOH	252(3.04)	13-0801-64
Gona-1,3,5(10),8-tetraen-17β-ol, 13β-ethyl-17α-ethynyl-3-methoxy-	n.s.g.	278(4.20)	39-4472-64
Gona-1,3,5(10),9(11)-tetraen-17β-ol, 13β-ethyl-17α-ethynyl-3-methoxy-	n.s.g.	263(4.30)	39-4472-64
19-Norandrost-4-en-3-one, 17α-butadiynyl-17β-hydroxy-	EtOH	240(4.23)	78-1197-65
Pregna-4,6,14,16-tetraene-3,20-dione, 16-methyl-	MeOH	282(4.38),300(4.35)	73-2513-64
$C_{22}H_{26}O_3$			
Chrysene, 8α-acetoxy-2-methoxy-10a-methyl-5,6,6a,7,8,9,10,10a-octahydro-, cis	EtOH	228(4.82),254(3.64), 266(3.73),276(3.75), 287(3.58),308(3.06), 315s(3.11),321(3.29), 329(3.22),335(3.40)	23-0591-64

Compound	Solvent	$\lambda_{max}(\log \epsilon)$	Ref.
Estra-1,3,5(10),8,14-pentaene, 3-methoxy-17,17-trimethylenedioxy-	EtOH	310(4.45)	87-0755-64
Estra-1,3,5(10)-trien-17-one, 3-(allyloxy)-16-(hydroxymethylene)-	pH 1	222(4.28),268(4.02)	44-2234-65
	pH 13	222(3.93),305(3.29)	44-2234-65
	MeOH	222(3.93),265(3.91), 305(3.40)	44-2234-65
Gona-1,3,5(10),8,14-pentaen-17-one, 3-methoxy-13-methyl-, cyclic ethylene acetal	EtOH	312(4.49)	39-4472-64
$C_{22}H_{26}O_4$			
9,10-Anthracenedipropanol, 2,6-dimethoxy-	EtOH	233(4.27),260(4.79), 269(5.19),302(3.13), 318(3.40),333(3.65), 351(3.68),373(3.41), 394(3.72),418(3.78)	39-4565-64
Cyclohexene, 4-(benzyloxy)-1-(2,3,4-trimethoxyphenyl)-	EtOH	240(5.0)	39-5189-65
Estra-1,3,5(10),6-tetraen-17-one, 3-acetoxy-, cyclic ethylene acetal	EtOH	262(3.93)	87-0755-64
16,21-Methanopregna-1,4,16-triene-3,20-dione, 11β,21-dihydroxy-	EtOH	244(4.40)	5-0218-65E
19-Nor-5ξ-carda-14,20(22)-dienolide, 3,11-dioxo-	MeOH	211.4(4.26)	44-0527-64
$C_{22}H_{26}O_5$			
Costatolide	EtOH	230(4.33),285(4.38), 328(4.12)	44-3604-64
Costatolide, dihydrooxo-	EtOH	221(4.49),289(4.36), 315(4.22)	44-3604-64
Estra-1,3,5(10)-triene-3,17β-diol, 6β,7β-epoxy-, diacetate	EtOH	215s(3.81),272(2.87), 278(2.90),287s(2.55)	44-1325-64
Estra-1,3,5(10)-trien-17-one, 1,4-diacetoxy-	EtOH	265(2.57),273(2.49)	24-1940-64
Estra-1,3,5(10)-trien-17-one, 3,15α-diacetoxy-	EtOH	268(2.90),275(2.87)	44-2731-64
Stilbene, 2-ethoxy-4,4',5,5'-tetramethoxy-2'-vinyl-	MeOH	230(4.36),261(4.36), 310s(4.27),345(4.48)	35-2177-64
$C_{22}H_{26}O_6$			
Chaparrol, acetate	EtOH	270(2.66),278(2.61)	23-2996-65
Pregna-1,4-diene-2-carboxaldehyde, 17,21-dihydroxy-3,11,20-trioxo-	EtOH-HCl	242(4.15)	44-3481-64
	EtOH	215(4.18)	44-3481-64
	EtOH-NaOH	241(4.18),349(3.99)	44-3481-64
$C_{22}H_{26}O_7$			
Neogmelinol	EtOH	232(3.58),279(3.03)	39-2709-64
Podocarpic acid, 13-hydroxy-7-oxo-, methyl ester, diacetate	EtOH	209(4.35),253(4.08), 293(3.33)	39-0361-65
$C_{22}H_{26}O_8$			
[Bicyclohexyl]-2,2',4,4',6,6'-hexone, 5,5'-diacetyl-1,1',3,3,3',3'-hexamethyl-	EtOH	240(4.33),275(4.31)	39-6543-65
	EtOH-base	275(4.50)	39-6543-65
$C_{22}H_{26}S_2$			
Anthracene, 9,10-bis(butylthio)-	n.s.g.	222(4.03),244s(4.32), 253s(4.57),264(4.88), 352s(3.36),370(3.69), 387(3.87),409(3.90)	39-2571-64

Compound	Solvent	λ_{max}(log ε)	Ref.
$C_{22}H_{27}BrO_4$			
Estra-1,3,5(10)-triene-3,17β-diol, 4-bromo-, diacetate	EtOH	269(2.59),277(2.56)	44-2195-64
	EtOH-base	244(3.86),304(3.87)	44-2195-64
$C_{22}H_{27}BrO_5$			
Estra-1,3,5(10)-triene-3,6α,17-triol, 7α-bromo-, 3,17-diacetate	EtOH	267(2.79),275(2.76)	44-1325-64
Estra-1,3,5(10)-triene-3,6β,17-triol, 7α-bromo-, 3,17-diacetate	EtOH	268(2.77),276(2.76)	44-1325-64
$C_{22}H_{27}ClN_2O_3$			
Macusine C, chloride	H_2O	222(4.63),272(3.88), 278(3.87),289(3.76)	39-4419-64
$C_{22}H_{27}ClO_3$			
9β,10α-Androsta-1,4,6-trien-3-one, 17β-acetoxy-6-chloro-17α-methyl-	EtOH	228(4.08),252(4.04), 300(4.04)	39-0989-65
$C_{22}H_{27}FO_4$			
18-Nor-17α-pregna-1,4,13-triene-3,20-dione, 9-fluoro-11β,21-dihydroxy-16α,17-dimethyl-	MeOH	238(4.18)	44-3486-64
Pregna-4,17(20)-dien-21-oic acid, 4-fluoro-3,11-dioxo-, methyl ester, cis	EtOH	237(4.31)	44-2982-64
Pregna-4,17(20)-dien-21-oic acid, 6α-fluoro-3,11-dioxo-, methyl ester, cis	EtOH	229(4.45)	44-2982-64
6β-isomer	EtOH	227(4.36)	44-2982-64
Pregna-5,17(20)-dien-21-oic acid, 4α-fluoro-3,11-dioxo-, methyl ester, cis	EtOH	222(4.11)	44-2982-64
$C_{22}H_{27}NOS$			
3-Pentanone, 1-phenyl-1-phenylthio-5-piperidino-, hydrochloride	H_2O	260(3.51)	7-1093-65
	EtOH	262(3.58)	7-1093-65
$C_{22}H_{27}NO_2$			
D-Homoandrosta-4,16-diene-16-carbonitrile, 17-methyl-3,17a-dioxo-	MeOH	240.5(4.43)	44-1272-65
$C_{22}H_{27}NO_2S$			
3-Pentanone, 5-morpholino-1-phenyl-1-(p-tolylthio)-, hydrochloride	H_2O	260(3.68)	7-0652-65
	EtOH	262(3.64)	7-0652-65
$C_{22}H_{27}NO_4$			
Pavine methine, N-methyl-	EtOH-HCl	290.5(3.98)	23-2183-65
	EtOH	290.5(3.98)	23-2183-65
$C_{22}H_{27}NO_5$			
Prednisolone, 2-cyano-	EtOH	244(4.00)	44-3300-64
$C_{22}H_{27}NO_6$			
Benzoin, 2-(2-acetamidoethyl)-3',4,4',5-tetramethoxydeoxy-	EtOH	230(4.51),278(4.15)	39-4014-65
Isoquinoline, 1-(α-hydroxy-2'-methyl-3',4'-dimethoxybenzyl)-2-methyl-8-methoxy-6,7-(methylenedioxy)-1,2,3,4-tetrahydro-, α-form	EtOH	282(3.42)	94-1080-64
β-form	EtOH	282.5(3.36)	94-1080-64

Compound	Solvent	λ_{max}(log ε)	Ref.
$C_{22}H_{27}N_3O$			
3-Indazolinone, 1-benzyl-2-(3-piperidinopropyl)-	EtOH-HCl	218(4.32),242(4.08), 319(3.69)	33-1986-64
3-Indazolinone, 2-benzyl-1-(3-piperidinopropyl)-	EtOH-HCl	221(4.28),243(4.03), 314(3.59)	33-1986-64
$C_{22}H_{27}N_5O_6$			
Benzamide, N,N'-[(butylimino)-diethylene]bis[4-nitro-	MeOH	261(4.34)	87-0107-65
$C_{22}H_{27}N_5O_{10}$			
Pteridine, 4-(dimethylamino)-7-(tetra-acetyl-β-D-glucopyranosyloxy)-	MeOH	238(4.19),255(4.07), 352(4.05)	4-0023-64
7(8H)-Pteridinone, 4-(dimethylamino)-8-(tetraacetyl-β-D-glucopyranosyl)-	MeOH	212(4.24),252(4.19), 295(3.78),365(3.91)	4-0023-64
$C_{22}H_{27}N_9O_4$			
Distamycin A	EtOH	237(4.5),303(4.57)	50-1064-64C
$C_{22}H_{28}$			
Benzene, [4-methyl-6-(2,6,6-trimethyl-1-cyclohexen-1-yl)-2,4-cyclohexa-dien-1-yl]-	n.s.g.	259(4.05)	39-2892-65
Biphenyl, 4-methyl-2-(2,2,6-tri-methylcyclohexyl)-	n.s.g.	245(3.84)	39-2892-65
1H-Cyclopent[a]anthracene, 3-ethyl-2,3,3a,4,5,11b-hexahydro-3a,7,11b-trimethyl-	EtOH	234(4.76),269s(3.65), 278(3.77),287(3.80), 296s(3.65),325(2.68)	35-4414-64
1-Decene, 1,1-diphenyl-	C_6H_{12}	251(4.04)	22-0693-64
1,3,5-Hexatriene, 3-methyl-6-phenyl-1-(2,6,6-trimethylcyclohex-1-enyl)-, 3-cis-1,5-ditrans	n.s.g.	338(4.58)	39-2892-65
all trans	n.s.g.	323s(4.59),337(4.69), 345(4.69)	39-2892-65
$C_{22}H_{28}Br_2O_2$			
9β,10α-Pregna-4,6-diene-3,20-dione, 6-(dibromomethyl)-	MeOH	258(3.95),293(3.85)	54-0904-65
9β,10α-Pregn-4-ene-3,20-dione, 6-(dibromomethylene)-	MeOH	250(4.02),285(3.77)	54-0904-65
$C_{22}H_{28}Br_2O_3$			
9β,10α-Androst-4-en-3-one, 17β-acetoxy-6-(dibromomethylene)-	MeOH	250(3.99),285(3.77)	54-0904-65
$C_{22}H_{28}ClNOS$			
3-Pentanone, 1-phenyl-1-phenylthio-5-piperidino-, hydrochloride	H_2O	260(3.51)	7-1093-65
	EtOH	262(3.58)	7-1093-65
$C_{22}H_{28}ClNO_2S$			
3-Pentanone, 5-morpholino-1-phenyl-1-(p-tolylthio)-, hydrochloride	H_2O	260(3.68)	7-0652-65
	EtOH	262(3.64)	7-0652-65
$C_{22}H_{28}Cl_2O_2$			
9β,10α-Pregna-4,6-diene-3,20-dione, 6-(dichloromethyl)-	MeOH	253(3.89),285(3.86)	54-0904-65
9β,10α-Pregn-4-ene-3,20-dione, 6-(dichloromethylene)-	MeOH	241(3.97),272(3.83)	54-0904-65

Compound	Solvent	λ_{max}(log ε)	Ref.
$C_{22}H_{28}Cl_2O_3$			
9β,10α-Androst-4-en-3-one, 17β-acetoxy-6-(dichloromethylene)-	MeOH	242(3.98),271(3.83)	54-0904-65
$C_{22}H_{28}Cl_4O_2$			
Benzo[1",2":3,4;4",5":3',4']dicyclobuta[1,2:1',2']dicyclooctene-7,14-dione, 6b,7a,13b,14a-tetrachloro-eicosahydro-	C_6H_{12}	295(1.65)	88-3471-64
1,12-Cyclodocosa-2,10,13,21-tetraene-dione, 2,11,13,22-tetrachloro-	n.s.g.	246(4.30)	88-3471-64
$C_{22}H_{28}INO_3$			
Atherosperminine, methoxy-, methiodide	EtOH	217(4.52),259(4.80), 284(4.03),296(4.05), 308(4.11)	12-1997-65
Trimethyl[2-(2,3,4-trimethoxy-1-phenanthryl)ethyl]ammonium iodide	EtOH	259(4.80),282(4.09), 294(4.09),306(4.13)	12-1997-65
$C_{22}H_{28}N_2O$			
2-Indolinone, 3-benzyl-3-(2-diethylaminoethyl)-1-methyl-	EtOH-HCl	206s(4.49),255(3.85), 285s(3.15)	87-0626-65
Stilbene, 4-[2-(4-ethyl-1-piperazinyl)ethoxy]-	EtOH	304(4.5),319(4.5)	87-0511-64
$C_{22}H_{28}N_2O_2$			
1H-Cyclopenta[5,6]naphtho[1,2-g]quinazolin-1-one, tetradecahydro-2-(hydroxymethylene)-11a,13a-dimethyl-	EtOH	257(3.94),278(3.98)	32-0338-65
Cyclopenta[7,8]phenanthro[2,3-c]pyrazol-1(2H)-one, dodecahydro-2-(hydroxymethylene)-7,10a,12a-trimethyl-	EtOH	251(3.68),299(3.70)	32-0257-65
$C_{22}H_{28}N_2O_3$			
Alstophyllinol	EtOH	233(4.57),282(3.71), 298(3.78)	33-1349-65
2-Indolinone, 3-benzyl-5,6-dimethoxy-3-(2-dimethylaminoethyl)-1-methyl-	MeOH-HCl	209(4.49),281(3.77), 296s(3.72)	87-0626-65
Kopsinoline	EtOH	205(4.43),244(3.87), 295(3.47)	33-1002-65
Mycelianamide, deoxy-	EtOH	227(4.16),320(4.36)	32-1301-64
$C_{22}H_{28}N_2O_4$			
Catharosine	MeOH	252(3.94),306(3.49)	25-1260-65
Isovenenatine	EtOH	226(4.57),271(3.92), 281s(3.84),293(3.81)	78-2951-65
Isovoacristine	EtOH	227(4.51),297(3.71)	25-2064-65
Rhyncophylline	n.s.g.	243(4.3),280s(3.1)	28-3872-64B
Venenatine	EtOH	226(4.57),271(3.92), 281s(3.84),293(3.81)	78-2951-65
	n.s.g.	226(4.57),271(3.92), 281s(3.84),293(3.81)	88-0901-64
Vincine	50% MeOH-HCl	223(4.44),269(3.84), 291(3.67)	73-0447-64
Vindorosine, deacetyl-	MeOH	252(3.94),306(3.49)	73-1913-64
Voaluteine	EtOH	228(4.46),407(3.56)	12-1279-65
Yohimbine pseudoindoxyl, 1-methyl-	EtOH	232(4.41),260(3.85), 419(3.62)	35-2229-65

Compound	Solvent	$\lambda_{max}(\log \epsilon)$	Ref.
$C_{22}H_{28}N_2O_5$			
Mycelianamide	EtOH	234(4.11),324(4.47)	32-1301-64
Venoxidine	EtOH	224(4.43),268(3.77), 291(3.60)	88-0159-65
$C_{22}H_{28}N_4$			
[17,16-d]Pyrimidino-5α-androstano-[3,2-c]pyrazole	EtOH	223(3.71),251(3.60)	32-0338-65
[17,16-d]Pyrimidino-5α-estrano-1'-methyl[3,2-c]pyrazole	EtOH	231(3.87),251(3.73)	32-0338-65
$C_{22}H_{28}O$			
Cyclohexanol, 2,2,6-trimethyl-1-(4-methylbiphenyl-2-yl)-, cis	n.s.g.	261(2.97),269(2.92), 279(2.77)	39-2892-65
trans	n.s.g.	263(2.79),269(2.79), 279(2.69)	39-2892-65
Unknown diphenylcarbinol from geijerene	EtOH	220(2.98),254(2.78), 260(2.78),280(1.50)	12-0075-64
$C_{22}H_{28}O_2$			
Androst-4-en-3-one, 17α-ethynyl-17-hydroxy-2-methylene-	EtOH	260(4.14)	44-0307-65
Androst-4-en-3-one, 17β-propiolyl-	MeOH	240(4.23)	87-0537-64
5α-Androst-2-en-17-one, 2-propiolyl-	EtOH	223(3.76),259(3.95)	13-0001-64
11H-Benzo[a]fluorene-9-acetaldehyde, 1,2,3,4,6,6aβ,11aα,11b-octahydro-3β-hydroxy-10,11bβ-dimethyl-	EtOH	254(2.64),267(2.53), 276(2.71)	13-0463-64
8α,10α-Bisnorchola-4,6,17(20)-trien-22-al, 3-oxo-	MeOH	258s(4.31),285(4.47)	54-0841-65
9β-Bisnorchola-4,6,17(20)-trien-22-al, 3-oxo-	MeOH	280(4.46)	54-0889-65
Estra-1,3,5(10)-trien-17β-ol, 17α-ethynyl-3-methoxy-2-methyl-	EtOH	282(3.43),286(3.42)	94-0196-64
Gona-4,9-dien-3-one, 13β-ethyl-17β-hydroxy-17-propynyl-	n.s.g.	306(4.29)	39-4472-64
2-Hexene, 3,4-bis(4-methoxy-3-methylphenyl)-	EtOH	232(4.23),277(3.67), 284s(3.63)	39-6509-65
2-Hexene, 3,4-bis(3,5-dimethyl-4-hydroxyphenyl)-	EtOH	235s(4.09),276(3.55)	39-6509-65
2-Hexene, 3-(3,5-dimethyl-4-methoxyphenyl)-4-(4-methoxyphenyl)-	EtOH	230s(4.24),277(3.39), 285(3.26)	39-6509-65
	EtOH	232s(4.16),278s(3.56)	39-6509-65
Pregna-4,14,16-triene-3,20-dione, 16-methyl-	MeOH	237(4.22),307(4.10)	73-2513-64
9β,10α-Pregna-1,4,6-triene-3,20-dione, 6-methyl-	MeOH	227(4.15),250(3.99), 309(4.05)	54-0904-65
Stilbene, α,α'-diethyl-4,4'-dimethoxy-3,3'-dimethyl-	EtOH	210(4.48),239(4.23), 275s(3.82)	39-6509-65
4,4'-Stilbenediol, α,α'-diethyl-3,3',5,5'-tetramethyl-	EtOH	215(4.48),242(4.16), 280s(3.71)	39-6509-65
$C_{22}H_{28}O_2S$			
Estr-4-en-3-one, 17β-hydroxy-17-(2-thienyl)-	EtOH	239(4.31)	78-1197-65
$C_{22}H_{28}O_3$			
Androsta-1,4,6-trien-3-one, 17β-hydroxy-, propionate	EtOH	223(4.04),258(3.99), 300(4.13)	94-0687-65
Estra-1,3,5(10),16-tetraen-17-ol, 3-methoxy-1-methyl-, acetate	EtOH	280(3.21),287(3.22)	94-0687-65

Compound	Solvent	λ_{max}(log ε)	Ref.
Estra-1,3,5(10)-trien-17β-ol, 2-acetyl-, acetate	EtOH	210(4.35),259(4.17)	24-0140-64
Estra-1,3,5(10)-trien-17-one, 3-methoxy-16-methylene-, cyclic ethylene acetal	EtOH	278(3.33),288(3.30)	87-0755-64
Etiojerva-12,14,16-trien-3-one, 17-acetyl-16-methoxy-	MeOH	259(3.44),288(3.44)	44-4220-65
Gona-1,3,5(10),8-tetraen-17-one, 13β-ethyl-3-methoxy-, cyclic ethylene acetal	EtOH-NaOH	278(4.18)	39-4472-64
16,21-Methanopregna-4,21-diene-3,20-dione, 21-hydroxy-	EtOH	243(4.27)	5-0218-65E
19-Norandrost-4-en-6-one, 17α-ethynyl-17β-hydroxy-, acetate	MeOH	245(3.86)	78-2509-65
19-Norpregna-4,9,11-trien-3-one, 20-acetoxy-	EtOH	237(3.72),340(4.39)	28-4545-65A
Pregna-1,4-diene-2-carboxaldehyde, 3,20-dioxo-	EtOH-HCl	247(4.02)	44-3481-64
	EtOH	221(4.08),247(4.14)	44-3481-64
	EtOH-NaOH	242(4.15),348(3.99)	44-3481-64
Pregna-4,6-diene-3,20-dione, 17α-hydroxy-16-methylene-	EtOH	283(4.51)	73-2351-64
Pregna-1,4-diene-3,11,20-trione, 1-methyl-	n.s.g.	244(4.25)	44-0163-64
Pregn-4-ene-3,20-dione, 16α,17α-epoxy-6-methylene-	EtOH	261(4.07)	78-0597-64
$C_{22}H_{28}O_3S$			
Cyclopenta[7,8]phenanthro[10,1-bc]thiophen-3(2H)-one, 1,6,6a,6b,7,8,9,9a,-10,11,11a,11b-dodecahydro-9-hydroxy-4,9a-dimethyl-, acetate	EtOH	222(4.09),268(4.05), 303(3.34)	94-1433-64
$C_{22}H_{28}O_4$			
Androsta-4,16-diene-17α-propionic acid, 3,15-dioxo-	MeOH	236.5(4.42)	94-0859-64
Androsta-1,4,6-trien-3-one, 4,17β-dihydroxy-17α-methyl-, 4-acetate	EtOH	225(4.11),252(3.56), 305(4.08)	7-0288-65
Androst-4-ene-17α-propionic acid, 17β-hydroxy-1,3-dioxo-, lactone	MeOH	243(4.04)	94-0859-64
Androst-4-ene-17α-propionic acid, 17β-hydroxy-3,12-dioxo-, lactone	MeOH	238.5(4.20)	94-0859-64
Androst-4-ene-17α-propionic acid, 17β-hydroxy-3,15-dioxo-, lactone	MeOH	238(4.20)	94-0859-64
Costatolide, dihydrodeoxy-	EtOH	213(4.5),254(3.84), 263(3.96),331(4.17)	44-3604-64
4,4'-Diphenoquinone, 3,3'-di-tert-butyl-5,5'-dimethoxy-	EtOH	219(4.29),261s(3.84), 449(4.43)	39-2914-65
Guaiareic acid, dimethyl-	n.s.g.	211(4.45),259(4.20)	39-4011-64
3-Hexene-1,5-diyne, 3,4-bis(4,4-ethylenedioxycyclohexyl)-, 3-cis	ether	260(4.25),272(4.16)	5-0024-65D
Isoguaiacin, dimethyl-	n.s.g.	283(3.83)	39-4011-64
Pregna-1,4-diene-2-carboxaldehyde, 17β-hydroxy-3,20-dioxo-	EtOH-HCl	247(4.12)	44-3481-64
	EtOH	220(4.05),247(4.12)	44-3481-64
	EtOH-NaOH	242(4.09),348(3.93)	44-3481-64
Pregna-4,16-diene-3,11,20-trione, 21-methoxy-	MeOH	238(4.37)	44-0513-64
$C_{22}H_{28}O_5$			
Cassa-8,11,13-triene-16,19-dioic acid, 7-oxo-, dimethyl ester	EtOH	219(4.3),255(3.95), 305(3.32)	39-0403-65

Compound	Solvent	λ_{max}(log ε)	Ref.
Costatolide, dihydro-	EtOH	215(4.59),251(3.84), 261(3.86),330(4.16)	44-3604-64
Estra-1,3,5(10)-triene-1,3,17β-triol, 1,17-diacetate	EtOH	280s(3.26),283(3.26)	24-1940-64
	NaOH	297(3.49)	24-1940-64
Estra-1,3,5(10)-triene-3,6α,17β-triol, 3,17-diacetate	EtOH	261s(2.75),268(2.85), 276(2.82)	44-1325-64
	EtOH	263s(2.82),267(2.88), 275(2.81)	44-1325-64
6-β-isomer	EtOH	215s(3.99),267(2.70), 275(2.65)	44-1325-64
Estr-4-en-3-one, 10β-acetoxyperoxy-17α-ethynyl-17β-hydroxy-	MeOH	233(4.16)	88-0663-64
Fluorene-8a-acetic acid, 4b,5,6,7,8,8a-hexahydro-1-methyl-7,9-dioxo-, cyclic ethylene acetal, tert-butyl ester	EtOH	248(4.09),297(3.36)	44-2528-65
Galgravin	n.s.g.	231(4.27),279(3.80)	39-4011-64
A-Norpregn-3(5)-ene-1,2,20-trione, 21-acetoxy-	EtOH	283(3.79)	94-0050-65
Pregna-1,4-diene-3,20-dione, 9α,11α-epoxy-17α,21-dihydroxy-16β-methyl-	MeOH	238(4.15)	31-0208-64
Saunderolide, dehydro-, acetate	C_6H_{12}	233(3.78),330(1.70)	33-2330-64
	EtOH	236(3.84),324(1.82)	33-2330-64
$C_{22}H_{28}O_6$			
Estra-1,3,5(10)-triene-3,6β,7α,17β-tetrol, 3,17-diacetate	EtOH	215s(3.97),267(2.76), 275(2.74)	44-1325-64
Pregna-1,4-diene-3,11-dione, 17α,20α,21-trihydroxy-, 21-formate	EtOH	241(4.15)	78-0179-65
$C_{22}H_{28}O_7$			
2-Butanone, 3,3-bis(2,4,5-trimethoxyphenyl)-	EtOH	292.0(4.006)	39-3040-65
Gibb-2-ene-10-carboxylic acid, 2-acetyl-1-carboxy-4a,7-dihydroxy-1-methyl-8-methylene-, methyl ester	n.s.g.	227(3.99)	39-1835-64
A-Norpregn-3(5)-ene-1,2,20-trione, 11β,17α,21-trihydroxy-, 21-acetate	EtOH	282(3.75)	94-0050-65
$C_{22}H_{29}BrN_2O_4$			
5H-Pyrrole-3-carboxylic acid, 5-[4-(4-carbethoxy-3,5-dimethyl-2-pyrrolyl)-but-3-en-2-ylidene]-2,4-dimethyl-, ethyl ester, hydrobromide	EtOH	289(4.05),356(3.74), 562s(4.49),600(5.28)	39-7001-65
$C_{22}H_{29}BrO_3$			
9β,10α-Androsta-4,6-dien-3-one, 17β-acetoxy-6-bromo-17α-methyl-	EtOH	289(4.30)	33-0989-65
A-Homoestra-2,4,5(10)-trien-17β-ol, 3-bromo-4-methoxy-, acetate	EtOH	205(4.27),296(3.60)	22-0906-64
Totaryl acetate, 6α-bromo-7-oxo-	MeOH	258(3.80),290(3.36), 292s(3.34)	44-0501-65
$C_{22}H_{29}BrO_4$			
Podocarpic acid, 6α-bromo-13-isopropyl-O-methyl-7-oxo-, methyl ester	MeOH	235(4.09),257(3.49), 305(3.94)	44-0501-65
	EtOH	242(3.39),306(3.86)	44-0501-65
$C_{22}H_{29}Br_3O_2$			
9β,10α-Pregn-4-ene-3,20-dione, 6β-(tribromomethyl)-	MeOH	241(4.14)	54-0904-65

Compound	Solvent	λ_{max}(log ϵ)	Ref.
$C_{22}H_{29}Br_3O_3$			
9β,10α-Androst-4-en-3-one, 17β-acetoxy-6β-(tribromomethyl)-	MeOH	240(4.11)	54-0904-65
$C_{22}H_{29}ClO_2$			
Androsta-1,3,5-trien-17β-ol, 3-chloro-17α-methyl-, acetate	C_6H_{12}	306(3.77),317(3.79)	44-2495-64
	MeOH	306(3.75),316(3.76)	44-2495-64
Estra-1,3,5(10)-trien-17β-ol, 1-chloro-4,17α-dimethyl-, acetate	MeOH	271(2.40)	44-2495-64
Estra-1,3,5(10)-trien-17β-ol, 1-chloro-4-methyl-, propionate	MeOH	271(2.42)	44-2495-64
Testosterone, 17α-chloroethynyl-3-methoxy-	EtOH	240(4.27)	78-0597-64
$C_{22}H_{29}ClO_3$			
Androsta-5,16-diene-16-carboxaldehyde, 3β-acetoxy-17-chloro-	EtOH	261(4.08)	88-1839-64
	ether	254(4.13),259(4.06)	39-0788-65
9β,10α-Androsta-4,6-dien-3-one, 17β-acetoxy-6-chloro-17α-methyl-	EtOH	286(4.34)	33-0989-65
Pregn-4-ene-3,20-dione, 2β-chloro-1α,11α-epoxy-2α-methyl-	EtOH	246(4.01)	32-0151-65
Testosterone, 4-chloro-17α-methyl-, acetate	MeOH	254(4.15)	94-1217-64
$C_{22}H_{29}ClO_4$			
Androsta-1,4-dien-3-one, 2-chloro-11α,17β-dihydroxy-17α-methyl-, 11-acetate	EtOH	251(4.22)	32-0138-65
$C_{22}H_{29}ClO_4S_3$			
Spiro[benzofuran-2(3H),1'-[2]cyclohexene]-3,4'-dione, 7-chloro-4,6-dimethoxy-6'-methyl-, 4'-(diethyl mercaptole)	EtOH	288(4.35)	87-0705-64
$C_{22}H_{29}ClO_5$			
Pregn-4-ene-3,20-dione, 2β-chloro-1α,11α-epoxy-17α,21-dihydroxy-2α-methyl-	EtOH	245(3.99)	32-0151-65
$C_{22}H_{29}Cl_3O_2$			
9β,10α-Pregn-4-ene-3,20-dione, 6-(trichloromethyl)-	MeOH	238(4.14)	54-0904-65
$C_{22}H_{29}Cl_3O_3$			
9β,10α-Androst-4-en-3-one, 17β-acetoxy-6β-(trichloromethyl)-	MeOH	238(4.16)	54-0904-65
$C_{22}H_{29}IN_2O_3$			
N-Methylvincamininium iodide	MeOH	223(4.61),269(3.91)	73-0433-64
$C_{22}H_{29}NO$			
Estra-1,3,5(10)-trien-17-one, 3-(N-pyrrolidinyl)-	MeOH	258(4.26),312(3.45)	44-2047-65
3-Heptanone, 6-(dimethylamino)-4-phenyl-4-p-tolyl-, hydrochloride	H_2O	289(2.89)	35-5277-64
Stilbene, 4-diisopropylaminoethoxy-	EtOH	304(4.5),319(4.5)	87-0511-64
$C_{22}H_{29}NO_2$			
Estra-1,3,5(10)-triene, 17β-(2-cyanoethoxy)-3-methoxy-	EtOH	279(3.27),287(3.22)	13-0557-65

Compound	Solvent	λ_{max}(log ϵ)	Ref.
3-Heptanone, 6-(dimethylamino)-4-(p-methoxyphenyl)-4-phenyl-, hydrochloride	H_2O	291(3.13)	35-5277-64
D-Homopregna-5,16-diene-16-carbonitrile, 3β-hydroxy-17-methyl-17a-oxo-	MeOH	247(4.03)	44-1272-65
2α-Phenanthrenecarboxamide, N-cyclohexyl-1,2,3,9,10,10aβ-hexahydro-7-methoxy-	EtOH	264(4.31),298(3.51)	44-2849-65
$C_{22}H_{29}NO_3$			
Estra-1,3,5(10)-trien-17β-ol, 2-acetamido-, acetate	EtOH	208(4.50),249(4.19), 288(3.06)	24-0140-64
Estra-1,3,5(10)-trien-17β-ol, 2-acetyl-, acetate, oxime	EtOH	211(4.45),252(4.13)	24-0140-64
Pregn-4-ene-16α-carbonitrile, 17α-hydroxy-3,20-dioxo-	MeOH	240(4.20)	44-1272-65
Pregn-4-ene-17α-carbonitrile, 16α-hydroxy-3,20-dioxo-	MeOH	240(4.19)	5-0228-65E
$C_{22}H_{29}NO_4$			
Laudanosomethine base	MeOH	295s(4.3),334(4.5)	83-0129-64
Pavine methine, dihydro-N-methyl-	EtOH-HCl	225s(4.22),286(3.78)	23-2183-65
$C_{22}H_{29}NO_5$			
17a-Aza-D-homo-19-norandrosta-3,5-dien-17-one, 3,17a-diacetoxy-	MeOH	234(4.30)	78-0743-65
O-Ethomostephanoline methine	EtOH	270(3.08)	95-0584-65
$C_{22}H_{29}NO_6$			
Heteratisone, oxo-	n.s.g.	310(1.48)	88-0669-64
$C_{22}H_{29}N_3O_2$			
Dipyrromethene-5'-carboxylic acid, 3,3'-diethyl-4,4'-dimethyl-5-(1-pyrrolin-2-yl)-, ethyl ester	EtOH-HCl	275(4.26),345(4.47)	39-2614-65
	EtOH	224(4.20),255(4.36), 261(4.43),446(4.42)	39-2614-65
$C_{22}H_{29}N_3O_2S$			
Estra-1,3,5(10)-triene-2-carboxaldehyde, 3-methoxy-17-oxo-, 16-(2-methyl-3-thiosemicarbazone)	pH 1	222(4.08),280(3.75), 288(3.71)	44-2234-65
	pH 13	222(4.11),280(3.68), 288(3.68)	44-2234-65
	MeOH	230(4.05),242(3.98), 281(4.32),300s(4.15)	44-2234-65
	MeCN	280(4.50)	44-2234-65
	$CHCl_3$	284(4.52)	44-2234-65
$C_{22}H_{29}N_3O_3$			
16,21-Methanopregna-1,4-diene-3,20,21-trione, trioxime	EtOH	241.5(4.38)	5-0218-65E
$C_{22}H_{29}N_3O_4$			
16,21-Methanopregna-1,4-diene-3,20,21-trione, 11β-hydroxy-, trioxime	EtOH	246(4.35)	5-0218-65E
$C_{22}H_{29}N_7$			
Isobiguanide, 1,2-bis(isopropylideneamino)-3,4-di-p-tolyl-	EtOH	240(4.48),274(4.50)	39-3912-65
$C_{22}H_{30}$			
Decane, 1,1-diphenyl-	C_6H_{12}	256(2.83),259(2.83), 262(2.84),269(2.71)	22-0587-65

Compound	Solvent	λ_{max}(log ε)	Ref.
$C_{22}H_{30}B_2N_2$			
1,4-Diaza-2,5-diborine, 2,3,5,6-tetraethyl-1,4-diphenyl-	EtOH	230(4.23)	88-0703-65
$C_{22}H_{30}ClNO$			
3-Heptanone, 6-(dimethylamino)-4-phenyl-4-p-tolyl-, hydrochloride	H_2O	289(2.89)	35-5277-64
$C_{22}H_{30}ClNO_2$			
3-Heptanone, 6-(dimethylamino)-4-(p-methoxyphenyl)-4-phenyl-, hydrochloride	H_2O	291(3.13)	35-5277-64
$C_{22}H_{30}ClNO_8$			
4,5-Dimethoxy-2-(3,4-dimethoxy-α-methylstyryl)-α-benzyldimethylammonium perchlorate	MeOH	265(4.8)	83-0129-64
$C_{22}H_{30}Cl_2O_3$			
Testosterone, 2α,4-dichloro-17α-methyl-, acetate	MeOH	261(4.05)	94-1217-64
$C_{22}H_{30}F_2O_2$			
Vitamin A acid, 10,12-difluoro-, ethyl ester	MeOH	353(4.52)	5-0021-65A
Vitamin A acid, 12,13'-difluoro-, ethyl ester	MeOH	354(4.50)	5-0021-65A
$C_{22}H_{30}N_2$			
Aniline, N-isopentylidene-, dimer	MeOH	242(4.25),276(4.00), 298s(3.51)	65-0356-64
$C_{22}H_{30}N_2O_2$			
Estrone, 16-[(2,2-dimethylhydrazinomethylene]-, methyl ether	pH 1	220(3.97),278(3.96), 287(3.96)	44-2234-65
	pH 13	220(3.97),309(4.31)	44-2234-65
	MeOH	220(3.97),309(4.33)	44-2234-65
1,4-Naphthalenediol, 2,3-bis-(piperidinomethyl)-	3N HCl	218(4.45),245(4.42), 338(3.62)	39-0042-64
	ether	252(4.47),333(3.76), 346(3.79)	39-0042-64
$C_{22}H_{30}N_2O_3$			
Echitinolide, α,β-dihydro-N^a-methyl-	0.05N HCl	243(3.96),299(3.53)	33-1598-65
	EtOH	257(4.08),315(3.62)	33-1598-65
Isoschizogamine, tetrahydro-N-methyl-	MeOH	213(4.50),262(4.13), 291(3.84)	100-0406-64
Schizozygine methine, hexahydro-N,N-dimethyl-	MeOH	265(4.04),311(3.99)	33-0308-65
$C_{22}H_{30}N_2O_4$			
Bicyclo[3.2.1]octane-2,4-dione, 3,3'-azobis[1,8,8-trimethyl-	CH_2Cl_2	265(3.96),426(4.30)	5-0214-65G
Erininediol, 19,20-dihydro-	0.05N HCl	244(3.90),301(3.47)	33-1598-65
	EtOH	256(4.05),316(3.61)	33-1598-65
Pyrrole-2-carboxylic acid, 5,5'-vinylenebis[4-ethyl-3-methyl-, diethyl ester	EtOH	273(4.34),383(4.32)	39-4385-65
Spiro[5.5]undecane-2,4-dione, 3,3'-azobis-	CH_2Cl_2	264(3.97),428(4.31)	5-0214-65G

Compound	Solvent	λ max(log ε)	Ref.
$C_{22}H_{30}N_4$			
2H-Pyrazolo[3",4":3',4']cyclopenta-[1',2':5,6]naphth[1,2-f]indazole, tetradecahydro-2,6a,8-trimethyl-	MeOH	231(4.04)	32-0257-65
$C_{22}H_{30}N_4O_4$			
3,5-Hexadien-2-one, 6-p-3-menthyl-, 2,4-dinitrophenylhydrazone	$CHCl_3$	399(4.69)	28-4712-64B
$C_{22}H_{30}N_4O_7S$			
6-Purinethiol, 9-(2,3,5-tri-O-butyryl-β-D-ribofuranosyl)-	pH 1	320(4.35)	87-0200-64
	pH 11	237(4.02),309(4.32)	87-0200-64
6-Purinethiol, 9-(2,3,5-tri-O-isobutyryl-β-D-ribofuranosyl)-	pH 1	320(4.47)	87-0200-64
	pH 11	236(4.20),310(4.45)	87-0200-64
$C_{22}H_{30}N_5O_4P$			
Phosphorane, trimorpholino[(1-oxo-2(1H)-naphthylidene)hydrazono]-	MeOH	330(3.0),470(3.2)	5-0056-64I
$C_{22}H_{30}O$			
Estra-1,3,5(10)-trien-17β-ol, 17-allyl-4-methyl-	MeOH	263(2.37)	87-0409-65
$C_{22}H_{30}O_2$			
Androsta-1,4-dien-3-one, 17β-hydroxy-17α-methyl-19-vinyl-	EtOH	248(4.13)	33-0094-65
5α-Androstan-17-one, 3α-hydroxy-2-propynylidene-	EtOH	228(4.22)	13-0001-64
Androsta-3,5,7-trien-17-one, 3-ethoxy-6-methyl-	EtOH	323(4.27)	78-1753-65
Androst-4-ene-3,17-dione, 16α-allyl-	EtOH	240(4.21)	32-0351-65
Androst-4-ene-3,17-dione, 16β-allyl-	EtOH	241(4.21)	32-0351-65
5α-Androst-2-en-17β-ol, 2-propynoyl-	EtOH	259(3.97)	13-0001-64
8α,10α-Bisnorchola-4,6-dien-22-al, 3-oxo-	MeOH	288(4.40)	54-0841-65
9β-Bisnorchola-4,6-dien-22-al, 3-oxo-	MeOH	284(4.42)	54-0889-65
9β-Bisnorchola-4,17(20)-dien-22-al, 3-oxo-	MeOH	251.5(4.44)	54-1069-64
Estra-1,3,5(10)-trien-17β-ol, 17α-cyclopropyl-3-methoxy-	MeOH	206(4.24),218s(3.94), 228s(3.70),273s(3.29), 278(3.34),287(3.29)	24-1470-65
Gona-1,3,5(10),8-tetraen-17β-ol, 13β,17α-diethyl-3-methoxy-	n.s.g.	278(4.20)	39-4472-64
Gona-1,3,5(10),9(11)-tetraen-17β-ol, 13β,17α-diethyl-3-methoxy-	n.s.g.	265(4.20)	39-4472-64
19-Norpregna-1,3,5(10)-trien-20-one, 3-ethoxy-	EtOH	280(3.34),287(3.28)	13-0013-64
9β,10α-Pregna-1,4-diene-3,20-dione, 6β-methyl-	MeOH	242(4.19)	54-0904-65
9β,10α-Pregna-4,6-diene-3,20-dione, 6-methyl-	MeOH	291(4.36)	54-0904-65
Pregna-5,14,16-trien-20-one, 3-hydroxy-16-methyl-	MeOH	310(4.13)	73-2513-64
5,10-Seco-5,19-cycloandrosta-1(10),2,4-trien-17β-ol, 3-methyl-, acetate	EtOH	216(4.35),257(3.79)	35-3727-65
$C_{22}H_{30}O_2S$			
Estra-1,3,5(10)-trien-17β-ol, 3-methoxy-16-S,16-O-isopropylidene-16β-mercapto-	EtOH	279(3.33),287(3.30)	94-0905-64

Compound	Solvent	λ_{max}(log ε)	Ref.
$C_{22}H_{30}O_3$			
Androsta-1,4-dien-3-one, 17β-hydroxy-, propionate	EtOH	245(4.23)	94-0050-65
Androsta-1,4-dien-3-one, 17β-hydroxy-17α-methyl-, acetate	MeOH	244(4.21)	13-0183-64
Androsta-1,5-dien-3-one, 17β-hydroxy-17α-methyl-, acetate	MeOH	226(4.04)	13-0183-64
Androsta-3,5-dien-7-one, 17β-hydroxy-17α-methyl-, acetate	MeOH	280(4.36)	31-0249-64
Androsta-3,5-dien-17-one, 3-hydroxy-6-methyl-, acetate	EtOH	245(4.28)	35-1977-64
Androsta-5,15-dien-17-one, 3β-hydroxy-16-methyl-, acetate	EtOH	240(3.85)	73-1173-64
9β,10α-Androsta-4,6-dien-3-one, 17β-hydroxy-6-methyl-, acetate	MeOH	290(4.37)	54-0904-65
Androst-4-en-3-one, 17α-ethynyl-17-hydroxy-2α-(hydroxymethyl)-	EtOH	242(4.16)	44-0307-65
Chrysene, 8α-acetoxy-4b,5,6,6a,7,8,-9,10,10a,10b,11,12-dodecahydro-2-methoxy-10a-methyl-, cis-syn-cis	EtOH	221(3.89),280(3.29), 287(3.29)	23-0591-64
Desoxycorticosterone, 6-methylene-	EtOH	260(4.05)	78-0597-64
Estra-1,3,5(10)-trien-17-one, 3-ethoxy-, cyclic ethylene acetal	EtOH	280(3.23),288(3.18)	87-0755-64
Etiojerva-12,14,16-triene-3β,16-diol, 17-acetyl-, 16-methyl ether	MeOH	287(3.44),360(3.44)	44-4220-65
Gona-1,3,5(10)-triene, 13β-ethyl-17,17-ethylenedioxy-3-methoxy-	n.s.g.	278(3.30)	39-4472-64
3-Hexanol, 3,4-bis(4-methoxy-3-methylphenyl)-	EtOH	209(4.36),230(4.37), 277(3.62)	39-6509-65
3-Hexanol, 3-(3,5-dimethyl-4-methoxyphenyl)-4-(4-methoxyphenyl)-	EtOH	225(4.39),276(3.47), 285(3.32)	39-6509-65
D-Homo-1,3,5(10)-estratrien-17-one, 3-methoxy-, cyclic ethylene acetal	EtOH	279(3.33),287(3.29)	87-0755-64
16,21-Methanopregn-5-en-20-one, 3β,21-dihydroxy-	EtOH	262(3.86)	5-0218-65E
Phenanthrene, 1,2,3,4-tetrahydro-7-isopropyl-5,6,7-trimethoxy-1,1-dimethyl-	C_6H_{12}	234s(4.60),243(4.66), 292s(3.77),303(3.82), 317(3.68),331(3.49)	33-1234-64
9β,10α-Pregna-4,6-diene-3,20-dione, 6-methoxy-	MeOH	247(3.88),307(4.19)	54-0918-65
Pregna-3,5-dien-20-one, 16α,17α-epoxy-3-methoxy-	EtOH	240(4.30)	78-0597-64
9β,10α-Pregna-4,6-dien-3-one, 20α-hydroxy-, 20-formate	MeOH	287(4.42)	54-0853-65
Pregna-5,14,16-trien-20-one, 3β,21-dihydroxy-16-methyl-	MeOH	310(4.05)	73-2513-64
5α-Pregn-1-ene-2-carboxaldehyde, 3-oxo-	EtOH-HCl	236(3.95)	44-3481-64
	EtOH	241(3.89)	44-3481-64
	EtOH-NaOH	304(4.23)	44-3481-64
	dioxan	245(3.98)	44-3481-64
Pregn-4-ene-3,20-dione, 2-(hydroxymethylene)-	EtOH	253(4.08),307(3.80)	44-3481-64
Pregn-4-ene-3,20,21-trione, 21-methyl-	EtOH	241(4.17)	94-1181-64
5,10-Seco-5,19-cycloandrosta-1(10),2,4-trien-17β-ol, 3-methoxy-, acetate	EtOH	257(3.78)	35-3727-65
Testosterone, 6-methylene-, acetate	EtOH	260(4.06)	78-0597-64
$C_{22}H_{30}O_4$			
Androsta-1,4-dien-3-one, 4,17β-dihydroxy-17α-methyl-, 4-acetate	EtOH	246(4.11)	7-0288-65

Compound	Solvent	$\lambda_{max}(\log \epsilon)$	Ref.
Androsta-4,6-dien-3-one, 4,17β-dihydroxy-17α-methyl-, 4-acetate	EtOH	289(4.33)	7-0288-65
Androst-4-ene-3,11-dione, 17β-(acetoxymethyl)-	MeOH	239(4.20)	39-0586-64
Androst-4-ene-17α-propionic acid, 1β,17β-dihydroxy-3-oxo-, lactone	MeOH	240(4.17)	94-0859-64
Androst-4-ene-17α-propionic acid, 2β,17β-dihydroxy-3-oxo-, lactone	MeOH-KOH	242(4.11)	94-0859-64
Androst-4-ene-17α-propionic acid, 6β,17β-dihydroxy-3-oxo-, lactone	MeOH	235.5(4.14)	94-0859-64
Androst-4-ene-17α-propionic acid, 9α,17β-dihydroxy-3-oxo-, lactone	MeOH	241(4.20)	94-0859-64
Androst-4-ene-17α-propionic acid, 12β,17β-dihydroxy-3-oxo-, lactone	MeOH	240(4.22)	94-0859-64
Androst-4-ene-17α-propionic acid, 14α,17β-dihydroxy-3-oxo-, lactone	MeOH	240.5(4.19)	94-0859-64
Androst-4-ene-17α-propionic acid, 15β,17β-dihydroxy-3-oxo-, lactone	MeOH	240(4.20)	94-0859-64
Biphenyl, 2,2'-dihydroxy-3,3'-dimethoxy-5,5'-di-tert-butyl-	EtOH	218(4.73),281(3.70)	39-2914-65
Biphenyl, 4,4'-dihydroxy-3,3'-dimethoxy-5,5'-di-tert-butyl-	EtOH	218(4.63),272(4.22), 289s(4.15)	39-2914-65
Cassa-8,11,13-triene-16,19-dioic acid, dimethyl ester	EtOH	266(2.7)	39-0403-65
Guairetic acid, dihydrodimethyl-	n.s.g.	229(4.30),281(3.79)	39-4011-64
19-Nor-5α-androsta-1(10),2-diene, 3,17β-diacetoxy-	MeOH	277(3.89)	24-3165-65
19-Norpregn-4-ene-3,20-dione, 10β-acetoxy-	EtOH	243(4.18)	13-0013-64
Podocarpic acid, 13-isopropyl-O-methyl-7-oxo-, methyl ester	n.s.g.	230(4.13),280(4.10)	78-0409-64
Pregn-4-ene-3,20-dione, 8,19-epoxy-19-methoxy-	EtOH	242(4.15)	44-0351-64
Tauranin	MeOH	266(4.07),415(3.07)	94-0796-64
Testosterone, 1-dehydro-7α-hydroxy-17α-methyl-, 7-acetate	EtOH	244(4.19)	24-3363-64
$C_{22}H_{30}O_5$			
Cortexone, 8,19-epoxy-19-methoxy-	EtOH	242(4.23)	44-0345-64
Cortisone, 6α-methyl-	EtOH	236.5(4.16)	39-0148-65
Estr-4-en-3-one, 1β,17β-diacetoxy-	EtOH	240(4.22)	54-0626-65
Eti-4-enic acid, 8,19-epoxy-19-methoxy-3-oxo-, methyl ester	EtOH	242(4.21)	44-0357-64
Hydrocortisone, 2-methylene-	EtOH	261(4.16)	87-0528-64
A-Norandrost-3(5)-en-2-one, 1β,17β-diacetoxy-	EtOH	235(4.18)	94-0156-65
A-Norandrost-3-en-2-one, 3,17β-diacetoxy-	EtOH	240(4.18)	44-1325-65
19-Norcortexone, 8-acetoxy-	EtOH	242(4.23)	44-0345-64
Pregn-4-ene-3,11,20-trione, 17-hydroxy-21-methoxy-	MeOH	238(4.21)	44-0513-64
5α-Pregn-17(20)-en-21-oic acid, 3β,16β-dihydroxy-11-oxo-, γ-lactone, 3-acetate	EtOH	217(4.16)	87-0348-64
Saunderolide, acetate	EtOH	284(1.43)	33-2330-64
$C_{22}H_{30}O_6$			
4-Etiocholenic acid, 17α-acetoxy-14α-hydroxy-3-oxo-	MeOH	240(4.20)	44-2776-65
Unsaturated ketolactone, acetate	EtOH	255(4.16)	78-3091-65

Compound	Solvent	λ_{max}(log ε)	Ref.
$C_{22}H_{30}O_6S$			
5α-Pregnane-5α-carboxylic acid, 11β,17α,21-trihydroxy-3,20-dioxo-9α-mercapto-, γ-(thio lactone)	EtOH	241(3.55)	13-0305-64
$C_{22}H_{31}BrO_3$			
9β,10α-Androst-4-en-3-one, 17β-acetoxy-6β-bromo-17α-methyl-	EtOH	236(4.11)	33-0989-65
Pregna-3,5-dien-20-one, 16β-bromo-17α-hydroxy-3-methoxy-	EtOH	240(4.31)	78-0597-64
$C_{22}H_{31}BrO_4$			
Testosterone, 6β-bromo-7α,17-dihydroxy-, 17-propionate	EtOH	248(4.17)	24-3363-64
Testosterone, 2α-bromo-4-hydroxy-17α-methyl-, 4-acetate	EtOH	251(4.15)	7-0288-65
$C_{22}H_{31}BrO_5$			
16α-Kaurane-17,19-dioic acid, 2β-bromo-1-oxo-, dimethyl ester	EtOH	292(1.49)	12-2005-65
$C_{22}H_{31}ClN_4O_2$			
2-Octanone, 8-[p-[[3-(2-amino-4-hydroxy-5-methyl-5-pyrimidinyl)-propyl]amino]phenyl]-1-chloro-	pH 1	265(3.89)	87-0035-65
	pH 7	239(4.05)	87-0035-65
	pH 13	279(3.92)	87-0035-65
$C_{22}H_{31}ClN_4O_3$			
2-Hexanone, 6-[p-[(2-amino-4-hydroxy-6-methyl-5-pyrimidinyl)-3-propyl]-amino]phenyl-1-chloro-, ethylene ketal	pH 1	265(3.94)	87-0035-65
	pH 7	241(4.22),292s(--)	87-0035-65
	pH 13	241(4.26),278(3.96)	87-0035-65
$C_{22}H_{31}ClO_3$			
5α-Androst-2-ene-2-carboxaldehyde, 3-chloro-17β-hydroxy-, acetate	EtOH	260(4.02)	88-1839-64
5α-Androst-3-ene-4-carboxaldehyde, 3-chloro-17β-hydroxy-, acetate	EtOH	255(3.81)	88-1839-64
Androst-4-en-3-one, 4-chloro-17β-(hydroxymethyl)-, acetate	n.s.g.	256(4.1)	31-0688-65
Androst-5-en-7-one, 6-chloro-17β-hydroxy-17α-methyl-, acetate	MeOH	254(4.11)	31-0249-64
9β,10α-Androst-4-en-3-one, 6β-chloro-17β-hydroxy-17α-methyl-, acetate	EtOH	235(4.18)	33-0989-65
Pregn-4-en-3-one, 2β-chloro-1α,11α-epoxy-20β-hydroxy-2α-methyl-	EtOH	247(3.98)	32-0151-65
$C_{22}H_{31}ClO_4$			
Testosterone, 6α-chloro-4-hydroxy-17α-methyl-, 4-acetate	EtOH	242(4.06)	7-0288-65
Testosterone, 6β-chloro-4-hydroxy-17α-methyl-, 4-acetate	EtOH	245(4.07)	7-0288-65
$C_{22}H_{31}FO_2$			
Vitamin A, 10-fluoro-, acetate	MeOH	324(4.52)	5-0020-64I
Vitamin A, 14-fluoro-, acetate	MeOH	326(4.68)	5-0020-64I
Vitamin A acid, 10-fluoro-, ethyl ester	MeOH	352(4.58)	5-0020-64I
Vitamin A acid, 12-fluoro-, ethyl ester	MeOH	352(4.55)	5-0021-65A
Vitamin A acid, 13'-fluoro-, ethyl ester	MeOH	355(4.60)	5-0021-65A
Vitamin A acid, 14-fluoro-, ethyl ester	MeOH	356(4.52)	5-0020-64I

Compound	Solvent	λ_{max}(log ε)	Ref.
$C_{22}H_{31}FO_3$			
9β,10α-Androst-4-en-3-one, 6α-fluoro-17β-hydroxy-17α-methyl-, acetate	EtOH	230(4.12)	33-0989-65
6β-fluoro isomer	EtOH	233(4.23)	33-0989-65
B-Homoestr-4-en-3-one, 7-fluoro-17β-hydroxy-, propionate	EtOH	238(4.20)	44-4160-65
$C_{22}H_{31}IN_2O$			
Normacusine B, dihydro-N[a],O-dimethyl-, methiodide	EtOH	223(4.73),276(3.87), 282(3.88),291s(3.78)	39-4419-64
$C_{22}H_{31}NO$			
Androstano[3,2-b]pyridine, 17β-hydroxy-	EtOH	263(3.67),269(3.76), 277(3.64)	94-0077-64
Cona-4,6-dienin-3-one	EtOH	286(4.36)	22-2169-64
Estra-1,3,5(10)-trien-17β-ol, 3-(N-pyrrolidinyl)-	MeOH	256(4.25),312(3.44)	44-2047-65
Estra-1,3,5(10)-trien-17-one, 3-(diethylamino)-	MeOH	263(4.16),310(3.33)	44-2047-65
Funtudienine	EtOH	280(4.34)	22-2169-64
$C_{22}H_{31}NOS$			
Androstano[3,2-b]pyridine-6'(1'H)-thione, 17β-hydroxy-	EtOH	282(4.10),373(3.83)	94-0077-64
$C_{22}H_{31}NO_2$			
Androstano[3,2-b]pyridin-6'(1'H)-one, 17β-hydroxy-	EtOH-HCl	219(3.75),297(4.02)	94-0077-64
	EtOH	230(3.96),316(3.96),	94-0077-64
	EtOH-KOH	233(4.03),307(3.90)	94-0077-64
5α-Androst-2-ene-2-carbonitrile, 17β-acetoxy-	C_6H_{12}	208.5(3.96)	44-3300-64
Atisone	EtOH	229(3.98)	35-0777-65
1H,7H-6a,9-Ethano-4,10b-propanobenz-[h]isoquinolin-7-one, 2-acetyl-decahydro-4-methyl-8-methylene-	EtOH	228s(3.96)	35-0777-65
Isoatisone	EtOH	227.5(3.91)	35-0777-65
$C_{22}H_{31}NO_3$			
3H,7H-6a,9-Ethano-4,10b-propanobenz-[h]isoquinoline-3,7-dione, decahydro-2-(2-hydroxyethyl)-4-methyl-8-methylene-	EtOH	227s(3.93)	35-0777-65
$C_{22}H_{31}NO_4$			
2H-1,3-Benzoxazine-2,4(3H)-dione, 3-myristoyl-	MeOH	206(4.56),238(3.94), 290(--)	4-0037-65
Isopyroheteratisine	EtOH	327(2.06)	88-0215-65
Pyroheteratisine	EtOH	238(4.03)	88-0215-65
$C_{22}H_{31}NO_5$			
Heteratisone	n.s.g.	270(1.84)	88-0669-64
Pregn-4-ene-3,11-dione, 17,21-dihydroxy-20-methoxyimino-	EtOH	235.5(4.22)	39-0148-65
$C_{22}H_{31}N_3O_2$			
Dipyrromethane-5'-carboxylic acid, 3,3'-diethyl-4,4'-dimethyl-5-(1-pyrrolin-2-yl)-, ethyl ester	EtOH-HCl	278(4.32),339(4.48)	39-2614-65
	EtOH	246(3.98),280(4.36), 299(4.30),339(3.72)	39-2614-65

Compound	Solvent	λ max(log ε)	Ref.
$C_{22}H_{31}N_3O_2S$			
Naphth[2',1':4,5]indeno[1,2-c]pyrazole-7(2H)-carboxamide, tetradecahydro-6b-hydroxy-N,4a,6a-trimethyl-2-oxothio-	MeOH	243(4.36),271(4.22)	44-2234-65
	pH 1	242(4.35),271(4.15)	44-2234-65
	pH 13	240(4.29),366(4.45)	44-2234-65
$C_{22}H_{31}N_3O_4$			
A-Norandrost-3(5)-ene-1,2-dione, 17β-hydroxy-, 2-semicarbazone, 17-propionate	EtOH	245(3.84),322(4.33)	94-0156-65
$C_{22}H_{31}N_3O_{14}S$			
Anhydroepitetrodotoxin, tetra-acetyl-, p-toluenesulfonate	H_2O	220(4.08),260(2.96)	78-2059-65
$C_{22}H_{31}N_5O_7S$			
6-Purinethiol, 2-amino-9-(2,3,5-tri-butyryl-β-D-ribofuranosyl)-	pH 1	264(3.93),342(4.25)	87-0200-64
	pH 11	252(4.12),318(4.30)	87-0200-64
6-Purinethiol, 2-amino-9-(2,3,5-tri-isobutyryl-β-D-ribofuranosyl)-	pH 1	264(3.90),344(4.38)	87-0200-64
	pH 11	250(4.12),317(4.33)	87-0200-64
$C_{22}H_{31}O_5P$			
Estrone, diethyl phosphate	MeOH	269(2.99),277(2.96)	87-0409-65
$C_{22}H_{32}$			
Naphthalene, 2,6-dihexyl-	C_6H_{12}	230(5.24),274(3.79), 303(2.78),309(3.00), 317(2.78),324(3.11)	39-2324-65
$C_{22}H_{32}N_2O$			
5α-Androstano[3,2-d]-2'-methyl-pyrimidine, 17β-hydroxy-	EtOH	260(3.65)	87-0238-64
5α-Androstano[3,2-d]pyrimidine, 17β-hydroxy-17α-methyl-	EtOH	254(3.61)	87-0238-64
Cona-4,6-dienin-3-one, oxime	EtOH	278(4.37)	22-2169-64
Funtudienine, oxime	EtOH	278(4.30)	22-2169-64
$C_{22}H_{32}N_2O_2$			
Androst-5-eno[16,17-c]-5'-methyl-pyrazole, 3β-acetoxy-	MeOH	225(3.82)	44-1142-64
$C_{22}H_{32}N_2O_3$			
Androstano[2,3-c]furazan, 17β-acetoxy-17α-methyl-	EtOH	217(3.66)	94-1445-65
Androstano[2,3-c]furazan, 17β-pro-pionyloxy-	EtOH	217(3.69)	94-1445-65
2-Indolinone, 3-[1-methyl-1-(nitro-methyl)-2-(2,2,6-trimethylcyclo-hexyl)ethyl]-	EtOH	250(3.86),283(3.13)	87-0626-65
$C_{22}H_{32}N_2O_4$			
Androstano[2,3-c]furazan, 17-acetoxy-17-methyl-, N-oxide	EtOH	264(3.84)	94-1445-65
Androstano[2,3-c]furazan, 17β-pro-pionyloxy-, N-oxide	EtOH	264(3.83)	94-1445-65
Pyrrole-2-carboxylic acid, 5,5'-ethyl-enebis[4-ethyl-3-methyl-, diethyl ester	EtOH	290(4.54)	39-4385-65

Compound	Solvent	$\lambda_{max}(\log \epsilon)$	Ref.
$C_{22}H_{32}N_2O_5$			
11bH-Benzo[a]quinolizine, 2-acetoxy-3-diethylcarbamoyl-9,10-dimethoxy-1,2,3,4,6,7-hexahydro-	MeOH-HCl	284(3.59)	87-0635-64
$C_{22}H_{32}N_4O_2$			
p-Benzoquinone, tetrapyrrolidino-	methyl cellosolve	253(4.18),359(4.10), 565s(2.80)	78-1889-64
$C_{22}H_{32}N_4O_6$			
p-Benzoquinone, tetramorpholino-	methyl cellosolve	241(4.25),399(3.99), 595s(2.64)	78-1889-64
$C_{22}H_{32}O_2$			
Androsta-3,5-dien-17β-ol, propionate	EtOH	228(4.33),234(4.36), 243(4.16)	44-0068-64
Androsta-4,6-dien-17β-ol, propionate	EtOH	230(4.32),238(4.35), 246(4.14)	44-0068-64
Androsta-1,4-dien-3-one, 19-ethyl-17β-hydroxy-17α-methyl-	EtOH	247(4.20)	33-0094-65
Androsta-3,5-dien-17-one, 3-ethoxy-6-methyl-	EtOH	248(4.32)	78-0569-65
5α-Androstane-3α,17β-diol, 2-(propynylidene)-	EtOH	228(4.20)	13-0001-64
5α-Androstano[3,2-b]-5'-methylfuran, 17β-hydroxy-	EtOH	222(3.88)	13-0001-64
Androsta-3,5,7-trien-17β-ol, 3-ethoxy-6-methyl-	EtOH	322(4.26)	78-1753-65
5α-Androst-1-en-17β-ol, 3-methylene-, 17-acetate	EtOH	228s(--),235(4.37), 244s(--)	87-0345-64
9β-Bisnorchol-4-en-22-al, 3-oxo-	MeOH	247(4.14)	54-1069-64
Etiojerva-12,14,16-triene-3β,16-diol, 17-ethyl-, 16-methyl ether	MeOH	279(3.53),284(3.52)	44-4220-65
Gona-1,3,5(10)-trien-17β-ol, 13β,17α-diethyl-3-methoxy-	n.s.g.	278(3.30)	39-4472-64
A-Homo-B-norpregn-3-ene-4a,20-dione, 5β-methyl-	n.s.g.	226.5(3.61)	78-3185-65
19-Norprogesterone, 17-ethyl-	MeOH	239(4.25)	78-0357-64
Pregna-3,5-dien-20-one, 3-methoxy-	EtOH	240(4.28)	78-0597-64
5β-Pregnane-3,20-dione, 6-methylene-	EtOH	285(1.88)	78-3185-65
9β,10α-Pregn-4-ene-3,20-dione, 6α-methyl-	MeOH	246(4.16)	54-0904-65
9β,10α-Pregn-4-ene-3,20-dione, 6β-methyl-	MeOH	241(4.21)	54-0904-65
Pregn-4-en-3-one, 20β-hydroxy-6-methylene-	EtOH	257(4.05)	78-0597-64
Testosterone, 16α-allyl-	EtOH	242(4.19)	32-0351-65
Testosterone, 16β-allyl-	EtOH	243(4.20)	32-0351-65
Testosterone, 17α-cyclopropyl-	MeOH	241(4.22)	24-1470-65
Vitamin A, acetate	MeOH	326(4.73)	5-0020-64I
Vitamin A acid, ethyl ester	MeOH	350(4.62)	5-0020-64I
$C_{22}H_{32}O_2S$			
Estra-1,3,5(10)-triene, 3-methoxy-17β-[(2-methoxyethyl)thio]-	EtOH	279(3.27),287(3.23)	13-0557-65
$C_{22}H_{32}O_2S_3$			
5α-Androstan-17β-ol, 2β,3α-dimercapto-, acetate, cyclic 2,3-trithiocarbonate	EtOH	320(4.23)	78-0329-65

Compound	Solvent	λ_{max}(log ϵ)	Ref.
$C_{22}H_{32}O_3$			
Androsta-3,5-dien-17β-ol, 3-methoxy-, acetate	EtOH	240(4.28)	78-0597-64
Androstan-3-one, 17β-hydroxy-2-(hydroxymethylene)-, acetate	EtOH	285(3.92)	88-2161-64
	EtOH-NaOH	314(4.28)	88-2161-64
Androst-2-en-1-one, 17β-acetoxy-3-methyl-	ether	232(4.09)	24-1770-64
Androst-4-en-3-one, 17β-(acetoxymethyl)-	MeOH	241(4.2)	31-0499-65
Androst-4-en-3-one, 17β-hydroxy-17α-methyl-, acetate	EtOH	241(4.22),312(1.86)	44-3300-64
Androst-5-en-7-one, 17β-hydroxy-17α-methyl-, acetate	MeOH	238(4.11)	31-0249-64
Androst-5-en-17-one, 3β-propionoxy-, p-hydroxybenzaldehyde complex	HOAc-$HClO_4$	600(4.23)	96-0134-65
resorcylic aldehyde complex	HOAc-$HClO_4$	540(3.70)	96-0134-65
vanillin complex	HOAc-$HClO_4$	600(4.10)	96-0134-65
9β,10α-Androst-4-en-3-one, 17β-hydroxy-6α-methyl-, acetate	MeOH	246(4.16)	54-0904-65
9β,10α-Androst-4-en-3-one, 17β-hydroxy-6β-methyl-, acetate	MeOH	242.5(4.18)	54-0904-65
9β,10α-Androst-4-en-3-one, 17β-hydroxy-17α-methyl-, acetate	EtOH	240(4.21)	33-0989-65
Biphenyl, 6-acetyl-1,2,5,6-tetrahydro-2',6'-dimethoxy-4'-pentyl-3-methyl-	MeOH	210(4.55),230s(3.84), 272(3.16),278(3.16), 325(2.78)	5-0122-65E
Epitauranin, dihydrodeoxy-	EtOH	267(4.11),420(3.03)	94-0796-64
Estra-3,5-dien-17β-ol, 3-methoxy-6-methyl-, acetate	EtOH	246(4.22)	78-0569-65
Estra-1,3,5(10)-trien-17β-ol, 3-methoxy-2-(methoxymethyl)-17α-methyl-	EtOH	287(3.45)	94-0196-64
Estra-1,3,5(10)-trien-17β-ol, 3-methoxy-4-(methoxymethyl)-17α-methyl-	EtOH	282(3.38),286(3.36)	94-0196-64
Etiojerva-12,15,17-triene-3,17-diol	MeOH	280(3.34)	44-0123-65
A-Homo-5α-androst-1-en-4-one, 17β-acetoxy-	MeOH	210(3.4),280(2.15)	5-0047-64G
Pregna-1,4-dien-3-one, 11α,20β-dihydroxy-2-methyl-	EtOH	250(4.18)	32-0151-65
Pregna-3,5-dien-20-one, 17-hydroxy-3-methoxy-	MeOH	239(4.30)	78-0357-64
$C_{22}H_{32}O_3S$			
1-Naphthalenesulfonic acid, 3,7-dihexyl-	EtOH	232(4.90),280(3.78), 313(3.28),319(3.08), 327(3.38)	39-2324-65
19-Nortestosterone, 16β-(ethylthio)-, acetate	EtOH	240.5(4.28)	94-0905-64
$C_{22}H_{32}O_4$			
p-Benzoquinone, 2-[(1α,2,3,4,4aα,5,6,-7,8,8a-decahydro-2,5,5,8aβ-tetramethyl-1-naphthyl)methyl]-3,6-dihydroxy-5-methyl-	EtOH	298(4.28),445(2.32)	94-0796-64
Deoxycorticosterone, 21-(hydroxymethyl)-	EtOH	240(4.23)	94-1180-64
1,3,5-Hexatriene, 3,4-bis(4,4-ethylenedioxycyclohexyl)-	MeOH	239.5(3.81)	5-0024-65D
Picrosalvin	EtOH	232(3.85),272(2.79)	33-1234-64
Pregn-4-ene-3,20-dione, 17α,21-dihydroxy-19-methyl-	n.s.g.	244(4.21)	13-0371-65

Compound	Solvent	λ_{max}(log ε)	Ref.
Testosterone, 7α-hydroxy-, 17-propionate	EtOH	242(4.19)	24-3363-64
Testosterone, 7α-hydroxy-17α-methyl-, 7-acetate	EtOH	238(4.23)	24-3363-64
Testosterone, 16α-hydroxy-17α-methyl-, acetate	MeOH	243(4.23)	7-0205-65
$C_{22}H_{32}O_5$			
Andromedienol, acetate	EtOH	205.5(4.00)	44-2756-64
Androst-1-en-3-one, 4α,5α,17β-tri-hydroxy-, 17-propionate	EtOH	231(3.97)	94-0050-65
Androst-4-en-3-one, 1α,2α,17β-tri-hydroxy-, 17-propionate	EtOH	240(4.14)	94-0050-65
Cortexolone, 2β-hydroxy-16α-methyl-	EtOH	242(4.16)	13-0057-65
16α-Kaurane-17,19-dioic acid, 1-oxo-, dimethyl ester	EtOH	295(1.40)	12-2005-65
Octanorebelo-16,22-dioic acid, 17-hydroxy-3-oxo-, 16,17-lactone	EtOH	277(2.08)	12-1451-65
Pregn-4-ene-3,20-dione, 17α,21-di-hydroxy-21-(hydroxymethyl)-	EtOH	241(4.24)	94-1180-64
Pregn-4-ene-3,20-dione, 7β,17α,21-tri-hydroxy-16α-methyl-	MeOH	242(4.19)	13-0459-65
Pregn-4-ene-3,20-dione, 11α,17α,21-tri-hydroxy-16α-methyl-	MeOH	242(4.21)	13-0459-65
Pregn-4-ene-3,20-dione, 11β,17α,21-tri-hydroxy-19-methyl-	n.s.g.	246(4.17)	13-0371-65
$C_{22}H_{32}O_5S$			
Estra-1,3,5(10)-triene, 17β-(2-hydroxy-ethoxy)-3-methoxy-, methanesulfonate	EtOH	278(3.32),286(3.27)	13-0557-65
$C_{22}H_{33}BrO_3$			
A-Homo-5α-androstan-4-one, 3-bromo-17β-hydroxy-, acetate	MeOH	302(2.2)	78-2937-64
$C_{22}H_{33}NO_2$			
Estra-3,5-dien-17-one, 6-(dimethyl-aminomethyl)-3-methoxy-	EtOH	250(4.31)	78-0569-65
Estradiol, 2-(dimethylaminomethyl)-17α-methyl-	EtOH	286(3.46)	94-0196-64
Estradiol, 4-(dimethylaminomethyl)-17α-methyl-	EtOH	286(3.32)	94-0196-64
Pregn-4-ene-3,20-dione, 17β-(methyl-amino)-, hydrochloride	MeOH	240(4.22)	44-0579-65
$C_{22}H_{33}NO_2S$			
Cyclopenta[7,8]phenanthro[3,2-b][1,4]-thiazin-8(9H)-one, hexadecahydro-1-hydroxy-1,11a,13a-trimethyl-	EtOH	232(3.58),298(3.38)	87-0555-64
$C_{22}H_{33}NO_3$			
11-Aza-5α,8ξ-pregn-16-en-20-one, 11-acetyl-3β-hydroxy-	EtOH	234(3.92)	13-0595-64
$C_{22}H_{33}NO_4$			
Androstane-2,3-dione, 17β-hydroxy-17α-methyl-, 2-O-acetyloxime	ether	220(3.87)	94-1445-65
Androstane-2,3-dione, 17β-propionyl-oxy-, 2-oxime	EtOH	243(3.86)	94-1445-65
Pyroheteratisine, dihydro-	EtOH	312(1.38)	88-0215-65

Compound	Solvent	λ_{max}(log ε)	Ref.
$C_{22}H_{33}NO_5$			
Octanorebelo-16,22-dioic acid, 17-hydroxy-3-oxo-, 16,17-lactone, oxime	EtOH	277(2.07)	12-1451-65
$C_{22}H_{34}$			
Anthracene, octahydrooctamethyl-	C_6H_{12}	270(2.98),278(3.04)	23-0579-64
$C_{22}H_{34}ClNO_2$			
Pregn-4-ene-3,20-dione, 17β-(methyl-amino)-, hydrochloride	MeOH	240(4.22)	44-0579-65
$C_{22}H_{34}N_2O$			
2-Indolinone, 3-[1-methyl-1-(amino-methyl)-3-(2,2,6-trimethylcyclo-hexyl)propyl]-	EtOH	237(3.88),291(3.43)	87-0626-65
$C_{22}H_{34}N_2O_2$			
1,3-Cyclohexadiene-1,4-dicarboxamide, N,N'-bis(cyclohexylmethyl)-	EtOH	307(4.17)	23-2852-64
$C_{22}H_{34}N_2O_4$			
Androstane-2,3-dione, 17β-hydroxy-, 17-propionate, dioxime	EtOH	238(3.8)	94-1445-65
$C_{22}H_{34}N_2S_2$			
4H,10H-Dithieno[3,4-c:3',4'-h][1,6]-diazecine, 5,6,11,12-tetrahydro-5,11-diisopropyl-1,3,7,9-tetra-methyl-	n.s.g.	220(4.1),239(4.1), 244s(4.1)	70-2182-64
$C_{22}H_{34}O$			
Androst-4-en-17β-ol, 17α-ethyl-3-methylene-	EtOH	238(4.42)	87-0345-64
$C_{22}H_{34}O_2$			
Androst-5-en-17-one, 3β-methoxy-16,16-dimethyl-	MeOH	298(1.56)	24-1799-64
Communic acid, ethyl ester	EtOH	232(4.48)	22-0348-64
Gon-4-en-3-one, 13β-ethyl-17β-hydroxy-17α-propyl-	n.s.g.	240(4.20)	39-4472-64
A-Homo-B-norpregnane-3,4a-dione,	EtOH	263(2.49),292s(2.11)	78-3185-65
5β-methyl-	EtOH-KOH	298.5(4.29)	78-3185-65
19-Nortestosterone, 17α-ethyl-18,18-dimethyl-	EtOH	240(4.08)	94-1294-65
$C_{22}H_{34}O_3$			
5α-Androstan-3-one, 2α-acetonyl-17β-hydroxy-	EtOH	284(1.95)	13-0001-64
A-Homo-5α-androstan-4-one, 17β-acetoxy-	MeOH	282(1.5)	5-0047-64G
5α-Pregnane-3,20-dione, 16α-methoxy-	EtOH	284(1.87)	1-0750-64
Pregn-4-en-3-one, 11α,20β-di-hydroxy-2α-methyl-	EtOH	242(4.11)	32-0151-65
$C_{22}H_{34}O_4$			
Andromedenol, 10-deoxyacetyl-	EtOH	203(3.82)	44-2756-64
Kolavic acid, dimethyl ester	n.s.g.	217(4.33)	88-3751-64
19-Norandrosta-3,5-diene-6-carboxalde-hyde, 17β-acetoxy-3-methoxy-	EtOH	220(4.00),322(4.21)	78-0597-64

Compound	Solvent	$\lambda_{max}(\log \epsilon)$	Ref.
$C_{22}H_{34}O_5$			
Hexanoic acid, 6-[2-(4-carbethoxy-butyl)-5-methoxyphenyl]-, ethyl ester	EtOH	221(3.91),275(3.39)	39-3872-65
$C_{22}H_{34}O_6$			
$\Delta^{10(18)}$-Andromedenol, acetate	EtOH	204(3.63)	44-2756-64
3,5-Hexadiene-1,2-diol, 3,4-bis-(4,4-ethylenedioxycyclohexyl)-	MeOH	231.5(3.74)	5-0024-65D
$C_{22}H_{35}N_3O_7$			
Griseoviridin, octahydrodethio-	EtOH	212(4.15)	23-0371-64
$C_{22}H_{35}O_4P$			
Pregn-5-en-20-one, 3β-hydroxy-, hydrogen methylphosphonate	MeOH	284(1.82)	22-0933-65
$C_{22}H_{36}$			
1,3,12,14-Cyclodocosatetraene	hexane	229(4.70),235(4.70)	39-6674-65
$C_{22}H_{36}N_2O_2$			
Hydroquinone, 2,3-bis(cyclohexyl-aminomethyl)-5,6-dimethyl-	dioxan	258(3.53),303(3.61)	39-0042-64
$C_{22}H_{36}N_7O_{14}P$			
6-Azauridine, 6-azauridylyl-(5'→5'), triethylamine salt	n.s.g.	260(4.05)	73-2567-64
$C_{22}H_{36}O_3$			
p-Benzoquinone, 2-hexadecyl-5-hydroxy-	EtOH	209(4.04),268(4.12), 383(2.85)	94-0236-64
$C_{22}H_{36}O_4$			
3-Hexene, 3,4-bis(4,4-ethylenedioxy-cyclohexyl)-	hexane	187(4.30)	5-0024-65D
$C_{22}H_{37}NO$			
5α-Androstan-3α-ol, 17-(propylimino)-	MeOH	235(2.41)	39-4508-65
$C_{22}H_{38}$			
Benzene, 1,2,3,5-tetra-tert-butyl-	n.s.g.	307(2.64)	88-3803-64
Benzene, 1,2,4,5-tetra-tert-butyl-	n.s.g.	273(2.57)	88-3803-64
$C_{22}H_{38}O_3$			
Labd-8(20)-ene-3β,15-diol, 3-acetate	EtOH	206.0(3.69)	39-3648-64
$C_{22}H_{40}N_4O_2$			
p-Benzoquinone, tetrakis(di-ethylamino)-	methyl cellosolve	240(4.20),407(3.89), 625s(2.56)	78-1889-64
$C_{22}H_{40}Si_5$			
Pentasilane, 1,1,2,2,3,3,4,4,5,5-deca-methyl-1,5-diphenyl-	C_6H_{12}	257.5(4.40)	25-1063-64
$C_{22}H_{42}N_4O_8S_2$			
D-Pantethine	EtOH	247s(2.70)	94-0180-65
$C_{22}H_{44}N_2O$			
2-Imidazoline-1-ethanol, 2-heptadecyl-	EtOH	232(3.78)	4-0188-64

Compound	Solvent	$\lambda_{max}(\log \epsilon)$	Ref.
$C_{22}H_{66}Si_{10}$			
Decasilane, docosamethyl-	C_6H_{12}	215s(4.45),230s(4.32), 255s(4.39),279(4.63)	101-0176-65B

Compound	Solvent	λ_{max}(log ϵ)	Ref.
$C_{23}H_{12}O_4$			
5,8,13,14-Pentaphenetetrone, 6-methyl-	dioxan	245(4.40),275(4.54)	65-3484-64
$C_{23}H_{13}N_5O_2$			
Benzene, 1-(2-phenyl-1,3,4-oxadiazol-5-yl)-3-[2-(3-cyanophenyl)-1,3,4-oxadiazol-5-yl]-	DMF	286.5(4.68)	24-2966-65
$C_{23}H_{14}Br_2O_2$			
1,3-Indandione, 2-(α,β-dibromostyryl)-2-phenyl-	MeOH	229(4.70)	44-1930-65
$C_{23}H_{14}N_4O$			
3,6-Ethanocyclopenta[ef]heptalene-1-carboxaldehyde, 11,11,12,12-tetracyano-3,6-dihydro-3,5-dimethyl-	DMF	549(2.79)	5-0194-64B
isomer	DMF	628(2.83),660(2.84), 720(2.58)	5-0194-64B
$C_{23}H_{14}O_2$			
12H-Benzo[b]xanthen-12-one, 6-phenyl-	EtOH	240(4.46),263(4.79), 313(3.80),324(3.84), 400(3.66)	35-5424-65
12H-Benzo[b]xanthen-12-one, 11-phenyl-	EtOH	238(4.39),262(4.83), 311(3.81),323(3.86), 390(3.63)	35-5424-65
1,3-Indandione, 2-phenyl-2-(phenylethynyl)-	MeOH	227(4.65)	44-1930-65
$C_{23}H_{14}O_3$			
12H-Benzo[b]xanthen-12-one, 11-hydroxy-6-phenyl-	C_6H_{12}	229(4.40),274(4.83), 290(4.42),316(4.07), 326s(3.50),438(3.74)	35-5424-65
9H-Furo[3,4-b][1]benzopyran-9-one, 1,3-diphenyl-	MeOH	264(4.38),294(4.18), 320s(3.99),422(3.87)	35-5424-65
$C_{23}H_{14}O_4$			
Chromone, 2,3-dibenzoyl-	MeOH	250(4.38),300s(3.97)	35-5424-65
Isocoumarin, phenanthrenequinone adduct	dioxan	244(4.54),281(3.91), 329(3.51)	24-3102-65
$C_{23}H_{15}ClO_2$			
1,3-Indandione, 2-(α-chlorostyryl)-2-phenyl-	MeOH	227(4.69),252(4.46)	44-1930-65
$C_{23}H_{15}NO$			
Naphth[1,2-d]oxazole, 2-(4-biphenylyl)-	heptane	225(4.58),300(4.32), 310(4.39),345(4.60)	65-0427-64
Naphth[2,1-d]oxazole, 2-(4-biphenylyl)-	heptane	285(4.58),320(4.55), 340(4.50),355(4.38)	65-0427-64
Naphth[2,3-d]oxazole, 2-(4-biphenylyl)-	heptane	287(4.21),336(4.25)	65-0427-64
$C_{23}H_{15}NO_5$			
Chromone, 2-benzyl-3-(p-nitrobenzoyl)-	MeOH	227(4.49),265(4.38), 293(4.16),303(4.06)	35-5417-65
$C_{23}H_{16}$			
Fluorene, 9-(inden-3-ylmethylene)-	ether	324(4.02)	5-0001-65J
	DMSO-tert-BuOK	300(4.35),344(3.94), 384s(3.52),543(4.97)	5-0050-65J

Compound	Solvent	λ_{max}(log ε)	Ref.
Fluorene, 9-(inden-3-ylmethylene)-, anion	DMSO	543(4.97)	5-0001-65J
$C_{23}H_{16}ClNO_4$			
5-Phenylbenzo[a]phenanthridizinium perchlorate	EtOH	202(4.52),216s(4.50), 242(4.32),286(4.62), 360(4.02),375(4.19), 394(4.27)	44-1846-65
$C_{23}H_{16}Cl_2N_2O_3$			
Barbituric acid, 5-benzyl-1,3-bis(p-chlorophenyl)-	MeOH-HCl	218(4.54),264(3.18)	49-1352-65
	MeOH	219(4.50),274(4.10)	49-1352-65
	MeOH-NaOH	218(4.50),274(4.31)	49-1352-65
	CH_2Cl_2	219(4.57),264(3.07)	49-1352-65
$C_{23}H_{16}Cl_2O_5$			
2-(o-Chlorophenyl)-4,6-diphenylpyrylium perchlorate	MeOH	314(3.90)	89-0437-64
2-(p-Chlorophenyl)-4,6-diphenylpyrylium perchlorate	MeOH	320(3.98)	89-0437-64
$C_{23}H_{16}N_2$			
Pyridazine, 3-fluoren-2-yl-6-phenyl-	EtOH	318(4.64)	39-7005-65
$C_{23}H_{16}N_2O$			
Benzo[a]phenazin-6-ol, 5-benzyl-	n.s.g.	236(4.68),309(4.77), 362(3.77),380(3.94), 430(3.66)	24-0666-65
Pyridazine, 3-phenyl-6-xanthen-2-yl-	EtOH	309(4.45)	39-7005-65
$C_{23}H_{16}N_2O_2$			
3H-Benz[f]indazole-4,9-diol, 3,3-diphenyl-	EtOH	243(4.4),308(3.5)	23-2665-64
$C_{23}H_{16}N_2O_3$			
1,7-Dioxa-2,8-diazaspiro[4.4]nona-2,8-dien-6-one, 3,4,9-triphenyl-	EtOH	233(4.00),262(4.37)	32-0127-65
$C_{23}H_{16}N_4$			
s-Triazolo[1,5-a]pyrazine, 2,5,6-triphenyl-	C_6H_{12}	204(4.66),220s(4.31), 269(4.69),292(4.16)	44-2542-64
s-Triazolo[4,3-a]pyrazine, 3,5,6-triphenyl-	EtOH	208(4.83),223s(4.35), 254(4.31),318s(3.71)	44-2542-64
$C_{23}H_{16}O$			
Acrylophenone, 3-(1-phenanthryl)-	EtOH	317s(3.85)	23-2580-64
Acrylophenone, 3-(9-phenanthryl)-	EtOH	352(4.12)	23-2580-64
2-Propen-1-one, 1-(9-anthryl)-	EtOH	251(5.14),267(4.27), 291(4.36),299(4.36), 304(4.36),345(3.77), 363(3.83),380(3.75)	39-5666-64
2-Propen-1-one, 1-(2-phenanthryl)-3-phenyl-	$CHCl_3$	316(4.31)	23-2580-64
2-Propen-1-one, 1-(3-phenanthryl)-3-phenyl-	$CHCl_3$	328(4.30)	23-2580-64
$C_{23}H_{16}O_2$			
4-Cyclopentene-1,3-dione, 2,4,5-triphenyl-	EtOH-acid	232(4.31),330(4.00)	65-0813-64
Indene, phenanthrenequinone adduct	dioxan	246(4.53),278(3.98), 328(3.51)	24-3102-65

Compound	Solvent	λ max(log ε)	Ref.
$C_{23}H_{16}O_2S$			
2H-Naphtho[2,3-b]thiete, 3,8-diphenyl-, 1,1-dioxide	EtOH	242(4.80),315(4.16), 338(4.05)	44-0629-65
	MeCN	241(4.72),315(4.08), 338(3.96)	88-3809-64
$C_{23}H_{16}O_3$			
12H-Benzo[b]xanthen-12-one, 5a,6-dihydro-11-hydroxy-6-phenyl-	C_6H_{12}	244(4.16),272(4.30), 294(4.00),306(4.04), 318(4.08),330s(3.95), 394(4.16)	35-5424-65
Benzoic acid, o-(1-oxo-3-phenylinden-2-yl)-, methyl ester	ether	245(4.47),278s(4.05), 412(3.09)	6-0359-64
Chromone, 3-benzoyl-2-benzyl-	MeOH	225(4.40),250(4.36), 295(3.93),303(3.89)	35-5417-65
1,3-Indandione, 2-phenacyl-2-phenyl-	MeOH	226(4.68),246(4.44)	44-1930-65
$C_{23}H_{16}O_4$			
Benzoic acid, o-(1-oxo-4-phenyl-1H-2-benzopyran-3-yl)-, methyl ester	EtOH	230(4.37),280(3.76), 306(3.75),324s(3.69)	6-0359-64
Chromone, 3-benzoyl-2-(p-hydroxybenzyl)-	MeOH	226(4.50),251(4.35), 295(4.00),304(3.92)	35-5417-65
$C_{23}H_{16}O_5$			
Benzoic acid, o-[(o-benzoylphenyl)-glyoxyloyl]-, methyl ester	EtOH	250(4.48),450(2.29)	6-0359-64
10H-o-Dioxino[4,5-b][1]benzopyran-10-one, 1,4-dihydro-1-hydroxy-1,4-diphenyl-	MeOH	263s(3.75),296(3.66), 303(3.66)	35-5424-65
$C_{23}H_{16}O_5S$			
Chromone, 3-benzoyl-2-[(phenyl-sulfonyl)methyl]-	MeOH	232(4.45),257(4.34), 298(4.01),307(3.99)	35-5417-65
$C_{23}H_{17}BrO_{11}$			
Flavone, 8-bromo-3,3',4',5,7-penta-hydroxy-, 3,3',4',7-tetraacetate	EtOH	272(4.48),302(4.21), 340(3.97)	44-0897-65
$C_{23}H_{17}ClO_4$			
7-[2-(9-Anthryl)vinyl]tropylium perchlorate	MeCN	223(4.52),255(5.13), 351(4.16),365s(4.12), 385s(3.89),586(4.08)	24-1349-64
7-[2-(9-Phenanthryl)vinyl]tropylium perchlorate	MeCN	224s(4.50),240(4.65), 244(4.65),299(3.68), 335(3.69),500(4.40)	24-1349-64
$C_{23}H_{17}ClO_5$			
Triphenylpyrylium perchlorate	MeOH	234(4.24),242(4.27), 255(4.33),324(4.03)	65-3979-64
$C_{23}H_{17}IN_2O_2$			
6-Hydroxy-5-methylindolizino[1,2-c]-quinolinium iodide, benzoate	MeOH	255(4.68),315(4.25), 344(3.84),428(4.30)	5-0196-65H
$C_{23}H_{17}N$			
Pyridine, 2,4,6-triphenyl-	MeOH	253(4.64),312(3.94)	65-3979-64
$C_{23}H_{17}NO_2$			
4(1H)-Quinolone, 3-benzoyl-2-benzyl-	MeOH	239(4.59),318(4.15), 328(4.06)	35-5417-65

Compound	Solvent	λ_{max}(log ε)	Ref.
$C_{23}H_{17}N_2PS_2$			
[[(2,2-Dicyano-1-mercaptovinyl)thio]-methyl]triphenylphosphonium hydroxide, inner salt	MeCN	275(3.99),342(4.26)	44-0497-64
$C_{23}H_{18}$			
Anthracene, 9-(1-phenylpropenyl)-	MeOH	255(5.05),333(4.36), 348(4.66),367(4.81), 386(4.77)	39-6710-65
Butatriene, 1-methyl-1,4,4-triphenyl-	benzene	393(4.53)	56-0763-65
Naphthalene, 7-phenyl-1-o-tolyl-	EtOH	252(4.72),294(4.08)	39-5110-64
Phenanthrene, 1-cinnamyl-	MeOH	213(4.75),248(4.80), 256(4.89),276(4.33), 287(4.25),299(4.31)	39-6710-65
Phenanthrene, 9-cinnamyl-	MeOH	246(4.82),253(4.90), 285(4.18),296(4.17)	39-6710-65
Phenanthrene, 1-(3-phenylpropenyl)-	MeOH	218(4.78),259(4.67), 300(4.26)	39-6710-65
Phenanthrene, 2-(3-phenylpropenyl)-	MeOH	260s(4.92),270(5.03), 291(4.55),304(4.44)	39-6710-65
Phenanthrene, 3-(3-phenylpropenyl)-	MeOH	254(4.75),262(4.80), 281(4.47),290(4.21), 302(4.45),316(4.43)	39-6710-65
Phenanthrene, 9-(3-phenylpropenyl)-	MeOH	258(4.72),273(4.32), 301(4.18)	39-6710-65
$C_{23}H_{18}BrNO$			
2-Phenacyl-1-phenylisoquinolinium bromide	EtOH	235(4.76),258s(3.86), 274s(3.81),280s(3.79), 342(3.78)	44-1846-65
$C_{23}H_{18}BrNO_2$			
2,3-Dimethoxydibenzo[h,j]acridizinium bromide	EtOH	203(4.69),220(4.65), 247(4.57),258(4.60), 267(4.62),283s(4.72), 295(4.75),315s(4.57), 345s(4.35),407(4.20)	54-0593-64
$C_{23}H_{18}Br_2N_4$			
6,6'-Methylenebis[dipyrido[1,2-c:-2',1'-e]imidazol-5-ium bromide]	n.s.g.	243(4.98),300(4.21), 327s(4.35),340(4.60), 356(4.56),373(4.06), 395(3.91),417(3.83)	12-1819-65
$C_{23}H_{18}Br_2O_{11}$			
Flavanone, 6,8-dibromo-3,3',4',5,7-pentahydroxy-, 3,3',4',7-tetraacetate	EtOH	278(4.13),359(3.68)	44-0897-65
$C_{23}H_{18}ClNO_5$			
2,4-Diphenyl-6-p-tolyl-1,3-oxazin-1-ium perchlorate	n.s.g.	354(4.41),419(4.51)	24-3892-65
1-(o-Hydroxyphenyl)-2-styryl-quinolinium perchlorate	EtOH	416(3.98)	65-3373-64
$C_{23}H_{18}ClNO_8$			
7,8-Diacetoxy-10-phenylbenzo[b]quino-lizinium perchlorate	MeCN	240(4.64),250(4.56), 366(4.12),390(3.92), 410(3.88)	44-0252-65

Compound	Solvent	$\lambda_{max}(\log \epsilon)$	Ref.
$C_{23}H_{18}Fe_2O$			
Iron, [(oxopropynylene)dicyclopentadienylene]bis[cyclopentadienyl-	$CHCl_3$	304(4.14),376(3.60), 486(3.45)	49-1750-64
$C_{23}H_{18}I_2O_{11}$			
Flavanone, 3,3',4',5,7-pentahydroxy-6,8-diiodo-, 3,3',4',7-tetraacetate	EtOH	286(3.99),366(3.65)	44-0897-65
$C_{23}H_{18}N_2O_2$			
Fluorene, 2-nitro-9-(3-methylanilino-2-propen-1-ylidene)-	MeCN	245(4.51),272(4.22), 428(4.63)	24-3331-64
$C_{23}H_{18}N_2O_4$			
Cinnamanilide, p-nitro-N-phenacyl-	EtOH	242(4.35),313(4.39)	78-0449-65
Ethane, 1-(N-acetyl-2-oxo-3-indolinylidene)-2-(N-propionyl-2-oxo-3-indolinylidene)-	benzene	395(4.20),416(4.24), 470(4.31)	39-1455-65
$C_{23}H_{18}O_2$			
9-Anthracenemethanol, α-phenyl-, acetate	n.s.g.	255.5(5.13)	1-0756-65
Anthraquinone, 2,3-dimethyl-1-p-tolyl-	EtOH	263(4.67),333(3.64)	65-2740-64
$C_{23}H_{18}O_3$			
Anthraquinone, 1-(p-methoxyphenyl)-2,3-dimethyl-	EtOH	263(4.67),333(3.65)	65-2740-64
1,4-Butanedione, 2-benzoyl-1,4-diphenyl-	n.s.g.	248(4.7)	23-2822-64
$C_{23}H_{18}O_4$			
Isoflavone, 7-(benzyloxy)-4'-methoxy-	EtOH	255(3.99),305s(3.56)	78-1141-64
$C_{23}H_{18}O_6$			
Pentane, 1,3,5-tris(p-benzoquinonyl)-	95% THF	249(4.65),315(3.44)	44-2602-65
$C_{23}H_{18}O_7$			
Benzo[b]naphtho[2,3-d]furan-6,10,11-triol, 2-methyl-, triacetate	EtOH	260(4.75),271(5.05), 281(4.08),293(3.76), 312(3.97),324(4.11), 341(3.93),357(3.97)	33-1459-64
$C_{23}H_{18}Pb$			
Lead, (1,3-pentadiynyl)triphenyl-	hexane	242(3.71),250(3.45), 257(3.31),264(3.20)	22-3518-65
$C_{23}H_{18}S$			
2-Naphthalenemethanethiol, 1,4-diphenyl-	EtOH	237(4.76),296(4.08)	44-0629-65
$C_{23}H_{18}Sn$			
Tin, (1,3-pentadiynyl)triphenyl-	EtOH	252(3.15),258(3.04), 265(3.00)	22-0035-65
$C_{23}H_{19}BO_{10}$			
Rubrocurcumin	EtOH-HOAc	540(4.76)	83-0660-64
	EtOH-buffer	645(4.75)	83-0660-64
$C_{23}H_{19}BrO_{11}$			
Flavone, 8-bromo-3,3',4',5,7-pentahydroxy-, 3,3',4',7-tetraacetate	EtOH	275(4.11),348(3.63)	44-0897-65

Compound	Solvent	$\lambda_{max}(\log \epsilon)$	Ref.
$C_{23}H_{19}ClO_4$			
1-[1-(1-Azulenyl)propylidene]dihydro-azulenylium perchlorate	HOAc	639(4.47)	65-0894-64
7-(4,4-Diphenyl-1,3-butadien-1-yl)-tropylium perchlorate	MeCN	230(4.39),294(4.13), 333(3.84),515(4.64)	24-1590-64
$C_{23}H_{19}ClO_6S$			
6,7-Dimethoxy-2,4-diphenyl-1-benzo-thiopyrylium perchlorate	HOAc	248(4.38),292(4.40), 351(3.90),444(4.38)	5-0136-64C
$C_{23}H_{19}FN_2O_6S$			
Uridine, 2'-deoxy-5-fluoro-5'-thio-, 3',5'-dibenzoate	pH 1	268(4.16)	44-0554-64
	pH 7	267(3.90)	44-0554-64
	pH 13	270(4.00)	44-0554-64
$C_{23}H_{19}IO_{11}$			
Flavanone, 3,3',4',5,7-pentahydroxy-8-iodo-, 3,3',4',7-tetraacetate	EtOH	279(4.02),355(3.58)	44-0897-65
$C_{23}H_{19}N$			
Aniline, N-(3-fluoren-9-ylidene-propenyl)-N-methyl-	MeCN	245s(4.73),251(4.76), 303(3.79),419(4.84)	24-3331-64
Crotononitrile, 2-benzyl-3,4-diphenyl-	EtOH	237(4.08)	28-1146-64B
2-Naphthylamine, N-methyl-1,4-diphenyl-	EtOH	220(4.43),253(5.64), 298(3.96),308(3.95), 366(3.72)	44-0190-65
$C_{23}H_{19}NO$			
3H-Benzazepine, 3-benzoyl-4,5-dihydro-1-phenyl-	$CHCl_3$	245(4.14),298(4.21)	6-0213-65
2-Indolinone, 3-(α-benzyl-phenethylidene)-	EtOH	255(4.45),298(3.93), 361(3.27)	87-0626-65
Indone, 3-(N-ethylanilino)-2-phenyl-	MeOH	278(4.34),454(3.71)	65-0448-64
$C_{23}H_{19}NO_2$			
Aziridine, 2,3-dibenzoyl-1-benzyl-, cis	EtOH	251(4.34)	35-1050-65
trans	EtOH	254(4.39)	35-1050-65
Cyclohepta[b]pyrrole-2,8-dione, 3,3-dibenzyl-1,3-dihydro-	EtOH	235(4.13),264(4.08), 345(3.94),370(3.93), 386(3.91),405(3.67)	94-0473-65
Cyclohepta[b]pyrrol-2(1H)-one, 1,3-dibenzyl-8-hydroxy-	EtOH	233(4.29),278(4.33), 297(4.50),369(3.69), 391(3.92),413(4.01)	94-0473-65
Vinylamine, benzoyl-α-acetyl-β,β-diphenyl-	EtOH	235(4.47),290(4.19)	32-1115-65
$C_{23}H_{19}NO_3$			
2-Buten-1-one, 4-acetoxy-1-(2,6-di-phenyl-4-pyridyl)-	MeOH	238(4.32),270(4.35), 300(4.35)	65-3979-64
3-Indolineacetic acid, 1-benzyl-2-oxo-3-phenyl-	EtOH	256(3.88),283s(3.24)	44-2973-65
$C_{23}H_{19}NO_4$			
Indole-2-carboxylic acid, 5,6-bis(benzyloxy)-	EtOH	220(4.35),316(4.20)	4-0387-65
$C_{23}H_{19}N_2OP$			
Phospholo[2,3-c]pyrazole, 3,3a,6,6a-tetrahydro-5,6,6a-triphenyl-, 6-oxide	EtOH	215(4.20),260(3.84), 325(2.48),365(1.88)	39-2184-65

Compound	Solvent	λ_{max}(log ϵ)	Ref.
Phosphorane, (furfurylidenehydrazono)-triphenyl-	$CHCl_3$	258s(3.48),262s(3.56), 265(3.63),274(3.66), 328(3.77)	44-0417-65
$C_{23}H_{19}N_3O_9$			
Uridine, 2',3'-O-benzylidene-, 5'-p-nitrobenzoate	MeOH	258(4.40)	22-2489-65
$C_{23}H_{19}N_5O_9$			
4-Ethoxy-1-[2-(3-indolyl)-2-oxo-ethyl]pyridinium picrate	MeOH	244(4.46),309(4.28)	44-3407-64
$C_{23}H_{19}OP$			
2-Phosphabicyclo[3.1.0]hex-3-ene, 1,2,3-triphenyl-, 2-oxide	EtOH	215(4.44),273(4.06), 373(2.24)	39-2184-65
$C_{23}H_{20}$			
Naphthalene, 1,2-dihydro-6-phenyl-4-o-tolyl-	EtOH	248(4.68)	39-5110-64
$C_{23}H_{20}ClNO_3$			
1-Pyrazolin-3-one, 2-(4-oxo-8-chloro-5-methoxy-1,2,3,4-tetrahydronaphth-2-yl)ethylidene-5-phenyl-	MeCN	222(4.52),242(4.22), 255(4.01),299(4.42)	35-0933-65
$C_{23}H_{20}ClNO_4S$			
7-(Dimethylamino)-2,4-diphenylbenzo-[b]thiopyrylium perchlorate	EtOH	220(4.50),230(4.49), 262(4.64),333(4.65), 565(4.43)	39-0032-65
$C_{23}H_{20}ClN_3O$			
3H-1,4-Benzodiazepine, 2-(benzyl-methylamino)-7-chloro-5-phenyl-, 4-oxide	EtOH-acid	253(4.52),307s(3.98)	87-0235-64
$C_{23}H_{20}Cl_2N_4$			
Phenazine, 2-[N-[p-[bis(2-chloroethyl)-amino]phenyl]formimidoyl]-	C_6H_{12}	232(4.31),268(4.66), 314(4.37),342(4.42)	65-1969-64
$C_{23}H_{20}Fe_2O$			
2-Propen-1-one, 1,3-diferrocenyl-	n.s.g.	314(4.16),384(3.64), 498(3.60)	101-0398-64B
$C_{23}H_{20}IN_3O_5$			
Cytidine, N-benzoyl-2',5'-dideoxy-, 3'-benzoate	MeOH	230(4.39),260(4.42), 302(4.02)	44-3067-65
$C_{23}H_{20}IP$			
1-Methyl-1,2,5-triphenyl-phospholium iodide	EtOH	222(4.56),405(4.10)	39-2184-65
$C_{23}H_{20}N_2$			
3H,5H-Pyrido[1,2,3-de]quinoxaline, 6,7-dihydro-2,3-diphenyl-	MeOH	250s(4.26),276(4.36), 318(3.81),427(3.62)	44-2589-65
$C_{23}H_{20}N_2O_3$			
Phthalimide, N-(1,2,3,4,5,10-hexahydro-1,3-dimethyl-4-oxobenzo[f]quino-lin-6-yl)-	EtOH-HCl	330(3.97)	94-1405-64

Compound	Solvent	$\lambda_{max}(\log \epsilon)$	Ref.
$C_{23}H_{20}N_2S_2$			
Benzothiazoline, 2-phenyl-2,2'-methylidynebis[3-methyl-	hexane	330(4.31)	22-2879-64
	EtOH	308(3.97)	22-2879-64
	HOAc	294(4.27)	22-2879-64
$C_{23}H_{20}N_4$			
Phenazine, 3-(1-benzimidazolyl)-1-tert-butyl-	EtOH	215(4.84),260(4.99), 370(4.29)	12-1241-65
$C_{23}H_{20}N_4O_5$			
Formamide, N-[2-(2,4-dinitroanilino)-1,2-diphenylvinyl]-N-ethyl-	EtOH	364(4.41)	65-0269-64
	$CHCl_3$	368(4.27)	65-0269-64
$C_{23}H_{20}O$			
Cyclopenta[e,f]heptalene-3-ethanol, 5-methyl-α-phenyl-	EtOAc	376(4.05),393(4.11), 452(2.43),735(2.01), 809(2.10),915(2.04), 1050(1.64)	5-0194-64B
Fluoranthene, 1,2,3,10b-tetrahydro-4-methoxy-10b-phenyl-	EtOH	282(4.18),312(3.79)	39-5110-64
Fluorene, 9-benzoyl-9-isopropyl-	EtOH	213(4.51),258(4.32), 292(3.85),305(3.72)	22-2345-65
isomer	EtOH	236(4.73),298(4.18), 330(4.25)	22-2345-65
Ketone, isopropyl 9-phenyl-9-fluorenyl-	EtOH	217(4.41),269(4.19), 303(3.50)	22-2345-65
$C_{23}H_{20}O_2$			
Phenanthrene, 9-methoxy-10-(p-methylbenzyloxy)-	dioxan	292(4.00),305(4.08), 324(2.95),341(3.08), 358(3.08)	44-2362-64
$C_{23}H_{20}O_3$			
Biphenyl-2-carboxylic acid, 2'-(2,5-dimethylbenzoyl)-, methyl ester	MeOH	210(4.62),255s(4.08), 290s(3.64)	44-3333-64
Biphenyl-2-carboxylic acid, 2'-(p-tolylacetyl)-, methyl ester	MeOH	220(4.48),285(3.51)	44-3333-64
Chalcone, 2'-(benzyloxy)-4-methoxy-	EtOH	238s(--),340(4.24)	22-3350-65
Chalcone, 2'-(benzyloxy)-4'-methoxy-	EtOH	233s(--),311(4.32)	22-3350-65
Phenanthrene, 1-(2-furfurylideneacetyl)-9,10-dihydro-7-methoxy-2-methyl-	EtOH	282(4.52),324(4.55)	70-1051-65
$C_{23}H_{20}O_4$			
2-Anthraceneacetic acid, 5-(benzyloxy)-1,2,3,4-tetrahydro-4-oxo-	EtOH	220(4.49),236(4.20), 261(4.47),307(3.70), 320(3.59),370(3.57)	65-0662-65
Chalcone, 2'-(benzyloxy)-4-methoxy-, epoxide	EtOH	258(4.04),314(3.41)	22-3350-65
	EtOH	258(4.09),313(3.69)	78-1141-64
Chalcone, 2'-(benzyloxy)-4'-methoxy-, epoxide	EtOH	235(4.24),279(4.08), 314(4.01)	22-3350-65
	EtOH	234(4.30),279(4.14), 312(4.04)	78-1141-64
1,2-Propanedione, 1-(2-benzyloxy-4-methoxyphenyl)-3-phenyl-	EtOH	233(4.07),280(3.96), 319(3.93)	22-3350-65
	EtOH-NaOH	379(3.92)	22-3350-65
1,2-Propanedione, 1-(2-benzyloxyphenyl)-3-(4-methoxyphenyl)-	EtOH	259(3.88),323(3.55)	22-3350-65
	EtOH-NaOH	381(4.11)	22-3350-65

Compound	Solvent	$\lambda_{max}(\log \epsilon)$	Ref.
$C_{23}H_{20}O_5$			
4,9-Anthracenedione, 10β-acetoxy-5-(benzyloxy)-1,4aα,9aα,10-tetrahydro-	EtOH	224(4.58),256(3.90), 316(3.56)	65-2558-64
	EtOH	225(4.57),258(3.92), 320(3.51)	65-2570-64
Anthraquinone, 4β-acetoxy-5-(benzyloxy)-1,4,4aα,9aα-tetrahydro-	EtOH	229(4.38),338(3.72)	65-2558-64
Anthrone, 4β-acetoxy-5,10β-(benzylidenedioxy)-1,4,4aα,9aα-tetrahydro-	EtOH	229(4.17),260(3.39), 315(3.33)	65-2563-64
1-Naphthacenecarboxylic acid, 2-ethyl-6,11-dihydro-5,7-dimethoxy-11-oxo-	EtOH	245(5.1),301(4.63), 427(4.40)	94-0797-65
1,3-Propanedione, 1-[3-(benzyloxy)-4-methoxyphenyl]-3-(o-hydroxyphenyl)-	EtOH	265(4.16),295(3.88), 380(4.14)	2-0351-65
$C_{23}H_{20}O_6$			
6,12-Methano-12H-dibenzo[d,g][1,3]dioxocin-1,3-diol, 6-(p-hydroxyphenyl)-8-methoxy-13-methyl-	EtOH	273(3.64),280s(3.58)	78-1471-65
$C_{23}H_{20}O_7$			
Amorphigenin, 6a,12a-dehydro-	EtOH	238(4.45),279(4.36), 309(4.25)	39-6023-64
Ichthynone	EtOH	232(4.53),262(4.39), 309(4.15),331(4.04), 345(3.97)	78-1317-64
$C_{23}H_{20}O_{11}$			
Flavanone, 3,3',4',5,7-pentahydroxy-, 3,3',4',7-tetraacetate	EtOH	277(4.13),340(3.57)	44-0897-65
$C_{23}H_{20}Sn$			
Tin, (3-methyl-3-buten-1-ynyl)-triphenyl-	EtOH	242.2(4.20)	22-0035-65
$C_{23}H_{21}BrN_2S_2$			
4-[(2,3-Dihydro-1H-pyrido[2,1-b]benzothiazol-4-yl)methylene]-1,2,3,4-tetrahydropyrido[2,1-b]benzothiazolium bromide	EtOH	582(5.09)	65-2455-64
$C_{23}H_{21}ClO_4$			
1-Propanone, 1-(2-benzyloxy-4-methoxyphenyl)-3-chloro-2-hydroxy-3-phenyl-	EtOH	228(4.24),272(4.04), 305(3.89)	22-3350-65
$C_{23}H_{21}IN_4OS$			
8-Ethyl-7-[3-(1-ethyl-4(1H)-quinolylidene)propenyl]oxazolo[4,5-e]-2,1,3-benzothiadiazolium iodide	EtOH	610(4.93)	17-0048-64
$C_{23}H_{21}NO_2$			
Isoquinoline, 7-(benzyloxy)-3,4-dihydro-6-methoxy-1-phenyl-	EtOH	235(5.45),284(3.87), 311(3.85)	95-0960-65
$C_{23}H_{21}N_2OP$			
Phospholo[2,3-c]pyrazole, 1,2,3,3a,6,6a-hexahydro-5,6,6a-triphenyl-, 6-oxide	EtOH	221(4.45),255(3.69), 333(2.08)	39-2184-65
$C_{23}H_{21}N_3O_2$			
Acetanilide, 2-benzoyl-2-[[p-(dimethylamino)phenyl]imino]-	n.s.g.	251(4.26),311(3.83), 368(3.79),424(4.11)	20-0609-65

Compound	Solvent	$\lambda_{max}(\log \epsilon)$	Ref.
Acetanilide, 2-benzoyl-2-[[p-dimethylamino)phenyl]imino]- (cont.)	n.s.g.	253(4.35),430(4.17)	20-0843-64
$C_{23}H_{21}N_3O_6$			
Cytidine, N-benzoyl-2'-deoxy-, 3'-benzoate	EtOH	232(4.33),260(4.39), 305(4.00)	44-3067-65
Cytidine, N-benzoyl-2'-deoxy-, 5'-benzoate	H_2O	230(4.27),260(4.31), 305(3.94)	44-3067-65
$C_{23}H_{22}ClNO_4$			
2-(p-Chlorophenyl)-6,7-dihydro-9,10-dimethoxybenzo[a]quinolizinium acetate	EtOH-NaOH	303(4.5),383(4.06)	33-1852-64
7-[p-(Dimethylamino)-β-phenylstyryl]-tropylium perchlorate	MeCN	277(4.17),343(3.94), 434(3.85),670(4.73)	24-1590-64
$C_{23}H_{22}N_2O$			
2-Indolinone, 3-benzyl-1-(dimethylamino)-3-phenyl-	MeOH	257(3.79)	44-3451-65
$C_{23}H_{22}N_2O_2$			
p-Propionotoluidide, 3-anilino-3-benzoyl-	EtOH	246(4.47)	39-2040-65
$C_{23}H_{22}N_2O_2S$			
5-Pyrimidinecarboxylic acid, 1,4-dihydro-6-methyl-1-phenyl-2-(2-phenylcyclopropyl)-4-thioxo-, ethyl ester	EtOH	249(4.07),338(4.41)	44-1115-64
$C_{23}H_{22}N_2O_{10}$			
1-Propene-1,2,3-tricarboxylic acid, 3-cyano-3-(2,3-dicarboxy-1-indolizinyl)-, 3-ethyl tetramethyl ester	MeOH-ether	228(4.39),238(4.38), 250(4.31),325(4.08), 339(4.07)	24-3537-65
$C_{23}H_{22}N_4O$			
Formanilide, 2'-[(4-tert-butyl-2-phenazinyl)amino]-	EtOH	235(4.62),286(4.60), 357(3.71),456(3.97)	12-1241-65
$C_{23}H_{22}N_4O_7$			
Fluorene-8aβ(4bβH)-acetic acid, 5,6,7,8-tetrahydro-1-methyl-7,9-dioxo-, methyl ester, 7-(2,4-dinitrophenylhydrazone)	EtOH	252(4.38),360(4.35), 276s(3.98),303s(3.72)	44-2528-65
$C_{23}H_{22}OS_2$			
3-Pentanone, 1,5-bis(phenylthio)-1-phenyl-	EtOH	256(4.09)	7-1093-65
$C_{23}H_{22}O_4$			
Acetophenone, 2',4'-bis(benzyloxy)-2-methoxy-	EtOH	273(4.11),305(3.90)	23-3572-65
6,12-Methano-1H-dibenzo[d,g][1,3]dioxocin-1-one, 2,3,4,12-tetrahydro-9-hydroxy-3,3-dimethyl-6-phenyl-	EtOH	224s(--),255s(--), 272(3.85)	78-3697-65
6,12-Methano-1H-dibenzo[d,g][1,3]dioxocin-1-one, 2,3,4,12-tetrahydro-6-(p-hydroxyphenyl)-3,3-dimethyl-	EtOH	226(4.37),268(4.03)	78-3697-65
5(12H)-Naphthacenone, 8-ethyl-1,4,11-trimethoxy-	EtOH	228(4.97),266(4.78), 303(3.96),315(4.02), 364(3.88)	94-0797-65
Xanthen-1(2H)-one, 3,4-dihydro-9-(p-hydroxyphenacyl)-3,3-dimethyl-	EtOH	218(--),281(4.36)	78-3707-65
	NaOEt	332(4.41)	78-3707-65

Compound	Solvent	λ_{max}(log ϵ)	Ref.
$C_{23}H_{22}O_5$			
Anthrone, 4β-acetoxy-5-(benzyloxy)-1,4,4aα,9aα-tetrahydro-10β-hydroxy-	EtOH	224(4.37),253(3.71), 308(3.44)	65-2558-64
Anthrone, 10β-acetoxy-5-(benzyloxy)-1,4,4aα,9aα-tetrahydro-4β-hydroxy-	EtOH	224(4.28),256(3.78), 311(3.29),314(3.29)	65-2558-64
	EtOH	224(4.52),257(3.88), 315(3.54)	65-2570-64
$C_{23}H_{22}O_6$			
Macluraxanthone	EtOH	242(4.31),283(4.64), 338(4.28)	44-0689-64
Xanthen-9-ol, 9-(2,6-dimethoxyphenyl)-1,8-dimethoxy-	EtOH	215(4.77),276(4.03), 285s(--)	35-2252-64
carbonium ion from	HOAc	288(4.62),357(4.34), 450(4.00),525s(3.62)	35-2252-64
$C_{23}H_{22}O_7$			
Amorphigenin	EtOH	236(4.14),242s(4.08), 293(4.22)	39-6023-64
Ichthynone, dihydro-	EtOH	220s(4.52),233s(4.39), 256s(4.20),312(4.27)	78-1317-64
$C_{23}H_{22}O_8$			
Podophyllotoxin, 4'-demethyldeoxy-, acetate	MeOH	293(3.69)	33-1203-64
$C_{23}H_{22}O_{10}$			
Flavone, 3',5-diacetoxy-3,4',6,7-tetramethoxy-	MeOH	232(4.32),260(4.19), 325(4.41)	24-2857-64
Flavone, 3',5-diacetoxy-3,4',7,8-tetramethoxy-	EtOH	265(4.25),318s(4.13), 342(4.23)	12-0934-64
Flavone, 4',5-diacetoxy-3,3',6,7-tetramethoxy-	MeOH	241(4.33),320(4.37)	24-0548-65
Glaucophanic acid, diethyl ester	EtOH	242(3.90),260(3.84), 290(3.64),330(3.72), 385(3.87),425(3.81), 515(3.92),672(5.13)	88-2313-64
$C_{23}H_{23}ClN_2O_4S$			
2,3-Dihydro-3-[2-(1,3,3-trimethyl-2-indolinylidene)ethylidene]-1H-pyrrolo-[2,1-b]benzothiazolium perchlorate	EtOH	543(5.01)	65-2455-64
4-(1-Ethyl-6-methyl-2(1H)-quinolylidene)-1,2,3,4-tetrahydropyrido[2,1-b]-benzothiazolium perchlorate	EtOH	550(4.49)	65-2455-64
$C_{23}H_{23}ClO_8$			
2,2',4,4'-Tetramethoxytriphenylcarbonium perchlorate	HOAc-$HClO_4$	565(4.40)	27-0339-65
2,2',5,5'-Tetramethoxytriphenylcarbonium perchlorate	HOAc-$HClO_4$	725(3.70)	27-0339-65
$C_{23}H_{23}IN_2O_2$			
2-Benzyl-1-(2-carboxyethyl)-9-methyl-9H-pyrido[3,4-b]indolium iodide, methyl ester	EtOH	269(4.50),315(4.28), 393(3.69)	94-1231-65
$C_{23}H_{23}N$			
5H-Dibenzo[a,d]cycloheptene, 5-(p-dimethylaminophenyl)-10,11-dihydro-	EtOH	260(4.23)	87-0457-64

Compound	Solvent	λ_{max}(log ε)	Ref.
$C_{23}H_{23}NO_3$			
1-Cyclopentene, 1-morpholino-5-benzoyl-, exo-trans enol benzoate	EtOH	232(4.27),293(4.08)	1-2139-65
Nicotinic acid, 5-acetyl-1,4-dihydro-6-methyl-2,4-diphenyl-, ethyl ester	MeOH	263(4.33),375(4.04)	44-3102-64
$C_{23}H_{23}NO_4$			
1-Pyrroline-5,5-dicarboxylic acid, 3-(diphenylmethylene)-, diethyl ester	EtOH	234(4.22),305(4.26)	95-0971-65
$C_{23}H_{23}NO_6$			
Acetamide, N-[2-(6-hydroxy-4-methoxy-phenanthro[3,4-d]-1,3-dioxol-5-yl)-ethyl]-N-methyl-, acetate	EtOH	252(4.57),284s(4.31), 324(3.87),364(3.10)	44-0432-65
$C_{23}H_{23}N_3O$			
5H-Dibenzo[b,e]-1,4-diazepine, 10-(N-benzyl-N-methylamino-ethyl)-10,11-dihydro-	EtOH	222(4.47)	33-1590-65
$C_{23}H_{23}N_3O_7S$			
Cytidine, N-benzoyl-2'-deoxy-, 5'-p-toluenesulfonate	EtOH	223(4.34),260(4.35), 305(3.98)	44-3067-65
$C_{23}H_{23}N_4O_3$			
Verdazyl, 1,3,5-tris(p-methoxyphenyl)-	benzene	750(3.65)	49-0457-64
$C_{23}H_{23}N_5$			
1,2,4-Triazole, 3,5-di-p-toluidino-4-p-tolyl-	EtOH	263(4.41)	39-3912-65
$C_{23}H_{24}$			
Azulene, 7-isopropyl-1,4-dimethyl-3-styryl-	C_6H_{12}	268(4.33),328(4.43), 645(2.62)	44-0270-65
	EtOH	231(4.20),269(3.43), 329(4.46),647(2.65)	44-0270-65
$C_{23}H_{24}ClNO_4$			
p-Diethylaminotriphenylcarbonium perchlorate	acetone	462.5(4.59)	28-6933-65A
$C_{23}H_{24}Cl_2N_2O_4$			
Malachite Green, m-chloro-, perchlorate	acid	441(4.53)	1-0157-64
	buffer	425(3.27),628(5.00)	1-0157-64
$C_{23}H_{24}IN$			
1,2,3,4-Tetrahydro-1,1-dimethyl-2,3-di-phenylquinolinium iodide, cis	EtOH	258(3.05),263(3.05), 270(2.94)	39-0249-64
$C_{23}H_{24}N_2O_3$			
Yohimban, 10-acetyl-15,16,17,18,19,20-hexadehydro-17,18-dimethoxy-	EtOH	257(4.67),287(4.17)	87-0141-64
$C_{23}H_{24}N_2O_5S$			
5-Pyrimidinecarboxylic acid, 1,4-di-hydro-6-methyl-1-phenyl-4-thioxo-2-(3,4,5-trimethoxyphenyl)-, ethyl ester	EtOH	230(4.25),340(4.40)	44-1115-64

Compound	Solvent	$\lambda_{max}(\log \epsilon)$	Ref.
$C_{23}H_{24}N_2O_6$			
3H-Phenoxazine-1,9-dicarboxylic acid, 2-(cyclopentylamino)-4,6-dimethyl-3-oxo-, dimethyl ester	MeOH	228(4.20),252(4.34), 428(4.33),446(4.37)	44-3185-65
Pleiocarpine, 10,11-dioxo-	n.s.g.	243(4.23),274(3.78)	88-3239-65
$C_{23}H_{24}N_2O_7$			
Anhydrotetracycline, methyl betaine	pH 1	272(4.70),432(3.54)	35-1795-65
Spiro[cyclopentane-1,2'(3'H)-[11aH]-oxazolo[4,5-b]phenoxazine]-4',6'-dicarboxylic acid, 11a-hydroxy-9',11'-dimethyl-, dimethyl ester	MeOH	222(4.09),244(4.33), 270(4.15),312(3.98), 376(4.31)	44-3185-65
$C_{23}H_{24}N_2O_9$			
4H-Quinolizine-1,7,9-tricarboxylic acid, 3-cyano-8-carboxymethyl-4-oxo-, tetraethyl ester	EtOH	213(4.4),225(4.41), 266(4.15),427(4.35)	39-1629-65
$C_{23}H_{24}N_4O_2$			
m-Benzanisidide, 4'-[[p-(dimethylamino)-phenyl]azo]-6'-methyl-	EtOH	232(4.15),310(3.83), 430(4.47)	65-3602-64
o-Benzanisidide, 4'-[[p-(dimethylamino)-phenyl]azo]-5'-methyl-	EtOH	245(--),260(4.13), 290s(3.91),425(4.53)	65-3602-64
$C_{23}H_{24}N_4O_3$			
Alloxazine, 5-acetyl-3-benzyl-5,10-di-hydro-1,7,8,10-tetramethyl-	MeOH	214(4.54),252s(4.24), 310(3.96)	33-1354-64
$C_{23}H_{24}N_4O_5$			
2(3H)-Phenanthrone, 4,4a,9,10-tetra-hydro-6-methoxy-1,4a-dimethyl-, 2,4-dinitrophenylhydrazone	$CHCl_3$	262(4.25),287s(4.11), 391(4.47)	39-4521-64
$C_{23}H_{24}N_6$			
Biguanide, 1-(isopropylideneamino)-2,4,5-triphenyl-	EtOH	277s(4.45)	39-0932-65
$C_{23}H_{24}N_6O_{11}$			
Atropine, 4-nitro-, nitric acid ester, picrate	MeOH	263(4.92)	83-0826-65
$C_{23}H_{24}N_8O_{11}$			
Adenine, 9-[3,4,6-tri-O-acetyl-2-deoxy-2-(2,4-dinitroanilino)-α-D-gluco-pyranosyl]-	EtOH	264(4.25),337(4.24)	44-1556-65
β-isomer	EtOH	265(4.12),341(4.11)	44-1556-65
$C_{23}H_{24}O_2$			
Cyclohexanone, 2-(3-oxo-1,5-di-phenyl-4-pentenyl)-	EtOH	290(4.37)	44-1409-65
Spiro[5.5]undecane-1,9-dione, 7,11-diphenyl-	EtOH	249(2.42),253(2.54), 259(2.64),265(2.56), 298(2.01)	44-1409-65
$C_{23}H_{24}O_5$			
Anthracene-4β,9β-diol, 10β-acetoxy-5-(benzyloxy)-1,4,4aα,9,9aα,10-hexahydro-	EtOH	209(4.51),278(3.39), 284(3.38)	65-0662-65
2H-1-Benzopyran-7-ol, 4-ethyl-3-(p-hy-droxyphenyl)-2,2-dimethyl-, diacetate	MeOH	270(3.98),309(3.90)	44-4114-65

Compound	Solvent	$\lambda_{max}(\log \epsilon)$	Ref.
$C_{23}H_{24}O_6$			
Alvaxanthone	EtOH	257(4.88),280(3.94), 332(4.38)	44-0689-64
	EtOH	257(4.88),280(3.94), 332(4.38)	44-1088-65
Macluraxanthone, dihydro-	EtOH	243(4.38),286(4.63), 338(4.32)	44-0692-64
Pentane, 1,3,5-tris(2,5-dihydroxy-phenyl)-	95% THF	299(4.06)	44-2602-65
$C_{23}H_{24}O_8$			
2-Naphthoic acid, 1,2,3,4-tetrahydro-3-(hydroxymethyl)-6,7-dimethoxy-4-oxo-1-(3,4,5-trimethoxyphenyl)-, γ-lactone	EtOH	235(4.45),278(4.03), 315(3.86)	2-0190-64
Picropodophyllone, 6,7-O-demethylene-6,7-di-O-methyl-	MeOH	237(4.41),282(4.09), 318(3.89)	33-1529-64
Podophyllotoxin, O-methyl-	MeOH	292(3.63)	33-1529-64
Podophyllotoxone, 6,7-O-demethylene-6,7-di-O-methyl-	MeOH	234(4.47),278(4.04), 313(3.80)	33-1529-64
$C_{23}H_{24}O_9$			
2-Naphthalenemalonic acid, 3-(1-carb-oxyacetonyl)-1,4-dihydro-1,4-di-oxo-, triethyl ester	MeCN	247(4.40),253(4.43), 260(4.31),335(3.49)	44-3819-65
$C_{23}H_{25}BrN_2O$			
Malachite Green, m-bromo-	acid	444(4.49)	1-0157-64
	buffer	425(4.24),628(4.95)	1-0157-64
$C_{23}H_{25}ClN_2$			
Malachite Green	$POCl_3$	618(5.0)	49-0678-64
$C_{23}H_{25}ClN_2O_5$			
1H-Azepino[5,4,3-cd]indole-2-methanol, 9-chloro-3,4,5,6-tetrahydro-6-methyl-, 3,4,5-trimethoxybenzoate	EtOH	215(4.73),275(4.26)	87-0200-65
$C_{23}H_{25}ClO_3$			
9β,10α-Androsta-1,4,6-trien-3-one, 17β-acetoxy-6-chloro-17α-ethynyl-	EtOH	226(4.05),256(4.02), 296(4.05)	33-0989-65
$C_{23}H_{25}FO_5$			
Pregna-1,4,16-triene-3,11,20-trione, 21-acetoxy-9-fluoro-	MeOH	236(4.38)	44-3486-64
$C_{23}H_{25}IN_2O$			
Malachite Green, m-iodo-	acid	446(4.45)	1-0157-64
	buffer	425(4.24),628(4.93)	1-0157-64
$C_{23}H_{25}NO$			
4-Piperidone, 3,5-dibenzylidene-N-tert-butyl-	EtOH	228(4.21),326(4.54)	39-2270-65
$C_{23}H_{25}NO_3$			
Alkaloid from Vincetoxicum officinale	n.s.g.	256(4.70),284(4.51), 341(3.18),354(3.00)	49-1094-65

Compound	Solvent	λ_{max}(log ϵ)	Ref.
$C_{23}H_{25}NO_5$			
Malonic acid, (3,3-diphenylallyl)-formamido-, diethyl ester	EtOH	253(4.25)	95-0971-65
Proline, 1-[(3,6,7-trimethoxy-9-phenanthryl)methyl]-, hydrochloride	EtOH	240(4.46),260(4.73), 284(4.50),316s(3.93), 342(3.14),358(2.78)	78-2573-65
$C_{23}H_{25}NO_6S$			
Zephyranthine, p-toluenesulfonate	EtOH	290(3.58)	95-0200-65
$C_{23}H_{25}NO_7$			
8H-Dibenzo[a,g]quinolizin-13-ol, 5,6,13,13a-tetrahydro-1,9,10-trimethoxy-2,3-(methylenedioxy)-, acetate, α-form	EtOH	282(3.44)	94-1080-64
β-form	EtOH	281(3.42)	94-1080-64
$C_{23}H_{25}NO_{10}$			
Carbostyril, 1-D-glucopyranosyl-, tetraacetate	n.s.g.	232(4.34),257(3.59), 305(3.46)	83-0824-65
$C_{23}H_{25}N_5O_7$			
L-Tryptophan, N-[N-(dinitrophenyl)-L-leucyl]-	HOAc	339(4.17)	35-1848-64
$C_{23}H_{26}ClNO_4$			
$\Delta^{1,\beta}$-Isoquinolinepropionic acid, α-(p-chlorophenyl)-6,7-dimethoxy-2-methyl-1,2,3,4-tetrahydro-, ethyl ester	EtOH-HCl	219(4.23),247(4.11), 308(3.72),360(3.89)	87-0259-65
	EtOH-base	258(3.60)	87-0259-65
$C_{23}H_{26}ClNO_5$			
Proline, 1-[(3,6,7-trimethoxy-9-phenanthryl)methyl]-, hydrochloride	EtOH	240(4.46),260(4.73), 284(4.50),316s(3.93), 342(3.14),358(2.78)	78-2573-65
$C_{23}H_{26}N_2O$			
Quinoline, 2-[p-(diethylamino)-ethoxy]styryl]-	EtOH	354(4.69)	54-0441-65
$C_{23}H_{26}N_2O_3$			
Vincolidine	EtOH	244(4.27),302(4.41)	100-0203-64
$C_{23}H_{26}N_2O_4$			
8-Azabicyclo[4.3.1]decan-10-ol, 8-methyl-10-phenyl-, p-nitrobenzoate	EtOH	258(4.21)	44-3634-65
Neblinine	EtOH	221(4.57),250s(3.80), 256(3.78),295(3.20)	35-2451-64
$C_{23}H_{26}N_2O_5$			
7H-Ajmalicine, 7-acetoxy-	EtOH	219(4.46),237(4.13), 248(4.06)	35-2229-65
Isovoacarpine, N-acetyl-	MeOH-HCl	240(4.20),315(4.25)	20-0170-65
	MeOH	240(4.20),315(4.25)	20-0170-65
	MeOH-KOH	240(4.20),315(4.25)	20-0170-65
$C_{23}H_{26}N_2O_6$			
α-D-Glucoside, methyl 4,6-O-benzylidene-3-deoxy-3-(o-tolylazo)-, 2-acetate	EtOH	272(4.03),398(2.45)	39-1045-64

Compound	Solvent	λ_{max}(log ε)	Ref.
α-D-Glucoside, methyl 4,6-O-benzylidene-3-deoxy-3-(p-tolylazo)-, 2-acetate	EtOH	276(4.22),390(2.35)	39-1045-64
$C_{23}H_{26}N_2O_7$			
α-D-Glucoside, methyl 4,6-O-benzylidene-3-deoxy-3-(p-methoxyphenylazo)-, 2-acetate	EtOH	305(4.14),396(2.60)	39-1045-64
$C_{23}H_{26}N_2O_8$			
Tetracycline methyl betaine	MeOH-HCl	271(4.24),358(4.17)	23-1382-65
$C_{23}H_{26}N_4O_2$			
Acrylonitrile, 3-[p-(diethylaminoethoxy)phenyl]-2-(p-ureidophenyl)-	EtOH-HCl	351(4.49)	87-0511-64
$C_{23}H_{26}N_4O_4$			
3H-Benz[e]inden-3-one, 1,2,3a,4,5,9b-hexahydro-7-isopropyl-9b-methyl-, 2,4-dinitrophenylhydrazone	EtOH	366(4.40)	44-1604-65
$C_{23}H_{26}N_4O_9$			
Flexuosin A, dehydro-, 2,4-dinitrophenylhydrazone	$CHCl_3$	294(3.44),363(4.42)	78-0979-64
$C_{23}H_{26}O$			
Estra-1,3,5(10)-trien-17β-ol, 17α-butadiynyl-4-methyl-	EtOH	241(2.61),255(2.59)	78-1197-65
Spiro[5.5]undecan-1-one, 7,11-diphenyl-	EtOH	249(2.44),253(2.57), 259(2.67),265(2.57), 302(1.80)	44-1409-65
$C_{23}H_{26}O_2$			
Estra-1,3,5(10)-trien-17β-ol, 17α-butadiynyl-3-methoxy-	MeOH	278(3.29),287(3.26)	13-0801-64
	EtOH	256(2.72),279(3.29), 288(3.26)	78-1197-65
$C_{23}H_{26}O_4$			
2H-1-Benzopyran-7-ol, 2,2-diethyl-3-(p-methoxyphenyl)-4-methyl-, acetate	MeOH	274(3.94),312(3.95)	44-4114-65
2H-1-Benzopyran-7-ol, 3-(p-methoxyphenyl)-2,2-dimethyl-4-propyl-, acetate	MeOH	270(3.98),307(3.95)	44-4114-65
$C_{23}H_{26}O_5$			
Carda-4,14,20(22)-trienolide, 19-hydroxy-3,11-dioxo-	MeOH	217(4.31),237s(4.26)	44-0527-64
Estra-1,3,5(10),6-tetraen-17-one, 1-acetoxy-4-acetoxymethyl-	EtOH	222(4.37),266(4.04)	39-3621-64
Furan, 2-(3,4-dimethoxyphenyl)-3,4-dimethyl-5-O-ethylisovanillyl-	n.s.g.	250(4.14),325(4.48)	39-1572-65
Furan, 2-(3,4-dimethoxyphenyl)-3,4-dimethyl-5-O-ethylvanillyl-	n.s.g.	251(4.10),327(4.49)	39-1572-65
Furoguaiacin, ethylmethyl-	n.s.g.	252(4.09),325(4.47)	39-4011-64
Naphthalene, 4-(3,4-dimethoxyphenyl)-2,6,7-trimethoxy-1-ethyl-	n.s.g.	248(4.8),298(3.9), 348(3.8)	49-0369-65
Pregna-1,4,8(14),9(11)-tetraene-3,20-dione, 21-acetoxy-17α-hydroxy-	MeOH	241(4.27)	35-2309-64
Siphulin, decarboxy-	EtOH	242s(4.48),249s(4.42), 287(4.36)	1-1677-65

Compound	Solvent	$\lambda_{max}(\log \epsilon)$	Ref.
$C_{23}H_{26}O_6$			
Amorphigenin, 12-deoxy-6',7'-dihydro-	EtOH	287s(3.81),290(3.83)	39-6023-64
12β,18-Cyclopregna-1,4-diene-3,11-dione, 17,20:20,21-bis(methylenedioxy)-	MeOH	240(4.22)	35-3394-64
Macluraxanthone, tetrahydro-	EtOH	254(4.60),287(4.07), 332(4.31)	44-0692-64
$C_{23}H_{26}O_7$			
Podophyllotoxin, 6,7-O-demethylenedeoxy-6,7-di-O-methyl-	MeOH	280(3.64)	33-1529-64
$C_{23}H_{26}O_8$			
Epipicropodophyllin, 6,7-O-demethylene-6,7-di-O-methyl-	MeOH	280(3.61)	33-1529-64
2-Naphthoic acid, 1β,2β,3α,4α-tetrahydro-4-hydroxy-3-(hydroxymethyl)-6,7-dimethoxy-1-(3,4,5-trimethoxyphenyl)-, lactone	MeOH	279(3.56)	33-1529-64
4β-isomer	MeOH	280(3.55)	33-1529-64
Neopodophyllotoxin, 6,7-O-demethylene-6,7-di-O-methyl-	MeOH	281.5(3.55)	33-1529-64
Picropodophyllic acid, deoxy-O-methyl-	EtOH	295.5(3.69)	44-1594-64
$C_{23}H_{26}O_9$			
Epipicropodophyllic acid, 3-O-methyl-	EtOH	292(3.64)	44-1594-64
Picropodophyllic acid, 3-O-methyl-	EtOH	293(3.65)	44-1594-64
$C_{23}H_{26}O_{16}$			
2,4-Cycloheptadiene-1,1,2,3,4,5,6,7-octacarboxylic acid, octamethyl ester	MeOH	228(3.94),264(3.60)	44-0423-64
3,5-Cycloheptadiene-1,1,2,3,4,5,6,7-octacarboxylic acid, octamethyl ester	MeOH	211(3.94),284(3.62)	44-0423-64
$C_{23}H_{27}BrO_6$			
Pregna-1,4-diene-3,11,20-trione, 18-bromo-17,21-dihydroxy-, 21-acetate	MeOH	239(4.21)	35-3394-64
$C_{23}H_{27}ClF_2O_6$			
Pregna-1,4-diene-3,20-dione, 16α-chloro-6α,9α-difluoro-11β,17α,21-trihydroxy-, 21-acetate	EtOH	238(4.23)	87-0751-64
$C_{23}H_{27}ClN_2O$			
2-Pentenenitrile, 2-[p-(2-diethylaminoethoxy)phenyl]-3-p-chlorophenyl-, cis, (as citrate)	EtOH	298(4.13)	87-0511-64
trans	EtOH-HCl	284(4.16)	87-0511-64
2-Pentenenitrile, 3-[p-(2-diethylaminoethoxy)phenyl]-2-p-chlorophenyl-, cis, (as citrate)	EtOH	305(4.09)	87-0511-64
trans	EtOH-HCl	294(4.17)	87-0511-64
$C_{23}H_{27}ClO_3$			
9β,10α-Androsta-4,6-dien-3-one, 17β-acetoxy-6-chloro-17α-ethynyl-	EtOH	283(4.34)	33-0989-65
$C_{23}H_{27}FN_2O$			
2-Pentenenitrile, 3-[p-(2-diethylaminoethoxy)phenyl]-2-p-fluorophenyl-, trans	EtOH-HCl	292(4.14)	87-0511-64

Compound	Solvent	λ_{max}(log ϵ)	Ref.
$C_{23}H_{27}FO_3$			
9β,10α-Androsta-1,4-dien-3-one, 17β-acetoxy-6α-fluoro-17α-ethynyl-	EtOH	242(4.21)	33-0989-65
$C_{23}H_{27}FO_5$			
Pregna-1,4,16-triene-3,20-dione, 9α-fluoro-11β,21-dihydroxy-, 21-acetate	MeOH	239(4.40)	44-3486-64
$C_{23}H_{27}FO_6$			
Pregna-1,4-diene-3,20-dione, 16α-acetoxy-17α,21-epoxy-9α-fluoro-11β-hydroxy-	MeOH	238(4.16)	13-0615-65
$C_{23}H_{27}IN_2O$			
1-[(7-Benzyloxy)-2,3-dihydro-1H-pyrrolo[1,2-a]indol-1-yl]-1-methylpyrrolidinium iodide	MeOH	278(4.03),306(3.63)	44-2904-65
$C_{23}H_{27}IO_6$			
Prednisone, 17,20:20,21-bis(methylenedioxy)-18-iodo-	MeOH	239(4.21)	35-3394-64
Prednisone, 18-iodo-, acetate	MeOH	239(4.12)	35-3394-64
$C_{23}H_{27}NO_3$			
Androsta-1,4-diene-2-carbonitrile, 17β-acetoxy-3-oxo-	EtOH	248(4.05)	44-3300-64
$C_{23}H_{27}NO_6$			
1-Demicolcine, acetate	MeOH	242(4.53),350(4.26)	5-0122-64D
Lumidemicolcine, acetate	MeOH	222(4.49),268(4.33), 290s(4.11),336(3.33), 347s(3.28)	5-0122-64D
$C_{23}H_{27}N_3O$			
Acrylonitrile, 2-[p-[2-(4-ethyl-1-piperazinyl)ethoxy]phenyl]-3-phenyl-	EtOH-HCl	330.5(4.36)	87-0511-64
$C_{23}H_{27}N_3O_2$			
Acrylonitrile, 2-[p-(diethylaminoethoxy)phenyl]-3-(p-acetamidophenyl)-	EtOH	351(4.55)	87-0511-64
Hydrouracil, 5,5-dimethyl-1,3-diphenyl-6-piperidino-	EtOH	240(3.89)	44-0812-65
$C_{23}H_{27}N_3O_3$			
Indole-3-carbonitrile, 2-[p-[2-(diethylamino)ethoxy]phenyl]-5,6-dimethoxy-	MeOH	233(4.23)	44-2253-65
$C_{23}H_{27}N_3O_5$			
Acrylonitrile, 3-[p-[2-(diethylamino)ethoxy]phenyl]-2-(4,5-dimethoxy-2-nitrophenyl)-	MeOH	230(4.27),311(4.46)	44-2253-65
$C_{23}H_{27}N_3O_6$			
Lysergic acid 2'-propanolamide, bimaleate	MeOH	321(3.91)	33-0756-64
$C_{23}H_{27}N_3O_7$			
Lysergic acid 2'-propanolamide, 12-hydroxy-, bimaleate	MeOH	325(4.04)	33-0756-64

Compound	Solvent	$\lambda_{max}(\log \epsilon)$	Ref.
$C_{23}H_{28}$			
Spiro[5.5]undecane, 1,5-diphenyl-	EtOH	249(2.46),254(2.60), 259(2.69),265(2.59), 268(2.43)	44-1409-65
$C_{23}H_{28}ClFO_4$			
Pregna-4,6-diene-3,20-dione, 17α-acetoxy-6-chloro-21-fluoro-	MeOH	284(4.33)	87-0386-65
Pregna-1,4,17(20)-trien-3-one, 16α-chloro-6α-fluoro-11β,21-dihydroxy-, 21-acetate	EtOH	242(4.24)	87-0751-64
$C_{23}H_{28}ClFO_6$			
Pregna-1,4-diene-3,20-dione, 16α-chloro-6α-fluoro-11β,17α,21-trihydroxy-, 21-acetate	EtOH	241(4.22)	87-0751-64
Pregna-1,4-diene-3,20-dione, 16α-chloro-9α-fluoro-11β,17α,21-trihydroxy-, 21-acetate	EtOH	238(4.20)	87-0751-64
$C_{23}H_{28}F_2O_6$			
Pregna-1,4-diene-3,20-dione, 16α-acetoxy-9α,21-difluoro-11β,17α-dihydroxy-	MeOH	238(4.10)	13-0615-65
$C_{23}H_{28}IN_2P$			
Bis(p-dimethylaminophenyl)methylphenylphosphonium iodide	MeOH	304(4.65)	88-2729-64
$C_{23}H_{28}N_2O$			
2-Pentenenitrile, 2-[p-(2-diethylaminoethoxy)phenyl]-3-phenyl-, cis (as citrate)	EtOH	291(4.04)	87-0511-64
trans	EtOH-HCl	277(4.13)	87-0511-64
2-Pentenenitrile, 3-[p-(2-diethylaminoethoxy)phenyl]-2-phenyl-, cis	EtOH-HCl	300(4.07)	87-0511-64
trans	EtOH-HCl	290.5(4.14)	87-0511-64
Propionanilide, N-[1-methyl-2-(1,2,3,6-tetrahydro-4-phenyl-1-pyridyl)-ethyl]-, as oxalate	MeOH	244.5(4.16)	87-0721-64
Propionanilide, N-[2-(1,2,3,6-tetrahydro-4-phenyl-1-pyridyl)propyl]-, as oxalate	MeOH	240.8(4.23)	87-0721-64
$C_{23}H_{28}N_2OS$			
Acrylonitrile, 2-[p-(diethylaminoethoxy)phenyl]-3-(p-ethylthiophenyl)-	EtOH-HCl	359(4.53)	87-0511-64
$C_{23}H_{28}N_2O_3$			
Acrylonitrile, 2-[p-(diethylaminoethoxy)phenyl]-3-[p-(2-hydroxyethoxy)phenyl]-	EtOH-HCl	346(4.48)	87-0511-64
	EtOH	346(4.47)	87-0511-64
Limatine, acetate	EtOH	216(4.35),255(4.01), 285s(3.43)	33-0822-65
Macroline, acetate	EtOH	229(4.68),284(3.91), 294s(3.86)	33-0689-65
$C_{23}H_{28}N_2O_4$			
Δ^6-Codeine, 6-deoxy-7-formyl-14-hydroxy-6-(1-pyrrolidinyl)-	EtOH	347(4.15)	78-1407-64

Compound	Solvent	λ_{max}(log ε)	Ref.
Echitovenine	EtOH	228(3.94),298(3.91), 328(4.08)	88-2239-65
Ochropine	EtOH	236(4.16),258(3.90), 337(4.35)	12-0246-64
	EtOH	236(4.16),258(3.90), 337(4.35)	100-0374-64
Secoaspidodasycarpine, N-acetyl-	EtOH	224(4.40),277s(3.75), 284(3.79),292(3.73)	78-1717-65
Voacoline, O-methyl-	MeOH	230(4.54),285(4.86)	20-0534-65
$C_{23}H_{28}N_2O_5$			
Isoreserpiline	EtOH	228(4.5),303(4.0)	28-0597-64B
Neoreserpiline	EtOH	227(4.50),304(4.02)	100-0220-64
Reserpiline	EtOH	228(4.5),303(4.0)	28-0597-64B
$C_{23}H_{28}N_2O_6$			
Carapanaubine	EtOH	218(4.44),280(3.70)	28-0597-64B
Isoreserpiline pseudoindoxyl	EtOH	224(4.36),283(4.07), 405(3.75)	35-2229-65
Rauvoxine	EtOH	218(4.44),280(3.70)	28-0597-64B
Rauvoxinine	EtOH	218(4.44),280(3.70)	28-0597-64B
$C_{23}H_{28}N_2O_8$			
Tryptamine, N,5-diacetyl-2,α,α-tricarbethoxy-	EtOH	268(4.75),311(3.99)	87-0141-64
$C_{23}H_{28}N_4O_7$			
Oxycolchicine, semicarbazone	EtOH	295(4.28)	78-1449-64
$C_{23}H_{28}O$			
Gon-4-en-3-one, 13-phenyl-	dioxan	241(4.22)	13-0409-65
$C_{23}H_{28}O_2$			
Androsta-1,4-dien-3-one, 17α-ethynyl-17β-hydroxy-19-vinyl-	EtOH	247(4.15)	33-0094-65
Androst-4-en-3-one, 17α-butadiynyl-17β-hydroxy-	MeOH	241(4.21)	13-0801-64
	EtOH	241(4.22)	78-1197-65
Benzo[fg]cyclopent[a]anthracene-3,8-dione, dodecahydro-3a,5b,11-trimethyl-	EtOH	268(4.14)	13-0023S-65
Estra-2,5(10)-dien-17β-ol, 17α-butadiynyl-3-methoxy-	EtOH	240s(2.70),252(2.60), 255(2.60),263s(2.30)	78-1197-65
$C_{23}H_{28}O_2S$			
Estra-1,3,5(10)-trien-17β-ol, 3-methoxy-17α-(2-thienyl)-	EtOH	231(4.14),278(3.32), 288(3.30)	78-1197-65
$C_{23}H_{28}O_3$			
Estra-1,3,5(10)-trien-17-one, 3-(2-propynyloxy)-, cyclic ethylene acetal	EtOH	280(3.18),288(3.20)	87-0755-64
$C_{23}H_{28}O_4$			
Cannabinolic acid, methyl ester	EtOH	220s(4.45),266(4.62), 295s(4.05),326(3.79)	78-1223-65
$C_{23}H_{28}O_5$			
Androst-4-en-3-one, 2α-acetoxy-1α,17β-dihydroxy-17α-ethynyl-	MeOH	240(4.17)	13-0345-65
Digitoxigenone, 12β-hydroxy-4,5:6,7-didehydro-	EtOH	217(4.22),283(4.40)	94-1143-64

Compound	Solvent	λ_{max}(log ϵ)	Ref.
Estra-1,3,5(10)-trien-17-one, 1-acetoxy-4-acetoxymethyl-	EtOH	267(3.53)	39-3621-64
Pregna-4,17(20)-dien-21-al, 20-acetoxy-3,11-dioxo-	MeOH	243(4.36)	39-0586-64
$C_{23}H_{28}O_6$			
Carda-4,20(22)-dienolide, 14,19-dihydroxy-3,11-dioxo-	MeOH	218(4.30),244s(4.16)	44-0527-64
Cortisone, 8(9)-dehydro-, acetate	EtOH	235(4.27)	39-0156-65
Digoxigenone, 7β-hydroxy-4,5-dehydro-	EtOH	225(4.36)	94-1143-64
Pregna-1,5-diene-3,11-dione, 17,20:20,21-bis(methylenedioxy)-	MeOH	225(4.04)	13-0183-64
5α-Pregna-1,8(9)-diene-3,11,20-trione, 21-acetoxy-17-hydroxy-	EtOH	226.5(4.11)	39-0156-65
Pregna-1,4-dien-3-one, 11β,18-epoxy-17,20:20,21-bis(methylenedioxy)-	MeOH	242(4.21)	35-3394-64
$C_{23}H_{28}O_7$			
Pregna-1,4-diene-3,11,20-trione, 21-acetoxy-16α,17-dihydroxy-	EtOH	237.5(4.20)	39-0130-65
$C_{23}H_{28}O_9$			
Picropodophyllic acid, 6,7-O-demethylene-6,7-di-O-methyl-	MeOH	281(3.54)	33-1529-64
$C_{23}H_{29}BrO_6$			
Cortisone, 9α-bromo-, acetate	EtOH	237(4.20)	39-0156-65
$C_{23}H_{29}ClO_3$			
9β,10α-Androst-4-en-3-one, 17β-acetoxy-6β-chloro-17α-ethynyl-	EtOH	234(4.20)	33-0989-65
$C_{23}H_{29}ClO_4$			
9β,10α-Pregna-4,6-diene-3,20-dione, 17α-acetoxy-6-chloro-	MeOH	286(4.32)	54-0918-65
Pregna-1,4,16-trien-3-one, 20α-chloro-11β,21-dihydroxy-, 21-acetate	EtOH	242(4.18)	87-0751-64
Pregna-1,4,17(20)-trien-3-one, 16α-chloro-11β,21-dihydroxy-, 21-acetate, trans	EtOH	241.5(4.22)	87-0751-64
$C_{23}H_{29}ClO_6$			
Pregna-1,4-diene-3,20-dione, 16α-chloro-11β,17α,21-trihydroxy-, 21-acetate	EtOH	243(4.18)	87-0751-64
$C_{23}H_{29}FO_3$			
9β,10α-Androst-4-en-3-one, 17β-acetoxy-6α-fluoro-17α-ethynyl-	EtOH	233(4.13)	33-0989-65
6β-fluoro isomer	EtOH	235(4.22)	33-0989-65
$C_{23}H_{29}FO_4$			
Pregna-4,6-diene-3,20-dione, 17α-acetoxy-21-fluoro-	MeOH	282(4.40)	87-0386-65
9β,10α-Pregna-4,6-diene-3,20-dione, 17α-acetoxy-6-fluoro-	MeOH	285(4.38)	54-0918-65
$C_{23}H_{29}FO_5$			
Pregn-4-ene-3,20-dione, 17α-acetoxy-6α,7α-epoxy-21-fluoro-	MeOH	240(4.17)	87-0386-65

Compound	Solvent	$\lambda_{max}(\log \epsilon)$	Ref.
$C_{23}H_{29}IO_6$			
Prednisolone, 18-iodo-, acetate	MeOH	242(4.19)	35-3394-64
$C_{23}H_{29}NOS$			
3-Pentanone, 1-phenyl-5-piperidino-1-(p-tolylthio)-, hydrochloride	H_2O	260(3.69)	7-0652-65
	EtOH	262(3.65)	7-0652-65
$C_{23}H_{29}NO_2S$			
3-Pentanone, 1-(p-methoxyphenylthio)-1-phenyl-5-piperidino-, hydrochloride	H_2O	232(4.13),258s(3.72)	7-1093-65
	EtOH	236(4.18),260s(3.72)	7-1093-65
$C_{23}H_{29}NO_4$			
Isoxazolino[17,16-d]pregn-4-ene-3,11,20-trione, 3'-methyl-	n.s.g.	237(4.22)	5-0139-64G
Pregn-4-ene-19-nitrile, 21-hydroxy-3,20-dioxo-, 21-acetate	EtOH	232(4.22)	87-0681-64
$C_{23}H_{29}NO_5$			
Isoxazolino[17,16-d]pregn-4-ene-3,20-dione, 3'-carboxy-	MeOH	240(4.35)	5-0228-65E
	MeOH	240(4.27)	44-1272-65
Isoxazolino[17,16-d]pregn-4-ene-3,11,20-trione, 21-hydroxy-3'-methyl-	n.s.g.	237(4.23)	5-0139-64G
$C_{23}H_{29}N_3O$			
Acrylonitrile, 2-[p-(diethylamino-ethoxy)phenyl]-3-(p-dimethylamino-phenyl)-	EtOH-HCl	395(4.58)	87-0511-64
$C_{23}H_{29}N_3O_3$			
Acetamide, N-[[2-[p-(diethylamino)-phenyl]-5,6-dimethoxyindol-3-yl]-methyl]-	MeOH	220(4.49),280(4.09), 332(4.57)	44-2253-65
$C_{23}H_{29}N_3O_7$			
Lysergic acid 2'-propanolamide, 9,10-dihydro-12-hydroxy-, bimaleate	MeOH	277(3.83),301(3.70)	33-0756-64
$C_{23}H_{29}N_5O_{10}$			
7(8H)-Pteridinone, 4-(dimethylamino)-6-methyl-8-(tetraacetyl-β-D-gluco-pyranosyl)-	MeOH	211(4.23),252(4.22), 296(3.81),356(3.94)	4-0023-64
$C_{22}H_{30}$			
1-Undecene, 1,1-diphenyl-	C_6H_{12}	250.5(4.15)	22-0693-64
$C_{23}H_{30}Br_2O_2$			
9β,10α-Pregn-4-ene-20α-carboxaldehyde, 6-(dibromomethylene)-3-oxo-	MeOH	250(3.99),285(3.83)	54-0904-65
$C_{23}H_{30}ClFO_4$			
Pregna-4,17(20)-dien-3-one, 16ξ-chloro-6α-fluoro-11β,21-dihydroxy-, 21-acetate, trans	EtOH	216s(4.12),235(4.22)	87-0751-64
$C_{23}H_{30}ClNOS$			
3-Pentanone, 1-phenyl-5-piperidino-1-(p-tolylthio)-, hydrochloride	H_2O	260(3.69)	7-0652-65
	EtOH	262(3.65)	7-0652-65

Compound	Solvent	$\lambda_{max}(\log \epsilon)$	Ref.
$C_{23}H_{30}ClNO_2S$			
3-Pentanone, 1-(p-methoxyphenylthio)-1-phenyl-5-piperidino-, hydrochloride	H_2O	232(4.13),258s(3.72)	7-1093-65
	EtOH	236(4.18),260s(3.72)	7-1093-65
$C_{23}H_{30}Cl_4O_4$			
Androst-4-en-3-one, 4-chloro-17β-(1-acetoxy-2,2,2-trichloroethoxy)-	MeOH	254(4.13)	83-0124-64
$C_{23}H_{30}N_2$			
5α-Androstane-$\Delta^{3,\alpha:17,\alpha'}$-diacetonitrile	EtOH	219(4.38)	44-0505-65
$C_{23}H_{30}N_2O$			
Propionanilide, N-[2-(4-phenylpiperidion)propyl]-, (as oxalate)	MeOH	258(3.02),264(2.92)	87-0721-64
$C_{23}H_{30}N_2OS$			
Androstano[3,2-b]pyridine-5'-carbonitrile, 1',6'-dihydro-17β-hydroxy-6'-thioxo-	EtOH	241s(3.79),309(4.30), 410(3.69)	94-0077-64
$C_{23}H_{30}N_2O_3$			
Androstano[3,2-b]pyridinecarbonitrile, 1',6'-dihydro-17β-hydroxy-6'-oxo-	EtOH	238(3.92),347(4.06)	94-0077-64
$C_{23}H_{30}N_2O_3$			
Lochrovine	EtOH	229(4.06),301(4.10), 344(4.25)	100-0203-64
Pleiocarpolinine	EtOH	207(4.42),251(4.00), 300(3.53)	33-1002-65
$C_{23}H_{30}N_2O_4$			
1H-Azecino[5,4-b]indole-6-acetic acid, 5-ethyl-2,3,4,6,7,8,9-octahydro-α-(methoxymethylene)-3-methyl-8-oxo-, methyl ester	ether	307(4.17)	35-5362-64
Fendlerine	EtOH	225(4.28),258(3.55)	25-0235-64
$C_{23}H_{30}N_2O_5$			
invert-Isoreserpic acid, methyl ester	EtOH	227(4.55),267(3.64), 297(3.71)	35-2229-65
Vindoline, deacetyl-	MeOH	252(3.91),305(3.76)	73-1913-64
$C_{23}H_{30}N_2O_6$			
7H-Reserpic acid, 7-hydroxy-, methyl ester	EtOH	233s(4.23),283(3.45)	35-2229-65
$C_{23}H_{30}N_4$			
Dipyrazolo[3,2-c][17,16-c]androst-4-ene, 1",2'-dimethyl-	EtOH	218(4.19),274(4.04)	32-0257-65
Pyrimidino[17,16-d]-5α-androstano-[3,2-c]pyrazole, 1'-methyl-	EtOH	232(3.87),251(3.72)	32-0338-65
$C_{23}H_{30}OS$			
Estra-1,3,5(10)-trien-17-one, 3-(cyclopentylthio)-	EtOH	261(3.95)	44-0295-65
$C_{23}H_{30}O_2$			
Androst-5-ene-3β,17β-diol, 17α-butadiynyl-	EtOH	230(2.50),241(2.53), 254(2.33)	78-1197-65

Compound	Solvent	λ max(log ε)	Ref.
Benzo[fg]cyclopent[a]anthracene-3,8-dione, tetradecahydro-3aα,5bα,11-trimethyl-	EtOH	318(4.03)	13-0023S-65
1H-Cyclopenta[a]chrysene-3,9(2H,3aH)-dione, dodecahydro-3aα,5bα-dimethyl-	EtOH	304(4.30)	13-0023S-65
3H-Cyclopenta[a]chrysen-3-one, tetradecahydro-9-hydroxy-3aα,5bα-dimethyl-	EtOH	281(3.33)	13-0023S-65
Gona-1,3,5(10),8-tetraen-17β-ol, 17α-allyl-13β-ethyl-3-methoxy-	n.s.g.	280(4.26)	39-4472-64
Pregn-4-ene-3,20-dione, 6;16α,17-dimethylene-	EtOH	260(4.05)	78-0597-64
Pregn-4-ene-3,20-dione, 4-ethynyl-	EtOH	268(4.10)	22-0331-64
$C_{23}H_{30}O_2S$			
Androst-4-en-3-one, 17β-hydroxy-17α-(2-thienyl)-	EtOH	239(4.38)	78-1197-65
$C_{23}H_{30}O_2S_2$			
Pregna-3,5-dien o[3,4-b]dithiane, 16α,17-epoxy-20-oxo-	EtOH	239(4.03),293(4.10)	94-1078-65
$C_{23}H_{30}O_3$			
9β,10α-Androst-4-en-3-one, 17β-acetoxy-17α-ethynyl-	EtOH	240(4.22)	33-0989-65
Estra-1,3,5(10)-trien-17-one, 3-(allyloxy)-, cyclic ethylene acetal	EtOH	280(3.30),288(3.26)	87-0755-64
Etiojerva-5,12,17(20)-triene-3,11-dione, 17-ethyl-, 3-ethylene ketal	EtOH	302(4.38)	44-0755-64
A(2)-Norhexanor-1(10),16-cucurbitadiene-3,11,20-trione	EtOH	235(4.44)	78-2665-64
A(2)-Norhexanorelatericin A, dehydro-16-deoxy-	EtOH	240(4.00)	78-2665-64
Pregna-5,14,16-trien-20-one, 3β-acetoxy-	EtOH	309(4.05)	44-1658-65
Pregna-5,14,16-trien-20-one, 3β-hydroxy-16-methyl-, formate	MeOH	310(4.14)	73-2513-64
$C_{23}H_{30}O_3S$			
Estra-1,3,5(10)-trien-17β-ol, 16β-mercapto-4-methyl-, diacetate	EtOH	234(3.72),263(2.62), 270(2.45)	94-0687-65
Pregna-3,5-dieno[3,4-b]oxathiane, 16α,17-epoxy-20-oxo-	EtOH	218(3.99),271(3.94)	94-1078-65
Thieno[4',3',2'-4,5,6]pregn-5-ene-3,20-dione, 17α-hydroxy-5'-methyl-	EtOH	221(4.11),269(4.07), 304(3.39)	94-1433-64
$C_{23}H_{30}O_4$			
Estra-1,3,5(10)-triene-1,17β-diol, 4-methyl-, diacetate	EtOH	268(2.49)	39-3621-64
Pregna-4,17(20)-dien-21-al, 20-acetoxy-3-oxo-	MeOH	244(4.33)	39-0586-64
Pregna-1,4-diene-3,20-dione, 11α-acetoxy-	MeOH	244(4.23)	24-1974-65
Pregna-1,4-diene-3,20-dione, 21-hydroxy- (acetate?)	EtOH	245(4.21)	94-0050-65
Pregna-4,6-diene-3,20-dione, 17-acetoxy-	MeOH	283(4.36)	87-0804-64
Pregna-3,5-dien-20-one, 16α,17α-epoxy-6-formyl-3-methoxy-	EtOH	217(4.08),320(4.19)	78-0597-64
5,10-Seco-5,19-cycloandrost-1(10),2,4-triene-3,17β-diol, acetate	EtOH	285(3.72)	35-3727-65

Compound	Solvent	$\lambda_{max}(\log \epsilon)$	Ref.
$C_{23}H_{30}O_4S$			
Thieno[4',3',2'-4,5,6]pregn-5-ene-3,20-dione, 17α,21-dihydroxy-5'-methyl-	EtOH	221(4.15),269(4.09), 302(3.40)	94-1433-64
$C_{23}H_{30}O_5$			
13α-Androsta-5,15-diene-3β,16-diol, 17-oxo-, diacetate	MeOH	235(3.92)	44-3304-64
5β-Card-20(22)-enolide, 19-hydroxy-3,11-dioxo-	MeOH	215(4.23)	44-0527-64
Digitoxigenone, 4,5-dehydro-7β-hydroxy-	EtOH	223.5(4.36)	94-1143-64
Digitoxigenone, 4,5-dehydro-	EtOH	226(4.32)	94-1143-64
12β-hydroxy-	EtOH	229(4.36)	94-1143-64
Estra-1,3,5(10)-triene-6β,17β-diol, 3-methoxy-, diacetate	EtOH	268(3.92)	44-2731-64
Estra-1,3,5(10)-trien-17-one, 16α-acetoxy-3-methoxy-, cyclic ethylene acetal	EtOH	280(3.31),288(3.31)	87-0755-64
16β-isomer	EtOH	281(3.18),287(3.18)	87-0755-64
A-Norandrosta-2,5-dien-1-one, 2,17β-dihydroxy-, 2-acetate 17-propionate	EtOH	297(3.97)	94-0156-65
Pregna-1,4-diene-3,20-dione, 21-acetoxy-17α-hydroxy-	EtOH	244(4.14)	13-0645-64
Pregn-4-ene-3,20-dione, 6β,19-epoxy-17α-hydroxy-, acetate	MeOH	237(4.15)	35-1528-64
Pregn-4-ene-3,20-dione, 8,19-epoxy-21-hydroxy-, acetate	EtOH	240(4.21)	35-4655-65
Pregn-4-ene-3,11,21-trione, 21-acetoxy-, vanillin complex	$HOAc-HClO_4$	560(3.48)	96-0134-65
Pregn-4-ene-3,12,20-trione, 11α-acetoxy-	EtOH	238(4.23)	44-2169-65
Pregn-4-ene-3,11,20-trione, 21,21-ethylenedioxy-	MeOH	238(4.15)	44-0640-64
Pregn-4-en-18-oic acid, 11α-acetoxy-20β-hydroxy-3-oxo-	MeOH	238(4.12)	24-1974-65
Pregn-4-en-20-yn-3-one, 1,2,17-trihydroxy-, 2-acetate	MeOH	240(4.17)	13-0403-65
Testosterone, 1-dehydro-7α-hydroxy-, 7,17-diacetate	EtOH	242(4.24)	24-3363-64
$C_{23}H_{30}O_6$			
Androst-5-ene-7,17-dione, 3β,x-diacetoxy-	EtOH	236(4.14)	95-0816-65
Card-4-enolide, 14,19-dihydroxy-3,11-dioxo-	MeOH	242(4.16)	44-0527-64
Citreoviridin	EtOH	204(4.23),234(4.01), 286s(4.39),294(4.43), 388(4.68)	88-1825-64
Cortexone, 8,19-epoxy-19-hydroxy-, 21-acetate	EtOH	243(4.15)	44-0345-64
Digitoxigenone, 4,5-dehydro-7β,12β-dihydroxy-	EtOH	226(4.32)	94-1143-64
Estra-1,3,5(10)-triene-3,7α,17β-triol, 6β-methoxy-, 3,17-diacetate	EtOH	267(2.79),275(2.76)	44-1325-64
Estra-1,3,5(10)-triene-3,7β,17β-triol, 6α-methoxy-, 3,17-diacetate	EtOH	267(2.75),273(2.72)	44-1325-64
Hydrocortisone, 2-methylene-, 21-formate	EtOH	261(4.16)	87-0528-64

Compound	Solvent	λ_{max}(log ε)	Ref.
Pregna-1,4-diene-3,11-dione, 17α,20α,21-trihydroxy-, 21-acetate	EtOH	240(4.18)	78-0179-65
Pregna-1,5-dien-3-one, 11β-hydroxy-17,20;20,21-bis(methylenedioxy)-	MeOH	226(4.05)	13-0183-64
Pregn-4-ene-3,20-dione, 11β,19-epoxy-17α,21-dihydroxy-, 21-acetate	n.s.g.	244(4.22)	13-0371-65
Pregn-4-ene-1,3,20-trione,	EtOH	245(3.56),282(3.71)	13-0645-64
21-acetoxy-17α-hydroxy-	MeOH-NaOH	245s(3.85),347(3.85)	13-0645-64
Pregn-4-en-21-oic acid, 20α-acetoxy-3,11-dioxo-	MeOH	238(4.19)	44-2559-64
$C_{23}H_{30}O_7$			
Card-1-enolide, 5β,14,19-trihydroxy-3,11-dioxo-	MeOH	232(4.04)	44-0527-64
$C_{23}H_{30}O_8$			
Helicobasidin, leucotetraacetate	EtOH	268(2.62)	94-0236-64
$C_{23}H_{31}BrO_4$			
Pregna-3,5-diene-6-carboxaldehyde, 16β-bromo-17α-hydroxy-3-methoxy-	EtOH	220(4.04),324(4.15)	78-0597-64
5α-Pregn-16-ene-12,20-dione, 3β-acetoxy-15β-bromo-	MeOH	232(4.00)	24-1188-65
$C_{23}H_{31}ClO_2$			
Androst-4-ene-3,17-dione, 4-(3-chloro-2-butenyl)-	EtOH	250(4.14)	13-0023S-65
$C_{23}H_{31}ClO_3$			
9β,10α-Androsta-4,6-dien-3-one, 17β-acetoxy-6-chloro-17α-ethyl-	EtOH	286(4.27)	33-0989-65
$C_{23}H_{31}ClO_4$			
Androsta-1,4-dien 3-one, 4-chloro-11α,17β-dihydroxy-2,17α-dimethyl-, 11-acetate	EtOH	254(4.19)	32-0159-65
9β,10α-Pregn-4-ene-3,20-dione, 17α-acetoxy-6α-chloro-	MeOH	240.5(4.13)	54-0863-65
6β-chloro isomer	MeOH	236(4.16)	54-0863-65
$C_{23}H_{31}ClO_5$			
Androst-4-ene-3,11-dione, 2β-chloro-1α,17β-dihydroxy-2α,17α-dimethyl-, 1-acetate	EtOH	241(4.02)	32-0159-65
Pregn-4-en-3-one, 19-chloro-17α,20:20,21-bis(methylenedioxy)-	n.s.g.	240(4.21)	13-0371-65
$C_{23}H_{31}FN_2O_4$			
Pregn-4-eno[3,2-c]pyrazole, 9α-fluoro-11β,17,21-trihydroxy-16α-methyl-20-oxo-	MeOH-HCl	260(4.06)	87-0352-64
$C_{23}H_{31}FO_3$			
Pregna-4,6-diene-3,20-dione, 21-fluoro-6,16α-dimethyl-17α-hydroxy-	EtOH	290(4.37)	87-0548-64
$C_{23}H_{31}FO_4$			
9β,10α-Pregn-4-ene-3,20-dione, 17α-acetoxy-6α-fluoro-	MeOH	234(4.10)	54-0863-65
6β-fluoro-	MeOH	236(4.22)	54-0863-65

Compound	Solvent	λ max(log ε)	Ref.
$C_{23}H_{31}FO_5$			
B-Homo-19-norpregn-4-en-3-one, 7ξ-fluoro-17,20;20,21-bis-(methylenedioxy)-	EtOH	240(4.17)	44-4160-65
$C_{23}H_{31}FO_6$			
Dexamethazone, 21-hydroxymethyl-	EtOH	239(4.19)	94-1181-64
Pregn-4-ene-3,20-dione, 4-fluoro-11β,17α,21-trihydroxy-, 21-acetate	EtOH	248(4.17)	44-2982-64
$C_{23}H_{31}IN_2O_4$			
Vincine, methiodide	MeOH	222(4.62),269(3.87), 292(3.80)	73-0447-64
$C_{23}H_{31}N$			
18-Norandrost-13-eno[3,2-b]pyridine, 17,17-dimethyl-	EtOH	269(3.77),277(3.66)	94-0092-64
$C_{23}H_{31}NO$			
Estra-1,3,5(10)-trien-17-one, 3-piperidino-	MeOH	251(4.04),322(3.21)	44-2047-65
5,10-Seco-5,19-cycloandrosta-1(10),2,4-trien-17-one, 3-(N-pyrrolidinyl)-	EtOH	222(4.13),283(3.91)	35-3727-65
$C_{23}H_{31}NO_2$			
Estradiol, 2-(dimethylaminomethyl)-17α-ethynyl-	EtOH	286(3.48)	94-0196-64
Pregna-3,5-diene-16α-carbonitrile, 3-methoxy-20-oxo-	EtOH	240(4.27)	78-0597-64
Pregna-5,17(20)-diene-21-nitrile, 3β-acetoxy-	EtOH	219(4.18)	44-0505-65
$C_{23}H_{31}NO_3$			
Androstano[3,2-b]pyridine-5'-carboxylic acid, 17β-hydroxy-	EtOH	276(3.81),283(3.75)	94-0077-64
Isoxazolino[17,16-d]pregn-4-ene-3,20-dione, 3'-methyl-	n.s.g.	239(4.28)	5-0139-64G
Isoxazolino[17,16-d]pregn-5-ene-3,20-dione, 3'-methyl-	MeOH	239(4.23)	4-0280-64
Pregn-4-ene-16α-carbonitrile, 17α-hydroxy-6α-methyl-3,20-dioxo-	EtOH	240(4.20)	44-3476-64
$C_{23}H_{31}NO_3S$			
Androstano[3,2-b]pyridine-5'-carboxylic acid, 1',6'-dihydro-17β-hydroxy-6'-thioxo-	EtOH	242s(3.65),305(4.30), 394(3.77)	94-0077-64
$C_{23}H_{31}NO_4$			
Androstano[3,2-b]pyridine-5'-carboxylic acid, 1',6'-dihydro-17β-hydroxy-6'-oxo-	EtOH	237(3.89),343(4.00)	94-0077-64
Pregn-4-ene-3,16-dione, 17,20-epoxy-, 16-O-acetyloxime	n.s.g.	237.5(4.27)	35-2451-65
$C_{23}H_{31}NO_4S$			
5 -Androstano[3,2-e]-1',4'-thiaz-5'-en-3'-one, 17β-acetoxy-11-oxo-	EtOH	230(3.60),298(3.41)	87-0555-64
Pregn-4-ene-[3,2-d]thiazole, 11β,17α,21-trihydroxy-16α-methyl-20-oxo-	MeOH	263(3.93)	87-0584-64

Compound	Solvent	λ max(log ε)	Ref.
$C_{23}H_{31}NO_5$			
Laudanosine methylmethine, 6'-methoxy-	MeOH	282(4.28),343(4.48)	35-2177-64
$C_{23}H_{31}N_3O_2$			
Androst-4-eno[3,2-b]pyridine-5'-carboxamide, 6'-amino-17β-hydroxy-	EtOH	223(4.32),242(4.27), 259(4.28),284s(3.99), 370(4.11)	94-0077-64
$C_{23}H_{31}N_3O_{13}S$			
D-Glucose, 2-deoxy-2-(2,4-dinitroanilino)-, S-ethyl O-methyl monothioacetal, 3,4,5,6-tetraacetate	EtOH	266(3.64),348(4.28)	44-3280-64
$C_{23}H_{32}$			
Undecane, 1,1-diphenyl-	C_6H_{12}	255(2.76),260(2.81), 262(2.82),269(2.74)	22-0587-65
$C_{23}H_{32}Br_2O_4$			
Estran-17β-ol, 3,3-(ethylenedioxy)-5α,6α-(dibromomethylene)-, acetate	EtOH	215(3.36)	22-3516-65
5β,6β-isomer	EtOH	212(3.46)	22-3516-65
$C_{23}H_{32}INO_3$			
Laudanosine methiodide, 6'-methoxy-	MeOH	289(3.88)	35-2177-64
$C_{23}H_{32}N_2O_2$			
Neblininane, 6,7-dihydro-	EtOH	221(4.48),261(3.70), 296(3.23)	35-2451-64
Pyrimidino[17,16-d]-5α-androstan-3β-ol, acetate	EtOH-HCl	252(3.72)	32-0338-65
	EtOH	250(3.75)	32-0338-65
$C_{23}H_{32}N_2O_3$			
Androstano[3,2-b]pyridinecarboxamide, 1',6'-dihydro-17β-hydroxy-6'-oxo-	EtOH	239(3.98),340(4.11)	94-0077-64
$C_{23}H_{32}N_2O_4$			
Acetic acid, (17β-hydroxy-1-oxo-A-norandrost-3(5)-en-2-ylidene)hydrazide, propionate	EtOH	245(3.88),322(4.32)	94-0156-65
Ochropine, tetrahydro-	EtOH	230(4.43),298(3.90)	12-0246-64
Secoaspidodasycarpine, N-acetyltetrahydro-	EtOH	224(4.23),273(3.67), 282(3.67),292(3.59)	78-1717-65
$C_{23}H_{32}N_2O_5$			
Androstano[2,3-c]furazan, 17β-hydroxy-, 17-hydrogen succinate	EtOH	219(3.60)	94-1445-65
$C_{23}H_{32}N_4$			
2H-Pyrazolo[3',4":3',4']cyclopenta-[1',2':5,6]naphth[1,2-f]indazole, tetradecahydro-2,4a,6a,8-tetramethyl-	MeOH	231(4.04)	32-0257-65
$C_{23}H_{32}N_4O_4S_4$			
5-Thia-1-azabicyclo[4.2.0]oct-2-ene-2-carboxylic acid, 3-(mercaptomethyl)-8-oxo-7-[2-(2-thienyl)acetamido]-, dimethyl dithiocarbamate, cyclohexylamine salt	EtOH	269(4.38)	39-5015-65

Compound	Solvent	$\lambda_{max}(\log \epsilon)$	Ref.
$C_{23}H_{32}O_2$			
Androsta-1,4-dien-3-one, 17β-hydroxy-19-(β,β-dimethylvinyl)-	EtOH	245(4.16)	33-0094-65
1H-Cyclopenta[a]chrysene-3,9(2H,3aH)-dione, tetradecahydro-3aα,5bβ-dimethyl-	EtOH	242(4.17)	13-0023S-65
9H-Cyclopenta[a]chrysen-9-one, hexadecahydro-3α-hydroxy-3aα,5bβ-dimethyl-	EtOH	305(4.27)	13-0023S-65
9,19-Cyclopregn-15-ene-3,20-dione, 4α,14-dimethyl-	EtOH	243(3.89)	35-4424-64
Estra-1,3,5(10)-trien-17-one, 3-pentoxy-	EtOH	279(3.34),288(3.29)	13-0013-64
Estra-1,3,5(10)-trien-17-one, 1,3,4-trimethyl-, cyclic ethylene acetal	EtOH	271(2.52)	94-0393-65
Gona-1,3,5(10)-trien-17β-ol, 17α-allyl-13β-ethyl-3-methoxy-	n.s.g.	278.5(3.30)	39-4472-64
Gon-4-en-3-one, 17β-hydroxy-13β-propyl-17α-propynyl-	n.s.g.	240(4.22)	39-4472-64
Pregna-4,16-diene-3,20-dione, 21-ethyl-	MeOH	238(4.43)	78-0357-64
Pregna-3,5-dien-20-one, 3-methoxy-16α,17α-methylene-	EtOH	240(4.30)	78-0597-64
Pregna-3,5,16-trien-20-one, 3-methoxy-6-methyl-	MeOH	242(4.34)	78-0357-64
Pregn-4-ene-3,20-dione, 4-vinyl-	EtOH	261(3.97)	22-0331-64
Testosterone, 2α-methyl-6-methylene-, acetate	EtOH	259(4.05)	78-0597-64
$C_{23}H_{32}O_2S$			
Androst-5-ene-3β,17β-diol, 17α-(2-thienyl)-	EtOH	237.5(3.87)	78-1197-65
Estra-1,3,5(10)-trien-17-one, 16β-(butylthio)-3-methoxy-	EtOH	222(3.98),278(3.34), 287(3.31),321(2.28)	94-0905-64
$C_{23}H_{32}O_2S_2$			
Androsta-3,5-dieno[3,4-b]dithiane, 17β-acetoxy-	hexane	240(4.11),294(4.16)	94-0383-64
	EtOH	240(4.11),294(4.16)	94-0769-65
$C_{23}H_{32}O_3$			
Androst-5-ene-3β,17β-diol, 17α-(3-oxo-1-butynyl)-	EtOH	227(3.86)	78-1197-65
Androst-4-ene-3,17-dione, 4-(3-oxobutyl)-	EtOH	250(4.12)	13-0023S-65
5β,17α-Carda-14,20(22)-dienolide, 3β-hydroxy-	EtOH	214(4.19)	94-0312-65
5β,14β,17α-Card-20(22)-enolide, 3-oxo-	EtOH	218(4.17)	94-0312-65
17β-Estradiol, 3-tetrahydropyranyl ether	n.s.g.	277(3.19)	13-0423-64
17β-Estradiol, 17-tetrahydropyranyl ether	n.s.g.	282(3.32)	13-0423-64
Estra-1,3,5(10)-trien-17-one, 3-propoxy-, cyclic ethylene acetal	EtOH	281(3.33),288(3.33)	87-0755-64
Etiojerva-5,12,17(20)-trien-3-one, 17-ethyl-11β-hydroxy-, 3-ethylene ketal	EtOH	243(4.19),250(4.26), 258(4.02)	35-0701-64
D-Homoandrosta-5,16-dien-17a-one, 3β-acetoxy-17-methyl-	EtOH	230(4.03)	73-1178-64
16,21-Methanopregna-5,16-dien-20-one, 3β-hydroxy-21-methoxy-	EtOH	246(3.98)	5-0218-65E

Compound	Solvent	λ max(log ε)	Ref.
16,21-Methanopregna-5,21-dien-20-one, 3β-hydroxy-21-methoxy-	EtOH	259(3.82)	5-0218-65E
A(2)-Norhexanor-1(10)-cucurbitene-3,11,20-trione	EtOH	236(4.21)	78-2665-64
19-Norpregn-9-en-20-one, 3β-hydroxy-5,6β-methylene-, acetate	EtOH	218(3.89)	44-4160-65
Pregna-4,6-diene-3,20-dione, 17α-hydroxy-6,16α-dimethyl-	EtOH	290(4.37)	87-0540-64
5β-Pregna-1,9(11)-diene-3,20-dione, cyclic 20-(ethylene acetal)	EtOH	226(4.31)	44-0163-64
Pregna-3,5-dien-20-one, 16α,17α-epoxy-3-methoxy-6-methyl-	EtOH	247(4.29)	78-0569-65
Pregna-3,5-dien-20-one, 16α,17α-epoxy-3-methoxy-16β-methyl-	EtOH	240(4.29)	78-0597-64
Pregna-4,17(20)-dien-3-one, 16α-acetoxy-, trans	MeOH	241(4.23)	44-1142-64
16β-acetoxy isomer	MeOH	241(4.23)	44-1142-64
Pregna-5,6-dien-20-one, 3β-acetoxy-	n.s.g.	239(3.91)	78-0387-64
Pregna-5,16-dien-20-one, 3β-acetoxy-	EtOH	239(4.01)	44-3915-65
Pregna-5,17-dien-16-one, 3β-acetoxy-, cis	EtOH	243(3.97)	94-1184-64
trans	EtOH	243(3.96)	94-1184-64
5α-Pregna-2,16-dien-20-one, 3β-acetoxy-	EtOH	192(3.88),240(3.99), 318(1.80)	1-0750-64
5α-Pregna-8,14-dien-20-one, 3β-acetoxy-	n.s.g.	250(4.41)	44-0315-64
5α-Pregna-14,16-dien-20-one, 3β-acetoxy-	MeOH	310(4.15)	24-1188-65
9β,10α-Pregna-3,5-dien-20-one, 3-acetoxy-	MeOH	236(4.28)	54-0863-65
Pregn-4-ene-3,20-dione, 17,16α-(epoxymethylene)-6α-methyl-	EtOH	240(4.21)	44-3476-64
Testosterone, 17α-methyl-6-methylene-, acetate	EtOH	260(4.07)	78-0597-64
$C_{23}H_{32}O_3S$			
Androsta-3,5-dieno[3,4-b]oxathiane, 17β-acetoxy-	hexane	222(3.96),270(3.93)	94-0383-64
	EtOH	222(3.96),270(3.93)	94-0769-65
Estradiol, 16β-ethyl-3-methoxy-, 17-acetate	EtOH	208(4.38),279(3.33), 288(3.30)	94-0905-64
Estra-1,3,5(10)-triene, 17-[(2-hydroxyethyl)thio]-3-methoxy-, acetate	EtOH	278(3.33),288(3.28)	13-0557-65
Pregn-4-ene-3,20-dione, 16α,17-epoxy-4-(ethylthio)-	EtOH	247(4.03),314(3.25)	94-1078-65
Pregn-4-ene-3,20-dione, 16β-mercapto-, acetate	EtOH	239(4.30)	87-0531-64
Thieno[4',3',2'-4,5,6]pregn-5-ene-3,20-dione, 7-hydroxy-5'-methyl-	EtOH	222(4.21),270(4.06), 295s(3.55)	94-1433-64
$C_{23}H_{32}O_4$			
Androsta-3,5-diene-6-carboxaldehyde, 17β-acetoxy-3-methoxy-	EtOH	220(4.01),320(4.17)	78-0597-64
5α-Androsta-1,3-diene-3,17β-diol, diacetate	MeOH	207(3.61)	24-3165-65
9β,10α-Androsta-3,5-diene-3,17β-diol, diacetate	MeOH	236(4.26)	54-0863-65
5α-Androst-1-en-3-one, 2-formyl-17β-hydroxy-, 17-propionate	EtOH-HCl	236(3.87)	44-3481-64
	EtOH	240(3.84)	44-3481-64
	EtOH-NaOH	305(4.16)	44-3481-64
	dioxan	245(3.98)	44-3481-64

Compound	Solvent	λ_{max}(log ε)	Ref.
Androst-5-en-3-one, 4-acetyl-17β-hydroxy-, acetate	MeOH	227(4.20),297(3.68), 324(3.72)	35-5213-64
	MeOH-base	249(4.08),316(4.01)	35-5213-64
5β,14α-Card-20(22)-enolide, 3β-hydroxy-15-oxo-	EtOH	221(4.06)	13-0593-64
Estra-1,3,5(10)-triene, 17β-(2-hydroxy-ethoxy)-3-methoxy-, acetate	EtOH	279(3.31),287(3.26)	13-0557-65
Isopregn-5-en-16β-ylacetic acid, 3β-acetoxy-20-oxo-, enol lactone	EtOH	210(3.73)	44-3578-65
A-Norpregn-3-ene-2,20-dione, 16α,17α-isopropylidenedioxy-	EtOH	234(4.19)	13-0493-64
Pregn-4-ene-16α-carboxaldehyde, 17α-hy-droxy-6α-methyl-3,20-dioxo-	EtOH	240(4.20)	44-3476-64
Pregn-4-ene-3,20-dione, 11α-acetoxy-	MeOH	240(4.09)	24-1974-65
5β-Pregn-16-ene-11,20-dione, 3α-acetoxy-	EtOH	235(3.9)	13-0805-65
9β,10α-Pregn-4-ene-3,20-dione, 2β-acetoxy-	MeOH	242.5(4.20)	54-0863-65
9β,10α-Pregn-4-ene-3,20-dione, 6β-acetoxy-	MeOH	238(4.19)	54-0863-65
Pregn-4-ene-3,20-dione, 17α-acetyl-16α-hydroxy-	MeOH	240(4.20)	5-0228-65E
Pregn-4-ene-3,6,20-trione, cyclic 20-(ethylene acetal)	EtOH	253(4.0)	39-3388-64
Taxinol, anhydroisopropylidene-	EtOH	227(3.95)	95-0404-65
Testosterone, 4-acetyl-, acetate	EtOH	241(4.18)	35-5213-64
Testosterone, 6β-acetyl-, acetate	EtOH	246.5(4.11)	35-5213-64
$C_{23}H_{32}O_4S$			
5α-Androstano[2,3-e]-2,3-dihydro-1',4'-oxathiin-2'-one, 17β-acetoxy-	EtOH	281(3.15)	87-0555-64
Pregn-4-ene-3,20-dione, 7α-hydroxy-6β-mercapto-, 6-acetate	EtOH	244(4.19)	94-1433-64
$C_{23}H_{32}O_5$			
5α-Androst-1-en-3-one, 4α,17β-diacetoxy-	EtOH	230(3.95)	44-1348-64
5α-Androst-1-en-3-one, 4β,17β-diacetoxy-	EtOH	235(4.05)	44-1348-64
5α-Androst-8-en-11-one, 3β,17β-diacetoxy-	EtOH	253(3.96)	39-2933-65
9β,10α-Androst-4-en-3-one, 2β,17β-diacetoxy-	MeOH	242(4.20)	54-0863-65
9β,10α-Androst-4-en-3-one, 6α,17β-diacetoxy-	MeOH	236(4.12)	54-0863-65
6β,17β-diacetoxy isomer	MeOH	237(4.19)	54-0863-65
5β-Card-20(22)-enolide, 14,15β-epoxy-3β,12β-dihydroxy-	EtOH	214(4.17)	33-2217-64
Etiojerva-5,12-diene-3,11-dione, 17-ethyl-17,20-dihydroxy-, 3-ethylene ketal	EtOH	251(4.15)	44-0755-64
Etiojerva-5,12-dien-11-one, 17-ethyl-3β,17ξ,20ξ-trihydroxy-, 3-acetate	EtOH	252(4.16)	35-0701-64
Pacholide, acetate	EtOH	291(1.79)	33-2330-64
Pregn-4-ene-3,20-dione, 11α-acetoxy-12β-hydroxy-	EtOH	238(4.23),310(2.55)	44-2169-65
Pregn-4-ene-3,20-dione, 21-acetoxy-11β-hydroxy-, vanillin complex	HOAc-HClO	560(4.08)	96-0134-65
Pregn-4-ene-3,20-dione, 17α,19-di-hydroxy-, 17-acetate	MeOH	241(4.21)	35-1528-64

Compound	Solvent	$\lambda_{max}(\log \epsilon)$	Ref.
9β,10α-Pregn-4-ene-3,20-dione, 6α,17α-dihydroxy-, 17-acetate	MeOH	239(4.11)	54-0863-65
6β,17α-isomer	MeOH	241(4.20)	54-0863-65
Pregn-4-ene-3,11,20-trione, 16α,21-dimethoxy-	MeOH	238(4.22)	44-0513-64
Pregn-4-ene-3,11,20-trione, 21,21-dimethoxy-	MeOH	239(4.20)	39-0586-64
17α-Pregn-4-ene-3,11,20-trione, 16α,21-dimethoxy-	MeOH	238(4.20)	44-0513-64
17α-Pregn-4-ene-3,11,20-trione, 21,21-dimethoxy-	MeOH	238(4.19)	44-0513-64
Pregn-4-en-21-oic acid, 20β-acetoxy-3-oxo-	MeOH	242(4.22)	44-2559-64
Pregn-4-en-3-one, 11α-acetoxy-18,20-epoxy-20-hydroxy-	MeOH	240(4.13)	24-1974-65
Testosterone, 7α-hydroxy-, 7,17-diacetate	EtOH	238(4.25)	24-3363-64
$C_{23}H_{32}O_6$			
Hexanedioic acid, 2-[2-(2,3,4,6,7,8-hexahydro-6-oxo-1-naphthyl)ethyl]-2-methyl-3-oxo-, diethyl ester	EtOH	301(4.21)	70-1131-64
19-Nortestosterone, 4,16α-diacetoxy-17α-methyl-	MeOH	246(2.59)	7-0221-65
Pregna-1,4-dien-3-one, 11β,17α,20α,21-tetrahydroxy-, 21-acetate	EtOH	245(4.15)	78-0179-65
Pregn-1-ene-3,20-dione, 4α,5α,21-tri-hydroxy-, 21-acetate	EtOH	231(3.96)	94-0050-65
Pregn-4-ene-3,20-dione, 1α,2α,21-tri-hydroxy-, 21-acetate	EtOH	240(4.14)	94-0050-65
Pregn-4-ene-3,20-dione, 1β,17α,21-tri-hydroxy-, 21-acetate	EtOH	241(4.18)	13-0645-64
17α-Strophanthidin	EtOH	218(4.17),284s(1.57)	33-1634-65
Strophanthidin, 16-dehydro-	EtOH	226s(3.6),268(4.2)	33-1634-65
Strophanthidol, 16-dehydro-	EtOH	218(3.71),269(4.17)	33-1634-65
$C_{23}H_{32}O_7$			
Isoandrographolide, 3-dehydro-, 19-monocathylate	EtOH	214(3.89),220(3.72), 230(3.13)	78-2617-65
Strophadogenin	EtOH	216(4.11),268s(3.0)	33-1634-65
	H_2SO_4	230(4.16),305(4.13), 415(4.32),543(3.90)	65-2463-64
$C_{23}H_{32}O_8$			
Pregn-1-ene-3,20-dione, 4α,5α,11β,17α-21-pentahydroxy-, 21-acetate	EtOH	231(3.98)	94-0050-65
Pregn-4-ene-3,20-dione, 1α,2α,11β,17α,-21-pentahydroxy-, 21-acetate	EtOH	241(4.12)	94-0050-65
$C_{23}H_{33}BF_2O_4$			
5α-Androstan-3-one, 17β-acetoxy-2-acetyl-, borofluoride complex	CH_2Cl_2	307(4.09)	44-2059-64
$C_{23}H_{33}BrO_4$			
5β-Card-20(22)-enolide, 14α-bromo-3β,15α-dihydroxy-	EtOH	211(4.06)	13-0593-64
$C_{23}H_{33}BrO_5$			
5α-Androstan-11-one, 3β,17β-diacetoxy-9α-bromo-	EtOH	317.0(1.99)	39-2933-65

Compound	Solvent	$\lambda_{max}(\log \epsilon)$	Ref.
$C_{23}H_{33}ClO_2$			
Pregna-4,6-dien-20-one, 6-chloro-17-ethyl-3β-hydroxy-	MeOH	237(4.27),244(4.34), 252(4.16)	87-0684-64
$C_{23}H_{33}ClO_3$			
5α-Androst-2-ene-2-carboxaldehyde, 17β-acetoxy-3-chloro-17α-methyl-	ether	249(4.14)	39-0788-65
9β,10α-Androst-4-en-3-one, 17β-acetoxy-6β-chloro-17α-ethyl-	EtOH	235(4.18)	33-0989-65
Pregn-4-en-3-one, 20β-hydroxy-, chloroacetate	MeOH	241(4.2)	13-0195-65
$C_{23}H_{33}ClO_5$			
Cerberigenin, chlorohydrin	n.s.g.	217(4.16)	12-1079-65
$C_{23}H_{33}FO_2$			
Pregn-4-ene-3,20-dione, 4-fluoro-6α,16α-dimethyl-	EtOH	247(4.15)	13-0337-64
$C_{23}H_{33}FO_3$			
9β,10α-Androst-4-en-3-one, 17β-acetoxy-17α-ethyl-6α-fluoro-	EtOH	233(4.12)	33-0989-65
6β-fluoro isomer	EtOH	234(4.22)	33-0989-65
$C_{23}H_{33}N$			
Androstano[3,2-b]pyridine, 17β-methyl-	EtOH	269(3.75),277(3.65)	94-0092-64
$C_{23}H_{33}NO$			
9β,10α-Androsta-3,5-dien-17-one, 3-pyrrolidino-	EtOH	281(4.33)	33-0989-65
Androstano[3,2-b]pyridine, 17β-hydroxy-17α-methyl-	EtOH	263s(3.63),269(3.76), 277(3.66)	94-0077-64
1-Dodecanone, 12-(methylamino)-1-(1-naphthyl)-	n.s.g.	215(4.60),238s(4.69), 300(3.74)	73-3111-65
24-Norchola-5,20(22)-diene-23-nitrile, 3-hydroxy-	EtOH	222(4.07)	44-0505-65
$C_{23}H_{33}NOS$			
Androstano[3,2-b]pyridine, 17β-hydroxy-6'-(methylthio)-	EtOH	249(4.05),303(3.86)	94-0077-64
$C_{23}H_{33}NO_2$			
Androstano[3,2-b]pyridine, 17β-hydroxy-17α-methyl-, 1'-oxide	EtOH	220(4.43),265(4.08)	94-0077-64
	ether	226(4.39),230(4.40), 277(4.13)	94-0077-64
Androstano[3,2-b]pyridin-6'(1'H)-one, 17β-hydroxy-17α-methyl-	EtOH	231(3.93),316(3.92)	94-0077-64
Androst-2-ene-2-carbonitrile, 17β-acetoxy-17α-methyl-	C_6H_{12}	208.5(3.99)	44-3300-64
$C_{23}H_{33}NO_2S$			
5α-Pregnano[3,2-e]-1',4'-thiaz-5'-en-3'-one, 20-oxo-	EtOH	230(3.59),299(3.38)	87-0555-64
$C_{23}H_{33}NO_3$			
5α-Androstan-3-one, 17β-acetoxy-2-acetyl-, isoxazole from	EtOH	227(3.84)	44-2059-64
Pregna-5,16-dien-20-one, 3β-acetoxy-, oxime	EtOH	215(3.90),236(4.19)	33-0608-65
17α-Pregn-4-ene-3,20-dione, 17β-acetamido-	MeOH	240(4.16)	44-0579-65

Compound	Solvent	λ_{max}(log ϵ)	Ref.
$C_{23}H_{33}NO_3S$			
5α-Androstano[3,2-e]-1',4'-thiaz-5'-en-3'-one, 17β-acetoxy-	EtOH	230(3.61),299(3.42)	87-0555-64
$C_{23}H_{33}NO_5$			
Pregn-4-ene-3,11-dione, 17,21-dihydroxy-20-methoxyimino-6β-methyl-	EtOH	237.5(4.21)	39-0148-65
Pregn-4-ene-3,20-dione, 2α-nitro-, cyclic 20-(ethylene acetal)	pH 1	252(4.21)	78-0373-64
	pH 13	242(3.94),269(3.93), 353(4.11)	78-0373-64
	MeOH	244(4.20),370(3.26)	78-0373-64
$C_{23}H_{33}NO_6$			
Pregn-4-ene-3,20-dione, 21-hydroxy-2α-nitro-, cyclic 20-(ethylene acetal)	pH 1	252(4.18)	78-0373-64
	pH 13	245(3.89),268(3.90), 354(4.08)	78-0373-64
	MeOH	245(4.19),375(2.80)	78-0373-64
1H-4,10b-Propanobenz[h]isoquinoline-6a,9(7H)-dicarboxylic acid, 2-acetyl-decahydro-8-hydroxy-10a,12a-dimethyl-, acetate	EtOH	253(3.99)	35-0291-64
$C_{23}H_{33}N_3O_2$			
Androstano[3,2-b]pyridine-5'-carboxamide, 6'-amino-17β-hydroxy-	EtOH	252(4.06),338(3.88)	94-0077-64
Dipyrromethane-5'-carboxylic acid, 3,3'-diethyl-4,4',α-trimethyl-5-(1"-pyrrolin-2"-yl)-, ethyl ester	EtOH-HCl	279(4.32),349(4.45)	39-2614-65
	EtOH	284(4.36),305(4.36)	39-2614-65
$C_{23}H_{34}$			
Anthracene, 1,2,5,6,7,8-hexahydro-4-isopropyl-1,1,5,5,8,8-hexamethyl-	C_6H_{12}	262(4.17)	23-0579-64
$C_{23}H_{34}Br_2O_3$			
5α-Pregnan-20-one, 3β-acetoxy-16β,17α-dibromo-	MeOH	303(2.04)	24-1188-65
$C_{23}H_{34}ClNO_5$			
1-Dodecanone, 12-(methylamino)-1-(1-naphthyl)-, perchlorate	n.s.g.	218(5.04),242s(4.69), 300(4.22)	73-3111-65
free base	n.s.g.	215(4.60),238s(4.69), 300(3.74)	73-3111-65
$C_{23}H_{34}N_2O$			
5α-Androstano[3,2-d]-2'-methylpyrimidine, 17β-hydroxy-17α-methyl-	EtOH	260(3.65)	87-0238-64
$C_{23}H_{34}O$			
Pregna-3,5-dien-20-one, 17-ethyl-	MeOH	228(4.31),234(4.33), 243(4.14)	87-0684-64
$C_{23}H_{34}O_2$			
Biphenyl, 1,2,5,6-tetrahydro-2',6'-dimethoxy-6-isopropylene-3-methyl-4'-pentyl-	MeOH	210s(4.49),230(3.91), 270(3.00),278(2.91)	5-0122-65E
Biphenyl, 1,4,5,6-tetrahydro-2',6'-dimethoxy-6-isopropylene-3-methyl-4'-pentyl-	EtOH	271(3.05),278s(3.01)	35-3273-65
Cannabidiol, dimethyl ether	EtOH	271(3.06),277(3.01)	35-3273-65

Compound	Solvent	λ max(log ε)	Ref.
Gona-1,3,5(10)-trien-17β-ol, 13β-ethyl-3-methoxy-17α-propyl-	n.s.g.	279(3.30)	39-4472-64
Gon-4-en-3-one, 17α-allyl-17β-hydroxy-13β-propyl-	n.s.g.	241.5(4.25)	39-4472-64
D-Homogona-1,3,5(10)-trien-17aβ-ol, 13β,17aα-diethyl-3-methoxy-	n.s.g.	280(3.14)	39-4472-64
Pregna-3,5-dien-20-one, 3-methoxy-17-methyl-	MeOH	239(4.33)	78-0357-64
Pregna-4,6-dien-20-one, 17-ethyl-3β-hydroxy-	MeOH	232(4.37),239(4.42), 247(4.23)	87-0684-64
Testosterone, 16α-allyl-17α-methyl-	EtOH	242(4.19)	32-0351-65
Testosterone, 16β-allyl-17α-methyl-	EtOH	242(4.19)	32-0351-65
$C_{23}H_{34}O_2S$			
Estra-1,3,5(10)-trien-17β-ol, 16β-(butylthio)-3-methoxy-	EtOH	279(3.35),287(3.30), 308(1.92)	94-0905-64
$C_{23}H_{34}O_3$			
Androsta-3,5-dien-17β-ol, 3-methoxy-6-methyl-, acetate	EtOH	246.5(4.29)	78-0569-65
Androsta-3,5-dien-17β-ol, 3-methoxy-17α-methyl-, acetate	EtOH	240(4.28)	78-0597-64
Androsta-3,5,7-trien-17-one, 3-methoxy-6-methyl-, dimethyl acetal	EtOH	320(4.29)	78-1753-65
9β,10α-Androst-4-en-3-one, 17β-acetoxy-17α-ethyl-	EtOH	239(4.24)	33-0989-65
Androst-4-en-3-one, 17β-(hydroxy-methyl)-, 20-acetonide	MeOH	242(4.2)	31-0499-65
Androst-5-en-7-one, 3β-hydroxy-4,4-dimethyl-, acetate	EtOH	238(4.11)	35-2832-64
5α,14α,17α-Card-20(22)-enolide, 3β-hydroxy-	EtOH	217.5(4.15)	13-0357-65
5β,14β,17α-Card-20(22)-enolide, 3β-hydroxy-	EtOH	218(4.17)	94-0312-65
Digitoxigenin, 14-deoxy-	EtOH	217(4.23)	94-0312-65
13β-Etiojerv-16-en-3β-ol, 17-acetyl-, acetate	MeOH	236(4.06)	44-2545-64
A-Homo-5α-androst-1-en-4-one, 17β-acetoxy-1-methyl-	MeOH	210(3.6),281(2.4)	5-0047-64G
Pregna-3,5-dien-20-one, 17-hydroxy-3-methoxy-6-methyl-	MeOH	245(4.27)	78-0357-64
Pregna-3,5-dien-20-one, 6-(hydroxy-methyl)-3-methoxy-	EtOH	250(4.28)	78-0597-64
Pregn-4-ene-3,20-dione, 17α-hydroxy-6α,16α-dimethyl-	EtOH	242(4.19)	87-0540-64
Pregn-4-ene-3,20-dione, 17-(2-hydroxyethyl)-	MeOH	241(4.29)	78-0357-64
Pregn-4-en-3-one, 20,20-(ethylene-dioxy)-	EtOH	241(4.15)	39-3388-64
5α-Pregn-16-en-20-one, 3β-acetoxy-, p-hydroxybenzaldehyde complex	$HOAc-HClO_4$	510(3.30)	96-0134-65
5α-Pregn-17(20)-en-16-one, 3β-acetoxy-, cis	EtOH	243(3.98)	94-1184-64
trans	EtOH	243(3.95)	94-1184-64
$C_{23}H_{34}O_3S$			
Androst-4-en-3-one, 17β-acetoxy-4-(ethylthio)-	EtOH	247(4.14)	94-0383-64
	EtOH	247(4.14),314(3.51)	94-0769-65

Compound	Solvent	λ_{max}(log ε)	Ref.
$C_{23}H_{34}O_4$			
Androsta-3,5-diene-6-methanol, 17β-acetoxy-3-methoxy-	EtOH	250(4.30)	78-0597-64
5α-Androstan-3-one, 2-acetyl- 17β-hydroxy-, acetate	EtOH	289(3.95)	35-5207-64
	EtOH	290(3.98)	44-2059-64
5α-Androstan-3-one, 4α-acetyl- 17β-hydroxy-, acetate	C_6H_{12}	288(2.25)	35-5213-64
	MeOH-NaOH	315.5(4.08)	35-5213-64
5α-Androstan-3-one, 6β-acetyl- 17β-hydroxy-, acetate	EtOH	285(1.98)	35-5213-64
5β-Androstan-3-one, 6α-acetyl- 17β-hydroxy-, acetate	EtOH	284(2.18)	35-5213-64
5α-Androstan-3-one, 17β-hydroxy- 2-(hydroxymethylene)-, propionate	EtOH	283(3.97)	44-3481-64
Pregn-4-ene-3,20-dione, 17α-hydroxy- 16α-(hydroxymethyl)-6α-methyl-	EtOH	241(4.20)	44-3476-64
5β-Pregn-1-ene-3,20-dione, 11β-hydroxy-, 20-(ethylene acetal)	n.s.g.	227(3.97)	44-0163-64
Pregn-4-en-3-one, 20α,21-dihydroxy-, 20-acetate	MeOH	241(4.23)	44-2559-64
Pregn-4-en-3-one, 20β,21-dihydroxy-, 20-acetate	MeOH	242(4.23)	44-2559-64
Pregn-4-en-3-one, 20α,21-dihydroxy-, 21-acetate	MeOH	241(4.24)	44-2559-64
Pregn-4-en-3-one, 20β,21-dihydroxy-, 21-acetate	MeOH	242(4.23)	44-2559-64
Pregn-16-en-20-one, 3β,12ξ-dihydroxy-, 3-acetate	EtOH	243(3.95)	88-2281-64
5α-Pregn-16-en-20-one, 3β,15α-di- hydroxy-, 3-acetate	MeOH	244(3.95)	24-1188-65
5α-Pregn-16-en-20-one, 3β,15β-di- hydroxy-, 3-acetate	MeOH	231.5(3.87)	24-1188-65
Uzarigenin	EtOH	217.5(4.23)	33-1775-64
$C_{23}H_{34}O_5$			
5α-Androstan-11-one, 3β,17β-diacetoxy-	EtOH	297(1.40)	39-2933-65
Androst-2-ene-2-carboxaldehyde, 17β-hy- droxy-3-(2-hydroxyethoxy)-, formate	EtOH	277(4.12)	88-2161-64
Coroglaucigenin	EtOH	218(4.20)	33-1775-64
Deoxycorticosterone, 21,21-bis- (hydroxymethyl)-	EtOH	242(4.20)	94-1181-64
Pregn-4-en-3-one, 11α,20β,21-tri- hydroxy-, 20-acetate	MeOH	242(4.20)	44-2559-64
Pregn-4-en-3-one, 17α,20α,21-tri- hydroxy-, 21-acetate	EtOH	241(4.19)	78-0179-65
$C_{23}H_{34}O_5S$			
5α-Pregnane-3,20-dione, 5α,17α-di- hydroxy-6β-mercapto-, 6-acetate	EtOH	234(3.73),305(2.20)	94-1433-64
$C_{23}H_{34}O_6$			
Antiogenin	EtOH	217(4.19)	33-2164-64
Hydrocortisone, 1α-methyl-, acetate	n.s.g.	247(4.12)	44-0163-64
Pregn-4-en-3-one, 11β,17α,20α,21- tetrahydroxy-, 21-acetate	EtOH	244(4.18)	78-0179-65
$C_{23}H_{34}O_7$			
Isoandrographolide, cathylate	EtOH	214(3.85),220(3.67), 230(3.08)	78-2617-65

Compound	Solvent	$\lambda_{max}(\log \epsilon)$	Ref.
$C_{23}H_{35}NO_2$			
Androsta-3,5-dien-17-one, 6-(dimethylaminomethyl)-3-methoxy-	EtOH	250(4.28)	78-0569-65
17α-Pregn-4-ene-3,20-dione, 17β-(dimethylamino)-	MeOH	240(4.22)	44-0579-65
hydrochloride	MeOH	239(4.18)	44-0579-65
$C_{23}H_{35}NO_2S$			
Eti-5-enamide, 3β-acetoxy-N-methylthiono-	MeOH	265(4.12),333(1.71)	35-0051-65
$C_{23}H_{35}NO_3$			
3-Aza-A-homopregn-4a-en-3-one, 20,20-(ethylenedioxy)-	EtOH	220(4.22)	39-3388-64
13β-Etiojerv-16-en-20-one, 17-acetyl-3β-hydroxy-, 3-acetate, 20-oxime	MeOH	234(4.18)	44-0123-65
5ζ-Pregn-16-ene-3β,16-diol, 20-imino-, 3-acetate	n.s.g.	322(4.12)	65-3565-64
Pregn-4-en-3-one, 20,20-(ethylenedioxy)-, oxime	EtOH	241(4.33)	39-3388-64
$C_{23}H_{35}N_5O_8$			
Bicarbamic acid, diethyl ester, reaction product with N-benzylidenebutylamine	MeCN	244(3.97)	44-2226-64
$C_{23}H_{36}$			
Anthracene, 1,2,3,4,5,6,7,8-octahydro-4-isopropyl-1,1,5,5,8,8-hexamethyl-	C_6H_{12}	271(3.09),280(3.13)	23-0579-64
$C_{23}H_{36}ClNO_2$			
17α-Pregn-4-ene-3,20-dione, 17β-(dimethylamino)-, hydrochloride	MeOH	239(4.18)	44-0579-65
$C_{23}H_{36}O_2$			
Pregna-3,5-dien-20β-ol, 3-ethoxy-	EtOH	241(4.29)	78-0597-64
Testosterone, 16α-tert-butyl-	EtOH	240(4.20)	35-0078-64
$C_{23}H_{36}O_3$			
5α-Pregnan-20-one, 3β-acetoxy-, p-hydroxybenzaldehyde complex	$HOAc-HClO_4$	520(4.26)	96-0134-65
vanillin complex	$HOAc-HClO_4$	600(4.18)	96-0134-65
$C_{23}H_{36}O_3S$			
5α-Pregnan-20-one, 3β-hydroxy-16α-mercapto-, 16-acetate	EtOH	233(3.67)	87-0531-64
5β-Pregnan-20-one, 3β-hydroxy-16α-mercapto-, 16-acetate	EtOH	233(3.67)	87-0531-64
5β-Pregnan-20-one, 3β-hydroxy-16β-mercapto-, 16-acetate	EtOH	235(3.72)	87-0531-64
$C_{23}H_{36}O_4$			
Grevillol, diacetate	EtOH	259(3.38)	12-2015-65
$C_{23}H_{36}O_5$			
Taxinine, isopropylidenedihydrodecinnamoyltrisdeacetyl-	n.s.g.	273(3.75)	95-0762-64
$C_{23}H_{37}N$			
Indole, 5,7-diisopentyl-2-tert-pentyl-	EtOH	224(4.54),273(3.96),283(3.89),294(4.54)	32-1221-64

Compound	Solvent	$\lambda_{max}(\log \epsilon)$	Ref.
$C_{23}H_{37}NO_2$			
Androsta-3,5-dien-17β-ol, 6-(dimethylaminomethyl)-3-methoxy-	EtOH	249(4.31)	78-0569-65
$C_{23}H_{38}O_2$			
Cannabidiol, dimethyl ether, reduction product	MeOH	210(4.55),271(3.12), 279(3.03)	5-0122-65E
$C_{23}H_{39}NO$			
Butylamine, N-(3α-hydroxy-5α-androstan-17-ylidene)-	hexane	248(2.24)	39-4508-65
	EtOH	238(2.34)	39-4508-65
Butylamine, N-(3β-hydroxy-5α-androstan-16-ylidene)-	MeOH	235(2.35)	39-4508-65

Compound	Solvent	$\lambda_{max}(\log \epsilon)$	Ref.
$C_{24}F_{20}Sn$			
Tin, tetrakis(pentafluorophenyl)-	C_6H_{12}	267(3.58)	39-4782-64
	MeOH	265(3.49)	39-4782-64
$C_{24}H_5F_{15}Sn$			
Tin, tris(pentafluorophenyl)phenyl-	C_6H_{12}	265(3.48)	39-4782-64
	MeOH	263(3.49)	39-4782-64
$C_{24}H_{10}F_{10}Sn$			
Tin, bis(pentafluorophenyl)-diphenyl-	C_6H_{12}	260(3.32),265(3.36)	39-4782-64
	MeOH	259(3.42),264(3.43), 268s(3.38)	39-4782-64
$C_{24}H_{12}$			
Coronene, 1,3,5-trinitrobenzene complex	$CHCl_3$	415(4.38)	38-0166-64A
$C_{24}H_{13}Br$			
1,2:4,5-Dibenzopyrene, 8-bromo-	benzene	298(4.82),310(4.93), 335(3.80),349(4.15), 366(4.42),386(4.48), 400(3.55)	24-0597-65
$C_{24}H_{13}I_2NO_9$			
Fluorescein, 4',5'-diiodo-4-nitro-, diacetate	EtOH	204(4.83)	44-0490-64
$C_{24}H_{14}$			
1,2:4,5-Dibenzopyrene	benzene	294(4.76),306(4.88), 330(3.82),344(4.10), 360(4.32),380(4.36), 399(3.22)	24-0597-65
Naphtho[2',3':11,12]fluoranthene	benzene	360(3.8),385(4.0), 405(4.2),440(4.3)	12-1138-64
$C_{24}H_{14}N_2O$			
16H-Dibenz[3,4:5,6]azepino[1,2-a]perimidin-16-one	$C_{10}H_7Cl$	390(3.81)	33-1999-65
$C_{24}H_{14}N_3PS_3$			
1-Propene-1,1,2-tricarbonitrile, 3-phenyl-3-(tri-2-thienylphosphoranylidene)-	dioxan	406(4.28)	49-1967-65
$C_{24}H_{14}O_4$			
5,8,13,14-Pentaphenetetrone, 6-ethyl-	dioxan	245(4.40),275(4.51)	65-3484-64
$C_{24}H_{15}F_5Sn$			
Tin, (pentafluorophenyl)triphenyl-	C_6H_{12}	253(3.08),260(3.20), 265(3.18)	39-4782-64
	MeOH	254(3.30),260(3.37), 264(3.36),269s(3.29), 253s?(4.30)	39-4782-64
$C_{24}H_{15}N$			
5H-Anthra[2,3-b]carbazole	$C_6H_3Cl_3$	322(5.00),364(4.16), 398(3.54),422(3.54), 485(3.37),517(3.64), 556(3.65)	24-0212-64

Compound	Solvent	$\lambda_{max}(\log \epsilon)$	Ref.
5H-Anthra[2,3-c]carbazole	EtOH	278(4.76),304(4.66), 322(4.54),337(4.48)	24-0212-64
	$C_6H_3Cl_3$	388(3.58),412(3.62), 446(3.42),475(3.60), 508(3.56)	24-0212-64
6H-Benzo[b]naphtho[2,3-h]carbazole	$C_6H_3Cl_3$	335(3.70),348(4.84), 373(4.20),394(4.06), 435(3.24),458(3.42), 492(3.45)	24-0588-65 24-2814-65
$C_{24}H_{15}NO_3$			
Benzonitrile, p-[(2-benzyl-4-oxo-4H-1-benzopyran-3-yl)carbonyl]-	MeOH	227(4.40),250(4.43), 287(4.00),304(3.91)	35-5417-65
$C_{24}H_{15}NO_9$			
Fluorescein, 4-nitro-, diacetate	EtOH	218(4.82)	44-0490-64
Fluorescein, 5-nitro-, diacetate	EtOH	219(4.85)	44-0490-64
$C_{24}H_{16}B_2O_3$			
Phenoxaborin, 10,10'-oxydi-	EtOH	229(4.8),315(2.8)	83-0524-64
$C_{24}H_{16}Br_4N_2$			
Hydrazine, tetrakis(p-bromophenyl)-	dioxan	257(4.37),294(4.49)	24-0844-65
$C_{24}H_{16}Cl_2N_2O_6$			
6,13-Triphenodioxazinedicarboxylic acid, 3,10-dichloro-, diethyl ester	DMF	482(4.71),515(4.84)	27-0242-65
$C_{24}H_{16}F_4N_2$			
Hydrazine, tetrakis(p-fluorophenyl)-	dioxan	256(4.21),289(4.37)	24-0844-65
$C_{24}H_{16}Fe_2$			
Bis(as-indacenyliron)	CH_2Cl_2	395s(2.92),509(2.51)	35-3169-64
$C_{24}H_{16}N_2O_2$			
6,11-o-Benzenonaphtho[2,3-g]quinoxaline-5,12-dione, 6,11-dihydro-2,3-dimethyl-	MeCN	251(4.65),273(4.59), 292(4.46)	44-2583-65
Pyrazine, 2,5-di-2-furyl-3,6-diphenyl-	$CHCl_3$	273(4.35),374(4.33)	22-3476-65
$C_{24}H_{16}N_2O_3$			
4-Pentenoic acid, 5-(2-benzimidazolyl)-4-hydroxy-3-oxo-2,5-diphenyl-, γ-lactone	MeOH	265(4.28)	78-0017-64
$C_{24}H_{16}N_3P$			
1-Propene-1,1,2-tricarbonitrile, 3-(triphenylphosphoranylidene)-	dioxan	404(4.22)	49-1967-65
$C_{24}H_{16}N_4O_2$			
7H,16H-Diquino[1,2-b:1',2'-b']benzo-[1,2-d:4,5-d']dipyrazole-7,16-dione, 5,6,14,15-tetrahydro-	EtOH	295.0(3.78)	39-5871-65
$C_{24}H_{16}O$			
Benzanthrone, 6-benzyl-	MeOH	232(4.67),237(4.62), 255(4.36),272(4.15), 282(4.02),304(3.94), 340(3.65),357(3.90), 389(3.97)	35-5417-65

Compound	Solvent	λ_{max}(log ϵ)	Ref.
$C_{24}H_{16}O_2$			
Dibenz[a,h]anthracen-7-ol, acetate	$CHCl_3$	287(4.91),297(5.00), 325(4.08),338(4.11), 353(4.09)	44-0686-64
$C_{24}H_{16}O_2S_2$			
Acetic acid, thio-, S,S'-3,10-perylenylene ester	$CHCl_3$	256s(4.39),260(4.49), 346s(3.50),380s(3.71), 409(4.18),433(4.47), 461(4.56)	39-2571-64
$C_{24}H_{16}O_5$			
Acenaphthenone, 2,2-bis-(3,4-dihydroxyphenyl)-	EtOH	250s(4.26),285(3.92), 318(3.73),343(3.71)	39-3040-65
$C_{24}H_{17}NO_2$			
Benzene, 1-(p-nitrophenyl)-2,3-diphenyl-	heptane	230(4.6),292(4.1)	56-0545-65
	heptane	230s(4.5),250s(4.3), 293(4.12)	56-0557-65
Benzene, 2-(o-nitrophenyl)-1,3-diphenyl-	heptane	252(4.6),300(3.3)	56-0545-65
Benzene, 2-(p-nitrophenyl)-1,3-diphenyl-	heptane	235(4.5),301(4.0)	56-0545-65 56-0557-65
Benzene, 4-nitro-1,2,3-triphenyl-	heptane	245s(4.3),260s(4.0)	56-0545-65 56-0557-65
Benzene, 5-nitro-1,2,3-triphenyl-	heptane	253(4.4),308(3.9)	56-0545-65
$C_{24}H_{17}N_3O_2$			
3,5-Pyrazolidinedione, 4-(indol-3-ylmethylene)-1,2-diphenyl-	MeOH	247(4.21),274(4.09), 430(4.56)	32-0320-65
$C_{24}H_{18}$			
Benzene, 1,2,4-triphenyl-	C_6H_{12}	250(4.67)	78-0519-64
Benzene, 1,3,5-triphenyl-	C_6H_{12}	252.5(4.76)	78-0519-64
Biphenyl, 2,2'-diphenyl-	C_6H_{12}	228.6(4.53)	78-0519-64
Biphenyl, 2,3'-diphenyl-	C_6H_{12}	231.5(4.65)	78-0519-64
Biphenyl, 2,4'-diphenyl-	C_6H_{12}	248(4.42),276(4.34)	78-0519-64
Biphenyl, 3,3'-diphenyl-	C_6H_{12}	247(4.77)	78-0519-64
Biphenyl, 3,4'-diphenyl-	C_6H_{12}	267.5(4.57)	78-0519-64
Biphenyl, 4,4'-diphenyl-	C_6H_{12}	294(4.66)	78-0519-64
1,2,9,10,17,18-Dehydro[2.2.2]paracyclophane	EtOH	245(4.3),300s(3.7)	12-0070-65
Naphth[2,1-a]anthracene, 3,12-dimethyl-	benzene	294(5.24),314(4.81), 339(3.96),348s(3.96), 357(3.96),368(3.97), 378(4.02),384s(3.92), 400(3.81)	39-0724-64
Picene, 2,9-dimethyl-	$CHCl_3$	261(4.77),278(4.85), 289(4.99),306(4.58), 319(4.27),334(4.36), 360(3.03),378(3.00)	39-0724-64
$C_{24}H_{18}Cl_2N_2O_{12}$			
7,8-Dihydroxydipyrido[2,1-b:1',2'-j]-[3,8]phenanthrolinediium diperchlorate, diacetate	MeOH	218(4.63),253(4.47), 300(4.06),345(4.32), 360(4.31),367(4.31), 386(4.41)	44-0252-65

Compound	Solvent	λ_{max}(log ε)	Ref.
$C_{24}H_{18}Cl_2O$			
3,8-Epoxycyclobuta[b]naphthalene, 1,2-dichloro-1,2,2a,3,8,8a-hexahydro-3,8-diphenyl-, cis	C_6H_{12}	247(2.82),252(2.92), 258(3.02),265(3.04), 272(2.88)	24-0732-64
$C_{24}H_{18}Fe_2$			
Diacetylene, 1,4-diferrocenyl-	MeOH	287(4.26),450(3.23)	49-1750-64
$C_{24}H_{18}N_2$			
Fluorene-2-carbonitrile, 9-[3-(N-methylanilino)allylidene]-	MeCN	258(4.83),303(4.22), 314(4.25),445(4.84)	24-3331-64
$C_{24}H_{18}N_2O$			
Pyridazine, 3-(2-fluorenyl)-6-(p-methoxyphenyl)-	EtOH	323(4.70)	39-7005-65
$C_{24}H_{18}N_2O_2$			
Pyridazine, 3-(p-methoxyphenyl)-6-(2-xanthenyl)-	EtOH	313(4.66)	39-7005-65
$C_{24}H_{18}N_2O_2S$			
Phenol, 4-[2'-(2"-phenoxyphenylmercapto)phenylazo]-	HOAc	351(4.35)	7-0821-64
Phenol, 4-[4'-(2"-phenoxyphenylmercapto)phenylazo]-	HOAc	366(4.50)	7-0821-64
Phenol, 4-[4'-(3"-phenoxyphenylmercapto)phenylazo]-	HOAc	366(4.48)	7-0821-64
Phenol, 4-[2'-(2"-phenylmercaptophenoxy)phenylazo]-	HOAc	355(4.40)	7-0821-64
Phenol, 4-[4'-(2"-phenylmercaptophenoxy)phenylazo]-	HOAc	354(4.49)	7-0821-64
Phenol, 4-[4'-(3"-phenylmercaptophenoxy)phenylazo]-	HOAc	353(4.51)	7-0821-64
$C_{24}H_{18}N_2O_4S$			
Phenol, 4-[2'-(2"-phenoxyphenylsulfonyl)phenylazo]-	HOAc	358(4.33)	7-0821-64
Phenol, 4-[4'-(2"-phenoxyphenylsulfonyl)phenylazo]-	HOAc	359(4.45)	7-0821-64
Phenol, 4-[4'-(3"-phenoxyphenylsulfonyl)phenylazo]-	HOAc	360(4.44)	7-0821-64
Phenol, 4-[2'-(2"-phenylsulfonylphenoxy)phenylazo]-	HOAc	357(4.31)	7-0821-64
Phenol, 4-[4'-(2"-phenylsulfonylphenoxy)phenylazo]-	HOAc	353(4.49)	7-0821-64
Phenol, 4-[4'-(3"-phenylsulfonylphenoxy)phenylazo]-	HOAc	352(4.48)	7-0821-64
$C_{24}H_{18}N_2O_6$			
6,13-Triphenodioxazinedicarboxylic acid, diethyl ester	DMF	475(4.68),507(4.81)	27-0242-65
$C_{24}H_{18}N_2Zn$			
Zinc, (1,10-phenanthroline)diphenyl-	toluene	361(3.01)	101-0222-65A
$C_{24}H_{18}N_4$			
Benzo[c]cinnoline, 2-(4,4'-diamino-3-biphenylyl)-	22N H_2SO_4	258(4.6),395(4.2)	12-0190-65
Phenazine, 2,3-dianilino-	EtOH	285(--),480(--)	12-1241-65
Phenazine, 2,7-dianilino-	EtOH	285(--),490(--)	12-1241-65

Compound	Solvent	λ_{max}(log ε)	Ref.
$C_{24}H_{18}O$			
Bicyclo[3.1.0]hex-3-en-2-one, 4,5,6-triphenyl-	n.s.g.	281(4.14)	77-0243-65
3,8-Epoxycyclobuta[b]naphthalene, 2a,3,8,8a-tetrahydro-3,8-diphenyl-	C_6H_{12}	247(2.74),253(2.89), 258(3.02),265(3.06), 272(2.91)	24-0372-64
Tricyclo[2.1.1.0^{5,6}]hexan-2-one, 4,5,6-triphenyl-	dioxan	243(4.23)	35-2091-64
$C_{24}H_{18}O_4$			
Benzene, p-bis(3,4-methylene-dioxystyryl)-	dioxan	246(4.28),373(4.85)	38-2839-64A
Chromone, 3-benzoyl-2-(p-methoxybenzyl)-	MeOH	226(4.50),252(4.34), 295(3.99),305(3.90)	35-5417-65
$C_{24}H_{18}O_6S_3$			
1,4:7,10:13,16-Triepithio[18]annulene-5,11,18-tricarboxylic acid, trimethyl ester	EtOH	233(4.29),250(4.29), 325(4.17),390(4.11)	12-0070-65
$C_{24}H_{18}O_8$			
Benzo[1,2-b:4,5-b']bisbenzofuran-6,12-diol, 3,9-dimethoxy-, diacetate	dioxan	232(4.62),246s(4.43), 269(4.28),307s(4.41), 320(4.78),336(4.96)	1-1063-65
Benzo[1,2-b:5,4-b']bisbenzofuran-6,12-diol, 3,9-dimethoxy-, diacetate	dioxan	251(4.77),261(4.63), 296s(4.50),300(4.51), 305(4.49),325(4.34), 335s(3.75)	1-1063-65
$C_{24}H_{18}O_9$			
Spiro[naphthalene-2(1H),1'-phthalan]-1,3'-dione, 4,7',8-trihydroxy-4'-methyl-, triacetate	CH_2Cl_2	238(4.65),290(3.82), 342(3.44)	33-1459-64
$C_{24}H_{19}ClN_4O_5$			
Purine, 6-chloro-9-(2,5-di-O-benzoyl-3-deoxy-β-D-ribofuranosyl)-	EtOH	232(4.45),265(4.00), 281(3.44)	87-0659-65
$C_{24}H_{19}ClO_5$			
2,4-Diphenyl-6-p-tolylpyrylium perchlorate	MeOH	325(4.09),415(2.89)	89-0437-64
$C_{24}H_{19}NO$			
2(1H)-Pyridone, 3-benzyl-1,6-diphenyl-	EtOH	322(4.05)	78-3255-65
$C_{24}H_{19}NO_2$			
Quinoline, 2-benzyl-3-benzoyl-4-methoxy-	MeOH	226(4.72),250(4.37), 317(3.32)	35-5417-65
4(1H)-Quinolone, 2-benzyl-3-benzoyl-1-methyl-	MeOH	242(4.54),250(4.52), 325(4.16),338(4.10)	35-5417-65
$C_{24}H_{19}N_3$			
Indole, 2-methyl-3-[2-methyl-3H-indol-3-ylidene)-4-pyridylmethyl]-	N HCl	282(4.15),550(3.52)	50-0077-64D
	EtOH	280(4.03),460(3.47)	50-0077-64D
$C_{24}H_{19}N_3O_2$			
3-Pyridinemethanol, 5-hydroxy-4-(indol-3-yl-3H-indol-3-ylidenemethyl)-6-methyl-	N HCl	282(4.10),530(3.32)	50-0077-64D
	EtOH	280(4.07),440(3.08)	50-0077-64D

Compound	Solvent	λ_{max}(log ε)	Ref.
$C_{24}H_{19}OPS$			
Ketone, 2-thienyl (triphenyl- phosphoranylidene)methyl	EtOH	248(4.11),253(4.12), 325(4.16)	65-2738-64
$C_{24}H_{19}O_2P$			
Ketone, 2-furyl (triphenyl- phosphoranylidene)methyl	EtOH	255(4.23),260(4.19), 320(4.27)	65-2738-64
$C_{24}H_{19}O_4P$			
Phosphinic acid, bis(p-phenoxyphenyl)-	EtOH	246.5(4.47)	87-0891-65
$C_{24}H_{20}$			
Benzo[j]fluoranthene, 4,6,9,11-tetramethyl-	EtOH	232(4.63),246(4.74), 321(4.54),342(4.05), 358(3.67),377(4.00), 397(4.12)	39-2380-64
1,2-Benzophenanthreno[9',10':3,4]- tetracene	benzene	304(4.92),328(5.12), 345(4.74),362(4.76), 388(3.78),412(3.94), 438(4.12),466(4.04)	78-2107-64
	C_6H_{12}	254(5.00),268(4.91)	78-2107-64
Perylene, 1-butyl-	C_6H_{12}	252(4.52),258(4.71), 385(4.08),404(4.51), 428(4.60)	44-2469-64
Perylene, 3-butyl-	n.s.g.	227(4.28),247(4.54), 255(4.75),265(4.04), 373s(3.70),392(4.08), 413(4.58),437(4.73), 443(4.83)	44-2469-64
Styrene, β,β-distyryl-	EtOH	237(4.21),299(4.47), 331(4.48)	44-2335-64
$C_{24}H_{20}Fe_2$			
Butatriene, 1,4-diferrocenyl-	THF	370(4.43),510(3.72)	101-0355-65A
$C_{24}H_{20}Fe_2S$			
Thiophene, 2,5-diferrocenyl-	$CHCl_3$	329(4.20),455(3.23)	49-1750-64
$C_{24}H_{20}Ge$			
Germane, tetraphenyl-	heptane	252(3.32),259(3.23), 263(3.09),265(3.10), 269(2.93)	28-3931-65A
$C_{24}H_{20}IP$			
Tetraphenylphosphonium iodide	EtOH	233(4.36),263(3.55), 269(3.68),276(3.60)	59-1143-64
$C_{24}H_{20}N_2$			
Aniline, 2-(p-biphenylylamino)- 4-phenyl-	EtOH	336(4.64)	77-0294-65
Hydrazine, tetraphenyl-	dioxan	258(4.19),294(4.31)	24-0844-65
$C_{24}H_{20}N_2O_2P_2$			
Diimide, bis(diphenylphosphinyl)-	dioxan	569(1.36)	24-2273-65
$C_{24}H_{20}N_2O_4$			
2-Indolinone, 3,3'-ethanediylidene- bis[1-propionyl-	benzene	395(4.24),415(4.28), 469(4.34)	39-1455-65
isomer	benzene	389(4.17),410(4.19), 461(4.28)	39-1455-65

Compound	Solvent	λ max(log ε)	Ref.
$C_{24}H_{20}N_2P_2S_2$			
Diimide, bis(diphenyl-phosphinothioyl)-	EtOH	538(0.7)	24-2273-65
$C_{24}H_{20}N_4O_4$			
Bicyclo[3.1.0]hexan-2-one, 5,6-di-phenyl-, 2,4-dinitrophenylhydrazone	EtOH	226(4.15),370(4.42)	35-4036-64
isomer	EtOH	368(4.40)	35-4036-64
$C_{24}H_{20}N_4O_5$			
Purine, 9-(2,5-di-O-benzoyl-3-deoxy-β-D-ribofuranosyl)-	EtOH	231(4.45),263(3.91), 280s(3.37)	87-0659-65
$C_{24}H_{20}N_4O_5S$			
6-Purinethiol, 9-(2,5-di-O-benzoyl-3-deoxy-β-D-ribofuranosyl)-	EtOH	230(4.53),278(3.60), 285(3.69),324(4.37)	87-0659-65
$C_{24}H_{20}N_4O_{10}$			
2-Xantheneacetaldehyde, 1-hydroxy-3,5,6-trimethoxy-, 2,4-dinitro-phenylhydrazone	MeOH	246(4.58),283(3.88), 316(4.26)	78-2653-65
$C_{24}H_{20}O$			
Cyclopentene, 1-benzoyl-2,3-diphenyl-	EtOH	246.5(4.33)	22-0722-65
Cyclopentene, 3-benzoyl-1,2-diphenyl-	EtOH	246.5(4.42)	22-0722-65
$C_{24}H_{20}O_2$			
2-Cyclobutene-1-acetic acid, 1,2,4-triphenyl-	EtOH	257(4.12)	35-2091-64
o-Toluic acid, α-(1,4-dimethyl-2-anthryl)-	50% EtOH	260(4.97),340(3.58), 367(3.66),386(3.56)	78-0467-65
$C_{24}H_{20}O_4Si$			
Phenyl silicate	C_6H_{12}	266(3.55)	12-1579-65
$C_{24}H_{20}O_5$			
Flavone, 2'-(benzyloxy)-5,7-dimethoxy-	EtOH	262(4.41),320(4.22)	78-2633-65
Xanthorrhoeol, methyl ether, piperonylidene derivative	EtOH	223(4.61),301(4.11), 348(4.35)	12-0575-65
$C_{24}H_{20}O_6$			
Flavone, 3',4',5'-trimethoxy-7-phenoxy-	n.s.g.	272(3.86),310(4.41)	28-6930-65A
$C_{24}H_{20}O_8$			
Viridin, diacetate	$CHCl_3$	242(4.50),295(3.97), 310(4.02),317(4.02)	39-3803-65
	n.s.g.	317(4.0)	77-0343-65
$C_{24}H_{20}O_{14}$			
s-Indacene-1,2,3,5,6,7-hexacarboxylic acid, 4,8-dihydroxy-, hexamethyl ester	aq. KOH	254(4.64),323(5.62), 403(3.98),465s(3.20)	88-1161-64
	CH_2Cl_2	232(4.36),287(4.68), 305s(4.59),345(4.81), 386(3.87),451s(3.84), 479(4.29),513(4.63), 700(2.59)	88-1161-64
$C_{24}H_{20}Pb$			
Lead, tetraphenyl-	isooctane	260(3.00)	101-0265-64B

Compound	Solvent	λ_{max}(log ε)	Ref.
$C_{24}H_{20}Si$			
Silane, tetraphenyl-	$CHCl_3$	254(3.04),260(3.11), 265(3.15),272(3.04)	39-4690-65
$C_{24}H_{20}Sn$			
Tin, tetraphenyl-	C_6H_{12}	228(4.9),258f(4.2)	5-0001-64G
	heptane	247(2.90),252(3.06), 259(3.16),262(2.96), 265(3.05),268(2.74)	28-3931-65A
	isooctane	260(2.90)	101-0265-64B
$C_{24}H_{21}ClO_4$			
Methyldi(1-methyl-3-azulenyl)-methene perchlorate	HOAc	670(4.49)	65-0894-64
$C_{24}H_{21}ClO_5$			
7-[4-(p-Methoxyphenyl)-4-phenyl-1,3-butadien-1-yl]tropylium perchlorate	MeCN	230(4.45),255s(4.32), 300s(4.03),350(3.65), 557(4.67)	24-1590-64
$C_{24}H_{21}N$			
1,2-Benzofluorene, 9-(5-dimethylamino-2,4-pentadien-1-ylidene)-	MeCN	240s(4.42),254s(4.52), 267(4.69),285s(4.51), 296s(4.38),494(4.70)	24-3331-64
2,3-Benzofluorene, 9-(5-dimethylamino-2,4-pentadien-1-ylidene)-	MeCN	256s(4.70),274(4.86), 288s(4.73),310s(4.25), 473(4.87)	24-3331-64
3,4-Benzofluorene, 9-(5-dimethylamino-2,4-pentadien-1-ylidene)-	MeCN	240(4.48),255(4.48), 275(4.52),323(3.89), 337(3.93),490(4.79)	24-3331-64
Indene, 1-(dimethylaminomethylene)-2,3-diphenyl-	MeCN	243(4.45),254s(4.41), 289(4.36),389(4.40)	24-3331-64
$C_{24}H_{21}NO_2S$			
Crotonic acid, 2-amino-3-(hydroxymethyl)-4-(tritylthio)-, γ-lactone	HCl	267(4.00)	44-3560-64
	EtOH	267(4.04)	44-3560-64
$C_{24}H_{21}NO_3$			
3-Indolineacetic acid, 1-benzyl-2-oxo-3-phenyl-, methyl ester	EtOH	256(3.87),266s(3.76), 280s(3.27)	44-2973-65
$C_{24}H_{21}N_3O_5S$			
Spiro[indoline-3,3'-pyrrolidin]-2-one, 2'-(p-nitrophenyl)-1'-(p-tolylsulfonyl)-	EtOH	229(4.27),262(4.16), 272s(4.11)	78-1327-65
$C_{24}H_{22}$			
1,1'-Binaphthyl, 5,5',7,7'-tetramethyl-	EtOH	228(5.00),292(4.16), 297(4.16)	39-2380-64
Distyrylbenzene, 2,2'-dimethyl-	dioxan	244(4.18),280s(4.34), 350(4.72)	38-2839-64A
Distyrylbenzene, 3,3'-dimethyl-	dioxan	242(4.15),250s(4.13), 280s(4.10),357(4.83)	38-2839-64A
Distyrylbenzene, 4,4'-dimethyl-	dioxan	359(4.76)	38-2839-64A
Ethane, 1,2-bis(4-methyl-1-naphthyl)-	C_6H_{12}	213s(4.78),230(5.03), 269s(3.87),283(4.11), 293(4.23),303(4.13), 316(3.34)	44-0963-64

Compound	Solvent	λ_{max}(log ϵ)	Ref.
$C_{24}H_{22}Br_2O_4$			
Propiophenone, 2'-(benzyloxy)-2,3-dibromo-4'-methoxy-3-(p-methoxyphenyl)-	EtOH	228(4.25),281(4.02), 315(3.96)	22-3350-65
$C_{24}H_{22}ClNO_4$			
6-Methyl-2-phenyl-1-p-tolyllepidinium perchlorate	EtOH	258(4.96),335(3.92)	65-0506-65
$C_{24}H_{22}Cl_2Ti$			
Titanium, dichlorobis(1-methyl-3-phenylcyclopentadienyl)-	CH_2Cl_2	246(4.44),288(4.30), 316(4.22),462(3.24), 550(2.98)	44-2205-65
$C_{24}H_{22}Fe$			
1,3,5,7-Octatetraene, 1-ferrocenyl-8-phenyl-	$CHCl_3$	386(4.83)	5-0088-64F
$C_{24}H_{22}Fe_2$			
Butadiene, 1,4-diferrocenyl-	C_6H_{12}	328(--),466(--)	5-0088-64F
	$CHCl_3$	335(4.44),468(3.48)	5-0088-64F
	THF	330(4.48),463(3.56)	101-0355-65A
$C_{24}H_{22}Fe_2O_2$			
Biferrocenyl, 1',6'-diacetyl-	EtOH	227(4.4),284(4.2)	44-0323-64
$C_{24}H_{22}NO_4P$			
Succinic acid, imino(triphenylphosphoranylidene)-, dimethyl ester	n.s.g.	289(3.70)	88-1263-64
$C_{24}H_{22}N_2O_2$			
3H-Pyrrolo[1,2,3-de]quinoxaline, 5,6-dihydro-2,3-bis(p-methoxyphenyl)-	MeOH	258s(4.19),284(4.36), 322s(4.14),419(3.68)	44-2589-65
$C_{24}H_{22}N_2O_6S$			
Thymidine, 5'-thio-, 3',5'-dibenzoate	pH 1	268.5(4.16)	44-0554-64
	pH 7	268.5(4.19)	44-0554-64
	pH 13	265(3.99)	44-0554-64
$C_{24}H_{22}N_2O_6S_2$			
Cephalosporadesic acid, O-p-toluoyl-7-phenylmercaptoacetamido-, sodium salt	n.s.g.	240(4.41)	87-0022-65
$C_{24}H_{22}O_2$			
Benzene, p-bis(o-methoxystyryl)-	dioxan	246(4.36),366(4.76)	38-2839-64A
Benzene, p-bis(p-methoxystyryl)-	dioxan	240(4.07),360(4.56)	65-2073-65
	dioxan	358(4.86)	38-2839-64A
$C_{24}H_{22}O_4$			
2-Anthraceneacetic acid, 5-(benzyloxy)-1,2,3,4-tetrahydro-9-methyl-4-oxo-	EtOH	223(4.51),260(4.60), 309(3.74),320(3.64), 372(3.74)	65-0670-65
Chalcone, 2'-(benzyloxy)-3,4-dimethoxy-	EtOH	252s(--),353(4.30)	22-3350-65
Chalcone, 2'-(benzyloxy)-4,4'-dimethoxy-	EtOH	237(4.17),345(4.30)	22-3350-65
$C_{24}H_{22}O_5$			
Chalcone, 2'-(benzyloxy)-3,4-dimethoxy-, epoxide	EtOH	256(3.98),291(3.75), 315s(--)	22-3350-65
Chalcone, 2'-(benzyloxy)-4,4'-dimethoxy-, epoxide	EtOH	231(4.32),278(4.16), 312(4.01)	22-3350-65

Compound	Solvent	λ_{max}(log ε)	Ref.
Chalcone, 4'-(benzyloxy)-2'-hydroxy-4,5'-dimethoxy-	EtOH	241(3.94),367(4.29)	2-0369-65
1,2-Propanedione, 1-(2-benzyloxy-4-methoxyphenyl)-3-(4-methoxyphenyl)-	EtOH	228(4.22),280(3.97), 320(3.89)	22-3350-65
	EtOH-NaOH	379(3.87)	22-3350-65
$C_{24}H_{22}O_7$			
Toxicarol isoflavone methyl ether	n.s.g.	265(4.63),294s(4.10)	39-4203-65
$C_{24}H_{22}O_{11}$			
Flavone, 3',4',5-triacetoxy-3,7,8-trimethoxy-	EtOH	264(4.49),308(4.28), 333(4.26)	78-3219-65
Flavone, 3',5,5'-triacetoxy-3,4',7-trimethoxy-	EtOH	255(4.23),317(4.34)	12-0934-64
Jaceidin, triacetate	EtOH	248(4.29),325(4.17)	39-5651-65
$C_{24}H_{23}Br_2NO_3$			
Morpholine, 4-[6,6-bis(o-bromobenzoyl)-1-cyclohexen-1-yl]-	n.s.g.	230s(4.36),265(4.00)	44-3711-65
$C_{24}H_{23}ClN_2O_2$			
Propionanilide, 4'-chloro-3-p-toluidino-3-p-toluoyl-	EtOH	248(4.51)	39-2040-65
$C_{24}H_{23}Cl_2NO_3$			
Morpholine, 4-[6,6-bis(o-chlorobenzoyl)-1-cyclohexen-1-yl]-	n.s.g.	230s(4.36),268(3.90)	44-3711-65
$C_{24}H_{23}F_2NO_3$			
Morpholine, 4-[6,6-bis(o-fluorobenzoyl)-1-cyclohexen-1-yl]-	n.s.g.	229(4.34),270(4.11)	44-3711-65
$C_{24}H_{23}I_2NO_3$			
Morpholine, 4-[6,6-bis(o-iodobenzoyl)-1-cyclohexen-1-yl]-	n.s.g.	220s(4.42),265(3.95)	44-3711-65
$C_{24}H_{23}NO_4S$			
6-Azaestra-1,3,5(10),8,14-pentaen-17-one, 3-methoxy-6-(phenylsulfonyl)-	n.s.g.	255(4.24),319(4.34)	31-0418-64
$C_{24}H_{23}N_3O_8S$			
Cytidine, N-benzoyl-2'-deoxy-, 5'-benzoate 3'-methanesulfonate	MeOH	228(4.45),259(4.40), 302(4.01)	44-3067-65
$C_{24}H_{23}N_5O_6$			
Aniline, N,N-bis(p-nitrobenzamidoethyl)-	MeOH	256(4.56)	87-0107-65
$C_{24}H_{24}Br_2N_2S_2$			
2,2'-Vinylenebis[3,4-dimethyl-5-phenylthiazolium bromide]	n.s.g.	415(4.1)	65-0075-64
$C_{24}H_{24}ClNO_4$			
Pseudoberberine, tetrahydro-, β-methochloride	EtOH	235(4.03),288(3.93)	78-2971-64
$C_{24}H_{24}FeO_4$			
Malonic acid, 9-ferrocenylnona-2,4,6,8-tetraen-1-ylidene-, dimethyl ester	$CHCl_3$	422(4.71)	5-0088-64F

Compound	Solvent	λ max(log ε)	Ref.
$C_{24}H_{24}N_2$			
1H-2,6-Methanobenzo[e][1,3]diazocine, 4-benzyl-3,4,5,6-tetrahydro-3-phenyl-	pH 1	250(4.08),284(3.11)	94-1225-65
	EtOH	254(2.86),260(2.92), 266(2.93),274(2.72)	94-1225-65
$C_{24}H_{24}N_2O$			
4H-1,3,4-Oxadiazine, 2-(diphenylmethyl)-5,6-dihydro-4,5-dimethyl-6-phenyl-	$CHCl_3$	244(3.79)	44-0668-64
m-Toluamide, N-[m-methyl-α-[(m-methylbenzylidene)amino]benzyl]-	EtOH	238(4.13)	35-1701-64
o-Toluamide, N-[o-methyl-α-[(o-methylbenzylidene)amino]benzyl]-	EtOH	253(4.15)	35-1701-64
p-Toluamide, N-[p-methyl-α-[(p-methylbenzylidene)amino]benzyl]-	EtOH	257(4.43)	35-1701-64
$C_{24}H_{24}N_2O_4$			
p-Anisamide, N-[p-methoxy-α-[(p-methoxybenzylidene)amino]benzyl]-	EtOH	275(4.57)	35-1701-64
$C_{24}H_{24}N_2O_6$			
Carbonic acid, phenyl ester, ester with ethyl 1-(2-cyanoethyl)-3-(hydroxymethyl)-5-methoxy-6-methylindole-2-carboxylate	MeOH	212(4.55),305(4.30)	44-2897-65
$C_{24}H_{24}N_2O_8S$			
Thymidine, 3'-benzoate 5'-p-toluenesulfonate	pH 1	266(3.83)	44-0554-64
	pH 7	263(4.06)	44-0554-64
	pH 13	266(3.93)	44-0554-64
$C_{24}H_{24}N_2O_9$			
Neobetanidine, 6-methyl ether, trimethyl ester, 5-acetate	MeOH-HCl	265(4.20),306(3.79), 496(4.66)	33-0252-65
	MeOH	223(4.40),263(4.21), 311(3.89),393(4.54)	33-0252-65
$C_{24}H_{24}N_4O_2$			
Formazan, N,N'-bis(4-ethoxyphenyl)-C-benzoyl-	EtOH	243(4.26),351(4.30), 468(4.42)	5-0099-65B
$C_{24}H_{24}N_4O_4S$			
4-Pyrimidinol, 2-amino-6-(2-furyl)-5-[N-(p-tolylsulfonyl)anilinopropyl]-	pH 1	231(4.33),316(4.19)	87-0283-65
	pH 7	234(4.38),280(4.11), 321(3.99)	87-0283-65
	pH 13	310(3.90)	87-0283-65
$C_{24}H_{24}N_6$			
Melamine, 1,2,6-tri-p-tolyl-	EtOH	263(4.36)	39-3459-64
$C_{24}H_{24}N_6O_{11}$			
Hydratropic acid, β-cyano-p-nitro-, 3α-tropanyl ester, picrate	acetone-KOH	539(4.66)	83-0826-65
	NaOMe	540(4.57)	83-0826-65
	NaCN	535(4.40)	83-0826-65
	NaSPr	542(4.15)	83-0826-65
$C_{24}H_{24}O_2$			
Gona-1,3,5(10),8-tetraen-17-one, 3-methoxy-13-phenyl-	dioxan	285(4.18)	13-0409-65

Compound	Solvent	λ_{max}(log ε)	Ref.
$C_{24}H_{24}O_3$			
1,3-Cyclopentanedione, 2-[2-(3,4-dihydro-6-methoxy-1(2H)-naphthylidene)-ethyl]-2-phenyl-	dioxan	267.5(4.32)	13-0409-65
$C_{24}H_{24}O_4$			
2-Anthraceneacetic acid, 5-(benzyloxy)-1,2,3,4-tetrahydro-4-hydroxy-9-methyl-	EtOH	238(4.67),292(3.79), 303(3.85),316(3.68), 330(3.26)	65-0670-65
Anthrone, 5-(benzyloxy)-1,4,4aα,9aβ-tetrahydro-4β,10β-(isopropylidene-dioxy)-	EtOH	223(4.51),255(3.90), 313(3.51)	65-2563-64
$C_{24}H_{24}O_5$			
Anthrone, 4β-acetoxy-5-(benzyloxy)-1,2,3,4,4aα,9aα-hexahydro-10β-hydroxy-	EtOH	255(3.95),309(3.32)	65-2563-64
Xanthen-1(2H)-one, 3,4-dihydro-9-(4-hydroxy-3-methoxyphenacyl)-3,3-dimethyl-	EtOH	225(4.38),280(4.27)	78-3707-65
	NaOEt	252(4.07),287(4.00), 352(4.39)	78-3707-65
Xanthen-1(2H)-one, 3,4-dihydro-9-(p-hydroxyphenacyl)-5-methoxy-3,3-dimethyl-	EtOH	225(4.33),266(4.00)	78-3697-65
$C_{24}H_{24}O_9$			
2-Butanone, 3,3-bis(3,4-diacetoxy-phenyl)-	EtOH	267(3.20),272(3.19), 288s(2.87)	39-3040-65
2,3,6,7-Dibenzofurantetrol, 1,4,5,8-tetramethyl-, tetraacetate	MeOH	232(4.68),253(4.11), 262(4.23),287(4.36)	5-0010-64F
$C_{24}H_{25}NO$			
Hexanophenone, 4'-(diphenylamino)-	MeOH	232(4.06),290(4.04), 350(4.37)	44-3536-64
$C_{24}H_{25}NO_2$			
9a(5aH)-Dibenzofuranpropionitrile, 6-benzylidene-4-ethoxy-6,7,8,9-tetrahydro-	EtOH	243(4.34)	94-1012-64
$C_{24}H_{25}NO_3$			
2-Cyclohexene-$\Delta^{1,\alpha}$-methanol, 2-morpholino-α-phenyl-, benzoate	hexane	229(4.42),268(4.14)	1-2139-65
	MeOH	232(4.37),265(4.11)	1-2139-65
Morpholine, 4-(2,6-dibenzoyl-1-cyclohexen-1-yl)-	hexane	243(4.38),280(3.52)	1-2139-65
	n.s.g.	232(4.40),265(4.15)	44-3711-65
$C_{24}H_{25}NO_4$			
Nicotinic acid, 5-(α-ethoxybenzyl)-1,6-dihydro-1-methyl-6-oxo-2-phenyl-, ethyl ester	EtOH	267(4.07),310(4.00)	78-3255-65
Pyridine-3,5-dicarboxylic acid, 1,4-dihydro-6-methyl-2,4-di-phenyl-, diethyl ester	MeOH	245(4.28),360(3.88)	44-3102-64
$C_{24}H_{25}NO_4S$			
6-Azaestra-1,3,5(10),8-tetraen-17-one, 3-methoxy-6-(phenylsulfonyl)-	n.s.g.	242(4.33),279(4.24)	31-0418-64
$C_{24}H_{25}NO_5$			
2-Naphthacenecarboxamide, N-tert-butyl-4,4a,5,12-tetrahydro-1,3,10-tri-hydroxy-6-methyl-12-oxo-	EtOH	246(4.43),265(4.44), 320(4.00),385(4.53)	65-0670-65

Compound	Solvent	$\lambda_{max}(\log \epsilon)$	Ref.
$C_{24}H_{25}NO_5S$			
Quinoline, 1,2,3,4-tetrahydro-7-methoxy-4-[2-(2-methyl-1,3-dioxocyclopentyl)ethylidene]-1-(phenylsulfonyl)-	n.s.g.	255(4.23)	31-0418-64
$C_{24}H_{25}N_5S$			
Acetone, 3-benzyl-4-(N,N'-diphenylamidino)-3-thiosemicarbazone	EtOH	272s(4.38),304(4.29)	39-0932-65
$C_{24}H_{25}P$			
Phosphorane, (3-methylcyclopentylidene)triphenyl-	isooctane	416.5(3.34)	77-0540-65
$C_{24}H_{26}$			
Bicyclo[2.2.2]oct-2-ene, 5-(2,2-diphenylvinyl)-1,3-dimethyl-	$CHCl_3$	258(4.19)	35-5646-64
Xylene, di-p-xylyl-	EtOH	222s(4.51),260s(3.14), 269(3.31),275(3.32), 280s(3.14)	5-0152-64D
$C_{24}H_{26}BF_4N_2P$			
1-Ethyl-2-[(1-ethyl-6-methyl-2(1H)-quinolylidene)phosphino]-6-methylquinolinium tetrafluoroborate	n.s.g.	587(4.61)	89-0433-64
$C_{24}H_{26}N_2O_6$			
Phenoxazine-1,9-dicarboxylic acid, 2-(cyclohexylamino)-4,6-dimethyl-3-oxo-, dimethyl ester	MeOH	224(4.30),250(4.40), 426(4.35),446(4.42)	44-3185-65
$C_{24}H_{26}N_2O_7$			
Spiro[cyclohexane-1,2'(3'H)-[11aH]oxazolo[4.5-b]phenoxazine]-4',6'-dicarboxylic acid, 11'a-hydroxy-9',11'-dimethyl-, dimethyl ester	MeOH	222(4.19),240(4.4), 268(4.24),310(4.08), 376(4.4)	44-3185-65
$C_{24}H_{26}N_4Ni_2O_4$			
Nickel, [1,1,2,2-ethanetetracarboxaldehydato(2-)]bis[N-methyl-7-(methylimino)-1,3,5-cycloheptatrien-1-ylaminato]di-	$CHCl_3$	268(4.77),306(4.59), 330(4.58),405(4.49), 463(4.10),501(4.06), 660(3.06)	44-3046-64
$C_{24}H_{26}N_4O_6$			
Phenanthrene, 6,7,8,9,10,14-hexahydro-3,4-dimethoxy-6-oxo-14-ethyl-, 2,4-dinitrophenylhydrazone	EtOH	220(4.38),265s(4.06), 305(4.15),400(4.50)	78-2553-65
$C_{24}H_{26}N_6$			
Mesoxaldehyde, tris(methylphenylhydrazone)	EtOH	255(4.35),335s(4.35), 370(4.35)	35-0732-64
$C_{24}H_{26}N_8NiO_2$			
Nickel, bis[3-imino-2-oxobutyronitrile (p-ethoxyphenyl)hydrazonato]-	EtOH	223(4.46),281(4.41), 296s(4.36),345(4.13), 488(3.91),600(2.65)	5-0033-65J
$C_{24}H_{26}O_2$			
Gona-1,3,5(10)-trien-17-one, 3-methoxy-13-phenyl-	dioxan	278(3.33)	13-0409-65

Compound	Solvent	λ_{max}(log ε)	Ref.
8α-Gona-1,3,5(10)-trien-17-one, 3-methoxy-13-phenyl-	dioxan	278(3.44)	13-0409-65
2(1H)-Phenanthrone, 1α-(benzyloxy-methyl)-3,4,9,12-tetrahydro-1β,12β-dimethyl-	n.s.g.	265(2.74)	39-4521-64
$C_{24}H_{26}O_6$			
Averythrin, tetra-O-methyl-	EtOH	228(4.48),293(4.61), 374(4.00)	39-3666-65
$C_{24}H_{26}O_7$			
Archangelicin	EtOH	216s(4.56),246s(3.67), 258(3.55),301s(3.95), 322(4.14)	1-0932-64
Siphulin	EtOH	243s(4.47),251(4.46), 264(4.32),293(4.36)	1-1677-65
$C_{24}H_{26}O_{11}$			
Luteolin, 4'-β-D-glucoside, tri-O-methyl ether	EtOH	241(4.31),267(4.28), 330(4.36)	94-0841-64
	EtOH-NaOAc	242(4.34),267(4.33), 329(4.35)	94-0841-64
	EtOH-NaOEt	242(4.34),267(4.33), 329(4.37)	94-0841-64
	EtOH-$AlCl_3$	243(4.32),267(4.33), 330(4.36)	94-0841-64
8-Hydroxy-1-naphthyltetra-O-acetyl-β-D-glucoside	EtOH	226(4.66),290s(3.68), 300(3.78),308s(3.68), 316(3.73),323s(3.52), 331(3.74)	39-4363-65
$C_{24}H_{26}O_{13}$			
Centaurein	MeOH	258(4.30),349(4.31)	24-1666-64
$C_{24}H_{26}S_4$			
Dispiro[m-dithiane-2,5'(4'H)-dibenzo-[ef,kl]heptalene-11'(6'H),2"-[m]-dithiane], 10',12'-dihydro-	dioxan	255(4.15),290(3.02)	35-1710-64
$C_{24}H_{27}ClN_2O_6$			
2-Naphthacenecarboxamide, 4-amino-N-tert-butyl-7-chloro-1,4,4a,5,5a,6,-11,11a-octahydro-3,12-dihydroxy-10-methoxy-1,11-dioxo-	MeOH-borax	240(4.10),272(4.03), 300(4.05),464(4.64), 490(4.54)	35-0933-65
$C_{24}H_{27}ClN_2O_9$			
Chlorotetracycline, N-(methoxymethyl)-	pH 1	230(4.24),268(4.26), 370(3.96)	87-0870-65
$C_{24}H_{27}ClO_4$			
Pregna-1,4,6,9(11)-tetraene-3,20-dione, 21-chloro-16 ,17-dihydroxy-, cyclic acetal with acetone	MeOH	226(4.08),245(3.96), 303(4.00)	13-0615-65
$C_{24}H_{27}Cl_3O_4$			
Pregna-1,4,6-triene-3,20-dione, 9,11β,21-trichloro-16α,17-dihydroxy-, cyclic acetal with acetone	MeOH	222(4.08),247(3.97), 296(4.11)	13-0615-65

Compound	Solvent	$\lambda_{max}(\log \epsilon)$	Ref.
$C_{24}H_{27}FO_5$			
Pregna-1,4,14,16-tetraene-3,20-dione, 9α-fluoro-11β,21-dihydroxy-16-methyl-, 21-acetate	MeOH	236(4.21),307(4.09)	44-3486-64
$C_{24}H_{27}NO_2$			
1H-Dibenzofuran-9b-propionitrile, 4-benzyl-6-ethoxy-2,3,4,4a-tetrahydro-	EtOH	280(3.38)	94-1012-64
$C_{24}H_{27}NO_5$			
Malonic acid, acetamido(3,3-diphenyl-allyl)-, diethyl ester	EtOH	253(4.21)	95-0971-65
Proline, 1-[(3,6,7-trimethoxy-9-phen-anthryl)methyl]-, methyl ester	EtOH	240s(4.57),268(4.81), 284(4.55),298s(4.28), 308s(3.93)	78-2573-65
$C_{24}H_{27}NO_6$			
Prednisone, 2-cyano-, bis(methylene-dioxy)-	EtOH	240(4.04)	44-3300-64
Prednisone, 2-cyano-, 21-acetate	EtOH	240(4.04)	44-3300-64
$C_{24}H_{27}N_3O_6$			
Elliotin, 1,3,5-trinitrobenzene derivative	EtOH	231(5.03)	44-0429-65
$C_{24}H_{27}N_5O_6$			
Purine, 6-morpholino-9-(2',3'-O-iso-propylidene-5'-O-benzoyl-β-D-ribofuranosyl)-	H_2O	237(3.96),279(4.20)	94-0267-64
$C_{24}H_{28}$			
3,4,5-Octatriene, 2,2,7,7-tetra-methyl-3,6-diphenyl-, cis	hexane	246(4.13),332(4.21)	24-3218-65
trans	hexane	321(4.31)	24-3218-65
$C_{24}H_{28}Br_2CoN_4$			
Cobalt, dibromotetrakis(4-picoline)-	$CHCl_3$	650(2.93)	12-0507-65
$C_{24}H_{28}Cl_2CoN_4$			
Cobalt, dichlorotetrakis(4-picoline)-	$CHCl_3$	615(2.79)	12-0507-65
$C_{24}H_{28}CoI_2N_4$			
Cobalt, diiodotetrakis(4-picoline)-	$CHCl_3$	660(2.95)	12-0507-65
$C_{24}H_{28}FIO_6$			
Pregna-1,4-diene-3,11,20-trione, 9-fluoro-17,21-dihydroxy-18-iodo-16α-methyl-, 21-acetate	EtOH	233(4.18)	35-3394-64
$C_{24}H_{28}NO_6P$			
2,2'-Spirobi[1,3,2-benzodioxaphosphole], 2-(o-hydroxyphenoxy)-, compound with triethylamine	MeCN	227s(4.17),280s(4.14), 284(4.17),290s(4.07)	35-2591-64
$C_{24}H_{28}N_2O_3$			
2,4-Diazaestra-3,5(10)-dien-1-one, 17β-acetoxy-3-phenyl-	MeOH	240(4.12),295(4.04)	13-0631-64

Compound	Solvent	$\lambda_{max}(\log \epsilon)$	Ref.
$C_{24}H_{28}N_2O_4$			
Voachalotine, acetate	MeOH	229(4.38),285(3.61)	20-0253-65
$C_{24}H_{28}N_2O_5$			
Obscurinervidine	EtOH	219(4.5),253(3.77), 310(3.42)	35-2451-64
$C_{24}H_{28}N_2O_6$			
Compactinervine, diacetate	EtOH	230(4.14),294(4.11), 324(4.16)	78-1141-65
$C_{24}H_{28}N_2S$			
Thieno[3,4-f][1,4]diazocine, 5,8-dibenzyl-4,5,6,7,8,9-hexahydro-1,3-dimethyl-	n.s.g.	238(3.9),254s(3.8)	70-2182-64
$C_{24}H_{28}N_6O_9$			
L-Glutamic acid, N-[p-[N-[2-[N-(2-acetamido-4-hydroxy-5-pyrimidinyl)formamido]ethyl]formamido]benzoyl]-, dimethyl ester	EtOH	255(4.30),304s(3.91)	35-5664-64
$C_{24}H_{28}O$			
Gona-1,3,5(10)-triene, 3-methoxy-13-phenyl-	dioxan	277(3.5)	13-0409-65
$C_{24}H_{28}O_2$			
Phenanthrene, 1α-benzyloxymethyl-1,2,3,4,11,12-hexahydro-1β,12β-dimethyl-	n.s.g.	219(4.39),265(3.97)	39-4521-64
$C_{24}H_{28}O_3S$			
Cyclopenta[7,8]phenanthro[10,1-bc]thiophen-3(2H)-one, 9-ethynyldodecahydro-9-hydroxy-4b,9a-dimethyl-, acetate	EtOH	222(4.16),268(4.10), 303(3.37)	94-1433-64
$C_{24}H_{28}O_4$			
2H-1-Benzopyran-7-ol, 4-butyl-3-(p-methoxyphenyl)-2,2-dimethyl-, acetate	MeOH	271(4.04),308(3.93)	44-4114-65
Guaiaretic acid, dehydrodiethyl-	n.s.g.	239(4.87),284(4.00), 315(3.52),329(3.58)	39-4011-64
Unknown phenol	n.s.g.	304(3.73)	39-1552-65
$C_{24}H_{28}O_5$			
Carda-4,14,20(22)-trienolide, 11,19-epoxy-11α-methoxy-3-oxo-	MeOH	217(4.31),241s(4.26)	44-0527-64
2,8-Dibenzofurandiol, 4,6-di-tert-butyl-, diacetate	C_6H_{12}	228(4.50),240s(4.17), 248(4.23),282(4.24), 296s(3.89),302s(3.73)	39-2904-65
Furan, 2,5-bis(4-ethoxy-3-methoxyphenyl)-3,4-dimethyl-	n.s.g.	250(4.10),326(4.41)	39-1572-65
Furan, 2-(3,4-dimethoxyphenyl)-5-(3,4-diethoxyphenyl)-3,4-dimethyl-	n.s.g.	249(4.14),326(4.49)	39-1572-65
Furoguaiacin, diethyl-	n.s.g.	252(4.08),326(4.49)	39-4011-64
16,21-Methanopregna-1,4,21-triene-3,20-dione, 21-acetoxy-11β-hydroxy-	EtOH	241(4.36)	5-0218-65E
$C_{24}H_{28}O_6$			
Prednisone, 16-methylene-, 21-acetate	MeOH	238(4.15)	44-3486-64

Compound	Solvent	$\lambda_{max}(\log \epsilon)$	Ref.
$C_{24}H_{28}O_7$			
Prednisone, 2-formyl-, bis(methylenedioxy)-	EtOH-HCl	241(4.18)	44-3481-64
	EtOH	219(4.09)	44-3481-64
	EtOH-NaOH	241(4.18),351(3.99)	44-3481-64
$C_{24}H_{28}O_8$			
Epipodophyllotoxin, 6,7-O-demethylene-tri-O-methyl-	MeOH	280(3.60)	33-1529-64
Flavone, 3',4',5-triethoxy-3,7,8-trimethoxy-	EtOH	254(4.40),272(4.36), 354(4.37)	78-3219-65
Podophyllotoxin, 6,7-O-demethylene-tri-O-methyl-	MeOH	280(3.56)	33-1529-64
$C_{24}H_{29}BrO_6$			
Estra-1,3,5(10)-triene-3,6,17β-triol, 7α-bromo-, triacetate	EtOH	214s(4.07),263s(2.75), 268(2.86),276(2.82)	44-1325-64
isomer	EtOH	217s(4.03),267(2.76), 275(2.75)	44-1325-64
$C_{24}H_{29}ClO_4$			
Pregna-1,4,9(11)-triene-3,20-dione, 21-chloro-16α,17-dihydroxy-, cyclic acetal with acetone	MeOH	238(4.19)	13-0615-65
$C_{24}H_{29}ClO_5$			
Pregna-1,4,6-triene-3,20-dione, 21-chloro-11β,16α,17-trihydroxy-, cyclic 16,17-acetal with acetone	MeOH	221(4.08),254(4.00), 298(4.09)	13-0615-65
$C_{24}H_{29}Cl_3O_4$			
Pregna-1,4-diene-3,20-dione, 9,11β,21-trichloro-16α,17-dihydroxy-, cyclic acetal with acetone	MeOH	236(4.17)	13-0615-65
$C_{24}H_{29}Cl_5O_4$			
Pregn-4-ene-3,20-dione, 2α,4,9α,11β,21-pentachloro-16α,17α-dihydroxy-, cyclic acetal with acetone	MeOH-dioxan	257(4.11)	13-0615-65
$C_{24}H_{29}FO_4$			
2H-Naphth[2',1':4,5]indeno[1,2-b]furan-2-one, 6b-acetyl-4b-fluorododecahydro-5-hydroxy-4a,6a,12-trimethyl-	EtOH	239(4.2)	44-3476-64
$C_{24}H_{29}FO_5$			
18-Nor-17α-pregna-1,4,13-triene-3,20-dione, 9-fluoro-11β,21-dihydroxy-16α,17-dimethyl-, 21-acetate	MeOH	236(4.22)	44-3486-64
Pregna-1,4,16-triene-3,20-dione, 9α-fluoro-11β,21-dihydroxy-16-methyl-, 21-acetate	MeOH	243(4.36)	44-3486-64
$C_{24}H_{29}FO_6$			
Prednisolone, 16-methylene-, 21-acetate	MeOH	238(4.20)	44-3486-64
Pregna-1,4-diene-3,20-dione, 16α,17α-epoxy-11β,21-dihydroxy-16β-methyl-, 21-acetate	MeOH	237(4.18)	44-3486-64
Pregna-1,4,15-triene-3,20-dione, 9α-fluoro-11β,17α,21-trihydroxy-16-methyl-, 21-acetate	MeOH	238(4.18)	44-3486-64

Compound	Solvent	$\lambda_{max}(\log \epsilon)$	Ref.
$C_{24}H_{29}N$			
Indene, 1-[4-(dibutylamino)-benzylidene]-	MeOH	420(4.6)	87-0390-65
	HOAc	340(4.1),420(4.2)	87-0390-65
$C_{24}H_{29}NO_2S$			
Acetophenone, 4'-[α-(2-oxo-4-piperidinobutyl)benzyl]thio]-, hydrochloride	H_2O	230(3.96),240s(3.92), 304(4.12)	7-1093-65
	EtOH	230(3.94),240s(3.92), 304(4.15)	7-1093-65
$C_{24}H_{29}NO_6$			
Prednisolone, 2-cyano-, bis(methylenedioxy)-	EtOH	245(3.99)	44-3300-64
$C_{24}H_{29}N_3O_8S$			
Griseoviridin, dihydrodehydro-, acetate	EtOH	216(4.33)	23-0371-64
$C_{24}H_{30}$			
Benzene, hexaisopropenyl-	n.s.g.	225s(4.41),290s(2.45)	35-2050-65
1,3-Butadiene, 1,4-di-tert-butyl-1,4-diphenyl-	ether	247(4.40)	24-3218-65
form m. 125-130°	ether	253(4.36)	24-3218-65
Coronene, dodecahydro-	n.s.g.	257(4.14),267(4.15), 278(3.60),288(3.47), 300(3.47),316(2.95), 332(3.08),347(3.20)	12-0055-64
$C_{24}H_{30}BrFO_4$			
17α-Pregna-1,4-diene-16α-acetaldehyde, 17β-bromo-9α-fluoro-11β-hydroxy-3,20-dioxo-6α-methyl-	EtOH	238(4.2)	44-3476-64
$C_{24}H_{30}Cl_2N_2$			
Bicyclo[2.2.0]octane, 1,4-bis-(2-chlorobenzylaminomethyl)-	EtOH	263(4.68)	23-2852-64
$C_{24}H_{30}Cl_2O_4$			
Pregna-1,4-diene-3,20-dione, 9,11β-dichloro-16α,17-dihydroxy-, cyclic acetal with acetone	MeOH	237(4.19)	13-0615-65
$C_{24}H_{30}FIO_6$			
Pregna-1,4-diene-3,20-dione, 9-fluoro-11β,17,21-trihydroxy-18-iodo-16α-methyl-, 21-acetate	EtOH	236(4.10)	35-3394-64
$C_{24}H_{30}FNaO_9S$			
Pregna-1,4-diene-3,20-dione, 9α-fluoro-16α,17α-isopropylidenedioxy-, 21-(hydrogen sulfate), sodium salt	H_2O	243(4.19)	13-0247-65
$C_{24}H_{30}F_2O_6$			
Pregna-1,4-diene-3,20-dione, 9α,16α-difluoro-11β,17α,21-trihydroxy-6α-methyl-, 21-acetate	EtOH	239(4.19)	87-0748-64
$C_{24}H_{30}N_2O$			
Acrylonitrile, 3-p-cumenyl-2-[p-(diethylaminoethoxy)phenyl]-	EtOH-HCl	335(4.41)	87-0511-64

Compound	Solvent	$\lambda_{max}(\log \epsilon)$	Ref.
$C_{24}H_{30}N_2OS$			
Acrylonitrile, 2-[p-(diethylamino-ethoxy)phenyl]-3-(p-isopropyl-thiophenyl)-	EtOH-HCl	356(4.51)	87-0511-64
	EtOH	355(4.48)	87-0511-64
$C_{24}H_{30}N_2O_4$			
Voacoline, O-ethyl-	MeOH	230(4.54),285(4.86)	20-0534-65
$C_{24}H_{30}N_2O_5$			
Obscurinervidine, dihydro-	EtOH	219(4.52),255(3.80), 312(3.43)	35-2451-64
Pleiocarpoline	EtOH	208(4.41),245(4.13), 281(3.33),287(3.32)	33-1002-65
Rindline	EtOH	222(4.4),253(3.9), 290(3.55)	39-2812-65
$C_{24}H_{30}N_3OP$			
Phosphine oxide, tris(p-dimethyl-aminophenyl)-	MeOH	284(4.88)	88-2729-64
$C_{24}H_{30}N_3P$			
Phosphine, tris(p-dimethylamino-phenyl)-	MeOH	285(4.85)	88-2729-64
$C_{24}H_{30}N_4O$			
2-Pyrazolin-5-one, 3-tert-butyl-4-[[4-(diethylamino)-o-tolyl]-imino]-1-phenyl-	n.s.g.	460(4.4),495(4.3)	46-1501-64
$C_{24}H_{30}N_4O_2$			
Benzoquinone, 2,5-dibenzylamino-3,6-diacetidino-	methyl cellosolve	228(4.22),251(4.12), 379(4.05),632(2.51)	78-1889-64
$C_{24}H_{30}O_2$			
Androsta-1,4-diene-3,17-dione, 4-allyl-19-vinyl-	EtOH	247(4.07)	33-0094-65
Estrone, 2,4-diallyl-	EtOH	335(3.69)	33-0094-65
Ethane, 1,2-bis[(5,6,7,8-tetra-hydro-4-methyl-1-naphthyl)oxy]-	MeOH	275(3.34)	78-2297-65
$C_{24}H_{30}O_3$			
D-Homoestra-1,3,5(10),8-tetraen-17a-one, 3-[(tetrahydropyran-2-yl)oxy]-	EtOH	276(4.18)	70-1814-64
$C_{24}H_{30}O_4$			
Estra-1,3,5(10),6-tetraene-3,17β-diol, 1-methyl-, 3-acetate 17-propionate	EtOH	224(4.50),229(4.46), 266(3.98)	94-0687-65
16,21-Methanopregna-1,4,21-triene-3,20-dione, 11β-hydroxy-21-ethoxy-	EtOH	249(4.29)	5-0218-65E
Pregna-4,6-diene-3,20-dione, 17α-acetoxy-16-methylene-	EtOH	283(4.50)	73-2351-64
$C_{24}H_{30}O_4S$			
19-Nor-17α-pregn-4-en-20-yn-3-one, 17-hydroxy-6α-mercapto-, diacetate	EtOH	235.5(4.24)	94-1433-64
$C_{24}H_{30}O_5$			
Mogoltin	EtOH	216(4.12),324(4.21)	65-1013-64
Norethindrone, hemisuccinate	EtOH	240(4.23)	13-0441-65

Compound	Solvent	λ_{max}(log ε)	Ref.
Pregna-1,4,9(11)-triene-3,20-dione, 21-hydroxy-16α,17α-isopropylidene-dioxy-	MeOH	238(4.13)	13-0615-65
$C_{24}H_{30}O_6$			
1,4-Butanedione, 1,4-bis(4-ethoxy-3-methoxyphenyl)-2,3-dimethyl-	n.s.g.	230(4.55),277(4.41), 305(4.34)	39-1572-65
Cortisone, 6-methylene-17α,20:20,21-bis(methylenedioxy)-	EtOH	259(4.06)	78-0597-64
Cortisone, 6-methylene-, 21-acetate	EtOH	258(4.08)	78-0597-64
Estra-1,3,5(10)-triene-1,3,17β-triol, triacetate	EtOH	267(2.71),274s(2.68), 285s(2.39)	24-1940-64
Estra-1,3,5(10)-triene-3,15α,17-triol, triacetate	EtOH	257(2.89),266(2.88)	44-1333-64
Hydrocortisone, 6-dehydro-16-methylene-, 21-acetate	EtOH	284(4.45)	5-0218-65E
Pregna-1,4-diene-3,20-dione, 9α,11α-epoxy-17α,21-dihydroxy-16β-methyl-, 21-acetate	MeOH	238(4.2)	31-0208-64
Pregna-1,4-diene-3,11,20-trione, 21-hy-droxy-16α,17α-isopropylidenedioxy-	MeOH	238(4.18)	13-0345-65
	EtOH	237(4.19)	39-0130-65
$C_{24}H_{30}O_7$			
Hypophyllanthin	EtOH	231(4.56),280(2.23)	88-1557-64
Prednisolone, 2-formyl-, 21-acetate	EtOH	220(4.18),244(4.08)	44-3481-64
	EtOH-HCl	247(4.14)	44-3481-64
	EtOH-NaOH	244(4.14),347(4.01)	44-3481-64
Prednisolone, 2-formyl-, bis(methylenedioxy)-	EtOH	221(4.16),245(4.07)	44-3481-64
	EtOH-HCl	246(4.14)	44-3481-64
	EtOH-NaOH	243(4.16),348(4.01)	44-3481-64
$C_{24}H_{30}O_8$			
Desaspidin	C_6H_{12}	231(4.49),274(4.38)	1-1292-64
$C_{24}H_{30}O_9$			
Podophyllic acid, 6,7-O-demethylene-6,7-di-O-methyl-, methyl ester	MeOH	280(3.56)	33-1529-64
$C_{24}H_{30}O_{11}$			
Harpagoside	EtOH	216(4.19),222(4.12), 276(4.36)	88-0835-64
$C_{24}H_{30}O_{14}$			
Swertiamarin, tetraacetate	n.s.g.	234(3.88)	22-0403-64
$C_{24}H_{31}BrO_6$			
Pregn-4-ene-3,11-dione, 2α-bromo-16α-methyl-17α,20:20,21-bis-(methylenedioxy)-	MeOH	240(4.13)	87-0584-64
$C_{24}H_{31}ClO_4$			
Pregna-4,9(11)-diene-3,20-dione, 21-chloro-16α,17-dihydroxy-, cyclic acetal with acetone	MeOH	238(4.28)	13-0615-65
Pregna-1,4,17(20)-trien-3-one, 16α-chloro-11β,21-dihydroxy-6α-methyl-, 21-acetate	EtOH	242(4.19)	87-0751-64

Compound	Solvent	$\lambda_{max}(\log \epsilon)$	Ref.
$C_{24}H_{31}ClO_5$			
Pregna-1,4-diene-3,20-dione, 21-chloro-11β-hydroxy-16α,17α-isopropylidene-dioxy-	MeOH	242(4.14)	13-0615-65
$C_{24}H_{31}ClO_6$			
Pregna-1,4-diene-3,20-dione, 16α-chloro-11β,17α,21-trihydroxy-6α-methyl-, 21-acetate	EtOH	243(4.15)	87-0751-64
Pregn-4-en-3-one, 2β-chloro-2α-methyl-17α,20:20,21-bis(methylenedioxy)-1α,11α-oxido-	EtOH	245(4.03)	32-0151-65
$C_{24}H_{31}Cl_3O_4$			
Pregn-4-ene-3,20-dione, 9,11β,21-trichloro-16α,17-dihydroxy-, cyclic acetal with acetone	MeOH	238(4.23)	13-0615-65
$C_{24}H_{31}FO_4$			
Pregna-1,4,17(20)-trien-3-one, 16ξ-fluoro-11β,21-dihydroxy-6α-methyl-, 21-acetate	EtOH	242(4.18)	87-0748-64
Pregn-4-ene-3,20-dione, 17α-acetoxy-21-fluoro-6-methylene-	EtOH	260(4.03)	78-0597-64
$C_{24}H_{31}FO_6$			
Pregna-1,4-diene-3,20-dione, 16α-fluoro-11β,17α,21-trihydroxy-6α-methyl-, 21-acetate	EtOH	242(4.17)	87-0748-64
$C_{24}H_{31}FO_9S$			
Pregna-1,4-diene-3,20-dione, 9-fluoro-11β,16α,17,21-tetrahydroxy-, 16-acetate, 21-methanesulfonate	MeOH	239(4.20)	13-0615-65
$C_{24}H_{31}NO_3$			
D-Homoandrosta-5,16-diene-16-carbonitrile, 3β-acetoxy-17-methyl-17a-oxo-	MeOH	246(4.03)	44-1272-65
Pregna-5,16-diene-16-carbonitrile, 3β-acetoxy-20-oxo-	EtOH	253(3.91),332(1.91)	22-1276-64
$C_{24}H_{31}NO_6$			
Hydrocortisone, 2α-cyano-, bis(methylenedioxy)-	EtOH	243(4.15)	44-3300-64
$C_{24}H_{31}NO_7$			
Pregn-4-ene-3,11-dione, 2-hydroxyimino-16α-methyl-17α,20:20,21-bis-(methylenedioxy)-	MeOH	260(4.14)	87-0584-64
$C_{24}H_{31}N_3O_5$			
Δ^6-Codeine, 6-(3-butylureido)-6-deoxy-7-formyl-14-hydroxy-	EtOH	324(4.23)	78-1407-64
Pregn-4-ene[3,2-d]-1',2',3'-triazole, 17α,20:20,21-bis(methylenedioxy)-16α-methyl-11-oxo-	MeOH	258(4.09)	87-0584-64
$C_{24}H_{31}N_3O_{14}S$			
D-Glucose, 2-deoxy-2-(2,4-dinitroanilino)-, S-ethyl monothiohemiacetal, 1,3,4,5,6-pentaacetate	EtOH	265(4.15),399(4.40)	44-1096-65

Compound	Solvent	$\lambda_{max}(\log \epsilon)$	Ref.
$C_{24}H_{31}N_3O_{15}$			
D-Glucose, 2-deoxy-2-(2,4-dinitroanilino)-, ethyl hemiacetal, 1,3,4,5,6-pentaacetate	EtOH	264(4.00),340(4.45)	44-1096-65
$C_{24}H_{31}O_6P$			
Ketone, 2,2-dihydro-2,2,2-triisopropoxy-5-phenyl-1,3,2-dioxaphosphol-4-yl phenyl	n.s.g.	250(4.28),346(3.90)	44-2575-65
$C_{24}H_{32}$			
1-Dodecene, 1,1-diphenyl-	C_6H_{12}	251(4.06)	22-0693-64
[2,2]Paracyclophane, octamethyl-	isooctane	233(4.21),257s(3.69), 308(2.84)	44-3245-64
$C_{24}H_{32}Cl_4O_4$			
Androst-4-en-3-one, 4-chloro-17β-(2,2,2-trichloro-1-hydroxyethoxy)-, propionate	MeOH	254(4.15)	83-0124-64
$C_{24}H_{32}N_2OS$			
1H-Cyclopenta[5,6]naphtho[1,2-g]quinoline-9-carbonitrile, hexadecahydro-1-hydroxy-1,11a,13a-trimethyl-8-thioxo-	EtOH	241s(3.85),310(4.34), 410(3.75)	94-0077-64
$C_{24}H_{32}N_2O_2$			
Androstano[3,2-b]pyridine-5'-carbonitrile, 1',6'-dihydro-17β-hydroxy-17α-methyl-6'-oxo-	EtOH	238(3.90),347(4.06)	94-0077-64
Androst-4-eno[3,2-d]-2'-methylpyrimidine, 17β-acetoxy-	EtOH	256(3.72),302(4.10)	87-0238-64
$C_{24}H_{32}N_2O_3$			
Androsta-5,16-dieno[16,17:4',5']-6'H-1',2'-diazepin-6'-one, 3β-acetoxy-1',7'-dihydro-	EtOH	220(4.02),253(3.56), 321(3.60),405(3.45)	44-0336-64
	EtOH-NaOH	283(3.53),323(3.32), 410(3.68)	44-0336-64
$C_{24}H_{32}N_2O_5$			
Pregn-4-eno[3,2-c]pyrazol-11β-ol, 17,20:20,21-bis(methylenedioxy)-	MeOH	261(4.02)	35-1520-64
$C_{24}H_{32}N_2S_2$			
2,2'-Bi-2H-1,3-thiazine, octahydro-4,4,4',4'-tetramethyl-2,2'-diphenyl-	EtOH-HCl	212(4.39),258(4.11)	44-3314-64
	EtOH	211(4.05)	44-3314-64
$C_{24}H_{32}O_2$			
Androst-5-ene-3β,17β-diol, 17α-butadiynyl-6-methyl-	EtOH	241(2.58),249(2.40), 272(1.72),289(1.70), 307(1.59)	78-1197-65
Estra-1,5(10)-dien-3-one, 4,4-diallyl-17β-hydroxy-	EtOH	335(3.67)	33-0094-65
2-Hexene, 3,4-bis(4-methoxy-3,5-xylyl)-	EtOH	278s(3.15)	39-6509-65
Pregna-1,4,6-triene-3,20-dione, 17-ethyl-6-methyl-	MeOH	228(4.18),255(4.00)	78-0357-64
Stilbene, α,α'-diethyl-4,4'-dimethoxy-3,3',5,5'-tetramethyl-	EtOH	216(4.41),236s(4.12)	39-6509-65

Compound	Solvent	λ_{max}(log ϵ)	Ref.
$C_{24}H_{32}O_3$			
Estra-1,3,5(10)-trien-17-one, 3-(3-butenyloxy)-, cyclic ethylene acetal	EtOH	281(3.09),288(3.10)	87-0755-64
D-Homoestrone, 3-tetrahydropyranyl ether	EtOH	276(3.32)	70-1814-64
Pregna-3,5-diene-6-carboxaldehyde, 3-methoxy-16α,17α-methylene-	EtOH	215(4.10),321(4.19)	78-0597-64
Pregna-4,17(20)-dien-21-oic acid, 6-methylene-3-oxo-, ethyl ester	EtOH	259(4.06)	78-0597-64
Pregna-3,5,17(20)-trien-11-one, 21-acetoxy-3-methoxy-	EtOH	239(4.24)	78-0597-64
Pregna-5,14,16-trien-20-one, 3β-acetoxy-16-methyl-	MeOH	310(4.08)	73-2513-64
Pregn-4-ene-16α-acetaldehyde, 17α-hydroxy-6α-methyl-3,20-dioxo-, 16b,17-cyclic enol ether	EtOH	241(4.2)	44-3476-64
$C_{24}H_{32}O_3S$			
Thieno[4',3',2'-4,5,6]androst-5-en-3-one, 5'-methyl-17β-propionyloxy-	EtOH	221(4.10),269(4.05), 304(3.36)	94-1433-64
$C_{24}H_{32}O_4$			
Androst-4-en-3-one, 2α-(acetoxymethyl)-17α-ethynyl-17-hydroxy-	EtOH	241(4.10)	44-0307-65
Androst-4-en-3-one, 17α-ethynyl-17β-hydroxy-1α,2α-isopropylidenedioxy-	MeOH	244(4.14)	13-0345-65
$\Delta^{8(14)}$-Anhydrogamabufotalin	MeOH	205(4.20)	95-0077-64
Δ^{14}-Anhydrogamabufotalin	MeOH	203(3.95)	95-0077-64
Desoxycorticosterone, 6-methylene-, acetate	EtOH	260(4.07)	78-0597-64
Estra-1,3,5(10)-triene-3,17β-diol, 1-methyl-, 3-acetate 17-propionate	EtOH	270(2.59),277(2.53)	94-0687-65
Guaiacin, diethyl-	n.s.g.	283(3.86)	39-4011-64
Isoguaiacin, diethyl-	n.s.g.	283(3.85)	39-4011-64
Pregna-3,5-diene-6-carboxaldehyde, 17α-hydroxy-3-methoxy-16-methylene-	EtOH	322(4.17)	78-0597-64
Pregna-5,16-diene-21-carboxaldehyde, 3β-hydroxy-20-oxo-, acetate	EtOH	271(4.01)	88-1839-64
Pregna-4,6-diene-3,20-dione, 17-acetoxy-6-methyl-	MeOH	289(4.38)	87-0804-64
Pregna-4,7-diene-3,20-dione, 17α-acetoxy-6α-methyl-	EtOH	238(4.19)	78-1753-65
5α-Pregna-6,8(14)-diene-3,20-dione, 17α-acetoxy-6-methyl-	EtOH	257.5(4.28)	78-1753-65
Pregna-1,4-diene-3,11,20-trione, 1-methyl-, cyclic 20-(ethylene acetal)	n.s.g.	244(4.23)	44-0163-64
Pregn-4-ene-16α-acetic acid, 17α-hydroxy-6α-methyl-3,20-dioxo-, lactone	EtOH	240(4.21)	44-3476-64
Pregn-4-ene-3,20-dione, 17α-acetoxy-6-methylene-	EtOH	260(4.15)	78-0597-64
$C_{24}H_{32}O_5$			
Androst-4-ene-17α-propionic acid, 6β-acetoxy-17β-hydroxy-3-oxo-, lactone	MeOH	235.5(4.12)	94-0859-64
Androst-4-ene-17α-propionic acid, 7α-acetoxy-17β-hydroxy-3-oxo-, lactone	MeOH	238(4.22)	94-0859-64
7β-acetoxy isomer	MeOH	238(4.21)	94-0859-64
Asperugin	EtOH	209(4.43),254(4.21), 315(3.84)	39-4672-65
	EtOH-NaOH	215(4.65),246(4.47), 280(4.15),327(4.01), 415(3.97)	39-4672-65

Compound	Solvent	λ_{max}(log ε)	Ref.
5β-Bufa-20,22-dienolide, 14,15β-epoxy-3β,12β-dihydroxy-	EtOH	298(3.76)	33-2217-64
5β-Card-20(22)-enolide, 11,19-epoxy-11α-methoxy-3-oxo-	MeOH	216(4.22)	44-0527-64
Pregna-1,4-diene-2-carboxaldehyde, 17β-hydroxy-3-oxo-, cyclic 20-(ethylene acetal)	EtOH	223(4.06),248(4.13)	44-3481-64
	EtOH-HCl	248(4.17)	44-3481-64
	EtOH-NaOH	243(4.12),347(3.96)	44-3481-64
Pregna-1,4-diene-3,11-dione, 17α,20α,21-trihydroxy-, cyclic 17,21-acetal with acetone	EtOH	240(4.16)	78-0179-65
Pregna-1,4-diene-3,20-dione, 11β,16α,17α-trihydroxy-, cyclic 16,17-acetal with acetone	MeOH-methyl cellosolve	242(4.19)	13-0615-65
9β,10α-Pregna-4,6-diene-3,20-dione, 17α-acetoxy-6-methoxy-	MeOH	246(3.89),306(4.20)	54-0918-65
Pregna-3,5,9(11)-trien-20-one, 21-acetoxy-17α-hydroxy-3-methoxy-	EtOH	241(4.29)	78-0597-64
Pregn-4-ene-16α-carboxaldehyde, 17-hydroxy-6α-methyl-3,20-dioxo-, formate	EtOH	240.5(4.21)	44-3476-64
Pregn-4-ene-3,20-dione, 17α,21-dihydroxy-19-methylene, 21-acetate	n.s.g.	239(4.19)	13-0371-65
Pregn-4-ene-3,20,21-trione, 17-acetoxy-21-methyl-	EtOH	240(4.25)	94-1181-64
Pregn-4-en-3-one, 17α,20:20,21-bis-(methylenedioxy)-19-methylene-	n.s.g.	239(4.21)	13-0371-65
$C_{24}H_{32}O_6$			
Cortexone, 8,19-epoxy-19-methoxy-, 21-acetate	EtOH	241(4.19)	44-0345-64
Cortisone, 1α-methyl-, acetate	EtOH	242(4.15)	44-0163-64
Cortisone, 6α-methyl-, 21-acetate	EtOH	238(4.19)	39-0148-65
Marinobufagin, 12β-hydroxy-	EtOH	195(4.08),298(3.69)	33-2226-64
Phenanthrene-12-carboxylic acid, 5,6-diacetoxy-1,2,3,4,9,10,11,12-octahydro-7-isopropyl-1,1-dimethyl-	EtOH	221s(4.03),268(2.64), 276(2.60)	33-1234-64
Pregna-1,4-diene-3,11-dione, 17α,20α,21-trihydroxy-, 17,21-(ethyl orthoformate)	EtOH	241(4.14)	78-0179-65
Pregna-1,4-diene-3,11-dione, 17α,20α,21-trihydroxy-, 17,21-(methyl orthoacetate)	EtOH	240(4.15)	78-0179-65
Pregna-3,5-diene-11,20-dione, 3-methoxy-17α,21-(methoxymethylenedioxy)-	EtOH	240(4.18)	78-0597-64
Pregna-1,4-dien-3-one, 11α-hydroxy-2-methyl-17α,20:20,21-bis-(methylenedioxy)-	EtOH	249(4.21)	32-0151-65
Pregn-4-ene-3,20-dione, 9α,11α-epoxy-17α,21-dihydroxy-16β-methyl-	MeOH	238(4.1)	31-0208-64
Pregn-4-ene-3,20-dione, 17α,21-dihydroxy-19-methylenoxy-,21-acetate	n.s.g.	244(4.16)	13-0371-65
Pregn-4-ene-3,11,20-trione, 21-acetoxy-17-hydroxy-6β-methyl-	EtOH	238(4.21)	39-0148-65
Pregn-4-ene-3,11,20-trione, 21-acetoxy-17α-hydroxy-16α-methyl-	MeOH	236(4.24)	13-0459-65
Pregn-4-ene-3,11,20-trione, 21-acetoxy-17α-hydroxy-19-methyl-	n.s.g.	240(4.16)	13-0371-65
Pregn-4-en-21-oic acid, 20β-acetoxy-3,11-dioxo-, methyl ester	MeOH	238(4.21)	44-2559-64
Pregn-4-en-3-one, 11β-hydroxy-2-methylene-17α,20:20,21-bis(methylenedioxy)-	EtOH	261(4.12)	87-0528-64

Compound	Solvent	λ_{max}(log ε)	Ref.
Schizandrin, deoxy-	EtOH	248(4.22),273s(3.49)	70-1036-64
$C_{24}H_{32}O_7$			
Pregn-4-en-3-one, 11β-hydroxy-2-(hydroxymethylene)-17α,20:20,21-bis(methylenedioxy)-	EtOH	255(4.03),308(3.74)	87-0528-64
$C_{24}H_{32}O_7S$			
5α-Pregnane-5α-carboxylic acid, 11β,17α,21-trihydroxy-3,20-dioxo-9α-mercapto-, γ-(thio lactone), 21-acetate	EtOH	241(3.54)	13-0305-64
$C_{24}H_{32}O_8$			
Melicopol, methyl-, diacetyl-	EtOH	228(4.22),267(3.83), 308(3.68)	12-2021-65
$C_{24}H_{32}O_{13}$			
Monotropein, dideoxydihydro-, acetate	EtOH	232(4.05)	94-0888-64
$C_{24}H_{32}O_{14}$			
Swertiamarin, dihydro-, tetraacetate	n.s.g.	234(3.99)	22-0403-64
$C_{24}H_{33}BrO_3$			
17α-Pregn-4-ene-16α-acetaldehyde, 17β-bromo-6α-methyl-3,20-dioxo-	EtOH	240(4.23)	44-3476-64
$C_{24}H_{33}ClO_3$			
Androsta-2,5-diene-2-carboxaldehyde, 17β-acetoxy-3-chloro-4,4-dimethyl-	MeOH-KOH	256(3.99)	39-0788-65
	ether	254(4.04)	39-0788-65
$C_{24}H_{33}ClO_4$			
5α-Androst-2(or 3)-ene-2,4-dicarboxaldehyde, 17β-acetoxy-3-chloro-17α-methyl-	MeOH-KOH	254(3.70),408(4.60)	39-0788-65
	ether	250(3.98),312(3.53)	39-0788-65
Pregna-4,16-dien-3-one, 20ξ-chloro-11β,21-dihydroxy-6α-methyl-, 21-acetate	EtOH	241.5(4.19)	87-0748-64
$C_{24}H_{33}ClO_5$			
Pregn-4-ene-3,20-dione, 21-chloro-11β,16α,17-trihydroxy-, cyclic 16,17-acetal with acetone	MeOH	241(4.21)	13-0615-65
$C_{24}H_{33}FO_4$			
Pregna-3,5-dien-20-one, 17α-acetoxy-21-fluoro-3-methoxy-	EtOH	240(4.29)	78-0597-64
Pregna-4,17(20)-dien-3-one, 16ξ-fluoro-11β,21-dihydroxy-6α-methyl-, 21-acetate	EtOH	241.5(4.19)	87-0748-64
$C_{24}H_{33}FO_6$			
Pregnene-3,20-dione, 2β,16α,17α,21-tetrahydroxy-6α-fluoro-, cyclic 16,17-acetal with acetone	pH 13	225(3.96),250(3.70), 361(3.93)	13-0057-65
	EtOH	237(4.14)	13-0057-65
$C_{24}H_{33}IO_5$			
Pregn-4-ene-3,20-dione, 11β,16α,17α-trihydroxy-21-iodo-, cyclic 16,17-acetal with acetone	MeOH	241(4.21)	13-0615-65

Compound	Solvent	λ_{max}(log ε)	Ref.
$C_{24}H_{33}NO_2$			
Androstano[3,2-b]pyridine, 17β-hydroxy-	EtOH	264s(3.68),269(3.75), 277(3.65)	94-0077-64
Estra-1,3,5(10)-trien-17β-ol, 3-(N-pyrrolidinyl)-, acetate	MeOH	258(4.22),312(3.40)	44-2047-65
Pregna-5,17(20)-diene-20-carbo- nitrile, 3β-acetoxy-	MeOH	222(4.17)	35-3737-65
isomer	MeOH	222(4.15)	35-3737-65
$C_{24}H_{33}NO_2S$			
Androstano[3,2-b]pyridine-6'(1'H)- thione, 17β-hydroxy-, 17-acetate	EtOH	281(4.03),373(3.80)	94-0077-64
$C_{24}H_{33}NO_3$			
Androstano[3,2-b]pyridine-1'-oxide, 17β-acetoxy-	EtOH	220(4.44),264(4.09)	94-0077-64
Androstano[3,2-b]pyrid-6'(1'H)-one, 17β-acetoxy-	EtOH	231(3.93),316(3.93)	94-0077-64
Estra-1,3,5(10)-trien-17-one,	pH 1	222(3.98),272(3.89)	44-2234-65
2-diethylaminoethyl-3-hydroxy-	pH 13	303(4.38)	44-2234-65
16-(hydroxymethylene)-	MeOH	222(3.98),292(3.99)	44-2234-65
Isoxazolino[17,16-d]pregn-4-ene- 3,20-dione, 3',6α-dimethyl-	n.s.g.	240(4.21)	5-0139-64G
$C_{24}H_{33}NO_3S$			
Androstano[3,2-b]pyridine-5'-carboxylic acid, 1',6'-dihydro-17β-hydroxy- 17α-methyl-6'-thioxo-	EtOH	244s(3.74),303(4.24), 392(3.73)	94-0077-64
$C_{24}H_{33}NO_4$			
Androstano[3,2-b]pyridine-5'-carboxylic acid, 1',6'-dihydro-17β-hydroxy- 17α-methyl-6'-oxo-	EtOH	236(3.84),340(3.90)	94-0077-64
Tembetarine methine, O,O'-diethyl-	EtOH	215(4.45),333(4.52)	5-0200-65E
$C_{24}H_{33}NO_6$			
Cortisone, 3-O-methyloxime, 21-acetate	EtOH	247.5(4.32)	39-0130-65
Cortisone, 20-O-methyloxime, 21-acetate	EtOH	236(4.25)	39-0130-65
$C_{24}H_{33}NO_6S$			
5α-Pregnane-5α-carboximidic acid, 11β,17α,21-trihydroxy-9α-mercapto- 3,20-dioxo-, γ-(thio lactone), 21-acetate	pH 1	259(3.83)	13-0305-64
$C_{24}H_{33}NS$			
18-Norandrost-13-eno[3,2-b]pyridine, 17,17-dimethyl-6'-(methylthio)-	EtOH	250(4.13),302(3.94)	94-0092-64
$C_{24}H_{33}N_3O_2$			
Androst-4-eno[3,2-b]pyridine-5'-carb- oxamide, 6'-amino-17β-hydroxy- 17α-methyl-	EtOH	224(4.33),242(4.29), 259(4.30),284s(3.98), 370(4.13)	94-0077-64
$C_{24}H_{33}N_3O_8S$			
Griseoviridin, hexahydrodehydro-, acetate	EtOH	214(4.31)	23-0371-64

Compound	Solvent	λ_{max}(log ε)	Ref.
$C_{24}H_{33}N_3O_{12}S_2$			
D-Glucose, 2-deoxy-2-(2,4-dinitroanilino)-, diethyl mercaptal, 3,4,5,6-tetraacetate	EtOH	263(3.90),345(4.26)	44-3280-64
$C_{24}H_{33}N_3O_{13}S$			
D-Glucose, 2-deoxy-2-(2,4-dinitroanilino)-, diethyl monothioacetal, 3,4,5,6-tetraacetate	EtOH	266(3.78),346(4.20)	44-3280-64
$C_{24}H_{34}$			
Dodecane, 1,1-diphenyl-	C_6H_{12}	255(2.80),260(2.84), 262(2.86),269(2.77)	22-0587-65
$C_{24}H_{34}FK_2O_9P$			
5α-Pregnane-3,20-dione, 9α-fluoro-11β,21-dihydroxy-16α,17α-isopropylidenedioxy-, 21-(dihydrogen phosphate), dipotassium salt	H_2O	240(0.30)	13-0247-65
$C_{24}H_{34}FNaO_9S$			
5α-Pregnane-3,20-dione, 9α-fluoro-11β,21-dihydroxy-16α,17α-isopropylidenedioxy-, 21-(hydrogen sulfate), sodium salt	H_2O	242(0.43)	13-0247-65
$C_{24}H_{34}INO_4$			
Tembetarine, O,O'-diethyl-, iodide	EtOH	282(3.75)	5-0200-65E
$C_{24}H_{34}N_2O_3$			
Androstano[3,2-b]pyridine-5'-carboxamide, 1',6'-dihydro-17β-hydroxy-17α-methyl-6'-oxo-	EtOH	239(3.94),340(4.08)	94-0077-64
$C_{24}H_{34}N_2O_4$			
Obscurinervidinol, dihydro-	EtOH	219(4.50),259(3.77), 309(3.50)	35-2451-64
$C_{24}H_{34}O_2$			
Androst-4-ene-3,17β-diol, 3-(cyclopentadienyl)-	MeOH	327(4.14)	83-0244-64
2H-1-Benzopyran-6-ol, 2-(4,8-dimethylnona-3,7-dienyl)-2,5,7,8-tetramethyl-	hexane	235(4.28),263s(--), 273(3.93),284(3.90), 339(3.53)	39-5060-65
Benzoquinone, trimethyl(3,7,11-trimethyl-2,6,10-dodecatrienyl)-	EtOH	259(4.25),267(4.26)	39-5060-65
Pregna-1,4-diene-3,20-dione, 17-ethyl-6α-methyl-	MeOH	244(4.18)	78-0357-64
$C_{24}H_{34}O_3$			
Androsta-3,5,7-trien-17β-ol, 3-ethoxy-, 17-propionate	EtOH	320.5(4.29)	78-1753-65
Androsta-3,5,7-trien-17β-ol, 3-ethoxy-6-methyl-, 17-acetate	EtOH	321(4.28)	78-1753-65
Estra-1,3,5(10)-trien-17-one, 3-butoxy-, cyclic ethylene acetal	EtOH	281(3.32),288(3.29)	87-0755-64
Pregn-4-en-3-one, 20β-acetoxy-6-methylene-	EtOH	256(4.04)	78-0597-64

Compound	Solvent	$\lambda_{max}(\log \epsilon)$	Ref.
$C_{24}H_{34}O_3S$			
Estra-1,3,5(10)-trien-17β-ol, 16β-(ethylthio)-3-methoxy-1-methyl-, acetate	EtOH	227(3.98),279(3.21), 286(3.23)	94-0687-65
Pregn-4-ene-3,20-dione, 16β-mercapto-, propionate	EtOH	239(4.33)	87-0531-64
Pregn-4-ene-3,20-dione, 16α-mercapto-6α-methyl-, 16-acetate	EtOH	238(4.29)	87-0531-64
$C_{24}H_{34}O_4$			
Androsta-3,5-diene-6-carboxaldehyde, 17β-acetoxy-3-ethoxy-	EtOH	220(4.03),323(4.20)	78-0597-64
Androsta-3,5-diene-6-carboxaldehyde, 17β-acetoxy-3-methoxy-17α-methyl-	EtOH	220(4.02),322(4.19)	78-0597-64
9β,10α-Androsta-3,5-diene-3,17β-diol, 17α-methyl-, diacetate	EtOH	235(4.28)	33-0989-65
Androst-4-ene-3,17-dione, 19-tetrahydropyranyloxy-	EtOH	242(4.21)	33-1961-64
Biphenyl, 5,5'-di-tert-butyl-2,2',3,3'-tetramethoxy-	EtOH	218(4.73),281(3.70)	39-2914-65
Chola-1,4-dienoic acid, 12α-hydroxy-3-oxo-, methyl ester	EtOH	246(4.19)	37-0094-64
Epitauranin, dihydrodeoxy-, acetate	EtOH	260(4.09),355(2.82)	94-0796-64
Pregna-3,5-dien-20-one, 17α-acetoxy-3-methoxy-	EtOH	240(4.30)	78-0597-64
Pregna-3,5-dien-20-one, 21-acetoxy-3-methoxy-	EtOH	240(4.29)	78-0597-64
Pregn-4-ene-3,20-dione, 11α,12α-dihydroxy-, cyclic acetal with acetone	EtOH	239(4.23)	44-2169-65
Pregn-4-ene-3,20-dione, 11β,12β-dihydroxy-, cyclic acetal with acetone	EtOH	238(4.23)	44-2169-65
Pregn-4-ene-3,20-dione, 17α-hydroxy-6β-methyl-, acetate	EtOH	241(4.20)	78-1619-65
Tauranin, deoxydihydro-, acetate	EtOH	260(4.18),348(2.78)	94-0796-64
Testosterone, 6β-acetyl-4-methyl-, acetate	EtOH	252(4.13)	35-5213-64
	MeOH-KOH	410(3.96)	35-5213-64
$C_{24}H_{34}O_4S$			
Testosterone, 6α-acetylthio-, propionate	EtOH	235.5(4.21)	94-1433-64
$C_{24}H_{34}O_5$			
5α-Androst-9(11)-ene-5-carboxaldehyde, 3,17-dioxo-, cyclic 3,17-bis-(ethylene acetal)	EtOH	211(3.43)	13-0305-64
Bufalin, 12β-hydroxy-	EtOH	301(3.72)	95-1092-65
Digifologenin, dihydro-, cyclic acetal with acetone	MeOH	306(1.57)	78-1469-64
Pregna-1,4-dien-3-one, 11β,17α,20α,21-tetrahydroxy-, 17α,21-acetonide	EtOH	245(4.16)	78-0179-65
Pregn-4-ene-3,20-dione, 11β-hydroxy-16α,17α-isopropylidenedioxy-	MeOH	240(4.19)	13-0615-65
Pregn-4-ene-3,20-dione, 17α-hydroxy-16α-(1,2,2-trihydroxyethyl)-6α-methyl-16b,17-oxido-	EtOH	241(4.20)	44-3476-64
Pregn-4-en-21-oic acid, 20α-acetoxy-3-oxo-, methyl ester	MeOH	242(4.21)	44-2559-64
Pregn-4-en-21-oic acid, 20β-acetoxy-3-oxo-, methyl ester	MeOH	242(4.23)	44-2559-64

Compound	Solvent	λ_{max}(log ε)	Ref.
Pregn-4-en-3-one, 19-methyl-17α,20:20,21-bis(methylenedioxy)-	n.s.g.	244(4.21)	13-0371-65
$C_{24}H_{34}O_6$			
Androst-2-ene-2-carboxaldehyde, 17β-hydroxy-3-(2-hydroxyethoxy)-, diformate	EtOH	275(4.07)	88-2161-64
17a-Oxa-5α-D-homoandrostan-17-one, 3β-acetoxy-16-(acetoxymethylene)-	MeOH	237(4.12)	44-0229-64
Phyllanthin	EtOH	230(4.33),280(1.89)	88-1557-64
Pregna-1,4-dien-3-one, 11β,17α,20α,21-tetrahydroxy-, 21-propionate	EtOH	244(4.15)	78-0179-65
Pregn-4-ene-3,11-dione, 20,20-ethylenedioxy-17,21-dihydroxy-6β-methyl-	EtOH	238(4.18)	39-0148-65
Pregn-4-ene-3,20-dione, 11β,17α,21-trihydroxy-19-methyl-, 21-acetate	n.s.g.	246(4.17)	13-0371-65
Pregn-4-en-3-one, 11α-hydroxy-17α,20:20,21-bis(methylenedioxy)-2α-methyl-	EtOH	242(4.15)	32-0151-65
Pregn-4-en-3-one, 11β-hydroxy-17,20:20,21-bis(methylenedioxy)-16α-methyl-	MeOH	242(4.20)	35-1520-64
Pseudostrophanthidin methylal	EtOH	217(4.24)	44-0342-64
Testosterone, 4,16α-dihydroxy-17α-methyl-, diacetate	MeOH	246(4.16)	7-0205-65
$C_{24}H_{34}O_6S$			
Androstane-5α-carboxylic acid, 3,3:17,17-bis(ethylenedioxy)-11β-hydroxy-9α-mercapto-, (thio lactone)	EtOH	240(3.59)	13-0305-64
$C_{24}H_{34}O_7$			
Isoandrographolide, diacetate	EtOH	214(3.95),220(3.74), 230(3.18)	78-2617-65
5α-Pregnan-3-one, 11β-hydroxy-2-(hydroxymethylene)-17,20:20,21-bis(methylenedioxy)-	MeOH	288(3.87)	35-1520-64
	MeOH-KOH	314.5(4.30)	35-1520-64
$C_{24}H_{34}O_8$			
Pregn-4-ene-3,20-dione, 1α,2α,11β,21-tetrahydroxy-16α,17α-isopropylidenedioxy-	MeOH	238(4.10)	13-0345-65
$C_{24}H_{34}Si_4$			
Trisilane, 2-(dimethylphenylsilyl)-1,1,3,3-tetramethyl-1,3-diphenyl-	C_6H_{12}	237(4.48)	101-0163-65B
$C_{24}H_{35}BrO_3$			
Androst-4-en-3-one, 17β-[(3-bromotetrahydropyran-2-yl)oxy]-	n.s.g.	241(4.22)	13-0397-65
$C_{24}H_{35}BrO_5$			
5α-Pregn-16-en-20-one, 3β-acetoxy-15β-bromo-12β-hydroxy-12α-methoxy-	MeOH	246(4.06)	24-1188-65
$C_{24}H_{35}ClN_4O_3$			
2-Octanone, 1-chloro-8-[p-[(2-amino-4-hydroxy-6-methyl-5-pyrimidinyl)-3-propyl]amino]phenyl-, ethylene ketal	pH 1	265(3.89)	87-0035-65
	pH 7	241(4.23),293(3.89)	87-0035-65
	pH 13	240(4.25),281(3.89)	87-0035-65

Compound	Solvent	λ max(log ε)	Ref.
$C_{24}H_{35}NO_2$			
Androst-4-en-3-one, 17α-(3-dimethyl-amino-1-propynyl)-17β-hydroxy-	EtOH	240(4.21)	78-2295-64
Cona-3,5-dienin, 7β-acetoxy-	EtOH	235(4.30)	22-2169-64
Pregn-4-en-3-one, 20β-(2-cyanoethoxy)-	EtOH	242(4.22)	13-0585-65
$C_{24}H_{35}NO_3$			
Pregna-5,16-dien-20-one, 3β-acetoxy-16-methyl-, oxime m. 163°	EtOH	237(3.76)	33-0608-65
isomer m. 168°	EtOH	239(3.76)	33-0608-65
isomer m. 182°	EtOH	238(3.73)	33-0608-65
isomer m. 214°	EtOH	234(3.54)	33-0608-65
$C_{24}H_{35}NO_3S$			
5α-Androstano[3,2-e]-4'-methyl-1',4'-thiaz-5'-en-3'-one, 17β-acetoxy-	EtOH	228(3.66),291(3.37)	87-0555-64
$C_{24}H_{35}NO_4$			
5α-Androst-9(11)-ene-3,17-dione, 5-formimidoyl-, cyclic bis-(ethylene acetal)	EtOH	207(3.51)	13-0305-64
11-Aza-5α,8ξ,9α-pregn-16-en-20-one, 3β,11-diacetoxy-	EtOH	235(3.99)	13-0595-64
$C_{24}H_{35}NO_5$			
Androst-4-en-3-one, 2α-nitro-17β-(2-tetrahydropyranyloxy)-	pH 1	253(4.18)	78-0373-64
	pH 13	246(3.86),268(3.88), 355(4.05)	78-0373-64
	MeOH	245(4.15),375(3.02)	78-0373-64
$C_{24}H_{35}N_3O_2$			
Androstano[3,2-b]pyridine-5'-carbox-amide, 6'-amino-17β-hydroxy-17α-methyl-	EtOH	252(4.05),338(3.88)	94-0077-64
$C_{24}H_{36}$			
Cyclohexane, hexapropylidene-	C_6H_{12}	263s(3.78)	33-1289-65
$C_{24}H_{36}Br_6$			
Benzene, hexakis(1-bromopropyl)-	C_6H_{12}	245(4.60)	33-1289-65
$C_{24}H_{36}O_2$			
Pregna-3,5-dien-20-one, 17-ethyl-3-methoxy-	MeOH	239(4.20)	78-0357-64
Pregn-4-ene-3,20-dione, 17-ethyl-6α-methyl-	MeOH	240(4.22)	78-0357-64
$C_{24}H_{36}O_3$			
Androsta-3,5-dien-17β-ol, 3-ethoxy-2α-methyl-, acetate	EtOH	243(4.29)	78-0597-64
Androsta-3,5-dien-17β-ol, 3-ethoxy-6-methyl-, acetate	EtOH	247.5(4.27)	78-0569-65
Androst-4-en-3-one, 16β,17β-dihydroxy-16α,17α-dimethyl-, cyclic acetal with acetone	MeOH	240(4.22)	7-0205-65
5α-Androst-7-en-6-one, 3β-[(tetra-hydropyranyl)oxy]-	EtOH	244(4.07)	35-2837-64
Chol-4-enic acid, 3-oxo-	EtOH	242(4.19)	44-3495-64
5β,19-Cycloandrost-6-en-3β-ol, 17-[(tetrahydropyranyl)oxy]-	EtOH	213(3.49)	13-0001-64

Compound	Solvent	λ_{max}(log ε)	Ref.
Gon-4-en-3-one, 13β-ethyl-17β-hydroxy-, isovalerate	n.s.g.	240(4.19)	39-4472-64
$C_{24}H_{36}O_4$			
Androsta-3,5-diene-6-methanol, 17β-acetoxy-3-ethoxy-	EtOH	251(4.30)	78-0597-64
5α-Pregnane-3,20-dione, 2-(hydroxymethylene)-, cyclic 20-(ethylene acetal)	EtOH	282(3.91)	44-3481-64
Pregn-4-en-3-one, 11β,12β,20α-trihydroxy-, cyclic 11,12-acetal with acetone	EtOH	239(4.21)	44-2169-65
Pregn-4-en-3-one, 11β,12β,20β-trihydroxy-, cyclic 11,12-acetal with acetone	EtOH	238(4.28)	44-2169-65
5α-Pregn-7-en-20-one, 3β-hydroxy-16α-methoxy-, 3-acetate	EtOH	236(3.57),302(2.64)	78-2455-64
Testosterone, 7α-methoxy-17α-methyl-, 17-propionate	EtOH	242.5(4.19)	24-3363-64
$C_{24}H_{36}O_5$			
Androst-9(11)-ene-3,17-dione, 5α-(hydroxymethyl)-, bis-(cyclic ethylene acetal)	EtOH	207(3.49)	13-0305-64
3α,5α-Epoxymethano-5α-androst-9(11)-en-17-one, 3β-(2-hydroxyethyloxy)-, cyclic ethylene acetal	EtOH	206(3.57)	13-0305-64
Octanorebelan-16-oic 16,17-lactone, 3β-acetoxy-17-hydroxy-22-oxo-	EtOH	296(1.38)	12-1451-65
$C_{24}H_{36}O_6$			
Carda-4,20(22)-dienolide, 11,19-epoxy-14-hydroxy-11α-methoxy-	MeOH	219(4.33),237s(4.28)	44-0527-64
$C_{24}H_{36}O_7$			
18-Norandromedol, 10-oxo-, cyclic acetal with acetone, acetate	EtOH	288(1.52)	44-2756-64
$C_{24}H_{37}NO_2$			
Androsta-3,5-dien-17-one, 6-dimethylaminomethyl-3-ethoxy-	EtOH	249(4.33)	78-0569-65
$C_{24}H_{37}NO_3$			
Androst-4-en-3-one, 17β-acetoxy-6α-dimethylaminomethyl-	EtOH	239(4.15)	78-0569-65
$C_{24}H_{38}N_2O$			
Conessine, 7-oxo-	MeOH	236(4.12)	31-0256-64
$C_{24}H_{38}N_2S_2$			
4H,10H-Dithieno[3,4-c:3',4'-h][1,6]-diazecine, 5,6,11,12-tetrahydro-5,11-diisobutyl-1,3,7,9-tetramethyl-	n.s.g.	220(4.1),239(4.1), 245(4.1)	70-2182-64
$C_{24}H_{38}O_3S$			
5β-Pregnan-20-one, 3β-hydroxy-16β-mercapto-, 16-propionate	EtOH	234(3.68)	87-0531-64

Compound	Solvent	λ_{max}(log ϵ)	Ref.
$C_{24}H_{38}O_4$			
Docosa-10,12-diynedioic acid, dimethyl ester	EtOH	226(2.67),230(2.36), 254(2.36)	78-1773-64
$C_{24}H_{39}NO_2$			
5α-Androstan-2-one, 3-(diethylamino-methylene)-17β-hydroxy-	n.s.g.	333(4.36)	44-3786-65
$C_{24}H_{40}$			
1,3,13,15-Cyclotetracosatetraene	hexane	235(4.65)	39-6674-65
$C_{24}H_{40}BNO_2$			
Androsta-3,5-dien-17-one, 6-dimethyl-aminomethyl-3-ethoxy-, compound with BH_3	EtOH	257(4.32)	78-0569-65
$C_{24}H_{40}O_3$			
Lithocholic acid, p-hydroxy-benzaldehyde complex	HOAc-$HClO_4$	530(4.08)	96-0134-65
resorcylic aldehyde complex	HOAc-$HClO_4$	540(3.48)	96-0134-65
vanillin complex	HOAc-$HClO_4$	600(4.06)	96-0134-65
$C_{24}H_{40}O_4$			
5β-Cholanic acid, 3α,12α-dihydroxy-, p-hydroxybenzaldehyde complex	HOAc-$HClO_4$	540(4.11)	96-0134-65
resorcylic aldehyde complex	HOAc-$HClO_4$	540(4.08)	96-0134-65
vanillin complex	HOAc-$HClO_4$	600(4.10)	96-0134-65
$C_{24}H_{40}O_5$			
5β-Cholanic acid, 3α,7α,17α-tri-hydroxy-, p-hydroxybenzaldehyde complex	HOAc-$HClO_4$	530(4.04)	96-0134-65
resorcylic aldehyde complex	HOAc-$HClO_4$	540(4.00)	96-0134-65
vanillin complex	HOAc-$HClO_4$	600(4.04)	96-0134-65
$C_{24}H_{42}$			
Benzene, hexaisopropyl-	C_6H_{12}	222s(4.40),273(2.34)	33-0509-65
	n.s.g.	273(2.27)	35-4729-64
Benzene, hexapropyl-	EtOH	213(4.68),274(2.37)	33-0509-65
$C_{24}H_{42}Cl_2Pd_2$			
Palladium, dichlorodi-π-cyclododecen-yldi-	dioxan	238(4.33),328(3.34)	24-1753-65
$C_{24}H_{44}N_2O_{20}$			
Lactose, azine	MeOH	260(4.02)	24-1404-65
Maltose, azine	MeOH	260(4.02)	24-1404-65
$C_{24}H_{46}Si_6$			
Hexasilane, dodecamethyl-1,6-diphenyl-	C_6H_{12}	265.0(4.48)	25-1063-64
$C_{24}H_{52}IN$			
Tetrahexylammonium iodide	MeOH	295(4.0)	60-0488-64
$C_{24}H_{54}Sn_2$			
Tin, hexabutyldi-	C_6H_{12}	none	101-0265-64B

Compound	Solvent	λ_{max}(log ε)	Ref.
$C_{25}H_{14}N_4$			
Benzo[b]tricycloquinazoline	$CHCl_3$	255(4.67),270(4.68), 290(4.73),303(4.63), 344(4.27),355(4.30), 370(4.30),394(4.29), 420(4.21),445(4.12)	39-3670-64
Benzo[c]tricycloquinazoline	$CHCl_3$	273(4.61),298(4.46), 323(4.32),339(4.48), 383(4.29),397(4.35), 401(4.34),419(4.21), 450(3.73),479(3.44)	39-3670-64
$C_{25}H_{14}O$			
2,3-Benzonaphtho[1",2":5,6]phenalen-1-one	heptane	242(4.66),257(4.54), 268(4.56)	24-0743-65
	benzene	292(4.37),324(4.38), 405(4.20),424(4.26)	24-0743-65
2,3-Benzonaphtho[2",1":7,8]phenalen-1-one	heptane	254(4.78),260(4.80)	24-0743-65
	benzene	295(4.25),308(4.24), 316(4.11),382(4.08), 400(4.22),425(4.20)	24-0743-65
$C_{25}H_{14}O_5$			
Psoralen, phenanthrenequinone adduct	dioxan	246(4.60),282(4.19), 324(4.02)	24-3102-65
$C_{25}H_{15}N_4P$			
1-Propene-1,1,2,3-tetracarbonitrile, 3-(triphenylphosphoranylidene)-	dioxan	406(4.28)	49-1967-65
$C_{25}H_{16}$			
2,3-Benzonaphtho[1",2":5,6]phenalene	heptane	250(4.51),264(4.57), 270(4.61),282(4.72), 291(4.62),308(4.07), 332(4.22),348(4.41), 368(4.40)	24-0743-65
7,8-Benzonaphtho[1",2":2,3]phenalene	heptane	250(4.60),278(4.68), 343(4.26),356(4.10)	24-0743-65
7,8-Benzonaphtho[2",1":2,3]phenalene	heptane	232(5.08),272(4.65), 288(4.52),302(4.14), 328(4.45),344(4.40), 375(3.60)	24-0743-65
$C_{25}H_{16}ClN_3O$			
Nicotinonitrile, 2-benzamido-4-(p-chlorophenyl)-6-phenyl-	n.s.g.	295(4.42),392(4.35)	24-3892-65
Nicotinonitrile, 2-benzamido-6-(p-chlorophenyl)-4-phenyl-	n.s.g.	295(4.42),387(4.30)	24-3892-65
Nicotinonitrile, 2-(p-chlorobenzamido)-4,6-diphenyl-	n.s.g.	295(4.47),389(4.38)	24-3892-65
$C_{25}H_{16}N_3OP$			
3-Butenal, 3,4,4-tricyano-2-(triphenylphosphoranylidene)-	dioxan	416(4.23)	49-1967-65
$C_{25}H_{17}ClO_4$			
7-[2-(3-Pyrenyl)vinyl]tropylium perchlorate	MeCN	234(4.65),275(4.49), 330s(4.10),346(4.23), 430(3.99),581(4.53)	24-1349-64

Compound	Solvent	λ max (log ε)	Ref.
$C_{25}H_{17}Cl_2N$			
Acridan, 9,9-bis(m-chlorophenyl)-	EtOH	248(3.73),294(4.02)	65-3472-64
Acridan, 9,9-bis(p-chlorophenyl)-	EtOH	248(3.89),296(4.02)	65-3472-64
$C_{25}H_{17}Cl_3OSi$			
Silane, benzoyltris(p-chlorophenyl)-	C_6H_{12}	229(4.58),257s(4.29), 387s(2.17),403(2.44), 422(2.56),442(2.36)	23-1175-65
	EtOH	259(4.24),400(2.46), 411(2.51),425(2.34)	23-1175-65
$C_{25}H_{17}N_3O$			
Benzamide, N-(4,4-dicyano-1,3-diphenyl-1,3-butadienyl)-	MeCN	290(4.31),384(4.33)	24-3892-65
Nicotinonitrile, 2-benzamido-4,6-diphenyl-	n.s.g.	293(4.43),386(4.37)	24-3892-65
$C_{25}H_{17}N_3O_2$			
3,5-Pyrazolidinedione, 1,2-diphenyl-4-(2-quinolylmethylene)-	MeOH	247(4.30),336(3.96), 440(3.97)	32-0320-65
$C_{25}H_{17}N_3O_4$			
Acrylonitrile, 3-[p-(diethylaminoethoxy)phenyl]-2-(p-maleylaminophenyl)-	EtOH-HCl	352(4.54)	87-0511-64
$C_{25}H_{18}$			
Propane, 1-fluoren-9-ylidene-3-inden-1-ylidene-	DMSO-tert-BuOK	630(5.30)	5-0050-65J
$C_{25}H_{18}NOPS$			
2-Thiophenepropionitrile, β-oxo-α-(triphenylphosphoranylidene)-	EtOH	255(4.05),275(3.97), 314(4.08)	65-2205-65
$C_{25}H_{18}NO_2P$			
2-Furanpropionitrile, β-oxo-2-(triphenylphosphoranylidene)-	EtOH	240(4.00),263(4.13)	65-2205-65
$C_{25}H_{18}N_4O$			
Cycloheptimidazol-2(1H)-one, 1,3-dibenzyl-6-dicyanomethylene-3,6-dihydro-	EtOH	264(4.14),455(4.49)	94-0810-65
$C_{25}H_{18}N_4O_7$			
p-Aminotriphenylcarbonium picrate	EtCOMe	440(4.60)	28-6933-65A
$C_{25}H_{19}$			
Methyl, m-biphenylyldiphenyl-	ether	253(4.48),330s(3.60), 490(4.51)	60-0264-64
Methyl, p-biphenylyldiphenyl-	ether	254(4.95),425(4.23), 488s(4.54),545(4.78)	60-0264-64
Methyl, bis-p-biphenylyl-	ether	258(4.85),335s(3.60), 395(3.95),545(5.20)	60-0264-64
$C_{25}H_{19}BrOSi$			
Silane, (p-bromobenzoyl)triphenyl-	C_6H_{12}	271(4.34),390s(2.07), 407(2.33),428(2.43), 448(2.20)	23-1175-65
	EtOH	271(4.33),410(2.43), 421(2.46),442(2.22)	23-1175-65

Compound	Solvent	λ_{max}(log ε)	Ref.
$C_{25}H_{19}Br_2NSi$			
Phenazasiline, 2,8-dibromo-5,10-dihydro-5-methyl-10,10-diphenyl-	$CHCl_3$	221(4.52)	44-3248-65
$C_{25}H_{19}ClOSi$			
Silane, (p-chlorobenzoyl)triphenyl-	C_6H_{12}	268(4.26),391s(2.07), 408(2.34),427(2.44), 448(2.20)	23-1175-65
$C_{25}H_{19}Cl_2NO$			
Methanol, (o-anilinophenyl)bis-(m-chlorophenyl)-	EtOH	252(3.71),298(4.29)	65-3472-64
Methanol, (o-anilinophenyl)bis-(p-chlorophenyl)-	EtOH	252(3.66),297(4.26)	65-3472-64
$C_{25}H_{19}Cs$			
Cesium, (p-biphenylyldiphenylmethyl)-	$C_6H_{11}NH_2$	573(4.65)	35-0384-65
$C_{25}H_{19}FOSi$			
Silane, (p-fluorobenzoyl)triphenyl-	C_6H_{12}	260(4.14),384s(2.08), 400(2.35),419(2.46), 440(2.27)	23-1175-65
$C_{25}H_{19}NO_3$			
Bicyclo[2.2.2]oct-5-ene-2,3-dione, 7-(1-naphthyl)-5(or 6)-benzamido-	EtOH	226(4.45),275(3.6), 285(3.6),295(3.5), 313(3.25)	12-1775-65
Nicotinic acid, 5-benzyl-1,6-dihydro-6-oxo-1,2-diphenyl-	EtOH	266(4.05),311(3.96)	78-3255-65
$C_{25}H_{19}NO_3Si$			
Silane, (p-nitrobenzoyl)triphenyl-	C_6H_{12}	266(4.25),411s(2.12), 434(2.28),448(2.28), 478(1.95)	23-1175-65
	EtOH	267(4.19),430(2.33)	23-1175-65
$C_{25}H_{19}N_3$			
Indole, 3,3',3"-methylidynetri-	n.s.g.	285(4.44),292(4.47)	28-0609-64A
$C_{25}H_{20}ClNO_6$			
1-(o-Acetoxyphenyl)-2-styryl-quinolinium perchlorate	EtOH	419(3.99)	65-3373-64
$C_{25}H_{20}ClN_2P$			
Phosphorane, [(p-chlorobenzylidene)-hydrazono]triphenyl-	$CHCl_3$	262s(3.85),268s(3.91), 276s(3.98),315s(4.23), 338(4.29)	44-0417-65
$C_{25}H_{20}Fe_2O$			
1-Penten-4-yn-3-one, 1,5-diferrocenyl-	$CHCl_3$	330(4.28),391(3.61), 508(3.71)	49-1750-64
$C_{25}H_{20}N_2O$			
Azobenzene, 2-methoxy-2',6'-diphenyl-	EtOH	236(4.49),274(4.05), 344(3.84),446(2.88)	7-0685-64
$C_{25}H_{20}N_2O_2$			
Aniline, N-methyl-N-[5-(2-nitrofluoren-9-ylidene)-1,3-pentadienyl]-	MeCN	246(4.53),280(4.32), 345(4.24),488(4.75)	24-3331-64

Compound	Solvent	λ_{max}(log ε)	Ref.
$C_{25}H_{20}N_3O_2P$			
Phosphorane, [(p-nitrobenzylidene)-hydrazono]triphenyl-	$CHCl_3$	257s(4.18),261(4.19), 265s(4.18),273s(4.11), 292s(3.90),419(4.38)	44-0417-65
$C_{25}H_{20}N_4O$			
Cycloheptimidazolemalononitrile, 1,3-dibenzyl-1,2,3,6-tetrahydro-2-oxo-	EtOH	295(3.80)	94-0810-65
$C_{25}H_{20}N_4O_2$			
2(1H)-Cycloheptimidazolone, 6-(carbamoylcyanomethylene)-1,3-dibenzyl-3,6-dihydro-	EtOH	264(4.09),452(4.48)	94-0810-65
$C_{25}H_{20}N_4O_7S$			
3-(2,4-Dinitrophenyl)-1-methyl-1H-naphth[1,2-d]imidazolium p-toluenesulfonate	EtOH	348(4.05)	65-0269-64
$C_{25}H_{20}N_4S_2$			
1H-Imidazo[4,5-d]pyridazine, 4,7-bis-(benzylthio)-2-phenyl-	EtOH	207(4.68),268(4.50), 289s(4.35),299(4.44), 309s(4.41)	4-0182-64
$C_{25}H_{20}N_8$			
Benzoic acid, (2-phenyl-1H-imidazo-[4,5-d]pyridazine-4,7-diyl)-dihydrazone	EtOH	204(4.33),229(4.30), 253s(4.24),296s(4.28), 334(4.36)	4-0182-64
$C_{25}H_{20}O$			
Methanol, m-biphenylyldiphenyl-	$HClO_4$	246(4.32),324s(3.34), 423(4.63),490s(3.47)	60-0264-64
Methanol, p-biphenylyldiphenyl-	$HClO_4$	257(4.15),286s(3.92), 342(3.79),423(4.48), 507(4.74)	60-0264-64
Methanol, bis(p-biphenylyl)-	$HClO_4$	255(4.04),333(3.38), 424(3.90),579(5.06)	60-0264-64
$C_{25}H_{20}OSi$			
Silane, benzoyltriphenyl-	C_6H_{12}	257(4.21),388s(2.07), 405(2.35),424(2.47), 440(2.25)	23-1175-65
	EtOH	258(4.19),403s(2.42), 417(2.48)	23-0298-64
	EtOH	258(4.19),403(2.42), 417(2.48)	23-1175-65
$C_{25}H_{20}O_2S$			
2H-Naphtho[2,3-b]thiete, 2,3-dimethyl-3,8-diphenyl-, 1,1-dioxide	EtOH	241(4.83),310(4.04), 336(3.78)	44-0629-65
$C_{25}H_{20}O_3$			
3-Chromene, 2,2-dimethyl-, phenanthrenequinone adduct	dioxan	244(4.48),281(3.92), 329(3.45)	24-3102-65
$C_{25}H_{20}O_4$			
Tetraphenyl orthocarbonate	C_6H_{12}	264(3.29)	12-1579-65

Compound	Solvent	λ_{max}(log ε)	Ref.
$C_{25}H_{20}O_5$			
9-Anthrol, 10-(4-hydroxy-3-methoxyphenyl)diacetal	benzene	280(5.2),340(4.6), 360(4.8),380(4.8), 400(4.8)	56-0245-65
$C_{25}H_{20}S$			
2H-Naphtho[2,3-b]thiete, 2,2-dimethyl-3,8-diphenyl-	EtOH	221(4.62),232(4.59), 258(4.54),296(4.01)	44-0629-65
$C_{25}H_{21}ClN_2O_5S$			
2-[(3-Ethyl-2-benzothiazolinylidene)-methyl]-1-(o-hydroxyphenyl)quinolinium perchlorate	EtOH	496(4.48)	65-3373-64
$C_{25}H_{21}ClO_4$			
7-(6,6-Diphenyl-1,3,5-hexatrien-1-yl)-tropylium perchlorate	MeCN	230s(4.36),268(4.27), 314(4.12),365(3.92), 560(4.71)	24-1590-64
$C_{25}H_{21}N$			
Aniline, N-(5-fluoren-9-ylidene-1,3-pentadienyl)-N-methyl-	MeCN	249(4.71),291(4.09), 457(4.88)	24-3331-64
$C_{25}H_{21}NO_2$			
3-Butenoic acid, 2-benzyl-2-cyano-3,4-diphenyl-, methyl ester	EtOH	258(4.26)	28-1146-64B
4(1H)-Pyridone, 2-phenyl-6-(2,2-diphenyl-2-hydroxyethyl)-	MeOH	242(4.38)	44-4263-65
4(1H)-Quinolone, 3-benzoyl-1-methyl-2-(α-methylbenzyl)-	MeOH	243(4.53),250(4.51), 325(4.09),337(4.09)	35-5417-65
$C_{25}H_{21}NO_3$			
Chromone, 3-benzoyl-2-benzyl-7-(dimethylamino)-	MeOH	264(4.54),293(3.93), 354(4.24)	35-5417-65
$C_{25}H_{21}N_2P$			
Phosphorane, (benzylidenehydrazono)-triphenyl-	$CHCl_3$	262s(3.99),269s(4.06), 277s(4.11),307s(4.25), 330(4.30)	44-0417-65
$C_{25}H_{22}N_2O_2$			
2-Quinolinecarbonitrile, 1,2,5,6,7,8-hexahydro-2-oxo-8-(α-phenacylbenzyl)-	EtOH	242(4.28),351(4.09)	35-5198-65
$C_{25}H_{22}N_2S$			
Benzothiazoline, 3-methyl-2-phenyl-2-[(1-methyl-1,2-dihydroquinolin-2-ylidene)methyl]-	hexane	312(4.06),404(3.90)	22-2879-64
	EtOH	312(4.16)	22-2879-64
	HOAc	307(4.31)	22-2879-64
Benzothiazoline, 3-methyl-2-phenyl-2-[(1-methyl-1,4-dihydroquinolin-4-ylidene)methyl]-	hexane	265(4.19),314(3.82), 395(4.08)	22-2879-64
	EtOH	311(4.14)	22-2879-64
	HOAc	305(4.35)	22-2879-64
$C_{25}H_{22}N_4O_2$			
6-Cycloheptimidazoleacetamide, 1,3-dibenzyl-α-cyano-1,2,3,6-tetrahydro-2-oxo-	EtOH	295(3.71)	94-0810-65
$C_{25}H_{22}N_4O_8$			
Rufocromomycin	1% $NaHCO_3$	227(4.57),245(4.59), 367(4.20)	28-4911-65B

Compound	Solvent	$\lambda_{max}(\log \epsilon)$	Ref.
$C_{25}H_{22}O$			
Cyclohexanone, 6-benzylidene-2,2-diphenyl-, cis	MeOH	268(3.89)	78-2201-64
trans	MeOH	292(4.23)	78-2201-64
$C_{25}H_{22}O_3$			
Limocitrol, 4'-benzyl ether	EtOH	256(4.33),279(4.16), 349(4.05),376(4.09)	78-2977-64
$C_{25}H_{22}O_6$			
2-Anthracenemalonic acid, 5-(benzyloxy)-1,2,3,4-tetrahydro-9-methyl-	EtOH	223(4.53),260(4.57), 308(3.68),321(3.58), 371(3.65)	65-0670-65
9-Anthracenone, 4,10-diacetoxy-5-(benzyloxy)-1,4,4a,9a-tetrahydro-	EtOH	225(4.48),258(4.02), 313(3.57)	30-0229-64B
$C_{25}H_{22}O_7$			
1-Naphthacenecarboxylic acid, 2-ethyl-6,11-dihydro-5,7,10-trimethoxy-6,11-dioxo-, methyl ester	EtOH	247(4.78),300(4.4), 428(4.17)	94-0797-65
$C_{25}H_{22}O_8$			
Amorphigenin, 6a,12a-dehydro-, acetate	EtOH	238(4.47),280(4.37), 309(4.24)	39-6023-64
$C_{25}H_{22}O_{12}$			
Flavone, 3',4',5,7-tetraacetoxy-3,8-dimethoxy-	EtOH	259(4.57),304(4.24), 328(4.26)	78-3219-65
$C_{25}H_{22}Pb$			
Lead, (1-cyclopenten-1-ylethynyl)-triphenyl-	hexane	218(4.69)	22-3518-65
$C_{25}H_{23}ClN_2O_5$			
2-p-Dimethylaminostyryl-1-(o-hydroxyphenyl)quinolinium perchlorate	EtOH	524(4.69)	65-3373-64
$C_{25}H_{23}ClO_6$			
7-[4,4-Bis(p-methoxyphenyl)-1,3-butadien-1-yl]tropylium perchlorate	MeCN	234(4.44),268(4.35), 360(3.84),425(3.91), 587(4.67)	24-1590-64
$C_{25}H_{23}NO_2$			
Isoquinoline, 1,4-dibenzyl-6,7-dimethoxy-	MeOH	243(4.79),270(3.71), 284(3.73),315(3.64), 328(3.69)	83-0879-65
$C_{25}H_{23}NO_5$			
Galanthaminone, piperonylidene-	EtOH	263(3.89),360(3.64)	94-0696-64
$C_{25}H_{23}N_5O_5$			
Adenosine, 3'-deoxy-N-methyl-2',5'-dibenzoate	EtOH	225(4.40),231(4.47), 266(4.23),270s(4.22)	87-0659-65
$C_{25}H_{23}OP$			
2-Cyclohexen-1-one, 3-methyl-6-(triphenylphosphoranylidene)-	MeOH	218(4.42),265(3.74)	5-0079-65J

Compound	Solvent	$\lambda_{max}(\log \epsilon)$	Ref.
$C_{25}H_{24}$			
Naphth[2,1-a]anthracene, 10,11,12,13-tetrahydro-3,12,12-trimethyl-	EtOH	227s(4.50),235(4.55), 250s(4.26),269(5.09), 278(5.18),292s(4.12), 305(4.08),311(4.08), 321(4.09),333(4.04), 351(2.80),357s(2.75), 370(2.58)	39-0724-64
Picene, 1,2,3,4-tetrahydro-2,2,9-trimethyl-	EtOH	227(4.53),233(4.53), 265(4.84),275(5.30), 289s(4.30),303(4.26), 316(4.27),331(3.28), 351(3.24),357s(3.06), 368(3.15)	39-0724-64
$C_{25}H_{24}BrNO_3$			
Isoquinoline, 1-(p-benzyloxybenzyl)-8-bromo-3,4-dihydro-6,7-dimethoxy-	EtOH	296(4.234)	65-0550-64
$C_{25}H_{24}BrNO_4$			
1-[2,3-Dimethoxy-9-phenanthryl)methyl]-2-(1,3-dioxolan-2-yl)pyridinium bromide	EtOH	212(4.49),260(4.81), 275(4.54),278(4.52), 302(4.06),312(4.04), 332(3.47),353(3.30)	54-0593-64
Piperidine, 4-(3-bromo-5H-dibenzo-[a,d]cyclohepten-5-ylidene)-1-methyl-, maleate	MeOH	220(4.67),244s(--), 291(4.12)	87-0829-65
$C_{25}H_{24}ClNO_4$			
1-Pyrrolin-4-one, 5-[2-(5'-chloro-3',4'-dihydro-8'-methoxyspiro[1,3-dioxolane-2,1'(2'H)-naphthalen]-3'-yl)-ethylidene]-2-phenyl-	MeCN	233(4.30),299(4.41)	35-0933-65
$C_{25}H_{24}N_2$			
3H,5H-Pyrido[1,2,3-de]quinoxaline, 6,7-dihydro-2,3-di-p-tolyl-	MeOH	254s(4.29),278(4.42), 320(3.97),420(3.69)	44-2589-65
$C_{25}H_{24}N_2O_2$			
3H,5H-Pyrido[1,2,3-de]quinoxaline, 6,7-dihydro-2,3-bis(p-methoxyphenyl)-	MeOH	282(4.43),320(4.09), 413(3.75)	44-2589-65
$C_{25}H_{24}N_2O_3$			
Quinolino[6,7-b]-14-hydroxy-dihydrodeoxycodeine	EtOH	280(3.72),298(3.63), 311(3.66),325(3.70)	78-1407-64
$C_{25}H_{24}OS$			
6H-Dibenzo[b,e]thiocine, 12-(3-benzyl-oxypropylidene)-7,12-dihydro-	MeOH	227(4.28),265(3.83)	24-0685-65
$C_{25}H_{24}O_2$			
Equilenin, 3-benzyl ether	MeOH	233(4.79),267(3.76), 278(3.72),289(3.52), 321(3.20),336(3.31)	44-0316-65
Naphth[2,1-a]anthracene-6,13-dione, 3,4,5,6,10,11,12,13-octahydro-3,12,12-trimethyl-	EtOH	236s(4.58),250(4.55), 258(4.63),268(4.58), 321(4.29),333(4.27), 358(3.69),378(3.63)	39-0724-64

Compound	Solvent	λ_{max}(log ε)	Ref.
Picene-1,2-dione, 1,2,3,4,9,10,11,12-octahydro-2,2,9-trimethyl-	EtOH	227(4.56),256s(4.53),264(4.62),313(4.25),333s(4.46),370s(3.45)	39-0724-64
$C_{25}H_{24}O_3$			
4a(2H)-Naphthalenecarboxylic acid, 3,4,5,6-tetrahydro-2-oxo-5,7-diphenyl-, ethyl ester	EtOH	330(4.38)	78-0195-64
$C_{25}H_{24}O_4$			
2-Anthraceneacetic acid, 1,2,3,4-tetrahydro-9-methyl-4-oxo-, methyl ester	EtOH	222(4.50),260(4.10),308(3.24),319(2.95),370(3.01)	65-0670-65
$C_{25}H_{24}O_5$			
Chalcone, 2'-(benzyloxy)-α,4,4'-trimethoxy-	EtOH	234(4.20),336(4.27)	22-3572-65
Chalcone, 2'-(benzyloxy)-3,4,4'-trimethoxy-	EtOH	246(4.07),359(4.33)	22-3350-65
Mammeigin	EtOH	234(4.45),286(4.52),365(4.11)	25-1065-64
	EtOH	234(4.45),286(4.52),365(4.11)	44-2342-65
	EtOH-NaOH	251(4.38),312(4.41),438(3.84)	25-1065-64
	EtOH-NaOH	218(4.33),251(4.38),312(4.41),438(3.84)	44-2342-65
$C_{25}H_{24}O_6$			
2-Anthracenemalonic acid, 5-(benzyloxy)-1,2,3,4-tetrahydro-4-hydroxy-9-methyl-	EtOH	236(4.71),302(3.87),315(3.74),329(3.45)	65-0670-65
Anthrone, 4β,10β-diacetoxy-5-(benzyloxy)-1,4,4a,9a-tetrahydro-	EtOH	226(4.46),254(3.90),311(3.45)	65-2558-64
4aα,9aβ-isomer	EtOH	226(4.44),258(3.93),318(3.54)	65-2558-64
Chalcone, 2'-(benzyloxy)-3,4,4'-trimethoxy-, epoxide	EtOH	233(4.34),280(4.14),311(4.04)	22-3350-65
1-Naphthacenecarboxylic acid, 2-ethyl-6,11-dihydro-5,7,10-trimethoxy-11-oxo-, methyl ester	EtOH	227(4.86),267(4.58),304(4.14),317(4.17),364(4.07)	94-0797-65
1,2-Propanedione, 1-[2-(benzyloxy)-4-methoxyphenyl]-3-(3,4-dimethoxyphenyl)-	EtOH	232(4.23),280(4.04),319(3.91)	22-3350-65
	EtOH-NaOH	389(4.05)	22-3350-65
$C_{25}H_{24}O_8$			
Amorphigenin, acetate	EtOH	235(4.17),241s(4.12),292(4.25)	39-6023-64
$C_{25}H_{25}ClN_2$			
Pinacyanol chloride	H_2O	590(4.8),600(4.9)	46-1894-65
	MeOH	560(4.9),605(5.2)	46-1894-65
$C_{25}H_{25}ClN_2O_4S_2$			
3,8,3',10-Bis(tetramethylene)thiacarbocyanine perchlorate	EtOH	586(4.94)	65-2455-64
$C_{25}H_{25}ClN_2S_2$			
3-Ethyl-2-[7-(3-ethyl-2-benzothiazolinylidene)-1,3,5-heptatrienyl]-benzothiazolium chloride	H_2O	690s(4.9),753(5.4)	46-1894-65

Compound	Solvent	$\lambda_{max}(\log \epsilon)$	Ref.
$C_{25}H_{25}NO_4$			
Isocarbostyril, 2-benzyl-8-(benzyloxy)-3,4-dihydro-6,7-dimethoxy-	iso-PrOH	217(4.64),264(4.10)	33-2089-64
$C_{25}H_{25}NO_5$			
Thebainehydroquinone	dioxan	310(3.75)	88-0275-65
$C_{25}H_{25}N_3O_7S$			
L-Tryptophan, p-nitrobenzyl ester, p-toluenesulfonate	80% EtOH	280(4.14)	44-2272-64
$C_{25}H_{25}N_3O_8$			
Cytidine, N-benzoyl-2',3'-O-(2,4-dimethoxybenzylidene)-	MeOH	260(4.43),302(4.02)	5-0156-64I
$C_{25}H_{25}N_3O_8S$			
Cytidine, N-benzoyl-2'-deoxy-, 3'-acetate, 5'-p-toluenesulfonate	MeOH	224(4.38),260(4.39), 302(4.02)	44-3067-65
$C_{25}H_{25}N_5O_6S$			
L-Glutamic acid, N-[p-[3-(2-amino-4-mercapto-6-phenyl-5-pyrimidinyl)propionamido]benzoyl]-	pH 1	272(4.44),347(4.24)	4-0263-64
	pH 8.4	270(4.49),360(4.25)	4-0263-64
	pH 13	269(4.46),327(4.20)	4-0263-64
$C_{25}H_{25}N_5O_7$			
L-Glutamic acid, N-[p-[3-(2-amino-4-hydroxy-6-phenyl-5-pyrimidinyl)-propionamido]benzoyl]-	pH 1	272(4.43)	4-0263-64
	pH 8.4	272(4.41)	4-0263-64
	pH 13	272(4.35)	4-0263-64
$C_{25}H_{26}BrNO_3$			
2-Benzyl-8-(benzyloxy)-3,4-dihydro-6,7-dimethoxyisoquinolinium bromide	pH 1	246(4.12),335(4.28)	33-2089-64
	iso-PrOH	248(4.11),341(4.31)	33-2089-64
$C_{25}H_{26}BrNO_4$			
Acetamide, 2-[p-(benzyloxy)phenyl]-N-(3-bromo-4,5-dimethoxyphenethyl)-	EtOH	280(3.556)	65-0550-64
$C_{25}H_{26}N_2O$			
2-Indolinone, 1-benzyl-3-(2-dimethylaminoethyl)-3-phenyl-	EtOH-HCl	256(3.84),264s(3.76), 283s(3.19)	87-0626-65
$C_{25}H_{26}N_2O_4$			
Δ^6-Codeine, 6-anilino-6-deoxy-7-formyl-14-hydroxy-	EtOH	376(4.28)	78-1407-64
$C_{25}H_{26}N_4O_5$			
Spiro[2-cyclohexene-1,1'-inden]-4-one, 2',3'-diethyl-6'-methoxy-, 2,4-dinitrophenylhydrazone	CH_2Cl_2	267(4.39),386(4.40)	5-0036-64D
isomer m. 182°	CH_2Cl_2	266(4.33),385(4.43)	5-0036-64D
$C_{25}H_{26}N_4O_6$			
Cytidine, N-benzoyl-2',3'-O-(p-dimethylaminobenzylidene)-	MeOH	262(4.64)	5-0156-64I
$C_{25}H_{26}N_8O_{12}$			
Purine, 6-acetamido-9-[(3,4,6-tri-O-acetyl-2-deoxy-2-(2,4-dinitroanilino)-α-D-glucopyranosyl]-	EtOH	262(4.37),341(4.26)	44-1556-65
β-D- isomer	EtOH	262(4.16),342(4.06)	44-1556-65

Compound	Solvent	λ_{max}(log ϵ)	Ref.
$C_{25}H_{26}OS_2$			
3-Pentanone, 1-phenyl-1,5-bis(p-tolylthio)-	EtOH	258(4.05)	7-0652-65
$C_{25}H_{26}O_2$			
Estra-1,3,5(10),6-tetraen-17-one, 3-(benzyloxy)-	MeOH	262(3.93),272(3.90)	44-0316-65
$C_{25}H_{26}O_3S_2$			
3-Pentanone, 1,5-bis(p-methoxyphenylthio)-1-phenyl-	EtOH	232(4.40),256s(4.12)	7-1093-65
$C_{25}H_{26}O_4$			
3-Cyclohexene-1-carboxylic acid, 2-oxo-1-(3-oxobutyl)-4,6-diphenyl-, ethyl ester	EtOH	292(4.25)	78-0195-64
Estra-1,3,5(10)-trien-17-one, 3,15-dihydroxy-, 3-benzoate	EtOH	231(4.28)	44-2731-64
[3.2]Metacyclophane, 2,2-dicarbethoxy-	EtOH	269(2.59)	88-2571-64
$C_{25}H_{26}O_5$			
2,4,6-Cycloheptatrien-1-one, 4-[3-[3-(benzyloxy)-4,5-dimethoxyphenyl]-propyl]-2-hydroxy-	EtOH	230s(4.56),326(3.89), 348s(3.85)	78-3605-65
	EtOH-base	241s(4.60),335(4.19), 392(4.19)	78-3605-65
Mammeigin, dihydro-	EtOH	287(4.54),341(4.05)	44-2342-65
	EtOH-NaOH	293(4.07),321(4.12), 420(3.89)	44-2342-65
$C_{25}H_{26}O_6$			
Anthracene, 9β-hydroxy-4a,10-diacetoxy-5-(benzyloxy)-1,4,4aβ,9,9aα,10-hexahydro-	EtOH	277(3.19),283(3.16)	65-0662-65
Indene, 5,7-diacetoxy-2-(4-acetoxyphenyl)-1-ethyl-3,3-dimethyl-	MeOH	269(4.24)	44-4120-65
Macluraxanthone, O,O'-dimethyl-	EtOH	247(4.22),289(4.57), 333(4.26),365s(3.66)	44-0692-64
$C_{25}H_{26}O_9$			
Podophyllotoxin, 6,7-O-demethylenedeoxy-, diacetate	MeOH	268(3.29),275(3.26)	33-1529-64
$C_{25}H_{27}ClN_2O_4$			
7-[2,2-Bis(p-dimethylaminophenyl)-vinyl]tropylium perchlorate	MeCN	252(4.13),330(4.22), 485(4.53),648(4.72)	24-1590-64
	MeCN-HCl	495(4.40)	24-1590-64
$C_{25}H_{27}ClO_{10}$			
2,2',2",4,4',4"-Hexamethoxytriphenylcarbonium perchlorate	HOAc-$HClO_4$	559(4.50)	27-0339-65
2,2',2",5,5',5"-Hexamethoxytriphenylcarbonium perchlorate	HOAc-$HClO_4$	669(3.91)	27-0339-65
$C_{25}H_{27}NO_2$			
Isoquinoline, 1,4-dibenzyl-1,2,3,4-tetrahydro-6,7-dimethoxy-, hydrochloride	MeOH	284(3.59)	83-0879-65
$C_{25}H_{27}NO_4$			
Estradiol, 3-benzoate, 17-nitrite	MeOH	229(4.34)	78-0743-65

Compound	Solvent	$\lambda_{max}(\log \epsilon)$	Ref.
Naphtho[2,1-f]quinolin-2(1H)-one, 3,4,4a,4b,5,6,10b,11,12,12a-decahydro-1,8-dihydroxy-12a-methyl-, 8-benzoate	MeOH	229(4.33),266(3.59), 274(3.51)	78-0743-65
isomer	MeOH	228(4.30),266(3.57), 274(3.51)	78-0743-65
$C_{25}H_{27}NO_4S$			
6-Azaestra-1,3,5(10),8,14-pentaen-17β-ol, 3-methoxy-6-(p-tolylsulfonyl)-	EtOH	254(4.24),316(4.26)	78-2517-65
6-Azaestra-1,3,5(10),8-tetraen-17-one, 3-methoxy-6-(p-tolylsulfonyl)-	EtOH	223(4.30),241(4.34), 280(4.09)	78-2517-65
$C_{25}H_{27}NO_5$			
Lyfoline	MeOH	282(4.15)	100-0084-65
$C_{25}H_{27}NO_7$			
Dibenzo[a,g]quinolizidine, 13-acetoxy-9,10-dimethoxy-2,3-(methylenedioxy)-8-(2-oxopropyl)-	EtOH	283(3.83)	95-0146-64
$C_{25}H_{27}N_3O_2$			
5H-Dibenzo[b,e][1,4]diazepin-11-one, 8-(benzyloxy)-10-(2-dimethylaminoethyl)-10,11-dihydro-5-methyl-	EtOH	211(4.64),225(4.54), 302(3.58)	33-1590-65
$C_{25}H_{27}N_3O_4$			
Estra-1,3,5(10)-trien-17-one, 3-methoxy-1-[(p-nitrophenyl)azo]-	MeOH	279(3.98),350(4.21)	35-2943-64
$C_{25}H_{27}N_7O_5$			
Benzoic acid, p-[N-[4-[[2-amino-6-hydroxy-5-(phenylazo)-4-pyrimidinyl]amino]-3-oxobutyl]acetamido]-, ethyl ester, hydrochloride	pH 13	382(4.25)	44-3404-65
$C_{25}H_{28}$			
Picene, 1,2,3,4,9,10,11,12-octahydro-2,2,9-trimethyl-	EtOH	230(4.41),256(4.76), 265(4.84),285(4.26), 296(4.24),310(4.30), 327s(3.10),342(3.03), 357(2.87)	39-0724-64
$C_{25}H_{28}ClNO_2$			
Isoquinoline, 1,4-dibenzyl-1,2,3,4-tetrahydro-6,7-dimethoxy-, hydrochloride	MeOH	284(3.59)	83-0879-65
$C_{25}H_{28}N_2$			
Yohimban, 3ξ-phenyl-	EtOH	228(4.58),284(3.99), 291(3.94)	44-0105-65
$C_{25}H_{28}N_2O_6$			
Voacarpine, N-acetyl-, acetate	MeOH-HCl	241(4.23),314(4.28)	20-0170-65
	MeOH	241(4.23),314(4.28)	20-0170-65
	MeOH-KOH	241(4.23),314(4.28)	20-0170-65
$C_{25}H_{28}N_4O_4$			
Estra-1,3,5(10)-trien-17-one, 4-amino-3-methoxy-1-[(p-nitrophenyl)azo]-	MeOH	280(4.00),406(4.39)	35-2943-64

Compound	Solvent	$\lambda_{max}(\log \epsilon)$	Ref.
$C_{25}H_{28}N_6$			
Biguanide, 1-(isopropylideneamino)-2-phenyl-4,5-di-p-tolyl-	EtOH	240(4.33),280s(4.52)	39-0932-65
$C_{25}H_{28}OS$			
Estra-1,3,5(10)-trien-17-one, 3-(benzylthio)-	EtOH	258(3.94)	44-0295-65
$C_{25}H_{28}O_3$			
Estra-1,3,5(10)-trien-17-one, 15β-(benzyloxy)-3-hydroxy-	MeOH	222(3.88),282(3.28)	44-0214-64
2(1H)-Phenanthrone, 1α-(benzyloxymethyl)-1,2,3,4,9,12-hexahydro-6-methoxy-1β,12β-dimethyl-	n.s.g.	232(4.23),280(3.49)	39-4521-64
$C_{25}H_{28}O_4$			
9-Anthrol, 5-(benzyloxy)-1,4,4aα,9,-9aβ,10-hexahydro-4β,10β-(isopropylidenedioxy)-9β-methyl-	EtOH	275(3.39),281(3.38)	65-2563-64
Estr-4-en-3α-ol, 9α,10α-epoxy-17β-hydroxy-, benzoate	n.s.g.	230(4.20)	39-4492-64
$C_{25}H_{28}O_6$			
Furo[2,3-a]xanthen-11-one, 2,3-dihydro-7-hydroxy-4,8-dimethoxy-10-(3-methyl-2-butenyl)-	MeOH	257(4.69),276(4.06), 314(4.29)	78-2653-65
Macluraxanthone, dihydro-, O,O'-dimethyl-	EtOH	248(4.32),289(4.53), 333(4.27),369s(3.57)	44-0692-64
Xanthen-9-one, 1,5-dihydroxy-3,6-dimethoxy-4,8-bis(3-methyl-2-butenyl)-	MeOH	242s(4.48),258(4.67), 286(3.86),335(4.30)	78-2653-65
Xanthen-9-one, 3,6-dimethoxy-1,5-bis-[(3-methyl-2-butenyl)oxy]-	MeOH	247(4.80),287s(4.22), 307(4.44)	78-2653-65
$C_{25}H_{28}O_7$			
Methanol, tris(2,6-dimethoxyphenyl)-carbonium ion from	EtOH	277(3.66),284(3.65)	35-2252-64
	0.5N HCl	272(4.04),522(4.25)	35-2252-64
	MeOH-HCl	522(4.29)	35-2252-64
	EtOH-HCl	522(4.30)	35-2252-64
	dioxan-HCl	522(4.28)	35-2252-64
Siphulin, methyl ester	EtOH	243s(4.43),252(4.43), 268(4.33),295(4.35)	1-1677-65
$C_{25}H_{28}O_9$			
2-Naphthoic acid, 1,2,3,4-tetrahydro-3-(hydroxymethylene)-6,7-dimethoxy-4-oxo-1-(3,4,5-trimethoxyphenyl)-, ethyl ester	EtOH	233(4.4),282(4.03), 355(4.16)	2-0190-64
$C_{25}H_{29}ClN_2O_9$			
Chlorotetracycline, N-(1-methoxyethyl)-	pH 1	230(4.25),268(4.26), 370(4.01)	87-0870-65
$C_{25}H_{29}NO_4S$			
6-Azaestra-1,3,5(10),8-tetraen-17-ol, 3-methoxy-6-(p-tolylsulfonyl)-	EtOH	226(4.31),241(4.37), 279(4.18)	78-2517-65
$C_{25}H_{29}NO_5$			
Lyfoline, dihydro-	MeOH	295.5(3.90)	100-0084-65

Compound	Solvent	$\lambda_{max}(\log \epsilon)$	Ref.
$C_{25}H_{30}$			
Naphthalene, 1β-(2,2-diphenylvinyl)-decahydro-4aβ-methyl-	EtOH	252(4.21)	78-2641-65
$C_{25}H_{30}ClN_3$			
Crystal violet	H_2O	590(5.0)	49-0678-64
	$POCl_3$	590(5.0)	49-0678-64
complex with two molecules $SbCl_5$	$POCl_3$	425(4.2),640(4.9)	49-0678-64
complex with four molecules $SbCl_5$	$POCl_3$	430(4.4)	49-0678-64
$C_{25}H_{30}N_2O_3S$			
Vincanidine, dihydro-, derivative	n.s.g.	246(3.58),276(2.96), 303(3.32)	70-1992-65
$C_{25}H_{30}N_2O_4$			
Secoaspidodasycarpine, N,O-diacetate	EtOH	223(4.61),276s(3.95), 283(3.96),292(3.91)	78-1717-65
$C_{25}H_{30}N_2O_5$			
Obscurinervine	EtOH	220(4.49),253(3.81), 312(3.43)	35-2451-64
$C_{25}H_{30}N_2O_6$			
Aspidodasycarpine, N-acetyl-, acetate	EtOH	238(3.78),295(3.30)	78-1717-65
$C_{25}H_{30}N_2O_9$			
Tetracycline, N-(1-methoxy)ethyl-	pH 1	218(4.17),270(4.28), 360(4.08)	87-0870-65
$C_{25}H_{30}OS$			
Estra-1,3,5(10)-trien-17β-ol, 3-(benzylthio)-	EtOH	258(3.92)	44-0295-65
$C_{25}H_{30}O_2$			
Phenanthrene, 1α-benzyloxymethyl-1,2,3,4,11,12-hexahydro-6-methoxy-1β,12β-dimethyl-	n.s.g.	274(4.1)	39-4521-64
$C_{25}H_{30}O_3S$			
Estra-1,3,5(10)-trien-3-ol, p-toluenesulfonate	EtOH	221(4.30),256s(3.18), 263(3.22),267(3.25), 274(3.16),326s(1.34)	24-0140-64
$C_{25}H_{30}O_5$			
Macluraxanthone, tetrahydro-, dimethyl ether	EtOH	248(4.57),262(4.28), 289(3.91),327(4.30), 355s(3.79)	44-0692-64
16,21-Methanopregna-1,4,21-triene-3,20-dione, 11β,21-dihydroxy-, 21-propionate	EtOH	241(4.36)	5-0218-65E
Siphulin, decarboxy-, dimethyl ether	EtOH	249s(4.32),283(4.27)	1-1677-65
$C_{25}H_{30}O_6S$			
Thieno[4',3',2'-4,5,6]pregn-5-ene-3,11,20-trione, 5'-methyl-17α,20:20,21-bis(methylenedioxy)-	EtOH	219(4.11),267(4.06), 298(3.39)	94-1433-64
$C_{25}H_{30}O_7$			
Estra-1,3,5(10)-trien-17-one, 1,6α-diacetoxy-4-(acetoxymethyl)-	EtOH	269(2.71)	39-3621-64

Compound	Solvent	λ_{max}(log ε)	Ref.
Phenanthrene-12-carboxylic acid, 5,6-diacetoxy-1,2,3,4,9,12-hexahydro-7-isopropyl-1,11-dimethyl-9-oxo-, methyl ester	EtOH	257(4.14),272s(4.06)	33-1234-64
$C_{25}H_{30}O_8$			
Flavone, 3',4',5,7-tetraethoxy-3,8-dimethoxy-	EtOH	254(4.42),272(4.40), 354(4.39)	78-3219-65
Orthodesaspidin	C_6H_{12}	230(4.40),293(4.38)	1-1292-64
Pregna-1,4-diene-3,11,20-trione, 16α,21-diacetoxy-17-hydroxy-	EtOH	237.5(4.20)	39-0130-65
$C_{25}H_{30}O_{16}$			
3,5-Cycloheptadiene-1,1,2,3,4,5,6,7-octacarboxylic acid, 1,1-diethyl hexamethyl ester	EtOH	215s(3.96),285(3.61)	44-0423-64
4,6-Cycloheptadiene-1,1,2,3,4,5,6,7-octacarboxylic acid, 1,1-diethyl hexamethyl ester	EtOH	215s(4.08),263(3.77), 275s(3.67)	44-0423-64
$C_{25}H_{31}FN_2O_5$			
Pregn-4-eno[3,2-c]pyrazole, 9α-fluoro-16α-methyl-17,20:20,21-bis(methylenedioxy)-11-oxo-	MeOH	260(4.0)	87-0352-64
$C_{25}H_{31}FO_7$			
Pregn-4-ene-3,11-dione, 9α-fluoro-2-(hydroxymethylene)-16α-methyl-17,20:20,21-bis(methylenedioxy)-	MeOH-NaOH	239(4.29),360(4.08)	87-0352-64
$C_{25}H_{31}NO_5S$			
Pregn-4-ene[3,2-d]thiazole, 17α,20:20,21-bis(methylenedioxy)-16α-methyl-11-oxo-	$CHCl_3$	271(3.92)	87-0584-64
$C_{25}H_{31}NO_5S_2$			
Pregn-4-ene[3,2-d]thiazole, 17α,20:20,21-bis(methylenedioxy)-2'-mercapto-16α-methyl-11-oxo-	MeOH	248(4.05),327(4.12), 337(4.13)	87-0584-64
	MeOH-NaOH	246(4.30),293(4.03)	87-0584-64
$C_{25}H_{31}N_3O_3$			
Urea, 1,3-dicyclohexyl-1-[(1,3,4,5-tetrahydro-5-oxobenz[cd]indol-3-yl)carbonyl]-	EtOH	242(4.22),320(3.68), 360(3.63)	44-0843-64
$C_{25}H_{32}N_2OS$			
Acrylonitrile, 2-[p-(diethylamino-ethoxy)phenyl]-3-(p-isobutylthio-phenyl)-	EtOH-HCl	360(4.54)	87-0511-64
$C_{25}H_{32}N_2O_5$			
Obscurinervine, dihydro-	EtOH	220(4.46),256(3.81), 313(3.43)	35-2451-64
$C_{25}H_{32}N_2O_6$			
Aspidodasycarpine, N^b-acetyldihydro-, acetate	EtOH	244(3.68),293(3.34)	78-1717-65
Vindoline	EtOH	250(3.71),304(2.76)	100-0470-64

Compound	Solvent	λ_{max}(log ε)	Ref.
$C_{25}H_{32}N_2O_{10}$			
Tetracycline, N-(methoxymethyl)-	pH 1	218(4.17),270(4.30), 360(4.16)	87-0870-65
$C_{25}H_{32}N_4O_5$			
D-Homo-C-norgon-13(17a)-en-17-one, 3β-hydroxy-10,17a-dimethyl-, 2,4-dinitrophenylhydrazone	EtOH	258(4.16),388(4.39)	78-1215-65
$C_{25}H_{32}O_2S$			
Androsta-5,16-dien-3β-ol, 17-(2-thienyl)-, acetate	EtOH	279(4.06)	78-1197-65
$C_{25}H_{32}O_4$			
16,21-Methanopregna-1,4,21-triene-3,20-dione, 11β-hydroxy-21-isopropoxy-	EtOH	248(4.26)	5-0218-65E
Pregna-4,6-diene-3,20-dione, 17α-acetoxy-6-methyl-16-methylene-	EtOH	287.5(4.35)	78-1753-65
Pregna-1,4,6-triene-3,20-dione, 17α-acetoxy-6,16α-dimethyl-	EtOH	228(4.08),256(3.93), 302(4.06)	87-0540-64
$C_{25}H_{32}O_4S$			
Thieno[4',3',2'-4,5,6]pregn-5-ene-3,20-dione, 17α-acetoxy-5'-methyl-	EtOH	221(4.09),269(4.05), 302(3.38)	94-1433-64
$C_{25}H_{32}O_5$			
5α-Carda-16,20(22)-dienolide, 3β-acetoxy-14β,15β-epoxy-	EtOH	276(4.15)	13-0645-65
5β,14β-Carda-16,20(22)-dienolide, 3β-acetoxy-14β,15β-epoxy-	EtOH	276(4.17)	94-0308-65
Norethindrone, carbomethoxypropionate	EtOH	239(4.22)	13-0441-65
5,14,16-Pregnatrien-20-one, 3β,21-dihydroxy-16-methyl-, 21-acetate 3-formate	MeOH	313(4.11)	73-2513-64
$C_{25}H_{32}O_5S$			
Thieno[4',2',3'-4,5,6]pregn-5-en-3-one, 5'-methyl-17α,20:20,21-bis-(methylenedioxy)-	EtOH	221(4.10),269(4.06), 304(3.36)	94-1433-64
$C_{25}H_{32}O_6$			
Pregna-3,5,9(11)-triene-6-carboxaldehyde, 21-acetoxy-17α-hydroxy-3-methoxy-	EtOH	220(4.08),322(4.19)	78-0597-64
$C_{25}H_{32}O_7$			
Estra-1,3,5(10)-triene-3,7α,17β-triol, 6β-methoxy-, triacetate	EtOH	263s(2.70),267(2.79), 275(2.75)	44-1325-64
Phenanthrene-12-carboxylic acid, 5,6-diacetoxy-1,2,3,4,9,10,11,12-octahydro-7-isopropyl-1,1-dimethyl-9-oxo-, methyl ester	EtOH	212(4.45),257(4.08)	33-1234-64
Pregna-3,5-diene-6-carboxaldehyde, 21-acetoxy-17 -hydroxy-3-methoxy-11,20-dioxo-	EtOH	218(4.04),322(4.18)	78-0597-64
Pregna-3,5-diene-6-carboxaldehyde, 3-methoxy-, cyclic methyl orthoformate	EtOH	219(4.07),322(4.16)	78-0597-64

Compound	Solvent	$\lambda_{max}(\log \epsilon)$	Ref.
Pregn-4-ene-3,11-dione, 2-formyl-16α-methyl-17α,20:20,21-bis-(methylenedioxy)-	MeOH	241(4.09)	87-0584-64
$C_{25}H_{32}O_7S$			
Pregn-4-ene-3,11-dione, 6α-mercapto-17,20:20,21-bis(methylenedioxy)-, acetate	EtOH	233(4.20)	94-1433-64
$C_{25}H_{32}O_8$			
Aspidin	C_6H_{12}	230(4.43),292(4.30)	1-1292-64
Isoaspidin	C_6H_{12}	228(4.48),292(4.33)	1-1294-64
Paraaspidin	C_6H_{12}	228(4.43),271(4.33)	1-1292-64
5α-Pregn-1-ene-3,11,20-trione, 16α,21-diacetoxy-17-hydroxy-	EtOH	226.5(4.04)	39-0130-65
$C_{25}H_{32}O_9$			
Melicopol, triacetate	EtOH	250(3.60),300(3.47)	12-2021-65
$C_{25}H_{33}ClO_6$			
Androst-4-ene-3,11-dione, 2β-chloro-1α,17β-dihydroxy-2α,17α-dimethyl-, diacetate	EtOH	240(4.05)	32-0159-65
$C_{25}H_{33}FN_2O_5$			
Pregn-4-eno[3,2-c]pyrazole, 9α-fluoro-11β-hydroxy-16α-methyl-17,20:20,21-bis(methylenedioxy)-	MeOH-HCl	267(4.02)	87-0352-64
$C_{25}H_{33}FO_4$			
Pregna-4,6-diene-3,20-dione, 21-fluoro-17α-hydroxy-6,16α-dimethyl-, acetate	EtOH	290(4.36)	87-0548-64
$C_{25}H_{33}FO_5$			
Pregna-3,5-diene-6-carboxaldehyde, 17α-acetoxy-21-fluoro-3-methoxy-	EtOH	220(4.01),322(4.16)	78-0597-64
$C_{25}H_{33}NO_2$			
Cyclohexanecarboxylic acid, 1-(4-biphenylyl)-, 2-(diethylamino)ethyl ester	EtOH-HCl	256(4.39)	87-0504-64
$C_{25}H_{33}NO_4$			
Pregn-4-ene-16α-carbonitrile, 17α-acetoxy-6α-methyl-3,20-dioxo-	EtOH	239.5(4.22)	44-3476-64
$C_{25}H_{33}NO_5$			
Isoxazolino[17,16-d]pregn-4-ene-3,20-dione, 3'-carbethoxy-	MeOH	240(4.35)	5-0228-65E
	MeOH	240(4.32)	44-1272-65
$C_{25}H_{33}NO_5S$			
Pregn-4-ene[3,2-d]thiazole, 11β-hydroxy-16α-methyl-17α,20:20,21-bis-(methylenedioxy)-	MeOH	263(3.99)	87-0584-64
$C_{25}H_{33}NO_6$			
p-Benzoquinone, 2-(4-methoxy-3-nitro-6-tert-butylphenoxy)-3,6-di-tert-butyl-	EtOH	229(4.26),263(4.24)	39-2921-65

Compound	Solvent	$\lambda_{max}(\log \epsilon)$	Ref.
$C_{25}H_{33}N_3O$			
Acrylonitrile, 2-[p-(diethylamino-ethoxy)phenyl]-3-(p-diethylamino-phenyl)-	MeOH-HCl	234(4.01),329(4.21), 405(3.65)	87-0511-64
$C_{25}H_{33}N_3O_6S_2$			
Cephalosporanic acid, 7-(benzylthio-carboxamido)-, cyclohexylamine salt	pH 6.0	258(3.98)	39-5015-65
$C_{25}H_{34}ClFO_5$			
Pregn-4-ene-3,20-dione, 21-chloro-9-fluoro-11β,16α,17-trihydroxy-2α-methyl-, cyclic 16,17-acetal with acetone	MeOH	236(4.26)	13-0615-65
$C_{25}H_{34}N_2O_3$			
Androst-4-eno[3,2-b]pyridine-5'-carb-oxylic acid, 6'-amino-17β-hydroxy-, ethyl ester	EtOH	225(4.25),242s(4.18), 259(4.26),282(4.07), 371(4.11)	94-0077-64
$C_{25}H_{34}N_2O_5$			
Hydrazine, 1,1-diacetyl-2-(17β-hy-droxy-1-oxo-A-norandrost-3(5)-en-2-ylidene)-, propionate	EtOH	243(3.98)	94-0156-65
Pregn-4-eno[3,2-c]pyrazole, 11β-hy-droxy-16α-methyl-17,20:20,21-bis-(methylenedioxy)-	MeOH	261(4.01)	35-1520-64
$C_{25}H_{34}N_4O_4S_3$			
5-Thia-1-azabicyclo[4.2.0]oct-2-ene-2-carboxylic acid, 3-(mercapto-methyl)-8-oxo-7-(2-phenylacetamido)-, dimethyldithiocarbamate, cyclohexyl-amine salt	pH 6.0	268(4.37)	39-5015-65
$C_{25}H_{34}N_6O_2$			
5α-Androstano[2,3-g]-2',4'-diamino-pteridine, 17β-acetoxy-	EtOH	257(4.41),373(4.00)	87-0678-64
$C_{25}H_{34}N_7O_2P$			
2H-Naphtho[2,3-d]triazole-4,9-dione, 2-[(tripiperidylphosphoranylidene)-amino]-	$CHCl_3$	317(4.51)	39-1003-65
$C_{25}H_{34}O$			
12'-Apo-α-carotenal	hexane	390(4.48),409(4.81), 431(4.77)	39-2019-65
12'-Apo-β-carotenal, all trans	hexane	414(4.90)	28-1840-64A
12'-Apolycopenal	$CHCl_3$	418(4.79),443(4.91), 468(4.81)	39-2019-65
$C_{25}H_{34}O_2$			
24-Norchola-5,17(20)-dien-22-yn-3β-ol, acetate (isomers)	MeOH	233(4.18)	35-3737-65
	MeOH	233(4.18)	35-3737-65
24-Norchola-5,20-dien-22-yn-3β-ol, acetate	MeOH	223(4.18)	35-3737-65
1,4,6,8,10,12-Tridecahexaen-3-one, 1-hydroxy-2,7,11-trimethyl-13-(2,6,6-trimethylcyclohex-1-en-1-yl)-	hexane	435(4.85)	28-1840-64A

Compound	Solvent	λ_{max}(log ε)	Ref.
$C_{25}H_{34}O_3$			
Estra-1,3,5(10)-trien-17-one, 3-(cyclopentyloxy)-, cyclic ethylene acetal	EtOH	281(3.30),289(3.25)	87-0755-64
Testosterone, 6-methylene-17α-vinyl-, propionate	EtOH	260(4.05)	78-0597-64
$C_{25}H_{34}O_4$			
Androsta-3,5,7-triene-6-carboxaldehyde, 3-ethoxy-17β-hydroxy-, propionate	EtOH	383(4.12)	78-1753-65
5β,17α-Carda-14,20(22)-dienolide, 3β-acetoxy-	EtOH	213(4.21)	94-0308-65
Pregna-1,4-diene-3,20-dione, 17α-acetoxy-6α,16α-dimethyl-	EtOH	245(4.20)	87-0540-64
Pregna-4,6-diene-3,20-dione, 17α-acetoxy-6,16α-dimethyl-	EtOH	288(4.40)	87-0540-64
Pregna-4,7-diene-3,20-dione, 17α-acetoxy-6α,16α-dimethyl-	EtOH	238(4.16)	78-1753-65
Pregna-3,5-dien-20-one, 17α-acetoxy-3-methoxy-16-methylene-	EtOH	241(4.29)	78-0597-64
Pregna-3,5,7-trien-20-one, 17α-acetoxy-3-ethoxy-	EtOH	215(3.87),322(4.27)	78-1753-65
Pregna-3,5,7-trien-20-one, 17α-acetoxy-3-methoxy-6-methyl-	EtOH	322(4.29)	78-1753-65
Pregna-5,16,21-trien-20-one, 3β-acetoxy-21-(methoxymethylene)-	EtOH	279(4.15)	88-1839-64
Pregn-4-ene-3,20-dione, 16α,17α-isopropylidenedioxy-6-methylene-	EtOH	260(4.05)	78-0597-64
$C_{25}H_{34}O_5$			
5β-Card-20(22)-enolide, 3β-acetoxy-14,15β-epoxy-	EtOH	212(4.10)	13-0593-64
5β,14α-Card-20(22)-enolide, 3β-acetoxy-15-oxo-	EtOH	212(4.07)	13-0593-64
5β,14β-Card-20(22)-enolide, 3β-acetoxy-15-oxo-	EtOH	212.5(4.09)	13-0593-64
5β,14β,17α-Card-20(22)-enolide, 3β-acetoxy-14β,15β-oxido-	EtOH	216(4.16)	94-0308-65
5β,14α,17α-Card-20(22)-enolide, 3β-acetoxy-15-oxo-	EtOH	214(4.18)	94-0308-65
5β,14β,17α-Card-20(22)-enolide, 3β-acetoxy-15-oxo-	EtOH	215(4.14)	94-0308-65
Pregna-3,5-diene-6-carboxaldehyde, 21-acetoxy-3-methoxy-20-oxo-	EtOH	219(4.01),320(4.19)	78-0597-64
Pregna-3,5-diene-11,20-dione, 17α,21-isopropylidenedioxy-3-methoxy-	EtOH	238(4.31)	78-0597-64
Pregna-3,5-dien-20-one, 3,15α-diacetoxy-	n.s.g.	230(4.33)	78-0107-64
Pregna-14,16-dien-20-one, 3,12-diacetoxy-	n.s.g.	304(4.16)	13-0271-64
5α-Pregna-14,16-dien-20-one, 3β,12β-diacetoxy-	EtOH	305(4.29)	94-1332-65
$C_{25}H_{34}O_5S$			
Pregn-4-ene-3,20-dione, 21-hydroxy-16α-mercapto-, diacetate	EtOH	238(4.29)	87-0531-64
$C_{25}H_{34}O_6$			
Androsta-3,5-diene-6-carboxaldehyde, 11α,17β-diacetoxy-3-methoxy-	EtOH	220(4.02),320(4.16)	78-0597-64
Cortexone, 2β-hydroxy-, 2,21-diacetate	EtOH	242(4.17)	13-0057-65

Compound	Solvent	λ_{max}(log ε)	Ref.
5β,19-Cyclopregn-6-en-3β-ol, 17α,20:20,21-bis(methylene-dioxy)-, acetate	n.s.g.	215(3.68)	13-0371-65
17α-Digitoxigenin, 16β,17β-epoxy-, 3-acetate	EtOH	218.5(4.13)	5-0246-65E
Etiojerva-5,12-diene-3,11-dione, 17-ethyl-17,20-dihydroxy-, 3-ethylene ketal, 21-acetate	EtOH	251(4.15)	44-0755-64
Phenanthrene-12-carboxylic acid, 5,6-diacetoxy-1,2,3,4,9,10,11,12-octahydro-7-isopropyl-1,1-dimethyl-, methyl ester	EtOH	220s(4.02),267(2.72), 274s(2.63)	33-1234-64
Pregna-3,5-diene-11,20-dione, 21-acetoxy-17α-hydroxy-3-methoxy-6-methyl-	EtOH	246(4.25)	78-0569-65
Pregna-3,5-dien-3-ol, 17α,20:20,21-bis(methylenedioxy)-, acetate	n.s.g.	235(4.26)	13-0371-65
Pregna-3,5-dien-11-one, 3-ethoxy-17α,20:20,21-bis(methylenedioxy)-	EtOH	241(4.27)	78-0597-64
5β-Pregn-17(20)-en-21-al, 3α,20-diacetoxy-11-oxo-	ether	246(4.18)	44-0521-64
Pregn-4-ene-3,20-dione, 11α,18-diacetoxy-	MeOH	238(4.16)	24-1974-65
5α-Pregn-16-ene-11,20-dione, 3β,21-diacetoxy-	EtOH	235(3.93)	39-0130-65
Pregn-16-en-20-one, 14,15β-epoxy-3β,12β-dihydroxy-, diacetate	n.s.g.	244(3.65)	13-0271-64
5α-Pregn-16-en-20-one, 14β,15β-epoxy-3β,12β-dihydroxy-, diacetate	EtOH	243(3.63)	94-1332-65
$C_{25}H_{34}O_7$			
14α-Digacetigenin, 3-acetate	EtOH	288(1.7)	39-3611-64
Pregna-3,5-diene-11,20-dione, 21-acetoxy-17α-hydroxy-6-(hydroxymethyl)-3-methoxy-	EtOH	249(4.20)	78-0597-64
Pregna-1,4-dien-3-one, 11β,17α,20α,21-tetrahydroxy-, 20,21-diacetate	EtOH	244(4.17)	78-0179-65
Pregn-4-ene-3,20-dione, 1β,21-diacetoxy-17α-hydroxy-	EtOH	241(4.17)	13-0645-64
Pregn-4-ene-3,20-dione, 15,21-diacetoxy-17-hydroxy-	EtOH	241(4.23)	13-0713-64
5α-Pregn-8(14)-ene-11,20-dione, 3β,21-diacetoxy-17-hydroxy-	EtOH	208s(3.85)	39-0156-65
Pregn-4-en-3-one, 11β-hydroxy-2-(hydroxymethylene)-16α-methyl-17,20:20,21-bis(methylenedioxy)-	MeOH-NaOH	242(4.17),357(4.00)	35-1520-64
Taxinol, acetyldehydroisopropylidene-oxo-	EtOH	237(3.88)	95-0404-65
	EtOH-NaOH	357(3.61)	95-0404-65
$C_{25}H_{34}O_8$			
Strophanthidin, 16-acetoxy-	EtOH	213(4.12),268s(2.8)	33-1634-65
$C_{25}H_{35}BrO_4$			
17α-Pregn-4-ene-16α-acetic acid, 17β-bromo-6α-methyl-3,20-dioxo-, methyl ester	EtOH	239(4.21)	44-3476-64
$C_{25}H_{35}ClO_4$			
9β,10α-Pregna-3,5-dien-20-one, 17α-acetoxy-6-chloro-3-ethoxy-	MeOH	253(4.29)	54-0918-65

Compound	Solvent	λ_{max}(log ε)	Ref.
$C_{25}H_{35}ClO_4S_3$			
Spiro[benzofuran-2(3H),1'-[2]cyclo-hexen]-3-one, 7-chloro-4,6-dimethoxy-6'-methyl-2',4',4'-tris(propylthio)-	EtOH	288(4.37)	87-0705-64
$C_{25}H_{35}ClO_5$			
5β-Card-20(22)-enolide, 3β-acetoxy-15α-chloro-14β-hydroxy-	EtOH	213(4.03)	13-0593-64
Pregn-4-ene-3,20-dione, 21-chloro-11β,16α,17-trihydroxy-6α-methyl-, cyclic 16,17-acetal with acetone	MeOH	241.5(4.17)	13-0615-65
$C_{25}H_{35}ClO_6$			
Cerberigenin, chlorohydrin, acetate	isooctane	217(4.17)	12-1079-65
$C_{25}H_{35}NO_3$			
Androsta-1,4-dien-3-one, 17β-acetoxy-2-(N-pyrrolidinyl)-	EtOH	243(4.23),377(3.70)	7-0288-65
Estr-4-en-3-one, 17β-acetoxy-17α-(3-dimethylaminoprop-1-ynyl)-	EtOH	239(4.19)	78-2295-64
$C_{25}H_{35}NO_4$			
Androstano[3,2-b]pyridine-5'-carboxylic acid, 1',6'-dihydro-17β-hydroxy-6'-oxo-, ethyl ester	EtOH	241(3.88),346(4.04)	94-0077-64
$C_{25}H_{35}NO_5$			
Phenethylamine, 2-(2-ethoxy-4,5-di-methoxystyryl)-N-ethyl-4,5-dimeth-oxy-N-methyl-	MeOH	290(4.08),336(4.10)	35-2177-64
$C_{25}H_{35}NO_5S$			
5α-Androstano[3,2-e]-1',4'-thiaz-5'-en-3'-one, 17β-hydroxy-, 17-(hydrogen succinate)	EtOH	230(3.61),298(3.42)	87-0555-64
$C_{25}H_{35}NO_6$			
Cortisone, 20-methoxyimino-6α-methyl-, 21-acetate	EtOH	234(4.20)	39-0148-65
$C_{25}H_{36}I_2N_4$			
4,4a,4b,5,6,6a,8,10,10a,10b,11,12-Do-decahydro-1,2,4a,6a,7,8-hexamethyl-1H-pyrazolo[3",4":3',4']cyclopenta-[1',2':5,6]naphth[1,2-f]indazolium diiodide	H_2O	225(4.55),280(4.11)	32-0257-65
$C_{25}H_{36}N_2O_3$			
Androstano[3,2-b]pyridine-5'-carboxylic acid, 6'-amino-17β-hydroxy-, ethyl ester	EtOH	252(4.05),342(3.99)	94-0077-64
$C_{25}H_{36}N_2O_4$			
Cyclopenta[7,8]phenanthro[2,3-c]pyra-zol-1-ol, hexadecahydro-1,10a,12a-trimethyl-, hydrogen succinate	EtOH	223(3.71)	13-0441-65
$C_{25}H_{36}N_4O_7S$			
6-Purinethiol, 9-(2,3,5-tri-O-valeryl-β-D-ribofuranosyl)-	pH 1	321(4.36)	87-0200-64
	pH 11	237(4.10),309(4.37)	87-0200-64

Compound	Solvent	λ_{max}(log ϵ)	Ref.
$C_{25}H_{36}O_2$			
19-Norpregna-1,3,5(10)-trien-20-one, 3-(pentyloxy)-	EtOH	278(3.37),287(3.32)	13-0013-64
Pregna-4,6-diene-3,20-dione, 6-methyl-17-propyl-	MeOH	290(4.36)	78-0357-64
$C_{25}H_{36}O_2S$			
Androst-5-ene, 3β,17β-dimethoxy-17α-(2-thienyl)-	EtOH	236(3.67)	78-1197-65
$C_{25}H_{36}O_2S_2$			
Pregna-3,5-dien-20-one, 16α,17-epoxy-3,4-bis(ethylthio)-	EtOH	292(4.09)	94-1078-65
$C_{25}H_{36}O_3$			
Pregna-5,20(21)-dien-3β-ol, 20-acetyl-, acetate	MeOH	227(3.98)	35-3737-65
isomer	MeOH	249(3.48)	35-3737-65
	MeOH	253(3.95)	35-3737-65
Pregna-4,6-dien-20-one, 17-ethyl-3β-hydroxy-, acetate	MeOH	232(4.39),239(4.43), 247(4.24)	87-0684-64
$C_{25}H_{36}O_3S$			
Pregn-4-ene-3,20-dione, 16α-mercapto-, butyrate	EtOH	239(4.31)	87-0531-64
$C_{25}H_{36}O_4$			
Androsta-3,5-diene-6-carboxaldehyde, 17β-acetoxy-3-ethoxy-2α-methyl-	EtOH	221(4.01),323(4.18)	78-0597-64
Androsta-3,5,7-triene-6-methanol, 3-ethoxy-17β-hydroxy-, 17-propionate	EtOH	320.5(4.23)	78-1753-65
5α,14α,17α-Card-20(22)-enolide, 3β-acetoxy-	EtOH	217.5(4.13)	13-0357-65
5β,14β,17α-Card-20(22)-enolide, 3β-acetoxy-	EtOH	218(4.20)	94-0312-65
Pregna-5,20-diene-21-carboxaldehyde, 3β-hydroxy-20-methoxy-, acetate	MeOH	260(4.27)	35-3908-64
Pregna-3,5-dien-20-one, 17α-acetoxy-3-ethoxy-	EtOH	240(4.31)	78-0597-64
Pregna-3,5-dien-20-one, 21-acetoxy-3-ethoxy-	EtOH	241(4.29)	78-0597-64
Pregna-3,5-dien-20-one, 17α-acetoxy-3-methoxy-6-methyl-	EtOH	247(4.29)	78-0569-65
Pregna-3,5-dien-20-one, 17α-acetoxy-3-methoxy-16α-methyl-	EtOH	240(4.29)	78-0597-64
Pregna-3,5,17(20)-triene-11β,21-diol, 3-ethoxy-, 21-acetate	EtOH	240(4.30)	44-2982-64
Pregn-4-ene-3,20-dione, 17α-acetoxy-6α,16α-dimethyl-	EtOH	242(4.20)	87-0540-64
Propionic acid, 2-(16β-hydroxy-3-oxo-13β-etiojerv-1-en-17β-yl)-, lactone	MeOH	233(3.98)	44-2545-64
$C_{25}H_{36}O_4S$			
Pregn-5-en-20-one, 3β-hydroxy-16α-mercapto-, diacetate	EtOH	234(3.73)	87-0531-64
Pregn-5-en-20-one, 3β-hydroxy-16β-mercapto-, diacetate	EtOH	234(3.72)	87-0531-64
$C_{25}H_{36}O_5$			
5β-Card-20(22)-enolide, 3β-acetoxy-14β-hydroxy-	EtOH	216(4.05)	13-0593-64

Compound	Solvent	λ_{max}(log ε)	Ref.
5β,14α-Card-20(22)-enolide, 3β-acetoxy-15ξ-hydroxy-	EtOH	215(4.05)	13-0593-64
Pregna-3,5-dien-20-one, 17α-acetoxy-6-(hydroxymethyl)-3-methoxy-	EtOH	250(4.30)	78-0597-64
Pregna-3,5-dien-20-one, 21-acetoxy-6-(hydroxymethyl)-3-methoxy-	EtOH	249(4.26)	78-0597-64
Pregn-4-ene-3,20-dione, 17α,21-dihydroxy-6α,16α-dimethyl-, 21-acetate	EtOH	241(4.23)	87-0548-64
Pregn-4-ene-3,20-dione, 17α,21-dihydroxy-6β,16α-dimethyl-, 21-acetate	EtOH	243(4.21)	87-0548-64
5β-Pregn-17(20)-en-21-oic acid, 3,11-dioxo-, methyl ester, cyclic 3-(ethylene acetal)	n.s.g.	233(4.07)	44-0163-64
Pregn-4-en-3-one, 20α,21-diacetoxy-	MeOH	241(4.23)	44-2559-64
Pregn-4-en-3-one, 20β,21-diacetoxy-	MeOH	242(4.24)	44-2559-64
5α-Pregn-16-en-20-one, 3β,12β-diacetoxy-	EtOH	236(3.97)	94-1332-65
5α-Pregn-17(20)-en-11-one, 3β,20-diacetoxy-	EtOH	298.0(1.46)	39-1161-64
Uzarigenin, O-acetyl-	EtOH	217.5(4.12)	33-1775-64
$C_{25}H_{36}O_6$			
Androsta-3,5-diene-11α,17β-diol, 6-(hydroxymethyl)-3-methoxy-	EtOH	250(4.28)	78-0597-64
5α-Androstan-3-one, 17β-hydroxy-2-(hydroxymethylene)-17α-methyl-, 17-(hydrogen succinate)	EtOH	282(3.96)	13-0441-65
	NaOH	315(4.26)	13-0441-65
Androst-2-ene-2-carboxaldehyde, 17β-hydroxy-3-(2-hydroxyethoxy)-, 17-acetate 3-formate	EtOH	277(4.11)	88-2161-64
5α-Androst-3-ene-1α,3, 17β-triol, triacetate	MeOH	208(3.23)	24-3165-65
Pregn-5-ene-7,20-dione, 3β,11α-diacetoxy-	EtOH	236(4.08)	70-2016-64
$C_{25}H_{36}O_7$			
Pregn-4-en-20-one, 21-acetoxy-3-(ethylenedioxy)-11β,17α-dihydroxy-	MeOH	241(2.90)	35-2183-64
Pregn-4-en-3-one, 11β,17α,20α,21-tetrahydroxy-, 20,21-diacetate	EtOH	243(4.18)	78-0179-65
$C_{25}H_{36}O_9$			
Melicopol, tetrahydro-, acetate	EtOH	252(3.60),303(3.47)	12-2021-65
$C_{25}H_{37}BrO_5$			
5α-Pregnan-11-one, 3β,20β-diacetoxy-9α-bromo-	EtOH	319.0(1.94)	39-2933-65
5α-Pregnan-11-one, 3β,20β-diacetoxy-12α-bromo-	EtOH	317.0(2.16)	39-2933-65
$C_{25}H_{37}NO$			
Buxpiine, anhydro-	EtOH	241(4.02)	88-3579-65
Cyclomicrobuxine, dehydration product	EtOH	243(3.99)	39-6688-65
$C_{25}H_{37}NO_3$			
Pregna-3,5-dien-20-one, 6-(dimethylaminomethyl)-16α,17α-epoxy-3-methoxy-	EtOH	250(4.30)	78-0569-65
$C_{25}H_{37}NO_4$			
Piericidin A	n.s.g.	239(4.6)	35-2066-65

Compound	Solvent	λ max(log ε)	Ref.
$C_{25}H_{38}Br_2N_4$			
Tetradecahydro-1,2,4a,6a,7,8-hexamethyl-2H-pyrazolo[3",4":3',4']-cyclopenta[1',2':5,6]naphth[1,2-f]-indazolium dibromide	EtOH	237(4.13)	32-0257-65
$C_{25}H_{38}Cl_2N_4$			
Dichloride corresponding to preceding dibromide	EtOH	237(4.11)	32-0257-65
$C_{25}H_{38}I_2N_4$			
Diiodide corresponding to preceding dichloride	H_2O	227.5(4.55)	32-0257-65
$C_{25}H_{38}O_2$			
Androsta-3,5-diene-16β,17β-diol, 3,16α,17α-trimethyl-, cyclic acetal with acetone	MeOH	239.5(4.23)	7-0205-65
Pregn-4-ene-3,20-dione, 6α-methyl-17-propyl-	MeOH	240(4.21)	78-0357-64
$C_{25}H_{38}O_2S_2$			
Androsta-3,5-dien-17β-ol, 3,4-bis-(ethylthio)-, acetate	EtOH EtOH	292(4.24) 292(4.24)	94-0383-64 94-0769-65
$C_{25}H_{38}O_3$			
Androsta-3,5-dien-17β-ol, 3-ethoxy-2α,6-dimethyl-, acetate	EtOH	249(4.30)	78-0569-65
Chol-4-enic acid, 3-oxo-, methyl ester	EtOH	241(4.17)	44-3495-64
Estr-4-en-3-one, 17β-hydroxy-, heptanoate	MeOH	240(4.20)	32-0675-64
Pregna-3,5-dien-20β-ol, 3-ethoxy-, acetate	EtOH	241(4.30)	78-0597-64
Pregn-5-ene-3,20-dione, 17α-hydroxy-4,4,6,16α-tetramethyl-	EtOH	none	87-0552-64
$C_{25}H_{38}O_4$			
Androsta-3,5-diene-6-methanol, 17β-acetoxy-3-ethoxy-2α-methyl-	EtOH	252(4.30)	78-0597-64
5β-Cholanic acid, 3,4-dioxo-, methyl ester	EtOH	271(3.84)	44-3495-64
$C_{25}H_{38}O_4S$			
5β-Pregnan-20-one, 3β-hydroxy-16α-mercapto-, diacetate	EtOH	232(3.65)	87-0531-64
$C_{25}H_{38}O_4S_2$			
5α-Androstan-17β-ol, 2β,3α-dimercapto-, triacetate	EtOH	232(3.95)	78-0329-65
$C_{25}H_{38}O_5$			
Andromedenol, 10-deoxy-, acetate, cyclic acetal with acetone	EtOH	203(3.81)	44-2756-64
5α-Pregnan-11-one, 3β,20β-diacetoxy-	EtOH	296.5(1.46)	39-2933-65
$C_{25}H_{38}O_6$			
$\Delta^{10(18)}$-Andromedenol, acetate, cyclic acetal with acetone	EtOH	205(3.75)	44-2756-64

Compound	Solvent	λ_{max}(log ε)	Ref.
$C_{25}H_{38}O_7$			
Andromedol, 3-oxo-, acetate, cyclic acetal with acetone	EtOH	290(1.43)	44-2756-64
$C_{25}H_{38}O_{12}S$			
Glucofuranose, 1,2:5,6-di-O-isopropylidene-, thiocarbonate	MeOH	233(3.98),305(1.53)	44-3071-65
$C_{25}H_{39}NO$			
9,19-Cyclo-5α,9β-pregn-2-en-16β-ol, 20α-(dimethylamino)-14-methyl-4-methylene-	EtOH	225s(4.21),230(4.22), 238s(4.02)	35-4414-64
$C_{25}H_{39}NO_2$			
Androstan-3-one, 17β-hydroxy-2-(piperidinomethylene)-	EtOH	334(4.33)	94-0077-64
$C_{25}H_{39}NO_3$			
Androsta-3,5-dien-17β-ol, 6-(dimethylaminomethyl)-3-methoxy-, acetate	EtOH	250(4.30)	78-0569-65
5α-Androstan-3-one, 17β-acetoxy-2-(dimethylaminomethylene)-17α-methyl-	MeOH	333(4.33)	39-0788-65
$C_{25}H_{39}NO_6$			
Erythrophleguine	EtOH	221(4.48)	88-4203-65
$C_{25}H_{40}O_4$			
5β-Pregnane-3α,20α-diol, diacetate, p-hydroxybenzaldehyde complex	HOAc-$HClO_4$	560(4.28)	96-0134-65
resorcylic aldehyde complex	HOAc-$HClO_4$	540(4.32)	96-0134-65
vanillin complex	HOAc-$HClO_4$	600(4.20)	96-0134-65
$C_{25}H_{41}N_3O_2$			
Hydroquinone, methyltris(piperidinomethyl)-	ether	309(3.76)	39-0042-64
$C_{25}H_{42}BNO_3$			
Androsta-3,5-dien-17β-ol, 6-(dimethylaminomethyl)-3-methoxy-, acetate, compound with BH_3	EtOH	255(4.29)	78-0569-65
$C_{25}H_{42}N_2$			
Buxenine G	EtOH	215s(--),230s(--), 238(4.43),247(4.46), 255(4.27),290s(--)	88-3145-64
Norbuxamine	MeOH	215s(3.84),230s(4.27), 238(4.44),246(4.47), 254(4.26),290s(2.18)	33-0968-64
$C_{25}H_{42}O_3$			
5β-Cholanic acid, 3α-hydroxy-, methyl ester, p-hydroxybenzaldehyde complex	HOAc-$HClO_4$	540(4.18)	96-0134-65
resorcylic aldehyde complex	HOAc-$HClO_4$	540(3.48)	96-0134-65
vanillin complex	HOAc-$HClO_4$	600(4.06)	96-0134-65
$C_{25}H_{42}O_5$			
5β-Cholanic acid, 3α,7α,17α-trihydroxy-, methyl ester, p-hydroxybenzaldehyde complex	HOAc-$HClO_4$	540(4.08)	96-0134-65
resorcylic aldehyde complex	HOAc-$HClO_4$	540(4.00)	96-0134-65
vanillin complex	HOAc-$HClO_4$	600(4.00)	96-0134-65

Compound	Solvent	$\lambda_{max}(\log \epsilon)$	Ref.
$C_{25}H_{42}O_8$			
2,4-Nonadienoic acid, 8-[6-(formyl-methyl)tetrahydro-4-hydroxy-3,5-di-methylpyran-2-yl]-7-hydroxy-2,6-di-methyl-, methyl ester, dimethyl acetal, acetate	n.s.g.	268(4.48)	31-0336-64
$C_{25}H_{43}NS$			
Cholanamide, N-methylthio-	MeOH	262(4.08),309s(1.83)	35-0051-65
$C_{25}H_{44}O_2$			
Resorcinol, 5-nonadecyl-	EtOH	275(3.25),282(3.24)	44-0435-64
$C_{25}H_{44}O_9$			
Lankolide, acetyl-	EtOH	288(1.54)	33-0078-64

Compound	Solvent	λ_{max}(log ϵ)	Ref.
$C_{26}H_8Cl_{12}$			
Ethylene, tetrakis(2,4,6-trichlorophenyl)-	C_6H_{12}	262(4.57),306(4.23)	35-2149-64
$C_{26}H_{10}Cl_{12}$			
Ethane, 1,1,2,2-tetrakis(2,4,6-trichlorophenyl)-	C_6H_{12}	244(4.53)	35-2149-64
$C_{26}H_{12}O_4$			
Dinaphtho[1,2-d:1',2'-d']benzo-[1,2-b:4,5-b']difuran-8,16-dione	dioxan	299(4.49),323(4.53), 434(3.95)	1-1063-6
Dinaphtho[1,2-d:1',2'-d']benzo-[1,2-b:5,4-b']difuran-8,16-dione	dioxan	296(4.48),316(4.29), 410(3.79),526(3.79)	1-1063-65
$C_{26}H_{14}$			
Benzene, o-bis(4-phenylbuta-1,3-diynyl)-	C_6H_{12}	232(4.69),243(4.67), 250(4.68),263(4.65), 271(4.54),286(4.46), 304(4.57),327(4.44), 349(4.36)	39-1147-64
11,12-(peri-Naphthylene)fluoranthene	benzene	325(4.72),394(4.10), 417(4.24)	78-1559-64
	C_6H_{12}	236(4.58)	78-1559-64
$C_{26}H_{16}$			
Anthraceno[2',1':1,2]anthracene	benzene	294(5.11),307(5.44), 329(4.40),358(3.71), 377(3.90),396(4.06), 420(4.03)	78-2107-64
1,2-Benzopentacene	benzene	290(4.80),318(5.00), 331(5.22),344(4.78), 360(4.04),418(3.08), 446(3.32),477(3.64), 510(3.94),550(4.04)	24-0212-64
1,2:7,8-Dibenzochrysene	EtOH	270(4.72),288(4.72), 301(4.90),336(4.16), 351(4.14),383(3.10)	78-2107-64
Fulminene	benzene	295(5.04),312(4.60), 322(4.30)	24-0494-64
	$C_6H_3Cl_3$	342(4.30),364(3.12), 372(2.90),384(3.03)	24-0494-64
Naphtho[2',1':1,2]tetracene	benzene	304(4.98),314(5.20), 342(4.20),376(3.21), 393(3.46),416(3.76), 442(3.98),472(3.98)	78-2107-64
$C_{26}H_{16}ClNO_2$			
Anthracene, 9-(p-chlorophenyl)-10-(p-nitrophenyl)-	CH_2Cl_2	258(5.00),323s(3.35), 340s(3.61),357(3.88), 375(4.07),395(4.08)	44-3695-64
$C_{26}H_{16}N_2O_2$			
Benzoic acid, p-(phenylazo)-, 7-phenyl-2,4,6-heptatriynyl ester	ether	240(4.85),252(5.07), 275(4.25),293(4.49), 312(4.66),334(4.61)	24-2135-64
$C_{26}H_{16}N_2O_4$			
Anthracene, 9,10-bis(m-nitrophenyl)-	CH_2Cl_2	251s(4.99),259(5.22), 320s(3.46),338s(3.68), 356(3.95),375(4.10), 395(4.10)	44-3695-64

Compound	Solvent	λmax(log ε)	Ref.
Anthracene, 9,10-bis(o-nitrophenyl)-	CH_2Cl_2	251s(4.90),260(5.09), 322s(3.37),340s(3.64), 357(3.90),376(4.04), 396(3.99)	44-3695-64
Anthracene, 9,10-bis(p-nitrophenyl)-	CH_2Cl_2	256(5.05),322s(3.44), 339s(3.62),357(3.87), 376(4.07),395(4.10)	44-3695-64
Anthracene, 9-(m-nitrophenyl)-10-(o-nitrophenyl)-	CH_2Cl_2	252s(4.96),260(5.15), 322s(3.28),339(3.64), 357(3.92),375(4.11), 395(4.04)	44-3695-64
Anthracene, 9-(m-nitrophenyl)-10-(p-nitrophenyl)-	CH_2Cl_2	259(5.12),323s(3.43), 339s(3.65),357(3.90), 375(4.08),394(4.09)	44-3695-64
Anthracene, 9-(o-nitrophenyl)-10-(p-nitrophenyl)-	CH_2Cl_2	258(4.96),322s(3.36), 339s(3.62),357(3.87), 376(4.04),396(4.04)	44-3695-64
$C_{26}H_{16}O_3S_2$			
Anthraquinone, 1-(phenylsulfinyl)-4-(phenylthio)-	EtOH	247(4.51),317(4.10), 445(3.54)	22-1648-65
$C_{26}H_{16}O_4S_2$			
Anthraquinone, 1,4-bis(phenylsulfinyl)-	EtOH	258(4.63),405(2.80)	22-1648-65
$C_{26}H_{16}O_6$			
Bergapten, phenanthrenequinone adduct	dioxan	247(4.61),286(4.16), 310(4.20)	24-3102-65
Xanthotoxin, phenanthrenequinone adduct	dioxan	245(4.65),282(4.23), 316(4.15)	24-3102-65
$C_{26}H_{16}O_6S_2$			
Anthraquinone, 1,4-bis(phenylsulfonyl)-	EtOH	255(4.67),313s(3.79)	22-1648-65
$C_{26}H_{17}N$			
9b-Azaphenalene, 3,9-diphenyl-	EtOH	207(4.5),262(4.68), 285(4.70),342(4.14), 366(4.23),467(4.03)	77-0011-65
	THF	602(2.65),655(2.71), 723(2.44)	77-0011-65
Phenanthridine, 6-(9-fluorenyl)-	hexane	250s(--),255(4.70), 261s(--),267s(--), 273s(--),279s(--), 290(4.03),301(4.05), 314(3.42),323(3.13), 338(2.98),345(3.49)	23-2919-65
Spiro[fluoren-9,3'-[3H]indole], 2'-phenyl-	EtOH	267(4.19),282(3.94), 295(3.72),306(3.98)	23-2919-65
$C_{26}H_{17}NO_2$			
Anthracene, 9-(m-nitrophenyl)-10-phenyl-	CH_2Cl_2	254s(4.89),261(5.09), 322s(3.16),340s(3.56), 357(3.90),375(4.10), 395(4.07)	44-3695-64
Anthracene, 9-(o-nitrophenyl)-10-phenyl-	CH_2Cl_2	254s(4.90),261(5.07), 324s(3.32),340s(3.62), 357(3.90),376(4.08), 396(4.03)	44-3695-64

Compound	Solvent	λ_{max}(log ϵ)	Ref.
Anthracene, 9-(p-nitrophenyl)-10-phenyl-	CH_2Cl_2	259(5.07),322s(3.46),339s(3.61),357(3.89),375(4.07),395(4.08)	44-3695-64
$C_{26}H_{18}$			
Anthracene, 1,4-diphenyl-	CH_2Cl_2	260(5.02),320(3.19),337(3.52),354(3.82),371(3.99),391(3.93)	44-1981-65
Anthracene, 1,9-diphenyl-	CH_2Cl_2	260(5.21),319(3.29),336(3.56),353(3.89),371(3.98),391(3.93)	44-1981-65
Anthracene, 1,10-diphenyl-	CH_2Cl_2	259(5.12),322(3.27),336(3.66),354(3.91),372(4.10),392(4.05)	44-1981-65
Anthracene, 2,9-diphenyl-	CH_2Cl_2	265(4.77),283(4.88),322(3.18),339(3.55),356(3.81),374(3.86),395(3.88)	44-1981-65
Anthracene, 2,10-diphenyl-	CH_2Cl_2	266(5.08),282(5.17),321(3.19),337(3.51),355(3.76),373(3.81),393(3.83)	44-1981-65
Anthracene, 9,10-diphenyl-	CH_2Cl_2	259(5.02),323(3.19),338(3.54),354(3.88),372(4.09),392(4.11)	44-1981-65
	CH_2Cl_2	262(5.02),323s(3.19),340s(3.54),358(3.88),376(4.09),396(4.11)	44-3695-64
Phenanthrene, 9,10-diphenyl-	EtOH	210(4.63),257(4.77),288(4.06),300(4.09),334(2.74),342(2.55),351(2.69)	39-5544-64
$C_{26}H_{18}N_2$			
Quinoline, 5,7-diphenyl-6-(2-pyridyl)-	n.s.g.	242(4.61),320(3.66)	88-1513-64
$C_{26}H_{18}N_2O$			
2,2'-Azoxyfluorene	EtOH	248(4.23),382(4.52)	31-0128-64
Nicotinonitrile, 4-(2,2-diphenylvinyl)-1,2-dihydro-2-oxo-6-phenyl-	EtOH	276(4.25),340(4.34),385(4.16)	35-5198-65
$C_{26}H_{18}N_3OP$			
1-Pentene-1,1,2-tricarbonitrile, 4-oxo-3-(triphenylphosphoranylidene)-	dioxan	437(4.00)	49-1967-65
$C_{26}H_{18}N_4$			
Acridine, 9,9'-hydrazodi-	pH 3.0	255(5.10),324(3.50),338(3.99),354(4.34),386(3.50),402(3.48)	39-4653-65
	pH 8.0	209(4.21),249(5.25),324s(3.61),338(3.83),346(3.85),354(4.02),366s(3.66),380s(3.44)	39-4653-65
	MeOH-KOH	235(4.92),299(4.18),468(4.37)	39-4653-65
$C_{26}H_{18}N_4O_4$			
Phenazine, 1,4-bis(p-nitrobenzyl)-	C_6H_{12}	210(4.81),360(4.52),388(4.54)	65-1969-64

Compound	Solvent	λ_{max}(log ϵ)	Ref.
$C_{26}H_{18}N_6$			
5,10-Dihydro-5,10-dimethyl-5-phenazinylium tetracyanoquinodimethan salt	MeCN	258(4.76),394(4.63), 842(4.42)	23-1448-65
$C_{26}H_{18}O$			
Cyclohept[f]inden-7(1H)-one, 6,8-diphenyl-	EtOH	266(4.57),300(4.64)	44-0891-65
Fluorene, 9-benzoyl-9-phenyl-	EtOH	219(4.51),255(4.31), 304(3.65)	22-2345-65
$C_{26}H_{18}O_4$			
5,8,13,14-Pentaphenetetrone, 2,3,10,11-tetramethyl-	dioxan	255(4.50),285(4.57)	65-3484-64
$C_{26}H_{18}O_8$			
Dianellinone	dioxan	223(4.72),275s(4.29), 429(3.89)	12-0218-65
$C_{26}H_{18}O_{10}$			
Benzo[1,2-b:4,5-b']bisbenzofuran, 3,6,9,12-tetraacetoxy-	dioxan	234(4.70),263(4.22), 300s(4.52),311(4.84), 328(4.86)	1-1063-65
Benzo[1,2-b:5,4-b']bisbenzofuran, 3,6,9,12-tetraacetoxy-	dioxan	229(4.51),248(4.79), 258(4.73),287(4.63), 307s(4.33),323(4.34)	1-1063-65
$C_{26}H_{19}BrSn$			
Tin, [(p-bromophenyl)ethynyl]-triphenyl-	EtOH	258(4.58),270(4.52)	22-0035-65
$C_{26}H_{19}N$			
Isoindole, 1,2,3-triphenyl-	dioxan	235(4.5),280(4.3), 320(4.0),380(4.3)	5-0096-64C
1H-Isoindole, 1,1,3-triphenyl-	dioxan	250s(4.2)	5-0096-64C
$C_{26}H_{19}N_3O$			
Nicotinonitrile, 2-benzamido-6-phenyl-4-p-tolyl-	n.s.g.	295(4.37),388(4.28)	24-3892-65
p-Toluamide, N-(3-cyano-4,6-diphenyl-2-pyridyl)-	n.s.g.	295(4.43),388(4.33)	24-3892-65
$C_{26}H_{19}OP$			
Phosphine oxide, 1-anthryldiphenyl-	EtOH	258(4.70),354(3.48), 372(3.60),392(3.51)	49-0285-65
Phosphine oxide, 2-anthryldiphenyl-	EtOH	264(5.25),353(3.53), 368(3.57),387(3.55)	49-0285-65
$C_{26}H_{19}P$			
Phosphine, 1-anthryldiphenyl-	EtOH	258(4.92),335(3.54), 350(3.8),369(4.0), 388(3.96)	49-0285-65
Phosphine, 2-anthryldiphenyl-	EtOH	264(5.12),346(3.52), 363(3.57),384(3.53)	49-0285-65
$C_{26}H_{20}BrN_4$			
Verdazyl, 1,3,5-triphenyl-6-(p-bromophenyl)-	benzene	700(3.64)	49-0457-64
$C_{26}H_{20}BrOP$			
Phosphoran, triphenylbromobenzoylmethylene-	EtOH	278(3.77),325(4.12)	65-2738-64

Compound	Solvent	λ_{max}(log ε)	Ref.
$C_{26}H_{20}ClOP$			
Phosphoran, triphenylchlorobenzoyl-methylene-	EtOH	240(4.34),274(3.93), 320(4.21)	65-2738-64
$C_{26}H_{20}Fe$			
Butatriene, 1,1-diphenyl-4-ferrocenyl-	THF	390(4.35),525(3.49)	101-0355-65A
$C_{26}H_{20}NO_3P$			
Acetophenone, 4'-nitro-2-(triphenyl-phosphoranylidene)-	EtOH	262(4.23),268(4.25), 360(4.92)	65-2738-64
$C_{26}H_{20}N_2$			
Fluorene-2-carbonitrile, 9-[5-(N-meth-ylanilino)-2,4-pentadien-1-ylidene]-	MeCN	259(4.75),306(4.24), 317(4.13),483(4.86)	24-3331-64
Phenazine, 1,4-dibenzyl-	C_6H_{12}	217(4.48),240(4.33), 276(4.42),360(4.22)	65-1969-64
$C_{26}H_{20}N_2O$			
1,3,4-Oxadiazole, 4,5-dihydro-2,4,5,5-tetraphenyl-	EtOH	250(4.08),305s(3.85), 348(4.07)	32-0033-65
$C_{26}H_{20}N_2O_2$			
Nicotinonitrile, 1,2-dihydro-4-(2-hy-droxy-2,2-diphenylethyl)-2-oxo-6-phenyl-	EtOH	246(4.40),351(4.24)	35-5198-65
$C_{26}H_{20}N_2O_3$			
2,4-Pentadienoic acid, 5-benzamido-2-cyano-3,5-diphenyl-, dimethyl ester	MeCN	360(3.80)	24-3892-65
$C_{26}H_{20}N_2O_4$			
4-Pentenoic acid, 4-hydroxy-5-(p-meth-oxyphenyl)-5-(1-methyl-2-benzimida-zolyl)-3-oxo-2-phenyl-, γ-lactone	MeOH	233(4.45),274(4.35), 284(4.34),299(4.24), 323(4.39)	78-0017-64
$C_{26}H_{20}N_6O_4$			
1,2-Diazetidine, 3,4-bis(m-nitrophenyl)-1-phenyl-4-(phenylazo)-	$CHCl_3$	275(4.7),400(2.8)	28-2113-64B
Benzil, m,m'-dinitro-, bis(phenyl-hydrazone)	EtOH	233s(4.5),250(4.8), 350(4.7)	28-2113-64B
$C_{26}H_{20}O$			
Acetophenone, 2,2,2-triphenyl-	EtOH	253(4.06),293(2.89), 329(2.48)	23-0298-64
Cyclohept[f]inden-7(1H)-one, 2,3-dihydro-6,8-diphenyl-	EtOH	255(4.50),292(4.64)	44-0891-65
$C_{26}H_{20}O_4S_2$			
Acetic acid, (benzo[a]pyren-3,6-ylene-dithio)di-, dimethyl ester	$CHCl_3$	261(4.55),272(4.62), 287(4.36),299(4.51), 311(4.57),353s(3.69), 376(4.10),397(4.43), 419(4.53)	39-2571-64
$C_{26}H_{20}Pb$			
Lead, triphenyl(phenylethynyl)-	hexane	250(4.41),262(4.37), 281(2.88)	22-3518-65
$C_{26}H_{20}S$			
Dibenzo[c,e]thiepin, 5,7-dihydro-3,9-diphenyl-	n.s.g.	286(4.51)	4-0181-65

Compound	Solvent	λ_{max}(log ϵ)	Ref.
$C_{26}H_{20}Se$			
Dibenzo[c,e]selenepin, 5,7-dihydro-3,9-diphenyl-	n.s.g.	288(4.55)	4-0181-65
$C_{26}H_{20}Sn$			
Tin, triphenyl(phenylethynyl)-	EtOH	249(4.45),260(4.38)	22-0035-65
$C_{26}H_{21}$			
Bis(p-biphenylyl)methyl carbanion	ether	258(4.85),315(4.18), 440(4.48),590(5.18)	60-0264-64
$C_{26}H_{21}AsO$			
Acetophenone, triphenylarsenic derivative	n.s.g.	255(4.32),310(3.72)	30-0424-64B
$C_{26}H_{21}BF_4$			
Tetrahydro-6,8-diphenylcyclohept[f]-indenylium tetrafluoroborate	H_2SO_4	249(4.39),327(4.81), 473(3.58)	44-0891-65
$C_{26}H_{21}ClN_2O_5$			
1-(o-Hydroxyphenyl)-2-[(1-methyl-2(1H)-quinolylidene)methyl]quino-linium perchlorate	EtOH	530(4.81)	65-3373-64
1-(o-Hydroxyphenyl)-2-[(1-methyl-4(1H)-quinolylidene)methyl]quino-linium perchlorate	EtOH	566(4.24)	65-3373-64
$C_{26}H_{21}NO_5$			
2-Naphthacenecarboxamide, 10-(benzyl-oxy)-4,4a,5,12-tetrahydro-1,3-dihydroxy-12-oxo-	n.s.g.	245(4.36),259(4.39), 382(4.52)	70-0945-64
$C_{26}H_{21}NO_{12}$			
Laccaic acid A_1	MeOH	225(4.7),298(4.7), 500(4.1)	39-6067-65
$C_{26}H_{21}N_3$			
Indole, 2-methyl-3,3',3"-methyli-dynetri-	n.s.g.	284(4.28),291(4.24)	28-0609-64A
1,2,4-Triazole, 4,5-dihydro-1,3,4,5-tetraphenyl-	EtOH	255(4.2),360(4.0)	24-2174-65
$C_{26}H_{21}N_4$			
Verdazyl, 1,3,5,6-tetraphenyl-	benzene	700(3.66)	49-0457-64
$C_{26}H_{21}N_5O$			
Adenine, N^6-benzoyl-3,7-dibenzyl-	pH 1	225s(4.32),303(4.26)	4-0115-64
	pH 7	238(4.15),336(4.32)	4-0115-64
	pH 13	238(4.14),336(4.28)	4-0115-64
$C_{26}H_{21}OP$			
Acetophenone, 2-(triphenyl-phosphoranylidene)-	EtOH	268(3.88),275(3.89), 317(4.13)	65-2738-64
$C_{26}H_{22}$			
Pentaphene, 2,3,10,11-tetramethyl-	dioxan	251(4.85),260(4.81), 295(4.37),307(4.74), 320(4.93),350(4.48), 361(4.23),382(3.22), 402(2.73),414(2.27), 424(2.47)	65-0130-65

Compound	Solvent	λ_{max}(log ε)	Ref.
p-Quaterphenyl, 2",3'-dimethyl-	n.s.g.	265(4.63)	4-0181-65
$C_{26}H_{22}Fe$			
1,3-Butadiene, 1,1-diphenyl-4-ferrocenyl-	THF	338(4.45),460(3.30)	101-0355-65A
$C_{26}H_{22}N_2$			
Quinoline, 5,6,7,8-tetrahydro-5,7-diphenyl-6-(2-pyridyl)-	EtOH	262(3.90),269(3.89), 277(3.66)	88-1513-64
$C_{26}H_{22}N_2O_2$			
1H-Indazol-5-ol, 4,5,6,7-tetrahydro-1,3-diphenyl-, benzoate	C_6H_{12}	277(4.50)	44-1582-64
$C_{26}H_{22}N_2O_8$			
6,13-Triphenodioxazinedicarboxylic acid, 3,10-dimethoxy-, diethyl ester	DMF	504(4.70),538(4.85)	27-0242-65
$C_{26}H_{22}N_4$			
Benzil, bis(phenylhydrazone)	C_6H_{12}	300(4.30),342(4.58)	7-0100-65
	EtOH	241(4.6),297(4.3), 347(4.6)	28-2113-64B
Benzoic acid, benzylidenephenylhydrazide, phenylhydrazone	EtOH	288s(4.4),335(4.6)	28-2113-64B
Bibenzyl, α,α'-bis(phenylazo)-	$CHCl_3$	277(4.34)	7-0100-65
1,2-Diazetidine, 1,3,4-triphenyl-4-(phenylazo)-	EtOH	276(4.3),335(3.7), 422s(2.6)	28-2113-64B
	$CHCl_3$	276(4.2),403(2.4)	28-2113-64B
$C_{26}H_{22}N_4O_7$			
Cytidine, N-benzoyl-2',5'-dideoxy-5'-phthalimido-, 3'-acetate	MeOH	258(4.47),300(4.16)	44-3067-65
$C_{26}H_{22}O$			
Ethanol, 1,1-bis(p-biphenylyl)-	$HClO_4$	252(4.18),283s(3.86), 308s(3.34),416(3.99), 544(4.84)	60-0264-64
	H_2SO_4	515(--)	60-0264-64
$C_{26}H_{22}OSi$			
Acetophenone, 2-(triphenylsilyl)-	EtOH	243(4.12),271(3.49), 310s(2.52)	23-0298-64
$C_{26}H_{22}O_2$			
Anthraquinone, 1,4,4a,5,8,9a-hexahydro-1,5-diphenyl-	$CHCl_3$	245(3.81),272(3.72), 367(2.31)	39-5110-64
$C_{26}H_{22}O_2Si$			
Silane, p-anisoyltriphenyl-	C_6H_{12}	290(4.32),298s(4.24), 380s(2.12),396(2.39), 414(2.53),434(2.37)	23-1175-65
$C_{26}H_{22}O_6$			
Phthalide, 3,3-bis(α-hydroxybenzyl)-, diacetate	EtOH	228(4.02),258(3.07), 262(3.05),264(3.10), 268(3.11),276(3.16), 283(3.16)	44-2778-64

Compound	Solvent	$\lambda_{max}(\log \epsilon)$	Ref.
$C_{26}H_{22}O_6S$			
D-Arabinitol, 1,5-anhydro-, 3,4-dibenzoate 2-thiobenzoate	EtOH	282(3.99),292(3.97)	44-1282-65
$C_{26}H_{22}O_{10}$			
Benzo[1,2-b:4,5-b']bisbenzofuran-6,12-diol, 1,3,7,9-tetramethoxy-, diacetate	dioxan	229(4.72),249(4.51), 272(3.99),298(4.39), 312(4.73),321(4.69), 337(4.96)	1-1063-65
Benzo[1,2-b:4,5-b']bisbenzofuran-6,12-diol, 3,4,9,10-tetramethoxy-, diacetate	dioxan	234(4.72),249s(4.53), 269(4.32),304s(4.41), 320(4.76),336(4.83)	1-1063-65
Benzo[1,2-b:5,4-b']bisbenzofuran-6,12-diol, 2,3,9,10-tetramethoxy-, diacetate	dioxan	254(4.67),263s(4.61), 311(4.62),315s(4.61), 322s(4.58)	1-1063-65
Benzo[1,2-b:5,4-b']bisbenzofuran-6,12-diol, 3,4,8,9-tetramethoxy-, diacetate	dioxan	255(4.87),288(4.57), 315(4.24),329(4.44)	1-1063-65
$C_{26}H_{22}Pb$			
Lead, triphenyl(p-vinylphenyl)-	$CHCl_3$	258(4.46)	46-0300-64
$C_{26}H_{22}Sn$			
Tin, triphenyl(p-vinylphenyl)-	$CHCl_3$	258(4.36)	46-0300-64
$C_{26}H_{23}N$			
1,3-Hexadienylamine, 5-fluoren-9-ylidene-N-methyl-N-phenyl-	MeCN	256(4.70),295(4.14), 471(4.72)	24-3331-64
Indene, 1-(3-dimethylamino-2-propen-1-ylidene)-2,3-diphenyl-	MeCN	255(4.52),298(4.17), 306s(4.15),440(4.70)	24-3331-64
$C_{26}H_{23}NO_3$			
4(1H)-Pyridone, 2-(p-methoxyphenyl)-6-(2,2-diphenyl-2-hydroxyethyl)-	MeOH	254(4.37)	44-4263-65
$C_{26}H_{23}NO_{12}$			
Laccaic acid A_1, dihydro-	MeOH	225(4.53),275(4.52), 347(4.25),447(3.80), 493(3.69)	39-6067-65
$C_{26}H_{23}N_3O_2$			
1,3-Azulenedicarboxamide, 2-amino-N,N'-dibenzyl-	EtOH	252(4.45),322(4.63), 390(3.83)	94-0473-65
3-Pyridinemethanol, 5-hydroxy-6-methyl-4-[(2-methylindol-3-yl)(2-methyl-3H-indol-3-ylidene)methyl]-	N HCl	260(4.06),540(3.74)	50-0077-64D
	EtOH	282(4.03),430(3.33)	50-0077-64D
$C_{26}H_{23}N_5$			
Adenine, N,3,7-tribenzyl-	pH 1	289(4.33)	4-0115-64
	pH 7	289(4.32)	4-0115-64
$C_{26}H_{23}N_5O_7$			
Hypoxanthine, 7-(2,5-di-O-benzoyl-3-deoxy-β-D-ribofuranosyl)-2-acetamido-	EtOH	232(4.45),254(4.22), 260(4.18),276(4.08), 282(4.08)	44-2851-65
Hypoxanthine, 9-(2,5-di-O-benzoyl-3-deoxy-β-D-ribofuranosyl)-2-acetamido-	EtOH	225(4.58),267(4.17), 280s(4.08)	44-2851-65

Compound	Solvent	$\lambda_{max}(\log \epsilon)$	Ref.
$C_{26}H_{23}O_2P$			
Phosphole, 1-(carboxymethylene)-1,1-dihydro-1,2,5-triphenyl-, ethyl ester	EtOH	223(4.49),239(4.29), 385(4.17)	39-2184-65
$C_{26}H_{24}$			
1H-Benz[e]indene, 2,3-dihydro-1,3,3-trimethyl-1-(2-naphthyl)-	EtOH	222(5.44),265s(4.03), 275(4.13),278(4.12), 283(4.09),291s(3.97), 307(3.37),313(3.23), 321(3.43)	44-3883-65
Butatriene, 1-tert-butyl-1,4,4-triphenyl-	C_6H_{12}	258(4.36),369(4.41)	56-0763-65
1,2,3-Hexatriene, 5,5-dimethyl-1,1,4-triphenyl-	benzene	393(4.53)	56-0763-65
Perylene, 3-hexyl-	EtOH	227(4.26),247(4.45), 254(4.59),266(3.88), 372(3.63),392(4.06), 414(4.42),441(4.54)	44-2469-64
$C_{26}H_{24}Fe$			
1,3,5,7,9-Decapentaene, 1-ferrocenyl-10-phenyl-	$CHCl_3$	407(4.92)	5-0088-64F
$C_{26}H_{24}Fe_2$			
Hexatriene, 1,6-diferrocenyl-	$CHCl_3$	355s(--),368(4.65), 474(3.78)	5-0088-64F
$C_{26}H_{24}Fe_2N_2$			
Iron, (cyclopentadienyl)[(2-formylvinyl)cyclopentadienyl]-, azine	$CHCl_3$	358(4.47),498(3.90)	5-0088-64F
$C_{26}H_{24}N_2O$			
Pyrrole, 1-morpholino-2,3,5-triphenyl-	EtOH	210(3.8),255(2.7), 290(2.7)	6-0073-64
$C_{26}H_{24}N_2O_3$			
1,3-Cyclohexanedione, 5,5-dimethyl-2-[(α-2-naphthoylbenzyl)azo]-	CH_2Cl_2	295(4.21),384(4.09)	89-0920-64
$C_{26}H_{24}O_2$			
Phenanthrene, 1-(3,4-dihydro-6-methoxy-1-naphthyl)-9,10-dihydro-7-methoxy-	EtOH	279(4.24)	70-2021-64
$C_{26}H_{24}O_4$			
2H-1-Benzopyran-7-ol, 3-(p-methoxyphenyl)-2,2-dimethyl-4-phenyl-, acetate	MeOH	284(4.00),310(4.00)	44-4114-65
p-Dioxan, [[10-[(p-methylbenzyl)oxy]-9-phenanthryl]oxy]-	dioxan	280s(4.02),293(4.02), 305(4.04),324s(2.72), 340(2.91),357(2.92)	44-2362-64
$C_{26}H_{24}O_7$			
1,3,5-Cycloheptatriene-1-carboxylic acid, 2-[3-[3-(benzyloxy)-4,5-dimethoxyphenyl]-2-hydroxypropyl]-6-hydroxy-7-oxo-, δ-lactone	pH 12	270(4.10),278s(4.06), 348(3.93),428(4.19)	78-3605-65
	EtOH	257(4.20),320s(3.66), 329(3.71),378(3.86)	78-3605-65

Compound	Solvent	$\lambda_{max}(\log \epsilon)$	Ref.
$C_{26}H_{24}O_8$			
Flavone, 3'-(benzyloxy)-3-hydroxy-4',5,6,7-tetramethoxy-	MeOH	254(4.36),358(4.38)	24-2857-64
Flavone, 4'-(benzyloxy)-3-hydroxy-3',5,6,7-tetramethoxy-	MeOH	254(4.31),359(4.34)	24-0548-65
$C_{26}H_{24}Pb$			
Lead, (1-cyclohexen-1-ylethynyl)-triphenyl-	hexane	218(4.71)	22-3518-65
$C_{26}H_{25}ClN_4O_2S$			
Pyrimidine, 2-amino-4-chloro-6-phenyl-5-[N-(p-tolylsulfonyl)-3-anilinopropyl]-	pH 1	237(4.47),322(3.89)	87-0283-65
	pH 7, 13	237(4.52),315(3.81)	87-0283-65
$C_{26}H_{25}NO_8$			
3,4,5,6-Pyridinetetracarboxylic acid, 1-benzyl-1,2-dihydro-2-phenyl-, tetramethyl ester	EtOH	206(4.40),233(4.21), 281(4.12),388(3.85)	5-0098-65H
$C_{26}H_{26}$			
Benzene, p-bis(2,5-dimethylstyryl)-	dioxan	246(4.36),358(4.72)	38-2839-64A
Benzene, p-bis(3,4-dimethylstyryl)-	dioxan	236(4.35),246(4.27), 252(4.22),363(4.76)	38-2839-64A
$C_{26}H_{26}INO_2$			
1,4-Dibenzyl-6,7-dimethoxy-2-methyl-isoquinolinium iodide	MeOH	258(4.78),319(4.04)	83-0879-65
$C_{26}H_{26}N_2O_5S$			
Spiro[indoline-3,3'-pyrrolidin]-2-one, 2'-(3,4-dimethoxyphenyl)-1'-(p-tolyl-sulfonyl)-	EtOH	230(4.04),279(3.64)	78-1327-65
$C_{26}H_{26}N_2O_9S_2$			
Cephalosporadesic acid, O-3,4,5-tri-methoxybenzoyl-7-phenylmercapto-acetamido-, as sodium salt	n.s.g.	262(4.26)	87-0022-65
$C_{26}H_{26}N_4O_2S_2$			
4-Pyrimidinethiol, 2-amino-6-phenyl-5-[N-(p-tolylsulfonyl)-3-anilino-propyl]-	pH 1	348(4.12)	87-0283-65
	pH 7	365(4.12)	87-0283-65
	pH 13	330(4.03)	87-0283-65
$C_{26}H_{26}O_3S_2$			
3-Pentanone, 1-phenyl-1-(p-acetyl-phenylthio)-5-(p-methoxyphenylthio)-	EtOH	230(4.28),238(4.21), 304(4.54)	7-1093-65
$C_{26}H_{26}O_3Si_3$			
Cyclotrisiloxane, 2,2-dimethyl-4,4,6,6-tetraphenyl-	$CHCl_3$	265(3.22)	46-1066-65
$C_{26}H_{26}O_4$			
Benzene, p-bis(2,3-dimethoxystyryl)-	dioxan	255(4.26),360(4.77)	38-2839-64A
Benzene, p-bis(2,4-dimethoxystyryl)-	dioxan	242(4.70),375(4.75)	38-2839-64A
Benzene, p-bis(3,4-dimethoxystyryl)-	dioxan	246(4.38),374(4.72)	38-2839-64A
$C_{26}H_{26}O_6$			
2,3:6,7-Anthracenetetracarboxylic di-anhydride, 1,2,3,4,5,6,7,8-octahydro-1,4,5,8-tetramethyl-9,10-divinyl-	$CHCl_3$	249(3.56),277(2.26)	33-0437-65

Compound	Solvent	λ_{max}(log ε)	Ref.
6,12-Methano-12H-dibenzo[d,g][1,3]dioxocin, 1,3,8-trimethoxy-6-(p-methoxyphenyl)-13-methyl-	EtOH	272(3.61),279(3.56)	78-1471-65
$C_{26}H_{26}O_7$			
Chalcone, 3-(benzyloxy)-2'-hydroxy-4,4',5',6'-tetramethoxy-	MeOH	374(4.39)	24-2857-64
Chalcone, 4-(benzyloxy)-2'-hydroxy-3,4',5',6'-tetramethoxy-	MeOH	377(4.40)	24-0548-65
$C_{26}H_{26}O_9$			
Coumarin, 4-(benzyloxy)-7-(2,3-O-carbonyl-α-noviosyloxy)-8-methyl-	EtOH	284(4.09),306(4.15)	33-0390-64
$C_{26}H_{26}O_{14}$			
Aphloiol, hexaacetate	EtOH	237(4.42),250(4.43), 287(3.95),360(3.73)	22-0376-64
$C_{26}H_{26}Si_2$			
Disilane, 1,1-dimethyl-1,2,2,2-tetraphenyl-	C_6H_{12}	239.0(4.38)	101-0369-64B
Disilane, 1,2-dimethyl-1,1,2,2-tetraphenyl-	C_6H_{12}	239.0(4.41)	101-0369-64B
$C_{26}H_{27}ClO_7$			
Propiophenone, 2'-(benzyloxy)-3-chloro-3-(3,4-dimethoxyphenyl)-2-hydroxy-3',4'-dimethoxy-	EtOH	279(4.0)	78-0963-65
$C_{26}H_{27}NO_5$			
Lythrine, reduction product	EtOH	294(3.88)	100-0015-64
$C_{26}H_{27}NO_7$			
2-Anthraceneacetic acid, 5-(benzyloxy)-1,2,3,4,9,9aα-hexahydro-9β-hydroxy-9α-methyl-4-oxo-α-nitro-	EtOH	239(3.89),351(4.03)	65-0655-65
$C_{26}H_{27}NO_8$			
2-Anthraceneacetic acid, 5-(benzyloxy)-1,2,3,9,9aα,10-hexahydro-4,9β-dihydroxy-9α-methyl-10-oxo-α-nitro-	EtOH	277(3.76),340(4.16)	65-0655-65
isomer	EtOH	278(3.77),340(4.17)	65-0655-65
$C_{26}H_{27}NS$			
6H-Dibenzo[b,e]thiocine, 12-(3-methylbenzylaminopropylidene)-7,12-dihydro-	MeOH-HCl	230(4.23),261(3.80)	24-0685-65
$C_{26}H_{27}N_5O_2S$			
Pyrimidine, 2,4-diamino-6-phenyl-5-[N-(p-tolylsulfonyl)anilinopropyl]-	pH 1	232(4.41),288(3.86)	87-0283-65
	pH 7	233(4.43),293(3.83)	87-0283-65
	pH 13	300(3.83)	87-0283-65
$C_{26}H_{28}CoN_6S_2$			
Cobalt, dithiocyanatotetrakis-(4-picoline)-	$CHCl_3$	630(2.83)	12-0507-65
$C_{26}H_{28}INO_2$			
1,4-Dibenzyl-3,4-dihydro-6,7-dimethoxy-2-methylisoquinolinium iodide	MeOH	249(4.24),314(3.86), 361(3.95)	83-0879-65

Compound	Solvent	λ_{max}(log ϵ)	Ref.
$C_{26}H_{28}N_2$			
Benzene, p-bis(p-dimethylamino-styryl)-	dioxan	255(5.17),261(5.08), 398(4.71)	38-2839-64A
	$C_2H_4Cl_2$	248(4.22),325(4.15), 400(4.70)	65-2073-65
$C_{26}H_{28}N_2O_4$			
Δ^6-Codeine, 6-deoxy-14-hydroxy-7-(N-phenylformimidoyl)-	EtOH	283(4.22)	78-1407-64
Pyrrolo[2,1-a]isoquinoline, 5,6-dihydro-8,9-dimethoxy-3-(6,7-dimethoxy-3,4-dihydro-1-isoquinolyl)-2-methyl-	pH 2	300(4.25),450(4.25)	94-0775-65
$C_{26}H_{28}N_6O_2S_2$			
Indole-2-carboxylic acid, 3,3'-dithio-bis[1-methyl-, bis(isopropylidene-hydrazide)	EtOH	219(4.67),295(4.57), 320s(4.51),372s(3.78)	44-0178-64
$C_{26}H_{28}O_6$			
4-Cyclohexene-1,3-dicarboxylic acid, 6-(p-methoxyphenyl)-2-(p-methoxy-styryl)-, dimethyl ester	EtOH	264(4.17)	70-1911-64
Macluraxanthone, O,O',O"-trimethyl-	EtOH	254(4.60),272(4.53), 314(4.25)	44-0692-64
$C_{26}H_{28}O_9$			
Coleon A, triacetate	EtOH	266(4.48),354(3.52)	33-0471-65
$C_{26}H_{29}ClN_2O_9$			
Tetracycline, N-tert-butyl-7-chloro-6-deoxy-6-peroxydehydro-	MeOH-HCl	253(4.31),375(3.60)	44-2746-64
$C_{26}H_{29}NO_3$			
Isoquinoline, 2-benzyl-8-(benzyloxy)-1,2,3,4-tetrahydro-6,7-dimethoxy-1-methyl-, as oxalate	pH 1	258s(2.98),264s(3.04), 269(3.08),281s(3.04)	33-2089-64
	pH 13	259(3.02),265s(3.07), 274s(3.19),280(3.20)	33-2089-64
	iso-PrOH	258(3.00),264(3.06), 269(3.10),274(3.12), 277s(3.12),280(3.12)	33-2089-64
$C_{26}H_{29}NO_5$			
Cryogenine	EtOH	260(4.08),285(4.17)	100-0015-64
Lythrine	EtOH	260(4.04),285(4.14)	100-0015-64
Nesodine	MeOH	281(4.09)	100-0084-65
Sinicuichine	EtOH	285(4.01)	100-0015-64
$C_{26}H_{29}NO_6$			
Heimine	EtOH	260(4.09),285(4.15)	100-0015-64
$C_{26}H_{29}N_5S$			
Acetone, 3-benzyl-4-(N,N'-di-p-tolyl-amidino)-3-thiosemicarbazone	EtOH	273s(4.36),302(4.31)	39-0932-65
$C_{26}H_{29}N_6O_2P$			
2H-Naphtho[2,3-d]triazole-4,9-dione, 2-(P,P-dipiperidyl-P-phenyl-phosphoranylideneamino)-	$CHCl_3$	269s(4.22),276s(4.24), 317(4.53)	39-1003-65

Compound	Solvent	$\lambda_{max}(\log \epsilon)$	Ref.
$C_{26}H_{30}$			
Benzene, o-bis(4-phenylbutyl)-	C_6H_{12}	263(3.04),266(3.00), 269(2.98)	39-1147-64
17H-Cyclopenta[a]phenanthrene, 17-(1,5-dimethylhexyl)-1-methyl-	EtOH	221(4.83),242(4.56), 265(4.66),295(4.05), 307(4.19),320(4.18), 345(3.07),362(2.98)	5-0152-64D
	EtOH	222(4.77),242(4.52), 268(4.62),295(4.03), 308(4.16),320(4.14), 346(3.03),363(2.93)	5-0152-64D
$C_{26}H_{30}F_6N_2$			
Bicyclo[2.2.2]octane-1,4-bis(methylamine), N,N'-bis[o-(trifluoromethyl)benzyl]-	EtOH-HCl	265(3.31),272(3.23)	23-2852-64
$C_{26}H_{30}N_2$			
Yohimban, 3ξ-benzyl-	EtOH	225(4.59),240(3.85), 275s(3.87),283(3.90)	44-0105-65
$C_{26}H_{30}N_2O_9$			
L-Glutamic acid, colchicidyl-	EtOH	253(4.50),357(4.38), 411(4.21)	65-0620-64
$C_{26}H_{30}N_4O_6$			
Carbamic acid, [(2,2'-dioxo[3,3'-biindoline]-3,3'-diyl)diethylene]-di-, diethyl ester	EtOH	250(4.19),285(3.59)	78-0565-64
$C_{26}H_{30}N_8O_{13}$			
Glucitol, 1-(6-amino-9H-purin-9-yl)-2-deoxy-2-(2,4-dinitroanilino)-1-O-methyl-, 3,4,5,6-tetraacetate	EtOH	261(4.36),340(4.23)	44-1096-65
$C_{26}H_{30}N_{10}O_5$			
Benzoic acid, p-[N-[4-[[2-amino-6-hydroxy-5-(phenylazo)-4-pyrimidinyl]-amino]-3-oxobutyl]acetamido]-, ethyl ester, semicarbazone	pH 13	388(4.27)	44-3404-65
$C_{26}H_{30}O_2$			
Androsta-1,4-dien-3-one, 17β-benzoyl-	MeOH	252(4.45)	87-0537-64
Androsta-4,9(11)-dien-3-one, 17β-benzoyl-	MeOH	240(4.48)	87-0537-64
$C_{26}H_{30}O_3$			
Estra-1,3,5(10)-trien-17-one, 15β-(benzyloxy)-3-methoxy-	MeOH	222(3.94),278(3.30), 288(3.30)	44-0064-64
Estrone, 2-[(benzyloxy)methyl]-	EtOH	286(3.51)	94-0196-64
Sugiol, benzoate	EtOH	230(4.35),300(3.37)	78-0409-64
$C_{26}H_{30}O_4$			
Benzene, p,p'-bis(4-ethoxy-3-methoxystyryl)-	dioxan	244(4.48),375(4.74)	38-2839-64A
$C_{26}H_{30}O_4P_2$			
(3,6-Dihydroxy-p-benzoquinon-2,5-ylene)-bis[diethylphenylphosphonium hydroxide], bis(inner salt)	CH_2Cl_2	267(4.24),286(4.37), 296(4.40),307(4.28), 370(2.70)	78-1941-65

Compound	Solvent	λmax(log ε)	Ref.
$C_{26}H_{30}O_5$			
4α,10α-Anthracenediol, 5-(benzyloxy)-1,4,4aβ,9,9aα,10-hexahydro-9β-(tetrahydropyranyloxy)-	EtOH	277(3.33),284(3.32)	65-0662-65
$C_{26}H_{30}O_6$			
Alvaxanthone, trimethyl ether	EtOH	249(4.64),276s(4.01),323(4.39)	44-0689-64
Xanthone, 1-hydroxy-3,5,6-trimethoxy-2,4-bis(3-methyl-2-butenyl)-	MeOH	245(4.57),316(4.24),430(4.0)	78-2653-65
$C_{26}H_{30}O_8$			
Chaparrol, triacetate	EtOH	244(4.10),276(3.01)	23-2996-65
p-Terphenyl, 2,2',2",4,4",5,5',5"-octa-methoxy-	dioxan	308(4.33)	1-0540-65
$C_{26}H_{31}ClN_2O_9$			
Chlorotetracycline, N-(1-methoxypropyl)-	pH 1	230(4.25),266(4.26),370(4.01)	87-0870-65
$C_{26}H_{31}FO_3$			
Androst-4-en-3-one, 17β-benzoyl-9α-fluoro-11β-hydroxy-	MeOH	240(4.44)	87-0537-64
$C_{26}H_{31}FO_6$			
18-Nor-17α-pregna-1,4,13-triene-3,20-dione, 9-fluoro-11β,21-di-hydroxy-16α,17-dimethyl-, diacetate	MeOH	235(4.07)	44-3486-64
$C_{26}H_{31}NO_5$			
Cryogenine, dihydro-	EtOH	293(3.88)	100-0015-64
Nesodine, dihydro-	MeOH	286(3.69)	100-0084-65
$C_{26}H_{31}N_3O_2$			
Phthalamide, N-butyl-N'-[1-(1,3,4,5-tetrahydro-2-methylbenz[cd]indol-3-yl)ethyl]-	EtOH	227(4.66),274(3.92),278f(3.91),290f(3.74)	44-0843-64
$C_{26}H_{31}N_5O_{11}S$			
Hydratropic acid, p-nitro-β-(propyl-thio)-, 3α-tropanyl ester, picrate	acetone-KOH	556(4.26)	83-0826-65
	NaOMe	563(4.54)	83-0826-65
	NaSPr	560(4.40)	83-0826-65
$C_{26}H_{31}O_5P$			
Estradiol, 3-benzoate 17-(hydrogen phosphonate)	MeOH	230.5(4.15)	22-0933-65
$C_{26}H_{32}ClNO_6$			
2-Benzyl-1-(2-carbethoxyethyl)-3-carb-ethoxy-3,4-dihydro-6,7-dimethoxy-isoquinolinium chloride	EtOH	254(4.19),310(3.92)	94-1478-64
$C_{26}H_{32}INO_6$			
2-Benzyl-1-(2-carbethoxyethyl)-3-carb-ethoxy-3,4-dihydro-6,7-dimethoxy-isoquinolinium iodide	EtOH	253(4.23),317(3.99)	94-1478-64
isomer	EtOH	253(4.22),317(3.98)	94-1478-64

Compound	Solvent	$\lambda_{max}(\log \epsilon)$	Ref.
$C_{26}H_{32}N_2O_4$			
1H-Cyclopenta[7,8]phenanthro[2,3-c]-[1,2,5]oxadiazol-1-ol, tetradecahydro-10a,12a-dimethyl-, benzoate, N-oxide	EtOH	230(4.23),264(3.9)	94-1445-65
$C_{26}H_{32}N_2O_7$			
Compactinervine, N-acetyl-2,16-dihydro-, diacetate	EtOH	250(4.03),280(3.47), 289(3.44)	78-1141-65
$C_{26}H_{32}N_2O_8$			
Tetracycline, N-tert-butyl-	MeOH-HCl	267(4.34),359(4.1)	44-2746-64
	MeOH-NaOH	240(4.31),265s(4.21), 378(4.16)	44-2746-64
$C_{26}H_{32}N_2O_8S_4$			
2,2'-Ethylenebis[3,4-dimethyl-5-phenylthiazolium methyl sulfate]	n.s.g.	290(4.0)	65-0075-64
$C_{26}H_{32}N_2O_9$			
Tetracycline, N-(1-methoxypropyl)-	pH 1	218(4.19),270(4.28), 360(4.14)	87-0870-65
$C_{26}H_{32}N_5O_4P$			
Phosphorane, trimorpholino[(10-oxo-9(10H)-phenanthrylidene)hydrazono]-	MeOH	380(2.9),430(3.1)	5-0056-64I
$C_{26}H_{32}O_2$			
Androsta-3,5-dien-17-one, 3-(benzyloxy)-	EtOH	241(4.35)	78-0597-64
Androst-4-en-3-one, 17β-benzoyl-	MeOH	242(4.45)	87-0537-64
$C_{26}H_{32}O_3$			
Androst-4-en-3-one, 17β-benzoyl-11β-hydroxy-	MeOH	241(4.44)	87-0537-64
Androst-4-en-17-one, 3β-hydroxy-, benzoate	EtOH	229(4.13),273(2.95), 282(2.85)	44-1421-65
Estradiol, 2-benzyloxymethyl-	EtOH	285(3.49)	94-0196-64
$C_{26}H_{32}O_5$			
Pregna-3,5,7-trien-20-one, 3,17α-diacetoxy-16-methylene-	EtOH	301(4.32),313(4.41), 328(4.27)	73-2351-64
Siphulin, decarboxy-, trimethyl ether	EtOH	250s(4.30),282(4.26)	1-1677-65
$C_{26}H_{32}O_5S$			
Androst-4-ene-3,17-dione, 19-hydroxy-, p-toluenesulfonate	EtOH	228(4.36),291(2.00)	13-0001-64
$C_{26}H_{32}O_6$			
Macluraxanthone, tetrahydro-O,O',O"-trimethyl-	EtOH	249(4.73),282(3.97), 317(4.38)	44-0692-64
$C_{26}H_{32}O_7$			
Cortisone, 6-methylene-, 17α,21-diacetate	EtOH	257(4.04)	78-0597-64
Epiisoobacunoic acid, deoxy-	n.s.g.	215(4.11)	78-2985-64
Pregna-1,4-diene-3,11,20-trione, 21-acetoxy-16α,17-isopropylidenedioxy-	EtOH	237.5(4.18)	39-0130-65

Compound	Solvent	$\lambda_{max}(\log \epsilon)$	Ref.
$C_{26}H_{32}O_8$			
Estradiol, 6α,7β-dihydroxy-, tetraacetate	EtOH	260s(2.77),267(2.89), 276(2.86)	44-1325-64
isomer	EtOH	214s(4.09),269(2.87), 276(2.84)	44-1325-64
$C_{26}H_{32}O_9$			
Chaparrin A, anhydro-, triacetate	EtOH	223(4.21),230(4.25), 239(4.04)	23-2996-65
$C_{26}H_{33}NO_6$			
Erythroskyrine	pH 13	260(4.14),392(4.78)	25-0419-64
	EtOH	260(3.95),409(4.45)	25-0419-64
	EtOH	260(3.95),409(4.45)	94-1240-65
$C_{26}H_{33}N_3O_3$			
Ferruginol, 11-p-nitrophenylazo-	MeOH	283(2.31),379(3.97), 479(2.54)	44-2293-64
$C_{26}H_{34}$			
19-Norcholesta-1,3,5(10),6,8,14-hexaene	EtOH	238(4.39),247(4.55), 255(4.66),264(4.61), 283(4.15),293(4.26), 305(4.20),328(2.73), 345(2.59)	5-0109-64E
$C_{26}H_{34}O_2$			
5α-Androstan-3-one, 2-benzylidene-17β-hydroxy-	n.s.g.	291(4.23)	5-0139-64G
$C_{26}H_{34}O_3$			
Androst-4-en-3-one, 17-hydroxy-6-methylene-17-(1-propynyl)-, propionate	EtOH	260(4.03)	78-0597-64
$C_{26}H_{34}O_4$			
Anthracene, 9,10-bis(3-ethoxypropyl)-2,6-dimethoxy-	EtOH	232(4.08),260(4.68), 269(5.05),302(2.48), 318(3.23),333(3.56), 350(3.52),373(3.23), 394(3.56),417(3.63)	39-4565-64
Pregna-5,14,16,20-tetraene-3β,20-diol, 16-methyl-, diacetate	MeOH	295(4.34)	73-2513-64
Pregn-4-ene-3,20-dione, 17α-acetoxy-16-ethylidene-6-methylene-	EtOH	261(4.07)	78-0597-64
$C_{26}H_{34}O_5$			
16,21-Methanopregna-5,16-dien-20-one, 3β,21-diacetoxy-	EtOH	245(4.03)	5-0218-65E
Pregna-3,5-diene-6-carboxaldehyde, 17α-acetoxy-3-methoxy-16-methylene-	EtOH	220(4.05),321(4.22)	78-0597-64
Pregna-1,4-diene-3,11-dione, 17α,20α,21-trihydroxy-, cyclic 17,21-acetal with cyclopentanone	EtOH	240(4.16)	78-0179-65
Pregna-3,5-dien-20-one, 3,17α-diacetoxy-16-methylene-	EtOH	243(4.32)	73-2351-64
Pregna-3,5,7-triene-6-carboxaldehyde, 17α-acetoxy-3-ethoxy-	EtOH	218(4.04),272(4.03), 383(4.10)	78-1753-65
Pregna-5,14,16-trien-20-one, 3β,21-diacetoxy-16-methyl-	MeOH	312(4.11)	73-2513-64

Compound	Solvent	$\lambda_{max}(\log \epsilon)$	Ref.
$C_{26}H_{34}O_5S$			
5α-Androstane-3,17-dione, 19-hydroxy-, p-toluenesulfonate	EtOH	226(4.12),264(2.82), 273(2.74)	44-2198-65
$C_{26}H_{34}O_6$			
Alvaxanthone, tetrahydro-, trimethyl ether	EtOH	249(4.65),261s(4.41), 277s(4.02),323(4.40), 357s(3.68)	44-1088-65
Pregna-3,5-diene-6-carboxaldehyde, 17α,21-isopropylidenedioxy-3-methoxy-11,20-dioxo-	EtOH	219(4.11),320(4.18)	78-0597-64
$C_{26}H_{34}O_7$			
Pregna-3,5-diene-6-carboxaldehyde, 3-ethoxy-17α,20:20,21-bis-(methylenedioxy)-	EtOH	219(4.05),321(4.19)	78-0597-64
Pregna-3,5-diene-6-carboxaldehyde, 3-ethoxy-17α,21-dihydroxy-11,20-dioxo-, 21-acetate	EtOH	218(4.05),320(4.17)	78-0597-64
Pregna-3,5-diene-11,20-dione, 17α,21-diacetoxy-3-methoxy-	EtOH	240(4.27)	78-0597-64
Pregn-4-ene-3,11-dione, 2-methoxymethylene-16α-methyl-17,20:20,21-bis-(methylenedioxy)-	MeOH	252(4.08),301(3.91)	35-1520-64
$C_{26}H_{34}O_8$			
Pregna-3,5-diene-6-carboxaldehyde, 21-acetoxy-17α-hydroxy-3-(2-hydroxyethoxy)-11,20-dioxo-	EtOH	219(4.02),321(4.14)	78-0597-64
Pregna-1,4-diene-3,20-dione, 21-acetoxy-2,11β-dihydroxy-16α,17α-isopropylidenedioxy-	MeOH	253(4.17)	13-0345-65
$C_{26}H_{34}O_{10}$			
Chaparrin A, triacetate	$CHCl_3$	315(1.40)	23-2996-65
$C_{26}H_{34}O_{16}$			
Monotropein, dihydro-, pentaacetate	EtOH	232(4.12)	94-0888-64
$C_{26}H_{35}BrN_3P$			
Tris(p-dimethylaminophenyl)ethylphosphonium bromide	MeOH	301(4.91)	88-2729-64
$C_{26}H_{35}ClO_4$			
Pregna-5,16,20-triene-21-carboxaldehyde, 20-(2-chloroethoxy)-3β-hydroxy-, acetate	EtOH	270(4.15)	88-1839-64
$C_{26}H_{35}NO_3$			
5β-Pregnane-3,20-dione, 11β-hydroxy-11α-(2-pyridyl)-	EtOH	263(3.59)	44-2095-65
$C_{26}H_{35}NO_4$			
Androstano[3,2-b]pyridine, 6',17β-diacetoxy-	C_6H_{12}	265s(3.67),270(3.73), 273s(3.73),278s(3.62)	94-0077-64
$C_{26}H_{35}NO_5$			
Isoxazolino[17,16-d]pregn-4-ene-3,20-dione, 3'-carbethoxy-6α-methyl-	MeOH	240(4.32)	5-0228-65E

Compound	Solvent	$\lambda_{max}(\log \epsilon)$	Ref.
$C_{26}H_{36}$			
19-Norcholesta-1,3,5(10),6,8-pentaene	EtOH	213(4.54),231(4.99), 277(3.72),283(3.73), 287(3.74),293(3.65), 308(3.13),315(2.91), 323(3.05)	5-0109-64E
$C_{26}H_{36}N_2O_3$			
Androst-4-en-3-one, 16α,17α-isopropylidenedioxy-17β-(3-methylpyrazol-5-yl)-	EtOH	240(4.22)	78-1927-64
Androst-4-eno[3,2-b]pyridine-5'-carboxylic acid, 6'-amino-17β-hydroxy-17α-methyl-, ethyl ester	EtOH	225(4.29),259(4.29), 371(4.15)	94-0077-64
$C_{26}H_{36}N_2O_5$			
Pregn-4-eno[3,2-c]pyrazole, 11β-hydroxy-1',16α-dimethyl-17,20:20,21-bis(methylenedioxy)-	MeOH-HCl	283(4.19)	35-1520-64
	MeOH	266.5(4.12)	35-1520-64
Pregn-4-eno[3,2-c]pyrazole, 11β-hydroxy-2',16α-dimethyl-17,20:20,21-bis(methylenedioxy)-	MeOH-HCl	282.5(4.12)	35-1520-64
	MeOH	277.5(4.01)	35-1520-64
Pregn-4-eno[3,2-c]pyrazole, 11β,17,21-trihydroxy-1',16α-dimethyl-20-oxo-, 21-acetate	MeOH	267(4.05)	35-1520-64
Pregn-4-eno[3,2-c]pyrazole, 11β,17,21-trihydroxy-2',16-dimethyl-20-oxo-, 21-acetate	MeOH-HCl	282.5(4.15)	35-1520-64
	MeOH	276.5(4.03)	35-1520-64
$C_{26}H_{36}O_2$			
Pregn-4-en-20-one, 3-(2,4-cyclopentadien-1-yl)-3-hydroxy-	MeOH	327(4.18)	83-0244-64
$C_{26}H_{36}O_4$			
5α-Androstane-3α,17β-diol, 2-(2-propynylidene)-, diacetate	EtOH	230(4.22)	13-0001-64
Pregna-3,5-dien-20-one, 17α-acetoxy-3-methoxy-16-methylene-6-methyl-	EtOH	246(4.30)	78-0569-65
Pregna-5,16,20-triene-3β,20-diol, 16-methyl-, diacetate	MeOH	243.5(3.84)	73-2513-64
Pregna-3,5,17(20)-trien-21-oic acid, 3-ethoxy-6-formyl-, ethyl ester	EtOH	223(4.42),323(4.17)	78-0597-64
Pregna-3,5,7-trien-20-one, 17α-acetoxy-3-ethoxy-6-methyl-	EtOH	322.5(4.28)	78-1753-65
$C_{26}H_{36}O_5$			
Pregna-3,5-diene-6-carboxaldehyde, 17α-acetoxy-3-methoxy-16α-methyl-20-oxo-	EtOH	218(4.05),320(4.18)	78-0597-64
Pregna-3,5-diene-11,20-dione, 17α,21-dihydroxy-3-methoxy-6-methyl-, cyclic acetal with acetone	EtOH	244.5(4.26)	78-0569-65
Pregna-1,4-dien-3-one, 11β,17α,20α,21-tetrahydroxy-, cyclic 17α,21-acetal with cyclopentanone	EtOH	244(4.15)	78-0179-65
Pregna-3,5-dien-20-one, 17α-acetoxy-6-(hydroxymethyl)-3-methoxy-16-methylene-	EtOH	247(4.28)	78-0597-64
Pregna-4,6-dien-20-one, 17α-acetoxy-3,3'-ethylenedioxy-6-methyl-	EtOH	242(4.36)	78-1753-65

Compound	Solvent	λ_{max}(log ϵ)	Ref.
5α-Pregna-6,8(14)-dien-20-one, 17α-acetoxy-3,3'-ethylenedioxy-6-methyl-	EtOH	256(4.37)	78-1753-65
$C_{26}H_{36}O_5S$			
Pregn-4-ene-3,20-dione, 21-hydroxy-16β-mercapto-, 21-acetate 16-propionate	EtOH	238(4.27)	87-0531-64
$C_{26}H_{36}O_6$			
9,10-Anthracenediol, 9,10-bis(3-ethoxypropyl)-9,10-dihydro-2,6-dimethoxy-	EtOH	275(3.65),286(3.63), 343(2.71)	39-4565-64
Prednisolone, pivalate	EtOH	243(4.18)	13-0339-65
Pregna-3,5-diene-11,20-dione, 21-acetoxy-17α-hydroxy-6-methyl-3-ethoxy-	EtOH	246.5(4.26)	78-0569-65
Pregna-3,5-diene-11,20-dione, 6-(hydroxymethyl)-17α,21-isopropylidenedioxy-3-methoxy-	EtOH	248(4.23)	78-0597-64
$C_{26}H_{36}O_7$			
Cortexolone, 2β-hydroxy-16α-methyl-, 2,21-diacetate	EtOH	242(4.27)	13-0057-65
Pregna-3,5-dien-11-one, 3-ethoxy-6-(hydroxymethyl)-17α,20:20,21-bis(methylenedioxy)-	EtOH	249(4.28)	78-0597-64
Pregn-4-en-3-one, 11β-hydroxy-2-methoxymethylene-16α-methyl-17α,20:20,21-bis(methylenedioxy)-	MeOH	255(4.10),298(3.94)	35-1520-64
$C_{26}H_{37}ClO_4$			
Pregna-5,20-diene-21-carboxaldehyde, 20-(2-chloroethoxy)-3β-hydroxy-, acetate	EtOH	265(4.30)	88-1839-64
$C_{26}H_{37}FO_8S$			
Pregn-4-ene-3,20-dione, 9-fluoro-11β,16α,17,21-tetrahydroxy-2α-methyl-, cyclic 16,17-acetal with acetone, 21-methanesulfonate	MeOH	237(4.19)	13-0615-65
$C_{26}H_{37}NO_6$			
5α-Pregn-16-en-11-one, 21-acetoxy-3,3-ethylenedioxy-20-methoxyimino-	EtOH	246(4.16)	39-0130-65
5α-Pregn-16-en-11-one, 3β,21-diacetoxy-20-methoxyimino-	EtOH	246.5(4.16)	39-0130-65
$C_{26}H_{38}Cl_4N_6O_{10}$			
Cytidine, N-mustard derivative	H_2O	216(4.26),294(4.56)	73-0635-64
$C_{26}H_{38}INO_5$			
[2-(2-Ethoxy-4,5-dimethoxystyryl)-4,5-dimethoxyphenethyl]ethyldimethylammonium iodide	MeOH	291(4.04),336(4.05)	35-2177-64
$C_{26}H_{38}N_2O_3$			
Androstano[3,2-b]pyridine-5'-carboxylic acid, 6'-amino-17β-hydroxy-17α-methyl-, ethyl ester	EtOH	252(3.99),342(3.93)	94-0077-64
Androst-4-en-3-one, 16α,17α-isopropylidenedioxy-6α-methyl-17β-(3-methylpyrazol-5-yl)-	EtOH	240(4.19)	78-1927-64

Compound	Solvent	$\lambda_{max}(\log \epsilon)$	Ref.
$C_{26}H_{38}O$			
Acetophenone, 2',4',6'-tricyclohexyl-	C_6H_{12}	219(3.67),241(3.14), 245(3.06),249(2.91), 255(2.56),270(2.35), 300(2.08)	73-3462-65
$C_{26}H_{38}O_3$			
Pregna-3,5,17(20)-trien-21-oic acid, 3-ethoxy-6-methyl-, ethyl ester	EtOH	232(4.35)	78-0569-65
$C_{26}H_{38}O_4$			
5β,19-Cycloandrost-6-en-3β-ol, 17-tetrahydropyranyloxy-, 3-acetate	EtOH	213(3.78)	13-0001-64
Estra-1,3,5(10)-triene, 3-methoxy-17β-(tetrahydro-2-pyranyloxyethoxy)-	EtOH	279(3.22)	13-0557-65
Pregna-3,5-diene-6-carboxaldehyde, 20β-acetoxy-3-ethoxy-	EtOH	223(4.04),322(4.13)	78-0597-64
Pregna-5,20-diene-21-carboxaldehyde, 3β-acetoxy-20-ethoxy-	MeOH	261(4.29)	35-3908-64
Pregna-3,5-dien-20-one, 17α-acetoxy-3-ethoxy-6-methyl-	EtOH	247.5(4.30)	78-0569-65
Pregna-3,5-dien-20-one, 17α-acetoxy-3-methoxy-6,16α-dimethyl-	EtOH	247(4.26)	78-0569-65
Pregna-3,5-dien-20-one, 16α,17α-dihydroxy-3-methoxy-6-methyl-, cyclic acetal with acetone	EtOH	247(4.29)	78-0569-65
$C_{26}H_{38}O_5$			
5α-Androstan-3-one, 17β-hydroxy-2-(hydroxymethylene)-, dipropionate	EtOH	256(4.08)	44-3481-64
Pregna-5,16-diene-21-carboxaldehyde, 3β-hydroxy-20-oxo-, 21-(dimethyl acetal), acetate	EtOH	243(3.95)	88-1839-64
Pregna-3,5-dien-20-one, 17α-acetoxy-6-(hydroxymethyl)-3-methoxy-16α-methyl-	EtOH	247(4.29)	78-0597-64
Pregn-4-en-3-one, 11β,12β,20α-trihydroxy-, cyclic 11,12-acetal with acetone, acetate	EtOH	238(4.28)	44-2169-65
Pregn-5-en-20-one, 21-acetyl-3β-hydroxy-16α,17α-isopropylidenedioxy-	pH 13	300(4.38)	78-1927-64
	EtOH	280(4.12)	78-1927-64
$C_{26}H_{38}O_6$			
Androst-9(11)-ene-3,17-dione, 5-(hydroxymethyl)-, cyclic bis(ethylene acetal), acetate	EtOH	206(3.51)	13-0305-64
3α,5α-Epoxymethano-5α-androst-9(11)-ene, 3β-(2-hydroxyethyloxy)-17,17-(ethylenedioxy)-, acetate	EtOH	207(3.57)	13-0305-64
Oxymetholone, carbomethoxypropionate	pH 13	217(4.42),314(4.25)	13-0441-65
	EtOH	283(3.96)	13-0441-65
Oxymetholone, 17-hemiglutarate	pH 13	315(4.23)	13-0441-65
	EtOH	283(3.96)	13-0441-65
Pregna-1,4-dien-3-one, 11,17α,20α,21-tetrahydroxy-, 21-valerate	EtOH	244(4.15)	78-0179-65
$C_{26}H_{38}O_7$			
Taxinine, dihydro-decinnamoyloxy-	n.s.g.	269(3.80)	95-0762-64

Compound	Solvent	λ_{max}(log ε)	Ref.
$C_{26}H_{38}O_8S$			
Pregn-4-ene-3,20-dione, 11β,16α,17,21-tetrahydroxy-6α-methyl-, cyclic 16,17-acetal with acetone, 21-methanesulfonate	MeOH	242(4.22)	13-0615-65
$C_{26}H_{38}O_9$			
Carbonic acid, ethyl ester, diester with isoandrographolide	EtOH	214(3.87),220(3.70), 230(3.11)	78-2617-65
$C_{26}H_{38}O_{12}S_4$			
Galactopyranose, 1,2:3,4-di-O-isopropylidene-, 5,5'-[dithiobis(thioformate)]	MeOH	240(4.27),288(3.95)	44-3071-65
$C_{26}H_{39}NO_2$			
5α-Androst-2-ene-2-carbonitrile, 17β-hydroxy-, hexanoate	C_6H_{12}	208(4.05)	44-3300-64
Estra-1,3,5(10)-triene, 3-methoxy-17β-(2-piperidinoethoxy)-	EtOH	278(3.30),287(3.25)	13-0557-65
$C_{26}H_{39}NO_4$			
Pregn-4-ene-3,20-dione, 17α-acetoxy-6α-dimethylaminomethyl-	EtOH	239.5(4.15)	78-0569-65
$C_{26}H_{39}N_3O_7$			
5α-Pregnane-11,20-dione, 21-acetoxy-3,3-ethylenedioxy-17-hydroxy-, 20-semicarbazone	EtOH	237.5(4.08)	39-0130-65
$C_{26}H_{40}$			
Naphthalene, 2,6-bis(1,1,3,3-tetramethylbutyl)-	C_6H_{12}	232(5.13),272(3.72), 302s(2.60),308(2.70), 316(2.48),323(2.78)	39-2324-65
Naphthalene, 2,6-dioctyl-	C_6H_{12}	230(5.23),274(3.83), 303(3.00),309(3.04), 317(2.85),324(3.18)	39-2324-65
$C_{26}H_{40}B_2N_2$			
1,3,2,4-Diazadiboretidine, 2,4-bis-(1,1-diethylpropyl)-1,3-diphenyl-	EtOH	253(2.29),259(2.27), 265(2.14)	88-0703-65
$C_{26}H_{40}N_2O_3$			
Cyclopentanone, ethyl tetradecahydro-10a,12a-dimethyl-1H-cyclopenta-[7,8]phenanthro[2,3-c][1,2,5]oxadiazol-1-yl acetal	EtOH	217(3.68)	94-1445-65
$C_{26}H_{40}N_4O_2$			
p-Benzoquinone, tetrapiperidino-	methyl cellosolve	243(4.22),404(3.97)	78-1889-64
$C_{26}H_{40}O_2$			
A-Norcholest-3(5)-ene-1,2-dione	EtOH	284(3.73)	94-0050-65
$C_{26}H_{40}O_3$			
Pregn-4-en-3-one, 17α,20β-isopropylidenedioxy-, 6α,16α-dimethyl-	EtOH	242(4.22)	87-0540-64

Compound	Solvent	λ_{max}(log ϵ)	Ref.
$C_{26}H_{40}O_3S$			
1-Naphthalenesulfonic acid, 3,7-bis-(1,1,3,3-tetramethylbutyl)-	EtOH	233(4.87),279(3.72), 312(3.11),318s(2.95), 326(3.15)	39-2324-65
1-Naphthalenesulfonic acid, 3,7-dioctyl-	EtOH	232(4.90),280(3.76), 313(3.26),318(3.08), 327(3.34)	39-2324-65
$C_{26}H_{40}O_4$			
24-Nor-5α-cholanic acid, 4,4,14-trimethyl-2,3-dioxo-	EtOH	270(3.97)	78-1755-64
23-Nor-5α,17α-chol-20(22)-enic acid, 3β-acetoxy-, methyl ester	EtOH	229(4.16)	13-0357-65
Pregna-3,5-diene-6-methanol, 20β-acetoxy-3-ethoxy-	EtOH	249(4.28)	78-0597-64
$C_{26}H_{41}NO_2S$			
23-Norchol-5-enamide, 3β-acetoxy-N-methylthio-	MeOH	262(4.08),317(1.72)	35-0051-65
$C_{26}H_{41}NO_3$			
Androsta-3,5-dien-17β-ol, 6-dimethylaminomethyl-3-ethoxy-, acetate	EtOH	251(4.31)	78-0569-65
$C_{26}H_{42}BNO_3$			
Androsta-3,5,7-trien-17β-ol, 6-dimethylaminomethyl-3-ethoxy-, acetate, compound with BH_3	EtOH	325(4.26)	78-1753-65
$C_{26}H_{42}N_2O$			
A-Norcholestan-1-one, 2-diazo-	EtOH	255(4.22),300(3.51)	44-3775-65
$C_{26}H_{42}N_2O_2$			
1,12-Diazatricyclo[10.10.4.2^{23,26}]octacosa-25,28-diene-24,27-dione	EtOH	233(4.13),360(3.96)	24-3439-65
$C_{26}H_{42}O$			
19-Norcholestenone	MeOH	240(4.15)	35-1528-64
B-Norcholest-4-en-3-one	n.s.g.	240(4.17)	39-6117-65
B-Nor-5α-cholest-1-en-3-one	n.s.g.	231(3.93)	39-6117-65
$C_{26}H_{42}O_3$			
24-Nor-5β-cholest-9(11)-en-12-one, 3α,25-dihydroxy-	EtOH	237(4.09)	24-2361-65
$C_{26}H_{42}O_4$			
p-Benzoquinone, 2,5-dihydroxy-3-methyl-6-nonadecenyl-	EtOH	292(4.37)	94-0236-64
p-Benzoquinone, 2,5-dihydroxy-3-methyl-6-(14-nonadecenyl)-	EtOH	209(4.24),294(4.36), 440(2.47)	94-0236-64 94-0511-65
24-Nor-5β-cholest-9(11)-en-12-one, 3α,7α,25-trihydroxy-	EtOH	238(4.02)	24-2361-65
$C_{26}H_{42}O_6$			
Cortisol, 17α,21-benzylidene-, 11-(α-methoxybenzyl) ether	EtOH	241(4.21)	78-0179-65
$C_{26}H_{43}NO_2$			
4-Azacholest-5-en-3-one, 4-hydroxy-	EtOH	238(4.17)	25-0648-65
A-Norcholestane-1,2-dione, 2-oxime	EtOH	236(3.87)	44-3775-65

Compound	Solvent	$\lambda_{max}(\log \epsilon)$	Ref.
$C_{26}H_{43}NO_6$			
Erythoskyrine, decahydro-	pH 13	246(4.15),288(4.15)	25-0419-64
	EtOH	225(3.86),284(4.04)	25-0419-64
$C_{26}H_{43}N_3O_9$			
Isoperhydrodethiogriseoviridin, diacetate	EtOH	385(4.46)	23-0371-64
$C_{26}H_{44}BNO_3$			
Androsta-3,5-dien-17β-ol, 6-dimethylaminomethyl-3-ethoxy-, acetate, compound with BH_3	EtOH	255(4.33)	78-0569-65
$C_{26}H_{44}N_2$			
Buxamine	MeOH	230s(4.28),238(4.42), 246(4.45),254(4.24), 277s(2.39),288s(2.24)	33-0968-64
$C_{26}H_{44}N_2O$			
4-Azacholest-5-en-3-one, 4-amino-	EtOH	244(4.12)	4-0212-65
Buxaminol	MeOH	238(4.45),245(4.49), 254(4.28),277(2.68), 287(2.58)	33-0968-64
$C_{26}H_{44}N_8O_2$			
p-Benzoquinone, tetrakis(4-methyl-1-piperazinyl)-	methyl cellosolve	240(4.26),396(4.02)	78-1889-64
$C_{26}H_{44}O_4$			
p-Benzoquinone, 2,5-didecyl-3,6-dihydroxy-	EtOH	209(4.14),295(4.29), 429(2.30)	94-0236-64 94-0511-65
p-Benzoquinone, 2,5-dihydroxy-3-methyl-6-nonadecyl-	EtOH	292(4.37)	78-2319-64
Maesaquinone, dihydro-	EtOH	294(4.37),440(2.58)	94-0511-65
$C_{26}H_{46}Cl_2Pd_2$			
Palladium, dichlorobis(1-methyl-π-cyclododecenyl)di-	dioxan	235s(4.22),332(3.34)	24-1753-65

Compound	Solvent	$\lambda_{max}(\log \epsilon)$	Ref.
$C_{27}H_{16}$			
Methane, difluoren-9-ylidene-	THF	440(4.44),456(4.45)	24-2975-64
$C_{27}H_{16}O_2$			
2,5-Cyclohexadiene-$\Delta^{1,\alpha}$-acetic acid, 4-fluoren-9-ylidene-2-hydroxy-α-phenyl-, γ-lactone	EtOH	258(4.11),516(4.00)	44-4333-65
$C_{27}H_{17}N_3O_{12}$			
2-Heptenedioic acid, 2-hydroxy-3-(α-hydroxy-m-nitrobenzyl)-5-(m-nitrobenzylidene)-4-(m-nitrophenyl)-6-oxo-, γ-lactone	MeOH	253(4.46)	44-1800-65
$C_{27}H_{18}$			
Fluorene, 9-(diphenylvinylidene)-	THF	314(4.30)	24-2975-64
Methane, fluoren-9-yl-fluoren-9-ylidene-	dioxan	231(4.74),238(4.62), 249(4.66),258(4.80), 270(4.43),290(4.37), 303(4.40),317(4.33)	49-0003-64
	DMSO-KO-tert-Bu	300(4.50),344(4.16), 450(3.34),559(5.05)	5-0050-65J
$C_{27}H_{18}O$			
Fluoren-9-ol, 9-(fluoren-9-ylidenemethyl)-	ether	248(4.67),257(4.71), 277(4.31),289(4.29), 305(4.24),317(4.20)	5-0050-65J
$C_{27}H_{18}O_6$			
Visnagin, phenanthrenequinone adduct	dioxan	244(4.80),278(4.24), 329(3.55)	24-3102-65
$C_{27}H_{19}BrNOP$			
Acetonitrile, (p-bromobenzoyl)(triphenylphosphoranylidene)-	EtOH	267(3.94),275(3.93), 296(3.92)	65-2205-65
$C_{27}H_{19}ClNOP$			
Acetonitrile, (p-chlorobenzoyl)(triphenylphosphoranylidene)-	EtOH	268(3.92),275(3.95), 299(3.95)	65-2205-65
$C_{27}H_{19}NO_3$			
Anthracene, 9-(m-methoxyphenyl)-10-(m-nitrophenyl)-	CH_2Cl_2	253s(4.96),260(5.06), 323s(3.28),339s(3.61), 357(3.92),375(4.11), 396(4.08)	44-3695-64
Anthracene, 9-(m-methoxyphenyl)-10-(o-nitrophenyl)-	CH_2Cl_2	253s(4.90),261(5.08), 323s(3.28),340s(3.62), 358(3.91),377(4.08), 396(4.05)	44-3695-64
Anthracene, 9-(m-methoxyphenyl)-10-(p-nitrophenyl)-	CH_2Cl_2	260(5.09),323s(3.28), 340s(3.60),357(3.88), 375(4.08),395(4.08)	44-3695-64
Anthracene, 9-(o-methoxyphenyl)-10-(m-nitrophenyl)-	CH_2Cl_2	253(4.95),261(5.16), 322s(3.24),339s(3.61), 357(3.93),375(4.13), 396(4.10)	44-3695-64
Anthracene, 9-(o-methoxyphenyl)-10-(o-nitrophenyl)-	CH_2Cl_2	253s(4.84),261(5.05), 323s(3.22),340s(3.56), 357(3.86),376(4.04), 396(4.00)	44-3695-64

Compound	Solvent	λ_{max}(log ε)	Ref.
Anthracene, 9-(o-methoxyphenyl)-10-(p-nitrophenyl)-	CH_2Cl_2	260(5.07),321s(3.21),339s(3.55),357(3.86),375(4.05),395(4.06)	44-3695-64
Anthracene, 9-(p-methoxyphenyl)-10-(m-nitrophenyl)-	CH_2Cl_2	253s(4.87),261(5.10),322s(3.17),340s(3.56),358(3.89),376(4.09),396(4.06)	44-3695-64
Anthracene, 9-(p-methoxyphenyl)-10-(o-nitrophenyl)-	CH_2Cl_2	254s(4.81),262(5.04),323s(3.13),341s(3.55),358(3.85),377(4.04),397(4.00)	44-3695-64
Anthracene, 9-(p-methoxyphenyl)-10-(p-nitrophenyl)-	CH_2Cl_2	261(4.98),324s(3.29),340s(3.54),358(3.83),376(4.01),396(4.02)	44-3695-64
$C_{27}H_{19}N_2O_3P$			
Acetonitrile, (p-nitrobenzoyl)(triphenylphosphoranylidene)-	EtOH	260(4.23),270(4.29)	65-2205-65
$C_{27}H_{19}N_4P$			
Malononitrile, [p-(triphenylphosphonazo)phenyl]-, inner salt	MeCN	262(3.84),318(3.64),337(3.64),563(4.56),600(4.49)	35-2174-64
$C_{27}H_{20}$			
1H-Cycloprop[a]anthracene, 1a,9b-dihydro-1,1-diphenyl-	dioxan	252s(4.57),261(4.58),303(4.15),316(4.12),343(3.31),360(3.28)	78-1657-65
Cyclopropene, 1,2,3,3-tetraphenyl-	EtOH	229(4.52),303(4.32),316(4.36),333(4.32)	35-3896-64
Fluorene, 9-(2,2-diphenylethylidene)-	DMSO-KO-tert-Bu	291(4.41),373(4.18),400s(3.99),515(4.48),540s(4.43)	5-0050-65J
Fluorene, 9-(2,2-diphenylvinyl)-	THF	259(4.56),292(3.83),303(3.94)	24-2975-64
Propadiene, tetraphenyl-	THF	265(4.46)	24-2975-64
$C_{27}H_{20}NOP$			
Acetonitrile, benzoyl(triphenylphosphoranylidene)-	EtOH	269(3.94),275(3.97),290(3.91)	65-2205-65
$C_{27}H_{20}N_2$			
Pyrazole, 1,3,4,5-tetraphenyl-	C_6H_{12}	238s(4.42),273s(4.26)	44-1582-64
$C_{27}H_{20}N_2O_2$			
1,3,4-Oxadiazole, 4-benzoyl-4,5-dihydro-2,5,5-triphenyl-	EtOH	300(4.25)	32-0033-65
$C_{27}H_{20}N_3O_2P$			
3-Butenoic acid, 3,4,4-tricyano-2-(triphenylphosphoranylidene)-, ethyl ester	dioxan	416(4.18)	49-1967-65
$C_{27}H_{20}N_4O_2$			
1,3-Propanedione, 1,3-diphenyl-2,2-bis(phenylazo)-	MeOH	245(4.36),315(4.09),420(4.31)	44-2959-64
$C_{27}H_{20}O$			
Fluoren-9-ol, 9-(fluoren-9-ylmethyl)-	ether	237(4.39),267(4.48),289(4.11),302(4.06)	5-0050-65J

Compound	Solvent	λ_{max}(log ε)	Ref.
$C_{27}H_{20}O_4$			
5,8,13,14-Pentaphenetetrone, 2,3,6,10,11-pentamethyl-	dioxan	255(4.49),285(4.55)	65-3484-64
$C_{27}H_{20}O_6$			
2,3-Diphenylindenone oxide, dimethyl acetylenedicarboxylate adduct	MeCN	230s(4.25),300s(3.27), 360s(2.20),368(2.45), 385s(2.26)	35-3814-64
$C_{27}H_{21}NOSn$			
Tin, (8-quinolinolato)triphenyl-	C_6H_{12}	245(4.5),260(4.3), 400(3.4)	101-0159-65B
$C_{27}H_{22}$			
Propene, 1,1,2,3-tetraphenyl-	EtOH	227(4.30),270(4.06)	88-2951-65
1-Propene, 1,1,3,3-tetraphenyl-	DMSO-KO-tert-Bu	455(4.43),559(4.67)	5-0050-65J
Propene, 1,2,3,3-tetraphenyl-, trans	EtOH	220(4.34),305(4.15)	88-2951-65
Propene, 1,3,3,3-tetraphenyl-	EtOH	220(4.37),253(4.22)	88-2951-65
$C_{27}H_{22}BrOP$			
Phosphorane, (α-bromobenzoyl-ethylidene)triphenyl-	EtOH	275(3.85),315(3.85)	65-2738-64
$C_{27}H_{22}ClOP$			
Phosphorane, (α-chlorobenzoyl-ethylidene)triphenyl-	EtOH	275(3.81),315(3.78)	65-2738-64
$C_{27}H_{22}N_3P$			
1-Hexene-1,1,2-tricarbonitrile, 3-(triphenylphosphoranylidene)-	dioxan	414(4.22)	49-1967-65
$C_{27}H_{22}O$			
2-Propanone, 1,1,1,3-tetraphenyl-	EtOH	253(3.27),259(3.18), 300(2.61)	23-0298-64
Propiophenone, 3,3,3-triphenyl-	EtOH	241(4.13),270s(3.32), 323(1.98)	23-0298-64
$C_{27}H_{22}O_7$			
1,10-Anthracenediol, 9-(4-hydroxy-3-methoxyphenyl)-, triacetate	benzene	262(5.1),341(4.3), 360(4.5),376(4.6), 398(4.6)	56-0847-65
$C_{27}H_{22}O_7S_2$			
Benzophenone, 2,2'-dihydroxy-, di-p-toluenesulfonate	EtOH	228(4.32),255(3.91), 283(3.71),323(2.53)	39-5074-64
$C_{27}H_{23}ClN_2O_5S$			
2-[3-(3-Ethyl-2-benzothiazolinylidene)-propenyl]-1-(o-hydroxyphenyl)-quinolinium perchlorate	EtOH	587(5.09)	65-3373-64
$C_{27}H_{23}ClN_2O_6S$			
2-[(3-Ethyl-2-benzothiazolinylidene)-methyl]-1-(o-hydroxyphenyl)quinol-inium perchlorate, acetate	EtOH	503(4.85)	65-3373-64
$C_{27}H_{23}ClO_4$			
7-(8,8-Diphenyl-1,3,5,7-octatetraen-1-yl)tropylium perchlorate	MeCN	234(4.37),271(4.27), 303(4.28),343(4.17), 597(4.81)	24-1590-64

Compound	Solvent	λ_{max}(log ε)	Ref.
$C_{27}H_{23}N$			
Acridan, 9,9-di-m-tolyl-	EtOH	248(3.67),291(4.11)	65-3472-64
Acridan, 9,9-di-p-tolyl-	EtOH	248(3.61),292(4.09)	65-3472-64
5H-Dibenz[c,e]azepine, 6,7-dihydro-6-methyl-3,9-diphenyl-	n.s.g.	290(4.55)	4-0181-65
$C_{27}H_{23}NO$			
Acridan, 9,9-di-p-anisyl-	EtOH	254(3.84),288(4.11)	65-3472-64
$C_{27}H_{23}NO_3$			
Nicotinic acid, 5-benzylidene-1,2,5,6-tetrahydro-6-oxo-1,2-diphenyl-, ethyl ester	EtOH	323(4.41)	78-3255-65
$C_{27}H_{23}NO_5$			
2-Naphthacenecarboxamide, 10-(benzyloxy)-4,4a,5,12-tetrahydro-1,3-dihydroxy-6-methyl-12-oxo-	EtOH	246(4.40),260(4.43), 316(3.93),382(4.56)	65-0670-65 70-0945-64
$C_{27}H_{23}N_3$			
Indole, 2,2'-dimethyl-3,3',3"-methylidynetri-	n.s.g.	285(4.35),292(4.13)	28-0609-64A
$C_{27}H_{23}N_3O_3$			
2(1H)-Cycloheptimidazolone, 6-[cyano-(carbethoxy)methylene]-1,3-dibenzyl-3,6-dihydro-	EtOH	265(4.16),465(4.66)	94-0810-65
$C_{27}H_{23}OP$			
Phosphorane, (α-benzoylethylidene)-triphenyl-	EtOH	265(3.81),310(3.76), 320(3.77)	65-2738-64
Phosphorane, (p-toluoylmethylene)-triphenyl-	EtOH	275(3.96),322(4.18)	65-2738-64
2-Propanone, 1-phenyl-3-(triphenylphosphoranylidene)-	EtOH	267(3.88),273(3.86), 287(3.81)	44-3327-64
$C_{27}H_{23}O_2P$			
Phosphorane, (anisoylmethylene)-triphenyl-	EtOH	248(4.14),320(4.26)	65-2738-64
$C_{27}H_{24}$			
Cyclohexane, 1,2,3-tribenzylidene-	EtOH	275(4.44)	44-2335-64
$C_{27}H_{24}Cl_2N_2O_4$			
Phthalonitrile, 4,5-dichloro-3-[3,17-dioxoandrosta-4,6-dien-1-yl)oxy]-6-hydroxy-	EtOH-HCl	224(4.37),235s(4.31), 292(4.29),345s(3.71)	44-0601-64
$C_{27}H_{24}N_2O_4$			
Benzoic acid, p-(phenylazo)-, 2-(2,4-hexadiynylidene)-1,6-dioxaspiro[4.5]dec-4-yl ester	ether	221(4.52),261(4.28), 275(4.40),291(4.43), 321(4.42)	24-1179-64
$C_{27}H_{24}N_2O_7$			
3,3-Pyrrolidinedicarboxylic acid, 1,2-diphenyl-4-(p-nitrobenzyl)-5-oxo-, dimethyl ester	EtOH	220s(4.18),242s(4.01), 270(4.10)	78-0449-65
$C_{27}H_{24}OSi$			
Propiophenone, 3-(triphenylsilyl)-	EtOH	242(4.09),270(3.24), 311s(2.12)	23-0298-64

Compound	Solvent	λ_{max}(log ε)	Ref.
$C_{27}H_{24}O_3$			
Spiro[cyclohexane-1,3'(1'H)-[1,4]epoxy-[2]benzoxepin]-5'(4'H)-one, 1',4'-diphenyl-	MeCN	255(4.02),263s(3.93), 293(3.29),301(3.24), 332s(2.05),344(2.18), 358(2.17),372s(1.88)	35-3814-64
$C_{27}H_{25}BO_{13}$			
Curcumin, boric acid citric acid complex	EtOH-buffer	630(4.82)	83-0660-64
	iso-PrOH-acid	510(4.88)	83-0660-64
	iso-PrOH-NH_3	652(4.74)	83-0660-64
$C_{27}H_{25}ClN_2O_6$			
1-o-Acetoxyphenyl-2-(p-dimethylamino-styryl)quinolinium perchlorate	EtOH	558(4.71)	65-3373-64
$C_{27}H_{25}NO$			
Methanol, (o-anilinophenyl)di-m-tolyl-	EtOH	254(3.57),296(4.25)	65-3472-64
Methanol, (o-anilinophenyl)di-p-tolyl-	EtOH	252(3.88),297(4.45)	65-3472-64
$C_{27}H_{25}NO_3$			
Methanol, (o-anilinophenyl)bis-(p-methoxyphenyl)-	EtOH	254(3.95),297(4.38)	65-3472-64
$C_{27}H_{25}NO_5$			
Benzophenone, 5-nitro-2-[5,6,7,8-tetrahydro-6,6,8,8-tetramethyl-7-oxo-2-naphthyl)oxy]-	EtOH	205(4.60),256(4.28)	39-0361-65
$C_{27}H_{25}NO_6$			
Benzophenone, 5-nitro-2-[5,6,7,8-tetrahydro-3-hydroxy-6,6,8,8-tetra-methyl-7-oxo-2-naphthyl)oxy]-	EtOH	208(4.67),258(4.22), 291(4.14)	39-0361-65
$C_{27}H_{26}N_4O_7$			
Hydrocinnamic acid, β-acetonyl-α-benzoyl-, ethyl ester, 2,4-dinitrophenylhydrazone	$CHCl_3$	367.5(4.26)	78-0195-64
$C_{27}H_{26}N_6O_4$			
Oxamic acid, (3-carboxy-1,4-diphenyl-Δ^2-1,2,4-triazolin-5-ylidene)-, diethyl ester, 2-(phenylhydrazone)	EtOH	240(4.4),375(4.3)	24-2185-65
Oxanilic acid, ethyl ester, 2-[(3-carb-oxy-1-phenyl-1H-1,2,4-triazol-5-yl)-phenylhydrazone], ethyl ester	EtOH	235(4.4),345(4.3)	24-2185-65
$C_{27}H_{26}O_3Si_3$			
Cyclotrisiloxane, 2-methyl-4,4,6,6-tetraphenyl-2-vinyl-	$CHCl_3$	265(3.22)	46-1066-65
$C_{27}H_{26}O_5$			
2H-1-Benzopyran-7-ol, 3,4-bis(p-meth-oxyphenyl)-2,2-dimethyl-, acetate	MeOH	284(4.02),308(4.01)	44-4114-65
$C_{27}H_{26}O_8$			
Flavone, 3'-(benzyloxy)-3,4',5,6,7-pentamethoxy-	MeOH	240(4.34),333(4.38)	24-2857-64

Compound	Solvent	$\lambda_{max}(\log \epsilon)$	Ref.
Flavone, 4'-(benzyloxy)-3,3',5,6,7-pentamethoxy-	MeOH	243(4.33),335(4.39)	24-0548-65
Macluraxanthone, diacetate	EtOH	244(3.94),272s(4.25), 294s(4.52),300(4.55), 337(4.06)	44-0692-64
$C_{27}H_{27}ClNP$			
Benzyl(p-dimethylaminophenyl)diphenylphosphonium chloride	MeOH	303(4.44)	88-2729-64
$C_{27}H_{27}ClN_4O_2S$			
Pyrimidine, 2-amino-6-benzyl-4-chloro-5-[N-(p-tolylsulfonyl)-3-anilinopropyl]-	pH 1, 7	236(4.42),309(3.62)	87-0283-65
	pH 13	305(3.66)	87-0283-65
$C_{27}H_{27}NO_3$			
3H-Benzazepine, 3-benzoyl-7,8-diethoxy-4,5-dihydro-1-phenyl-	$CHCl_3$	258(4.25),310(4.15)	6-0213-65
$C_{27}H_{27}NO_5$			
Benzo[a]heptalene-9-carbonitrile, 5,6,7,8,11,12-hexahydro-10-hydroxy-1,2,3-trimethoxy-, benzoate	EtOH	220(4.62),255s(4.17)	44-1752-65
$C_{27}H_{27}NO_8$			
3,4,5,6-Pyridinetetracarboxylic acid, 1,2-dihydro-1-phenethyl-2-phenyl-, tetramethyl ester	EtOH	206(4.40),231(4.18), 283(4.08),386(3.82)	5-0098-65H
$C_{27}H_{27}N_3$			
Pyrrole, 1-(1-methyl-4-piperazinyl)-2,3,5-triphenyl-	EtOH	210(3.0),255(2.7), 285(2.7)	6-0073-64
$C_{27}H_{28}N_2$			
3H,5H-Pyrido[1,2,3-de]quinoxaline, 2,3-bis(p-ethylphenyl)-6,7-dihydro-	MeOH	258s(4.24),278(4.35), 318(3.84),420(3.58)	44-2589-65
$C_{27}H_{28}N_2O_3$			
Ajmalol B, benzoate	EtOH	229(4.73),275s(3.91), 282(3.93),292s(3.84)	33-1349-65
$C_{27}H_{28}O_3Si_3$			
Cyclotrisiloxane, 2-ethyl-2-methyl-4,4,6,6-tetraphenyl-	$CHCl_3$	265(3.22)	46-1066-65
$C_{27}H_{28}O_7$			
5H-Tribenzo[a,d,g]cyclononen-5-one, 10,15-dihydro-2,3,7,8,12,13-hexamethoxy-	EtOH	206(4.81),240(4.51), 286(4.15),326(4.16)	39-1685-65
$C_{27}H_{28}O_8$			
Macluraxanthone, dihydro-, diacetate	EtOH	244(4.20),267(4.28), 295s(4.41),302(4.46), 334(4.08)	44-0692-64
$C_{27}H_{28}O_{11}$			
Podophyllotoxin, 6,7-O-demethylene-, triacetate	MeOH	268(3.24)	33-1529-64

Compound	Solvent	$\lambda_{max}(\log \epsilon)$	Ref.
$C_{27}H_{29}BrN_2O_4$			
Bisnorrubremetinium bromide	n.s.g.	260(4.3),300(4.3), 435(4.48)	94-0775-65
$C_{27}H_{29}NO_4$			
Isohypognavinone	EtOH	230(4.31)	94-1124-64
$C_{27}H_{29}NO_5$			
2-Pyrroline-3,4-dicarboxylic acid, 5-benzoyl-1-cyclohexyl-2-phenyl-, dimethyl ester	EtOH	255(4.19),303(4.21)	88-4363-65
$C_{27}H_{29}N_5O_2S$			
Pyrimidine, 2,4-diamino-6-benzyl-5-[N-(p-tolylsulfonyl)anilino-propyl]-	pH 1	276(3.93)	87-0283-65
	pH 7	280(3.94)	87-0283-65
	pH 13	293(3.99)	87-0283-65
$C_{27}H_{29}O_2P$			
(2-Hydroxy-3,3,5-trimethyl-4-oxo-1-hexenyl)triphenylphosphonium hydroxide, inner salt	EtOH	225s(4.40),266(3.80), 273(3.79),293(3.78)	44-2957-64
$C_{27}H_{30}FeN_2O$			
Methanol, bis(p-dimethylaminophenyl)-ferrocenyl-	EtOH-HCl	458(4.26),608(4.72)	39-5759-65
$C_{27}H_{30}N_2O$			
Benzamide, N-[α-[(2,4-dimethylbenzylidene)amino]-2,4-dimethylbenzyl]-2,4-dimethyl-	EtOH	260(4.38)	35-1701-64
Benzamide, N-[α-[(2,5-dimethylbenzylidene)amino]-2,5-dimethylbenzyl]-2,5-dimethyl-	EtOH	256(4.30)	35-1701-64
Benzamide, N-[α-[(3,4-dimethylbenzylidene)amino]-3,4-dimethylbenzyl]-3,4-dimethyl-	EtOH	248(4.33)	35-1701-64
$C_{27}H_{30}N_2O_2$			
6bH-Naphth[2',1':4,5]indeno[2,1-d]isoxazole-6b-carbonitrile, tetradecahydro-4a,6a-dimethyl-2-oxo-β-phenyl-	MeOH	246(4.37)	4-0280-64
$C_{27}H_{30}N_2O_3$			
Ajmalol B, benzoate	EtOH	229(4.73),276s(3.88), 283(3.91),292s(3.82)	33-1349-65
$C_{27}H_{30}N_2O_3S$			
Voachalotinol, de(hydroxymethyl)-, p-toluenesulfonate	MeOH	229(4.70),286(3.87)	20-0253-65
$C_{27}H_{30}N_2O_5$			
Pyrrolo[2,1-a]isoquinoline-2-ethanol, 3-(6,7-dimethoxy-3,4-dihydro-1-isoquinolyl)-5,6-dihydro-8,9-dimethoxy-	pH 2	255(4.18),293(4.27), 300(4.27),440(4.24)	94-0775-65
$C_{27}H_{30}N_4O_5$			
Androsta-1,4-diene-3,17-dione, 19-vinyl-, 2,4-dinitrophenylhydrazone	EtOH	395(4.46)	33-0094-65

Compound	Solvent	λ_{max}(log ε)	Ref.
$C_{27}H_{30}O_3$			
Ferruginol, 5,6-dehydro-7-oxo-, benzoate	EtOH	234(4.30),260(4.18)	39-0361-65
$C_{27}H_{30}O_5$			
5H-Tribenzo[a,d,g]cyclononene, 10,15-dihydro-3,7,8,12,13-pentamethoxy-2-methyl-	EtOH	205(4.87),233(4.52), 291(3.96)	39-1685-65
$C_{27}H_{30}O_6$			
12H-Benzo[b]xanthen-12-one, 1-heptyl-11-hydroxy-3,8,10-trimethoxy-	EtOH	253(4.67),292(4.69), 326(4.06),341(4.10), 420(4.00)	1-1677-65
Cyclotriveratrylene	EtOH	205(4.82),233(4.49), 292(3.99)	39-1685-65
$C_{27}H_{30}O_7$			
Nimbinic acid, acid treatment product	EtOH	351(4.16)	2-0108-64
$C_{27}H_{30}O_8$			
Macluraxanthone, tetrahydro-, diacetate	EtOH	243(4.43),266(4.30), 324(4.17),366(3.62)	44-0692-64
Phthalide, 3-(5-heptyl-7-methoxy-4-oxo-4H-1-benzopyran-2-yl)-3-hydroxy-5,7-dimethoxy-	EtOH	249s(4.48),296(4.34)	1-1677-65
$C_{27}H_{30}O_{10}$			
2-Naphthoic acid, 1,2,3,4-tetrahydro-3-(hydroxymethylene)-6,7-dimethoxy-4-oxo-1-(3,4,5-trimethoxyphenyl)-, ethyl ester, acetate	EtOH	235(4.35),282(3.95), 355(4.09)	2-0190-64
$C_{27}H_{30}O_{12}$			
Podophyllotoxin β-D-glucoside, 4-demethyldeoxy-	MeOH	292(3.66)	33-1203-64
$C_{27}H_{30}O_{14}$			
Lanceolarin	EtOH	262(4.55),335(4.06)	88-3191-65
Rhoifolin	EtOH	268(4.34),336(4.42)	31-0562-64
Spherobioside	EtOH	261(4.59),326(3.73)	24-2193-65
$C_{27}H_{31}BrO_3$			
Sugiol, 6α-bromo-, benzoate	isooctane	224(4.38),261(4.09), 299s(3.34)	44-0501-65
	MeOH	227(4.34),264(4.08), 302s(3.61)	44-0501-65
Totaryl benzoate, 6α-bromo-7-oxo-	EtOH	340(4.10)	44-0501-65
$C_{27}H_{31}ClN_4O_6$			
Estra-3,5-diene-6-carboxaldehyde, 3-chloro-17β-hydroxy-, 2,4-dinitrophenylhydrazone, acetate	EtOH	270(4.12),313(4.03), 398(4.51)	88-0137-65
$C_{27}H_{31}NO_3$			
19-Nor-17α-pregn-4-en-20-yn-3-one, 17β-hydroxy-, carbanilate	MeOH	237(4.55)	4-0207-65
$C_{27}H_{31}NO_4$			
Isohypognavine, hydrochloride	EtOH	230(4.09)	94-1124-64

Compound	Solvent	$\lambda_{max}(\log \epsilon)$	Ref.
$C_{27}H_{31}NO_5$			
Lyfoline, O,O-dimethyl-	MeOH	278(4.22)	100-0084-65
Lythrine, O-methyl-	EtOH	280(4.23)	100-0015-64
$C_{27}H_{31}NO_8$			
Dibenzo[a,g]quinolizidine, 8-(2-hydroxypropyl)-13-hydroxy-2,3-(methylenedioxy)-9,10-dimethoxy-	EtOH	286(3.73)	95-0146-64
$C_{27}H_{31}NO_{12}S$			
Galactopyranose, 2-(carboxyamino)-2-deoxy-, benzyl ester, 1,3,4-triacetate, 6-p-toluenesulfonate	$CHCl_3$	258(2.81),262(2.85), 267(2.79),273(2.69)	44-3654-64
$C_{27}H_{32}$			
17H-Cyclopenta[a]phenanthrene, 17-(1,5-dimethylhexyl)-1,17-dimethyl-	EtOH	224(4.74),245s(4.46), 264(4.69),297(4.06), 309(4.25),322(4.25), 347(2.94),364(2.70)	5-0152-64D
$C_{27}H_{32}ClNO_4$			
Isohypognavine, hydrochloride	EtOH	230(4.09)	94-1124-64
$C_{27}H_{32}N_2$			
Indeno[1,2-b]azirine, 1-cyclohexyl-6-(cyclohexylimino)-1a,6a-dihydro-1a-phenyl-	isooctane	252(4.25)	35-2510-65
$C_{27}H_{32}N_4O_8$			
Cassamic acid, dehydro-, 2,4-dinitrophenylhydrazone	EtOH	388(4.3)	39-0403-65
$C_{27}H_{32}N_8O_8$			
1-Cyclohexene-1-valeraldehyde, β,2,6,6-tetramethyl-δ-oxo-, bis(2,4-dinitrophenylhydrazone)	n.s.g.	244(4.36),354(4.64)	70-1197-65
$C_{27}H_{32}N_8O_{12}S$			
Glucitol, 1-(6-amino-9H-purin-9-yl)-2-deoxy-2-(2,4-dinitroanilino)-1-S-ethyl-1-thio-, 3,4,5,6-tetraacetate	EtOH	264(4.19),342(4.07)	44-3280-64
$C_{27}H_{32}N_8O_{13}$			
Glucitol, 1-(6-amino-9H-purin-9-yl)-2-deoxy-2-(2,4-dinitroanilino)-1-O-ethyl-, 3,4,5,6-tetraacetate	EtOH	260(4.48),340(4.34)	44-1096-65
$C_{27}H_{32}O_3$			
Androsta-3,5-diene-6-carboxaldehyde, 3-(benzyloxy)-17-oxo-	EtOH	322(4.21)	78-0597-64
Ferruginol, 7-oxo-, benzoate	EtOH	240s(4.12),257(4.14), 293s(3.47)	39-0361-65
$C_{27}H_{32}O_7$			
Mexicanolide	EtOH-NaOH	287(4.50)	77-0162-65
Siphulin, methyl ester, dimethyl ether	EtOH	252(4.45),266(4.37), 292(4.34)	1-1677-65
$C_{27}H_{32}O_8$			
Muconomycin B	n.s.g.	221(4.34),261(4.34)	44-0746-65

Compound	Solvent	λ_{max}(log ε)	Ref.
Verrucarin J	EtOH	196(4.19),219(4.30), 262(4.16)	33-1079-65
$C_{27}H_{32}O_9$			
Verrucarin A, dehydro-	EtOH	260(4.36)	33-0157-65
$C_{27}H_{32}O_{10}$			
Echiodin, penta-O-methyl-	EtOH	260(4.41),310(4.20)	78-3715-65
$C_{27}H_{32}O_{15}$			
Naringenin, 5,7-di-β-D-glucosyl-	EtOH	278(4.23)	22-2937-65
$C_{27}H_{33}NO_4$			
17a-Aza-D-homoandrost-5-en-17-one, N-benzoyl-3β-hydroxy-16-(hydroxymethylene)-	MeOH	245(4.09)	13-0291-64
	MeOH-NaOH	279(4.16)	13-0291-64
$C_{27}H_{33}NO_5$			
Lyfoline, dihydro-O,O-dimethyl-	MeOH	293(3.89)	100-0084-65
Nesodine, dihydro-O-methyl-	MeOH	285(3.59)	100-0084-65
$C_{27}H_{33}N_3O_8$			
Reverin	EtOH	266(4.2),366(4.2)	83-0034-65
$C_{27}H_{34}$			
15H-Cyclopenta[a]phenanthrene, 17-(1,5-dimethylhexyl)-16,17-dihydro-1,17-dimethyl-	EtOH	213(4.45),227(4.26), 257(4.89),282(4.13), 293(4.08),306(4.18), 337(2.75),353(2.60), 364(2.64)	5-0152-64D
$C_{27}H_{34}N_2OS$			
Acrylonitrile, 3-(p-cyclohexylthiophenyl)-2-[p-(diethylaminoethoxy)-phenyl]-	EtOH	358(4.53)	87-0511-64
$C_{27}H_{34}N_2O_3$			
Pregn-4-en-21-al, 17-hydroxy-3,20-dioxo-, 21-phenylhydrazone	MeOH	241(4.42),365(4.32)	44-0521-64
5β-Pregn-16-en-21-al, 3α-hydroxy-11,20-dioxo-, 21-phenylhydrazone	MeOH	245(4.10),385(4.23)	44-0521-64
$C_{27}H_{34}N_4O_5$			
Progesterone, dinitrophenylhydrazone	DMF	395(4.45)	13-0255-64
	DMF-NaOH	483(4.56)	13-0255-64
$C_{27}H_{34}N_8O_{10}$			
Undecanoic acid, 6,10-dimethyl-5,9-dioxo-, bis(2,4-dinitrophenylhydrazone)	$CHCl_3$	363(4.62)	39-2340-65
$C_{27}H_{34}O_3$			
Androst-4-en-3-one, 17β-o-anisoyl-	MeOH	242(4.31),297(3.45)	87-0537-64
Estradiol, 2-(benzyloxymethyl)-17α-methyl-	EtOH	286(3.51)	94-0196-64
$C_{27}H_{34}O_4$			
2H-1-Benzopyran-7-ol, 2,2-dibutyl-3-(p-methoxyphenyl)-4-methyl-, acetate	MeOH	274(3.94),312(3.95)	44-4114-65
16,21-Methanopregna-1,4,21-triene-3,20-dione, 11β-hydroxy-21-cyclopentyloxy-	EtOH	249(4.25)	5-0218-65E

Compound	Solvent	λ_{max}(log ε)	Ref.
$C_{27}H_{34}O_6$			
Mogoltavin	EtOH	217(4.15),322(4.22)	65-1013-64
$C_{27}H_{34}O_7$			
Veprisone, deoxy-	n.s.g.	214(4.10)	78-2985-64
$C_{27}H_{34}O_8S$			
Melicopol, methyl-, 3-p-toluene-sulfonate	EtOH	218s(3.53),223s(3.45), 227s(3.42),236s(4.28), 290(4.35),330(3.64)	12-2021-65
$C_{27}H_{34}O_9$			
Verrucarin A	EtOH	260(4.25)	33-0157-65
$C_{27}H_{34}O_{15}$			
Monotropein, anhydrodihydro-, pentaacetate, methyl ester	EtOH	234(4.08)	94-0888-64
$C_{27}H_{35}NO_{10}$			
Glucopyranosiduronamide, 17-hydroxy-3,11,20-trioxopregna-1,4-dien-21-yl-	EtOH	242(3.97)	94-0450-64
$C_{27}H_{35}N_3O_3$			
Ferruginol, 11-(p-nitrophenylazo)-, methyl ether	MeOH	281(3.87),335(2.62), 490(2.09)	44-2293-64
Progesterone, p-nitrophenylhydrazone	DMF	424(4.54)	13-0255-64
	DMF-NaOH	543(4.71)	13-0255-64
$C_{27}H_{35}N_3O_7$			
Glycine, N-[N-(N-carboxy-3-phenyl-L-alanyl)-L-tyrosyl]-, N-tert-butyl ethyl ester	n.s.g.	278(3.20)	31-0490-64
$C_{27}H_{36}$			
19-Norcholesta-1,3,5(10),6,8,14-hexaene, 1-methyl-	EtOH	252(4.60),259(4.63), 268(4.52),288(4.15), 298(4.25),311(4.16), 332(3.00),349(2.87)	5-0109-64E
Tetracyclo[9.9.5.1^{3,19}.1^{9,13}]heptacosa-1,3(26),9,11,13(27),19-hexaene	hexane	199(4.98),204s(4.83), 217s(4.18),260(2.53), 264s(2.59),266(2.65), 269(2.61),273(2.61), 276(2.42)	39-3160-65
$C_{27}H_{36}F_2O_6$			
Pregna-1,4-diene-3,20-dione, 6α,9-difluoro-11β,17,21-trihydroxy-16α-methyl-, 21-pivalate	EtOH	239(4.21)	13-0339-65
$C_{27}H_{36}N_2O_3$			
5β-Pregnan-21-al, 3α-hydroxy-11,20-dioxo-, 21-phenylhydrazone	MeOH	238(4.08),295(3.63), 350(4.34)	44-0521-64
$C_{27}H_{36}N_2O_4$			
5β-Pregnan-21-al, 3α,17-dihydroxy-11,20-dioxo-, 21-phenylhydrazone	MeOH	241(4.01),298(3.36), 365(4.32)	44-0521-64
$C_{27}H_{36}N_4O_2$			
Acrylonitrile, 2-[p-(diethylaminoacetamido)phenyl]-3-[p-(diethylaminoethoxy)phenyl]-	EtOH-HCl	347.5(4.50)	87-0511-64

Compound	Solvent	λ_{max}(log ε)	Ref.
$C_{27}H_{36}O_3$			
Pregn-4-en-3-one, 20,21-dihydroxy-20-phenyl-	MeOH	241(4.23)	87-0537-64
$C_{27}H_{36}O_4$			
Pregna-3,5,7-trien-20-one, 17α-acetoxy-3-ethoxy-6-methyl-16-methylene-	EtOH	324(4.27)	78-1753-65
$C_{27}H_{36}O_5$			
Gratiogenin, CrO_3 oxidation product	MeOH	247(3.87)	5-0196-64D
Pregna-3,5-diene-6-carboxaldehyde, 17α-acetoxy-16-ethylidene-3-methoxy-	EtOH	320(4.19)	78-0597-64
$C_{27}H_{37}FO_6$			
Dexamethasone, pivalate	EtOH	240(4.18)	13-0339-65
Paramethasone, pivalate	EtOH	244(4.2)	13-0339-65
$C_{27}H_{37}NO_3$			
5β-Pregnane-3,20-dione, 11β-hydroxy-11α-(2-pyridylmethyl)-	EtOH	263(3.59)	44-2095-65
$C_{27}H_{37}NO_5$			
Androst-5-ene-3β,17β-diol, 16α,17α-[3-phenyl-3,1-(2-isoxazolino)]-, diacetate	MeOH	254(4.07)	4-0280-64
$C_{27}H_{37}NO_6$			
Pregn-5-en-20-one, 3β-acetoxy-16α,17α-[3-carbethoxy-3,1-(2-isoxazolino)]-	MeOH	249(3.69)	4-0280-64
	MeOH	249(3.70)	44-1272-65
$C_{27}H_{37}NO_7$			
Androst-5-ene-3β,17β-diol, 16α,17α-[3-carbethoxy-3,1-(2-isoxazolino)]-	MeOH	230(3.77)	4-0280-64
$C_{27}H_{37}NO_{10}$			
Glucosiduronamide, 17-hydroxy-3,11,20-trioxopregn-4-en-21-yl	EtOH	241(3.73)	94-0450-64
$C_{27}H_{37}N_3O_2$			
Acrylonitrile, 2,3-bis[p-(diethylaminoethoxy)phenyl]-	EtOH-HCl	343.5(4.48)	87-0511-64
Acrylonitrile, 3-[o-(diethylaminoethoxyphenyl]-2-[p-(diethylaminoethoxy)-phenyl]-	EtOH-HCl	342.5(4.25)	87-0511-64
Acrylonitrile, 3-[p-(diethylaminoethoxy)phenyl]-2-[p-(3-dimethylamino-2-methylpropoxy)phenyl]-	EtOH-HCl	343(4.44)	87-0511-64
$C_{27}H_{37}N_3O_3$			
Pregnenolone, p-nitrophenylhydrazone	DMF	406(4.38)	13-0255-64
$C_{27}H_{38}$			
19-Norcholesta-1,3,5(10),6,8-pentaene, 1-methyl-	EtOH	233(4.81),279(3.63), 284(3.67),296(3.58), 313(3.05),320(2.89), 328(3.22)	5-0109-64E
$C_{27}H_{38}N_2O_2$			
5β-Pregnan-21-al, 3α-hydroxy-20-oxo-, 21-phenylhydrazone	MeOH	238(4.08),295(3.66), 350(4.34)	44-0521-64

Compound	Solvent	$\lambda_{max}(\log \epsilon)$	Ref.
$C_{27}H_{38}N_2O_3$			
5β-Pregnan-21-al, 3α,11β-dihydroxy-20-oxo-, 21-phenylhydrazone	MeOH	238(4.08),295(3.65), 350(4.33)	44-0521-64
$C_{27}H_{38}N_2O_6$			
Pregn-4-eno[3,2-c]pyrazole, 11β-hydroxy-2'-(2-hydroxyethyl)-16α-methyl-17,20:20,21-bis(methylenedioxy)-	MeOH	277(4.01)	35-1520-64
$C_{27}H_{38}O_3$			
D-Homo-C-nor-5α-furosta-12,14,16,20(22)-tetraene-3β,26-diol	MeOH	216(4.43),255(4.18), 263(4.18),284(3.85), 289(3.82),294(3.86)	44-4220-65
D-Homo-C-nor-5α-spirosta-12,14,16-trien-3β-ol	MeOH	289(3.64)	44-4220-65
$C_{27}H_{38}O_4$			
Pregna-3,5,7-trien-20-one, 17α-acetoxy-3-ethoxy-6,16α-dimethyl-	EtOH	323(4.30)	78-1753-65
$C_{27}H_{38}O_4S_2$			
5β,14α,17α-Card-20(22)-enolide, 3β-acetoxy-15-oxo-, ethylenethio ketal	EtOH	215(4.14)	94-0308-65
5β,14β,17α-Card-20(22)-enolide, 3β-acetoxy-15-oxo-, ethylenethio ketal	EtOH	213(4.24)	94-0308-65
$C_{27}H_{38}O_5$			
Manogenin, 9,11,25,27-tetradehydro-	EtOH	237(4.07)	78-2089-65
Pregna-3,5-diene-6-carboxaldehyde, 3-ethoxy-16α,17α-isopropylidenedioxy-20-oxo-	EtOH	220(4.07),323(4.20)	78-0597-64
$C_{27}H_{38}O_6$			
Digitoxigenin, 16,17-dehydro-21-ethoxy-, 3-acetate	EtOH	212(3.65),276(4.25)	5-0246-65E
Pregna-3,5-diene-11,20-dione, 21-acetoxy-17α-hydroxy-6-methyl-3-propoxy-	EtOH	247.5(4.30)	78-0569-65
$C_{27}H_{38}O_6S$			
Pregn-5-en-20-one, 3β,21-dihydroxy-16α-mercapto-, triacetate	EtOH	232(3.67)	87-0531-64
$C_{27}H_{38}O_7$			
Pregn-4-en-3-one, 11α,20β,21-triacetoxy-	MeOH	240(4.21)	44-2559-64
Pregn-5-en-20-one, 3β,7α,11α-triacetoxy-	EtOH	280(2.45)	70-2016-64
$C_{27}H_{38}O_8$			
17α-Digitoxigenin, 16α,17β-dihydroxy-, 3,16-diacetate	EtOH	223(4.03)	5-0246-65E
Drevogenin, dehydro-, triacetate	EtOH	292(1.50)	33-0857-65
Pregn-4-ene-3,20-dione, 11β,21-dihydroxy-1α,2α,16α,17α-diisopropylidenedioxy-	MeOH	245(4.19)	13-0345-65
Taxinine, diacetyldecinnamoyldihydroisopropylidenoxo-, acetate	EtOH	295(3.56)	95-0404-65

Compound	Solvent	$\lambda_{max}(\log \epsilon)$	Ref.
$C_{27}H_{39}Br$			
5,7,8,14-Anthracholestatetraene, 15-bromo-	isooctane	222(4.33),227(4.34), 270(4.29),306(3.21)	44-4384-65
$C_{27}H_{39}NO$			
8α,10α-Bisnorchola-4,6,20(22)-trien-3-one, 22-(1-piperidino)-	MeOH	220(3.87),288(4.39)	54-0841-65
9β-Bisnorchola-4,6,20(22)-trien-3-one, 22-(1-piperidino)-	MeOH	220(3.91),284(4.43)	54-0889-65
Verarine	C_6H_{12}	263(3.09)	73-2570-64
$C_{27}H_{39}NO_2$			
Androst-4-en-3-one, 17β-hydroxy-17α-(3-piperidinoprop-1-ynyl)-	EtOH	240(4.21)	78-2295-64
Δ^4-Tomatillidine, 3-oxo-	EtOH	240(4.25)	44-0754-65
$C_{27}H_{39}NO_3$			
Isojervine, isomer	EtOH	220(3.88),252s(3.55), 316(2.26)	44-2282-64
$C_{27}H_{39}NO_5$			
Pregna-3,5,9(11)-trien-20-one, 21-acetoxy-6-dimethylaminomethyl-17α-hydroxy-3-methoxy-	EtOH	250.5(4.29)	78-0569-65
$C_{27}H_{39}NO_6$			
Pregna-3,5-diene-11,20-dione, 21-acetoxy-6-dimethylaminomethyl-17α-hydroxy-3-methoxy-	EtOH	249.5(4.29)	78-0569-65
$C_{27}H_{39}NO_9$			
Glucopyranosiduronamide, 17-hydroxy-3,20-dioxopregn-4-en-21-yl	EtOH	240(4.01)	94-0450-64
$C_{27}H_{39}NO_{10}$			
Glucopyranosiduronamide, 11,17-dihydroxy-3,20-dioxopregn-4-en-21-yl	EtOH	244(4.08)	94-0450-64
$C_{27}H_{40}O_2$			
D-Homo-26,C-dinor-5α-furosta-12,14,16-trien-3β-ol	MeOH	290(3.69)	44-4220-65
$C_{27}H_{40}O_3$			
Cholest-4-en-19-oic acid, 6β-hydroxy-3-oxo-, γ-lactone	MeOH	238(4.09)	35-1528-64
Estra-1,3,5(10)-trien-17-one, 3-heptyloxy-, cyclic ethylene acetal	EtOH	281(3.28),288(3.28)	87-0755-64
D-Homo-C-norcholesta-12,14,16-triene-3,16-diol, 22,26-epoxy-	MeOH	287(3.54)	44-4220-65
D-Homo-C-nor-5α,20β-furosta-12,14,16-triene-3β,26-diol	MeOH	289(3.68)	44-4220-65
Isonuatigenin, 3-anhydro-	n.s.g.	234(4.30)	78-0387-64
Nuatigenin, anhydro-	n.s.g.	234(4.22)	78-0387-64
Pregna-5,16-dien-20-one, 6-methyl-3β-(2-tetrahydropyranyloxy)-	MeOH	240(3.94)	78-0357-64
$C_{27}H_{40}O_4$			
Hecogenin, 9-dehydro-	EtOH	238(4.13)	94-0779-64

Compound	Solvent	$\lambda_{max}(\log \epsilon)$	Ref.
Pregna-3,5-dien-20-one, 3-ethoxy-16α,17α-dihydroxy-6-methyl-, cyclic acetal with acetone	EtOH	248(4.29)	78-0569-65
Pregn-5-ene-3,20-dione, 17α-acetoxy-4,4,6,16α-tetramethyl-	EtOH	none	87-0552-64
25D-Spirost-4-en-3-one, 2α-hydroxy-	EtOH	242.5(4.14)	78-3633-65
$C_{27}H_{40}O_5$			
Manogenin, 9-dehydro-	EtOH	237(4.12)	94-0779-64
Pregna-3,5-dien-20-one, 3-ethoxy-6-(hydroxymethyl)-16α,17α-isopropylidenedioxy-	EtOH	251(4.30)	78-0597-64
Pregn-5-en-20-one, 21-acetyl-3β-hydroxy-16α,17α-isopropylidenedioxy-6-methyl-	EtOH	279(4.10)	78-1927-64
25D-Spirost-1-en-3-one, 4α,5α-dihydroxy-	EtOH	230(3.91)	94-0050-65
25D-Spirost-4-en-3-one, 1α,2α-dihydroxy-	EtOH	238(4.06)	94-0050-65
$C_{27}H_{40}O_5S$			
Pregn-5-ene-3β,20β-diol, 16α-mercapto-, triacetate	EtOH	235(3.70)	44-2009-65
$C_{27}H_{40}O_5S_2$			
5α-Pregnan-20-one, 3β-hydroxy-6β,16α-dimercapto-, triacetate	EtOH	234(3.99)	44-2009-65
5α-Pregnan-20-one, 3β-hydroxy-6β,16β-dimercapto-, triacetate	EtOH	234(4.01)	44-2009-65
$C_{27}H_{41}BrCl_2O$			
Cholest-4-en-3-one, 4-bromo-2,2-dichloro-	C_6H_{12}	268(4.39)	39-4992-64
$C_{27}H_{41}NO$			
3,5-Tomatidadiene	MeOH	234(4.40)	5-0187-65A
$C_{27}H_{41}NO_3$			
Androst-4-en-3-one, 17β-acetoxy-6α-(piperidinomethyl)-	EtOH	239.5(4.16)	78-0569-65
3-Aza-A-homo-22β-spirost-4a-en-4-one	EtOH	217(4.40)	2-0522-65
Isojervine, 5α-dihydro-	EtOH	239(3.97),326(2.32)	44-2282-64
Isojervine, 5β-dihydro-	EtOH	236(4.04),337(2.40)	44-2282-64
Isojervine, 5,6-dihydro-	EtOH	238(3.98),333(2.32)	44-0262-64
Isojervine, 8,9-dihydro-	EtOH	312(2.48)	44-2282-64
Isojervinol	EtOH-HCl	311(4.08)	44-2282-64
15H-Naphth[2',1':1,2]indeno[5,4-b]indolizin-15-one, eicosahydro-3,10-dihydroxy-9,12,14a,15b-tetramethyl-	EtOH	239(3.93)	44-2282-64
22β-Spirost-4-en-3-one, oxime	EtOH	234(4.36)	2-0522-65
$C_{27}H_{41}NO_4$			
Cassine, piperonylidene-	EtOH	247(4.00),297(4.01), 338(4.27)	44-0471-64
Pregna-3,5-dien-20-one, 17α-acetoxy-6-dimethylaminomethyl-3-methoxy-	EtOH	251(4.31)	78-0569-65
$C_{27}H_{42}$			
Cholesta-2,6,8(14)-triene	EtOH	247(4.19)	39-1160-65
19-Norcholesta-1,3,5(10)-triene, 1-methyl-	EtOH	266(2.43)	5-0152-64D

Compound	Solvent	λmax(log ε)	Ref.
$C_{27}H_{42}Br_2O$			
Cholest-4-en-3-one, 4,6β-dibromo-	EtOH	272(3.99)	12-1049-65
$C_{27}H_{42}ClNO_2$			
Jervine, N-chlorotetrahydro-	MeOH	252(2.79)	78-0727-65
Tomatid-5-en-3β-ol, N-chloro-	MeOH	266(2.51)	78-0727-65
	dioxan	270(2.47)	78-0727-65
$C_{27}H_{42}Cl_2O$			
Cholest-4-en-3-one, 2,2-dichloro-	C_6H_{12}	256(3.99)	39-4992-64
$C_{27}H_{42}N_2O$			
Cholest-3-en-1-one, 2-diazo-	dioxan	260(4.23),324(3.72)	44-3775-65
$C_{27}H_{42}N_2O_3$			
Tomatid-5-en-3β-ol, N-nitroso-	MeOH	233(3.90),363(1.80)	5-0187-65A
$C_{27}H_{42}O$			
Cholesta-1,4-dien-3-one	EtOH	247(4.2)	94-0050-65
Cholesta-4,6-dien-3-one	EtOH	308(4.18)	12-0661-64
$C_{27}H_{42}O_2$			
Cholest-2-ene-1,7-dione	EtOH	225(3.89)	94-0971-64
Cholest-4-ene-3,6-dione	EtOH	252(4.00)	95-0390-64
5α-Cholest-7-ene-3,6-dione	EtOH	245(4.13)	24-2361-65
Cholest-4-en-3-one, 6β,19-oxido-	MeOH	240(4.17)	31-0500-65
	EtOH	240(4.15)	5-0152-64D
$C_{27}H_{43}Br$			
Cholesta-3,5-diene, 3-bromo-	EtOH	240(4.6)	44-2784-64
$C_{27}H_{43}BrO$			
Cholest-4-en-3-one, 4-bromo-	EtOH	261(4.02)	44-3469-65
5α-Cholest-1-en-3-one, 2-bromo-	EtOH	256(3.97)	44-3498-64
$C_{27}H_{43}BrO_2$			
5α-Cholestan-3-one, 6β-bromo-4α,5-epoxy-	C_6H_{12}	302(1.38)	12-1049-65
5β-Cholestan-3-one, 6β-bromo-4β,5-epoxy-	C_6H_{12}	284(1.90),307(1.93), 317(1.93)	12-1049-65
$C_{27}H_{43}ClO$			
Cholest-4-en-3-one, 2α-chloro-	C_6H_{12}	235(3.31)	39-4992-64
Cholest-4-en-3-one, 4-chloro-	EtOH	258(4.17)	78-0733-65
$C_{27}H_{43}NO$			
Cholesta-3,5-dien-16β-ol, 22,26-imino-	MeOH	234(4.31)	5-0196-65A
$C_{27}H_{43}NO_2$			
Cholest-3-ene-1,2-dione, 2-oxime, anti	EtOH	260(4.03)	44-3775-65
	base	308(4.13)	44-3775-65
isomer	EtOH	205(3.59),225(3.69)	44-3775-65
	base	225(3.58),295(3.84)	44-3775-65
$C_{27}H_{43}NO_3$			
Cholest-4-en-3-one, 6α-nitro-	EtOH	232(4.20)	23-2919-64
	EtOH	234(4.15)	39-2830-64
	n.s.g.	230.0(4.17)	39-3210-64

Compound	Solvent	$\lambda_{max}(\log \epsilon)$	Ref.
Cholest-4-en-3-one, 6β-nitro-	EtOH	233(4.10)	23-2919-64
	EtOH	235(4.06)	39-2830-64
	n.s.g.	230.0(4.08)	39-3210-64
Isojervine, 5α-tetrahydro-	EtOH	310(2.38)	44-2282-64
Isojervine, 5β-tetrahydro-	EtOH	302(2.30)	44-2282-64
Isojervine, 5,6,8,9-tetrahydro-	EtOH	230s(3.36),310(2.35), 320(2.34),332s(2.16)	44-0262-64
$C_{27}H_{44}$			
Cholesta-3,5-diene	hexane	229(4.39),236(4.42), 244(4.24)	94-1415-64
	EtOH	228(4.26),235(4.27), 245(4.13)	44-3469-65
$C_{27}H_{44}BNO_3$			
Androsta-3,5,7-trien-17β-ol, 6-dimethylaminomethyl-3-ethoxy-, propionate, borane adduct	EtOH	323.5(4.27)	78-1753-65
$C_{27}H_{44}BNO_4$			
Pregna-3,5-dien-20-one, 17α-acetoxy-6-dimethylaminomethyl-3-methoxy-, borane adduct	EtOH	256(4.28)	78-0569-65
$C_{27}H_{44}Br_2O_2$			
5α-Cholestan-3-one, 4α,6β-dibromo-5-hydroxy-	$CHCl_3$	276(2.00)	12-1049-65
5α-Cholestan-3-one, 4β,6β-dibromo-5-hydroxy-	$CHCl_3$	283(2.30)	12-1049-65
$C_{27}H_{44}N_2O$			
Cholestan-1-one, 2-diazo-	dioxan	276(3.95)	44-3775-65
$C_{27}H_{44}O$			
Cholest-1-en-3-one	EtOH	232(4.02)	44-3469-65
Cholest-4-en-3-one	EtOH	240(4.22)	23-0464-64
	EtOH	241(4.26)	28-4901-65B
Cholest-5-en-3-one	C_6H_{12}	196(3.83),212s(3.30), 296(1.77)	60-0285-64
	MeOH-HCl	242(4.23)	35-5670-65
	EtOH	198(3.74),217s(3.11), 289(1.94)	60-0285-64
Cholest-17(20)-en-16-one	C_6H_{12}	250(4.06)	35-0269-64
5α-Cholest-1-en-3-one	EtOH	230(4.00)	12-0440-64
	EtOH	230(3.99),242(3.85)	44-3498-64
5α-Cholest-2-en-6-one	EtOH	284(1.67)	39-2349-65
1β,5-Cyclo-5 ,10 -cholestan-2-one	EtOH	212(3.90),285(1.84)	78-2973-65
B-Norcholest-4-en-3-one, 4-methyl-	n.s.g.	247(4.29)	39-6117-65
$C_{27}H_{44}O_2$			
5α-Cholestane-1,3-dione	EtOH	255(4.20)	88-1387-64
	EtOH-NaOH	285(4.28)	88-1387-64
Cholest-1-en-3-one, 7β-hydroxy-	EtOH	230(4.02)	94-0971-64
Cholest-4-en-3-one, 2α-hydroxy-	EtOH	243(4.14)	12-0661-64
	EtOH	243(4.08)	78-0733-65
Cholest-4-en-3-one, 19-hydroxy-	MeOH	242(4.19)	31-0500-65
	EtOH	244(4.20)	5-0152-64D
5α-Cholest-7-en-6-one, 3β-hydroxy-	EtOH	244(4.14)	24-2361-65
B-Nor-5α-cholestan-3-one, 2-(hydroxymethylene)-	n.s.g.	284(3.88)	39-6117-65

Compound	Solvent	λ_{max}(log ε)	Ref.
$C_{27}H_{44}O_3$			
Cholest-1-en-3-one, 4α,5α-dihydroxy-	EtOH	231(3.95)	94-0050-65
Cholest-4-en-3-one, 1α,2α-dihydroxy-	EtOH	241(4.1)	94-0050-65
Cholest-4-en-3-one, 7α,12α-dihydroxy-	EtOH	243(4.23)	37-2396-65
5β-Cholest-9(11)-en-12-one, 3α,25-dihydroxy-	EtOH	239(4.01)	24-2361-65
Clerodolone	n.s.g.	202(3.89)	78-0797-65
$C_{27}H_{44}O_6$			
Ecdysone	EtOH	242(4.09)	24-2361-65
$C_{27}H_{45}BrO$			
Cholestanone, 2α-bromo-	EtOH	280(1.71)	44-1423-65
	dioxan	285(1.48)	44-1423-65
Cholestanone, 4β-bromo-	EtOH	308(2.12)	44-1423-65
	dioxan	312(2.06)	44-1423-65
$C_{27}H_{45}BrO_2$			
Cholestan-3-one, 6β-bromo-5α-hydroxy-	n.s.g.	283(1.30)	22-1538-65
5α-Cholestan-3-one, 6β-bromo-5-hydroxy-	$CHCl_3$	280(1.34)	12-1049-65
$C_{27}H_{45}ClO_2$			
Cholestan-3-one, 6β-chloro-5α-hydroxy-	n.s.g.	284(1.00)	22-1538-65
$C_{27}H_{45}NO$			
7a-Aza-B-homocholest-5-en-7-one	EtOH	221(4.2)	39-3392-64
Cholest-5-en-7-one, oxime	EtOH	235(4.13)	39-3392-64
$C_{27}H_{45}NO_2$			
Cholestane-1,2-dione, 2-oxime, anti	EtOH	236(3.80)	44-3775-65
	base	290(4.09)	44-3775-65
Cholest-4-ene, 4-nitro-	EtOH	262.0(3.43)	39-2601-65
Cholest-5-ene, 6-nitro-	EtOH	261.0(3.44)	39-2601-65
$C_{27}H_{45}NO_3$			
Isojervine, hexahydro-	EtOH	215(3.49)(end absorption)	44-2282-64
$C_{27}H_{45}N_3O_2S$			
1H-Cyclopenta[5,6]naphtho[2,3-d]thiazole-6-propionic acid, 3-(1,5-dimethylhexyl)-8-hydrazino-2,3,3a,4,-5,5a,6,10,10a,10b-decahydro-3a,6-dimethyl-	EtOH	286(3.84)	13-0399-65
$C_{27}H_{46}$			
Cholestene	MeOH	210(2.61)	13-0487-64
$C_{27}H_{46}BNO_3$			
Androsta-3,5-dien-17β-ol, 6-dimethylaminomethyl-3-methoxy-17α-methyl-, 17-propionate, borane adduct	EtOH	255.5(4.32)	78-0569-65
$C_{27}H_{46}N_2O_2$			
Cyclomicrophylline	hexane	196(4.23)	22-0657-65
Cyclomicrophylline B	EtOH	204(3.76)	39-4512-65
Cyclomicrophylline C	EtOH	204(3.75)	39-4512-65
$C_{27}H_{46}O$			
Cholestanone	EtOH	280(1.40)	44-1423-65
	dioxan	287(1.18)	44-1423-65

Compound	Solvent	$\lambda_{max}(\log \epsilon)$	Ref.
$C_{27}H_{46}OS$			
Cholestan-3β-ol, 5β,6β-epithio-	EtOH	263(1.66)	78-0329-65
$C_{27}H_{46}O_2$			
Cholestan-3-one, 5α-hydroxy-	n.s.g.	284(1.30)	22-1538-65
19-Norcholest-9-ene-3α,4α-diol, 5β-methyl-	C_6H_{12}	200(4.13)	13-0111-65
Peniocerol	EtOH	210(3.59),220(3.13) (end absorptions)	39-1160-65
$C_{27}H_{46}O_3$			
3,5-Seco-A-norcholestan-3-oic acid, 5-oxo-, methyl ester	iso-PrOH	292(1.45)	44-1513-65
$C_{27}H_{46}S$			
Cholestane, 2α,3α-epithio-	isooctane	265(1.55)	78-0583-65
Cholestane, 2β,3β-epithio-	isooctane	246(1.58)	78-0583-65
Cholestane, 3β,4β-epithio-	dioxan	259(1.98)	78-0583-65
Cholestane, 5β,6β-epithio-	dioxan	263(2.11)	78-0583-65
5α-Cholestane, 2α,3α-epithio-	EtOH	262(1.68)	78-0329-65
$C_{27}H_{48}O_2$			
Nonadecane, 1-(3,5-dimethoxyphenyl)-	EtOH	273(3.19),280(3.20)	44-0435-64
Resorcinol, 5-heneicosyl-	EtOH	275(3.24),282(3.24)	44-0435-64
$C_{27}H_{48}O_3$			
Cholestane-3β,5α,8α-triol	n.s.g.	none	37-4176-65
1-Nonadecanol, 1-(3,5-dimethoxyphenyl)-	EtOH	274(3.27),280(3.27)	44-0435-64

Compound	Solvent	λ_{max}(log ε)	Ref.
$C_{28}H_{14}O_4$			
1,1'-Bianthraquinone	n.s.g.	255(4.98),275(4.66), 340(4.02)	25-0088-65
2,2'-Bianthraquinone	n.s.g.	255(4.53),272(4.75), 290s(4.71),345(4.12), 350s(4.10)	25-0088-65
$C_{28}H_{14}O_8$			
2,2'-Biquinizaryl	o-chloro-phenol	305(4.30),500(4.36)	56-1251-64
$C_{28}H_{15}ClN_2O_4$			
Naphth[2,3-c]acridan-5,8,14-trione, 6-(m-chlorobenzamido)-	$C_6H_3Cl_3$	604(4.00)	65-1574-64
Naphth[2,3-c]acridan-5,8,14-trione, 6-(o-chlorobenzamido)-	$C_6H_3Cl_3$	601(4.00)	65-1574-64
Naphth[2,3-c]acridan-5,8,14-trione, 6-(p-chlorobenzamido)-	$C_6H_3Cl_3$	605(3.95)	65-1574-64
$C_{28}H_{15}Cl_4N_4O_2P$			
1,4-Naphthoquinone, 2-azido-5,6,7,8-tetrachloro-3-(tri-phenylphosphoranylideneamino)-	$CHCl_3$	263s(3.98),267(3.99), 274(4.03),286(4.17), 298(4.32),319(3.15), 334(3.29),347(3.25)	39-1003-65
2H-Naphtho[2,3-d]triazole-4,9-dione, 5,6,7,8-tetrachloro-2-(triphenyl-phosphoranylideneamino)-	$CHCl_3$	261(4.50),335(4.57)	39-1003-65
$C_{28}H_{15}FN_2O_4$			
Naphth[2,3-c]acridan-5,8,14-trione, 6-(m-fluorobenzamido)-	$C_6H_3Cl_3$	609(3.95)	65-1574-64
Naphth[2,3-c]acridan-5,8,14-trione, 6-(o-fluorobenzamido)-	$C_6H_3Cl_3$	597(4.00)	65-1574-64
Naphth[2,3-c]acridan-5,8,14-trione, 6-(p-fluorobenzamido)-	$C_6H_3Cl_3$	605(3.95)	65-1574-64
$C_{28}H_{15}N_3O_6$			
Naphth[2,3-c]acridan-5,8,14-trione, 6-(m-nitrobenzamido)-	$C_6H_3Cl_3$	604(3.95)	65-1574-64
Naphth[2,3-c]acridan-5,8,14-trione, 6-(o-nitrobenzamido)-	$C_6H_3Cl_3$	600(3.95)	65-1574-64
Naphth[2,3-c]acridan-5,8,14-trione, 6-(p-nitrobenzamido)-	$C_6H_3Cl_3$	606(3.95)	65-1574-64
$C_{28}H_{16}$			
Benzonaphthofluoranthene	benzene	380(3.8),400(3.9), 432(4.4),460(4.5)	12-1138-64
2,3-Benzonaphtho[2',3':11,12]fluor-anthene	benzene	420(4.2),450(4.5), 475(4.6),510(4.7)	12-1138-64
1,2-Benzonaphtho[1",2":4,5]pyrene	dioxan	260(4.68),282(4.67), 292(4.82),304(4.74), 318(4.83),342(4.26), 358(4.50),377(4.62), 409(3.18)	24-0743-65
1,2-Benzonaphtho[1",2":6,7]pyrene	dioxan	270(4.57),285(4.50), 298(4.56),312(4.67), 350(4.01),370(4.04), 386(4.06),395(4.03), 410(4.00)	24-0743-65
Butatriene, bisbiphenylene-	THF	484(4.70)	24-2975-64

Compound	Solvent	$\lambda_{max}(\log \varepsilon)$	Ref.
Phenanthro[9',10':1,2]pyrene	C_6H_{12}	240(4.65),248(4.65), 275(4.48),290(4.42), 301(4.68),314(4.83), 360(4.26),374(4.28), 398(3.45)	78-2107-64
$C_{28}H_{16}Br_2$			
9,9'-Bianthryl, 2,2'-dibromo-	$CHCl_3$	268(5.27),325s(3.53), 340(3.75),357(3.93), 376(4.11),397(4.10)	88-4129-65
$C_{28}H_{16}Cl_2$			
9,9'-Bianthryl, 2,2'-dichloro-	$CHCl_3$	263(5.32),325s(3.60), 340(3.80),357(3.99), 376(4.16),397(4.23)	88-4129-65
$C_{28}H_{16}F_2$			
9,9'-Bianthryl, 2,2'-difluoro-	$CHCl_3$	257(5.16),320s(3.47), 336(3.75),355(3.96), 374(4.15),395(4.25)	88-4129-65
$C_{28}H_{16}INO_4$			
Alkaloid B of Doryphora sassafras	EtOH	251(4.71),310(4.44), 340s(4.37)	100-0237-65
$C_{28}H_{16}N_2O_2$			
Anthraquinone, azine	CH_2Cl_2	233(4.80),260(4.44), 313(4.44),381(4.23)	28-2259-65A
$C_{28}H_{16}N_2O_4$			
Naphth[2,3-c]acridan-5,8,14-trione, 6-benzamido-	$C_6H_3Cl_3$	610(4.00)	65-1575-64
$C_{28}H_{16}N_6O$			
11H-Isoindolo[2,1-a]benzimidazol-11-one, 2,3-bis(2-benzimidazolyl)-	$CHCl_3$	272(4.34),324(4.62), 380s(4.17)	33-2211-64
$C_{28}H_{16}O_2$			
6,13-o-Benzenopentacene-17,20-dione, 6,13-dihydro-	acetone	213(4.9),238(5.0)	70-1470-64
$C_{28}H_{16}O_4$			
2,6-Naphthalenediacrylic acid, 1,5-dihydroxy-β,β'-diphenyl-, di-δ-lactone	MeOH	230(4.55),292(4.19), 330(3.55)	44-0502-64
$C_{28}H_{17}F_9OSi$			
Silane, benzoyltris(p-trifluoromethylphenyl)-	C_6H_{12}	266(4.24),384s(2.12), 401(2.38),418(2.50), 438(2.30)	23-1175-65
$C_{28}H_{17}N$			
Dinaphtho[2,3-b:2',3'-g]carbazole	dioxan	250(4.90),266s(4.66), 310(4.72)	24-2814-65
	$C_6H_3Cl_3$	323(4.78),354(4.16), 370s(3.98),390(4.16), 413(4.48),453(3.56), 480(3.78),510(3.77)	24-2814-65
Dinaphtho[2,3-b:2',3'-h]carbazole	$C_6H_3Cl_3$	365(4.66),378(4.88), 403(4.36),426(4.18), 465(3.26),498(3.37), 535(3.38)	24-0588-65 24-2814-65

Compound	Solvent	λ_{max}(log ϵ)	Ref.
$C_{28}H_{17}NO_2$			
9,9'-Biphenanthrene, 10-nitro-	THF	226(4.75),238(5.00), 255(5.02),273(4.52), 286(4.34),298(4.29)	39-0387-64
$C_{28}H_{17}N_3$			
5H-Dibenzo[b,h]quino[2,3,4-de][1,6]-naphthyridine, 5-phenyl-	EtOH	254(4.6),332(4.4), 380(3.4),402(3.5)	56-1215-65
$C_{28}H_{18}$			
Butatriene, 1,1-diphenyl-4,4-bi-phenylene-	THF	455(4.70)	24-2975-64
Dianthracene, 9,10-dehydro-	CH_2Cl_2	269(3.62),275(3.61), 284(3.48)	35-1389-64
$C_{28}H_{18}Br_2N_2O_2$			
8,9-Dihydroxydiisoquino[3,2-b:2',3'-j]-[3,8]phenanthrolinediium dibromide	H_2O	248(4.52),269(4.54), 300s(4.34),419(4.35), 440(4.43)	44-0252-65
$C_{28}H_{18}N_2$			
9,9'-Azoanthracene	$CHCl_3$	252(5.1),288s(4.3), 281(3.7),400(3.7), 476(4.0),592s(3.7)	28-4799-64A
$C_{28}H_{18}N_5O_4P$			
2H-Naphtho[2,3-d]triazole-4,9-dione, 5-nitro-2-(triphenylphosphoranyl-ideneamino)-	$CHCl_3$	269(4.14),276(4.15), 322(4.45)	39-1003-65
$C_{28}H_{18}O_2$			
6,13-o-Benzenopentacene-17,20-diol, 6,13-dihydro-	EtOH	225(5.0),248(4.9)	70-1470-64
5,14-o-Benzenopentacene-1,4-dione, 4a,5,14,14a-tetrahydro-	acetone	206(4.8),238(5.0), 258(3.68),323(3.40), 338(3.65),354(3.72), 375(3.57)	70-1470-64
6,13-o-Benzenopentacene-17,20-dione, 6,13,15,16-tetrahydro-	acetone	206(4.9),238(5.0)	70-1470-64
$C_{28}H_{18}O_3$			
Coumarone, 2-phenyl-, phenanthrene-quinone adduct	dioxan	243(4.57),281(3.99), 329(3.51)	24-3102-65
$C_{28}H_{19}Br$			
Naphthalene, 2-bromo-1,3,4-triphenyl-	C_6H_{12}	241(4.81),293(4.00)	24-3637-65
$C_{28}H_{19}ClNO_2P$			
1,2-Naphthoquinone, 3-chloro-4-(tri-phenylphosphoranylideneamino)-	$CHCl_3$	282(4.46),339s(3.75)	39-1003-65
$C_{28}H_{19}ClN_2$			
Phenanthridine Red	EtOH	250f(4.8),280f(4.5), 320(4.5),350f(4.1), 490(4.5),510(4.6)	27-0325-65
$C_{28}H_{19}NO_2$			
9,9'-Biphenanthrene, 9,10-dihydro-10-nitro-	THF	225(4.68),299(4.06)	39-2587-65

Compound	Solvent	λ_{max}(log ε)	Ref.
9,9'-Biphenanthrene, 9,10-dihydro-10'-nitro-	THF	249(5.05),256(5.17),277(4.48),288(4.51),299(4.35)	39-2587-65
1,4-Epoxy-5H-2-benzazepin-5-one, 1,4-dihydro-1,3,4-triphenyl-	MeCN	247s(4.24),259(4.26),358s(2.29),370(2.31),380s(2.11)	35-3814-64
$C_{28}H_{19}NO_5$			
Isocoumarin, 4-(N-benzoyl-1,4-dihydro-4-pyridyl)-3-hydroxy-, 3-benzoate	dioxan	235(4.70),268(4.37),313(4.01)	83-0722-65
	$CHCl_3$	235(4.70),268(4.37),313(4.01)	83-0722-65
$C_{28}H_{19}N_3$			
Dibenzo[b,h][1,6]naphthyridine, 7,12-dihydro-6-phenyl-7-(phenylimino)-	EtOH	272(4.7),328(4.3),372(3.6),390(3.8),440(3.9)	56-1215-65
$C_{28}H_{19}N_4O_2P$			
2H-Naphtho[2,3-d]triazole-4,9-dione, 2-(triphenylphosphoranylideneamino)-	$CHCl_3$	270s(4.18),276(4.21),318(4.45)	39-1003-65
$C_{28}H_{20}$			
9,9'-Biphenanthrene, 9,10-dihydro-	dioxan	246(4.69),252(4.94),257(5.06),300(4.20),326(2.57),334(2.72),342(2.54),350(2.70)	39-0387-64
Butatriene, tetraphenyl-	THF	420(4.58)	24-2975-64
Fluorene, 9-(2,3-diphenylcyclopropylidene)-	MeCN	238s(4.74),242(4.77),262s(4.49),296s(4.03),363s(4.48),375(4.50),437(3.60)	35-1608-65
	MeCN-HBF_4	252(4.53),262(4.50),273s(4.47),296(4.50),309(4.50)	35-1608-65
Naphthalene, 1,2,3-triphenyl-	C_6H_{12}	251(4.64)	32-0252-64
Naphthalene, 1,3,6-triphenyl-	EtOH	270(4.75),301(4.20)	78-0195-64
1-Propene, 1,3-bisbiphenylene-2-methyl-	dioxan	234(4.77),250(4.65),260(4.74),276(4.65),290(4.10),303(4.39),321(4.34)	49-0003-64
	DMF-NaOH	345(3.94),572(4.32)	49-0003-64
	DMSO-KO-tert-Bu	302(4.53),347(4.11),572(4.81)	5-0050-65J
$C_{28}H_{20}Br_2O_2$			
2-Furanol, 2,5-bis(p-bromophenyl)-2,5-dihydro-3,4-diphenyl-	EtOH	227(4.24),255(3.83)	78-2181-64
$C_{28}H_{20}Cl_2$			
9,9'-Biphenanthrene, 9,10-dichloro-9,9',10,10'-tetrahydro-	dioxan	268(4.01)	39-0387-64
$C_{28}H_{20}Ge_2I_4$			
1,4-Digermacyclohexa-2,5-diene, 1,1,4,4-tetraiodo-2,3,5,6-tetraphenyl-	n.s.g.	260(4.38),285(4.32),356(3.67)	101-0233-65A
$C_{28}H_{20}N$			
Pyrrolyl, 2,3,4,5-tetraphenyl-	toluene	370(4.2),570(3.7)	24-3124-65

Compound	Solvent	λ_{max}(log ε)	Ref.
$C_{28}H_{20}NO_2P$			
1,2-Naphthoquinone, 4-(triphenyl-phosphoranylideneamino)-	$CHCl_3$	278(4.31),340s(3.78)	39-1003-65
1,4-Naphthoquinone, 2-(triphenyl-phosphoranylideneamino)-	EtOH	280(4.50)	39-1003-65
$C_{28}H_{20}N_2O_2S$			
2-Naphthol, 1-[o-(m-phenoxyphenyl-mercapto)phenylazo]-	HOAc	487(4.25)	7-0821-64
2-Naphthol, 1-[o-(o-phenoxyphenyl-mercapto)phenylazo]-	HOAc	489(4.25)	7-0821-64
2-Naphthol, 1-[p-(m-phenoxyphenyl-mercapto)phenylazo]-	HOAc	495(4.35)	7-0821-64
2-Naphthol, 1-[p-(o-phenoxyphenyl-mercapto)phenylazo]-	HOAc	493(4.34)	7-0821-64
2-Naphthol, 1-[o-(m-phenylmercapto-phenoxy)phenylazo]-	HOAc	493(4.29)	7-0821-64
2-Naphthol, 1-[o-(o-phenylmercapto-phenoxy)phenylazo]-	HOAc	501(4.25)	7-0821-64
2-Naphthol, 1-[p-(m-phenylmercapto-phenoxy)phenylazo]-	HOAc	494(4.24)	7-0821-64
2-Naphthol, 1-[p-(o-phenylmercapto-phenoxy)phenylazo]-	HOAc	500(4.28)	7-0821-64
$C_{28}H_{20}N_2O_4$			
9,9'-Biphenanthrene, 9,9',10,10'-tetrahydro-9,10'-dinitro-	THF	248.5(4.75)	39-2587-65
$C_{28}H_{20}N_2O_4S$			
2-Naphthol, 1-[o-(m-phenoxyphenyl-sulfonyl)phenylazo]-	HOAc	476(4.22)	7-0821-64
2-Naphthol, 1-[o-(o-phenoxyphenyl-sulfonyl)phenylazo]-	HOAc	476(4.24)	7-0821-64
2-Naphthol, 1-[p-(m-phenoxyphenyl-sulfonyl)phenylazo]-	HOAc	484(4.39)	7-0821-64
2-Naphthol, 1-[p-(o-phenoxyphenyl-sulfonyl)phenylazo]-	HOAc	484(4.39)	7-0821-64
2-Naphthol, 1-[o-(m-phenylsulfonyl-phenoxy)phenylazo]-	HOAc	490(4.27)	7-0821-64
2-Naphthol, 1-[o-(o-phenylsulfonyl-phenoxy)phenylazo]-	HOAc	488(4.23)	7-0821-64
2-Naphthol, 1-[p-(m-phenylsulfonyl-phenoxy)phenylazo]-	HOAc	490(4.25)	7-0821-64
2-Naphthol, 1-[p-(o-phenylsulfonyl-phenoxy)phenylazo]-	HOAc	490(4.28)	7-0821-64
$C_{28}H_{20}N_2O_5$			
[9,9'-Biphenanthren]-10-ol, 9,9',10,10'-tetranitro-, nitrate	THF	271.0(4.48)	39-2587-65
$C_{28}H_{20}N_5O_2P$			
2H-Naphtho[2,3-d]triazole-4,9-dione, 5-amino-2-(triphenylphosphoranyl-ideneamino)-	$CHCl_3$	269s(4.24),278s(4.28), 306(4.41),321(4.40)	39-1003-65
2H-1,2,3-Triazoledicarboximide, N-phenyl-2-(triphenylphosphor-anylideneamino)-	$CHCl_3$	270s(4.18),276(4.23), 303(4.37)	39-1003-65
$C_{28}H_{20}N_8O_{17}S_3$			
Thymidine, tris(2,4-dinitrobenzenesul-fenate)	dioxan	264(4.56),310(4.45)	44-2615-64

Compound	Solvent	λ_{max}(log ε)	Ref.
$C_{28}H_{20}O$			
Furan, tetraphenyl-	MeOH	258(4.26),325(4.37)	32-0252-64
$C_{28}H_{20}O_2$			
2-Butene-1,4-dione, 1,2,3,4-tetraphenyl-	MeOH	255(4.50),307s(3.93)	32-0252-64
trans	MeCN	254(4.52)	44-0636-65
p-Dioxin, tetraphenyl-	EtOH	237(4.38)	23-1928-64
$C_{28}H_{20}O_2S$			
1,4-Butanedione, 2,3-epithio-1,2,3,4-tetraphenyl-	MeCN	330(3.83)	44-0636-65
$C_{28}H_{20}O_2S_2$			
Anthraquinone, 1,4-bis(benzylthio)-	benzene	340(4.08),497(3.83)	22-1648-65
$C_{28}H_{20}O_3S_2$			
Anthraquinone, 1-(benzylsulfinyl)-4-(benzylthio)-	benzene	277(4.19),325(4.17), 463(3.58)	22-1648-65
$C_{28}H_{20}O_4S_2$			
Anthraquinone, 1,4-bis(benzylsulfinyl)-	benzene	277(4.18),322(3.96), 436(3.20)	22-1648-65
$C_{28}H_{20}O_6S_2$			
Anthraquinone, 1,4-bis-(benzylsulfonyl)-	benzene	277(4.06)	22-1648-65
$C_{28}H_{20}O_7$			
Khellin, phenanthrenequinone adduct	dioxan	244(4.78),278(4.23), 333(3.60)	24-3102-65
$C_{28}H_{20}Pb$			
Lead, triphenyl(phenylbutadiynyl)-	hexane	209(4.78),220(4.81), 232(4.84),252(3.96), 265(4.29),281(4.50), 298(4.44)	22-3518-65
$C_{28}H_{20}Sn$			
Tin, triphenyl(phenylbutadiynyl)-	EtOH	266(4.34),281(4.52), 299(4.48)	22-0035-65
$C_{28}H_{21}BrN_2$			
5-Benzyl-x-phenylindolizino[1,2-c]-quinolinium bromide	MeOH	260(4.65),310(4.36), 342(3.80),440(4.24)	5-0196-65H
$C_{28}H_{21}N$			
9,9'-Biphenanthrene, 10-amino-9,10-dihydro-	THF	227(4.91),255(4.92), 283(2.54),284(4.44), 296(4.34)	39-2587-65
$C_{28}H_{21}NO$			
Indone, 3-anilino-2-(diphenylmethyl)-	MeOH	265(4.48),445(3.59)	65-0448-64
2H-Pyrrol-2-ol, 2,3,4,5-tetraphenyl-	ether	242(4.41),310s(3.78)	28-3133-65B
$C_{28}H_{21}NO_2$			
Hydroperoxide, 2,3,4,5-tetraphenyl-2H-pyrrol-2-yl-	ether	242(4.34),310s(3.78)	28-3133-65B
$C_{28}H_{21}NO_3$			
[9,9'-Biphenanthren]-10-ol, 9,9',10,10'-tetrahydro-10'-nitro-	THF	269.5(4.35)	39-2587-65

Compound	Solvent	λ_{max}(log ε)	Ref.
$C_{28}H_{21}N_2O_2P$			
1,4-Naphthoquinone, 2-amino-3-(triphenylphosphoranylideneamino)-	C_6H_{12}	262(4.39),303s(4.24), 315(4.26),350s(3.84), 582(3.42)	39-1003-65
$C_{28}H_{22}$			
Anthracene, 9,10-dibenzyl-	$CHCl_3$	265(4.82),343(3.47), 360(3.65),380(3.85), 400(3.85)	39-4396-65
Biphenyl, 2'-methyl-2-(9-phenanthrylmethyl)-	dioxan	256(4.81),287(4.04), 297(4.11),318(2.53), 326(2.47),333(2.61), 342(2.42),350(2.64)	39-0387-64
Butadiene, tetraphenyl-	THF	341(4.57)	24-2975-64
1,3-Butadiene, 1,1,3,4-tetraphenyl-	C_6H_{12}	328(4.28)	24-3637-65
9,10-Ethanoanthracene, 9,10-dihydro-9,10-diphenyl-	n.s.g.	259(3.06),264(3.05), 271(2.87)	23-1754-65
Fluorene, 9-(2,2-diphenylvinyl)-9-methyl-	THF	260(4.44),293(3.83), 305(3.87)	24-2975-64
$C_{28}H_{22}Br_4N_2Si$			
10,10'(5H,5'H)-Spirobiphenazasiline, 2,2',8,8'-tetrabromo-5,5'-diethyl-	$CHCl_3$	321(4.33)	44-3248-65
$C_{28}H_{22}NOP$			
Acetonitrile, p-toluoyl(triphenylphosphoranylidene)-	EtOH	268(3.96),275(3.98), 293(3.95)	65-2205-65
$C_{28}H_{22}NO_2P$			
Acetonitrile, p-anisoyl(triphenylphosphoranylidene)-	EtOH	261(4.17),275(4.13), 295(4.11)	65-2205-65
$C_{28}H_{22}N_2$			
Dibenzo[e,g][1,4]diazocine, 1,2-dimethyl-6,7-diphenyl-	EtOH	255(4.51),327s(3.62)	39-2326-64
$C_{28}H_{22}N_2O_2$			
Nicotinonitrile, 1,2-dihydro-2-oxo-4-(β-phenacylphenethyl)-6-phenyl-	EtOH	246(4.25),356(4.21)	35-5198-65
Syphilobine A	n.s.g.	221(4.67),254(4.64), 275(4.46),321s(3.33), 337(3.47),353(3.51)	78-2885-64
Unknown compound, m. 218-220°	THF	250(4.38),292(3.99), 353(3.94)	44-0043-65
$C_{28}H_{22}O$			
3-Butenophenone, 2,3,4-triphenyl-	C_6H_{12}	332(4.56)	32-0252-64
	n.s.g.	248(4.32),255s(4.20)	35-5046-64
3-Butenophenone, 2,4,4-triphenyl-	n.s.g.	250(4.45),255(4.42)	44-3660-64
Cyclopropane, 1-benzoyl-1,2,2-triphenyl-	n.s.g.	227(4.32),243s(4.2)	35-5046-64
Furan, 4,5-dihydro-2,3,5,5-tetraphenyl-	n.s.g.	245(4.01),303(3.99)	35-5046-64
$C_{28}H_{22}O_2$			
Benzofuran, 2-phenyl-, irradiation product	CH_2Cl_2	283(3.88),292s(3.79)	35-5277-64
5,11-Epoxy-1,4-methano-5H-dibenzo-[a,d]cyclohepten-10(1H)-one, 4,4a,11,11a-tetrahydro-5,11-diphenyl-	C_6H_{12}	248(4.06),288(3.28), 298(3.31),337s(2.06), 349(2.18),365(2.16), 379s(1.87)	35-3814-64

Compound	Solvent	$\lambda_{max}(\log \epsilon)$	Ref.
$C_{28}H_{22}O_4$			
Anthracene, 1,4-dimethoxy-9,10-diphenyl-, 1,4-photooxide	ether	244(4.6),274(4.0), 283(4.0),292s(3.9), 317(3.0),330(3.0)	28-5031-65A
5,8,13,14-Pentaphenetetrone, 6-ethyl-2,3,10,11-tetramethyl-	dioxan	255(4.51),285(4.58)	65-3484-64
$C_{28}H_{22}O_8$			
2,2'-Binaphthyl, 1,1',4,4'-tetra-acetoxy-	dioxan	217(4.75),235(4.82), 247s(4.77),286s(4.18)	12-0218-65
Dianellinone, di-O-methyl-	dioxan	218(4.63),251(4.44), 275s(4.32),365(3.77)	12-0218-65
Isodianellinone, di-O-methyl-	dioxan	236s(4.41),273(4.76), 299s(4.08),332s(3.65), 354s(3.45),480(3.90)	12-0218-65
$C_{28}H_{23}ClN_2O_6$			
1-(o-Hydroxyphenyl)-2-[(1-methyl-2(1H)-quinolylidene)methyl]quinolinium perchlorate, acetate	EtOH	532(4.68)	65-3373-64
1-(o-Hydroxyphenyl)-2-[(1-methyl-4(1H)-quinolylidene)methyl]quinolinium perchlorate, acetate	EtOH	567(3.92)	65-3373-64
$C_{28}H_{23}NO$			
[9,9'-Biphenanthren]-10-ol, 10'-amino-9,9',10,10'-tetrahydro-	THF	255(4.6),341(3.83)	39-2587-65
$C_{28}H_{23}N_3O_3$			
Maleanilic acid, 2'-(2,4-diindol-3-ylethyl)-	MeOH	276s(4.15),282(4.21), 291(4.16)	39-0526-64
	MeOH-acid	274s(4.38),281(4.42), 289(4.36)	39-0526-64
$C_{28}H_{24}$			
Anthracene, 9,10-dibenzyl-9,10-dihydro-	EtOH	208(4.57),260(3.37), 263(3.40),272(3.03)	39-4396-65
Cyclobutane, 1,2,3,4-tetraphenyl-	hexane	250(2.92),256(3.00), 261(3.07),263(3.05), 271(2.92)	88-3937-65
$C_{28}H_{24}Cl_2CoN_4O_8$			
1,10-Phenanthroline, 2,9-dimethyl-, compound with $Co(ClO_4)_2$	n.s.g.	535(1.52)	12-0691-65
$C_{28}H_{24}CoN_6O_6$			
1,10-Phenanthroline, 2,9-dimethyl-, compound with $Co(NO_3)_2$	n.s.g.	505(1.52)	12-0691-65
$C_{28}H_{24}N_2O_4$			
Uridine, 2',3'-didehydro-2',3'-dideoxy-5'-O-trityl-	EtOH	242(3.78),261(3.99)	35-1896-64
$C_{28}H_{24}N_2O_5$			
Uracil, 2,2'-anhydroarabinosyl-5'-O-trityl-	EtOH	248s(3.83)	44-0564-64
$C_{28}H_{24}O$			
3-Buten-1-ol, 1,2,4,4-tetraphenyl-	n.s.g.	255(4.23)	44-3660-64

Compound	Solvent	λ_{max}(log ε)	Ref.
Butyrophenone, 4,4,4-triphenyl-	EtOH	235(4.19),270s(3.24), 314s(1.92)	23-0298-64
$C_{28}H_{24}O_2$			
2-Furanol, tetrahydro-2,3,5,5-tetra-phenyl-	MeCN	243(4.13)	35-5046-64
$C_{28}H_{24}O_4$			
Anthrone, 4β,10β-(benzylidenedioxy)-5-(benzyloxy)-1,4,4aα,9aβ-tetrahydro-	EtOH	223(4.48),255(3.84), 314(3.45)	65-2563-64
4H-Pyran-4-one, 2-methyl-6-styryl-, dimer	EtOH	253(4.36)	65-2766-64
$C_{28}H_{24}O_5$			
1-Acenaphthenone, 2,2-bis-(3,4-dimethoxyphenyl)-	EtOH	236s(4.46),255s(4.28), 280(3.86),319(3.74), 346(3.73)	39-3040-65
$C_{28}H_{25}FN_2O_5$			
Uridine, 2'-deoxy-2'-fluoro-5'-O-trityl-	EtOH	259(3.95)	44-0564-64
$C_{28}H_{25}N$			
Indene, 1-(5-dimethylamino-2,4-penta-dien-1-ylidene)-2,3-diphenyl-	MeCN	259(4.45),300s(3.95), 487(4.69)	24-3331-64
$C_{28}H_{25}N_3$			
Indole, 3,3',3"-methylidyne-tris[2-methyl-	n.s.g.	284(2.91),292(2.89)	28-0609-64A
$C_{28}H_{25}OP$			
Phosphorane, triphenyl(2,4-dimethyl-benzoyl)methylene-	EtOH	275(4.02),302(4.07)	65-2738-64
Phosphorane, triphenyl(α-toluoyl-ethylidene)-	EtOH	265(3.88),275(3.85), 320(3.80)	65-2738-64
$C_{28}H_{25}O_2P$			
Phosphorane, triphenyl(α-anisoyl-ethylidene)-	EtOH	275(3.96),320(3.79)	65-2738-64
$C_{28}H_{26}$			
Butane, 1,2,3,4-tetraphenyl-	EtOH	254(2.79),259(2.77), 262(2.83),265(2.89), 269(2.69)	39-5951-64
p-Quaterphenyl, 2",3',5',6"-tetra-methyl-	n.s.g.	264(4.94)	4-0181-65
$C_{28}H_{26}Fe$			
1,3,5,7,9,11,13-Tetradecaheptaene, 1-ferrocenyl-14-phenyl-	$CHCl_3$	437(5.0)	5-0088-64F
$C_{28}H_{26}Fe_2$			
1,3,5,7-Octatetraene, 1,8-diferrocenyl-	$CHCl_3$	372s(--),389(4.76), 478(3.98)	5-0088-64F
$C_{28}H_{26}N_2O_3$			
Syphilobine F	n.s.g.	274(4.14),280(4.14)	78-2885-64
$C_{28}H_{26}N_2O_5$			
6(1H)-Pyrimidinone, 1-(5-trityl-β-D-ribofuranosyl)-	EtOH	270(3.62)	94-0828-64

Compound	Solvent	λ max(log ε)	Ref.
Uracil, 1-(2-deoxy-5-O-trityl-β-D-lyxosyl)-	EtOH	262(4.03)	87-0385-64
$C_{28}H_{26}N_2O_6$			
Uracil, 5'-O-tritylarabinosyl-	EtOH	261.5(3.99)	44-0564-64
$C_{28}H_{26}N_2O_8$			
6,13-Triphenodioxazinedicarboxylic acid, 3,10-diethoxy-, diethyl ester	DMF	507(4.70),542(4.82)	27-0242-65
$C_{28}H_{26}N_2Si$			
10,10'(5H,5'H)-Spirobiphenazasiline, 5,5'-diethyl-	$CHCl_3$	223(4.74)	44-3248-65
$C_{28}H_{26}N_4O_2$			
p-Anisil, bis(phenylhydrazone)	EtOH	245(4.6),297(4.6), 344(4.7)	28-2113-64B
Uretidine, 3,4-bis(p-methoxyphenyl)-1-phenyl-4-(phenylazo)-	$CHCl_3$	279(4.6),407(2.8)	28-2113-64B
$C_{28}H_{26}N_4O_4S_2$			
5-Pyrimidinecarboxylic acid, 4,4'-dithiobis[6-methyl-2-phenyl-, diethyl ester	EtOH	273(4.77)	44-1115-64
$C_{28}H_{26}N_4O_5$			
Spiro[4,5]decane-1,8-dione, 6,10-diphenyl-, 2,4-dinitrophenylhydrazone	EtOH	249(2.42),253(2.52), 259(2.62),265(2.56), 299(2.09)	44-1409-65
$C_{28}H_{26}OSi$			
Silane, benzoyltris(p-tolyl)-	C_6H_{12}	263(--),404(--), 423(--),446(--)	23-1175-65
$C_{28}H_{26}O_2$			
Cyclopentadiene, 2,3-dibenzoyl-5-(dipropyl-2-cyclopropen-1-ylidene)-	C_6H_{12}	255(4.25),346(4.34)	35-1609-65
	MeOH	260(4.32),361(4.35)	35-1609-65
cation	12N HCl	402(4.37),430(4.36)	35-1609-65
$C_{28}H_{26}O_4$			
Anthracene, 4-oxo-5,9β-di(benzyloxy)-10α-hydroxy-1,4,4aβ,9,9aα,10-hexahydro-	EtOH	258(3.06),276(3.33), 284(3.33)	65-0662-65
$C_{28}H_{26}O_8$			
[2,2'-Binaphthalene]-7,7',8,8'-tetrone, 1,1',6,6'-tetrahydroxy-5,5'-diisopropyl-3,3'-dimethyl-	EtOH	310(4.11),445(4.23), 550(3.97)	44-3617-64
$C_{28}H_{26}O_{10}$			
Averythrin, tetraacetate	EtOH	213(4.39),272(4.50), 348(3.68)	39-3666-65
$C_{28}H_{27}NO_6$			
Malonamic acid, 2-[[5-(benzyloxy)-1,2,3,4-tetrahydro-4-oxo-2-anthryl]-acetyl]-, ethyl ester	n.s.g.	220(4.56),264(4.62), 308(3.70),320(3.61), 372(3.63)	70-0945-64

Compound	Solvent	λ_{max}(log ε)	Ref.
$C_{28}H_{27}NSi$			
Phenazasiline, 5-ethyl-5,10-dihydro-2,8-dimethyl-10,10-diphenyl-	$CHCl_3$	223(4.56)	44-3248-65
$C_{28}H_{27}N_3O_2$			
Hydrouracil, 6-(dimethylamino)-5,5-dimethyl-1,3-di-1-naphthyl-	$CHCl_3$	282(4.15)	44-0812-65
$C_{28}H_{28}CuN_4O$			
Copper, [α-hydroxy-1,2,3,4,5,6,7,8-octamethylporphinato(2-)]-	$CHCl_3$	403(5.24),407s(5.22), 498s(3.74),529(4.09), 563(3.94)	12-1835-65
$C_{28}H_{28}N_2$			
Hydrazine, tetra-p-tolyl-	dioxan	264(4.33),299(4.46)	24-0844-65
$C_{28}H_{28}N_2O_2$			
4(1H)-Pyridone, 2,2'-hexamethylenebis-6-phenyl-	MeOH	239(4.69)	44-4263-65
$C_{28}H_{28}N_2O_4$			
[2,2'-Bipyrrole]-4,4'-dicarboxylic acid, 3,3',5,5'-tetramethyl-, dibenzyl ester	$CHCl_3$	260(4.19)	39-3315-64
[2,2'-Bipyrrole]-5,5'-dicarboxylic acid, 3,3',4,4'-tetramethyl-, dibenzyl ester	$CHCl_3$	325(4.39)	39-3315-64
Hydrazine, tetrakis(p-methoxyphenyl)-	dioxan	264(4.34),300(4.47)	24-0844-65
$C_{28}H_{28}N_4O_5$			
Bicyclo[3.3.1]nonan-9-one, 2-(p-methoxyphenyl)-4-phenyl-	EtOH	227(4.1),280(3.35)	2-0561-65
$C_{28}H_{28}N_6O_6S_2$			
Hydrazine, 1,1'-[dithiobis[(1-methylindole-2,3-diyl)carbonyl]]bis-[1,2-diacetyl-	EtOH	212(4.72),278(4.30), 291(4.29),348s(3.88)	44-0178-64
$C_{28}H_{28}O$			
5,7-Octadien-3-one, 2,2-dimethyl-5,7,8-triphenyl-	EtOH	273(3.92)	5-0183-64H
$C_{28}H_{28}O_4$			
Anthracene, 5,9β-di(benzyloxy)-4α,10α-dihydroxy-1,4,4aβ,9,9aα,10-hexahydro-	EtOH	276(3.42),284(3.41)	65-0662-65
$C_{28}H_{28}O_6$			
2-Anthracenemalonic acid, 5-(benzyloxy)-1,2,3,4-tetrahydro-4-oxo-, diethyl ester	EtOH	219(4.52),237(4.15), 263(4.57),309(3.69), 320(3.58),372(3.64)	65-0662-65
$C_{28}H_{28}O_{10}$			
Averythrin, dihydro-, tetraacetate	EtOH	212(4.47),264(4.63), 342(3.80)	39-3666-65
$C_{28}H_{29}FN_4O_7$			
Pregna-4,6,17(20)-trien-21-oic acid, 4-fluoro-3,11-dioxo-, 3-(2,4-dinitrophenylhydrazone), methyl ester	$CHCl_3$	269(4.21),296(4.10), 309(4.19),392(4.60)	44-2982-64
$C_{28}H_{29}NO_{11}$			
Glucosiduronic acid, 7-acetamidofluoren-2-yl-, methyl ester, triacetate	MeOH	290(4.49)	37-1011-65

Compound	Solvent	λ_{max}(log ϵ)	Ref.
Glucuronic acid, 1-O-(N-fluoren-2-yl-acetamido)-, methyl ester, triacetate	MeOH	276(4.35),303(4.16)	37-1011-65
$C_{28}H_{30}$			
Benzene, p-bis(isopropylstyryl)-	dioxan	247(4.30),362(4.86)	38-2839-64A
Ethane, 1,2-bis(2,3,4-trimethyl-1-naphthyl)-	C_6H_{12}	219s(4.82),233s(5.07), 235(5.09),275s(3.88), 287(4.07),298(4.18), 311(4.08),323(3.32)	44-0963-64
$C_{28}H_{30}N_2$			
Quinoxaline, 6,7-di-tert-butyl-2,3-diphenyl-	EtOH	254(4.71),353(4.25)	35-5281-64
$C_{28}H_{30}N_2OS$			
Acrylonitrile, 3-(p-benzylthiophenyl)-2-[p-(diethylaminoethoxy)phenyl]-	EtOH-HCl	357.5(4.54)	87-0511-64
$C_{28}H_{30}N_4O$			
Porphine, N,α-dihydro-1,2,3,4,5,6,7,8-octamethyl-α-oxo-	5N HCl	406(5.09),522s(3.12), 556(3.79),607(3.80)	12-1835-65
	HOAc	408(4.97),677(4.14)	12-1835-65
	$CHCl_3$	404(5.00),546s(3.66), 583(3.95),630(4.19)	12-1835-65
$C_{28}H_{30}O_2$			
1H-Benz[e]indene-3,7(2H,3aH)-dione, 6β-(2,2-diphenylvinyl)-4,5,5a,6,-8,9,9a,9b-octahydro-3a-methyl-	EtOH	250(4.26)	78-2473-64
$C_{28}H_{30}O_3$			
Dispiro[5.2.5.2]hexadecane-3,7,16-trione, 1,5-diphenyl-	EtOH	253(2.62),260(2.68), 266(2.60),290(2.18)	78-2553-64
$C_{28}H_{30}O_3Si_3$			
Cyclotrisiloxane, 2-methyl-4,4,6,6-tetraphenyl-2-propyl-	$CHCl_3$	265(3.22)	46-1066-65
$C_{28}H_{30}O_6$			
Benzene, p-bis(3,4,5-trimethoxy-styryl)-	dioxan	255(4.16),280(4.18), 369(4.67)	38-2839-64A
2',2'''-Bispiro[cyclopropane-1,5'-[5H]-indene]-3',3''',7',7'''(2'H,2'''H,-6'H,6'''H)-tetrone, 6',6'''-dihydroxy-2',2''',4',4''',6',6'''-hexamethyl-	EtOH	302(4.42)	94-0853-64
	EtOH	302(4.42)	78-1231-65
	EtOH	305(4.11)	78-2671-65
$C_{28}H_{31}NO_3$			
19-Nor-17α-pregna-1,3,5(10)-trien-20-yn-17β-ol, 3-methoxy-, carbanilate	MeOH	232(4.24),277(3.29)	4-0207-65
$C_{28}H_{31}NO_4$			
Isoxazolino[17,16-d]pregn-4-ene-3,11,20-trione, 3'-phenyl-	n.s.g.	212(4.33),218(4.28), 242(4.38)	5-0139-64G
$C_{28}H_{31}N_2P$			
Phosphorane, [(2-bornylidene)hydra-zono]triphenyl-	$CHCl_3$	256s(3.68),260(3.70), 266(3.70),273(3.63), 290s(3.46)	44-0417-65
Phosphorane, triphenyl[1,3,3-trimethyl-2-norbornylidene)hydrazono]-	$CHCl_3$	258s(3.79),264s(3.76), 272s(3.68),298(3.60)	44-0417-65

Compound	Solvent	λ_{max}(log ε)	Ref.
$C_{28}H_{31}N_3O_{10}$			
D-Fructosone, pentaacetyl-, 1-diphenylhydrazone 2-oxime	EtOH	237(4.15),349(4.37)	24-0725-64
$C_{28}H_{32}N_2O$			
Phenanthro[1',2':5,6]pentaleno[1,2-b]-quinoxalin-2(3H)-one, tetradecahydro-4a,6a-dimethyl-	EtOH	243(4.63),324(4.09)	5-0218-65E
$C_{28}H_{32}O_4Si_4$			
Cyclotetrasiloxane, 1,1,3,3-tetramethyl-5,5,7,7-tetraphenyl-	$CHCl_3$	265(3.17)	46-1066-65
Cyclotetrasiloxane, 1,1,5,5-tetramethyl-3,3,7,7-tetraphenyl-	$CHCl_3$	265(3.17)	46-1066-65
$C_{28}H_{32}O_6$			
12H-Benzo[b]xanthen-12-one, 1-heptyl-3,8,10,11-tetramethoxy-	EtOH	253(4.62),272(4.73), 341(4.00),389(3.85)	1-1677-65
Pregna-1,4-diene-3,11-dione, 17α,20α,21-trihydroxy-, 21-benzoate	EtOH	233(4.42)	78-0179-65
5H-Tribenzo[a,d,g]cyclononene, 2-ethoxy-10,15-dihydro-3,7,8,12,13-pentamethoxy-	EtOH	204(4.86),233(4.52), 292(4.00)	39-1685-65
$C_{28}H_{32}O_9$			
Chaparrol, dihydro-, tetraacetate	EtOH	244(4.11),276(3.01)	23-2996-65
$C_{28}H_{32}O_{10}$			
Dihydrocoleon-γ-lactone, leucoacetate	EtOH	241(5.10),288(3.95), 298(3.91),322(3.36), 337(3.38)	33-0471-65
Glaucanol, tetraacetate	EtOH	215(4.43),244(4.00)	22-1818-64
$C_{28}H_{33}BrO_7$			
Siphulin, bromo-, methyl ester, trimethyl ether	EtOH	250s(4.39),290(4.31)	1-1677-65
$C_{28}H_{33}IN_2O_3$			
Ajmalol B, benzoate, methiodide	EtOH	224(4.45),285(3.58)	33-1349-65
$C_{28}H_{33}NO_3$			
Isoxazolino[17,16-d]pregn-4-ene-3,20-dione, 3'-phenyl-	MeOH	245(4.34)	4-0280-64
	n.s.g.	212(4.23),219(4.18), 245(4.34)	5-0139-64G
17α-Pregn-4-en-20-yn-3-one, 17β-hydroxy-, carbanilate	MeOH	238(4.16)	4-0207-65
Spiro[androst-4-ene-17,5'(1'β)-oxazolidine]-2',3-dione, 4'-methylene-3'-phenyl-	MeOH	232(4.32)	4-0207-65
$C_{28}H_{33}NO_4$			
5-Androsten-3-one, 17β-acetoxy-16α,17α-[3-phenyl-3,1-(2-isoxazolino)]-	MeOH	245(4.42)	4-0280-64
$C_{28}H_{33}NO_7$			
Sinine	EtOH	292.5(3.85)	100-0015-64
$C_{28}H_{33}N_9O_{14}$			
Pyrrolidine, 2,2'-pyrrole-2,5-diylbis[5,5-dimethyl-, dipicrate	EtOH	223(4.04)	23-1073-64

Compound	Solvent	λ max(log ε)	Ref.
$C_{28}H_{34}N_2O_4$			
Pregn-4-eno[3,2-c]pyrazole, 11β,17,21-trihydroxy-20-oxo-2'-phenyl-	MeOH	260(4.20)	35-1520-64
$C_{28}H_{34}N_2O_8S$			
1-(5,6-Dihydro-8,9-dimethoxy-2-methyl-pyrrolo[2,1-a]isoquinolin-3-yl)-3,4-dihydro-6,7-dimethoxy-2-methyliso-quinolinium methyl sulfate	EtOH	300(4.3),370(3.9), 450(4.07)	94-0775-65
$C_{28}H_{34}N_4Ni_2O_4$			
Nickel, [1,1,2,2-ethanetetracarboxalde-hydato(2-)]bis[N-ethyl-7-(ethylimino)-1,3,5-cycloheptatrien-1-ylaminato]di-	$CHCl_3$	279(4.70),319(4.52), 338s(4.48),412(4.44), 460s(4.14),499(4.04), 655(2.77)	44-3046-64
$C_{28}H_{34}N_4O_7$			
5α-Pregn-8-ene-20-carboxylic acid, 3,11-dioxo-, 3-(2,4-dinitrophenylhydra-zone)	$CHCl_3$	363(4.38)	39-0156-65
$C_{28}H_{34}N_4O_8$			
Cassamic acid, dehydro-, methyl ester, 2,4-dinitrophenylhydrazone	EtOH	392(4.3)	39-0403-65
$C_{28}H_{34}O_4$			
Estra-1,3,5(10)-trien-17β-ol, 15β-(benzyloxy)-3-methoxy-, acetate	MeOH	222(3.94),278(3.34), 288(3.30)	44-0064-64
$C_{28}H_{34}O_5$			
Pregn-4-ene-3,11-dione, 17α,21-benzyl-idenedioxy-20α-hydroxy-	EtOH	239(4.18)	78-0179-65
$C_{28}H_{34}O_7$			
Siphulin, methyl ester, trimethyl ether	EtOH	250s(4.49),286(4.40)	1-1677-65
$C_{28}H_{34}O_{12}$			
Renifolin, pentaacetate	EtOH	276(2.81)	94-0533-64
$C_{28}H_{34}O_{15}$			
Chalcone, 2',4',6'-trihydroxy-4-meth-oxy-, 2',4'-di-β-D-glucopyranoside	EtOH	368(4.48)	22-2937-65
Isosakuranetin, 5,7-β-D-diglucosyl-	EtOH	280(4.04)	22-2937-65
	EtOH	280(4.04)	28-0402-64B
$C_{28}H_{34}O_{16}$			
Hesperetin, 5,7-di-β-D-glucosyl-	EtOH	280(4.10)	22-2937-65
$C_{28}H_{35}NO_2$			
17α-Pregn-4-ene-3,20-dione, 17β-(benzylideneamino)-	MeOH	248(4.54),288(3.34)	44-0579-65
$C_{28}H_{35}NO_3$			
Pregn-4-en-3-one, 20ξ-hydroxy-16α,17α-[3-phenyl-3,1-(2-isoxazolino)]-	MeOH	245(4.34)	4-0280-64
Pregn-5-en-20-one, 3β-hydroxy-16α,17α-[3-phenyl-3,1-(2-isoxazolino)]-	MeOH	262(4.11)	4-0280-64

Compound	Solvent	λ_{max}(log ε)	Ref.
$C_{28}H_{35}NO_7$			
Erythroskyrine, acetate	EtOH	260(4.14),402(4.65)	94-1240-65
$C_{28}H_{35}N_3O_4$			
Pregn-4-eno[3,2-d]-2'H-1',2',3'-triazole, 11β,17α,21-trihydroxy-16α-methyl-20-oxo-2'-phenyl-	MeOH	304(4.52),310(4.51)	87-0584-64
$C_{28}H_{36}N_2O_4$			
Hydrocinnamic acid, tetradecahydro-10a,12a-dimethyl-1H-cyclopenta[7,8]-phenanthro[2,3-c][1,2,5]oxadiazol-1-yl ester, N-oxide	EtOH	208(3.98),264(3.84)	94-1445-65
$C_{28}H_{36}N_4O_7$			
5α-Pregnane-20-carboxylic acid, 3,11-dioxo-, 2,4-dinitrophenylhydrazone	$CHCl_3$	367.5(4.37)	39-0156-65
$C_{28}H_{36}O_3$			
Pregn-4-en-21-al, 20-(p-methoxyphenyl)-3-oxo-	MeOH	240(4.32)	87-0537-64
$C_{28}H_{36}O_4S_2$			
Estra-1,3,5(10)-triene, 17β-[(2-hydroxy)thio]-3-methoxy-, p-toluenesulfonate	EtOH	225(4.29),278(3.53), 287(3.44)	13-0557-65
$C_{28}H_{36}O_5$			
Ramanone, benzoyl-	EtOH	233(4.10),278(3.13)	94-1332-65
$C_{28}H_{36}O_5S$			
Estra-1,3,5(10)-triene, 17β-(2-hydroxyethoxy)-3-methoxy-, p-toluenesulfonate	EtOH	224(4.31),274(3.30), 287(3.31)	13-0557-65
$C_{28}H_{36}O_6$			
Pregna-1,4-diene-3,11-dione, 17α,20α,21-trihydroxy-, 17,21-acetal with cyclopentanone, 20-acetate	EtOH	240(4.16)	78-0179-65
Withaferin A, didehydro-	EtOH	223(4.17)	35-5805-65
$C_{28}H_{36}O_7$			
Resibufogenin	EtOH	300(3.71)	95-1092-65
$C_{28}H_{36}O_{11}$			
Chaparrin, tetraacetate	$CHCl_3$	308(1.58)	23-2996-65
$C_{28}H_{37}FO_8$			
Pregn-4-ene-3,20-dione, 6α-fluoro-2β,16α,17α,21-tetrahydroxy-, 2,21-diacetate, cyclic 16,17-acetal with acetone	EtOH	236(4.18)	13-0057-65
$C_{28}H_{37}NO_2$			
17α-Pregn-5-en-20-one, 17β-(benzylideneamino)-3β-hydroxy-	MeOH	249(4.31),281(3.31)	44-0579-65
$C_{28}H_{37}N_3O_2$			
Acrylonitrile, 3-[p-(diethylaminoethoxy)phenyl]-2-[p-(2-piperidinoethoxy)phenyl]-	EtOH-HCl	342.5(4.43)	87-0511-64

Compound	Solvent	$\lambda_{max}(\log \epsilon)$	Ref.
$C_{28}H_{38}N_2OS$			
Acrylonitrile, 2-[p-(diethylaminoethoxy)phenyl]-3-(p-heptylthiophenyl)-	EtOH-HCl	360(4.53)	87-0511-64
$C_{28}H_{38}O_2$			
Ergosta-4,6,8(9),14,22-pentaen-3-one, 4-hydroxy-	EtOH	208(4.12),270(4.30), 371(3.97),623(0.71)	39-6991-65
	NaOH	229(4.09),287(4.22), 348(4.04)	39-6991-65
Unknown dimer	MeOH	304(3.80),317(3.96), 335(3.89)	5-0062-65D
$C_{28}H_{38}O_3$			
C,26-Dinor-D-homo-5α-furosta-12,14,16,20(22)-tetraen-3β-ol, acetate	MeOH	255(4.18),263(4.18), 283(3.79),289(3.76), 294(3.81)	44-4220-65
5β-Pregnan-11-one, 21-benzylidene-3α,20β-dihydroxy-	MeOH	252(4.32)	44-3158-64
$C_{28}H_{38}O_4$			
5β-Pregnane-3,20-dione, 11β-hydroxy-11-(o-methoxyphenyl)-	EtOH	271(3.30),278(3.28)	44-2095-65
Pregn-4-en-3-one, 20,21-dihydroxy-20-(o-methoxyphenyl)-	MeOH	227(4.16),242(4.19), 277(3.30)	87-0537-64
$C_{28}H_{38}O_5S$			
Pregnane-3,20-dione, 12α-hydroxy-, p-toluenesulfonate	isooctane	224(4.08)	23-0189-64
$C_{28}H_{38}O_6$			
Pregna-3,5-dien-20-one, 3β-acetoxy-21-acetyl-16α,17α-isopropylidenedioxy-	EtOH	235(4.28),279(4.11)	78-1927-64
Withaferin A	EtOH	214(4.24)	35-5805-65
	EtOH	214(4.25)	44-1774-65
$C_{28}H_{39}Br_3O$			
18-Norergosta-8,11,13-trien-3-one, 2α,22,23-tribromo-12-methyl-	EtOH	222s(4.17),262s(2.51), 267(2.53),275s(2.45)	78-0929-64
$C_{28}H_{39}FO_9$			
5α-Pregnane-3,20-dione, 9α-fluoro-16α,17α-isopropylidenedioxy-11β,21-dihydroxy-, 21-(hydrogen succinate), sodium salt	H_2O	238(0.6)	13-0247-65
$C_{28}H_{39}NO_4$			
Androsta-3,5-diene-4,6-dicarboxaldehyde, 17β-hydroxy-3-pyrrolidinyl-, propionate	EtOH	230(4.22),315s(--), 343(4.00),392(4.08)	88-1839-64
$C_{28}H_{40}Br_2O$			
18-Nor-5α-ergosta-8,11,13-trien-3-one, 22,23-dibromo-12-methyl-	EtOH	222s(4.05),261(2.43), 268(2.54),275s(2.40)	78-0929-64
$C_{28}H_{40}Br_2O_2$			
18-Nor-5α-ergosta-8,11,13-trien-7-one, 22,23-dibromo-3β-hydroxy-12-methyl-	EtOH	219(4.42),264(4.10), 309(3.45)	78-0929-64
$C_{28}H_{40}N_2O$			
5α-Pregnan-20-one, 3β-(benzylideneamino)-, oxime	EtOH	247(4.27)	22-0761-64

Compound	Solvent	λ_{max}(log ε)	Ref.
$C_{28}H_{40}O_2$			
Diphenoquinone, 3,3',5,5'-tetra-tert-butyl-	CCl_4	400(3.95)	70-1717-64
Ergosta-4,6,8(14),22-tetraen-3-one, 4-hydroxy-	EtOH	205(3.96),262(3.82), 370(4.32)	39-6991-65
	NaOH	225(4.56),277(3.82), 390(4.16)	39-6991-65
$C_{28}H_{40}O_3$			
Diosgenin, 6-methylene-	EtOH	260(4.05)	78-0597-64
Ergosta-4,6,8(9),22-tetraen-3-one, 6,14α-dihydroxy-	EtOH	206(3.89),240(3.91), 266(3.95),375(4.19)	39-6991-65
	NaOH	263(3.46),444(4.05)	39-6991-65
$C_{28}H_{40}O_4$			
Androsta-3,5-diene-6-methanol, 3-ethoxy-17α-(1-propynyl)-, propionate	EtOH	251(4.24)	78-0597-64
Androst-4-en-3-one, 6α,7α-epoxy-17β-hydroxy-, cyclohexanepropionate	EtOH	241(4.17)	24-3363-64
$C_{28}H_{40}O_5$			
7H-Indeno[2,1-a]phenanthren-7-one, octadecahydro-2,6-dihydroxy-4a,6a,9-trimethyl-, diacetate	MeOH	235(4.10)	44-0604-64
isomer	MeOH	234(4.09)	44-0604-64
Withaferin A, deoxydihydro-	EtOH	226(3.90)	44-1774-65
$C_{28}H_{40}O_6$			
Pregn-5-en-20-one, 3β-acetoxy-21-acetyl-16α,17α-isopropylidenedioxy-	pH 13	300(4.38)	78-1927-64
	EtOH	279(4.12)	78-1927-64
Tauranin, deoxydihydro-, leucotriacetate	EtOH	267(2.65)	94-0796-64
Withaferin A, dihydro-	EtOH	210(4.00)	44-1774-65
$C_{28}H_{40}O_9$			
Taxinol, tetraacetyl-	EtOH	210(3.36),271(3.54)	95-0404-65
$C_{28}H_{41}BrO_9$			
Taxinol, tetraacetylbromodihydro-	EtOH	280(2.58)	95-0404-65
$C_{28}H_{41}Br_2Cl$			
18-Norergosta-8,11,13-triene, 22,23-dibromo-3ξ-chloro-12-methyl-	EtOH	215(4.40),258(2.68), 267(2.68),276(2.60)	78-0929-64
$C_{28}H_{41}ClO_4S_3$			
2'-Grisen-3-one, 2',4',4'-tris(butylthio)-7-chloro-4,6-dimethoxy-6'-methyl-	EtOH	289.5(4.42)	87-0705-64
$C_{28}H_{41}NO_3$			
Isojervine, N-methyl-	EtOH	245s(3.64),327(2.54)	44-2282-64
$C_{28}H_{41}NO_4$			
Pregna-3,5-dien-20-one, 17α-acetoxy-6-dimethylaminomethyl-3-methoxy-16-methylene-	EtOH	249.5(4.32)	78-0569-65
$C_{28}H_{41}NO_5$			
Pregna-3,5-diene-11,20-dione, 6-dimethylaminomethyl-17α,21-dihydroxy-3-methoxy-, cyclic acetal with acetone	EtOH	249.5(4.30)	78-0569-65

Compound	Solvent	λ_{max}(log ϵ)	Ref.
$C_{28}H_{41}NO_6$			
Pregna-3,5-dien-11-one, 6-dimethyl-aminomethyl-3-ethoxy-17α,20:20,21-bis(methylenedioxy)-	EtOH	250(4.29)	78-0569-65
$C_{28}H_{42}Br_2O$			
18-Norergosta-8,11,13-trien-3β-ol, 22,23-dibromo-12-methyl-	EtOH	268(2.60),276(2.42)	78-0929-64
$C_{28}H_{42}O$			
9β-Ergosta-4,6,22-trien-3-one	MeOH	285(4.41)	54-1069-64
9β-Ergosta-4,7,22-trien-3-one	EtOH	244(4.12)	54-1069-64
9β-Ergosta-5,7,22-trien-3-one	MeOH	274(3.98),284(3.98)	54-1069-64
$C_{28}H_{42}O_2$			
Ergosta-4,6,22-trien-3-one, 4-hydroxy-	EtOH	228(4.23),288(4.16)	39-6991-65
	NaOH	220(3.88),257(4.24), 300(4.00)	39-6991-65
$C_{28}H_{42}O_3$			
9β-Ergosta-7,22-diene-3,6-dione, 5α-hydroxy-	EtOH	250(4.19)	39-2054-65
9α-Lumista-7,22-diene-3,6-dione, 5β-hydroxy-	EtOH	249(4.20)	39-2054-65
5β,25D-Spirostan-3-one, 6-methylene-	EtOH	279(1.46)	78-3185-65
$C_{28}H_{42}O_4$			
Androst-4-en-3-one, 7α,17β-dihydroxy-, 17-(cyclohexanepropionate)	EtOH	242(4.20)	24-3363-64
Cholest-5-en-7-one, 3β,4β-dihydroxy-, 3,4-carbonate	EtOH	225(4.04)	24-2383-65
Pregn-4-en-3-one, 3-ethoxy-17α,20α,21-trihydroxy-, cyclic 17,21-acetal with cyclopentanone	EtOH	241(4.29)	78-0179-65
25D-Spirost-4-en-3-one, 2α-methoxy-	EtOH	241.6(4.16)	78-3633-65
$C_{28}H_{43}NO$			
5α-Cholest-1-ene-1-carbonitrile, 3-oxo-	EtOH	236(4.04)	88-1387-64
$C_{28}H_{43}NO_3$			
Androsta-3,5-dien-17β-ol, 3-methoxy-6-(piperidinomethyl)-, acetate	EtOH	250(4.29)	78-0569-65
$C_{28}H_{43}NO_4$			
Pregna-3,5-dien-20-one, 17α-acetoxy-6-dimethylaminomethyl-3-ethoxy-	EtOH	250.5(4.31)	78-0569-65
$C_{28}H_{44}O$			
Cholesta-4,6-dien-3-one, 4-methyl-	EtOH	292(4.36)	39-2633-64
Cholest-4-en-3-one, 6-methylene-	MeOH	260(4.05)	39-2285-64
5α-Cholest-1-en-3-one, 4-methylene-	EtOH	245(4.0),346(1.77)	44-2925-65
Ergosta-7,22-dien-3-one	EtOH	206(3.70),210(3.65)	39-1142-64
9β-Ergosta-4,22-dien-3-one	MeOH	247(4.14)	54-1069-64
$C_{28}H_{44}O_3$			
Cholest-4-en-19-oic acid, 3-oxo-, methyl ester	EtOH	242(4.20)	5-0152-64D
9β-Ergosta-7,22-dien-6-one, 3β,5α-dihydroxy-	EtOH	252(4.09)	39-2054-65
A-Nor-10α-cholest-3-ene-3-carboxylic acid, 2-oxo-, methyl ester	EtOH	244(4.11)	78-2973-65

Compound	Solvent	λ_{max}(log ε)	Ref.
Pregn-5-en-3-one, 17α,20β-isopropyl-idenedioxy-4,4,6,16α-tetramethyl-	EtOH	none	87-0106-64 87-0552-64
$C_{28}H_{44}O_5$			
p-Benzoquinone, 2,5-dihydroxy-3-methyl-6-(14-nonadecenyl)-, acetate	EtOH	275(4.0),410(2.8)	94-0511-65
$C_{28}H_{45}BrO$			
Cholest-4-en-3-one, 6β-bromo-4-methyl-	EtOH	262(4.09)	39-2633-64
	EtOH	262(4.11)	39-2476-65
$C_{28}H_{45}FO$			
Cholesta-3,5-diene, 6-fluoro-3-methoxy-	EtOH	239(4.35)	23-2919-64
$C_{28}H_{45}NO$			
Cholest-5-ene-6-carbonitrile, 3β-hydroxy-	MeOH	220(4.10)	22-1538-65
$C_{28}H_{45}NOS$			
Thiazolo[5',4':2,3]-5α-cholest-2-ene, 2'-oxo-	EtOH	251(3.72)	78-0329-65
Thiocyanic acid, 3-oxo-5α-cholest-an-2α-yl ester	EtOH	231(3.67)	78-0329-65
$C_{28}H_{45}NO_2$			
Cholestane-6β-carbonitrile, 5α-hydroxy-3-oxo-	n.s.g.	283(1.34)	22-1538-65
$C_{28}H_{45}NO_2S$			
1H-Cyclopenta[5,6]naphtho[2,3-d]thia-zole-6-propionic acid, 3-(1,5-di-methylhexyl)-2,3,3a,4,5,5a,6,10,10a,-10b-decahydro-3α,6,8-trimethyl-	EtOH	251(3.68)	13-0399-65
$C_{28}H_{45}NO_3$			
Cholesta-3,5-diene, 3-methoxy-6-nitro-	EtOH	233(4.15)	23-2919-64
$C_{28}H_{46}$			
Cholesta-3,5-diene, 3-methyl-	EtOH	232(4.26),239(4.29), 248(4.16)	23-2695-64
	n.s.g.	232(4.26),239(4.29), 248(4.16)	23-2153-64
Cholesta-3,5-diene, 4-methyl-	C_6H_{12}	232s(4.20),239(4.26), 247(4.04)	39-2633-64
	EtOH	232(4.20),238(4.20), 245s(4.10)	39-1850-64
Cholesta-3,5-diene, 6-methyl-	EtOH	236(4.31),243(4.32), 251(4.19)	78-0559-65
$C_{28}H_{46}BNO_3$			
Pregna-3,5,17(20)-trien-21-oic acid, 6-dimethylaminomethyl-3-ethoxy-, ethyl ester, borane adduct	EtOH	224(4.31),255(4.34)	78-0569-65
$C_{28}H_{46}BNO_4$			
Pregna-3,5-dien-20-one, 17α-acetoxy-6-dimethylaminomethyl-3-ethoxy-, borane adduct	EtOH	256(4.34)	78-0569-65

Compound	Solvent	λ max(log ε)	Ref.
$C_{28}H_{46}N_2$			
Buxenine-G, isopropylideneimine	EtOH	230s(--),238(4.39), 247(4.43),256(4.23), 280s(--)	88-3145-64
Norbuxamine, N-isopropylidene-	MeOH	230s(4.30),238(4.45), 246(4.48),254(4.28), 280s(2.06)	33-0968-64
$C_{28}H_{46}N_2O_2$			
5α-Pregn-8-ene-7,11-dione, 3β,20α-bis-(dimethylamino)-4,4,14α-trimethyl-	EtOH	270(3.8)	28-4139-65B
$C_{28}H_{46}O$			
Cholesta-3,5-diene, 3-methoxy-	EtOH	239(4.30)	23-0079-64
Cholesta-4,6-dien-3β-ol, 6-methyl-	EtOH	242(4.35)	39-3106-64
A-Homo-B-nor-5β-cholest-3-en-4a-one, 5-methyl-	EtOH	226.5(3.78)	78-0559-65
A-Norcholest-3-ene, 3-acetyl-	EtOH	257.5(4.14)	39-2633-64
$C_{28}H_{46}OS_2$			
5α-Cholestan-2β-ol, 3α-mercapto-, cyclic O,S-dithiocarbonate	dioxan	280(4.2),370(2.0)	78-1581-65
5α-Cholestan-2β-ol, 3β-mercapto-, cyclic O,S-dithiocarbonate	dioxan	232(3.69),283(4.25), 373(1.91)	78-0583-65
	dioxan	232(3.70),267s(3.9), 283(4.23),374(1.91)	78-1581-65
5α-Cholestan-3α-ol, 2α-mercapto-, cyclic O,S-dithiocarbonate	dioxan	229(3.68),283(4.22), 375(1.87)	78-1581-65
5α-Cholestan-3α-ol, 2β-mercapto-, cyclic O,S-dithiocarbonate	dioxan	282(4.24),370(2.00)	78-1581-65
5α-Cholestan-3α-ol, 4α-mercapto-, cyclic O,S-dithiocarbonate	dioxan	281(4.01),369(1.73)	78-1581-65
5α-Cholestan-3β-ol, 2α-mercapto-, cyclic O,S-dithiocarbonate	dioxan	227(3.47),247(3.72), 282(4.15),380(1.89)	78-0583-65
	dioxan	227(3.72),272s(4.0), 282(4.15),380(1.89)	78-1581-65
	dioxan	229(3.68),283(4.22), 375(1.87)	78-0583-65
5α-Cholestan-3β-ol, 4α-mercapto-, cyclic O,S-dithiocarbonate	dioxan	284(4.26),374(2.00)	78-1581-65
$C_{28}H_{46}O_2$			
Cholestan-3-one, 2-(hydroxymethylene)-	EtOH	283(3.93)	88-2161-64
	EtOH-NaOH	314(4.27)	88-2161-64
Cholest-4-en-3-one, 6β-hydroxy-6α-methyl-	EtOH	236(4.08)	78-0559-65
Clerosterol	EtOH	203.6(3.99)	78-0797-65
A-Norcholest-3-ene-3-carboxylic acid, methyl ester	EtOH	237(4.17)	39-5416-65
$C_{28}H_{46}O_5$			
p-Benzoquinone, 2,5-dihydroxy-3-methyl-6-nonadecyl-, acetate	EtOH	275(4.0),410(2.8)	94-0511-65
$C_{28}H_{46}S_3$			
Cholestane-2β,3α-dithiol, cyclic trithiocarbonate	EtOH	242(3.02),317(4.13), 450(1.98)	78-0329-65
	dioxan	295s(4.2),318(4.27), 455(1.94)	78-1581-65
	dioxan	225s(3.34),295s(4.2), 318(4.27),455(1.94)	78-0583-65

Compound	Solvent	$\lambda_{max}(\log \epsilon)$	Ref.
$C_{28}H_{48}N_2O$			
5α-Pregn-8-en-11-one, 3β,20α-bis(dimethylamino)-4,4,14α-trimethyl-	EtOH	253(3.9)	28-4139-65B
5α-Pregn-9-en-12-one, 3β,20α-bis(dimethylamino)-4,4,14α-trimethyl-	EtOH	240(3.5)	28-4139-65B
$C_{28}H_{48}N_2O_2$			
Buxaminol, N-acetyldehydrotetrahydro-	MeOH	230s(2.74),300(1.59), 325s(1.33)	33-0968-64
Cyclomicrophylline A	EtOH	204(3.76)	39-4512-65
$C_{28}H_{48}O$			
Cholestanone, 2α-methyl-	EtOH	280(1.60)	44-1423-65
	dioxan	283(1.54)	44-1423-65
Cholest-7-en-3β-ol, 7-methyl-	EtOH	203(3.81)	39-3106-64
$C_{28}H_{48}O_4$			
Polygonaquinone	EtOH	295(4.28)	78-2319-64
$C_{28}H_{50}N_2O_2$			
Buxaminol, N-acetyltetrahydro-	MeOH	235s(2.60)	33-0968-64
$C_{28}H_{50}N_8O_{17}P_2$			
P^1,P^2-Di(6-azauridylyl-5')pyrophosphate, triethylamine salt	n.s.g.	260(4.10)	73-2567-64

Compound	Solvent	$\lambda_{max}(\log \epsilon)$	Ref.
$C_{29}H_{18}N_2O_4$			
m-Toluamide, N-(5,14-dihydro-5,8,14-trioxonaphth[2,3-c]acridan-6-yl)-	$C_6H_3Cl_3$	607(3.95)	65-1575-64
o-Toluamide, N-(5,14-dihydro-5,8,14-trioxonaphth[2,3-c]acridan-6-yl)-	$C_6H_3Cl_3$	603(3.95)	65-1575-64
p-Toluamide, N-(5,14-dihydro-5,8,14-trioxonaphth[2,3-c]acridan-6-yl)-	$C_6H_3Cl_3$	607(4.28)	65-1575-64
$C_{29}H_{18}O$			
Anthrone, 10-(diphenyl-2-cyclopropen-1-ylidene)-	$CHCl_3$	242(4.59),275(4.47), 315s(--),368s(--), 405(4.10)	89-0784-64
	H_2SO_4	251(4.68),287(4.26), 379(4.11),432(4.04), 540(4.15)	89-0784-64
Fluoren-9-ol, 9-(3-fluoren-9-ylidene-1-propynyl)-	ether	257(4.56),266(4.66), 279(4.19),288(4.17), 312(4.11),328(4.41), 344(4.53)	5-0050-65J
$C_{29}H_{18}O_2$			
12H-Benzo[b]xanthen-12-one, 6,11-diphenyl-	EtOH	240(4.49),264(4.86), 313(3.82),326(3.83), 403(3.84)	35-5424-65
$C_{29}H_{18}O_3$			
7,8-Benzonaphtho[1",2":2,3]phenalen-1-ylsuccinic anhydride, K salt	50% EtOH	260(4.57),270(4.52), 280(4.52),292(4.44), 298(4.30),330(4.10), 346(4.22),363(4.11)	24-0743-65
7,8-Benzonaphtho[2",1":2,3]phenalen-1-ylsuccinic anhydride, K salt	50% EtOH	234(5.03),274(4.63), 287(4.44),332(4.42), 350(4.30),372(3.53)	24-0743-65
$C_{29}H_{18}O_4$			
Isocoumarin, 3-phenyl-, phenanthrenequinone adduct	dioxan	244(4.58),281(3.90), 326(3.48)	24-3102-65
$C_{29}H_{18}O_7$			
Flavone, 5',6-dibenzoyl-2',4',7-trihydroxy-	MeOH	295(4.31),328(4.33)	44-3445-64
$C_{29}H_{20}$			
Pentatetraene, tetraphenyl-	THF	335(4.90),420(3.70)	24-2975-64
Propane, 1,3-difluoren-9-ylidene-	DMSO-KO-tert-Bu	305(4.37),350(4.24), 590s(4.62),634(5.36)	5-0050-65J
$C_{29}H_{20}O$			
Fluoren-9-ol, 9-(3-fluoren-9-ylidenepropenyl)-	ether	257(4.54),268(4.58), 300(4.18),311(4.24), 338(4.54),354(4.50)	5-0050-65J
$C_{29}H_{20}O_2$			
6-Oxabicyclo[3.1.0]hex-3-en-2-one, 1,3,4,5-tetraphenyl-	EtOH	233(4.23),338(3.85)	88-0505-64
$C_{29}H_{20}O_3$			
Chromone, 3-benzoyl-2-(diphenylmethyl)-	MeOH	248(4.33),295(3.90), 305(3.85)	35-5417-65

Compound	Solvent	λ_{max}(log ϵ)	Ref.
$C_{29}H_{20}O_5$			
10H-o-Dioxino[4,5-b][1]benzopyran-10-one, 1,4-dihydro-1-hydroxy-1,4,4-triphenyl-	MeOH	261s(3.96),293(3.83), 299s(3.82)	35-5424-65
$C_{29}H_{22}$			
Fluorene, 9-(4,4-diphenyl-3-buten-ylidene)-	DMSO-KO-tert-Bu	290(4.38),375(4.16), 418s(3.95),542s(4.67), 570(4.80)	5-0050-65J
Naphthalene, 1-benzhydryl-4-phenyl-	n.s.g.	294(4.13)	24-2959-64
$C_{29}H_{22}NO_3P$			
1,4-Naphthoquinone, 2-methoxy-3-(tri-phenylphosphoranylideneamino)-	C_6H_{12}	285(4.46)	39-1003-65
$C_{29}H_{22}N_2O$			
Fulvene, 2-(α-hydroxy-α-2-pyridyl-benzyl)-6-phenyl-6-(2-pyridyl)-	MeOH	324(4.37)	57-0412-64B
$C_{29}H_{22}O_4$			
Benzoic acid, o-(5-hydroxy-2,4,6-tri-phenyl-m-dioxan-5-yl)-, γ-lactone	EtOH	250s(2.96),257(3.00), 261s(3.03),263(2.99), 267(2.99),278(3.16), 286(3.16)	44-2778-64
$C_{29}H_{22}O_{11}$			
Duclauxin	EtOH	233(3.95),318(3.21)	88-1287-65
$C_{29}H_{23}N$			
Indene, 1-(methylanilinomethylene)-2,3-diphenyl-	MeCN	251(4.60),294(3.30), 404(4.50)	24-3331-64
$C_{29}H_{24}$			
1,4-Pentadiene, 1,1,5,5-tetraphenyl-	DMSO-KO-tert-Bu	475(4.17),614(5.06)	5-0050-65J
$C_{29}H_{24}N_2O_6S$			
Uridine, 5-O-trityl-, cyclic 2',3'-thiocarbonate	EtOH	234(4.38)	44-4353-65
$C_{29}H_{24}O$			
Cyclopenta[ef]heptalene-3-ethanol, 5-methyl-α,α-diphenyl-	EtOAc	372(4.14),392(4.20), 422(2.94),449(2.54), 745(2.09),810(2.20), 915(2.16),1050(1.78), 1090(1.76)	5-0194-64B
Phthalan, 1,1-dibenzyl-3-benzylidene-	EtOH	232(4.01),240(3.99), 250s(3.81),307s(4.27), 318(4.40),333(4.44), 350(4.20)	44-0886-64
$C_{29}H_{24}O_3$			
Chalcone, 2',4-bis(benzyloxy)-	EtOH	240s(--),340(4.37)	22-3350-65
Chalcone, 2',4'-bis(benzyloxy)-	C_6H_{12}	310(4.33)	22-3350-65
$C_{29}H_{24}O_4$			
Propiophenone, 2'-(benzyloxy)-3-[p-(benzyloxy)phenyl]-2,3-epoxy-	EtOH	260(4.04),315(3.58)	22-3350-65
Propiophenone, 2',4'-bis(benzyloxy)-2,3-epoxy-3-phenyl-	EtOH	235(4.25),279(4.12), 314(4.01)	22-3350-65

Compound	Solvent	λ_{max}(log ε)	Ref.
$C_{29}H_{24}O_8$			
Isodianellinone, tri-O-methyl-	dioxan	272(4.73),295s(4.16), 358(3.62),470(3.79)	12-0218-65
$C_{29}H_{25}BO_6S_2$			
Curcumin dithienylboric acid complex	iso-PrOH-HOAc	440(4.4),505(4.2)	83-0617-64
	iso-PrOH-NH	460(3.9),610(4.6)	83-0617-64
$C_{29}H_{25}ClN_2O_6S$			
2-[3-Ethyl-2-benzothiazolinylidene)-propenyl]-1-(o-hydroxyphenyl)quinolinium perchlorate, acetate	EtOH	593(5.14)	65-3373-64
$C_{29}H_{25}N$			
Azuleno[1,8-cd]azepine, 7-isopropyl-9-methyl-1,3-diphenyl-	n.s.g.	265(4.3),300(4.5), 400(4.2),470(3.2), 500(3.3),800(2.0)	88-1877-65
$C_{29}H_{25}N_3O$			
3-Butenophenone, 2,4,4-triphenyl-, semicarbazone	n.s.g.	254(4.33)	44-3660-64
$C_{29}H_{26}N_2O_4$			
Uridine, 2',3'-didehydro-2',3'-dideoxy-3-methyl-5'-O-trityl-	EtOH	244(3.71),260(3.86)	44-4353-65
$C_{29}H_{26}O_4$			
m-Dioxan-5-ol, 5-(α-hydroxy-o-tolyl)-2,4,6-triphenyl-	EtOH	250(2.72),257(2.85), 261(2.74),264(2.78), 267s(2.56)	44-2778-64
$C_{29}H_{26}O_9$			
6,12-Methano-12H-dibenzo[d,g][1,3]dioxocin-1,3-diol, 6-(p-hydroxyphenyl)-8-methoxy-13-methyl-, triacetate	EtOH	271(3.49)	78-1471-65
$C_{29}H_{27}ClN_2O_5$			
1-(o-Hydroxyphenyl)-2-[3-(1,3,3-trimethyl-2-indolinylidene)propenyl]-quinolinium perchlorate	EtOH	570(5.03)	65-3373-64
$C_{29}H_{27}NO$			
7H-Benzocyclononen-7-one, 10,11-dihydro-6,8-diphenyl-9-(1-pyrrolidinyl)-	EtOH	227(4.40),275(4.15), 315(4.23)	35-1404-65
$C_{29}H_{27}NO_5$			
2(1H)-Pyridone, 1-(5-O-trityl-β-D-ribofuranosyl)-	EtOH	303(3.74)	94-0828-64
$C_{29}H_{28}OSi$			
Silane, (p-tert-butylbenzoyl)-triphenyl-	C_6H_{12}	269(4.30),386s(2.08), 403(2.36),421(2.48), 442(2.30)	23-1175-65
$C_{29}H_{28}O_4$			
9-Anthrol, 4β,10β-(benzylidenedioxy)-5-(benzyloxy)-1,4,4aα,9,9aβ,10-hexahydro-9β-methyl-	EtOH	276(3.30),282(3.30)	65-2563-64

Compound	Solvent	λ_{max}(log ε)	Ref.
$C_{29}H_{28}O_9$			
Macluraxanthone, triacetate	EtOH	245s(3.85),269(4.84), 295s(4.21),312(3.98)	44-0692-64
$C_{29}H_{29}NO_6$			
Malonamic acid, [5-(benzyloxy-1,2,3,4-tetrahydro-9-methyl-4-oxo-2-anthryl)-acetyl]-, ethyl ester	EtOH	224(4.77),262(4.89), 308(3.98),320(3.88), 373(3.92)	65-0670-65
$C_{29}H_{29}N_3O_6S$			
Cytidine, 2'-deoxy-5'-O-trityl-, 3'-methanesulfonate	pH 1	270(4.03)	44-3067-65
	pH 13	270(3.94)	44-3067-65
	EtOH	265(3.87)	44-3067-65
$C_{29}H_{30}N_2O_6$			
Henningsoline, benzoate, hydrochloride	EtOH	280(4.6)	39-2818-65
	EtOH-KOH	305(4.6)	39-2818-65
$C_{29}H_{30}O_6$			
2-Anthracenemalonic acid, 5-benzyloxy-1,2,3,4-tetrahydro-9-methyl-4-oxo-, diethyl ester	EtOH	222(4.53),261(4.63), 308(3.76),319(3.66), 372(3.76)	65-0655-65
$C_{29}H_{30}O_9$			
Alvaxanthone, triacetate	EtOH	238(4.62),258(4.56), 312(4.20),367(3.72)	44-0689-64
Macluraxanthone, dihydro-, triacetate	EtOH	245s(4.23),267(4.45), 318(4.10)	44-0692-64
$C_{29}H_{31}N_3O_7$			
Brefeldin A, mono-O-[p-(p-nitro-phenylazo)benzoyl]-	MeOH	328(4.54),462(2.85)	33-1401-64
$C_{29}H_{32}N_2O_2$			
Benzoic acid, p-(phenylazo)-, 7,12,14-hexadecatrien-10-ynyl ester	ether	264(4.60),273s(4.50), 320(4.36)	24-0872-65
$C_{29}H_{32}O_7$			
2α-Anthracenemalonic acid, 5-benzyloxy-1,2,3,4,9,9aα-hexahydro-9β-hydroxy-9α-methyl-4-oxo-, diethyl ester	EtOH	242(3.93),352(4.16)	65-0655-65
2β-isomer	EtOH	242(3.84),352(4.12)	65-0655-65
$C_{29}H_{32}O_8$			
Cyclohexanemalonic acid, 3-[7-(benzyl-oxy)-3-methylphthalidyl]-5-oxo-, diethyl ester	EtOH	236(4.11),298(3.87)	65-0655-65
$C_{29}H_{32}O_9$			
Macluraxanthone, tetrahydro-, triacetate	EtOH	243(4.64),267s(3.89), 279(3.89),313(4.23)	44-0692-64
$C_{29}H_{33}BrN_4O_9$			
5α-Pregn-1-ene-3,11,20-trione, 21-acet-oxy-9α-bromo-17-hydroxy-, 3-(2,4-dinitrophenylhydrazone)	$CHCl_3$	381(4.47)	39-0156-65

Compound	Solvent	$\lambda_{max}(\log \epsilon)$	Ref.
$C_{29}H_{33}N_3O_3$			
A β-carboline derivative	EtOH	208(4.52),232(4.49), 245s(4.30),260s(4.17), 286s(4.09),291(4.10), 298(4.22)	77-0317-65
$C_{29}H_{34}N_2O_3$			
Androst-5-ene-17β-carbonitrile, 3β-acetoxy-16α,17α-[3-phenyl-3,1-(2-isoxazolino]-	MeOH	259(4.14)	4-0280-64
$C_{29}H_{34}N_2O_7$			
Pyrrolo[2,1-a]isoquinoline-2-acetic acid, 3-[(3,4-dimethoxyphenethyl)-carbamoyl]-5,6-dihydro-8,9-dimethoxy-, ethyl ester	EtOH	318(4.45)	94-0775-65
$C_{29}H_{34}N_2$			
Corrole, 8,12-diethyl-2,3,7,13,17,18-hexamethyl-	$CHCl_3$	346s(4.22),396(5.09), 408s(5.00),474s(3.52), 502s(3.83),525s(3.92), 536(4.26),550(4.26), 593(4.33)	39-1620-65
	H_2SO_4	292(4.29),358(4.70), 671(4.46)	39-1620-65
hydrobromide	$CHCl_3$	406(5.14),504s(3.40), 544s(3.94),577(4.39)	39-1620-65
copper complex	$CHCl_3$	396(3.91),502s(3.85), 549(4.00)	39-1620-65
nickel complex	$CHCl_3$	349(4.65),419s(4.44), 652(3.96)	39-1620-65
$C_{29}H_{34}N_4O_4$			
ac-Bisnorbilene-b-6,7-dicarboxylic acid, 1',8'-dideoxy-1,2,3,4,5,8-hexamethyl-, diethyl ester	EtOH-HBr	238(4.28),292(4.28), 338(4.36),422(3.88), 621(4.73)	39-1460-65
$C_{29}H_{34}O_6$			
5H-Tribenzo[a,d,g]cyclononene, 2,3-diethoxy-10,15-dihydro-7,8,12,13-tetramethoxy-	EtOH	209(4.84),234(4.50), 292(4.01)	39-1685-65
$C_{29}H_{34}O_8$			
Siphulin, methyl ester, dimethyl ether, acetate	EtOH	245s(4.48),279(4.22)	1-1677-65
$C_{29}H_{34}O_9$			
Alvaxanthone, tetrahydro-, triacetate	EtOH	238(4.53),259(4.49), 308(4.16),367(3.66)	44-1088-65
$C_{29}H_{35}ClN_2O_4$			
Emetine, 1,3,5(11b)-tridehydro-, chloride	EtOH-HCl	235s(4.38),267s(4.07), 285(4.26),366(4.16)	33-1117-64
$C_{29}H_{35}NO_3$			
Isoxazolino[17,16-d]pregn-4-ene-3,20-dione, 6α-methyl-3'-phenyl-	n.s.g.	212(4.22),218(4.15), 244(4.29)	5-0139-64G
$C_{29}H_{35}N_5O_5$			
Ergovaline	MeOH	241(4.32),312(3.97)	33-1911-64

Compound	Solvent	λ_{max}(log ε)	Ref.
Ergovalinine	MeOH	239(4.32),311(3.93)	33-1911-64
$C_{29}H_{36}N_2O_4$			
Pregn-4-ene[2,3-d]imidazole, 11β,17α,21-trihydroxy-16α-methyl-20-oxo-1'-phenyl-	MeOH	225(4.35),288(3.99), 293(4.00)	87-0584-64
Pregn-4-ene[3,2-c]pyrazole, 11β,17,21-trihydroxy-16α-methyl-20-oxo-1'-phenyl-	MeOH	298(4.50)	35-1520-64
$C_{29}H_{36}N_4$			
Biladiene-ac, 1',8'-dideoxy-4,5-diethyl-1,2,3,6,7,8-hexamethyl-, hydrobromide	$CHCl_3$	288(3.43),373(4.20), 458(4.30),522(5.28)	39-1620-65
$C_{29}H_{36}O_4$			
Androsta-3,5-diene-6-carboxaldehyde, 17β-acetoxy-3-(benzyloxy)-	EtOH	322(4.21)	78-0597-64
$C_{29}H_{36}O_6$			
Pentane, 1,3,5-tris(2,5-dimethoxyphenyl)-	95% THF	232(4.20),292(4.06)	44-2602-65
$C_{29}H_{36}O_8$			
Verrucarin H	EtOH	195(4.2),223(4.35), 259(4.21)	33-1079-65
$C_{29}H_{36}O_{10}$			
Verrucarin A, acetate	EtOH	260(4.41)	33-0157-65
$C_{29}H_{36}O_{16}$			
Olivine 4'-diglucoside	n.s.g.	251(4.39),268(4.36), 343(4.48)	73-1484-64
$C_{29}H_{37}NO_4$			
Phthalimide, N-(3β-hydroxy-5β-androstan-17β-yl)-, acetate	MeOH	292(3.27),300s(3.24)	24-0533-64
$C_{29}H_{37}N_3O_3$			
Alkaloid AL 64	EtOH	226(4.28),283(4.18)	2-0468-64
Tubulosine	MeOH	223(4.60),279(4.16), 310s(3.67)	100-0212-65
	base	215(4.73),279(4.09), 320(3.64)	100-0212-65
	n.s.g.	225s(4.55),281(4.16)	35-1895-64
$C_{29}H_{37}N_5O_5$			
Ergovaline, 9,10-dihydro-	MeOH	223(4.51),280(3.84), 291(3.75)	33-1911-64
$C_{29}H_{38}ClNO_4$			
Dihydrometoxazine chloride	EtOH	247(4.40),372(2.41)	44-0270-64
$C_{29}H_{38}N_2O_4$			
5β-Pregnan-21-al, 3-acetoxy-11,20-dioxo-, 21-phenylhydrazone	MeOH	238(4.09),295(3.63), 350(4.35)	44-0521-64
$C_{29}H_{38}O_4$			
Androsta-3,5-diene-6-methanol, 17β-acetoxy-3-(benzyloxy)-	EtOH	251(4.33)	78-0597-64

Compound	Solvent	λ_{max}(log ϵ)	Ref.
$C_{29}H_{38}O_5$			
Heptanoic acid, 8-ester with dodeca-hydro-5,8-dihydroxy-4a,6a-dimethyl-pentaleno[2,1-a]phenanthrene-2,7-dione	EtOH	241(4.35)	5-0218-65E
$C_{29}H_{38}O_7S$			
5,8,10a-(Epoxymetheno)-10aH-cyclohepta-[b]naphthalen-1(2H)-one, dodecahydro-3,9-dihydroxy-2,2,5,9-tetramethyl-, 9-acetate 3-p-toluenesulfonate	EtOH	227(4.07)	78-3091-65
$C_{29}H_{38}O_8$			
Roridin E	EtOH	195(4.2),223(4.40), 263(4.30)	33-1079-65
$C_{29}H_{38}O_9$			
Roridin D	EtOH	260(4.33)	33-1079-65
$C_{29}H_{39}NO_2$			
Androsta-3,5-dien-17-one, 3-(benzyl-oxy)-6-dimethylaminomethyl-	EtOH	250.5(4.33)	78-0569-65
Pregna-5,20-dien-3β-ol, 20-(5-methyl-2-pyridyl)-, acetate	hexane	234(3.92)	78-1707-64
$C_{29}H_{39}NO_3$			
Phthalimide, N-(3β-hydroxy-5α-pregnan-20β-yl)-	dioxan	292(3.24),300s(3.21)	24-0533-64
$C_{29}H_{39}NO_4$			
Isojervine, N-acetyl-5,6-dihydro-, 3,11,23-triketone from	EtOH	230s(4.06),305(2.45)	44-0262-64
Δ^4-Isojerv-3-one, N-acetyl-	EtOH	230(4.35),331(2.30)	44-0262-64
$C_{29}H_{39}NO_5$			
Δ^4-Isojerv-3-one, N-acetyl-, 17,17a-epoxide	EtOH	230(4.33),315(2.44)	44-0270-64
Δ^4-Jervisin-3-one, 8,9-dehydro-, 17-acetate	EtOH	232(4.33),399(2.28)	44-0270-64
hydrochloride	EtOH	232(4.40)	44-0270-64
$C_{29}H_{40}ClNO_5$			
A dihydrometoxazine chloride	EtOH	229(4.38),315(2.04)	44-0270-64
$C_{29}H_{40}N_2O_3$			
Urea, 1,3-dicyclohexyl-1-[(1,2β,3,9,-10,10aβ-hexahydro-7-methoxy-2-phenanthryl)carbonyl]-	EtOH	263(4.31),298(3.51)	44-2849-65
$C_{29}H_{40}N_4O_2$			
Acrylonitrile, 3-[p-(diethylaminoeth-oxy)phenyl]-2-(N-ethyl-N'-piperazin-ylethoxy)phenyl]-	EtOH-HCl	342.5(4.47)	87-0511-64
$C_{29}H_{40}O_4$			
C-Nor-D-homo-5α-spirosta-12,14,16-tri-en-3β-ol, acetate	MeOH	289(3.67)	44-4220-65
$C_{29}H_{40}O_6$			
Gratiogenin, oxidized acetate	MeOH	242(3.76)	5-0196-64D

Compound	Solvent	λ_{max}(log ε)	Ref.
Pregna-3,5-dien-20-one, 3-acetoxy-21-acetyl-16α,17α-isopropylidenedioxy-	EtOH	245(4.23),278(4.14)	78-1927-64
$C_{29}H_{40}O_6S$			
Androst-5-en-3-one, 17β-hydroxy-17-(hydroxymethyl)-, cyclic ethylene acetal, 17-p-toluenesulfonate	MeOH	225(4.15)	31-0499-65
$C_{29}H_{40}O_8S$			
5a,8-Methano-5aH-cyclohepta[b]naphthalen-4(1H)-one, dodecahydro-2,7,11,12-tetrahydroxy-3,3,7,11-tetramethyl-, 12-acetate 2-p-toluenesulfonate	EtOH	227(4.04)	78-3091-65
$C_{29}H_{40}O_{10}$			
Calotoxin	EtOH	217(4.17),309(1.76)	39-2187-64
$C_{29}H_{41}NO_4$			
Isojervine, N-acetyl-8,9-dihydro-	EtOH	309(1.78),319(1.78), 330s(1.59)	44-0528-65
Isojerv-3-one, N-acetyl-4,5-dihydro-	EtOH	237(3.99),332(2.27)	44-0262-64
Δ^4-Jerv-3-one, N-acetyl-13,17a-dihydro-	EtOH	234(4.22),295(2.12)	44-0262-64
$C_{29}H_{41}NO_6$			
9,19-Cyclopregn-4-en-3-one, 4,16-dihydroxy-14-methyl-20-(N-methylacetamido)-, diacetate	EtOH	277(4.10)	35-4414-64
Pregna-4,?-dien-3-one, 4,16α-dihydroxy-14-methyl-20α-(N-methylacetamido)-, diacetate	EtOH	247(4.10)	35-4414-64
Pregn-4-en-3-one, 11β-hydroxy-17α,20:20,21-bis(methylenedioxy)-2-(N-piperidinomethylene)-	EtOH	250(4.23),373(4.17)	87-0528-64
$C_{29}H_{41}N_3O_2$			
2-Pentenenitrile, 2,3-bis[p-(2-diethylaminoethoxy)phenyl]-, trans	EtOH-HCl	292.5(4.17)	87-0511-64
$C_{29}H_{42}Br_2O$			
9β,10α-Ergosta-4,22-dien-3-one, 6-(dibromomethylene)-	MeOH	250(3.99),285(3.77)	54-0904-65
$C_{29}H_{42}ClNO_4$			
A dihydrometoxazine chloride	EtOH	254(4.17),338(3.16)	44-0270-64
$C_{29}H_{42}O_2$			
2H-1-Benzopyran-6-ol, 2,5,7,8-tetramethyl-2-(4,8,12-trimethyltrideca-3,7,11-trienyl)-	hexane	235(4.27),264s(--), 274(3.94),284(3.91), 340(3.53)	39-5060-65
$C_{29}H_{42}O_4$			
p-Benzoquinone, 2-(2,5-di-tert-butyl-4-methoxyphenoxy)-3,6-di-tert-butyl-	CCl_4	266(4.0),291s(3.5)	39-2921-65
$C_{29}H_{42}O_5$			
25D-Spirost-4-en-3-one, 2 -acetoxy-	EtOH	242.5(4.22)	78-3633-65
$C_{29}H_{42}O_6$			
Pregn-5-en-20-one, 3β-acetoxy-21-acetyl-16α,17α-isopropylidenedioxy-6-methyl-	pH 13	300(4.38)	78-1927-64
	EtOH	279(4.12)	78-1927-64

Compound	Solvent	$\lambda_{max}(\log \epsilon)$	Ref.
Senegenic acid, dehydro-	EtOH	237(4.09),246(4.17), 254(4.06)	44-4234-65
	EtOH	237(4.10),246(4.17), 254(4.06)	88-3065-64
$C_{29}H_{42}O_8$			
Cholanic acid, 3α,9α-diacetoxy-11,12-dioxo-, methyl ester	EtOH	289(2.04),297(2.04), 357(1.95)	13-0445-64
$C_{29}H_{42}O_{10}$			
Erygypsoside	H_2SO_4	235(4.21),303(3.84), 410(4.24),483(3.98), 540(3.82)	65-2463-64
Strophalloside	EtOH	216(4.23),299(1.63)	33-2164-64
$C_{29}H_{42}O_{11}$			
Antialloside	EtOH	217(4.22),297(1.67)	33-2164-64
$C_{29}H_{43}Br_3O$			
9β,10α-Ergosta-4,22-dien-3-one, 6β-(tribromomethyl)-	MeOH	241(4.15)	54-0904-65
$C_{29}H_{43}NO_2$			
Solaso-3,5-diene, N-acetyl-	EtOH	234(4.42)	73-1178-64
$C_{29}H_{43}NO_4$			
Isojervine, N-acetyl-5,6-dihydro-	EtOH	238(3.97),333(2.32)	44-0262-64
$C_{29}H_{43}NO_5$			
Isojervine, N-acetyl-5,6-dihydro-, 17,17a-epoxide	EtOH	237(4.03),319(1.83)	44-0270-64
Jervisine, 8,9-dehydro-5,6-dihydro-, 17-acetate	EtOH	237(3.93)	44-0270-64
$C_{29}H_{44}$			
24-Nor-D:C-friedo-B':A'-neogamma-cera-3,5,9(11)-triene	EtOH	234(3.98),241(4.03), 249(3.86)	94-0986-65
$C_{29}H_{44}ClNO_4$			
A dihydrometoxazine chloride	EtOH	243(4.17),312(2.80)	44-0262-64
$C_{29}H_{44}N_2O_4S$			
4-Morpholinecarboxamide, N-(3β-hydroxy-20-methylpregn-5-en-21-oyl)thio-, acetate	n.s.g.	285(4.23),345(2.70)	32-1438-64
isomer	n.s.g.	285(4.20),345(2.60)	32-1438-64
$C_{29}H_{44}O$			
Cholest-4-en-3-one, 4-ethynyl-	EtOH	269(4.09)	22-0321-64
Ergosta-4,6,8(14),22-tetraene, 3β-methoxy-	EtOH	287(4.52)	24-2383-65
$C_{29}H_{44}O_2$			
5β-Lumista-7,22-dien-3-one, 2-(hydroxymethylene)-	EtOH	287(3.76)	39-5064-65
$C_{29}H_{44}O_3$			
$\Delta^{3,5}$-Diosgenin, 3-ethoxy-	EtOH	241(4.32)	78-0597-64
A(1)-Noroleanan-28-oic acid, 19-hydroxy-3-oxo-, γ-lactone	n.s.g.	296(1.49)	12-0141-64

Compound	Solvent	λ_{max}(log ε)	Ref.
$C_{29}H_{44}O_4$			
25α-Furosta-5,20(22)-diene-3β,27-diol, 23-acetyl-	EtOH	290(2.88)	44-3915-65
25α-Furosta-5,22(23)-diene-3β,27-diol, 23-acetyl-	EtOH	279(4.09)	44-3915-65
29-Nor-8α,9β,13α,14β-dammar-1-en-21-oic acid, 11α,16β-dihydroxy-3-oxo-, γ-lactone	EtOH	238(3.84)	78-3505-65
	dioxan	238(3.84)	31-0344-64
isomer	EtOH	238(3.96)	78-3505-65
	dioxan	238(3.96)	31-0344-64
Pregn-4-ene-3,20-dione, 17α-hydroxy-6α,16α-dimethyl-, hexanoate	EtOH	241(4.12)	87-0540-64
$C_{29}H_{44}O_6$			
Polygalic acid	EtOH	200(3.91)	88-2567-64
Senegenic acid	EtOH	205(3.70)	44-4234-65
	EtOH	205(3.70)	88-3065-64
$C_{29}H_{44}O_8$			
Ascleposide	EtOH	216.5(4.22)	33-1775-64
Digitoxigenin allomethyloside	n.s.g.	216(4.21)	31-0575-65
$C_{29}H_{44}O_9$			
Peripalloside	EtOH	217(4.20)	33-2164-64
Periplorhamnoside	EtOH	216(4.16)	33-2164-64
$C_{29}H_{44}O_{10}$			
Antiogoside	EtOH	217(4.17)	33-2164-64
Antioside	EtOH	218(4.19)	33-2164-64
Periplogenin, β-D-glucoside	EtOH	217(4.22)	33-0799-64
Uposide	EtOH	218(4.11)	33-2164-64
$C_{29}H_{44}O_{12}$			
Antiosemoside	EtOH	217(4.26)	33-2164-64
$C_{29}H_{45}N$			
Cholest-4-ene-$\Delta^{3,\alpha}$-acetonitrile	EtOH	225(3.91)	44-0505-65
$C_{29}H_{45}NO_3$			
Cholest-4-en-3-one, 16β-hydroxy-22,26-acetimino-	EtOH	241(4.25)	5-0196-65A
$C_{29}H_{45}NO_4$			
Isojervine, N-acetyl-5,6,7,8-tetrahydro-	EtOH	311(2.23),322(2.18), 331(2.03)	44-0262-64
3-Piperidinol, 1-acetyl-2-[1-(3β-hydroxy-10,17a-dimethyl-11-oxo-D-homo-C-nor-5α-gon-13(17a)-en-17β-yl)-ethyl]-5-methyl-	EtOH	255(4.19),352(1.96)	78-0779-65
Pregna-3,5-dien-20-one, 6-dimethylaminomethyl-3-ethoxy-16α,17α-dihydroxy-, cyclic acetal with acetone	EtOH	251(4.27)	78-0569-65
$C_{29}H_{45}NO_5$			
Isojervine, N-acetyl-5,6,8,9-tetrahydro-, 17,17a-epoxide	EtOH	235s(3.01),308(1.34)	44-0262-64
Jervisine, 5,6-dihydro-, 17-acetate	EtOH	235s(2.73)	44-0262-64

Compound	Solvent	λ_{max}(log ϵ)	Ref.
3-Piperidinol, 1-acetyl-5-methyl-2-[1-(2,3,4,4a,5,6,6b,7,8,9,10,10a,11,11b-tetradecahydro-3,9-dihydroxy-10,11b-dimethyl-11-oxo-1H-benzo-[a]fluoren-9-yl)ethyl]-	EtOH	239(4.03),290(3.97)	44-0270-64
$C_{29}H_{46}$			
19-Norcholesta-1,3,5(10)-triene, 1,3,4-trimethyl-	EtOH	271(2.80)	94-0393-65
$C_{29}H_{46}ClNO_5$			
A dihydrometoxazine chloride	EtOH	303(1.85)	44-0262-64
$C_{29}H_{46}O$			
Cholesta-3,5-diene, 6-acetyl-	MeOH	221(3.97),281(3.78)	39-2285-64
Cholest-4-en-3-one, 6-ethylidene-	MeOH	246(3.79),279(3.82)	39-2285-64
Cholest-4-en-3-one, 4-vinyl-	EtOH	260(3.99)	22-0321-64
Cholest-5-en-3-one, 4-ethylidene-	EtOH	242.5(4.20)	22-0321-64
Ergosta-7,22-dien-3-one, 4α-methyl-	EtOH	206(3.66),210(3.63)	39-1142-64
9β,10α-Ergosta-4,22-dien-3-one, 6α-methyl-	MeOH	247.5(4.17)	54-0904-65
$C_{29}H_{46}OS$			
Cholesta-3,5-dieno[3,4-b]oxathiane	hexane	223(3.97),270(3.94)	94-0383-64
$C_{29}H_{46}O_2$			
Cholesta-4,6-dien-3-one, 6-ethoxy-	n.s.g.	308(4.18)	12-0661-64
Cholest-4-en-3-one, 6β-acetyl-	MeOH	247(4.16)	39-2285-64
Clerodone	n.s.g.	205(3.54)	78-0797-65
$C_{29}H_{46}O_3$			
Cholest-4-en-3-one, 2α-acetoxy-	EtOH	242(4.19)	12-0661-64
	EtOH	243(4.16)	78-0733-65
	EtOH	243(4.17)	94-0383-64
Cholest-4-en-6-one, 3β-acetoxy-	EtOH	236(3.83)	44-3495-64
5α-Cholest-7-en-6-one, 3β-acetoxy-	EtOH	244(4.15)	24-2361-65
5,19-Cyclocholestan-6-one, 3β-acetoxy-	MeOH	208(3.54)	35-1528-64
Gon-4-en-3-one, 13β-ethyl-17β-hydroxy-, decanoate	n.s.g.	239(4.21)	39-4472-64
19-Norcholest-8-en-6-one, 7-(hydroxy-methylene)-3β-methoxy-5β-methyl-	EtOH	296(3.79)	5-0167-65F
	EtOH-base	322(4.20)	5-0167-65F
$C_{29}H_{46}O_4$			
Cholest-1-en-3-one, 4α,5α-dihydroxy-, 4-acetate	EtOH	231(3.99)	94-0050-65
16-Epideacetyl-24,25-dihydro-fusidic acid lactone	EtOH	230(4.11)	78-3505-65
$C_{29}H_{46}O_5$			
29-Nor-8α,9β,13α,14β-dammar-17(20)-en-21-oic acid, 11α,16α-dihydroxy-3-oxo-	EtOH	228(4.03)	78-3505-65
	dioxan	228(4.03)	31-0344-64
$C_{29}H_{46}S_2$			
Cholesta-3,5-dieno[3,4-b]dithiane	hexane	240(4.08),292(4.14)	94-0383-64
	EtOH	240(4.08),292(4.14)	78-0733-65
$C_{29}H_{47}BrO$			
Cholest-5-en-2-one, 3α-bromo-4,4-dimethyl-	C_6H_{12}	310(2.02)	22-2602-65
3β-bromo isomer	C_6H_{12}	297(1.43)	22-2602-65

Compound	Solvent	λ max(log ε)	Ref.
$C_{29}H_{47}N$			
5α-Cholestane-Δ$^{3,\alpha}$-acetonitrile	EtOH	221(4.09)	44-0505-65
$C_{29}H_{47}NO_2$			
Cholest-5-ene-2,3-dione, 4,4-dimethyl-, 2-oxime	EtOH	236(3.89)	13-0041-64
	EtOH-base	292(4.11)	13-0041-64
$C_{29}H_{48}$			
Cholesta-3,5-diene, 4-ethyl-	EtOH	232s(4.24),239(4.29), 247s(4.12)	22-0321-64
$C_{29}H_{48}BNO_4$			
Pregna-3,5-dien-20-one, 6-dimethyl-aminomethyl-3-ethoxy-16α,17α-dihydroxy-, cyclic acetal with acetone, borane adduct	EtOH	256(4.31)	78-0569-65
$C_{29}H_{48}Br_2O$			
5α-Cholestan-2-one, 1α,3β-dibromo-4,4-dimethyl-	C_6H_{12}	306(1.93)	22-2236-64
5α-Cholestan-2-one, 3,3-dibromo-4,4-dimethyl-	C_6H_{12}	306(1.93)	22-2236-64
$C_{29}H_{48}ClNO_4$			
5α-Cholestan-3β-ol, 5α-chloro-6β-nitro-, acetate	MeOH	264(1.59),283(1.70)	44-1350-64
	MeOH-KOH	265(3.16)(changing)	44-1350-64
$C_{29}H_{48}N_2$			
Buxamine, N-isopropylidene-	MeOH	230s(4.30),239(4.45), 247(4.48),256(4.29), 279(2.52),290s(2.37)	33-0968-64
Indeno[5,4-f]pyrimido[1,2-a]quinoline, 9-(1,5-dimethylhexyl)hexadecahydro-6a,8a-dimethyl-	EtOH	252(4.07)	25-0650-65
$C_{29}H_{48}N_2O$			
Buxaminol, N-isopropylidene-	MeOH	230s(4.33),238(4.46), 246(4.49),255(4.29), 278(2.27),290(2.13)	33-0968-64
$C_{29}H_{48}N_2O_2$			
Spiro[5α-cholestane-3,4'-imidazolidine]-2',5'-dione	dioxan	252(3.8)	94-1377-65
$C_{29}H_{48}N_4O_2$			
Hydroquinone, piperidinotris-(piperidinomethyl)-	dioxan	310(3.78)	39-0042-64
$C_{29}H_{48}O$			
Cholesta-3,5-diene, 3-ethoxy-	EtOH	239(4.25)	23-0079-64
Cholesta-3,5-diene, 3-methoxy-6-methyl-	EtOH	247(4.39)	23-2919-64
Cholest-4-en-3-one, 4-ethyl-	EtOH	251(4.14)	22-0321-64
	EtOH	250(4.25)	23-0464-64
Cholest-5-en-2-one, 4,4-dimethyl-	C_6H_{12}	292(1.52)	22-2602-65
5α-Cholest-1-en-3-one, 4,4-dimethyl-	C_6H_{12}	224(3.89)	22-2588-65
5α-Cholest-2-en-1-one, 4,4-dimethyl-	C_6H_{12}	219(3.97)	22-2588-65
3α,5-Cyclocholestan-6-one, 4,4-dimethyl-	EtOH	203(3.67)	23-0456-64

Compound	Solvent	λ_{max}(log ϵ)	Ref.
$C_{29}H_{48}OS$			
Cholest-4-en-3-one, 4-(ethylthio)-	EtOH	248(4.11),316(3.30)	78-0733-65
	EtOH	248(4.11),316(3.30)	94-0383-64
$C_{29}H_{48}O_2$			
2H-1-Benzopyran-6-ol, 2,5,7,8-tetra-methyl-2-(4,8,12-trimethyltridecyl)-	hexane	235(4.28),265s(--), 273(3.92),283(3.87), 339(3.51)	39-5060-65
p-Benzoquinone, trimethylphytyl-	EtOH	261(4.27),267(4.27)	39-5060-65
Cholestan-3-one, 2-acetyl-	EtOH	289(3.98)	35-5207-64
Cholest-4-en-3-one, 6β-(1-hydroxyethyl)-	MeOH	244(4.15)	39-2285-64
Clionast-4-en-3-one, 6β-hydroxy-	EtOH	238(4.15)	100-0040-64
5α,14α-Lumistan-3-one, 4-(hydroxy-methylene)-	n.s.g.	293(3.94)	39-7199-65
$C_{29}H_{48}O_3$			
Cholest-4-en-3-one, 4-(1,2-di-hydroxyethyl)-	EtOH	231.5(3.70)	22-0979-64
$C_{29}H_{48}O_4$			
Diels' acid, dimethyl ester	EtOH	216(3.88)	12-1049-65
Resorcinol, 5-nonadecyl-, diacetate	EtOH	261(2.46)	44-0435-64
$C_{29}H_{48}O_5$			
16-Epideacetyl-24,25-dihydrofusidic acid	EtOH	229(3.95)	78-3505-65
$C_{29}H_{49}BrO$			
5α-Cholestan-1-one, 2α-bromo-4,4-dimethyl-	C_6H_{12}	290(1.48)	22-2588-65
2β-bromo isomer	C_6H_{12}	290(1.79)	22-2588-65
5α-Cholestan-2-one, 3α-bromo-4,4-dimethyl-	C_6H_{12}	312(2.01),320(2.00)	22-2236-64
3β-bromo isomer	C_6H_{12}	290(1.34),297(1.34)	22-2236-64
$C_{29}H_{49}N$			
4-Azacholesta-2,5-diene, 3-ethyl-4-methyl-	EtOH	271(3.95)	4-0126-65
$C_{29}H_{49}NO_2$			
Cholestan-3-one, 2α-acetamido-	EtOH	279(2.11)	23-0712-64
$C_{29}H_{49}NO_2S_2$			
(22R:25S)-N-Dithiocarbomethoxy-22,26-imino-5α-cholestane-3β,16β-diol	dioxan	278(4.05),340s(1.95)	78-0407-65
(22S:25R)- form	dioxan	281(4.07),341(1.91)	78-0407-65
(22S:25S)- form	dioxan	281(4.08),338(1.94)	78-0407-65
$C_{29}H_{50}INO_2$			
Cholestane-2β-carbamic acid, 3α-iodo-, methyl ester	MeOH	260(2.70)	44-1748-65
	MeOH	260(2.70)	44-3640-64
$C_{29}H_{50}O$			
5α-Cholestan-1-one, 4,4-dimethyl-	C_6H_{12}	298(1.43)	22-2588-65
5α-Cholestan-2-one, 4,4-dimethyl-	C_6H_{12}	299(1.42)	22-2236-64
Clionasterol	heptane	190(4.07)	100-0040-64
$C_{29}H_{50}OS$			
Cholestan-3-one, cyclic ethylene monothioacetal (isomers)	dioxan	235s(1.70)	78-1581-65
	dioxan	242(1.65)	78-1581-65

Compound	Solvent	$\lambda_{max}(\log \epsilon)$	Ref.
$C_{29}H_{50}O_2S$			
5α-Cholestan-5α-ol, 4β-mercapto-, 4-acetate	EtOH	236(3.74),310(2.30)	78-0309-65
$C_{29}H_{50}O_3$			
3,5-Seco-A-norcholestan-3-oic acid, 5-oxo-, isopropyl ester	iso-PrOH	293(1.39)	44-1513-65
$C_{29}H_{50}S_2$			
Cholestan-1-one, cyclic ethylene mercaptole	dioxan	244(2.48)	78-1581-65
Cholestan-2-one, cyclic ethylene mercaptole	dioxan	240(2.60)	78-1581-65
5α-Cholestan-3-one, cyclic ethylene mercaptole	dioxan	243(2.57)	78-1581-65
5β-Cholestan-3-one, cyclic ethylene mercaptole	dioxan	243(2.57)	78-1581-65
$C_{29}H_{51}O_5P$			
α-Tocopheryl phosphate	H_2O	286(3.29)	37-1374-65
$C_{29}H_{52}O$			
Cholestane, 3β-ethoxy-	MeOH	210(2.25)	13-0487-64
$C_{29}H_{52}O_2$			
Benzene, 1-heneicosyl-3,5-dimethoxy-	EtOH	273(3.20),280(3.21)	44-0435-64
$C_{29}H_{52}O_3$			
1-Heneicosanol, 1-(3,5-dimethoxyphenyl)-	EtOH	274(3.28),280(3.29)	44-0435-64

Compound	Solvent	λ_{max}(log ϵ)	Ref.
$C_{30}H_{14}N_4O_2$			
Terephthaloylenebisnaphthimidazole	$C_{10}H_7Cl$	540(4.24),583s(4.10)	33-2211-64
$C_{30}H_{16}$			
Butatriene, difluoren-9-ylidene-	THF	542(4.93)	24-2975-64
2,3:6,7-Di(peri-naphthylene)-naphthalene	benzene	285(4.10),329(5.04), 394(4.09),418(4.54), 445(4.74)	78-1559-64
	C_6H_{12}	232(4.80)	78-1559-64
Pyreno[1',2':1,2]pyrene	C_6H_{12}	226(4.52),270(4.40), 311(4.75),326(5.01), 380(4.32),400(4.34)	78-2107-64
$C_{30}H_{18}$			
2-Butyne, 1,4-difluoren-9-ylidene-	CH_2Cl_2	233(4.89),269(4.75), 420(4.61),447(4.63)	5-0001-65I
$C_{30}H_{18}Br_2$			
Propene, 1,1-dibromo-3-fluoren-9-ylidene-2-(fluoren-9-ylidenemethyl)-	CH_2Cl_2	233(4.80),253(4.79), 261(4.89),342(4.55)	5-0001-65I
$C_{30}H_{18}N_2$			
Cyclobuta[1,2-1:3,4-1']diphenanthrene-8b,8c-dicarbonitrile, 16b,16c-dihydro-	EtOH	208(4.68),275(4.35)	39-5544-64
$C_{30}H_{18}N_6O_3$			
1,3,4-Oxadiazole, 2,5-bis[3-(2-phenyl-1,3,4-oxadiazol-5-yl)phenyl]-	DMF	287.2(4.89)	24-2966-65
$C_{30}H_{18}O_4$			
Benzene, 1-(1,3-dioxoindan-2-yl)-2-(1,3-dioxo-2-phenylindan-2-yl)-	EtOH	226(4.78),243(4.40), 288s(3.50),342(2.80)	6-0359-64
[2,2'-Biindan]-1,1',3,3'-tetrone, 2,2'-diphenyl-	EtOH	230(4.84)	6-0359-64
Phthalide, 3-[α-(1,3-dioxo-2-phenyl-2-indanyl)benzylidene]-	ether	317(3.80)	22-3061-65
$C_{30}H_{18}O_6$			
Benzoic acid, o-[[o-(1,3-dioxo-2-phenyl-2-indanyl)phenyl]glyoxyloyl]-	ether	226(4.40),282s(3.58), 360s(2.60)	6-0359-64
$C_{30}H_{19}I$			
Propene, 3-fluoren-9-ylidene-2-(fluoren-9-ylidenemethyl)-1-iodo-	CH_2Cl_2	234(4.86),253(4.78), 261(4.87),346(4.58)	5-0001-65I
$C_{30}H_{19}NO_2$			
5,10-Ethanocyclopenta[cd]pleiadene-13,14-dicarboximide, 5,10-dihydro-N-phenyl-	dioxan	218(4.48),242(4.46), 266s(3.60),274s(3.54), 285s(3.42),314s(3.81), 328(4.03),340(4.04), 350(3.86),355(3.92)	78-3051-65
$C_{30}H_{20}$			
Hexapentaene, tetraphenyl-	THF	490(4.78)	24-2975-64
$C_{30}H_{20}Br_2O_9$			
Erythroaphin-fb, dibromohydroxy-	$CHCl_3$	263(4.65),331(3.67), 345(3.68),452(4.41), 478(4.46),526(4.12), 568(4.22)	39-0062-64

Compound	Solvent	λ_{max}(log ε)	Ref.
$C_{30}H_{20}Br_2O_{10}$ Erythroaphin-fb, dibromodihydroxy-	80% dioxan	261(4.73),346(3.80), 478(4.60),526(4.17), 568(4.24)	39-0062-64
$C_{30}H_{20}Cl_2O_8$ Erythroaphin-tt, dichloro-	$CHCl_3$	262(4.22),345(3.59), 450(4.43),525(4.13), 567(4.26)	39-0072-64
$C_{30}H_{20}FeO_3$ Iron, tricarbonyl(tetraphenyl-propadiene)-	n.s.g.	468(2.91)	101-0007-65A
$C_{30}H_{20}I_2O_8$ Erythroaphin-fb, diiodo-	$CHCl_3$	268(4.64),337(3.68), 453(4.38),488(4.37), 526(4.23),569(4.30)	39-0062-64
Erythroaphin-sl, diiodo-	$CHCl_3$	267(4.64),337(3.68), 453(4.40),486(4.36), 527(4.22),569(4.30)	39-0062-64
$C_{30}H_{20}N_2O_2$ 4,4'-Biisoxazole, tetraphenyl-	n.s.g.	240(4.52),268(4.58)	32-0393-64
$C_{30}H_{20}N_3P$ 1-Propene-1,1,2-tricarbonitrile, 3-phenyl-3-(triphenylphosphor-anylidene)-	dioxan	414(4.22)	49-1967-65
$C_{30}H_{20}N_6O_9$ 3H-2,3-Benzoxazepine, 4-(2,4-dinitro-anilino)-1,4-epoxy-4,5-dihydro-3-[5-nitro-2-(1-oxo-1,2-dihydro-2-isoquinolyl)phenyl]-	EtOH	206(4.53),232s(4.41), 338(4.13)	24-1013-65
$C_{30}H_{20}O_2$ Benzo[b]fluoren-11-one, 10-(p-methoxy-phenyl)-5-phenyl-	EtOH	288(4.83)	39-5473-65
Cyclopropenone, diphenyl-, dimer	EtOH	198(4.8),222(4.57), 228(4.58),284s(4.55), 297(4.6),313(4.46)	35-1320-65
$C_{30}H_{20}O_4$ Cyclobuta[1,2-1:3,4-1']diphenanthrene-8b,8c-dicarboxylic acid, 16b,16c-dihydro-	EtOH	209(4.76),277(4.35)	39-5544-64
$C_{30}H_{20}O_9$ Sennidin C	MeOH	260(4.31),369(4.26)	33-1911-65
$C_{30}H_{21}Br$ Fluorene, 9-[2-(bromomethylene)-4,4-diphenyl-3-butenylidene]-	CH_2Cl_2	234(4.77),252(4.64), 261(4.69),303(4.57)	5-0001-65I
$C_{30}H_{21}ClO_4$ 1-(2,3-Diphenyl-2-cyclopropen-1-yl)-2,3-diphenylcyclopropenylium perchlorate	EtOH	278(4.32),288(4.34), 295(4.40),303(4.31)	35-5139-65

Compound	Solvent	λ_{max}(log ϵ)	Ref.
$C_{30}H_{21}NO_2$			
5,10-Ethanocyclopenta[cd]pleiadene-13,14-dicarboximide, 1,2,5,10-tetrahydro-N-phenyl-	dioxan	218s(4.70),232(4.78), 270s(3.58),282(3.79), 293(4.03),304(4.16), 318(4.00),328(3.65)	78-3051-65
$C_{30}H_{22}$			
Benzene, p-bis[2-(1-naphthyl)vinyl]-	dioxan	245(4.65),371(4.76)	38-2839-64A
Benzene, p-bis[2-(2-naphthyl)vinyl]-	dioxan	242(4.8),373(4.6)	38-2839-64A
Bi-2-cyclopropen-1-yl, 2,2',3,3'-tetraphenyl-	n.s.g.	326(4.48)	35-5139-65
5,5'-Bi-5H-dibenzo[a,d]cycloheptene	$CHCl_3$	296(4.31)	87-0088-64
1,4-Pentadiene, 1,5-bisbiphenylene-2-methyl-	DMSO-KO-tert-Bu	669(--)	5-0050-65J
1,4-Pentadiene, 1,5-bisbiphenylene-3-methyl-	DMSO-KO-tert-Bu	352(4.0),649(4.96), 704(4.82)	5-0050-65J
$C_{30}H_{22}Cl_2N_2O_8$			
Erythroaphin-fb, diaminodichloro-, pyridine adduct	$CHCl_3$	261(4.62),333(3.66), 347(3.67),477(4.45), 526(4.11),568(4.21)	39-0072-64
$C_{30}H_{22}Cl_4O_{10}$			
Xanthoaphin-fb, tetrachloro-	$CHCl_3$	265(4.91),284(4.43), 363(4.09),383(4.34), 408(3.83),432(4.09), 461(4.14)	39-0080-64
$C_{30}H_{22}N_2$			
Cyclobutanedicarbonitrile, 1,2,3,4-tetraphenyl-	EtOH	255(2.91),261(3.14), 267(3.95),271(3.78)	39-5544-64
$C_{30}H_{22}N_2O_3S$			
Benzamide, N-[4-cyano-1,3-diphenyl-4-(phenylsulfonyl)-1,3-butadienyl]-	MeCN	372(4.06)	24-3892-65
$C_{30}H_{22}N_3OP$			
2H-Pyrrole-3-carbonitrile, 5-amino-2-oxo-4-[α-(triphenylphosphoranylidene)benzyl]-	dioxan	455(4.10)	49-1967-65
$C_{30}H_{22}O$			
Benzo[b]fluorene, 10-(p-methoxyphenyl)-5-phenyl-	EtOH	270(4.84),314(4.29), 323(4.33)	39-5473-65
$C_{30}H_{22}O_3$			
Acrylic acid, 2,3-diphenyl-, anhydride, cis	hexane	292(4.45)	88-3085-65
trans	hexane	291(4.47)	88-3085-65
$C_{30}H_{22}O_4$			
3-Hexynedioic acid, tetraphenyl-	EtOH	207(4.51),260(4.00)	44-1973-65
$C_{30}H_{22}O_8$			
Erythroaphin-tt	$CHCl_3$	253(4.49),422(4.39), 448(4.50),488(3.70), 524(4.03),565(4.22), 590(3.84)	39-0072-64

Compound	Solvent	λ_{max}(log ε)	Ref.
$C_{30}H_{22}O_9$			
Erythroaphin-sl, 2-hydroxy-	$CHCl_3$	264(4.63),349(3.62), 420s(4.17),442(4.33), 490s(4.26),516(4.31), 561(4.22)	39-6923-65
	dioxan	226(4.69),264(4.32), 348(3.67),439(4.32), 512(4.33),550(4.21)	39-6923-65
	50% acetone-NaOH	418(--),442s(--), 584(--)	39-6923-65
$C_{30}H_{22}O_{14}$			
Benzo[1,2-b:4,5-b']bisbenzofuran, 1,3,6,7,9,12-hexaacetoxy-	dioxan	239(4.69),266(4.04), 300s(4.43),310(4.75), 326(4.73)	1-1063-65
Benzo[1,2-b:4,5-b']bisbenzofuran, 3,4,6,9,10,12-hexaacetoxy-	dioxan	235(4.70),265(4.10), 297s(4.48),308(4.78), 324(4.77)	1-1063-65
Benzo[1,2-b:5,4-b']bisbenzofuran, 2,3,6,9,10,12-hexaacetoxy-	dioxan	250(4.80),260(4.79), 292(4.72),324(4.23)	1-1063-65
$C_{30}H_{23}BrO_{11}$			
Chrysoaphin-sl-1, bromodihydroxy-	$CHCl_3$	270(4.77),390(4.50), 410(4.62),457(4.22), 487(4.32)	39-6923-65
$C_{30}H_{23}N_2O_3P$			
1,4-Naphthoquinone, 2-acetamido-3-(triphenylphorphoranylideneamino)-	$CHCl_3$	290(4.46)	39-1003-65
$C_{30}H_{23}OP$			
2'-Acetonaphthone, 2-(triphenylphosphoranylidene)-	EtOH	268(4.25),275(4.23), 317(4.27),324(4.31)	65-2738-64
$C_{30}H_{24}$			
Cyclobutane, 1,3-bis(diphenylmethylene)-	EtOH	267(4.50)	88-0913-65
Dibenzo[a,e]cyclooctene, 5,6-dimethyl-11,12-diphenyl-	n.s.g.	285(3.50)	35-0086-65
$C_{30}H_{24}CoN_6S_2$			
Cobalt, diisothiocyanatobis(2,9-dimethyl-1,10-phenanthroline)-	n.s.g.	510(--)	12-0691-65
$C_{30}H_{24}N_2O_8$			
4(3H)-Pyrimidone, 3-β-D-ribofuranosyl-, tribenzoate	MeOH	228(4.63),273(3.81)	24-1511-65
$C_{30}H_{24}O$			
2H-Pyran, 2-benzyl-2,4,6-triphenyl-	EtOH	257(4.35),335(4.05)	5-0183-64H
4H-Pyran, 4-benzyl-2,4,6-triphenyl-	EtOH	248(4.37)	5-0183-64H
$C_{30}H_{24}O_2$			
1,4-Cyclohexanedione, 2,3,5,6-tetraphenyl-	EtOH	260(3.14),286(2.59)	39-2009-65
$C_{30}H_{24}O_2P_2$			
Phosphine oxide, p-phenylenebis-[diphenyl-	EtOH	225(4.59),260(3.44)	49-0285-65

Compound	Solvent	λ_{max}(log ε)	Ref.
$C_{30}H_{24}O_4$			
1,5-Pentanedione, 3-benzoyl-2-hydroxy-1,2,5-triphenyl-	n.s.g.	250(4.46)	23-2822-64
Phenanthro[9,10-l]phenanthrene, 2,3,6,7-tetramethoxy-	EtOH	226(4.54),252s(4.77),259(4.81),285(4.28),301(4.22),325s(4.47),333(4.51),363(3.52)	39-3040-65
$C_{30}H_{24}O_6$			
Kaempferol, 4',7-dibenzyl-3-methyl-	EtOH	269(4.37),350(4.30)	22-0779-65
	EtOH-$AlCl_3$	279(4.35),304(4.17),342(4.30),400(4.14)	22-0779-65
$C_{30}H_{24}O_9$			
Isodianellinone, di-O-methyl-, acetate	dioxan	269(4.77),295s(4.22),373(3.61),440(3.72)	12-0218-65
$C_{30}H_{24}O_{11}$			
Chrysoaphin-sl-1, dihydroxy-	$CHCl_3$	268(4.66),381(4.39),403(4.52),456(4.20),487(4.23)	39-6923-65
$C_{30}H_{24}S$			
2H-Thiopyran, 2-benzyl-2,4,6-triphenyl-	EtOH	258(4.32),347(3.66)	5-0183-64H
4H-Thiopyran, 4-benzyl-2,4,6-triphenyl-	EtOH	233(4.48)	5-0183-64H
$C_{30}H_{25}BrNO_4P$			
Succinic acid, [(p-bromophenyl)imino]-(triphenylphosphoranylidene)-, dimethyl ester	n.s.g.	222s(4.52),274(4.03),313s(3.92)	77-0087-64
	n.s.g.	321(4.05)	88-1263-64
$C_{30}H_{25}ClSi_2$			
Disilane, chloropentaphenyl-	C_6H_{12}	241.5(4.23)	25-1492-64
	C_6H_{12}	241.5(4.23)	39-4690-65
$C_{30}H_{25}NO_2$			
Cyclohepta[b]pyrrole-2,8(1H,3H)-dione, 1,3,3-tribenzyl-	EtOH	234(4.20),268(3.93),347(3.95),378(3.90),397(3.71)	94-0473-65
Cyclohepta[b]pyrrol-2(1H)-one, 1,3-dibenzyl-8-(benzyloxy)-	EtOH	238(4.28),285(4.41),306(4.47),395(3.87),418(3.96)	94-0473-65
$C_{30}H_{25}N_2O_6P$			
Succinic acid, [(p-nitrophenyl)imino]-(triphenylphosphoranylidene)-, dimethyl ester	n.s.g.	377(4.20)	88-1263-64
$C_{30}H_{25}O_3P$			
Acrylic acid, 2-[α-(triphenylphosphoranylidene)phenacyl]-, methyl ester	EtOH	267(3.54),274(3.54),305(3.43)	44-3312-65
$C_{30}H_{26}$			
7,7'-Bicyclo[4.2.0]octa-1,3,5-triene, 8,8'-dimethyl-7,7'-diphenyl-	isooctane	256s(3.39),261(3.52),267(3.66),274(3.62),289(2.37),298(2.43)	35-0086-65
$C_{30}H_{26}Cl_2O_8$			
2,2'-(1,3-Azulenylene)bis[1,2,3,?-tetrahydroazulenylium perchlorate	H_2SO_4	234(5.00),268(4.43),302(4.06),360(4.13)	88-1497-65

Compound	Solvent	λ_{max}(log ε)	Ref.
$C_{30}H_{26}Cl_4O_{10}$			
Xanthoaphin-fb, tetrachloro-tetrahydro-	$CHCl_3$	271(4.72),278(4.82), 344(3.84),363(3.91), 380(3.95),405(3.94), 427(3.91)	39-0080-64
$C_{30}H_{26}NO_4P$			
Tetraphenylphosphine imine, dimethyl acetylenedicarboxylate adduct	n.s.g.	313(3.92)	88-1263-64
$C_{30}H_{26}O$			
1,3-Butadiene, 1-ethoxy-1,2,3,4-tetraphenyl-	C_6H_{12}	228(4.36),324(4.25)	32-0252-64
9(10H)-Phenanthrone, 10,10-bis-(p-methylbenzyl)-	MeOH	245(4.54),253s(4.40), 301s(3.40),339(3.52)	44-2362-64
$C_{30}H_{26}O_2$			
2',2'''-Biacetophenone, 2,2"-di-p-tolyl-	MeOH	210(4.63),290(3.49)	44-2362-64
$C_{30}H_{26}O_4$			
Chalcone, 2',4'-bis(benzyloxy)-α-methoxy-	EtOH	230(4.19),307(4.25)	22-3572-65
Chalcone, 2',4-bis(benzyloxy)-3-methoxy-	EtOH	255s(--),356(4.27)	22-3350-65
Chalcone, 2',4-bis(benzyloxy)-4'-methoxy-	EtOH	245s(--),349(4.40)	22-3350-65
5,8,13,14-Pentaphenetetrone, 6-octyl-	dioxan	275(4.54)	65-3484-64
Tectol	n.s.g.	275(4.50),348(3.78), 363(3.65)	24-0588-64
$C_{30}H_{26}O_5$			
Chalcone, 2',4-bis(benzyloxy)-3-methoxy-, epoxide	EtOH	255(4.03),291(3.80), 318s(--)	22-3350-65
Chalcone, 2',4-bis(benzyloxy)-4'-methoxy-, epoxide	EtOH	231(4.35),278(4.15), 313(3.99)	22-3350-65
6,12-Methano-1H-dibenzo[d,g][1,3]dioxocin-1-one, 2,3,4,12-tetrahydro-6-(p-hydroxyphenyl)-3,3-dimethyl-, benzoate	EtOH	230(4.44),263s(4.08)	78-3707-65
1,2-Propanedione, 1-(2-benzyloxy-4-methoxyphenyl)-3-(4-benzyl-oxyphenyl)-	EtOH	228(4.42),280(4.00), 317(3.89)	22-3350-65
	EtOH-NaOH	383(3.97)	22-3350-65
1,2-Propanedione, 1-[2,4-bis(benzyl-oxy)phenyl]-3-(p-methoxyphenyl)-	MeOH	282(3.37),315(3.37)	78-1141-64
Xanthen-1(2H)-one, 3,4-dihydro-9-(p-hydroxyphenacyl)-3,3-di-methyl-, benzoate	EtOH	231(4.42),251(4.45)	78-3707-65
$C_{30}H_{26}O_{10}$			
[2,2'-Binaphthalene]-8,8'-dicarbox-aldehyde, 6,6',7,7'-tetrahydroxy-5,5'-diisopropyl-3,3'-dimethyl-1,1',4,4'-tetraoxo-	EtOH	223(4.5),272(4.4), 315s(4.2),408s(3.7)	44-4111-65
Xanthoaphin-fb	$CHCl_3$	258(4.66),283(4.74), 359(4.03),379(4.33), 406(3.78),430(4.04), 459(4.08)	39-0080-64

Compound	Solvent	λ_{max}(log ε)	Ref.
$C_{30}H_{26}O_{14}$			
Ergoflavin	6N HCl	246(4.29),270(4.18), 360(3.82)	78-1417-65
	pH 13	237(4.31),286(4.13), 414(4.07)	78-1417-65
	EtOH	242(4.33),278(4.33), 384(3.90)	39-4130-65
	EtOH	207(4.40),241(4.29), 279(4.30),380(3.86)	78-1417-65
$C_{30}H_{26}Si$			
Silacyclopenta-2,4-diene, 1,1-dimethyl-2,3,4,5-tetraphenyl-	C_6H_{12}	247(4.41),357(3.99)	35-1596-64
$C_{30}H_{27}BrN_2S_2$			
1-Ethyl-2-[3-(1-ethylnaphtho[1,2-d]-thiazolin-2-ylidene)-2-methyl-propenyl]naphtho[1,2-d]thia-zolium bromide	H_2O	510(4.79)	46-1896-64
	EtOH	575(4.98)	46-1896-64
$C_{30}H_{27}ClO_5$			
Propiophenone, 2',4'-bis(benzyloxy)-3-chloro-2-hydroxy-3-(p-methoxy-phenyl)-	EtOH	270(3.56),305(3.39)	78-1141-64
$C_{30}H_{27}N_3O_8S$			
Cytidine, N-benzoyl-2'-deoxy-, 3'-benzoate 5-p-toluenesulfonate	MeOH	227(4.54),260(4.43), 302(4.05)	44-3067-65
$C_{30}H_{28}Cl_2O_6$			
3-Cyclohexene-1-carboxylic acid, 6,6'-[1,2-bis(2-chloro-5-methoxy-phenyl)-1,2-dihydroxyethylene]di-, di-γ-lactone	EtOH	205(4.75),230(4.34), 284(3.64)	70-0806-65
$C_{30}H_{28}CoN_2O_2$			
Ethylamine, α-phenyl-N-salicylidene-, cobalt chelate	MeOH	253(4.51),317(3.94)	65-3055-64
$C_{30}H_{28}CuN_2O_2$			
Ethylamine, α-phenyl-N-salicylidene-, copper chelate	MeOH	240(4.69),365(4.02)	65-3055-64
$C_{30}H_{28}Fe_2$			
1,3,5,7,9-Decapentaene, 1,10-diferrocenyl-	$CHCl_3$	390s(--),406(4.86)	5-0088-64F
$C_{30}H_{28}N_2NiO_2$			
Ethylamine, α-phenyl-N-salicylidene-, nickel chelate	MeOH	255(4.49),315(3.94), 395(3.28)	65-3055-64
$C_{30}H_{28}N_2O_2$			
Benzoic acid, p-(phenylazo)-, 2,8,10,16-heptadecatetraene-4,6-diynyl ester	ether	251(4.54),268(4.51), 298s(4.63),316(4.74), 339(4.69)	24-3010-65
2-cis	ether	249(4.52),267(4.49), 298(4.62),316(4.77), 337(4.68)	24-1225-65

Compound	Solvent	λ_{max}(log ε)	Ref.
$C_{30}H_{28}N_4O_{12}$			
3,10-Triphenodioxazinedicarbamic acid, 6,13-dicarboxy-2,9-dimethoxy-, 6,13-diethyl 3,10-dimethyl ester	DMF	552(4.72),591(4.91)	27-0242-65
$C_{30}H_{28}O_2$			
9,10-Phenanthrenediol, 9,10-dihydro-9,10-bis(p-methylbenzyl)-	MeOH	210(4.56),273(4.18)	44-2362-64
$C_{30}H_{28}O_5$			
2-Anthraceneacetic acid, 5,9β-bis-(benzyloxy)-1,2,3,4,9,9aα-hexahydro-4-oxo-	EtOH	207(4.55),284(3.29), 349(3.66)	65-0662-65
$C_{30}H_{28}O_6$			
Lactic acid, 2-[2,4-bis(benzyloxy)-phenyl]-3-(p-methoxyphenyl)-	EtOH	276(3.81),282(3.79)	78-1141-64
$C_{30}H_{28}O_8$			
Rottlerin	EtOH	228(4.40),295(4.41), 355(4.18)	32-0725-65
$C_{30}H_{28}O_9$			
Rottlerin, 4-hydroxy-	EtOH	298(4.31),387(4.16)	32-0725-65
$C_{30}H_{28}O_{13}$			
Θ-Rhodomycinone, tetraacetate	EtOH	260(4.35),350(3.73)	39-3927-64
$C_{30}H_{30}Br_2N_4O_4$			
Byssochlamic acid, bis[(p-bromo-phenyl)hydrazide]	EtOH	236(4.61),292(3.49)	39-1787-65
$C_{30}H_{30}N_2O_7S$			
Thymine, 1-(2-deoxy-5-O-trityl-βD-lyxosyl)-, 3'-methanesulfonate	EtOH	266(4.00)	44-2076-64
$C_{30}H_{30}N_4$			
Phenazine, 1,4-bis(p-dimethylamino-benzyl)-	C_6H_{12}	260(4.48)	65-1969-64
$C_{30}H_{30}O_2$			
peri-Xanthenoxanthene, 1,7-dipentyl-	n.s.g.	277(4.17),289(4.13), 314(3.56),370(3.75), 390(3.68),414(4.01), 442(4.14)	78-2095-65
$C_{30}H_{30}O_4$			
Tectol, tetrahydro-	n.s.g.	252(4.76),328(3.91), 340(3.94)	24-0588-64
$C_{30}H_{30}O_7$			
Flavone, 6-allyl-3,3',4',7-tetrakis-(allyloxy)-5-hydroxy-	EtOH	258(4.3),274(4.3), 361(4.2)	24-0114-65
	EtOH-NaOAc	258(4.3),274(4.3), 361(4.2)	24-0114-65
	EtOH-NaOEt	287(4.3),383(3.8)	24-0114-65
	EtOH-$AlCl_3$	284(4.2),351(4.1), 411(4.0)	24-0114-65

Compound	Solvent	$\lambda_{max}(\log \epsilon)$	Ref.
$C_{30}H_{30}O_7S$			
2,9-Epoxyanthracen-10(1H)-one, 2α,3,4,4aα,9,9aα-hexahydro-3α-hydroxy-9α-methyl-, sulfite	EtOH	247(4.31),286(3.40)	70-1024-64
$C_{30}H_{30}O_8$			
[2,2'-Binaphthalene]-7,7',8,8'-tetrone, 1,1'-dihydroxy-5,5'-diisopropyl-6,6'-dimethoxy-3,3'-dimethyl-	EtOH	265(4.44),303(3.99), 450(4.13),535s(3.90)	44-3617-64
$C_{30}H_{30}O_{14}$			
Trifolirhizin, tetraacetate	MeOH	285(3.58),311(3.84)	94-0093-65
$C_{30}H_{30}O_{16}$			
p,p'-Bitolyl, octaacetoxy-	EtOH	272(3.43)	57-0152-65A
$C_{30}H_{31}BrN_2O_7$			
7H-Isoreserpiline, 7-hydroxy-, m-bromobenzoate	EtOH	239(4.68),294(3.76)	35-2229-65
$C_{30}H_{31}N_3O_5$			
Cytosine, 1-(2,3,5-tri-O-benzyl-β-D-arabinofuranosyl)-	MeOH	233(3.87),273(3.95)	44-0835-65
$C_{30}H_{32}N_2O_4$			
4(1H)-Pyridone, 2,2'-hexamethylene-bis[6-(p-methoxyphenyl)]-	n.s.g.	240(4.80)	44-4263-65
$C_{30}H_{32}N_4O_2$			
α-Porphinol, 1,2,3,4,5,6,7,8-octa-methyl-, acetate	5N HCl	403(5.57),516s(3.20), 550(4.02),594(3.33)	12-1835-65
	CHCl	397(5.00),499(4.01), 532(3.63),570(3.62), 625(3.03)	12-1835-65
$C_{30}H_{32}O_{10}$			
Siphulin, acetate	EtOH	264s(4.04),299(3.93)	1-1677-65
$C_{30}H_{33}NO_6S$			
Estradiol, 3-benzoate 17-(hydrogen sulfate), compound with pyridine	MeOH	230(4.31)	13-0845-65
$C_{30}H_{34}F_2N_2O_4$			
Pregna-4,6-dieno[3,2-c]pyrazole, 2'-(2,4-difluorophenyl)-11β,17,21-trihydroxy-6,16α-dimethyl-20-oxo-	MeOH	270(4.14),313(4.30)	87-0355-64
$C_{30}H_{34}N_2O_{19}$			
Amaranthine	H_2O	536(4.75)	7-0963-64
$C_{30}H_{34}N_4$			
Porphyrin, 3,17-diethyl-2,7,8,12,13,18-hexamethyl-	$CHCl_3$	269(3.66),330s(4.17), 400(5.11),498(3.87), 535(3.75),568(3.58), 597(2.95),624(3.45)	39-1620-65
$C_{30}H_{34}N_5O_4P$			
Phosphorane, trimorpholino[(5-oxo-6(5H)-chrysenylidene)hydrazono]-	MeOH	400(3.1),450s(3.0)	5-0056-64I

Compound	Solvent	λ max(log ε)	Ref.
$C_{30}H_{34}O_7$			
Anthracene, 4a,10a-diacetoxy-5-(benzyloxy)-1,4,4aβ,9,9aα,10-hexahydro-9β-(tetrahydropyranyloxy)-	EtOH	278(3.37),284(3.36)	65-0662-65
Pregna-1,4-diene-3,11-dione, 17α,20α,21-trihydroxy-, 20-acetate 21-benzoate	EtOH	233(4.42)	78-0179-65
$C_{30}H_{35}N_3O_5$			
Dispiro[cyclopenta[7,8]phenanthro[2,3-d]-triazole-1(11H),4'-[1,3]dioxolane-5',4"-[1,3]dioxolan]-11-one, dodecahydro-2,10a,12a-trimethyl-8-phenyl-	MeOH	304(4.51),310(4.51)	87-0584-64
$C_{30}H_{35}N_3O_6$			
5α-Pregnano[3,2-d]-3'H-1',2',3'-triazole, 17α,20:20,21-bis(methylenedioxy)-16α-methyl-6,11-dioxo-3'-phenyl-	MeOH	228(4.03)	87-0584-64
Pregn-4-eno[3,2-d]-2'H-1',2',3'-triazole, 17α,20:20,21-bis(methylenedioxy)-16α-methyl-11-oxo-2'-phenyl-, N-1'-oxide	MeOH	227(4.19),252(4.22), 297(4.27)	87-0584-64
$C_{30}H_{36}N_2O_4$			
Pregn-4-eno[3,2-d]pyrimidine, 11β,17α,21-trihydroxy-16α-methyl-20-oxo-2'-phenyl-	MeOH-HCl	276(4.44)	87-0584-64
	MeOH	258(4.61),307(3.98)	87-0584-64
$C_{30}H_{36}N_2O_5$			
Pregn-4-eno[3,2-c]pyrazole, 11β-hydroxy-17,20:20,21-bis(methylenedioxy)-2'-phenyl-	MeOH	261(4.22)	35-1520-64
$C_{30}H_{36}N_4$			
Corrole, 8,12-diethyl-2,3,7,13,17,18,21-heptamethyl-	$CHCl_3$	400(4.94),502s(3.73), 541(4.20),559(4.20), 600(4.20)	39-1620-65
Corrole, 8,12-diethyl-2,3,7,13,17,18,22-heptamethyl-	$CHCl_3$	413(4.88),510(3.87), 527(3.92),547(4.18), 625(3.88)	39-1620-65
$C_{30}H_{36}O_6$			
Pregn-4-ene-3,11-dione, 20α-acetoxy-17α,21-benzylidenedioxy-	EtOH	239(4.18)	78-0179-65
$C_{30}H_{36}O_9$			
Isonimbin	EtOH	230(4.28),330(1.73)	2-0108-64
$C_{30}H_{36}O_{11}$			
Glaucanol, dihydro-, pentaacetate	EtOH	244(4.11)	22-1818-64
$C_{30}H_{37}ClN_2O_8$			
Tetracycline, 7-chloro-9,N-di-tert-butyl-dehydro-	MeOH-HCl	255(4.37),395(3.80)	44-2746-64
	MeOH-NaOH	245(4.41),345(3.70), 420(4.15)	44-2746-64
$C_{30}H_{37}NO_4$			
Pregn-4-en-20-one, 3β-acetoxy-16α,17α-[3-phenyl-3,1-(2-isoxazolino)]-	MeOH	262(4.10)	4-0280-64
	n.s.g.	212(4.19),218(4.00), 262(4.19)	5-0139-64G

Compound	Solvent	λ_{max}(log ε)	Ref.
$C_{30}H_{37}N_3O_5$			
Pregn-4-eno[3,2-d]-2'H-1',2',3'-triazole, 11β-hydroxy-17,20:20,21-bis-(methylenedioxy)-16α-methyl-2'-phenyl-	MeOH	305(4.51),310(4.51)	87-0584-64
$C_{30}H_{37}N_3O_6$			
Pregn-4-ene-2,3,11-trione, 17α,20:20,21-bis(methylenedioxy)-16α-methyl-, 2-oxime 3-phenylhydrazone	MeOH	249(4.15),366(4.21)	87-0584-64
$C_{30}H_{38}N_2O_5$			
5α-Pregnano[3,2-c]pyrazole, 11β-hydroxy-17,20:20,21-bis(methylenedioxy)-1'-phenyl-	MeOH	269.5(4.30)	35-1520-64
5α-Pregnano[3,2-c]pyrazole, 11β-hydroxy-17,20:20,21-bis(methylenedioxy)-2'-phenyl-	MeOH	251(4.04)	35-1520-64
5α-Pregnano[3,2-c]pyrazole, 11β,17,21-trihydroxy-20-oxo-2'-phenyl-, 21-acetate	MeOH	251(3.96)	35-1520-64
$C_{30}H_{38}O_4$			
Pregna-5,16-dien-20-one, 3β-hydroxy-, (p-methoxyphenyl)acetate	MeOH	229(4.14),277(3.31), 283(3.25)	35-0995-65
$C_{30}H_{38}O_5Si_5$			
Cyclopentasiloxane, 2,2,4,4,6,6-hexamethyl-8,8,10,10-tetraphenyl-	$CHCl_3$	265(3.14)	46-1066-65
$C_{30}H_{38}O_6$			
Pregn-4-en-3-one, 20α-acetoxy-17α,21-benzylidenedioxy-11β-hydroxy-	EtOH	243(4.16)	78-0179-65
$C_{30}H_{38}O_8$			
2,3,6,7-Anthracenetetracarboxylic acid, 1,2,3,4,5,6,7,8-octahydro-1,4,5,8-tetramethyl-9,10-divinyl-, tetramethyl ester	EtOH	215(4.55),272(2.84)	33-0437-65
$C_{30}H_{38}O_9$			
Isonimbin, dihydro-	EtOH	233(3.95)	2-0108-64
$C_{30}H_{39}ClN_2O_8$			
Tetracycline, 9,N-di-tert-butyl-7-chloro-	MeOH-HCl	258(4.32),342(3.89), 373(3.96)	44-2746-64
	MeOH-NaOH	243(4.33),270(4.23), 391(4.11)	44-2746-64
$C_{30}H_{39}NO_2$			
Cyclopenta[5,6]naphth[1,2-f]indol-1-ol, 7-benzylhexadecahydro-10a,12a-dimethyl-, acetate	EtOH	209(4.18)	44-3612-64
$C_{30}H_{39}N_3O_7S$			
5β-Pregnane-11,20-dione, 17α-azido-3β,16α-dihydroxy-, 3-acetate 16-p-toluenesulfonate	EtOH	226(4.1)	13-0805-65

Compound	Solvent	λ_{max}(log ε)	Ref.
$C_{30}H_{40}N_2O_8$			
Tetracycline, 9,10-di-tert-butyl-	MeOH-HCl	271(4.35),341(4.06), 366(4.07)	44-2746-64
	MeOH-NaOH	269(4.26),380(4.23)	44-2746-64
$C_{30}H_{40}N_{10}$			
Bis(tetraethylammonium) 1,1,4,4-tetra-cyano-2,3-bis(dicyanomethyl)-butadienediide	MeCN	328(4.30),420(4.30)	35-2898-64
$C_{30}H_{40}O$			
2,4,6,8,10,12,14,16-Heptadecaoctaenal, 2,6,11,15-tetramethyl-17-(2,6,6-tri-methylcyclohex-1-en-1-yl)-, all trans	hexane	457(5.04)	28-1840-64A
$C_{30}H_{40}O_2$			
1,4,6,8,10,12,14,16-Heptadecaoctaen-3-one, 1-hydroxy-2,6,11,15-tetra-methyl-17-(2,6,6-trimethyl-cyclohex-1-en-1-yl)-	hexane	470(4.98)	28-1840-64A
$C_{30}H_{40}O_2S_3$			
2,5-Cyclohexadien-1-one, 4,4'-(1,2,4-trithiolane-3,5-diylidene)bis-[2,6-di-tert-butyl-	MeOH	433(4.65)	5-0037-65D
$C_{30}H_{40}O_6S$			
Pregn-4-ene-3,6-dione, 20β-(2-hydroxy-ethoxy)-, p-toluenesulfonate	EtOH	226(4.25),252(4.06)	13-0585-65
Pregn-4-ene-3,20-dione, 17-hydroxy-16α-(hydroxymethyl)-6α-methyl-, 16-p-toluenesulfonate	EtOH	227(4.34),242(4.20), 272(2.85)	44-3476-64
$C_{30}H_{40}O_9$			
Isonimbin, tetrahydro-	EtOH	210(3.81),290(1.86)	2-0108-64
$C_{30}H_{41}Br_3O_3$			
18-Norergosta-8,11,13-trien-7-one, 3β-acetoxy-6α,22,23-tribromo-12-methyl-	EtOH	220(4.30),274(4.08), 306s(3.49)	78-0929-64
$C_{30}H_{41}NO_4$			
5α-Pregn-20-ene, 3β,16β-diacetoxy-20-(2-pyridyl)-	hexane	236(3.83),267(3.61)	24-1961-65
$C_{30}H_{41}NO_7S$			
5β-Pregnane-11,20-dione, 3α-acetoxy-17α-amino-16α-(p-toluenesulfonoxy)-	EtOH	225(4.05)	13-0805-65
$C_{30}H_{42}Br_2O_6$			
6,7-Seco-18-norergosta-8,11,13-triene-6,7-dioic acid, 3β-acetoxy-22,23-dibromo-12-methyl-	EtOH	257(4.04),300(2.48)	78-0929-64
$C_{30}H_{42}N_2O_2S$			
Neothiobinupharidine	acid	284(2.95)	88-0129-64
$C_{30}H_{42}N_4O_4$			
Ceanothine, dihydro-	n.s.g.	232(4.0),280(3.0)	23-2594-65

Compound	Solvent	λ_{max}(log ϵ)	Ref.
$C_{30}H_{42}O_2$			
3-Apolycopenal, 1,2-dihydro-1-hydroxy-	benzene	489(5.16),516s(--)	1-1739-64
Ergosta-3,5,7,9(11),22-pentaene, 3-acetoxy-	EtOH	206(3.80),236(4.09), 338(4.01),354(4.14), 373(4.02)	39-6991-65
Lumista-3,5,7,9(11),22-pentaene, 3-acetoxy-	EtOH	234(4.02),320(4.09), 334(4.09),353(3.99)	39-6991-65
$C_{30}H_{42}O_3$			
Ergosta-4,6,8(14),22-tetraen-3-one, 4-acetoxy-	EtOH	250(3.67),362(4.25)	39-6991-65
	NaOH	218(4.12),398(4.05)	39-6991-65
$C_{30}H_{42}O_4$			
Ergosta-4,6,8(9),22-tetraen-3-one, 6-acetoxy-14α-hydroxy-	EtOH	345(4.42)	39-6991-65
	NaOH	265(3.46),440(3.51)	39-6991-65
Pregna-3,5,7-trien-20-one, 17α-acetoxy-3-(cyclohexyloxy)-6-methyl-	EtOH	324(4.31)	78-1753-65
$C_{30}H_{42}O_6$			
Withaferin A, deoxydihydro-, acetate	EtOH	226(3.91)	44-1774-65
$C_{30}H_{42}O_8$			
2,3,6,7-Anthracenetetracarboxylic acid, 9,10-diethyl-1,2,3,4,5,6,7,8-octahydro-1,4,5,8-tetramethyl-, tetramethyl ester	EtOH	215(3.61),285(2.76)	33-0437-65
$C_{30}H_{42}O_9$			
14ξ-Card-20(22)-enolide, 1β,5β,19-trihydroxy-3β,11α-diacetoxy-, cyclic 1,19-acetal with acetone	MeOH	215(4.21)	44-0527-64
Nimbin, hexahydro-	EtOH	208(3.93),290(1.67)	2-0108-64
$C_{30}H_{43}NO_5$			
5β-Pregnane-3,20-dione, 11β-hydroxy-11α-(2-pyridyl)-, diethylene ketal	EtOH	264(3.56)	44-2095-65
$C_{30}H_{44}Br_2O_2$			
18-Norergosta-8,11,13-trien-3β-ol, 22,23-dibromo-12-methyl-, acetate	isooctane	220s(4.13),261(2.39), 268(2.45),277s(2.33)	78-0929-64
$C_{30}H_{44}O_2$			
9β-Ergosta-3,5,7,22-tetraen-3-ol, 3-acetate	MeOH	314(4.22)	54-1069-64
$C_{30}H_{44}O_3$			
22,25,27-Ebelotrien-16-oic acid, 17-hydroxy-3-oxo-, 16,17-lactone	n.s.g.	269(4.61),278(4.71), 290(4.59)	12-1451-65
Ergosta-4,6,22-trien-3-one, 4-acetoxy-	EtOH	214(4.16),267(3.99), 307(4.05)	39-6991-65
	NaOH	260(4.18),302(4.01)	39-6991-65
$C_{30}H_{44}O_4$			
Colupone	EtOH-acid	245s(--),285(3.98)	39-6542-65
	EtOH-base	240s(--),280(4.29)	39-6542-65
Ergosta-3,6,22-trien-3-ol, 5α,8α-epidioxy-, acetate	EtOH	212(3.89)	39-6991-65

Compound	Solvent	λ_{max}(log ϵ)	Ref.
$C_{30}H_{44}O_5$			
Androst-4-en-3-one, 7α,17β-dihydroxy-, 7-acetate 17-(β-cyclohexane-propionate)	EtOH	238(4.22)	24-3363-64
Ceanothic acid, epiketo acid from	n.s.g.	290(1.85)	12-0141-64
Ergosta-3,7,22-trione, 3-acetoxy-5α,8α-epidioxy-	EtOH	240(4.02)	39-6991-65
25D-Spirost-4-en-3-one, 1α,2α-dihy-droxy-, cyclic acetal with acetone	EtOH	244.5(4.18)	94-0050-65
$C_{30}H_{44}O_5S$			
Pregn-5-en-3β-ol, 20β-(2-hydroxy-ethoxy)-, 20-p-toluenesulfonate	EtOH	225(4.09),262(2.77), 274(2.64)	13-0585-65
$C_{30}H_{44}O_6$			
Cyclosenegenin	EtOH	209(3.83)	35-2065-65
$C_{30}H_{44}O_7$			
Cucurbitacin L	EtOH	270(3.90)	39-0529-64
	50% EtOH-KOH	313(3.76)	39-0529-64
$C_{30}H_{44}O_8$			
Apocannoside	EtOH	216(4.10)	87-0803-64
Cucurbitacin J	EtOH	270(3.93)	39-0529-64
	EtOH-KOH	313(3.79)	39-0529-64
Cucurbitacin K	EtOH	270(3.90)	39-0529-64
	EtOH-KOH	312(3.77)	39-0529-64
$C_{30}H_{44}O_9$			
Cymarin	EtOH	216(3.99)	87-0803-64
$C_{30}H_{44}O_{10}$			
Strophanthojavoside	EtOH	216(4.18),299(1.41)	33-2164-64
Taxinol, pentaacetyldihydro-	EtOH	210(3.43)	95-0404-65
$C_{30}H_{44}O_{11}$			
Antiarojavoside	EtOH	218(4.21),299(1.61)	33-2164-64
$C_{30}H_{45}ClO_6$			
Senegenin	MeOH	205(3.84)	23-0491-64
$C_{30}H_{45}ClO_9$			
Cerbertin, deacetyl-, chlorohydrin	n.s.g.	217(4.12)	12-1079-65
$C_{30}H_{45}NO_3S$			
5α-Androstano[3,2-e]-1',4'-thiaz-5'-en-3'-one, 17β-(3-cyclohexylpropionoxy)-	EtOH	230(3.63),298(3.36)	87-0555-64
$C_{30}H_{45}NO_4$			
Pregna-5,20-diene, 21-acetyl-3β-hydroxy-16α,17α-isopropylidene-dioxy-20-(N-pyrrolidyl)-	EtOH	313(4.51)	78-1927-64
$C_{30}H_{45}N_3S_3$			
4H,10H,16H-Trithieno[3,4-c:3',4'-h:-3",4"-m][1,6,11]triazacyclopenta-decine, 5,11,17-triethyl-5,6,11,12,-17,18-hexahydro-1,3,3,9,13,15-hexa-methyl-	n.s.g.	220(4.3),239(4.3), 248s(4.3)	70-2182-64

Compound	Solvent	$\lambda_{max}(\log \epsilon)$	Ref.
$C_{30}H_{46}$			
7,9(11)-Fernadiene	EtOH	232(4.13),240(4.17), 248(3.99)	88-3413-64
$C_{30}H_{46}Br_2O_2$			
Ergosta-7,14-dien-3β-ol, 22,23-dibromo-, acetate	EtOH	240(4.08)	13-0637-65
18α-Oleanan-1-one, 2,2-dibromo-19β,28-epoxy-	n.s.g.	215(3.2),265(3.1), 315(1.9)	24-2837-65
$C_{30}H_{46}Br_2O_4$			
Ergost-8(14)-en-15-one, 3β-acetoxy-22,23-dibromo-17-hydroxy-	EtOH	253(4.11)	13-0637-65
$C_{30}H_{46}Br_4O_2$			
Ergost-8-en-3β-ol, 7,11,22,23-tetrabromo-, acetate	isooctane	236(3.75),261(3.86)	13-0637-65
Ergost-14-en-3β-ol, 7,8,22,23-tetrabromo-, acetate	isooctane	253(3.88)	13-0637-65
$C_{30}H_{46}N_2O$			
Lup-20(29)-en-3-one, 2-diazo-	MeOH	292(4.0)	24-2284-65
$C_{30}H_{46}N_2O_2$			
18α-Oleanan-1-one, 2-diazo-19β,28-epoxy-	MeOH	264(4.0)	24-2291-65
18α-Oleanan-3-one, 2-diazo-19β,28-epoxy-	n.s.g.	291(4.0)	24-1837-65
$C_{30}H_{46}O$			
Ergosta-4,6,8(14),22-tetraene, 3ξ-ethoxy-	EtOH	287(4.48)	13-0539-65
	EtOH	287(4.52)	24-2383-65
Lumista-5,7,22-trien-3-one, 4,4-dimethyl-	n.s.g.	287(3.97)	39-7199-65
$C_{30}H_{46}O_2$			
Allo-2-betulene, 1α-hydroxy-	ether	219(4.0)	24-2291-65
Ergosta-7,14,22-trien-3β-ol, acetate	EtOH	242(4.00)	13-0637-65
18α-Olean-12-ene-3,11-dione	EtOH	246(4.07)	44-1698-65
$C_{30}H_{46}O_3$			
Ebelo-22,25,27-trien-16-oic acid, 3β,17-dihydroxy-, 16,17-lactone	n.s.g.	269(4.65),278(4.76), 290(4.63)	12-1451-65
Gon-4-en-3-one, 13β-ethyl-17β-hydroxy-, 10-undecenoate	n.s.g.	240(4.23)	39-4472-64
9α-Lumista-6,8,22-triene-3β,5α-diol, 3-acetate	EtOH	275(3.64)	39-2054-65
18α-Oleanane-2,3-dione, 19β,28-epoxy-	n.s.g.	270(3.8)	24-1837-65
	base	305(3.5)	24-1837-65
$C_{30}H_{46}O_4$			
Ether, 2,5-di-tert-butyl-6-hydroxy-3-methoxyphenyl 2,5-di-tert-butyl-4-methoxyphenyl	n.s.g.	233(4.1),295(3.8)	5-0061-64A
29-Nor-8α,9β,13α,14β-dammara-15,17(20)-dien-21-oic acid, 11α-hydroxy-3-oxo-, methyl ester	EtOH	273(4.28)	78-3505-65
	dioxan	272(4.28)	31-0344-64

Compound	Solvent	λ_{max}(log ϵ)	Ref.
$C_{30}H_{46}O_5$			
29-Nor-8α,9β,13α,14β-dammaran-21-oic acid, 11α,16β-dihydroxy-2-(hydroxy-methylene)-3-oxo-, γ-lactone	EtOH	292(3.88)	78-3505-65
	dioxan	292(3.88)	31-0344-64
$C_{30}H_{46}O_6$			
Isosenegenin, dechloro-	MeOH	199(3.82)	23-0491-64
$C_{30}H_{46}O_8$			
Solanoside	EtOH	216(4.15)	33-0065-65
Vallaroside	EtOH	217(4.16)	33-0065-65
$C_{30}H_{46}O_{10}$			
Bipindaloside	EtOH	217(4.19)	33-0799-64
Callengoside	EtOH	217(4.20)	33-0799-64
$C_{30}H_{47}BrO_2$			
18α-Oleanan-1-one, 2α-bromo-19β,28-epoxy-	n.s.g.	230(3.3),295(1.5)	24-2837-65
$C_{30}H_{47}NO_2$			
20(29)-Lupene-2,3-dione, 2-oxime, anti	MeOH	240(4.0)	24-2284-65
	NaOH	292(4.2)	24-2284-65
$C_{30}H_{47}NO_3$			
18α-Oleanane-1,2-dione, 2-oxime, anti	MeOH	234(3.6)	24-2291-65
	NaOH	284(3.8)	24-2291-65
18α-Oleanane-2,3-dione, 2-oxime, anti	n.s.g.	240(4.0)	24-1837-65
	NaOH	291(4.2)	24-1837-65
$C_{30}H_{48}$			
Naphthalene, 2,6-didecyl-	C_6H_{12}	230(5.22),274(3.79), 303(2.70),309(2.95), 317(2.78),324(3.11)	39-2324-65
Neo-11,13(18)-hopadiene	C_6H_{12}	247(4.37),256(4.45), 267(4.25)	88-3337-65
$C_{30}H_{48}N_2O_3$			
18α-Oleanane-2,3-dione, 19β,28-epoxy-, 2,3-amphi-dioxime	n.s.g.	225(3.7)	24-1837-65
	NaOH	278(4.0)	24-1837-65
$C_{30}H_{48}N_4O_2$			
p-Benzoquinone, tetrakis(hexahydro-1H-azepin-1-yl)-	methyl cellosolve	238(4.09),421(3.77), 575s(2.84)	78-1889-64
$C_{30}H_{48}O$			
Cholest-4-en-3-one, 4-allyl-	hexane	242(4.10)	22-0227-65
	EtOH	250(4.26)	22-0227-65
Cholest-4-en-3-one, 6ξ-allyl-	EtOH	243.5(4.23)	22-0227-65
Dustaninone A, anhydro-	C_6H_{12}	237(4.10)	88-3337-65
9β,10α-Ergosta-3,5,22-triene, 3-ethoxy-	MeOH	244(4.26)	54-0904-65
3-Filicen-2-one	EtOH	237(4.11)	88-3413-64
Lupenone	EtOH	290(1.76)	95-0887-64
$C_{30}H_{48}O_2$			
Cholesta-3,5-dien-3-ol, 4-methyl-, acetate	EtOH	230s(4.28),236(4.30), 243s(4.20)	39-1850-64
Clerodol	EtOH	203.5(3.95)	78-0797-65
12,18-Cyclolanostane-3,11-dione	EtOH	217(3.85)	33-0704-65

Compound	Solvent	$\lambda_{max}(\log \epsilon)$	Ref.
$C_{30}H_{48}O_3$			
Cholest-4-en-3-one, 1α,2α-dihydroxy-, cyclic acetal with acetone	EtOH	245(4.16)	94-0050-65
Cholest-2-en-4-one, 3-hydroxy-5-methyl-, acetate	MeOH	231(3.72)	78-2531-64
5α-Cholest-2-en-4-one, 3-hydroxy-5-methyl-, acetate	MeOH	240(3.72)	78-2531-64
Ebelin, lactone	n.s.g.	269(4.65),278(4.76), 290(4.64)	12-1451-65
Ergosta-6,8(14),22-trien-5α-ol, 3β-acetoxy-	EtOH	253.5(4.38)	13-0539-65
5α-Ergost-8(9)-en-11-one, 3β-acetoxy-	EtOH	254(3.83)	39-0156-65
5α,14β-Ergost-8(9)-en-11-one, 3β-acetoxy-	EtOH	249(3.92)	39-0156-65
Mangiferolic acid	n.s.g.	217(4.05)	88-2377-65
$C_{30}H_{48}O_3S$			
1-Naphthalenesulfonic acid, 3,7-didecyl-	EtOH	232(4.91),280(3.78), 313(3.26),318(3.08), 327(3.36)	39-2324-65
$C_{30}H_{48}O_4$			
29-Nor-8α,9β,13α,14β-dammar-17(20)-en-21-oic acid, 11α-hydroxy-3-oxo-, methyl ester	dioxan	234(4.03)	31-0344-64
Saikogenin A	n.s.g.	242(4.43),250(4.48), 260(4.29)	88-3783-65
$C_{30}H_{48}O_5$			
29-Nor-8α,9β,13α,14β-dammar-17(20)-en-21-oic acid, 11α,16α-dihydroxy-3-oxo-, methyl ester	EtOH	231(4.03)	78-3505-65
	dioxan	231(4.03)	31-0344-64
$C_{30}H_{48}S_2$			
Cholesta-4,6-dien-3-one, 4-methyl-, cyclic ethylene mercaptole	EtOH	253(4.41)	39-2476-65
$C_{30}H_{49}BrO$			
Lanost-8-en-2-one, 3α-bromo-	dioxan	311(2.05),320(2.03)	22-2245-64
Lanost-8-en-2-one, 3β-bromo-	dioxan	287(1.38),293(1.38)	22-2245-64
Lanost-8-en-3-one, 2α-bromo-	dioxan	290(1.70)	22-2245-64
Lanost-8-en-3-one, 2β-bromo-	dioxan	282(1.63),288(1.61), 297(1.54),305(1.18)	22-2245-64
$C_{30}H_{50}$			
Cholesta-3,5-diene, 3-ethyl-4-methyl-	C_6H_{12}	238(4.29),245(4.33), 253s(4.13)	23-0464-64
Cholesta-3,5-diene, 4-ethyl-3-methyl-	C_6H_{12}	237(4.29),245(4.30), 253s(4.11)	23-0464-64
γ-Lupene	EtOH	214(3.60),216(3.49), 218(3.37),220(3.25) (end absorptions)	12-1451-65
$C_{30}H_{50}N_2O_2$			
Baleabuxine	EtOH	219(3.79)	22-0657-65
Isobaleabuxine	EtOH	252(4.01)	22-0657-65
$C_{30}H_{50}N_2O_4$			
Baleabuxidine	EtOH	219(3.85)	28-4139-65B

Compound	Solvent	λ_{max}(log ε)	Ref.
$C_{30}H_{50}N_4O_2$			
Hydroquinone, tetrakis-(piperidinomethyl)-	ether	313(3.84)	39-0042-64
$C_{30}H_{50}O$			
Cholestan-3-one, 2-isopropylidene-	EtOH	254(3.92)	35-0085-64
Lanost-8-en-2-one	dioxan	287(1.40),295(1.40)	22-2245-64
Moretenol	EtOH	210(2.73)	12-0226-65
Shionone	$CHCl_3$	289(1.5)	95-0318-64
	dioxan	293(1.70)	22-0584-64
$C_{30}H_{50}O_2$			
Aglaiol	EtOH	210(3.07)(end absorption)	78-0917-65
12,18-Cyclolanostane, 3β-hydroxy-11-oxo-	EtOH	222(3.95)	33-0704-65
$C_{30}H_{50}O_2S_2$			
Cholestan-2-one, 3β-mercapto-, ethylxanthate	dioxan	223(3.90),279(4.06), 354(1.75)	78-0583-65
$C_{30}H_{50}O_3$			
Cholest-2-ene-2-carboxaldehyde, 3-(2-hydroxyethoxy)-	EtOH	277(4.12)	88-2161-64
Ebelan-16-oic acid, 17-hydroxy-3-oxo-, γ-lactone	n.s.g.	271(2.29)	12-1451-65
Ebelo-22,25,27-triene-3β,16,17-triol	n.s.g.	269(4.61),278(4.72), 290(4.60)	12-1451-65
Ebelo-22,25,27-trien-3β-ol, 16,17-oxide, acetate	n.s.g.	269(4.60),278(4.71), 290(4.59)	12-1451-65
$C_{30}H_{50}O_4$			
29-Nor-8α,9β,13α,14β-dammar-17(20)-en-21-oic acid, 3β,11α-dihydroxy-, methyl ester	EtOH	234(4.08)	78-3505-65
	dioxan	234(4.08)	31-0344-64
Spergulagenin A	n.s.g.	280(1.17)	2-0339-64
$C_{30}H_{51}BrO$			
Shionone, α-bromodihydro-	$CHCl_3$	308(2.04)	95-0325-64
Shionone, α'-bromodihydro-	$CHCl_3$	311.5(2.19)	95-0325-64
$C_{30}H_{52}N_2$			
Buxamine, N-isopropyl-N-methyl-	MeOH	221s(4.10),230s(4.31), 238(4.45),245(4.49), 255(4.29),280(1.80)	33-0968-64
$C_{30}H_{52}O$			
Shionan-3-one	dioxan	290(1.60)	22-0584-64
$C_{30}H_{52}OS$			
Cholestan-2β-ol, 3α-mercapto-, cyclic monothioacetal with acetone	heptane	249(1.49)	78-1581-65
Cholestan-2β-ol, 3β-mercapto-, cyclic monothioacetal with acetone	heptane	243(1.42)	78-1581-65
Cholestan-3α-ol, 2α-mercapto-, cyclic monothioacetal with acetone	heptane	245(1.44)	78-1581-65
Cholestan-3α-ol, 2β-mercapto-, cyclic monothioacetal with acetone	heptane	248(1.45)	78-1581-65
Cholestan-3α-ol, 4α-mercapto-, cyclic monothioacetal with acetone	heptane	252(1.85)	78-1581-65
Cholestan-3β-ol, 2α-mercapto-, cyclic monothioacetal with acetone	heptane	250(1.48)	78-1581-65

Compound	Solvent	λ_{max}(log ε)	Ref.
Cholestan-3β-ol, 4α-mercapto-, cyclic monothioacetal with acetone	heptane	255(1.78)	78-1581-65
$C_{30}H_{52}O_2$			
Dustanin	C_6H_{12}	none	88-3337-65
Ebelan-16-oic acid, 17-hydroxy-, 16,17-lactone (end absorptions)	EtOH	214(2.50),216(2.39), 218(2.27),220(2.16), 222(2.10),224(2.03)	12-1451-65
$C_{30}H_{52}O_2S_2$			
Cholestan-3β-ol, 2α-mercapto-, ethylxanthate	dioxan	227(3.69),285(4.08), 355(1.88)	78-0583-65
$C_{30}H_{52}O_3$			
Ebelin, hexahydro-, lactone (end absorptions)	EtOH	214(2.36),216(2.23), 218(2.06),220(1.91), 222(1.79),224(1.67)	12-1451-65
Isoebelin, hexahydro-, lactone (end absorptions)	EtOH	214(1.92),216(1.87), 218(1.85),220(1.80), 222(1.77),224(1.75)	12-1451-65
$C_{30}H_{52}O_4$			
Poligonaquinone, dimethyl ether	EtOH	287(4.06)	78-2319-64
	$CHCl_3$	288(4.12),398(2.60)	78-2319-64
$C_{30}H_{53}NO_2$			
Abietic acid, compound with diisopentylamine	EtOH	241(4.28)	35-0096-64

Compound	Solvent	λ_{max}(log ε)	Ref.
$C_{31}H_{18}$			
Methane, (4H-cyclopenta[def]phenanthren-4-yl)(4H-cyclopenta[def]phenanthren-4-ylidene)-	DMSO-KO-tert-Bu	310(4.34),330s(4.16), 523s(4.51),558(5.15)	5-0050-65J
$C_{31}H_{20}$			
11H-Benzo[b]fluorene, 11-(fluoren-9-ylidenemethyl)-	dioxan	302(4.67),324(4.19), 341(3.82),368(3.24), 385(3.21)	5-0001-65I
anion	DMSO	573(5.01)	5-0001-65I
$C_{31}H_{20}CrO_3$			
Chromium, tricarbonyl(tetraphenylbutatriene)	THF	402(4.43),493(4.14)	101-0355-65A
$C_{31}H_{20}O_2$			
1-Naphthylideneacetic acid, 4-(diphenylmethylene)-2-hydroxy-α-phenyl-, γ-lactone	EtOH	264(4.04),447(4.46)	44-4333-65
$C_{31}H_{20}O_6$			
Benzoic acid, o-[[o-(1,3-dioxo-2-phenyl-2-indanyl)phenyl]glyoxyloyl]-, methyl ester	MeOH	224(4.59),280s(3.90), 360s(2.76)	6-0359-64
$C_{31}H_{20}O_{10}$			
Hinokiflavone, methyl ether	EtOH	272(4.49),335(4.63)	25-2020-64
	EtOH-NaOEt	281(4.67),304(4.67), 375(4.46)	25-2020-64
$C_{31}H_{21}Cl_3N_4O_7S$			
9H-Purine-6-thiol, 9-[2,3,5-tris-(p-chlorobenzoyl)-β-D-ribofuranosyl]-	pH 1	245(4.61),322(4.31)	87-0200-64
	pH 11	238(4.60),310(4.21)	87-0200-64
$C_{31}H_{21}N_7O_{13}S$			
9H-Purine-6-thiol, 9-[2,3,5-tris-(p-nitrobenzoyl)-β-D-ribofuranosyl]-	pH 1	264(4.52),320(4.44)	87-0200-64
	pH 11	234(4.33),305(4.52)	87-0200-64
$C_{31}H_{22}Cl_3N_5O_7S$			
9H-Purine-6-thiol, 2-amino-9-[2,3,5-tris(p-chlorobenzoyl)-β-D-ribofuranosyl]-	pH 1	246(4.65),340(4.30)	87-0200-64
	pH 11	244(4.62),323(4.16)	87-0200-64
$C_{31}H_{22}NOP$			
Acetonitrile, (2-naphthoyl)(triphenylphosphoranylidene)-	EtOH	268(4.11),275(4.11), 305(4.03)	65-2205-65
$C_{31}H_{22}N_2O_2$			
Benzamide, N-(4-benzoyl-4-cyano-1,3-diphenyl-1,3-butadienyl)-	MeCN	351(3.96)	24-3892-65
$C_{31}H_{22}N_8O_{13}S$			
9H-Purine-6-thiol, 2-amino-9-[2,3,5-tris(p-nitrobenzoyl)-β-D-furanosyl]-	pH 1	263(4.55),344(4.38)	87-0200-64
	pH 11	253(4.50),273(4.50), 314(4.44)	87-0200-64
$C_{31}H_{23}$			
Bis(p-biphenylyl)phenyl, carbanion	ether	260(5.20),435(4.23), 512(4.99),578(5.04)	60-0264-64

Compound	Solvent	$\lambda_{max}(\log \epsilon)$	Ref.
$C_{31}H_{23}N_2P$			
Phosphorane, (fluoren-9-ylidene-hydrazono)triphenyl-	$CHCl_3$	288s(3.88),297(3.95), 309(3.95),372s(4.35), 386(4.37)	44-0417-65
$C_{31}H_{24}N_3OP$			
2H-Pyrrole-3-carbonitrile, 5-(methyl-amino)-2-oxo-4-[α-(triphenylphos-phoranylidene)benzyl]-	dioxan	440(4.25)	49-1967-65
$C_{31}H_{24}N_4O_7S$			
9H-Purine-6-thiol, 9-(2,3,5-tri-O-benzoyl-β-D-ribofuranosyl)-	pH 1	228(4.62),320(4.40)	87-0200-64
	pH 11	310(4.38)	87-0200-64
$C_{31}H_{24}O$			
Methanol, bis(m-biphenylyl)phenyl-	$HClO_4$	256(4.53),333(3.80), 422(4.52),505s(3.68)	60-0264-64
	H_2SO_4	490s(--)	60-0264-64
Methanol, bis(p-biphenylyl)phenyl-	$HClO_4$	271(4.34),352(3.94), 460(4.71),538(4.86)	60-0264-64
	H_2SO_4	500(--)	60-0264-64
$C_{31}H_{24}O_2$			
10H-Dibenzo[a,d]cyclohepten-10-one, 11,11'-methylenebis[5,11-dihydro-	MeOH	207(3.13),256(2.61), 278(2.51)	49-0182-65
$C_{31}H_{24}O_3$			
4-Hexenoic acid, 5-hydroxy-4-methyl-3-oxo-2,2,6,6-tetraphenyl-, δ-lactone	MeOH	283(3.95)	78-2735-64
$C_{31}H_{25}N$			
Indene, 1-(3-methylanilino-2-propen-1-ylidene)-2,3-diphenyl-	MeCN	255(4.47),289(3.79), 438(4.72)	24-3331-64
$C_{31}H_{25}NO_8$			
Pyridine, 2-(D-ribofuranosyloxy)-, tribenzoate	EtOH	230(4.58),268(3.66)	94-0828-64
2(1H)-Pyridone, 1-D-ribofuranosyl-, tribenzoate	EtOH	230(4.68),276(3.71), 283(3.76),303(3.73)	94-0828-64
$C_{31}H_{25}N_2P$			
Phosphorane, [(diphenylmethylene)-hydrazono]triphenyl-	$CHCl_3$	262s(3.79),269s(3.81), 275(3.83),321(3.94)	44-0417-65
$C_{31}H_{25}N_5$			
Adenine, N,3-bis(diphenylmethyl)-	pH 1	280(4.41),294s(4.32)	4-0115-64
	pH 7	283(4.47),292s(4.46)	4-0115-64
	pH 13	281(4.45),291s(4.39)	4-0115-64
$C_{31}H_{25}N_5O_7S$			
9H-Purine-6-thiol, 2-amino-9-(2,3,5-tribenzoyl-β-D-ribofuranosyl)-	pH 1	227(4.60),266(4.02), 344(4.34)	87-0200-64
	pH 11	250(4.22),317(4.32)	87-0200-64
$C_{31}H_{25}OP$			
Phosphorane, triphenyl-α-(2-naphthyl)-ethylidene-	EtOH	275(4.16),315(3.89)	65-2738-64

Compound	Solvent	λ_{max}(log ε)	Ref.
$C_{31}H_{26}BrP$			
(Diphenylmethyl)triphenylphosphonium bromide	EtOH	262(3.35),269(3.38), 277(3.27)	59-1143-64
$C_{31}H_{26}O_4$			
1,5-Pentanedione, 3-benzoyl-2-methoxy-1,2,5-triphenyl-	n.s.g.	252(4.32)	23-2822-64
$C_{31}H_{26}O_6$			
Kaempferol, 4',7-dibenzyl-3,5-dimethyl-	EtOH	265(4.37),335(4.33)	22-0779-65
	EtOH-$AlCl_3$	265(4.37),335(4.33)	22-0779-65
$C_{31}H_{26}O_{14}$			
s-Indacene-1,2,3,5,6,7-hexacarboxylic acid, 4,8-dihydroxy-, heptamethyl ester, tropylium salt	CH_2Cl_2	234(4.32),287(4.64), 300s(4.57),342s(3.61), 387(3.73),452s(3.83), 480(4.25),515(4.80), 618(3.02),800s(2.33)	88-1161-64
$C_{31}H_{27}BO_6S$			
Borinic acid, phenyl-2-thienyl-, ester with curcumin	iso-PrOH-acid	425(4.4)	83-0617-64
	iso-PrOH-NH_3	460(4.1),600(4.8)	83-0617-64
$C_{31}H_{27}ClN_2O_4$			
4-(p-Dimethylaminostyryl)-1,2-diphenylquinolinium perchlorate	n.s.g.	578(4.60)	65-2008-64
$C_{31}H_{27}IN_2O_2$			
4-(p-Dimethylaminostyryl)-6-hydroxy-1-(p-hydroxyphenyl)-2-phenylquinolinium iodide	n.s.g.	565(4.47)	65-2008-64
$C_{31}H_{28}$			
1,1'-Spirobiindan, 3,3-dimethyl-3',3'-diphenyl-	C_6H_{12}	261s(3.21),266(3.31), 273(3.19)	23-0025-64
$C_{31}H_{28}O$			
1,4-Pentadiene, 3-ethoxy-1,1,5,5-tetraphenyl-	dioxan	276(4.45)	5-0050-65J
$C_{31}H_{28}O_5$			
Chalcone, 2',4-bis(benzyloxy)-3,4'-dimethoxy-	EtOH	246(4.08),357(4.34)	22-3350-65
$C_{31}H_{28}O_6$			
Chalcone, 2',4-bis(benzyloxy)-3,4'-dimethoxy-, epoxide	EtOH	233(4.30),279(4.06), 311(3.97)	22-3350-65
$C_{31}H_{28}O_{14}$			
Ergochrysin A	6N HCl	251(4.17),268(4.11), 360(3.75)	78-1417-65
	pH 13	229(4.28),361(4.15), 400s(3.98)	78-1417-65
	EtOH	208(4.30),243(4.21), 270(4.22),336(4.21), 374(3.86)	78-1417-65
	n.s.g.	244(4.25),269(4.29), 338(4.25)	39-4144-65

Compound	Solvent	$\lambda_{max}(\log \epsilon)$	Ref.
Ergochrysin B	6N HCl	252(4.19),268(4.12), 359(3.79)	78-1417-65
	pH 13	227(4.29),362(4.20), 400s(3.99)	78-1417-65
	EtOH	208(4.30),241(4.22), 266(4.23),336(4.23), 374(3.90)	78-1417-65
Ergoxanthin	EtOH	209(4.40),268(4.37), 373(3.90)	78-1417-65
Isoergochrysin A	n.s.g.	212(4.44),264(4.35), 334(4.25)	39-4144-65
$C_{31}H_{28}Si_2$			
Disilane, methylpentaphenyl-	C_6H_{12}	242.5(4.45)	101-0369-64B
$C_{31}H_{29}ClN_2O_6$			
1-(o-Hydroxyphenyl)-2-[3-(1,3,3-trimethyl-2-indolinylidene)propenyl]-quinolinium perchlorate, acetate	EtOH	578(5.05)	65-3373-64
$C_{31}H_{30}O_3$			
4a(2H)-Naphthalenecarboxylic acid, 3,4,5,6-tetrahydro-2-hydroxy-2,5,7-triphenyl-, ethyl ester	EtOH	275(4.45),343(4.18)	78-0195-64
$C_{31}H_{31}ClN_2O_7$			
2-Naphthacenecarboxamide, 4-benzamido-N-tert-butyl-7-chloro-1,4,4a,5,5a,-6,11,11a-octahydro-3,12-dihydroxy-10-methoxy-1,11-dioxo-	MeOH-$Na_2B_4O_7$	220(4.49),303(3.97), 463(4.68),487(4.59)	35-0933-65
$C_{31}H_{31}NO_5$			
2-Naphthacenecarboxamide, 10-(benzyloxy)-N-tert-butyl-4,4a,5,12-tetrahydro-1,3-dihydroxy-6-methyl-12-oxo-	EtOH	247(4.30),262(4.34), 316(3.76),382(4.39)	65-0670-65
$C_{31}H_{32}N_2O_6$			
2(1H)-Pyrimidinone, 1-(2,3,5-tri-O-benzoyl-β-D-arabinofuranosyl)-	MeOH	276(3.81)	44-0835-65
$C_{31}H_{32}O_4$			
Anthracene, 5,9β-bis(benzyloxy)-1,4,4aβ,9,9aα,10-hexahydro-4α,10α-(isopropylidenedioxy)-	EtOH	276(3.32),283(3.33)	65-0662-65
$C_{31}H_{32}O_7$			
Glutaric acid, 2-[5-(benzyloxy)-1,2,3,4-tetrahydro-9-methyl-4-oxo-2-anthryl]-3-oxo-, diethyl ester	EtOH	228(4.63),242(4.70), 294(3.79),305(3.84), 317(3.71),331(3.48), 374(2.55)	65-0655-65
$C_{31}H_{33}N_3O_{14}S$			
Glucopyranuronic acid, 1-[N^4-carboxy-N^1-(5-methyl-3-isoxazolyl)sulfanilamido]-1-deoxy-, N^4-benzyl methyl ester, 2,3,4-triacetate	MeOH	263(4.45)	95-1104-64

Compound	Solvent	λ_{max}(log ε)	Ref.
$C_{31}H_{34}O_8$			
Glutaric acid, 2-[5-(benzyloxy)-1,2,3,4,9,9aα-hexahydro-9β-hydroxy-9α-methyl-4-oxo-2-anthryl]-3-oxo-, diethyl ester	EtOH	230(4.25),264(4.07), 307(3.70),317(3.66), 346(3.11)	65-0655-65
$C_{31}H_{34}O_9$			
Glutaric acid, 2-[3-(7-(benzyloxy)-3-methylphthalidyl]-5-oxocyclohexyl]-3-oxo-, diethyl ester	EtOH	239(4.14),244(4.15), 298(3.79)	65-0655-65
$C_{31}H_{34}O_{10}$			
Siphulin, methyl ester, triacetate	EtOH	263s(4.08),298(4.00)	1-1677-65
$C_{31}H_{35}FN_2O_5$			
Pregn-4-eno[3,2-c]pyrazole, 9α-fluoro-2'-(p-fluorophenyl)-16α-methyl-17,20:20,21-bis(methylenedioxy)-	MeOH	261(4.20)	87-0352-64
$C_{31}H_{36}F_2N_2O_5$			
Pregn-4-eno[3,2-c]pyrazole, 9α-fluoro-2'-(p-fluorophenyl)-11β,17,21-trihydroxy-16α-methyl-20-oxo-, 21-acetate	MeOH	260(4.20)	87-0352-64
$C_{31}H_{36}N_2O_5$			
Pregn-4-eno[2,3-d]imidazole, 17,20:20,21-bis(methylenedioxy)-16α-methyl-11-oxo-1'-phenyl-	MeOH	225(4.23),290(4.00), 295(4.00)	87-0584-64
Pregn-4-eno[3,2-c]pyrazole, 17,20:20,21-bis(methylenedioxy)-16α-methyl-11-oxo-1'-phenyl-	MeOH	297(4.49)	35-1520-64
Pregn-4-eno[3,2-c]pyrazole, 17,20:20,21-bis(methylenedioxy)-16α-methyl-11-oxo-2'-phenyl-	MeOH	262(4.22)	35-1520-64
$C_{31}H_{36}N_2O_5S$			
Pregn-4-eno[2,3-d]imidazole, 17,20:20,21-bis(methylenedioxy)-2'-mercapto-16α-methyl-11-oxo-1'-phenyl-	MeOH	270(3.94),321(4.14)	87-0584-64
	MeOH-NaOH	331(4.07)	87-0584-64
$C_{31}H_{36}N_2O_8S$			
p-Toluenesulfonamide, N-[3,4-dimethoxy-α-methyl-α-(4-methyl-5-oxo-4-veratryl-2-oxazolin-2-yl)phenethyl]-	MeOH	230(4.45),278(3.75)	44-1424-64
$C_{31}H_{37}ClN_2O_5$			
Pregn-4-eno[3,2-c]pyrazole, 2'-(p-chlorophenyl)-11β-hydroxy-17,20:20,21-bis(methylenedioxy)-16α-methyl-	MeOH	231(4.05),262(4.31)	35-1520-64
$C_{31}H_{37}ClN_2O_9$			
Emetine, N-acetyl-1,3,5(11b)-tridehydro-, perchlorate	EtOH	232s(4.24),268s(4.08), 283(4.22),368(4.05)	33-1117-64
$C_{31}H_{37}FN_2O_5$			
Pregn-4-eno[3,2-c]pyrazole, 2'-(p-fluorophenyl)-11β-hydroxy-17,20:20,21-bis(methylenedioxy)-16α-methyl-	MeOH-HCl	252(4.07),289(4.14)	35-1520-64
	MeOH	261.5(4.20)	35-1520-64

Compound	Solvent	$\lambda_{max}(\log \epsilon)$	Ref.
$C_{31}H_{37}N_3O_5$			
Pregna-4,6-dieno[3,2-d]-3'H-1',2',3'-triazole, 11β,17α,21-trihydroxy-6,16α-dimethyl-20-oxo-3'-phenyl-	MeOH	263(4.02),318(4.35)	87-0584-64
Pregn-5-eno[3,2-d]-3'H-1',2',3'-triazole, 17α,20:20,21-bis(methylenedioxy)-6,16α-dimethyl-11-oxo-3'-phenyl-	MeOH	224s(4.05)	87-0584-64
$C_{31}H_{37}N_3O_6$			
Pregn-5-eno[3,2-d]-3'H-1',2',3'-triazole, 4β-hydroxy-17α,20:20,21-bis-(methylenedioxy)-6,16α-dimethyl-11-oxo-3'-phenyl-	MeOH	228(4.03)	87-0584-64
$C_{31}H_{38}N_2O_5$			
Pregn-4-eno[2,3-d]imidazole, 11β-hydroxy-17α,20:20,21-bis(methylenedioxy)-16α-methyl-1'-phenyl-	MeOH-HCl	222(4.02),278(4.04), 283(4.05)	87-0584-64
	MeOH	227(4.20),288(3.97), 293(3.98)	87-0584-64
Pregn-4-eno[3,2-c]pyrazole, 11β-hydroxy-17,20:20,21-bis(methylenedioxy)-16α-methyl-1'-phenyl-	MeOH	298(4.47)	35-1520-64
Pregn-4-eno[3,2-c]pyrazole, 11β-hydroxy-17,20:20,21-bis(methylenedioxy)-16α-methyl-2'-phenyl-	MeOH	261(4.23)	35-1520-64
$C_{31}H_{38}N_2O_9S$			
Alanine, 3-(3,4-dimethoxyphenyl)-N-[3-(3,4-dimethoxyphenyl)-2-methyl-N-(p-tolylsulfonyl)alanyl]-2-methyl-	MeOH	230(4.42),279(3.74)	44-1424-64
$C_{31}H_{39}NO_5$			
3-Piperidinol, 1-acetyl-5-methyl-2-[1-(2,3,5,6,6b,7,11,11b-octahydro-10,11b-dimethyl-3,11-dioxo-1H-benzo-[a]fluoren-9-yl)ethyl]-, acetate	EtOH	233(4.30),328(3.99)	44-0270-64
$C_{31}H_{39}N_3O_6$			
5α-Pregnano[3,2-d]-3'H-1',2',3'-triazole, 6β-hydroxy-17,20:20,21-bis-(methylenedioxy)-6α,16α-dimethyl-11-oxo-3'-phenyl-	MeOH	229(4.05)	87-0584-64
$C_{31}H_{40}N_2O_5$			
5β-Pregnan-21-al, 3α-acetoxy-11,20-dioxo-, 21-(N-acetylphenylhydrazone)	MeOH	274(4.23)	44-0521-64
5β-Pregn-17(20)-en-21-al, 3α,20-diacetoxy-11-oxo-, phenylhydrazone	MeOH	248(4.00),304s(4.17), 336(4.47)	44-0521-64
$C_{31}H_{40}N_2O_{19}$			
2(1H)-Pyrimidone, 1-(4-O-β-D-galactopyranosyl-β'-D-glucopyranosyl)-4-methoxy-, heptaacetate	EtOH	274(3.74)	44-2723-65
$C_{31}H_{40}N_4O_9$			
5β-Pregnan-21-al, 3α,20β-diacetoxy-11-oxo-, 21-(2,4-dinitrophenylhydrazone)	$CHCl_3$	261(4.04),351(4.35)	44-3158-64

Compound	Solvent	λ_{max}(log ε)	Ref.
$C_{31}H_{40}O_6$			
Polygalic acid, ethyl ester	EtOH	200(3.86)	88-2567-64
$C_{31}H_{40}O_9$			
Andirobin, tetrahydro-, diacetate	n.s.g.	209(3.96)	88-2607-64
$C_{31}H_{41}NO_4$			
Phthalimide, N-(3β-hydroxy-5α-pregnan-20β-yl)-, acetate	dioxan	292(3.24),300s(3.21)	24-0533-64
$C_{31}H_{41}NO_5$			
Isojervone, diacetyl-	EtOH	234(4.43),320(2.38)	44-2282-64
$C_{31}H_{41}NO_6$			
Δ^4-Isojerv-3-one, 23,N-diacetyl-, 17,17a-epoxide	EtOH	233(4.33),318(2.06)	44-0270-64
$C_{31}H_{42}ClNO_6$			
Octadecahydro-5',10-dihydroxy-1',4',7',10,11b-pentamethyl-3,11-dioxospiro[9H-benzo[a]fluorene-9,3'-[3H]pyrido[1,2-c][1,3]oxazinium chloride, 5'-acetate	EtOH	299(4.37),308(2.29)	44-0270-64
$C_{31}H_{42}O_4$			
Estra-1,3,5(10),8,14-pentaen-17β-ol, 3-methoxy-, 1-(menthyloxy)acetate	EtOH	310(4.53)	94-1285-65
$C_{31}H_{42}O_5$			
D:C-Friedo-B':A'-neogammacera-5,8-dien-24-oic acid, 7,11,12-trioxo-, methyl ester	EtOH	215(3.87),283(3.9)	94-0986-65
$C_{31}H_{43}NO_3$			
Androsta-3,5-dien-17β-ol, 3-ethoxy-6-[(N-methylanilino)methyl]-, acetate	EtOH	255(4.50)	78-0569-65
$C_{31}H_{43}NO_5$			
Isojervine isomer, diacetate	EtOH	220(4.00),252s(3.65)	44-2282-64
$C_{31}H_{44}O_5$			
29-Nor-8,9,14-dammara-12,17(20)-dien-21-oic acid, 3,16-dihydroxy-11-oxo-, γ-lactone, acetate	EtOH	280(4.24)	78-3505-65
$C_{31}H_{44}O_6$			
Helvolic acid, deacetoxydihydro-	EtOH	219(3.95)	94-0121-64
29-Nor-9β,13α,14β-dammar-7-en-21-oic acid, 3α,16β-dihydroxy-6,11-dioxo-, γ-lactone, acetate	dioxan	238(4.00)	31-0344-64
$C_{31}H_{44}O_7$			
11,12-Seco-D:C-friedo-B':A'-neogammacera-5,8-diene-11,12,24-trioic acid, 7-oxo-, 24-methyl ester	EtOH	252(3.94)	94-0986-65
$C_{31}H_{45}NO_4$			
Δ^{22}-Tomatillidine, N-acetyl-, acetate	EtOH	222(3.68),275(3.55)	44-0754-65

Compound	Solvent	λ_{max}(log ε)	Ref.
$C_{31}H_{45}NO_5$			
5β-Pregnan-11β-ol, 3,3:20,20-bis-(ethylenedioxy)-11α-(2-pyridyl-methyl)-	EtOH	264(3.58)	44-2095-65
$C_{31}H_{45}NO_6$			
3-Piperidinol, 1-acetyl-5-methyl-2-[1-(tetradecahydro-9-hydroxy-10,11b-dimethyl-3,11-dioxo-1H-benzo[a]fluor-en-9-yl)ethyl]-, 3-acetate	EtOH	238(3.96),317s(1.87)	44-0270-64
$C_{31}H_{46}Br_2O_2$			
Ergosta-7,22-dien-3β-ol, 5α,6α-(di-bromomethylene)-, acetate	C_6H_{12}	218(3.95)	44-1737-65
$C_{31}H_{46}O_4$			
D:C-Friedo-B':A'-neogammacer-8-en-24-oic acid, 7,11-dioxo-, methyl ester	EtOH	271(3.87)	94-0986-65
29-Nordammara-9(11),17(20)-dien-21-oic acid, 3α,11α-dihydroxy-, γ-lactone, acetate	EtOH	221(4.19)	78-3505-65
$C_{31}H_{46}O_5$			
Cholesta-5,8(14)-diene-$\Delta^{7,\alpha}$-succinic acid, 3β-hydroxy-	EtOH	210(4.00),224(3.98), 282(4.18)	78-3641-65
Epidehydroceanothic acid, methyl ester	n.s.g.	290(1.70)	12-0141-64
Maleic acid, (3β-hydroxycholesta-5,8-dien-7α-yl)-	EtOH	217(3.97)	78-3641-65
29-Nordammar-17(20)-en-21-oic acid, 3,16-dihydroxy-11-oxo-, γ-lactone, acetate	EtOH	222(4.14)	78-3505-65
$C_{31}H_{46}O_6$			
Senegenic acid, dehydro-, dimethyl ester	EtOH	238(4.01)	44-4234-65
$C_{31}H_{46}O_7$			
Ceanothic acid, acetyltrinorketone from	n.s.g.	293(1.39)	12-0141-64
$C_{31}H_{47}NO_3$			
Δ^{22}-Tomatillidine, N-acetyl-deoxo-, acetate	EtOH	235(3.90)	44-0754-65
$C_{31}H_{48}O_2$			
Cycloeucalenone, 2-(hydroxymethylene)-	EtOH	290.0(3.86)	39-1692-65
Ergosta-7,22-dien-3β-ol, 5α,6α-meth-ylene-, acetate	C_6H_{12}	227(3.78)	44-1737-65
D:C-Friedo-B':A'-neogammacera-7,9(11)-dien-24-oic acid, methyl ester	EtOH	233(3.32),240(3.38), 249(3.15)	94-0986-65
$C_{31}H_{48}O_3$			
Abieslactone	EtOH	207.5(4.30)	95-0453-64
D:C-Friedo-B':A'-neogammacer-9(11)-en-24-oic acid, 12-oxo-, methyl ester	EtOH	246(3.97)	94-0986-65
$C_{31}H_{48}O_4$			
Cholesta-5,7-diene-3β,4α-diol, diacetate	EtOH	272(4.07),282(4.09), 294(3.84)	24-2383-65

Compound	Solvent	λ_{max}(log ε)	Ref.
$C_{31}H_{48}O_5$			
Cholesta-5,8-diene-7α-succinic acid, 3β-hydroxy-	EtOH	209(3.72)	78-3641-65
Cholesta-5,8(14)-diene-7α-succinic acid, 3β-hydroxy-	EtOH	211(3.85)	78-3641-65
$C_{31}H_{48}O_6$			
29-Nordamma-17(20)-en-21-oic acid, 11,16-dihydroxy-3-oxo-, 16-acetate	EtOH	220(3.91)	78-3505-65
	dioxan	220(3.91)	31-0344-64
isomer	EtOH	218(3.90)	78-3505-65
$C_{31}H_{48}O_{10}$			
Cerbertin, deacetyl-	n.s.g.	218(4.15)	12-1423-64
$C_{31}H_{48}O_{11}$			
Cerbertatin, deacetyl-	n.s.g.	217(4.15),290(1.78)	12-1423-64
$C_{31}H_{49}NO_3$			
Δ^{22}-Tomatillidine, N-acetyldeoxo-5α,6-dihydro-, acetate	EtOH	235(3.86)	44-0754-65
$C_{31}H_{50}N_2O_2$			
Pyrrolidine, 1-(6-nitrocholesta-3,5-dien-3-yl)-	EtOH	260(4.26),479(4.48)	39-2830-64
$C_{31}H_{50}O$			
Cycloartanone, 24-methylene-	EtOH	280.0(1.58)	39-1692-65
$C_{31}H_{50}O_2$			
Davallic acid, methyl ester	EtOH	207(3.71)	94-0986-65
Isodavallic acid I, methyl ester	EtOH	207(3.74)	94-0986-65
$C_{31}H_{50}O_3$			
5α-Cholestane-$\Delta^{3,\alpha}$-acetic acid, 7-oxo-, ethyl ester	EtOH	223(4.10)	44-0505-65
$C_{31}H_{50}O_4$			
2-Cholestene-2-carboxaldehyde, 3-(2-hydroxyethoxy)-, formate	EtOH	275(4.10)	88-2161-64
Peniocerol, diacetate	EtOH	210(3.56),220(3.13) (end absorptions)	39-1160-65
$C_{31}H_{51}NO_4$			
Cholestan-22-one, 3β-acetoxy-26-acetamido-	EtOH	280(2.03)	44-0754-65
$C_{31}H_{52}$			
Cholesta-3,5-diene, 3,4-diethyl-	EtOH	238(4.14),245(4.17), 254(3.98)	23-0792-65
$C_{31}H_{52}O_2$			
5α-Cholestane-$\Delta^{3,\alpha}$-acetic acid, ethyl ester	EtOH	226(4.22)	44-0505-65
$C_{31}H_{52}O_2S_2$			
5α-Cholestane-2β,3α-dithiol, diacetate	EtOH	232(3.95)	78-0329-65

Compound	Solvent	λ_{max}(log ϵ)	Ref.
$C_{31}H_{52}O_3$			
Bicyclo[25.3.1]hentriaconta-27,30-diene-29,31-dione, 28-hydroxy-	EtOH	269(4.02),413(2.90)	89-0567-64
catenane with N-acetyl-14-azacyclohexacosanone	EtOH	266(3.97),406(2.87)	89-0567-64
$C_{31}H_{52}O_3S$			
5α-Cholestan-3β-ol, 6β-mercapto-, diacetate	EtOH	234(3.74)	78-0309-65
$C_{31}H_{52}O_4$			
Resorcinol, 5-heneicosyl-, diacetate	EtOH	262(2.48)	44-0435-64
$C_{31}H_{52}S_2$			
Cholesta-3,5-diene, 3,4-bis-(ethylthio)-	hexane	292(4.17)	94-0383-64
	EtOH	292(4.17)	78-0733-65

Compound	Solvent	λ_{max}(log ε)	Ref.
$C_{32}H_{16}N_8O_2U$			
Uranyl phthalocyanine	$C_{10}H_7Cl$	410(4.7),915(4.62)	23-2201-64
$C_{32}H_{18}$			
Butadiyne, di-1-anthryl-	THF	260(5.27),285(4.39), 297(4.42),317(3.80), 332(3.89),349s(3.90), 367(4.14),391(4.36), 410(4.50),430(4.46)	88-0719-64
1,2:4,5-Dibenzo[naphtho-2''',3''':8,9-pyrene]	benzene	282(3.88),306(4.69), 317(4.89),341(4.68), 358(4.83),385(4.16), 406(4.38),432(4.48), 452(3.66)	24-0597-65
6:7-(peri-Naphthylene)-1,2:3,4-dibenzanthracene	benzene	295(4.74),320(4.78), 334(4.76),350(4.92), 369(4.58),386(4.00), 410(4.14),436(4.31)	78-1559-64
	C_6H_{12}	254(4.66)	78-1559-64
$C_{32}H_{18}N_2O_4$			
Quino[2,3-b]acridine-6,7,13,14(5H,12H)-tetrone, 4,11-diphenyl-	H_2SO_4	381(3.95),486(3.91)	27-0242-65
$C_{32}H_{18}N_2O_6$			
Quino[2,3-b]acridine-6,7,13,14(5H,12H)-tetrone, 2,9-diphenoxy-	H_2SO_4	485(4.18)	2--0242-65
$C_{32}H_{18}O_2$			
6,13[2',3']-Naphthalenopentacene-5,14-dione, 6,13-dihydro-	n.s.g.	225(4.9),243(5.1)	70-1260-64
$C_{32}H_{19}N_6O_6PS_2$			
5H-Triazolo[4',5':2,3]anthra[1,9-de:-4,10-d'e']bis[1,2,3]oxathiazine, 5-triphenylphosphoranylidene-amino-, 2,2,8,8-tetraoxide	$CHCl_3$	260s(4.42),270s(4.44), 278s(4.47),290(4.52), 369(4.45)	39-1003-65
$C_{32}H_{20}Cl_2N_2O_{10}$			
7,15-Dihydro-7,15-dioxo-8,16-diphenylbenzo[1,2-b:4,5-b']diquinolizinediium perchlorate	MeOH-HCl	235s(4.52),255(4.56), 382s(4.17),390(4.20)	44-0061-64
	MeOH	238(4.51),360(4.02), 375(4.09)	44-0061-64
$C_{32}H_{20}N_2O_4S$			
1,4-Naphthoquinone, 2,2'-thiobis[3-anilino-	dioxan	240(4.58),282(4.84), 396(4.03),524(4.07)	5-0151-64E
$C_{32}H_{20}N_2O_4S_2$			
1,2-Naphthoquinone, 3,3'-dithiobis[4-anilino-	dioxan	253(4.68),323(4.44), 359(4.45),514(4.28)	5-0151-64E
$C_{32}H_{20}N_4$			
A tetraazaheterocycle from 6-methoxy-6,5-borazarochrysene and nitrous acid	EtOH	257(4.65),264(4.69), 276s(4.52),306(3.85), 343(4.05),361s(3.90), 380(3.75)	44-1757-64
	H_2SO_4	268(4.98),354(3.85), 449(4.42)	44-1757-64

Compound	Solvent	λ max(log ε)	Ref.
$C_{32}H_{20}O_2$			
6,15-o-Benzenohexacene-5,16-dione, 5α,6,15,15a-tetrahydro-	n.s.g.	215(4.8),240(5.1)	70-1260-64
6,13[2',3']-Naphthalenopentacene-5,14-dione, 5α,6,13,13a-tetrahydro-	n.s.g.	254(5.04),320(3.42), 339(3.60),355(3.68), 375(3.57)	70-1260-64
$C_{32}H_{22}$			
Anthracene, 1,9,10-triphenyl-	CH_2Cl_2	268(4.97),328(3.11), 345(3.52),364(3.86), 383(4.06),403(4.03)	44-1981-65
Triptycene, 2,5-diphenyl-	C_6H_{12}	241(4.52),253(4.43), 262(4.38),277(4.10)	44-1959-65
	EtOH	241(4.50),252(4.44), 261(4.41),277s(4.14)	44-1959-65
$C_{32}H_{22}N_2O_6$			
Dibenzo[a,l]triphenodioxazine-8,17-dicarboxylic acid, diethyl ester	DMF	544(4.64),575(4.84)	27-0242-65
Dibenzo[c,n]triphenodioxazine-8,17-dicarboxylic acid, diethyl ester	DMF	539(4.71),580(4.90)	27-0242-65
$C_{32}H_{22}O$			
Isobenzofuran, 2,3,6,7-tetraphenyl-	$CHCl_3$	324(3.40),410(4.04)	78-0195-64
$C_{32}H_{22}O_2$			
Anthracene, 9,10-dicinnamoyl-	$CHCl_3$	251(4.88),257(4.97), 292(4.66),349(3.86), 370(3.97),390(3.86)	39-5666-64
p-Terphenyl, 2',3'-dibenzoyl-	$CHCl_3$	254(4.50)	78-0195-64
$C_{32}H_{22}O_4$			
6,13-o-Benzenopentacene-17,20-diol, 6,13-dihydro-, diacetate	acetone	225(5.0),248(5.1)	70-1470-64
$C_{32}H_{22}O_6$			
[2,2'-Biindan]-1,1',3,3'-tetrone, 2,2'-bis(p-methoxyphenyl)-	ether	228(4.86)	22-3061-65
Phthalide, 3-[p-methoxy-α-[2-(p-methoxyphenyl)-1,3-dioxo-2-indanyl]-benzylidene]-	ether	225(4.90),274(4.38), 320(4.08)	22-3061-65
$C_{32}H_{22}O_{10}$			
Hinokiflavone, dimethyl ether	EtOH	271(4.51),333(4.59)	25-2020-64
$C_{32}H_{23}Cl_3N_4O_4$			
3,5-Pyrazolidinedione, 4,4'-(2,2,2-trichloroethylidene)bis[1,2-diphenyl-	MeOH	258(4.68)	32-0320-65
$C_{32}H_{23}N_2OP$			
Phosphorane, [(10-oxo-9(10H)-anthrylidene)hydrazono]triphenyl-	$CHCl_3$	246(4.62),300(4.02), 350(4.01),448(4.27)	28-2259-65A
$C_{32}H_{24}$			
Cyclooctatetraene, 1,2,4,7-tetraphenyl-	heptane	263(4.66),325(3.81)	35-0453-64
	n.s.g.	263(4.71)	35-4212-64
Cyclooctatetraene, 1,3,5,7-tetraphenyl-	EtOH	261(4.73)	35-0453-64
Tricyclo[4.2.0.0^{2,5}]octa-3,7-diene, 1,2,4,7-tetraphenyl-	ether	261(4.20)	35-0453-64

Compound	Solvent	$\lambda_{max}(\log \epsilon)$	Ref.
Tricyclo[4.2.0.0^{2,5}]octa-3,7-diene, 1,3,5,7-tetraphenyl-	ether	262(4.46)	35-0453-64
$C_{32}H_{24}N_2O_2$			
Bicyclo[2.2.2]oct-5-ene, 2-(1-naphthyl)-7,8-dibenzimido-	EtOH	225(4.65),274(3.7), 283(3.5),294s(3.4), 313(3.35)	12-1775-65
$C_{32}H_{24}N_6O_{11}$			
3H-2,3-Benzoxazepine, 4-(2,4-dinitroanilino)-7-methoxy-1,4-epoxy-3-[5-nitro-2-(1,2-dihydro-6-methoxy-1-oxo-2-isoquinolyl)phenyl]-4,5-dihydro-	EtOH	205(4.59),253(4.58), 335(4.21)	24-1023-65
$C_{32}H_{24}O_4$			
Cyclobuta[1,2-1:3,4-1']diphenanthrenedicarboxylic acid, 8b,8c,16b,16c-tetrahydro-, dimethyl ester	EtOH	206(4.68),270(4.29)	39-5544-64
$C_{32}H_{24}O_9$			
Acenaphthenone, 2,2-bis(3,4-diacetoxyphenyl)-	EtOH	256s(4.45),281(4.03), 313(3.88),347(3.82)	39-3040-65
$C_{32}H_{24}O_{11}$			
Isodianellinone, triacetate	dioxan	268(4.82),295s(4.24), 435(4.35)	12-0218-65
$C_{32}H_{25}ClO_4$			
7-Methoxy-4-(7-methoxyflav-2-enyl)-flavylium chloride	EtOH	264(4.18),305(4.15), 421(4.45)	78-0657-65
$C_{32}H_{25}N$			
Ethylene, 1-dimethylamino-2,2-bis-(9-fluorenylidenemethyl)-	MeCN	247s(4.98),251(5.00), 294(4.30),303s(4.28), 457(4.62)	24-3331-64
$C_{32}H_{25}N_2O_4P$			
Diacetamide, N-[1,4-dihydro-1,4-dioxo-3-[(triphenylphosphoranylidene)-amino]-2-naphthyl-	$CHCl_3$	282(4.50)	39-1003-65
$C_{32}H_{26}$			
Amitriptyline	dioxan	318(4.52)	87-0555-65
$C_{32}H_{26}Fe$			
1,3-Butadiene, 4-ferrocenyl-1,1,4-triphenyl-	THF	352(4.44),470(3.40)	101-0355-65A
$C_{32}H_{26}N_2O_{11}$			
Barbituric acid, 5-acetyl-1-(2,3,5-tri-O-benzoyl-β-D-ribofuranosyl)-	EtOH	232(4.58),274(4.26)	94-0459-64
$C_{32}H_{26}O_2$			
Cyclohexene, 4,5-dibenzoyl-3,6-diphenyl-	$CHCl_3$	250(4.30),282(3.28), 308(2.23)	78-0195-64
$C_{32}H_{26}O_{10}$			
Xylindein, dihydroxy-	n.s.g.	328(4.17),339(3.48), 422(3.74),448(3.81)	78-2095-65

Compound	Solvent	λ_{max}(log ϵ)	Ref.
$C_{32}H_{26}O_{11}$			
Chrysoaphin-sl-1, acetoxy-	$CHCl_3$	268(4.69),382(4.42), 402(4.54),460(4.10), 486(4.20)	39-6923-65
$C_{32}H_{27}BrN_2O_8$			
3-Carbamoyl-1-(2,3,5-tri-O-benzoyl-D-arabinofuranosyl)pyridinium bromide	MeOH	267(3.86)	39-0610-65
	MeOH-KCN	323(3.83)	39-0610-65
$C_{32}H_{27}N$			
Cyclopentadiene, 5-(dimethylamino-methylene)-1,2,3,4-tetraphenyl-	MeCN	290(4.46),368(4.50)	24-3331-64
$C_{32}H_{27}N_5O_6$			
Adenosine, N-benzoyl-3'-deoxy-N-methyl-, 2',5'-dibenzoate	EtOH	228(4.57),281(4.15)	87-0659-65
$C_{32}H_{27}O_5P$			
Fumaric acid, [α-(triphenylphosphoran-ylidene)phenacyl]-, dimethyl ester	EtOH	225(4.55),267(3.88), 274(3.84),307(3.75), 367(3.77)	44-3312-65
Maleic acid, [α-(triphenylphosphoran-ylidene)phenacyl]-, dimethyl ester	EtOH	228(4.53),266(3.90), 274(3.86),305(3.74), 403(3.51)	44-3312-65
Succinic acid, phenacylidene(triphenyl-phosphoranylidene)-, dimethyl ester	EtOH	224(4.73),259(4.10), 263(4.11),405(4.34)	44-3312-65
$C_{32}H_{28}$			
1,5-Cyclooctadiene, 1,2,5,6-tetra-phenyl-	hexane	260(4.23)	88-3937-65
2,3,4-Hexatriene, 2,5-dibenzyl-1,6-diphenyl-	C_6H_{12}	274.5(4.40)	89-0095-65
Tricyclo[4.2.0.0^{2,5}]octane, 1,2,5,6-tetraphenyl-	hexane	228(4.46),255(3.15), 262(3.13),268(2.93), 272(2.86)	88-3937-65
$C_{32}H_{28}BF_4O_5P$			
[α-(1,2-Dicarboxyethylidene)phenacyl]-triphenylphosphonium tetrafluoro-borate, dimethyl ester	EtOH	228(4.5),261(4.02), 268(4.02),274(3.99)	44-3312-65
$C_{32}H_{28}Br_2O_3$			
Furan, 2,5-bis(p-bromophenyl)-2,5-di-ethoxy-2,5-dihydro-3,4-diphenyl-	EtOH	229(4.25),260(3.92)	78-2181-64
$C_{32}H_{28}NO_5P$			
Succinic acid, [(p-acetylphenyl)imino]-(triphenylphosphoranylidene)-, dimethyl ester	n.s.g.	346(4.22)	88-1263-64
$C_{32}H_{28}NO_6P$			
Succinic acid, [(p-carboxyphenyl)-imino]-(triphenylphosphoranylidene)-, trimethyl ester	n.s.g.	340(4.02)	88-1263-64
$C_{32}H_{28}O_7$			
Flavone, 4',7-bis(benzyloxy)-5,6,8-trimethoxy-	EtOH	272(4.49),325(4.65)	39-6255-64

Compound	Solvent	λ_{max}(log ε)	Ref.
$C_{32}H_{28}O_{12}Pb$			
Lead tetraanisate	EtOH	256(4.46)	35-2229-65
$C_{32}H_{29}NO_4$			
Pyrryl, 2,3,4,5-tetrakis(p-methoxyphenyl)-	toluene	395(4.5),615(4.3)	24-3124-65
	ether	395(4.6),600(4.3)	
$C_{32}H_{29}NO_{15}$			
Xantholaccaic acid A_1, acetate	MeOH	254(4.48),294(4.44)	39-6067-65
$C_{32}H_{29}O_5P$			
Succinic acid, phenacyl(triphenylphosphoranylidene)-, dimethyl ester	EtOH	228(4.53)	44-3312-65
$C_{32}H_{30}$			
2,4-Hexadiene, 2,5-dibenzyl-1,6-diphenyl-	C_6H_{12}	253(4.51)	89-0095-65
$C_{32}H_{30}O_4$			
1,2-Butadiene, 1,1,4,4-tetrakis-(p-methoxyphenyl)-	n.s.g.	234(4.51),276(4.39)	24-2959-64
5,8,13,14-Pentaphenetetrone, 6-decyl-	dioxan	275(4.53)	65-3484-64
Tectol, O,O'-dimethyl-	n.s.g.	228(4.64),265(4.61), 273(4.67),368(3.85)	24-0588-64
$C_{32}H_{30}O_7$			
Propiophenone, 2',6'-bis(benzyloxy)-2,3-dihydroxy-3-(p-methoxyphenyl)-, 3-acetate	MeOH	272(4.02),305(3.82)	78-1141-64
$C_{32}H_{30}O_{14}$			
Secalonic acid B	6N HCl	254(4.21),268(4.19), 357(3.81)	78-1417-65
	pH 13	247(4.35),364(4.46)	78-1417-65
	EtOH	214(4.28),242(4.20), 262s(4.15),340(4.48), 377s(3.88)	78-1417-65
Secalonic acid C	6N HCl	356(3.80)	78-1417-65
	pH 13	362(4.47)	78-1417-65
	EtOH	210(4.30),242(4.27), 261s(4.18),341(4.53), 375s(3.89)	78-1417-65
$C_{32}H_{32}CuN_2O_4$			
Ethylamine, α-phenyl-N-(3-methoxysalicylidene)-, copper chelate	MeOH	270(4.15),373(3.53)	65-3055-64
$C_{32}H_{32}Ge_2$			
1,4-Digermacyclohexa-2,5-diene, 1,1,4,4-tetramethyl-2,3,5,6-tetraphenyl-	n.s.g.	227(4.67),270(3.98)	101-0233-65A
$C_{32}H_{32}N_4O_{10}$			
6,13-Triphenodioxazinedicarboxylic acid, 2,9-diethoxy-, diethyl ester	DMF	555(4.73),596(4.92)	27-0242-65
$C_{32}H_{32}O_{15}$			
Echioidin, pentaacetate	EtOH	248s(4.37),303(4.32)	78-3715-65

Compound	Solvent	$\lambda_{max}(\log \epsilon)$	Ref.
$C_{32}H_{33}CuN_7O_6$			
Etioporphyrin I, trinitro-, copper complex	$CHCl_3$	401(5.21),538(4.09), 575(4.24)	39-4303-65
$C_{32}H_{33}N_3O_6$			
Cytosine, N-acetyl-1-(2,3,5-tri-O-benzyl-β-D-arabinofuranosyl)-	MeOH	248(4.15),300(3.87)	44-0835-65
$C_{32}H_{34}CuN_6O_4$			
Etioporphyrin I, dinitro-, copper complex	$CHCl_3$	402(5.21),536(4.10), 570(4.29)	39-4303-65
$C_{32}H_{34}N_2O_4$			
Pyrrole-2-carboxylic acid, 5,5'-vinylenebis[4-ethyl-3-methyl-, dibenzyl ester	EtOH	265(4.15),387(4.44)	39-4385-65
$C_{32}H_{34}N_4$			
Benzil, p,p'-diisopropyl-, bis(phenylhydrazone)	EtOH	244(4.6),297(4.4), 344(4.7)	28-2113-64B
1,2-Diazetidine, 3,4-bis(p-cumenyl)-1-phenyl-4-(phenylazo)-	$CHCl_3$	275(4.6),410(2.6)	28-2113-64B
$C_{32}H_{34}N_8$			
Erythrose, N-phenyl-, N-methylalkazone	EtOH	220s(4.42),242s(4.38), 307s(4.33),357(4.72)	35-0732-64
$C_{32}H_{34}N_8O_{10}$			
3-Cyclohexene-$\Delta^{1,\alpha}$-acetaldehyde, 6-[2-[6-hydroxy-6-(hydroxymethyl)-bicyclo[3.2.1]oct-2-en-1-yl]ethyl]-5-methyl-2-oxo-, bis(2,4-dinitrophenylhydrazone)	$CHCl_3$	386(4.68)	24-2652-64
$C_{32}H_{34}O_4$			
Tectol, tetrahydro-O,O'-dimethyl-	n.s.g.	248(4.84),328(3.99)	24-0588-64
$C_{32}H_{35}CuN_5O_2$			
Etioporphyrin I, nitro-, copper complex	$CHCl_3$	401(5.22),530(4.11), 566(4.31)	39-4303-65
$C_{32}H_{35}N_5O_{12}$			
β-Narcotinediol, picrolonate	50% EtOH	283(3.54)	39-1087-65
$C_{32}H_{35}N_7O_6$			
Etioporphyrin I, trinitro-	$CHCl_3$	403(4.96),509(4.06), 537(3.78),585(3.69), 637(3.46)	39-4303-65
$C_{32}H_{36}F_2N_2O_5$			
Pregna-4,6-dieno[3,2-c]pyrazole, 2'-(2,4-difluorophenyl)-11β-hydroxy-6,16α-dimethyl-17,20:20,21-bis-(methylenedioxy)-	MeOH	269(4.15),312(4.31)	87-0355-64
$C_{32}H_{36}N_2O_4$			
Pyrrole-2-carboxylic acid, 5,5'-ethylenebis[4-ethyl-3-methyl-, dibenzyl ester	EtOH	290(4.59)	39-4385-65

Compound	Solvent	λ_{max}(log ε)	Ref.
$C_{32}H_{36}N_6O_4$			
Etioporphyrin I, dinitro-	$CHCl_3$	380(5.00),397(4.99), 504(4.07),537(3.86), 575(3.73),629(3.60)	39-4303-65
$C_{32}H_{37}CuN_5$			
Etioporphyrin I, amino-, copper complex	$CHCl_3$	417(5.36),541(4.02), 572(3.67),592(3.71)	39-4303-65
$C_{32}H_{37}N_5O_2$			
Etioporphyrin I, nitro-	$CHCl_3$	399(5.07),504(4.08), 538(3.90),573(3.77), 627(3.69)	39-4303-65
$C_{32}H_{38}$			
Ethane, 1,2-bis(2,3-diethyl-4-methyl-1-naphthyl)-	C_6H_{12}	219s(4.76),233(5.01), 239(5.09),266s(3.60), 276s(3.82),289(4.02), 300(4.13),310(4.03), 324(3.37)	44-0963-64
$C_{32}H_{38}N_2O_5$			
Pregneno[3,2-d]pyrimidine, 11β-hydroxy-16α-methyl-17α,20:20,21-bis(methylenedioxy)-2'-phenyl-	MeOH-HCl	276(4.41)	87-0584-64
	MeOH	258(4.59),307(3.93)	87-0584-64
$C_{32}H_{38}N_2O_6$			
Pregn-4-eno[3,2-c]pyrazole, 11β-formyloxy-16α-methyl-17,20:20,21-bis-(methylenedioxy)-2'-phenyl-	MeOH	261(4.21)	35-1520-64
$C_{32}H_{38}N_4$			
Etioporphyrin I	$CHCl_3$	397(5.24),496(4.15), 531(4.01),564(3.83), 595(3.13),615(3.65)	39-4303-65
	n.s.g.	506(5.14),537(5.03), 572(4.87),625(4.74)	65-3352-64
$C_{32}H_{38}N_4O_2$			
4-Piperidone, 1,1'-[ethylenebis-(indole-2,3-diylethylene)]di-	MeOH	283(4.21),290(4.21)	24-2579-65
$C_{32}H_{38}O$			
Ether, bis[(4-methyl-2,3-diethyl-1-naphthyl)methyl]	EtOH	218s(4.80),230(5.14), 236(5.19),263s(3.65), 274s(3.91),284(4.09), 294(4.18),306(4.07), 323(3.17),326s(3.07)	44-0963-64
$C_{32}H_{38}O_4$			
Dispiro[1,3-dioxolane-2,3'-[3H]benz-[e]indene-7'(3'aH),2"-[1,3-dioxolane]-6' -(2,2-diphenylvinyl)decahydro-3'aα-methyl-	EtOH	250(4.27)	78-2473-64
Progesterone, 21-benzylidene-16α,17α-isopropylidenedioxy-6-methylene-	EtOH	299(4.26)	78-0597-64
$C_{32}H_{39}N_5$			
Etioporphyrin I, amino-	$CHCl_3$	416(5.20),519(4.07), 553(3.73),587(3.56), 646(3.80)	39-4303-65

Compound	Solvent	$\lambda_{max}(\log \epsilon)$	Ref.
$C_{32}H_{40}N_6$			
Hydrazine, tetrakis(p-dimethylaminophenyl)-	dioxan	285(4.49+),318(4.46+)	24-0844-65
$C_{32}H_{40}O_6$			
Dibenzo[d,f]dioxepin-6-spiro-1'-cyclohex-3'-ene-2',5'-dione, 3,3',9'-tri-tert-butyl-2,10-dimethoxy-	C_6H_{12}	218(4.58),233s(4.47), 258s(4.23),303(4.04)	39-3660-65
Dibenzo[d,f]dioxepin-6-spiro-1'-cyclohex-3'-ene-2',5'-dione, 3,4',9'-tri-tert-butyl-2,10-dimethoxy-	C_6H_{12}	218(4.55),233s(4.44), 259s(4.20),303(4.01)	39-3660-65
$C_{32}H_{41}N_3O_{19}$			
2(1H)-Pyrimidone, 4-acetamido-1-(hepta-O-acetyl-β-cellobiosyl)-	EtOH	251(3.82)	44-2723-65
2(1H)-Pyrimidone, 4-acetamido-1-(hepta-O-acetyl-β-lactosyl)-	MeOH	249(4.26),298(3.82)	44-2723-65
$C_{32}H_{42}N_2O_6$			
5β-Pregn-17(20)-en-21-al, 3α,20-diacetoxy-11-oxo-, 21-(N-acetylphenylhydrazone)	MeOH	269(4.12)	44-0521-64
$C_{32}H_{42}N_2O_{19}$			
2(1H)-Pyrimidone, 4-ethoxy-1-(hepta-O-acetyl-β-cellobiosyl)-	EtOH	275(3.78)	44-2723-65
2(1H)-Pyrimidone, 4-ethoxy-1-(hepta-O-acetyl-β-lactosyl)-	EtOH	275(3.81)	44-2723-65
$C_{32}H_{42}N_4Ni_2O_4$			
Nickel, [1,1,2,2-ethanetetracarboxaldehydato(2-)]bis[N-propyl-7-(propylimino)-1,3,5-cycloheptatrien-1-ylaminato]di-	$CHCl_3$	296(4.75),309(4.57), 338(4.52),413(4.48), 500(4.08),659(2.78)	44-3046-64
$C_{32}H_{42}N_4O_8$			
Pregn-5-en-3β-ol, 21-acetyl-, 20-(2,4-dinitrophenylhydrazone)	EtOH	225(4.20),354(4.34)	78-1927-64
$C_{32}H_{42}O_5$			
5β-Pregnan-11-one, 3α,20β-diacetoxy-21-benzylidene-	MeOH	253(4.35)	44-3158-64
$C_{32}H_{42}O_7$			
Condurangogenin A	MeOH	218(4.19),223(4.16), 280(4.35)	78-1777-65
$C_{32}H_{42}O_8$			
Withaferin A, diacetate	EtOH	214(4.26)	35-5805-65
	EtOH	214(4.25)	44-1774-65
$C_{32}H_{42}O_9S$			
Pregn-5-ene-3β,19-diol, 17α,20:20,21-bis(methylenedioxy)-, 3-acetate 19-p-toluenesulfonate	n.s.g.	226(4.08),262(2.74), 273(2.60)	13-0371-65
$C_{32}H_{43}NO_8S$			
5β-Pregnane-11,20-dione, 17α-acetamido-3α,16α-dihydroxy-, 3-acetate 16-p-toluenesulfonate	EtOH	224(4.1)	13-0805-65

Compound	Solvent	λ_{max}(log ε)	Ref.
$C_{32}H_{44}O_6Si_6$			
Cyclohexasiloxane, 2,2,4,4,8,8,10,10-octamethyl-6,6,12,12-tetraphenyl-	$CHCl_3$	265(3.14)	46-1066-65
$C_{32}H_{44}O_8$			
Withaferin A, dihydro-, diacetate	EtOH	214(3.99)	44-1774-65
$C_{32}H_{46}N_2O$			
Buxenine-G, N^{20}-salicylidene-	C_6H_{12}	227(4.21),238(4.25), 245(4.40),253(4.26), 280(3.67),314(3.51), 400(3.23)	88-3145-64
$C_{32}H_{46}O_5$			
Senegenic acid, diosphenol acetate	EtOH	249(3.99)	44-4234-65
$C_{32}H_{46}O_6$			
5β-Pregnan-11β-ol, 3,3;20,20-bis(ethylenedioxy)-11α-(o-methoxyphenyl)-	EtOH	219(3.86),272(3.29), 279(3.28)	44-2095-65
$C_{32}H_{46}O_6S$			
Pregn-5-en-3β-ol, 20β-(2-hydroxyethoxy)-, 3-acetate 20β-(2-p-toluenesulfonate)	EtOH	225(4.06),262(2.74), 274(2.62)	13-0585-65
$C_{32}H_{46}O_7S$			
Pregnane-3,20-dione, 12α-(p-toluenesulfonyloxy)-, bis(ethylene ketal)	isooctane	224(4.11)	23-0189-64
$C_{32}H_{46}O_{10}$			
Cerbertin	n.s.g.	218(4.22)	12-1423-64
$C_{32}H_{46}O_{11}$			
Cerbertatin	n.s.g.	217(4.18),290(1.88)	12-1423-64
$C_{32}H_{47}ClO_7$			
Senegenin, acetate	MeOH	205(3.81)	23-0491-64
$C_{32}H_{48}Cl_2O_3$			
18ξ-Olean-12-en-11-one, 15,18-dichloro-3β-hydroxy-, acetate	C_6H_{12}	242(4.01)	44-1698-65
$C_{32}H_{48}N_2$			
1H-Cyclopenta[7',8']phenanthro-[2',3':4,5]imidazo[1,2-a]pyridine, 1β-(1,5-dimethylhexyl)tetradecahydro-13aβ,15aβ-dimethyl-	n.s.g.	232(4.20),238(4.21), 275(3.44),283(3.46), 310(3.44)	32-0485-64
$C_{32}H_{48}O_3$			
Multiflora-7,9(11)-dien-15-one, 3β-acetoxy-	EtOH	233(4.13)	44-1698-65
$C_{32}H_{48}O_4$			
Ebelin, lactone, acetate	n.s.g.	269(4.62),278(4.72), 290(4.59)	12-1451-65
Taraxer-14-en-16-one, 3β-acetoxy-, 11α,12α-epoxide	EtOH	242(4.04)	44-1698-65
$C_{32}H_{48}O_5$			
Lup-1-ene-27,28-dioic acid, 3-oxo-, methyl ester	n.s.g.	231(3.71)	78-1529-65

Compound	Solvent	$\lambda_{max}(\log \epsilon)$	Ref.
27-Norurs-12(14)-en-28-oic acid, 3β-acetoxy-13-oxo-, methyl ester	MeOH	254(3.99)	5-0159-65C
27-Norurs-13(15)-en-28-oic acid, 3β-acetoxy-14-oxo-, methyl ester	MeOH	256(4.04)	5-0159-65C
$C_{32}H_{48}O_6$			
Senegenin, dechlorodehydro-, dimethyl ester	MeOH	243(4.12),249(4.17), 256(4.07)	23-0491-64
$C_{32}H_{48}O_9$			
Solanoside, acetate	EtOH	217(4.16)	33-0065-65
Swietenin, octahydro-	n.s.g.	202(3.80)	39-6935-65
Vallaroside, acetate	EtOH	217(4.19)	33-0065-65
$C_{32}H_{48}O_{10}$			
Acoschimperoside P	EtOH	216(4.10)	33-0065-65
Vallarosolanoside	EtOH	215(4.20)	33-0065-65
$C_{32}H_{49}ClO_3$			
Multiflor-7-en-3β-ol, 15ξ-chloro-11α,12α-epoxy-, acetate	EtOH	206(3.57)	44-1698-65
Olean-13(18)-en-11-one, 3β-acetoxy-15-chloro-	C_6H_{12}	207(3.92),297(1.79)	44-1698-65
$C_{32}H_{49}NO_4$			
18α-Oleanane-2,3-dione, 19β,28-epoxy-, 2-O-acetyloxime	n.s.g.	213(3.8)	24-1837-65
$C_{32}H_{50}N_2O_2$			
Urea, 1,3-dicyclohexyl-1-(3-oxo-androst-4-en-17-yl)-	MeOH	240(4.20)	35-5670-65
$C_{32}H_{50}O_2$			
Multiflora-7,9(11)-dien-3β-ol, acetate	EtOH	232(4.15),238(4.18), 247(3.99)	44-1698-65
$C_{32}H_{50}O_3$			
Ebelo-22,25,27-trien-3β-ol, 16,17-epoxy-, acetate	EtOH	269(4.64),278(4.75), 290(4.62)	12-1451-65
Oxidation product of α-amyrin acetate	EtOH	251(3.93)	100-0040-64
Urs-12-en-11-one, 3β-acetoxy-	EtOH	250(4.08)	44-1698-65
	n.s.g.	250(4.08)	32-1378-64
$C_{32}H_{50}O_4$			
Lanost-7(11)-ene-8,9-dione, 3β-acetoxy-, 9(8→7),8(9→11)-diabeo-	MeOH	263(3.95)	5-0159-65C
Lanost-8-ene-7,11-dione, 3β-acetoxy-	MeOH	272(3.93)	5-0159-65C
Oleanolic acid, acetyl- (end absorptions)	EtOH	214(3.39),216(3.28), 218(3.12),220(2.94), 222(2.74),224(2.53)	12-1451-65
$C_{32}H_{50}O_6$			
Medicagenic acid, dimethyl ester (end absorptions)	EtOH	214(3.48),216(3.38), 218(3.27),220(3.13), 222(2.98),224(2.82)	12-1451-65
$C_{32}H_{52}O_2$			
α-Amyrin, acetate	heptane	195(4.04)	100-0040-64
Cholesterol, 7-dehydro-, 3-(2-tetra-hydropyranyl) ether	EtOH	272(4.06),282(4.08), 293(3.83)	44-1131-64

Compound	Solvent	λ_{max}(log ϵ)	Ref.
Lupenol, acetate	heptane	196(3.97)	100-0040-64
$C_{32}H_{52}O_3$			
12,18-Cyclolanostan-11-one, 3β-acetoxy-	EtOH	222(3.74)	33-0704-65
Lanost-8-en-7-one, 3β-acetoxy-	MeOH	255(4.0)	5-0159-65C
24-Oxocycloartanyl acetate	hexane	280(1.78)	39-1692-65
$C_{32}H_{52}O_6$			
Polygonaquinone, diacetate	EtOH	269(4.02)	78-2319-64
	$CHCl_3$	269(4.20),343(2.61)	78-2319-64
$C_{32}H_{54}O_3S_2$			
Cholestan-3α-ol, 2α-mercapto-, 3-acetate 2-ethylxanthate	dioxan	225s(3.86),280(4.09), 365(1.73)	78-0583-65
Cholestan-3β-ol, 2α-mercapto-, 3-acetate 2-ethylxanthate	dioxan	223s(3.91),281(4.12), 357(1.76)	78-0583-65
$C_{32}H_{54}O_4$			
Ebelin, hexahydro-, lactone, acetate (end absorptions)	EtOH	214(2.46),216(2.41), 218(2.36),220(2.00), 222(2.20),224(2.06)	12-1451-65

Compound	Solvent	$\lambda_{max}(\log \epsilon)$	Ref.
$C_{33}H_{16}Br_4Cl$ Fluoren-9-yl, 2,7-dibromo-9-[p-chloro-α-(2,7-dibromofluoren-9-ylidene)-benzyl]-	dioxan	255(4.81),285(4.75), 293(4.79),412(3.90), 491(4.41),874(3.16)	49-0003-64
$C_{33}H_{17}Br_4Cl$ Methane, (p-chlorophenyl)(2,7-dibromo-fluoren-9-yl)(2,7-dibromofluoren-9-ylidene)-	dioxan	225(4.78),248(4.62), 258(4.70),268(4.82), 282(4.69),306(4.55)	49-0003-64
	DMF-NaOH	268(4.91),300(4.70), 337(4.30),597(4.64)	49-0003-64
$C_{33}H_{20}Br$ Fluoren-9-yl, 9-(p-bromo-α-fluoren-9-ylidenebenzyl)-	dioxan	244(4.80),286(4.47), 321(4.18),492(4.43), 860(3.14)	49-0003-64
$C_{33}H_{20}Cl$ Fluoren-9-yl, 9-(p-chloro-α-fluoren-9-ylidenebenzyl)-	dioxan	245(4.80),286(4.48), 320(4.17),382(3.65), 491(4.43),860(3.14)	49-0003-64
$C_{33}H_{21}$ Fluoren-9-yl, 9-(α-fluoren-9-ylidene-benzyl)-	dioxan	245(4.82),265(4.65), 287(4.51),317(4.11), 380(3.65),485(4.47), 860(3.20)	49-0003-64
$C_{33}H_{21}Br$ Methane, (p-bromophenyl)fluoren-9-yl-fluoren-9-ylidene-	dioxan	224(4.78),251(4.63), 260(4.71),275(4.42), 291(4.32),303(4.36), 320(4.34)	49-0003-64
	DMF-NaOH	307(4.43),344(4.25), 610(4.60)	49-0003-64
$C_{33}H_{21}Cl$ Methane, (p-chlorophenyl)fluoren-9-yl-fluoren-9-ylidene-	dioxan	224(4.79),252(4.64), 259(4.72),274(4.44), 291(4.32),303(4.37), 319(4.34)	49-0003-64
$C_{33}H_{22}$ Methane, fluoren-9-ylfluoren-9-ylidenephenyl-	dioxan	223(4.78),252(4.61), 260(4.69),276(4.41), 292(4.32),304(4.36), 320(4.33)	49-0003-64
	DMF-NaOH	307(4.53),346(4.25), 600(4.64)	49-0003-64
	DMSO-KO-tert-Bu	306(4.55),345(4.28), 600(4.65)	5-0050-65J
$C_{33}H_{25}ClN_2O_6$ 1-[(o-Hydroxyphenyl)-2-[3-]1-(o-hy-droxyphenyl)-2(1H)-quinolylidene]-propenyl]quinolinium perchlorate	EtOH	618(5.25)	65-3373-64
$C_{33}H_{25}CoO$ Cobalt, carbonylcyclopentadienyl-(tetraphenylpropadiene)-	n.s.g.	266(4.435)	101-0007-65A

Compound	Solvent	$\lambda_{max}(\log \epsilon)$	Ref.
$C_{33}H_{25}NO$			
2-Indolinone, 1-(diphenylmethyl)-3,3-diphenyl-	EtOH	260(4.02)	88-1553-64
$C_{33}H_{25}NO_{12}$			
Coumarin, 3-[4-acetoxy-3-isopenten-2'-ylbenzamido]-7-(2,3-O-carbonyl-α-noviosyloxy)-4-hydroxy-8-methyl-	EtOH	321(4.28)	33-0390-64
$C_{33}H_{25}N_3O_3$			
5-Norbornene-2,3-dicarboximide, 5-(α-hydroxy-α-2-pyridylbenzyl)-7-(α-2-pyridylbenzylidene)-	MeOH	250(4.24)	57-0412-64B
$C_{33}H_{26}$			
Indene, 3a,7a-dihydro-4,5,6,7-tetraphenyl-	EtOH	312(3.98)	39-1881-65
Unknown hydrocarbon	EtOH	278(4.10)	39-1881-65
$C_{33}H_{27}BO_7$			
10-Boraxanthydrol, curcumin complex	iso-PrOH-HOAc	430(4.7)	83-0617-64
	iso-PrOH-NH_3	370(3.8),470(4.4),600(4.7)	83-0617-64
$C_{33}H_{27}ClN_2O_4$			
4-[(1-Ethyl-4(1H)-quinolylidene)-methyl]-1,2-diphenylquinolinium perchlorate	n.s.g.	594(4.57)	65-2008-64
$C_{33}H_{27}IN_2O_2$			
4-[(1-Ethyl-4(1H)-quinolylidene)-methyl]-6-hydroxy-1-(p-hydroxyphenyl)-2-phenylquinolinium iodide	n.s.g.	612(3.70)	65-2008-64
$C_{33}H_{27}N$			
Indene, 1-(5-methylanilino-2,4-pentadien-1-ylidene)-2,3-diphenyl-	MeCN	259(4.53),305(4.16),475(4.87)	24-3331-64
$C_{33}H_{27}NO_3$			
Nicotinic acid, 5-(diphenylmethylene)-1,2,5,6-tetrahydro-6-oxo-2,2-diphenyl-, ethyl ester	EtOH	248s(4.13),340(4.20)	78-3255-65
$C_{33}H_{27}N_3$			
2-Naphthylamine, N-methylene-, trimer	DMF	290s(4.4),341(3.8)	7-0969-64
$C_{33}H_{28}$			
Indan, 3a,7a-dihydro-4,5,6,7-tetraphenyl-	EtOH	320(4.12)	39-1881-65
$C_{33}H_{28}O_4$			
Pentatetraene, tetrakis-(p-methoxyphenyl)-	C_6H_{12}	353(4.9),433(3.7)	24-1760-64
$C_{33}H_{29}BO_6$			
Borinic acid, diphenyl-, curcumin complex	iso-PrOH-HOAc	380(4.1),450(4.5),470(4.7),495(4.7)	83-0617-64
	iso-PrOH-NH_3	360(3.8),450(4.1),600(4.8)	83-0617-64

Compound	Solvent	λ_{max}(log ε)	Ref.
$C_{33}H_{29}ClN_2O_5$			
4-(o-Hydroxyphenyl)-3-[3-(1,3,3-trimethyl-2-indolinylidene)propenyl]-benzo[f]quinolinium perchlorate	EtOH	568(4.95)	65-3377-64
$C_{33}H_{29}NO_5$			
Benzoic acid, o-(benzyloxy)-, ester with 1-[o-(benzyloxy)benzoyl]-1,4,5,6-tetrahydro-2-pyridinol	dioxan	218(4.57),285(3.88)	70-0774-64
$C_{33}H_{29}N_5O_8$			
Benzophenone, 5-nitro-2-[(5,6,7,8-tetrahydro-6,6,8,8-tetramethyl-7-oxo-2-naphthyl)oxy]-, 2,4-dinitrophenylhydrazone	90% EtOH-10% $CHCl_3$	221(4.59),295s(4.23), 380(4.45)	39-0361-65
	+NaOH	313s(4.20),496(4.43)	39-0361-65
$C_{33}H_{30}O_4$			
5H-Dibenzo[a,c]cycloheptene-6,6-dicarboxylic acid, 6,7-dihydro-3,9-diphenyl-, diethyl ester	n.s.g.	296(4.65)	4-0181-65
Naphthalene, 7-methoxy-1-(4,4'-dimethoxybenzhydryl)-4-(p-methoxyphenyl)-	n.s.g.	288(4.15),338(3.75)	24-2959-64
$C_{33}H_{30}O_8$			
Flavone, 4',7-bis(benzyloxy)-3',5,6,8-tetramethoxy-	EtOH	251(4.32),271(4.31), 334(4.45)	39-6255-64
$C_{33}H_{30}O_{10}$			
Isodianellinone, tri-O-methyl-, diacetate	dioxan	241(4.43),280(4.85), 347(3.92),370(4.11), 389(4.25)	12-0218-65
$C_{33}H_{31}ClO_8$			
1,1,5,5-Tetrakis(p-methoxyphenyl)-1,4-pentadien-3-yl perchlorate	n.s.g.	481(4.48),673(5.13)	24-2959-64
$C_{33}H_{31}Cl_3O_4$			
3-Chloro-1,1,5,5-tetrakis(p-methoxyphenyl)-1,4-pentadien-3-yl perchlorate, hydrochloride	n.s.g.	522(4.26),723(4.56)	24-2959-64
$C_{33}H_{31}IN_2$			
4-(p-Dimethylaminostyryl)-6-methyl-2-phenyl-1-p-tolylquinolinium iodide	n.s.g.	570(4.47)	65-2008-64
$C_{33}H_{32}$			
Methane, bis(hexahydropyrenyl)-	EtOH	244(5.23),293(4.15), 300(4.21),319(3.93), 334(3.75)	39-5783-65
$C_{33}H_{32}N_2O_2$			
Estra-1,3,5(10)-trien-17-one, 4-methyl-1-[(2-phenyl-4-quinazolinyl)oxy]-	MeOH	257(4.53),286(4.23)	44-1893-64
4(3H)-Quinazolinone, 3-(4-methyl-17-oxo-estra-1,3,5(10)-trien-1-yl)-2-phenyl-	MeOH	282(4.15)	44-1893-64

Compound	Solvent	$\lambda_{max}(\log \epsilon)$	Ref.
$C_{33}H_{32}O_6$			
1,5-Pentanedione, 1,5-bis(2,4-dimethoxyphenyl)-2,4-diphenyl-	MeOH	276(4.33)	78-0555-64
$C_{33}H_{37}NO_6$			
Malonamic acid, 2-[(4-oxo-5-benzyloxy-1,2,3,4-tetrahydro-9-methyl-2-anthryl)acetyl]-N-tert-butyl-, ethyl ester	EtOH	224(4.54),262(4.66), 308(3.73),320(3.62), 372(3.69)	65-0670-65
$C_{33}H_{38}N_2O_8$			
Grevillol, bis(p-nitrobenzoate)	EtOH	260(3.46)	12-2015-65
$C_{33}H_{38}N_8O_8$			
Progesterone, bis(2,4-dinitrophenylhydrazone)	DMF	390(4.70)	13-0255-64
	DMF-NaOH	478(4.76)	13-0255-64
$C_{33}H_{38}O_5$			
Androst-5-ene-3β,17β,19-triol, 3,17-dibenzoate	EtOH	230(4.45),273(3.34)	33-1961-64
$C_{33}H_{38}O_{15}$			
Chromomycinoquinone, hexaacetate	EtOH	259(4.33),276(4.04), 355(3.56)	88-2355-64
$C_{33}H_{40}N_2O_9$			
invert-Isoreserpine	EtOH	266(4.20)	35-2229-65
invert-Reserpine	EtOH	267(4.19)	35-2229-65
$C_{33}H_{40}N_2O_{10}$			
Isoreserpine, pseudoindoxyl	EtOH	232(4.33),249(4.39), 276(4.29),381(3.63)	35-2229-65
$C_{33}H_{40}N_4O_4$			
Biladiene-ac, 1',8'-dideoxy-4,5-bis-(2'-carbomethoxyethyl)-1,2,3,6,7,8-hexamethyl-, hydrobromide	$CHCl_3$	291(3.45),373(4.18), 457(4.40),520(5.20)	39-1620-65
$C_{33}H_{40}N_4O_4S_2$			
p-Toluenesulfonic acid, androsta-1,4-diene-3,17-diylidenehydrazide	EtOH	227(4.43),287(4.27)	78-1611-65
$C_{33}H_{40}N_6O_4$			
Progesterone, bis(p-nitrophenylhydrazone)	DMF	417(4.64)	13-0255-64
	DMF-NaOH	535(4.90)	13-0255-64
$C_{33}H_{40}O_4$			
Grevillol, dibenzoate	EtOH	233(4.45)	12-2015-65
$C_{33}H_{41}N_5$			
5-Azaporphine, 2,3,7,8,13,17-hexaethyl-12,18-dimethyl-	$CHCl_3$	273(3.92),380(4.96), 480(3.62),506(3.88), 538(4.38),560(3.90), 613(4.39)	39-1620-65
$C_{33}H_{42}N_2O_6$			
5β-Pregn-17(20)-en-21-al, 3α,20-diacetoxy-11-oxo-, 21-(N-acetylphenylhydrazone)	MeOH	281(4.53)	44-0521-64

Compound	Solvent	λ_{max}(log ε)	Ref.
$C_{33}H_{42}N_2O_{10}$			
Isoreserpine, pseudoindoxyldihydro-	EtOH	260(4.10)	35-2229-65
$C_{33}H_{42}N_4$			
Corrole, 2,3,8,12,17,18-hexaethyl-7,13-dimethyl-	$CHCl_3$	346(4.21),396(5.06),408(5.00),503s(3.85),516s(3.95),537(4.24),552(4.26),592(4.36)	39-1620-65
copper complex	$CHCl_3$	397(3.95),502s(3.86),549(4.06)	39-1620-65
$C_{33}H_{42}O_4$			
Pregna-3,5-dien-20-one, 21-benzylidene-3-ethoxy-16α,17α-isopropylidenedioxy-	EtOH	240(4.32)	78-0597-64
$C_{33}H_{42}O_6$			
Dibenzo[d,f]dioxepin-6-spiro-2'-cyclohexa-3',5'-dienone, 2,4',10-tri-tert-butyl-4,6',8-trimethoxy-	EtOH	220(4.82),255(4.27),287(3.71)	39-2914-65
Dibenzo[d,f]dioxepin-6-spiro-2'-cyclohexa-3',5'-dienone, 3,3',9-tri-tert-butyl-2,5',10-trimethoxy-	C_6H_{12}	220(4.63),260(4.22),307(4.12),340s(3.34)	39-3660-65
5H-Tribenzo[a,d,g]cyclononene, 2,3,7,8,12,13-hexaethoxy-10,15-dihydro-	EtOH	207(4.85),233(4.51),289(3.97)	39-1685-65
$C_{33}H_{44}N_4$			
Biladiene-ac, 1',8'-dideoxy-1,2,4,5,7,8-hexaethyl-3,6-dimethyl-, dihydrobromide	$CHCl_3$	288(3.43),373(4.22),458(4.32),523(5.23)	39-1620-65
$C_{33}H_{44}N_4O_{22}$			
Urea, 1,1'-malonylbis[3-β-D-glucopyranosyl-, octaacetate	EtOH	295(3.88)	94-0459-64
$C_{33}H_{44}O_2$			
1,4-Naphthoquinone, 3-geranylgeranyl-2,6,7-trimethyl-	hexane	253(4.36),259(4.40),275(4.16)	73-0197-64
$C_{33}H_{44}O_4$			
1,4-Naphthoquinone, 3-geranylgeranyl-6,7-dimethoxy-2-methyl-	hexane	268(4.50),270(4.42),277(4.45)	73-0197-64
$C_{33}H_{44}O_6$			
5α-Androstan-3-one, 17β-hydroxy-2-(hydroxymethylene)-17-methyl-, 2-(ethyl carbonate) 17-hydrocinnamate	EtOH	208(3.99),251(4.03)	13-0441-65
Biphenyl, 2-hydroxy-2'-(2-hydroxy-3-methoxy-5-tert-butylphenoxy)-3,3'-dimethoxy-5,5'-di-tert-butyl-	EtOH	223(4.87),252s(4.41),298(4.03)	39-2914-65
$C_{33}H_{44}O_8$			
Helvolic acid	EtOH	231(4.24)	94-0121-64
$C_{33}H_{45}NO_6$			
3-Piperidinol, 1-acetyl-2-[1-(2,3,4,4a-5,6,6b,7,11,11b-decahydro-3-hydroxy-10,11b-dimethyl-11-oxo-1H-benzo[a]-fluoren-9-yl)ethyl]-5-methyl-, diacetate	EtOH	233s(3.58),294(3.66),325(3.82)	44-0270-64

Compound	Solvent	λ_{max}(log ε)	Ref.
$C_{33}H_{46}N_4O_5$			
Cholest-4-ene-3,6-dione, 2,4-dinitrophenylhydrazone	DMF	398(4.51)	95-0390-64
	DMF-NaOH	562(4.67)	95-0390-64
$C_{33}H_{46}O_5$			
Cholesta-5,8(14)-diene-Δ$^{7,\alpha}$-succinic anhydride, 3β-acetoxy-	EtOH	213(3.99),224(3.97), 284(4.13)	78-3641-65
5,8-Etheno-15H-cyclopenta[a]phenanthrene-6,7-dicarboxylic anhydride, 17-(1,5-dimethylhexyl)dodecahydro-3-hydroxy-10,13-dimethyl-, acetate	C_6H_{12}	214(3.67)	78-3651-65
$C_{33}H_{47}NO$			
A-Nor-5α-cholest-1-eno[2,1-b]quinolin-4'(1'H)-one	EtOH	242(4.48),248(4.45), 317(4.02),330(4.04)	94-1073-65
$C_{33}H_{47}NO_6$			
Isojervine, 5α-dihydro-, triacetate	EtOH	234(4.15),332(2.26)	44-2282-64
Isojervine, 5β-dihydro-, triacetate	EtOH	236(4.04),328(2.26)	44-2282-64
Isojervine, 5,6-dihydro-, triacetate	EtOH	238(4.01),332(2.13)	44-0262-64
Isojervine, 8,9-dihydro-, triacetate	EtOH	310(2.43),320(2.42)	44-0262-64
Isojervine, 5,6-dihydro-triacetyl-	EtOH	237(3.93)	44-0528-65
Isojervine, 8,9-dihydro-triacetyl-	EtOH	309(1.81),319s(1.77), 330(1.58)	44-0528-65
3-Piperidinol, 1-acetyl-2-[1-(dodecahydro-3,9-dihydroxy-10,11b-dimethyl-11-oxo-1H-benzo[a]fluoren-9-yl)-ethyl]-5-methyl-, diacetate	EtOH	227s(3.95),260s(3.72), 317(4.02)	44-0262-64
$C_{33}H_{47}NO_7$			
Isojervine, 5,6-dihydro-triacetyl-, 17,17a-epoxide	EtOH	237(4.06),316(1.85)	44-0270-64
3-Piperidinol, 1-acetyl-2-[1-(dodecahydro-3,9-dihydroxy-10,11b-dimethyl-11-oxo-1H-benzo[a]fluoren-9-yl)-ethyl]-5-methyl-, acetate	EtOH	273(4.05),337s(2.68)	44-0270-64
$C_{33}H_{47}N_3O_3$			
Cholest-4-ene-3,6-dione, (p-nitrophenylhydrazone)	DMF	428(4.71)	95-0390-64
	DMF-NaOH	612(4.81)	95-0390-64
$C_{33}H_{48}N_4O_4$			
Cholest-4-en-3-one, 2,4-dinitrophenylhydrazone	DMF	405(4.39)	95-0390-64
	DMF-NaOH	490(4.53)	95-0390-64
Cholest-5-en-3-one, 2,4-dinitrophenylhydrazone	DMF	380(4.30)	95-0390-64
	DMF-NaOH	480(4.46)	95-0390-64
$C_{33}H_{48}O$			
Cholest-4-en-3-one, 4-phenyl-	EtOH	244(4.10)	22-0321-64
$C_{33}H_{48}O_6$			
25a-Furosta-5,22(23)-diene-3β,27-diol, 23-acetyl-, diacetate	EtOH	278(4.17)	44-3915-65
$C_{33}H_{49}N$			
4-Azacholesta-2,5-diene, 4-methyl-3-phenyl-	EtOH	248(4.06),296(3.56)	4-0126-65
$C_{33}H_{49}NO$			
2,3-Secocholest-4-eno[3,4-b]indole, 2-oxo-	EtOH	241(3.82),290(3.31)	94-1073-65

Compound	Solvent	λ max(log ε)	Ref.
$C_{33}H_{49}NO_6$			
Isojervine, tetrahydro-, triacetyl derivative	EtOH	307(2.40)	44-2282-64
5β-isomer	EtOH	308(2.40)	44-2282-64
3-Piperidinol, 1-acetyl-2-[1-(3-hydroxy-10,17a-dimethyl-11-oxo-D-homo-C-norgon-16-en-17-yl)ethyl]-5-methyl-, diacetate	EtOH	254(4.21),350(2.01)	78-0779-65
$C_{33}H_{49}N_3O_2$			
Cholest-4-en-3-one, (p-nitrophenylhydrazone)	DMF	425(4.52)	95-0390-64
	DMF-NaOH	541(4.67)	95-0390-64
Cholest-5-en-3-one, (p-nitrophenylhydrazone)	DMF	418(4.48)	95-0390-64
	DMF-NaOH	537(4.64)	95-0390-64
$C_{33}H_{50}O_2$			
1,4-Naphthoquinone, 2,6,7-trimethyl-3-phytyl-	hexane	252(4.41),258(4.45), 276(4.24)	73-0197-64
$C_{33}H_{50}O_4$			
1,4-Naphthoquinone, 6,7-dimethoxy-2-methyl-3-phytyl-	hexane	267(4.58),270(4.55), 278(4.64)	73-0197-64
Sanguisorbigenin, methyl ester, acetate	EtOH	215(3.92)(end absorption)	95-0477-64
Ursa-9(11),12-dien-24-oic acid, 3α-hydroxy-, methyl ester, acetate	n.s.g.	282(4.0)	12-0477-64
$C_{33}H_{50}O_5$			
Cholesta-5,8-diene-$\Delta^{7,\alpha}$-succinic acid, 3β-hydroxy-, dimethyl ester	C_6H_{12}	217(4.10)	78-3641-65
Urs-12-en-24-oic acid, 3α-hydroxy-11-oxo-, methyl ester, acetate	EtOH	250(4.06)	32-0328-64
$C_{33}H_{50}O_6$			
Ceanothic acid, acetyl-, dimethyl ester	EtOH	214(2.97),216(2.81), 218(2.67),220(2.57) (end absorptions)	12-1451-65
$C_{33}H_{50}O_7$			
29-Nordammar-17(20)-en-21-oic acid, 11β,16β-dihydroxy-3-oxo-, diacetate	EtOH	218(3.92)	78-3505-65
$C_{33}H_{51}N$			
4-Aza-5α-cholest-2-ene, 4-methyl-3-phenyl-	EtOH	216(4.05),280(3.53)	4-0126-65
$C_{33}H_{52}O_4$			
Urs-12-en-24-oic acid, 3α-hydroxy-, methyl ester, acetate	EtOH	244(2.96),251(2.99), 260(2.90)	32-0328-64
$C_{33}H_{52}O_7$			
29-Nordammar-17(20)-en-21-oic acid, 11,16-dihydroxy-3-oxo-, cyclic ethylene acetal, 16-acetate	EtOH	221(3.92)	78-3505-65
	dioxan	221(3.92)	31-0344-64
$C_{33}H_{54}O_5S$			
5α-Cholestane-3β,5α-diol, 6β-mercapto-, triacetate	EtOH	232(3.67),306(2.69)	78-0329-65

Compound	Solvent	λ_{max}(log ϵ)	Ref.
$C_{33}H_{64}O_2$			
16,18-Tritriacontanedione	hexane	273(4.20)	12-0464-64
	EtOH	276(4.16)	12-0464-64
	EtOH-KOH	297(4.54)	12-0464-64

Compound	Solvent	λ max(log ε)	Ref.
$C_{34}H_{16}O_2$			
4,11-Dioxa-5,6;12,13-dibenzo-bisanthene	benzene	312(4.84),326(4.95)	24-2860-64
	$C_6H_3Cl_3$	360(4.04),410(4.20), 423(4.26),453(4.24), 484(4.28)	24-2860-64
$C_{34}H_{18}$			
5,6-Benzo-7,10-(o-phenylene)naphtho-[2''',3''':2,3]fluoranthene	benzene	318(4.84),354(4.50), 361(4.56),394(3.82), 400(3.84),420(4.02), 436(3.94),446(4.24), 468(3.66),498(3.74), 524(3.57),584(3.46)	78-2107-64
2,3:7,8-Dibenzo-1,12-(o-phenylene)-perylene	benzene	289(4.94),340(4.69), 372(4.18),416(4.06), 441(4.24),462(4.18), 490(4.10)	78-2107-64
2,3:6,7-Di-(peri-naphthylene)-anthracene	benzene	338(4.96),354(5.22), 403(4.00),416(4.12), 442(4.46),473(4.66)	78-1559-64
	C_6H_{12}	257(4.98),270(4.80)	78-1559-64
Hexatriyne, 1,1'-dianthryl-	THF	256(5.20),263(5.17), 287(4.45),304(4.55), 320s(4.10),380s(4.24), 406s(4.50),420(4.53), 440(4.62)	88-0719-64
1,12-o-Phenylene-2,3:10,11-di-benzoperylene	benzene	301(4.67),313(4.82), 326(4.87),358(3.95), 382(4.11),402(4.37), 426(4.43)	78-0507-64
	C_6H_{12}	245(4.80)	78-0507-64
$C_{34}H_{20}$			
Tetrapheno[6',5':5,6]tetraphene	benzene	304(4.81),322(4.84), 336(4.54),400(4.14), 420(4.14)	78-2107-64
	C_6H_{12}	240(4.80),272(4.65)	78-2107-64
$C_{34}H_{22}$			
Dianthracene, 9',10'-dehydro-9-phenyl-	CH_2Cl_2	268(3.69),275(3.67), 285(3.54)	44-2126-65
$C_{34}H_{23}O$			
Fluoren-9-yl, 9-(α-fluoren-9-ylidene-p-methoxybenzyl)-	dioxan	240(4.79),334(4.15), 432(4.10),503(4.40), 882(3.21)	49-0003-64
$C_{34}H_{24}$			
Naphthalene, 1,4,5,8-tetraphenyl-	C_6H_{12}	248(4.50),334(4.18)	78-0195-64
	C_6H_{12}	250(4.7),337(4.3)	28-5447-64A
$C_{34}H_{24}Br_2O_{10}$			
Erythroaphin-sl, diacetyldibromo-	$CHCl_3$	264(4.37),275s(4.28), 328(4.01),361(3.92), 450(4.41)	39-0072-64
$C_{34}H_{24}N_2O_4S$			
1,4-Naphthoquinone, 2,2'-thio-bis[3-p-toluidino-	dioxan	242(4.47),282(4.64), 403(3.77),495(3.73)	5-0151-64E

Compound	Solvent	λ_{max}(log ϵ)	Ref.
$C_{34}H_{24}N_2O_6S$			
1,4-Naphthoquinone, 2,2'-thiobis-[3-p-anisidino-	dioxan	240(4.52),282(4.67), 413(3.95),550(3.95)	5-0151-64E
$C_{34}H_{24}O$			
Anisole, p-(fluoren-9-ylfluoren-9-ylidenemethyl)-	dioxan	225(4.80),251(4.65), 260(4.72),275(4.44), 291(4.31),304(4.35), 320(4.30)	49-0003-64
anion	DMF-NaOH	306(4.51),344(4.26), 592(4.69)	49-0003-64
$C_{34}H_{24}O_4$			
α-Toluic acid, α,α'-6,12-chrysenylenedi-, potassium salt	50% EtOH	264(4.90),274(5.12), 292(4.12),303(4.17), 316(4.28),330(4.30), 347(3.48),365(3.39)	24-2860-64
$C_{34}H_{24}O_6$			
1,2-Naphthoquinone, 4-benzyl-3-hydroxy-, dimer	n.s.g.	259(4.38),301(3.51)	24-0666-65
$C_{34}H_{24}O_8$			
Phenanthro[9,10-l]phenanthrene, 2,3,6,7-tetraacetoxy-	EtOH	253(4.78),261s(4.70), 281s(4.28),298(4.25), 312(4.36),324(4.39), 340(3.80),357(3.72)	39-3040-65
$C_{34}H_{24}O_9$			
Chrysoaphin-fb	$CHCl_3$	268(4.64),380(4.30), 402(4.51),457(4.19), 485(4.21)	39-0080-64
$C_{34}H_{25}P$			
Phosphole, pentaphenyl-	n.s.g.	248(4.51),320(3.94), 357(3.95)	89-1109-65
$C_{34}H_{26}$			
Benzene, p-bis(p-phenylstyryl)-	dioxan	360(--),376(--)	38-2839-64A
1,3-Butadiene, 1,1,2,4,4-pentaphenyl-	EtOH	240(4.39),342(4.23)	44-3660-64
$C_{34}H_{26}Br_2N_2O_{10}$			
Erythroaphin-fb, diacetamido-dibromo-	$CHCl_3$	264(4.60),340(3.67), 483(4.49),532(4.14), 574(4.21)	39-0072-64
$C_{34}H_{26}Cl_2N_2O_{10}$			
Erythroaphin-fb, diacetamido-dichloro-	$CHCl_3$	263(4.63),333(3.75), 478(4.49),529(4.14), 571(4.19)	39-0072-64
$C_{34}H_{26}Fe_2$			
Bi(2,3-ferrocoindenyl)	EtOH	218(4.35),242(4.01), 289(3.89),345(2.79), 459(2.37)	35-5607-65
$C_{34}H_{26}O_2P_2$			
Phosphine oxide, 1,4-naphthylenebis[diphenyl-	EtOH	220s(4.82),226s(4.72), 245(4.51),266(3.68), 274(3.65),307(3.91), 326(3.75)	49-0285-65

Compound	Solvent	λ_{max}(log ε)	Ref.
Phosphine oxide, 1,5-naphthylene-bis[diphenyl-	EtOH	224(4.82),262s(3.63), 268s(3.77),274(3.86), 284(3.98),293(4.11), 304(3.98),324(3.38)	49-0285-65
Phosphine oxide, 2,6-naphthylene-bis[diphenyl-	EtOH	226(4.79),245(5.17), 268(4.14),272(4.14), 280(4.08),292s(3.75), 322(3.28),334(3.36)	49-0285-65
Phosphine oxide, 2,7-naphthylene-bis[diphenyl-	EtOH	225(4.82),245(5.09), 266(3.97),273(3.95), 285s(3.77),313(2.82), 321(2.83),329(2.92)	49-0285-65
$C_{34}H_{26}O_{10}$			
3''',8-Biflavone, 5,5''-dihydroxy-4',4''',7,7''-tetramethoxy-	EtOH	270(4.65),328(4.60)	12-1491-65
	0.02N NaOEt	290(4.81),380(4.20)	12-1491-65
	0.002N NaOEt	290(4.74),360(4.20)	12-1491-65
	0.0002N NaOEt	271(4.65),326(4.59)	12-1491-65
	EtOH-$AlCl_3$	225s(4.79),282(4.69), 302(4.62),346(4.66), 382s(4.49)	12-1491-65
$C_{34}H_{26}P_2$			
Phosphine, 1,4-naphthylene-bis[diphenyl-	EtOH	276s(4.13),332(4.15)	49-0285-65
Phosphine, 1,5-naphthylene-bis[diphenyl-	EtOH	220(4.74),318(4.02)	49-0285-65
Phosphine, 2,6-naphthylene-bis[diphenyl-	EtOH	234(4.82),268(4.6), 315s(4.24)	49-0285-65
Phosphine, 2,7-naphthylene-bis[diphenyl-	EtOH	236(4.78),270(4.48), 290s(4.38),330s(3.36)	49-0285-65
$C_{34}H_{28}Br_2N_2S_2$			
2,2'-Vinylenebis[3-methyl-4,5-di-phenylthiazolium bromide]	n.s.g.	435(4.0)	65-0075-64
$C_{34}H_{28}CuN_2O_2$			
Copper, bis[1-(α-aminobenzyl)-2-naphtholato]-	90% EtOH-10% THF	237s(5.01),243(5.02), 263s(4.58),272s(4.45), 283(4.40),292s(4.30), 380s(3.23),450s(2.65), 490s(2.53)	7-0696-64
$C_{34}H_{28}N_2NiO_2$			
Nickel, bis[1-(α-aminobenzyl)-2-naphtholato]-	90% EtOH-10% THF	233(4.80),257(4.88), 273s(4.42),286(4.28), 300s(4.13),343s(3.83), 353(3.87),390s(2.86), 480(2.34),575(2.30)	7-0696-64
$C_{34}H_{28}N_2O_2Pd$			
Palladium. bis[1-(α-aminobenzyl)-2-naphtholato]-	90% EtOH-10% THF	245s(4.99),253(5.02), 264s(4.46),273s(4.36), 284(4.28),294s(4.10), 342s(3.86),349(3.87), 393(3.11)	7-0696-64

Compound	Solvent	λ_{max}(log ε)	Ref.
$C_{34}H_{28}N_2O_{10}$			
Erythroaphin-fb, diacetamido-	$CHCl_3$	255(4.56),322(3.61), 433(4.40),460(4.54), 529(4.05),570(4.22), 600(3.81)	39-0072-64
$C_{34}H_{28}N_6O_2S_2$			
Indole-2-carboxylic acid, 3,3'-dithiobis[1-methyl-, bis(benzylidenehydrazide)	EtOH	214(4.79),303(4.75), 324s(4.67),376s(3.87)	44-0178-64
$C_{34}H_{28}O$			
Furan, 2,5-dihydro-2,2-diphenyl-5-(6,6-diphenylhexatrienyl)-	n.s.g.	239(4.22),326(4.62)	35-0136-65
$C_{34}H_{28}O_6$			
1,4-Ethenoanthracene-2,3-dicarboxylic acid, 1,4-dihydro-1,4-dimethoxy-9,10-diphenyl-, dimethyl ester	ether	245(4.7),274(4.1), 283(4.1),292s(4.0), 319(3.0),333(2.9)	28-5031-65A
$C_{34}H_{28}O_{10}$			
Xylindein, dimethyl-	$CHCl_3$	281(4.13),403(4.05), 453(3.74),485(4.12), 521(4.51),566(4.66)	78-2095-65
$C_{34}H_{29}NO_{16}$			
Laccaic acid A , tetraacetate	MeOH	251(4.49),295(4.44)	39-6067-65
$C_{34}H_{30}$			
Bi-2-cyclopropen-1-yl, 2,2'-diethyl-1,1',3,3'-tetraphenyl-	n.s.g.	271(4.23)	35-5139-65
p-Xylene, α,α'-bis(o-benzylphenyl)-	hexane	260(2.31),266(2.31), 276(2.15)	70-0124-65
$C_{34}H_{30}Cl_2O_8$			
Erythroaphin-fb, dichlorodihydrotetramethyl-	$CHCl_3$	287(4.61),367(3.19), 416(3.49),440(3.93), 470(4.37),503(4.52)	39-0072-64
Erythroaphin-sl, dichlorodihydrotetramethyl-	$CHCl_3$	287(4.60),367(3.14), 442(3.93),470(4.37), 504(4.52)	39-0072-64
$C_{34}H_{30}N_4O_4$			
3,5-Pyrazolidinedione, 4,4'-isobutylidenebis[1,2-diphenyl-	MeOH	258(4.18)	32-0320-65
$C_{34}H_{30}N_4O_{10}S$			
6-Purinethiol, 9-[2,3,5-tris(p-methoxybenzoyl)-β-D-ribofuranosyl]-	pH 1	261(4.61),323(4.34)	87-0200-64
	pH 11	236(4.39),258(4.67), 310(4.37)	87-0200-64
$C_{34}H_{30}O_2$			
2,4,6,8-Decatetraene-1,10-diol, 1,1,10,10-tetraphenyl-	n.s.g.	294(4.72),306(4.85), 320(4.82)	35-0136-65
$C_{34}H_{30}O_4$			
1,2,3,5-Hexatetraene, 1,1,6,6-tetrakis(p-methoxyphenyl)-	n.s.g.	288(4.46),456(4.58)	24-2959-64
1,2,4,5-Hexatetraene, 1,1,6,6-tetrakis(p-methoxyphenyl)-	n.s.g.	252(4.75)	24-2959-64

Compound	Solvent	λ_{max}(log ϵ)	Ref.
$C_{34}H_{30}O_6$			
Tectol, diacetate	n.s.g.	228(4.72),264(4.76), 272(4.81),345(4.03), 363(3.96)	24-0588-64
$C_{34}H_{31}ClN_2O_5$			
2-(4-Benzamido-2,4-diphenyl-1,3-butadienyl)-1,3,3-trimethyl-3H-indolium perchlorate	MeCN	343(4.09),455(4.57)	24-3892-65
$C_{34}H_{31}NO_{16}$			
Laccaic acid A_1, dihydro-, tetraacetate	MeOH	261(4.52),311(4.20), 359s(3.98),459(3.54)	39-6067-65
$C_{34}H_{31}N_5O_7S$			
9H-Purine-6-thiol, 2-amino-9-(2,3,5-tritoluoyl-β-D-ribofuranosyl)-	pH 1	245(4.78),340(4.32)	87-0200-64
	pH 11	244(4.63),318(4.39)	87-0200-64
$C_{34}H_{31}N_5O_{10}S$			
9H-Purine-6-thiol, 2-amino-9-[2,3,5-tris(p-methoxybenzoyl)-β-D-ribofuranosyl]-	pH 1	261(4.65),344(4.30)	87-0200-64
	pH 11	256(4.70),318(4.30)	87-0200-64
$C_{34}H_{32}CoN_4O_4$			
Cobalt protoporphyrin	pH 7.9	416(5.0)	37-1018-64
$C_{34}H_{32}CuN_4O_4$			
Copper protoporphyrin	pH 7.9	388(4.81)	37-1018-64
$C_{34}H_{32}FeN_4O_4$			
Heme	pH 7.9	385(4.75)	37-1018-64
$C_{34}H_{32}N_4NiO_4$			
Nickel protoporphyrin	pH 7.9	385(4.72)	37-1018-64
$C_{34}H_{32}O_4$			
Tricyclo[8.2.2.2^{4,7}]hexadeca-4,6,10,12,13,15-hexaenecarboxylic acid, [p-(p-carboxyphenethyl)-α-methylbenzyl]-	EtOH	227(4.48),282(3.91)	44-1815-65
isomer?	EtOH	227(4.48),280(3.89)	44-1815-65
$C_{34}H_{32}O_8$			
Erythroaphin-fb, dihydrotetramethyl-	$CHCl_3$	279(4.54),347(3.50), 363(3.72),390(3.34), 415(3.56),440(4.03), 470(4.38),505(4.57)	39-0072-64
Erythroaphin-sl, dihydrotetramethyl-	$CHCl_3$	265(4.50),277(4.44), 347(3.52),363(3.74), 390(3.32),413(3.55), 439(4.02),468(4.39), 503(4.54)	39-0072-64
Erythroaphin-tt, dihydrotetramethyl-	$CHCl_3$	265(4.44),278(4.50), 347(3.57),363(3.74), 390(3.40),415(3.57), 440(4.01),470(4.38), 504(4.52)	39-0072-64

Compound	Solvent	λ_{max}(log ϵ)	Ref.
$C_{34}H_{33}ClO_8$			
1,1,5,5-Tetrakis(p-methoxyphenyl)-3-methyl-1,4-pentadien-3-yl perchlorate	n.s.g.	500(4.50),687(4.95)	24-2959-64
$C_{34}H_{34}Cl_4N_4$			
Phenazine, 1,4-bis[p-bis(2-chloroethyl)aminobenzyl]-	C_6H_{12}	264(4.53)	65-1969-64
$C_{34}H_{34}Cl_4N_4O_2$			
Phenazine, 1,4-bis[p-bis(2-chloroethyl)-aminobenzyl]-, 5,10-dioxide	C_6H_{12}	252(4.53)	65-1969-64
$C_{34}H_{34}N_4O_4$			
Protoporphyrin	pH 7.9	373(4.87)	37-1018-64
$C_{34}H_{34}O_4$			
5,8,13,14-Pentaphenetetrone, 2,3,10,11-tetramethyl-6-octyl-	dioxan	255(4.48),287(4.55)	65-3484-64
$C_{34}H_{34}O_{14}$			
Ergochrysin A, 1,1',9-tri-O-methyl-	n.s.g.	255(4.44),319(3.20)	39-4144-65
Secalonic acid B, dimethyl ether	EtOH	210(4.33),265(4.15), 328(4.42),376s(3.81)	78-1417-65
Secalonic acid C, dimethyl ether	EtOH	265(4.11),327(4.45), 374s(3.82)	78-1417-65
$C_{34}H_{36}FeN_4O_4$			
Mesoheme IX	pH 7.9	386(4.73)	37-1018-64
$C_{34}H_{36}MgN_4O_4$			
Magnesium mesoporphyrin	pH 7.9	365(4.91),459(4.50)	37-1018-64
$C_{34}H_{36}N_2O_7S$			
Bisnorrubremetinium p-toluenesulfonate	H_2O	255(4.28),285(4.28), 300(4.29),435(4.48)	94-0775-65
$C_{34}H_{36}N_6O_2$			
1,9-Nonanedianilide, 3-methyl-4',4"-bis(phenylazo)-	EtOH	237(4.55),349(4.86)	94-0073-65
$C_{34}H_{36}O_6$			
Tectol, tetrahydro-, diacetate	n.s.g.	248(4.87),305(4.14), 330(4.01)	24-0588-64
$C_{34}H_{37}N_5O_5$			
Ergostine	MeOH	315(3.95)	33-1921-64
$C_{34}H_{38}N_4$			
Phenazine, 1,4-bis(p-diethylaminobenzyl)-	C_6H_{12}	268(4.48)	65-1969-64
$C_{34}H_{38}N_4O_4$			
Phenazine, 1,4-bis[p-bis(2-hydroxyethyl)aminobenzyl]-	C_6H_{12}	260(4.59)	65-1969-64
$C_{34}H_{40}N_4O_{12}S_2$			
5H,11H-Bisthiazolo[3,2-a:3',2'-e][1,5]-diazocine-3,9-dicarboxylic acid, 6,12-bis(2,6-dimethoxybenzamido)-octahydro-2,2,8,8-tetramethyl-5,11-dioxo-	MeOH	284(3.83)	44-1826-64

Compound	Solvent	λ_{max}(log ϵ)	Ref.
$C_{34}H_{40}O_{16}$			
Amorphin	EtOH	236(4.19),240s(4.15), 293(4.23)	39-6023-64
$C_{34}H_{41}N_5O$			
Etioporphyrin I, acetamido-	$CHCl_3$	406(5.20),505(4.14), 537(3.91),575(3.80), 633(3.64)	39-4303-65
$C_{34}H_{42}Br_2N_4O_4$			
18-Norergosta-1,8,11,13-tetraen-3-one, 22,23-dibromo-12-methyl-, 2,4-dinitrophenylhydrazone	$CHCl_3$	256(4.33),286s(4.10), 382(4.55)	78-0929-64
$C_{34}H_{42}N_2O_6$			
Pregn-4-eno[3,2-d]pyrimidine, 11β-(methoxymethyleneoxy)-16α-methyl-17α,20:20,21-bis-(methylenedioxy)-2'-phenyl-	MeOH-HCl	276(4.46)	87-0584-64
	MeOH	258(4.61),307(3.98)	87-0584-64
$C_{34}H_{42}O_5$			
Pregna-3,5-dien-20-one, 21-benzylidene-3-ethoxy-16α,17α-isopropylidenedioxy-	EtOH	224(4.28),298(4.37)	78-0597-64
$C_{34}H_{42}O_7$			
Taxinin, dideacetyl-, cyclic acetal with acetone (benzene solvate)	n.s.g.	218(4.24),224(4.19), 280(4.46)	95-0762-64
$C_{34}H_{42}O_{14}$			
Glucopyranosiduronic acid, 17-hydroxy-3,11,20-trioxopregna-1,4-dien-21-yl-, methyl ester, 2,3,4-triacetate	EtOH	238(4.3)	94-0450-64
$C_{34}H_{42}O_{18}$			
β-Cellobioside, p-vinylphenyl-, heptaacetate	MeOH	253.5(4.29)	87-0242-64
β-Maltoside, p-vinylphenyl-, heptaacetate	MeOH	253.5(4.29)	87-0242-64
$C_{34}H_{44}Br_2N_4O_4$			
18-Norergosta-8,11,13-trien-3-one, 22,23-dibromo-12-methyl-, 2,4-dinitrophenylhydrazone	$CHCl_3$	260s(4.00),367(4.32)	78-0929-64
$C_{34}H_{44}N_2O_2$			
Succinonitrile, α,β-bis(3,5-di-tert-butyl-4-hydroxybenzylidene)-	MeOH	255(4.3),380(4.5)	5-0141-65A
$C_{34}H_{44}O_{14}$			
Glucosiduronic acid, 17-hydroxy-3,11,20-trioxopregn-4-en-21-yl, methyl ester, triacetate	MeOH	239(4.1)	44-2462-64
	EtOH	237(4.20)	94-0450-64
$C_{34}H_{46}Cl_2O_2$			
Cholesta-2,4-dien-6β-ol, 2,6-dichloro-, benzoate	C_6H_{12}	203(4.69),267(3.71)	44-2549-65

Compound	Solvent	λ_{max}(log ϵ)	Ref.
$C_{34}H_{46}O_4$			
Succinaldehyde, bis(2,5-di-tert-butyl-4-hydroxybenzylidene)-	MeOH	240(4.3),340(4.5)	5-0141-65A
$C_{34}H_{46}O_8$			
Helvolic acid, methyl ester	EtOH	231(4.24)	94-0121-64
$C_{34}H_{46}O_9$			
Cucurbitacin I, diacetate	EtOH	233(4.35)	39-0529-64
$C_{34}H_{46}O_{13}$			
Glucopyranosiduronic acid, 17-hydroxy-3,20-dioxopregn-4-en-21-yl, methyl ester, 2,3,4-triacetate	EtOH	242(4.02)	94-0450-64
$C_{34}H_{46}O_{14}$			
Glucopyranosiduronic acid, 11,17-dihydroxy-3,20-dioxopregn-4-en-21-yl, methyl ester, 2,3,4-triacetate	EtOH	242(4.25)	94-0450-64
$C_{34}H_{46}O_{18}$			
Eluteroside E	EtOH	234(4.2),271(3.6)	70-2065-65
$C_{34}H_{47}ClO_{11}$			
Cerbertin, acetyl-, chloroketone	n.s.g.	217(4.16),292(1.92)	12-1079-65
$C_{34}H_{47}N_{11}O_{13}$			
Itorine	H_2O	277(3.1)	20-0329-65
	pH 13	295(3.43)	20-0329-65
$C_{34}H_{48}Cl_2O_3$			
5α-Cholest-2-ene-5,6β-diol, 6-(2,6-dichlorobenzoate)	C_6H_{12}	203(4.62),272(2.53), 279(2.58)	44-2549-65
$C_{34}H_{48}N_4O_4$			
Cholesta-4,6-dien-3-one, 4-methyl-, 2,4-dinitrophenylhydrazone	MeOH	266(4.22),305(4.16), 314(4.21),407(4.57)	78-2531-64
$C_{34}H_{48}O$			
1β,5-Cyclo-5β,10α-cholestan-2-one, 3-benzylidene-	EtOH	224(3.98),230(3.99), 300(4.47)	78-2973-65
$C_{34}H_{48}O_8$			
Senegenin, dechlorodehydro-, dimethyl ester, diformate	MeOH	240(4.08),248(4.14), 256(3.97)	23-0491-64
$C_{34}H_{49}ClO_{11}$			
Cerbertin, chlorohydrin, acetate	n.s.g.	217(4.14)	12-1079-65
$C_{34}H_{50}N_2O_2S$			
p-Toluenesulfonic acid, cholesta-1,4-dien-3-ylidenehydrazide	EtOH	226(4.18),244(4.18), 288(4.26)	78-1611-65
$C_{34}H_{50}O$			
A-Homocholest-4a-en-3-one, 4ξ-phenyl-	MeOH	208(5.1)	5-0047-64G
$C_{34}H_{50}O_6$			
Spergulagenic acid, acetyl-, dimethyl ester, oxidation product of	n.s.g.	242(4.27),250(4.32), 260(4.13)	2-0283-65

Compound	Solvent	λ max(log ε)	Ref.
$C_{34}H_{50}O_7$			
Gratiogenin, oxidized diacetate	MeOH	242(3.76)	5-0196-64D
$C_{34}H_{50}O_{11}$			
Acoschimperoside P, acetate	EtOH	215(4.21)	33-0065-65
$C_{34}H_{50}O_{14}$			
Gypsobioside	EtOH	219(4.21),275s(2.00), 310s(1.82)	65-2463-64
	H_2SO_4	235(4.12),315(4.34), 408(4.28),485(3.92), 540(3.92)	65-2463-64
$C_{34}H_{52}O_5$			
Serratenone, diacetoxy-	EtOH	256(3.97)	88-1303-64
$C_{34}H_{54}N_2O_4$			
Pseudourea, 1,3-dicyclohexyl-2-(3-oxo-androst-4-en-17-yl)-, acetate	MeOH	237(4.19)	35-5670-65
$C_{34}H_{54}O_6$			
29-Nordammar-17(20)-en-21-oic acid, 3β,11α-dihydroxy-, methyl ester, diacetate	EtOH	234(4.08)	78-3505-65
	dioxan	234(4.08)	31-0344-64
$C_{34}H_{54}O_{14}$			
Neutramycin	n.s.g.	216(4.37)	31-0372-65
$C_{34}H_{56}$			
Naphthalene, 2,6-didodecyl-	C_6H_{12}	230(5.18),274(3.77), 303(2.95),309(3.00), 317(2.78),324(3.11)	39-2324-65
$C_{34}H_{56}O_3S$			
1-Naphthalenesulfonic acid, 3,7-didodecyl-	EtOH	232(4.94),280(3.80), 313(3.28),318(3.11), 327(3.38)	39-2324-65

Compound	Solvent	λ_{max}(log ε)	Ref.
$C_{35}H_{22}$			
7H-Dibenzo[c,g]fluorene, 7-(fluoren-9-ylidenemethyl)-	dioxan	259(4.87),287(4.44), 302(4.36),317(4.46), 333(4.33),349(4.42), 365(4.35),420(2.0)	5-0001-65I
anion	DMF	584(4.98)	5-0001-65I
13H-Dibenzo[a,i]fluorene, 13-(fluoren-9-ylidenemethyl)-	dioxan	258(4.98),268(5.02), 280(4.73),291(4.50), 320(4.35),330(4.33), 346(4.15),358(3.64)	5-0001-65I
anion	DMF	611(4.54)	5-0001-65I
$C_{35}H_{24}$			
Dianthracene, 9',10'-dehydro-9-p-tolyl-	CH_2Cl_2	268(3.55),273(3.57), 283(3.51)	44-2126-65
Propane, 1,3-difluoren-9-ylidene-2-phenyl-	dioxan	250(4.79),259(4.84), 291(4.41),304(4.41), 320(4.48)	5-0001-65I
anion	DMF	684(4.93)	5-0001-65I
	DMSO-KO-tert-Bu	308s(4.35),354(4.24), 372(4.22),435(4.19), 684(4.93)	5-0050-65J
$C_{35}H_{24}O$			
2-Naphthaldehyde, 1,4,5,8-tetraphenyl-	$CHCl_3$	280(4.45),330(4.06), 384(3.72)	78-0195-64
$C_{35}H_{25}N_3O_3S$			
2-Naphthol, 3-N-phenylcarbamyl-1-[2'-(2"-phenoxyphenylmercapto)-phenylazo]-	HOAc	513(4.35)	7-0821-64
2-Naphthol, 3-N-phenylcarbamyl-1-[2'-(3"-phenoxyphenylmercapto)-phenylazo]-	HOAc	513(4.32)	7-0821-64
2-Naphthol, 3-N-phenylcarbamyl-1-[4'-(2"-phenoxyphenylmercapto)-phenylazo]-	HOAc	526(4.42)	7-0821-64
2-Naphthol, 3-N-phenylcarbamyl-1-[4'-(3"-phenoxyphenylmercapto)-phenylazo]-	HOAc	525(4.48)	7-0821-64
2-Naphthol, 3-N-phenylcarbamyl-1-[2'-(2"-phenylmercaptophenoxy)-phenylazo]-	HOAc	518(4.36)	7-0821-64
2-Naphthol, 3-N-phenylcarbamyl-1-[2'-(3"-phenylmercaptophenoxy)-phenylazo]-	HOAc	517(4.37)	7-0821-64
2-Naphthol, 3-N-phenylcarbamyl-1-[4'-(2"-phenylmercaptophenoxy)-phenylazo]-	HOAc	524(4.42)	7-0821-64
2-Naphthol, 3-N-phenylcarbamyl-1-[4'-(3"-phenylmercaptophenoxy)-phenylazo]-	HOAc	523(4.44)	7-0821-64
$C_{35}H_{25}N_3O_5S$			
2-Naphthol, 3-N-phenylcarbamyl-1-[2'-(2"-phenoxyphenylsulfonyl)-phenylazo]-	HOAc	500(4.34)	7-0821-64
2-Naphthol, 3-N-phenylcarbamyl-1-[2'-(3"-phenoxyphenylsulfonyl)-phenylazo]-	HOAc	499(4.24)	7-0821-64

Compound	Solvent	λ_{max}(log ϵ)	Ref.
2-Naphthol, 3-N-phenylcarbamyl-1-[4'-(2"-phenoxyphenylsulfonyl)-phenylazo]-	HOAc	504(4.42)	7-0821-64
2-Naphthol, 3-N-phenylcarbamyl-1-[4'-(3"-phenoxyphenylsulfonyl)-phenylazo]-	HOAc	505(4.42)	7-0821-64
2-Naphthol, 3-N-phenylcarbamyl-1-[2'-(2"-phenylsulfonylphenoxy)-phenylazo]-	HOAc	515(4.36)	7-0821-64
2-Naphthol, 3-N-phenylcarbamyl-1-[2'-(3"-phenylsulfonylphenoxy)-phenylazo]-	HOAc	513(4.35)	7-0821-64
2-Naphthol, 3-N-phenylcarbamyl-1-[4'-(2"-phenylsulfonylphenoxy)-phenylazo]-	HOAc	514(4.42)	7-0821-64
2-Naphthol, 3-N-phenylcarbamyl-1-[4'-(3"-phenylsulfonylphenoxy)-phenylazo]-	HOAc	516(4.44)	7-0821-64
$C_{35}H_{26}N_2O_6$			
3-Buten-1-ol, 1,2,4,4-tetraphenyl-, 2,4-dinitrobenzoate	n.s.g.	250(4.43)	44-3660-64
$C_{35}H_{26}O_2$			
3-Cyclopenten-1-one, 2-hydroxy-2,3,4,5,5-pentaphenyl-	EtOH	248(4.165)	28-4061-64B
2-Naphthaldehyde, 1,4-epoxy-1,2,3,4-tetrahydro-1,4,5,8-tetraphenyl-	EtOH	244s(4.20)	78-0195-64
2,4-Pentadienoic acid, 2,3,4,5,5-pentaphenyl-, trans-cis	EtOH	242(4.42),290(4.26), 328s(4.03)	28-4061-64B
isomer	EtOH	300(4.095)	28-4061-64B
$C_{35}H_{30}$			
5H-Benzocycloheptene, 6,7,8,9-tetrahydro-1,2,3,4-tetraphenyl-	isooctane	236(4.62)	35-1326-65
$C_{35}H_{30}N_4O_9S$			
Thymidine, 5'-O-trityl-, 3'-(2,4-dinitrobenzenesulfenate)	dioxan	264(4.30),330(4.04)	44-2615-64
$C_{35}H_{31}IN_2$			
4-[(1-Ethyl-4(1H)-quinolylidene)-methyl]-6-methyl-1-p-tolyl-2-phenyl-quinolinium iodide	n.s.g.	601(4.58)	65-2008-64
$C_{35}H_{31}IN_2O_4$			
6-Acetoxy-1-p-acetoxyphenyl-4-(p-dimethylaminostyryl)-2-phenyl-quinolinium iodide	n.s.g.	590(4.47)	65-2008-64
$C_{35}H_{32}N_4O_4$			
3,5-Pyrazolidinedione, 4,4'-isopentylidenebis[1,2-diphenyl-	MeOH	248(4.38),400(3.45)	32-0320-65
$C_{35}H_{32}O_2$			
Indan, 3a,7a-dihydro-4,7-bis(p-methoxyphenyl)-5,6-diphenyl-	EtOH	333(4.14)	39-1881-65

Compound	Solvent	λ_{max}(log ε)	Ref.
$C_{35}H_{34}O_4$			
4-Hexenoic acid, 5-hydroxy-4-methyl-3-oxo-2,2,6,6-tetraphenyl-, δ-lactone, diethyl acetal	MeOH	260s(3.30)	78-2735-64
5-Hexenoic acid, 5-hydroxy-4-methyl-3-oxo-2,2,6,6-tetraphenyl-, δ-lactone, diethyl acetal	MeOH	252(4.11)	78-2735-64
$C_{35}H_{36}N_2O_6$			
Atherospermoline (as $CHCl_3$ adduct)	EtOH	284(3.97)	25-0694-65
$C_{35}H_{36}N_2O_8S_2$			
Voacarpine, N-(p-tolylsulfonyl)-, p-toluenesulfonate	MeOH-HCl	228(4.46),315(4.18)	20-0170-65
	MeOH	228(4.46),315(4.18)	20-0170-65
$C_{35}H_{40}N_4$			
Malonamidine, N,N',N'',N'''-tetraphenethyl-, hydrochloride	MeOH	308(4.33)	44-0308-64
$C_{35}H_{42}O_9$			
Taxinin	n.s.g.	217(4.31),223(4.20), 278(4.46)	95-0762-64
$C_{35}H_{44}O_9$			
Taxinin, dihydro-	n.s.g.	212(3.94),268(3.84)	95-0762-64
$C_{35}H_{46}O_9$			
Taxinin, tetrahydro-	n.s.g.	212(3.94),269(3.85)	95-0762-64
$C_{35}H_{48}O_6$			
Biphenyl, 5,5'-di-tert-butyl-2'-(2,3-dimethoxy-5-tert-butylphenoxy)-2,3,3'-trimethoxy-	EtOH	217(4.83),280(3.78)	39-2914-65
$C_{35}H_{48}O_{10}$			
27-Norolean-13-ene-23,28-dioic acid, 2β,3β-dihydroxy-12,15-dioxo-, dimethyl ester, diacetate	EtOH	270(3.98)	44-4234-65
$C_{35}H_{50}N_4O_5$			
Clionast-4-ene-3,6-dione, 2,4-dinitrophenylhydrazone	$CHCl_3$	265(4.09),387(4.59)	100-0040-64
$C_{35}H_{50}O_2$			
Cholesta-4,6-dien-3β-ol, 6-methyl-, benzoate	EtOH	238(4.49)	39-3106-64
$C_{35}H_{50}O_6$			
5,8-Etheno-15H-cyclopenta[a]phenanthrene-6,7-dicarboxylic acid, 17-(1,5-dimethylhexyl)decahydro-3-hydroxy-10,13-dimethyl-, dimethyl ester, acetate	C_6H_{12}	215(4.13),256s(4.05)	78-3651-65
$C_{35}H_{50}O_7$			
25α-Furosta-5,20(22)-diene-3β,27-diol, 23-(1-hydroxyethylidene)-, triacetate	EtOH	253(3.67)	44-3915-65

Compound	Solvent	λ_{max}(log ε)	Ref.
$C_{35}H_{50}O_{10}$			
Taxinin, decahydrooxo-	EtOH	299(3.59),450(2.48)	95-0404-65
$C_{35}H_{51}NO_2$			
Phthalimide, N-5α-cholestan-3α-yl-	dioxan	291(3.22),298s(3.20)	24-0533-64
$C_{35}H_{52}N_2O_3$			
Cyclomicrophyllidine A	EtOH	265(2.34),273(2.38), 280(2.30)	39-4512-65
$C_{35}H_{52}O$			
Cholesta-3,5-diene, 3-(p-ethoxy-phenyl)-	EtOH	227(4.01),234s(--), 288(4.41)	23-2919-64
$C_{35}H_{52}O_2$			
Cholest-5-en-3β-ol, 7β-methyl-, benzoate	EtOH	231(4.22)	39-3106-64
Cholest-7-en-3β-ol, 7-methyl-, benzoate	EtOH	231(4.21)	39-3106-64
$C_{35}H_{52}O_3$			
Cholest-6-ene-3β,5α-diol, 6-methyl-, 3-benzoate	EtOH	231(4.21)	39-3106-64
Cholest-6-ene-3β,7α-diol, 6-methyl-, 3-benzoate	EtOH	231(4.22)	39-3106-64
$C_{35}H_{52}O_6$			
Cholest-9(11)-en-3β-ol, 5α,8α-(1,2-di-carbomethoxyvinylene)-, acetate	EtOH	215(3.78),244s(3.45)	78-3651-65
Maleic acid, (3β-hydroxycholesta-5,8-dien-7α-yl)-, dimethyl ester, acetate	EtOH	218.5(4.03)	78-3641-65
Maleic acid, (3β-hydroxycholesta-5,8(14)-dien-7α-yl)-, dimethyl ester, acetate	EtOH	219(3.99)	78-3641-65
$C_{35}H_{52}O_9$			
Taxinin, decahydro-	n.s.g.	268(3.84)	95-0762-64
$C_{35}H_{54}N_2O_3$			
Cyclomicrophyllidine A, dihydro-	EtOH	265(2.34),273(2.38), 280(2.30)	39-4512-65
$C_{35}H_{54}O_3$			
Cholestane-3β,7α-diol, 7β-methyl-, 3-benzoate	EtOH	231(4.21)	39-3106-64
$C_{35}H_{54}O_{12}$			
Glucoevatromonoside	MeOH	217(4.20)	5-0137-64H
$C_{35}H_{54}O_{13}$			
Digitoxigenin, glucosido-glucomethyloside	n.s.g.	216(4.20)	5-0216-65H
Glucodigifucoside	n.s.g.	216(4.22)	31-0575-65
Glucogitoroside	MeOH	217(4.21)	5-0137-64H
Neoglucodigifucoside	n.s.g.	216(4.21)	31-0575-65
$C_{35}H_{56}OS$			
Cycloeucalenone, butylthio-methylene-	EtOH	311.5(4.2)	39-1692-65

Compound	Solvent	$\lambda_{max}(\log \epsilon)$	Ref.
$C_{35}H_{56}O_{14}$ Chalcomycin	EtOH	218(4.36)	35-2724-64

Compound	Solvent	$\lambda_{max}(\log \varepsilon)$	Ref.
$C_{36}H_{18}$			
Octatetrayne, di-1-anthryl-	THF	250(5.20),299(4.67), 315(4.71),352(4.09), 370s(4.12),390s(4.34), 422(4.62),456(4.61)	88-0719-64
1,12:2,3:6,7:8,9-Tetrabenzanthanthrene	$C_6H_3Cl_3$	309(4.88),324(4.89), 362(4.35),380(4.69), 404(4.86)	78-0507-64
1,2:3,4:5,6-Tribenzocoronene	benzene	272(4.60),293(4.62), 340(4.79),354(5.00), 384(4.34),407(4.25), 436(2.98),464(2.62)	78-0507-64
	C_6H_{12}	236(4.81)	78-0507-64
$C_{36}H_{18}N_2O_8S_4$			
2,2'-Ethylenebis[3-methyl-4,5-di-phenylthiazolium methyl sulfate]	n.s.g.	295(4.0)	65-0075-64
$C_{36}H_{22}Cl_2N_4O_8$			
Quino[2,3-b]acridine-6,7,13,14(5H,12H)-tetrone, 2,9-dibenzamido-1,8-di-chloro-4,11-dimethoxy-	H_2SO_4	425(3.96),528(4.05)	27-0242-65
$C_{36}H_{22}N_2O_8$			
Quino[2,3-b]acridine-6,7,13,14(5H,12H)-tetrone, 1,8-dibenzoyl-4,11-dimethoxy-	H_2SO_4	407(4.31),484(4.01)	27-0242-65
$C_{36}H_{24}$			
7H-Dibenzo[c,g]fluorene, 7-(fluoren-9-ylidenemethyl)-7-methyl-	dioxan	260(4.83),284(4.41), 302(4.31),317(4.38), 337(4.20),355(4.33), 372(4.30)	5-0001-65I
Hexa-m-phenylene	$CHCl_3$	251.2(5.14)	88-0319-64
$C_{36}H_{24}Br_6O_6W$			
Tungsten, hexakis(p-bromophenoxy)-	C_6H_{12}	234(4.85),398(4.54)	12-1579-65
$C_{36}H_{24}Cl_6O_6W$			
Tungsten, hexakis(o-chlorophenoxy)-	C_6H_{12}	229(4.57),256(4.45), 384(4.38)	12-1579-65
Tungsten, hexakis(p-chlorophenoxy)-	C_6H_{12}	235(4.84),420(4.49)	12-1579-65
$C_{36}H_{26}$			
Biphenyl, 3,3'-bis(biphenylyl)-	C_6H_{12}	249.0(5.01)	39-0114-64
Biphenyl, 2,2',4,4'-tetraphenyl-	C_6H_{12}	253(4.76),291(4.46)	39-0114-64
Naphtho[2',3':1,2]-4,5,8,9-tetrahydro-pyrene, 1',4'-diphenyl-	benzene	319(4.41)	78-0507-64
p-Sexiphenyl	C_6H_{12}	308.0(4.78)	39-0114-64
1,1':2',1'':2'',1''':2''',1'''':-2'''',1'''''-Sexiphenyl	C_6H_{12}	235.0(4.60)	39-0114-64
1,1':4',1'':2'',1''':2''',1'''':-4'''',1'''''-Sexiphenyl	C_6H_{12}	273.0(4.67)	39-0114-64
1,1':4',1'':3'',1''':3''',1'''':-4'''',1'''''-Sexiphenyl	C_6H_{12}	274.0(4.86)	39-0114-64
1,1':4',1'':4'',1''':3''',1'''':-4'''',1'''''-Sexiphenyl	C_6H_{12}	291.0(4.80)	39-0114-64
$C_{36}H_{26}O_2$			
2-Naphthoic acid, 1,4,5,8-tetra-phenyl-, methyl ester	$CHCl_3$	255(4.49),335(4.14)	78-0195-64

Compound	Solvent	λ_{max}(log ε)	Ref.
$C_{36}H_{27}N$			
Cyclopentadiene, 5-(dimethylamino-methylene)-2,3-bis(9-fluorenyli-denemethyl)-	MeCN	249(4.92),288s(4.35), 413(4.63),453(4.64)	24-3331-64
$C_{36}H_{30}Ge_2$			
Digermane, hexaphenyl-	C_6H_{12}	239.0(4.48)	39-4690-65
$C_{36}H_{30}N_3O_6P_3$			
1,3,5,2,4,6-Triazatriphosphorine, 2,2,4,4,6,6-hexahydro-2,2,4,4-tetraphenoxy-6,6-diphenyl-	MeCN	206s(4.70),259(3.23), 266(3.31),272(3.18)	35-2591-64
$C_{36}H_{30}OSi_2$			
Disiloxane, hexaphenyl-	C_6H_{12}	253(3.28),265(3.35), 272(3.24)	39-4690-65
$C_{36}H_{30}O_3Si_3$			
Cyclotrisiloxane, 1,1,3,3,5,5-hexa-phenyl-	$CHCl_3$	265(3.40)	46-1066-65
$C_{36}H_{30}O_6W$			
Tungsten, hexaphenoxy-	C_6H_{12}	228(4.70),395(4.44)	12-1579-65
$C_{36}H_{30}Pb_2$			
Lead, hexaphenyldi-	C_6H_{12}	245s(4.53),293(4.54)	39-4690-65
	C_6H_{12}	293(4.54)	101-0265-64B
$C_{36}H_{30}Si_2$			
Disilane, hexaphenyl-	C_6H_{12}	246.5(4.51)	39-4690-65
	C_6H_{12}	246.5(4.51)	101-0369-64B
	n.s.g.	246.5(4.51)	25-1492-64
	n.s.g.	246.5(4.50)	101-0176-65B
$C_{36}H_{30}Sn_2$			
Tin, hexaphenyldi-	C_6H_{12}	220(4.8),247(4.5)	5-0001-64G
	C_6H_{12}	247.5(4.53)	39-4690-65
	C_6H_{12}	247.5(4.53)	101-0265-64B
	EtOH	246.5(4.51)	39-4690-65
$C_{36}H_{31}N$			
Cyclopentadiene, 5-(5-dimethylamino-2,4-pentadien-1-ylidene)-1,2,3,4-tetraphenyl-	MeCN	250s(4.25),256s(4.30), 276(4.42),510(4.88)	24-3331-64
$C_{36}H_{31}NSi_2$			
Disilazane, 1,1,1,3,3,3-hexaphenyl-	C_6H_{12}	254(3.12),260(3.24), 265(3.20),271(3.06)	39-4690-65
$C_{36}H_{33}NO_{18}$			
Laccaic acid A_1, tetraacetate	MeOH	251(4.49),295(4.44)	39-6067-65
$C_{36}H_{36}O_2$			
Acetophenone, 4'-[p-[1-(acetyltri-cyclo[8.2.2.2^{4,7}]hexadeca-4,6,10,12,13,15-hexaen-5-yl)-ethyl]phenethyl]-	EtOH	204(4.70),253(4.37), 290(3.92)	44-1815-65
isomer	EtOH	253(4.37),290(3.90)	44-1815-65

Compound	Solvent	λ_{max}(log ε)	Ref.
$C_{36}H_{36}O_{13}$			
Hemiergoflavin, 2-[2,4-dimethoxy-3-(2-methoxy-6-carbomethoxy-4-methylbenzoyl)phenyl]-1,9-di-O-methyl-	n.s.g.	252(4.53),319s(3.88)	39-4144-65
$C_{36}H_{38}N_2O_6$			
Isochondrodendrine	EtOH	278(3.76),284(3.76)	100-0191-65
$C_{36}H_{38}N_4O_5$			
Pheophorbide a, methyl ester	HOAc	410(5.0),508(4.0),535(4.0),608(4.0),670(4.4)	50-0343-64B
$C_{36}H_{38}O_4$			
5,8,13,14-Pentaphenetetrone, 6-decyl-2,3,10,11-tetramethyl-	dioxan	255(4.50),287(4.56)	65-3484-64
$C_{36}H_{38}O_{16}$			
Protoaphin-fb	aq.Na_2HPO_4	234(4.75),298(4.26),342(3.73),356(3.75),512(3.72)	39-0051-64
	70% EtOH	223(4.82),273(4.19),296(4.06),310(3.97),343(3.73),357(3.78),449(3.69)	39-0051-64
Protoaphin-sl	aq.Na_2HPO_4	233(4.74),299(4.23),340(3.59),354(3.63),516(3.71)	39-0051-64
	75% EtOH	223(4.73),274(4.14),297(3.98),311(3.91),343(3.67),356(3.72),450(3.64)	39-0051-64
$C_{36}H_{40}N_4O_{13}$			
α-D-Glucopyranoside, 2-(3-formyl-1-phenylpyrazol-5-yl)-2-hydroxyethyl-, acetylphenylhydrazone, pentaacetate	EtOH	282(4.46)	44-3072-64
$C_{36}H_{41}N_5O_6$			
Mesoporphyrin, nitro-, dimethyl ester	$CHCl_3$	399(5.02),503(4.01),536(3.83),571(3.72),625(3.62)	44-2791-65
$C_{36}H_{42}N_4O_5$			
6,7-Porphinedipropionic acid, 2,4-diethyl-β-hydroxy-1,3,5,8-tetramethyl-, dimethyl ester	n.s.g.	404(5.08),585(4.30),635(4.48)	35-0676-65
tautomer	n.s.g.	404(5.00),700(4.48)	35-0676-65
$C_{36}H_{42}O_6$			
Pregna-1,4-dien-3-one, 17α,21-benzylidenedioxy-11β,20α-dihydroxy-11-(α-methoxybenzyloxy)-	EtOH	244(4.17)	78-0179-65
$C_{36}H_{43}NO_4$			
Pregn-5-en-20-one, 3β-hydroxy-16α,17α-(2,3-diphenyl-3,1-isoxazolidino)-, acetate	MeOH	250(3.80)	4-0280-64

Compound	Solvent	λ_{max}(log ε)	Ref.
$C_{36}H_{43}N_5NiO_2$			
Nickel, [1,2,3,4,5,6,7,8-octaethyl-α-nitroporphinato(2-)]-	$CHCl_3$	397(5.16),525(3.99), 560(4.33)	44-2791-65
$C_{36}H_{43}N_7O_6$			
Porphine, octaethyl-α,β,γ-trinitro- (Soret bands not listed)	$CHCl_3$	514s(4.01),537s(3.73), 590(3.60),638(3.28)	44-2791-65
$C_{36}H_{43}O_6P$			
Estrone, hydrogen phosphate	MeOH	270(3.47),276(3.41)	22-0029-65
$C_{36}H_{44}N_4O_{12}S_2$			
5H,11H-Bisthiazolo[3,2-a:3',2'-e]-[1,5]diazocine-3,9-dicarboxylic acid, 6,12-bis(2,6-dimethoxybenzamido)-octahydro-2,2,8,8-tetramethyl-5,11-dioxo-, dimethyl ester	MeOH	283(3.94)	44-1826-64
$C_{36}H_{44}N_6O_4$			
Porphine, octaethyl-α,β-dinitro- (Soret bands not listed)	$CHCl_3$	508(4.00),538(3.75), 578(3.67),628(3.47)	44-2791-65
Porphine, octaethyl-α,γ-dinitro-	$CHCl_3$	505(4.10),538(3.96), 576(3.77),628(3.82)	44-2791-65
$C_{36}H_{44}O_6$			
Pregn-4-en-3-one, 17α,21-benzylidenedioxy-11β,20α-dihydroxy-11-(α-methoxybenzyloxy)-	EtOH	243(4.16)	78-0179-65
$C_{36}H_{44}O_{22}$			
Phloroacetophenone, 2,4-bis-(tetraacetyl-β-D-glucosyl)-	EtOH	275(4.22)	22-2937-65
$C_{36}H_{45}N_5O_2$			
Porphine, octaethyl-α-nitro-	$CHCl_3$	400(5.10),504(4.09), 538(3.92),571(3.8), 623(3.72)	44-2791-65
$C_{36}H_{46}N_4$			
Porphine, octaethyl-	$CHCl_3$	401(5.22),499(4.12), 534(4.01),566(3.82), 594(3.08),618(3.69)	44-2791-65
Porphine, tetramethyltetrapropyl-	benzene	400(5.29),498(4.14), 532(4.00),566(3.84), 620(3.69)	39-2262-64
$C_{36}H_{46}N_4O_4$			
4-Piperidinol, 1,1'-[ethylenebis-(indole-2,3-diylethylene]di-, diacetate	MeOH	228(4.71),282(4.16), 291(4.16)	24-2579-65
$C_{36}H_{46}N_6O_4$			
Chlorin, octaethyl-γ,δ-dinitro-	$CHCl_3$	394(4.93),499(4.09), 530s(3.45),560(3.11), 617s(3.59),626(3.61), 662(4.63)	44-2791-65
$C_{36}H_{46}O_2S_2$			
Estra-1,3,5(10)-trien-17β-ol, 3,3'-dithiodi-	EtOH	238(4.34)	44-0295-65

Compound	Solvent	$\lambda_{max}(\log \epsilon)$	Ref.
$C_{36}H_{47}N_5$			
Porphine, α-aminooctaethyl-	$CHCl_3$	519(4.05),555(4.72), 588(3.56),644(4.76)	44-2791-65
$C_{36}H_{47}N_5O_2$			
Chlorin, octaethyl-γ-nitro-	$CHCl_3$	392(4.95),497(4003), 536(3.30),605s(3.53), 621s(3.57),650(4.52)	44-2791-65
$C_{36}H_{48}N_4$			
Chlorin, octaethyl-	$CHCl_3$	392(5.19),491(4.04), 498(4.06),525(3.56), 544(3.45),592(3.56), 617(3.54),644(4.68)	44-2791-65
$C_{36}H_{50}O_9$			
29-Nordammara-3,5,17(20)-trien-21-oic acid, 3,6,16β-trihydroxy-7-oxo-, methyl ester, triacetate	EtOH	219(4.05),288(4.16)	94-0121-64
$C_{36}H_{52}N_4O_4$			
Cholest-4-en-3-one, 6-allyl-, 2,4-dinitrophenylhydrazone	$CHCl_3$	391(4.45)	22-0227-65
Cycloeucalenone, 2,4-dinitrophenylhydrazone	EtOH	374.0(4.38)	39-1692-65
$C_{36}H_{52}N_6O_8S$			
Methioninamide, N-carbo-tert-butyloxyphenylalanyltyrosylglycylleucyl-	n.s.g.	278(3.23)	31-0490-64
$C_{36}H_{52}O_3$			
Cholest-8-en-7-one, 3-hydroxy-4,4-dimethyl-, benzoate	C_6H_{12}	229(4.31),245s(4.03)	22-1295-65
Cholest-8(14)-en-7-one, 3-hydroxy-4,4-dimethyl-, benzoate	MeOH	229(4.30),263(4.21)	22-1295-65
$C_{36}H_{54}O_3$			
Cholest-5-en-3β-ol, 4,4-dimethyl-, benzoate	dioxan	277(3.27),283(3.21)	35-0995-65
$C_{36}H_{54}O_8$			
Sulphurenic acid, dihydro-7,11-dioxo-, methyl ester, diacetate	EtOH	270(3.99)	78-2297-64
$C_{36}H_{54}O_{14}$			
Glucolanadoxin	MeOH	217(4.16)	5-0137-64H
$C_{36}H_{56}O_4S$			
Cholestan-6-one, 3β-hydroxy-4,4-dimethyl-, p-toluenesulfonate	C_6H_{12}	225(4.03)	23-0456-64
$C_{36}H_{56}O_7$			
Ketotriacetate from ceanothic acid	n.s.g.	292(1.65)	12-0141-64
$C_{36}H_{56}O_8$			
Cocarcinogen A	EtOH	232(3.73),333(1.86)	89-0225-64
Compound from croton oil	n.s.g.	233(3.74),333(3.74)	25-0350-65
$C_{36}H_{56}O_{11}$			
Bulloside	EtOH	217(4.12)	33-0649-65

Compound	Solvent	λ_{max}(log ε)	Ref.
$C_{36}H_{56}O_{13}$			
Neo-odorobioside G	MeOH	217(4.19)	5-0216-65H
Odorobioside G	MeOH	217(4.17)	5-0137-64H
$C_{36}H_{56}O_{14}$			
Neodigitalinum verum	EtOH	217(4.14)	5-0216-65H
Sargenoside	EtOH	217(4.20)	33-0799-64
$C_{36}H_{56}O_{15}$			
Kisantoside	EtOH	219(4.23)	33-0799-64
$C_{36}H_{58}OS$			
Cycloartanone, 2-butylthio-methylene-24-methylene-	EtOH	311.5(3.91)	39-1692-65
$C_{36}H_{58}O_8$			
Polygonaquinone, tetraacetate	EtOH	263(2.58)	78-2319-64
$C_{36}H_{66}Pb_2$			
Lead, hexacyclohexyldi-	C_6H_{12}	254(4.51)	101-0265-64B

Compound	Solvent	λ_{max}(log ε)	Ref.
$C_{37}H_{23}$			
Fluoren-9-yl, 9-(fluoren-9-yl-idene-1-naphthylmethyl)-	dioxan	223(4.93),287(4.48), 330(4.14),456(4.33), 845(3.10)	49-0003-64
$C_{37}H_{24}$			
7H-Dibenzo[c,g]fluorene, 7-(3-fluor-en-9-ylidenepropylidene)-	dioxan	258(4.79),289(4.74), 301(4.68),313(4.53), 330(4.24),345(4.29), 361(4.37),420(3.28)	5-0001-65I
anion	n.s.g.	659(5.33)	5-0001-65I
Indene, 3-(fluoren-9-ylidenemethyl)-1-(fluoren-9-ylmethylene)-	dioxan	260(4.68),302(4.32), 329(4.38),390(4.00)	5-0001-65I
anion	DMF	694(4.86)	5-0001-65I
Methane, fluoren-9-ylfluoren-9-ylidene-1-naphthyl-	dioxan	225(5.07),252(4.70), 261(4.75),290(4.41), 303(4.38),321(4.32)	49-0003-64
anion	DMF-NaOH	306(4.46),346(4.19), 481(3.87),602(4.61)	49-0003-64
$C_{37}H_{28}N_4O_4$			
3,5-Pyrazolidinedione, 4,4'-benzyli-denebis[1,2-diphenyl-	MeOH	248(4.55)	32-0320-65
$C_{37}H_{28}O$			
Methanol, tris(3-biphenylyl)-	$HClO_4$	259(4.71),332(3.77), 422(4.65),510s(3.97)	60-0264-64
	H_2SO_4	500s(--)	60-0264-64
carbanion	ether	252(4.90),322(3.95), 460(4.68),485s(4.65)	60-0264-64
Methanol, tris(4-biphenylyl)-	$HClO_4$	248(4.45),281(4.18), 384(3.64),541(4.93)	60-0264-64
	H_2SO_4	510(--)	60-0264-64
carbanion	ether	260(5.42),315s(3.0), 435(3.0),580(3.86)	60-0264-64
$C_{37}H_{28}O_3$			
3-Cyclopenten-1-one, 2-acetoxy-2,3,4,5,5-pentaphenyl-	EtOH	268(4.105)	28-4061-64B
$C_{37}H_{29}ClN_2O_8$			
1-(o-Hydroxyphenyl)-2-[3-[1-(o-hy-droxyphenyl)-2(1H)-quinolylidene]-propenyl]quinolinium perchlorate, diacetate	EtOH	622(5.19)	65-3373-64
$C_{37}H_{29}N$			
Cyclopentadiene, 5-(methylanilino-methylene)-1,2,3,4-tetraphenyl-	MeCN	265(4.43),399(4.50)	24-3331-64
$C_{37}H_{30}O_6$			
2-Cyclohexen-1-one, 3-hydroxy-2-[2-(p-hydroxyphenyl)-4H-benzo-pyran-4-yl]-5,5-dimethyl-, dibenzoate	EtOH	238(4.74),271s(--)	78-3707-65
$C_{37}H_{30}O_7$			
Quercetin, 3',4',7-tribenzyl-3-methyl-	EtOH	256(4.37),356(4.30)	22-0779-65
	EtOH-$AlCl_3$	263(4.29),280(4.32), 355(4.23),402(4.17)	22-0779-65

Compound	Solvent	λ_{max}(log ε)	Ref.
$C_{37}H_{31}BF_4P_2$			
Triphenyl[(triphenylphosphoranylidene)methyl]phosphonium tetrafluoroborate	EtOH	267(3.87)	44-2427-64
$C_{37}H_{31}IP_2$			
Triphenyl[(triphenylphosphoranylidene)methyl]phosphonium iodide	EtOH	268(3.89)	44-2427-64
$C_{37}H_{31}O_4P$			
Succinic acid, (diphenylmethylene)-(triphenylphosphoranylidene)-, dimethyl ester	n.s.g.	365(3.48)	88-1263-64
$C_{37}H_{33}N$			
Cyclopentadiene, 5-(3-piperidino-2-propen-1-ylidene-1,2,3,4-tetraphenyl-	MeCN	257s(4.21),270s(4.37), 276(4.38),450(4.81)	24-3331-64
$C_{37}H_{34}N_2O_4$			
1H-Pyrrolo[1,2-a]indol-1-ol, 9,9'-methylenebis[7-(benzyloxy)-2,3-dihydro-	MeOH	283(4.26),300s(4.04), 315s(3.89)	35-4612-64
$C_{37}H_{35}BP_2$			
Triphenyl[(triphenylphosphoranylidene)-methyl]phosphonium tetrahydroborate	EtOH	267(3.89)	44-2427-64
$C_{37}H_{38}MgN_4O_5$			
Chlorophyllide a, ethyl ester	DMSO	340(4.4),390(4.6), 410(4.8),435(4.9), 540(3.7),580(3.9), 625(4.0),660(4.6)	50-0637-64C
$C_{37}H_{38}N_2O_6$			
Thalmethine, O-methyl-	MeOH	280(4.13),314(3.87)	25-1595-65
$C_{37}H_{38}O_{19}$			
Pigment A, octaacetate	n.s.g.	300(4.33)	83-0838-65
$C_{37}H_{39}ClO_8$			
1,1,5,5-Tetrakis(p-methoxyphenyl)-3-tert-butyl-1,4-pentadien-3-yl perchlorate	n.s.g.	500(4.50),673(3.59)	24-2959-64
$C_{37}H_{42}N_2O_6$			
Isoliensinine	EtOH	286(4.04)	88-2637-64
	EtOH	257(3.41),286(4.05)	94-0039-65
$C_{37}H_{44}O_{17}$			
Chromomyciquinone, "reductive octaacetate"	EtOH	235(4.80),285(3.27)	88-2355-64
$C_{37}H_{45}N_3O_{18}$			
2(1H)-Pyrimidone, 1-[2,3,4-tri-O-acetyl-6-(2-deoxy-2-benzamido-3,4,6-tri-O-acetyl-β-D-glucopyranosyl)-β-D-glucopyranosyl]-4-ethoxy-	EtOH	274(3.69)	44-2723-65

Compound	Solvent	$\lambda_{max}(\log \varepsilon)$	Ref.
$C_{37}H_{50}N_4O_{10}$			
Acetyltrinorketone from ceanothic acid, 2,4-dinitrophenylhydrazone	n.s.g.	365(4.33)	12-0141-64
$C_{37}H_{54}N_4O_4$			
Cycloartanone, 24-methylene-, 2,4-dinitrophenylhydrazone	EtOH	370.0(4.38)	39-1692-65
$C_{37}H_{56}O_{15}$			
Glucoverodoxin	MeOH	217(4.13)	5-0137-64H
$C_{37}H_{66}O_6$			
Myristin, 2-trans,trans-sorbo-1,3-di-	EtOH	261(4.38)	39-5651-65

Compound	Solvent	$\lambda_{max}(\log \epsilon)$	Ref.
$C_{38}H_{18}$			
Decapentayne, di-1-anthryl-	THF	252(5.30),265(5.16), 312(4.86),328(4.88), 354(4.26),380s(4.30), 410(4.60),440(4.79), 473(5.61)	88-0719-64
$C_{38}H_{20}$			
Naphtho[2",3":4,5]pyreno- [1',2':1,2]pyrene	benzene	317(4.66),326(4.63), 342(4.82),363(4.10), 385(4.08),414(4.24), 456(4.02),485(4.10)	78-2107-64
	C_6H_{12}	257(4.60),283(4.60)	78-2107-64
$C_{38}H_{22}N_8O_4$			
1,3,4-Oxadiazole, 2,2'-m-phenylene- bis[5-[m-(5-phenyl-1,3,4-oxadiazol- 2-yl)phenyl]-	DMF	287.6(5.01)	24-2966-65
$C_{38}H_{24}Cl_4N_4O_8$			
6,13-Triphenodioxazinedicarboxylic acid, 2,9-dichloro-3,10-bis(p-chloro- benzamido)-, diethyl ester	DMF	465(--),548(--)	27-0242-65
$C_{38}H_{25}IN_4$			
[4-(4-Quinolyl)-5-(4-quinolylmethyl)- pyrrolo[1,2-a:3,4-c']diquinolinium iodide	MeOH	250(4.61),315(4.48), 358(3.96),450(4.29)	5-0196-65H
$C_{38}H_{26}$			
Anthracene, 1,4,9,10-tetraphenyl-	CH_2Cl_2	277(4.86),408(4.08)	44-1981-65
$C_{38}H_{26}Cl_2N_4O_8$			
6,13-Triphenodioxazinedicarboxylic acid, 3,10-dibenzamido-2,9-di- chloro-, diethyl ester	DMF	467(4.70),551(4.79)	27-0242-65
$C_{38}H_{26}N_2O_8$			
6,13-Triphenodioxazinedicarboxylic acid, 2,9-dibenzoyl-, diethyl ester	DMF	475(4.79),508(4.90)	27-0242-65
$C_{38}H_{26}N_4O_4$			
3,5-Pyrazolidinedione, 4,4'-(p-phenyl- enedimethylidyne)bis[1,2-diphenyl-	MeOH	393(4.73)	32-0320-65
	$CHCl_3$	413(4.67)	32-0320-65
$C_{38}H_{26}O_2$			
1,4-Anthracenedione, 2,3-dihydro- 5,8,9,10-tetraphenyl-	$CHCl_3$	245(4.52),284(4.72), 365(4.09)	22-2342-65
$C_{38}H_{26}S_2$			
1,4-Dithiafulvalene, 2,5,6,7,8-penta- phenyl-	CF_3COOH	542(4.28)	88-2121-65
	THF	455(4.52)	88-2121-65
$C_{38}H_{28}O_2P_2$			
Phosphine oxide, 1,5-anthrylene- bis[diphenyl-	EtOH	265(4.98),340(3.52), 355(3.83),374(4.06), 394(4.03)	49-0285-65
$C_{38}H_{28}P_2$			
Phosphine, 1,5-anthrylenebis- [diphenyl-	EtOH	266(4.79),360(3.06), 378(3.82),396(3.78)	49-0285-65

Compound	Solvent	$\lambda_{max}(\log \epsilon)$	Ref.
$C_{38}H_{30}Cl_2O_{12}$			
Erythroaphin-fb, tetraacetyl-dichlorodihydro-	$CHCl_3$	285(4.64),363(3.27), 436(4.03),467(4.39), 498(4.53)	39-0072-64
Erythroaphin-sl, tetraacetyl-dichlorodihydro-	$CHCl_3$	283(4.63),363(3.18), 411(3.52),437(4.01), 466(4.38),498(4.51)	39-0072-64
$C_{38}H_{30}Cl_4O_{14}$			
Xanthoaphin-fb, tetraacetyl-tetrachloro-	$CHCl_3$	255(4.44),279(4.84), 365(3.77),385(4.10), 424(3.96)	39-0080-64
$C_{38}H_{30}N_6O_7$			
Purine, 2,6-dibenzamido-9-(2,5-di-O-benzoyl-3-deoxy-β-D-ribofuranosyl)-	EtOH	232(4.68),272s(4.35), 280s(4.09),292(4.27)	87-0659-65
$C_{38}H_{32}O_2$			
p-Terphenyl, 2',3'-dihydro-4,4''-di-methoxy-2',5',6'-triphenyl-	EtOH	335.6(4.25)	39-1881-65
$C_{38}H_{32}O_{12}$			
Erythroaphin-tt, tetraacetyldihydro-	$CHCl_3$	281(4.55),357(3.55), 413(3.70),438(4.03), 465(4.31),502(4.43)	39-0072-64
$C_{38}H_{34}O_{14}$			
Dianellinone, hexaacetate	dioxan	233s(4.89),246(4.96), 288s(4.36)	12-0218-65
Gossypolone, tetraacetate	EtOH	262(4.6),340s(3.8)	44-4111-65
$C_{38}H_{36}O_4Si_4$			
Cyclotetrasiloxane, 2,2-dimethyl-4,4,6,6,8,8-hexaphenyl-	$CHCl_3$	265(3.34)	46-1066-65
$C_{38}H_{44}$			
Ethylene, tetramesityl-	C_6H_{12}	253s(4.40),257(4.42), 301s(4.08),311(4.15)	35-2149-64
Stilbene, 2,2'-bis(mesitylmethyl)-3,3',5,5'-tetramethyl-	C_6H_{12}	215(4.49),295(4.23)	35-2149-64
$C_{38}H_{44}N_2$			
Benzophenone, 2,2',4,4',6,6'-hexa-methyl-, azine	C_6H_{12}	233s(4.42),317(4.27)	35-2149-64
$C_{38}H_{44}N_2O_6$			
Dauricine	EtOH	279(4.29)	39-6141-64
Phenanthrene-1-carboxylic acid, 1,2,3,4,4a,9-hexahydro-6-methoxy-1,4a-dimethyl-9-oxo-, methyl ester, azine	EtOH	243(4.38),296(4.27), 368(4.27)	39-0361-65
$C_{38}H_{44}O_4$			
Ethylene, tetrakis(2,6-dimethyl-4-methoxyphenyl)-	C_6H_{12}	216s(4.72),261(4.56), 323(4.32)	35-2149-64
Stilbene, 2,2'-bis(2,6-dimethyl-4-methoxybenzyl)-3,3'-dimethyl-5,5'-dimethoxy-	C_6H_{12}	216s(4.71),287(4.26), 295s(4.17)	35-2149-64

Compound	Solvent	λ_{max}(log ε)	Ref.
$C_{38}H_{44}O_7$			
Pregna-1,4-dien-3-one, 17α,21-benzylidenedioxy-11β,20α-dihydroxy-11-(α-methoxybenzyloxy)-, 20-acetate	EtOH	244(4.18)	78-0179-65
$C_{38}H_{44}O_8$			
Gambogic acid	EtOH	217(4.56),280(4.22), 291(4.23),362(4.17)	78-1453-65
$C_{38}H_{46}N_4O_6S$			
Indole, 3-(2-aminobutyl)-6-benzyloxy-, sulfate	EtOH	223(4.89),258s(3.88), 265s(3.94),274(3.97), 292(4.03)	87-0274-64
$C_{38}H_{46}O_7$			
Pregn-4-en-3-one, 17α,21-benzylidenedioxy-11β,20α-dihydroxy-11-(α-methoxybenzyloxy)-, 20-acetate	EtOH	243(4.20)	78-0179-65
$C_{38}H_{48}O_4$			
Androsta-3,5-diene-7,17-dione, dimer	n.s.g.	248(4.38)	88-1075-64
$C_{38}H_{52}O_2$			
Androsta-4,6-dien-3-one, photodimer	EtOH	256(4.01)	44-0068-64
Cholesterol, 1-naphthoate	dioxan	295(3.88)	35-0995-65
$C_{38}H_{52}O_4$			
Androsta-4,6-dien-3-one, 17β-hydroxy-, photodimer	EtOH	257(4.00)	44-0068-64

Compound	Solvent	λ_{max}(log ϵ)	Ref.
$C_{39}H_{25}$			
Fluoren-9-yl, 9-(α-fluoren-9-ylidene-p-phenylbenzyl)-	dioxan	249(4.83),330(4.37), 501(4.41),870(3.19)	49-0003-64
$C_{39}H_{26}$			
Methane, 4-biphenylylfluoren-9-ylfluoren-9-ylidene-	dioxan	252(4.78),262(4.84), 291(4.41),304(4.40), 320(4.38)	49-0003-64
	DMF-NaOH	297(4.53),345(4.26), 614(4.54)	49-0003-64
$C_{39}H_{27}NO_3$			
Benzophenone, 4,4'',4''''-nitrilotri-	CHCl	248(4.59),382(4.64)	44-3536-64
$C_{39}H_{27}N_3O_3S$			
2-Naphthol, 3-N-(1-naphthylcarbamyl)-1-[2'-(2''-phenoxyphenylmercapto)-phenylazo]-	HOAc	515(4.34)	7-0821-64
2-Naphthol, 3-N-(1-naphthylcarbamyl)-1-[2'-(3''-phenoxyphenylmercapto)-phenylazo]-	HOAc	513(4.32)	7-0821-64
2-Naphthol, 3-N-(1-naphthylcarbamyl)-1-[4'-(2''-phenoxyphenylmercapto)-phenylazo]-	HOAc	526(4.48)	7-0821-64
2-Naphthol, 3-N-(1-naphthylcarbamyl)-1-[4'-(3''-phenoxyphenylmercapto)-phenylazo]-	HOAc	526(4.48)	7-0821-64
2-Naphthol, 3-N-(1-naphthylcarbamyl)-1-[2'-(2''-phenylmercaptophenoxy)-phenylazo]-	HOAc	524(4.35)	7-0821-64
2-Naphthol, 3-N-(1-naphthylcarbamyl)-1-[2'-(3''-phenylmercaptophenoxy)-phenylazo]-	HOAc	519(4.36)	7-0821-64
2-Naphthol, 3-N-(1-naphthylcarbamyl)-1-[4'-(2''-phenylmercaptophenoxy)-phenylazo]-	HOAc	526(4.45)	7-0821-64
2-Naphthol, 3-N-(1-naphthylcarbamyl)-1-[4'-(3''-phenylmercaptophenoxy)-phenylazo]-	HOAc	522(4.44)	7-0821-64
$C_{39}H_{27}N_3O_5S$			
2-Naphthol, 3-N-(1-naphthylcarbamyl)-1-[2'-(2''-phenoxyphenylsulfonyl)-phenylazo]-	HOAc	503(4.33)	7-0821-64
2-Naphthol, 3-N-(1-naphthylcarbamyl)-1-[2'-(3''-phenoxyphenylsulfonyl)-phenylazo]-	HOAc	499(4.30)	7-0821-64
2-Naphthol, 3-N-(1-naphthylcarbamyl)-1-[4'-(2''-phenoxyphenylsulfonyl)-phenylazo]-	HOAc	505(4.44)	7-0821-64
2-Naphthol, 3-N-(1-naphthylcarbamyl)-1-[4'-(3''-phenoxyphenylsulfonyl)-phenylazo]-	HOAc	504(4.43)	7-0821-64
2-Naphthol, 3-N-(1-naphthylcarbamyl)-1-[2'-(2''-phenylsulfonylphenoxy)-phenylazo]-	HOAc	519(4.34)	7-0821-64
2-Naphthol, 3-N-(1-naphthylcarbamyl)-1-[2'-(3''-phenylsulfonylphenoxy)-phenylazo]-	HOAc	517(4.36)	7-0821-64

Compound	Solvent	λ_{max}(log ε)	Ref.
2-Naphthol, 3-N-(1-naphthylcarbamyl)-1-[4'-(2''-phenylsulfonylphenoxy)-phenylazo]-	HOAc	519(4.45)	7-0821-64
2-Naphthol, 3-N-(1-naphthylcarbamyl)-1-[4'-(3''-phenylsulfonylphenoxy)-phenylazo]-	HOAc	516(4.43)	7-0821-64
$C_{39}H_{28}OS_2$			
1,4-Dithiafulvalene, 2-(p-methoxy-phenyl)-5,6,7,8-tetraphenyl-	$CHCl_3$	461(4.45)	88-2121-65
	CF_3COOH	550(4.31)	88-2121-65
$C_{39}H_{30}N_6$			
1,2,4,6,7,9-Hexaazaspiro[4,4]nona-2,7-diene, 1,3,4,6,8,9-hexaphenyl-	EtOH	252(4.5),333(4.4)	24-2174-65
$C_{39}H_{31}N$			
Cyclopentadiene, 5-(3-methylanilino-2-propen-1-ylidene)-1,2,3,4-tetra-phenyl-	MeCN	273(4.52),451(4.81)	24-3331-64
$C_{39}H_{35}ClO_8$			
1,1,5,5-Tetrakis(p-methoxyphenyl)-3-phenyl-1,4-pentadien-3-yl perchlorate	n.s.g.	505(4.45),715(4.25)	24-2959-64
$C_{39}H_{36}O_4Si_4$			
Cyclotetrasiloxane, 2-methyl-4,4,6,6,8,8-hexaphenyl-2-vinyl-	$CHCl_3$	265(3.35)	46-1066-65
$C_{39}H_{36}O_{18}$			
Ergoxanthin, tetraacetate	EtOH	214(4.31),242(4.42), 330(3.66)	78-1417-65
$C_{39}H_{38}O_4Si_4$			
Cyclotetrasiloxane, 2-ethyl-2-methyl-4,4,6,6,8,8-hexaphenyl-	$CHCl_3$	265(3.34)	46-1066-65
$C_{39}H_{46}O_6$			
Pregna-5,15,17(20)-triene-3β,20-diol, bis[(p-methoxyphenyl)acetate]	MeOH	231(4.31),275(3.51), 282(3.43)	35-0995-65
$C_{39}H_{46}O_8$			
Gamboginic acid, methyl ester	EtOH	217(4.62),361(4.22)	78-1453-65
$C_{39}H_{50}N_2O_{13}$			
Rifamycin B amide	pH 7.38	223(4.62),305(4.32), 428(4.21)	87-0596-64
$C_{39}H_{52}N_4O_4$			
Cholest-4-en-3-one, 4-phenyl-, 2,4-dinitrophenylhydrazone	$CHCl_3$	375(4.45)	22-0321-64
$C_{39}H_{52}N_6O_4$			
Cholest-4-ene-3,6-dione, bis-(p-nitrophenylhydrazone)	DMF-NaOH	535(4.90),670(4.64)	95-0390-64
$C_{39}H_{54}O$			
3-Oxa-5α-cholestane, 2-(diphenyl-methylene)-	EtOH	212(4.24),265(4.19)	12-0440-64

Compound	Solvent	$\lambda_{max}(\log \epsilon)$	Ref.
$C_{39}H_{56}O$			
5,6-Secocholest-6-en-3β-ol, 6,6-diphenyl-	C_6H_{12}	254(4.23)	24-0958-64
$C_{39}H_{56}O_4$			
Androst-4-en-3-one, 17β,17'β-(methylenedioxy)di-	EtOH	241(4.21)	44-2925-65

Compound	Solvent	λ_{max}(log ε)	Ref.
$C_{40}H_{18}$			
Dodecahexayne, di-1-anthryl-	THF	254(5.29),273(5.05), 307(5.00),325(4.96), 373(4.38),397(4.53), 425(4.75),457(4.81), 494(4.50)	88-0719-64
$C_{40}H_{24}$			
Acepleiadylene, dimer	dioxan	221(4.89),248(4.54), 306(3.73),320(4.06), 335(4.26),352(4.37), 369(4.41)	78-3051-65
$C_{40}H_{24}O_{10}$			
1,2-Naphthoquinone, 3-hydroxy-4-(2,3-dihydroxy-1-naphthyl)-, dimer	dioxan	238(5.12),302(4.05), 312(4.02),327(3.90)	24-0666-65
$C_{40}H_{26}N_2$			
9,9'-Azotriptycene	heptane	271(3.32),280(3.42)	24-0428-65
$C_{40}H_{28}$			
Acepleiadylene dimer, tetrahydro-	dioxan	220(4.90),317(4.31), 334(4.31)	78-3051-65
$C_{40}H_{29}N$			
Ethylene, 1-(dimethylamino)-2,2-bis-(3,4-benzofluorenyl-9-idenemethyl)-	MeCN	239(4.81),259(4.89), 320s(4.42),332(4.44), 486(4.65)	24-3331-64
$C_{40}H_{30}Sn_2$			
Tin, butadiynylenebis[triphenyl-	EtOH	212(4.85),224(4.84), 249(3.15),259(3.11), 263(3.18),278(2.60)	22-0035-65
$C_{40}H_{32}N_4O_{10}$			
6,13-Triphenodioxazinedicarboxylic acid, 3,10-dibenzamido-2,9-dimethoxy-, diethyl ester	DMF	556(4.60),597(4.80)	27-0242-65
$C_{40}H_{34}N_2O_9$			
D-Glucose, 3,4,5,6-tetrabenzoyl-, phenylhydrazone	dioxan	274(4.32),281(4.32)	24-2409-64
azo-form	dioxan	273(4.13),410(1.12)	24-2409-64
$C_{40}H_{34}N_4O_4$			
[2,2'-Bibenzo[a]phenazine]-1,1',6,6'-tetrol, 5,5'-diisopropyl-3,3'-dimethyl-	EtOH	290(4.84),315(4.66), 430(4.10),490(4.25)	44-3617-64
$C_{40}H_{34}O_9$			
5,7:9,15-Dimethano-5H,15H-benzo-[1,2-d:3,4-d']bis[1,3]benzodioxocin-6-ol, 9,17-bis(p-hydroxyphenyl)-1,11-dimethoxy-19,20-dimethyl-	EtOH	274(3.88),280s(3.83)	78-1471-65
triacetate	EtOH	273(3.64),280s(3.59)	78-1471-65
trimethylated product	EtOH	272(3.73),279(3.70)	78-1471-65
$C_{40}H_{34}O_{14}$			
Erythroaphin-sl, 2-acetoxy-tetraacetyldihydro-	$CHCl_3$	278(4.53),357(3.34), 408(3.62),434(3.93), 463(4.25),496(4.35)	39-6923-65

Compound	Solvent	λ_{max}(log ε)	Ref.
Xylindein, tetraacetyldihydro-	$CHCl_3$	311(4.25),338(3.70), 358(3.86),375(4.11), 396(3.86),419(4.04), 446(4.05)	78-2095-65
$C_{40}H_{36}O_2$			
Unknown compound, m. 224-5°	EtOH	319.5(4.00)	39-1881-65
$C_{40}H_{38}Cl_2N_2O_8$			
Erythroaphin-fb, dichloro-dipiperidino-	$CHCl_3$	264(4.56),494(4.32), 533(4.28),578(4.33)	39-0072-64
$C_{40}H_{39}IN_2O_8$			
Erythroaphin-fb, iodopiperidino-	$CHCl_3$	264(4.57),448(4.28), 470s(4.23),537(4.24), 579(4.31)	39-0062-64
$C_{40}H_{40}O_4Si_4$			
Cyclotetrasiloxane, 2-methyl-4,4,6,6,8,8-hexaphenyl-2-propyl-	$CHCl_3$	265(3.34)	46-1066-65
$C_{40}H_{42}N_{10}$			
Glutaraldehyde, trioxo-, pentakis-(methylphenylhydrazone)	EtOH	251(4.61),289s(4.45), 360(4.61)	35-0732-64
$C_{40}H_{42}O_5Si_5$			
Cyclopentasiloxane, 2,2,4,4-tetra-methyl-6,6,8,8,10,10-hexaphenyl-	$CHCl_3$	265(3.33)	46-1066-65
$C_{40}H_{44}$			
p-Xylylenecyclophane, dimer	isooctane	228(4.30),258s(3.84), 294(3.45),330s(2.65)	35-3898-64
$C_{40}H_{46}ClFeN_6O_8S_2$			
Hemin c	aq. C_5H_5N	521(4.16),551(4.37)	24-1647-65
$C_{40}H_{48}N_4O_3$			
Villalstoninol	EtOH	232(4.63),249(4.14), 286(3.98),293(3.97)	33-0689-65
$C_{40}H_{48}N_6O_8S_2$			
Porphyrin c	90% C_5H_5N	504(4.25),538(4.03), 572(3.89),627(3.63)	24-1647-65
$C_{40}H_{48}O_8$			
Gambogic acid, O-methyl-, methyl ester	EtOH	224(4.56),299(4.13)	78-1453-65
$C_{40}H_{52}$			
Chlorobactene	pet ether	435(5.06),461(5.21), 491(5.16)	1-1739-64
Isochlorobactene	pet ether	448s(--),474(5.12), 504(--)	1-1739-64
$C_{40}H_{52}N_2O_4$			
Benzilic acid, hydrazide, hydrazone with testosterone heptanoate	EtOH	282(4.53)	87-0573-64
$C_{40}H_{52}O_2$			
25,26,27,28-Tetranorcycloeucal-23-en-3β-ol, 24,24-diphenyl-, acetate	EtOH	252(4.14)	35-4424-64

Compound	Solvent	λ_{max}(log ϵ)	Ref.
$C_{40}H_{52}O_8$			
Gambogic acid, tetrahydro-O-methyl-, methyl ester	EtOH	225(4.47),289(4.14)	78-1453-65
$C_{40}H_{54}N_2O_{12}$			
Rifamycin SV, N,N-dimethylamino-methyl-	pH 7.38	314(4.26),447(4.16)	87-0790-65
$C_{40}H_{54}N_2O_{13}$			
Rifamycin SV, N,N-dimethylamino-methyl-, N-oxide	pH 7.38	316(4.22),452(4.09)	87-0790-65
$C_{40}H_{54}O_6$			
Ceanothic acid, phenylbutadiene derivative	n.s.g.	225(4.03),231(4.03), 239(3.90),296(4.49), 304(4.49),320(4.30)	12-0141-64
$C_{40}H_{54}O_7$			
30-Norlupane-27,28-dioic acid, 3 -acetoxy-20-oxo-, methyl ester, benzylidene derivative	n.s.g.	290(4.36)	78-1529-65
$C_{40}H_{56}$			
α-Carotene	pet ether	422(5.02),444(5.17), 473(5.13)	27-0294-65
β-Carotene	pet ether	453(5.15),481(5.09)	27-0294-65
γ-Carotene	pet ether	437(5.05),462(5.23), 494(5.16)	27-0294-65
δ-Carotene	pet ether	431(5.04),456(5.15), 489(5.20)	27-0294-65
	hexane	280(4.60),431(5.04), 456(5.24),488(5.22)	39-2019-65
ε-Carotene	pet ether	416(5.01),440(5.23), 470(5.23)	27-0294-65
	hexane	266(4.56),416(5.02), 505(5.23)	39-2019-65
Lycopene	pet ether	446(5.08),472(5.28), 505(5.23)	27-0294-65
$C_{40}H_{58}N_2O_6$			
Bicarbamic acid, (3β-hydroxycholesta-5,8-dien-7-yl)-, diethyl ester, benzoate	n.s.g.	230(4.13)	94-0517-65
3b,5a-Etheno-1H-indeno[4,5-c]cinnoline-4,5-dicarboxylic acid, 1-(1,5-di-methylhexyl)-2,3,3a,6,7,8,9,9a,9b,-10,11,11a-dodecahydro-7-hydroxy-9a,11a-dimethyl-, diethyl ester, benzoate	n.s.g.	230(4.20)	94-0517-65
$C_{40}H_{60}N_2O_6$			
3b,5a-Ethano-1H-indeno[4,5-c]cinnoline-4,5-dicarboxylic acid, 1-(1,5-di-methylhexyl)dodecahydro-7-hydroxy-9a,11a-dimethyl-, diethyl ester, benzoate	n.s.g.	229(4.13)	94-0517-65
$C_{40}H_{64}$			
Phytaene	pet ether	285f(4.7)	10-0291-65B

Compound	Solvent	$\lambda_{max}(\log \epsilon)$	Ref.
$C_{41}H_{29}ClN_2O_4$			
1-(1-Naphthyl)-2-[3-[1-(1-naphthyl)-2(1H)-quinolylidene]propenyl]quinolinium perchlorate	EtOH	619(5.15)	65-3366-64
4-Phenyl-3-[3-(4-phenylbenzo[f]quinolin-3(4H)-ylidene)propenyl]benzo-[f]quinolinium perchlorate	EtOH	645(5.16)	65-3366-64
$C_{41}H_{33}N$			
Cyclopentadiene, 5-(5-methylanilino-2,4-pentadien-1-ylidene)-1,2,3,4-tetraphenyl-	MeCN	271(4.52),498(4.94)	24-3331-64
$C_{41}H_{48}N_4O_3$			
Epivoacamine, decarbomethoxy-	EtOH	227(4.66),287(4.16), 294(4.18)	35-4631-64
$C_{41}H_{48}N_4O_4$			
Villamine	EtOH	231(4.56),250s(3.96), 286(3.97),294s(3.93)	33-0689-65
Villalstonine	EtOH	231(4.57),250s(4.00), 286(3.96),294s(3.93)	33-0689-65
$C_{41}H_{50}N_4O_4$			
Villalstonine, dihydro-	EtOH	230(4.62),285(3.95), 292s(3.91)	33-0689-65
$C_{41}H_{50}O_2$			
Phenol, 4,4'-fluoren-9-ylidene-bis[2,6-di-tert-butyl-	C_6H_{12}	299(2.6),310(2.8)	35-0117-64
$C_{41}H_{56}I_2N_2O_2$			
Aquiloginine, iodide	EtOH	223(4.68),285(4.48)	100-0111-64
$C_{41}H_{56}N_2O_{13}$			
Rifamycin SV, N-methyl-N-hydroxy-ethylaminomethyl-	pH 7.38	314(4.28),447(4.16)	87-0790-65
$C_{41}H_{58}O_3$			
Trinorshionanoic acid, 3β-acetoxy-, methyl ester, benzhydryl derivative	EtOH	253(2.59),259(2.67), 265(2.58)	95-0888-65
$C_{41}H_{60}O_2$			
Spiro[5α-androstane-2,8'(1'H)-cyclopenta[7,8]phenanthro[2,3-b]pyran-3-one, hexadecahydro-11'a,13'a-dimethyl-4-methylene-	EtOH	239(3.88)	44-2925-65
Spiro[5α-androstane-4,8'(1'H)-cyclopenta[7,8]phenanthro[2,3-b]pyran-3-one, hexadecahydro-11'a,13'a-dimethyl-2-methylene-	EtOH	251(4.02)	44-2925-65
$C_{41}H_{64}O_{15}$			
Glucodigitoxigenin bisdigitoxoside	MeOH	217(4.18)	5-0216-65H

Compound	Solvent	λ_{max}(log ε)	Ref.
$C_{42}H_{20}N_6$			
Butadiene, hexacyano-, pyrene complex	CH_2Cl_2	605(3.42)	35-2898-64
$C_{42}H_{25}N$			
Dinaphtho[2',3':3,4;2'',3'':5,6]carbazole, 9-(1-anthryl)-	benzene	300(4.60),312(4.64)	24-0304-64
	$C_6H_3Cl_3$	360(4.26),387(4.10), 411(4.14),438(4.22)	24-2814-65
			24-0304-64
$C_{42}H_{28}$			
Allene, 1-phenyl-3,3-biphenylene-, dimer, colorless form	dioxan	261(4.60),336(4.47), 351(4.46),367(4.18)	24-2611-65
yellow form	dioxan	273(4.64),388(4.49)	24-2611-65
	DMSO-KO-tert-Bu	361(4.67),584(4.10)	24-2611-65
red form	dioxan	245(4.86),271(4.67), 280s(4.62),463(4.47)	24-2611-65
$C_{42}H_{28}O_4$			
Indone, 2,3-diphenyl-, oxide, dimer	MeCN	245s(3.97),292(3.24), 300s(3.19),332s(1.90), 343(2.02),355(2.00), 372s(1.70)	35-3814-64
	MeCN	257(4.02),295(3.27), 305s(3.19),328s(2.08), 343(2.16),358(2.14), 375(1.82)	35-3814-64
$C_{42}H_{30}$			
Bi-2-cyclopropen-1-yl, hexaphenyl-	n.s.g.	306(4.51)	35-5139-65
$C_{42}H_{30}N_8O_2P_2$			
Benzo[1,2-d:4,5-d']bistriazole-4,8(2H,6H)-dione, 2,6-bis(triphenylphosphoranylideneamino)-	$CHCl_3$	267(4.35),274s(4.33), 337(4.72)	39-1003-65
$C_{42}H_{30}O_3$			
Diphenylketene trimer	$CHCl_3$	237(4.65),340(3.38)	54-0965-65
$C_{42}H_{30}Sn_2$			
Tin, hexatriynylenebis[triphenyl-	EtOH	221(5.10),231(5.10), 243(5.22),290(3.26), 308(3.20)	22-0035-65
$C_{42}H_{34}$			
9,9'-Bianthryl, 10,10'-dibenzyl-9,9',10,10'-tetrahydro-	EtOH	243(3.46),248(3.48), 251(3.53),260(3.14)	39-4396-65
$C_{42}H_{35}N_3O_6$			
Cytidine, N-benzoyl-2'-deoxy-5'-O-trityl-, 3'-benzoate	EtOH	226(4.55),256(4.46), 306(3.97)	44-3067-65
$C_{42}H_{36}N_4O_{10}$			
6,13-Triphenodioxazinedicarboxylic acid, 3,10-dibenzamido-2,9-diethoxy-, diethyl ester	DMF	558(4.52),598(4.70)	27-0242-65
$C_{42}H_{38}N_6O_{10}$			
Brefeldin A, di-O-[p-(p-nitrophenylazo)benzoyl]-	MeOH	333(4.77),466(3.16)	33-1401-64

Compound	Solvent	λ_{max}(log ϵ)	Ref.
$C_{42}H_{39}BO_{16}$			
Rosocyanin	EtOH-HOAc	555(4.72)	83-0660-64
	EtOH	660(4.96)	83-0660-64
	iso-PrOH-HOAc	555(4.73)	83-0660-64
	iso-PrOH-NH_3	680(4.82)	83-0660-64
$C_{42}H_{40}O_{19}$			
Rhein aloe emodin dianthrone, 8,8'-diglucoside (or sennoside C)	70% MeOH	268(4.29),333(4.17)	33-1911-65
$C_{42}H_{41}NO_5$			
Indole, 1-(2,3,4,6-tetra-O-benzyl-β-D-glucopyranosyl)-	EtOH	265(3.86)	65-0896-65
$C_{42}H_{42}O_6W$			
Tungsten o-cresoxide	C_6H_{12}	228(4.64),255(4.51), 398(4.76)	12-1579-65
Tungsten p-cresoxide	C_6H_{12}	230(4.77),415(4.50)	12-1579-65
$C_{42}H_{42}Sn_2$			
Tin, hexabenzyldi-	C_6H_{12}	246.5(4.70)	39-4690-65
Tin, hexa-p-tolyldi-	C_6H_{12}	247.5(4.63)	39-4690-65
$C_{42}H_{43}NO_5$			
Indoline, 1-(2,3,4,6-tetra-O-benzyl-β-D-glucopyranosyl)-	EtOH	248(4.08),298(3.4)	65-0896-65
$C_{42}H_{50}N_2O_9$			
Turumiquirensine	EtOH	220(4.80),265(4.27), 300(4.19)	50-0296-64C
	EtOH-NaOH	225(4.84),275(4.14), 315(4.15)	50-0296-64C
$C_{42}H_{50}N_4O_5$			
Gabunine	EtOH	226(4.83),287(4.24), 295(4.22)	88-0931-65
$C_{42}H_{50}N_8O_{18}$			
Isopilocereine, dipicrate	MeOH	220(4.36),285(3.77)	5-0207-65E
$C_{42}H_{52}N_4O_5$			
Voacamine, methine	EtOH	225(4.80),286(4.29), 294(4.29)	35-4631-64
$C_{42}H_{54}O_2$			
Cholesterol, 9-anthroate	dioxan	330(3.49),346(3.79), 363(3.96),383(3.92)	35-0995-65
$C_{42}H_{56}N_2O_{12}$			
Rifamycin SV, N-pyrrolidinomethyl-	pH 7.38	314(4.25),448(4.13)	87-0790-65
$C_{42}H_{56}N_2O_{13}$			
Rifamycin SV, N-morpholinomethyl-	pH 7.38	316(4.30),450(4.15)	87-0790-65
Rifamycin SV, N-pyrrolidinomethyl-, N-oxide	pH 7.38	316(4.28),453(4.13)	87-0790-65

Compound	Solvent	$\lambda_{max}(\log \epsilon)$	Ref.
$C_{42}H_{56}O_4$			
Cholest-5-ene-3β,7α-diol, 6-methyl-, dibenzoate	EtOH	231(4.49)	39-3106-64
$C_{42}H_{56}O_6$			
6-Chromanol, 2,2,5,7,8-pentamethyl-, trimer, reduction product	isooctane	219(4.42),300(3.83)	44-3601-64
$C_{42}H_{60}N_2O_{12}$			
Rifamycin SV, N,N-diethylaminomethyl-	pH 7.38	315(4.24),445(4.15)	87-0790-65

Compound	Solvent	λ_{max}(log ϵ)	Ref.
$C_{43}H_{28}$			
Methane, bis(biphenylenevinyl)-	dioxan	342(4.54)	88-0383-64
biphenyleneethylidene-	DMF-NaOH	647(5.0)	88-0383-64
Methane, tris(biphenylenevinyl)-	dioxan	249(4.97),258(5.06), 289(4.61),305(4.63), 319(4.72)	5-0001-65I
	dioxan	319(4.72)	88-0383-64
	DMF-NaOH	647(5.0)	88-0383-64
$C_{43}H_{30}$			
Cyclopentadiene, 1,2,3,5-tetraphenyl-4-biphenylenevinyl-	dioxan	259(4.69),298s(4.22), 310s(4.15),397(4.10)	5-0001-65I
anion	DMSO	535(4.65)	5-0001-65I
	DMSO-KO-tert-Bu	308(4.43),330(4.44), 535(4.65)	5-0050-65J
Methane, (2,2-diphenylvinyl)bis-(biphenylenevinyl)-	dioxan	250(4.86),259(4.94), 283(4.53),291(4.55), 306(4.53),321(4.49)	5-0001-65I
anion	DMF	675(4.99)	5-0001-65I
1,3-Pentadiene, 5-(biphenylenevinyl)-5,5-biphenylene-1,1-diphenyl-	dioxan	251(4.77),260(4.83), 307(4.78),320(4.61)	5-0001-65I
1,3-Pentadiene, 1,1;5,5-bis(biphenylene)-3-(2,2-diphenylvinyl)-	CH_2Cl_2	234(4.85),252(4.76), 262(4.79),295(4.67), 305(4.68)	5-0001-65I
$C_{43}H_{32}$			
Fluorene, 9-[2-(2,2-diphenylvinyl-4,4-diphenyl-3-butenylidene]-	dioxan	249(4.68),257(4.72), 279(4.50),309(4.28), 322(4.32)	5-0001-65I
	dioxan	320s(4.30)	88-0383-64
anion	DMF	619(4.76)	5-0001-65I
$C_{43}H_{34}O_7$			
Quercetin, 3,3',4',7-tetrabenzyl-	EtOH	257(4.43),355(4.29)	22-0779-65
	EtOH-$AlCl_3$	280(4.39),355(4.21), 402(4.16)	22-0779-65
$C_{43}H_{36}O_5$			
Chalcone, 2',3,4,4'-tetrakis-(benzyloxy)-	C_6H_{12}	244s(--),340(4.32)	22-3350-65
$C_{43}H_{38}O_{19}$			
Ergochrysin A, acetylated	EtOH	210(4.59),238(4.52), 264(4.27),294(3.74), 333(3.30)	78-1417-65
$C_{43}H_{50}N_4O_5$			
Villamine, acetate	EtOH	231(4.60),286(3.95), 294s(3.92)	33-0689-65
$C_{43}H_{58}N_2O_{12}$			
Rifamycin SV, N-piperidinomethyl-	pH 7.38	315(4.26),450(4.14)	87-0790-65
$C_{43}H_{58}N_2O_{13}$			
Rifamycin B, diethylamide	pH 7.38	222(4.63),302(4.32), 421(4.21)	87-0596-64
$C_{43}H_{59}N_3O_{12}$			
Rifamycin SV, (N'-methyl-N-piperazino)methyl-	pH 7.38	314(4.32),450(4.15)	87-0790-65

Compound	Solvent	λ_{max}(log ε)	Ref.
$C_{43}H_{61}NO_4$			
Bis-5α-androstano[3,2-b:2',3'-e]pyridine-17β,17β'-diol, diacetate	EtOH	281(3.98)	44-2922-65

Compound	Solvent	λ_{max}(log ϵ)	Ref.
$C_{44}H_{24}Cl_4N_4Zn$			
Zinc, [α,β,γ,δ-tetrakis(o-chlorophenyl)porphinato(2-)]-	benzene	424(5.61),512(3.77), 549(4.21),588(3.32)	12-1028-64
$C_{44}H_{26}Cl_4N_4$			
Porphine, α,β,γ,δ-tetrakis-(o-chlorophenyl)-	benzene	418(5.57),478(3.49), 513(4.21),543(3.58), 589(3.78),645(3.04)	12-1028-64
Porphine, α,β,γ,δ-tetrakis-(p-chlorophenyl)-	benzene	421(5.71),485(3.60), 515(4.32),550(3.95), 592(3.77),647(3.56)	12-1028-64
$C_{44}H_{28}Cl_2F_2$			
Cyclobutadiene, fluorotriphenyl-, dimer	n.s.g.	235(4.70),242(4.72), 293(4.48)	35-0449-64
$C_{44}H_{28}N_4Zn$			
Zinc, [(α,β,γ,δ-tetraphenylporphinato(2-)]-	benzene	424(5.75),514(3.52), 550(4.34),590(3.61)	12-1028-64
$C_{44}H_{30}$			
Triapentafulvalene, hexaphenyl-	C_6H_{12}	263(4.54),377(4.47)	77-0512-65
	MeCN	266(4.55),371(4.51)	77-0512-65
$C_{44}H_{30}F_2$			
Tricyclo[4.2.0.0]octa-3,7-diene, 1,2-difluoro-3,4,5,6,7,8-hexaphenyl-, anti	n.s.g.	227(4.73),288(4.45)	35-0449-64
isomer	n.s.g.	250(4.46),373(4.13)	35-0449-64
$C_{44}H_{30}N_4$			
Meso-tetraphenylporphin	benzene	419(5.67),485(3.53), 514(4.27),549(3.89), 591(3.73),647(3.53)	12-1028-64
$C_{44}H_{36}Br_2P_2$			
Bicyclo[4.2.0]octa-1,3,5-trien-7,8-ylenebis[triphenylphosphonium bromide]	EtOH	230s(4.07),232s(4.06), 266(3.81),272(3.92), 279(3.86)	35-5041-64
$C_{44}H_{36}O_7$			
Quercetin, 3,3',4',7-tetrabenzyl-5-methyl-	EtOH	252(4.46),340(4.34)	22-0779-65
	EtOH-$AlCl_3$	252(4.47),340(4.34)	22-0779-65
$C_{44}H_{36}O_8$			
Myricetin, 3',4',5',7-tetrabenzyl-3-methyl-	EtOH	269(4.32),349(4.29)	22-0779-65
	EtOH-$AlCl_3$	281(4.32),302(4.10), 345(4.25),401(4.12)	22-0779-65
$C_{44}H_{38}N_4O_3$			
2(1H)-Cycloheptimidazolone, 6,6'-oxybis[1,3-dibenzyl-3,6-dihydro-	EtOH	220(4.45),290(3.75)	94-0810-65
carbon tetrachloride adduct	EtOH	220(4.74),290(4.08)	94-0810-65
$C_{44}H_{38}O_{18}$			
Secalonic acid B, acetylated product	EtOH	211(4.52),234(4.56), 289(3.92)	78-1417-65

Compound	Solvent	λ_{max}(log ε)	Ref.
$C_{44}H_{42}O_{20}$			
Secalonic acid C, acetylated product	EtOH	211(4.31),254(4.57), 298(4.52),305(4.51), 333(4.27)	78-1417-65
$C_{44}H_{56}N_6NiO_4$			
Dipyrromethene-5'-carboxylic acid, 3,3'-diethyl-4,4'-dimethyl-5-(1-pyrrolin-2-yl)-, ethyl ester, nickel complex	EtOH	268(4.17),279(4.16), 286(4.13),470(4.30), 533(4.58),558(4.64)	39-2614-65
$C_{44}H_{58}N_8O_{10}S_2$			
Mesoporphyrin IX, 2,4,α,α'-bis-[S-glycyl-L-cysteinyl]-	90% C_5H_5N	504(4.10),538(4.00), 572(3.84),627(3.61)	24-1647-65
$C_{44}H_{58}O_7$			
Unnamed compound, m. 242-244°	isooctane	217(4.39),300(3.79)	44-3601-64
$C_{44}H_{60}N_2O_{12}$			
Rifamycin SV, (2-methyl-N-piperidino-methyl)-	pH 7.38	314(4.25),448(4.14)	87-0790-65
Rifamycin SV, (4-methyl-N-piperidino-methyl)-	pH 7.38	315(4.24),446(4.14)	87-0790-65
$C_{44}H_{60}N_2O_{14}$			
Rifamycin SV, [2,5-bis(hydroxymethyl)-N-pyrrolidinomethyl]-	pH 7.38	314(4.25),447(4.13)	87-0790-65
$C_{44}H_{60}O_6$			
Androsta-4,6-dien-3-one, 17β-hydroxy-, propionate, dimer	EtOH	256(4.03)	44-0068-64
4,5-(6',7'-Bicyclo[3.2.0]heptano)-androst-6-en-3-one, 17β-hydroxy-, propionate	n.s.g.	285(2.10)	88-1075-64

Compound	Solvent	$\lambda_{max}(\log \epsilon)$	Ref.
$C_{45}H_{22}Cl_2N_4O_{18}$			
4H-Pyran-4-one, 2,6-bis(5-chloro-2,4-dihydroxyphenyl)-, tetrakis-(p-nitrobenzoate)	MeOH	264(3.5),333(4.0)	44-2482-64
$C_{45}H_{24}N_4O_{18}$			
4H-Pyran-4-one, 2,6-bis(2,4-dihydroxyphenyl)-, tetrakis(p-nitrobenzoate)	MeOH	284(3.3),327(3.5)	44-2482-64
$C_{45}H_{28}$			
Propane, 1,3-bis(7H-dibenzo-[c,g]fluoren-9-ylidene)-	dioxan	290(4.89),301(4.84), 331(4.38),346(4.51), 362(4.69),420(3.55)	5-0001-65I
anion	DMSO	694(5.46)	5-0001-65I
$C_{45}H_{30}$			
1,3-Pentadiene, 1-fluoren-9-yl-5-fluoren-9-ylidene-4-(fluoren-9-ylidene-methyl)-	THF	303(4.32),365(4.59), 384(4.50)	5-0001-65I
anion	DMF	652(4.66)	5-0001-65I
$C_{45}H_{32}$			
Fluorene, 9-[3-(2,3,4,5-tetraphenyl-1,3-cyclopentadien-1-yl)allylidene]-	DMSO-KOEt	313(4.36),323(4.35), 582s(4.80),622(5.11)	5-0050-65J
$C_{45}H_{33}ClN_2O_4$			
4-[3-(1,2-Diphenyl-4(1H)-quinolylidene)propenyl]-1,2-diphenyl-quinolinium perchlorate	n.s.g.	724(5.09)	65-2008-64
$C_{45}H_{33}IN_2O_4$			
6-Hydroxy-4-[3-[6-hydroxy-1-(p-hydroxy-phenyl)-2-phenyl-4(1H)-quinolylidene]propenyl]-1-(p-hydroxyphenyl)-2-phenylquinolinium iodide	n.s.g.	743(5.00)	65-2008-64
$C_{45}H_{62}N_2O_{12}$			
Rifamycin SV, (2,6-dimethyl-N-piperidino)methyl-	pH 7.38	314(4.29),448(4.17)	87-0790-65
$C_{45}H_{62}N_2O_{16}$			
Rifamycin SV, N-methyl-N-cyclohexyl-aminomethyl-	pH 7.38	314(4.25),446(4.15)	87-0790-65
$C_{45}H_{65}N_3O_6$			
Pilocereine	MeOH	220(4.77),285(3.96)	5-0207-65E
$C_{45}H_{68}O_6$			
2,5-Cyclohexadien-1-one, 4-methoxy-4,6-bis(4-methoxy-2,5-di-tert-butyl-phenoxy)-2,5-di-tert-butyl-	n.s.g.	240(4.5),288(3.9), 333s(3.2)	5-0061-64A

Compound	Solvent	$\lambda_{max}(\log \epsilon)$	Ref.
$C_{46}H_{26}N_{10}O_5$			
1,3,4-Oxadiazole, 2,5-bis[m-[5-[m-(5-phenyl-1,3,4-oxadiazol-2-yl)-phenyl]-1,3,4-oxadiazol-2-yl]-phenyl]-	DMF	287.6(5.09)	24-2966-65
$C_{46}H_{30}Cl_4N_2O_2P_2$			
1,4-Naphthoquinone, 5,6,7,8-tetra-chloro-2,3-bis[(triphenylphos-phoranylidene)amino]-	$CHCl_3$	252(4.62),305(4.45), 318(4.40)	39-1003-65
$C_{46}H_{33}N_3O_4P_2$			
1,4-Naphthoquinone, 5-nitro-2,3-bis-[(triphenylphosphoranylidene)-amino]-	$CHCl_3$	271s(4.49)	39-1003-65
$C_{46}H_{34}N_2O_2P_2$			
1,4-Naphthoquinone, 2,3-bis[(tri-phenylphosphoranylidene)amino]-	$CHCl_3$	269s(4.35),273(4.36), 314(4.22),367(3.94)	39-1003-65
$C_{46}H_{34}O_6$			
[2,2'-Bichroman]-4,4'-dione, 2,2'-di-benzyl-3,3'-bis(α-hydroxybenzyli-dene)-	C_6H_{12}	234(4.72),264s(4.23), 271(4.23),280(4.25), 306(4.32)	35-5424-65
$C_{46}H_{36}O_4Si$			
7-Silabicyclo[2.2.1]hepta-2,5-diene-2,3-dicarboxylic acid, hexaphenyl-, dimethyl ester	C_6H_{12}	273(4.25),285s(4.21), 379(3.78)	35-5584-64
$C_{46}H_{42}B_2N_2O_8$			
Tetracycline-triphenylboron reaction product	EtOH	280(4.0),375(4.1), 410s(3.9)	83-0034-65
$C_{46}H_{50}Cl_2P_2$			
Tetramethylenebis[tribenzyl-phosphonium chloride]	EtOH	254(3.06),260(3.15), 266(3.06)	59-1143-64
$C_{46}H_{61}NO_2$			
5,6-Secocholest-6-en-3β-ol, 6,6-diphenyl-, carbanilate	C_6H_{12}	234(4.22),275(2.97), 282(2.88)	24-0958-64
$C_{46}H_{65}N_3O_{13}$			
Rifamycin B, dimethylpentylhydrazide	pH 7.38	222(4.62),302(4.33), 421(4.21)	87-0596-64
$C_{46}H_{66}CuO_8$			
5α-Androstan-3-one, 2-acetyl-17β-acetoxy-, copper chelate	EtOH	243(3.61),311(4.05)	44-2059-64
5α-Androstan-17-one, 16-acetyl-3β-acetoxy-, copper chelate	EtOH	260(4.09),308(4.31)	35-5207-64
$C_{46}H_{78}N_4O_7S$			
9H-Purine-6-thiol, 9-(2,3,5-tri-O-myristyl-β-D-ribofuranosyl)-	pH 1	320(4.38)	87-0200-64
	pH 11	234(4.15),310(4.35)	87-0200-64

Compound	Solvent	$\lambda_{max}(\log \epsilon)$	Ref.
$C_{47}H_{28}N_4O_{18}$			
4H-Pyran-4-one, 2,6-bis(4,6-dihydroxy-o-tolyl)-, tetrakis(p-nitrobenzoate)	MeOH	263(3.70),311(4.03)	44-2482-64
$C_{47}H_{39}FN_2O_5$			
Uridine, 3',5'-di-O-trityl-2'-fluorodeoxy-	EtOH	259(3.98)	44-0564-64
$C_{47}H_{40}N_2O_6$			
Uracil, 1-(3,5-di-O-tritylarabinosyl)-	EtOH	261.5(4.03)	44-0564-64
$C_{47}H_{41}N_2O_9P$			
Uridine, 2',5'-di-O-trityl-, 3'-phosphate, ammonium salt	EtOH	247(3.7),263(3.86)	94-1503-64
$C_{47}H_{42}OSi_2$			
Valerophenone, 3,5-bis-(triphenylsilyl)-	$CHCl_3$	263(3.56),270(3.40), 314s(2.07)	23-0298-64
$C_{47}H_{66}O_6$			
1H-7,17b-Ethano-as-indaceno[2,1-a:-7,6-a']diphenanthren-21-one, triacontahydro-2,13-dihydroxy-7-methoxy-4a,6a,15a,17a-tetramethyl-, diacetate	dioxan	295(1.66)	35-0995-65
$C_{48}H_{24}$			
Hexabenzo[a,d,g,j,m,p]coronene	benzene	357(4.83),376(5.07), 402(4.63),456(3.01), 486(2.64)	78-0467-65
$C_{48}H_{36}N_4O_4Zn$			
Zinc, [α,β,γ,δ-tetrakis(o-methoxyphenyl)porphinato(2-)]-	benzene	424(5.71),513(3.46), 550(4.34),588(3.32)	12-1028-64
Zinc, [α,β,γ,δ-tetrakis(p-methoxyphenyl)porphinato(2-)]-	benzene	427(5.65),515(3.52), 553(4.28),592(3.75)	12-1028-64
$C_{48}H_{37}N_3O_3P_2$			
1,4-Naphthoquinone, 5-acetamido-2,3-bis[(triphenylphosphoranylidene)amino]-	$CHCl_3$	261(4.47),274s(4.34), 360(4.31)	39-1003-65
$C_{48}H_{38}N_4O_4$			
Porphine, α,β,γ,δ-tetrakis-(o-methoxyphenyl)-	benzene	420(5.54),513(4.18), 546(3.66),590(3.62), 647(3.18)	12-1028-64
Porphine, α,β,γ,δ-tetrakis-(p-methoxyphenyl)-	benzene	424(5.61),488(3.53), 518(4.13),556(3.95), 595(3.74),652(3.65)	12-1028-64
$C_{48}H_{39}N_{12}$			
s-Tetrazin-1(2H)-yl, 6,6',6''-s-phenenyltris[3,4-dihydro-2,4-diphenyl-	DMF	721(4.10)	89-0043-65
$C_{48}H_{40}O_4Si_4$			
Cyclotetrasiloxane, octaphenyl-	$CHCl_3$	265(3.47)	46-1066-65
$C_{48}H_{40}Si_3$			
Trisilane, octaphenyl-	C_6H_{12}	254(4.51)	101-0163-65B
	n.s.g.	255(4.51)	101-0176-65B

Compound	Solvent	λ_{max}(log ε)	Ref.
$C_{48}H_{40}Si_4$			
Cyclotetrasilane, octaphenyl-	n.s.g.	234(4.81),270(4.54)	101-0176-65B
$C_{48}H_{40}Sn_3$			
Tin, octaphenyltri-	C_6H_{12}	220(5.0),248(4.7), 275(4.7)	5-0001-64G
$C_{48}H_{48}Cl_6O_6W$			
Tungsten, hexakis(4-chloro-3,5-dimethylphenoxy)-	C_6H_{12}	235(4.79),265(4.48), 398(4.40)	12-1579-65
$C_{48}H_{68}O_{20}$			
Sargenoside, hexaacetate	EtOH	217(4.17)	33-0799-64
$C_{49}H_{28}N_4O_{20}$			
4H-Pyran-4-one, 2,6-bis(5-acetyl-2,4-dihydroxyphenyl)-, tetrakis-(p-nitrobenzoate)	MeOH	283(4.42),318(4.5)	44-2482-64
$C_{49}H_{39}N_5O_4S$			
Adenosine, N-trityl-5'-O-trityl-, cyclic 2',3'-thiocarbonate	MeCN	228s(4.55),270(4.36), 273(4.35),283s(4.17)	44-2854-65
$C_{49}H_{41}IN_2$			
6-Methyl-4-[3-(6-methyl-2-phenyl-1-p-tolyl-4(1H)-quinolylidene)propenyl]-2-phenyl-1-p-tolylquinolinium iodide	n.s.g.	732(5.20)	65-2008-64
$C_{50}H_{34}$			
Anthracene, 1,4,5,8,9,10-hexa-phenyl-	$CHCl_3$	301(4.81),439(4.21)	22-2342-65
isomer	$CHCl_3$	293(4.28),315(4.10), 317(4.10)	22-2342-65
$C_{50}H_{38}O_2$			
Anthracene-1,4-diol, 1,2,3,4-tetra-hydro-1,4,5,8,9,10-hexaphenyl-	$CHCl_3$	262(4.50),330(3.94)	22-2342-65
$C_{50}H_{46}Si_5$			
Cyclopentasilane, 1,1-dimethyl-2,2,3,3,4,4,5,5-octaphenyl-	n.s.g.	249s(4.61)	101-0176-65B
$C_{50}H_{50}N_2O_{10}$			
Thebainehydroquinone, dimer	dioxan	309(4.04)	88-0275-65
$C_{52}H_{36}N$			
Pyrrolyl, 2,3,4,5-tetrakis-(p-biphenylyl)-	toluene	425(4.8),640(4.2)	24-3124-65
$C_{52}H_{38}Cl_6Pt$			
Bis(dihydro-6,8-diphenylcyclohept[f]-indenylium) hexachloroplatinate	96% H_2SO_4	260(4.43),344(4.59), 500(3.70)	44-0891-65
$C_{52}H_{52}O_6Si_6$			
Cyclohexasiloxane, 2,2,8,8-tetramethyl-4,4,6,6,10,10,12,12-octaphenyl-	$CHCl_3$	265(3.45)	46-1066-65

Compound	Solvent	$\lambda_{max}(\log \epsilon)$	Ref.
$C_{53}H_{41}IN_2O_8$ 6-Hydroxy-4-[3-[6-hydroxy-1-(p-hydroxy-phenyl)-2-phenyl-4(1H)-quinolylidene]propenyl]-1-(p-hydroxyphenyl)-2-phenylquinolinium iodide, diacetate	n.s.g.	740(5.01)	65-2008-64
$C_{54}H_{38}$ 9,9'-Bifluorene, 9,9'-bis(2,2-di-phenylvinyl)-	THF	256(4.50),291(3.79), 310(3.80)	24-2975-64
$C_{54}H_{45}Cl_9P_3Re_3$ Rhenium, nonachlorotris(tri-phenylphosphine)tri-	acetone	516(3.23),755(2.82)	39-5683-64
$C_{54}H_{46}GeSi$ Silane, tris(triphenylgermyl)-	C_6H_{12}	240s(4.69)	101-0163-65B
$C_{54}H_{46}Si_4$ Trisilane, 1,1,1,3,3,3-hexaphenyl-2-(triphenylsilyl)-	C_6H_{12}	240s(4.62)	101-0163-65B
$C_{54}H_{70}N_{12}O_{12}S_2$ Triostin C	MeOH	243(4.87),321(4.13)	78-2931-65
$C_{54}H_{84}O_2$ Cholesta-4,6-dien-3-one, dimer	C_6H_{12}	243(4.03)	44-0068-64
$C_{55}H_{45}BP_2$ Triphenyl[(triphenylphosphonia)(tri-phenylphosphoranylidene)-methyl] borate	MeCN	267(3.98)	44-2427-64
$C_{55}H_{72}MgN_4O_5$ Chlorophyll a	CCl_4	410(4.8),430s(5.0), 620(4.1),660(4.8)	10-0251-64D
$C_{56}H_{32}Cl_8$ Cyclobutadiene, tetrakis(p-chloro-phenyl)-, dimer	n.s.g.	238(4.62),273(4.53)	23-0470-65
$C_{56}H_{42}O_{12}$ Hopeaphenol	EtOH	281(4.16)	39-0406-65
$C_{56}H_{52}N_4Zn$ Zinc, [α,β,γ,δ-tetramesityl-porphinato(2-)]-	benzene	424(5.78),512(3.39), 547(4.32),587(3.25)	12-1028-64
$C_{56}H_{54}N_4$ Porphine, α,β,γ,δ-tetramesityl-	benzene	420(5.57),483(3.45), 515(4.19),548(3.72), 593(3.58),649(3.36)	12-1028-64
$C_{56}H_{58}O_{26}$ Protoaphin-fb, decaacetate	EtOH	240(4.86),294(4.03), 306(4.04),338(3.84)	39-0051-64
$C_{56}H_{86}O_4$ Ethane, 1,1,2,2-tetrakis(3,5-di-tert-butyl-4-hydroxyphenyl)-	C_6H_{12}	276(3.89),283(3.81)	35-1263-64

Compound	Solvent	λ max(log ε)	Ref.
$C_{57}H_{48}N_4O_{12}$			
Malondiamide, N,N'-bis(2,3,5-tri-O-benzoyl-β-D-ribofuranosylcarbamoyl)-	EtOH-Et_3N	237(4.87),296(4.57)	94-0459-64
$C_{58}H_{40}Cl_6N_4Zn$			
Zinc meso-o-chlorophenyl-5,5'-di-o-chlorobenzylpyrromethene	CCl_4	303(3.36),351(3.14), 481(4.10),502(4.62)	12-1028-64
$C_{58}H_{73}N_7O_{12}$			
Proline, N-[3-(benzyloxy)picolinoyl]-L-threonyl-D-α-aminobutyryl-L-prolyl-N-methyl-L-phenylalanyl-, tert-butyl ester, ester with N-carboxy-L-2-phenylglycine tert-butyl ester	EtOH	291(3.72)	35-4373-65
$C_{58}H_{82}O_4$			
2,5-Cyclohexadien-1-one, 4,4'-[bis-(3,5-di-tert-butyl-4-hydroxyphenyl)-ethanediylidene]bis[2,6-di-tert-butyl-	C_6H_{12}	385(3.48),478(4.26)	35-1263-64
$C_{58}H_{82}O_{26}$			
Odorotrioside G, octaacetate	EtOH	217(4.17)	33-0799-64
Sarmentocymarin, diglucoside, octaacetate	EtOH	217(4.17)	33-0799-64
$C_{58}H_{88}O_6$			
Coenzyme Q, hydroquinone diacetate	EtOH	268(3.88)	37-0514-65
$C_{59}H_{30}N_6O_{26}$			
4H-Pyran-4-one, 2,6-bis(2,4,6-trihydroxyphenyl)-, hexakis(p-nitrobenzoate)	MeOH	273(3.7),329(3.9)	44-2482-64
$C_{59}H_{32}N_4O_{20}$			
4H-Pyran-4-one, 2,6-bis(5-benzoyl-2,4-dihydroxyphenyl)-, tetrakis-(p-nitrobenzoate)	MeOH	265(4.2),295(4.5), 330(4.5)	44-2482-64
$C_{59}H_{90}O_4$			
Coenzyme Q_{10}	EtOH	275(5.15)	37-0514-65
$C_{60}H_{36}$			
Fluoren-9-yl, 9,9'-[p-phenylenebis-(fluoren-9-ylidenemethylene)]di-	dioxan	240(5.04),253(5.02), 347(4.39),498(4.51), 865(3.34)	49-0003-64
$C_{60}H_{38}$			
p-Xylene, α,α'-difluoren-9-yl-α,α'-difluoren-9-ylidene-	dioxan	223(5.10),252(4.97), 261(5.05),292(4.69), 304(4.69),321(4.66)	49-0003-64
anion	DMF-KOMe-dioxan	295(4.68),355(4.45), 402(4.21),484(4.28), 626(4.74)	49-0003-64
$C_{60}H_{50}Si_4$			
Tetrasilane, decaphenyl-	n.s.g.	255s(4.52),288(4.36)	101-0176-65B

Compound	Solvent	λ_{max}(log ε)	Ref.
$C_{60}H_{50}Si_5$			
Cyclopentasilane, decaphenyl-	n.s.g.	251s(4.78)	101-0176-65B
$C_{60}H_{64}O_{28}$			
Protoaphin-fb, dihydro-, dodecaacetate	EtOH	238(5.06),309(4.21), 335(3.99)	39-0051-64
$C_{60}H_{78}O_6W$			
Tungsten, hexa(p-tert-butylphenoxy)-	C_6H_{12}	230(4.85),256(4.58), 387(4.56)	12-1579-65
$C_{60}H_{82}N_{12}O_{16}$			
Actinomycin C_1, 1,8-didemethyl-	MeOH	236(4.50),438(4.34)	88-4803-65
$C_{60}H_{86}N_{12}O_{18}$			
Actinomycinic acid	MeOH	446(4.43)	88-3523-64
Actinomycinic acid C_1, 1,8-di-demethyl-	MeOH	234(4.60),422(4.34), 441(4.34)	88-4803-65
$C_{61}H_{44}$			
Cyclopentadiene, 2-[2-(2,3,4,5-tetra-phenylfulven-6-yl)vinyl]-1,3,4,5-tetraphenyl-	ether	275(4.53),325(4.22), 473(4.66)	5-0001-65I
anion	n.s.g.	655(5.18)	5-0001-65I
	DMSO-KOEt	313(4.60),655(5.18)	5-0050-65J
$C_{61}H_{51}BP_2$			
Triphenyl[(triphenylphosphoranylidene)-methyl]phosphonium tetraphenylborate	MeCN	267(4.03)	44-2427-64
$C_{61}H_{78}O_8$			
Acetic acid, (p-methoxyphenyl)-, di-ester with triacontahydro-2β,13β-di-hydroxy-7-methoxy-4a,6a,15a,17a-tetramethyl-1H-7,17b-ethano-as-indaceno[2,1-a:7,6-a']diphenan-thren-21-one	dioxan	277(3.60),283(3.54)	35-0995-65
$C_{62}H_{56}Si_6$			
Cyclohexasilane, 1,1-dimethyl-decaphenyl-	n.s.g.	250s(4.75)	101-0176-65B
$C_{62}H_{92}O_4$			
Vitamin K_1, dimerization acid	C_6H_{12}	249s(--),253(4.77), 324(3.90),338(3.92)	22-2513-65
$C_{63}H_{49}ISn$			
Tin,(2''-1-4',4'',5',5'',6',6''-hexa-phenyl-m-quaterphenyl-2'-yl)tri-methyl-	heptane	247(3.68)	101-0417-64B
$C_{64}H_{49}N$			
Ethylene, 1-(dimethylamino)-2,2-bis-(1,2,3,4-tetraphenylcyclopentadien-5-ylidenemethyl)-	MeCN	279(4.73),470s(4.60), 522(4.76)	24-3331-64
$C_{64}H_{108}N_4O_7S$			
9H-Purine-6-thiol, 9-(2,3,5-tri-O-oleyl-β-D-ribofuranosyl)-	pH 1	321(4.43)	87-0200-64
	pH 11	235(4.21),310(4.38)	87-0200-64

Compound	Solvent	$\lambda_{max}(\log \epsilon)$	Ref.
$C_{64}H_{114}N_4O_7S$			
9H-Purine-6-thiol, 9-(2,3,5-tri-O-stearyl-β-D-ribofuranosyl)-	pH 1	322(4.44)	87-0200-64
	pH 11	234(4.28),311(4.43)	87-0200-64
$C_{65}H_{60}Si_6$			
1,2,3,4,5,6-Hexasilaspiro[5.5]undecane, decaphenyl-	n.s.g.	250s(4.70)	101-0176-65B
$C_{65}H_{74}O_6$			
1-Naphthoic acid, diester with triacontahydro-2,13-dihydroxy-7-methoxy-4a,6a,15a,17a-tetramethyl-1H-7,17b-ethano-as-indaceno[2,1-a:7,6-a']diphenanthren-21-one	dioxan	295(4.16)	35-0995-65
$C_{66}H_{62}O_{12}$			
Hopeaphenol, deca-O-methyl-	EtOH	283(4.18)	39-0406-65
$C_{68}H_{44}MgN_4$			
Magnesium octaphenylporphine	DMF	550(4.28),590(4.11)	44-0859-65
$C_{68}H_{46}N_4$			
Porphine, octaphenyl-	$CHCl_3$	423(5.14),514(4.23), 550(4.04),583(3.95), 635(3.48)	44-0859-65
	$CHCl_3$-HCl	442(5.15),573(4.08), 590s(3.74),615(3.70)	44-0859-65
$C_{70}H_{47}FeN_4O_2$			
Ferrioctaphenylporphine, acetate	pyridine	432(5.14),530(4.04), 562(4.20)	44-0859-64
$C_{72}H_{54}O_6W$			
Tungsten, hexa(p-phenylphenoxy)-	C_6H_{12}	265(4.99),412(4.65)	12-1579-65
$C_{72}H_{60}GePb_4$			
Germane, tetrakis(triphenylplumbyl)-	$CHCl_3$	328(4.80)	101-0265-64B
$C_{72}H_{60}GeSn_4$			
Germane, tetrakis(triphenylstannyl)-	$CHCl_3$	276(4.86)	101-0265-64B
$C_{72}H_{60}Pb_4Sn$			
Tin, plumbanetetrayltetrakis-[triphenyl-	benzene	319(4.83)	101-0265-64B
	$CHCl_3$	298(4.77)	101-0265-64B
$C_{72}H_{60}Pb_5$			
Lead, tetrakis(triphenylplumbyl)-	benzene	358(4.75),444(4.46)	101-0265-64B
$C_{72}H_{60}Si_5$			
Pentasilane, dodecaphenyl-	n.s.g.	250s(4.56),297(4.44)	101-0176-65B
$C_{72}H_{60}Si_6$			
Cyclohexasilane, dodecaphenyl-	n.s.g.	248s(4.78)	101-0176-65B
$C_{72}H_{60}Sn_5$			
Tin, tetrakis(triphenylstannyl)-	$CHCl_3$	277(4.90)	101-0265-64B
$C_{72}H_{60}Sn_6$			
Cyclohexastannane, dodecaphenyl-	C_6H_{12}	215(5.0),250f(4.8)	5-0001-64G

Compound	Solvent	$\lambda_{max}(\log \epsilon)$	Ref.
$C_{76}H_{60}MgN_4O_8$			
Magnesium octa(p-methoxy-phenyl)porphine	pyridine	438(5.45),562(4.40), 600(4.18)	44-0859-65
$C_{76}H_{62}N_4O_8$			
Porphine, octa(p-methoxyphenyl)-	$CHCl_3$	424(5.52),518(4.36), 555(4.23),583(4.08), 638(3.78)	44-0859-65
	$CHCl_3$-HCl	446(5.45),535s(4.15), 582(4.62),603s(4.23), 625(4.18)	44-0859-65
$C_{80}H_{154}O_2$			
Smegmamycolic acid, α-anhydro-, methyl ester	n.s.g.	218(4.09)	22-0868-64
$C_{84}H_{70}Si_6$			
Hexasilane, tetradecylphenyl-	n.s.g.	255s(4.58),312(4.49)	101-0176-65B
$C_{96}H_{80}Si_7$			
Heptasilane, hexadecaphenyl-	n.s.g.	255s(4.68),324(4.58)	101-0176-65B

1- -64, Acta Chem. Scand., 18 (1964)
0157 R. Cigen and C.-G. Ekström
0191 N.S. Hjelte and T. Agback
0421 R. Magnusson
0441 M. Nilsson
0447 G. Bengtsson
0483 I. Könyves and A. Olsson
0750 C. Keller-Juslen
0815 A. Biezais and G. Bergson
0843 J. Bjerrum
0871 J. Sandström
0932 B.E. Nielsen and J. Lemmich
1059 B. Persson and J. Sandström
1292 A. Penttilä and J. Sundman
1368 F. Merenyi and M. Nilsson
1379 B.E. Nielsen and J. Lemmich
1389 O. Buchardt
1739 R. Bonnet, A.A. Spark and B.C.L. Weedon
1806 T. Hase
1984 H. Lund
2000 G. Bergson and C. Frisell
2303 G. Pettersson

1- -65, Acta Chem. Scand., 19 (1965)
0540 M. Lounasmaa
0612 K. Nilsson
0653 D. Dyrssen and D. Petkovic
0756 K.G. Flynn and G. Bergson
0766 P. Friis
0839 T. Bruun, D.P. Hollis and R. Ryhage
0913 C.R. Enzell and B.R. Thomas
1051 J. Gripenberg and J. Martikkala
1063 J. Gripenberg and M. Lounasmaa
1088 W.O. Godtfredsen and S. Vangedal
1113 M.L. Shankaranarayana and C.C. Patel
1120 O. Buchardt, J. Becher and Chr. Lohse
1188 G. Bendz et al.
1191 E. Akerblom and M. Sandberg
1239 K.A. Jensen et al.
1271 S. Gronowitz and A. Bugge
1607 B. Lüning and K. Leander
1677 T. Bruun
1875 C.R. Enzell and B.R. Thomas
1897 E. Plahtz, J. Grundnes and P. Klaeboe
1951 K. Nilsson
2139 R. Helmers
2432 J. Sandström and B. Uppström

2- -64, Indian J. Chem., 2 (1964)
0017 S.C. Agarwal and T.R. Seshadri
0108 C.R. Narayanan et al.
0169 T.R. Govindachari and K. Nagarajan
0182 N.R. Krishnaswamy, T.R. Seshadri and B.R. Sharma
0190 H.D. Shroff, A.B. Diwadkar and A.B. Kulkarni
0319 G. Chakravarthy and T.R. Seshadri
0339 P. Chakrabarthi and A.K. Barua
0399 S.C. Bhrara et al.
0417 S. Husai and R. Vaidyeswaran
0423 S. Swaminathan and K. Narasimhan
0449 B.R. Pai et al.
0464 B.K. Gupta et al.
0468 S.C. Pakrashi
0491 B.R. Pai et al.

2- -65, Indian J. Chem., 3 (1965)
0071 T.R. Govindachari et al.
0091 G.D. Joshi and S.N. Kulkarni
0143 A.S.N. Murthy
0162 S. Naqui and V.R. Srinivasan
0251 S.S. Kanhere, R.S. Shah and S.L. Bafna
0283 P. Chakrabarti et al.
0351 A.C. Jain, S.K. Mathur and T.R. Seshadri
0354 T.R. Seshadri and M.S. Sood
0369 A.C. Jain, P.D. Sarpal and T.R. Seshadri
0422 G.D. Bhatia, S.K. Mukherjee and T.R. Seshadri
0522 H. Singh and T.K. Kaw
0524 A.K. Bhatnagar, S. Bhattacharki and S.P. Popli
0561 J.R. Merchant, J.B. Mehta and V.B. Desai

4- -64, J. Heterocyclic Chem., 1 (1964)
0001 W.H. Nyberg and C.C. Cheng
0006 E.F. Elslager et al.
0013 A.I. Meyers et al.
0023 W. Pfleiderer and D. Soll
0030 C.K. Bradsher and J.C. Parham
0034 C.W. Noell and R.K. Robins
0042 T. Kuraishi and R.N. Castle
0079 B.R. Baker, B. Ho and T. Neilson
0088 B.R. Baker, B. Ho and G.B. Cheda
0113 A.D. Broom and R.K. Robins
0115 J.A. Montgomery and H.J. Thomas
0121 C.K. Bradsher and J.C. Parham
0130 G.D. Daves, Jr., et al.
0158 B.J. Bergot and L. Jurd
0159 J.E. Pike, L. Slechta and P.F. Wiley
0168 C.K. Bradsher and E.F. Litzinger
0171 M.R.W. Levy and M.M. Joullié
0182 M. Malm and R.N. Castle
0188 B.G. Harnsberger and J.L. Riebsomer
0208 C.K. Bradsher and R.W.L. Kimber
0213 J.A. Montgomery and K. Hewson
0215 J.A. Montgomery, S.J. Clayton and W.E. Fitzgibbon, Jr.
0229 B.G. Harnsberger and J. Riebsomer
0247 G.A. Gerhardt and R.N. Castle
0263 B.R. Baker and P.I. Almaula
0275 L. Bauer, C. Nambury and D. Dhawan
0280 T.P. Culbertson, G.W. Moersch and W.A. Neuklis
0288 V.J. Bauer and S.R. Safir
0295 R.E. Jensen and R.T. Pflaum

4- -65, J. Heterocyclic Chem., 2 (1965)
0001 W.J. Haggerty, Jr., R.H. Springer and C.C. Cheng
0015 L.H. Klemm and A. Weisert

0021 B.R. Baker and J.H. Jordaan
0026 F. Johnson and W.A. Nasutavicus
0037 J.D. Crum and J.A. Franks, Jr.
0049 R.H. Springer, W.J. Haggerty, Jr., and C.C. Cheng
0053 J.P. Paolini and R.K. Robins
0063 R.R. Shoup and R.N. Castle
0097 R.W. Balsiger, J.A. Montgomery and T.P. Johnston
0100 E. Campaigne and E.S. Neiss
0105 K.T. Potts, S.K. Roy and D.P. Jones
0110 W.H. Nyberg, C.W. Noell and C.C. Cheng
0113 A.S. Dey and M.M. Joullie
0120 A.S. Dey and M.M. Joullie
0126 N.J. Doorenbos and K.A. Kerridge
0130 H.J. Brabander and W.B. Wright, Jr.
0140 L.H. Klemm and A. Weisert
0144 W. Korytnyk and B. Paul
0162 B.R. Baker and H. Jordaan
0181 R.L. Taber et al.
0196 R.J. Rousseau and R.K. Robins
0207 G.W. Moersch and P.L. Creger
0212 N.J. Doorenbos and M.T. Wu
0228 C.K. Bradsher, J.C. Parham and J.D. Turner
0231 E. Campaigne and E.S. Neiss
0242 G. DiModica, E. Barni and F.D. Monache
0247 G.A. Gerhardt, D.L. Aldous and R.N. Castle
0253 T. Ichikawa, T. Kato and T. Takenishi
0272 J.S. Driscoll and R.H. Nealey
0291 N.J. Leonard et al.
0306 G.A. Gerhardt and R.N. Castle
0308 M.D. Coburn and H.E. Ungnade
0310 A. Brossi and E. Wenis
0313 J.A. Montgomery and K. Hewson
0315 C.D. Slater and D.L. Heywood
0318 E.J. Seus
0323 J. Iacobelli et al.
0329 A.I. Meyers and J.C. Sircar
0330 T. Kametani et al.
0331 C.K. Bradsher et al.
0340 B.R. Baker and B.-T. Ho
0371 R.A. Mitsch et al.
0385 G. Casini, F. Gualteri and M.L. Stein
0387 J.D. Benigni and R.L. Minnis
0453 H.F. Ridley et al.
0475 L.J. Chinn
0495 T.Y. Shen, W.V. Ruyle and R.L. Bugianesi

5- -64A, Ann. Chem. Liebigs, 671 (1964)
0010 W.R. Roth
0031 K. Hafner and K.D. Asmus
0061 E. Müller, H. Kaufmann and A. Rieker
0092 D. Schulte-Frohlinde and P. Erhardt

5- -64B, Ann. Chem. Liebigs, 672 (1964)
0055 G. Wittig and H. Dürr
0194 K. Hafner and G. Schneider

5- -64C, Ann. Chem. Liebigs, 673 (1964)
0096 W. Theilacker et al.
0136 N. Engelhard and A. Kolb

5- -64D, Ann. Chem. Liebigs, 674 (1964)
0001 H. Machleidt and R. Wessendorf
0018 G. Kresze, H.-G. Hankel and H. Goetz
0028 H.H. Inhoffen et al.
0036 H.H. Inhoffen et al.
0062 H. Wolf et al.
0079 H. Dickmann, D. Hofmann and K. Wellenfels
0122 O.A. Neumüller et al.
0129 E. Buchta and H. Maar
0152 H. Dannenberg et al.
0196 R. Tschesche, G. Biernoth and G. Snatzke

5- -64E, Ann. Chem. Liebigs, 675 (1964)
0083 G. Ohloff, J. Seibl and E. Kovats
0109 H. Dannenberg and H.-G. Neumann
0151 R. Gompper, H. Euchner and H. Kast

5- -64F, Ann. Chem. Liebigs, 676 (1964)
0010 H. Musso et al.
0021 G. Wittig and G. Steinhoff
0066 H. Machleidt
0088 K. Schlögl and H. Egger
0101 M. Regitz
0168 H. Morimoto and H. Oshio

5- -64G, Ann. Chem. Liebigs, 677 (1964)
0001 W.P. Neumann and K. König
0021 R. Huisgen, G. Seidl and I. Wimmer
0047 E. Müller, B. Zech and R. Heischkeil
0139 W. Fritsch, G. Seidl and H. Ruschig

5- -64H, Ann. Chem. Liebigs, 678 (1964)
0039 K. Hafner, G. Schulz and K. Wagner
0137 F. Kaiser, E. Haack and H. Spingler
0183 K. Dimroth, K. Wolf and H. Kroke

5- -64I, Ann. Chem. Liebigs, 679 (1964)
0020 H. Machleidt and R. Wessendorf
0026 H. Disselnkötter and P. Kurtz
0042 C. Grot et al.
0056 W. Ried and H. Appel
0100 G. Hesse and B. Wehling
0109 F. Boberg
0156 F. Cramer et al.

5- -64J, Ann. Chem. Liebigs, 680 (1964)
none

5- -65A, Ann. Chem. Liebigs, 681 (1965)
0021 H. Machleidt and G. Strehlke
0039 H.P. Kaufmann and A.K. SenGupta
0084 H. Reinshagen
0089 S. Bodforss
0123 B. Eistert and G. Heck
0141 E. Muller, H.-D. Spanagel and A. Ricker

0187 K. Schreiber and H. Rönsch
0196 K. Schreiber and H. Rönsch

5- -65B, Ann. Chem. Liebigs, 682 (1965)
0001 E. Vogel et al.
0058 H. Oediger and K. Eiter
0062 K. Eiter and H. Oediger
0099 B. Hirsch and E.-A. Jauer
0123 H. Gehlen and J. Schmidt
0188 H. Behringer and A. Grimm
0212 H. Morimoto and H. Oshio

5- -65C, Ann. Chem. Liebigs, 683 (1965)
0159 G. Snatzke and A. Nisar

5- -65D, Ann. Chem. Liebigs, 684 (1965)
0014 H. Kabbe, E. Truscheit and K. Eiter
0024 H.H. Inhoffen et al.
0037 R. Gompper, R.R. Schmidt and E. Kutter
0062 G. Snatzke and G. Zanati
0092 G. Opitz, H. Schempp and H. Adolph
0146 G. Henseke and R. Jacobi
0200 H. Dannenberg, J. Sonnenbichler and H.J. Gross

5- -65E, Ann. Chem. Liebigs, 685 (1965)
0010 K.-H. Büchel, H. Röchling and F. Korte
0122 F. Korte, E. Hackel and H. Sieper
0134 M. El-Dakhakhny
0167 O. Dann and G. Volz
0200 S.M. Albonico, A.M. Kuck and V. Deulofeu
0207 B. Franck, G. Blaschke and K. Lewejohann
0218 F. Werder et al.
0228 U. Stache, W. Fritsch and H. Ruschig
0246 D. Satoh et al.

5- -65F, Ann. Chem. Liebigs, 686 (1965)
0064 G. Hesse and P. Thiene
0134 F. Lingens and H.Schneider-Bernlohr
0145 R. Huisgen and H. Blaschke
0167 G. Snatzke

5- -65G, Ann. Chem. Liebigs, 687 (1965)
0191 C. Grundmann et al.
0214 M. Regitz and D. Stadler
0236 W. Lüttke and V. Schabacker

5- -65H, Ann. Chem. Liebigs, 688 (1965)
0028 W.R. Roth and J. König
0098 R. Huisgen and K. Herbig
0196 G. Niederdallmann and F. Kröhnke
0216 F. Kaiser, E. Haack and H. Spingler

5- -65I, Ann. Chem. Liebigs, 689 (1965)
0001 R. Kuhn et al.
0040 R. Köster and G.W. Rotermund
0093 H. Musso and D. Maassen
0109 H. Suhr
0197 C. Schiele
0202 F. Hein and K.-H. Vogt
0221 M. Goto et al.

5- -65J, Ann. Chem. Liebigs, 690 (1965)
0009 S. Hünig et al.
0033 B. Hirsch
0050 R. Kuhn and D. Rewicki
0079 H. Machleidt and W. Grell
0138 E. Fahr, K. Königsdorfer and F. Scheckenbach
0170 G. Pfleiderer and C. Woenckhaus

6- -64, Ann. Chim. (Paris), 9 (1964)
0073 F. Erb-Debruyne
0359 P. Aubrun

6- -65, Ann. Chim. (Paris), 10 (1965)
0213 M. Hamon
0583 R. Bloch

7- -64, Ann. chim. (Rome), 54 (1964)
0080 A. Fravolini, G. Grandolini and G. Monzali
0128 A. Arcoria and G. Scarlata
0170 R. Nicoeletti and L. Baiocchi
0180 L. Pentimalli and P. Bruni
0206 M. DiFonzo, R. Giuliano and R. Maddalena
0462 D. DalMonte, E. Sandri and C. Brizzi
0476 C. Brizzi, D. DalMonte and E. Sandri
0530 G. DiModica and E. Barni
0685 A. Risaliti and S. Bozzini
0696 V. Carassiti and A. Seminara
0805 M. Covello, E. Abignente and A. Dini
0821 F. Bottino and A. Compagnini
0963 M. Piattelli, L. Minale and G. Prota
0969 R. Carpignano and R. Antonuccio
0987 F. Russo and M. Ghelardoni
1165 P. Vita-Finzi and M. Arbasino
1364 M. Ruccia, V. Sprio and S. Cusmano

7- -65, Ann. chim. (Rome), 55 (1965)
0100 N. Gallo
0143 R. Andrisano, A.S. Angeloni and M. Tramontini
0205 R. Modelli
0221 R. Modelli
0239 M. Covello, E. Abignente and A. Dini
0277 R. Sciaky and A. Consonni
0288 B. Camerino and R. Sciaky
0310 R. Modelli
0329 V. Carassiti et al.
0441 L. Pentimalli and S. Bozzini
0452 L. Baiocchi
0485 E. Perrotti et al.
0576 M. Grifantini and M.L. Stein
0583 G. Palazzo and G. Corsi
0652 R. Andrisano et al.

0763 T. Pozzo-Balbi, G. Romano and M.T. Bassi
0935 G. Palazzo and L. Baiocchi
0968 R. Andrisano et al.
1069 L. Chierici, C. Dell'Erba and D. Spinelli
1093 R. Andrisano, A.S. Angeloni and M. Tramontini
1154 M. Tramontini
1223 G. Tacconi and A. Perotti
1223 P.V. Finzi, P.L. Caramella and P. Grunanger
1252 D. Spinelli, C. Dell"Erba and G. Guanti

9- -65, Appl. Spectroscopy, 19 (1965)
0091 J.D. Margerum and J.A. Sousa

10- -64A, Arch. Biochem. Biophys., 104 (1964)
0111 P. Rainford
0156 M.C. Rebstock

10- -64B, Arch. Biochem. Biophys., 105 (1964)
None

10- -64C, Arch. Biochem. Biophys., 106 (1964)
0015 J.R. Pasqualini
0379 S.P. Bag, Q. Fernando and H. Freiser

10- -64D, Arch. Biochem. Biophys., 107 (1964)
0251 A.F.H. Anderson and M. Calvin

10- -64E, Arch. Biochem. Biophys., 108 (1964)
0334 G. Jommi, P. Manitto and M.A. Silanos
0375 J.H. Baxter
0510 F.L. Rodkey

10- -65A, Arch. Biochem. Biophys., 109 (1965)
None

10- -65B, Arch. Biochem. Biophys., 110 (1965)
0133 G. Cilento and D.L. Sanioto
0291 F.B. Jungalwala and J.W. Porter
0373 A. Kobayashi and H. Matsumoto
0444 M. Goto et al.

10- -65C, Arch. Biochem. Biophys., 111 (1965)
0713 H.T. Shigeura et al.

10- -65D, Arch. Biochem. Biophys., 112 (1965)
0076 G.L. Tong et al.
0313 R.A. Vitali et al.

11- -64, Arkiv Kemi, 22 (1964)
0211 A.B. Hörnfeldt
0281 B. Holmström

12- -64, Australian J. Chem., 17 (1964)
0047 R.F.C. Brown
0055 J.D. Brooks, R.A. Durie and H. Silberman
0066 P.K. Grant and N.R. Hill
0075 M.D. Sutherland
0109 A.L.J. Beckwith and L.B. See
0119 W.D. Crow and J.H. Hodgkin
0141 R.A. Eade, G. Kornis and J. Simes
0154 R.F.C. Brown
0246 B. Douglas et al.
0294 A.H. White et al.
0337 G.D.F. Jackson and W.H.F. Sasse
0353 G.M. Badger, P. Cheuychit and W.H.F. Sasse
0379 R.G. Cooke and I.D. Rae
0390 R.M. Carman and N. Dennis
0393 R.M. Carman
0440 J.E. Bolliger and J.L. Courtney
0447 R.F.C. Brown and I.D. Rae
0455 R.A. Jones and A.R. Katritzky
0461 L.H. Briggs and T. Cebalo
0464 D.H. Horn, Z.H. Kranz and J.A. Lamberton
0477 D.H. Horn and J.A. Lamberton
0489 H.H. Hatt
0567 D.J. Brown and T. Teitei
0578 G.V. Baddeley et al.
0661 D.J. Collins and J.J. Hobbs
0692 J.A. Lamberton
0765 R.B. Johns and A.B. Kriegler
0877 J.W. Clark-Lewis et al.
0894 R.A. Jones
0901 B.P. Moore
0934 C.A. Hanrick and P.R. Jefferies
1028 G.M. Badger, R.A. Jones and R.L. Laslett
1036 G.M. Badger, R.J. Drewer and G.E. Lewis
1138 G.M. Badger, J.K. Donnelly and T.M. Spotswood
1164 J.W. Clark-Lewis and V. Nair
1170 J.W. Clark-Lewis and I. Dainis
1174 L.K. Dalton
1245 G.W.K. Cavill and F.B. Whitefield
1260 G.W.K. Cavill and F.B. Whitefield
1270 J.H. Gough and M.D. Sutherland
1309 C.J. Moye
1406 J.L. Huppatz and W.H.F. Sasse
1418 R.O. Hellyer
1423 J. Cable, R.G. Coombe and T.R. Watson

12- -65, Australian J. Chem., 18 (1965)
0061 R.F.C. Brown, I.D. Rae and S. Sternhell
0070 G.M. Badger, J.A. Elix and G.E. Lewis
0108 G.S. Chandler, R.A. Jones and W.H.F. Sasse
0168 A. Meisters and J.M. Swan
0182 T.F. Low et al.
0190 G.M. Badger, N.C. Jamieson and G.E. Lewis
0199 D.J. Brown and T. Teitei
0206 J.L. Huppatz and W.H.F. Sasse
0218 R.G. Cooke and L.G. Sparrow
0226 M.N. Galbraith et al.
0337 R.W. Hay
0363 R.W. Guy and R.A. Jones
0379 G.M. Badger and R.P. Rao
0507 D.P. Graddon and E.C. Watton

0543 J.T. Pinhey and S. Sternhell
0559 D.J. Brown and T. Teitei
0575 H. Duewell
0691 D.B. Fox, J.R. Hall and R.A. Plowman
0731 R.F.C. Brown, S. Sternhell and R.N. Warrener
0763 D.D. Perrin and I.H. Pitman
0837 E.F.L.J. Anet
0875 R.A. Jones and J.A. Lindner
1049 D.J. Collins and J.J. Hobbs
1071 R.F.C. Brown and I.D. Rae
1079 J. Cable, R.G. Coombe and T.R. Watson
1111 M.J. Gallagher and M.D. Sutherland
1211 R.F.C. Brown, I.D. Rae and S. Sternhell
1221 L.M. Jackman, Q.N. Porter and G.R. Underwood
1227 J.M. Edwards and L.M. Jackman
1241 F.R. Hewgill, D.G. Hewitt and P.B. Langley
1279 G.B. Guise, E. Ritchie and W.C. Taylor
1331 J.R. Hall, M.R. Litzow and R.A. Plowman
1451 R.A. Eade et al.
1491 R. Hodges
1579 P.I. Mortimer and M.I. Strong
1775 T.G. Corbett, W. Davies and Q.N. Porter
1819 I.C. Calder and W.H.F. Sasse
1835 P.S. Clezy and A.W. Nichol
1865 J.T. Craig
1871 R.M. Dawson et al.
1977 P.S. Clezy and A.W. Nichol
1997 I.R.C. Bick and G.K. Douglas
2005 C.A. Henrick and P.R. Jefferies
2015 E. Ritchie et al.
2021 E. Ritchie et al.

13- -64, Steroids, 3 (1964)
0001 J.C. Orr et al.
0013 F. Alvarez
0183 E.L. Shapiro et al.
0189 J.A. Hogg, B.J. Magerlein and J. Korman
0271 H. Mitsuhashi and T. Nomura
0337 S. Nakanishi
0391 W.R. Slaunwhite, Jr., L. Neely and A.A. Sandberg
0487 Y. Inouye and K. Nakanishi
0493 A.E. Hydron, J.N. Korzun and J.R. Moetz
0593 C.R. Engel and G. Bach
0631 D.M. Piatak and E. Caspi
0639 C.H. Robinson et al.

13- -64, Steroids, 4 (1964)
0001 O. Halpern et al.
0031 A.V. Zakharychev et al.
0041 M.P. Cava and P.M. Weintraube
0255 T. Nishina, Y. Sakai and M. Kimiura
0291 S. Rakhit and M. Gut
0305 K. Takeda et al.
0423 A.D. Cross et al.
0445 K.B. Bharucha and H.M. Schrenk
0463 R.W. Frank, G.P. Rizzi and W.S. Johnson
0595 J.P. Kutney and I.J. Vlattas
0645 V. Schwarz, M. Ulrich and K. Syhora
0713 K. Tori and E. Kondo
0729 N.N. Gaidamovich and I.V. Torgov
0801 J.N. Gardner et al.

13- -65, Steroids, 5 (1965)
0057 A.I. Laskin et al.
0345 K.J. Sax et al.
0399 N.J. Doorenbos and A.P. Schroff
0403 K.J. Sax, R.H. Evans, Jr., and C.E. Holmlund
0441 D.D. Evans and P.J. Palmer
0459 C.E. Holmlund et al.
0539 K.M. Sivanandiah and W.R. Nes
0557 A.D. Cross et al.
0585 A.D. Cross et al.
0615 M. Heller, R.H. Lenhard and S. Bernstein
0637 C.F. Hammer and R. Stevenson

13- -65, Steroids, 6 (1965)
0111 N.D. Hall and G. Just
0195 H.J.V. Molan et al.
0239 C. Djerassi and A.R. Vanhorn
0247 A.E. Hydorn, L.J. Lerner and J. Schwatz
0339 P.A. Desaulles et al.
0357 M. Okada and Y. Saito
0371 J.A. Edwards et al.
0397 A.D. Cross and I.T. Harrison
0409 T.B. Windholz, R.D. Brown and A.A. Patchett
0645 M. Okada and Y. Saito
0651 M. Okada and Y. Saito
0805 F. Winternitz and C.R. Engel
0845 R.Y. Kirdani

13- -65, Steroids, Mika Hyano suppl. (1965)
0023 J.M.H. Graves and H.J. Ringold
0185 Y. Kurosawa, H. Shimojima and Y. Osawa

15- -64, Biochem. Z., (1964)
None

15- -65, Biochem. Z., 342 (1965)
0040 H. Eggerer and I. Grumm
0190 T. Hopner and J. Knappe

17- -64, Boll. sci. fac. chim. ind. Bologna, 22 (1964)
0033 D. Dal Monte and E. Sandri
0041 D. Dal Monte and E. Sandri
0048 E. Animali, D. Dal Monte and E. Sandri

17- -65, Boll. sci. fac. chim. ind. Bologna, 23 (1965)
0021 I. Degani et al.

0079 P.E. Todesco, A. Trombetti and P. Vivarelli
0151 I. Degani, R. Fochi and G. Spunta
0165 I. Degani, R. Fochi and G. Spunta
0203 G. Adembri and P. Tedeschi
0255 G. Adembri et al.
0405 R. Danielli and G. Maccagnani

20- -64, Bull. soc. chim. Belges (1964)
0076 L. De Borger et al.
0081 N. Schamp and M. Verzele
0275 A. Lepoivre et al.
0483 M. Renson
0491 M. Renson and R. Collienne
0532 G. Rosseels
0585 I. Flament, R. Promel and R.H. Martin
0628 M. Vandewalle
0703 M. Barbieux and R.H. Martin
0741 M. Mandel and P. Decroly
0782 P. Van Brandt, E. de Hoffmann and A. Bruylants
0843 P. Van Brandt and A. Bruylants

20- -65, Bull. soc. chim. Belges (1965)
0091 G. Rosseels
0136 J. Indeherbergh and A. Bruylants
0170 M. Denayer-Tournay et al.
0253 J.C. Braekman et al.
0329 L. Delcambe
0344 V. de Smedt and A. Bruylants
0518 J. Charette et al.
0534 G. Lhoest et al.
0591 J. D'Souza and A. Bruylants
0609 E. de Hoffmann and A. Bruylants
0629 M. Anteunis et al.

22- -64, Bull. soc. chim. France, Series 5, (1964)
0101 S. David and P. Regent
0161 P. Caubère
0204 A. Casadevall, G. Cauquil and R. Corriu
0225 N. Thoai
0301 R. Granger and J.P. Girard
0321 S. Julia and C. Moutonnier
0331 S. Julia nd P. Simon
0348 A. Ahond et al.
0376 M.S. Adjangba
0392 W.I. Taylor et al.
0403 M. Koch et al.
0500 J. Bourdais et al.
0550 F. Weiss and R. Rusch
0584 Y. Tanahashi et al.
0651 M. Julia and J. Pilard
0693 A. Jung and M. Brini
0729 S. Munavalli and G. Ourisson
0761 J. Schmitt et al.
0817 M. Pais et al.
0844 B. Waegell and C.W. Jefford
0868 A.H. Etemadi, R. Okuda and E. Lederer
0906 A.B. Font
0979 S. Julia and C. Moutonnier
0989 P. Cagniant, M. Mennrath and D. Cagniant
1038 J. Chopin et al.
1276 J. Romo et al.
1818 J. Polonsky et al.
1957 J.-M. Conia and J. Salaun
1968 J.-M. Conia and J. Gore
1976 J.-M. Conia et al.
2020 J. Levisalles and H. Rudler
2124 J. Bourdais
2169 Q. Khuong-Huu et al.
2236 A. Lablache-Combier and J. Levisalles
2245 B. Lacoume and J. Levisalles
2258 E. Elkik
2533 M. Julia et al.
2541 M. Julia and C. Descoins
2683 J. Poisson et al.
2868 J. Metzger et al.
2953 J.-J. Godfroid
3218 S. Julia, M. Julia and P. Graffin

22- -65, Bull. soc. chim. France, Series 5, (1965)
0018 J. Riess
0029 J. Riess
0035 M. LeQuan and P. Cadiot
0111 S. Labadum et al.
0227 B. Decouvelaere et al.
0315 C. Lumbroso and P. Maitte
0393 P. Lochon
0525 A. Resplandy
0587 A. Jung and M. Brini
0657 D. Herlem-Gaulier et al.
0704 J. Kossanyi
0714 J. Kossanyi
0722 J. Kossanyi
0779 H. Pacheco and A. Grouiller
0933 J. Riess and G. Ourisson
1295 F. Lederer and G. Ourisson
1464 P. Demerseman et al.
1518 L. Vo-Quang and P. Cadiot
1525 L. Vo-Quang and P. Cadiot
1538 J.-C. Jacquesy and J. Levisalles
1648 R. Panico and O. Pouchot
1895 Y. Israeli
2198 J.E. Dubois and C. Moulineau
2306 S. David and J.-C. Fischer
2342 Y. Lepage and O. Pouchot
2345 J. Rouzaud et al.
2489 J.-P. Coat et al.
2513 P. Mamont et al.
2588 P. Francois et al.
2602 B. Lacoume and J. Levisalles
2635 M. Sy and M. Maillet
2724 L. Lang et al.
2747 J.-M. Conia and J. Salaun
2751 J.-M. Conia and J. Salaun
2755 J.-L. Ripoll and J.-M. Conia
2793 J. Polonsky and N. Bourguignon-Zylber
2937 H. Pacheco and A. Grouiller
2988 J.-J. Basselier et al.
3047 J. Rigaudy and P. Derible
3061 J. Rigaudy and P. Derible
3136 R. Ramasseul and A. Rassat
3273 R. Briere, H. Lemaire and A. Rassat

3290 R. Briere et al.
3350 J. Chopin and P. Durual
3476 M.J. Wiemann et al.
3511 H. Piotrowska et al.
3516 A.B. Font
3518 J.-C. Masson and P. Cadiot
3572 J. Chopin, P. Durual and M. Chadenson
3643 R.-M. Dupeyre, A. Rassat and P. Rey
3718 M. Fetizon and M. Moreau

23- -64, Can. J. Chem., 42 (1964)
0010 A.F. McKay et al.
0025 L.R.C. Barclay and R.A. Chapman
0079 G. Just and C.C. Leznoff
0113 J.M. Pepper and M. Saha
0189 P. Ziegler and A.A. Amos
0190 J. Parrick
0298 A.G. Brook and J.B. Pierce
0371 M.C. Fallona et al.
0456 G. Just and K. St.C. Richardson
0464 G. Just and K. St.C. Richardson
0491 J.J. Dugan, P. de Mayo and A.N. Starratt
0579 L.R.C. Barclay, B.A. Ginn and C.E. Milligan
0591 J.P. Kutney and A. By
0698 J.P. Kutney et al.
0712 O.E. Edwards and K.K. Purushothaman
0724 T.S. Sorensen
0764 Y. Tsuda and L. Marion
0836 B.T. Newbold and D. Tong
0856 P.R. Murthy and C.C. Patel
1073 R. Bonnett, K.S. Chan and I.A.D. Gale
1279 H.J. Anderson and L.C. Hopkins
1500 M.R. Kamal and J.E. Wicklatz
1524 R.S. Atkinson and E. Bullock
1595 T.C. McMorris and M. Anchel
1599 C.J. Abshire and L. Berlinguet
1605 B.M. Lynch and Y.-Y. Hung
1664 E.W. Warnhoff
1681 R. Stewart and J.P. O'Donnell
1760 K.H. Palmer
1901 P. Singh and L. Berlinguet
1917 W.A.E. McBryde
1928 S. Jerumanis and J.M. Lalancette
1957 K. Yates, J.B. Stevens and A.R. Katritzky
2153 G. Just and V. DiTullio
2201 J.E. Bloor et al.
2362 H. Morita
2456 S.N. Alam, K.A.H. Adams and D.B. MacLean
2580 O.H. Wheeler et al.
2665 W.I. Awad and A. Boulos
2674 A. Balasubramanian, J.B. Capindale and W.F. Forbes
2695 G. Just and V. DiTullio
2768 T.S. Sorensen
2781 T.S. Sorensen
2806 G.W. Kosicki, S.N. Lipovac and R.G. Annett
2822 G. Kornis and P. de Mayo
2852 L.G. Humber et al.
2900 B. Robinson
2919 C.C. Leznoff and G. Just

23- -65, Can. J. Chem., 43 (1965)
0190 H.H. Baer and F. Kienzle
0306 J.S. Chadha
0332 E. Bullock and B. Gregory
0356 R.J. Crawford and R. Raap
0409 H.J. Anderson and S.-F. Lee
0470 P.M. Maitlis et al.
0679 X.A. Dominguez et al.
0700 R.J. Alaimo and D.G. Farnum
0792 D.N. Gupta, G. Schilling and G. Just
0862 E. Buncel and B.T. Lawton
1175 A.G. Brook, R. Kivisikk and G.E. LeGrow
1225 R. Stewart, J.P. O'Donnell and K. Bowden
1345 J.L.R. Williams et al.
1382 R.K. Blackwood and C.R. Stephens
1448 L.R. Melby
1527 N.A. Nelson and J.M. Schuck
1607 C. Ainsworth
1754 L.R.C. Barclay and R.A. Chapman
1835 P.V. Divekar et al.
2180 R.H.F. Manske and K.H. Shin
2183 R.H.F. Manske et al.
2306 P. Yates and J.R. Lynch
2328 K. Yates and J.C. Riordan
2408 A.M. El-Abbady and S.H. Doss
2426 R.P. Mariella and T. Conway
2512 S.S. Kulp, V.B. Fish and N.R. Easton
2594 E.W. Warnhoff, S.K. Pradhan and J.C.N. Ma
2603 A. Balasubramanian et al.
2685 V. Krishnan and C.C. Patel
2717 R. Bonnett, S.C. Ho and J.A. Raleigh
2744 T.S. Sorensen
2919 P.M.G. Bavin
2996 T.A. Davidson et al.

24- -64, Chem. Ber., 97 (1964)
0005 L. Hainisch, W. Ozegowski and M. Mühlstädt
0029 C. Jutz and F. Voithenleitner
0140 H. Dannenberg and T. Köhler
0172 G. Neurath and E. Doerk
0186 T. Severin and M. Adam
0212 M. Zander and W.H. Franke
0218 K.F. Lang and M. Zander
0304 M. Zander and W.H. Franke
0307 J.-C. Salfeld and E. Bawme
0363 E. Biekert and T. Funck
0372 M. Avram et al.
0494 K. Lang, H. Buffleb and J. Kalowy
0520 F. Bohlmann and U. Hinz
0533 H. Wolf, E. Bunnenberg and C. Djerassi
0549 H. Oediger and K. Eiter
0557 H.-J. Teuber et al.

0588 W. Sandermann and M.H. Simatupang
0598 H.-G. Viehe
0602 H.-G. Viehe and E. Franchimont
0654 R. Mayer et al.
0725 G. Henseke and H.-W. Pelz
0794 F. Bohlmann et al.
0801 F. Bohlmann et al.
0809 F. Bohlmann et al.
0815 H. Marxmeier and E. Pfeil
0958 G. Quinkert et al.
1127 E. Uhlig and K. Doering
1176 F. Bohlmann and K. Prezewowsky
1179 F. Bohlmann et al.
1266 G. Opitz and F. Zimmermann
1298 R. Mayer and B. Gebhardt
1318 W. Tochtermann et al.
1329 W. Tochtermann et al.
1337 C. Jutz and F. Voithenleitner
1349 C. Jutz
1354 F. Bohlmann and O. Schmidt
1373 E. Tenor and C.-F. Kröger
1414 H. Bredereck and R. Bangert
1453 P.R. Shah and N.M. Shah
1482 M. Regitz and G. Heck
1548 W. Flitsch
1590 C. Jutz and F. Voithenleitner
1625 S. Dähne and H. Paul
1631 G. Neurath, B. Pirmann and M. Dünger
1662 L. Hörhammer et al.
1666 L. Farkas et al.
1732 H. Behringer, M. Ruff and R. Wiedenmann
1760 R. Kuhn, H. Fischer and H. Fischer
1770 F. Bohlmann and C. Rufer
1799 G. Quinkert, E. Blanke and F. Homburg
1811 W. Adam
1839 F. Bohlmann and W. Sucrow
1846 F. Bohlmann and W. Sucrow
1910 F. Fischer and W. Arlt
1926 E. Hecker and E. Meyer
1940 E. Hecker
1952 E. Winterfeldt
1959 E. Winterfeldt
1963 F. Korte and F.-F. Wiese
1970 F. Korte and H. Wamhoff
2037 R. Hüttel, H. Dietl and H. Christ
2050 C. Jutz
2109 F. Bohlmann and E. Bresinsky
2118 F. Bohlmann, R. Enkelmann and W. Plettner
2125 F. Bohlmann, K.-M. Kleine and C. Arndt
2135 F. Bohlmann, H. Bornowski and K.-M. Kleine
2409 F. Micheel and I. Dijong
2463 E. Winterfeldt
2652 H.P. Kaufmann and A.K. SenGupta
2689 D. Martin
2700 E. Klein and W. Rojahn
2713 G. Neurath and M. Dünger
2732 R.M. Srivastava and M.P. Khare
2785 E. Graf and E. Dahlke
2857 L. Hörhammer et al.
2860 K.F. Lang and M. Zander
2884 R. Huisgen, G. Binsch and H. Konig
2949 R. Criegee and H. Furrer
2953 D. Seebach
2959 H. Fischer and H. Fischer
2975 H. Fischer and H. Fischer
3131 G. Schröder
3140 G. Schröder
3322 A. Roedig, F. Hagedorn and G. Märkl
3331 C. Jutz and H. Amschler
3363 K. Irmscher et al.
3452 J. Knabe and H. Roloff
3469 F. Bohlmann, K.-M. Kleine and C. Arndt

24- -65, <u>Chem. Ber.</u>, <u>98</u> (1965)
0024 K. Eckardt
0036 M. Regitz
0046 A. Mondon et al.
0114 W. Heimann and H. Bar
0140 W. Lüttke and J. Grussdorf
0155 F. Bohlmann et al.
0164 L. Farkas and M. Nogradi
0369 F. Bohlmann, K.-M. Kleine and H. Bornowski
0387 R. Criegee and M. Krieger
0428 W. Theilacker, K. Albrecht and H. Uffmann
0471 G. Wittig and J. Weinlich
0548 L. Hörhammer et al.
0557 C. Szantay and J. Rohaly
0567 H. Gnichtel
0588 M. Zander and W.H. Franke
0593 K.F. Lang et al.
0597 K.F. Lang and M. Zander
0623 P. Grunanger, P.V. Finzi and C. Scotti
0653 F. Bohlmann et al.
0666 H.-J. Teuber and G. Steinmetz
0685 F. Bickelhaupt, K. Stach and M. Thiel
0743 R.K. Erünlü
0764 H.-D. Scharf and F. Korte
0781 E. Däbritz and A.I. Virtanen
0829 R. Mayer and S. Scheithauer
0844 F.A. Neugebauer and P. Fischer
0864 G. Hilgetag, H. Teichmann and M. Krüger
0872 F. Bohlmann and K.-M. Kleine
0883 F. Bohlmann and E. Berger
1013 C. Szantay and L. Szabo
1023 C. Szantay and L. Szabo
1031 W. Kampe
1060 G. Nübel and W. Pfleiderer
1164 T. Okuda and H. Zahn
1188 R. Tschesche et al.
1225 F. Bohlmann et al.
1228 F. Bohlmann et al.
1246 L. Horner and K.-H. Weber
1282 O. Christmann
1374 R. Gompper, E. Kutter and R.R. Schmidt
1404 H.H. Stroh, A. Arnold and H.-G. Scharnow

1411 F. Bohlmann et al.
1416 F. Bohlmann and C. Arndt
1427 W. Ziegenbein
1470 H.-G. Lehmann, H. Müller and R. Wiechert
1476 R. Huisgen, E. Aufderhaar and G. Wallbillich
1511 W. Pfleiderer and R.K. Robins
1514 B. Franck and I. Zimmer
1556 J. Goerdeler and K. Stadelbauer
1562 H. Musso and H. Schröder
1581 E. Winterfeldt
1588 H. Stroh and H. Scharnow
1616 F. Bohlmann et al.
1647 H. Gnichtel and W. Lautsch
1727 T. Wieland and D. Grimm
1736 F. Bohlmann et al.
1753 R. Hüttel and H. Dietl
1774 H. Plieninger et al.
1837 S. Huneck
1858 H. Dürr, G. Ourisson and B. Waegell
1949 W. Heimann and A.N. Sagredos
1961 K. Schreiber et al.
1974 R. Tschesche et al.
1988 G. Etzold and P. Langen
2111 H.-J. Teuber and D. Cornelius
2174 R. Huisgen et al.
2185 R. Huisgen and E. Aufderhaar
2193 H. Rösler, T.J. Mabry and J. Kagan
2201 H. Prinzbach et al.
2236 F. Bohlmann, H. Bornowski and C. Arndt
2273 H. Bock and G. Rudolph
2284 S. Huneck
2291 S. Huneck
2327 R. Criegee, H. Hofmeister and G. Bolz
2331 R. Criegee et al.
2339 R. Criegee et al.
2361 H. Hoffmeister et al.
2383 C. Rufer et al.
2438 G. Maier
2579 E. Winterfeldt and P. Strehlke
2596 F. Bohlmann et al.
2605 F. Bohlmann et al.
2608 F. Bohlmann and G. Grau
2611 R. Kuhn and D. Rewicki
2643 H.-J. Teuber and O. Glosauer
2738 J.N. Chatterjea and K.D. Banerji
2742 G. Quinkert, A. Moschel and G. Buhr
2762 C.H. Krauch and W. Metzner
2774 H. Musso and D. Bormann
2797 H. Brockmann and E. Wimmer
2814 M. Zander and W.H. Franke
2822 K.B. Prasad and S.C. Shaw
2825 R. Gompper and E. Kutter
2837 S. Huneck
2844 H. Bock and E. Baltin
2859 L. Hörhammer et al.
2939 H.-J. Teuber and O. Glosauer
2954 J. Goerdeler and H. Schenk
2966 R. Huisgen, C. Axen and H. Seidl
3010 F. Bohlmann et al.
3015 F. Bohlmann and A. Seyberlich
3020 U. Türck and H. Behringer
3045 E. Klein and W. Rojahn
3087 F. Bohlmann, D. Bohm and C. Rybak
3095 L. Heinisch, W. Ozegowski and M. Mühlstädt
3102 C.H. Krauch, S. Farid and G.O. Schenck
3124 K. Schilffarth and H. Zimmermann
3145 H. Brockmann, J. Niemeyer and W. Rode
3165 R. Wiechert and G. Schulz
3196 H. Behringer and K. Leiritz
3218 R. Kuhn and B. Schulz
3385 J.F.M. Oth et al.
3439 G. Schill
3508 H. Zinner and K. Peseke
3515 H. Zinner and K. Peseke
3537 E. Winterfeldt
3571 K. Gewald
3637 G. Köbrich and H. Fröhlich
3672 H.-D. Scharf and F. Korte
3680 E. Vogel and E.-G. Wyes
3838 R. Criegee and F. Zanker
3892 R.R. Sc-midt
4033 M. Zander and W.H. Franke

25- -64, Chem. and Ind.(London) (1964)
0033 R.H. Burnell, J.D. Medina and W.A. Ayer
0195 W.F. Little et al.
0235 R.H. Burnell, J.D. Medina and W.A. Ayer
0319 F. Fish, M. Qaisuddin and J.B. Stenlake
0366 D. McHale and J. Green
0368 T.A. Liss
0419 J. Shoji and S. Shibata
0459 B. Robinson and G. Spiteller
0542 R.F. Childs and A.W. Johnson
0546 J.R. Holker
0622 B. Meek, S.S. Szinai and D. Wallis
0709 W.H. Chang
0839 P.T. Izzo and A.S. Kende
0931 J.M. Bobbitt et al.
1063 H. Gilman, W.H. Atwell and G.L. Schwebke
1065 R.A. Finnegan and W.H. Mueller
1261 M. Barash
1264 L.A. Summers and D.J. Shields
1313 P. Bamfield et al.
1364 E.J. Reist et al.
1492 D.N. Hague and R.H. Prince
1524 R. Mukherjee and A. Chatterjee
1580 S.M. Albonico, A.M. Kuck and V. Deulofeu
1686 M. Vondracek and Z. Vanek
1800 S. Ghosal and B. Mukherjee
1801 C.K. Bradsher and D.F. Lohr, Jr.
1837 F.J. Allan and G.G. Allan
1861 M.O. Bagley, C.R. Smith, Jr., and I.A. Wolff
1865 J.S.E. Holker and J.G. Underwood
1917 A.J. Birch et al.
1954 G. Vogel

2020 M. Kawano et al.
2059 L.J. Gough

25- -65, Chem. and Ind. (London) (1965)
0088 S. Matsueda et al.
0182 A.M. Roe and J.B. Harbridge
0184 S. Masamune and N.J. Castellucci
0186 I. Iwai and H. Mishima
0270 M.M. Coombs
0350 E.R. Arroyo and J. Holcomb
0383 C.W.L. Bevan and D.E.U. Ekong
0424 P.K. Freeman and D.G. Kuper
0467 S. Sarel and E. Breuer
0562 S.H. Harper, A.D. Kemp and W.G.E. Underwood
0563 R. Waack and M.A. Doran
0648 N.J. Doorenbos and M.T. Wu
0650 N.J. Doorenbos and M.T. Wu
0694 I.R.C. Bick and G.K. Douglas
0767 R. Filler and F.N. Miller
0782 W.R. Boon
0794 E.J. Kupchik and J.A. Ursino
0795 X.A. Dominguez et al.
0847 D.H. Corr and E.E. Glover
0899 H. Minato and J. Nagasaki
1074 G. Aguilar-Santos
1183 B. Halpern and A.D. Cross
1260 B.K. Moza and J. Trojanek
1379 I. Maclean and R. Stevenson
1382 G.W.H. Cheeseman and B. Tuck
1425 R. Dowbenko
1561 E.J. Reist, D.F. Calkins and L. Goodman
1595 N.M. Mollov et al.
1629 B. Wilhelm, M. Stoll and A.F. Thomas
1729 R. Seltzer and W.J. Considine
1767 S. Swaminathan et al.
1837 R.A. Finnegan and D. Knutson
2064 M.P. Cava et al.
2065 M. Kimura, M. Kawata and M. Nakadate

27- -64, Chimia, 18 (1964)
0174 D. Felix, M. Stoll and A. Eschenmoser
0244 G. Schetty
0358 C.H. Eugster and P. Kuser

27- -65, Chimia, 19 (1965)
0208 G.M. Blackburn et al.
0242 A. Pugin and J.V.D. Crone
0294 U. Schwieter et al.
0325 E. Daltrozzo, G. Scheibe and J. Smits
0333 G. Eigenmann
0339 R. Wizinger
0538 D. Felix, P. Jacober and A. Eschenmoser

28- -64A, Compt. rend., 258 (1964)
0196 C. Hélène and P. Douzou
0207 J. Armand
0237 P. Miginiac and L. Miginiac
0243 J. Martel et al.
0585 J.-M. Saveant
0597 R. Granger et al.
0600 C. Faget, J.-M. Conia and E.H. Eschinazi
0609 A. Calvaire and R. Pallaud
0954 C. Prévost, J. Filippi and P. Grammaticakis
1259 J.-P. Quillet and J. Dreux
1262 P. Grammaticakis
1512 M. Fleury
1537 C. Charrier et al.
1541 H. Christol, D. Lafont and F. Plenat
1840 J. Redel and J. Roch
1844 H. Christol, M. Levy and Y. Pietrasanta
3059 S. David and A. Veyrières
3705 M. Mousseron et al.
3728 M. Julia and A. Guy-Roualt
3865 S. Deswarte and J. Armand
4579 H. Najer, J. Menin and J.-F. Giudicelli
4795 V-K. Long, N.P. Buu-Hoï and N.D. Xuong
4799 J. Rigaudy and J. Barcelo
5087 J. Loeb
5228 P. Rouiller and J. Dreux
5447 C. Dufraisse and Y. Lepage
5470 J. Tempé, H. Heslot and G. Morel
5614 R. Martin and F. Krausz
5669 R. Joly et al.
5873 C. Prévost and M. Fleury
5895 J. Royer and J. Dreux

28- -64B, Compt. rend., 259 (1964)
0402 H. Pacheco, A. Grouiller and A. Hourfar
0404 M. Bertrand and J. LeGras
0583 Y. Wormser and R. Contant
0594 M. Bertrand and C. Rouvier
0597 J.-L. Pousset and J. Poisson
0827 M. Bertrand, H. Reggio and G. Leandri
1146 G. Morel
1418 H. Gault and R. Bloch
1530 C. Rouvier and M. Bertrand
1649 R. Rambaud and B. Cheminat
2113 P. Grammataticakis
2453 L. Pichat, P. Dufay and Y. Lamorre
2466 J. Redel, J. Boch and S.-Y. Tchen
2841 O. Bagno and P. Bonet-Maury
2859 R. Carrié, R. Bougot and B. Potteau
2868 H. Najer et al.
3780 J.-P. Quillet and J. Dreux
3872 Raymond-Hamet
4054 M. de Botton
4035 C. Dufraisse and Y. Lepage
4061 G. Rio and A. Ranjon
4295 P. Grammaticakis
4387 C. Hélène and P. Douzou
4712 M. de Botton
4853 C. Hélène and P. Douzou

28- -65A, Compt. rend., 260 (1965)
0209 M. Bertrand and C. Rouvier
0337 J. Parello and S. Munavalli
0606 J. Elguero and R. Jacquier
0610 J. Brigando and D. Colaitis
1440 S. Julia and C. Papantoniou
2259 G. Cauquis, G. Reverdy and M. Rastoldo
2827 P. Souchay and M. Fleury
2833 R. Dabard et al.
2839 M. Hedayatullah and L. Denivelle
2847 G. Pourcelot
2851 M. Mousseron-Canet and J.-P. Boca
3102 J. Hamelin et al.
3212 L.-A. Pradel et al.
3425 J.-P. Bégué and M. Fétizon
3688 H. Bouas-Laurent
3931 J. Marrot, J.-C. Maire and J. Cassan
4538 J.-F. Giudicelli, J. Menin and H. Najer
4545 G. Nomie et al.
4783 M. de Botton
5031 C. Dufraisse et al.
5582 J. Chopin and G. Dellamonica
5783 F. Terrier, P. Pastour and R. Schaal
6379 P. Souchay and S. Deswarte
6479 A. Resplaudy
6922 H. Bouas-Laurent and R. Lapouyade
6930 A.V. Goudou and M. Blanchecotte
6933 R. Reynaud

28- -65B, Compt. rend., 261 (1965)
0464 O. Roussel-Périn et al.
0759 A. Kirrmann and C. Wakselman
0766 J. Menin, J.-F. Giudicelli and H. Najer
0972 J. Lemerle and S. Valladas-Dubois
1015 E. Elkik and P. Vaudescal
1026 A. Pavia et al.
1332 J. Hamelin and R. Carrié
1339 M. Pesson and D. Richer
1343 J. Elguero, R. Jacquier and C. Marzin
1551 P. Courtemanche and J.-C. Merlin
1695 A. Bezaguet and M. Bertrand
1983 J. Moulines and R. Lalande
1987 A. Foucaud and A. Robert
2339 A. Macadré and C. Moncuit
2374 Dang Quoc Quan
2676 J.P. Almauge
3133 C. Dufraisse et al.
3420 F. Larèze
4139 F. Khuong-Huu-Lainé et al.
4776 J. Hamelin
4901 M. Devys and M. Barbier
4911 M. Dubost et al.
5237 N. Bisset
5487 C. Leibovici and J. Deschamps
5538 J.-J. Pousset et al.

29- -63A, Discussions Faraday Soc., 35 (1963)
0043 R.E. Ballard, S.F. Mason and G.W. Vane
0175 J.C.D. Brand et al.
0184 J.C.D. Brand and D.G. Williamson
0192 K.K. Innes and L.E. Giddings

29- -63B, Discussions Faraday Soc., 36 (1963)
0153 E. Collinson, J.J. Conlay and F.S. Dainton

30- -64, Doklady Akad. Nauk S.S.S.R. (1964) (English translation pagination)
0277 V.I. Litvinenko
0424 N.A. Nesmeyanov et al.
0734 F.P. Sidel'kovskaya and A.A. Avetisyan
1101 V.V. Ershov and G.A. Nikiforov
1135 N.N. Vorozhtsov et al.
1164 M.E. Perel'son et al.

30- -65, Doklady Akad. Nauk S.S.S.R. (1965) (English translation pagination)
0031 V.O. Lukashevich and E.S. Lisitsyna
0644 P.K. Yuldashev and S.Y. Yunusov
0815 Y.I. Naumov and V.A. Izmail'skii
0833 S.A. Hillers et al.
0939 G.Y. Kondrat'eva and H. Chih-heng
0969 R.G. Kostyanovskii and A.K. Prokof'ev

31- -64, Experientia, 20 (1964)
0128 H.R. Gutman
0208 C. Coronelli, D. Kluepfel and P. Sensi
0249 H.J. Siemann, W. Pohnert and S. Schwarz
0256 E.L. Patterson, W.W. Andres and R.E. Hartman
0336 W. Oppolzer, V. Prelog and P. Sensi
0344 D. Arigoni et al.
0363 M. Ohashi et al.
0380 K. Bernauer
0418 H. Smith, G.H. Douglas and C.R. Walk
0490 L. Bernardi et al.
0562 J.D. Coussio
0668 K. Fukui and M. Nakayama

31- -65, Experientia, 21 (1965)
0162 F. Barbatschi et al.
0189 J. Koziol
0372 M.P. Kunstmann and L.A. Mitscher
0374 K. Bernauer, G. Englert and W. Vetter
0425 D.R. Babin et al.
0432 K.L. Agarwal and M.M. Dhar
0434 G.E. Risinger, P.N. Parker and H.H. Hsieh
0499 A. Schubert and S. Schwarz
0500 S. Schwarz, H.J. Siemann and W. Pohnert
0563 J. Tadanier
0566 R.R. Arndt and C. Djerassi
0575 F. Kaiser
0617 R. Zelnik and F. Strehlau
0688 A. Schubert and S. Schwarz

32- -64, Gazz. chim. ital., 94 (1964)
0091 G.F. Bettinetti, G. Desmoni and F. Grunanger
0203 I. Degani, R. Fochi and C. Vincenzi
0252 L.M. Vallarino and G. Santarella
0328 S. Corsano and C. Iavarone
0393 G. Gaudiano, P. Bravo and A. Ricca
0485 L. Caglioti, G. Cianelli and A. Selva
0584 G. DiMaio and P.A. Tardella
0590 G. DiMaio and P.A. Tardella
0675 C. Gandolfi and P. De Ruggieri
0902 L. Pentimalli
0915 V. Bertini et al.
1054 L. Musajo et al.
1073 G. Rodighiero et al.
1108 L. Mangoni and M. Delardini
1137 N. Marziano and R. Passerini
1221 G. Casnati et al.
1248 F. Piozzi and A. Umani-Ronchi
1287 G. Favini
1301 C. Gallina et al.
1342 F. D'Angeli, C. Di Bello and V. Giormani
1378 S. Corsano and A. Piantelli
1438 V. Tortorella and T. Toscano

32- -65, Gazz. chim. ital., 95 (1965)
0033 G.F. Bettinetti and L. Capretti
0083 R. Nicoletti and M.L. Forcellese
0127 G. Lo Vecchio et al.
0138 L. Canonica et al.
0151 G. Jommi, P. Manitto and C. Scholastico
0159 P. Manitto, F. Pelizzoni and C. Scholastico
0206 G. Lo Vecchio et al.
0257 P. De Ruggieri et al.
0311 C. Cardani et al.
0320 G. Cardillo, L. Merlino and R. Mondelli
0338 P. De Ruggieri
0351 R. Gardi and R. Vitali
0455 P. De Ruggieri, C. Gandolfi and U. Guzzi
0513 G. Caporale et al.
0533 S. Cabani and G. Conti
0546 F. Sparatore and G. Pirisino
0693 F. Bottino and G. Purrello
0699 G. Purrello
0725 G. Cardillo et al.
0735 F. D'Angeli, C. Di Bello and V. Giorman
0786 R. Cameroni et al.
0814 F. Piozzi, A. Umano-Ronchi and L. Merlini
0831 G. Cignorella, L. Mariani and E. Testa
0948 A. Corbella et al.
1078 G. Purrello
1115 G. Cainelli, S. Morrocchi and A. Quilico
1130 V. Balzani
1322 G. Galiazzo
1371 R. Mondelli and L. Merlini
1478 G. Renzi and V. Dal Piaz

33- -64, Helv. Chim. Acta, 47 (1964)
0033 H.U. Daeniker
0051 Y.-R. Naves and P. Ochsner
0066 J. Gmunder and A. Lindenmann
0078 W. Keller-Schierlein and G. Roncari
0185 H. Hauth and D. Stauffacher
0319 E. Demole
0358 C. Kump, J. Seibl and H. Schmid
0390 B.P. Vaterlaus et al.
0408 E. Sundt, B. Wilhelm and M. Stoll
0558 C.A. Grob and P.W. Schiess
0567 P.-Y. Blanc, A. Perret and F. Teppa
0725 P.-Y. Blanc, A. Perret and F. Teppa
0756 P.A. Stadler et al.
0769 P. Bosshard et al.
0799 K. Brenneisen et al.
0827 C. Djerassi et al.
0873 E. Felder, D. Pitre and E.B. Grabitz
0878 M. Hesse et al.
0942 C. Moussebois and J.F.M. Oth
0968 D. Stauffacher
1022 M. Neunschwander, D. Menche and H. Schaltegger
1047 P. Baudet, M. Calin and E. Cherbuliez
1052 H. Kobel, E. Schreier and J. Rutschmann
1117 M. Gerecke and A. Brossi
1147 B.W. Bycroft et al.
1203 A. von Wartburg, M. Kuhn and H. Lichti
1211 D. Meuche et al.
1234 H. Linde
1265 W. Ludwig and G. Wittmann
1354 K.H. Dudley et al.
1385 C.A. Grob, W. Schwarz and H.P. Fischer
1401 H.P. Sigg
1424 H. Dahn and H. Hauth
1459 E. Simonitsch et al.
1484 F. Gerson
1497 W.G. Kump et al.
1529 E. Schreier
1581 P. Dietrich et al.
1766 E. Demole
1775 J.M. do Nascimento et al.
1852 H. Bruderer et al.
1860 H. Dahn and H. Moll
1911 P.A. Stadler et al.
1921 W. Schlientz et al.
1961 D. Hauser et al.
1986 J. Schmutz, F. Hunziker and W. Michaelis
2017 E. Hardegger et al.
2072 A. Walser and C. Djerassi
2089 A. Brossi, F. Schenker and W. Leimgruber
2119 K. Bernauer
2122 K. Bernauer
2164 P. Mühlradt, E. Weiss and T. Reichstein
2186 D. Stauffacher and H. Tscherter

2195 M. Viscontini et al.
2211 B.K. Manukian
2217 M. Schüpbach and Ch. Tamm
2226 M. Schüpbach and Ch. Tamm
2234 J. Gutzwiller et al.
2330 U. Meyer, E. Weiss and T. Reichstein

33- -65, Helv. Chim. Acta, 48 (1965)
0010 G. Ohloff and G. Uhde
0065 H. Kaufmann, W. Wehrli and T. Reichstein
0094 R. Barner et al.
0119 J. Winkler and W. Jenny
0157 J. Gutzwiller and Ch. Tamm
0252 M.E. Wilcox et al.
0308 U. Renner and H. Fritz
0361 H. Wyler, M.E. Wilcox and A.S. Dreiding
0433 H. Sigel and H. Brintzinger
0437 H. Hopff and G. Komany
0460 R.D. Studer and W. Lergier
0471 D. Karanatsios and C.H. Eugster
0509 H. Hopff and A. Gati
0517 M. Syz and H. Zollinger
0538 E. Widmer et al.
0608 M. Roth et al.
0649 G.R. Duncan, E. Weiss and T. Reichstein
0689 M. Hesse et al.
0704 E. Altenburger, H. Wehrli and K. Schaffner
0764 M. Viscontini and S. Huwyler
0791 C.A. Grob and H.J. Lutz
0799 C.A. Grob and H.R. Kiefer
0822 M. Pinar et al.
0857 H.H. Sauer, E. Weiss and T. Reichstein
0927 R. Good, G.F.R. Mueller and C.H. Eugster
0955 H. Schaltegger et al.
0962 H.P. Sigg et al.
0989 H. Els et al.
1002 C. Kump, J. Seibl and H. Schmid
1079 B. Böhner et al.
1113 E.E. Flury, E. Weiss and T. Reichstein
1289 H. Hopff and A. Gati
1322 A. Hofmann, W. von Philipsborn and C.H. Eugster
1349 T. Kishi et al.
1395 H. Göth, P. Cerutti and H. Schmid
1494 F. Gerson et al.
1590 F. Hunziker and O. Schindler
1598 B.W. Bycroft, M. Hesse and H. Schmid
1634 R. Berthold, W. Wehrli and T. Reichstein
1725 W. Acklin et al.
1800 L. Chardonnens, B. Laroche and G. Gamba
1822 H.H.A. Linde
1911 W. Schmid and E. Angliker
1922 M.E. Wilcox, H. Wyler and A.S. Dreiding
1933 R.M. Dodson et al.
1945 H. Bruderer and A. Brossi
1957 Z.M. Khan, M. Hesse and H. Schmid
1988 M. Dvolaitzky and A.S. Dreiding
1999 B.K. Manukian

35- -64, J. Am. Chem. Soc., 86 (1964)
0078 C. Djerassi, P.A. Hart and E.J. Warawa
0085 C. Djerassi, P.A. Hart and C. Beard
0096 A.W. Burgstahler and L.R. Worden
0107 J.B. Hendrickson et al.
0117 A. Chandross and R. Kreilick
0249 T.J. Katz, M. Rosenberger and R.K. O'Hara
0269 C. Beard et al.
0288 S. Masamune
0291 S. Masamune
0433 M.J.S. Dewar and R.C. Dougherty
0449 K. Nagarajan, M.C. Caserio and J.D. Roberts
0453 E.H. White and H.C. Dunathan
0465 C. Djerassi et al.
0471 G. Stork and M. Tomasz
0478 E.J. Corey et al.
0498 O.L. Chapman and E.D. Hoganson
0503 M.S. Newman and J. Blum
0521 Y. Gaoni and F. Sondheimer
0525 R. Anet and F.A.L. Anet
0658 C.G. Overberger and J.P. Anselme
0661 J.K. Stille and R. Ertz
0673 D.G. Farnum, M.A.T. Heybey and B. Webster
0694 B. Gilbert et al.
0701 S.M. Kupchan and S.D. Levine
0708 A.G. Anderson, Jr., and W.F. Harrison
0720 J.A. Carbon
0729 M.F. Bartlett, B.F. Lambert and W.I. Taylor
0732 O.L. Chapman et al.
0736 E. Kondo and K. Tori
0746 J.C. Sheehan and I. Lengyel
0905 K.F. Bangert and V. Boekelheide
0908 A.L. Goodman and R.H. Eastman
0935 G. Stork and R. Borch
0936 G. Stork and R. Borch
0942 M.A. Battiste
0944 W.M. Jones and J.M. Denham
0951 E.C. Taylor and R.W. Hendess
0953 J.P. Kutney and E. Piers
0962 P.E. Eaton and T.W. Cole, Jr.
1125 J.C. Culling, M.J.S. Dewar and P.A. Marr
1127 B. Miller
1242 R.E. Holmes and R.K. Robins
1251 M.J. Robins, W.A. Bowles and R.K. Robins
1252 W.A. Bowles and R.K. Robins
1263 E.A. Chandross
1264 P.E. Brenneisen, T.E. Acker and S.W. Tannenbaum
1270 P.R. Story and S.R. Fahrenholtz
1389 D.E. Applequist and R. Searle

1434 H.E. Zimmerman and G.L. Grunwald
1444 J.M. Rice
1454 A.J. Waring and H.J. Hart
1457 P. Kohn, R.H. Samaritano and L.M. Lerner
1520 R. Hirschmann et al.
1528 M. Akhtar and D.H.R. Barton
1596 H. Gilman, S.G. Cottis and W.H. Atwell
1600 T.J. Katz and E.H. Golad
1626 G.I. Drummond et al.
1636 P.W. Wigler and H.U. Choi
1640 E.J. Corey and M. Chaykovsky
1641 Y. Gaoni and R. Mechoulam
1701 H. Schehter, S.S. Rawalay and M. Tubis
1710 K. Mislow et al.
1755 C. Djerassi and J.E. Gurst
1761 G. Stork and S.D. Darling
1830 E.C. Taylor and G.W.H. Cheeseman
1835 M.S. Newman and J. Blum
1848 S. Ishii and B. Witkop
1869 T.K. Liao, E.G. Podrebarac and C.C. Cheng
1876 N.A. LeBel, A.G. Phillips and R.N. Liiesemer
1884 H.S. Corey, Jr., et al.
1890 L.J. Dolby and S. Sakai
1891 W.A. Sheppard and J. Diekmann
1895 P. Brauchli et al.
1896 J.P. Horwitz et al.
1918 G.S. Hammond and J.R. Fox
1959 W.S. Johnson et al.
1966 W.S. Johnson et al.
1972 W.S. Johnson, W.H. Lunn and K. Fitzi
1997 S.K. Malhotra and H.J. Ringold
2025 P.A.S. Smith, L.O. Kibecheck and W. Reeseman
2038 E. Wenkert et al.
2073 G. Bozzato et al.
2083 E. Smith et al.
2086 W.E. Noland and R.F. Modler
2088 A.S. Kende and P.T. MacGregor
2091 A. Small
2095 J. Tsuji, M. Morikawa and N. Iwamoto
2149 H.E. Zimmerman and D.H. Paskovich
2174 H.D. Hartzler
2177 S.M. Kupchan and N. Yokoyama
2183 J.J. Brown, R.H. Lenhard and S. Bernstein
2252 J.C. Martin and R.G. Smith
2286 J.P. Kutney, R.T. Brown and E. Piers
2309 M. Heller, R.H. Lenhard and S. Bernstein
2315 R. Lovrien and J.C.B. Waddington
2384 W.D. Closson and P. Hang
2392 L.F. Fieser and M.J. Haddadin
2395 D.M. Lemal, F. Menger and E. Coats
2419 D.L. Webb and H.H. Jaffe
2451 K.S. Brown, Jr., and C. Djerassi
2490 W.G. Dauben and R.M. Coates
2591 H.R. Allcock
2623 H. Powell et al.
2645 S.L. Manatt et al.
2660 O.L. Chapman et al.
2724 P.W.K. Woo, H.W. Dion and Q.R. Bartz
2726 P.W.K. Woo, H.W. Dion and Q.R. Bartz
2736 R.K. Blackwood and C.R. Stephens
2738 J.J. Bloomfield and W.T. Quinlin
2811 N.L. Allinger and M.A. Miller
2819 R.F. Heck
2825 R.H. Shapiro and C. Djerassi
2832 R. Bengelmans et al.
2837 R.H. Shapiro et al.
2861 J.M. Ross and W.C. Smith
2884 G. Buchi and J.D. White
2898 O.W. Webster
2943 E.W. Cantrall, R.B. Conrow and S. Bernstein
2948 R. Shapiro
2957 K. Mislow, P. Schneider and A.L. Ternay, Jr.
2959 C.W.J. Chang, R.E. Moore and P.J. Scheuer
3056 H. Jadamus, Q. Fernando and H. Freiser
3068 W. Reusch and R. LeMahieu
3085 G. Smolinsky and B.I. Feuer
3137 L.L. Replogle
3142 W.C. Baid, Jr., and R.L. Shriner
3162 R.B. Woodward, T. Fukunaga and R.C. Kelly
3168 F. Sondheimer and A. Shani
3169 T.J. Katz and J. Schulman
3265 G.N. Schranzer and K.C. Dewhirst
3329 D.J. Bertelli, C. Golino and D.L. Dreyer
3375 W.L. Alworth, A.A. Liebman and H. Rapoport
3394 M. Akhtar, D.H.R. Barton and P.G. Samme
3397 G.I. Glover and H. Rapoport
3398 H.W. Moore
3796 R.L. Hinman and J. Lang
3814 E.F. Ullman and J.E. Milks
3874 R.C. Esse et al.
3877 G.R. Allen, Jr., J.F. Poletto and M.J. Weiss
3878 G.R. Allen, Jr., J.F. Poletto and M.J. Weiss
3891 Q.E. Thompson, M.M. Crutchfield and M.W. Dietrich
3896 J.A. Berson and M. Pomerantz
3898 D.T. Longone and H.S. Chow
3908 J.P. Dusza, J.P. Joseph and S. Bernstein
3973 W.H. Knoth et al.
4018 M.J. Kamlet, H.G. Adolph and J.C. Hoffsommer
4036 H.E. Zimmerman and J.W. Wilson
4053 P.J. Kropp
4074 J. Meinwald et al.
4080 J.E. Hodgkins et al.
4089 L.A. Paquette
4092 L.A. Paquette

4096 L.A. Paquette
4152 D.F. Veber and W. Lwowski
4162 R. Paul and A.S. Kende
4211 P.G. Gassman, D.H. Aue and D.S. Patton
4212 A. Padwa and R. Hartmann
4373 G.J. Karabatsos and R.A. Taller
4406 S.A. Bernhard, Y. Shabitin and Z.H. Tashjian
4414 K.S. Brown, Jr., and S.M. Kupchan
4424 K.S. Brown, Jr., and S.M. Kupchan
4434 S.M. Laiho and H.M. Fales
4438 G. Buchi, W.D. Macleod, Jr., and J. Padilla
4506 F.D. Marsh and M.E. Hermes
4507 S. Marumo, K. Sasaki and S. Suzuki
4509 J.E. Balwin et al.
4608 W.A. Remers
4612 W.A. Remers, R.H. Roth and M.J. Weiss
4628 L.A. Strait et al.
4631 G. Buchi, R.E. Manning and S.A. Monti
4729 E.M. Arnett and J.M. Bollinger
4876 T.J. Katz and P.J. Garratt
4898 J.L. Kice and N.E. Pawlowski
4917 V.J. Traynelis and P.L. Pacini
4928 W.R. Vaughan and R. Caple
4942 C.K. Hancock and A.D.H. Cague
4976 A.L. Burlingame, H.M. Fales and R.J. Highet
5032 R.F. Bleiholder and H. Schechter
5041 A.T. Blomquist and V.J. Hruby
5046 D.W. Boykin, Jr., and R.E. Lutz
5194 T.J. Katz and P.J. Garratt
5202 D. Valentine et al.
5207 A. Yogev, M. Gorodetzky and Y. Mazur
5213 M. Gorodetzky and Y. Mazur
5218 M. Gorodetzky, D. Amar and Y. Mazur
5244 T.L. Jacobs and R.A. Meyers
5277 J.C. Sheehan and R.M. Wilson
5281 A.W. Burgstahler et al.
5293 J. Blake, C.D. Willson and H. Rapoport
5307 N.J. Leonard and G.E. Wilson, Jr.
5320 L.B. Townsend et al.
5342 W.H. Urry, M.H. Pai and C.Y. Chen
5360 G.C. Brumlik, A.I. Kosak and R. Pitcher
5362 L.J. Dolly and S.I. Sakai
5364 C.G. Overberger, N. Weinshanker and J.P. Anselme
5511 K.M. Harmon, A.B. Harmon and F.E. Cummings
5515 E.M. Kosower and E.J. Pozionek
5542 J.E. Gurst and C. Djerassi
5548 R.T. LaLonde, S. Emmi and R.R. Fraser
5584 H. Gilman, S.G. Cottis and W.H. Atwell
5589 H. Gilman and W.H. Atwell
5593 W.S. Johnson and R. Owyang
5600 M.S. Newman and J. Blum
5637 K.K. Andersen et al.
5646 G. Buchi, E. Koller and C.W. Perry
5654 G. Buchi and G. Lukas
5664 G.T. Long, W.W. Lee and L. Goodman

35- -65, J. Am. Chem. Soc., 87 (1965)
0011 L.B. Clark and I. Tinoco, Jr.
0051 J.V. Burakevich and C. Djerassi
0086 A.T. Blomquist and C.G. Bottomley
0093 M. Gorman, N. Neuss and N.J. Cone
0134 M. Schach von Wittenau et al.
0136 T.S. Cantrell and H. Schechter
0275 G. Stork et al.
0321 W.E. Parham et al.
0384 A. Streitwieser, Jr., et al.
0528 T.W.G. Solomons, F.W. Fowler and J. Calderazzo
0539 K.M. Harmon and F.E. Cummings
0580 J. Karliner, H. Budziekievicz and C. Djerassi
0652 E. Ciganek
0676 A.H. Jackson et al.
0777 S.W. Pelletier and P.C. Parthasarathy
0817 C. Djerassi et al.
0863 D.Y. Curtin et al.
0874 D.Y. Curtin et al.
0882 T. Asao et al.
0933 H. Muxfeldt and W. Rogalski
0995 S.A. Latt, H.T. Cheung and E.R. Blout
1050 A.B. Turner et al.
1145 A.D. Broom and R.K. Robins
1149 E. Ciganek
1320 R. Breslow et al.
1326 R. Breslow et al.
1353 E.J. Corey and M. Chaykovsky
1364 D.B. Straus and J.R. Fresco
1381 H.G. Richey, Jr., J.C. Phillips and L.E. Rennick
1385 C.J. Sih et al.
1386 C.J. Sih et al.
1397 J. Blake, J.R. Tretter and H. Rapoport
1403 R.M. Pike and R.R. Luongo
1404 J. Ciabattoni and G.A. Berchtold
1410 G.W. Griffin et al.
1573 W.L. Meyer et al.
1580 E. Wenkert and B. Wickberg
1594 T.C. McMorris and M. Anchel
1608 W.M. Jones and R.S. Pyron
1609 A.S. Kende and P.T. Izzo
1623 P.R. Story and S.R. Fahrenholtz
1627 M.S. Silver
1739 H. Wynberg and H.J. Kooreman
1757 M.E. Warren, Jr., and H.E. Smith
1772 R.E. Holmes and R.K. Robins
1794 J.R.D. McCormick and E.R. Jensen
1795 J.J. Hlavka, P. Bitha and J.H. Boothe
1818 M.F. Hawthorne, D.C. Young and P.A. Wegner

1819 F.D. Marsh and F.E. Hermes
1857 H.E. Zaugg and A.D. Schaefer
1925 M.J. Goldstein and G.L. Thayer, Jr.
1941 E.A. LaLancette and R.E. Benson
1958 K. Mislow et al.
1976 E.C. Taylor and R.W. Morrison Jr.
1980 E.C. Taylor and R.W. Hendess
1995 E.C. Taylor and R.W. Hendess
2003 G.I. Glover, R.B. Smith and H. Rapoport
2050 E.M. Arnett, J.M. Bollinger and J.C. Sanda
2065 Y. Shimizu and S.W. Pelletier
2066 N. Takahashi, A. Suzuki and S. Tamura
2229 N. Finch et al.
2451 A.L. Nussbaum et al.
2510 A.E. Pohland et al.
2887 J.A. Berson et al.
2908 E.M. Evleth, Jr., et al.
3004 T. Money, I.H. Qureshi and G.B. Webster
3135 H. Hart and J.L. Corbin
3158 C.L. Osborn et al.
3186 T.M. Harris, S. Boatman and C.R. Hauser
3269 T.N. Margulis et al.
3273 R. Mechoulam and Y. Gaoni
3275 J. Tsuji and H. Takahashi
3283 N.C. Cook and J.E. Lyons
3365 P. Beak and J. Bonham
3484 G. Buchi, J.D. White and G.N. Wogan
3510 T. Hanafusa et al.
3530 B. Frydman, M.E. Despuy and H. Rapoport
3532 J. Meinwald, A. Eckell and K.L. Erickson
3638 R. Wolovsky
3651 W.J. Linn, O.W. Webster and R.E. Benson
3657 W.J. Linn and R.E. Benson
3665 W.J. Linn
3719 D.J. Bertelli and C.C. Ong
3727 L.H. Knox, E. Velarde and A.D. Cross
3737 N.K. Chaudhuri and M. Gut
3752 J.F. Gerster and R.K. Robins
4301 N.A. Lebel and R.N. Liesemer
4341 F.S. Fawcett and W.A. Sheppard
4365 A. Padwa
4373 M.A. Ondetti and P.L. Thomas
4393 S. Trofimenko
4506 L. Skattebol and S. Solomon
4533 N.C. Deno et al.
4576 B.T. Gillis and J.D. Hagarty
4601 N. Akhtar, D.H.R. Barton and P.G. Sammes
4629 R. Ginsig and A.D. Cross
4655 E. Kondo, T. Mitsugi and K. Tori
4656 M.G. Burdon and J.G. Moffatt
4794 H. Tanida and R. Muneyuki
4912 R.W. King, C.F. Murphy and W.C. Wildman
4934 M.J. Robins and R.K. Robins
4940 C.P. Whittle and R.K. Robins
4944 H. Achenbach and K. Biemann
4971 E.W. Garbisch, Jr.
4972 A.J. Hortmann
5075 T.S. Sorenson
5115 B. Miller
5132 R. Breslow et al.
5139 R. Breslow et al.
5148 W.S. Johnson, P.J. Neustaedter and K.K. Schmiegel
5186 L.A. Paquette
5198 S. Boatman, T.M. Harris and C.R. Hauser
5218 J. Meinwald and R.A. Schneider
5237 S.C. Cascon et al.
5256 T.W.G. Solomons and C.F. Voigt
5262 W.A. Remers and M.J. Weiss
5264 M.J. Jorgenson and C.H. Heathcock
5417 K.R. Huffmann et al.
5424 W.A. Henderson, Jr., and E.F. Ullman
5442 J.A. Montgomery and H.J. Thomas
5461 E. Wenkert, K.G. Dave and F. Haglid
5512 A.G. Anastassiou
5515 B. Miller
5607 M. Cais, A. Modiano and A. Raveh
5661 K.E. Pfitzner and J.G. Moffatt
5670 K.E. Pfitzner and J.G. Moffatt
5716 M. Rosenblum et al.
5720 R. Wolovsky and F. Sondheimer
5728 E.J. Corey and S. Nozoe
5736 E.J. Corey and A.G. Hortman
5791 W. Leimgruber et al.
5793 W. Leimgruber et al.
5804 L. A. Carpino and L.V. McAdams III
5805 S.M. Kupchan et al.

37- -64, J. Biol. Chem., 239 (1964)
0094 S. Hayakawa and B. Samuelsson
0560 C.C. Levy
1018 J. Ozols and P. Slrittmattev
1237 R.B. Martin and J.G. Hull
1284 S. Dagley and D.T. Gibson
1381 M.E. Dempsey et al.
2189 S.A. Narrod and W.B. Jakoby
2259 T. Shiota et al.
2267 B. Levenberg
2285 J.C. Ensign and S.C. Rittenberg
2865 M.D. Lane et al.
3407 W.M. Dairman and W.S. McNutt
3493 H. Wacker et al.
3964 W.S. Zaugg
3981 H. McKennis et al.
4272 W.S. McNutt and S.P. Damle

37- -65, J. Biol. Chem., 240 (1965)
0514 R.E. Olson et al.
0699 J.H. Phillips, S.A. Robrish and C. Bates
0722 H.R. Horton, H. Kelly and D.E. Koshland
0740 A. Ichiyama et al.
1011 C.C. Irving
1165 K.A. Schellenberg

1374 P.M. Scott
1941 K. Schimizu
2396 O. Berseus, H. Danielsson and A. Kallner
2491 J. Fridovich
3123 S.H. Cohen and K.L. Yielding
3264 P.S. Marfey et al.
3580 E.J. Williams and M. Laskowski
4176 M.E. Dempsey

38- -64A, J. Chem. Phys., 40 (1964)
0166 N. Christodouleas and S.P. McGlynn
0890 P.L. Kronick, H. Scott and M.M. Labes
2839 A. Heller
3749 H. Kuroda and H. Akamatu

38- -64B, J. Chem. Phys., 41 (1964)
0895 H. Baba, A. Matsuyama and H. Kokubun
0979 I.A. Taub et al.
1082 T. Pavlopoulos and M.A. El-Sayed
2003 R.B. Aust, G.A. Samava and H.G. Drickamer

38- -65B, J. Chem. Phys., 43 (1965)
2548 J.A. Millins, A.D. Adler and R.M. Hochstrasser
3393 J.M. Goodenow and M. Tamnes

39- -64, J. Chem. Soc. (1964)
0026 R. Hodges et al.
0042 D.W. Cameron et al.
0051 D.W. Cameron et al.
0062 D.W. Cameron et al.
0072 D.W. Cameron et al.
0080 A. Calderbank et al.
0114 J.A. Cade and A. Pilbeam
0141 G.B. Barlow and A.J. MacLeod
0249 D.A. Archer, H. Booth and P.C. Crisp
0289 J. Martin, W. Parker and R.A. Raphael
0387 M. Asgar Ali et al.
0411 M. Dean et al.
0438 E.A. Clarke and M.F. Grundon
0446 M.P.L. Caton et al.
0515 E.K. Weisburger and R.E. Boyd
0526 R.M. Acheson, R.S. Feinberg and A.R. Hands
0529 P.R. Enslin and K.B. Norton
0531 G. Valkanas and E.S. Waight
0543 S.T.D. Gough and S. Trippett
0586 E. Caspi and H. Zajac
0594 W.J. Conradie et al.
0688 D.H. Dewar et al.
0724 W. Carruthers and D.A.M. Watkins
0766 D.M. Green et al.
0751 H. Bauer et al.
0783 G.C. Barrett, V.V. Kane and G. Lowe
0788 G.C. Barrett et al.
0842 M.L. Dhar, V. Thalles and M.C. Whiting
0865 J.S. Pizey and W.E. Truce
0868 G.B. Barlow
0888 M. Elliott
0893 J.H. Atkinson, R. Grigg and A.W. Johnson
0906 W.V. Farrar
0915 D.J. Rabiger and M.M. Joullie
0928 C.W. Rees and C.E. Smithen
0938 C.W. Rees and C.E. Smithen
0952 J.S. Burton and R. Stevens
0991 R.J.S. Beer and J. Hollowood
1029 T.G. Halsall, D.W. Theobald and K.B. Walshaw
1045 G.J.F. Chittenden and R.D. Guthrie
1067 W. Baker, N.J. McLean and J.F.W. McOmie
1142 R. O'Dorachai, P.G. Flanagan and J.B. Thomson
1147 O.M. Behr et al.
1151 O.M. Behr et al.
1154 J. Eglinton et al.
1161 S. Binns et al.
1184 D.D. Evans, D.E. Evans and R.W.J. Williams
1190 A.G. Long and A. Tulley
1197 D.J. Austin and M.B. Meyers
1265 J.W. Barton and J.F. Thomas
1334 Z. Rappoport, P. Greenzaid and A. Horowitz
1423 J.M. Cox, J.A. Elvidge and D.E.H. Jones
1476 E.R.H. Jones, B.E. Lowe and G. Lowe
1500 J.M. Burgess and M.S. Gibson
1507 R.G. Micetich and J.C. MacDonald
1511 M.A. Kazi et al.
1622 J.W. Barton et al.
1632 E. Bullock et al.
1637 A.D. Campbell and A.R. Keen
1666 A. Albert and J. Clarke
1835 D.F. Jones, J.F. Grove and J. MacMillan
1850 A.D. Tait
1857 D.E.A. Rivett
1858 J.P. Dickinson, J. Harley-Mason and J.H. New
2146 D. Nasipuri et al.
2150 G.B. Barlin
2153 R. Robson, P.W. Grubb and J.A. Barltrop
2165 I. Flemin and J. Harley-Mason
2175 R.A. Abramovitch and J.G. Saha
2187 D.H.G. Crout et al.
2222 M.V. Sargent and C.J. Timman
2244 R.R. Arndt and W.H. Baaschers
2262 A.H. Jackson, P. Johnston and G.W. Kenner
2285 B.C. Elmers, M.P. Hartshorn and D.N. Kirk
2289 R.W. Bycroft, J.C. Roberts and P.M. Baker
2306 E. El Khadem et al.
2319 T.R. Emerson and C.W. Rees
2326 D.M. Hall and J.M. Insole
2380 M. Crawford and V.R. Supanekar

2421 D.A. Denton et al.
2465 D.F. Schneider and C.F. Garbess
2518 D.H.R. Barton, J.T. Pinkey and R.G. Wells
2571 A.L.J. Beckwith and L. Beng See
2579 M.K.A.Khan and K.J. Morgan
2633 S. Julia et al.
2640 A.J. Birch et al.
2657 C.H. Hassall and E.M. Wilson
2676 R.M. Acheson and A.O. Plunkett
2699 R.S. Dickson and G. Wilkinson
2709 A.J. Birch et al.
2760 A. Fozard and G. Jones
2763 A. Fozard and G. Jones
2816 I.A. Kaye, R.S. Matthews and A.A. Scala
2829 D.W. Russel
2830 M. Davis
2932 A.J. Birch, D.N. Butler and J.B. Siddall
2941 A.J. Birch, D.N. Butler and J.B. Siddall
3001 A.C. Day
3005 M.J. Perkins
3030 A. Fozard and G. Jones
3035 D.R. Sayers, R. Stephens and J.C. Tatlow
3043 R.C. Cookson, R.R. Hill and J. Hudec
3062 R.C. Cookson et al.
3097 R.I. Fryes, B.Brust and L.H. Sternbach
3106 R.K. Callow and G.A. Thompson
3114 D.H. Jones et al.
3126 J.S. Davier, C.H. Hassall and J.A. Schofield
3204 D.J. Brown and T. Teitei
3210 W.A. Harrison et al.
3221 J. Clarke et al.
3225 R.M. Acheson and D.M. Goodall
3229 R.M. Acheson et al.
3234 J.F. Grove
3239 D.C. Aldridge and J.F. Grove
3315 R. Grigg and A.W. Johnson
3357 A. Albert and E.P. Sergeant
3366 E.E. Glover and G.H. Morris
3388 C.W. Shoppee et al.
3392 C.W. Shoppee et al.
3459 F. Kurzer and E.D. Pitchfork
3484 J. Klein and E.D. Bergmann
3554 S.L. Mukherjee and P.-C. Dutta
3577 K. Takeda et al.
3611 C.W. Shoppee and R.E. Lack
3621 R. Villotti et al.
3635 A.F. Casy, N.J. Harper and J.R. Dimmock
3648 G. Chaudra et al.
3663 M.W. Partridge and M.F.G. Stevens
3670 M.W. Partridge et al.
3673 M.W. Partridge et al.
3816 J.S. Burton, J.A. Elvidge and R. Stevens
3856 T. Kametani and H. Sugahara
3822 B.R. Brown and J. MacBride
3841 R.E. Bowman et al.
3927 J.H. Bowie and A.W. Johnson
4004 J.L. Garraway
4008 J.L. Garraway
4011 F.E. King and J.G. Wilson
4035 A.J. Bellamy and G.H. Witham
4077 W. Carruthers and A.G. Douglas
4122 T. Kametani and K. Ogasawara
4154 M.S.R. Nais et al.
4157 J.A. Elvidge et al.
4167 A.J. Birch and D.N. Butler
4190 E.A. Clarke and M.F. Grundon
4212 W.J. Bowyer et al.
4254 K.J. Crowley
4257 A.R. Battersby
4274 G.R. Proctor
4310 R. Hodges et al.
4315 H. Hermann, R. Hodges and A. Taylor
4320 G.W. Kershaw and A. Taylor
4419 A.R. Battersby and D.A. Yeowell
4472 H. Smith et al.
4492 D. Hartley and H. Smith
4521 C.L. Graham and F.G. McQuillin
4565 D.W. Cameron et al.
4578 K. Takeda et al.
4591 W.A.W. Cummings and A.C. Davis
4613 E.P. White
4636 E.D. Andrews and W.E. Harvey
4769 A. Stuart, D.W. West and H.C.S. Wood
4782 R.D. Chambers and T. Chivers
4907 R.S. Davidson et al.
4920 J. Clark
4941 R. Bryant, G.H. Hassall and J. Weatherston
4972 W. Cocker and P.H. Boyle
4978 W. Davey and J.A. Hearne
4992 C.W. Shoppee, P.J. Havlicek and R.E. Lack
5002 S.D. Robinson and B.L. Shaw
5017 J. Burdon et al.
5074 J.R. Lewis and B.H. Warrington
5110 A.S. Bailey et al.
5130 A.F. Casy and H. Birnbaum
5135 M. Gawlk and R.F. Robbins
5161 J.W. Barton
5225 M. Elliott
5243 E.P. White
5302 T.A. Nour et al.
5317 P.B.D. De La Mare et al.
5343 A.J. Birch et al.
5378 B.J. Calvert and J.D. Hobson
5382 R.F. Curtis, P.C. Harries and C.H. Hassall
5488 J.M.H. Graves et al.
5503 H. Minato, S. Nosaka and I. Horibe
5510 A.H. Jackson and A.E. Smith
5544 M.V. Sarjent and C.J. Timmons
5569 D.W. Cameron and P.M. Scott
5573 A.B.A. Jansen, J.M. Johnson and J.R. Surtees
5617 E. Wenkert et al.
5640 L. Crombie and D.A. Mitchard
5646 A.K. Kiang and S.F. Tau
5666 P.H. Gore and J.A. Hoskins
5683 B.H. Robinson and J. Fergusson

5704 E.R. Clark
5748 P. Robson et al.
5815 K. Mori, S.K. Roy and D.M.S. Wheeler
5819 H.O. Larson et al.
5822 P.S. Gray and J.S. Mills
5888 V. Gold, G. Socrates and M.R. Crampton
5907 E.R.H. Jones
5911 E.R.H. Jones
5916 P.E. Cross and E.R.H. Jones
5919 P.E. Cross and E.R.H. Jones
5951 R.L. Huang, H.H. Lee and M.S. Malhotra
5957 R.L. Huang and K.H. Lee
5963 R.L. Huang and K.H. Lee
5969 R.R. Arndt, A. Jordaan and V.P. Joynt
5991 C.W.L. Bevan et al.
5999 J.H. Atkinson, R.S. Atkinson and A.W. Johnson
6023 J. Claisse, L. Crombie and R. Peace
6061 A.J. Nunn, D.J. Chadbourne and J.T. Ralph
6072 J.A. Elvidge and V.A. Moss
6076 A.D. Beveridge and G.S. Harris
6090 P.F. Holt and R. Oakland
6095 P.F. Holt and A.E. Smith
6141 T. Kametani and K. Fukumoto
6185 J.N. Murrell and S. Carter
6255 H.H. Lee and C.H. Tan

39- -65, J. Chem. Soc. (1965)
0027 A. Albert and J. Clark
0032 D. Leaver et al.
0130 S. Eardley and A.G. Long
0148 S. Eardley, G.F.H. Green and A.G. Long
0156 S. Earley, A.G. Long and C.H. Robinson
0181 D.H.R. Barton et al.
0194 C.E. Berkoff et al.
0208 D.M. Brown and P. Schell
0361 J.A. Hill et al.
0403 G.T. Chapman et al.
0406 P. Coggon et al.
0459 P. Hodge and R.W. Rickards
0480 T.J. King and C.E. Newall
0575 R.E. Banks et al.
0594 R.E. Banks et al.
0610 M.R. Atkinson, R.K. Morton and R. Naylor
0676 C.H. Rochester
0755 D.J. Brown and T. Teitei
0788 G.W. Moersch and W.A. Neuklis
0826 T.C. Owen and A.C. Wilbraham
0833 G.L. Buchanan et al.
0932 F. Kurzer and K. Dourahi-Zadeh
0940 H.J. Clase and E.A.V. Ebsworth
0948 R.M. Acheson, R.S. Feinberg and J.M.F. Gagan
0954 H.D. Cossey, J. Judd and F.F. Stephens
0974 T.J. King and C.E. Newall
0990 H.J.E. Loewenthal and S.K. Malhotra
1003 W.L. Mosby and M.L. Silva
1020 A. Bhati
1028 J. Dale
1034 W. Locker et al.
1051 M.W. Austin et al.
1080 R.E. Bowman et al.
1087 A.R. Battersby and H. Spencer
1137 D. Hucke, I.M. Lockhart and M. Wright
1149 G.I.H. Hanania, D.H. Irvine and F. Shurayh
1160 C. Djerassi et al.
1175 D.J. Brown and N.W. Jacobsen
1219 H. Dorn et al.
1224 A.R. Forrester and R.H. Thomson
1243 G. Eglinton, J. Martin and W. Parker
1258 W.L.F. Armarego and R.E. Willette
1262 S.H. Eggers, V.V. Kane and G. Lowe
1276 J.S. Burton, J.A. Elvidge and R. Stevens
1298 M. Kyaw and L.N. Owen
1338 D. Becker and H.J.E. Loewenthal
1344 P. Doyle et al.
1356 P.N. Rao and L.R. Axelrod
1390 G.M. Iskander and F. Stansfield
1455 A.W. Johnson and A.S. Katner
1460 D. Dolphin et al.
1515 J.A. Hill and W.J. LeQuesne
1518 R. Bonnett et al.
1530 D.J. Brown and B.T. England
1552 A.J. Birch and J.B. Siddall
1558 J.D. Hepworth and E. Tittensor
1572 F.E. King and J.G. Wilson
1605 P.J. Keay, J.S. Moffatt and T. Mulholland
1620 A.W. Johnson and I.T. Kay
1629 D. Leaver and J.D.R. Vass
1642 J.H. Dewar and G. Shaw
1648 R.J.S. Beer and R.W. Turner
1653 W. Carruthers and R.A.W. Johnstone
1685 A.S. Lindsey
1692 W. Cocker, T. McMurry and M.S. Ntamila
1700 H.M. Frey and I.D.R. Stevens
1761 J.M. Roberts
1772 D.H.R. Barton et al.
1779 D.H.R. Barton et al.
1787 J.E. Baldwin, D.H.R. Barton and J.K. Sutherland
1881 R.C. Cookson and D.W. Jones
1951 R.D. Gillard, J.A. Osborn and G. Wilkinson
2009 R.C. Cookson and M.J. Nye
2019 P.S. Mauchaud et al.
2040 W.I. Awad et al.
2054 G.M.L. Cragg and G.D. Meakins
2072 G.B. Arrowsmith, G.H. Jeffery and A.I. Vogel
2096 M. Bellas and H. Suschitzky
2141 H.A. Anderson, J. Smith and R.H. Thomson
2184 I.G.M. Campbell et al.

2251 J.A. Elvidge and R. Stevens
2258 C.H. Williams
2260 G.B. Barlin
2270 J.B. Robinson and J. Thomas
2281 B. Weinstein et al.
2283 A.K. Kiang and S.F. Tan
2285 W.H. Hui, S.N. Loo and H.R. Arthur
2305 W.-H. Chang
2313 R. Bonnett
2324 D.C.Garbutt, K. Pachler and J.R. Parrish
2340 D.J. Baisted and J.S. Whitehurst
2349 C.W. Shoppee et al.
2355 A.G. Brown, J.C. Lovie and R.H. Thomson
2361 R. Bryant and D.L. Haslam
2372 S.J. Moss and H. Steiner
2411 M. Anderson and A.W. Johnson
2423 D.H.R. Barton et al.
2476 C.W. Shoppee et al.
2492 L.H. Briggs et al.
2543 K. Anderton and R.W. Rickards
2549 A.J. Boulton and J.F.W. McOmie
2587 H. Heaney et al.
2601 J.R. Bull, E.R.H. Jones and G.D. Meakins
2614 J.H. Atkinson and A.W. Johnson
2633 R.M. Acheson et al.
2720 R.E. Banks, J.E. Burgess and R.N. Haszeldine
2723 B.J. Calvert and J.D. Hobson
2743 H.H. Lee and C.H. Tan
2778 W.L.F. Armarego
2788 A.J. Bellamy and R.D. Guthrie
2812 J.S. Grossert et al.
2818 K. Biemann et al.
2844 C.H. Hassall and J. Weatherston
2892 F.S. Edmunds and R.A.W. Johnstone
2898 F.S. Edmunds and R.A.W. Johnstone
2904 F.R. Hewgill et al.
2914 F.R. Hewgill and B.S. Middleton
2921 F.R. Hewgill and B.R. Kennedy
2933 E.R.H. Jones and D.A. Wilson
2955 A.R. Pinder and B.W. Staddon
2983 B.W. Nash et al.
2988 D.A. Thomas and W.K. Warburton
3001 A. Jordaan, V.P. Joynt and R.R. Arndt
3007 A.J. Birch and G.S.R. Subba Rao
3017 G.B. Barlin and N.B. Chapman
3032 B.R.T. Keene and P. Tissington
3037 P.L. Pauson et al.
3040 I.M. Davidson et al.
3052 E. Caspi et al.
3075 P.S. Steyn et al.
3090 S. Golding, A.R. Katritzky and H.Z. Kucharska
3093 A.R. Katritzky, H.Z. Kucharska and J.D. Rowe
3097 M. Elliott
3101 H.M. Frey and I.D.R. Stevens
3154 J.M. Greenwood et al.
3160 A.J. Hubert and J. Dale
3200 R.M. Acheson, M.W. Foxton and G.R. Miller
3312 L.A. Summers, P. Freeman and D.J. Shields
3319 W. Cocker and D.M. Sainsbury
3336 B. Robinson
3342 F.G. Badelar, A. El-Habashi and A.K. Fateen
3357 G.W. Miller and F.L. Rose
3369 W. Broadbent, G.W. Miller and F.L. Rose
3379 B. Coffin and R.F. Robbins
3456 L.A. Cort
3550 J.R. Hanson and T.P.C. Mulholland
3563 D.H.R. Barton and D.W. Jones
3610 I.M. Lockhart and E.M. Tanner
3660 F.R. Hewgill and D.G. Hewitt
3666 J.C. Roberts and P. Roffey
3678 G.W.H. Cheeseman and B. Tuck
3690 I.R.C. Bick et al.
3770 D.J. Brown and N.W. Jacobsen
3785 D. Lloyd et al.
3803 J.F. Grove, J.S. Moffatt and E.B. Vischer
3811 P. McCloskey
3872 V. Askam and D. Bailey
3885 E.E. Glover and G.H. Morris
3887 E.N. Morgan et al.
3912 F. Kurzer and K. Douraghi-Zadeh
3928 A. Fredga et al.
3987 J.M. Carpenter and G. Shaw
4004 L.H. Roach and D.G. Neilson
4007 T.R. Emerson et al.
4014 I. Baxter and G.A. Swan
4130 J.W. Apsimon et al.
4144 J.W. Apsimon et al.
4203 S.H. Harper and W.G.E. Underwood
4226 P.J. Brignell, U. Eisner and H. Williams
4292 A.G. Brown and R.H. Thomson
4303 A.W. Johnson and D. Oldfield
4348 C.S.L. Baker et al.
4355 P. Bamfield et al.
4363 D.W. Cameron et al.
4385 A. Hayes et al.
4396 K.C. Bass and P. Nababsing
4399 R.B. Horner and R.B. Moodie
4426 B.R.T. Keene and P. Tissington
4448 F. Kurzer and K. Douraghi-Zadeh
4456 D.S. Wulfman, J.J. Korst and R.W. Franck
4503 Z. Badr et al.
4508 R. Bonnett and T.R. Emerson
4512 T. Nakano and S. Terao
4546 J.F. Cavalla et al.
4599 N.H.P. Smith
4603 C.H. Rochester
4646 K. Mackenzie
4653 A. Albert
4659 C.S.L. Baker, P.D. Landor and S.R. Landor
4672 J.A. Ballantine, C.H. Hassall and G. Jones
4690 D.N. Hague and R.H. Prince
4744 W.-H. Chang
4765 A. Gandini and P.H. Plesch
4773 J.A. Joule et al.

4831 J.I.G. Cadogan et al.
4900 R. Clarkson
4930 J.M. Blatchly et al.
4939 T.P.C. Mulholland et al.
5015 J.D. Cocker et al.
5060 D. McHale and J. Green
5064 W.T. Pike, G.H.R. Summers and W. Klyne
5134 R.F. Curtis and G.T. Phillips
5137 A.J. Birch et al.
5139 A.J. Birch and G.S.R. Subba Rao
5140 R. Bryant
5182 J.A. Barltrop and B. Hesp
5189 J.D. Hardstone and K. Schofield
5245 P.F. Holt and A.E. Smith
5311 D.J.W. Bullock, C.W.N. Cumper and A.I. Vogel
5360 W.L.F. Armarego and J.I.C. Smith
5377 J.D. Scribner and J.A. Miller
5391 D.E. Ames et al.
5414 A.J. Boulton and D.P. Clifford
5416 G.H. Whitham and J.A.F. Wickramusinghe
5473 N. Campbell and H.G. Heller
5518 K.D. Warren and J.R. Yandle
5537 J.W. Barton, A.M. Rogers and M.E. Barney
5542 D.J. Brown and J.S. Harper
5551 J.W. Clark-Lewis and J.A. Edgar
5556 J.W. Clark-Lewis and J.A. Edgar
5625 S.R. Landor and P.F. Whiter
5651 J.H. Bowie and D.W. Cameron
5707 F.D. Schlosser and F.L. Warren
5744 P.H. Gore and J.A. Hoskins
5759 C.C. Barker et al.
5760 D.A. Crombie and S. Shaw
5762 C.W. Bird
5772 A.K. Hiscock and J.S. Whitehurst
5783 W. Carruthers and D.A.M. Watkins
5868 M. Akhtar
5871 D.L. Hammick and D.J. Voaden
5877 S.A. Procter and G.A. Taylor
5920 D.H. Reid and W. Bonthrone
5927 J.J. Armstrong and W.B. Turner
5958 A.J. Boulton, A.C.G. Gray and A.R. Katrizky
5976 J.C. Hanson et al.
5984 J.C. Hanson et al.
6036 D.E. Ames and A.C. Lovesey
6057 R. Grigg, J.A. Knight and M.V. Sargent
6061 J. Burgess and R.H. Prince
6067 R. Burwood et al.
6117 Y.M.Y. Haddad et al.
6125 T.L.V. Ulbricht and G.T. Rogers
6130 T.L.V. Ulbricht and G.T. Rogers
6221 W. Carruthers and H.N.M. Stewart
6296 F. Kurzer and E.D. Pitchfork
6421 H.S. Turner and R.J. Warne
6464 J. Lewis et al.
6509 E.R. Clark and S.R. O'Donnell
6531 J.H. Dunlop and R.D. Gillard
6542 D.R.J. Laws
6543 P.R. Ashurst et al.
6570 R.S. Nyholm et al.
6587 G. Read
6629 M.J. Mays and G. Wilkinson
6658 W.B. Turner
6674 A.J. Hubert and J. Dale
6688 T. Nakano and M. Hasegawa
6710 D.R. Ross and E.S. Waight
6784 D.K. Black and S.R. Landor
6851 M. Brickman, J.H.P. Utley and J.H. Ridd
6923 D.W. Cameron et al.
6930 A. Albert and J.J. McCormack
6935 J.D. Connolly et al.
6960 R.F. Curtis, C.H. Hassall and D.W. Jones
6972 B.T. Newbold
6984 J.W. Bayles and B. Evans
6991 P. Bladon and T. Sleigh
7001 P. Bamfield, A.W. Johnson and J. Leng
7005 F.G. Baddar, N. Latif and A.A. Nada
7018 E. Jones and I.M. Moodie
7109 R.E. Atkinson, R.F. Curtis and G.T. Phillips
7165 B. Heath-Brown and P.G. Philpott
7199 W.T. Pike, G.H.R. Summers and W. Klyne
7246 G.A. Ellestad et al.
7348 J. Hill and G.R. Ramage
7358 A.B. Clayton et al.

42- -64, J. Indian Chem. Soc., 41 (1964)
0093 J.N. Chatterjee, N. Prasad and K.D. Banerjee
0163 A. Chatterjee, R. Mukherjee and B. Das and S. Ghosal
0242 D. Nasipuri and M. Guha
0479 B.K. Bhattachrya et al.
0643 A. Chatterjee and S. Banerjee
0821 H.K. Desai and R.N. Usgaonkar

43- -64, J. Optical Soc. Am., 54 (1964)
0817 V.M. Bhucar and S.R. Das

44- -64, J. Org. Chem., 29 (1964)
0016 R.L. Rowland et al.
0021 H.J. Shine, C.F. Davis and R.J. Small
0051 V. Georgian and L.L. Skaletzky
0061 C.K. Bradsher and M.W. Barker
0064 E.W. Cantrall, K. Littell and S. Bernstein
0068 M.B. Rubin, G.E. Hipps and D. Glover
0074 H.O. House and R.G. Carlson
0087 J.P. Ferris, C.E. Sullivan and B.G. Wright
0105 J.H. Brewster and H.O. Bayer
0160 P.G. Gassman and P.G. Pape
0163 W.J. Wechter
0178 J. Szmuszkovicz
0214 E.W. Cantrall, R. Littell and S. Bernstein
0229 S. Rakhit and M. Gut
0243 R.L. Letsinger and J.A. Gilpin

0262 O. Wintersteiner and M. Moore
0270 O. Wintersteiner and M. Moore
0279 R.M. Scribner
0305 I.W. Elliott
0308 W.J. Fanshawe, V.J. Bauer, E.F. Ullman and S.R. Safir
0311 T.E. Stevens
0315 P.-L. Chien et al.
0323 S.I. Goldberg and R.L. Matteson
0336 J.A. Moore and L.J. Pandya
0339 I. Kagawa and Y. Sato
0342 T. Kubota and M. Ehrenstein
0345 T. Kubota and M. Ehrenstein
0351 T. Kubota and M. Ehrenstein
0357 T. Kubota and M. Ehrenstein
0366 V.J. Traynelis and R.F. Love
0370 C.F. Howell et al.
0382 D.N. Kevill, G.A. Coppens, M. Coppens and N.H. Cromwell
0385 W.R. Benson and A.E. Pohland
0415 W.B. Lutz et al.
0423 E. Legoff and R.B. Lacouny
0435 E. Wenkert et al.
0445 F.M. Beringer and S.J. Huang
0452 C.K. Bradsher and M.W. Barker
0471 R.J. Highet
0476 D.J. Rabiger
0490 A.J. Verbiscar
0497 D.C. Dittmer, H.E. Simmons and R.D. Vest
0499 D.N. Kevill and N.H. Cromwell
0502 L.L. Woods and J. Sterling
0513 M.L. Lewbart and V.R. Mattox
0521 M.L. Lewbart and V.R. Mattox
0527 J.S. Baran
0554 E.J. Reist, A. Benitez and L. Goodman
0558 J.F. Codington, I.L. Doerr and J.J. Fox
0564 J.F. Codington, I.L. Doerr and J.J. Fox
0574 W. Korytnyk, E.J. Kris and R.P. Singh
0579 E. Brill and H.P. Schultz
0582 M. Uskokovic et al.
0601 S.K. Pradhan and H.J. Ringold
0604 M.E. Wall et al.
0636 A.L. Bluhm, J.A. Sousa and J. Weinstein
0640 E. Caspi, H. Zajac and T.A. Wittstruck
0650 W.P. Norris and R.A. Henry
0660 W.R. Hatchard
0665 W.R. Hatchard
0668 D.L. Trepanier, V. Sprancmanis and K.G. Wiggs
0678 C. Grundmann and V. Mini
0681 T. Masamune et al.
0686 B.L. Van Duuren, I. Bekerskey and M. Lefar
0689 M. L. Wolfrom et al.
0692 M. L. Wolfrom et al.
0707 A.D. Josey
0734 J.A. Montgomery and N.F. Wood
0738 R.W. Addor
0751 R.J. Cotter and W.F. Beach
0755 S.M. Kupchan, T. Masamune and G.W.A. Milne
0759 H.H. Takimoto and G.C. Denault
0776 B.E. Fischer and J.E. Hodge
0782 T.A. Spencer, M.A. Schwartz and K.B. Sharpless
0787 T.A. Spencer, M.D. Newton and S.W. Baldwin
0794 G.H. Alt and A.J. Speziale
0798 G.H. Alt and A.J. Speziale
0843 J. Szmuszkovicz
0856 C.K. Bradsher and J.C. Parham
0862 G.G. Gallo et al.
0883 H. Rapoport, N. Castagnoli, Jr., and K.G. Holden
0886 P.R. Jones, G. Visser and R.M. Stinson
0942 W.J. Farnshawe, V.J. Bauer and S.R. Safir
0943 D.D. Bly
0947 W.E. Noland and K.R. Rush
0963 E. Wolthuis and D.L. Vander Jagt
0975 J.C. Goan, E. Berg and H.E. Podall
0988 W.H. Tallent
0996 S.I. Goldberg and J.S. Crowell
1022 W. Herz and N. Viswanathan
1092 V.J. Traynelis and J.R. Livingston, Jr.
1097 T. Taguchi et al.
1110 E.D. Weil
1115 G. De Stevens, B. Smolinsky and L. Dorfman
1120 M. Nussim, Y. Mazur and F. Sondheimer
1131 M. Nussim, Y. Mazur and F. Sondheimer
1142 W.R. Benn and R.M. Dodson
1158 J.B. Hester, Jr.
1180 D.L. Ross and J.J. Chang
1194 H.D. Hartzler
1206 R.L. Hinman and C.P. Bauman
1270 M.E. Kuehne and T. Kitagawa
1276 D.N. Kevill et al.
1296 L.H. Zalkow and D.R. Brannon
1307 V. Boekhelheide and G.R. Wenzinger
1325 O. Wintersteiner, M. Moore and A.I. Cohen
1333 A.I. Laskin et al.
1341 I.A. Kaye and R.S. Mathews
1348 S. Kaufmann
1350 A. Hassner and C. Heathcock
1379 J.P. Freeman
1391 T.M. Harris and C.R. Hauser
1419 T. Masamune et al.
1424 H.L. Slates, D. Taub, C.H. Kuo and N.L. Wendler
1435 A.I. Meyers and G. Garcia-Munoz
1445 J.D. Park and W.S. Frank
1449 R.L. Hinman and J. Lang
1453 C.A. Giza and R.L. Hinman
1508 K.E. Pfitzner and J.G. Moffatt
1537 G.W. Stacy, B.V. Ettling and A.J. Papa
1543 E. Campaigne et al.

1549 R.A. Lucas et al.
1575 W.E. Parham and M.D. Bhavsar
1582 M.E. Kuehne, S.J. Weaver and P. Franz
1594 W.J. Gensler et al.
1621 W. Metlesics et al.
1623 A. Prosen, B. Stanovic and M. Tisler
1645 W.B. Lutz et al.
1647 E.M. Fry
1650 F.J. Dinan and H. Tieckleman
1673 D.M. Lemal and A.J. Fry
1677 H.R. Nace and D.H. Nelander
1700 W. Herz, G. Hogenauer and A. Romo de Vivar
1703 E. Campaigne and N.W. Jacobsen
1708 E. Campaigne, R.D. Hamilton and N.W. Jacobsen
1711 E. Campaigne and R.D. Hamilton
1720 C.G. Overberger and H.A. Friedmann
1723 W.E. Rosen, L. Dorfman and M.P. Linfield
1740 A. Takamizawa, K. Hirai, Y. Sato and K. Tori
1751 P. Coad, R.A. Coad and J. Hyepock
1757 M.J.S. Dewar and W.H. Poesche
1762 T. Ueda and J.J. Fox
1770 T. Ueda and J.J. Fox
1772 N. Miller and J.J. Fox
1776 D. Horton
1782 A.L. Clingman and N.K. Richtmeyer
1800 F. Scotti and E.J. Fraza
1812 A. Nishinaga and T. Matsuura
1821 J.C. Dearden
1826 D.A. Johnson and C.A. Panetta
1834 A.L. Logothetis
1893 D.F. Morrow and M.E. Butler
1919 H. Wynberg and D.J. Zwannenberg
1932 M.B. Rubin and E.C. Blossey
1961 L.H. Klemm, E. Huber and C.E. Klopfenstein
1983 L.D. Freedman and G.O. Doak
1994 C.A. Aufdermarsh, Jr.
2003 W. Garner and H. Tiekelmann
2009 P. Tarrant, J. Savory and E.S. Iglehart
2018 W.W. Zorbach and S. Saeki
2024 D.L. Ross and J.J. Chang
2028 T.I. Bieber and M.T. Dorsett
2030 T.E. Young and M.F. Mizianty
2044 C.J. Thiman et al.
2053 E.W. Tristram et al.
2059 G.I. Fujimoto and R.W. Ledeen
2064 G. De Stevens and V.P. Arya
2068 H. Zinnes, R.A. Comes and J. Shavel, Jr.
2077 J.P. Horwitz, J. Chua and M. Noel
2085 Y. Mizuhara
2088 E.W. Crandell and J. Olguin
2101 R.A. Lucas, R.G. Smith and L. Dorfman
2101 R.G. Hiskey and R.L. Smith
2109 R.W. Griffin, Jr., et al.
2117 E.C. Taylor and E.E. Garcia
2121 E.C. Taylor and E.E. Garcia
2124 E.C. Taylor and E.E. Garcia
2128 R.K. Bly, E.C. Zoll and J.A. Moore
2135 Y.F. Shealy and C.A. O'Dell
2141 Y.F. Shealy and J.D. Clayton
2146 W.J. Humphlett and R.W. Lamon
2187 L.H. Knox et al.
2195 A.D. Cross et al.
2211 W.E. Parham et al.
2214 W.E. Parham and S.H. Groen
2226 W.A. Thaler and B. Franzus
2256 L.A. Kaplan
2265 H.E. Smith, S.L. Cook and M.E. Warren, Jr.
2272 D. Theodoropoulos and J. Tsangaris
2282 T. Masamune et al.
2293 C.H. Brieskorn et al.
2298 J.M. Bobbitt and R.E. Doolittle
2331 A.T. Blomquist and E.A. LaLancette
2335 G. Witshard and C.E. Griffin
2351 K.J. Sax et al.
2362 M.B. Rubin and P. Zwitkovits
2368 S.C. Bell, G.L. Conklin and S.J. Childress
2382 G.O. Doak, L.D. Freedman and J.B. Levy
2427 J.S. Driscoll et al.
2431 R.L. Hinman and C.P. Bauman
2452 M.N. Appelbaum, R.W. Fish and M. Rosenblum
2455 H. Wynberg and A. Kraak
2462 W.W. Zorbach and G.D. Valiaveedan
2467 D.K. Chatterjee, R.M. Chatterjee and K. Sen
2469 H.E. Zieger and J.E. Rosenkranz
2471 D.S. Tarbell and T. Parasaran
2482 L.L. Woods and J.B. Sapp
2486 G.C. Morrison and J. Shavel, Jr.
2493 C.M. Orlando, Jr., and K. Weiss
2495 G.W. Moersch et al.
2501 J.A. Marshall and W.I. Fanta
2527 Z.G. Hajos, K.J. Doebel and M.W. Goldberg
2534 E. Wenkert, D.B.R. Johnston and K.G. Dave
2542 G.M. Badger, B.J. Nelson and K.T. Potts
2545 W.F. Johns
2553 T.A. Geissman and R.J. Turley
2559 M.L. Lewbart and J.J. Schneider
2582 F.F. Ebetino
2595 H.J. Schaeffer and V.K. Jain
2598 J.W. Huffmann and L.E. Browder
2602 L. Jurd
2611 Y. Mizuno, I. Itoh and K. Saito
2615 R.L. Letsinger et al.
2623 B. Klein, E. O'Donnell and J.M. Gordon
2658 R.K. Griffith and H.J. Harwood
2663 C.J. Argoudelis and F.A. Kummerow
2670 J.D. Fissekis, A. Myles and G.B. Brown
2674 R.B. Trattner et al.
2693 C. Postmus, Jr., et al.
2727 H. Rapoport and J. Bordner
2731 P. Crabbe and C. Casa-Campillo

2746 M. Schach von Wittenau
2756 W.H. Tallent
2766 R. Dowbenko
2771 G.C. Morrison, W. Cetenko and J. Shavel, Jr.
2775 R.E. Schaub, H.M. Kissman and M.J. Weiss
2778 H. Zimmer, R.D. Barry and F. Kaplan
2784 J.A. Ross and M.D. Martz
2785 D.L. Roberts
2805 L.L. Raplogle
2824 L.A. Carpino and S. Gowecke
2860 J.A. Moore and E.C. Capaldi
2864 J.B. Hester, Jr.
2877 E. Campaigne and R.D. Hamilton
2881 A. Rosowsky et al.
2895 M. Freifelder
2898 A.T. Nielsen, D.W. Moore, J.H. Mazur and K.H. Berry
2903 D.M. Mulvey, S.G. Cottis and H. Tieckelmann
2947 M.P. Cava, D. Mangold and K. Muth
2951 L. Skattebøl
2957 E.A. LaLancette
2959 H.C. Yao
2982 B.J. Magerlein et al.
2986 R.H. Wiley, C.E. Staples and T.H. Crawford
3014 A. Nickon and B.R. Aaronoff
3028 J.D. Edwards, Jr., S.E. McGurie and C. Hignite
3032 D.J. Bertelli
3036 L. Jurd
3046 S. Trofimenko
3049 A.L. Logothetis
3061 T. Shiba et al.
3070 H.D. Becker
3072 H. El Khadem
3074 M.L. Wolfrom, H. El Khadem and H. Alfes
3087 E.R. Altwicker and C.D. Cook
3092 E.D. Bergmann, P. Bracha, J. Blum and M. Engelbrath
3102 K.L. Marsi and K. Torre
3108 D. Caine and J.B. Dawson
3110 P.J. Kropp
3146 C.L. Stevens et al.
3151 L.R. Caswell and P.C. Atkinson
3158 V.R. Mattox and W. Vrieze
3161 E.E. Smissman and A.N. Voldeng
3165 M. Masaki and M. Ohta
3180 H.H. Baer and B. Achmatowicz
3206 S.C. Bell, P.H.L. Wei and S.J. Childress
3216 R.H. Barker, J.P. Collman and R.L. Marshall
3229 R.R. Haynes and H.R. Snyder
3234 R.A. Finnegan and R.S. McNees
3241 R.A. Finnegan and R.S. McNees
3245 D.T. Longone and L.H. Simanyi
3252 F. Johnson and J.P. Heeschen
3280 M.L. Wolfrom, H.G. Garg and D. Horton
3300 J.C. Orr, A. De La Roz and A. Bowers
3304 L.J. Chinn
3314 D.S. Tarbell et al.
3327 H.O. House, V.K. Jones and G.A. Frank
3333 M.B. Rubin
3370 J.J. McCarmack and H.G. Mautner
3401 R.J. Baumgarten and M.C. Henry
3403 D.L. Garmaise and J. Komlossy
3407 K.T. Potts and I.D. Nasri
3416 C.F. Spencer and J.G. Michels
3430 R.L. Schaff and C.T. Lenk
3436 J.A. Montgomery and K. Hewson
3438 W. Herz and Y. Sumi
3441 T. Nakano, T.H. Yang and S. Terao
3445 L.L. Woods and J. Sapp
3447 L.A. Paquette
3467 A.C. Cope and R.J. Cotter
3469 J. Meinwald et al.
3476 J.E. Pike
3481 J.A. Edwards et al.
3486 D. Taub, R.D. Hoffsommer and N.L. Wendler
3495 R.N. Iacona, A.T. Rowland and H.R. Nace
3498 H.R. Nace and R.N. Iacona
3503 C.H. DePuy et al.
3520 D.J. Goldsmith and J.A. Hartman
3524 D.J. Goldsmith and J.A. Hartman
3527 W.F. Forbes, R. Shilton and A. Balasubramanian
3536 C.J. Fox and A.L. Johnson
3560 E. Galantay et al.
3574 C.L. Stevens, K.G. Taylor and M.E. Munk
3577 E.J. Moriconi and J.J. Murray
3584 C.K. Bradsher and E.F. Litzinger
3587 T.N. Hall
3591 G.A. Reynolds and J.A. Van Allan
3596 J. Wolinsky, R. Novak and R. Vasileff
3601 W.A. Skinner and R.M. Parkhurst
3604 G.H. Stout and K.L. Stevens
3610 C.M. Baugh and E. Shaw
3612 T.C. Miller and R.G. Christiansen
3617 E.W. Scheiffel and D.A. Shirley
3640 A. Hassner and C. Heathcock
3654 U. Zehari and N. Sharon
3660 R.E. Lutz, R.G. Bass and D.W. Boykin, Jr.
3687 R.G. Hiskey and J. Hollander
3695 S.C. Dickermann, M. Klein and G.B. Vermont
3739 J. Hannah and J.H. Fried
3740 J. Wolinsky, M.R. Slabaugh and T. Gibson

44- -65, J. Org. Chem., 30 (1965)
0010 R.S. Bly and R.T. Swindel
0043 E. Negishi and A.R. Day
0091 C.E. Griffin et al.
0097 C.E. Griffin et al.
0105 H. Zinnes, R.A. Comes and J. Shavel, Jr.
0112 O.L. Galmarini and F.H. Stodola
0115 B.J. Whitlock, S.H. Lipton and F.M. Strong

0118 W. Hey, A. Romo de Vivar and M.V. Lakshmikantham
0123 W.F. Johns and I. Laos
0131 A.G. Anderson, Jr., et al.
0144 M.L. Wolfrom, F. Komitsky, Jr., and J.H. Looker
0162 W.M. Doane et al.
0184 R.K. Olsen and H.R. Snyder
0190 E. Wolthuis et al.
0194 T. Sherdasky and R.M. Dodson
0199 V. Papesch and R.M. Dodson
0203 G. Vogel
0212 D.F. Morrow et al.
0243 A.L. Borror and A.F. Haeberer
0252 D.L. Fields, J.B. Miller and D.D. Reynolds
0257 L.J. Chinn and J.S. Michina
0259 V.J. Grenda et al.
0270 R.N. Macdonald and W.S. Stewart
0277 R. Brown, C. Kelley and S.E. Wiberley
0285 A. Rosowsky, H.K. Protopapa and E.J. Modest
0295 R. Gardi and C. Pedrali
0307 A.J. Manson, R.E. Sjogren and M. Riano
0316 W.M. Hoehn, C.R. Dorn and B.A. Nelson
0354 H.E. Ungnade and L.W. Kissinger
0394 R.E. Lyle and W.E. Krueger
0396 E. Campaigne and W.L. Rolofs
0408 R.M. Creswell, H.K. Maurer, T. Strauss and G.B. Brown
0417 G. Singh and H. Zimmer
0429 N.M. Joye, Jr., et al.
0432 W.M. Harris and T.A. Geissman
0467 I.L. Doerr, J.F. Codington and J.J. Fox
0476 J.F. Codington, I.L. Doerr and J.J. Fox
0501 A.K. Bose, M.S. Manhas and R.C. Cambie
0505 A.K. Bose and R.T. Dahill, Jr.
0518 T. Matsuura and T. Suga
0526 P.A. Duke, A. Fozard and G. Jones
0528 O. Wintersteiner and M. Moore
0560 T.K. Liao, F. Barochi and C.C. Cheng
0563 R.B. Carlin and J.W. Harrison
0567 W.G. Finnegan and R.A. Henry
0579 D.F. Morrow, M.E. Butler and E.C.Y. Huang
0603 C.H. Mao and L. Anderson
0610 R.G. Powell et al.
0629 L.A. Paquette
0636 D.C. Dittmer and G.C. Levy
0639 S. Rakhit and M. Gut
0644 T.K. Mukherjee and L.A. Levasseur
0650 H.O. House, J.J. Reihl and C.G. Pitt
0669 T. Suga, K. Mori and T. Matsuura
0695 T. Hayashi et al.
0702 H. Bader et al.
0707 H. Bader
0722 E. Wenkert, L.H. Liu and D.B.R. Johnston
0728 W.E. Parham and S.H. Groen
0746 J.S. Vittiberga and B.M. Vittiberga
0754 E. Bianchi et al.
0807 R.L. Letsinger et al.
0812 A.K. Bose and G. Mina
0829 C. Temple, Jr. et al.
0835 T.Y. Shen, H.M. Lewis and W.V. Ruyle
0859 M. Friedman
0891 D.J. Bertelli
0897 H. Aft
0910 T. Kato and H. Yamanaka
1012 D.E. Machiele et al.
1020 H.M. Blatter, H. Lukaszewski and G. DeStevens
1038 J. Meinwald et al.
1050 J.L. E. Erickson and F.E. Collins
1058 H. Wynberg and W.E. Wiersum
1061 H.O. House and T.H. Cronin
1088 M.L. Wolfrom, H.G. Garg and D. Horton
1096 M.L. Wolfrom, H.G. Garg and D. Horton
1104 J.L. Pinkus, G.G. Woodyard and T. Cohen
1107 J.D. Albright and L. Goodman
1110 W.C. Colburn, Jr., et al.
1118 S. Takahashi and H. Kano
1126 W.R. Benson and A.E. Pohland
1129 W.R. Benson and A.E. Pohland
1213 Z.G. Hajos, D.R. Parrish and M.W. Goldberg
1242 H.W. Davies and M. Schwartz
1247 J.R. Piper and T.P. Johnston
1251 F.H. Greenberg
1255 E.E. Royals and J.C. Leffingwell
1272 G.W. Moersch et al.
1277 S.C.J. Fu, M. Reiner and T.L. Loo
1278 W.J. Fanshawe, V.J. Bauer
1279 J.W. Schulenberg and S. Archer
1282 E.J. Hedgley and H.G. Fletcher,Jr.
1292 J.A. Marshall and N.H. Andersen
1294 T.A. Spencer, S.W. Baldwin and K.K. Schmiegel
1325 S.D. Levine and P.A. Diassi
1333 R.A. Finnegan
1398 W.J. Middleton and C.G. Krespan
1407 G.H. Alt and A.J. Speziale
1409 H.A.P. De Jongh et al.
1416 G.N. Walker
1421 M.G. Ward, J.C. Orr and L.L. Engel
1423 N.L. Allinger et al.
1431 J.C. Kauer, R.E. Benson and G.W. Parshall
1470 A. Streitwieser, Jr., R.G. Lawler and D. Schwaab
1473 O.H. Wheeler and H.N. Battle de Pabon
1513 E.C. Pesterfield
1523 A. Fozard and G. Jones
1528 J.A. Montgomery and K. Hewson
1539 C.K. Bradsher, R.W.L. Kimber and S.D. Mills
1542 E.J. Moriconi and A.J. Fritsch

1550 L.J. Dolby and D.L. Booth
1556 M.L. Wolfrom, H.G. Garg and D. Horton
1604 J.W. Huffmann and P.G. Arapakos
1626 T.A. Spencer, K.K. Schmiegel and W.W. Schmiegel
1629 R.F. Stockel, M.T. Beachem and F.H. Megson
1657 G.G. Gallo, C.R. Pasqualucci and A. Diena
1658 A.J. Solo and B. Singh
1661 H. Takahashi, S. Tai and M. Yamaguchi
1690 K.L. Stevens, R.E. Lundin and R. Teranishi
1687 A.A. Griswold and P.S. Starcher
1693 W.G. Dauben, W.E. Thiessen and P.R. Resnick
1698 I. Agata et al.
1723 R.H. Hesse and M.M. Pechet
1737 M.H. Nazer
1744 P. Resnelle and G. Ourisson
1748 A. Hassner and C. Heathcock
1752 J. Martel, E. Toromanoff and C. Huynh
1769 W.H. Pirkle and M. Gates
1774 D. Lavie, E. Glotter and Y. Shvo
1800 E.D. Stecher, A. Waldman and D. Fabiny
1815 D.J. Cram and H.P. Fischer
1832 A. Rosowsky and E.J. Modest
1837 E.J. Modest, S. Chatterjee and H.K. Protopapa
1840 J. Diamond, W.F. Bruce and F.T. Tyson
1846 C.K. Bradsher and R.W.L. Kimber
1859 K.A. Schellenberg and F.H. Westheimer
1881 W. Herz and H.J. Wahlborg
1887 J.A. Moore and W.J. Theuer
1889 J.A. Moore and C.L. Habraken
1916 S.C.J. Fu, E. Chinoporos and H. Terzian
1926 C.G. Overberger and H.A. Friedman
1930 F.M. Beringer and S.A. Galton
1959 C.F. Wilcox, Jr., and F.D. Roberts
1973 R. Nahon and A.R. Day
1981 S.C. Dickermann, D.D. Sousa and P. Wolf
1986 Y. Makisumi
1989 Y. Makisumi
2009 R.M. Dodson, P.B. Sollman and J.R. Deason
2037 R.W. Kluiber
2047 F.S. Alvarez and A.B. Rinz
2086 L.A. Spurlock and P.E. Newallis
2095 G.S. Fonken
2107 L.A. Paquette
2109 E.W. Garbisch, Jr.
2126 D.E. Applequist et al.
2130 H.J. Shine and E.E. Mach
2165 N.L. Allinger and E.S. Jones
2169 E.J. Becker, R.M. Palmere, A.I. Cohen and P.A. Diassi
2198 L.H. Knox et al.
2205 R.F. Heck
2218 C.W. Spangler and G.F. Woods
2222 W.I. Awad et al.
2228 D.L. Trepanier et al.
2234 R.E. Schaub, J.H. Van Den Hende and M.J. Weiss
2241 H. Zinnes et al.
2251 R.D. Reynolds and R.J. Conboy
2253 J.T. Shuh and B.M. Puma
2259 H. Tanida et al.
2264 D.J. Goldsmith and C.J. Cheer
2272 J. Diekmann
2290 A. Takamizawa and K. Hirai
2330 P.F. Wiley et al.
2334 J.J. Berreboom et al.
2342 R.A. Finnegan and W.H. Mueller
2344 E.H. White et al.
2353 A.L. Livingstone et al.
2356 B.A. Parkin, Jr., and G.W. Hedrick
2359 E.J. McWhorter and M. Anchel
2371 R.G. Binder, L.A. Goldblatt and T.H. Applewhite
2384 W.J. Middleton et al.
2395 C. Temple, Jr. et al.
2398 J.L. Wong, M.S. Brown and H. Rapoport
2403 J.G. Lombardino
2407 W.P. Norris and J. Osmundsen
2410 G.B. Butler and M.A. Raymond
2420 L.J. Dolby and R.H. Iwamoto
2425 M. Schwarz, A. Besold and E.R. Nelson
2457 D. Horton and M.J. Miller
2488 Y.F. Shealy and C.A. O'Dell
2502 H.O. House and B.M. Trost
2513 H.O. House et al.
2519 H.O. House, S.G. Boots and V.K. Jones
2528 H.O. House and R. Darns
2534 J.C. Powers
2549 K.D. McMichael and G.A. Selter
2575 F. Ramirez, A.V. Patwardhan and C.P. Smith
2583 W.Gum, Jr., and M.M. Jouillie
2589 A. Richardson, Jr.
2593 D.F. O'Brien and J.W. Gates, Jr.
2602 R.E. Moser and H.G. Cassidy
2610 E. Campaigne and W.R. Roelofs
2642 G.A. Berchtold, G.R. Harvey and G.E. Wilson, Jr.
2660 W.D. Crow and N.J. Leonard
2674 F.T. Williams, Jr., et al.
2678 H.N. Simpson, C.K. Hancock and E.A. Meyers
2715 L.L. Replogle, R.M. Arluck and J.R. Maynard
2723 C.L. Stevens and P. Blumbergs
2754 C.T. Mathew et al.
2763 L.D. Heustis, M.L. Walsh and N. Hahn
2766 T.J. Delia, M.J. Olsen and G.B. Brown
2776 M. Tanabe and D.F. Crowe
2784 A. Winston et al.

2791 R. Bonnett and G.F. Stephenson
2837 T.L. Loo and R.L. Dion
2838 R.E. Beyler and G. Ourisson
2840 E.D. Bergmann and A.M. Meyer
2849 Z.G. Hajos et al.
2851 S.R. Jenkins, F.W. Holly and E. Walton
2854 G.L. Tong, W.W. Lee and L. Goodman
2862 W.J. Fanshawe, V.J. Bauer and S.R. Safir
2875 R.C. Bertelson
2882 S.G. Smith and J.P. Petrovich
2897 G.R. Allen, Jr., J.F. Poletto and M.J. Weiss
2904 G.R. Allen, Jr., and M.J. Weiss
2910 W.A. Remers, R.H. Roth and M.J. Weiss
2918 R.R. Johnson and J.A. Nicholson
2922 T.C. Miller
2925 W.H.W. Lunn
2942 H.O. House and R.W. Bashe, II
2956 J.W. Huffman and T.W. Bethea
2967 C.L. Stevens et al.
2973 G.N. Walker, D. Alkalay and R.T. Smith
3000 J.U. Lowe, Jr., and L.N. Ferguson
3031 S. Wawzonek and R.C. Gueldner
3067 E. Benz, N.F. Elmore and L. Goldman
3071 W.M. Doane et al.
3105 M.S. Newman and G. Kaugars
3111 M. Uskokovic et al.
3166 T.K. Mukherjee and A. Golubovic
3185 S.G. Levine and M.C. Wani
3190 W. Herz et al.
3205 A.T. Botlini and W. Schear
3207 J. Wolinsky, M. Senyek and S. Cohen
3209 B. Weinstein and A.H. Fenselau
3211 N. Muramatsu and T. Takenishi
3215 K.V. Scherer, Jr., and R.S. Lunt
3225 E. Wolthuis and A. De Boer
3231 J.C. Winter et al.
3242 C.G. Pitt
3248 D. Wasserman et al.
3250 O.W. Webster, M. Brown and R.E. Benson
3295 M.S. Newman and G. Kaugars
3312 J.B. Hendrickson et al.
3333 T.A. Spencer, T.D. Weaver and W.J. Greco, Jr.
3377 S. Portnoy
3401 E.J. Reist et al.
3404 J.I. Degraw et al.
3427 T.S. Splitter and M. Calvin
3436 C.H. Stammer and J.D. McKinney
3451 R.F. Meyer
3454 S. Klutchko, H.V. Hansen and R.I. Meltzer
3457 W.E. Noland, L.R. Smith and K.R. Rush
3469 D. Levy and R. Stevenson
3552 S. Noguchi and D.K. Fukushima
3561 L. Prakash et al.
3564 J.S. Baran
3566 R.A. Kloss and D.A. Clayton
3569 J.A. Berson and M.R. Willcott
3573 H. Muxfeldt, M. Weigele and V. Van Rheenen
3578 J.L. Mateos, A. Dosal and C. Carbajal
3593 S. Boatman, T.M. Harris and C.R. Hauser
3597 J.W. Marsico and L. Goldman
3601 C. Temple, Jr. et al.
3604 R.J. Sundberg
3610 P.L. Creger
3613 T.E. Young and P.H. Scott
3618 J.J. D'Amieco et al.
3634 H.O. House and W.M. Bryant, III
3642 J.A. Marshall and D.J. Schaeffer
3647 R.L. Cargill et al.
3650 A.T. Nielsen
3657 R.M. Scribner
3667 W. Sobotka et al.
3679 G.A. Berchtold, J. Ciabattoni and A.A. Tunick
3698 E.T. McBee et al.
3705 M.E. Munk and Y. Ki Kim
3711 R.D. Campbell and J.A. Jung
3739 H.C. Brown and R. Pater
3775 M.P. Cava and B.R. Vogt
3781 L.L. Smith, T.J. Foell and D.M. Teller
3786 R.L. Clarke and S.J. Daum
3792 S.M. Kupchan and H.C. Wormser
3819 G.A. Reynolds, J.A. Van Allan and R.E. Adel
3878 N. Ishikawa, M.J. Namkung and T.L. Fletcher
3883 L.A. Paquette and T.R. Phillips
3895 B. Miller and H. Margulies
3913 P.K. Chang
3915 F.C. Uhle
3933 S.M. Kupchan and H.C. Wormser
3935 S.M. Kupchan, H.C. Wormser and M. Sesso
3941 J.W. Huffman and R.L. Asbury
3955 W.W. Zorbach, H.R. Munson and K.V. Bhat
3957 G.F. Field, W.J. Zally and L.H. Sternabach
3960 J.B. Wright
3991 E.N. Marvell and J. Tashiro
3993 W.F. Johns
4008 E.E. Smissman and J.R.J. Sorenson
4066 Y. Mizuno et al.
4074 G.W. Stacy, F.W. Villaescusa and S.F. Silver
4078 H.J. Richter, R.L. Dressler and S.F. Silver
4085 J.P. Paolini and R.K. Robins
4107 J.H. Day and A. Joachim
4111 R.H. Haas and D.A. Shirley
4114 C.E. Cook, R.C. Corley and M.E. Wall
4120 C.E. Cook, R.C. Corley and M.E. Wall
4122 M.S. Newman and C.K. Dalton

4125 M.S. Newman and C.K. Dalton
4145 R.A. Finnegan and P.L. Bachman
4154 H. Carpio et al.
4160 L.H. Knox et al.
4175 J.C. Martin et al.
4180 J.J. Looker
4188 W.R. Brasen et al.
4198 E. Ciganek
4220 W.F. Johns and I. Laos
4230 J.J. Beereboom
4234 S.W. Pelletier et al.
4263 K.G. Hampton, T.M. Harris, C.M. Harris and C.R. Hauser
4293 P.R. Jones and P.J. Desio
4303 A.J. Speziale, L.R. Smith and J.E. Fedder
4333 R.D. Kimbrough, Jr., et al.
4344 K.D. Kaufmann, D.W. McBride and D.C. Eaton
4353 W.V. Ruyle, T.Y. Shen and A.A. Patchett
4366 E. Ciganek
4381 W.A. Remers, R.H. Roth and M.J. Weiss
4384 A.W. Burgstahler and C.P. Kulier
4387 E. Wenkert and B.L. Mylari

46- -64, J. Phys. Chem., 68 (1964)
0300 S.R. Sandler and K.C. Tsou
0752 J.R. Totter, W. Stevenson and G.E. Philbrook
1205 K. Shinzawa and I. Tanaka
1501 W.F. Smith
1768 E. Rutner and S.H. Bauer
1786 G.O. Pritchard, M. Venugopalan and T.F. Graham
1793 D.P. Chong and G.B. Kistiakowsky
1842 D.G. Herries, W. Bishop and F.M. Richards
1896 R.E. Kay, E.R. Walwick and C.K. Gifford
1999 R.H. Linnell, F. Raab and R. Clifford
3225 H.E. Ungnade, E.M. Roberts and L.W. Kissinger

46- -65, J. Phys. Chem., 69 (1965)
0053 J. Rabani, W.A. Mulac and M.S. Matheson
0457 T. Urbaski and W.K. Mathews
0641 S. Chaberek, A. Shepp and R.J. Allen
0647 S. Chaberek and R.J. Allen
0821 G.R. Seely
0978 B.J. Litman and J.A. Schellman
1001 G.O. Pritchard and R.L. Thommason
1066 C.R. Sporck and A.E. Coleman
1466 M. Abu-Hamdiyyah and K.J. Mysels
1588 D. Rosenthal et al.
1758 H.E. Ungnade et al.
1773 J. Fajer
1894 W. West and S. Pearce
1992 D. Verdin, S.M. Hyde and F. Neighbour
2004 N.M. Trieff and B.R. Sundheim
2475 B.C. Roquitte
2545 K. Koyano and I. Tanaka
3225 R.L. Alumbaugh, G.O. Pritchard and B. Rickborn
3615 L.B. Clark, G.G. Peschel and I. Tinoco
3791 J.S. Brinen et al.
3872 M.E. Lamm and D.M. Neville

49- -64, Monatsh. Chem., 95 (1964)
0003 R. Kuhn and F.A. Neugebauer
0402 E. Zbiral et al.
0415 T. Kappe and E. Ziegler
0457 R. Kuhn and H. Trischmann
0485 F. Bickelhaupt, K. Stach and M. Thiel
0512 E. Zbiral, O. Saiko and F. Wessely
0576 K. Schlogl, M. Fried and H. Falk
0649 F. Wessely, J. Swoboda and V. Guth
0678 V. Gutmann et al.
1068 G. Doleschall and K. Lempert
1228 M. Spitelier-Friedmann et al.
1283 G. Swoboda, J. Swoboda and F. Wessely
1698 H. Tuppy and E. Kuchler
1750 H. Egger and K. Schlogl

49- -65, Monatsh. Chem., 96 (1965)
0025 A. Brossi, M. Baumann and R. Borer
0077 E. Ziegler and T. Kappe
0182 J.O. Jilek et al.
0212 E. Ziegler and E. Steiner
0220 G. Mixich and A. Zinke
0285 H. Schindlbauer and H. Hagen
0369 A. Muller and M. Lempert-Sreter
0450 S.H. Dandegaonker and G.R. Revankar
0614 S.H. Dandegaonker and D. Shastri
0631 D. Braun and M. Kiessel
0888 E. Ziegler and T. Kappe
0909 R. Kaschnitz and G. Spiteller
1094 M. Pailer and W. Streicher
1173 V. Gutmann and A. Steininger
1214 S.H. Dandegaonker and S.G. Shet
1314 G. Kunesch and F. Wessely
1324 M. Pailer et al.
1352 G. Kleinberg and E. Ziegler
1409 A. Brossi and R. Borer
1512 W. Fleischhacker and F. Viebock
1520 K. Schlogl and W. Steyrer
1793 H. Schindlbauer
1967 E. Zbiral

50- -64A, Nature, 201 (1964)
0378 D.G. O'Sullivan, D. Pantic and A.K. Wallis

50- -64B, Nature, 202 (1964)
0343 P.A. Loach and M. Calvin

50- -64C, Nature, 203 (1964)
0296 R.H. Burnell and D.D. Casa
0523 M.M. Coombs and H.R. Roderick

0637 R. Felton, G.M. Sherman and H. Linschitz
0970 S. Chatterjee, D.H. Trites and E.J. Modest
1064 F. Arcamone et al.
1065 S.F. Chang and J.E. Liener

50- -64D, Nature, 204 (1964)
0077 T.A. Scott
0186 B.S. Shasha et al.

50- -65B, Nature, 206 (1965)
0630 F.F. Elslager and D.F. Worth

54- -64, Rec. trav. chim., 83 (1964)
0031 M.J.D. Van Dam
0039 M.J.D. Van Dam and F. Kogl
0081 J.H.S. Weiland
0154 T.H. Van Der Meulen and G.J.M. Van Der Kerk
0249 J.S. Wieczorek and E. Plazek
0364 H. De Koning et al.
0593 C.K. Bradsher and R.B. Desai
0711 R. Foster
0949 A.P. De Jonge, A. Verhage and B. Van Der Ven
0995 J.C. Overeem and G.J.M. Van Der Kerk
1005 J.C. Overeem and G.J.M. Van Der Kerk
1069 P. Westerhof
1160 F.A. Buiter et al.
1173 J.L.M.A. Schlatmann, J. Pot and E. Havinga
1215 H. Dolman et al.

54- -65, Rec. trav. chim., 84 (1965)
0137 D.S. Deorha and S.B. Sareen
0193 H. Dolman, J. Van Der Goot and H.D. Moed
0245 A.P. Ter Borg and H. Kloosterziel
0289 O. Korver, J.U. Veenland and T.J. de Boer
0314 H.I.X. Mager and W. Berends
0334 N.P. Buu-Hoi et al.
0389 A.G.M. Willems, U.K. Pandit and H.O. Huisman
0441 R.F. Shuman and E.D. Amstutz
0516 R. Foster, C.A. Fyfe and J.W. Morris
0521 J. Van Dijk et al.
0581 A.P. de Jonge et al.
0626 C.C. Bolt et al.
0648 G.J.N. Egmond et al.
0806 O. Korver et al.
0841 R. Van Moorselaar, S.J. Halkes and E. Havinga
0853 H. van Kamp
0863 P. Westerhof, J. Hartog and S.J. Halkes
0889 S.J. Halkes and E. Havinga
0904 H. van Kamp and S.J. Halkes
0918 P. Westerhof and J. Hartog
0965 H. Das and E.C. Kooyman
1094 N. Maoz and S. Vromen

1113 H.A.M. Jacobs et al.
1233 H.C. Volger and W. Brackman
1478 W.H. Laarhoven and R.J.F. Nivard

56- -64, Roczniki Chem., 38 (1964)
0385 A. Uzarewicz
0515 S. Goszczynski
0591 M. Zaidlewicz et al.
0599 A. Uzarewicz
0789 L. Skulski, G.C. Palmer and M. Calvin
0893 S. Goszczynski
1251 L. Czuchajewski and A. Erndt
1523 T. Lesiak
1533 J. Gronowska
1709 T. Lesiak
1767 J. Gronowska
1807 J. Ziolkowski and K. Guminski

56- -65, Roczniki Chem., 39 (1965)
0007 A. Hendrich and H. Kuczynski
0237 J. Gronowska
0245 J. Gronowska
0375 J. Gronowska
0545 D. Buza and W. Polaczkowa
0557 D. Buza and W. Polaczkowa
0589 T. Lesiak
0639 T. Lesiak and J. Liesiecki
0681 T. Lesiak
0757 T. Lesiak
0763 W. Jasiobecki
0847 J. Gronowska
0931 C. Parkanyi and A. Vystrcil
0939 T. Lesiak
1019 J. Gronowska
1215 J. Moszew and Zankowska-Jasinska
1423 W.E. Hahn et al.
1625 J. Izdebski
1713 W.E. Hahn, J. Epsztajn and B. Rybczynski

57- -64B, Science, 144 (1964)
0412 A.P. Roszkowski, G.I. Poos and R.J. Mohrbacher
0540 J.W. Wheeler et al.

57- -65A, Science, 147 (1965)
0152 H. Takeshita and M. Anchel

59- -64, Spectrochim. Acta, 20 (1964)
0299 J. Fabian and R. Mayer
0397 K.M. Sancier, A.P. Brady and W.W. Lee
0597 J. Charette, G. Falthansl and P. Teyssie
0993 M. Koyanagi and Y. Kanda
1143 H. Schindlbauer
1227 C. Dijkgraaf
1437 C.V. Berney
1665 J.F. Corbett
1709 D.W. Ellis and L.B. Rogers

59- -65, Spectrochim. Acta, 21 (1965)
0529 W.P. Hayes and C.J. Timmons
0931 K.L. Wierzchowski and D. Shugar

1229 O. Popovych and L.B. Rogers
1625 K.K. Chatterjee and B.E. Douglas
1881 B. Ellis and P.J.F. Griffiths

60- -64, Trans. Faraday Soc., 60 (1964)
0062 A.K. Chandra and D.C. Mukherjee
0264 R. Grinter and S.F. Mason
0274 R. Grinter and S.F. Mason
0285 R. Grinter, S.F. Mason and G.W. Vane
0386 G.R. Haugen and W.H. Melhuish
0465 S.K. Chakrabarti and S. Basu
0476 M.E. Peover and J.D. Davies
0488 M.J. Blandamer, T.E. Gough and M.C.R. Symons
1053 Z.L. Ernest and F.G. Herring
1131 Z.R. Grabowski and A. Bylina
1424 D. Gill, J. Jagur-Gradzinski and M. Szware
2177 A.A. Burr, E.J. Llewellyn and G.F. Lothian
2189 R. Foster and P. Hanson

60- -65, Trans. Faraday Soc., 61 (1965)
0408 W. Slough
0597 R.F. Schaufele and L. Goodman
0891 A.G. Evans and B.J. Tabner
1406 G. Aloisi et al.
1437 B.W. Brooks, F.S. Dainton and K.J. Ivin
1787 R.B. McKay
1800 R.B. McKay and P.J. Hillson
1981 K. Tickle and F. Wilkinson
2097 B. Ghosh and S. Basu

61- -64, Ber. Bunsengesellschaft Phys. Chem., 68 (1964)
0296 H. Berg and K. Kramarczyk
0973 G. Kortum and H. Rau

61- -65, Ber. Bunsengesellschaft Phys. Chem., 69 (1965)
0448 B. Ziolkowsky and F. Dorr
0716 E. Merkel

65- -64, Zhur. Obshchei Khim., 34 (1964)*
0075 A.I. Kiprianov and M.Y. Kornilov
0149 V.A. Koptyug et al.
0159 K.A. Chkhikvadze et al.
0190 A.V. Dombrovskii and M.I. Shevchuk
0204 Y.A. Levin et al.
0269 A.D. Garnovskii et al.
0278 S.G. Fridman and D.K. Golub
0282 L.A. Tsoi et al.
0326 V.I. Bliznyukov et al.
0345 I.A. Bessonova et al.
0356 V.I. Minkin and L.E. Nivorozhkin
0411 Y.P. Shvachkin and M.T. Azarova
0427 L.N. Pushkina and I.Y. Postovskii
0448 Y.F. Freimanis and G.Y. Vanag
0489 A.I. Artemenko et al.
0495 L.N. Yakhontov and M.V. Rubtsov

* English translation edition pagination

0504 Y.A. Levin and V.A. Kukhtin
0541 V.A. Zagorevskii et al.
0550 O.N. Tolkachev et al.
0620 V.V. Kiselev
0632 A.F. Pozharskii
0674 A.E. Lipkin et al.
0710 Z.N. Nasarova et al.
0796 Y.A. Berlin et al.
0808 E.Y. Gren and G.Y. Vanag
0813 E.Y. Gren and G.Y. Vanag
0818 E.Y. Gren and G.Y. Vanag
0841 Y.K. Yur'ev et al.
0894 F.N. Stepanov and A.G. Yurchenko
0949 E.R. Zakhs and L.S. Efros
0955 E.R. Zakhs and L.S. Efros
1013 G.K. Nikonov and V.B. Kuvaev
1353 G.K. Nikonov et al.
1454 L.N. Yakhontov et al.
1474 V.M. Naidan and A.V. Dombrovskii
1575 N.S. Dokunikhin et al.
1582 A.M. Simonov and A.F. Pozharskii
1588 M.V. Gorelik et al.
1642 E.R. Zakhs and L.S. Efros
1686 M.K. Yusupov and A.S. Sadykov
1786 I.S. Berdinskii
1924 V.A. Zagorevskii et al.
1969 Z.P. Penyugalova et al.
1985 G.S. Grinenko et al.
2008 G.T. Pilyugin et al.
2064 V.G. Yashunskii et al.
2171 Y.P. Shvachkin and L.A. Syrtsova
2179 Y.P. Shvachkin and M.T. Azarova
2201 Y.K. Yur'ev and N.K. Sadovaya
2212 N.K. Kochetkov et al.
2258 M.I. Shevchuk et al.
2328 S.V. Tsukerman et al.
2371 I.M. Mishina and L.S. Efros
2400 N.S. Vul'fson and V.E. Kolchin
2447 F.S. Babichev et al.
2455 F.S. Babichev and E. Shchetsinskaya
2463 R.N. Tursunova and N.K. Abubokirov
2558 M.N. Kolosov et al.
2563 M.N. Kolosov et al.
2599 K.A. Chkhikvadze and O.Y. Magidson
2632 M.V. Rubtsov
2699 N.S. Vul'fson et al.
2738 M.I. Shevchuk and A.V. Dombrovskii
2740 N.I. Ganushchak et al.
2749 A.E. Lutskii et al.
2766 N.S. Vul'fson et al.
2774 V.D. Lyashenko et al.
2790 I.A. Romadan and E.A. Kochetkova
2799 F.T. Pozharskii et al.
2843 N.P. Kir'yalov and S.V. Serkerov
2848 G.K. Nikonov
2914 S.V. Tsukerman et al.
2994 V.K. Daukshas and B.A. Puodzhyunaite
3025 A.N. Kost et al.
3049 A.P. Terent'ev et al.
3055 A.P. Terent'ev et al.

3135 V.M. Dashunin et al.
3150 N.O. Pastushak and A.V. Dombrovskii
3182 G.M. Kheifets and N.V. Khromov-Borisov
3340 N.B. Tarusova et al.
3352 A.F. Mironov et al.
3360 M.I. Rogovik et al.
3366 M.I. Rogovik and G.T. Pilyugin
3373 G.T. Pilyugin et al.
3377 G.T. Pilyugin et al.
3393 A.V. Dobrovskii and K.G. Tashchuk
3472 M.E. Konshin and P.A. Petyunin
3484 V.L. Florent'ev
3487 A.N. Kost, L.G. Yudin and C. Yu-Chou
3565 N.N. Suvorov and L.I. Klimova
3602 V.A. Izmail'skii and A.V. Malygina
3645 S.V. Tsukerman et al.
3788 A.V. El'tsov et al.
3819 N.S. Dokunikhin and V.Y. Fain
3837 B.M. Krasovitskii et al.
3904 B.I. Stepanov and A.I. Bokanov
3912 G.K. Nikonov and D.I. Barnan-skaite
3979 G.N. Dorofeenko et al.
4106 A.N. Kost et al.
4171 A.P. Prokopenko
4184 B.A. Ivin and V.G. Nemets
4193 A.S. Sadykov et al.

65- -65, Zhur. Obshchei Khim., 35 (1965)*
0049 V.M. Stepanov and V.F. Krivtsov
0058 N.S. Zefirov et al.
0094 V.F. Larrushin et al.
0130 V.L. Floren'ev
0244 Y.S. Shabarov et al.
0280 R.N. Elizareva and A.D. Kuzovkov
0298 V.B. Leont'ev et al.
0372 M.G. Imaev et al.
0428 G.L. Ryzhova et al.
0453 N.A. Domnin, T.A. Rakova and V.A. Cherkasova
0502 V.M. Dziomko and V.M. Ostrov-skaya
0506 G.T. Pilyugin et al.
0519 M.A. Mostoslavskii and V.A. Izmail'skii
0542 K.M. Kirillova and V.A. Kukhtin
0554 V.F. Krivtsov and V.M. Stepanov
0561 Y.P. Shvachkin and M.T. Azarova
0631 O.D. Strizhakov et al.
0655 A.I. Gurevich et al.
0662 M.N. Kolosov et al.
0670 A.I. Gurevich et al.
0675 V.M. Berezovskii et al.
0886 N.M. Turkevich and A.F. Minka
0896 M.N. Preobrazhenskaya and N.N. Suvorov
0988 V.M. Stepanov and V.F. Krivtsov
1304 V.G. Nemets and B.A. Ivin
1308 A.M. Shkrob et al.
1433 V.G. Nemets, B.A. Ivin and V.I. Slesaev
1724 S.V. Tsukerman et al.
1930 Y.I. Mushkin and A.I. Finkel'shtein
2073 A.E. Lutskii et al.
2080 A.E. Lutskii et al.
2088 A.E. Lutskii and A.F. Soldatova
2094 A.E. Lutskii et al.
2205 M.I. Shevchuk et al.
2240 Y.P. Shvachkin and I.K. Shprunka

67- -65, J. Structural Chem., 6 (1965) (English translation)
0375 Y.L. Erolov, A.V. Kalihina and A.K. Filippova
0379 N.N. Magdesieva et al.

70- -64, Izvest. Akad. Nauk S.S.S.R.(1964)
0111 L.F. Matyash and V.M. Stepanov
0121 S.V. Svetozarskii et al.
0174 A.V. Bogdanova et al.
0164 M.V. Mavrov and V.F. Kucherov
0197 A.N. Nesmeyanov, T.V. Nikitina and E.G. Perevalova
0293 G.A. Nikiforov and V.V. Ershov
0310 Y.A. Arbusov et al.
0371 G.A. Zlobina and V.V. Ershov
0482 Y.A. Arbusov et al.
0492 Y.P. Volkov et al.
0512 E.A. Mistryukov et al.
0576 V.P. Parini, E.L. Frankevich and M.V. Deichmeister
0680 R.M. Khomutov, M.Y. Karpeiskii and E.S. Severin
0774 A.M. Shkrov et al.
0776 M.L. Khidekel' et al.
0860 L.D. Bergel'son and Y.G. Molotovskii
0864 V.A. Mironov et al.
0934 B.M. Mikhailov and L.C. Povarov
0936 V.N. Setkina and S.D. Sokolov
0945 A.I. Gurevich et al.
0996 L.K. Freidlin et al.
1013 M.M. Shemyakin et al.
1024 M.M. Shemyakin et al.
1036 N.K. Kochetkov, A.Y. Khorlin and O.S. Chizhov
1054 L.P. Vinograova et al.
1131 N.N. Gaidamovich and I.V. Torgov
1236 E.I. Budovskii et al.
1241 I.A. Gurvich and V.F. Kucherov
1245 I.I. Nazarova, L.A. Yanovskaya and V.F. Kucherov
1260 L.V. Antik et al.
1318 B.P. Gusev and V.F. Kucherov
1349 L.P. Larina, A.M. Sladkov and A.G. Makhsumov
1355 A.A. Akhrem, Y.A. Titov and Z.A. Kravchenko
1453 L.D. Bergel'son et al.
1456 V.F. Kucherov, I.A. Gurvich and B.A. Rudenko
1470 L.V. Antik et al.
1475 Y.A. Levin, R.N. Platonova and V.A. Kukhtin
1481 Y.A. Levin, N.A. Shvink and V.A. Kukhtin
1515 V.F. Mironov and T.K. Gar

1640 S.G. Batrakov and L.D. Bergel'son
1648 G.B. Kondrat'eva et al.
1653 M.V. Mavrov and V.F. Kucherov
1666 G.A. Zlobina and V.V. Ershov
1703 M.I. Gugeshashvili et al.
1717 V.V. Karpov et al.
1814 V.E. Limanov, S.N. Ananchenko and I.V. Torgov
1820 M.V. Mavrov and V.F. Kucherov
1911 A.A. Akhrem, Y.A. Titov and I.S. Levina
2003 L.D. Bergel'son, E.V. Dyatlovitskaya and M.M. Shemyakin
2008 L.M. Kogan et al.
2016 L.M. Kogan et al.
2021 T.I. Sorkina, I.I. Zaretskaya and I.V. Torgov
2055 Y.L. Gol'dfarb et al.
2064 F.P. Sidel'kovskaya and A.A. Avetisyan
2086 S.I. Zav'yalov et al.
2093 L.A. Yanovskaya, R.N. Stepanova and V.F. Kucherov
2097 L.A. Yanovskaya and V.F. Kucherov
2175 K.S. Bokarev et al.
2182 Y.L. Gol'dfarb et al.
2203 V.A. Petukhov, V.F. Mironov and P.P. Shorygin
2241 Y.L. Gol'dfarb, F.D. Alashev and V.K. Zvorykina
2246 A.A. Akhrem, Y.A. Titov and I.S. Levina

70- -65, Izvest. Akad. Nauk S.S.S.R. (1965)
0110 Z.A. Krasnaya and V.F. Kucherov
0124 M.A. Chel'tsova et al.
0240 I.P. Beletskaya, A.E. Myshkin and O.A. Revtov
0322 E.D. Vasil'eva et al.
0336 V.V. Ershov and A.A. Volod'kin
0371 F.P. Sidel'kovskaya and F.L. Kolodkin
0466 Z.G. Isaeva et al.
0502 S.Y. Yunusov et al.
0510 Y.L. Gol'dfarb, V.P. Litvinov and S.A. Ozolin'
0544 A.R. Derzhinskii, M.V. Mavrov and V.F. Kucherov
0546 M.V. Mavrov and V.F. Kucherov
0669 B.A. Arbusov et al.
0684 L.A. Yanovskaya, B.G. Kovalev and V.F. Kucherov
0702 F.P. Sidel'kovskaya et al.
0716 V.A. Golubev and E.G. Rozantsev
0729 I.I. Nazarova, B.P. Gusev and V.F. Kucherov
0747 V.I. Gunar and S.I. Zav'yalov
0760 A.V. Zakharychev et al.
0800 L.A. Kalashmkova, E.G. Rozantsev and A.M. Chaikin
0806 Y.A. Arbuzov et al.
0818 L.D. Bergel'son and S.G. Batrakov
0843 L.N. Ivanova, T.A. Severina and V.F. Kucherov
0851 B.P. Kusev and V.F. Kucherov
0914 N.K. Kochetkov et al.
1039 I.G. Bolesov et al.
1051 I.I. Zaretskaya et al.
1058 I.I. Zaretskaya. T.I. Sorkina and I.V. Torgov
1066 S.S. Novikov et al.
1070 Z.A. Krasnaya and V.F. Kucherov
1076 V.I. Gunar et al.
1107 V.P. Mamaev and V.M. Ignat'ev
1197 B.M. Mikhailov and G.S. Ter-Sarkisyan
1237 A.R. Derzhinskii, M.V. Mavrov and V.P. Kucherov
1281 Y.L. Gol'dfarb et al.
1283 S.S. Novikov, L.A. Nikonova and V.I. Slovetskii
1290 B.A. Arbuzov and A.I. Konovalov
1382 A.A. Nesmeyanov et al.
1413 A.V. Zakharychev, S.N. Ananchenko and I.V. Torgov
1434 S.N. Godovikova and Y.L. Gol'd-farb
1460 M.V. Mavrov, A.R. Derzhinskii and V.F. Kucherov
1491 A.I. Ivanov et al.
1675 V.V. Ershov and G.A. Zlobina
1798 A.T. Prudchenko et al.
1809 A.V. Zakharychev et al.
1885 V.I. Gunar, D.F. Ovechkina and and S.I. Zav'yalov
1887 S.I. Zav'yalov et al.
1992 K.U. Ubaev, P.K. Yuldashev and and S.Y. Yunusov
2006 E.A. Mistryukov et al.
2063 V.I. Slovetskii et al.
2065 Y.S. Ovodov et al.
2152 M.A. Kuchenkova, P.K. Yuldashev and S.Y. Yunusov
2155 E.A. Mistryukov
2206 L.I. Zakharkin and V.N. Kalinin
2215 E.A. El'perina, B.P. Gusev and V.F. Kucherov
2216 N.I. Shuikin et al.
2220 G.Y. Legin et al.

73- -64, Coll. Czech. Chem. Comm., 29 (1964)
0143 J. Kuthan, E. Janeckova and M. Havel
0197 J. Weichet, J. Hodrova and L. Blaha
0214 S. Chladek and J. Smrt
0400 I. Jirkovsky and M. Protiva
0433 J. Trojanek et al.
0447 O. Strouf and J. Trojanek
0635 J. Zemlicka, J. Smrt and F. Sorm
1029 B. Pelc
1173 Z. Koblicova and K. Syhora
1178 H. Bockova, V. Schwarz and K. Syhora
1394 J. Gut, J. Jonas and J. Pitha
1484 H. Bockova, J. Holubek and Z. Cekan
1495 E. Janeckova and J. Kuthan
1654 J. Kuthan and E. Janeckova
1689 J.L. Kaul et al.
1913 B.K. Moza et al.

2060 A. Piskala and F. Sorm
2182 L. Novotny, V. Herout and F. Sorm
2189 L. Novotny, V. Herout and F. Sorm
2351 K. Syhora and R. Mazac
2513 R. Mickova and K. Syhora
2567 A. Holy, N.C. Spasovska and J. Smrt
2570 J. Tomko and S. Bauer
2576 J. Pliml and F. Sorm
2607 V. Zatka
2956 M. Prystas and F. Sorm
2980 Z. Budesinsky, J. Prikryl and E. Svatek

73- -65, Coll. Czech. Chem. Comm., 30 (1965)
0195 R. Zahradnik and C. Parkanyi
0940 O. Exner and J. Holubek
1158 M. Prochazka
2609 J. Kuthan and R. Bartonickova
2783 Z. Arnold
3016 R. Zahradnik and C. Parkanyi
3102 J. Klienar and F. Kosek
3111 O. Cervinka and L. Hub
3479 F. Santavy et al.
3575 B. Pele and J. Hodkova
3697 J. Slavik et al.
3711 J. Kuthan and E. Janeckova
3730 Z. Budesinsky, F. Roubinek and E. Svatek
3744 J. Pliml and F. Sorm
3895 Z. Budesinsky et al.

77- -64, Proc. Chem. Soc. (1964)
0017 K.J. Crowley
0082 G.M. Badger, J.A. Elix and G.E. Lewis
0087 G.W. Brown et al.
0108 A. Ledwith and N. McFarlane
0120 H. Minato, S. Nosaka and I. Horibe
0144 R.C. Cookson et al.
0195 J.A. Barltrop and B. Hesp
0217 R.L. Jones, C.W. Rees and C.E. Smithen
0232 C. Dickinson, J.R. Holden and M.J. Kamlet
0368 J. Harley-Mason and T.J. Leeney
0370 W.D. Bannister et al.

77- -65, Chemical Communications (1965)
0001 H. Wynberg and U.E. Wiersum
0011 W.K. Gibson and D. Leaver
0015 M.L. Burstall
0040 J.A. Miller and H.C.S. Wood
0057 R. Dietz
0064 A. Ledwith and M. Sambhi
0098 R.C. Cookson et al.
0114 S.D. Ibekwe and M.J. Newlands
0149 D. Becker and H.J.E. Loewenthal
0151 J.M. Locke and E.W. Duck
0162 J.D. Connolly, R. McCrindle and K.H. Overton
0192 C.D. Campbell and C.W. Rees
0193 C.W. Rees and R.C. Storr
0197 J.E.D. Barton and J. Harley-Mason
0243 A.M. Small
0248 J. Barrett et al.
0269 G.M. Badger et al.
0272 J.C. Emmett, D.F. Veber and W. Lwowski
0288 A. Fozard and C.K. Bradsher
0294 H.J. Shine and J.P. Stanley
0309 S.H. Harper et al.
0313 H. van Zwet and E.C. Kooyman
0314 B. Miller and H. Margulies
0317 H. Monteiro et al.
0321 P.M. Greaves et al.
0343 J.F. Grove et al.
0347 G. Pattenden et al.
0353 R.M. Dodson and A.G. Zielske
0377 H. Minato and T. Nagasaki
0391 P.G. Gassman and K. Mansfield
0422 C.H. Fawcett et al.
0439 H.H. Hatt and A.C.K. Triffett
0447 K.A. Muszkat, D. Gegiou and E. Fischer
0453 R. Ramasseul and A. Rassat
0464 G.O. Dudek and E.P. Dudek
0479 W. Cocker and D.M. Sainsbury
0492 G.M. Badger et al.
0509 R.P.A. Sneeden and H.P. Throndsen
0512 E.D. Bergmann and I. Agranat
0524 J.E. Baldwin and N.H. Rogers
0539 J. King and D. Leaver
0540 S.F. Mason and G.W. Vane
0569 S.A. Procter and G.A. Taylor
0574 R.N. Pratt, S.A. Procter and G.A. Taylor
0607 J.F. Munshi and M.M. Joullie
0622 R.A. Moss
0640 G. Ferguson et al.

78- -64, Tetrahedron, 20 (1964)
0017 S.C. Agarwal and T.R. Seshadri
0033 V. Boekelheide et al.
0079 J. Romo et al.
0087 Y.N. Sharma, A. Zaman and A.R. Kidwai
0107 G. Snatzke, H. Piper and R. Tschesche
0159 H. Kano and E. Yamazaki
0177 K. Pilgram and F. Korte
0189 O.H. Wheeler and D. Gonzalez
0195 E.D. Bergmann, S.H. Blumberg, P. Bracha and S. Epstein
0211 M. Kamel, S. Sherif and M.M. Kamel
0215 N.A. Lebel and L.A. Spurlock
0231 R. Huisgen and G. Seidl
0299 A.R. Katritzky and F.W. Maine
0315 A.R. Katritzky and F.W. Maine
0333 A. Manjarrez, T. Rios and A. Guzman
0341 W. Herz and S. Inayama
0357 M.J. Weiss et al.
0373 R.E. Schaub, W. Fulmor and M.J. Weiss
0387 R. Tsechche and K.H. Richert
0409 R.C. Cambie, L.N. Mander, A.K. Bose and M.S. Manhas

0437 S. Oae, W. Takagi and A. Ohno
0461 H. Kano and E. Yamazaki
0507 E. Clar and A. McCallum
0515 B. Robinson
0519 J.A. Cade and A. Pilbeam
0555 S.K. Grover, A.C. Jain and T.R. Seshadri
0565 J.B. Hendrickson, R. Goschke and R. Rees
0597 D. Burn et al.
0717 R.C. Cookson, N.S. Isaacs and M. Szelke
0791 A.L. Beckwith and R.J. Leydon
0803 M.V. Bhatt
0831 P. Beak
0861 J.E. Bloor and A. Burawoy
0871 L.H. Klemm et al.
0929 C.F. Hammer et al.
0979 W. Herz, Y. Kishida and M.V. Lakshmikantham
0991 K.R. Markham
0999 L. Ramachandra Row et al.
1037 A. Hassner and C. Heathcock
1051 B.S. Thyagarajan and P.V. Gopalakrishnan
1057 G. Cignarella et al.
1119 S. Swaminathan, S. Ramachandran and S.K. Sankarappa
1141 S.C. Bhrara, A.C. Jain and T.R. Seshadri
1185 B. Willhalm, A.F. Thomas and F. Gautschi
1289 K.S. Kulkarni, S.K. Parnikar and S.C. Bhattacharyya
1317 S.P. Schwarz et al.
1331 S.F. Dyke, W.D. Ollis, M. Sainsbury and J.S. Schwarz
1381 B.R. Letchford, C.R. Patrick and J.C. Tatlow
1397 G. Berti, A. Da Settimo and O. Livi
1407 M.G. Lester, V. Petrow and O. Stephenson
1435 R. Tschesche, P. Welzel, R. Moll and G. Legler
1449 G.L. Buchanan et al.
1455 D.W. Theobald
1469 R. Tschesche and G. Brugmann
1547 A.J. Ryan
1555 M.J.Y. Foley, N.H.P. Smith and P. Watts
1559 E. Clar and J.E. Stephen
1593 K. Hunger, U. Hasserodt and F. Korte
1613 H. Goldwhite, M.S. Gibson and C. Harris
1625 S.A. Fuqua, R.M. Parkhurst and R.M. Silverstein
1707 K. Schreiber and G. Adam
1725 B. Weinstein and T.A. Hylton
1729 A. Mondon and H.U. Menz
1737 H. Fritz and O. Fischer
1755 G.R. Pettit, D.S. Alkalay, P. Hofer and P.A. Whitehouse
1773 W. McCrae
1781 W.H. Tallent
1789 G. Lucas et al.
1889 K. Wallenfels and W. Draber
1927 M.T. Davies et al.
1963 A.L. Livingston et al.
1971 H. Mitsuhashi and T. Muramatsu
1987 R.H. Shapiro and C. Djerassi
2047 H. Fritz and O. Fischer
2091 J.C. Anderson, D.G. Lindsay and C.B. Reese
2107 E. Clar, J.F. Guye-Vuilleme and J.F. Stephen
2177 T.L. Jacobs, D. Dankner and S. Singer
2181 H.H. Freedman and G.A. Doorakian
2185 J.B.F. Lloyd and P.A. Ongley
2201 A. Hassner and T.C. Mead
2217 M. Ballester and J. Riera
2295 D. Burn and V. Petrov
2297 J. Fried, P. Grabowich, E.F. Sabo and A.I. Cohen
2319 H. Nakata, K. Sasaki, I. Morimoto and Y. Hirata
2331 J. Romo and P.J. Nathan
2455 P. Crabbe et al.
2473 F. Sondheimer, R. Mechoulam and M. Sprecher
2487 D.K. Banerjee et al.
2531 J.M. Coxon, M.P. Hartshorn and D.N. Kirk
2553 H.A.P. De Jongh and H. Wynberg
2593 D.W. Theobald
2605 S.N. Shanbhag et al.
2617 P.S. Kalsi, K.K. Chakravarti and S.C. Bhattacharyya
2639 G.H. Kulkarni, G.R. Kelkar and S.C. Bhattacharyya
2655 K. Takeda et al.
2665 D. Lavie and B.S. Benjaminov
2701 C. Belil, J. Pascual and F. Serratosa
2735 R. Scarpati, D. Sica and C. Santacroce
2815 T.M. Jacob, P.A. Vatakencherry and S. Dev
2821 T.M. Jacob, P.A. Vatakencherry and S. Dev
2835 A.J. Boulton et al.
2885 R. Tschesche, D. Kloden and H.W. Fehlhaber
2911 B.N. Joshi et al.
2921 V.K. Howard and A.S. Rao
2927 K.R. Varma and S.C. Bhattacharyya
2937 G. Snatzke, B. Zeeh and E. Muller
2951 M.G. Gianturco et al.
2971 D. Giacopello, V. Deulofeu and J. Comin
2977 D.L. Dreyer, S. Tabata and R.M. Horowitz
2985 T.R. Govindachari et al.

78- -65, <u>Tetrahedron</u>, <u>21</u> (1965)
0031 I. Maclean and R.P.A. Sneeden
0035 E.H. Hoffmeister and D.S. Tarbell
0089 W.L. Stanley et al.

0093 A.K. Ganguly, T.R. Govindachari and P.A. Mohamed
0115 K.R. Varma et al.
0179 R. Gardi, R. Vitali, A. Ercoli and W. Klyne
0245 J.B.F. Lloyd and P.A. Ongley
0261 K. Bowden and R. Stewart
0267 L.A. Mitscher, W. McCrae and S.E. Devoe
0273 M.P. Hartshorn and A.F.A. Wallis
0309 K. Tori and T. Komeno
0329 K. Takeda et al.
0407 H. Ripperger and K. Schreiber
0449 A.K. Bose, M.S. Manhas and R.M. Ramer
0467 E. Clar and J.E. Stephen
0515 H.A.P. De Jongh and H. Wynberg
0559 J.W. Blunt, M.P. Hartshorn and D.N. Kirk
0569 D. Burn et al.
0583 D.A. Lightner and C. Djerassi
0607 V.H. Kapadia et al.
0657 B.J. Bergot and L. Jurd
0681 D.P. Chakraborty, B.K. Barman and P.K. Bose
0727 H. Ripperger, K. Schreiber and G. Snatzke
0733 M. Tomoeda et al.
0743 C.H. Robinson et al.
0759 K. Yoshida and T. Kubota
0779 O. Wintersteiner and M. Moore
0791 D.W. Theobald
0797 M. Manzoor-I-Khuda and S. Sarela
0817 G. Bianchi and P. Grunanger
0823 A. Da Settimo and M.E. Saettone
0909 R.A. Corral and O.O. Orazi
0917 D. Shiengthong et al.
0927 J. Burdon, D. Harrison and R. Stephens
0945 B.S. Thyagarajan and P.V. Gopalakrishnan
0963 S.C. Bhrara, A.C. Jain and T.R. Seshadri
0989 A.H. Jackson and A.E. Smith
1001 K.J. Crowley
1015 R.A. Finnegan and J.J. Mattice
1067 N. Isenberg and H.F. Herbrandson
1141 B. Gilbert et al.
1167 K.S. Kulkarni and A.S. Rao
1197 C. Burgess et al.
1215 H. Mitsuhashi and N. Kawahara
1223 R. Mechoulam and Y. Gaoni
1231 K. Nakanishi et al.
1299 A.J. Neale, T.J. Rawlings and E.B. McCall
1323 A. Pollak and M. Tisler
1327 A.B.A. Jansen and C.G. Richards
1357 L. Skattebøl
1369 N. Suciu et al.
1411 A. Kamal et al.
1417 D.J. Aberhart et al.
1441 B.P. Chaliha, G.P. Sastry and P.R. Rao
1449 K.R. Markham
1453 W.D. Ollis et al.
1471 L. Jurd and A.C. Waiss, Jr.
1489 A.H. Beckett, G. Kirk and A.J. Sharpen
1495 D. Kumari, S.R. Mukerjee and T.R. Seshadri
1509 T.R. Govindachari, B.S. Joshi and V.N. Kamat
1521 A.S. Bawdekar and G.R. Kelkar
1529 C.S. Chopra et al.
1581 D.A. Lightner et al.
1611 H. Dannenberg and H.J. Gross
1619 D. Burn, D.N. Kirk and V. Petrow
1657 H. Nozaki, M. Yamabe and R. Noyori
1681 A.R. Katritzky et al.
1693 A.R. Katritzky, F.W. Maine and S. Golding
1711 W. Herz and M.V. Lakshmikantham
1717 J.A. Joule et al.
1741 A. Romo De Vivar and H. Jiminez
1753 G. Cooley, B. Ellis and V. Petrow
1777 R. Tschesche, P. Welzel and G. Snatzke
1889 R.D. Stolow and K. Sachdev
1917 A. Butt and I.A. Akhtar
1923 A. Da Settimano and M.F. Saettone
1931 P.A. Argabright, H.D. Rider and M.W. Hanna
1941 F. Ramirez, D. Rhum and C.P. Smith
2059 T. Goto et al.
2089 K. Takeda et al.
2095 R.L. Edwards and N. Kale
2117 H. Immer et al.
2133 A. Tahara, K. Hirao and Y. Hamazaki
2155 W.H. Baarschers and R.R. Arndt
2183 P.J. Kropp
2205 R. Eisenthal and A.R. Katritzky
2281 J.B.F. Lloyd and P.A. Ongley
2289 B.S. Thyagarajan, K.K. Balasubramanian and R. Bhima Rao
2297 M. Los and A.D. Mighell
2313 S. McClean and P. Haynes
2413 E. Jones and I.M. Moodie
2501 Y. Yamato and H. Kaneko
2509 C.H. Robinson, O. Gnoj and F.E. Carlon
2517 W.N. Speckamp et al.
2529 G. Jones and J. Wood
2553 K.W. Bentley, S.F. Dyke and A.R. Marshall
2573 T.R. Govindachari et al.
2579 B.R. Pai and G. Shanmugasundarm
2593 V.K. Honwad and A.S. Rao
2605 H. Ishii, T. Tozyo and H. Minato
2617 M.P. Cava et al.
2633 T.R. Govindachari et al.
2641 N.K. Basu et al.
2653 E.D. Burling, A. Jefferson and F. Scheinmann
2671 T. Matsumato et al.
2683 W.B. Eyton et al.
2697 W.B. Eyton et al.
2707 M.F. Barnes et al.

2735 C.E. Griffin, N.F. Hepfinger and B.L. Shapiro
2857 E.H. Hoffmeister and D.S. Tarbell
2865 E.H. Hoffmeister and D.S. Tarbell
2931 H. Otsuka and J. Shoji
2939 A.G. McInnes, S. Yoshida and G.H.N. Towers
2951 T.R. Govindachari et al.
2957 T.R. Govindachari et al.
2961 G. Jones and J. Wood
2973 B.A. Shoulders et al.
2997 S.F. Campbell, R. Stephens and J.C. Tatlow
3019 D.S. Kemp and R.B. Woodward
3051 M.P. Cava and R.H. Schlessinger
3073 M.P. Cava and R.H. Schlessinger
3091 H. Kakisawa et al.
3185 M.T. Davies et al.
3205 S.C. Agarwal and T.R. Seshadri
3219 C.A. Henrick and P.R. Jefferies
3237 T.R. Govindachari et al.
3255 M. Shamma et al.
3289 K.D. Bartle et al.
3305 B.S. Thyagarajan and P.V. Gopalakrishnan
3325 D.L.E. Bronnert and B.C. Saunders
3401 I.O. Sutherland and M.V.J. Ramsey
3505 W.O. Godtfredsen et al.
3537 M.M. Shemyakin et al.
3573 B.R. Pai, P.S. Subramaniam and V. Subramanyam
3591 S.N. Shanbhag et al.
3599 P.K. Grant and M.G.H. Munro
3605 A.I. Scott et al.
3633 T. Kawasaki and K. Miyahara
3641 A. Van Dergen et al.
3651 A. Van Dergen et al.
3697 L. Jurd and B.J. Bergot
3707 L. Jurd
3715 T.R. Govindachari et al.
3721 R.E. Taylor-Smith
3727 D. Adinarayana and T.R. Seshadri
3731 W. Jordan and P.J. Scheuer

82- -64B, J. Mol. Spect., 13 (1964)
0001 S. Nagakura, K. Kaya and H. Tsubomura
0174 S. Nagakura, M. Kojima and Y. Maruyama

83- -64, Arch. Pharm., 297 (1964)
0124 R. Huttenrauch
0129 J. Knabe and J. Kubitz
0146 K.W. Merz and G. Grafe
0182 K.E. Schulte and G. Rucker
0244 R. Huttenrauch
0321 H. Bohme, G. Berg and H. Schneider
0362 W. Wiegrebe
0367 F. Eiden and B.S. Nagar
0474 H. Mohrle
0488 F. Eiden and B.S. Nagar
0524 H.J. Roth and B. Miller
0529 J. Knabe and G. Grund
0617 H.J. Roth and B. Miller
0623 R. Neidlein
0660 H.J. Roth and B. Miller
0703 G. Schenk, M. Huke and T. Naumann

83- -65, Arch. Pharm., 298 (1965)
0034 H.J. Roth and R. Brandes
0124 R. Neidlein
0209 M. Maturova, L. Hruban, F. Santavy and W. Wiegrebe
0262 H. Bohme and L. Kreutzig
0326 H.J. Roth, C. Schwenke and G. Dvorak
0385 S. Pfeifer and S.K. Banerjee
0411 J. Schnekenburger
0561 J. Knabe and H. Roloff
0672 T. Beyrich
0704 W. Dopke
0715 J. Schnekenburger
0722 J. Schnekenburger
0824 A. Mustafa and A.K. Mansour
0826 G. Schwenker
0838 H. Rimpler and R. Hansel
0879 J. Knabe, H. Roloff and U.R. Shukla
0885 H.J. Roth, K. Jager and R. Brandes

87- -64, J. Med. Chem., 7 (1964)
0001 C. Heidelberger, D.G. Parsons and D.C. Remy
0005 M. Israel et al.
0010 E. Dyer and H.S. Bender
0024 B.R. Baker et al.
0088 M.A. Davis et al.
0094 H.H. Keasling, R.E. Willette and J. Szmuszkovicz
0106 S. Nakanishi
0108 S. Nakanishi
0141 M. von Strandtmann, C. Puchalski and J. Shavel, Jr.
0150 J. Lehrfeld, A.M. Burkman and J.E. Gearien
0200 L.R. Lewis, R.K. Robins and C.C. Cheng
0207 I. Wempen and J.J. Fox
0235 S. Farber, H.M. Wuest and R.I. Meltzer
0238 J.H. Ackerman et al.
0242 A.L. Clingman
0269 J.G. Topliss et al.
0274 J.B. Hester et al.
0310 W.E. Kreighbaum and H.C. Scarborough
0337 C. Piantadosi et al.
0345 K. Irmscher et al.
0348 J.E. Pike et al.
0352 R. Hirschmann et al.
0355 R.G. Strachan et al.
0385 J.P. Horwitz et al.
0389 J. DeGraw and L. Goodman
0392 R.D. Hoffsommer, D. Taub and N. Wendler
0399 R.P. Buhs et al.
0415 G.A. Youngdale et al.
0439 M.A. Davis et al.
0446 R.B. Moffett
0457 F.J. Villani et al.
0504 P. Morand et al.

0508 S.K. Figdor et al.
0511 G.M.K. Hughes, P.F. Moore and R.B. Stebbins
0519 R.E. Juday, D.P. Page and G.A. DuVall
0528 J.A. Edwards, M.C. Calzada and A. Bowers
0531 L.L. Smith and D.M. Teller
0537 D.F. Morrow et al.
0540 R.P. Graber et al.
0548 M.B. Meyers, R.P. Graber and D.A. Jones
0552 S. Nakanishi and R.P. Graber
0555 P.E. Shaw et al.
0567 H.J. Minnemeyer et al.
0573 C.H. Gleason
0584 H. Mrozik et al.
0590 D.M. Pistek et al.
0596 S. Furesz, R. Pallanza and V. Arioli
0614 V.P. Shah and R. Ketcham
0632 A.L. Davis et al.
0635 B.K. Koe and R. Pinson, Jr.
0655 L.M. Lerner and P. Kohn
0678 S.P. Raman et al.
0681 M.E. Wolff and W.Ho
0684 G.R. Allen, Jr., and M.J. Weiss
0689 E.L. Patterson et al.
0695 R.J. Fessenden et al.
0705 B.K. Coe and W.D. Celmer
0716 J.F. Cavalla, J.P. Marshall and R.A. Selway
0721 D.E. Fancher et al.
0726 N.J. Harper, C.F. Chignell and G. Kirk
0748 B.J. Magerlein et al.
0751 F. Kagan, R.D. Birkenmeyer and B.J. Magerlein
0755 G.C. Buzby, Jr., et al.
0792 M. Israel et al.
0801 P.K. Joseph and M.M. Joullie
0803 S.M. Kupchan, R.J. Hemingway and R.W. Doskotch
0804 M.J. Weiss et al.
0806 E.G. Podrebarac and C.C. Cheng
0819 K. Hayes
0826 L.G. Humber

87- -65, <u>J. Med. Chem.</u>, <u>8</u> (1965)
0001 C.L. Stevens et al.
0022 E. Van Heyningen
0033 H.J. Schaeffer and R. Vince
0035 B.R. Baker and J.H. Jordaan
0048 P.D. Klimstra and R.E. Counsell
0107 P, Thyrum and A.R. Day
0112 W. Korytnyk
0129 P. Coad et al.
0137 C.T. Bahner, H. Kinder and T. Rigdon
0139 T.L. Loo
0140 G.J. Durr
0147 P.S. Portoghese
0174 E. Van Heyningen and C.N. Brown
0182 D.E. O'Brien et al.
0187 A.P. Martinez, W.W. Lee and L. Goodman
0190 G.B. Brown et al.
0200 M. von Strandtmann, M.P. Cohen and J. Shavel, Jr.
0204 E. Walton et al.
0253 G.J. Durr
0259 G.C. Morrison and J. Shavel, Jr.
0268 R.Y. Kirdani and R.I. Dorfman
0283 B.R. Baker et al.
0350 D.L. Smith, A.A. Forist and W.E. Dulin
0368 T. Kappe and M.D. Armstrong
0386 R.F.R. Church and M.J. Weiss
0390 C.T. Bahner et al.
0397 C.T. Bahner, H. Kinder and L. Gutman
0409 A.H. Goldkamp et al.
0474 W.J. Wechter, W.A. Phillips and F. Kagan
0486 R. Thedford, M.H. Fleysher and R.H. Hall
0502 H.J. Schaeffer, D. Vogel and R. Vince
0531 W.H. Nyberg and C.C. Cheng
0536 J. Hannah and J.H. Fried
0554 G.W.E. Plant
0555 R.D. Hoffsommer, D. Taub and N.L. Wendler
0559 B.C. Joshi et al.
0583 G.N. Walker
0626 G.N. Walker, R.T. Smith and B.N. Weaver
0659 E. Walton et al.
0667 A. Giner-Sorolla and A. Bendich
0700 W.A. Remers and M.J. Weiss
0708 J.A. Montgomery and K. Hewson
0710 H.J. Schaeffer and R. Vince
0713 T.L. Loo et al.
0722 R. Littell and D.S. Allen, Jr.
0727 J.A. Montgomery, K. Hewson and J.R. Piper
0732 J.A. Skorcz and F.E. Kaminski
0737 J.A. Montgomery and K. Hewson
0790 M. Maggi, V. Anoli and P. Sensi
0797 W.J. Haggerty, Jr., R.H. Springer and C.C. Cheng
0821 K.S. Cheah and J.C. Watkins
0829 C.L. Stone et al.
0866 G. Gough and M.H. Maguire
0870 C.R. Tamorria and R.C. Esse
0884 P.K. Chang
0886 H.L. Slates and N.L. Wendler
0891 L.D. Freedman and G.O. Doak

88- -64, <u>Tetrahedron Letters</u> (1964)
0029 D.R. Perrin
0043 W.A. Gibbons and H. Fischer
0061 A.W. Burgstahler et al.
0103 S.W. Pelletier and P.C. Partasarathy
0129 O. Achmatowicz and J.T. Wrobel
0137 R. Bognar, M. Rakosi and J. Balint
0171 A.V. Zakharichev, S.N. Anachenko and I.V. Torgov
0189 P.G. Farrell and J. Newton

0223 S. Rakhit and M. Gut
0245 D.M. Lemal and E.H. Banitt
0253 A. Pollak and M. Tisler
0319 H.A. Staab and F. Binnig
0365 R.N. Mirrington et al.
0383 R. Kuhn and D. Rewicki
0499 L. Bernardi and A. Leone
0505 J.M. Dunston and P. Yates
0659 J.S. Shannon, H. Silberman and S. Sternhell
0663 E.L. Shapiro, T. Legatt and E.P. Oliveto
0669 R. Aneja and S.W. Pelletier
0719 S. Akiyama and M. Nakagawa
0773 G. Schroder, R. Mirenyi and J.F.M. Oth
0813 A. Padwa
0827 D.E. Bublitz and K.L. Rinehart, Jr.
0835 H. Lichti and A. von Wartburg
0849 A.I. Scott et al.
0901 T.R. Govindachari et al.
0961 W. von E. Doering and M. Pomerantz
0973 G. van Binst et al.
1075 M.B. Rubin, D. Glover and R.G. Parker
1107 A. Roedig and R. Kohlhaupt
1151 I.J. Borowitz and G. Gonis
1161 E. LeGoff and R.B. LaCount
1263 G.W. Brown, R.C. Cookson and I.D.R. Stevens
1275 H.O. Huisman et al.
1303 Y. Inubushi, T. Sano and Y. Tsuda
1337 Y.-L. Chow
1345 J. Tadanier and W. Cole
1375 D.I. Schuster, M.J. Nash and M.L. Kantor
1387 A.T. Gleu and J. McLean
1403 M. Regitz
1465 N.J. Leonard and G.E. Wilson, Jr.
1471 N.J. Leonard and G.E. Wilson, Jr.
1477 W.D. Crow and N.J. Leonard
1513 M.M. Robison et al.
1525 E. Muller, H. Fricke and H. Kessler
1553 S. Sarel et al.
1557 L.R. Row et al.
1623 A.J. Birch, M. Salahud-Din and D.C.C. Smith
1629 I.R.C. Bick and C.K. Douglas
1635 Y. Makisumi
1671 P. Potier, C. Kan and J. LeMen
1719 W.C. Howell, M. Ktenas and J.M. MacDonald
1733 K. Hafner, D. Zinser and K.-L. Moritz
1743 K.B.L. Mathur and K.P. Sarbhai
1781 R.M. Dupeyre, H. Lemaire and A. Rassat
1797 D.H. Reid and F.S. Skelton
1825 N. Sakabe, T. Goto and Y. Hirata
1839 R. Sciaky and U. Pallini
1897 R.E. Bowman, M.D. Closier and P.J. Islip
1903 R.E. Corbett and S.G. Wyllie
1961 A. Chatterjee and S. Sen Gupta
2049 A.T. Balaban, C.N. Rentea and M. Mocanu
2109 M.P. Cava and R.H. Schlessinger
2117 J. Jones and J. Jones
2124 H. Saito, K. Nukuda and M. Ohno
2131 H.E. Zimmerman and L. Craft
2137 D. Bryce-Smith, G.I. Fray and A. Gilbert
2151 M. Ohno and N. Naruse
2157 G. Bianchi, P. Grunanger and A. Perotti
2161 R.D. Youssefyeh
2185 K. Hafner and W. Kaiser
2201 W.A. Ayer et al.
2211 A.J. Birch and M. Salahuddin
2281 H. Misuhashi and K. Shibata
2313 L. Crombie, D.E. Games and M.H. Knight
2355 M. Miyamoto et al.
2411 C.J. Moyer and S. Sternhell
2419 Y. Kanaoka et al.
2427 F. Kaplan and G.K. Meloy
2443 S.I. Sallay
2489 E. Campaigne and C.D. Blanton, Jr.
2531 D. Horton and W.N. Turner
2567 J.J. Dugan, P. de Mayo and A.N. Starratt
2571 R.W. Griffin, Jr., and R.A. Coburn
2587 A. Piskala
2607 W.D. Ollis, A.D. Ward and R. Zelnik
2637 M. Tomita et al.
2643 L. Mangoni and M. Belardini
2659 L.P. Volodarsky and V.A. Koptyug
2711 M. Khaimova et al.
2729 G.P. Schiemenz
2791 J.M. Conia and P. LePerchec
2797 J.M. Conia and P. Briet
2855 K. Hunger and F. Korte
2881 S.P. Gubin, A.Z. Rubezhov and B.L. Winch
2919 C.R. Ganellin
2955 M. Tanabe and D.F. Crowe
2987 H.M. Buck et al.
3039 E.W. Muller and F. Korte
3043 C.A. Bishop and L.K.J. Tong
3065 S.W. Pelletier et al.
3145 S.M. Kupchan and W.L. Asbun
3159 R.E. Atkinson, R.F. Curtis and G.T. Phillips
3173 H. Takahasi, T. Kimata and M. Yamaguchi
3189 J.M. Conia, P. Leriverend and J.L. Bouket
3201 F. Cramer and G. Schlingloff
3363 S.G. Smith and J.P. Petrovich
3367 T.R. Seshadri et al.
3413 H. Ageta, K. Iwata and S. Natori
3431 S. Senoh et al.
3437 S. Senoh et al.
3459 W.M. Jones and M.E. Stowe
3471 D. Bryce-Smith and A. Gilbert
3523 H. Brockmann and H. Lockner

3557 C.W.J. Chang, R.E. Moore and P.J. Scheuer
3575 N.N. Vorozhtsov et al.
3605 M. Tomita et al.
3617 M. Tomita, T. Ibuka and Y. Inubushi
3653 H. Morrison
3671 K.K. Pivnitsky and I.V. Torgov
3751 R. Misra, R.C. Pandey and S. Dec
3775 A.F. Thomas and B. Willhalm
3803 E.M. Arnett and J.M. Bollinger
3809 D.C. Dittmer and N. Takashina
3815 M.P. Cava and R.H. Schlessinger
3819 S. McLean and M.S. Lin
3841 N.J. McCorkindale
3871 I.C. Calder and W.H.F. Sasse
3899 M. Ohashi, J.A. Joule and C. Djerassi
3927 T.R. Govindachari, B.S. Joshi and V.N. Kamat
3935 P. Grafen and R.B. Turner
3983 J. Polonsky and J.-L. Fourrey

88-65, Tetrahedron Letters (1965)
0023 R.T. LaLonde and R.I. Akentijevich
0029 A.J. Birch et al.
0037 H. Nozaki, Z. Yamaguti and R. Noyori
0041 S.W. Pelletier, R.L. Chappell and P.C. Parthasarathy
0051 H.A. Staab and F. Vogtle
0073 M. Inatome and L.P. Kuhn
0137 R. Sciaky and F. Mancini
0153 R. Nicoletti and M.L. Forcellese
0159 A. Chatterjie et al.
0171 F. Bohlmann and R. Mayer-Mader
0215 R. Aneja and S.W. Pelletier
0251 H. Nozaki, K. Kondo and M. Takaku
0275 Z.J. Barneis, D.M.S. Wheeler and T.H. Kinstle
0297 R.E. Atkinson and R.F. Curtis
0359 S.M. Kupchan and H.C. Wormser
0365 R. Breslow, W. Vitale and K. Wendel
0377 D.S. Glass, J.W.H. Watthey and S. Winstein
0385 E.N. Marvell, G. Caple and B. Schatz
0391 E. Vogel, W. Grimme and E. Dinne
0423 E. Muller, H. Kessler and H. Suhr
0479 W.M. Jones and R.S. Pyron
0485 P. Yates and A.G. Szabo
0609 E. Vogel, W.A. Bole and H. Gunther
0703 J. Casanova, Jr., et al.
0721 A. Hochrainer and F. Wessely
0821 S. Sarel and J. Rivlin
0913 M. Simonetta and S. Carra
0931 M.P. Cava et al.
0945 S. Masamune
0979 J.-L. Ripoll and J.-M. Conia
1031 K.H. Scheit
1041 T. Nozoe, T. Mukai and K. Sakai
1049 A. Padwa
1053 S. Sarel and R. Ben-Shoshan
1059 K. Kitahonoki et al.
1071 D. Misiti, H.W. Moore and K. Folkers
1091 T. Irie, M. Suzuki and T. Masamune
1117 G. Giacomello
1175 J. Elguero and R. Jacquier
1261 H. Inouye et al.
1287 S. Shibata et al.
1359 J.F. Harris, Jr.
1389 H.W. Whitlock, Jr., and G.L. Smith
1405 N. Otake et al.
1411 N. Otake et al.
1433 K. Nakagawa and H. Onoue
1465 I.C. Calder and W.H.F. Sasse
1477 U. Biethan, U. v. Gizycki and H. Musso
1497 P.C. Myrhe and R.D. Andersen
1509 M. Shamma and W.A. Slusarchyk
1539 A. Chatterjie et al.
1565 L.B. Volodarsky et al.
1577 F. Walls et al.
1599 K. Wiedhaup et al.
1625 A.W. Burgstahler, D.J. Malfer and M.O. Abdel-Rahman
1717 W.F. Erman and H.C. Kretschmar
1723 C.K. Bradsher et al.
1757 H. Behringer and F. Scheidl
1761 H.A. Lloyd
1829 F. Piozzi et al.
1877 L.L. Replogle, K. Katsumoto and T.C. Morrill
1987 S.R. Johns, J.H. Russel and M.L. Heffernan
2059 B. Shimizu and M. Miyaki
2075 H. Fritz et al.
2121 A. Luttringhaus, H. Berger and H. Prinzback
2175 L. Skattebøl
2181 T. Axenrod, E. Biering and L.H. Schwartz
2233 S.D. Levine
2239 B. Das et al.
2285 D. Beck et al.
2317 E.V. Dehmlow
2365 K. Hideg and H.O. Hankovszky
2377 S. Corsano and E. Micione
2399 I.R.C. Bick and G.K. Douglas
2433 F. Bohlmann and D. Schumann
2473 Y.-L. Chow
2559 K. Fukui and M. Nakayama
2597 M. Kurihara and N. Yoda
2655 H. Nakata and Y. Hirata
2673 E. Muller and H. Kessler
2817 I. Moritani and N. Obata
2831 J.A. Miller, M.H. Durand and J.E. Dubois
2847 K. Heusler, H. Labhart and H. Loeliger
2903 R.W. Murray and M.L. Kaplan
2941 E. Molenaar and J. Strating
2951 G.W. Griffin et al.
2967 M. Ueno, I. Murata and Y. Kitahara
2983 J. Wiemann et al.

3085 C.G. Kratchanov and B.J. Kurtev
3095 T. Hashizume and H. Iwamura
3145 U.R. Ghatek, J. Chakravarty and A.K. Banerjee
3151 J.M. Conia and J.C. Limasset
3175 R. Haller
3191 A. Malhotra, V.V.S. Murti and T.R. Seshadri
3239 H. Achenbach and K. Biemann
3301 P.K. Freeman et al.
3337 Y. Tsuda and K. Isobe
3391 T.H. Applewhite
3397 H.E. Miller et al.
3439 M.L. Scarpati, M. Guiso and L. Panizzi
3505 J. Fried and N.A. Abraham
3569 B.D. Tilak, R.B. Mitra and C.V. Deshpande
3579 Z. Voticky and J. Tomko
3595 M. Shamma, M.A. Greenberg and B.S. Dudock
3613 E. Vogel, W. Pretzer and W.A. Boll
3619 T. Irie et al.
3625 E. Vogel, W. Grimme and S. Korte
3783 S. Shibata, I. Kitagawa and H. Fujimoto
3873 T.R. Govindachari et al.
3937 E.H. White and J.P. Anhalt
4003 E.V. Dehmlow
4053 S.B. Nerali et al.
4097 T. Sa-an et al.
4129 H. Weiler-Feilchenfeld, E.D. Bergmann and A. Hirschfeld
4167 K. Munakata et al.
4191 T. Okuda and T. Yoshida
4203 O. Lindwall et al.
4259 N.J. Leonard and A.E. Yethon
4363 A. Padwa and L. Hamilton
4413 M.J. Goldstein and A.H. Gevirtz
4487 T.D.J. D'Silva and H.J. Ringold
4537 H.A. Lloyd
4545 L. Caglioti, P. Grasselli and G. Rosini
4557 K. Hata and M. Kozawa
4569 K. Bott
4585 D. Caine and J.F. DeBardeleben
4603 P. de Ruggieri, C. Gandolfi and U. Guzzi
4647 S. Masamune and K. Fukumoto
4655 I.R.C. Bick and G.K. Douglas
4675 S. Nozoe et al.
4737 I. Morimoto et al.
4803 H. Brockmann and F. Seela
4833 R.P. Rebman and N.H. Cromwell
4837 M. Masaki et al.
4857 G.B. Marini Bettolo, C.G. Casinovi and C. Galeffi

89- -64, Angew. Chem., 76 (1964)
0076 E. Bayer
0145 E. Vogel and H.D. Roth
0157 R. Mayer, J. Morgenstern and J. Fabian
0225 E. Hecker, H. Bresch and Ch. v. Szczepanski
0393 E. Bertele et al.
0432 E. Vogel, H. Kiefer and W.R. Roth
0433 K. Dimroth and P. Hoffmann
0437 R. Schmidt
0535 E. Vogel, R. Schubart and W.A. Boll
0567 G. Schill and A. Luttringhaus
0569 S. Masamune and N.T. Castellucci
0684 E. Fahr and H.-D. Rupp
0784 B. Fohlisch and P. Burgle
0784B E. Vogel and W.A. Boll
0785 E. Vogel et al.
0786 E. Vogel, W. Meckel and W. Grimme
0862 E. Sturm and K. Hafner
0920 M. Regitz and D. Stadler
0922 H.G. Viehe et al.

89- -65, Angew. Chem., 77 (1965)
0042 R. Hafner, R. Fleischer and K. Fritz
0043 R. Kuhn, F.A. Neugebauer and H. Trischmann
0095 G. Kobrich and W. Drischel
0171 A. Messmer and A. Gelleri
0258 H. Prinzbach, D. Seip and U. Fischer
0260 W. Mack
0261 R. Mayer and J. Wehl
0262 V. Bertini and P. Pino
0345 Y. Kitahara, I. Murata and S. Katagiri
0346 H. Prinzbach and V. Freudenberger
0348 W. Grimme, H. Hoffmann and E. Vogel
0453 H. Prinzbach, H. Berger and A. Luttringhaus
0621 H. Prinzbach and U. Fischer
0768 H.G. Viehe
0810 R.W. Hoffmann and W. Sieber
0814 K.H. Buchel and F. Korte
0861 K. Wallenfels and W. Hanstein
1077 W. Rohr and H.A. Staab
1109 G. Markl
1136 W. Pfleiderer and W. Hutzenlaub

94- -64, Chem. Pharm. Bulletin (Japan), 12 (1964)
0018 T. Kato and H. Yamanaka
0023 J. Kinugawa and M. Ochiai
0077 M. Shimizu et al.
0087 G. Ohta, K. Ueno and M. Shimizu
0092 K. Ueno
0112 H. Mori, V.S. Gandhi and E. Schwenk
0121 S. Okuda et al.
0127 Y. Kanaoka et al.
0196 H. Kaneko, M. Hashimoto and A. Kobayashi
0204 Y. Makisumi, H. Watanabe and K. Tori
0214 T. Tanaka
0228 T. Itai and S. Natsume
0236 S. Natori, H. Nishikawa and H. Ogawa
0249 S. Yamada and A.M.E. Omar

0253 S. Ozeki
0267 M. Ikehara, H. Uno and F. Ishikawa
0312 S. Fukushima, A. Ueno and Y. Akahori
0316 S. Fukushima, Y. Akahori and A. Ueno
0357 T. Nishimura and I. Iwai
0383 M. Tomoeda et al.
0386 Y. Yamamoto, S. Uyeo and K. Ueda
0393 A. Takamizawa and K. Hirai
0408 S. Uyeo, T. Kitagawa and Y. Yamomoto
0450 Y. Nitta, M. Shindo and K. Takamura
0454 T. Ukita, A. Hamada and M. Yoshida
0459 T. Ukita, M. Yoshida, Y. Kato and A. Hamada
0473 H. Hasegawa and K. Tsuda
0489 S. Uyeo et al.
0533 H. Inoyue and T. Arai
0558 A. Takamizawa et al.
0595 R. Dohmori
0696 J. Koizumi, S. Kobayashi and S. Uyeo
0725 K. Okumura, K. Kotera and I. Inouye
0744 T. Miyasake
0747 I. Murakoshi et al.
0752 T. Wada and D. Satoh
0755 H. Hikino, Y. Hikino and I. Yosioke
0773 Y. Kanaoka et al.
0779 K. Takeda et al.
0789 Y. Makisumi
0796 K. Kawashina et al.
0804 A. Takamizawa and K. Hirai
0813 I. Iwai and T. Hiraoka
0820 A. Ogiso and I. Iwai
0828 T. Ukita, R. Funakoshi and Y. Hirose
0841 M. Aritomi
0853 M. Tada et al.
0859 R.C. Tweit et al.
0866 Y. Mizuno, T. Itoh and K. Saito
0888 H. Inoyue, T. Arai and Y. Miyoshi
0905 K. Takeda et al.
0951 T. Narito, K. Ueno and F. Ishikawa
0971 T. Nakano and M. Hasegawa
0984 A. Tahara and K. Hirao
0987 H. Tsukamoto et al.
1012 S. Minami and S. Uyeo
1030 K. Imai
1072 M. Ohta, H. Tami and S. Morizumi
1080 M. Ohta et al.
1094 I. Iwai and J. Ide
1111 E. Hayashi and T. Higashino
1117 T. Wada
1118 Z. Horrii et al.
1124 T. Okamoto, M. Natsume and S. Kamata
1129 G. Kobayashi and S. Furukawa
1143 T. Okumura et al.
1181 S. Noguchi, F. Nakayama and K. Morita
1184 S. Noguchi, M. Imanishi and K. Morita
1217 K. Yasuda
1232 T. Miyazaki and S. Mihashi
1240 H. Mizuno et al.
1259 T. Kishikawa and H. Yuki
1296 Y. Ban et al.
1329 T. Sasaki and K. Minamoto
1338 T. Miyadera and I. Iwai
1344 T. Miyadera and I. Iwai
1351 F. Yoneda, T. Ohtaka and Y. Nitta
1375 K. Sugimoto and S. Ohki
1378 Y. Ban and M. Seo
1381 Y. Ban and I. Inoue
1405 Z. Horii et al.
1415 T. Momose et al.
1418 A. Takamizawa and K. Hirai
1424 Y. Makisumi
1433 K. Takeda, T. Komeno and S. Ishihara
1439 K. Tori
1446 I. Iwai et al.
1471 T. Nishimura et al.
1478 N. Yomeda
1495 E. Ochiai and H. Mitarashi
1503 T. Ukita, Y. Takeda and H. Hayatsu
1510 C. Kaneko, S. Hayashi and M. Ishikawa
1520 S. Saito et al.

94- -65, <u>Chem. Pharm. Bulletin (Japan), 13</u> (1965)

0007 K. Imai and M. Honjo
0039 M. Tomita et al.
0043 T. Wada
0050 T. Kubota et al.
0058 S. Nakajima, K. Kinoshita and S. Shibata
0064 S. Nakajima
0069 S. Nakajima
0073 S. Nakajima
0088 S. Yamada et al.
0093 Y. Fujise, T. Toda and S. Ito
0118 I. Iwai, K. Tomita and J. Ide
0142 A. Takamizawa and Y. Hamashima
0156 K. Yoshida and T. Kubota
0180 M. Shimizu et al.
0253 H. Hagiwara et al.
0291 E. Hayashi and T. Higashino
0308 T. Wada and D. Satch
0312 T. Wada
0389 K. Harada and S. Emoto
0393 Y. Sato et al.
0420 Z. Horii et al.
0427 S. Uyeo et al.
0443 G. Sunagawa and H. Nakao
0450 G. Sunagawa and H. Nakao
0457 N. Soma et al.
0465 H. Nakao and G. Sunagawa
0473 H. Nakao et al.
0503 T. Miyadera
0511 H. Ogawa and S. Natori
0517 M. Tomoeda, R. Kikuchi and M. Urata
0580 T. Ohtaka and Y. Nitta
0622 N. Yoneda
0628 H. Hikino, K. Aota and T. Takemoto

0633 S. Natori, Y. Kumada and H. Nishi-kawa
0651 Z. Hori, T. Momose and Y. Tamura
0663 I. Iwai and J. Ide
0681 A. Takamizava and K. Hirai
0687 K. Takeda et al.
0695 M. Tomita et al.
0717 H. Minato et al.
0769 M. Tomoeda et al.
0775 Y. Ban and M. Terashima
0786 S. Saito et al.
0797 Z. Horii, T. Momose and Y. Tamura
0803 T. Nishimura and B. Shimizu
0810 H. Nakao
0828 H. Nakao, N. Soma and G. Sunagawa
0895 M. Shimizu et al.
0912 M. Hamana and H. Noda
0931 Y. Ban, R. Sakaguchi and M. Nagai
0935 Y. Yamamoto et al.
0942 K. Takeda et al.
0951 K. Mitsuhashi and K. Nomura
0963 T. Kato and T. Niitsuma
0986 Y. YengLin et al.
1017 T. Fujii, T. Itaya and S. Yamada
1073 Y. Ban and Y. Sato
1078 M. Tomoeda, T. Furuta and T. Koga
1084 S. Minami et al.
1138 P. Bey, F. Lederer and G. Ourisson
1168 T. Sasaki and K. Minamoto
1207 A. Takamizawa and Y. Hamashima
1225 T. Kametani, K. Kigaswa and T. Hayasaka
1231 N. Yoneda
1240 J. Shoji et al.
1247 Y. Arata and T. Ohashi
1285 T. Miki, K. Hiraga and T. Asako
1289 K. Hiraga
1294 K. Hiraga, T. Asako and T. Miki
1300 K. Hiraga
1307 Z. Horii et al.
1319 M. Sekiya and Y. Osaki
1332 H. Mitsuhashi and T. Nomura
1377 Y. Maki, M. Sato and K. Obata
1408 H. Hikino et al.
1430 I. Yosioka and T. Kimura
1435 T. Nambara and M. Kato
1445 G. Ohta et al.
1472 S. Natori and Y. Kumada
1484 H. Hikino et al.

95- -1964, J. Pharm. Soc. Japan, 84 (1964)
0001 A. Takai and I. Saikawa
0009 A. Takai and I. Saikawa
0016 A. Takai and I. Saikawa
0047 T. Sasaki
0052 S. Mizukami and Y. Kanaya
0057 S. Mizukami and Y. Kanaya
0077 M. Komatsu
0109 I. Saikawa, A. Takai and Y. Kodama
0121 I. Saikawa
0131 I. Saikawa and T. Maeda
0146 Y. Kondo
0195 T. Sasaki
0207 I. Saikawa
0212 I. Saikawa
0219 I. Saikawa, A. Takai and Y. Kodama
0225 I. Saikawa, A. Takai and Y. Kodama
0318 W. Kamisako and M. Takahashi
0325 W. Kamisako
0337 H. Inouye, T. Arai and Y. Yaoi
0360 M. Aritomi
0362 M. Tomita and M. Kozuka
0381 T. Sasaki et al.
0390 T. Nishina and M. Kimura
0399 T. Kametani and H. Sugahara
0412 T. Kametani and H. Yagi
0453 S. Uyeo, J. Okada and S. Matsunaga
0477 H. Wada, H. Nakata and Y. Hirata
0548 S. Uyeo et al.
0555 S. Uyeo and T. Shingu
0562 Y. Deguchi and K. Miura
0566 I. Saikawa
0646 I. Saikawa and Y. Suzuki
0671 I. Murakoshi et al.
0674 I. Murakoshi et al.
0721 T. Takemoto, K. Kondo and Y. Kondo
0762 S. Uyeo et al.
0874 T. Kato and T. Kitagawa
0887 Y. Tanabe et al.
0895 M. Aritomi, M. Shimojo and T. Mazaki
0930 H. Nishimura
0955 H. Taguchi and I. Imaseki
0971 T. Yamana et al.
1061 T. Fujita, T. Fujii and A. Ide
1072 J. Taga
1080 H. Otomasu, H. Takahashi and K. Yoshida
1085 S. Tagami and D. Shiho
1104 M. Ueda and K. Kuribayashi
1113 A. Takamizawa and Y. Hamashima
1194 S. Ozeki
1201 T. Kato, H. Yamanaka and H. Moriya
1220 Z. Horii
1227 M. Kimura et al.

95- -65, J. Pharm. Soc. Japan, 85 (1965)
0055 A. Nitta
0077 M. Tomita and M. Kozuka
0095 H. Inoue et al.
0158 A. Takamizawa, S. Hayashi and H. Sato
0200 S. Ozeki
0206 S. Ozeki
0271 Y. Kuwayama and S. Kataoka
0310 G. Kobayashi et al.
0314 S. Uyeo, T. Shingu and H. Harada
0339 M. Yanai et al.
0344 M. Yanai and T. Kinoshita
0374 N. Morita, M. Shimizu and M. Fukuta
0387 Y. Kuwayama and S. Kataoka
0391 Y. Kuwayama and S. Kataoka
0399 K. Takiura et al.
0404 S. Uyeo et al.
0429 Y. Maki, M. Sato and K. Yamane
0437 S. Uyeo et al.
0442 S. Hayashi

0451 T. Kato and Y. Goto
0472 H. Furukawa, T.-H. Yang and T.-J. Lin
0507 S. Toyoshima et al.
0553 M. Yasue et al.
0565 A. Fujita et al.
0584 Y. Watanabe, M. Matsui and K. Ido
0615 S. Uyeo and Y. Yamato
0667 T. Kametani et al.
0699 S. Ozeki
0731 Y. Kuwayama and Y. Matsuda
0757 K. Murayama et al.
0812 T. Kato, T. Atsumi and H. Sasaki
0816 M. Okada, A. Yamada and M. Ishidate
0827 M. Tomita, S.-T. Lu and S.-J. Wang
0839 T. Onaka
0857 Y. Sasaki and H. Kurokawa
0858 Y. Sasaki and Y. Nakamura
0871 T. Kametani et al.
0875 S. Umemoto and H. Takamatsu
0888 W. Kamisako and M. Takahashi
0948 I. Saikawa and A. Takai
0960 T. Kametani et al.
0971 S. Ohki and F. Hamaguchi
0975 H. Takamatsu et al.
0981 T. Amano and S. Mizukami
0985 T. Kametani and K. Ogasawara
0998 N. Oi and I. Umeda
1035 T. Amano and S. Mizukami
1042 T. Amano, T. Sakano and S. Mizukami
1049 T. Amano
1092 M. Okada, M. Hasunuma and Y. Saito

96- -64, The Analyst, 89 (1964)
0730 D.W. Andrews
0735 J.E. de Souza and M. Scherbak
0788 G.C. Jayson, T.C. Owen and A.C. Wilbraham
0803 L. Mosconi and A. Lama

96- -65, The Analyst, 90 (1965)
0134 J.D. Few
0155 J. Nabney and B.F. Nosbitt
0161 T.R. Andrew and P.N.R. Nichols
0409 J. Metcalfe
0432 N. Wahba

100- -64, Lloydia, 27 (1964)
0015 R.N. Blomster, A.E. Schwarting and J.M. Bobbitt
0040 R. Hanna
0111 C.L. Winek, J.L. Beal and M.P. Cava
0203 G.H. Svoboda, M. Gorman and R.H. Tust
0220 A.H. Kiang et al.
0374 J.A. Weisbach and B. Douglas
0406 U. Renner
0456 J. Lemen
0470 W.D. Loub, N.R. Farnsworth, R.N. Blomster and W.W. Brown

100- -65, Lloydia, 28 (1965)
0073 M.P. Cava, T.A. Reed and J.L. Beal
0084 H.G. Appel et al.
0090 A. Rother et al.
0095 P.J. Schever and T.R. Pattabhiraman
0191 J.R. Boissier et al.
0199 A. Brossi and R. Borer
0203 K.L. Euler et al.
0212 J.D. Albright et al.
0237 S.A. Gharbo et al.
0359 T.B. Tjio, T. Sproston and R. Tomlinson

101- -64B, J. Organometallic Chem., 2 (1964)
0265 W. Drenth et al.
0336 G.N. Schrauzer and G. Kratel
0369 H. Gilman, W.H. Atwell and G.L. Schwebke
0398 H. Egger and K. Schlogl
0417 D. Seyferth, C. Sarafidis and A.B. Evnin

101- -65A, J. Organometallic Chem., 3 (1965)
0007 A. Nakamura, P.J. Kim and N. Hagihara
0097 E.P. Blanchard, Jr., et al.
0107 J.H. Osiecki, C.J. Hoffman and D.P. Hollis
0222 J.G. Noltes and J.W.G. Van den Hurk
0233 F. Johnson, R.S. Gohlke and W.A. Nasutavicus
0269 M. Cais and N. Narkis
0355 A. Nakamura, P.J. Kim and N. Hagihara

101- -65B, J. Organometallic Chem., 4 (1965)
0067 W.H. Nelson and D.F. Martin
0159 M. Wada et al.
0163 H. Gilman et al.
0176 H. Gilman and W.H. Atwell
0261 R.A. Finnegan et al.
0271 C.H. Langford and J.P. Aplington
0430 R.E. Bailey and R. West